Communications Standard Dictionary

Third Edition

Martin H. Weik, D.Sc.

formerly on staff of:
U.S. Forces European Theater
Columbia University
University of Delaware
U.S. Army Ballistic Research Laboratories
U.S. Army Office of the Chief of Research and Development
Defense Communications Agency
National Communications System

former consultant to:
Naval Sea Systems Command
Space and Naval Warfare Systems Command

consulting affiliations:
Dynamic Systems, Incorporated, Alexandria, Virginia
Science Applications International Corporation, Sierra Vista, Arizona

former:
Chairman, Glossary Subcommittee
Federal Telecommunication Standards Committee
National Communications System

Chairman, Glossary Subcommittee
Computer Society Standards Committee
Institute of Electrical and Electronic Engineers

Chairman, Technical Committee X3K5
Vocabulary for Information Systems
American National Standards Institute

Chief U.S. Delegate to TC97/SC1
Vocabulary for Information Systems
International Organization for Standardization

Communications Standard Dictionary

Third Edition

Martin H. Weik, D.Sc.

New York • Albany • Bonn • Boston • Cincinnati • Detroit • London • Madrid • Melbourne
Mexico City • Pacific Grove • Paris • San Francisco • Singapore • Tokyo • Toronto • Washington

™ A division of International Thomson Publishing, Inc.
The ITP logo is a trademark under license

Printed in the United States of America

For more information, contact:

Van Nostrand Reinhold
115 Fifth Avenue
New York, NY 10003

International Thomson Publishing GmbH
Königswinterer Strasse 418
53227 Bonn
Germany

International Thomson Publishing Europe
Berkshire House 168-173
High Holborn
London WCIV 7AA
England

International Thomson Publishing Asia
221 Henderson Road #05-10
Henderson Building
Singapore 0315

Thomas Nelson Australia
102 Dodds Street
South Melbourne, 3205
Victoria, Australia

International Thomson Publishing Japan
Hirakawacho Kyowa Building, 3F
2-2-1 Hirakawacho
Chiyoda-ku, 102 Tokyo
Japan

Nelson Canada
1120 Birchmount Road
Scarborough, Ontario
Canada M1K 5G4

International Thomson Editores
Campos Eliseos 385, Piso 7
Col. Polanco
11560 Mexico D.F. Mexico

1 2 3 4 5 6 7 8 9 10 HAM 01 00 99 98 97 96

To order this or any other Chapman & Hall book, please contact **International Thomson Publishing, 7625 Empire Drive, Florence, KY 41042.** Phone: (606) 525-6600 or 1-800-842-3636. Fax: (606) 525-7778, e-mail: order@chaphall.com.

For a complete listing of Chapman & Hall titles, send your requests to
Chapman & Hall, Dept. BC, 115 Fifth Avenue, New York, NY 10003

To my wife, Helen,

for the love and encouragement she has given me

during the preparation of this and prior editions.

And God said, Let there be light:
and there was light. And God saw the light,
that it was good: and God divided the light from the darkness.

Genesis 1:3–4

Now the whole earth had one language and few words.
And the Lord said, "Behold, they are one people,
and they have all one language;
and this is only the beginning of what they will do;
and nothing that they propose to do
will now be impossible for them."

Genesis 11:1, 6

In the beginning was the Word,
and the Word was with God,
and the Word was God.

John 1:1

Contents

Preface

The Early Days

The modern age of communications began in the latter half of the nineteenth century with the advent of the telegraph and the telephone. Three-quarters of a century later, about the middle of the twentieth century, electronic computers made their debut. The computer age was introduced by gargantuan building-size computing machines, such as (a) the ENIAC *(Electronic Numerical Integrator and Calculator)*, a decimal machine with 20 words of decade ring counter storage, 19,000 full-size vacuum (radio) tubes, a 400-ms (millisecond) operation time, and programming based on external wiring which had to be rewired each time the program was changed, (b) the EDVAC *(Electronic Discrete Variable Automatic Computer)*, a serial binary machine with 1024 words of mercury tube storage, a 48-μs (microsecond) operation time, and internally stored programming, and (c) the ORDVAC *(Ordnance Variable Automatic Computer)*, a parallel digital machine with an analog adder that had an operation time of the order of several microseconds, depending on the nature of the operands, 1024 words of cathode ray tube storage (Williams tubes), and internally stored programming. The ENIAC and EDVAC were developed by the University of Pennsylvania and the ORDVAC by the University of Illinois. All were installed, further developed, and operated at the U.S. Army's Aberdeen Proving Ground primarily for ballistic computations, numerical analyses, and other scientific applications. Along with a pair of Bell relay computers, a Busch Mechanical Differential Analyzer, and punch card facilities, these machines constituted the largest single concentration of computer capability in the world. Their combined capacity and computing power was a small fraction of the power of a single modern electronic computer.

No Interconnection and No Networking

Visitors from around the world, including Japan, came to witness the Aberdeen machines in operation. The year was 1950. There were no commercially produced electronic computers. Teams of government engineers and technicians spent years to complete the development and connect input/output and peripheral devices to these machines. Though all the machines were installed in the same Ballistic Research Laboratories building, not a word was uttered about connecting these machines directly to each other, not to mention connection to other machines via telephone lines. The state of the art was not ready for the interconnection of computers. There were no standards, no protocols, no computer interconnection, and no need or incentive for such interconnection.

Computers and Communications

During the almost half century that followed these early computing stalwarts, a myriad of companies developed and marketed commercial electronic computers, always increasing computer capabilities and decreasing size. Gradually, the electronic computer changed the face of communications systems. Computer-to-computer, database-to-database, and terminal-to-terminal communications capabilities were added to the conventional person-to-person, multicast, and broadcast communications systems. Up until the 1990s, the communications community considered computers as just another end instrument on their communications lines. The computer community considered communications lines simply as links between computers and work stations. Multimedia interconnection became essential. Homes and offices with personal computers required video and graphics. Broader bandwidths, higher data signaling rates, and more channels were required. The communications community began to use computers to control their communications systems, and the communications systems were used to interconnect computers. Finally, with remote job entry, computer networking, and database interconnection, the boundary between the two communities began to disappear.

The Superhighway

The computer and communications communities began supporting each other. Data signaling rates and computer operational speeds were matched. Computers needed the higher computer-to-computer transmission rates and broader bandwidths. Communications systems needed faster communications traffic control capabilities. Initially there was no real amalgamation of communications, computer, data processing, and control systems. Only during the last decade of the twentieth century has meaningful integration occurred, giving rise to the modern mix of communications concepts, such as multimedia, cyberspace, the information superhighway, Internet, servers, World Wide Web®, clients, B-ISDN, SONET, cellular telephones, personal communications, open systems architecture, the OSI-RM, and E-mail, to name but a few. All of these fast-moving and accelerated developments and applications of communications systems occurred with equal rapidity in the military, civilian, and commercial sectors. Integration of computer and communications systems became commonplace.

Real Time

Modern communications systems enable (a) real time acquisition and distribution of information, (b) remote control of systems and devices, (c) telemetering of data from any remote place to any other place, (d) search of remote databases, (e) conversational operation of geographically widely separated personal computers, (f) remote job entry and access into computers from and to nearly anywhere around the world, and (g) worldwide rapid message and packet transmission. News reaches homes and offices as it occurs. House-to-house street fighting in civil and national wars, the devastating effects of earthquake, flood, and fire, and the investigations of legislative, judicial, and law enforcement authorities may be witnessed in billions of homes and offices around the world in the same moment in which they occur. Communications has in fact tied the elements of the world together, though the elements may be far from united in purpose.

Security

The establishment of Internet, cyberspace, and the information superhighway has given rise to a generation of predator hackers, crackers, and larcenists who invade privacy, violate copyright laws, break security codes, steal trade and industrial secrets, illegally manipulate financial accounts, illegally obtain information for private profit, distribute pornographic literature, promote drug transactions, and entice unsuspecting citizens into illegal activities. Cyberspace is difficult to police, though many traps have been set and many predators have been caught, arrested, and charged. With little precedence for such cases, courts are ill prepared to handle them. In many cases, legal procedures are not clear. Identification of exactly what is considered criminal is difficult. Shortly after arrest and conviction, most crackers and criminal hackers are given small fines, suspended sentences, and the freedom to continue their clandestine illegal and illicit operations. Much needs to be done in the area of secure communications systems to provide adequate protection against the criminal hacker before the enormous benefits of the superhighway and the Internet are further jeopardized. The situation is similar to that of the early days of the telephone when everyone was concerned about wiretapping. However, the current situation goes beyond wiretapping. It includes illegal tapping of information resources. Further enhancement of secure systems and firewalls is becoming more and more essential.

MARTIN H. WEIK, D.Sc.

Introduction

Data, Information, and Communications

Simply stated, data are the representation of information, information is the meaning assigned to data, and communications is the transfer of data, and hence the transfer of the information, from one or more persons, places, or machines to one or more other persons, places, or machines. Among the desires are that (a) the data adequately and accurately represent the information, (b) the sender and receiver agree on the meaning assigned to the data, and (c) nothing is lost, changed, or added during the transfer. If all these criteria are met, effective communication can occur. Thus, strictly speaking, communications systems primarily accept, store, transport, and furnish data.

The New Edition

This *Communications Standard Dictionary, Third Edition,* is written with improved clarity, style, format, accuracy, and precision over prior editions. Every definition has been recast to (a) reflect these qualities, (b) include the latest trends in communications, computer, data processing, and control systems hardware, software, and firmware, (c) obtain closer alignment with international, national, federal, military, technical society, and industrial standards, (d) improve internal consistency among entries, and (e) maintain compatibility with the thousands of terms, definitions, cross references, illustrations, examples, and explanations that have been added to update the Second Edition.

Scope

Every effort has been made to ensure that this dictionary covers all aspects of (a) communications applications, such as computer networking, data exchange, database management, telemetering, multicasting, broadcasting, and point-to-point communications, (b) all communications media, such as telephone, telegraph, radio, television, facsimile, wirephoto, radar, spread spectrum, navigation, military, and related systems and components, (c) all propagation media, i.e., transmission media, such as open wire, metallic cable, fiber optic cable, satellite, microwave relay, coaxial cable, and visual systems, (d) the complete communications system life cycle, such as design, development, testing, fabrication, installation, operation, application, maintenance, and salvage of communications, computer, data processing, and control systems networks, equipment, and components, and (e) most of the codes, protocols, procedures, computer programs, and arrangements associated with all of these aspects. Specifically, this dictionary covers, among many others, the following topics:

acoustic communications
aeronautical communications
active systems
air-air communications
air-ground communications
air-sea communications
analog communications
antennas
battlefield surveillance
broadcasting
buffering
carriers
checking systems
circuitry
coding schemes
command guidance
communications devices
communications engineering
communications security
communications systems
communications traffic
communication theory
compaction
computer engineering
computer programming
computing systems
control systems
cybernetics
cyberspace
data conversion
data integrity
data processing
data transmission
database management
detection
digital systems
direction finding
display devices
display systems
distortion
documentation
electromagnetic theory
electronic warfare communications
emanation security
enciphering
encryption
error control
facsimile
fiber optics
fixed communications
gating
information management
information retrieval
information storage
information theory
information systems
interactive systems
interference
jamming
key
keying
layering
lightwave communications
local area networks
message switching
microwave communications
military communications
mobile communications
modulation
multiplexing
navigation communications
networking
noise
office machines
open systems
optical communications
optics
packet switching
passive systems
personal computers
polarization
program management
protocols
pulsed systems
radar
radiation control
radio
radionavigation
receivers
reception
repeaters
satellite communications
secure systems
sensing systems
sensors
servers
signal processing
signaling
software engineering
space communications

spread spectrum systems
strategic communications
surveillance
switching systems
synchronization
tactical communications
telegraphy
telephony
timing
transmission
transmission security
transmitters
video telephony
visual communications
wave propagation communications
word processing

Application

The terms defined and referenced in this dictionary are those that (a) are written and spoken by designers, developers, manufacturers, vendors, users, managers, administrators, operators, and maintainers of communications, computer, data processing, and control systems and components, (b) are used by educators, students, members of standards organizations, and government personnel, (c) are used by individuals to communicate with each other, particularly in written and spoken form in seminars, letters, conversations, calls, and messages, and (d) are used by communications systems personnel, such as administration, operation, and maintenance personnel, to effect the communications process itself, particularly in the areas of transmission control, routing, protocols, networking, and repair.

Sources

Sources of material for this dictionary are many. This edition subsumes the entire Second Edition, which, of course, subsumed the First Edition. Further, definitions are based on international, national, federal, military, industrial, communications carrier, and technical society standards. In order to overcome delays in standards development, the latest literature in communications, computer, data processing, and control systems was screened for new terms and definitions, until diminishing returns precluded further search for new terms. Many definitions from various sources had to be edited to ensure technical accuracy, precise wording, consistency in format, and compatibility with existing standards. In thousands of cases, examples, explanations, and clarifications were added. In about half of the entries, several definitions for the same term had to be included, not only because usage of the same term is different in different fields, but also because different sources used a different approach to describe the concept represented by the term, such as the use of a verbal definition in comparison to a mathematical expression or the use of a description of functions of an entity rather than its composition. The reader has the benefit of all of these variations.

Organization of Entries

Ordering

Entries are arranged in natural spoken English alphabetical word order, letter by letter. Punctuation and special signs and symbols, such as spaces, hyphens, slashes, ampersands, parentheses, superscripts, and subscripts, are ignored in ordering. No attempt is made to embed Greek letters, Roman letters, and Arabic numerals. Numerals follow Z, and Greek letters are at the very end. Except for a few isolated cases, such as *laser* and *radar*, acronyms and abbreviations are defined at the spelled-out version. Cross references to the spelled-out version are at the acronym or abbreviation.

Cross Referencing

Every significant word in a multiple-word entry is also entered in the main listing along with a reference to the fully expressed entry. For example, **near-field diffraction pattern** is defined. Following the definition is *Synonym* **Fresnel diffraction pattern.** There is an entry **diffraction pattern** with a definition. Following the definition is the entry *See* **near-field diffraction pattern,** along with the names of other types of diffraction patterns. Also, there is an entry **pattern:** *See* **near-field diffraction pattern,** along with the names of other types of patterns. This makes all significant words in a multiword entry accessible. There also is an entry **Fresnel diffraction pattern:** *Synonym* **near-field diffraction pattern.** The definition will always be found at the preferred term as recommended by standards bodies. The practice of arbitrarily cross referencing all significant words in a multiword entry creates a useful tool. Quite often readers will seek a multiword entry without knowing the exact words in the entry, but the correct entry will be recognized when seen. For example, the name and the definition of a specific type of network are being sought. Under the entry **network** all the types of networks are listed. Scanning the list usually brings to mind the sought name, from which the definition can be found.

Format

Usually each entry consists of a fixed set of parts presented in the same sequence. The parts are (a) the term being defined, i.e., the term name in boldface followed by a colon, (b) a defining phrase, which is the generally accepted definition based on current standards, (c) notes, usually devoted to units, examples, equations, and explanations, (d) synonyms, and (e) cross references to other entries, following the words *See, See also, Refer to* **Fig.**, and *Refer to* **Table.** *See* is used to refer the reader to various types of the entity being defined. For example, the definition of **modulation** is followed by *See* **amplitude modulation, frequency modulation, pulse code modulation.** *See also* is used to refer the reader to the definitions of closely related or contrasting concepts. For example, the definition of **optical fiber** is followed by *See also* **cladding, core.** *Refer to* **Fig.** identifies figures that illustrate the term being defined. The numerals following the figure numbers are the page numbers on which the referenced figure may be found. If there is no numeral, the referenced figure is on the same page or is one page away. *Refer to* **Table** identifies tables in Appendix B that further explain the term being defined as well as indicate its relationship with other terms.

Restructuring of Terms

Except where absolutely required by published standards, prepositional phrases are reduced to modified nouns. For example, **index of refraction** is reduced to **refractive index, angle of incidence** to **incidence angle, angle of reflection** to **reflection angle,** and **period of performance measurement** to **performance measurement period.** The reader is also cautioned in the use of abbreviations of international terms and organizations. For example, **ISO** is the official abbreviation for the **International Organization for Standardization,** and **BIH** is the official abbreviation for the **International Time Bureau.** Most of these variances occur because international standards bodies adopted some French titles, thus making the letters, their sequencing, or both, different in English. However, in the vast majority of cases, most international standards bodies draft and approve vocabulary definitions in American English. These may then be translated into other languages.

Optical *versus* *Fiber Optic* *versus* *Optical Fiber*

The standards organizations have taken different positions in regard to these three terms. One international standards organization defines **optical cable** as consisting of optical fibers and other components surrounded by a jacket. Other standards organizations define **fiber optic cable** in the same way that **optical cable** is defined. Then, there are different words for the same thing, such as **optical coupler, optical fiber coupler, fiber optic coupler,** and **fiber coupler,** further contributing to the difficulty of preparing definitions and indexing of terms. One international standards body uses **optical regenerator section** because optical pulses are generated and dispatched by the section. Another standards body uses **fiber optic section** because optical fibers are used instead of wires or coaxial cables. This author has adopted a kind of compromise. Optical is generic. Thus, "optical system" is not synonymous with "fiber optic system," even in the restricted area of communications. A system of mirrors and lenses, such as is found in a microscope or a telescope, is an optical system rather than a fiber optic system. Thus, a fiber optic system is an optical system, but not vice versa. "Fiber optic" is more specific than "optical." If the optical fiber itself is addressed, "optical fiber" is used, as in "optical fiber core" and "optical fiber cladding." Thus, "fiber optic cladding" would be inappropriate. If an optical fiber is combined with other components to create a device, the device is a fiber optic device, such as a fiber optic cable, fiber optic coupler, and fiber optic data link. This is consistent with most standards. Finally, "optical" is always used to modify terms that are not hardware, such as **optical signal, optical pulse,** and **optical power.** *If a term being sought is not found under one of these terms, the reader should check the others.* In this dictionary, most cases of the various forms have been cross referenced.

Communication *versus* *Communications*

Among the vocabulary standards and the technical literature, the author was unable to find consistency in the use of "s" on the end of "communications" or "telecommunications," particularly when used as a modifier. For example, in the same standard, one can find "communications system" and "communication system" used in the same sense, and often in the same entry of a given standard. Even the names of bodies responsible for communications standards are not consistent in this regard. For example, there is the *International Telecommunication Union* and the *National Communications System.* Except in rare instances, such as **communication theory,** the author has elected to use the more prevalent final "s" when **communications** and **telecommunications** are used as modifiers. If uniformity in this regard is NOT maintained, alphabetization becomes erratic. Terms containing "communications" become alphabetically widely separated from those containing "communication." Look-up becomes difficult. The reader is placed in the position of having to guess which word to search for and then possibly having to search for both in two different areas of the alphabetic sequencing. Cross referencing all of these would create excessive redundancy.

Earth *versus* *earth*

In accordance with federal, national, and international standards, references to the planets are made with uppercase initial letters, such as Earth, Mars, and Venus. There are numerous occurrences of these terms in the entries concerning satellite and deep space communications. In terrestrial communications systems, reference is often made to a ground connection directly to the soil, i.e., to the earth. Thus, earth-moving equipment does not imply that the Earth is being moved.

Encyclopedia

This dictionary may be used as an encyclopedia and therefore as a handy reference manual. The *See* and *See also* references serve as linkages to related entries, enabling the reader to develop complete concepts in specialized areas without having to guess which concepts relate to the definition being sought. Finally, technical terms used in definitions are also defined. There is little value in a dictionary in which the definitions contain complex, unknown, and undefined terms that are more complex than the term being defined.

Acknowledgments

The author extends his appreciation and gratitude to the various persons he has worked with during the years since the printing of the Second Edition. Much of their expertise has found its way into the pages of this edition. In particular, special thanks go to Mr. James H. Davis, Fiber Optics Program Office, the Naval Sea Systems Command; Mr. Lonnie Benson, Dynamic Systems, Incorporated; Mr. Thomas F. Blizzard, the Logicon Eagle Technology Corporation; and Ms. Evelyn Gray, Institute for Telecommunication Sciences, the U.S. Department of Commerce.

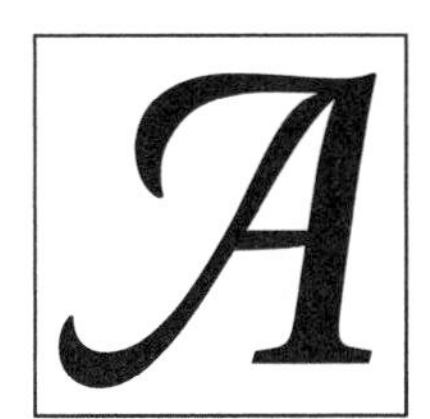

a: atto (10^{-18}).

A: ampere.

A AND NOT B gate: A two-input, binary, logic coincidence circuit or device that performs the logic operations of A AND NOT B, i.e., if A is a statement and B is a statement, the result is true only if A is true and B is false, and for the other three combinations of A and B, the result is false. *Synonyms* **A except B gate, AND NOT gate, exclusion gate, negative A implies B gate, NOT if then gate, sine function gate, subjunction gate.** *Refer to* **Fig. A-1.** *Refer also to* **Fig. B-1 (66).**

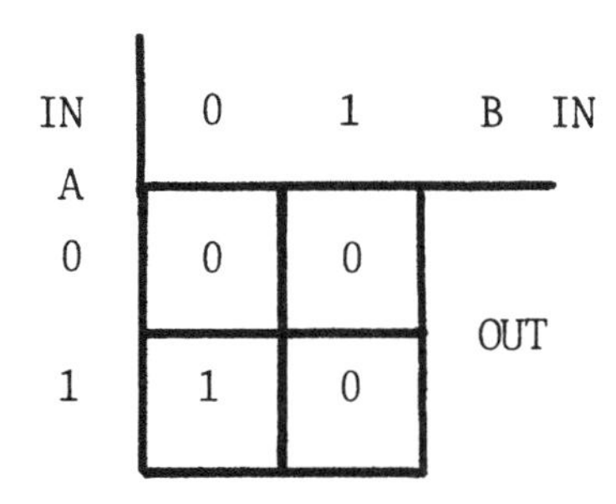

Fig. A-1. A truth table that shows the input and output digits of an **A AND NOT B gate.**

abandoned call: A call in which the call originator cancels the call after a connection has been made, but before the call is completed, such as before the call receiver goes off-hook. *See also* **access phase, call, call processing, clearing, disconnect.**

Abbe constant: A mathematical expression for determining the correction for chromatic aberration of an optical system, usually expressed by the relation $V = (n_d - 1)/(n_f - n_c)$ where n_d, n_f, and n_c are the refractive indices for light of the wavelengths of the D line of sodium and the E and C lines of hydrogen, respectively. *Note:* The Abbe constant is also considered as the refractivity/dispersion ratio. *Synonyms* **Nu value, Vee value.**

abbreviated address: An address that has fewer characters than the full address, usually for special communications and other services or for certain users. *Note:* Examples of abbreviated addresses are (a) a four-digit telephone number for a user calling another user connected to the same switching exchange and (b) message addresses that have only the addressee name and station code or number.

abbreviated address calling: Calling that enables a user to employ an address having fewer characters than the full address when initiating a call. *Note:* Communications network users may be allowed to designate a given number of abbreviated address codes. The allocation of the abbreviated address codes to a destination or group of destinations may be changed as required by means of a suitable procedure. *See* **group abbreviated address calling.**

abbreviated answer: In radiotelegraph communications, a response to a preliminary call (a) in which the call sign of the calling station is omitted and (b) that usually is used after good communications have been established.

abbreviated dialing: A network-provided service feature that permits the user to dial fewer digits to access the network than are required under the nominal numbering plan. *See also* **repertory dialer, service feature, speed calling.**

abbreviated title: *Synonym* **short title.**

abbreviation: *See* **aeronautical communications system abbreviation.**

aberration: 1. In an optical system, a systematic departure from an idealized path of light rays that form an image, causing the image to be imperfect. **2.** In physical optics, a systematic departure of a wavefront from its ideal plane or spherical form. *Note:* Common aberrations include spherical aberration, chromatic aberration, coma, distortion of image, curvature of field, and astigmatism. *See* **chromatic aberration.**

abort: 1. In data transmission, a function invoked by a primary or secondary sending station that causes the recipient to discard or ignore all bit sequences transmitted by the sender since the preceding flag sequence. **2.** In a computer system, to terminate, usually in a controlled manner, a processing activity because it is impossible or undesirable for the activity to proceed. *See also* **disengagement attempt, flag sequence, lost call, unsuccessful call.**

abrasive: A hard material, such as aluminum oxide, silicon carbide, and diamond, that (a) is in powder form, (b) is graded according to particle size, and (c) is used to shape and finish optical elements, such as lenses and optical fiber endfaces. *Note:* For finishing optical fiber endfaces, (a) the abrasive is made to adhere to a substrate of plastic film, resulting in a material similar to fine sandpaper, (b) the film is supported by a hard, flat plate, (c) the connector is supported by a fixture that holds it securely in the proper position and orientation for finishing, and (d) the grinding and polishing motion may be performed manually or by a machine. *See also* **optical fiber.**

absent user service: A user-provided service feature in which the user automatically advises all callers that the called terminal is not available.

absolute address: In communications, computer, and data processing systems, an address that directly identifies a storage location without the use of an intermediate reference, such as a base address or a relative address.

absolute cell address: In a personal computer spreadsheet formula, a cell address that has column and row specifiers that are preceded by a special symbol, such as a dollar sign, to indicate that the cell reference in the formula will always refer to the same cell in the spreadsheet, even if the formula is copied to another location on the spreadsheet.

absolute coordinate: In display systems, a coordinate, expressed in absolute coordinate data, that (a) identifies an addressable point in the display space on the display surface of a display device or image storage space and (b) indicates the displacement of the given addressable point from the origin of the particular coordinate system in which it lies. *See also* **absolute coordinate data, relative coordinate.**

absolute coordinate data: In display systems, such as computer interactive monitors, display terminals, and fiberscope faceplates, values that specify the actual coordinates in the display space on the display surface of a display device or image storage space. *Note 1:* The absolute coordinate data may be contained in a computer program, stored in a storage unit, such as a memory or a buffer with a display device, or be recorded on a hard copy document, such as a sheet of paper, usually called "graph paper" or "coordinate paper." *Note 2:* Absolute coordinate data (a) may be linear, logarithmic, or polar and (b) may be single or multidimensional. *See also* **absolute coordinate, relative coordinate data.**

absolute delay: The time between transmission and reception of a signal. *Note:* Absolute delay may be expressed in any suitable units, such as units of time or units of phase. *See also* **delay distortion, delay equalizer, delay line, phase delay.**

absolute error: 1. The algebraic difference obtained when (a) a true, specified, theoretically correct, or calculated value is subtracted from (b) an observed or measured value, both usually expressed in the same units. **2.** The absolute value of the difference between (a) a true, specified, theoretically correct, or calculated value and (b) the observed or measured value, both usually expressed in the same units. *See also* **normalized error, relative error.**

absolute gain: 1. In an antenna and for a given direction, the ratio of (a) the power that would be required at the input of an ideal isotropic radiator to (b) the power actually supplied to the given antenna so that the radiation intensity in the far field region in the given direction would be the same. *Note:* If no direction is given, the direction corresponding to maximum radiation is assumed. *Synonym* **isotropic gain. 2.** The ratio of (a) the signal level at the output of a device to (b) the signal level at the input to the device under specified operating conditions. *Note 1:* Absolute gain may be determined under various operating conditions, such as no load, full load, and small signal conditions. *Note 2:* Absolute gain is usually expressed in dB. *See also* **antenna, gain, level, loss.**

absolute loader: A loader that performs its loading function by beginning at the assembled origin.

absolute luminance threshold: The lowest limit of luminance necessary for visual perception to occur in a person with normal or average vision.

absolute luminosity curve: The plot of spectral luminous efficiency versus optical wavelength.

absolute machine code: A machine code (a) that is loaded into a computer fixed storage location at each use and (b) that may not be relocated during execution of a computer program. *See also* **relocatable machine code.**

absolute magnetic permeability: The ratio of the magnetic flux density, B, to the magnetic field strength, H, at a point in a material medium, given by the relation $\mu_{abs} = B/H$. *See also* **incremental magnetic permeability, magnetic permeability, relative magnetic permeability.**

absolute magnification: The magnification produced by a lens placed in front of a normal eye at such a distance from the eye that either (a) the rear focal point of the lens coincides with the center of rotation of the eye or (b) the front focal point of the eye coincides with the second principal point of the lens. *Note 1:* The absolute magnification is numerically equal to the distance of distinct vision divided by the equivalent focal length of the lens, with both distances expressed in the same units. *Note 2:* The object must be located close to the front focal point of the lens.

absolute order: In display systems, a display command that causes a display device to interpret the data following the order as absolute coordinate data rather than relative coordinate data. *Note:* A display command can occur in a segment, a display file, a computer program, or an instruction repertory. *See also* **relative order.**

absolute potential: 1. A voltage, i.e., a potential difference, between (a) a specified point and (b) ground, i.e., a zero potential reference level. *Note:* An electric circuit element functions with other circuit elements in accordance with potential differences across its terminals and in accordance with relative potential differences, i.e., with biases, maintained between its terminals and other terminals of the same circuit. However, for the circuit to exist and operate in its environment, the absolute potential may be significant, requiring the circuit to be insulated from its environment. **2.** The spatial line integral of all the electric field gradients from earth to a specified point.

absolute power: The real power developed, dissipated, transferred, or used in a circuit. *Note:* Absolute power must be expressed in real power units, such as watts, milliwatts, microwatts, dBm, and dBW. *See also* **reactive power, real power, volt•amperes.**

absolute refractive index: *See* **refractive index.**

absolute signal delay: The time difference between (a) the generation or the occurrence of a point on a signal waveform, such as the leading edge, at the beginning of a propagation medium, such as a waveguide, and (b) the arrival or the reception of the corresponding point of the same signal at the end of the propagation medium, i.e., the transit time from one place to another for a specified point on a signal waveform.

absolute temperature scale: *See* **Kelvin temperature scale.**

absolute unit: The actual value of a unit of measure of a quantity. *Note:* Examples of absolute units are volts, amperes, watts, meters, hours, and dBm (dB referred to 1 mW (milliwatt)).

absolute vector: In display systems with display devices that have display surfaces, such as computer monitors, fiberscope faceplates, or light-emitting diode panels, a vector with starting and ending points that are specified by vectors that originate from a designated point, usually from the point designated as the origin. *See also* **relative vector.**

absorbed dose: *See* **radiation absorbed dose.**

absorptance: *See* **spectral absorptance.**

absorption: In the transmission of electrical, electromagnetic, or acoustic signals, the conversion of the transmitted energy into other forms of energy, such as heat. *Note 1:* Absorption causes signal attenuation. *Note 2:* In an optical fiber, intrinsic absorption is caused by tails of ultraviolet and infrared bands in the propagating lightwave. Extrinsic absorption is caused by impurities and defects. *Synonyms* **material absorption, radiation absorption.** *See* **optical fiber absorption, phonon absorption.** *See also* **absorption modulation, attenuation, ionospheric absorption,**

precipitation attenuation. *Refer to* **Figs. H-1 (430), S-15 (931).**

absorption coefficient (AC): In the propagation of an electromagnetic or acoustic wave in a material, a measure of the attenuation caused by absorption of the radiant or sound energy that results from passage of the wave through the material. *Note 1:* The absorption coefficient is expressed in units of reciprocal distance. *Note 2:* The absorption coefficient is represented by b in the exponent of the absorption equation that expresses Bouger's law, given by the relation $F = F_0 e^{-bx}$ where F is the power level at the point x and F_0 is the initial value of power level, i.e., the power level at $x = 0$. *Note 3:* A typical value of b for optical fibers is 0.023/km (kilometers^{-1}), which corresponds to an attenuation rate of 0.2 dB/km, both figures for a wavelength of 1.31 μm (microns), the standard wavelength. *Note 4:* The sum of the absorption coefficient and the scattering coefficient is the attenuation coefficient.

absorption electronic countermeasure: In electronic warfare, an electronic countermeasure in which devices and materials that reduce the electromagnetic reflectivity of a target are used.

absorption factor: *See* **radiation absorption factor, solar absorption factor.**

absorption index: 1. A measure of the attenuation caused by the absorption of energy per unit of distance that occurs in an electromagnetic wave of given wavelength propagating in a material medium of given refractive index. *Note:* The value of the absorption index is given by the relation $K = b\lambda/(4\pi n)$, where K is the absorption index, b is the absorption coefficient, λ is the wavelength in vacuum, and n is the refractive index of the absorptive material. **2.** The functional relationship between the sun angle, at any latitude and local time, and the ionospheric absorption. *Note:* The sun angle is the angle between the horizontal plane and the sun at the point where the absorption index is measured. *See also* **radiation scattering, refractive index.**

absorption loss: In the propagation of an electromagnetic or acoustic wave in a material, the part of the total transmission loss caused by the dissipation, i.e., the conversion of electromagnetic or acoustic energy into other forms of energy, that results from passage of the wave through the medium. *See also* **absorption coefficient, absorption index, adjacent channel interference, attenuation, crosstalk.**

absorption modulation: At the output of a radio transmitter, amplitude modulation in which a variable impedance circuit is used to absorb carrier power in accordance with the modulating signal, i.e., the signal that is controlling the variable impedance. *See also* **absorption, amplitude modulation, carrier (cxr), modulation.**

absorption peak: In lightwave transmission media, the specific wavelength at which a particular impurity absorbs the most power, i.e., causes a maximum attenuation of propagated lightwaves. *Note:* Absorption by these impurities at other wavelengths is less than that of the absorption peak. Glass, quartz, silica, and plastics used in optical fibers, slab dielectric waveguides, optical integrated circuits (OICs), and similar media, usually display absorption peaks. Impurities that cause absorption peaks include copper, iron, nickel, chromium, manganese, and hydroxyl ions.

absorptivity: In a material propagation medium, the internal absorptance per unit of thickness, equal to unity minus the transmissivity. *See also* **absorption coefficient, transmissivity.**

abstraction: In communications, computer, and data processing systems, a concept, problem, problem description, or system that allows relevant information to be considered and nonrelevant information to be suppressed or ignored.

abstract machine: A representation of the characteristics of a machine. *Note:* The user of the abstract machine may be a person or a computer program.

abstract syntax: In open systems architecture, the specification of Application Layer data or application protocol control information by using notation rules that are independent of the encoding technique used to represent the information.

abstract syntax notation one (ASN.1): A standard, flexible, formal, precise, unambiguous notation system that (a) describes data structures for representing, encoding, transmitting, and decoding data, (b) provides a set of formal rules that is used for describing the structure of objects, (c) is independent of machine-specific encoding schemes, (d) is a formal network management Transmission Control Protocol/Internet Protocol (TCP/IP) language that uses human-readable notation, and (e) is a compact, encoded representation of the same information used in communications protocols.

ac: alternating current.

AC: access charge, alternating current.

accelerator board: A printed circuit (PC) board that has sufficient buffer storage, high speed capability, and other sophisticated high technology features that enable it significantly to speed up the operations of a computer, such as by (a) avoiding repeated references to disk storage, (b) performing operations in parallel, (c) performing calculations at higher speeds, and (d) releasing the computer to the operator while certain operations, such a printing, are being performed. *Note:* The accelerator board does not speed up the disk drive on a personal computer (PC). However, it can buffer large blocks of data, thus making the disk drive appear to be faster. *Synonym* **turboboard.**

accentuated contrast: In the transmission of images, such as in facsimile or fiber optic systems, the contrast, in a picture, drawing, or text on a document, obtained when (a) elements with a luminance less than a specified value are transmitted as nominal black, (b) elements with a luminance greater than a specified value are transmitted as nominal white, and (c) all values in between are transmitted at their respective levels.

accept: In data transmission, the condition assumed by a primary or secondary station upon receipt of a correct frame. *See also* **acknowledge character, frame, primary station, secondary station.**

acceptance: Pertaining to the condition that exists when a functional unit, such as a facility or a system, meets specified requirements, such as technical performance and security requirements.

acceptance angle: In fiber optics, half the vertex angle of that cone within which optical power may be coupled into bound modes of an optical fiber. *Note 1:* For an optical fiber, the acceptance angle is the maximum angle, measured from the longitudinal axis or centerline of the fiber to an incident ray, within which the ray will be accepted for transmission along the fiber, i.e., total (internal) reflection of the incident ray will occur for long distances within the fiber core. If the acceptance angle is exceeded, optical power in the incident ray will be coupled into leaky modes or rays, or lost by scattering, diffusion, or absorption in the cladding. For a cladded fiber, the sine of the acceptance angle is given by the square root of the difference of the squares of the refractive indices of the fiber core and the cladding, i.e., by the relation $\sin A = (n_1^2 - n_2^2)^{1/2}$ where A is the acceptance angle and n_1 and n_2 are the refractive indices of the core and the cladding, respectively. If the refractive index is a function of distance from the center of the core, then the acceptance angle at a given distance from the center is given by the relation $\sin A_r = (n_r^2 - n_2^2)^{1/2}$ where A_r is the acceptance angle at a point on the entrance face of the fiber at a distance r from the center, and n_2 is the minimum refractive index of the cladding. Quantities Sin A and sin A_r are the numerical apertures (NA). Acceptance angles and numerical apertures for optical fibers usually are given for the center of the endface of the fiber, i.e., where the refractive index, and hence the NA, is the highest. Optical power may be coupled into leaky modes at angles exceeding the acceptance angle, that is, at internal incidence angles less than the critical angle. *Note 2:* The axis of the cone is collinear with the fiber axis, the vertex of the cone is on the fiber endface, and the base of the cone faces the optical power source. *Synonyms* **acceptance one-half angle, collection angle.** *See also* **acceptance cone, bound mode, launch angle, launch numerical aperture, maximum acceptance angle, numerical aperture.** *Refer to* **Fig. A-2.** *Refer also to* **Fig. A-3 (6).**

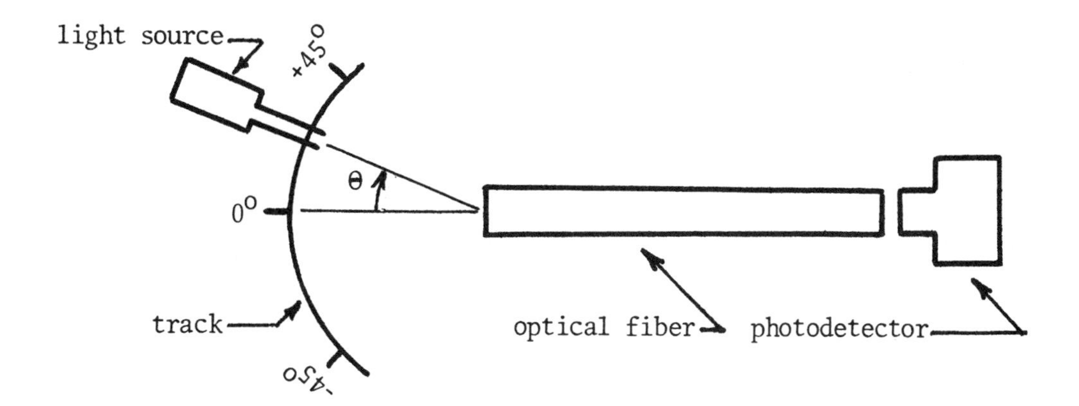

Fig. A-2. An acceptance angle plotter used to obtain the **acceptance cone** of an optical fiber. In addition to setting the source angles and plotting the optical output power corresponding to the angles, the fiber can be rotated to obtain the three-dimensional cone.

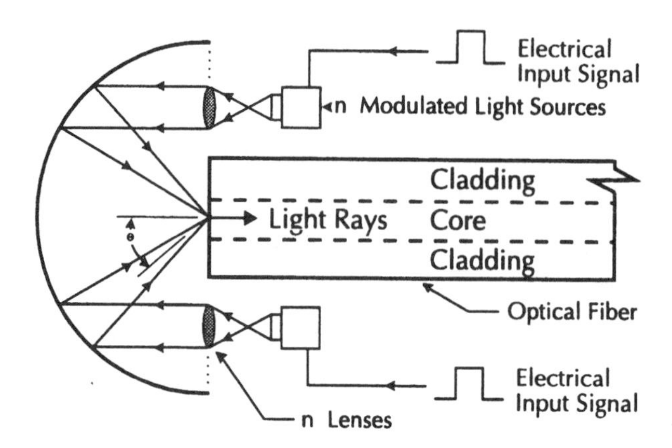

Fig. A-3. Within the acceptance cone of an optical fiber, using the **acceptance angle** and a focusing lens to time division multiplex or wavelength division multiplex each of *n* modulated input signals into a single optical fiber.

acceptance angle plotter: A device capable of varying the incidence angle of a narrow beam that is incident upon a surface, such as the endface of an optical fiber, and measuring the intensity of the transmitted light, i.e., the light coupled into the surface, such as that of the optical fiber, for each angular position of the source relative to the incident surface, i.e., for each incidence angle. *Refer to* **Fig. A-2 (5).**

acceptance cone: In fiber optics, the cone within which optical power may be coupled into the bound modes of an optical waveguide. *Note:* The acceptance cone is derived by rotating the acceptance angle, i.e., the maximum angle within which light will be coupled into a bound mode, about the fiber axis. The acceptance cone for a round optical fiber is a solid angle with an included apex angle that is twice the acceptance angle. Rays of light that are within the acceptance cone can be coupled into the end of an optical fiber and be totally internally reflected as it propagates along the core. An acceptance cone apex angle typically is 40°. For noncircular waveguides, the acceptance cone transverse cross section is not circular, but is similar to the cross section of the fiber. *See also* **acceptance angle, bound mode, coupling, fiber optics, numerical aperture.** *Refer to* **Fig. A-3.** *Refer also to* **Fig. A-2 (5).**

acceptance criterion: A criterion that a product must meet to successfully complete a specified test. *Note:* The criterion may take one of many forms, such as performance in a given environment, size, color, shape, elasticity, strength, and weight.

acceptance one-half angle: *Synonym* **acceptance angle.**

acceptance pattern: 1. For an antenna, a distribution plot of the off-axis power relative to the on-axis power as a function of angle or position with respect to the antenna. *Note:* The acceptance pattern is the equivalent of a horizontal antenna pattern. *See also* **antenna lobe, directive gain, directivity pattern. 2.** For an optical fiber or fiber bundle, a curve of total transmitted power plotted against the launch angle. *See also* **acceptance cone.**

acceptance test: 1. A test of a system or a functional unit, usually performed by the user on the user premises after installation, with the participation of the vendor to ensure that the contractual requirements are met. **2.** In communications, to operate and test a communications system, subsystem, or component, to ensure that the specified performance characteristics have been met. *See also* **performance parameter, quality assurance, reliability, test and validation.**

acceptance testing: Operating and testing of a communications system, subsystem, or component, to ensure that the specified performance characteristics have been met. *See also* **performance parameter, quality assurance, reliability, test and validation.**

acceptance trial: In military systems, a trial carried out by representatives of the eventual military users of the weapon or equipment to determine if the specified performance characteristics have been met.

accepted interference: Interference that (a) is at a higher level than that defined as permissible interference and (b) has been agreed upon by two or more administrations without prejudice to other administrations. *See also* **interference, permissible interference.**

accepted signal: *See* **call not accepted signal.**

acceptor: In an intrinsic semiconducting material, such as germanium or silicon, a dopant, such as gallium, that has nearly the same electronic bonding structure as the intrinsic material, but with one fewer electron among its valence electrons than the number required to complete the intrinsic bonding structural pattern. *Note:* The structural pattern has a "space," i.e., a "hole," for one electron for each dopant atom in the structure. The dopant atoms are relatively few and far apart and hence do not interfere with the electrical conductivity of the intrinsic material. An electron from a neighboring intrinsic material atom can fill the hole at the dopant site, leaving a hole from whence it came. Thus, the hole can appear to move or wander about, although with less mobility than the electrons that are free and that are present in excess to donor atoms. *See also* **donor, hole.**

access: 1. The ability and the means necessary to store data in, retrieve data from, or communicate with a system. **2.** To obtain the use of a resource. **3.** In communications security (COMSEC), the capability and the opportunity to gain detailed knowledge of, or to alter, information or material. *See* **carrier sense multiple access, controlled access, customer access, demand assignment multiple access, dial dictation access, direct access, dual access, frequency division multiple access, multiple access, remote access, serial access, time division multiple access.** *See also* **attendant access loop, disengagement attempt, dual use access line, maximum access time, medium access control sublayer, multiple homing, service access point, special grade access line.**

access and cross connect system: *See* **digital access and cross connect system.**

access arrangement: *See* **data access arrangement.**

access attempt: In a telecommunications system, an attempt (a) to enable initiation of user information transfer, i.e., an attempt to accomplish an information transfer transaction, (b) that begins with an issuance of an access request by an access originator, and (c) that ends either in access or in failure to obtain access, i.e., in access failure. *See also* **abandoned call, access request, access time, call receiver camp on, call originator camp on, interface.**

access barred signal: In a communications system, a signal sent in the backward direction indicating that a call will not be completed because of a call originator or a call receiver facility requirement. *Note:* An access barred signal may occur for many reasons, such as the failure of a closed user group validation check or an incoming calls barred facility. *See also* **backward direction.**

access card: *See* **free-access card, secure-access card.**

access category: In an automated information system, a class to which a user, such as a person, program, process, or equipment, may be assigned, based on the resources each user is authorized to use. *See also* **classmark, user.**

access charge (AC): 1. A charge that (a) is made by a local exchange carrier, i.e., local operating company, for use of its local exchange facilities, such as for the origination or the termination of calls to or from a distant exchange via an interexchange carrier and (b) may be (i) paid directly by local end users, (ii) paid by interexchange carriers, or (iii) shared by both. **2.** A surcharge levied in accordance with the *Code of Federal Regulations*, Title 47, part 69, on each line or circuit that can access, or be accessed by, a public exchange network. *See also* **tariff.**

access code (AC): The preliminary digits that a user must dial to be connected to a particular outgoing trunk group or line. *See also* **code, NXX code.**

access contention: In Integrated Services Digital Network (ISDN) applications, *synonymous with* **contention.** *See* **contention.**

access control: 1. A network-provided service feature that allows or denies the use of the components of a communications system. **2.** The defining or the restricting of the rights of individuals or application programs to obtain data from, or place data into, a storage device. **3.** The definition or the restriction of the rights of individuals or application programs to obtain data from, or to place data into, a storage unit. **4.** The limiting of access to the resources of a communications, computer, data processing, control, or automated information system (AIS) to authorized users, programs, processes, or other systems. **5.** In a system, the function performed by the system resource controller that allocates system resources to satisfy user requests. *Synonym* **controlled accessibility.** *See also* **authenticate, classmark, controlled access, restricted access, service feature.**

access control key: *Synonym* **pass key.**

access control mechanism: A hardware, software, or firmware feature, operating procedure, or management procedure that (a) permits authorized access to a system, such as a communications, computer, and data processing system, (b) prevents unauthorized access to the system, and (c) is considered to have failed when unauthorized access is permitted or when authorized access is prevented.

access control message: A message that pertains to the control of access to a communications, computer, data processing, or control system, such as a user request, a resource controller response, or a request/response between resource controllers.

access control sublayer: *See* **medium access control sublayer.**

access controller: *See* **terminal access controller.**

access coupler: In fiber optics, a device placed between two fiber ends to allow signals to be extracted from, or injected into, one of the fibers. *See also* **fiber optic coupler.**

access data processing: *See* **remote access data processing.**

access denial: **1.** Failure of an access attempt caused by the issuing of a system blocking signal by a communications system that does not include provision for call originator camp on. **2.** To exceed the maximum access time and nominal system access time fraction during an access attempt. *Synonym* **system blocking.**

access denial probability: *Synonym* **access denial ratio.**

access denial ratio: The ratio of (a) the number of access attempts that result in access denial, i.e., system blocking, to (b) the number of access attempts, both counted during a measurement period. *Note 1:* The access denial ratio is 1 minus the access success ratio. *Note 2:* The access denial ratio may be used as an access denial probability. *Synonym* **access denial probability.**

access denial time: The time between (a) the instant an access attempt starts and (b) the instant an access failure signal is received. *Note:* Access denial time is measured only on access attempts that result in access denial.

access digit: In automatic telephone direct outward dialing, a digit that (a) enables access to outside a local switchboard, local exchange, or local central office (C.O.), (b) is dialed before the long distance number, (c) usually is dialed before an area code, and (d) is often a 1 or a 9.

access failure: In a communications system, an unsuccessful access attempt that results in termination of the access attempt in any manner other than initiation of user information transfer between the intended source and destination (sink) within the specified maximum access denial time. *Note:* Access failure can be the result of access denial, user blocking, or incorrect access.

access line: A transmission path between user terminal equipment and a switching center. *See also* **line, local exchange loop, loop.**

access list: In communications, computer, and data processing systems, a catalog of users and the access categories to which each is assigned. *See also* **access category.**

access loop: *See* **attendant access loop.**

access memory: *See* **random access memory.**

access memory disk: *See* **random access memory disk.**

access node: In packet switching, the switching point for customer traffic to and from a communications network. *Note:* Protocol conversion may occur at the access node. *See also* **originating user, source user, user.**

access originator: In a telecommunications system, the functional unit that initiates a particular access attempt. *Note:* An access attempt can be initiated by a source user, a destination user, or the telecommunications system. *See also* **originating user, source user, user.**

access path: The steps required to obtain the use of a system or device. *Note:* Examples of access paths are (a) the operations required of a database management system to obtain access to a database and (b) the sequence of steps required to reach a file.

access period: The period during which access rights prevail.

access phase: In an information transfer transaction, the phase during which an access attempt is made. *Note:* The access phase is the first phase of an information transfer transaction. *See also* **access request, disengagement phase, information transfer phase, information transfer transaction, phase, successful disengagement.**

access plan: *See* **demand assignment access plan, preassignment access plan.**

access point: **1.** A junction point that (a) is in a dedicated outside plant, (b) is usually a semipermanent splice point at a junction between a branch feeder cable and distribution cables, and (c) is a point where connections may be made for testing or using particular communications circuits. **2.** The point at which a user interfaces with a circuit or a network. *See also* **connection, demarcation point, outside plant, point of presence.**

Access Procedure B: *See* **Link Access Procedure B.**

Access Procedure D: *See* **Link Access Procedure D.**

access provider: *See* **competitive access provider.**

access request: A control message issued by an access originator for the purpose of initiating an access attempt. *See also* **access attempt, access phase.**

access rights terminal: *See* **multiple access rights terminal.**

access service: *See* **restricted access service.**

access success: In a communications system, the termination of an access attempt in such a manner that an information transfer transaction can occur, i.e., user information can be transferred between the call originator and the call receiver, i.e., between the message originator and the addressee, within the specified maximum access time. *See also* **access attempt, access time, access failure, information transfer transaction.**

access success ratio: The ratio of (a) the number of access successes to (b) the total number of access attempts, both counted during a specific performance measurement period. *Note 1:* The access success ratio is 1 minus the access denial ratio. *Note 2:* The access success ratio may be used as an access success probability. *See also* **performance measurement period.**

access system: In communications, computer, and data processing systems, a program that (a) allows an operator to call up different parts of the program package and (b) usually allows functions to be selected from menus in the same way as other commands. *Note:* An example of an access system is the program supplied with the Lotus 1-2-3 program package that (a) allows the computer to shift between Lotus 1-2-3 and Printgraph and (b) provides access to various support functions.

access time: **1.** In a telecommunications system, the time between (a) the instant an access attempt is started and (b) the instant successful access occurs. *Note:* Access time can be measured only on access attempts that result in access. **2.** In a computer, the time between (a) the instant at which an instruction control unit initiates a call for data and (b) the instant at which delivery of the data is completed. **3.** The time between (a) the instant at which storage of data is requested and (b) the instant at which storage is started. **4.** In magnetic disk devices, the time for the access arm to reach the desired track and the delay for the rotation of the disk to bring the required sector under the read-write mechanism. *See* **maximum access time.** *See also* **access attempt, call setup time.**

access unit: *See* **medium access unit.**

access unit interface (AUI): In a local area network (LAN), the interface between the medium access unit (MAU) and the data terminal equipment (DTE) within a data station. *Synonym* **attachment unit interface.**

access with collision avoidance: *See* **carrier sense multiple access with collision avoidance.**

access with collision detection: *See* **carrier sense multiple access with collision detection.**

accommodation: A function of the human eye whereby the eye's total refractive power is varied in order for the eye to see clearly objects at different distances. *See also* **light adaptation.**

accommodation limit: The distance from an observer to the nearest point, or to the farthest point, at which an image of an object can be clearly focused on the retina of the eye of the observer. *Note:* Accommodation limits usually are (a) between 4 in. (inches) and 5 in. and (b) infinity.

accountability: In communications system security, the quality or the state that enables violations or attempted violations of system security to be traced to specific persons who then may be held responsible.

accounting: *See* **automatic message accounting, centralized automatic message accounting.**

accounting legend: In communications system security, a number that (a) is assigned to items listed on accounting reports, (b) is assigned by the producing organization, and (c) usually identifies the minimum accounting controls required for the item.

accounting management: In network management, the set of functions that (a) enables network service use to be measured and the costs for such use to be determined and (b) includes all the resources consumed, the facilities used to collect accounting data, the facilities used to set billing parameters for the services used by customers, maintenance of the databases used for billing purposes, and the preparation of billing reports on resource usage.

accounting symbol: In a communications system, a combination of letters used in message headings to identify the organization that is financially responsible for the message.

accumulator: 1. A register in which one operand can be stored and subsequently be replaced by the result of the execution of an instruction. **2.** A storage register. **3.** A storage battery (U.K.). *See also* **buffer, register.**

accuracy: 1. The degree of conformity of (a) a measured or calculated value to (b) the actual or specified value. **2.** A quality of that which is free of error. **3.** A qualitative assessment of freedom from error, a high assessment corresponding to a small error. *See also* **precision.**

accuracy control character: A control character appended to data to indicate an error-related characteristic, such as whether the data contain errors, are to be ignored, cannot be represented, or cannot be processed by a particular device. *Synonym* **error control character.**

ACD: automatic call distributor.

ac-dc ringing: Telephone ringing that makes use of both ac and dc voltages and currents. *Note:* An alternating current may be used to operate a ringer and direct current to aid the relay action that stops the ringing when the call receiver telephone is answered, i.e., when the telephone goes off-hook. *See also* **ringdown.**

achromat: A compound lens corrected to have the same focal length for at least two different wavelengths of light passing through the lens. *See also* **achromatic, achromatic lens.**

achromatic: The quality of being free from color or hue, such as free from chromatic aberration. *See also* **achromat.**

achromatic lens: A lens, consisting of two or more elements or parts, usually made of crown and flint glass, that has been corrected so that light of at least two selected wavelengths is focused at a single point on the optical axis. *See also* **achromat.**

ACK: acknowledge character.

ACK/NAK: acknowledge/negative acknowledge.

acknowledge character (ACK): A transmission control character that (a) is transmitted by the receiving station as an affirmative response to the sending station and (b) may also be used as an accuracy control character. *See* **negative acknowledge character.** *See also* **accept, compelled signaling, control character, delivery confirmation.**

acknowledgment: 1. A response sent by a receiver to indicate successful receipt of a transmission. *Note:* An example of an acknowledgment is a protocol data unit, or element thereof, between peer entities to indicate the status of data units that have been successfully received. **2.** A message from a message addressee informing the message originator that the originator message has been received and understood.

acknowledgment delay period: *Synonym* **sliding window.**

acknowledgment signal: A signal that is (a) generated at the receiving end of a communications circuit to confirm the receipt of signaling digits from the sending equipment and (b) sent back to the sending equipment. *See* **circuit released acknowledgment signal.**

A condition: In start-stop teletypewriter systems, the significant condition of the signal element that immediately precedes a character signal or a block signal and prepares the receiving equipment for the reception of the code elements. *See also* **start signal.**

acoustic alarm: *Synonym* **audio alarm.**

acoustic conduction: The conduction of sound in which (a) longitudinal compressional waves in an elastic material medium are propagated through the material, and (b) energy exchanges with phonons affect electrical conductive, magnetic, and quantum electronic properties of the material medium.

acoustic coupler: 1. A device that couples electrical signals, by acoustic means, usually into and out of a telephone instrument. **2.** A terminal device used to link data terminals and radio receivers with the telephone network. *Note:* The link is achieved through acoustic (sound) signals rather than through direct electrical connection. *See also* **coupling, modem.**

acoustic delay line: A delay line in which (a) operation is based on the transmission of sound waves in a material medium, such as a column of mercury, ferrite rods, or carbon rods, and (b) digital data may be serially stored between transmitter and receiver transducers by means of continuous recirculation.

acoustic noise: Noise that (a) is in the audio frequency range, (b) usually is an undesirable disturbance, and (c) usually is audible. *See also* **ambient noise level, interference, noise.**

acoustics: The branch of science and technology that is devoted to the production, transmission, control, processing, transformation, reception, and effects of sound, particularly as vibration, pressure, or elastic waves and shock phenomena in material media. *See also* **phonon.**

acoustic sensor: *See* **optical fiber acoustic sensor.**

acoustic storage: A device that stores data in the form of sound waves propagating in acoustic delay lines.

acoustic transducer. *See* **optoacoustic transducer.**

acoustic wave: A longitudinal wave that (a) consists of a sequence of pressure pulses or elastic displacements of the material, whether gas, liquid, or solid, in which the wave propagates, (b) in gases consists of a sequence of compressions (dense gas) and rarefactions (less dense gas) that travel through the gas, (c) in liquids consists of a sequence of combined elastic deformation and compression waves that travel through the liquid, and (d) in solids consists of a sequence of compression and expansion waves that travel through the solid. *Note 1:* The speed of an acoustic wave in a material medium is determined by the temperature, pressure, and elastic properties of the medium. In air, acoustic waves propagate at 332 $m \cdot s^{-1}$ (meters per second) at 0°C at sea level. In air, acoustic wave speed increases about 0.6 $m \cdot s^{-1}$ (2 $ft \cdot s^{-1}$) for each kelvin above 0°C. *Note 2:* Acoustic waves audible to the normal human ear are called sound waves. *Note 3:* Acoustic waves may resonate in closed chambers if the length of the chamber is a multiple of the wavelength. *See also* **sound wave.**

acoustooptic: Pertaining to the interaction of optical and acoustic waves. *See also* **acoustooptic effect.**

acoustooptic effect: A variation of the refractive index of a material caused by acoustic energy, such as the energy in a sound wave or shock pulse. *Note 1:* The acoustooptic effect is used in devices that modulate or deflect light. *Note 2:* The changes in refractive index are also produced in diffraction gratings or phase patterns produced in a propagation medium in which a lightwave is propagating when the medium is subjected to a sound wave that causes photoelastic changes to occur in the material composing the propagation medium. The acoustic waves may be created by a force developed by an impinging sound wave, such as by the piezoelectric effect or by magnetostriction. The acoustooptic effect can be used to modulate a light beam in a material because of the changes that occur in light velocities, reflection and transmission coefficients, acceptance angles, critical angles, and transmission modes resulting from changes in the refractive index caused by the acoustic wave. *See also* **modulation.**

acoustooptic modulator: A modulator that uses the acoustooptic effect to modulate a lightwave carrier.

acoustooptics: The branch of science and technology devoted to interactions between acoustic waves and optical waves, such as lightwaves, in a material medium. *Note:* Acoustic waves can be made to modulate, deflect, and focus lightwaves by causing a variation in the refractive index of the material in which the lightwaves are propagating. *See also* **fiber optics.**

acoustooptic transducer: A device that converts a sound tone into an audio frequency modulated lightwave by causing a crystal to vibrate at the audio modulation frequency of the lightwave. *See also:* **modulation, optoacoustic transducer.**

acquisition: 1. In satellite communications, the locking of tracking equipment on a signal from a communications satellite. **2.** The achieving of synchronization. **3.** In servo systems, the entering of the boundary conditions that will allow the loop to capture the signal and achieve lock-on. *See* **sweep acquisition.** *See also* **phase locked loop.**

acquisition and logistics support: *See* **joint computer-aided acquisition and logistics support system.**

acquisition and tracking orderwire (ATOW): A downlink orderwire that provides a terminal with information regarding uplink satellite acquisition and synchronization status. *See also* **downlink, uplink.**

acquisition radar: Radar equipment that (a) is used to detect and locate targets to be tracked by the tracking radar, (b) scans volumes of space, (c) maintains continuous surveillance of a sector or an area, (d) obtains position data, such as height, azimuth, and direction of moving objects, such as targets and aircraft, and (e) passes these data to the tracking radar. *Synonym* **surveillance radar.** *See also* **tracking radar.**

acquisition time: 1. In a communications system, the time required to attain synchronism. *See also* **synchronization. 2.** In satellite communications control, the time required to lock tracking equipment on a signal from a communications satellite. **3.** In a device that emits a signal in response to an input signal, such as a fiber optic transmitter or receiver, the time between (a) the instant of application of the leading edge of an input signal and (b) the stabilization of the corresponding output signal. *See also* **satellite.**

actinometry: The science of measurement of radiant energy, particularly that of the Sun.

action: In quantum mechanics, the product of the total energy, E, in a stream of photons and the time during which the flow occurs, expressed by the relation $E = h\sum f_i n_i t_i$ where f_i is the ith frequency, n_i is the number of photons of the ith frequency, t_i is the time duration of the ith frequency summed over all the frequencies, photons, and time durations of each in a given light beam or beam pulse, i.e., from i=1 to i=m, and h is Planck's constant. *See also* **electrooptics, magnetooptics, optooptics, photonics.**

action addressee: The organization, activity, or person required by a message originator to take the action specified in the message. *See also* **exempted addressee, information addressee.**

action office: The office or the action addressee required to take the action specified in a message.

activated chemical vapor deposition process: *See* **plasma activated chemical vapor deposition (PACVD) process.**

active: In communications, computer, and data processing systems, pertaining to a state of readiness to perform a function, usually pertaining to a connected, energized, loaded, or ready to run system, component, or module. *See* **mark active, optically active, space active.**

active atom: *See* **radioactive atom.**

active connector: *See* **optical fiber active connector.**

active detection system: A system that (a) emits electromagnetic radiation to determine the existence, location of source, nature, and type of received radiation and (b) usually consists of early warning radar equipment, height finders, acquisition radar, i.e., surveillance radar, and tracking radar.

active device: **1.** A device that (a) contains a source of energy and (b) has an output that is a function of present and past input signals that modulate the output of the energy source. **2.** In electronic circuits, a device (a) that contains an energy source, or that requires a source of energy, other than that contained in the input signals and (b) that modulates the output of the energy source in accordance with an input signal. *Note:* Examples of active devices include controlled power supplies, transmitters, operational amplifiers, repeaters, oscillators, phototransistors, lasers, optical masers, photomultipliers, and photodetectors. *See also* **passive device.**

active electronic countermeasure: In electronic warfare, an electronic countermeasure that (a) produces detectable electromagnetic radiation and (b) usually includes electronic jamming and electronic deception.

active fiber optic connector: *Synonym* **fiber optic active connector.**

active file: In communications, computer, and data processing systems, a file (a) to or from which data may be currently and readily entered or retrieved, (b) that has an expiration date that is later than the date of the database to which it belongs, and (c) that is not an archival file. *See also* **inactive file.**

active filter: A filter that requires power in addition to the power in that which is being filtered to perform its function. *See also* **passive filter.**

active homing: Pertaining to a missile to target guidance system in which (a) the missile tracks radiation that is originated by the missile itself and that is reflected by the target, (b) the missile contains the radar transmitter and receiver, and (c) the missile rides the maximum gradient of the reflected radar waves. *See also* **passive homing, semiactive homing.**

active infrared device: An infrared (IR) device that (a) contains a source of IR radiation, (b) is subject to jamming by other IR sources, such as fires, furnaces, and the Sun, and (c) is relatively immune to electronic deception countermeasures. *Note:* Examples of active infrared devices are sniperscopes, night vision surveillance devices, IR searchlights, and IR flash cameras. *See also* **passive infrared device.**

active laser medium: In a laser, the material that emits radiation or exhibits gain as the result of stimulated electronic or molecular transitions to lower energy states. *Note 1:* Radiation from a laser usually is coherent, i.e., has a high coherence degree, and results from stimulated electronic, atomic, or molecular energy transitions from higher to lower energy levels. The action is maintained by causing population inversion. *Note 2:* Examples of active laser media include certain crystals, gases, glasses, liquids, and semiconductors. *Synonym* **laser medium.** *See also* **laser, optical cavity.**

active line: In a telecommunications system, a transmission line that is currently available for use. *See also* **inactive line.**

active link: In telecommunications and telemetry systems, a link that is connected or is available for connection to the system. *See also* **inactive system.**

active material: *See* **optically active material.**

active matrix liquid crystal display (AMLCD): A liquid crystal display image source that provides display images in full color.

active network: A network that requires a source of power other than that contained in signals. *See also* **communications satellite, satellite.**

active node: In a telecommunications network, a node that is energized and connected, or available for connection, to the network. *See also* **inactive node.**

active optical countermeasure: An optical countermeasure that makes use of an emission of signals or substances that obscure vision, produce illusions, or produce psychological effects by optical means. *Note:* Examples of active optical countermeasures are the use of smoke screens, unidirectional aerosols, i.e., micron-size particle clouds, high intensity flares, and flashing lights on aircraft.

active optical device: A device (a) that operates with, or performs specific operations on, electromagnetic waves having wavelengths that are in the optical spectrum, i.e., in the optical region of the electromagnetic spectrum, usually within the visible and near visible spectra and (b) in which operations are performed with the use of input energy in addition to that contained in the waves themselves. *Note:* Examples of active optical devices are fiber optic transmitters, receivers, repeaters, switches, active multiplexers, and active demultiplexers. *See also* **passive optical device.**

active optical fiber: An optical fiber in which the active laser medium is the fiber itself, i.e., an optical fiber laser that serves as an optical fiber amplifier.

active optics: The development and the use of optical components with characteristics that are controlled during their operational use in order to modify the characteristics of lightwaves propagating within them. *Note:* Controlled lightwave characteristics include wavefront direction, wave polarization, modal power distribution, electromagnetic field strength, and the path the waves take. *See also* **fixed optics.**

active particle: *See* **radioactive particle.**

active program: 1. In communications, computer, and data processing systems, a computer program that is loaded and ready to be executed. **2.** A system development program (a) that is currently being executed and (b) in which resources, such as labor, materials, funds, and information, are being consumed. *See also* **inactive program.**

active satellite: 1. A satellite that (a) receives signals, (b) performs signal processing functions, such as amplification, regeneration, frequency translation, and link switching, (c) renders the signals suitable for transmission, and (d) transmits the signals. **2.** A satellite with a station that transmits or retransmits radiocommunications signals. *See also* **communications satellite, satellite.** *Refer to* **Fig. W-4 (1087).**

active sensor: 1. A sensor that requires input energy from a source other than that which is being sensed. **2.** In surveillance, a sensor that emits energy capable of being detected. *Note:* An example of an active sensor is a measuring instrument that generates a signal, transmits it to a target, and receives a reflected signal from the target. Information concerning the target is obtained by comparison of the received signal with the transmitted signal. **3.** In the Earth exploration satellite service or in the space research service, a measuring instrument used to obtain information by transmission and reception of radio waves. *See also* **passive sensor, sensor.**

active sonar: Sound navigation and ranging (sonar) that relies on reception of a sound wave reflected by an object, such as a target, against which a sound wave was launched. *See also* **passive sonar.**

active station: In a telecommunications network, a station that is currently able to enter or accept messages to and from the network. *See also* **inactive station.**

active time: Time that (a) is spent in the information transfer phase within the user service time interval of an information transfer transaction and (b) excludes all the time spent in the access phase, the disengagement phase, the idle state, the exit state, and any other time outside the service time interval.

active wiretapping: The unauthorized attaching of a device, such as a telephone, transmitter-receiver, playback recorder, or computer terminal, to a communications circuit for the purpose of obtaining access to data by generating false messages, falsifying control signals, or altering user messages.

activity factor: During a specified time interval, such as the busy hour, in a communications channel, the ratio of (a) the time that a signal is present in the channel in either direction to (b) the measurement interval. *Note:* The activity factor may be expressed as a fraction or in percent. *See also* **channel, erlang, station message detail recording.**

activity loading: The storing of files or records such that minimum time is consumed in searching for the files or the records. *Note:* Examples of activity loading are (a) storing most frequently sought files in the fastest-accessed area of storage and (b) minimizing the overall number of reads by storing most often needed files or records in first read locations.

activity ratio: *See* **file activity ratio.**

ACTS: advanced communications technology satellite.

actual key: A key that directly identifies the physical location of a record (a) in or on a storage medium or (b) in a storage device.

actual transfer rate: The average number of binary digits, characters, blocks, or frames transferred per unit time between two points whether accepted or not accepted as valid at the receiving end.

actuation method: In switching systems, the way in which a motive force must be applied to a switch to place it into its various states.

actuator: A device that provides the motive force that must be applied to a switch to place it into its various states.

ACU: automatic calling unit.

A/D: analog to digital.

A-D: analog to digital.

Ada®: A high level computer programming common language that (a) is the official computer programming language of the U.S. Department of Defense (DoD) for embedded computer real time applications, as defined in MIL-STD-1815, (b) is designed for diverse applications with capabilities offered by classical programming languages, such as PASCAL, (c) has capabilities found only in specialized languages, (d) is a modern algorithmic language with the usual control structures, (e) has the ability to define data types and subprograms, (f) serves the need for modularity, whereby the data types and subprograms can be packaged, and (g) is the result of a collective effort to design a common language for programming large-scale and real time defense systems. *Note:* Ada is a registered trademark of the U.S Government Ada Joint Program Office.

ADAPT: architectures design, analysis, and planning tool.

adaptability: *See* **software adaptability.**

adaptation: *See* **dark adaptation, light adaptation.**

adapter: A device that permits two parts to be joined or mated when they were not designed and constructed to be joined or mated. *See* **homing adapter, integrated communications adapter, panoramic adapter, right angle adapter, terminal adapter.**

adapter circuit: *See* **line adapter circuit.**

adaptive channel allocation: Communications channel allocation in which information-handling capacities of channels are not predetermined and are assigned on demand. *Note:* Adaptive channel allocation may be accomplished by multiplexing. *See also* **adaptive routing, channel, multiplexing, spill forward.**

adaptive communications system: A communications system, or portion thereof, that automatically uses feedback information obtained from the system itself, or from the signals carried by the system, to dynamically modify one or more of the system operational parameters to improve system performance or to resist degradation. *Note:* The modification of a system parameter may be discrete, as in hard switched diversity reception, or may be continuous, as in a predetection combining algorithm. *See also* **adaptive system, communications.**

adaptive differential pulse code modulation (ADPCM): Differential pulse code modulation in which the prediction algorithm is modified in accordance with one or more aspects of the incoming signal. *See also* **differential pulse code modulation, modulation, signal, subband adaptive differential pulse code modulation.**

adaptive equalization: Equalization that is automatically accomplished while signals that represent user or test traffic are being transmitted. *Note:* Adaptive equalization dynamically adjusts signal characteristics to compensate for changing transmission path characteristics. *See also* **equalization.**

adaptive high frequency (AHF) radio: A radio that (a) operates in the high frequency (HF) band of the electromagnetic spectrum, (b) monitors its own performance, (c) monitors the path quality through sounding or polling, (d) automatically varies operating characteristics, such as frequency, power, and data rate, (e) uses closed loop action to optimize its performance by automatically selecting frequencies or channels, and (f) usually is used for automatic link establishment (ALE) calls, such as query calls. *See also* **query call.**

adaptive maintenance: 1. In communications, computer, and data processing systems engineering, maintenance in which software, hardware, or both are changed so as to make them usable in a changed environment. **2.** To maintain a system in such a manner that its performance is not affected by changes in its physical or operating environment.

adaptive predictive coding (APC): Coding that uses (a) narrowband analog to digital conversion and (b) a one-level or multilevel sampling system in which the value of the signal at each sample time is adaptively predicted to be a linear function of the past values of the quantized signals. *Note:* Adaptive predictive coding (APC) is related to linear predictive coding (LPC) in that both use adaptive predictors. However, APC uses fewer prediction coefficients, thus requiring a higher bit rate than LPC. *See also* **code, level, linear predictive coding, signal.**

adaptive radio: A radio that (a) monitors its own performance, (b) monitors the path quality through sounding or polling, (c) varies operating characteristics, such as frequency, power, and data rate, and (d) uses closed loop action to optimize its performance by automatically selecting frequencies or channels. *See also* **adaptive system.**

adaptive routing: Routing that is automatically adjusted to compensate for network changes, such as changing traffic patterns, channel availability, and equipment failures. *Note:* The experience used for adaptation comes from the traffic being carried. *See* **dynamically adaptive routing.** *See also* **adaptive channel allocation, alternate routing, automatic route selection, directionalization, line load control, time assignment speech interpolation.**

adaptive system: 1. A system that (a) has a means of monitoring its own performance, (b) has a means of varying its own parameters, usually by closed loop action, and (c) is able to improve its own performance. **2.** A system that continues to perform the functions it was designed to perform by making adjustments to compensate for environmental changes, i.e., the system adapts to its environment. *See also* **adaptive communications system.**

adaptive technique: *See* **coherent optical adaptive technique.**

activity: In communications, an organization that performs a communications service, function, or operation. *See* **mean circuit activity, radioactivity, sun spot activity.**

ADC: analog to digital conversion, analog to digital converter.

ADCCP: Advanced Data Communications Control Procedure.

added bit: A bit delivered to the addressee in addition to intended user information bits and delivered user and system overhead bits. *Synonym* **extra bit.** *See also* **binary digit, character count and bit count integrity, deleted bit.**

added block: A block delivered to the addressee in addition to intended user information bits and delivered user and system overhead bits. *Synonym* **extra block.** *See also* **binary digit, block, deleted block.**

added block probability: *Synonym* **added block ratio.**

added block ratio: The ratio of (a) the number of added blocks to (b) the number of blocks received by a specified destination user, both counted during a measurement period. *Note:* The added block ratio may be used as an added block probability. *Synonym* **added block probability.**

added data unit: A data unit, such as a bit, byte, character, block, or other delimited bit group, delivered to the intended destination user in addition to intended user information bits and delivered overhead bits. *Synonym* **extra data unit.** *See also* **binary digit, block, character count integrity, deleted block.**

addend: A number that is to be added to another number, the augend. *Note:* The augend usually is on hand first, perhaps already in an accumulator or a register from a previous operation, or it may written first and the addend is then added to the augend. The addend usually is written under the augend. Thus, the augend is augmented by the addend to produce the sum. However, in an adder, the distinction between addend and augend usually is lost.

adder: 1. A device that produces an output that is a representation of the sum of the numbers represented by its input data. **2.** A device that produces an output that is a representation of the sum of the quantities represented by its input data. *Note:* An adder may add entities other than representations of numbers, such as voltages and currents. Analog adders are not limited to summing representations of numbers. An adder may operate on digital or analog code elements. *See* **binary adder, binary half-adder, modulo-two adder, parallel adder, quarter adder, serial adder.** *Refer to* **Figs. B-5 (77), B-6 (78).**

adder-subtracter: A device that performs addition or subtraction depending upon the received control signal. *Note 1:* The adder-subtracter may be constructed so as to yield a sum and a difference at the same time. *Note 2:* An arithmetic adder-subtracter yields arithmetic sums and differences. A logical adder-subtracter yields logical sums and differences.

add-in board: One of a large family of printed circuit (PC) boards that may be plugged into a spare connector slot in a computer, communications, or data processing system, such as a personal computer (PC) or a microprocessor, in order to expand its capability. *Note:* Examples of add-in boards are accelerator boards, communications boards, emulator boards, expanded memory boards, gateway boards, multifunction boards, and video graphic boards.

addition: *See* **logical addition.**

additive white Gaussian noise (AWGN): *Synonym* **white noise.**

add mode: In addition and subtraction operations, a mode in which the decimal marker is placed at a predetermined location with respect to the last digit entered. *Note 1:* In the United States, the decimal marker and the binary marker are each a period, called a decimal point and a binary point, respectively. In many foreign countries, the decimal marker and the binary marker are each a comma. *Note 2:* To avoid confusion, especially when expressing monetary values, documents intended for use in foreign countries should specify whether the decimal marker is a period or a comma. For example, in the United States, the value of π is approximately 3.1416, whereas in Germany it is 3,1416.

add-on conference: A network-provided service feature that allows another user to be added to an established call without attendant assistance. *Note:* A common implementation allows a call originator or a call receiver to add at least one more user. *See also* **computer conferencing, conference call, service feature.**

add-on security measure: An additional protective measure, hardware or software, adopted after a system, such as a communications, computer, or data processing system, has become operational.

address: **1.** In communications, the coded representation of the source or the destination of a message. **2.** In data processing, a character or a group of characters that identifies a data source or a destination, such as a register or a particular part of storage. **3.** An expression, usually an alphanumeric expression, that identifies a location. **4.** To refer to a device or a data item by means of an identifying label. **5.** To assign to a device or a data item a label to identify its location. **6.** The part of a selection signal that indicates the destination of a call. *See* **directed broadcast address, global address, group address, subnet address.** *See also* **area code, en bloc signaling.**

addressability: **1.** In computer graphics, the number of addressable points on a display surface or in storage. **2.** In micrographics, the number of addressable points within a specified film frame. *Note 1:* The addressability may be expressed as (a) the number of addressable points along a specified dimension, such as in Cartesian coordinates, the number of addressable horizontal points, i.e., abscissa points, times (b) the number of addressable vertical points, i.e., ordinate points. *Note 2:* Examples of expressing addressability in Cartesian coordinates are 3,000 by 4,000, 3000 x 4000, and 3K by 4K (K = 1024). An example in polar coordinates is the number of addressable points along a radius times the number of addressable angular positions, i.e., angular displacements.

addressable horizontal positions: *Synonym* **display line.**

addressable point: In computer graphics, any point in a device that can be addressed.

addressable vertical positions: *Synonym* **display column.**

address calling: *See* **group abbreviated address calling.**

address calling facility: *See* **multiaddress calling facility.**

address code: In communications, computer, and information processing systems, a group of characters that express or identify an address. *Note:* Examples of address codes are (a) cable address codes and (b) the addresses of storage locations in computer program instructions. *See* **cable address code.**

address designator: A combination of characters or pronounceable words designated for use in message headings to (a) identify an entity, such as an organization, authority, unit, or communications facility, and (b) assist in the transmission and delivery of messages. *Note:* Classes of address designators include call signs, address groups, plain language address designators, and routing indicators. *Synonym* **station designator.**

addressee: The organization, activity, or person to whom a message is directed by the message originator. *Note:* Addressees are often indicated as action, information, or exempted addressees. *See also* **action addressee, exempted addressee, information addressee.**

address field: The portion of a message header that contains the source user address and the destination user address. *Note:* In a communications network, the transmitted signal for the message consists of a message header, the user data, and a trailer. *See also* **data transmission, header, signal.**

address field extension: In data transmission, an enlargement of the address field to include more addressing information.

address format: In communications, computer, and data processing systems, the arrangement of the parts of an address, such as the arrangement of (a) the characters of an address that identify the band, channel, track, or cylinder on magnetic disks, (b) the several addresses in a computer instruction word, and (c) the parts of the address of a message.

address group: 1. In radiotelephone communications systems, the word that means that the group that follows is an address indicating group. **2.** A station or address designator that usually consists of a group of four letters assigned to represent entities, such as organizations, authorities, activities, units, and geographic locations, and primarily used for the forward and backward addressing of messages. *See* **collective address group, conjunctive address group, geographic address group.**

address group allocation: The assignment of individual address groups to specified elements of an entity, such as elements of an organization, activity, unit, or geographic location.

address indicating group (AIG): A station or address designator, used to represent a set of specific and frequently recurring combinations of action or information addresses. *Note:* The identity of the message originator may also be included in the AIG. An address group is assigned to each AIG for use as an address designator. *Synonym* **address indicator group.**

address indicating group (AIG) allocation: The assignment of address indicating groups to specified groups of elements of an entity, such as elements of an organization, activity, unit, or geographic location.

address indicator group: *Synonym* **address indicating group.**

address message: A message sent in the forward direction that contains (a) address information, (b) the signaling information required to route and connect a call to the called line, (c) class of service information, (d) information relating to user and network facilities, and (e) call originator identity or call receiver identity.

address message sequencing: In common channel signaling, a procedure for ensuring that address messages are processed in the correct order when the order in which they are received is incorrect. *See also* **queue, routing indicator.**

address multiple access: *See* **pulse address multiple access.**

address part: The part of a computer instruction word that (a) usually contains only an address or a part of an address and (b) specifies (i) the storage locations where operands are to be obtained, (ii) the storage location where the results of the instruction are to be placed, or (iii) the storage location where the next instruction is to be obtained. *See also* **address.**

address pattern: A prescribed data structure used to represent the destination of a delimited set of data, such as a block, message, or packet. *See also* **address, frame synchronization pattern.**

address resolution protocol (ARP): The Internet protocol that (a) is used in Transmission Control Protocol/Internet Protocol (TCP/IP) data networks, (b) dynamically binds a high level IP address to a low level physical hardware address, (c) retrieves information on the physical addresses of network workstations, and (d) requires a single physical network that supports hardware broadcasts.

address separator: The character that separates the different addresses in a selection signal. *See also* **character.**

address signal: A signal that represents one element of the address of an entity, such as an organization, activity, unit, geographic location, or person. *Note:* Examples of address signals are a decimal digit or an end of number indicator in an address.

address space: In communications, computer, and data processing system operations, the range of addresses available to a programmer or an operator. *See* **virtual address space.**

address trace: A record of the contents of specified registers and control storage areas of a communications, computer, or data processing system. *Note:* Entries usually are made in the record at each interrupt as an aid to servicing.

address track: A track that contains addresses that are used to locate data on other tracks of the same storage medium. *Note 1:* An example of an address track is the track on a storage medium, such as an optical disk or a magnetic disk, card, drum, or tape, that is used to store addresses that identify storage locations on other tracks. *Note 2:* Addresses may be recorded on an address track when a storage medium is formatted.

add without carry gate: *Synonym* **EXCLUSIVE OR gate.**

ADH: automatic data handling.

adjacent channel: A channel that is contiguous to another channel in the time, frequency, or spatial domain. *Note:* Examples of adjacent channels are (a) in time division multiplexing, the channel occupying the time frame immediately following or immediately preceding the time frame of a given channel, (b) in frequency division multiplexing, the channel that occupies the next frequency band immediately higher or lower in frequency than that occupied by a given channel, (c) the channel that uses an optical fiber proximate to the fiber used by another channel in a fiber optic cable, and (d) in wavelength division multiplexing in an optical fiber, an adjacent color in the wavelength spectrum.

adjacent channel interference: Extraneous power from a signal in an adjacent channel. *Note 1:* Adjacent channel interference may be caused by inadequate filtering, such as incomplete filtering of unwanted modulation products in frequency modulation (FM), improper tuning or poor frequency control, in either the reference channel or the interfering channel, or both. *Note 2:* Adjacent channel interference is distinguished from crosstalk that results from undesired capacitive, inductive, or conductive coupling. *See also* **channel, co-channel interference, interference.**

adjacent channel selectivity: The degree to which a receiver is capable of discriminating between the signals in the desired channel and the signals in an adjacent channel.

adjacent domain: In distributed data processing systems, one of two domains that are directly connected to each other by a data link with no intervening domains.

adjacent node: In distributed data processing systems, one of two nodes that are directly connected to each other by a data link with no intervening nodes. *See also* **node.**

adjunct service point (ASP): An intelligent network feature that resides at the intelligent peripheral equipment and responds to service logic interpreter requests for service processing. *See also* **intelligent network.**

adjusting: *See* **self-adjusting.**

adjust text mode: In word processing systems, a mode (a) in which text is reformatted to accommodate specified line lengths and page lengths and (b) that is used to assist the operator in adjusting line lengths automatically or manually with the aid of word processing control functions implemented with function keys, such as the F8 key in Volkswriter.

administration: 1. In network management, a group of network support functions that ensures that (a) services are performed, (b) the network is used efficiently, and (c) prescribed service quality objectives are met. **2.** Any governmental department or service responsible for discharging the obligations undertaken in the convention of the International Telecommunication Union and the *Regulations.* **3.** In international telecommunications, the governmental agency, in a given country, assigned responsibility for the implementation of telecommunications standards, regulations, recommendations, practices, and procedures. **4.** The internal management functions and resources of an organization. **5.** In military units, the management and the execution of all military matters not included in tactics and strategy. *Note:* Military administration is primarily in the fields of logistics and personnel management. *See* **telecommunications administration.**

administrative management complex (AMC): In network management, a complex that (a) is controlled by a network provider, and (b) is responsible for, and performs, network management functions, such as network maintenance.

administrative network: A communications network that is used for (a) the exchange of messages of a nonoperational nature, such as network operational personnel assignment and payroll data, (b) operational messages of a nonurgent nature, (c) messages that contain administrative information in contrast to network real time control and operational information, and (d) messages that contain long-term background and policy information needed by the operators of a communications network.

administrative security: In a communications, computer, or data processing system, security implemented as management controls, such as management constraints, operational procedures, and accountability procedures, to provide an acceptable level of protection against unauthorized access to the system.

ADP: automatic data processing.

ADPCM: adaptive differential pulse code modulation.

ADPE: automatic data processing equipment.

ADPS: automatic data processing system.

ADP system: *Synonym* **computer system.**

ADS: automatic dependent surveillance.

ADSU: automatic dependent surveillance unit.

advanced communications technology satellite (ACTS): A satellite test bed for satellite communications systems that (a) is a National Aeronautics and Space Administration (NASA) satellite for communications experimentation, (b) is structured to validate on-orbit, high-risk systems in multiple frequency bands for supporting NASA, other government agencies, academia, and industry, and (c) has onboard capabilities, such as baseband digital processing, storage and switching using multiple hopping spot beams to reach isolated areas, time division multiple access (TDMA), a microwave switching matrix for high volume traffic, and adaptive signal fade compensation.

advanced intelligent network (AIN): A proposed intelligent network (IN) architecture that includes both IN/1 and IN/2 concepts. *See also* **intelligent network.**

advanced satellite system for evaluation and test: *See* **Functional Advanced Satellite System for Evaluation and Test.**

advanced tactical optical fiber (ATOF): Optical fiber used in high performance avionic systems for providing (a) a high speed bus to interconnect avionics subsystems, (b) point-to-point links that transmit sensor data to signal processors and video data to cockpit controls and displays, and (c) network interface units that allow point-to-point communications for the aircraft common integrated processor.

advanced television (ATV): A family of television systems that (a) are improvements over distribution quality television and (b) include extended definition television (EDTV), high definition television (HDTV), and improved definition television (IDTV). *See also* **enhanced definition television, high definition television, improved definition television.**

advanced television system: A television system in which (a) an improvement in performance has been made to an existing television system to create the advanced system, (b) the improvement may or may not result in a system that is compatible with the original system or with other present systems, and (c) some features and technical characteristics of the original system on which the advanced system is based may be included. *See also* **enhanced definition television (EDTV) system, high definition television (HDTV) system, improved definition television (IDTV) system.**

Advanced Data Communications Control Procedure (ADCCP): A bit-oriented Data Link Layer protocol used to provide point-to-point and point-to-multipoint transmission of a data frame with error control. *Note:* Advanced Data Communications Control Procedure (ADCCP) closely resembles high level data link control (HDLC) and synchronous data link control (SDLC). *See also* **binary synchronous communications, frame, high level data link control, link, synchronous data link control.**

Advanced Mobile Phone Service (AMPS): A first-generation cellular radio service pioneered by Bell Laboratories during the 1970s.

Advanced Mobile Phone System (AMPS) standard: A standard that is intended to provide compatibility among cellular radio systems in North America, particularly in the areas of frequencies, channel identification, and use of time division multiple access (TDMA) versus code division multiple access (CDMA).

Advanced Project for Information Exchange (APEX): A system that (a) is operated by the European Research Cooperative Action (EUREKA) and (b) handles the electronic transfer of documents, correspondence, designs, drawings, photos, graphics, and imagery.

Advanced Research Projects Agency Network (ARPANET): A packet-switching network that (a) was developed and used by the U.S. Department of Defense (DoD) and (b) evolved into the Internet. *See also* **Internet.**

advantage: *See* **range advantage.**

advisory station: *See* **aeronautical advisory station.**

AECS: Aeronautical Emergency Communications System.

AECSP: Aeronautical Emergency Communications System Plan.

aerial cable: A communications cable designed for installation on, or suspension from, a pole or other overhead structure. *See also* **cable, direct buried cable, outside plant, underground cable.**

aerial fiber optic cable: A fiber optic cable designed for use in overhead suspension devices, such as towers or poles.

aerial insert: In a direct buried or underground cable run, a cable rise to a point high above ground, followed by an overhead run on poles, followed by a drop back into the ground, in places where it is not possible or practical to remain underground, such as might be encountered in crossing a deep ditch, canal, river, or subway line. *Refer to* **Fig. A-4 (20).**

aerodrome control service: *Synonym* **airport control service.**

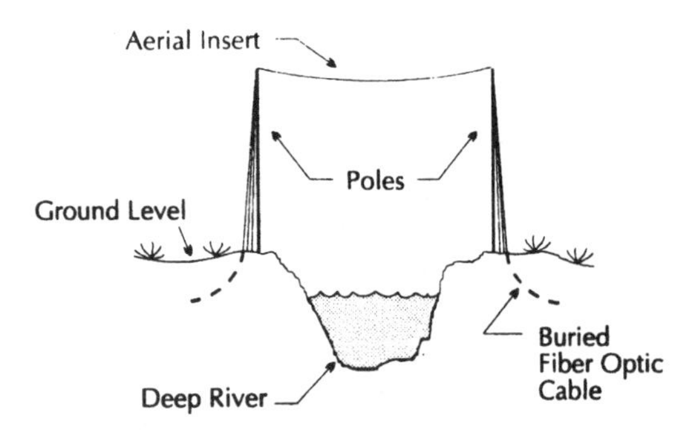

Fig. A-4. An **aerial insert** for a direct buried cable.

aerodrome control station: *Synonym* **airport control station.**

aeronautical advisory station (AAS): An aeronautical station used for advisory and civil defense communications primarily with private aircraft stations. *Synonym* **UNICOM station.**

aeronautical broadcast station (ABS): An aeronautical station that makes scheduled meteorological information broadcasts and Notice to Airmen (NOTAM) broadcasts. *Note:* An aeronautical broadcast station may also be on board a ship.

aeronautical communications system abbreviation: An abbreviation of frequently used words and expressions that may be used, when authorized, by operators in the aeronautical mobile service to facilitate copying of voice transmissions.

aeronautical Earth station (AES): In the fixed satellite service or in the aeronautical mobile satellite service, an Earth station at a specified fixed point on land to provide a feeder link for the aeronautical mobile satellite service. *See also* **Earth station, fixed satellite service.**

Aeronautical Emergency Communications System (AECS) Plan: A plan that, in an emergency, (a) provides for the operation of aeronautical communications stations, (b) is operated on a voluntary organized basis, and (c) provides (i) the president of the United States, (ii) the federal government, (iii) the heads of state and local governments, or their designated representatives, and (iv) the aeronautical industry with an expeditious means of communications during the emergency.

aeronautical emergency frequency: *See* **international aeronautical emergency frequency.**

aeronautical fixed service (AFS): A radiocommunications service between specified fixed points provided primarily for the safety of air navigation and for the regular, efficient, and economical operation of air transport. *See also* **aeronautical radionavigation service, mobile service.**

aeronautical fixed station (AFS): A station in the aeronautical fixed service.

aeronautical marker beacon station: A radionavigation land station in the aeronautical radionavigation service that provides a signal to designate a small area above the station. *Note:* In certain instances, an aeronautical marker beacon station may be on board a ship or aircraft.

aeronautical mobile satellite (off-route) service (AMS(OR)S): An aeronautical mobile satellite service intended for communications with aircraft outside national and international civil air routes. *Note:* The aeronautical mobile satellite (off-route) service includes communications relating to flight coordination. *See also* **aeronautical mobile satellite (route) service (AMSS).**

aeronautical mobile satellite (route) service (AMSS): An aeronautical mobile satellite service reserved for communications with aircraft primarily along national or international civil air routes. *Note:* The aeronautical mobile satellite (route) service includes communications relating to safety and regularity of flight. *See also* **aeronautical mobile satellite (off-route) service (AMS(OR)S).**

aeronautical mobile service (AMS): A mobile service (a) that is between aeronautical stations and aircraft stations, or between aircraft stations and (b) in which survival craft stations and emergency position-indicating radio beacon stations may participate on designated distress and emergency frequencies. *See also* **aeronautical station, aircraft station, mobile Earth station, mobile service.**

aeronautical multicom land station: A land station operating in the aeronautical multicom service.

aeronautical multicom mobile station: A mobile station operating in the aeronautical multicom service.

aeronautical multicom service (AMS): A mobile service that (a) is used to provide communications essential to the conduct of activities being performed by, or directed from, private aircraft and (b) is not open to general public communications.

aeronautical radio beacon station: A radionavigation land station in the aeronautical radionavigation service that provides signals that enable mobile service elements, such as aircraft, ships, or land mobile stations, to determine their bearings or their direction in relation to the aeronautical radio beacon station. *Note:* In certain instances, an aeronautical radio beacon station may be on board a ship or aircraft.

aeronautical radionavigation satellite service (ARNSS): A radionavigation satellite service in which Earth stations are on board aircraft.

aeronautical radionavigation service (ARNS): A radionavigation service intended for the benefit and the safe operation of aircraft.

aeronautical station: A land station in the aeronautical mobile service. *Note 1:* In certain instances, an aeronautical station may be on board a ship or on a platform at sea. *Note 2:* An aeronautical station may also serve as a guard, secondary, or standby station during the flight of an aircraft. *See also* **aeronautical mobile service, aircraft station.**

aeronautical station master log: A brief record of major or significant events that occur at an aeronautical station during a radio day, including, but not necessarily limited to, station identification, time of station opening and closing, date, pertinent actions, service interruptions, troubles, system failures, and supervisor signatures.

aeronautical telemetering land station: A telemetering land station used (a) in the flight testing of manned or unmanned aircraft or missiles and (b) in the flight testing of major components of manned or unmanned aircraft or missiles.

aeronautical telemetering mobile station: A telemetering mobile station used (a) in the flight testing of manned or unmanned aircraft or missiles and (b) in the flight testing of major components of manned or unmanned aircraft or missiles.

aeronautical utility land station: A land station that (a) is located in aerodrome or airport control towers and (b) is used for control of ground vehicles and aircraft on the ground at aerodromes and airports.

aeronautical utility mobile station: A mobile station that (a) communicates with ground vehicles, aircraft on the ground, and the aeronautical utility land station in the aerodrome or airport control tower, (b) transmits when authorized by the aerodrome or airport control station, (c) is subject to the control of the aerodrome or airport controller, and (d) discontinues transmission immediately when requested to do so by the aerodrome or airport controller.

AES: aircraft Earth station.

A except B gate: *Synonym* **A AND NOT B gate.**

AF: audio frequency.

affinity routing: In communications systems operations, the routing of messages based on a temporary relationship between (a) a data source, such as an originating office, a source user, or a message originator and (b) a data sink, such as a destination office, a destination user, or a message addressee.

AFNOR: Association Française de Normalisation.

AFRS: Armed Forces Radio Service.

AGC: automatic gain control.

agency: *See* **communications agency, operating agency, telecommunications private operating agency.**

aggregate bit rate: *See* **multiplex aggregate bit rate.**

agile radar: *See* **frequency agile radar.**

agility: *See* **frequency agility.**

aging: *See* **frequency aging.**

AHF radio: adaptive high frequency radio.

AI: absorption index, artificial intelligence.

aid: *See* **radio landing aid, radionavigation aid, runway approach aid.**

aid to navigation: *See* **long range aid to navigation, short range aid to navigation.**

aided software engineering: *See* **computer aided software engineering.**

aided tracking: Tracking of a target in azimuth, elevation, range, or any combination of these variables simultaneously, in which a constant rate of motion of the tracking mechanism is maintained.

aids service: *See* **Meteorological Aids service.**

AIG: address indicating group.

A IGNORE B gate: A two-input binary, logic coincidence circuit or device in which (a) normal operation can be interrupted by a control signal that enables the gate to function so as to pass the A input signal and completely disregard the B input signal, and (b) the output is the same as the A input signal and completely independent of the B input signal. *Refer to* **Fig. A-5 (22).** *Refer also to* **Fig. B-4 (76).**

AIM: amplitude intensity modulation.

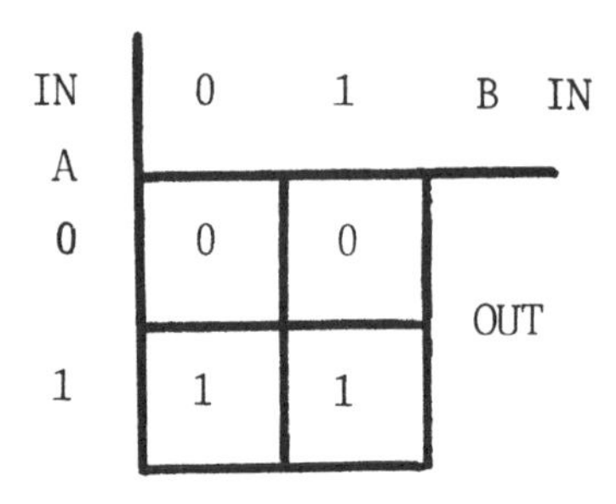

IN
A
0
1
0
1
B IN
0
0
OUT
1
1

Fig. A-5. A truth table that shows the input and output digits of an **A IGNORE B gate.**

aiming circle: An aiming symbol in the shape of a circle.

aiming field: On the display surface of a display device, the area covered or bounded by an aiming symbol.

aiming symbol: A pattern of light used to guide the positioning of a light pen or to indicate the area in the display space of a display device within which the presence of a light pen can be detected at a given time. *Note:* Examples of aiming symbols are a circle, square, triangle, or pair of brackets of light on a cathode ray tube (CRT) screen, fiberscope faceplate, light-emitting diode (LED) panel, or gas panel.

A IMPLIES B gate: *Synonym* **B OR NOT A gate.**

AIN: advanced intelligent network.

AIOD: automatic identified outward dialing.

air-air communications: Communications methods, systems, and equipment used for the transmission of messages (a) containing information about air to air, i.e., aircraft station to aircraft station, operations or facilities, (b) to or from air units in flight, and (c) for the purpose of using, directing, or coordinating air units in flight. *Synonym* **air to air communications.** *See also* **air-ground communications.**

airborne command post: A suitably equipped aircraft used by a commander, coordinator, or director for the control of units engaged in an operation, such as a military, rescue, fire fighting, crop dusting, police, evacuation, disaster, or emergency medical operation.

airborne direction finding recorder: A direction finding recording device that records all intercepts so that detailed analysis can be made by processing units to ascertain signal parameters. *Note:* Typical manual airborne direction finding recordings include operator written logs, audio and video magnetic recordings, and photographs.

airborne direction finding unit: A direction finding unit (a) that uses a portion of the receiver output to visually display relative bearings from an aircraft station to another transmitter, (b) in which the emitter location is determined through computation of a series of relative bearings versus aircraft headings and through polarization of the received signal, and (c) in which the polarization of the received signal is determined using specialized antenna circuitry.

airborne early warning: Pertaining to systems, equipment, devices, personnel, and procedures used to detect and signal the approach of airborne or space vehicles.

airborne early warning net: A network of airborne radar equipment that provides long range detection, identification, and relaying of radar signals to land, aircraft, and ship stations.

airborne interception: 1. Pertaining to the location, identification, and maintenance of contact with airborne objects, such as missiles and aircraft. **2.** The interception of a mobile object or vehicle by means of an airborne vehicle, such as a missile or aircraft.

airborne intercept radar: In military communications systems, intercept radar that (a) is on board aircraft on an airborne intercept mission, (b) is in contrast to ground-based radar, (c) is vectored, i.e., is guided, toward a strike attack force acquired by ground radar, (d) tracks the force so interceptor aircraft can be guided toward it, and (e) performs search, i.e., scanning, and track, i.e., lock-on, functions that may be combined into one track while scan airborne intercept radar set. *See* **track while scan airborne intercept radar.**

airborne radar: *See* **side looking airborne radar.**

airborne radar jammer: A jammer that (a) is on board an aircraft and (b) is used to jam ground, ship, and airborne radar equipment.

airborne radar warning system: A wideband system that (a) is on board an aircraft, (b) detects radio detection and ranging (radar) signals in several frequency bands, (c) performs an analysis of the threat, and (d) usually is capable of providing video information to the air crew.

airborne radio relay (ARR): 1. Airborne equipment used to relay radio transmissions from selected originating transmitters. **2.** The use of aircraft fitted with radio relay stations for the purpose of increasing the range, flexibility, or physical security of communications systems.

air circuit breaker switchgear: *See* **low voltage enclosed air circuit breaker switchgear.**

air communications: *See* **air-air communications, airway/air communications, maritime air communications, surface to air communications.**

air conditioning: The process of simultaneously controlling the characteristics of air, such as temperature, humidity, cleanliness, motion, and pollutant concentration, in a given space to meet the requirements of the occupants, a process, or equipment in that space. *Note:* In the U.S. Department of Defense, **environmental control** is preferred over the synonym air conditioning. *See also* **critical areas.**

air control authority: *See* **maritime air control authority.**

aircraft: *See* **automatic radio relay aircraft.**

aircraft call sign: Aircraft signal letters or identification numbers used as an international call sign.

aircraft communications standard plan: A table listing standard communications equipment recommended as minimum equipment for long range maritime patrol aircraft.

aircraft control warning system: A system that consists of (a) observation facilities, such as radar, passive electronic, and visual equipment, (b) control centers, and (c) necessary communications established to control and report the movement of aircraft.

aircraft Earth station (AES): In the aeronautical mobile satellite service, a mobile Earth station on an aircraft. *See also* **mobile Earth station.**

aircraft emergency frequency (AEF): An international aeronautical emergency frequency (a) for aircraft stations, (b) for stations concerned with safety and regulation of flight along national or international civil air routes, and (c) for maritime mobile service stations authorized to communicate for safety purposes. *Note:* An aircraft emergency frequency is 121.5 MHz (megahertz) or 243.0 MHz.

aircraft signature identification: The identification of aircraft by (a) analyzing and cataloging the way the compressor and turbine blades of jet engines modulate radar signals and (b) identifying and discriminating between different aircraft, discriminating between decoys and warhead missiles, and discriminating among other moving objects when sufficient signatures of the modulation are available. *See also* **radar signature.**

aircraft station: In the aeronautical mobile service, a mobile station, other than a survival craft station, on board an aircraft. *See also* **aeronautical mobile service, aeronautical station.**

air defense command-control-communications (C^3) network: All the facilities and equipment used to integrate the operations of the passive detection systems, the active detection systems, and the weapon systems of an air defense system.

air defense communications: Communications (a) in which methods, systems, and equipment are used for transmission of messages that contain information about military air defense situations, (b) in which messages are transmitted to or from air defense units, facilities, or communications activities, and (c) that are used for directing, coordinating, and conducting air defense operations.

air defense communications system: A communications system established among ground, air, and sea units to permit recognition of aircraft by various means, such as visual, optical, radio, and radar, and to transmit the information obtained to appropriate units.

air dielectric coaxial cable: A coaxial cable that (a) has air between the inner and outer conductors, (b) has insulating spacers at intervals to maintain separation of the conductors, and (c) may be used as a transmission line from a transmitter to an antenna.

airdrome control service: *Synonym* **airport control service.**

airdrome control station: *Synonym* **airport control station.**

air-ground communications: 1. Two-way communications between aircraft stations and land stations. **2.** Communications systems, equipment, devices, and procedures that (a) are used for the transmission of messages from air to ground and from ground to air stations, (b) contain information about air and ground units, and (c) are used for directing or coordinating air and ground units. *See also* **air-air communications.**

air-ground communications emergency frequency: A frequency primarily used during emergencies for contact purposes rather than as a working frequency for handling the emergency situation itself. *Note:* Aeronautical communications systems use 121.5 MHz (megahertz) or 243.0 MHz as emergency frequencies.

air-ground radiotelephone service (AGRS): A public radio service between a base station and airborne mobile stations.

air-ground worldwide communications system (AGWCS): A worldwide military network of ground agencies and stations that (a) provide a two-way communications link between aircraft and ground stations for navigation and control, including air route traffic control, and (b) may also provide support for special functions, such as for civil aircraft providing assistance to military missions and for meeting communications requirements for aircraft flying distinguished visitors.

air net: *See* **ship-air net, ship to air net.**

air operations communications net: *See* **maritime tactical air operations communications net.**

air portable: Pertaining to material that is suitable for transport by an aircraft internally or externally loaded, with no more than minor dismantling by the shipper and minor reassembly by the destination user. *Note:* The extent of air portability should be specified, such as by specifying the packaged size and weight of the items being shipped by air.

airport control service: Communications activities and services used for the control of operations in and around an airport or aerodrome. *Synonyms* **aerodrome control service, airdrome control service.**

airport control station (ACS): An aeronautical station that provides communications services between an airport control tower and aircraft. *Synonyms* **aerodrome control station, airdrome control station.**

airport land mobile radio base station (ALMRBS): A radio base station that (a) is used by the control tower, (b) is located at or near an airport, and (c) is used to control airport ground support vehicles, such as runway and grounds maintenance, aircraft maintenance, aircraft supply services, and emergency services vehicles.

airport land mobile radio station (ALMRS): A mobile radio station used by airport ground support vehicles, such as runway and grounds maintenance, aircraft maintenance, aircraft supply, and emergency services, to communicate with each other and with the control tower, over the Land Mobile Radio Trunking System.

airport surface surveillance device: A device that uses electromagnetic waves to detect motion on the surface of an airport.

air radio organization: *See* **maritime air radio organization.**

air reporting net: *See* **maritime patrol air reporting net.**

air-sea rescue frequency: *See* **scene of air-sea rescue frequency.**

air sounding: Measuring atmospheric phenomena or determining atmospheric conditions, especially by means of apparatus carried by balloons, rockets, or satellites. *See also* **ionosphere, virtual height.**

air-spaced doublet: In optics, a compound lens of two elements or parts with air or empty space between them.

air-supported optical fiber: An optical fiber that relies on air-filled space between the core and cladding to provide a refractive index that is less than that of the core in order to ensure total internal reflection of lightwaves in the core. *Refer to* **Fig. A-6.**

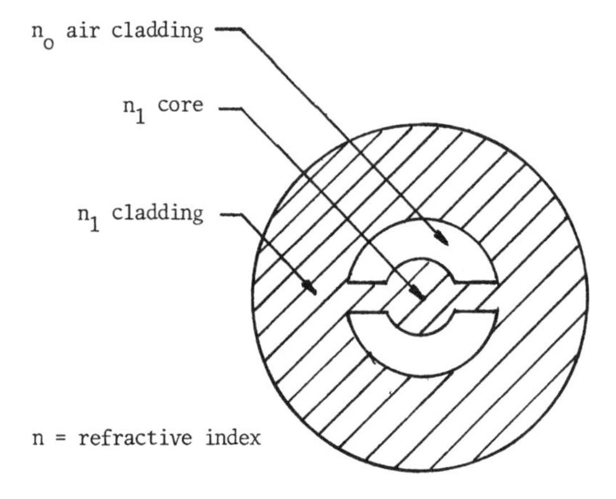

Fig. A-6. A cross section of an **air-supported optical fiber.** The numerical aperture for this case is $n_1{}^2 - n_0 p^2)^{1/2}$. The coupling efficiency approaches 100%.

air surface detection equipment: 1. A radiolocation device used for air landing surface surveillance to determine the distribution of equipment on airport surfaces. **2.** A device for locating the landing surfaces of an airport.

air surveillance: The systematic observation of airspace by any means, such as by radio, radar, optical, and visual means, primarily to identify and determine the movements of aircraft and missiles in the airspace under observation.

air telecommunications organization: *See* **maritime air telecommunications organization.**

air terminal: 1. The lightning rod or conductor placed on or above a building, a structure, or external conductors to intercept lightning. *See also* **facility grounding system, lightning down conductor, lightning protection subsystem. 2.** An airport for commercial airline ground and air route interconnection.

air to air communications: *Synonym* **air-air communications.**

air traffic control communications: Communications systems, equipment, and procedures used to provide services for (a) preventing collisions between aircraft, (b) preventing collisions between aircraft and ground obstructions, (c) providing landing instructions and advice, and (d) expediting and maintaining an orderly flow of air traffic.

airway/air communications: Communications methods, systems, and equipment that are used for transmission of messages that (a) contain information about air routes, terminals, and facilities, (b) are sent to or from air operations and communications facilities, and (c) are used to conduct air-related military or civilian operations.

AIS: automated information system.

alarm: *See* **audible alarm, audio alarm, emergency signal alarm.**

alarm center: A location that receives local and remote alarms. *Note:* An alarm center is usually in a technical control facility.

alarmed cable: *Synonym* **intrusion resistant communications cable.**

alarm indicator: A device that responds to a signal from an alarm sensor. *Note:* Examples of alarm indicators include bells, lamps, horns, gongs, and buzzers.

alarm rate: *See* **constant false alarm rate.**

alarm sensor: 1. In communications systems, a device that can sense an abnormal condition within a system and either locally or remotely provide a signal indicating the presence or the nature of the abnormal condition. *Note:* The signal may be in any desired form ranging from a simple contact opening or closure to a time phased automatic shutdown and restart cycle. **2.** In a physical security system, an approved device used to indicate a change in the physical environment of a facility or a part thereof. *Note:* Alarm sensors may also be redundant or chained, such as when one alarm sensor is used to protect the housing, cabling, or power protected by another alarm sensor. *See also* **communications security, variation monitor.**

alarm signal: 1. A signal that draws the attention of a user or an operator. *Note:* Examples of alarm signals are bells, blinking lights, and sirens. **2.** In distress and rescue communications, a signal that precedes the distress call to operate automatic alerting equipment. *Note 1:* In radio telegraphy, the alarm signal usually is followed, in order, by the distress call, an interval of 2 min. (minutes), the distress call again, and then the distress message. *Note 2:* In radio telephony, the alarm signal usually is followed, in order, by the distress call and the distress message. *Note 3:* In international radio telegraphic Morse code (cw), the alarm signal usually is a series of 12 dashes sent in 1 min. (minute). Each dash is 4 s (seconds) long with a 1-s interval between dashes. *Note 4:* In voice transmission, the alarm signal consists of two sinusoidal audio frequency tones, one of 2200 Hz (hertz) and the other of 1300 Hz, which are transmitted alternately. Each tone is 250 ms (milliseconds) long. *See* **ship alarm signal.** *See also* **distress traffic.**

a-law: A non–North American encoding methodology of sampling the audio waveforms used in 2048-kbps, 30-channel, pulse code modulation (PCM) systems.

a-law algorithm: A standard signal-compression algorithm that (a) is used in digital communications systems of the European digital hierarchy to modify, such as to optimize, the dynamic range of an analog signal prior to digitizing and (b) effectively reduces the dynamic range of the signal, thereby increasing the coding efficiency and resulting in a signal to distortion ratio that is greater than that obtained by linear encoding for a given number of bits. *Note:* The wide dynamic range of speech is not susceptible to efficient linear digital encoding. *See also* **linear encoding.**

ALBAM: air-land battle assessment model.

albedo: A measure of the surface reflective capability of a body that is not self-radiating. *Note:* The Earth albedo is 0.39, the Moon albedo is 0.15, Mars is 0.15, and Venus is 0.59. The albedo is similar to the reflection coefficient applied to the body as a whole. For Earth satellites, the albedo usually refers to the infrared and visible wavelengths. The incident light and heat at the satellite consist of energy directly received from the Sun and also indirectly from the Earth and the Moon. The greater the albedo of the satellite itself, the less it is heated by solar rays.

ALE: automatic link establishment.

alert-dispatch system: *See* **group alert-dispatch system.**

alerting and dispatching system: *See* **group alerting and dispatching system.**

algebra: *See* **Boolean algebra.**

algebraic expression: 1. An expression that specifies an operation in conventional algebra to be performed on abstract variables. **2.** An expression that specifies a set of arithmetic or algebraic operations that consists of operators and operands and that can be reduced to a single numeric result when numeric values are substituted for the operands and the specified operations are performed. *Note:* Examples of algebraic expressions are $x + y$, $x^2 + y^2$, $\sin (x + y)$, $mx + b$, and $e^x - e^y$.

ALGOL: A standard block structured high level computer programming common language used primarily for (a) arithmetic and logic operations and computations, (b) string handling, and (c) record handling. *Note:* The name is derived from Algorithmic Language.

algorithm: 1. A finite set of well-defined rules for the solution of a problem in a finite number of steps and to a stated precision. *Note:* An example of an algorithm is a complete specification of a sequence of arithmetic operations for evaluating π, $\sin \Theta$ given Θ, or e^x given x, to a given precision. **2.** A finite set of well-defined rules that specifies a sequence of operations for performing a specific task. *See* **a-law algorithm, mu-law algorithm, traveling salesman algorithm, type 3 algorithm, type 4 algorithm.**

algorithmic language: An artificial language established for expressing a given class of algorithms.

alias: 1. An alternate label, name, or key for a set of data, a file, or a record in a database. **2.** An additional name or label by which an item, such as a data item, storage location, database file, channel, software, hardware, or person is known or identified. **3.** In pulse code modulation (PCM), a spurious signal that results from beats caused by the combining of frequencies that are used in accomplishing the modulation, such as a signal that may result from beats between signal and sampling rates. *Synonym* **alternate name.**

aligned bundle: A bundle of optical fibers in which the relative spatial coordinates of each fiber are the same at the two ends of the bundle, i.e., a fixed relationship exists between the optical fibers at one end of the bundle relative to the fibers at the other end, such that each fiber has the same relative position with respect to all other fibers at both ends. *Note:* Aligned bundles may be used for the transmission of images because an image focused on one end of the bundle will appear, with only limited distortion, at the other end of the bundle. *Synonym* **coherent bundle.** *See also* **fiber bundle, fiber optics, spatial coherence.**

alignment: *See* **frame alignment, level alignment, message alignment, octet alignment.**

alignment connector: *See* **fiber optic grooved alignment connector.**

alignment indicator: *See* **message alignment indicator.**

alignment sensor: *See* **fiber optic axial alignment sensor.**

alignment signal: *See* **bunched frame alignment signal, distributed frame alignment signal, frame alignment signal, multiframe alignment signal.**

alignment time: *See* **out of frame alignment time.**

Allan variance: One half of the time average over the sum of the squares of the differences between successive readings of the frequency deviation sampled over the sampling period. *Note:* The Allan variance is conventionally expressed by $\sigma_y^2(\tau)$. The samples are taken with no dead time between them. *Synonym* **two-sample variance.**

allcall: In adaptive high frequency (HF) radio automatic link establishment (ALE), a broadcast in which (a) all stations are called, (b) no automatic response is required, and (c) all stations receiving the call must stop scanning to receive the message. *See also* **anycall.**

Allen belt: *See* **Van Allen belt.**

all-glass optical fiber: An optical fiber with a core and cladding that consist entirely of glass. *Note:* Most optical fibers are of silica glass core and silica glass cladding.

allocation: *See* **adaptive channel allocation, address group allocation, address indicating group allocation, dynamic allocation, dynamic storage allocation, frequency allocation, frequency band allocation, indicator allocation, resource allocation, storage allocation.**

allocation plan: *See* **call sign allocation plan, frequency allocation plan.**

allotment: *See* **radio frequency allotment, radio frequency channel allotment.**

all-plastic optical fiber: An optical fiber with a core and cladding that consist entirely of plastic. *Note:* Most optical fibers have a glass core and one or more glass claddings.

all-silica optical fiber: An optical fiber that (a) is composed of a silica-based core and one or more silica-based claddings and (b) may have a protective polymer overcoat or tight buffer.

all trunks busy (ATB): An equipment condition in which all trunks in a given trunk group are busy. *Note:* All-trunks-busy registers usually do not indicate subsequent attempts to reach trunk groups. *See also* **busy hour, erlang.**

alphabet: 1. An ordered set of all the letters used in a language, including letters with diacritical signs where appropriate, but not including punctuation marks. **2.** An ordered set of all the symbols used in a language, including punctuation marks, numeric digits, nonprinting control characters, and other symbols. *Note:* Examples of alphabets include the Roman alphabet, the Greek alphabet, the Morse code, and the 128 characters of the American Standard Code for Information Interchange (ASCII) (IA No. 5). *See also* **alphanumeric, American Standard Code for Information Interchange, character, character set, code, coded set, digit, digital alphabet, extended binary coded decimal interchange code, language.** *Refer to* **Fig. P-5 (718).**

alphabetic character set: A character set that contains letters and may contain control characters, special characters, and the space character, but not digits, i.e., not numerals or bit patterns that represent numbers.

alphabetic character subset: A character subset that contains letters and may contain control characters, special characters, and the space character, but not numerals, i.e., not digits.

alphabetic code: A code according to which data are represented through the use of an alphabetic character set.

alphabetic coded character set: A coded character set that has a character set that is an alphabetic character set.

alphabetic flag: A flag used on an international basis in visual communications systems to represent a letter of the alphabet. *Note 1:* An example of an alphabetic flag is a square flag on which a pattern of colors is used to represent a letter of the alphabet. *Note 2:* Alphabetic flags are usually strung in sequence on a flaghoist, lanyard, or mast of a ship to convey a message.

alphabetic string: 1. A string that (a) solely consists of letters and (b) only contains letters from the same alphabet. **2.** A character string that (a) solely consists of letters and associated special characters, (b) only contains letters and associated special characters from the same alphabet, and (c) does not contain digits, i.e., does not contain numerals or bit patterns that represent numbers.

alphabetic telegraphy: 1. Telegraphy, applicable to text, in which coded signals are used, each signal or group of signals corresponding to a character, such as a letter, symbol, or punctuation mark. **2.** Telegraphy in which the received signals represent characters, such as letters, symbols, or punctuation marks, that are automatically recorded as printed characters. *Synonym* **printing telegraphy.**

alphabetic word: 1. A word that (a) solely consists of letters and (b) only contains letters from the same alphabet. **2.** A word that (a) solely consists of letters and associated special characters, (b) only contains letters and associated special characters from the same alphabet, and (c) does not contain digits, i.e., does not contain numerals or bit patterns that represent numbers. *See also* **word.**

alphabet number: *See* **international telegraph alphabet number 1, 2, 3, 4, 5.**

alphabet translation: *Deprecated. Use* **alphabet transliteration.**

alphabet transliteration: The conversion of the characters of one alphabet to characters of a different alphabet. *Note 1:* Alphabet transliteration is usually accomplished on a character by character basis. For example, if the Roman letters a, b, and p are replaced with the Greek letters α, β, and π, respectively, the transliteration would be from Roman to Greek. *Note 2:* Alphabet transliteration is reversible. *Note 3:* Alphabet transliteration often becomes necessary in telecommunications systems because of the different alphabets and codes used worldwide. *Note 4:* In alphabet transliteration, no consideration is given to the meaning of the individual characters or their combinations. *See also* **code, language, transliterate, translator.**

alphanumeric: 1. Pertaining to a character set that contains letters, digits, and sometimes other characters, such as punctuation marks. **2.** Pertaining to a set of unique bit patterns that are used to represent letters of an alphabet, decimal digits, punctuation marks, and other special signs and symbols used in grammar, business, and science, such as those displayed on conventional typewriter keyboards. *See also* **alphabet, character set, code, language.**

alphanumeric character set: A character set that contains letters; digits, i.e., numerals or bit patterns that represent numbers; special characters; and the space character.

alphanumeric character subset: A character subset that contains both letters and digits and that may contain control characters, special characters, and the space character.

alphanumeric code: 1. A code derived from an alphanumeric character set. **2.** A code that, when used, results in a code set with elements taken from an alphanumeric character set.

alphanumeric coded character set: A coded character set that has a character set that is an alphanumeric character set.

alphanumeric data: Data that consist of letters; digits, i.e., numerals or bit patterns that represent numbers; usually the space character; and sometimes special characters.

alpha particle: A particle that consists of the nucleus of a helium atom, which has two protons, two neutrons, and a positive electric charge equivalent to the two protons.

alpha profile: *See* **power law index profile.**

altazimuth mount: An antenna mount, such as a mount for a directional antenna, that allows slewing in (a) a horizontal plane, i.e., a plane tangent to the Earth surface, and (b) elevation angles above and below that plane. *See also* **antenna, directional antenna.**

alternate binary coding: *Synonym* **bipolar coding.**

alternate code: *Synonym* **paired disparity code.**

alternate communications: *See* **two-way alternate communications.**

alternate communications net: A communications net used when a primary net fails for any reason, such as adverse atmospheric conditions, overloading, destruction, saturation, or lack of operators.

alternate mark inversion code: *See* **modified alternate mark inversion code.**

alternate mark inversion (AMI) signal: A pseudoternary signal, representing binary digits, in which (a) successive "marks" are of alternately positive and negative polarity and the absolute values of their amplitudes are usually equal and (b) "spaces" are of zero amplitude. *See also* **alternate mark inversion violation, bipolar signal, modified AMI, paired disparity code, return to zero code.**

alternate mark inversion violation: A "mark" that has the same polarity as the previous "mark" in the transmission of alternate mark inversion (AMI) signals. *Note:* In some transmission protocols, alternate mark inversion (AMI) violations are deliberately introduced to aid synchronization or to signal a special event. *See also* **alternate mark inversion signal, error, paired disparity code.**

alternate name: *Synonym* **alias.**

alternate party: In multilevel precedence and preemption systems, the call receiver, i.e., the destination user, to which a precedence call will be diverted. *Note 1:* Diversion occurs when the response timer expires, when the call receiver is busy on a call of equal or higher precedence, or when the call receiver is busy with access resources that are nonpreemptable. *Note 2:* Alternate party diversion is an optional terminating feature subscribed to by the call receiver. Thus, the alternate party is specified by the call receiver at the time of subscription.

alternate routing: The routing of a call or a message over a substitute route when the primary route is unavailable. *See also* **adaptive routing, call, dispersion, dual access, dual homing, heuristic routing, multiple access, multiple homing, routing.**

alternating current (ac): Electric current that (a) is continuously changing in magnitude, (b) periodically reverses its polarity, and (c) usually is sinusoidal. *Note:* The abbreviation "ac" usually is also used when referring to a voltage, such as in "115 V (volts) ac." *See also* **direct current.**

alternation by line: *See* **phase alternation by line.**

alternation by line (modified): *See* **phase alternation by line (modified).**

alternation gate: *Synonym* **OR gate.**

alternative: *Synonym* **variant.** *See* **typing alternative.**

alternative denial gate: *Synonym* **NAND gate.**

alternative frequency: A frequency, or a group of frequencies, that may be assigned (a) for use on any channel, or on a particular channel, (b) at a certain time, or (c) for a certain purpose to replace or supplement the frequencies usually used on that channel.

alternative key (Alt Key): On a printer, computer, or communications keyboard, a key used in conjunction with another key to designate a function different from the function designated by the other key when it is used alone.

alternative routing: A process in which substitute routes are used for transmitting messages when (a) circuit failures occur on operating transmission paths or (b) message backlogs occur. *See also* **primary route.**

alternative routing indicator: Information that (a) is sent in the forward direction, and (b) that indicates that the associated call or message has been subjected to an alternative routing.

alternative test method (ATM): In fiber optics, a test method in which a given characteristic of a specified class of fiber optic devices, such as optical fibers, fiber optic cables, connectors, photodetectors, and light sources, is measured in a manner that is consistent with the definition of this characteristic, gives reproducible results that are relatable to the reference test method, and is relatable to practical use. *Synonym* **practical test method.** *See also* **reference test method.**

altimeter: *See* **radio altimeter.**

altimeter station: A radionavigation mobile station that (a) is in the aeronautical radionavigation service, (b) is on board an aircraft, and (c) is used to determine the aircraft altitude above the Earth's surface.

altimetry area: *See* **radar area.**

altitude: *See* **apogee altitude, azimuth over altitude, perigee altitude.**

altitude of the apogee or perigee: *See* **apogee altitude, perigee altitude.**

altitude over azimuth: Pertaining to an antenna mount (a) in which the azimuth axis is the primary axis fixed to earth, and the altitude, i.e., the elevation, drive is the secondary axis, and (b) that is used in moderate latitudes for tracking drifting equatorial satellites where considerable azimuth movement is required. *Synonym* **elevation over azimuth.** *See also* **azimuth over altitude.**

ALU: arithmetic and logic unit.

aluminum garnet source: *See* **YAG/LED source.**

ALWAC: An airborne scanning, acquisition, tracking, and track telling radar system. *See also* **acquisition radar, airborne radar warning system, tracking radar, track telling, track while scan airborne intercept radar.**

AM: amplitude modulation, angle modulation, ante meridian.

AMA: automatic message accounting.

amateur satellite service (ASS): A radiocommunications service that uses space stations on Earth satellites for the same purposes as those of the amateur service.

amateur service: A radiocommunications service used for self-training, intercommunications, and technical investigations by licensed qualified persons who have a personal interest in radio techniques and operations as a pastime rather than as a profession and who are not seeking to derive a profit from operating their amateur radio stations.

amateur station: A station in the amateur service.

Amazon surveillance system: A surveillance system that (a) includes (i) an integrated network of telecommunications, remote satellite sensing and imagery, and ground-based and airborne sensor systems, (ii) control by regional and national coordination centers, and (iii) various components, such as air traffic control radars, automation and flight certification aircraft, air- and ground-based surveillance and environmental sensors, weather radar systems, and image analysis systems and (b) in the Amazon River basin is used for (i) reducing deforestation and wildlife destruction, (ii) combating illegal mining and drug trafficking, (iii) protecting indigenous tribes, (iv) strengthening border security, and (v) monitoring air and waterways traffic.

ambient light susceptibility: The optical power that enters a device from ambient illumination incident upon the device. *Note 1:* Ambient light susceptibility may be measured in absolute power units, such as microwatts, or in dB relative to the incident ambient optical power. *Note 2:* An example of ambient light susceptibility in a fiber optic connector or rotary joint is the ambient optical power that leaks into the optical path in the component.

ambient noise level: The level of acoustic noise existing at a given location, such as in a room, in a compartment, or at a place out of doors. *Note 1:* Ambient noise level is measured with a sound level meter. *Note 2:* Ambient noise level is usually measured in dB above a reference pressure level of 0.00002 Pa (20 μPa) in International System (SI) units. A pascal is a newton per square meter. *Note 3:* In the centimeter-gram-second (cgs) system of units, the reference level for measuring ambient noise level is 0.0002 dyne•cm^{-2}. *See also* **acoustic noise, background noise, level, noise.**

ambient temperature: The temperature of air or other media in a designated space or area, particularly the area surrounding equipment. *See also* **technical area.**

ambiguity: *See* **time ambiguity, time code ambiguity.**

AME: amplitude modulation equivalent, automatic message exchange.

American National Standards Institute (ANSI): The U.S. standards organization that establishes procedures for the development and coordination of American National Standards. *Note:* American National Standards are voluntary standards.

American Standard Code for Information Interchange (ASCII): The standard code used for information interchange among data processing systems, data communications systems, and associated equipment in the United States. *Note 1:* The ASCII character set contains 128 coded characters. *Note 2:* Each ASCII character is a 7-bit coded unique character, 8 bits when a parity check bit is included. *Note 3:* The ASCII character set consists of control characters and graphic characters. *Note 4:* When considered simply as a set of 128 unique bit patterns, or 256 with a parity bit, disassociated from the character equivalences in national implementations, the ASCII may be considered as an alphabet used in machine languages. *Note 5:* The ASCII is the U.S. implementation of International Alphabet No. 5 (IA No. 5) as specified in International Telegraph and Telephone Consultative Committee (CCITT) Recommendation V.3. *See also* **binary digit, bit pairing, code, data, Extended Binary Coded Decimal Interchange Code.**

AMI: alternate mark inversion (signal).

AMI violation: alternate mark inversion violation.

AMLCD: active matrix liquid crystal display.

ampere: The International System (SI) unit of electric current, equivalent to 1 C (coulomb) of electric charge passing though a point in 1 s (second), i.e., equivalent to approximately 6.242 x 10^{18} electrons per second. *Note:* The charge equivalent to one electron is approximately 1.602 x 10^{-19} C.

amplification by stimulated emission of radiation: *See* **laser, maser.**

amplification jammer: *See* **direct noise amplification jammer.**

amplifier: A device that has an output signal that (a) is a function of its input signal, (b) is at a higher power level than the input signal, and (c) has a gain expressed as a transfer function and often in positive dB. *Note:* If the gain is less than 1, or negative dB, the device is an attenuator. *See* **fiber optic amplifier, linear-logarithmic intermediate frequency amplifier, low noise amplifier, operational amplifier, optical fiber amplifier, parametric amplifier, photodetector transimpedance amplifier.** *See also* **amplitude, amplitude distortion, dB, distortion, feedback, frequency level, noise, signal.** *Refer to* **Fig. T-7 (1024).**

amplifying message: A message that contains information that is in addition to information contained in a previous message.

amplifying prefix: In communications net operations, a prefix or word that may be used to modify the basic functional word that describes or names a type of communications net. *Note:* Examples of amplifying prefixes are the words navigation, search, safety, police, broadcast, medical emergency, amateur, and rescue that precede and describe a communications net, land station, ship station, aircraft station, or other communications entity that is a part of a net.

amplifying suffix: In communications net operations, a suffix or a word that may be used to modify the basic functional word that describes or names a type of communications net. *Note:* An example of an amplifying suffix is the letters "UHF" following the call sign of a radio station.

amplifying switch message: A message that contains detailed information in addition to that contained in a previously sent switch message.

amplitude: The magnitude or ordinate value of a periodic variation or excursion of a parameter, such as electrical voltage, current, or power. *Note 1:* An example of amplitude is one half of the peak to peak value, i.e., trough to crest value, measured in a direction transverse to the direction of propagation of a wave or measured perpendicular to the time axis of a time plot of the wave. *Note 2:* "Amplitude" should not be applied to frequency or phase. Thus, amplitude, frequency, phase, their transitions, and their shifts have magnitudes. *See* **pulse amplitude.**

amplitude-amplitude distortion: *Synonym* **amplitude distortion.**

amplitude distortion: Distortion of a signal that occurs in a system, subsystem, or device when the output amplitude is not a linear function of the input amplitude under specified conditions. *Note:* Amplitude distortion is measured with the system operating under steady state conditions with a sinusoidal input signal. When other frequencies are present, the term "amplitude" refers to that of the fundamental frequency only. *Synonym* **amplitude-amplitude distortion.** *See* **phase-amplitude distortion.** *See also* **distortion, insertion loss versus frequency characteristic.**

amplitude equalizer: A network that can be used to modify the amplitude characteristics of a circuit or a system over a desired frequency range. *Note:* Amplitude equalizers may be fixed, manually adjustable, or operated automatically. *See also* **amplitude versus frequency distortion, equalization.**

amplitude-frequency distortion: *Synonym* **frequency distortion.**

amplitude frequency response: *Synonym* **insertion loss versus frequency characteristic.**

amplitude hit: In a data transmission channel, a momentary disturbance caused by a sudden change in amplitude of a signal. *See also* **hit.**

amplitude intensity modulation (AIM): *See* **intensity modulation.**

amplitude keying: Keying in which the amplitude of a signal is varied between a set of discrete values. *See also* **keying, modulation, signal.**

amplitude limiting: *See* **video amplitude limiting.**

amplitude modulation (AM): Modulation in which the amplitude of a carrier wave is varied in accordance with some characteristic of the modulating signal, usually performed by (a) modulating a coherent carrier wave by mixing it in a nonlinear device with the modulating signal, (b) producing discrete upper and lower sidebands that are the sum and difference frequencies of the carrier and the signal, (c) creating an envelope of the resultant modulated wave, i.e., creating an analog of the modulating signal, the values of which are the vector sums of the corresponding instantaneous values of the carrier wave, upper sideband, and lower sideband, and (d) recovering the modulating signal by direct detection or by heterodyning. *See* **balanced amplitude modulation, pulse amplitude modulation, quadrature amplitude modulation.** *See also* **absorption modulation, balanced modulator, carrier (cxr), intensity modulation, modulation, signal.**

amplitude modulation equivalent (AME): *Synonym* **compatible sideband transmission.**

amplitude quantized control: *Synonym* **amplitude quantized synchronization.**

amplitude quantized synchronization: In a signal transmission control system, synchronization in which (a) the functional relationship between the actual phase error and the derived error signal includes discontinuities, (b) the working range of phase error is divided into a finite number of subranges, and (c) a unique signal is derived for each subrange whenever the error falls within a subrange. *Synonym* **amplitude quantized control.**

amplitude system: *See* **frequency and amplitude system.**

amplitude versus frequency distortion: 1. In a transmission system, distortion caused by the nonuniform attenuation, or gain, in the system with respect to frequency under specified operating conditions. *Note:* The response to different frequencies is different. Resonance causes certain frequencies to be passed and others to be blocked. Multimode effects add and remove certain frequencies. In most instances, the impedance to different frequencies is different. Except for the theoretical pure resistance, impedance is a function of frequency. **2.** An undesired variation, with respect to frequency, of the ratio of (a) the magnitude of the fundamental component of the response to a sinusoidal excitation to (b) the magnitude of the excitation. *Synonyms* **amplitude-frequency distortion, attenuation-frequency distortion, frequency distortion, gain-frequency distortion.** *See also* **amplitude equalizer, dispersion, distortion, equalization, frequency, group delay frequency distortion, insertion loss versus frequency characteristic, phase-frequency distortion, response.**

AMPS: Advanced Mobile Phone Service.

AMPS standard: Advanced Mobile Phone System standard.

AMS: aeronautical mobile service, automatic message processing system.

AMTS: automated maritime telecommunications system.

analog: 1. Pertaining to the representation of data or physical quantities in the form of a continuous signal. *Note 1:* Usually the instantaneous magnitude is a function of the value of the data or physical quantity being represented. *Note 2:* An example of an analog signal is the baseband signal in the form of a continuous current produced by a telephone transmitter before any digitization occurs. **2.** A system or device that is used to simulate, model, or represent another system or device. *See* **network analog.** *See also* **analog network, digital, discrete.**

analog adder: *Synonym* **summer.**

analog compression: *Synonym* **analog speech interpolation.**

analog computer: A device that performs operations on data that are represented within the device by continuous variables having a physical resemblance to the quantities being represented. *Note:* The earliest analog computers were constructed with purely mechanical components, such as levers, cogs, cams, disks, gears, and balls. These components represented the quantities being manipulated or the operator-inserted values.

Modern analog computers usually employ electrical parameters, such as voltages, resistances, or currents, to represent the quantities being manipulated. *See also* **computer, digital computer.**

analog control: *Synonym* **analog synchronization.**

analog converter: *Synonym* **digital to analog converter.**

analog data: Data represented by a physical quantity that (a) is a continuously changing parameter or variable and (b) has a magnitude directly proportional to the data or to a suitable function of the data. *Note:* Examples of analog data are the representations of (a) the amplitude, phase, or frequency of a voltage, (b) the amplitude or duration of a pulse, (c) the angular position of a shaft, and (d) the pressure of a fluid, before any digitization of the data occurs or after conversion in a digital to analog (D-A) converter. *See also* **data, digital data.**

analog data channel: A one-way path for data signals that includes a voice frequency channel, an associated data modulator, and a demodulator.

analog decoding: Decoding in which an analog signal is reconstructed from a digital signal that represents the original analog signal. *See also* **analog encoding, signal.**

analog-digital converter: *Synonym* **analog to digital converter.**

analog-digital encoder: *Synonym* **analog to digital encoder.**

analog divider: A device that has an output analog variable that is proportional to the quotient of two input analog variables.

analog encoding: 1. Encoding in which digital signals are generated that represent samples taken of an analog signal at a given instant or consecutive instants. **2.** In analog to digital conversion, the generation of digital signals resulting from sampling analog signals. *See also* **analog decoding, signal, uniform encoding.**

analog facsimile equipment: In facsimile systems, equipment in which analog techniques are used to encode the image detected by the scanner and in which the output is an analog signal. *Note:* Examples of analog facsimile equipment are International Telegraph and Telephone Consultative Committee (CCITT) Group 1 and CCITT Group 2 facsimile equipment.

analog intensity modulation: In an optical modulator, the variation of the intensity of a light source in accordance with an intelligence-bearing signal or continuous wave, with the resulting envelope usually being detectable at the other end of a lightwave propagation medium or system.

analog interpolation: *Synonym* **analog speech interpolation.**

analog multiplier: 1. A device that has an output analog variable that is proportional to the product of two input analog variables. **2.** A device that can perform two or more multiplication operations at the same time. *Note:* An example of an analog multiplier is a servomultiplier.

analog network: An electrical network that can simulate, represent, or serve as a model for some nonelectrical system, such as a mechanical, chemical, or theoretical system. *Note 1:* An example of an analog network is a network of resistors, capacitors, and inductors that can oscillate and thus simulate a swinging pendulum. *Note 2:* Applicable electrical equations can be used to predict the performance of the system simulated by the analog network.

analog radio: Radio in which transmissions consist of continuously varying signals, such as continuous phase and frequency modulated signals rather than discontinuous signals, i.e., discrete signals, such as pulse code modulated (PCM) signals as in radiotelegraph transmission and digital data transmission.

analog representation: 1. A representation of data or physical quantities in the form of a continuous signal. *Note 1:* Usually the instantaneous magnitude of the representation is a function of the value of the data or physical quantity being represented. *Note 2:* An example of an analog representation is the baseband signal in the form of a continuous current produced by a telephone transmitter before any digitization occurs. **2.** A representation of a system or a device that is used to simulate, model, or represent another system or device. *See also* **analog, analog network, digital, discrete.**

analog signal: 1. A signal that has a continuous nature rather than a pulsed or discrete nature. *Note:* Electrical or physical analogs, such as continuously varying voltages, frequencies, and distances, may be used as analog signals. **2.** A nominally continuous electrical signal that varies in some direct correlation with another signal impressed on a transducer. *Note:* Examples of analog signals are signals that vary in frequency, phase, or amplitude in response to changes in physical phenomena, such as sound, light, heat, position, or pressure. *See also* **analog, digital signal, signal.** *Refer to* **Fig. F-2 (339).**

analog speech interpolation: The utilization of the transient inactivity periods that always exist in both transmission directions of a telephone conversation. *Note 1:* The interpolation is used in order to increase the efficiency and lower the cost of long distance conversations. Analog signals with quiescent periods or long vowels are compressed, such as by time assignment speech interpolation (TASI), transmitted, and expanded by inverse compression, i.e., by expansion, at the receiving end. *Note 2:* The principle of analog speech interpolation is being used for digitized speech. When pulse code modulation (PCM) is used, the process is called "digital speech interpolation (DSI)." *Synonyms* **analog compression, analog interpolation.** *See also* **digital speech interpolation.**

analog switch: Switching equipment that connects circuits between users for real time transmission of analog signals. *See also* **circuit, digital switch, switch, switching center.**

analog switching: The interconnecting of input and output terminations in a switching matrix for the purpose of transferring analog signals between them.

analog synchronization: Synchronization in which the relationship between the actual phase error between clocks and the error signal device is a continuous function over a given range. *Synonym* **analog control.** *See also* **synchronization.**

analog telephone: A telephone that has (a) an electrical output that is a nominally continuous electrical signal correlated with the sound pressure variation on its transmitter, i.e., its microphone or sound-electrical transducer, and (b) a sound output that is a nominally continuous sound signal correlated with the electrical variation on its receiver, i.e., its speaker or electrical-sound transducer.

analog to digital (A-D) coder: *Synonym* **analog to digital converter.**

analog to digital converter (ADC): A device that converts an analog signal to a digital signal that represents equivalent information. *Synonyms* **analog to digital coder, analog to digital encoder.** *See also* **digital to analog converter, digital voice transmission, digitizer.**

analog transmission: Transmission of an analog signal, i.e., a continuously varying signal, as opposed to transmission of a digital signal, i.e., a discretely varying signal.

analog variable: A continuously changing phenomenon that represents either a mathematical variable in a mathematical function or a physical quantity. *Note:* Examples of analog variables are (a) a continuously varying electric current that represents speech, (b) light intensity or color that represents the temperature of an incandescent body, (c) the position of a servomechanism, and (d) a baseband television video or audio signal before any digitization or analog to digital conversion occurs.

analysis: *See* **design analysis, dynamic analysis, electromagnetic compatibility analysis, error analysis, Fourier analysis, numerical analysis, path quality analysis, requirements analysis, risk analysis, signal analysis, signature analysis, spectrum analysis, static analysis, system analysis, traffic analysis.**

analysis and planning tool: *See* **architectures design, analysis, and planning tool.**

analyzer: *See* **dynamic analyzer, integrated optical spectrum analyzer, jammer analyzer, light analyzer, lightwave spectrum analyzer, multichannel optical analyzer, program analyzer, pulse analyzer, static analyzer, wave analyzer.**

anamorphic: In optical systems, pertaining to a configuration of optical components, such as lenses, mirrors, and prisms, that produce different effects on an image in different directions or different effects on different parts of the image. *Note:* Examples of anamorphic effects are (a) the production of different magnifications in different directions and (b) the conversion of a point on an object to a line in its image.

AND circuit: *Synonym* **AND gate.**

AND element: *Synonym* **AND gate.**

AND gate: A device that is capable of performing the logical AND operation, such that if P, Q, and R are statements, then the AND of P, Q, and R is true if all statements are true, and false if any statement is false. P AND Q is often represented as P•Q, $P \cap Q$, or simply PQ. *Synonyms* **AND circuit, AND element, AND unit, coincidence gate, coincidence unit, conjunction gate, intersection gate, logic product gate, positive AND gate.** *Refer to* **Fig. A-7 (34).** *Refer also to* **Figs. B-6 (78), K-1 (493), L-6 (529), V-3 (1064).**

AND NOT gate: *Synonym* **A AND NOT B gate.**

AND unit: *Synonym* **AND gate.**

angle: *See* **acceptance angle, borescope articulation angle, borescope axial viewing angle, Brewster angle, convergence angle, critical angle, departure angle, deviation angle, elevation angle, exit angle, horizon angle, incidence angle, launch angle, limiting angle, look angle, maximum acceptance angle,**

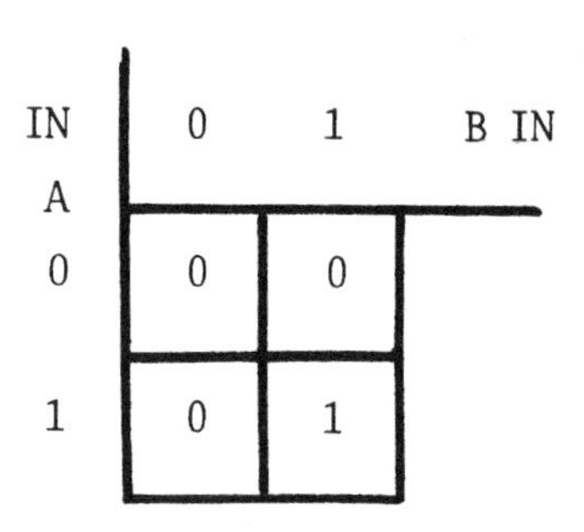

Fig. A-7. A truth table that shows the input and output digits of an **AND gate.**

pointing angle, radiation angle, reflection angle, refraction angle, total acceptance angle, vertex angle.

angle between half-power points: *See* **emission beam angle between half-power points.**

angle break lock: In radar systems, a loss of the ability of tracking radar to continue tracking an object in azimuth or elevation but not necessarily in range.

angle diversity: In tropospheric scattering propagation of electromagnetic waves, the difference created by two or more feeders that produce multiple beams from the same reflector at slightly different launch angles.

angle measuring equipment (AME): In aeronautical aids to navigation on an aircraft, a radio aid to navigation for measuring the vertical angle to a ground-based or sea-based transmitter by determining the phase difference between signals received by antennas located at different positions on the aircraft.

angle modulation (AM): Modulation in which the phase or the frequency of a sine wave carrier is varied in accordance with an information-bearing signal. *Note:* Examples of angle modulation are phase modulation (PM) and frequency modulation (FM). *See also* **deviation ratio, digital phase modulation, frequency modulation, modulation, modulation index, phase deviation, phase modulation.**

angle modulation noise improvement: An improvement in the output signal to noise ratio (S/N) obtained in a receiver through the use of threshold extension demodulators.

angle of deviation: *Synonym* **deviation angle**.

angle of incidence: *Synonym* **incidence angle.**

angle sensor: *See* **critical angle sensor.**

angle tracker: *See* **infrared angle tracker.**

angstrom (Å): A unit of length. *Note 1:* 1 Å = 10^{-10} m (meters) or 10^{-4} μm (microns or micrometers) or 10^{-1} nm (nanometers). *Note 2:* The angstrom is not an SI (International System) unit. *Note 3:* The angstrom is, and historically has been, used in the field of optics.

angular frequency: *Synonym* **angular velocity.**

angular magnification: The ratio of (a) the apparent size of an image, seen through an optical system, element, or instrument, to (b) that of the object viewed by the unaided eye, when both the object and the image are considered to be at infinity, which is the case for telescopes, or when both the object and the image are considered to be at the distance of distinct vision, which is the case for microscopes. *Synonym* **magnifying power.**

angular misalignment loss: In fiber optic systems, a power loss caused by the deviation from optimum angular alignment of the optical axes at a connector or a splice, such as at a light source to optical fiber junction, a fiber to fiber junction, or a fiber to photodetector junction. *Note 1:* Angular misalignment loss is considered to be extrinsic to the fiber. *Note 2:* Angular misalignment loss does not include lateral offset loss and longitudinal displacement loss. *See also* **coupling efficiency, extrinsic joint loss, intrinsic joint loss, lateral offset loss, longitudinal displacement loss**.

angular resolution: In optics and display systems, a measure of the capability of a device to distinguish two points as separate at a given distance, i.e., to distinguish between two divergent directions from a given point. *Note:* The smallest angle that enables two points to be distinguished as separate points is the limiting resolution angle. *See also* **limiting resolution angle.**

angular velocity: 1. The rate of rotation of a rotating body, given by the relation $\varpi = 2\pi f$, where ϖ is the angular velocity, π is approximately 3.1416, and f is the number of revolutions per unit time. *Note 1:* Angular velocity is usually expressed in radians per second. *Note 2:* Angular velocity is a vector quantity, the direction being parallel to the axis of rotation and in the direction of a right-hand screw. **2.** In a sinusoidal electromagnetic wave, 2π times the number of complete cycles of the wave per unit time. *Synonym* **angular frequency.**

ANI: automatic number identification.

anisochronous: Pertaining to transmission in which the time interval separating any two significant instants in sequential signals is not necessarily related to the time interval separating any other two significant instants. *Note:* Isochronous and anisochronous are characteristics, whereas synchronous and asynchronous are relationships. *See also* **heterochronous, homochronous, isochronous, mesochronous, plesiochronous.**

anisochronous signal: A signal in which the interval separating two significant instants is not necessarily a multiple of the unit interval. *Note:* Isochronous and anisochronous are characteristics. Synchronous and asynchronous are relationships. *See also* **heterochronous, homochronous, isochronous, mesochronous, plesiochronous, signal.**

anisochronous transmission: Transmission in which the time interval separating any two significant instants in sequential signals is not necessarily related to the time interval separating any two other significant instants. *See also* **asynchronous transmission, isochronous burst transmission, synchronous transmission.**

anisotropic: 1. Pertaining to a material with properties, such as density, electrical conductivity, electric permittivity, magnetic permeability, refractive index, or composition, that vary with distance or direction. **2.** Pertaining to a material with magnetic, electrical, or electromagnetic properties that vary with the direction of static or propagating magnetic, electrical, or electromagnetic fields within the material. *Note:* Anisotropy is exhibited in (a) certain crystals that have different refractive indices for light rays propagating in different directions or with different polarizations and (b) a piece of iron that demonstrates different magnetic permeabilities to differently directed magnetic fields. *See also* **birefringence, isotropic.**

anisotropic propagation medium: Pertaining to a propagation medium in which properties and characteristics of the medium are not the same in reference to a specific phenomenon. *Note:* An example of an anisotropic propagation medium is a medium in which electromagnetic properties, such as the refractive index, at each point are different for different directions of propagation or for different polarizations of a wave propagating in the medium.

annex storage: *Synonym* **associative storage.**

annunciator: 1. A visual or audible signaling device that indicates the conditions of associated circuits. *Note:* Early annunciators were operated by relays. **2.** A visual or audible signaling device used to make announcements, post messages, or create sound signals, such as the "five o'clock whistle" or a sign made with letters or numbers that can be changed to announce train, bus, or aircraft arrivals and departures.

anode: A positively charged collector of negatively charged particles, such as negative ions and electrons. *See also* **cathode.** *Refer to* **Fig. E-1 (285).**

anomalous propagation (AP): Abnormal propagation caused by discontinuities in the propagation medium. *Note:* In many instances, anomalous propagation (AP) results in the reception of signals well beyond the distances usually expected. *See also* **ionosphere, propagation, sporadic E propagation.**

ANSI: American National Standards Institute.

ANSI/EIA/TIA-568 cabling standard: A U.S. industry standard that (a) specifies a generic telecommunications cabling system, (b) will support a multiproduct, multivendor environment for commercial buildings, (c) specifies performance characteristics for unshielded twisted pair telecommunications cabling, including categories that allow data signaling rates (DSRs) up to 100 Mb•s^{-1} (megabits per second), designated as categories 3, 4, and 5, while categories 1 and 2 have not been defined, and (d) is adopted as FIPS PUB 174.

answer: 1. The transmission made by the call receiver station in response to the call received. **2.** To remove a telephone handset from the cradle of hook, i.e., to go off-hook, in response to a ringing signal. *See* **abbreviated answer, call forwarding don't answer.**

answer back: A signal sent by receiving equipment to the sending equipment to indicate that the receiving equipment is ready to accept transmission from the sending equipment. *See also* **acknowledge character, call control signal, handshaking, signaling.**

answer back code: A unique sequence of characters that identify a particular terminal or data station.

answer back unit: The part of a telegraph station or data terminal equipment (DTE) that automatically transmits its answer back code upon receipt of a who are you (WRU) signal.

answering: A call receiving station responding to a call originating station to complete a connection between the stations. *Refer to* **Fig. S-22 (977).**

answering frequency: The frequency used to respond to a received message. *See also* **calling frequency, crossband frequency, working frequency.**

answering net: A communications net that (a) is usually associated with ship–shore communications systems and (b) is used for answering calls from ships.

answering plug: Of the two plugs associated with any cord circuit in a telephone switchboard, the plug that (a) is nearer to the face of the switchboard and (b) is usually used for answering calling signals.

answering sign: In semaphore communications systems, the flag position that (a) is used as an answer to a call and (b) may be preceded by a call sign to denote the station answering.

answer lamp: A lamp bulb mounted on a telephone switchboard that lights up when a connecting plug of the call receiver is inserted into the corresponding jack of the call originator, goes out when the call receiver answers, and lights up again when the call is finished.

answer list: In communications systems operations, a list of network station names or call signs, along with required control information, used to ensure that a switched line is connected only to authorized users.

answer signal: A supervisory signal returned by the called telephone to the originating switch when the call receiver answers, i.e., the call receiver telephone goes off-hook. *Note 1:* The answer signal stops the ringback signal from being returned to the call originator. *Note 2:* The answer signal is usually returned by means of a closed loop. *See also* **call control signal, loop, signal.**

antenna: 1. A structure or a device used to collect or radiate electromagnetic waves. **2.** A transducer that (a) converts electrical energy from a source, such as a transmitter or a light source, to electromagnetic waves in free space or in a propagation medium, or (b) converts electromagnetic energy from free space or a transmission medium to electrical energy for a receiver or a detector. *Note 1:* A transmitting antenna usually carries the currents that launch the transmitted waves. A receiving antenna usually carries the currents that are induced by the waves that impinge upon it. *Note 2:* Light sources and photodetectors may be considered as antennas. *See also* **aperiodic antenna, biconical antenna, billboard antenna, buoyant antenna, cassegrain antenna, dielectric guide feed antenna, dipole antenna, directional antenna, electrically despun antenna, fan beam antenna, global beam satellite antenna, half-wave dipole antenna, helical antenna, high gain antenna, Hogg horn antenna, horn antenna, image antenna, isotropic antenna, jamming antenna, light antenna, log periodic antenna, lossless antenna, narrow beam antenna, notch antenna, omnidirectional antenna, parabolic antenna, periodic antenna, periscope antenna, reference antenna, reflective array antenna, rhombic antenna, single polarized antenna, slot antenna, telescopic antenna, test antenna, unidirectional antenna, wide beam antenna, Yagi antenna.** *Refer to* **Figs. R-2 and R-3 (799).** *Refer to* **Table 9, Appendix B.**

antenna aperture: The physical area of the main reflector of an antenna, i.e., the radiating area of the source of the transmitted electromagnetic beam, such as a light beam or a radio beam. *Note:* The effective antenna aperture may be less than the physical area because of the blockage caused by intervening parts, such as a subreflector, antenna supports, or waveguide feed supports. In front fed systems, the area is measured in a plane normal to the beam axis. *Synonym* **antenna capture area.**

antenna array: Antenna or light source elements arranged so that the resulting electromagnetic radiation pattern has its main lobe such that the launched waves have maximum radiance, or propagate with maximum irradiance, in the desired direction. *See* **broadside antenna array, collinear antenna array.**

antenna beam rotation: The movement of an antenna beam which may be accomplished in many different ways, such as by mechanical movement of the antenna, phasing of different elements of an array, reflection, tuning, beam switching, holding, lobe switching, and searchlighting.

antenna blind cone: The volume of space, usually approximately conical with its vertex at the antenna, that cannot be scanned by the antenna owing to limitations of the antenna mount and antenna radiation pattern. *Note:* An example of an antenna blind cone is that of an air route surveillance radar (ARSR). The horizontal radiation pattern of the antenna is very narrow. The vertical radiation pattern is fan-shaped, reaching to about 70° of elevation above the horizontal. As the antenna is rotated about a vertical axis, it can illuminate objects, i.e., targets, only if they are 70° or less from the horizontal. Above that elevation angle they are in the antenna blind cone, the apex angle of the antenna blind cone being about 40°.

antenna blockage: The portion of the antenna aperture that is blocked, i.e., shaded, by equipment and structures mounted in front of the antenna, such as subreflectors and subreflector supports.

antenna capture area: *Synonym* **antenna aperture.**

antenna design: The physical, electrical, and magnetic characteristics of an antenna, including its geometrical shape and spatial orientation.

antenna dissipative loss: 1. Signal power absorbed by an antenna and lost by thermal dissipation. *Note:* Antenna dissipative losses usually consist of electrical conductor ohmic losses and insulator dielectric losses. **2.** A power loss calculated as the difference between (a) the loss calculated from the measured impedance of a practical antenna and (b) the power loss calculated for a theoretically perfect antenna.

antenna diversity: *See* **spaced antenna diversity.**

antenna dynamics: The electrical, magnetic, and structural characteristics of an antenna, including light sources, during operating conditions, such as current, voltage, power, impedance, cross section, efficiency, average power, irradiance, instantaneous power, peak power, luminous intensity, directive gain, radiation pattern, shape, size, and orientation.

antenna effective area: The area from which an antenna directed toward the source of the received signal gathers or absorbs the energy of an incident electromagnetic wave. *Note 1:* Antenna effective area is usually expressed in square meters. *Note 2:* In the case of parabolic and horn parabolic antennas, the antenna effective area is about 0.35 to 0.55 of the geometric area of the antenna aperture.

antenna efficiency: The ratio of the total radiated power to the total input power. *Note:* The total radiated power is the total input power to the antenna less antenna dissipative losses.

antenna electrical beam tilt: In a transmitting antenna, the shaping of the radiation pattern in the vertical plane by electrical means so that maximum radiation occurs at an angle below the horizontal plane.

antenna equation: *See* **Hertzian dipole antenna equation.**

antenna feed: The conductor, cable, or waveguide that conveys the energy that is to be transmitted from the transmitter to the antenna. *See also* **antenna, transmitter.**

antenna feed system: A system that (a) transports signals from the output terminal of the transmitter to the antenna or to the antenna radiator, (b) usually consists of (i) the transmitter-antenna feed connector, (ii) the antenna feed, and (iii) the antenna mechanical structure, such as a tower, mast, and radiator support, and (c) does not include (i) the antenna or (ii) the radiator and reflector, if any. *Note:* The antenna, radiator, or antenna reflector is what is being "fed," and therefore is not part of the antenna feed system. *See* **monopulse antenna feed system.** *See also* **antenna, antenna feed, antenna reflector, transmitter.**

antenna gain (AG): The ratio of (a) the power required at the input of a loss-free reference antenna to (b) the power supplied to the input of the given antenna to produce, in a given direction, the same field strength, or the same irradiance, at the same distance. *Note 1:* Antenna gain is usually expressed in dB. *Note 2:* Unless otherwise specified, the antenna gain refers to the direction of maximum radiation. *Note 3:* Antenna gain may be considered for a specified polarization. *Note 4:* Depending on the choice of the reference antenna, a distinction is made between (a) absolute or isotropic gain (G_i) when the reference antenna is an isotropic antenna isolated in space, (b) gain relative to a half-wave dipole (G_d) when the reference antenna is a half-wave dipole isolated in space and with an equatorial plane that contains the given direction, and (c) gain relative to a short vertical antenna (G_r) when the reference antenna is a linear conductor, much shorter than one quarter of the wavelength, normal to the surface of a perfectly conducting plane that contains the given direction. *Synonyms* **antenna power gain, gain of an antenna, power gain of an antenna.** *See also* **antenna, aperture to medium coupling loss, directive gain, gain, radio transmitter mean power.**

antenna gain contour: *See* **effective antenna gain contour.**

antenna gain to noise temperature (G/T): A figure of merit, given by the relation *G/T* where *G* is the antenna gain in dB at the receive frequency and *T* is the equivalent noise temperature of the receiving system in kelvins. *See also* **antenna gain, noise.**

antenna height above average terrain: The antenna height above the terrain elevation averaged over the distance from 3.2 to 16 kilometers, i.e., 2 to 10 miles, from the antenna and averaged over eight directions evenly spaced in azimuth. *Note 1:* The directional antenna height above average terrain is usually obtained for each of the eight directions evenly spaced by 45° in azimuth. *Note 2:* Usually a different directional antenna height above average terrain will be obtained for each of the eight directions from the antenna. *Note 3:* The average of the eight directional antenna heights above average terrain is considered the antenna height above average terrain. *Note 4:* In some cases, fewer than eight directions may be used. *See also* **antenna.**

antenna interference: The mutual interference caused by electromagnetic coupling that occurs when two or more antennas are closely located.

antenna length: *See* **effective antenna length.**

antenna lobe: A three-dimensional section of the radiation pattern of a directional antenna bounded by one or more cones of nulls, i.e., by one or more regions of diminished field strength or diminished irradiance. *See also* **fan beam antenna, main beam, main lobe, radiation pattern, side lobe.**

antenna matching: The adjusting of the impedance of an antenna so that its input impedance equals or approximates its transmission line characteristic impedance over a specified range of frequencies. *Note:* The impedance of the transmission line, the antenna, or both may be adjusted to effect the match. *See also* **balun, impedance matching.**

antenna mount: The structure that supports an antenna. *See* **nonorthogonal antenna mount.**

antenna multiplier: A device connected to an antenna that permits several pieces of equipment to be connected to one antenna. *Note:* Though the use of an antenna multiplier reduces the number of antennas, the channel frequency spacing will have to be greater than when a separate antenna is used for each frequency in the channel in order to avoid excessive interference.

antenna noise temperature (ANT): The temperature of a hypothetical passive resistor at the input of an ideal noise free receiver that would generate the same output noise power per unit bandwidth as that at the antenna output at a specified frequency. *Note:* The antenna noise temperature depends on (a) the antenna coupling to all noise sources in its environment and (b) the noise generated within the antenna. *See also* **noise.**

antenna pattern: *Synonym* **radiation pattern.**

antenna power gain: *Synonym* **antenna gain.**

antenna reflector: The portion of a directional antenna array that reduces the field strength behind the array and increases it in the forward direction.

antenna rotation: Antenna movement which (a) may be accomplished in many different ways, such as by mechanical movement of the antenna, by phasing of different elements of an array, or by reflection methods, and (b) may assume different forms, such as irregular motion, tuning, beam switching, holding, lobe switching, searchlighting, or phase control.

antenna rotation period: The time required for one complete revolution or movement cycle of an antenna. *Note:* The antenna rotation period is usually expressed in seconds. The sweep rate is usually expressed in revolutions per minute.

antenna sweep: The space angle within which a mutating or oscillating antenna moves relative to its base or other fixed or moving reference. *Note:* In addition to a mutating or oscillating motion, an antenna may be sweeping in a vertical or horizontal direction. Antenna sweep may be designated as fixed, continuous, or sector.

anticlash key: On a typewriter, a device that (a) is used to restore jammed type bars to their resting position and (b) usually is in the form of a key or a lever.

anticlockwise polarized wave: *Synonym* **left-hand polarized wave.**

anticoincidence gate: *Synonym* **EXCLUSIVE OR gate.**

anticoincidence unit: *Synonym* **EXCLUSIVE OR gate.**

anticyclone: A distribution of atmospheric pressure that (a) has a pressure that increases toward a center of high pressure, (b) has winds that circulate in a clockwise direction in the Northern Hemisphere and in a counterclockwise direction in the Southern Hemisphere as viewed from above, (c) gives rise to clear, calm weather conditions although in winter fog is likely to occur, (d) usually advances at 30 $km{\bullet}hr^{-1}$ (kilometers per hour) to 50 $km{\bullet}hr^{-1}$, (e) has a diameter of 2500 km to 4000 km, and (f) often brings cool, dry weather.

antielectronic jamming: Organizational, tactical, strategic, and technical measures taken to overcome the effects of electronic jamming. *See also* **electronic jamming.**

antiinterference: In radio communications, pertaining to equipment used or actions taken to reduce the effect of natural and man-made noise.

antijam: In radio communications, pertaining to equipment used and actions taken to reduce the effects of jamming on a desired signal. *See also* **electronic countercountermeasure, electronic warfare, frequency hopping, spread spectrum.**

antijamming: 1. Pertaining to the ability of a device or system to withstand jamming or interference without seriously degrading its performance. **2.** Pertaining to measures that are taken to reduce the effects of attempts to deliberately cause interference in a system.

antijamming margin: The maximum jamming to signal power, voltage, or current ratio at which a device will still continue to satisfactorily perform its intended function. *See also* **jamming to signal ratio.**

antijamming measure: A measure taken to minimize the effect of jamming or to reduce the effectiveness of jamming efforts.

antilog: The number that corresponds to a logarithm. *Note 1:* An example of an antilog is a gain of 50, which corresponds to a gain of about 1.70 B (bel), which is the log to base 10 of 50. Also, 1.70 B is equivalent to 17.0 dB if the gain is a power gain and 34.0 if the gain is a voltage gain. *Note 2:* Antilogs, and logarithms, can be had in many number systems, such as number systems to base *e*, 2, 10, and 16.

antinode: A point in a standing wave, i.e., in a stationary wave, at which the amplitude is a maximum, i.e., there is a crest. *Note:* The type of wave should be identified, such as a voltage wave or a current wave. *See also* **node, standing wave ratio.**

antireflection coating: A thin, dielectric or metallic film, or several such films, applied to an optical surface to reduce its reflectance and thereby increase its transmittance. *Note:* For minimum reflection, the ideal value of the refractive index of a single layer antireflection coating is the square root of the product of the two refractive indices, one on each side of the coating. For minimum reflectance, the ideal thickness of the coating is one quarter of the wavelength of the incident lightwave. A precise thickness of coating will cancel the reflection from the interface surface by destructive interference. *See also* **dichroic filter, Fresnel reflection, index matching material, reflectance, transmittance.**

antisinging device: A device that prevents singing, i.e., prevents resonant oscillations, in a circuit. *Note:* Resonant oscillations often occur in circuits, such as audio circuits, when positive feedback occurs. *See also* **singing suppression circuit, volcas circuit.**

antispoof: Pertaining to measures taken to prevent (a) participation in a telecommunications network or (b) operation or control of a cryptographic or communications security (COMSEC) system.

anycall: In adaptive high frequency (HF) radio automatic link establishment (ALE), a broadcast in which (a) the called stations are unspecified, (b) stations receiving the call stop scanning, and (c) each station automatically responds in pseudorandom time slots. *See also* **allcall.**

a-o: acoustooptic.

A OR NOT B gate: A two-input binary logic coincidence circuit or device capable of performing the logic operation of A OR NOT B, i.e., if A is a statement and B is a statement, the result is false only if A is false and B is true, and for the other three combinations of A and B the result is true. *Synonyms* **B implies A gate, if B then A gate, implication gate, inclusion gate.** *Refer to* **Fig. A-8.** *Refer also to* **Fig. B-8 (92).**

AP: anomalous propagation.

APC: adaptive predictive coding.

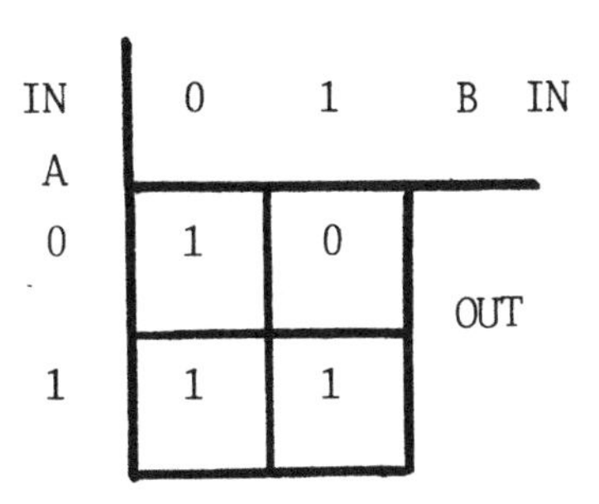

Fig. A-8. A truth table that shows the input and output digits of an **A OR NOT B gate.**

a.p.d.: avalanche photodiode.

apd: avalanche photodiode.

APD: avalanche photodiode.

APD coupler: avalanche photodiode coupler.

aperiodic antenna: An antenna designed to have an approximate constant input impedance over a wide range of frequencies. *Note:* Examples of aperiodic antennas include terminated rhombic antennas and wave antennas. *Synonym* **nonresonant antenna.** *See also* **antenna.**

aperture: For a directional antenna, the portion of a plane surface that (a) is near the antenna, (b) is normal to the direction of maximum irradiance, and (c) conveys the major part of the radiation. *Note:* In the movement or passage of any entity, such as a fluid, an electromagnetic wave, a sound wave, or time, an aperture is an opening or window in the path or period of time through which or during which the entity is constrained to flow or occur. Apertures can be controlled so as to perform coupling, guidance, and control functions. The size of the aperture is a function of its physical dimension, the direction of flow, the boundary conditions, and parameters of the moving entity, such as the viscosity of fluids or the wavelength of light. The dimensions of an aperture may be expressed in terms of length, width, area, planar angle, solid angle, and time units as well as in terms of dimensionless or normalized numbers. The numerical values may be relative, that is, normalized, or they may be absolute. Apertures usually occur at interfaces or transitional points, such as at a radiating antenna or at the entrance to an optical fiber. Apertures usually introduce losses and restrict passage. For example, the aperture of a lens depends on the size of the uncovered portion of the lens, the refractive index of the glass, and the color (wavelength) of the incident light. The aperture presented by a device may also depend on the incidence angle of the entering light. The aperture of an electromagnetic wave antenna is limited by the wavelength and the dimensions of the feeder, antenna array, or dish. For a phased array antenna, the aperture is limited by the dimensions of the array, the launch angle, and the wavelength. In an optical fiber, the aperture is the acceptance cone angle as determined by the numerical aperture. For a sound wave, the aperture may be determined by the size of the hole in an inelastic material and the wavelength. For fluids, the aperture might be the size of a hole in the side of a container or a constriction in a pipe, in which case the size of the aperture may depend on the dimensions of the hole or the constriction, the turbulence, and the viscosity. For granular substances, the hole must be large enough to sustain flow, or else the effective aperture is zero. *See also* **launch numerical aperture, numerical aperture, output aperture, tone control aperture.** *Refer to* **Fig. N-21 (641).**

aperture distortion: In facsimile transmission systems, the effect caused by the finite shape and dimensions of the scanning spot of the transmitter and the recording spot at the receiver. *Note 1:* Aperture distortion may occur in one or more attributes of the recorded image, such as in resolution, density, or shape. *Note 2:* Aperture distortion causes contours of images to become blurred, and details that are smaller than the scanning spot are suppressed. *See also* **facsimile, X dimension of recording spot, X dimension of scanning spot, Y dimension of recording spot, Y dimension of scanning spot.**

aperture loss: *See* **numerical aperture loss.**

aperture ratio: The value of R_a in the relation $R_a = 2n \sin A$, where R_a is the aperture ratio, n is the refractive index of the image space, and A is the maximum angular opening of the axial bundle of refracted rays. *Note:* The speed, i.e., the energy per unit area of images, of an objective is proportional to the square of its aperture ratio. When the angular opening is small, when $n = 1$, as for air, and when the object distance is great, (a) $n \sin A = D/2F$ and (b) F/D = F-number or $1/(2 \sin A)$ or 1/(aperture ratio). *See also* **F-number.**

aperture stop: The physical diameter that limits the size of the cone of radiation that an optical system will accept from an object point on the optical axis of a lens system. *See also* **axial bundle, exit pupil, field stop.**

aperture tagging: In image restoration, wavefront control in which the distortion of the wavefront of an electromagnetic wave, introduced when passing through an aperture, is compensated by prior or after-the-fact methods, i.e., trial perturbations are made in the outgoing wavefront, and the resulting variations in power reflected from an object are analyzed to optimize the irradiance in specific portions of the wavefront.

aperture to medium coupling loss: The difference between (a) the theoretical gain of a very large antenna, such as those used in beyond-the-horizon microwave links, and (b) the gain that can be realized in operation. *Note 1:* Aperture to medium coupling loss is related to the ratio of the scatter angle to the antenna beamwidth. *Note 2:* Very large antennas are referred to in wavelengths. Aperture to medium coupling loss also applies to line-of-sight systems. *See also* **antenna, antenna gain, coupling, coupling loss, loss.**

APEX: Advanced Project for Information Exchange.

API: application program interface.

APL: A computer programming language primarily used in a conversational mode, i.e., interactive mode, for mathematical applications with multidimensional arrays. *Note:* The name is derived from "A Programming Language" or from "Array Programming Language."

aplanatic lens: A lens that has been corrected for spherical aberration, coma, color, and departure from the sine condition.

apogee: In an orbit of a satellite orbiting the Earth, the point that is farthest from the gravitational center of the Earth. *See also* **apogee altitude, geostationary orbit, perigee, perigee altitude, satellite.**

apogee altitude: The vertical distance from the apogee to a specified reference surface serving to represent the surface of the Earth. *See also* **apogee, perigee, perigee altitude.**

apparent power: In alternating current (ac) power generation, transmission, and distribution, the product of the root mean square (rms) voltage and the rms amperage. *Note 1:* The apparent power is equal to the square root of the sum of the squares of the effective power, i.e., the real power, and the reactive power. *Note 2:* The apparent power is only the effective power when the voltage and the current are in phase, in which case the reactive power is 0. When they are 90° out of phase, the apparent power is only the reactive power, in which case the effective power is 0. *Note 3:* Apparent power is expressed in volt•amperes, not watts. *Note 4:* The apparent power times the cosine of the phase angle between the voltage and the current is the effective power. The apparent power times the sine of the phase angle between the voltage and the current is the reactive power. *Synonym* **volt•amperes.**

application: *See* **spatial application, temporal application.**

application language: *Synonym* **application oriented language.**

application layer: In open systems architecture, the layer that (a) is directly accessible to, visible to, and usually explicitly defined by users and (b) provides all the functions and services needed to execute their programs, processes, and data exchanges. *See also* **layer, Open Systems Interconnection—Reference Model.**

Application Layer: *See* **Open Systems Interconnection—Reference Model.**

application network: A network of interconnected data processing equipment, such as computers, processors, controllers, and terminals, used for processing and exchanging user data. *Note:* An application network may include public or private communications facilities provided by private organizations, common carriers, or recognized private or public operating administrations or agencies. *Synonym* **computer network.**

application-oriented language: 1. A computer programming language that has capabilities or notation schemes specially suited for solving problems in specific classes of applications, such as communications, scientific, business, engineering, design, or simulation applications. **2.** A language in which statements contain the terminology of the occupation, profession, or other interests or pursuits of the user. *Synonyms* **application language, problem-oriented language.**

application parameter: In the design of an optical station/regenerator section, a performance parameter specified by the manufacturer or the user, such as the application type (aerial, buried, underwater (salt, fresh, or brackish), or underground) temperature range, cabled optical fiber reel length, nominal central wavelength, central wavelength range, type of splice, splice insertion loss, cable designation, maximum cable cutoff wavelength, maximum cable attenuation rates and expected increases, dispersion parameters, interconnection parameters (cladding diameter, cladding ovality, core/cladding concentricity errors, mode field diameter), and global fiber parameters (standard and extended).

application program: A computer program that directly meets the specific need of a user. *Note:* Examples of application programs are payroll, inventory control, database, and traffic control programs.

application program interface (API): A formalized set of software-controlled calls and routines that can be referenced by an application program in order to access supporting network services.

application software: Software designed for a specialized use. *Note:* Examples of application software are computer programs for navigation, gunfire control, payroll preparation, telemetry, message routing, database management, and communications network control.

application-specific integrated circuit (ASIC): An integrated circuit that is designed and built to perform a particular set of functions, such as (a) execute a specific computer program using particular sets of input data, (b) control an aircraft or missile using input from distributed sensors, (c) direct operations on a satellite using onboard sensors and instructions from ground stations, and (d) control traffic on railroad, highway, and communications systems.

applique: Circuit components added to an existing system to provide additional, alternate, or modified functions. *Note:* An example of applique is a circuit used to modify carrier telephone equipment designed for ringdown manual operation to allow its use between points having dial equipment. *See also* **switching center.**

appointment directory: A telephone directory in which (a) each telephone number designates the title, appointment, assignment, or function of that user, (b) personnel changes do not affect the directory, (c) a reorganization of the user group, with new titles and assignments, will affect the directory, (d) entries usually are alphabetical by title, appointment, assignment, or function, and (e) there are no names of individual persons. *See also* **fixed directory.**

approach: *See* **carrier controlled approach.**

approach aid: *See* **radio approach aid, runway approach aid.**

approach control service: A communications system used to communicate with aircraft stations that are approaching an airport.

approved circuit: A circuit that has been authorized by responsible authority for the transmission of information. *Note:* Approved circuits usually relate only to those wireline or optical fiber systems to which electromagnetic and physical safeguards have been applied to permit transmission of unencrypted information. The circuit includes all the individual electrical conductors and optical fibers in the path. From a security or privacy standpoint, radio circuits usually are not considered as approved circuits. *See also* **protected distribution system, protected optical fiber distribution system, protected wireline distribution system.**

APT: A computer programming language that is used primarily for the numerical control of machine tools. *Note:* The name is derived from "Automatically Programmed Tools."

aramid yarn: A tough synthetic yarn that is often used in fiber optic cable construction, such as for strength members, protective braids, or rip cords for jacket removal.

arbitrary sequence computer: A computer in which each instruction explicitly specifies or determines the location of the next instruction to be executed. *Note:* An example of an arbitrary sequence computer is a computer that is capable of executing multiple address instructions in which one of the addresses is the address of the next instruction to be executed.

Archie: A distributed system server that (a) searches indices of files available on public servers in the Internet, (b) may provide access via Telnet, E-mail, or a special client, and (c) requires a user to be familiar with where the indices are located, i.e., the user must provide the domain name or the internet protocol (IP) address. *See also* **Internet, server.**

architecture: In communications, computer, and data processing systems, the organizational structure of an entity, such as a telephone network, a satellite communications system, or a radio network. *See* **client-server architecture, computer architecture, information system architecture, network architecture, open network architecture, open system architecture, systems architecture, telecommunications system architecture.**

architectures design, analysis, and planning tool (ADAPT): 1. A software tool that (a) was developed by the U.S. government, (b) is geotechnical modeling software, and (c) is a shell around (i) AutoCAD, a personal computer-based computer aided design (CAD) tool, (ii) Oracle, a relational database, and (iii) a geographic database. **2.** Network modeling software that (a) documents connectivity of networks, (b) enables network components to be described graphically, (c) allows description of geographic location, (d) uses icons to represent equipment, and (e) allows past, present, budgeted, and future representation of networks to be overlaid in different colors.

archive: *See* **Automated Recourse to Electronic Negotiations Archive.**

archiving: In database management, the storing of records, such as files and journals, usually for a long time in a relatively slow-access, inexpensive data medium for backup or historical purposes. *Note:* An example of archiving is the storing of former employee personnel records on magnetic tape.

area: *See* **antenna effective area, blanketing area, blind area, broadcast area, cladding tolerance area, coherence area, controlled area, coordination area, core area, core tolerance area, critical area, data numbering plan area, effective boresight area, elemental area, exchange area, input area, local service area, maximum call area, maximum calling area, output area, primary service area, radar altimetry area, restricted area, save area, storage area, technical area, white area.**

area broadcast shift: The changing from (a) listening to transmissions intended for one broadcast area to (b) listening to transmissions intended for another broadcast area. *Note 1:* An area broadcast shift may occur when a ship or an aircraft crosses the boundary between listening areas. *Note 2:* Shift times on the date that a ship or an aircraft is expected to pass into another area must be strictly observed, or the ship or the aircraft will miss messages intended for it. *Synonym* **radio watch shift.**

area broadcast station: 1. A land station responsible for radio broadcasting to a specific geographical area of the Earth. **2.** A radio station responsible for broadcasting to one of the 12 numbered areas into which the world has been divided for operating the merchant ship broadcast system.

area code: A telephone system number that has been assigned to a specific geographical area. *See* **access code, code, country code, NXX code.**

area communications organization: The communications organization that provides communications between radio-equipped organizations in a given area, including ship stations, shore stations, and aircraft stations, particularly maritime patrol aircraft on a mission, ships at sea, and aircraft in flight.

area coverage: *See* **jammer area coverage.**

area information server: *See* **Wide Area Information Server.**

area loss: When the end faces of optical fibers are butted, i.e., are joined, by a splice or by a pair of mated connectors, the power loss that (a) is caused by a mismatch in size or shape of the cross section of the cores of the mating fibers, (b) may consist of light from the core of the transmitting optical fiber, i.e., the source fiber, that enters the cladding of the receiving fiber, where it is lost within a very short distance from the joint, and (c) may be dependent on the direction of propagation, such that in coupling a signal from an optical fiber having a smaller core to an otherwise identical one having a larger core, there will be no loss, but in the opposite direction, there will be a loss. *See also* **fiber optic splice, optical fiber.**

area network: *See* **local area network, metropolitan area network, wide area network.**

area radio station: *See* **ship broadcast area station.**

area service: *See* **extended area service.**

area signaling service: *See* **custom local area signaling service.**

ARENA: Automated Recourse to Electronic Negotiations Archive.

argument: 1. An independent variable. **2.** Any value of an independent variable. *Note:* Examples of arguments include search keys, numbers that identify the location of data in a table, and the Θ in sin Θ.

arithmetic: *See* **modulo-n arithmetic.**

arithmetic check: *Synonym* **mathematical check.**

arithmetic expression: An expression that specifies a simple arithmetic operation to be performed on numbers. *Note:* Examples of arithmetic expressions are 2 + 3, 6 − 9, and 13 ÷ 4. *See also* **algebraic expression.**

arithmetic-logic unit (ALU): The part of a computer that performs arithmetic, logic, and related operations.

arithmetic operation: An operation performed according to the rules of arithmetic, in contrast to a logical operation. *See* **binary arithmetic operation.** *See also* **logical operation.**

arithmetic overflow: 1. In a digital computer, the condition that occurs when a calculation produces a result that is larger than a given register or storage location can store or represent. **2.** In a digital computer, the amount that a calculated value is larger than a given register or storage location can store or represent. *Note:* The overflow may be placed at another storage location. *See also* **arithmetic underflow.**

arithmetic register: A register that holds the operands or the results of operations, such as arithmetic operations, logic operations, and shifts.

arithmetic shift: A shift, with respect to a reference, of the representation of a number in a fixed radix numeration system. *Note 1:* In a floating point representation of a number, an arithmetic shift applies only to the fixed point part. *Note 2:* An arithmetic shift is usually equivalent to multiplying the number by a positive or a negative integral power of the radix, except for the effect of any rounding. *Note 3:* A representation of a number is a numeral.

arithmetic underflow: 1. In a digital computer, the condition that results when a calculation produces a nonzero result that is smaller than the smallest nonzero quantity that the computer storage unit can store or represent. **2.** In an arithmetic operation, a result that has an absolute value too small to be represented within the range of the numeration system in use, particularly when a floating point number representation system is used. *See also* **arithmetic overflow.**

arithmetic unit: In a processor, the part (a) that performs arithmetic operations and (b) that may perform both arithmetic and logical operations.

Armed Forces Radio Service (AFRS): A radio broadcasting service that is operated by and for the personnel of the armed services in the area covered by the broadcast. *Note:* An example of an AFRS is the radio service operated by the U.S. Army for U.S. and allied military personnel on duty in overseas areas.

armor: In a communications cable, a component that (a) protects the critical internal components, such as buffer tubes, optical fibers, or electrical conductors, from damage from external environmental conditions, such as mechanical abuse, rodent attack, fish bite, and abrasion, (b) usually consists of a steel or aluminum tape wrapped about an inner jacket that covers the critical internal components, and (c) usually is covered by an outer jacket. *Note:* Armor usually is used on cables installed in harsh environments, such as beach and river crossings, ship-to-shore tethers, and aerial inserts. *See* **overarmor.**

Army Special Operations Command Network (ASOCNet): A U.S. Army worldwide information network that (a) links forces in the field with commanders and staff and (b) combines existing equipment with new communications equipment and architecture.

ARN: air reporting net. *See* **maritime patrol air reporting net.**

ARP: Address Resolution Protocol.

ARPANET: Advanced Research Projects Agency Network.

ARQ: automatic recovery quotient, automatic repeat request.

arrangement: *See* **basic serving arrangement, common control switching arrangement, data access arrangement, remote trunk arrangement.**

array: 1. An arrangement of elements in one or more dimensions. **2.** In a programming language, an aggregate that consists of data objects with identical attributes, each of which may be uniquely referenced. *See* **antenna array, broadside antenna array, collinear antenna array, phased array, planar array, programmable logic array, sensor array, solar array, uniform linear array.**

array antenna: *See* **reflective array antenna.**

array processor: A processor that executes instructions in which the operands may be arrays rather than data items. *Synonym* **vector processor.**

arrester: A device that protects hardware, such as systems, subsystems, circuits, and equipment, from voltage or current surges that may be produced by lightning or an electromagnetic pulse (EMP). *Note:* If the hardware is adequately protected, associated software may also be adequately protected. *See also* **air terminal, lightning down conductor, protector.**

arrow key: One of the four keys on a standard keyboard that is used to move a cursor or pointer up, down, right, or left on the display surface of a display device.

ARS: automatic route selection.

Artemis: A vehicle license plate and traffic bottleneck recognition, imaging, and identification system developed by Siemens Plessey.

arteriovenous oximeter: *See* **fiber optic arteriovenous oximeter.**

articulation angle: *See* **borescope articulation angle.**

articulation index: A measure of the intelligibility of voice signals. *Note 1:* The articulation index is usually expressed as a percentage of speech units that are understood by the listener when heard out of context. *Note 2:* The articulation index is affected by noise, interference, and distortion.

artificial intelligence (AI): Pertaining to equipment that performs functions that (a) are usually associated with human intelligence or (b) are similar to functions performed by humans, such as reasoning, learning, decision making, and self-improvement. *Note:* Artificial intelligence (AI) is the branch of computer science that attempts to approximate the results of human reasoning by organizing and manipulating factual and heuristic knowledge. Areas of AI activity include expert systems, natural language understanding, speech recognition, vision, and robotics.

artificial language: A language in which symbols, conventions, and rules for its use are explicitly established before its use. *Note:* Examples of artificial languages are computer programming languages, such as PL/1, ALGOL, FORTRAN, BASIC, COBOL, and Ada. *Synonym* **formal language.** *See also* **natural language.**

artificial pupil: A beam-limiting device, such as a diaphragm, that confines a beam of light to a cone with a smaller apex angle than the cone would have without the limiting device.

artificial transmission line: A four-terminal electrical network, i.e., an electrical circuit, that has the characteristic impedance, transmission time delay, phase shift, or other parameters of a real transmission line and therefore can be used to simulate a real transmission line in one or more of these respects. *Note:* An artificial transmission line can be used to study, and predict the performance of, a real transmission line. *Synonym* **art line.**

art line: *Synonym* **artificial transmission line.**

ARU: audio response unit.

ASAS: all-source analysis system.

ASCII: American Standard Code for Information Interchange.

ASE: automatic switching equipment.

ASIC: application-specific integrated circuit.

ASOCNet: Army Special Operations Command Network.

ASP: adjunct service point.

aspect: *See* **image aspect.**

aspect ratio: In facsimile or television systems, the ratio of the width to the height of a recorded copy, object, picture, document, or scanning field. *See also* **facsimile.**

ASS: automatic switching system.

assemble: To translate a computer program expressed in an assembly language into a machine language.

assemble and go: Pertaining to computer programming and operation in which there are no stops or breaks (a) between assembling and loading and (b) between loading and execution of a computer program.

assembler: A computer program that is used to assemble. *Synonym* **assembly program.** *See* **cross assembler.** *See also* **compiler, translator.**

assembler/disassembler: *See* **packet assembler/disassembler.**

assembly: An item that (a) usually consists of replaceable parts, (b) is usually a portion of a larger piece of equipment, subsystem, or system, and (c) can be provisioned and replaced as an entity. *See* **cable assembly, fiber optic cable assembly, fiber optic harness assembly, multiple bundle cable assembly, multiple fiber optic cable assembly.** *See also* **communications subsystem, component.**

assembly-disassembly facility: *See* **packet assembly-disassembly facility.**

assembly language: A computer-oriented language (a) in which instructions are symbolic and usually in one-to-one correspondence with sets of machine language instructions and (b) that may provide various facilities, such as the use of macroinstructions. *Synonym* **computer-dependent language.** *See also* **compile, computer language, computer-oriented language, high level language, language, machine language.**

assembly phase: 1. In the execution of a computer program, the logical subdivision of a computer run that includes the execution of an assembler. **2.** In computer operations, the logical subdivision of a run that includes the execution of an assembler. **3.** In the production of an item, such as a system, the phase in which the subsystems, component, or parts are connected to form the whole item.

assembly program: *Synonym* **assembler.**

assembly time: The elapsed time required to execute an assembler. *See also* **assembler, compiler.**

assertions method: *See* **inductive assertions method.**

assessment: *See* **operations security assessment, reliability assessment.**

assigned frequency: 1. The center of the assigned frequency band assigned to a station. *See also* **assigned frequency band. 2.** The frequency of the center of the radiated bandwidth. *Note:* The frequency of the radio frequency (rf) carrier, whether suppressed or radiated, is usually given in parentheses following the assigned frequency, and is the frequency appearing in the dial settings of rf equipment intended for single sideband or independent sideband transmission. *See also* **authorized frequency, bandwidth, carrier, center frequency, frequency, frequency allocation, frequency assignment, frequency tolerance.**

assigned frequency band: The frequency band authorized for the emissions of a station. *Note 1:* The width of the assigned frequency band is equal to the necessary bandwidth plus twice the absolute value of the frequency tolerance. *Note 2:* For space stations, the assigned frequency band also includes twice the maximum Doppler shift that may occur in relation to any point of the Earth's surface. *See also* **authorized bandwidth, frequency allocation, radio frequency allotment.**

assignment: In telecommunications systems for National Security Emergency Preparedness (NS/EP), the designation of priority levels. *See* **demand assignment, frequency assignment, priority level assignment, routing indicator assignment.** *See also* **priority level.**

assignment access plan: *See* **demand assignment access plan, preassignment access plan.**

assignment multiple access: *See* **demand assignment multiple access.**

assignment plan: *See* **frequency assignment plan.**

assistance: *See* **dial service assistance.**

assistance switchboard: *See* **dial service assistance switchboard, service assistance switchboard.**

assisted call: *See* **operator-assisted call.**

associated common channel signaling (ACCS): Common channel signaling in which (a) the signal channel is associated with a specific trunk group and terminates at the same pair of switches as the trunk group and (b) the signaling is usually accomplished by using the same facilities as the associated trunk group uses. *See also* **common channel signaling, nonassociated common channel signaling.**

Association Français Normal (AFNOR): The national standards-setting organization of France. *Note:* The Association Français Normal (AFNOR) provides the Secretariat for International Organization for Standardization Technical Committee 97, Subcommittee 1 (ISO TC97/SC 1), Information Technology Vocabulary, which includes vocabulary for computers, communications, information processing systems, and office machines.

associative storage: 1. Storage in which locations are identified by their contents, or by a part of their contents, rather than by their names or positions. *Synonyms* **annex storage, content addressable storage. 2.** Storage that supplements another storage.

assumed value: A value that (a) lies within a specified range of values, parameters, or levels, (b) is assumed for use in a mathematical model, hypothetical circuit, or network, (c) is used as the basis for analyses, additional estimates, or calculations, (d) is not measured, (e) represents the best engineering judgment, (f) usually is derived from values found or measured in real circuits or networks of the same generic type, and (g) includes projected improvements. *See also* **design objective.**

assurance: *See* **quality assurance.**

astigmatism: An aberration of a lens or a lens system that causes an off-axis point on an object to be imaged as two separated lines perpendicular to each other.

astronomy: *See* **radio astronomy.**

astronomy service: *See* **radio astronomy service.**

astronomy station: *See* **radio astronomy station.**

asymmetrical modulator: *Synonym* **unbalanced modulator.**

asymmetry: *See* **bidirectional asymmetry.**

asynchronous: Pertaining to a lack of any time relationship among the occurrences of events, such as signals, operations, or spatial movements, i.e., the time relationships of these events are unpredictable.

asynchronous communications system (ACS): A data communications system in which operations are performed asynchronously. *Note 1:* An example of an asynchronous communications system is one in which signal elements are appended to the data for synchronizing individual characters or blocks. *Note 2:* The interval between successive characters or blocks may be of arbitrary duration. *Synonym* **start-stop system.** *See also* **asynchronous operation, block, character, communications.**

asynchronous data transmission channel: A data transmission channel in which separate timing information is not transferred between the data terminal equipment (DTE) and the data circuit terminating equipment (DCE). *Synonym* **nonsynchronous data transmission channel.** *See also* **channel, clock, network.**

asynchronous digital computer: A digital computer in which (a) each event, such as the execution of each operation, starts as a result of a signal generated by the completion of the previous event and (b) the next event may proceed when the parts or components required for the event are available. *See also* **synchronous digital computer.**

asynchronous network: A network in which the clocks do not need to be synchronous or mesochronous. *Synonym* **nonsynchronous network.** *See also* **clock, network.**

asynchronous operation: 1. A sequence of operations in which operations are executed out of time coincidence with any event. **2.** An operation that occurs without a regular or predictable time relationship to a specified event. *Note:* An example of an asynchronous operation is the calling of an error diagnostic routine that may receive control at any time during the execution of a computer program. *Synonym* **asynchronous working.**

asynchronous satellite: A satellite that revolves about its parent body independently of the rotation of the parent body about its axis. *Synonym* **nonsynchronous satellite.** *See also* **satellite.**

asynchronous signal: A signal in a group of signals, such as occurs in bit stream transmission, arranged in such a manner that (a) between any two significant instants in the same group there is always an integral number of unit intervals, (b) between two significant instants in different groups there is not always an integral number of unit intervals, and (c) each group begins with a start signal, such as the start signal of each character in start-stop telegraphy. *Note 1:* In data transmission, a group is a block or a character; in telegraphy, a group is a character. *Note 2:* Asynchronous signals occur in an asynchronous counter in which the change of output state of each flip-flop is dependent on the previous flip-flops having changed state, i.e., the flip-flops do not all change state at the same time because it takes a finite time for changes to propagate throughout the bank of flip-flops. *See also* **synchronous signal.**

asynchronous system: 1. A system in which operations are executed out of time coincidence with any event. *Note:* In an asynchronous system, transmission is such that (a) between similar significant instants in the same data group there is always an integral number of unit intervals, (b) there is not always an integral number of unit intervals between two significant instants located in different groups, (c) the group usually is a block or a character, (d) the time of occurrence of the start of each character, or block of characters, is arbitrary, and (e) once started, the time of occurrence of each significant instant, in the signal representing a bit that is within the character or the block, has the same relationship to the significant instants of a fixed time frame. **2.** A system in which operations occur without a regular or predictable time relationship with specified events. *Note:* An example of an asynchronous system is a system in which an error diagnostic routine assumes control during the execution of a computer program. *Synonym* **nonsynchronous system.** *See also* **block, character, intercharacter interval, isochronous, plesiochronous, synchronous system, synchronous transmission.**

asynchronous time division multiplexing (ATDM): Time division multiplexing in which asynchronous transmission is used. *See also* **multiplexing, synchronous time division multiplexing, time division multiplexing.**

Asynchronous Transfer Mode (ATM): A communications protocol that (a) enables the transmission of voice, data, image, and video signals over wide-area, high-bandwidth communications systems; (b) provides fast packet switching in which information is inserted in small, fixed-size cells (32 to 120 octets) that are multiplexed and switched in a slotted operation, based upon header content, over a virtual circuit established immediately upon a request for service; (c) has been chosen as the switching standard for broadband integrated services digital networks (BISDNs); (d) has transmission rates from 45 to 150 Mbps (megabits per second) that may go up to 600 Mbps; (e) offers bandwidth on-demand service, and (f) supports multiple concurrent connections over single communications lines. *See also* **cell relay, fast packet switching.**

asynchronous transmission (AT): 1. Transmission such that between similar significant instants in the same data group there is always an integral number of unit intervals. *Note:* There is not always an integral number of unit intervals between two significant instants located in different groups. The group usually is a block or a character. **2.** Data transmission in which the time of occurrence of the start of each character, or block of characters, is arbitrary. Once started, the time of occurrence of each significant instant, in the signal representing a bit that is within the character or block, has the same relationship to the significant instants of a fixed time frame. *See also* **block, character, intercharacter interval, isochronous, plesiochronous, synchronous transmission.**

asynchronous transmission (AT) command: A command in a set of de facto standard commands for a modem that (a) is connected to a central processing unit (CPU) and (b) is used in most 1200 bps (bit per second) and 2400 bps modems. *See also* **modem.**

asynchronous working: *Synonym* **asynchronous operation.**

ATB: all trunks busy.

ATC: air traffic control.

ATDM: asynchronous time division multiplexing.

ATF: advanced tactical optical fiber.

ATM: alternative test method, asynchronous transfer mode.

atmosphere: 1. The envelope of air that (a) surrounds the Earth, (b) is bound to the Earth by the Earth's gravitational field, (c) extends from the solid or liquid surface of the Earth to an indefinite height, and (d) has a density that asymptotically approaches that of interplanetary space. **2.** A unit of pressure equal to 101,325 $N\bullet m^{-2}$ (newtons per square meter) or approximately 14.70 $lbs\bullet in^{-2}$ (pounds per square inch), representing the atmospheric pressure at mean sea level under standard conditions.

atmosphere laser: A laser in which the active laser medium is a gas. *See* **longitudinally excited atmosphere laser, transverse excited atmosphere laser.**

atmospheric duct: In the lower atmosphere, a layer that (a) occasionally is of great horizontal extent and (b) has vertical refractive index gradients of the air such that radio signals are (i) guided or focused within the duct, (ii) tend to follow the curvature of the Earth, and (iii) experience less attenuation in the ducts than they would if the ducts were not present. *Note:* The reduced refractive index of the air at the higher altitudes bends the signals back toward the Earth. Signals in a higher refractive index layer, i.e., in a duct, tend to remain in that layer because of the reflection and the refraction encountered at the boundary with lower refractive index air. *See also* **ducting, hop, ionosphere, total internal reflection, troposphere.**

atmospheric noise: Radio noise caused by natural atmospheric processes, primarily by lightning discharges in thunderstorms. *See also* **interference, noise.**

atom: *See* **radioactive atom.**

atomic defect absorption: In lightwave transmission media, such as optical fibers and integrated circuits made of glass, silica, quartz, and plastic, the absorption of light energy caused by atomic changes that are introduced into the media, during or after their fabrication, by exposure to high radiation levels. *Note:* Titanium doped silica can develop losses of several thousand dB per kilometer when optical fibers are drawn under high temperature, and conventional fiber optic glasses can develop losses of 20,000 dB•km^{-1} (decibels per kilometer) during and after exposure to gamma radiation of 3000 rads.

atomic laser: A laser in which the active laser medium is an element in atomic form, such as sodium or boron. *See also* **molecular laser.**

atomic time: *See* **International Atomic Time.**

atomic time scale: *See* **International Atomic Time scale.**

ATOW: acquisition and tracking orderwire.

attachment unit interface (AUI): In a local area network, the interface between the propagation medium attachment unit and the data terminal equipment (DTE) within a data station. *Synonym* **access unit interface.**

attack: *See* **NAK attack.**

attack time: The time between (a) the instant that a signal at the input of a device or a circuit exceeds the activation threshold of the device or the circuit and (b) the instant that the device or the circuit reacts in a specified manner, or to a specified degree, to the input. *Note:* Attack time occurs in various electronic circuits, such as clippers, peak limiters, compressors, and voice-operated circuits (voxes).

attack time delay: *See* **receiver attack time delay, transmitter attack time delay.**

attempt: *See* **access attempt, block transfer attempt, call attempt, disengagement attempt, unsuccessful call attempt.**

attendance: *See* **centralized attendance.**

attendant access loop: A switched circuit that provides a manual means for call completion and control. *Note:* An attendant access loop might be given a specific telephone number. *See also* **call, circuit, fixed loop, loop.**

attendant conference: A network-provided service feature that allows an attendant to establish a conference connection of three or more users. *See also* **conference call, service feature.**

attendant position: The part of a switching system used by an attendant, i.e., an operator, to assist users in call completion and the use of special services. *See also* **busy verification, call, switching center.**

attendant service: *See* **centralized attendant service.**

attention key: A function key on computer and communications system terminal keyboards that is used to cause an input-output interruption in the processing or control unit.

attention sign: In semaphore communications, the flag positions used as a preliminary call and to establish communications.

attention signal: The signal to be used by amplitude modulation (AM), frequency modulation (FM), and television (TV) broadcast stations to actuate muted receivers for interstation receipt of emergency cuing announcements and broadcasts during a range of emergency contingencies that pose a threat to the safety of life and the integrity of property.

attenuate: To decrease the power of an entity, such as a signal, light beam, or lightwave propagating in a device, the decrease usually being the reduction that occurs between the input and the output of the device, such as over the entire length of an electrical cable, an optical fiber, or a connector. *Note:* A signal may be attenuated by absorption, reflection, scattering, deflection, dispersion, or diffusion rather than by geometric spreading.

attenuation: **1.** In a signal, beam, or wave propagating in a material medium, the decrease in field strength, or the irradiance, caused (a) by the absorption of energy by the medium, (b) by the scattering caused by the scattering centers of the material medium, but not (c) by the reduction caused by geometric spreading, i.e., the inverse square of distance effect for the irradiance. *Note 1:* Attenuation is usually expressed in dB. *Note 2:* Attenuation is not the same as the attenuation rate, which is expressed in dB per kilometer. *Note 3:* A distinction must be made as to whether the attenuation is that of signal power, usually expressed in watts/square meter, or signal electric field strength, usually expressed in volts/meter. **2.** The loss of power that occurs between the input and the output of a device. *Note:* Attenuation usually is expressed in dB. *See* **cloud attenuation, differential mode attenuation, echo attenuation, induced attenuation, light attenuation, macrobend attenuation, optical dispersion attenuation, path attenuation, precipitation attenuation, transient attenuation.** *See also* **absorption, attenuation constant, attenuation rate, coupling loss, damping, differential mode attenuation, equilibrium mode power distribution, extrinsic joint loss, flat fading, insertion loss, intrinsic joint loss, leaky mode, macrobend loss, material scattering, microbend loss.** *Refer to* **Fig. A-9.** *Refer also to* **Figs. H-1 (430), S-8 (874), V-1 (1060).**

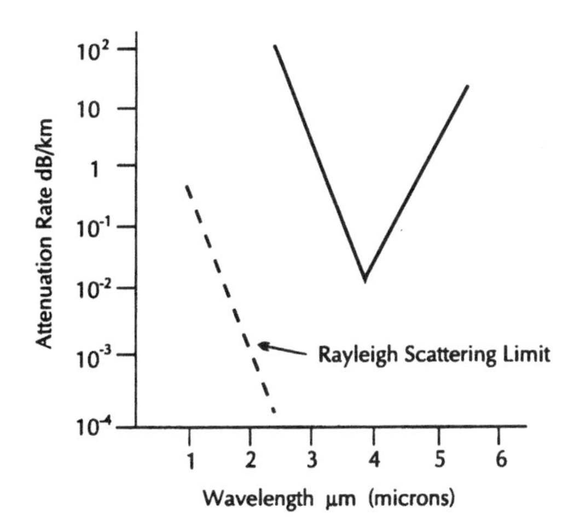

Fig. A-9. Idealized **attenuation** versus wavelength curve for a typical ultralow loss optical fiber. Beryllium fluoride glass optical fibers demonstrate losses at the indicated levels and wavelengths.

attenuation coefficient: *Synonym* **attenuation rate.**

attenuation constant: *Synonym* **attenuation term.**

attenuation factor: *See* **dispersion attenuation factor.**

attenuation-frequency distortion: *Synonym* **frequency distortion.**

attenuation-limited operation: 1. The condition that prevails in a device or a system when the magnitude of the power at a point in the device or the system limits its performance or the performance of the device or the system to which it is connected. *Note:* An example of attenuation-limited operation is the condition that prevails in an optical or radio link when the power level at the receiving end is the predominant mechanism that limits link performance. When a link is attenuation-limited, the power level at the receiver is below the sensitivity threshold of the receiver because the transmitter power is not high enough or the attenuation between transmitter and receiver is too high, or both. The attenuation is the sum of all the component insertion losses and the propagation medium losses. **2.** The condition that occurs when the received signal amplitude, rather than the received signal distortion, limits performance. *See also* **bandwidth-limited operation, dispersion-limited operation, distortion-limited operation, quantum-noise-limited operation.**

attenuation rate: 1. The rate of diminution of average power with respect to distance along a transmission path. *Note 1:* The attenuation rate usually is expressed in dB per kilometer (dB/km). *Note 2:* The attenuation rate for typical optical fibers is 0.1 dB•km^{-1}. **2.** In an electromagnetic wave that is propagating in a waveguide, such as a lightwave in an optical fiber, the attenuation in signal power that occurs per unit of distance along its length. *Note:* The attenuation rate, α, in dB/km, is given by the relation $\alpha = [10 \log_{10}(P_i/P_o)]/L$ where L is a given length of the waveguide in kilometers, P_i is the input power for the length L, and P_o is the output power for the length L. In this form, the attenuation rate will be a positive value because P_i is the larger of the two powers. *Synonym* **attenuation coefficient.** *See* **transient attenuation rate.** *See also* **attenuation, attenuation constant, axial propagation constant.** *Refer to* **Fig. A-10.** *Refer also to* **Figs. A-9, H-1 (430), S-15 (931), S-16 (932).**

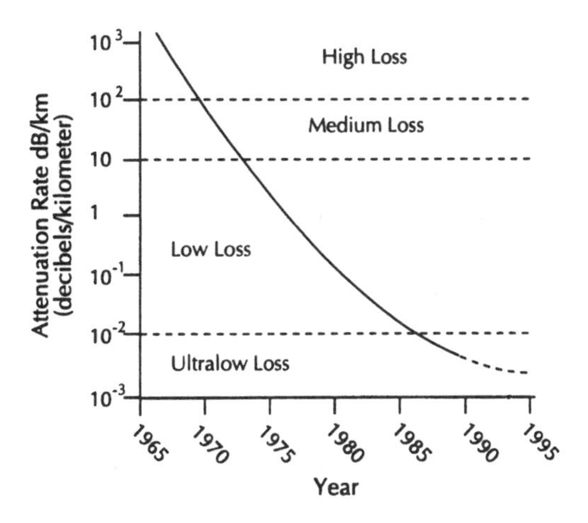

Fig. A-10. The reduction in **attenuation rate** in optical fibers achieved by industry and research laboratories brought about primarily by (a) improvement in purification and forming methods and (b) production of specially doped glasses and special glasses, such as halide glasses. Causes of the lower limit include Rayleigh scattering and infrared, overtone, and ultraviolet absorption.

attenuation rate characteristic: *See* **optical fiber global attenuation rate characteristic, spectral attenuation rate characteristic, wavelength-dependent attenuation rate characteristic.**

attenuation ratio: *See* **echo attenuation ratio.**

attenuation term: In the propagation of an electromagnetic wave in a uniform waveguide, such as an optical fiber or metal waveguide, the parameter that (a) is the real part of the axial propagation constant for a particular mode, (b) indicates the attenuation per unit distance of the wave at any point along the waveguide, and (c) is represented by a in the expression for the exponential variation of the electric field strength of the propagating wave, given by the relation $E = E_0e^{-pz} = E_0e^{-ihz-az}$ where E_0 is the initial electric field strength at $z = 0$, $p = h + a$ is the propagation constant where h is the phase term, a is the attenuation term caused by absorption and scattering, z is the distance from where E_0 is taken in the direction of propagation, and $i = (-1)^{1/2}$ is the complex indicator that identifies the quadrature phase component of the exponent. *Note 1:* The attenuation term usually is expressed as a pure number per unit length. It can be converted to dB per kilometer. *Note 2:* The attenuation term, a, is dependent upon the absorptive qualities of the material composing the waveguide, the frequency of the electromagnetic wave, and the modal characteristics of the wave. *Note 3:* The attenuation term may also be applied to the magnetic field, in which case H may be substituted for E, and H_0 may be substituted for E_0. *Note 4:* The attenuation term is supplied by the manufacturer. It can be measured experimentally. *Synonym* **attenuation constant.** *See also* **axial propagation constant, mode, phase, phase term, propagation constant.**

attenuation test: In an optical fiber, fiber bundle, or fiber optic cable, a test that (a) is used to measure the attenuation rate or the dispersion attenuation factor, (b) is conducted at a specified wavelength, such as 1.31 µm (microns), and (c) usually is performed by measuring the attenuation in an entire reel or spool of fiber of known length.

attenuator: 1. In electrical systems, a network that reduces the amplitude of a signal without appreciably distorting its waveform. *Note 1:* Attenuators are usually passive and continuously or incrementally adjustable. Fixed attenuators are often called pads, especially in telephony. *Note 2:* The input and output impedances of an attenuator are usually matched to the impedances of the signal source and load, respectively. **2.** In optical systems, an optical device that reduces the amplitude of a signal without appreciably distorting its waveform. *Note:* Optical attenuators usually are passive devices. *See* **continuously variable optical attenuator, fiber optic attenuator, fixed attenuator, longitudinal displacement attenuator, optical attenuator, pad, stepwise variable optical attenuator.**

at the dip: In flaghoist communications, the signaling condition that exists when the flaghoist is raised three-fourths of the way up toward the point of hoist. *See also* **close up, flaghoist, hauled down.**

attitude: *See* **satellite attitude.**

attribute: 1. In a database, a property inherent in an entity, or associated with an entity, in the database. **2.** In display systems, a property or a characteristic of any displayed entity. *Note:* Examples of attributes are the intensity, color, shape, or size of display elements, groups, or segments. **3.** In communications, computer, and data processing systems, a characteristic or a property of the system. *Note:* Examples of attributes are the data signaling rate (DSR), the propagation medium used, and the power level used. *See* **data attribute.**

audible alarm: An alarm that (a) can be heard by the normal human ear and (b) is activated when predetermined events occur that require operator attention or intervention. *See also* **audio alarm.**

audible ringing tone: The tone that (a) is received by the call originator telephone indicating that the call receiver telephone is ringing and (b) is not generated by the call receiver telephone. *Synonym* **ringback tone, ringing tone.** *See also* **ringback signal.**

audio: In communications systems, pertaining to a capability or a characteristic enabling detection, transmission, or processing of signals having frequencies that lie within the range that can be heard by the human ear. *Note:* Audio frequencies range from 30 Hz (hertz) to 15,000 Hz.

audio alarm: An alarm, i.e., a device, that generates a sound signal. *Synonym* **acoustic alarm.** *See also* **audible alarm.**

audio frequency (AF): A frequency that lies in the band of frequencies that, when transmitted as sound waves of sufficient power level, can be heard by the normal human ear. *Note:* Audio frequencies range from approximately 30 Hz (hertz) to 15 kHz (kilohertz). *See also* **voice frequency.**

audio frequency wave: A wave with a frequency that lies in the band of frequencies that, when transmitted as sound waves of sufficient power level, can be heard by the normal human ear. *Note 1:* Audio frequencies range from approximately 30 Hz (hertz) to 15 kHz (kilohertz). *Note 2:* Audio frequency waves can be transmitted as analog electrical signals in wires. *Note 3:* Radio wave and lightwave carriers can be modulated at audio frequencies.

audio response unit (ARU): A device that provides synthesized voice responses to dual tone multifrequency signaling input by processing calls based on the call originator input, information received from a host database, and information in the incoming call, such as the time of day. *Note:* Audio response units (ARUs) are used to increase the rate at which information calls are handled and to provide consistent quality in information retrieval.

audiovisual: Pertaining to the application and use of electrical, chemical, mechanical, and optical media to record or reproduce audible signals and visual images.

audit: To conduct an independent review and examination of system records and activities in order to (a) test the adequacy and effectiveness of data security and data integrity procedures, (b) ensure compliance with established policy and operational procedures, and (c) recommend any changes.

auditing: *See* **configuration auditing.**

audit review file: A file created by executing statements included in a program for the explicit purpose of providing data for auditing.

audit trail: 1. In information systems, a record of access attempts, accesses, and services. **2.** Data in the form of a logical path that links a sequence of events and that is used to trace the transactions that have affected the contents of a record. **3.** A chronological record of system activities that is sufficient to enable the reconstruction, review, and examination of the sequence of environments and activities surrounding or leading to an operation, a procedure, or an event in a transaction from its inception to final results. *Note:* Audit trails may apply to information in an automated information system, to the routing of messages in a communications system, or to material exchange transactions. *See also* **automatic message accounting, call record, station message detail recording, trace packet.**

augend: A number to which another number, the addend, is to be added to produce the sum. *See also* **addend.**

AUI: attachment unit interface.

aurora: The sporadic radiant emission from the upper atmosphere that usually occurs over the middle and upper latitudes. *Note:* The aurora is most intense at times of magnetic storms and sunspot activity when it is also nearest to the equator. The aurora interferes with radio communications. The distribution of aurora with altitude shows a pronounced maximum near 100 km (kilometers) above the Earth's surface.

aurora australis: Aurora in the Southern Hemisphere. *Synonym* **southern lights.** *See also* **aurora.**

aurora borealis: Aurora in the Northern Hemisphere. *Synonym* **northern lights.** *See also* **aurora.**

auroral display: A particular visible occurrence of an aurora in one or more of its many shapes and colors, including (a) arcs, i.e., bands of light extending across the sky, the highest point of the arc being in the direction of the magnetic meridian, (b) rays, i.e., single lines, like a searchlight beam or bundle of lines, (c) draperies or curtains, i.e., a curtain-like appearance, sharp at the bottom and tenuous at the top, billowing or blowing areas of brightness running across the northern or southern night sky, (d) crowns or corona, i.e., rays that appear to spread out from a single point in the sky, (e) bands, i.e., streaks similar to arcs, but less structured, (f) streamers extending to great heights, and (g) other shapeless, dynamic, and sunlit forms. *Note 1:* The auroral display can last from several seconds to several hours, with various levels of brightness and colors. *Note 2:* Auroral displays often pulsate at frequencies of the order of 0.2 Hz (hertz) to 2 Hz. *Note 3:* High levels of radio communications interference are encountered during auroral displays.

auroral zone: 1. The areas of the Earth that include (a) the latitudes of the Arctic, extending around the Earth from central Alaska, through James Bay, southern Greenland, Iceland, and northern Norway, and along the Siberian coast, in which the aurora borealis, i.e., the northern lights, occurs about 250 times per year and (b) the latitudes of the Antarctic where the aurora australis, i.e., the southern lights, occurs somewhat less often. **2.** A region in which a particular characteristic of an aurora occurs, such as extreme height, specific shapes, increased brightness, or longer-lasting effects, particularly in areas farther south, or areas in which particularly more frequent or particularly less frequent auroral displays occur.

authenticate: 1. To establish, usually by challenge and response, that a transmission attempt is authorized and valid. **2.** To verify the identity of a user, a device, or other entity in a computer system, or to verify the integrity of data that have been stored, transmitted, or otherwise exposed to possible unauthorized modification. **3.** A challenge given by voice or electrical means to attest to the authenticity of a message or a transmission. *See also* **access control, fetch protection, password, recognition.**

authentication: 1. A security measure designed to protect a communications system against acceptance of a fraudulent transmission or fraudulent simulation by establishing the validity of a transmission, message, or originator. **2.** A means of identifying individuals and verifying their eligibility to receive specific categories of information. **3.** Evidence by proper signature or seal that a document is genuine or official. **4.** A security measure designed to protect a communications system against fraudulent transmission. *See also* **authenticator, fetch protection, password, recognition.**

authentication equipment: Equipment that (a) provides protection against fraudulent transmission and communications deception or (b) establishes the authenticity of a transmission, message, station, originator, user, or equipment.

authentication system: 1. A system that (a) authenticates, i.e., serves as a secure means of establishing the authenticity of a transmission or message, or (b) challenges the identity of a station. **2.** A cryptosystem or cryptographic process used for authentication. **3.** A technique designed to provide effective authentication.

authenticator: 1. A symbol or a group of symbols, or a series of bits, selected or derived in a prearranged manner and usually inserted at a predetermined point within a message or a transmission for the purpose of attesting to the validity of the message or a transmission. **2.** A letter, numeral, group of letters or numerals, or any combinations of these, attesting to the authenticity of a message or a transmission. **3.** In information security (INFOSEC), a means used to confirm the identity or eligibility of a station, a message originator, or a person.

authority: *See* **maritime air control authority, post-telegraph-telephone authority.**

authorization: 1. In computer and communications systems, the right granted to a user to access, read, modify, insert, or delete certain data, or to execute certain programs. **2.** In computer and communications systems, the granting of access rights to a user, a program, or a process.

authorized bandwidth: In the transmission and the reception of radio signals, the necessary bandwidth excluding allowance for transmitter drift and Doppler shift. *See also* **assigned frequency, assigned frequency band, bandwidth, necessary bandwidth.**

authorized frequency: A frequency allocated and assigned by a competent authority to a specific user for a specified purpose. *See also* **assigned frequency, frequency, frequency assignment, frequency band allocation.**

authorized station: A station legitimately provided with all the current keys, procedures, and time information necessary to communicate with another station.

autocontrol: *See* **radio autocontrol.**

autocorrelation: In communications, the time integral of the product of a given signal and a time-delayed replica of itself, expressed by the relation

$$AC = \int_0^t s(t) \cdot s(t - \tau) dt$$

where AC is the autocorrelation, $s(t)$ is the signal magnitude as a function of time, t is the time, and $s(t - \tau)$ is a replica of $s(t)$ delayed by the time τ. *Note:* In the process of correlating two or more interrelated signals or other phenomenon, autocorrelation is the special case of cross correlation in which the functions are identical in shape but occur at different times. *See also* **cross correlation.**

autodialer: A communications device (a) into which a call originator may insert often called or important telephone numbers so that when one of these numbers is to be dialed only one or two keys need to be actuated, (b) that may dial again when a busy signal is received until the call receiver answers or until a specified time out is reached, whichever occurs first, and (c) that may display access attempt status on a display panel or screen.

AUTODIN: Automatic Digital Network.

AUTOEXEC.BAT: The name of the file (a) that a personal computer (PC) usually reads after it loads the disk operating system (DOS) and (b) in which the user usually lists the commands that are to be automatically executed when electrical power is applied to the PC.

automated data medium: *Synonym* **machine-readable medium.**

automated information system (AIS): 1. An assembly of computer hardware, software, firmware, or any combination of these, configured to accomplish specific information-handling operations, such as communicate, compute, disseminate, process, store, or control information represented by data. **2.** In information security (INFOSEC), a system, subsystem, or equipment that (a) is used in the automatic handling of data, such as the automatic acquisition, storage, manipulation, management, movement, control, display, switching, interchange, transmission, or reception of data, (b) includes computer software, firmware, and hardware, and (c) includes associated electronic and optical equipment, such as computers, word-processing systems, storage units, and networks. *See also* **data, information.**

automated information systems security (AISS): Measures and controls that (a) protect an automated information system against service denial or against unauthorized disclosure, modification, or destruction of the systems or the data in the system, (b) may be accidental or intentional, and (c) include consideration of (i) all hardware and software functions, characteristics, and features, (ii) operational and accountability procedures, (iii) access controls at the central computer facility, remote computer, and terminal facilities, (iv) management constraints, (v) physical structures and devices, (vi) personnel and communications controls needed to provide an acceptable level of risk for the automated information system and for the data and information contained in the system, and (vii) the totality of security safeguards needed to provide an acceptable protection level for an automated information system and for the data handled by an automated information system.

automated maritime telecommunications system (AMTS): An automatic, integrated, and interconnected maritime communications system that serves ship stations on specified inland and coastal waters of the United States.

automated radio: A radio that can be automatically controlled by electronic devices and that requires little or no human intervention.

Automated Recourse to Electronic Negotiations Archive (ARENA): A database that (a) contains the records of the arms control talks and discussions in which the United States has participated, including the Strategic Arms Limitation Talks (SALT), Conference on Disarmament (CD) discussions, Strategic Arms Reduction Talks (START), Intermediate range Nuclear Forces (INF) negotiations, and Nonproliferation Negotiations (NN) and (b) is operated by the Arms Control and Disarmament Agency.

automated software verification: Proving or disproving that a computer program performs according to its specification by using a software tool as a means of verifying the correctness of the program. The software tool may perform both static and dynamic analysis in the code verification process.

automated tactical command and control system: A command and control system, or part thereof, that can manipulate the movement of information from source to user without human intervention. *Note:* In automated tactical command and control systems, automated execution of a decision may occur with human intervention.

automatic: Pertaining to a process or device that, under specified conditions, can function without human intervention.

automatic answer back code exchange: Automatic identification of stations in which each communications terminal or data terminal equipment (DTE) transmits its own answer back code.

automatic answering: A network-provided service feature in which the called terminal automatically responds to the calling signal and the call may be established whether or not the called terminal is attended by a human operator. *See also* **call, data terminal equipment, facility, service feature.**

automatic branch exchange: *Synonym* **private branch exchange.**

automatic call distributor (ACD): A device that distributes incoming calls to a specific group of terminals. *Note:* If the number of active calls is less than the number of terminals, the next call will be routed to the terminal that has been in the idle state the longest. If all terminals are busy, the incoming calls are held in a first in, first out queue until a terminal becomes available. *See also* **call, proration.**

automatic callback: A network feature that permits a user, when encountering a busy condition, to instruct the system to retain the called and calling numbers and to establish the call upon the availability of a clear circuit. *Note:* Automatic callback may be implemented in the terminal or in the switching system, or shared between them. *Colloquial synonym* **camp on.** *See also* **automatic calling unit, call, card dialer, circuit, data terminal equipment, service feature.**

automatic calling: Calling in which the elements of the selection signal, generated by the data terminal equipment (DTE), are contiguously entered into the data network at the full data signaling rate. *Note:* A limit may be imposed by the network to prevent more than a permitted number of unsuccessful call attempts to the same address within a specified period.

automatic calling unit (ACU): A device that permits business machines, such as computers and card dialers, to originate calls automatically over a telecommunications network. *See also* **automatic callback, call, card dialer.**

automatic charge indication: The automatic indication of the charge of a call given by a communications system to the paying terminal either prior to the release of the paying terminal or by recall at a convenient time.

automatic check: A check performed by equipment specifically built into a system for checking purposes. *Note:* Examples of automatic checks are (a) checks performed by equipment used to count the bits of a character to perform parity checking and (b) echo checks. *See also* **hardware check, programmed check.**

automatic continuous tape relay switching: *See* **semiautomatic continuous tape relay switching.**

automatic data handling (ADH): 1. Data handling that includes automatic data processing and data transfer. **2.** The combining of data processing and data transfer, i.e., the combining of (a) data processing, transmission, and reception or (b) data processing and communications.

automatic data processing (ADP): 1. An interacting assembly of procedures, processes, methods, personnel, and equipment that automatically performs a series of data processing operations on data. *Note:* The data processing operations may result in a change in the semantic content of the data. **2.** Data processing by means of one or more devices that (a) use common storage for all or part of a computer program or for all or part of the data necessary for execution of the program, (b) execute user-written or user-designated programs, (c) perform user-designated symbol manipulation, such as arithmetic operations, logic operations, or character string manipulations, and (d) execute programs that modify themselves during their execution. *Note:* Automatic data processing may be performed by a stand-alone unit or by several connected units. **3.** Data processing largely performed by automatic means. **4.** The branch of science and technology concerned with methods and techniques related to data processing largely performed by automatic means.

automatic data processing equipment (ADPE): 1. Equipment, or interconnected systems or subsystems of equipment, that is used in the automatic acquisition, storage, manipulation, management, movement, control, display, switching, interchange, transmission, or reception of data. **2.** In U.S. government operations, any equipment or interconnected system or subsystems of equipment that is used in the automatic acquisition, storage, manipulation, management, movement, control, display, switching, interchange, transmission, or reception of data or information (a) by a federal agency or (b) under contract with a federal agency that (i) requires the use of such equipment or (ii) requires the performance of a service or the furnishing of a product that is performed or produced making significant use of such equipment. *Note 1:* Automatic data processing equipment (ADPE) includes (a) computers, (b) ancillary equipment, (c) software, firmware, and related procedures, (d) services, including support services, and (e) related resources as defined by regulations issued by the Administrator for General Services. *Note 2:* This definition is based on Public Law 99-500, Title VII, Sec. 822 (a) Section 111(a) of the Federal Property and Administrative Services Act of 1949 (40 U.S.C. 759(a) revised).

automatic data processing system (ADPS): In communications and teleprocessing, automatic data processing equipment (ADP) linked together by communications and data transmission equipment to form an integrated system for the processing and the transfer of data.

automatic date and time indication: In a communications system, the automatic indication of the date and the time of the commencement of a call or a message to (a) the call or a message originating terminal, (b) the call or message receiver terminal, or (c) both the call or message originating terminal and the call or message receiver terminal.

automatic dependent surveillance (ADS) system: For flights over remote ocean areas, an aircraft automatic position reporting system that (a) at regular intervals reports aircraft progress, including position, vector, and meteorological data, via satellite link, (b) is used by air traffic services, and (c) is based on (i) satellite air and ground communications data link relays and (ii) information derived from onboard navigation sensors transmitted from the aircraft to a communications satellite, such as the international maritime satellite (INMARSAT). *See also* **oceanic display and planning system.**

automatic dependent surveillance unit (ADSU): In an aircraft, a device that (a) is in the avionics air/ground data link subsystem of the integrated automatic dependent surveillance (ADS) system, (b) is an onboard black box router and data link processor that collects relevant data from the aircraft, reformats the data, and transfers the data to the aircraft satellite communications unit, and (c) accepts uplink messages that set reporting rates, selects the data fields from which data are to be transmitted, and performs functions pertaining to establishing, maintaining, updating, and terminating communications connections.

automatic dialing: Dialing in which the dialing pulses or signals are generated automatically, usually by means of an automatic calling unit (ACU). *See also* **automatic calling unit.**

Automatic Digital Network (AUTODIN): The worldwide data communications network of the U.S. Department of Defense (DoD) Defense Communications System (DCS) which has been replaced by the Defense Switched Network (DSN). *See also* **Automatic Secure Voice Communications Network, Automatic Voice Network, Defense Data Network, Federal Telecommunications System.**

Automatic Digital Network (AUTODIN) operation: Mode 1: A duplex synchronous message transmission network operation with automatic error and channel controls that allows independent and simultaneous two-way message transmission. **Mode 2:** A duplex asynchronous network operation, without automatic error and channel controls, that allows simultaneous two-way message transmission. **Mode 3:** A duplex synchronous network operation, with automatic error and channel controls, that allows only one-way message transmission. **Mode 4:** A unidirectional asynchronous network operation with a send-only or a receive-only capability without error control and without channel coordination. The Mode 4 channel is nonreversible. **Mode 5:** A duplex synchronous network operation that allows independent and simultaneous two-way message transmission. Control characters are used to acknowledge receipt of messages and perform limited channel coordination.

Automatic Digital Network (AUTODIN) user: A person, an installation, an activity, or equipment that has access to the Automatic Digital Network (AUTODIN) through an AUTODIN switching center.

automatic digital switching system: *See* **tactical automatic digital switching system.**

automatic duration indication: The automatic indication of the chargeable time of a call given by a communications system to the paying terminal either prior to the release of the paying terminal or by recall at a convenient time.

automatic error correction: 1. The automatic correction of an error in a transmitted signal. **2.** In a communications system, pertaining to the correction of errors in messages by automatic means. *See also* **error correcting code.**

automatic error detection: 1. The automatic detection and indication of an error in a transmitted signal. **2.** In a communications system, pertaining to the detection of errors in messages by automatic means.

automatic error detection and correction system: 1. A system that uses an error-detecting code so conceived that any received false or incorrect signal initiates a retransmission of the character. **2.** A system in which errors that occur in transmission are detected and corrected automatically, through the use of error-detection equipment and error-detection codes without initiating a request for repetition. *Note:* Error-correcting codes are usually redundant codes.

automatic exchange: A telephone exchange in which communications among users are effected by means of switches set in operation by the originating users' equipment without human intervention. *See also* **data switching exchange, exchange, switching center.**

automatic function: A machine function controlled by a computer program and carried out without human intervention. *Note:* A computer may execute a series of functions as if the series were a single function.

automatic gain control (AGC): The automatic adjustment of the gain of a device in a specified manner and as a function of a specified parameter, such as input level. *Note:* Automatic gain control (AGC) usually is applied to amplifiers, antennas, and light sources to maintain a constant output level. *See* **instantaneous automatic gain control.** *See also* **compressor, level, limiter circuit.**

automatic identification: The automatic transmission of the identification of the call-originating terminal or station to the call-receiving terminal or station, or vice versa, or the identification of terminals and stations to one another. *Note:* Automatic identification may be established when a connection is made. *See* **terminal automatic identification.**

automatic identified outward dialing (AIOD): A network-provided service feature that provides the user with an itemized statement of usage on directly dialed calls. *Note 1:* Automatic identified outward dialing (AIOD) is provided by automatic number identification (ANI) equipment installed in an exchange and connected via a data link to the serving automatic message accounting (AMA) central office. *Note 2:* The automatic identified outward dialing (AIOD) capability is often referred to as AMA/ANI. *See also* **call, service feature.**

automatic identified outward dialing (AIOD) lead: A terminal equipment lead used solely to transmit automatic identified outward dialing (AIOD) data from a private branch exchange (PBX) to the public switched telephone network or to switched service networks, such as enhanced private switched communications systems (EPSCSs), so that a vendor can provide a detailed monthly bill identifying long distance usage by individual PBX stations, tie trunks, or attendants.

automatic link establishment (ALE): 1. The capability of a high frequency (HF) radio station to make contact, or initiate a circuit, between itself and another specified radio station, without human intervention and usually under processor control. *Note:* Automatic link establishment (ALE) techniques include automatic signaling, selective calling, and automatic handshaking. Other automatic techniques that are related to ALE are channel scanning and selection, link quality analysis (LQA), polling, sounding, message store and forward, address protection, and antispoofing. **2.** In high frequency (HF) radio, a link control system that (a) includes automatic scanning, selective calling, sounding, and transmit channel selection using link quality analysis data, (b) includes the optional automatic link functions of polling and the exchange of orderwire commands and messages, and (c) allows users to enter the address of the station they are calling so that the radio automatically routes the call to that destination in a manner similar to a telephone. *See also* **controller.**

automatic message accounting (AMA): A network-provided service feature that automatically records data on user-dialed calls. *See* **centralized automatic message accounting.** *See also* **audit trail, automatic number identification, call, call record, service feature.**

automatic message exchange (AME): 1. In message exchange in an adaptive high frequency (HF) radio network, the automatic transmission by one node and reception by another node when a direct link between the two nodes is established. **2.** In an adaptive high frequency (HF) radio network, the actions that include (a) the automatic transfer of a message from message insertion to addressee reception without human intervention, (b) the use of machine-addressable transport guidance information usually contained in the message header, and (c) the automatic routing of the message through an online direct connection through single or multiple propagation media, i.e., transmission media.

automatic message processing system (AMPS): An organized assembly of resources and methods used to collect, process, and distribute messages usually by automatic means.

automatic message switching: The automatic handling of messages through a switching center, either from local users or from other switching centers, in which (a) a distant electrical connection is established between the call originating station and a call receiving station or (b) a store and forward data transmission system is used.

automatic multilevel precedence: The automatic sensing and application of precedence procedures in accordance with established criteria to include several levels of preemption and override capability in an automatic data transmission or telephone switching network.

automatic number identification (ANI): A network-provided service feature in which the directory number or equipment number of a calling station is automatically obtained. *Note:* Automatic number identification (ANI) is used in message accounting. *See also* **automatic message accounting, call, service feature.**

automatic numbering equipment: Equipment that automatically generates and transmits the transmission identification number for each message.

automatic numbering transmitter: A transmitter that automatically transmits a serial number before each message.

automatic operation: The functioning of systems, equipment, or processes in a desired manner at the proper time under control of mechanical or electronic devices that operate without human intervention.

automatic preemption: In a telephone network or data transmission system, a preemption implementation that (a) permits a user to enter the precedence assigned to a call at the time the call is placed, dialed, or keyed, (b) processes the call with the assigned precedence, and (c) automatically preempts circuits with lower-precedence calls when required or as necessary in accordance with the preemption criteria.

automatic process control: The control of operations or processes using automatic devices, such as computers, servomechanisms, feedback loops, and autopilots, rather than measuring devices coupled with human observation and manual intervention.

automatic programming: Using a programmed computer to assist in generating, i.e., preparing, a computer program.

automatic radio relay aircraft: An aircraft that contains the necessary equipment to relay automatically radio messages intended for other aircraft.

automatic receiver: A radio receiver that (a) performs automatic tuning, (b) slowly sweeps the frequency spectrum, (c) alerts the operator when a signal is picked up, and (d) records the pickup. *See also* **panoramic receiver.**

automatic recovery quotient (ARQ): An automatic error detection and correction scheme (a) that is used in telegraph systems and (b) in which the backward channel is used to obtain repetition of corrupted signals until they are received uncorrupted.

automatic relay equipment: *See* **telegraph automatic relay equipment.**

automatic relay system: *See* **semiautomatic relay system.**

automatic remote rekeying: 1. Rekeying in which distant equipment is electronically rekeyed without specific actions by the receiving terminal and without human intervention. A system that (a) contains switching equipment and (b) causes automatic equipment to record and retransmit messages. **2.** In information security (INFOSEC), the electronic rekeying of distant cryptoequipment or noncryptoequipment without specific actions by the receiving terminal operator. *See also* **encryption.**

automatic remote reprogramming: Reprogramming in which distant equipment is electronically reprogrammed without specific actions by the receiving terminal and without human intervention.

automatic repeat request (ARQ): In data transmission, error control in which the receive terminal is arranged to detect a transmission error and automatically transmit a repeat request (RQ) signal to the transmit terminal, whereupon the transmit terminal then retransmits the character, code block, or message until either it is correctly received or the error persists beyond a predetermined number of transmittals. *Synonyms* **decision feedback system, error detecting and feedback system, repeat request system.** *See also* **block, block parity, character, cyclic redundancy check, echo check, error, error control, error correcting system.**

automatic retransmission: A recording of received signals immediately followed by their automatic transmission.

automatic retransmitter: Equipment that automatically records and retransmits received signals.

automatic ringdown: *Synonym* **verified off-hook.**

automatic route selection (ARS): Electronically or mechanically controlled selection and routing of outgoing calls without human intervention. *See also* **adaptive routing, call, proration.**

automatic search jammer: A jammer that (a) consists of an intercept receiver, i.e., a search receiver, and a jamming transmitter; (b) searches for signals that have specific characteristics; (c) automatically jams the signals; and (d) operates in one or more domains, such as the frequency, time, spatial, and polarization domains.

Automatic Secure Voice Communications Network (AUTOSEVOCOM): A worldwide, switched, secure voice network developed to meet U.S. Department of Defense (DoD) long-haul, secure voice communications requirements. *See also* **Automatic Digital Network Automatic Voice Network, communications, Federal Telecommunications System.**

automatic send-receive teletypewriter: A teletypewriter arranged for automatic transmission and reception of messages. *See also* **keyboard send-receive teletypewriter, receive-only teletypewriter.**

automatic sequential connection: 1. A network-provided service feature that automatically connects, in a prearranged sequence, the data terminal equipment (DTE) at each of a set of specified addresses to a single DTE at a specified address. **2.** A network-provided service feature in which the terminals at each of a set of specified addresses are automatically connected, in a predetermined sequence, to a single terminal at a specified address. *See also* **data terminal equipment, proration.**

automatic signaling service: *Synonym* **off-hook service.**

automatic sounding: The testing of selected channels or paths by providing a very brief beacon-like identifying broadcast that may be used by other stations to evaluate connectivity, propagation, and availability, and to identify known working channels for possible later use for communications or calling. *Note 1:* Automatic soundings are primarily intended to increase the efficiency of automatic link establishment (ALE) functions, thereby increasing system throughput. *Note 2:* Sounding information is used for identifying the specific channel to be used for a particular automatic link establishment (ALE) connectivity attempt. *See also* **automatic link establishment, path.**

automatic spot jammer: A spot jammer that (a) has a noise generator with a frequency bandwidth that is small in comparison to the operating band, (b) automatically tracks frequency, (c) sets the jammer center frequency to the frequency being tracked, (d) may be a single or multiple spot jammer, (d) uses a look-through period with rapidly sweeping receivers to detect and analyze signals, (e) has a jamming transmitter that operates between the look-through intervals, and (f) is limited by fixed transmission times, time sharing, and signal delay in reception and transmission.

automatic switching: Communications system operation that (a) effects the automatic interconnection of channels, circuits, and trunks and (b) handles traffic through a switching facility.

automatic switching equipment (ASE): Switching equipment that can automatically process communications traffic without human intervention, such as (a) recognize dial signals and pushbutton signals, i.e., keyed signals, (b) switch from line to line, trunk to trunk, line to trunk, and trunk to line, and (c) automatically route calls, handle precedences, and dispose of incomplete calls.

automatic switching system (ASS): 1. A switching system in which all the operations required to execute the three phases of information transfer transactions are automatically executed in response to signals from a user end instrument. *Note:* In an automatic switching system, the information transfer transaction is performed without human intervention, except for initiation of the access phase by a user. **2.** A telephone system in which all the operations required to set up, supervise, and release connections required for calls are automatically performed in response to signals from a calling device. *See also* **call, switching center.**

automatic system: *See* **Wheatstone automatic system.**

automatic tape relay system: A system of tape relay that includes automatic switching.

automatic telegraph transmitter: A telegraph transmitter in which the forming of signals is actuated step by step from a signal storage device, usually a perforated tape and a tape reader, rather than by an operator.

automatic telegraphy: Telegraphy in which manual operations are effectively reduced or eliminated by the use of automatic equipment.

automatic telephone system: A switching system through which telephone calls are automatically routed under control of the call originator instrument, i.e., the originating data terminal equipment (DTE).

automatic traffic overload protection: A communications system operating procedure in which automatic equipment is used to reduce the volume of line, trunk, or switch traffic in accordance with specified line load control categories when demands for service exceed the capacity to meet the demand. *Note 1:* Examples of automatic traffic overload protection are (a) the automatic delaying of dial tone and (b) the denying of originating user call privileges on certain lines.

automatic voice network: An automatically switched telephone network. *See also* **Automatic Voice Network.**

Automatic Voice Network (AUTOVON): The principal long-haul, nonsecure voice communications network within the U.S. Department of Defense (DoD) Communications System, now replaced by the Defense Switched Network (DSN). *See also* **Automatic Digital Network, Automatic Secure Voice Communications Network, communications.**

Automatic Voice Network (AUTOVON) user: A person, an installation, an activity, or equipment that, through an AUTOVON switching center, has access to the Automatic Voice Network (AUTOVON), which has now been replaced by the Defense Switched Network (DSN).

automation: 1. The implementation of processes by automatic means. **2.** The investigation, design, development, and application of methods of rendering processes automatic, self-moving, or self-controlling. **3.** The conversion of a procedure, a process, or equipment to automatic operation.

autonomous system: In communications, a collection of gateways and networks that (a) falls under one administrative entity, (b) promotes network connectivity, and (c) propagates routing information using a specified interior gateway protocol.

AUTOSEVOCOM Network: Automatic Secure Voice Communications Network.

AUTOVON: Automatic Voice Network, now replaced by the **Defense Switched Network (DSN).**

autumnal equinox: The line of intersection of the celestial equatorial and ecliptic planes at which the Sun passes from the Northern Hemisphere into the Southern Hemisphere. *Note:* The autumnal equinox occurs about September 22, bringing autumn to the Northern Hemisphere and spring to the Southern Hemisphere. *See also* **vernal equinox.**

auxiliary channel: In data transmission, a secondary channel that (a) has a direction of transmission that is independent of the primary channel and (b) is controlled by an appropriate set of secondary control interchange circuits.

auxiliary operation: An offline operation performed by equipment not under control of the processing unit.

auxiliary power: An alternate source of electric power, serving as backup for the primary power source at the station main bus or prescribed subbus. *Note 1:* An offline unit provides electrical isolation between the primary power source and the critical technical load, whereas an online unit does not. *Note 2:* A Class A power source is a primary power source, i.e., a source that assures an essentially continuous supply of power. *Note 3:* Types of auxiliary power sources include Class B, a standby power plant to cover extended outages of the order of days; Class C, a 10- to 60-second quick start unit to cover short-term outages of the order of hours; and Class D, an uninterruptive nonbreak unit using stored energy to provide continuous power within specified voltage and frequency tolerances. *See also* **power, primary power, station battery.**

auxiliary station: *See* **fixed microwave auxiliary station.**

auxiliary storage: 1. Storage that is available to a processor only through its input/output channels. **2.** In a computer, any storage that is not internal memory, i.e., random access memory (RAM). *Note:* Examples of auxiliary storage are diskettes, compact disks, optical disks, and magnetic tape cassettes. *See also* **main storage, memory**.

availability: 1. During a specified time interval, the ratio of (a) the total time a functional unit is capable of being used to (b) the duration of the interval. *Note 1:* If a unit is capable of being used, or performs satisfactorily, for 100 hours in a week, the availability of the unit is 100/168 ≈ 0.595 ≈ 59.5%. *Note 2:* The conditions determining operability and committability must be specified. **2.** A measure of the degree to which a system, subsystem, or equipment is operable and in a committable state at the start of a mission, when the mission is called for at an unknown point in time, i.e., at a random point in time. *Note 1:* The conditions determining operability and committability must be specified. *Note 2:* The availability is 1 minus the unavailability. If the availability is 100/168, the unavailability is 68/168. These values will always sum to 1. *See also* **idle state, maintainability, mean time between failures, mean time between outages, mean time to repair, mean time to service restoral, reliability, unavailability.**

available at action office time: *See* **message available at action office time.**

available for delivery time: *See* **message available for delivery time.**

available line: 1. In voice, video, or data communications, a circuit between two points that is ready for service but is in the idle state. *Synonym* **open line.** *See also* **idle state, line. 2.** In facsimile transmission, the portion of the scanning line that can be specifically used for image signals. *Synonym* **useful line.** *See also* **facsimile, scanning line.**

available line length: In facsimile transmission, the portion of the scanning line that can be used specifically for picture signals. *Synonym* **useful line.**

available point: In the display space on the display surface of a display device, an addressable point at which one or more display elements, display groups, or display image characteristics, such as color, intensity, on/off condition, shape, and orientation may be changed. *Note:* The available point characteristics that may be changed are usually changed by the operator or by a program-controlled computer.

available time: From the point of view of a user, the time during which a functional unit can be used. *Note:* From the point of view of operating and maintenance personnel, the available time is the same as the uptime, i.e., the time during which a functional unit is fully operational. *See also* **downtime, uptime.**

avalanche multiplication: In semiconductors, the sudden or rapid increase in the number or density of hole-electron pairs that is caused when the semiconductor is subjected to high, i.e., near breakdown, electric fields. *Note:* The increase in charge carriers causes a further increase in their numbers and density. Incident photons of sufficient energy can further multiply the carriers. These effects are used in avalanche photodiodes.

avalanche photodetector (APD): A photodetector that uses the avalanche multiplication effect to increase the photocurrent. *See also* **avalanche multiplication, avalanche photodiode.**

avalanche photodiode (APD): A photodiode that operates with a reverse bias voltage that causes the primary photocurrent to undergo amplification by cumulative multiplication of charge carriers. *Note 1:* As the reverse bias voltage increases toward the breakdown voltage, hole-electron pairs are created by absorbed photons. An avalanche effect occurs when the hole-electron pairs acquire sufficient energy to create additional pairs when the incident photons collide with the ions, i.e., collide with the holes and electrons. Thus, a signal gain is achieved. *Note 2:* Silicon and germanium avalanche photodiodes (APDs) are inherently electrically noisy. Gallium indium arsenide phosphide on indium phosphide substrates and gallium antimonide APDs are being made. *See also* **photodiode, positive-intrinsic-negative photodiode.**

avalanche photodiode (APD) coupler: A device that couples light energy from an optical fiber onto the photosensitive surface of an avalanche photodetector (APD) at the receiving end of a fiber optic data link. *Note:* The avalanche photodiode (APD) coupler may be only an optical fiber pigtail epoxied to the APD photosensitive surface.

avalanching: The amplification of an electrical signal within a device by electron multiple collision and ionization.

average block length: The average value of the total number of bits in blocks transferred across a user-communications interface. *Note:* The average block length is specified by the communications system operator. It is used to determine values for the block-oriented performance parameters. In systems where the information transferred across the functional interface is not delimited into blocks, the data signaling rate (DSR) is used instead of the average block length.

average diameter: In a round optical fiber, the mean value of a measurable diameter, such as the core, cladding, or reference surface diameter.

average error rate: In a transmission system, the ratio of (a) the number of incorrect data units, such as bits, characters, words, and frames, to (b) the total number of the same data units sent, both counted during the same measurement period. *Note:* A data unit is counted as an erroneous unit when one or more bits are different from the bits that are sent. Thus, if a block is sent with an erroneous bit, and the block is received exactly as it was sent, with the erroneous bit, it is not counted as an erroneous block.

average information content: *Synonym* **character mean entropy.**

average information rate: The mean entropy per character per unit of time. *Note:* The average information rate may be expressed in a unit such as a shannon per second. *See also* **shannon.**

average picture level (APL): In video systems, the average level of the picture signal during active scanning, integrated over a frame period. *Note:* The average picture level (APL) usually is expressed in percent of the range between blanking and the reference white level. *See also* **blanking, reference white level.**

average power: **1.** The total energy generated, radiated, or transmitted during a period divided by the period. **2.** In pulsed transmission, the energy per pulse times the pulse repetition rate. *Note:* The average power is usually expressed in watts and identified as average power. A joule per second is 1 watt.

average rate of transmission: *Synonym* **effective transmission rate.**

average ratio: *See* **peak to average ratio.**

average terrain: *See* **antenna height above average terrain.**

average transfer rate: *Synonym* **effective transfer rate.**

average transfer speed: *Synonym* **effective transfer rate.**

average transinformation rate: The mean transinformation content per character per unit of time. *Note:* The average transinformation rate may be expressed in a unit such as a shannon per second.

average transmission rate: *Synonym* **effective transmission speed.**

aviation instructional station: A land or mobile station in the aeronautical mobile service used for radiocommunications for instructions to students or pilots while actually operating aircraft or engaged in training flights.

avionics: The branch of science and technology devoted to the design, development, and operation of electronic and photonic equipment used in aviation, such as (a) the microwave landing system (MLS) used for runway approach guidance and (b) the electronic equipment on board aircraft for communications and flight control.

avoidance: *See* **carrier sense multiple access with collision avoidance.**

avoidance routing: The assignment of circuits and paths to avoid certain critical or trouble-prone circuits and nodes. *See also* **alternate routing, directionalization, dynamically adaptive routing, line load control, route diversity.**

avoidance system: *See* **terrain avoidance system.**

AVPO: axial vapor phase oxidation.

AVPO process: *See* **axial vapor phase oxidation (AVPO) process.**

awareness beacon with reply: *See* **situation awareness beacon with reply.**

AWGN: additive white Gaussian noise. *See* **white noise.**

AWSIM: air warfare simulation.

axial alignment sensor: *See* **fiber optic axial alignment sensor.**

axial bundle: A cone of electromagnetic rays that emanates from an object point that is on the optical axis of a lens system. *See also* **aperture stop.**

axial deposition process: *See* **vapor phase axial deposition process.**

axial displacement sensor: *See* **fiber optic axial displacement sensor.**

axial interference microscopy: *Synonym* **slab interferometry.**

axial misalignment loss: *Synonym* **angular misalignment loss.**

axial propagation constant: In an optical fiber, the propagation constant evaluated along the optical axis of the fiber in the direction of transmission. *Note:* The real part of the axial propagation constant is the attenuation term. The imaginary part is the phase term. *See also* **attenuation, attenuation coefficient, attenuation term, phase term, propagation constant.**

axial ratio: In an electromagnetic wave with elliptical polarization, the ratio of (a) the length of the major axis to (b) the length of the minor axis of the ellipse described by the tip of the electric field vector.

axial ratio polarization: *Synonym* **elliptical polarization.**

axial ray: **1.** A light ray that propagates along and is coincident with the axis of an optical waveguide. **2.** In an optical fiber, a ray that travels along a path that is parallel to the fiber axis. *Note:* Along with meridional rays, axial rays lie in meridional planes. *See also* **geometric optics, meridional ray, paraxial ray, skew ray.** *Refer to* **Fig. R-5 (817).**

axial slab interferometry: *Synonym* **slab interferometry.**

axial vapor phase oxidation (AVPO) process: A vapor phase oxidation (VPO) process for making graded index (GI) optical fibers in which (a) the glass preform is grown radially rather than longitudinally as in other processes, (b) the refractive index profile is controlled in a spatial domain rather than in a time domain, and (c) the chemical gases are burned in an oxyhydrogen flame, as in the outside vapor phase oxidation (OVPO) process, to produce a stream of soot particles to create the graded refractive index profile.

axial viewing angle: *See* **borescope axial viewing angle.**

axis: *See* **optic axis, optical axis, optical fiber axis, primary axis, secondary axis, signal communications axis.**

axis of signal communications: *Synonym* **signal communications axis.**

axis paraboloidal mirror: *See* **off-axis paraboloidal mirror.**

azimuth: **1.** A given horizontal angular direction from a specified reference direction, measured in a clockwise direction, when looking vertically downward, from the reference direction. **2.** A direction expressed as a horizontal angle, usually expressed in degrees or mils from either true, grid, or magnetic north. **3.** The angle at the zenith between the observer's celestial meridian and the vertical circle through a heavenly body. *See also* **altitude over azimuth, back azimuth, jamming azimuth, reciprocal bearing, reciprocal heading.** *Refer to* **Figs. I-7 (476), R-11 (848).**

azimuth over altitude: Pertaining to an antenna mount (a) in which the altitude axis, i.e., the elevation axis, is the primary axis, (b) that is fixed to earth, i.e., fixed to the ground, and (c) that primarily is used in equatorial regions for communicating with and controlling drifting equatorial satellites. *Synonym* **azimuth over elevation.** *See also* **altitude over azimuth.**

azimuth over elevation: *Synonym* **azimuth over altitude.**

b: bit. *See* **binary digit.**

B: bel, byte.

babble: In transmission systems, the aggregate of crosstalk induced in a given line by all other lines.

back: *See* **callback, ringback.**

back azimuth: A direction opposite to an azimuth, i.e., 180° from a given azimuth. *See also* **azimuth, reciprocal bearing, reciprocal heading.** *Refer to* **Fig. R-11 (848).**

back bias gain control circuit: *See* **detector back bias gain control circuit.**

backbone: 1. The high traffic density connectivity portion of a communications network. **2.** In a communications network, a primary forward direction path traced sequentially through two or more major relay or switching stations. *Note:* A backbone consists primarily of switches and interswitch trunks. **3.** The route between two principal terminal microwave stations. **4.** A distribution core that brings electrical power and various materials, such as gases and chemicals, to various parts of a chip in an automated fabrication and processing system. *See* **Multicast Backbone.** *See also* **network.**

back focal length (BFL): The distance from the vertex of the back surface of a lens to its rear focal point.

background: 1. The area on a document or display surface that surrounds the subject matter represented on the document or display surface. **2.** In multiprogramming, the environment in which computer programs, usually of low priority, are executed. *See* **display background.** *See also* **foreground.**

background-limited infrared detector: An infrared detector that is background-radiation-noise-limited in its sensitivity when viewing an ambient temperature background. *Note:* The limitation is primarily caused by the inherent inability of the detector to shield itself entirely from all ambient infrared noise. The background radiation tends to saturate the detector, i.e., to create such a large signal that infrared signals that do occur cannot be detected. The infrared from an aircraft approaching a detector from the direction of a brightly visible Sun cannot be detected because the infrared from the Sun obliterates the infrared from the aircraft. An infrared personnel detector may not function when exposed to the radiation of the noonday Sun.

background noise: In a communications system, the total system noise in the absence of information transmission. *See also* **ambient noise level, noise.**

background processing: In computer systems, the automatic execution of lower-priority computer programs when higher-priority programs are not using the system resources. *See also* **batch processing.**

backing storage: *Synonym* **auxiliary storage.**

backlog: *See* **significant backlog.**

back lobe: In an antenna radiation pattern, including that of a light source, a lobe—or the radiant energy density corresponding to the lobe, i.e., the radiance—that is directed opposite to the main lobe. The back lobe usually is extremely small compared to the main lobe, i.e., the irradiance is far lower than that of the main lobe at a given distance from the source.

backoff: *See* **truncated binary exponential backoff.**

backscatter: 1. The deflection, by reflection, refraction, or diffraction, of an incident electromagnetic wave or signal in such a manner that a component of the deflected wave or signal is in a direction of propagation opposite to that of a component of the incident wave or signal. **2.** In an electromagnetic wave or signal that is deflected by reflection, refraction, or diffraction, a component of the deflected wave or signal that is in a direction of propagation opposite to that of a component of the incident wave or signal. **3.** To deflect, by reflection, refraction, or diffraction, an incident electromagnetic wave or signal in such a manner that a component of the deflected wave or signal is in a direction of propagation opposite to that of a component of the incident wave or signal. *Note:* The term "scatter" can be applied to reflection or refraction by uniform media and diffraction by certain apertures, but it usually is applied to a modification of the wavefront and the direction of propagation in a random or disorderly fashion. **4.** The components of electromagnetic waves that are directed back toward their source when resolved along a line from the source to the point of scatter. *Note:* Backscatter occurs in optical fibers because of the reflecting surfaces of particles or occlusions in the propagation medium, resulting in attenuation. Backscatter of lightwaves and radio waves also occurs in the atmosphere and the ionosphere. *See also* **backscattering, forward scatter, propagation.**

backscattering: 1. The deflecting by reflection, refraction, or diffraction, or any combination of these, of an electromagnetic wave in such a manner that a component of the wave is deflected opposite to the direction of propagation of a component of the incident wave or signal. *Note:* The term "scatter" can be applied to reflection, refraction, or diffraction by relatively uniform media, but it usually is taken to mean wave propagation in which the wavefront and its direction are modified in a relatively disorderly fashion. **2.** Radio wave propagation in which the incident and scattered waves, resolved along a reference direction, are oppositely directed. *Note 1:* A signal received by backscattering is often referred to as "backscatter." *Note 2:* In radio propagation, backscatter usually is considered to occur in the horizontal plane, i.e., back toward the transmitter. **3.** In optics, the scattering of a lightwave into a direction generally opposite to the original direction. *See also* **forward scatter, propagation, Rayleigh scattering, reflectance, reflection, scatter.**

backscattering technique: *Synonym* **optical time domain reflectometry.**

backshell: In a fiber optic connector, the portion of the shell (housing) that (a) is immediately to the rear of the fastening mechanism, which is considered the front, (b) contains the optical fibers, and (c) attaches to the bend limiter.

backspace: To move a data medium, pointer, or cursor a specified distance opposite to the forward direction. *Note:* Examples of backspacing are (a) to move a paper in a typewriter or printer backward by means of an escapement without a character being typed or printed and (b) to move a cursor on a computer monitor backward one space while obliterating the character occupying the space to which the cursor was moved. *See also* **backspace key.**

backspace key: On keyboards, such as on printers, computers, and communications equipment, a key that, when actuated, (a) causes the device under control of the keyboard to erase the previously entered character, (b) causes the device to step back in space or position, (c) cancels a previously entered command, or (d) returns the device to a previous state or condition. *Note:* In a personal computer, actuation of the backspace key may resemble an eraser during entry or editing by (a) removing previously entered characters, i.e., characters behind the cursor, (b) canceling a stored range in a command that requires ranges to be recalled, or (c) causing a pointer to return to the cell where the pointer was located before its present position. *See also* **backtab key.**

back surface mirror: An optical mirror in which (a) the highly reflective coating or reflecting surface is applied to the back surface of the substrate that forms the mirror, i.e., not to the surface of first incidence of light, and (b) the reflected light must pass through the substrate twice, once as part of the incident light and once as the reflected light. *See also* **front surface mirror.**

backtab key: On keyboards, such as on printers, computers, and communications equipment, a key that, when actuated, will cause the device to step backward a specified number of increments, such as positions, columns, screens, or storage locations. *Note:* In a personal computer (PC), actuation of the backtab key may cause the computer to display an entire screen of columns to the left on a spreadsheet. On a printer, it might cause the printer to step back along a line to a preset position. The action of the backtab key might be equivalent to (a) actuating the backspace key a requisite number of times on a printer keyboard, (b) actuating the left arrow key a requisite number of times on a PC keyboard, or (c) effecting a margin release. *See also* **backspace key, tab key.**

back tell: Information transfer from facilities at a higher to a lower operational level or echelon of command or administration.

back-to-back connection (BBC): A direct connection between (a) the output of a device, such as a transmitter and (b) the input of another device, such as an associated receiver, thereby eliminating the effects of transmission channels or transmission media. *Note:* The insertion of a passive element, such as a pad, i.e., an attenuator, between the devices is still considered as a direct connection, i.e., a back-to-back connection. *See also* **loop back.**

backup: In communications, computer, and data processing systems, the provisions that are made for the recovery of data or software, the restart of operations, or the use of alternative equipment after failure or interruption of system operations occurs.

backup copy: In database management, a copy of a file or a set of data that is retained for reference in the event that part or all of the original is lost or destroyed.

backup file: A file that (a) is a copy of an original file, (b) is made for the express purpose of possible later reconstruction of the original file should this become necessary, (c) may be used for preserving the integrity of the original file, and (d) may be recorded on any suitable medium from which it can be recovered. *Synonym* **job recovery control file.** *See also* **backward recovery, data integrity.**

Backus-Naur form (BNF): A metalanguage used to specify or describe the syntax of a language in which each unique symbol represents a single set of strings of symbols.

backward channel: 1. In data transmission, a secondary channel in which the direction of transmission is constrained to be opposite to that of the primary channel, i.e., the forward or user information channel. *Note:* The backward channel direction of transmission is restricted by the control interchange circuit that controls the direction of transmission in the primary channel. **2.** In a data circuit, the channel that passes data in a direction opposite to that of its associated forward channel. *Note 1:* The backward channel usually is used for transmission of supervisory, acknowledgment, or error control signals. The direction of flow of these signals is opposite to that in which user information is being transferred. *Note 2:* The backward channel bandwidth usually is less than that of the primary channel, i.e., the forward or user information channel. *See also* **backward signal, data transmission, forward channel, forward signal, information bearer channel.**

backward current: In a device, such as a vacuum tube, transistor, or diode, the current that flows when the device is driven in reverse to the forward direction. *Note:* In the case of a diode or a transistor, the backward resistance is high, allowing only a very small backward current magnitude when the cathode or the emitter is driven positive, compared to the forward current magnitude when the anode or the collector is driven positive. *See also* **forward current.**

backward direction: 1. In a communications system, the direction from a data sink to a data source. **2.** The direction from an addressee to a message originator. **3.** The direction from a call receiver to a call originator.

backward ionospheric scatter: *Synonym* **backward propagation ionospheric scatter.**

backward propagation ionospheric scatter: Ionospheric scatter in the backward direction. *See also* **backward direction.** *Synonym* **backward ionospheric scatter.**

backward recovery: The reconstruction of an earlier version of a file by using a later version and known, usually documented, changes.

backward signal: A signal sent from the called station back to the calling station, i.e., from the original data sink to the original data source. *Note:* Backward signals usually are sent in a backward channel and usually consist of supervisory, acknowledgment, or error control signals. *See also* **backward channel, communications sink, communications source, forward channel, forward signal, signal.**

backward supervision: The use of supervisory signal sequences from a secondary to a primary station. *See also* **control station, secondary station, supervisory signals.**

bait tube: The basic structure upon which a fiber optic preform is built.

balance: In electrical circuits and networks, to adjust the impedance to achieve specific objectives, such as to reach specified return loss objectives at junctions of two-wire and four-wire circuits. *See* **Earth radiation balance, hybrid balance, line balance, line filter balance, longitudinal balance.** *See also* **balancing network.**

balanced: Pertaining to electrical symmetry.

balanced amplitude modulation: Suppressed carrier modulation in which (a) the modulator suppresses the carrier by means of a balance circuit, such as a modulator in which the modulating voltage enters a transformer primary with an amplifier connected to each secondary end and the carrier signal is fed through the center tap of the transformer in a push-pull connection, (b) the resulting signal can be either a single sideband or a double sideband amplitude-modulated signal, (c) all the intelligence is contained in one of the sidebands and none is contained in the carrier because either sideband can be suppressed as well as the carrier, and (d) there is no direct current (dc) component in the antenna feed signal. *See also* **phase exchange keying, suppressed carrier modulation.**

balanced code: 1. In pulse code modulated (PCM) systems, a code constructed so that the frequency spectrum resulting from the transmission of any code word has no direct current (dc) component. **2.** In pulse code modulation (PCM), a code that has a finite digital sum variation. *See also* **code, pulse code modulation.**

balanced error: A set of errors that has a mean value of zero.

balanced line: A transmission line that (a) consists of two conductors in the presence of ground and (b) is capable of being operated in such a way that (i) the voltages of the two conductors at all transverse planes are equal in magnitude and opposite in polarity with respect to ground, (ii) the currents in the two conductors are equal in magnitude and opposite in direction. *Note:* A balanced line may be operated in an unbalanced condition. *Synonym* **balanced signal pair.** *See also* **balance return loss, ground return circuit, hybrid balance, line, line balance, longitudinal balance, metallic circuit, symmetrical pair, unbalanced line, unbalanced wire circuit.**

balanced link level operation: At the data link level, the operation of data links in which balanced lines are used. *See also* **balanced line.**

balanced modulation: Suppression of the carrier in an amplitude-modulated wave so that (a) only sideband signals appear in the wave, and (b) there is approximately equal power in each sideband, i.e., in the upper and lower sidebands.

balanced modulator: A modulator constructed so that the carrier is suppressed and any associated carrier noise is balanced out. *Note 1:* The balanced modulator output contains only the sidebands. *Note 2:* Balanced modulators are used in amplitude-modulated (AM) transmission systems. *See also* **amplitude modulation, sideband transmission, single sideband suppressed carrier transmission.**

balanced signal pair: *Synonym* **balanced line.**

balanced station: A station that (a) performs balanced link level operations, (b) generates commands, (c) interprets responses, (d) interprets received commands, and (e) generates responses.

balanced to unbalanced: *Synonym* **balun.**

balance return loss (BRL): 1. A measure of the degree of balance between two impedances connected to two conjugate sides of a hybrid set, network, or junction. **2.** A measure of the effectiveness with which a balancing network simulates the impedance of a two-wire circuit at a hybrid coil. *See also* **balanced line, balancing network, hybrid balance, hybrid junction, hybrid network, hybrid set, return loss.**

balancing: *See* **power balancing.**

balancing mechanism: *See* **bandwidth balancing mechanism.**

balancing network: 1. A circuit used to simulate the impedance of a uniform two-wire cable or open wire circuit over a selected range of frequencies. **2.** A device used between a balanced device or line and an unbalanced device or line in order to transform the operating condition from balanced to unbalanced or vice versa. *Deprecated synonym* **balun.** *See also* **balance return loss, balun, characteristic impedance, circuit, impedance matching.**

ball: *See* **control ball, type ball.**

balsam: *See* **Canada balsam.**

balun: A device used to couple a balanced device or line and an unbalanced device or line. *Note:* "Balun" is derived from "balanced to unbalanced." *Synonym* **balanced to unbalanced.** *Deprecated synonym* **balancing network.** *See also* **antenna, antenna matching, balancing network, Pawsey stub.**

band: 1. In communications, the frequency spectrum between two defined frequency limits. **2.** A group of tracks on a magnetic drum or on one side of a magnetic disk. **3.** A geographical area defined by common carriers for purposes of communications system management. **4.** In an atom, the difference between two electron energy levels. *See* **baseband, emergency frequency band, frequency band, narrow band, octave band, stop band, wideband.** *See also* **common carrier, communications system management.**

band compaction: *See* **variable tolerance band compaction.**

band edge absorption: In a lightwave, absorption of energy that (a) occurs in the visible region of the electromagnetic frequency spectrum and extends from the ultraviolet to the infrared regions and (b) in glass usually is caused by (i) oxides of certain elements, such as silicon, sodium, boron, calcium, and germanium and (ii) hydroxyl ions.

band elimination filter: *Synonym* **band stop filter.**

band gap energy: The difference between any pair of allowable energy levels of the electrons of an atom. *Note:* An electron that absorbs a photon absorbs energy equivalent to one or more band gaps. When an electron loses energy, a photon is emitted with energy equivalent to the energy of one or more band gaps.

banding: *See* **rubber banding.**

B AND NOT A gate: A two-input, binary, logic coincidence circuit or device that performs the logic operations of B AND NOT A, i.e., if A is a statement and B is a statement, the result is true only if B is true and A is false, and for the other three combinations of A and B, the result is false. *Synonyms* **B except A gate, negative B implies A gate.** *Refer to* **Fig. B-1.** *Refer also to* **Fig. A-1 (1).**

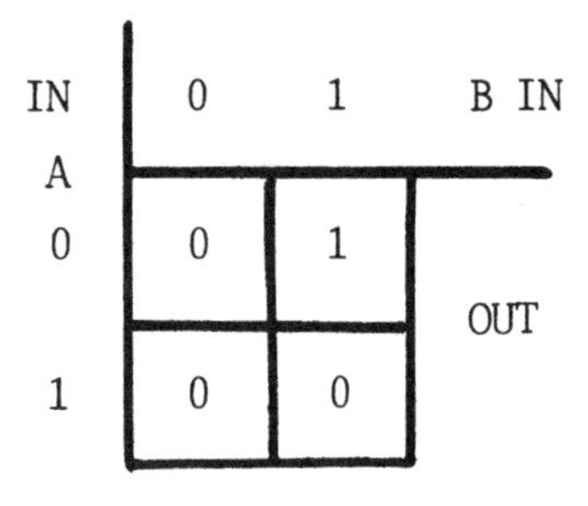

Fig. B-1. A truth table that shows the input and output digits of a **B AND NOT A gate.**

bandpass: Pertaining to a device that will allow or support a fixed band of frequencies.

bandpass filter: 1. A filter that passes all frequencies between two nonzero finite frequency limits and bars all frequencies not within the limits. **2.** A filter that transmits a single, continuous band of frequencies, the lower and upper cutoff frequencies being nonzero and finite. *Note:* The cutoff frequencies are usually considered to be at the 3 dB level. *See also* **band stop filter, filter, high pass filter, low pass filter.**

bandpass limiter: A device that (a) imposes hard limiting on a signal and (b) contains a filter that suppresses the unwanted frequencies, such as transients and harmonics, caused by the limiting process. *See also* **band stop filter, filter.**

band rejection filter: *Synonym* **band-stop filter.**

bandspreading: Electromagnetic wave transmission in which (a) signals are artificially spread over a wider bandwidth than that of the original signals, (b) modulation usually is performed according to a prearranged coded high speed digital process, (c) resulting signals have a noise-like wideband structure and can be received and decoded only by the use of a correlation process, and (d) synchronization must be maintained with the prearranged code. *See also* **code division multiple access, spread spectrum multiple access.**

band stop filter: A filter that attenuates a single, continuous band of frequencies, the lower and upper cutoff frequencies being nonzero and finite, respectively, and passes all frequencies not within the limits. *Note:* A band stop filter is designed to stop the specified band of frequencies, but usually only attenuates them to below a specified level. *Synonyms* **band elimination filter, band rejection filter, band suppression filter, notched filter.** *See also* **bandpass filter, bandpass limiter, filter, frequency.**

band suppression filter: *Synonym* **band stop filter.**

bandwidth (BW): 1. The frequency band within which the performance of a device with respect to some characteristic, such as gain, falls within specified limits of the characteristic. **2.** The difference between the limiting frequencies of a continuous frequency band. *Note:* The particular performance characteristic, such as the gain of an amplifier, the responsivity of a device, or the bit error ratio (BER), and the limiting frequencies must be specified. **3.** A range of frequencies, in a given frequency spectrum, usually specifying the number of hertz of the band or of the upper and lower limiting frequencies. **4.** The range of frequencies that a device is capable of generating, handling, passing, or allowing, usually the range of frequencies in which the responsivity is not reduced greater than 3 dB from the maximum response. **5.** In fiber optics, the value numerically equal to the lowest modulation frequency at which the magnitude of the baseband transfer function of an optical fiber decreases to a specified fraction, usually to one-half of the zero frequency value. *Note 1:* In multimode fibers, bandwidth is limited mainly by multimode distortion and material dispersion, which cause intersymbol interference at the receiving end of an optical fiber and thus a limit is placed on the bit error ratio (BER). In single-mode fibers, bandwidth is limited mainly by material and waveguide dispersion. *Note 2:* In fiber optics, the concept of bandwidth, as applied to electronic systems, is not entirely useful in describing the data signaling rate capability or the transmission capacity of an optical fiber because the optical frequencies involved range over a spectrum about a decade wide near 10^{15} Hz, or 1 PHz (petahertz). The bandwidth•distance factor, i.e., the bit-rate•length product (BRLP), is a better measure of the signal-carrying capacity of an optical fiber than the bandwidth. However, the modulation scheme must also be specified in order to make the value of the BRLP meaningful. *See* **authorized bandwidth, Carson bandwidth, closed loop noise bandwidth, facsimile bandwidth, necessary bandwidth, nominal bandwidth, Nyquist bandwidth, occupied bandwidth, optical fiber bandwidth, phase bandwidth, radio frequency bandwidth.**

bandwidth balancing mechanism: In a distributed queue dual bus (DQDB) network, a procedure in which a node occasionally skips the use of empty queued arbitrated slots and thereby more effectively shares the occupied bandwidth. *See also* **occupied bandwidth.**

bandwidth compression: 1. The reduction of the bandwidth needed to transmit a given amount of data in a given time. **2.** The reduction of the time needed to transmit a given amount of data in a given bandwidth. *Note:* Bandwidth compression implies a reduction in normal bandwidth of an information-carrying signal (a) without reducing the information content of the signal and (b) usually without increasing the bit error ratio (BER). *See also* **biternary transmission, data compression, necessary bandwidth.**

bandwidth•distance factor (BWDF): In fiber optics, a figure of merit used to express the signal-carrying capacity of a fiber optic cable over various specified distances. *Note 1:* The bandwidth•distance factor usually is expressed in megahertz•kilometers. *Note 2:* The bandwidth•distance factor implies that bandwidth and distance are reciprocally related. However, the trade-off is not necessarily linear. Thus, fiber optic cable characteristics, such as the bandwidth, length, and allowable bit error ratio (BER), should be separately specified. *Synonyms* **bandwidth•distance product, bandwidth•length factor, bandwidth•length product.** *See also* **bandwidth, bit-rate•length product, fiber bandwidth.**

bandwidth•distance product: *Synonym* **bandwidth•distance factor.**

bandwidth•length factor: *Synonym* **bandwdith•distance factor.**

bandwidth•length product: *Synonym* **bandwidth•distance factor.**

bandwidth-limited: 1. In transmission circuits, pertaining to an ability to pass signals with frequency components that lie only within certain limits. *Note:* An optical filter that passes only a narrow frequency band, such as blue or red frequencies only, is bandwidth-limited because of the characteristics of the materials used in its construction. **2.** In a communications system, pertaining to a signal with spectral components that extend only between specified wavelengths. *Note:* If a signal is passed through a filter with a transfer function of 1 for frequencies equal to or less than the maximum frequency, the signal will be transmitted without distortion. Signals that are not themselves originally bandwidth-limited may be passed through filters that introduce band limiting at either high or low frequencies. Hence, distortion is introduced. **3.** In a communications system, pertaining to the situation that exists when the total traffic capacity is limited by the allocated available bandwidth or assigned frequency bandwidth.

bandwidth-limited operation: The condition that prevails when the system bandwidth, rather than a signal parameter, such as signal amplitude, power, phase shift, or frequency shift, limits performance. *Note:* Bandwidth-limited operation is reached when the system distorts the signal waveform beyond specified limits. For linear systems, bandwidth-limited operation is equivalent to distortion-limited operation. *See also* **attenuation-limited operation, bandwidth, bit-rate•length product-limited operation, dispersion-limited operation, distortion-limited operation, linear element, quantum-noise-limited operation.**

bandwidth product: *See* **gain•bandwidth product.**

bandwidth rule: *See* **Carson bandwidth rule.**

bank: *See* **channel bank, data bank.**

bar: *See* **reverse video bar, type bar, writing bar.**

bar code: A code that represents characters by sets of parallel bars of different thickness and separation that are optically read by transverse scanning. *Note:* Bar codes are often used to identify equipment, components, and merchandise. *Refer to* **Fig. B-2.**

Fig. B-2. Two UPC codes that are examples of a **bar code.**

Barker code: A binary code that (a) is suitable for pulse code modulation and synchronization, (b) has optimal correlation properties, (c) is relatively immune to phase displacement caused by random pulses immediately adjacent to the code patterns when compared to other codes, such as the pure binary numeration system code, and (d) is relatively immune to phase displacement errors caused by the transmitter.

bar printer: An impact printer in which type characters are mounted on elongated strips of metal.

barrage jammer: A jammer that (a) spreads its radiated electromagnetic energy over a wide frequency band, (b) may jam several transmitters simultaneously, (c) discourages escape simply by shifting the transmission frequency of the victim transmitter, i.e., the transmitter being jammed, (d) usually is tunable over a wide frequency spectrum, and (e) has a disadvantage in that spreading the energy over a wide frequency spectrum results in less power at a particular frequency where higher power jamming is needed.

barrage jamming: Jamming that (a) makes it possible to jam emitters on different frequencies simultaneously, (b) reduces the need for operator assistance or complex control equipment, (c) has advantages that are gained at the expense of reduced jamming power at any given frequency, and (d) is accomplished by (i) transmitting a band of frequencies that is large with respect to the bandwidth of a single emitter, (ii) presetting multiple jammers on adjacent frequencies, or (iii) using a single wideband transmitter.

barrage jamming radar display: *See* **spot barrage jamming radar display.**

barred facility: *See* **calls barred facility.**

barred signal: *See* **access barred signal.**

barrel distortion: In display systems, a distortion of the image of an object in such a manner that the display element, display group, or display image of an otherwise straight-sided square or rectangular object has its sides bowed out, i.e., has a convex appearance, relative to the object. *See also* **pincushion distortion.**

barrier layer: In the manufacture of optical fibers, a layer that prevents hydroxyl ion diffusion into the fiber core.

barrier layer cell: *Synonym* **photovoltaic cell.**

base: 1. In the numeration system commonly used in scientific work, the number that is raised to a power denoted by the exponent and then multiplied by the coefficient to determine the value of the same number represented without the use of exponents. *Note:* An example of a base is the number 6.25 in the expression $2.70 \times 6.25^{1.5} \approx 42.19$. The 2.70 is the coefficient, and the 1.5 is the exponent. In the decimal numeration system, the base is 10, and in the binary numeration system, the base is 2. The value $e \approx 2.718$ is the natural base. **2.** A reference value. **3.** A number that is multiplied by itself as many times as indicated by an exponent.

base address: 1. An address that is used as the origin in the calculation of addresses in the execution of a computer program. **2.** A given address from which an absolute address is derived by combination with a relative address. *Note:* Base addresses are primarily used by computer programmers rather than computer users. *See also* **absolute address, relative address.**

baseband (BB): 1. The band of frequencies contained in the original signal that is produced by a transducer, or other signal-initiating device, such as a microphone, telegraph key, or sensor, prior to initial modulation. *Note 1:* Demodulation recreates the baseband signal. *Note 2:* In transmission systems, the baseband signal usually is used to modulate a carrier. *Note 3:* The baseband describes the signal state prior to modulation, prior to multiplexing, following demultiplexing, and following demodulation. *Note 4:* Baseband frequencies usually are characterized by being much lower in frequency than the frequencies that result when the baseband signal is used to modulate a carrier or a subcarrier. *Note 5:* The baseband signal usually is the original information-bearing signal. **2.** In fiber optics, the spectral band occupied by a signal that (a) neither requires carrier modulation nor describes the signal state prior to modulation or following demodulation and (b) usually is characterized by being much lower in the frequency spectrum than the resultant signal after it is used to modulate a carrier or a subcarrier. **3.** In facsimile systems, the bandwidth of a signal equal in width to that between zero frequency and maximum keying frequency. **4.** The spectral band occupied by an unmodulated signal. *See also* **baseband signaling, carrier, frequency, modulation, multiplex baseband, multiplexing.**

baseband local area network (BBLAN): A local area network in which information is encoded, multiplexed, and transmitted without modulation of carriers. *See also* **local area network.**

baseband receive terminal: *See* **multiplex baseband receive terminal, radio baseband receive terminal.**

baseband send terminal: *See* **multiplex baseband send terminal, radio baseband send terminal.**

baseband signal: 1. A signal as it was originally generated prior to any modulation, conversion, or translation. **2.** In a digital data stream, signals that are represented in the form of voltage levels or current levels that are continuous rather than discrete. *Note:* Examples of baseband signals are (a) the voice signal generated in an ordinary telephone transmitter, (b) the signal generated by a microphone, and (c) the signal generated by a transducer, such as a piezoelectric crystal when subjected to pressure variations.

baseband signaling: Transmission of a digital or analog signal at its original frequencies, i.e., a signal in its original form, not changed by modulation. *See also* **baseband, modulation, signal.**

basecom: base communications.

base communications (basecom): Communications services (a) that are provided for the operation of a military installation, such as a post, camp, or station and (b) that usually includes the installation, operation, maintenance, augmentation, modification, and rehabilitation of communications networks, systems, facilities, and equipment that provide local and intrabase communications, including off-installation extensions. *Synonym* **communications base section.**

base Earth station (BES): An Earth station that (a) is in the fixed satellite service or in the land mobile satellite service, (b) is at a specified fixed point or within a specified area on land, and (c) provides a feeder link for the land mobile satellite service.

base equipment: *See* **embedded base equipment.**

base group: *Synonym* **channel basic group.**

baseline: 1. Pertaining to a specification, product, or system that (a) has been formally reviewed and agreed upon, (b) serves as the basis for further development, such as redesign or improvement, and (c) can be changed only through a formal change control procedure. **2.** In a display device, a line that represents a reference level, usually zero, on which a scale, such as a time, frequency, or distance scale, is placed, for the display of signals, waves, or other events. *Note:* Display devices that exhibit a baseline include the cathode ray tube (CRT), vidicon, reflectometer, spectroscope, and fiberscope. **3.** On a map, a defined straight line (a) that serves as a reference line for a coordinate system in lieu of the existing map coordinates and (b) the map coordinates of which are sent to those in need of them. **4.** On a map, a line that (a) is determined by two known fixed points, (b) is used to define positions of points relative to the line and relative to one or more fixed points on the line, and (c) usually is used in direction finding, triangulation, intersection, and resection. *See also* **direction finding, triangulation.** *Refer to* **Fig. I-7 (476).**

base node record: *Synonym* **root record.**

base station: A land station that (a) is in the land mobile service, (b) primarily communicates with mobile stations, and (c) communicates with other base stations incidental to communications with land mobile stations. *See* **airport land mobile radio base station.** *See also* **mobile service, mobile station.**

BASIC: A computer programming language that (a) has a simple syntax, such as simplified grammar, and simplified semantics, such as mnemonic instruction codes, (b) is designed for ease in learning, (c) is used for conversational mode, numeric computations, and string handling, and (d) has many variations, i.e., dialects, that are in use. *Note:* The name is derived from Beginner's All-purpose Symbolic Instruction Code.

basic exchange telecommunications radio service (BETRS): A service that can extend telephone service to rural, outlying, and remote areas by replacing the local loop with radio communications. *Note:* In the basic exchange telecommunications radio service (BETRS), (a) the ultrahigh frequency (UHF) and very high frequency (VHF) common carrier frequencies and (b) the private radio frequencies are shared.

basic group: 1. In frequency division multiplexing of wideband systems, a group of voice channels that (a) is a separate group or a group within a supergroup, (b) usually consists of up to 12 voice channels, (c) occupies the frequency band from 60 kHz (kilohertz) to 108 kHz, (d) may be divided into four 3-channel pregroups, (e) is designated CCITT Basic Group B, and (f) in carrier telephone systems, is an assembly of 12 channels that occupies upper sidebands in the 12 kHz to 60 kHz band and that is designated as CCITT Basic Group A. **2.** In signaling, a group of characters to which suffixes or prefixes may be added to form signal groups for specific signaling purposes. *Synonym* **primary group.** *See* **channel basic group.** *See also* **group.**

basic mode: *See* **laser basic mode.**

basic mode link control: In an open system, the control of a data link at the data link level rather than at higher layers, such as in high level data link control (HDLC).

Basic Mode Link Control (BMLC): Control of data links by use of the 7-bit control characters defined in International Standards Organization (ISO) Standard 646-1983 and International Telegraph and Telephone Consultative Committee (CCITT) Recommendation V.3-1972.

basic rate interface (BRI): In communications systems, a user interface standard that describes the capability of simultaneous voice and network-provided data services.

Basic Rate Interface (BRI): An International Telegraph and Telephone Consultative Committee (CCITT) Integrated Services Digital Network (ISDN) multipurpose interface standard that defines the network user access arrangement for simultaneous voice and data services provided over two clear 64-kb/s (kilobits per second) channels and one clear 16-kb/s channel (2B+D). *See also* **Integrated Services Digital Network.**

basic service: 1. A transmission service provided over a communications path that is essentially transparent to user information. **2.** A transmission service that offers sufficient capacity between two or more points to satisfy user transmission needs. *Note:* The basic service must meet acceptable performance criteria, such as fidelity, distortion, and availability. *See also* **enhanced service.**

basic service element (BSE): 1. An optional unbundled network-provided service feature that (a) usually is associated with the basic serving arrangement (BSA), (b) an enhanced service provider (ESP) may require or find useful for providing an enhanced service, and (c) allows a user to select services. **2.** A fundamental, i.e., a basic, communications network service that (a) usually is an optional network capability associated with a basic serving arrangement and (b) constitutes an optional network capability to which the user may subscribe or decline to subscribe. *See also* **basic serving arrangement, unbundling.**

basic serving arrangement (BSA): 1. The fundamental tariffed communications services, such as switching and transmission, that a common carrier must provide to an enhanced service provider (ESP) to interconnect its customers. **2.** In open network architecture, the fundamental underlying connection of an enhanced service provider (ESP) to and through a common carrier network. *Note 1:* The basic serving arrangement (BSA) includes (a) an enhanced service provider (ESP) access link along with the features and functions associated with the link at the central office serving the ESP and other offices and (b) the dedicated or switched transport within the common carrier network that completes the connection from the ESP (i) to the central office serving the ESP customers or (ii) to the equipment associated with the customer complementary network services. *Note 2:* Each basic serving arrangement (BSA) component, such as an enhanced service provider (ESP) access link, transport, and usage, may have a number of categories of network characteristics. Within these categories of network characteristics are options the customer may choose. *See also* **basic service element, unbundling.**

basic status: In data transmission, the status of the capability of a secondary station to send or receive a frame containing an information field.

basis weight: The weight in pounds of a ream, i.e., 500 sheets, of paper of a given standard size, such as 25 in. (inches) by 38 in. for book papers and 17 in. by 22 in. for bond papers. *Note:* Continuous forms, such as for computer and telecommunications printers, are given the same basis weights as for bond papers.

basket grip: A device for gripping a cable so that the pulling force is transferred to the sheath.

bastion host: In a firewall, the combination of hardware, software, and firmware that (a) provides and maintains the protective environment, (b) contains a screening router with the intelligence needed to provide an effective firewall, and (c) contains a rules table for acceptability determination. *See also* **firewall.**

batch: In communications, computer, and data processing systems, a grouping of items. *Note:* Examples of batches are (a) an accumulation of data to be processed, (b) a group of records or jobs assembled for processing, transmission, or storage, and (c) a set of punched cards or tickets having a characteristic in common.

batched communications: *Synonym* **batched transmission.**

batched transmission: The transmission of two or more messages from one station to another without intervening responses from the receiving station. *Synonym* **batched communications.**

batch entry: *See* **remote batch entry.**

batch file: In personal computer (PC) operation, a disk operating system file that (a) contains a list of commands to be executed automatically and usually regularly when setting up a sequence of steps and (b) is convenient for handling various PC startup procedures. *Note:* An example of a batch file is the DOS AUTOEXEC.BAT file.

batch processing: 1. In a computer, the processing of data, the running of programs, or the accomplishment of jobs accumulated in advance and entered into the computer such that the user cannot further influence the processing while it is in progress. **2.** The processing of data accumulated over a period of time. **3.** The execution of computer programs in sequence, i.e., one at a time. **4.** Pertaining to the execution of a set of computer programs, usually in a single run, such that each program is completed before the next program of the set is started. **5.** Pertaining to the sequential input of computer programs or data. *See also* **background processing, job, remote batch processing.**

bathtub curve: A curve that (a) is shaped like the longitudinal cross-sectional profile of a bathtub, (b) characterizes failure rates of large populations of components, such as are used in communications, computer, and data processing systems, as a function of time, and (c) shows (i) an early failure period with a high failure rate that weeds out defective items, (ii) followed by a nearly constant failure rate period with a low failure rate based on statistical failures, and (iii) followed by a rapidly rising failure rate as the life of each component expires. *See also* **component life.** *Refer to* **Fig. B-3.**

battery: *See also* **common battery, local battery, station battery.**

battery signaling: *See* **common battery signaling.**

battery signaling exchange: *See* **common battery signaling exchange.**

battlefield: *See* **digitized battlefield.**

baud (Bd): 1. For a modulated signal, the reciprocal of the unit interval measured in seconds. **2.** For a communications channel, the maximum data signaling rate (DSR) supported. **3.** A unit of modulation rate. *Note:* One baud corresponds to a rate of one unit interval per second. The baud, i.e., the modulation rate, is expressed as the reciprocal of the duration in seconds of the unit interval. If the duration of the unit interval is 20 milliseconds, the signaling speed is 50 baud. If the signal transmitted during each unit interval can take on any one of M discrete states, the bit rate is equal to the rate in baud times $\log_2 M$. The technique used to encode the allowable signal states may be any combination of amplitude, frequency, or phase modulation, but it cannot use further time division multiplexing technique to subdivide the unit intervals into multiple subintervals. In some signaling systems, non-information-carrying signals may be inserted to facilitate synchronization, e.g., in certain forms of binary modulation coding, there is a forced inversion of the signal state at the center of the bit interval. In these cases, the synchronization signals are included in the calculation of the rate in baud but not in the computation of bit rate. *See also* **bit rate, data signaling rate, unit interval.**

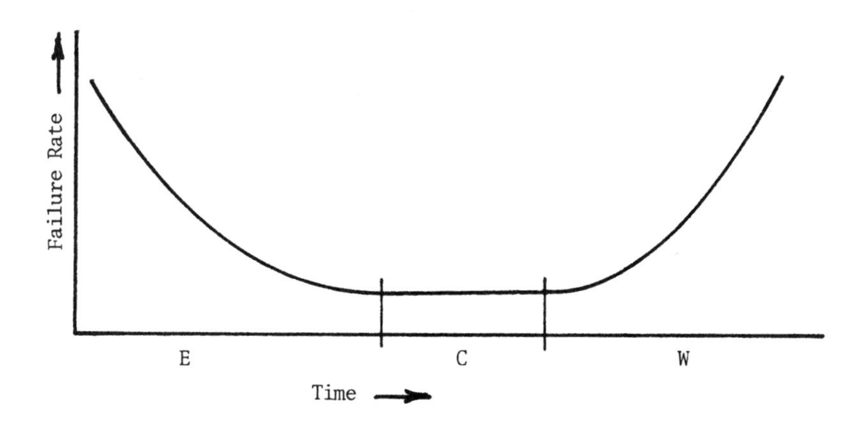

Fig. B-3. The shape of a typical **bathtub curve** showing the failure rate, *i.e.,* the number of components that fail in a system per given number of components per unit of time, such as the number that fail per thousand per month, as a function of time over the life of the system. *E* is the early failure period, *C* is the constant failure period, and *W* is the wearout failure period.

Baudot code: Developed about 1880 for the transmission of data, a synchronous code in which five equal-length bits represent one character. *Note 1:* The Baudot code has been replaced by the start-stop asynchronous International Telegraph Alphabet 2 (ITA-2). *Note 2:* The International Telegraph Alphabet 2 (ITA-2) should not be identified as "Baudot code." *Note 3:* The Baudot code has been widely used in teletypewriter systems that use five-hole punched paper tape. *See also* **code, International Telegraph Alphabet 2.**

bay: *See* **coaxial patch bay, digital circuit patch bay, digital primary patch bay, direct current (dc) patch bay, D patch bay, group patch bay, K patch bay, MM patch bay, M patch bay, patch bay, U link patch bay, voice frequency primary patch bay.**

BCC: block check character.

BCD: binary coded decimal.

B channel: The International Telegraph and Telephone Consultative Committee (CCITT) designation for a clear channel that (a) provides 64-kbps (kilobits/second) service to a customer in the Integrated Services Digital Network (ISDN) and (b) is intended for transport of user information, as opposed to signaling information. *See also* **bearer channel, Integrated Services Digital Network.**

BCH code: Bose-Chaudhuri-Hocquenghem code.

BCI: bit count integrity.

Bd: baud.

beacon: A light or electronic source that emits a distinctive or characteristic signal used for the determination of bearings, courses, or locations. *See* **crash locator beacon, fan marker beacon, marker beacon, nondirectional beacon, personnel locator beacon, radar beacon, radar homing beacon, radio radar beacon, satellite beacon, search and rescue beacon, Z-marker beacon.**

beacon frequency: *See* **search and rescue beacon frequency.**

beacon navigation system: A navigation system (a) that uses electromagnetic energy to establish a signal pattern that can be followed or used as a guide and (b) in which the direction of motion relative to a line from the present position of a mobile receiver to the beacon transmitter can be determined.

beacon precipitation gage station: *See* **radio radar beacon precipitation gage station.**

beacon with reply: *See* **situation awareness beacon with reply.**

beam: In communications, a shaft or a column of electromagnetic radiation that consists of parallel, converging, or diverging rays. *See* **diverging beam, fan beam, Gaussian beam, laser beam, radio beam, spot beam, telecommunications laser beam.** *Refer to* **Fig. N-21 (641).**

beam angle between half-power points: *See* **emission beam angle between half-power points.**

beam antenna: *See* **fan beam antenna, narrow beam antenna, wide beam antenna.**

beam connector: *See* **expanded beam connector.**

beam diameter: The distance between two diametrically opposed points on a plane perpendicular to the beam axis at which the irradiance is a specified fraction of the beam peak irradiance. *Note 1:* Beam diameters can only be determined for beams that are circular or nearly circular in cross section. *Note 2:* The specified fraction of the beam peak irradiance usually is stated as being 1/2, $1/e$, $1/e^2$, or 1/10 of the peak irradiance at the point where the beam diameter is measured. *Note 3:* The beam profile of the emission from a multimode optical fiber is assumed to be Gaussian, and the beam diameter is given by the relation $D = (2d \tan(\sin^{-1} NA))/1.73 \approx 2d(NA)/1.73$ where D is the beam diameter, d is the distance from the end of the fiber, and NA is the numerical aperture measured at the 5% of peak irradiance points. *Note 4:* Single mode optical fiber beam diameters are given by the relation $D = 2d\lambda/\pi d_o$ where d_o is the mode field (core) diameter, and λ is the wavelength. The numerical aperture is defined at the 5% of peak irradiance points. To calculate the axial beam irradiance of a Gaussian beam, the beam cross-sectional area at the $1/e$ points is used. The value 1.73 accounts for the difference between single mode fibers and multimode fibers. *See also* **antenna, beam divergence, beamwidth.**

beam divergence: 1. In an electromagnetic beam, such as a radio or light beam, the increase in diameter with increase in distance from the appropriate aperture in the direction of propagation along the beam axis. **2.** In an electromagnetic beam that is circular or nearly circular in cross section, the angle, with its vertex at the source, subtended by the far field beam diameter. **3.** In an electromagnetic beam that is not circular in cross section, the far field angle subtended by two diametrically opposite points in a plane perpendicular to the beam axis, at which points the irradiance is a specified fraction of the maximum irradiance. *Note:* Usually, for beams not circular in cross section, the maximum and minimum beam divergences, corresponding to the angles subtended by the major and minor diameters of the far field, must be specified. **4.** The far field angle subtended by two diametrically opposed points in a plane perpendicular to the beam axis, at which points the irradiance, i.e., the power density, is a specified fraction of the beam peak irradiance. *Note 1:* The divergence, usually expressed in milliradians, is measured at specified points along the beam and radially from the beam central axis, such as where the energy density is 1/2 or $1/e$ of the maximum value at that point along the beam. The divergence must be specified as the half angle or as the full angle. Usually, only the maximum and minimum divergences, corresponding to the major and minor diameters of the far field irradiance, need by specified. *Note 2:* The emission from a multimode optical fiber is assumed to be Gaussian, in which case the beam divergence, Φ_m, is given by the relations $\Phi_m = (\tan(\sin^{-1} NA))/1.73 \approx NA/1.73$ where Φ_m is the beam divergence, and NA is the numerical aperture measured at the 5% of peak irradiance points. *Note 3:* The beam divergence, Φ_s, for a single mode fiber is given by the relation $\Phi_s = 2\lambda/\pi d_o$ where Φ_s is the beam divergence, d_o is the mode field (core) diameter, and λ is the wavelength.

beaming: *See* **multiple beaming.**

beam lobe switching: The determination of the direction to a remote object by comparison of the signals from the object corresponding to two or more successive beam angles that differ slightly. *Note:* Beam lobe switching may be either (a) continuous or (b) discontinuous, i.e., may be periodic.

beam radar: *See* **narrow beam radar, wide beam radar.**

beam riding guidance system: A guidance system in which an object is made to follow a directionally controlled radio beam.

beam rotation: *See* **antenna beam rotation.**

beam satellite antenna: *See* **global beam satellite antenna.**

beam split radar: *See* **conical beam split radar.**

beam splitter: A device that divides a beam, such as a light beam, into two or more separate beams with a total power that is less than or equal to the original beam. *Note:* Examples of beam splitters are (a) a partially reflecting mirror and (b) a flat parallel transparent plate coated on one side with a dielectric or metallic material that reflects a portion and transmits the remaining portion of a light beam. *Synonyms* **beamsplitter, optical splitter.** *See also* **fiber optic coupler, partial mirror.**

beamsplitter: *Synonym* **beam splitter.**

beam station: *See* **radar beam station.**

beam steering: Changing the direction of the major lobe of a radiation pattern. *Note:* In radio systems, beam steering is usually accomplished by switching or by controlling the phase of signals in antenna elements. In optical systems, beam steering may be accomplished (a) by changing the refractive index of the medium through which the beam is propagating or (b) by using mirrors and lenses. *See also* **antenna, beam diameter, collimation, decollimation, diffraction-limited, far field region.**

beam tilt: The angular tilt of the main lobe of the vertical radiation pattern above or below 0° elevation. *Note:* The tilt must be referred to the free space pattern unless the actual vertical radiation pattern approaches free space conditions, i.e., unless it is free from ground reflection effects. *See* **antenna electrical beam tilt.**

beamwidth: **1.** In an electromagnetic beam, the angle, measured from the antenna aperture, subtended by the beam diameter. **2.** In a radio frequency antenna pattern, the angle subtended by the diametrically opposite half-power points, i.e., the 3-dB points, of the main lobe of the antenna pattern when referenced to the maximum irradiance. *Note:* The beamwidth usually is measured in the horizontal plane and expressed in degrees. **3.** The angle between the directions, on either side of the axis, at which the intensity of the radio frequency field drops to 1/2, $1/e$, $1/e^2$, or 1/10 of the value it has on the axis, i.e., of the peak irradiance. *See also* **antenna, aperture, beam diameter, directive gain, directivity pattern.**

bearer: In communications systems, a facility or propagation medium for messages. Examples of bearers are land lines, cables, microwaves between towers, radio relays, high frequency radios, ionospheric reflections, tropospheric scatter, and satellites.

bearer channel: A communications channel used for the transmission of an aggregate signal generated by multichannel transmitting equipment. *See* **information bearer channel.** *See also* **B channel, channel, group, multiplexing.**

bearer service: In integrated services digital networks (ISDNs), a telecommunications service that allows transmission of user information signals between user-network interfaces. *See also* **interface.**

bearing: The direction of a point from another point, usually the direction from true north of a station requesting its bearing measured from the direction finding station. *Note:* Bearings are usually expressed in degrees, from 001 to 360, or from 000 to 359. *See* **gyro bearing, jamming bearing, magnetic bearing, reciprocal bearing, relative bearing, true bearing, uncoordinated bearing.** *See also* **azimuth.** *Refer to* **Figs. I-7 (476), R-11 (848).**

bearing classification: An evaluation of the precision with which the bearing of a station is measured by a direction-finding station or stations. *Note:* Bearings are (a) categorized in classes, such as Class A, B, and C or (b) are not categorized, depending upon the precision with which the bearing is measured. Class A is the most precise. The class of a position determination or a bearing is specified by the direction-finding station. Thus, a Class A bearing might be estimated to be accurate within ±2°, a Class B bearing within ±5°, and a Class C bearing within ±10°.

bearing cut: In direction finding, the angle or intersection of two bearings, the angle being opposite, i.e., facing, the baseline, with each bearing being taken from two different locations on the baseline. *Note:* Unless bearing cuts are between 30° and 150°, the location results will be inaccurate.

bearing discrimination: A measure of the precision with which a bearing, i.e., direction or azimuth, can be measured. *Note:* Radar has high range discrimination because the yardstick is a radar pulse. Radar and radio have a low bearing discrimination because the yardstick is the antenna aperture, and because precision to only a few degrees of bearing corresponds to large distances at the large operating ranges that occur in direction-finding and positioning operations. *See also* **direction discrimination, range discrimination.**

bearing intersection: Determination of the location of a stationary or moving signal source by means of direction finding from two more locations. *See also* **running fix.**

beat frequency oscillator: An oscillator that produces a desired frequency by combining two other frequencies, such as (a) an audio frequency produced by combining two radio frequencies and (b) a desired radio frequency, such as the intermediate frequency produced by a superheterodyne circuit. *Synonym* **heterodyne oscillator.**

beat: A pulse or signal that results when two or more signals of different frequency are combined. *Note:* Beat frequencies usually are the sum and difference frequencies of two combined frequencies.

beating: The phenomenon in which two or more periodic quantities having slightly different frequencies produce a resultant having periodic variations in amplitude. *Note:* Beating two frequencies together results in a frequency, i.e., a beat frequency, that is equal to the absolute value of the difference between the two frequencies. *See also* **heterodyne.**

beat length: *See* **polarization beat length.**

bed: *See* **plotter bed, test bed, vacuum bed.**

beeper: *Synonym* **radio pager.**

beeping: *Synonym* **radio paging.**

Beer's law: In the transmission of electromagnetic radiation through a liquid solution, such as a nonabsorbing nonscattering solvent containing an absorbing or scattering solute, the attenuation, reduction, decay, or diminution of the electromagnetic or optical power density, i.e., the irradiance, is given by the relation $I = I_0 e^{-cax}$ where I is the power density at distance x, I_0 is the power density at $x = 0$, c is the concentration of the solute, a is the spectral absorption/scattering coefficient per unit of concentration per unit of distance, and x is the distance the radiation travels from the $x = 0$ reference point, i.e., the thickness of the solution from $x = 0$ to the point x. *See also* **Bouger's law, Lambert's law.**

beginning of tape mark: A mark on a magnetic tape used to indicate the beginning of the permissible recording area. *Note:* Examples of beginning of tape marks are photoreflective strips and transparent sections of tape.

bel (B): A unit of measure of ratios of power levels, i.e., relative power levels. *Note 1:* The number of bels for a given ratio of power levels is calculated by taking the logarithm to the base 10 of the ratio. The number of bels is given by the relation $B = \log_{10}(P_1/P_2)$ where P_1 and P_2 are power levels. *Note 2:* The dB (decibel), equal to 0.1 B, is a more commonly used unit. *See also* **dB.**

bell: *See* **extension bell, margin bell.**

Bell: 1. Alexander Graham Bell, the person credited with inventing the telephone. **2.** The name applied to the former Bell System of Telephone Companies and to the current Bell Operating Companies (BOCs).

bell (BEL) character: A transmission control character that is used when there is a need to call for user or operator attention in a communications system, and that usually activates an audio or visual alarm or other attention-getting device.

bell curve: A curve that (a) is somewhat bell-shaped, (b) is symmetrical, and (c) follows a Gaussian distribution, i.e., a normal distribution. *See also* **Gaussian distribution, Gaussian function.** *Refer to* **Fig. P-13 (759).**

bellfast: A teletypewriter switching system formerly available by leasing from the former American Telephone and Telegraph Company (now AT&T).

Bell integrated optical device: An optical integrated circuit (OIC) (a) that has interconnected active and passive elements used as logic components and control components in optical devices, such as memories, pulse shapers, optical switches, differential amplifiers, optical amplifiers, and logic gates and (b) has a configuration that is dependent upon on how the elements are interconnected during fabrication.

Bell Operating Company (BOC): Any of the 22 operating companies that were divested from AT&T by court order. *Note:* The Cincinnati Bell Telephone Company and the Southern New England Bell Telephone Company are not included.

belt: *See* **Van Allen belt.**

benchmark problem: A problem that (a) has a solution that can be used to comparatively evaluate the performance of the functional unit used in obtaining the solution and (b) may be used in a benchmark test. *See also* **benchmark test.**

benchmark test: A test that (a) is designed to comparatively evaluate the performance of functional units, such as communications links, computer hardware, computer software, and data processing systems, and (b) usually makes use of a benchmark problem. *Note:* An example of a benchmark test is a test designed to check the bit error ratio (BER) of a data link and compare it with the BER of another data link. *See also* **benchmark problem.**

bend: *See* **cold bend, E bend, H bend, macrobend, microbend.**

bending: *See* **macrobending, microbending.**

bending loss: In an optical fiber, the radiation that (a) is emitted at bends in the fiber and (b) consists of discrete modes of the transmitted light. *Note:* The bends may be (a) microbends that occur along the core-cladding interface, causing incidence angles to occur that are greater than the critical angle for total reflection or (b) macrobends with radii of curvature less than the critical radius, in which case evanescent waves can no longer remain coupled to bound modes and hence radiate laterally away from the fiber. The outside edges of the evanescent waves will radiate because their wavefronts cannot exceed the velocity of light as they sweep around the outside of the bend, i.e., they cannot maintain a constant phase relationship with the wave inside the fiber. Thus, they become uncoupled or unbound and hence radiate away. *See also* **bound mode, critical angle, curvature loss, evanescent wave, microbend, optical fiber, radiation.**

bend limiter: In fiber optic cable systems, a device, such as a rod, bracket, fixture, or tube, that (a) reduces the tendency of a fiber optic cable to bend when bending forces are applied and (b) assists in preventing the cable from bending at a radius of curvature less than the critical radius or the minimum bend radius, whichever is greater.

bend loss: 1. *Synonym* **curvature loss. 2.** *See* **microbend loss.**

bend radius: *See* **minimum bend radius.**

bend sensor: *See* **microbend sensor.**

BEP: block error probability.

BER: bit error ratio, block error ratio.

Bernoulli box disk drive: A personal computer (PC) disk drive that can accept a disk mounted in a cartridge with capacities up to 50 Mb (megabytes).

BERT: bit error ratio tester.

beta particle: *Synonym* **electron.**

beta switch: *See* **delta-beta switch.**

BETRS: Basic Exchange Telecommunications Radio Service.

between failures: *See* **mean time between failures.**

between outages: *See* **mean time between outages.**

between the lines entry: Unauthorized access to a momentarily inactive terminal of a legitimate user assigned to a communications channel. *Note:* The between the lines entry is obtained through the use of active wiretapping by an unauthorized user.

BEX: broadband exchange.

B EXCEPT A gate: *Synonym* **B AND NOT A gate.**

BFL: back focal length.

bias: 1. A systematic deviation of a value from a reference value. **2.** The amount by which the average of a set of values departs from a reference value. **3.** An electrical, mechanical, magnetic, or other force field applied to a device to establish a reference level to operate the device. **4.** An effect on telegraph signals produced by the electrical characteristics of the line and the terminal equipment. *Note:* An example of telegraph bias is a positive or negative direct current (dc) voltage at a point that should remain at zero.

bias distortion: 1. Distortion of a signal resulting from a shift in the bias. **2.** In two-condition signaling, i.e., in binary signaling, distortion of the signal in which all the significant intervals corresponding to one of the two significant conditions have uniformly longer or shorter durations than the corresponding theoretical durations. *Note:* The magnitude of the distortion is expressed in percent of a perfect unit pulse duration. *See also* **bias, cyclic distortion, distortion, end distortion, internal bias, marking bias, spacing bias.**

bias gain control circuit: *See* **detector back bias gain control circuit.**

biconditional gate: *Synonym* **EXCLUSIVE NOR GATE.**

biconical antenna: An antenna that (a) consists of two conical conductors that have a common axis and vertex, (b) is excited at the common vertex, and (c) if one of the cones is reduced to a plane, is called a discone. *See also* **antenna, discone.**

bid: In communications, computer, data processing, and control systems, an attempt by a system, or equipment in the system, to gain control of a device, such as a line, circuit, circuit group, or computer, that is also available to other communications, computer, data processing, or control systems. *Note:* Depending on the form of the invitation protocol, i.e., the selection protocol, the bid may be successful, be unsuccessful, or result in contention.

bidirectional asymmetry: In two-way data transmission, the condition that exists when information flow characteristics are different for both directions. *See also* **bidirectional symmetry.**

bidirectional coupler: In fiber optic transmission systems, a directional coupler that (a) permits signals to propagate in both directions or (b) has terminals or connections for sampling waves in both directions of transmission. *See also* **directional coupler, unidirectional coupler.**

bidirectional fiber optic cable: A fiber optic cable that (a) can handle signals simultaneously in both directions, (b) has the necessary components to operate successfully, (c) avoids crosstalk between the oppositely directed signals, and (d) usually consists of fiber optic cable, beam splitters, entrance and exit ports, mixing rods, couplers, and interference filters.

bidirectional flow: In a channel or a circuit, flow in either direction, i.e., flow that may be in one direction or the other at any given time or in both directions simultaneously. *Note:* Bidirectional flow may be represented by a flowline on a flowchart by arrows pointing in both directions on the same flowline.

bidirectional symmetry: In two-way data transmission, the condition that exists when information flow characteristics are the same for both directions. *See also* **bidirectional asymmetry.**

bidirectional transmission: In an optical waveguide, such as an optical fiber, simultaneous transmission of signals in both directions.

bifocal: In optics, pertaining to a system or a component that has, or is characterized by, two or more optical focuses.

bifurcation connector: *Synonym* **tee coupler.**

B IGNORE A gate: A two-input, binary, logic coincidence circuit or device in which (a) normal operation can be interrupted by a control signal that enables the gate to function so as to pass the B input signal and completely disregard the A input signal, (b) the output signal is therefore the same as the B input signal and independent of the A input signal during the controlled period, and (c) operation usually is temporary, normal operation being that of an OR gate. *Refer to* **Fig. B-4.**

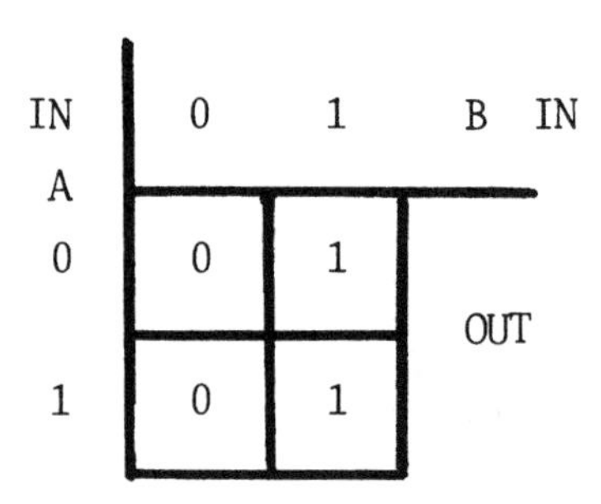

Fig. B-4. A truth table that shows the input and output digits of a **B IGNORE A gate.**

BIH: International Time Bureau. *See* **International Atomic Time (TAI).**

bilateral control: *Synonym* **bilateral synchronization.**

bilateral synchronization: 1. A synchronization control system between exchanges A and B in which the clock at exchange A controls the data received at exchange B and the clock at exchange B controls the data received at exchange A. *Note:* Bilateral synchronization usually is implemented by deriving the timing from the incoming bit stream. **2.** A synchronization control system between exchanges A and B in which the clock at exchange A controls the clock at exchange B, and the clock at exchange B controls the clock at exchange A. *Synonym* **bilateral control.** *See also* **clock, double ended synchronization, single ended synchronization, synchronization.**

billboard antenna: An antenna that (a) consists of a group of parallel dipoles with flat reflectors placed in a single, usually straight line, (b) has a dipole spacing that depends on the wavelength, and (c) has a main lobe direction that can be controlled, within limits, by appropriate phasing of the signals to each dipole of the array. *Synonym* **broadside antenna.** *See also* **antenna.**

B IMPLIES A gate: *Synonym* **A OR NOT B gate.**

binary: 1. Pertaining to a selection, choice, or condition that has only two possible different values or states. **2.** Pertaining to a fixed radix numeration system that has a radix of two.

binary adder: A device that obtains the sum of two or more binary numbers. *Note:* A full binary adder may be constructed using two binary half-adders and an OR gate, the output being the final binary sum and a final carry bit, if any. *Refer to* **Fig. B-5.**

binary arithmetic operation: An operation in which the operands and the results are represented in the pure binary numeration system.

binary card: A punched card (a) in which the punched and unpunched positions represent numbers in the binary numeration system and (b) that may be in column binary or in row binary form.

binary cell: A storage unit in which (a) one binary digit, i.e., one bit, a 0 or a 1, may be stored, and (b) a stored bit may be represented in many ways, such as two voltage levels, two phases, magnetic polarization in one direction or the other, or a hole or no hole at a punch position on a card.

binary code: A code composed by selection and configuration of an entity that can assume either one of two possible states. *See* **dense binary code, fixed weight binary code, modified reflected binary code, nonreturn to zero binary code, pseudobinary code, return to zero binary code, symmetrical binary code, unit disparity binary code.** *See also* **binary digit, code.**

binary coded decimal (BCD): Pertaining to a numeration system in which each of the decimal digits 0 through 9 is represented by a binary numeral. *Note:* In binary coded decimal notation that uses the weights 8-4-2-1, the decimal number 93 is represented by 1001 0011. In binary coded decimal (BCD) with the 8-4-2-1 weighting, six of the sixteen possible bit patterns go unused. However, their occurrence can be used for error detection. The same decimal number in the pure binary numeration system is 1011101, and all the bit patterns are used. *See also* **binary digit, binary notation, code.**

binary coded decimal code: *Synonym* **binary coded decimal notation.**

binary coded decimal interchange code: A binary code used for representing decimal numerals in information interchange circuits. *See* **extended binary coded decimal interchange code.**

binary coded decimal (BCD) notation: Notation in which each of the decimal digits 0 through 9, and each of the digits in a decimal numeral, is represented by a binary numeral, i.e., by a bit pattern. *Note 1:* In one binary coded decimal (BCD) notation system, the decimal numeral 58 is represented by the binary numerals 0101 1000. *Note 2:* In pure binary, i.e., in straight, or-

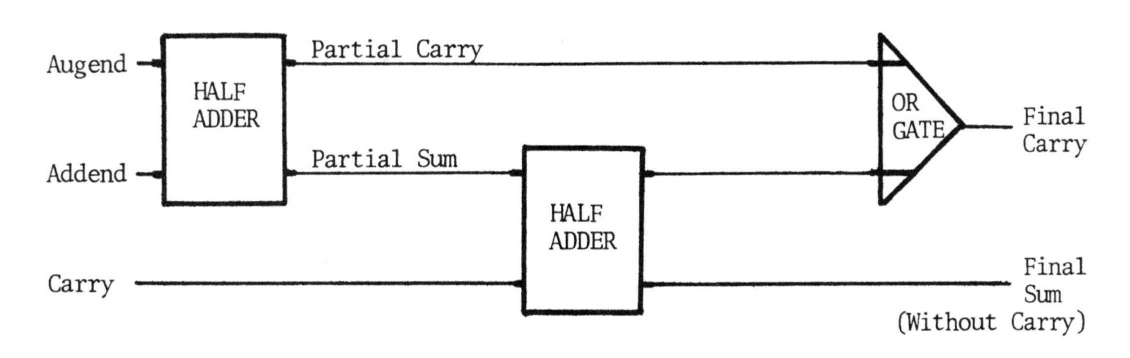

Fig. B-5. A full **binary adder.**

dinary, regular, or plain binary, decimal 58 would be represented as binary 111010. This is not considered as binary coded decimal notation. *Synonyms* **binary coded decimal code, binary coded decimal representation.**

binary coded decimal representation: *Synonym* **binary coded decimal notation.**

binary counter: A counter that displays its results in the form of a binary numeral. *See also* **modulus counter, ring counter.**

binary digit (bit): 1. In binary notation, either of the characters 0 or 1. **2.** A unit of information equal to one binary decision or the designation of one of two possible and equally likely states of anything used to store, convey, or represent information. **3.** A character used to represent one of the two digits in the numeration system with a base of two, each digit representing one of two, and only two, possible states of a physical entity or system. *See also* **byte, code element, digital signal, octet alignment.**

binary digit interval: *Synonym* **bit interval.**

binary element: A constituent element of data that can have either of two different values, i.e., either of two different states, at any given instant.

binary exponential backoff: *See* **truncated binary exponential backoff.**

binary half-adder: A device or a circuit composed of combinational circuits, such as gates, that form the logical sum bit, i.e., modulo-two sum bit, and the carry bit, one at each of its two output terminals when bits are simultaneously present at each of its two input terminals, one bit representing the augend and the other the augend. *Note:* A binary full adder can be built from two binary half-adders and an OR gate to produce the final sum and a carry, if any. *See also* **combinational circuit, gate.** *Refer to* **Fig. B-6.** *Refer also to* **Figs. B-5 (77), L-6 (529).**

binary modulation: The varying of a parameter of a carrier as a function of two finite, discrete states. *See also* **carrier, modulation.**

binary notation: Notation that (a) uses two different characters, usually the binary digits 0 and 1, (b) may be used to encode data, and (c) may be in many different forms, such as pure binary, binary coded decimal (BCD), Gray code, biquinary, or two out of five code form. *See also* **binary coded decimal, binary coded decimal (BCD) notation, binary digit, pure binary numeration system.**

binary number: A number that (a) is expressed in binary notation and (b) is usually characterized by the arrangement of bits in sequence, with the understanding that successive bits are interpreted as coefficients of successive powers of the base 2. *Note:* A distinction must be made between a number and a numeral. Thus, 365, CCCLXV, and 101101101 are numerals, each in a different numeration system but each representing the number of days in a year, excluding leap year. *See also* **binary notation, pure binary numeration system.**

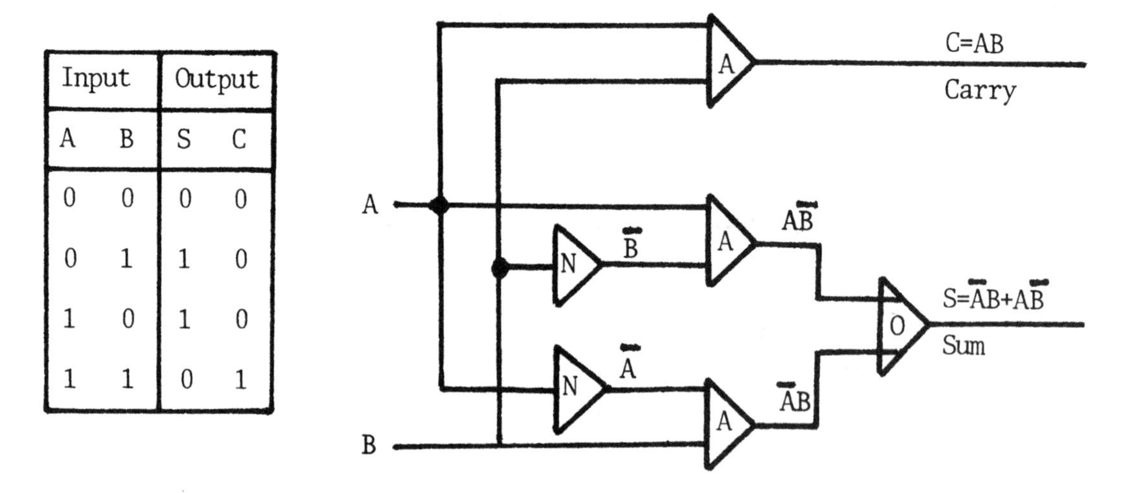

Input		Output	
A	B	S	C
0	0	0	0
0	1	1	0
1	0	1	0
1	1	0	1

Fig. B-6. The combinational logic circuits that compose the **binary half-adder** and the corresponding truth table of input and output digits.

binary numeral: A numeral that consists only of digits that are either 0 or 1. *See also* **binary number.**

binary numeration system: *See* **pure binary numeration system.**

binary phase-shift keying (BPSK): Phase-shift keying (PSK) in which there are only two phase significant conditions, such as an in-phase condition and a 180° out-of-phase condition. *See also* **phase-shift keying, significant condition.**

binary search: A dichotomized search, in which the two parts of the set being searched are equal, with special arrangements of sets with an odd number of items. *See also* **dichotomized search.**

binary sequence: *See* **pseudorandom binary sequence.**

binary serial signaling rate: In the particular case of serial two-state transmission, the reciprocal of the unit interval measured in seconds and expressed in bits per second.

binary signaling: Signaling in which there are two significant conditions, such as voltage or current on versus off, up versus down, or open versus closed, in which one condition represents a 0 and the other represents a 1. *Note 1:* Binary signaling is the most fundamental and widely used signaling method for digital signaling. However, binary signaling has limitations. In some environments, coding techniques must be used that take into account signal spectral characteristics, the requirements or limitations of channels, signal synchronization capabilities and requirements, error detection capabilities and requirements, interference, noise immunity, cost, and complexity. *Note 2:* Analog voice, ternary, quaternary, i.e., any *n*-ary signals other than binary, and fiberscope transmission are examples of nonbinary signaling.

binary synchronous communications (bi-sync): A character-oriented, Data Link Layer protocol. *Note:* The binary synchronous communications (bi-sync) protocol is being phased out of computer communications networks in favor of bit-oriented protocols, such as the Synchronous Data Link Control (SDLC), High level Data Link Control (HDLC), and Advanced Data Communications Control Procedure (ADCCP) protocols. *See also* **Advanced Data Communications Control Procedure, high level data link control, synchronous data link control.**

binder: In fiber optic systems, a string or a tape used to tie together a number of fiber optic cable components (FOCCs), such as unbuffered fibers, buffered fibers, jackets, strength members, and buffers.

binding: In communications, computer, data processing, and control systems, the assigning of a value or a reference to an identifier. *Note:* Examples of binding include assigning a value to a parameter, assigning an absolute address to a virtual or relative address, and assigning a device identifier to a symbolic address or label. *See also* **identifier, value.**

binocular instrument: An optical instrument that makes use of, or allows a person to use, two eyes at the same time by providing for convergence of the optical axis of each of its two optical systems on the same object in order to produce the effect of a single image in the brain. *Note:* Examples of binocular instruments are most field glasses, range finders, and aerial photograph interpretation devices. Hand-held telescopes, compasses, and magnifying glasses are monocular instruments.

BIOD: Bell integrated optical device.

bionics: The branch of science and technology devoted to the study and application of the functions, characteristics, and phenomena of living organisms to the design, development, and manufacture of systems, components, and equipment, such as electronic, optical, electrooptical, electromechanical, and mechanical systems, components, and equipment.

Biot-Savart law: The magnetic field intensity, H, and hence the magnetic flux density, B, at a point in the space surrounding an electric current, is directly proportional to the magnitude of the electric current. *Note:* Mathematically, the Biot-Savart law is given by the relation $dB = \mu dH = \mu I dl \sin \beta / 4\pi r^2$ where μ is the magnetic permeability of the surrounding medium, I is the electric current in the elemental conductor dl, β is the angle between the elemental conductor, i.e., the direction of I, and a vector from the elemental conductor to the point at which the magnetic field is being determined, and r is the distance from the elemental conductor to the point.

biphase coding: Coding in which there are two phases, i.e., plus and minus 90° phase shift keying (PSK). *Note:* If a 1 is represented by a +90° phase shift from a reference, a 0 is represented by a −90° phase shift from the same reference. Thus, the 0s and 1s are sine waves or pulses shifted 180° (electrical) from each other. Therefore, they are opposite in electrical polarity.

biphase level coding: Biphase signal coding in which a change of state, i.e., a change in amplitude significant condition, in the original input binary baseband signal is represented by a change of phase in the output biphase signal.

biphase mark coding: Biphase signal coding that has a signal transition at the beginning of every bit period, a 1 producing a second transition in the biphase signal a significant interval later, such as one-half of a bit period later, and a 0 producing no second transition. *Note:* Biphase mark coding may be generated by using conditioned signals, i.e., conditioned on 0, and passing the conditioned signals through an EXCLUSIVE OR gate with the system clock. The advantage of conditioning the biphase signals is that any ambiguity in the detected signal is removed. The spectrum for conditioned biphase signaling is the same as that for conventional biphase signaling. *Synonym* **conditioned biphase coding.**

biphase signaling: Signaling in which binary digits are represented by sine waves that are 180° out of phase with each other, i.e., shifted by 180 electrical degrees. *Note 1:* If a 1 is represented by sin *x*, a 0 is represented by $-\sin x$. *Note 2:* Biphase signaling has many forms, such as biphase level signaling, biphase mark signaling, and biphase space signaling. Biphase signaling is one of the source formats, such as nonreturn to zero (NRZ), biphase, and delay modulation, usually used in radio applications. Biphase signaling may be generated by passing the binary signal and the system clock through an EXCLUSIVE OR gate. It is decoded by passing the biphase signal sequence and the recovered clock through an EXCLUSIVE OR gate. The biphase signal spectrum has a low direct current (dc) content, which is the major advantage of the code. It has high energy levels up to a frequency of twice the bit rate. Therefore, biphase signaling in not unduly affected by high-pass filtering, but has a low tolerance to low-pass filtering. Biphase signaling requires a wide band, usually between about 0.08 and 0.75 times the bit rate. This is a major disadvantage. For a 16-kb•s^{-1} (kilobit per second) data signaling rate (DSR), the bandwidth is 10.8 kHz (kilohertz), extending from about 1.2 kHz (kilohertz) to 12 kHz. Nonreturn to zero (NRZ) signaling requires only an 8-kHz bandwidth.

biphase space coding: Biphase signal coding that, like biphase mark coding, has a signal transition at the beginning of every bit period, a 0 producing a second transition in the biphase signal a significant interval later, such as one-half of a bit period later, and a 1 producing no second transition. *Note:* Generation of biphase space coding is the same as for biphase mark coding except that the data are conditioned on 1s.

bipolar: Pertaining to the transmission of signals in which both positive and negative voltages and currents are used in the signals. *Note:* A mark could be represented by a voltage or a current in one direction and a space by a voltage or a current in the opposite direction, with or without the use of the zero condition. *See also* **unipolar.**

bipolar coding: Binary signal coding in which each "off" bit is represented by a 0 and each "on" bit is represented by a polarity opposite from the previous one, thus allowing three significant conditions, i.e., positive, zero, and negative. *Note:* Bipolar coding can be either full-baud bipolar (FBB), or half-baud bipolar (HBD) coding. An FBB coding process consists of representing binary 0s by zero levels and binary 1s by alternate positive or negative levels for the entire baud interval, i.e., pulse period, whereas HBB pulses return to the zero level at the midpoint of the baud interval. Bipolar coding is a ternary signal coding scheme and thus has a greater susceptibility to noise errors than does binary coding. The bipolar frequency spectrum has an energy null at direct current (dc), i.e., at 0 Hz (hertz), as does biphase coding, but requires less bandwidth than biphase coding. Bipolar coding is an inherent error-detecting code in that if a pulse is decoded improperly, the bipolar phase alternation rule will be violated, if not compensated by a pair of errors, and the error will be detected. Bipolar coding may be generated by conditioning nonreturn to zero signals, delaying this sequence one bit period, and adding the delayed version and the undelayed version. *Synonyms* **alternate mark inversion coding, alternating binary coding.** *See* **autocorrelation, full-baud bipolar coding, half-baud bipolar coding.**

bipolar signal: A signal that (a) has pulses of either one of two polarities, neither of which is zero, (b) may have a two-state nonreturn to zero (NRZ) or a three-state return to zero (RZ) binary coding scheme, and (c) usually is symmetrical with respect to zero amplitude, i.e., the plus polarity and minus polarity pulse amplitudes have the same absolute value. *See also* **alternate mark inversion signal, nonreturn to zero code, polar operation, return to zero code, signal.**

bipolar telegraphy: *Synonym* **double current signaling.**

bipolar transmission: *Synonym* **double current signaling.**

bipolar with eight-zero substitution (B8ZS): A T carrier line code (a) in which bipolar violations are deliberately inserted if user data contains a string of eight or more consecutive 0s and (b) that is used in the European hierarchy at the T1 rate.

bipolar with six-zero substitution (B6ZS): A T carrier line code (a) in which bipolar violations are deliberately inserted if user data contains a string of six or more consecutive 0s and (b) that is used in the European hierarchy at the T2 rate.

bipolar with three-zero substitution (B3ZS): A T carrier line code (a) in which bipolar violations are deliberately inserted if user data contains a string of three or more consecutive 0s and (b) that is used in the European hierarchy at the T3 rate.

biquinary code: A code in which (a) a decimal digit, *n,* is represented by a pair of numerals, **a** and **b**, where $n = 5\mathbf{a} + \mathbf{b}$ and (b) the **a** may be a 0 or a 1 and the **b** may be a 0, 1, 2, 3, or 4, which may also be represented in a binary code. *Refer to* **Fig. B-7.**

Decimal Digit	BIQUINARY Binary Form	BIQUINARY Decimal Form
0	01 00001	00
1	01 00010	01
2	01 00100	02
3	01 01000	03
4	01 10000	04
5	10 00001	10
6	10 00010	11
7	10 00100	12
8	10 01000	13
9	10 10000	14

Fig. B-7. Two forms of a **biquinary code** for decimal digits.

birefringence: 1. The refraction of a light ray such that (a) two rays are formed, and (b) the two rays travel in different directions. *Note:* Birefringence literally means double refraction. **2.** The property of a material that causes a light beam passing through the material to be split into two beams with orthogonal polarizations. **3.** The splitting of a light beam into two divergent components upon passage into and through a doubly refracting propagation medium, with the two components propagating at different velocities in the medium. *Note:* The medium is characterized by having two different refractive indices in the same direction, which causes the existence of the different velocities of propagation for the different orthogonal polarizations. **4.** In a transparent material, anisotropy of the refractive index, which varies as a function of orientation with respect to the incident ray, and also with the polarization of the incident ray. *Note 1:* All crystals except those of cubic lattice structure exhibit some degree of anisotropy with regard to their physical properties, including refractive index. Other materials, such as glasses or plastics, become birefringent when subjected to mechanical strain. *Note 2:* Birefringent materials, including crystals, have the ability to refract an unpolarized incident ray into two separate orthogonally polarized rays, which usually take different paths, depending on orientation with respect to the incident ray. The refracted rays are referred to as the "ordinary" or "O" ray, which apparently obeys Snell's Law, and the "extraordinary," or "E" ray, which does not. In certain anisotropic materials the refractive index is a vector, i.e., the refractive index has a different magnitude in different directions, thus causing the dependence on the relative directions of the incident ray and the refractive index in the direction of the incident ray. *Synonym* **double refraction.** *See also* **birefringent medium, fiber optics, refraction.**

birefringence fiber: *See* **high birefringence fiber, low birefringence fiber.**

birefringence noise: *See* **polarization noise.**

birefringence sensor: A fiber optic sensor in which (a) light entering an optical fiber made of birefringent material, i.e., material demonstrating two refractive indices, is split into two beams, each propagating in a different path of different optical path length, (b) the two beams are recombined producing interference patterns, (c) external stimuli to be sensed are used to alter the optical path lengths to produce reinforcement or cancellation of the lightwaves at a photodetector, and (d) the output signal of the photodetector is a function of the type and degree of applied stimuli, such as force, pressure, electric field, and temperature. *Note:* Examples of birefringence sensors are the Pockels effect sensor and the Kerr effect sensor.

birefringent material: *Synonym* **birefringent medium.**

birefringent medium: An anisotropic material that exhibits different refractive indices depending on the orientation of the material with respect to the polarization of the incident light. *Note 1:* Some materials exhibit a specific anisotropism intrinsically. Other materials exhibit the same anisotropism only when induced, for example by mechanical stress, i.e., by applied pressure. *Note 2:* The phase velocity of a wave in a birefringent medium depends on the polarization of the wave. *Note 3:* If light propagating in a medium is decomposed into two perpendicular directions of polarization that travel at different velocities, the medium is said to be birefringent. *Note 4:* Optical fibers may exhibit birefringence. *Synonym* **birefringent material.** *See also* **anisotropic, birefringence, fiber optics, isotropic, refraction.**

B-ISDN: broadband ISDN.

bistable: Pertaining to a device capable of assuming either one of two stable states. *Note:* Usually a bistable device will remain in a given state until application of a suitable stimulus, at which time the device will revert to the other state. *See also* **flip-flop.**

bistable circuit: *Synonym* **flip-flop.**

bistable multivibrator: *Synonym* **flip-flop.**

bistable trigger circuit: *Synonym* **flip-flop.**

bi-sync: binary synchronous communications.

bit: binary digit. *See* **signal to noise ratio per bit.**

bit by bit asynchronous operation: In data transmission systems, operation in which rapid manual, semiautomatic, or automatic shifts in the data modulation rate are accomplished by gating or slewing the clock modulation rate. *Note:* The equipment may be operated at 50 b•s^{-1} (bits per second) one moment and at 1200 b•s^{-1} the next moment. *See also* **synchronous transmission.**

bit byte: *See* **n-bit byte.**

bit configuration: 1. The order for encoding the bits of information that define a character. **2.** The sequencing of the bits used to encode a character. *See also* **binary digit, bit pattern.**

bit count integrity (BCI): The preservation of the exact number of bits that (a) were originated in a message in the case of message communication or (b) occur per unit time in the case of user-to-user connection. *Note:* Bit count integrity is not to be confused with bit integrity, which requires that the bits delivered are, in fact, as they were originated. *See also* **added bit, binary digit, bit error ratio, bit integrity, bit inversion, bit slip, deleted bit, digital error, error.**

bit density: The number of bits recorded per unit length, area, or volume. *See also* **packing density, recording density.**

biternary transmission: Transmission in which two binary pulse trains are combined for transmission over a system in which the available bandwidth is sufficient for transmission of only one of the two pulse trains when in binary form. *Note:* The biternary signal may be generated from two synchronous binary signals operating at the same bit rate. The two binary signals are adjusted in time to have a relative time difference of one-half the binary interval and are combined by linear addition to form the biternary signal. Each biternary signal element can assume any one of three possible states, corresponding to +1, 0, and −1. Each binary signaling element contains information on the state of the two binary signaling elements as defined in the following table:

Bit 1	Bit 2	Biternary
0	0	−1
0	1	0
1	0	0
1	1	+1

The method of addition of Bit 1 and Bit 2 as described above does not permit the biternary signal to change from −1 to +1 or from +1 to −1 without an intermediate binary signal of 0. Because there is a half a unit interval time difference between the binary signals for Bit 1 and Bit 2, only one of them can change its state during the binary unit interval. This makes it possible in the decoding process to ascertain the state of the binary signal that has not changed its state, and thus avoid ambiguity in decoding the biternary signal of 0. *See also* **code, pulse, pulse train.**

bit error rate: *Deprecated synonym for* **bit error ratio.**

bit error ratio (BER): The ratio of (a) the number of erroneous bits to (b) the total number of bits transmitted, received, or processed over some stipulated period. *Note 1:* Examples of bit error ratios (BERs) are (a) transmission BER, i.e., the ratio of (i) the number of erroneous bits received to (ii) the total number of bits transmitted, and (b) information BER, i.e., the ratio of (i) the number of erroneous decoded (corrected) bits to (ii) the total number of decoded (corrected) bits. *Note 2:* The bit error ratio (BER) is usually expressed as a coefficient and a power of 10. Thus, 2.5 erroneous bits out of 100,000 bits transmitted would be expressed as 2.5 out of 10^5 or 2.5 x 10^{-5}. If there is no coefficient, such as a BER of 10^{-7}, a coefficient of unity is implied. *Note 3:* The bit error ration (BER) is a dimensionless number. *Deprecated synonym* **bit error rate.** *See* **pseudo bit error ratio.** *See also* **binary digit, bit count integrity, character count integrity, error, error budget, error burst, error control, error ratio, undetected error ratio.**

bit error ratio tester (BERT): A testing device that compares a received data pattern with a known transmitted pattern to determine the level of transmission quality.

bit insertion: *See* **zero bit insertion.**

bit integrity: 1. The assurance that the bits delivered are, in fact, as they were originated. **2.** Preservation of the value of a bit during processing, storage, and transmission. *See also* **bit count integrity, bit inversion.**

bit interval: 1. In a bit stream, the time or the distance between corresponding points on two consecutive signals, each representing a binary digit (bit). **2.** In the transmission of a binary digit, the time or the distance allowed or required to represent the bit. *See* **binary digit, character interval, unit interval.**

bit inversion: The changing of the state that represents a given bit, i.e., a 0 or a 1, to the state that represents the opposite value. *Note:* An example of bit inversion is that if a 1 is represented by a given polarity or phase at one stage in a circuit, the 1 is represented by the opposite polarity or phase at the next stage and then back to the original polarity or phase in the following stage. Thus, whether a 0 or a 1 is represented by a given polarity or phase at a given stage is a matter of bookkeeping. *See also* **bit count integrity, character count integrity.**

bit misdelivery probability: 1. *Synonym* **bit misdelivery ratio. 2.** The probability that a bit will be misdelivered, i.e., delivered as a wrong bit to the proper addressee, as a correct bit to the wrong addressee, or not delivered at all, the probability usually being based on the bit misdelivery ratio. *See also* **bit misdelivery probability.**

bit misdelivery ratio (BMR): The ratio of (a) the number of misdelivered bits to (b) the total number of bit transfer attempts during a specified period. *Note:* The bit misdelivery ratio may be used as a bit misdelivery probability. *Synonym* **bit misdelivery probability.** *See also* **block error ratio, block loss ratio, lost bit, misdelivered block.**

bitoric lens: A lens with both surfaces ground and polished in a cylindrical form or in a toroidal shape.

bit pairing: The practice of establishing, within a code set, a number of subsets that have an identical bit representation except for the state of a specified bit. *Note:* An example of bit pairing in the International Alphabet No. 5 and the American Standard Code for Information Interchange (ASCII) is that the upper case letters are related to their respective lower case letters by the state of bit 6 in the 7-bit string that represents ASCII characters. *See also* **ASCII, binary digit.**

bit pattern: The arrangement of the bits used to encode a character. *See also* **bit configuration.**

bit position: A character position in a word in binary notation.

bit rate (BR): In a bit stream at a point on a line, the number of bits occurring per unit time. *Note 1:* The bit rate usually is expressed in bits per second, such as 4.2 Gb/s (gigabits per second). *Note 2:* An example of the bit rate for a two-condition serial transmission in a single channel in which each significant condition represents a bit is that the bit rate is $1/T$, where T is the unit interval, i.e., the minimum interval between significant instants. In this case, the bit rate in bits per second and the rate in baud have the same numerical value. *See also* **baud, binary digit, bits per second, data signaling rate, modulation rate, multiplex aggregate bit rate.**

bit-rate•length product (BRLP): For an optical fiber or cable, the product of a given length of fiber or cable and the bit rate, i.e., the data signaling rate (DSR), that the fiber or the cable can handle for specified input conditions, tolerable dispersion, acceptable attenuation, and a given bit error ratio (BER). *Note 1:* The bit-rate•length product (BRLP) usually is stated in units of megabit•kilometers/second. A typical BLRP value for graded index fibers with a numerical aperture (NA) of 0.2 is 1000 Mb•km•s^{-1}. High performance fibers have a higher BRLP than this. Higher values are expected in the future. The value of the BRLP is a good indicator of fiber performance in terms of transmission capability. Another useful measure of optical fiber performance or transmission capacity is the bit rate at which the full wave half-power point occurs at the receiving end of a given length of the fiber. *Note 2:* For digital signals, the bit-rate•length product (BRLP) is a better measure of fiber performance than the bandwidth•distance factor. *Note 3:* In stating the bit-rate•length product (BRLP) for a fiber, either the modulation scheme or the signaling rate in baud should also be stated. *See also* **bandwidth, bandwidth•distance factor, fiber bandwidth.**

bit-rate•length product-limited operation: The condition that prevails in a device or a system when distortion of the signal caused by dispersion, i.e., waveguide and material dispersion, limits its performance or the performance of the device or the system to which it is connected; for example, the condition that prevails in an optical link when intersymbol interference at the receiving end is the predominant mechanism that limits link performance. *See also* **bandwidth-limited operation.**

bit robbing: 1. *Synonym* **speech digit signaling. 2.** Using the least significant bit in a time slot or channel for conveying voice-related signaling or supervisory information.

bit sequence independent: Pertaining to digital data transmission systems in which no restrictions or modifications are imposed on the transmitted bit sequence. *Note:* Bit sequence independent protocols are in contrast to protocols that reserve certain bit sequences for special meanings, such as the flag sequence, 01111110, for high level data link control (HDLC), synchronous data link control (SDLC), and Advanced Data Communications Control Procedure (ADCCP) protocols. *See also* **binary digit, bit stream transmission.**

bit sequential: Pertaining to data transmission in which the signal elements are successive in time.

bit slip: The insertion or the deletion of bits by a device to accommodate accumulated variations and differences between the clock reference for the received signal and the clock of the device. *See also* **binary digit, bit count integrity, character count integrity, clock, error.**

bits per inch (BPI): A unit used to express the linear density of data in storage. *See also* **bit density, packing density.**

bits per second (b•s^{-1}, bps, or b/sec.): In a transmission system, a unit used to express the number of bits occurring at or passing through a point per second, i.e., to express the bit rate. *Note 1:* For two-condition serial transmission in a single channel in which each significant condition represents a bit, i.e., a 0 or a 1, the bit rate in bits per second and the baud have the same numerical value only if each bit occurs in a unit interval. In this case, the data signaling rate in bits per second is $1/T$, where T is the unit interval, i.e., the minimum interval between significant instants. *Note 2:* Values of modulation rate and transmission rate in baud and in bits per second are numerically equal if, and only if, (a) all pulses, i.e., all bits, are the same duration, (b) all pulses are equal to the unit interval, and (c) binary operation is used. In n-ary operation, the bits per second equals the modulation rate, transmission rate, or line rate in baud multiplied by the logarithm to the base two of N, where N is the number of distinct states used in the digital modulation or transmission process. A distinct state might be any significant condition, such as amplitude, phase, or frequency. *See also* **baud, binary digit, bit rate, data signaling rate, data transfer rate, modulation rate.**

bit stepped: Pertaining to the control of digital equipment such that operation is incremented one step at a time at the applicable bit rate. *See also* **character stepped.**

bit stream: An uninterrupted sequence of pulses that represent binary digits transmitted in a propagation medium. *Note:* An example of a bit stream is a continuous sequence of bits in a wireline or optical fiber. *See* **mission bit stream.** *See also* **data stream.** *Refer to* **Figs. N-17 (635), N-18 (636).**

bit stream signaling: *See* **random bit stream signaling.**

bit stream transmission: The continuous transmission of characters represented by bit patterns. *Note:* In bit stream transmission, the bits usually occur at fixed time intervals, start and stop signals are not used, and the bit patterns follow each other in sequence without interruption. *See also* **binary digit, bit sequence independent, data stream.**

bit string: A sequence of binary digits, i.e., bits, that (a) usually is in a bit stream, (b) has a beginning and an end identified by data delimiters and (c) is treated as a unit. *See* **bit string, character string.** *See also* **binary digit, byte, data delimiter, packet, pulse train, word.**

bit stuffing: The insertion of extraneous, i.e., noninformation, bits into data, such as into a bit stream, a block, or a frame. *Note:* In data transmission, bit stuffing is used for various purposes, such as for synchronizing bit streams that do not necessarily have the same or rationally related bit rates, or to fill buffers or frames. The location of the stuffing bits is communicated to the receiving end of the data link, where these extra bits are removed to return the bit streams to their original bit rates or form. Bit stuffing may be used to synchronize several channels before multiplexing or to rate-match two single channels to each other. *Synonyms* **justification, positive justification, positive pulse stuffing.** *See also* **binary digit, destuffing, digital multiplexer, idle character, interframe time fill, maximum stuffing rate, multiplexing, nominal bit stuffing rate, pulse stuffing, synchronization.**

bit stuffing rate: *See* **nominal bit stuffing rate.**

bit synchronism: In a digital system, synchronism in which each bit is in exact synchronism with a clock pulse.

bit synchronization: The process in which the decision instant at which a receiver selects the state to be assigned to a received bit or basic signaling element is brought into alignment with the received bit or basic signaling element. *See also* **binary digit, decision instant, frame synchronization, synchronization, synchronization bit.**

bit synchronous operation (BSO): Operation in which data circuit-terminating equipment (DCE), data terminal equipment (DTE), and transmitting circuits are all operated in bit synchronization with a clocking system. *Note 1:* Clock timing is delivered at twice the modulation rate. One bit is transmitted or received during each clock cycle. *Note 2:* Bit synchronous operation is sometimes erroneously referred to as "digital synchronization." *See also* **binary digit, clock, data circuit-terminating equipment, data terminal equipment, synchronization, terminal.**

BIU: bus interface unit. *See* **network interface device.**

black: *See* **noisy black, nominal black, picture black.**

BLACK: A designation applied to telecommunications circuits, components, systems, equipment, and areas in which only encrypted or unclassified signals occur. *See also* **BLACK signal, communications security, RED/BLACK concept, RED signal.**

black and white facsimile telegraphy: *Synonym* **document facsimile telegraphy.**

blackbody: A body that (a) totally absorbs all the radiation that is incident upon it, (b) does not reflect radiation, (c) can absorb and radiate energy, (d) in thermal equilibrium, absorbs and radiates energy at the same rate, (e) radiates at a rate expressed by the Stefan-Boltzmann law, and (f) has a spectral distribution of radiation as expressed by Planck's radiation formula. *Synonym* **ideal blackbody.** *See also* **emissivity.**

black concept: *See* **red-black concept.**

black designation: A security designation applied to (a) all wirelines, fiber optic cables, components, equipment, and systems that handle only encrypted or security unclassified signals and (b) physical areas in which no unencrypted security classified signals occur. *See also* **red-black concept, red designation.**

black facsimile transmission: 1. In facsimile systems using amplitude modulation, transmission in which the maximum transmitted power corresponds to the maximum density, i.e., the maximum darkness, of the object. **2.** In facsimile systems using frequency modulation, transmission in which the lowest transmitted frequency corresponds to the maximum density, i.e., the maximum darkness, of the object. *See also* **facsimile, object, white facsimile transmission.**

black noise: Noise that has a frequency spectrum of predominately zero power level over all frequencies except for a few narrow bands or spikes. *Note:* An example of black noise in facsimile transmission systems is the spectrum that might be obtained when scanning a black area in which there are a few white spots. Thus, in the time domain, a few random pulses occur while scanning. *See also* **blue noise, pink noise, white noise.**

blackout: In communications systems, a complete disruption of communications capability caused by conditions that occur in the transmission media. *Note:* Examples of blackouts are (a) the complete disruption of radiotelephone communications caused by sunspots, solar flares, or ionospheric disturbances and (b) the disruption of radio communications caused by a nuclear burst.

black recording: 1. In facsimile systems using amplitude modulation, that form of recording in which the maximum received power corresponds to the maximum density, i.e., the maximum darkness, of the record medium. **2.** In a facsimile system using frequency modulation, recording in which the lowest received frequency corresponds to the maximum density, i.e., the maximum darkness, of the record medium. *See also* **facsimile.**

black signal: In facsimile, the signal resulting from scanning a maximum density area, i.e., a maximum darkness area, of the object. *See also* **facsimile, signal.**

BLACK signal: In cryptographic systems, a signal that represents only unclassified or encrypted information. *See also* **communications security, RED signal.**

blank: A thick rod of glass with a specific uniform refractive index used to make optical fiber preforms. *See* **optical blank, optical fiber blank.** *See also* **optical fiber preform, refractive index.**

blanket: *See* **thermal blanket.**

blanketing: In amplitude modulation (AM) broadcasting, the interference caused by the presence of an AM signal with an irradiance, i.e., a field intensity, of 1 V/m (volt per meter) or greater in the area adjacent to the transmitting antenna. *Note:* The 1 V/m contour is referred to as the blanket contour, and the area within this contour is referred to as the blanket area. *See also* **white area.**

blanketing area: In the vicinity of a transmitting antenna, the area in which the signal from the antenna interferes with the reception of signals from other transmitters. *Note:* The blanketing area for a given transmitting antenna depends on the selectivity and the quality of the receiver as well as on the signal levels from the other transmitters. *See also* **interference.**

blanking: 1. In graphic display devices, the suppression of the display of one or more display elements, display groups, or display images in the display space on the display surface of the display device. *Note:* Display surfaces that are particularly capable of blanking are cathode ray tube (CRT) screens, fiberscope faceplates, light-emitting diode panels, and gas panels. **2.** In radio reception, the receipt of a signal of such magnitude as to saturate, i.e., to overdrive, circuits and therefore prevent signal resolution, frequency discrimination, and direction finding. **3.** In infrared scanning systems, the receipt of a pulse, such as a laser pulse, of such magnitude as to saturate, i.e., to overdrive, circuits and therefore prevent signal resolution, frequency discrimination, and direction finding. *See* **pulse interference suppression and blanking, sidelobe blanking.** *See also* **clipping.**

blemish: In optical fibers, an area in the optical fiber or fiber bundle that has a reduced light transmission capability for any reason, such as lodged foreign substances, enlarged microcracks, breaks, or cuts.

blind: In a communications system, to render a device nonreceptive to certain transmissions that reach it, such as to prevent reception of arriving data by recognition of certain characters that precede the data. *See also* **lockout, poll, select.**

blind area: A region of space not reached, i.e., not illuminated, by a radar signal. *Note:* Blind areas may be caused by directional antennas that radiate so as not to reach certain regions, by shadows caused by obstructions or land features, or by ground clutter in which an object, i.e., a target, may be obscured. Because a blind area cannot be seen by radar or direction finding equipment, high frequency line of sight radiation cannot be obtained from sources in these areas, nor can objects be identified in them. *Synonym* **shadow area.**

blind cone: *See* **antenna blind cone.**

blind range: *See* **radar blind range.**

blind speed: *See* **radar blind speed.**

blind transmission: Transmission without obtaining a receipt from the intended receiving station. *Note:* Blind transmission may occur or be necessary when (a) security constraints, such as radio silence, are imposed, (b) technical difficulties with receivers or transmitters occur, or (c) a lack of time precludes the delay caused by waiting for receipts.

blinking: In graphic display devices, an intentional periodic change in the intensity of one or more display elements or display groups. *See also* **flicker, wink pulsing.**

blip: 1. In display systems, a mark, spot, or symbol that appears in the display space on the display surface of a display device. *Note:* An example of a blip, on a radar screen, is a bright spot that represents or identifies the polar coordinate position of an aircraft, target, or missile. **2.** An optical mark used for counting film frames by automatic means. *See also* **document mark.**

block: 1. A group of bits or digits that is transmitted as a unit and that may be encoded for error control purposes. **2.** A string of data units, such as records, words, and characters, that for technical or logical purposes is treated as a unit. *Note 1:* Blocks (a) are separated by interblock gaps, (b) are delimited by an end of block (EOB) signal, and (c) may contain one or more records. *Note 2:* A block is usually subjected to some type of block processing, such as multidimensional parity checking. **3.** In computer programming, a subdivision of a program that performs various specific functions, such as group related statements, delimit routines, specify storage allocations, delineate the applicability of labels, and segment parts of the program for other purposes. **4.** In flaghoist communications, the pulley or the loop at the tip of a mast or a pole for raising a flag, pennant, or panel on a lanyard. *See* **added block, deleted block, delivered overhead block, digital block, erroneous block, error block, horizontal terminating block, incorrect block, lost block, misdelivered block, program block, self-delineating block, time block, transmission block, user information block, vertical terminating block.**

blockage: *See* **antenna blockage.**

block cancel character: A cancel character used to indicate that the portion of the block back to the last block mark, i.e., the most recently occurring block mark, is to be ignored or deleted.

block character: *See* **end of transmission block character.**

block check: A check used to determine whether a block of data is structured according to given rules. *See also* **block, block check character, block code, block parity, error control.**

block check character (BCC): A character added to a message or transmission block to facilitate error detection. *Note:* In longitudinal redundancy checking and cyclic redundancy checking, block check characters are computed for, and added to, each message block transmitted. The computed block check character is compared with a second block check character computed by the receiver to determine if the transmission is error-free. *See also* **block check, block parity, character, cyclic redundancy check, error control.**

block code: In a block consisting of information and check characters, an error detection or correction code constructed so that errors can be detected or corrected. *See also* **block, block parity, convoluted code, error control, error correcting code, forward error correction.**

block correction efficiency factor: 1. In data transmission systems that use error control, a performance evaluation parameter given by the ratio of (a) the number of blocks that have been received without error during a given period to (b) the total number of blocks received during the same period. *Note:* The block correction efficiency factor usually is converted to a percentage by multiplying the calculated block correction efficiency factor by 100. **2.** In data transmission systems, the ratio of (a) the total number of erroneous blocks corrected by a receiver to (b) the total number of erroneous blocks that arrived at the receiver, both counted during the same measurement period.

block delivery: *See* **successful block delivery.**

block diagram: A diagram of a system, a computer, or a device in which the principal parts are represented by suitably annotated geometrical figures to show both the basic functions of the parts and their functional relationships. *See also* **flowchart.**

block distortion: In video systems, distortion characterized by the appearance of an underlying block encoding structure in the received image. *See also* **distortion, image.**

block efficiency: In a block, the ratio of (a) the number of user information bits to (b) the total number of bits. *Note:* For a given block scheme, block efficiency represents the maximum possible efficiency for a given block transmitted over a perfect transmission link.

block error probability (BEP): The expected block error ratio. *See also* **block error ratio.**

block error ratio (BER): 1. The ratio of (a) the number of incorrect blocks, i.e., erroneous blocks, received to (b) the total number of blocks received, both counted during the same measurement period. **2.** The ratio of (a) the number of incorrectly received or missing blocks to (b) the total number of blocks transmitted during a measurement period. *Note:* The block error ratio may be used as a block error probability. *See also* **block error probability, block loss ratio, block misdelivery ratio, incorrect block.**

blocking: 1. The formatting of data into blocks in order to perform various functions, such as transmission, storage, and checking functions. **2.** Denying access to, or use of, an entity, such as a facility, system, or component. *See also* **classmark, lost call, system blocking signal.**

blocking cement: An adhesive used to bond optical elements to blocking tools. *Note:* Examples of blocking cements are thermoplastic materials such as resin, beeswax, pitch, or shellac.

blocking criterion: In telephone traffic engineering, a criterion that specifies the maximum number of calls or service demands that fail to receive immediate service. *Note:* The blocking criterion is usually expressed in a probabilistic notation, such as *P*.001.

blocking factor: The number of records per block, i.e., the ratio of (a) the block length to (b) the length of each record contained in the block. *Note 1:* The block length and the record length must be expressed in the same data unit. *Note 2:* Each record in the block must have the same length. *Synonym* **grouping factor.**

blocking formula: A specific probability distribution function that approximates the call pattern of a user when failing to find available facilities.

blocking network: In telecommunications, a network that (a) has fewer transmission paths than would be required if all users were to communicate simultaneously, (b) is used to reduce equipment requirements, (c) is used because not all users require service simultaneously, and (d) makes use of certain statistical distributions that apply to the patterns of user demand. *Note:* Certain events that encourage simultaneous use often cause blocking networks to become saturated.

blocking probability: In a communications system, the probability that an access attempt will result in an access failure. *Note:* When expressed as a percentage, the blocking probability implies the number of calls that can be expected not to be completed out of 100 calls that are attempted. *See also* **nonblocking.**

blocking signal: *See* **system blocking signal.**

block length: The number of data units, such as bits, bytes, characters, or records, in a block.

block loss probability: The expected block loss ratio. *See also* **block loss ratio.**

block loss ratio (BLR): The ratio of (a) the number of lost blocks to (b) the total number of block transfer attempts during a specified period. *Note:* The block loss ratio may be used as a block loss probability. *See also*

block error ratio, block loss probability, block misdelivery ratio.

block misdelivery probability: The expected block misdelivery ratio. *See also* **block misdelivery ratio.**

block misdelivery ratio (BMR): The ratio of (a) the number of misdelivered blocks to (b) the total number of block transfer attempts during a specified period. *Note:* The block misdelivery ratio may be used as a block misdelivery probability. *See also* **block error ratio, block loss ratio, block misdelivery probability, lost block, misdelivered block.**

block parity: The designation of one or more bits in a block as parity bits to ensure that a designated parity occurs, either odd or even. *Note:* Block parity is used to assist in error detection or correction. *See also* **binary digit, block code, cyclic redundancy check, error control, error correcting code, error detecting code, parity, parity check.**

block probability: *See* **added block probability.**

block rate efficiency: The ratio of (a) the product of the block transfer rate and the average block length to (b) the signaling rate.

block ratio: *See* **added block ratio.**

block reek: A chain-like scratch accidently occurring during the polishing of optical elements. *See also* **optical element.**

block transfer: The transfer of one or more blocks of data. *Note:* A block transfer is initiated by a single action.

block transfer attempt: A coordinated sequence of user and telecommunications system actions undertaken to effect the transfer of a single block from a source user to a destination user. *Note:* A block transfer attempt begins the instant the first bit of the block crosses the functional interface between the source user and the telecommunications system. A block transfer attempt ends either in successful block transfer or in block transfer failure. *See also* **block transfer time, interface, successful block transfer.**

block transfer efficiency: In a successfully transferred block, the ratio of (a) the number of user information bits in the block to (b) the total number of bits in the block. *Note:* The block transfer efficiency may be expressed as a common fraction, a decimal fraction, or a percent. *See also* **overhead information, throughput.**

block transfer failure: Following a block transfer attempt, a failure to deliver a block successfully. *Note:* The principal block transfer failures result in lost blocks, misdelivered blocks, and added blocks. *See also* **added block, deleted block, failure, incorrect block, lost block, successful block delivery, successful block transfer.**

block transfer rate (BTR): The rate at which blocks are successfully transferred. *Note 1:* The block transfer rate (BTR) should be qualified, such as the instantaneous rate, average rate, minimum rate, or maximum rate. *Note 2:* The average block transfer rate (BTR) during a period may be obtained by dividing (a) the number of blocks successfully transferred during a performance measurement period by (b) the duration of the period. *See also* **data transfer rate, data transfer time, error ratio, maximum block transfer time.**

block transfer success: *Synonym* **successful block transfer.**

block transfer time: The time between (a) the instant a block transfer attempt is initiated and (b) the instant successful block transfer is achieved. *Note:* A block transfer attempt is successful if (a) the transmitted block is delivered to the intended destination user within the maximum allowable performance period, and (b) the contents of the delivered block are correct. *See also* **block, block transfer attempt, maximum allowable transfer time, successful block transfer.**

bloodflow meter: *See* **fiber optic bloodflow meter.**

blooming: The formation of trapped gas pockets in an optical fiber.

blowback: *Synonym* **reenlargement.**

blowback ratio: The ratio of (a) a linear dimension of an enlarged image of an object to (b) the linear dimension of the original image of the same object before enlargement.

BLP: block loss probability.

BLR: block loss ratio.

blue noise: Noise power with a spectral density that is directly proportional to the noise frequency over a specified frequency range. *Note:* When noise power is plotted with respect to frequency, blue noise has a positive slope, white noise has a zero slope, and pink noise has a negative slope. *See also* **noise, pink noise, white noise.**

blue screening: *Synonym* **chroma keying.**

blurring: In video display systems, a global distortion that is characterized by (a) reduced sharpness of image edges and (b) limited spatial detail. *See also* **resolution.**

BNF: Backus Naur form.

board: *See* **accelerator board, add-in board, bulletin board, communications board, daughter board, dial service board, emulator board, gateway board, Joint Telecommunications Resources Board, long board, mother board, multifunction board, printed circuit board, secure voice cord board, short board, test board, video graphic board.**

boating: *See* **motorboating.**

bobbin-wound sensor: A distributed fiber sensor in which the optical fiber is wound on a bobbin in such a manner as to be differentially, sequentially, or selectively stimulated by the parameter to be measured, such as a spatially shaded winding or a segmented in-line winding of many coils on a single bobbin, the bobbin serving only to support the distributed fiber but playing no other role, such as stretching or squeezing the fiber.

BOC: Bell Operating Company.

body: *See* **blackbody.**

body language: A language in which messages may be created, transmitted, received, and understood by positions, movements, and appearances of the body and its parts through the use of established conventions. *Note 1:* Military tactical control may be exercised by arm and hand signals. *Note 2:* Examples of body language are (a) a wave of an arm and hand to indicate "go away" or "come here," depending on the direction of motion, (b) a pivoting of the head vertically to indicate "yes" and horizontally to indicate "no," (c) a shrug of the shoulders to indicate "I don't know," (d) a smile to indicate agreement or approval, (e) a frown or scowl to indicate disagreement, disapproval, or deep thought, (f) a raising of the hand to indicate "stop," (g) a spreading of the hands to indicate size, (h) shaking hands to indicate friendship, and (i) saluting to indicate obedience, respect, subordination, greeting, and recognition. *Note 3:* One can transmit messages unintentionally by body language, such as whistling to indicate joy or fear, slouching to indicate sadness, or clapping hands to indicate joy or approval. Thumbing one's nose, putting one's tongue in the cheek, winking an eye, and giving the Bronx cheer are other examples of body language messages that can be as precise, imprecise, meaningful, explicit, and implicit as the written and spoken natural and artificial languages. It has been said that changes observed in the iris of an eye may serve as a crude lie detector. Sign language is a precise form of body language.

body problem: *See* **n-body problem, restricted circular three-body problem, three-body problem, two-body problem.**

body signal: *See* **ground to air visual body signal.**

Boehme equipment: 1. Transmitting equipment used for sending international Morse code characters by passing Wheatstone tape through a keying head. **2.** Receiving equipment used for recording international Morse code characters on a moving paper tape using ink syphon equipment.

Boehme tape: *Synonym* **Wheatstone tape.**

bolometer: A device that (a) is used to measure the radiant energy from an emitter by measuring the change in resistance of a temperature-sensitive element exposed to the radiation and (b) usually is used to measure temperatures high above ambient, such as that of a furnace or other body heated to incandescence.

bolometry: The study, design, development, or application of electrical instrumentation to measure radiant energy by measuring the changes in electrical conductivity or resistance of a blackened temperature-sensitive device exposed to the radiation.

Boltzmann's constant (*k*): A physical constant that defines the ratio of the universal gas constant to Avogadro's number. *Note 1:* Boltzmann's constant usually is represented by the symbol *k*. Boltzmann's constant is approximately equal to 1.38042×10^{-16} erg/K (ergs per kelvin) or 1.38042×10^{-23} J/K (joules per kelvin). *Note 2:* Boltzmann's constant relates the average energy of a molecule to the absolute temperature of the environment. *Note 3:* Boltzmann's constant is applicable in areas such as photoemission, thermal noise power, and signal to noise ratio in photodetectors. *Note 4:* Examples of the use of Boltzmann's constant (*k*) are that (a) in a photoemissive photodetector, the dark current, I_d, is given by the relation $I_d = AT^2e^{qw/kT}$ where I_d is the dark current, A is the surface area constant, T is the absolute temperature, q is the electron charge, w is the work function of the photoemissive surface material, and k is Boltzmann's constant, (b) the signal to noise ratio, SNR, for a photodetector is given by the relation $SNR = (N_p/B)(1 - e^{-hf/kT})$ where SNR is the signal to noise ratio, N_p/B is the number of incident photons per unit of bandwidth, h is Planck's constant, f is the frequency of the incident photons, k is Boltzmann's constant, and T is the absolute temperature, and (c) for the noise generated by thermal agitation of electrons in a conductor, the noise power, P, is given by the relation $P = kT(\Delta f)$ where P is the noise power in watts, k is Boltzmann's constant, T is the conductor temperature in kelvins, and Δf is the bandwidth in hertz.

Boltzmann's emission law: The ratio of (a) the number of electrons in energy level *m* to (b) the number of electrons in level *n* is an exponential function of the energy level difference between the levels *m* and *n,* the temperature, and a proportionality constant, given by the relation $N_m/N_n = e^{(Em - En)/kT}$ where N_m is the number of electrons in energy level *m,* N_n is the number of electrons in energy level *n,* E_m is the energy of level *m,* E_n is the energy of level *n,* *k* is Boltzmann's constant, i.e., a proportionality constant, and *T* is the absolute temperature.

bombardment resistance: *See* **neutron bombardment resistance.**

bond: 1. In electrical circuits, an electrical connection that provides a low resistance path between two metallic surfaces. **2.** The established degree of electrical continuity between two or more conductive surfaces. *See* **direct bond, permanent bond.** *See also* **ground.**

bond frequency: The natural frequency of vibration caused by thermal excitation, i.e., by infrared radiation, of the electronic bonds of molecules in a material. *Note:* Molecular vibrations give rise to electric and magnetic fields that interact with the electric and magnetic fields of electromagnetic waves propagating in the material, such as an optical fiber, resulting in an absorption of energy from the propagating waves. In glass, the thermal vibrations of the oxides of silicon result in absorption of specific frequencies. *See also* **infrared absorption.**

bonding: 1. In electrical circuits, the connecting of metal parts so that they make low resistance electrical contact for direct current (dc) and lower frequency alternating currents (ac). **2.** The establishing of the required degree of electrical continuity between two or more conductive surfaces. *See also* **direct bond, ground.**

book: *See* **code book.**

booked call: In telephone switchboard operation, a call that is recorded so that the operator can call the calling party when the call has been set up, i.e., when the connections are made. *Note:* Calls are booked when delays are encountered so as to preclude a call originator's remaining on the line for extended periods. *See also* **delayed call.**

book message: A message that (a) is destined for two or more addressees, (b) is of such a nature that the originator considers that no addressees need to be informed of any of the other addressees, and (c) usually has each addressee indicated as an action addressee or an information addressee.

Boolean algebra: An algebra that (a) is an algebra of classes and propositions, like ordinary algebra, but dealing with truth values as variables and having basic operators, such as AND, OR, NOT, EXCEPT, and IF . . . THEN, (b) is a branch of symbolic logic, named after George Boole (1815–1864), an English mathematician, (c) allows variables that may assume either of only two different states, and (d) readily lends itself to adaptation and implementation of logic propositions and statements, in the form of on-off circuits and coincidence gates, permitting fabrication of many logic circuits and switching circuits to automatically perform the arithmetic and logic operations that form the basis of electronic digital computers. *Note:* Examples of Boolean algebra axioms are (a) B OR 0 = B, (b) B AND 1 = B, and B OR 1 = 1. Examples of Boolean propositions are (a) NOT (A AND B) = (NOT A) OR (NOT B) and (b) A AND B = NOT {(NOT A) AND (NOT B)}. *See also* **Boolean function, Boolean operation.**

Boolean function: 1. In the Boolean algebra, a mathematical function that describes a Boolean operation. **2.** A switching function in which the number of possible values of the function and each of its independent variables is two. *See also* **Boolean algebra, Boolean operation.** *Refer to* **Figs. A-7 (34), B-6 (78), K-1 (493), N-1 (609), N-2 (615), N-19 (637), O-4 (681), T-11 (1042), V-2 and V-3 (1064).**

Boolean operation: 1. An operation in which each of the operands and the result may have one of two different possible values, such as either "0 or 1," "on or off," "open or closed," or "present or absent." **2.** An operation that follows the rules of Boolean algebra. *See also* **Boolean algebra, Boolean function.** *Refer to* **Figs. A-7 (34), B-6 (78), K-1 (493), O-4 (681).**

Boolean operator: An operator that (a) has operands and produces results that can have either of two possible values at any given instant and (b) may be physically realized by means of gates. *See also* **gate, operand.**

Boolean table: A table that (a) lists permissible combinations of Boolean operators and operands, (b) lists all possible values of operands, (c) indicates the result for each of the Boolean functions listed, and (d) may be used to prove the equivalence or nonequivalence of Boolean functions, such as NOT (A AND B) = NOT A inclusive OR NOT B. *Refer to* **Fig. T-11 (1042).**

boot: 1. In a personal computer (PC) or a microcomputer system, to load a program or an operating system into the computer internal or main memory from an external or auxiliary storage, such as from a diskette, magnetic tape, or magnetic card. *Note:* In nearly all PCs or microcomputers, the intrinsic operating system, such as DOS for IBM and IBM-compatible systems, will permit the operator to boot a specific program. Most systems can be configured to boot automatically when electrical power is applied to the system. **2.** To bootstrap.

boot diskette: A diskette that (a) contains a copy of the disk operating system and all the files necessary to activate the hardware in a personal computer (PC), (b) is used to communicate with the other parts of the PC, and (c) may contain application software.

bootstrap: 1. A technique or device designed to bring about a desired state by means of its own action. **2.** A part of a computer program that may be used to establish another version of the computer program. **3.** An automatic procedure whereby the basic operating system of a processor is reloaded following a complete shutdown or loss of memory. **4.** A set of instructions that cause additional instructions to be loaded until the complete computer program is in storage. **5.** To use a technique or a device designed to bring about a desired state by means of its own action. *See also* **computer.**

bootstrap loader: In computer programming, an input routine (a) that contains instructions that, when executed, cause further instructions to be loaded into memory until the complete computer program is stored in the computer and (b) that may part of the program being stored. *See also* **computer program.**

border: *See* **display border.**

borescope: A device that (a) may be inserted into the interior of a system and (b) is used to observe internal functions and structures of the system. *See* **fiber optic borescope, flexible borescope, rigid borescope.**

borescope articulation angle: In a flexible borescope, the angle between (a) the axis of the articulating tip of the borescope at its maximum deflection and (b) the axis of the tip at the undeflected position, i.e., when the tip is collinear with the straight-line working length. *Note:* A typical articulation angle is 90°, though borescopes with larger articulation angles are available.

borescope axial viewing angle: The angle by which the axis of the view field deviates from the nominal straightforward axis of the working section of a borescope.

borescope magnification: The ratio of (a) the apparent size of an object seen through a borescope to (b) that of the size of the same object viewed directly by the unaided eye, with the object-borescope distance equal to the object-eye distance.

borescope overall length: The length of a borescope, including (a) the working length, which is the length of fiber optic cable to the tip, (b) the tip length, and (c) the eyepiece assembly length.

borescope target resolution: The resolving power of a borescope, usually given in line pairs per millimeter of a back-lighted high contrast resolution object, i.e., target, placed a selected nominal operating distance from the objective end, i.e., from the distal tip.

borescope tip length: The distance between the center objective window and the objective end, i.e., the distal tip, of a flexible or rigid borescope. *Note:* The borescope tip length determines how close to the bottom of a cavity an image can be observed when the borescope is completely inserted. The tip length should be compatible with the diameter of the probe.

borescope view field: The area or solid angle that can be viewed through a borescope. *Note:* The solid angle is determined by the aperture. The area is determined by the aperture and the focal length.

borescope working diameter: In a borescope, the diameter of the working section, i.e., the diameter of (a) the fiber optic cable and tip of a flexible borescope or (b) the diameter of the probe of a rigid borescope. *Note:* The diameter must be sufficiently smaller than the smallest-size opening it must pass through so that it reaches the object to be viewed without undue friction or binding.

borescope working length: In a borescope, the distance between the distal end of the hand-held body portion and the optical axis of the tip. *Note:* The borescope working length does not include the tip length.

boresight area: *See* **effective boresight area.**

B OR NOT A gate: A two-input, binary, logic coincidence circuit or device that (a) performs the logic operation B OR NOT A, (b) is the same as A OR NOT B gate except that the variables are reversed, and (c) if A is a statement and B is a statement, produces the result of false only when A is true and B is false, while for the other three combinations of A and B it produces a result of true. *Synonyms* **A implies B gate, if A then B gate, implication gate, inclusion gate.** *Refer to* **Fig. B-8.**

Bose-Chaudhuri-Hochquenghem (BCH) code: A multilevel, cyclic, error-correcting, variable-length digital code that (a) is used to correct errors up to

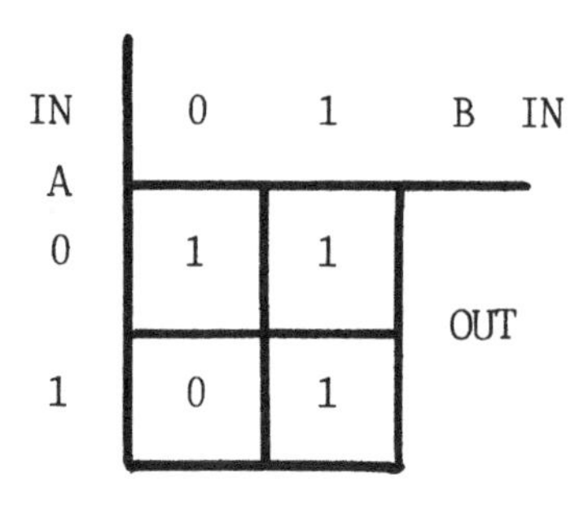

Fig. B-8. A truth table that shows the input and output digits of a **B OR NOT A gate.**

about 25% of the total number of digits and (b) is not limited to binary codes but may be used with multi-level phase shift keying whenever the number of levels is a prime number or a power of a prime number, such as 2, 3, 4, 5, 7, 8, 11, and 13. *Note:* A BCH code in 11 levels has been used to represent the 10 decimal digits plus a sign digit.

both way communications: *Synonym* **two-way simultaneous communications.**

bottom-up: In communications, computer, automatic data processing, and control systems engineering, pertaining to an approach to systems design, development, or fabrication that starts with the lowest level components and progressively proceeds through higher level components until the system is completely designed, developed, or fabricated. *See* **bottom-up programming, bottom-up designing, bottom-up testing.** *See also* **top-down.**

bottom-up designing: The designing of a system by using a bottom-up approach, i.e., by starting with the most basic or primitive components and proceeding to higher level components or modules by using the lower level components as building blocks, until the system design is completed. *See also* **top-down designing.**

bottom-up programming: Creating software, such as a computer program, by using a bottom-up approach, i.e., by starting with the most basic or primitive instructions and subroutines and redesigning them until they perform satisfactorily, then building more complex and sophisticated higher level routines and programs, while using the lower level ones as building blocks, until the software is completed. *See also* **top-down programming.**

bottom-up testing: Testing software or hardware by using a bottom-up approach, i.e., by starting with tests of the most basic or primitive lower level parts and redesigning them, if necessary, until they perform satisfactorily, then building more complex and sophisticated higher level parts and modules, components, and subsystems with these tested lower level parts, and testing the higher level parts, until the entire system is assembled and tested as a whole. *See also* **top-down testing.**

Bouger's law: In the transmission of electromagnetic radiation through a propagation medium, the attenuation of the irradiance, i.e., the electromagnetic field strength or optical power density, is an exponential function of the product of a constant coefficient, which is dependent upon the material and its thickness, that is given by the relationship $I = I_0 e^{-ax}$ where I is the irradiance at distance x, I_0 is the irradiance at $x = 0$, and a is a material constant coefficient that depends upon the scattering and absorptive properties of the propagation medium, such that (a) if only absorption takes place, a is the spectral absorption coefficient and is a function of wavelength, (b) if only scattering takes place, it is the scattering coefficient, and (c) if both absorption and scattering occur, it is the extinction coefficient, being then the sum of the absorption and scattering coefficients. *Note:* The irradiance, or optical power density, usually is measured in watts per square meter.

boule: An assembly of fused optical fibers from which fiber faceplates may be sliced. *See also* **optical fiber faceplate.**

bounce: **1.** In transmission systems, to reflect a signal, or the reflected signal itself. *Note:* Bounce may occur at the end of a line, a point of sudden impedance difference, i.e., impedance mismatch, a dish, a target, the ionosphere, or the surface of the Earth or the Moon. **2.** The instantaneous momentary reopening of relay contacts caused by the elasticity of the colliding materials. *See also* **switch bounce time, switch operating bounce time, switch release bounce time.**

bounce time: *See* **switch bounce time, switch operating bounce time, switch release bounce time.**

boundary layer photocell: *Synonym* **photovoltaic cell.**

boundary limit marker: *See* **drop zone boundary limit marker.**

bound mode: In an optical waveguide, a mode in which (a) the electromagnetic field strength or optical power density decays monotonically in the transverse direction everywhere external to the core, and (b) power is not lost by radiation. *Note 1:* For an optical fiber in which the refractive index decreases with distance from the optical axis, i.e., a graded index fiber, and where there is no central refractive index dip, a bound mode is governed by the relation $n_a k < \beta < n_0 k$ where β is the imaginary part, i.e., the phase term, of the axial propagation constant, n_a is the refractive index at $r = a$, i.e., the core radius, n_0 is the refractive index at $r = 0$, i.e., the core axis, k is the angular free space wavenumber, i.e., $2\pi/\lambda$, and λ is the free space wavelength, i.e., the wavelength of the source. *Note 2:* Bound modes correspond to guided rays in geometric optics. *Note 3:* When more than one mode is propagating in a fiber, the optical power in bound modes is largely confined to the core. *Synonyms* **guided mode, trapped mode.** *See also* **acceptance angle, acceptance cone, cladding mode, guided wave, leaky mode, leaky ray, mode, normalized frequency, radiation mode, unbound mode.** *Refer to* **Fig. E-4 (314).**

bound ray: *Synonym* **guided ray.**

bounds checking: *See* **storage bounds checking.**

bounds register: A register that holds an address that designates a storage boundary.

box: *See* **breakout box, fiber optic interconnection box, optical fiber distribution box, optical fiber interconnection box, optical mixing box, stunt box, television set top box.**

bpi: bits per inch.

BPI: bits per inch.

bps: bits per second.

BPSK: binary phase-shift keying.

BR: bit rate.

braid: **1.** In a fiber optic cable, a layer of woven yarn. *Note:* In single-fiber loose-buffered fiber optic cables or two-fiber zip-cord loose-buffered fiber optic cables, the braid is situated between the buffer tube and the jacket. In cables having multiple buffer tubes, the braid usually is situated between the inner jacket and the outer jacket. **2.** An unwoven parallel bundle of yarn that (a) is situated around the tight buffer of a single-fiber cable or a two-fiber zip-cord fiber optic cable, (b) adds tensile strength to the cable, (c) may be anchored to an optical connector or splice organizer assembly to secure the end of the cable, and (d) is often an aramid yarn.

branch: **1.** In a computer program, a conditional jump or departure from the implicit or declared order in which instructions are being executed. **2.** In a computer program, to select a conditional jump or departure from the implicit or declared order in which instructions are being executed. **3.** In networks, a direct path joining two adjacent nodes of a network. **4.** In a power distribution device, such as a power panel, one of the output circuits. *Note:* The branch usually has a lower power-handling capability than that of the input circuit. *See also* **node.**

branched cable: A multiple-wire, multiple-optical fiber, or multiple-bundle cable that contains one or more breakouts or divergences, i.e., one or more branches. *Note:* An example of a branched cable is a cable that up to a point has eight conductors, at which point the cable is divided into two separate cables of four conductors each that continue on in divergent directions.

branch exchange: *See* **private branch exchange, satellite private branch exchange.**

branch exchange tie trunk: *See* **private branch exchange tie trunk.**

branch feeder cable: A cable that conducts signals to or in a branch in a communications network.

branching device: *See* **fiber optic branching device, optical branching device.**

branching network: A network used for the transmission or the reception of signals over two or more channels.

branching repeater: A repeater that (a) drives several output channels or circuits and (b) has input from only one input channel or circuit. *See also* **repeater.**

branchpoint: **1.** The point in the sequence of instructions of a computer program at which a branch could or does occur. **2.** In a branched cable, the point at which a branch occurs. *See also* **branch.**

breadboard: **1.** An assembly of circuits or parts used to prove the feasibility of a device, circuit, system, or principle with little or no regard to the final configuration or packaging of the parts. **2.** To create an assembly of circuits or parts used to prove the feasibility of a device, circuit, system, or principle with little or no regard to the final configuration or packaging of the parts.

break: **1.** In a communications circuit, an action in which the destination user interrupts the source user and takes control of the circuit. *Note:* A break is performed when operating half-duplex telegraph circuits and two-way telephone circuits equipped with voice-operated devices. **2.** In radiotelephone systems, the separation of the text, i.e., the main body, from other portions of a message. **3.** An interruption in transmission. **4.** In personal computers (PCs) and microcomputers, to cancel a current command, entry, or operation, such as a printing operation.

breaker: *See* **circuit breaker, molded case circuit breaker, power circuit breaker.**

break-in procedure: In radiotelegraph communications, a procedure whereby a receiving station interrupts a transmission to request the transmitting station to perform some specified function, such as to wait, shift frequency, repeat, or receive. *Note:* A series of dashes is usually used to break in.

break lock: To lose frequency lock, phase lock, or synchronism. *See* **angle break lock.**

breakout box: In an interface cable, a device that permits a user to access individual leads for various purposes, such as to monitor, break, switch, interconnect, or test selected circuit leads. *See also* **distribution frame.**

breakout cable: *See* **fiber optic breakout cable.**

breakout chain: In a relay circuit of a telephone exchange dialing system, a set of relays operated in such a manner that when the first relay of the set is energized, any others in the set are prevented from being energized.

breakout kit: *See* **fiber optic breakout kit.**

breakout point: The point where a branch meets, merges, or joins with or diverges from the main cable or harness run. *Note:* An example of a breakout point is the point of convergence or divergence of the fibers in a fiber optic cable.

bremsstrahlung: A pulsed ray or beam, literally, from the German, a braked ray, i.e., an interrupted ray.

brevity code: A code that (a) makes use of brief codes, such as prosigns and prowords, to represent a complete message or instruction, (b) is used to save time, energy, and money, (c) requires the originator and the addressee to have a priori knowledge of the code, and (d) is not used to maintain communications security.

brevity list: A list of the brevity codes in use, usually arranged for lookup access to entries in the list.

Brewster angle: The angle, measured with respect to the normal, at which an electromagnetic wave incident upon an interface surface between two dielectric media of different refractive indices is totally transmitted from one propagation medium into another. *Note 1:* The transmittance will be unity and the reflectance will be zero only for an electromagnetic wave that has its electric field vector in the plane defined by (a) the direction of propagation, i.e., the direction of the Poynting vector, and (b) the normal to the surface. Thus, the magnetic field vector of the incident wave must be parallel to the interface surface. *Note 2:* The Brewster angle, Θ_B, for propagation from medium 1 to medium 2, is given by the relation $\Theta_B = \arctan(n_2/n_1) = \arctan(\varepsilon_2/\varepsilon_1)^{1/2}$ where Θ_B is the Brewster angle, n_2 and n_1 are the refractive indices of medium 2 and medium 1, respectively, and ε_2 and ε_1 are their electric permittivities. *Note 3:* The Brewster angle is a convenient angle for transmitting all the energy in an optical fiber to a photodetector, or from a source to a fiber. There is no Brewster angle when the electric field component of the incident wave is parallel to the interface, except when the electric permittivities are equal, in which case there is no interface. For entry into a more dense medium, such as from air into an optical fiber, Θ_B is less than 45°. From a more dense medium into a less dense medium, such as fiber to air, Θ_B is greater than 45°. *See also* **Brewster's law, incidence angle, reflectance, refractive index.**

Brewster's law: When an electromagnetic wave is incident upon an interface surface, i.e., an incidence plane, and the angle between the refracted and reflected ray is 90°, (a) maximum polarization occurs in both rays, (b) the reflected ray has its maximum polarization in a direction normal to the interface surface, and (c) the refracted ray has its maximum polarization parallel to the interface surface, i.e., incidence plane. *See also* **Brewster angle.**

BRI: basic rate interface.

brick: In the mobile service, a station that consists of a hand-held radiotelephone unit that (a) is licensed under a site authorization and (b) is capable of operation while being hand-carried.

bridge: 1. In communications networks, a means that (a) links or routes signals from one ring or bus to another or from one network to another, (b) extends the distance span and capacity of a single local area network (LAN), (c) does not modify packets or messages, (d) operates at the Data link Layer of the Open Systems Interconnection—Reference Model (OSI-RM) (Layer 2), and (e) reads packets and messages and passes only those with addresses on the same segment of the network as the originating user. **2.** A functional unit that interconnects two local area networks (LANs) that use the same logical link control procedure but may use different medium access control procedures. *See also* **gateway, hybrid coil, port, router.**

bridged ringing: The part of a signaling system in which ringers associated with a particular line are connected to that line.

bridge lifter: A device that electrically or physically removes bridged telephone pairs. *Note:* Bridge lifters may be in various forms, such as relays, saturable inductors, and semiconductors.

bridge to bridge station (BBS): A ship station that (a) operates in the port operations service, (b) transmits messages restricted to navigational communications, (c) is capable of operation from the ship navigational bridge or, in the case of a dredge, from its main control station, and (d) operates in the 156 MHz (megahertz) to 162 MHz band.

bridge transformer: *Synonym* **hybrid coil.**

bridging: The use of two or more telephone instruments connected in parallel across a single line.

bridging connection: A parallel connection used to extract some of the signal energy from a circuit, usually with negligible effect on the normal operation of the circuit. *See also* **branching network, circuit, monitor jack.**

bridging loss: At a given frequency, the loss that results when an impedance is connected across a transmission line, measured as the ratio of (a) the signal power delivered to a given point in a system downstream from the bridging point to (b) the signal power delivered to the given point after bridging. *Note:* Bridging loss usually is expressed in dB (decibels). *See also* **loss.**

brightness: An attribute of visual perception in which a source appears to emit a given amount of light. *Note 1:* "Brightness" should be used only for nonquantitative references to physiological sensations and perceptions of light. *Note 2:* "Brightness" was formerly used as a synonym for the photometric term "luminance" and (incorrectly) for the radiometric term "radiance." *See also* **irradiance, radiance, radiant intensity.**

brightness conservation: *Synonym* **radiance conservation law.**

brightness theorem: *Synonym* **radiance conservation law.**

Brillouin diagram: A diagram that shows allowable and unallowable frequencies or energy levels of electromagnetic waves that can pass through certain materials that have periodic microstructures. *Note:* When an electric field is applied to the material, electrons experience periodic attractive and repulsive forces as they approach and depart from atomic nuclei in their path as they move through the material. The resulting electron vibration results in a field that interacts or resonates with electromagnetic waves, causing energy from the waves to be absorbed by the electrons at the resonant frequencies. The net result is that electromagnetic waves of the resonant frequencies are absorbed, while the others are passed. Crystals and certain glasses have the periodic microstructure, and therefore Brillouin diagrams can be drawn for them. *See also* **Brillouin scattering.**

Brillouin scattering: The scattering of lightwaves in a propagation medium caused by thermally driven density fluctuations that cause frequency shifts of several gigahertz at room temperature. *See also* **Brillouin diagram, Mie scattering, Raman scattering, Rayleigh scattering.**

BRLP: bit-rate•length product.

broadband: *Synonym* **wideband.**

broadband exchange (BEX): A switched communications system that has interconnection bandwidths greater than voice bandwidth. *See also* **bandwidth, group, switching center.**

Broadband Integrated Services Digital Network (B-ISDN): An International Telegraph and Telephone Consultative Committee (CCITT) Integrated Services Digital Network (ISDN) offering broadband capabilities, features, and services that include (a) access to and from 150 Mbps (megabits/second) to 600 Mbps interfaces, (b) use of the Asynchronous Transfer Mode (ATM) to carry all services over a single, integrated, high speed packet switched network, (c) local area network (LAN) interconnection, (d) the ability to connect LANs at different locations, (e) access to a remote, shared disk server, (f) voice/video/data teleconferencing from a user desk, (g) transport for programming services, such as cable television (TV), (h) user-controlled access to remote video sources, (i) voice/video telephone calls, and (j) access to shop at home and other information services. *Note:* Techniques used in the Broadband Integrated Services Digital Network (B-ISDN) include code conversion, information compression, multipoint connections, and multiple connection calls. B-ISDN uses a service-independent call structure that allows modular control of access and transport, the service components of a connection that (a) can provide each user in a connection with independent control of access features and (b) allows a simplified control structure for multipoint and multiconnection calls. B-ISDN offers a variety of ancillary information processing functions. *See also* **Integrated Services Digital Network.**

broadband system: *Synonym* **wideband system.**

broadcast: Transmission in which (a) any number of receivers and stations, such as home receivers and organization, unit, ship, aircraft, vehicle, and network stations may receive the transmissions, and (b) transmission usually is in the form of radio, television, or radiotelephone signals. *See* **routine meteorological broadcast, special meteorological broadcast, time signal standard frequency broadcast.** *Refer to* **Fig. B-9.**

broadcast address: *See* **directed broadcast address.**

broadcast area: 1. One of the 12 numbered areas in which the Earth has been divided for purposes of operating the merchant ship broadcast system. **2.** The geographical area covered by the signals from a radio station or a television station.

broadcast area radio station: *See* **ship broadcast area radio station.**

broadcast communications method: 1. A method of transmitting messages or information to a number of receiving stations that make no receipt. **2.** A method of communications that (a) broadcasts messages for which the addressee does not furnish a receipt, (b) allows the receiver to maintain radio silence, and (c) is used by shore stations to transmit messages to ships at sea, to aircraft in flight, or to units in the field. *Synonym* **broadcast method.** *See also* **intercept communications method, receipt communications method, relay communications method.**

broadcast communications net: *See* **maritime broadcast communications net.**

broadcast day: *See* **radio broadcast day, television broadcast day.**

broadcast frequency: The frequency used to broadcast messages and programs by radio. *Note:* An example of a broadcast frequency is the frequency used to

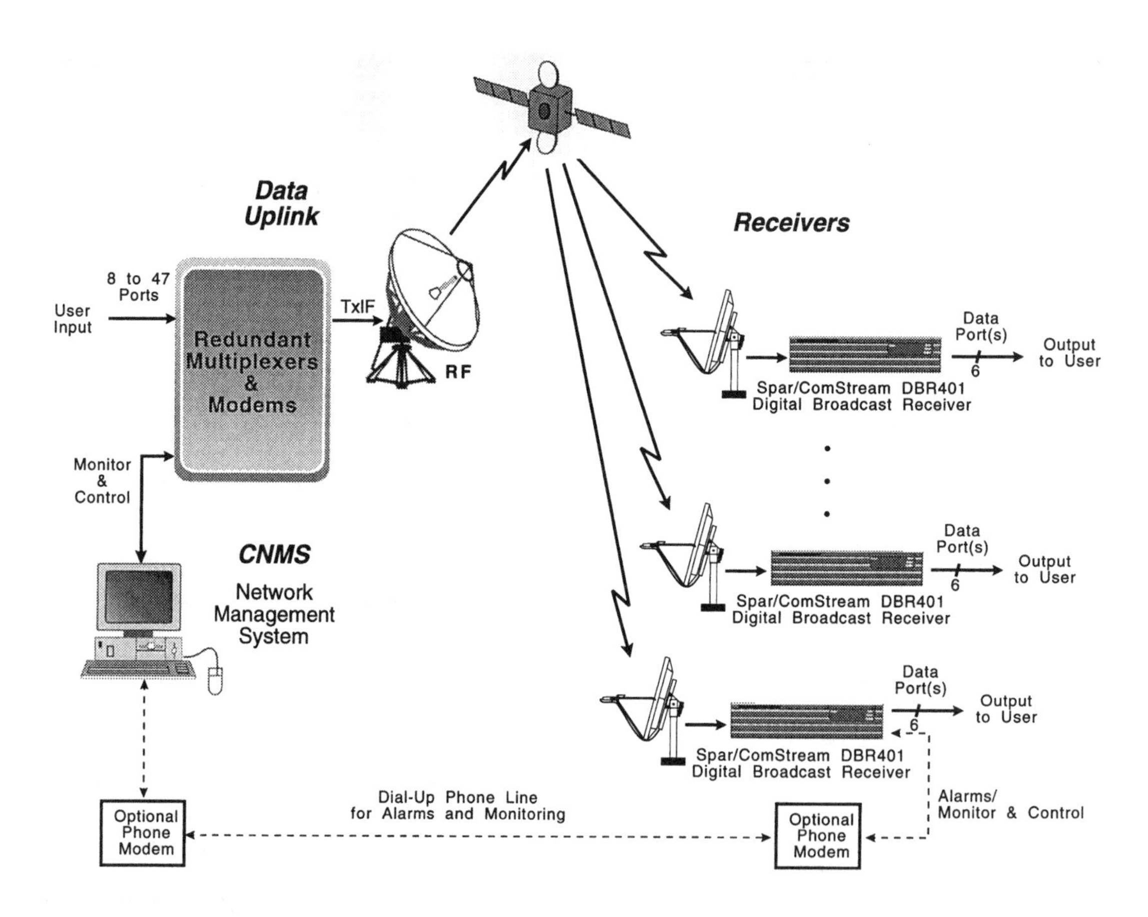

Fig. B-9. A typical satellite data **broadcast** network. (Courtesy ComStream Company.)

broadcast messages from a shore station to all ship stations.

broadcasting satellite service (BSS): A radiocommunciations service in which signals transmitted or retransmitted by space stations are intended for direct reception by the general public. *Note:* In the broadcasting satellite service, the term "direct reception" encompasses both individual reception and community reception. *See also* **individual reception.**

broadcasting satellite space station: A space station in the broadcasting satellite service primarily used for sound broadcasting.

broadcasting service (BS): A radiocommunications service in which the transmissions are intended for direct reception by the general public. *Note:* The broadcasting service may include several types of transmissions, such as sound transmissions and television transmissions.

broadcasting station: A station in the broadcasting service that transmits messages and programs by radio or television. *See* **international broadcasting station.**

broadcast method: *Synonym* **broadcast communications method.**

broadcast net: *See* **maritime patrol air broadcast net.**

broadcast operation: The operation of a transmission system in which information may be simultaneously received by stations that usually make no acknowledgment. *See also* **antenna, point to point transmission.**

broadcast repeater: A repeater that connects one incoming channel to several outgoing channels. *See also* **branching repeater, radio relay system, repeater.**

broadcast satellite: *See* **direct broadcast satellite.**

broadcast schedule: *See* **ship broadcast schedule.**

broadcast shift: *See* **area broadcast shift.**

broadcast station: *See* **aeronautical broadcast station, area broadcast station, maritime broadcast station.**

broadcast system: *See* **Emergency Broadcast System, merchant ship broadcast system.**

broadcast television satellite: *See* **direct broadcast television satellite.**

broadcast translator: *See* **television broadcast translator.**

broadening: *See* **pulse broadening, root-mean-square pulse broadening.**

broadside antenna: *Synonym* **billboard antenna.**

broadside antenna array: A group of parallel vertical dipole antennas that are placed in a single usually straight horizontal line.

brouter: A combination of a bridge and a router that (a) operates without protocol restrictions, (b) routes data using a protocol it supports, and (c) bridges data it cannot route.

browser: Any computer software program that (a) is used for reading hypertext, (b) usually is associated with the Internet and the World Wide Web (WWW), and (c) usually is able to access information in many formats using various services and protocols, such as the Hypertext Transfer Protocol (HTTP), File Transfer Protocol (FTP), Gopher, and Archie. *See also* **hypertext.**

browsing: The searching through an information system storage, such as a database or an automated information storage system, to locate or acquire information without necessarily knowing of the existence or the format of the information being sought.

b/s: bits per second.

b•s^{-1}: bits per second.

BSA: basic serving arrangement.

BSE: basic service element.

b•sec^{-1}: bits per second.

BSI: British Standards Institution.

bubble: In a propagation medium or an optical element, such as a lens or an optical fiber, a minute quantity of trapped vacuum or free gas that (a) usually consists of air, carbon dioxide, nitrogen, or water vapor, (b) usually is formed when the medium, such as glass or plastic, is in the molten state, (c) usually is spherical because, in accordance with Pascal's principle, pressure is exerted equally in all directions against the surface tension of the molten medium, and (d) causes dispersion, reflection, deflection, diffusion, absorption, and scattering of lightwaves.

bubble sort: An exchange sort in which the sequence of examination of pairs is reversed when an exchange of the position of items in a pair is made. *Synonym* **sifting sort.**

budget: *See* **error budget, loss budget, optical power budget, power budget.**

budget constraint: *See* **loss budget constraint, statistical loss budget constraint.**

buffer: 1. A computer routine or a storage device used to (a) compensate for a difference in the rate of flow of data, or time of occurrence of events, when transferring data from one device to another, (b) interconnect two digital circuits operating at different rates, (c) hold data for use at a later time, (d) allow timing corrections to be made on a data stream, (e) collect binary data bits into groups that can then be operated on as a unit, and (f) delay the transit time of a signal in order to allow other operations to occur. **2.** To allocate and schedule the use of buffers. **3.** An isolating circuit, usually an amplifier, used to prevent a driven circuit from influencing the driving circuit. *Synonym* **buffer amplifier. 4.** In a fiber optic cable, a component used to encapsulate an optical fiber, thus providing mechanical isolation and protection from physical damage. *Note:* Fiber optic cable buffers include (a) cushioning materials, such as Kevlar®, and (b) tubes in which the optical fiber is loose, or in a gel. These buffers reduce external pressures transferred to the fibers. The buffer material is used to cushion and protect the optical fibers, particularly from mechanical damage, such as microbends and macrobends, caused during manufacture, cabling, spooling, and subsequent handling, and when pressure is applied during use. Buffers may be bonded to the cladding. *See* **data buffer, elastic buffer, optical fiber buffer, variable length buffer.** *See also* **data; first in, first out; optical fiber cable; queuing; queuing delay; queue traffic.** *Refer to* **Figs. C-1 (102), F-2 (339).**

buffer amplifier: *See* **buffer.**

buffered fiber: An optical fiber that has a coating over the cladding for protection, increased visibility, and ease of handling.

buffer gate: *Synonym* **OR gate.**

buffering: *See* **dynamic buffering, message buffering, simple buffering, static buffering.**

buffer storage: A functional unit that stores data and usually is used to compensate for a difference in the flow rate of data or for a disparity in time of occurrence of events when data are being transferred from one device to another. *Note:* A buffer storage may be used when transferring data from a high speed computer to a slow speed display device or from a high speed communications channel to several slow speed terminal devices.

bug: 1. A concealed audiosurveillance device, such as a microphone or a listening device. **2.** An error in software, such as in a computer program or in a protocol. **3.** To install a means for audiosurveillance. **4.** A semiautomatic telegraph key. **5.** A malfunction of hardware. *See also* **communications security.**

building out: The adding of a combination of electrical inductance, capacitance, and resistance to a cable pair to (a) increase its electrical length by a desired amount and (b) control its impedance characteristics. *Synonym* **line buildout.** *See also* **balancing network, impedance matching.**

buildout: *See* **line buildout.**

bulk encryption: 1. The use of a single encryption device to encrypt the output signal from a multiplexer on a multichannel telecommunications trunk. **2.** The simultaneous encryption of all the data in all the channels of a multichannel telecommunications trunk. *See also* **communications security, cryptology.**

bulkhead connector: A fiber optic connector that (a) enables the connection of a cable on one side of a barrier to a cable on the other side, usually with a butted, epoxied, optical fiber interface between the fiber ends and (b) usually is used to pass through the walls, sides, or skins of various entities, such as cabinets, aircraft, missiles, compartments, and houses.

bulk material absorption: The electromagnetic wave power absorption that occurs per unit volume or per unit thickness of basic materials of construction, such as those used in optical fibers and lead chambers. *Note 1:* Measurement of bulk material absorption usually is made prior to use of the material. *Note 2:* Bulk material absorption usually is expressed in rads or in dB per kilometer, i.e., as an attenuation rate.

bulk material scattering: The electromagnetic wave power that (a) is scattered per unit volume or per unit thickness of basic materials of construction, such as those used in optical fibers and building materials, (b) is measured prior to use of the material, and (c) follows a Rayleigh distribution, which is characteristic of a propagation medium with a refractive index that fluctuates over small distances compared to the wavelength of the incident light. *Note:* Bulk material scattering usually is expressed in dB per kilometer, i.e., as an attenuation rate. *Synonym* **material scattering.**

bulk optical glass: Large quantities of glass suitable in purity for making optical elements, such as lenses, prisms, mirrors, and particularly optical fibers.

bulk optical plastic: Large quantities of plastic suitable in purity for making optical elements, such as lenses, prisms, mirrors, and particularly optical fibers.

bulk storage: *Synonym* **mass storage.**

bulletin board: An electronic mail system in which addressed messages are entered by message originators into a network consisting of remote stations connected to one or more host or central computers and in which addressees obtain, at their convenience and request, messages addressed to them that were sent to the host computer by the originators. *Note:* Originators and addressees need only log-on, insert, or call for their messages, using their own addresses, and operate their own station, using telephone lines, modems, printers, and usually personal computers.

bunched frame alignment signal: A frame alignment signal in which the signal elements occupy consecutive digit positions. *See also* **distributed frame alignment signal, frame, frame alignment, frame alignment signal, signal.**

bunching strip: In telephone switchboard operation, a strip of jacks connected in parallel in some types of switchboards to facilitate the connection of multiple calls. *See also* **multiple call.**

bundle: A group of associated (a) electrical conductors, such as wires and coaxial cables, (b) optical fibers, aligned or unaligned, or (c) both conductors and fibers, all usually contained within a single jacket along with buffers and strength members. *See* **aligned bundle, cable bundle, fiber optic bundle, ray bundle, unaligned bundle.** *See also* **cable, fiber optics.**

bundle cable: *See* **multiple bundle cable, multichannel bundle cable, single channel single bundle cable.**

bundle cable assembly: *See* **multiple bundle cable assembly.**

bundle jacket: The outer protective covering applied over a fiber optic bundle. *See* **cable bundle jacket.**

bundle of rays: *Synonym* **ray bundle.**

bundle resolving power: The ability of an aligned fiber bundle to transmit the details of an image. *Note:* Bundle resolving power usually is expressed in lines per millimeter. *See also* **aligned bundle.**

bundle transfer function: *See* **fiber optic bundle transfer function.**

bundpack: In fiber optics, an optical fiber wound in a spherical or conical shape, after the fiber drawing process, in such a manner that it can be payed out, i.e., unwound, from either the inside or the outside of the winding. *Note:* The conical shape permits easy removal of the collapsible winding mandrel after winding. The bundpack can be wrapped in a tight plastic skin for protection during storage, shipping, and handling. There are no spools or reels involved during payout, though one may be used for outside payout if it is convenient to do so. A light bonding cement is used to control payout so that payout is steady and without bunching. A twist is placed in the fiber when winding so that payout is without strain and hockels, i.e., kinks do not occur. *Synonym* **bundpaket.**

bundpaket: *Synonym* **bundpack.**

buoy: A tethered floating device that may be used for various purposes, such as marking waterways and channels, holding communications equipment, or marking underwater obstacles. *See* **fog buoy, sonobuoy.**

buoyant antenna: An antenna designed to operate when floating on the surface of the sea, such as on a buoy, or when streaming submerged at listening depth when towed by a submerged submarine and, therefore, not visible from above the surface.

bureau: *See* **International Time Bureau.**

buried cable: *See* **direct buried cable.**

burn through range: The maximum range from a given radar being jammed to an object at which a radar jammer can no longer mask the echoes from the object to the given radar. *Note:* Consider the relative position of the given radar, its object, i.e., its target, and the jammer. At long range, the jammer is at an advantage because it can launch a pulse stronger than the object echo and thus easily masks the echo on the screen of the radar being jammed. As the range to the object decreases, the echo becomes stronger while the jamming signal remains the same, unless jamming power is increased or the jammer moves closer to the radar being jammed. As the range diminishes, the echo is no longer masked, i.e., it propagates through the masking and is detected, i.e., it "burns through" to the radar being jammed. The object at that range is no longer masked by the jamming signals and becomes visible to the radar that until then was being jammed and now is no longer considered jammed, even though interference may still be visible on the radar screen.

burn through range equation: *Synonym* **self-screening range equation.**

Burrus diode: *Synonym* **surface-emitting light-emitting diode.**

Burrus light-emitting diode: *Synonym* **surface-emitting light-emitting diode.**

burst: **1.** In data communications, a sequence of signals, noise, or interference counted as a unit in accordance with specified criteria or measurements. *See also* **burst transmission.** **2.** To separate continuous form or multipart paper into discrete sheets. *See also* **data burst, error burst.**

burst communications: *See* **meteor burst communications.**

burst isochronous: *Deprecated synonym for* **isochronous burst transmission.**

burst switching: In a packet switched network, switching in which each network switch extracts routing instructions from an incoming packet header to establish and maintain the appropriate switch connection for the duration of transmission of the packet, following which the connection is automatically released. *Note:* Burst switching is similar to connectionless mode transmission, but it differs in that burst switching implies an intent to establish the switch connection in near real time so that only minimum buffering is required at the node switch. *See also* **network, packet, packet switching.**

burst transmission: 1. Transmission in which a very high data signaling rate is combined with very short transmission times. **2.** Operation of a data network by interrupting, at intervals, the data being transmitted. *Note:* Burst transmission enables communications between data terminal equipment (DTE) and a data network that operate at different data signaling rates (DSRs). *See also* **data burst.**

bursty transmission: Transmission characterized by long periods of idleness between relatively short periods of transmission. *Note:* An example of bursty transmission is transmission in which 3-s (second) to 5-s bit streams are separated by 2-min. (minute) to 3-min. periods of idleness, i.e., periods of transmission of idle characters or no transmission of any characters.

bus: One or more conductors or optical fibers that serve as a common connection for a group of related devices. *See* **data bus, dual bus, fiber optic bus, fiber optic data bus, infinite bus, interface bus, open dual bus, power bus.** *See also* **highway.** *Refer to* **Figs. N-5 (622), P-15 (763).**

busback: *Synonym* **feedback.**

bus coupler: *See* **data bus coupler.**

bus driver: A circuit that amplifies a signal on a bus to assure reception of the signal at the destination. *Note:* An example of a bus driver is an amplifier that energizes a communications bus.

bus interface unit (BIU): *See* **network interface device.**

bus network: *See* **network topology, token bus network.** *Refer to* **Fig. N-5 (622).**

bus topology: *See* **network topology.**

busy: In communications systems, the condition that exists when a system component is fully engaged and cannot undertake an additional task. *Note:* In telephone systems, if an access attempt is made when a circuit, line, or channel is busy, a busy signal is generated and sent to the call originator. *See* **all trunks busy.** *See also* **busy signal.**

busy back: *Deprecated synonym for* **busy signal.**

busy circuit: *See* **pilot make busy circuit.**

busy hour: In a communications system, the 60-min. (minute) period during which the traffic load in a given 24-hour period is at a maximum. *Note 1:* The busy hour is measured by fitting a horizontal line segment equivalent to one hour under the traffic load curve at the peak load point. If the service time interval is less than 60 min., the busy hour is the 60-min. interval that contains the service time interval in the center. *Note 2:* In cases where more than one busy hour occurs in a 24-hour period, i.e., saturation is occurring, the busy hour, or hours, most applicable to the particular situation is used. *Synonym* **peak busy hour.** *See also* **all trunks busy, erlang, group busy hour, switch busy hour, traffic capacity, traffic intensity, traffic load.** *Refer to* **Fig. B-10.**

busying: *See* **forward busying.**

busy line: *See* **call forwarding busy line.**

busyness: *See* **edge busyness.**

busy party: *Synonym* **busy user.**

busy season: During a 1-year cycle, the 3-consecutive-month period that has the highest busy hour traffic. *See also* **busy hour.**

busy signal: 1. In telephony, an audible or visual signal that (a) is sent in the backward direction and (b) indicates that the call receiver line is engaged or the transmission path to the call receiver line is unavailable. **2.** In telephony, an audible or visual signal that indicates that no transmission path to the called number is available. *Synonym* **reorder signal.** *See* **camp on busy signal.** *See also* **backward direction, busy tone, call, engaged tone, signal, terminal engaged signal.**

busy test: In telephony, a test made to determine whether certain facilities, such as a customer line or a central office trunk, are available for use.

busy tone: An audible busy signal. *See also* **busy signal.**

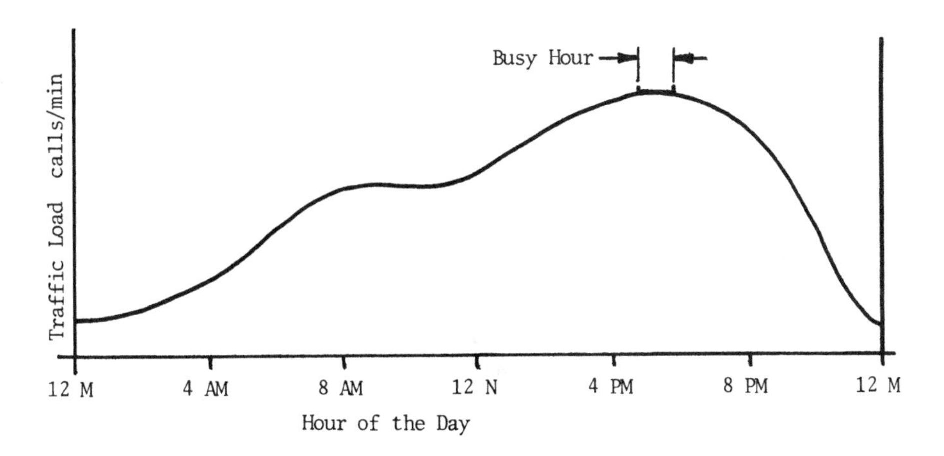

Fig. B-10. A communications traffic load for a given day, including the **busy hour.**

busy user: A user of a communications system that is actively using a terminal station or an end instrument, such as a telephone or facsimile machine. *Note:* A call originator will usually receive a busy signal when the call receiver is a busy user. *Synonym* **busy party.** *See also* **busy signal.**

busy verification: In a public switched telephone network, a network-provided service feature that permits an attendant to (a) verify the busy or idle state of station lines and (b) break into a conversation. *Note:* A 440-Hz tone is applied to the line for 2 seconds, followed by a 0.5-second burst every 10 seconds, to alert both parties that the attendant is connected to the circuit. *See also* **attendant position, call.**

butt coupling: In fiber optics, the coupling of one optical element, such as an optical fiber, to another by placing the endface of one against the endface of the other so that an electromagnetic wave can be transmitted with a minimum loss of power and a maximum transmission coefficient at the interface. *Note:* The interface usually is transverse to the direction of propagation. The Brewster angle can be used for maximum coupling and total transmission, i.e., for zero reflection, when lightwaves pass from one propagation medium to another, each medium having a different refractive index. However, in butt coupling optical fibers, the endfaces are perpendicular to the optical axis because they are easy to make and the joint can be rotated and shifted laterally for maximum coupling. Fresnel reflection usually is about 4% of the incident irradiance, i.e., 4% of the incident optical power. *Synonym* **face to face coupling.** *Refer to* **Figs. L-5 (511), L-7 (535), P-11 (749).**

button: *See* **virtual pushbutton.**

BW: bandwidth.

bypass: 1. The use of any telecommunications facilities or services that circumvent those of the local exchange common carrier. *Note:* The alternative facilities or services may be either customer-provided or vendor-provided. **2.** An alternate circuit around equipment or a system component. *Note:* A bypass is often used to allow system operation to continue when the bypassed equipment or system component is inoperable, malfunctioning, or unavailable.

byte (B): A sequence of *n* consecutive binary digits, i.e., of adjacent binary digits, usually treated as a unit, where *n* is a nonzero integral number. *Note:* Before 1970, "byte" was defined as a variable length bit string, such as a 4-bit byte, 6-bit byte, and 8-bit byte. Since then, byte has come to mean an 8-bit string, i.e., an octet. This usage predominates in computer and data transmission literature, especially in referring to block length, frame size, and storage capacity. Thus, a 1.44-Mb diskette has a capacity of 1.44 megabytes. *See* **six-bit byte.** *Synonym* **octet.** *See also* **binary digit, bit string, block, computer word, octet, octet alignment, word.**

byte serial transmission: Transmission of successive bytes in sequence, in which (a) the bits of the bytes may be serial on one line and the bytes are serial on the same line, i.e., serial by bit and byte, or the bits of the same byte may be parallel on several lines with the bytes occurring serially on the multiple lines, i.e., parallel by bit, serial by byte, i.e., the bytes occur serially but the bits of the same byte are on separate lines, with a line for each information bit and usually for a parity bit.

B3ZS: bipolar with three-zero substitution.

B6ZS: bipolar with six-zero substitution.

B8ZS: bipolar with eight-zero substitution.

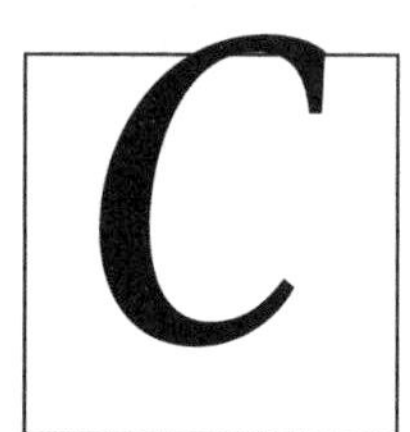

c: The symbol used for the speed of an electromagnetic wave, such as a lightwave, in a vacuum, i.e., 3 x 10^8 m/s (meters per second). The speed in a material propagation medium is given by the relation $v = c/n$ where n is the refractive index of the medium relative to that of a vacuum.

cable: 1. An assembly that (a) consists of one or more insulated conductors or optical fibers, or a combination of both, within an enveloping jacket or sheath and (b) is constructed so that the conductors or fibers may be used singly or in groups. **2.** In fiber optic systems, an assembly of one or more optical fibers, strength members, buffers, and perhaps other components all within an enveloping sheath or jacket, constructed so as to permit access to the fibers singly or in groups. **3.** A message sent by any telegraphic means. *See* **aerial cable, aerial fiber optic cable, air dielectric coaxial cable, bidirectional fiber optic cable, branched cable, branch feeder cable, buried cable, Category 3 cable, Category 4 cable, Category 5 cable, central strength member fiber optic cable, coaxial cable, composite cable, direct buried cable, distribution cable, duct fiber optic cable, fiber optic breakout cable, fiber optic cable, fiber optic station cable, fiber optic telecommunications cable, filled cable, flat fiber optic cable, foam dielectric coaxial cable, grooved cable, house cable, hybrid cable, intrusion-resistant communications cable, loose tube cable, monofilament cable, multichannel bundle cable, multichannel cable, multichannel single fiber cable, multipaired cable, multiple bundle cable, optoelectrical cable, paired cable, peripheral strength member fiber optic cable, plow-in fiber optic cable, quadded cable, ribbon cable, single channel single bundle cable, single channel single fiber cable, special fiber optic cable, spiral four cable, station fiber optic cable, submerged cable, submersible cable, symmetrical pair cable, TEMPEST proofed fiber optic cable, tight jacketed cable, twin cable, underground cable, wire cable.** *See also* **bundle, message.** *Refer to* **Fig. C-1.**

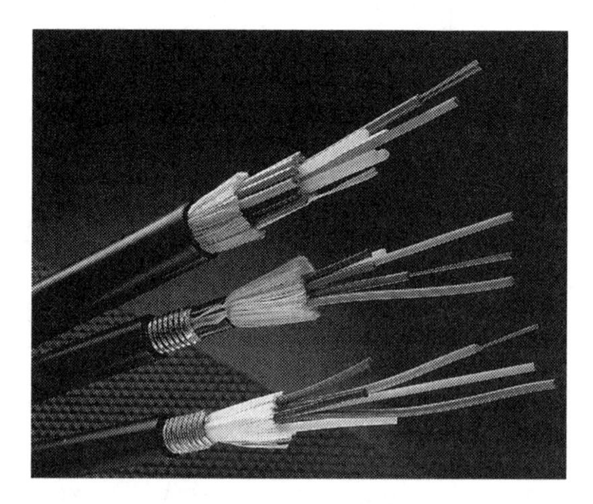

Fig. C-1. Fiber optic **cables.** (Courtesy Siecor Corporation, Hickory, NC.)

cable address code: A code, usually registered with all commercial telecommunications carriers, used as the address for messages to specific organizations in foreign countries. *Note:* Examples of cable address codes are 23 887/ISO CH for Telex and ISORGANIZ for telegrams, both for the International Organization for Standardization in Geneva, Switzerland.

cable assembly: A cable that (a) is ready for installation in specific applications and (b) usually is terminated with connectors. *Note:* A cable assembly usually is terminated by the manufacturer, though connectors are often installed by the user. *See also* **cable, fiber optic cable assembly, multiple bundle cable assembly, multiple fiber optic cable assembly.**

cable bundle: A number of optical fibers, buffered optical fibers, ribbons, or fiber optic cable components (OFCCs) grouped together in the cable core within a common protective layer.

cable bundle jacket: The material that forms a protective layer around a fiber optic bundle, buffered optical fibers, ribbons, or fiber optic cable components (FOCCs).

cablecasting: The distribution via cables of video, radio, audio, and data signals from a point of reception directly to the public or to other distribution points. *Note:* The cables are usually coaxial or fiber optic cables. *See also* **broadcasting.**

cable code: A variation of the Morse code, usually used in submerged cables, in which dots, dashes, and spaces all have equal durations but differ in electrical polarity. *See also* **submersible cable.**

cable component: A component that (a) is ready for cabling or (b) is in an existing cable. *Note:* Examples of cable components are wires, optical fibers, strength members, buffers, and jackets. *See* **fiber optic cable component.** *See also* **cabling.**

cable core: **1.** The portion of a cable inside a common covering, i.e., inside the cable jacket. **2.** The signal-bearing elements of a cable, such as conducting wires, individual optical fibers, and ribbon cables. *See* **fiber optic cable core.**

cable core component: A cable core part, such as a buffered optical fiber, fiber optic cable component (FOCC), cable bundle, and ribbon.

cable cutoff wavelength (λ_{cc}): **1.** For a cabled single-mode optical fiber under specified length, bend, and deployment conditions, the wavelength at which the fiber second-order mode is attenuated a measurable amount when compared to (a) a multimode reference fiber or (b) a tightly bent single-mode fiber. **2.** For a cabled optical fiber, the wavelength region above which the fiber supports the propagation of only one mode and below which multiple modes are supported. *Note 1:* Operation of the optical fiber below the cutoff wavelength usually results in modal noise, modal distortion, increased pulse broadening, and unsatisfactory operation of connectors, splices, and wavelength division multiplexing (WDM) couplers. The operating wavelength range, as determined by the transmitter nominal central wavelength and the transmitter spectral width, must be less than the maximum cutoff wavelength and greater than the minimum cutoff wavelength to ensure that the cable operates entirely in the fiber single-mode region. Usually, the highest value of cabled fiber cutoff wavelength occurs in the shortest cable length. The maximum and minimum cutoff wavelengths should be specified. *Note 2:* A criterion that will ensure that a system is free from high cutoff wavelength problems is given by the relation $\lambda_{cc\max} > \lambda_{t\min}$ where $\lambda_{cc\max}$ is the maximum cutoff wavelength and $\lambda_{t\min}$ is the minimum value of the transmitter central wavelength range, which is also caused by the worst-case variations introduced by manufacturing, temperature, aging, and any other significant factors determined when the cable is operated under standard or extended operating conditions. *See* **maximum cable cutoff wavelength, minimum cable cutoff wavelength.** *See also* **cutoff wavelength, optical fiber cutoff wavelength.**

cable driver: *See* **fiber optic cable driver.**

cable facility: *See* **fiber optic cable facility.**

cable facility loss: *See* **fiber optic cable facility loss, statistical fiber optic cable facility loss.**

cable feed through: *See* **fiber optic cable feed through.**

cable interconnect feature: *See* **fiber optic cable interconnect feature.**

cable jacket: The outer protective covering applied over the internal cable elements for (a) protection of the internal elements and (b) bearing the cable identification markings. *Synonym* **sheath.** *See* **fiber optic cable jacket.** *Refer to* **Fig. C-1.**

cable pigtail: *See* **fiber optic cable pigtail.**

cable retention: In a fiber optic cable, the ability of a connector to hold or retain the fiber optic cable against a specified pulling force on the cable when the connector is held fixed.

cable run: The portion of a branched cable or harness where the cross-sectional area is largest. *Synonym* **harness run.**

cable splice: In fiber optic systems, a connection that (a) is made between two cables, consisting of one or more optical fiber splices within a cable splice closure, (b) provides optical continuity between cables, (c) protects against environmental conditions, (d) provides mechanical strength to the cable joint, and (e) primarily is used to complete a cable span or to repair a cable.

cable splice closure: In a cable, such as a fiber optic cable or metallic conductor cable, the portion of a splice that (a) covers the optical fiber splice or conductor housings, (b) seals against the outer jackets of the joined cables, (c) provides protection against the environment, and (d) provides mechanical strength to the joint.

cable television (CATV): Television transmission in which signals from distant stations are received, amplified, and then retransmitted to local users, usually by means of coaxial cables, fiber optic cables, or microwave links. *Note 1:* In most countries cable television (CATV) originated in areas where good reception of direct broadcast television (TV) was not possible. CATV also provides a highly sophisticated cable distribution system to large metropolitan areas. *Note 2:* The abbreviation "CATV" originally applied to "community antenna television." However, "CATV" usually is used to mean "cable television." *Synonyms* **cable TV, community antenna television.**

cable television relay service (CARS) station: A fixed or mobile station used for (a) transmitting television and related audio signals, signals of standard and FM broadcast stations, and signals of instructional television fixed stations and (b) cablecasting from the point of reception to a terminal point from which the signals are distributed to the public. *See also* **cablecasting.**

cable TV: *Synonym* **cable television.**

cabling: **1.** The sequence of operations required to manufacture a cable, usually by serially combining the component parts. **2.** In the production of fiber optic cables, a sequence of operations that includes (a) paying an optical fiber out of a spool, reel, or bundpack, (b) passing the fiber through dipping troughs for applying buffers or coatings, (c) wrapping the fiber with additional buffers and strength members, (d) winding additional fibers, applying additional strength members and adhesives, (e) applying a jacket, (f) adding markings, (g) winding the finished cable on a reel or a spool, and (h) storing or shipping the wound cables. *Note:* Cabling can introduce microbends in the optical fibers because of (a) the requirement to pass the fibers around pulleys and capstans, (b) twisting the fibers around other cable elements, and (c) paying out the fibers initially and winding the finished cables. *See also* **bundpack, cable component.**

cabling process: In the manufacture of fiber optic cables, starting with one or more reels of optical fiber or fiber ribbon, usually with appropriate coatings and buffers already applied, the (a) assembling of a fiber optic cable by wrapping additional buffers, strength members, jackets, and perhaps other materials, such as overarmor and harnesses, (b) placing of coatings on the jacket, adding markings to indicate the type of cable, (c) winding the finished cable on spools or other forms for internal or external payout, (d) ensuring that microbends are not produced in the optical fibers, and (e) ensuring the optical fibers are not bent with a radius of curvature less than the minimum bend radius.

cabling standard: *See* **ANSI/EIA/TIA-568 cabling standard.**

CAC^2: combined arms command and control.

cache memory: A buffer that (a) is smaller and faster than main storage and (b) is used to hold a copy of instructions and data in main storage that are likely to be needed next by the processor and that have been obtained automatically from main storage. *Synonym* **cache storage.** *Refer to* **Fig. P-15 (763).**

cache storage: *Synonym* **cache memory.**

CAD: computer aided dispatching.

CAD/CAM: computer aided design/computer aided manufacturing.

calculate key (Calc Key): In personal computer (PC) systems with standard keyboards, a key that will cause the PC to calculate the value or the result of a function for the parameters it has been given. *Note:* An example of a calculate key is the F9 key when using Lotus 1-2-3. Pressing the F9 key when in the ready mode will cause a recalculation of all the function-driven cell values on the worksheet, i.e., the spreadsheet.

calculation: *See* **recalculation.**

calculator: A device that (a) performs arithmetic and algebraic operations, (b) usually does not have a stored program, (c) requires human intervention for manually inserting each operator and operand one at a time, and (d) obtains and usually displays the result of each operation after each such insertion. *See* **handheld calculator.** *See also* **computer.**

calendar: *See* **Gregorian calendar, Julian calendar.**

calendar time: The time on the Coordinated Universal Time (UTC) scale. *See also* **Coordinated Universal Time.**

calibration factor: *See* **transmitter calibration factor.**

call: **1.** In communications, a request to set up a connection. **2.** The sequence of events begun when a user causes a calling signal to reach an exchange and asks for a connection, and concluded when conversation is finished and all the connections have been severed. **3.** A unit of traffic measurement. **4.** The actions performed by a call originator. **5.** The operations required to establish, maintain, and release a connection. **6.** To use a connection between two stations. **7.** To cause a computer program or routine to execute. *See* **abandoned call, allcall, anycall, booked call, completed call, conference call, data call, direct call, distress call, emergency call, external call, finished call, follow-on call, function call, internal call, junction call, load-on call, local call, long distance call, lost call, meet-me conference call, multiple call, nonprecedence call, number call, operator-assisted call, person to person call, precedence call, preempting call, preliminary call, private call, progressive conference call, query call, radiotelephone call, radio Telex® call, redirected call, single call, station to station call, telephone call, ticketed call, toll call, trunk call, virtual call.** *See also* **message.**

call abandoned: *See* **abandoned call.**

call accepted signal: A call control signal sent by the called terminal to indicate that it accepts the incoming call. *Note 1:* The call accepted signal is sent in the backward direction, indicating that the call can be completed if the call receiver answers. *Note 2:* Usually the call will be automatically charged to the call originator. In the United States, if the called number is a 1-800 number followed by the call receiver number, the call will be charged to the call receiver. *See also* **call, call control signal, data terminal equipment, signal.**

call address: The part of the call selection signal that indicates the destination of the call. *See also* **call selection signal.**

call area: *See* **maximum call area.**

call attempt: In a telecommunications system, a request by a user for a connection to another user. *Note:* In telephone traffic analysis, call attempts are counted during a specific period. The count usually includes all completed calls and all unsuccessful calls, such as overflowed, abandoned, and lost calls. *See* **unsuccessful call attempt.** *See also* **access attempt.**

callback: **1.** The positive identification of a calling terminal by disconnecting and reestablishing the connection by having the called terminal, i.e., the call receiver, dial the number of the calling terminal, i.e., the call originator. *Note:* An example of a callback is a computer system dialing of the number of the call originator. **2.** The identifying of a call originator remote terminal, such as an automated information system remote terminal, whereby the host system, i.e., the called terminal, disconnects the call originator remote terminal and then dials the authorized telephone number of the remote terminal to reestablish the connection, thereby verifying the authenticity of the remote terminal. *See* **automatic callback.**

call barred facility: *Synonym* **calls barred facility.**

call capability: *See* **virtual call capability.**

call collision: **1.** The contention that occurs when a terminal and data circuit-terminating equipment (DCE) specify the same channel to transfer a call request and handle an incoming call. *Note:* The data circuit-terminating equipment (DCE) proceeds with the call request and cancels the incoming call. **2.** The condition that occurs when a trunk or a channel is seized at both ends simultaneously, thereby blocking a call. *See also* **blocking, call, clear collision, collision, data circuit-terminating equipment, data terminal equipment, head-on collision.**

call completion objective: Policy established within a telephone system or network concerning the manner in which an operator-assisted call will be handled. *Note:* Examples of call completion objectives include (a) processing and completing a call without releasing the call originator and (b) not holding a call originator when the call receiver line is busy or the call destination is unavailable.

call completion rate: *Deprecated. Use* **call completion ratio.**

call completion ratio: During a specified period, the ratio of (a) the number of completed calls to (b) the total number of call attempts. *Note:* The call completion ratio is usually expressed as either a percentage or a decimal fraction.

call control character (CCC): One of a set of control characters used in call control signaling. *Note:* The signals that represent call control characters may be used under defined conditions on interchange circuits other than the originating circuit. *See also* **call, call control signal, character, control character.**

call control procedure: The entire set of interactive procedures required to establish, maintain, and release a call.

call control signal: A member of the set of interactive signals required to establish, maintain, and release a call.

call delay: 1. The delay that occurs when a call arrives at an automatic switching device and no channel or facility is immediately available to process the call. **2.** The time between (a) the instant a communications system receives a call attempt and (b) the instant of initiation of ringing at the call receiver end instrument. *See also* **call attempt.**

call detail recording (CDR): A network-provided service feature in which call data on a specific telephone extension or group of customers are collected and recorded for cost of service accounting purposes. *See also* **automatic message accounting.**

call directly code (CDC): A code contained in the first part of calls and messages so that all stations on a line can listen to all codes but a receiver responds only to its own code. *Note:* If broadcast or group call directly codes (CDCs) are used, called stations usually answer before the transmitting station begins sending the message. *See also* **transmitter start code.**

call distributor: *See* **automatic call distributor.**

call duration: The time between (a) the instant a connection is established between a call originator and the call receiver, i.e., the call receiver goes off-hook, and (b) the instant the call originator or the call receiver terminates the call, i.e., either party hangs up. *Note:* The call duration corresponds to the duration of the information transfer phase of an information transfer transaction. *See also* **call release time, call second, call setup time, information transfer phase, information transfer transaction.**

called line identification facility: A network-provided service feature that enables a calling data terminal equipment (DTE), i.e., a call originator, to be notified by the network of the number, i.e., the address, to which the calling DTE has been connected. *See also* **calling line identification facility, calling line identification signal, data terminal equipment, facility, service feature.**

called line identification request indicator: An indicator, i.e., a signal, sent in the forward direction indicating whether or not the called line identity should be included in the response message.

called line identification signal: A sequence of characters that (a) is transmitted to the calling data terminal equipment (DTE), i.e., the call originator, (b) is transmitted in the backward direction, and (c) identifies the called line, i.e., the call receiver. *See also* **call control signal, call receiver, character, data terminal equipment, signal, terminal.**

called line identity: Information that (a) is sent in the backward direction and (b) consists of a number of signals that indicate the address of the called line.

called party: *Deprecated synonym for* **call receiver.**

called party camp on: *Deprecated synonym for* **call receiver camp on.**

called station: The station to which a message is routed, a transmission is directed, or a call is placed.

called user: In telephone switching systems, the user end instrument or destination data terminal equipment (DTE) to which a call originator has been, or is to be, connected.

caller: *Synonym* **call originator.**

caller ID: *Synonym* **caller identification.** *Refer to* **Fig. S-19 (954).**

caller identification: A network-provided service feature that permits the call receiver to determine, before answering, the number of the call originator.

call facility: *See* **virtual call facility.**

call failure signal: A signal that (a) is sent in the backward direction, (b) indicates that a call cannot be completed because of a condition, such as a time out or a fault, and (c) does not correspond to, i.e., is not identical to, any other particular signal.

call filing time: The time at which a call is accepted by an operator for placement. *Note:* The call filing time and date may be recorded on the call ticket for various purposes, such as call tracing, call charging, and callback.

call forward: In a call that was completed only up to the user switching equipment, such as a private branch exchange (PBX), the operation that enables a user to forward an incoming call to another terminal on the user switching equipment. *Note:* The call forward usually is preceded by a call hold to allow the user to establish whether or not the call receiver is available.

call forwarding: 1. A network-provided service feature in which calls may be rerouted automatically from one line, i.e., station number, to another line or to an attendant. **2.** A network-provided service feature that sends all received calls to any local or long-distance number where the user plans to be. *Note:* Call forwarding may be implemented in many different ways. *See also* **service feature, switching center.**

call forwarding busy line: A network-provided service feature in which an incoming call automatically is routed to an attendant when both the called and alternate lines are busy. *See also* **call forwarding.**

call forwarding don't answer: A network-provided service feature in which an incoming call automatically is rerouted to an attendant after a predetermined number of rings or seconds. *See also* **call forwarding.**

call hold: A service feature in which a user may retain an existing call connection while accepting or originating another call using the same facilities, such as the same telephone. *See also* **service feature.**

call identification: *See* **incoming call identification.**

call identifier: A network utility that consists of a name assigned by the originating network for each established or partially established virtual call. *Note:* When a call identifier is used in conjunction with the calling data terminal equipment (DTE) address, it uniquely identifies the virtual call. *See also* **data terminal equipment, virtual call.**

call indicator: *See* **redirected call indicator.**

calling: The transmitting of a call selection signal to establish a connection between stations. *See* **automatic calling, digital selective calling, double call sign calling, group abbreviated address calling, offnet calling, selective calling, single call sign calling, three-way calling.** *See also* **call selection signal.**

calling area: *See* **maximum calling area.**

calling facility: *See* **manual calling facility, multi-address calling facility.**

calling frequency: The frequency that (a) a radio station uses to call another station, (b) may not be the frequency used to respond to, i.e., to answer, the call, and (c) may not be the frequency used for the transmission of messages. *See also* **answering frequency, cross-band radiotelegraph procedure, radio telegraphy distress frequency, working frequency.**

calling line identification facility: **1.** A network-provided service feature in which called data terminal equipment (DTE), i.e., the call receiver, is notified by the network of the number, i.e., the address, from which the call originated. **2.** A network-provided service feature in which called data terminal equipment (DTE), i.e., the call receiver, is notified by the network of the number, i.e., the address, of the calling terminal, which may be an intermediate terminal, i.e., not necessarily the terminal from which the call originated. *See also* **called line identification facility, data terminal equipment, facility, incoming call identification, service feature.**

calling line identification request indicator: Information that (a) is sent in the backward direction and (b) indicates whether or not the calling line identity should be sent forward in a calling line identity message.

calling line identification signal (CLIS): A signal that (a) consists of a sequence of characters transmitted to the called terminal and (b) identifies the address from which the call originated. *See also* **call, call control signal, character, data terminal equipment, signal.**

calling line identity: Information that (a) is sent in the forward direction and (b) consists of a number of address signals that indicate the complete identity of the call originator, i.e., of the calling data terminal equipment (DTE). *See also* **calling line identity message.**

calling line identity message: A message that (a) is sent in the forward direction, (b) contains the identity of the call originator, i.e., of the calling data terminal equipment, and (c) is sent subsequent to an address message that does not contain the identity of the call originator, i.e., the calling DTE. *See also* **calling line identity.**

calling net: A communications network that (a) usually is associated with ship-shore communications and (b) is used by ship stations to call shore stations.

calling party: *Deprecated synonym for* **call originator.**

calling party camp on: *Deprecated synonym for* **call originator camp on.**

calling plug: In telephone switchboard operation, of the two plugs usually associated with any cord circuit, the plug that (a) is nearer the operator and (b) usually is used to call another switchboard or user.

calling rate: The number of telephone calls originated during a specified performance measurement period, such as a specified hour.

calling sequence: A sequence of instructions together with any associated data necessary to service a call.

calling signal: **1.** A call control signal transmitted over a circuit to indicate that a connection is desired. **2.** In telephone switchboard operation, a visible or audible signal at a switchboard that indicates that (a) a user wishes to speak to the operator, (b) a user wishes to speak to another user, or (c) the operator wishes to speak to a user. *See also* **call control signal, signal.**

calling station: 1. The station that initiates a transmission. **2.** The station preparing a tape for transmission.

calling unit: *See* **automatic calling unit.**

call intensity: *Synonym* **traffic intensity.**

call message: *See* **multiple call message, single call message.**

call not accepted signal: A call control signal that (a) is sent by the called data terminal equipment (DTE) to indicate that it does not accept the incoming call, (b) is sent in the backward direction indicating that the call cannot be completed, and (c) usually indicates there will be no charge to the call originator or the call receiver. *See also* **call accepted signal, call control signal, data terminal equipment, signal.**

call originator: A user, such as a person, equipment, or a computer, that originates a call, i.e., that places a call or dials a call. *Synonym* **caller.** *Deprecated synonym* **calling party.** *See also* **call receiver, source user.**

call originator camp on: A network-provided service feature that enables the network to (a) complete an access attempt in spite of the issuance of a user blocking signal, (b) monitor the busy user until the user blocking signal ends, (c) complete the access request, (d) hold an incoming call until the call receiver is free, and (e) issue a blocking signal to inform the call originator of the access delay. *Deprecated synonym* **calling party camp on.** *See also* **access attempt, call, call receiver camp on, queue traffic, service feature.**

call pending indication: *Synonym* **call pending signal.**

call pending signal: A signal that indicates to a busy call receiver that another call originator is waiting. *Synonym* **call pending indication.**

call pickup: A service feature that enables a user, by dialing a predetermined code, to answer incoming calls that are directed to another user in a preselected call group. *See also* **service feature.**

call processing: 1. The sequence of operations performed by a switching system from the acceptance of an incoming call through the final disposition of the call. **2.** The sequence of operations performed by a network from the instant a call attempt is initiated until the instant the call is finished. **3.** The operations required to complete all three phases of an information transfer transaction. *See also* **completed call, finished call, information transfer transaction, switching center.**

call progress signal (CPS): 1. A call control signal transmitted by the called data circuit-terminating equipment (DCE) to the calling data terminal equipment (DTE) to report (a) the progress of a call by using a positive call progress signal or (b) the reason why a connection could not be established by using a negative call progress signal. **2.** For packet services, a call control signal for virtual call service to inform the calling and called data terminal equipments (DTEs) of the reason why a call has been cleared. **3.** In permanent virtual circuit service, a call control signal to inform the data terminal equipment (DTE) of the reason why the permanent virtual circuit has been reset. **4.** For datagram service, a call control signal to inform the source data terminal equipment (DTE) about the delivery or nondelivery of a specific datagram or about general operation of the DTE-datagram interface or service. *See also* **call control signal, data circuit-terminating equipment, data terminal equipment, signal.**

call progress tone: An audible signal returned by a network to a call originator to indicate the status of a call. *Note:* Examples of call progress tones are dial tones and busy signals. *See also* **call control signal.**

call receiver: A user, such as a person, equipment, or a computer program, to which a call is directed. *Deprecated synonym* **called party.** *See also* **call originator, destination user.**

call receiver camp on: A network-provided service feature in which the network (a) completes an access attempt in spite of the issuance of a user blocking signal, (b) monitors the busy user until the user blocking signal ends, (c) holds an incoming call until the call receiver line is free, and (d) completes the requested access. *Deprecated synonym* **called party camp on.** *See also* **access attempt, call, call originator camp on, queue traffic, service feature.**

call record: Recorded data pertaining to a single call. *See also* **audit trail, automatic message accounting.**

call release time: In communications systems, the time between (a) the instant a clearing signal is initiated by the called data terminal equipment (DTE) and (b) the instant the available circuit condition occurs at the originating DTE. *See also* **call, call duration, call setup time, disengagement time, terminal.**

call request time: In the establishment of a connection or the placement of a call, the time between (a) the instant a calling signal is initiated by a call originator and (b) the instant a proceed to select signal, such as a dial tone, is obtained by the call originator.

call restriction: A network-provided service feature that prevents selected data terminal equipment (DTE) from using one or more service features otherwise available from the network. *See also* **classmark, class of service, service feature.**

calls barred facility: A network-provided service feature that permits data terminal equipment (DTE) to either make outgoing calls or receive incoming calls, but not both. *Synonym* **call barred facility.** *See also* **classmark, data terminal equipment, facility, service feature.**

call-second: A unit used to measure communications traffic. *Note 1:* A call-second is equivalent to 1 call of 1 second duration. *Note 2:* One user making two 75-second calls is equivalent to two users each making one 75-second call. Each case produces 150 call-seconds of traffic. *Note 3:* The CCS, equivalent to 100 call-seconds, is often used. The initial "C" in "CCS" represents 100, as does the Roman numeral "C." *Note 4:* 3600 call-seconds = 36 CCS = 1 call-hour. *Note 5:* 3600 call-seconds per hour = 36 CCS per hour = 1 call-hour per hour = 1 erlang = 1 traffic unit. *See also* **call duration, connections per circuit hour, erlang, holding time, traffic intensity.**

call selection signal: A signal that (a) usually consists of the electrical representation of a set of characters, (b) contains all the information needed to establish a call, (c) often consists of the facility request, the address, or both, (d) usually has the facility request ahead of the address, and (e) may contain several facility requests and several addresses.

call selection time: In the establishment of a connection or the placement of a call, the time between (a) the instant the call originator receives a proceed to select signal, such as a dial tone, and (b) the instant all the selection signals have been transmitted, such as dialing has been completed.

call setup: The establishment of a circuit switched connection between two users or between two sets of data terminal equipment (DTE).

call setup time: 1. The length of time required to establish a circuit switched call between users, i.e., the time between (a) the instant the final digit of the call receiver number is entered by the call originator data terminal equipment (DTE) and (b) the instant at which the calling signal starts at the call receiver DTE. **2.** For data communications, the overall length of time required to establish a circuit switched call between data terminal equipments (DTEs), i.e., the time between (a) the instant a call request is initiated and (b) the instant the call message begins. *Note:* Call setup time is the summation of (a) the call request time, i.e., the time between the instant a calling signal is initiated by the call originator and the instant the call originator receives a proceed to select signal, (b) selection time, i.e., the time between the instant a proceed to select signal is received by the call originator and the instant all the selection signals have been transmitted, and (c) post selection time, i.e., the time between the instant the last selection signal element is transmitted and the instant the call connected signal is received by the originating data terminal equipment (DTE), i.e., when the call receiver goes off-hook. *See also* **access time, call, call duration, call release time, data terminal equipment.**

call sign: A station or address designator that (a) usually is represented by a combination of characters or pronounceable words, (b) is used to introduce or identify a communications facility, station, command, authority, activity, or unit, and (c) is primarily needed for the establishment and maintenance of communications. *See* **aircraft call sign, collective call sign, indefinite call sign, individual call sign, international call sign, military call sign, military ship international call sign, net call sign, permanent voice call sign, ship call sign, ship collective call sign, visual call sign.**

call sign allocation plan: The table of allocation of international call sign series contained in the current edition of the *International Telecommunication Union (ITU) Radio Regulations. Note:* In the table of allocation, the first two characters of each call sign, whether two letters or one number and one letter, in that order, identify the country in which the station is located. In instances where the complete alphabetical block is allocated to a single nation, the first letter is sufficient for national identity. Individual assignments are made by appropriate national assignment authorities from the national allocation.

call sign calling: *See* **double call sign calling, single call sign calling.**

call sign linkage: The association of a new call sign and address group of an activity or an organization with the old call sign and address group of the same activity or organization.

call spillover: In common channel signaling, the effect on a user traffic circuit of the arrival at a switching center of an abnormally delayed call control signal relating to a previous call, while a subsequent call is being set up on the circuit. *See also* **call control signal, circuit, lockout.**

call splitting: A network-provided service feature that allows an attendant to talk privately with either the call originator or the call receiver. *See also* **cord circuit, service feature.**

call tape: In radiotelegraph communications systems that use automatic broadcast equipment, a tape on which the standard signals for initiating a broadcast are recorded for automatic transmission.

call ticket: A record of a call, usually indicating the time and the duration of the call, the call originator number, the call receiver number, the date, the charges, and the operator initials.

call tracing: A facility that (a) permits an entitled user to be given the routing data for an established connection, (b) identifies the entire route from the call originator to the call receiver, and (c) provides the routing data on a permanent basis covering all calls or on a demand basis in which only a specific call is traced immediately after it is disestablished. *See* **on-demand call tracing, permanent call tracing.**

call transfer: 1. A network-provided service feature that allows a call originator or receiver to instruct the local switching equipment or attendant to transfer an existing call to another data terminal equipment (DTE), usually one that is connected to the same switching exchange, switchboard, or private branch exchange. *Note:* Call transfer may be available on a call by call basis or on a semipermanent basis. **2.** A network-provided service feature in which a user is able to arrange call selections personally. *See also* **service feature.**

callup signal: One of the members of a set of signals used by a radio station to establish contact with another station. *Note:* When confidential call signs are used in the callup signals, their meanings in a message heading should not be used. *See* **preliminary callup signal.**

call waiting: 1. A network-provided service feature that provides an indication to data terminal equipment (DTE) already engaged in an established call that one or more calls are awaiting connection. **2.** A network-provided service feature that informs a call originator or a call receiver that another caller is trying to reach the user and that allows the user to switch back and forth between the two callers. *See also* **call control signal, signal.**

call waiting signal: A call control signal that indicates to a call receiver or a called data terminal equipment (DTE) that one or more calls are awaiting completion. *See also* **completed call.**

call without selection signal: *Synonym* **direct call.**

CALS: continuous acquisition and lifecycle support.

CAMA: Centralized Automatic Message Accounting.

camcorder: A video camera that (a) makes a full-motion record of the scanned scene and (b) usually is hand-held. *Note:* With appropriate equipment, the record made by the camcorder can be played back on a television set for viewing. *Synonym* **camera-recorder.** *See also* **video cassette, video cassette recorder.**

camera-recorder: *Synonym* **camcorder.**

camp on: 1. A network-provided service feature whereby (a) an incoming call is held until the call receiver is free, (b) the busy call receiver hears a tone indicating that a call originator is waiting, and (c) when the call receiver hangs up on the original call, the phone will ring and will automatically be connected to the call originator that has been waiting. **2.** The holding by a network of a call attempt that was unsuccessful because the called terminal is busy, with subsequent connection as soon as the called terminal becomes free. *Synonym* **connect when free.** *See* **automatic callback, call originator camp on, call receiver camp on, queue traffic.**

camp on busy signal: 1. A telephone signal that informs a busy user that another call originator is waiting for a connection. **2.** A teleprinter exchange facility signal that automatically causes a calling station to retry the call receiver number after a given interval when the call receiver teleprinter is occupied or the circuits are busy. *Synonym* **speedup tone.**

camp on with recall: A camp on in which (a) the call originator terminal is released until the call receiver terminal becomes free, and (b) the call originator can establish other calls until the recall signal is obtained rather than simply wait until the call receiver line is free.

CAN: cancel character.

Canada balsam: An adhesive used to cement optical elements.

cancel character (CAN): 1. A control character used to indicate that the data with which it is associated are in error or are to be disregarded. **2.** An accuracy control character used to indicate that the data with which it is associated are in error, are to be disregarded, or cannot be represented on a particular device. *See also* **control character.**

cancellation: *See* **echo cancellation, side lobe cancellation.**

candela (cd): The SI unit (Système International d'Unités) of light flux. *Note:* 1 cd (candela) (a) has a light emission capacity equivalent to that of 1/600,000 m^2 (square meters) of a blackbody radiator at the temperature of solidification of platinum, i.e., 2045 K (kelvin) and (b) emits 4π lumens of light flux. *See also* **blackbody, candlepower, lumen.**

candlepower: A unit of measure of the illuminating power of a light source, equal to the number of international candles, i.e., candela, of the source. *Note 1:* A flux density of 1 lumen of light flux per steradian of solid angle measured from a source is produced by a point source of 1 cd (candela) emitting equally in all directions. Thus, a 1-cd light source emits 4π lumens. *Note 2:* Formerly, 1 cp (candlepower) was the illuminating power of the international candle, a candle built according to precise specifications for materials and dimensions. *See also* **candela.**

capability: *See* **direct interface capability, expansion capability, multilevel precedence capability, virtual call capability, virtual circuit capability.**

capacitance: 1. The ability of material media to store an electrical charge. *Note:* The SI unit (Système International d'Unités) unit for capacitance is the farad. The microfarad (μF) and the picofarad (pF or μμF) are also often used. **2.** The property of a capacitor that determines the amount of electrical energy that can be stored in it by applying a given voltage, given by the relation $E = (1/2)CV^2$ where E is the energy, C is the capacitance, and V is the voltage. If C is in farads and V is in volts, E will be in joules. Also, the electrical charge that can be stored in a capacitor is given by the relation $Q = CV$, where Q is the stored charge, C is the capacitance, and V is the voltage. If C is in farads and V is in volts, the stored charge will be in coulombs. *Synonym* **capacity.** *See also* **inductance, resistance.** *Refer to* **Table 3, Appendix B.**

capacitive coupling: 1. The transfer of energy from one circuit to another by means of the mutual capacitance between the circuits. *Note 1:* Capacitive coupling may be deliberate, as in the case of a capacitive coupled amplifier circuit, or inadvertent, as in the case of crosstalk. *Note 2:* Capacitive coupling favors transfer of higher frequency components of a signal, inductive coupling favors transfer of lower frequency components, and conductive coupling favors neither higher nor lower frequency components. **2.** The coupling of signals or noise in which one conductor is linked to another by means of the electrical capacitance between them. *Note:* Electric charge concentrations on one side of the capacity induce charge concentrations on the other side. The accumulated charges produce voltages, and their migrations constitute currents. *See also* **conductive coupling, coupling, inductive coupling.**

capacitive reactance: The opposition to the flow of alternating current (ac) when an alternating voltage is applied to a capacitor, given by the relation $X_c = 1/2\pi fC$ where X_c is the capacitive reactance, f is the frequency, and C is the capacity. *Note 1:* If f is in hertz and C is in farads, X_c will be in ohms. The capacity is a function of geometry and the materials of construction of the capacitor. *Note 2:* When a voltage is applied to the capacitor, electric charges accumulate on it, causing its voltage to rise to the level of the applied voltage. When the capacitor is charged to its maximum charge, depending on the voltage and the capacitance, no more current will flow. Thus, the current is reduced to zero. When the applied voltage reverses, the capacitor will discharge and recharge with the opposite polarity. The applied ac voltage reverses again, and the process is repeated. For a pure capacitor, the current will lead the voltage by 90°, i.e., be in phase quadrature. Thus, for a given capacitance and a given applied ac voltage, there will be a resulting ac current. The current in the capacitor is given by the relation $I_c = V/X_c$ where I_c is the current, V is the applied voltage, and X_c is the capacitive reactance. If the sinusoidal V is in volts (rms) and X_c is in ohms, the sinusoidal I_c will be in amperes (rms). *See also* **impedance, inductive reactance.**

capacitor: An electrical device that can store and hold an electrostatic charge and therefore can store electric energy. *See also* **capacitive coupling, capacitive reactance.**

capacitor storage: Storage in which electrical capacitance is used to store data, such as storage where a charged capacitor might represent a 1 and a discharged capacitor might represent a 0. *Synonyms* **condenser storage, dicap storage.** *See also* **capacitor.**

capacity: 1. *See* **channel capacity, traffic capacity. 2.** *Synonym* **capacitance.**

Capstone: A system that (a) provides a software key escrow encryption capability for communications, computer, data processing, and control systems and (b) contains classified algorithms. *Note:* Clipper is a reduced, i.e., a stripped-down, version of Capstone. *See also* **Clipper, encryption.**

capture: *See* **data capture.**

capture effect: In the reception of frequency-modulated signals in which two signals are received on or near the same frequency, phenomenon where only the stronger of the two will appear in the radio output. *Note 1:* The complete suppression of the weaker signal occurs at the receiver limiter, where it is treated as noise and rejected. *Note 2:* When both signals are nearly equal in strength, or are fading independently, the receiver may switch from one to the other, back and forth, depending on path characteristics, such as fluctuating circuit parameters, atmospheric conditions, ionospheric conditions, multipath, and fading. *Synonym* **frequency modulation capture effect.** *See also* **carrier, frequency modulation.**

capture range: The range of frequency differences over which a given phase-locked loop can accomplish synchronism and maintain lock-on at a specified phase relationship.

carbon ribbon: A ribbon, usually designed for one-time use and usually of plastic, paper, or cloth, with a substance on one side that will adhere to paper when the ribbon is pressed against the paper by the type of an impact printer or typewriter to form characters on the paper. *See also* **inked ribbon, type.**

card: *See* **binary card, edge notched card, flash card, free-access card, key card, magnetic card, memory card, microprocessor card, punch card, punched card, secure-access card, smart card.**

card code: A code in which the set of coded characters is a set of unique patterns of holes punched in a punch card, each pattern corresponding to a character of an alphabet, such as the letters of the English alphabet, i.e., the Roman alphabet, and the decimal numerals 0 through 9. *Note 1:* The standard punch card has 80 columns with 12 punch positions, designated 0 through 9, x, and y, in each column. Each column is used to represent a character. Each different character corresponds to a different hole pattern in the column. *See also* **punch card.**

card dialer: An automatic dialer that (a) is combined with a terminal and (b) dials telephone numbers coded on a card by various means, such as punched holes, magnetic recording, or printed bar codes. *See also* **automatic callback, automatic calling unit, repertory dialer, speed calling.**

card interpreter: A card punch that prints, usually at the top of each column, the graphic character represented by the hole pattern in each column. *See also* **card punch, printer.**

card punch: A device that punches patterns of holes in cards to represent information.

card reader: A device that senses data in the form of patterns of holes in cards and converts the patterns to electrical signals for various purposes, such as to serve as input to a computer, printer, tape perforator, or card punch.

card storage: A container that holds punched cards. *See* **magnetic card storage.**

cardinal radial: **1.** In radio transmission systems, one of the eight radials at 0°, 45°, 90°, 135°, 180°, 225°, 270°, and 315° of azimuth with respect to true north. **2.** In navigation, one of the four radials at 0°, 90°, 180°, and 270° of azimuth with respect to true north. *Note:* The cardinal radials are equivalent to true north, west, south, and east.

carriage control character: On a printer or a typewriter, a character that (a) specifies the number of lines that are to be skipped before the next line is printed and (b) usually is the first character of an output line of characters.

carriage control tape: A tape that contains data used to control vertical tabulation, i.e., indexing, of printing or display positions. *Note:* An example of a carriage control tape is a tape that contains line feed control data for printers.

carriage return: **1.** A machine function that (a) controls horizontal movement of the carriage of typewriter, teletypewriter, and teleprinter equipment by moving it to the left margin of the paper and (b) is accomplished by (i) pressing the carriage return (CR) character key or the ENTER key on a keyboard or (ii) reading the corresponding perforation in punched paper tape. **2.** The return to the starting position of a line on a page.

carriage return (CR) character: A format effector that causes the print or display position to move to the first position on the current or next line of print. *See also* **format effector.**

carriage return key: On a typewriter, teletypewriter, or teleprinter, a key that returns the paper carrier or the type carrier to the first position on the current or next printing line.

carrier: 1. An electromagnetic wave, such as a sine wave or a pulse train, suitable for modulation by an information-bearing signal that is to be transmitted over a communications system. *Note:* Types of carrier modulation include frequency, phase, amplitude, pulse code, and wavelength. A carrier wave usually has a particular frequency, and may be a radio wave, lightwave, microwave, or sound wave. **2.** An unmodulated emission. *Note:* The carrier (cxr) usually is a sinusoidal electromagnetic wave or a series of electromagnetic pulses. **3.** An organization that provides a transport service to the public, such as a railroad company, an airline company, a steamship company, a communications company, or a trucking company. *Synonyms* **carrier system, carrier wave.** *See* **charge carrier, commercial communications common carrier, common carrier, communications common carrier, continuous carrier, digital loop carrier, double sideband suppressed carrier, full carrier, local exchange carrier, other common carrier, pulse carrier, radio common carrier, reduced carrier, resale carrier, single sideband suppressed carrier, specialized common carrier, subcarrier, suppressed carrier, T carrier, telecommunications carrier, two-independent-sideband carrier, type carrier, value-added carrier, virtual carrier, wireline common carrier.**

carrier controlled approach (CCA): A communications and control system used (a) for directing the approach of aircraft to an aircraft carrier in search and rescue operations and military operations and (b) to assist in homing.

carrier dropout: A temporary loss of a carrier signal. *Note:* Some causes of carrier dropout are noise, maintenance activity, system malfunction, system degradation, and temporary loss of power.

carrier frequency: 1. The frequency of a carrier wave. **2.** The frequency of an unmodulated wave capable of being modulated with an information-bearing signal. *Note:* In frequency-modulated systems, the carrier frequency is also the center frequency. *In frequency modulation, synonym* **center frequency.** *See also* **carrier, center frequency, frequency spectrum designation, modulation.** *Refer to* **Fig. W-3 (1082).**

carrier frequency shift: *See* **subcarrier frequency shift.**

carrier frequency stability: A measure of the ability of a transmitter to maintain, i.e., to remain on, an assigned frequency. *See also* **assigned frequency.**

carrier hole: One of the holes in a vertical column of holes in the side margins of continuous-form paper in which tractor pins are inserted and used to control paper alignment, print line registration, and paper movement.

carrier interrupt signaling: *See* **digital carrier interrupt signaling.**

carrier leak: In a suppressed carrier transmission system, the carrier that remains after carrier suppression. *Note:* The carrier leak may be used as a reference for an automatic frequency control system. *See also* **carrier.**

carrier level: The level of a carrier signal at a specified point in a communications system. *Note:* The carrier level may be expressed (a) in dB (decibels) relative to a specified reference level or relative to a specified point in the system or (b) in absolute power units, such as kilowatts, watts, milliwatts, or microwatts. *See also* **carrier power, level.**

carrier multiplex: *Synonym* **frequency division multiplex.**

carrier noise level (CNL): The noise level that results from undesired variations of a carrier in the absence of any intended modulation. *Synonym* **residual modulation.** *See also* **carrier, level, noise.**

carrier pigeon: A homing pigeon trained to carry messages from a given site to its home location.

carrier power (PC): In a radio transmitter, the average power supplied to the antenna feeder when only the unmodulated carrier is transmitted. **2.** The average power supplied to the antenna feeder by a transmitter during one radio frequency cycle taken under the condition of no modulation. *Note:* These concepts do not apply to pulse modulation or frequency shift keying transmission systems. *See also* **carrier, carrier level, peak envelope power.**

carrier sense: *Synonym* **carrier sensing.**

carrier sense multiple access (CSMA): A network control feature in which a transmitter checks for a clear channel before transmitting. *See also* **carrier, code division multiple access, collision, local area network.**

carrier sense multiple access with collision avoidance (CSMA/CA): A protocol in which (a) carrier sense is used, (b) a data station that intends to transmit sends a jam signal, (c) after waiting a sufficient time for all stations to receive the jam signal, the data station transmits a frame, and (d) while transmitting, if the data station detects a jam signal from another station, it stops transmitting for a designated time and then tries again. *See also* **local area network.**

carrier sense multiple access with collision detection (CSMA/CD): A protocol in which (a) carrier sense is used, and (b) a transmitting data station that detects another signal while transmitting a frame stops transmitting that frame, transmits a jam signal, and then waits for a specified time interval before trying to send that frame again. *See also* **local area network.**

carrier sensing: In a local area network (LAN), an ongoing activity of a data station to detect whether another station is transmitting. *Synonym* **carrier sense.** *See also* **local area network.**

carrier shift: 1. In the transmission of binary or teletypewriter signals, keying in which the carrier frequency is changed in one direction for marking signals and in the opposite direction for spacing signals. **2.** In amplitude modulation, a condition that (a) results from imperfect modulation such that the positive and negative excursions of the modulating envelope are unequal in amplitude, (b) results in a change in the power associated with the carrier, and (c) may be positive or negative, i.e., an increased or a decreased frequency. *See also* **carrier, frequency, frequency shift keying, modulation.**

carrier single sideband emission: *See* **reduced carrier single sideband emission.**

carrier suppression: *See* **suppressed carrier transmission.**

carrier synchronization: In a radio receiver, the generation of a reference carrier that has a phase closely matched to that of a received signal. *See also* **carrier, phase locked loop, synchronization.**

carrier system (CS): A multichannel telecommunications arrangement in which a number of individual data or voice circuits are multiplexed for use in transmission between the nodes of a network. *Note 1:* In carrier systems, many different forms of multiplexing are used, such as time division and frequency division multiplexing. *Note 2:* Multiple layers of multiplexing may be imposed on the same carrier. *See also* **channel, multiplexing, T carrier.** *Refer to* **Table 10, Appendix B.**

carrier to noise (C/N or CNR) power ratio: The ratio of (a) the power of the carrier wave to (b) the power of the noise measured at a receiver of defined bandwidth before any nonlinear signal processing is performed. *Note 1:* The carrier to noise (C/N) power ratio usually is expressed in dB. *Note 2:* Nonlinear signal processing includes (a) amplitude limiting, i.e., clipping, and (b) detection, i.e., rectification and demodulation. In lightwave communications systems, noise is introduced by light sources, coupling with adjacent channels, the propagation media, and various sources of interference.

carrier to noise ratio (C/N or CNR): 1: In radio receivers, the ratio of (a) the amplitude of the carrier to (b) the amplitude of the noise in the intermediate frequency (IF) bandwidth before any nonlinear process occurs, such as amplitude limitation and detection. **2.** The ratio of the received signal level to the receiver noise level. *Note 1:* If power values are used, 10 times the logarithm to the base 10 of the carrier to noise ratio (C/N or CNR) converts the ratio to dB. If current or voltage values are used, 20 times the logarithm of C/N converts the ratio to dB. *Note 2:* The carrier to noise ratio (CNR) usually is expressed in dB. *See also* **carrier, signal to noise ratio.**

carrier to receiver noise density (*C/kT*): In satellite communications systems, the ratio of (a) the received carrier power to (b) the receiver noise power density. *Note 1:* The carrier to receiver noise power density is usually expressed in dB (decibels). *Note 2:* The carrier to receiver noise density is given by the relation C/kT where C is the received carrier power in watts, k is Boltzmann's constant in joules per kelvin, and T is the receiver system noise temperature in kelvins. *Note 3:* The receiver noise power density, kT, is the receiver noise power per hertz. *See also* **carrier, signal to noise ratio.**

carrier transmission: *See* **full carrier transmission, reduced carrier transmission, suppressed carrier transmission.**

carrier wave: *See:* **carrier.**

carry: In arithmetic operations performed in positional numeration systems, to forward a digit produced at one digit position to another digit position for processing there, or the digit itself. *Note:* A carry usually occurs when the sum of digits in a column exceeds the value of the radix minus 1 of the numeration system, such as exceeds 9 in the decimal numeration system. The digit is transferred to the digit position with the next higher weight for processing there. *See also* **numeration system, positional numeration system.**

CARS: cable television relay service.

Carson bandwidth: The transmission bandwidth required for a frequency-modulated signal as determined by the Carson bandwidth rule. *See also* **Carson bandwidth rule.**

Carson bandwidth rule: A rule that (a) defines the approximate bandwidth requirements of communications system components for a carrier signal that is frequency-modulated by a continuous or broad spectrum of frequencies rather than a single frequency and (b) is given by the relation $BW = 2(\Delta f + f_m)$ where BW is the required bandwidth, Δf is the carrier peak deviation frequency, and f_m is the highest modulating frequency. *Note 1:* Bandwidth requirements for modulated carriers and equipment capabilities impose limits on the extent of multiplexing that can be accommodated. *Note 2:* The Carson bandwidth rule is often applied to communications systems components, such as transmitters, antennas, light sources, receivers, and photodetectors.

Cartesian lens: A lens with one surface that is a Cartesian oval, which thus can be an aplanatic lens. *See also* **aplanatic lens.**

cartridge: In communications, computer, and data processing systems, a container that holds a material until it is drawn out and used. *Note:* An example of a cartridge is a plastic container that holds a reel of magnetic tape or carbon ribbon. *See also* **cassette.**

cartridge drive: A device that (a) usually is used in conjunction with personal computers (PCs) to provide additional storage area and (b) consists of a driving mechanism and a control device. *Note:* A cartridge drive with an inserted cartridge in a personal computer (PC) might provide storage capacity equivalent to over 30 high density double-sided diskettes. *See also* **hard disk, hard drive, diskette.**

CAS: centralized attendant service.

cascading: The forming of a linear string of entities, each entity interacting only with its neighbors. *Note:* Examples of cascading are forming (a) a single sequence of nodes forming a one-path network, (b) a series of amplifiers with each output connected to the input of another, except the last, and (c) a series of circuits connected end to end. *Synonym* **concatenation.** *See also* **tandem.**

CASE: computer aided software engineering, computer assisted software engineering.

case dispersion coefficient: *See* **worst-case dispersion coefficient.**

case shift: 1. In data equipment, the change from letters to other characters, or vice versa. **2.** In typewriting, typesetting, or keyboard operation, the change from lower case letters to upper case letters, or vice versa. **3.** The changeover of the mechanism of telegraph receiving equipment from letter case to figure case, or vice versa.

cash: *See* **digital cash.**

Cassegrain: *See* **uniform illumination Cassegrain.**

Cassegrain antenna: An antenna in which (a) the feed radiator is mounted at or near the surface of a concave main reflector and is aimed at a convex secondary reflector slightly inside the focus of the main reflector, (b) energy from the feed unit illuminates the secondary reflector which reflects it back to the main reflector, which then forms the desired forward beam, (c) operation and technology are adapted from optical telescope technology, and (d) the feed radiator is easily supported. *See also* **antenna.**

cassette: A device that (a) consists of two reels that can be driven by a spindle for repeated payout and take-up of tape or film, (b) can be used for recording and playback of analog or digital data, (c) has tape or film that can be moved, in both directions, past an external read/record head, (d) can be reversed in direction, i.e., played and rewound, and (e) may be placed in a record/playback device. *See* **magnetic tape cassette.** *See also* **cartridge.**

catastrophic degradation: The rapid reduction of the ability of a system, subsystem, component, equipment, or software to perform its intended function, usually total failure to perform any function. *Note:* An example of catastrophic degradation is the sudden reduction of optical power of a laser caused by facet failure when intense fields bring about local disassociation of material to the point of fracturing or cracking. Catastrophic degradation can occur in an entire system, including entire communications, computer, data processing, control, and power systems, and usually results in shutdown of the system. *See also* **graceful degradation.**

category: *See* **access category, emergency category.**

Category 3 cable: The ANSI/EIA/TIA-568 designation for 100-Ω (ohm) unshielded twisted-pair cables, and associated connector hardware, that are specified for data transmission up to 16 Mbps (megabits per second).

Category 4 cable: The ANSI/EIA/TIA-568 designation for 100-Ω (ohm) unshielded twisted-pair cables, and associated connector hardware, that are specified for data transmission up to 20 Mbps (megabits per second).

Category 5 cable: The ANSI/EIA/TIA-568 designation for 100-Ω (ohm) unshielded twisted-pair cables, and associated connector hardware, that are specified for data transmission up to 100 Mbps (megabits per second).

catenation: *Synonym* **tandem.**

catheter: *See* **fiber optic catheter.**

cathode: An emitter of negatively charged particles, i.e., a source of electrons or negative ions. *Note:* Examples of cathodes are the heated filaments in vacuum tubes, photomultipliers, and cathode ray tubes (CRTs) and the negative terminal of batteries. *Note:* Conventional positive current direction is opposite to the flow of negative particles, i.e., from the anode to the cathode through the inside of the device, such as a vacuum tube, cathode ray tube (CRT), photomultiplier, or battery. Electron flow is from the cathode to the anode. For a battery, the positive direction of external current is from the anode, i.e., from the positive terminal, through the external circuit being driven, and back to the negative terminal, which usually is grounded. *See also* **anode.**

cathode ray tube (CRT): A vacuum tube in which (a) an electron gun produces a concentrated stream of high speed electrons, i.e., an electron beam or "cathode ray," that impinges on an inside surface with a phosphorescent coating that can be viewed from the outside, (b) the electron bombardment of the surface produces a spot of light at the point of impact of the beam on the surface, (c) the intensity, i.e., the brightness, of the spot of light is controlled by regulating the electron flow rate, and (d) deflection of the beam is achieved by passing the beam through a magnetic field or an electric field that can be varied in accordance with an input signal so that displays can be produced on the phosphorescent screen in display devices such as are used in personal computer monitors and radar equipment.

cathode ray tube (CRT) storage: A cathode ray tube (CRT) used as a storage device by placing charged spots on the inner surface of the tube with an electron beam to store data and interrogating the spots with the electron beam to read the data. *Note:* Cathode ray tube (CRT) storage is volatile storage. *See also* **volatile storage.**

CATV: cable television, community antenna television.

cavity: A volume that (a) is defined by conductor-dielectric or dielectric-dielectric reflective boundaries, or a combination of both, (b) has dimensions that produce specific interference effects, such as constructive and destructive interference effects, when excited by an electromagnetic wave of appropriate wavelength, and (c) is used as a source of high-powered electromagnetic radiation with specific characteristics, such as narrow spectral widths and narrow beams. *See also* **optical cavity, resonant cavity, selective cavity.**

cavity combiner: A combiner that (a) has one or more resonant bandpass selective cavities that are used to combine two or more frequencies and (b) is placed in the transmission lines leading from a number of transmitters to a single antenna. *See also* **combiner, resonant cavity, selective cavity.**

cavity mirror sensor: *See* **optical cavity mirror sensor.**

C band: *Obsolete. Note 1:* C band was used to designate the frequency band between 4 GHz (gigahertz) and 6 GHz used in satellite communications. *Note 2:* Radio frequency bands should be specified by citing the lower and upper limits of the band. *Note 3:* Letter designators of radio frequency bands are imprecise, deprecated, and legally obsolete.

CBR: Carson bandwidth rule.

CBS: corps battle simulation.

CBX: computerized branch exchange.

cc: cubic centimeter.

CCA: carrier controlled approach.

CCD: charge coupled device.

CCH: connections per circuit-hour, hundred circuit-hours.

CCI: character count integrity.

CCIR: International Radio Consultative Committee.

CCIS: common channel interoffice signaling.

CCITT: International Telegraph and Telephone Consultative Committee, a predecessor organization of the ITU-T.

CCS: hundred call-seconds.

CCSA: common control switching arrangement.

CDF: combined distribution frame.

CDM: color division multiplexing.

CDMA: code division multiple access.

CDPSK: coherent differential phase-shift keying.

CDR: call detail recording.

CD-ROM: compact disk read only memory.

C-E: communications-electronics.

CE: communications electronics.

CEI: comparably efficient interconnection.

ceilometer: In air-ground communications, an optoelectronic instrument used to measure by triangulation the height of the cloud base, i.e., the ceiling, above the surface of the Earth. *Note:* Low ceilings reduce pilot visibility, requiring that radio landing aids, such as instrument landing systems (ILS) and carrier controlled approach (CCA) systems, be used almost all the way to the ground runway or aircraft carrier deck, at least until runway markers become visible. Most major airports and large aircraft are equipped with ground controlled approach (GCA) landing equipment for use in low ceilings. Most airports close at ceiling zero. Different protocols are used for landing during different ceilings. *See also* **ceilometry.**

ceilometry: In air-ground communications, the study of the height of a cloud base, i.e., the ceiling, above the surface of the Earth. *Note:* Ceilometry usually is accomplished with an optoelectronic instrument to measure the cloud base by triangulation. *See also* **ceilometer.**

celestial equator: The great circle of the celestial sphere that (a) is in a plane perpendicular to the Earth's axis of rotation, (b) divides the celestial sphere into Northern and Southern Hemispheres, and (c) lies in the same plane as the Earth's equatorial plane.

celestial latitude: The angle measured north or south from the ecliptic along a secondary great circle perpendicular to it. *See also* **ecliptic, great circle.**

celestial longitude: The angle measured from the vernal equinox along the ecliptic to the secondary great circle passing though the point at which the longitude is being determined. *See also* **ecliptic, great circle, vernal equinox.**

celestial mechanics: The branch of astronomy that is devoted to the study of (a) the translational, rotational, and deformational motions of bodies under the influence of forces, such as gravitational forces, forces caused by the resistance of media to the motion of bodies, and light pressure, (b) the motion of artificial celestial bodies in unpowered phases of flight, and (c) with special emphasis, the translational motion of celestial bodies, such as stars, planets, satellites, and meteors under mutual attraction according to the law of gravitation.

celestial meridian: The great circle on the celestial sphere that (a) passes through the north and south points, the observer's zenith and nadir, and the celestial poles and (b) lies in a plane that is parallel to that of the Earth's meridian that passes through the observation points on the Earth. *See also* **celestial pole, celestial sphere.**

celestial pole: One of the points on the celestial sphere at which it is intersected by the celestial axis. Consequently, there are a north and a south celestial pole. *See also* **celestial sphere.**

celestial sphere: In celestial mechanics, a sphere that (a) is of near-infinite radius, (b) has a point of observation at its center, (c) has a radius so large that any point on Earth may serve as its center, (d) has a maximum radius approaching that of the known universe, and (e) has a minimum radius equal to the distance to the nearest star. *See also* **celestial mechanics.**

cell: 1. In cellular telephone systems, a specified geographic area that (a) is defined for a specific mobile communications system and (b) has its own base station and a single controller interconnected with the public telephone network. **2.** In communications, a string of bits consisting of a header string and an information string. *Note 1:* A cell is dedicated to one user for one session. Cells for a given system usually are of fixed length and smaller than a frame, such as 424 bits for a cell compared to 1024 bits for a frame. *Note 2:* In asynchronous transfer mode (ATM) systems, a cell consists of 53 bytes (5-byte header field and 48-byte information field). The 5-byte headers for the user-network interface (UNI) and network-network interface (NNI) have (a) a 4-bit general flow control (GFC) field to control the flow of information in the cell for different qualities of service (only for UNI), (b) 8 to 12 bits (UNI) or 12 bits (NNI) virtual path identifier (VPI) for explicit path identification, (c) 12 to 16 bits (UNI) or 16 bits (NNI) virtual channel identifier (VCI) for explicit channel identification (VPI and VCI total bits at UNI is 24, but only 20 are active on a given UNI), (d) a 2-bit payload type (PT) to indicate whether the cell contains user or network information, (e) a 4-bit reserved (RES) field, and (f) an 8-bit header error check (HEC) for limited header error correction. *Note 3:* A cell does not have error correction capability and is therefore suited for low error communications systems, such as digital fiber optic systems. **3.** In the Open Systems Interconnection—Reference Model (OSI-RM), a fixed-length block labeled at the Physical Layer. **4.** In computer systems, an addressable, internal hardware location. **5.** In computer systems, a single location on a spreadsheet. *Note:* An example of a cell in Lotus 1-2-3 is the cell, designated by M18, that lies in row 18 and column M of the spreadsheet. *See* **binary cell, function driven cell, index matching cell, Kerr cell, photoemissive cell, photovoltaic cell, Pockels**

cell, refractive index matching cell, radar resolution cell, storage cell.

cell address: The symbols, usually letters, numerals, or a combination of letters and numerals, used to designate a particular cell, such as a cellular radio cell or a spreadsheet cell. *See* **absolute cell address, mixed cell address, relative cell address.** *See also* **cell.**

cellar: *Synonym* **push down storage.**

cell pointer: In personal computer spreadsheet software systems, such as Lotus 1-2-3, the reverse video bar that can be moved to a location, i.e., a cell, on a worksheet, i.e., a spreadsheet, to indicate the cell that is to be affected by the next operation. *See also* **cell, location, reverse video bar.**

cell relay: A statistically multiplexed interface protocol for packet switched data communications that (a) uses fixed sized packets, i.e., cells, to transport user packets, (b) has transmission rates that are between (i) 56 kb/s (kilobits per second) to 64 kbps and (ii) 1.544 Mbps (megabits per second), i.e., the T-1/DS1 rate, (c) has neither flow control nor error correction capability, (d) is information-content-independent, (e) corresponds only to Level 1 and Level 2 of the Open System Interconnection—Reference Model (OSI-RM), (f) encloses variable-sized user packets within fixed-sized packets, i.e., cells, that add addressing and verification information, (g) has a frame size that is fixed in hardware and based on time delay and user packet size considerations, (h) has a user packet that may be segmented over many cells, (i) is an implementation of fast packet technology that is used in connection-oriented broadband integrated services digital networks (B-ISDN) and connectionless IEEE 802.6, SMDS networks, and (j) is used for time-sensitive traffic, such as voice and video traffic. *See also* **cell, fast packet switching, frame relay.**

cellular mobile service: A communications service that (a) allows a user to access the telephone network from a stationary or moving vehicle, (b) is based on a combination of radio transmission and telephone switching, (c) provides a communications link to the user by segmenting a large geographical area into many smaller areas, i.e., into cells, each of which has its own base station and a single controller interconnected with the public switched telephone network, and (d) allows a call in progress to be handed over without interruption to the adjacent base station as the user passes from cell to cell. *Note 1:* The same frequency may be used in noncontiguous cells because transmitted power levels and signal directions are controlled. *Note 2:* In the United States, a total bandwidth of 50 MHz (megahertz) is allocated to the cellular mobile service, the 50 MHz distributed (a) between 824 MHz and 849 MHz and (b) between 869 MHz and 894 MHz of the frequency spectrum. *See also* **cellular telephone, frequency spectrum designation.**

cellular phone: *Synonym* **cellular telephone.**

cellular radio: *Synonym* **cellular telephone.**

cellular radiotelephone: *Synonym* **cellular telephone.**

cellular telephone: Based on a combination of radio transmission and telephone switching, a communications system that permits telephone communications to and from mobile users within a specified area. *Note 1:* In cellular telephone systems, large geographical areas are segmented into many smaller areas, i.e., cells, each of which has its own base station and a single controller interconnected with the public switched telephone network. The same frequency may be used in noncontiguous cells because transmitted power levels and signal directions are controlled. *Note 2:* In the United States, a total bandwidth of 50 MHz is allocated to the cellular mobile service, the 50 MHz distributed (a) between 824 MHz and 849 MHz and (b) between 869 MHz and 894 MHz of the frequency spectrum. *See also* **cellular mobile service, frequency spectrum designation.** *Synonyms* **cellular phone, cellular radio, cellular radiotelephone.**

CELP: code excited linear prediction.

Celsius temperature scale: A temperature scale in which the freezing point of water is set at 0°C (degrees Celsius) and the boiling point of water at 100°C, both at sea level. *Note:* Fahrenheit temperatures may be converted to Celsius temperatures by the relation $C = 5(F - 32)/9$ where F is the Fahrenheit temperature. Both scales have the same numerical value of temperature at −40°. *See also* **Celsius temperature scale, Kelvin temperature scale, temperature.**

cement: An adhesive used to bond items together. *Note:* An example of cement is optical cement used to bond optical elements to each other or to holding devices, fixtures, or frames. The three general types of cement used in the optical industry are blocking cements, mounting cements, and optical cements. *See* **blocking cement, mounting cement, optical cement, thermoplastic cement, thermosetting cement.**

cemented doublet: A compound lens that consists of two lenses joined together with optical cement. *See also* **compound lens, optical cement.**

center: *See* **alarm center, circuit switching center, cladding center, communications center, core center, filter center, gateway switching center, internal message center, National Coordinating Center, network control center, nodal switching center, processing center, reference surface center, regional center, scattering center, signal center, signal communications center, storage center, store and forward switching center, switching center, tandem center, telecommunications center, test center.**

center frequency: 1. In frequency modulation, the rest frequency, i.e., the frequency of the unmodulated carrier. **2.** In facsimile systems, the frequency midway between the picture black and the picture white frequencies, i.e., the arithmetic mean of the picture black and picture white frequencies. *See also* **assigned frequency, carrier, carrier frequency, facsimile, frequency, modulation.**

center sampling: Sampling a digital data stream at the center of each signal element.

centiliter (cl): A unit of volumetric measure equal to one-hundredth of a liter, i.e., 10^{-2} l (liter), equivalent to about 0.338 fluid ounces. *See also* **milliliter.**

centimetric wave: A wave, usually an electromagnetic wave, with a wavelength between 1 cm (centimeter) and 10 cm, corresponding to a frequency range between 3 GHz (gigahertz) and 30 GHz, i.e., an electromagnetic wave in the superhigh frequency (SHF) range. *See also* **frequency spectrum designation.**

central battery exchange: *Synonym* **common battery exchange.**

central computer: *Synonym* **host computer.**

centralized attendance: The monitoring and controlling of a number of local or remote switches in a communications system by a centrally located attendant console.

centralized attendant service (CAS): In communications systems, a service provided by an attendant who serves many switches. *Note:* The attendant console is usually centrally located, and the switches may be remote.

centralized automatic message accounting (CAMA): An automatic message accounting system that serves more than one switch from a central location. *Note:* When using centralized automatic message accounting (CAMA), human intervention may be required. *See also* **automatic message accounting.**

centralized computer network: A computer network in which all of the data processing capability required by users is located at one node in the network.

centralized operation: Operation of a communications network in which transmission may occur between the control station and any tributary station, but not between tributary stations. *See also* **communications, distributed control, distributed switching, network, system signaling and supervision.**

centralized ordering group (COG): An organization provided by some communications service providers to coordinate services between the customer companies and the vendors.

central office (C.O.): A common carrier switching center in which trunks and loops are terminated and switched. *Note:* In the U.S. Department of Defense (DoD) "common carrier" is called "commercial carrier." *Synonyms* **exchange, local central office, local exchange, local office, switch, switching center (except in DoD AUTOVON usage), switching exchange, telephone exchange.** *See also* **regional switching center, tandem exchange, tandem office, tandem switch, trunk.**

central office connecting facility: A connection facility that (a) is between a central office (C.O.) and a private branch exchange (PBX) and (b) is not between central offices. *Synonym* **central office trunk.** *See also* **trunk.**

central office side: *Synonym* **equipment side.**

central office trunk: *Synonym* **central office connecting facility.**

central primary control: The selection and the management of the control station of a communications network from a given single node in the network.

central processing unit (CPU): 1. The portion of a computer that includes circuits that control the interpretation and the execution of instructions. **2.** The portion of a computer that executes programmed instructions, performs arithmetic and logical functions on data, and controls input/output functions. *Note:* CPU is also used as an abbreviation for "communications processor unit." *Synonyms* **central processor, main frame.** *See also* **communications processor unit, computer, multiprocessing.**

central processor: *Synonym* **central processing unit.**

central strength member fiber optic cable: A cable that (a) contains optical fibers wrapped around a high tensile strength material, such as stranded steel, Kevlar, and Nylon, and (b) has a crush-resistant jacket wrapped or extruded around the outside. *See also* **peripheral strength member fiber optic cable.**

central wavelength: In optical communications, the wavelength at which the transmitted power is considered to be concentrated, usually determined by a peak mode power or a power weighted measurement. *See* **nominal central wavelength.** *See also* **peak wavelength.**

central wavelength measurement: For the transmitter terminal/regenerator facility of a fiber optic station/regenerator section, the measurement of the wavelength that identifies where the effective optical power of the transmitter resides using either the peak mode method or the power weighted method.

central wavelength range: *See* **transmitter central wavelength range.**

Centrex® (CTX) service: A service offered by Bell Operating Companies that provides functions and features comparable to those provided by a private branch exchange (PBX) or a private automatic branch exchange (PABX). *Note:* "Centrex® C.O." indicates that all equipment except the attendant position and station equipment is located on the central office (C.O.) premises. "Centrex® C.U." indicates that all equipment, including the dial switching equipment, is located on the customer premises.

certification: For an automated information system, a comprehensive evaluation of the technical and nontechnical security features and safeguards made in support of the accreditation process that establishes the extent to which a particular design and implementation meets a specified set of security requirements.

cesium clock: A clock that contains a cesium standard as a frequency-determining element. *See also* **cesium standard, coordinated clock, Coordinated Universal Time, DoD master clock, precise frequency, precise time, primary frequency standard, primary time standard, second.**

cesium clock second: *See* **second.**

cesium standard: A primary frequency standard in which a specified hyperfine transition of cesium-133 atoms is used to control the output frequency of a clock. *Note:* The accuracy and the precision of the cesium standard are intrinsic and achieved without calibration. *See also* **cesium clock, Coordinated Universal Time, frequency, primary frequency standard, primary time standard, second.**

CFAR: constant false alarm rate.

chad: The material separated from a punched tape or a punched card in forming a hole. *See also* **reperforator.**

chadless perforation: Perforation of a tape in which the cuts are only partial and do not form a hole, the chads or cuttings remaining attached, i.e., remaining hinged, to the tape.

chadless tape: 1. Punched tape that has been punched in such a way that chad is not formed. **2.** A punched tape in which only partial perforation is performed so that the chad remains attached to the tape. *Note:* The use of chadless tape is a deliberate process and should not be confused with imperfect production of chad. *Note:* Chadless tape can be read by a mechanical perforated tape reader but not by an optical perforated tape reader. *See also* **reperforator, tape relay.**

chad tape: Punched tape that (a) is used in telegraph and teletypewriter operations, (b) is punched with hole patterns to represent data, usually in transverse rows to represent characters sequentially in the longitudinal direction, (c) is punched in such a manner that chad is formed, and (d) usually does not have printed characters on it. *See also* **reperforator, tape relay.**

chaff: A resonant piece of metal, or a substrate with an electrically conductive coating, used to reradiate, i.e., to reflect, a radar pulse and thereby create a radar echo. *Note:* Reflection is the highest when the chaff length is one-half the radar wavelength. Chaff is used to screen objects, i.e., screen targets, confuse search radar, and cause radar to break lock-on and lock on the chaff rather than a desired target. *See also* **lock-on, radar echo.**

chain: 1. In communications, computer, and data processing systems, the sequence of binary digits (bits) used to generate a chain code. **2.** *Synonym* **tandem.** *See* **breakout chain, Markov chain, visual responsibility chain.** *See also* **chain code, tandem.**

chain code: A code derived from a given binary digit sequence in such a fashion that the code groups are related to their neighbors and are based on the given sequence. *Note:* An example of a chain code is the code set derived by taking a binary digit sequence and deriving *n*-bit bytes by taking *n*-bits at a time from the sequence, each time moving, i.e., displacing, one bit position to the right. If the given binary digit sequence is 011010, the codes would be 011, 110, 101, 010, 100, and 001. All the codes are different, and the combination 000 does not occur. If 000 did occur, it could be interpreted as an error in an error detection scheme. Chain codes usually are made with unit distance

Decimal	00010111 Chain Code
0	000
1	001
2	010
3	101
4	011
5	111
6	110
7	100

Fig. C-2. A three-bit **chain code** for a given binary sequence and possible octal digit equivalents. The codes could be used to gate spread spectrum frequencies or scramble sequences of digits that can be unscrambled using the same sequence.

codes. *See also* **code groups, n-bit byte, unit distance code.** *Refer to* **Fig. C-2.**

chained list: A list of data items in which each item contains an identifier for locating the next item in the list, in which case the items may be scattered in storage and still be found. *See also* **identifier.**

chaining: *See* **daisy chaining.**

chain search file: A file in which each item in the file contains a means for locating the next item to be considered when searching the file.

challenge: To ascertain the identity of a person, place, or thing, by citing a code that should trigger a specific coded response, i.e., an appropriate reply. *Note:* If the challenge and reply combination is secret, the combination can be used to establish the relationship between the challenger and the challenged entity. The challenge and reply can be easily compromised by eavesdropping. *See also* **challenge and reply authentication.**

challenge and reply: 1. A prearranged procedure in which one station requests authentication of another station, i.e., the challenge, and the latter station, by a proper reply, establishes its authenticity. **2.** In communications, computer, data processing, and control systems, calling in which (a) the calling station identifies itself, i.e., the challenge, and requests the identity of another station, i.e., the called station, and (b) the called station identifies itself, i.e., the reply, in accordance with a prearranged code in which the reply must match the challenge.

challenge and reply authentication: In the transmission of messages, a prearranged procedure in which (a) a message originator sends a request for authentication, i.e., a challenge, to a potential message receiver, and (b) the proper reply to the challenge establishes the validity or the identity of the receiver.

change: *See* **nonreturn to zero change (NRZ-C).**

changeable storage: A storage unit in which the user or the operator can remove parts containing data from the unit and replace them by parts containing other data or no data. *Note:* Examples of changeable storage units are (a) the storage units of personal computers (PCs) in which the diskettes can be removed from the computers and replaced by others, (b) a magnetic tape storage unit in which reels of tape in magnetic tape cartridges can be changed, and (c) reels of magnetic tape that can be changed on magnetic tape drives. From the point of view of the user or the operator, the hard disk of a PC is not changeable. From the point of view of the computer maintenance shop person or factory repair person, the hard disk can be replaced, but this is not considered changeable storage because the purpose is to repair rather than to remove or store data. *See also* **diskette, magnetic tape cartridges.**

change coefficient: *See* **phase change coefficient.**

change control: The formal control of changes to a system, or its specification, such as changes that are proposed, evaluated, approved, rejected, scheduled, tracked, and effected.

changed address interception: The interception, usually by automatic means, of a call to an obsolete address, i.e., obsolete number, in which the call originator data terminal equipment (DTE) is advised of the new address, followed by either call redirection to the new address or by release of the call originator DTE.

change dump: A dump of only those storage locations in which changes to their contents have been made. *See also* **dump.**

change in operational control (CHOP): In the course of a voyage, flight, or vehicular trip, control of (a) the transfer of a ship, aircraft, or vehicle from one operational control authority (OCA) to another, (b) the shift in radio guard that may take place at the same time, and (c) the change and shift times, usually according to prescribed sailing or departure orders, that enable the new OCA to communicate with any ship, aircraft, or vehicle at any time. *See also* **radio guard.**

change in transmittance: *Synonym* **induced attenuation.**

changer: *See* **gender changer.**

channel: **1.** A connection between initiating and terminating nodes of a circuit. **2.** A single path provided by a propagation medium either by physical separation, such as (a) separation by electrical insulation in a multiwire cable or (b) separation by other means, such as separation by frequency division or time division multiplexing. **3.** A single unidirectional or bidirectional path for transmitting and receiving electrical or electromagnetic signals. *Note:* Each channel is usually distinguished from other parallel paths. **4.** A specified radio or television (TV) frequency usually used in conjunction with a predetermined letter, number, or codeword to identify the radio or TV frequency. **5.** A path along which signals may be sent, such as a data channel or an output channel. **6.** The portion of a storage medium, such as a track or a band, that is accessible to a given reading or writing station or head. **7.** The part of a communications system that connects the message source to the message sink. **8.** In data transmission, a means of performing simplex transmission in one preassigned direction. **9.** A connection between two nodes in a network. **10.** A complete facility for telecommunications on a system or circuit, such as a television channel, radio channel, or facsimile channel. *Note:* The number of independent channels available in a system depends on the number of separate communications facilities that can be provided by them. *See also* **adjacent channel, analog data channel, asynchronous data transmission channel, auxiliary channel, backward channel, clear channel, data channel, data transmission channel, D channel, deep channel, digital data channel, drop channel, engineering channel, forward channel, frequency-derived channel, idle channel, information bearer channel, information channel, logical channel, network control channel, one-way-only channel, positioned channel, primary channel, radio channel, rescue control channel, restricted channel, secondary channel, status channel, subvoice grade channel, symmetrical channel, symmetric binary channel, synchronous data transmission channel, telegraph channel, time-derived channel, transmission channel, transmission service channel, voice frequency channel, wideband channel.** *Refer to* **Fig. S-21 (967).** *Refer to* **Tables 6 and 7, Appendix B.**

channel allocation: *See* **adaptive channel allocation.**

channel allotment: *See* **radio frequency channel allotment.**

channel-associated signaling: Signaling in which the signals necessary for controlling the traffic carried by a single channel are transmitted in the channel itself or in a signaling channel permanently associated with it. *See also* **channel, common channel signaling, inband signaling, out-of-band signaling, signal.**

channel bank: The part of a carrier multiplex terminal that performs the first step of modulation by multiplexing a group of channels into a higher bandwidth analog channel or higher bit rate digital channel and, conversely, demultiplexes these aggregates back into individual channels. *Note:* D1, D1A, D2, . . . D*n* are often used to denote individual configurations of digital channel banks containing A-D converters. *See also* **channel, common channel signaling, group, multiplexing, wideband.**

channel basic group: In frequency division multiplexed wideband systems, a group of channels, usually 12 channels in the United States and Canada. *Synonym* **base group.**

channel bundle cable: *See* **multichannel bundle cable.**

channel cable: *See* **multichannel cable.**

channel capacity: The maximum possible data transfer rate through a channel, subject to specified constraints. *Note:* Channel capacity usually is measured as the ability of a given channel, subject to constraints, to transmit messages from a specified message source. It is expressed either at the maximum possible mean transinformation content per character or as the maximum possible average transinformation rate that can be achieved. An arbitrarily small probability of error is allowed, and an appropriate code is used. *See also* **average transinformation content, channel, mean transinformation content.**

channel gate: A device that connects a channel to a highway, or a highway to a channel, at specified times. *See also* **channel, highway.**

channel group: One of the assemblies of a number of channels. *Note:* Examples of channel groups are basic groups, supergroups, master groups, and jumbo groups.

channel increment: *See* **radio frequency channel increment.**

channeling equipment: *Synonym* **channel translation equipment.**

channel interference: *See* **adjacent channel interference, cochannel interference.**

channel interoffice signaling: *See* **common channel interoffice signaling.**

channelization: Using a single wideband facility, i.e., a single high capacity facility, to create many relatively narrowband channels, i.e., many relatively lower capacity channels, by subdividing the frequency spectrum of the wideband facility. *See also* **bandwidth, channel, frequency division multiplexing, narrowband, time division multiplexing, wideband.**

channel jumbo group: In frequency division multiplexed wideband systems in the United States and Canada, a group of master groups, usually six 600-channel master groups. *See also* **master group.**

channel loading: *See* **idle channel loading.**

channel master group: In frequency division multiplexed wideband systems, a group of supergroups. In most communications systems in the United States and Canada, a master group is composed of ten 60-channel supergroups, i.e., 600 voice channels. Some master groups that are used in terrestrial telephone systems carry 300 or 900 channels, which are formed by multiplexing five or fifteen 60-channel supergroups, respectively. *See also* **supergroup.**

channel mode: *See* **satellite channel mode.**

channel noise: *See* **idle channel noise, total channel noise.**

channel noise level: 1. The ratio of (a) the channel noise level at any point in a transmission system to (b) an arbitrary level chosen as a reference. *Note 1:* The channel noise level usually is expressed in dBrn (decibels above reference noise), dBrnC (decibels above reference noise with C message weighting), or dBa (adjusted decibels). *Note 2:* Each unit used to measure channel noise level reflects a circuit noise reading of a specialized instrument designed to account for different interference effects that occur under specified conditions. **2.** The noise power density spectrum in the frequency range of interest. **3.** The average noise power in the frequency range of interest. **4.** The indication, on a specified instrument, of the noise in a channel, the characteristics of the instrument being determined by the type of noise being measured and the use being made of the channel. *See also* **channel, circuit noise level, dBa, dBa0, dBm(psoph), dBm0, dBm0p, dBrn, dBrnC, level, noise, signal plus noise to noise ratio, signal to noise ratio.**

channel offset: The constant frequency difference between a channel frequency and a reference frequency which may frequency hop. *See also* **frequency hopping.**

channel operation: *See* **drop channel operation.**

channel packing: Maximizing the use of voice frequency channels for data transmission by multiplexing a number of lower data signaling rate (DSR) channels into a single DSR voice frequency channel. *See also* **channel, data signaling rate, digital multiplexer, multiplexing.**

channel patching jackfield: A patching jackfield that consists of a rack of telephone jack patch panels that facilitate the connection of user discrete channels to channeling equipment by jack and plug connections. *Note:* A channel patching jackfield may be used in Earth stations to (a) enable patching of traffic channels to multiplex equipment on a day-to-day basis for fault location purposes and (b) allow access to traffic paths for monitoring and test purposes.

channel reliability (ChR): The percentage of time a channel was available for use in a specified period of scheduled availability. *Note 1:* Channel reliability is given by the relation $ChR = 100(1 - (T_o/T_s)) = 100T_a/T_s$ where T_o is the channel total outage time, T_s is the channel total scheduled time, and T_a is the channel total available time. *Note 2:* $T_s = T_a + T_o$. *See also* **channel, circuit reliability.**

channel restoration: The repair or the reconnection of an existing channel.

channel selectivity: A measure of the extent to which a channel can be specifically selected from a group of channels by a given set of equipment. *See* **adjacent channel selectivity.**

channel service unit (CSU): A line bridging device that is the last signal regeneration point on the loop side, coming from the central office (C.O.), before the regenerated signal reaches a multiplexer or data terminal equipment (DTE). *Note:* The channel service unit (a) is used to perform loop back testing, (b) may perform bit stuffing, and (c) provides a framing and formatting pattern compatible with the network.

channel signaling: *See* **associated common channel signaling, common channel signaling.**

channel single bundle cable: *See* **single channel single bundle cable.**

channel single fiber cable: *See* **multichannel single fiber cable, single channel single fiber cable.**

channel supergroup: In frequency division multiplexed wideband systems, a group of basic groups, usually five 12-channel basic groups in the United States and Canada. *See* **group.**

channel supervisory signaling: *See* **separate channel supervisory signaling.**

channel-surf: To rapidly browse through many television channels looking for a desirable program.

channel time slot: A time slot that starts at a particular instant in a frame and is allocated to a channel for transmitting data, such as a character or an in-slot signal. *Note:* The type of channel time slot depends on the type of channel, such as a telephone channel time slot for multiplexing calls. *See also* **channel, digital multiplexer, multiplexing, time division multiple access, time division multiplexing.**

channel translation equipment: Equipment used for (a) the frequency translation of audio channels, (b) assembling the channels into channel groups, and (c) performing the reverse process, i.e., for demultiplexing into single audio channels. *See* **channel groups, frequency translation.** *Synonym* **channeling equipment.**

channel vocoder. *Synonym* **channel voice encoder.**

channel voice encoder: A voice encoding and decoding device that uses speech spectrum compression of up to 90% by separation of the speech frequency band into a number of subband channels by rectifying the signal in each channel. *Note:* The channel voice encoder direct current (dc) signal output represents a time-varying average signal for the subband. These outputs are multiplexed with a separate channel conveying the fine-structure characteristics of the original speech needed to reconstitute the original voice by decoding at the receiver. *See also* **bandwidth compression.** *Synonym* **channel vocoder.**

character: 1. A letter, digit, or other symbol that is used as part of the organization, control, or representation of data. **2.** One of the units of an alphabet. *Note 1:* In most communications systems, characters are represented as strings, i.e., as sequences of pulses or pulse trains that may be transmitted, stored, and retrieved as a unit. *Note 2:* Characters may be letters, digits, punctuation marks, control signals, special signs, and special symbols that may be represented in various forms, such as (a) a spacial arrangement of adjacent or connected strokes, (b) temporal or spatial sequences of electrical pulses, (c) patterns of magnetized spots, and (d) patterns of holes in cards and tapes. *See* **accuracy control character, acknowledge character, bell character, binary character, block cancel character, block check character, call control character, data link escape character, delete character, device control character, dummy character, editing character, end of medium character, end of selection character, end of text character, end of transmission character, end of transmission block character, enquiry character, escape character, facility request separator character, fill character, font change character, form feed character, gap character, graphic character, horizontal tabulation character, idle character, illegal character, inactive character, line feed character, negative acknowledge character, new line character, nonlocking code extension character, null character, print control character, repeating character, shift-in character, shift-out character, space character, special character, start of heading character, start of text character, start-stop character, stuffing character, substitute character, synchronous idle character, transmission control character, vertical tabulation character, who are you character, wildcard character.**

character check: In error detection systems, a check in which (a) preset rules are used for the formation of characters, and (b) characters that are not formed in accordance with the rules are considered to be erroneous. *See also* **character, cyclic redundancy check, error control, parity.**

character count integrity (CCI): The preservation of the exact number of characters that (a) were originated in a message in the case of message communications or (b) occur per unit time in the case of a user-to-user connection, in both cases whether the characters are correct or incorrect. *Note:* Character count integrity is not to be confused with character integrity, which requires that the characters delivered are, in fact, as they were originated. *See also* **added bit, binary digit, bit count integrity, bit error ratio, bit inversion, bit slip, character, deleted bit, digital error, error.**

character display device: A display device that displays data only in the form of characters. *See also* **display device.**

character-fill: To add meaningless characters to a data unit, such as a fixed-length word, block, or frame, solely for the purpose of achieving a specified total, such as a specified number of characters (a) stored in a given storage medium or frame, (b) transmitted in a given time frame, or (c) contained in a fixed-length string, when the meaningful data do not complete the data unit. *See also* **block, character, storage, string, word.**

character filter: Software that is capable of selectively removing characters from a data stream, such as software that removes communications control characters so that they will not be printed. *Note:* Some personal computer (PC) software packages allow the user to turn the filter on and off. *See also* **data stream, software package.**

character generator: **1.** A device that controls a display writer during the generation of a graphic character in the display space of a display device. **2.** A functional unit that converts the code for a graphic character into a character suitable for display in accordance with prescribed conventions. *See* **dot matrix character generator, stroke character generator.** *See also* **display writer.**

character integrity: **1.** Assurance that the characters delivered are, in fact, as they were originated. **2.** Preservation of a character during one or more operations, such as data processing, storage, and transmission operations. *See also* **character count integrity.**

character interval: In a communications system, the total number of unit intervals required to transmit a given character, (a) including intervals for synchronizing information, error checking, and control bits and (b) not including extra signals that are not associated with the individual character. *Note:* An extra signal that is excluded is a character added in the time interval between the end of the stop element and the beginning of the next start element for various reasons, such as for a change in the data signaling rate (DSR) and for buffering. This additional time is a part of the intercharacter interval. *See also* **character, intercharacter interval, start element, stop element, unit interval.**

characteristic: **1.** A quality, attribute, or value of an entity. **2.** The whole-number part of a logarithm. *Note:* An example of a characteristic for the decimal numeration system is the 2 in the decimal logarithm 2.5440. The .5440 is called the "mantissa." The number, i.e., the antilog, for the logarithm 2.5440 is $10^{2.544}$ = 349.945. *See* **fiber global attenuation rate characteristic, fiber global dispersion characteristic, halftone characteristic, loading characteristic, wavelength-dependent attenuation rate characteristic.**

characteristic distortion: Distortion that (a) is caused by signal transients that are present in the transmission channel as a result of modulation, (b) produces effects that are not consistent, (c) usually occurs during transition from one significant condition to another, and (d) is dependent upon the remnants of signal transients from previous signal elements. *See also* **cyclic distortion, data, distortion, signal element, significant condition.**

characteristic frequency: In a given emission or signal, a frequency that can be easily identified, measured, and associated with the emission or the signal. *Note:* An example of a characteristic frequency is a carrier frequency. *See also* **emission, frequency, reference frequency, signal.**

characteristic impedance (Z_o): **1.** At any point along a uniform transmission line, such as a coaxial cable, twin lead, and twisted pair, where there are no reflections present, the ratio of (a) the voltage across the line to (b) the current in the line. *Note 1:* In order that there be no reflections, the total impedance connected across the output end of the line must be the line characteristic impedance. This impedance consists of the electrical loads being driven by the line and impedance added, if any, to create the characteristic impedance. *Note 2:* The characteristic impedance of a uniform line can be calculated by measuring the input impedance when the line is short-circuited at the output end and when the line is open-circuited at the output end, and then taking the square root of the product of the measurements, i.e., $Z_o = (Z_{sc}Z_{oc})^{1/2}$. *Note 3:* When a uniform transmission line is terminated in its characteristic impedance, the line will appear, from the input end, to be infinitely long because there will be no reflected waves. However, the characteristic impedance is a function of frequency. *Note 4:* A line terminated in its characteristic impedance will have no standing waves, no reflections from the end, and a constant ratio of voltage to current at a given frequency at every point on the line. *Note 5:* If the transmission line is not uniform, the iterative impedance must be used. **2.** From Maxwell's equations, the impedance of a linear, homogeneous, isotropic, dielectric, and electric charge-free propagation medium, given by the relation $Z = (\mu/\varepsilon)^{1/2}$ where μ is the magnetic permeability and ε is the electric permittivity. *Note:* In free space, and approximately in air, $\mu = 4\pi \times 10^{-7}$ H/m (henries per meter) and $\varepsilon = (1/36\pi) \times 10^{-9}$ F/m (farads per meter), from which 120π, or 377 ohms, is obtained. For dielectric media, the permittivity is the dielectric constant, which for glass ranges from about 2 to 7. However, the refractive index is the square root of the dielectric constant. Thus, the characteristic impedance of dielectric materials is $377/n$ ohms, where n is the refractive index. **3.** In a uniform electromagnetic plane wave propagating in free space, or in dielectric propagation media, the ratio of (a) the electric field strength to (b) the magnetic field strengths, given by the relation $Z = E/H$ where E and H are orthogonal and are in a direction perpendicular to the direction of propagation, i.e., perpendicular to the Poynting vector, which is the vector obtained from the cross (vector) product of the electric and magnetic field vectors, with the direction of a right-hand screw obtained when rotating the electric vector into the magnetic vector over the smaller angle. *Note:* If E is expressed in volts per meter and H is expressed in oersteds, i.e., 4π ampere•turns, the characteristic impedance will be in ohms. *See also* **impedance, iterative impedance.**

character mean entropy: The mean entropy per character of all possible messages from a stationary source, given by the relation $H = \lim \rightarrow \infty H_m/m$ where H is the character mean entropy and H_m is the entropy of the set of all sequences of m characters from the source. *Note 1:* The character mean entropy may be expressed in an information content unit, such as the shannon per character. *Note 2:* The limit of H_m/m may not exist if the source is not stationary. *Synonyms* **average information content, information rate, mean information content.** *See also* **character mean transinformation content, shannon.**

character mean transinformation content: The mean transinformation content per character for all possible messages from a stationary message source, given by the relation $CMTC = \lim \rightarrow \infty T_m/m$ where $CMTC$ is the character mean transinformation content, T_m is the mean transinformation content for all pairs of corresponding input and output sequences of characters, and m is the number of characters. *Note:* The character mean transinformation content may be expressed in an information content unit, such as the shannon per character. *See also* **average transinformation rate, transinformation content.**

character recognition (CR): The identification of characters by automatic means. *Note:* Examples of character recognition are optical character recognition (OCR), magnetically recorded character reading, and magnetic ink character recognition (MICR).

character set: 1. A finite set of different characters that is complete for a given purpose. *Note:* A character set may or may not include punctuation marks or other symbols. **2.** An ordered set of characters, i.e., a set of unique representations. *Note:* Examples of character sets are (a) the 26 letters of the English alphabet, (b) Boolean 0 and 1, (c) the 128 American Standard Code for Information Interchange (ASCII) characters, and (d) the International Telegraph Alphabet 5 (ITA-5), published as International Telephone and Telegraph Consultative Committee (CCITT) Recommendation V.3 and International Organization for Standardization (ISO) 646. *See also* **alphabet, alphanumeric, binary digit, character, code, coded character set, coded set, digital alphabet, language.**

character signal: A set of signal elements that represent a character. *See also* **signal element.**

characters per inch (cpi): A unit of character linear packing density on a recording medium, such as the number of characters recorded per inch of magnetic tape, paper tape, disk track, microfilm, and printed line. *Note:* At any given point in a record medium, the number of characters per inch is the reciprocal of the space, expressed in inches, occupied by a character, including the space between characters, at that point. *See also* **packing density.**

characters per second (cps): A unit of data signaling rate (DSR), i.e., data signaling speed used to express the number of characters passing a designated point per second. *See also* **characters per second, data signaling rate.**

character stepped: Pertaining to control of start-stop teletypewriter equipment in which a device is stepped one character at a time. *Note:* The step interval is equal to or greater than the character interval at the applicable signaling rate. *See also* **bit stepped, character.**

character string: A sequence of characters, such as bits, letters, and numerals usually of the same type, that (a) has a beginning and an end identified by data delimiters, and (b) is treated as a unit. *See also* **bit string, character, data delimiter, pulse train, string.**

character subset: A selection of characters from a character set, such as a group of characters, in which each member of the group has a characteristic in common with all the other members of the group. *Note:* An example of a character subset is the digits 0 through 9 in the American National Code for Information Interchange or in the International Telegraph Alphabet Number 5 (ITA-5). *See* **alphabetic character subset, alphanumeric character subset, numeric character subset.** *See also* **numeric character set.**

charge: *See* **access charge, electronic charge.**

charge carrier: An electron or a hole in a semiconducting material that results when some electrons have been thermally excited into the conduction band from the valence band. *Note:* Thermal excitation creates hole-electron pairs. One hole is created for each free electron, i.e., each excited electron. Both the hole and the electron are capable of transporting, i.e., of carrying, a charge. Hence, they are called "charge carriers," or simply "carriers." The moving carriers constitute electric currents in devices such as lasers, light-emitting diodes, photodetectors, diodes, and transistors. Electrons excited from a lower to a higher energy level cause photon absorption. When electrons change from a higher to a lower energy level, photons, i.e., lightwaves, are emitted. *See also* **conduction band, semiconductor, valence band.**

charge indication: In telephone system operations, the indication by a network, upon request of the paying station, of the monetary charge of a call. *Note:* The charge indication is issued prior to the release of the paying station or after release by recall at a convenient time. *See* **automatic charge indication.**

chart: *See* **radio organization chart, routing chart, Smith chart, time conversion chart.**

Chaudhuri-Hocquenghem code: Bose-Chaudhuri-Hocquenghem code.

check: To verify a characteristic of an entity, such as to verify the accuracy of a measurement or verify an operating condition. *See* **automatic check, block check, character check, continuity check, cyclic redundancy check, echo check, hardware check, longitudinal check, marginal check, mathematical check, modulo-n check, parity check, programmed check, range check, reasonableness check, redundancy check, summation check, transverse parity check, transverse redundancy check.**

check bit: A binary digit (bit) appended to a bit string for later use in error detection. *Note:* An example of a check bit is a parity bit. *See also* **binary digit, error, error control, overhead information, parity check, redundancy check.**

check character: A single character that (a) is derived from and appended to data items, such as blocks and frames, in accordance with specified rules, (b) is used to detect errors in processing or transmitting the data items, (c) is derived from the data items after processing, such as transmission and storage, using the same specified rules, and (d) is checked to assure that the characters derived before and after processing are identical. *See* **block check character.** *See also* **character, overhead information, parity check.**

check coding: *See* **parity check coding.**

check digit: A single digit, derived from and appended to digitized data items, that can be used to detect errors in processing or transmitting the data items. *Note:* An example of a check digit is a 7 for the numeral 106242370. The sum of the digits in the numeral is 25. The sum of the digits in 25 is 7. If the digits of the numeral are summed, modulo 9, the result is 7 again. Also, the sum of the four digits from the left is 9, the next three digits sum to 9, and only 7 and 0 remain, which, when summed, is 7. Thus, "casting out nines" is a modulo-9 check. The 7 could then be appended to the numeral as a check digit at a specified position when transmitted or stored and used to check reception or retrieval. The check digit could be appended to any specified group of numerals or digitized letters to ensure a high probability of accurate reception. Arithmetic operations may also be checked using check digits for the operands and results and performing the same operations on the check digits as on the operands. Thus, $544 \times 496 = 269{,}824$. The modulo-9 check, i.e., casting out nines, digits are 4 and 1 for the operands. Their product is 4. The modulo-9 check for the product of the operands is also 4, i.e., the sum of the digits excluding the 9 is 22, and the sum of these digits is also 4. *See also* **overhead information, parity check.**

check group: A transmitted group of digits or characters, that, by transmitting the information in a second form, or as a repetition, serves as a check on the proper reception of the original group.

checking: *See* **storage bounds checking.**

checksum: 1. The sum of a group of data items that is used for checking purposes. *Note 1:* A checksum is stored or transmitted with the group of data items. *Note 2:* The checksum is calculated by treating the data items as numeric values. *Note 3:* Checksums are used for detecting and correcting errors that may occur during storage, transmission, and processing of data. **2.** The value that (a) is computed, via a parity or hashing algorithm, on data that require protection against error or manipulation, (b) is stored or transmitted with the data, and (c) is intended to detect data integrity problems.

chemical vapor deposition (CVD): In optical fiber manufacturing, the production of deposits, by heterogeneous gas-solid and gas-liquid chemical reactions, on the surface of a substrate. *Note:* Chemical vapor deposition (CVD) is often used in fabricating optical fiber preforms by causing gaseous materials to react and deposit glass oxides. The preform may be processed further in preparation for pulling into an optical fiber. *See also* **preform.**

chemical vapor deposition (CVD) process: A process for making optical fibers in which silica and other glass-forming oxides and dopants are deposited at high temperatures on the inner wall of a fused silica tube, called a bait tube, which is then collapsed into a short, thick, preform from which a long thin optical fiber is pulled at a high softening temperature using an optical fiber pulling machine. *See* **modified chemical vapor deposition (CVDP) process, plasma activated chemical vapor deposition (PACVD) process.** *Refer to* **Fig. C-3 (128).** *Refer also to* **Fig. S-12 (917).**

Fig. C-3. Intel engineers monitor the **chemical vapor deposition process** during the manufacturing of computer integrated circuit chips. (Courtesy Intel Corporation.)

chemical vapor phase oxidation (CVPO) process: A process for making low loss (less than 10 dB/km), high bit-rate•length product, i.e., greater than 300 Mb•km•s^{-1} (megabit-kilometers per second), multimode, graded index (GI), optical fiber using either the inside vapor phase oxidation (IVPO) process, the outside vapor phase oxidation (OVPO) process, the modified chemical vapor deposition (MCVD) process, the plasma-activated chemical vapor deposition (PCVD) process, the axial vapor phase oxidation (AVPO) process, or a combination or variation of these, by soot deposition on a glass substrate followed by oxidation and pulling of the optical fiber.

chicken wire: In an optical element, such as an optical fiber or lens, a blemish that appears as a grid of lines along the boundaries, particularly along fiber boundaries in a multifiber bundle. *See also* **multifiber bundle, optical element, optical fiber.**

chief operator: *Synonym* **chief supervisor.**

chief ray: The central ray of a ray bundle. *See also* **ray bundle.**

chief supervisor: The person responsible for (a) directing, assisting, and instructing switchboard operators and attendants in the performance of their duties and (b) ensuring that maintenance is effectively performed, duty schedules are prepared and met, operating policies are adhered to, and telephone traffic is effectively routed at all times. *Synonym* **chief operator.**

CHINAPAC: The Chinese nationwide data communications packet-switched network that serves (a) 28 provinces and autonomous regions and (b) three municipalities.

chip: 1. In spread spectrum systems, a single frequency output from a frequency hopping signal generator. **2.** In micrographic and display systems, a relatively small and separate piece of microform that (a) contains microimages and (b) contains coded information for search, identification, and retrieval of the microimages. **3.** A minute piece of material cut from a single crystal on which electronic or optical circuit active and passive elements are mounted, usually by etching, deposition, and diffusion processes, to form an integrated circuit. *See also* **wafer. 4.** In satellite communications systems, the smallest element of data in an encoded signal. *See* **clipper chip, competitive chip, flip chip large-scale integrated chip, microimage chip.** *Refer to* **Figs. C-3, I-3 (457), P-14 (762), P-15 (763), P-16 (764), W-1 and W-2 (1078).**

chip frequency: In spread spectrum systems, one of the frequencies that (a) is generated by a frequency hopping generator and (b) that occurs during chip time of the spread spectrum code sequence generator. *See also* **chip time, frequency hopping, frequency hop-**

ping generator, spread spectrum, spread spectrum code sequence generator.

chip rate: In direct sequence modulation spread spectrum systems, the rate at which the information signal bits are transmitted as a pseudorandom sequence of chips. *Note:* The chip rate usually is several times the information bit rate.

chip time: In spread spectrum systems, the duration of one of the frequencies, i.e., a chip frequency, of a frequency hopping generator. *See also* **chip frequency, frequency hopping generator, spread spectrum.**

chirp modulation: *Synonym* **pulse frequency modulation.**

chirp: *Synonym* **pulsed frequency modulated signal.**

chirping: 1. A rapid change, as opposed to a long-term drift, of the frequency of an electromagnetic wave. *Note:* Chirping is most often observed in generating pulses. **2.** In the wavelengths emitted by a source, a sudden shift from one set of spectral lines to another set of spectral lines. **3.** Pulse compression in which usually linear frequency modulation is used during the pulse. *See also* **frequency fluctuation, frequency instability.**

choledochoscope: *See also* **fiber optic choledochoscope.**

CHOP: change in operational control.

chopper: *See* **optical chopper.**

chopper bar: *Synonym* **writing bar.**

chroma keying: In color television systems, instantaneous switching that (a) is between multiple video signals, based on the state, i.e., on the phase, of the color (chroma) signal to form a single composite video signal and (b) is used to create an overlay effect in the final picture, such as to insert a false background consisting of a weather map or a scenic view, behind the principal subject being photographed. *Note:* In chroma keying, the principal subject is photographed against a background having a single color or a relatively narrow range of colors, usually in the blue region of the visible spectrum. When the phase of the chroma signal corresponds to the preprogrammed state or states associated with the background color, or range of colors, behind the principal subject, the signal from the alternate background, i.e., from the false background, is inserted into the composite signal and presented at the output. When the phase of the chroma signal deviates from that associated with the background colors behind the principal subject, video signals associated with the principal subject are presented at the output. *Synonyms* **blue screening, color keying.** *See also* **composite video, television.**

chromatic aberration: Image imperfections that are caused by lightwaves consisting of different wavelengths, which each follow different paths through an optical system, i.e., image imperfections that are caused by the dispersion introduced by the optical elements of the system.

chromatic dispersion: *Synonym* **material dispersion.** *See also* **dispersion.**

chromatic dispersion coefficient [D(λ)]: In an electromagnetic wave propagating in an optical fiber, the derivative, with respect to wavelength, of the normalized group delay, $\tau(\lambda)$, i.e., $D(\lambda) = d\tau(\lambda)/d\lambda$ where $D(\lambda)$ is the chromatic dispersion coefficient, τ is a function of wavelength and is the normalized group delay, and λ is the source wavelength. *See also* **Sellmeier equation.**

chromatic dispersion slope [S(λ)]: In an electromagnetic wave propagating in a waveguide, such as an optical fiber, the derivative, with respect to wavelength, of the chromatic dispersion coefficient, $D(\lambda)$ of the waveguide, i.e., $S(\lambda) = dD(\lambda)/d\lambda$ where $S(\lambda)$ is the chromatic dispersion slope, $D(\lambda)$ is the chromatic dispersion coefficient and is a function of wavelength, and λ is the source wavelength. *See also* **zero dispersion wavelength.**

chromatic distortion: *Synonym* **intramodal distortion.**

chromaticity: In the visible region of the spectrum, the frequency, or wavelength, composition of electromagnetic waves that (a) usually is characterized by the dominant frequencies or wavelengths, which describe the quality of color, and (b) affects propagation in filters, attenuation in optical fibers, and the sensitivity of photographic film.

chromatic radiation: *See* **polychromatic radiation.**

chromatic resolving power: 1. The ability of an instrument to separate two electromagnetic waves of different wavelengths. *Note:* For two wavelengths, the chromatic resolving power is equal to the ratio of (a) the shorter of the two wavelengths to (b) the difference between the two wavelengths. **2.** A measure of the ability of an optical component to separate or differentiate two or more object points that are close together relative to the distance between the optical component and the object points.

chrominance: In color television, the wavelength difference between a reproduced color and a standard reference color having the same radiance. *Note 1:* Chrominance is used in describing the fidelity of color reproduction. *Note 2:* Color component coding techniques may be used to represent the spectral composition of a color. *See also* **luminance.**

chrominance signal: In color television systems, the signal, or the portion of the composite signal, that bears the color information. *See also* **composite video signal.**

CIF: common intermediate format.

cifax: 1. The application of cryptography or ciphony to facsimile signals. **2.** Facsimile signals that have been enciphered to preserve the confidentiality of the transmitted material. *Synonym* **ciphax.** *See also* **ciphony, civision, cryptoinformation, cryptology, cryptomaterial.**

cifony: *Synonym* **ciphony.**

CIM: corporate information management.

cine oriented mode: A mode of recording images on strip or roll film in which the top edge of each image, when viewed for reading, is perpendicular to the long edge, i.e., to the longitudinal axis, of the film. *Note:* Most motion picture films, i.e., movie films, are made in cine oriented mode. The films are made and played past camera and projector lens in a vertical direction. *See also* **comic strip oriented mode.**

ciphax: *Synonym* **cifax.**

cipher: 1. A cryptographic system in which (a) arbitrary symbols, or groups of symbols, are used to represent units of plain text of regular length, usually single letters, or (b) units of plain text are rearranged, or (c) both, all in accordance with certain predetermined rules. **2.** The result obtained when (a) arbitrary symbols, or groups of symbols, are used to represent units of plain text of regular length, usually single letters, or (b) units of plain text are rearranged, or (c) both, in accordance with certain predetermined rules. *See also* **cryptology, cryptosystem, plain text.**

cipher equipment: Equipment that (a) converts plain text to cipher text, and vice versa, (b) may be operated manually or automatically, and (c) may be operated with and without a power source. *See also* **cipher text, plain text.** *See* **literal cipher equipment.**

cipher group: A group of letters or numbers that result from conversion into cipher text, the cipher text being arranged in cipher groups with a fixed number of characters in each group for ease of transmission and for ease in performing encryption-decryption processes. *See also* **cipher text, decryption, encryption.**

cipher mode: In secure radiotelephone communications systems, a mode of operation in which voice transmissions are automatically enciphered and deciphered by ciphony equipment. *See also* **ciphony, secure, secure communications, secure transmission.**

cipher system: A cryptosystem that requires the use of a key to convert, unit by unit, (a) plain or encoded text and (b) plain or encoded signals into an unintelligible form for secure transmission. *Note:* The capability to decipher must be available at the receiving site.

cipher text: 1. Enciphered information. **2.** The product resulting from the encryption of plain text. *See also* **cipher, encipher, encrypt, plain text.**

ciphony: 1. The application of cryptography to telephone communications. **2.** Enciphered speech signals. **3.** The enciphering of audio information, resulting in encrypted speech. *Synonym* **cifony.** *See also* **civision, cryptographic information, cryptology, cryptomaterial.**

ciphony communications system: A radiotelephone, wire, or fiber optic communications system in which messages are enciphered to maintain confidentiality of the transmitted information. *See also* **ciphony protected net, civision, cryptographic information, cryptology, cryptomaterial, secure, secure communications, secure transmission.**

ciphony protected net: A radiotelephone system in which communications are secured with equipment that automatically enciphers and deciphers voice messages when synchronism has been established between sending and receiving equipment. *See also* **ciphony communications system, civision, cryptographic information, cryptology, cryptomaterial, secure, secure communications, secure transmission.**

circle: *See* **aiming circle, great circle.**

circuit: 1. The complete path that provides one-way or two-way communications between two terminals. **2.** An electrical path between two or more points, capable of providing a number of channels. **3.** A number of conductors connected together for the purpose of carrying an electrical current. **4.** An electrical closed loop path among two or more points used for signal transfer. **5.** A number of electrical components, such as resistors, inductances, capacitors, transistors, and power sources, connected together in one or more closed loops. *See* **application-specific integrated circuit, approved circuit, clamping circuit, closed circuit, combinational circuit, common user circuit,**

composite circuit, conditioned circuit, conditioned voice grade circuit, conference circuit, cord circuit, data circuit, data transmission circuit, decision circuit, dedicated circuit, deemphasis circuit, detector back bias gain control circuit, direct current restoration circuit, disturbed circuit, disturbing circuit, divider circuit, driving circuit, duplex circuit, engineering circuit, equivalent four-wire circuit, exclusive user circuit, express engineering circuit, external liaison circuit, four-wire circuit, gating circuit, ground return circuit, half-duplex circuit, homogeneous multiplexed circuit, hybrid fiber optic circuit, integrated circuit, integrated fiber optic circuit, integrated optical circuit, integrating circuit, interchange circuit, line adapter circuit, link engineering circuit, lithium niobate integrated circuit, local engineering circuit, maintenance control circuit, monolithic integrated circuit, monostable circuit, multipoint circuit, nailed-up circuit, open circuit, optical circuit, orderwire circuit, permanent circuit, permanent virtual circuit, phantom circuit, pilot make busy circuit, point to point circuit, preemphasis circuit, printed circuit, radio circuit, reference circuit, ringdown circuit, self-repairing circuit, self-testing circuit, sequential circuit, side circuit, signal return circuit, simplex circuit, simplexed circuit, singing suppression circuit, sole user circuit, solid state circuit, squelch circuit, switched circuit, switched hot line circuit, tail circuit, telecommunications circuit, telegraph circuit, telephone cord circuit, temporary circuit, unbalanced wire circuit, video circuit, video discrimination circuit, virtual circuit, voice grade circuit, voice operated relay circuit, volcas circuit. *See also* **transmission channel.**

circuit activity: *See* **mean circuit activity.**

circuit board: *See* **printed circuit board.**

circuit breaker: A device that (a) opens and closes a circuit between separable contacts under both normal and abnormal conditions and (b) may (i) automatically open when an overload condition occurs, (ii) automatically reclose to test if the overload condition persists, (iii) be used to disconnect or isolate a circuit, and (iv) be used to apply power to a circuit. *Note:* Circuit breakers may be of many types and sizes, and usually are classified according to (a) the medium in which the interruption takes place, such as (i) liquid, such as oil, or (ii) gas, such as air or helium, (b) the number of pairs of contacts, and (c) the manner of suppressing arc-over when a circuit is interrupted, such as by magnetic blowout or capacitive storage. *See* **molded case circuit breaker, power circuit breaker.** *See also* **circuit, disconnect switch, protector, switch.**

circuit breaker switchgear: *See* **low voltage enclosed air circuit breaker switchgear.**

circuit capability: *See* **virtual circuit capability.**

circuit connection: *See* **data circuit connection.**

circuit current: *See* **metallic circuit current.**

circuit designator: A set of standard characters, such as letters, numerals, and special symbols, that represent information concerning a circuit, such as its operating frequency, application, and channel types. *See also* **emission designator, net glossary.**

circuit discipline: In a communications network, the operational discipline that results in (a) the maintenance of message security and integrity, (b) the proper use of communications equipment, (c) the strict adherence to authorized frequencies and operating procedures, (d) the use of remedial action, (e) the exercise of network control, (f) the proper monitoring of the network, and (g) the proper training of operators and maintenance personnel.

circuit filter-coupler-switch-modulator: *See* **integrated optical circuit filter-coupler-switch-modulator, optical integrated circuit filter-coupler-switch-modulator.**

circuit group congestion signal: *See* **national circuit group congestion signal.**

circuit-hour: A unit of communications traffic volume. *Note:* One circuit-hour is equivalent to a circuit occupied for one hour. Thus, two circuits occupied for a half-hour each would be equivalent to a traffic volume of 1 ch (circuit-hour). Fifteen circuits occupied for 2 min. (minutes) each and 20 circuits for 3 min. each would be a traffic volume of 1.5 ch. *See* **connections per circuit-hour, hundred circuit-hours.** *See also* **erlang.**

circuit log: In a radiotelephone, telephone, or telegraph net, a record of transmitted traffic, received traffic, and operating conditions for each day on each circuit or channel, including data on significant events, such as operating and closing time, delays, frequency changes, procedural errors, and security violations.

circuit multiplex set: *See* **engineering circuit multiplex set.**

circuit noise level: At any point in a transmission system, the ratio of (a) the circuit noise to (b) some arbitrary amount of circuit noise chosen as a reference. *Note:* The circuit noise level usually is expressed in dBrn, signifying the reading of a circuit noise meter, or in dBa, signifying circuit noise meter reading adjusted to represent an interfering effect under specified conditions. *See also* **channel noise level, circuit, dBa, dBrn, image rejection ratio, level, noise, signal plus noise to noise ratio.**

circuit patch bay: *See* **digital circuit patch bay.**

circuit quality control: The actions taken to assure proper performance of circuits, such as circuit evaluation, reporting of poor quality circuits, and repairing of poor quality circuits.

circuit release: To discontinue the use of a circuit, usually by a user going off-hook, thus allowing the circuit to be used by other users. *Note:* A circuit that has been released usually remains in the idle state until used again. *See also* **circuit released signal.**

circuit released acknowledgment signal: A signal sent in the forward direction in response to a circuit released signal indicating that a circuit has been released. *See also* **circuit, circuit released signal, forward direction.**

circuit released signal: A signal sent in the forward direction and in the backward direction indicating that a circuit has been released. *See also* **backward direction, circuit, forward direction, signal.**

circuit reliability (CR): The percentage of time a circuit was available for use in a specified period of scheduled availability. *Note 1:* Circuit reliability is given by the relation $CR = 100(1 - (T_o/T_s)) = 100T_a/T_s$ where T_o is the circuit total outage time, T_s is the circuit total scheduled time, and T_a is the circuit total available time. *Note 2:* $T_s = T_a + T_o$. *See also* **circuit, channel reliability.**

circuit restoration: The restoration of communications between two users usually by using another communications circuit after disruption or loss of the original circuit. *Note:* Circuit restoration usually is performed in accordance with planned procedures and priorities.

circuitry: 1. A complex of circuits within or between systems. **2.** A complex of circuits used for establishing connections among components of a system and among systems. *See also* **circuit.**

circuit security: *See* **telephone circuit security.**

circuit service: *See* **reserved circuit service.**

circuit signaling: *See* **telephone circuit signaling.**

circuit switched connection: A circuit that is established and maintained, usually on demand, between two or more stations or sets of data terminal equipment, in order to allow the exclusive use of the circuit until the connection is released.

circuit switched data transmission service: A data transmission service that (a) requires establishment of a circuit switched connection before data can be transferred from the source data terminal equipment (DTE) to the sink DTE and (b) uses a connection-oriented network.

circuit switching: 1. The handling of communications traffic through a switching center, either from local users or from other switching centers, whereby a connection is established between the originating and receiving stations until the connection is released by the originating or the receiving station. **2.** The establishing of a connection, usually on demand, between two or more data terminal equipments (DTEs) for the exclusive use of a circuit between users until the connection is released. *See also* **circuit, message switching, packet switching, switching system.**

circuit switching center (CSC): A complex of communications circuits and supporting facilities that can be interconnected and released for establishing and disestablishing connections between users. *See also* **central office, circuit switching, switching center.**

circuit switching unit: Equipment used for routing messages over common user circuits that interconnect the source data terminal equipment (DTE) to the sink DTE for information interchange.

circuit-terminating equipment: *See* **data circuit-terminating equipment.**

circuit transfer mode (CTM): In integrated services digital network (ISDN) applications, a transfer mode in which there is a permanent allocation of channels or bandwidth between connections. *See also* **connection, transfer mode.**

circuit transparency: *See* **data circuit transparency.**

circuit unavailability procedure: A procedure to be followed by telephone operators and attendants when circuits are busy, such as when (a) backlogs occur, (b) other facilities must be used, (c) callbacks are requested, and (d) precedence calls are to be handled.

circular dielectric waveguide: A waveguide that consists of a long circular cylinder of concentric dielectric materials, capable of sustaining and guiding an electromagnetic wave in one or more propagation modes. *Note:* Most optical fibers are circular dielectric waveguides that are capable of transmitting, or transporting, lightwaves over lang distances. Some polarization-

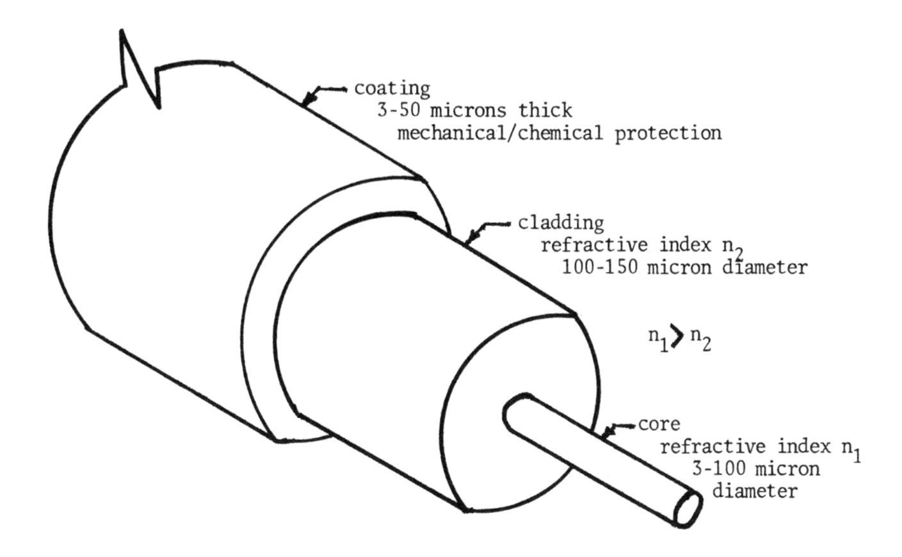

Fig. C-4. A **circular dielectric waveguide,** such as an optical fiber.

maintaining (PM) fibers are somewhat elliptical or rectangular in cross section, especially the core. *Refer to* **Fig. C-4.**

circularity: *See* **cladding noncircularity, core noncircularity.**

circularly symmetric optical fiber: An optical fiber of circular cross section that has material parameters, such as a refractive index profile, that are independent of (a) the angular displacement of the central angle of the fiber and (b) the longitudinal coordinate of the fiber, i.e., the transmission media parameters are the same in all radial directions at any cross section, although parameters are a function of the radial distance from the fiber axis.

circular polar diagram: In radio communications antennas, a circular plot indicating that the electric field strength, in volts per meter, or the irradiance, in watts per square meter, is about the same in all horizontal directions. *Note:* A completely uniform omnidirectional antenna would exhibit a circular polar diagram, in contrast to a directional antenna, which concentrates power in certain directions.

circular polarization: Polarization of the electric and magnetic field vectors in a uniform plane polarized electromagnetic wave, such as some lightwaves and radio waves, in which the two arbitrary sinusoidally varying rectangular components of the electric field vector are equal in amplitude but are 90° out of phase, causing the electric field vector to rotate, the direction of rotation depending on which component leads or lags the other. *Note 1:* The tip of the electric field vector rotates in a circle. *Note 2:* The magnetic field vector is at the center of the circle described by the tip of the electric field vector and is perpendicular to the plane of the circle. *See* **left-hand circular polarization, right-hand circular polarization.** *See also* **antenna, clockwise polarized wave, elliptical polarization, linear polarization.**

circular scan: 1. In satellite communications systems, an Earth station directional antenna search pattern in which the antenna usually is rotated about a vertical or a horizontal axis. *Note:* Circular scan is used after initial pointing to the predicted satellite position. This enables acquisition of the downlink signals prior to automatic tracking. **2.** In radar systems, the continuous rotation of a directional antenna, usually about a vertical axis. *Note:* The rotation may be a full circle or only a sector.

circular shift: In computer and data processing systems, the movement of a computer word such that vacant positions at one end of the word are filled in strict sequence by characters dropped off the other end. *Note:* In a register containing a numeral 32 bits long, a circular shift of 32 places results in a zero net change in its contents. The circulation may be to the right or to the left. No bits are omitted, and no bits are added to the contents of the register during the circular shift. If the numeral 106242370 in a register were right-circular-shifted four places, the contents of the register would be 237010624. *Synonym* **cyclic shift, end around shift.**

circular three-body problem: *See* **restricted circular three-body problem.**

circulating storage: Storage in which data are moved in a closed loop, such as (a) serially into one end of a register, out the other end, and back to the beginning in a continuous manner, (b) into a delay line, such as a mercury, magnetic, or electromagnetic delay line, out the other end, and back to the beginning, and (c) out from a read head and back to a write head on a magnetic disk, diskette, drum, or tape. *Note:* In a circulating storage, data can be gated in for storage and out for retrieval. If data are in the storage, data gated in replace data that were in. Data gated out can remain in or be removed. *See also* **circular shift, gate.**

circulator: 1. A passive junction of three or more ports in which the ports can be accessed in such a sequence that power applied to any port is transferred to the next port, the first port being counted as following the last in sequence. **2.** In radar, a device that switches an antenna alternately between the transmitter and the receiver.

C^{3}I reusable software system: *See* **command, control, communications, and intelligence reusable software system.**

CIS: communications and information system.

civil communications network: A nonmilitary communications network that (a) is owned and operated by a public authority or a private organization and (b) usually is operated for public or private use.

civil day: A 24-hour day of 86,400 s (seconds), i.e., 236 s longer than a sidereal day, that relates to an imaginary, i.e., an ideal, Sun and Earth with uniform relative motion. *Synonym* **mean solar day.** *See also* **second, sidereal day, true solar day.**

civil fixed telecommunications network: A network of commercial and government telecommunications facilities for transmission of messages on a worldwide basis by radio, cable, and satellite links that sends the traffic to specific addressees, usually via their nearest station. *See also* **military fixed telecommunications network.**

civision: 1. The application of cryptography to television (TV) signals. **2.** Enciphered television (TV) signals, i.e., TV signals that have been enciphered to preserve the confidentiality of the transmitted information. *See also* **cifax, ciphony, cryptoinformation, cryptology, cryptomaterial.**

C/kT: carrier to receiver noise density.

clad: *Synonym* **cladding.**

cladded optical fiber: *See* **doped silica cladded optical fiber.**

cladded slab dielectric waveguide: *See* **doubly cladded slab dielectric waveguide.**

cladding: 1. In a dielectric waveguide, such as an optical fiber, slab dielectric waveguide, or optical integrated circuit (OIC), a layer of material of lower refractive index in intimate contact with a core material of higher refractive index. *Note 1:* The cladding confines lightwave modes to the core by means of total internal reflection, provides some protection to the core, and transmits evanescent waves that are usually bound to waves in the core. *Note 2:* Evanescent waves will propagate in the cladding and even extend beyond the cladding. They will not radiate away if the cladding is sufficiently thick. They will radiate away at a bend if the dielectric waveguide is bent too sharply, i.e., is bent at a radius less than the critical radius. **2.** In a metal cable, a covering of one metal with another metal, usually achieved by pressure rolling, extruding, drawing, or swaging, until a bond is achieved. *Synonym* **clad.** *See* **depressed cladding, extramural cladding, homogeneous cladding, matched cladding, optical fiber cladding.** *See also* **cable, core, core diameter, deeply depressed cladding optical fiber, depressed cladding optical fiber, doubly clad optical fiber, fiber optics, multimode optical fiber, normalized frequency, optical fiber, single mode optical fiber, tolerance field, total internal reflection.** *Refer to* **Figs. A-6 (24), D-4 (264), L-8 (536), L-9 (536), P-12 (751), R-1 (796), R-10 (836), S-18 (951).**

cladding center: In a circular cross section optical fiber, the center of the circle that circumscribes the outer surface of the homogeneous cladding. *Note:* The cladding center may not be the same as the cladding reference surface center. *See also* **cladding, tolerance field.**

cladding concentricity: *See* **core-cladding concentricity.**

cladding diameter: 1. In an optical fiber, the diameter of the circle that best fits the outer limit of the cladding. *Note:* The center of this circle is the cladding center. **2.** In an optical fiber, the outer diameter of the low refractive index region that surrounds the core. *Note:* The cladding diameter is determined only for optical fibers with a circular cross section. *See also* **cladding, core diameter, fiber optics, tolerance field.**

cladding eccentricity: The ratio of (a) the minimum cladding thickness to (b) the maximum cladding thickness. *See also* **concentricity error.**

cladding guided mode: *Synonym* **cladding mode.**

cladding mode: In an open waveguide, i.e., in an all-dielectric waveguide such as an optical fiber, a transmission mode supported by the cladding and the core, i.e., a mode in addition to the modes supported by the core alone. *Note 1:* Cladding modes are confined to the cladding and the core because of the lower refractive index propagation medium surrounding the outermost cladding. *Note 2:* Cladding modes may be attenuated by using lossy media in the cladding to absorb the energy in the propagating modes, thus preventing reconversion of energy to core guided modes, and thereby reducing dispersion. *Synonyms* **cladding guided mode, cladding ray.** *See also* **bound mode, leaky mode, mode, radiation mode, unbound mode.**

cladding mode stripper: 1. In fiber optic systems, a device in which (a) cladding modes are converted to radiation modes, and (b) a material usually is applied to the cladding of a short section of optical fiber to allow electromagnetic energy propagating in the cladding to radiate from the cladding by refraction, by having the refractive index of the applied material greater than that of the cladding. *Note:* Certain buffers, coatings, and jackets applied for other purposes may also perform the function of a cladding mode stripper. **2.** An optical element that (a) supports only certain modes in a dielectric waveguide, such as an optical fiber, planar waveguide, or optical integrated circuit (OIC), (b) is placed in the optical system in which cladding modes are to be removed, (c) does not support the cladding modes in the optical system, and (c) removes the cladding modes without disturbing the core supported modes. *See also* **cladding, cladding mode.**

cladding noncircularity: 1. In the cross section of a round optical fiber, the ratio of (a) the difference between (i) the diameter of the smallest circle that can circumscribe the core area and (ii) the diameter of the largest circle that can be inscribed within the cladding, both circles concentric with the cladding center, to (b) the cladding diameter. **2.** In an optical fiber, the percentage of deviation from a circle of the cladding cross section. *Synonym* **cladding ovality.** *See also* **concentricity error, core noncircularity, ovality.**

cladding optical fiber: *See* **deeply depressed cladding optical fiber.**

cladding ovality: *Synonym* **cladding noncircularity.**

cladding power distribution: *See* **core-cladding power distribution.**

cladding ray: *Synonym* **cladding mode.**

cladding tolerance area: In the cross section of a round optical fiber, the region between (a) the smallest circle that can circumscribe the core area and (b) the largest circle that can be inscribed within the cladding, both circles being concentric with the cladding center. *See also* **cladding center.**

clad optical fiber: *See* **doubly clad optical fiber, quadruply clad optical fiber.**

clad silica optical fiber: *See* **hard clad silica optical fiber, low-loss FEP clad silica optical fiber, plastic clad silica optical fiber.**

clad switchgear: *See* **metal clad switchgear.**

clamper: *Synonym* **clamping circuit.**

clamping circuit: A circuit that prevents the voltage level of another circuit from going above or below predetermined levels. *Note:* A diode and a constant voltage source can be used as a clamping circuit, provided that power, voltage, and current levels do not exceed the capability of the clamping circuit. Clamping circuits may be used in almost any circuit, such as those found in lasers, photodetectors, amplifiers, gates, optical integrated circuits, computer circuits, and communications circuits. Clamping circuits serve as protective devices. *Synonym* **clamper.** *See also* **clipper.**

C language: A general-purpose, high level, structured computer programming language that supports a wide range of data types. *Note:* C language was originally designed for and implemented on the UNIX™ operating system.

CLASS: custom local area signaling services.

class d address: *In Internet Protocol, synonym* **multicast address.**

classification: *See* **bearing classification, office classification, position classification.**

classified: *See* **security classified.**

classified information: 1. Official information or matter in any form or of any nature that requires protection in the interests of national security. **2.** Information that must be handled according to specific rules in order to prevent compromise, i.e., prevent unauthorized disclosure. *Note:* In the U.S. Department of Defense, one form of classification is to categorize information into unclassified, confidential, secret, and top secret information. Unauthorized disclosure of confidential, secret, and top secret information affects the security of the United States.

class indicator: *See* **user class indicator.**

classmark: A designator used to describe the service feature privileges, restrictions, and circuit characteristics for lines or trunks that access a switch. *Note:* Examples of classmarks are precedence levels, conference privileges, security levels, and zone restrictions. *Synonym* **class of service mark.** *See also* **blocking, call restriction, calls barred facility, class of service, code restriction, controlled access, line load control, precedence, restricted access, service feature.**

class of emission: The set of characteristics of an emission, such as the type of (a) modulation of the main carrier, (b) modulating signal, (c) information to be transmitted, and (d) any additional signal characteristics, each designated by a standard symbol.

class of office: A ranking, assigned to each switching center in a communications network, determined by the center switching functions, interrelationships with other offices, and transmission requirements. *Synonym* **office class.** *See also* **office classification.**

class of service: 1. A designation assigned to describe the service treatment and privileges given to a particular terminal. **2.** A subgrouping of telephone users for the sake of rate distinction. *Note:* Examples of class of service subgrouping include distinguishing between (a) individual and party lines, (b) government and nongovernment lines, (c) lines that permit unrestricted international dialed calls and those that do not, (d) business, residence, and coin, (e) flat rate and message rate, and (f) restricted and extended area service. **3.** A category of data transmission provided by a public data network in which the data signaling rate, the terminal operating mode, and the code structure, if any, are standardized. *Note:* Class of service is defined in International Telegraph and Telephone Consultative Committee (CCITT) Recommendation X.1. *Synonyms* **service class, user service class.** *See also* **call restriction, classmark.**

class of service mark: *Synonym* **classmark.**

clear: 1. In telephone switching systems, to cause a clearing signal to be given at an exchange, either by replacing the handset on its cradle or hook, i.e., going on-hook by hanging up, or by ring-off by the call originator or the call receiver, thus indicating that the call is finished. *See also* **finished call, off-hook, ring-off. 2.** In telephone system operations, to take down, i.e., to remove, connections at an exchange, thereby freeing the circuit at all points for other calls. *Synonym* **clear down. 3.** To cause one or more storage locations to be in a prescribed state, usually the state that corresponds to zero or to the space character. **4.** In computer systems, to cancel or erase a previously specified condition or value, such a tab or range setting value. *Note:* In most computer systems, clearing may be accomplished in a variety of ways, such as by erasing, deleting, resetting, escaping, initiating a new operation, exiting from the program, and simply turning the power off. *See* **double clear, false clear, single clear.**

clearance: *See* **path clearance.**

clear channel: A signal path that provides its full bandwidth for user services. *Note:* Clear channels are not used for system control signaling. *See also* **bandwidth.**

clear collision: Contention that occurs when data terminal equipment (DTE) and data circuit-terminating equipment (DCE) simultaneously transfer a clear request packet and a clear indication packet, both specifying the same logical channel. *Note:* The data circuit-terminating equipment (DCE) usually performs as though the clearing is completed and will not transfer a DCE clear confirmation packet. *See also* **call collision, clear confirmation signal, collision, data circuit-terminating equipment, data terminal equipment.**

clear confirmation: 1. An interrogation signal or message to determine if a circuit has been cleared. **2.** The signal or the message, in response to an interrogation, confirming that a circuit has in fact been cleared.

clear confirmation signal (CCS): A call control signal used to acknowledge reception of the data terminal equipment (DTE) clear request by the data circuit-terminating equipment (DCE) or the reception of the DCE clear indication by the DTE. *See also* **call control signal, clear collision, data circuit-terminating equipment, data terminal equipment, signal.**

clear down: *Synonym* **clear.**

clear forward signal: In semiautomatic and automatic telephone systems, a signal transmitted in a forward direction on termination of a call by a telephone operator or when a user replaces the handset on its cradle or hook, i.e., goes on-hook by hanging up, thus freeing the circuit at all points for other calls. An exception may occur where the signaling system includes a switching release guard signal. *Synonym* **forward clearing signal.** *See also* **finished call, forward direction.**

clear indicator: A signal or a message indicating that a circuit has just been cleared.

clearing: 1. A sequence of actions taken to disconnect a call and return to the ready state. **2.** A procedure used to erase classified or sensitive information stored on a magnetic medium. *Note:* Clearing creates a product that may be reused within, but not outside of, a secure facility. Clearing does not produce a declassified or desensitized product by itself, but is the first step in the declassification or desensitization process. *See* **negative clearing, positive clearing.** *See also* **call, disengagement time.**

clearing indicator shutter: *See* **fallen clearing indicator shutter.**

clearing signal: A signal sent by a user or an attendant to indicate that the temporary use of a line, loop, or trunk is ended, and the line, loop, or trunk is available for further use. *See also* **clear indicator.**

clear message: 1. A message that (a) is sent in the forward direction and the backward direction, (b) contains a circuit released signal or circuit released acknowledgment signal, and (c) usually contains an indication of whether the message is in the forward or the backward direction. **2.** A message in plain language, i.e., a message that is not enciphered. *See* **modified clear message.**

clear request signal: A signal that (a) is transmitted in the forward direction and the backward direction in the data channels between exchanges and (b) usually is sent by the user end instrument or the user data terminal equipment (DTE). *See also* **backward direction, data terminal equipment clear request, forward direction.**

clear signal: A signal that initiates clearing actions. *See* **data circuit-terminating equipment clear signal.** *See also* **clearing.**

clear text: *Synonym* **plain text.**

clear to send signal: A signal generated by a device to indicate that it is ready to transmit data. *See also* **request to send signal.**

clear traffic: Communications traffic that has not been encrypted, scrambled, or in any way deliberately made unintelligible.

cleave: 1. In an optical fiber, a break that (a) is deliberate and controlled, (b) has a smooth, flat endface perpendicular to the fiber axis, and (c) is made by (i) creating a microscopic fracture, nick, or groove in the fiber with a cleaving tool that has a sharp blade of hard material, such as diamond, sapphire, or tungsten carbide, and (ii) applying tension to the fiber as the fracture is made or immediately afterward, either with the cleaving tool or manually, causing the fracture to propagate through the cross section of the fiber and thereby creating the smooth, flat, perpendicular endface. **2.** To break an optical fiber in a controlled precise manner. *Note:* A good cleave is required for a successful low-attenuation joint made by mechanical splice, fusion splice, or connector. Some connectors do not require the use of abrasives and polishers for the endface. They use a cleaving technique that trims the fiber to its proper length and produces a smooth, flat, perpendicular endface.

client: In communications networks, a software application that allows the user to access a service from a server computer, such as a server computer on the Internet.

client-server: Any hardware/software combination that (a) usually is associated with a client-server architecture and (b) is independent of the type of application. *See also* **client, client-server architecture, server.**

client-server architecture: Any network-based software system that uses (a) a client to request a specific service and (b) a corresponding server to provide the service from another computer on the network. *See also* **client, server.**

clipper: A circuit or a device that limits the instantaneous output signal amplitude to a predetermined maximum value, regardless of the amplitude of the input signal. *See also* **clamping circuit, companding, compandor, compressor, expander, limiter, peak limiting, vogad.**

clipper chip: 1. An integrated circuit (IC) designed for secure communications. **2.** A protective device that (a) is used to conduct legalized wiretaps of digital circuitry that would be difficult to tap without a device and (b) is used with a private key that (i) is furnished to the protective agency, such as the National Security Agency (NSA) or the U.S. Department of Justice, and (ii) is held in protective custody by the protective agency.

Clipper chip: A U.S. government-mandated escrowed encryption standard chip for use in telephone encryption systems.

clipping: 1. In telephony, the loss of the initial or final parts of words or syllables, usually caused by the nonideal operation of voice-actuated devices. **2.** The deliberate limiting of the instantaneous signal amplitude to a predetermined maximum value. **3.** The limitation in amplitude of a signal by a circuit component, such as an amplifier, photodetector, or light source. *Note:* In pulse circuits, clipping is a means of reducing the amplification of frequencies below a specified frequency or removing the tail of a pulse after a fixed time. In telephone circuits, clipping is the perceptible mutilation of signals or speech syllables during transmission, resulting in the loss of the beginnings or endings of signal elements or speech syllables, called initial clipping and final clipping. In analog systems, clipping is an intentional signal processing action to reduce the dynamic range of an analog signal. **4.** In a display device, the removal of those parts of display elements that lie outside of a given boundary. *Synonym* **scissoring.** *See also* **clipper, clamping circuit, hard limiting, limiting.**

clock: 1. A reference source of timing information, especially for components, equipment, or systems. **2.** A device that provides signals used in a transmission system to control the timing of certain functions, such as to control the duration of signal elements or to control the sampling rate. **3.** A device that generates periodic, accurately spaced signals used for various purposes, such as timing, synchronization, regulation of the operations of a processor, or the generation of interrupts. *See also* **cesium clock, coordinated clock, Department of Defense master clock, equipment clock, false clock, independent clock, master clock, principal clock, quartz clock, reference clock, remote clock, rubidium clock, slave clock, station clock, synchronized clock.** *See also* **Coordinated Universal Time, precise time.**

clock calendar driver: A software package for a communications, computer, or data processing system, such as a personal computer (PC), used to control the system clock and calendar for displaying or recording time, date, place, and other time-related events.

clock control: *See* **direct clock control, indirect clock control.**

clock difference: *Synonym* **clock time difference.**

clock driver: In a communications, computer, data processing, or control system, software used to control the system clock. *Refer to* **Fig. P-15 (763).**

clock error: At any given instant, the algebraic difference between (a) the time or the value, i.e., the indication, of a given clock and (b) the time or the value, i.e., the indication, of a designated reference clock, determined by subtracting the time or the value, i.e., the indication, of the reference clock from the time or the value of the given clock. *Note:* Subtracting the clock error from the given clock reading will produce the reference clock reading. *See also* **error, local clock, reference clock.**

clock phase slew: The rate of relative phase change between a given clock signal and a stable reference signal. *Note:* The two signals are approximately at or near the same clock pulse rate or have an integral multiple clock pulse rate relationship. *Synonym* **clock pulse slew.** *See also* **clock, phase.**

clock pulse duration modulation: *See* **suppressed clock pulse duration modulation.**

clock pulse slew: *Synonym* **clock phase slew.**

clock rate: The rate at which a clock issues timing pulses. *Note:* Clock rates are usually expressed in pulses per second, such as 4.96 Mpps (megapulses per second). *Note:* Timing pulses are usually used to (a) synchronize communications transmission systems and (b) gate or synchronize logic operations or data transfers in digital computers. *Synonym* **clock speed.**

clock second: *See* **cesium clock second.**

clock speed: *Synonym* **clock rate.**

clock stability: A measure of the ability of a device to produce the same number of equally spaced timing pulses during equal and relatively large periods. *Note:* Clock stability usually is expressed in parts per million, i.e., in parts per 10^6. Thus, the stability could be cited as correct to within 2.4 parts per million, such as $+2.4 \times 10^{-6}$, indicating that there are 2.4 excess timing pulses per million pulses of a standard reference clock.

clock time difference (CTD): 1. The difference between the readings of two clocks taken at the same instant. *Note:* In order to avoid confusion in sign, algebraic quantities should be given, applying the convention that at an instant T, if a denotes the reading of a reference time scale A and b denotes the reading of a given time scale B, then the time scale difference is expressed as $a - b$ at the instant T. The same convention applies to the case where A and B are clocks. **2.** A measure of the separation between the respective time marks of two clocks. *Note:* Clock differences are reported as algebraic quantities on a given time scale. The date of the measurement should be given. An example of clock time difference might be stated as 1645 UTC, 5 October 1995; UTC(USNO)−UTC(USAF Primary #1) = 0.9 μs ± 0.2 μs (microseconds). The local clock time must be subtracted from the reference clock time to get the proper arithmetic sign for the clock time difference. *Synonym* **clock difference.**

clock tolerance: The maximum permissible departure of a clock indication from a designated time reference, such as from Coordinated Universal Time (UTC).

clock track: A track, such as on a magnetic or optical disk, used to record a pattern of signals to provide a timing reference for reading and recording data. *See also* **track.**

clockwise helical polarization: *Synonym* **right-hand helical polarization.**

clockwise polarized electromagnetic wave: *Synonym* **right-hand polarized electromagnetic wave.**

clockwise polarized wave: *Synonym* **right-hand polarized electromagnetic wave.**

close: In display systems, to restore continuity in a specific sequence of graphic or display commands after inserting additional commands. *See also* **open.**

close confinement junction: *Synonym* **single heterojunction.**

closed circuit: **1.** In radio and television (TV) transmission, pertaining to an arrangement in which programs are transmitted directly to specific users and not broadcast to the general public. **2.** In telecommunications, a circuit dedicated to specific users. **3.** A completed electrical circuit. *See also* **circuit, dedicated circuit.** *Refer to* **Fig. F-3 (342).**

closed code: *Synonym* **cyclic code.**

closed loop noise bandwidth: The integral, over all frequencies, of the absolute value of the closed loop transfer function of a phase locked loop. *Note:* The closed loop noise bandwidth multiplied by the noise spectral density equals the output noise power in a phase locked loop. *See also* **phase locked loop.**

closed loop transfer function: In a closed loop feedback circuit, a mathematical function, expressed as an algorithm, that describes the net result of the effects of a closed loop feedback of a function of the output signal to the input via the circuits that form the loop. *Note 1:* The closed loop transfer function is measured at the output. *Note 2:* The output signal waveform can be calculated from the closed loop transfer function and the input signal waveform. *See also* **feedback, loop, regeneration.**

closed orbit: A simple elliptical orbit, i.e., an orbit that the orbiting body repeats. *Note:* Parabolic and hyperbolic orbits are not elliptical orbits. A circular orbit is a special case of an elliptical orbit. *Synonym* **elliptical orbit.**

closed shop operation: In computer and data processing operations, the operation of a computing center such that originators describe the problems they want solved to the programming specialists on the center staff, who then write the programs, run them, and give the solutions to the problem originator. *Note:* Often discussions ensue, changes are made, and the revised program is run again, until a satisfactory solution is obtained. *See also* **originator.**

closed source intelligence (CSCINT): Intelligence derived by gathering, processing, and analyzing information that (a) is derived or obtained from sources that (i) do not make the information available to the public and (ii) apply protective measures that ensure that the information is made available only to select persons or systems, (b) is not available via multimedia, (c) usually is security-classified by the source, (d) may be proprietary, but includes copyrighted information, (e) may be derived from sensitive contracting, and (f) usually is obtained by clandestine or covert means. *Note:* An example of closed source intelligence (CSCINT) is information that, if known to certain persons or systems, would place the national security of a nation in jeopardy. *See also* **open source intelligence.**

closed user group: In a network, a group of specified users assigned to a facility that permits them to communicate with each other but precludes communications with all other users of the network service or services, i.e., with users outside the group. *Note:* A user data terminal equipment (DTE) may belong to more than one closed user group. *See also* **data, facility, special network service.** *Refer to* **Fig. C-5.**

closed user group with outgoing access: A closed user group in which at least one member of the group has a facility that permits communications with one or more users external to the closed user group, such as a user that (a) is not in the closed user group and (b) has data terminal equipment (DTE) that is connected to a public data network. *Refer to* **Fig. C-5.**

closed waveguide: An electromagnetic waveguide (a) that has the shape of a tube or pipe, (b) that has electrically conducting walls, i.e., conducting sides, or optically reflecting walls, (c) that may be hollow or filled with a dielectric material, (d) that can support an infinite number of discrete propagating modes, though only a few are practical, (e) in which each discrete mode has its own propagation constant, (f) in which the field at any point is describable in terms of these modes, (g) in which there is no radiation field, i.e., there is no lateral radiation from the guide, and (g) in which discontinuities and bends cause mode conversion but not radiation. *Note 1:* The two cross section dimensions and the polarization determine the modal distribution in the waveguide. *Note 2:* Examples of closed waveguides are (a) metallic rectangular cross section pipes, much like an ordinary rectangular downspout for rain from a gutter, and (b) optical fibers with metallic reflective coating that may be used as

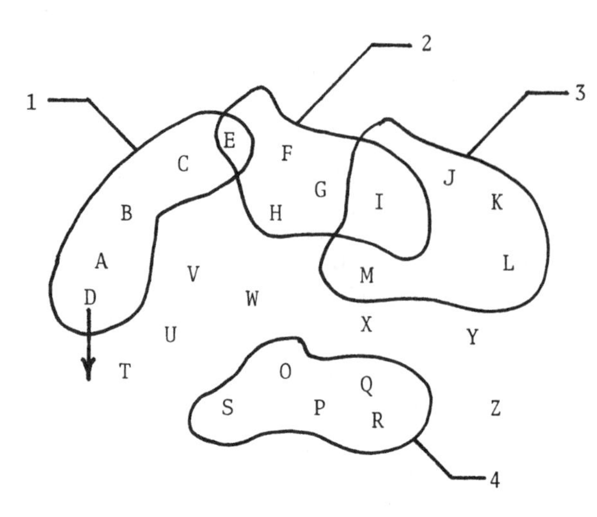

Fig. C-5. Users in and out of **closed user groups** with and without outgoing access in a data network. Each letter represents a user in the network.

light pipes for guiding optical energy for various purposes, such as for powering optical devices and for illumination. *See also* **open waveguide, optical waveguide.**

closedown: *See* **flush closedown.**

closeup: In flaghoist communications systems on board ships and on land, the condition that exists when the flaghoist top is touching the block at the point of hoist, i.e., when it is raised all the way up. *Note:* The signal originating station, i.e., the transmitting station, usually raises flags, pennants, and panels at closeup. *See also* **at the dip, flaghoist, hauled down.**

Clos network: A switching subnetwork that (a) has a high degree of connectivity, (b) provides a large number of routing alternatives, and (c) requires messages or packets to pass through several queues before reaching the ultimate destination. *See also* **connectivity, subnetwork, switched network, switching, switching system.**

closure: *See* **cable splice closure, splice closure.**

cloud attenuation: In the transmission of electromagnetic signals, attenuation caused by absorption and scattering by water or ice particles in clouds. *Note:* The amount of cloud attenuation depends on many factors, including (a) the density, particle size, and turbulence of the clouds and (b) the propagation path length in the clouds.

cluster: A pyrotechnic signal in which all the stars or flares of a group burn at the same time. *See also* **pyrotechnic signal.**

clutter: In radar, random noise that (a) appears on a screen and (b) obscures part of the display space. *Note:* An example of clutter is the noise caused by objects close to the ground, such as trees, buildings, and hills. *See* **ground clutter.**

cm: centimeter.

C-message weighting: A noise weighting used in a noise power measuring set to measure noise power on a line that would be terminated by a 500-type or similar instrument. *Note:* The instrument is calibrated in dBrnC. *See also* **flat weighting, F1A line weighting, HA1 receiver weighting, message, 144-line weighting, 144-receiver weighting.**

CMOS: complementary metal oxide semiconductor.

CMRR: common mode rejection ratio.

CMW: compartmented mode workstation.

C/N: carrier to noise ratio.

CNR: carrier to noise ratio, combat net radio.

C/N ratio: carrier to noise ratio.

CNS: complementary network services.

CO: central office.

C.O.: central office.

COAM: customer owned and maintained equipment. *Deprecated. Use* **customer premises equipment.**

coast Earth station (CES): An Earth station in the fixed satellite service or, in some cases, in the maritime mobile satellite service, located at a specified fixed point on land to provide a feeder link for the maritime mobile satellite service.

coasting mode: In timing-dependent systems, a free-running operational timing mode in which (a) continuous or periodic measurement of clock error, i.e., timing error, is not made, and (b) operation may be enhanced for a period by using clock error or clock correction data obtained during a prior tracking mode to estimate clock corrections. *See also* **tracking mode.**

coast station: A land station that (a) is in the maritime mobile service and (b) usually communicates with ship stations but may also communicate with other coast stations in connection with communicating with ship stations.

coat: *See* **overcoat.**

COAT: coherent optical adaptive technique.

coated optical fiber sensor: A fiber optic sensor in which a special coating is used to apply stress to an optical fiber resulting in fiber strain. *Note 1:* An example of a coated optical fiber sensor consists of (a) a narrow-spectral-line-width light source, (b) a length of optical fiber, serving as a sensing leg, with a bonded coating sensitive to the parameter being measured, (c) a shielded length of optical fiber, serving as a reference leg, that is not exposed to the parameter being measured, (d) a fiber optic splitter to split the same wave into both legs, (e) a coupler to combine the waves from both legs, and (f) a photodetector responsive to the light intensity produced by the two interfering waves, all connected together. *Note 2:* A magnetostrictive coating may be used to elongate an optical fiber proportionally to an applied longitudinal magnetic field. The variations in length can be measured by using interferometric techniques. If the magnetic field is made proportional to an electric current, the electric current can be measured. If the coating is thermally sensitive, the coefficient of thermal expansion can be used to measure temperature, or the thermal expansion effect of an electric current in the coating can be used to measure the electric current.

coated optics: The branch of optics devoted to the study, development, and use of optical elements or components that have optical refractive and reflective surfaces that have been coated with one or more layers of dielectric or metallic material for reducing or increasing reflection from the surfaces, either totally or for selected wavelengths, and for protecting the surfaces from abrasion and corrosion. *Note:* Some antireflection coating materials that can serve as optical coatings are magnesium fluoride, silicon monoxide, titanium oxide, and zinc sulfide.

coating: In fiber optics, special materials bonded to optical fibers to make them sensitive to a specific physical variable that is to be measured, such as magnetostrictive materials to measure magnetic fields, metals to measure electric currents, and thermally sensitive materials to measure temperature. *See* **antireflection coating, optical fiber coating, optical protective coating, primary coating, secondary coating.** *See also* **buffer, cladding, jacket, overarmor, sheath.**

coating offset ratio: *See* **optical fiber/coating offset ratio.**

coax: coaxial cable.

coaxial cable (coax): A cable that (a) consists of a center conductor surrounded by an insulating material, a concentric outer conductor, and an outer sheath and (b) is used as a waveguide primarily for wideband, video, and radio frequency applications. *Synonyms* **coaxial line, concentric cable, concentric line.** *See* **air dielectric coaxial cable, foam dielectric coaxial cable.** *See also* **cable.** *Refer to* **Fig. C-6.** *Refer also to* **Fig. S-20 (959).**

coaxial line: *Synonym* **coaxial cable.**

coaxial patch bay: A patch bay that is served by coaxial cable circuits. *See also* **patch bay.**

COBOL: A high level general-purpose computer programming language that (a) makes use of English words rather than mathematical notation and (b) is principally used for record and file handling. *Note:* COBOL is derived from Common Business Oriented Language.

cochannel interference: Interference that results from two or more transmissions occurring simultaneously in the same channel. *See also* **adjacent channel interference, channel, interference.**

code: 1. A set of unambiguous rules that specifies the manner in which data may be represented in a discrete form. *Note:* Use of a code provides a means of converting information into a form suitable for communications, processing, or encryption. **2.** Any communications system in which arbitrary groups of symbols represent units of plain text of varying length. *Note:*

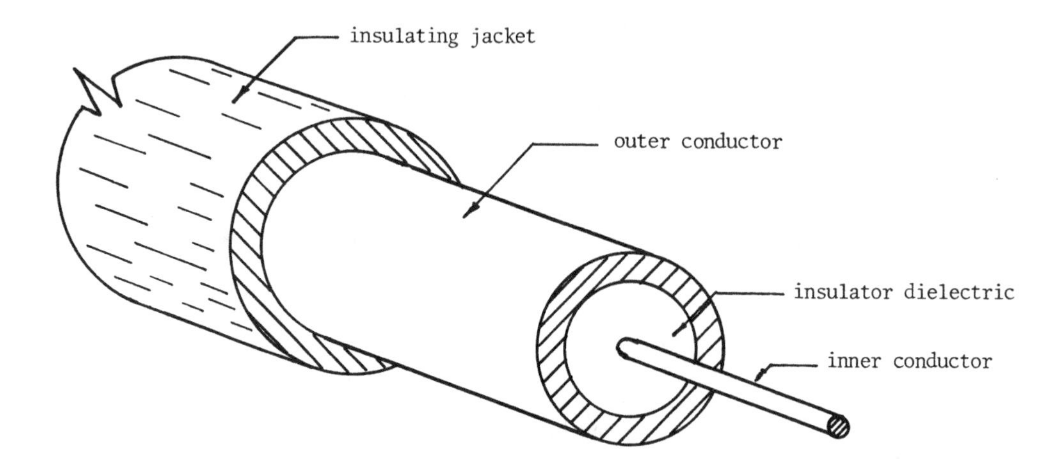

Fig. C-6. A **coaxial cable** that serves as a waveguide, *i.e.,* a uniform transmission line. The outer conductor usually is braided for ease of bending and often serves as ground. The inner conductor is insulated and usually is solid.

Codes may be used to (a) achieve brevity, (b) provide security, (c) convert information into a form suitable for communications or encryption, (d) reduce the length of time required to transmit information, (e) describe the instructions that control the operation of a computer, and (f) convert plain text into combinations of letters or numbers that do not reveal their meaning, and vice versa. **3.** A cryptosystem in which the cryptographic equivalents, i.e., the code groups, usually consisting of letters or digits in otherwise meaningless combinations, are substituted for plain text elements that are primarily words, phrases, or sentences. **4.** A set of rules that maps the elements of one set, the coded set, onto the elements of another set, the code element set. *Synonym* **coding scheme.** **5.** A set of items, such as abbreviations, that represent corresponding members of another set. **6.** To represent data or a computer program in a symbolic form that can be accepted by a processor. **7.** To write computer programs or routines. *See* **absolute machine code, access code, address code, alphabetic code, alphanumeric code, answer back code, area code, balanced code, bar code, Barker code, Baudot code, binary code, biquinary code, cable address code, cable code, call directly code, card code, chain code, comma-free code, composite code, convoluted code, country code, cyclic code, data code, data country code, data network identification code, dense binary code, equal length code, error correcting code, error detecting code, excess three code, extended binary coded decimal interchange code, fixed weight binary code, gold code, Gray code, Hagelbarger code, Hamming code, Hollerith code, illegal code, interlock code, international Morse code, international signal code, international visual signal code, instruction code, language code, linear sequence code, line code, lock code, machine instruction code, Manchester code, microcode, minimum distance code, mnemonic code, modified alternate mark inversion code, modified nonreturn to zero level code, modified reflected binary code, Morse code, m-out-of-n code, n-ary code, National Electrical Code, network code, nonreturn to zero binary code, numeric code, NXX code, object code, operation code, optical line code, paired disparity code, panel code, prearranged code, privacy code, pseudobinary code, pseudocode, pulse code, pyrotechnic code, redundant code, relocatable machine code, return to zero binary code, self-demarcating code, self-synchronizing code, signaling panel code, source code, stop code, symmetrical binary code, synchronization code, syncopated code, telegraph code, time code, transmitter start code, two out of five code, unit disparity binary code, unit distance code, walking code, ZIP code.** *See also* **decode, encode.** *Refer to* **Figs. B-2 (68), B-7 (81), C-2 (121), E-3 (310), E-5 (315), G-3 (402), I-6 (472), M-1 (584), P-15 (763), S-17 (940).**

code alphabet: *See* **n-unit code alphabet.**

code ambiguity: *See* **time code ambiguity.**

code audit: In computer programming, an independent review of computer program coding that (a) verifies compliance with design documentation, standards, and organizational policy, (b) evaluates coding accuracy and efficiency, and (c) may be accomplished by one or more persons or by a software tool. *See also* **software tool.**

code book: A book that contains (a) codes arranged in systematic order, (b) data that are arranged so that messages can be encoded and decoded, and (c) a vocabulary (i) that consists of arbitrary meanings, such as letters, syllables, words, phrases, and sentences, each accompanied by one or more groups of symbols, and (ii) that may be used as equivalents to the plain text messages.

codec: 1. coder-decoder. 2. *Synonym* **coder-decoder.**

code character: The representation of a discrete value or symbol in accordance with a code. *See also* **digital alphabet.**

code combination: A set of significant conditions formed by means of an *n*-unit code in which each code element assumes a given significant condition. *See also* **code element, n-unit code, significant condition.**

code conversion: 1. Conversion of signals, or groups of signals, in one code into corresponding signals, or groups of signals, in another code. **2.** A process for converting a code with some predetermined bit structure, such as 5, 7, or 14 bits per character interval, to a second code with a different structure, such as a different number of bits per character interval. *Note:* In code conversion, alphabetical order is not significant. *See also* **binary digit, character, code, line code, pulse code modulation, signal.**

code converter: A data converter that changes the representation of data using one code, or coded character set, into data representing the same information but using a different code, or a different coded character set. *Note:* Examples of code converters are (a) a device that converts data encoded in international Morse code into data encoded in the American National Code for Information Interchange (ASCII) and (b) a device that converts a nonreturn to zero (NRZ) code to a return to zero (RZ) code.

coded character set: A character set established in accordance with unambiguous rules that define the character set and the one-to-one relationships between the characters of the set and their coded representations. *See also* **character, character set, code, digital alphabet.**

coded decimal interchange code: *See* **extended binary coded decimal interchange code.**

coded decimal notation: *See* **binary coded decimal notation.**

code-dependent: Pertaining to a communications, computer, data processing, or control system in which correct and proper functioning depends on the use of a specific coded representation or coded character set. *See also* **coded character set, coded representation, code-independent.**

code-dependent system: A system that depends for its correct functioning upon the character set or code required for transmission of data. *Synonym* **code-sensitive system.** *See also* **code-transparent system.**

coded image: A representation of a display image in a form suitable for transmission, storage, and processing.

coded image space: *Synonym* **image storage space.**

code division multiple access (CDMA): 1. Modulation that (a) independently codes data in multiple channels for transmission over a single wideband channel, (b) may be used as an access method that permits carriers from different stations to use the same transmission equipment by using a wider bandwidth than the individual carriers do, whereupon, on reception, each carrier can be distinguished from the others by means of a specific modulation code, thereby allowing for the reception of signals that were originally overlapping in frequency and time, (c) permits several transmissions to occur simultaneously within the same bandwidth, with the mutual interference reduced by the degree of orthogonality of the unique codes used in each transmission, and (d) permits a more uniform distribution of energy in the emitted bandwidth. **2.** A coding scheme in which (a) digital information is encoded in an expanded bandwidth format, (b) several transmissions can occur simultaneously within the same bandwidth with the mutual interference reduced by the degree of orthogonality of the unique codes used in each transmission, and (c) a high degree of energy dispersion occurs in the emitted bandwidth. *See also* **carrier sense multiple access, code, modulation, spread spectrum.**

coded modulation: *See* **differential trellis coded modulation, trellis coded modulation.**

coded representation: 1. The representation of an item of data by a code. **2.** The representation of a character established by a coded character set. *Note:* Examples of coded representation using coded character sets are (a) the three-letter codes used to identify international airports, such as IAD for the Washington, DC Dulles International Airport, and (b) the seven-binary-digit codes that represent the characters of the American National Standard Code for Information Interchange (ASCII). *Synonym* **code value.** *See also* **coded character set.**

coded set: A set of elements onto which another set of elements has been mapped according to a code. *Note:* Examples of coded sets include the list of names of airports that is mapped onto a set of corresponding three-letter representations of airport names, the list of classes of emission that is mapped onto a set of corresponding standard symbols, and the names of the months of the year mapped onto a set of two-digit decimal numbers. *See also* **alphabet, character set, code, digital alphabet.**

coded speech: The output of any device that converts a signal derived from plain speech into another type of signal.

code element: In the set of characters of a code, one of a set of parts used to compose or construct the characters. *See also* **binary digit, character, mark, pulse code modulation, space.**

code exchange: *See* **automatic answer back code exchange.**

code excited linear prediction (CELP) coding: Digital coding of voice signals by a specialized method of analog to digital conversion. *See also* **linear predictive coding.**

code extension character: A transmission control character used to indicate that one of the succeeding coded representations is to be interpreted according to a different code or according to a different coded character set. *See* **nonlocking code extension character.** *See also* **coded character set, coded representation.**

code generator: In communications, computer, data processing, and control systems software, a functional unit, such as a part of a computer, that transforms a computer program from an intermediate level of representation, such as the output of a parser, into a lower level representation, such as assembly language code or machine language code. *See also* **assembly language, machine language, parser.**

code group: In a code system, a group of letters, numbers, or a combination of these, used to represent a plain text element, such as a symbol, word, phrase, or sentence. *See also* **code, group.**

code identifier: **1.** A code or a signal used to identify another particular code from a set of available codes for use in a given situation or application. **2.** In visual signaling systems, an identifier used to identify an object. *Note:* An example of a code identifier is a configuration of panels used to identify a drop zone or a landing zone. *See also* **drop zone, panel, visual signaling system.**

code-independent: Pertaining to communications, computer, or data processing systems in which correct and proper functioning does not depend upon any particular coded representations or coded character sets. *Note:* Limitations or dependency on mechanical, electrical, and pulse configurations may exist. *Synonym* **code-transparent.** *See also* **coded character set, code-dependent, coded representation.**

code-independent data communications: Data communications in which protocols are used that do not depend for their correct functioning upon the data character set or data code used. *Synonym* **code-transparent data communications.** *See also* **code, data, data communications, data transmission, source.**

code-independent system: A system in which correct functioning does not depend upon a particular coded character set or a particular coded representation of data. *Synonyms* **code-insensitive system, code-transparent system.** *See also* **coded character set, code-dependent system, coded representation.**

code indicator: *See* **destination code indicator, Morse code indicator.**

code-insensitive system: *Synonym* **code-independent system.**

code line index: In micrographics, a visual index that (a) consists of a pattern of transparent and opaque bars parallel to the long edge of a microfilm roll and (b) is located between the frames.

code modulation: *See* **adaptive differential pulse code modulation, differential pulse code modulation, code modulation, pulse code modulation, subband adaptive differential pulse code modulation.**

code modulation multiplexing: *See* **pulse code modulation multiplexing.**

code multiplexing equipment: *See* **pulse code multiplexing equipment.**

coder: **1.** An analog to digital converter, i.e., a device for converting continuously varying signals into discrete signals. **2.** An encoder or a decoder. *See* **pattern matching voice coder, vocoder, voice coder.** *See also* **analog to digital encoder, channel voice encoder.**

coder-decoder (codec): **1.** An assembly consisting of an encoder and a decoder in one piece of equipment. **2.** A circuit that converts analog signals to digital code and vice versa. **3.** An electronic device that converts analog signals, such as video and voice signals, into digital form and compresses them to conserve bandwidth on a transmission path. *Note:* Codec is used in this sense for video conferencing systems. *Synonym* **codec.** *See also* **code, pulse code modulation.**

code repertory: *Synonym* **instruction code.**

code repertoire: *Synonym* **instruction code.**

code restriction: A network-provided service feature in which certain data terminal equipment (DTE) and user end instruments are prevented from having access to certain features of the network. *See also* **classmark, restricted access, service feature.**

code-sensitive system: *Synonym* **code-dependent system.**

code sequence: *See* **maximal code sequence.**

code sequence generator: *See* **Mersenne code sequence generator, modular spread spectrum code sequence generator, spread spectrum code sequence generator.**

code sequence register: In a spread spectrum code sequence generator, the bank of series-connected flip-flops used to generate the spread spectrum code sequence. *See also* **flip-flop, spread spectrum code sequence generator.**

code set: 1. The complete set of representations of a code or coded character set. **2.** The set of representations defined by or used in a specific code or language. *Note:* An example of a code set is the set of two-letter codes used to represent the states of the United States, such as NY for New York State. *See also* **alphabet, character set, coded character set, code, digital alphabet.**

code-transparent data communications: *Synonym* **code-independent data communications.**

code-transparent system: *Synonym* **code-independent system.**

code value: *Synonym* **coded representation.**

code word: 1. In a code, a word that consists of a sequence of symbols assembled in accordance with the specific rules of the code and assigned a unique meaning. *Note:* Examples of code words are error detecting or correcting code words and communications code words, such as SOS, MAYDAY, ROGER, TEN-FOUR, and OUT. **2.** A cryptonym used to identify sensitive intelligence data. **3.** A word that has been assigned a classification and a classified meaning to safeguard intentions and information regarding a classified plan or operation. *Note:* An example of a code word is OVERLORD for the June 6, 1944 Allied landing in Normandy, France, during World War II. *See* **inactive code word.** *See also* **code, comma-free code, word.**

coding: 1. In communications systems, the altering of the characteristics of a signal to make the signal more suitable for an intended application, such as for making the signal more suitable for transmission, improving transmission quality and fidelity, modifying the signal spectrum, increasing the information content, providing error detection and correction, and providing data security. *Note:* A single coding scheme usually does not provide more than one or two specific capabilities. Each resulting code has distinct advantages and disadvantages. **2.** In communications and computer systems, implementing rules that are used to map the elements of one set onto the elements of another set, usually on a one-to-one basis. *See* **adaptive predictive coding, biphase coding, biphase level coding, biphase mark coding, biphase space coding, bipolar coding, code excited linear prediction coding, dipulse coding, duobinary coding, full-baud bipolar coding, half-baud bipolar coding, linear predictive coding, nonreturn to zero coding, parity check coding, return to zero coding, three-condition coding, transcoding.**

coding compaction: *See* **floating point coding compaction, variable precision coding compaction.**

coding scheme: *Synonym* **code.**

codress message: In military communications systems, a message in which the entire address is encrypted with the message text.

codress message address: An address that is encrypted in the text of a message.

coefficient: A number that multiplies another number or expression. *Note:* An example of a coefficient is the numeral 5.44 in the expression $5.44e^{4.96}$. The 4.96 is the exponent. The *e* is the base of natural logarithms, equal to about 2.718. The value of the expression is about 1252.9. *See* **absorption coefficient, chromatic dispersion coefficient, dispersion coefficient, electrooptic coefficient, extinction coefficient, global dispersion coefficient, phase change coefficient, reference coefficient, reflection coefficient, scattering coefficient, spectral absorption coefficient, transmission coefficient, worst-case dispersion coefficient.**

COG: centralized ordering group.

COGO: A computer programming language primarily used for topological problems, such as geometric problems and network topological problems. *Note:* COGO is derived from coordinate geometry.

coherence: 1. The phenomenon pertaining to a correlation between the phase points of two waves at the same instant or at the same point in space. **2.** The phenomenon pertaining to a correlation between the phase points of one wave at two different instants of time at the same point in space or at two points in space at the same instant. *Note:* In practice, coherence is a matter of degree. If coherence were perfect, the phase relationships within a wave, and the phase relationships among waves, would be always and everywhere precisely predictable. **3.** Pertaining to the similarity of two signal wave forms in regard to a given particular feature, such as the same polarization or the same frequency. *Synonyms* **space coherence, spatial coherence.** *See* **spatial coherence, coherence degree, partial coherence, time coherence.** *See also* **coherent, correlation.** *Refer to* **Fig. O-2 (671).**

coherence area: 1. In an electromagnetic wave, the area of a surface (a) every point in which the surface is perpendicular to the direction of propagation, (b) over which the electromagnetic wave maintains a specified coherence degree, and (c) usually over which the coherence degree exceeds 0.88. **2.** In optical communications, the area in a plane perpendicular to the direction of propagation over which light may be considered highly coherent, i.e., the area over which the coherence degree exceeds 0.88. *Synonym* **coherent area.** *See also* **coherence degree, coherence length, coherence time, coherent.**

coherence degree: A unit used to indicate the extent of coherence of light. *Note 1:* Coherence degree is expressed as a ratio. *Note 2:* The magnitude of the coherence degree is equal to the visibility, V, of the fringes of a two-beam interference test, as given by the relation $V = (I_{max} - I_{min})/(I_{max} + I_{min})$ where V is the visibility, I_{max} is the intensity at a maximum of the interference pattern, and I_{min} is the intensity at a minimum of the interference pattern. *Note 3:* Light is considered to be highly coherent when the coherence degree exceeds 0.88, partially coherent for values less than 0.88 but more than nearly zero values, and incoherent for nearly zero and zero values. *Synonym* **degree of coherence.** *See also* **coherence area, coherence length, coherence time.**

coherence length: The propagation distance from a coherent source to a point where an electromagnetic wave no longer maintains a specified degree of coherence. *Note 1:* In long distance transmission systems, the coherence length and the coherence time usually are reduced by propagation factors, such as dispersion, scattering, and diffraction. *Note 2:* In optical communications, the coherence length is approximately given by the relation $L_c = \lambda^2/(n\Delta\lambda)$ where L_c is the coherence length, λ is the central wavelength of the source, n is the refractive index of the medium in which the electromagnetic wave is propagating, and $\Delta\lambda$ is the spectral width of the source. *See also* **coherence area, coherence time, degree of coherence, spectral width.**

coherence of frequency: *Synonym* **phase coherence.**

coherence of phase: *Synonym* **phase coherence.**

coherence time: The time required for an electromagnetic wave to propagate from a coherent source to a point where the wave no longer maintains a specified coherence degree. *Note 1:* In long distance transmission systems, the coherence time and the coherence distance usually are reduced by propagation factors, such as dispersion, scattering, and diffraction. *Note 2:* In optical communications, coherence is calculated by dividing the coherence length by the phase velocity of the light in the medium in which the wave is propagating. Thus, coherence time is approximately given by the relation $\tau_c = \lambda^2/c\Delta\lambda$ where τ_c is the coherence time, λ is the central wavelength of the source, $\Delta\lambda$ is the spectral width of the source, and c is the velocity of light in vacuum. In long distance transmission systems, the coherence time may be degraded, i.e., reduced, by many other propagation factors, such as refractive index, temperature, impurity, and pressure variations at various points in the propagation medium. *See also* **coherence area, coherence degree, coherence length, fiber optics, multimode distortion.**

coherency: In spread spectrum communications systems, a synchronized and phase-matched condition between a receiver reference and the desired signal, i.e., the received signal.

coherent: 1. Pertaining to a fixed phase relationship between corresponding points everywhere on an electromagnetic wave. *Note 1:* A completely coherent wave would be perfectly coherent at all points in space. In practice, however, the region of high coherence may extend over only a finite distance because of various propagation factors, such as dispersion, diffraction, and scattering. *Note 2:* An electromagnetic wave is said to be coherent at distances from the source less than the coherent length, i.e., where the coherence degree is greater than 0.88. **2.** Pertaining to an electromagnetic wave in which the electric and magnetic field vectors are uniquely and specifically definable always and everywhere within specified tolerances. *Note:* An electromagnetic wave of single frequency, i.e., a purely monochromatic wave, would be coherent at all points in free space and in a homogeneous isotropic propagation medium. Thus, the electric and magnetic field values could be precisely predicted at every point in space and time using the solutions to Maxwell's equations. In actual practice monochronism is not perfect. The region of high coherence may extend only a short distance from the source. The area on the surface of a practical wavefront over which the wave may be considered highly coherent is the coherent area or is a coherence patch. The distance in the direction of propagation in which the wave is highly coherent is the coherence length, in which case the wave is phase- or length-coherent. This coherence length divided by the velocity of the light in the propagation medium is the coherence time. Thus, phase and time coherence are related. Phase coherence may also be called time coherence because both are taken in the direction of propagation rather than transverse to the direction of propagation as area coherence is. Area coherence is taken at an instant of time. Time coherence, i.e., temporal coherence, and phase coherence of a wave occur over the period of time that corresponds to its coherent length. *Synonyms* **spatially coherent, spatially coherent radiation.** *See also* **phase, phase coherent.**

coherent area: *Synonym* **coherence area.**

coherent bundle: *Synonym* **aligned bundle.**

coherent detection: In fiber optics, optical detection in which a low power phase modulated lightwave is coherently mixed with an optical signal to produce an amplitude-modulated signal that can be directly detected downstream by conventional means.

coherent detector: A signal detector in which (a) both phase and amplitude information in the received signal are used, and (b) a phase reference signal is used to remove out-of-phase noise. *Note:* Examples of coherent detectors are specialized photodetectors and demodulators.

coherent differential phase-shift keying (CDPSK): In digital data transmission, phase-shift keying in which (a) the phase of the carrier is discretely modulated in relation to the phase of a reference signal, (b) the modulated carrier is of constant amplitude and frequency, (c) a phase comparison is made of successive pulses, and (d) information is recovered by examining the phase transitions between the carrier and successive pulses rather than by the absolute phases of the pulses relative to a synchronizing signal. *Synonym* **differentially coherent phase-shift keying.** *See also* **keying, modulation, phase-shift keying.**

coherent frequency shift keying (CFSK): Frequency shift keying (FSK) in which the instantaneous frequency is shifted between two discrete values, called the "mark" and "space" frequencies, such that there is no phase discontinuity in the output signal when the frequencies are shifted. *See also* **frequency shift keying, incoherent frequency shift keying.**

coherent light: Light in which all parameters are predictable and correlated at any point in time or space, particularly over an area in a plane perpendicular to the direction of propagation or over time at a point in space. *See* **space coherent light, time coherent light.**

coherent moving target indicator: In radar systems, a moving target indicator (MTI) in which a reference frequency from an oscillator is used in the receiver for phase comparison with the return signals for determining the radial component of the velocity of an object. *See also* **incoherent moving target indicator, moving target indicator.**

coherent optical adaptive technique (COAT): A technique that (a) is used to improve the optical power density of electromagnetic wavefronts propagating through a turbulent atmosphere, using various approaches, such as aperture tagging, compensating phase shift, image sharpening, and phase conjugation and (b) is used in wavefront control. *See also* **aperture tagging, compensating phase shift, image sharpening, phase conjugation.**

coherent pulse operation: In pulsed carrier transmission, a method of operation in which a fixed phase relationship of the carrier wave is maintained from one pulse to the next. *See also* **fixed reference modulation, phase.**

coherent radiation: At a given point, electromagnetic radiation that has a coherence degree greater than 0.88. *See* **time coherent radiation.** *See also* **coherence, coherence degree, coherent.**

coherent repeater jammer: A repeater jammer that (a) produces a jamming signal that maintains a fixed relationship with the signal that it receives, (b) transmits a signal that is made to be identical to a signal that would be reflected from a target if there were a target, and (c) has a signal strength, shape, pulse repetition rate (PRR), and frequency that are adjusted for deception.

coherent signal: A signal that is locked in phase or bears a fixed time relationship with another signal.

coil: *See also* **hybrid coil, loading coil, repeating coil.**

coil optical fiber sensor: *See* **flat coil optical fiber sensor.**

coincidence circuit: *Synonym* **AND gate.**

coincidence gate: *Synonym* **AND gate.**

coincidence unit: *Synonym* **AND gate.**

cold bend: In testing fiber optic cabling, the alternate cyclic bending of an optical fiber, fiber optic bundle, or fiber optic cable using a specified device at a specified temperature, usually room temperature.

cold boot: *Synonym* **cold restart.**

cold restart: The restarting of equipment after a usually unintentional, quick, sudden, or flush closedown or shutdown in which all previously initialized input data, programs, and output queues in storage have been lost and thus a warm restart cannot occur. *Synonyms* **cold boot, cold start.** *See also* **warm restart.**

cold standby: Pertaining to electronic equipment that (a) is available for substitute use, (b) usually is connected or is ready for connection, (c) does not have power turned on, and (d) is not warmed up and therefore is not ready for immediate use. *See also* **hot standby.**

cold start: *Synonym* **cold restart.**

collate: In data systems, to place a set of items in a prescribed single sequence. *Note:* An example of collate is to place a set of words in alphabetical order, numerical order, or order of length. *See also* **merge, sort.**

collect: In a communications system, a prefix in a message indicating that the charges for the message were not paid at the source and are to be collected at the destination, such as from the addressee. *See also* **prepaid.**

collection: *See* **data collection.**

collection angle: *Synonym* **acceptance angle.**

collection facility: *See* **data collection facility.**

collective address: *Synonym* **group address.**

collective address group: An address group that (a) represents two or more entities, such as organizations, authorities, activities, units, persons, and any combination of these, and (b) usually is a list of the names of the heads of the entities. *See also* **address group.**

collective call sign: **1.** A call sign that represents two or more facilities, commands, authorities, or units. **2.** In radio communications systems, a call sign that identifies a predetermined group of stations. *See* **ship collective call sign.** *See also* **individual call sign, net call sign.**

collective lens: A lens of positive power, used in an optical system to refract the chief rays of image-forming bundles of rays, so that these rays will pass through subsequent optical elements of the system. *Note 1:* If all the rays do not pass through an optical element, a loss of light ensues, known as vignetting. *Note 2:* Any lens of positive power is not necessarily a collective lens. *See also* **chief ray, converging lens, positive power, vignetting.**

collective routing: Routing in which a switching center automatically delivers messages to a specified list of destinations, thus avoiding the need to list each single address in the message heading. *Note:* Major relay stations usually transmit messages bearing collective routing indicators to tributary, minor, and other major relay stations.

collective routing indicator: A group of symbols, usually letters, that identifies a group of stations, such as all the stations of a given relay network in a specified geographical area or all the minor and tributary stations of a major relay station. *Note:* Major relay stations usually transmit messages bearing collective routing indicators to tributary, minor, and other major relay stations.

collimated light: A bundle of light rays in which the rays emanating from any single point on the object are parallel to one another. *Note:* Examples of collimated light are (a) light from an extremely distant real source, such as from a star, and (b) light from an apparent source, such as a collimator reticle. *Synonym* **parallel light.** *See also* **collimation, collimator, light, object, ray.**

collimated transmittance: Transmittance of an optical waveguide in which the lightwave has the coherency at the output related to the coherency at the input. *See also* **coherency, transmittance.**

collimation: **1.** In a beam with divergent or convergent rays, the conversion of the rays into a beam with the minimum possible ray divergence or convergence. *Note 1:* If collimation is perfect, a collimated beam would consist of a bundle of parallel rays. *Note 2:* Collimation is used in many applications, such as radio, radar, sonar, cathode ray tubes, and optical communications equipment. **2.** The making of electromagnetic rays, such as light rays, parallel. *Note 1:* An example of collimation is the use of a lens system to make divergent or convergent rays parallel. Light rays from a very distant and not necessarily point source are practically parallel. **3.** The aligning of the optical axis of optical systems to the reference mechanical axis or surfaces of an instrument. **4.** The aligning of two or more optical axes with respect to each other. *See also* **beam divergence, decollimation.**

collimator: A device that (a) renders divergent or convergent rays more nearly parallel and (b) may be used to simulate a distant target or object, align the optical axes of instruments, or prepare rays for entry into the end of an optical fiber. *Note:* The degree of collimation, i.e., the maximum divergence angle or the maximum convergence angle, should be stated.

collinear antenna array (CAA): An array of dipole antennas in which (a) the dipole antennas are all mounted in a single line that is an extension of their long axes, (b) the dipole antennas usually are in a vertical line so as to produce an antenna gain in the horizontal direction at the expense of vertical gain, and (c) when stacking the dipole antennas, doubling their number produces a 3-dB increase in directive gain.

collision: 1. In a data transmission system, the simultaneous occurrence of two or more requests on equipment that can handle only one request at any given instant. *Note:* A protocol is used to resolve the contention that immediately follows the instant of collision. **2.** In a computer, the occurrence of attempts to simultaneously store two different data items at a given address that can hold only one of the items. *See* **call collision, clear collision, head-on collision.** *See also* **carrier sense multiple access, data transmission.**

collision avoidance: *See* **carrier sense multiple access with collision avoidance.**

collision detection: *See* **carrier sense multiple access with collision detection.**

color: 1. The sensation produced by light of a given wavelength, or group of wavelengths, in the visible region of the electromagnetic spectrum. *Note:* Color-related properties of light include chromatic aberration, color shade, and the number of Newton's rings present when two optical surfaces are placed together multiples and submultiples of wavelength apart. *See also* **chromatic aberration, Newton's rings, visible spectrum. 2.** A given wavelength, or set of wavelengths, in the visible spectrum.

color division multiplexing (CDM): In optical communications systems, the multiplexing of channels on a single propagation medium, such as using each color in a transmitted polychromatic light beam as a channel in one optical fiber or fiber optic bundle. *Note:* Color division multiplexing in the visible region of the electromagnetic frequency spectrum is the same process as frequency division multiplexing in the nonvisible region of the spectrum. Each color corresponds to a different frequency and a different wavelength. Demultiplexing can be accomplished with prisms that separate the colors by dispersion. *See also* **wavelength division multiplexing.**

colored light transmission: Transmission in which an arrangement of colored lights is displayed in accordance with a prearranged code.

color error: In color video systems, a distortion of hues in all or a portion of the received image.

colorimeter: An optical instrument used to compare the color of a sample with a source reference or synthesized stimulus. *Note:* In a three-color colorimeter, the synthesized stimulus is produced by a mixture of three colors of fixed chromaticity but variable luminance.

color keying: *Synonym* **chroma keying.**

colors: *See* **false colors.**

color temperature: For a given body, the thermal temperature of a blackbody that emits light of the same color as the given body. *Note:* The color temperature is also expressed in kelvins, as is the thermal temperature.

column: 1. A string of items that has a spatial orientation perpendicular to the interpupillary line of the person viewing the string. **2.** In data processing, a vertical arrangement of items, such as a vertical list of characters on a page. *See also* **row.**

column binary: Pertaining to the representation of binary numerals in which the successive digits are placed in a vertical column, usually with the least significant digit at the top, such as using a column of holes on a punched card to represent a binary numeral, with the hole at the top of the column representing the least significant digit. *Synonym* **Chinese binary.** *See also* **row binary.**

column width: The maximum number of characters permitted for an item in a column without obscuring any portion of the item, such as the maximum number of characters that may be placed in one row of a column in a spreadsheet. *Note:* Most spreadsheet software allows column widths to be individually varied or globally varied by means of appropriate commands.

COM: computer output microfilm, computer output microfilmer, computer output microfilming, computer output microform.

coma: An aberration of a lens that causes oblique bundles of light rays from a point on an object or a point source to be imaged as a comet-shaped blur, i.e., a smear. *See also* **aberration, object, image.**

combat net radio (CNR): A radio operating in a military network that (a) usually provides a half-duplex circuit, (b) employs either (i) a single radio frequency or (ii) a discrete set of radio frequencies in a frequency hopping mode, and (c) primarily is used for command and control of combat, combat support, and combat service support operations among military ground, sea, and air forces. *See also* **tactical communications.**

combination: *See* **code combination.**

combinational circuit: 1. A circuit that performs the function of a combinational logic element. **2.** A circuit that (a) consists of an interconnected set of logic elements, i.e., gates, (b) executes logic functions, such as those of the Boolean algebra, (b) may perform arithmetic operations, (c) usually reaches decisions based on input pulse patterns and circuit configuration, (d) usually consists of interconnected transistors, optical elements, or memory elements, and (e) has input and output ports. *See also* **circuit, combinational logic element, gate.** *Refer to* **Figs. B-6 (78), L-6 (529), P-15 (763).**

combinational logic element: A device that (a) has at least one output channel and one or more input channels, all characterized by discrete states, such that at any instant the state of each output channel is completely determined by the states of the input channels at the same instant, and (b) is used to construct digital control components, such as (i) half-adders, (ii) arithmetic and control units, and (iii) transmission control devices. *Note:* Examples of combinational logic elements are AND, OR, and NEGATION gates. *See also* **combinational circuit.**

combined communications: The common use of communications facilities by two or more military services, each belonging to a different nation. *Note:* Such use might be specified by a combined communications-electronics agency or board. *See also* **joint communications.**

combined distribution frame (CDF): A distribution frame that (a) combines the functions of main and intermediate distribution frames, (b) contains both vertical and horizontal terminating blocks, (c) uses the vertical blocks to terminate the permanent outside lines entering the station, (d) uses the horizontal blocks to terminate inside plant equipment, (e) permits the association of any outside line with any desired terminal equipment, (f) usually makes connections either with twisted-pair wire, usually referred to as jumper wire, or with optical fiber cables, usually referred to as jumper cables, (g) in technical control facilities, may use the vertical side to terminate equipment as well as outside lines, and (h) may then use the horizontal side for jackfields and common battery terminations. *See also* **distribution frame, frame, horizontal terminating block, intermediate distribution frame, main distribution frame, technical control facility, vertical terminating block.**

combined station: In high level data link control (HDLC) operation, the station that (a) usually is responsible for performing balanced link level operations and (b) generates commands, interprets responses, interprets received commands, and generates responses.

combiner: In antenna systems, a device that (a) enables two or more transmitters operating at different frequencies to simultaneously use a single antenna, (b) eliminates the need for separate antennas for each frequency or for each transmitter, and (c) should minimize both insertion loss and intermodulation noise. *See* **cavity combiner, diversity combiner, equal gain combiner, fiber optic combiner, linear combiner, maximal ratio combiner, postdetection combiner, selective combiner.**

combining: Using a single device, such as a transmission line or antenna, to handle several different signals either concurrently or simultaneously, or both, usually by using frequency division multiplexing or time division multiplexing. *See* **digital combining, linear diversity combining, predetection combining, pregroup combining, quadruple diversity combining, radio frequency combining.** *See also* **combiner.**

combining method: *See* **maximal ratio square diversity combining method.**

comic strip oriented mode: A mode of recording images on strip or roll film in which the top edge of each image, when viewed for reading, is parallel to the long edge, i.e., to the longitudinal axis, of the film. *Note:* Pictures taken in most personal cameras, most comic strips, sequences of photos on a page, and rows of frames on microfiche are made in the comic strip oriented mode. *See also* **cine oriented mode.**

COMINT: communications intelligence.

COMJAM: communications jamming.

comma-free code: A code (a) that is constructed so that any partial code word, beginning at the start of a code word but terminating prior to the end of that code word, is not a valid code word and (b) that permits the proper framing of transmitted code words when (i) external synchronization is provided to identify the start of the first code word in a sequence of code words and (ii) no uncorrected errors occur in the symbol stream. *Note:* Examples of comma-free codes are the variable-length Huffman codes. *Synonym* **prefix-free code.** *See also* **code, self-synchronizing code.**

command: **1.** An order for an action to take place. **2.** In data transmission, an instruction sent by the primary station instructing a secondary station to perform a specific function or operation. *Note:* An example of a command is, in high level data link control (HDLC) operation, the content of the control field of a command frame sent by the primary station or combined station to perform a specific link level function. **3.** In signaling systems, a control signal. **4.** The part of a computer instruction word that specifies the operation to be performed. **5.** A mathematical or logic operator that operates on an operand. *See* **asynchronous transmission command, copy command, data command, disconnect command, display command, editing command, file command, graph command, move command, print command, quit command, range command, spreadsheet command, telemetry command, theater area command, unnumbered command.** *Refer to* **Fig. S-22 (977).**

command and control (C^2): In military organizations and systems, (a) the exercise of authority and direction by an authorized commander over assigned forces in the accomplishment of an assigned mission and (b) the performance of functions through an arrangement of personnel, equipment, communications, facilities, and procedures used in planning, directing, coordinating, and controlling forces and operations in the accomplishment of the mission.

command and control (C^2) system: The personnel, facilities, communications, equipment, and procedures essential to a commander for planning, directing, and controlling operations. *See* **automated tactical command and control system, semiautomated tactical command and control system.**

command, control, and communications (C^3): **1.** In military systems, the capabilities required by commanders to accomplish their assigned missions. **2.** The use of communications systems in the exercise of command and control. *See also* **command and control, communications.**

command, control, communications, and intelligence (C^3I) reusable software system: A library database that (a) provides commonly used reusable software for Operations Support System (OSS) applications, (b) has a user interface and tool set for library management, storage, retrieval, and distribution, (c) provides text and diagrams for computer aided software engineering (CASE), and (d) serves as a text editor that can display and manipulate text and conduct keyword searches.

command-control-communications network: *See* **air defense command-control-communications network.**

command, control, computers, and communications (C^4): The use of computers and communications in the exercise of command and control. *See also* **command and control, communications, computer.**

command, control, computers, communications, and intelligence (C^4I): The use of computers, communications, and information in the exercise of command and control. *See also* **command and control, communications, computer.**

command frame: In data transmission, a frame that (a) contains a command and (b) is transmitted by a primary station for reception by a secondary station. *Note:* An example of a command frame, in high level data link control (HDLC), is a frame that (a) is transmitted by a primary station or a combined station and (b) has the remote, i.e., the receiving, combined station or secondary station address on it. *See also* **command, frame.**

command fuzing: Fuzing in which a device is detonated only when directed to detonate. *Note:* An example of command fuzing is fuzing in which a computer is used to determine relative position between a missile and a target. When the computed distance is a minimum or below a threshold distance, the fuze is ordered to detonate.

command guidance: Guidance in which intelligence transmitted to a missile from an outside source causes the missile to take a directed flight path.

command guidance system: A guidance system (a) that is based on signals transmitted from the ground, the air, or the sea to a controllable object, such as a spacecraft, missile, aircraft, ship, or land vehicle, and (b) in which the surface station or the aircraft station transmits to the mobile station that converts the command signals to maneuvering control signals.

command language: A source language that primarily consists of procedural operators that identify or invoke operations or functions that are to be executed. *Note:* An example of a command language is the set of terms in a command menu along with the rules for their use. *See also* **command menu, source language.**

command list: *Synonym* **command menu.**

command menu: A list of all the different commands that may be given to a computer or communications system by an operator. *Note:* Commands on a command menu may be selected by the operator usually by using (a) an indicator, such as a light pen, a cursor in conjunction with a keyboard, a reverse video bar followed by depression of the enter key, or depression of one or more keys on a keyboard, or (b) a computer program instruction. *Synonym* **command list.** *See also* **nested command menu.**

command net: In military communications systems, a network that connects an echelon of command with some or all of its subordinate echelons for the purpose of exercising command and control, usually from a command center or a command post. *See also* **network.**

command post: The point or the place from which command is exercised via communications systems and networks. *See* **airborne command post.** *See also* **command.**

command protocol data unit: In communications networks, a protocol data unit (PDU) transmitted by a logical link control (LLC) sublayer in which the PDU command/response (C/R) bit is equal to 0.

command station: *See* **telemetry command station, tracking telemetry command station.**

comm center: *Synonym* **communications center.**

commercial carrier: *Synonym* **common carrier.**

commercial communications common carrier: A privately owned organization, usually a corporation, that provides communications services to the public. Examples of commercial communications common carriers are AT&T, MCI, and SPRINT.

commercial communications service: A specific class of service available from a commercial communications common carrier. *Note:* Examples of commercial communications services are user-to-user service, user-to-nonuser service, public message service, full-rate facsimile service, and overnight telegram service. *See also* **class of service, common carrier, user.**

commercial refile: In military communications systems, the processing of a message from (a) a given military network, such as a tape relay network, a point-to-point telegraph network, a radiotelegraph network, or the Automatic Voice Network (AUTOVON), to (b) a commercial communications network. *Note:* Commercial refiling of a message usually requires a reformatting of the message, particularly the heading.

commission: *See* **public utility commission.**

committee: *See* **International Radio Consultative Committee, International Telegraph and Telephone Consultative Committee.**

common address multiple lines: A facility that permits a user to receive calls to a single address on more than one circuit.

commonality: **1.** A quality that applies to materiel or systems that (a) possess like and interchangeable characteristics enabling the materiel or systems to be used, operated, or maintained by personnel trained on other materiel or systems without additional specialized training, (b) have interchangeable repair parts or components, and (c) use consumable items interchangeably without adjustment. **2.** Pertaining to equipment or systems that have the quality of possessing like and interchangeable parts with another equipment or system. **3.** Pertaining to system design in which a given part can be used in more than one place in the system, i.e., subsystems and components of the system have parts in common. *Note:* Examples of commonality include the use of a firing pin that fits in many different weapons and the use of a light source that fits in many different types of fiber optic transmitters. *See also* **compatibility, interchangeability, interface, interoperability, transparency, transparent interface.**

common battery: **1.** Pertaining to the use of a single electrical power source to energize more than one circuit, component, equipment, or system. **2.** An electrical power source that serves as a central source of energy for more than one circuit. *Note 1:* In many telecommunications applications, the common battery is at a nominal −48 volts dc, relative to ground. *Note 2:* A telephone switch common battery supplies power to operate all directly connected instruments. *See also* **local battery.**

common battery exchange: In a manual or automatic telephone system, an exchange in which the power needed for (a) supervisory signals, (b) user calling signals, and (c) usually user voice, data, and video signals is supplied from a power source, such as a battery or power supply, located at the exchange. *Synonym* **battery exchange.**

common battery signaling (CBS): In telephone systems, signaling in which the signaling power of telephones is supplied by the serving switchboard or automatic switch. *Note 1:* In common battery signaling, "talking power" may be supplied by common or local battery.

common battery signaling exchange: A manual or automatic telephone exchange in which the power needed for (a) supervisory signals and (b) user calling signals is supplied from a power source, such as a battery or a power supply, located at the exchange while power needed for user voice, data, and video signals may be supplied from a power source, such as a battery or a power supply, located on the user premises. *Synonym* **battery exchange.**

common business oriented language (COBOL). *See* **COBOL.**

common carrier: 1. An organization that provides telecommunications facilities, services, or classes of service to the public for hire. **2.** An individual person, corporation, partnership, association, joint stock company, business, trust, or any other organized group, or any receiver or trustee engaged for hire, in interstate or foreign communications by any means, such as by wire, fiber optic cable, or radio or in interstate or foreign radio transmission of energy. *Note:* Radio broadcasting networks and systems are not considered as common carriers because of the one-way nature of their transmissions. *Synonyms* **carrier, commercial carrier.** *See also* **commercial communications common carrier, communications common carrier, other common carrier, radio common carrier, specialized common carrier, wireline common carrier.** *See also* **divestiture, resale carrier.**

common channel interoffice signaling (CCIS): In multichannel switched networks, signaling for a group of trunks in which the signaling information is encoded and transmitted over a separate voice channel using time division digital techniques. *See also* **channel, signal.**

common channel signaling: Signaling in which one of the channels on a multichannel link is used for controlling, accounting, and managing traffic on all channels of the link. *Note:* On a given link, the channel used for common channel signaling does not carry user information. *Synonym* **separate channel supervisory signaling.** *See* **associated common channel signaling.** *See also* **channel, in-band signaling, out-of-band signaling, signal, Signaling System No. 7, signal transfer point.**

common control: An automatic switching arrangement in which (a) the control equipment necessary for the establishment of connections is shared by being associated with a given call only during the period required to accomplish the control function for the given call and (b) the channels that are used for signaling, whether frequency bands or time slots, are not used for message traffic. *See also* **switching system.**

common control switching arrangement (CCSA): A switching arrangement in which switching for a private network is provided by one or more common control switching systems. *Note:* The switching systems may be shared by several private networks and also may be shared with the public telephone networks. *See also* **common control switching system, switching system.**

common control switching system: An automatic switching system in which common equipment is used to establish a connection. *Note:* In common control switching systems, after a connection is released, the common equipment is made available to establish other connections. *See also* **common control switching arrangement (CCSA), common equipment, communications system.**

common emergency frequency: *See* **military common emergency frequency.**

common equipment: 1. Equipment used by more than one system, subsystem, component, or other equipment, such as a channel or a switch. **2.** Items used by more than one channel or equipment function.

common frequency: *See* **convoy common frequency.**

common intermediate format (CIF): A video signal format that (a) is defined in International Telegraph and Telephone Consultative Committee Recommendations and (b) has four times as many pixels as the quarter common intermediate format (QCIF). *Synonym* **full common intermediate format.** *See* **quarter common intermediate format.**

Common Management Information Protocol (CMIP): A protocol used by an application process to exchange information and commands for the purpose of managing remote computer and communications resources. *See also* **common management information service, Simple Network Management Protocol.**

common management information service (CMIS): In communications networks, a service that specifies the service interface to the Common Management Information Protocol (CMIP). *Note:* In order to transfer management information between open systems by using common management information service (CMIS)/Common Management Information Protocol (CMIP), peer connections, i.e., associations, must be established. This requires the establishment of an application association, a session connection, a transport connection, and, depending on the supporting communications technology, network and link connections. *See also* **Common Management Information Protocol.**

common mode interference: 1. Interference that appears (a) between a signal lead and ground or (b) between a terminal of a measuring circuit and ground. **2.** Coherent interference that affects two or more elements of a network in a similar manner as distinct from locally generated noise or interference that is statistically independent between pairs of network elements. *See also* **differential mode interference, interference, mode.**

common mode rejection ratio (CMRR): The ratio of (a) the common mode interference voltage at the input of a circuit to (b) the corresponding interference voltage at the output. *See also* **interference, mode.**

common mode voltage: 1. In a two-terminal input device, the voltage common to both input terminals. **2.** In a differential amplifier, the unwanted part of the voltage between each input connection point and ground that is added to the voltage of each original signal. **3.** An uncompensated combination of (a) generator or receiver ground potential difference, i.e., voltage, (b) generator common return offset voltage, and (c) longitudinally coupled peak random noise voltage measured between the receiver circuit ground and receiver cable with the generator ends of the cable short-circuited to ground. **4.** In a two-terminal input receiver, the algebraic mean of the two voltages appearing at the receiver input terminals with respect to the receiver circuit ground. *See also* **mode.**

common return: A return path that is common to two or more circuits and that serves to return currents to their source or to ground. *See also* **circuit, ground return circuit, neutral, unbalanced wire circuit.**

common return offset: *Synonym* **common return offset voltage.**

common return offset voltage: The direct current (dc) voltage above ground of the common return of a line. *Synonym* **common return offset.** *See also* **balanced, unbalanced line, unbalanced wire circuit.**

common trunk: In telephone systems having a grading arrangement, a trunk accessible to all groups of the grading. *See also* **trunk.**

common trunk line: In fiber optic communications systems, a fiber optic transmission channel that (a) consists of a series of fiber optic transmitter and receiver pairs connected in tandem, (b) has a transmitter and a receiver located at each station or node, and (c) is capable of receiving, transmitting, and bypassing wavelength division or time division multiplexed signals at various points along the channel length. *See also* **bus.**

common user: In communications systems, pertaining to communications facilities and services provided to essentially all users in the area served by the system, rather than to one or a relatively small number of users, such as a closed user group with outgoing access.

common user circuit: A circuit designated to furnish a communications service to two or more users. *See also* **circuit.**

common user communications: Communications in which communications services are provided to many users.

common user communications service: A communications service established to provide communications services and support to a group of users that have a common interest or relationship, such as a group of users that are (a) in a single organization, (b) at a single command post, and (c) in the same building.

common user network: A system of circuits or channels allocated to furnish communications paths between switching centers to provide communications services on a common basis to all connected stations or users. *Note:* In military communications, a common user network is often called a "general purpose network." *See also* **channel, circuit, network.**

common user service: A communications service that is provided by a common user network. *See also* **common user network.**

communication intelligence: *Synonym* **communications intelligence.**

communications: 1. The branch of science and technology that is concerned with the process of representing, transferring, interpreting, or processing data among persons, places, or machines usually without loss of the meaning assigned to the data. **2.** The transfer of information among entities, such as persons, places, processes, and machines. **3.** The transfer of information between a source and a sink over one or more channels in accordance with a protocol and in a manner suitable for interpretation or comprehension by the receiver. **4.** Pertaining to the transfer of information among entities, such as persons, places, processes, and machines. **5.** Pertaining to the transfer of information between a source and a sink over one or more channels in accordance with a protocol and in a manner suitable for interpretation or comprehension by the receiver. *See* **air-air communications, air defense communications, air-ground communications, air traffic control communications, airways/air communications, base communications, binary synchronous communications, code-dependent communications, code-independent data communications, combined communications, common user communications, control communications, convoy communications, convoy internal communications, data communications, dedicated communications, distress communications, fiber optic communications, duplex communications, free-space optical communications, global system for mobile communications, ground to air communications, intelligence communications, joint communications, lateral communications, lightwave communications, long-haul communications, maritime air communications, meteor burst communications, military communications, one-way communications, public relations communications, radiotelegraph communications, radiotelephone communications, record communications, rescue communications, satellite communications, scene of action communications, search and rescue communications, secure communications, ship communications, ship-shore communications, signal communications, sound communications, surface to air communications, tactical communications, teletypewriter communications, terrestrial radio communications, two-way alternate communications, two-way simultaneous communications, weather communications, wireline communications.** *See also* **data, sink, source.**

communications adapter: *See* **integrated communications adapter.**

communications agency: A facility that (a) uses personnel and equipment to provide communications services to private organizations or the general public and (b) performs other communications-related functions, such as allowing charges to be appended to telephone bills, such as for the sending of telegrams or flowers or the purchasing of theater tickets. *See* **Defense Communications Agency.**

communications axis: *See* **signal communications axis.**

communications base section: *Synonym* **base communications.**

communications board: **1.** A printed circuit (PC) board that (a) is placed in each of two or more computer systems, such as mainframes, microprocessors, and personal computer (PC) systems, and (b) is used to enable the systems to share capabilities, such as share storage, printers, and monitors, thus forming a kind of local area network. **2.** A printed circuit (PC) board that enables one computer to communicate with another.

communications cable: *See* **intrusion-resistant communications cable.**

communications center: **1.** A facility that (a) is charged with the responsibility for handling and controlling communications traffic and (b) includes transmitting and receiving facilities. **2.** A facility that (a) serves as a node for a communications network, (b) is equipped for technical control and maintenance of the circuits originating, transitting, or terminating at the node, (c) may contain message center facilities, and (d) may serve as a gateway. *See* **filing communications center.** *See also* **information processing center, signal center.**

communications channel: *See* **channel.**

communications common carrier: An organization, agency, or system that operates as a common carrier and that provides communications services to public and private organizations and to the general public. *See* **commercial communications common carrier.**

communications control character: *Synonym* **transmission control character.**

communications control procedure: **1.** *Synonym* **data communications control procedure.** **2.** *See* **Advanced Data Communications Control Procedure.**

communications deception: In military communications systems, the use of devices, operations, and techniques with the intent of confusing or misleading the user of a communications link or a navigation system. *Note:* Communications deception may be accomplished by the radiation, reradiation, reflection, absorption, and alteration of signals with the intent of (a) confusing operating personnel, (b) misleading operating personnel, (c) introducing errors in transmissions, and (d) causing misinterpretation of messages. *See* **manipulative communications deception.** *See also* **electronic deception.**

communications device: *See* **low power communications device.**

communications-electronics (C-E): The specialized field that (a) is concerned with the use of electronic devices and systems for the acquisition, acceptance, processing, storage, display, analysis, protection, disposition, and transfer of information and (b) includes the wide range of responsibilities and actions relating to (i) electronic devices and systems used in the transfer of ideas and perceptions, (ii) electronic sensors and sensory systems used in the acquisition of information devoid of semantic influence, and (iii) in military systems, electronic devices and systems intended to allow friendly forces to operate in hostile environments and to deny to hostile forces the effective use of electromagnetic resources. *See* **theater director of communications-electronics.** *See also* **communications.**

communications emergency frequency: *See* **air-ground communications emergency frequency.**

communications equipment: Equipment used in communications systems and networks. *See* **data communications equipment, high performance communications equipment, low performance communications equipment, mobile communications equipment, transportable communications equipment.**

communications exercise: The transmission, reception, or processing of information specifically to evaluate the efficiency of communications facilities, procedures, personnel, and training.

communications facility: An installation or equipment that provides communications services. *See* **internal communications facility, record communications facility.**

communications guard: *See* **radio communications guard.**

communications intelligence (COMINT): Technical and intelligence information that (a) is derived from foreign communications by other than the intended recipients and (b) is obtained by using communications systems, procedures, and equipment to intercept transmissions. *Synonym* **communication intelligence.** *See also* **communications deception, electronic deception, electronic intelligence, signal intelligence.**

communications intercept: A reception of a signal in which the operator is able to read words, letters, or numbers, or understand codes, such as the international Morse code, that were not intended for the operator.

communications jamming (COMJAM): 1. The portion of electronic jamming that is directed against communications circuits and systems. **2.** The prevention of successful radio communications by the use of electromagnetic signals, i.e., the deliberate radiation, reradiation, or reflection of electromagnetic energy with the objective of impairing the effective use of electronic communications systems. *Note:* The aim of communications jamming is to prevent communications by electromagnetic means, or at least to degrade communications sufficiently to cause delays in transmission and reception. Jamming may be used in conjunction with deception to achieve an overall electronic countermeasure (ECM) plan implementation.

communications line: A land, sea, or air route that (a) connects an operating unit with one or more bases of operations and along which supplies and reinforcements may move, (b) consists of the path, means, and media of communications maintained in support of operations, and (c) is maintained between higher and lower echelons of control, between supporting and supported elements, and among other elements involved in a given situation or operation, such as rescue, fire fighting, disaster area control, medical evacuation, and military operations. *Synonym* **line of communications.**

communications link: *See* **link.**

communications means: A system or mode of communications usually identified by the propagation medium, coding method, class of service, or operating organization. *Note:* Examples of communications means are telephone, telegraph, wire, fiber optic, radio, television, visual, messenger, teletypewriter, military, and commercial means.

communications method: *See* **broadcast communications method, intercept communications method, receipt communications method, relay communications method.**

communications net: An organization, i.e., a network, of stations capable of direct communications on a common channel or frequency. *See* **alternate communications net, distress communications net, maritime amphibious communications net, maritime broadcast communications net, maritime distress communications net, maritime ship-shore communications net, maritime tactical air operations communications net.** *See also* **communications network.**

communications net operation: *See* **net operation.**

communications network: 1. An organization of stations capable of intercommunications but not necessarily on the same channel or frequency. **2.** Two or more interrelated circuits, switches, and terminals for communications. **3.** A combination of switches, circuits, and terminals that serve a given purpose. **4.** A combination of terminals and circuits in which transmission facilities interconnect the user station directly without the use of switching, control, or message processing centers. **5.** A combination of circuits and ter-

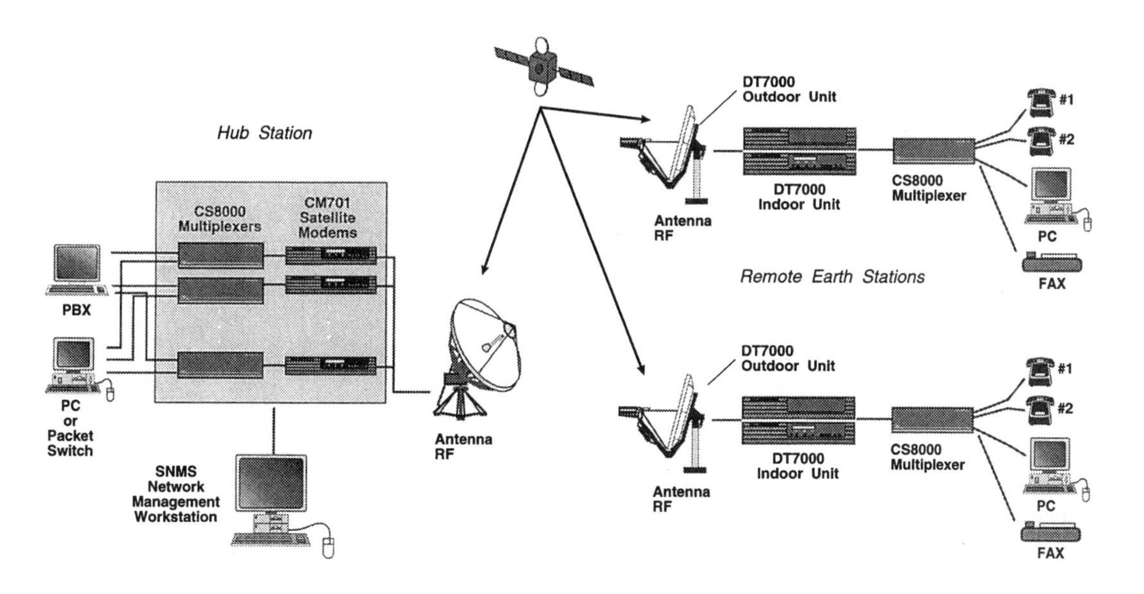

Fig. C-7. A typical single-channel-per-carrier (SCPC) satellite **communications network.** (Courtesy ComStream Company.)

minals serviced by a single switch or message processing center. *See* **air defense command-control-communications network, hybrid communications network, long-haul communications network, military fixed communications network, short-haul communications network.** *Synonym* **net.** *See also* **communications, network.** *Refer to* **Fig. C-7.** *Refer also to* **Fig. T-5 (1012).**

communications operating instruction: *Synonym* **signal operations instruction.**

communications organization: *See* **area communications organization.**

communications parameter: A transmission characteristic, feature, capability, or attribute of a system or a device, such as a host computer or a modem. *Note:* Examples of communications parameters are baud rate, bits per word, data signaling rate (DSR), bit rate, buffer capacity, and automatic checking capability.

communications plan: A plan that covers various aspects of communications systems development, installation, operations, and maintenance, such as communications requirements, equipment, communications networks, personnel, facilities, policies, and procedures. *See also* **communications plan format.**

communications plan format: The arrangement, layout, data items, and sequence of topics in a communications plan. *See also* **communications plan.**

communications precedence system: *See* **Joint Uniform Telephone Communications Precedence System.**

communications processor unit (CPU): A computer embedded in a communications system. *Note 1:* An example of a communications processor unit (CPU) is the message data processor of an Automatic Digital Network (AUTODIN) switching center. *Note 2:* "CPU" is also an abbreviation for "central processing unit" of a computer. *See also* **central processing unit, communications, computer.**

communications program: In a computer network, a program used to control the transmission of data or programs over a communications link. *Note:* An example of a communications program is a program that directs a personal computer to find, retrieve, and dispatch the data to be transmitted over a telephone line via a modem.

communications protection: The application of communications security (COMSEC) measures to telecommunications systems in order to (a) deny unauthorized persons access to sensitive unclassified information, (b) prevent disruption of telecommunications services, and (c) ensure the authenticity of infor-

communications publication: A publication that contains information concerning (a) communications personnel, intelligence, operations, and logistics and (b) communications systems, networks, and equipment, including their acquisition, operation, and maintenance.

communications reliability: The probability that information transmitted from (a) the communications station that serves a source user, such as a message originator or a call originator, to (b) the communications station that serves the destination user, such as the addressee or the call receiver, will arrive in a timely manner without loss of content.

communications satellite: An orbiting vehicle that relays signals between communications stations. *Note:* Active communications satellites receive, regenerate, and retransmit signals between stations. Passive communications satellites reflect signals between stations. *See* **active communications satellite, passive communications satellite.** *See also* **communications, satellite.** *Refer to* **Figs. P-10 (740), T-4 (994), W-4 (1087).**

communications satellite Earth station: An Earth station in the communications satellite service. *See also* **communications satellite service, Earth station.** *Refer to* **Fig. C-8.** *Refer also to* **Figs. N-16 (624), P-10 (740), T-4 (994).**

communications satellite service: 1. A communications service between Earth stations that use active or passive satellites for the exchange of messages in a fixed or mobile service. **2.** A communications service between an Earth station and stations on active satellites for the exchange of messages in a mobile service with a view to their retransmission to or from stations in the mobile service, such as from train to train, truck to truck, ship to ship, aircraft to aircraft, and ship to aircraft and all other combinations.

Fig. C-8. A single-channel-per-carrier (SCPC) **communications satellite Earth station.** (Courtesy ComStream Company.)

communications satellite space station: A space station on an Earth satellite in the communications satellite service. *See also* **communications satellite service, Earth station, space station.** *Refer to* **Fig. R-4 (800).**

communications saturation: *Synonym* **communications system saturation.**

communications security (COMSEC): The protection that results from all measures designed to deny unauthorized persons information of value that might be derived from the possession and study of telecommunications, or to mislead unauthorized persons in their interpretation of the results of such possession and study. *Note:* Communications security includes (a) cryptosecurity, (b) transmission security, (c) emission security, and (d) physical security of communications security materials and information. *See also* **alarm sensory, BLACK signal, bug, bulk encryption, communications, compromise, compromising emanations, controlled access, controlled area, cryptology, cryptosecurity, electronic security, emission control, emission security, failure access, information systems security, limited protection, physical security, RED/BLACK concept, RED signal, signal security, transmission security, vocoder.**

communications security (COMSEC) equipment: Equipment that (a) provides security to telecommunications systems by converting information to a form unintelligible to an unauthorized interceptor and by reconverting such information to its original form for authorized recipients, (b) specifically aids in, or is an essential element of, the conversion process, and (c) includes cryptoequipment, cryptoancillary equipment, cryptoproduction equipment, and authentication equipment. *See* **integrated communications security equipment.** *See also* **communications security, security.**

communications security information: Information that concerns communications security, communications security material, and communications security systems. *See also* **cryptoinformation.**

communications security (COMSEC) material: 1. An item that is used to secure or authenticate telecommunications. *Note:* Examples of communications security material are key, equipment, devices, documents, firmware, and software that embody or describe cryptographic logic and other items that per-

tions security material are key, equipment, devices, documents, firmware, and software that embody or describe cryptographic logic and other items that perform communications security functions. **2.** All documents, devices, equipment, or systems, including cryptomaterial, used in establishing or maintaining secure communications. *See also* **communications security, key.**

communications security policy: The overall policy applicable to all areas of communications security, including topics such as communications security monitoring, control systems, analyses, equipment, materials, information, custodianship, codes, aids, surveillance, and surveys.

communications security profile: An identification of all communications security measures and materials available for a given operation, system, or organization, including a determination of the amount and the type of use of these measures.

communications service: A service or a facility provided to users of a communications system that enables them to communicate with each other. *Note 1:* Examples of communications services are telephone, telegraph, and facsimile services. *Note 2:* Communications services are provided by common carriers, such as AT&T, MCI, and SPRINT. *See* **commercial communications service, common user communications service, dedicated communications service, personal communications service.**

communications silence: The avoidance of any type of transmission, emission, or radiation by any means, including radiation from receiving equipment. *Note:* An example of communications silence is the maintaining of a listening watch only if the receivers do not radiate. *See also* **radio silence.**

communications sink: 1. A device that receives data, control, timing, or other signals from communications sources. *Note:* Examples of communications sinks are telephone receivers and radio receivers. **2.** A place where signal energy from one or more sources is collected, absorbed, or dissipated. **3.** A signal receiving device, such as a photodetector in a fiber optic link or the load in a communications circuit. *See also* **communications source, sink, source.**

communications source: 1. A device that transmits data, control, timing, or other signals to communications sinks. *Note:* Examples of communications sources are telephone transmitters and radio transmitters. **2.** A device that generates information, control, or other signals destined for communications sinks. **3.** A place where energy originates for transmission to and collection, absorption, or dissipation at a sink. **4.** A signal generating device, such as a modulated laser in a fiber optic link. *See also* **communications source, sink, source.**

communications standard plan: *See* **aircraft communications standard plan, ship-fitting standard plan.**

communications subsystem: A functional unit or operational assembly that is smaller than the assembly it is a part of. *Note:* Examples of communications subsystems are (a) a satellite link with one Earth terminal in the continental United States (CONUS) and one in Europe, (b) the interconnect facilities at each Earth terminal of a satellite link, and (c) a fiber optic cable with its driver and receiver in either of the interconnect facilities. *See also* **assembly, communications, component, link.**

communications survivability (CS): The ability of communications systems to continue to operate effectively under adverse conditions, though portions of the system may be damaged or destroyed, by using various methods for maintaining communications services, such as by using alternate routing, different transmission media or methods, redundant equipment, and sites and equipment that are radiation-hardened. *See also* **electromagnetic survivability.**

communications system: A system (a) that consists of an integrated collection of individual communications networks, transmission systems, relay stations, tributary stations, data circuit-terminating equipments (DCEs), and data terminal equipments (DTEs) capable of interconnection and interoperation and (b) that has parts that serve a common purpose, are technically compatible, employ common procedures, respond to controls, and operate in unison. *See* **adaptive communications system, air defense communications system, air-ground worldwide communications system, asychronous communications system, Automatic Secure Voice Communications system, ciphony communications system, fiber optic communications system, flashing light communications system, hybrid communications system, integrated communications system, optical communications system, satellite communications system, space communications system, strategic military communications system, tactical communications system, visual communications system, wideband communications system.** *See also* **common control system, communications, error correcting system, error detecting system, hybrid communications network, link, neutral direct current telegraph system, personal communications system, polarential tele-**

graph system, protected distribution system, radio communications system, switching system, visual communications system, wideband communications system. *Refer to* **Figs. S-3 through S-6 (869), T-1 and T-2 (983).**

communications system abbreviation: *See* **aeronautical communications system abbreviation.**

communications system consolidation: A combination of two or more existing autonomous communications facilities into a single entity with all or most of the original capability and some of the original autonomy remaining. *Note:* Communications system consolidation is a facet of system rationalization.

communications system engineering: System engineering that includes (a) the translation of user communications requirements for the exchange of information into cost-effective and low-risk technical solutions, (b) the design and development of equipment, subsystems, and systems that meet these requirements, and (c) the integration of these parts into a complete entity at minimum cost over the entire system lifecycle. *See also* **communications, communications system.**

communications system plan: *See* **Aeronautical Emergency Communications System Plan.**

communications system saturation: The condition that exists in a communications system handling traffic at the maximum capacity of the system. *Synonym* **communications saturation.**

communications systems management (CSM): The planning, organizing, coordinating, directing, controlling, and supervising of communications systems and networks, including (a) the authority and responsibility for executing all applicable network planning functions, (b) the formulation, review, and revision of the applicable management information systems, (c) the issuance and the enforcement of standing operating procedures, standing communications operating instructions, and technical directives, (d) the development of routine logistical processes, including the supply, maintenance, and replacement of equipment, (e) the development of procedures for advice and assistance visits by representatives of other authorities, (f) the programming for acquisition of equipment or modifications to improve system efficiency and reliability, (g) the development of manning and training criteria for systems personnel, and (h) the preparation of budgetary policies within established rules. *Note 1:* Though communications systems management is a long-term support function, it may respond instantaneously to unforeseen or emergency situations. *Note 2:* Communications systems management functions may be performed by using the telecommunications management network and the communications management information system. *See also* **telecommunications management network.**

communications systems survivability: *See* **survivability.**

communications technology satellite: *See* **advanced communications technology satellite.**

communications test: Pertaining to the transmission and reception of information specifically intended to evaluate the efficiency of communications media or facilities.

communications theory: *Synonym* **communication theory.**

communications traffic: 1. The totality of transmitted and received messages and calls. **2.** The streams of messages transmitted and calls placed within and among networks, including user and system operator messages and calls.

communications watch: The (a) monitoring of one or more communications lines, frequencies, or channels to obtain information by listening to or receiving all transmissions on them and (b) transmitting and receiving messages as required. *See* **continuous communications watch.** *See also* **listening watch.**

communications zone: In a combat theater of operations, the rear part of the theater of operations that (a) is behind but contiguous to the combat zone and (b) contains the lines of communications, facilities for supply and evacuation, and other agencies required for the immediate support and maintenance of the field forces. *See also* **communications.**

communication theory: Theory that (a) is devoted to the probabilistic characteristics of the transmission of data in the presence of noise and (b) is used to advance the design, development, and operation of communications systems. *Note:* The mathematics of communication theory was considerably developed and advanced by many persons, including (a) Claude Elwood Shannon, who developed theories in combinational logic and contributed to advances in information theory, and (b) Norbert Wiener, who contributed to advances in cybernetics, i.e., in control theory. *Synonym* **communications theory.**

community antenna television (CATV): *Synonym* **cable television.**

community of interest: A group of users who (a) have a common interest and (b) generate a majority of their traffic in calls to each other. *Note:* Examples of communities of interest are groups of communications system users in a given geographical area, users in a given organization, or users that have a common goal. *Synonym* **special interest group.** *See also* **call, closed user group.**

community reception: In the broadcasting satellite service, the reception of emissions from a space station in the broadcasting satellite service by receiving equipment that (a) may be complex, (b) usually has antennas larger than those used for individual reception, and (c) is intended for use (i) by a group of the general public at one location or (ii) through a distribution system covering a limited area.

commutation: 1. In communications, the sampling of two or more channels, circuits, sources, or quantities in a cyclic or repetitive manner for transmission over a single channel using a multiplexing method for transmission. **2.** The process of converting alternating current (ac) to direct current (dc) in a direct current generator.

compact: *See* **data compaction.**

compact disk read-only memory (CD-ROM): A disk memory that (a) can only be read after data have been stored on it, i.e., the data on the disk cannot be changed, and (b) can store up to 600 Mb (megabytes) of computer-accessible data on a thin 4.625-in.-diameter metal plate for storing large-volume items, such as encyclopedias, dictionaries, and complex games.

compaction: In communications, computer, data processing, and control systems, the reduction of the spatial and temporal requirements for information by using an alternative manner or mode for representing the information. *Note 1:* Compaction does not change the information content. *Note 2:* Compaction is not the same as compression. *See* **curve fitting compaction, data compaction, fixed tolerance band compaction, floating point coding compaction, frequency analysis compaction, incremental compaction, probability analysis compaction, sample change compaction, slope keypoint compaction, variable precision coding compaction, variable tolerance band compaction.** *See also* **compression, data compaction.** *Refer to* **Figs. F-4 (356), P-13 (759).**

compander: A device that consists of a compressor and an expander, each of which may be used independently. *Synonym* **compandor.** *See also* **clipper, companding, compression ratio, compressor, expander, peak limiting, vogad.**

companding: An operation in which the dynamic range of signals is compressed before transmission and expanded to the original value at the receiver. *Note:* The use of companding allows signals with a large dynamic range to be transmitted over facilities that have a smaller dynamic range capability. Companding reduces the noise and crosstalk levels at the receiver. *See* **instantaneous companding, logarithmic companding, syllabic companding.** *See also* **clipper, compandor, compressor, image processing, expander.**

compandor: *Synonym* **compander.**

comparably efficient interconnection (CEI): In network interconnections, an equal-access concept, developed by the Federal Communications Commission (FCC), stating, in part, in FCC *Report and Order* June 16, 1986, that ". . . if a carrier offers an enhanced service, it should be required to offer network interconnection (or collocation) opportunities to others that are comparably efficient to the interconnection that its enhanced service enjoys. Accordingly, a carrier would be required to implement comparably efficient interconnection (CEI) only as it introduces new enhanced services." *See also* **unbundling.**

comparator: 1. In analog computing, a functional unit that compares two analog variables and indicates the result of the comparison. **2.** A device that compares two items of data and indicates the result of the comparison. **3.** A device for determining the dissimilarity of two items, such as the difference between two pulse patterns or words.

compartmentalization: *Synonym* **compartmentation.**

compartmentation: 1. The isolation of components, programs, and information in order to provide protection against compromise, contamination, or unauthorized access. **2.** The segregation of information into separate groups or categories for various purposes, such as access control and security classification. *Synonym* **compartmentalization.**

compatibility: 1. The capability of two or more items or components of equipment or material to exist or function in the same system or environment without mutual interference. **2.** In computing, the ability to execute a given program on different types of computers without modification of the program or of the computers. **3.** In command, control, communications, computer, and intelligence (C^4I) systems, the characteristic of systems, equipment, and components such that (a) information can be exchanged directly in usable form, (b) signals can be exchanged between them without the addition of devices for the specific purpose of achieving workable interface connections, and (c) the systems, equipment, and components being interconnected possess comparable radiation characteristics. *See* **electromagnetic compatibility.** *See also* **commonality, fully intermateable connectors, interchangeability, interoperability, mobile service, mobile station, portability, transportability.**

compatibility analysis: *See* **electromagnetic compatibility analysis.**

compatible data encryption: Data encryption that provides for secure transmission among networks with differing operating protocols. *See also* **encryption, protocol.**

compatible sideband transmission (CST): Independent sideband transmission (a) in which the carrier is deliberately reinserted at a lower power level after its normal suppression to permit reception by conventional amplitude modulation (AM) receivers and (b) that usually is single sideband (SSB) amplitude modulation equivalent (AME) transmission consisting of the emission of the carrier plus the upper sideband. *Synonyms* **amplitude modulation equivalent, compatible single sideband transmission.** *See also* **sideband transmission.**

compatible single sideband transmission: *Synonym* **compatible sideband transmission.**

compelled signaling: Signaling in which the transmission of each signal in the forward direction from an originating terminal is inhibited until an acknowledgment of the satisfactory receipt of the previous signal is received by the originating terminal. *See also* **acknowledge character, error control, negative acknowledge character, packet switching, signal.**

compensated optical fiber: An optical fiber with a refractive index profile that has been adjusted so that light rays (a) propagating in the higher refractive index portion of the core, i.e., near the core center, thus propagating a shorter distance at lower speeds because of the higher index material, and (b) propagating in the outer lower index material, thus undergoing total (internal) reflection, or bending back and forth, and thus traversing a longer path but propagating faster because of the lower index, arrive at the end of the optical fiber at the same time as the skew rays that travel a helical path, thus reducing modal dispersion to nearly zero. *Note:* Higher order modes have higher eigenvalues in the wave equation solutions, higher frequencies, and shorter wavelengths than lower order modes. *See also* **overcompensated optical fiber, undercompensated optical fiber.**

compensating equalizer: *See* **line temperature compensating equalizer.**

compensating phase shift: A coherent optical adaptive technique used to improve the power density, i.e., the irradiance, of electromagnetic wavefronts by changing the shape, phase, or character of the wavefront with optical systems or fields in order to adjust for unwanted variations of these parameters introduced by transmission media and system components, such as optical fibers, connectors, couplers, lenses, and the atmosphere.

compensation: *See* **luminance compensation, wavefront compensation.**

compensator: *See* **signal level compensator.**

competitive access provider (CAP): An organization that provides exchange access services in competition with an established local exchange carrier.

competitive chip: In time assignment speech interpolation (TASI) or digital speech interpolation (DSI), truncation of the initial part of a speech spurt, caused when all channels in a given direction of transmission are busy, and the transmission of the spurt must wait for an available channel.

compile: **1.** To translate a computer program expressed in a high level language into a program expressed in a lower level language, such as an intermediate language, assembly language, or a machine language. **2.** To prepare a machine language program, i.e., a target program, from a computer program, i.e., a source program, written in another programming language, by (a) making use of the overall logic structure of the source program, (b) generating more than one machine language instruction for each symbolic source program statement, or (c) performing the function of an assembler. *See also* **Ada®, assembly language, computer, computer language, computer oriented language, cross compiler, high level language, machine language, programmer, root compiler.**

compiler: **1.** A computer program that compiles. **2.** A computer program that converts computer programs written in a procedure-oriented language into a form suitable for execution on a computer. **3.** A computer program that, when executed, produces another computer program with new features, such as a capability to be directly executed by a computer as a machine language program. **4.** An artificial language used to generate, express, or write computer programs. *Synonym* **compiling program.** *See also* **compile.**

compiling program: *Synonym* **compiler.**

complement: *See* **nines complement, ones complement, radix complement, radix minus one complement, tens complement, twos complement.**

complementary network service (CNS): A communications service that (a) provides a means for an enhanced service provider customer to connect to a network and to the enhanced service provider and (b) usually consists of the customer local service, such as a business or a residence, and several associated service options, such as a call forwarding service. *See also* **call forwarding.**

complementary wave: A wave with a polarization that (a) is obtained or derived from the polarization or the modulation of another wave, (b) is derived in order to avoid interference with other waves, and (c) is at right angles to the given wavefront if the given wavefront is circularly or elliptically polarized.

completed call: **1.** A call in which connections have been established and the call originator and the call receiver are in communication with each other, i.e., are in the information transfer phase. **2.** A call in which the access phase of an information transfer transaction has ended, and the information transfer phase has begun. *See also* **finished call.**

completion objective: *See* **call completion objective.**

completion ratio: *See* **call completion ratio.**

complex: *See* **administrative management complex, customer management complex, Earth terminal complex, orderwire complex.**

complex pulse radar: A radar that has an output signal shape that may be described, for precise analytical purposes, in terms of a complex mathematical function, such as a complex function of time with the general equation $s(t) = x(t) + jy(t)$ where $s(t)$ is the complex signal function of time, $x(t)$ is the real part, $y(t)$ is the imaginary part, and j is the imaginary component operator, i.e., the quadrature component operator, where $j = (-1)^{1/2}$.

complex transfer function: **1.** A transfer function that (a) has an imaginary component, i.e., a quadrature component, and (b) may also have a real component. **2.** In signal transmission, a transfer function that includes the phase transformation, as well as the attenuation transformation, that occurs to a signal propagating in a medium. *Note:* A complex transfer function may be determined from the two Fourier analyses of a 100-ps (picosecond) rectangular pulse entered into and received at the end of a length of optical fiber.

component: **1.** An assembly, or a part thereof, that (a) is essential to the operation of some larger assembly and (b) is an immediate subdivision of the assembly to which it belongs. *Note:* A radio receiver may be a component of a complete radio set consisting of a combined transmitter-receiver, i.e., a transceiver. The same radio receiver could also be a subsystem of the combined transmitter-receiver, in which case the intermediate frequency (IF) amplifier section would be a component of the receiver but not of the radio set. Similarly, within the IF amplifier section, items, such as resistors, capacitors, vacuum tubes, and transistors, are components of that section. **2.** A part, or a combination of parts, that (a) has a specified function, (b) is usually installed or replaced as a whole, and (c) is usually expendable. *See* **cable component, cable core component, electric field component, fiber optic cable component, firewall component, functional component, magnetic field component, optical path component, quadrature component.** *See also* **assembly, communications subsystem.** *Refer to* **Fig. B-3 (71).**

component life: The period of acceptable usage of a component, part, or device after the expiration of which the likelihood of failure sharply increases. *Note:* In the interest of achieving a high level of reliability, components are removed from service prior to the expiration of their component life. *See* **indefinite component life, inphase component life, out-of-phase component life.** *See also* **bathtub curve, wear-out failure period.**

composite cable: A communications cable that (a) has both optical and metallic signal-carrying components and (b) is not a cable that has only one type of optical fiber and metallic strength members or armor. *See also* **hybrid cable.**

composite circuit: *Synonym* **composited circuit.**

composite code: In spread spectrum systems, a code that is obtained from two or more other distinct codes. *Note:* An example of a composite code is the gold code.

composited circuit: A circuit (a) that can be used simultaneously for either telephony and direct current (dc) telegraphy or for signaling and (b) in which separation between the two is accomplished by frequency division multiplexing. *Synonyms* **composite circuit, voice-plus circuit.** *See also* **circuit, speech-plus-duplex operation, speech-plus signaling, speech-plus telegraph.**

composite signaling (CX): Signaling that (a) provides direct current (dc) signaling and dial pulsing beyond the range of conventional loop signaling and (b) like DX signaling, permits duplex operation, i.e., it provides simultaneous two-way signaling. *Synonym* **CX signaling.** *See also* **dial signaling, direct current signaling, pulse, signal.**

composite two-tone test signal: A test signal that (a) is composed of two different frequencies of the same power level and (b) is used for intermodulation distortion measurements in which the intermodulation products of the two frequencies, f_1 and f_2, are $2f_1 - f_2$ and $2f_2 - f_1$. *See also* **intermodulation distortion, signal.**

composite video signal: In television systems, a video signal in which synchronizing information, i.e., synchronizing pulses, and picture information, including chroma information, i.e., color information, if any, are combined.

compound: *See* **flooding compound.**

compound glass process: *Synonym* **double crucible process.**

compound lens: A lens that (a) is composed of two or more separate pieces of optical material, such as glass, plastic, or mineral, and (b) may be cemented together. *Note:* An example of a common compound lens is a two-element objective, one element being a converging lens, i.e., a convex lens, of crown glass and the other a diverging lens, i.e., a concave lens, of flint glass. The combination of suitable optical materials, properly ground and polished, reduces aberrations usually present in a single lens. *See also* **aberration, converging lens, crown glass, diverging lens, flint glass, single lens.**

compound signal: In alternating current (ac) signaling, a signal consisting of the simultaneous transmission of more than one frequency. *Note:* An example of a compound signal is a dual tone multifrequency (DTMF) signal. *See also* **dual tone multifrequency signaling, signal.**

compress: *See* **data compaction, data compression, signal compression.**

compressed dialing: Dialing that makes use of a shortened directory with shorter telephone numbers to obtain frequently called numbers.

compression: 1. In signaling, the reduction of the dynamic range of a signal by controlling it as a function of the inverse relationship of its instantaneous value relative to a specified reference level. *Note 1:* Instantaneous values of the input signal that are low relative to the reference level are increased, and those that are high are decreased. *Note 2:* Compression is usually accomplished by separate devices called compressors and is used for many purposes, such as improving signal to noise ratios, preventing overload of succeeding elements of a system, or matching the dynamic ranges of two devices. *Note 3:* The amount of compression, usually expressed in dB, may be a linear or nonlinear function of the signal level across the frequency band of interest and essentially may be instantaneous or have fixed or variable delay times. *Note 4:* Compression always introduces distortion, which usually is not objectionable, provided that the compression is limited to a few dB. **2.** In facsimile systems, operation in which the number of pels scanned on the object is larger than the number of encoded bits of picture information transmitted. *See* **bandwidth compression, data compression, facsimile data compression, luminance range compression, optical pulse compression, radar pulse compression, signal compression.** *See also* **compander, compression ratio, compressor, expander, expansion, level, redundancy.**

compression and expansion: *See* **linked compression and expansion.**

compression radar: *See* **pulse compression radar.**

compression ratio: 1. The ratio of (a) the dynamic range of compressor input signals to (b) the dynamic range of the compressor output signals. *Note:* The compression ratio usually is expressed in dB (decibels). Thus, a 40-dB input range compressed to a 25-dB output range would be equivalent to a 15-dB compression. **2.** In digital facsimile systems, the ratio of (a) the total pels scanned for the object to (b) the total encoded bits sent for picture information. **3.** The ratio of (a) the gain of a device at a low power level to (b) the gain at some higher level. *Note:* The compression ratio is usually expressed in dB (decibels). *See also* **compandor, compression, expander, level.**

compression sensor: *See* **fiber optic longitudinal compression sensor, fiber optic transverse compression sensor.**

compressor: A device that (a) has a nonlinear gain characteristic, (b) has a lower gain at higher input levels than it does at lower input levels, and (c) is usually used to allow signals with a large dynamic amplitude range to be sent through devices and circuits with a smaller amplitude range capability. *Note:* An example of a compressor is an amplifier that (a) is usually part of a signal transmitter and (b) has a gain that always decreases with increasing input signal level in order to partly compensate for amplitude variation. The process may be reversed by an expander at the receiver. *See also* **attack time, automatic gain control,**

clipper, companding, compandor, expander, limiter, peak limiting, vogad.

compromise: 1. The known or suspected exposure of clandestine personnel, installations, or other assets, or of classified information or material, to one or more unauthorized persons. **2.** The disclosure of cryptographic information to one or more unauthorized persons. **3.** The recovery of plain text of encrypted messages by one or more unauthorized persons through cryptanalysis methods. **4.** The disclosure of any information or data to one or more unauthorized persons. **5.** A violation of the security policy of an automated information system such that an unauthorized disclosure of sensitive information may have occurred. **6.** The availability, exposure, or transfer of information to unauthorized persons through any means, such as loss, photography, theft, and breaking of codes. *See also* **communications security.**

compromise equalizer: *See* **fixed compromise equalizer.**

compromising emanation: 1. An unintentional intelligence-bearing signal that, if intercepted and analyzed, discloses the information transmission received, handled, or otherwise processed by communications and information processing equipment. **2.** In a circuit that is transmitting information, radiation from the circuit that can disclose information to unauthorized persons or equipment. *See also* **communications security, RED/BLACK concept, TEMPEST.**

COMPUSEC: computer security.

compute mode: The operating mode of an analog computer during which the solution is present or in progress at the output. *See also* **set mode.** *Synonym* **operate mode.**

computer: 1. A device that (a) accepts data, processes the data in accordance with a stored program, and supplies the results and (b) usually consists of input, output, storage, arithmetic, logic, and control units. **2.** A functional unit that can perform substantial computation, including numerous arithmetic operations or logic operations, without human intervention during a run. *Note 1:* A computer is distinguished from similar devices, such as hand-held calculators and certain types of control devices that are not considered as computers. *Note 2:* Computers may be classified into microcomputers, minicomputers, and main frame computers, based on their size. These distinctions are rapidly disappearing as the capabilities of even the smaller units have increased. Microcomputers now usually are more powerful and versatile than the minicomputers and the main frame computers were a few years ago. *See* **analog computer, arbitrary sequence computer, asynchronous digital computer, digital computer, host computer, hybrid computer, laptop computer, minicomputer, multiterminal computer, optical computer, palmtop computer, parallel computer, personal computer, satellite computer, self-adapting computer, self-organizing computer, sequential computer, serial digital computer, special purpose computer, stored program computer, synchronous digital computer.** *See also* **assembly language, central processing unit, communications processor unit, compile, computer language, concentrator, front end processor, high level language, machine language, multiprocessing, overlay, programmer, software.** *Refer to* **Fig. T-10 (1041).**

computer aided acquisition and logistics support: *See* **joint computer aided acquisition and logistics support system.**

computer aided manufacturing (CAM): The use of computers and computer programs in the design and the construction of systems and components, such as in the design and the manufacture of automobiles, ships, and chips. *Synonym* **computer assisted manufacturing.** *See also* **manufacturer automation protocol.**

computer aided software engineering (CASE): The branch of science and technology that is devoted to the use of computers to perform software engineering, i.e., to design, develop, and produce software. *Note:* Design and performance criteria include accuracy, efficiency, flexibility, comprehensibility, maintainability, time, and cost. *Synonym* **computer assisted software engineering.** *See also* **software, software engineering.**

computer aided software engineering (CASE) technology: Technology that makes use of computer aided software engineering (CASE) to enhance the development of systems design and development in order to improve accuracy, efficiency, flexibility, comprehensibility, maintainability, time, and cost of systems. *Synonym* **computer assisted software engineering (CASE) technology.** *See also* **computer aided software engineering, software, software engineering.**

computer architecture: 1. In a computer, (a) the physical configuration, logical structure, formats, protocols, and operational sequences for processing data, (b) the controls used to manage the configuration and the operations, and (c) possibly the word lengths, instruction codes, and interrelationships among the main parts of the computer or a group of computers. **2.** The documentation describing the principles, physical configuration, functional organization, operational procedures, and data formats used as the basis for the design, construction, modification, and operation of a computer. *See also* **network architecture, systems architecture.**

computer assisted manufacturing: *Synonym* **computer aided manufacturing.**

computer assisted software engineering (CASE): *Synonym* **computer aided software engineering.**

computer assisted software engineering technology: *Synonym* **computer aided software engineering technology.**

computer base: *See* **trusted computer base.**

computer-based office work: *Synonym* **office automation.**

computer code: *Synonym* **machine code.**

computer conferencing: 1. Teleconferencing supported by one or more computers. **2.** An arrangement in which access, by multiple users, to a common database is mediated by a controlling computer. **3.** The interconnection of two or more computers working in a distributed manner on a common application process.

computer-dependent language: *Synonym* **assembly language.**

computer environment: *See* **portable operating system interface for computer environment.**

computer graphics: 1. Graphics implemented through the use of computers. **2.** Methods and techniques for converting data to or from graphic displays via computers. **3.** The branch of science and technology devoted to methods and techniques for converting data to or from visual presentations by using computers.

computer instruction: 1. A instruction in a computer program. **2.** An instruction in machine code that (a) is directly executable by a computer and (b) usually is derived from a computer program instruction using a programming language, such as ADA, COBOL, or BASIC, from which the computer compiles a new program consisting of instructions written in machine language using machine codes. *Note:* "Computer instruction" is not synonymous with "machine instruction." "Machine" is more general than "computer." *See also* **machine instruction.**

computer interface: *See* **host computer interface.**

computer language: A language used to program a computer. *Note:* A computer language may be a high level language, an assembly language, or a machine language. *See also* **assembly language, compile, high level language, language, machine language.**

computer net: A communications net designed for the transfer of computer-generated data taken directly from computer output terminals or devices.

computer network: 1. A network of data processing nodes that are interconnected for communications purposes. **2.** A communications network in which the end instruments are computers, i.e., two or more computers connected together by a communications network. *Synonym* **application network.** *See also* **computer, network.**

computer oriented language: 1. A programming language in which words and syntax are designed for use on a specific computer or class of computers. **2.** A programming language that is dependent upon or reflects the structure of a given computer or class of computers. *Synonym* **machine oriented language.** *See also* **assembly language, compile, high level language.**

computer output microfilm (COM): In display systems, a microfilm that contains data that are recorded directly from computer-generated signals that may be used to control a display device, such as a cathode ray tube (CRT), an array of light-emitting diodes (LEDs), a gas panel, an array of lasers, or a fiber optic aligned bundle that leads to the faceplate of a fiberscope, the display being used to expose microfilm. *Note:* The abbreviation "COM" is also used for the process of microfilming, for the microfilmer that performs the process, and for microform. *See also* **aligned bundle, computer output microform, fiberscope, microfilm, microform.**

computer output microform (COM): In display systems, a microform that contains data that are recorded directly from computer-generated signals that may be used to control a display device, such as a cathode ray tube (CRT), an array of light-emitting diodes (LEDs), a gas panel, an array of lasers, or a fiber optic aligned bundle that leads to the faceplate of a fiberscope, the display being used to expose microfilm. *Note:* The "M" in the abbreviation "COM" is also used for the process of microfilming, for the microfilmer that performs the process, and for microfilm. *See also* **computer output microfilm, fiberscope, microfilm, microform.**

computer patch: *Synonym* **computer program patch.**

computer peripheral: Pertaining to peripheral equipment connected to, or associated with, a computer, i.e., equipment that is online or offline with respect to a computer, or the equipment itself.

computer peripheral device: An auxiliary device under direct or indirect control of a computer, i.e., a device that is online or offline with respect to a computer. *Note:* Examples of devices that might serve as computer peripheral devices are card punches, card readers, printers, magnetic tape units, facsimile machines, display devices, and other computers, such as front end processors.

computer program: **1.** A plan or a routine for solving a problem on a computer. **2.** A sequence of instructions used by a computer to do a particular job or solve a given problem. **3.** A sequence of instructions that (a) are suitable for processing by a computer, (b) may include the use of an assembler, a compiler, an interpreter, or a translator to prepare and execute the sequence of instructions, and (c) may include statements and necessary declarations. **4.** To design, write, and test programs. *See also* **assembler, compiler, computer, job.**

computer programming: **1.** The planning, designing, writing, and testing of computer programs. **2.** The art and science of planning the solution of problems and using a computer for assisting in solving these problems. **3.** The process of setting up a computer, such as an analog computer, for specific automatic operation or series of operations.

computer program origin: The address assigned to the initial storage location of a computer program in main storage. *See also* **address, computer program, main storage.**

computer program patch: In a computer program, routine, or subroutine, a temporary change that (a) usually is accomplished by adding a section of new instructions to the existing program, routine, or subroutine, and using transfer or branch instructions to go to or from the added section, (b) allows the use of old programs to handle new situations or changes in existing situations, and (c) eventually, along with other added sections, may become a permanent part of the program. *Note:* If there are too many computer program patches in a program, the program may have to be rewritten to reduce loading time, execution time, and the probability of errors. *Synonym* **program patch.**

computer routine: A sequence of computer instructions that (a) has general use, (b) is usually called by another, usually larger, program, (c) may have to be executed repeatedly, and (d) accomplishes a specific task required by the calling program, such as (i) solve a specific mathematical function, such as a series expansion of sin x, (ii) retrieve a file from bulk storage, (iii) update a payroll record, or (iv) select a communications channel. *See also* **computer program, computer subroutine.**

computer science: The branch of science and technology that is concerned with methods and techniques relating to computing and data processing performed by automatic means.

computer security (COMPUSEC): **1.** Measures that control the confidentiality, integrity, and availability of the information processed and stored by a computer. **2.** The application of hardware, firmware, and software security features and measures to a computer system in order (a) to protect against, or prevent, the unauthorized disclosure, manipulation, or deletion of information and (b) to provide continuity of service. **3.** The security that results from the measures that control the confidentiality, integrity, and availability of the information processed and stored by a computer. *See* **automated information systems security.**

computer simulation: The use of a programmed computer to represent or model the features or behavior of a physical or abstract system so as to make the computer produce the results or permit analysis, as would the system being simulated. *Note:* Computer simulation may be performed on one computer by using different software for each of the simulated systems, such as the behavior of a high traffic communications system, the flight of a missile or a satellite, the economy of a country, war games, a railroad network, or another computer. *See also* **emulator, simulator.** *Refer to* **Fig. T-10 (1041).**

computer subroutine: A set of sequenced computer instructions or statements that may be used in one or more programs and at one or more points in a program. *Note:* Open subroutines are integrated into a program. Closed subroutines are arranged so that program control is shifted to them for execution. When execution is completed, program control is returned to the program in which the subroutine is embedded. *Note 2:* Subroutines may be dynamic, recursive, or reentrant.

computer system: A functional unit that (a) consists of one or more computers and associated software, (b) uses common storage for all or part of a program and also for all or part of the data necessary for the execution of the program, (c) executes user written or user designated programs, (d) performs user designated data manipulation, including arithmetic and logic operations, (e) may execute programs that modify themselves during their execution, and (f) may be a stand-alone system or may consist of several interconnected systems. *Synonym* **computing system.** *See* **heterogeneous computer system, homogeneous computer systems, process computer system, trusted computer system.**

computer system fault tolerance: The ability of a computer system to continue to operate correctly even though one or more of its components are malfunctioning. *Note:* System performance, such as diminished speed and throughput, may not be normal until the faults are corrected. *Synonym* **computer system resilience.**

computer system resilience: *Synonym* **computer system fault tolerance.**

computer tape: A flat strip of material that (a) may be punched or coated with magnetic material or optically sensitive material, (b) may be used by a computer to store data, and (c) may be read by a computer.

computer terminal: A device that enables its operator to interact with a computer. *See also* **data processing terminal.**

computer time: The time scale that (a) is used by a computer for the solution of a problem or for the execution of a computer program and (b) is usually different from calendar time, problem time, and real time. *Note:* In the computation of a parameter as a function of time, such as the points on the trajectory of a moving body, such as a satellite, missile, or artillery round, if a point, such as the ordinate, is computed for say 2-s (second) intervals of flight time (problem time), and if the value of 2 is inserted by the program into the equation for the ordinate as a function of time every 5 ms (milliseconds, real time), the motion is computed 400 times faster than the body actually moves in space (calendar time). Thus, a point on the trajectory at a future time can be calculated in less time than it takes for the moving body to reach the point. In this case, the computer is said to be operating in real time because there is time to influence the motion of the body before it moves much farther, i.e., to guide it.

computer word: In computing, a word that occupies one or more storage locations and is treated by computer circuits as a unit. *Synonym* **machine word.** *See also* **binary digit, byte, character, word.**

computing system: *Synonym* **computer system.**

COMSEC: communications security.

Com21™ system: An interactive international telecommunications system that (a) provides broadband multimedia services to homes and offices, (b) integrates fiber optic, cable television (CATV), and wireless (radio) technologies, (c) uses asynchronous transfer mode (ATM) and fast packet switching using 424-bit standard cells routed through the system, (d) provides a standard 2 Mb•s^{-1} (megabit per second) data signaling rate (DSR) and a 45 Mb•s^{-1} to 155 Mb•s^{-1} DSR for business applications, (e) has options for installations where certain facilities, such as CATV, are not available, and (f) may be accessed via existing telephone instruments and wiring.

concatenation: 1. To place in tandem. *See also* **tandem. 2.** The linking of components such that (a) the output of a given component is the input to the next component, and (b) each component interacts only with its immediate neighbor or neighbors. *Note:* Examples of concatenation are (a) the connection of two or more lengths of cable end to end and (b) the connection of the output of an amplifier to the input of another amplifier. *Synonym* **cascading.**

concave: Pertaining to a hollow curved surface, usually of a given entity, material, or representation of a given entity or material, such as a lens or a drawing of a lens. *Note:* If a concave item is embedded in a material, the material in which the item is embedded is convex.

concave lens: *Synonym* **diverging lens.**

concentrator: 1. In data transmission, a functional unit that (a) lies within a path, (b) permits a common path to handle more data sources than there are channels currently available within the path, (c) usually provides a communications capability between (i) many low speed, usually asynchronous, channels and (ii) one or more high speed, usually synchronous channels, and (d) on the low speed side, usually accommodates different speeds, codes, and protocols that usually operate in contention and require buffering. **2.** A device that connects a number of circuits, which are not all used at once, to a smaller group of circuits for economy. *See* **fiber optic concentrator.**

concentricity: *See* **core-cladding concentricity, core reference surface concentricity.**

concentricity error: When used in conjunction with a tolerance field to specify the core/cladding geometry of an optical fiber, the distance between (a) the center of the two concentric circles that specify the cladding diameter and (b) the center of the two concentric circles that specify the core diameter. *Synonyms* **core to cladding concentricity, core to cladding eccentricity, core to cladding offset, optical fiber concentric-

ity error. *See also* **cladding, cladding diameter, core, core diameter, tolerance field.**

concentric lens: A lens in which the centers of curvature of the surfaces coincide and thus have a constant radial thickness in all zones.

concentric line: *Synonym* **coaxial cable.**

concept: *See* **red-black concept.**

concurrent operation: 1. An operation that occurs within a time interval during which another operation shares a common resource. *Note:* For example, several operations are concurrent when they are executed by multiprogramming techniques in a computer having a single instruction control unit. **2.** In data link operations, an operation in which two or more data links are used during a single, usually short, time interval, while adhering to the protocols of each link without performing data forwarding among the links. *Synonym* **multitasking.**

concurrent process: A process that occurs during the same period as another process, such as in parallel on multiple processors or asynchronously on a single processor. *Note:* Concurrent processes may interact with one another. One process may suspend execution pending receipt of data from another process, or two or more processes may start and stop in accordance with the occurrence of external events. *See also* **sequential process.**

condenser storage: *Synonym* **capacitor storage.**

condensing lens: A lens or a lens system of positive lens power used for condensing, i.e., for converging, radiant energy from a source onto an object or a surface.

condition: *See* **A condition, decibel-above-isotropic condition, exception condition, extended operating condition, initial condition, launch condition, modulation condition, not-ready condition, prelasing condition, ready condition, reeling condition, significant condition, standard operating condition, steady state condition, trunk-free condition, trunk seized condition.**

conditional control program: A computer program that allows alternate branching of program flow depending upon whether or not specified conditions are met. *Note:* An example of a conditional control program is a program in which IF . . . THEN statements are used. *See also* **branch, computer program, control program, IF . . . THEN statement.**

conditional information content: The information content conveyed by the occurrence of an event of definite conditional probability, given the occurrence of another event. *See* **conditional information entropy.**

conditional information entropy: In information theory, the mean of the measure of the information content conveyed by the occurrence of one of a finite set of mutually exclusive and jointly exhaustive events, each with definite conditional probability, given the occurrence of another set of mutually exclusive events. *Synonyms* **average conditional information content, mean conditional information content.**

conditional jump: In computer programming, a jump that takes place when the program instruction that specifies it is executed, and specified conditions have been met. *See also* **computer programming, jump, unconditional jump.**

conditioned baseband representation: *Synonym* **nonreturn to zero mark.**

conditioned biphase coding: *Synonym* **biphase mark coding.**

conditioned circuit: A circuit that has conditioning equipment to obtain the desired characteristics for voice, video, or data transmission. *See also* **conditioning equipment, equalization.**

conditioned diphase modulation (CDM): Modulation in which diphase modulation and signal conditioning are used to obtain desired signal characteristics, such as by eliminating the direct current (dc) component, enhancing timing recovery, and facilitating transmission over voice frequency (VF) circuits or coaxial cables. *See also* **modulation.**

conditioned loop: A loop that (a) has conditioning equipment to obtain the desired line characteristics for voice, video, or data transmission, (b) is used to improve the amplitude versus bandwidth characteristics of the circuit, and (c) provides for impedance matching. *See also* **conditioning equipment, equalization, loop.**

conditioned voice grade circuit: A voice grade circuit that has signal conditioning equipment to equalize envelope or phase delay response so as to improve data transmission through the circuit. *See also* **voice grade circuit.** *Synonym* **data grade circuit.**

conditioning: *See* **air conditioning.**

conditioning equipment: 1. At junctions of circuits, equipment used to obtain desired circuit characteristics, such as matched transmission power levels, matched impedances, and equalization between facilities. **2.** In a desired frequency range, corrective networks used to improve data transmission, such as equalization of the insertion loss versus frequency characteristic and equalization of the envelope delay distortion. *See also* **building out, conditioned circuit, conditioned loop, equalization.**

condition mode: *See* **initial condition mode.**

condition number: *See* **significant condition number.**

conductance: A measure of the ability of a material to conduct or carry an electrical current in relation to the magnitude of the applied voltage, equivalent to the reciprocal of resistance. *Note 1:* Conductance usually is expressed in mhos, from ohm spelled backward. The greater the conductance is, the greater the current that can be carried for a given potential difference, i.e., given voltage, whereas the greater the resistance is, the less the current. *Note 2:* The current in a component is given by the relation $I = \delta V$, where I is the current, δ is the conductance, and V is the voltage across the terminals of the component. If δ is in mhos and V is in volts, I will be in amperes. *Note 3:* Conductance should not be confused with the electrical conductivity of a material. The current density in a medium is given by the relation $J = \sigma E$, where J is the current density, σ is the electrical conductivity, and E is the electrical field strength. If σ is in mhos per meter and E is in volts per meter, J will be in amperes per square meter. *See also* **resistance.**

conducted coupling: *Synonym* **direct coupling.**

conducted interference: 1. Interference resulting from noise or unwanted signals entering a device by conductive coupling, i.e., by direct coupling. **2.** An undesired voltage or current generated within, or conducted into, a receiver, transmitter, or associated equipment, and appearing at the antenna terminals. *See also* **antenna, coupling, interference.**

conduction: *See* **acoustical conduction, optical conduction.**

conduction band: 1. In a semiconductor, the range of electron energy, higher than that of the valence band, sufficient to make the electrons free to move from atom to atom under the influence of an applied electric field and thus constitute an electric current. **2.** In the atomic structure of a material, a partially filled or empty energy level in which electrons are free to move, thus allowing the material to conduct an electrical current upon application of an electric field by means of an applied voltage.

conductive coupling: Energy transfer by means of direct physical contact, such as by means of a wire, i.e., coupling achieved without the use of inductive or capacitive coupling. *Note 1:* Conducted coupling may be achieved by wire, resistor, or common terminal, such as a binding post or metallic bonding. *Note 2:* Conductive coupling passes the full spectrum of frequencies, including direct current (dc). *Synonym* **direct coupling.** *See also* **capacitive coupling, coupling, inductive coupling.**

conductivity: *See* **electrical conductivity, photoconductivity.**

conductor: 1. In electric circuits, a material that readily permits a flow of electrons through itself upon application of an electric field. *Note:* Examples of good electrical conductors are copper, aluminum, lead, gold, silver, and platinum. Examples of poor electrical conductors are carbon, iron, water, and air. Examples of nonconductors are dielectric materials, such as glass, plastics, and natural and synthetic fibers, such as cotton and nylon. **2.** In fiber optics, a transparent medium that is capable of transmitting or conveying lightwaves for useful distances. *Note 1:* Examples of optical conductors are pure silica glass and transparent plastic. *Note 2:* To avoid confusion in practice, the term "conductor" should be used only in relation to electrical currents, perhaps because optical conductors usually are electrical nonconductors, and many optical nonconductors are electrical conductors, such as the metals. Some materials, such as black plastic and black paper, are nonconductors of electrical currents and lightwaves. *See* **electrical conductor, fortuitous conductor, lightning down conductor, optical conductor.** *See also* **conductance, electrical conductivity.**

conductor loss: *See* **connector-induced optical conductor loss.**

conduit: A protective tube that (a) usually is installed externally to walls and foundations and (b) houses and provides protection for wires or fiber optic cables that may be preinstalled within it or pulled through it after installation. *See also* **incoherent bundle.**

cone: *See* **acceptance cone, antenna blind cone.**

cone of silence: In an antenna radiation pattern, the cone-shaped region of space with its apex at an antenna, in which signals from the antenna are greatly reduced in amplitude. *Note:* The near-zero or zero portions of the antenna radiation pattern may be inverted cone-shaped space directly over the antenna, particularly if the antenna is a vertical dipole or vertical stacked dipole antenna. The cone of silence occurs directly over the antenna towers of some forms of radio beacons. *See also* **dipole antenna, radiation pattern, radio beacon.**

CONEX: connectivity exchange.

conference: *See* **add-on conference, attendant conference, preset conference, random conference, telecommunications conference.**

conference call: 1. A network-provided service feature that allows a call to be established among three or more stations in such a manner that each of the stations is able to communicate with all other stations. **2.** A call in which more than two access lines are connected. *See* **meet-me conference call, progressive conference call.** *See also* **add-on conference, attendant conference, bridging connection, call, computer conferencing, group alerting and dispatching system, multiaddress calling facility, multiple call, service feature, teleconference.**

conference circuit: A circuit that (a) allows simultaneous communications between two or more stations for conference purposes and (b) allows all stations to originate and receive messages when connected to the circuit.

conference operation: In a communications network, operation that allows a call to be established among three or more stations in such a manner that each of the stations is able to communicate directly with all the other stations. *Note 1:* In radio systems, the stations may receive simultaneously but must transmit one at a time. *Note 2:* During conference operation, the common operational modes are "push to talk" for telephone operation and "push to type" for telegraph and data transmission. *See also* **computer conferencing, data conferencing repeater.**

conference repeater: A repeater that (a) connects several circuits, (b) receives telephone or telegraph signals from any one of the circuits, and (c) automatically retransmits the signals over all the other circuits. *See also* **data conferencing repeater.**

conferencing: *See* **computer conferencing.**

CONFIG.SYS: In a disk operating system (DOS), such as the IBM DOS or MS DOS for a personal computer (PC), the file that (a) is used to determine the various device drivers that must be loaded and (b) usually is stored in the boot disk or diskette. *See also* **boot, disk, diskette, disk operating system.**

configurable station: In high level data link control (HDLC) operation, a station that (a) is a logical station and (b) has a mode-setting capability of being a different type of station at different times, such as a primary station, a secondary station, or a combined station at different times. *See also* **combined station, high level data link control, logical station, primary station, secondary station.**

configuration: 1. A specific interconnection of the hardware and software components of a system, such as (a) the interconnection of storage, buffer, input, output, main frame, monitor, and printing units of a general-purpose computer system and (b) the connection of a specific disk drive to the computer so that various files are always retrieved from and stored on a specific disk as a matter of convention or default in a personal computer (PC). **2.** In communications, computer, data processing, and control systems, an arrangement of the functional units that compose a system according to their nature, number, and main characteristics. *Note 1:* Configuration pertains to hardware, software, firmware, and documentation. *Note 2:* The configuration will affect system performance. *See* **bit configuration, default configuration, network configuration, reference configuration, relay configuration.**

configuration auditing: In system manufacturing, verifying that (a) all required components of a system have been produced, (b) the current system configuration meets specified requirements, and (c) the accompanying documentation completely and accurately describes the system components and their interrelationships.

configuration control: 1. After establishing a configuration, such as that of a telecommunications or computer system, the evaluating and approving of changes to the configuration and to the interrelationships among system components. **2.** In distributed queue dual bus (DQDB) networks, the function that ensures the resources of all nodes of a DQDB network are configured into a correct dual bus topology. *Note:* Examples of configuration control functions include the head of bus, external timing source, and default slot generator functions. *See also* **configuration management.**

configuration identification: The identification and the definition of the configuration items in a system and the documentation that identifies and describes the configuration. *See also* **configuration, configuration item.**

configuration item: An item of hardware or software that may consist of many parts but is treated as a unit to effect configuration management and configuration control. *See also* **configuration control, configuration management.**

configuration item control: Evaluating, approving, recording, and reporting changes to, and the status of, a configuration item for the purposes of configuration management and configuration control. *See also* **configuration control, configuration item, configuration management.**

configuration management (CM): 1. The controlling and the recording of changes that are made to system hardware, software, firmware, and documentation throughout the lifecycle of a system. **2.** In acquisitions, the technical and administrative direction and surveillance actions taken to (a) identify and document functional and physical characteristics of items, such as communications and computer systems, (b) control changes to the items and their characteristics, and (c) record and report the changes and the implementation status of the items. **3.** In network management, (a) a set of primary functions that (i) control, identify, and collect data from network elements and (ii) provide data to network elements and (b) a set of subfunctions that include (i) setting the parameters that control operations, (ii) associating names with managed objects and sets of managed objects, (iii) initializing the closing down of managed objects, (iv) collecting information on the current network status, (v) obtaining reports on changes, and (vi) directing configuration changes. **4.** In system security, the management of security features and assurances through the controlling and the recording of changes that are made to system hardware, software, firmware, tests, test fixtures, and documentation throughout the lifecycle of the system. *See also* **configuration control.**

configuration scattering: The scattering of electromagnetic radiation caused by variations in the configuration of a propagation medium. *Note:* Examples of configuration scattering include scattering caused by (a) variations in geometry of an optical fiber, (b) variations in the refractive index profile of an optical fiber, and (c) discontinuities in a waveguide cross section. *See also* **optical fiber scattering.**

confinement: In database management systems, the prevention of unauthorized access to specified information in a file during authorized file access. *See also* **integrity, protection.**

confinement factor: For a given guided mode of an electromagnetic wave propagating in a waveguide, the ratio of (a) the power within the guiding layer, such as the core of an optical fiber or the inner layer of a planar waveguide, to (b) the total guided power.

confirmation: *See* **clear confirmation, delivery confirmation.**

confirmation signal: *See* **clear confirmation signal.**

confirmation signaling: On some Automatic Voice Network (AUTOVON) intertoll trunks, signaling that ensures error-free transmission of dialed information by returning a unique digit-dependent signal from the far end as each digit is sent over the trunk. *See also* **signal.**

confirmation to receive: In facsimile systems, a signal from an International Telegraph and Telephone Consultative Committee (CCITT) Group 1, 2, or 3 facsimile receiver, indicating that it is ready to receive picture signals. *See also* **facsimile.**

conformance statement: *See* **protocol implementation conformance statement.**

conformance statement pro forma: *See* **protocol implementation conformance statement pro forma.**

conformance test: A test performed by an independent body to determine if a particular system or piece of equipment satisfies the criteria stated in a specified controlling document, such as an American National Standard, a federal standard, a military standard, or a military specification.

congestion: 1. In a communications system, a state or a condition that occurs when more users attempt to use a service than the system is capable of handling. **2.** In a communications system, a state that exists when a call is placed and all circuits capable of routing the call are busy, i.e., there are no idle circuits to handle the call. **3.** In a communications system, a state that exists when a call is placed and at least one circuit required to route the call is busy. **4.** In a saturated communications system, the condition that exists when one additional demand for service occurs. *See* **frequency spectrum congestion, network congestion, reception congestion.** *See also* **collision, contention, saturation.**

congestion signal: *See* **international congestion signal, national circuit group congestion signal, national switching equipment congestion signal.**

congruency: In facsimile systems, a measure of the ability of a facsimile transmitter or receiver to perform in a manner identical to the manner in which the corresponding equipment of another facsimile system performs.

conical beam split radar: A radar that (a) has an antenna with an output beam that is spun about an axis so as to describe a cone in space, (b) distributes energy in two or more directions at any given instant, and (c) combines the advantages of conical scanning and multiple beaming. *See also* **antenna, conical scanning, multiple beaming, radar.**

conical optical fiber: *Synonym* **optical taper.**

conical scanning: In radar systems, object scanning, i.e., target scanning, in which (a) the lobing consists of movement of the antenna axis and sampling of the echo signal at various points in the image plane, (b) the scanning is accomplished by rotating the antenna axis so as to describe a cone with its apex at the radar antenna site and its base at the object, i.e., at the target, thus providing an echo signal, (c) the echo signal can be made symmetrical during a single rotation, thus indicating that the antenna beam is on the object and the direction of the object can be determined, and (d) the scanning permits automatic tracking and eliminates the necessity of manual tracking by an operator who must otherwise maintain a balance of two pulses that become unequal when the antenna moves off the object. *See also* **conical beam split radar, lobing, paired lobing, radar, sequential lobing.**

conjugation: *See* **phase conjugation.**

conjugation gate: *Synonym* **AND gate.**

conjunction: The situation in which two or more satellites are in line with their parent body. *Note:* Examples of conjunction occur when (a) two or more Earth satellites and the Earth's center are in one straight line, i.e., are collinear, and (b) the Sun, Earth, and Mars are collinear.

conjunctive address group: An address group that has an incomplete meaning and must be used in combination with one or more other address groups.

connected network: *See* **fully connected network.**

connected station: *See* **directly connected station.**

connected topology: *See* **fully connected topology.**

connecting arrangement: In public switched telephone network operation, the arrangement in which equipment provided by the network, i.e., the common carrier, establishes the interconnection between (a) the equipment provided by the user, i.e., the customer or the subscriber, and (b) the facilities of the common carrier. *See also* **common carrier.**

connection: **1.** In communications systems, a provision for a signal to propagate from one point to another, such as from one circuit, line, subassembly, or component to another. **2.** In communications systems, an association established between functional units for conveying information. *See* **automatic sequential connection, back to back connection, bridging connection, circuit switched connection, cross connection, data circuit connection, data connection, dedicated connection, fiber optic cross connection, fiber optic interconnection, laser service connection, logical connection, multipoint connection, point to point connection, ring connection, star connection.** *See also* **circuit, connector, cross section, splice.**

connection in progress signal: A call control signal at the data circuit-terminating equipment/data terminal equipment (DCE/DTE) interface that indicates to the DTE that the establishment of the data connection is in progress, and that the ready for data signal will follow. *See also* **call, call control signal, data circuit-terminating equipment, interface, signal.**

connectionless data transfer: *See* **connectionless mode transmission.**

connectionless mode transmission (CMT): In a packet-switched network, a mode of transmission in which each packet is encoded with a header containing a destination address sufficient to permit the independent delivery of the packet without the aid of additional instructions. *Note 1:* A packet transmitted in a connectionless mode is frequently called a "datagram." *Note 2:* In connectionless mode transmission of a packet, the service provider usually cannot guarantee that there will be no loss, error insertion, misdelivery, duplication, or out-of-sequence delivery of the packet. However, the risk of these hazards occurring may be reduced by providing a reliable transmission service at a higher protocol layer, such as the Transport Layer of the Open Systems Interconnection—Reference Model. *Synonym* **connectionless transmission.** *See also* **connection oriented mode transmission, datagram, Open Systems Interconnection—Reference Model, packet switching.**

connectionless transmission: *Synonym* **connectionless mode transmission.**

connection mode transmission: *Synonym* **connection oriented mode transmission.**

connection oriented data transfer protocol: A data transfer protocol in which a logical connection is established between end points. *See also* **virtual circuit.**

connection oriented mode transmission: In a packet-switched network, a mode of transmission in which (a) there is a complete information transfer transaction for each packet or group of packets, i.e., the information transfer phase is preceded by an access phase and followed by a disengagement phase, (b) during the information transfer phase of the transmission, more than one packet may be transmitted, (c) the header of each information packet contains a sequence number and an identifier field that associates the packet with the connection that was established during the access phase before the beginning of the information transfer phase, and (d) detection of lost, erroneous, duplicated, or out-of-sequence packets usually occurs because a connection is established from end to end before transmission begins. *Note:* The International Telegraph and Telephone Consultative Committee (CCITT) X.25 protocols are widely used to implement connection oriented mode transmission on packet-switched public data networks. The protocols are implemented at Layers 1, 2, and 3 of the Open Systems Interconnection—Reference Model. *Synonym* **connection mode transmission.** *See also* **connectionless mode transmission, Open Systems Interconnection—Reference Model, switching center, X-series Recommendations.**

connections per circuit hour (CCH): A unit of traffic measurement, i.e., the rate at which connections are being established by a switch. *Note:* The magnitude of the number of connections per circuit hour (CCH) is an instantaneous value subject to change as a function of time, i.e., from moment to moment. *See also* **call-second, erlang, traffic intensity.**

connectivity: The level or the degree to which users are interconnected, usually in terms of (a) the availability of multimedia and (b) the use of modems, personal computers, main frames, and interconnected networks, such as the Internet and the World Wide Web. *See* **network connectivity.** *See also* **Clos network.**

connectivity exchange (CONEX): In an adaptive or manually operated high frequency (HF) radio network, the automatic or manual exchange of information concerning routes, such as (a) link quality and availability information and (b) identification information on indirect paths and possible relay stations to stations that are not directly reachable by the information exchange originator.

connector: A device for coupling and decoupling propagation media and devices, such as electrical conductors, optical fibers, and fluid flow lines. *Note:* A connector is different from a splice in that a splice is a permanent joint. *See* **bulkhead connector, dimpled connector, double eccentric connector, expanded beam connector, female connector, fiber optic active connector, fiber optic connector, fiber optic grooved alignment connector, fiber optic passive connector, field installable connector, fixed connector, fixed fiber optic connector, free fiber optic connector, fully intermateable connector, grooved fiber alignment connector, heavy duty connector, hybrid connector, interchangeable connector, laser connector, light duty connector, male connector, mechanically intermateable connector, medium interface connector, optical waveguide connector, receiver fiber optic connector, tapered plug connector, transmitter fiber optic connector, wet mateable connector.** *See also* **splice.**

connector-induced optical loss: The part of connector insertion loss caused by optical fiber end contamination or structural changes to the optical fiber introduced by termination or handling within the connector. *Note:* Connector-induced optical power loss is usually expressed in dB.

connector insertion loss: The power loss sustained by a signal in a propagation medium, such as a wire, coaxial cable, fiber optic cable, or optical integrated circuit (OIC), caused by the insertion of a connector between two elements of the medium, a loss that would not occur if the medium were continuous without the connector, i.e., if there were no reflected, absorbed, dispersed, or scattered power caused by the connector. *Note 1:* Connector insertion loss usually is expressed in dB. *Note 2:* Connector insertion loss often and invariably occurs when a mated connector is inserted into a cable. In a fiber optic cable, the insertion loss usually is about 0.1 dB per connector.

connector set: *See* **fiber optic connector set.**

connector variation: *See* **fiber optic connector variation.**

connect when free: *Synonym* **camp on.**

conservation: *See* **radiance conservation.**

conservation law: *See* **radiance conservation law.**

conservation of radiance: *Synonym* **radiance conservation.**

conservation of radiance law. *Synonym* **radiance conservation law.**

CONSOL: A specific long-range radio navigational aid that (a) emits signals that enable bearings to be determined by means of the audio frequency modulation characteristics of the signals, (b) is operated such that one line of bearing is obtained from one station on any low frequency receiver, (c) requires at least two stations for position fixing by obtaining crossed bearings, i.e., by resection, and (d) requires certain charts for aid in interpreting the information received.

console: **1.** In communications systems, a control device that (a) has access to a communications network, usually to a switch, and (b) is installed so as to control network access lines and instruments in a local area. **2.** In automatic data processing systems, the part of a system, such as a computer, process controller, or operations controller, that (a) is used for communications between the system and the operating or service personnel, (b) usually contains display registers, counters, indicator lights, keys, switches, and related circuits, (c) may be a desk-like unit, permitting the seating of persons, (d) may be an upright panel for standing personnel, (e) may be used for performing various functions, such as error control, error detection, error correction, revising storage constants, checking data flow, diagnosing faults, correcting malfunctions, monitoring program execution, and supervising operations. *See* **display console.**

consolidated local telecommunications service: A local communications service provided by the General Services Administration (GSA) to all federal agencies located in a given building, complex, or geographical area.

consolidation: *See* **communications system consolidation.**

constant: *See* **Abbe constant, axial propagation constant, Boltzmann's constant, fast time constant, ground constant, logarithmic fast time constant, Planck's constant, propagation constant, time constant, transverse propagation constant, Verdet's constant.**

constant current modulation: Amplitude modulation in which (a) a source of constant electrical current supplies a radio frequency generator and a modulation amplifier in parallel, and (b) current variations in one can cause equal and opposite variations in the other, resulting in modulation of the carrier output.

constant failure period: In the life of a device with a large number of components, the period that (a) follows the early failure period and (b) is characterized by a lower and fairly constant failure rate as each component statistically approaches the end of its useful life. *See also* **bathtub curve.** *Refer to* **Fig. B-3 (71).**

constant false alarm rate (CFAR): In radar systems, the rate at which objects, i.e., targets, that do not exist are indicated as being present when a constant noise level is maintained at the radar data processor input. *Note:* False alarms usually are caused by noise, interference, or jamming. An alarm is the indication of an identified object. Constant false alarm rate techniques in radar systems (a) maintain a constant noise level at the input of an automatic data processor, (b) usually do not permit the indication of an object if the signal from the object is weaker than noise, interference, or jamming, and (c) remove some of the confusing effects caused by noise, interference, or jamming. *Note:* Constant false alarm rate techniques are also applicable to systems other than radar, particularly sensing and telemetering systems.

constant watch ship: In radiotelegraph operations at sea, a ship station that is on listening watch, is transmitting, or is receiving transmissions on a continuous basis.

constant watch station: In radiotelegraph operations at sea, particularly in ship-shore operations, a station that is on listening watch, is transmitting, or is receiving transmissions on a continuous basis.

constitutive relation: One member of the set of three relations pertaining to the properties of a propagation medium in which electric and magnetic fields, electric currents, and electromagnetic waves exist and propagate, given by the relations $\boldsymbol{D} = \varepsilon \boldsymbol{E}$ where $\boldsymbol{D}$ is the electric flux density or electric displacement vector, ε is the electric permittivity, i.e., the dielectric constant, and $\boldsymbol{E}$ is the electric field strength, $\boldsymbol{B} = \mu \boldsymbol{H}$ where $\boldsymbol{B}$ is the magnetic flux density, μ is the magnetic permeability, and $\boldsymbol{H}$ is the magnetic field strength, and $\boldsymbol{J} = \sigma \boldsymbol{E}$ where $\boldsymbol{J}$ is the electric current density, σ is the electrical conductivity, i.e., the reciprocal of the resistivity, and $\boldsymbol{E}$ is the electric field strength as before, and where $\boldsymbol{D}, \boldsymbol{E}, \boldsymbol{B}, \boldsymbol{H},$ and $\boldsymbol{J}$ are always vector quantities because they have direction. *Note:* In some substances, the constitutive relations assume a complex form and are also represented as vectors or tensors. The constitutive relations are used in conjunction with Maxwell's equations for electromagnetic wave propagation. In dielectric materials, such as optical fibers, $\sigma = 0$ and $\mu = 1$. *See also* **electrical conductivity, electric permittivity, magnetic permeability, Maxwell's equations.**

constraint: *See* **loss budget constraint, statistical loss budget constraint.**

contact: 1. *Synonym* **terminus. 2.** *See* **optical contact.**

content: *See* **character mean transinformation content, conditional information content, decision content, information content, joint information content, mean transinformation content, mutual information content, natural unit of information content, signaling information content, transinformation content.**

content addressable storage: *Synonym* **associative storage.**

contention: 1. A condition that arises when two or more data stations attempt to transmit at the same time over a shared channel, or when two data stations attempt to transmit at the same time in two-way alternate communications. *Note:* A contention can occur in data communications when no station is designated as a master station. In contention, each station must monitor the signals and wait for a quiescent condition before initiating a bid for master station status. **2.** Competition by users of a system for use of the same facility at the same time. *See* **access contention.** *See also* **carrier sense multiple access, collision, data communications, link, master station.**

continuity check: 1. A check made of a circuit to determine whether a communications or power path exists. **2.** A check made to determine if conduction can occur between two points. *See also* **circuit.**

continuity failure signal: A signal sent in the backward direction indicating that a call cannot be completed because of the failure of the continuity check in the forward direction. *See also* **backward direction, call, continuity check, forward direction.**

continuous carrier: 1. Pertaining to a signal in which the transmission of the carrier is continuous, i.e., is not pulsed on and off. **2.** A radio wave that (a) has a constant amplitude and constant frequency, (b) may be amplitude, phase, or frequency modulated, and (c) is transmitted without interruption by a modulating signal, i.e., is not turned on and off in accordance with modulating pulses, such as in pulse code modulation and Morse code transmission. *See also* **continuous emission, continuous wave.**

continuous communications watch: A communications watch on a 24-hours-a-day basis, usually accomplished when there are at least three operators available at a station to maintain the watch. *See also* **single operator period watch, two-operator period watch.**

continuous emission: The transmission of an uninterrupted signal. *Note:* Examples of continuous emissions are an unending dash in radiotelegraphy, a 1000-Hz (hertz) audio signal in voice transmission, a continuous carrier transmission as a standard frequency or timing signal, and the continuous emission of the radio transmitter that is set before abandonment of a ship or an aircraft. *See also* **continuous carrier, continuous wave.**

continuous form: Pertaining to a series of pages of paper joined at the tops and bottoms with other pages so they may be fed into and out of devices, such as printers, typewriters, printing calculators, facsimile machines, and cash registers, usually by means of sprockets that engage feed holes or by nonslip rollers. *Note:* The use of sprockets and feed holes allows for positive page positioning. In some devices, precise positioning is not required.

continuously variable optical attenuator: 1. In an optical system, a device that (a) attenuates the irradiance, i.e., the optical density, of lightwaves over a continuous range rather than in discrete steps and (b) may be inserted in an optical link to control the irradiance of light at the receiving photodetector. **2.** In fiber optic systems, a device that (a) is inserted into a fiber optic link, (b) attenuates the intensity of lightwaves over a continuous rage of attenuation, and (c) is controlled by a fixed setting or by a control signal. *See also* **stepwise variable optical attenuator.**

continuously variable slope delta (CVSD) modulation: Delta modulation in which the size of the steps of the approximated signal is progressively increased or decreased as required to make the approximated signal closely match the input analog wave. *See also* **delta modulation, differential modulation, differential pulse code modulation, modulation.**

continuous operation: 1. Operation in which certain components, such as nodes, facilities, circuits, or equipment, are in an operational state at all times. *Note:* Continuous operation usually requires that there be a fully redundant configuration, or at least a sufficient X out of Y degree of redundancy for compatible equipment, where X is the number of spare components and Y is the number of operational components. **2.** In data transmission, operation in which the master station need not stop for a reply from a slave station after transmitting each message or transmission block. *See also* **cutover, degraded service state, downtime, dynamically adaptive routing, fail safe operation, graceful degradation, operational service state, outage, redundancy, survivability.**

continuous phase modulation: Modulation in which the phase of the modulated carrier is not discontinuous, i.e., the transitions from one phase-significant condition to another are smooth so as not to introduce increased bandwidth requirements. *See also* **modulation, phase, phase modulation, significant condition.**

continuous presence: In teleconferencing, the concurrent presence of two or more video images, such as two images that may appear on a single monitor on a split screen or on two separate monitors.

continuous receiver: In facsimile systems, equipment that records line by line on a data medium, such as paper, that moves with a constant pitch between consecutive lines so as to record several messages in succession without the need for an operator to change the medium between consecutive messages. *See also* **record medium.**

continuous spectrum: In a band, a frequency spectrum in which all frequencies are present. *See also* **blue noise, frequency spectrum, pink noise, white noise.**

continuous tape relay switching: *See* **semiautomatic continuous tape relay switching.**

continuous time system: A system in which (a) input and output signals can change at any instant, and (b) operations may be modeled by the use of differential equations. *See also* **discrete time system.**

continuous tone copy: In facsimile systems, an object or a recorded copy that contains shades of gray, i.e., contains densities between black and white, such as a photographic print. *See also* **facsimile, halftone.**

continuous wave (CW): A wave of constant amplitude and constant frequency. *See* **Morse continuous wave.** *See also* **continuous carrier, continuous emission, frequency, interrupted continuous wave.**

continuous wave jamming: *See* **unmodulated continuous wave jamming.**

continuous wave radar: A radar system in which a continuous flow of radio frequency energy is transmitted in the direction of the object, i.e., the target, which scatters and reflects some of the incident energy and returns a small fraction of it to a receiving antenna.

contour: *See* **coordination contour, effective antenna gain contour.**

contouring: In digital facsimile systems, density step lines in the recorded copy resulting from analog to digital conversion when the object has observable shades of gray between the smallest density steps of the digital system. *See also* **facsimile.**

contract developmental station: *See* **experimental contract developmental station.**

contrast: The comparative difference between two entities. *Note:* Examples of contrast are (a) the difference between two colors, two shades of the same color, or two shades of gray, ranging from white to black, (b) the relative luminance of two objects, and (c) the difference between two modes of communication. *See* **refractive index contrast.** *See also* **contrast ratio.**

contrast ratio (CR): 1. In systems that have display devices, such as video and facsimile systems, the ratio of (a) the maximum reflectivity of a spot on an object or recorded copy to (b) the reflectivity of a spot at the point of interest, i.e., the point at which the contrast is observed. **2.** The ratio of (a) the luminance at one point to (b) the luminance at another point. *Deprecated synonym* **brightness ratio. 3.** In optical character recognition, the difference between color or shading of the printed material on a document and the background on which it is printed. **4.** In computer graphics, the difference in luminance between a display image and the area in which it is displayed. **5.** In display systems, the relation between (a) the intensity of color, brightness, or shading of an area occupied by a display element, display group, or display image in the display surface of a display device and (b) the intensity of an area not occupied by elements, groups, or images. *Note:* The contrast ratio is always greater than unity. *See* **print contrast ratio.** *See also* **contrast, refractive index contrast.**

contribution: In broadband integrated services digital network (B-ISDN) applications, the use of broadband transmission of audio or video information to the user for postproduction processing and distribution. *See also* **broadband ISDN, distribution, postproduction processing.**

control: *See* **access control, automatic gain control, automatic process control, basic mode link control, central primary control, change control, change in operational control, circuit quality control, command and control, common control, configuration control, configuration item control, direct clock control, distributed control, electromagnetic radiation control, emission control, error control, failure control, fast automatic gain control, flow control, frequency control, function control, guided missile control, high level control, high level data link control, indirect clock control, instantaneous automatic gain control, line load control, link control, low level control, numerical control, operational network control, port radio transmission control, process control, quality control, radio control, radio transmission control, pulse jet control, receiver intermediate frequency gain control, remote control, right-through control, sensitivity time control, sequential control, supervisory control, synchronous data link control, telecontrol, transmit flow control, wavefront control.**

control aperture: *See* **tone control aperture.**

control authority: *See* **maritime air control authority.**

control ball: A ball that (a) usually is embedded in a slot in a surface, such as that of a mouse or a control console, (b) can be rotated about its center, and (c) may be used as an input device, usually for inserting locator or coordinate data. *Synonym* **trackball.**

control center: *See* **network control center.**

control channel: *See* **network control channel, rescue control channel.**

control character: 1. In a particular context, a character that, when it occurs, initiates, modifies, or stops a function, event, or operation. **2.** In a particular context, a character that specifies a control operation. *Note:* A control character may be recorded for use in a subsequent action. Control characters are not graphic characters, but they may have graphic representations in some circumstances. *See also* **acknowledge character, call control signal, character, data link escape character, end of selection character, end of text character, end of transmission block character, end of transmission character, enquiry character, idle character, negative acknowledge character, start of heading character, start of text character, stop signal.**

control circuit: *See* **maintenance control circuit.**

control communications: The branch of science and technology devoted to the design, development, and application of communications facilities used specifically for control purposes, such as for controlling (a) industrial processes, (b) movement of resources, (c) electric power generation, distribution, and utilization, (d) communications networks, and (e) transportation systems, such as airlines, railroads, truck lines, bus lines, taxi systems, and steamship lines. *See* **air traffic control communications.**

control communications network: *See* **air defense command-control-communications network.**

control designation: *See* **line load control designation.**

control equipment: *See* **process control equipment, remote control equipment.**

control facility: *See* **technical control facility.**

control field: 1. In high level data link control (HDLC) operation, the sequence of 8 bits, or 16 bits if extended, that (a) immediately follow the address field of a frame, (b) is interpreted by the receiving secondary station designated by the address field as a command instructing the performance of some specific function, (c) is interpreted by the receiving primary station as a response from the secondary station designated by the address field to one or more commands, (d) is also interpreted by the receiving combined station as a command instructing the performance of some specific function if the address field designates the receiving station as a combined station, and (e) is also interpreted by the receiving combined station as a response to one or more transmitted commands if the address field designates a remote combined station. **2.** In a protocol data unit (PDU), the field that (a) contains data interpreted by the receiving destination logical link controller (LLC) and (b) may be the field immediately following the destination service access point (DSAP) and source service access point (SSAP) address fields of the PDU. *Note:* The destination service access point (DSAP) address field designates the destination logical link controllers (LLC), and the source service access point (SSAP) address field designates the source logical link controllers (LLD). *See also* **address field, combined station, control field extension, frame, primary station, secondary station.**

control field extension: In standard high level data link control (HDLC) operations, an enlargement of the control field to include additional control information. *See also* **control structure.**

control flow: The sequence in which instructions, routines, and subroutines are executed when a computer program is executed.

control function: *Synonym* **control operation.**

control information: *See* **protocol control information.**

control key: 1. A key that (a) is used to cause, initiate, or control the execution of an operation in a system and (b) may be located on any component of the system, such as on a console, control panel, or remote controller. **2.** On a standard keyboard, such as is used in communications, computer, and data processing systems, a key that is used, usually in conjunction with other keys, to initiate operations, such as (a) in WordPerfect, to delete an entire word when used in conjunction with the delete key, or (b) in Volkswriter, to delete an entire line when used in conjunction with the F4 key.

controllable coupler: *See* **electronically controllable coupler.**

control language: *See* **job control language.**

controlled access: Access in which the resources of an area or a system are limited to authorized personnel, users, programs, processes, or other systems, and denied to all others. *See also* **access code, access control, classmark, communications security, restricted access.**

controlled accessibility: *Synonym* **access control.**

controlled approach: *See* **carrier controlled approach.**

controlled approach landing: *See* **ground controlled approach landing.**

controlled approach system: *See* **ground controlled approach system.**

controlled area: **1.** An area (a) in which uncontrolled movement does not result in compromise of classified information, (b) that provides administrative control and safety, and (c) that serves as a buffer for controlling access to limited-access areas. **2.** An area to which security controls have been applied to protect information processing system equipment and communications lines, equivalent to that protection required for the information transmitted through the system. *See also* **communications security.**

controlled not-ready signal: A signal sent in the backward direction indicating that a call cannot be completed because the called line is not in a ready condition, but is under control, rather than not being in a ready condition and not under control.

controlled security mode: *Synonym* **controlled security operation.**

controlled security operation: In communications, computer, data processing, and control systems, a mode of system operation in which (a) internal security controls prevent inadvertent disclosure of information, (b) personnel, physical, and administrative controls prevent deliberate, malicious attempts to gain unauthorized access, (c) service may be provided to both cleared and uncleared users, and (d) service may be provided to both secured and unsecured terminals in remote areas. *Synonym* **controlled security mode.** *See also* **access control, classmark, communications security.**

controlled space: The three-dimensional space (a) that surrounds a communications, computer, data processing, or control system or component and (b) within which unauthorized persons are denied unrestricted access and are either escorted by authorized persons or are under continuous physical or electronic surveillance. *Synonym* **restricted area.**

controller: In an automated radio, the device that commands the radio transmitter and receiver, and that performs processes, such as automatic link establishment (ALE), channel scanning and selection, link quality analysis, polling, sounding, message store and forward, address protection, and antispoofing. *See* **network controller, resource controller, system controller, terminal access controller.** *See also* **automatic link establishment.**

control link: A dedicated data link (a) in which the transmitted data are used to control the operation of a device, such as a ship or aircraft propulsion system or subsystem, and (b) that usually is used (i) to transmit data from the device being controlled to a control point, such as a control panel, or (ii) to transmit data, usually in the form of a control signal, to the device, where the signal is converted to an electric current or mechanical actuating device to control the operation of that device and hence the system or the subsystem.

control mechanism: *See* **access control mechanism.**

control message: *See* **access control message.**

Control Message Protocol: *See* **Internet Control Message Protocol.**

control net: A communications net that is used for the direction and control of activities, systems, organizations, operations, or processes.

control of electromagnetic radiation: *Synonym* **electromagnetic radiation control.**

control operation: In communications systems, an operation that affects the recording, processing, transmission, or interpretation of data. *Note:* Examples of control operations are (a) starting and stopping a process, (b) executing a carriage return, a font change, or a rewind, and (c) transmitting an end of transmission (EOT) control character. *Synonym* **control function.**

control panel: **1.** A board or a panel on which are mounted (a) instruments for indicating the status of the various components of a system and (b) devices that enable an operator to enter data or signals that will influence the operation of the system. **2.** In personal computers, a part of the display area on a display device, such as a monitor, reserved for use by an operator to enter and record decisions, queries, and results, such as in Lotus 1-2-3, where the control panel displays operating messages, such as (a) data concerning the cell of current interest, (b) the current mode, (c) operator entries as they are typed or edited, (d) menus and their explanations, and (e) command prompts and the operator reply to these prompts.

control phase: *See* **network control phase.**

control point: *See* **secondary control point.**

control procedure: *See* **Advanced Data Communications Control Procedure, call control procedure, data communications control procedure, flow control procedure.**

control program: A computer program that (a) retrieves programs, i.e., calls programs, from storage in accordance with schedules built into the program and (b) supervises their execution. *See* **conditional control program.**

control protocol: *See* **transmission control protocol.**

Control Protocol/Internet Protocol Suite: *See* **Transmission Control Protocol/Internet Protocol Suite.**

control service: *See* **airport control service, approach control service.**

control signal: *See* **call control signal.**

control signaling: **1.** Signaling used for effecting control of a communications system. **2.** The use of directory numbers of users, operators, and maintenance personnel for effecting control of an automatic telephone system.

control station: **1.** In communications networks, the station that (a) selects the master station, (b) supervises operational procedures, such as polling, selecting, and recovery, (c) is responsible for the orderly operation of the entire network, (d) initiates recovery in the event of an abnormal condition on the network, and (e) is designated in accordance with control procedures. *Note:* The designation of a particular station as a control station usually is not affected by the control procedures. **2.** In a multipoint or point-to-point connection and using basic mode link control procedures, the terminal installation that nominates the master station and supervises polling, selecting, interrogating, and recovery. *See* **master net control station.** *See also* **backward supervision, data communications, master station, network, primary station, secondary station, slave station, supervisory program, tributary station.** *Refer to* **Figs. N-5 through N-15 (622, 623).**

control structure: In computer programming, a construct that determines, or is used to manage, the control flow through a computer system. *See also* **control flow.**

control sublayer: *See* **logical link control sublayer, medium access control sublayer.**

control switching arrangement: *See* **common control switching arrangement.**

control switching system: *See* **common control switching system.**

control system: *See* **automated tactical command and control system, command and control system, common control system, process control system, semiautomated tactical command and control system.**

control tape: *See* **carriage control tape.**

control tower: In an airport, a facility that (a) provides for the control of aircraft and vehicles operating on and around the airport, landing zone, or landing area, (b) is equipped with communications facilities, and (c) is high enough to afford direct visibility of the surrounding spatial volume.

control unit: The part of a system, such as a communications, computer, or data processing system, that performs the control functions of the system, such as, for a computer system, obtain instructions, send results to storage, and control input and output devices. *See* **teletypewriter control unit.** *Refer to* **Fig. L-6 (529).**

control warning system: *See* **aircraft control warning system.**

control zone: In communications, computer, data processing, and control systems, the space that (a) surrounds equipment used to process sensitive information and (b) is under sufficient physical and technical control to prevent unauthorized entry or compromise of sensitive information, such as security classified information, private information, and industrial proprietary information.

convention: *See* **fifteen ones convention.**

convergence: In optics, the bending of light rays toward each other, as by a converging lens, i.e., a convex lens. *See also* **converging lens, divergence.**

convergence angle: **1.** In optical systems, the angle, i.e., the smaller interior angle, at which two light rays converge with each other. **2.** The angle formed by the lines of sight of both eyes when focusing on any line, corner, surface, point, or part of an object. *See also* **accommodation, convergent light, converging lens, ray.** *Synonym* **convergent angle.**

convergent angle: *Synonym* **convergence angle.**

convergent lens: *Synonym* **converging lens.**

convergent light: Light consisting of a bundle of rays, each of which is propagating in such a direction as to be approaching every other ray, i.e., they are not parallel to each other and are propagating toward a line, or are heading toward a point of intersection or focus. *Note:* The wavefront of convergent light is somewhat spherical, i.e., concave on the trailing side. If a collimated beam enters a converging lens, i.e., a concave lens, the emerging light is convergent to the focal point and divergent after the focal point. *See also* **divergent light.**

converging lens: A lens that (a) adds convergence to an incident light ray bundle and (b) may have (i) one surface spherically convex and the other plane, i.e., a planoconvex lens, (ii) both surfaces convex, i.e., a double convex or biconvex lens, and (iii) one surface convex and the other concave, i.e., a converging meniscus. *Synonyms* **convergent lens, convex lens, crown lens, plus lens, positive lens.** *See also* **collective lens, diverging lens.**

conversation: *See* **telegraph conversation.**

conversational: Pertaining to a mode of system operation similar to a conversation between two persons, i.e., a mode of operation similar to a dialogue. *Note:* Communications and computer systems may operate in a conversational mode with each other and with persons. Communications systems are used to support the dialogue and often make it possible for the participants in the dialogue to be at great distances from each other while they converse in real time. Under such arrangements, each user entry usually engenders a response from the system and vice versa. *See also* **interactive mode.**

conversational mode: In communications and computer systems operation, a mode of system operation similar to a conversation between two persons, i.e., the communications or computer system responds immediately to queries and commands. *See also* **duplex operation, Hamming code, interactive data transaction, push to talk operation.**

conversational service: In telecommunications, a service that provides two-way, interactive, real time, end-to-end information exchange.

conversion: *See* **code conversion, double conversion, mode conversion, serial to parallel conversion.**

conversion chart: *See* **time conversion chart.**

convert: In communications, computer, data processing, and control systems, to change the representations of information, i.e., data, from one form to another without changing the information they convey. *Note:* Examples are to convert one radix to another, one code to another, analog data to digital data, and data on cards to data on tape, each without loss of meaning, accuracy, or precision. *See* **translation.**

converter: In information and related systems, a device that performs the function of converting an entity from one form to another, i.e., a device that converts. *Note:* Examples of converters are (a) a section of a superheterodyne radio receiver that converts the desired incoming radio frequency (rf) signal to a lower frequency signal, i.e., to the intermediate frequency (if) signal, (b) a machine that consists of an electric motor driving an electric generator to change alternating current (ac) to direct current (dc) or to change alternating current at one frequency to alternating current at a different frequency, (c) a facsimile device that changes the type of modulation delivered by its scanner, (d) a facsimile device that changes amplitude modulation to audio frequency shift modulation, often called a remodulator, (e) a device that changes audio frequency shift modulation to amplitude modulation, usually called a discriminator, (f) a transducer in which the output frequency is the sum of the input frequency and a local oscillator frequency, and (g) a device that changes the representation of data from one form to another without changing the meaning they convey, such as a card to printer converter or a tape to printer converter. *See* **analog to digital converter, code converter, digital to analog converter, down converter, facsimile converter, parallel to serial converter, serial to parallel converter, signal converter, upconverter.**

convex: Pertaining to a continuous surface (a) to which more than one tangent plane can be drawn and (b) that, if reflective, causes incident parallel rays to diverge. *Note:* Examples of convex surfaces are (a) the surface of a solid sphere and (b) the outside surface of an intact orange rind. *See also* **concave.**

convex lens: *Synonym* **converging lens.**

convoluted code: An error correction code in which (a) each m-bit information symbol, i.e., each m-bit string, to be encoded is transformed into an n-bit symbol where $n > m$, and (b) the transformation is a function of the last k information symbols where k is the constraint length of the code. *Note 1:* An example of a convoluted code is a code that has a parity bit that is the modulo-2 sum of several past data and parity bits. Logical analysis of successive parity bits allows identification of the erroneous bit and facilitates correction of the data. *Note 2:* Convoluted codes are often used to improve the performance of radio and satellite links. *Synonym* **convolutional code.** *See also* **block code, error control, error correcting code, forward error correction, Hagelbarger code, Hamming code.**

convolutional code: *Synonym* **convoluted code.**

convoy common frequency: A frequency that can be used by all ship stations of a convoy. *Note:* Examples of convoy frequencies are (a) a convoy radiotelegraph frequency of 500 kHz (kilohertz) and (b) a voice radiotelephone frequency for use within a convoy, both as might be prescribed by the convoy commander or commodore.

convoy communications: The branch of communications devoted to (a) methods, systems, and equipment used for transmission of messages to, from, and within land and maritime convoys, (b) transmission of information for controlling convoy formations, routes, and procedures, and (c) transmission and reception of messages and calls to, from, and within convoys, particularly between fixed land stations and mobile stations and stations within the convoy.

convoy internal communications: In a land or maritime convoy, communications (a) that are among the elements of the convoy, (b) that include radio, visual, and sound transmissions, (c) in which radio intervehicle and intership communications usually are by radiotelephone using the receipt method of operation, (d) in which the convoy commander vehicle or ship usually has the net control station on board, and (e) that usually use a single assigned frequency.

convoy radiotelephone frequency: The radio frequency (rf) common to all vehicles or ships in a convoy, usually for radiotelephone transmissions.

cooperation factor: In facsimile systems, the product of the total scanning length and the scanning density, given by the relation $CF = L\sigma$, where L is the scanning line length and σ is the scanning line density, both in compatible units. *Note:* For a 20-cm length line and a line density of 6 scanning pitches per centimeter, the cooperation factor would be 120 scanning pitches. *See also* **cooperation index.**

cooperation index (CI): In facsimile systems, the quotient obtained by dividing the cooperation factor by π. *Note 1:* In the case of a drum transmitter or receiver, the cooperation index is equal to the product of the drum diameter and the scanning density. *Note 2:* The cooperation index, CI, is given by the relation $CI = D\sigma$, where D is the drum diameter and σ is the scanning density, both in compatible units. For a drum diameter of $20/\pi$ cm, and a σ of 6 scanning pitches per cm, the cooperation index would be $120/\pi$, and the cooperation factor would be 120. The relationship is based on the line length being the circumference of the drum. Thus, the line length is π times the drum diameter. *Synonyms* **diametral cooperation index, index of cooperation.** *See also* **cooperation factor, drum factor, facsimile.**

coordinate: *See* **absolute coordinate, device coordinate, real world coordinate, relative coordinate, screen coordinate, world coordinate.**

coordinate data: Data that (a) represent positions or locations on display surfaces, such as cathode ray tube (CRT) screens, gas panels, fiberscope faceplates, light-emitting diode panels, and plotter beds, of a display device, (b) may be absolute or relative, (c) may be used to control the operations of a system, such as a communications, computer, or data processing system, and (d) may be used to control or perform operations on the display surface, such as moving display elements, moving display groups, moving images, rubberbanding, scissoring, clipping, zooming, or scrolling. *See* **absolute coordinate data, relative coordinate data.**

coordinated clock: In a set of clocks distributed over a spatial region, the clock that produces time scales that are synchronized to the time scale of a reference clock at a specified location. *See also* **cesium clock, clock, Coordinated Universal Time, DoD master clock, master clock, primary time standard, reference clock.**

coordinated time scale: A time scale synchronized within given tolerances to a reference time scale. *See also* **Coordinated Universal Time, DoD master clock, Greenwich Mean Time, International Atomic Time (TAI), time, time scale.**

Coordinated Universal Time (UTC): The time on a scale based on the second, an SI (Système International d'Unités) time unit, as defined and recommended by the International Radio Consultative Committee (CCIR) and maintained by the Bureau International de l'Heure (BIH) (International Time Bureau) of the Bureau International des Poids et Mesures (BIPM). *Note 1:* For most practical purposes associated with the Radio Regulations, Coordinated Universal Time (UTC) is equivalent to mean solar time at the prime meridian (0° longitude), formerly expressed as Greenwich Mean Time (GMT). *Note 2:* The full definition of UTC is contained in International Radio Consultative Committee (CCIR) Recommendation 460-4. *Note 3:* The maintenance of UTC by BIPM includes cooperation among various national laboratories around the world. *Note 4:* The international second was formerly based on observations of astronomical phenomena. When this practice was abandoned in order to take advantage of atomic resonance, i.e., atomic time, to define the second more precisely, it became necessary to make occasional adjustments in the atomic time scale (UTC) to coordinate it with the workday mean solar time scale, UT-1, which is based on the somewhat irregular angular velocity of the Earth. Rotational irregularities usually cause a net decrease in the Earth's average angular velocity, resulting in lags of UT-1 with respect to UTC. *Note 5:* Adjustments to the atomic time scale, i.e., the UTC time scale, consist of an occasional addition or deletion of one full second, i.e., the SI (Système International d'Unités) second. The added or deleted second is called a leap second. Twice yearly, during the last minute of the day on June 30 and December 31, Universal Time (UT), adjustments may be made to ensure that the accumulated difference between UTC and UT-1 will not exceed 0.9 s (seconds) before the next scheduled adjustment. Historically, adjustments, when necessary, usually have consisted of adding an extra second to the UTC time scale in order to allow the rotation of the Earth to catch up. Thus, the last minute of the UTC time scale, on the day when an adjustment is made, will have 61 seconds. *Synonyms* **World Time, Z time, Zulu Time.** *See also* **coordinated time scale, dating format, DoD master clock, Greenwich Mean Time, International Atomic Time, leap second, precise frequency, precise time, second, time scale, time standard.**

Coordinated Universal Time (*i*) [UTC(*i*)]: Coordinated Universal Time (UTC), as kept by the *"i"* laboratory, where *i* is any laboratory cooperating in the determination of UTC. *Note:* In the United States, the official Coordinated Universal Time (UTC) is kept by the U.S. Naval Observatory (USNO) and is referred to as UTC (USNO). *See also* **Coordinated Universal Time.**

Coordinated Universal Time (UTC) U.S. Naval Observatory (USNO): For the United States, the official Coordinated Universal Time (UTC) kept by the U.S. Naval Observatory.

coordinate graphics: In display systems and devices, the generation of display elements, display groups, and display images in the display area on the display surface of a display device using display commands and coordinate data. *See also* **coordinate data, display command, display device, display image.**

coordinate position: In display systems, a single location in the display surface of a display device specified by absolute or relative coordinate data. *Note:* In a fiberscope, each optical fiber or fiber optic bundle in an aligned bundle can be assigned a coordinate position, and each position on a cathode ray tube (CRT) screen can be assigned a device coordinate. *See* **coordinate data, device coordinate, display system.**

coordinating center: *See* **National Coordinating Center.**

coordination: *See* **frequency coordination.**

coordination area: 1. In the satellite service, the area in which an Earth station and terrestrial stations sharing the same frequency band must coordinate their transmissions in order to avoid mutual interference above an acceptable level. **2.** The area associated with an Earth station outside of which a terrestrial station sharing the same frequency band neither causes nor is subject to interfering emissions greater than a permissible level. *See also* **accepted interference, blanketing area, interference.**

coordination contour: The perimeter of a coordination area. *See also* **blanketing, coordination area**.

coordination distance: 1. On a given azimuth, the distance from an Earth station beyond which a terrestrial station sharing the same frequency band neither causes nor is subject to interfering emissions greater than a permissible level. **2.** In satellite communications systems, the distance between (a) an Earth station and (b) a fixed or mobile service station within which there is a possibility that (i) the use of a given transmitting frequency at the Earth station will cause harmful interference to stations in the fixed or mobile service that share the same frequency band, or (ii) the use of a given frequency for reception at the Earth station will cause reception of harmful interference from stations in the fixed or mobile service. *See also* **accepted interference, blanketing, Earth station, frequency band, interference.**

coordination net: A communications net used for the free and rapid exchange of information usually associated with the control of specific distributed entities, such as processes, activities, organizations, programs, and communications systems. *See* **communications net, communications system.**

coordinator: *See* **line traffic coordinator.**

cop: *See* **cybercop.**

copier: *See* **microform reader-copier.**

coplanar self-assembled monolayers (SAM): Molecular layers that (a) have feature sizes ranging from 1 μm (micron) to 100 μm, (b) spatially control adhesion and growth of biological cells, and (c) are used to develop biological neuronal networks that react to stimuli.

coprocessor: In a personal computer (PC), a single chip or a printed circuit (PC) board that is plugged into the computer to perform some of the processing assigned to the main microprocessor, thus yielding higher overall performance of the computer. *Note:* Use of a coprocessor usually requires that the associated software be modified. *See* **chip, microprocessor, personal computer, software.**

copy: 1. To receive a message. **2.** A recorded message or a duplicate of it. **3.** In radio communications, to maintain a continuous radio receiver watch and to keep a complete radio log. **4.** To read data from a source, leaving the source data unchanged at the source, and to write elsewhere data that represent the same information as the original data. *Note:* The copy data may be in a form different from that of the source data. **5.** To understand a transmitted message. *See* **backup copy, continuous tone copy, edited copy, hard copy, record copy, recorded copy, soft copy.** *See also* **read.**

copy command: In personal computers, a command that directs the system to transfer the contents of a cell, group of cells, storage area, or storage unit to another cell, group of cells, storage area, or storage unit, usually while retaining the contents at the original location, depending on the software. *Note:* In the disk operating system (DOS) disk copy command and in the WordPerfect copy command, the data will also remain at the original location.

copy-protected: Pertaining to data, computer programs, or files on a data medium that either (a) should not be copied because of copyright laws, (b) cannot be copied because of secret or embedded codes that prevent copying by conventional means, or (c) both, even though the data, programs, or files can be accessed and used. *Note 1:* Many information files, computer programs, and software packages on diskettes available for purchase are copy-protected to prevent a user from buying only one and making as many copies as desired for use throughout a large organization or community of users. *Note 2:* A simple copyright statement accompanying the diskette does not copy-protect the data on the diskette.

copy watch: A radio communications watch or a video communications watch in which an operator is required to maintain a continuous receiver watch and to keep a complete log. *See also* **receiver watch.**

cord: *See* **plug-ended cord, zip cord.**

cord board: *See* **secure voice cord board.**

cord circuit: A switchboard circuit in which (a) a plug-terminated cord is used to establish connections manually between user lines or between trunks and user lines, (b) a number of cord circuits are furnished as part of the manual switchboard position equipment, (c) the cords may be referred to as front cord and rear cord or trunk cord and station cord, and (d) in modern cordless switchboards, the cord circuit is switch-operated. *See also* **call splitting, circuit, switchboard, switched loop.**

cord lamp: The lamp associated with a cord circuit that indicates supervisory conditions for the respective part of the connection.

cordless switchboard: A telephone switchboard in which manually operated keys, rather than circuit cords, are used to make connections. *See also* **circuit cord, private branch exchange, switchboard.**

cordless telephony: Telephony that is devoted to the use of a handset, i.e., a receiver and a transmitter, that (a) is linked by radio to a local base station on the user premises, (b) usually operates at maximum distances between 50 m (meters) and 100 m from the base station, the maximum range depending on the make and the model and local transmission conditions, (c) has a base station that is either (i) wired to the public switched telephone network (PSTN) via a loop or (ii) linked to the PSTN via a cellular radio system, and (d) eliminates the need for an extension cord for the handset.

core: 1. A piece of magnetic material, usually toroidal in shape, used for computer storage. **2.** The material at the center of an electromechanical relay or coil winding. **3.** In an optical fiber, the center region, i.e., the region surrounded by the innermost cladding. *Note 1:* The core contains the bound modes propagating in an optical fiber. *Note 2:* Usually a significant fraction of the light energy in a bound mode propagates in the cladding. *Note 3:* The refractive index of the core must be higher than that of the cladding for the light energy propagating in the core to remain confined to the core. *See* **fiber optic cable core.** *See also* **cladding, core diameter, fiber optics, normalized frequency, optical fiber, optical fiber core.** *Refer to* **Figs. A-6 (24), D-4 (264), L-5 (511), L-8 (536), L-9 (536), P-1 (695), P-12 (751), R-1 (796), R-10 (836), S-18 (951), T-9 (1037).**

core area: The part of the cross-sectional area of a dielectric waveguide that (a) has a refractive index that is everywhere greater than that of the adjacent, i.e., the innermost, homogeneous cladding by a given fraction of the difference between the maximum refractive index in the core and the refractive index of the homogeneous cladding, excluding any refractive index dip, (b) is the smallest cross-sectional area within which the refractive index is given by the relation $n_3 = n_2 + m(n_1 - n_2)$ where n_2 is the refractive index of the homogeneous cladding adjacent to the core, n_1 is the maximum refractive index of the core, and m is a fraction less than 1, usually not greater than 0.05, and (c) is enclosed by the locus of all points nearest to the optical axis on the periphery of the core where the refractive index exceeds that of the homogeneous cladding by m times the difference between the maximum refractive index in the core and the refractive index of the homogeneous cladding, where m is a positive or negative constant such that $|m| < 1$. *See also* **cladding, core, homogeneous cladding, tolerance field.**

core center: In an optical fiber with a circular cross section, the center of the circle that best fits the outer limit of the core area. *Note:* The core center may not be the same as the cladding and reference circle centers.

core-cladding concentricity: 1. In a multimode round optical fiber, (a) the distance between the core and cladding centers divided by (b) the core diameter. **2.** In a single mode optical fiber, the distance between the core and cladding centers.

core-cladding power distribution. For a given mode of an electromagnetic wave propagating in an optical fiber, the fraction of the total power in the mode that is propagating within the core, the remaining fraction being in the cladding. *Note:* The fraction of power in the core is given by the relation $P_{\text{core}}/P_T = 1 - P_{\text{clad}}/P_T$ where P_{core} is the power in the core, P_T is the total power in the mode, and P_{clad} is the modal power in the cladding. The power distribution for a given mode can change with distance along the guide, especially for the higher order modes in which the cladding power can be lost to lateral radiation.

core component: *See* **cable core component.**

core diameter: In a dielectric waveguide that is circular in cross section, the diameter of the circle that best fits the outer limit of the core area. *Note 1:* The center of the circle that best fits the outer limit of the core area is the core center. *Note 2:* The refractive index of the core is higher than that of the cladding that surrounds it. *See also* **cladding, cladding diameter, core, core area, fiber optics, optical fiber core diameter.**

core eccentricity: *See* **optical fiber concentricity error.**

core mode filter: A device, such as one turn of an optical fiber wrapped around a mandrel, that removes high order propagation modes from the core of the optical fiber.

core noncircularity: 1. For a round optical fiber, the percent that the shape of the core cross section deviates from a circle. *Note 1:* Core noncircularity is not the same as core eccentricity. *Note 2:* In an optical fiber that is circular in cross section, core noncircularity is calculated by dividing (a) the absolute difference between the diameters of the smallest circle that can circumscribe the core area and the largest circle that can be inscribed within the core area by (b) the core diameter. This is accomplished by first circumscribing the smallest circle around the outside of the core area. This circle is used to define the core center. This center is then used to inscribe the largest circle that fits within the core area. **2.** In the cross section of a round optical fiber, (a) the difference between the diameters of the smallest circle that can circumscribe the core area and the largest circle that can be inscribed within the core area divided by (b) the core diameter. *Synonym* **core ovality.** *See also* **cladding, core area, concentricity error, core center, core-cladding concentricity, ovality.**

core ovality: *Synonym* **core noncircularity.**

core reference surface concentricity: 1. In a multimode round optical fiber, the distance between the core and reference surface centers, divided by the core diameter. **2.** In a single mode round optical fiber, the distance between the core and reference surface centers.

core-rope storage: Fixed storage, i.e., permanent storage, that (a) consists of an array of magnetic cores and wires constructed so that coded data are stored as structured wiring, rather than in the cores themselves, as in a core storage unit, (b) has wires that pass through the core in one direction or another, or bypass the cores, resulting in a rope-like appearance and permitting the selection of a single core for a given pattern of pulses, and (c) allows selection of a particular core that will cause a set of pulses to appear on a set of readout lines from that core, representing the stored numeral such that the number of digits in the numeral depends upon the number of readout lines, rather than on the number of cores, resulting in more wires but fewer cores than other types of core storage units. *Synonyms* **Diamond switch, Olsen memory, Rajchman selection switch, rope storage, selection switch.**

core storage: *See* **magnetic core storage.**

core to cladding concentricity: *Synonym* **concentricity error.**

core to cladding eccentricity: *Synonym* **concentricity error.**

core to cladding offset: *Synonym* **concentricity error.**

core tolerance area: In the cross section of a round optical fiber, the region between the smallest circle that can circumscribe the core area and the largest circle that can be inscribed within the core area, both circles concentric with the core center.

core wrap: In a fiber optic cable, a material placed around the lightwave conducting and electrical conducting elements of the cable for mechanical protection, dielectric insulation, heat insulation, and reduction of strain in flexure.

Corguide™: A fiber optic cable produced by Siecor.

corner cube reflector: *Synonym* **triple mirror.**

corner reflector: 1. A device that (a) usually consists of three metallic surfaces or screens perpendicular to one another and (b) is designed to serve as a radar target or marker. **2.** In radar, an object that, by means of multiple reflections from smooth surfaces, produces a radar return of greater magnitude than might be expected from the physical size of the object. **3.** A reflector that (a) consists of three mutually perpendicular intersecting conducting flat surfaces and (b) returns a reflected electromagnetic wave to its point of origin. **4.** A passive optical mirror that (a) consists of three mutually perpendicular flat intersecting reflecting surfaces and (b) returns an incident light beam in the opposite direction. **5.** A directional antenna that uses two mutually intersecting conducting flat surfaces. *See* **dihedral corner reflector, trihedral corner reflector.** *See also* **triple mirror.**

corner reflector antenna: An antenna that consists of a feed and one or more corner reflectors. *See also* **dihedral corner reflector, trihedral corner reflector.**

corporate information management: A program for organizing and streamlining federal government information management systems, such as defense, lifecycle, functional improvement, data administration, functional economic analysis, technical architectural framework, and graphic interface information management systems.

Corporation for Open Systems (COS): A nonprofit research and development consortium that (a) consists of communications companies and federal agencies, such as the Defense Information Systems Agency (DISA) and the General Services Administration (GSA), (b) operates a test bed for telecommunications systems and components, and (c) develops and tests communications systems for (i) end-to-end interoperability, i.e., user-to-user interoperability, (ii) viability of protocols, and (iii) fragmentation in existing systems, such as in the Integrated Services Digital Network (ISDN) standard and implementations.

corrected lens: A lens is relatively free from one or more aberrations. Examples of corrected lenses are (a) simple lenses with aspheric surfaces, i.e., nonspherical surfaces, and (b) compound lenses that consist of several optical elements and different types of glass. *See also* **aberration, lens, optical element.**

correcting code: *See* **error correcting code.**

correcting system: *See* **error correcting system.**

correction: *See* **forward error correction, synchronous correction.**

correction bit: *See* **error correction bit.**

correction efficiency factor: *See* **block correction efficiency factor.**

correction request: A service message requesting correction of all or part of a message that has errors, such as incorrect groups, incorrect code groups, omitted portions, wrong character count, and wrong block count, that the communications system is capable of detecting. *Note:* Certain types of errors, such as complete garbling and mutilation, may not be correctable because the system is not designed to detect them.

correction signal: *See* **synchronization correction signal.**

correction system: *See* **automatic error detection and correction system.**

corrective maintenance: 1. Maintenance actions carried out to restore a defective item to a specified condition. **2.** Tests, measurements, and adjustments made to remove or correct a fault. *See also* **maintenance, preventive maintenance.**

correlation: The comparing of two characteristics that interrelated phenomena have in common. *Note:* In signaling systems, a phase shift, i.e., a time difference, between two signal sequences may be correlated with their spatial separation. Thus, the time correlation function may be compared with the spatial correlation function, yielding a temporal and spatial correlation function. *See* **autocorrelation, cross correlation, track correlation.** *See also* **coherence.**

correlator: In a spread spectrum system, the equipment that compares a received signal with a local reference to check for agreement.

correspondence: *See* **public correspondence.**

corruption: *See* **data corruption.**

COS: Corporation for Open Systems.

cosine emission law: *Synonym* **Lambert's cosine law.**

cosine law: *Synonym* **Lambert's cosine law.**

cosmic noise: 1. Noise or interference coupled into a communications system, originating from sources outside the solar system, mainly from certain areas of the Milky Way. **2.** Noise that (a) is measured by antennas pointed toward the center of the Milky Way and pointed in the direction of the galactic arm, (b) is random, (c) is similar to thermal noise, (d) occurs at frequencies above 15 MHz (megahertz), and (e) may be extragalactic, coming from emission nebulae in other galaxies and from remnants of supernovae. *Synonyms* **galactic noise, galactic radio noise, radio noise.**

cosmic velocity: *See* **third cosmic velocity.**

Costas loop: 1. A phase-locked loop demodulator circuit that (a) does not require that symbol synchronization be known, (b) has a tracking loop in which the integrators are replaced by arbitrary filters, (c) is used particularly in spread spectrum systems, (d) usually is used in pulse amplitude modulated (PAM) and binary phase shift key (PSK) loop circuits, and (e) has a variation that consists of a squaring loop that includes a square law device. **2.** A phase-locked loop that (a) is used for carrier phase recovery from suppressed carrier modulation signals, such as from double sideband suppressed carrier signals, (b) in the usual implementation, has a local voltage-controlled oscillator that provides quadrature outputs, one to each of two phase detectors, i.e., product detectors, (c) applies the same phase of the input signal to both phase detectors, (d) passes the output of each phase detector through a low-pass filter, and (e) applies the outputs of these low-pass filters to another phase detector, the output of which passes through a loop filter before being used to control the voltage-controlled oscillator. *See also* **loop, phase-locked loop, square law device, squaring loop.**

coulomb (C): The International System (SI) (Système International d'Unités) practical MKS (meter-kilogram-second) unit of electric charge, equal to (a) the quantity of electricity transferred by a current of 1 A (ampere), equivalent to about 6.2418×10^{18} $e \cdot s^{-1}$ (electrons per second), flowing for one second and (b) 6.2418×10^{18} electrons. *Note:* The charge of an electron is about 1.6021×10^{-19} C (coulombs).

Coulomb's law: The universal law of attraction and repulsion of electric charges, given by the relation $F = q_1 q_2 / \pi \varepsilon r^2$, where F is the force of attraction, if the charges are of opposite polarity, or the force of repulsion, if they are of like polarity, q_1 and q_2 are the magnitudes of the charges, ε is the electric permittivity of the surrounding medium in which the charges are embedded or located, i.e., is the constant of proportionality, and r is the distance of separation of q_1 and q_2. *Note 1:* The charges are assumed to be points compared to their distance of separation. *Note 2:* The variables may be expressed in any compatible system of units. *Refer to* **Fig. I-1 (436).**

count: *See* **optical fiber count, peg count, raster count.**

counter: A device that is capable of representing or displaying the number of pulses that it has received since it was last set. *See* **binary counter, modulo-n counter, modulus counter, reversible counter, ring counter, synchronous counter, zero-crossing counter.**

counterclockwise polarized electromagnetic wave: *Synonym* **left-hand polarized electromagnetic wave.**

countercountermeasure: *See* **electronic counter-countermeasure, infrared jamming electronic countercountermeasure, radio relay electronic countercountermeasure.**

countermeasure: *See* **absorption electronic countermeasure, active electronic countermeasure, active optical countermeasure, electronic countercountermeasure, electronic countermeasure, infrared jamming electronic countercountermeasure, optical countermeasure, radio relay electronic counter-countermeasure.**

counterpoise: A conductor or a system of conductors used as a substitute for earth or ground in an antenna system. *See also* **earth, ground.**

count integrity: *See* **bit count integrity, character count integrity.**

count register: *See* **peg count register.**

country code: 1. In international direct telephone dialing, a code that consists of one-, two-, or three-digit numbers in which the first digit designates the region, and succeeding digits, if any, designate the country. **2.** In international record carrier transmissions, a code consisting of two- or three-letter abbreviations of the country names, or two- or three-digit numbers that represent the country names, that follow the geographical place names. *See* **data country code.** *See also* **access code, NXX code.**

country network identity: In a communications system, information sent in the backward direction consisting of address signals that indicate the identity of a country or a network to which a call has been switched. *See also* **backward direction.**

coupled mode: 1. In an electromagnetic wave propagating in a waveguide, such as an optical fiber, a coaxial cable, or a metal waveguide, a mode that shares energy with one or more other modes in such a manner that the coupled modes propagate together, and the propagating energy is distributed among the coupled modes. *Note 1:* Coupled modes (a) coexist, (b) have fields that are interrelated, and (c) mutually interchange their energies. *Note 2:* Modulation may be common or different among the modes. *Note 3:* The energy share of each mode changes with propagation distance in the waveguide. *Note 4:* The energy share of each mode does not change after the equilibrium length has been reached. **2.** In microwave transmission, a condition in which energy is transferred from the fundamental mode to higher order modes. *Note:* Energy transferred to coupled modes usually is undesirable in microwave transmission in a waveguide. The frequency is held low, i.e., the wavelength is held high, so that propagation in the waveguide is limited to the fundamental mode. *See also* **differential mode attenuation, fiber optics, mode, multimode optical fiber, single mode optical fiber.**

coupled power: *See* **source to optical fiber coupled power.**

coupled reperforator tape reader: A perforated tape retransmitter that ensures the retransmission of all signals recorded during perforation, including the last one. *Synonym* **fully automatic reperforator transmitter distributor.**

coupler: 1. In fiber optics, a device that enables the transfer of optical energy from one fiber optic waveguide to another. **2.** A passive optical device that distributes or combines optical power among two or more ports, including protective housing and pigtails. **3.** In electronics, the means by which energy is transferred from one conductor, including a fortuitous conductor, to another. *Note 1:* Types of electronic coupling are (a) capacitive coupling, i.e., the linking of one conductor with another by means of capacitance, also known as electrostatic coupling, (b) inductive coupling, i.e., the linking of one conductor with another by means of magnetic inductance, also known as magnetic coupling, and (c) conductive coupling, i.e., hard wire or resistive connection of one conductor to another, also known as direct coupling or resistive coupling. *Note 2:* Electronic coupling is frequency-sensitive. *See* **access coupler, acoustic coupler, antenna multicoupler, bidirectional coupler, data bus coupler, directional coupler, electronically controllable coupler, fiber optic coupler, fiber optic multiport coupler, fiber optic rod coupler, hybrid coupler, laser diode coupler, light-emitting diode coupler, nonreflective star coupler, nonuniformly distributive coupler, optical directional coupler, optical multiport coupler, optical waveguide coupler, optoelectronic directional coupler, positive-intrinsic-negative photodiode coupler, reflective star coupler, star coupler, tee coupler, unidirectional coupler, uniformly distributive coupler, wye coupler.**

coupler excess loss: Fiber optic coupler power loss, given by the relation $P_{ex} = \Sigma P_j - \Sigma P_{ij(id)}$ where P_{ex} is the coupler excess loss, i is the number of input ports, P_j is the output power at port *j,* and $P_{ij(id)}$ are the transmittances that would be obtained from an ideal coupler in which all of the input power is coupled to the nonisolated output ports.

coupler loss: In a coupler, the insertion loss, i.e., the power loss, that occurs when power is transferred from a given input port to a given output port with all other ports properly terminated. *See* **source-coupler loss.**

coupler-switch modulator: *See* **integrated optical circuit filter-coupler-switch modulator.**

coupler transmittance: In a fiber optic coupler, the optical transmittance measured between an input port and an output port of the coupler by launching a known optical power level into each input port *i,* one at a time, and measuring the power at each of the *j* output ports for each of the *i* input ports. *Note:* In a 2×4 coupler there are six ports. For each of the two input ports, the power at each of the four output ports would be measured to determine the eight coupler transmittances. *See also* **coupler transmittance matrix.**

coupler transmittance matrix: For a fiber optic coupler, a matrix of $L \times L$ dimensions, where L is the total number of ports, no distinction being made between input and output ports. *Note:* In a 2×4 coupler L would be 6; there would be six ports, numbered 1 to 6. Each of the transmittances, T_{ij}, in dB, in the matrix is represented by the relation $T_{ij} = 10 \log_{10}(P_i / P_j)$ where T_{ij} is the transmittance in dB from the input port i to the output port j, P_i is the optical power launched into port i, P_j is the output power at port j, and $i \neq j$. The subscripts of the elements of the transmittance matrix correspond to the numbers labeled on the coupler, such as $T_{12}, T_{13}, T_{14} \ldots T_{21}, T_{23}, T_{24} \ldots T_{31} \ldots T_{n(n-1)}$ where n is the number of ports.

coupling: The transfer of energy from one conductive or dielectric medium, such as an optical waveguide or wire, to another, including fortuitous transfer. *Note 1:* Coupling may be desirable or undesirable. Types of electrical coupling include (a) capacitive coupling, i.e., electrostatic coupling, (b) inductive coupling, i.e., magnetic coupling, and (c) conductive coupling, i.e., resistive or hard wire coupling. *Note 2:* Undesirable coupling between optical fibers is very effectively prevented by a polymer coating, which reduces the propagation of cladding modes and provides some degree of physical protection. Desirable coupling between optical fibers, such as in fiber optic couplers, is achieved by stripping the cladding from a section of the fibers and placing the cores against each other. **2.** In fiber optics, the part of a fiber optic connector that aligns the optical termini. *See* **butt coupling, capacitive coupling, conductive coupling, cross coupling, crosstalk coupling, direct coupling, end fire coupling, evanescent field coupling, fiber-detector coupling, free-space coupling, furcation coupling, inductive coupling, lens coupling, mode coupling, optical fiber detector coupling, proximity coupling, reference coupling, resistance coupling, source-fiber coupling, tangential coupling.** *See also* **aperture to medium coupling loss, conducted interference, cross modulation, crosstalk, directional coupler, interaction crosstalk, mode scrambler.** *Refer to* **Figs. E-2 (291), L-5 (511), L-7 (535), P-17 (767).**

coupling coefficient: A number that (a) expresses the degree of electrical coupling that exists between two circuits, (b) is calculated as the ratio of (i) the mutual impedance to (ii) the square root of the product of the self-impedances of the coupled circuits, all impedances being of the same kind, such as all being inductive reactances, and (c) is given by the relations $\Gamma = Z_m/(Z_1Z_2)^{1/2}$ where Γ is the coupling coefficient, Z_m is the mutual impedance, and Z_1 and Z_2 are the self-impedances of the coupled circuits, again all impedances being of the same kind, i.e., capacitive, inductive, or resistive, such as $\Gamma_L = M/(L_1L_2)^{1/2}$ where Γ_L is the inductive coupling coefficient, M is the mutual inductance, and L_1 and L_2 are the self-inductances of the coupled circuits. *See also* **coupling, coupling efficiency.**

coupling efficiency: In fiber optics, the efficiency with which optical power is transferred between two components. *Note 1:* Coupling efficiency usually is expressed as the percentage of power output from the transmitting component that is received at the receiving component. *Note 2:* If the coupling efficiency is expressed in dB, it corresponds to the coupling loss. *Note 3:* The maximum light coupling efficiency, η_c, from a Lambertian radiator, i.e., a light source, such as a light-emitting diode (LED) and a graded index optical fiber, is given by the relation $\eta_c = A_c(NA)^2/2A_sn_0^2$ where η_c is the coupling efficiency, A_c is the core area, A_s is the source area, n_0 is the refractive index of the propagation medium between the source and optical fiber, and NA is the numerical aperture. *See* **source coupling efficiency.** *See also* **coupling, coupling coefficient, numerical aperture.** *Refer to* **Fig. A-6 (24).**

coupling loss: 1. The power loss that occurs when energy is transferred from one circuit element or propagation medium to another. *Note 1:* Coupling loss usually is expressed in the same units as used for the originating circuit, such as watts or dBm. *Note 2:* Coupling loss may be expressed in absolute units, such as watts, or as a ratio, i.e., a fraction or a percentage of the power in the originating circuit. If expressed as 10 times the logarithm to base 10 of the ratio, i.e., in dB or dBm, it is called coupling efficiency. **2.** In fiber optics, the power loss that occurs when coupling light from one optical device or propagation medium to another. *See* **aperture to medium coupling loss, extrinsic coupling loss, intrinsic coupling loss.** *See also* **angular misalignment loss, coupling, extrinsic joint loss, insertion loss, intrinsic joint loss, lateral offset loss, longitudinal displacement loss, path loss, transmission loss.**

couriergram: A message that usually would be sent by electrical means, but, if circuits are not available, is carried over all or part of its route by a messenger or an agency that is not necessarily a part of the communications system that accepted the message for transmission.

courier service: Transmission based on the use of persons who carry messages between persons, stations, commands, organizations, and places. *Note:* Examples of courier services are (a) a truck message delivery system in which a fleet of trucks and drivers carry messages directly from the message originator to the message receiver and (b) a daily two-way transfer of messages between a subordinate unit and its higher headquarters using a motorcycle, a rider, and a message pouch.

course: *Synonym* **heading.**

cover: 1. To convert, by means of transmission security (TRANSEC) and cryptographic techniques, a transmitted waveform into a form that may be used only by authorized persons. **2.** To alter characteristic communications patterns to conceal information that could be of value to unauthorized persons or systems. **3.** The act of maintaining a continuous receiver watch with transmitter calibrated and available, but not necessarily available for immediate use. **4.** The measures necessary to give protection to a person, plan, operation, formation, or installation from hostile intelligence efforts and leakage of information. *See* **horizontal cover, radio communications cover, vertical cover.** *See also* **continuous receiver watch.**

coverage: *See* **Earth coverage, jammer area coverage.**

coverage diagram: *See* **radio coverage diagram, vertical coverage diagram.**

covering: *See* **protective covering.**

cover watch: A copy watch in which the maintenance of a complete station log is optional. *See also* **copy watch, station log.**

CPE: customer premises equipment.

cpi: characters per inch.

CP/M: An operating system used on 8-bpb (bits per byte) microprocessors. *See also* **disk operating system.**

cps: characters per second, cycles per second. *Note:* The use of "cycles per second" for designating frequency is obsolete. The hertz (Hz), the SI unit (Système International d'Unités), is the proper unit for frequency. However, "cycles per second" and "cps" may be used in mechanical systems, such as internal combustion engines, and in cyclical industrial processes, such as machine tool and automatic screw machine repetitive operations. "Cycles per minute, hour, or day" may also be used in these instances.

CPU: central processing unit, communications processor unit.

CR: carriage return, channel reliability, circuit reliability.

crack: *See* **microcrack.**

cracker: A malevolent hacker, i.e., a hacker who has turned to crime, such as a hacker who violates copyright laws, steals personal and private information in violation of the Privacy Act, distributes pornographic materials, launches worms and viruses that destroy computer programs and databases, steals and sells industrial and defense secrets, causes illegal transfer of funds, uses communications systems without authorization, or illegally breaks key. *See also* **cybercop, firewall, hacker, key.**

CRADA: cooperative research and development agreement.

craft frequency: *See* **emergency craft frequency.**

crash locator beacon: An automatic radio beacon that can identify the location of a crash site by means of direction finding.

crash protected: Pertaining to a hardware or software capability that prevents the loss, or allows recovery, of software in the event of a total system failure.

CRC: cyclic redundancy check.

creeping code: *Synonym* **walking code.**

crime: *See* **cybercrime.**

criterion: In testing, a condition or performance requirement that a system must meet. *See* **acceptance criterion, blocking criterion, Nyquist criterion.** *See also* **test.**

critical angle: In geometric optics, at a refractive boundary between two media each with a different refractive index, the smallest incidence angle at which total internal reflection occurs. *Note 1:* The incidence angle is measured with respect to the normal at the refractive boundary. *Note 2:* The critical angle is given by the relation $\Theta_c = \arcsin(n_2/n_1)$ where Θ_c is the critical angle, n_1 is the refractive index of the more dense medium, i.e., the medium with the higher refractive index, and n_2 is the refractive index of the other medium. *Note 3:* For there to be total internal reflection and therefore a critical angle, the incident ray must be in the denser medium, i.e., higher refractive index medium. *Note 4:* If the incident ray is precisely at the critical angle, the refracted ray is tangent to the boundary at the incidence point and propagates along the interface surface. There is no reflected ray. *See*

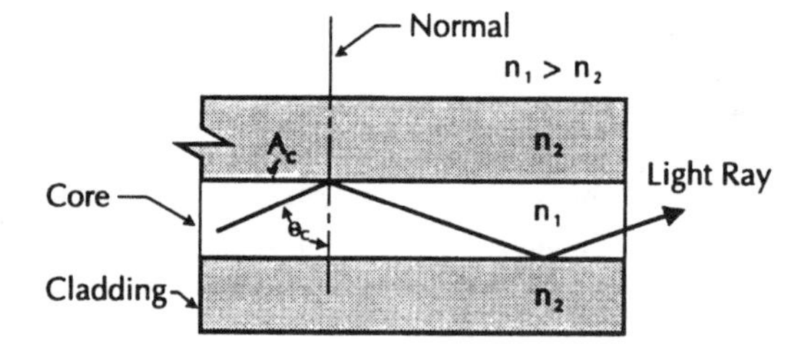

Fig. C-9. For a step index optical fiber with a core refractive index of 1.500 and a cladding index of 1.485, the **critical angle** Θ_c, given by the relation $\Theta_c = \sin^{-1}$ 1.485/1500, is about 84°, making the complement about 6°. Total internal reflection occurs when the incidence angle of the light rays striking the core-cladding interface surface is greater than the critical angle, in this case less than 6° with the interface surface.

also **incidence angle, fiber optics, total internal reflection.** *Refer to* **Fig. C-9.** *Refer also to* **Figs. E-4 (314), L-8 (536), L-9 (536), S-18 (951).**

critical angle sensor: A fiber optic sensor in which (a) the optical power coupled into an optical fiber is a function of the angle at which the light enters the fiber endface, and (b) the optical power, i.e., the irradiance, that reaches a photodetector at the end of the fiber will depend on the entry angle, permitting the angle between the direction of the source beam and the fiber axis to be measured.

critical area: An operational area that requires specific environmental control, such as air conditioning, because of the equipment or information contained in the area. *See also* **air conditioning.**

critical frequency (CF): 1. In radio wave propagation by way of the ionosphere, the limiting frequency at or below which a wave component is reflected by, and above which it penetrates through, an ionospheric layer. **2.** At vertical incidence, the limiting frequency at or below which a wave component is reflected by, and above which it penetrates through, an ionospheric layer. *Note:* The existence of the critical frequency is the result of electron limitation, i.e., the inadequacy of the existing concentration of free electrons, i.e., free-electron density, to support reflection at higher frequencies. *See also* **critical wavelength, cutoff frequency, frequency, ionosphere.**

critical radius: As the radius of curvature of an optical fiber is decreased from infinity, i.e., from that of a straight line, the radius at which the fields of an electromagnetic wave propagating in the fiber begin to become detached from the fiber and radiate out into space, i.e., bound modes begin to cease being bound modes. *Note 1:* At radii less than the critical radius, radiation and attenuation occur because (a) the evanescent waves on the outside of the curve must exceed the speed of light of the bound modes if they are to remain bound, and (b) the propagating waves will no longer meet the critical angle requirement for maintaining total internal reflection. Thus, the velocity of the portion of the wavefront on the outside part of the curve must increase beyond the speed of light in order to maintain the proper phase relationship with the wave inside the fiber and so no longer can remain bound to the inside wave. When the velocity of an evanescent wavefront element farthest from the central axis of the fiber begins to exceed the velocity of light as it sweeps around the bend, the wavefront can no longer keep up with the part of the wave inside the fiber, and the energy begins to radiate out from the fiber, resulting in attenuation of the energy in the propagating wave. If the radius is less than the critical radius, the incidence angles at the bend become less than the critical angle, causing part of the energy in the wave to enter the cladding. *Note 2:* The attenuation drops off sharply at the critical radius. At radii below the critical radius, nearly all the energy radiates from the fiber as mode conversion takes place at the bend. Care must be taken in laying fiber optic cables so as not to bend them at radii approaching the value of either the critical radius or the minimum bend radius, whichever is the larger. However, the phenomenon can be used in displacement sensors and switches. *See also* **critical radius sensor, minimum bend radius.**

critical radius sensor: A fiber optic sensor in which an optical fiber is bent in such a manner that increased energy in a lightwave propagating within it is radiated out of the fiber as the bend radius is decreased. *Note:* The optical power radiated away at the bend increases sharply from zero to 100% as the radius of the bend is decreased to and passes through the critical value. The loss is caused by mode conversion in the transition from the straight fiber to the curved fiber and the separation of the evanescent wave into space when the outer edge of the wave cannot keep up with the internal wave, i.e., the outer edge has reached the speed of light as it swings around the bend. At the end of the fiber, a photodetector is used to detect the variation in optical power, which is a function of the bend radius. Thus, the displacement that is causing the bend can be measured after the device is calibrated. *See also* **critical radius.**

critical technical load (CTL): That part of the total technical power load required for synchronous communications and automatic switching equipment. *See also* **load, technical load.**

critical wavelength: The free-space wavelength that corresponds to the critical frequency. *Note 1:* The critical wavelength is obtained by dividing the free-space speed of light by the critical frequency. *Note 2:* The critical wavelength usually is expressed in meters, which will be obtained when the free-space speed of light is expressed in meters per second, i.e., 3×10^8 $m \cdot sec^{-1}$, and the critical frequency is expressed in hertz. *See also* **critical frequency, ionosphere.**

cross assembler: An assembler that can run symbolic language input on one type of computer and produce machine language output for another type of computer.

cross band radiotelegraph procedure: A radiotelegraph network operational procedure in which (a) calling stations, such as ship stations, call other stations, such as shore stations, using one frequency, (b) the calling stations shift to another frequency to transmit their messages, and (c) the called stations answer using a third frequency. *See also* **answering frequency, calling frequency, simplex radiotelegraph procedure, working frequency.**

crossbar switch: A switch that has (a) a plurality of vertical paths, (b) a plurality of horizontal paths, and (c) electromagnet-operated mechanical means, i.e., electromechanical relays, for interconnecting any one of the vertical paths with any one of the horizontal paths. *Note:* The same logical function may also be achieved by using combinational circuits, i.e., gates, made from solid-state semiconductor devices. *See also* **combinational circuit, electronic switching system, gate, step by step switching system, switch, switching center.**

crossbar system: In a communications system, such as a telephone network, an automatic switching system that (a) consists of (i) crossbar switches, (ii) selecting mechanisms, (iii) common circuits to select and test the switching paths and control the operation of the selecting mechanisms, and (iv) control mechanisms that receive the switching information, store the switching information, and determine the operations necessary to establish a connection, and (b) makes connections (i) between circuits, i.e., between loops, (ii) between circuits and trunks, and (iii) between trunks. *See also* **connection, crossbar switch, switch.**

cross compiler: A compiler that can be executed on one computer and compiles a program that can be executed on a different computer. *Note:* An example of a cross compiler is a compiler that, using one computer, generates assembly language code or object language code for execution by a different computer, i.e., a different make or model computer.

cross connect: *Synonym* **cross connection.**

cross connection: In a distribution frame, a connection (a) between terminal blocks on the two sides of the distribution frame or (b) between terminals on the same terminal block. *Note:* A connection between terminals on the same block is also called a "strap." *Synonyms* **cross connect, jumper, strap.** *See* **fiber optic cross connection.** *See also* **connection, distribution frame, strap.**

cross connect system: *See* **digital access and cross connect system.**

cross correlation: The time integral of the product of one signal and another related, but time-delayed, signal, given by the relation

$$CC = \int_o^t f(t)g(t-\tau)dt$$

where CC is the cross correlation, $f(t)$ and $g(t)$ are the two signals, or arbitrary functions of time, and τ is the time $g(t)$ is delayed in relation to the occurrence of $f(t)$. *Note:* Correlating two or more interrelated phenomena includes finding the similarity between two different functions. *See also* **autocorrelation, correlation.**

cross coupling: The coupling of a signal from one channel, circuit, or conductor to another, where it becomes an undesired signal, such as crosstalk. *See also* **circuit, coupling, interference.**

crossfire: In telegraph systems, electrical current interference in one channel caused by signaling currents in another channel. *See* **receiving end crossfire, sending end crossfire.**

crossing counter: *See* **zero crossing counter.**

cross level: Pertaining to one axis of a three-axis antenna mount, one axis in each Cartesian direction, i.e., orthogonal direction. *Note:* Cross level antenna mounts are often used by Earth stations on board ships.

crosslink: In a satellite communications system, a direct link between two geostationary satellites that (a) are separated by approximately 90° of longitude, equivalent to about 60,000 km (kilometers), and (b) carry narrowbeam antennas pointed at each other. *Note:* The cross link allows long-haul communications traffic to avoid the signal attenuation and distortion caused by atmospheric conditions associated with a double hop, i.e., from satellite to Earth and Earth to satellite. *See also* **attenuation, geostationary satellite, hop, long-haul communications.**

cross linking: In satellite communications, direct transmission between two geostationary satellites via a crosslink. *See also* **attenuation, crosslink, geostationary satellite, hop, long-haul communications.**

cross modulation: Intermodulation caused by the modulation of the carrier of a desired signal by an undesired signal. *See also* **coupling, interference, modulation.**

cross office trunk: A trunk that has its terminations within a single facility. *See also* **facility, trunk.**

crosspoint: In a switch, a single element in the array of elements that compose the switch, consisting of a set of physical or logical contacts that operate together to extend the speech and signaling channels in a switched network. *See also* **switch, switching center.**

cross polarized operation: The operation of two transmitters on the same frequency, with one transmitter-receiver pair operating with vertically polarized waves and the other pair with horizontally polarized waves, i.e., the polarization planes of the waves are orthogonal. *See also* **antenna, polarization diversity, space diversity.**

cross product: *Synonym* **vector product.**

cross section: *See* **radar cross section, scattering cross section.**

cross site link: In a satellite communications system, the signal power and control connections between the components of an Earth station. *Note:* Examples of cross site links are (a) links between transmitters and antennas and (b) links between control consoles and transmitters.

cross strapping: Connecting two or more points in a circuit or a device with a short piece of wire or metal. *Note:* Examples of cross strapping are (a) connecting terminals on the same block of a distribution frame and (b) connecting resonator segments that have the same polarity in a multicavity magnetron to suppress undesired modes of oscillation. *See also* **cross connection, distribution frame.**

crosstalk (*XT*): 1. Undesired capacitive, inductive, or conductive coupling from one circuit, part of a circuit, or channel to another. **2.** Any phenomenon by which a signal transmitted on one circuit or channel of a transmission system creates an undesired effect in another circuit or channel. **3.** In fiber optics, the undesired optical power that is transferred from, i.e., leaks from, one optical waveguide to another. *Note 1:* The crosstalk is often a measure of the optical power picked up by an optical fiber from an adjacent energized fiber. *Note 2:* The crosstalk, XT, is given by the relation $XT = 10 \log_{10}(P_d/P_o)$ where XT is the crosstalk in dB (decibels), P_o is the optical power at the receiving end of the energized fiber, and P_d is the power in the disturbed fiber at the corresponding end. P_d is always less than P_o. Thus, the crosstalk, XT, is always negative. The negative sign is commonly ignored. The crosstalk may be given simply in dB, such as 90 dB. *See* **far end crosstalk, intelligible crosstalk, interaction crosstalk, near end crosstalk, optical fiber crosstalk, unintelligible crosstalk.** *See also* **adjacent channel interference, channel, crosstalk coupling, interference, stereophonic crosstalk, susceptiveness.**

crosstalk coupling: The ratio of (a) the power in a disturbing circuit to (b) the induced power in the disturbed circuit observed at specified points in the circuits under specified terminal conditions. *Note:* Crosstalk coupling usually is expressed in dB (decibels). *Synonym* **crosstalk coupling loss.** *See also* **circuit, coupling, crosstalk, interference, loss.**

crosstalk coupling loss: *Synonym* **crosstalk coupling.**

crosstalk resistance: In a multiplexed communications system or component, the ability to prevent transfer of noise or unwanted signals from one channel to another.

cross tell: *Synonym* **lateral tell.**

crown glass: An optical glass that (a) is made of alkaline lime silicate, (b) has a relatively low refractive index, (c) has a low dispersion attenuation factor, and (d) has a low absorption coefficient, compared to other types of glass. *See also* **absorption coefficient, dispersion attenuation factor, refractive index.**

crown lens: *Synonym* **converging lens.**

CRSS: C^{3}I reusable software system.

CRT: cathode ray tube.

crucible process: *See* **double crucible process.**

crunching: *See* **number crunching.**

crush: 1. A small scratch, or series of small scratches, on the surface of an optical element, usually caused by mishandling, scuffing, scraping, or rubbing the element. **2.** In communications systems, the damage done by a compressive force to a cable or its components, such as twisted pairs, coaxial insulation, optical fibers, fiber optic bundles, or cable cores. *See also* **fiber optic cable, cable, coaxial cable, optical element.**

cryogenics: The branch of science and technology devoted to the behavior and application of materials at or near absolute zero temperature, i.e., at or near 0 K (kelvin). *Note:* Examples of cryogenic applications are (a) the reduction of thermal noise in materials at very low temperatures, such as the cooling of preamplifiers to reduce intrinsic noise by circulating liquid helium or liquid nitrogen through or around the preamplifiers, (b) the design, development and use of cryogenic storage devices, and (c) the use of superconductivity for switching. *See also* **cryogenic storage, Kelvin temperature scale, superconductivity.**

cryogenic storage: Storage in which (a) the superconductive property of materials is used to store data and (b) use is made of the phenomenon that superconductivity is destroyed in the presence of a magnetic field, thus enabling a binary 1 to be stored and removed. *See also* **cryogenics, storage, superconductivity.**

cryptanalysis: 1. The steps, operations, and processes that are performed to convert encrypted messages into plain text without initial knowledge of the key employed in the encryption. **2.** The study of encrypted texts. *Synonym* **cryptographic analysis.** *See also* **cryptography, cryptology.**

crypto: Pertaining to cryptography, cryptanalysis, cryptology, and the materials, systems, procedures, information, and operations related to these areas.

CRYPTO: A marking or designator that identifies communications security (COMSEC) keying material used to secure or authenticate classified or sensitive information, the loss of which could adversely affect the national security of the United States. *See also* **cryptomaterial, key.**

cryptochannel: A complete crypto communications system, such as the basic unit for U.S. Navy cryptographic communications, that includes (a) the prescribed cryptographic aids, (b) the holders of the cryptographic aids, (c) the indicators or other means of identification, (d) the effective or applicable areas, (e) the special purpose, if any, for which the channel is provided, and (f) pertinent information, such as information concerning distribution, operation, and usage. *Synonym* **cryptographic channel.** *See also* **channel, cryptology.**

cryptoequipment: Equipment used to encrypt messages. *Synonym* **cryptographic equipment.**

cryptogram: A message, such as a telegram, that has been encrypted.

cryptographer: 1. A person who encrypts or decrypts messages. **2.** A person who creates components for cryptosystems. *See also* **cryptanalysis.**

cryptographic analysis: *Synonym* **cryptanalysis.**

cryptographic channel: *Synonym* **cryptochannel.**

cryptographic equipment: *Synonym* **cryptoequipment.**

cryptographic information: 1. Information that (a) is descriptive of cryptographic systems and equipment, including their functions and capabilities, and (b) includes all cryptomaterial. **2.** Information that (a) can be used for the cryptanalytic conversion of encrypted text, (b) would make a significant contribution to the cryptanalytic conversion of encrypted text, or (c) would contribute to the operation of a cryptosystem. *Synonym* **cryptoinformation.** *See also* **cifax, ciphony, civision, cryptology, cryptomaterial, key.** *Refer to* **Fig. C-10.**

Fig. C-10. The modular AN/PSC-5 VHF/UHF manpack military line-of-sight and satellite communications equipment that has an embedded **cryptographic information** capability and is operated from a simple front panel keypad. (Courtesy Magnavox Systems Company, Fort Wayne, IN.)

cryptographic key: *Deprecated. See* **key.**

cryptographic net: *Synonym* **cryptonet.**

cryptographic operation: *See* **offline cryptographic operation, online cryptographic operation.**

cryptographic security: *Synonym* **cryptosecurity.**

cryptographic system: *Synonym* **cryptosystem.**

cryptography: 1. The branch of cryptology that (a) is devoted to the principles, means, and methods of designing, developing, and using cryptosystems and (b) includes the transformation of plain text to encrypted text and vice versa. **2.** The principles, means, and methods for (a) transforming plain information into an unintelligible form and (b) restoring encrypted information to intelligible form. *See* **key escrow cryptography.** *See also* **cryptanalysis, cryptology.**

cryptoinformation: *Synonym* **cryptographic information.** *Refer to* **Fig. C-10.**

crypto key: *Deprecated. See* **key.**

cryptologic: Pertaining to cryptology. *See also* **cryptology.**

cryptologic: Pertaining to cryptology. *See also* **cryptology.**

cryptology: The branch of science and technology devoted to (a) hidden, disguised, or encrypted communications, (b) communications security, and (c) communications intelligence. *See also* **bulk encryption, ciphony, communications security, cryptanalysis, cryptography, cryptologic, encode, encrypt, traffic flow security.**

cryptomaterial: 1. All material, including documents, devices, equipment, and apparatus, essential to the encryption, decryption, or authentication of telecommunications information. *Note:* Classified cryptomaterial is designated CRYPTO and is subject to special safeguards. **2.** All material, including documents, devices, or equipment that (a) contains cryptographic information and (b) is essential to the encryption, decryption, or authentication of telecommunications information. *See also* **ciphony, cryptographic information, cryptology, key.**

cryptonet: A net that consists of stations that (a) hold a specific key for use, (b) usually communicate directly with each other, and (c) usually use a common cryptochannel. *Note 1:* A cryptonet may be formed by two or more activities that have a cryptosystem in common and possess a means of communications. *Note 2:* Activities that hold key for a purpose other than use, such as cryptologistic depots, are not cryptonet members for that key. Controlling authorities are de facto members of the cryptonets that they control. *Synonym* **cryptographic net.** *See also* **key.**

cryptooperation: *See* **synchronous cryptooperation.**

cryptosecurity: 1. Communications security that results from the provision of technically sound cryptosystems and their proper use. **2.** The portion of communications security concerned with the application of cryptology to communications systems, such as cryptographic information, cryptomaterials, cryptosystems, and cryptoequipment. *Synonym* **cryptographic security.** *See also* **communications security, cryptoequipment, cryptographic information, cryptomaterials, cryptosystems, key.**

cryptosystem: The interaction of associated cryptomaterials that (a) are used as a unit and (b) provide a single means of encryption and decryption. *See also* **cipher, cryptanalysis, cryptochannel, cryptology, cryptomaterial, decrypt, key, link encryption.** *Synonym* **cryptographic system.**

crystal: *See* **doubly refracting crystal, multirefracting crystal.**

crystal display: *See* **active matrix liquid crystal display.**

crystal optics: The branch of optics devoted to the study and application of the propagation of radiant energy through crystals, especially anisotropic crystals, and their effects on polarization of electromagnetic waves, particularly lightwaves.

crystal oscillator (XO): An oscillator in which the frequency is controlled by a piezoelectric crystal. *Note 1:* Crystal oscillators often require a controlled temperature because their operating frequency may be a function of temperature. *Note 2:* Types of crystal oscillators include voltage controlled crystal oscillators (VCXO), temperature compensated crystal oscillators (TCXO), oven controlled crystal oscillators (OCXO), temperature compensated voltage controlled crystal oscillators (TCVCXO), oven controlled voltage controlled crystal oscillators (OCVCXO), microcomputer compensated crystal oscillators (MCXO), and rubidium crystal oscillators (RbXO).

CSCINT: closed source intelligence.

CSMA: carrier sense multiple access.

CSMA/CA: carrier sense multiple access with collision avoidance.

CSMA/CD: carrier sense multiple access with collision detection.

CSU: channel service unit, circuit switching unit, customer service unit.

CTX: Centrex®.

cube: magic cube.

cubic centimeter: *Synonym* **milliliter.**

cue: *Synonym* **prompt.**

current: *See* **alternating current, backward current, dark current, direct current, disturbance current, electrical current, forward current, metallic circuit current, noise current, photocurrent, pulsating direct current, stationary direct current, sneak current, stray current, threshold current, vector current.**

current disk: *Synonym* **data disk.**

current modulation: *See* **constant current modulation.**

current moment: At a point in space and at a given instant, the integral of all of the elemental current moments that (a) are in the space surrounding the point and (b) contribute to the total magnetic field strength in both magnitude and direction at the given point and given instant. *See* **elemental current moment.** *See also* **effective height.**

current patch bay: *See* **direct current patch bay.**

current position: 1. In display systems, the present position of a display element, display group, or display image, such as a display group that represents an object in space, in the display space of a display device. *Note:* The current position usually is expressed in user coordinates, device coordinates, or real world coordinates. **2.** The present position of an object, such as a target, in space. *Note:* The current position is expressed in spatial real world coordinates, such as slant range, elevation angle, and azimuth. *See also* **device coordinate, display element, display group, real world coordinate.**

current spreadsheet: In a personal computer with spreadsheet software in effect, the specific version of a spreadsheet that (a) has been called up from a diskette or a disk, (b) now resides in the computer, (c) is displayed on a display device, i.e., on a monitor, (d) can be worked on, i.e., is ready to be changed, edited, deleted, or printed, and (e) indicates the latest entries, calculations, and commands.

current telegraph system: *See* **neutral direct current telegraph system.**

current telegraph transmission: *See* **polar direct current telegraph transmission.**

cursor: A movable, visible mark that (a) is used to indicate a position of interest on a display surface, (b) has a specific fixed or controllable shape, such as an underline, a rectangle, or a pointer, and (c) usually indicates where (i) the next character or graphic will be entered or revised or (ii) the next selection is to be made.

curvature: In the measurement or specification of lenses, the amount of departure from a flat surface, i.e., a plane surface. *Note:* Curvature is expressed as the reciprocal of the radius of curvature.

curvature loss: In an optical fiber or a fiber optic cable, the light energy loss attributed to macrobending of the fiber or the cable. *Note:* Power will radiate away from an optical fiber if the radius of the bend is less than the critical radius. *Synonyms* **bend loss, macrobending loss, macrobend loss.** *See also* **fiber optics, macrobending, microbending, microbend loss, modal loss.**

curve: *See* **absolute luminosity curve, bathtub curve, bell curve, luminosity curve, spectral loss curve.**

curve-fitting compaction: Data compaction accomplished by substituting an analytical expression for the data to be stored or transmitted. *Note:* Examples of curve fitting compaction are (a) the breaking of a continuous curve into a series of straight-line segments and specifying the slope, intercept, and range for each segment and (b) using a mathematical expression, such as a polynomial or a trigonometric function, and a single point on the corresponding curve instead of storing or transmitting the entire graphic curve or a series of points on it. *See also* **compaction, data compaction.**

curve generator: A functional unit, such as a display device or a computer program, that generates curves, usually consisting of a connected sequence of dots or of very short vectors or line segments, that are either tangent to the smoothed curve or are cords to the smoothed curve. *Note:* A vector generator might produce the tangent line segments or short cords. *See also* **functional unit, vector generator.**

custodian: *See* **security custodian.**

custom local area signaling service (CLASS): In a communications network, a network-provided service feature that is one of an identified group of network-provided enhanced services. *Note:* Examples of custom local area signaling service (CLASS) offerings are incoming call identification, call trace, call blocking, automatic return of the most recent incoming call, call redial, and selective forwarding and programming to permit distinctive ringing for incoming calls. *See also* **call forwarding, calling line identification signal.**

customer: In communications systems, the ultimate entity for which a communications system was installed, i.e., the ultimate user of a communications service. *Note 1:* Customers include individuals, activities, organizations, positions, functions, offices, or units that are associated with a communications system directory entry and that are approved by the proper authority. *Note 2:* Customers use end instruments, such as telephones, modems, facsimile machines, computers, and remote terminals, that are connected via a loop to a central office. *Note 3:* Customers usually are subject to tariff. *Note 4:* Customers do not include communications systems management, administrative, operating, and maintenance personnel except for their personal terminals. *Synonym* **subscriber.**

customer access: In Integrated Services Digital Networks (ISDNs), the portion of the ISDN access that a network provider supplies to connect the customer installation, i.e., the subscriber installation, to the network. *Note:* Customer access includes those network elements or portions of elements that extend from the access switch to the network interface.

customer identification: *Synonym* **user identification.**

customer installation: At a customer premises, the equipment and the facilities that are on the customer side of the network interface.

customer line: *Synonym* **loop.**

customer management complex (CMC): In network management, a complex that (a) consists of the facilities and the equipment controlled by a customer and (b) is responsible for, and performs, maintenance for the customer installation. *See also* **administrative management complex.**

customer office terminal (COT): 1. Termination equipment that (a) is located on the customer premises and (b) performs a function that may be integrated into the common carrier equipment. *Note:* An example of a customer office terminal (COT) is a stand-alone multiplexer located on the customer premises. **2.** The digital loop carrier (DLC) multiplexing function that (a) is near the exchange termination (ET) and (b) is provided by a stand-alone multiplexer. *Note:* The customer office terminal (COT) may be integrated into the exchange termination (ET).

customer owned and maintained equipment (COAM): *Deprecated synonym for* **customer premises equipment.**

customer premises equipment (CPE): Equipment, such as data terminal equipment (DTE), i.e., telephone sets, facsimile transmitters and receivers, modems, personal computers, and any other associated equipment located on customer premises and connected to one or more carrier communications channels at the network interface, such as the network data circuit-terminating equipment (DCE) at the customer premises. *Note:* Customer premises equipment (CPE) does not include overvoltage protection equipment, inside wiring, and pay telephones. *Deprecated synonyms* **customer owned and maintained equipment, customer provided equipment.** *See also* **embedded customer premises equipment, new customer premises equipment.**

customer provided equipment: *Deprecated synonym for* **customer premises equipment.**

customer service unit (CSU): At a user location, a device that provides access to either switched or point-to-point data circuits at a specified data signaling rate (DSR). *Note:* A customer service unit (CSU) may provide local loop equalization, transient protection, isolation, and central office loop-back testing capabilities. *See also* **data service unit.**

customer set: *Synonym* **user set.**

cut: *See* **bearing cut, runner cut, scene cut.**

cutback technique: In an optical fiber, a technique for measuring and calculating certain transmission characteristics, such as the attenuation rate, insertion loss, bit-rate•length product, bandwidth, and dispersion, by performing two transmission measurements, measuring applicable lengths, and using the results to calculate the characteristic. *Note 1:* The transmission measurements to be made depend on the characteristic to be determined. *Note 2:* One measurement is made at the output of a length of the optical fiber, and the other is made at the output of a shorter length of the same fiber without changing the launching conditions. Access to the shorter length is obtained by cutting off a length of the exit end of the fiber. The cut should not be made less than 1 meter from the launch end. The two transmission measurements and the measured lengths can be used to calculate the desired transmission characteristic. *Note 3:* Several characteristics may be determined by using the same test optical fiber. *Note 4:* A variation of the cutback technique is to make measurements before and after a component to be measured is spliced into a fiber without changing the launch and exit conditions. *See also* **launch angle, launching optical fiber, launch numerical aperture.**

cutoff: *See* **low frequency cutoff.**

cutoff attenuator: A waveguide of adjustable length that may be used to change the attenuation of signals passing through it.

cutoff frequency: 1. The frequency either above which or below which the output level of a circuit, such as a line, amplifier, or filter, is reduced to a specified level. **2.** The frequency below which a radio wave fails to penetrate a layer of the ionosphere at the incidence angle required for transmission between two specified points by reflection from the layer. **3.** In an optical fiber, the frequency below which a specified propagation mode fails to exist. *Note:* As the frequency is decreased, the wavelength increases until it is too long for the dimensions of an optical fiber to support the wave. *See also* **critical frequency, cutoff wavelength, frequency, ionosphere.**

cutoff mode: The highest-order mode that will propagate in a given waveguide at a given frequency. *See also* **frequency, mode.**

cutoff wavelength: 1. In a waveguide, the wavelength corresponding to the cutoff frequency, i.e., the free-space wavelength above which a given bound mode cannot exist in the guide. **2.** The wavelength greater than which a particular mode ceases to be a bound mode, i.e., the mode ceases to propagate within the guide. *Note:* In a single mode waveguide, concern is with the cutoff wavelength of the second-order mode in order to ensure single mode operation. For single mode operation of an optical fiber, the cutoff wavelength can be determined from the relation $V = (2\pi a/\lambda)(n_1^2 - n_2^2)^{1/2}$ where V is the normalized frequency, a is the core radius, λ is the operational wavelength, n_1 is the core refractive index, and n_2 is the cladding refractive index. For single mode operation $V = 2.405$, and the solution for λ will be the cutoff wavelength. For a short uncabled single mode optical fiber with a specified large radius of curvature, i.e., a macrobend radius very much larger than the critical radius, the cutoff wavelength is the wavelength at which the presence of a second-order mode introduces a significant attenuation increase when compared to a fiber with differential mode attenuation that is not changing at that wavelength. Because the cutoff wavelength of a fiber is dependent upon length, bend, and cabling, the cabled fiber cutoff wavelength usually is a more useful value for cutoff wavelength from a systems point of view. In general, the cable cutoff wavelength is shorter than the optical fiber cutoff wavelength. *See* **cable cutoff wavelength, maximum cable cutoff wavelength, minimum cable cutoff wavelength, optical fiber cutoff wavelength.** *See also* **cable cutoff wavelength, cutoff frequency, fiber optics, frequency, mode.**

cutover: 1. The changing of circuits or lines from one physical configuration to another. **2.** To change circuits or lines from one physical configuration to another. *See also* **continuous operation.**

cut response: *See* **scene cut response.**

cutting tool: *See* **optical fiber cutting tool.**

CVD: chemical vapor deposition.

CVD process: chemical vapor deposition process.

CVPO: chemical vapor phase oxidation.

CVPO process: chemical vapor phase oxidation process.

CVSD: continuously variable slope delta.

CVSD modulation: continuously variable slope delta modulation.

cw or CW: carrier wave, continuous wave.

CX: composite signaling.

cxr: carrier.

CX signaling: composite signaling.

cybercop: A person specially trained and equipped, or a system specially designed and built, to search for, locate, and identify persons and systems that (a) operate among legitimate computer networks and servers, such as Internet, Oracle, World Wide Web (WWW), Archie, Hypertext Transfer Protocol (HTTP), File Transfer Protocol (FTP), Gopher, E-mail, and Bulletin Board, (b) illegally seek to break cryptographic codes through cryptanalysis and theft of key, (c) steal and illegally use software, services, and credit, (d) violate copyright, privacy, and proprietary rights, and (e) commit other information-related crimes, such as dealing in pornography, illegal drug trafficking, and stalking using the cyberspace. *See also* **cracker, cyberspace, hacker.**

cybercrime: A crime, such as larceny, copyright violation, privacy invasion, drug trafficking, or pornographic distribution, committed through the use of communications, computer, data processing, and control systems, i.e., committed in cyberspace. *See also* **cracker, cybercop, cyberspace, hacker.**

cyberia: *Synonym* **cyberspace.**

cybernetics: The branch of science and technology devoted to the theories, study, design, development, and application of communications and control in machines and living organisms.

cyberrevolution: The change from (a) traditional station-to-station or person-to-person call and message switching with inherent circuit establishment and disestablishment to (b) the complete interconnection of all stations ready for online and real time use at any arbitrary moment, such as that provided by Internet, the World Wide Web, and electronic mail. *See also* **electronic mail, information superhighway, Internet, World Wide Web.**

cyberspace: 1. The domain of integrated computer and communications systems. **2.** The world of information that (a) includes information generation, storage, access, exchange, and use and (b) usually includes the use of computers and databases connected by communications systems on a worldwide basis. *Synonym* **cyberia.** *See also* **domain, information superhighway.** *Refer to* **Fig. S-10 (880).**

cycle: A sequence of events that usually is repeated periodically, such as amplifying, clocking, shaping, and gating a serial set of similar pulses. *See* **display cycle, duty cycle, lifecycle, software lifecycle, storage cycle, sunspot cycle, system lifecycle.**

cycle of Saros: Approximately 223 lunar months, equivalent to about 18 years and 11 days, after which both the Moon and the line of nodes return practically to their original positions relative to the Sun. *Note:* Eclipses tend to recur after the cycle of Saros interval. Accurate astronomical measurements show that there is a difference of 0.46 day between the time of return of the Moon nodes to the same position relative to the Sun and the return of the Moon to the same point in its orbit after 223 lunar months. A complete cycle of Saros lasts about 122 years before the Moon and the line of nodes return exactly to their original positions relative to the Sun. *See also* **eclipse, orbit.**

cyclic binary unit distance code: *Synonym* **Gray code.**

cyclic code: A unit distance code of finite length in which the representation of the largest quantity is one unit distance away from the code that represents zero. *Note:* Cyclic codes include reflected codes. *Synonym* **closed code.** *See also* **Gray code, reflected code, unit distance code.**

cyclic distortion: In telegraphy, distortion (a) that is periodic and (b) that is neither characteristic, bias, nor fortuitous. *Note:* Causes of cyclic distortion include (a) irregularities in the duration of contact time of the brushes of a transmitter distributor and (b) interference by fixed frequencies, such as a disturbing alternating current (ac). *See also* **bias distortion, characteristic distortion, distortion, end distortion, fortuitous distortion.**

cyclic redundancy check (CRC): A check in which errors are detected by using parity bits generated by polynomial encoding and decoding algorithms that detect the errors generated in transit. *Note:* Error correction, if required, usually is accomplished through the use of an automatic repeat request (ARQ) system. *See also* **automatic repeat request (ARQ), block, block check character, character check, error control, error correcting code, error detecting system.**

cyclic storage: *Synonym* **circulating storage.**

cyclone: *See* **anticyclone.**

cylinder: On a coaxial set of magnetic or optical disks, a set of corresponding tracks on two or more disks, the corresponding tracks all being at the same radial distance from the common axis of the disks. *See* **type cylinder.**

cylindrical lens: A lens that (a) has a cylindrical surface, (b) is used in range finders to introduce astigmatism in order that a point-like source may be imaged as a line of light, and (c) may be combined with spherical surfaces to produce an optical system that gives a certain magnification in a given azimuth of the image and a different magnification at right angles in the same image plane, the resulting system being designated as anamorphic. *See* **microcylindrical lens.**

cylindrical microlens: *See* **microcylindrical lens.**

cyphony: *Synonym* **ciphony.**

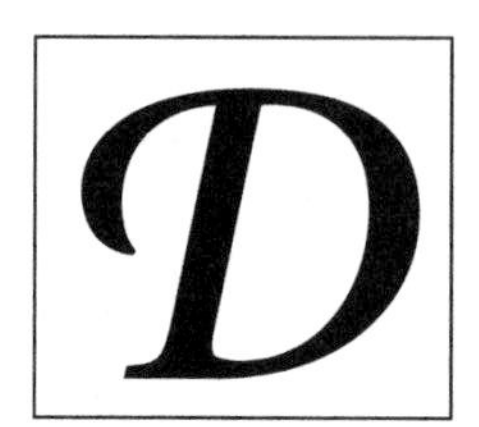

D: specific detectivity.

D*: D star. *Synonym* **specific detectivity.**

D/A: digital to analog.

D-A: digital to analog.

DACS: digital access and cross connect system.

daisy chaining: In communications, the transmitting of signals along a bus in which (a) a series of devices are sequentially connected so that devices not requesting a given signal allow it to pass on to the first device that either requested the signal or that is designated to receive the signal, and (b) the device responds by performing the requested action and breaking the continuity to the devices that are on the bus beyond the device. *Note:* Daisy chaining permits the assignment of priorities based on electrical position on the bus.

DAMA: demand assignment multiple access.

damage: *See* **optical fiber radiation damage.**

damping: 1. The progressive reduction with time of certain quantities characteristic of a phenomenon. **2.** The progressive decay with time in the amplitude of the free voltage, current, power, or energy oscillations in a circuit.

dancer: In testing the tensile strength of optical fibers, a device that (a) applies a given controlled tensile force at a given controlled application rate to an optical fiber as it passes from a payout spool, around the pulleys and capstans of the device, to a takeup spool, and (b) halts if any segment of the fiber breaks because it is unable to withstand the applied tensile force at the prescribed application rate for a given length of time. *Note:* A broken optical fiber may be spliced at the breakpoint and the test continued, including the spliced segment, or the fiber may be discarded. If the fiber does not break throughout its length, the fiber is considered to be "proofed" at the given conditions.

dark adaptation: The ability of the human eye to adjust itself to low levels of illumination. *Note:* The pupil dilates at low levels of illumination. *See also* **accommodation.**

dark current: The external current that, (a) under specified biasing conditions, flows in a photodetector when there is no incident radiation, (b) usually increases with increased temperature for most photodetectors, and (c) in a photoemissive photodetector, is given by the relation $I_d = AT^2e^{qw/kT}$ where I_d is the dark current, A is the surface area constant, T is the absolute temperature, q is the electron charge, w is the work function of the photoemissive surface material, and k is Boltzmann's constant. *See also* **photosensitive recording, quantum noise.** *Refer to* **Fig. P-6 (721).**

dark resistance: The electrical resistance of the photoconductive material of a photodetector (a) that exists when there is no radiant energy incident upon its photosensitive surface, and (b) that usually decreases with increases in temperature for most photodetectors. *See* **infrared detector dark resistance.**

dark trace: Pertaining to a display device, such as a cathode ray tube (CRT), light-emitting diode (LED) panel, gas panel, or fiberscope faceplate, that creates display elements, display groups, and display images by means of dark points, lines, or areas on the display surface of its brightened screen. *See also* **display element, display group, display image.**

dark-trace tube: A cathode ray tube (CRT) that creates display elements, display groups, and display images by means of dark points, lines, or areas on the display surface of its brightened screen. *See also* **display element, display group, display image.**

darlington: *See* **photodarlington.**

darlington pair: A base driver or emitter-follower transistor circuit that (a) usually consists of a voltage amplifier followed by a power amplifier connected in tandem, (b) has high input impedance, increased power gain, and high stability, and (c) is widely used in electronic and integrated circuits.

dash: In the international Morse code, a signal element of mark condition having a duration of three unit intervals followed by a signal element of space condition having duration of one unit interval. *Note:* The duration of a dash is about equal to three times the duration of a dot. *See also* **dot, international Morse code, mark, signal element, significant condition, space, unit interval.** *Refer to* **Figs. I-6 (472), T-3 (991).**

data: 1. A representation, such as characters or analog quantities, to which meaning is assigned. *Note:* Examples of data are bit strings, groups of alphanumeric characters, numerals, and hole patterns in punched cards that have been assigned a meaning. **2.** A representation of information. **3.** A representation of facts, concepts, or instructions in a formalized manner suitable for communication, interpretation, or processing by humans or by machines. *Note 1:* Data may be transferred, i.e., be transported, from place to place, such as from storage unit to storage unit, city to city, computer to computer, and person to person. Data may assume many different forms, such as hole patterns in punched cards, magnetized spots on disks, drums, tapes, and cards, electrical pulses in wires, optical pulses in optical fibers, and modulated electromagnetic waves in free space. Data may be displayed on the display surfaces of display devices, such as cathode ray tube (CRT) screens. *Note 2:* Arbitrary sets or strings of characters to which meaning has not been assigned are not considered to be data. *See* **absolute coordinate data, alphanumeric data, analog data, coordinate data, digital data, formatted data, immediate data, numeric data, output data, relative coordinate data, reliability data, source data, static data, user data.** *See also* **information.**

data access arrangement: 1. In public switched telephone networks, a single item or group of items present at the customer side of the network interface for data transmission purposes, including all equipment that may affect the characteristics of the interface. **2.** Data circuit-terminating equipment (DCE) supplied or approved by a common carrier that permits a DCE or data terminal equipment (DTE) to be attached to the common carrier network. *Note:* Data access arrangements are now an integral part of all modems built for the public telephone network. *Synonym* **data arrangement.** *See also* **data, data circuit-terminating equipment, data terminal equipment, data transmission, interface.**

data alphabet: A set of correspondences established by convention that indicates the relationship between data characters and the coded signals that represent them. *Note:* An example of a data alphabet is the set of alphabetic characters and their bit pattern equivalents in the American Standard Code for Information Interchange (ASCII).

data arrangement: *Synonym* **data access arrangement.**

data attribute: A characteristic of data, such as length, value, or method of representation.

data bank: 1. A set of data related to a given subject and organized in such a way that it can be consulted by users. **2.** A data repository that (a) is accessible by local and remote users, (b) may contain information on single or multiple subjects, (c) may be organized in any rational manner, (d) may contain more than one database, (e) may be geographically distributed, and (f) may be used to build a comprehensive database. *See also* **database.**

database: A set of data that (a) is required for a specific purpose or fundamental to a system, project, enterprise, or business, (b) may consist of one or more data banks, and (c) may be geographically distributed among several repositories. *Note 2:* Examples of databases are (a) all the data concerning items stored in a warehouse, whether or not the data are organized or automated, (b) all the distributed data concerning the economy of a country, and (c) all the data in a set of diskettes. *See* **distributed database, integrated database, relational database.** *See also* **data bank, data command, field, record.** *Refer to* **Fig. P-18 (769).**

database management system (DBMS): 1. A software system that facilitates (a) the creation and the maintenance of a database and (b) the execution of computer programs using the database. **2.** A computer-based system used to establish, make available, and maintain the integrity of a database. **3.** An integrated set of computer programs that collectively provide all of the capabilities required for centralized management, organization, and access control of a database that is shared by many users. **4.** A special-purpose data processing system, or a part thereof, that facilitates creating, storing, manipulating, reporting, managing, and controlling data. *Synonym* **data management system.** *See also* **data dictionary, facility.**

data buffer: A data storage device that can (a) compensate for a difference in the rate of flow of data or compensate for the difference of time of occurrence of events, such as the occurrence of control signals in a communications system, (b) accept short bursts of high speed data flow, i.e., high data signaling rates (DSRs), and emit a continuous lower speed data flow, (c) accept data at a low rate of flow and emit the data in short higher speed bursts, perhaps interleaved, i.e., multiplexed, with high speed bursts from other buffers, and (d) in any case, store data temporarily. *Synonym* **speed buffer.**

data burst: An uninterrupted sequence of digital data that (a) usually is transmitted at the maximum data signaling rate (DSR) of a channel and (b) may be of any finite nonzero length. *Synonym* **burst transmission.**

data bus: A bus used to transfer data within, or to and from, a processing unit or a storage device. *See* **fiber optic data bus.** *See also* **bus, power bus.**

data bus coupler: A coupler that connects a device to a data bus. *Note:* In a fiber optic communications system, an example of a data bus coupler is a coupler that interconnects a number of fiber optic waveguides and provides an inherently bidirectional system by mixing and splitting all signals within the coupler. *See also* **bus, coupler.**

data call: A call established and used for the purpose of transferring digital data or analog data but not analog voice signals. *See also* **analog data, call, digital data.**

data capture: To store data streaming into a device, such as a storage unit at a communications terminal or station. *Note:* An example of data capture is to place data on a disk or to print data that are entering a computer from a communications channel via a modem. *Synonym* **download.** *See also* **upload.** *Refer to* **Fig. T-10 (1041).**

data carrier: A unitized piece of material in or on which a limited amount of data may be stored or recorded to facilitate transport of the data. *Note:* Examples of data carriers are diskettes, lengths of paper tape, sheets of paper, magnetic cards, identification badges, and magnetic tape cassettes. *See also* **data medium.**

data channel: A channel that allows one-way transmission of data. *See* **analog data channel, digital data channel.**

data circuit: A pair of associated transmit and receive channels that (a) provides a means of two-way data communication, with transmit and receive capabilities at both ends, (b) may include data circuit-terminating equipment (DCE) between data switching exchanges, depending on the type of interface used at the data switching exchange or concentrator, (c) includes data circuit-terminating equipment (DCE) at the data terminal installation, (d) includes equipment similar to data circuit-terminating equipment at the exchange or concentrator location, (e) may consist of associated data channels that permit transmission in both directions between two locations at the same time, and (f) may be either a physical or a virtual circuit. *See also* **channel, circuit, concentrator, data, data circuit-terminating equipment, data terminal equipment, data transmission.** *Refer to* **Fig. D-1.**

data circuit connection: The interconnection of two or more data links or trunks, on a tandem basis, by means of switching equipment to enable data transmission to take place among data terminal equipments (DTEs). *See also* **circuit, data, data circuit-terminating equipment, data link, data terminal equipment, data transmission, data transmission circuit, link, tandem.**

data circuit-terminating equipment (DCE): 1. In a data station, the equipment that (a) provides signal conversion, coding, and other functions, (b) is located at the customer end of the line between the data terminal equipment (DTE) and the line, and (c) may be a separate or an integral part of the DTE or of the intermediate equipment. **2.** The interfacing equipment that may be required to couple the data terminal equipment (DTE) into a transmission circuit or channel and from a transmission circuit or channel into the DTE. *Note:* The term "data communications equipment (DCE)" is not to be used for "data circuit-terminating equipment (DCE)" and vice versa. *Deprecated synonyms:* **data communications equipment, data set.** *See also* **data communications equipment, data terminal equipment, interface, modem.** *Refer to* **Fig. D-1.**

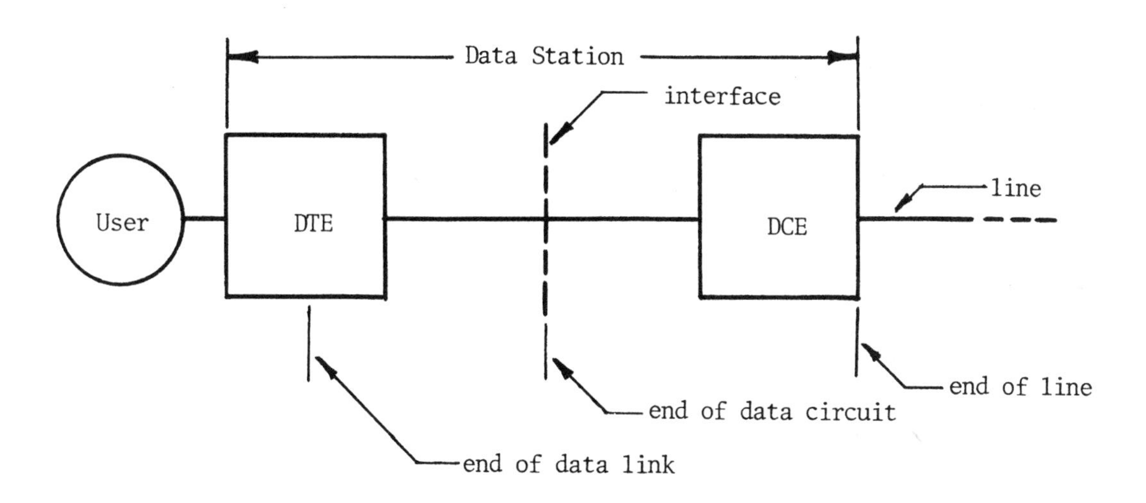

Fig. D-1. Components of a **data link.**

data circuit-terminating equipment (DCE) clear signal: A call control signal transmitted by data circuit-terminating equipment (DCE) to indicate that it is clearing the associated circuit after a call is finished. *See also* **call, call control signal, data circuit-terminating equipment, finished call, signal.**

data circuit-terminating equipment (DCE) waiting signal: A call control signal at the data circuit-terminating equipment/data terminal equipment (DCE/DTE) interface that indicates that the DCE is ready for another event in the call establishment procedure. *See also* **call, call control signal, data circuit-terminating equipment, data terminal equipment, signal.**

data circuit transparency: The capability of a data circuit to transmit all data without changing the information content or structure. *Note:* An example of data circuit transparency is a measure of the ability of a circuit to handle many alphabetic codes or to function at a wide range of pulse repetition rates. *See also* **data circuit, transparency.**

data code: A code used to represent data items. *Note:* Examples of data codes are 03 for March and 06 for colonel, where "March" is the data item for the data element "month" and "colonel" is the data item for the data element "rank." *See also* **data element, data item.**

data collection facility: A facility for gathering and organizing data from a group of sources. *See also* **data, database management system, facility.**

data collection station: *Synonym* **data input station.**

data command: One of a set of commands that allows an operator to manipulate data in a database. *See also* **command, database.**

data communications: 1. The transfer of data among functional units by means of data transmission according to a protocol. *Note:* The transfer of data from one or more sources to one or more sinks may be accomplished over one or more data links. **2.** Two-way data transfer between data source and data sink that is accomplished (a) via one or more data channels or data links and (b) according to a protocol. *See also* **code-independent data communications, communications, contention, control station, data, data transmission, link, master station, primary station, secondary station, sink, source, throughput, tributary station.**

data communications control character: *See* **control character.**

data communications control procedure: 1. A procedure used to control the orderly communication of information among stations in a data communications network. **2.** *Synonym* **communications control procedure.** *See* **Advanced Data Communications Control Procedure.** *See also* **data, data transmission, serial access, serial transmission.**

data communications equipment (DCE): Equipment that is (a) used in a data communications system or (b) stand-alone equipment used for data communications purposes. *Note:* The term "data communications equipment (DCE)" is not to be used for "data circuit-terminating equipment (DCE)" and vice versa. *See* *also* **data circuit-terminating equipment, data communications.**

data compaction: The reduction of space, bandwidth, cost, and time for the generation, transmission, and storage of data without loss of information by eliminating redundancy, removing irrelevancy, and using various forms of special coding. *Note 1:* Examples of data compaction methods are the use of fixed tolerance bands, variable tolerance bands, slope-keypoints, sample changes, curve patterns, curve fitting, variable precision coding, frequency analysis, and probability analysis. *Note 2:* Simply squeezing noncompacted data into a smaller space, such as by increasing packing density or by transferring data on punched cards to magnetic tape, is not considered to be data compaction. Whereas data compaction reduces the amount of data used to represent a given amount of information, data compression does not. *Note 3:* Both data compaction and data compression result in a reduction of the physical space required to store a given amount of information. *See also* **compaction, compression, curve fitting compaction, data compression, fixed tolerance band compaction, floating point coding compaction, frequency analysis compaction, incremental compaction, probability analysis compaction, sample change compaction, slope-keypoint compaction, variable precision coding compaction, variable tolerance band compaction.**

data compression: 1. Increasing the amount of data that (a) can be stored in a given domain, such as the spatial, temporal, or frequency domain, or (b) can be contained in a given message length. **2.** Reducing the amount of storage space required to store a given amount of data, or reducing the length of message required to transfer a given amount of information. *Note 1:* Data compression may be accomplished by simply squeezing a given amount of data into a smaller space, such as by increasing packing density or by transferring data on punched cards onto magnetic tape. *Note 2:* Data compression does not reduce the amount of data used to represent a given amount of information, whereas data compaction does. Both data compression and data compaction result in the use of less space for a given amount of information. **3.** Without changing the basic mode of representation, (a) increasing the amount of data that can be stored in a given physical space or (b) reducing the amount of physical space for storing a given amount of data, such as (a) transferring data from punched cards to magnetic tape, (b) increasing the bit packing density by transferring data from magnetic disk to optical disk, and (c) placing full-size paper documents and photographs on microfilm. *See also* **data compaction.**

data concentrator: 1. A functional unit that permits a common propagation medium to serve more data sources than there are channels currently available within the propagation medium. **2.** Switching equipment that enables connections between circuits and stations to be established and disestablished as required. *Note:* The number of data circuit-terminating equipment (DTE) units always exceeds the number of stations. *See also* **concentrator, multiplexer, switching equipment.**

data conferencing repeater: A device that enables any one user of a group of users to transmit a message to all other users in that group. *Synonym* **technical control hubbing repeater.** *See also* **conference operation, conference repeater, data, data transmission, network, teleconferencing.**

data connection: A connection in which (a) a number of data circuits are connected on a tandem basis by means of switching equipment that enables data transmission to take place between data terminal equipment (DTE) units, (b) one or more of the interconnected data circuits may be a virtual circuit, and (c) the overall connection includes the data circuit-terminating equipment (DCE) at the respective data station locations. *See also* **data circuit, data circuit-terminating equipment, data terminal equipment, switching equipment, virtual circuit, virtual connection.**

data contamination: *Synonym* **data corruption.**

data corruption: 1. A violation of data integrity. **2.** The process in which data integrity is violated. *Synonym* **data contamination.** *See also* **data, data integrity, data transmission.**

data country code: A 3-digit numerical country identifier that (a) is part of the 14-digit network terminal numbering plan and (b) is a prescribed numerical designation that further constitutes a segment of the overall 14-digit X.121 numbering plan for an International Telegraph and Telephone Consultative Committee (CCITT) X.25 network.

data definition language: *Synonym* **data description language.**

data delimiter: *Synonym* **delimiter.**

data-dependent protection: The addition of protective data elements to a data stream in such a manner that the composition of the data stream determines the amount or the type of protective elements to be added. *See also* **data integrity.**

data description language (DDL): A high level language that (a) may be used to describe relationships among different records in a database and (b) usually is different from and independent of computer programming languages, such as BASIC, COBOL, ADA®, ALGOL, and PL/1. *Note:* An example of a data description language (DDL) is a set of declarative statements that describe the contents or the structure of a database. *Synonyms* **data definition language, description language.** *See also* **ADA®, ALGOL, BASIC, COBOL, database, high level language, PL/1, record, statement.**

data dictionary: 1. A database used for data that refer to the use and the structure of other data, i.e., a database for the storage of metadata, such as information about the meaning, relationships, origin, usage, and format of data. **2.** A dictionary that (a) consists of an inventory that describes, defines, and lists all of the data elements that are stored in a database, (b) may be part of a database management system, and (c) usually is a system database rather than a user database. *Synonym* **data element dictionary.** *See also* **database management system, data directory, data element.**

data directory: A directory, i.e., an inventory, that specifies the source, location, ownership, usage, and destination of all of the data elements that are stored in a database.

data disk: A disk on which data are stored and from which data are retrieved. *Note:* In a personal computer (PC), the data on the disk are grouped into files, each file having its own name for storage and retrieval purposes. *Synonyms* **current disk, file disk, storage disk.** *See also* **program disk.**

data element: 1. A named unit of data that may be indivisible or may consist of data items. **2.** A named identifier of each of the entities and their attributes that are represented in a database. **3.** A basic unit of information that has a unique meaning and subcategories, called data items, of distinct units or values. *Note:* Examples of data elements are personnel grade, gender, race, month, and day. Data items for these data elements might be captain, male, black, June, and Tuesday, respectively. Examples of data codes for these items might be 03 for captain and 06 for June. *See also* **data code, data item.**

data element dictionary: *Synonym* **data dictionary.**

data encryption: *See* **compatible data encryption.**

data encryption standard (DES): A standard procedure that (a) is used for enciphering plain text and (b) may not be approved for protection of national security classified information. *Note:* An example of a data encryption standard is the algorithm designed to encipher and decipher 8-byte blocks of data using a 64-bit key, as specified in Federal Information Processing Standard Publication 46-1. *See also* **algorithm, block, byte, cipher, cipher text, data, decipher, encipher, key, plain text.**

Data Encryption Standard (DES): A cryptographic algorithm for the protection of unclassified computer data, issued as Federal Information Processing Standard Publication (FIPS PUB) 46-1. *Note 1:* The Data Encryption Standard (DES), which was adopted as a standard by the National Institute of Standards and Technology (NIST), formerly the National Bureau of Standards (NBS), is intended for public and government use. *Note 2:* The Data Encryption Standard (DES) is not approved for protection of national security classified information.

data flowchart: A flowchart that (a) shows the path through various pieces of equipment and devices, from input to output, to be taken by data, (b) defines the major phases of processing, and (c) indicates the various data media, data carriers, and equipment used to store and process the data. *See also* **data carrier, data medium, flowchart.**

data format: *Synonym* **format.**

data forwarder: A device that (a) receives data from one data link and retransmits data representing the same information, using proper format and link protocols, to another data link and (b) may forward data between (i) links that are identical, such as from TADIL B to TADIL B, (ii) links that are similar, such as from TADIL A to TADIL B, or (iii) links that are dissimilar, such as from TADIL A to TADIL J. *See also* **data link, link protocol.**

data forwarding: Receiving data from one data link and retransmitting data representing the same information, using proper format and link protocols, on another data link. *See also* **data link, link protocol.**

data fusion: The gathering and processing of related data from more than one source.

data grade circuit: *Synonym* **conditioned voice grade circuit.**

datagram: In packet switching, a self-contained packet, independent of other packets, that contains information sufficient for routing from the originating data terminal equipment (DTE) to the destination DTE without relying on prior exchanges between the equipment and the network. *Note:* Unlike virtual call service, when datagrams are sent there are no call establishment or clearing procedures. Thus, the network may not be able to provide absolute protection against loss, duplication, or misdelivery. *See also* **connectionless mode transmission, data, data transmission, packet switching, virtual call capability.**

datagram protocol: *See* **User Datagram Protocol.**

datagram service: A transmission service in which (a) a datagram is routed to the destination identified in its address field without reference by the data network to another datagram previously sent or likely to follow, (b) datagrams may be delivered to a destination address in a different order from that in which they entered the network, (c) it may be necessary for users to provide procedures to ensure delivery of datagrams to the destination user, i.e., the addressee, and (d) in packet mode operation, a datagram usually is conveyed as a single packet. *See also* **address field, datagram, packet mode operation, transmission service, user.**

data handling: *See* **automatic data handling.**

data input station: A user terminal primarily used for the insertion of data into a data processing system or a communications service. *Synonym* **data collection station.** *See also* **communications service, data processing system, terminal, user.**

data integrity: 1. The condition in which data are maintained without loss of information content during any operation, such as transfer, storage, and retrieval. **2.** The state that exists when data (a) are handled as intended and (b) are not exposed to accidental or malicious modification, destruction, or disclosure. **3.** The preservation of data for their intended use. **4.** The property of data such that the data meet an a priori expectation of quality. **4.** A condition of data such that the data remain current, consistent, and accurate. *See also* **data, data corruption, data security, data transmission.**

data interchange: *See* **electronic data interchange.**

data interface standard: *See* **Fiber Distributed Data Interface standard.**

data interrogating station: *See* **oceanographic data interrogating station.**

data item: 1. A named component of a data element, usually the smallest component. **2.** A subunit of descriptive information or value classified under a data element. *Note 1:* Examples of data items are captain, male, black, June, and Tuesday for the data elements personnel grade, gender, race, month, and day, respectively. *Note 2:* Data items may be coded, such as 01 for January for the data element month. In a structured file, only the data codes may need to be stored and transmitted. *See also* **data code, data element.**

data keyboard: A device that (a) consists of an array of keys and (b) is used to control a data source, such as a telegraph transmitter.

data label: A label that is associated with a given set of data. *Note:* In some systems, such as personal computer (PC) systems, a label may be (a) placed above, below, at the center, left, or right of the associated data or (b) may be embedded in the associated data. *See also* **label.**

data link: 1. The means of connecting facilities or equipment at one location to facilities or equipment at another location for the purpose of transmitting and receiving data. **2.** An assembly, consisting of parts of two data terminal equipments (DTEs) and the interconnecting data circuit, that is controlled by a link protocol enabling data to be transferred from a data source to a data sink. *See* **dedicated data link, fiber optic data link.** *See also* **data, data terminal equipment, data transmission, link, Open Systems Interconnection—Reference Model, tactical data information link.** *Refer to* **Fig. D-1 (203).** *Refer also to* **Figs. F-2 (339), S-5 (869), S-6 (869), S-7 (870).**

Data Link: A digital communications system that (a) provides communications between aircraft pilots and air traffic controllers, (b) reduces the need for voice communications between pilots and controllers by using displayed messages, (c) makes use of an interlocking system of computers, satellites, sensors, and communications software that integrates ground automation systems with aircraft computers, (d) makes use of technical advances in computers, electronic and optical sensors, and communications systems, and (e) was developed by the Federal Aviation Administration (FAA). *See also* **Mode S radar.**

data link control: *See* **Advanced Data Communications Control Procedures, binary synchronous communication, high level data link control, Open Systems Interconnection—Reference Model, synchronous data link control.**

data link escape character (DLE): A transmission control character that changes the meaning of a limited number of contiguously following characters or coded representations. *See also* **character, control character.**

data link layer: In open systems architecture, the layer that (a) provides the functions, procedures, and protocol needed to establish, maintain, and release data link connections between user end instruments and switching centers and between two switching centers of a network, (b) is a conceptual level of data processing or control logic existing in the hierarchical structure of the station that is responsible for maintaining control of the data link, and (c) provides an interface between the station high level logic and the data link, including (i) bit injection at the transmitter and bit extraction at the receiver, (ii) address and control field interpretation, (iii) command-response generation, transmission, and interpretation, and (iv) frame check sequence computation and interpretation. *See also* **data link, interface, protocol, switching center.** *See also* **Open Systems Interconnection—Reference Model.**

Data Link Layer: *See* **Open Systems Interconnection—Reference Model.**

data logging: The dating, time labeling, and recording of data.

data management: The controlling of data handling operations, such as acquisition, analysis, translation, coding, storage, retrieval, and distribution of data, but not necessarily the generation and the use of data. *See also* **coding, retrieval, storage, translation.**

data management system (DMS): *Synonym* **database management system.**

data medium: 1. The material in or on which information can be represented to facilitate storage or transport. **2.** The material in or on which one or more characteristics of the material may be used to statically or dynamically represent information. *Note:* Examples of data media are the materials used in roll films, wires, cards, disks, drums, papers, optical fibers, cathode ray tube (CRT) screens, light-emitting diode (LED) displays, and gas panels. *Note:* Free space is the data medium for modulated electromagnetic waves propagating in free space. **3.** The physical quantity, material, or substance that may be varied to represent data. *See also* **data carrier, machine readable medium, medium, propagation medium.**

data mode: In a communications network, the state of data circuit-terminating equipment (DCE) when connected to a communications channel and ready to transmit data, usually digital data. *Note:* When in the data mode, the data circuit-terminating equipment (DCE) is not in a talk or dial mode. *See also* **data, data circuit-terminating equipment, data transmission, mode.**

data multiplexer: A functional unit that permits two or more data sources to share a common propagation medium so that each data source has its own channel. *See also* **propagation medium.**

data network: 1. The assembly of functional units that establishes data circuits between data terminal equipments (DTEs) at data stations. **2.** A network that transfers data among a group of interconnected stations. *See* **Defense Data Network, public data network, synchronous data network.** *See also* **data circuits, data station, data terminal equipment, functional unit, integrated services digital network.** *Refer to* **Figs. C-5 (140), D-1 (203), S-7 (870).**

data network identification code (DNIC): In the International Telegraph and Telephone Consultative Committee (CCITT) X.121 code, the first four digits of the international data number, consisting of the three digits that may represent the data country code (DCC), and the one-digit network code, i.e., the network digit. *See also* **data country code.**

data numbering plan area (DNPA): In the U.S. implementation of an International Telegraph and Telephone Consultative Committee (CCITT) X.25 network, the first three digits of a network terminal number (NTN). *Note:* The ten-digit network terminal number (NTN) is the specific addressing information for an end point terminal in an X.25 network.

data path: In a personal computer (PC), the part of the interface bus (a) on which data are transferred among different parts of the computer and (b) which usually is eight bits wide, i.e., consists of eight parallel lines that transmit the eight bits of a byte simultaneously. *See also* **bus, interface, personal computer.**

data phase: A phase of a data call during which data may be transferred between data terminal equipments (DTEs) that are interconnected via the network. *Note:* The data phase of a data call corresponds to the information transfer phase of an information transfer transaction. *See also* **data terminal equipment, data transmission circuit, phase.**

data portability: A measure of the ease with which specific data can be transferred from one location to another. *See also* **data value.** *Synonym* **data transportability.**

data processing: 1. The performing of one or more operations on data. *Note:* While data processing may provide new information, the information content of the original input data should not be changed by the processing. **2.** The performance of operations upon data, such as transmitting, merging, sorting, and storing data. *Note:* The semantic content of the processed data may be changed, such as might occur when data are merged or associated with other data to produce new information represented by different data. *Synonym* **information processing.** *See* **automatic data processing, optical data processing, remote access data processing.** *See also* **computer system, data.**

data processing node: A node in a network at which processors, controllers, data terminals, and other data processing equipment and their associated software are located. *See also* **network, node, processor.**

data processing station: 1. In a network, a station at a data processing node that has input-output devices that serve the needs of the user. **2.** An input-output device at a data processing node that serves the needs of users. *See also* **network, node.**

data processing system: *See* **automatic data processing system, distributed data processing system.**

data processing terminal: A device that enables an operator to interact with a data processing system. *See also* **computer terminal.**

data rate: In data communications, the rate at which data units, such as bits, bytes, characters, symbols, or frames, are transmitted or pass though a specific point per unit time. *Note:* An example of a data rate is 1.41 Gbps (gigabits per second). *See* **data signaling rate.**

data scrambler: A device used in digital transmission systems to convert digital signals into a pseudorandom sequence that is free from many repetitions of simple patterns, such as marks and spaces. *Note:* The data scrambler facilitates timing, reduces the accumulation of jitter, and prevents baseline drift. *See also* **randomizer, scrambler.**

data security: 1. The protection of data from accidental or intentional (a) modification, (b) destruction, or (c) disclosure to unauthorized personnel. **2.** The protection of data from unauthorized (accidental or intentional) modification, destruction, or disclosure. **3.** The procedures and the actions designed to prevent the unauthorized disclosure, transfer, modification, or destruction of data, whether accidental or intentional. *See also* **communications security, data.**

data service: A communications service devoted to the handling of large volumes of data, such as in facsimile, digital data, and telemetry transmission systems, rather than voice or video transmission systems, such as telephone or television systems. *See* **switched multimegabit data service.** *See also* **voice-data service.**

data service unit (DSU): 1. A device used for connecting data terminal equipment (DTE) to the public telephone network. **2.** A type of short-haul, synchronous data line driver, usually installed at a user location, that connects user synchronous equipment over a four-wire circuit at a preset transmission rate to a servicing central office. *Note:* The data service unit (DSU) can be used for point-to-point or multipoint connection in a digital data network. *See also* **customer service unit.**

data set: 1. A set or a group of data items, usually stored, retrieved, and organized as a unit of data. *Note:* An example of a data set is all the data constituting a file or a record concerning one person, one contract, one project, or one account. **2.** *Deprecated synonym for* **data circuit-terminating equipment.**

data shift: To spatially justify or move data with respect to a reference, such as a margin or one end of a register. *Note 1:* Examples of data shifts are (a) a shift that eliminates all zeros at the right-hand or left-hand end of a data word and (b) a shift that is equivalent to multiplication or division by the radix of a number system; e.g., a two-place shift to the left of a decimal numeral is the same as multiplication by 100, and a three-place shift to the left of a binary numeral is the same as multiplication by eight. Division is accomplished by shifting to the right. In multiplying two binary numbers, the only operations that need be done are shift and add the multiplicand, according to the digits in the multiplier. *Note 2:* A data shift may be accomplished in a register.

data signaling rate (DSR): The aggregate rate at which data pass a point in the transmission path of a data transmission system. *Note 1:* The data signaling rate (DSR) usually is expressed in bits per second. *Note 2:* The data signaling rate (DSR) is given by the relation $DSR = \Sigma(\log_2 n_i)/T_i$ summed from $i = 1$ to $i = m$ where DSR is the data signaling rate, m is the number of parallel channels, n_i is the number of significant conditions of the modulation in the ith channel, and T_i is the unit interval, expressed in seconds, for the ith channel. *Note 3:* For serial transmission in a single channel, the data signaling rate (DSR) reduces to $(1/T)\log_2 n$, and with a two-condition modulation, i.e., $n = 2$, the DSR is $1/T$. *Note 4:* For parallel transmission with equal unit intervals and an equal number of significant conditions on each channel, the DSR is $(m/T)\log_2 n$, and in the case of a two-condition modulation, this reduces to m/T. *Note 5:* The DSR may be expressed in baud, in which case, the factor $\log_2 n_i$ in the above summation formula should be deleted when calculating baud. *Note 6:* In synchronous binary signaling, the data signaling rate (DSR) in bits per second may be numerically the same as the modulation rate expressed in baud. Signal processors, such as four-phase modems, cannot change the DSR, but the modulation rate depends on the line modulation scheme. Thus, in a 2400 bps four-phase sending modem, the signaling rate is 2400 bps on the serial input side, but the modulation rate is only 1200 baud on the four-phase output side. *Synonym* **data transmission rate.** *See also* **baud, binary digit, bit rate, bits per second, data, data signaling rate (DSR) range designation, data transfer rate, data transmission, effective data transfer rate, effective speed of transmission, efficiency factor, maximum user signaling rate, multiplex aggregate bit rate.**

data signaling rate (DSR) range designation: The systematic designation of the ranges for data signaling rates, such that (a) the designator is a two- or three-letter abbreviation for the name, (b) each range is one decade wide, (c) the range limits for each decade have coefficients of 3, (d) the logarithmic midrange value of each range, i.e., each decade, is close to one times a power of ten, and (e) each range is designated by a number that is the power of ten of the midrange value. *Note 1:* For example, the logarithmic midrange value of the 30–300 Gbps EHR (Extremely High Data Signaling Rate) is 10^{11}. Therefore, the numerical designator for this range is 11. *Note 2:* The data signaling rate (DSR) describes the rate at which data pass through a point, usually expressed in data units per unit time, such as bits, characters, or words per second. It is not the speed at which pulses or lightwaves propagate in a line. This speed might be expressed in meters per second. *Synonym* **range designation of data signaling rate.** *See also* **data signaling rate, spectrum designation of frequency.**

data signaling rate transparency: A communications network characteristic that enables the transfer of data between one user and another without placing any restriction in regard to a specific data signaling rate (DSR) except that it must lie within certain limits. *Note:* Data signaling rate transparency may be achieved in fixed rate systems by using stuffing characters. *See also* **stuffing character, data signaling rate, transparency.**

data sink: In a communications system, equipment that accepts data after transmission, usually checks the data, and may originate error control and system control signals. *See also* **data source, sink, source.**

data source: In a communications system, equipment that supplies data to be transmitted, usually participates in checking actions, and may originate error control and system control signals. *See also* **data sink, sink, source.**

data station: Data terminal equipment (DTE), data circuit-terminating equipment (DCE), and any intermediate equipment connected together at one location. *Note:* The data terminal equipment (DTE) may be connected directly to a data processing system, or it may be a part of the data processing system. *See* **oceanographic data station.** *Synonym* **data terminal installation.** *See also* **data, data circuit-terminating equipment, data terminal equipment, data transmission.** *Refer to* **Fig. D-1 (203).**

data stream: In data transmission, a sequence of digitally encoded signals used to represent information. *See also* **bit stream transmission, code, data, data transmission, serial transmission.**

data subscriber terminal equipment (DSTE): In the Automatic Digital Network (AUTODIN), a general-purpose terminal device that consists of (a) all the equipment necessary to provide interface functions, perform code conversion, and transform messages on various data media, such as punched cards, magnetic tapes, and paper tapes, to electrical signals for transmission, and vice versa, and (b) all the equipment necessary to convert received electrical signals into data stored or recorded on various data media. *See also* **data medium, terminal.**

data switching: *See* **digital data switching.**

data switching exchange (DSE): The equipment that (a) is installed at a single location, such as a switching center, to switch data traffic and (b) may provide only circuit switching, only packet switching, or both. *See also* **circuit switching, packet switching, switching center, traffic.**

data system: A system that processes data, transfers data, or both. *Note:* Examples of data systems are automatic data processing systems, computer systems, and communications systems that handle data.

data terminal equipment (DTE): 1. Digital end instruments that convert user information into signals for transmission or reconvert the received signals into user information. **2.** The functional unit of a data station that serves as a data source or a data sink and provides for the data communications control function to be performed in accordance with link protocol. *Note 1:* The data terminal equipment (DTE) may be a single piece of equipment or an interconnected subsystem of multiple pieces of equipment that performs all the required functions necessary to permit users to intercommunicate. *Note 2:* A user interacts with the data terminal equipment (DTE), or the DTE may be the user. The DTE interacts with the data circuit-terminating equipment (DCE). *See also* **data circuit-termi-**

nating equipment, data sink, data source, data station, link protocol. *Refer to* **Fig. D-1 (203).**

data terminal equipment (DTE) clear request signal: A call control signal sent by data terminal equipment (DTE) to initiate clearing. *See also* **call, call control signal, clear request signal, data terminal equipment, signal.**

data terminal equipment (DTE) waiting signal: A call control signal, sent by the data circuit-terminating equipment/data terminal equipment (DCE/DTE) interface, that indicates that the DTE is waiting for a call control signal from the DCE. *See also* **call, call control signal, data circuit-terminating equipment, data terminal equipment, signal.**

data terminal installation: *Synonym* **data station.**

data transaction: *See* **interactive data transaction.**

data transfer: *See* **request data transfer.**

data transfer network (DTN): A network that transports data in any form among a group of interconnected stations or users. *Note:* The stations or the users may be switching centers, switchboards, devices being controlled, or other systems, subsystems, or user end-instruments, such as telephones, computers, video terminals, and control panels. *See* **fiber optic data transfer network.**

data transfer phase: In a call, the phase (a) during which user data are transferred between a source and a destination user, and (b) that occurs between the call setup phase and the call termination phase. *See also* **information transfer phase, information transfer transaction.**

data transfer protocol: *See* **connection oriented data transfer protocol.**

data transfer rate (DTR): 1. The number of data units transferred or passing a point per unit time. *Note:* Examples of data rate units are bits per second (b/s), characters per second (ch/s), and words per minute (w/m). **2.** The average number of data units, such as bits, characters, and words, passing between corresponding equipment per unit time in a data transmission system. *Note:* The corresponding equipment may be any piece of transmission equipment, such as modems, data sources, data sinks, data stations, and facsimile equipment. *See also* **baud, binary digit, block, block transfer rate, character, data signaling rate, data transmission, effective data transfer rate, throughput.**

data transfer request signal (DTRS): At a data station, a call control signal transmitted by the data circuit-terminating equipment (DCE) to the data terminal equipment (DTE) indicating that a request signal, originated by a distant DTE, has been received from a distant DCE to exchange data with the station. *Synonym* **request data transfer signal.** *See also* **call, call control signal, data, data circuit-terminating equipment, data terminal equipment, data transmission, signal.**

data transfer time: The time between (a) the instant at which a user data unit, such as a character, word, block, or message, is made available to a network by a transmitting data terminal equipment (DTE) and (b) the receipt of that complete data unit by a receiving DTE. *See also* **block transfer rate, data, data transmission, throughput, transmission time, transmit flow control.**

data transmission: 1. The sending of data from one place to another by means of signals over a channel. **2.** The conveying of data from one place for reception elsewhere by telecommunications means. *See* **code-independent data transmission, store and forward data transmission.** *See also* **data communication.**

data transmission channel: A channel that (a) consists of the transmission media and related equipment used in the transfer of data between data terminal equipments (DTEs) at different locations, (b) includes signal conversion equipment, and (c) may support the transfer of information in one direction only, in either direction alternately, or simultaneously in both directions. *Note:* When the data terminal equipment (DTE) has more than one speed capability associated with it, such as 1200 bps (bits per second) in the forward direction and 600 bps in the backward direction, a data transmission channel is defined for each speed capability. *See* **asynchronous data transmission channel, synchronous data transmission channel.** *See also* **backward direction, forward direction.**

data transmission circuit: A circuit that (a) includes the transmission media and the intervening equipment used for the transfer of data between data terminal equipments (DTEs), (b) includes any required signal conversion equipment, and (c) may transfer information in one direction only, in either direction but one way at a time, or in both directions simultaneously. *See also* **channel, circuit, data circuit, data phase, data terminal equipment, data transmission.**

data transmission control character: *Synonym* **transmission control character.**

data transmission interface: In a data transmission system, a shared boundary identified by a distinguishing feature or transition characteristic, such as a functional change, physical interconnection, signal transformation, change in data medium, or change in propagation medium. *See also* **data medium, interface, propagation medium.**

data transmission rate: *Synonym* **data signaling rate.**

data transmission service: A service, usually provided to the public by a recognized private operating agency (RPOA) or an administration, that transmits data from one user to another. *See* **circuit switched data transmission service, leased circuit data transmission service, packet switched data transmission service, public data transmission service.** *See also* **administration.**

data transportability: *Synonym* **data portability.**

data unit: A quantity or group of data that (a) is treated as a single entity, i.e., is processed, stored, and transmitted as a single entity, and (b) is bounded by data delimiters. *Note:* Examples of data units are bits, bytes, characters, words, blocks, and frames. *See* **added data unit, command protocol data unit, expedited data unit, protocol data unit, service data unit.** *See also* **data delimiter.**

data validation: The checking of data to ensure correctness and compliance with applicable standards, rules, and conventions for the representation, organization, structuring, and formatting of data.

data value: In a database or data processing system, one of the allowable attributes of a data item. *Note:* An example of a data value is green for the data item color of eyes. The data values for green might be dark, medium, light, pea, grass, and sea. *See also* **database, data element, value.**

data volatility: Pertaining to the rate of change in the values of stored data over a period of time. *See also* **volatile storage.**

datawire: In a communications system, the channel or the path that is used to transmit data, i.e., perform the primary function for which the system was built. *See also* **channel, orderwire, path.**

date: An instant in the passage of time, identified with specific precision by a clock and a calendar. *Note 1:* An example of a date is 23 seconds after 3:14 P.M. on February 9, 1998. This date might be represented as 1998FEB091514.23. *Note 2:* In a message, the date is represented by a date-time group. *See* **epoch date.** *See also* **date-time group.**

date and time indicator: *See* **automatic date and time indicator.**

date format: The arrangement of numeric and alphabetic characters that define the date, such as October 5, 1922, 5 OCT 22, 05 October 1922, 10/5/22, and 1922-10-05. *Note:* In using the date format 10/5/22, the month, in U.S. communications, must always come first, followed by day and year, or confusion will result. Europeans and the U.S. military use a day-month-year date format, such as 05 OCT 22. The 1922-10-05 format is widely used in automatic data processing systems because the units are in descending order of magnitude, which can be followed by hour, minute, and second in that order. Thus, files may be more easily sorted by date. In many personal computer (PC) software systems, (a) the date may be stored as the number of days since the beginning of the current century and (b) the date is tracked and, in a conventional date format, is placed on the directory of file names when a document is stored. *See also* **date-time group.**

date-time group (DTG): In a message, a set of characters that (a) usually is in a prescribed format, (b) is used to express the day of the month, the hour of the day, the minute of the hour, the time zone, and the year, (c) usually is placed in the header of the message, (d) may indicate either the date and the time that a message was dispatched by a transmitting station or the date and the time that it was handed in to a transmission facility by the originator or the user for dispatch, (e) may be used as a message identifier if it is unique for each message, and (f) should not be used to compute message processing time, delivery time, or delay time. *Note:* An example of a date-time group (DTG) is 081330Z AUG 98, indicating 1330 hours or 1:30 P.M. Universal Coordinated Time (UTC) on the 8th day of August of the year 1998. The DTG usually is expressed by two digits for the day of the month and four digits for the time on a 24-hour clock, followed by a letter indicating the time zone suffix, the first three letters of the month to express the month, and two or four digits to express the year. Often the year is not included. Seconds or fractional minutes should be clearly indicated after the minute digits, such as 1330.8 to indicate eight-tenths of a minute after 1330. *See* **true date-time group.**

dating format: A format used to express a moment in time, such as a format used for expressing a moment in the occurrence an event, such as the start of an event. *Note 1:* Dating formats usually are precisely prescribed for date-time groups on messages. *Note 2:* An example of a dating format is that used to express a moment of time on the Coordinated Universal Time (UTC) Time Scale, given in the sequence hour and minute, day, month, and year, such as 1201UT04Jul96, i.e., one minute after noon on the 220th anniversary of U.S. Independence Day. The hour is designated by the 24-hour system. *Note 3:* For increased precision, a decimal point may be placed after the four hour and minute digits, followed by either two-digit second indicators or two-digit decimal fraction indicators, with a common understanding of which indicators are being used in a given system. *See also* **Coordinated Universal Time (UTC), date-time group.**

daughter board: A printed circuit (PC) board that (a) is plugged into another printed circuit board, i.e., into a mother board and (b) is used to add or expand system capabilities, especially when physical space is extremely limited, as it is in microcomputers and personal computers (PCs). *See also* **mother board, printed circuit, printed circuit board.**

day: Nominally, the duration of one complete revolution of the Earth about its own axis. *See* **civil day, radio broadcast day, sidereal day, television broadcast day, true solar day, undisturbed day.**

day of year: In each year, the sequential day count starting at 001 on the first day of January and ending at 365 on the last day of December, 366 in leap years. *See also* **Coordinated Universal Time, day.**

dB: A unit of relative electrical or acoustical power equal to 0.1 B (bel). *Note 1:* The dB (decibel) is the conventional relative power unit, rather than the B (bel), for expressing power ratios because the dB is smaller and therefore more convenient than the B. The number of dB is given by the relation $dB = 10\log_{10}(P_1/P_2)$ where P_1 and P_2 are the actual powers. Power ratios may be expressed in terms of voltage and impedance, E and Z, or current and impedance, I and Z. Thus, dB is also given by the relations

$$dB = 10\log_{10}\{(E_1{}^2/Z_1)/(E_2{}^2/Z_2)\}$$
$$= 10\log_{10}\{(I_1{}^2Z_1)/(I_2{}^2Z_2)\}$$

If $Z_1 = Z_2$, the result is given by the relation $dB = 20\log(E_1/E_2) = 20\log(I_1/I_2)$. Thus, if the ratio of the optical power at the end of an optical fiber to the power at the beginning of the fiber is 0.50, the loss would be expressed as 10 log 0.50 = −3.01, i.e., about 3 dB down. *Note 2:* The reference level usually is placed in the denominator and the level to be calculated in the numerator. If the logarithm is negative, a power loss is indicated. If the logarithm is positive, a gain is indicated. If a label is used, such as loss, down, gain, or up, the number of dB is stated without a sign. Thus, "−3 dB down" would not be used. *Note 3:* The dB is used rather than ratios or percentages because when circuits are connected in sequence, i.e., in tandem, the changes in power level at different times or places may simply be added and subtracted. Thus, in an optical link, if the level at a source is known and the required threshold level at a photodetector is known, the available number of dB can be calculated. The sum of the losses in intervening connectors, splices, and fibers cannot exceed the available dB. Also, there should be a safety margin. *Synonym* **decibel.** *See also* **power budget.**

dBa: A unit of weighted absolute noise power, measured and expressed in dBa, referenced to approximately 3.16 pW (picowatts), which is 0 dBa, i.e., −85 dBm. *Note:* When measuring and recording dBa, whether F1A line or HA1 receiver weighting was used must be indicated in parentheses. A 1000-Hz (hertz), 1-mW (milliwatt) tone will read +85 dBa, but the same power as white noise, randomly distributed over a 3-kHz band, nominally 300 Hz to 3300 Hz, will read +82 dBa because of the frequency weighting. *Synonym* **dBrn adjusted.** *See also* **channel noise level, circuit noise level, dB, noise level, noise weighting.**

dBa(F1A): A unit of weighted absolute noise power, measured and expressed in dBa(F1A), referenced to a measurement by a noise-measuring set with F1A line weighting. *Note:* F1A weighting is no longer used in U.S Department of Defense (DoD) applications. *See also* **dB, F1A line weighting.**

dBa(HA1): A unit of weighted absolute noise power, measured and expressed in dBa(HA1), referenced to a measurement across the receiver of a 302-type, or similar subscriber set, made by a noise-measuring set with HA1 receiver weighting. *Note:* HA1 weighting is no longer used in U.S Department of Defense (DoD) applications.

dbase: *Synonym* **database.**

dBa0: A unit of absolute noise power, measured and expressed in dBa0, referenced to or measured at zero transmission level point (0TLP), i.e., a point of zero relative transmission level (0 dBr). *Note:* Noise readings should be converted from dBa to dBa0 so that the relative transmission level at the point of actual measurement need not be known or stated. *See also* **channel noise level, circuit noise level, dB, level, noise level.**

dBc: A unit of relative power, expressed in dBc, relative to the carrier power.

dBm: A unit of absolute power measured and expressed in dBm referenced to 1 mW (milliwatt). *Note 1:* 0 dBm equals 1 mW (milliwatt). *Note 2:* In the U.S. Department of Defense (DoD), unweighted measurement usually is assumed and understood to be applicable to a stated or implied bandwidth. *Note 3:* In European practice, psophometric weighting may be implied by context. However, this is equivalent to dBm0p, which is preferred in the United States. *See also* **dB, dBmV, level, neper, signal level.**

dBm(psoph): A unit of absolute noise power, measured and expressed in dBmp, measured with psophometric weighting given by $dBm(psoph) = (10\log_{10}pWp) - 90 = dBa - 84$ where pWp is power in picowatts psophometrically weighted and dBa is the weighted noise power in dB referenced to 3.16 pW (picowatts). *See also* **channel noise level, circuit noise level, dB, noise level.**

dBm transmission level point: *See* **zero dBm transmission level point.**

DBMS: database management system.

dBmV: A unit of absolute power, measured and expressed in dBmV, referenced to 1 mV (millivolt) across 75 ohms. *Note:* The dBmV reference level is not equal to the dBm reference level. 1 dBmV is equal to approximately 1.33×10^{-5} mW (milliwatts). *See also* **dB, dBm.**

dBm0: A unit of absolute power, measured and expressed in dBm, referenced to or measured at a zero transmission level point (0TLP), i.e., a point of zero relative transmission level (0 dBr0). *Note:* In some international documents, dBm0 means noise power in dBm0p, i.e., psophometrically weighted dBm0. In the United States, dBm0 is not used that way. *See also* **dBm, level, zero transmission level point.**

dBm0p: A unit of absolute noise power, measured and expressed in dBm0, i.e., in dBm, measured by a psophometer, i.e., a noise-measuring set, having psophometric weighting. *See also* **channel noise level, circuit noise level, dB, dBm, noise level, noise weighting.**

dBp: A unit absolute of power, expressed as dB referenced to 1 pW (picowatt). *Note:* The dBp power unit is used in communications when measuring very low signal levels and low noise levels and, in acoustics, sound levels. 0 dBp is equal to 1 pW (picowatt), i.e., 10^{-12} W (watts). *See also* **pascal.**

dBr: A unit of relative power, expressed in dBr, measured between any point and a reference point selected as the zero relative transmission level point. *Note:* Any power expressed in dBr does not specify the absolute power. It is a relative measurement only. *See also* **dB, level, relative transmission level, transmission level, transmission level point, zero transmission level point.**

dBrn: A unit of absolute noise power, measured and expressed in dBrn, above reference noise. *Note 1:* Weighted noise power is expressed in dB (decibels) referenced to 1 pW (picowatt). Thus, 0 dBrn = −90 dBm = −120 dBfW (dBfemtowatts) = −150 dBaW (dBattowatts). The use of 144-line, 144-receiver, C message, and flat weighting must be indicated in parentheses as required. *Note 2:* With C message weighting, a 1000-Hz (hertz), 1-mW (milliwatt) tone will read +90 dBrn, but the same power as white noise, randomly distributed over a 3-kHz (kilohertz) band will read approximately +88.5 dBrn because of the frequency weighting. The +88.5 dBrn usually is rounded off to +88 dBrn. *Note 3:* With 144-line weighting, a 1000-Hz, 1-mW white noise tone will also read +90 dBrn, but the same 3-kHz power will only read +82 dBrn because of the different frequency weighting. *See also* **channel noise level, circuit noise level, dB, noise level, noise weighting.**

dBrn adjusted: *Synonym* **dBa.**

dBrnC: A unit of weighted absolute noise power, measured and expressed in dBrnc, measured by a noise-measuring set with C-message weighting. *See also* **circuit noise level, dB, level, noise level.**

dBrnC0: A unit of absolute noise power, expressed in dBrnC, referenced to or measured at zero transmission level point (0TLP). *See also* **level, noise level, zero transmission level point.**

dBrn(f_1-f_2): A unit of absolute noise power, measured and expressed in dBrn(f_1-f_2), measured over an amplitude-frequency characteristic that is flat between frequencies f_1 and f_2. *See also* **dB, flat weighting, frequency, noise level, noise weighting.**

dBrn(144 line): A unit of absolute noise power, measured and expressed in dBrn(144 line), measured by a noise-measuring set with 144-line weighting. *See also* **dB, noise level, noise weighting.**

DBS: direct broadcast satellite.

DBTVS: direct broadcast television satellite.

dBV: dB relative to 1 volt peak to peak. *Note:* The dBV is usually used for television video signal level measurements.

dBW: A unit of absolute power, measured and expressed in dBW, referenced to 1 W (watt). *Note:* 0 dBW equals 1 W. *See also* **dB, bel, rated output power.**

dBx: dB above reference coupling. *Note:* dBx is used to express the amount of crosstalk coupling in telephone circuits. dBx is measured with a noise measuring set.

dc: direct current.

DC: direct current, double crucible.

DCE: data circuit-terminating equipment.

DCE clear signal: *See* **data circuit-terminating equipment clear signal.**

DCE waiting signal: *See* **data circuit-terminating equipment waiting signal.**

D channel: A channel that (a) consists of a 16-kbps (kilobit per second) segment of a 144-kbps full-duplex subscriber service channel that is subdivided into conventional 2B+D channels, i.e., two 64-kbps clear channels and one 16-kbps channel, (b) usually is used for out-of-band signaling, (c) has two 64-kbps clear channels used for customer voice and data services, and (d) conforms to and is part of the International Telegraph and Telephone Consultative Committee (CCITT) customer accessing standard that is part of the Integrated Services Digital Network (ISDN) standard. *See also* **Integrated Services Digital Network.**

dc patch bay: *Synonym* **direct current patch bay.**

DC process: double crucible process.

DC restoration circuit: *See* **direct current restoration circuit.**

DC signaling: *See* **direct current signaling.**

DCPSK: differentially coherent phase-shift keying.

DDD: direct distance dialing.

DDL: data description language.

DDN: Defense Data Network.

DDR: disengagement denial ratio.

DDS: doped deposited silica.

deadlock: 1. Unresolved contention for the use of a system or a component. **2.** In computer and data processing systems, an error condition such that processing cannot continue because each of two components or processes is waiting for an action or a response from the other. **3.** A permanent condition in which a system cannot continue to function unless some corrective action is taken. *See also* **interlock.**

dead reckoning: To calculate a present position using the last known position and the movements that have been made since leaving the last known position. *Note:* An example of dead reckoning is to use the integrated velocity with respect to time and the initial coordinates to obtain new coordinates. In radionavigation, dead reckoning may have to be used if radio transmission or reception is lost.

dead reckoning position: In navigation, the position calculated by dead reckoning.

dead sector: 1. In facsimile systems, the time between the end of scanning of one object line and the start of scanning of the following line. **2.** In facsimile systems, the portion of the surface of a drum transmitter in which the scanning time cannot be used for picture signal transmission. *See also* **facsimile, scanning, scanning line length.**

dead space: In radio transmission and reception, the area, zone, or volume of space (a) that is otherwise within range of a transmitter, such as a radio or radar transmitter, and (b) in which a signal is not detectable by a given receiver.

dead zone: 1. The region in an optical waveguide, such as an optical fiber, in which a measurement cannot be made with a given optical time domain reflectometer (OTDR). *Note 1:* In an optical fiber, the dead zone usually extends from immediately beyond the light source, or the entrance face of the fiber to which the optical time domain reflectometer (OTDR) is connected for the fiber portion of the dead zone, to a point, some distance into the fiber, where the measurement is 3 dB (decibel) down from measurements made a much greater distance into the fiber where measurements are their normal values. *Note 2:* Dead zones are of the order of 1 m (meter). **2.** The range over which an input analog variable may vary without resulting in a change in the output analog variable.

dead zone unit: A device that has an output analog variable that is constant over a specified range of the input analog variable.

deaf: *See* **teletypewriter/telecommunications device for the deaf.**

debug: 1. To detect, trace, and eliminate mistakes in a computer program. **2.** To detect, locate, and eliminate faults in equipment. *See also* **bug.**

debugging model: A model that may be used to determine, predict, assess, or evaluate the number, distribution, location, category, or probability of detection of mistakes, errors, malfunctions, failures, or faults, i.e., bugs, in the software or hardware components of a system. *See also* **bug, error, failure, fault, malfunction, mistake, model.**

decametric wave: An electromagnetic wave in the high frequency (HF) range of the electromagnetic frequency spectrum, having a frequency between 3 MHz (megahertz) and 30 MHz and a wavelength between 10 m (meters) and 100 m.

decay time: *See* **pulse decay time.**

DECCA: A radio phase comparison system that (a) uses a master and several slave stations to establish a hyperbolic radio wave lattice, (b) provides accurate position-fixing services, (c) is a long-range navigational system that operates in the 70 kHz (kilohertz) to 130 kHz frequency band, (d) is a continuous wave system in which the receiver measures and integrates the relative phase difference between the signals received from two or more synchronized ground stations, (e) usually has one master station and three slave stations arranged in a fixed formation, and (f) has an operational range of about 250 mi (miles) or about 400 km (kilometers).

decentralized computer network: A computer network in which the data processing and computing capability required by users is distributed over two or more nodes. *See also* **computer network, node.**

deception: *See* **communications deception, electronic deception, manipulative communications deception, manipulative electronic deception, radar scan rate modulation deception, radio deception.**

deception jammer: A jammer that induces false indications in the system that it jams. *See also* **jammer.**

deception repeater: A device that (a) receives signals, (b) amplifies, delays, reshapes, or otherwise manipulates the signals, and (c) retransmits the signals solely for creating deception. *See* **radar deception repeater.**

decibel (dB): *Synonym* **dB.**

decibel above isotropic (DBI) condition: An antenna gain unit of measure that (a) is used instead of dB and (b) has the same numerical value as the dB. *See also* **antenna gain.**

decimal: Pertaining to a numeration system in which the radix is 10. *See* **sexadecimal.** *See also* **numeration system.**

decimal digit: One of the digits in the decimal numeration system, i.e., one of the digits 0, 1, 2, 3, 4, 5, 6, 7, 8, and 9.

decimal interchange code: *See* **extended binary coded decimal interchange code.**

decimal notation: *See* **binary coded decimal notation.**

decimal numeration system: A positional numeration system in which the radix, i.e., the base, is 10. *See also* **numeration system, positional numeration system.** *Refer to* **Figs. B-7 (81), E-5 (315).**

decimal unit of information content: *Synonym* **hartley.**

decimetric wave: An electromagnetic wave in the ultrahigh frequency (UHF) range of the electromagnetic frequency spectrum, having a frequency between 300 MHz (megahertz) and 3000 MHz and a wavelength between 0.1 m (meter) and 1 m.

decimillimetric wave: An electromagnetic wave in the tremendously high frequency (THF) range of the electromagnetic frequency spectrum, having a frequency between 300 GHz (gigahertz) and 3000 GHz, i.e., 3 THz (terahertz), and a wavelength between 0.0001 m (meter) and 0.001 m. *Note:* Decimillimetric waves have a wavelength between 10- and 100-thousandths of a centimeter, hence the name decimillimetric.

decipher: To convert enciphered text to the equivalent plain text by means of a cipher system. *Note:* Deciphering does not include deriving plain text by cryptanalysis. *See also* **cipher, communications security, decode, decrypt.**

decision circuit: **1.** A circuit that measures the probable value of an input signal significant condition and derives a corresponding output signal significant condition based on the measured value of the input signal significant condition and predetermined criteria. **2.** A circuit that generates an output signal that represents a decision based on past or present, or both, input signals. *Note:* An example of a decision circuit is a combinational circuit, such as an AND gate. *See also* **circuit, gate, gating.**

decision content: A measure of the number of possible decisions needed to select a given event from among a finite number of mutually exclusive events, given by the relation $H = \log n$, where H is the decision content and n is the number of events. *Note 1:* The base of the logarithm determines the unit used. *Note 2:* The decision content is independent of the probabilities of occurrence of the events. However, in some applications, it is assumed that these probabilities are equal. *See also* **hartley, natural unit of information content, shannon.**

decision feedback: *Synonym* **error detection and feedback.**

decision feedback system: *Synonym* **automatic request repeat.**

decision instant: In the reception of a digital signal, the instant at which a decision is made by a receiving device as the probable value of a significant condition. *Note:* The decision instant occurs during a signal transition, usually at or near a significant instant. *Note:* An example of a decision instant is the instant at which a decision is reached as to whether a given significant condition represents a 0 or a 1. *Synonym* **selection position.** *See also* **bit synchronization, significant instant.**

decision support system: *See* **global decision support system.**

deck: *See* **virtual reality deck.**

deck log: *See* **ship deck log.**

declaration: 1. In a computer programming language, an explicit specification of the computing environment or of the characteristics, attributes, or aspects of one or more identifiers in a computer program. *Note:* Examples of declarations are (a) a definition of a collating sequence, (b) a description of the attributes of a named variable or parameter, and (c) the establishment of an identifier as a name or the construction of a set of attributes of the named item. **2.** In a computer programming language or a computer program, an expression that influences the interpretation of other expressions in that language or program. *Synonym* **declarative statement.**

declarative statement: *Synonym* **declaration.**

declination parallel: Any small circle of the celestial sphere that has a plane parallel to the plane of the celestial equator and perpendicular to the celestial axis. *See also* **celestial equator, celestial sphere.**

decode: 1. To convert data, i.e., to return data to their original form, by reversing the effect of previous encoding. **2.** To interpret a code. **3.** To convert encoded text into equivalent plain text by means of a code, usually contained in a code book in which the code groups are in a systematic order, such as an alphabetical, numerical, or structured order. *Note:* Decoding does not include deriving plain text by cryptanalysis. *See also* **analog decoding, code, decipher, decrypt, encipher, encode.**

decoder: 1. A device that decodes data. **2.** A device (a) that has a number of input lines, of which any number may carry signals, and a number of output lines, of which not more than one may carry a signal at any given instant, and (b) in which there usually is a one-to-one correspondence between the combinations of the input signals and the output signals. *See* **zero level decoder.** *See also* **coder, coder-decoder, encoder, signal.**

decoding: *See* **analog decoding.**

decollimation: 1. In a beam, increasing the divergence or the convergence of the rays in the beam. **2.** In a beam with a given degree of convergence or divergence of rays, the causing of the rays to converge or diverge farther from parallelism. *Note 1:* Other than deliberate decollimation, any of a large number of factors may cause decollimation, such as refractive index inhomogeneities, occlusions, scattering, deflection, diffraction, reflection, and refraction. *Note 2:* Decollimation occurs in many applications, such as radio, radar, sonar, and optical communications applications. *Note 3:* Decollimation may be desirable or undesirable. *See also* **beam divergence, collimation, diffraction, refraction, scattering.**

decomposition: *See* **hierarchical decomposition, modular decomposition.**

decrypt: To convert encrypted text into equivalent plain text by means of a cryptosystem. *Note 1:* Decryption does not include solution by cryptanalysis. *Note 2:* To decrypt includes to decipher and to decode. *See also* **decipher, decode, garble.**

DECT: Digital European Cordless Telecommunications, digital European cordless telephone, digital European cordless telephony.

dedicated circuit: 1. A circuit that (a) is designated for exclusive use by specified users and (b) usually serves a designated purpose. **2.** A circuit designated for exclusive use by two and only two users. *See also* **circuit, closed circuit, hot line, leased circuit, off-hook service, point to point link.**

dedicated communication: Communication that is (a) between specific users, (b) for specific purposes, and (c) under the direct control of the users.

dedicated communications service: A communications service provided for the exclusive use of one user, user group, system, or subsystem. *See also* **common user, common user communications group, communications service.**

dedicated connection: A user-to-user connection that (a) is automatically set up on some simple command, such as by a key depression or by short dialing, by one user and (b) usually includes a preemption facility to ensure an adequate grade of service. *See also* **command, dialing, grade of service, preemption.**

dedicated data link: A data link that (a) is used for one and only one purpose, (b) connects two points, and (c) has no connections to other points or components. *Note:* Examples of dedicated data links are (a) links from sensors to display panels and (b) data links between control panels and motor control actuators. *See also* **dedicated link.**

dedicated mode: In communications, computer, data processing, and control systems, a mode of operation of a system or component by a specific group of users, for a specific application or use, or for a specific class of information.

dedicated service: **1.** A service provided by a communications network that (a) is devoted to a single purpose or group of users, (b) usually is provided in a specified format, such as voice, digital data, facsimile, or video format, and (c) may be a subset of a larger network. **2.** In a communications system, a specified set of services provided to designated users. *Note:* An example of a dedicated service is the Automatic Voice Network (AUTOVON) used by the military.

dedicated storage: Storage, or an area of storage, that has been committed, allocated, obligated, or otherwise set aside, earmarked, or assigned for a specified use, problem, or user and none other. *Note:* Examples of dedicated storage are (a) a set of tracks on a magnetic disk assigned to a division of a corporation for its sole use and (b) the track on a disk used for addresses or timing pulses. *See also* **storage.**

deducible directory: A directory constructed in such a manner that (a) users can deduce the directory numbers of other users using defined rules and fixed lookup tables, (b) users can form a directory number for most users in a closed user group, and (c) the rules are simple to implement and easy to memorize. *See also* **closed user group, directory.**

deemphasis: **1.** In a receiver, the intentional alteration of the amplitude, phase, frequency, or shape characteristics of a signal for the purpose of restoring the signal to its original condition. *Note:* In frequency modulation (FM), an example of deemphasis is the reduction in amplitude of the higher frequency components of the received signal, thus restoring them to their original relative levels. **2.** In frequency modulation (FM) transmission, the restoration, after detection, of the amplitude versus frequency characteristics of the received signal to those of the original signal. *See also* **emphasis, preemphasis, preemphasis improvement.**

deemphasis circuit: A circuit that restores the original amplitudes of the various frequencies that were in a signal prior to their modification by a preemphasis circuit, usually by decreasing the relative amplitude of the higher frequency signal components that were emphasized at the input. *Note:* A deemphasis of the lower frequency components and their restoration would have similar effects. *See also* **preemphasis, preemphasis circuit, preemphasis improvement.**

deemphasis network: A network inserted into a system in order to decrease the amplitude of electromagnetic waves or signals in one frequency range with respect to the amplitude in another frequency range. *See also* **amplitude, network, preemphasis network.**

deep channel: A channel used for communication into, out of, or within deep space. *See also* **channel, communication, deep space.**

deeply depressed cladding optical fiber: An optical fiber that has an outer cladding with approximately the same refractive index as the core and an inner cladding with a refractive index such that the refractive index contrast is greater than 1%. *Note 1:* The depressed cladding is between the outer cladding and the core. *Note 2:* The outer cladding is used to protect the inner cladding, reduce microbending, and remove cladding modes usually to reduce dispersion and thereby obtain increased data signaling rates (DSRs). *Note 3:* Deeply depressed cladding optical fibers usually are used for single mode transmission. *See also* **cladding, depressed cladding optical fiber, doubly clad optical fiber, fiber optics, quadruply clad optical fiber, single mode optical fiber.**

deep space: **1.** Space at distances from the Earth approximately equal to or greater than the distance between the Earth and the Moon. **2.** Space at distances from the Sun greater than that of the farthest planet from the Sun. *See also* **outer space.**

de facto standard: A standard that (a) is widely accepted and used and (b) lacks formal approval by a recognized standards organization. *See also* **standard.**

default: **1.** Pertaining to the predefined initial, original, or specific setting, condition, value, or action that a system will assume, use, or take in the absence of instructions from the user. **2.** The predefined initial, original, or specific setting, condition, value, or action that a system will assume, use, or take in the absence of instructions from the user.

default configuration: In communications, computer, and data processing systems, the configuration that a system will automatically assume when the system configuration options are not specified by the operator or the program. *Note:* Examples of default configurations are (a) in communications systems, the trunks and lines that will be interconnected when specific connections have not been specified, (b) in computer systems, the peripheral equipment that will be connected before any selections are made, (c) in personal computers (PCs), the disk drive that the computer will use when a specific drive has not been selected, and (d) in data processing systems, the output printer or line that will be used when an output printer or line has not been selected. *See also* **configuration.**

defect: *See* **interstitial defect, vacancy defect.**

defect absorption: *See* **atomic defect absorption.**

defense command-control-communications network: *See* **air defense command-control-communications network.**

defense communications: *See* **air defense communications.**

Defense Communications Agency: *See* **Defense Information Systems Agency.**

defense communications system: *See* **air defense communications system.**

Defense Data Network (DDN): The U.S. Department of Defense (DoD) integrated packet-switching network capable of worldwide multilevel secure and nonsecure data transmission. *Note 1:* The unclassified MILNET portion of the Defense Data Network (DDN) is part of Internet. *Note 2:* The classified Defense Integrated Secure Network (DISNET) and the Worldwide Military Command and Control System (WWMCCS) Intercomputer Network (WIN) are essentially the same network hardware and software as MILNET. However, for security reasons, DISNET and WIN are not part of the Internet. *See also* **Internet, MILNET.**

Defense Information Systems Agency (DISA): The U.S. Department of Defense (DoD) agency responsible for (a) policy pertaining to acquisition, operation, and deployment of defense communications systems and (b) operation of DoD-wide and worldwide information systems. *Note:* The Defense Information Systems Agency (DISA) was formerly called the Defense Communications Agency (DCA).

Defense Information Systems Network (DISN): A U.S. Department of Defense (DoD) network that (a) interconnects several hundred DoD networks at military installations in the United States into a single international broadband communications network, i.e., a single platform, and (b) is divided into three regions, i.e., the eastern, western, and southern regions, of the United States.

defense master clock: *See* **Department of Defense master clock.**

Defense Network Security Information Exchange (DNSIX): A standardized security arrangement that (a) provides security for information passing across gateways to various interconnected distributed open system architecture networks, each with different operating protocols and compartmentation schemes, (b) affords protection to network data by various security procedures and mechanisms, such as identity checks, data encryption controls, authentication procedures, and fail-safe procedures, (c) provides network-level security traps, alarms, and journals of events, and (d) consists primarily of a network-level module and a revised interconnection protocol security option (RIPSO). *See also* **compartmentation, revised interconnection protocol security option.**

definition: 1. A figure of merit for image quality. *Note:* In images, such as in video displays, definition usually is expressed in terms of the smallest resolvable element, such as lines per inch or pels per square inch. **2.** In facsimile systems. the distinctness or clarity of detail or outline on a record sheet. *See also* **facsimile, resolution.**

deflection: 1. A change in the direction of a traveling particle, usually without loss of particle kinetic energy, representing a velocity change without a speed change. **2.** A change in the direction of a wave, beam, electron, or other entity, such as might be accomplished by an electric or magnetic field. *Note:* If the deflection is caused by a prism (refraction), a mirror (reflection), or an optical grating (diffraction), the specific applicable term should be used.

degauss: To remove the magnetization from a magnetic data medium, usually by raising and then gradually lowering an alternating magnetic field in the magnetic material. *Note:* A thoroughly degaussed magnetic tape, card, or disk should not produce signals or noise when scanned by a magnetic head. *See also* **data medium, magnetic field, magnetic head.**

degenerate waveguide mode: One of a set of waveguide modes that have the same propagation constant for all frequencies of interest. *See also* **frequency, mode, propagation constant, waveguide.**

degradation: 1. The deterioration in quality, level, or standard of performance of a functional unit. **2.** In communications, the condition in which one or more of the required performance parameters fall outside predetermined limits, resulting in a lower quality of service. *See* **motion response degradation.** *See also* **degraded service state, tolerance.**

degraded service state: 1. The condition that exists when one or more of the required service performance parameters fall outside specified limits, resulting in a lower quality of service. **2.** The condition in which degradation prevails in a communications link. *Note:* For some applications, such as automatic switching to a nondegraded standby link, the degraded service state must persist for a specified period before a degraded service state is considered to exist. *See also* **catastrophic degradation, continuous operation, degradation, graceful degradation, outage.**

degree: *See* **coherence degree, electrical degree, spatial degree.**

degree of coherence: *Synonym* **coherence degree.**

degree of isochronous distortion: *Synonym* **isochronous distortion ratio.**

degree of start-stop distortion: *Synonym* **start-stop distortion ratio.**

dejitterizer: A device that reduces jitter in a digital signal. *Note 1:* A dejitterizer usually consists of an elastic buffer in which the signal is temporarily stored and then retransmitted at a rate based on the average rate of the incoming signal. *Note 2:* The dejitterizer usually is ineffective in dealing with low frequency jitter, such as waiting time jitter. *See also* **jitter, regeneration, signal regeneration.**

DEL: delete character.

delay: 1. The amount of time by which an event is retarded. *Note:* In radar, an example of delay is the delay of the start of the time base used to select a particular segment of the total delay time. **2.** The time between (a) the instant at which a given event occurs and (b) the instant at which a related aspect of that event occurs. *Note:* The events, relationships, or aspects of the entity being delayed must be specified. *Synonym* **delay time. 3.** In communications systems, the lost or waiting time introduced because a call cannot be connected immediately or because a message or signal cannot be transmitted immediately, usually caused by a queue on the required line or circuit. *See* **absolute delay, absolute signal delay, call delay, dial tone delay, differential mode delay, group delay, multimode group delay, near real time delay, operator delay, phase delay, propagation delay, queuing delay, receive after transmit time delay, receiver attack time delay, receiver release time delay, round trip delay, transit delay, transmit after receive time delay, transmit delay, transmitter attack time delay, transmitter release time delay.** *See also* **delay distortion, delay line, total delay time, transmission time.**

delay distortion: In a waveform or signal that consists of two or more different frequencies, distortion caused by the difference in arrival times of the frequencies at the output of a transmission system. *Note 1:* The output signal wave shape is not the same as the input signal wave shape for the same signal. *Note:* In an optical fiber, the spectral width gives rise to delay distortion because the different wavelengths in an optical pulse arrive at the end of a fiber at different times. *Synonyms* **phase distortion, time delay distortion.** *See* **group delay distortion, waveguide delay distortion.** *See also* **absolute delay, distortion.**

delayed delivery facility: In a communications network, a facility that stores data, destined for delivery to one or more addressees, for subsequent delivery at a later time. *See also* **store and forward.**

delay element: A device that yields an output signal essentially similar to a previously introduced input signal, the time between the signals being the delay. *See also* **input, output, signal.**

delay encoding: The encoding of binary data in which a two-level signal is formed such that (a) a binary 0 causes no change of signal level unless it is followed by another zero, in which case a transition to the other level takes place at the end of the first bit period, and (b) a binary 1 causes a transition from one level to the other in the middle of the bit period. *Note:* Delay encoding is used primarily for encoding radio signals because the frequency spectrum of the encoded signal contains less low frequency energy than a conventional nonreturn to zero (NRZ) signal and less high frequency energy than a biphase signal. *See also* **modulation, delay modulation, nonreturn to zero code.**

delay equalizer: A corrective circuit or network that (a) makes the phase delay or envelope delay of another circuit or network substantially constant over a desired frequency range, (b) compensates for group delay frequency distortion, and (c) makes the transmission time substantially constant for all frequency components within the specified frequency band for all signals in the circuit or the network. *Synonyms* **delay frequency equalizer, phase equalizer.** *See also* **absolute delay, circuit, equalization, group delay, phase, phase delay, phase-frequency equalizer.**

delay line: 1. A device, such as a transmission line, that (a) is designed and operated to deliberately delay a signal for timing purposes and (b) is used to delay a signal. **2.** A single input channel device, such as a single input sequential logic element, in which the output channel state at a given instant, *t*, is the same as the input channel state at the instant $t-n$, where *n* is a number of time units, i.e., the input sequence undergoes a delay of *n* time units, such as *n* attoseconds, femtoseconds, nanoseconds, microseconds, or milliseconds. **3.** A sequential logic element or device that (a) has one and only one input channel, (b) has an output channel state at a given instant, *t,* that is the same as the input channel state at the instant $t-n$, i.e., the input sequence undergoes a delay of *n* units, and (c) may have additional taps yielding output channels with smaller values of *n. See* **acoustic delay line, fiber optic delay line, magnetic delay line, magnetostrictive delay line.** *See also* **absolute delay.**

delay line storage: Storage in which (a) a delay line is used to store strings of binary digits temporarily for the time length of the line or permanently if circulated indefinitely, (b) from the end of the line the pulses representing the bits may be amplified, shaped, timed, and gated back into the beginning of the line, and (c) the pulses may be (i) acoustic, using quartz crystal transducers at each end to launch and sense the sonic pulses, (ii) electromagnetic, using continuous or lumped capacitors and inductors, (iii) moving magnetic, using a pair of read and write heads spaced on the same track on a magnetic disk, drum, or tape loop arranged so that a pulse is written back behind the read head and held until it reaches the read head, or (iv) fiber optic, using an optical fiber with a fiber optic transmitter at the beginning and a fiber optic receiver at the end. *See also* **delay line, fiber optic receiver, fiber optic transmitter, gate, read head, storage, write head.**

delay modulation (DM): Modulation that (a) is achieved by introducing different forms of delay in a signal element, (b) is a form of signal modulation that is often used in radio applications, (c) provides a compromised waveform between nonreturn to zero (NRZ) and biphase, and (d) produces a spectrum that has less low frequency power than NRZ and less high frequency power than biphase. *Note:* An example of delay modulation (DM) is to encode binary data into one of four symbols such that (i) a 1 (mark) is represented by a transition from either level, i.e., from either high or low level to the other level in the center of the bit interval, and thus the DM 1 can be encoded into either of the biphase symbols, (ii) a 0 (space) is represented by no transition, except in the case of two or more consecutive 0s (spaces), in which case transitions occur at the ends of the bit interval, and thus the DM 0 can be encoded in either of the two NRZ symbols, and (iii) there is no transition at the end of a bit interval when a 0 (space) is followed by a 1 (mark), or vice versa. *See also* **delay encoding.**

delay operator: In telephone switchboard operations, an operator specifically assigned to deal with all booked calls or lines over which delay working is in progress. *See also* **booked calls, delay working.**

delay spread: *See* **multimode group delay spread.**

delay time: 1. *Synonym* **delay. 2.** *See* **group delay time, round trip delay time, total delay time.**

delay working: In telephone switchboard operations, operations intended to ensure fair distribution of the time of one or more lines among groups of call originators. *Note:* An example of delay working is the withdrawing of one or more lines from general use and placing them under the control of a delay operator so that when other operators book call demands on tickets, the tickets can be passed to the delay operator for connection in the order in which they are booked. *See* **trunk delay working.**

DEL character: *Synonym* **delete character.**

delete character: A transmission control character primarily used to obliterate an erroneous character. *Note:* On perforated tape, the delete character consists of code holes at all punch positions. *Synonyms* **DEL character, erase character, rub-out character.**

deleted bit: A bit not delivered to the intended destination. *See also* **added bit, binary digit, bit count integrity, character count integrity, error.**

deleted block: A block not delivered to the intended destination. *See also* **added block, block, block transfer failure.**

delimiter: 1. A mark, signal, or character that is used to separate data in a data stream into data units or groups of data units. *Note:* Examples of delimiters are end of text (EOT) characters between messages, end of frame signals between frames, and spaces between words. **2.** A character used to indicate the beginning and the end of a character string. **3.** A flag that separates and organizes items of data. *Synonym* **data delimiter.**

delineating block: *See* **self-delineating block.**

delineation map: *See* **routing indicator delineation map.**

delineation table: *See* **routing indicator delineation table.**

delivered block: A successfully transferred block. *See also* **binary digit, block.**

delivered overhead bit: A bit that (a) is successfully transferred to a destination user, (b) performs its primary function within the telecommunications system, and (c) does not represent user information. *See also* **binary digit, overhead information, user information.**

delivered overhead block: A block that (a) is successfully transferred to a destination user, (b) performs its primary function within the telecommunications system, and (c) does not contain user information bits. *See also* **block, overhead information, user information.**

delivery: *See* **successful block delivery.**

delivery confirmation: Information returned to the originator indicating that a given unit of information has been delivered to the intended addressees. *See also* **acknowledge character.**

delivery facility: *See* **delayed delivery facility.**

delivery time: In a communications system, the date and the time at which a message is delivered to the addressee, i.e., to the destination user. *See also* **user.**

Dellinger effect: An effect that (a) lasts from 10 minutes to several hours and (b) causes electromagnetic sky wave signals to disappear rapidly as a result of greatly increased ionization in the ionosphere caused by increased noise from solar storms. *Note:* An example of the Dellinger effect is the fading of shortwave radio transmissions because of the formation of a highly absorbing D layer, which is lower in altitude than the regular E layer and F layer of the ionosphere, caused by a burst of hydrogen particles from eruptions associated with sunspot activity. *Synonyms* **Dellinger fade out, Dellinger fading.** *See also* **flutter.**

Dellinger fade out: *Synonym* **Dellinger effect.**

Dellinger fading: *Synonym* **Dellinger effect.**

delta-beta switch: A fiber optic switch (a) in which a two-branch Y-splitter lithium niobate waveguide is made so that one path is made of titanium lithium niobate and the other of just lithium niobate, (b) in which an applied voltage can cause one path to shift the wavelength of light passing through it relative to the other path, (c) that acts as a fast wavelength division multiplexer, and (d) in which electrons line up preferentially along the length of the titanium doped channel, which changes its refractive index to a different, nonlinear refractive index, thus shifting the wavelength of light down that channel. *Note 1:* If a specific voltage is guided to one waveguide, the polarization state of light in that waveguide is shifted. The shifted mode is coupled instantly with the wave going down the second waveguide, thus shutting off the first channel. Because this switch operates by change of beta, the nonlinear coefficient of the waveguide, and because a flow of electrons, i.e., beta particles, takes place, it is referred to as a delta-beta switch. *Note 2:* In another variety of lithium niobate electrooptic behavior, the device can be made into a picosecond-speed on-off switch. *Note 3:* By replacing the Y-splitters on a common Mach-Zehnder intensity modulator with 3-dB couplers, a 2-by-2 high speed balanced bridge switch can be mass-produced to perform the same function as the delta-beta switch. *Synonym* **Δ-β switch.**

delta modulation (DM): In analog to digital conversion, modulation in which (a) the analog signal is approximated by a series of straight-line segments, (b) the approximated signal is compared to the original analog signal to determine whether there is an increase or a decrease in relative amplitude, (c) the decision process for establishing the state of successive binary digits is determined by this comparison, (d) the change of information, i.e., an increase or a decrease of the signal amplitude from the previous sample, is sent, and (e) if there is no change, the output condition remains at the same, i.e., 0 or 1, state as the previous sample. *Note:* There are several variations of delta modulation (DM). *See* **continuously variable slope delta modulation.** *See also* **delta-sigma modulation, differential modulation, differential pulse code modulation, pulse code modulation, quantized feedback.**

delta-sigma modulation: Signal modulation in which (a) the information-bearing signal, $s(t)$, is first integrated and then subjected to delta modulation in such a manner that the input to the modulator, $E(t)$, is the difference between the integrated information-bearing signal, i.e., $\int s(t)dt$, and the integrated output pulses, $\int p(t)dt$, given by the relations $E(t) = \int s(t)dt - \int p(t)dt = \int [s(t) - p(t)]dt$ where $E(t)$ is the input signal to the modulator, and s and p are signal values, such as voltage, as a function of time, (b) the integral of the input signal is encoded rather than the signal itself, resulting in a flat load capacity spectrum with increasing signal frequency rather than the 6 dB per octave reduction that is characteristic of delta modulation, and (c) modulation may be achieved by preceding a conventional delta modulation encoder by an integrating network. *Synonym* **Δ-Σ modulation.** *See also* **delta modulation, modulation, signal.**

demand: In telephone switchboard operation, any request made to an operator by a user or by another operator. *See also* **user.**

demand assignment: The sharing among several users of a communications channel on a real time basis such that (a) a user needing to communicate with another user of the network activates the required circuit, (b) when the call is finished, the circuit is deactivated so that its capacity is made available to other users, and (c) the arrangement is similar to conventional telephone switching that provides common trunking for many users, on a demand basis, through a limited-size trunk group. *See also* **time division multiple access.**

demand assignment access plan: In satellite communications systems operations, a variable communications channel access plan in which the allocation of accesses or the number of channels per access is varied by demand, as opposed to a preassigned, i.e., a fixed, access plan. *See also* **preassignment access plan.**

demand assignment multiple access (DAMA): The allocation of communications system capacity based on demand for access by the user. *Note:* Demand assignment multiple access (DAMA) may be implemented by using (a) frequency division multiplexing (FDM) or time division multiplexing (TDM) and (b) modulation techniques. *See also* **frequency division multiplexing, time division multiplexing.**

demand factor: **1.** The ratio of (a) the maximum real power consumed by a system to (b) the maximum real power that would be consumed if the entire load connected to the system were to be activated at the same time. *Note:* The maximum real power usually is integrated over a specified time interval, such as 15 or 30 minutes, and usually is expressed in kilowatts. The real power that would be consumed if the entire load connected to the system were to be activated at the same time is obtained by summing the power required by all the connected equipment. This load is expressed in kilowatts if the consumed real power is expressed in kilowatts. **2.** The ratio of (a) the maximum power, integrated over a specified time interval such as 15 or 30 minutes and usually expressed in kilowatts, consumed by a system to (b) the maximum volt-amperes, expressed in kilovolt-amperes (kVA) if the power is expressed in kilowatts (kW), integrated over a time interval of the same duration, though not necessarily during the same interval, the kVA always being greater than the kW. *Note:* Charges for electrical power may be based on the demand factor as well as the kilowatt-hours of electrical energy consumed, where the lower the demand factor the higher the charges. *See also* **demand load.**

demand load: **1.** In communications, the total power required by a specified unit, such as a specified facility, equal to the sum of (a) the operational loads, including any tactical load, and (b) the nonoperational demand loads. *Note:* The demand load is determined by applying the proper demand factor to each of the connected loads and a diversity factor to the sum total. **2.** At a communications center, the power required by (a) all automatic switching, synchronous, asynchronous, and terminal equipment operated simultaneously online or in standby, (b) control and keying equipment, and (c) lighting, ventilation, and air conditioning equipment required to maintain full continuity of communications services. **3.** The power required for ventilating equipment, shop lighting, and other support items that may be operated simultaneously with the technical load. **4.** The sum of the technical demand and nontechnical demand loads of an operating facility. **5.** At a receiver facility, the power required for all (a) receivers and auxiliary equipment that may be operated on prime or spare antennas simultaneously, including those in standby condition, (b) multicouplers, (c) control and keying equipment, and (d) lighting, ventilation, and air conditioning equipment required to maintain full continuity of communications services. **6.** At a transmitter facility, the power required for all (a) transmitters and auxiliary equipment that may be operated on prime or spare antennas simultaneously, including dummy loads and including those in standby condition, (b) multicouplers, (c) control and keying equipment, and (d) lighting, ventilation, and air conditioning equipment required for full continuity of communications services. *See also* **demand factor, load, nontechnical load, technical load.**

demand service: In Integrated Services Digital Network (ISDN) applications, a telecommunications service that, in response to a user request made through user-network signaling, establishes an immediate communications path.

demarc: *Synonym* **demarcation point.**

demarcation point: In a communications system, the point (a) at which operational control or ownership changes from one organizational entity to another, (b) that usually is the interface point (i) between customer-premises equipment and external network service provider equipment or (ii) between two distinct networks, and (c) that usually is easily identifiable and readily accessible, such as distribution frame terminals, cable heads, or microwave transmitter terminals. *Synonyms* **demarc, network terminating interface.** *See also* **interface point, presence point.**

democratically synchronized network: A mutually synchronized network in which (a) all clocks in the network exert equal amounts of control on the others, and (b) the network operating clock pulse repetition rate is the average of the natural, i.e., the uncontrolled, clock pulse repetition rates of the population of clocks. *See also* **frequency averaging, hierarchically synchronized network, master-slave timing, mutually synchronized network, mutual synchronization, oligarchically synchronized network.**

demodulate: To recover an intelligence-bearing signal from a wave that has been modulated by the signal, i.e., the reverse of modulate. *See also* **modulate.**

demodulation: The recovery, from a modulated signal, of a signal having substantially the same characteristics as the original modulating signal. *See* **isochronous demodulation, synchronous demodulation.** *See also* **detection, modulation, restitution.**

demodulation linearity: *See* **modulation-demodulation linearity.**

demodulator: A device that (a) recovers the baseband signal, i.e., the information-bearing signal, from the modulated signal, thereby performing demodulation, and (b) produces an output signal suitable for further processing. *See* **modulator-demodulator, telegraph demodulator.** *See also* **remodulator.**

demonstration software: In personal computer (PC) systems, software programs that (a) only represent, describe, demonstrate, or picture actual software programs and hardware capabilities and (b) are not the original unrestricted or protected operational software programs. *Synonym* **demo software.**

demo software: *Synonym* **demonstration software.**

demultiplex (deMUX): The inverse of multiplex, i.e., to separate two or more signals that previously were combined, i.e., were multiplexed, by a compatible multiplexer and transmitted over a single channel for subsequent demultiplexing. *See also* **demultiplexing.**

demultiplexer: A device that performs demultiplexing. *See* **fiber optic demultiplexer (active), fiber optic demultiplexer (passive), optical demultiplexer (active), optical demultiplexer (passive).** *See also* **demultiplex, demultiplexing.**

demultiplexing (deMUXing): The separation of two or more channels previously multiplexed, i.e., the reverse of multiplexing. *See* **time division demultiplexing.** *See also* **frequency division multiplexing, multiplexing, space division multiplexing, time division multiplexing, wavelength division multiplexing.** *Refer to* **Fig. W-3 (1082).**

deMUX: demultiplex, demultiplexer.

deMUXing: demultiplexing.

denial: *See* **access denial, disengagement denial.**

denial probability: *See* **access denial probability, disengagement denial probability.**

denial ratio: *See* **access denial ratio, disengagement denial ratio.**

denial time: *See* **access denial time.**

dense binary code (DBC): A binary code in which all possible bit patterns that can be made from a fixed number of bits in a pattern are used in the code. *Note:* Examples of dense binary codes are (a) a pure binary representation for sexadecimal digits using all 16 possible patterns and (b) an octal representation using all 8 patterns. A binary representation of decimal numbers using 4 binary digits of which only 10 of the possible 16 patterns are used is not a dense binary code. If a binary code is not dense, the unused patterns can be used to detect errors because if they occur, an error has been made.

density: 1. In a facsimile system, a measure of the light transmission or reflection properties of an area of an object. *Note 1:* Density usually is expressed as the logarithm to the base 10 of the ratio of incident to transmitted or reflected irradiance. *Note 2:* There are many types of density that usually will have different numerical values for a given material, such as diffuse density, double diffuse density, and specular density. The relevant type of density depends upon the type of optical system and the materials that are used. **2.** The mass per unit volume of a substance. *See* **bit density, carrier to receiver noise density, diffuse density, electromagnetic energy density, internal optical density, luminous density, magnetic flux density, noise power density, optical density, optical energy density, optical power density, packing density, power density, raster density, reflectance density, scanning density, spectral density, track density, transmittance density, uniform density.** *See also* **facsimile.**

Department of Defense (DoD) master clock: The master clock to which time and frequency measurements for the U.S. Department of Defense are referenced, i.e., are traceable. *Note 1:* The U.S. Naval Observatory (USNO) master clock is designated as the Department of Defense (DoD) Master Clock. *Note 2:* The U.S. Naval Observatory (USNO) master clock is one of the two standard time and frequency references for the U.S. government in accordance with federal Standard 1002. The other standard time and frequency reference for the U.S. government is the National Institute for Standards and Technology (NIST) master clock. *Synonym* **DoD master clock.** *See also* **cesium clock, clock, coordinated clock, coordinated time scale, Coordinated Universal Time, International Atomic Time, precise frequency, precise time, primary time standard, reference clock, remote clock, second, standard frequency and time signal service.**

departure: *See* **phase departure.**

departure angle: The angle between (a) the axis of the main lobe of an antenna pattern and (b) the horizontal plane at the transmitting antenna. *Synonym* **takeoff angle.** *See also* **launch angle, radiation pattern.**

dependent: *See* **code-dependent.**

dependent attenuation rate characteristic: *See* **wavelength-dependent attenuation rate characteristic.**

dependent protection: *See* **data-dependent protection.**

dependent surveillance system: *See* **automatic dependent surveillance system.**

dependent surveillance unit: *See* **automatic dependent surveillance unit.**

dependent system: *See* **code-dependent system.**

depolarization: 1. The reduction or the randomization of the polarization of an electromagnetic wave. *Note:* Depolarization may be caused by transmission through a nonhomogeneous medium or a depolarizer. *See also* **polarization, polarization diversity. 2.** Prevention of polarization in an electric cell or battery.

deposited silica process: *See* **doped deposited silica process.**

deposition: *See* **inside vapor deposition, outside vapor deposition.**

deposition process: *See* **chemical vapor deposition process, modified chemical vapor deposition process, plasma activated chemical vapor deposition process, vapor phase axial deposition process.**

depressed cladding: Cladding in which the region adjacent to the core has a refractive index less than that of regions not adjacent to the core.

depressed cladding optical fiber: An optical fiber that has two claddings, the outer cladding having a refractive index intermediate between that of the core and that of the inner cladding. *Note 1:* The depressed cladding is between the outer cladding and the core. *Note 2:* The outer cladding is used to (a) protect the inner cladding, (b) reduce microbending, (d) remove cladding modes, (d) reduce dispersion, and (e) obtain increased data signaling rates. *Note 3:* Depressed cladding optical fibers usually are used for single mode transmission. *See* **deeply depressed cladding optical fiber.** *See also* **cladding, doubly clad optical fiber, fiber optics, single mode optical fiber.**

depressed inner cladding optical fiber: *Synonym* **doubly clad optical fiber.**

depression angle: An angle, in a vertical plane, between a horizontal plane and a line from (a) an observation point, i.e., an origin or a vertex, in the horizontal plane to (b) a point below the horizontal plane.

depth: *See* **focus depth, listening depth.**

deregulation: In telecommunications, the reduction of regulation of (a) tariffs, (b) market entry and exit, and (c) facilities of public telecommunications services.

derivative: In the calculus, the result obtained when an algebraic function is differentiated. *See* **differentiation.**

derived channel: *See* **time-derived channel, frequency-derived channel.**

DES: data encryption standard, Data Encryption Standard.

descrambler: The inverse of a scrambler. *Note:* The descrambler output is a signal restored to the state it had when entering the associated scrambler, provided that no errors have occurred. *See also* **randomizer, scrambler.**

description language: *Synonym* **data description language.**

descriptor: 1. *Synonym* **keyword. 2.** *See* **receiver information descriptor, transmitter information descriptor.**

desensitization: *See* **receiver desensitization.**

design: *See* **antenna design, detailed design, functional design, system design.**

design analysis: The evaluation of a design, or alternative designs, to determine the extent of conformance with stated criteria, such as performance requirements, design standards, and system efficiency.

design, analysis, and planning tool: *See* **architectures design, analysis, and planning tool.**

designation: *See* **black designation, data signaling rate range designation, frequency spectrum designation, line load control designation, precedence designation, red designation.**

designation of data signaling rates: *See* **range designation of data signaling rates.**

designator: *See* **address designator, circuit designator, emission designator, precedence designator, special handling designator.**

designing: *See* **bottom-up designing, top-down designing.**

design margin: The additional performance capability above required standard basic system parameters that may be specified by a system designer to compensate for uncertainties. *See also* **fade margin, radio frequency power margin.**

design objective (DO): In communications, computer, data processing, and control systems, a desired performance characteristic for circuits and equipment that is based on engineering judgment but is not considered feasible to establish as a system standard at the time when the standard is written. *Note 1:* Examples of reasons for designating a performance characteristic as a design objective (DO) rather than as a standard are that (a) the characteristic may be an advancement in the state of the art, (b) the requirement may not have been fully confirmed by measurement or experience with operating circuits, and (c) it may not have been demonstrated that the requirement can be met, considering other constraints, such as cost and size. *Note 2:* A design objective (DO) should be considered as guidance in preparation of specifications for development or procurement of new equipment or systems, and should be used, if technically and economically practical, at the time when such specifications are written. *See also* **assumed values, specification, standard, system standard.**

design phase: In a system lifecycle, the period during which the various system conditions and requirements, such as descriptions of configuration management criteria, performance requirements, system architecture, software components, hardware components, and interfaces, are prepared, documented, and verified.

design requirement: A requirement that impacts or places a constraint upon the design of a system hardware or software component, such as might be imposed by a requirement to adhere to a standard. *See also* **functional requirement, performance requirement, physical requirement, standard.**

design review: A review conducted by any means, such as a meeting or through distributed documentation, in which the design of a system is presented to interested persons, such as the production personnel, user, or customer, for comment or approval.

despotically synchronized network: A synchronized network in which a unique master clock has the full power to control all other clocks in the network. *See also* **democratically synchronized network, hierarchically synchronized network, oligarchically synchronized network.**

despun antenna: In a rotating communications satellite, an antenna with a main beam that is continuously adjusted in direction with respect to the satellite so that the antenna illuminates a given area on the surface of the Earth, i.e., the footprint does not move with respect to the Earth's surface, the Earth's rotation and the satellite's rotation or oscillation notwithstanding. *See also* **electrically despun antenna, mechanically despun antenna.** *See also* **satellite.**

destination address: In a communications system, data that (a) are usually in the header of a message, (b) are sent in the forward direction, and (c) consist of signals that indicate the complete address of the destination user, i.e., the addressee. *See also* **address, forward direction, user.**

destination code indicator: In a communications system, data sent in the forward direction indicating whether or not the destination code that refers to the destination country or network is included in the destination address. *See also* **code, forward direction.**

destination dialing: A network-provided service feature in which (a) a facility is provided by an automatically switched communications system that uses automatic trunk routing, and (b) source users, i.e., call originators, are not required to define or have knowledge of the route to the destination user, i.e., the call receiver. *See also* **call originator, call receiver, route.**

destination office: The office, such as a switching center or a central office (C.O.), that routes a call or a message directly to a destination user, i.e., a call receiver. *See also* **call, call receiver, central office, destination user, message, route, switching center, user.**

destination routing: In communications system operations, the routing of messages based on the name of the destination office, the destination user, or the address on the message, i.e., the addressee.

destination user: The intended recipient of information that (a) is from the originating user and (b) is usually transferred during an information transfer transaction. *See also* **communications sink, communications source, data sink, data source, information transfer transaction, sink, source, source user.**

destructive cursor: A cursor that erases the symbols or the characters at its current display position or that erases the symbols or the characters at all the display positions through which it has passed. *See also* **display position.**

destructive storage: Storage in which (a) the contents at a storage location are deleted when the data in that location are read, (b) the data that are read have to be regenerated and restored after each reading, i.e., after each sensing, if it is desired that the data be retained in storage, and (c) regeneration usually is automatically performed after reading and within the read cycle. *Note:* Examples of destructive storage are cathode ray tube (CRT) storage, ferroelectric storage, some types of magnetic core storage, and electrostatic storage. *See also* **nondestructive storage, storage location.**

destuffing: The controlled deletion of stuffing bits from a stuffed digital signal in order to recover the original signal that existed prior to stuffing. *Note:* The deleted data may be transmitted via a separate low traffic capacity time slot to be compared with the stuffing data. *Synonyms* **negative justification, negative pulse stuffing.** *See also* **binary digit, bit stuffing, maximum stuffing rate, nominal bit stuffing rate.**

detail recording: *See* **call detail recording.**

detailed design: The act, or the result, of refining and expanding a preliminary design so as to contain more explicit descriptions of the hardware and software components of a system, such as the processing logic, data structures, data definitions, and hardware parts, to a level of specificity such that the design is sufficiently complete so that fabrication of the system may proceed.

detectable element: A display element that can be detected by a pointer. *See also* **display element, detectable group, pointer.**

detectable group: A display group that can be detected by a pointer. *See also* **display group, detectable element, pointer.**

detecting system: *See* **error detecting system.**

detection: **1.** The recovery of information from an electrical or electromagnetic signal. *Note:* Radio waves usually are detected by heterodyning, i.e., by coherent reception and detection, in which the received signal is mixed, in a nonlinear device, with the output of a local oscillator to produce an intermediate frequency (IF) from which the modulating signal is recovered. The inherent instabilities of available optical sources have prevented the practical use of coherent reception and detection in optical communications receivers. Optical receivers use direct detection, i.e., the received optical signal impinges directly onto a photodetector. Direct detection is less sensitive than coherent detection. Progress is being made in the area of optical coherent detection. **2.** In a modulated signal, the restoration of the modulating signal to its original state. **3.** In radar, the separation of a target echo from the background noise. **4.** In surveillance, the determination and the transmission of a signal by a surveillance system, indicating that an event has occurred. **5.** The discovery by any means of the presence of a person, object, or phenomenon. *See* **automatic error detection, carrier sense multiple access with collision detection, coherent detection, fiber optic illumination detection, homodyne detection, light pen detection, low probability of detection, passive detection, pitch detection, radio detection, wild point detection.** *See also* **demodulation, discriminator, identification friend or foe, modulation, recognition, restitution.**

detection and correction system: *See* **automatic error detection and correction system.**

detection and feedback: *See* **error detection and feedback.**

detection and ranging: *See* **light detection and ranging.**

detection combiner: *See* **postdetection combiner, predetection combiner.**

detection combining: *See* **predetection combining.**

detection discrimination: The ability of a receiver to distinguish, separate, identify, and detect a given signal or signal source from all other incident signals, noise, interference, or environmental effects. *See also* **bearing discrimination, range discrimination.**

detection equipment: *See* **airport surface detection equipment.**

detection range: *See* **maximum detection range.**

detection resolution: In communications systems, the precision or the accuracy with which signal parameters, such as frequency, duration, phase, amplitude, or shape, can be measured or determined. *Note:* Resolution may be expressed in terms of temporal or spatial aspects. *See also* **range resolution.**

detection system: *See* **active detection system, passive detection system.**

detection threshold: *Synonym* **sensitivity.**

detection tracking: *See* **passive detection tracking.**

detectivity: The reciprocal of noise equivalent power (NEP), i.e., $D = 1/(NEP)$, where NEP is the noise equivalent power. *Note:* Detectivity usually is expressed in reciprocal watts, i.e., the noise equivalent power usually is expressed in watts. *See* **specific detectivity.** *See also* **noise equivalent power, signal to noise ratio.**

detector: **1.** A device that is responsive to the presence or the absence of a stimulus. **2.** In an amplitude modulated (AM) radio receiver, a circuit or a device that recovers the signal of interest from the modulated wave. **3.** A transducer. **4.** A signal conversion device that converts power from one form to another, such as from optical power to electrical power, from sound power to electrical power, or from mechanical power to electrical power, while preserving the information in, or extracting the information from, the input signal. *Note:* An example of a detector is a device that accepts an amplitude modulated carrier and emits only the modulation signal. *See* **background-limited infrared detector, coherent detector, external photoeffect detector, internal photoelectric effect detector, fiber optic photodetector, infrared detector, infrared point source flash detector, internal photoeffect detector, optical detector, optoelectronic detector, phase detector, photodetector, photoelectromagnetic photodetector, photoemissive photodetector, photon detector, photovoltaic photodetector, pitch detector, video optical detector.**

detector back bias gain control circuit: A circuit that performs instantaneous response automatic gain control (AGC) by operating on the detector rather than on the intermediate frequency (IF) amplifier. *See also* **automatic gain control, detector.**

detector bandwidth: *See* **postdetector bandwidth.**

detector coupling: *See* **optical fiber-detector coupling.**

detector dark resistance: *See* **infrared detector dark resistance.**

detector-emitter (DETEM or detem): An optoelectronic transducer in which the functions of an optical detector and an optical emitter are combined in a single device or module. *See also* **transducer.**

detector noise-limited operation: In fiber optic communications system operations, the situation in which (a) the amplitude of pulses, rather than their duration, limits the distance between source and detector, and (b) the losses are sufficient to attenuate the amplitude of the pulse to such an extent, in relation to the detector noise level, that a proper decision based on the presence or the absence of a pulse in the signal is not possible. *See also* **dispersion-limited operation.**

detector signal to noise ratio: *See* **photodetector signal to noise ratio.**

detector type: *See* **optical detector type.**

detem: detector/emitter.

DETEM: detector/emitter.

determination: *See* **isochrone determination, orbit determination, radio determination, radio position line determination.**

determination service: *See* **radio determination service.**

determination station: *See* **radio determination station.**

deterministic routing: **1.** Routing within a network in which the routes between pairs of nodes are preprogrammed, i.e., are determined in advance of transmission. **2.** In telephone systems, routing (a) that uses an algorithm in which a subset of the seven-digit U.S. telephone numbering plan represents the switching center, and the remaining digits identify the customer associated with that switching center, and (b) in which the path used to complete a call through a network is decided, i.e., is determined, in advance by routing tables maintained in each switching center database. *Note:* The tables assign the trunks to reach each switching center code, area code, and International Access Prefix (IAP), usually with one or two alternate routes. *See also* **database, path, routing, switching center.**

deterministic transfer mode: An asynchronous transfer mode in which the maximum information transfer capacity of a telecommunications service is provided throughout a call. *See also* **asynchronous transfer mode, transfer mode.**

detonation: *See* **proximity fuze detonation.**

Deutsches Institut für Normung (DIN): The organization that (a) sets standards for Germany, (b) performs functions similar to those of the American National Standards Institute (ANSI), and (c) is the member body for Germany in the International Organization for Standardization (ISO). *Note:* The American National Standards Institute (ANSI) is the member body for the United States in the International Organization for Standardization (ISO). *See also* **American National Standards Institute.**

developmental station: *See* **experimental contract developmental station, experimental developmental station.**

development process: *See* **hardware development process, software development process.**

deviation: *See* **frequency deviation, phase deviation, root mean square deviation, two-sample deviation.**

deviation angle: 1. In radio and radar systems, the angular change in direction of a beam after crossing the interface between two different transmission media. **2.** At a point in a transmission medium, the angle through which a light ray is bent by reflection, refraction, or diffraction. **3.** In optics, the net angular deflection experienced by a light ray after one or more reflections, refractions, diffractions, or combinations of these. *Synonym* **angle of deviation.** *See also* **launch angle, reflection, refraction.**

deviation ratio: In a frequency modulation (FM) system operating under specified conditions, the ratio of (a) the maximum frequency deviation of the carrier to (b) the maximum modulating frequency. *See also* **angle modulation, frequency, frequency deviation, frequency modulation, modulation index, phase modulation.**

device: *See* **active device, active infrared device, active optical device, airport surface surveillance device, antisinging device, Bell integrated optical device, character display device, computer peripheral device, display device, fiber optic branching device, fiber optic display device, fiber optic interface device, incidental radiation device, infrared device, intelligent display device, linear device, logic device, low power communications device, network interface device, nonlinear device, optical branching device, optoelectronic device, passive device, passive infrared device, passive optical device, pattern recognition device, photoconductive device, radio-wire integration device, square law device, telephone traffic metering device, trunk encryption device.**

device control character: A character used for the control of ancillary, auxiliary, and peripheral devices associated with a communications, computer, data processing, or control system. *Note:* Examples of device control characters are (a) a character used to select a device from a given set of devices, such as a printer or tape station, and (b) a character used to switch a device on or off. *See also* **character.**

device coordinate: In display systems, a coordinate in a coordinate system, such as Cartesian or polar coordinate system, that (a) is physically marked on the display surface of a display device and (b) identifies a physical location on the display surface. *See also* **real world coordinate, user coordinate.**

device driver: A combination of software and hardware that allows the selection and the control of peripheral equipment of a communications, computer, data processing, or control system. *Note:* An example of a device driver for a personal computer (PC) is a file that (a) is used by the disk operating system (DOS) to describe a nonstandard device to the PC, (b) usually is used via a special file, such as the CONFIG.SYS file, and (c) is specifically written for most peripheral equipment options. *See also* **hardware, software, peripheral equipment.**

device for the deaf: *See* **teletypewriter/telecommunications device for the deaf.**

device interface: The interface between a device and the system to which it is connected. *See* **interface, virtual device interface.**

devitrification: The changing of glass from the vitreous state to a crystalline state, thus greatly changing most of its optical properties, usually for the worst for most applications, such as reduced light transmission in optical fibers and optical integrated circuits (OICs).

dewpoint: As the temperature of a mixture of air and a given amount of water vapor, i.e., water in gaseous form, usually expressed in partial pressure, the temperature at which (a) the water vapor begins to condense into droplets visible as fog, (b) the relative humidity is 100%, and (c) the air cannot hold moisture in gaseous form, and thus dew forms, so that, in effect, it rains. *Note:* In a warm area, moisture forms on a cold component because the air immediately surrounding the component has reached the dewpoint.

DF: direction finder, direction finding.

DFB laser: distributed feedback laser.

DFSK: double frequency shift keying.

DGPS: differential global positioning system.

diad: *Synonym* **dibit.**

diagnostic: Pertaining to the detection, location, isolation, and possibly correction of errors, faults, failures, malfunctions, and mistakes in the firmware, hardware, and software of a system, equipment, device, or component. *See also* **error, fault, failure, firmware, hardware, malfunction, mistake, software.**

diagnostic imagery system: *See* **medical diagnostic imagery system.**

diagnostic program: 1. A computer program that determines the cause or the nature of conditions or problems within specified elements of a system. **2.** A computer program that recognizes, locates, or explains (a) faults in equipment, networks, or systems, (b) predefined errors in input data, or (c) syntax errors in other computer programs. **3.** A computer program that detects, locates, or identifies faults in equipment, mistakes in input data, or errors in other computer programs.

diagram: *See* **block diagram, Brillouin diagram, circular polar diagram, logic diagram, phase diagram, polar diagram, radio coverage diagram, routing diagram, state diagram, Veitch diagram, Venn diagram, vertical coverage diagram.**

dial: 1. A device that generates signals and is used as a calling device to select and establish connections. **2.** To use a calling device to select and establish connections. **3.** In telephone systems, a calling device that (a) is on a user end instrument or on a switchboard, (b) consists of a rotatable return spring-loaded disk, and (c) when rotated an amount corresponding to a desired digit and released, generates signals that control equipment at an automatic exchange to enable the selection of trunks or loops without the assistance of an attendant. *Note:* The term "dial" is used to designate or refer to all calling devices, such as mechanical keyboards, pushbutton keyboards, touch keyboards, and switch panels, that generate signals used for establishing connections. *See also* **calling, calling signal, signal, signaling.**

dial mode: A mode of operating data circuit-terminating equipment (DCE) so that circuitry associated with call origination is directly connected to a communications channel. *Synonym* **talk mode.** *See also* **data circuit-terminating equipment, data mode.**

dial pulse: A direct current (dc) pulse produced by a telephone instrument that interrupts a steady current at a sequence and a rate determined by the selected digit and the operating characteristics of the instrument. *See also* **dial signaling, dial tone, multifrequency signaling, pulsing, rotary dial.**

dial pulsing: Pulsing (a) in which a direct current (dc) pulse train is produced by interrupting a steady signal according to a fixed or formatted code for each digit and (b) at a standard pulse repetition rate. *See also* **pulse train, rotary dial service.**

dial service: *See* **rotary dial service.**

dial service assistance (DSA): A network-provided service feature, associated with the switching center equipment, in which services, such as directory assistance, call interception, random conferencing, and precedence calling assistance, are performed by an attendant. *See also* **service feature.**

dial service assistance (DSA) switchboard: A dial central office switchboard that handles special assistance calls, such as intercepted calls, calls from miscellaneous lines and trunks, certain toll calls, international calls, and precedence calls.

dial service board (DSB): In a dial telephone system, a switchboard for completing incoming calls received from manual offices. *See also* **dial service assistance, dial service assistance board.**

dial signaling: Signaling in which dual tone multifrequency (DTMF) signals or direct current (dc) pulse trains are transmitted to a switching center. *Note:* Rotary dials and keypad instruments may produce either dual tone multifrequency (DTMF) signals or pulse trains. *See also* **composite signaling, dial pulsing, direct current signaling, dual tone multifrequency signaling, pulse, pulse train, pulsing, rotary dial, signal.**

dial switching equipment: Switching equipment actuated by electrical pulses generated by rotary dial or keypad pulsing equipment. *See also* **keypad, rotary dial.**

dial through: A communications network access arrangement that permits outgoing routine calls to be dialed by the private branch exchange (PBX) user after the PBX has established the initial connection. *See also* **service feature.**

dial tone: A tone employed in a dial telephone system to indicate to the call originator that the equipment is ready to receive dial signaling. *See also* **call control signal.**

dial tone delay: The time between a user going off-hook and the receipt of a dial tone from a central office (C.O.).

dial dictation access: A network-provided service feature that permits dialing a special number to access centralized dictation equipment. *See also* **service feature.**

dialer: *See* **autodialer.**

dialing: In a communications system, using a device that generates signals for selecting and establishing connections. *Note:* A dialing device is any device that performs the dialing function. Dialing devices may consist of (a) rotatable disks that, when rotated an amount corresponding to a desired digit and released, generate signals that control equipment at an automatic exchange, thus allowing any user to enter the number of another user without the assistance of an operator, (b) touch keyboards that generate groups of tones corresponding to selected digits and symbols, (c) mechanical keyboards that generate pulse trains, corresponding to selected digits, and (d) pushbutton dialing pads or keypads. The term "dialing" is used to designate or refer to all calling devices used for inserting data used to establish connections. *See* **abbreviated dialing, automatic dialing, automatic identified outward dialing, compressed dialing, destination dialing, direct dialing service, direct distance dialing, direct outward dialing, network indialing, pushbutton dialing, route dialing, speed dialing.**

dialing directory: A list of telephone numbers stored in such a manner that they are available to an automatic dialer.

dialing key: In most manually operated telephone switchboards, a key that (a) must be actuated before dialing commences in order that the dial tone can be heard, (b) must not be restored until the dial has come to rest after dialing the last digit, (c) if it becomes necessary to clear and redial on the same line, must be restored before redialing, or else the dial tone will not be heard in most systems, and (d) may be part of the switch in a telephone cradle or hook operated by the handset when going off-hook or on-hook.

dialing lead: *See* **automatic identified outward dialing lead.**

dialing pad: *See* **pushbutton dialing pad.**

dialing system: *See* **long-distance direct current dialing system.**

dialup: 1. A network-provided service feature that allows a user to initiate service on a previously arranged trunk or to transfer from an active trunk to a standby trunk without human intervention. **2.** A network-provided service feature that allows a computer terminal to use telephone systems to initiate and effect communications with other computers. *Refer to* **Figs. S-5 (869), S-6 (869), S-7 (870).**

diamagnetic: Pertaining to a material, such as bismuth or silver, that (a) always is repelled by a magnet, (b) has a magnetic permeability slightly less than that of air, and (c) tends to take a position such that the longitudinal axis is at right angles to the lines of magnetic flux when its placed in a magnetic field, whereas paramagnetic material aligns its longitudinal axis parallel to the lines of magnetic flux, similar to a magnetic compass needle. *See also* **ferromagnetic, paramagnetic.**

diameter: *See* **average diameter, beam diameter, borescope working diameter, cladding diameter, core diameter, mode field diameter, optical fiber core diameter, optical fiber diameter.**

diameter tolerance: In a round optical fiber, the maximum allowable deviation from the nominal values of the core, cladding, or reference surface diameters.

diametral cooperation index (DCI): *Synonym* **cooperation index.**

diametral index of cooperation: *Synonym* **cooperation index.**

Diamond switch: *Synonym* **core-rope storage.**

diathermy equipment: *See* **medical diathermy equipment.**

dibit: A group of two bits. *Note 1:* The four possible states for a dibit are 00, 01, 10, and 11. *Note 2:* A dibit may be used in a quaternary numeration system. *Synonym* **diad.** *See also* **binary digit, numeration system.**

dicap storage: *Synonym* **capacitor storage.**

dichotic: Pertaining to the condition in which the sound stimulus presented at one ear differs physically or perceptually from the stimulus presented at the other ear. *Note:* The sound stimuli may (a) differ physically in sound pressure, frequency, phase, time, duration, or bandwidth, or (b) differ perceptually in pitch or loudness. *Note:* The dichotic condition permits perception of the direction to the source of a sound and hence the direction of the sound wave.

dichotomizing search: In a search for a specific item in a given ordered set of items, the division of the set into two parts, one of which is rejected, the process being repeated on the accepted part until the sought item is found or it is determined that the specific item is not in the given set.

dichroic: Pertaining to the quality of dichroism. *See also* **dichroism.**

dichroic filter: An optical filter that (a) transmits light selectively according to wavelength, (b) usually is a high-pass or a low-pass filter, (c) usually is used as a beam splitter, (d) is capable of transmitting all frequencies in an electromagnetic wave above a certain cutoff frequency and reflecting all lower frequencies, being either a high-pass or a low-pass filter, depending on whether the transmitted or the reflected wave is used, and (e) separates radiation into two spectral bands. *See also* **high-pass filter, low-pass filter, mode filter.** *Refer to* **Fig. O-2 (671).**

dichroic mirror: A mirror that selectively reflects light according to the wavelength and not according to the polarization plane. *Refer to* **Fig. O-2 (671).**

dichroism: 1. In anisotropic crystalline propagation media, the absorption of light rays propagating in only one particular plane relative to the crystalline axes of the material media. **2.** The effect that occurs in an isotropic propagation medium in which (a) the selective reflection and transmission of light is a function of wavelength regardless of the direction of the polarization plane, and (b) the color of the medium, as seen by transmitted light, varies with the thickness of the medium. *Synonym* **dichromatism, polychromatism.**

dichromatic radiation: *Synonym* **polychromatic radiation.**

dichromatism: *Synonym* **dichroism.**

dictation access: *See* **dial dictation access.**

dictation machine: A machine that (a) consists of at least a microphone, recorder, playback controller, and headphone and (b) records spoken text that can be played back at a later time. *Note:* On playback, the recording may be keyed into a typewriter, printer, or computer.

dictation system: *See* **centralized dictation system.**

dictionary: *See* **data dictionary.**

DID: direct inward dialing.

dielectric: 1. A substance in which an electric field may be maintained with zero or near zero power dissipation, i.e., the electrical conductivity is zero or near zero. **2.** Pertaining to a substance that has a zero or near zero electrical conductivity and therefore an extremely high, or nearly infinite, electrical resistivity. *Note 1:* Optical elements, optical fibers, and insulators are dielectric. A transient polarization current may occur in a dielectric material only when an electric field is applied or removed. The transient current results from dipole rotation and alignment, and from polarization upon application of the electric field and from depolarization upon removal of the field. Electric charges placed on dielectric materials will remain in place for relatively long periods compared to charges placed on conducting materials. *Note 2:* Polarization and polarization currents are specified in Maxwell's equations by the electric permittivity, ε, of dielectric materials. **3.** Pertaining to material composed of atoms with electrons so tightly bound to their atomic nuclei that electric currents are negligible even when high electric field strengths, i.e., near breakdown voltages, are applied to the material. *Note:* One of the constitutive relations for any material is given by $J = \sigma E$ where J is the electric current density, σ is the electrical conductivity, and E is the electric field strength. In dielectric materials, $\sigma = 0$. Therefore, $J = 0$ for dielectric materials. *See also* **electrical conductivity, electric field, electric permittivity, magnetic permeability, optical element, optical fiber.**

dielectric coaxial cable: *See* **air dielectric coaxial cable, foam dielectric coaxial cable.**

dielectric constant: *Synonym* **electric permittivity.**

dielectric film: *See* **multilayer dielectric film.**

dielectric filter: *Synonym* **interference filter.**

dielectric guide feed antenna: An antenna that has a cone of dielectric foam between the antenna horn feeder and the subreflector. *Note:* The dielectric material confines the rays by total internal reflection resulting in low spillover, and the dielectric cone can support the subreflector and thus eliminate the need for the usual support legs. *See also* **dielectric, horn antenna, spillover, subreflector, total internal reflection.**

dielectric lens: A lens made of dielectric material that refracts radio waves in the same manner that an optical lens refracts light waves.

dielectric optical waveguide: *See* **slab dielectric optical waveguide.**

dielectric strength: 1. The minimum voltage gradient, i.e., field strength, at which the insulating property of a material no longer performs an insulating function. *Note 1:* The dielectric strength of a given material is an intrinsic property of the material and is not dependent on the configuration of the material. *Note 2:* The dielectric strength usually is expressed in volts per meter or kilovolts per centimeter. **2.** For a given dielectric material, a given configuration of the dielectric material and electrodes, and a given application rate of applied voltage, the minimum voltage that causes an arcover or produces a breakdown, rupture, or destruction of the material. *Note 1:* At breakdown, the increased electric field frees bound electrons, causing the material to become a conductor. *Note 2:* The specific voltage at which breakdown occurs in a given case is dependent on (a) the intrinsic dielectric material, (b) the respective geometries of the dielectric material and the electrodes with which the electric field is applied, and (c) the rate of increase of the applied voltage. Thus, such conditions should be specified. *See also* **dielectric.**

dielectric waveguide: An open waveguide that (a) is constructed of dielectric materials and (b) does not contain electrical conducting materials. *Note 1:* Examples of dielectric waveguides are optical fibers, fiber optic bundles, and slab dielectric waveguides. *Note 2:* A metallic waveguide filled with a dielectric material is not a dielectric waveguide. **2.** A waveguide that consists of a dielectric material surrounded by another dielectric material, such as air, glass, or plastic, with a lower refractive index. *See* **circular dielectric waveguide, slab dielectric waveguide.** *See also* **open waveguide, optical fiber.** *Refer to* **Figs. C-4 (133), P-17 (767), S-12 (917).**

difference: The number obtained when one number, the subtrahend, is subtracted from another number, the minuend. *See* **clock difference, clock time difference, phase difference, potential difference.** *See also* **contrast, delta modulation, delta-sigma modulation, refractive index contrast.**

differential encoding: Encoding in which signal significant conditions represent binary data, such as "0" and "1," "off" and "on," and "no" and "yes," are represented as changes to succeeding values rather than with respect to a given reference. *Note:* An example of differential encoding is phase-shift keying (PSK), in which the information is not conveyed by the absolute phase of the signal with respect to a reference but by the difference between phases of successive symbols, thus eliminating the requirement for a phase reference at the receiver. *See also* **relative phase telegraphy.**

differential global positioning system (DGPS): 1. A global positioning system for use as an autocoupled precision approach landing system for aircraft guidance. **2.** An enhanced global positioning system (GPS) that (a) improves the accuracy of the GPS, (b) calculates the combined error in satellite range data, and (c) uses the error data to correct positioning measurements in the same general area. *See also* **global positioning system.**

differentially coherent phase-shift keying: *Synonym* **coherent differential phase-shift keying.**

differentially encoded baseband (DEB): *Synonym* **nonreturn to zero mark.**

differential Manchester encoding: Encoding in which (a) data and clock signals are combined to form a single self-synchronizing data stream, (b) one of the two bits, i.e., 0 or 1, is represented by no transition at the beginning of a pulse period and a transition in either direction at the midpoint of a pulse period, and (c) the other is represented by a transition at the beginning of a pulse period and a transition at the midpoint of the pulse period. *Note:* In differential Manchester encoding, if a 1 is represented by one transition, a 0 is represented by two transitions, or vice versa. *See also* **Manchester encoding.**

differential mode attenuation (DMA): In an electromagnetic wave propagating in a waveguide, such as an optical fiber, the differences in attenuation that occur to each of the modes that compose the wave. *Note:* In a given optical fiber, higher-order modes are attenuated more than lower-order modes. *See also* **coupled modes, fiber optics, mode.**

differential mode delay: In a lightwave propagating in a dielectric waveguide, such as an optical fiber, the variation in propagation delay that occurs because of the different group velocities of the modes in the lightwave. *Synonym* **multimode group delay.** *See also* **fiber optics, mode, multimode distortion.**

differential mode interference: 1. Interference that causes a change in electric potential, i.e., a change in voltage, of one side of a signal transmission path relative to the other side. **2.** Interference that results from an interference current path coinciding with the signal path. *See also* **common mode interference, interference, mode.**

differential modulation: Modulation in which the choice of the significant condition for any signal element is dependent on the choice for the previous signal element. *Note:* An example of differential modulation is delta modulation. *See also* **continuously variable slope delta modulation, delta modulation, modulation, significant instant.**

differential phase-shift keying (DPSK): Phase-shift keying used for digital transmission in which the phase of the carrier is discretely varied in relation to the phase of the immediately preceding signal element, in accordance with the data being transmitted. *Note 1:* Differential phase-shift keying (DPSK) is a version of phase-shift keying (PSK) in which the phase reference for each signaling element is the phase state of the preceding signaling element. *Note 2:* Differential phase-shift keying (DPSK) systems usually are designed so that the carrier can assume only two different phase angles. Each change of phase of the signal element carries one bit of information, i.e., the bit rate equals the modulation rate. However, if the number of recognizable phase angles is increased to four, then two bits of information, i.e., a dibit, can be encoded into each signal element. The two bits can count from 0 to 3. Likewise, eight distinguishable phase angles can encode three bits, i.e., can encode a triad. *See* **filtered symmetric differential phase-shift keying.** *See also* **double frequency shift keying, frequency shift keying, phase, phase deviation, phase modulation, phase shift, phase-shift keying.**

differential pulse code modulation (DPCM): Pulse code modulation in which an analog signal is sampled, and the difference between the actual value of each sample and its predicted value, derived from the previous sample or samples, is quantized and converted, by encoding, to a digital signal. *Note:* There are several variations of differential pulse code modulation. *See* **subband adaptive differential pulse code modulation.** *See also* **adaptive differential pulse code modulation, code, continuously variable slope delta modulation, delta modulation, modulation.**

differential quantum efficiency: 1. In a quantum device, the derivative, or slope, of the characteristic graph or equation that defines the countable elementary events at the output as a function of the countable events at the input where the resulting function is the device transfer function for the countable events. **2.** In an optical source or detector, the slope of the curve relating output quanta to input quanta. *Note:* In a photodetector, the differential quantum efficiency is the slope at a given point in the curve in which the number of electrons generated per unit time for the photocurrent is plotted against the number of incident photons per unit time. *Synonym* **incremental quantum efficiency.** *See also* **optoelectronic, response quantum efficiency.**

differential trellis coded modulation (DTCM): Trellis coded modulation based on the use of phase difference rather than absolute phase position. *See also* **modulation, phase modulation, trellis coded modulation.**

differentiating circuit: *Synonym* **differentiating network.** *Refer to* **Fig. D-2.**

differentiating network: A network, or circuit, that (a) has one input and one output, (b) produces an instantaneous output waveform that is directly proportional to the derivative, with respect to time, of the input waveform, and (c) at any instant, has an output signal value that is directly proportional to the rate at which the input signal value is changing. *Note 1:* If the input to a differentiating network is a series of rectangular pulses, the output will be a series of high amplitude short spikes of alternating polarity. If the input is a series of triangular sawtooth waves of constant slope and rapid return, the output will be a wave consisting of a series of rectangular pulses, each followed by a spike of opposite polarity. *Note 2:* Differentiating networks are used in signal processing, such as for producing short timing pulses from square waves. *Synonym* **differentiating circuit.**

differentiation: In the calculus, a mathematical operation that (a) determines the rate at which a function is changing with respect to a variable in the function, such as a signal value as a function of time, and (b) is the reverse of integration, such that if a function is differentiated, integrating the result produces the original

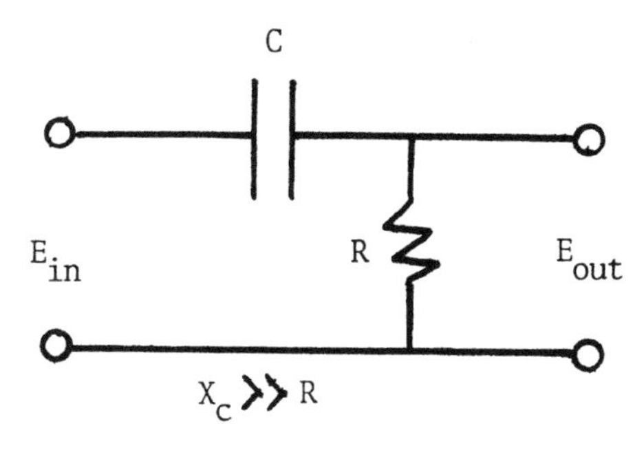

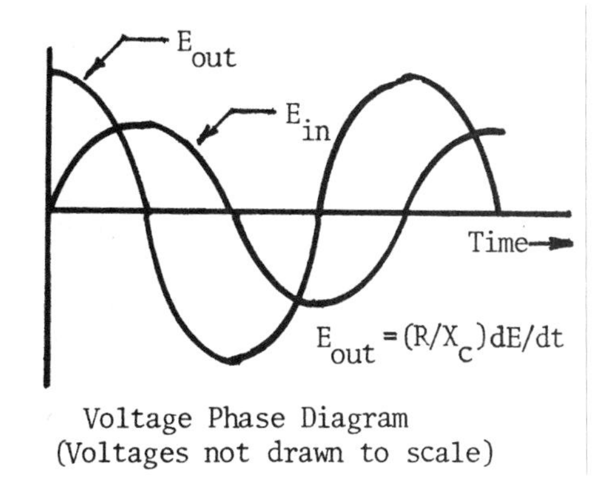

Fig. D-2. In the **differentiating circuit,** the phase angle of the output voltage leads the phase angle of the input voltage by nearly 90° (electrical).

function, except for a constant, whereas the derivative of an integral of a function always yields the original function. *Note:* If a signal voltage waveform is expressed by the relation $v = v_0 + at^2$ where v_0 is the voltage at $t = 0$, a is a constant, and t is time, the time derivative of v is 2 at, i.e., $dv/dt = 2at$. Thus, the slope of the signal waveform at $t = 0$ is 0, and the slope at $t = t_1$ is $2at_1$. *See also* **derivative, function, integration, signal, value, variable.**

differentiator: A device that performs differentiation. *Note:* In analog systems, an example of a differentiator is a device that has an output analog variable, such as the voltage of a signal form, that is the derivative, with respect to a specified variable, such as time, of an input analog variable. *See also* **differentiation, integrator.**

diffraction: The modification of the propagation of waves as they interact with objects in their path, such as the bending of radio, sound, or lightwaves around an object, barrier, or aperture edges. *Note 1:* Waves are modified by their interaction with obstacles or objects. Some of the rays in a beam are deviated from their path while others are not. As the objects become small in comparison to the wavelength, the concepts of reflection and refraction become useless, and diffraction plays a dominant role in determining the redistribution of rays following incidence upon the objects. Thus, even with a very small distant source of light, some light, in the form of bright and dark bands, is found within a geometrical shadow because of the diffraction of the light at the edge of the object forming the shadow. Diffraction gratings, with spacings of the order of the wavelengths of the incident light, cause diffraction that results in the formation of light and dark areas called "diffraction patterns." Such gratings can be ruled grids, spaced spots, or crystal lattice structures. *Note 2:* When a wavefront or a ray is restricted by an opening or an edge, the deviation of the wavefront or the ray from the direction of incidence may be predicted by physical optics or geometric optics, respectively. *Note 3:* Diffraction usually is most noticeable for openings of the order of a wavelength. However, diffraction may still be significant for apertures many orders of magnitude larger than the wavelength. *Note 4:* Knife edges and diffraction gratings cause diffraction of incident waves to occur. *See also* **diffraction limited, far field radiation pattern, fiber optics, knife edge effect, propagation, refraction.**

diffraction grating: An array of fine, parallel, equally spaced reflecting or transmitting lines on a substrate that mutually enhances the effects of diffraction at the edges of each line so as to concentrate the diffracted light very close to a few directions, depending on the spacing of the lines and the wavelength of the incident light. *Note 1:* Diffraction gratings produce diffraction patterns of many orders, each numbered. The spectrum order number is given by the relation $N = (s/\lambda)(\sin \Theta_i + \sin \Theta_d)$ where N is the spectrum order number, s is the center-to-center distance between successive rulings, λ is the wavelength of the incident light, Θ_i is the incidence angle, and Θ_d is the diffraction angle. *Note 2:* Conversely, for each spectrum order number, the wavelength, λ, is given by the relation $\lambda = (s/N)(\sin \Theta_i + \sin \Theta_d)$ where Θ_i is the incidence angle, Θ_d is the diffraction angle, s is the center-to-center distance between successive lines, and N is the spectrum order number. *Note 3:* If there is a large number of narrow, close, equally spaced opaque lines upon a transparent or reflecting substrate, the grating will be capable of dispersing incident light into its wavelength components, i.e., its various colors. Though the action is different, the diffraction grating will produce an effect similar to that of a prism that disperses white light into its constituent colors.

diffraction grating spectral order: The consecutive integers that (a) label the different directions of each member of a family of light rays emerging from a diffraction grating, such as occurs when a beam of parallel rays of monochromatic light passes through the diffraction grating, (b) start with the emergent rays that have remained undeviated, i.e., the zero spectral order, and (c) continue with the rays in the family of deviated rays that (i) emerge after diffraction at the grating, (ii) exhibit pronounced maxima and minima, and (iii) have well-defined and enumerable directions, on each side of the undeviated rays, such that the integers that are assigned to distinguish these directions mark the spectral orders, beginning with the zero spectral order. *See also* **diffraction grating, grating chromatic resolving power.**

diffraction-limited: **1.** In ordinary optics, pertaining to a light beam in which the far field beam divergence is equal to that predicted by diffraction theory. **2.** In focusing optics, pertaining to a light beam in which the impulse response or resolution limit is equal to that predicted by diffraction theory. *See also* **beam divergence, diffraction, refraction.**

diffraction pattern: **1.** When incident light is diffracted and allowed to be incident on a reflective surface, such as a screen, the pattern of light and dark areas caused by the diffraction of the incident light, such as the patterns caused by gratings that may be ruled grids, spaced spots, or crystal lattice structures. **2.** The pattern of light and dark areas that (a) is formed when incident light passes through diffracting media, such as diffraction gratings or knife edges, and (b) may be visualized on a screen placed in, and transverse to, the diffracted beam. *See* **far field diffraction pattern, Fraunhofer diffraction pattern, Fresnel diffraction pattern, near field diffraction pattern.** *See also* **diffraction, diffraction grating, diffraction grating spectral order.**

diffraction region: 1. In radio propagation, a region outside the line-of-sight region in which radio reception is made possible or is enhanced by the diffraction of the radiated waves. **2.** In radio transmission, the region beyond the radio horizon. *See also* **diffraction, knife edge effect, line-of-sight propagation.**

diffusant: A dopant introduced into a material by diffusion. *See also* **diffusion.**

diffuse density: The logarithm to the base 10 of the reciprocal of the diffuse transmittance.

diffused optical waveguide: An optical waveguide that (a) consists of a substrate into which one or more dopants have been diffused to a depth of a few microns, thus producing a lower refractive index material on the outside, as in an optical fiber, (b) has a graded refractive index profile, and (c) can be made from single crystals of zinc selenide or cadmium sulfide and dopants that include cadmium for the zinc selenide and zinc for the cadmium sulfide. *See* **strip loaded diffused optical waveguide.**

diffused waveguide: *See* **planar diffused waveguide.**

diffuse reflectance: 1. The ratio of (a) the total amount of light, i.e., the optical power, reflected diffusely in all directions to (b) the total incident optical power, specular reflection excluded. **2.** The reflectance of a sample relative to a perfectly diffusing and a perfectly reflecting standard, with a 45° incidence angle and observation along the perpendicular to the surface. *Synonym* **total diffuse reflectance.** *See also* **diffuse reflection, specular reflection.**

diffuse reflection: Reflection of an incident, but not necessarily parallel, coherent, or collimated, group of light rays from a rough surface, such that (a) different parts of the same group are reflected in different directions as Snell's laws of reflection and refraction are microscopically obeyed at each undulation of the rough surface, and (b) formation of a clear image of the light source or of the illuminated object is impossible. *See also* **diffuse reflectance, reflection, Snell's law, specular reflection.**

diffuse transmission: The propagation of lightwaves in a propagation medium in which (a) there is a high level of attenuation caused by diffusion, (b) clear images are not transmitted, and (c) signal attenuation is excessive. *Note:* Diffuse transmission can be observed when a narrow light beam is passed through fog, murky water, and particles in suspension in a liquid. *See also* **attenuation, diffusion, propagation medium, signal.**

diffuse transmittance: 1. The transmittance measured with diffusely incident lightwaves. **2.** The ratio of (a) the total amount of light, i.e., the optical power, diffusely transmitted in all directions to (b) the total incident optical power. *See also* **diffuse reflectance, transmittance.**

diffusion: 1. Electromagnetic wave propagation in which rays, bundles, or beams (a) are deviated by obstacles in their path, (b) have deviated components that propagate forward, backward, and sideways, depending on the incidence angles with the scattering surfaces and their transmission and reflection coefficients at the macroscopic and microscopic levels, and (c) are deviated by granular surfaces and materials, such as fog, frosted glass, rough surfaces, the ionosphere, dust on surfaces and suspended in air, smoke, occlusions in a propagation medium, and propagation medium discontinuities. *Note:* Because of diffusion, automobile headlight, flashlight, and spotlight beams can be seen in fog, smog, and smoke without the source of the light being seen. **2.** The controlled migration of dopant atoms and molecules into a material, such as the dopants used to produce (a) semiconductor junctions and (b) refractive index profiles of optical fibers. *See also* **beam, diffuse transmission, dopant, junction, propagation medium, ray.** *Refer to* **Figs. R-8 (834), R-9 (835).**

digit: A symbol, numeral, or graphic character that (a) represents an integer, (b) usually is a member selected from a finite set of whole numbers, i.e., set of integers, (c) might be represented by a signal element, and (d) in storage, might be represented by a specified physical condition, such as the direction of polarization of magnetic material. *Note 1:* Examples of digits include any one of the decimal numerals 0 through 9 and either of the binary numerals 0 or 1. *Note 2:* In a given fixed radix numeration system, the number of allowable different digits, including zero, is always equal to the radix. *Synonym* **numeric character.** *See* **access digit, binary digit, check digit, decimal digit, fractional digit, n-ary digit, significant digit.** *See also* **alphabet, binary digit, character, character set.**

digital: Pertaining to discrete states, particularly the discrete states used to represent data, such as pulse trains. *Note:* Examples of digital data are (a) pulses that represent bits in an electrical or fiber optic transmission line and (b) magnetized polarized spots on a magnetic medium, such as magnetic tapes, cards, disks, and drums. *See also* **analog, pulse train.**

digital access and cross connect system (DACS): In communications, a system in a category of systems in which (a) T-1 hardware architecture is used in private and public networks with centralized switching, and (b) D3/D4 framing is used for switching digital-signal-0 (DS-0) channels to other DS-0 channels. *See also* **digital signal 0.**

digital alphabet: 1. A coded character set in which the characters of an alphabet have a one-to-one relationship with their digitally coded representations. **2.** A table of correspondences between characters, or functions, and the binary digit patterns, i.e., bit patterns, that are used to represent them. *See also* **alphabet, character, character set, code, coded character set, coded set.**

digital block: In communications systems hardware, a set of multiplexed equipment that (a) includes one or more data channels and associated circuits and (b) usually is designated in terms of signaling speed, such as a 134.5-baud digital block. *See also* **channel, data, data channel, signaling speed.**

digital carrier interrupt signaling: Telephone system signaling in which direct current (dc) pulses are used to key a carrier on and off. *See also* **carrier, key, signaling.**

digital cash: Pertaining to systems that transfer credit among accounts via information networks.

digital circuit patch bay: A patch bay in which low level digital data circuits may be patched, monitored, and tested. *Note:* The digital circuit patch bay may be either (a) a D type, i.e., an unbalanced type, or (b) a K type, i.e., a balanced type. *See also* **circuit, facility, level, patch bay.**

digital combining: The interleaving, i.e., interlacing or interspersing, of digital data signals that are either synchronous or asynchronous without converting the data into an analog signal. *Note:* The combining is accomplished by interleaving baseband, independently clocked digital data into one combined data stream containing the information from the individual independently clocked sources. Digital data containing timing corrections for these sources are added to the data stream. *See also* **digital alphabet, diversity combiner, interface, multiplexing.**

digital compression: *See* **digital speech compression.**

digital computer: A computer that (a) consists of associated processing units and peripheral equipments that are controlled by internally stored programs, (b) performs operations on data that are represented by discrete values only, (c) usually uses electrical signals having two permissible states or levels, which represent the two possible characters, i.e., the two possible numerals, in the binary numeration system, and (d) usually is structured with input channels, output channels, arithmetic and logical units, control units, and internal memories. *See* **serial digital computer.** *See also* **analog computer, computer.**

digital converter: *See* **analog to digital converter.**

digital data: 1. Data represented in digital form. *Note:* Examples of digital data are (a) groups of pulses that represent digits, i.e., groups of bits, in an electrical or fiber optic transmission line and (b) magnetized polarized spots that represent bits on a magnetic medium, such as magnetic tapes, cards, disks, and drums. **2.** Data represented by discrete values or conditions, as opposed to analog data. **3.** A discrete representation of a quantized value of a variable, such as the representation of a number by digits that are represented by bit patterns, perhaps with special characters and the space character. *See also* **analog data, data, data transmission, digital.**

digital data channel: A one-way path for data signals that includes a digital channel and associated interface equipment at each end. *See also* **channel, digital data, signals.**

digital data switching: Switching in which connections are established by operations directly on digital signals without converting them to analog signals.

digital encoder: *Synonym* **analog to digital converter.**

digital equipment: *See* **facsimile digital equipment.**

digital error: An inconsistency between the existing representation of a digit and the representation that it should be. *Note:* Examples of digital errors are (a) a single digit or multiple digit inconsistency between the signal actually received and the signal that should have been received, (b) an inconsistency between a digit stored at a given location and the digit read from that location, (c) a dropout of one or more bits, and (d) a lost bit. *See also* **bit count integrity, character count integrity, dropout, error, error control, lost bit.**

digital facsimile: Facsimile in which the scanned picture elements are encoded into a digital form for transmission by a facsimile transmitter and decoded upon reception by the facsimile receiver. *See also* **facsimile, picture element.**

digital facsimile equipment: *Synonym* **facsimile digital equipment.**

digital filter: 1. A filter that (a) substitutes a programmed digital process for an analog filter network or device, such as a filter in an optical system, (b) has the advantages of accuracy, stability, reliability, and flexibility, although not simplicity, over normal analog filters, (c) in its simplest form, consists of sampling circuits, analog to digital converters, digital pattern filters or comparators, digital to analog converters, and first-order hold circuits, (d) has an accuracy that depends on filter word length and not on electric circuit RLC components which are subject to drifting and temperature changes, and (e) can be changed simply by changing parameters in the program, such as the filter word pattern. **2.** A filter that (a) usually is linear, (b) operates in discrete time, (c) usually is implemented by means of digital electronic computation, (d) differs from continuous-time filters primarily in application, (e) usually is more stable than the parameters of most analog filters, i.e., of continuous filters, and (f) can be used as an optimal estimator. *Note:* Examples of digital filters are finite impulse response (FIR) filters and infinite impulse response (IIR) filters. *See also* **discrete time filtering system, Kalman filter.**

digital frequency modulation (DFM): The transmission of digital data by frequency modulation of a carrier, as in binary frequency shift keying (FSK). *See also* **angle modulation, carrier, differential phase-shift keying, frequency modulation, frequency shift keying, modulation.**

digital group: *Synonym* **digroup.**

digital grouping: A basic digital multiplexing grouping. *Note:* In the United States the digital grouping is based on a data signaling rate (DSR) of 1.544 Mbps (megabits per second). In Europe the digital grouping is usually based on 2.048 Mbps. *Synonym* **digrouping.** *See also* **binary digit, digital multiplex hierarchy, digital transmission group, group, T-carrier.**

digital information link: *See* **tactical digital information link.**

digital interface: *See* **high level digital interface.**

digital interpolation: *Synonym* **digital speech interpolation.**

digital loop carrier (DLC): 1. Equipment and facilities, including the lines, that (a) are used for multiplexing telephone circuits and (b) are provided by the network as part of the subscriber access. **2.** Equipment and facilities that (a) multiplex 24 telephone circuits onto a single pair of wires, (b) sample and digitize analog input signals, and (c) time division multiplex the resulting digital signals from the 24 user circuits into a single bit stream.

digital mark-space signaling: Telephone system binary signaling in which signaling pulses or bits are time division multiplexed into the output data stream of a digital vocoder. *See also* **signaling, vocoder.**

digital milliwatt: 1. In digital telephony, a test signal that (a) consists of eight 8-bit words that correspond to one cycle of a sinusoidal signal approximately 1 kHz (kilohertz) in frequency and 1 mW (milliwatt), rms, in power, (b) is stored in digital form in read-only memory (ROM), (c) enables a continuous signal of arbitrary length, i.e., an indefinite number of cycles, to be realized by continually reading and concatenating the stored information to create a digital data stream that is to be converted into analog form, (d) is used in lieu of separate test equipment, and (e) has the advantage of being tied in frequency and amplitude to the relatively stable digital clock signal and power supply, respectively, that are used by the digital channel bank. **2.** A digital signal that is the coded representation of a 0-dBm, 1000-Hz (hertz) sine wave. *Synonym* **digital milliwatt signal.**

digital milliwatt signal: *Synonym* **digital milliwatt.**

digital modulation: The varying of one or more parameters of a carrier wave as a function of two or more finite and discrete states of a signal. *Note:* Examples of digital modulation are (a) coding of data by delaying the data two bit intervals and then adding and (b) ternary signal coding modulation in which (i) two independent bipolar coding sequences are interleaved, (ii) the digital modulation power spectrum nulls at a direct current (dc) level and again at one-half the bit rate, (iii) less bandwidth is required than nonreturn to zero (NRZ) signals, (iv) better low frequency response is provided than return to zero (RZ) signals, and (v) there is a low tolerance of poor signal to noise ratios. *See also* **carrier, digital data, frequency shift keying, modulation, phase-shift keying.**

digital multiplexer: A multiplexer that multiplexes digital signals usually by interleaving bits, in rotation, from several digital bit streams, either with or without the addition of extra framing, control, or error detection bits. *See also* **binary digit, bit stuffing, channel packing, frame, interleaving, multiplexing.**

digital multiplex hierarchy: An ordered hierarchy in which digital signals are combined by the repeated application of digital multiplexing. *Note 1:* Digital multiplexing hierarchies may be implemented in many different configurations, depending upon the number of channels desired, the signaling system to be used, and the bit rate allowed by the communications propagation media. *Note 2:* Some currently available multiplexers have been designated as D1, DS, or M series, all of which operate at T-carrier rates. *Note 3:* Care must be exercised in selecting equipment for a specific system to ensure interoperability because there are incompatibilities among equipment produced by different manufacturers in accordance with different national standards. *See also* **digital transmission group, digital transmission system, digroup, multiplex hierarchy, multiplexing.**

digital network: A communications network in which the information contained in messages is represented by digital data. *See* **Automatic Digital Network, Broadband Integrated Services Digital Network, integrated digital network, integrated services digital network.** *See also* **digital data.**

digital phase-locked loop: A phase-locked loop in which the reference signal, the controlled signal, or the controlling signal, or any combination of these, is in digital form, i.e., is represented by digital data. *See also* **digital data, loop, phase, phase-locked loop.**

digital phase modulation (DPM): Modulation in which the instantaneous phase of the modulated wave is shifted between a set of predetermined discrete values in accordance with the significant conditions of the modulating signal. *See also* **angle modulation, differential phase-shift keying, frequency shift keying, modulation, phase, phase-shift keying.**

digital primary patch bay (DPPB): A patch bay that provides (a) the first access to most local user digital circuits in a technical control facility and (b) patching, monitoring, and testing capabilities for both high level and low level digital circuits. *See also* **circuit, D patch bay, facility, K patch bay, patch bay.**

digital selective calling (DSC): A synchronous calling system used to automatically establish contact with a station or a group of stations by means of radio. *Note 1:* Digital selective calling (DSC) was developed by the International Radio Consultative Committee (CCIR). *Note 2:* The operational and technical characteristics of DSC are contained in International Radio Consultative Committee (CCIR) Recommendation 493.

digital signal (DS): 1. A signal characterized by discrete states. *Note 1:* Digital signals usually are binary signals. *Note 2:* Analog signals may be converted to digital signals by sampling and then quantizing. **2.** A nominally discontinuous electrical signal that changes from one significant condition to another in one or more discrete steps. *Note:* Examples of digital signals are (a) electromagnetic wave pulses, phase shifts, or frequency shifts that represent zeros or ones and (b) electrical output pulses from a photodetector caused by light input pulses. **3.** A signal that is timewise discontinuous, i.e., is timewise discrete, and can assume a limited set of values, i.e., of significant conditions. *Note:* Digital signals may be *n*-ary signals, such as binary, ternary, and quaternary signals. *See* **n-ary digital signal.** *See also* **analog signal, binary digit, character, signal, significant condition.**

digital signal 0 (DS0): A signal with a basic digital signaling rate of 64 kbps (kilobits per second), corresponding to the capacity of one voice frequency equivalent channel. *Note:* Digital signal 0 (DS0) (a) forms the basis for the North American digital multiplex transmission hierarchy and (b) may consist of (i) twenty 2.4-kbps (kilobits/second) channels, (ii) ten 4.8-kbps channels, (iii) five 9.67-kbps channels, (iv) one 56-kbps channel, or (v) one 64-kbps clear channel.

digital signal 1 (DS1): A signal with a digital signaling rate of 1.544 Mbps (megabits per second), corresponding to the North American and Japanese T1 designator. *See also* **T-carrier.**

digital signal 1C (DS1C): A signal with a digital signaling rate of 3.152 Mbps (megabits/second), corresponding to the North American T1C designator. *See also* **T-carrier.**

digital signal 2 (DS2): A signal with a digital signaling rate of 6.312 Mbps (megabits per second), corresponding to the North American and Japanese T2 designator. *See also* **T-carrier.**

digital signal 3 (DS3): 1. A signal with a digital signaling rate of 44.736 Mbps (megabits per second), corresponding to the North American T3 designator. **2.** A signal with a digital signaling rate of 32.064 Mbps, corresponding to the Japanese T3 designator. *See also* **T-carrier.**

digital signal 4 (DS4): 1. A signal with a digital signaling rate of 274.176 Mbps (megabits per second), corresponding to the European T4 designator. **2.** A signal with a digital signaling rate of 97.728 Mbps, corresponding to the Japanese T4 designator. *See also* **T-carrier.**

digital signaling: The transmitting of signaling information that (a) is sent over a digital transmission system and (b) usually is transmitted as characters coded as bit strings. *See also* **bit string, digital, signaling.**

digital slip: 1. When the phases of two bit streams that are essentially synchronous are compared with each other, the gaining or losing of all or part of one digit interval, i.e., a pulse period, in one bit stream relative to the other bit stream. **2.** When the phases of two bit streams that are asynchronous are compared, the rate at which the phase of one bit stream gains or loses digit intervals, i.e., pulse periods, relative to the other bit stream. *See also* **bit stream.**

digital speech interpolation (DSI): In pulse code modulation signals, digital speech transmission in which (a) periods of inactivity or constant signal level are used to interleave, i.e., to intersperse, a different signal and thereby increase the transmission efficiency, and (b) the constant periods of the pulse code modulation signals are compressed, transmitted, and reconstituted using complementary algorithms on both ends. *Synonyms* **digital compression, digital interpolation.** *See also* **algorithm, analog speech compression, time assignment speech interpolation.**

digital subscriber line (DSL): In Integrated Services Digital Networks (ISDNs), equipment that provides full-duplex service on a single metallic twisted pair at a rate sufficient to support ISDN basic access and additional framing, timing recovery, and operational functions. *Note:* The physical termination of the digital subscriber line (DSL) at the network end is the line termination. The physical termination at the customer end is the network termination. *See also* **Integrated Services Digital Network.**

digital switch: Equipment designed, designated, or used to perform time division multiplexed switching of digital signals. *Note 1:* When it is used with analog inputs, an analog-digital/digital-analog (A-D/D-A) conversion is required. *Note 2:* Usual implementation of a digital switch is accomplished by interchanging time slots between input and output ports on a regular sequential basis under the direction of control systems. The control systems may be automatic, semiautomatic, or manual. *See also* **circuit, digital to analog converter, digital transmission group, digital transmission system, digital voice transmission, digitize, digitizer, digit time slot, switching center.**

digital switching: Switching in which digital signals are switched without converting them to or from analog signals. *See also* **digital switch, time division switching.**

digital switching system: *See* **tactical automatic digital switching system.**

digital synchronization: Synchronization (a) in which the start of a transmitted digital entity, such as a bit, character, word, or frame, is defined, and (b) that may consist of several hierarchical stages, such as bit synchronization, symbol synchronization, and frame synchronization. *Note:* Digital synchronization is often confused with bit synchronous operation. *See also* **bit synchronous operation.**

digital to analog (D-A) converter: 1. A device that converts a digital input signal to an analog output signal that represents the same information as the input signal. **2.** A functional unit that converts data from a digital representation to an analog representation. *See also* **analog to digital converter, digital voice transmission, digitize.**

digital transmission group: A group of voice and data channels with bit streams that are combined into a single digital bit stream for transmission over communications propagation media. *Note:* In a digital transmission group there usually is no specific number of channels. However, a digital transmission group usually contains no more channels than can be accommodated by a multiplexer with an output rate that is not greater than that equivalent to about 24 voice channels. *See also* **digital multiplex hierarchy, digroup, group, multiplexing, transmission.**

digital transmission system: A transmission system in which (a) all circuits carry digital signals, and (b) the signals are combined into one or more serial bit streams that include all framing and supervisory signals. *Note:* Analog-digital/digital-analog (A-D/D-A) conversion, if required, must be accomplished external to the system. *Synonym* **digital transport system.** *See also* **binary digit, circuit, digital multiplex hierarchy, frame, multiplexing, T-carrier.**

digital transport system: *Synonym* **digital transmission system.**

digital video disk (DVD): A disk that (a) has about 15 times the storage capacity of conventional compact disks (CDs) and compact disk read-only memories (CD-ROMs) and (b) stores 135 minutes of broadcast-quality video.

digital voice transmission: The transmission of analog voice signals that have been converted into digital signals. *Note:* An example of digital voice transmission is transmission of pulse code modulated (PCM) analog voice signals. *Refer to* **Fig. S-7 (870).**

digital voltage level: The peak-to-peak amplitude excursion of a digital signal. *Note:* Digital voltage levels usually are expressed in volts or millivolts.

digitize: To convert an analog signal to a digital signal that represents the same information as the analog signal. *See also* **analog to digital converter, digital to analog converter.**

digitized battlefield: The use of digital communications systems and components in tactical environments. *See also* **survivable adaptive system.**

digitized speech: *Synonym* **digitized voice.**

digitized voice: Digital signals that have been derived from, i.e., converted from, electrical or optical analog voice signals, particularly for transmission and storage, through the use of a digitizer. *Note:* Pulse code modulation may be used to digitize analog speech signals. *Synonym* **digitized speech.** *See also* **analog, digital, digitizer, pulse code modulation.** *Refer to* **Fig. S-7 (870).**

digitizer: 1. A device that converts an analog signal into a digital representation of the analog signal. *Note:* A digitizer usually samples the analog signal at a constant sampling rate and encodes each sample into a numeric representation proportional to the amplitude value of the sample. **2.** A device that converts the position of a point on a surface into digital coordinate data. *See also* **analog to digital converter.**

digit signaling: *See* **speech digit signaling.**

digit time slot: In a digital data stream, a time interval that (a) is allocated to or occupied by a single digit and (b) can be uniquely recognized and defined. *See also* **signaling time slot, time division multiplexing, time slot.**

digroup: In telephony, a basic digital multiplexing group. *Note 1:* In the North American and Japanese T-carrier digital hierarchies, each digroup supports 12 pulse code modulated (PCM) channels or their equivalent in other services. The DS1 line rate for two digroups plus overhead bits is 1.544 Mbps (megabits per second), supporting 24 voice channels or their equivalent in other services. *Note 2:* In the European hierarchy, each digroup supports 15 pulse code modulated (PCM) channels or their equivalent in other services. The DS1 line rate for two digroups plus overhead bits is 2.048 Mbps, supporting 30 voice channels or their equivalent in other services. *Synonym* **digital group.**

digrouping: *Synonym* **digital grouping.**

dimensional stability: The capability of an entity, such as an optical fiber, bundle, or cable, to retain its original dimensions when stimulated by environmental factors, such as heat, humidity, electric fields, magnetic fields, and water immersion. *See* **optical fiber dimensional stability.**

dimension of recorded spot: *See* **x dimension of recorded spot, y dimension of recorded spot.**

dimpled connector: A fiber optic cable connector (a) in which the optical fiber end surfaces lie in a concave surface at the cable ends, (b) in which divergent rays from a driving light source can enter each optical fiber within its individual, and perhaps small, acceptance angle, and (c) that has an increased coupling efficiency over other connectors. *See also* **connector, coupling, coupling efficiency, fiber optic cable, optical fiber.**

DIN: Deutsches Institut für Normung.

DINA jammer: direct noise amplification jammer.

diode: *See* **double heterojunction diode, edge-emitting light-emitting diode, five-layer four-heterojunction diode, four-heterojunction diode, large-optical-cavity diode, laser diode, injection laser diode, light-emitting diode, monorail double heterojunction diode, photodiode, positive-intrinsic-negative diode, positive-intrinsic-negative photodiode, restricted edge-emitting diode, stripe laser diode, superluminescent light-emitting diode.**

diode coupler: *See* **avalanche photodiode coupler, laser diode coupler, light-emitting diode coupler, positive-intrinsic-negative photodiode coupler.**

diode laser: *Synonym* **laser diode.**

diode photodetector: *Synonym* **photodiode.**

diopt: diopter.

diopter: 1. A unit of measure of curved surfaces and refractive powers, such as those of lenses, equal to the reciprocal of the radius of curvature when expressed in meters. *Note:* The diopter is expressed in reciprocal meters. Thus, a radius of curvature of 0.5 m (meter) is equivalent to 2 diopters. **2.** A unit of refractive power of optical elements, such as lenses and prisms, equal to the reciprocal of the focal length in meters. *Note:* "Diopter" is abbreviated as "diopt." *See also* **focal length, optical element.**

dioptric power: The refractive power of an optical element expressed in diopters. *See also* **diopter, optical element.**

dip: *See* **at the dip, refractive index dip.**

DIP: dual inline package.

diphase modulation: *See* **conditioned diphase modulation.**

diplexer: 1. A three-port frequency-dependent device that may be used as a separator or a combiner of signals. *Note:* Duplex transmission through a diplexer is not possible. **2.** A multicoupler that (a) permits the connection and the simultaneous use of several devices, such as several transmitters or several receivers, to a common or single device, such as an antenna or a channel, (b) does not permit simultaneous transmission and reception, (c) permits the simultaneous transmission of two or more signals using the same circuit, such as an antenna feed and a transmitting antenna, without interference, and (d) permits the simultaneous reception of two or more signals using the same circuit, such as an antenna lead, without interference. *See also* **diplex operation, duplexer, duplex operation, multicoupler.**

diplexing: *See* **radar diplexing.**

diplex operation: Simultaneous one-way transmission or reception of two independent signals using a common element, such as a single antenna or channel. *Note:* An example of diplex operation is the use of one antenna for two or more radio transmitters on different frequencies. *See also* **duplexer, duplex operation.**

diplex telegraphy: *See* **four-frequency diplex telegraphy.**

dipole antenna: An antenna that usually is a straight, center-fed, one-half-wavelength antenna. *See* **half-wave dipole antenna, multielement dipole antenna.** *See also* **antenna, lossless half-wave antenna.** *Refer to* **Fig. R-2 (799).**

dipole antenna equation: *See* **Hertzian dipole antenna equation.**

dipulse coding: The coding of 1s and 0s in a message in which (a) one full cycle of a square wave, i.e., a positive pulse followed by a negative pulse in the same bit period is transmitted when the message bit is a 1 and (b) nothing is transmitted when the bit is a 0, or vice versa. *Note:* A dipulse signal can be generated by encoding the data into 50% return to zero (RZ) unipolar data and sending the bits through an AND gate with the system clock pulse. This RZ bit stream is then delayed one half-bit period and then added to the undelayed RZ stream. This produces the final dipulse waveform. The dipulse power spectrum is similar to that of the biphase coding power spectrum except that dipulse coding produces a pulse repetition rate equal to the bit rate. *See also* **bit stream, pulse repetition rate.**

direct access: 1. To access data in a storage device or to enter data into a storage device in a sequence independent of the relative locations of the data by means of addresses that indicate the physical location of the data to be retrieved or entered. **2.** Pertaining to the organization and access method that must be used for a storage structure in which locations of records are determined by their keys, without reference to an index or to other records that may have been previously accessed. *See also* **browsing, random access, serial access.**

direct access storage: Data or program storage in which the access time to the data files or computer program instructions is independent of the physical location of the data or the instruction in the storage unit, i.e., access to each storage location is direct with no requirement to pass over or pass by other locations to reach a desired location. *Note:* In most personal computer (PC) systems, the data and programs in the computer memory are stored in direct access storage, i.e., the random access memory (RAM). Diskette storage is not direct access storage because time is required for the disk to rotate to the desired location, thus passing over undesired locations. However, the content of the direct access storage is lost when power is removed, whereas the content of diskettes is not lost when power is removed. *See also* **random access memory, read only memory, volatile storage.**

direct address: In computer and data processing systems, an address that designates the storage location of an item of data that is to be treated as an operand and not as another address.

direct bond: In electrical circuits, an electrical connection in which there is continuous metal-to-metal contact between the members being joined. *See also* **bond, bonding.**

direct broadcast satellite (DBS): 1. A satellite that (a) receives signals that are to be broadcast, such as radio and television signals, and (b) broadcasts the signals directly to Earth-based antennas that are within its footprint. **2.** Pertaining to the broadcasting of radio and television programs from satellites directly to terrestrial receiving antennas. *Note:* The programs may be (a) rebroadcast on Earth to conventional receiving antennas at homesites, (b) broadcast directly from satellites to homesite receiving antennas, and (c) distributed locally via coaxial cable television (CATV). *See also* **footprint.**

direct broadcast television satellite (STVS): A satellite that broadcasts television programs directly to customer-owned receiving antennas rather than via an intermediate system, such as a community or cable television (CATV) system.

direct buried cable: A communications cable that is buried in direct contact with the earth. *See also* **cable, underground cable.** *Refer to* **Fig. A-4 (20).**

direct call: A call that (a) is handled by a facility that avoids the use of addresses or address selection signals by having the network interpret the call request signal as an instruction to immediately establish a connection based on previously designated user information, (b) reduces call setup time, (c) has no special real or implied priority over other users of the network in establishing connections, and (d) usually can be placed only during an agreed period. *Synonym* **call without selection signal.** *See also* **call.**

direct clock control: **1.** In digital data transmission, the use of a clock rate that is twice the modulation rate. **2.** In digital data transmission, clock control in which a clock at twice the modulation rate is used, rather than a higher modulation rate, such as 4, 8, or 128 times the modulation rate, as is done in indirect clock control. *Synonym* **direct control.** *See also* **clock, indirect clock control, modulation, modulation rate.**

direct coupling: *Synonym* **conducted coupling.**

direct current (DC): An electric current of constant direction, i.e., of constant polarity. *Note:* If direct current is of constant level, it is called stationary direct current. If it changes periodically, it is called pulsating direct current. *See* **level, pulsating direct current, stationary direct current.**

direct current dialing system: *See* **long-distance direct current dialing system.**

direct current (dc) patch bay: A patch bay in which direct current (dc) circuits are grouped. *Synonym* **dc patch bay.** *See also* **patch bay.**

direct current restoration circuit: A circuit that forces a voltage level back to its normal value after it has deviated from the normal level because of an abnormal condition, such as a disturbance, interference, malfunction, or excessive signal overdrive. *Note:* Infrared scanning systems usually use a direct current restoration circuit to reduce blanking by laser pulses.

direct current signaling (DX): In telephone systems, signaling in which (a) the signaling circuit E&M leads use the same pair of conductors as the voice circuit, and (b) no filter is required to separate the signaling currents from the voice currents during simultaneous transmission because of the great disparity in frequency. *Synonym* **DX signaling.** *See also* **dial signaling, E&M leads, composite signaling, signal.**

direct current telegraph system: *See* **neutral direct current telegraph system.**

direct current telegraph transmission: *See* **polar direct current telegraph transmission.**

direct dialing service (DDS): A network-provided service feature that allows a user to place certain kinds of information, such as information concerning credit card calls, collect calls, and special billing calls, into the public telephone network without human intervention. *See also* **service feature.**

direct distance dialing (DDD): In public switched telephone networks, a network-provided service feature in which a call originator may, without human intervention, call any other user outside the local calling area. *Note 1:* Direct distance dialing (DDD) extends beyond the boundaries of the national public telephone network. *Note 2:* Direct distance dialing (DDD) requires more digits in the number dialed than are required for use within the local area. *Synonym* **subscriber trunk dialing.** *See also* **service feature.**

directed beam scan: *Synonym* **directed scan.**

directed broadcast address (DBA): An Internet Protocol address that specifies all hosts on a specified network. *Note:* A single copy of a directed broadcast is routed to the specified network where it is broadcast to all terminals on that network.

directed net: A radio net in which no station other than the net control station may communicate with any other station, except for the transmission of emergency, urgent, or distress messages, without first obtaining permission of the net control station. *Note:* The directed net is established by the net control station. The net control station may restore the net to a free net. *See also* **distress message, emergency message, free net, net control station, urgent message.**

directed scan: In display systems, a scan in which display elements, display groups, and display images are generated, recorded, and displayed in any sequence, as directed by an operator, a computer program, or a program stored in the display device. *Synonyms* **directed beam scan, random scan.** *See also* **display element, display group, display image, scan.**

direct interface capability: The capability of equipment to interface directly with equipment that conforms to other interfacing standards without the need for external modems.

direct inward dialing (DID): **1.** A service feature that allows inward-directed calls to a private branch exchange (PBX) to reach a specific PBX station without attendant assistance. **2.** In a switching system, a network-provided service feature that permits incoming trunk calls to be forwarded directly to a user without attendant intervention. *Synonym* **network inward dialing.** *See also* **direct outward dialing, private branch exchange, service feature.**

direction: *See* **backward direction, forward direction, scanning direction.** *See also* **heading.**

directional antenna: An antenna in which (a) the radiation pattern has at least one lobe, (b) the radiation pattern is not omnidirectional, (c) the radiation pattern is anisotropic, and (d) the bulk of radiated power, a vector quantity, is confined to a given solid angle, i.e., power is radiated in, or received from, some directions more than others. *See also* **antenna, directive gain, directivity pattern, rhombic antenna, unidirectional antenna.**

directional coupler: 1. A transmission coupling device for separately sampling (a) the forward wave, i.e., the incident wave or (b) the backward wave, i.e., the reflected wave, in a transmission line. *Note 1:* The directional coupler loss usually is known. *Note 2:* A unidirectional coupler has terminals or connections for sampling waves in only one direction of transmission. A bidirectional coupler has terminals or connections for sampling waves in both directions. **2.** In fiber optics, a coupler in which optical power applied to certain input ports is transferred to only one or more specified output ports. *Note:* An example of an optical directional coupler is a fiber optic coupler in which a portion of the energy fed into one end of either optical fiber is transferred to only one end of the other fiber. A portion of the energy fed into the other end of the first fiber is transferred to the other end of the second fiber. *See* **optoelectronic directional coupler.** *See also* **fiber optic branching device, fiber optic coupler, TEE coupler.**

directional signaling system: A signaling system in which the signals are transmitted with increased intensity or power density, i.e., radiance, within the range of a small solid angle, such as transmitted by means of a narrow beam searchlight, a spotlight, or a directional antenna. *See also* **irradiance, omnidirectional signaling system.**

directional transmission: The concentration of the electromagnetic wave energy of an antenna or a light source in a given direction instead of its being spread in many or all directions. *Note:* Directional transmission may be accomplished by various devices, such as antenna arrays, lasers, collimators, parabolic reflectors, and mirrors, in order to make maximum use of the radiation by confining it to a specific direction, such as into an optical fiber, toward the sea, to a microwave repeater, or toward a densely populated area.

directionalization: The temporary conversion of a portion or all of a two-way trunk group to one-way trunks that favor traffic flowing away from a congested switching center. *Note:* Adjacent nodes must cooperate to accomplish directionalization. *See also* **adaptive routing, avoidance routing, dynamically adaptive routing, line load control, minimize.**

direction finder (DF): A device that indicates or measures the bearing of a source of electromagnetic radiation from a given reference direction and from a given point. *Note 1:* If the true, magnetic, grid, or relative bearings of an electromagnetic source are measured from two or more points whose locations are known, the position of the source can be determined, provided that the source is not on the same line that the known points are on. This method of direction finding is known as intersection. *Note 2:* If the bearings to two or more sources of known location are measured from a point whose position is unknown, the position of the unknown point can be determined, provided that the sources are not on the same line that the point of measurement is on. This method of direction finding is known as resection. *Note 3:* Direction finding by either intersection or resection is considered as triangulation. *See also* **bearing, direction finding, intersection, relative bearing, resection, triangulation.**

direction finding: 1. Determining the bearing of an object from a given location, such as the bearing of a source of electromagnetic radiation from a given location. **2.** The obtaining of bearings of radio frequency emitters by using a highly directional antenna and a display unit on an intercept receiver or ancillary equipment. *Note:* Two or more bearings can be used for fixing the point from which the bearings are taken or for fixing a source of radiation. Coupled with time and distance, the velocity and the acceleration of sources and points of measure can also be determined. If a station's position is to be determined, the station may request its bearing from two or more other stations of known location and thereby fix its location. Direction finding may make use of any vector quantities that can be received or transmitted, such as electromagnetic waves, electrostatic fields, magnetic fields, sound waves, pressure gradients, water waves, vibrations, and gravitational fields. *See* **radio direction finding, simultaneous direction finding.** *See also* **bearing, direction finder.** *Refer to* **Fig. R-11 (848).**

direction finding base line: A line that (a) joins two direction finding stations and (b) is used for determining bearings, i.e., for direction finding. *Note:* If more than one base line is available, a base line is chosen that provides maximum triangulation precision and reception advantage. Groundwave, very high frequency (VHF), ultrahigh frequency (UHF), and superhigh frequency (SHF) direction finding stations, i.e., receiving stations, need to be as close as possible to the area of the transmitter whose location is to be determined. Skywave and high frequency (HF) stations can be farther away from the transmitter whose location is to be determined. UHF and SHF stations practically need line of sight (LOS) contact with the transmitter whose location is to be determined. *See also* **direction finder, direction finding.**

direction finding equipment: *See* **radio direction finding equipment.**

direction finding message: In distress and rescue communications, a message that (a) is sent immediately after the distress message, (b) permits direction finding stations to fix the position of the requesting station, (c) is repeated as frequently as required, and (d) is sent according to specific procedures established by international standards; e.g., in radiotelegraph and radiotelephone procedure, two dashes of approximately 10 s (seconds) are sent, each followed by the station call sign or other identification. *See also* **direction finding, distress message, distress communications, rescue communications.**

direction finding picket: A ship, aircraft, or vehicle with direction finding (DF) equipment on board that is stationed at a distance from an area, such as a heavily populated area or an area occupied by a military force, to provide a DF capability to elements within the area or to elements outside the area that need the DF capability to reach elements in the area. *Note:* Examples of direction finding (DF) pickets are (a) direction finding (DF) aircraft stations near an air armada in flight and (b) DF ship stations near a fleet or convoy on the high seas. *See also* **aircraft station, direction finding equipment, ship station.**

direction finding recorder: *See* **airborne direction finding recorder.**

direction finding service: A service that (a) provides bearing, direction, course, or position information to a transmitting or a receiving station, (b) operates direction finding stations that may operate singly or in pairs, and (c) provides the information to requesting stations, such as ship stations, aircraft stations, or mobile land stations.

direction finding station: *See* **radio direction finding station.**

direction finding unit: *See* **airborne direction finding unit.**

direction of scanning: *Synonym* **scanning direction.**

directive gain: 1. At an antenna aperture, the ratio of (a) 4π times the radiance, i.e., the power radiated per unit area per unit solid angle, such as watts per square meter per steradian, in a given direction to (b) the power radiated to 4π steradians. *Note 1:* The directive gain usually is expressed in dB. *Note 2:* The directive gain is relative to an isotropic antenna. *Note 3:* The power radiated to 4π steradians is the total power radiated by the antenna because 4π steradians constitute an entire sphere. **2.** In an antenna, for a given direction, the ratio of the radiance, i.e., the radiation intensity, produced in the given direction to the average value of the radiance in all directions. *Note 1:* If the direction is not specified, the direction of maximum radiance is assumed. *Note 2:* The directive gain usually is expressed in dB. *See also* **antenna, antenna gain, beamwidth, effective radiated power, gain, high gain antenna, main beam, mean power, power.**

directivity pattern: *Synonym* **radiation pattern.**

direct line: In telephone system operation, a line that (a) connects two switchboards and (b) does not pass through an intervening exchange, switchboard, or switching center.

directly connected station: In a communications network, a station connected to another station by means of a dedicated circuit or by means of a direct line. *See also* **dedicated circuit, direct line.**

direct mode: In a radionavigation system, the operation or the use of the several transmitters of the radionavigation service as a single system without consideration of the value of the individual transmitters or radio beacons for direction finding. *See also* **direction finding, indirect mode, radionavigation service.**

direct noise amplification (DINA) jammer: A barrage jammer in which noise is directly amplified and transmitted. *See also* **amplification, barrage jammer, jammer, noise.**

director: *See* **telephone exchange director.**

direct orbit: Of a satellite orbiting the Earth, an orbit in which the projection of the satellite on the equatorial plane revolves around the Earth in the same direction as the rotation of the Earth. *Note:* If a satellite orbit inclination angle with the equatorial plane lies between east and north, the orbit is a direct orbit, and if between north and west, the orbit is a retrograde orbit. *See also* **equatorial orbit, geostationary orbit,**

inclined orbit, polar orbit, retrograde orbit, satellite, synchronous orbit.

director of communications-electronics: *See* **theater director of communications-electronics.**

directory: 1. In communications system operations, a list that (a) contains essential information to assist in identifying the users, such as their titles, appointments, and addresses, that have end instruments connected to an exchange, (b) contains the telephone number, or numbers, associated with each user, (c) usually is alphabetical by user identification, such as (i) by person name or title, (ii) by area, such as city or town, or (iii) by organization, such as division or department, (d) usually is distributed or circulated to the users, and (e) is also kept at the exchange. *Note 1:* In a fixed directory, the number assigned to each user remains fixed regardless of the location of the user, thus allowing freedom of movement of users without directory changes and as an aid to call originators. *Note 2:* In an appointment directory, each number designates the appointment, title, function, or assignment of that user. Personnel changes do not affect the updating of such a directory, though reorganizations do affect it. **2.** In personal computer (PC) systems, a listing of the names of the files that are stored on a disk or a diskette. *See* **appointment directory, data directory, deducible directory, dialing directory, field directory, fixed directory, inverted directory, routing directory, telephone directory.**

direct outward dialing (DOD): 1. A network-provided service feature that allows outgoing calls from a private branch exchange (PBX) terminal to be directly dialed without human intervention. **2.** A network-provided service feature that allows outgoing trunk calls to be directly dialed without human intervention. *Synonym* **network outward dialing.** *See also* **direct inward dialing, private branch exchange, service feature.**

direct ray: A ray of electromagnetic radiation that follows the path of least possible propagation time between transmitting and receiving antennas. *Note:* The least time path is not always the shortest distance path. In a graded index optical fiber, a light ray propagating in the lower refractive index regions propagates faster than light rays propagating in higher index regions, but the higher speed ray might travel a longer distance, particularly if it is a skew ray and takes a helical path around the fiber axis. Thus, the skew ray may be a direct ray even if its path is geometrically longer than an axial ray because the propagation time of the skew ray may be less than that of the axial ray. *See also* **antenna, direct ray, ground wave, line of sight propagation, propagation, propagation mode, skip zone, sky wave.**

direct recording: In facsimile systems, recording in which a visible record is produced, without subsequent processing, in response to received signals. *See also* **electrochemical recording, electrolytic recording, electrostatic recording, facsimile.**

direct sequence modulation: In spread spectrum systems, modulation in which a sequence of binary pulses directly modulates a carrier, usually by phase-shift keying. *Synonym* **direct spread modulation.** *See* **frequency hopper direct sequence modulation.**

direct sequence spread spectrum (DSSS): 1. Pertaining to a system (a) that is used for generating spread spectrum transmissions by phase-modulating a sine wave pseudorandomly with a continuous string of pseudonoise code symbols, each of duration much smaller than a bit, and (b) that may be time-gated, where the transmitter is keyed-on periodically or randomly within a specified time interval. **2.** Pertaining to a signal structuring system in which (a) a digital code sequence having a chip rate much higher than the information signal bit rate is used, and (b) each information bit of the digital signal is transmitted as a pseudorandom sequence of chips. *See also* **chip, spread spectrum.**

direct spread modulation (DSM): *Synonym* **direct sequence modulation.**

direct wave: 1. A wave that arrives at a point via a direct path through the propagation medium without reflection, refraction, diffraction, or scattering from objects, including the Earth, ionosphere, or surrounding objects, except that it may be refracted by the atmosphere. **2.** A wave that travels directly between (a) the transmitting antenna, i.e., the source, and (b) the receiving antenna, i.e., the sink, without reflecting from any object. The direct wave usually consists of direct rays. *See also* **antenna, direct ray, indirect wave, propagation medium, surface wave.**

direct working: In radio communications net operation, the direct two-way transmission of messages between pairs of stations or from any one station to a group of stations in the same net, i.e., the local net, or another net. *See also* **net.**

direct working local net: In radio communications net operation, the direct two-way transmission of messages between pairs of stations, or from any one station to a group of stations, in the same net, i.e., in the local net. *See also* **net.**

directory service: In telephone systems operations, a service in which (a) calls may be placed and answered not necessarily by number, and (b) the call originator describes the call receiver to the operator or the attendant, referring to such matters as location, function, title, name, organization, or services needed. *Note:* An example of directory service is a service that responds to calls for police, fire, medical, or rescue service, or other important persons or services, by completing the call without referring the call originator to the information operator, who would only furnish a number. *See also* **call originator, call receiver, information operator, operator.**

DISA: Defense Information Systems Agency.

disabling signal: A signal that prevents the occurrence of a specific event that would or could occur if the signal were not present or did not occur. *See also* **enabling signal.**

disabling tone: A tone, transmitted over a communications path, used to control equipment along, or at the end of, the path. *Note:* A disabling tone usually is used to place an echo suppressor in a nonoperative condition during data transmission over a telephone circuit. *See also* **call, call control signal.**

disassembler: *See* **packet assembler/disassembler.**

disaster communications: *Synonym* **distress communications.**

disaster dump: A dump that is made when a nonrecoverable computer malfunction or a program error occurs. *See also* **dump, recoverable error.**

DISC: disconnect command.

discernible signal: *See* **minimum discernible signal.**

discipline: *See* **circuit discipline.**

discone: The antenna obtained when one of the cones of a biconical antenna is reduced to a plane. *See also* **biconical antenna.**

disconnect: In telephone systems, the disassociation, i.e., the release, of (a) a switched circuit between two stations or (b) a connection between a call originator and a call receiver after a call is finished. *See also* **abandoned call, circuit, completed call, disconnect signal, disengagement attempt, disengagement phase, finished call, information transfer transaction, successful disengagement.**

disconnect command (DISC or disc): 1. In a data communications network, an unnumbered command (a) that is used to terminate the operational mode previously established, (b) that is used to inform the secondary station that the primary station is suspending the current operation, (c) that does not have an associated information field, (d) prior to the institution of which the secondary station confirms its acceptance by the transmission of a response, and (e) after the issuance of which previously transmitted frames that are unacknowledged remain unacknowledged. **2.** In link layer protocols, such as (a) high level data link control (HDLC) protocols, (b) synchronous data link control (SDLC) protocols, and (c) in Advanced Data Communication Control Procedure (ADCCP), an unnumbered command used to terminate the operational mode previously set. *See also* **frame, LAP B, network, Open Systems Interconnection—Reference Model, primary station, secondary station, X series Recommendations.**

disconnecting switch: *See* **fuze disconnecting switch.**

disconnection: *See* **line disconnection.**

disconnect signal: In a switched telephone network, a signal transmitted from one end of a user line or trunk to indicate at the other end that the established connection should be disconnected. *See also* **call, call control signal, disconnect, disengagement attempt, information transfer transaction, signal.**

disconnect switch: In power systems, a switch that (a) is used to close, open, or change connections in a circuit or a system, (b) usually is used for isolating purposes and therefore may not have an interrupting capacity rating because it is not used to interrupt a circuit that is bearing, i.e., is conducting, an electric current, and (c) is intended to be operated only after the circuit has been opened by some other means, such as by reducing the voltage to zero, individually removing loads, or using an interrupting switch. *See also* **circuit breaker, interrupting switch, operational load, station load.**

disconnect system: *See* **loop disconnect system.**

discontinuity: 1. In communications systems, a sudden or step change of the value of a parameter or variable, such as a signal level. **2.** A break in an electrical conductor or an optical fiber that causes a step change in power level of a signal at the break. *See* **optical impedance discontinuity.**

discrete: Pertaining to distinct, separable elements. *Note:* Discrete data are data represented by discrete elements, such as pulses, bits, bytes, characters, and words. Quantities that have distinct integral values are discrete quantities. *See also* **analog, digital, integral, value.**

discrete time filtering system: A filtering system (a) in which input and output signals change only at discrete instants, (b) in which signals do not vary continuously, and (c) that may be modeled by the use of difference equations. *See also* **continuous time system, filtering.**

discrete time system: A system (a) in which input and output signals change only at discrete instants, (b) in which signals do not vary continuously, and (c) that may be modeled by the use of difference equations. *See also* **continuous time system, signals.**

discrimination: The recognition of the distinction between, i.e., the difference between, two separate entities, or the difference itself. *See* **bearing discrimination, detection discrimination, range discrimination.**

discrimination circuit: *See* **video discrimination circuit.**

discriminator: A circuit, usually a part of a frequency modulation (FM) receiver, that extracts the desired signal from an incoming FM wave by changing frequency variations into amplitude variations. *See* **telegraph discriminator.** *See also* **detection, frequency modulation improvement factor, frequency modulation threshold effect.**

disengagement: In a communications system, the termination of an access, usually after the information transfer phase of an information transfer transaction or after a call is finished. *See also* **completed call, finished call, information transfer transaction.**

disengagement attempt: In a communications system, an attempt on the part of one or more users, or the system itself, to interact with the system in order to terminate an access. *See also* **disconnect, disconnect command, disconnect signal, information transfer transaction, disengagement failure, disengagement success.**

disengagement denial: In a communications system, after a disengagement attempt, a continuation of access. *Note 1:* Disengagement denial usually is caused by (a) a malfunction in the communications system, (b) excessive delay by the telecommunications system, and (c) a valid system action. Disengagement failure is a result of system malfunction only. *Note 2:* A specified time, i.e., the maximum disengagement time, must elapse before disengagement denial is considered to have occurred. *See also* **disengagement failure, disengagement time, failure, maximum disengagement time.**

disengagement denial probability: The probability that disengagement denial will occur, based on the disengagement ratio measured using acceptable statistical sampling techniques and criteria. *See also* **disengagement denial ratio, disengagement success ratio.**

disengagement denial ratio (DDR): The ratio of (a) the number of disengagement attempts that result in disengagement denial to (b) the total disengagement attempts counted during a measurement period. *Note 1:* The disengagement denial ratio may be used to establish a disengagement denial probability. *Note 2:* The disengagement denial ratio (DDR) must equal 1 minus the disengagement success ratio (DSR). *See also* **disengagement denial probability, disengagement success ratio.**

disengagement failure: In a communications system, a continuation of access, i.e., a failure of the system to return to the idle state for a given user, or for the system, within a specified maximum disengagement time following a disengagement attempt. *See also* **disengagement denial, disengagement request, failure, maximum disengagement time, successful disengagement.**

disengagement originator: The user or the functional unit that initiates a disengagement attempt. *Note 1:* A disengagement originator may be the originating user, the destination user, or the communications system. *Note 2:* The communications system may deliberately originate the disengagement because of preemption or inadvertently because of system malfunction. *See also* **access originator.**

disengagement phase: In an information transfer transaction, the phase during which successful disengagement occurs. *Note:* The disengagement phase is the third phase of an information transfer transaction. *See also* **access phase, disconnect, information transfer phase, information transfer transaction, disengagement success.**

disengagement request: A control or overhead signal issued by a disengagement originator for the purpose of initiating a disengagement attempt. *See also* **disengagement originator.**

disengagement request signal (DRS): A control or overhead signal issued by a disengagement originator for the purpose of initiating a disengagement attempt. *See also* **disengagement attempt, disengagement failure, disengagement originator, disengagement success.**

disengagement success: The termination of user information transfer between a source user and a destination user in response to a disengagement request. *Synonym* **successful disengagement.** *See also* **disengagement failure, disengagement request.**

disengagement success ratio (DSR): The ratio of (a) the number of disengagement attempts that result in disengagement to (b) the total number of disengagement attempts counted during a measurement period. *Note 2:* The disengagement success ratio may be used to establish a disengagement success probability. *Note 2:* The disengagement success ratio (DSR) must equal 1 minus the disengagement denial ratio (DDR). *See also* **disengagement denial ratio.**

disengagement time: 1. The average time between (a) the instant a disengagement attempt is started on the part of the originating user, the destination user, or the system and (b) the instant the disengagement occurs. **2.** The time between (a) the instant a disengagement attempt is started and (b) the instant disengagement occurs, i.e., disengagement success occurs. *See* **maximum disengagement time.** *See also* **call release time, clearing.**

dish: A bowl-shaped antenna that (a) usually is a paraboloid of revolution used to transmit or receive electromagnetic waves, (b) usually consists of sheet metal or a grid-like wire basket and a source of radiation or a sensor at the focus of the paraboloid, and (c) is used in satellite, microwave, radar, and radio astronomy systems. *Refer to* **Figs. C-7 (157), R-4 (800).**

disjoint signal: A signal that does not overlap, is not coincident with, or is not superimposed upon any other signal in time or in space. *Note:* Frequency division multiplexed signals and time division multiplexed signals are disjoint signals in the time domain. Space division signals and signals in a delay line or en route to a satellite are disjoint in time and space. *See also* **signal, space domain, time domain.**

disjunction gate: *Synonym* **OR gate.**

disk: *See* **data disk, digital video disk, diskette, external hard disk, hard disk, internal hard disk, magnetic disk, optical disk, optical video disk, program disk, random access memory disk, shared disk, type disk.**

disk drive: A device (a) that writes data on or reads data from one or both sides of a rotating disk, usually a disk that is coated with magnetizable or optically sensitive material, and (b) in which one or more disks in cartridges may be inserted or removed, or in which a disk is permanently installed as far as the user is concerned although the disk may be replaced at a specially equipped shop or factory. *Note 1:* In personal computer (PC) system configurations, the number and the type of disk drives are selected and installed by the manufacturer. One or two diskette drives, such as floppy disk drives, plus a hard drive is a usual configuration. These disk drives might be designated the A drive and the B drive for the two diskette drives and the C drive for the hard disk drive, i.e., the hard drive. *Note 2:* Computer programs, data, and system configuration information may be stored on diskettes or the hard drive. *See* **Bernoulli box drive, virtual disk drive.**

diskette: A small magnetic disk enclosed in a rigid or a semirigid jacket. *Note:* A typical high density 3.5-in. formatted rigid diskette has a capacity of approximately 1.44 Mb (megabytes). A typical high density (HD) double-sided 5.25-in. formatted floppy diskette has a capacity of approximately 1.2 Mb (megabytes). *Note:* A floppy diskette may be considered as a diskette, but all diskettes are not necessarily floppy diskettes. *See* **boot diskette, formatted diskette, unformatted diskette.** *See also* **floppy disk, hard disk.**

diskette exchange format: A format for transferring data on diskettes among systems or devices. *Note:* An example of a diskette exchange format is the IBM 128-byte format used on the basic exchange diskette.

disk operating system (DOS): Software that (a) is loaded when a computer, particularly a personal computer (PC) with one or more disk drives, is powered, (b) provides the computer with the data and programs it needs to handle higher level functions, such as disk drive access, (c) usually is a standard part of most computer software systems, (d) enables the operator to direct the computer and thus control the flow of data and instructions among the computer system components, such as between the computer and a disk drive, the computer and a printer, or a disk drive and a printer, and (e) is not a word processor, a spreadsheet program, or an application program, such as WordPerfect, Lotus 1-2-3, Microsoft Money, Microsoft Publisher, and SAMS EXCEL 4 for WINDOWS. *Note:* In IBM, and IBM-compatible computers, the disk operating system (DOS) is labeled PC DOS. There are other versions, such as PC DOS 2, DOS 2.1, DOS 3.0, MS DOS, and CP/M. *See also* **disk drive, operating system.**

disk pack: An assembly of magnetic disks that can be removed as a whole from a disk drive, together with a container from which the assembly must be separated when operating.

disk read-only memory: *See* **compact disk read-only memory.**

disk storage: Storage that consists of a flat, circular plate that (a) has a surface of magnetic or photosensitive material, (b) may be used to store and retrieve data, and (c) is rotated in conjunction with a read, write, or read/write head. *See* **optical disk storage.** *See also* **disk, diskette, disk drive, storage.**

DISN: Defense Information Systems Network.

DISNET: Defense Integrated Secure Network.

disparity: In pulse code modulation (PCM), the digital sum, i.e., the algebraic sum, of a set of signal elements. *Note:* In electronic systems, the disparity will be zero, and there will be no cumulative or drifting electrical polarization if there are as many positive elements, say those that represent 1, as there are negative elements, say those that represent 0. *See also* **paired disparity code, pulse code modulation.**

disparity binary code: *See* **unit disparity binary code.**

disparity code: *See* **paired disparity code.**

dispatching system: *See* **group alerting and dispatching system.**

dispatch system: *See* **group alert dispatch system.**

dispatch time: *See* **message dispatch time.**

dispersal: *See* **frequency dispersal.**

dispersion: 1. The separation of electromagnetic radiation into components that have different characteristics, such as different frequencies, energies, and velocities. **2.** In communications, any phenomenon in which electromagnetic wave propagation parameters are dependent upon wavelength. **3.** In an electromagnetic signal propagating in a material medium, the degrading of the signal because the various wavelength components of the signal have different propagation characteristics within the medium. **4.** In classical optics, the wavelength dependence of the refractive index in matter, i.e., the $dn/d\lambda$ where n is the refractive index and λ is the wavelength, caused by interaction between the matter and light. **5.** In an optical fiber, pulse broadening, i.e., the increasing of pulse duration (temporal) and pulse length (spatial), as a result of traversing a length of the fiber. *Note 1:* In an optical fiber, there are several significant dispersion effects, such as material dispersion, profile dispersion, and waveguide dispersion, that degrade the signal. *Note 2:* If a lightwave pulse of given pulse duration and given spectral line width is launched into an optical fiber, the different wavelengths experience different refractive indices and therefore take different paths as well as propagate at different speeds, causing pulse broadening as the pulse propagates along the fiber because the different wavelengths that make up the spectral line width arrive at the far end at different times (material dispersion). At the same time, energy in the different wavelengths is distributed among different modes, also causing pulse broadening (modal dispersion). Dispersion imposes a limit on the bit-rate•length product (BRLP) for digital data and bandwidth for analog data transmitted in an optical fiber for a given bit error ratio (BER) at the limit of intersymbol interference. *Note 3:* In an optical fiber, dispersion usually is expressed in picoseconds per kilometer of fiber for each nanometer of spectral width of the lightwave launched into the fiber, i.e., $ps{\bullet}nm^{-1}{\bullet}km^{-1}$. *Note 3:* The term "dispersion" as used in fiber optic communications should not be confused with the term "dispersion" as used by optical lens designers or with "dispersion" as applied to alternate routing in communications networks. **6.** The spreading of a signal in time and space as a result of traversing transmission paths of different physical, electrical, or optical lengths. *Note:* Dispersion does not include geometric spreading, i.e., spreading the signal power over wider areas only because of increasing distances from the source. *See* **chromatic dispersion, intermodal dispersion, material dispersion, maximum transceiver dispersion, modal dispersion, optical fiber dispersion, optical multimode dispersion, profile dispersion, waveguide dispersion, zero dispersion.** *See also* **distortion, intramodal distortion, dispersion coefficient, multimode distortion, profile dispersion parameter, route dispersion, zero dispersion.** *Refer to* **Fig. W-3 (1082).**

dispersion attenuation: *See* **optical dispersion attenuation.**

dispersion attenuation factor: A factor that (a) when multiplied by an initial value of optical power, yields the value of optical power at another point in an optical system when the reduction in power is caused only by dispersion and (b) is given by the relation

$$\delta = e^{-df^2}$$

where δ is the dispersion attenuation factor, e is the natural exponential function base, i.e., $e \approx 2.718$, d is a material constant that includes substance and geometry, and f is the frequency component of the signal being attenuated by dispersion.

dispersion characteristic: *See* **optical fiber global dispersion characteristic.**

dispersion coefficient: For a lightwave propagating in an optical fiber, a value that expresses the amount of dispersion as a function of spectral width and distance along the fiber, i.e., the pulse broadening per unit length of fiber and per unit of spectral width of the source of the propagating lightwave. *Note 1:* The dispersion coefficient usually is expressed in ps•km^{-1}•nm^{-1} (picoseconds/(kilometer•nanometer). *Note 2:* The dispersion coefficient is a measure of the change in group index, (ΔN), with respect to wavelength, given by the relations $M(\lambda) = (1/c)dN/d\lambda = -(1/c)(d^2n/d\lambda^2)$ where $M(\lambda)$ is the dispersion coefficient, c is the velocity of light in vacuum, N is the group index expressed as $n-\lambda(dn/d\lambda)$, λ is the wavelength of interest, i.e., the wavelength of the source, and n is the refractive index of the material. *Note 3:* For many optical fiber materials, M equals zero at a specific wavelength, λ_0, between 1.2 μm (microns) and 1.5 μm. Below λ_0, the dispersion coefficient, M, is negative and increases with increasing wavelength. It is positive above λ_0. A function that passes from negative to positive, or vice versa, must pass through 0. *Note 4:* Pulse broadening caused by dispersion in a unit length of optical fiber is given by the product of the dispersion coefficient, M, and the spectral width ($\Delta\lambda$), except at $\lambda = \lambda_0$, where terms proportional to $(\Delta\lambda)^2$ are more important. *Synonyms* **dispersion parameter, material dispersion coefficient, material dispersion parameter.** *See* **chromatic dispersion coefficient, global dispersion coefficient, worst-case dispersion coefficient.** *See also* **dispersion, group index.**

dispersion equation: An equation that indicates the dependence of the refractive index of a propagation medium on the free space wavelength of the light transmitted by the medium. *Note:* The adjustment of the refractive index for wavelength permits more accurate calculation of reflection and refraction angles, phase shifts, and propagation paths that are dependent upon the refractive index and the refractive index gradient. Often it is necessary to obtain a value of the rate of change of refractive index with respect to the wavelength. There are several useful dispersion equations. In all the following equations, n is the refractive index, λ is the wavelength, n_0 is the nominal refractive index measured with the free space wavelength, λ_0 is the free space wavelength, and the other symbols are material-dependent and empirically determined. In one, the refractive index, n, is given by the relation

$$n = n_0 + C/(\lambda - \lambda_0),$$

which is attributed to Hartmann. In another, the refractive index is given by the relation

$$n = A + B/\lambda^2 + C/\lambda^4$$

which is attributed to Cauchy. In yet another, derived by Sellmeier, the refractive index is given by the relation

$$n^2 = 1 + \Sigma_{i=o}^{i=m} A_i\lambda/(\lambda^2 - \lambda_i^2).$$

An extension of the Sellmeier equation is useful for covering more than one absorption region. Helmholtz included an additional term to the Sellmeier equation, given by the relation

$$B_i/(\lambda^2 - \lambda_i^2).$$

The additional term is also useful within absorption regions. Usually, some of the terms in the Sellmeier summations are replaced by a constant. In applications, one of the equations is selected, and then a more accurate fit is derived by an appropriate curve-fitting technique, such as the method of least squares. *See also* **dispersion, refractive index, propagation medium.**

dispersion flattened optical fiber: An optical fiber that has a reduced dispersion coefficient over the wavelength region from 1.29 μm (microns) to 1.57 μm.

dispersion gate: *Synonym* **NAND gate.**

dispersion-limited: In fiber optic transmission systems, pertaining to the limitations placed on the pulse repetition rate (PRR) caused by dispersion and resulting in intersymbol interference at the end of the transmission line. *Note:* In certain step index (SI) optical fibers with a core refractive index of 1.51 and a cladding index of 1.49, the dispersion-limited pulse repetition rate (PRR) is 18 Mbps (megabits per second) for a 1-km length, 9 Mbps for a 2-km length, and so on. *See also* **dispersion, pulse repetition rate.**

dispersion-limited length: The sheathed length of fiber optic cable, such as might be in the cable facility of a station/regenerator section, that is limited by dispersion. *Note 1:* The dispersion arises from the combined effects of chromatic dispersion and modal noise. *Note 2:* The dispersion limited length, l_D, is given by the relation $l_D = D_{TR}/D_{\max}$ where l_D is the dispersion limited length, D_{TR} is the maximum transceiver dispersion in psec•nm^{-1}, and $D_{\max}$ is the absolute value of the worst case chromatic dispersion coefficient in ps•nm^{-1}•km^{-1} over the range of transmitter wavelength and over the specified variation in cable dispersion. *Note 3:* A regenerator section will not be limited by dispersion as long as the relation $D(\lambda_t)\bullet l < D_{TR}$ holds, where $D(\lambda_t)$ is the optical fiber chromatic dispersion coefficient in ps•nm^{-1}•km^{-1} evaluated at λ_t, l is the fiber length in km, and D_{TR} is the maximum transceiver dispersion in ps•nm^{-1}. Generally for single mode systems operating at bit rates, i.e., data signaling rates (DSRs), under 0.5 Gbps (gigabits per second), the length of the fiber optic cable facility of a regenerator section can be expected to be limited by attenuation (loss) rather than by dispersion. At higher bit rates the optical link length can be expected to be limited by dispersion. In the design of an optical station/regenerator section, the effect of dispersion is accounted for in the equation for the terminal/regenerator system gain, *G*, via the dispersion power penalty, P_D, in dB. *See also* **sheathed length, terminal/regenerator system gain.**

dispersion-limited operation: Operation of a communications link under specific conditions and limits in which signal waveform degradation attributable to the dispersive effects of the propagation medium is the dominant mechanism that limits circuit performance. *Note 1:* The amount of allowable degradation is directly dependent on the quality of the receiver. *Note 2:* In fiber optic communications, dispersion-limited operation is often confused with distortion-limited operation. *See also* **attenuation-limited operation, bandwidth-limited operation, circuit, detector-noise-limited operation, distortion-limited operation, fiber optics, multimode distortion, quantum-noise-limited operation.**

dispersion parameter: *Synonym* **dispersion coefficient.**

dispersion power penalty: In an optical transmitter, the worst-case increase in receiver input optical power needed to compensate for the total pulse distortion caused by intersymbol interference and mode partition noise at the specific bit rate, i.e., data signaling rate (DSR), bit error ratio (BER), and maximum transceiver dispersion specified by the manufacturer (a) when operated under standard or extended operating conditions and (b) as measured under standard test procedures.

dispersion-shifted (DS) optical fiber: 1. An optical fiber that has its minimum dispersion wavelength shifted to coincide with its minimum attenuation wavelength. *Note:* The shifting is accomplished by adjusting the refractive index contrast, the profile dispersion parameter, and the profile parameter by controlling the concentration and the distribution of dopants. **2.** A single mode optical fiber that (a) has a nominal zero dispersion wavelength of 1.55 μm (microns) with applicability in the 1.5-μm to 1.6-μm range and (b) has a dispersion coefficient that is a monotonically increasing function of wavelength. *Note:* The refractive index profile is adjusted for minimum dispersion at the operating wavelength. The optical fiber refractive index profile is designed in such a manner that because of the variation in refractive index with various wavelengths in the spectral width of the pulse and because of the various lengths of the paths taken by the various propagation modes, at a certain median wavelength and particular refractive index profile, all modes can be made to have the same end-to-end propagation time, thus reducing signal pulse distortion caused by material dispersion to an absolute minimum, actually nearly zero. A fiber can be designed such that the zero material dispersion wavelength is also at the minimum attenuation wavelength, i.e., is at a spectral window, or trough, in the attenuation rate versus wavelength curve. *Synonym* **EIA Class IVb optical fiber.** *See also* **dispersion-unshifted optical fiber, single mode optical fiber.**

dispersion slope: *See also* **chromatic dispersion slope, minimum dispersion slope, zero dispersion slope.**

dispersion specification: *See* **pulse dispersion specification.**

dispersion-unmodified optical fiber: *Synonym* **dispersion-unshifted optical fiber.**

dispersion-unshifted (DU) optical fiber: A single mode optical fiber that (a) has a nominal zero dispersion wavelength in the 1.3-μm (microns) transmission window and (b) has a dispersion curve that increases monotonically, with a single crossing of the zero dispersion axis in the vicinity of 1.30 μm (microns). *Synonyms* **dispersion-unmodified optical fiber, dispersion-unshifted single mode optical fiber, EIA Class IVa optical fiber, nonshifted optical fiber.** *See also* **dispersion-shifted optical fiber, single mode optical fiber.**

dispersion-unshifted single mode optical fiber: *Synonym* **dispersion-unshifted optical fiber.**

dispersion wavelength: *See* **zero dispersion wavelength, zero material dispersion wavelength.**

dispersion window: *See* **minimum dispersion window.**

dispersive lens: *Synonym* **diverging lens.**

dispersive medium: In the propagation of electromagnetic waves in material media, a medium in which the phase velocity varies with frequency. *Note:* If an optical pulse consisting of more than a single wavelength is passed through a dispersive medium, the different wavelengths will arrive at a given distant point at different times. Thus, the phase velocity and the group velocity are not the same. All materials are dispersive to some extent. *See also* **nondispersive medium.**

displacement: 1. In the spatial domain, the linear movement, usually of the center of gravity, of an object from one position to another. *Note 1:* The SI unit (Système International d'Unités) for linear displacement is the meter. *Note 2:* Examples of displacement are (a) the movement of a satellite from one point on its orbit to another point and (b) the movement of a cursor on the display surface of a display device. *In a spatial sense, synonym* **translation. 2.** The spatial distance from one point to another, usually indicated as a vector that is equal to the difference between two position vectors. **3.** The change in angular orientation of an object with respect to a fixed reference direction. *Note:* The International System (SI) unit for angular displacement is the radian. There are 2π radians in a full circle. Other units for angular displacement are the degree, equal to 1/360 of a full circle, and the mil, equal to an angle whose tangent is 1/1000. **4.** The weight of water displaced by a ship.

displacement attenuator: *See* **longitudinal displacement attenuator.**

displacement sensor: *See* **fiber optic axial displacement sensor.**

display: 1. A visual presentation of data. **2.** To present data visually using a display device, such as a cathode ray tube, a fiberscope face plate, a film projector, or a light-emitting diode panel. *See* **active matrix liquid crystal display, auroral display, flat panel display, radar display, raster display, spot-barrage jamming radar display, spot-continuous-wave jamming radar display.**

display background: In display systems, the fixed portion of a display image that cannot be changed by an operator or a user during a particular application. *Note:* Examples of display backgrounds are form flashes and form overlays. *See also* **display foreground, form flash, form overlay**.

display border: On the display surfaces of display devices, the area that (a) is along the top, sides, or bottom of the display space in the display surface and (b) may contain data, such as headers, footers, column headings, and line or row numbers. *Note:* In Lotus 1-2-3 programs, there are several types of borders, such as (i) column headings A through IV on the upper horizontal border, (ii) numbers 1 through 2048 on a vertical border, and (iii) the repeating borders that are used with the \Print Printer Options Borders command. *See also* **command, display device, display space, display surface.**

display column: In a display system, the column of display positions that constitutes a full-length sequence of vertical positions on the display surface of a display device. *Note 1:* In display systems, "vertical" is defined as being perpendicular to the direction of right to left when a display image is viewed by an observer. *Note 2:* The positions in a display column usually are addressable. *See also* **addressable vertical position, display device, display image, display line, display surface.**

display command: In display systems, a command that can be used to control a display device. *Synonym* **display order.** *See also* **command.**

display console: A control device that (a) has one or more input units, such as an alphanumeric keyboard, function keys, a joy stick, a control ball, a light pen, or a mouse, (b) has a display device with a display surface, such as a cathode ray tube screen, a light-emitting diode panel (LED), a gas panel, a fiberscope faceplate, or a plotter bed or tablet, and (c) usually is equipped for use by an operator. *See also* **display device, display surface.**

display cycle: In a display device, the sequence of events needed for a one-time regeneration of a display element, display group, or display image on the display surface of the device, if and when such regeneration is necessary to preserve and continue the presentation of the element, group, or image. *See also* **display generation.**

display device: An output unit that (a) gives a visual presentation of data and (b) has a display surface within which there is a display space. *Note:* In personal computer (PC) systems, the display device is called a "monitor." *See* **character display device, intelligent display device.** *See also* **display element, display group, display image, display space, display surface, fiber optic display device.**

display element: A basic graphic element that can be placed in the display space of a display device to produce or construct display groups and display images, such as graphs, drawings, pictures, and words, that can be manipulated, such as translated, rotated, enlarged, and shrunk, as a unit. *Note:* Examples of display elements are dots, line segments, arcs, squares, crosses, and corners. *See also* **display group, display image, display device, display surface.**

display field: A specified part of the storage space of a display buffer or a specified area on a display surface that contains or displays a set of characters that can be manipulated or operated upon as a unit. *See also* **display surface, storage space.**

display field attribute: An attribute, i.e., a characteristic, of a display field. *Note:* Examples of display field attributes are (a) protected or unprotected against operator manual input or copy, (b) having alphabetic, numeric, or alphanumeric input control, and (c) displayed or nondisplayed, intensified or nonintensified, pen-detectable or nondetectable, modified or unmodified. *See also* **attribute, display field.**

display file: In display systems, a sequence of instructions that (a) is similar to a computer program and (b) is to be executed by the processor in a display device. *See also* **computer program, display device, display system, processor.**

display foreground: In display systems, the variable portion of a display image that can be changed by an operator or a user during a particular application. *Note:* A computer program or a pointer may be used to change the display foreground. *See also* **computer program, display background, display image, display system, pointer.**

display format: The arrangement of data in the display space of a display device. *Note:* In word processing and spreadsheet software systems, such as WordPerfect, Lotus 1-2-3, and Volkswriter, the format of the displayed data on the display device screen, the format of the corresponding data that are printed, and the format of the corresponding data stored on a diskette may be different, though they all represent the same information. *See also* **diskette, display device, display space, format, monitor, screen.**

display frame: In display devices, a specified portion of the area of the display surface available for viewing. *Note:* Examples of display frames are (a) a specified portion of the area of a microform available for viewing a microimage, (b) the display space used on a computer monitor, and (c) a page of a book. *See also* **microform, microimage.**

display framing: On the display surface of a computer monitor, controlling display coverage by moving the display. *Note:* Examples of display framing include paging, scrolling, offsetting, and expanding. *See also* **display, display surface, monitor.**

display generation: On a display device connected to a computer, the means by which output data may be specified. *Note 1:* The display generation specification may be entered manually or by the computer to which the display device is connected. *Note 2:* A computer may also be considered as a user.

display group: A group of display elements that (a) can be manipulated or processed as a unit, such as transferred, converted, translated, rotated, or tumbled as a unit, and (b) can be used to create display images. *Synonym* **display segment.** *See also* **display element, display image.**

display image: A collection of display elements or a collection of display groups associated and displayed together at one time in the display space of a display device. *Note:* Examples of images are assembly drawings, pictures, graphs, and text that are displayed by a display device. *See also* **display device, display element, display space.**

display line: In a display device, a row of display positions that constitutes a full-length horizontal line on the display surface of the display device. *Note 1:* In display systems, "horizontal" is defined at the direction of right to left or left to right, or the direction of reading of lines of English text, when a display image is viewed by an observer. *Note 2:* The positions in a display line usually are addressable. *Synonym* **display row.** *See also* **addressable horizontal position, display column, display device, display image, display surface.**

display order: *Synonym* **display command.**

display position: In the display space on the display surface of a display device, an addressable location that may be occupied by a symbol, such as a character or a display element. *See also* **character, display device, display element, display space, display surface.**

display row: *Synonym* **display line.**

display segment: *Synonym* **display group.**

display space: In a display device, the part of the display surface (a) in which a display element, display group, or display image may be placed, (b) which may be equal to, but usually is smaller than, the display surface. *Note:* Examples of display spaces are (a) the portion of a cathode ray tube (CRT) screen that can be used to display a display image, (b) the portion of a sheet of drawing paper that can be accessed by the writing stylus of a plotter, and (c) the portion of the faceplate of a fiberscope upon which an image can be formed because of the limitations imposed by a frame at the source, i.e., at the opposite end of an aligned bundle of optical fibers. *See also* **aligned bundle, display element, display group, display image, display device, display surface, faceplate, fiberscope, optical fiber, stylus.**

display surface: The data medium of a display device on which display elements, display groups, and display images may be placed. *Note 1:* The entire display surface may not be available for display, i.e., the display is limited to the display space. *Note 2:* Examples of display surfaces are (a) the entire screen of a cathode ray tube (CRT), (b) the sheet of drawing paper in a plotter, and (c) the entire faceplate of a fiberscope. *See also* **data medium, display device, display element, display group, display image, display space, faceplate, fiberscope.**

display system: A system that presents data in a form that is visible or legible to the human eye. *Note:* Examples of display systems are (a) a monitor connected to a computer, (b) a fiberscope, complete with light source, objective lens, aligned bundle, and faceplate, and (c) a plotter attached to a computer. *See also* **fiberscope, monitor.**

display tailoring: On a display device, designing displays to meet the specific requirements of a user, rather than providing a general display that can be used for many purposes. *See also* **display, display device.**

display unit: *See* **visual display unit.**

display update: On a display device, the generation of a new display to show current status, based on changes to the previously displayed data. *Note:* The display update can be accomplished by user request or by automatic means.

display writer: In a display device, the component used to create visible marks or traces in the display space on the display surface of the device. *Note:* Examples of display writers are light pens, laser beams, electron guns, and plotter pens. *See also* **display space, display surface, electron, laser beam, light pen.**

dissector: *See* **image dissector.**

dissipative loss: *See* **antenna dissipative loss.**

distance: *See* **coordination distance, downwind distance, Hamming distance, signal distance, skip distance, transmission loss distance.**

distance code: *See* **minimum distance code, unit distance code.**

distance dialing: *See* **direct distance dialing.**

distance gate: *See* **EXCLUSIVE OR gate.**

distance learning: *Synonym* **teletraining.**

distance measuring equipment (DME): 1. In radio location systems, equipment that ascertains the distance between an interrogator and a transponder by (a) measuring the time of transit of a signal to and from the transponder, (b) applying a known signal propagation rate, (c) calculating the round-trip distance, and (d) dividing by 2. **2.** A radionavigation aid in the aeronautical radionavigation service that determines the distance of an interrogator from a transponder by measuring the transit time of a pulse to and from the transponder. *See also* **aeronautical radionavigation service, interrogator, pulse, radionavigation aid, transit time, transponder.**

distance resolution: In an optical time domain reflectometer (ODTR), (a) a rating based on the shortest distance along the length of an optical waveguide, such as an optical fiber, that the OTDR can distinguish on its display screen, usually a cathode ray tube and (b) a measure of how close together two events, or faults, can be distinguished as separate events. *Synonym* **spatial resolution.** *See also* **dynamic range, measurement range.**

distance-square law: *See* **inverse distance-square law.**

distance training: *Synonym* **teletraining.**

distortion: 1. In an entity, any departure of the value of an attribute from a specified value. **2.** The extent to which the value of an attribute falls outside a specified range. **3.** The extent to which the value of an attribute exceeds specified tolerances. *Note:* An example of distortion is the change that occurs in the physical parameters of a radiation pattern during propagation through material media. **4.** In communications, any departure of an output signal shape from the corresponding input signal shape, over a range of frequencies, amplitudes, or phase shifts, during a time interval. **5.** Any departure from a specified input-output signal relationship over a range of frequencies, amplitudes, or phase shifts, during a time interval. *Note:* Signal degradation can occur in a multimode optical fiber from multimode distortion. In addition, several dispersive mechanisms can cause signal distortion in any optical fiber, such as waveguide dispersion, material dispersion, and profile dispersion. **6.** In start-stop teletypewriter signaling, the shifting of the transition points of the signal pulses from their proper positions relative to the beginning of the start pulse. *Note:* The magnitude of the distortion usually is expressed in percent of a perfect pulse duration (temporal) or pulse length (spatial). **7.** In a signal transmission system, the undesirable amount that an output wave shape or pulse differs from the input form, usually expressed as the undesirable difference in amplitude, frequency composition, phase, shape, or other attribute between an input signal and its corresponding output signal. *See* **amplitude distortion, amplitude versus frequency distortion, aperture distortion, barrel distortion, bias distortion, block distortion, characteristic distortion, cyclic distortion, delay distortion, early distortion, end distortion, envelope delay distortion,**

fortuitous distortion, group delay distortion, group delay frequency distortion, harmonic distortion, intermodal distortion, intermodulation distortion, intramodal distortion, isochronous distortion, late distortion, mode distortion, modulation frequency harmonic distortion, modulation frequency intermodulation distortion, multimode distortion, nonlinear distortion, optical distortion, optical pulse distortion, phase-frequency distortion, pincushion distortion, pulse distortion, pulse duration distortion, quantization distortion, radial distortion, significant instant distortion, single harmonic distortion, start-stop distortion, telegraph distortion, teletypewriter signal distortion, total harmonic distortion, waveguide delay distortion. *See also* **dispersion, distortion-limited operation, eye pattern, pulse, signal.**

distortion-limited: Pertaining to the operation of a transmission system in which the distortion of a received signal rather than its amplitude or power limits performance under stated conditions. *See also* **amplitude, distortion, distortion-limited operation, operation, power, signal, system, transmission.**

distortion-limited operation: Operation in which the distortion of a received signal rather than its attenuated amplitude or power limits performance of a system under stated operational conditions and limits. *Note 1:* In distortion-limited operation, a condition is reached such that the system distorts the shape of the waveform beyond specified limits. *Note 2:* In linear systems, distortion-limited operation is equivalent to bandwidth-limited operation. *Note 3:* An example of distortion-limited operation is the condition that prevails in an optical link when a combination of attenuation, dispersion, and delay distortion, rather than attenuation, contributes to limiting link performance. *See also* **attenuation-limited operation, bandwidth-limited operation, dispersion-limited operation, multimode distortion, quantum-noise-limited operation.**

distortion ratio: *See* **signal plus noise plus distortion to noise plus distortion ratio, start-stop distortion ratio.**

distortion to noise plus distortion ratio: *See* **signal plus noise plus distortion to noise plus distortion ratio.**

distraction: An extraneous audible signal appearing in a circuit during a telephone conversation, the signal being caused by a disturbance, such as crosstalk, echo, noise, interference, or any combination of these. *See also* **disturbance.**

distress call: In distress and rescue communications, a call that (a) immediately precedes a distress message, (b) combines the distress signal with the call sign of the calling station or the call originator, (c) usually has priority over all other transmissions, (d) is not addressed to a particular station, and (e) is not acknowledged before the distress message that follows it is sent. *Note 1:* Stations that hear a distress call usually immediately cease any transmission that might interfere with the distress message and subsequent distress traffic. Stations usually continue to listen on the frequency used for transmission of the distress call. *Note 2:* In continuous wave (cw) transmission, the distress call is SOS followed by the call sign of the mobile station in distress. In voice transmission, the distress signal is MAYDAY followed by the call sign or other identification of the mobile station in distress. MAYDAY usually is said in groups of three. *Refer to* **Fig. D-3.**

distress communications: Communications (a) in which methods, systems, and equipment are used for transmission of messages that contain information about emergency or distress situations, (b) in which distress messages are sent to or from a specific emergency or disaster area for the purpose of directing, controlling, reporting, or assisting in survival or recovery operations, (c) that include the transmission of

Mode	Distress Call: Distress Signal	Distress Call: Distressed Station	Message
Voice	Mayday (3 times)	Name (3 times)	Distress Message
Code	SOS (3 times)	Call Sign (3 times)	Distress Message

Fig. D-3. The structure of **distress calls** for voice and code transmission.

messages related to the immediate assistance required by the station that is in distress, and (d) that include alarm signals, distress signals, distress calls, distress messages, and direction finding transmissions. *Synonym* **emergency communications.** *See also* **distress call, distress message, distress signal.**

distress communications net: A communications net exclusively used for the transmission, control, and coordination of distress calls, distress messages, and distress signals. *See* **maritime distress communications net.** *See also* **communications net, distress call, distress message, distress signal.**

distress frequency: A radio frequency that is exclusively used for distress traffic, such as distress calls, distress messages, and distress signals. *Note:* The following are examples of distress frequencies that have been established by international agreement for emergency and distress purposes:

500 kHz	International distress calls (MF)
2182 kHz	International distress calls and messages (MF)
8364 kHz	International lifeboat, lifecraft, and survival (HF)
121.5 MHz	Aeronautical emergency messages
243.0 MHz	Survival emergency messages
156.8 MHz	International distress calls and safety messages

These radio frequencies are designated for use by mobile stations or survival craft for handling distress traffic. Thus, the first transmission by an aircraft in distress usually is made on the frequency in use by that aircraft. If no response is obtained, 121.5 kHz (kilohertz) or 243.0 MHz (megahertz) should be used for distress traffic, i.e., for emergency traffic. For radiotelegraphy, 500 kHz should be used. *Synonym* **emergency frequency.** *See* **international radiotelegraph distress frequency, international radiotelephone distress frequency, radiotelegraph distress frequency, radiotelephone distress frequency.** *See also* **distress call, distress message, distress signal, distress traffic.**

distress message: In distress and rescue communications, the message that (a) contains the details of the distress situation, (b) follows the distress call as soon as possible, (c) follows the distress signal, such as SOS or MAYDAY, and (d) contains the name or other identification of the ship, aircraft, or unit in distress, the particulars of its position, the nature of the distress, the kind of assistance needed, and any other information that might facilitate search and rescue. *Note:* A common practice is that the distress message is repeated at one-quarter before and one-quarter after the hour in some areas at 500 kHz (kilohertz), and on the hour and the half-hour in other areas at 2182 kHz. As a rule, a ship signals its position in latitude and longitude using degrees and minutes suffixed by north, south, east, or west. In carrier wave (cw) transmission, the signal AAA is used to separate the degrees from the minutes. If possible, the true north bearing and the distance in nautical miles from a known geographic point should be given. Conventions are published by the International Telecommunication Union (ITU). *See also* **bearing, distress call, distress communications, distress signal, emergency message.** *Refer to* **Fig. D-3 (255).**

distress period: One of the specified periods during which a continuous listening watch is to be maintained on the international distress and calling frequencies and during which communications silence should be maintained. *Note:* Examples of international distress periods are (a) 15 to 18 min. (minutes) and 45 to 48 min. past the hour for the 500-kHz (kilohertz) distress calling frequency and (b) 00 to 03 min. and 30 to 33 min. past the hour for the 2182-kHz distress message and distress calling frequency. *See also* **calling frequency, distress call, distress frequency, distress message, distress signal.**

distress signal: In distress and rescue communications, a signal that (a) indicates that the unit, i.e., the station, sending the signal is threatened by a grave or imminent danger and requests immediate assistance, (b) is sent before the distress call and at the beginning of the distress message, (c) in carrier wave (cw) transmission is SOS, and (d) in voice transmission is MAYDAY. *See* **pyrotechnic distress signal.** *See also* **distress call, distress communications, distress message.** *Refer to* **Fig. D-3 (255).**

distress traffic: All communications, such as alarm signals, distress signals, distress calls, distress messages, and direction finding transmissions, relevant to the assistance needed by a mobile station that is in distress or is experiencing an emergency. *Note:* The distress signals SOS or MAYDAY should accompany all communications for as long as the emergency category is that of distress. *See also* **distress call, distress message, distress signal, emergency category, mobile station.**

distributed control: In communications networks, control of a network from multiple points. *Note:* Each distributed control point controls a portion of the network using local information or information transmitted over the network from distant points. *See also* **centralized operation, distributed switching, hier-**

archical computer network, network architecture, routing table, system signaling and supervision.

distributed database: A database that is stored at more than one location, in more than one device, or both. *Note:* Examples of distributed databases are databases that are (a) dispersed among the nodes of a network, (b) placed in several storage units under the control of one or more data processors, and (c) controlled by a single database management system, but are contained in storage units that are not all connected to or under the control of the same processor. *See also* **database, database management system.**

distributed data interface standard: *See* **Fiber Distributed Data Interface standard.**

distributed data processing: Data processing in which the data processing functions, such as program execution, storage, control, and input-output functions, are performed at more than one node of a network. *See also* **data processing, network, node.**

distributed data processing system: In open systems architecture, a data processing system in which (a) two or more computer systems or data processing systems are interconnected for a specific purpose, and (b) the interconnections may be made at corresponding layers in each of the interconnected systems. *See also* **computer, data processor, layer, open systems architecture.**

distributed feedback (DFB) laser: A laser that (a) has part of its output fed back to its input using more than one propagation mode in the feedback path for controlling mode generation and mode conversion, (b) usually uses periodically inhomogeneous thin films, periodically inhomogeneous substrate guides, or thin film waveguides with periodic surfaces, and (c) operates more efficiently when feedback and direct input waves are in the same mode. *See* **surface emitting distributed feedback laser.** *See also* **feedback, laser, mode.**

distributed frame alignment signal: A frame alignment signal in which the signal elements occupy digit positions that are not consecutive. *See also* **bunched frame alignment signal, frame, frame alignment, frame alignment signal, multiframe.**

distributed network: A network in which (a) the network resources, such as switching equipment and processors, are at different locations within the geographical area being served by the network and (b) control may be centralized or distributed. *See also* **network, network architecture, network connectivity.**

distributed optical fiber sensor: A fiber optic sensor that (a) uses a spatial distribution of optical fibers to sense a parameter, such as a planar distribution, a bobbin wound with spatial shading, a series of windings in line, two or more parallel coils, or an array of wound bobbins, (b) can sense one or more of many parameters, such as a pressure point at a Cartesian coordinate, a formed beam direction, a pressure gradient, and an electric field gradient, and (c) usually uses optical time domain reflectometry (OTDR), interferometry, and other optical techniques to modulate irradiance incident upon a photodetector. *See also* **fiber optic sensor, optical time domain reflectometry.**

distributed parameter: A component in a communications, power, or control circuit in which (a) the electrical components, such as resistances, inductances, and capacitances (R-L-C) are structurally distributed in space such that they cannot be considered as being located at a single point in the circuit, i.e., they are not lumped, and (b) an input electrical signal affects various interconnected components at different times depending on the propagation speed through the components. *Note:* An example of a distributed parameter is the resistance, capacitance, and inductance in (a) a pair of parallel wires, (b) a coaxial cable, and (c) a rectangular metallic waveguide. In these examples, the values of the distributed parameters are expressed per unit length, such as ohms, henries, and farads per meter. *See also* **lumped parameter.**

distributed processing: Processing in which an integrated set of information processing functions is performed within two or more physically separated devices. *See also* **distributed network.**

distributed-queue dual-bus (DQDB) network: A distributed multiaccess network that (a) supports integrated communications using a dual bus and distributed queuing, (b) provides access to local area networks (LANs), metropolitan area networks (MANs), or wide area networks (WANs), and (c) supports connectionless data transfer, connection oriented data transfer, and isochronous communications, such as voice communications.

distributed radar system: A system in which (a) radar detectors are placed at more than one location, and (b) all the gathered radar data are processed at one location.

distributed sensor system: A system in which (a) sensors are placed at more than one location, and (b) all the sensed data are telemetered to and processed at one location.

distributed star coupled bus: *Synonym* **star-mesh network.** *See* **network topology.**

distributed star coupled network: *See* **network topology.**

distributed switching: Switching in which many processor-controlled switching units are distributed, usually close to concentrations of users, and operated in conjunction with a host switch. *Note:* Distributed switching provides improved communications services for concentrations of users remote from the host switch and reduces the transmission requirements, i.e., the traffic requirements, between such concentrations and the host switch. *See also* **centralized operation, concentrator, distributed control, spill forward, switching center, system signaling and supervision.**

distributed thin film waveguide: *See* **periodically distributed thin film waveguide.**

distribution: In integrated services digital network (ISDN) applications, the use of broadband transmission of audio or video information to the user without applying any postproduction processing to the information. *Note:* Distribution is in contrast to contribution. *See* **core-cladding power distribution, equilibrium modal power distribution, exceedance distribution, fading distribution, Gaussian distribution, Lambertian distribution, mean power of the talker volume distribution, modal distribution, nonequilibrium modal power distribution, normal distribution, Poisson distribution, Rayleigh distribution, uniform Lambertian distribution.** *See also* **contribution, integrated services digital network, postproduction processing.**

distribution box: *See* **fiber optic distribution box.**

distribution cable: In a communications system, a cable that leads from a distribution frame to a destination user terminal. *See also* **distribution frame.**

distribution center: *See* **internal message distribution center, switching and message distribution center.**

distribution frame: In communications systems, a structure with terminations for connecting the permanent wiring of a facility in such a manner that interconnection by cross connections may readily be made. *See* **combined distribution frame, fiber optic distribution frame, group distribution frame, high frequency distribution frame, intermediate distribution frame, main distribution frame, supergroup distribution frame.** *See also* **group.**

distribution length: *See* **nonequilibrium modal power distribution length.**

distribution panel: *See* **optical fiber distribution panel.**

distribution quality television (DQTV): Television (TV) that conforms to the National Television Standards Committee (NTSC) standard, the Système Electronique Couleur avec Mémoire (SECAM) standard, the Phase Alternation by Line (PAL) standard, or the Phase Alternation by Line Modified (PAL-M) standard. *Synonym* **existing quality television.** *See also* **enhanced quality television, high definition television.**

distribution service: In Integrated Services Digital Network (ISDN) applications, a telecommunications service that allows one-way flow of information from one point in the network to other points in the network with or without user individual presentation control.

distribution substation: A substation that (a) transforms electrical power usually from higher to lower voltage levels and (b) distributes the output over many local lines.

distribution system: *See* **local distribution system, multifunction information distribution system, primary distribution system, protected fiber optic distribution system, protected distribution system, protected wireline distribution system.**

distribution voltage drop: In a power distribution system, the voltage drop, i.e., the decrease in voltage that occurs over the distance between any two specified points.

distributive coupler: *See* **nonuniformly distributive coupler, uniformly distributive coupler.**

distributor: In data transmission, a device that (a) accepts a data stream from one line and (b) places a sequence of signals, one or more at a time, on several lines, thus performing a spatial multiplexing of the original stream. *Note:* Examples of a distributor are (a) a mechanical unit with input to a rotor and output through many contacts wiped by the rotor and (b) a set of combinational circuits, such as a series of AND gates, that are sequentially enabled by a set of pulses and that are all connected to a common bus carrying the input signals. *See* **automatic call distributor.**

disturbance: In communications systems, an unwanted perturbation, or the cause of the perturbation, such as a voltage induced by lightning or interference visible on a television (TV) screen when a vacuum cleaner is turned on nearby. *See* **ionospheric distur-**

bance, sudden ionospheric disturbance. *See also* **disturbance power, disturbance voltage.**

disturbance current: An unwanted extraneous current induced in a system by natural or man-made sources. *Note:* In telecommunications systems, the disturbance currents limit or interfere with the interchange of information. Disturbance currents may produce false signals in a telephone, noise in a radio receiver, or distortion in a received television (TV) signal. *See also* **interference.**

disturbance power: The unwanted extraneous power that (a) originates from natural or man-made power sources, (b) is associated with transmission, propagation, and reception of signals, (c) tends to limit or interfere with the interchange of information, (d) degrades or otherwise interferes with normal operation of a communications system, (f) may produce false signaling causing circuit disruptions and outages, and (g) produces noise and distortion of signals in telephone, radio, and television receivers.

disturbance sensor: *See* **field disturbance sensor.**

disturbance voltage: An unwanted extraneous voltage induced in a system by natural or man-made sources. *Note:* In telecommunications systems, the disturbance voltage creates currents that limit or interfere with the interchange of information. Disturbance voltages may produce false signals in a telephone, noise in a radio receiver, or distortion in a received television (TV) signal. *See also* **interference.**

disturbed circuit: 1. In telephone sytems, a circuit in which crosstalk is generated by disturbance power from a disturbing circuit. **2.** A circuit that generates noise in itself. *Note 1:* Examples of a disturbed circuit are (a) a component that vibrates in the Earth's magnetic field and induces a disturbance voltage and current in itself, (b) a circuit that is experiencing microphonics, and (c) a circuit in which there is positive feedback. *Note 2:* A disturbed circuit may also be a disturbing circuit. *See also* **crosstalk, disturbance power, disturbing circuit, microphonics.**

disturbing circuit: A circuit that generates crosstalk in another circuit by transferring disturbance power to it. *Note 1:* The power from the disturbing circuit may be transferred to the disturbed circuit by various means, such as by (a) conduction, i.e., current leakage, (b) induction, i.e., inductive coupling by magnetic fields, (c) electrostatic coupling, i.e., capacitive coupling, (d) electromagnetic coupling, i.e., the disturbing circuit behaving as a transmitting antenna and the disturbed circuit behaving as a receiving antenna, and (e) optical coupling, i.e., by light leakage. *Note 2:* A disturbing circuit may also be a disturbed circuit. *See also* **disturbed circuit.**

dither: In spread spectrum systems, a phase shift of the clock signal that (a) is supplied to the receiver by a controlled amount so as to generate an error signal for clock control and (b) usually is only for the time corresponding to a fraction of a binary digit, i.e., a fraction of a bit. *Synonyms* **tau jitter, τ jitter.** *See* **frequency multidither, polystep dither.** *See also* **clock, phase shift, binary digit, spread spectrum system.**

diurnal: 1. Pertaining to the variation with time of day of an environmental condition or physical quantity, such as temperature, humidity, atmospheric pressure, magnetic declination, magnetic inclination, or height of the ionosphere. **2.** Pertaining to a daily recurring phenomenon, such as the rising sun, falling dew, and dusk. *Note:* Diurnal environmental conditions significantly affect communications by causing changes in propagation media. *See also* **diurnal phase shift, propagation medium.**

diurnal phase shift: The phase shift of electromagnetic signals associated with daily changes in environmental conditions, particularly changes in the ionosphere. *Note 1:* The major changes usually occur during the periods when sunrise or sunset is present at critical points along the path. *Note 2:* Significant phase shifts may occur on paths in which a reflection area of the path is subject to a large tidal range. *Note 3:* In cable systems, significant phase shifts can be caused by diurnal temperature variances. *See also* **diurnal, fading, ionosphere, phase, phase interference fading.**

divergence: The propagation or bending of waves away from each other, such as water waves from the bow of a moving ship, shock waves at the forward wing tip of a moving aircraft, lightwaves emerging from a concave lens with collimated wave input, sound waves emanating from a conical loudspeaker or horn, and electromagnetic waves radiating from a dipole antenna. *See* **beam divergence, convergence, divergent light.**

divergent lens: *Synonym* **diverging lens.**

divergent light: Light that (a) consists of a bundle of rays such that each ray is propagating in such a direction as to be departing from every other ray, i.e., they are not parallel to each other but are spreading and hence are not collimated or convergent, and (b) has a wavefront that is somewhat spherical or concave on the trailing side. *Note:* If a collimated beam enters a diverging lens, i.e., a concave lens, the emerging light is divergent light. *See also* **convergence, convergent light, divergence.**

divergent meniscus lens: A lens that (a) consists of one convex and one concave surface, the concave surface having the smaller radius of curvature and hence the greater dioptric power, and (b) behaves like a concavoconvex lens. *Synonyms* **diverging meniscus lens, negative meniscus.** *See also* **concavoconvex lens, dioptric power, lens.**

diverging beam: A beam of light (a) whose rays are directed away from each other, (b) that has a concave wavefront on the trailing side, and (c) that is not collimated. *See also* **beam, divergent light.**

diverging lens: A lens that (a) causes (i) parallel light rays to spread out, (ii) diverging light rays to diverge farther, and (iii) converging light waves to be less convergent, i.e., more divergent, (b) may have (i) one surface concavely spherical and the other plane, i.e., planoconcave, (ii) both surfaces concave, i.e., double-convergent, or (iii) one surface concave and the other convex, i.e., concavoconvex or divergent meniscus, and (c) has a negative focal length, measured from the focal point toward the object. *Synonyms* **concave lens, dispersive lens, divergent lens, minus lens, negative lens.** *See also* **converging lens.**

diverging meniscus lens: *See* **divergent meniscus lens.**

diverse routing: A survivability feature of a communications system (a) in which access lines or channels serving the same station or facility are routed over geographically separated circuits, and (b) that also may include the use of two or more radio nets in addition to the land lines.

diversion: *See* **toll diversion.**

diversity: The property of being made up of two or more different elements, media, or methods. *Note:* In communications, diversity is usually used to provide robustness, reliability, and security. *See* **angle diversity, dual diversity, frequency diversity, order of diversity, polarization diversity, quadruple diversity, route diversity, spaced antenna diversity, space diversity, time diversity, tone diversity.** *See also* **avoidance routing, diversity combiner, diversity reception.**

diversity combiner: A circuit or a device for combining two or more signals that represent the same information but are received via separate paths or channels with the objective of providing a single resultant signal that is superior in quality to any of the contributing signals. *See also* **digital combining, equal gain combiner, linear combiner, maximal ratio combiner, postdetection combiner, predetection combiner, selective combiner.**

diversity combining: *See* **quadruple diversity combining.**

diversity combining method: *See* **maximal-ratio-square diversity combining method.**

diversity factor: *Synonym* **diversity ratio.**

diversity gain: In high frequency communications, the ratio of (a) the signal field strength or irradiance obtained by diversity combining to (b) the signal strength or irradiance obtained by a single path or channel. *Note:* The diversity gain usually is expressed in dB.

diversity gate: *Synonym* **EXCLUSIVE OR gate.**

diversity order: *Synonym* **order of diversity.**

diversity ratio: The ratio of (a) either (i) the sum of the individual maximum actual real power demands of the various parts of a power distribution system, (ii) the sum of the individual maximum possible demands of the various parts of a power distribution system, or (iii) the sum of the individual component rated power of the various electrical components connected to a power distribution system to (b) the maximum demand of the whole system. *Note:* The diversity ratio is always greater than unity. *Synonym* **diversity factor.** *See also* **load, load factor.**

diversity reception: Reception in which a resultant signal is obtained by combining signals (a) that originate from two or more independent sources that have been modulated with identical information-bearing signals and (b) that may vary in their transmission characteristics at any given instant. *Note 1:* Diversity transmission and reception are used to obtain reliability and signal improvement by overcoming the effects of fading, outages, and circuit failures. *Note 2:* In using diversity transmission and reception, the amount of received signal improvement depends on the independence of the fading characteristics of the signal as well as circuit outages and failures. *Synonym* **diversity system.** *See also* **diversity transmission, dual diversity, frequency diversity, order of diversity, quadruple diversity, space diversity, tone diversity.**

diversity system: *Synonym* **diversity reception.**

diversity transmission: Transmission in which two or more independent sources are modulated with identical information-bearing signals, and the modulated signals are independently transmitted over different paths or media in order that their recombination can result in an improved signal. *Note 1:* Diversity transmission and reception are used to obtain reliability and signal improvement by overcoming the effects of fading, outages, and circuit failures. *Note 2:* In using diversity transmission and reception, the amount of received signal improvement depends on the independence of the fading characteristics of the signal as well as circuit outages and failures. *Note 3:* On reception, a resultant signal is obtained by combining the signals from the two or more independent sources. *See* **time diversity transmission.** *See also* **diversity reception, dual diversity, frequency diversity, order of diversity, quadruple diversity, space diversity, tone diversity.**

divestiture: The court-ordered separation of the Bell Operating Telephone Companies from the American Telephone and Telegraph Company. *Note:* The Bell Operating Telephone Companies, along with other operating telephone companies, provide loops and central office (C.O.) services, i.e., local exchange services, directly to the user. *See also* **common carrier, local access and transport area, network interface, other common carrier, resale carrier, specialized common carrier.**

dividend: A number, i.e., a numerator, that is to be divided by another number, i.e., a divisor or a denominator. *Note:* Examples of a dividend is the 496 in the expressions (a) 496/74, (b) 496÷74, (c) 496 divided by 74, (d) $496 \cdot 74^{-1}$, and (e) 496×74^{-1}. *See also* **divisor.**

divider: *See* **analog divider, frequency divider.**

divider circuit: A filtering circuit that provides for the separation of simultaneous multiple frequency signals, arriving on a single transmission path, into separate paths for each frequency or group of frequencies. *See also* **filtering circuit, path, transmission.**

division: An arithmetic operation that determines how many times a number, i.e., a divisor, can be subtracted from another number, i.e., a dividend, before the resulting differences change sign, where the result is the quotient, and a remainder also is produced if the dividend, after successive subtraction by the divisor, becomes smaller than the divisor. *Note:* The number 74 can be subtracted 6 times from 496 with 52 left over. Thus, 74 is the divisor, 6 is the quotient, 496 is the dividend, and 52 is the remainder. The remainder, when calculated decimally, is combined with the quotient, to obtain a net quotient of about 6.70. *See* **frequency division, space division, time division, wavelength division.**

division demultiplexing: *See* **time division demultiplexing.**

division multiple access: *See* **code division multiple access.**

division multiplexing: *See* **color division multiplexing, frequency division multiplexing, optical space division multiplexing, space division multiplexing, synchronous time division multiplexing, time division multiplexing, wavelength division multiplexing.**

divisor: A number, i.e., a denominator, used to divide another number, i.e., a numerator or a dividend. *See also* **dividend, division.**

D layer: In the D region of the atmosphere, a layer of increased free electron density caused by the ionizing effect of solar radiation in the daylight hemisphere. *Note 1:* When present, the D layer is the lowest layer of ionized air above the Earth. *Note 2:* The D layer reflects frequencies at or below the critical frequency, approximately 50 kHz (kilohertz), and partially absorbs higher frequencies. *Note 3:* Radio waves in the 1 MHz (megahertz) to 100 MHz band propagated via any of the higher reflecting layers pass through the absorbing D region, encountering there the principal attenuation. *Note 4:* The D layer does not exist at night because the free electron density is diminished, and attenuation of radio waves becomes negligible. *See also* **critical frequency, D region, ionosphere.**

DLE: data link escape transmission control character.

DLE character: data link escape character.

DM: delay modulation, delta modulation, differential modulation.

DMA: differential mode attenuation.

DMW: digital microwave.

DNSIX: Defense Network Security Information Exchange.

DO: design objective.

document: 1. A data medium that has data recorded on it, usually has permanence, and can be read by a person or a machine. **2.** To place data on a data medium that usually has permanence and can be read by a person or a machine. *See* **source document.** *See also* **documentation.**

documentation: 1. The management of documents, including the actions of identifying, acquiring, processing, storing, and disseminating them. **2.** A collection of documents on a given subject, usually grouped by some criteria, such as (a) for a given organization and purpose and (b) in a given container and data medium. **3.** The collecting, organizing, storing, citing, indexing, retrieving, and disseminating of documents or their contents, including the techniques necessary for the orderly presentation, organization, and communication of recorded information. **4.** The management of documents, including the identification, acquisition, process, storage, and dissemination of the documents or their contents. **5.** A collection of documents, usually pertaining to a given subject. *See* **system documentation.** *See also* **document.**

document facsimile equipment: *See* **group one (two, three) document facsimile equipment.**

document facsimile telegraphy: Facsimile telegraphy intended primarily for the transmission of documents that have only two levels of density, usually black and white. *Synonym* **black and white facsimile telegraphy.** *See also* **facsimile system.**

document mark: In micrographics, an optical mark used for automatically counting display frames. *Synonym* **blip.**

DoD: Department of Defense (U.S.).

DOD: Department of Defense (U.S.), direct outward dialing.

DoD master clock: *Synonym* **Department of Defense master clock.**

domain: 1. In distributed networks, all the hardware, firmware, and software under the control of a specified set of host processors. **2.** The independent variable used to express a function. *Note 1:* Examples of domains are the complex domain, frequency domain, spatial domain, and time domain. *Note 2:* Domains usually have names. *See* **adjacent domain, complex domain, frequency domain, spatial domain, time domain.** *See also* **Domain Name System.**

domain name server: A server that retains the addresses and routing information for Transmission Control Protocol/Internet Protocol (TCP/IP) local area network (LAN) users. *See also* **domain, server.**

Domain Name System (DNS): The online distributed database system that (a) is used to map human readable addresses into Internet Protocol addresses, (b) has servers throughout the Internet to implement hierarchical addressing that allows a site administrator to assign machine names and addresses, (c) supports separate mappings between mail destinations and IP addresses, and (d) uses domain names that (i) consist of a sequence of names, i.e., labels, separated by periods, i.e., by dots, (ii) usually are used to uniquely name Internet host computers, (iii) are hierarchical, and (iv) are read from right to left. *Note:* For example, the Domain Name System (DNS) for a specific host might be nic.ddn.mil, in which nic (Network Information Center) is the name of the host, ddn (Defense Data Network) is the subdomain, and mil (MILNET) is the primary domain. *See also* **domain.**

domain network: *See* **multiple domain network.**

domain reflectometry: *See* **optical time domain reflectometry.**

domestic fixed public service: A fixed service, whose stations are open to public correspondence, for radio communications originating and terminating solely at points that all lie within the entire United States and certain other geographical areas specified in the *Code of Federal Regulations* (CFR), Section 47, Telecommunications.

domestic public radio service (DPRS): A radio service that (a) provides domestic land mobile and fixed communications services to the public and (b) consists of (i) individual radio stations or (ii) networks of radio stations.

dominant mode: In a waveguide that can support more than one propagation mode, the mode that propagates with minimum degradation, i.e., the mode with the lowest cutoff frequency. *Note:* Designations for the dominant mode are TE_{10} for rectangular waveguides and TE_{11} for circular waveguides. *See also* **lowest-order transverse mode, mode, mode conversion, multimode distortion, waveguide.**

donor: In an intrinsic semiconducting material, such as germanium or silicon, a dopant that has nearly the same electronic bond structure as the intrinsic material, but with one more electron among its valence band electrons than that number required to complete the intrinsic bonding pattern, thus leaving one "extra" or "excess" electron for each impurity atom, i.e., each dopant atom, in the structure. *Note:* The dopant atoms, i.e., the donor atoms, are relatively few and far apart, hence do not interfere with the electrical conductivity of the intrinsic material. Arsenic can serve as a donor for intrinsic germanium and silicon. The extra electron moves or wanders from atom to atom more freely than the bound electrons that are required to complete the bonding structure, although interchanges do occur with the bound electrons. The extra electrons move about more freely than the holes created by acceptors. Hence, the electrons are more mobile than the holes. Under the influence of electric fields, the electrons and the holes move in the direction of the field according to their sign. *See also* **acceptor, dopant, electron, hole.**

do nothing instruction: *Synonym* **no-op instruction.**

don't answer: *See* **call forwarding don't answer.**

door: *See* **trap door.**

dopant: A material mixed, fused, amalgamated, crystallized, or otherwise added to another in order to achieve desired characteristics of the resulting material. *Note:* Examples of dopants are the germanium tetrachloride or the titanium tetrachloride added to pure glass to control the refractive index of the glass for optical fibers and (b) the gallium or the arsenic added to germanium or silicon to produce p-type or n-type semiconducting material for making diodes and transistors. *See also* **acceptor, donor, electron, hole.** *Refer to* **Figs. R-8 (834), R-9 (835).**

doped deposited silica (DDS) process: A process for making optical fibers in which dopants are deposited in inside or outside walls of silica glass tubes or cylinders and fused to produce desired refractive index profiles.

doped silica cladded optical fiber: An optical fiber that (a) consists of a doped silica core with doped silica cladding, (b) usually is produced by the chemical vapor deposition process (CVDP), (c) usually exhibits very low loss and moderate dispersion, and (d) usually is a step indexed fiber.

doped silica graded optical fiber: A silica optical fiber in which (a) the dopant varies radially so as to produce a decreasing refractive index from the center toward the outside, thus eliminating the need for cladding, (b) the refractive index profile is graded and tailored to reduce multimode dispersion, (c) the non-axial rays of light, although traveling farther than the axial rays, travel faster in the outer propagation medium where the refractive index is lower, and (d) the axial rays arrive at the end of the fiber about the same time as the nonaxial rays. *See also* **axial rays, cladding, dopant, paraxial ray, refractive index.** *Refer to* **Fig. R-9 (835).**

Doppler: *See* **radio Doppler.**

Doppler effect: The difference between the observed frequency and the transmitted frequency of a source that has a radial component of velocity toward or away from the observer. *Note 1:* The observed frequency is higher if the source is moving with a radial component of velocity toward the observer and lower if away from the observer. *Note 2:* For a constant-frequency source, the observed change in frequency is proportional to the magnitude of the radial component of velocity between the observer and the source. *Note 3:* Examples of the Doppler effect are (a) the increase in observed frequency of a radio, light, or sound wave from a source of constant frequency speeding toward an observer, (b) the red shift of the stars, and (c) the relatively gradual transition from a higher pitch to a lower pitch on approach and departure of a constant-speed, constant-frequency train whistle one observes while standing a significant distance from a straight track. A detector on the track bed would sense a sudden frequency shift as the train whistle passed over the sensor.

Doppler navigation system: A navigation system that uses electromagnetic waves to measure the radial component of velocity by measuring the phase and frequency shifts of transmitted or reflected signals.

Doppler pulse repetition rate: In Doppler radar systems, the difference between (a) the transmitted radar pulse repetition rate and (b) the received, i.e., the radar return or the reflected, pulse repetition rate. *Note:* The Doppler pulse repetition rate is directly proportional to the radial component of velocity of the illuminated object, i.e., the target, toward or away from the radar site. If the received pulse repetition rate is higher than the transmitted pulse repetition rate, the radial component of the velocity is toward the radar site. *See also* **Doppler radar, radar blind speed, radar return.**

Doppler radar: Radar that detects the radial component of the velocity of a distant object, i.e., a distant target, relative to the radar antenna by means of the Doppler effect. *See* **pulse Doppler radar.** *See also* **Doppler effect, Doppler shift.**

Doppler shift: The change in observed or measured frequency of a wave from a source that is moving with a radial component of velocity relative to the observer or the antenna making the measurement. *See also* **Doppler effect, Doppler radar.**

Doppler spread: 1: The nominal width of the frequency spectrum of a sine wave when it is passed through a channel that causes fading. **2.** The change in the nominal width of the frequency spectrum of a sine wave when it is passed through a channel that causes fading.

DOS: disk operating system.

dose: *See* **radiation absorbed dose.**

dosimeter: An instrument that measures the cumulative exposure to high energy radiation, such as X-ray and gamma radiation. *Note 1:* A unit of measure of high energy exposure is the rad•hour for a given material. *Note 2:* A specific dosimeter for a particular application, such as a dosimeter worn by persons working in radiation hazardous areas, must be calibrated specifically for the purpose. *See* **fiber optic dosimeter.** *See also* **hazards of electromagnetic radiation to fuel, hazards of electromagnetic radiation to ordnance, hazards of electromagnetic radiation to personnel, rad.**

dot: 1. In the international Morse code, a short interval, i.e., a short on-key interval, that (a) has many different physical representations, such as (i) a short burst of carrier energy, (ii) a short pulse of applied modulating voltage, (iii) a short burst of audio, visual, or optical energy, and (iv) a short period between clicks, (b) is often used as the unit of duration to specify the duration of other signaling intervals, such as the duration of dashes and spaces, and (c) is basic, so that its duration or spatial length is arbitrary but usually constant in any given system or mode of operation. *Note:* In manual telegraph transmission, the duration of dots and dashes may vary according to the speed and the "wrist" of the operator. However, even in manual transmission, a fair amount of proportionality is maintained for the duration of dots, dashes, spaces between characters, spaces between words, and spaces between sentences. **2.** In Morse code, a signal element of mark condition and of duration of one unit interval followed by a signal element of space condition having the same duration. *See also* **international Morse code, mark, Morse code, signal element, space, unit interval.** *Refer to* **Figs. I-6 (472), T-3 (991).**

dot frequency: 1. In facsimile transmission systems, half the number of contiguous picture elements transmitted in a second. **2.** The fundamental frequency of the signal representing alternately black and white picture elements along a scanning line. *See also* **facsimile system, picture element.**

dot matrix: A pattern of dots in two-dimensional space (a) that can be used for constructing (i) display images, either directly or through the use of display groups or display elements, (ii) display groups, either directly or through the use of display elements, and (iii) display elements, (b) that can be transmitted by scanning or by one or more optical fibers, and (c) that can be used to form characters, such as letters, numerals, and special signs and symbols. *See also* **display element, display group, display image, dot matrix character generator.**

dot matrix character generator: A character generator that generates characters that (a) are composed of dots for presentation in the display space on the display surface of a display device, (b) usually are formed from an array of dot positions such that the presence or the absence of a dot at each position is used to form the character, and (c) can be formed and transmitted (i) in bit patterns to control a display device, such as a cathode ray tube (CRT) or a dot matrix printer, or (ii) in aligned bundles of optical fibers, in which case the display surface is the faceplate of a fiberscope on the end of an aligned bundle of optical fibers. *See also* **aligned bundle, character generator, display device, display space, display surface, dot matrix, dot matrix printer.**

dot matrix printer: A high speed printer that (a) prints characters composed of dots, (b) is well suited for printing characters but not well suited for printing graphic displays, (c) uses character codes to govern which dots in an array on the print head are to be printed, and (d) may have a laser jet print head. *See also* **daisy wheel printer, ink jet printer, laser jet printer.**

double call sign calling: Calling to establish and conduct communications by transmitting (a) the call sign of the called station, (b) a letter group or word meaning "this is," and (c) the call sign of the calling station, in that order.

double clear: In telephone switchboard operation, a double supervisory working signal that indicates a clearing condition.

double converter: In satellite communications systems, a two-stage up-converter or two-stage down-converter for frequency translation. *See also* **down converter, frequency translation, up-converter.**

double crucible (DC) process: In the production of optical fibers, an optical fiber drawing process in which (a) two concentric crucibles are used, one for the core

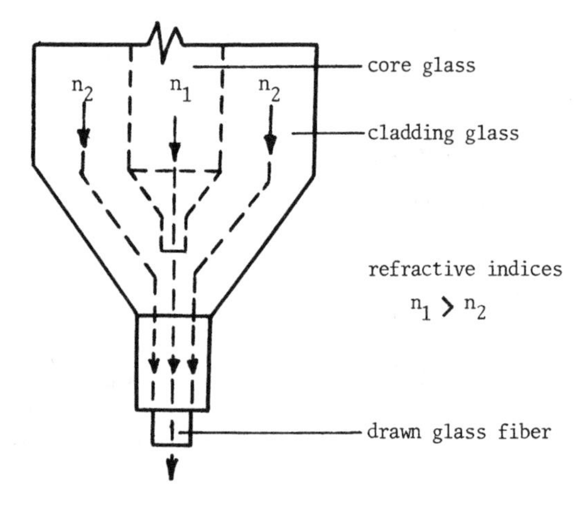

Fig. D-4. The **double crucible process** for drawing optical fibers.

glass and one for the cladding glass, (b) the cladded fiber is pulled out of the bottom at the convergence or the apex of the two crucibles, and (c) diffusion of the dopant materials in the glasses produces a graded refractive index profile. *Synonyms* **compound glass process, ion exchange process.** *Refer to* **Fig. D-4.**

double current signaling: Telegraph signaling in which (a) marks and spaces are represented by electric currents of opposite polarities rather than by current and no current conditions, and (b) transmission is binary in which positive and negative direct currents (dc) denote the significant conditions. *Note:* Some advantages of double current signaling are (a) the fast operation of sensitive polarized receiving relays, (b) the fast cutoff, i.e., fast-acting disabling combinational circuits, (c) that no bias is required, and (d) that the system operates in a neutral condition when there are no signals. *Synonyms* **bipolar telegraphy, bipolar transmission, double current telegraphy, double current transmission, mark-space signaling, polar direct current telegraph transmission, polar transmission.** *See also* **mark, space, unipolar.**

double current telegraphy: *Synonym* **double current signaling.**

double current transmission: *Synonym* **double current signaling.**

double eccentric connector: A fiber optic connector with an eccentric center in each mating pair so that relative rotation between the mating pairs compensates for eccentricity of the optical fiber cores and permits core-to-core alignment for improved coupling.

double ended control: *Synonym* **double ended synchronization.**

double ended synchronization (DES): For two connected exchanges in a communications network, a synchronization control system in which the phase error signals used to control the clock at one exchange are derived by comparison with the phase of the incoming digital signal and the phase of the internal clock at both exchanges. *Synonym* **double ended control.**

double frequency shift keying (DFSK): Frequency shift keying in which two telegraph signals are multiplexed and transmitted simultaneously by frequency shifting among four frequencies. *See also* **differential phase-shift keying, frequency shift keying, keying, phase-shift keying.**

double heading: *Synonym* **message readdressal.**

double heterojunction: In a laser diode, two heterojunctions in close proximity, resulting in full carrier and radiation confinement as well as improved control of recombinations. *See also* **heterojunction, laser diode, recombination.**

double heterojunction diode: A laser diode that (a) has two different heterojunctions, the difference being primarily in the stepped changes in refractive indices of the material in the vicinity of the p-n junctions, and (b) is widely used for pulse code modulation (PCM) operation. *See* **monorail double heterojunction diode.** *See also* **heterojunction, laser diode, p-n junction, refractive index.**

double image: 1. Two images of the same object, the two images caused by imperfections in the imaging system, such as an optical system. **2.** Pertaining to the doubling of an image caused by optical imperfections in an optical system. *See also* **image, optical system.**

double modulation: Modulation in which (a) a subcarrier is modulated with an information-bearing signal, and (b) the resulting modulated subcarrier is then used to modulate another carrier that has a higher frequency. *See also* **carrier (cxr), modulation.**

double refraction: *Synonym* **birefringence.**

double sideband: In an amplitude modulated carrier wave, the frequencies that (a) are above and below the carrier frequency and (b) are caused by the modulation of the carrier wave by a baseband signal, i.e., the information-bearing signal. *See also* **amplitude modulation, baseband, carrier.**

double sideband reduced carrier (DSB-RC) transmission: Reduced carrier transmission in which (a) the frequencies produced by amplitude modulation are symmetrically spaced above and below the carrier, and (b) the carrier level is reduced for transmission at a fixed level below that which is provided to the modulator. *Note:* In double sideband reduced carrier (DSB-RC) transmission, the carrier usually is transmitted at a level suitable for use as a reference by the receiver, except for the case in which it is reduced to the minimum practical level, i.e., the carrier is suppressed. *See also* **double sideband suppressed carrier transmission, double sideband transmission, level, modulation, single sideband suppressed carrier transmission, single sideband transmission.**

double sideband suppressed carrier: The carrier that (a) is used in double sideband suppressed carrier transmission systems and (b) after amplitude modulation by an intelligence-bearing signal, i.e., a baseband signal, is fully or partially suppressed, so that only both of the symmetrical sidebands are transmitted. *See also* **amplitude modulation, baseband, carrier, modulation, sideband.**

double sideband suppressed carrier (DSB-SC) transmission: Suppressed carrier transmission in which (a) frequencies produced by amplitude modulation are symmetrically spaced above and below the carrier frequency, and (b) the carrier level is reduced to the lowest practical level, ideally completely suppressed. *Note:* Double sideband suppressed carrier (DSB-SC) transmission is a special case of double sideband reduced carrier transmission. *See also* **double sideband reduced carrier transmission, level, modulation, sideband transmission, single sideband suppressed carrier transmission, single sideband transmission.**

double sideband suppressed carrier (DSB-SC) transmission system: A transmission system in which (a) amplitude modulation of a carrier is used, (b) the modulated electromagnetic waves produced by modulation are symmetrically spaced in the frequency domain above and below the carrier frequency, and (c) the carrier power level is reduced to a predetermined level below that of the transmitted sidebands. *See also* **amplitude modulation, carrier, frequency domain.**

double sideband (DSB) transmission: Sideband transmission in which (a) both sidebands are transmitted, (b) unless otherwise indicated, the sidebands are symmetrically spaced in the frequency domain above and below the carrier frequency, and (c) the two sidebands and the carrier are all transmitted at full power. *See also* **double sideband reduced carrier transmission, independent sideband transmission, sideband transmission, single sideband suppressed carrier transmission, single sideband transmission.**

double supervisory working: In telephone switchboard operation, a supervisory system in which (a) two supervisory signal keys or switches are provided for each cord circuit, (b) the keys or switches are arranged in a switchboard keyshelf or panel to correspond with the arrangement of the cords, (c) the back signal of each pair, i.e., the answering supervisory signal, is associated with the answering plug of a cord, thus constituting the answer back unit, (d) the front signal of each pair, i.e., the calling supervisory signal, is associated with the calling plug, i.e., the front plug, of a cord, and (e) either the front or the back signal indicates that (i) the user has left the circuit by restoring the telephone instrument receiver to its rest, cradle, or on-hook condition, i.e., has hung up, or (ii) the exchange, i.e., the central office (C.O.) has disconnected.

doublet: 1. A byte composed of two binary digits or pulses. **2.** In optics, a compound lens that consists of two optical elements. *See* **air-spaced doublet, cemented doublet.** *See also* **byte, compound lens, optical element.**

doubly cladded slab dielectric waveguide: A slab dielectric waveguide with a core surrounded by two layers of dielectric material, the core having the highest refractive index, the inner cladding having the lowest refractive index, and the outermost cladding having an intermediate refractive index. *See also* **cladding, core, refractive index, slab dielectric waveguide.**

doubly clad optical fiber (DCF): An optical fiber with (a) a core surrounded by an inner cladding with a refractive index lower than that of the core and (b) an outer cladding surrounding the inner cladding, the outer cladding having a refractive index higher than that of the inner cladding. *Note 1:* Doubly clad optical fibers are usually used for single mode transmission. *Note 2:* Doubly clad optical fibers are in contrast to fibers having only one cladding surrounding the core. *Note 3:* The doubly clad optical fiber construction is used for single mode transmission to (a) reduce bending losses and remove cladding modes, if any, and (b) reduce dispersion and thereby obtain higher data signaling rates (DSRs). The inner cladding modes are stripped off because the outer cladding refractive index is higher than the inner cladding refractive index so that when rays are incident upon the intercladding interface surface, they are bent toward the normal to the intercladding interface surface and hence escape. *Note 4:* A symmetrical plot of the refractive index profile of a doubly clad optical fiber superficially resembles the letter W. *Synonyms* **depressed inner cladding optical fiber, W-profile optical fiber.** *See also* **cladding, deeply depressed cladding optical fiber, depressed cladding optical fiber, fiber optics, single mode optical fiber.**

doubly refracting crystal: A transparent crystalline substance that (a) is anisotropic with respect to the velocity of light propagating within it in two different directions, (b) demonstrates two different light velocities in different directions because the refractive index is different in one direction from that in the other, and (c) demonstrates that light travels more slowly in a material with a higher refractive index than in a material with a lower refractive index. *See also* **refractive index, birefringence.**

down-converter: A device for performing frequency translation in such a manner that the output signals are lower in frequency and longer in wavelength than the input signals. *Synonym* **down translator.** *See also*

erect position, frequency, frequency translation, inverted position, up-converter.

downlink: The portion of a communications link used for transmission of signals from a satellite or an airborne platform to a surface terminal. *Note:* A downlink is the converse of an uplink. *See also* **Earth terminal complex, link, satellite, uplink.**

download: *Synonym* **data capture.**

downrange site: An area that (a) lies in a direction away from a launch site of a missile or projectile, (b) if not a target area, usually has sensing and tracking equipment, observation stations, and communications equipment, and (c) if a target area, may have test objects and other instrumentation for terminal ballistic testing.

downstream: 1. The direction of data flow. **2.** In area networks, the direction of data flow along a bus, i.e., in the direction away from the head of bus function. *See also* **upstream.**

down the hill radio relay: A short range, high capacity radio relay station for extending radio relay trunks from a complex radio relay site to the local communications center.

downtime: The time during which a functional unit is inoperable. *Note:* Downtime can result for a fault within the functional unit or from an environmental condition. In the latter case, the functional unit itself may be operable except for the environmental condition that prevents it from being operated. The condition might be that there is no power or no operator. The time during which a functional unit is simply turned off because there is no need to operate it is not considered downtime. *See also* **continuous operation, failure, fault, mean time between failures, mean time between outages, mean time to repair, mean time to service restoral, uptime.** *Refer to* **Fig. D-5.**

Fig. D-5. The AN/ARC-222 VHF transceiver unit's high reliability assures minimum downtime. (Courtesy Magnavox Electronic Systems Company, Fort Wayne, IN.)

down-translator: *Synonym* **down-converter.**

downwind distance: In communications, the distance in the direction of the wind from the point of a nuclear explosion to a point where fallout from the exploded nuclear device would no longer be dangerous, though communications might still be disrupted. *Note:* One common standard specifies the downwind distance of a 1-Mt (megaton) detonated nuclear bomb as a unit of measure.

D patch bay: A patch bay designed for patching and monitoring unbalanced digital data circuits at data signaling rates (DSR), i.e., at bit rates, up to 1 Mbps (megabits per second). *See also* **K patch bay, M patch bay, MM patch bay, patch bay.**

DPCM: differential pulse code modulation.

DPSK: differential phase-shift keying.

DQDB: distributed queue dual bus.

drafter: *See* **message drafter.**

dragging: In display systems, picking a display element, display group, or display image in the display space on the display surface of a display device and moving it so that all points on the element, group, or image continuously move in a path parallel to that taken by the picking device. *See also* **display element, display group, display space, display systems, picking.**

drawing: *See* **optical fiber drawing.**

drawn glass optical fiber: A glass optical fiber formed by pulling a strand from a heat-softened preform through a die or from a mandrel with an optical fiber pulling machine.

D region: The region of the atmosphere between approximately 50 and 95 km above the surface of the Earth. *Note 1:* The D region is below the E and F regions. *Note 2:* The D layer of the ionosphere forms in the D region. In the D layer, attenuation of radio waves, caused by ionospheric free electron density generated by cosmic rays from the Sun, is pronounced during daylight hours. At night the cosmic rays are gone, ionization ceases, and hence attenuation of radio waves ceases. *See also* **D layer, ionosphere.**

drift: A comparatively long-term change in any attribute or value of any system or equipment operational parameter. *Note 1:* The drift should be characterized, such as "diurnal frequency drift" or "output level drift." *Note 2:* Drift usually is undesirable and unidirectional, but may be bidirectional, cyclic, or of such long-term duration and low excursion rate as to be insignificant and negligible. *Note 3:* Examples of drift are (a) carrier frequency drift of a transmitter, (b) voltage drift of a communications line as charges build up on the line, (c) amplifier gain drift caused by changing environmental conditions, (d) ship drift caused by ocean currents and wind, and (e) aircraft drift caused by the wind. *Note 4:* Jitter, swim, wander, and drift have increasing periods of variation in that order. *See also* **frequency drift.** *See also* **jitter, swim, wander.**

drive: *See* **Bernoulli box disk drive, cartridge drive, disk drive, magnetic tape drive, streaming tape drive, virtual disk drive, Winchester drive.**

drive circuit: In fiber optic transmission systems, the electrical circuit that drives the light-emitting source, modulating it in accordance with an intelligence-bearing signal, i.e., an information-bearing baseband signal.

driver: A functional unit or a criterion that exercises or directs the activity of another functional unit, such as a system or a component. *See* **bus driver, clock driver, device driver, fiber optic cable driver.** *See also* **functional unit.**

drone: A pilotless radio-controlled aircraft.

droop: *See* **signal droop.**

drop: 1. In a communications network, the portion of a device directly connected to internal station facilities, such as toward an Automatic Voice Network (AUTOVON) four-wire switch, a switchboard, or a switching center. **2.** The central office (C.O.) side of test jacks. **3.** A wire or a cable from a pole or cable terminal to a building. **4.** To intentionally or unintentionally delete or lose part of a signal, such as removing bits from a bit stream. *See* **distribution voltage drop, fiber optic drop, hyphen drop.** *See also* **local exchange loop, loop.**

drop and insert: At an intermediate point in a multichannel transmission system, the termination of one or more channels and the initiation of other channels in the corresponding multiplexed positions of the terminated channels. *Note 1:* Part of the data in the transmission system is demodulated or filtered out, i.e., is dropped, at the intermediate point, and data that represent different information are entered, i.e., are inserted, for subsequent transmission in the same position, such as the time, frequency, or phase, previously occupied by the terminated data. *Note 2:* Information not of interest at the drop and insert location is not demodulated or filtered out. *See also* **drop repeater, radio relay system.**

drop channel: In a multichannel transmission path or branch, a channel that (a) is terminated at an intermediate point between the beginning and the end of the path or the branch and (b) contains data that need not go beyond the intermediate point. *See also* **drop channel operation.**

drop channel operation: Operation in which one or more channels of a multichannel system are terminated, i.e., are dropped, at any intermediate point between the end terminals of the system. *See also* **channel, drop and insert, drop repeater, radio relay system.**

drop message: A message that is dropped from a higher to a lower altitude, such as a message dropped from an aircraft in flight to the ground.

dropout: 1. In a communications system, a momentary loss of signal. *Note:* Dropouts usually are caused by noise, propagation anomalies, or system malfunctions. **2.** A failure to properly read a binary character from data storage. *Note:* Dropouts usually are caused by a defect in the storage media or by a malfunction of the read mechanism. **3.** In magnetic tape, disk, card, or drum systems, a recorded signal with an amplitude less than a predetermined percentage of a reference signal. *Note:* An example of a dropout is the reading of a 0 when a 1 should have been read, because of a speck of dirt, an inclusion, or a blemish on the surface of a magnetic disk. *See* **carrier dropout.** *See also* **character, failure, hit.**

drop repeater: In a multichannel communications system, a repeater that has the necessary equipment for the local termination, i.e., for the dropping, of one or more channels. *See also* **channel, drop and insert, drop channel operation.**

drop zone: An area designated to serve as a receiving area for dropping objects, such as messages, equipment, supplies, or persons, from aircraft. *See also* **drop message.**

drop zone boundary limit marker: A marker, such as panels, lights, contrasting objects, and footprints, used to indicate the boundary of the usable area of a drop zone.

drop zone impact point: The point in a drop zone where the first of a series of dropped objects or persons should land.

drop zone indicator: A previously agreed upon signal used to mark the boundary of a drop zone.

drum: In data storage systems, a storage device that (a) is in the shape of a right circular cylinder, (b) has a surface, usually coated with a magnetic material, on which data may be recorded, and (c) is rotated so that data may be read or recorded by using one or more read-write heads. *See also* **magnetic head, read-write head.**

drum factor: In facsimile systems, the ratio of drum length to the drum diameter. *Note:* In nondrum systems, the drum factor is the ratio of the equivalent dimensions. *See also* **channel, drop and insert, drop channel operation.**

drum receiver: In facsimile systems, a receiver in which the recording medium, such as sensitized or photosensitive paper, is fastened to a rotating drum and scanned helicoidally by a recording head that moves in a straight line along an element of the drum cylinder as it rotates.

drum speed: In facsimile systems, the rotation rate of the facsimile transmitter or receiver drum. *Note:* Drum speed usually is expressed in revolutions per minute (rpm or RPM). *See also* **facsimile.**

drum transmitter: In facsimile systems, a transmitter in which the object, i.e., the original document, whose contents are to be transmitted is fastened to a rotating drum and scanned helicoidally by a recording head that moves in a straight line along an element of the drum cylinder as it rotates.

dry glass: Glass from which as much water as possible has been driven out. *Note 1:* An example of dry glass is glass with a water concentration less than 1 ppm (part per million). *Note 2:* Drying is done primarily to reduce hydroxyl ion absorption loss. *Note 3:* Drying can be accomplished by many days of drying of starting powders in a 250°C (celsius) vacuum oven, followed by many hours of dry gas melt bubbling to dry out water to extremely low levels.

DS: digital signal.

DSA: Defense Security Agency, dial service assistance.

DSA board: dial service assistance board.

DSB: dial service board, double sideband.

DSB-SC: double sideband suppressed carrier.

DSC: digital selective calling.

DSE: data switching exchange.

DSI: digital speech interpolation.

DSM: delta-sigma modulation.

DSR: data sampling rate, data signaling rate, digital signaling rate.

D star: *Synonym* **specific detectivity.**

D*: *Synonym* **specific detectivity.**

DSTE: data subscriber terminal equipment.

DSU: data service unit.

DS0: digital signal 0.

DS1 . . . DS4: digital signal 1 . . . digital signal 4.

DS1C: digital signal 1C.

DTE: data terminal equipment.

DTE clear signal: data terminal equipment clear signal.

DTE waiting signal: data terminal equipment waiting signal.

DTG: date-time group.

DTMF: dual tone multifrequency.

DTN: data transfer network.

dual access: 1. The connection of a user to two switching centers by separate access lines using a single message routing indicator or telephone number. **2.** In satellite communications, the simultaneous transmission of two carriers through a single communications satellite repeater. **3.** Pertaining to the availability of voice transmission and digital transmission from the same station or terminal. *See also* **alternate routing, dual homing, multiple access, multiple homing, pulse address multiple access, split homing.**

dual access tributary station: A tributary station that has (a) a user access line usually used for voice communications and (b) special circuits or equipment for digital transmission. *Synonyms* **voice-data tributary station, voice-graphics tributary station.** *See also* **tributary station.**

dual bus: A pair of parallel buses arranged so that the direction of data flow in one bus is opposite to the direction of data flow in the other bus. *See* **open dual bus.** *Refer to* **Figs. N-5 and N-6 (622).**

dual-bus network: *See* **distributed-queue dual-bus network.**

dual coding: *Synonym* **dual programming.**

dual combining: *Synonym* **dual diversity.**

dual diversity: Pertaining to the simultaneous combining, selecting, and detecting of two independently fading signals through the use of space, frequency, angle, time, or polarization characteristics. *Synonym* **dual combining.** *See also* **diversity reception, frequency diversity, order of diversity, polarization diversity, quadruple diversity, space diversity, tone diversity.**

dual homing: The connection of a terminal such that (a) the terminal is served by either of two switching centers, and (b) a single directory number or a single routing indicator is used. *See also* **alternate routing, dual access, multiple access, multiple homing.**

dual inline package (DIP): An integrated circuit package with a rectangular housing and a row of pins along two opposite sides that are compatible with standard integrated circuit sockets. *Note:* An example of a dual inline package (DIP) is a microcircuit package with two rows of seven vertical leads that is specially designed for mounting on a printed circuit board.

dual inline package switch: A subminiature switch, or group of subminiature switches, that (a) is mounted in a dual inline package, (b) usually is used to select the operational options of a device, and (c) usually is compatible with standard integrated circuit sockets.

dual precedence message (DPM): A message that contains two precedence designations. *Note:* Usually the higher precedence message is for all action addressees and the lower for all information addressees. *See also* **single precedence message.**

dual programming: Computer program development in which two functionally identical programs (a) are developed by different programmers, (b) are based on the same requirements, (c) are functionally identical, (d) may be written in different programming languages, and (e) may be used for several purposes, such as to (i) reduce running time, (ii) reduce operating cost, (iii) improve error detection, (iv) increase reliability, (v) improve documentation, and (vi) reduce programming errors. *Synonym* **dual coding.**

dual-ring network: *See* **network topology.**

dual tone multifrequency (DTMF) signaling: Multifrequency signaling that makes use of standard fixed combinations of two specific voice band frequencies, one from a group of four lower frequencies and the other from a group of four higher frequencies. *Note 1:* Dual tone multifrequency (DTMF) signals, unlike dial pulses, can pass through the entire connection to the destination user, and therefore lend themselves to various schemes for remote control after access, i.e., after the connection is established. *Note 2:* Telephones using dual tone multifrequency (DTMF) signaling usually have 12 keys, each corresponding to a pair of frequencies that represent the ten decimal digits plus the symbols "#" and "*," the "*" being reserved for special purposes. *Note 3:* The standard signal frequency pairs transmitted by switching equipment used by the public exchange carriers for the indicated keys are, in Hz (hertz), 697/1209 for 1, 697/1336 for 2, 697/1477 for 3, 770/1209 for 4, 770/1336 for 5, 770/1477 for 6, 852/1209 for 7, 852/1336 for 8, 852/1477 for 9, 941/1336 for 0, 941/1209 for the asterisk, and 941/1477 for the pound sign. *Note 4:* Automatic Voice Network (AUTOVON) telephones have 16 keys, the extra 4 being used for precedence. For AUTOVON, the signal frequency pairs transmitted for the ten decimal digits, the asterisk, and the pound sign are the same as those used by the public exchange carriers. In AUTOVON, the frequency pairs for the indicated key and precedence level are 697/1633 Hz (hertz) for FO (Flash Override), 770/1633 Hz for F (Flash), 852/1633 Hz for I (Immediate), and 941/1633 Hz for P (Priority). *See also* **dial pulse, dial signaling, key pulsing, multifrequency signaling, signal.**

dual use access line: A user access line that (a) usually is used for voice communications and (b) has special signal-conditioning equipment that allows its use as a digital transmission line.

duct: 1. A device that confines, guides, or routes materials, such as air, cables, and liquids. *Note:* An example of a duct is a permanent tube or tunnel, usually installed inside walls and foundations or underground, through which cables may be pulled or threaded for installation or replacement. **2.** Any configuration of material that serves as a passive guide for materials or phenomena. *Note:* Examples of ducts are (a) atmospheric ducts for radio waves, (b) underwater ducts for sound waves, (c) aquaducts, and (d) chutes. *See* **atmospheric duct.** *See also* **ducting.**

duct fiber optic cable: A fiber optic cable designed for use in underground ducts. *See also* **underground cable.**

ducting: 1. The propagation of radio waves within an atmospheric duct. **2.** The trapping of a wave, caused by a layering of the propagation medium, in which a layer of material of a given refractive index is bounded on both sides by layers of material of lower refractive index, thus trapping a wave in the higher refractive index medium. *Note:* For the case of sound waves, the surrounding wall of the duct can reflect and refract sound waves rather than absorb them, thus trapping the waves. Layers of the atmosphere can trap electromagnetic waves the same way. Ducting occurs in optical fibers to keep electromagnetic signals, i.e., optical signals, confined to the fiber. *See also* **atmospheric duct, ionosphere, surface refractivity, total internal reflection, trapped electromagnetic wave.**

dumb terminal: An asynchronous terminal that (a) does not use a transmission control protocol, (b) sequentially only sends or receives data one character at a time, and (c) usually handles only one coded character set, such as the American Standard Code for Information Interchange (ASCII) characters. *See also* **programmable terminal, smart terminal.**

dummy character: A character that (a) is inserted into stored data or a data stream to accomplish a specific purpose, such as fill a buffer or a frame, and (b) does not affect the meaning of the data into which it is inserted. *See also* **null character.**

dummy group: In cryptographic systems, a code group that has the appearance of a valid code group or cipher group, but has no plain text significance. *See also* **cipher group, code group.**

dummy load: A dissipative impedance matched network, usually used at the end of a transmission line, arranged so that all incident energy is absorbed. *Note 1:* The incident energy in the dummy load usually is converted to heat. *Note 2:* The actual useful load plus the dummy load usually is made equal to the characteristic impedance of the line. If the line has no useful load, the dummy load is made equal to the characteristic impedance of the line so as to avoid reflections.

dump: 1. To transfer all or part of the contents of a storage unit or a data medium to another storage unit or data medium, such as to transfer all the data from an internal storage, i.e., memory, of a computer to an external storage unit or data medium, usually for a specific purpose, such as (a) to allow other use of the storage, (b) to safeguard the stored data against faults, errors, or power outages, and (c) to assist in debugging. **2.** The data transferred from one storage unit or data medium to another storage unit or data medium, such as from internal storage to external storage or to a printer. *See* **change dump, disaster dump, dynamic dump, formatted dump, postmortem dump, selective dump, snapshot dump, static dump.**

duobinary coding: Binary coding in which (a) a 0 in the input sequence of binary digits is represented by a 0 in the output sequence, whereas a 1 in the input sequence might cause a change in the output sequence pulse level, this change depending on the number of 0s since the last 1 occurred, (b) if the number of 0s since the last 1 in the input sequence is even, no change occurs in the output level, and (c) if the number of 0s since the last 1 is odd, a change occurs in the output signal level to the appropriate level at which it was last positioned, either positive or negative. *Note 1:* The duobinary coding bandwidth is about half of that required for nonreturn to zero (NRZ) coding. Duobinary coding requires about 6.5 kHz (kilohertz) to pass 16-kbps (kilobits per second) data. *Note 2:* Duobinary is a ternary signal and thus has a poorer signal to noise ratio than regular binary. However, this loss is offset by its increased data-handling capacity. *Note 3:* Duobinary coding may be accomplished by conditioning the data, delaying the data one bit interval, inverting it, and then adding it to the undelayed version. *See also* **duobinary signal.**

duobinary signal: A pseudobinary coded signal in which a 0 bit is represented by a zero level electric current or voltage, and a 1 bit is represented by a positive level current or voltage if the quantity of 0 bits since the last 1 bit is even, and by a negative level current or voltage if the quantity of 0 bits since the last 1 bit is odd. *Note 1:* Duobinary signals require less bandwidth than nonreturn to zero (NRZ) signals. *Note 2:* Duobinary signals also permit the detection of some errors without the addition of error checking bits, such as parity bits. *See also* **code, duobinary signal, nonreturn to zero code, return to zero code, signal.**

duplex: In communications, computer, data processing, and control systems, pertaining to a double arrangement, condition, or state, usually of two mutually associated entities. *See* **semiduplex, two-frequency half-duplex.**

duplex cable: A fiber optic cable that contains two optical fibers.

duplex circuit: A circuit that permits transmission in both directions. *Note:* A single optical fiber can handle lightwaves in both directions at the same time, but a single wire cannot handle direct currents (dc) both ways at the same time. *See also* **circuit, duplex operation, full-duplex circuit, half-duplex circuit, half-duplex operation, simplex circuit, simplex operation.**

duplex communication: Communication between two points in both directions simultaneously. *Synonym* **full duplex communication.** *See also* **two-way simultaneous communication.** *Refer to* **Fig. T-3 (991).**

duplex equipment: Equipment that consists of transmitting and receiving components arranged for duplex transmission. *See also* **half-duplex equipment.**

duplexer: 1. A device that allows the simultaneous use of a transmitter and a receiver in connection with a common element, such as an antenna system, thus allowing the same antenna to be used for transmission and reception at the same time. **2.** In radar, a transmitter-receiver switch that automatically couples an antenna to a receiver during the receiving period and to a transmitter during the transmitting period. *See also* **diplexer, diplex operation, duplex operation.**

duplexing: time division duplexing.

duplex operation: Operation of a telecommunications channel in which transmission is simultaneously permitted in both directions. *Note 1:* Duplex operation is not limited to radio transmission systems. *Note 2:* In radio transmission systems, duplex operation and semiduplex operation usually require two frequencies. Simplex operation may use either one or two frequencies. *Note 3:* A single optical fiber can handle lightwaves in both directions at the same time, but a single wire connot handle direct currents (dc) both ways at the same time. *Synonyms* **full duplex operation, two-way simultaneous operation.** *See also* **conversational mode, diplexer, diplex operation, duplex circuit, duplexer, half-duplex operation, simplex circuit, simplex operation.**

duplex transmission: Transmission between two points in both directions simultaneously. *Synonym* **full duplex transmission.** *See also* **two-way simultaneous communication.**

duplicating: Producing copies. *See* **stencil duplicating.**

duplicator: A device that produces copies. *See* **stencil duplicator.**

duration: *See* **call duration, frame duration, pulse duration, pulse half-duration, root mean square pulse duration, significant interval theoretical duration.**

duration discrimination: *See* **pulse duration discrimination.**

duration discriminator: *See* **pulse duration discriminator.**

duration distortion: *See* **pulse duration distortion.**

duration indication: In telephone system operation, the indication, upon request by the network to the paying station, that (a) the time of a call is chargeable, and (b) the chargeable time is to be indicated (i) prior to the release of the paying station or (ii) by recall at a more convenient time. *See* **automatic duration indication.**

duration modulation: *See* **pulse duration modulation, suppressed clock pulse duration modulation.**

duress signal: In communications systems, a prearranged signal used to indicate to the addressee or the destination user that the operator is being forced to transmit the particular message in which the signal occurs.

duty cycle: 1. The ratio of (a) the duration of a phenomenon occurring in a stated interval to (b) the duration of the interval. *Note:* The limiting conditions must be specified. **2.** In a periodic phenomenon, the ratio of (a) the duration of the phenomenon in a given period to (b) the period. *Note:* The duty cycle is 0.25 for a pulse train in which the pulse duration is 1 μs (microsecond) and the pulse period is 4 μs. **3.** In pulse code transmission, the ratio of (a) the sum of all pulse durations during a specified performance period to (b) the total specified measurement period. **4.** In a continuously variable slope delta (CVSD) modulation converter, the mean proportion of binary 1 digits at the converter output in which each 1 indicates a run of a specified number of consecutive bits of the same polarity in the digital output signal. **5.** In radio station operations, the daily schedule of transmissions. **6.** In each cycle of operation of equipment to be operated intermittently, the ratio of (a) the length of time the equipment is to operate to (b) the period of the cycle. *Note:* The maximum period of the cycle, or its equivalent, must be stated. If the allowable duty cycle is unity, continuous operation is allowable. *See also* **continuously variable slope delta modulation, operating time, performance measurement period.**

duty factor: In a wave that consists of periodic pulses, the product of pulse duration and pulse repetition rate (PRR).

DVD: digital video disk.

dwell time: The period of time during which a dynamic process is halted in order that another process may occur.

DX: direct current signaling.

DX signaling: *Synonym* **direct current signaling.**

DX signaling unit: A duplex signaling unit that (a) repeats E and M lead signals into a cable via A and B leads and (b) transmits these signals on the same cable that transmits the message. *Note:* The cable usually consists of cable pairs. *See also* **E and M lead signaling, E&M signaling, signal.**

dyadic operator: An operator that can operate on two operands. *Note:* Examples of dyadic operators are the addition, subtraction, multiplication, division, AND, INCLUSIVE OR, EXCLUSIVE OR, NOR, NAND, and COMPARE operators. *See also* **monadic operator.**

dynamic allocation: Resource allocation in which components are assigned to users or systems during operation, such as while a computer program is being executed. *See also* **resource allocation.**

dynamically adaptive routing (DAR): In route determination for packet-switched networks, adaptive routing in which an algorithm is used that (a) automatically routes traffic around congested, damaged, or destroyed switches and trunks, and (b) allows the network to continue to function over the remaining portions of the network. *See also* **adaptive routing, avoidance routing, communications, continuous operation, directionalization, routing, traffic.**

dynamic analysis: Examining, evaluating, and analyzing the performance of a functional unit while it is running or operating. *Note:* An example of dynamic analysis is evaluating the performance of a computer program based on check printouts and display register contents while the program is being executed.

dynamic analyzer: A software tool that aids in the evaluation of computer programs by monitoring the execution of programs. *Note:* Examples of dynamic analyzers are instrumentation tools, software monitors, and tracers. *See also* **software monitor, static analyzer, tool, tracer.**

dynamic binding: Binding performed during execution of the computer program in which the binding is being performed. *See also* **binding, static binding.**

dynamic buffering: Allocating buffer storage capacity to incoming or outgoing data as needed during the performance of a job, running of a program, or execution of a routine. *See also* **static buffering.**

dynamic dump: A dump that (a) is made during the execution of a computer program and (b) usually is controlled by that program. *See also* **dump.**

dynamicization: *Synonym* **parallel to serial conversion.**

dynamicizer: *Synonym* **parallel to serial converter.**

dynamic programming: Optimizing the solution of a multistage problem in which a number of options are available for decision at each stage of the solution.

dynamic range: **1.** In a system or a device, the ratio of (a) a specified maximum level of a parameter, such as power, current, voltage, frequency, or floating point number representation, to (b) its minimum detectable or positive value. *Note 1:* The dynamic range is usually expressed in dB. *Note 2:* Floating point number representations are not expressed in dB. **2.** In a transmission system, the ratio of (a) the overload level, i.e., the maximum signal power that the system can handle without distortion of the signal, to (b) the noise level of the system. *Note:* The dynamic range of transmission systems usually is expressed in dB. **3.** In an optical time domain reflectometer (OTDR), a measure of how far into an optical waveguide, such as an optical fiber, a measurement can be made. *Note:* Thus, the dynamic range may be (a) a measure of the optical power rating of an optical time domain reflectometer (OTDR) based on the inherent loss in dB/km (decibels per kilometer) of the waveguide, the pulse duration, the spectral width of the light source, the operational wavelength, the length of the waveguide being tested, the number of times the source pulse is launched, and whether the rating is based on a one-way or a round trip of the pulse, and (b) the dynamic range may be the loss through the waveguide that can be measured, based on the waveguide inherent attenuation rate. If a multimode optical fiber has an attenuation rate of 0.8 dB/km and the goal is to measure a 40-km (kilometer) piece of fiber, or it is 40 km to a break, the OTDR has to have a dynamic range of at least 32 dB in order to measure the length of the fiber or the distance to the break. *See* **optical dynamic range.** *See also* **distance resolution, level, measurement range.**

dynamic restructuring: Changing the configuration of a system while it is operating or being used. *Note:* Examples of dynamic restructuring are (a) changing hardware or software components of a communications, computer, data processing, or control system while the system is operating and (b) rearranging data in a database during its use. *See also* **configuration, system.**

dynamics: *See* **antenna dynamics.**

dynamic scanning: In fiber optic transmission systems, the vibration of a fiber optic aligned bundle about a fixed point with reference to the impressed image in order to render the optical fiber pattern less visible at the output end. *See also* **aligned bundle, image, optical fiber pattern.**

dynamic storage: Storage in which (a) the data medium used for storage must move during reading and writing, or (b) the stored data must move relative to the storage medium. *Note:* Examples of dynamic storage are magnetic disks, drums, cards, and tapes. They, or their read/write heads, must be moved. These are nonvolatile storage media. Acoustic delay lines, electromagnetic delay lines, and cathode ray tubes are also examples of dynamic storage. The data moves with respect to the data medium, or the data must be regenerated. These are volatile storage media. *See also* **data medium, nonvolatile storage, storage, volatile storage.**

dynamic storage allocation: In communications, computer, and data processing systems, the assignment of storage capacity to programs and data while the programs are being executed. *See also* **storage, storage capacity.**

dynamic variation: In communications systems, a short time variation in the characteristics of power delivered to communications equipment. *Note:* Dynamic variations do not include steady state conditions. *See also* **steady state condition, transient.**

D1: *See* **channel bank.**

D2: *See* **channel bank.**

D3: *See* **channel bank.**

D4: *See* **channel bank.**

D-4: A framing standard for traditional time division multiplexing that describes (a) user channels multiplexed onto a trunk that has been segmented or framed into 24 bytes of eight bits each and (b) the performance of the multiplexing function in this framing structure by interleaving bits of consecutive bytes as they are presented from individual circuits into each frame. *See also* **digital signal, time division multiplexing, trunk.**

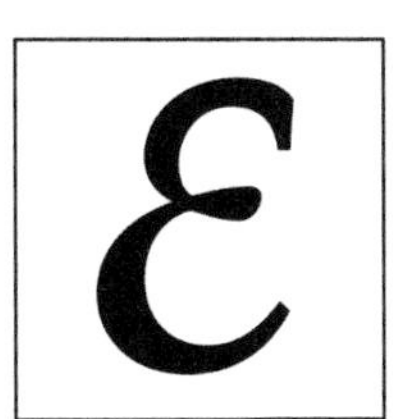

e: The base of natural logarithms, approximately equal to 2.718. *Note 1:* To convert base 10 logarithms to base *e* logarithms, multiply the base 10 logarithm by 2.302, i.e., ln 10 = 2.302 and log 10 = 1. Thus, $\log_e$ 1,000,000 = ln 1,000,000 = 13.815, and $\log_{10}$ 1,000,000 = log 1,000,000 = 6. *Note 2:* Examples of the use of *e* are: (a) the absorption coefficient is represented by *b* in the exponent of the absorption equation that expresses Bouger's law, given by the relation $F = F_0 e^{-bx}$ where *F* is the power level at the point *x*, F_0 is the initial value of power level, i.e., the power level at $x = 0$, and a typical value of *b* for optical fibers is 0.023/km (kilometer^{-1}), which corresponds to an attenuation rate 0.2 dB/km, both figures for a wavelength of 1.31 μm (microns), the standard wavelength; and (b) in an electromagnetic wave that is propagating in a waveguide, such as a lightwave in an optical fiber, the amplitude, *A*, at any point *z* along the guide, is given by the relation $A = A_0 e^{-pz} = A_0 e^{-jb-az}$ where *A* is the amplitude, A_0 is the amplitude at the point $z = 0$, *p* is the propagation constant, *j* expresses the imaginary part, *b* is the phase term, *a* is the attenuation term, and *z* is the propagation distance at which *A* occurs. *See also* **absorption coefficient, attenuation term, Bouger's law, phase term.**

E: 1. exa. 2. *See* **sporadic E.**

E and M lead signaling: 1. Signaling in which separate paths are used for signaling and voice signals. *Note:* The E was originally derived from ear (receiver) and the M from mouth (transmitter). **2.** In telephone systems, signaling in which separate leads, respectively called the E and M leads, are used for signaling and supervisory purposes. *Note 1:* In one implementation of E & M signaling, the near end signals the far end by applying −4 V (volts) direct current (dc) to the M lead, which results in a ground being applied to the far-end E lead. When −48 V direct current (dc) is applied to the far-end M lead, the near-end E lead is grounded. *Note 2:* The E originally stood for "ear," i.e., when the near-end E lead was grounded, the far end was calling and "wanted your ear." The M originally stood for "mouth" because when the near end wanted to call, i.e., to speak to, the far end, −48 V direct current (dc) was applied to that lead. *Synonyms* **E and M signaling, E&M lead signaling, E&M signaling.** *See also* **circuit, direct current signaling, DX signaling unit, pulse link repeater, signal.**

early distortion: Distortion that is caused by the advanced arrival of parts of a signal. *Note:* Signal parts are transmitted in a given sequence, each part in a specific time slot. A part that arrives ahead of the proper instant causes early distortion. *Synonym* **negative distortion.** *See also* **late distortion.**

early failure period: In a device or a system with a large population of components, the period that (a) immediately follows the initial use or turn-on of the device or the system, (b) is characterized by a component failure rate that is higher than the rate that occurs during the following period, called the constant failure period, and (c) primarily is caused by defects in the failing components. *Synonym* **infant mortality period.** *See also* **bathtub curve, constant failure period, failure rate.** *Refer to* **Fig. B-3 (71).**

early warning: *See* **airborne early warning.**

early warning net: *See* **airborne early warning net.**

early warning radar: Long range radar that provides (a) range, direction, and elevation information on objects in space, i.e., on targets, such as missiles, aircraft, and satellites, as far as the radar horizon and at slant ranges beyond the atmosphere and (b) provides the earliest possible data on the position and the movement of the objects. *See also* **long range radar.**

earphone: A receiver mounted near the ear for the purpose of audio reception, such as reception via telephone, radio, radiotelephone, telegraph, or radiotelegraph. *See also* **headphone.**

earth: In an electric circuit, a point in the circuit that is connected by a zero resistance path directly into or under the ground. *Note 1:* An example of earth is a point connected to a thick copper stake driven deeply into a salted and wetted area in the ground, or to a metal water pipe that is buried in the ground for a significant distance. The earth can also be used as one part of a circuit as in a ground return. A perfect earth is at zero electric potential. It never rises above zero potential no matter how much current is driven into it or drawn from it. It is of infinite capacity. *Note 2:* Earth is often erroneously used as a synonym for ground. A distinction should be made. *See* **smooth Earth.** *See also* **earth ground, ground.**

Earth coverage (EC): In satellite communications, the coverage that occurs when the satellite-to-Earth beam is wide enough to cover all of the surface of the Earth exposed to the satellite, i.e., the footprint is as large as it possibly can be. *See also* **footprint, satellite.**

earth electrode subsystem: A network of zero resistance electrically interconnected materials, such as rods, plates, mats, or grids, installed for the purpose of establishing a contact with the earth that is as close to zero resistance as possible. *See also* **equipotential ground plane, facility grounding system, fault protection subsystem, ground potential, lightning protection subsystem, signal reference subsystem.**

Earth exploration satellite service (EESS): A radiocommunications service between Earth stations and one or more space stations (a) that may include links between space stations, (b) in which information relating to the characteristics of the Earth and its natural phenomena is obtained from active or passive sensors on Earth satellites, (c) in which similar information is collected from airborne or Earth-based platforms, and in which such information may be distributed to Earth stations within the system concerned, (d) that may include platform interrogation, and (e) that may include necessary feeder links.

earth ground: Ground that is connected to the earth via a zero or a near zero resistance path. *See also* **ground.**

Earth radiation: The total electromagnetic radiation that emanates from the Earth.

Earth radiation balance: The balance of the Earth's thermal radiation, i.e., infrared radiation, as determined by the difference between (a) solar radiation absorbed by the Earth's surface-atmosphere system and (b) the Earth's surface-atmosphere system radiation at a given point. *Note 1:* The annual average radiation balance of the whole surface of the Earth equals zero. The principal factors that determine the Earth radiation balance at various points on the Earth's surface are cloudiness, albedo, and the temperature of the underlying surface. The radiation balance at a given point on the Earth's surface varies with the time of day, the time of year, and the latitude. Near the equator the balance is always positive, whereas in the high latitudes it is nearly always negative. *Note 2:* Examples of the average annual value of the Earth's radiation balance at two points are (a) $40.3 - 34.7 = 5.6$ $cal \cdot m^{-2} \cdot min^{-1}$ (calories per square meter per minute) in the 0° to 10°N latitude zone and (b) $10.6 - 24.5 = -13.9$ $cal \cdot m^{-2} \cdot min^{-1}$ in the 80°N to 90°N latitude zone. *See also* **albedo, radiation.**

Earth satellite: A satellite that orbits the Earth. *See* **synchronous Earth satellite.**

Earth station (ES): A station that (a) is located either on the Earth's surface or within the major portion of the Earth's atmosphere, such as on board a ship, an aircraft, or a land vehicle, or fixed to the ground, (b) is intended for communications with (i) one or more space stations or (ii) one or more stations of the same kind by means of one or more reflecting satellites or other objects in space, (c) is in the space service, (d) contains all the radio and support equipment for communications with, and control of, a satellite, and (e) may not include all the baseband signal processing equipment necessary for channeling and interfacing with terrestrial communications networks and user terminal equipment. *Synonym* **satellite Earth station.** *See* **aeronautical Earth station, aircraft Earth station, base Earth station, coast Earth station, communications satellite Earth station, meteorological**

satellite Earth station, mobile Earth station, radionavigation satellite Earth station, ship Earth station, space research Earth station, space telecommand Earth station, space telemetering Earth station, space tracking Earth station.

Earth station engineering control center: The part of an Earth station that contains control consoles for switching, monitoring, patching, and supervising transmission and reception equipment.

Earth terminal (ET): 1. In a satellite link, one of the nonorbiting communications stations that (a) is on the Earth's surface or within the major portion of the Earth's atmosphere and (b) receives, processes, and transmits signals between itself and a satellite. *Note:* Earth terminals may be at mobile, fixed, airborne, and waterborne Earth terminal complexes. **2.** The portion of a satellite link that receives, processes, and transmits signals between the Earth and a satellite. *Note:* Earth terminals includes satellite terminals on board ships, aircraft, and ground vehicles. *See* **satellite Earth terminal.** *See also* **Earth station, Earth terminal complex, facility, link, satellite.** *Refer to* **Figs. C-7 (157), S-3 through S-5 (869), T-4 (994), T-5 (1012).**

Earth terminal complex (ETC): In satellite communications systems, the equipment and facilities necessary to integrate an Earth terminal into a communications network. *Note:* The Earth terminal complex (a) includes the Earth terminal and its support equipment, (b) includes any required interconnect facilities and their support equipment, and (c) does not include facilities at the site that are not necessary to establish and integrate the satellite links with the network. *See also* **downlink, Earth terminal, facility, link, satellite.**

Earth-width pulse: In a satellite, an electromagnetic pulse that (a) is generated by a device that can sense or detect radiation from the Earth, (b) is used to determine satellite attitude, i.e., orientation with respect to the Earth, and (c) for a spin stabilized satellite, has a duration depending on the time that a portion of the Earth is in the sensor field of view as the satellite spins on its axis.

EAS: extended area service.

eavesdropping: The unauthorized interception of emanations from which information may be derived, usually through the use of methods other than wiretapping.

EBCDIC: Extended Binary Coded Decimal Interchange Code.

E bend: For a transverse electromagnetic (TEM) wave propagating in a waveguide, a smooth change, i.e., a gradual change, in the direction of the axis of the waveguide, such that throughout the change the axis remains in a plane parallel to the direction of the electric field vector, **E,** i.e., parallel to the electric field transverse polarization. *Synonym* **E plane bend.** *See also* **H bend, macrobending, microbending, waveguide.**

EBS: Emergency Broadcast System.

EC: Earth coverage.

eccentric connector: *See* **double eccentric connector.**

eccentricity: *See* **cladding eccentricity.**

ECCM: electronic countercountermeasure.

echo: 1. A wave that has been reflected or otherwise returned with sufficient magnitude and delay to be perceived. *Note 1:* Echoes are frequently measured in dB relative to the directly transmitted wave. *Note 2:* Echoes may be desirable, as in radar systems, or undesirable, as in telephone systems. **2.** In computing, to print or display characters (a) as they are entered from an input device, (b) as instructions are executed, or (c) as retransmitted characters are received from a remote terminal. **3.** For an interactive computer graphics display, the immediate notification of the current value of a graphics parameter or operation as selected by the user. **4.** In computer graphics, the immediate notification of the current values provided by an input device to the operator at the display console. *See also* **echo attenuation, echo suppressor, feeder echo noise, forward echo, ghost, return loss.**

echo area: *Synonym* **scattering cross section.**

echo attenuation: The attenuation that (a) has occurred in a signal in the time between the instant it is launched and the instant an echo is received, usually at the point of launch, and (b) is based on the strength of the launched signal and the strength of the received echo. *See also* **echo attenuation ratio.**

echo attenuation ratio (EAR): 1. The ratio of (a) the strength of launched energy to (b) the strength of a received echo of the launched energy. **2.** In a four-wire or a two-wire communications circuit in which the two directions of transmission can be separated from each other, the ratio of (a) the level of the echo signal that returns to the input of the circuit to (b) the transmitted signal level. *Note:* The echo attenuation ratio is usually expressed in dB. **3.** In a fiber optic link, the ratio of (a) the level of the echo signal that returns to the input of the link to (b) the transmitted signal level. *Note:* The echo attenuation ratio is usually expressed in dB. *See also* **attenuation, echo, echo suppressor, reflection loss, return loss.**

echo canceler: *See* **echo suppressor.**

echo cancellation: **1.** In a system, the reduction of the power level of an echo. **2.** In a system, the elimination of an echo. *Note:* Echo cancellation usually is an active process in which echo signals are measured and either cancelled or eliminated by combining the inverted echo signal with the upright echo signal. *See also* **echo suppressor.**

echo chamber effect: *Synonym* **rain barrel effect.**

echo check: A check in which received data are returned to the source for comparison with the originally transmitted data. *Synonym* **loop check.** *See also* **ARQ, data transmission, feedback, information feedback.**

echo effect: In facsimile or video transmission, a defect in reproduction caused by transmission anomalies consisting of the appearance of a second outline, or several other outlines, of the scanned object. *Note:* The image outlines are displaced in the scanning direction from the outline of the normal picture. An example of the echo effect in television (TV) transmission is the extra outlines caused by reflections from objects in or near the transmission path from the transmitter to the receiver.

echoing: *Synonym* **rain barrel effect.**

echo noise: *See* **feeder echo noise.**

echoplex: An echo check used in public switched networks operating in the duplex transmission mode, i.e., the two-way simultaneous mode. *See also* **network.**

echo return loss (ERL): *See* **return loss.**

echo singing suppression circuit: *See* **volcas circuit.**

echo sounder: Equipment used to measure the depth of a body of water or the distance to an object in a body of water by measuring the time required for sound, elastic, or electromagnetic waves of known velocity to reflect from the bottom of the water body or from the distant object. *Note:* In echo sounding, damped continuous wave (cw) transmission usually is used.

echo sounding: The measurement of the depth of a body of water or the distance to an object in a body of water by measuring the time required for sound, elastic, or electromagnetic waves of known velocity to reflect from the bottom of the water body or from the distant object. *Note:* In echo sounding, damped continuous wave (cw) transmission usually is used.

echo suppressor: In a two-way circuit, a device that attenuates echo energy originating from signals transmitted in the opposite direction. *See also* **echo, echo attenuation, return loss.**

eclipse: **1.** An obscuration of light from a source by an intervening body. **2.** In celestial mechanics, the obscuration of light from one celestial body by another body that lies between the source and an observer or a photodetector. *Note:* Examples of eclipses are (a) an eclipse of the Sun by a planet, i.e., a transit of the planet, which occurs when the planet prevents light from the Sun from reaching an observer, usually on Earth, (b) an eclipse of the Sun by the Moon, i.e., a solar eclipse, and (c) an eclipse of the Moon by the Earth, i.e., a lunar eclipse. *See* **lunar eclipse, solar eclipse, transit.** *See also* **celestial mechanics, transit.**

ecliptic: The great circle traced on the celestial sphere annually by the Sun. The inclination of the ecliptic plane to the celestial equatorial plane is about 23.5°. *See also* **celestial sphere, ecliptic plane, equinox, great circle.**

ecliptic obliquity: The inclination of the ecliptic plane to the equatorial plane of the Earth, about 23.5°. *Note:* The precise value of the ecliptic obliquity for any date is obtainable from almanacs.

ecliptic plane: The plane that (a) is perpendicular to the Earth's axis of rotation and (b) contains the Earth's center. *See also* **ecliptic, ecliptic obliquity.**

ECM: electronic countermeasure.

EDACS™: Enhanced Digital Access Communications System.

edge: *See* **reference edge, stroke edge.**

edge absorption: *See* **band edge absorption.**

edge busyness: In video systems, distortion that (a) is concentrated at the edges of images and (b) is characterized by spatially varying noise or temporally varying sharpness.

edge-emitting diode: *See* **restricted edge-emitting diode.**

edge-emitting light-emitting diode (ELED): A light-emitting diode (LED) that (a) has a spectral output that emanates from between the heterogeneous layers, (b) emits light parallel to the plane of the junction, and (c) has a higher output radiance and a greater coupling efficiency to an optical fiber or integrated optical circuit than the surface-emitting LED, but not as great as the injection laser diode. *Note:* Edge-emitting and surface-emitting light-emitting diodes (LEDs) provide several milliwatts of optical power output in the 0.8-µm (micron) to 1.2-µm wavelength range at drive currents of 100 mA (milliamperes) to 200 mA. Diode lasers at these currents provide tens of milliwatts.

edge-notched card: 1. A card into which notches that represent data are punched around the edges. **2.** A card that is punched with hole patterns in tracks along the edges. *Note:* Usually the hole patterns are in the same codes that are used in punched paper tape.

edge response: The ability of a fiber optic aligned bundle to form, maintain, and resolve an image of a sharply outlined object, such as a knife edge.

edge test: *See* **Foucault knife edge test.**

EDI: electronic data interchange.

edit: To intentionally change data for a specific purpose and in accordance with prescribed rules. *Note:* Examples of editing are to (a) prepare data for a specific operation to follow, (b) correct errors, (c) delete data that do not conform to specified rules, (d) rearrange data, (e) change data format, (f) convert data to prescribed codes, and (g) apply standard processes, such as zero suppression.

edited copy: A copy of text that has been marked with amendments in the form of deletions, corrections, and additions, usually made in accordance with specified rules using standard symbols for indicating the changes to be made.

editing: *See* **full-screen editing, line editing.**

editing character: 1. A character, or a combination of characters, used to control the format of text. *Note:* Examples of editing characters are an end of line character, a skip a line character, a font change character, and a centering character. **2.** A sign or symbol used to indicate changes to be made in text. *Note:* Examples of editing characters are an insertion character, a delete character, an uppercase character, a lowercase character, a straight underscore to indicate italics, and a wavy underscore to indicate boldface.

editing command: In personal computer (PC) systems, a command, i.e., an imperative instruction, that (a) an operator or a program may give to a computer to require the computer to perform a specified operation on specified data in a selected file, such as insert a character, delete a character, start a new paragraph, insert a line, or delete a word or a line, and (b) usually is limited to the part of a specific file that is being worked on, whereas a file command is devoted to those commands that pertain to an entire file, such as retrieve, save, or delete file commands. *See* **command, file command.**

editing session: In a computer or data processing system, the period that (a) starts when an editor is invoked and ends when the editor has completed its processing and (b) includes the time the system is in the edit mode. *See also* **edit mode, editor.**

editing statement: A computer program instruction, command, or statement that specifies the editing that is to be done, such as a statement that specifies syntactic and formatting operations to be performed on data. *See also* **command, computer program, editing, format, instruction, statement.**

edit key: In personal computer (PC) systems, a key that, when actuated, enables an operator to change text, data, formulas, or format, i.e., to perform an editing action.

edit mode: 1. The status of a system when it is executing, or can execute, editing statements and editing commands during an editing session. **2.** In personal computer (PC) systems, the state or the condition of the PC that allows changes to be made in text, data, or formulas in a file or on a spreadsheet. *Note 1:* In most word processing and spreadsheet software systems, a mode indicator on the monitor, such as EDIT, will appear. In this mode, the pointer movement keys may be used to position the cursor for deleting or inserting characters as necessary. *Note 2:* Some software programs, such as Lotus 1-2-3, automatically place the system in the edit mode when an error is detected. *See also* **editing, editing commands, editing statements.**

editor: 1. A computer program, routine, or subroutine that (a) is capable of executing editing commands and editing statements, (b) usually contains editing commands or editing statements, and (c) is executed during editing sessions. **2.** A person who ensures that prepared text is correct, accurate, and logically consistent according to prescribed rules, such as grammatical, syntactical, factual, and policy rules. *Note:* An example of an editor is a person who directs the publication of a newspaper, magazine, periodical, or book.

EDTV: enhanced definition television, extended definition television.

EDTV system: enhanced definition television system, extended definition television system.

EDTV-Wide: *See* **extended definition television.**

EEP: electromagnetic emission policy.

effect: *See* **acoustooptic effect, capture effect, Dellinger effect, Doppler effect, echo effect, electrooptic effect, fluid immersion effect, flywheel effect, frequency modulation threshold effect, internal photoelectric effect, intrinsic internal photoelectric effect, Kendall effect, Kerr effect, knife edge effect, magnetooptic effect, night effect, photoconductive effect, photoelastic effect, photoelectric effect, photoelectromagnetic effect, photoemissive effect, photovoltaic effect, piezoelectric effect, Pockels effect, rain barrel effect, skin effect, speckle effect, Stark effect, thermoelectric effect, trim effect, Zeeman effect.**

effective antenna gain contour: In a steerable satellite beam, an envelope of antenna gain contours that is obtained by moving the boresight of the beam along the limits of the effective boresight area.

effective antenna length: In an antenna, the ratio of (a) the open circuit voltage, usually expressed in volts, to (b) the electric field intensity, usually expressed in volts per unit length, such as volts per meter, in which case the effective antenna length will be expressed in meters.

effective area: *See* **antenna effective area.**

effective boresight area: In a steerable satellite beam, the footprint on the surface of the Earth within which the boresight of the beam is pointed. *Note:* There may be more than one unconnected effective boresight area to which a single steerable satellite beam can be pointed, i.e., can be aimed.

effective data transfer rate (EDTR): The average number of data units, such as bits, characters, blocks, or frames, transferred per unit time from a data source and accepted as valid by a data sink. *Note:* The effective data transfer rate usually is expressed in bits, characters, blocks, or frames per second, minute, or hour. *See also* **data signaling rate, data transmission, effective speed of transmission, efficiency factor, throughput.**

effective Earth radius (EER): The radius of a hypothetical Earth for which the distance to the radio horizon, assuming rectilinear propagation, is the same as that for the actual Earth with an assumed uniform vertical gradient of refractive index. *Note:* For the standard atmosphere, the effective Earth radius is 4/3 that of the actual Earth radius. *See also* **Fresnel zone, k factor, path clearance, path profile, path survey, propagation path obstruction.**

effective height: **1.** The height of the center of radiation of an antenna above the effective ground level. **2.** In low frequency applications using loaded or non-loaded vertical antennas, the current moment in the vertical section divided by the input current. *Note:* For an antenna with symmetrical current distribution, the center of radiation is the center of distribution. For an antenna with asymmetrical current distribution, the center of radiation is the center of current moments when viewed from directions near the direction of maximum radiation. *See also* **antenna, antenna height above average terrain.**

effective input noise temperature: In a two-port network or amplifier, the source noise temperature that (a) will result in the same output noise power when connected to a noise-free network or amplifier as (b) that of the actual network or amplifier connected to a noise-free source. *Note:* The effective noise temperature is given by the relation $T_n = 290(F - 1)$ where T_n is the effective noise temperature in kelvin, F is the noise figure, and 290 K (kelvin) is the standard noise temperature. *See also* **effective noise temperature, noise, noise figure, thermal noise.**

effective isotropically radiated power (e.i.r.p.): **1.** The arithmetic product of (a) the power supplied to an antenna and (b) the antenna gain. **2.** The arithmetic product of (a) the power supplied to an antenna and (b) the antenna gain relative to an isotropic antenna. **3.** At a point in the far field of an antenna, the arithmetic product of (a) the power supplied to the antenna and (b) the antenna gain in the direction from the antenna to the point. *See also* **antenna gain, effective radiated power, isotropic antenna.**

effective mode volume: In a waveguide, such as an optical fiber, the square of the product of the diameter of the near field pattern and the sine of the radiation angle of the far field pattern. *Note 1:* The diameter of the near field radiation pattern is taken as the full-width half-maximum irradiance, and the radiation angle is taken at half maximum irradiance. *Note 2:* In a multimode fiber, the effective mode volume is proportional to the breadth of the relative distribution of optical power among the modes. *Note 3:* The effective mode volume is not a spatial volume but rather an optical volume that has the units of the product of area and solid angle. *See also* **mode volume, radiation pattern.**

effective monopole radiated power (e.m.r.p.): The product of (a) the power supplied to an antenna and (b) its gain, in a given direction, relative to a short vertical antenna. *See also* **effective radiated power.**

effectiveness: *See* **shielding effectiveness.**

effective noise temperature: The temperature that corresponds to a given noise level from all sources, including thermal noise, source noise, induced noise, and internal noise. *Note:* If F is the noise figure (NF) numerical value and 290 K the standard noise temperature, then the effective noise temperature is given by the relation $T_n = 290(F - 1)$. *See also* **effective input noise temperature, noise, noise figure, thermal noise.**

effective power: In steady state sinusoidal voltages and currents in a circuit, the product of (a) the root mean square (rms) voltage, i.e., the effective voltage, (b) the rms current, i.e., the effective current, and (c) the power factor, i.e., the cosine of the phase angle between the voltage and the current. *Note 1:* Only the effective power, i.e., the real or true power, delivered to or consumed by the load, is expressed in watts. The apparent power is expressed in volt•amperes. The reactive power is expressed in vars (volt•amperes, reactive). *Note 2:* In vector notation, the effective power is expressed as the dot product of the vector voltage and the vector current. *Note 3:* Effective power is a scalar quantity, not a vector quantity, i.e., not a phasor quantity. *Synonyms* **real power, true power.** *See also* **absolute power, reactive power, relative power, volt•amperes.**

effective radiated power (e.r.p. or ERP): 1. The product of (a) the power supplied to an antenna and (b) the antenna gain in a given direction. **2.** The product of (a) the power supplied to an antenna and (b) its gain relative to a half-wave dipole in a given direction. *Note 1:* If the direction is not specified, the direction of maximum gain is assumed. *Note 2:* The ratio of (a) the power gain referred to an isotropic antenna to (b) the power gain referred to a standard half-wave dipole antenna is equivalent to 2.15 dB. *See also* **antenna, directive gain, effective isotropically radiated power, equivalent monopole radiated power, mean power.**

effective speed of transmission: *Synonym* **effective transmission rate.**

effective transfer rate: The average number of data units, such as binary digits, characters, or blocks, transferred between two points and accepted as valid at the receiving point per unit of time, i.e., the average number of error-free data units received per unit of time. *Synonym* **average transfer rate.**

effective transmission rate (ETR): The rate at which information is processed by a transmission facility. *Note 1:* The effective transmission rate is calculated as (a) the measured number of units of data, such as bits, characters, blocks, or frames, transmitted during a significant performance measurement period divided by (b) the measurement period. *Note 2:* The effective transmission rate is usually expressed as a number of units of data per unit time, such as bits per second or characters per second. *Synonyms* **average rate of transmission, average transmission speed, effective speed of transmission, effective transmission speed.** *See also* **data signaling rate, data transmission, effective data transfer rate, efficiency factor, instantaneous transmission rate, maximum user signaling rate, speed of service, throughput.**

effective transmission speed (ETS): *Synonym* **effective transmission rate.**

effector: *See* **format effector.**

effects on electronics: *See* **transient radiation effects on electronics.**

efficiency: *See* **antenna efficiency, block efficiency, block rate efficiency, coupling efficiency, differential quantum efficiency, luminous efficiency, luminous radiation efficiency, optical power efficiency, quantum efficiency, radiant efficiency, radiation efficiency, response quantum efficiency, source coupling efficiency, source efficiency, source power efficiency, transmission efficiency.**

efficiency factor: In data communications, the ratio of (a) the time needed to automatically transmit a text at a specified modulation rate to (b) the time actually required to receive the same text at a specified maximum error rate. *Note 1:* All of the communications facilities are assumed to be in the normal condition of adjustment and operation. *Note 2:* Telegraph communications may have different temporal efficiency factors for the two directions of transmission. *Note 3:* The practical conditions of measurement should be specified, especially the duration of the performance measurement period and the error detection criteria. *Synonym* **efficiency ratio.** *See* **block correction efficiency factor.** *See also* **data signaling rate, data transmission, effective data transfer rate, effective transmission rate, instantaneous transmission rate, speed of service, throughput.**

efficiency ratio: *Synonym* **efficiency factor.**

efficiency test: *See* **termination efficiency test.**

efficient interconnection: *See* **comparably efficient interconnection.**

EFIS: electronic flight information system.

EFL: equivalent focal length.

EHF: extremely high frequency.

EIA Class IVa optical fiber: *Synonym* **dispersion-unshifted optical fiber.**

EIA Class IVb optical fiber: *Synonym* **dispersion-shifted optical fiber.**

EIA interface: Electronic Industries Association interface.

EIA standard: Electronic Industries Association standard.

eigenvalue: A constant in an equation or an expression that can assume any one of a set of finite values and still have the equation or the expression be a solution of an integral, a differential, an equation, or a function. *Note:* In the solution of the wave equation, each eigenvalue produces another propagation mode that is allowable for a particular waveguide. The allowable eigenvalues are determined by physical factors, such as waveguide dimensions, and for a given guide may be simply multiples of π or of wavelength. *See also* **mode, propagation mode, wave equation.**

eight hundred (800) service: A network-provided service feature that allows call originators within specified rate areas to place toll telephone calls to subscribers of the service without charge to the call originator. *Note:* The charges are billed to the call receiver. *Synonyms* **800 service, Inward Wide Area Telephone Service.**

eight-zero substitution: *See* **bipolar with eight-zero substitution.**

either-way communications: *Synonym* **two-way alternate communications.**

eject key: A key that controls the release, movement, or removal of data media from communications, computer, data processing, and control systems. *Note:* Examples of eject keys are (a) a key that, when actuated, causes a video cassette recorder (VCR) to move the cassette so that it can be removed from the VCR and (b) a key that, when actuated, enables a cassette to be removed from a tape recorder.

EKMS: electronic key management system.

elastic: *See* **photoelastic.**

elastic buffer: A buffer that has an adjustable capacity for data or that introduces an adjustable delay of signals. *See also* **buffer; first in, first out; queue traffic; variable length buffer.**

elastic wave: A wave that consists of a series of pulses of elastic deformation traveling in a propagation medium. *Note 1:* The propagation speed and attenuation of an elastic wave depend on the elastic properties of the medium. *Note 2:* Examples of elastic waves are sonar pulses in water, sound waves in air, and ultrasound pulses in mercury.

E layer: In the E region, a layer of increased free electron density caused by the ionizing effect of solar radiation. *Note 1:* The E layer, because of its low altitude and low nighttime ion density, provides useful, mainly daytime, radio wave reflections for distances up to 2000 km. *Note 2:* The E layer is important for daytime high frequency (HF) propagation and for nighttime medium frequency (MF) and low frequency (LF) propagation. *Note 3:* The noontime minimum sun spot activity free electron density is roughly 10^5 free electrons per cubic centimeter. It is increased approximately 50% during maximum sun spot activity. *Note 4:* The E layer exhibits sporadic characteristics. *Synonym* **Heaviside layer, Kennelly-Heaviside layer.** *See also* **critical frequency, D layer, E region, F layer, ionosphere, sporadic E.**

electric: Pertaining to the natural phenomena of the motion, effects, and characteristics of charged particles, particularly electrons, that exhibit attraction and repulsion in accordance with Coulomb's law. *See* **transverse electric.** *See also* **Coulomb's law, electron.**

electrical beam tilt: *See* **antenna electrical beam tilt.**

electrical cable: *See* **optoelectrical cable.**

electrical charge: *Synonym* **electronic charge,**

electrical conductivity: In a material, (a) a measure of the ability of the material to conduct an electric current when under the influence of an applied electric field and (b) the constant of proportionality in the constitutive relation between the electrical current density and the applied electric field strength at a point in the material, given by the relation $\mathbf{J} = \sigma\mathbf{E}$, where $\mathbf{J}$ is the vector current density, σ is the electrical conductivity, and $\mathbf{E}$ is the vector electric field strength. *Note:* If $\mathbf{J}$ is in amperes per square meter and $\mathbf{E}$ is in volts per meter, the electrical conductivity is given as $\sigma = \mathbf{J}/\mathbf{E}$ amperes/volt•meter, $(\text{ohm}\cdot\text{meter})^{-1}$, or mhos/meter.

electrical conductor: A material (a) in which there are large numbers of atoms with electrons that are easily excited from the valence band into the conduction band and hence are easily moved in large quantities upon application of small electric fields, (b) in which it is difficult to maintain a large potential difference, i.e., voltage, between two points because of the large currents, which prevent the buildup or segregation of charges at any specific point that is required to maintain a potential difference, and (c) that is used to provide equipotential surfaces between points or terminals because of its high conductivity, i.e., low resistance to current flow. *Note:* Usually, good electrical conductors are poor light conductors, and vice versa. Some materials, such as opaque rubber, ceramics, and plastics are poor electrical and light conductors. *See also* **conduction band, equipotential surface, valence band.**

electrical degree: 1/360 of a cycle of a wave. *Note:* A complete cycle of a wave is measured from a point in one wave to the corresponding point in the next wave, such as from crest to crest or from a zero crossing from negative to positive value to the corresponding next zero crossing from negative to positive. *See also* **spatial degree.** *Refer to* **Figs. P-4 (712), Q-1 (786).**

electrical delay line: *Synonym* **electromagnetic delay line.**

electrical interface: In a communications, computer, or data processing system, a point (a) that serves as the boundary between two electrical systems, subsystems, or components, (b) that enables the systems, subsystems, or components to interoperate, and (c) is precisely specified, the specifications usually including (i) the type, quantity, and function of the interconnection circuits and (ii) the type and form of signals to be interchanged via those circuits. *See also* **interoperability.**

electrical length: **1.** A length, such as a point-to-point measure of phase shift of a wave, expressed in wavelengths, radians, or degrees. *Note:* When electrical length is expressed in angular units, it is the length, in wavelengths, multiplied by 2π to give radians, or by 360 to give degrees. **2.** In a periodic phenomenon, such as a wave, a length, i.e., a measurement of transit time or a distance between specified points, expressed in units such as wavelengths, radians, degrees, or seconds. *Note 1:* Electrical length is a convention established to express length in terms of time durations and spatial distances in a representation of any periodic phenomenon, not just an electrical phenomenon. *Note 2:* A whole wavelength, i.e., one complete cycle of the wave, whether in absolute distance or in time units, is considered to have 360° or 2π radians. These dimensionless relative units can then be used to express lengths or locations of points within a wave. **3.** In an antenna or a transmission line, the effective length as it affects transmission performance, such as the radiation pattern of an antenna or the location of nodes in a transmission line. *Note:* The electrical length differs from the physical length because of many factors, such as geometric size, the distribution of resistances and reactances, and the different units of measure used for expressing physical and electrical length. *See also* **wavelength.**

electrically despun antenna: An antenna that (a) has no moving mechanical parts and (b) can electrically direct its main lobe in a given fixed direction while the platform on which it is mounted pitches, rolls, and sways. *Note:* An example of an electrically despun antenna is a phased array antenna that is mounted on a spinning satellite and that can fix its main beam in the direction of the Earth at all times. *See also* **mechanically despun antenna.**

electrically powered telephone: A telephone in which the operating power is obtained either from batteries located at the telephone, i.e., a local battery, or from a central office, i.e., a common battery. *See also* **sound-powered telephone.**

electrical requirement: A requirement that defines a specific electrical characteristic that a system or a system component must have, such as operating voltage, current, power factor, input signal level, output power level, and duty cycle. *See also* **requirement.**

electrical signal: *Synonym* **electronic signal.**

electrical transmission: **1.** The transmission of messages by any means that (a) includes, in whole or in part, an electrical circuit and (b) may include other components, such as radio and fiber optic components. *Note:* Examples of electrical transmission include transmission by telephone, telegraph, teletype, facsimile, radio, and television. **2.** A message transmitted by electrical means, such as a telegram and a teletype message.

electric charge: *Synonym* **electronic charge.**

electric code. *See* **National Electric Code.**

electric current: The amount of electronic charge that (a) passes through a given cross-sectional area per unit time and (b) consists of the net drift of charges, such as electrons, past the point per unit time. *Note:* The SI (Système International d'Unités) unit of current is the ampere, equal to 1 coulomb per second. *See also* **coulomb, electronic charge, second.**

electric field: The effect produced by the existence of an electric charge, such as an electron, ion, or proton, in the volume of space or material medium that surrounds the charge. *Note 1:* Each electric charge in a distribution of charges contributes to the whole field at a point, on the basis of superposition. *Note 2:* An electric charge placed in an electric field has a force exerted upon it by the field. The magnitude of the force is the product of the electric field strength and the magnitude of the charge. The force on the charge is a measure of the electric field strength. Thus, $\mathbf{E} = \mathbf{F}/q$, where **E** is the electric field strength, **F** is the force the field exerts on the charge, and q is the charge. The direction of the force is in the same direction as the field for positive charges, and in the opposite direction for negative charges. Moving an electric charge against the field requires work, i.e., requires energy, and increases the potential energy of the charge. *Note 3:* Electric field strength is measurable as (a) force per unit charge, (b) potential difference per unit distance, i.e., a voltage gradient, such as volts/meter, and (c) a number of electric lines of flux per unit of cross-sectional area. *Note 4:* The force and the electric field are vector quantities. The charge is a scalar quantity. *Refer to* **Fig. I-1 (436).**

electric field component: The part of an electromagnetic wave that (a) consists of a time-varying electric field that interacts with a magnetic field to produce propagation of a field, i.e., of electromagnetic energy, in a direction perpendicular to both fields and (b) the direction and magnitude of which, along with the magnetic field component relative to the direction of the interface surface as well as the refractive indices, electric permittivities, magnetic permeabilities, and electrical conductivities of the transmission media on both sides of the interface surface, governs the reflection, refraction, and transmission that occur at media interfaces. *See also* **electric field, magnetic field component, Maxwell's equations, wave equation.**

electric field force: 1. The force exerted by an electric field upon an electronic charge placed in the field, given by the relation $\mathbf{E} = \mathbf{F}/q$, where **E** is the electric field strength, **F** is the force the field exerts on the charge, and q is the amount of charge. *Note:* The direction of the force is in the same direction as the field for positive charges, and in the opposite direction for negative charges. **2.** The force of attraction or repulsion between two electric charges, each subject to the other's electric field, such that like charges, either positive or negative charges, repel each other, whereas opposite charges, one negative and one positive, attract each other. *See also* **Coulomb's law, electronic charge.**

electric field intensity: *Synonym* **electric field strength.**

electric field strength: 1. The intensity, i.e., the amplitude or the magnitude, of an electric field at a given point. *Note 1:* In radiometric measurements, the electric field strength usually refers to the root mean square (rms) value of the electric field, expressed in volts per meter. *Note 2:* An instantaneous electric field strength usually is given as a gradient, and therefore is a vector quantity, because it has both magnitude and direction. *Note 3:* The units of electric field strength may be volts per meter, kilograms per coulomb, or electric lines of flux per square meter divided by the electric permittivity of the medium at the point where it is measured. **2.** The field strength produced at a point because of the existence of an electronic charge, given by the relation $\mathbf{E} = (q/4\pi\varepsilon_0 r^2)\mathbf{a_r}$ where **E** is the electric field strength, q is the electronic charge that produces the electric field, ε_0 is the electric permittivity of the intervening space between the point and the charge, r is the distance from the charge to the point, and $\mathbf{a_r}$ is the unit vector in the direction of the line from the charge to the point. *Synonym* **electric field intensity.** *See also* **electric permittivity, magnetic field strength.**

electric field vector: 1. At a point, a vector that represents the magnitude and the direction of the electric field strength at that point. **2.** In an electromagnetic wave, such as a radio wave or a lightwave, the vector that represents the instantaneous electric field strength in magnitude and direction at any point in a medium in which the wave is propagating. *See also* **magnetic field, magnetic vector.**

electricity: *See* **static electricity.**

electric mode: *See* **transverse electric mode.**

electric permittivity: A parameter of free space or a material that serves as the constant of proportionality between the magnitude of the force exerted between two point electric charges of known magnitude separated by a given distance. *Note 1:* The magnitude of the force is defined by the relation $F = q_1 q_2/4\varepsilon\pi d^2$ where q_1 and q_2 are the point electric charges, F is the force between them, d is their distance of separation, ε is the electric permittivity of the medium in which they are embedded, and π is approximately 3.1416. The direction of the force is along the line joining the two point charges, being one of attraction if the charges are of opposite polarity and repulsion if they are of the same polarity. The electric permittivity, magnetic permeability, and electrical conductivity determine the refractive index of a material. *Note 2:* For dielectric materials, i.e., electrically nonconducting media, such as the glass and the plastic used to make optical fibers, the electrical conductivity is zero, the relative velocities of electromagnetic waves, such as lightwaves, propagating within them, are given by the relation $v_1/v_2 = \{(\mu_2\varepsilon_2)/(\mu_1\varepsilon_1)\}^{1/2}$ where v is the velocity, and μ and ε are the magnetic permeability and the electric permittivity, respectively, of media 1 and 2. If propagation medium 1 is free space, the equation becomes the relation $c/v = \{(\mu\varepsilon)/(\mu_0\varepsilon_0)\}^{1/2}$ where c is the velocity of light in a vacuum, v is the velocity in a material medium, the nonsubscripted values of the constituent constants are for the material medium, and the subscripted values are for free space. However, c/v is the definition of the refractive index of a medium relative to free space. From this is obtained the relation $n = (\mu_r\varepsilon_r)^{1/2}$ where n is now a dimensionless number and relative to free space. The magnetic permeability and electric permittivity are also relative to free space, or air. As stated above, for dielectric materials, μ_r is approximately 1, being very much greater than 1 only for ferrous substances and slightly greater than 1 for a few metals, such as nickel and cobalt. For a few diamagnetic materials, μ_r is slightly less than unity. Hence, for the glass and plastics used to make optical fibers and other dielectric waveguides, the refractive index, relative to a vacuum, or air, is given by the relation $n \approx \varepsilon_r^{1/2}$ where ε_r is the electric permittivity of the propagation medium. Refractive indices usually are given relative to a vacuum, or relative to 1.0003 for air near the Earth's surface and closer to 1, i.e., closer to unity, as the air gets thinner in the upper atmosphere and outer space. In absolute units for free space, the electric permittivity is $\varepsilon = 8.854 \times 10^{-12}$ F/m (farads per meter), and the magnetic permeability is $\mu = 4\pi \times 10^{-7}$ H/m (henries per meter). The reciprocal of the square root of the product of these values yields approximately 2.998×10^8 m/s (meters per second), the speed of light in a vacuum. Because the refractive index for a propagation medium is defined as the ratio between the velocity of light in a vacuum and the velocity of light in the medium, the ratio becomes unity when the propagation medium is a vacuum. *See* **relative electric permittivity.** *See also* **refractive index.** *Refer to* **Fig. I-1 (436).**

electric vector: A representation of the magnitude and direction of an electric field, such as that associated with an electromagnetic wave or an electrostatic field. *See also* **electric field, electromagnetic wave, electrostatic field, magnetic vector, Poynting vector.**

electric wave: A wave that (a) is launched by a source, such as an accumulated and variable electric charge, that produces an electric field, such as the field between the plates of a condenser, (b) does not travel very far from the source, (c) consists of an electric field, i.e., electric lines of force, that emanates from the source, (d) when varied in magnitude or reversed in polarity, may be used to induce currents in conductors, and (e) is not necessarily associated with a magnetic wave and not necessarily part of an electromagnetic wave. *See* **transverse electric wave.** *See also* **magnetic wave, electromagnetic wave.**

electrochemical recording: Recording in which a chemical reaction is caused by the passage of a signal-controlled current through the sensitized portion of the surface on which the recorded copy is made. *Note:* Electrochemical recording may be used in facsimile systems. *See also* **direct recording, electrolytic recording, facsimile, recording.**

electrochemical series: A sequence of metal elements and alloys that indicates their relative chemical activity or affinity for replacing one another in electrolytic chemical action and thus corroding each other. *Note 1:* Electrical currents through joints of two different metals enhance the chemical action, particularly in the presence of moisture. *Note 2:* In constructing electrical equipment, such as communications, computer, data processing, and control equipment, particularly antennas and structural members, metals in contact should be as close together as possible in the electrochemical series, thus reducing the tendency to replace one another. For example, copper or brass should be zinc-plated before connection to aluminum, i.e., copper and brass should not be placed in contact with aluminum. Platings or coatings should be used to prevent physical contact of metals far apart in the electrochemical series. An iron nail placed in a solution of copper sulfate will develop a coating of copper because iron is much higher than copper in the electrochemical series and

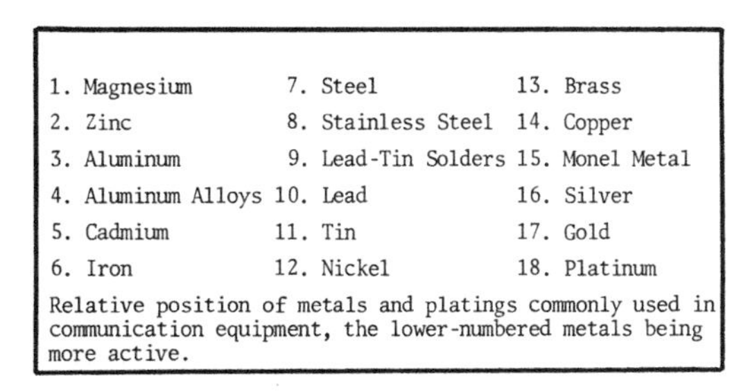

1. Magnesium	7. Steel	13. Brass
2. Zinc	8. Stainless Steel	14. Copper
3. Aluminum	9. Lead-Tin Solders	15. Monel Metal
4. Aluminum Alloys	10. Lead	16. Silver
5. Cadmium	11. Tin	17. Gold
6. Iron	12. Nickel	18. Platinum

Relative position of metals and platings commonly used in communication equipment, the lower-numbered metals being more active.

Fig. E-1. The **electrochemical series,** *i.e.,* Galvanic series. The metals toward the top of the series are chemically more active than those toward the bottom. When metals are in contact, the lower-numbered metal becomes the anode of an electocytic cell, which (a) causes corrosion by reaction with environmental substances, (b) produces salts and various other compounds, and (c) reduces the strength of the remaining metal. Metals that must be placed in physical contact should be as close together as possible in the series because this proximity lowers the potential difference between them and thus reduces the corrosion rate.

replaces the copper in the copper sulfate solution. *Synonym* **Galvanic series.** *Refer to* **Fig. E-1.**

electrode subsystem: *See* **earth electrode subsystem.**

electrographic recording: *See* **electrostatic recording.**

electroluminescence: The direct conversion of electrical energy into light. *Note 1:* Electroluminescence does not produce incandescence in the emitting material. *Note 2:* An example of electroluminescence is photon emission resulting from electron-hole recombination in a positive-negative (pn) junction, such as occurs in light-emitting diodes (LEDs). The emitted optical radiation is in excess of the radiation caused by thermal emission that would result from the application of electrical energy. *See also* **injection laser diode, light-emitting diode.**

electrolytic recording: 1. Electrochemical recording in which the chemical change is made possible by the presence of an electrolyte. **2.** Electrochemical facsimile recording in which the recorded copy is made by the passage of a signal-controlled current through an electrolyte which causes metallic ions to be deposited, thus forming an image of the object, i.e., forming the recorded copy. *See also* **direct recording, electrochemical recording, facsimile recorder.**

electromagnetic: Pertaining to the interaction and interdependence of coupled electric and magnetic fields that (a) interchange their energies at various rates, i.e., various frequencies, (b) may propagate in vacuum and material media over long distances, (c) may develop standing wave patterns, and (d) may be reflected and refracted as well as absorbed and scattered. *See also* **electromagnetic wave, Maxwell's equations, Poynting vector, wave equation.**

electromagnetic compatibility (EMC): 1. The condition that exists when telecommunications equipment (a) performs its intended function in a common electromagnetic environment without causing or experiencing unacceptable degradation because of unintentional electromagnetic interference to or from other equipment in the same environment, (b) includes engineering design characteristics or features in all electromagnetic radiating and receiving components in order to eliminate or reject undesired signals and enhance operating compatibilities, (c) is designed with sufficient flexibility to ensure interference-free operation, (d) is designed under frequency management concepts and doctrines that maximize operational effectiveness, (e) includes full flexibility in order that operators may deal with interference if and when it occurs, and (f) is designed to operate with a defined margin of safety. *See also* **compatibility, electromagnetic compatibility analysis, electromagnetic interference.**

electromagnetic compatibility (EMC) analysis: 1. Analysis of a system, a subsystem, a facility, or equipment to determine its electromagnetic compatibility (EMC) status. *Note 1:* The electromagnetic compatibility (EMC) analysis may be a projected analysis before construction or an analysis after construction. *Note 2:* In a historic note, in 1952, during the continued development of the Electronic Discrete Variable Automatic Computer (EDVAC) at the U.S. Army Aberdeen Proving Ground in Maryland, a phenomenon occurred in which the EDVAC would malfunction for no known reason. The electromagnetic environment became a prime suspect. An electromagnetic compatibility analysis revealed that every time a local radar, part of the antiaircraft Skysweeper development program, aimed its beam at the EDVAC, the EDVAC would malfunction. The radar induced extraneous pulses and transients and caused pulse dropouts, which immediately brought the computer to a halt because illegal characters would occur. **2.** An examination of equipment, subsystems, or systems (a) that is conducted to determine their electromagnetic compatibility status and (b) that can be conducted at any time during the lifecycle. *See also* **electromagnetic compatibility, electromagnetic radiation hazard, illegal character.**

electromagnetic delay line: A sequence of distributed or lumped electric circuit parameters, i.e., circuit elements, that introduce a delay in a signal because of the propagation time of the signal from one point to another in the sequence of circuit elements. *Note 1:* The time constants of the circuit elements introduce the delay in each element as the signal propagates. A rectangular pulse introduced at one end usually has the higher frequencies attenuated more than the lower frequencies, so that the output signal is less square than the input signal. *Note 2:* The electromagnetic delay line can be used to bring about (a) synchronism at any point and (b) time coincidence and synchronism at combinational circuits by delaying the earlier signals in order to obtain proper time coincidence gating. *Synonym* **electrical delay line.** *See also* **combinational circuit, delay line, distributed constant, gating, lumped constant.**

electromagnetic effect: *See* **photoelectromagnetic effect.**

electromagnetic emission control (EEC): 1. The control of electromagnetic emissions, such as radio, radar, and sonar transmissions, for the purpose of preventing or minimizing their use by unintended recipients. **2.** Emission control applicable to electromagnetic radiation, signal transmission, and related communications-electronics emissions only. *See also* **electromagnetic interference, electronic warfare, emanations security, emission control, TEMPEST.**

electromagnetic emission policy (EEP): Policy that (a) governs the operation of equipment that emits electromagnetic, electrostatic, magnetostatic, and electromagnetostatic field energy, and (b) is concerned with the bands of frequencies and the use of the equipment on which transmission is permitted. *Synonym* **electronic emission policy.** *See also* **electromagnetic emission control, emission control.**

electromagnetic energy density: In a standing or propagating electromagnetic wave at a point in free space or a material medium, the energy contained per unit volume. *Note 1:* The electromagnetic energy density is (a) a scalar quantity, (b) a function of the product of the electric and magnetic field strengths, and (c) usually expressed in joules per cubic meter. *Note 2:* In a propagating electromagnetic wave, the energy flow rate is the product of the energy density and the velocity, resulting in watts per square meter, i.e., the irradiance, if the energy density is expressed in joules per cubic meter and the velocity in meters per second. *See also* **irradiance.** *Refer to* **Fig. R-3 (799).**

electromagnetic environment (EME): 1. For a telecommunications system, the time-varying, frequency-dependent, spatial and temporal distribution of electromagnetic energy surrounding a given site. *Note:* The electromagnetic environment may be expressed in terms of the spatial and temporal distribution of electric field strength (volts/meter), irradiance (watts/square meter), or energy density (joules/cubic meter). **2.** The resulting product of the power and time distribution, in various frequency ranges, of the radiated or conducted electromagnetic emission levels that may be encountered by an organization, a system, or a platform when performing its assigned mission in its intended operational environment. **3.** At a specified point, or in the volume immediately surrounding given equipment, the sum of (a) electromagnetic interference (EMI), (b) electromagnetic pulse (EMP), (c) hazards of electromagnetic radiation to personnel, ordnance, and volatile materials, such as fuels and gases, and (d) natural phenomena effects, such as lightning and precipitation static (p-static). *See also* **electronic reconnaissance, electronics intelligence, frequency, interference, precipitation static.**

electromagnetic field: A field characterized by electric and magnetic fields that (a) vary with time at a point and, except when confined to a cavity, propagate as a wave from its source, and (b) are interdependent, remain coupled, interchange their energies, and interact with one another rather than exist independently of each other, as do the electrostatic, staticmagnetic, and electromagnetostatic fields, which decay rapidly with distance from their sources even when they vary with time, position, orientation, or magnitude. *Note 1:* Except for geometric spreading, electromagnetic fields are not affected by free space, i.e., unbounded vacuum that does not contain any other field. *Note 2:* Electromagnetic fields are affected by the characteristics of the material media through which they propagate. *See also* **electric field, magnetic field, propagation medium.**

electromagnetic hazard: *See* **hazards of electromagnetic radiation to fuel, hazards of electromagnetic radiation to ordnance, hazards of electromagnetic radiation to personnel.**

electromagnetic interference (EMI): 1. An electromagnetic disturbance that (a) interrupts, obstructs, or otherwise degrades or limits the effective performance of electronic or electrical equipment, (b) is coupled into a circuit by electromagnetic radiation from sources outside the circuit, (c) is coupled by any means, such as conductive, capacitive, or inductive coupling, (d) may be from a wide range of sources and frequencies, such as those of lightwaves, radio waves, radar pulses, gamma rays, high energy neutrons, X rays, and microwaves, (e) may endanger the functioning of radionavigation, (f) may constitute a safety hazard, and (g) may be unintentional or deliberate. *Note:* Electromagnetic interference is considered to have occurred when an unacceptable or undesirable response, malfunction, degradation, or interruption of the intended operation or performance of equipment has occurred as a result of the interference. **2.** An electromagnetic disturbance that interrupts, obstructs, or otherwise degrades or limits the effective performance of electronic or electrical equipment. *Note:* Electromagnetic interference (EMI) can be induced intentionally, as in some forms of electronic warfare, or unintentionally, as a result of various emissions, such as transmissions, spurious emissions and responses, and intermodulation products. *See also* **electromagnetic compatibility, electromagnetic emission control, electromagnetic vulnerability, electronic counter-countermeasure, electronic countermeasure, electronic reconnaissance, electronics intelligence, electronic warfare, interference, interference emission, radio frequency interference.**

electromagnetic interference (EMI) control: The control of radiated and conducted energy so that the emissions unnecessary for system, subsystem, or equipment operation (a) are reduced, minimized, or eliminated and (b) are controlled regardless of their origin within the system, subsystem, or equipment. *Note 1:* Effective electromagnetic interference (EMI) control with effective susceptibility control enables electromagnetic compatibility (EMC) to be achieved. *Note 2:* Electromagnetic radiated and conducted emissions, regardless of their origin within the system, subsystem, or equipment, are held to a level of acceptability for specific purposes, such as prevention of interference, compromise, and loss of energy. *See also* **compromise, electromagnetic interference, electronic countercountermeasure, electronic countermeasure, electronics security, electronic warfare, interference, interference emission.**

electromagnetic intrusion: The intentional insertion of electromagnetic energy into transmission paths in any manner, with the objective of deceiving operators or of causing confusion. *See also* **electronic deception, electronic warfare, interference.**

electromagnetic mode: *See* **transverse electromagnetic mode.**

electromagnetic photodetector: *See* **photoelectromagnetic photodetector.**

electromagnetic pulse (EMP): 1. The electromagnetic radiation from a nuclear explosion caused by Compton recoil electrons and photoelectrons from photons scattered in the materials of the nuclear device or in a surrounding medium. *Note 1:* The electric and magnetic fields resulting from electromagnetic pulses (EMPs) may couple with electrical/electronic systems to produce damaging current and voltage surges. *Note 2:* Electromagnetic pulses (EMPs) may be caused by nonnuclear means. **2.** A broadband, high intensity, short duration burst of electromagnetic energy. *Note 1:* In the case of a nuclear detonation, the resulting electromagnetic pulse (EMP) consists of a continuous frequency spectrum, with most of the energy distributed throughout the lower frequencies between 3 Hz and 30 kHz. *Note 2:* The electric and magnetic fields caused by an electromagnetic pulse (EMP) may couple with electrical and electronic systems to produce damaging current and voltage surges. *See also* **high altitude electromagnetic pulse, pulse.**

electromagnetic pulse resistance: The ability of a system or a component, such as a communications, computer, data processing, or control system, or component, to withstand the effects of a high energy burst of electromagnetic energy, such as might be produced by a nuclear detonation.

electromagnetic radiation (EMR): Radiation that (a) consists of electromagnetic waves, i.e., oscillating electric and magnetic fields, (b) propagates with the speed of light, and (c) includes cosmic rays, gamma rays, X rays, ultraviolet light, visible light, infrared light (heat), radar pulses, microwaves, and radio waves. *Note:* Electromagnetic radiation is propagated with a phase velocity, v, in the propagation medium, given by the relation $v = \lambda f = c/n$ where v is the phase velocity, λ is the wavelength in the propagation medium, f is the frequency generated by the source, c is the velocity of light in a vacuum, i.e., approximately 3×10^8 m•s^{-1} (meters per second), and n is the refractive index of the propagation medium. *See also* **far field radiation pattern, far field region, intermediate field region, near field region, radiation pattern, radiation scattering.** *Refer to* **Fig. R-2 (799).**

electromagnetic radiation control (ERC): 1. Measures taken to (a) minimize undesired electromagnetic radiation emanating from a system or a component, (b) control all electromagnetic radiation, useful and nonuseful, aiming toward its reduction, redistribution, minimization, elimination, frequency change, modulation, power level, spreading, and repolarization, (c) control electromagnetic radiation for purposes of security and the reduction of interference, especially on board ships and aircraft, (d) minimize electromagnetic interference, and (e) control electromagnetic radiation when the main purpose is security, particularly against missile homing, interception, jamming, and navigation. **2.** A national operational plan to minimize the use of electromagnetic radiation in the United States and its possessions and the Panama Canal Zone in the event of an attack, an imminent attack, or a threat of an attack, and thereby preclude the use of the radiation as an aid to the navigation of devices and systems, such as hostile aircraft, ships, and guided missiles. *See also* **electronic warfare.**

electromagnetic radiation hazard (RADHAZ or EMR hazard): A hazard caused by a transmitter and antenna installation that (a) generates electromagnetic radiation in the vicinity of ordnance, personnel, or fueling operations in excess of established safe levels, (b) increases the existing radiation levels to a level hazardous to personnel, fueling, or ordnance installations located in an area illuminated by electromagnetic radiation at a level that is hazardous to the planned operations or occupancy, and (c) exists when an electromagnetic field of sufficient strength is generated to (i) induce or otherwise couple currents or voltages of magnitudes large enough to potentially initiate electroexplosive devices or other sensitive explosive components of weapon systems, ordnance, or explosive devices, (ii) cause harmful or injurious effects to humans and wildlife, and (iii) create sparks of sufficient magnitude to ignite flammable mixtures of materials in the affected area. *See also* **electromagnetic vulnerability, electronic warfare, hazards of electromagnetic radiation to fuel, hazards of electromagnetic radiation to ordnance, hazards of electromagnetic radiation to personnel.**

electromagnetic spectrum: 1. The range of frequencies, or wavelengths, of electromagnetic radiation that extends from near zero to infinity. *Note 1:* Light is the visible portion of the electromagnetic spectrum. Heat is the infrared portion. **2.** The distribution of frequencies in a specified band of electromagnetic radiation. *Note:* A particular electromagnetic spectrum could include a single frequency or a wide range of frequencies, depending on the amount of electromagnetic power in each of the frequencies. *See also* **band, frequency, frequency spectrum, frequency spectrum designation, optical spectrum, visible spectrum, visual spectrum.**

electromagnetic surveillance: A search for, and identification of, electromagnetic signals in a specified area, at a specified location, or perhaps in a specified frequency range. *See* **exploratory electromagnetic surveillance.**

electromagnetic survivability: The ability of a system, subsystem, or equipment to resume functioning without evidence of degradation following temporary exposure to an adverse electromagnetic environment, such as might be caused by (a) excessive sunspot activity, (b) high-intensity jamming, infrared, or ultraviolet radiation bursts, or (c) a nuclear detonation. *Note:* System, subsystem, or equipment performance may be degraded during exposure to adverse electromagnetic environment, but if the electromagnetic level of survivability is high enough, the system will not experience permanent damage, such as component burnout, that will prevent proper operation when the adverse electromagnetic environment subsides or is removed. *See also* **survivability.**

electromagnetic theory: The theory that (a) concerns the propagation of energy by coupled electric and magnetic fields and (b) is described by Maxwell's equations and the relations derived from them. *See also* **Maxwell's equations.**

electromagnetic vulnerability (EMV): 1. A measure of the extent to which the characteristics of a system cause the system to degrade to a level such that it is incapable of performing its designated mission (a) as a result of being subjected to a certain level of electromagnetic environmental effects or conditions or (b) as a result of electromagnetic interference. *See also* **electromagnetic pulse, high altitude electromagnetic pulse, pulse.**

electromagnetic wave (EMW): The effect obtained when time-varying electric and magnetic fields interact by exchanging their energies, causing electrical and magnetic energy to be propagated in a direction that is dependent upon the spatial relationship of the two interacting fields. *Note 1:* In an energized antenna, the electric charges produce an electric field. When these same charges move, they produce a magnetic field. The charges and their movement are interlocked. The energies in the two fields they produce are also interlocked and can thus propagate over long distances at high speeds. Only certain material media can significantly influence their propagation. *Note 2:* In telecommunications, including radio frequency (rf) and fiber optic systems, the most commonly used electromagnetic wave (EMW) consists of a time-varying electric field and a time-varying magnetic field that have spatial vectors at right angles to each other, thus defining a plane, i.e., the polarization plane, in which they both lie. *Note 3:* The direction of energy propagation is perpendicular to the polarization plane, and the wave is called plane polarized. A plane polarized wave may be linearly, circularly, or elliptically polarized, depending on the phase relationship between the varying electric and magnetic fields. *Note 4:* The electric and magnetic fields define the wavefront at a point as well as the polarization plane at that point. The cross product of the two fields, with the electric field vector rotated into the magnetic field vector, defines a vector, called the Poynting vector, which indicates the direction of propagation, and defines a ray, which is perpendicular to the wavefront. *Note 5:* If the magnitude as a function of time and the spatial distribution of the oscillating currents in a source are given, the electromagnetic field strengths, power flow rates, and energy densities can be determined for the electromagnetic wave everywhere in space, provided also that the parameters of the materials in the space are known. *Note 6:* The units of measure for electromagnetic waves are frequency, wavelength, radiance, irradiance, and field strength. *Note 7:* Electromagnetic waves are also produced by excited materials when electrons that have been raised to higher energy levels return to the lower levels, such as occurs in lasers and lamps. *Note 8:* Examples of electromagnetic waves (EMW) are radio waves, lightwaves, X rays, gamma rays, and cosmic rays. Lightwaves can travel in free space or transparent materials, such as optical fibers, where they can be trapped, guided, and made to energize photodetectors at large distances from the light source. *See* **horizontally polarized electromagnetic wave, left-hand polarized electromagnetic wave, plane electromagnetic wave, plane wave, plane polarized electromagnetic wave, polarized electromagnetic wave, right-hand polarized electromagnetic wave, transverse electromagnetic wave, trapped electromagnetic wave, uniform plane polarized electromagnetic wave, vertically polarized electromagnetic wave.** *See also* **elliptical polarization, Hertzian wave.** *Refer to* **Figs. P-2 and P-3 (708).**

electromagnetic wave transmission: 1. The dispatching of an electromagnetic wave, such as a radio wave, radar pulse, or lightwave, by an antenna, such as (a) a radio antenna, (b) a radar antenna, or (c) a light source. **2.** The transmission, propagation, and usually reception of an electromagnetic wave. *Note:* In a radio wave, transmission occurs at the transmitting antenna, propagation occurs between the transmitting antenna and a receiving antenna, and reception occurs at a receiving antenna. *See* **low detectability electromagnetic wave transmission, propagation medium.**

electromagnetic wave velocity: The velocity, including magnitude and direction, of an electromagnetic wave in a given propagation medium. *Note:* The magnitude of the electromagnetic wave velocity in a vacuum is about 3×10^8 m•s^{-1} (meters per second), usually represented by *c*. In material media, the magnitude of the electromagnetic wave velocity is given by the relation $v = c/n$, where v is the magnitude of the velocity, c is the magnitude in a vacuum, and n is the refractive index of the medium in which the wave is propagating. The refractive index is about 1.002 for air, 1.33 for seawater, 9 for distilled water, and 1.5 to 2.0 for certain types of glasses and plastics.

electromagnetostatic field: 1. A field of force in which (a) both electrostatic fields and magnetostatic fields exist at the same point and are produced by independent charges and currents in a conducting region with an electric conductivity that is so low that the electric field intensity in the conducting region cannot be neglected, (b) each field vector is nonzero, (c) the electrostatic and magnetostatic fields are coupled by the electric current density, given by the relation $\mathbf{J} = \sigma\mathbf{E}$ where $\mathbf{J}$ is the vector current density, σ is the electrical conductivity, and $\mathbf{E}$ is the vector electric field strength, (d) the current gives rise to the magnetic field in accordance with Ampere's law, and (e) the fields are coupled time-invariant fields, while the electromagnetic wave has coupled time-variant electric and magnetic fields. **2.** An independent time-invariant electrostatic field and an independent time-invariant magnetostatic field coexisting at the same point. *Note:* Slowly varying fields are considered time-invariant. *See* **electrostatic field, magnetostatic field.**

electromechanical recording: Recording by means of a signal actuated mechanical device, such as an impact printer. *See also* **facsimile, recording.**

electromechanics: The branch of science and technology devoted to the design, development, fabrication, testing, and use of systems or devices in which mechanical and electrical components are integrated, usually the electrical components for control and the mechanical components for motion and force development, such as relays, heliograph mechanisms, and solenoid-operated power switches. *See also* **mechatronics, relay.**

electromotive force (emf or EMF): 1. The driving force of an electric field equal to the electric field strength, which is a gradient, integrated over any path between two points. **2.** The potential difference between two points, or the difference between the absolute potentials of the two points. **3.** The voltage at a point with respect to another point, i.e., the voltage between two specified points, one of which usually is ground. *See also* **electric field, electric field strength, potential difference.**

electron: A basic negatively charged particle that (a) carries a charge of 1.6021×10^{-19} C (coulomb), (b) has a mass of 9.1091×10^{-31} kg (kilogram), (c) exists outside the nuclei of chemical element atoms, (d) exists with different discrete energy levels in a given chemical element atom, (e) distinguishes the elements by its population outside the nucleus of each atom, and (e) is the moving matter that contributes most to the formation of electric currents and voltages. *Synonym* **beta particle.** *See also* **charge.** *Refer to* **Fig. L-2 (502).**

electron beam recording: Recording on film in which a beam of electrons impinging on the screen of a cathode ray tube (CRT) is used to expose film. *See also* **fiberscope recording.**

electron energy: *See* **emitted electron energy.**

electronic: Pertaining to the effects, motion, and behavior of electrons and other charged particles under the influence of electric and magnetic fields, chemical reaction, physical force (sound, vibration, shock, gravity), bombardment by atomic particles and photons, and other forms of excitation. *Note:* Control of the flow of electrons allows for power transmission and signal transmission. Moving electrons, particularly those that form oscillating currents, produce electromagnetic waves. Thus, electromagnetic wave propagation is also included in electronics. Changes in electron distribution cause changes in the electric fields produced by the electrons. Moving electrons constitute an electric current, which produces magnetic fields. When these occur in concert, electromagnetic waves are produced. *See* **optoelectronic.** *See also* **electronics.**

electronically controllable coupler: An optical element that enables other optical elements to be coupled to or uncoupled from each other, in accordance with an applied electrical signal. *Note:* An example of an electronically controllable coupler is two parallel slab dielectric waveguides with an optical material between them with a refractive index that can be altered by the application of an electronic signal, thus turning the coupling of the waveguides on or off according to the applied signal. *See also* **coupler, fiber optic coupler.**

electronically controlled coupling (ECC): The coupling of a lightwave from one dielectric waveguide into another dielectric waveguide upon the application of an electrical signal. *Note:* Devices that perform electronically controlled coupling (ECC) can be used as switches. *Refer to* **Fig. E-2.**

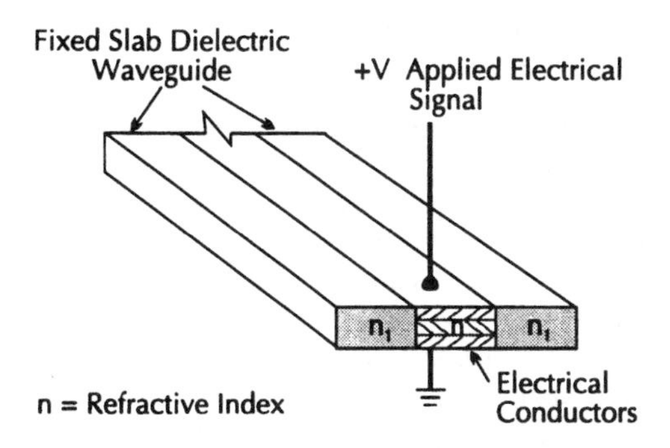

Fig. E-2. An **electronically controlled coupler (ECC)** between two slab dielectric (planar) waveguides.

electronic charge: The quantity of charge carried by one electron, equal to 1.6021×10^{-19} C (coulomb).

electronic classroom: *Synonym* **teletraining.**

electronic countercountermeasure (ECCM): In electronic warfare, an action taken to (a) ensure friendly effective use of the electromagnetic spectrum despite the use of electronic warfare by hostile forces and (b) render electronic countermeasures ineffective. *Note:* Electronic countercountermeasures (ECCM) is a subdivision of electronic warfare (EW). *See* **infrared jamming electronic countercountermeasure, radio relay electronic countercountermeasure.** *See also* **antijamming measure, electromagnetic interference, electromagnetic interference control, electronic warfare, frequency hopping.**

electronic countermeasure (ECM): An action taken to prevent or reduce the effective use of the electromagnetic spectrum by hostile forces. *Note:* Electronic countermeasures (ECM) is a subdivision of electronic warfare. *See also* **electromagnetic interference, electromagnetic interference control, electronic jamming, electronic warfare, susceptibility.**

electronic data interchange (EDI): The interchange of information that is (a) represented in electronic form and (b) transferred by electronic systems, such as radio, television, radar, radio, and wire systems.

electronic deception: **1.** The deliberate radiation, reradiation, alteration, suppression, absorption, denial, enhancement, or reflection of electromagnetic energy in a manner intended to convey misleading information and to deny valid information to hostile forces and their electronic systems. *Note:* Among the types of electronic deception are (a) manipulative electronic deception, i.e., actions taken to eliminate revealing or convey misleading telltale indicators that may be used by hostile forces, (b) simulative electronic deception, i.e., actions taken to represent friendly notional or actual capabilities to mislead hostile forces, and (c) imitative electronic deception, i.e., the introduction of electromagnetic energy that imitates hostile emissions into hostile systems. **2.** Deliberate activity designed to mislead hostile forces in the interpretation or the use of information received by their electronic systems. *See* **manipulative electronic deception.** *See also* **electromagnetic intrusion, electronic warfare.**

electronic device: *See* **optoelectronic device.**

electronic directional coupler: *See* **optoelectronic directional coupler.**

electronic emission policy: *Synonym* **electromagnetic emission policy.**

electronic emission security: The measures taken to protect all transmissions from interception and electronic analysis. *See also* **electromagnetic interference control, electronics security, electronic warfare, emanations security, TEMPEST.**

electronic exchange: An automatic exchange in which solid state switching and computing equipment are used. *See also* **automatic exchange.**

electronic flight information system (EFIS): A cockpit flight information system that contains (a) multifunction, navigation, and engine instrument displays, (b) data link functions, and (c) radio communications capabilities.

Electronic Industries Association (EIA): An association of organizations that are engaged in or concerned with the development, standardization, and manufacture of electrical, electronic, and fiber optic components.

Electronic Industries Association (EIA) interface: An interface compliant with standards developed by the Electronic Industries Association (EIA). *Note 1:* Most interface equipment is compliant with the voluntary industrial standards developed by Electronic Industries Association (EIA). Some of these standards have been adopted by the federal government as federal standards. *Note 2:* The telecommunications standards bodies of the Electronic Industries Association (EIA) are part of the Telecommunications Industry Association (TIA). The standards are designated EIA/TIA-XXX. *Synonym* **EIA interface.** *See also* **interface.**

Electronic Industries Association standard: A standard developed for the commercial world by the Electronic Industries Association (EIA) to define and standardize electrical, electronic, and fiber optic system parameters, components, interfaces, and connectors for all industrial applications, including communications, computer, information processing, and control systems. *Note:* Some Electronic Industries Association (EIA) standards have been adopted by the U.S. government as federal standards.

electronic intelligence (ELINT): Intelligence in electronic form. *See also* **electronics intelligence, intelligence.**

electronic jamming: An electronic countermeasure that (a) includes the deliberate radiation, reradiation, or reflection of electromagnetic energy for the purpose of disrupting the use of electronic devices, equipment, or systems by hostile forces and (b) can be accomplished in many ways, such as by radiating or reradiating electromagnetic energy at the frequency that a victim emitter uses, modulating the jamming emission so as to mask, interfere with, or otherwise prevent full use of the emitter. *See* **antielectronic jamming.** *See also* **electronic countermeasure, electronic warfare, interference, victim emitter, victim frequency.**

electronic key management system (EKMS): In communications, computer, data processing, and control security systems, a system for managing the generation, distribution, and use of electronic key in electronic communications systems. *See also* **key.**

electronic library: A database that (a) consists of data stored in electronic circuits, (b) may be fixed or programmable, and (c) is usually mounted on integrated circuit chips for high capacity and transportability. *Note:* An electronic library might contain and process all the data required to operate a passenger aircraft, including flight scheduling, flight path data, weather data, airport configuration data, passenger listings, announcements, ticketing data, and meal menus.

electronic line of sight: The path traversed by electromagnetic waves that are not reflected or refracted by material media, such as the atmosphere. *See also* **line of sight propagation.**

electronic line scanning: In facsimile systems, scanning in which the scanning spot is moved along the scanning line by electronic means. *See also* **facsimile, line.**

electronic mail (E-mail): An electronic means of communications in which (a) primarily text is transmitted, (b) operations include sending, storing, processing, and retrieving information, (c) users are allowed to communicate under specified conditions, and (d) messages are held in storage until called for by the addressee. *Synonym* **E-mail.** *See also* **data transmission, electronic message system, Integrated Services Digital Network, store and forward.**

electronic message system (EMS): Electronic mail with an additional feature in which the central facility ensures delivery of messages to the intended addressees without requests from the addressees. *Note:* In contrast to an electronic message system, electronic mail is passive in that it delivers messages only in response to requests from addressees. *See also* **electronic mail.**

Electronic Negotiations Archive: *See* **Automated Recourse to Electronic Negotiations Archive.**

electronic reconnaissance: The detection, identification, evaluation, and location of foreign electromagnetic radiations, such as those that emanate from communications, radar, and control guidance sources rather than from radioactive sources, such as nuclear detonations, radioactive waste, or radioactive materials used for medical diagnostic or therapeutic purposes. *See also* **electromagnetic environment, electromagnetic interference, electronic warfare, monitoring.**

electronics: The branch of science and technology devoted to the study and the application of the effects, motion, and behavior of electrons, and other charged particles, under the influence of electric and magnetic fields, chemical reaction, physical force (sound, vibration, shock, gravity), bombardment by atomic particles and photons, and other forms of excitation. *See* **communications-electronics, optoelectronics, theater director of communications-electronics, transient radiation effects on electronics.** *See also* **electronic.**

electronic search: An investigation of the distribution of signals in the electromagnetic spectrum, or portions thereof, in order to determine the existence, sources, and characteristics of the electromagnetic radiation. *See also* **electromagnetic spectrum.**

electronic security: Security that is obtained through the use of electronic equipment and systems, including fiber optic equipment and systems. *See also:* **electronics security.**

electronic signal: A signal that consists of an electrical current, i.e., a flow of electrons, that varies with time or space in accordance with specified conditions or parameters. *Synonym* **electrical signal.** *See also* **electron, signal.**

electronic silence: The prevention of any form of electromagnetic emission, including radio, video, radar, sonar, sounding, and signaling transmissions as well as emanations from equipment not intended for transmission, such as receivers, ignition systems, motors, generators, and power switching equipment. *Note:* When electronic silence is in effect, all equipment capable of electromagnetic radiation usually is kept inoperative. Frequency bands and types of equipment affected usually are specified.

electronics intelligence (ELINT): 1. Technical and intelligence information derived from foreign or hostile noncommunications electromagnetic radiations emanating from sources other than nuclear detonations or radioactive sources. **2.** Intelligence in any form obtained from activities engaged in the collection and the processing of electromagnetic radiation received from any and all sources. *See also* **electromagnetic environment, electromagnetic interference, electronic intelligence, electronic warfare, intelligence, intercept, interference, signal intelligence.**

electronics security (ELSEC): The protection that results from all measures taken to deny unauthorized persons information of value that might be derived from interception and study of electromagnetic radiations from electronic equipment, such as radio, video, wire, optical fiber, microwave, radar, sonar, and missile systems, rather than from interception and interpretation of the traffic, i.e., the calls and messages, in these systems. *See also* **communications security, electronic security, radiation, signal security, traffic.**

electronic surveillance: The receipt, analysis, and evaluation of signals obtained from scanning a designated volume of space with electronic systems, equipment, and devices, such as receivers, radars, and infrared scanners, that (a) are used to monitor, intercept, classify, and analyze all electromagnetic transmissions and radiations from or in a given area, (b) may be ground-based, shipborne, or airborne, and (c) may be side-looking, forward-looking, up-looking, or area-searching. *See also* **radiation, signal, transmission.**

electronic switch: A switch that (a) is composed of electronic parts, (b) is electronically operated, and (c) usually is constructed of solid state circuits. *See* **semielectronic switch.** *See also* **solid state circuit, switch.**

electronic switching system (ESS): A switching system with devices that use semiconductor components, such as transistor combinational circuits, i.e., logic elements, semiconductor diode matrices, and integrated circuits to select and switch circuits. *Note:* A switching system that has reed relays or crossbar matrices as well as semiconductor components is considered to be an electronic switching system (ESS). *See also* **combinational circuit, crossbar switch, integrated circuit, semiconductor, semielectronic switch, switch, switching system.**

electronic village: Pertaining to a computer, communications, and information system that (a) supports all forms of data and multimedia capabilities on a national scale and (b) supports the information and communications requirements of the homes and businesses in a typical village, such as E-mail, database access for medical data, switched video services, access to the 2.5 million files of Internet, university services, integrated services digital network (ISDN) services, and an installation capability that consists of four tiers of increasing data rate capability based on individual user needs.

electronic warfare (EW): All actions and supporting measures, such as electronic countermeasures (ECM) and electronic countercountermeasures (ECCM), taken to (a) determine, exploit, reduce, or prevent the use of the electromagnetic spectrum by hostile forces and (b) retain the effective use of the electromagnetic spectrum by friendly forces. *Note:* Uses of the electromagnetic spectrum include radio, video, radar, sonar, microwave, and control guidance. *See also* **electromagnetic spectrum, electronic countercountermeasure (ECM), electronic countermeasure.**

electronic warfare support measure (ESM): In electronic warfare (EW), a measure that (a) includes actions taken under direct control of an operational commander to search for, intercept, identify, and locate sources of radiated electromagnetic energy for the purpose of immediate threat recognition, (b) provides a source of information required for tactical employment of forces, such as immediate decisions concerning electronic countermeasures (ECM), electronic countercountermeasures (ECCM), avoidance, targeting, and homing, and (c) provides data that can be used to produce signals intelligence (SIGINT), i.e., both communications intelligence (COMINT) and electronics intelligence (ELINT). *Note:* The total of all electronic warfare support measures (ESMs) is a subdivision of electronic warfare (EW). *See also* **electronic warfare.**

electron lens: A lens that is capable of converging or diverging an electron beam by means of an electric field, a magnetic field, or both.

electron microscope: An electron-optical instrument that can produce an enlarged image of a small object on a fluorescent screen or photographic plate by focusing an electron beam by means of an electron lens. *See* **scanning electron microscope.** *See also* **electron lens.**

electron optics: The branch of science and technology devoted to the study and application of methodologies for using electric fields, magnetic fields, or both, to direct and guide electron beams in the same manner in which lenses influence light beams. *Note:* Devices developed and used in electron optics include image converter tubes, electron microscopes, Kerr cells, Pockels cells, and other devices whose operation depends on the modification of the refractive index by electric fields. *See also* **electron microscope, Kerr cell, lens, Pockels cell.**

electron•volt (eV): A unit of energy equal to the amount of energy released or acquired when an electronic charge equivalent to the charge of one electron moves through a potential difference of 1 V (volt). *Note:* The electron volt is convenient to use when expressing energies of charged particles under the influence of electric and magnetic fields. The electronic charge carried by one electron is 1.6021×10^{-19} C (coulomb), and the electron•volt (eV) is equivalent to 1.6021×10^{-19} J (joules). *See also* **electron, potential difference, volt.**

electrooptic: 1. Pertaining to the effect that an electric field has on a lightwave, such as rotating the direction of the polarization plane as the lightwave passes through the field, or, if the wave is propagating in material media, the effect the electric field has on the refractive index of the material. *Note:* "Electrooptic" is often erroneously used as a synonym for "optoelectronic." **2.** Pertaining to the conversion of electrical power or energy into optical power or energy, such as conversion of an electrical signal to an optical signal. *See also* **optoelectronic.**

electro-optical intelligence (ELECTRO-OPTINT): Intelligence, i.e., information, other than signals intelligence (SIGINT), derived from the monitoring of the optical portion of the electromagnetic spectrum from ultraviolet, 0.01 μm (microns, micrometers), through far infrared, 1000 μm. *Synonym* **electrooptical intelligence.**

electrooptical intelligence: *Synonym* **electro-optical intelligence.**

electrooptic coefficient (EC): A measure of the extent to which the refractive index changes with applied high electric fields, such as several parts per 10,000 for applied fields of the order of 20 $V \cdot cm^{-1}$ (volts per centimeter). *Note:* The phase shift of a lightwave is a function of the change of refractive index of the propagation medium in which it is propagating. Thus, the change in refractive index can be used to phase-modulate the lightwave by shifting its phase at a particular point along the guide by changing the propagation time to the point. *See also* **electrooptic phase modulation.**

electrooptic detector: *Deprecated synonym for* **photodetector.**

electrooptic effect: 1. Any one of a number of phenomena that occur when an electromagnetic wave in the optical spectrum interacts with an electric field or with matter under the influence of an electric field. **2.** The effect an electric field has on the polarization of a lightwave, such as rotating the polarization plane. *Note:* The applied electric field strength adds to the electric field strength of the electromagnetic wave, causing polarization plane rotation as the wave propagates. Pockels and Kerr effects are electrooptic effects that are linear and quadratic, respectively, in relation to the applied electric field. **3.** The change in the refractive index of a material when subjected to an electric field. *Note 1:* The electrooptic effect can be used to modulate a light beam in a material because many properties, such as propagation velocities, reflection and transmission coefficients at interfaces, acceptance angles, critical angles, and transmission modes, are dependent upon the refractive indices of the media in which the beam propagates. *Note 2:* Two electrooptic effects having application as modulation mechanisms in optical communications are the Kerr effect and the Pockels effect, in which birefringence is induced (Kerr effect) or modified (Pockels effect) in a material. *Note 3:* The terms "electrooptic" and electro-optic" are often misused as synonyms for "optoelectronic," and vice versa. *See also* **magnetooptic effect, optoelectronic.**

electrooptic frequency response: The response obtained when an electrooptic device is electronically driven by a signal that consists of many frequencies. *Note:* An example of an electrooptic frequency response is the output signal amplitude, plotted as a function of the frequency of a fixed-amplitude input signal, obtained from a Bragg-type surface interaction device that launches surface waves on a piezoelectric zinc oxide thin film photoelastic waveguide section using electromechanical coupling. *See also* **photoelastic, response, waveguide.**

electrooptic modulator: An optical device in which a signal-controlled element is used to modulate a light beam. *Note 1:* The modulation may be imposed on the phase, frequency, amplitude, or direction of the modulated beam. *Note 2:* Modulation bandwidths into the gigahertz range are possible using laser-controlled modulators. *See also* **electrooptic effect, fiber optics, optoelectronic device.**

electrooptic phase modulation: Modulation of the phase of a lightwave in accordance with an applied field serving as the modulating signal. *Note:* An example of electrooptic phase modulation is the changing of the refractive index and thus the velocity of propagation, and thus the phase at a point in the propagation medium in which the wave is propagating. *See also* **electrooptic coefficient.**

electrooptic receiver: A receiver that accepts an electrical signal and converts it to an optical signal for further processing or use, such as for transmission on a fiber optic cable. *See also* **fiber optic cable, receiver, signal.**

electrooptics: The branch of science and technology devoted to (a) the design, development, and use of components, devices, and systems that are designed to interact between the electromagnetic (optical) and the electric (electronic) state, (b) the study and application of the interaction between optics and electronics leading to the transformation of electrical energy into light, or vice versa, with the use of optical devices, particularly the generation and the control of lightwaves by electronic means, and (c) the design, development, and use of components, devices, and systems in which operation depends on modification of the refractive index of a material by an electric field. *Note 1:* In a Kerr cell, the refractive index change is proportional to the square of the electric field strength, and the material usually is a liquid. *Note 2:* In a Pockels cell, the refractive index change is proportional to the electric field strength, and the material usually is a crystal. *Note 3:* "Electrooptic" is often erroneously used as a synonym for "optoelectronic."

electrooptic transmitter: 1. A transmitter that accepts an electrical signal and converts it to an optical signal that represents the same information as the electrical signal. **2.** A light source that produces an output optical signal when modulated by an electrical input signal.

ELECTRO-OPTINT: electrooptical intelligence, electro-optical intelligence.

electrophotographic recording: Recording in which (a) light is used to produce a change in electrostatic charge distribution to form a photographic image, and (b) subsequent processing usually is required to make the image visible. *See also* **facsimile, recording.**

electrosensitive recording: Recording in which an electrical signal is directly impressed on the record medium. *See also* **electrophotographic recording, facsimile, recording.**

electrostatic field: 1. In an antenna, such as a dipole, that is radiating an electromagnetic wave, the near field. *See also* **near field. 2.** The electric field produced by a static, or slowly moving, electric charge. *Note:* An electrostatic field exists between the plates of a capacitor that has a constant voltage applied between the plates. *See also* **electric field, field strength.**

electrostatic recording: Recording in which (a) a signal-controlled electrostatic field is used, and (b) subsequent processing usually is required to make the image visible, i.e., make the recorded copy visible. *See also* **direct recording, facsimile, recording.**

electrostatic storage: Storage in which electrically charged areas on a dielectric surface are used to store data. *Note:* An example of electrostatic storage is a cathode ray tube (CRT) storage. *See also* **cathode ray tube storage.**

electrothermal recording: Recording in which signal-controlled thermal action is used. *See also* **facsimile, recording.**

ELED: edge-emitting light-emitting diode.

element: *See* **basic service element, code element, combinational logic element, data element, delay el-**

ement, detectable element, display element, message element, n-ary information element, optical element, parasitic element, picture element, receiving element, signal element, stop element, thin-film laser element, transmitting element, unit element.

elemental area: In facsimile transmission systems, any segment of a scanning line, the dimension of which along the line is (a) exactly equal to the nominal line width and (b) is not necessarily the same as that of the scanning spot. *See also* **facsimile, picture element, pixel, scanning spot.**

elemental current moment: The contribution to the magnetic field strength at a point made by an electric current in an infinitesimal length of conductor in an arbitrary direction and at an arbitrary distance from the point, given by the relation $IdsR^{-1} \sin \Theta$ where i is the instantaneous current in the elemental length of conductor, ds is the elemental length bearing the current I, R is the radial distance from the referenced point about which the moment is taken, and Θ is the smaller angle between ds and R. *Note:* When the elemental current moments are integrated over all current paths, the magnetic field strength at a point can be determined, i.e., ampere•turns can be calculated, such as the magnetic field strength produced by solenoids and transmitting antennas, given the geometrical configuration of the conductors and the currents in them. *See also* **current moment.**

elementary quantum of action: *Synonym* **Planck's constant.**

elementary signaling element: *See* **unit interval.**

element protocol: *See* **remote operations service element protocol.**

elevated duct: An atmospheric duct consisting of a high density air layer that starts at high altitudes and continues upward or remains at high altitudes, thus affecting primarily very high frequency (VHF) transmission. *See* **atmospheric duct.**

elevation: The distance from a reference point, such as a point on the Earth's surface, to another point radially away from the center of the Earth, i.e., the vertical component of distance between the points. *See also* **elevation angle.**

elevation angle: An angle, in a vertical plane, between a horizontal plane and a line extending from a point (such as an observation point, origin, or vertex) in the vertical plane to a point (such as an object) above or below the horizontal plane, the angle being measured from the horizontal plane to the line. *Note:* If the object is below the horizontal plane containing the observation point, the elevation angle is negative.

elevation over azimuth: *See* **altitude over azimuth.**

ELF: extremely low frequency.

ELINT: electronic intelligence, electronics intelligence.

elliptical orbit: *Synonym* **closed orbit.**

elliptical polarization: In a uniform plane polarized electromagnetic wave, polarization in which (a) the tip, i.e., the extremity, of the electric field vector describes an ellipse in a plane that is normal to the direction of propagation, (b) the electric vector varies in magnitude and angular velocity as it rotates, and (c) the two arbitrary sinusoidally varying rectangular components of the electric field vector are not equal in magnitude, and there is a phase angle between them, i.e., one leads or lags the other by an arbitrary phase angle such that (i) if the phase angle is not 90°, the ellipse will be inclined to both rectangular components, (ii) if the phase angle is 0, the ellipse flattens to a straight line and the wave is linearly polarized, and (iii) if the vectors are equal and 90° out of phase with each other, the wave is circularly polarized. *Note 1:* An elliptically polarized wave may be resolved into two linearly polarized waves in phase quadrature with their polarization planes at right angles to each other. *Note 2:* Circular and linear polarization are special cases of elliptical polarization. *Synonym* **axial ratio polarization.** *See also* **axial ratio, circular polarization, linear polarization, polarization.**

ELSEC: electronics security.

EM: 1. end of medium. 2. The end of medium transmission control character.

E-mail: *Synonym* **electronic mail.**

emanation: Undesirable electromagnetic radiation that (a) exits from a material medium, such as a wire or a device, and (b) usually travels a considerable distance from the medium or the device. *See* **compromising emanation.** *See* **radiation.**

emanations security (EMSEC): 1. The protection that results from all measures taken to deny unauthorized persons information of value that might be derived from intercept and analysis of compromising emanations from sources other than cryptoequipment and their message traffic. **2.** The security that results from all measures taken to deny unauthorized persons information of value that might be derived from intercept and analysis of all compromising emanations, including cryptoequipment and their message traffic. *See* *also* **electromagnetic emission control, electronic warfare, TEMPEST.**

embedded base equipment (EBE): All customer premises equipment that (a) had been provided by the Bell Operating Companies (BOCs) prior to January 1, 1984 and (b) was ordered transferred from the BOCs to the American Telephone and Telegraph (AT&T) Company by court order.

embedded customer-premises equipment (ECPE): Telephone-company-provided equipment on customer premises in use or in inventory of a regulated telephone utility as of December 31, 1982. *See also* **customer premises equipment, new-customer premises equipment.**

EMC: electromagnetic compatibility.

EMC analysis: electromagnetic compatibility analysis.

EMCON: emission control.

EMD: equilibrium mode distribution.

EME: electromagnetic environment.

emergence: Pertaining to the trigonometric relation between an emergent ray and the surface of a propagation medium. *See* **grazing emergence, normal emergence.** *See also* **propagation medium, ray.**

emergency: A sudden, generally unexpected event that does or could do harm to people, the environment, resources, property, or institutions. *Note:* Emergencies range from relatively local events to regional and national events and may be caused by natural or technological factors, human actions, or national-security-related events.

Emergency Broadcast System (EBS): A system that (a) is composed of (i) amplitude-modulated (AM), frequency-modulated (FM), and television (TV) broadcast stations, (ii) low power television (TV) stations, and (iii) nongovernment industry entities and (b) operates on a voluntary, organized basis during emergencies at national, state, or operational (local) area levels.

emergency call: 1. A call used to establish communications during an emergency, distress, or disaster. **2.** A call used to exchange information pertaining to an emergency. *Note:* An emergency call is not necessarily a call placed during an emergency. *See also* **emergency message, emergency signal.**

emergency category: A level, i.e., a degree, of emergency indicating the extent of danger, urgency, or probability of loss of life or property. *Note:* The emergency categories, in descending order of precedence, are distress, urgent, and safety. These three levels indicate the precedence or priority of messages dispatched during an emergency. *See also* **precedence designation.**

emergency communications: *Synonym* **distress communications.**

emergency communications system plan: *See* **Aeronautical Emergency Communications System Plan.**

emergency craft frequency: An international distress and emergency frequency for survival purposes, such as frequencies for emergency craft position indicating radio beacons, distress and emergency situations, and direction finding transmissions for search and rescue operations. *Note 1:* Examples of emergency craft frequencies are (a) 2182 kHz (kilohertz) for international distress and calling and (b) 8364 kHz for international lifeboat, lifecraft, and survival craft. *Note 2:* Frequencies for distress, safety, emergency, search, and rescue operations are assigned by the International Telecommunication Union (ITU), Geneva, Switzerland. *See also* **distress frequency.**

emergency frequency: A frequency that is to be used only for emergency situations, such as distress, search, rescue, and survival operations. *See* **aircraft emergency frequency, air-ground communications emergency, distress frequency, international aeronautical emergency frequency, military common emergency frequency.**

emergency frequency band: The band of frequencies that (a) is used for emergency messages and (b) lies between 118 MHz (megahertz) and 132 MHz in the electromagnetic frequency spectrum. *See also* **emergency message, frequency assignment, frequency spectrum designation, international aeronautical emergency frequency.**

emergency locator transmitter (ELT): A transmitter that (a) is on board an aircraft or a survival craft, (b) is manually or automatically actuated, and (c) is used as an alerting and locating aid for survival and rescue operations.

emergency maintenance: Corrective maintenance performed immediately upon detection of a fault or an error in order to ensure continued satisfactory operation of a functional unit.

emergency message: A message that (a) contains information pertaining to an emergency and (b) is transmitted to or from the station that is experiencing the emergency. *Note:* An example of an emergency message is a message that contains information concerning the nature of the emergency or the type of assistance that is required at the scene of the emergency. *See also* **distress message, emergency call, emergency signal.**

emergency meteorological warning: An urgent warning that (a) is of interest to all units, such as aeronautical, maritime, and land-based units, (b) usually covers a broad area, (c) usually is broadcast on several frequencies, (d) usually includes information on wind, rain, lightning, hail, tidal waves, earthquakes, and abnormal temperatures, (e) covers location, direction, area, and severity, (f) could be dispatched to specific units by specific means, and (g) is for the preservation of life, property, and communications continuity. *See also* **routine meteorological broadcast, special meteorological broadcast.**

emergency position-indicating radiobeacon: *See* **satellite emergency position indicating radiobeacon.**

emergency position-indicating radiobeacon station: A station in the mobile service whose emissions are intended to facilitate search and rescue operations. *See also* **mobile service.**

Emergency Preparedness (EP) telecommunications service: *See* **private National Security or Emergency Preparedness (NS/EP) telecommunications service.**

emergency signal: In communications systems operations, a signal used to alert listening stations to suspend operations to the degree necessary to hear the message that follows. *Note 1:* An example of an emergency signal is the alerting continuous-tone signal used by the Emergency Broadcast System, a signal that is followed by information, announcements, and instructions in an emergency. *Note:* In order of accepted descending precedence, the emergency categories are distress, urgent, and safety. In voice communications systems, the emergency signal may be sent by speaking three times, respectively for each category, the words MAYDAY, PAN, or SECURITÉ (pronounced say-coor-ee-tay). *See* **ground-air visual emergency signal.** *See also* **emergency call, emergency message.**

emergency signal alarm: In communications systems maintenance, a signal that is used for automatically attracting attention to some abnormal condition in an emergency, such as a lamp and a loud ringing bell that alerts central office (C.O.) personnel to serious troubles, such as above normal voltages, major fuse blowing, circuit breaker operation, or failure of the machine that sounds alarms.

emergency transmitter: *See* **ship emergency transmitter.**

emergent ray: A ray of light emerging from a propagation medium. *See also* **incident ray, propagation medium, ray.**

emf: electromotive force.

EMF: electromotive force.

EMI: electromagnetic interference.

emission: 1. Electromagnetic energy that (a) is propagated from a source by radiation or conduction, (b) may be either desired or undesired, and (c) may occur anywhere in the electromagnetic spectrum. **2.** Radiation produced, or the production of radiation, by a radio transmitting station. *Note:* The energy radiated by the local oscillator of a radio receiver would be radiation but not an emission. *See* **class of emission, continuous emission, interference emission, out of band emission, parasitic emission, primary emission, reduced carrier single sideband emission, secondary emission, single sideband emission, spurious emission, spontaneous emission, stimulated emission, suppressed carrier single sideband emission.** *See also* **electromagnetic interference control, electronic emission security, frequency, harmful interference, necessary bandwidth, radiation, radio frequency bandwidth, spurious radiation.**

emission class: *Synonym* **class of emission.**

emission control (EMCON): 1. The selective control of emitted electromagnetic or acoustic energy. **2.** In military systems, the selective and controlled use of electromagnetic, acoustic, or other emitters to (a) optimize command and control capabilities while minimizing, for operations security (OPSEC), detection by hostile sensors, (b) minimize mutual interference among friendly systems, (c) execute a military deception plan, (d) minimize the detection of emissions and exploitation of the information so gained by hostile forces, and (e) improve the performance of friendly sensors. *See* **electromagnetic emission control.** *See also* **communications security, electromagnetic emission policy, electronic warfare, emission policy, interference.**

emission designator: A set of standard characters, such as letters, numerals, and special symbols, that (a) is used to designate a type or a class of transmitting stations or communications networks and (b) usually consists of an approved basic functional word with approved prefixes and suffixes. *See also* **circuit designator, net glossary.**

emission intercept range: A range (a) at which a specified emission or transmitted signal can be received, interpreted, and analyzed, and (b) that has a maximum at which emissions may be intercepted that depends upon the emitter characteristics, propagation medium conditions, and receiver characteristics and location.

emission law: *See* **Boltzmann's emission law, cosine emission law.**

emission of radiation: *See* **microwave amplification by stimulated emission of radiation.** *See also* **laser.**

emission policy: Policy concerning (a) the controls and limitations imposed upon or used in an organization pertaining to electromagnetic, sonic, and other forms of emission, transmission, radiation, or emanation by any means, including conduction, and (b) the operation of radio receivers, radar receivers, and other devices that are not specifically designed to emit radiation when operated. *Note:* Forms of emission include radio, video, sonar, visible light, infrared, and any other electromagnetic, magnetostatic, electrostatic, magnetic, electric, or sonic means. *See* **electromagnetic emission policy.**

emission security: Communications security that results from all measures taken to deny unauthorized persons information of value that might be derived from intercept and analysis of compromising emanations from cryptoequipment and telecommunications systems. *See also* **communications security, emanation, emanation security, emission.**

emission wavelength: *See* **peak emission wavelength.**

emissivity: 1. The ratio of (a) the power radiated by a substance to (b) the power radiated by a blackbody at the same temperature. *Note 1:* Emissivity is a function of wavelength and temperature. *Note 2:* Power units and area units for both bodies must be the same or be normalized. **2.** The ratio of (a) the radiant emittance of, i.e., the radiated flux from, a source to (b) the radiant emittance of, i.e., the radiated flux from, a blackbody having the same temperature. *Note:* Emissivity is a function of wavelength and temperature. *See* **total emissivity.** *See also* **optical spectrum, radiance, radiant emittance.**

emit: In a given entity, the action of ejecting an entity, usually into free space, such as to eject (a) electrons from a heated cathode, (b) signals from an antenna, (c) electrons from a photosensitive surface, (d) gamma rays and X rays from a radioactive substance, and (e) lightwaves from a laser. *See also* **transmit.**

emittance: A measure of the amount of light energy emitted from a source. *See* **luminous emittance, radiant emittance, spectral emittance.**

emitted electron energy: In a photoemissive detector, the remaining energy of an electron that escapes from the emissive material, caused by the energy imparted to it by an incident photon, which is given by the relation $E_e = hf - qw$ where E_e is the emitted electron energy, h is Planck's constant, f is the photon frequency, q is the charge of an electron, and w is the work function of the emissive material, so that hf is the photon energy, and qw is the energy required for an electron to escape from the emissive material, i.e., to overcome the boundary effects. *See also* **work function.**

emitter: A device or a component that emits energy in any form, such as lightwaves, electrons, and sound waves. *See* **detector-emitter, optical emitter, victim emitter.**

emitting surface: A surface from which (a) waves emerge, such as lightwaves, radio waves, microwaves, X rays, gamma rays, and cosmic rays, and (b) particles emerge, such as electrons, ions, protons, neutrons, and alpha particles. *Note:* Light sources, such as light-emitting diodes (LEDs), lasers, lamps, and the ends of optical fibers from which optical signals emerge, have emitting surfaces.

E&M lead signaling: *Synonym* **E and M lead signaling.**

EMP: electromagnetic pulse.

emphasis: *Synonym* **preemphasis.**

emphasis circuit: *Synonym* **preemphasis circuit.**

emphasis network: *Synonym* **preemphasis network.**

EMP resistance: electromagnetic pulse resistance.

empty medium: A data medium that could hold data but does not contain data other than format control data or indicators, such as fixed frame, legend, or position marks. *Note:* Examples of empty media are preprinted paper blank forms, paper tapes with only sprocket feed holes, erased magnetic tapes, and diskettes that have been formatted but have not yet been used to store data, or that have been used and all files therein have been deleted. *See also* **data medium.**

EMR: electromagnetic radiation.

EMR hazards: electromagnetic radiation hazards.

e.m.r.p.: effective monopole radiated power.

EMS: electronic message system.

EMSEC: emanations security.

E&M signaling: *Synonym* **E and M lead signaling.**

emulate: To duplicate the functions of one system with a second, different system, so that the second system appears to behave like the first system. *Note:* For example, a computer emulates another different computer if it accepts the same data, executes the same programs, and achieves the same results as the computer it emulates. *See also* **emulator, simulate, simulation, simulator.**

emulator: A hardware, software, or firmware system that (a) emulates another, different hardware, software, or firmware system and (b) appears to have all the characteristics of the other system, such as to accept the same data, accept and execute the same programs, and produce the same set of outputs for the same set of inputs when running the same programs that the imitated system runs, although the internal programs run by each system may differ. *See* **terminal emulator.** *See also* **emulate, simulator.**

emulsion storage: *See* **laser emulsion storage.**

EMUT: enhanced manpack ultrahigh frequency terminal.

EMV: electromagnetic vulnerability.

EMW: electromagnetic wave.

E_bN_0: *See* **signal to noise ratio per bit.**

enabling signal: A signal that permits the occurrence of a specific event that would not occur if the signal were not present. *Note:* Examples of enabling signals are (a) a signal that would allow an AND gate to perform its normal function only when the signal is present, and (b) a signal that enables a ringing signal to be dispatched to a call receiver. *See also* **signal.**

en bloc signaling: Signaling in which (a) address digits are transmitted in one or more blocks, and (b) each block contains sufficient address information to enable switching centers to carry out progressive onward routing. *See also* **address, block, switching center.**

encapsulated packet: In open systems, an addressed packet that is contained in the data portion of a frame, along with other addressed packets, such that upon arrival at the frame destination the inner addresses on the packets in the frame can be used to route the packets to their ultimate destination. *See also* **tunnel.**

encapsulation: In open systems, the layering of protocols such that a lower-layer protocol accepts a message from a higher-layer protocol and places it in the data portion of a frame in the lower layer.

encipher: To convert plain text into an unintelligible form by means of a cipher system. *See also* **cipher, decode, encode, encrypt.**

enciphered telephony: *Synonym* **ciphony.**

enclosure: *See* **shielded enclosure.**

encode: 1. To convert data by the use of a code, frequently one consisting of binary numerals, such that reconversion to the original form is possible. **2.** To substitute numerals, letters, and characters for other numerals, letters, or characters, usually to hide the meaning of the message to everyone except the persons who know the encoding scheme. **3.** To append redundant check symbols to a message for the purpose of generating an error detection and correction code. **4.** To apply a code, usually one consisting of binary digits that represent individual characters or groups of characters in a message. **5.** To convert plain text into an unintelligible form by means of a code, i.e., to convert a plain text message into a coded message. *See also* **character, code, cryptology, decode, encipher, encrypt.**

encoded recording: *See* **phase encoded recording.**

encoder: 1. A device that encodes data. **2.** A device that has a number of input lines, of which not more that one may carry a signal at a given instant, and a number of output lines, of which any number may carry signals, there being a one-to-one correspondence between the input signals and the combinations of output signals. *See* **analog to digital encoder, channel voice encoder.** *See also* **analog to digital converter, decoder.**

encoding: *See* **analog encoding, delay encoding, differential encoding, differential Manchester encoding, Manchester encoding, run length encoding, segmented encoding, uniform encoding.**

encoding law: The law defining the relative values of the quantum steps used in quantizing and encoding signals. *See also* **code, segmented encoding law.**

encrypt: To convert plain text into unintelligible forms by means of a cryptosystem. *Note:* The term "encrypt" covers the meanings of "encipher" and "encode." *See also* **cryptology, encipher, encode, garble.**

encrypted voice: A voice transmission that is secured against compromise through the use of an approved ciphony system. *Synonym* **secure voice.**

encryption: The conversion of plain text into a disguised or unintelligible form by means of a cryptosystem. *See* **bulk encryption, compatible data encryption, end-to-end encryption, link encryption,**

multiplex link encryption, reencryption, superencryption, variable spacing encryption. *See also* **key, public key encryption system, private key encryption system.**

encryption algorithm: A set of mathematically expressed rules for rendering data unintelligible by effecting a series of conversions controlled by the use of key. *See also* **algorithm, key.**

encryption device: A device that converts plain text into a disguised or unintelligible form by means of a cryptosystem. *See* **trunk encryption device.**

encryption key: *See* **traffic encryption key.**

encryption standard: A standard devoted to an aspect of encryption. *Note:* Examples of encryption standards are (a) the format for expressing key and (b) a procedure for enciphering plain text. *See* **data encryption standard, Data Encryption Standard, key.**

end: *See* **local end.**

end around shift: The movement of data out of one end of a storage unit or register and reentry of the data into the other end. *Synonym* **circular shift.**

end crossfire: *Synonym* **sending end crossfire.**

end crosstalk: *See* **far end crosstalk.**

end distortion: In start-stop teletypewriter operation, the shifting of the end of all marking pulses, except the stop pulse, from their proper positions in relation to the beginning of the next start pulse. *Note 1:* Shifting of the end of the stop pulse is a deviation in character time and rate rather than an end distortion. *Note 2:* Spacing end distortion is the termination of marking pulses before the proper time. *Note 3:* Marking end distortion is the continuation of marking pulses past the proper time. *Note 4:* The magnitude of end distortion usually is expressed in percent of an ideal pulse duration. *Synonym* **spacing end distortion.** *See also* **bias distortion, characteristic distortion, cyclic distortion, distortion, start-stop distortion, teletypewriter signal distortion.**

ended cord: *See* **plug-ended cord.**

end exchange: An exchange that serves as an originating, gateway, or destination exchange. *Synonym* **end office**.

end finish: *See* **optical fiber end finish.**

end fire coupling: *Synonym* **butt coupling.**

ending: 1. In a message, the parts that follow the text. **2.** In radiotelephone system messages, the part of the message that (a) consists of the date-time group, final instructions, and ending sign and (b) usually follows the break that follows the text. *See* **transmission ending.**

end instrument: A device that (a) serves as the user interface to a communications circuit, such as data circuit-terminating equipment (DCE) or a loop, (b) is connected to the terminal of a circuit, (c) is used to convert information into electrical signals or electrical signals into useful information, (d) usually belongs to the user rather than to the communications agency or administration, such as a telephone company, (d) usually is on the user premises, and (e) is considered part of the communications system to which it is connected. *Note:* Examples of end instruments are telephones, teletypewriters, radio receivers, facsimile transmitters, and facsimile receivers. *See* **user end instrument.** *See also* **circuit, data terminal equipment, main station, terminal.**

end key: On standard keyboards, such as those used with personal computer (PC) systems, a key that is used separately or in conjunction with other keys, such as an arrow key, a control key, or a home key, to perform a specific function, such as delete text from a cursor position to the end of a line, move to the beginning or end of a document, move a cursor, or move a pointer. *See also* **cursor, pointer, key.**

end office (E.O. or EO): A central office (C.O.) at which user lines and trunks are interconnected. *Note:* An end office serves local customers. *See also* **central office, switching center, tandem center, trunk.**

end of medium (EM) character: A control character that may be used to identify (a) the physical end of a data medium, (b) the end of the usable or used portion of a data medium, or (c) the end of the wanted portion of data that are recorded on a data medium.

end of message function: In tape relay systems, the letter and key functions, including the end of message indicator, that constitute the last format line. *See also* **relay, reperforator, tape relay, torn tape relay.**

end of message indicator: In tape relay communications systems, an indicator that is used to signal the end of a message at the termination of a transmission. *Note:* An example of an end of message indicator is NNNN.

end of selection character: The character that indicates the end of the selection signal. *See also* **binary synchronous communications, character, selection signal.**

end of tape mark: A mark on a magnetic tape that indicates the end of the permissible recording area.

Note: Examples of end of tape marks are a photo reflection strip, a punched hole in the tape, a transparent section of tape, and a particular bit pattern, each of which excites a sensing mechanism.

end of text character (ETX): A transmission control character used to terminate text. *See also* **binary synchronous communications, character.**

end of transmission block character (ETB): A transmission control character used to indicate the end of a transmission block of data when the data are divided into such blocks for transmission purposes. *See also* **binary synchronous communications, block, character.**

end of transmission character (EOT): A transmission control character used to indicate the conclusion of a transmission that may have included one or more texts and any associated message headings. *Note:* The end of transmission character (EOT) is often used to initiate other functions, such as releasing circuits, disconnecting terminals, or placing receive terminals in a standby or ready condition. *See also* **binary synchronous communications, character.**

end of transmission function: In tape relay procedures, the letters and machine functions that (a) include the end of message indicator and (b) signify the termination of transmission.

end of transmission group: A standardized, uninterrupted sequence of characters and machine functions that is used to terminate a transmission, disconnect the circuit, and turn off the transmitting equipment.

endoscope: A device that (a) is used to observe specific surfaces, particularly areas that are relatively inaccessible to photographic and television cameras, such as interior parts of functioning equipment and internal organs of the human body, and (b) usually consists of (i) a lens that focuses an image on a fiber optic faceplate made from an endface of an aligned bundle of optical fibers, (ii) an aligned bundle of optical fibers that transmits the image to the fiber faceplate at the other end of the bundle, also made from the endface of the bundle, and (iii) perhaps another lens for displaying the image. *Note:* Arrangements must be made for the object to be illuminated so the image of the object can be transmitted. The illumination can be accomplished with additional optical fibers placed in the same fiber optic cable with the aligned bundle and energized by a light source to illuminate the object. *Synonym* **endoscopic device.** *See* **fiber optic endoscope.**

endoscopic device: *Synonym* **endoscope.**

endpoint functional group: *See* **terminal endpoint functional group.**

endpoint node: In network topology, a node connected to one and only one branch. *Synonym* **peripheral node.** *Refer to* **Figs. N-10 and N-11 (623).**

end system (ES): A system that (a) contains the application processes that are the ultimate source and sink of user traffic and (b) performs functions that can be distributed among two or more processors or computers. *See also* **sink, source, traffic.**

end-to-end encryption: The encryption of information at its origin and decryption at its intended destination without an intermediate decryption.

end-to-end security: The safeguarding of information in a secure telecommunications system by means of cryptographic or protected distribution from point of origin to point of destination of the information.

end-to-end separation sensor: *See* **optical fiber end-to-end separation sensor.**

end-to-end transport layer: In a layered communications system, the set of functions and protocols that pertain to the control of integrity and errors in the functions performed by the transmission layer to assure accuracy in the application layer.

endurability: The property of a system, subsystem, equipment, or process that enables it to continue to function within specified performance limits for an extended period of time, usually months, despite a severe natural or man-made disturbance, such as a nuclear attack or a loss of external logistic or utility support. *Note:* Endurability includes temporary failures when the local capability exists to restore and maintain the system, subsystem, equipment, or process to an acceptable performance level. *See also* **catastrophic degradation, endurable operation, fail safe operation, graceful degradation, survivability.**

endurable operation: An operation that (a) is performed by a system, subsystem, equipment, or process and (b) is continued to be performed within specified performance limits for an extended period of time, usually months, despite a severe natural or man-made disturbance, such as a nuclear attack or a loss of external logistic or utility support. *See also* **endurability.**

end user: In communications, computer, and data processing systems, (a) a customer, subscriber, or operator, such as an engineer, executive, secretary, or clerk, whom the system was established to serve, i.e., the beneficiary of a communications, computer, or data processing service, and (b) the ultimate source or consumer of the information represented by the data handled by the system. *See also* **destination user, source user.**

energy: *See* **band gap energy, emitted electron energy, luminous energy, radiant energy.**

energy band: A specified range of energy levels that a constituent particle or component of a substance may have. *Note 1:* The particles may be electrons, protons, ions, neutrons, atoms, molecules, mesons, or others. *Note 2:* Some energy bands are allowable, and others are unallowable. *Note 3:* Electrons of a given element at a specific temperature can occupy only certain energy bands, such as the conduction and valence bands. When radiation strikes an electron and imparts enough energy to the electron to cause its energy to rise to a higher energy band, energy is absorbed. If the electron energy level is reduced to that of a lower energy band, energy is emitted in the form of electromagnetic radiation, i.e., a photon is emitted. Statistical laws apply to the number of electrons in each of the possible bands, the density of these electrons in the material, and the energies of the emitted photons. *Refer to* **Fig. L-2 (502).**

energy density: The energy per unit volume, such as joules per cubic meter, at a spatial point. *See* **optical energy density, electromagnetic energy density.**

energy level: The discrete amount of kinetic and potential energy possessed by an orbiting electron. *Note:* A quantum of energy is absorbed or radiated, depending on whether an electron moves from a lower to a higher energy level or vice versa. *Refer to* **Fig. L-2 (502).**

energy level transition: In molecular structures, the passing of an electron (a) from a lower energy level to a higher energy level when a quantum of energy is absorbed, or (b) from a higher energy level to a lower energy level when a quantum of energy is radiated. *See also* **energy level, quantum.**

energy/noise ratio: *Synonym* **energy to noise ratio.**

energy to noise ratio (E/N): In the transmission of a pulse of an electromagnetic wave, representing a bit, the ratio of (a) the energy per bit, *E,* to (b) the noise energy density per hertz, *N. Note:* The energy per bit, *E,* usually is expressed in joules per bit, and the noise energy density per hertz, *N,* is expressed in watts (joules per second) per hertz. Thus, *E/N* is hertz•seconds/bit, i.e., "per bit." A joule is a watt•second and a hertz is a cycle per second. Thus, the ratio *E/N* actually is cycles per bit. However, if a cycle is a bit, then *E/N* is dimensionless and usually is considered to be dimensionless. *Synonym* **energy/noise ratio.**

engaged signal: *See* **terminal engaged signal.**

engaged signaling: *See* **visual engaged signaling.**

engaged test: In telephone switchboard operation, a test that (a) is used to determine whether a line is in use, (b) applies only to switchboards with two or more positions, (c) may be accomplished manually by touching the bust of the jack with the tip of the calling plug, and (d) creates an audible click if the line is in use, provided that the associated key is actuated. *See also* **calling plug, jack, position, switchboard.**

engaged tone: In telephone switchboard operation, an audible signal indicating that the called line, call receiver, or intermediate equipment required for setting up a particular connection is in use. *See also* **busy signal.**

engagement clip: *See* **quick engagement clip.**

engineering: *See* **communications system engineering, computer aided software engineering, software engineering, traffic engineering.**

engineering channel: A communications channel that (a) is used for voice, data, video information, or any combination of these, (b) is provided to facilitate the installation, maintenance, restoral, or deactivation of segments of a communications system by equipment operators, attendants, and controllers, and (c) in satellite communications systems, is set aside from user traffic circuits to provide engineering links between the technical managing organizations at each Earth station and switching center in the systems. *Synonyms* **order channel, service channel, speaker channel.** *See also* **channel, engineering circuit.**

engineering circuit: A voice or data circuit used by technical control, operating, maintenance, and attendant personnel for coordination and control activities related to activation, deactivation, change, rerouting, reporting, establishment, operation, and maintenance of communications system services, facilities, and equipment. *Synonyms* **engineering orderwire circuit, order circuit, orderwire circuit, service circuit, speaker circuit.** *See also* **circuit, engineering channel, engineering circuit multiplex set, express orderwire, link orderwire, local orderwire, maintenance control circuit, network management, supervisory signal.**

engineering circuit multiplex set: A multiplex carrier set specifically designed to carry engineering circuit traffic rather than user traffic, i.e., subscriber or customer traffic. *Synonym* **orderwire multiplex set.** *See also* **engineering circuit, traffic, user.**

engineering orderwire (EOW) circuit: *Synonym* **engineering circuit.**

engineering path: In communications networks, a path that (a) consists of engineering channels and engineering circuits, (b) is used for voice, data, or both, and (c) is provided to facilitate the installation, maintenance, restoral, or deactivation of segments of a communications system by operator, attendant, maintenance, and control personnel. *See also* **engineering channel, engineering circuit, path.**

engineering technology: *See* **computer aided software engineering technology.**

enhanced definition television (EDTV) system: A television system in which improvements are made that are not compatible with present receivers. *Note:* Some features and technical characteristics of the original system on which the enhanced definition television (EDTV) system is based may be included in the enhanced system. *See* **advanced television system, extended definition television system, high definition television system, improved definition television system.**

Enhanced Digital Access Communications System (EDACS™): A trunked mobile radio system that (a) provides fault-tolerant architecture, (b) provides continuous trunking, (c) provides point of failure recognition, (d) may be linked to existing very-high frequency (VHF), ultrahigh frequency (UHF), and 800-MHz systems, and (e) is manufactured by Ericsson GE.

enhanced manpack ultrahigh frequency terminal (EMUT): 1. A portable satellite-base communications terminal developed for military tactical communications systems. **2.** A highly portable secure communications system that (a) is designed for use by the military and (b) usually includes (i) modular construction, (ii) use of Demand Assigned Multiple Access (DAMA) compatibility, and (iii) a broadband frequency range, such as 40 MHz (megahertz) to 400 MHz.

enhanced quality television: *Synonym* **improved definition television.**

enhanced service (ES): 1. In commercial carrier transmission facilities used in interstate communications, a service that (a) uses computer processing to act on various aspects of transmitted user information, such as the format, content, code, or protocol, (b) provides the user with additional, different, or restructured information, and (c) allows the user to interact with stored information. **2.** A combination of basic service and computer processing applications that results in a modification of the transmitted information. **3.** An operational mode characterized by the restructuring by the transmission system of the data that represent the transmitted information. *See also* **basic service.**

enlargement: In display systems, an enlarged copy of a display element, display group, or display image. *Note:* Examples of enlargements are (a) an enlarged copy of a microimage and (b) a positive print larger than the photographic negative from which it was made. *See also* **reduction, scale, zooming.**

ENQ: enquiry character.

enquiry character (ENQ): A transmission control character used as a request for a response from the station with which a connection has been set up. *Note:* The response may include station identification, the type of equipment in service, and the status of the remote station. *See also* **binary synchronous communications, character.**

E/N ratio: energy to noise ratio.

E-N ratio: energy to noise ratio.

en route guard station: *Synonym* **secondary station.**

enter: 1. To insert data or a command into a communications, computer, or data processing system by one of many ways, such as via keyboard, mouse, light pen, voice, or switch. **2.** In a personal computer (PC), to depress the enter key to cause a previously selected keyed command or default to be executed. *See also* **enter key.**

enter key: In standard computer keyboards, the key primarily used to initiate an indicated operation. *Note 1:* Examples of the use of the enter key are that (a) in Lotus 1-2-3, depressing the enter key will cause data that have been typed, i.e., keyed, into the data entry area to be entered into the indicated cell, (b) in Volkswriter, depressing the enter key will initiate an indicated operation, such as designate the end of a paragraph and move the cursor to the beginning of the next line, (c) in IBM DOS, depressing the enter key after typing COPY A: will initiate copying a complete disk, and (d) in WordPerfect, depressing the enter key after blocking and move commands will recover the data at the position of the cursor. *Note 2:* The functions performed by the carriage return lever or key on typewriters are performed by the enter key on personal computer (PC) keyboards. *Synonyms* **return key, start key.** *See also* **carriage return key.**

entity: In database management systems, the real or actual person, event, object, place, or thing, or any tangible representation of that item, described by data in a database. Examples of entities are the actual structure called the Taj Mahal, a picture of the Taj Mahal, the words "Taj Mahal" written on this page, and the bit patterns in storage that represent the letters that spell Taj Mahal.

entrance facility: **1.** All network facilities between the network interface on the customer premises and the central office (C.O.) of a commercial carrier. **2.** The entrance to a building for both public and private network service cables, including (a) antennas where applicable, (b) the entrance point at the building wall or floor slab, and (c) the entrance room or entrance space. *See also* **entrance room, entrance space, facility, interface, network interface.**

entrance point: In a building, the point of emergence (entry) of telecommunications cables through an exterior wall or floor slab usually via a conduit.

entrance pupil: **1.** In an optical system, the image of the limiting aperture stop formed in the object space by all optical elements of the system preceding the limiting aperture stop. **2.** The aperture of the objective when there are no other limiting stops following it in an optical system. *See also* **aperture, aperture stop, image, input aperture, object, objective, optical element.**

entrance room: In a building, a space that (a) is used to join interbuilding or intrabuilding telecommunications backbone facilities and (b) may also be an equipment room. *See also* **backbone, equipment room.**

entrapment: In communications system security, the deliberate planting of apparent flaws in a system for the purpose of detecting attempted penetrations. *See also* **penetration, pseudoflaw.**

entropy: *See* **character mean entropy, conditional information entropy, information entropy, relative entropy.**

entry: In communications, computer, data processing, and control systems, any data, instructions, or signals that are entered into a device or system by any means, such as keyboard, mouse, or light pen. *See* **between the lines entry, negative entry, piggyback entry, remote batch entry, remote job entry.** *See also* **enter.**

entry point: The point in a system where an entry is made, such as an end instrument. *See* **reentry point.** *See also* **end instrument.**

envelope: **1.** In an electromagnetic wave, the boundary of the family of curves obtained by varying a parameter of the wave. **2.** In a representation of a modulated carrier wave, the function that describes the extent of variation of the carrier that is caused by the modulating signal, as a function of time. *Note:* An example of an envelope is the line that joins the peaks of the modulated carrier when a high frequency carrier is modulated by a low frequency sinusoidal wave, and the resulting signal is plotted as a function of time. The envelope does not indicate instantaneous signal levels before detection, i.e., before demodulation. The shape of the output of the demodulator will be approximately the same as the shape of the modulated carrier envelope. **3.** In digital data transmission, a group of pulses that represent the binary digits that form a byte, with the additional bits required for transmission system operation, such as start and stop pulses. *See* **permanently locked envelope.** *See also* **demodulation, modulation.**

envelope delay distortion: **1.** In a given passband of a device or a transmission facility, the maximum difference of the group delays of any two specified frequencies. **2.** The distortion that occurs when the rate of change of phase shift with frequency of a circuit or a system is not constant over the frequency range that is required for the transmission. *Note:* Envelope delay distortion usually is expressed as one-half of the difference of the group delays of the two extremes of frequency that define the limits of the channel used. *See also* **distortion, ghost, group delay, frequency, passband, phase shift.**

envelope power: *See* **peak envelope power.**

environment: **1.** In spatial communications systems, the total radiated electromagnetic power in all frequencies as a function of time surrounding a point in space. **2.** In communications, computer, data processing, and control systems, the surrounding natural and man-made conditions under which a system is required to function, including climatic, vibration, radiation, sound, power, and human factors. *See* **electromagnetic environment, stressed environment.**

environmental control: *See* **air conditioning.**

environmental file: *See* **system environmental file.**

environmental loss time: In the measurement of the loss time of a functional unit, the downtime caused by a condition that is outside of the unit. *Synonym* **external loss time.**

environmental security: **1.** The security that is inherent in the physical surroundings in which a facility or a functional unit is located, such as on board ships and aircraft and in underground vaults, where locations by their nature provide a certain amount of protection against exploitation of compromising emanation even before other protective measures are implemented. **2.** The application of electrical, acoustic, physical, and other safeguards to an area to minimize the risk of unauthorized interception of information from the area.

environments: *See* **portable operating system interface for computer environments.**

environment software: *See* **virtual environment software.**

e-o: electrooptic, electro-optic.

EO: electrooptic, end office.

E.O.: end office.

EOM: 1. End of message. 2. The end of message character.

EOMI: end of message indicator.

EOS: 1. End of selection. 2. The end of selection character.

EOT: 1. End of transmission. 2. The end of transmission character.

EOW: engineering orderwire.

ephemeris: 1. A tabulation of the location of celestial bodies at specific intervals. **2.** A tabulation of the location of artificial satellites at specific intervals.

epitaxial growth: A growth of crystals formed in layers, such as the growth of crystals for light-emitting diodes (LEDs) and lasers.

E plane bend: *Synonym* **E bend.**

epoch date: A date in history, chosen as the reference date from which time may be measured. *Note 1:* An example of an epoch date is January 1, 1900, Universal Time, for Transmission Control Protocol/Internet Protocol (TCP/IP). *Note 2:* Transmission Control Protocol/Internet Protocol (TCP/IP) programs exchange date or time of day information with time expressed as the number of seconds past the epoch date.

E propagation: *See* **sporadic E propagation.**

equal access: In a communications system, such as a telephone network, a system capability of a local telephone service that allows all its users, i.e., all its subscribers or customers, to select and use any of the authorized operating long-distance companies to which the local system can be connected via a gateway, bridge, or port.

equal gain combiner (EGC): A diversity combiner in which (a) the signals on each channel are added, and (b) the channel gains can be made to remain equal so that the resultant signal remains approximately constant. *See also* **diversity combiner, gain, maximal ratio combiner.**

equality gate: *Synonym* **EXCLUSIVE NOR gate.**

equalization: 1. In a circuit, the maintenance of signal characteristics within specified limits by modifying the electrical parameters of the circuit, such as by modifying resistance, inductance, or capacitance, in order to adjust signal parameters, such as amplitude, frequency, phase, timing, transport mechanisms, and transport relationships. **2.** In a circuit, reducing the frequency distortion or the phase distortion, or both, of signals by the introduction of electric circuits or networks to compensate for the difference in attenuation or delay that occurs to the various frequency components of the signals in the circuit. *See* **adaptive equalization.** *See also* **amplitude equalizer, amplitude versus frequency distortion, building out, conditioned circuit, conditioned loop, conditioning equipment, delay equalizer, loading, phase, slope equalizer.**

equalizer: *See* **amplitude equalizer, delay equalizer, fixed compromise equalizer, line residual equalizer, line temperature compensating equalizer, phase-frequency equalizer, slope equalizer.**

equal-length code: A telegraph or a data code in which (a) all the words or code groups are composed of the same number of unit elements, (b) each element has the same duration or spatial length, (c) each word or code group has the same time duration or spatial length, and (d) each word or code group usually has the same number of characters.

equal-level patch bay: An analog patching facility at which all input and output voice frequency levels are uniform. *Note:* An equal-level patch bay permits patching without making transmission level adjustments. *See also* **circuit, facility, patch bay.**

equation: *See* **dispersion equation, Fresnel equation, Helmholtz equation, Hertzian dipole antenna equation, Maxwell equation, one-way radio equation, password length equation, photoelectric equation, radar line of sight equation, self-screening range equation, Sellmeier equation, two-way radar equation, wave equation.**

equator: *See* **celestial equator.**

equatorial orbit: For the Earth, an orbit in the equatorial plane. *Note:* The inclination angle of an equatorial orbit is 0°. *See also* **direct orbit, geostationary orbit, inclined orbit, polar orbit, retrograde orbit, satellite, synchronous orbit.**

equilibrium coupling length: *Synonym* **equilibrium length.**

equilibrium distance: *Synonym* **equilibrium length.**

equilibrium length: In a multimode waveguide, the distance from the input end necessary to attain electromagnetic wave equilibrium modal power distribution for a specific excitation condition, i.e., for a specific set of launch conditions at the input end. *Note 1:* Beyond the equilibrium length the fraction of total radiant power in each mode remains constant with respect to distance. *Note 2:* The equilibrium length usually is the longest distance obtained from a worst-case, but undefined, excitation condition. *Synonyms* **equilibrium coupling length, equilibrium distance, equilibrium modal power distribution length.** *See also* **equilibrium modal power distribution, fiber optics, mode coupling.**

equilibrium modal power distribution (EMPD): In an electromagnetic wave propagating in a waveguide, such as a lightwave in an optical fiber, the distribution of radiant power among the various modes such that the fraction of total power in each mode is stable and does not redistribute as a function of distance. *Note 1:* Equilibrium modal power distribution will not occur at the beginning of the guide until the wave has propagated a distance called the equilibrium length, because the launch conditions, wavelength, and refractive indices cannot be controlled with sufficient precision. *Note 2:* If the entrance conditions could be perfectly controlled, the equilibrium length could be reduced to zero. *Synonyms* **equilibrium mode distribution, steady state condition.** *See also* **equilibrium length, fiber optics, mode, nonequilibrium modal power distribution.**

equilibrium modal power distribution length: *Synonym* **equilibrium length.**

equilibrium mode distribution: *Synonym* **equilibrium modal power distribution.**

equilibrium mode simulator: A system, subsystem, component, or equipment used to create an approximate equilibrium modal power distribution. *See also* **equilibrium modal power distribution, mode filter.**

equilibrium radiation pattern: The radiation pattern at the output end of an optical fiber that is long enough so that the equilibrium modal power distribution has been reached, i.e., the fiber is longer than the equilibrium length. *See also* **equilibrium length.**

equinox: Either of two points on the celestial sphere where the ecliptic and the celestial equatorial plane intersect. *Note:* The ecliptic is the great circle traced on the celestial sphere annually by the Sun. Its inclination to the celestial equator is about 23.5° The vernal equinox is the point where the Sun crosses the equator from south to north, on or about March 21, bringing spring to the northern hemisphere and autumn to the southern hemisphere. At the autumnal equinox, the Sun crosses the equator from north to south on or about September 22, at which time autumn begins for the northern hemisphere and spring begins for the southern hemisphere. When the Sun is at either point, the day is only slightly longer than the night. *See* **autumnal equinox, ecliptic, vernal equinox.**

equipment: *See* **airport surface detection equipment, analog facsimile equipment, angle measuring equipment, authentication equipment, automatic data processing equipment, automatic numbering equipment, automatic switching equipment, Boehme equipment, channel translation equipment, cipher equipment, common equipment, communications security equipment, concentrating equipment, conditioning equipment, customer premises equipment, data circuit-terminating equipment, data communications equipment, data subscriber terminal equipment, data terminal equipment, dial switching equipment, distance measuring equipment, duplex equipment, embedded base equipment, embedded customer premises equipment, facsimile digital equipment, Federal Information Processing equipment, grandfathered terminal equipment, half-duplex equipment, high performance communications equipment, high performance equipment, industrial radio frequency heating equipment, industrial-scientific-medical radio frequency equipment, intermediate equipment, limited protection voice equipment, literal cipher equipment, low performance communications equipment, medical diathermy equipment, mobile communications equipment, offline equipment, online equipment, peripheral equipment, process control equipment, process interface equipment, pulse code multiplexing equipment, radio direction finding equipment, remote control equipment, start-stop equipment, switching equipment, telegraph automatic relay equipment, telegraph error rate measuring equipment, terminal equipment, through group equipment, through supergroup equipment, time division multiplex equipment, transportable communications equipment, ultrasonic equipment.**

equipment clock: A clock that (a) satisfies the particular needs of equipment and (b) may control the flow of data at the equipment interface. *See also* **clock.**

equipment congestion signal: *See* **national switching equipment congestion signal.**

equipment identification: Information that uniquely characterizes a product and provides a traceable indicator for determining its specifications, features, issue or revision, and manufacturer, such as the information given in the Common Language Equipment Identification system.

equipment intermodulation noise: Intermodulation noise introduced into a system by a specific piece of equipment. *See also* **intermodulation noise, noise.**

equipment mobility: In communications, pertaining to the capability of communications equipment to be moved from place to place. *Note:* Fully mobile implies the ability to fulfill its primary mission, i.e., perform communications operations, such as transmit and receive, while on the move. Transportable does not imply the ability to fulfill its primary mission while on the move.

equipment reference level: *See* **single sideband equipment reference level.**

equipment room: In a building, a centralized space for telecommunications equipment that (a) serves the occupants of the building and (b) is considered as distinct from a telecommunications closet because of the nature or the complexity of the equipment housed in the equipment room. *See also* **entrance facility, entrance room.**

equipment side: The portion of a device that is directly connected to facilities internal to a station, such as the data terminal equipment (DTE) side of the DTE/data circuit-terminating (DCE) interface, switches, and user end instruments. *Synonyms* **station side, central office side.** *See also* **line side.**

equipotential ground plane: A mass, or bonded masses, of conducting material that (a) offers a negligible impedance to current flow and (b) either may be in direct contact with the earth or may be physically isolated from the earth but suitably connected to it. *See also* **earth electrode subsystem, facility grounding system, ground, ground plane.**

equipotential surface: A surface of a material on which ideally (a) every point has the same potential difference, i.e., has the same voltage, with respect to a given reference, (b) there is no potential difference between any two points on the surface, (c) there is no current flow between any two points on the surface, and (d) there may be a buildup of charges on the surface. *Note:* Examples of equipotential surfaces are communications buses, power buses, equipotential ground planes, facility ground systems, and earth. *See also* **bus, earth, equipotential ground plane, facility ground system.**

equivalence gate: *Synonym* **EXCLUSIVE NOR gate.**

equivalence operation: An identity operation in which there are exactly two operands. *See also* **identity operation, operand.**

equivalency number: *See* **ringer equivalency number.**

equivalent binary digit factor: The ratio of (a) the approximate average least number of binary digits to (b) the number of digits of a nonbinary numeration system required to express a pair of numbers, one in binary and the other in nonbinary, that are of equal value in both systems. *Note:* An example of an equivalent binary digit factor is the ratio of 10/3 for the decimal $10^3 = 1000$, a nonbinary number, and 2^{10} for the approximately equivalent binary numeral for 1024, which yields about 3.4 for the equivalent binary digit factor. Thus, especially for large numbers, it takes about 3.4 times as many binary digits to express the same value in decimal digits. When other systems are used, the equivalent binary digit factor is different, such as the alphanumeric system, which requires as many as seven binary digits to express a character, including the ten decimal digits. *See also* **binary digit, numeration system.**

equivalent circuit: *Synonym* **equivalent network.** *Refer to* **Fig. P-7 (723).**

equivalent focal length: 1. The distance from a principal point to its corresponding principal focus point. **2.** The focal length of the equivalent thin lens. *Note:* The size of the image of an object is directly proportional to the equivalent focal length of the lens forming it. *See also* **focal length, object, principal focus point.**

equivalent four-wire circuit: A transmission system in which frequency division multiplexing is used to obtain duplex operation, i.e., full-duplex operation, over one pair of wires. *See also* **duplex operation, frequency division multiplexing.**

equivalent isotropically radiated power: The product of (a) the power supplied to an antenna and (b) the antenna gain relative to an isotropic antenna. *See also* **antenna gain, isotropic antenna.**

equivalent network: 1. In a system, a network that may replace another network without altering the performance of the system. **2.** A network with external characteristics that are identical to those of another network. **3.** A theoretical representation of an actual network such that the theoretical representation can be used to analyze and predict the performance of the actual network. *Synonym* **equivalent circuit.** *See also* **network.**

equivalent noise resistance: A quantitative representation in resistance units of the spectral density of a noise voltage generator, given by the relation $R_n = (\pi W_n)/(kT_0)$ where W_n is the spectral density, k is Boltzmann's constant, and T_0 is the standard noise temperature (290 K). *Note 1:* $kT_0 = 4.00 \times 10^{-21}$ W•s (watt•second). *Note 2:* The equivalent noise resistance in terms of the mean square noise generator voltage, e^2, within a frequency increment, Δf, is given by the relation $R_n = e^2/4kT_0\Delta f$. *See also* **frequency, noise.**

equivalent noise temperature (ENT): The physical temperature, i.e., the thermal temperature, of a matched resistance at the input of an assumed noiseless device, such as a noiseless amplifier, that would account for the measured output noise. *Note:* The equivalent noise temperature usually is expressed in kelvins.

equivalent power: *See* **noise equivalent power.**

equivalent pulse code modulation (PCM) noise: On a frequency division multiplexed (FDM) or wire channel, the amount of thermal noise power necessary to approximate the same judgment, based on comparative tests, of speech quality created by quantizing noise in a pulse code modulation (PCM) channel. *Note:* The approximate equivalent pulse code modulation (PCM) noise of a seven-bit PCM system usually is considered to be 33.5 dBrnC ± 2.5 dB. *See also* **frequency division multiplexing, noise, pulse code modulation.**

equivalent satellite link noise temperature: In an Earth station of a satellite communications system, the noise temperature, referred to the output of the receiving antenna of the Earth station, that corresponds to the radio frequency noise power that produces the total observed noise at the output of the satellite link, excluding noise caused by interference from satellite links using other satellites and from terrestrial systems. *See also* **Earth station.**

equivalent step index (ESI) profile: The refractive index profile of a hypothetical step index optical fiber that (a) has the same propagation characteristics as a given single mode optical fiber and (b) has a constant refractive index throughout the cladding.

equivalent step index (ESI) profile refractive index difference: For an optical fiber, the difference between (a) the refractive index of the core and (b) the refractive index of the cladding in its equivalent step index (ESI) profile.

equivocation: The mean additional information content that must be supplied per message at the message sink to correct the received messages affected by a noisy channel, given the conditional information entropy of the occurrence of specific messages at the message source and given the occurrence of specific messages at a message sink that is connected to the message source by a specific channel. *See also* **channel, conditional information entropy, message sink, message source.**

erasable storage: Storage in which the contents can be changed by an operator or by a computer program without making hardware changes, in contrast to fixed storage in which changes can only be made by hardware replacement by a technician, usually at a shop or a factory. *Note:* Examples of erasable storage are the diskette storage, disk storage, and internal memory of a personal computer.

erase: 1. To obliterate information from a storage medium, such as to clear or to overwrite. **2.** In a magnetic storage medium, to remove all stored data by (a) changing the medium to an unmagnetized state or (b) changing the medium to a predetermined magnetized state. **3.** In paper tape and punched card storage, to punch a hole at every punch position. *See also* **read-only storage, storage.**

erase character: *Synonym* **delete character.**

erase head: A device that has only one function, that of deleting recorded data before new data are recorded.

erasure signal: A signal that is used for the purpose of invalidating a previous signal without actually erasing or deleting the original signal. *See also* **erase, signal.**

erect image: In an optical system, an image, either real or virtual, that has the same spatial orientation as the object. *Note:* The image obtained on the retina with the assistance of an optical system is said to be erect, in contrast to inverted, when the orientation of the image is the same as seen with the unaided eye. *See also* **image, inverted image, object.**

erect position: In frequency division multiplexing, a point in a translated channel at which an increasing signal frequency in the untranslated channel causes an increasing signal frequency in the translated channel. *Synonym* **upright position.** *See also* **down-converter, frequency division multiplexing, frequency translation, inverted position, up-converter.**

E region: In the atmosphere, the region (a) in which the E layer of the ionosphere forms, (b) that lies between approximately 95 km (kilometers) and 160 km above the surface of the Earth, and (c) that lies between the D region and the F region. *Synonyms* **Heaviside region, Kennelly-Heaviside region.** *See also* **D region, F region, ionosphere.**

ERL: echo return loss. *See* **return loss.**

erlang: An international, dimensionless unit that (a) indicates the average traffic intensity, i.e., the average occupancy, of a facility during a period, usually a busy hour, and (b) is indicative of how much a piece of equipment, such as a line, switch, or trunk, actually is used compared to how much it could have been used during a performance measurement period. *Note 1:* The number of erlangs is expressed as the ratio of (a) the time during which a facility is continuously or cumulatively occupied to (b) the time the facility is available for occupancy. *Note 2:* The facility, such as a switch register, trunk, or circuit, usually is shared by many users. *Note 3:* Communications traffic, measured in erlangs and offered to a group of shared facilities, such as a trunk group, is equal to the sum of the traffic intensity, in erlangs, of all individual sources, such as telephones, that share and are served exclusively by this group of facilities. *Note 4:* A single facility, such as a register, trunk, or circuit, cannot carry more than one erlang of traffic. *Synonym* **traffic unit.** *See also* **busy hour, call-second, communications traffic volume, group busy hour, traffic intensity, traffic load.**

erosion: *See* **facet erosion.**

erp: effective radiated power.

e.r.p.: effective radiated power.

ERP: effective radiated power.

erroneous bit: A bit that is not what it should be according to specified criteria. *Note:* An example of an erroneous bit is a 1 that should be a 0, such as a bit that was stored as a 1 but read as a 0. *See also* **binary digit.**

erroneous block: A block in which there are one or more erroneous bits. *See also* **binary digit, block, block error probability, block transfer failure, error, incorrect block.**

error: 1. The difference between a computed, estimated, or measured value and the true, specified, or theoretically correct value. *Note:* Examples of errors are (a) the letters "ta" in "thta" when the correct spelling is "that," (b) the letter "A" in the social security number 106-2A-2370, when the correct value for the "A" is "4," and (c) a wrong character in a computer program instruction word. **2.** A malfunction that is not reproducible. **3.** The result of a malfunction, failure, or mistake, usually a result that is not reproducible. *See* **absolute error, balanced error, clock error, color error, concentricity error, digital error, optical fiber concentricity error, recoverable error, relative error, time error, time interval error, tracking error, truncation error.** *Refer to* **Fig. E-3.**

Even Parity

```
|1 1 0 0|          |1 1 0 0|
|0 1 1 0|          |0 1 1 0|
|0 0 1 1|  = 6     |0[1]1 1|  = Error
|1 0 0 1|          |1 0 0 1|

Transmitted Array   Received Array
```

Fig. E-3. An **error-correcting code** used to represent decimal 6. Note that there are two 1s in each column and row. Other patterns that also meet this criterion are used to represent other decimal digits. Whenever a bit is wrong, the criterion in a row and column is violated, the erroneous bit in that row and column is identified, and the erroneous bit is corrected. The probability of the array for a given decimal digit being converted to the proper array for another decimal digit by the simultaneous occurrence of a group of errors is extremely small.

error analysis: The investigating of an observed or detected error, failure, fault, malfunction, or mistake, i.e., a bug, in the hardware, firmware, or software of a system with the purpose of tracing the bug to its source, i.e., to its cause. *See also* **bug, debug, error, failure, fault, malfunction, mistake.**

error block: In video systems, block distortion in which a block or blocks in the received image bear no resemblance to the current or previous scene and may contrast greatly with adjacent blocks.

error budget: The allocation of a bit error ratio (BER) requirement to the segments of a communications system, such as trunking, switching, access, and terminal devices, in a manner that satisfies the specified system end-to-end BER requirement for the transmitted traffic. *Note:* The circuit chosen for an error budget may be a postulated reference circuit, but usually it is a typical circuit in which the budgeted components are used. *See also* **binary digit, bit error ratio, error.**

error burst: In data communications, a group of signals that contains one or more errors but is considered as a unit, in regard to errors, in accordance with some criterion of measure, such as (a) if at least three correct bits follow an erroneous bit, the particular error burst is considered terminated, and (b) in a contiguous sequence of symbols received over a data transmission channel, (i) the first and last symbols are in error and there exists no contiguous subsequence of M correctly received symbols within the sequence, where M is an integer parameter referred to as the guard band, (ii) the last symbol in the sequence and the first symbol in the following sequence are accordingly separated by M correct bits or more, and (iii) the number M is specified. *See also* **binary digit, bit error ratio, burst, error, error control.**

error control: 1. Any technique that will detect or correct errors. **2.** To control the detection and the correction of errors. *Note 1:* The errors may be (a) in stored or transmitted data or (b) in computer programs. **3.** The improvement of communications, computer, data processing, and control systems through the use of procedures designed to reduce, eliminate, or prevent errors, such as the use of error detection schemes, forward-acting error correction procedures, block codes, error-detecting codes, error correcting codes, and parts of a protocol that controls the detection, and possibly the correction, of errors. *See also* **error-correcting code, error-detecting code, forward error correction.**

error control character: *See* **accuracy control character.**

error-correcting code (ECC): A code such that (a) each group of characters or signals conforms to specific rules of construction so that departures from this construction in the received signals can be detected automatically and some or all of the errors can be corrected automatically, (b) more signal elements are required than are necessary to convey the basic information, (c) redundant characters are used to assist in the restoration of the word that has been mutilated, (d) only certain kinds of errors may be corrected if the redundancy is less than 100%, (e) the redundancy usually is arranged so that the mutilated word when wholly or partially corrected resembles the original word more closely than any other valid representation of another possible word in the vocabulary, (f) if an error occurs that the code is not designed to correct, the adjustment that is made may be erroneous, and (g) in most applications, the code is in the form of additional digits appended to the word that is transmitted. *Note 1:* If the number of errors is equal to or less than the maximum correctable threshold of the error-correcting code, all errors will be corrected. *Note 2:* The two main classes of error-correcting codes are block codes and convoluted codes. *See also* **block code, code, convoluted code, error control, forward error correction, Hagelbarger code, Hamming code.** *See also* **code.** *Refer to* **Fig. E-3.**

error-correcting system: 1. A system in which (a) an error-correcting code is used, (b) some or all signals detected as being in error are automatically corrected at the receiving terminal before delivery to the data sink, and (c) in a packet-switched data service, the error correction might result in the retransmission of one or more complete packets should an error be detected. **2.** In digital data transmission, a system that uses either forward error correction (FEC) or automatic repeat-request (ARQ) techniques so that most transmission errors are automatically removed from the data prior to delivery to the destination facility. *See also* **ARQ, code, communications system, error, error control, error-correcting code, error-detecting code, forward error correction.**

error correction: *See* **forward error correction.**

error correction bit: One of a group of bits that is inserted into a bit stream at the end of a block of bits to enable receiving equipment to detect and correct the error that may exist in a block. *See also* **bit stream, block.**

error-detecting and feedback system: *Synonym* **automatic repeat-request.**

error-detecting code (EDC): A code (a) in which each telegraph or data signal conforms to specific rules of construction, so that departures from this construction in the received signal can be automatically detected, and (b) that requires more signal elements than are necessary to convey the basic information. *Note:* An example of an error-detecting code is the use of a parity bit, though it is possible that if two compensating errors occur, the errors can go undetected. *See also* **block parity, code, error, error control, error-correcting code, error-correcting system.**

error-detecting system: A system that uses an error-detecting code so that any data detected as being in error are either (a) deleted from the data delivered to the data sink, in some cases with an indication that such deletion has taken place, or (b) delivered to the data sink together with an indication that the data are in error. *See also* **code, cyclic redundancy check, error, error control.**

error detection: *See* **automatic error detection.**

error detection and correction system: *See* **automatic error detection and correction system.**

error detection and feedback: An error control technique that uses an error-detecting code and in which a signal detected as being in error automatically initiates a request for retransmission of the signal. *Synonyms* **decision feedback, request-repeat.**

error detection bit: One of a group of bits that (a) periodically is inserted into a bit stream and (b) enables the equipment at the receiving end to determine if errors are likely to exist in the data received immediately preceding these bits.

error function: The algebraic expression

$$\mathrm{erf}(x) = (1/2\pi)^{1/2}\int_0^x \exp(-y^2/2)dy,$$

where y is a function of x.

error indicator: 1. A device that indicates that an error, failure, fault, malfunction, or mistake has occurred. **2.** In a calculator, a visual indication that the operator has attempted to have the calculator carry out an impossible mathematical function, such as divide by 0 or take the square root of a negative number. *See also* **error, failure, fault, malfunction, mistake.**

error message: In communications, computer, data processing, and control systems, a visual or audio message that indicates (a) that an error has been made by the system or its operator, (b) the nature or the type of error that has been made, and (c) the next step to be taken by the operator. *Note:* In personal computer (PC) systems, the error message will appear on the screen, indicating that an error has been made, the nature of the error, and the action to be taken to clear the error in order to return to a previous condition or to a ready mode or to proceed properly. In communications systems, the error message will tell the operator the steps to be taken to return the system to a clear condition or the way to proceed to continue operations without loss. In many PC software packages, use of the escape or cancel key will cause the computer to return to a state in which operations may continue.

error model: A mathematical model used to predict or estimate the number of errors, failures, faults, malfunctions, or mistakes, i.e., bugs, that have occurred, or are likely to occur in the hardware, software, or firmware of a system. *See also* **bug, error, failure, fault, malfunction.**

error rate: The number of errors that occur per unit of time at a point in a propagation medium during data transmission independent of the data signaling rate (DSR). *See also* **data signaling rate, error ratio.**

error rate measuring equipment: *See* **telegraph error rate measuring equipment.**

error ratio: The ratio of (a) the number of bits, elements, characters, or blocks incorrectly received to (b) the total number of bits, elements, characters, or blocks sent during a performance measurement period. *Note:* Threshold, i.e., maximum allowable, error ratios for modern communications systems, before error correction, range from 10^{-9} to 10^{-7}, based on the ratio of erroneous bits received to the total bits sent during a given time interval, such as one hour. *Deprecated synonym* **error rate.** *See* **bit error ratio, block error ratio, pseudo bit error ratio, residual error ratio.** *See also* **binary digit, block, block transfer rate, character, error, error rate.**

error ratio tester: *See* **bit error ratio tester.**

error recovery: In a communications, computer, information processing, or control system, (a) correcting or bypassing errors, failures, faults, malfunctions, or mistakes, (b) restoring the system to a prescribed condition, and (c) continuing usual operation. *See also* **error, failure, fault, malfunction, mistake.**

error seeding: In a computer program, the intentional adding of a known number of errors to those that already might be in the program for the purpose of estimating the number of faults remaining in the program. *Synonym* **fault seeding.**

error signal: In a communications, computer, data processing, or control system, an audio or visual signal that indicates that an error has been made by the system or its operator. *Note:* In most systems, the error signal accompanies an error message and is used to draw operator attention to the error message. *See also* **error message.**

ES: end system.

ESC: escape character.

escape character (ESC): 1. In alphabet coding schemes, a specially designated character, the occurrence of which in the data signifies that one or more of the characters to follow are from a different character code, i.e., they have meanings other than normal. **2.** A code extension character used, in some cases with one or more succeeding characters, to indicate, by convention or agreement, that the coded representations following the character, or the group of characters, are to be interpreted according to a different code or according to a different coded character set. **3.** In a text control sequence of characters, a control character that indicates the beginning of the sequence and the end of any preceding text.

escape key: In most communications, computer, and data processing systems, especially personal computer

(PC) systems, a key on a standard keyboard that has the general effect of undoing what has just been done or ordered to be done, i.e., commanded to be done, usually to (a) correct or change a course of action, (b) enter a command different from the command that has just been entered, (c) cause the computer to back up one step, or (d) return to a ready mode.

escapement: On a printing device, such as a printer or a typewriter, the relative movement of one character space or unit space between the paper carrier and a print position on a line or a row.

escort jamming: Jamming that is accomplished with equipment that is on a ship, aircraft, or land vehicle that is accompanying the group that is being protected.

escrow cryptography: *See* **key escrow cryptography.**

ESF: extended superframe.

ESI: equivalent step index.

ESI profile: equivalent step index profile.

ESI refractive index difference: equivalent step index profile refractive index difference.

ESM: electronic warfare support measure.

ESS: electronic switching system.

essential service: A network-provided service feature (a) in which a priority dial tone is furnished, (b) that usually is provided to less than 10% of network users, and (c) that is used in conjunction with national security and emergency preparedness (NS/EP) telecommunications services. *Synonym* **critical service.** *See also* **national security and emergency preparedness telecommunications.**

establishment: *See* **automatic link establishment.**

estimated time of arrival (ETA): In radio communications, the time at which a mobile station, such as an aircraft, ship, or land vehicle station, is expected or calculated to arrive at a specified location, such as a ship arriving in port, an aircraft arriving on a runway, and a motor vehicle convoy arriving at a destination.

estimated time of departure (ETD): In radio communications, the time at which a mobile station, such as an aircraft, ship, or land vehicle station, is expected or calculated to depart from a specified location, such as a ship leaving port, an aircraft taking off, and a motor vehicle convoy departing.

estimation: *See* **reliability estimation.**

ETA: estimated time of arrival.

etalon: A device that (a) is used for spectral analysis of light beams, (b) operates from infrared to ultraviolet, (c) may be of fixed or tunable wavelength, (d) usually has piezoelectric crystals, cavities, mirrors, and gimbals mounting, and (e) uses interferometric techniques to accomplish the analysis.

ETB: 1. End of transmission block. 2. The end of block transmission control character.

ETD: estimated time of departure.

Ethernet: A communications network standard that (a) is for a 10-Mbps (megabit per second) baseband local area network (LAN) bus using carrier sense multiple access with collision detection (CSMA/CD) as the access method, (b) is implemented at the Physical Layer in the International Organization for Standardization (ISO) Open Systems Interconnection—Reference Model (OSI-RM), and (c) establishes (i) the physical characteristics of the network and (ii) the use of various cabling technologies, such as thick or thin coaxial cable, unshielded twisted pair cables, and fiber optic cables. *Note:* The Ethernet standard is designated as IEEE-802.3.

ETX: end of text, end of text transmission control character.

eV: electron volt.

evaluation: *See* **operational evaluation, performance evaluation, technical evaluation.**

evaluation and test: *See* **Functional Advanced Satellite System for Evaluation and Test.**

evanescent field: In a waveguide, a time-varying transverse electromagnetic field that (a) has an amplitude that decreases monotonically as a function of transverse distance, i.e., radial distance, from the waveguide, but without an accompanying phase shift, (b) is coupled, i.e., is bound, to an electromagnetic wave or mode propagating inside the waveguide, (c) is a surface wave, and (d) in fiber optics, may be used to provide coupling to another optical fiber by means of proximity coupling because the field is outside the core. *See also* **evanescent field coupling, evanescent wave, fiber optics, optical fiber.**

evanescent field coupling: Coupling between two waveguides achieved by allowing evanescent waves of one to penetrate, and be trapped by, the other. *Note 1:* Transfer of energy will take place if the guides are placed parallel and close together for even a short distance. *Note 2:* Evanescent field coupling may be achieved between optical fibers (a) by etching away the cladding of both fibers and placing their cores close together or (b) by locally modifying their refractive indices. *See also* **tangential coupling.**

evanescent mode: A mode of the evanescent field. *See also* **evanescent field, fiber optics, mode.**

evanescent wave: In a waveguide, a propagating transverse electromagnetic wave that (a) is on the outside of the guide, (b) is coupled, i.e., is bound, to an electromagnetic wave or mode propagating inside the waveguide, (c) radiates away at sharp bends in the guide if the radius of the bend is less than the critical radius, (d) usually has a frequency smaller than the cutoff frequency above which true propagation occurs and below which the wave decays exponentially with distance from the guide, (e) has a wavefront that may be perpendicular or at an angle less than 90° to the surface of the guide in the direction of propagation, and (f) will radiate away from the guide if the internal modes to which it is bound travel faster than the evanescent wave can travel in the medium it is in. *See also* **evanescent field, fiber optics, optical fiber.** *Refer to* **Fig. E-4.** *Refer also to* **Figs. L-8 (536), P-17 (767).**

evasion: *See* **frequency evasion.**

even parity: Parity in which the number of 1s, i.e., marks, in a group of binary digits, such as those that compose a character or a word, is always maintained as an even number, including the parity bit. *Note:* Consideration must be given to the fact that the electrical representations of 1s and 0s are often interchanged because of inversion and reversion that occurs in the electronic circuitry. When such inversion occurs, the parity bit is also inverted. *See* **parity, parity bit, parity check.** *Refer to* **Fig. E-3 (310).**

event: 1. An occurrence or happening, usually significant to the performance of a function, operation, or task. **2.** In Integrated Services Digital Networks (ISDNs), an instantaneous occurrence that changes at least one of the attributes of the global status of a managed object. *Note*: An event (a) may be persistent or temporary, thus allowing for functions, such as surveillance, monitoring, and performance measurement, (b) may generate reports, (c) may be spontaneous or planned, (d) may trigger other events, and (e) may be triggered by one or more other events.

evident code: *See* **self-evident code.**

EW: electronic warfare.

EWSM: electronic warfare support measure.

exa (E): A prefix that indicates multiplication by 10^{18}, i.e., by (a) a billion billion (U.S.) or (b) million trillion (U.S.). *Note:* An example of the use of "exa" is in 10 EHz (exahertz), equivalent to 10^{19} Hz.

exahertz (EHz): A unit of frequency that is equal to one million trillion (U.S.) hertz, i.e., 10^{18} Hz. *Note:* Care must be exercised in the use of billion and trillion. In some countries, such as the United States, a billion is a thousand million. In other countries, a billion is a million million, i.e., a bi-million.

exalted carrier reception: Reception of either amplitude-modulated or phase-modulated signals in which the carrier is separated from the sidebands, filtered, amplified, and then recombined with the sidebands at a higher power level prior to demodulation. *Synonym* **reconditioned carrier reception.** *See also* **carrier, sideband transmission.**

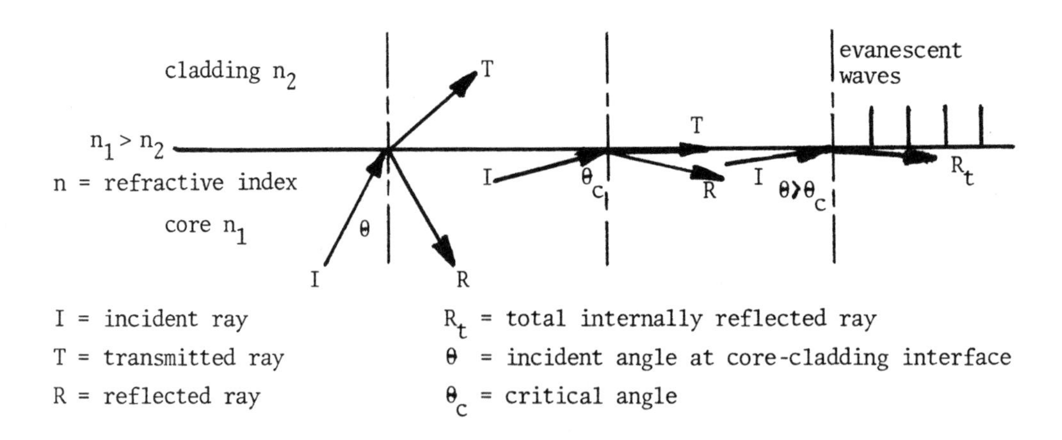

Fig. E-4. The production of **evanescent waves** at the core-cladding interface surface of an optical fiber, particularly when total internal reflection occurs. The evanescent waves remain bound, *i.e.,* remain coupled, to propagating modes in the core.

exceedance distribution: **1.** The probability that samples from a given set of values will exceed a specified value, given by the relation $E_X = P_r(X > x)$, where E_X is the exceedance function, given by the relation $E_X = 1 - FX$, where FX is the cumulative distribution function. **2.** In adaptive high frequency (HF) radio links, a time-related value that represents channel quality above a specified threshold value. *Note:* The exceedance distribution is usually expressed as a percentage of an epoch. *See also* **epoch.**

exception condition: **1.** In data transmission, the condition assumed by a device when it receives a command that it cannot execute. **2.** In data transmission, the condition assumed by a secondary station when it receives a command that it cannot execute. *Note:* In high level data link control (HDLC) operation, an example of an exception condition is the condition assumed by a station upon receipt of a control field that it cannot execute because an error has occurred either in transmission or in internal processing. *See also* **data transmission, error, failure, high level data link control, secondary station.**

excess insertion loss: *Deprecated synonym for* **insertion loss.** *Note:* "Excess insertion loss" was used to indicate that in a fiber optic coupler the loss occasioned by dividing the input power among the ports is not the total insertion loss.

excess loss: *See* **coupler excess loss.**

excess-three code: A code for representing binary coded decimal digits in which the decimal digit n is represented by the straight binary equivalent of $n + 3$. *Note:* In the excess three code, the 9s complement of a decimal digit is obtained simply by changing all 1s to 0s and all 0s to 1s in any given decimal digit. This is convenient because any bank of flip-flops that forms a register always has available a number and its complement, considering that both sides of each flop-flop are of opposite polarity. The complements are needed to perform arithmetic operations. *Refer to* **Fig. E-5.**

Excess-Three	Pure Binary	Decimal
0011	0000	0
0100	0001	1
0101	0010	2
0110	0011	3
0111	0100	4
1000	0101	5
1001	0110	6
1010	0111	7
1011	1000	8
1100	1001	9

Fig. E-5. The **excess-three code** equivalences to decimal and pure binary numbers.

exchange: **1.** A room or building containing equipment that (a) enables telephone lines terminating there to be interconnected as required and (b) may include manual or automatic switching equipment. **2.** In the telephone industry, pertaining to a geographical area, such as a city and its environs, established by a regulated telephone company for the provision of local telephone services. *Note:* An exchange area often includes the areas served by many different telephone central offices (COs). **3.** In the Modification of Final Judgment (MFJ), a local access and transport area (LATA). *See* **Advanced Project for Information Exchange, automatic answer back code exchange, automatic exchange, automatic message exchange, broadband exchange, common battery exchange, common battery signaling exchange, connectivity exchange, data switching exchange, Defense Network Security Information Exchange, electronic exchange, Internetwork Packet Exchange, main exchange, private branch exchange, satellite exchange, satellite private branch exchange, telephone exchange, terminal exchange, trunk exchange, zone exchange.**

exchange access: In telephone networks, access in which exchange services are provided (a) for originating or terminating interexchange telecommunications, and (b) by facilities in an exchange area for the transmission, switching, or routing of interexchange telecommunications originating or terminating within the exchange area. *See also* **telecommunications.**

exchange area: A geographical area (a) that is served by one or more central offices (COs) and (b) within which local telephone service is furnished under regulation. *See also* **extended area service, foreign exchange service, local access and transport area.**

exchange carrier: *See* **local exchange carrier.**

exchange director: *See* **telephone exchange director.**

exchange facility: A facility included within a local access and transport area (LATA). *See also* **local access and transport area.**

exchange format: *See* **diskette exchange format.**

exchange keying: *See* **phase exchange keying.**

exchange line: **1.** In telephone switchboard operation, a line that connects a user end instrument in a closed user group to an exchange. **2.** A line that connects a switchboard to an exchange. **3.** A line that connects one exchange to another. *See also* **closed user group, exchange, line, switchboard, user end instrument.**

exchange loop: *See* **local exchange loop.**

exchange service: *See* **foreign exchange service, Teletypewriter Exchange Service.**

exchange signaling: *See* **frequency exchange signaling.**

exchange sort: A sort in which (a) succeeding pairs of items in a set are placed in sequence according to the sort criterion, (b) the process is repeated until all the items in the set are sorted, and (c) only adjacent pairs are sorted in each step of the process.

exchange telecommunications radio service: *See* **basic exchange telecommunications radio service.**

exchange tie trunk: *See* **private branch exchange tie trunk.**

excitation: *See* **impulse excitation.**

excited atmosphere laser: *See* **longitudinally excited atmosphere laser, transversely excited atmosphere laser.**

excited linear prediction coding: *See* **code excited linear prediction coding.**

excited state: Any one of many orbital kinetic and potential energy states that an electron can have above the ground state. *Note:* An atom emits a quantum of energy when one of its electrons moves from (a) an excited state to the ground state and (b) a higher excited state to a lower excited state. *See also* **ground state.**

exclusion gate: *Synonym* **A AND NOT B** gate.

exclusive message: A message that is appropriately marked and that is to be delivered only to the person or authorized representative of the person, whose name or designator appears immediately following the appropriate marking in the address part of the message.

EXCLUSIVE NOR gate: A two-input, binary logic, combinational circuit or device that is capable of performing the logic operation of EXCLUSIVE NOR, such that if A is an input statement and B is an input statement, the output statement is true when both A and B are true or when both A and B are false, i.e., the result is true when A and B are the same and false when A and B are different. *Synonyms* **biconditional gate, equality gate, equivalence gate, identity gate, match gate.** *See also* **combinational circuit, equivalence, gate, identity.** *Refer to* **Fig. E-6.**

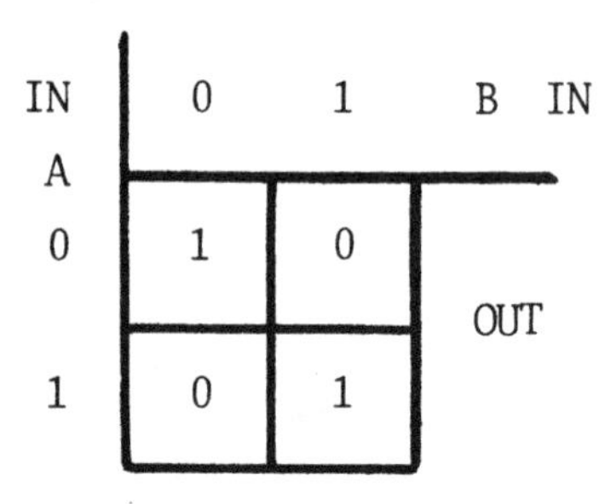

Fig. E-6. A truth table that shows the input and output digits of an **EXCLUSIVE NOR gate.**

EXCLUSIVE OR: A logic operator that has the property that if P is a statement and Q is a statement, then P EXCLUSIVE OR Q is true if either but not both statements are true, and false if both are true or both are false.

EXCLUSIVE OR gate: A two-input, binary logic, combinational circuit or device capable of performing the logic operation of EXCLUSIVE OR, such that if A is an input statement and B is an input statement, the output statement is true when A is true and B is false or when A is false and B is true, i.e., the result is false when A and B are both true and when A and B are both false. *Synonyms* **add without carry gate, anticoincidence gate, anticoincidence unit, difference gate, distance gate, diversity gate, exjunction gate, nonequality gate, nonequivalence gate, partial sum gate, symmetric difference gate.** *See also* **combinational circuit, gate.** *Refer to* **Fig. E-7.** *Refer also to* **Fig. S-17 (940).**

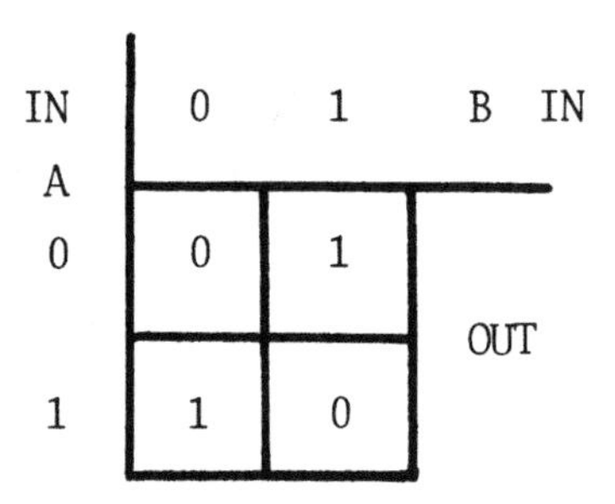

Fig. E-7. A truth table that shows the input and output digits of an **EXCLUSIVE OR gate.**

exclusive user circuit: A preset or programmed circuit between specified users or groups of users for their full-time use. *See also* **dedicated circuit, leased circuit data transmission service, sole user circuit, switched hot line circuit.**

execute: In communications, computer, data processing, and control systems, to perform an operation or a function, i.e., to carry out an instruction or a command. *Note 1:* In most systems, the instruction or the command to be executed is selected by the operator or a program, the operands are obtained or referenced, and a signal is given to commence the operation when ready or when the signal is given. *Note 2:* In some personal computer systems, if a command to print is given, the printing operation might start with the display of printing options, such as margin setting, line spacing, font style, character spacing, and the printer to be used. Actual printing may not start until all options have been selected or defaults are defined, power is on, paper is in the printer, a pointer is moved to a prescribed position, and certain keys are depressed.

executive program: *Synonym* **supervisory program.**

exempted addressee: An organization, activity, or person (a) usually included in the collective address group of a message, (b) deemed by the originator as having no need for the information in the message, and (c) possibly explicitly excluded from the collective address group for the particular message to which the exemption applies. *See also* **action addressee, collective address group, information addressee.**

exempt information: Information that (a) is exempted from public disclosure by law and (b) does not include (i) information that is security-classified in accordance with executive orders, (ii) proprietary information, and (iii) information covered by privacy statutes.

exemption: *See* **Warner exemption.**

exercise: *See* **communications exercise.**

existing quality television: *Synonym* **distribution quality television.**

exitance: *See* **radiant exitance.**

exit angle: When a light ray emerges from a surface of a material, i.e., passes from a medium with a given refractive index to a medium with a different refractive index, the angle between the ray and a normal to the surface at the point of emergence. *Note:* Examples of exit angles are (a) for an optical fiber, the angle between the fiber axis and the emerging ray, (b) for a light source, the launch angle, and (c) for an interface surface, the refraction angle.

exit instruction: An instruction in a computer program, routine, or subroutine that terminates the control exercised by that program, routine, or subroutine. *See also* **computer program, exit point.**

exit point: In a computer program, routine, or subroutine, the point (a) beyond which control is no longer exercised by that program, routine, or subroutine and (b) that usually is indicated by an exit instruction. *See also* **computer program, exit instruction.**

exit pupil: In an optical system, the image of the limiting aperture stop formed by all lenses following this stop. *Note:* In photographic objectives, this image is virtual and usually is not far from the iris diaphragm. In telescopes, this image is real and can be seen as a small, bright, circular disk by looking at the eyepiece of the instrument directed toward an illuminated area or light source. In telescopes, the image diameter is equal to the diameter of the entrance pupil divided by the magnification of the instrument. In Galilean telescopes, the exit pupil is a virtual image between the objective lens and the eyepiece and acts as an out-of-focus field stop. *See also* **aperture stop, field stop, objective, output aperture.**

exjunction gate: *Synonym* **EXCLUSIVE OR gate.**

expanded beam connector: A fiber optic connector in which (a) the diameter of the light beam and the launch angle are increased, so that the losses caused by longitudinal displacement, angular misalignment, and lateral offset are reduced to a minimum, (b) in many cases, special lenses and other passive optical devices are built into the connector to optimize coupling, and (c) in some cases, the glass in the optical fiber itself is used to form the optical elements in the connector. *Synonym* **lensed connector.**

expander: 1. A device that restores the dynamic range of a compressed signal to its original dynamic range. **2.** A device that (a) usually is an amplifier, (b) is part of a signal receiver, (c) has a gain that increases with increasing signal amplitude according to a specified law, function, or relationship, (d) may be used to undo the effects of a compressor at the transmitter, (e) compensates for transmission distortion, (f) provides for various image processing techniques, (g) has a nonlinear gain characteristic that acts to increase the gain more on larger input signals than it does on smaller input signals, and (h) usually is used to restore the original dynamic range of a signal that was compressed for transmission. *See also* **clipper, companding, compandor, compression, compression ratio, compressor, peak limiting, voice-operated gain adjusting device.**

expansion: The restoration of the dynamic range of a compressed signal to its original dynamic range. *See* **linked compression and expansion, luminance range expansion.** *See also* **compression.**

expansion capability: The inherent limit for increasing the capacity of a system beyond its installed capacity.

expedited data unit: In a connection in a layered system, a service data unit that is delivered to a peer entity in the destination open system before the delivery of any subsequent service data unit sent on that connection.

expendable jammer: An electronic jamming transmitter that (a) usually is designed for one-time use, (b) operates unattended, and (c) usually is placed in the vicinity of the equipment to be jammed.

experimental contract developmental station: A contractor-operated experimental station that is used for evaluation or testing of electronic equipment or systems that are in a design or development stage. *See also* **experimental station.**

experimental developmental station: An experimental station that is used for the evaluation or testing of electronic equipment or systems that are in a design or development stage. *See also* **experimental station.**

experimental export station: An experimental station that (a) is intended for export and (b) is to be used for the evaluation or testing of electronic equipment or systems that are in a design or development stage. *See also* **experimental station.**

experimental research station: An experimental station that is used in basic studies and scientific investigations for the advancement of radio communications.

experimental station: 1. In communications, a station that (a) conducts experiments using electromagnetic waves to further the development of science and technology, (b) conducts scientific investigations to promote the development of telecommunications techniques and systems, (c) operates primarily in the 10-MHz (megahertz) to 3000-MHz band, and (d) is not an amateur station. *See also* **amateur station.**

experimental testing station: An experimental station that is used for (a) evaluation or testing, (b) transmission site selection, and (c) transmission path surveys for electronic equipment or systems that have been developed for operational use. *See also* **experimental station.**

expert system (ES): 1. A system that (a) has artificial intelligence capabilities, (b) is designed to solve a specific class of problems or to process specific data in a certain manner, and (c) usually includes computer hardware and software. **2.** A computer system that facilitates solving problems in a given field or application by drawing inferences from a knowledge base developed from human experience. *Note 1:* The term "expert system" is sometimes used synonymously with "knowledge-based system," although it usually is taken to emphasize expert knowledge. *Note 2:* Some expert systems are able to improve their knowledge base and develop new inference rules based on their experience with previous problems.

exploration satellite service: *See* **Earth exploration satellite service.**

exploratory electromagnetic surveillance: Electromagnetic surveillance in an unknown environment. *See also* **electromagnetic surveillance.**

exponent: In certain number representation systems, the number to which another number, the base, is raised. *Note:* Examples of exponents are (a) the 2.5 in the expression $e^{2.5} \approx 12.18$, where e is the base of natural logarithms, about 2.71828, and (b) the 5 in the expression $10^5 = 100{,}000$. *See also* **base, natural logarithm.**

exponential backoff: *See* **truncated binary exponential backoff.**

export station: *See* **experimental export station.**

exposure: In a risk-bearing situation, the monetary value of an item multiplied by the probability of occurrence of its loss. *Note 1:* In a sequence of mutually exclusive risk-bearing situations, the exposure may be added as long as the total exposure remains small compared with the actual value. *Note 2:* If there is a probable salvage value in a given situation, then the monetary value is adjusted accordingly. *Note 3:* Examples of exposure are: (a) if the value of a mobile station is $\$10^6$ and the probability of its complete destruction in a given situation is 10^{-4}, i.e., one chance in 10,000, then the exposure is \$100 for this situation and (b) if an aircraft valued at $\$10^7$ has probable total losses of 2×10^{-6}, 4×10^{-6}, and 3×10^{-6} in three consecutive hops of a given flight, then the total exposure for the flight is \$90.

express: *See* **Pony Express.**

express engineering circuit: A permanently connected voice circuit between selected stations for technical control purposes. *Synonym* **express orderwire.** *See also* **circuit.**

expression: *See* **algebraic expression, arithmetic expression, logical expression.**

express orderwire: *Synonym* **express engineering circuit.**

extended area service (EAS): A network-provided service feature in which a user pays a higher flat rate to obtain wider geographical coverage without paying per-call charges for calls within the wider area. *See also* **exchange, exchange area, flat rate service, postalized rate.**

extended area telephone service: *Synonym* **wide area telephone service.**

Extended Binary Coded Decimal Interchange Code (EBCDIC): A code that (a) consists of a set of 256 different characters, each character represented by a unique eight-bit byte, i.e., an octet, (b) is used for data representation in storage and transmission, (c) has no parity bit, (d) has letters collated before numerals, and (e) is similar to the American Standard Code for Information Interchange (ASCII), except that the bit patterns that represent the alphanumeric characters are not the same for corresponding characters in each of the codes. *Note:* The Extended Binary Coded Decimal Interchange Code (EBCDIC) has been replaced by the American Standard Code for Information Interchange (ASCII). *See also* **American Standard Code for Information Interchange, code.**

extended definition television (EDTV): Television in which (a) improvements are made to the National Television Standards Committee (NTSC) standard television system, (b) the improvements are receiver-compatible with the NTSC television standard systems, and (c) the improvements modify the NTSC television system emission standards. *Note 1:* Extended definition television (EDTV) improvements may include (a) a wider aspect ratio, (b) a higher picture definition than distribution quality definition but lower than that of high definition television (HDTV), and (c) any of the improvements used in improved definition television (IDTV). *Note 2:* When extended definition television (EDTV) signals are transmitted in the 4:3 aspect ratio, it is referred to as "EDTV." When EDTV signals are transmitted in a wider aspect ratio, it is referred to as "EDTV-Wide." *See also* **advanced television, high definition television, improved definition television.**

extended operating condition: In the design of a fiber optic station/regenerator section, a condition that is more severe, such as (a) a higher or lower temperature condition than a standard operating condition, (b) a higher or lower humidity condition than a standard operating condition, or (c) any other environmental condition not included in the set of standard conditions, such as high pressures, corrosive atmospheres, severe abrasion, and high tensile stress, that may cause specified parameters to deviate from the performance specified for standard operating conditions. *See also* **standard operating condition.**

extended superframe: Pertaining to T-carrier framing in which (a) framing is provided for D-4 formatting and for online, real time testing of circuit capability and operating conditions without taking the circuit offline and (b) less frequent synchronization frees overhead bits for use in testing and monitoring.

extended time scale factor: In simulation, a time scale factor greater than 1; i.e., computer time is longer than problem, real, calendar, or any other measure of time or time scale. *Synonym* **slow time scale factor.** *See also* **calendar time, computer time, problem time, real time, simulation.**

extensible language: A computer programming language that readily permits a user to define new elements, such as new control structures, data types, and operational statements. *See also* **computer programming, language, statement.**

extension: 1. In telephone switchboard operation, a user end instrument, such as a telephone, that is directly connected to a switchboard or an exchange. **2.** Another telephone connected to an existing line and having the same call number as the other telephones connected to the same line, i.e., a telephone connected in parallel with another telephone on the same line. **3.** In most personal computer (PC) systems and software packages, such as PC DOS, MS DOS, Volkswriter, WordPerfect, Lotus 1-2-3, dBASE, and DR HALO, a period, i.e., a dot, followed by up to three characters that may be appended to the end of a file name to further identify the type of file, such as .EXE, .BAK, .WKS, and .CUT for executive, backup, worksheet, and pictorial/graphic, respectively. *See* **address field extension, control field extension, frequency modulation threshold extension, offnet extension, off-premises extension, program extension, threshold extension.**

extension bell: In telephone systems, an audible signal on a user end instrument, such as a subscriber telephone, that indicates there is an incoming call from a switchboard or an exchange. *Note:* The extension bell sounds when the switchboard or the exchange is ringing. *See also* **ringing.**

extension character: *See* **code extension character, nonlocking code extension character.**

extension facility: A facility that provides access to a user or a group of users isolated from a node in a communications network. *See also* **facility, multiple access, node, offline, online, on-premises extension, terminal.**

extension terminal: A terminal that is added to an existing terminal and that uses the same circuit and address, i.e., the same port and number, as the terminal to which it is added. *See also* **extension facility, node.**

external call: A call placed through a public exchange, i.e., placed beyond a private branch exchange (PBX) or local switchboard. *Synonym* **outside call.** *See also* **internal call.**

external hard disk: A hard disk that is mounted and operated in a container that is separate from the housing of the computer to which it is connected, i.e., is in a container of its own. *See also* **hard disk, hard drive.**

external heading: The parts of a message that precede the heading. *Note:* An example of an external heading is a special handling instruction given prior to the usual or standard heading, i.e., internal heading. *See also* **message heading, message text.**

external liaison circuit: An engineering circuit from a facility control, switching control, or Earth station to a remote facility control, switching control, or Earth station. *See also* **Earth station, engineering circuit, facility, remote control.**

external loss time: *Synonym* **environmental loss time.**

external noise: Noise that (a) originates outside a communications system, i.e., from external sources, such as galactic, sunspot, ionospheric, tropospheric, atmospheric, ground, and Earth noise sources, as well as artificial interference from man-made sources, and (b) in a given part of a system, does not include noise from sources in other parts of the system, such as thermal noise and circuit noise. *See also* **noise.**

external optical modulation: Modulation of a lightwave in a propagation medium by application of a stimulus, such as fields, forces, and waves, to a propagation medium conducting a light beam in such a manner that (a) a characteristic of the medium, the beam, or both is modulated in some way and (b) use is made of particular effects, such as the electrooptic, acoustooptic, magnetooptic, or absorptive effect.

external photoeffect: *Synonym* **photoemissive effect.** *Refer to* **Fig. P-7 (723).**

external photoeffect detector: *Synonym* **photoemissive effect detector.**

external photoelectric effect: *Synonym* **photoemissive effect.**

external photoelectric effect detector: *Synonym* **photoemissive effect detector.**

external security audit: A security audit of a system by an organization or an agency independent of the organization operating the system. *See also* **security audit.**

external sort: A sort that requires the use of auxiliary or external storage because of lack of storage capacity in the main or internal storage. *See also* **auxiliary storage, external storage, internal sort, internal storage, main storage, sort.**

external storage: Storage that (a) is accessible to a computer only through its input-output channels, (b) usually is in a separate storage unit, (c) is not internal storage, and (d) is not directly accessible to the computer arithmetic and logical unit (ALU).

external timing reference (ETR): In a given communications system, a timing reference obtained from a source, such as a navigation system, external to the given system. *Note:* External timing references usually are referenced to Coordinated Universal Time (UTC). *See also* **Coordinated Universal Time, local clock, master clock, time.**

extinction coefficient: The sum of the absorption coefficient and the scattering coefficient. *Note:* The extinction coefficient is expressed in units of reciprocal length. *See also* **absorption coefficient, Bouger's law, scattering coefficient.**

extinction ratio (r_e): **1.** In fiber optic systems, the ratio of (a) the average received optical energy in a logic 1 pulse to (b) the average received optical energy in a logic 0 pulse. **2.** The ratio of two optical power levels, P_1/P_2, of a digital signal generated by a light source, where P_1 is the optical power level generated when the light source is "on," and P_2 is the power level generated when the light source is "off." *Note:* The extinction ratio may be expressed as a fraction or in dB.

extra bit: *Synonym* **added bit.**

extra block: *Synonym* **added block.**

extra block probability: *Synonym* **added block probability.**

extra data unit: *Synonym* **added data unit.**

extramural absorption: The absorption of light that is transmitted radially through the cladding of an optical fiber. *Note:* The absorption may be accomplished by placing a light-absorbing dark, i.e., opaque, coating over the cladding. Coatings include dipped coatings, interstitial Kevlar fibers, and jackets.

extramural cladding: A layer of dark or opaque absorbent coating placed over the cladding of an optical fiber to (a) increase total (internal) reflection, (b) protect the smooth reflecting wall of the cladding, and (c) absorb scattered or escaped stray light rays that might penetrate the cladding. *See also* **cladding.**

extraordinary ray: A light ray that (a) has an anisotropic velocity in a doubly refracting crystal, (b) does not necessarily obey Snell's law upon refraction at the crystal surface, and (c) is caused by the different refractive indices in different directions in a doubly refracting crystal. *See also* **ordinary ray.**

extremely high frequency (EHF): A frequency that lies in the range from 30 GHz (gigahertz) to 300 GHz, the range having the alphabetic designator of EHF and a numeric designator of 11. *See also* **frequency, frequency spectrum designation.**

extremely low frequency (ELF): A frequency that lies in the range from 30 Hz (hertz) to 300 Hz, the range having the alphabetic designator of ELF and a numeric designator of 2. *See also* **frequency, frequency spectrum designation.**

extrinsic coupling loss: In fiber optic systems, the coupling loss that is caused when two optical fibers are imperfectly joined, such as by longitudinal displacement, angular misalignment, and lateral offset. *Synonym* **extrinsic joint loss.** *See also* **angular misalignment loss, intrinsic coupling loss, lateral offset loss, longitudinal displacement loss.**

extrinsic internal photoeffect: An internal photoeffect that makes use of dopants, or other impurities, in a basic material, i.e., an intrinsic material. *See also* **intrinsic internal photoeffect.**

extrinsic joint loss: *Synonym* **extrinsic coupling loss.**

eye pattern: An oscilloscope display (a) in which a pseudorandom digital data signal from a receiver repetitively is sampled and applied to the vertical input while the data rate is used to trigger the horizontal sweep of the display device, (b) from which system performance information can be derived by analyzing the display, (c) in which an open eye display corresponds to minimal signal distortion, (d) in which closure of the eye display indicates distortion of the signal waveform caused by intersymbol interference and noise, (e) that *qualitatively* describes the proper functioning of a digital system, whereas the bit error ratio (BER) *quantitatively* describes the proper functioning, (f) of which the quality, critical for achieving a low BER, is influenced by (i) noise, which is proportional to the receiver bandwidth, and (ii) the intersymbol interference, which arises from other bits interfering with the bit of interest, and (f) in which the lower level of the display and the zero level of the display area are not identical. *Note:* Intersymbol interference is inversely proportional to the bandwidth. A bandwidth compromise has to be made in order to obtain the best possible eye pattern aperture. *See also* **display area, display device, distortion.**

eyepiece: In a lens system, the lens that is (a) closest to the eye when using the system, (b) usually farthest from the object being viewed, (c) farthest from the objective lens, and (d) used to focus the image on a screen or to allow, in conjunction with the lens of the human eye, the image to be focused on the retina. *Note:* An objective lens, an aligned bundle of optical fibers, and an eyepiece can be used to form a flexible fiberscope, endoscope, or fiber optic choledochoscope.

eyes only: A message marker for a special category message that is intended for delivery only to a specific person, or authorized representative of that person, and thus to no one else.

FAADC²: forward area air defense command and control.

fabric ribbon: A ribbon that (a) is composed of a woven material, such as silk, cotton, linen, or a synthetic yarn, such as rayon, nylon, or Orlon, (b) may be inked or carbon-impregnated, (c) is used on impact printers and typewriters for forming characters, (d) may be used either repeatedly or only one time, and (e) usually is contained on a spool or in a cassette.

Fabry-Perot fiber optic sensor: A Fabry-Perot sensor (a) that can be used to measure short distances or varying distances, and (b) in which the ends of an optical fiber are made reflective such that moving one end relative to the other will produce an output optical signal to a photodetector proportional to the relative motion when monochromatic light, i.e., light with a narrow spectral width, is inserted into the fiber.

Fabry-Perot interferometer: A high resolution multiple beam interferometer that consists of two optically flat and parallel transparent plates, such as glass or quartz plates, that are held a short fixed distance apart, the adjacent surfaces of the plates or interferometer flats being made almost totally reflecting by means of a special coating, such as a thin silver film or a multilayer dielectric coating. *See also* **interferometer.**

Fabry-Perot sensor: A high resolution multiple beam interferometric sensor in which (a) two optically flat parallel transparent plates are held a short distance apart, and the adjacent surfaces of the plates, i.e., interferometric flats, are made almost totally reflective by a special coating, such as a thin silver film or a multilayer dielectric coating, (b) if one plate is moved relative to the other, interference patterns can be made to occur when a lightwave is directed at the plates, and (c) the interference patterns can be used (i) to measure the relative motion of the plates or (ii) if the plates are held fixed, to analyze the spectral composition of lightwaves.

face change character: *Synonym* **font change character.**

face to face coupling: *Synonym* **butt coupling.**

faceplate: *See* **fiber optic faceplate.**

facet erosion: In laser diodes, a phenomenon in which degradation of the mirror reflectivity decreases the internal quantum efficiency and increases the threshold current. *See also* **injection laser diode, quantum efficiency.**

facilities request signal: The part of the selection signal that identifies required facilities. *See also* **selection signal.**

facility: **1.** A fixed, mobile, or transportable structure, including (a) all installed electrical and electronic wiring, cabling, and equipment and (b) all supporting structures, such as utility, ground network, and electrical supporting structures. **2.** A network-provided service to users or the network operating administration. **3.** A transmission pathway and associated equipment.

4. In a protocol applicable to a data unit, such as a block or a frame, an additional item of information or a constraint encoded within the protocol to provide the required control. **5.** A real property entity consisting of (a) buildings, structures, utility systems, pavements, and land or (b) any combination of these. **5.** An operational capability or the means of providing a capability, including hardware, firmware, and software but not people. *See* **called line facility, calling line facility, calls barred facility, communications facility, data collection facility, delayed delivery facility, entrance facility, exchange facility, extension facility, fiber optic cable facility, glide slope facility, help facility, interconnect facility, information facility, internal communications facility, landline facility, mailbox facility, manual answering facility, manual calling facility, multiaddress calling facility, multimode facility, packet assembly-disassembly facility, patch and test facility, priority facility, station fiber optic facility, technical control facility, telecommunications facility, trouble desk facility, user facility, virtual call facility, whisper facility.**

facility ground system: *Synonym* **facility grounding system.**

facility grounding system (FGS): The electrically interconnected system of conductors and conductive elements that (a) provides multiple current paths to the earth electrode subsystem and (b) consists of the earth electrode subsystem, the lightning protection subsystem, the signal reference subsystem, and the fault protection subsystem. *Synonym* **facility ground system.** *See also* **air terminal, earth electrode subsystem, equipotential ground plane, facility, fault protection subsystem, ground, ground return circuit, lightning down conductor, lightning protection subsystem, neutral ground, signal reference subsystem.**

facility loss: *See* **fiber optic cable facility loss, statistical fiber optic cable facility loss.**

facility request separator character: The character that separates the different individual facility request signals in the facilities request signal of a selection signal. *See also* **facilities request signal, selection signal.**

facs: 1. facsimile, facsimile system. 2. *Synonym* **fax.**

facsimile (FAX or fax): 1. Pertaining to the transmission and the reception of data in graphic form. **2.** Telegraphy in which objects are scanned and the resulting signals are transmitted to a receiver where they are reproduced in permanent form, called "recorded copies." *Note 1:* The transmitter is considered a source and the receiver a sink. *Note 2:* The objects and recorded copies may be half-tones. **3.** A form of telegraphy for the transmission of images, with or without half-tones, with a view to their reproduction in a permanent form. *Note:* "Telegraphy" has the general meaning as defined by the Convention on Communications. **4.** A communications system for the transmission of images with a view to their reception in a permanent form. **5.** The process, or the result of the process, by which graphic images, such as printed text and pictures, are scanned and the information is converted into electrical signals that are transmitted over a communications system and used to create a recorded copy of the object, i.e., the original. *Note 1:* Wirephoto and telephoto are facsimile via wire circuits. Radiophoto is facsimile via radio. *Note 2:* Current facsimile systems are designated and defined as:

Group 1 Facsimile: The mode of black and white facsimile operation, defined in International Telegraph and Telephone Consultative Committee (CCITT) Recommendation T.2, that uses double sideband modulation without any special measures to compress the bandwidth. *Note 1:* A (216 × 279)-mm (millimeter) document (Size A4), i.e., an (8½ × 11)-inch document, may be transmitted in approximately 6 minutes via a telephone-type circuit. Additional modes in this group may be designed to operate at a lower resolution suitable for the transmission of (216 × 279)-mm documents in 3 to 6 minutes at about 4 scan lines per millimeter. *Note 2:* The CCITT frequencies used are 1300 Hz (hertz) for white and 230 Hz for black. The North American standard is 1500 Hz for white and either 2300 Hz or 2400 Hz for black. *Synonym* **group one document facsimile telegraph apparatus.**

Group 2 Facsimile: The mode of black and white facsimile operation, defined in CCITT Recommendation T.3, that accomplishes bandwidth compression by using encoding and vestigial sideband (VSB), but excludes processing of the document signal to reduce redundancy. *Note:* A (216 × 279)-mm document, i.e., an (8½ × 11)-inch document, may be transmitted in approximately 3 minutes using a 2100-Hz AM/PM/VSB over a telephone-type circuit. *Synonym* **group two document facsimile telegraph apparatus.**

Group 3 Facsimile: The mode of black and white facsimile operation, defined in CCITT Recommendation T.4, that incorporates means for reducing the redundant information in the signal by using a one-dimensional run-length coding scheme prior to the modulation process. *Note 1:* A (216 × 279)-mm document, i.e., an (8½ × 11)-inch document, may be transmitted in approximately 1 minute or less over a telephone-type circuit with twice the Group 2 horizontal resolution. The vertical resolution may also be doubled. *Note 2:* Group 3 Facsimile machines have integral digital modems. *Note 3:* An optional two-dimensional bandwidth compression scheme is also defined within the Group 3 Facsimile Recommendation. *Synonym* **group three document facsimile telegraph apparatus.**

Group 4 Facsimile: The mode of black and white facsimile operation defined in CCITT Recommendations T.5 and T.6. *Note:* Group 4 Facsimile uses bandwidth compression techniques to transmit, essentially without errors, a (216 × 279)-mm document, i.e., an (8½ × 11)-inch document, at a nominal resolution of 8 scan lines/mm in less than 1 minute over a public data network voice grade circuit. *Synonym* **group four document facsimile telegraph apparatus.** *Refer to* **Figs. C-7 (157), S-5 (869), S-6 (869), S-7 (870).**

facsimile bandwidth: In a given facsimile system, the difference between the highest and lowest frequency components that are required for adequate transmission of the facsimile picture signals.

facsimile converter: 1. In a facsimile receiver, a device that changes the signal modulation from frequency shift keying (FSK) to amplitude modulation (AM). **2.** In a facsimile transmitter, a device that changes the signal modulation from amplitude modulation (AM) to frequency shift keying (FSK).

facsimile data compression: The removal of redundancy from a facsimile picture signal before transmission to reduce the bandwidth required for transmission.

facsimile digital equipment: In facsimile systems, equipment in which (a) digital techniques are used to encode the picture signal, i.e., the baseband signal resulting from scanning the object, and (b) the output signal may be either digital or analog. *Note:* Examples of digital facsimile equipment are International Telegraph and Telephone Consultative Committee (CCITT) Group 3, CCITT Group 4, Standardization Agreement (NATO) (STANAG) 5000 Type I, and STANAG 5000 Type II equipment. *Synonym* **digital facsimile equipment.**

facsimile equipment: One of the four groups of facsimile equipment. *See* **analog facsimile equipment.** *See also* **facsimile.**

facsimile frequency shift: At any point in a frequency shift facsimile system, the numerical difference between (a) the frequency that corresponds to a white signal and (b) the frequency that corresponds to a black signal. *Note:* Facsimile frequency shift usually is expressed in hertz. *See also* **frequency shift telegraphy.**

facsimile picture signal: In a facsimile system, the baseband signal that results from scanning the object. *See also* **baseband signaling, facsimile, object, scanning.**

facsimile receiver: In a facsimile system, the portion of the system that converts the facsimile picture signals into a recorded copy. *See also* **facsimile, facsimile picture signal, facsimile transceiver, facsimile transmitter, recorded copy.**

facsimile receiver aperture: In a facsimile receiver, the final opening (a) through which the light from a local source that has been intensity modulated by the received facsimile picture signal passes, and (b) after which the light is focused on a photosensitive recording medium to produce a recorded copy. *See also* **facsimile transmitter aperture, x dimension of recorded spot, y dimension of recorded spot.**

facsimile recorder: In a facsimile receiver, the device that performs the final conversion of the facsimile picture signal to an image of the object, i.e., that makes the recorded copy. *See also* **facsimile, recorded copy.**

facsimile service: A public or private communications system that is devoted to the transmission and the reception of fixed images, such as photographs, charts, maps, drawings, and printed pages.

facsimile signal contrast: The ratio of (a) the facsimile signal level obtained when optically scanning a white area on the object to (b) the signal level obtained when scanning a black area on the object. *Note:* The facsimile signal contrast usually is expressed in dB. *See also* **object.**

facsimile signal level (FSL): In a facsimile system, the signal level at any point in the system. *Note 1:* The facsimile signal level is used to establish the operating levels. *Note 2:* The facsimile signal level usually is expressed in dB with respect to a standard value, such as 1 mW (milliwatt), i.e., such as 0 dBm. *See also* **signal, zero transmission level point.**

facsimile system: A communications system that (a) is capable of accepting fixed images, such as pictures, maps, drawings, and printed pages, transmitting them over long distances, receiving them at the distant location, and reproducing them in a permanent form, such as hard copy, (b) consists of at least a scanner, a propagation medium, and a receiver, (c) usually has a scanner that uses a spot of light so that the scanning signal can be amplified and then transmitted directly over fiber optic cables, and (d) may include a scanning system that has the capability of scanning halftones. *Note:* Wirephoto and telephoto are facsimile via wirelines. Radiophoto is facsimile via radio. *Synonyms* **facs, fax, FAX.**

facsimile telegraphy: Telegraphy in which (a) the surface of an object, such as a paper document, is systematically scanned on sending and synthesized on reception to produce on a recording medium an image that is geometrically similar to the object, such as the image on the paper document, and (b) the reproduction may be (i) in two density states, usually black on white, (ii) may have intermediate shades of black and white, i.e., halftones, or (iii) may be in color. *See* **document facsimile telegraphy, picture facsimile telegraphy.**

facsimile transceiver: In a facsimile system, the portion of the system that sends and receives facsimile signals, i.e., a facsimile transmitter and receiver in a single unit. *Note:* Full duplex facsimile transceivers can send and receive at the same time. Half-duplex facsimile transceivers cannot send and receive at the same time. *See also* **facsimile, facsimile picture signal, facsimile receiver, facsimile transmitter.**

facsimile transmission: *See* **black facsimile transmission, white facsimile transmission.**

facsimile transmitter: In a facsimile system, the portion of the system that converts the baseband facsimile picture signals resulting from scanning the object into signals suitable for transmission by a communications system. *See also* **facsimile, facsimile picture signal, facsimile receiver, facsimile transceiver.**

facsimile transmitter aperture: An opening through which a constant, controlled, unmodulated beam of light passes and which effectively controls the size of the scanning spot of a facsimile transmitter. *See also* **facsimile receiver aperture, x dimension of scanning spot, y dimension of scanning spot.**

factor: *See* **activity factor, bandwidth•distance factor, block correction efficiency factor, blocking factor, confinement factor, cooperation factor, demand factor, dispersion attenuation factor, drum factor, duty factor, efficiency factor, equivalent binary digit factor, extended time scale factor, fast time scale factor, feedback factor, frequency modulation improvement factor, intrinsic quality factor, K factor, load factor, modulation factor, noise factor, pulse duty factor, radiation absorption factor, reflection factor, scale factor, shape factor, solar absorption factor, spot noise factor, time scale factor, transmitter calibration factor, work factor.**

factorial: 1. A mathematical operator that produces the product of all natural numbers, excluding 0, up to a specified number. **2.** The product of all natural numbers, excluding 0, up to a specified number. **3.** Pertaining to the product of all the integers in the ascending series of all natural numbers, excluding 0, up to and including a specified number. *Note:* An example of a factorial is factorial $5 = 1 \times 2 \times 3 \times 4 \times 5 = 120 = 5!$. Factorial 5 may be written as (a) 5! and pronounced "5 factorial," or (b) as ⌊5 or ∠5 and pronounced "factorial 5."

fade: 1. In radar operations, to disappear from the radar display. **2.** In radio communications, at the receiving antenna input, a reduction of signal strength caused by (a) atmospheric propagation conditions, such as those that cause scattering, ducting absorption, dispersion, reflection, and refraction, and (b) ionospheric conditions that cause absorption, reflection, and refraction, which change continuously. *Note:* Changes in transmitted power are not considered as fading, though the effect at the receiver is the same.

fade margin: 1. In communications systems design, an allowance that (a) provides for sufficient system gain and sufficient system sensitivity to accommodate expected fading and (b) ensures that the required grade of service will be maintained for the specified percentage of time. **2.** The amount by which a received signal level may be reduced without causing system output, such as a channel output, to fall below a specified value, i.e., a specified threshold. *See also* **design margin, fading, radio frequency power margin.**

fadeout: *Synonym* **flutter.**

fading: In a received signal, the time variation of the level or the relative phase, or both, of one or more of the frequency components of the signal. *Note:* Fading is caused by changes in the characteristics of the propagation path with time. *See* **flat fading, multipath fading, phase interference fading, Rayleigh fading, selective fading.** *See also* **diurnal phase shift, fade margin, frequency, Rayleigh distribution, Rician fading.**

fading distribution: The probability that signal fading will exceed a given value relative to a specified reference level. *Note 1:* In the case of phase interference fading, the time distribution of the instantaneous field strength approximates a Rayleigh distribution a large part of the time when at least several signal components of equal amplitude are present. *Note 2:* The field strength usually is measured in volts per meter and expressed in dB. *Note 3:* The fading distribution may also be measured in terms of power level, in which case the unit of measure usually is watts per square meter, and the fading distribution is expressed in dB. *See also* **Rayleigh distribution, Rayleigh fading.**

Fahrenheit temperature scale: A temperature scale in which the freezing point of water is set at 32°F (degrees Fahrenheit) and the boiling point of water at 212°F, both at sea level. *Note:* Celsius temperatures may be converted to Fahrenheit temperatures by the relation $F = 32 + 9C/5$ where C is the Celsius temperature. Both scales have the same numerical value of temperature at −40°. *See also* **Celsius temperature scale, Kelvin temperature scale, temperature.**

fail safe: 1. Of a device, the capability to fail without detriment to other devices or danger to personnel. **2.** In computer and communications systems, the automatic protection of programs, transmissions, and components when a failure occurs. **3.** Pertaining to the structuring of a system such that either it cannot fail to accomplish its assigned mission regardless of environmental factors or that the probability of such failure is extremely close to zero. **4.** Pertaining to operation designed to ensure that a failure of equipment, process, or system does not propagate beyond the immediate environs of the failing entity. *See also* **fail safe operation.**

fail safe operation: 1. Operation that ensures that a failure of the equipment, process, or system does not propagate beyond the immediate environs of the failing entity. **2.** A control operation or function that prevents improper system functioning or catastrophic degradation in the event of (a) a malfunction within the system or (b) an operator error. **3.** The capability of a device to fail without creating danger to operating personnel, to itself, or to other devices. *See also* **catastrophic degradation, continuous operation, endurability, fail safe, graceful degradation.**

fail soft: The termination of selected lower-priority operations when a hardware or software failure is detected in a system in order to prevent catastrophic degradation. *See also* **catastrophic degradation, graceful degradation.**

failure: The temporary or permanent termination of the ability of an entity to perform its required function, the termination usually resulting from a fault or an environmental condition. *Note:* Failures may be (a) catastrophic, i.e., sudden and complete or (b) they may be graceful, i.e., gradual and partial. *Synonym* **malfunction.** *See* **access failure, block transfer failure, disengagement failure, mean time between failures, mean time to failure, radio failure.** *See also* **catastrophic degradation, disengagement denial, error, fault, graceful degradation, mistake.**

failure access: In an automated information system, an unauthorized or inadvertent access to data that resulted from a hardware or software failure. *See also* **communications security.**

failure control: In a communications, computer, data processing, or control system, control used to provide fail safe or fail soft recovery from hardware and software failures. *See also* **fail safe, fail soft.**

failure fraction automatically switched: In a communications system during a specified measurement period, the ratio of (a) the number of failures that are immediately bypassed by automatic switching to another component, such as another channel, line, or data terminal equipment, to (b) the total number of failures. *See also* **mean time between outages.**

failure period: *See* **constant failure period, early failure period, wear-out failure period.**

failure rate: 1. In a given system, the number of errors, failures, faults, malfunctions, and mistakes, i.e., bugs, that occur per unit time or per event, such as per transaction, run, pass, or iteration. **2.** The number of parts or components in a system that fail in a given quantity of the same part or component per unit time. *Note:* An example of a failure rate is the number of resistors that fail per million resistors of the same type per hour, such as $2 \times 10^{-7} \cdot hr^{-1}$ or 0.2 parts per million parts per hour. *See* **error, failure, fault, malfunction, mistake.** *Refer to* **Fig. B-3 (71).**

failures: *See* **mean time between failures.**

failure signal: *See* **call failure signal, continuity failure signal.**

failure transfer: *See* **system failure transfer.**

fair queuing: The controlling of congestion in gateways by restricting every host to an equal share of gateway bandwidth. *Note:* Fair queuing does not distinguish between small and large hosts or between hosts with few or many active connections. Thus, in these senses, fair queuing is not fair and is a misnomer.

fallen clearing indicator shutter: In a single-supervisory electromechanical telephone switchboard, a clear indicator that is in the form of a shutter or a lever that drops down to indicate that a user has finished a call. *See also* **clear indicator, clearing, finished call.**

fallout message: A message that contains data related to a nuclear detonation, such as (a) the time and the location of the detonation, (b) the direction, speed, and extent of the fallout zone, and (c) the expected radiation levels in affected areas. *See also* **preburst message.**

fall time: 1. The time required for a signal level, such as pulse amplitude, to decrease from a specified value, usually near the peak value in the case of a pulse, to another specified value, usually near the lowest value or the zero value. *Note:* The specified values for determining pulse fall time usually are 90% and 10% of the peak value. **2.** The time required for a signal at a given level to decrease to near zero, usually 10% of the original level. *Note:* Overshoot and oscillation, i.e., hunting, may need to be specified in determining fall time. *Synonym* **pulse decay time.** *See also* **drift, jitter, pulse fall time, rise time, root mean square pulse broadening, swim, wander.**

false alarm rate: *See* **constant false alarm rate.**

false character: *Synonym* **illegal character.**

false clear: In telephone switchboard operation, a signal that (a) produces the same effect as a clear signal, (b) is caused by (a) undesired, unintentional, or accidental means, such as interference, outages, and operator errors, and (c) is not caused by usual means of disconnection, such as by a ring-off or by a user deliberately replacing the telephone receiver on its hook, cradle, or switch, i.e., a user hanging up. *See also* **clearing, clear signal.**

false clock: A clock, controlled by a phase-locked loop, that is locked to a frequency other than the correct frequency. *Note 1:* A false clock can occur when there is excessive phase shift, usually as a function of frequency, in the loop. *Note 2:* A false clock often occurs when the clock locks on a harmonic of the correct frequency. *See also* **false lock, phase-locked loop.**

false colors: In visual signaling systems, a marking, such as a flag, insignia, or sign, of a country other than the country of registry of the ship, aircraft, spacecraft, or vehicle on which the marking is placed.

false lock: In a phase-locked loop, a lock on a frequency or phase other than the correct one. *See also* **false clock, phase-locked loop.**

family: *See* **frequency family.**

fan beam: A radar beam radiation pattern that (a) is broad in two Cartesian dimensions and narrow in the third and (b) usually is used in search radars. *Note:* An example of a fan beam is a beam with large vertical and narrow horizontal coverage. This fan beam would sweep by rotating about a vertical axis, i.e., sweep horizontally. *See also* **fan beam antenna, radiation pattern.**

fan beam antenna: A directional antenna that produces a main beam with a large ratio of major to minor dimension at any transverse cross section. *See also* **antenna, antenna lobe, fan beam.**

fan marker beacon: A radio beacon that (a) radiates with a radiation pattern that is a fan beam pattern and (b) may emit a keyed signal for identification purposes. *See also* **fan beam, radiation pattern, radio beacon.**

fan out: *Synonym* **break out.**

fanout cable: *Synonym* **fiber optic breakout cable.**

Faraday effect: A magnetooptic effect in which the polarization plane of an electromagnetic wave is rotated under the influence of a magnetic field parallel to the direction of propagation. *Note:* The Faraday effect may be used to modulate a lightwave. *See also* **magnetooptic effect.**

far end crosstalk (FEC): Crosstalk that is propagated in a disturbed channel in the same direction as the propagation of a signal in the disturbing channel. *Note:* The terminals of the disturbed channel, at which the far end crosstalk is present, and the energized terminals of the disturbing channel usually are remote from each other. *See also* **crosstalk, intelligible crosstalk, interference, near end crosstalk.**

far field: The field of an electromagnetic wave developed by an antenna or a source of electromagnetic radiation in which (a) the electric field strength varies primarily inversely as the distance from the antenna or the source, (b) other contributions to the field strength, such as those that vary inversely as the square or the cube of the distance, are negligible, though they are not negligible in the near and intermediate fields, and (c) the angular field distribution function essentially is independent of the distance from the antenna. *Note 1:* The far field exists at distances from the source greater than $\lambda/6$ or greater than d, given by the relation $d \rangle 2s^2/\lambda$ where s is the overall dimension of the source and is very much larger than the wavelength, i.e., $s \rangle\rangle \lambda$, and λ is the wavelength. *Note 2:* A region is a space, usually defined in terms of distances from a source. A field of force occupies the space. *Note 3:* The differential tangential component of the electric field vector produced by an ideal Hertzian dipole antenna, which most dipole antennas can be considered to be at the distances involved, is given by the relation

$$dE_\theta = Idl(\sin\theta)/4\pi\varepsilon)[-\omega(\sin\omega t)/rv^2 + (\cos\omega t)/r^2v + (\cos\omega t)/\omega r^3]$$

where dE_θ is the differential tangential component of the electric field vector, I is the antenna current, dl is the elemental length of antenna bearing the current I, θ is the angle between the antenna direction of dl and the line from the dipole center to the point of measurement, π is approximately 3.1416, ε is the electric permittivity of the intervening medium, ϖ is the angular velocity equal to $2\pi f$ where f is the frequency, t is time, r is the distance from the center of the dipole antenna to the point of measurement, and v is the propagation velocity equal to $(\mu\varepsilon)^{-1}$ where μ is the magnetic permeability. Note that for the expression in brackets, the terms are for the radiation field, the induction field, and the electrostatic field, in that order. Note that at great distances, i.e., large values of r, the radiation field, i.e., the far field, is dominant. Closer to the antenna the intermediate field is dominant, and just outside the antenna the near field is dominant. *Synonyms* **far zone, Fraunhofer region, radiation field.** *Deprecated synonym* **far field region.** *See also* **beam divergence, electromagnetic radiation, intermediate field, near field, transition zone.**

far field diffraction pattern: The diffraction pattern of a source, such as a light-emitting diode (LED), laser diode, or output end of an optical fiber, observed at a distance, d, from the source, given by the relation $d \gg s^2/\lambda$ where s is a characteristic dimension of the source, such as half the length or width, and λ is the wavelength. *Note 1:* If the source is a uniformly illuminated circle, then s is the radius of the circle. *Note 2:* The far field diffraction pattern of a source may be observed in the focal plane of a well-corrected lens. *Note 3:* The far field diffraction pattern of a diffracting screen illuminated by a point source may be observed in the image plane of the source. *Synonym* **Fraunhofer diffraction pattern.** *See also* **diffraction, diffraction-limited.**

far field pattern: *Synonym* **far field radiation pattern.**

far field radiation pattern: A radiation pattern in which (a) the electric and magnetic field strengths of an electromagnetic wave vary inversely as the distance from the antenna or source and (b) other contributions to the field strength, such as those that vary inversely as the square or the cube of the distance, are negligible. *Note 1:* The far field radiation pattern may be from any source, such as a radio antenna, a light source, or the end of an energized optical fiber. *Note 2:* The far field radiation pattern for an optical fiber describes the distribution of irradiance as a function of angle with the fiber axis at a distance from the fiber endface. *Synonym* **far field pattern.** *See also* **antenna, diffraction, electromagnetic radiation, intermediate field region, near field diffraction pattern, radiation pattern.**

far field region: In an electromagnetic wave, the spatial volume (a) that is occupied by the far field, (b) that lies farther from the antenna than the intermediate field region, and (c) in which the field strength varies as $1/d$ where d is the distance from the antenna and depends primarily on the interaction of the electric and magnetic fields in the wave. *Note:* Because all the points in the far field region are far from the source, the change in radiation pattern from point to point within the far field region is small. *Synonym* **far zone.** *See also* **intermediate field region, near field region, radiation pattern.**

far infrared: Pertaining to the region of the electromagnetic spectrum that lies between the longer wavelength end of the middle infrared region and the shorter wavelength end of the radio region, from about 30 μm (microns) to 100 μm. *Note:* The far infrared region is included in the optical spectrum. However, the near infrared region lies between the longer wavelength end of the visible spectrum and the shorter wavelength end of the middle infrared region, i.e., from about 0.8 μm (microns) to 3.0 μm. Thus, the near infrared is included in the near visible region, but the middle infrared and far infrared regions are not included in the near visible region. The infrared regions are not included in the visible spectrum.

far zone: *Synonym* **far field region.**

FASSET: Functional Advanced Satellite System for Evaluation and Test.

fast automatic gain control: Automatic gain control that (a) has a very short response time and (b) can be used as an electronic countercountermeasure. *See also* **automatic gain control, response time.**

fast packet switching (FPS): Increasing the speed of packet switching by eliminating overhead. *Note 1:* Overhead reduction is accomplished by allocating flow control and error correction functions to either the user applications or the network nodes that interface with the user. *Note 2:* Cell relay and frame relay are two implementations of fast packet switching. *See also* **cell relay, frame relay.**

fast select: In a call setup and call clearing procedure, an optional feature in the virtual call service that allows the inclusion of user data in the call-request-connected and call-clear-indication packets. *Note:* Fast select is an essential feature of the International Telegraph and Telephone Consultative Committee

(CCITT) X.25 (1984) Protocol. *See also* **virtual call capability, X series Recommendations.**

fast time constant (FTC): A time constant of extremely short duration compared to the time constant of conventional circuits. *Note 1:* In radar systems, fast time constants (a) are used in signal processing to reduce the effects of certain types of undesired signals, (b) are used to emphasize signals of short duration to produce discrimination against the low frequency components of clutter, and (c) protect against interference of either frequency-modulated or amplitude-modulated signals. *Note 2:* To obtain a fast time constant in an electric circuit, a low product of resistance and capacitance (RC), or a low ratio of inductance to resistance (L/R) is used. If these circuit elements are of proper magnitude and are properly connected in relation to input and output, the resulting circuit output will be the time derivative of its input. The differentiating circuit, i.e., the differentiator, inserted between the detector and the video amplifier, will pass individual pulses but not the modulation frequencies. Circuits of this type shorten the duration of the observed pulses and should be switched on only when interference makes it necessary. Such differentiating circuits are of real value as long as the intermediate frequency (IF) amplifier is not overloaded. *See* **differentiating circuit, differentiator, logarithmic fast time constant, time constant.**

fast time scale factor: In simulation, a time scale factor less than unity, i.e., computer time is shorter than any other measure of time or time scale, such as problem, real, clock, and calendar time.

fault: 1. An accidental condition that causes a functional unit to fail to perform its required function. **2.** A defect that causes a reproducible or catastrophic malfunction. *Note:* A malfunction is considered reproducible if it occurs consistently under the same circumstances. **3.** In power systems, an unintentional or partial short circuit between energized conductors or between an energized conductor and ground. *See also* **downtime, error, ground, preventive maintenance.**

fault dominance: If *x* and *y* are two faults that can occur in a system, then fault *x* dominates fault *y* if and only if every test to determine if *y* has occurred is also a test to determine if *x* has occurred.

fault management: In network management, the set of functions, based on examination and manipulation of database information, that (a) detect, isolate, and correct malfunctions in a telecommunications network, (b) compensate for environmental changes, (c) maintain and examine error logs, (d) accept and act on error detection notifications, (e) trace, localize, identify, and correct faults, (f) carry out sequences of diagnostic tests, and (g) report error conditions.

fault masking: If *x* and *y* are two faults that can occur in a system, then fault *x* masks fault *y* if and only if no test to determine if *y* has occurred is also a test for the joint occurrence of *x* and *y*.

fault protection subsystem (FPS): In a facility power distribution system, the subsystem that provides a direct path from each power sink to the earth electrode subsystem. *Note:* The fault protection subsystem usually is referred to as the "green wire." *Synonym* **green wire.** *See also* **earth electrode subsystem, facility, facility grounding system, ground, lightning protection subsystem, neutral, neutral ground, signal reference subsystem.**

fault rate threshold: A specified number of faults in a given category or class in a given period, that, if exceeded, results in a prescribed action, such as remedial action, notification of operators, calling diagnostic programs, and reconfiguration of system architecture to exclude a faulty unit from the system. *See also* **fault threshold.**

fault seeding: *Synonym* **error seeding.**

fault threshold: A specified number of faults in a given category or class, that, if exceeded, results in a prescribed action, such as remedial action, notification of operators, calling diagnostic programs, and reconfiguration of system architecture to exclude a faulty unit from the system. *See also* **fault rate threshold.**

fault tolerance: The extent to which a functional unit will continue to operate at a defined performance level when one or more of its components are malfunctioning. *See also* **catastrophic degradation, endurability, graceful degradation.**

fax: 1. facsimile, facsimile system. 2. To transmit or send a message via a facsimile system, as in the common expression "Fax me a copy." **3.** Pertaining to a facsimile system, as in the common expression "Do you have a fax machine?" *Synonym* **facs.**

FAX: facsimile, facsimile system.

F band: A band of frequencies that (a) includes frequencies in the range from 3 GHz (gigahertz) to 4 GHz and (b) is composed of ten numbered channels, each of which is 100 MHz (megahertz) wide.

FC: functional component.

FCC Registration Program: Federal Communications Commission Registration Program.

FCCRP: Federal Communications Commission Registration Program.

FCS: frame check sequence.

FDDI: fiber distributed data interface.

FDDI-2: *See* **fiber distributed data interface.**

FDHM: full duration half maximum.

FDM: frequency division multiplexing.

FDMA: frequency division multiple access.

FDX: full duplex.

FE: 1. format effector. 2. The format effector transmission control character.

feature: *See* **fiber optic cable interconnect feature, service feature, spill forward feature.**

FEC: forward error correction.

Federal Communications Commission (FCC): The U.S. government board of seven presidential appointees that has the authority to regulate all U.S. interstate communications systems as well as all international communications systems that originate or terminate in the United States. *Note:* The Federal Communications Commission (FCC) was established by the Communications Act of 1934.

Federal Communications Commission Registration Program (FCCRP): The Federal Communications Commission Program and associated directives that (a) require assurance that all connected data terminal equipment (DTE) and protective circuitry will harm neither the public switched telephone network nor certain private line services, (b) require the registration of terminal equipment and protective circuitry in accordance with Subpart C of part 68, Title 47 of the *Code of Federal Regulations,* (c) require the assignment of identification numbers to the equipment and the testing of the equipment, and (d) do not require that accepted terminal equipment be compatible or function with the network. *See also* **part 68.**

Federal Information Processing (FIP) equipment: For purposes relating to the U.S. government and its respective agencies, any equipment or interconnected system or subsystems of equipment used in the automatic acquisition, storage, manipulation, management, movement, control, display, switching, interchange, transmission, or reception of data or information.

Federal Information Processing Standard (FIPS): A standard in the field of computers and information processing that (a) is published and is for sale by the U.S. government, (b) is prepared, endorsed, or adopted by the U.S. government, and (c) is either voluntary or mandatory within U.S. government agencies.

Federal Information Processing Standard Publication (FIPS PUB): The document that contains a Federal Information Processing Standard (FIPS). *Note:* An example of a Federal Information Processing Standard Publication (FIPS PUB) is FIPS PUB 11-3, *American National Standard Dictionary for Information Systems.*

Federal Information Processing (FIP) system: Any organized collection of Federal Information Processing (FIP) equipment, software, services, support services, or related supplies.

Federal Law Enforcement Training Center (FLETC): The center that (a) is near Brunswick, Georgia (U.S.) for training law enforcement officers and (b) is engaged in training officers in all crime prevention areas, including cybercops to combat criminals operating in cyberspace. *See also* **cracker, cybercop, cyberspace, hacker.**

Federal Secure Telephone Service (FSTS): A full-service secure switched voice telecommunications service designed to protect sensitive and classified voice transmissions for the U.S. government worldwide. *Note 1:* The Government Services Administration provides information security (INFOSEC) system and equipment protection services to agencies within the 50 states, Puerto Rico, Bermuda, the Virgin Islands, Canada, Mexico, Europe, and the Far East. *Note 2:* Federal Secure Telephone Service (FSTS) currently uses the STU-III Low Cost Secure Voice/Data Terminal. *See also* **Automatic Digital Network, Automatic Secure Voice Communications Network, Automatic Voice Network, communications.**

Federal Telecommunications Standard (FTS): One of a series of standards that (a) is developed under the auspices of the Federal Telecommunications Standards Committee (FTSC) of the National Communications System (NCS-TS), Arlington, VA 22204 and (b) is published by the General Services Administration, Washington, DC 20405. Examples of Federal Telecommunications Standards (FTS) are (a) Federal Standard 1037C, *Telecommunications: Glossary of Telecommunication Terms* and (b) Federal Standard 1033, *Automatic Voice Network (AUTOVON).*

Federal Telecommunications System (FTS): The original umbrella contracted switched telecommunications network that (a) provided local and long-distance telecommunications analog and digital services, including FTS2000 long-distance services, (b) was operated, managed, or maintained by the General Services Administration, (c) was for the common use of all federal agencies and other authorized users, (d) was provided by leased facilities, and (e) has been replaced by Federal Telecommunications System 2000. *See also* **Automatic Digital Network, Automatic Secure Voice Communications Network, Automatic Voice Network, communications, Federal Telecommunications System 2000.**

Federal Telecommunications System 2000 (FTS 2000): A long-distance telecommunications service, including services such as switched voice service for voice or data up to 4.8 kbps (kilobits per second), switched data at 56 kbps and 64 kbps, switched digital integrated service for voice, data, image, and video transmission up to 1.544-Mbps, packet-switched service for data in packet form, video transmission for both compressed and wideband video transmission, and dedicated point-to-point private line for voice and data. *Note 1:* Use of the Federal Telecommunications System 2000 (FTS2000) contract services is mandatory for U.S. government agencies for all acquisitions subject to Title 40 U.S.C. 759. *Note 2:* No U.S. government information processing equipment or customer premises equipment other than that required to provide an FTS2000 service is furnished. *Note 3:* The FTS contractors are required to provide service directly to an agency terminal equipment interface. *Note 4:* The General Services Administration (GSA) awarded two ten-year, fixed-price contracts covering FTS2000 services on December 7, 1988. *Note 5:* The Warner Amendment excludes the mandatory use of FTS2000 in instances related to maximum security. *Note 6:* The ten-year contract for telecommunications services replaced the older Federal Telecommunications System (FTS), to provide enhanced, domestic, long-distance, transmission services for the U.S. government. *Note 7:* The Federal Telecommunications System 2000 (FTS2000) contract addresses, but is not limited to, (a) switched voice service (SVS), (b) switched data service (SDS), (c) packet-switched service (PSS), including dialup asynchronous or packet-switched-based access to E-mail services, (d) dedicated data transmission services (DTS), (e) video transmission services (VTS), (f) switched digital integrated services (SDIS), including Integrated Services Digital Network (ISDN) features, and (g) integrated customer network services (ICNS), a vehicle for modifying network architecture for fiber optic system components, such as interfaces and service nodes.

Federal Telecommunications System 2000 (FTS 2000) Service: A long-distance telecommunications service, including services such as (a) switched voice service for voice or data up to 4.8 kbps (kilobits per second), (b) switched data at 56 kbps and 64 kbps, (c) switched integrated services digital network (ISDN) services for voice, data, image, and video transmission up to 1.544 Mbps (megabits per second), (d) packet-switched service for data in packet form, (e) video transmission for both compressed and wideband video, and (f) dedicated point-to-point private lines for voice and data. *Note 1:* Use of Federal Telecommunications System 2000 (FTS 2000) Service contract services is mandatory for U.S. government agencies for all acquisitions subject to U.S.C. 759. *Note 2:* No U.S. government information processing equipment or customer premises equipment other than that required to provide Federal Telecommunications System 2000 (FTS 2000) Service is furnished. *Note 3:* The Federal Telecommunications System 2000 (FTS 2000) Service contractors are required to provide service directly to agency data terminal equipment (DTE) interfaces. For example, the FTS 2000 contractor might provide a terminal adapter to an agency in order to connect FTS 2000 Integrated Services Digital Network (ISDN) services to the agency data terminal equipment (DTE). *Note 4:* The General Services Administration (GSA) awarded two ten-year, fixed-price contracts covering Federal Telecommunications System 2000 (FTS 2000) Service on December 7, 1988. *Note 5:* The Warner Amendment excludes mandatory use of Federal Telecommunications System 2000 (FTS 2000) Service in instances related to maximum security as defined in the Federal Information Resources Management Regulations (FIRMRs).

FedWorld: A government-operated database that (a) allows the general public to call into federal computer files, programs, and databases, (b) allows browsing among federal and foreign engineering, research, and development files, (c) may be downloaded into both Macintosh and MS-DOS systems, and (d) may serve as a precursor and a contributor to the information superhighway.

feed: **1.** To supply a signal to the input of a system, subsystem, equipment, or component, such as a transmission line or an antenna. **2.** A coupling device between an antenna and its transmission line. *Note:* A feed may consist of a distribution network or a primary radiator. **3.** A transmission facility between (a) the point of origin of a baseband signal, such as is generated in a radio or television studio, and (b) the head-end of a distribution facility, such as a broadcasting or multicasting station in a network. **4.** Pertaining to the function of inserting one thing into another, such as in a feed horn, paper feed, card feed, and line feed. *See* **antenna feed, form feed, line feed, paper feed, prime focus feed.**

feed antenna: *See* **dielectric guide feed antenna.**

feedback: 1. The return of a portion of the output of an active device to the input. *Note:* Positive feedback adds to the input; negative feedback subtracts from the input. **2.** The return of a processed part of the output signal of a device or a system to the input. *Note 1:* The feedback signal will have a certain magnitude and phase relationship relative to the output signal or relative to the input signal. This magnitude and phase relationship can be used to influence the behavior, such as the gain and the stability, of the overall circuit. *Note 2:* If the feedback is regenerative, i.e., is additive, it is called "positive feedback," which (a) increases gain and distortion and (b) decreases linearity and stability of the overall circuit. *Note 3:* If the feedback is degenerative, i.e., is subtractive, it is called "negative feedback," which (a) reduces gain and distortion and (b) increases linearity and stability of the overall circuit. *Note 4:* Feedback can occur unintentionally or inadvertently. It can be harmful because of the huge gain and sustained high amplitude oscillations that can occur. **3.** Information returned as a response to an originating source. *Synonym* **bus back.** *See* **error detection and feedback, information feedback, message feedback, negative feedback, positive feedback, regenerative feedback, video integrator feedback.** *See also* **closed loop transfer function, echo check, lock-in frequency, regeneration.** *Refer to* **Figs. S-17 (940), T-7 (1024).**

feedback factor: In a device, the fraction, in magnitude and phase with respect to the input or the output signal, that is returned to the input terminal and combined with the input signal. *Note:* The feedback signal may be combined with the input signal in a subtractor. If there is no phase shift in the feedback signal, or if the phase shift is negligible, the feedback is positive. If the phase shift is 180° electrical, the feedback is negative. The feedback factor can be between these phase values and at any amplitude, usually a fraction of the device output amplitude, usually taken from a voltage divider circuit. *Refer to* **Fig. T-7 (1024).**

feedback laser: *See* **distributed feedback laser, surface emitting distributed feedback laser.**

feedback loop: In a device, the circuit components and the processes used to accomplish feedback.

feedback shift register: A shift register that (a) generates a sequence of binary combinations by feeding back to its input functions of the state of the register, and (b) generates binary combinations that do not necessarily follow the pure, i.e., the straight, binary counting sequence. *Synonym* **twisted ring counter.** *See* **linear feedback shift register, nonlinear feedback shift register.** *See also* **spread spectrum code sequence generator.**

feedback tap: In a spread spectrum code sequence generator, an output tap of one of the flip-flops of the code sequence generator register. *Note:* In a simple generator, the selected tap is connected to one of the inputs of an EXCLUSIVE OR gate, i.e., a modulo two summer. The output tap of the last flip-flop, i.e., the register output flip-flop, is connected to the other input of the gate, and the gate output is connected to the first flip-flop. More complex arrangements can be made to produce special effects, including a pseudorandom modification of the generator itself by gating back the outputs of arbitrarily selected flip-flops. *See also* **spread spectrum code sequence generator.** *Refer to* **Fig. S-17 (940).**

feed character: *See* **form feed character, line feed character.**

feeder: A means in which radio frequency (rf) energy can be transmitted between a radio transmitter amplifier output and a transmitting antenna input. *Note:* An example of a feeder is a rectangular metal waveguide, or a coaxial cable, that connects the transmitter amplifier output to the transmitting antenna input at a radio station. *See* **suction feeder.**

feeder echo noise (FEN): In a feed, such as an antenna feed or a line feed, reflected noise that distorts a signal, such as reflected noise in a transmission line that is many wavelengths long and is mismatched at both the source end, i.e., the generator end, and the sink end, i.e., the load end. *See also* **echo, noise.**

feeder link: At a specified fixed point, a radio link from an Earth station to a space station, or vice versa, that conveys information for a space radiocommunications service other than for the fixed satellite service. *See also* **Earth station.**

feed harness: The set of electrical and mechanical elements that (a) conducts electrical energy from a transmitter output to a transmitting antenna input terminal and from the antenna input terminal to the actual connections or terminals on the individual radiators of the antenna array, and (b) includes all impedance matching transformers and connectors.

feed hole: One of a series of holes punched in a data medium for moving or positioning the medium for reading or writing. *Note:* Examples of feedholes are (a) the holes down the center of paper tape that engage the feed sprocket and (b) the holes on the edges of continuous form printer paper that engage sprockets on both sides of a platen for feeding the paper, i.e., advancing the paper in a longitudinal direction. *See also* **data medium.**

feed horn: *See* **scalar feed horn.**

feed system: *See* **antenna feed system, five-horn feed system, four-horn feed system, monopulse antenna feed system.**

feed through: *See* **fiber optic cable feed through.**

female connector: The half of a connector that usually has elements or contacts in a cavity or is recessed so that the male connector half may be inserted into it to establish a connection. *See also* **gender changer, male connector.**

FEP: fluorinated ethylene propylene, front end processor, perfluorinated ethylene propylene.

FEP clad silica optical fiber: *See* **lowloss FEP clad silica optical fiber.**

Fermat's principle: A ray of light follows the path that requires the least time to travel from one point to another, including reflections and refractions that may occur. *Note:* The optical path length is an extreme path in the terminology of the calculus of variations. Thus, if all rays starting from point *A* and traveling via a medium arrive at some point *B,* the paths from *A* to *B* all have the same optical length, though their geometric length may be different. In optical fibers, because different rays take different paths that do not have the same path length, they arrive at the end at different points or at a given point at different times, giving rise to dispersion, distortion, and intersymbol interference. In radio systems, this form of multipath may cause distortion. *Synonym* **least time principle.**

ferrite switch: A passive switch that uses the phase-shifting properties of a high magnetic permeability ferrite at microwave frequencies to connect either of two outputs to either of two inputs by switching the electromagnetic wave fields.

ferromagnetic: Pertaining to material that (a) has a magnetic permeability very much greater than that of air or a vacuum, (b) acts like a good conductor of magnetic lines of flux when placed in a magnetic field, (c) tends to draw the lines of flux inside itself, and (d) tends to take a position with its long axes parallel to the lines of magnetic flux with a strong torque compared to other materials. *Note:* Examples of magnetic materials are iron, iron oxide, i.e., ferrous or ferric oxide, cobalt, and nickel. *See also* **diamagnetic, magnetic permeability, paramagnetic.**

ferrous shield: A low resistance, electric current conducting material that contains iron and therefore (a) prevents electric fields from being sustained, (b) provides a low reluctance, high permeability magnetic path that conducts magnetic lines of flux away from a shielded object, (c) causes incident electromagnetic waves to be absorbed, and (d) prevents electromagnetic waves from penetrating deeply and therefore prevents their transmission through the material.

ferrule: A mechanical device or fixture that (a) usually is a rigid tube, (b) is used to hold the stripped end of an optical fiber or fiber optic bundle consisting of individual optical fibers cemented together, (c) has a diameter designed to hold the fibers firmly with a maximum packing fraction, (d) may be made of nonrigid materials, such as shrink tubing, and (e) usually provides a means of positioning fibers within a connector by performing the function of a bushing.

FET: field-effect transistor.

fetch protection: An automated information system restriction that prevents a program from obtaining access to a segment of storage that contains data belonging to another user. *See also* **register, storage.**

FEXT: far end crosstalk.

FFL: front focal length.

fiber: 1. In fiber optics, an optical fiber. **2.** To install optical fiber and fiber optic cables, as in "to wire." *See* **active optical fiber, advanced tactical optical fiber, air supported optical fiber, all-glass optical fiber, all-plastic optical fiber, all-silica optical fiber, buffered optical fiber, circularly symmetric optical fiber, compensated optical fiber, deeply depressed cladding optical fiber, depressed cladding optical fiber, dispersion flattened optical fiber, dispersion-shifted optical fiber, dispersion-unshifted optical fiber, doped silica cladded optical fiber, doped silica graded optical fiber, doubly clad optical fiber, drawn glass optical fiber, fully filled optical fiber, graded index optical fiber, hard clad silica optical fiber, high birefringence optical fiber, high loss optical fiber, inhomogeneous optical fiber, jointed optical fiber, jumbo optical fiber, launching optical fiber, liquid core optical fiber, Lorentzian optical fiber, low birefringence optical fiber, lowloss FEP clad silica optical fiber, lowloss optical fiber, medium loss optical fiber, multimode optical fiber, optical fiber, overcompensated optical fiber, plastic clad silica optical fiber, polarization-maintaining optical fiber, quadruply clad optical fiber, self-focusing optical fiber, SELFOC^R optical fiber, SEL^R optical fiber, single optical fiber, single mode optical fiber, step index optical fiber, superfiber, tapered optical fiber, transition optical fiber, triangu-**

lar core optical fiber, ultralow-loss optical fiber, undercompensated optical fiber, uniform index profile optical fiber, weakly guiding optical fiber, W-type optical fiber.

fiber absorption: *See* **optical fiber absorption.**

fiber acoustic sensor: *See* **fiber optic acoustic sensor, optical fiber acoustic sensor.**

fiber active connector: *See* **fiber optic active connector, optical fiber active connector.**

fiber alignment connector: *See* **fiber optic alignment connector.**

fiber amplifier: *See* **optical fiber amplifier.**

fiber and splice organizer: *See* **optical fiber and splice organizer.**

fiber axial alignment sensor: *See* **fiber optic axial alignment sensor.**

fiber axial displacement sensor: *See* **fiber optic axial displacement sensor.**

fiber axis: *See:* **optical fiber axis.**

fiber bandwidth: *See* **optical fiber bandwidth.**

fiber blank: *See* **optical fiber blank.**

fiber borescope: *See* **fiber optic borescope.**

fiber branching device: *See* **fiber optic branching device.**

fiber buffer: *See* **optical fiber buffer.**

fiber bundle: *See* **fiber optic bundle.**

fiber bundle transfer function: *See* **fiber optic bundle transfer function.**

fiber cable assembly: *See* **fiber optic cable assembly, multiple optical fiber cable assembly.**

fiber cable: *See* **fiber optic cable.**

fiber cable component: *See* **fiber optic cable component.**

fiber choledochoscope: *See* **fiber optic choledochoscope.**

fiber circuit: *See* **hybrid optical fiber circuit.**

fiber cladding: *See* **optical fiber cladding.**

fiber coating: *See* **optical fiber coating.**

fiber/coating offset ratio: *See* **optical fiber/coating offset ratio.**

fiber communications system: *See* **fiber optic communications system.**

fiber concentrator: *See* **fiber optic concentrator.**

fiber concentricity error: *See* **optical fiber concentricity error.**

fiber connector set: *See* **optical fiber connector set.**

fiber core: *See* **optical fiber core.**

fiber core diameter: *See* **optical fiber core diameter.**

fiber count: *See* **optical fiber count.**

fiber coupled power: *See* **source to fiber coupled power.**

fiber coupling: *See* **source-fiber coupling.**

fiber crosstalk: *See* **optical fiber crosstalk.**

fiber cutoff wavelength: *See* **optical fiber cutoff wavelength.**

fiber cutting tool: *See* **optical fiber cutting tool.**

fiber delay line: *See* **fiber optic delay line.**

fiber-detector coupling: *See* **optical fiber-detector coupling.**

fiber diameter: *See* **optical fiber diameter.**

fiber dimensional stability: *See* **optical fiber dimensional stability.**

fiber dispersion: *See* **optical fiber dispersion.**

Fiber Distributed Data Interface (FDDI) standard: An optical fiber token-ring network standard that (a) is defined in four American National Standards Institute (ANSI) standards, with reliable data transfer, active link monitoring, station management, wideband capability, i.e., 100 Mbps (megabits per second) transmission rate, and survivability features, and (b) is a fiber optic communications network standard, developed by the American National Standards Institute (ANSI), in which a single bit stream format is used rather than a set of optical data signaling rates (DSRs) and formats using a byte format. *Note 1:* The four standards are (a) ANSI X3T9.5, containing Physical Media Dependent (PMD) specifications, (b) ANSI X3T9.5, containing Physical (PHY) specifications, (c) ANSI X3.139, containing Media Access Control (MAC) specifications, and (d) ANSI X39.5, containing Station Management (SMT) specifications. *Note 2:* Fiber Distributed Data Interface 2 (FDDI-2) is a second-generation FDDI network standard. *See also* **Synchronous Optical Network standard.**

fiber distribution box: *See* **optical fiber interconnection box.**

fiber distribution frame: *See* **fiber optic distribution frame.**

fiber distribution panel: *See* **optical fiber distribution panel.**

fiber distribution system: *See* **protected fiber distribution system.**

fiber dosimeter: *See* **fiber optic dosimeter.**

fiber drawing: *See* **optical fiber drawing.**

fiber drawing machine: *See:* **optical fiber pulling machine.**

fiber end-to-end separation sensor: *See* **fiber optic end-to-end separation sensor.**

fiber faceplate: *See* **fiber optic faceplate.**

fiber global attenuation rate characteristic: *See* **optical fiber global attenuation rate characteristic.**

fiber global dispersion characteristic: *See* **optical fiber global dispersion characteristic.**

fiber global loss characteristic: *See:* **optical fiber global attenuation rate characteristic.**

Fiberguide™: An optical fiber produced by Times Fiber Communications, Inc.

fiber gyroscope: *See* **fiber optic gyroscope.**

fiber hazard: *See* **optical fiber hazard.**

fiber identifier: *See* **optical fiber identifier.**

fiber jacket: *See* **optical fiber jacket.**

fiber jumper: *See* **fiber optic jumper.**

fiber junction: *See* **optical fiber junction.**

fiber laser: *See* **optical fiber laser.**

fiber light source: *See* **fiber optic light source.**

fiber lightguide: *Synonym* **optical fiber.**

fiber link: *See* **television fiber optic link.**

fiber longitudinal compression sensor: *See* **fiber optic longitudinal compression sensor.**

fiber loop: *See* **fiber optic loop.**

fiber loop multiplexer: *See* **fiber optic loop multiplexer.**

fiber loss: *See* **source to fiber loss.**

fiber mandrel: *See* **optical fiber mandrel.**

fiber merit figure: *See* **optical fiber merit figure.**

fiber mixer: *See* **fiber optic mixer.**

fiber multiplexer: *See* **fiber optic multiplexer (active).**

fiber net: *See* **fiber optic net.**

fiber nuclear hardening: *See* **optical fiber nuclear hardening.**

fiber optic: Pertaining to systems, subsystems, equipment, and components in which optical fibers are used. *Note:* The optical fibers (a) may be used for transmission, sensing, telemetering, amplification, illumination, and signaling and (b) may have cross sections other than round, such as elliptical, rectangular, and thin film cross sections. *See also* **fiber optics, optical fiber.**

fiber optic acoustic sensor: A sensor (a) that is highly sensitive, (b) that senses sound waves by allowing them to impinge upon a length of optical fiber, (c) in which the waves cause variations in the refractive indices of the materials in the fiber, (d) in which the variation of refractive indices is used to modulate lightwaves, usually from a laser, that are propagating in the fiber on their way to a photodetector, and (e) in which the spatial distribution of the fiber determines the directional capability of the sensor, such as (i) if the fiber is wound in the shape of a sphere, the sensor will be omnidirectional, (ii) if the fiber is distributed in a plane, it will be unidirectional, and (iii) if the fiber is shaped in the form of a rod, it will be cylindrically directional. *Synonym* **optical fiber acoustic sensor.**

fiber optic active connector: A fiber optic connector that (a) has a built-in active device, such as a light-emitting diode (LED) or a laser as a transmitter, a photodetector as a receiver, or both, usually built into one of the mating elements of the connector, usually in the form of a semiconductor chip with an optical fiber pigtail for connection to a fiber, (b) improves the fiber-to-fiber or fiber-to-pigtail coupling efficiency over that of a fiber optic passive connector, such as 50 times normal coupling efficiency at 30 Mbps (megabits per second) over a 1-km (kilometer) length and apertures of the semiconductor element of about 0.20 mm (millimeter) for a 0.20-mm-diameter core with a numerical aperture of 0.48, (c) is adaptable for bulkhead or printed circuit board mounting and electromagnetic interference (EMI) and radio frequency (rf) interference (RFI) shielding, and (d) serves as a repeater. *Synonyms* **active fiber optic connector, optical fiber active connector.**

fiber optic alignment connector: A fiber optic connector that holds optical fibers firmly and aligns them for effective optical coupling between the fibers. *See also* **grooved fiber optic alignment connector.**

fiber optic amplifier: An amplifier that (a) accepts an optical signal above threshold, (b) converts the optical signal to an electrical signal by means of a photodetector, (c) amplifies and reshapes the signal electronically, (d) either (i) modulates a light source, such as a laser or a light-emitting diode or (ii) modulates the light from a continuous light source, and (e) usually is used in a fiber optic link. *See also* **optical fiber amplifier.**

fiber optic arteriovenous oximeter: A device in which a fiber optic catheter is used for continuous prolonged *in vivo* measurement of blood oxygen content, blood pressure, and blood saturation.

fiber optic attenuator: In fiber optic systems, a device, such as a length of high loss optical fiber, that operates upon its input optical signal power level in such a way that its output signal power level is less than the input level, the reduction being caused by such means as absorption, reflection, diffusion, scattering, deflection, diffraction, and dispersion, and not a result of geometric spreading. *See also* **optical attenuator.**

fiber optic axial alignment sensor: A fiber optic sensor in which the amount of light that is coupled from (a) a source optical fiber mounted on one platform to (b) a photodetector optical fiber mounted on another platform decreases as the angular misalignment of the axes of the two fibers increases. *Note 1:* Axial alignment is considered to have occurred when the output signal of the photodetector becomes a maximum as the alignment is varied. *Note 2:* The fiber optic axial alignment sensor can be used to measure angular alignment of the two platforms, one attached to each optical fiber. *Note 3:* When the angle between the axes is greater than half the apex angle of the acceptance cone, light is no longer coupled from the source optical fiber to the detector fiber, at which point the output of the photodetector is reduced to zero. *See also* **fiber optic end-to-end separation sensor.**

fiber optic axial displacement sensor: A fiber optic sensor in which the amount of light that is coupled from (a) a source optical fiber mounted on one platform to (b) a photodetector fiber mounted on another platform decreases as the axes of the two fibers are moved laterally, i.e., transversely, farther apart, even though the axes remain parallel and the longitudinal displacement is held constant, thus causing increasing lateral offset loss. *Note:* The output signal from the photodetector is a function of the amount of displacement, varying from zero to a maximum. The time rate of change of the signal is proportional to the velocity at which the fibers are displaced laterally relative to each other. If the fibers oscillate with respect to each other, an oscillating signal will be produced by the photodetector. *See also* **lateral offset loss.**

fiber optic bloodflow meter: A bloodflow meter that (a) is mounted in a catheter, (b) is introduced *in vivo* into a vein, (c) has a single optical fiber to transmit light into the vein, (d) obtains reflected and scattered light from the moving erythrocytes, and (e) uses optical mixing spectroscopy and Doppler shift to measure the bloodflow rate.

fiber optic borescope: An optical device in which optical fibers are used to illuminate, view, and take pictures of the internal parts of a system, such as the interior of machines, e.g., turbine blade clearance, or the interior of internal organs of the human body, such as blood vessels and intestines. *See also* **flexible borescope, rigid borescope.**

fiber optic branching device: In fiber optic systems, an optical device that (a) possesses one input port and two or more output ports, (b) divides input signal optical power among its output ports in a predetermined manner without any modification, manipulation, or amplification of the input signal, and (c) reverses the function a fiber optic combiner. *Note:* Types of fiber optic branching devices include unidirectional, bidirectional, symmetrical, and asymmetrical branching devices. *Synonym* **passive optical fiber branching device.** *See also* **fiber optic combiner, fiber optic coupler.**

fiber optic breakout cable: A multifiber fiber optic cable that (a) consists of individual, usually tight-buffered, optical fibers surrounded by separate strength members and jackets, all of which are encased in a common jacket, (b) facilitates installation of fiber optic connectors, (c) may be prepared to receive connectors by removing the outer jacket, thus exposing what essentially are individual single-fiber cables, and (d) tends to induce bends in the fibers, usually resulting in higher transmission losses for a given fiber than loose-buffer cables produce. *Synonym* **fanout cable.**

fiber optic breakout kit: A kit of materials that (a) consists of an outer jacket containing a strength member usually made of a bundle of aramid yarn, (b) enables a fiber optic cable containing multiple loose-buffer tubes to receive connectors without the splicing of pigtails, and (c) allows (i) the jacket and yarn to be slipped over a loose-buffer tube containing a single fiber, thus converting the buffer tube and fiber to a complete single-fiber cable to which a fiber optic connector may be directly attached, and (ii) a heat-shrinkable plastic boot to be used for cosmetic purposes, strain relief, and sealing the point where the individual newly created cables diverge and merge.

fiber optic bundle: 1. An assembly of usually unbuffered optical fibers that (a) is used as a single transmission channel, as opposed to multifiber cables in which each fiber provides a separate channel, (b) is flexible and is unaligned if used only to transmit optical power as in illumination systems rather than signals as in fiber optic communications systems, and (c) is flexible and aligned if used to transmit and display images. **2.** Two or more optical fibers in a single protective sheath or jacket. *Note:* The number of optical fibers in a fiber optic bundle might range from a few to several thousand, depending on the application and the characteristics of the fibers. *Synonyms* **fiber bundle, optical fiber bundle.** *See also* **aligned bundle, buffer, bundle, cable, fiber optics, optical fiber.** *Refer to* **Figs. P-9 (739).**

fiber optic bundle transfer function: The function that produces the output signal waveform when it operates on the input signal waveform for a fiber optic bundle, given by the relation $T_b = A(f)e^{j\Theta(f)}$ where T_b is the fiber optic bundle transfer function, $A(f)$ is the amplitude portion, and $j\Theta(f)$ the phase portion of the transfer function.

fiber optic bus: A bus that consists of a fiber optic cable. *See also* **bus.**

fiber optic cable: 1. A cable that (a) has one or more optical fibers that are used as the propagation medium for lightwaves, (b) is capable of transmitting optical signals over long distances, (c) usually consists of optical fibers that are surrounded by buffers, strength members, and jackets for protection, stiffness, and strength, and (d) does not require the use of metals. *Note 1:* For a given signal transmission capacity, fiber optic cables are smaller and lighter than electrical cables. *Note 2:* Fiber optic cables are used for many applications, such as communications, illumination, optical power transmission, imaging, endoscopy, and telemetry. **2.** A cable in which optical fibers are used to achieve space division multiplexing. *See* **bidirectional fiber optic cable, duct fiber optic cable, special fiber optic cable, station fiber optic cable, TEMPEST-proofed fiber optic cable.** *See also* **wire cable.** *Synonyms* **fiber cable, optical cable, optical fiber cable.** *Refer to* **Fig. F-1.** *Refer also to* **Figs. F-2 (339), S-14 (927), S-20 (959), W-3 (1082).**

Fig. F-1. A **fiber optic cable** that outshines a copper cable. (Courtesy Siecor Corporation, Hickory, NC.)

fiber optic cable assembly: 1. A fiber optic cable that is terminated with fiber optic connectors, i.e., with one part of a two-part connector on each end. 2. A fiber optic cable that is terminated and ready for installation. *Synonyms* **optical cable assembly, optical fiber cable assembly.** *See also* **cable assembly.**

fiber optic cable component (FOCC): 1. A component or a part of a fiber optic cable. **2.** The waveguide portion of a fiber optic cable, such as (a) a 12-fiber ribbon complete with optical fibers and (b) a buffered fiber complete with strength members and jacket suitable for cabling with other similar components. **3.** A buffered optical fiber augmented with a concentric layer of strength members and an overall jacket.

fiber optic cable core: The portion of a fiber optic cable that (a) serves as the waveguide proper, such as an optical fiber, a fiber optic bundle, or a ribbon holding a group of optical fibers, and (b) usually is protected from environmental effects by various other components, such as additional buffers, strength members, gels, tubes, jackets, overall sheaths, or overarmor.

fiber optic cable driver: A device that (a) launches modulated lightwaves into optical fibers, fiber optic cables, or fiber optic bundles by (i) accepting electrical pulses from a propagation medium, such as a wire or coaxial cable, (ii) converting the pulses to optical pulses, i.e., to lightwave pulses, and (iii) coupling the optical pulses to an optical fiber, fiber optic cable, or fiber optic bundle and (b) usually consists of (i) an input channel for modulating the light source or for inserting modulated electrical pulses, (ii) a light source capable of being modulated by electrical pulses, and (iii) a means for coupling the output optical pulses into an optical fiber, fiber optic cable, or fiber optic bundle. *Synonym* **optical cable driver.** *See also* **coaxial cable, fiber optic bundle, optical pulse.**

fiber optic cable facility: 1. In a station facility of a fiber optic cable station/regenerator section, the device used to connect fiber optic station cables (inside plant) to a fiber optic cable facility (outside plant) via connectors and splices. **2.** The part of a fiber optic station/regenerator section that (a) consists of the fiber optic cable, (b) includes splices, (c) connects (i) the cable splice at the fiber optic cable interconnect feature at one fiber optic station facility to (ii) the splice at the fiber optic cable interconnect feature at another optic fiber station facility, and (d) serves as the outside plant between stations. *Note:* In designing a fiber optic station/regenerator section, the optical power losses in the fiber optic cable facility must be equal to or less than the terminal/regenerator system gain. *Synonyms* **fiber optic cable interconnect feature, fiber optic cable interconnect facility, optical cable facility.**

fiber optic cable facility loss: In a fiber optic cable station/regenerator section, the loss in the fiber optic cable facility, i.e., in the outside plant, given by the relation $L = l_t(U_c + U_{cT} + U_\lambda) + N_S(U_S + U_{ST})$ where L is the fiber optic cable facility loss, l_t is the total sheath (cable) length of spliced cable (km), U_c is the worst-case end-of-life cable attenuation rate (dB/km) at the transmitter nominal central wavelength, U_λ is the largest increase in cable attenuation rate that occurs over the transmitter central wavelength range, U_{cT} is the effect of temperature on the end-of-life cable attenuation rate at the worst-case temperature conditions over the cable operating temperature range, N_S is the number of splices in the length of the cable in the fiber optic cable facility, including the splice at the fiber optic station facility on each end and allowances for cable repair splices, U_S is the loss (dB/splice) for each splice, and U_{ST} is the maximum additional loss (dB/splice) caused by temperature variation. *Note 1:* The fiber optic cable facility loss, *L,* must be equal to or less than the fiber optic terminal/regenerator system gain, *G,* for the optical station/regenerator section to operate satisfactorily. *Note 2:* The fiber optic cable facility loss usually is expressed in dB. *Synonym* **optical cable facility loss.** *See* **statistical fiber optic cable facility loss.** *See also* **fiber optic global attenuation rate characteristic, terminal/regenerator system gain.**

fiber optic cable feed through: A mechanism that (a) provides strain relief to a fiber optic cable entering a fiber optic interconnection box and (b) may also be used to seal around the cable.

fiber optic cable interconnect facility: *Synonym* **fiber optic cable facility.**

fiber optic cable interconnect feature: *Synonym* **fiber optic cable facility.**

fiber optic cable jacket: In a fiber optic cable, the material that (a) is placed over or wrapped around buffered or unbuffered optical fibers, strength members, added buffers, fillers, and any other elements used in the construction of the cable, (b) forms the outside of the cable, except when overarmor is placed over it, and (c) has identification markings. *Note:* Overarmor may be placed over an optical fiber jacket to overcome abrasive forces in special applications, such as those that occur in sea-to-shore installations, tethered ship installations, and risers in aerial inserts. *Synonyms* **optical cable jacket, sheath.**

fiber optic cable link: *Synonym* **fiber optic link.**

fiber optic cable pigtail: A short length of fiber optic cable permanently fixed to a component and used to couple optical power between it and another fiber optic cable pigtail or between it and a fiber optic cable, such as a transmission cable.

fiber optic catheter: A catheter in which optical fibers are used to make measurements of physiological parameters and perform *in vivo* operations inside humans, animals, and other living organisms.

fiber optic choledochoscope: A device in which optical fibers are used for illuminating and viewing the interior of the bile duct.

fiber optic circuit: *See* **hybrid fiber optic circuit.**

fiber optic combiner (FOC): A passive device in which power from several optical fibers is distributed among fewer fibers. *Note:* The function of a fiber optic combiner is the reverse of that of a fiber optic branching device. *See also* **fiber optic branching device, fiber optic coupler, fiber optics, star coupler.**

fiber optic communications: The use of physical paths, such as lines, circuits, and links, that are made of optical fibers between components of a communications system, such as between (a) terminals, (b) terminals and user end instruments, and (c) exchanges and switching centers. *Synonym* **optical fiber communications.**

fiber optic communications system: A communications system in which fiber optic equipment and components, such as fiber optic transmitters and fiber optic cables, are used. *See* **integrated fiber optic communications system, optical communications system.**

fiber optic concentrator: A communications system component that multiplexes a number of separate incoming fiber optic communications channels. *Note:* Examples of fiber optic concentrators are (a) a distribution frame that (i) accepts 3000 fiber optic voice channels from home optical transceivers, (ii) time division multiplexes them using ten 10-to-1 multiplexers, (iii) sends the output of each multiplexer to 30 more 10-to-1 multiplexers, and (iv) produces a final output that modulates a lightwave carrier in an optical fiber, and (b) a device that (i) multiplexes 32 full duplex digital data channels and (ii) sends the output to a single fiber optic data link or cable by means of an interconnection box. *See also* **multiplexing, optical fiber.**

fiber optic connector (FOC): A device that (a) transfers optical power between two optical fibers, groups of fibers, or fiber optic bundles, (b) may be repeatedly connected and disconnected, and (c) usually consists of two mateable and demateable parts, one attached to each end of a fiber optic cable, or to a piece of fiber optic equipment or component, to provide connection and disconnection of fiber optic cables. *Note:* Fiber optic connectors are sometimes erroneously referred to as couplers. *Synonym* **optical connector.** *See* **fixed fiber optic connector, free fiber optic connector.** *See also* **fiber optic coupler, multifiber joint.** *Refer to* **Fig. F-2.**

fiber optic connector variation: The maximum value of the difference in insertion loss between mating fiber optic connectors of the same type and model and from the same manufacturer. *Synonym* **optical connector variation.**

fiber optic coupler: **1.** A device in which an optical fiber transfers optical power to other optical fibers without the use of splices or connectors. *Note:* The power transfer usually is accomplished by placing the transmitting and receiving cores in proximity and making use of evanescent waves and cladding modes. Thus, the coupler may transfer power to a number of ports in a predetermined manner. **2.** A device that transfers power between a light source and an optical fiber or between an optical fiber and a photodetector. *Note:* The fiber optic coupler may transfer optical power to a number of ports in a predetermined manner. The ports may be connected to fiber optic components, such as fiber optic light sources, waveguides, and photodetectors. *Synonym* **optical fiber coupler.** *See also* **fiber optic branching device, optical coupler, optical directional coupler, star coupler, tee coupler.**

fiber optic cross connection: Connection of two optical fibers by means of a third optical fiber, i.e., by means of a jumper, that serves as a link between the two fibers. *Note:* Although two connections have to be made rather than the one required for a direct connection between the two optical fibers, the jumpers can be mounted on a panel so connections can be made conveniently at one place. *See also* **fiber optic interconnection.**

fiber optic data bus: **1.** A data bus that (a) consists of fiber optic cables and (b) interconnects stations, nodes, or terminals that are connected to the bus. **2.** A single optical fiber or cable (a) to which stations and terminals are directly connected so that they can communicate with each other via the single bus, (b) in which messages intended for specific addressees must be addressed according to network protocol, and (c) all messages placed on the bus can be made available to any or all of the stations and terminals connected to the bus without having to pass through any specific station or terminal. *See also* **data bus.**

fiber optic data link: A data link that (a) usually consists of (i) an electronically modulated light source, (ii) a fiber optic cable, (iii) a photodetector, and (iv) associated fiber optic connectors, (b) is capable of transmitting data in the form of optical signals from one location to another, and (c) may include other components, such as amplifiers, time division multiplexers (TDM), time division demultiplexers (TDDM), clocks, and data recovery circuits. *See also* **data link.** *Refer to* **Fig. F-2.**

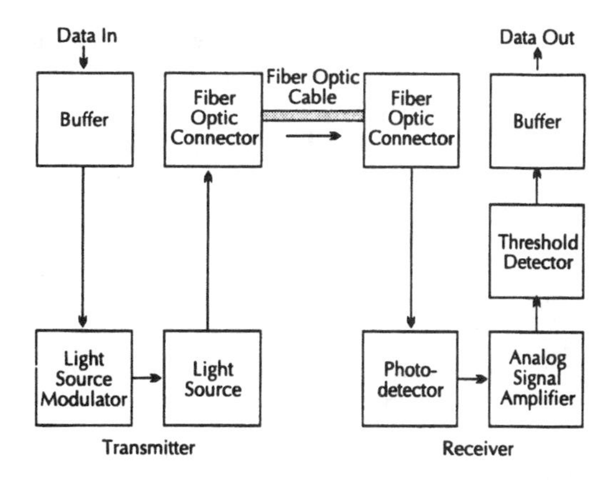

Fig. F-2. The components of a complete **fiber optic data link.**

fiber optic data transfer network: A data transfer network (DTN) in which fiber optic modules and components are used to sense, transmit, display, or otherwise process data.

fiber optic delay line: An optical fiber of precise length that (a) introduces a delay in a lightwave pulse equal to the time required for the pulse to propagate from beginning to end and (b) may be used for various purposes, such as phase adjustment, pulse positioning, pulse interval coding, or data storage. *Synonym* **optical fiber delay line.**

fiber optic demultiplexer (active): An electronically operated optical device, with input and output optical fibers, that (a) accepts information in the form of modulated lightwaves that have resulted from combining two or more optical signal streams in a multiplexer, (b) places each of the original streams on a separate output channel, (c) may have two or more output fibers, each operating at a different wavelength, and one input fiber bearing all the wavelengths, and (d) requires energy other than that contained in the input signals to operate. *Note:* An example of a fiber optic demultiplexer (active) is a device that uses the electrooptic effect to disperse a single channel input polychromatic lightwave into its separate constituent colors, each of which may have been separately modulated. *Synonym* **optical fiber demultiplexer (active).** *See also* **fiber optic multiplexer (active), optical demultiplexer (active), optical multiplexer (active).**

fiber optic demultiplexer (passive): A purely passive optical device that (a) accepts information in the form of modulated lightwaves that have resulted from combining two or more optical signal streams in a multiplexer, (b) places each of the original streams on a separate output channel, and (c) does not require energy to operate other than that contained in the input lightwave. *Note:* Examples of a fiber optic demultiplexer (passive) are (a) two or more output optical fibers, each operating at a different wavelength, and one input fiber bearing all the wavelengths and (b) a glass prism that disperses a single channel input polychromatic lightwave in an optical fiber into separate constituent colors, each of which may have been separately modulated and each of which is incident upon a separate photodetector. *Synonym* **optical fiber demultiplexer (passive).** *See also* **fiber optic multiplexer (passive), optical demultiplexer (passive), optical multiplexer (passive).**

fiber optic display device: A device that (a) is used to display pictorial, alphanumeric, or graphical data and (b) usually consists of (i) a fiber optic faceplate made of the endfaces of optical fibers in an aligned bundle, (ii) an aligned bundle of fibers used to transfer an image to the display surface, and (iii) perhaps a lens to enlarge or project the image.

fiber optic distribution frame: A structure that (a) has fiber optic cable terminations for interconnecting permanently installed fiber optic cables in such a way that interconnection by cross connections may be made, (b) may hold connectors, splice trays, or both, (c) usually is located at a central office (C.O.), switching center, or node in a communications network, and (d) is part of the inside plant. *Synonym* **fiber optic interconnection frame.** *See also* **fiber optic interconnection box.**

fiber optic distribution system: *See* **protected fiber optic distribution system.**

fiber optic dosimeter: An instrument that measures cumulative exposure to high energy radiation or bombardment, such as X rays, gamma radiation, cosmic rays, and high energy neutrons, by measuring the induced attenuation in an optical fiber from exposure to the radiation. *See also* **induced attenuation.**

fiber optic drop: A fiber optic transmission line that carries multiplexed signals from a central office (C.O.) inside plant to a distribution point where lines lead to taps for user end instruments, such as telephones and data terminals on user premises, e.g., individual offices and homes. *Note:* Current trends are to run optical fiber all the way to the taps for video, telephone, and data channels. Satellite dishes, coaxial cable, and fiber optic cables are competing signal transport media for television, telephone, and data transmission into home, office, and factory. *See also* **fiber optic loop.**

fiber optic endoscope: An optical device used to view and to take pictures of the internal parts of a system, such as the internal cavities and organs of the human body or the interior of machines. *Note:* In medicine, fiber optic endoscopes, such as bronchoscopes, gastroscopes, colonoscopes, choledochoscopes, cystoscopes, laparoscopes, and arthroscopes, are used for *in vivo* examination of internal body cavities, performing biopsies, and performing polypectomies. Industrial counterparts of the endoscope are used for remote nondestructive inspection of internal parts and cavities of machine systems while they are operating. *See also* **fiberscope.**

fiber optic end-to-end separation sensor: A fiber optic sensor in which the amount of light that is coupled from (a) a source optical fiber mounted on one platform to (b) a photodetector fiber mounted on another platform is a function of the longitudinal distance of separation of the endfaces of the two fibers; i.e., as the separation distance increases, the amount of coupled light decreases from a maximum to zero. *Note:* Operation of the fiber optic end-to-end separation sensor is dependent upon exit and acceptance cone angles, i.e., numerical aperture, core diameters, and wavelength. If the end-to-end separation distance oscillates, the photodetector output signal will oscillate. *See also* **fiber optic axial displacement sensor, fiber optic sensor.**

fiber optic faceplate: A plate that (a) is made by cutting a transverse slice from an aligned bundle or a boule, i.e., from a bundle of fused optical fibers, and (b) is used as a cover for a light source or for displaying an image of an object.

fiber optic facility: *See* **station fiber optic facility.**

fiber optic filter: A filter that (a) consists of one or more optical fibers and (b) operates on optical signals of various frequencies such that certain frequencies are blocked and others are passed. *Note:* A piece of blue glass will transmit only the blue wavelengths contained in incident white light and will block all the other wavelengths. *See also* **filter, optical filter.**

fiber optic flood illumination: Illumination of a relatively large area with a wide light beam emanating from the end of an optical fiber or fiber optic bundle. *See also* **fiber optic spot illumination.**

fiber optic global attenuation rate characteristic: *Synonym* **optical fiber global attenuation rate characteristic.**

fiber optic grooved alignment connector: An optical fiber connector that (a) holds the fibers in grooves cut in split blocks of material to assure proper alignment of the fibers between the mating pairs of the connector and (b) can be made of nearly any opaque plastic, such as polyvinyl chloride (PVC) and styrofoam. *Synonym* **grooved optical fiber alignment connector.** *See also* **connector, fiber optic alignment connector, fiber optic connector, optical fiber.**

fiber optic gyroscope (FOG): 1. A gyroscope in which (a) optical fibers, such as single mode or polarization-maintaining fibers, and (b) monochromatic light, such as that produced by lasers, are used to measure changes in rotational displacement, velocity, and acceleration, and hence changes in orientation and direction of the platform holding the sensor. *Note:* Interferometric techniques, such as are used in the Sagnac fiber optic sensor, are used to detect the changes in phase or polarization of the lightwaves. **2.** A gyroscope (a) in which optical fibers are used to conduct signals that represent tilt forces as part of the control system for maintaining stability and for conveying signals from perturbations, and (b) that is characterized by low power consumption, high sensitivity to perturbations, and high reliability.

fiber optic identifier: *See* **optical fiber identifier.**

fiber optic illumination detection: The use of an optical fiber or a fiber optic bundle to determine whether light is emanating from a given source.

fiber optic illuminator: A device (a) in which an optical fiber or a fiber optic bundle is used to convey light to and illuminate an area remote from the source of light, and (b) that may be used to shed light on specific and often inaccessible surfaces, such as surgical fields, instrument panels, and auxiliary equipment, particularly in atmospheres where electrical illumination might be hazardous.

fiber optic interconnection: Connection of two optical fibers by means of a direct connection from one to the other, thus requiring no fiber optic jumpers. *See also* **fiber optic cross connection.**

fiber optic interconnection box: A housing (a) that holds fiber optic splices, connectors, and couplers used to distribute signals on incoming cables to outgoing cables by means of connections, (b) in which connections may be made by interconnection or cross connection, and (c) that usually is a local box or housing at a user premises for holding connections to fiber optic cables from distribution frames and fiber optic splices, or splice trays that hold splices, to loops going to optical transceivers or other user end instruments. *Synonyms* **fiber optic distribution box, optical fiber interconnection box.** *See also* **fiber optic distribution frame.**

fiber optic interconnection frame: *Synonym* **fiber optic distribution frame.**

fiber optic interface device: A device that (a) accepts electrical signals and converts them into optical signals for transmission in an optical medium, such as an optical fiber, a fiber optic bundle, a fiber optic cable, or a slab dielectric waveguide, or (b) accepts optical signals and converts them into electrical signals for transmission in an electrical medium, such as a wire or a coaxial cable. *Note:* An example of a fiber optic interface device is a duplex asynchronous data transmission unit capable of handling a data signaling rate (DSR) of 0 to 20 kbps (kilobits per second) using a fiber optic cable for transmission and maintaining a bit error ratio (BER) of less than 10^{-9}. *Synonym* **fiber optic modem.**

fiber optic isolator: In a communications system, a fiber optic link that (a) consists of a fiber optic line driver, a fiber optic cable, and a photodetector, (b) provides electrical isolation between two or more parts of the system, (c) provides nonmetallic entry and exit from secure areas, (d) serves as a fiber optic data link as well as an electrical insulator, and (e) performs impedance transformation. *See also* **data link, optical isolator.**

fiber optic jumper: A jacketed short optical fiber that (a) has a fiber optic connector on both ends, (b) is used for effecting a fiber optic cross connection, and (c) usually is mounted, with others, on panels for easy access.

fiber optic junction: 1. A junction of fiber optic cables such that the cables that arrive at the junction are bound into a single sheath or jacket. **2.** A junction of fiber optic cables at which the individual fibers in the cables are coupled in such a manner that the signals in them are mixed in a fiber optic mixer. *See also* **fiber optic mixer, optical junction.**

fiber optic light source: A device that (a) emits lightwaves, i.e., radiation in the optical spectrum, i.e., in or near the visible region of the electromagnetic spectrum, (b) provides optical power to fiber optic systems, (c) may be modulated by an electronic input signal, (d) may have an integral optical fiber as a pigtail and one or more wires for input electrical power and modulation signals, and (e) may serve as a component in fiber optic transmitters for command, communications, control, and telemetry systems. *Note:* Examples of fiber optic light sources are light-emitting diodes (LEDs), lasers, and lamps.

fiber optic link (FOL): 1. A communications link that (a) transmits signals by means of modulated lightwaves that propagate in one or more optical fibers, (b) connects two end terminals or connects other links in series, i.e., in tandem, and (c) usually consists of a (i) fiber optic transmitter containing a modulated light source, (ii) a fiber optic cable, (iii) a fiber optic receiver containing a photodetector, (iv) associated fiber optic connectors, splices, penetrators, and cableways, and (v) perhaps one or more repeaters, amplifiers, transceivers, modulators, time division multiplexers (TDM), time division demultiplexers (TDDM), clocks, and data recovery circuits. **2.** An optical transmission channel designed to connect two end terminals or to be connected in series with other channels. *Note:* Terminal hardware, such as transmitter-receiver modules, may be part of a fiber optic link. **3.** A transmission channel that (a) transmits optical signals, usually in optical fibers, (b) may include regenerative repeaters, (c) connects two electronic or optoelectronic communications terminals, and (d) sometimes includes the terminal fiber optic transmitter and receiver, especially in the case of an all-electronic data link that has been retrofitted with fiber optic equipment. *Synonyms* **fiber optic cable link, optical fiber link, optical link.** *See* **repeatered fiber optic link, repeaterless fiber optic link, television fiber optic link.** *See also* **data link, fiber optic cable, fiber optic repeater, fiber optics, link, optical repeater, optical transmitter.** *Refer to* **Fig. F-3.** *Refer also to* **Fig. T-6 (1013).**

fiber optic longitudinal compression sensor: A fiber optic sensor in which (a) a force applied to a short length of optical fiber causes the fiber to shorten, and (b) by using interferometric techniques, the variation in monochromatic light irradiance at a photodetector can be made a function of the applied compressive force. *See also* **fiber optic transverse compression sensor.**

fiber optic loop: A loop that consists of a fiber optic link. *See also* **fiber optic link, loop.**

fiber optic loop multiplexer: A fiber optic multiplexer for providing multichannel (multiplexed) capability for loops to user end instruments in offices and homes. *See also* **loop.**

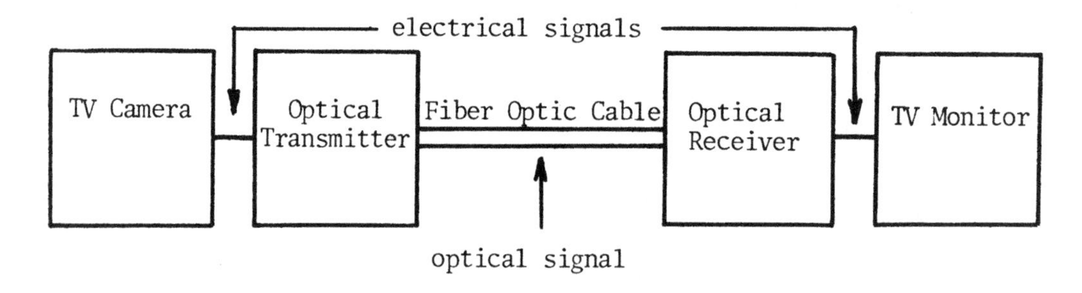

Fig. F-3. Closed-circuit television using a **fiber optic link.**

fiber optic mixer: A device that (a) accepts wavelength division multiplexed signals from two or more input optical fibers, (b) mixes the signals, and (c) transmits the mixed signals containing all the input wavelengths, i.e., transmits the resulting dichromatic or polychromatic light. *Note:* An example of a fiber optic mixer is a rod of transparent material with optical fiber input ports and internal mirrors for multiple reflection everywhere except at the output ports, at which optical fibers are attached. *See also* **optical mixing box.**

fiber optic modem: A modem that (a) consists of a fiber optic transmitter and receiver, (b) has appropriate arrangements for rotary dial and pushbutton signaling, and (c) is compatible with existing electronic equipment. *Synonym* **fiber optic interface device.** *See also* **modem.**

fiber optic mode stripper: Material applied to the cladding of an optical fiber, a slab dielectric waveguide, or an optical integrated circuit (OIC) that (a) allows optical energy propagating in the cladding to leave the cladding, (b) supports only certain propagating modes, (c) usually does not support the modes propagating in the cladding, and (d) does not disturb the modes propagating in the core. *Note:* Optical power removed from a waveguide by a stripper usually is not used.

fiber optic modulator: A device that (a) has at least one input or output lead that is an optical fiber and (b) is capable of (i) accepting a non-intelligence-bearing input signal, i.e., a carrier, (ii) accepting an intelligence-bearing input signal, and (c) mixing them in such a manner that the output signal consists of a parameter of the carrier varied in accordance with the intelligence-bearing input signal.

fiber optic multimode dispersion: Multimode dispersion that occurs in an optical fiber or fiber optic bundle. *See also* **dispersion, mode, multimode dispersion, optical multimode dispersion.**

fiber optic multiplexer (active): An optoelectronic device that (a) accepts information in the form of modulated lightwaves, i.e., optical signal streams, in optical fibers on two or more channels, (b) places all of the input information on a single output channel in such a way that the information on the input channels can be recovered at the end of the single output channel, (c) may consist of two or more input optical fibers, each operating at a different wavelength, and one output fiber containing all the wavelengths, (d) has a multiplexing component that may be constructed with optical fibers, such as fiber optic couplers, and (e) requires electrical energy or other forms of energy to operate, in addition to that contained in the input lightwaves. *Synonym* **optical fiber multiplexer (active).** *See also* **fiber optic demultiplexer (active), optical demultiplexer (active), optical multiplexer (active).**

fiber optic multiplexer (passive): A purely passive optical device that (a) accepts information in the form of modulated lightwaves in optical fibers, such as streams of optical pulses, on two or more channels, (b) places all of the input information on a single output channel in such a way that the information on the input channels can be recovered at the end of the single output channel, (c) may consist of two or more input optical fibers, each operating at a different wavelength, and one output fiber containing all the wavelengths, (d) may be constructed with optical components, such as optical fibers, fiber optic couplers, optical mixing rods, mixing boxes, or other passive mixing devices, and (e) does not require energy to operate other than that contained in the input lightwaves. *Synonym* **optical fiber multiplexer (passive).** *See also* **optical demultiplexer (passive), optical multiplexer (passive).**

fiber optic multiport coupler: An optical device that (a) has two or more output or input ports, (b) can be used to couple various sources to various receivers, (c) has optical fibers attached to the ports, (d) if there is only one input port and one output port, is a fiber optic connector or splice, (e) usually consists of a piece of transparent material, (f) depends on reflection, transmission, scattering, diffusion, or a combination of these, to operate, and (g) may be passive or active.

fiber optic myocardium stimulator: A device that (a) consists of a heart pacemaker, control circuits, a light source, optical fibers, and photodetectors and (b) delivers impulses directly into the heart muscles, i.e., the myocardium, to initiate periodic contraction at appropriate instants in strategic places in the heart muscles to overcome blocking of normal neuronal conduction.

fiber optic net: A communications network in which connecting cables contain optical fibers rather than copper wires. *Note:* An all-fiber net has optical fibers in trunk lines between switching centers and central offices (COs) as well as in local loops and local area networks (LANs).

fiber optic network: *See* **synchronous fiber optic network.**

fiber optic nuclear hardening: Reducing the degradation of communications, computer, data processing, and control systems in a nuclear environment produced by a nuclear detonation, i.e., raising the probability of survival in an environment that includes nuclear radiation and electromagnetic impulses (EMI). *Note 1:* The hardening may be in terms of either susceptibility or vulnerability to a nuclear environment. *Note 2:* The extent of performance degradation may be measured in terms of outage time, data lost, and equipment damage. The environment is defined in terms of radiation levels, overpressure, peak velocities, radiation energy absorbed, and electrical stress. *See also* **nuclear hardness, optical fiber nuclear hardening.**

fiber optic patch panel: A panel (a) on which fiber optic connectors and couplers are mounted in an organized array for easy access, (b) in which both sides usually are accessible with sufficient spacing between components to facilitate handling individual components, (c) in which individual connectors are easily identifiable, and (d) for which there usually is an individual connectorization and organization plan for the panel and for the interconnection box in which it is installed.

fiber optic pattern: In the fiber faceplate of a fiberscope, the arrangement of the ends of the individual optical fibers that comprise the faceplate. *Note:* Inspection of an image on the faceplate might reveal that the fiber pattern makes the image appear as a mosaic of individual picture elements. *Synonym* **optical fiber pattern.**

fiber optic penetrator: A device that allows an optical fiber or fiber optic cable to pass from one compartment to another, such as through a wall, partition, bulkhead, hull, fuselage, or cabinet.

fiber optic phase modulator: *See* **optical phase modulator.**

fiber optic photodetector: A device that (a) converts input optical signals into equivalent electronic output signals and (b) usually consists of (i) an optical fiber signal input pigtail, (ii) a light-sensitive element, such as a photodiode, (iii) an electrical signal output wire, and (iv) an electrical power input wire.

fiber optic polarizer: *See* **optical fiber polarizer.**

fiber optic preform: *See* **optical fiber preform.**

fiber optic probe: A flexible probe that consists of an aligned bundle of optical fibers for transmitting images.

fiber optic receiver: In fiber optic communications systems, a device that (a) detects an optical signal received via an optical fiber, (b) converts the signal to an electrical signal, and (c) processes the electrical signal for subsequent use. *See also* **optical receiver, photodetector.**

fiber optic reflective sensor: A sensor that (a) illuminates an object with a light source, (b) senses reflections from the object, (c) uses optical fibers to conduct the light both ways, (d) from a photodetector, has an electrical final output signal that (i) is proportional to the intensity or the color of the reflected light, (ii) may be sequential from each fiber, and (iii) may be the result of scanning an image of the object focused on a screen. *Note:* An example of a fiber optic reflective sensor is a hand-held optical scanner used to scan bar codes on merchandise.

fiber optic repeater: 1. An optical repeater in a fiber optic system. **2.** A repeater in which a fiber optic photodetector, a fiber optic transmitter, and associated electronic components, such as amplifiers and signal processors, are used. **3.** A repeater that consists only of an optical fiber amplifier and necessary connectors or splices. *See also* **optical repeater.**

fiber optic repeater power: 1. In a fiber optic communications system, the optical output power provided by a fiber optic repeater. **2.** The electrical or optical power delivered to a fiber optic repeater by means such as (a) a solar cell, (b) a local battery, (c) a commercial power outlet, and (d) a cable, such as (i) an electrical cable, (ii) a fiber optic cable, (iii) both electrical and fiber optic cables, or (iv) a hybrid cable. *Note:* If an optical signal channel cable is used to provide fiber optic repeater power, the optical power must be extracted without disturbing the information content of the signals. *See also* **fiber optic repeater, optical repeater, optical repeater power.**

fiber optic ribbon: A fiber optic cable consisting of optical fibers arrayed side by side and held in place by one or more materials, such as tapes, adhesives, or plastic strips. *Note:* An example of a fiber optic ribbon is a cable consisting of optical fibers laminated within a flat plastic strip. *Synonym* **fiber ribbon, optical fiber ribbon.** *See also* **fiber optic cable.**

fiber optic ringer: A signaling system for soliciting attention in which (a) a photodetector and a transmitter convert enough lightwave energy from information-bearing modulated carrier signals in optical fibers into an audio tone for sound signaling, i.e., for ringing, and (b) the ringer power may also be used for other applications, such as providing power for electronic circuits.

fiber optic rod coupler: A graded index, cylindrically shaped section of optical fiber or glass rod (a) that has a length that corresponds to the pitch of the undulation of lightwaves caused by the graded refractive index, (b) in which light beams are injected via optical fibers at an off-axis end point on the radius, (c) in which the undulations of the resulting wave vary periodically from one point to another along the fiber or the rod, and (d) in which half-reflection layers at the 1/4 pitch point of the undulations provide for coupling between the input and output fibers. *See also* **GRIN rod, half-reflection.**

fiber optic rod multiplexer-filter: A graded index, cylindrically shaped section of optical fiber or glass rod (a) that has a length that corresponds to the pitch of the undulation of lightwaves caused by the graded refractive index, (b) in which light beams are injected via optical fibers at an off-axis end point on the radius, (c) in which the undulations of the resulting wave vary periodically from one point to another along the fiber or the rod, and (d) in which interference layers at the 1/4 pitch point of the undulations provide for multiplexing or filtering as desired. *See also* **GRIN rod, interference.**

fiber optic rotary joint: A passive optical device made of optical fibers and connectors that allows signals from one or more fixed platforms to be coupled to one or more platforms that are rotating relative to the fixed platforms. *See also* **fiber optic rotor.**

fiber optic rotor: A fiber optic device that allows optical signals propagating in one optical element to pass across the axial interface into another element while they are axially rotating at different speeds, such as between an optical fiber rotating about its optical axis and a stationary fiber, which might occur when a fiber optic system fixed to a platform that is rotating with respect to the ground or the frame is connected to a system that is fixed to the ground or the frame. *Synonym* **optical rotor.** *See also* **fiber optic joint, fiber optic rotary joint, optical scanner.**

fiber optics (FO): The branch of science and technology devoted to (a) the transmission of radiant power through fibers made of transparent materials, such as glass, fused silica, and plastic, (b) the combining of the features of optical fibers with other dielectric waveguides, (c) the use of optical fibers in conjunction with other components, such as electronic and acoustic components, and (d) the development, fabrication, and application of optical fibers in various areas, such as communications, control, computers, data processing, sensing, telemetry, endoscopy, illumination, and display, using various configurations and modes of excitation and under various environmental conditions. *Note 1:* The optical fibers may be wound and bound in various shapes and distributions singly or in bundles, which may be aligned or unaligned bundles. The aligned bundles are used to transmit and display images as well as to scramble images. *Note 2:* In telecommunications applications, flexible low-loss optical fibers are used as a signal propagation medium. A single fiber might constitute a single channel, which could be time division or wavelength division multiplexed. Flexible and rigid fiber optic bundles are used in various industrial and medical applications. For image transmission, the individual fibers are spatially aligned at each end of a cable, such as in endoscopes and in faceplates on high speed oscilloscopes. Unaligned bundles are used for transmission of optical power for sensors, fiber amplifiers, signaling, and illumination. *Note:* In 1956, N. S. Kapany first defined fiber optics as the art of active and passive guidance of light as rays and waveguide modes in the ultraviolet, visible, and infrared spectra in transparent fibers along predetermined paths. *See* **woven fiber optics, ultraviolet fiber optics.** *See also* **fiber optic.**

fiber optic scrambler: A device that is similar to a fiberscope, except that the center section of the aligned bundle has its optical fibers randomly distributed, and when the randomized fibers are potted and sawed, each half is capable of coding a picture that can only be readily decoded by the other half.

fiber optic section: *See* **station/regenerator fiber optic section.**

fiber optic sensor: A device that (a) has an optical fiber sensor that responds to a physical, chemical, or radiant field input stimulus, such as heat, light, sound, pressure, temperature, strain, magnetic field, or electric field, and (b) transmits on an optical fiber a pulse, signal, or other representation indicating one or more characteristics of the stimulus. *Note:* Fiber optic sensors are used as thermometers, tachometers, hydrophones, microphones, and barometers that have an output lead consisting of an optical fiber. *See* **Fabry-Perot fiber optic sensor, Mach-Zehnder fiber optic sensor, Michelson fiber optic sensor, Sagnac fiber optic sensor.** *See also* **optical fiber sensor.**

fiber optic source: An electronically powered light source that (a) inserts radiant power into an optical fiber, usually via a pigtail for splicing or coupling to the fiber, (b) may be directly modulated by electronic signals before the light enters the fiber, (c) may have a constant output of radiant power that may be modulated after the light is coupled into the fiber, and (d) has a typical power output on the order of several milliwatts at optical wavelengths. *Note:* Examples of optical fiber sources are light-emitting diodes (LEDs) and lasers. *Synonym* **optical fiber source.**

fiber optic splice: An inseparable, permanent joint that (a) is made between two fiber optic cables or two optical fibers, (b) is mounted within a protective housing, such as a tube or sheath, (c) provides for minimal optical power loss at the junction, and (d) is used to couple optical signal power between the cables or the fibers. *Synonym* **optical splice.** *See also* **fusion splice, joint, mechanical splice, optical fiber splice.** *Refer to* **Figs. L-7 (535), P-11 (749).**

fiber optic splice housing: A housing, such as tapes, jackets, coatings, and other components necessary for attaching, supporting, and aligning optical fibers, applied over a fiber optic splice and proximate fiber for (a) their protection against the environment, (b) mechanical strength, (c) preservation of the integrity of the fiber coating, buffer, cladding, and core, and (d) mounting the aligned assembly in an interconnection box or a cable splice closure.

fiber optic splice tray: A flat, usually rectangular, pan-shaped device that (a) holds and protects the individual optical fibers from two fiber optic cables that have been spliced, and (b) holds the fibers even though they may not be sheathed, jacketed, or otherwise covered as a group, although the fiber optic splices may be individually covered.

fiber optic splitter: A device that (a) extracts a portion of the optical signal power propagating in an optical fiber, (b) consists of an optical tap, fiber optic coupler, or partial mirror, and (c) diverts the optical power to a destination different from that of the original optical fiber. *See also* **optical splitter.**

fiber optic spot illumination: Illumination of a relatively small area with a narrow beam of light emanating from the end of an optical fiber or fiber optic bundle. *See* **fiber optic bundle, fiber optic flood illumination.**

fiber optic station cable: *Synonym* **station fiber optic cable.**

fiber optic station facility: *Synonym* **station fiber optic facility.**

fiber optic station/regenerator section: A fiber optic transmission system consisting of (a) a station fiber optic facility, i.e., a fiber optic transmitter, connectors, station cables, and a fiber optic cable interconnect feature, (b) a cable facility, i.e., cable sections joined by splices between station facilities, and (c) another station facility, i.e., a fiber optic cable interconnect feature, station cables, connectors, and a receiver. *Synonym* **optical station/regenerator section.**

fiber optic strain-induced sensor: A fiber optic sensor in which (a) various methods, such as optical time domain reflectometry (OTDR), interferometry, birefringence, polarization and polarization plane rotation, wave coherency, special coatings, and microbending, are used, (b) a transverse, longitudinal, or torsional stress that produces a strain in an optical fiber is applied, and (c) a change in irradiance, i.e., light intensity, at a photodetector is produced that is a function of the applied stress that produced the strain.

fiber optic supporting structure: A structure that (a) is used to fasten, guide, and hold fiber optic components in place in operational systems and (b) without which a fiber optic system cannot be installed, held in place, and operated. *Note:* Examples of fiber optic supporting structures are splice trays, patch bays, bend limiters, raceways, harnesses, hangars, ducts, and housings.

fiber optic switch: A switch that (a) can selectively transfer optical signals from one optical fiber to another depending upon the application of an external stimulus, such as a force or an applied electric, magnetic, electromagnetic, acoustic, gravitational, or other force field, and (b) may be latching or nonlatching.

fiber optic system: *See* **integrated fiber optic system.**

fiber optic taper: *Synonym* **optical taper.**

fiber optic telecommunications cable: A long-haul fiber optic cable usually with groups of individual fibers or with fiber optic bundles, or both, strength members, stuffing, electrical conductors, insulation, jacketing, and protective sheathing for maximum protection, security, and reliability in rugged and wet overhead, aerial insert, direct buried, underground, conduit, and tethered installations.

fiber optic telemetry system: A telemetry system in which fiber optic components, such as fiber optic sensors, data links, light sources, photodetectors, and related components, are used to gather, transmit, and distribute data originating from one or more sensors.

fiber optic tension sensor: A fiber optic sensor in which (a) an increase in tension applied to an optical fiber results in an elongation of the fiber, (b) a reference fiber not subject to the increase in tension is used as a reference for interferometric comparison using coherent light, i.e., a phase shift is measured, and (c) the phase shift is used to measure the force producing the increase in tension. *Note:* If the force is applied directly, it can be measured directly. If the optical fiber is wound on a bobbin and pressure is applied to the inside of the bobbin to cause it to expand, the fiber will be subjected to tension and thereby elongated. The magnitude of the pressure and the variation in pressure as a function of time can be measured.

fiber optic terminus: The part of a fiber optic connector into which the optical fiber is inserted and attached by mechanical means or by adhesives. *See also* **terminus.**

fiber optic test method (FOTM): A general description of the overall approach to the testing of fiber optic systems and components, identifying (a) the system or the component to be tested, (b) the type and the nature of the test, (c) the parameters to be measured, and (d) the environmental conditions under which the test is to be conducted. *Note:* The fiber optic test method (FOTM) does not explicitly describe in every detail the exact steps to be taken during the conduct of the test. Engineering analysis is required, and one or more fiber optic test procedures (FOTPs) must be prepared before the FOTM can be executed. *See also* **fiber optic test procedure.**

fiber optic test procedure (FOTP): A detailed, highly explicit, unambiguous, stand-alone, step-by-step sequence of actions, or the document describing them, for testing fiber optic systems and components, each step precisely described, with all dimensions, instrument settings, apparatus, structures, parameters to be measured, environmental conditions, exact action to be taken at each step, and data to be taken clearly described for each part or run of the test. *Note:* A fiber optic test procedure (FOTP) can be executed without additional engineering analysis. FOTPs are usually prepared from or based on previously prepared fiber optic test methods (FOTMs). *See also* **fiber optic test method.**

fiber optic transceiver: A fiber optic receiver-transmitter, that consists of (a) a photodetector that receives optical signals and converts them to electrical signals, (b) a light source, such as a laser, that converts electrical signals to optical signals, and (b) is housed in a single unit with appropriate electrical and optical connectors. *Synonym* **optical transceiver.** *See also* **home fiber optic transceiver.**

fiber optic transimpedance: In a fiber optic transmission system, the ratio of (a) the output voltage at the photodetector to (b) the input current at the light source. *See also* **optical transimpedance.**

fiber optic transmission system (FOTS): A transmission system in which fiber optic links are used to transmit messages by means of modulated lightwaves. *See also* **fiber optic link.**

fiber optic transmitter: A device that (a) accepts an electrical signal as its input, (b) processes this signal, (c) uses the result of the processing to modulate an optoelectronic source, such as a light-emitting diode (LED) or an injection laser diode, to produce an optical signal, and (d) couples the optical signal into a fiber optic cable for transmission via the fiber optic cable. *See also* **fiber optics, optical link, optical transmitter.**

fiber optic transverse compression sensor: A fiber optic sensor in which (a) the optical attenuation is a function of the longitudinally applied pressure, i.e., the pressure is uniformly applied radially against the outside of an optical fiber, and (b) a length of cabled fiber, a not necessarily coherent or monochromatic light source, and a photodetector usually are required. *See* **fiber optic longitudinal compression sensor.**

fiber optic video trunk: A video signal trunk in which (a) optical fibers, fiber optic bundles, and fiber optic cables are used for the transmission of video signals, (b) color multiplexing is used over the full visible spectrum from infrared to ultraviolet inclusive, and (c) dispersion is used to separate the colors prior to detection and further processing in the video receiver. *Synonym* **optical fiber video trunk.** *See also* **dispersion, fiber optic bundle, multiplexing, trunk, video signal, visible spectrum.**

fiber optic visible/infrared spin-scan radiometer: A device that (a) measures and records the intensity of visible and infrared electromagnetic radiation from many directions sequentially, (b) uses optical fibers to conduct the received radiation from the scanned directions, (c) may be fixed or tilting, and (d) may be used in a satellite to create a terrestrial map based on infrared radiation received from the Earth.

fiber optic waveguide: A waveguide that (a) guides lightwaves and (b) consists of one or more optical fibers. *See also* **optical waveguide.**

fiberoptronics: The branch of science and technology devoted to exploiting the advantages obtained by combining optical fibers, other active and passive optical devices, and electronic circuits into useful systems. *Note 1:* An example of the application of fiberoptronics is electronically modulating a light source in order to impress information on its output light beam, guiding the beam through a fiber optic cable to a photodetector, and recovering the impressed information in electronic form. *Note 2:* "Fiberoptronics" is an abbreviation for fiber optics, optoelectronics, and electro-optics.

fiber organizer: *Synonym* **fiber and splice organizer.**

fiber parameter: *See* **global optical fiber parameter.**

fiber pattern: *See* **optical fiber pattern.**

fiber pigtail: *See* **optical fiber pigtail.**

fiber polarizer: *See* **optical fiber polarizer.**

fiber preform: *See* **optical fiber preform.**

fiber pulling machine: *See* **optical fiber pulling machine.**

fiber pulse compression: *See* **optical pulse compression.**

fiber radiation damage: *See* **optical fiber radiation damage.**

fiber retention: *See* **optical fiber retention.**

fiber ribbon: *See* **fiber optic ribbon.**

fiber ringer: *See* **optical fiber ringer.**

fiber scattering: *See* **configuration scattering.**

fiberscope: A device that (a) consists of (i) an aligned bundle of optical fibers with terminating fiber optic faceplates, (ii) an objective lens on one end for focusing the image of an object, and (iii) an eyepiece for viewing the transmitted image in full color on the fiber faceplate at the opposite end, and (b) is used for viewing objects that are otherwise inaccessible for direct viewing, such as the interior of machines and the internal organs of a body. *See* **hypodermic fiberscope.** *See also* **fiber optic endoscope.**

fiberscope recording: *Synonym* **fiberscopic recording.**

fiberscopic recording: Recording in which light emerging from the fiber optic faceplate of an aligned bundle of optical fibers, i.e., a fiberscope, is used to expose film. *Synonym* **fiberscope recording.**

fiber sensor: *See* **coated optical fiber sensor, distributed optical fiber sensor, flat coil optical fiber sensor.**

fiber source: *See* **optical fiber source.**

fiber splice: *See* **fiber optic splice, optical fiber splice.**

fiber splice housing: *See* **fiber optic splice housing.**

fiber strain: *See* **optical fiber strain.**

fiber strain-induced sensor: *See* **fiber optic strain-induced sensor.**

fiber tensile strength: *See* **optical fiber tensile strength.**

fiber tension sensor: *See* **fiber optic tension sensor.**

fiber transfer function: *See* **optical fiber transfer function.**

fiber transverse compression sensor: *See* **fiber optic transverse compression sensor.**

fiber trap: *See* **optical fiber trap.**

fiber video trunk: *See* **fiber optic video trunk.**

Fibonacci series: A series of integers in which the value of each integer is equal to the sum of the two preceding integers, starting with 0 and 1, given by the relation $x_i = x_{i-2} + x_{i-1}$ where x_i is the ith integer, $x_0 = 0$, and $x_1 = 1$. *Note 1:* The Fibonacci series starts with 0, 1, 1, 2, 3, 5, 8, 13, 21, and 34. *Note 2:* The Fibonacci series occurs in certain types of biological reproduction and statistical distributions.

fiche: *See* **microfiche.**

fictive temperature: The temperature at which glass will change from one state or condition to another, such as from a soft state to a hard state. *Note:* The fictive temperature is of critical concern in the pulling of optical fibers. *See also* **optical fiber pulling machine.**

fidelity: In a communications system, the degree to which a system, or a portion of a system, accurately reproduces, at its output, the essential characteristics of the signal impressed upon its input. *See also* **linearity, precision.**

field: **1.** A physical phenomenon that (a) is a source of force, (b) exists in a spatial domain, (c) usually is influenced by material media, (d) is a point function, (e) is a vector quantity, (f) is measurable, and (g) is subject to temporal changes. *Note 1:* Examples of fields are (a) the electric field of force emanating from an electric charge and (b) the electric and magnetic fields in a propagating electromagnetic wave. *Note 1:* Fields are usually expressed as vector quantities, i.e., they usually have both magnitude and direction. *Note 2:* In electromagnetic theory, force fields and their gradients are manifested by stationary or moving electric charges or magnetic poles because these charges or magnetic poles will have forces exerted on them if they are placed in or move in electric or magnetic fields. *Note 3:* The three main fields are the electric, magnetic, and gravitational fields. *Note 4:* In electromagnetic waves, such as radio waves and lightwaves, the electric and magnetic fields interact, and the waves can be influenced by a gravitational field. **2.** On a data medium or in storage, a specified area used for a particular class of data, such as a group of character positions used to enter or display wage rates on a screen or a specified part of a frame. **3.** Defined logical data that are part of a record. **4.** The elementary unit of a record that may contain a data item, a data aggregate, a pointer, or a link. **5.** A dimensional space, such as a defined line, surface, or volume. *See* **address field, aiming field, borescope view field, control field, display field, electric field, electromagnetic field, electrostatic field, evanescent field, far field, gravitational field, gravitational force field, information field, intermediate field, magnetic field, magnetostatic field, near field, protected field, reference surface tolerance field, scanning field, signaling information field, tolerance field, true field, unprotected field, view field.**

field attribute: *See* **display field attribute.**

field component: *See* **electric field component, magnetic field component.**

field coupling: *See* **evanescent field coupling.**

field curvature: In optics, an aberration of actions that causes a plane image to be focused onto a curved surface instead of a plane surface, i.e., a flat surface.

field diameter. *See* **mode field diameter.**

field diffraction pattern: *See* **far field diffraction pattern, near field diffraction pattern.**

field directory: 1. A directory specifically designed for use by communications system maintenance personnel outside central offices (COs). **2.** A directory specially designed for use by military personnel during a tactical situation. *See also* **directory.**

field disturbance sensor: A restricted radiation device that establishes a radio frequency (rf) field in its vicinity and detects changes in that field resulting from the movement of persons or objects within the rf field. *Note:* Examples of field disturbance sensors include microwave intrusion sensors and devices that use radio frequency (rf) energy for production line counting and sensing. *See* **restricted radiation device.**

field-effect transistor integrated receiver: *See* **positive-intrinsic-negative field-effect transistor integrated receiver.**

field-effect transistor (FET) photodetector: A photodetector that uses photogeneration of carriers in the channel region of a field-effect transistor structure to provide photodetection with electric current gain. *See also* **photocurrent, photodiode.**

field extension: *See* **address field extension, control field extension.**

field force: *See* **electric field force, gravitational field force, magnetic field force.**

field-installable connector: A connector that (a) simply and easily can be installed and operated in an outdoor environment without the use of special tools, without intensive training, and without the use of irreversible processes, such as molding, welding, fusing, brazing, soldering, and potting, (b) has halves, male and female, that are capable of being disconnected and reconnected under similar conditions, and (c) is considered to be a splice if irreversible processes are required to make the connection.

field intensity: The irradiance of an electromagnetic wave under specified conditions. *Note:* Field intensity usually is expressed in watts per square meter. *See also* **field strength, radio beam.**

field lens: In an optical system or instrument, a positive lens that (a) is used to collect the chief rays of image-forming ray bundles so that entire bundles, or sufficient portions of them, will pass through the exit pupil of the instrument, (b) usually is located at or near the focal point of the objective lens, and (c) increases the size of the field that can be viewed with any given eye lens diameter. *See also* **chief ray, exit pupil, focal point, objective, optical system, positive lens, ray bundle.**

field-programmable gate array: *Synonym* **programmable logic.**

field radiation pattern: *See* **near field radiation pattern.**

field ray: In a symmetrical optical system, a ray that (a) in the object space, intersects the optical axis at the center of the entrance pupil of the optical system, (b) in the image space, emerges, as the same ray, from the exit pupil, and (c) in a thick lens, is a principal ray. *See also* **entrance pupil, exit pupil, object, optical axis, optical system, principal ray, thick lens.**

field-replaceable unit: A component that (a) may be replaced by a user, operator, or repair service person in its entirety at its place of use when it fails, or any of its parts fails, and (b) can be replaced without the use of special tools, without a high level of training and experience, without intensive or detailed disassembly and reassembly, and without irreversible processes, such as molding, welding, fusing, brazing, soldering, and potting. *Note:* Examples of field-replaceable units are field-installable connectors, fiber optic transmitters, and fiber optic receivers.

field scanning technique: *See* **near field scanning technique.**

field splicing: The joining of the ends of two fiber optic cables in a field environment, such as on open terrain, on board ships, and under water, without the use of detachable connectors and in such a manner that continuity through the splice is preserved in all elements of the cable, such as optical continuity of optical fibers, electrical continuity, and continuity of jacketing, sheathing, buffers, stuffing, insulation, strength members, and waterproofing. *See also* **fiber optic cable, fiber optic splice.**

field stop: In an optical system, a limit or a boundary placed on an image at some point in the system in order to limit the size or the extent of the image and thus eliminate unwanted diffused, deflected, or divergent light. *See also* **aperture stop, entrance pupil, exit pupil, optical system.**

field strength: The magnitude of an electric, magnetic, electromagnetic, gravitational, or other force field at a given point. *Note:* In referring to electromagnetic waves, the field strength usually is expressed as the root mean square (rms) value of the electric field, expressed in volts per meter, or of the magnetic field, expressed in ampere•turns per meter. Field strength usually is given as a gradient, and therefore is a vector quantity, because it has both magnitude and direction. The field strength may also be given as the irradiance, i.e., the power per unit area incident upon a surface, such as watts per square meter. The radiance is the power per unit area per unit solid angle emitted by a source, such as watts per square meter per steradian. In referring to staticmagnetic and electrostatic fields, "field strength" is the proper term rather than "field intensity." *Synonym* **radio field intensity.** *See* **electric field strength, magnetic field strength.** *See also* **field intensity, radio beam.**

field switchboard: A telephone switchboard that is specially designed and packaged for use in open country, either on or off vehicles.

field template: *See* **four-concentric-circle near field template.**

field vector: *See* **electric field vector, magnetic field vector.**

field wire: A flexible insulated wire used in field telephone and telegraph systems. *Note 1:* WD-1 and WF-16 are types of field wire. *Note 2:* Field wire usually contains a strength member.

FIFO: first in, first out.

fifteen ones convention: In synchronous data link control (SDLC) and high level data link control (HDLC) transmission systems, a convention in which a series of fifteen consecutive 1s are transmitted to indicate that the station transmitting them is in the idle state.

figure: *See* **optical fiber merit figure.**

figure case: 1. In the teletypewriter code for tape relay systems, the machine function that (a) produces or interprets a hole pattern on tape or a pulse pattern on wires, optical fibers, or cables, as one of the Arabic numerals 0 through 9, a punctuation mark, or a special sign or symbol, and (b) is obtained by actuating a special function key. **2.** One of the groups (a) into which the characters and the machine functions of a code are placed, and (b) that contains the Arabic numerals 0 through 9. *See also* **function key, letter case.**

figure indicator: In visual signaling systems, a panel signaling indicator that consists of a single panel appropriately placed to mean that the characters that follow are to be interpreted as numerals rather than as letters. *See also* **visual signal.**

figure shift: In teletypewriter operations, a case shift that results in the translation of signals into another group of characters that is predominantly numerals and machine functions, rather than the letters group that was being used or interpreted before the figure shift occurred. *See also* **case shift, escape character, letter shift, machine function.**

figure shift signal: The signal that conditions a telegraph receiver to translate all received signals as the figures case, which is primarily numerals and machine functions, rather than as the letters case that was in use before the figure shift signal occurred.

filament: *See* **monofilament.**

file: 1. In a database of records, the largest unit of storage structure that consists of a named collection of all occurrences of a particular record type. **2.** A set of related records treated as a unit. *Note:* An example of a file is an invoice file that consists of a complete set of invoices on all received shipments. *Synonym* **logical file.** *See* **active file, audit review file, backup file, batch file, chain search file, display file, frequently asked question file, inactive file, inverted file, master file, system environmental file, tree file, work file.**

file activity ratio: The ratio of (a) the number of records in use in a file to (b) the total number of records in the file.

file application, transfer, access, and management (FATAM): An application protocol that (a) is based on the concept of virtual file storage, (b) allows remote access to various levels in a file structure, and (c) provides a comprehensive set of file management capabilities.

file command: In personal computer (PC) systems, a command, i.e., an instruction, that (a) an operator may give to a computer to perform an operation on an entire stored file and (b) usually may be selected from a menu displayed on a monitor when using a specific software package, such as Lotus 1-2-3, Volkswriter, dBase, and WordPerfect. *Note 1:* Examples of file commands are the save, retrieve, find, erase, and list commands. *Note 2:* File commands relate to an entire file, whereas editing commands pertain to the contents or the entries used to make changes, insertions, deletions, and corrections to the data in a particular file.

file disk: *Synonym* **data disk.**

file gap: In a data medium, an unrecorded area that (a) is between recorded files and (b) may be used to indicate (i) the end of a file, (ii) the beginning of a file, or (iii) the beginning or the end of a group of files, and (d) for dynamic media, permits time for specific operations or switching transients to occur. *See also* **dynamic medium.**

file layout: The organization and the arrangement of data in a file, including (a) format, structure, and sequencing schemes, (b) word, record, and file size, and (c) instructions for access, sensitivity, and use.

file maintenance: The action of keeping a file in a desired state, such as up-to-date, correct, and structured according to the specified file layout, usually by adding, deleting, changing, or rearranging the data in it. *See also* **file layout.**

file name: The name or the title that (a) is given to a file so that it can be found and retrieved, (b) is unique to the database in which it is stored, (c) is constructed according to the prescribed rules for the particular database, such as the number of characters are limited, characters must be contiguous, i.e., without spaces, and certain characters may not be allowed, such as @, #, $, and %. *Note:* Many personal computer (PC) programs allow up to eight characters, followed by a decimal point, followed by three more characters. Examples of file names for some PC software may be entered as Martin, Testdata, my_son, and Test0005.gec (For Test Number 5 of General Electric components). The file names usually will be recorded and displayed in all uppercase letters, though they need not be entered that way. Thus, lowercase and uppercase cannot be used to distinguish file names. Examples of file names that may not work are Mary Ann, six@two, and co-op. Computer software packages contain instructions on the rules for creating file names.

file protection: The prevention of unauthorized access, contamination, or elimination of a file.

file separator: A character that is used to mark logical boundaries between files in a database.

file server: 1. In a given computer, a process that allows programs running on other computers to access the files of the given computer. **2.** A computer that runs file control programs, such as network file server programs.

file system: *See* **Network File System.**

file time: *See* **message file time.**

file transfer, access, and management (FTAM): An application service and protocol that (a) is based on virtual file storage, (b) allows remote access to various levels in a file structure, and (c) provides a comprehensive set of file management capabilities.

File Transfer Protocol (FTP): The Transmission Control Protocol/Internet Protocol (TCP/IP) that (a) is a standard high level protocol for transferring files from one computer to another, (b) usually is implemented as an application level program, and (c) uses the TELNET and TCP protocols. *Note:* In conjunction with local software, the File Transfer Protocol (FTP) allows computers connected to the Internet to exchange files.

filing: In a communications system, the entering or the handing in of a message by a user, or a user representative, at a communications center for processing and transmission. *See also* **communications center, filing time, refile message.**

filing communications center: The communications center at which a message is first accepted from a message originator for transmission. *See also* **message originator.**

filing time: 1. In communications system operation, the date and the time a message is accepted by a communications center from a message originator for transmission. *Note 1:* The date-time group usually is placed on the message immediately upon acceptance at the message center. *Note 2:* For refile messages, the filing time is the data and the time the message is received by the communications center for refile. **2.** In telephone switchboard operation, the time at which a call is placed or booked. *Note:* For booked calls, the filing time is placed on the call ticket. *See* **call filing time.** *See also* **booked call, date-time group, filing.**

fill: In coupling optical power from (a) a source, such as a laser, a light-emitting diode (LED), or an optical fiber, to (b) a receiver, such as another fiber or a photodetector, the situation that occurs when the useful diameter of the receiver, such as the core of a fiber or the sensitive surface of a photodetector, is equal to the diameter, at the surface of the receiver, of the cone of light emanating from the source. *Note:* In coupling optical fibers, fill occurs when the spot of light incident on the endface of the fiber is equal to the core diameter, and the launch angle of the source equals the acceptance angle of the fiber. *See* **character fill, interframe time fill, overfill, underfill, zero fill.** *See also* **bit stuffing.**

fill character: 1. A character placed adjacent to an item of data to bring the item up to a specified size or length. *Note:* Examples of fill characters are characters that are (a) used to complete a field in storage or in transmission, (b) satisfy the character count of a fixed-length message or frame, and (c) fill a buffer so that it can be emptied. The fill character usually is a nondata character, such as a hyphen, so that it is not interpreted as meaningful data. **2.** A character used on a human-readable data medium, such as legal documents, a bank check, or a business form, to prevent any other characters from being added in the blank spaces of a field or group. *Note:* Examples of characters that have been used as fill characters are (a) dashes, dots, hyphens, and asterisks surrounding the amount figures on a check and (b) the letters NA on a business form to indicate the nonapplicability of the entry and prevent any subsequent entry that might be made if the field were left blank. *See also* **character, data medium, field.**

filled cable: A cable that has a nonhygroscopic material, usually a gel, inside the jacket or the sheath. *Note 1:* In filled cables a nonhygroscopic material fills the spaces between the interior parts of the cable, preventing moisture from entering minor leaks in the sheath and migrating inside the cable. *Note 2:* A metallic cable, such as a coaxial cable or a metal waveguide, filled with a dielectric material, is not considered to be a filled cable.

filled fiber: *See* **fully filled optical fiber.**

film: *See* **microfinishing film, optical thin film, photoconductive film.**

film frame: On a film, the area specifically dedicated to the recording of a display image. *Note:* Examples of film frames are (a) in a camera, the area on the film exposed each time the film is in place and the shutter is opened, and (b) the area exposed when microimages are placed on a film. *Synonym* **recording area.**

film optical modulator: *See* **thin film optical modulator.**

film optical multiplexer: *See* **thin film optical multiplexer.**

film optical switch: *See* **thin film optical switch.**

film optical waveguide: *See* **thin film optical waveguide.**

film waveguide: *See* **periodically distributed thin film waveguide.**

filter: 1. In electronic circuits and optical systems, a device that transmits only part of the incident energy and thereby has a spectral distribution of energy output that is different from the input. *Note:* Examples of filters include (a) high-pass filters that transmit energy above a certain frequency, (b) low-pass filters that transmit energy below a certain frequency, (c) bandpass filters that transmit energy of a certain bandwidth, and (d) band stop filters that transmit energy outside a certain frequency band. **2.** In optics, a device used to modify the spectral composition of light, such as to absorb certain wavelengths or transmit certain wavelengths. **3.** A functional unit that separates or removes data, signals, or material in accordance with specified criteria. *See also* **active filter, band pass filter, band stop filter, character filter, core mode filter, dichroic filter, fiber optic filter, fiber optic rod multiplexer filter, high-pass filter, interference filter, Kalman filter, low-pass filter, mode filter, notch filter, optical filter, passive filter, postdetection filter, preselection filter, roofing filter, security filter, tunable optical filter, video filter.**

filter balance: *See* **line filter balance.**

filter center: A communications center in which information from different data sources is checked and perhaps encrypted prior to further dissemination.

filter-coupler-switch modulator: *See* **integrated optical circuit filter-coupler-switch modulator.**

filtered symmetric differential phase-shift keying (FSDPSK): In digital transmission, the encoding of information by means of phase-shift keying in which (a) a binary 0 is encoded as a +90° change in carrier phase and a binary 1 is encoded as a −90° change in the carrier phase, and (b) abrupt phase transitions are smoothed by filtering or other functionally equivalent pulse shaping techniques. *See also* **phase-shift keying.**

filtering: The removal of specific components, features, or characteristics from an entity. *Note 1:* Examples of filtering are (a) removing the low frequencies from a radio signal, (b) removing all frequencies except a very narrow band of frequencies from a radar signal, (c) removing the color blue from sunlight, i.e., blue blocking, and (d) removing fine particles suspended in a liquid. *Note 2:* Processes that usually require some form of filtering are coupling, switching, modulating, image restoring, water purifying, and chemical and biological analyzing. *See* **inverse filtering, Kalman filtering, pseudoinverse filtering.**

filtering system: 1. An assembly of filters that perform a specified filtering function. **2.** A system that (a) removes specified ingredients from a material and (b) consists of one or more filters. *See* **discrete time filtering system.** *See also* **filter, filter center, filtering.**

final modulation: The last stage of the modulation of a carrier prior to transmission to the antenna to be radiated into space. *See also* **carrier, modulation.**

finder: *See* **direction finder, height finder.**

finding: *See* **direction finding, radio range finding, simultaneous direction finding.**

finding equipment: radio direction finding equipment.

finding station: *See* **radio direction finding station.**

finish: *See* **end finish, optical end finish.**

finished call: A call in which the call originator or the call receiver (a) no longer desires to communicate, and (b) signals the termination of the information transfer phase, such as rings off or returns the receiver to the cradle or hook, i.e., goes on-hook or hangs up.

finished lens: A lens with both surfaces ground and polished to a specific dioptric power or focal length. *See also* **diopter, dioptric power.**

finite state machine: In automata theory, a machine, or a model of a machine, that is capable of assuming only a finite number of states and transitions between these states. *Note:* An example of a finite state machine is a code sequence generator.

FIP: Federal Information Processing.

firewall: In secure communications systems, a device, procedure, or program that (a) provides trusted network security for a distributed computing and communications environment, (b) is designed to preclude unauthorized access to a computer or communications system component, such as a terminal, database, computer program, or transmission system, (c) accepts all traffic that passes from inside to outside and outside to inside the protected system, (d) allows only authorized traffic to pass according to specified policy, and (e) is immune to penetration. *Note:* A firewall is as secure as its bastion host. *See also* **bastion host, cracker, firewall principle, secure communications, secure line, secure transmission, security, transmission security.** *Refer to* **Figs. S-9 through S-11 (880).**

firewall component: A component of hardware, software, or firmware that is used to implement a firewall, such as (a) a screening router installed between the private network and the Internet for "permit" or "deny" determination—not very effective for preventing network penetration of and by itself, (b) a bastion host that provides the intelligence needed to determine what is acceptable and what is unacceptable, (c) a rules table that resides in the bastion host, (d) a labeled system in which all subjects and objects are labeled with group identifiers of categorized information, and (e) a combination of these. *See also* **cracker, firewall, firewall principle, hacker, secure communications, transmission security.**

firewall principle: In secure communications systems, (a) the effective blocking and checking of all incoming network traffic, (b) the permitting of authorized users to access and transmit privileged information, and (c) the denying of access to unauthorized users. *Note:* The two types of firewall implementation are those that (a) permit everything that is not expressly denied and (b) deny everything that is not expressly permitted. The latter assumes that what is not known cannot hurt. *See also* **cracker, secure communications, secure line, secure transmission, security, transmission security.**

FIRM: functionally integrated resource manager.

firmware: 1. Software that (a) is permanently stored as built-in hardware, (b) allows reading and executing the software represented by the hardware, (c) cannot be modified by software, (d) permits rapid execution of usually repetitive instructions, (e) usually is used where the operations always need to be performed regardless of the application or the use of the system, and (f) is used where only the operands, variables, or data are changed, and the sequence of operators is constant. *Note:* Hardware configurations, such as permanently connected semiconductor logic circuits and distributions of magnetic material, usually are used to implement, i.e., to create, the software. Originally, firmware was created by constructing a set of orthogonally intersecting wires with a magnetic core at each intersection and then removing the cores where the zeros of a binary numeral, representing an instruction word, are to be stored permanently, whereas the ones of the number are stored where the cores remain. The same effect can be created by thin film, evaporative deposition, printed, or integrated circuit techniques. In communications, computer, data processing, and control systems, firmware is permanently installed or wired into the system in either pluggable or nonremovable circuits capable of executing arithmetic and logic operations, such as executing specific built-in sequences of computer program instructions, performing code or language translations, compiling instructions, performing communications system store and forward operations, switching operations, and controlling operations automatically. Firmware in a computer can only be changed by changing the circuitry and not by computer programming. **2.** A computer program that is permanently implemented in hardware rather than stored as a sequence of coded instructions in a storage unit. *See also* **hardware, hardwire, read-only storage, software.**

first in, first out (FIFO): In a file, storage unit, or location in a system, pertaining to a queuing discipline in which entities leave in the same order, i.e., in the same sequence, in which they arrived. *Note 1:* Service is offered first to the entity that has been in the file the longest time. *Note 2:* First in, first out (FIFO) queuing is commonly used in message switching systems. *Synonym* **push-up.** *See also* **buffer; elastic buffer; last in, first out; push-up storage; queue traffic; variable length buffer.**

first out: *See* **first in, first out.**

first principal point: In an optical system, the first significant point in the system nearest the object. *Note:* An example of a first principal point is the geometric center of an objective lens. *See also* **object, objective lens, optical system, principal point, second principal point.**

first window: In silica-based optical fibers, the wavelength transmission window between about 0.83 μm (microns) and about 0.85 μm.

FISINT: foreign instrumentation signals intelligence.

five-horn feed system: In radar systems, a monopulse or pseudomonopulse antenna feed system in which four elements operate in the receive mode for signal tracking, and a separate element operates in the transmit mode. *See also* **antenna feed, four-horn feed system.**

five-hundred (500) service: A network-provided service feature in which individuals may receive, via a single number, telephone calls at various locations, such as home, office, or car phone, from call originators not necessarily using the same common carrier. *Synonym* **500 service.**

five-layer four-heterojunction diode: A four-heterojunction laser diode, that (a) consists of two pairs of heterojunctions, (b) has five layers of step-indexed material, i.e., five layers of material with a sudden transition of refractive index at the interfaces between layers, (c) confines the emitted light to a narrow beam for optimum coupling to an optical fiber, fiber optic bundle, or optical integrated circuit (OIC), (d) has only three different refractive indices because there may be a pair of identically indexed inside layers, on opposite sides of a center layer, each pair and center layer being of different refractive index, and (e) has the lower refractive indices toward the outside, resulting in a layered cross section with step indices of n_1:n_2:n_3:n_2:n_1, with $n_1<n_2<n_3>n_2>n_1$, thus confining almost all of the generated and emitted light to the center layer by internal reflection. *See also* **four-layer heterojunction diode.**

fix: A successful triangulation in direction finding, i.e., a determination of the location of a transmitter. *See* **radio fix, request fix, running fix.** *See also* **direction finding, triangulation.**

fixed: In communications, computer, data processing, and control systems, pertaining to installed equipment that is not portable, mobile, or transportable, such as a radio station or a main frame computer permanently installed within and attached to a building. *See also* **mobile, portable, transportable.**

fixed attenuator: An attenuator in which the amount of attenuation cannot be varied. *Note 1:* The attenuation is expressed in dB. *Note 2:* The operating wavelength for optical attenuators and fiber optic attenuators should be specified for the rated attenuation because optical attenuation of a material varies with wavelength. *See also* **pad.**

fixed communications network: *See* **military fixed communications network.**

fixed compromise equalizer: A circuit or network that (a) has fixed, i.e., nonvariable, circuit elements, (b) is connected to a transmission line to compensate for the different impedances encountered by the different frequencies that usually occur, (c) best fits all the frequencies that are encountered, and (d) is used because each frequency in a given signal requires a different balancing, matching, or terminating impedance, thus requiring a compromise circuit to accomplish specific purposes, such as to minimize reflections and obtain maximum power transfer to the load. *Note:* Impedance is a function of frequency. Thus, a given circuit cannot serve as a balancing, matching, or terminating impedance for all frequencies. The characteristic impedance of a transmission line depends on the parameters of the line and the frequency. *See also* **characteristic impedance.**

fixed connector: In fiber optic systems, a connector that has one of its mating halves mounted on a board, panel, casing, bulkhead, or other rigid object or surface. *See also* **free connector.**

fixed directory: A telephone directory in which (a) the telephone number assigned to each user, whether customer, subscriber, equipment, computer, program, or process, remains fixed regardless of where the user is connected to the systems within the area covered by the directory, (b) updating is required when personnel, equipment, and program changes are made, and (c) updating is not required when users change location within the area covered by the directory. *See also* **appointment directory, directory, directory service.**

fixed Earth station: An Earth station that (a) is intended to be used at a specified fixed ground location and (b) is an appurtenance to the ground and therefore is neither mobile nor portable. *See also* **Earth station.**

fixed fiber optic connector: A connector half that (a) is permanently attached to a device, such as a light-emitting diode (LED), laser, fiber optic transmitter, fiber optic repeater, fiber optic receiver, or photodetector, (b) that permits connection of the fiber optic component to which it is attached to be connected to other fiber optic components, such as those just mentioned as well as terminated fiber optic cables, i.e., connectorized fiber optic cables, and (c) is a mateable and demateable connector.

fixed focus: Pertaining to instruments that are not provided with a means of focusing, such as changing the focus or the location of the focal plane to compensate for distance to the object. *See also* **focus.**

fixed format message: A message (a) in which transmission line control characters must be inserted when the message is transmitted and deleted when it is received, and (b) that is intended for terminals and stations with dissimilar characteristics. *See also* **variable format message.**

fixed function generator: A function generator in which (a) the function that it generates is determined by construction and cannot be altered by the user, and (b) in analog systems, the generated function usually is an analog electric current or voltage that is a function of temporal or spatial coordinates. *See also* **function generator.**

fixed horizontal polarization: In a horizontally polarized electromagnetic wave propagating in free space or in a material propagation medium, such as an optical fiber, polarization of the electric field vector such that it remains horizontal always and everywhere and hence does not change from a horizontal direction with time or distance along the propagation path. *Note:* The magnetic field vector may change with time or distance, thus governing the angle of elevation with the horizontal at which the wave is propagating at a particular point in space or time. *See also* **fixed vertical polarization, horizontally polarized electromagnetic wave.**

fixed logic: A permanent, hard-wired, interconnected set of combinational circuits, i.e., logic elements, that can be changed by exchanging pluggable units, by resetting connectors or switches, or by rewiring, but not by software changes. *See also* **combinational circuit, hard wire, programmable logic.**

fixed loop: A network-provided service feature that permits an attendant on an assisted call to retain connection through the attendant position for the duration of the call. *Note:* The attendant usually will receive a disconnect signal when the call is finished, i.e., is terminated. *See also* **attendant access loop, finished call, loop, service feature.**

fixed magnetron: A fixed-frequency transmitter tube in which (a) pulses of ultrahigh frequency (UHF) energy are produced, (b) the flow of electrons is controlled by an applied magnetic field, (c) electrons are released from a large cathode and forced to gyrate in an axial magnetic field before reaching the anode, (d) electron energy is collected in a series of slot resonators in the face of the circular anode, (e) a radio frequency (rf) electromagnetic wave is generated, and (f) output power pulses are taken from a small coupling circuit and transmitted to an antenna, from which they are radiated.

fixed microwave auxiliary station: A fixed station used in connection with (a) the alignment of microwave transmitting and receiving antenna systems and equipment, (b) coordination of microwave radio survey operations, and (c) cue and contact control of television (TV) pickup station operations.

fixed optical attenuator: A device that attenuates the intensity of lightwaves a fixed, i.e., a given, number of dB when inserted into an optical waveguide link.

Note: An example of a set of fixed optical attenuators is the standard set that has fixed attenuators that attenuate 3, 6, 10, 20, and 40 dB. *See also* **attenuation, attenuator.**

fixed optics: The development and the use of optical components that have characteristics that cannot be changed or controlled during their operational use, except perhaps for minor adjustments in position, such as in focusing telescopes, binoculars, microscopes, and magnifying glasses to accommodate the human eye. *Synonyms* **inactive optics, passive optics.** *See also* **active optics.**

fixed reference modulation: Modulation in which the choice of the significant condition for any signal element is based on, or makes use of, a fixed reference, such as a timing pulse or a carrier fixed frequency. *See also* **coherent pulse operation, modulation, signal element, significant condition.**

fixed satellite service (FSS): A radiocommunications service between Earth stations at specified fixed points when one or more satellites are used. *Note 1:* In some cases fixed satellite service includes satellite-to-satellite links, which may also be used in the intersatellite service. *Note 2:* The fixed satellite service may also include feeder links for other space radiocommunications services. *See also* **aeronautical service, Earth station.**

fixed service (FX): A radiocommunications service between specified fixed points. *See* **aeronautical fixed service.** *See also* **mobile service.**

fixed station: A station that (a) is in the fixed service and (b) as a secondary service, may transmit to mobile stations on its usual frequencies. *See* **aeronautical fixed station, hydrological and meteorological fixed station, telemetering fixed station.** *See also* **fixed service.**

fixed storage: *Synonym* **read-only storage.**

fixed telecommunications network: *See* **civil fixed telecommunications network.**

fixed-tolerance-band compaction: Data compaction accomplished by storing or transmitting data only when the data deviate beyond prescribed limits. *Note:* An example of fixed tolerance band compaction in a telemetering system is the transmission of the temperature only when the temperature is above or below preestablished threshold limits. Thus, the recipient of the transmission is to assume that the value is in the prescribed range unless a signal to the contrary occurs. Because the value is in range most of the time, transmission requirements are considerably reduced. Redundant information is not transmitted continuously. Care must be taken to make the system fail safe, as is always the case when information is based on nontransmission. *See also* **data compaction.** *Refer to* **Fig. F-4.**

fixed vertical polarization: In a vertically polarized electromagnetic wave propagating in free space or in a material propagation medium, such as an optical fiber, polarization of the electric field vector such that it remains vertical always and everywhere and hence does not change from a vertical direction with time or dis-

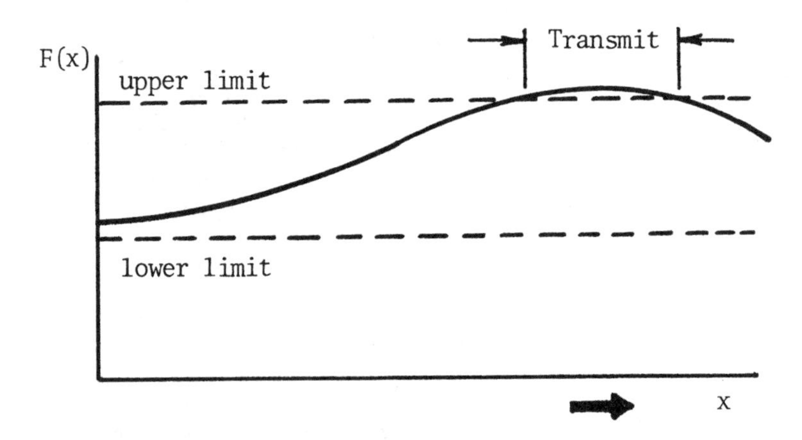

Fig. F-4. Fixed-tolerance-band compaction in which the value of the function **F(x),** such as pressure, temperature, linear speed, or angular rotation rate, is transmitted only when the function exceeds an upper limit or falls below a lower limit because the value of the variable, *x,* that determines the value of the function at every instant has caused the function to be outside the limit.

tance along the propagation path. *Note:* The magnetic field vector may change with time or distance, thus governing the horizontal direction, i.e., the azimuth, in which the wave is propagating at a particular point in space or time. *See also* **azimuth, fixed horizontal polarization, vertically polarized electromagnetic wave.**

fixed-weight binary code: A binary code that has a fixed number of 1s in each numeral. *Note:* An example of a fixed-weight binary code is one in which there are always two binary 1s out of 5 digits, such as 00011, 00101, 00110, 01001, 01010, or 01100. *See also* **binary code.**

fixer network: A combination of radio or radar direction finding installations that operate in conjunction and are capable of plotting the position, relative to the ground, of an aircraft that is in flight, either by obtaining fixes on its transmitter or by obtaining radar azimuths. *Note:* The radar can also confirm the range to an aircraft. *See also* **azimuth, direction finding, fix.**

fixing aid: *See* **radio fixing aid.**

flag: 1. In automated communications, computer, and data processing systems, an indicator used for identification. *Note:* Examples of flags are bytes, words, marks, group marks, or letters that signal the occurrence of some condition or event, such as the end of a word, block, or message. *See also* **block, character, indicator. 2.** In visual signaling systems, a symbol or an emblem usually made of colored cloth and raised by a halyard on a flaghoist, i.e., on a vertical structure, such as a pole or a mast, for signaling purposes. *See* **alphabetic flag, numeral flag, special flag, turning flag.** *See also* **halyard, pennant.**

flaghoist: A structure, such as a mast, stack, or pole, usually equipped with a pulley for mounting a halyard for raising flags and pennants. *See* **flag, halyard, inferior position flaghoist, pennant, superior position flaghoist.**

flaghoist signaling: A visual signaling system in which colored pennants and flags are hung from a hoist and used to represent letters, numerals, and words for the transmission of messages primarily from ship to ship, ship to shore, shore to ship, and ship to aircraft but also on shore to make announcements, such as indicating expected weather conditions, announcing beach closings, and issuing various warnings and allowances. *See also* **at the dip, close up, hauled down, inferior position flaghoist, superior position flaghoist.**

flaghoist transmission: Transmission of messages by means of flags or pennants that are displayed from halyards. *Note:* Although flaghoist transmission is a rapid and accurate method of visual transmission, its use usually is limited to daylight hours, short distances, and fixed coded messages, all in sharp contrast to telecommunications systems. *See also* **halyard, pennant.**

flag lockers: A container that (a) holds all the necessary flags and pennants for flaghoist signaling and (b) on board ships, is placed near the halyard mast for immediate hoisting when needed. *See also* **flag, flaghoist signaling, halyard, pennant.**

flag sequence: 1. In a bit-oriented link protocol, such as in the Advanced Data Communication Control Procedure (ADCCP), Synchronous Data Link Control (SDLC), and High Level Data Link Control (HDLC) protocols, a sequence of bits used to delimit, i.e., to mark, the beginning and the end of a frame. *Note:* An eight-bit sequence usually is used as the flag sequence. **2.** In data transmission, the sequence of bits used to delimit the beginning and the end of a frame. *Note:* For example, the eight-bit flag sequence 01111110 is used in some data transmission systems. *See also* **abort, binary digit, data transmission, flag, frame. 3.** In visual signaling systems, the arrangement of flags and pennants on a halyard for the transmission of messages. *See also* **flag, frame, halyard, pennant.**

flag signaling: In visual signaling systems, the sending of messages by means of flags, such as by displaying coded flags on a halyard and by positioning handheld flags to represent letters, numerals, transmission control characters, and data delimiters. *Note:* Examples of flag signaling are flaghoist signaling, semaphore signaling, and wigwag signaling. *See* **Morse flag signaling.** *See also* **data delimiter, flaghoist signaling, semaphore signaling, wigwag signaling.**

flag transmission: The transmission of data by means of positioned flags. *See also* **flag signaling.**

flare: A high intensity source of light that (a) is produced by chemical reaction, such as combustion, (b) may be any one of a number of different colors, (c) usually is launched skyward with a pistol or a small built-in rocket, and (d) is used for signaling or illumination.

flash: 1. In telephone switchboard operation, a signal that (a) is generated by momentarily and repeatedly depressing and releasing a switch, such as a manual key or telephone switchhook, (b) is used to obtain the attention of an attendant, and (c) usually is used to request additional services. *See* **form flash, index flash.**

See also **dual tone multifrequency signaling. 2.** In military communications systems, a precedence designation for high urgency messages, exceeded only by the flash override precedence. *See also* **precedence, precedence designation.**

flash card: 1. In micrographics, a document that is introduced during recording on microfilm to facilitate indexing. **2.** A hand-held card that (a) has graphic data on it, (b) is used as a teaching device, (c) usually is a member of a deck of such cards devoted to a particular subject, and (d) is "flashed" to the reader.

flash detector: *See* **infrared point source flash detector.**

flashing light communications system: A visual signaling system in which a light (a) is turned on and off, or blanked and unblanked, i.e., masked and unmasked, to create dots, dashes, and pauses, and (b) is used to transmit messages, usually by using the international Morse code.

flashing light ship signaling: Visual signaling to and from ships using a flashing light communications system. *See also* **flashing light transmission.**

flashing light transmission: In visual signaling systems, the transmission of messages by (a) turning a light on and off or by blanking and unblanking, i.e., masking and unmasking, a light that is on continuously, (b) confining the light to a narrow beam for security, reducing interception, and increasing distance at a given signal power, (c) distributing the light in a broad sector for transmitting to dispersed stations, (d) enciphering the messages, (e) using the optimum signal power to ensure reception and avoid undesired interception, and (f) usually using the international Morse code.

flash memory: A memory chip that (a) uses floating-gate storage cells, (b) uses high energy electrons to charge the cells, (c) uses applied electric fields to force electrons to tunnel away from the cells and thereby remove the charge, and (d) stores large amounts of data in the chip structures, thus supplanting bulkier conventional magnetic and optical data storage media. *Note:* Current technology is approaching 256-Mb (megabit) capacity using 0.5-μm (micron) flash memory chips, with gigabit capacities expected early in the twenty-first century when 0.25-μm chips are available.

FLASH message: In military communications systems, a brief message of extreme urgency. *Note 1:* FLASH messages are in a category of precedence messages. *Note 2:* Examples of FLASH messages are messages concerning initial contact with hostile forces or operational combat orders. *See also* **precedence, precedence designation.**

FLASH OVERRIDE message: In military communications systems, a message that (a) is of extreme importance, emergency, or criticality, greater than that of a FLASH message, and (b) can be sent only by the highest authority. *See also* **precedence designation.**

flash override precedence: The precedence designation that (a) is reserved for messages of greater importance, emergency, or criticality than flash messages and (b) can only be used when specifically authorized by the highest authority. *See also* **precedence, precedence designation.**

flash precedence: The precedence designation that is reserved for messages of extreme urgency, such as critical military action messages. *See also* **precedence, precedence designation.**

flatbed plotter: A plotter that (a) draws or creates an image on a display surface that is flat, and (b) usually uses a moving plotting head to draw or create the image. *See also* **display surface, image, plotter.**

flatbed transmitter: In facsimile systems, a transmitter that scans the object, such as a document, while it is lying on a flat surface, usually by scanning the object line by line with a moving optical reading head while the object remains stationary. *See also* **facsimile, facsimile system, object.**

flat coil fiber optic sensor: *Synonym* **flat coil optical fiber sensor.**

flat coil optical fiber sensor: A distributed optical fiber sensor in which the optical fiber is distributed in a plane in rows and columns, or in a flat spiral, so that the distance, usually measured by optical time domain reflectometry (OTDR), to a stimulated point on the plane, such as a pressure point or a hot point, can be used to calculate the Cartesian or polar coordinates of the point. *Synonym* **flat coil fiber optic sensor.**

flat fading: Fading in which all the frequency components of a radio signal simultaneously vary in the same proportion. *See also* **attenuation, fading, selective fading.**

flat fiber optic cable: In fiber optic systems, a cable that (a) has a group of optical fibers or slab dielectric waveguides arranged in a rectangular coordinate system, i.e., a Cartesian coordinate system, such that they are in rows and columns, (b) maintains a constant relative position of optical fibers and guides throughout the cable, (c) is of rectangular cross section, and

(d) has rectangular connectors on the ends. *Synonym* **flat optical cable.** *See also* **fiber optic ribbon.**

flat noise power: Noise power that is constant over a band of frequencies; i.e., the noise power per hertz is inversely proportional to the frequency, expressed by the relation $p/f = a/f$ where p is the power at any given frequency, f is the frequency, and a is a constant of proportionality, which implies that $p = a$ where a is a constant, and thus the power at any one frequency is the same as the power at any other frequency. *See also* **noise power.**

flat optical cable: *Synonym* **flat fiber optic cable.**

flat panel display (FPD): An electronic, thin, lightweight, display device that (a) is thin in depth, i.e., in thickness, compared to the dimensions of the display surface, such as 1 in. thick for an 8-in. × 11-in. surface, (b) is used to display data and images where space constraints are high, such as in laptop personal computers (PCs), airplane cockpits, portable maps, and military vehicles, and (c) may produce sharp detailed images suitable for many applications, such as high definition television (HDTV) and pocket-size computers. *Note:* Flat panel displays (FPDs) consist of an array of pixels made from transistors, liquid crystals, light-emitting diodes, and gas panels. A cathode ray tube (CRT) is not an FPD.

flat rate service: Telephone service in which a single rental payment for a specified period permits an unlimited number of local calls to be made without further charge. *See also* **extended area service, measured rate service, postalized rate, tariff.**

flattened fiber: *See* **dispersion-flattened optical fiber.**

flat weighting: In a noise measuring set, a noise weighting based on an amplitude-frequency characteristic, i.e., a frequency response curve, that is flat over a frequency range that must be stated. *Note 1:* Flat noise power is expressed in dBrn($f_1 - f_2$) or in dBm($f_1 - f_2$). *Note 2:* "3-kHz flat weighting" and "15-kHz flat weighting" are based on amplitude versus frequency characteristics that are flat between 30 Hz and the frequency indicated. *See also* **C message weighting, dBrn($f_1 - f_2$), F1A line weighting, HA1 receiver weighting, 144 line weighting, 144 receiver weighting.**

flaw: *Synonym* **loophole.**

F layer: In the F region of the atmosphere, a layer of increased free electron density caused by the ionizing effect of solar radiation. *Note 1:* The F layer reflects normal incidence frequencies at or below the critical frequency, approximately 10 MHz (megahertz), and partially absorbs higher frequencies. *Note 2:* The F_1 layer exists from about 160 to 250 km above the Earth's surface and only during the day. Though fairly regular in its characteristics, the F_1 layer is not observable everywhere or on all days. The F_1 layer is the principal reflecting layer during the summer for paths 2000 to 3500 km long. The F_1 layer has about 5×10^5 e•cm^{-3} (free electrons/cubic-centimeter) at noontime and minimum sunspot activity. The density increases to approximately 20×10^5 e•cm^{-3} during maximum sunspot activity. The density falls off to below 0.1×10^5 e•cm^{-3} at night. *Note 3:* The F_1 layer merges into the F_2 layer at night. *Note 4:* The F_2 layer exists from about 250 to 400 km above the Earth's surface. It is the principal reflecting layer for high frequency (HF) communications during both day and night. The F_2 layer has about 10×10^5 e•cm^{-3}; however, variations usually are large, irregular, and particularly pronounced during magnetic storms. *See also* **critical frequency, D layer, E layer, F region, ionosphere.**

Fleming's rule: If the thumb of the right hand points in the direction of an electric current, then the curled fingers point in the direction of the magnetic field that encircles the current; and further, if the curled fingers of the right hand describe the electric current in a solenoid, then the thumb points in the direction of the magnetic field inside the solenoid. *Synonym* **right-hand rule.**

flexible borescope: A portable fiber optic borescope that (a) has an aligned bundle of optical fibers in a fiber optic cable with sufficient flexibility to reach an inspection area via a multiturn path and sufficient rigidity to cross unsupported gaps in the path, (b) usually has an eyepiece or a fiber optic faceplate for viewing the image, (c) usually has a controllable articulating tip on the objective end to enable inspection at various angles, such as up to 90° from the cable optical axis, (d) is hand-held and energized via a fiber optic cable from a light source, and (e) usually comes in a set consisting of a basic borescope unit and the ancillary equipment required for operation, all packed in a portable container with instructions. *See also* **rigid borescope.**

flexible disk: *Synonym* **floppy diskette.**

flexible diskette: *Synonym* **floppy diskette.**

flexible spreadsheet: A spreadsheet that, by means of a special software package, can be varied in format and overall dimensions so that it may better fit a given application. *Note:* An example of a flexible spreadsheet is one in which the width, depth, and matrix dimensions can be varied by the operator to make more effective use of the available storage area. *See also* **software, software package, spreadsheet, storage area.**

flicker: An undesirable pulsation or temporary variation in the intensity of a display image on a display surface, usually caused by voltage or current variations or by low regeneration rates in those types of display images that must be regenerated to prevent loss of intensity by electric charge leakage or radiation decay. *See also* **blinking, wink pulsing.**

flight information system: *See* **electronic flight information system.**

flight management system (FMS): On board aircraft, a system that manages flight operations, such as (a) integrates inputs from a set of navigation systems, (b) usually is contained in a single avionics package, (c) allows for one cockpit control and display for all the integrated systems, (d) obtains input from a number of sensors, such as the very high frequency (VHF) omnidirectional range (VOR), distance measuring equipment (DME), Omega, very low frequency (VLF), Loran C, inertial, and global positioning system (GPS) sensors, (e) stores flight profile data, (f) provides fuel management data, and (g) performs frequency management for communications, navigation, and runway approach guidance systems.

flight telemetering land station: A telemetering land station that (a) is used for telemetering data from a launched object, such as a balloon, booster, or rocket, and (b) does not include (i) a booster or a rocket in orbit about the Earth or in deep space, (ii) a space ship in flight, (iii) an aircraft in flight, or (iv) a station that is used in the flight testing of an aircraft. *See also* **land station, telemetering land station.**

flight telemetering mobile station: A telemetering mobile station that (a) is used for telemetering data from a launched object, such as a balloon, booster, or rocket, and (b) does not include (i) a booster or a rocket in orbit about the Earth or in deep space, (ii) a space ship in flight, (iii) an aircraft in flight, or (iv) a station that is used in the flight testing of an aircraft. *See also* **mobile station, telemetering mobile station.**

flight test station: An aeronautical station that is used for the transmission of essential messages associated with the testing of aircraft or major components of aircraft. *See also* **aeronautical station.**

flint glass: A heavy brilliant glass that (a) contains lead oxide, (b) has a relatively high refractive index, and (c) is used more for optical structures, such as ornaments, figurines, beads, jewelry, and glassware, than for optical elements. *See also* **optical element, refractive index.**

flip chip: In an optical circuit, such as an optical integrated circuit (OIC), an optical switch that controls light paths into and out of an optical junction. *See also* **optical junction, optical switch.**

flip-flop: A device that (a) may assume either one of two reversible stable states and (b) usually is an active device. *Note 1:* The flip-flop is used as a basic control element in computer and communications systems. *Note 2:* A flip-flop usually has two output terminals that always maintain opposite polarities. The nature and the place of the input stimulus determine whether the output polarities are to be reversed or which of the two stable states the flip-flop is to assume. The polarity of the output terminals may represent a 0 or a 1. Thus, a flip-flop is capable of storing a 0, a 1, or both at the same time. *Note 3:* In a flip-flop, the transition from one stable state to another is unstable. *Synonym* **bistable circuit, bistable multivibrator, bistable trigger circuit.** *Refer to* **Figs. L-6 (529), S-17 (940).**

flip-flop storage: Storage that (a) consists of an array of flip flops, (b) usually is used as a bank of high speed registers, and (c) can control gates, change rapidly, store binary digits, and indicate its state, i.e., its contents. *See also* **flip-flop, register, storage.**

floating point coding compaction: Data compaction accomplished by using coefficients, a base, and exponents to specify the scale, range, or magnitude of numbers. *Note 1:* In a floating point number, the exponent indicates the location of the decimal point in the number. Thus, each floating point number has two parts, the coefficient, i.e., the fractional part, and the exponent part. The coefficient is to be multiplied by a power of 10 indicated by the exponent part. *Note 2:* An example of floating point coding compaction is using 119.8 x 10^6, 119.8(6), or 119.86 to represent 119,800,000. If the number is rounded to 120,000,000, it might be written as 1206 or 127 in which the last digit is the number of zeros to be appended to the preceding digits. Thus, only three positions are required instead of nine to represent the number in storage or in a message, which is only 33% of the original space and time requirement.

floating point operations per second (flops): The rate at which floating point operations are performed by a computer. *Note:* An example of a floating-point-operations-per-second (flops) rate is 15 Mflops, which equals 15 million floating point arithmetic operations per second, approximately the operating speed of minisupercomputers.

flood illumination: *See* **fiber optic flood illumination.**

flooding compound: In a fiber optic cable, a substance, such as a gel, that (a) surrounds the buffer tubes and (b) prevents fluids from entering the interstices within the cable, particularly at splices or at breaks in the jacket. *See also* **gel.**

flood projection: In facsimile systems, scanning in which the object is floodlighted, and the scanning spot is defined by a masked portion of the illuminated area. *See also* **facsimile.**

flood routing: *Synonym* **flood search routing.**

flood search routing: In a telephone network, routing (a) that is nondeterministic routing, and (b) in which a dialed number received at a switch is transmitted to all switches, i.e., is flooded, in the area covered by the same area code and directly connected to that switch, and, if the dialed number is not an affiliated customer at those switches, the number is then retransmitted to all directly connected switches, and then routed through the switch that has the dialed number corresponding to the particular user end instrument affiliated with it. *Note 1:* All digits of the numbering plan are used to identify a particular customer. *Note 2:* Flood search routing allows customers to have telephone numbers independent of switch codes. *Note 3:* Flood search routing provides the highest probability that a call will go through even though a number of switches and links fail or are in an outage state. *Note 4:* Flood search routing creates a larger trunk occupancy and more switched connections than deterministic routing. *Synonym* **flood routing.** *See also* **deterministic routing, hybrid routing.**

flop: floating point operation.

floppy: *Synonym* **floppy diskette.**

floppy disk: *Synonym* **floppy diskette.**

floppy diskette: A thin, flexible magnetic disk (a) that is permanently enclosed in a semirigid protective cover with openings for access by a read/write head, (b) on which data may be recorded on either or both sides, and (c) that usually is used for data and as a part of a software package. *Note:* A typical standard double-sided high density floppy diskette measures 5.25 in. by 5.25 in. and holds 1.2 Mb (megabytes) when IBM-formatted. *Synonyms* **flexible disk, flexible diskette, floppy.** *See also* **diskette, magnetic disk, read/write head, rigid diskette, software package.**

flops: floating point operations per second.

flow: *See* **bidirectional flow, control flow, pseudoflow.**

flowchart: A graphical or diagrammatic representation in which (a) various interconnected symbols are used to represent the sequence of steps to be taken, events to occur, operations to be performed, or decisions to be made to define, analyze, or solve a problem or represent a system, (b) special standardized symbols usually are used to represent various entities, such as operations, data, data media, and equipment, (c) flowlines with arrows are used to indicate the sequence of the interconnected symbols and the direction of flow of data, and (d) decisions, decision points, and options usually are indicated. *Synonym* **flow diagram.** *See* **data flowchart, programming flowchart.** *See also* **block diagram, flowline.**

flowcharter: In computer programming and software engineering, a software tool used to produce flowcharts. *See also* **software engineering, software tool.**

flow control: In communications, computer, and data processing systems, the control that regulates the data transfer rate, the data signaling rate (DSR), or both. *See* **packet flow control, transmit flow control.** *See also* **data signaling rate, data transfer rate.**

flow control procedure: In switching systems, the procedure for controlling the data transfer rate, the data signaling rate (DSR), or both, between two points in a data network, such as between two sets of data terminal equipment (DTE), between a DTE and a switching center, or between switching centers, usually to avoid overload. *See also* **data signaling rate, data terminal equipment, data transfer rate, data transmission, network, switching center.**

flow diagram: *Synonym* **flowchart.**

flowline: On a flowchart, a line that (a) has an indicated direction, (b) represents a connecting path between other symbols, and (c) indicates the sequence of operations, the transfer of control, or the flow of data. *See also* **flowchart.**

flow meter: *See* **fiber optic bloodflow meter.**

flow security: *See* **traffic flow security.**

fluerics: A branch of science and technology (a) that is a subdivision of fluidics, and (b) in which moving mechanical parts are not used. *See also* **fluidics.**

fluidics: The branch of science and technology devoted to the use of fluid dynamics and fluid phenomena to perform operations and execute functions, such as sensing, control, data transmission, information processing, logic, amplifier, and actuation functions.

fluid immersion effect: In optical fibers, fiber optic bundles, and fiber optic cables, the effect on propagation medium parameters when immersed in a given fluid for a specified time. *Note:* Examples of fluid immersion effects are, for a given period, (a) the ratio of transmitted power before and after immersion for the same input power and (b) the change in pulse dispersion as a result of immersion. *See also* **propagation medium, pulse dispersion.**

fluorescence: The emission of electromagnetic radiation by a material during absorption of electromagnetic radiation from another source. *See also* **incandescence, luminescence, phosphorescence.**

fluoride glass: A halide glass in which the halogen is fluorine. *See also* **halide glass.**

flush closedown: The shutdown or the closedown of a communications station or system in such a manner that incoming message traffic is suspended, but all messages currently in the outgoing message blacklog are transmitted, i.e., are dispatched, before closedown is complete.

flutter: Rapid variations of signal parameters, such as amplitude, phase, and frequency, particularly variations in significant conditions and significant instants. *Note:* Examples of flutter are (a) signal variations that may be caused by atmospheric variations, antenna movements in a high wind, or interaction with other signals, (b) in radio signal propagation, phenomena in which nearly all radio waves that usually are reflected by ionospheric layers in or above the E region experience partial or complete absorption, (c) in radio transmission, rapidly changing signal levels, together with variable multipath time delays, caused by reflection and possible partial absorption of the signal by aircraft flying through the radio beam or common scatter volume, (d) variations in the transmission characteristics of a loaded telephone circuit caused by the action of telegraph direct currents on the loading coils, (e) the distortions that are caused by the variation in loss that results from delay distortion, and (f) in recording and duplicating equipment, the deviations of frequency caused by irregular motion of the equipment during operation. *Synonyms* **Dellinger fadeout, fadeout, radio fadeout.** *See also* **delay distortion, distortion, phase interference fading, selective fading, significant condition, significant instant.**

flux: 1. *Obsolete in radiometric terminology. Use* **luminous power, radiant power. 2.** The lines of force of a magnetic field. *See* **magnetic flux.**

flux density: *See* **magnetic flux density.**

flux reversal: *See* **phase flux reversal.**

flux rise time: *Deprecated synonym for* **radiant power rise time.**

fly by light: Pertaining to the use of fiber optic systems on board aircraft for various purposes, such as for (a) sensing and monitoring various aircraft parameters, such as temperature, air speed, aircraft orientation, logging, fuel consumption, hydraulic pressure, and flap position, (b) controlling the flight of the aircraft, and (c) providing a local area network for communications on board the aircraft.

flying spot scan: In display systems, a scan by a spot of light such that the intensity of light from a beam reflected from any point in the display space on the display surface of a display device is measured at any instant, thus providing a means of reading a display image by light reflection rather than light transmission. *See also* **directed scan, display device, display space, display surface, image, raster scan.**

flying spot scanner: In display systems, a device that uses a moving beam of light to scan an object or an image so that a photodetector can generate signals from variations in intensity of transmitted or reflected light that represent the object. *See also* **display device, image, object.**

flying spot scanning: In display systems, scanning in which the intensity of light from a beam reflected from any point in the display space on the display surface of a display device is measured at any instant, thus providing a means of reading a display image by light reflection rather than light transmission. *See also* **display device, display space, display surface, image.**

flywheel effect: In an oscillator, the continuation of oscillations after removal of the applied stimulus. *Note 1:* The flywheel effect usually is caused by interacting inductive and capacitive circuits in the oscillator. *Note 2:* The flywheel effect may be desirable, such as in phase-locked loops used in synchronous systems, or undesirable, such as in voltage controlled oscillators that are turned on and off with a pulse. *Synonym* **flywheeling.**

flywheeling: *Synonym* **flywheel effect.**

FM: frequency modulation.

FM blanketing: frequency modulation (FM) blanketing.

FM broadcast translator: *See* **translator.**

FM capture effect: *Synonym* **capture effect.**

FM capture ratio: *See* **capture effect.**

FM improvement factor: frequency modulation improvement factor.

FMS: flight management system.

FM threshold effect: frequency modulation threshold effect.

FM threshold extension: frequency modulation threshold extension.

f-number: *Synonym* **F number.**

F number: A number that expresses the effectiveness of the aperture of a lens in relation to the brightness of the image of an object such that the smaller the number is, the brighter the image and therefore the shorter the exposure time required for a given amount of incident energy, such as to expose a given film emulsion. *Synonyms* **f-number, F-number.** *See also* **aperture ratio, brightness, image.**

F-number: *Synonym* **F number.**

FO: fiber optic, fiber optics.

foam dielectric coaxial cable: A coaxial cable with plastic foam insulation between the inner and outer conductors. *See also* **coaxial cable, dielectric.**

FOC: fiber optic communications.

focal length: The distance from a mirror, a lens, or some point therein, to the sharp image of a small, infinitely distant object or light source. *Note:* The location of the image is the focal point. *See* **back focal length, equivalent focal length, front focal length.** *See also* **image, light source, object.**

focal plane: In an optical system, such as a mirror or lens system, a plane through the focal point perpendicular to the principal axis of the system. *Note:* An example of a focal plane is the film plane in a camera when it is focused at infinity or when an image is focused on the film plane by moving the lens. *See also* **focus, optical system, principal axis.**

focal point: 1. In an optical system, the point at which a ray bundle forms a sharp image of an object. **2.** In an optical system, the point at which an object must be placed to obtain a sharp image. *Synonym* **principal focus.** *See also* **image, object, optical system, ray bundle.**

FOCC: fiber optic cable component.

focus: To make an adjustment in an optical system so that a sharp, distinct image is obtained. *Note:* An example of focus is to move the eyepiece or the objective of a telescope, microscope, or camera to obtain a sharp image of an object. The focused image will be in the image plane, which may be coincident with a given surface, such as (a) a screen or film or (b) when the eye is a part of the optical system, the retina. *See* **fixed focus.** *See also* **eyepiece, focal point, image, object, objective, optical system.**

focus depth: In coupling lightwaves in and out of an optical fiber, the distance the light source or photodetector is placed from the ends of the fiber, given by the relation $D = d/2 \tan \theta$ where D is the focus depth, d is the desired resolution diameter or image size, and θ is the acceptance angle. *See also* **acceptance angle, light source, optical fiber, photodetector, resolution.**

focus feed: *See* **prime focus feed.**

focusing optical fiber: *See* **self-focusing optical fiber.**

focus point: *See* **principal focus point.**

FOG: fiber optic gyroscope.

fog buoy: A device that (a) contains a radio transmitter and (b) is towed by a ship in a fog or heavy weather, i.e., thick weather, as a marker to assist ship and aircraft stations in maintaining their proper position, i.e., proper station, or their relative position in a convoy.

follow-on call: The establishment of a connection for a new call without completely releasing the call selection chain established for the preceding call. *Note:* An example of a follow-on call is a call established by performing line holding on an international circuit connection to establish a subsequent connection for another call receiver in the same called country. *See also* **call receiver, connection, line holding.**

followup: *See* **system followup.**

FOM: fiber optic modem.

FOMS: fiber optic myocardium stimulator.

font: A family, group, or assortment of graphic characters of given size and style, some of which are available for typewriters, printers, and display devices. *Note:* Examples of fonts are elite, pica, italic, script, Times Roman, and Courier. *See* **type font.**

font change character: A control character that (a) is used to select and make a change in the specific shape or size of the graphics for a set of characters and (b) does not change the character itself. *Note:* Examples of a font change character are (a) a character that denotes a change of a given character from Roman to italics, such as from "a" to "*a*" and (b) a character that denotes a change of a given character from italics to boldface italics, such as from "*a*" to "***a***". *Synonym* **face change character.** *See also* **character, control character, font.**

foot-candle: A unit of illuminance equal to 1 lumen (of light flux) incident per square foot, i.e., the illuminance of a curved surface, all points of which are placed 1 foot from a light source having a luminous intensity (candlepower) of 1 candle, i.e., 1 candela. *Note 1:* 4π lumens emanate from 1 candela. However, the foot-candle is not an SI (Système International d'Unités) unit. The SI unit is the lux, equal to 1 $lu{\bullet}m^{-2}$ (lumen per square meter). *Note 2:* A foot-candle is not a number of feet times a number of candles, such as is the case for foot•pound for energy and pound•feet for torque. *See also* **candela, illuminance, lumen.**

footprint: In satellite communications systems, the portion of the Earth's surface over which a satellite antenna delivers a specified signal power level, i.e., irradiance or field strength, under specified conditions. *Note:* The limiting case is somewhat less than one-half the Earth's surface, depending on the altitude of the satellite and a tangent cone with the satellite at the apex and the Earth as the base, which is further reduced by a limiting grazing angle. *See also* **Earth coverage, satellite.**

for: The indicator that (a) indicates the addressee of a message and (b) does not necessarily limit the distribution of the message to any other addressees. *See also* **indicator.**

forbidden character: *Synonym* **illegal character.**

forbidden code: *Synonym* **illegal code.**

forbidden landing point marker: In air to ground visual communications systems, a large cross, usually 45 ft. (foot) × 45 ft., 15 m (meter) × 15 m), displayed in the center of a landing strip, site, zone, or point that is not to be used for landing purposes. *Synonym* **no-landing marker.**

force: *See* **electric field force, electromotive force, gravitational field force, magnetic field force, magnetomotive force, psophometric electromotive force.**

forces radio service: *See* **Armed Forces Radio Service.**

forecast: *See* **long-term ionospheric forecast, short-term ionospheric forecast.**

forecasting: *See* **ionospheric forecasting.**

foreground: 1. In multiprogramming, the environment in which computer programs, usually of a high priority, are executed. *Note:* An example of foreground is the environment in which programs are moved in and out of main storage to allow processing time to be shared among several users. **2.** In a picture or a photograph, the part that is or is represented as nearest to the viewer. *See* **display foreground.** *See also* **background, computer program, main storage, multiprogramming.**

foreground processing: The processing of jobs of high priority, usually in real time, that (a) preempt the use of communications, computer, and data processing systems, equipment, or facilities and (b) are perceived by an operator as being executed immediately. *See also* **background processing.**

foreground program: In multiprogramming, a high priority program, i.e., a program executed in the foreground. *See also* **foreground, foreground processing, multiprogramming.**

foreign exchange (FX) service: A network-provided service feature in which a telephone in a given exchange area, instead of being connected directly to the central office (C.O.) serving that exchange area, is connected, via a private line, to a C.O. in another exchange area, i.e., in a "foreign" exchange area. *Note 1:* To call originators, it appears that the call receiver telephone is located in the exchange area served by the central office (C.O.) that serves the call originator. *Note 2:* An example of foreign exchange (FX) service is one in which a user end instrument, such as a telephone, is located in the area served by a central office (C.O.) in the 804 exchange area but is directly connected to a C.O. that serves a 703 area code exchange. *Synonyms* **long call, long local.** *See also* **exchange area.**

foreign instrumentation signals intelligence (FISINT): 1. Intelligence information derived from electromagnetic emissions associated with the testing and operational deployment of foreign aerospace, surface, and subsurface systems. **2.** Technical information and intelligence information derived from the interception of foreign instrumentation signals by other than the intended recipients. *Note 1:* Foreign instrumentation signals intelligence is a category of signals intelligence. *Note 2:* Foreign instrumentation signals include, but are not limited to, signals from telemetry, beaconry, electronic interrogators, tracking/fusing/arming/firing command systems, and video data links.

form: *See* **Backus-Naur form, continuous form.**

form feed: Paper movement to bring a specified area of a form to a particular position for printing. *Note:* An example of form feed is to advance a continuous form to a predetermined line on the following page.

form feed character: A format effector that causes the data medium to be moved so that the print or display position occurs at a predetermined spot on the next line, the next form, the next page, or the equivalent of these locations on any movable data medium. *See* **data medium, format effector.**

form flash: In display systems, the projection of a form overlay on the display surface of a display device. *Note:* An example of a form flash is the placement of a grid or a projection of a data form on the faceplate of a fiberscope or on the screen of a personal computer (PC) monitor. *See also* **display system, form overlay.**

form overlay: In display systems, a pattern used as a display background for a display image. *Note:* Examples of patterns that can be used as form overlays are blank forms, grids, maps, graphs, and outlines. *See also* **display background, display image, display surface, display system.**

formal language: *Synonym* **artificial language.**

formal logic: The rules of structure and form of valid argumentation, with no need for assigning particular meaning to the terms or the expressions used in the argument. *Note:* An example of formal logic is Boolean logic. *See also* **Boolean function.**

formal message: A message in which the format is rigidly controlled in order to facilitate transmission, handling, and distribution. *See also* **format, message, transmission.**

format: 1. The shape, size, and general makeup of a document. **2.** An arrangement of bits or characters within a group, such as a word, message, or language. **3.** The distribution, arrangement, or layout of data elements on or in a data medium. **4.** To distribute, arrange, or lay out data elements on or in a data medium. **5.** In personal computer (PC) systems, to process a diskette by placing a recording track and certain data on it in order that files may be stored and retrieved from it by specific software systems. **6.** To specify the manner in which data are to be arranged in a file or a data medium. *Synonym* **data format.** *See* **address format, communications plan format, date format, dating format, diskette exchange format, display format, ionospheric message format, packet format, quarter common intermediate format, radiotelegraph message format, signal format.** *See also* **packet format, software.**

format effector (FE): One of a group of control characters used to control the positioning of printed, displayed, or recorded data. *See also* **control character.**

formation radar jamming: Jamming a radar system by flying target aircraft in a formation that causes the radar system to experience difficulty in ascertaining azimuth and elevation angles. *Note:* An example of formation jamming is jamming by causing radar strobes to overlap on a radar scope so that the operator cannot determine where one strobe stops and another starts. *See also* **azimuth, elevation angle, jamming, radar strobe.**

format line: In a message, a line of printed characters that (a) starts with, or consists of, some format effector or indicator, (b) in some communications systems, is mandatory, and (c) is used to assist in switching and routing messages. *Note:* An example of a format line is a line that starts with to, from, information, or a break symbol. In some tape relay systems, there are as many as 16 possible format lines for messages. *See also* **break, format, format effector, indicator, message format.**

formatted data: Data that are arranged in a particular manner, structure, or scheme to facilitate access and processing.

formatted diskette: A diskette that has been subjected to the format process. *See also* **format.**

formatted dump: A dump in which specified data are identified and isolated. *See also* **dump.**

formula: 1. A description, usually in symbolic form, of a process, procedure, or set of operations that, when executed, usually produces a useful or desired result. *Note:* Examples of formulas are (a) a mathematical or algebraic expression, i.e., a mathematical or algebraic function, and (b) a description of a mixture or a compound, including the ingredients and instructions for their mixing or synthesis. **3.** In communications, computer, data processing, and control systems, an expression that (a) is placed at a storage location, and (b) is used to calculate a value from arguments, i.e., from parameters, that are placed in other storage locations, such that when the arguments in those storage locations are changed, the value will be calculated and entered at a specified location. *Note:* Formulas are used in spreadsheet applications. *See* **blocking formula.** *See also* **algebraic expression, argument, function, function-driven cell, value.**

FORTRAN: An international standard computer programming language that is used primarily for numeric, arithmetic, and algebraic operations. *Note 1:* Many versions, i.e., many dialects, of FORTRAN exist, most of which are extensions of the standard. *Note 2:* FORTRAN is derived from formula translation.

fortuitous conductor: Any conductor that may provide an unintended path for signals. *Note:* Examples of fortuitous conductors are metal water pipes, wires, metal cables, and metal building and equipment structural members. Fiber optic cables and polyvinylchloride (PVC) pipes are not fortuitous conductors.

fortuitous distortion: In a signal, distortion that results from causes usually subject to laws concerning random occurrences. *Note:* An example of fortuitous distortion is signal distortion caused by accidental voltage irregularities or current surges that occur in the operation of communications systems components and their moving parts. These random disturbances will affect transmission in any channel coupled to the source of the disturbance. *Synonym* **fortuitous jitter.** *See also* **cyclic distortion, distortion.**

fortuitous jitter: *Synonym* **fortuitous distortion.**

forward: *See* **call forward, spill forward, store and forward.**

forward busying: In a telecommunications system, a feature in which supervisory signals are forwarded in advance of address signals in order to seize assets of the system before attempting to establish a connection. *See also* **call, supervisory signals.**

forward channel: 1. The channel of a data circuit that transmits data from the originating user to the destination user. *Note:* The forward channel carries message traffic and some control information. **2.** In data transmission systems, a channel in which (a) the direction of transmission is the same as that in which data are being transferred, (b) the transmission of information is from the data source to the data sink, (c) the direction of transmission coincides with that in which user information is being transferred, and (d) if information is simultaneously transferred in both directions, the momentary data source to data sink determines the forward channel. *See also* **backward channel, backward signal, channel, circuit, data sink, data source, forward signal, information bearer channel.**

forward clearing signal: *Synonym* **clear forward signal.**

forward current: In a device, such as a vacuum tube, transistor, or diode, the conventional positive flow of electric current from the anode or the collector to the cathode or the emitter. *Note 1:* The actual flow of electrons is in the direction opposite to the conventional positive current direction. *Note 2:* In the case of a diode, the forward resistance is low, allowing a large forward current. The backward resistance is high, allowing a small backward current, given the same positive voltage driving source in each case.

forward data transmission: *See* **store and forward data transmission.**

forward direction: 1. In a communications system, the direction from a data source to a data sink. **2.** The direction from a message originator to the addressee. **3.** The direction from a call originator to the call receiver. *See also* **backward direction.**

forward echo: In a transmission line, an echo that (a) propagates in the same direction as the original wave and (b) consists of energy reflected back from one irregularity and then reflected forward again by another irregularity. *Note:* Forward echoes can occur at all irregularities in a length of cable. They can impair the performance of the cable, especially when they systematically add to or subtract from the desired signals. Forward echoes can be supported by reflections caused by splices or other discontinuities in the propagation media, such as optical fibers, twisted pairs, or coaxial cables. In metallic lines, forward echoes may be supported by impedance mismatches between source or load and the characteristic impedance of the propagation medium. In optical propagation media, forward echoes may be supported by refractive index mismatches between (a) the source or the detector and (b) the optical propagation media. *See also* **echo, return loss.**

forwarder: *See* **data forwarder.**

forward error correction (FEC): In data transmission and reception, error control in which the receiving device has the capability to detect and correct any character or code block that contains fewer than a predetermined number of erroneous symbols. *Note:* Forward echo correction (FEC) is accomplished by adding bits to each transmitted character or code block according to a predetermined algorithm. *See also* **binary digit, block code, character, code, convoluted code, data transmission, error, error control, error detecting code, information feedback.**

forward feature: *See* **spill forward feature.**

forwarding: *See* **call forwarding, data forwarding.**

forwarding busy line: *See* **call forwarding busy line.**

forwarding don't answer: *See* **call forwarding don't answer.**

forward ionospheric scatter: *Synonym* **forward propagation ionospheric scatter.**

forward propagation ionospheric scatter (FPIS): Ionospheric scatter in the forward direction. *Synonyms* **forward ionospheric scatter, ionospheric forward scatter.** *See also* **backward propagation ionospheric scatter, forward direction.**

forward scatter: **1.** The deflection, by reflection, refraction, or diffraction, of an incident electromagnetic wave or signal in such a manner that a component of the deflected wave or signal is in the same direction of propagation as a component of the incident wave or signal. **2.** In an electromagnetic wave or signal that is deflected by reflection, refraction, or diffraction, a component of the deflected wave or signal that is in the direction of propagation of a component of the incident wave or signal. **3.** To deflect, by reflection, refraction, or diffraction, an incident electromagnetic wave or signal in such a manner that a component of the deflected wave or signal is in the same direction of propagation as a component of the incident wave or signal. *Note:* The term "scatter" can be applied to reflection or refraction by uniform media and diffraction by certain apertures, but it is usually applied to a modification of the wavefront and the direction of propagation in a random or disorderly fashion. In some instances, particularly in radio transmission, the term "forward" refers to the resolution of field components when resolved along a line drawn from the source of radiation to the point of scatter. **4.** The component of an electromagnetic wave or signal that is deflected by reflection or refraction in the direction of propagation of the incident wave or signal. *See also* **backscatter, backscattering, forward direction, propagation.**

forward signal: A signal sent in the forward direction, i.e., from the calling to the called station, from the original data source to the original data sink, or from the call originator to the call receiver. *Note:* The forward signal is transmitted in the forward channel. *See also* **backward channel, backward signal, data transmission, forward channel, forward direction, signal.**

forward supervision: In communications systems, supervision in which the supervisory signals are sent from a primary or master station to a secondary or slave station. *See also* **backward supervision, supervisory signal, system supervision.**

forward switching center: *See* **store and forward switching center.**

forward tell: Information transfer from facilities at a lower operational level or echelon of command to a higher operational level or echelon of command.

FOT: optimum transmission frequency. *Note:* The abbreviation "FOT" was originally written for **frequency of optimum transmission.**

FOTM: fiber optic test method.

FOTP: fiber optic test procedure.

FOTS: fiber optic transmission system.

Foucault knife edge test: A test used (a) to determine the errors in an image of a point source by partially occluding the light from the image by means of a knife edge, and (b) to measure the errors in refracting or reflecting surfaces. *See also* **knife edge effect.**

four-address instruction: A computer program instruction that specifies four addresses, such as the addresses in storage of two operands, the address of the storage location at which the results of the operation specified in the instruction are to be placed, and the address of the next instruction to be executed. *See also* **multiple address instruction, one-address instruction, three-address instruction, two-address instruction.**

four-color theorem: Every planar map, such as a map of the United States or a map of the countries on a continent, can be colored with not more than four colors under the restriction that any two areas, such as states or countries, with a contiguous border must be colored differently.

four-concentric-circle near field template: A template that (a) consists of four concentric circles, (b) is applied to a near field radiation pattern radiating from the exit face of a round optical fiber, and (c) is used as an overall check of various geometrical properties of the fiber all at once.

four-concentric-circle refractive index template: A template that (a) consists of four concentric circles, (b) is applied to a complete refractive index profile of a round optical fiber, and (c) is used as an overall check of various geometrical properties of the fiber all at once.

four-frequency diplex telegraphy: Frequency shift telegraphy that (a) has four possible signals each of which is represented in two telegraph channels by a different frequency and (b) is used in radio telegraphy. *Synonym* **four-frequency duoplex telegraphy, twinplex telegraphy.**

four-frequency duoplex telegraphy: *Synonym* **four-frequency diplex telegraphy.**

four-heterojunction diode: A laser diode with two double heterojunctions, i.e., two pairs of heterojunctions, to provide improved control of the direction of radiation and radiative recombination. *Synonym* **symmetrical double heterojunction diode, five-layer four-heterojunction diode.**

four-horn feed system: In radar systems, a monopulse antenna feed system for signal tracking in which a diplexer allows simultaneous transmission and reception. *See also* **five-horn feed system.**

Fourier analysis: The definition of a periodic phenomenon or arbitrary wave shape, such as a square wave, in terms of a group of sine waves with specific amplitudes and frequencies that are multiples of the fundamental frequency of the periodic phenomenon, such as a square wave. *Note:* Each sine wave has an amplitude such that when all the waves are added, the shape of the periodic phenomenon is obtained. The representation is an approximation. A given circuit may discriminate against certain of the frequencies, resulting in distortion of the wave. A Fourier analysis is particularly well suited for communications equipment design and for predicting the performance of a given design.

four-wire circuit: A two-way circuit using two paths so arranged that the electrical signals are transmitted in one direction only in one path and in the other direction in the other path. *See also* **circuit, duplex circuit, duplex operation, four-wire terminating set, line adapter circuit, metallic circuit.**

four-wire repeater: A repeater that (a) consists of two amplifiers, one for each direction, and (b) is used in a four-wire circuit. *See also* **repeater.**

four-wire subset user service: In a telephone network, a telephone service in which (a) a user end instrument is connected directly into a switching center of the network without going through a private branch exchange (PBX), and (b) care usually is taken to ensure that overseas calls are controlled and precedence is not abused. *See also* **end instrument, switching center, user.**

four-wire terminating set: A set used to (a) terminate the transmit and receive channels of a four-wire circuit and (b) interconnect four-wire and two-wire circuits. *See also* **circuit, four-wire circuit, hybrid set, line adapter circuit.**

fox message: The standard test message that (a) includes all the English alphanumeric characters on a teletypewriter, (b) includes function characteristics, such as space, figures shift, and letters shift, and (c) consists of "THE QUICK BROWN FOX JUMPED OVER THE LAZY DOG'S BACK 1234567890."

FPD: flat panel display.

FPGA: field programmable gate array.

FPI: functional process improvement.

FPIS: forward propagation ionospheric scatter.

fraction: *See* **packing fraction.**

fractional: 1. The ratio of a given value of a parameter to its nominal value. *Synonym* **normalized. 2.** The ratio of a given value of a parameter to a specified reference value.

fractional digit: In a positional numeration system, a digit that occupies a position that has been assigned a weight of less than unity. *Note:* An example of a fractional digit in a numeral in the decimal numeration system is a digit to the right of the decimal point, such as the 4 in the mixed decimal numeral 6935.496. *See also* **positional numeration system.**

fractional frequency fluctuation: The deviation of the frequency of an oscillator from its nominal constant frequency, normalized to the nominal frequency.

fractional offset: *Synonym* **normalized offset.**

fraction automatically switched: *See* **failure fraction automatically switched.**

fraction loss: *See* **packing fraction loss.**

frame: 1. In data transmission, the sequence of contiguous bits that (a) is bracketed by, and includes, beginning and ending flag sequences, (b) usually consists of a specified number of bits between flags and contains an address field, a control field, and a frame check sequence, (c) may include an information field, (d) is usually associated with complex schemes, (e) is used with binary signaling, and (f) usually consists of (i) a representation of the original signal presented for transmission together with other signal elements used for error detection or error control, or both, (ii) routing information, (iii) synchronization information, and (iv) similar overhead functions not directly associated with the original signal. *Note 1:* The bit pattern and bit positions in a frame are used to specify certain purposes and meanings. *Note 2:* A frame could contain information from more then one source and could be sent to more than one destination. *Note 3:* Parts of information from a single call could be contained in more than one frame. Frame time slots are assigned for each call. *Note 4:* Frames are used in synchronous systems and are identifiable by reference to a clock. **2.** In the multiplex structure of pulse code modulated (PCM) systems, a set of consecutive digit time slots in which the position of each digit time slot can be identified by reference to a frame alignment signal. *Note:* The frame alignment signal does not necessarily occur, in whole or in part, in each frame. **3.** In a time division multiplexed (TDM) system, a repetitive group of signals that result from a single sampling of all channels, including any required system information, such as additional synchronizing signals. *Note:* In-frame is the condition that exists when there is a channel-to-channel and bit-to-bit correspondence, exclusive of transmission errors, between all inputs to a time division multiplexer (TDM) and the output of its associated demultiplexer. **4.** In facsimile systems, a rectangular area, whose width is the available line and whose length is determined by the service requirements. **5.** In the Integrated Services Digital Network (ISDN), a block of variable length, labeled at the Data Link Layer of the Open Systems Interconnection—Reference Model (OSI-RM). *See* **combined distribution frame, command frame, display frame, fiber optic distribution frame, film frame, freeze frame, group distribution frame, high frequency distribution frame, intermediate distribution frame, main distribution frame, loop frame, main frame, page frame, response frame, supergroup distribution frame, time frame, transmission frame.**

frame alignment: In receiving equipment, the extent to which the equipment frame is correctly phased with respect to that of the received frame. *See also* **bunched frame alignment signal, distributed frame alignment signal, frame, frame alignment signal, frame slip, framing, framing bit.**

frame alignment recovery time: *Synonym* **reframing time.**

frame alignment signal (FAS): 1. In data transmission, a distinctive signal that enables frame alignment to be accomplished. **2.** In facsimile systems, a signal used for adjustment of the picture signal to a desired position in the direct line of progression. *Synonym* **framing signal.** *See* **bunched frame alignment signal, distributed frame alignment signal.** *See also* **facsimile, frame, frame alignment, multiframe, signal, synchronous transmission.**

frame alignment time slot: A time slot that (a) starts at a particular phase or instant in each frame and (b) is allocated to the transmission of a frame alignment signal. *See also* **frame, synchronous transmission.**

frame check sequence (FCS): In high level data link control (HDLC) operation, the field that (a) immediately precedes the closing flag sequence of a frame and (b) contains the bit sequence that enables the receiving station to detect transmission errors. *See also* **cyclic redundancy check.**

framed interface: An interface through which the information flow is partitioned into physical periodic frames that consist of overhead information and an information payload. *See also* **overhead information, payload.**

frame duration: 1. The sum of all the unit time intervals of a frame. **2.** In a digital signal that consists of a bit pattern with an integral delimited number of bits called a "frame," the time between (a) the instant one frame starts and (b) the instant the next frame starts. *See also* **frame, framing, framing bit, synchronous transmission, unit interval.**

frame frequency: *Synonym* **frame rate.**

frame grabber: A device that can seize and record a single frame out of a sequence of many frames. *Note:* Frame grabbers have particular application in video transmission and reception.

frame multiplex structure: A set of consecutive digit time slots in which the position or timing of each digit time slot can be identified by reference to a frame alignment signal, which need not necessarily occur in whole or in part in each frame. *See also* **frame alignment signal, time slot.**

frame pitch: 1. The distance, time, or number of bits between corresponding points, i.e., significant instants, on two consecutive frames. *See also* **frame, significant instant. 2.** The distance between corresponding points on two contiguous or successive film frames on a recording or data medium, such as a microfilm or microfiche. *Synonym* **pulldown.**

frame rate: In a transmission system, the number of frames passing a point per unit time, such as the number of frames transmitted, received, or passing through a switch per unit time. *Note 1:* Frame rate usually is expressed in frames per second. *Note 2:* An example of a frame rate is 500 fps (frames per second) transmitted in a single channel. *Synonym* **frame frequency.** *See also* **frame, frame alignment, frequency, synchronous transmission.**

frame relay: A statistically multiplexed interface protocol for packet-switched data communications in which (a) variable-size packets, i.e., variable-size frames, are used that completely enclose the user packets they transport, (b) transmission rates are between 56/64 kbps (kilobits/second) and 1.544 Mbps (megabits/second) (T-1/DS1 rate multiples), (c) there is neither flow control nor error correction, (d) there is information content independence, (e) there is a correspondence only to the Open Systems Interconnection—Reference Model (OSI-RM) Layers 1 and 2, (f) variable-size user packets are enclosed in larger packets, i.e., larger frames, that add addressing and verification information, (g) frames may vary in length up to a design limit, usually 1 kb (kilobyte) or more, (h) one frame relay packet transports one user packet, (i) fast packet technology is used for connection-oriented services, and (j) time-delay-insensitive traffic, such as local area network (LAN) interworking and image transfer, can be handled. *Note:* Frame relay is referred to as the local management interface (LMI) standard and as Annex D to American National Standards Institute (ANSI) Standard T1.617.

frame slip: **1.** In a receiver, the loss of synchronization of a received frame with the receiver frame. **2.** The condition under which a received digital signal loses frame synchronization. **3.** The dropping or the repeating of a full frame by a transmission or switching facility without the loss of frame synchronization. *See also* **frame, frame alignment.**

frame synchronization: For a received signal, the alignment of a given digital channel, such as a given time slot, at the receiving end with the corresponding channel at the transmitting end. *Note:* In frame synchronization, extra bits, i.e., frame synchronization bits, usually are inserted at regular intervals to indicate the beginning of a frame and for use in frame synchronization. *See also* **binary digit, frame, synchronization, synchronization code, time slot.**

frame synchronization pattern: In digital communications, a prescribed recurring pattern of bits or pulses transmitted to enable the receiver to achieve frame synchronization. *Synonym* **address pattern.** *See also* **binary digit, frame.**

frame television: *See* **freeze frame television.**

framing: **1.** In time division multiplexed (TDM) reception, the process of adjusting the timing of the receiver to coincide with that of the received framing signals. **2.** In video reception, the process of adjusting the timing of the receiver to coincide with the received video synchronization pulse. **3.** In facsimile systems, the adjustment of the facsimile picture to a desired position in the direction of line progression. *See* **display framing.** *See also* **frame, frame alignment, frame duration, synchronous transmission, time division multiplexing.**

framing bit: **1.** A bit used in frame synchronization. **2.** In a bit stream, a bit that (a) is used to determine the beginning or the end of a frame, (b) is assigned to a specific time interval in the frame, and (c) is a noninformation bit used in grouping characters, such as separating them into lines, paragraphs, pages, and channels. *Note:* Framing in a digital signal usually is repetitive. *See also* **binary digit, frame, frame alignment, frame duration.**

framing signal: *Synonym* **frame alignment signal.**

franking: **1.** The stamping or the printing of a mark on an envelope indicating that postage requirements have been met. **2.** Making a mark on any document to indicate that the document has been processed in a specified manner, usually identified by the mark. *Note:* Examples of franking are printing a transaction number on a check, coupon, certificate, or postage stamp to indicate that the amount indicated has been honored in some way, such as paid, credited, or posted.

Fraunhofer diffraction pattern: *Synonym* **far field diffraction pattern.**

Fraunhofer region: *Synonym* **far field region.**

free-access card: A smart card that (a) has a variable storage capacity, (b) is reusable, and (c) has service functions that usually require frequent changing of stored data, such as portable data files and electronic marketing data. *See also* **smart card.**

free condition: *See* **trunk free condition.**

free connector: *Synonym* **free fiber optic connector.**

free fiber optic connector: In fiber optic systems, a connector that is not attached or associated with any other object or surface, i.e., neither of its mating halves is mounted on a board, panel, casing, bulkhead, or other rigid object or surface. *Synonym* **free connector.** *See also* **fixed connector.**

free line signaling: In telephone switchboard operations, signaling that (a) uses the lighting of a lamp to indicate that the first of a group of outgoing trunks is no longer engaged and (b) relieves the operator from performing the engage test. *See also* **engaged test, signaling, trunk.**

free net: A radio net in which any station may communicate with any other station in the net without first obtaining the permission of the net control station. *Note:* Permission to operate as a free net is granted by the net control station until such time as a directed net is established by the net control station. *See also* **net, net control station.**

free routing: The routing of messages in such a manner that they are forwarded toward their destination or addressee over any available channel without dependence upon predetermined routing. *See also* **routing.**

free running capability: In a synchronized oscillator, the capability to operate in the absence of a synchronizing signal.

free space: A theoretical concept of space that (a) is devoid of all matter, (b) implies remoteness from material objects that could influence the propagation of electromagnetic waves, (c) implies freedom from the influence of extraneous fields other than the fields of interest, and (d) is usually free of electric charges, except when a charge or charges in the space are of interest. *Note:* The interior of an optical fiber is not considered free space, but there might be some free space between fibers, fibers and light sources, and fibers and photodetectors. *See also* **free space loss, line of sight propagation.**

free space coupling: 1. Coupling by magnetic, electric, and electromagnetic fields that (a) are not confined to conductors, capacitors, or inductors and (b) do move across vacuum and air space, such as between (i) Earth stations and satellites and (ii) through the atmosphere between microwave relay towers. *See also* **coupling, free space, line of sight propagation.**

free space loss: The signal attenuation that (a) would result if all obstructing, scattering, absorbing, reflecting, refracting, and diffracting influences were sufficiently removed so as to have no effect on propagation, and (b) is primarily caused by geometric spreading, i.e., beam divergence, with signal energy spreading over larger areas as distance from the source increases. *Note:* In the far field of an electromagnetic wave from an omnidirectional antenna, the free-space electric field strength (volts per meter) decreases inversely with the distance from the antenna, while the irradiance (watts per square meter) decreases inversely with the square of the distance from the antenna. *See also* **geometric spreading, line of sight propagation, loss.**

free space optical communications: Communications using lightwaves that (a) propagate in unbounded propagation media, such as outer space, the atmosphere, the ionosphere, or the ocean, and (b) are not confined to waveguides, such as optical fibers, slab dielectric waveguides, and optical integrated circuits (OICs).

free storage: In communications, computer, and data processing systems, storage that (a) may be unprotected and thereby available for use by any operating system component, a supervisor, or a user in response to any request and (b) may be protected and thereby available to the system for its sole use, such as for control blocks, files, or supervisory programs. *See also* **operating system, supervisor, supervisory program.**

freeze frame: A frame of visual information that (a) is selected from a set of motion video frames, (b) is held in a buffer for continuous transmission through the video coder-decoder (codec) to remote sites, and (c) is different from a still image. *See also* **still image**.

freeze frame television: Television (TV) in which fixed images are sequentially transmitted. *Note 1:* The transmission of each image usually is performed every 30 seconds from a processing unit memory where each image is fixed prior to its transmission. *Note 2:* Freeze frame television lowers the bandwidth requirement below that of full-motion television (TV).

F region: The region of the atmosphere that (a) exists between approximately 175 km (kilometers) and 400 km above the Earth's surface, (b) is above the D and E regions, and (b) contains the F_1 layer and the F_2 layer of the ionosphere. *See also* **D region, E region, F layer, ionosphere.**

frequency: 1. For a periodic function, the number of cycles or events that occur per unit of time. *Note 1:* Frequency may be calculated (a) as the reciprocal of the period of a recurring phenomenon or (b) as (i) the number of cycles or events that occur during an interval of time divided by (ii) the time length of the interval. *Note 2:* When the period of a recurring phenomenon is expressed in seconds, the frequency will be expressed in hertz (Hz), i.e., seconds^{-1}. **2.** In an electromagnetic wave, the number of times per second the electric field vector reaches its peak value in a given direction. *See* **aircraft emergency frequency, air-ground communications emergency frequency, alternative frequency, answering frequency, assigned frequency, audio frequency, authorized frequency, bond frequency, broadcast frequency, calling frequency, carrier frequency, center frequency, characteristic frequency, ship frequency, convoy common frequency, convoy radiotelephone frequency, critical frequency, cutoff frequency, distress frequency, dot frequency, emergency craft frequency, extremely high frequency, extremely low frequency, frame frequency, fundamental frequency, guarded frequency, high frequency, image frequency, infralow frequency, intermediate frequency, international aeronautical emergency frequency, international radiotelegraph distress frequency, lock-in frequency, lowest usable frequency, lowest useful high frequency, low frequency, maritime mobile ra-**

diotelephone frequency, maximum keying frequency, maximum modulating frequency, maximum scanning frequency, maximum usable frequency, medium frequency, military common emergency frequency, natural frequency, normalized frequency, optimum traffic frequency, picture frequency, precise frequency, primary frequency, protected frequency, pump frequency, radio frequency, radiotelegraph distress frequency, radiotelephone distress frequency, reference frequency, resonant frequency, scene of air-sea rescue frequency, search and rescue beacon frequency, search and rescue frequency, secondary frequency, superhigh frequency, taboo frequency, telephone frequency, threshold frequency, transition frequency, transmission frequency, tremendously high frequency, ultrahigh frequency, very high frequency, very low frequency, victim frequency, video frequency, working frequency, voice frequency. *See also* **frequency spectrum designation.** *Refer to* **Fig. L-2 (502). Refer to Tables 1, 2, and 3, Appendix B.**

frequency accuracy: In a periodically recurring phenomenon, such as occurs in an electromagnetic or sound wave, the degree of conformity to a specified value of a frequency. *See also* **frequency, frequency lock, frequency stability, precise frequency.**

frequency agile radar: A radar (a) in which the operating radio frequency (rf) can be changed readily for various purposes, such as to avoid jamming, reduce mutual interference with other sources, enhance echoes from objects, i.e., from targets, and produce necessary patterns for electronic countermeasure and electronic countercountermeasure radiation, (b) that performs signal processing in the video portion of the receiver after the phase information has been removed from the signal, and (c) that acts on the amplitude information only as noncoherent signal processing, thereby reducing clutter and improving range resolution. *See also* **echo, electronic countermeasure, electronic countercountermeasure, frequency agility, interference, resolution.**

frequency agility: The capability to change the operating frequency of a device, such as a radio receiver, radio transmitter, radar receiver, radar transmitter, radiotelephone transmitter, or a radiotelegraph receiver, quickly and easily. *Note 1:* As an electronic countercountermeasure for radar, some radar systems may have the capability to shift frequency within their operating bands on a pulse-to-pulse basis. *Note 2:* For communications systems, frequency agility is used to avoid jamming, prevent interference, and reduce noise. Frequency shifting usually is performed by shifting the frequency above and below the assigned frequency. *See also* **assigned frequency, electronic countermeasure, interference, noise.** *Refer to* **Fig. F-5.**

frequency aging: In an oscillator, the change in frequency, over time, caused by internal changes in oscillator parameters even when external factors, such as environment and power supply characteristics, remain constant. *See also* **frequency drift.**

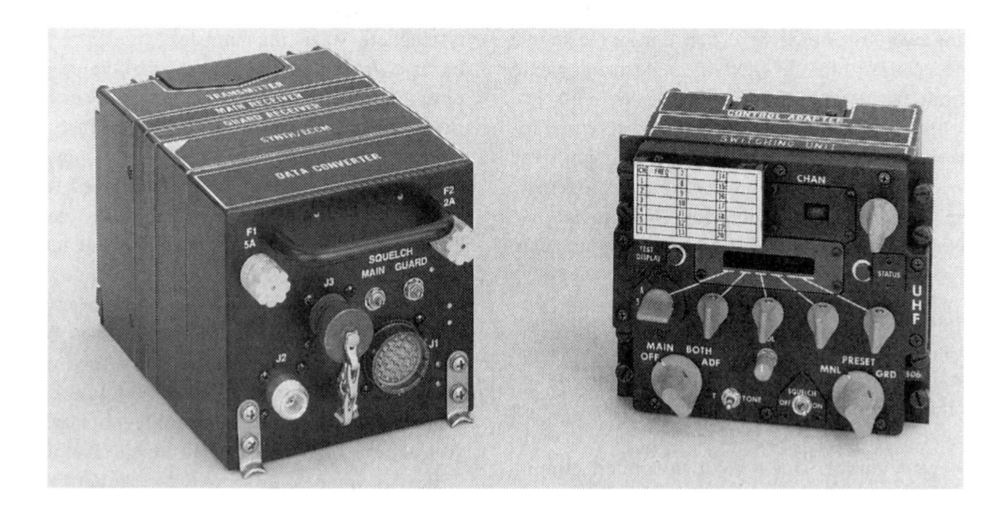

Fig. F-5. The AN/ARC-222 airborne communications system that provides **frequency agility** and frequency modular communications in the 30 MHz band and amplitude modulation communications in the high VHF band. (Courtesy Magnavox Electronic Systems Company, Fort Wayne, IN.)

frequency allocation: 1. The assignment of a part of the frequency spectrum to a specified category of users, such as the allocation of narrow bands in the 30-MHz (megahertz) to 3000-MHz very high frequency (VHF) and ultrahigh frequency (UHF) ranges to public safety, citizens radio, industrial, land transportation, and maritime mobile bands. **2.** The planning for, and establishing of, radio frequency bands for use by specific radio communications services. *Note:* Frequency allocation is accomplished by national and international agreements. International frequency allocation is conducted under the auspices of the International Telecommunication Union (ITU), Geneva, Switzerland, in accordance with current international agreements. National frequency allocation is accomplished by individual governments within their respective international frequency band allocation by the ITU. *See also* **frequency assignment, frequency assignment authority, frequency band allocation, frequency spectrum, frequency spectrum designation.**

frequency allocation plan: A plan that indicates the electromagnetic frequencies that are to be used in particular areas or by particular countries without specifying the stations to which the frequencies are to be assigned. *See also* **frequency assignment authority, frequency assignment plan.**

frequency analysis compaction: Data compaction accomplished by using an expression composed of a number of different frequencies of different magnitudes to represent a particular curve. *Note:* An example of frequency analysis compaction is the use of a Fourier analysis to represent an arbitrary curve, a periodic function, an aperiodic function, or a wave shape. Thus, the fundamental frequency, the amplitude of the fundamental frequency, and the amplitudes and the frequencies of the harmonics are all the data that are needed to reconstitute the function or the wave shape. The shape can thus be readily stored and transmitted in this compacted form. *See also* **data compaction.**

frequency and amplitude (FRENA) system: A system of transmission in which the frequency and amplitude components of a signal are transmitted separately and recombined at the receiver.

frequency assignment: *Synonym* **radio frequency assignment.**

frequency assignment authority (FAA): The power granted an administration, or its designated or delegated leader or agency via treaty or law, to specify frequencies, or frequency bands, in the electromagnetic spectrum for use in systems or equipment. *Note:* Primary authority for the United States is exercised by the National Telecommunications and Information Administration (NTIA) for the federal government and by the Federal Communications Commission (FCC) for nongovernment organizations. International frequency assignment authority is vested in the International Frequency Registration Board of the International Telecommunication Union (ITU).

frequency assignment plan: A plan that indicates the electromagnetic frequencies to be used by specified stations and organizations. *See also* **frequency allocation, frequency allocation plan, frequency assignment authority.**

frequency averaging: In the synchronization of a network, the achievement of synchronization at all nodes by (a) adjusting the frequency of oscillators to match the average frequency of the bit streams received from connected nodes, (b) assigning equal weight to all oscillators in determining the ultimate network frequency, and (c) not using a specified reference frequency oscillator for determining or influencing the ultimate network frequency. *See also* **democratically synchronized network, frequency.**

frequency band: A continuous and contiguous group of electromagnetic wave frequencies, usually defined by a lower and an upper limit of frequency. *Note:* In addition to specification of lower and upper limiting frequencies, frequency bands may be identified by frequency allocation, such as citizens band, police band, emergency band, travel information band, and land transportation band. *See* **assigned frequency band, emergency frequency band.** *See also* **frequency**

Fig. F-6. The AN/ARC-164 airborne ruggedized military ultrahigh frequency (UHF) radio that provides 225 MHz to 400 MHz **frequency band** communications in the presence of jamming. (Courtesy Magnavox Electronic Systems Company, Fort Wayne, IN.)

band allocation, frequency spectrum designation. *Refer to* **Fig. F-6.**

frequency band allocation: 1. In the Table of Frequency Allocations in the Radio Regulations of the International Telecommunication Union, the allocation of a given frequency band for use under specified conditions by (a) one or more terrestrial or space radiocommunications services and (b) the radio astronomy service. **2.** The frequency band allocated for use under specified conditions to (a) one or more terrestrial or space radiocommunications services and (b) the radio astronomy service. **3.** The process of designating radio frequency (rf) bands for use by specific radio services. *See also* **adaptive channel allocation, assigned frequency, authorized frequency, frequency, frequency allocation, frequency spectrum designation.**

frequency bandwidth: *See* **radio frequency bandwidth.**

frequency change signaling (FCS): Signaling (a) in which a specific set of frequencies corresponds to each desired significant condition of a code, (b) in which the transition from one set of frequencies to the other may be a continuous or a discontinuous change in frequency or phase, and (c) that includes supervisory signaling and user information signaling. *See also* **code, frequency, signal.**

frequency channel allotment: *See* **radio frequency channel allotment.**

frequency channel increment: *See* **radio frequency channel increment.**

frequency characteristic: *See* **insertion loss versus frequency characteristic.**

frequency code modulation: Digital frequency modulation in which (a) several discrete frequencies are used to represent binary data, (b) instructions or commands may be defined from a remote terminal, and (c) a number of such frequency-coded command bits may be sent in parallel. *See also* **digital frequency modulation.**

frequency coherence: *See* **phase coherence.**

frequency combining: *See* **radio frequency combining.**

frequency compatibility: A measure of the extent to which communications-electronic devices will operate in their intended operational environment at design levels of performance. *Note:* For example, a random selection of channel frequencies without consideration of equipment characteristics usually results in unacceptable interference on a number of channels. This usually results in a reduction of frequency compatibility.

frequency control: The regulation of the frequency generated by a source, i.e., the maintenance of the precise frequency of a source to within a desired degree of error at each instant.

frequency conversion: *Synonym* **frequency translation.**

frequency coordination: Coordination among public and private agencies for minimizing electromagnetic interference through cooperative use of the electromagnetic frequency spectrum. *Note:* To be effective, the coordination must extend through the planning, proposal, and actual use phases of the cooperative effort.

frequency cutoff: *See* **low frequency cutoff.**

frequency departure: An unintentional deviation from the nominal frequency value. *See also* **error, frequency, frequency offset.**

frequency-derived channel: 1. A channel that is obtained by frequency division multiplexing a line, circuit, or other channel. **2.** A channel that (a) is derived by dividing an allocated or available bandwidth into two or more portions, each separately usable, (b) is continuously available, and (c) may be further divided on either a frequency division or a time division basis. *See also* **bandwidth, channel, frequency, frequency division multiplexing, time-derived channel, time division multiplexing.**

frequency deviation: 1. The amount by which a frequency differs from a prescribed value. *Note:* Examples of frequency deviation are (a) the amount an oscillator frequency drifts from its nominal frequency and (b) the instantaneous value of the difference between the actual frequency at which a transmitter is transmitting and the assigned frequency. **2.** In frequency modulation, the maximum difference, during a specified period, between (a) the instantaneous frequency of the modulated wave and (b) the carrier frequency. *Note:* The carrier frequency may be present even when there is no modulation. **3.** In frequency modulation, the absolute difference between (a) the maximum permissible instantaneous frequency of the modulated wave or the minimum possible instantaneous frequency of the modulated wave and (b) the carrier frequency. *Synonym* **frequency settability.** *See also* **assigned frequency, carrier, deviation ratio, frequency, frequency modulation, frequency shift keying, frequency stability.**

frequency dispersal: An electronic countercountermeasure (ECCM) in which communications network operating frequencies are widely separated from each other, causing a requirement to spread jamming power over wider frequency bands and thus compelling a reduction of available jamming power on any single channel or frequency, or causing a requirement for more jamming power or more jamming equipment.

frequency displacement: The end-to-end shift in frequency that may result from independent frequency translation errors in a circuit or a sequence of circuits. *See also* **error, frequency, frequency translation.**

frequency distortion: *Synonym* **amplitude versus frequency distortion.**

frequency diversity: Diversity transmission and reception in which the same information signal is transmitted and received simultaneously on two or more independently fading carrier frequencies. *Note:* Frequency diversity is used (a) to overcome fading, provide reliability, and improve reception and (b) as an electronic countercountermeasure in which various types of equipment that operate in widely separated frequency bands are used. *See also* **carrier, diversity reception, diversity transmission, dual diversity, frequency, spread spectrum, time diversity.**

frequency divider: A device that produces output power at a frequency that is an exact integral submultiple of the input frequency, such as one-half, one-third, or one-fourth of the input frequency. *See also* **frequency translator.**

frequency division: In communications systems, the use of frequency to obtain separation between channels so that several messages can be handled by the same circuit at the same time, each message being transmitted at a different frequency. *See also* **space division, time division.**

frequency division multiple access (FDMA): The use of frequency division to provide multiple and simultaneous transmissions to a single transponder. *Note:* In satellite communications, frequency division multiple access (FDMA) is used to allow carrier waves, which are radiated from different Earth stations, to use the same satellite transponder. FDMA allows each carrier to be distinguished from the others when received at an Earth station. In FDMA, the available repeater bandwidth is divided into nonoverlapping frequency bands. Each carrier is assigned one of these bands. Modulation must occur within the band. *See also* **bandwidth, carrier, frequency, frequency guard band, multiplexing, time division multiple access.**

frequency division multiplex (FDM): To derive two or more simultaneous, continuous channels from a propagation medium that connects two points by (a) assigning separate portions of the available frequency spectrum to each of the individual channels, (b) dividing the frequency range into narrow bands, and (c) using each band as a separate channel. *Synonym* **carrier multiplex.** *See* **wavelength division multiplex.**

frequency division multiplex combining: *Synonym* **radio frequency combining.**

frequency division multiplexing (FDM): The multiplexing, i.e., the deriving, of two or more simultaneous, continuous channels from a propagation medium by assigning separate portions of the available frequency spectrum to each of the individual channels. *Note 1:* A broadband frequency spectrum allows for a large number of channels. *Note 2:* In optical communications, frequency division multiplexing (FDM) may be accomplished by wavelength division multiplexing (WDM), in which several distinct optical sources, such as lasers, are used, each having a distinct center frequency, i.e., each emitting a different optical wavelength or color and each modulated by a different information-bearing signal. Each separate wavelength provides a separate channel. All wavelengths may be simultaneously coupled into a single optical waveguide, such as an optical fiber. Each wavelength can be demultiplexed, i.e., can be separated and recovered at the far end of the optical fiber, as long as a combination of dispersive components, such as prisms, and photodetectors on the receiving end are wavelength-sensitive for demultiplexing. *Note 3:* In fiber optic systems, frequency division multiplexing (FDM) is usually called "wavelength division multiplexing (WDM)" because lightwaves and optical components are best and more often described in terms of wavelength. To be consistent, modulation of the instantaneous wavelength is called "wavelength modulation (WM)," and that avoids confusion with FDM which also may be in use in the same system. FDM signals can be modulated onto optical carriers by means of WM or by intensity modulation (IM). *See also* **carrier, channelization, frequency, frequency sharing, inverted position, multiplex hierarchy, multiplexing, time assignment speech interpolation, time division multiplexing, wavelength division multiplexing.**

frequency drift: A slow undesired progressive change in frequency with time. *Note 1:* Causes of frequency drift include component aging and environmental changes. *Note 2:* Frequency drift may be in either direction, i.e., increasing or decreasing frequency, and will not necessarily be linear. *Note 3:* Frequency drift may occur in devices that generate waves as well as devices that process them. *Note 4:* Frequency drift may occur in various devices, such as oscillators, clocks, transmitters, receivers, and repeaters. *See also* **drift, frequency, frequency aging, frequency instability, frequency lock, frequency offset, frequency stability, frequency tolerance, lock-in frequency.**

frequency equalizer: *See* **phase-frequency equalizer.**

frequency evasion: An electronic countercountermeasure (ECCM) that consists of changing the frequency of a transmitter, a receiver, or both, to avoid a jamming signal.

frequency exchange signaling (FES): Frequency change signaling in which the change from one significant condition to another is accompanied by decay in amplitude of one or more frequencies and by buildup in amplitude of one or more other frequencies. *Note:* Frequency exchange signaling is used for supervisory signaling and user information signaling. *Synonym* **two-source frequency keying.** *See also* **frequency, frequency change signaling, frequency shift, frequency shift keying, signal, two-tone keying.**

frequency family: A group of frequencies that are assigned to specific sites or geographical areas. *Note:* Frequency families have been developed for the ultrahigh frequency (UHF) range with 100-kHz (kilohertz), 50-kHz, and 25-kHz spacings. The number of usable channels at a given site or in a given area is limited to the channel separation that is required to prevent interference.

frequency fluctuation: A short-term variation, with respect to time, of the frequency of an oscillator. *Note:* Frequency fluctuation, $f(t)$, is given by the relation $f(t) = (1/2\pi)(d^2\Theta/dt^2)$ where Θ is the sinusoidal wave phase angle expressed as a function of time, and t is the time. *See also* **chirping, frequency instability, frequency stability.**

frequency frogging: 1. The interchanging of the frequency allocations of carrier channels to accomplish specific purposes, such as to prevent singing, reduce crosstalk, and correct for a high transmission line frequency response slope. *Note:* Frequency frogging is accomplished by having the modulators in a repeater translate a low frequency group to a high frequency group, and vice versa. A channel will appear in the low group for one repeater section and will then be translated to the high group for the next section. This results in nearly constant attenuation with frequency over two successive repeater sections, and eliminates the need for large slope equalization and adjustments. Singing and crosstalk are minimized because the high level output of a repeater is at a different frequency from the low level input to other repeaters. **2.** In line-of-sight microwave systems, the alternate use of two frequencies at repeater sites. *See also* **carrier, channel, frequency, frequency band allocation.**

frequency gain control: *See* **receiver intermediate frequency gain control.**

frequency generator: *See* **master frequency generator.**

frequency geographical sharing: Frequency sharing in which two or more stations do not transmit into certain sections or areas so as to avoid interference among their signals. *See also* **frequency time sharing.**

frequency ground: A dedicated, single-point network intended to serve as a reference for voltages and currents, whether signal, control, or power, from direct current (dc) to 30 kHz (kilohertz) and, in some cases, to 300 kHz. *Note 1:* Pulse and digital signals with rise and fall times greater than one microsecond are considered to be lower frequency signals. *Note 2:* The term "lower frequency ground" is no longer in common use. *Note 3:* The concept of a lower frequency ground is obsolete in military facilities because of the introduction of the equipotential ground plane system of signal grounding. *See also* **higher frequency ground.**

frequency guard band: 1. A frequency band deliberately left vacant between two channels to provide a margin of safety against mutual interference. **2.** A frequency band that (a) is deliberately left vacant, i.e., left unused, between two channels to provide a margin of safety against mutual interference and (b) has a width equal to the difference between the highest frequency used by a given channel and the lowest frequency used by the next-higher-frequency adjacent channel. *See also* **band, channel, frequency, time guard band.**

frequency harmonic distortion: *See* **modulation frequency harmonic distortion.**

frequency hopper: In spread spectrum communications systems, an electromagnetic wave signal source that generates a wideband signal by changing from one frequency to another over a large number of usually fixed frequency choices pseudorandomly selected by a code sequence generator. *See also:* **spread spectrum, spread spectrum code sequence generator.**

frequency hopper direct sequence modulation: In spread spectrum systems, a combination of frequency hopping modulation and direct sequence modulation of a signal. *See also* **direct sequence modulation, spread spectrum.**

frequency hopping: During radio transmission, the repeated changing of frequencies according to a specified algorithm, usually by means of a spread spectrum code sequence generator. *Note 1:* Frequency hopping is used to avoid or minimize unauthorized interception or jamming of telecommunications. *Note 2:* The overall bandwidth required for frequency hopping is much wider than the bandwidth of the carrier. *Note 3:* The receiver must use the same code to keep in synchronism with the hopping pattern. *See* **spread spectrum, spread spectrum code sequence generator.**

frequency hopping generator: A device that allows the gating of different frequencies in rapid succession for immediate transmission. *See also* **gate, spread spectrum code sequence generator.**

frequency hopping spread spectrum (FHSS): Pertaining to transmission signal structuring in which (a) the frequency is automatically changed according to a given algorithm, (b) selection of the frequency to be transmitted usually is made in a pseudorandom manner from a set of frequencies covering a wider bandwidth than the information bandwidth, and (c) the intended receiver frequency hops in synchronism with the transmitter in order to retrieve the information in the transmitted signal. *Note:* Frequency hopping spread spectrum transmission is used as an electronic countercountermeasure (ECCM). *See also* **antijam, electronic countercountermeasure, frequency, frequency hopping, M sequence, spread spectrum.**

frequency hour: One frequency used for one hour, regardless of the number of transmitters over which the frequency is simultaneously broadcast by a station during that hour.

frequency instability: 1. The variation of the frequency generated by a circuit or a device. *Note:* Frequency instability usually results from variations of oscillator circuit parameters caused by corrosion and changing environmental conditions, such as temperature, humidity, atmospheric pressure, and vibration. Often the cause of frequency instability is unknown. **2.** The undesired variation of the frequency response of a circuit. *See also* **chirping, frequency, frequency drift, frequency fluctuation, frequency offset, frequency response curve, frequency stability, jitter, swim.**

frequency interference: *See* **radio frequency interference, single frequency interference.**

frequency intermodulation: *See* **radio frequency intermodulation.**

frequency intermodulation distortion: *See* **radio frequency intermodulation distortion.**

frequency lock: The condition in which a frequency-correcting feedback loop maintains control of the output within the limits of one cycle. *Note:* Frequency lock does not imply phase lock, but phase lock does imply frequency lock. *See also* **frequency, frequency accuracy, frequency drift, frequency stability, lock-in frequency, lock-in range.**

frequency management: The control of functions, activities, policies, and procedures that govern the satisfaction of user requirements related to the assignment and the use of the electromagnetic frequency spectrum. *See also* **frequency spectrum.**

frequency meter: An instrument that (a) measures the frequency of an electromagnetic wave that falls within its range, and (b) is calibrated to indicate the frequency of the wave that it is tuned to receive and amplify.

frequency-modulated signal: *See* **pulsed frequency-modulated signal.**

frequency modulation (FM): Modulation in which the instantaneous frequency of a sine wave carrier, such as an electromagnetic, acoustic, or elastic wave, is caused to depart from the center frequency by an amount proportional to the instantaneous value of a modulating signal. *Note 1:* In frequency modulation (FM), the carrier frequency is called the "center frequency." *Note 2:* Frequency modulation (FM) and phase modulation are forms of angle modulation. Combinations of phase and frequency modulation are considered as frequency modulation. *Note 3:* In optical communications, even if the electrical baseband signal is used to frequency-modulate an electrical carrier wave before modulating the lightwave that is transmitted, i.e., a frequency-modulated lightwave transmission system, it is still the intensity of the lightwave that is modulated by the frequency-modulated carrier to convey the information. In this case, the information is represented by the electrical frequency-modulated carrier wave, and the lightwave is varied in intensity at an instantaneous rate corresponding to the instantaneous frequency of the electrical frequency-modulated carrier wave. The optical frequency, i.e., the wavelength, of the lightwave is not modulated. *See* **digital frequency modulation, pulse frequency**

modulation. *See also* **angle modulation, carrier, deviation ratio, digital frequency modulation, frequency, frequency deviation, frequency shift keying, modulation, modulation index, phase deviation, phase modulation, phase-shift keying.**

frequency modulation (FM) blanketing: At a transmitting station, blanketing caused by the presence of a frequency modulation (FM) broadcast signal field strength equal to or greater than 115 dBu, i.e., greater than 562 mV/m (millivolts per meter). *Note:* The 115-dBu contour is the blanket contour, and the area within this contour, is the blanketing area. *See also* **blanketing, blanketing area**.

frequency modulation (FM) improvement factor: In a frequency modulation (FM) receiver, the number obtained by (a) dividing the signal to noise (S/N) ratio at the output of the receiver by (b) the carrier to noise ratio (C/N) at the input of the receiver. *Note 1:* When the frequency modulation (FM) improvement factor is greater than unity, improvement in the signal to noise (S/N) ratio is always obtained at the expense of increased bandwidth in the receiver and in the transmission path. *Note 2:* An increase in the frequency modulation improvement factor is obtained at the same time as an increased bandwidth requirement for the receiver and the transmission path. *See also* **bandwidth, carrier to noise ratio, discriminator, frequency modulation, frequency modulation threshold effect, signal to noise ratio.**

frequency modulation (FM) improvement threshold: In a frequency modulation (FM) receiver, the point at which the peaks in the radio frequency (rf) signal equal the peaks of the thermal noise generated in the receiver. *Note:* A baseband signal to noise ratio (S/N) of about 30 dB is usual at the improvement threshold. This ratio improves 1 dB for each dB of increase in the signal above the threshold. *See also* **frequency modulation, frequency modulation threshold effect, signal to noise ratio, squelch.**

frequency modulation integrator: A device that (a) has an output signal that is the time integral of the frequency-modulated input signal, (b) provides a means of achieving integration of functions and of wave shapes, (c) is more sophisticated than the video integrator and (d) allows the use of higher feedback factors, i.e., higher backward gains. *See also* **frequency modulation, integration, integrator.**

frequency modulation (FM) threshold effect: In a frequency modulation (FM) receiver, the effect that (a) is produced when the desired signal gain just begins to produce a limiting of the desired signal and noise suppression, and (b) occurs at and above the point at which the FM signal to noise improvement factor is measured. *See also* **discriminator, frequency modulation, frequency modulation improvement factor, frequency modulation improvement threshold, frequency modulation threshold extension.**

frequency modulation (FM) threshold extension: In a frequency modulation (FM) receiver, a change in the value of the FM threshold that may be obtained by decreasing the operational bandwidth, thus decreasing the received noise power and allowing the threshold of the desired signal to occur at a lower signal input level. *See also* **bandwidth, frequency modulation, frequency modulation threshold effect, level, noise power, signal to noise ratio.**

frequency multidither: In an optical system, the obtaining and the tagging of perturbations consisting of an ensemble of sine waves of different amplitudes and frequencies. *See also* **polystep dither.**

frequency octave: The frequency interval between any two frequencies that have a ratio of 2 to 1. *Note:* The number of octaves between two given frequencies is calculated as $\log_2$, or $3.322 \times \log_{10}$, of the ratio of the upper frequency to the lower frequency. *Note 1:* Wideband receivers that can cover entire frequency octaves are used to augment normal tunable receivers for monitoring and recording transmissions in a wide band of frequencies. *Note 2:* Though the term "octave" is used, it is related to "eight" only in the sense that if one frequency is eight times another, there are three octaves between them, i.e., they are three octaves apart.

frequency offset: 1. The difference between the frequency of a source and a reference frequency. **2.** The fractional frequency deviation of one frequency with respect to another frequency, given by the relation $\delta = \Delta f/f_2 = (f_1 - f_2)/f_2$ where δ is the frequency offset, Δf is the difference between the two frequencies, f_2 is the reference frequency with respect to which the offset is taken, and f_1 is the frequency that is offset. *Note 1:* If the offset frequency is less than the reference frequency, the frequency offset is negative. If the offset frequency is greater than the reference frequency, the frequency offset is positive. *Note 2:* An example of frequency offset is the fractional difference between the frequency of the National Bureau of Standards and Technology (NIST) Frequency Standard and the quartz crystal controlled oscillators from which the NIST broadcast signal are derived. These frequency offsets may be on the order of one part in 10^{12}, i.e., one part in one trillion. *See also* **fractional, frequency, frequency departure, frequency drift, fre-**

quency instability, frequency stability, frequency tolerance, primary frequency standard.

frequency of optimum traffic (FOT): *Synonym* **optimum transmission frequency.**

frequency of optimum transmission (FOT): *Synonym* **optimum transmission frequency.**

frequency power margin: *See* **radio frequency power margin.**

frequency prediction: A prediction of the maximum usable frequency (MUF), the optimum transmission frequency, and the lowest usable frequency (LUF) for transmission between two specific locations or geographical areas during various times throughout a 24-hour period. *Note:* Frequency prediction usually is indicated by means of a graph for each frequency plotted as a function of time.

frequency priority: 1. The right of an organization to use a specific frequency for authorized purposes so as to remain potentially free from harmful interference that might be caused by signals from stations of other organizations in the same-priority area, i.e., the same jurisdiction. **2.** In the allocation of frequencies, the right of a station to occupy a specific frequency without harmful interference from other stations. *See also* **frequency, interference, signals.**

frequency pulse: *See* **radio frequency pulse.**

frequency range: A continuous range or spectrum of frequencies that extends from one limiting frequency to another. *Note 1:* The frequency range for given equipment specifies the frequencies at which the equipment is operable. For example, filters pass or stop certain bands of frequencies. The frequency range for propagation indicates the frequencies at which electromagnetic wave propagation in certain modes or paths is possible over given distances. Frequency allocation, however, is made in terms of bands of frequencies. There is little, if any, conceptual difference between a range of frequencies and a band of frequencies. *Note 2:* "Frequency band" usually identifies a specific band of frequencies, such as are found in the Tables of Frequency Allocations or contained in the specifications for specific equipment. *Refer to* **Tables 1, 2, Appendix B.**

frequency record: A record of frequency allocation agreements, frequency assignments, frequency regulations, and frequency management procedures for entities or areas, such as organizations and nations. *Note:* An example of a frequency record is the Radio Regulations published by the International Telecommunication Union, Geneva, Switzerland. *See also* **frequency allocation, frequency management.**

frequency register: *See* **Master International Frequency Register.**

frequency registration board: *See* **International Frequency Registration Board (IFRB).**

frequency response: In a device, the response, i.e., the behavioral characteristics, that occurs when different frequencies are applied to the device. *Note 1:* An example of frequency response is the gain of an amplifier as a function of frequency of applied signals. The frequency response can be adjusted for different purposes, such as raising or lowering the gain for higher or lower frequencies, or maintaining a uniform flat frequency response for all frequencies in the operating range. *Note 2:* The frequency response is determined by the various impedances, such as input impedance, internal impedance, and output impedance, of the device, all of which are a function of frequency. Capacitors have a low impedance at high frequencies and a high impedance at low frequencies, whereas the reverse is true for inductors. Pure resistance does not vary with frequency, but every structure has some capacitance and some inductance. At high frequencies, the skin effect increases the resistance of a structure, such as a wire, resistor, or coaxial cable. Some metallic cables are made hollow because the interior carries such a small part of the total current that the material is all but wasted, though it does contribute to strength. There is no skin effect from continuous direct current. *See* **electrooptic frequency response.** *See also* **frequency-response curve, insertion loss versus frequency characteristic, skin effect.**

frequency-response curve: 1. For a device, a plot of a characteristic or a parameter, such as the pulse rise time, 3-dB cutoff frequency, dispersion, gain, or attenuation, as a function of frequency. **2.** A plot of the gain or attenuation of a device, such as an amplifier or a filter, as a function of frequency. *Note:* A flat curve indicates a uniform gain or attenuation over the range of frequencies where the curve is flat. Most amplifiers will have a flat frequency-response curve up to a certain maximum frequency, at which the gain is reduced. A bandpass filter usually has a peak as a frequency-response curve, and a bandstop filter has a trough as a frequency-response curve. A low-pass filter has a falling frequency-response curve, and a high-pass filter has a rising frequency-response curve.

frequency-response function: *Synonym* **transfer function.**

frequency-response test: In a device, such as a filter, amplifier, transmission line, attenuator, coaxial cable, twisted pair, or optical fiber, a test to determine the effect of frequency on a parameter of the device, such as the pulse rise time, 3-dB cutoff frequency, dispersion, gain, or attenuation, as a function of frequency.

frequency scanning: Conducting a search for signals over a band or a range of frequencies by means of a manual or automatically tuned receiver. *Note:* The tuning rate, i.e., the frequency change rate, may be fixed or variable, or it may be performed mechanically at low speed or electronically at high speed. Frequency scanning may be used to enable radar to transmit on a clear frequency, i.e., no-interference frequency, by searching a frequency band and then tuning the system to a clear portion of that band.

frequency selective waveguide: A waveguide that passes only certain frequencies, i.e., certain wavelengths, and blocks all others. *Note:* An example of a selective waveguide is a slab dielectric center waveguide with slotted or periodically varying thicknesses on each side, one slotted with a pitch of λ_1/n and the other λ_2/n where the λs are the wavelengths. If the parallel waveguides are the proper distance apart, coupling will occur between them on a highly selective basis because only a narrow frequency band will propagate and transfer to the center waveguide and to the other slotted waveguide. Other wavelengths will not match the slots and therefore will not pass.

frequency series: A group of several harmonically related electromagnetic wave frequencies. *Note:* An example of a frequency series is a fundamental frequency and a group of higher harmonic frequencies on a transmission line. *See also* **Fourier series, harmonic.**

frequency service: *See* **standard frequency service.**

frequency settability: *Synonym* **frequency deviation.**

frequency sharing: The use of the same radio frequency by two or more users where interference may occur. *Note 1:* Frequency sharing is used for stations that have different objectives. For example, one station may use a frequency for voice communications, and another station may use the same frequency for telemetry. *Note 2:* Frequency sharing does not pertain to stations using the same frequency in a given communications net. *Note 3:* Frequency sharing may be accomplished by time sharing or by geographical sharing, i.e., by space sharing, such as by using directional antennas. *See also* **communications net.**

frequency shift: 1. A change in frequency. **2.** A change in the frequency of a device, such as a radio transmitter, oscillator, or circuit, to another operational frequency. **3.** In facsimile systems, a frequency modulation system in which (a) one frequency represents picture black and another frequency represents picture white, and (b) frequencies between these two limits may represent shades of gray. **4.** An intentional frequency change used for modulation purposes. **5.** A system of radioteletypewriter operation in which the mark and space signals are represented by different frequencies. *See* **facsimile frequency shift, radio frequency shift, signal frequency shift, subcarrier frequency shift.** *See also* **frequency, frequency assignment, frequency exchange signaling, frequency shift keying, frequency shift telegraphy, frequency tolerance, phase shift.**

frequency shift keying (FSK): 1. In a transmitter, frequency modulation (FM) in which the modulating signal shifts the output frequency between predetermined values. *Note 1:* Frequency shift keying (FSK) usually is incoherent, i.e., the instantaneous frequency is shifted between two discrete values, called the "mark" and "space" frequencies, without regard for phase discontinuity. *Note 2:* In coherent frequency shift keying (CFSK), there is no phase discontinuity in the output signal when the frequencies are shifted. **2.** Signaling in which different frequencies are used to represent different characters to be transmitted. *Note:* An example of frequency shift keying is keying in which one frequency represents a 0 and another frequency represents a 1, with a smooth phase transition between them if practical or necessary. Frequency shift keying is characterized by continuity of phase during the transition from one signaling condition to another. Frequency change signaling is similar to frequency shift keying but may have a discontinuous change in frequency and change in phase with significant condition transitions. *Synonyms* **frequency shift modulation, frequency shift signaling.** *See* **coherent frequency shift keying, double frequency shift keying, incoherent frequency shift keying, multiple frequency shift keying, narrow shift frequency shift keying, wide shift frequency shift keying.** *See also* **carrier shift, differential phase-shift keying, digital modulation, digital phase modulation, frequency, frequency deviation, frequency exchange signaling, frequency modulation, frequency shift, keying, phase-shift keying, significant condition, two-tone keying.**

frequency shift modulation: *Synonym* **frequency shift keying.**

frequency shift signaling: *Synonym* **frequency shift keying.**

frequency shift telegraphy (FST): Telegraphy by frequency modulation (FM) in which (a) each significant condition, under steady state conditions, is represented by a different frequency, (b) the telegraph signal shifts the frequency of the carrier between predetermined values, and (c) there usually is phase continuity during the shift from one frequency to another, i.e., during significant condition transition. *See also* **frequency modulation, significant condition.**

frequency signal: *See* **standard time-frequency signal.**

frequency source: *Synonym* **frequency standard.**

frequency spectrum: 1. In a wave, the array of frequency components present in the wave, such as an electromagnetic, elastic, or sound wave. **2.** A portion of the electromagnetic spectrum. **3.** The frequency components of a recurring phenomenon that could be plotted as a wave. *Note 1:* The frequency spectrum usually is displayed as a plot with frequency as the abscissa and the amplitude or the energy per hertz in the various frequencies or frequency bands present in a given wave as the ordinate. *Note 2:* The frequency spectrum for a given wave might be the result obtained by a Fourier analysis of the wave. *See also* **electromagnetic spectrum, Fourier analysis, frequency spectrum designation.**

frequency spectrum congestion (FSC): The situation that occurs when many stations transmit simultaneously using frequencies that are close together, i.e., with insufficient width of frequency guard bands or channel spacing. *Note:* Frequency spectrum congestion (FSC) causes (a) difficulty in discrimination by tuning, (b) overlap in sidebands and main signal, and (c) interference that occurs when frequencies shift slightly or are phase-shifted by ionospheric reflection.

frequency spectrum designation (FSD): The designation of ranges or bands of communications frequencies. *Note:* In the United States, the designator is a two-letter or three-letter abbreviation for the name of the range. In the International Telecommunication Union (ITU), the designator is a numeral that expresses the power of 10 of the midband frequency. The ranges are a decade of frequencies wide. Thus, 3 MHz (megahertz) to 30 MHz is the high frequency (HF) band. The International Telecommunication Union (ITU) designator for the HF band is 7. *Synonym* **spectrum designation of frequency.** *See also* **electromagnetic spectrum, frequency spectrum, radio frequency.** *Refer to* **Tables 1, 2, Appendix B.**

frequency stability: 1. An inverse measure of the undesired deviation of the frequency of a tuned or controlled source or circuit, such as an oscillator, transmitter, receiver, filter, or clock, from its mean frequency during a specified period, such that the less the deviation, the greater the stability. **2.** The variation of the frequency of an oscillator from its mean frequency during a specified performance measurement period. *See* **carrier frequency stability.** *See also* **error, frequency, frequency accuracy, frequency averaging, frequency deviation, frequency drift, frequency fluctuation, frequency instability, frequency lock, frequency offset.**

frequency standard: A stable oscillator used for frequency calibration or reference. *Note 1:* The frequency standard oscillator usually generates a fundamental frequency with a high degree of accuracy and precision. Harmonics of the fundamental are used to provide reference points. *Note 2:* Frequency standards in a network or a facility are sometimes administratively designated as "primary" or "secondary." A primary frequency standard, such as a cesium beam clock, is a standard that meets national standards for accuracy and operates without the need for calibration. *Synonym* **frequency source.** *See also* **frequency, frequency stability, primary frequency standard, secondary frequency standard.**

frequency standard accuracy: The degree to which the frequency of any given frequency standard agrees with a specified primary frequency standard or with a stated frequency value. *See also* **frequency standard, primary frequency standard, value.**

frequency standard precision: An exact measure, with a specified tolerance, of the degree to which a frequency standard reproduces the same frequency each time it is placed in operation, whether or not maintenance has been performed. *See also* **frequency standard, frequency standard accuracy.**

frequency standard stability: The degree to which a continuously operating frequency standard retains its initial frequency within a specified period. *See also* **frequency standard, frequency standard accuracy, frequency standard precision.**

frequency station: *See* **standard frequency station.**

frequency switch: *See* **laser frequency switch.**

frequency synthesizer: A device that, in addition to a basic frequency, produces one or more frequencies that are phase-coherent with a reference frequency. *Note:* The reference frequency usually is from an external standard but may be from an internal source. *See also* **frequency, phase coherence, station clock.**

frequency synthesizing: Generating multiple frequencies with high stability using only one stable oscillator.

frequency system: *See* **wired radio frequency system.**

frequency tight: *See* **radio frequency tight.**

frequency time sharing: Frequency sharing in which transmitting stations in the same geographical area operate on schedules in such a manner that any two stations do not use the same, or nearly the same, frequency at the same time. *Note:* Frequency time sharing avoids interference. However, application of frequency time sharing is limited because (a) most communities need full time use of their allocated and assigned frequencies, and (b) it usually becomes an unacceptable burden on the communications system. *See also* **frequency allocation, frequency assignment, frequency geographical sharing.**

frequency tolerance: 1. The maximum permissible departure by (a) the center frequency of the frequency band occupied by an emission from the assigned frequency or (b) the characteristic frequency of an emission from the reference frequency. *Note 1:* Frequency tolerances are established by international agreement. The tolerances may vary according to the assigned frequency, the type of transmitter, such as fixed versus mobile, and the power of the transmitting station. *Note 2:* By international agreement, frequency tolerance is expressed in (a) parts per million, i.e., parts per 10^6, or (b) in hertz. *Note 3:* Frequency tolerance includes both the initial setting tolerance and excursions related to short-term and long-term instability, such as might be caused by environmental conditions and aging. *Note 4:* In the United States, frequency tolerance is expressed in parts per 10^n, in hertz, or in percentage. **2.** The maximum departure of the characteristic frequency of an emission from a reference frequency, including both the initial setting tolerance and frequency excursions that are related to short-term frequency stability, long-term frequency stability, and aging. *See also* **assigned frequency, error, frequency, frequency drift, frequency offset, frequency spectrum designation, frequency stability.**

frequency translation: The transfer of signals that occupy a definite frequency band, such as a channel or a group of channels, from one position in the frequency spectrum to another, in such a way that the arithmetic frequency difference of signals within the band remains unaltered. *Synonym* **frequency conversion.** *See* **single frequency translation.** *See also* **channel, down-converter, erect position, frequency, frequency displacement, frequency spectrum, inverted position, up-converter.**

frequency translator: *See* **translator.**

frequency weighting: The adjustment that needs to be made to meter readings because of the frequency distribution in a set of signals or in noise when measuring signal or noise energy in an electrical circuit.

frequently asked question (FAQ) file: In a communications, computer, data processing, or control system, an online file that (a) contains frequently asked questions along with answers, (b) is provided to assist new users, (c) facilitates system use, and (d) avoids repetitive offline inquiries. *Note:* Examples of frequently asked question (FAQ) files are the Frequently Asked Question (FAQ) files usually created for Internet news groups.

Fresnel diffraction pattern: *Synonym* **near field diffraction pattern.**

Fresnel equation: One of a set of equations that (a) define the reflection and transmission coefficient at an optical interface when an electromagnetic wave is incident upon the interface surface, and (b) have coefficients that are functions of the refractive indices of the propagation media on both sides of the interface, the incidence angle, and the direction of polarization with respect to the interface surface.

Fresnel field: *Synonym* **intermediate field.**

Fresnel loss: *Synonym* **Fresnel reflection loss.**

Fresnel reflection: In optics, the reflection of a portion of incident light at an interface between two homogeneous propagation media having different reflective indices. *Note 1:* Fresnel reflection occurs at the air-glass interfaces at the entrance and exit ends of an optical fiber. Resultant transmission losses are on the order of 0.2 dB, or about 4%, and can be virtually eliminated by use of antireflection coatings or index matching materials. It also occurs at the core-cladding interface. *Note 2:* A Fresnel reflection is considered to be a coherent reflection. *Note 3:* The Fresnel reflection coefficient depends upon the refractive index difference across the interface and the incidence angle. In optical elements, a thin transparent film, called an "antireflection coating," may be used to give an additional Fresnel reflection that cancels the original one by destructive interference. *Note 3:* For incident electromagnetic waves with the magnetic field vector parallel to the interface, there is no Fresnel reflection at the Brewster angle. The reflection coefficient is zero, and the transmission coefficient is unity. *See also* **antireflection coating, Brewster angle, index matching material, reflectance, refractive index.**

Fresnel reflection coefficient: The reflection coefficient for a Fresnel reflection. *Note 1:* A Fresnel reflection is a coherent reflection. *Note 2:* The Fresnel reflection coefficient depends upon the refractive index difference across a smooth interface and the incidence angle. The Fresnel reflection coefficient is zero at the Brewster angle, i.e., there is no reflection. *See also* **Brewster angle, Fresnel reflection, Fresnel reflection loss, Fresnel reflection method.**

Fresnel reflection loss: 1. Loss caused by a refractive index difference across propagation media interfaces encountered by an electromagnetic wave. *Note:* The loss is the power in the Fresnel reflection. **2.** In fiber optic links, the optical power loss that occurs because of Fresnel reflection, particularly at the terminus interface. **3.** At an interface of two media with different refractive indices, such as at the surface of a glass mirror, a window pane, or the endface of an optical fiber, the loss that takes place where a fraction of the optical power is reflected back toward the source. *Note 1:* At normal incidence, the fraction of reflected power at the interface of two media is expressed by the relation $R = (n_1 - n_2)^2/(n_1 + n_2)^2$ where R is the fraction of incident power that is reflected, and n_1 and n_2 are the refractive indicies of the two media. *Note 2:* The reflection loss in dB is given by $10 \log_{10} R$. *Note 3:* Fresnel reflection loss for ordinary window pane at about normal incidence is about 0.2 dB, or about 4%. *Synonyms* **Fresnel loss, Fresnel reflectance loss, reflectance loss.** *See also* **Fresnel reflection, reflection coefficient, reflection loss, transmission coefficient.**

Fresnel reflection method: In fiber optics, a method for measuring the refractive index profile of an optical fiber by measuring the reflectance as a function of position on an endface of the fiber. *See also* **Fresnel reflection, index profile, reflectance.**

Fresnel region: *Synonym* **intermediate field region.**

Fresnel zone: In radio communications, (a) the zone that is bounded by a cigar-shaped shell of circular cross section, obtained by rotating an ellipse about its longitudinal axis, forming an ellipsoid of revolution that surrounds the direct path between a transmitter and a receiver, where (b) (i) for the first Fresnel zone, the distance from the transmitter to any point on this shell and on to the receiver is one half-wavelength longer than the direct path, (ii) for the second Fresnel zone, is two half-wavelengths longer, (iii) for the third Fresnel zone, is three half-wavelengths longer, and (iv) so on for higher-order Fresnel zones. *Note 1:* The cross section of the first Fresnel zone is circular. Subsequent Fresnel zones are annular in cross section, and concentric with the first. *Note 2:* Odd-numbered Fresnel zones have relatively intense field strengths, whereas even-numbered Fresnel zones are nulls. *Note 3:* Fresnel zones result from diffraction by the circular aperture. *See also* **effective Earth radius, Fresnel reflection, k factor, path clearance, path profile, path survey, propagation path obstruction.**

fringe: In optics, a light or dark band, region, or area caused by interference of two or more electromagnetic waves, usually lightwaves, so that bands, regions, or areas of wave reinforcement and cancellation occur. *See* **Newton's fringes.**

frogging: *See* **frequency frogging.**

front-emitting light-emitting diode. *Synonym* **surface-emitting light-emitting diode.**

front end computer: *Synonym* **front end processor.**

front end noise temperature: A measure of the thermal noise generated in the first stage of a receiver. *See also* **noise.**

front end processing: 1. The processing of information prior to insertion into a communications, computer, data processing, or control system or network. *Note 1:* Front end processing may include operations such as serial-to-parallel conversion, packetizing, multiplexing and concentration, network access signaling and supervision, protocol conversion, error control, and diagnostics. *Note 2:* Front end processing usually is performed by a computer. **2.** In a computer or a computer network, processing that relieves a host computer of certain nonapplication-oriented processing tasks, such as data formatting, data preprocessing, line control, message handling, code conversion, and error control. *Synonym* **preprocessing.**

front end processor (FEP): A programmable logic or stored program device that (a) performs operations for another processor and (b) serves as an interface with other functional units, such as data communications equipment, input-output buses, and the memory of a data processor or a computer. *Synonym* **front end computer.** *See also* **computer.** *Refer to* **Fig. T-5 (1012).**

front feed: *Synonym* **prime focus feed.**

front focal length (FFL): In a lens of an optical system, the distance measured from (a) the principal focus point, located in the front space, to (b) the first principal point on the lens. *See also* **first principal point, optical system, principal focus point.**

front sign: In semaphore signaling, the flag positions that are used after the attention sign to indicate other signs, such as the operating sign. *See also* **attention sign, flag position, operating signal, semaphore, semaphore signaling.**

front surface mirror: An optical mirror on which the reflecting surface is applied to the front surface, i.e., the viewing side or side of first incidence, instead of the back side.

front-to-back ratio: In a device, such as an antenna, rectifier, or transmission line, the ratio of (a) a parameter, such as signal strength, impedance, or power level, in one direction to (b) the same parameter in the opposite direction.

front velocity: *See also* **phase front velocity.**

frustrated total internal reflection sensor: A fiber optic sensor that (a) consists of two optical fibers with an end of each separated by an air gap produced by cutting a fiber and polishing the cut ends at a precise angle to produce total internal reflection in the source fiber when the fibers are widely separated, and in which (b) when the distance between the ends, i.e., the air gap, is sufficiently small, a large portion of the light in the source fiber is coupled to the fiber connected to a photodetector, (c) as the gap is widened, say by moving the fiber axes laterally away from one another, by moving the fiber endfaces longitudinally away from one another, or by increasing the angular misalignment of the fibers, the light coupled across the gap decreases, and (d) the irradiance, i.e., the light intensity, at the photodetector is a function of the lateral offset, longitudinal displacement, or angular misalignment of the fibers. *See also* **angular misalignment loss, fiber optic sensor, lateral offset loss, longitudinal displacement loss.**

FSDPSK: filtered symmetric differential phase-shift keying.

FSK: frequency shift keying.

FSTS: Federal Secure Telephone Service.

FTAM: file transfer, access, and management.

FTC: fast time constant.

FTP: File Transfer Protocol.

FTS: Federal Telecommunication Standard, Federal Telecommunications System.

FTS2000: Federal Telecommunications System 2000.

fulcrum: *See* **type bar fulcrum.**

full adder: *Synonym* **binary adder.**

full baud bipolar coding (FBBC): In a bipolar sequence of pulses, coding that (a) consists of representing a 0 by a zero-level voltage in the bipolar sequence of pulses and a 1 by alternate positive and negative pulses in the bipolar sequence of pulses, (b) has the pulses remain at their respective levels, i.e., significant conditions, for the entire baud length, i.e., a unit interval, and (c) has an inherent error detection capability because of the requirement that the polarity of the pulses alternate, so that, if the alternation does not occur, an error can be signaled by a circuit that is checking the alternation. *See also* **baud, bipolar, bipolar coding, half-baud polar coding, significant condition, unit interval.**

full carrier: A carrier wave that is transmitted at a power level that has not dropped more than a specified amount from the peak envelope power, such as not more than 3 dB or not more than 6 dB. *Note:* A full carrier usually has (a) double sideband and single sideband amplitude-modulated transmissions and (b) a power level that is 6 dB below the peak envelope power at 100% modulation.

full carrier single sideband emission: A single sideband emission without reduction, i.e., without suppression, of the carrier. *See also* **reduced carrier single sideband emission, single sideband emission.**

full carrier transmission: Transmission in which a modulated full carrier is transmitted, i.e., the amplitude of the carrier never drops more than a specified amount below the peak envelope power, such as not more 3 dB or not more than 6 dB.

full common intermediate format: *Synonym* **common intermediate format.**

full duplex (FDX) circuit: A circuit that permits simultaneous transmission in both directions. *Synonym* **duplex circuit.**

full duplex operation: *Synonym* **duplex operation.**

full duplex transmission: *Synonym* **duplex transmission.**

full duration at half maximum (FDHM): 1. Full width at half maximum in which the independent variable is time. **2.** Pertaining to the time interval during which a characteristic or a property of a variable, such as the amplitude of a pulse, has a value greater than 50% of its maximum value. *Synonym* **full duration half maximum.** *See also* **full width at half maximum.**

full duration half maximum: *Synonym* **full duration at half maximum.**

Fullerphone: An instrument that (a) uses a very low direct current (dc) in the line, (b) converts the dc into an intermittent current of audio frequency at the receiver, (c) enables hand-speed Morse telegraphy to take place over good or bad lines with the least chance of remote interception, (d) uses buzzer signals for keying and reception by listening, and (e) uses dc for telegraphic transmission, thus making the system less vulnerable to clandestine tapping. *See also* **audio frequency.**

full justify: To justify right and left at the same time. *See also* **justify, left justify, right justify.**

full modulation (FM): In an analog-digital converter, the condition in which the input signal amplitude has just reached the threshold at which clipping begins to occur.

full motion operation rate: In television, an encoded/decoded video frame update rate that provides the appearance of full motion without flicker or smear in the image. *Note:* Picture motion appears to be full at greater than 16 fps (frames per second). European television operates at 25 fps and North American television at 30 fps.

full screen editing: Editing on a display device in which an operator can access data displayed anywhere on the entire display space, i.e., the entire screen, using a device, such as a keyboard, cursor, stylus, light pen, or mouse, to enter commands. *See also* **command, display device, editing, line editing.**

full screen processing: Processing data using a display device in which the operator may enter data, such as by keying or by mouse clicking, into some or all unprotected fields on the screen before transferring the data into storage. *See also* **display device, editing, full screen editing, line editing, mouse, unprotected field.**

full wave half power point: In fiber optic systems, a measure of the information-carrying capacity of an optical fiber of given length in which the power or the amplitude of an optical pulse of a full wave, i.e., a 50% duty cycle, is reduced to half of the steady state level. *Note:* Dispersion and attenuation contribute to the reduction in pulse amplitude or power level.

full width at half maximum (FWHM): 1. Pertaining to a measure of the extent of a function, given by the difference between the two extreme values of the independent variable at which the dependent variable is equal to half of its maximum value. *Note:* Full width at half maximum (FWHM) usually pertains to the duration of pulse waveforms, the spectral extent of emission or absorption lines, and the angular or spatial extent of radiation patterns. "Full duration at half maximum" is preferred when the independent variable is time. **2.** Pertaining to the range over which a property of a variable has a value greater than 50% of its maximum value. *Note:* Full width at half maximum (FWHM) pertains to (a) characteristics, such as radiation patterns, optical pulse durations, spectral widths, spectral linewidths, beam diameters, and beam divergences, and (b) variables, such as wavelength, voltage, power, and current. *Synonym* **full width half maximum.** *See also* **full duration half maximum.**

full width half maximum: *Synonym* **full width at half maximum.**

fully automatic reperforator transmitter distributor: *Synonym* **coupled reperforator tape reader.**

fully connected network: *See* **network topology.**

fully connected topology: A network topology in which there is a direct path, i.e., branch, between any two nodes. *Note:* In a fully connected network with n nodes, there are $n(n-1)/2$ direct paths, i.e., branches. *Synonyms* **fully connected mesh network, fully connected network.** *See also* **branch, local area network, network topology, node.**

fully filled optical fiber: An optical fiber in which the optical power in a propagating lightwave is distributed equally among all the modes the fiber is capable of supporting in the equilibrium modal power distribution condition. *Note:* Leaking modes are lost in long optical fibers, i.e., leaky modes are filtered by being stripped off, and therefore are not supported by the fiber. Some light sources, such as lensed light-emitting diodes (LEDs) and edge-emitting LEDs, couple most of their output optical power into the lower-order modes, which tend not to leak, thus keeping the fiber fully filled. Connectors do some mode mixing by adding power into the higher-order modes. *See also* **equilibrium modal power distribution, filtering, leaky mode.**

fully intermateable connector: A connector from one source, i.e., from one manufacturer, that mates with complementary components from other sources (a) without mechanical damage and (b) with transmission properties that are maintained within specified limits. *See also* **compatibility.**

function: 1. In mathematics, a group of associated operators and operands, expressed symbolically and capable of being evaluated by humans or machines when parameters, i.e., values, are assigned to the operands. *Note:* An example of a function is $y = mx + b$; i.e., y, as a function of x, can be evaluated when m, x, and b are given. Examples of other functions are exponential functions, transfer functions, transcendental functions, hyperbolic functions, trigonometric functions, Bessel functions, LeGendre polynomials, and complex functions. **2.** The actual performance or execution of a capability, such as transmission, conversion, transformation, translation, or detection. **3.** A capability or a role. *See* **automatic function, Boolean function, closed loop transfer function, complex transfer function, control function, end of message function, end of transmission function, error function, fiber optic bundle transfer function, Gaussian function, harmonic function, impulse response function, machine function, mediation function, modulation transfer function, N function, optical fiber transfer function, recursive function, sectorial harmonic function, sinusoidal function, spherical harmonic function, step function, tesseral harmonic function, transfer function, work function, zonal harmonic function.** *See also* **argument, formula, function-driven cell.** *Refer to* **Figs. P-15 (763), P-19 (771), R-2 (799), T-7 (1024), T-11 (1042), V-2 (1064).**

Functional Advanced Satellite System for Evaluation and Test (FASSET): An extremely high frequency (EHF) functional advanced development model for evaluating satellite communications systems and concepts. *Note:* The Functional Advanced Satellite System for Evaluation and Test (FASSET) is a part of the Canadian Department of National Defence spacecraft program.

functional component (FC): In intelligent networks, an elemental call processing component that directs internal network resources to perform specific actions, such as collect dialed digits. *Note:* A functional component (FC) is unique to the intelligent network IN/2 architecture. *See also* **functional unit, intelligent network.**

functional design: A design that specifies the working relationships among the parts of a system.

functional group: 1. The functions performed by a specified set of equipment. **2.** In Integrated Services Digital Networks (ISDNs), the functions performed by specific equipment. *See* **terminal endpoint functional group.** *See also* **Integrated Services Digital Network.**

functionality: *See* **interface functionality.**

functionally integrated resource manager (FIRM): In fighter aircraft, an avionic programmable computer that (a) accepts, processes, and displays data from onboard sensors and external communications systems, (b) makes the data available to onboard avionic systems and to the air crew during critical situations, (c) relieves the crew of routine flight and fire control functions, and (d) allows the pilot to assume the role of a database manager.

functional profile: In a standard with options, the agreed-upon options and other variations allowed by the selected standards in order to (a) support a specific information transfer function, (b) provide for the development of uniform, recognized tests, and (c) ensure interoperability among different network elements and terminal equipment that implement the specific options.

functional requirement: A requirement that defines a specific function that a system, component, or functional unit must be capable of performing or executing. *Note:* Examples of functional requirements are a requirement to have the capability to (a) execute a specified computer program, (b) evaluate a specified algebraic expression, (c) interconnect 2500 lines, and (d) transmit 25 kW (kilowatts) of radio frequency (rf) energy at 1.4 MHz (megahertz). *See also* **functional unit.**

functional signaling: In an integrated services digital network (ISDN), signaling in which the signaling messages are unambiguous and have clearly defined meanings that are known to both the sender and the receiver of the messages. *Note:* Functional signaling is usually generated by the data terminal equipment (DTE).

functional signaling link: A combination of a communications link and the associated transfer control functions. *See also* **functional signaling, link, signal.**

functional specification: A documented function that a component, subsystem, or system is designed to perform and should be capable of performing.

functional unit (FU): A hardware, software, or firmware item that has a specific purpose or that is capable of performing a specific function. *Note:* Examples of functional units are transmitters, receivers, trunks, computers, computer programs, read-only memories, transponders, fiberscopes, repeaters, antennas, connectors, and data circuit-terminating equipment (DCE). *See also* **functional component.**

function call: In personal computer (PC) spreadsheet software systems, the citing or the entering of a function on a spreadsheet. *Note:* In most personal computer (PC) spreadsheet software, it is only necessary to enter a symbol to indicate that a function is being cited (not data per se), the name of the function, and the arguments, all in the proper format. An example of a function call entry in Lotus 1-2-3 is @SUM(A1..A6), in which @ indicates a function follows, SUM indicates that the operation to be performed is addition, and (A1..A6) indicates the locations, i.e., the cells, that contain the numbers to be added. The function call is accomplished simply by placing the function in the cell where the operator wishes the sum to appear, say A7. That cell will always display the sum. The function will appear in the top margin of the screen for checking or changing when the pointer is placed at A7, i.e., column A row 7. If any of the numbers in A1 through A6 is changed, the new sum in A7 will automatically be recalculated and displayed, whether or not the pointer is placed at A7.

function control: The control of an operation to be performed by a system, such as a telegraph, data recording, or data processing device, other than recording or printing a letter, numeral, punctuation mark, or the graphic symbols contained in a data stream, such as a message or a frame. *Note:* An example of function control is the control of line feed, carriage return, shifting, and spacing.

function-driven cell: In personal computer spreadsheet software systems, a cell on a spreadsheet that has a stored value that is the result of functions and values of parameters in those functions that are placed in other cells on the same spreadsheet. *See also* **argument, formula, function.**

function generator: In analog communications, computer, data processing, and control systems, a functional unit that has an output variable that is a function of its input analog variables. *See* **fixed function generator, variable function generator.**

functioning life: In a population of items or events, the total number of items or events times a specified measurement period. *Note:* If there are 10^4 transistors in a population of transistors and a performance measurement period is 100 s (seconds), then the functioning life for the transistors is $10^4 \cdot 10^2$, i.e., 10^6 transistor•seconds.

function key: **1.** A specific key on a keyboard that provides for function control. *Note:* Examples of function keys on a teletypewriter keyboard are keys that are marked LTRS, CR, LF, or FIGS. When operated, these keys cause the teletypewriter to perform the identified machine functions of letters shift, carriage return, line feed, or figures, i.e., numerals, shift, respectively. These keys enable a message to be transmitted, received, and recorded in proper form. **2.** In display systems, a button or a switch that may be used to generate and dispatch a signal that (a) identifies, or corresponds to, the button or the switch that is actuated, selected, or depressed, and (b) can be used to control a display device directly or through the use of a computer program. **3.** In communications, computer, and data processing systems, a key on a keyboard that (a) when actuated alone, or in concert with other keys, will initiate a specific action by the system, and (b) is marked with codes that indicate the function that will be performed when the key is actuated. *Note:* In a standard keyboard, the letter F and a numeral from 1 through 10 or 1 through 12 mark the function keys. In WordPerfect, the function keys, when actuated alone, perform the functions Cancel, Search, Help, Indent, List, Bold, Exit, Underline, End Field, and Save for the 10 function keys, in that order, and Reveal Codes and Block for F11 and F12, if provided. Other functions, such as Search, Switch, Center, and Print, require one of the 10 or 12 function keys and one other key, such as Shift, Control, or Alternate key. *See also* **function control.**

function keyboard: A keyboard that contains a set of function keys for initiating, activating, controlling, or causing control functions. *Note:* An example of a function keyboard is a keyboard, key set, key panel, or key pad mounted on a display device console to form a part of an input unit for the device. *See also* **function key, keyboard, keypad.**

function signal: A set of signal elements that is used to transmit or represent a function control character that actuates a control function, such as carriage return, line feed, letters shift, or figures shift, that is to be performed by communications devices, such as teletypewriters, teleprinters, and typewriters. *See also* **function control, function key, signal element.**

fundamental: *Synonym* **fundamental frequency.**

fundamental frequency: **1.** The basic or lowest frequency in a wave or periodic phenomenon, i.e., the repetition rate of the repeated phenomenon. **2.** In a square or triangular wave, the repetition rate of the complete square or triangular wave. *Note:* A pure sine wave has one and only one frequency. If it is distorted, it may be considered to consist of other higher frequencies that are multiples of the original undistorted wave. The fundamental frequency remains the same as long as the distortion is limited. In a Fourier analysis representation of an arbitrarily shaped periodic wave, the higher frequencies are multiples of the fundamental frequency. Each frequency has a specific amplitude, such that when all the frequencies are added, the original distorted or arbitrary wave is obtained. The multiples of the fundamental frequency are harmonics of the fundamental frequency, beginning with the second harmonic. The first harmonic is the fundamental frequency itself. *Synonym* **fundamental.** *See also* **Fourier analysis, frequency.**

fundamental keying frequency: *Synonym* **maximum keying frequency.**

fundamental mode: 1. The lowest-order mode in a waveguide, such as an optical fiber or a coaxial cable. *Note:* In optical fibers, the fundamental mode is designated as the LP_{01} or the HE_{11} mode. **2.** In a multimode optical fiber, the lowest-order mode that (a) the fiber with a given numerical aperture and set of launch conditions is capable of supporting, and (b) corresponds to the cutoff frequency, i.e., the longest wavelength the fiber is capable of supporting. *Note 1:* In a single-mode optical fiber, the one mode is the fundamental mode. *Note 2:* If the frequency of the source is increased, and consequently the wavelength decreased, so that the same optical fiber can support additional modes, the fundamental mode still remains the same mode, even though the same fiber is supporting more modes with wavelengths shorter than that of the fundamental mode. *See also* **mode.**

fundamental scanning frequency: *Synonym* **maximum scanning frequency.**

furcate: *Synonym* **breakout.**

furcation coupling: The mixing of signals from several separate optical fibers by passing them through a common single optical fiber, thus obtaining a signal containing all the components of the several signals, such as several colors contained in one pulse. *See also* **optical fiber, signal.**

fuse: In electrical circuits, an overcurrent protective device with a circuit-opening part that is heated and severed by the passage of a current in excess of a specified value. *Note:* Fuses are rated according to a current level in amperes, above which they will open the circuit in which they are connected. *See also* **fuse disconnecting switch, fuze, protector.**

fuse disconnecting switch: A disconnecting switch that (a) has a fuse that forms part of, or is connected to, a blade assembly, (b) provides circuit continuity when inserted into a circuit, (c) usually can be plugged into a circuit, and (d) when unplugged from the circuit, leaves the circuit open so that current flow is either interrupted or prevented. *See also* **fuse, switch.**

fused quartz: Glass that (a) is made by melting natural quartz crystals and (b) is not so pure as vitreous silica. *Note:* Fused quartz is used in the manufacture of some optical fibers. *See also* **vitreous silica.**

fused silica: *Synonym* **vitreous silica.**

fusion: *See* **data fusion.**

fusion splice: A fiber optic splice that (a) is accomplished by the application of heat sufficient to melt, fuse, and thus join two lengths of optical fiber, forming a single continuous fiber, (b) if properly made, results in a continuous length of material with low, or no, discontinuities at the splice and a near-zero attenuation, and (c) usually is made with commercially available fusion splicing equipment. *See also* **connection, splice.**

fusion splicing: In optical transmission systems using solid dielectric propagation media, the joining together of two media by butting them, forming an interface between them, and then removing the common surfaces by a melting process so that there is no interface between them. *Note:* A lightwave will be propagated from one propagation medium, such as an optical fiber, to another without reflection at the joint and therefore without an insertion loss. However, the joint is never perfect, and so some small loss will occur at the fiber optic splice.

fuze: 1. In ordnance devices, such as mortar and artillery rounds, bombs, and grenades, the assembly that detonates the device. *Note:* Examples of fuzes are point-detonating and proximity fuzes. **2.** A cord or a rope that (a) is impregnated with a combustible material that has its own supply of oxygen, (b) will burn progressively when ignited at any point, usually one end, (c) is used to carry a flame from one point to another, and (d) usually has a specific burning rate of progress. *See also* **command fuzing, proximity fuze detonation, fuse.**

fuzing: *See* **command fuzing.**

FWHM: full width at half maximum.

FX: fixed service, foreign exchange service.

F1A line weighting: A noise weighting used in a noise measuring set to measure noise on a line that is terminated by a 302-type or similar instrument. *Note 1:* The meter scale readings are in dBa(F1A). *Note 2:* F1A line weighting is obsolete for new U.S. Department of Defense (DoD) applications. *See also* **C message weighting, dBa(F1A), flat weighting, noise, noise weighting 144 line weighting.**

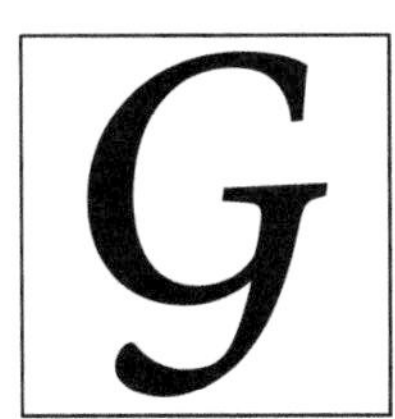

g: In fiber optics, the symbol used for the refractive index profile parameter. *See* **refractive index profile parameter.** *See also* **power law refractive index profile.**

G: giga.

gage station: *See* **radar beacon precipitation gage station.**

gain: 1. The ratio of output current, voltage, or power to input current, voltage, or power, respectively. *Note:* Gain usually is expressed in dB. **2.** In an active device or system, the ratio between (a) an output quantity and (b) an input quantity, the quantities being of the same nature, such as both voltage, both power, or both electric field strength, and expressed in the same units, such as both volts, both watts, or both volts per meter. *Note 1:* If the gain expressed in dB is positive, there is a gain between input and output. If the gain expressed in percent is greater than 100%, there is a gain. If the gain expressed per unit is greater than unity, there is a gain. *Note 2:* If the gain expressed in dB is negative, there is a loss between input and output. If the gain is expressed in percent is less than 100%, there is a loss. If the gain expressed per unit is less than unity, there is a loss. *See* **absolute gain, antenna gain, antenna power gain, directive gain, diversity gain, height gain, insertion gain, loop gain, net gain, pair gain, power gain, process gain, Q gain, signal processing gain, statistical terminal/regenerator system gain, terminal/regenerator system gain.** *See also* **dB, loss.** *Refer to* **Fig. T-7 (1024).**

gain antenna: *See* **high gain antenna.**

gain•bandwidth product: In an active device, the product of a measured gain and a specified bandwidth. *Note:* The gain•bandwidth product is expressed in hertz when the gain is expressed per unit and the bandwidth is expressed in hertz.

gain combiner: *See* **equal gain combiner.**

gain contour: *See* **effective antenna gain contour.**

gain control: In an active device, the ability to adjust the gain of the device to a desired value for a specific purpose. *See* **automatic gain control, instantaneous automatic gain control, receiver intermediate frequency gain control.**

gain control circuit: In an active device, a circuit that enables the gain of the device to be adjusted to a desired value for a specific purpose, such as to increase the gain when the input signal is weak so as to maintain a constant output level. *See* **detector back bias gain control circuit, level.**

gain factor: *See* **photoconductive gain factor.**

gain frequency distortion: *Synonym* **frequency distortion.**

gain hit: *See* **hit.**

gain medium: An active medium, device, or system in which amplification of input, such as electric power, current, voltage, or radiation, occurs with or without feedback. *Note:* Gain media include amplifiers, lasers, and avalanche photodiodes (APD).

gain of an antenna: *Synonym* **antenna gain.**

gain system: *See* **pair gain system.**

gain to noise temperature (G/T): *See* **antenna gain to noise temperature.**

galactic noise: *Synonym* **cosmic noise.**

galactic radio noise: *Synonym* **cosmic noise.**

Galite: An optical fiber produced by Galileo Electro-Optics Corporation available in single-fiber and multiple-fiber cables with small and large cores.

Galvanic series: *Synonym* **electrochemical series.**

gamma photon: A fundamental particle or quantum of gamma radiation that (a) has an energy equal to hf, where h is Planck's constant and f is the frequency of the radiation, i.e., of the photon, (b) has a frequency higher than that of X rays and optical spectra, which includes visible, infrared, and ultraviolet spectra, and (c) occupies the 10^{10} GHz (gigahertz) to 10^{12} GHz band of the electromagnetic spectrum. *Note:* At these frequencies even a single photon can be destructive to human tissues. *See also* **electromagnetic spectrum.**

gamma radiation: Electromagnetic waves that (a) have a frequency higher than that of lightwaves and X rays, (b) can only be stopped or absorbed by highly dense materials, such as lead and depleted uranium, (c) cannot be confined to or guided in optical fibers, (d) usually are produced by nuclear reaction and high-powered lasers when transitions occur from very high to very low electron energy levels in molecules, and (e) occupy the 10^{10} GHz (gigahertz) to 10^{12} GHz band of the electromagnetic spectrum. *Note:* At these frequencies even a single photon can be destructive to human tissues.

gamma ray: A ray of electromagnetic radiation from a source that (a) radiates in the 10^{10} GHz (gigahertz) to 10^{12} GHz frequency band, compared to 10^{6} GHz for lightwaves, (b) is undergoing high level to low level energy transitions, and (c) emits rays that have sufficient energy to destroy living tissue by changing its molecular structure.

gap: *See* **file gap, interblock gap.**

gap energy: *See* **band gap energy.**

gap loss attenuator: *Synonym* **longitudinal displacement attenuator.**

gap loss: In an optical fiber butted joint, such as is made by a splice or a connector, a loss of optical energy at the joint. *Note:* The three basic types of gap loss are angular misalignment loss, lateral offset loss, and longitudinal displacement loss. *Deprecated synonym* **longitudinal displacement loss.** *See* **angular misalignment loss, lateral offset loss, longitudinal displacement loss.**

garble: In communications systems, an effect that (a) usually is an undesired result of one or more errors in transmission, propagation, reception, encryption, or decryption that changes the text of a message or any portion thereof in such a manner that it is incorrect or undecryptable, and (b) usually appears as a random sequence of letters that may result from various causes, such as from (i) transposing, omitting, substituting, or adding characters, (ii) using wrong key when deciphering, (iii) dropout pulses on reading, recording, or transmitting, and (iv) equipment malfunctions, computer program errors, and operator mistakes. *See also* **decrypt, encrypt, error, key.**

garble table: An aid, such as a table, chart, spreadsheet, or list, that is used to assist in the correction or the interpretation of a garble. *See* **teletypewriter garble table.** *See also* **garble.**

garnet source: *Synonym* **YAG/LED source.**

gas laser: A laser in which the active medium is a gas. *Note:* Types of gas lasers are (a) atomic lasers, such as helium-neon lasers, (b) ionic lasers, such as argon, krypton, and xenon lasers, (c) metal vapor lasers, such as helium-cadmium molecular and helium-selenium lasers, and (d) molecular lasers, such as carbon dioxide, hydrogen cyanide, and water vapor lasers. *See* **mixed gas laser.** *See also* **active laser medium, laser.**

gas panel: In a display device, a part that consists of a display surface beneath which is a grid of electrodes or wires in a flat gas-filled envelope, in which (a) the energizing of a group of electrodes causes the electrons in the gas molecules to attain higher energy levels at specific locations, and (b) when the electrons return to normal levels, photons are emitted at those locations, thus creating an image of light. *Synonym* **plasma panel.**

gate: 1. A combinational circuit that (a) performs an elementary logic operation, i.e., such as the AND, OR, NOR, NAND, or NEGATION operation, (b) usually has one output channel and one or more input channels, and (c) has an output channel state that is completely determined by the input channel states, except during switching transients, i.e., during transition from one significant condition to another. *Note 1:* Inputs to gates may be 0 or 1. If there are two inputs, there are 16 different possible input arrangements, ranging from

0000 to 1111, giving rise to 16 different gating arrangements. *Note 2:* Strictly speaking, a device that emits a continuous series of 0s or 1s might be considered as a gate only if the output can be inhibited. **2.** To select portions of electrical pulses or electromagnetic waves that meet specified criteria, such as portions that lie between certain time or amplitude limits. **3.** To use combinational circuits, i.e., to use logic, for purposes such as to mix signals, execute logic functions, perform signal processing functions, and adjust the timing of signals. *See* **A AND NOT B gate, A IGNORE B gate, AND gate, A OR NOT B gate, B AND NOT A gate, B IGNORE A gate, B OR NOT A gate, channel gate, EXCLUSIVE NOR gate, GENERATOR gate, NAND gate, NEGATION gate, NEGATIVE B IGNORE A gate, NOR gate, NULL gate, OR gate, range gate.** *See also* **Boolean function, Boolean operation, Boolean operation table, combinational circuit, gating, gating circuit, logic, logical expression, significant condition.** *Refer to* **Figs. A-7 (34), B-6 (78), N-1 (609), N-2 (615), N-19 (637).**

gate stealing: *See* **range gate stealing, velocity gate stealing.**

gateway: 1. In a communications network, a functional unit that (a) is used for interfacing with a user or a network using different architectures or protocol suites, (b) may perform the functions of a router, (c) usually is located at, or may be part of, a node, and (d) is used for interconnecting (i) dissimilar local area networks (LANs), (ii) devices on the same LAN that use different high level protocols, and (iii) LANs with networks of different architectures. *Note 1:* A media conversion gateway interconnects networks that have similar high level protocols but different transmission level protocols, i.e., different physical layer or data link layer protocols. *Note 2:* A protocol translation/mapping gateway interconnects networks with different network and protocol technologies by performing the required protocol conversions. *Note 3:* A gateway may consist of various devices, such as protocol translators, impedance matching devices, rate converters, fault isolation equipment, and signal translators, as necessary, to provide system interoperability. Mutually acceptable administrative procedures must be established between the interconnected entities. **2.** A node at which two or more networks are connected, with the fundamental role of serving as the boundary between the internal protocols of the connected networks. **3.** The collection of hardware and software that is required to effect an interconnection between one communications network and another. **4.** The interface or the connection between one communications network and another. **5.** The path into or out of a communications network. *Note:* An example of a gateway is a connection between a local area network (LAN) of personal computer (PC) systems and a large-scale host computer. *See also* **bridge, communications, interface, internetworking, network, node, port, router.** *Refer to* **Figs. N-13 (623), S-9 and S-10 (880).**

gateway board: A printed circuit board (PCB) that performs the functions of a gateway, such as (a) connecting one network to another and (b) connecting a local area network (LAN) of personal computer (PC) systems to a host computer so that all the PCs can communicate with the host computer and with each other. *See also* **gateway, printed circuit board.**

gateway interface: An interface between two dissimilar communications systems in which (a) switching centers are provided that enable switching, signaling, supervisory, and transmission functions to be performed on calls from a user in one system to a user in another system, (b) switching functions are automatically performed under the control of a programmed computer, and (c) usually only a limited number of users in both systems are permitted access to the other system via the interface.

gateway switching center: An automatic telephone network switching center that is part of a gateway and provides interconnections among communications systems and networks. *Note:* An example of a gateway switching center is a switching center that provides connections to networks in foreign countries or overseas. *See also* **gateway, switching center.**

gating: 1. The permitting of an event to occur when specified conditions are met. **2.** The selecting of only those portions of a wave between specified instants or between specified amplitude limits. *See also* **limiting, synchronizing. 3.** The controlling of signals by means of combinational circuits. **4.** A process arranged such that when a predetermined set of conditions is met, a second process is allowed to occur. *See also* **combinational circuit, decision circuit, gate, gateway, gateway board, synchronizing.**

gating circuit: A circuit that is used to perform signal processing operations, such as limit input signals to another circuit, whether in time or space, prevent interference, commutate pulses, multiplex, perform logic decisions, limit output operations during receiving operations, prevent blanking, or prevent saturation. *Note:* Gating circuits are used (a) in infrared scanning systems, to reduce blanking by laser pulses by admitting pulses only during certain time intervals or when specified conditions are met, (b) in radars, to process echoes only at certain times after the launching of a signal, and (c) in computers, to perform arithmetic and logical operations. *See also* **blanking, combinational circuit, signal processing.**

Gaussian beam: A beam of light that has a Gaussian electric field strength distribution, i.e., a distribution resembling a bell curve. *Note:* When a Gaussian beam is circular in cross section, the electric field strength, E_r, is given by the relation $E_r = E_0e^{-x}$ where E_0 is the electrical field strength at the beam axis, i.e., at $r = 0$, $x = (r/r_a)^2$ where r is the distance from the center of the beam, and r_a is the value of r at which the intensity is $1/e$ of its value on the axis, where e is the base of natural logarithms, equal to approximately 2.718. *See also* **beam diameter.**

Gaussian distribution: 1. A normal distribution, given by the Gaussian function, i.e., by the relation $G(x) = (1/2\pi)^{1/2}\exp-x^2/2$. **2.** In a large population of entities, a distribution that (a) represents the number of entities that have a given value of an attribute, characteristic, or quality that matches the function $G(x) = A\exp-(x/a)^2$ where A is the ordinate value of $G(x)$ at $x = 0$, x is the independent value that represents the number in the population that have the specific attribute, characteristic, or quality of the entities, and a is the value of x where the value of the function $G(x)$ is $1/e$ of the maximum, (b) is symmetrical about a center maximum, and (c) has the shape of a bell curve when adjusted at the maximum and extremities. *Note 1:* Gaussian distributions occur in many areas, such as (a) the irradiance as a function of distance from the center of certain electromagnetic beams, (b) the distribution of pulse positions in a jittered pulse stream, and (c) the heights of persons in a large population of people. *Note 2:* Usually the top of the Gaussian distribution curve is artificially rounded at an appropriate level and the extremities are truncated and rounded, i.e., are cut off, at a reasonable point, depending on the extreme values encountered in the population. *See also* **bell curve, Gaussian beam, Gaussian function.** *Refer to* **Fig. P-13 (759).**

Gaussian function: The algebraic function given by the relation $G(x) = (1/2\pi)^{1/2}\exp-x^2/2$. *See also* **error function.**

Gaussian minimum shift keying (GMSK): Phase-shift keying that (a) uses constant envelope continuous phase modulation signaling and (b) is used in satellite communications systems, particularly in satellite station to mobile station links. *See also* **phase shift keying, satellite communications system.**

Gaussian pulse: A pulse that has the waveform of a Gaussian distribution, i.e., a distribution resembling a bell curve. *Note 1:* When the magnitude of the waveform is expressed as a function of time, the Gaussian pulse magnitude, A_t, at any time, t, is given by the relation $A_t = A_0e^{-x}$ where A_0 is the maximum magnitude, and $x = (t/\sigma)^2$ where t is time and σ is the pulse half duration, i.e., half of the pulse width, at the $1/e$ points, i.e., at the level where the magnitude is $1/e$ of its maximum value, expressed in the same time units as t, where e is the base of natural logarithms, equal to approximately 2.718. *Note 2:* On a conventional time plot of a Gaussian pulse, the equation yields only the right half of the pulse. The left half is a mirror image, thus producing a curve that nearly looks like a bell when the peaks and ends are slightly rounded. *Note 3:* Square pulses and pulses with discontinuities are not Gaussian pulses. *Note 4:* A Gaussian frequency distribution is obtained by replacing t with f, where f is the frequency in units compatible with σ, the pulse half duration. *See also* **full duration at half maximum, full width at half maximum.**

G band: Frequencies from 4 GHz (gigahertz) to 6 GHz, consisting of ten numbered frequency bands each 200 MHz (megahertz) wide, which overlap parts of the obsolete S, C, X, and J bands. *See also* **frequency spectrum, frequency spectrum designation.**

GBH: group busy hour.

Gbps: gigabits per second, 10^9 bits per second.

GCA landing: ground controlled approach landing.

GCA system: ground controlled approach system.

GCT: Greenwich Civil Time. *See* **Coordinated Universal Time.**

GDDS: global decision support system.

GDF: group distribution frame.

gel: A substance that (a) has the approximate viscosity of petroleum jelly, (b) surrounds a fiber, or multiple fibers, enclosed in a loose buffer tube, (c) lubricates and supports the fibers in the buffer tube, (d) prevents fluid intrusion into the interstices of the buffer tube in the event the tube is breached, and (e) may be in the form of an index matching material for use in fiber optic joints. *See* **matching gel.**

gender changer: A connection device that will convert a female connector or receptacle to a male connector or vice versa. *See also* **female connector, male connector.**

general communications: Two-way voice communications, through a base station, between (a) a common carrier land mobile or airborne station and a landline telephone station connected to a public message landline telephone system, (b) two common carrier land mobile stations, (c) two common carrier airborne stations, or (d) a common carrier land mobile station and a common carrier airborne station.

general message: A message that has a wide standard distribution, such as to all stations in a network, to all ships at sea, to all ships in a given convoy, to all ships under one command, to all ships in a given ocean area or of a given nation, to all aircraft, police patrols, or forest rangers, or to all of the civilian population of a country or an area.

general polling: Polling in which special characters are sent to solicit transmission from all attached devices that are ready to transmit. *See also* **character, polling.**

general purpose language: A computer programming language that is used in a broad class of applications. *See also* **artificial language, high level language, programming language.**

general purpose network: *Synonym* **common user network.**

general purpose operating system: An operating system that handles a wide variety of communications, computer, and data processing systems applications. *See also* **operating system.**

general purpose user service: A telephone service in which users are connected to a telephone network through a private branch exchange (PBX) or through a console with direct access to a switching center. *See also* **access, private branch exchange, switching center.**

generation: In recording and reproduction systems, a measure of the remoteness, in succession, of a copy from the original image, such that a first-generation copy is a recorded copy made from the object, i.e., from the original, a second-generation copy is made from the first-generation copy, a third-generation copy is made from the second-generation copy, and so on. *Note:* An example of generation is that if a microfilm is made from a display image on the display surface of a display device, the microfilm is a first-generation recorded copy. If another microfilm is made from the microfilm, or if a blowback is made on a screen, the second microfilm, or the image on the screen, would be the second generation. *See* **display generation.** *See also* **image, object, recorded copy.**

generator: *See* **character generator, code generator, curve generator, dot matrix generator, fixed function generator, frequency hopping generator, function generator, key generator, key variable generator, master frequency generator, Mersenne code sequence generator, modular spread spectrum code sequence generator, pseudorandom number generator, signal generator, shift register generator, spread spectrum code sequence generator, stroke character generator, variable function generator, vector generator.**

GENERATOR gate: A circuit or a device that (a) produces a continuous stream of 1 bits, pulses, or significant conditions that represent a sequence of 1s in a given system, (b) is considered as a special case of a combinational circuit that produces 1s regardless of the input or with no input, (c) is the reverse of a NULL gate, and (d) may be used to enable other gates that might otherwise be inhibited. *See also* **combinational circuit, inhibiting signal.** *Refer to* **Fig. G-1.**

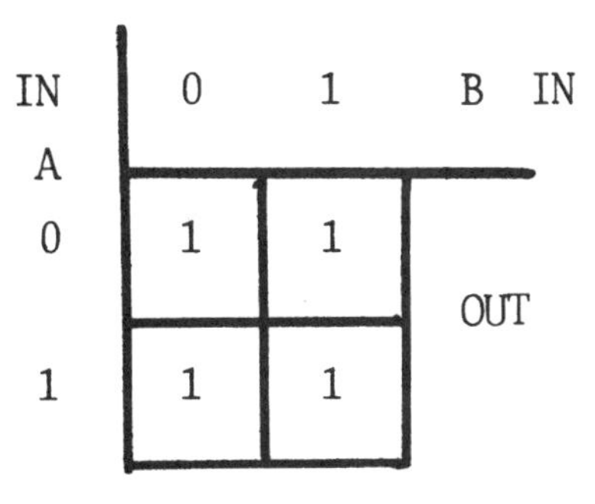

Fig. G-1. A truth table that shows the input and output digits of a **GENERATOR gate.**

geographic address group: In a message, an address group that represents a geographical location or area and that must be used in combination with another address group in order that a message can be routed from a source user, such as a message originator, to a destination user, such as an addressee. *See also* **address group.**

geographical frequency sharing: The sharing of the electromagnetic frequency spectrum among all users located in a given geographical area. *Note:* Geographical frequency sharing is one of the most effective methods of frequency sharing. In determining the necessary separation for the use of the same or adjacent frequencies, a number of factors must be considered, such as the receiver selectivity and sensitivity, transmitter and receiver separation distances, terrain features, propagation medium characteristics for the frequency ranges involved, environmental factors, transmitter and receiver siting, and emission bandwidths. Geographical frequency sharing can often be achieved or enhanced by the use of directional antennas. *Synonym* **geographical**

sharing. *See* **frequency sharing.** *See also* **frequency spectrum, frequency spectrum designation.**

geographical mile: The distance along the equator that corresponds to one minute of longitude, i.e., 6087.15 ft (feet) or 1855.4 m (meters). *Note:* The geographical mile should not be confused with (a) the nautical mile, which is 6076.1 ft (feet) or 1852 m (meters), or (b) the statute mile, which is 5280 ft or 1609.3 m.

geographical reference system: *See* **worldwide geographical reference system.**

geographical sharing: *Synonym* **geographical frequency sharing.**

geometric image: Pertaining to the location and the shape of the image of a particle (a) that is predicted by the use of geometric optics alone, (b) that is distinguished from the diffraction image determined from both physical optics and geometric optics, and (c) for which if the particle has two points, the image has two points, although the diffraction image (i) may suggest the presence of an object composed of two small particles, and (ii) more closely resembles the object. *See also* **geometric optics, image, object, physical optics.**

geometric length: *Synonym* **physical path length.**

geometric optics: 1. The branch of optics that describes light propagation in terms of rays. *Note 1:* Rays are bent at the interface between two dissimilar media or may be curved in a medium in which the refractive index is a function of position. *Note 2:* The ray in geometric optics is everywhere perpendicular to the wavefront in physical optics. **2.** The optics of light rays that follow mathematically defined paths when passing through optical elements, such as lenses, prisms, and other optical propagation media that refract, reflect, diffract, or transmit electromagnetic radiation. *See also* **axial ray, meridional ray, optical axis, paraxial ray, physical optics, skew ray.**

geometric spreading: In a wave propagating in a segment of a propagation medium in which there are no sources, the decrease in irradiance, as a function of distance in the direction of propagation, that is caused only by the divergence of rays rather than by absorption, scattering, diffusion, diffraction, or transverse radiation. *Note:* As a curved wavefront surface, such as for divergent electromagnetic waves, moves in the direction of propagation, the available power at one point must be spread over a larger area at the next point in space. In the case of a point source of radiation, the total transmitted power is spread over the surface of a sphere. Thus, the irradiance at a distance from the source is given by the relation $P_r = W/4\pi r^2$ where P_r is the irradiance at the distance r from the point source and W is the total radiated power, assuming there are no losses caused by absorption or scattering. If W is in watts and r is in meters, the irradiance will be in watts per square meter. The electric and magnetic field strengths will be proportional to the reciprocal of r because the irradiance is proportional to the square of the electric field strength or the square of the magnetic field strength. The electric field strength usually is expressed in volts per meter. Geometric spreading is not considered to take place in a collimated beam, such as a beam in a microwave link, though some spreading will occur because collimation is never perfect, and some light is scattered transverse to the beam by discontinuities in the propagation media, such as air molecules, smoke particles, dust, or water vapor. *See* **irradiance, propagation medium, wavefront.**

geostationary orbit: 1. The path that (a) is described by a satellite that always remains fixed with respect to all points on a rotating orbited body, albeit that the body and the satellite are rotating about their common center of gravity, (b) is circular, (c) lies in a plane, and (d) has points that revolve about the orbited body in the same direction and with the same period as the orbited body rotation. **2.** For the Earth, the orbit of a satellite that (a) always maintains a fixed position with respect to all points on Earth, (b) is circular, (c) lies in the Earth's equatorial plane, and (d) moves in the same direction as the Earth's rotation. *Note 1:* An object in a geostationary orbit will remain directly above a fixed point on the equator at a distance of approximately 42,164 km from the center of the Earth, i.e., approximately 35,787 km above mean sea level. *Note 2:* Strictly speaking, the orbit is not geostationary; the satellite in the orbit is geostationary. *See also* **direct orbit, equatorial orbit, inclined orbit, periapsis, perigee, polar orbit, retrograde orbit, synchronous orbit.**

geostationary satellite: 1. A satellite that (a) is located at a point on a geostationary orbit and (b) remains directly above a fixed point on the equator at a distance of approximately 42,164 km from the center of the Earth, i.e., approximately 35,787 km above mean sea level. **2.** A geosynchronous satellite in a circular and direct orbit that lies in the Earth's equatorial plane and that remains fixed relative to points on the Earth. **3.** A satellite that remains fixed relative to the Earth's surface.

geostationary satellite height: The height above the surface of a rotating celestial body, such as the Earth, at which a satellite in its equatorial plane remains directly above a given point on the equator without requiring internal forces, such as jet thrust, to remain in position, and such that the inward force of gravitational attraction is large enough to keep the satellite from drifting off into space but not so large as to cause the satellite to spiral inward to the Earth; that is, the centrifugal force, i.e., the outward force of acceleration, exactly equals the centripetal force, i.e., the force of gravitational attraction. *Note:* The Earth's geostationary satellite height is 35,787 km (kilometers) above the equator. The period of rotation of the satellite about the Earth is the same as the period of the Earth's rotation about its axis, about 24 hours. The satellite appears to be stationary from any point on Earth, but it is moving just as any point on Earth is moving as the Earth rotates about its axis. *Synonym* **synchronous equatorial satellite height.**

geostationary satellite orbit: The orbit in which a satellite must be placed to be a geostationary satellite.

geosynchronous satellite: An Earth satellite that has a period of revolution about the Earth's polar axis equal to that of the Earth.

germanium photodiode: A photodiode that (a) is used in fiber optic receivers, (b) consists of a germanium-based positive-negative (PN) junction or a germanium-based positive-intrinsic-negative (PIN) junction, (c) is used for direct detection of optical wavelengths from about 1 μm (micron) to several tens of microns, and (d) is noisier than silicon-based photodetectors, which are preferred for wavelengths less than 1 μm.

GFC: general flow control

ghost image: A secondary image or signal resulting from transmission anomalies, such as the presence of echoes, envelope delay distortion, or multipath reception. *Note:* Ghosts are often seen on television (TV) screens because of signal reflections from objects, such as buildings or mountains. *See also* **echo, envelope delay distortion, multipath.**

GHz: gigahertz.

giga (G): A prefix that indicates multiplication by 10^9, i.e., by (a) a thousand million or (b) in the United States, by a billion. *Note:* An example of the use of "giga" is in 10 GHz (gigahertz), equivalent to 10^{10} Hz.

gigabit per second signaling rate: A data signaling rate (DSR) of 10^9 bps (bits per second), i.e., one thousand million bits per second or 1 billion (U.S.) bits per second. *See also* **bit, data signaling rate.**

gigaflop: A billion (U.S.), i.e., 10^9, floating point operations. *Note 1:* The range of operational speeds of a supercomputer is between 1 Gflop/s (gigaflop per second) and 10 Gflops/s. *Note 2:* The unit for operational speeds, Gflops/s, has been shortened to Gflops (gigaflops), where the "ps" has a double role.

gigaflops: gigaflops per second. *See* **gigaflop.**

gigahertz: A unit of frequency that denotes a billion (U.S.) hertz, i.e., 10^9 Hz (hertz). *See also* **frequency, frequency spectrum designation, hertz, metric system.** *Refer to* **Table 4, Appendix B.**

GI optical fiber: *See* **graded index optical fiber.**

glare: *Deprecated synonym for* **call collision.**

glass: A somewhat vitreous, somewhat crystalline, somewhat amorphous compound that (a) is made of sand (silicon dioxide), soda, lime, and some other ingredients for special effects, such as lead oxide or barium oxide for increased sparkle, boric oxide for heat resistance, and various dopants for optical fibers, (b) is workable into many shapes when molten, (c) hardens when cooled to room temperature, and (d) except for certain refractory glasses, remains thermoplastic. *See* **bulk optical glass, crown glass, dry glass, flint glass, fluoride glass, halide glass.** *See also* **dopant, magnifier.**

glass laser: A solid state laser in which the active laser medium is glass. *See also* **active laser medium, laser.**

glass optical fiber: *See* **all-glass optical fiber, drawn glass optical fiber.**

glass powder: Pulverized glass that (a) is produced from raw materials by industrial processing and (b) when in the molten state, is purified and dried for making optical fiber preforms. *See also* **optical fiber preform.**

glass window: The interface between a light source and a glass optical element, such as a microcylindrical lens or an integrally grown lens. *See also* **integral lens light emitting diode, light source, microcylindrical lens.**

glide path: *See* **instrument landing system glide path.**

glide path indicator (GPI): *See* **instrument landing system glide path indicator.**

glide path slope station: A radionavigation land station that (a) is in the aeronautical radionavigation service, (b) provides vertical guidance in connection with an instrument landing system (ILS), and (c) in certain circumstances, may be located on board a ship, such as an aircraft carrier.

glide slope facility: In aeronautical navigation systems, instrument approach landing equipment that furnishes vertical guidance information to an aircraft from its approach altitude down to the surface of the runway. *See also* **glide slope indicator.**

glide slope indicator (GSI): In aeronautical navigation systems, a device that provides signals used for vertical guidance of an aircraft along an inclined surface or slope. *See also* **glide slope facility, instrument landing system, instrument landing system glide path.**

G line: Goubau line.

global: 1. Pertaining to, or involving, the entire world, i.e., pertaining to worldwide. **2.** Pertaining to that which is defined in one subsection of an entity and used in at least one other subsection of the same entity. **3.** In computer, data processing, and communications systems, pertaining to that which is applicable to an area beyond the immediate area of consideration. *Note:* Examples of global entities are (a) in computer programming, an entity that is defined in one subdivision of a computer program and used in at least one other subdivision of that program, and (b) in personal computer systems and their software packages, a setting, definition, or condition that applies to the entire software package, such as in Lotus 1-2-3, applies to the entire spreadsheet, and in WordPerfect, applies to an entire document rather than a line or a word. *See also* **local.**

global address: In a communications network, the predefined address (a) that is used as an all-users address and (b) that cannot be the address of an individual user or subgroup of users of the network. *See also* **group address.**

global attenuation rate characteristic: *See* **optical fiber global attenuation rate characteristic.**

global beam satellite antenna: An antenna on a satellite that emits a conical beam that provides, at the intended orbit height, complete coverage of all of the area of the Earth that faces the satellite, to the radio horizon in all directions. *Note:* The area of coverage of a global beam satellite antenna is the area inside the circle of tangency of a cone within which the Earth is fitted with the satellite at the apex, i.e., the footprint covers the Earth to the radio horizon. *See* **beam, footprint, radio horizon, satellite.**

global decision support system: A U.S. Air Force combination worldwide air traffic management, inventory tracking, and logistics support system that (a) provides multilevel security database management, (b) uses open systems architecture, (c) is a UNIX-based system, and (d) was developed by Informix Software.

global dispersion coefficient: In a fiber optic station/regenerator section, a value of an optical fiber chromatic dispersion coefficient that lies between the upper and lower limiting fiber global dispersion characteristics. *Note:* The upper limit of the global fiber dispersion coefficient, usually expressed in picoseconds per nanometer•kilometer ($ps \cdot nm^{-1} \cdot km^{-1}$), is given by the relation $D_1(\lambda) = S_{omax}(\lambda - \lambda_{omin}{}^4/\lambda^3)/4$, and the lower limit is given by the relation $D_2(\lambda) = S_{omax}(\lambda - \lambda_{omax}{}^4/\lambda^3)/4$ where λ is the operating wavelength, S_{omax} is the maximum worst-case zero-dispersion slope, usually expressed in picoseconds per (square nanometer)•kilometer ($ps \cdot nm^{-2} \cdot km^{-1}$), λ_{omin} is the minimum value of the zero dispersion wavelength, and λ_{omax} is the maximum value of the zero dispersion wavelength, the variation in values being caused by various factors, such as manufacturing tolerances, temperature, humidity, and aging. *See also* **dispersion coefficient.**

global optical fiber parameter: An optical fiber parameter, such as optical fiber attenuation rate, dispersion, or splice insertion loss, that is likely to vary as a result of temperature, humidity, and aging, and for which an allowance should be made in designing a terminal/regenerator section so as to permit unforeseen conditions and events, such as upgrading of terminal equipment, rerouting of traffic, restoration of services, and prevention of traffic saturation. *Note:* If a known upgrade is planned in the design of an optical station/regenerator section, the required (global) parameters should be specified at the outset.

global parameter: *Synonym* **global variable.**

global positioning inertial reference system (PIRS): An aircraft landing guidance system that (a) provides lateral positioning estimates for aircraft curved approach landing, (b) pseudo ranges, (c) time, and (d) Doppler range rate measurements.

global positioning system: *See* **differential global positioning system.**

Global Positioning System (GPS): A federal-government-sponsored system that enables the identification and determination of the position of a mobile transmitter anywhere on or near the Earth's surface. *See also* **signal intercept from low orbit.**

global search: In communications, computer, and data processing systems, a search that (a) is for a character or a character string wherever it appears in a document, storage area, buffer, or transmission, (b) usually is automatically performed with a single command or instruction, and (c) if successful, is followed by a specific action. *See also* **character string.**

global status: 1. The set of attributes of an entity, described at a particular time, when the set is extended to every occurrence of the entity within a prescribed boundary. **2.** The complete set of attributes necessary to describe an entity at a particular time.

global system for mobile communications: A European digital cellular communications system for mobile communications that (a) operates at 270 kb•s^{-1} (kilobits per second), (b) has each carrier divided into eight time slots, (b) has one channel, designated the primary carrier, serving as the signaling channel, (c) has the remaining seven channels, and each additional channel, devoted to voice, and (d) uses a smart card for subscriber identity and billing purposes.

global variable: A variable that is defined in one portion of one computer program and used in another portion of the same program or in other programs, but usually within a given operating system. *See also* **computer program, operating system.**

glossary: *See* **net glossary.**

GMSK: Gaussian minimum shift keying.

GMT: 1. Greenwich Mean Time. 2. Obsolete term. *See* **Coordinated Universal Time.**

gnomonic map: A map that (a) is made by projecting from the center of the Earth onto a plane that touches the Earth's surface at the point of tangency and (b) can show only a limited portion of the Earth's surface without severe distortion at the edges. *See also* **orthodromic map.**

go-ahead message: *Synonym* **go-ahead notice.**

go-ahead notice: In a tape relay communications system, a service message, usually sent to a relay station or to a tributary station, that contains a request to the operator to resume transmitting over a specified channel or channels. *Synonyms* **go-ahead message, start message, start notice.**

go-ahead tone (GAT): In communications systems, an audible signal transmitted by a system indicating that the system is ready to receive a message or a signal.

gold code: In spread spectrum systems, a code that is generated by adding, using modulo-two addition, the outputs of two spread spectrum code sequence generators. *See also* **modulo-two addition.**

goniometer: 1. In a radar system, a device for electrically shifting the characteristics of a directional antenna. **2.** The antenna system of a radio direction finding set. **3.** An electrical device that is used to determine the azimuth of a received signal by combining the outputs of individual elements of a receiving antenna array, because the radiating individual elements of the antenna have specific phase relationships among them, from which the direction of the received signal can be determined. *See also* **antenna array, directional antenna.**

Goos-Haenchen shift: The phase shift that occurs in a lightwave when it is reflected, the magnitude of the shift being a function of the incidence angle and the refractive index gradient at the reflecting interface surface. *Note:* The Goos-Haenchen shift occurs to some degree with every internal reflection that occurs in an optical fiber. When total (internal) reflection occurs, the amount of phase shift in the direction of propagation can be precisely determined. The ray trajectory is changed from that which would be predicted from simple ray theory because of the penetration of the ray into the lower refractive index material of the cladding a short distance before it is turned completely from the incidence angle to the reflection angle. The ray trajectory is shifted in the direction of the fiber axis. If the refractive indices at the core-cladding interface were a perfect step function, i.e., a step change in refractive index over zero distance, there would be no Goos-Haenchen shift. There simply would be Fresnel spectral reflection. *See also* **reflected ray.**

Gopher: A menu-based information-searching tool that (a) allows users to access various types of databases, such as File Transfer Protocol (FTP) and white pages databases, (b) is most often used as an Internet browser, and (c) has software that uses the client-server model. *See also* **client, client-server, Internet, server.**

GOS: grade of service.

GOSIP: Government Open Systems Interconnection Profile.

go to: Pertaining to a statement or an instruction in a computer program that unconditionally transfers control from one point in the program to some other designated point in the same program, thus enabling instructions to be either repeated or skipped. *See also* **computer program, instruction.**

go-to key: On a computer standard keyboard, a key that, when actuated, initiates the execution of an action by the computer related to a specific location, such as to display the contents of a storage location. *Note:* Examples of the use of the go-to key are: (a) in Lotus 1-2-3, the F5 (GoTo) function key causes the pointer to move to a specific location on the spreadsheet, i.e., the worksheet, by typing the specified cell in response to the prompt, such that if F5 is pressed, followed by H5, and <Enter> is pressed, the pointer will move to cell H5, column H, row 25, thus saving moving the cursor past all the columns and rows to get there from wherever it may be at the time, and (b) in WordPerfect, pressing Cntrl-Home keys will cause a prompt to enter the page number to go to, and pressing <Enter> will cause the specified page of the current document to appear on the screen, thus avoiding the necessity of pressing Pg Up or Pg Dn a number of times to reach a desired page from whichever page is displayed at the time.

Goubau line: A single wire open waveguide that guides an axial cylindrical surface wave. *See also* **surface wave, waveguide.** *Synonym* **G line.**

Government Open Systems Interconnection Profile (GOSIP): A definition of federal government functional requirements for open systems computer network products, including a common set of Open System Interconnection (OSI) data communications protocols that enables systems developed by different vendors to interoperate and enables the users of different applications on these systems to exchange information. *Note 1:* The Government Open Systems Interconnection Profile (GOSIP) is a subset of the Open Systems Interconnection (OSI) protocols and is based on agreements reached by vendors and users of computer networks participating in the National Institute of Standards and Technology (NIST) Implementors Workshop. *Note 2:* The Government Open Systems Interconnection Profile (GOSIP) is described in Federal Information Processing Standard (FIPS) Pub 146. *Note 3:* The Open Systems Interconnection (OSI) protocols were developed primarily by the International Organization for Standardization (ISO) and the International Telegraph and Telephone Consultative Committee (CCITT) of the International Telecommunication Union (ITU). *See also* **Open Systems Interconnection.**

GPIRS: global positioning inertial reference system.

GPS: global positioning system.

GPSS: A computer programming language based on block diagrams and used primarily for digital simulation. *See also* **block diagram, programming language, simulate.**

grabber: *See* **frame grabber.**

graceful degradation: Degradation of a system in such a manner that it continues to operate but provides degraded or reduced services rather than catastrophically failing, i.e., completely failing. *Note:* Examples of graceful degradation are that (a) in a communications network with a large number of available lines, channels and trunks continue to operate satisfactorily, though one or two lines fail each week, (b) in a light-emitting diode (LED), there is a reduction in the externally measured quantum efficiency, and (c) in a laser diode, there is a reduction in incremental quantum efficiency even though current density is increased, resulting in reduced power output for a given current density without evidence of facet damage, although the optical power output level can be restored to optimum level by increasing the current density. *Synonym* **gradual degradation.** *See also* **available line, catastrophic degradation, continuous operation, degraded service state, endurability, fail safe operation, failure.**

grade: *See* **subvoice-grade.**

graded index optical fiber: An optical fiber (a) that has a core refractive index that decreases with increasing radial distance from the fiber axis, (b) in which propagating light rays are continually refocused by refraction so as to remain within the core, and (c) that is distinguished from a step index fiber. *Synonym* **gradient index optical fiber.** *See also* **optical fiber, refractive index, step index optical fiber.** *Refer to* **Fig. G-2.** *Refer also to* **Figs. R-1 (796), R-5 (817).**

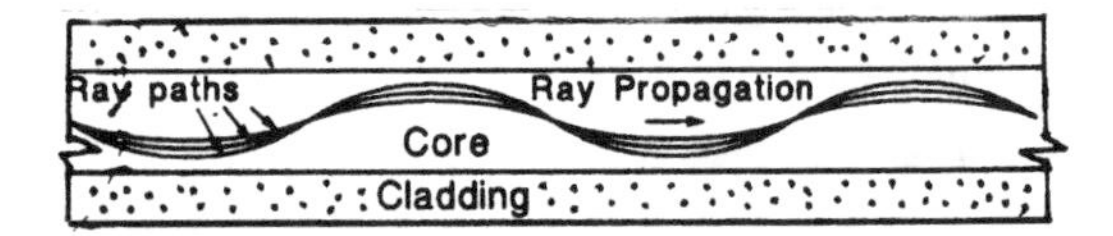

Fig. G-2. Propagating modes in a **graded index optical fiber.** Rays traveling longer physical path lengths travel faster in the lower index regions than the other rays travel. Thus, nearly all rays have the same optical path length and arrive at the same time at the end of the optical fiber.

graded index optical waveguide: An optical waveguide, usually a dielectric waveguide, such as an optical fiber, that has a graded refractive index profile. *See also* **optical waveguide, refractive index.**

graded index profile: 1. Any refractive index profile that varies with distance. **2.** The spatial variation of the refractive index. **3.** In the core of an optical fiber, the variation of refractive index such that the refractive index decreases with increasing radial distance from the fiber axis. **4.** In an optical fiber, any refractive index profile (a) in which the index in the core varies with radial distance from the optical fiber axis, and (b) that is distinguished from a step index profile. *See also* **dispersion, mode volume, multimode optical fiber, normalized frequency, optical fiber, parabolic profile, power law index profile, profile parameter, refractive index, refractive index profile, step index profile, uniform index profile.**

graded optical fiber: *See* **doped silica graded optical fiber.**

grade of service (GOS): 1. The probability of a call's being blocked or delayed for more than a specified time interval. *Note 1:* Grade of service usually is expressed as a decimal fraction, i.e., in parts per unit. A grade of service of P-03 could denote a grade of service in which three calls of each 100 access attempts would fail to complete, i.e., would result in access failure. *Note 2:* Grade of service may be applied to the busy hour or to some other specified period or set of traffic conditions. *Note 3:* Grade of service may be viewed independently from either end of the trunk and is not necessarily equal in each direction. **2.** In a communications channel or system, the quality specification that may be stated in terms of a performance parameter, such as a signal to noise ratio, bit error rate, message throughput rate, or call blocking probability. **3.** In communications systems, a measure of the traffic handling capability of a network from the point of view of sufficiency of equipment and trunking throughout a multiplicity of nodes. **4.** A subjective rating of communications quality and performance in which users judge a transmission or a function as excellent, good, fair, poor, or unsatisfactory. *Synonym* **service grade.** *See* **special grade of service.** *See also* **call, quality of service.**

gradient: *See* **refractive index gradient.**

gradual degradation: *Synonym* **graceful degradation.**

grandfathered system: In telephony, a telephone system, such as a private branch exchange (PBX) or a key telephone system, that (a) was directly connected to the public switched telephone network on June 1, 1978, and may remain permanently connected thereto without registration unless subsequently modified, or (b) is of the same type as those connected to the public switched telephone network on July 1, 1978, and was added before January 1, 1980, and may remain permanently connected thereto without registration unless subsequently modified. *See also* **grandfathered terminal equipment.**

grandfathered terminal equipment: In a telephone system, equipment that (a) is terminal equipment, other than private branch exchanges (PBXs) and key telephone systems, and (b) is protective circuitry, both being connected to the public switched telephone network before July 1, 1978. *Note:* Grandfathered terminal equipment may remain connected to the public switched telephone network for life without registration unless subsequently modified. *See also* **grandfathered systems.**

granular noise: *Synonym* **quantizing noise.**

graph command: In a computer system, a command to execute an operation related to creating, displaying, or printing graphs. *Note:* Examples of a graph command are (a) in Lotus 1-2-3, a command that creates and displays graphs using data contained in the spreadsheet, i.e., the Worksheet, or that stores graphs on a disk as picture files for printing, automatically identified as .PIC placed after a conventional file name, and (b) in WordPerfect, a command in which the Graphics key is used for performing graphics commands. *See also* **command.**

grapheme: A graphic representation of a semanteme. *Note:* Examples of graphemes are x-former for transformer, x-mission for transmission, and x-roads for crossroads. *See also* **semanteme.**

graphical user interface (GUI): A computer operating system, program, or environment that displays options on the screen as icons, i.e., as picture symbols, by which users enter commands by selecting an icon. *Note:* Icons may be selected in various ways, such as by pressing the <Enter> key on a computer keyboard, "clicking" the designated button on a computer mouse, or touching an icon on a touch panel.

graphic board: *See* **video graphic board.**

graphic character: 1. A visual representation of a character, other than a control character, that is usually produced by writing, printing, or displaying. **2.** In the American Standard Code for Information Interchange (ASCII) code, a character other than an alphanumeric character, intended to be written, printed, or otherwise displayed in a form that can be read by human beings. *Note 1:* Graphic characters are contained in rows 2 through 7 of the American Standard Code for Information Interchange (ASCII) code table. *Note 2:* The space and delete characters are also graphic characters.

graphics: 1. The branch of science and technology devoted to the study and the application of conveying information through the use of (a) arrays of symbols, such as graphs, letters, numerals, lines, drawings, and pictures, (b) display media, such as media that display graphs, letters, lines, drawings, and pictures, (c) non-voice analog information devices and systems, such as facsimile, television, wirephoto, and radiophoto systems, and (d) the storage, transmission, and display of coded images. **2.** The symbols that are formed by connected or adjacent strokes, such as the letter A. *See* **computer graphics, coordinate graphics, interactive graphics, passive graphics.** *See also* **facsimile, imagery.**

graphic service: A communications service or system in which graphics are used, i.e., a system that receives and transmits data in pictorial form, such as pictures, graphs, letters, numerals, lines, and drawings. *Note:* Examples of graphic services are facsimile, wirephoto, radiophoto, and television services.

graph key: In most personal computer (PC) software systems, a key on a standard keyboard that, when actuated, initiates the execution of an action by the PC related to producing a graph on the display area of a display device. *Note:* Examples of a graph key are (a) a function key that causes a graph to be redrawn using the most recently entered graph command specifications, and (b) a function key that can be used to view a selected graph prior to printing. *See also* **command.**

graph program: In personal computer (PC) systems, a program that allows graphs to be drawn, displayed, and printed. *Note:* An example of a graph program is the Lotus 1-2-3 PrintGraph program, which allows the printing of graph files created with the Lotus 1-2-3 program.

grass: In a radar receiver, the random noise that (a) is caused by various sources, such as atmospheric disturbances, cosmic noise, auroras, thermal excitation, and interference, (b) can be seen on a radar screen, and (c) appears as illumination displayed on a radar screen when the gain control is set so high that the screen displays the noise that occurs in the radar itself without receiving antenna input signals other than atmospheric disturbances, cosmic noise, auroras, thermal excitation, and interference.

grating: *See* **diffraction grating.**

grating chromatic resolving power: The resolving power that (a) determines the minimum wavelength difference for any spectral order that can be distinguished as separate, (b) for diffraction gratings, usually is stated for cases in which parallel rays of light are incident upon the grating, and (c) is numerically equal to the number of lines or ruled spacings per unit distance in the grating. *See also* **diffraction grating, diffraction grating spectral order, resolving power.**

grating sensor: *See* **moving grating sensor.**

grating spectral order: *See* **diffraction grating spectral order.**

gravitational field: The force field that (a) is established by the existence of a mass of material matter, such that two material masses attract each other, (b) exists in the space surrounding the material matter, (c) varies in magnitude inversely as the square of the distance from the material mass and extends to infinity, at which it is zero, (d) at a given point, is the sum, i.e., the integral, of all the contributions of the individual distributed masses, (e) is a vectorial field, having both magnitude and direction, and (f) can be measured by placing a small test element of mass in the field and measuring the magnitude and the direction of the force exerted on the element, i.e., measuring the force required to hold the element in place. *Note 1:* Mutual gravitational attraction is expressed by the relation $F = Gm_1m_2/d^2$ where F is the force of attraction, G is the universal gravitational constant, i.e., the proportionality constant, m_1 and m_2 are the masses of the attracting bodies, and d is the distance between their centers of gravity, which for homogeneous substances, are their geometric centroids. *Note 2:* The gravitational field established by the Earth and the Moon keeps the Moon in its orbit, and the gravitational field established by the Sun and the Earth keeps the Earth in its orbit. The gravitational field on the surface of a celestial body, such as the Earth, the Moon, or Mars, can be determined by measuring the force on a known mass. The force exerted by the Earth's gravitational field on 1 kg (kilogram) of mass on the Earth's surface is 9.8 N (newtons). *Synonym* **gravitational force field.**

gravitational force field: *Synonym* **gravitational field.**

Gray code: **1.** A special binary code used to minimize transmission errors, in which consecutive decimal numbers are represented by consecutive binary expressions that differ only in one bit position at a time. *Note 1:* In sequencing consecutive numerals in the Gray code, transitions from numeral to numeral are smoother, not like transitioning from 999 to 1000 in decimal, resulting in lower power fluctuations than in decimal. *Note 2:* The decimal numerals 0, 1, 2, 3, 4, 5, 6, 7, 8, 9, 10, 11, 12, 13, 14, and 15 are represented in pure binary as 0000, 0001, 0010, 0011, 0100, 0101, 0110, 0111, 1000, 1001, 1010, 1011, 1100, 1101, 1110, and 1111, respectively. In Gray code, the numerals are represented as 0000, 0001, 0011, 0010, 0110, 0111,

0101, 0100, 1100, 1101, 1111, 1110, 1010, 1011, 1001, and 1000, respectively. **2.** A cyclic binary unit distance code in which a positional binary code system for consecutive numbers is used such that in two consecutive numbers of the code the digits are the same in every digit place, i.e., digit position, except one, and in that place the digits differ by one unit, such that the signal distance, i.e., the Hamming distance, between consecutive numerals in the code is one. *Note 1:* The Gray code proves valuable for encoding devices because only one digit changes at a time. This provides for smooth operation and reduced ambiguities at change points. Instead of many digits changing at the same time, such as in counting in pure binary from 1111 to 10000, or in decimal from 999 to 1000, only one digit changes at these transition points. *Note 2:* In order to convert from pure binary to Gray code, starting from the left end of the pure binary numeral, copy any initial 0s and the first occurrence of a 1. If the next binary digit is different from 1, i.e., is a 0, write a 1. If it is the same, i.e., is a 1, write a 0. If the next digit is different from the previous digit, write a 1. If the next digit is different from the previous digit, write a 1, and each time the next digit is the same as the previous digit, write a 0 for the Gray code. Thus, each time the next digit in the pure binary numeral changes, write a 1, and each time the next digit is the same, write a 0 until the end of the pure binary numeral. Thus, 10111011 in pure binary is 1110110. *Note 3:* In order to convert a Gray code numeral to a pure binary numeral, start counting the Gray code 1s from the left. If the count is odd, write a 1. As long as the count remains odd as position by position, i.e., digit by digit, is passed, continue to write 1s for the pure binary numeral. Thus, write 1s for an odd count and 0s for an even count until the end of the numeral. Thus, given the Gray code numeral 1110110, the equivalent pure binary numeral will be 1 for the first 1 in the Gray code numeral, 0 for the second 1 because the count of 1s is now even, 1 for the third 1 because the count is now odd, 1 again because the count of 1s is still odd in the Gray code numeral, now 0 because the count is again even, and so on to yield 1011011 in pure binary. *Note:* The entire Gray code may be constructed in a reflected fashion. Begin with 0,1 written in a column, to obtain:

0
1

Then, reflect the 0 and 1 by writing its mirror image under it, i.e., in reverse sequence, to obtain:

0
1
1
0

Then, prefix the forward sequence with 0s and the reflected sequence, i.e., reverse sequence, with 1s, to obtain:

00
01
11
10

With this as the forward sequence, reflect it by writing the reverse sequence under it, to obtain:

00
01
11
10
10
11
01
00

Then, again prefix the forward sequence with 0s and the reflected sequence with 1s, to obtain:

000
001
011
010
110
111
101
100

This process may be continued indefinitely. To obtain the table of equivalences, simply number the Gray code sequence with pure binary numerals in pure binary sequence. *Note 4:* To construct an $(n + 1)$-bit Gray code from an n-bit Gray code, write the n-bit sequence forward and then in reverse, i.e., reflected. Prefix the forward sequence with 0s and the reverse with 1s. *Synonyms* **cyclic binary unit distance code, reflected binary unit distance code.** *See also* **binary digit, error, error control, pure binary numeration system.** *Refer to* **Fig. G-3.**

gray scale: 1. An optical pattern consisting of discrete steps between black and white, i.e., consisting of shades of gray. **2.** Apparent discrete light intensity levels between zero and a maximum level, usually when referring to diffused light from a rough or frosted surface. *Note:* The gray scale is not so much an exact measure of direct intensity, but more a measure of the proportionate distribution of black spots on a white surface or white spots on a black surface. In facsimile systems, the scale is made in discrete steps between all

Decimal	Pure Binary	Gray Code
0	0000	0000
1	0001	0001
2	0010	0011
3	0011	0010
4	0100	0110
5	0101	0111
6	0110	0101
7	0111	0100
8	1000	1100
9	1001	1101

Fig. G-3. The **Gray code** equivalences to decimal and pure binary numerals.

white and all black by introducing an array of small black dots on a white area and increasing their size in discrete steps, rather than their number, until they cover all the area. No dot at an array location is white. The gray scale is the steps in which the dot size, relative to the spacing between them, is increased, while their center-to-center spacing is held constant. *See also* **facsimile.**

grazing emergence: In optics, a condition in which an emergent ray makes an angle of 90° from the normal to the emergent surface of a propagation medium. *See also:* **emergence, emergent ray, grazing incidence, normal emergence.**

grazing incidence: In optics, a condition in which an incident ray makes an angle of 90° from the normal to the incident surface. *See also* **grazing emergence, normal incidence.**

great circle: 1. A circle on the surface of a sphere that (a) lies in a plane that passes through the center of the sphere, (b) has the same radius and center as the sphere itself, and (c) is the intersection of (i) a plane that passes through the center of the sphere and (ii) the surface of the sphere. *Note:* Any segment of a great circle is the shortest distance, along the surface, between the end points that define the segment, i.e., the shortest distance along the surface between any two points on the surface is along a great circle. In Earth measurements along great circles, the Earth is considered as a sphere. **2.** A circle defined by the intersection of the Earth's surface and any plane that passes through the center of the Earth. *Note 1:* The shortest distance, over the idealized surface of the Earth, between two points on the surface lies along a great circle. *Note 2:* Lines of longitude and the equator are special cases of great circles. The lines of longitude are great circles that pass through the poles. There are an infinite number of great circles that do not pass through the poles. *Note 3:* Except for the equator, lines of latitude are not great circles.

Green interferometer: *See* **Twyman-Green interferometer.**

Greenwich Civil Time (GCT): 1. *Synonym* **Greenwich Mean Time (GMT). 2.** Obsolete term. *See* **Coordinated Universal Time.**

Greenwich Mean Time (GMT): 1. Mean solar time at the meridian of Greenwich, England, i.e., the 0° meridian, formerly used as a basis for standard time throughout the world and usually expressed in 24-hour time. *Note 1:* Greenwich Mean Time is also called "Zulu Time" and "World Time." *Note 2:* "Greenwich Mean Time" is an obsolete term. Because the second is no longer defined in terms of astronomical phenomena, "Universal Time," which is defined in terms of cesium-133 transitions, is the most accurate measure of time, and the primary name for time is "Coordinated Universal Time (UTC)," which is synonymous with "Zulu Time." **2.** Time based on the time at Greenwich, England, counting 24 one-hour time zones from the Earth meridian, i.e., the Earth longitude, that is zero through Greenwich, England. *Note:* GMT is expressed in 24-hour time. By international agreement, Coordinated Universal Time (UTC) has replaced Greenwich Mean Time (GMT). *Synonyms* **Greenwich Civil Time, Universal Time, World Time, Zulu Time.** *See also* **Coordinated Universal Time.**

green wire: *Synonym* **fault protection subsystem.**

Gregorian calendar: The calendar that (a) was established by Pope Gregory XIII in 1582 and (b) is in worldwide use today. *See also* **Julian calendar.**

Griffith microcrack model: A microcrack model for glass, described by Griffith, in which the crack is described by the relation $s_t = s_f(1 - 2x/y)$ where s_t is the ultimate tensile strength stress of the material, s_f is the failure stress for a given crack, x is the crack depth, and y is the crack surface width. *See also* **microcrack.**

GRIN^R lens: *See* **GRIN^R rod.**

GRIN^R rod: A cylindrically shaped rod, such as a short length of optical fiber, that (a) has a graded refractive index profile parameter that will cause all light rays entering within a given acceptance cone to be focused at one point for a given operating wavelength, (b) collimates and focuses a light beam emanating from a point source, (c) has a profile parameter of approximately 2, making the refractive index profile close to that of a parabola, and (d) was developed by Bell Laboratories. *Synonym* **GRIN[R] lens.** *See also* **SELFOC[R] lens, self-focusing optical fiber.**

grip: *See* **basket grip.**

grooved alignment connector: *See* **fiber optic grooved alignment connector.**

grooved cable: A fiber optic cable in which the optical fibers are fitted into grooves in a cylindrical element. *Note:* Fiber optic cables with a large number of optical fibers can be produced by stranding two or more cylindrical elements together and sheathing or jacketing the whole assembly.

grooved optical fiber alignment connector: *Synonym* **fiber optic grooved alignment connector.**

grooved waveguide: A slab dielectric waveguide that (a) supports lightwaves in certain propagation modes, (b) confines the waves in grooves a few microns wide, and (c) is particularly used in optical integrated circuits (OICs). *See also* **mode, slab dielectric waveguide.**

groper: *See* **packet internet groper.**

ground: 1. The electrical connection to earth through an earth electrode subsystem. **2.** In an electrical circuit, a common return path that usually (a) is connected to an earth electrode subsystem and (b) is extended throughout a facility via a facility ground system consisting of (i) the signal reference subsystem, (ii) the fault protection subsystem, and (iii) the lightning protection subsystem. **3.** In an electrical circuit, a common return path, which (a) may be a chassis, cabinet, or frame and (b) may not necessarily be connected to earth ground. *Note:* The ground terminal may be connected to the earth via a low resistance path, or it may never be connected to the earth at all. If it is not connected to the earth, it is a floating ground, as in the case of an automobile, aircraft, small power supply, or portable air compressor. **4.** Ideally, a zero resistance, i.e., an infinite conductance, path to a capacitor with an infinite capacity. **5.** A point that has an absolute electric potential, i.e., an absolute voltage, of zero. *See* **frequency ground, neutral ground.** *See also* **bond, bonding, earth, earth electrode subsystem, equipotential ground plane, facility grounding system, fault, ground absorption, ground constants, ground plane, ground potential, ground return circuit, ground start, ground wave.** *Refer to* **Figs. C-6 (142), G-4 (404).**

ground absorption: In radio and television (TV) transmission, the dissipation of transmitted energy by earth, i.e., the ground, such as soil, rock, and water. *See also* **ground constants.**

ground-air net: A communications net that is established for one-way communications from Earth stations to aircraft stations.

ground-air visual emergency signal: 1. A pattern, figure, letter, or graphic symbol that (a) is used for ground-to-aircraft signaling in an emergency, (b) has an assigned meaning, and (c) usually is formed by straight lines or contrasting material on the ground. **2.** A symbol that (a) is made of straight lines of material that is used by survivors and search parties to send coded messages to aircraft and (b) may be made of sticks, stones, trampled sod, footprints in sand, or any other material. **3.** Any signal on the ground that can be used to attract attention in an emergency. *Note:* Examples of ground-air visual emergency signals are panels, flares, smoke, reflected light, flags, and fires. A specific signal, made by prearrangement, might be a single straight line on the ground made with anything available to indicate that medical attention is needed or two parallel lines to indicate that only food and water is needed.

ground clutter: A large distribution of radar echoes that (a) extends a few miles in all directions from a radar site, (b) is caused by radar signal reflections from the surface of the Earth and fixed objects, such as buildings and trees, and (c) creates a pattern on a radar screen, such as the planned position indicator or A-scope, that is approximately circular, is constant, and is near the origin of the screen coordinates. *See also* **echo, ground return signal.**

ground communications: *See* **air-ground communications.**

ground communications emergency frequency: *See* **air-ground communications emergency frequency.**

ground constant: The electrical parameters of earth, such as electrical conductivity, electric permittivity, and magnetic permeability of soil, rock, and water. *Note 1:* The values of these parameters vary with the chemical composition, density, and distribution of materials in earth and the frequency of the transmitted signal. *Note 2:* For a propagating electromagnetic wave, such as a surface wave propagating along the surface of the Earth, these parameters vary with the frequency of the wave and the direction of the propagating wave relative to the earth. *Note 3:* Ground constants influence the propagation of ground waves, surface waves, and evanescent waves. *See also* **facility grounding system, ground, ground absorption, ground potential.**

ground controlled approach (GCA) landing: The landing of an aircraft by a pilot who is given landing instructions by radiotelephone from a ground station that can determine the exact instantaneous position of the aircraft in relation to the landing strip, landing zone, or runway. *See also* **ground controlled approach system, instrument landing system.**

ground controlled approach (GCA) system: A ground radar system that (a) provides information needed by pilots of aircraft as they approach a landing zone, landing strip, or runway, (b) includes information transmitted to the pilots, during approach, by radiotelephone, signals from the ground station to the aircraft station, and radar, (c) is used to accomplish ground controlled approach landing with or without visibility, and (d) usually is accomplished by radar, radiotelephone, and signals from the ground station, though some systems can also control the aircraft with a pilot-override capability. *See also* **ground controlled approach system, instrument landing system.**

ground link subsystem: *See* **space ground link subsystem.**

ground loop: In an electrical system, a loop formed when two or more points that are nominally at earth potential are connected by a conducting path such that the points are at different "ground" electric potentials, i.e., there is a voltage between them. *Note:* Ground loops can be detrimental to the operation of an electrical system.

ground plane: The surface that serves as the near field reflection surface for an antenna. *See also* **antenna, equipotential ground plane, image antenna.**

ground potential: 1. In a system or a circuit, the zero reference level used to apply and measure voltages. *Note:* An electrical potential difference, i.e., a voltage, may exist between the zero voltage reference of a circuit and the ground potential of earth, which varies with locality, soil conditions, and meteorological phenomena. **2.** The absolute potential of earth that (a) usually is taken as a zero level of absolute potential and (b) is used as the reference level to express voltages, i.e., potential differences. *See also* **earth electrode subsystem, ground, ground constants.**

ground radiotelephone service: *See* **air-ground radiotelephone service.**

ground return: 1. In wire communications, a circuit arrangement in which one line of an insulated pair of conductors is replaced by a return path through earth, thus enabling the insulated wires to serve as a forward channel and a backward channel, two forward channels, or two backward channels. *Note:* Ground return requires that the transmitter and the receiver be connected to earth. **2.** In any electrical system, an arrangement of power and signal distribution in which all return paths for currents are via a common bus, such as a copper strip, cabinet, frame, or chassis of the equipment, or earth.

ground return circuit (GRC): 1. A circuit in which earth serves as one conductor. **2.** A circuit that uses a common return path that is at ground electrical potential. *Note:* Earth may serve as a portion of the ground return circuit. **3.** A circuit that consists of one insulated conductor from one point to another and that is connected to ground on both ends. **3.** A circuit in which there is a common return path, whether or not connected to earth. *Synonym* **ground return.** *See also* **circuit, common return, facility grounding system, ground, unbalanced line.** *Refer to* **Fig. G-4.**

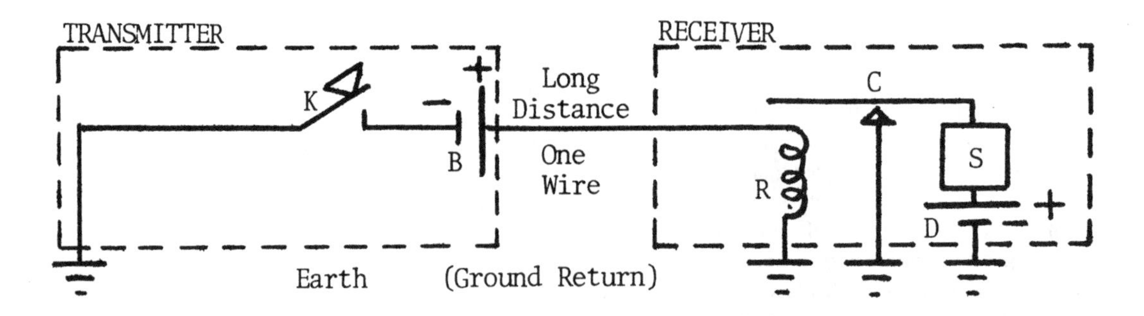

Fig. G-4. A **ground return circuit** in a single one-way telegraph channel. At the telegraph transmitter, depressing key *K* will cause local battery *B* to send current through relay *R*, closing contact *C* and sending current from battery *D* through sounder *S* at the telegraph receiver. The transmitted current returns through the earth to the transmitter. The local current at the receiver has its own loop, via wire, earth, or both.

ground return signal: In radar operations, a wave or an echo obtained by reflection from the ground, trees, buildings, or other objects fixed to the ground or moving on the ground. *See also* **ground clutter.**

ground segment: 1. In satellite communications, all of the Earth stations in a satellite communications system. **2.** In electrical circuits, all the ground return paths in a set of ground return circuits. *See also* **Earth station, ground return.**

ground start: A supervisory signal from a terminal to a switching center in which one side of the line is temporarily grounded. *See also* **ground, loop start.**

ground state: In a given element, the lowest total orbital kinetic and potential energy state that an electron can have. *Note:* An electron (a) absorbs a quantum of energy when it moves from the ground state to an excited state and (b) emits a quantum of energy when it moves from an excited state to the ground state. *See also* **excited state.**

ground system: *See* **facility ground system.**

ground to air communications: One-way communications from fixed or mobile ground stations to aircraft stations.

ground to air visual body signal: A signal that (a) is formed by arm, leg, and body position and movement and (b) is used for sending simple messages from the ground or sea surface to aircraft, such as signals for directing the landing of aircraft, sending emergency messages, and indicating that all is clear.

ground wave: 1. An electromagnetic wave that (a) propagates close to the Earth's land or sea surface, (b) is coupled to it, (c) interacts with it, (d) is absorbed by it, (e) travels above the surface and is coupled to modes beneath the surface, (f) can emerge from it at certain distances from the source, (g) is heavily attenuated, though other surface and evanescent waves are not, (h) usually is of low frequency, (i) does not depend on ionospheric reflection, and (j) is not a lightwave because lightwave frequencies are too high to function as a ground wave. **2.** In radio transmission, a surface wave that (a) propagates close to the surface of the Earth, (b) is affected by the presence of the Earth because the Earth has one refractive index and the atmosphere has another, the two thus constituting an interface surface, (c) is locked to earth, and (d) is affected by the absolute values, the gradients, and the temporal and spatial changes in the electrical conductivity, electric permittivity, and magnetic permeability of the ground and the lower atmosphere. *Note:* Ground waves do not include ionospheric and tropospheric waves. *See also* **antenna, direct ray, sky wave.**

ground worldwide communications system: *See* **air-ground worldwide communications system.**

group abbreviated address calling: In telephone operations, the establishment of a connection to a group of destinations, stations, addressees, or call receivers by using a single abbreviated address.

group: 1. In wideband frequency division multiplexed (FDM) communications systems, a specific number of voice channels that (a) is either within or separate from a supergroup, (b) consists of up to 12 voice channels, (c) occupies the frequency band from 60 kHz (kilohertz) to 108 kHz, and (d) is International Telegraph and Telephone Consultative Committee (CCITT) Basic Group B. *Note 1:* International Telegraph and Telephone Consultative Committee (CCITT) Basic Group A, for carrier telephone systems, consists of 12 channels, occupying upper sidebands in the 12-kHz (kilohertz) to 60-kHz band. Basic Group A is no longer mentioned in CCITT Recommendations. *Note 2:* A supergroup, i.e., a secondary group, (a) usually consists of 60 voice channels, or five groups of 12 voice channels each, and (b) occupies the frequency band from 312 kHz (kilohertz) to 552 kHz. *Note 3:* A master group consists of ten supergroups or 600 voice channels. *Note 4:* The International Telegraph and Telephone Consultative Committee (CCITT) standard master group consists of five supergroups. The U.S. commercial carrier standard master group consists of ten supergroups. *Note 7:* The terms "supermaster group" and "jumbo group" are sometimes used to refer to six master groups. *Note 8:* The preferred group hierarchy from lowest to highest is basic group, supergroup, master group, and jumbo group. **2.** A set of characters that form a unit for transmission or cryptographic treatment. *See* **address group, address indicating group, basic group, centralized ordering group, channel basic group, channel group, check group, cipher group, closed user group, code group, collective address group, conjunctive address group, date time group, detectable group, digital transmission group, digroup, dummy group, end of transmission group, functional group, graphic address group, high usage trunk group, international address indicating group, line group, national address indicating group, peer group, phantom group, signal group, synchronized transmitter group, terminal endpoint functional group, through group, true date-time group, trunk group.**

group address: In a communications network, a predefined address used for a specified set of users. *Synonym* **collective address.** *See also* **global address.**

group alerting and dispatching system: A network-provided service feature that (a) enables a controlling telephone to place a call to a specified number of telephones simultaneously, (b) enables the call to be recorded, (c) if any of the called lines is busy, enables the equipment to camp on until the busy line is free, (d) rings the free line, and (e) plays the recorded message. *See also* **conference call, multiaddress calling facility, service feature.**

group allocation: *See* **address group allocation, address indicating group allocation.**

group busy hour (GBH): The busy hour for a given trunk group. *See also* **busy hour, erlang, group, switch busy hour, traffic load.**

group congestion signal: *See* **national circuit group congestion signal.**

group delay: 1. In a group of waves of slightly different individual frequencies that is propagating through a device, system, or propagation medium, the rate at which the total phase shift, Θ, changes with respect to the angular frequency, ω, given by the relation $\tau = d\Theta/d\omega$ or $\tau = \Delta\Theta/\Delta\omega$ where τ is the group delay, Θ is the phase shift usually expressed in radians, and ω is the angular velocity, i.e., the angular frequency, equal to $2\pi f$, where ω is the angular velocity usually expressed in radians per second and f is the frequency usually expressed in hertz. *Note:* The group delay has the units of time, usually seconds, because Θ usually is expressed in radians and ω in radians/second. The phase changes occur over the length of the propagation medium in the direction of propagation. Thus, for a wave of a certain frequency in a group of waves with different frequencies, the group delay is the first derivative with respect to frequency of the phase shift between two given points in the propagating wave. In a transmission system where phase shift is proportional to frequency, the group delay is constant. **2.** In an optical fiber, the transit time required for optical power, propagating at a given mode group velocity, to travel a given distance. *Note:* In optical fiber dispersion measurements, the quantity of interest is group delay per unit length, which is the reciprocal of the group velocity of a particular mode. The measured group delay of a signal through an optical fiber exhibits a wavelength dependence caused by various dispersion mechanisms present in the fiber. *See* **multimode group delay.** *See also* **delay distortion, envelope delay distortion, group delay time, group velocity, Sellmeier equation.**

group delay distortion: In the propagation of a signal, the distortion that (a) is caused by the propagation medium because of the difference in propagation velocity, and hence propagation time, among the different frequencies that compose the signal, (b) results in dispersion that will make the shape of the received signal be different from that of the dispatched signal, (c) results in a spreading of the signal during propagation in the medium, as different parts of the signal arrive in a time sequence relatively different from the sequence in which they were dispatched, (d) can be reduced, or compensated for, by requiring the different frequencies to travel different path lengths, and (e) can be compensated for in an optical fiber and an optical integrated circuit (OIC) by adjusting the refractive index profile so that the higher velocity components of the signal travel a longer path than other components and thus will arrive at the end at the same time as the lower velocity components, which travel a shorter path. *See also* **dispersion, frequency, propagation medium, refractive index profile, signal.**

group delay frequency distortion: Distortion that (a) is an undesired variation of the rate of change of (i) the difference in phase angle between a sinusoidal excitation and the fundamental frequency component of the received signal with respect to (ii) the frequency, i.e., the first derivative with respect to frequency, and (b) consists of a combination of (i) group delay distortion and (ii) frequency distortion, both caused by undesired frequency and phase shift variations between the carrier and its modulating signal. *See* **carrier, distortion, frequency distortion, fundamental frequency, group delay distortion, phase angle.**

group delay spread: *See* **multimode group delay spread.**

group delay time: In a group of waves with different individual frequencies, the time that is required for a crest, or any other significant instant of the group, to propagate (a) through a component, device, or system or (b) from one specified point to another in a propagation medium. *See also* **envelope delay distortion, group delay, group velocity, propagation medium, Sellmeier equation, significant instant.**

group distribution frame (GDF): In frequency division multiplexing (FDM), a distribution frame that provides terminating and interconnecting facilities (a) for the modulator output and demodulator input circuits of the channel transmitting equipment and (b) for the modulator input and demodulator output circuits for the group translating equipment operating in the basic spectrum of 60 kHz (kilohertz) to 108 kHz. *See also* **distribution frame, group.**

group equipment: *See* **through group equipment.**

group four document facsimile equipment: *See* **facsimile.**

group four document facsimile telegraph apparatus: *See* **facsimile.**

group index: For a given mode in an electromagnetic wave propagating in an optical waveguide, the ratio of (a) the velocity of light in a vacuum to (b) the group velocity of the mode, given by the relation $N = c/v_g$ where N is the group index, c is the velocity of light in a vacuum, and v_g is the mode velocity in the waveguide. *Note:* For a plane wave, the group index, N, for a given mode is given by the relation $N = n - \lambda dn/d\lambda$ where n is the refractive index of the medium expressed as a function of wavelength, i.e., $dn/d\lambda$ is the rate of change of the refractive index with respect to wavelength, and λ is the source wavelength. *See also* **group velocity, mode.**

grouping: In facsimile systems, a periodic error in the spacing of recorded lines. *See* **digital grouping, line grouping.**

grouping factor: *Synonym* **blocking factor.**

group one document facsimile equipment: *See* **facsimile.**

group one document facsimile telegraph apparatus: *See* **facsimile.**

group patch bay: A patch bay that is used for a group of circuits, such as a number of basic groups, master groups, or jumbo groups. *See also* **patch bay.**

group selector: In an automatic telephone switching exchange, a selector that (a) responds to an appropriate control signal, such as a specific character, (b) connects one input path to one of a number of output paths, and (c) may be more specifically designated as a code, tandem, numeral, or final selector.

group three document facsimile equipment: *See* **facsimile.**

group three document facsimile telegraph apparatus: *See* **facsimile.**

group two document facsimile equipment: *See* **facsimile.**

group two document facsimile telegraph apparatus: *See* **facsimile.**

group velocity: 1. The velocity of propagation of an envelope produced when an electromagnetic wave is modulated by, or mixed with, other waves of different frequencies. *Note:* Ideally, the group velocity would be the velocity of a signal represented by two superimposed sinusoidal waves of equal amplitude and slightly different frequencies approaching a common limiting value. The group velocity is the velocity of information propagation and of energy propagation. The modulating signal is the information-bearing signal, and the signal that is to be modulated is the carrier. **2.** In fiber optic transmission of a particular mode, the reciprocal of the rate of change of the phase term with respect to angular velocity, i.e., the rate at which frequency changes with respect to the reciprocal of the wavelength. *Note:* The group velocity usually differs from the phase velocity. The group velocity, given by the relation $v_g = d\omega/d\beta$, is less than the phase velocity, represented by the relation $v_p = \omega/\beta$ where ω is the carrier angular velocity, $d\omega$ is the signal angular velocity, and β is the phase term at the carrier frequency. βL is the carrier phase delay for a line L meters long. The angular velocity, i.e., the angular frequency, is given by the relation $\omega = 2\pi f$ where f is the frequency. In nondispersive media, the group velocity and the phase velocity are equal if the phase term is a linear function of the angular velocity. In a waveguide, each mode has its own group velocity. The group velocity equals the phase velocity only if the phase term is a linear function of the angular velocity. **3.** In fiber optic transmission, the velocity of the modulated optical power. *See also* **differential mode delay, group index, phase velocity.**

group with outgoing access: *See* **closed user group with outgoing access.**

group 1, 2, 3, 4 facsimile: *See* **facsimile.**

growth: *See* **epitaxial growth, reliability growth.**

GRWSIM: ground warfare simulation.

GSM communications: global system for mobile communications.

G/T: antenna gain to noise temperature.

guard: In radio, radar, and sonar communications systems, to maintain a continuous receiver watch in which (a) the transmitter is ready for immediate use, (b) a complete radio log usually is kept, and (c) the radio receiver watch may be maintained on more than one frequency, thereby maintaining radio contact with specific fixed or mobile stations. *See also* **receiver watch.**

guard band: *See* **frequency guard band, time guard band.**

guarded frequency: A transmission frequency that is not to be jammed and for which interference is not to be created because of the value of the information being derived from it. *Note:* For example, a guarded frequency will not be jammed when the tactical, strategic, and technical information that can be obtained from the transmissions outweighs the potential operational gain achieved by jamming.

guard ring: In surface-dominated avalanche photodiodes (APDs), such as lattice-matched InGaAsP APDs, a specifically doped peripheral region that prevents leakage currents that cause gain saturation. *Note:* The guard ring improves performance in the 1.0 µm (micron) to 1.3 µm detection range for 0.01 mm to 0.02 mm (millimeter) diameter APDs operating at less than 5 µA (microamperes).

guard ship: A ship station that (a) is designated to maintain any one of the various types of receiver watches on a definite frequency on behalf of another ship station and (b) usually keeps a complete radio station log.

guard station: *See* **primary guard station.**

guard time: In time division multiplexing (TDM) and in time division multiple access (TDMA) systems, the time between the end of one burst of signals and the beginning of the next burst of signals.

guard watch: 1. A radio communications watch that (a) requires the operator to maintain a continuous receiver watch, (b) usually requires that a complete radio station log be kept, (c) keeps the transmitter ready for instant use, (d) may be any one of several types or a combination of several types of radio watch, (e) is performed by a guard station, and (f) may be performed by a ship station or a land station. **2.** In communications networks of any type, the monitoring, copying, and relaying of calls or messages that are intended for another station. *Note:* A guard watch is not the only type of watch that may be performed by a guard station. *See also* **copy watch, listening watch, radio communications cover, radio communications guard, receiver watch.**

guidance navigation system: *See* **terminal guidance navigation system.**

guidance radar: Radar that is used to track moving objects, such as aircraft, spacecraft, ships, missiles, and ground vehicles, in order to obtain position information that is to be used with other position information to calculate instructions that are needed to steer the object from its present position to a desired fixed or moving position.

guidance system: A system that guides objects, such as aircraft, spacecraft, ships, missiles, and ground vehicles, from their present position to desired fixed or moving positions. *See* **homing guidance system, inertial guidance system, noninertial guidance system.**

guided missile control: The control of guided missiles by (a) the use of communications equipment, i.e., flight control, (b) the control of guided missile organizational elements, i.e., command control, (c) the control of procurement, movement, receipt, storage, maintenance, and salvage of guided missiles, missile systems, and components, i.e., logistics control, and (d) the control of repair and replacement parts, i.e., supply control.

guided mode: 1. *Synonym* **bound mode. 2.** *See* **cladding guided mode.**

guided ray: In an optical fiber, a ray that (a) is completely confined to the core or (b) is bound to a mode in the core. *Note 1:* A guided ray satisfies the relation $0 \leq \sin \theta_r \leq (n_r^2 - n_a^2)^{1/2}$ where θ_r is the angle the ray makes with the fiber axis, r is the radial position, i.e., radial distance, of the ray from the fiber axis, n_r is the refractive index at the radial distance r from the fiber axis, and n_a is the refractive index at the core radius, a, i.e., at the core-cladding interface. Guided rays correspond to bound modes, i.e., to guided modes, in terms of modes rather than rays. Each individual ray corresponds to an individual mode. *Note 2:* The energy of a guided ray is confined between surfaces, or in the vicinity of surfaces, of materials because of specific properties of the materials. In guided electromagnetic waves, wave energy is concentrated near interfaces, i.e., near boundaries or between substantially parallel boundaries separating materials of different properties. The direction of propagation of guided rays effectively is parallel to these boundaries. A guided electromagnetic wave usually consists of many guided rays. *Synonyms* **bound ray, trapped ray.** *See also* **bound mode, ducting, fiber optics, guided wave, leaky ray, waveguide.**

guided wave: A wave in which (a) energy is concentrated near a boundary or between parallel boundaries separating materials of different properties, and (b) the direction of propagation is parallel to these boundaries. *Note:* Examples of guided waves are (a) waves propagating in a waveguide, (b) evanescent waves on the surface of a waveguide, (c) ground waves along the Earth's surface, (d) radio waves confined to layers of the atmosphere, and (e) sound waves guided by thermal layers of the ocean. *See also* **bound mode.**

guide edge: *Synonym* **reference edge.**

guide feed antenna: *See* **dielectric guide feed antenna.**

guiding optical fiber: *See* **weakly guiding optical fiber.**

gyro bearing: 1. A bearing or a direction that is determined or measured by means of a gyrocompass. **2.** The initial bearing or reference direction at which a gyrocompass initially or originally is set. **3.** The mechanical bearings on which a gyrocompass rotates.

gyroscope: *See* **fiber optic gyroscope.**

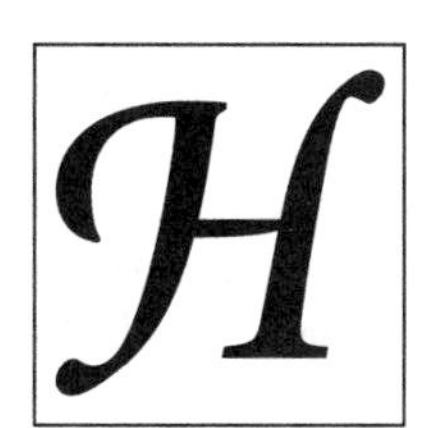

h: The symbol used to represent Planck's constant.

hacker: In computer operations, a person who makes a conscious effort to gain illicit access to a computer system owned by another person or a corporate entity. *Note:* Hackers break access codes by clandestine means, such as by studying material written on waste paper from trash bins, gaining access to computer rooms, talking to operating personnel, conducting trial-and-error routines, and using their knowledge of computer operating systems. Many hackers use their home or business personal computers (PCs) and the telephone networks to gain access to computers, computer systems, and computer networks. Most hackers engage in such activity for personal amusement and to gain knowledge. They consider the attempts to gain access as a challenge and actual access as a victory. A few engage in larceny by stealing computer time, gaining proprietary or private information, and obtaining personal financial credit, such as a competitor's production and delivery data or personal data that may be used to gain an unfair advantage. Many cases of such theft and several cases involving breaches in security have been reported, even though secret passwords were required to gain access. Private and government authorities are using advanced computer techniques to identify hackers and bring them to trial on criminal charges. *See also* **cracker, firewall, superhacker.** *Refer to* **Fig. P-18 (769).**

Hagelbarger code: A convoluted code that (a) is an efficient burst-correcting code used for data transmission through a telephone network, (b) enables error bursts to be corrected, provided that there are relatively long error-free intervals between the error bursts, and (c) has inserted parity check bits spread out in time so that an error burst is not likely to affect more than one of the groups in which parity is checked. *See also* **binary digit, code, convoluted code, error control, error-correcting code, group.**

half-adder: A digital device that (a) has two input lines to insert the addend and augend bits that are to be added, (b) consists of binary combinational circuits, such as AND, NAND, OR, NOR, or NEGATION gates, (c) produces a bit that represents the partial sum digit, i.e., the logical sum of the input bits, (d) produces a bit that represents the partial carry digit, and (e) has two output lines, one for the partial sum bit and one for the partial carry bit. *Note 1:* The gating stages in the half-adder each occur at successive clock pulse times. The clock pulses are used at each logic element to time and shape the output pulses for input to the next stage. *Note 2:* Two half-adders can be used to create a full digital adder. *Note 3:* Half-adders are used in digital computers and control circuits. *See also* **adder.**

half-baud bipolar (HBB) coding: A modified full-baud bipolar coding in which (a) a 0 is represented by a zero voltage level in the bipolar sequence of pulses, (b) a 1 is represented by alternate positive or negative pulses in the bipolar sequence, and (c) pulses return to the zero level at the midpoint of the baud, i.e., the midpoint of the unit interval. *Note:* Half-baud bipolar coding is similar to full-baud bipolar coding except for the pulse levels returning to the zero voltage level at the midpoint of each baud. Bipolar coding has an inherent error-detection capability because of the alternate-polarity pulses. The error occurs when the pulses do not alternate, and vice versa. *See also* **full-baud bipolar coding.**

half-duplex: *See* **two-frequency half-duplex.**

half-duplex (HD) circuit: A circuit that affords communications in either direction but only in one direction at a time. *Note:* If the transmission direction is reversed sufficiently rapidly, a half-duplex circuit can be used effectively to simulate full-duplex operation. *See also* **circuit, duplex circuit, duplex operation, half-duplex operation, push to talk operation, push to type operation, simplex circuit, simplex operation.**

half-duplex equipment: Equipment that (a) consists of a transmitting element and a receiving element and (b) is arranged to allow for transmission in both directions but not simultaneously. *See also* **duplex apparatus.**

half-duplex (HDX) operation: 1. Operation in which communications between two terminals may occur in either direction but only in one direction at a time. *Note:* Half-duplex operation may occur on a half-duplex circuit or on a duplex circuit, but not on a simplex circuit. **2.** Pertaining to an alternate, one-way-at-a-time transmission mode of operation. *Synonyms* **one-way reversible operation, two-way alternate operation.** *See also* **half-duplex circuit.**

half-duplex transmission: Transmission in either direction but only in one direction at a time. *See also* **push to talk operation.**

half-duration: *See* **pulse half-duration.**

half-echo suppressor: Telephone echo suppression equipment in which speech signals in one path control suppression of reflected waves, i.e., echoes, in another path, but the action is not reciprocated.

half-maximum: *See* **full-duration half-maximum (FDHM), full-width half-maximum (FWHM).**

half-power point: Any point at which power is determined to be one-half of what it is at some other reference point in a given system or component. *Note:* Examples of half-power points are (a) at a particular distance and direction from a source of radiation, the point at which the irradiance, or field strength squared, is one-half of the maximum irradiance, or field strength squared, at that distance but in some other direction, and (b) when moving radially outward from the center of a narrow circular cross-sectional beam, the point at which the optical power density, i.e., the irradiance, is one-half of the power density at the center of the beam, regardless of the distance from the source at which the cross section is taken. *See* **full-wave half-power point.**

half-power points: *See* **emission beam angle between half-power points.**

half-reflection: Reflection that occurs at the interface between one propagation medium and another when half of the incident power is reflected and half is transmitted. *Note:* In optical systems, a half-reflection coupler allows half of the light energy to be coupled out of an optical waveguide, such as an optical fiber, fiber optic bundle, or fiber optic cable, and half to be available for transmission beyond the coupler. The reflecting surface makes an angle of about 45° to the axis of the optical element, i.e., about 45° with the incident lightwave. *See also* **fiber optic rod coupler.**

halftone: 1. A photomechanical printing surface, or the impression therefrom, in which detail and tone values are represented by a series of evenly spaced dots in varying size and shape, varying in direct proportion to the intensity of tones, i.e., the shading, they represent. **2.** Simulated continuous tone recorded copy using variable-size dots, such as are used for creating pictures in magazines, periodicals, journals, pamphlets, and newspapers. *See also* **continuous tone copy, facsimile.**

halftone characteristic: 1. In facsimile systems, the relationship between the density of the recorded copy and the density of the object, i.e., the original. **2.** In facsimile systems, the relationship between the amplitude of the facsimile signal and either the density of the object or the density of the recorded copy when only a portion of the system is under consideration. **3.** The ratio of (a) the facsimile signal strength to (b) the density of the object or the density of the recorded copy. *Note:* In a frequency modulation (FM) facsimile system, an appropriate parameter other than the amplitude is used to define the halftone characteristic. *See also* **facsimile.**

halftone picture: A picture that has a range of tones or shades of gray that lie between the limits of picture black and picture white.

half-wave dipole antenna: An antenna that (a) is straight, (b) is one-half of the operating wavelength long, (c) is center-fed so as to have equal current distribution in both halves, (d) when mounted vertically, produces a volumetric doughnut-shaped radiation pattern that is circular in the horizontal plane and double-lobed in a vertical plane through the antenna long axis, (e) is easily constructed and mounted, (f) has some inherent losses, and (g) when used as an antenna gain reference, has a power gain of 0 dB. *Note:* The ratio of (a) the power gain when referred to an isotropic source of radiation, i.e., an isotropic antenna, to (b) the power gain when referred to the standard half-wave dipole antenna is equivalent to 2.15 dB. *See* **lossless half-wave dipole antenna.** *See also* **dipole antenna.**

halide glass: A special glass that (a) contains halogens, such as fluorine, in combination with heavy metals, such as zirconium, barium, or hafnium, (b) transmits light in the visible and infrared spectra of the electromagnetic spectrum, such as wavelengths up to 7 μm (microns), (c) has applications that include remote sensing systems, night vision devices, and fiber optic systems, and (d) has a potential ultralow-loss fiber optic attenuation rate of 10^{-3} dB/km (decibels per kilometer), making the glass particularly attractive for long-distance repeaterless communications links.

halyard: In flaghoist signaling systems, a rope, wire, or chain on which signaling flags or pennants are attached for hoisting up a pole or a mast.

Hamming code: An error-detecting and error-correcting binary code used in data transmission that can (a) detect all single bit and double bit errors and (b) correct all single bit errors. *Note:* Hamming codes must satisfy the relation $2^{(n-k)} \geq n+1$ where n is the total number of bits in the block, including the check bits, and k is the number of information bits in the block. The number of check bits is $n-k$. The information bits are all the bits in the block except the check bits. *See also* **binary digit, block, code, convoluted code, error-correcting code, error-detecting code.**

Hamming distance: *Synonym* **signal distance.**

Hamming weight: The number of nonzero symbols in a symbol sequence, such as a string. *Note:* For binary signaling, the Hamming weight is the number of 1 bits in a given sequence of binary digits. *See also* **binary code, binary digit, Hamming code.**

hand flag transmission: Message transmission in which (a) one or two flags held in the operator's hands are used, and (b) the position or the movement of the flags represents letters, numerals, special signs and symbols, and transmission control codes. *Note:* Examples of hand flag transmission are semaphore, wigwag, and Morse flag signaling.

hand-held calculator: A calculator that (a) can operate independently of electric power mains and (b) is light enough and small enough to be held in one hand while being operated with the other hand.

handler: *See* **magnetic tape handler.**

handling: *See* **automatic data handling.**

handoff: 1. In a cellular telephone system, the transfer of a phone call in progress from one cell transmitter and frequency to another cell transmitter using a different frequency without interrupting the call. **2.** In satellite communications, the process of transferring ground station control responsibility from one ground station to another without loss or interruption of service. *See also* **cellular radio.**

handset: A telephone transmitter and receiver in a single unit that (a) can be held in one hand in such a manner that the receiver portion can be placed against the ear and the transmitter portion can be placed in front of the mouth, (b) is placed on a cradle or a hook when not in use, and (c) is removed from the cradle or the hook when answering a call. *See also* **hang up.**

handshake: In communications system operation, error control in which data transmitted over a forward channel are returned over a backward channel and checked with the stored data originally transmitted, usually on an immediate and continuous basis. *See* **backward channel, forward channel.**

handshaking: 1. In a data communications system, a sequence of events governed by hardware or software that requires mutual agreement on the state of the operational conditions prior to information exchange. **2.** The actions that (a) are used to establish communications parameters between two stations, (b) follow the establishment of a circuit between the stations, (c) precede information transfer, and (d) are used to agree upon protocol features and parameters, such as information transfer rate, alphabet, parity, and interrupt procedure. **3.** An exchange of predetermined characters or signals between two stations that provides for control or synchronism after a connection is established, particularly during the information transfer phase of an information transfer transaction, such as a call or a message transfer. **4.** The establishment of communications parameters that (a) usually follows the establishment of a circuit between the stations, (b) usually precedes information transfer, and (c) is used to ensure that agreement has been reached on specified system parameters and protocols, such as signaling rates, alphabets, codes, parity, and interrupt procedures. *See also* **answer back, information transfer transaction, protocol.**

hangover: *Synonym* **tailing.**

hang up: To terminate a call by operating a switch on a telephone instrument, usually by placing the handset on a switch in a cradle or by hanging the handset on a hook switch, the weight of the handset holding the switch down against a spring, such that, in either case, the switch will operate when the handset is lifted from the cradle or the hook. *See also* **handset.**

hard clad silica (HCS) optical fiber: An optical fiber that has a silica core and a hard polymeric plastic cladding intimately bonded to the core.

hard copy: In computer graphics and in telecommunications, a permanent reproduction that (a) is on any medium suitable for direct use by a person and (b) consists of displayed or transmitted data. *Note 1:* Examples of hard copy include teletypewriter pages, continuous printed tapes, facsimile pages, computer printouts, and radiophoto prints. *Note 2:* Magnetic tapes, diskettes, and nonprinted punched paper tapes are not hard copy. *See also* **soft copy.**

hard disk: In a personal computer (PC) system, a permanently installed high capacity magnetic or optical disk and drive used to provide large amounts of storage capacity for the system, often 400 times the capacity of a diskette. *Note:* Although most PC system software can operate with only a hard disk, it is more convenient to have at least one diskette drive, even better to have two, one for each standard-size diskette, to provide a means of easily transferring files and programs from diskettes to hard disk and vice versa. Also, diskettes serve as an input-output channel because they can be removed, stored, updated, and transferred to other PCs. Some software programs require at least one diskette drive in addition to a hard disk. *See* **Bernoulli box disk drive, external hard disk, hard drive, internal hard disk.**

hard drive: In a personal computer (PC) system, an assembly that (a) consists of a hard disk, a controlled driving mechanism, a read-write head, and input-output gating, (b) is factory-installed or service-shop-replaced rather than removed and inserted by the operator as diskettes are, and (c) usually is used to hold all operating system, application, and utility software. *Note:* Hard drive capacities range from 10 Mb (megabytes) to 500 Mb, whereas diskette capacities usually are 1.2 Mb or 1.44 Mb. *See* **read-write head.**

hardened: Pertaining to the condition of a facility with protective features that enable it to withstand destructive environmental forces, such as lightning, explosions, shock, excessive vibration, excessive moisture and humidity, natural disasters, high intensity ionizing electromagnetic radiation, or any combination of these.

hardened optical fiber: *See* **radiation hardened optical fiber.**

hardening: *See* **fiber optic nuclear hardening, optical fiber nuclear hardening.**

hard hyphen: A hyphen that (a) is deliberately placed in text by an author or an operator, usually to hyphenate words that require hyphens, such as in "high-pass filter" and "time-division multiplexing," and (b) is retained when automatic end-of-line hyphenation is used in word processing system software. *See also* **soft hyphen.**

hard limiter: *Synonym* **limiter circuit.**

hard limiting: 1. In a device, limiting in which the limiting action produces (a) a negligible variation of the expected characteristic of an output signal of the device over the permitted dynamic range and (b) a steady state signal at the maximum permitted level for the duration of each period when the output would otherwise be required to exceed the permitted dynamic range in order to correspond precisely to the transfer function of the device. **2.** Limiting with negligible variation in output in the range where the output is limited, i.e., is controlled, when the input varies widely. *See also* **clipper, clipping, level, limiting, soft limiting, transfer function.**

hardness: *See* **limiter hardness, nuclear hardness, radiation hardness.**

hard sectoring: 1. The physical marking of sector boundaries on a magnetic or optical disk. **2.** Establishing sectors on a disk, such as a magnetic or optical disk, by making a physical mark on the disk, such as an index hole, that marks the beginning of the first sector recorded on the disk, such that when the presence of the physical mark is recognized by a reader, a reference signal is generated, and all sector locations can be measured from this signal. *See also* **soft sectoring.**

hard stop: An immediate or sudden cessation of operation or execution by a functional unit. *See also* **functional unit.**

hardware: 1. Physical equipment, as opposed to computer programs, protocols, procedures, rules, and associated documentation. *Note 1:* Examples of hardware are transmitters, receivers, wire, printed circuits, cables, transponders, and computers. *Note 2:* Firmware is considered as hardware because it is a configuration of tangible material. **2.** Physical items, as distinguished from their capability or function, such as equipment, tools, implements, instruments, devices, sets, fittings, trimmings, assemblies, subassemblies, components, and parts. *Note:* "Hardware" is often used in regard to the stage of development of equipment, as in the development of a device or a component from the design stage into a physical stage, such as a prototype stage, as a step toward the finished production item. **3.** In communications, computer, data processing, and control systems, the physical equipment or devices that compose the systems. *See* **off-the-shelf hardware.** *See also* **firmware, software.**

hardware check: 1. A check automatically performed by hardware components, such as a circuit used to perform a parity check or a circuit used to sense the on or off condition of a printer. **2.** A check made to determine the existence of faults or malfunctions in hardware. *Synonym* **automatic check.** *See also* **hardware.**

hardware reliability: 1. A measure of the ability of a hardware functional unit to continue to perform its intended function under either normal or adverse environmental conditions. **2.** The probability that a hardware system will function without failure over a specified period, i.e., the probability that the system will not fail in the time from time 0 to time t, given the assumed Poisson model that the probability of failure is given by the relation $P = e^{-\gamma t}$ where P is the probability of failure, γ is the constant component failure rate in parts per unit per second, such as parts per million per second, and t is time in seconds if γ is in reciprocal seconds. *See also* **bathtub curve, constant failure period, software reliability.**

hardwire: 1. To connect equipment or components permanently in contrast to using switches, plugs, or connectors. **2.** In communications, computer, data processing, and control systems, the wiring-in of fixed logic or read only storage that cannot be altered by program changes, i.e., by software. *See also* **firmware.**

harmful interference: 1. Any emission, radiation, or induction interference that endangers the functioning or seriously degrades, obstructs, or repeatedly interrupts a communications system, such as a telecommunications service, radio communications service, search and rescue service, or weather service, operating in accordance with approved standards, regulations, and procedures. *Note:* To be considered harmful interference, the interference must cause serious debilitating effects, such as circuit outages and message losses, as opposed to interference that is merely a nuisance or an annoyance that can be overcome by appropriate measures. **2.** Interference that endangers the functioning of a radionavigation service or of other safety services or seriously degrades, obstructs, or repeatedly interrupts a radiocommunications service operating in accordance with the Radio Regulations approved by the International Telecommunication Union (ITU). *See also* **accepted interference, interference.**

harmonic: 1. Of a sine wave, an integral multiple of the frequency of the wave. *Note 1:* The first harmonic is the fundamental frequency, the second harmonic is twice the fundamental frequency, the third harmonic is three times the fundamental frequency, and so on. *Note 2:* A pure sine wave has no harmonics above the first harmonic. **2.** Of a nonsinusoidal wave, such as a square wave, an integral multiple of the fundamental frequency of the wave. *Note 1:* The frequencies of the harmonics, including the fundamental frequency, and their amplitudes, may be identified by a Fourier analysis of the wave. *Note 2:* The fundamental frequency of a periodic phenomenon, such as a square wave, is the reciprocal of the period of the phenomenon. *See* **single harmonic.** *See also* **Fourier analysis, overtone.**

harmonic distortion: In the output signal of a device, distortion caused by the presence of frequencies that are not present in the input signal. *Note 1:* Harmonic distortion usually is caused by nonlinearities within the device. *Note 2:* In harmonic distortion measurements, the fundamental frequency, i.e., the first harmonic, of the input signal is the fundamental frequency of the output, regardless of what a Fourier analysis of the output signals identifies as the fundamental frequency. *See* **single harmonic distortion, total harmonic distortion.** *See also* **distortion.**

harmonic function: A mathematical function that (a) can be used to describe the curves used to represent waves and (b) has constant magnitudes at particular constant intervals of abscissa values, i.e., independent variable values, such as time, distance, and frequency. *Note 1:* An example of a harmonic function is a sine wave or a combination of sine waves, such as occurs in a Fourier analysis. *Note 2:* The oscillations of the electric and magnetic fields in plane polarized electromagnetic waves may be described by harmonic functions as well as by exponential functions. *See* **sectorial harmonic function, spherical harmonic function, tesseral harmonic function, zonal harmonic function.**

harness: In communications, computer, information processing, control, and power distribution systems, an assembly of wiring and cabling that conveys signals or power to a system consisting of distributed components. *See* **feed harness, optical harness, vehicle harness.**

harness assembly: *See* **optical harness assembly.**

harness run: *Synonym* **cable run.**

hartley: A unit of logarithmic measure of information content, 1 h (hartley) being equal to the decision content of a set of ten mutually exclusive events that are expressed as a logarithm to the base 10. *Note:* The decision content of a character set of eight characters equals 0.903 hartley, obtained from $\log_{10} 8$. *Synonym* **decimal unit of information content.** *See also* **decision content.**

hash total: The result obtained by subjecting a set of data to an algorithm for purposes of checking the data at the time the algorithm is applied or for use at a later time, such as after transmission or retrieval from storage. *Note:* Examples of hash totals are (a) sums produced by a modulo-*n* check and (b) summing numbers of different items, such as amplifiers, gates, and connectors, to obtain an item count that can be used later for checking purposes. *See also* **modulo-n check.**

hauled down: In flaghoist communications, the situation (a) that exists when a flaghoist is returned, i.e., is lowered, to the deck or the ground and (b) usually in which the moment of hauling down is the moment of execution of the instruction, command, or order that was specified by the flaghoist when it was raised. *See also* **at the dip, close up, flaghoist.**

hazard: *See* **electromagnetic radiation hazard, laser hazard, optical fiber hazard.**

hazards of electromagnetic radiation to fuel (HERF): The potential that electromagnetic radiation has to cause ignition or detonation of volatile combustibles, such as aircraft fuels, hydrogen, propane, carbon monoxide, powders, and dust. *See also* **radiation.**

hazards of electromagnetic radiation to ordnance (HERO): The potential that electromagnetic radiation has to adversely affect explosives, such as munitions, dynamite, trinitrotoluene (TNT), fuzes, and blasting caps, especially electroexplosive devices. *Note:* Exposure to electromagnetic radiation, such as is produced by radio transmitters and radar, can cause premature actuation of electroexplosive devices. Electromagnetic radiation can damage or trigger solid state circuits, cause erratic readings in test sets, and set off electrical blasting caps, any of which may be part of a munitions detonating device. Thus, there is a high potential for munitions or electroexplosive devices, such as proximity fuzes, to be adversely affected by electromagnetic radiation. *See also* **radiation.**

hazards of electromagnetic radiation to personnel (HERP): The potential that electromagnetic radiation has to produce harmful biological effects in humans. *Note:* Particular damage can occur to the retina of the eye and to abdominal tissue at microwave and higher frequencies. X-ray and gamma radiation can destroy living tissue cells and cause genetic disorders. *See also* **radiation.**

HA1 receiver weighting: A noise weighting used in a noise measuring set to measure noise across the HA1 receiver of a 302-type or similar instrument. *Note 1:* The meter scale readings are in dBa(HA1). *Note 2:* HA1 receiver weighting is obsolete for U.S. Department of Defense (DoD) applications. *See also* **C message weighting, flat weighting, 144-line weighting, 144-receiver weighting.**

H bend: For a transverse electromagnetic (TEM) wave propagating in a waveguide, a smooth change, i.e., a gradual change, in the direction of the axis of the waveguide, such that throughout the change the axis remains in a plane parallel to the direction of the magnetic field, i.e., the H field. *Synonym* **H plane bend.** *See also* **E bend, macrobending, microbending, waveguide.**

H channel: In Integrated Services Digital Networks (ISDN), a 384-kbps (kilobit per second), 1472-kbps, or 1536-kbps channel, designated as H_0, H_{10}, and H_{11}, respectively, accompanied by timing signals, used to carry a wide variety of user information. *Note:* User information representation forms carried by H channels include fast facsimile, video, high speed data, high quality audio, packet switched data, and bit streams at rates less than the respective H channel bit rate that have been rate-adapted or multiplexed together.

HCS: hard clad silica.

HCS optical fiber: *See* **hard clad silica optical fiber.**

HD: half-duplex.

HDLC: high level data link control.

HDSL technology: high-bit-rate digital subscriber line technology.

HDTV: high definition television.

HDTV system: *See* **high definition television system.**

head: A device that reads, writes, or erases data on a data medium. *See* **erase head, keying head, laser head, magnetic head, read head, read-write head, tape reading head.** *See also* **data medium.**

head end: **1.** In local area network/metropolitan area network (LAN/MAN) systems, a control device that provides centralized functions, such as remodulation, retiming, message accountability, contention control, diagnostic control, and access to a gateway. **2.** In cable television (CATV) systems, a control device that provides centralized functions, such as remodulation. *See also* **local area network.**

header: In a message, packet, or call, the portion that contains (a) information used to guide the message, packet, or call to the correct destination and (b) special requirements of the originator, such as security, precedence, and transmission speed requirements. *Note:* Examples of items that may be in a header are the addresses of the sender and the receiver, precedence level, routing instructions, user overhead information, system overhead information, and synchronizing pulses. *Synonym* **header part.** *See* **external header.** *See also* **address field, message heading, overhead information.**

header character: *See* **start of header character.**

header part: *Synonym* **header.**

heading: The direction in which a mobile object, such as a ship, aircraft, spacecraft, land vehicle, or missile, either is pointed or is moving with respect to a given reference direction, such as (a) true, magnetic, or grid north or (b) compass, baseline, or thrust line direction. *Note:* Headings and bearings usually are given as true or magnetic headings and bearings unless otherwise specified. *Synonym* **course.** *See* **massage heading.** *See also* **bearing, direction.**

heading sense: One of the two opposite directions that are associated with a given heading, i.e., a given course or direction, as determined by a direction finding station or stations. *Note:* Unless otherwise stated, the calling station, i.e., the station that desires to know its heading, will assume that sense was determined by the direction finding (DF) station and therefore the proper forward heading is given in the response, i.e., the 180° ambiguity has been removed. If the mobile station that is requesting the heading were to proceed in the direction of the bearing given by the DF station, it would arrive at the DF station. If two successive requests by a mobile station result in a determination of actual heading by a mobile station, the DF station will report the actual heading being taken by the mobile station. In normal usage, "heading" is the direction in which an object is moving relative to a reference direction, whereas "bearing" is the direction from an object to another object relative to a reference direction. Both must be established in the proper sense in regard to the 180° ambiguity. *See also* **bearing, direction finding, heading.**

head of bus function: In a data bus, the function that generates management information and empty bus slots at the point on each bus where data flow begins.

head-on collision: **1.** A collision that occurs on a communications channel when two or more users begin to transmit on the channel at approximately the same instant. **2.** A condition that can occur on a circuit, such as a data circuit, when two or more entities, such as nodes, switches, or exchanges, seize either end of the circuit and attempt to transmit at approximately the same instant. *See also* **call collision, contention, lockout.**

headphone: **1.** A telephone receiver and transmitter that (a) is attached to the head when in use by an operator and (b) usually is worn by persons, such as airplane pilots, telephone operators, airport traffic controllers, radio operators, and television studio operators, who need to communicate while keeping both hands free for other use, such as operating a computer keyboard or operating switches on a control console. **2.** A device used to record telephone conversations in a dictation machine. *See also* **earphones.**

hearing threshold: In audiometry, the sound energy level below which a person with normal hearing capability cannot hear sound and above which a person with normal heading capability can hear sound. *Note:* The hearing threshold varies with frequency. *See also* **audiometry, threshold.**

heating equipment: *See* **industrial radio frequency heating equipment.**

Heaviside layer: *Synonym* **E layer.**

Heaviside region: *Synonym* **E region.**

heavy duty connector: In fiber optic systems, a connector that is designed and intended for use outside of an interconnection box, i.e., outside a distribution box or cabinet.

heavy seeding: Pertaining to a condition in an optical propagation medium, such as glass, in which the fine and the coarse seeds are very numerous, such as over 25 seeds per square inch. *See also* **seed, propagation medium.**

heavy typewriter: By international convention, a typewriter that weighs more than 10 kg (kilograms).

HEC: header error check

hectometric wave: An electromagnetic wave that (a) is in the medium frequency (MF) subdivision of the frequency spectrum and (b) lies in the range between 300 kHz and 3000 kHz (kilohertz). *See also* **frequency spectrum, frequency spectrum designation.**

height: *See* **effective height, geostationary satellite height, synchronous height.**

height above average terrain: *See* **antenna height above average terrain.**

height finder: **1.** A device, such as radar, that (a) is located on the Earth's surface and (b) can measure the altitude of an object above ground or above the site at which it is located. **2.** A device, such as radar, that (a) is located in an object, such as an aircraft, spacecraft, satellite, or balloon, and (b) can measure its own height above a surface, such as that of the Earth or the Moon, usually the height above the surface directly below it.

height gain: For a given propagation mode of an electromagnetic wave, the ratio of (a) the field strength at a specified height to (b) the field strength at the Earth's surface. *Note:* The ratio usually is that of the electric field strengths, each expressed in the same units, such as volts per meter. The ratio may also be expressed in dB. *See also* **gain.**

height indicator: *See* **range-height indicator.**

helical antenna: An antenna that is shaped like a helix, i.e., like a coil. *Note:* When the helix circumference is much smaller than one wavelength, the antenna radiates at right angles to the axis of the helix. When the helix circumference is one wavelength, maximum radiation is along the helix axis. *Synonym* **helix antenna.**

helical polarization: *See* **left-hand helical polarization, right-hand helical polarization.**

helical ray: In a graded index optical fiber, a ray that (a) follows a path that winds around the fiber axis as it propagates along the fiber and (b) has its polarization plane rotated during propagation. *See also* **axial ray, meridional ray.** *Refer to* **Fig. R-5 (817).**

heliograph: In visual signaling systems, a device that (a) is used for signaling by means of reflected rays of the Sun, (b) contains a reflecting surface, such as a mirror, and (c) has a means of rapidly masking and unmasking the reflecting surface in accordance with a code, such as the international Morse code. *Note:* Heliographs are often used in ship-to-ship, ship-to-shore, and shore-to-ship signaling, particularly during periods of radio silence.

helix: In facsimile systems, a rotating part that is used in continuous receivers, consisting of a helicoidal rib whose intersection with the scanning line defines the position of the picture element on this line at a given instant. *See also* **picture element.**

helix antenna: *Synonym* **helical antenna.**

hellschreiber system: *Synonym* **hell system.**

hell system: Mosaic telegraphy in which (a) a telegraph code is used that represents each character by a fixed, unique number of unit elements, and (b) the received characters are formed with a rotating spiral that is continuously inked and that is brought into contact with a moving paper tape under the control of the unit elements. *Synonym* **hellschreiber system.** *See also* **mosaic telegraphy.**

Helmholtz equation: A member of a set of equations that describe uniform plane polarized electromagnetic waves in an unbounded, lossless, i.e., nonabsorptive, source-free propagation medium, such as an optical fiber, when (a) the wave equations are expressed in Cartesian coordinates, i.e., in rectangular coordinates, (b) the components of the electric and magnetic fields in the direction of propagation are identically equal to zero, and (c) the transverse first derivatives $\partial/\partial x$ and $\partial/\partial y$ when the wave is propagating in the z-direction are also identically zero. *Note:* The Helmholtz equations are given by the relations

$$d^2E_x/dz^2 - \Gamma^2 E_x = 0$$

$$d^2E_y/dz^2 - \Gamma^2 E_y = 0$$

$$d^2H_x/dz^2 - \Gamma^2 H_x = 0 \text{ and}$$

$$d^2H_y/dz^2 - \Gamma^2 H_y = 0$$

where E is the scalar electric field strength component and H is the scalar magnetic field strength component in the direction indicated by the subscript for a wave propagating in the z-direction. Γ is given by the relation $\Gamma^2 = j\omega\mu(\sigma + j\omega\varepsilon) = -\omega^2\mu\varepsilon + j\omega\mu\sigma$ where j is $-1^{1/2}$, $\omega = 2\pi f$ is the angular velocity, f is the frequency, μ is the magnetic permeability, ε is the electric permittivity, and σ is the electrical conductivity. Thus, Γ^2 is a complex function. For dielectric materials, such as the glass used in dielectric waveguides, $\mu = 1$ and $\sigma = 0$. These equations apply to propagation in optical elements under certain geometrical conditions. These equations are solutions of the vector Helmholtz equations given by the relations $\nabla^2\mathbf{E} - \Gamma^2\mathbf{E} = 0$ and $\nabla^2\mathbf{H} - \Gamma^2\mathbf{H} = 0$ where ∇ is the spatial derivative used in the vector algebra, and Γ is as above.

help facility: In most personal computer (PC) software systems, a special arrangement that provides information to an operator or a user on how to tell the PC to perform a desired operation. *Note 1:* Usually a special key is actuated to display such helpful information on the screen of the display device, i.e., the monitor. *Note 2:* In Lotus 1-2-3 pressing the F1 function key at any time while using Lotus 1-2-3 provides instant information about the operation currently in progress. In WordPerfect, pressing the F3 function key provides helpful information. Other software systems have similar arrangements. *See also* **help key, help screen.**

help key: In personal computer (PC) systems, a special key, or set of keys, that (a) is designated by the software program in use and (b) causes information to be displayed relevant to the operation currently in progress, so that (c) actuating the same, or another specially designated key, causes the system to return to the same point in the sequence of operations where it was when the help key was depressed. *Note:* In Lotus 1-2-3, pressing the F1 key will cause a special help screen to be displayed, and pressing the escape key will cause the personal computer (PC) to return to the same point in the spreadsheet, i.e., the Worksheet, where the operator left off when the F1 key was depressed. *See also* **help facility, help screen.**

help screen: In personal computer (PC) systems, information that (a) is displayed on a display device screen, i.e., on a monitor, (b) concerns the operation currently in progress, (c) enables the operator or the user to become more familiar with the operation, and (d) provides a basis for decisions concerning operations about to be performed by the PC. *See also* **help facility, help key.**

HE_{11} mode: In a lightwave propagating in an optical fiber, the designation for the fundamental hybrid mode. *See also* **fundamental mode, mode.**

HEMP: high altitude electromagnetic pulse.

HERF: hazards of electromagnetic radiation to fuel.

HERO: hazards of electromagnetic radiation to ordnance.

HERP: hazards of electromagnetic radiation to personnel.

hertz (Hz): 1. The SI unit (Système International d'Unités) for frequency. *Note:* 1 Hz (hertz) is equivalent to the frequency of a periodic phenomenon that has a period of one second. **2.** The frequency of a periodic phenomenon for which the period is one second. *Note:* One hertz corresponds to one cycle per second, which is the obsolete unit for electromagnetic wave frequency, but which may be used for other periodic or recurring phenomena, such as occur in internal combustion engines, machine tools, automatic screw machines, and assembly lines. *See also* **frequency, frequency spectrum designation, International System of Units.**

Hertzian dipole antenna equation: In a radiating Hertzian dipole antenna, the differential tangential component of the electric field vector is given by the relation $dE_\Theta = (Idl\,(\sin\Theta)/4\pi\varepsilon)[(-\omega\,(\sin\omega t)/rv^2 + (\cos\omega t)/r^2v + (\sin\omega t)/\omega r^3]$ where dE_Θ is the differential tangential component of the electric field vector, I is the instantaneous antenna current in the elemental length of antenna, dl is the elemental length of antenna bearing the current I, Θ is the angle between the antenna direction of dl and the line from the dipole center to the point of measurement of dE_Θ, π is approximately 3.1416, ε is the electric permittivity of the intervening medium, ω is the angular velocity which is equal to $2\pi f$ where f is the frequency of the radiation, t is time, r is the distance from the center of the dipole to the point of measurement of dE_Θ, and v is the propagation velocity. *Note:* The first term expresses the far field contribution, i.e., the radiation field contribution. It varies inversely with the distance from the antenna and therefore is dominant at large distances. The second term expresses the intermediate field contribution, i.e., the induction field contribution. It varies inversely with the square of the distance and is dominant closer in to the antenna. The third term expresses the near field contribution, i.e., the electrostatic field contribution. It varies inversely with the cube of the distance and is dominant immediately outside the antenna surface. The antenna length and the distances should be expressed in the same units. *See also* **far field, intermediate field, near field.**

Hertzian wave: An electromagnetic wave that (a) has a frequency lower than 3000 GHz (gigahertz), i.e., lower than 3×10^{12} Hz, and (b) may be propagated in free space, without a waveguide. *Note 1:* Examples of Hertzian waves are the full range of radio waves. *Note 2:* Lightwave frequencies are about 3×10^{14} Hz. *Note 3:* Hertzian waves, lightwaves, X rays, and gamma rays are electromagnetic waves.

heterochronous: A relationship between two signals such that their corresponding significant instants do not necessarily occur at the same time. *Note:* Two signals that (a) have different nominal signaling rates and (b) do not stem from the same clock or from homochronous clocks usually are heterochronous. *See also* **anisochronous, homochronous, isochronous, mesochronous, plesiochronous.**

heterodyne: 1. Pertaining to the generation of new frequencies by mixing two or more signals in a nonlinear device, such as a vacuum tube, transistor, or diode mixer. *Note:* A superheterodyne receiver converts selected incoming frequencies by heterodyne action to a common intermediate frequency (IF) for amplification and for improving selectivity, i.e., improving filtering. **2.** To generate new frequencies by mixing two or more signals in a nonlinear device, such as a vacuum tube, transistor, or diode mixer. **3.** Pertaining to a frequency produced by mixing two or more signals in a nonlinear device. **4.** The combining of two electromagnetic waves of different frequency in a nonlinear device in order to produce frequencies that are equal to the sum and the difference of the combining frequencies. **5.** A frequency produced by mixing two or more signals in a nonlinear device. *See* **self-heterodyne.** *See also* **beating, frequency, image frequency, intermodulation, intermodulation distortion.**

heterodyne oscillator: *Synonym* **beat frequency oscillator.**

heterodyne repeater: 1. In a radio receiver, a repeater in which the received radio frequency signals are converted to an intermediate frequency for further processing in the receiver, such as detection, i.e., demodulation, amplification, and conversion to sound. **2.** A repeater in which the received signals are converted, i.e., are translated, to an intermediate frequency, amplified, and reconverted, i.e., retranslated, to a new frequency for transmission over the next repeater section. *Synonym* **intermediate frequency repeater.** *See also* **frequency, repeater.**

heterodyning: The mixing of an electromagnetic wave of one frequency with an electromagnetic wave of another frequency to produce one or more additional frequencies. *Note 1:* The sum and difference frequencies are produced when the waves of two different frequencies are combined in a nonlinear device, such as a nonlinear amplifier. *Note 2:* Heterodyning is often performed in radio receivers to enable the receiver circuits to operate at a frequency lower than that of the carrier. *See also* **intermediate frequency.**

heteroepitaxial optical waveguide: An optical wavelength electromagnetic waveguide that (a) consists of an optical-quality crystal substrate upon which are deposited one or more layers of substances with different refractive indices, (b) has deposited layers that have closely matched lattice structures and indices of refraction less than that of the substrate so that the layers themselves do not act as ordinary waveguides with total internal reflection, and (c) propagates leaky modes, with attenuation losses inversely proportional to the square of the wavelength. *Note:* A heteroepitaxial optical waveguide may be made of (a) cubic zinc sulfide layered upon a calcium arsenide substrate or (b) zinc selenide on a gallium arsenide substrate. *See* **attenuation, leaky mode, refractive index, total internal reflection.**

heterogeneous computer system: A computer system that has characteristic differences, such as distinctive structural, electrical, and operational differences, from other computer systems.

heterogeneous multiplexing (HM): Multiplexing in which two or more information-bearer channels operate at different data signaling rates (DSRs), i.e., multiplexing such that the information-bearing channels resulting from multiplexing do not all operate at the same DSR. *See also* **homogeneous multiplexing, multiplexing.**

heterogeneous network: A network that consists of dissimilar or incompatible components, such as computers of different manufacture, different electrical characteristics, and different mechanical features.

heterojunction: A junction between two semiconductors that differ in their (a) doping levels, (b) electrical conductivities, and (c) atomic or alloy compositions. *Note:* In heterojunction transistors, diodes, and lasers, a sudden transition occurs in material composition across the junction boundary, such as a sudden change in refractive index or a transition from p-type, i.e., positively doped, to n-type, i.e., negatively doped, material. In most heterojunctions, a change in geometric cross section occurs, across which a voltage or a voltage barrier exists. Heterojunctions provide a controlled level and direction of radiation confinement. *See* **double heterojunction, single heterojunction.** *See also* **homojunction.**

heterojunction diode: *See* **double heterojunction diode, five-layer four-heterojunction diode, four-heterojunction diode, monorail double-heterojunction diode.**

heuristic: Pertaining to exploratory methods for solving problems by evaluating the progress made toward the final result, such as by empirical, untried, unproven methods or by methods that produce results that cannot be proved.

heuristic routing: In a communications network, routing in which certain data, such as time delay data, extracted from incoming messages, during specified periods and over different routes, are used to determine the optimum routing for transmitting data back to the sources. *Note:* Heuristic routing allows a measure of routine optimization based on recent empirical knowledge of the state of the network. *See also* **alternate routing, deterministic routing, flood search routing, routing.**

hexadecimal: 1. Pertaining to a selection, choice, or condition that has 16 possible different values or states. **2.** In a fixed-radix numeration system, a radix of 16. *Synonym* **sexadecimal.**

HF: high frequency.

HFAARS: High Frequency Adaptive Antenna Receiving System.

HFDF: high frequency distribution frame.

hidden line: In display systems in which there is a two-dimensional projection of a three-dimensional object in the display space on the display surface of a display device, a line that (a) is a representation of an object edge that is obscured by the object itself, or by another object, when the object is viewed, (b) is used to represent the shape of an object, (c) usually is in a format different from that of edges that are not obscured by the object when it is being viewed, (d) usually is represented by a sequence of short collinear line segments, (e) represents abrupt or discontinuous changes in surface direction, such as at intersections of surfaces, and (f) occurs as boundaries or perimeters of images of objects.

hierarchical computer network (HCN): A computer network in which processing and control functions are performed at several levels by computers specially suited for the functions performed, such as industrial process control, inventory control, database control, or hospital automation. *See also* **distributed control.**

hierarchical decomposition: Pertaining to the top down designing of a system by means of its top down subdivision into identifiable components or modules. *See also* **modular decomposition, top down designing.**

hierarchically synchronized network (HSN): A mutually synchronized network in which (a) some clocks exert more control than others and (b) the timing of all clocks is a weighted mean of the natural timing of the population of clocks. *See also* **democratically synchronized network, despotically synchronized network, master-slave timing, mutually synchronized network, mutual synchronization, oligarchically synchronized network.**

hierarchical network: A network in which processing and control functions are performed at several layers by computers that are specially designed to perform the functions at each layer.

hierarchical routing: In a network, routing that is based on a hierarchical, i.e., on a layered, addressing scheme. *Note:* Most Transmission Control Protocol/Internet Protocol (TCP/IP) routing is based on a two-level hierarchical routing in which an IP address is divided into a network portion and a host portion. Gateways use only the network portion until an IP datagram reaches a gateway that can deliver it directly. Additional levels of hierarchical routing indicators are introduced by the addition of subnetwork identifiers.

hierarchical structure: A system configuration in which groups of elements, i.e., of components, that are in a given class are layered above or below other groups of elements in a laminar fashion. *Note:* Examples of hierarchical structures are (a) structures that conform to the open systems architecture concept of a layered system and (b) structures of tree networks. *Refer to* **Fig. P-19 (771).**

hierarchy: *See* **digital multiplex hierarchy.**

high altitude electromagnetic pulse (HEMP): An electromagnetic pulse produced at an altitude effectively above the sensible atmosphere, i.e., higher than 120 km above the Earth's surface. *See also* **electromagnetic pulse.**

high birefringence optical fiber: An optical fiber in which the polarization beat length is 1 mm (millimeter) or less. *Note 1:* The beat length, L_B, is given by the relation $L_B = 2\pi/(\beta_x - \beta_y)$ where β_x and β_y are the propagation constants of the two polarized waves caused by the birefringence of the fiber. Thus, the propagation constant difference is large. *Note 2:* High birefringence optical fibers are designed for applications in which a stable polarization state is important and must be maintained. *See also* **birefringence, low birefringence optical fiber.**

high definition television (HDTV): Television that (a) has approximately twice the horizontal and twice the vertical emitted resolution specified by the National Television Standards Committee (NTSC) standard, (b) has 1125 horizontal scan lines, (c) has a wider than specified screen, (d) has over four times the total number of pixels that are specified in the NTSC standard, (e) may include any or all improved definition television (IDTV) and extended definition television (EDTV) improvements, (f) has a wide aspect ratio, (g) may use fiber optic cables that have the broader bandwidth required for over 100 high definition television (HDTV) station channels, with bandwidth to spare for an integrated services data network (ISDN) including optical fiber in the local loop, and (h) is not compatible with conventional systems and would render them obsolete. *See also* **advanced television, broadband integrated services digital network, distribution quality television, extended definition television, improved definition television.**

higher frequency ground: The interconnected metallic network intended to serve as a common reference for currents and voltages at frequencies above 30 kHz (kilohertz) and, in some cases, above 300 kHz. *Note:*

The term "higher frequency ground" is no longer in common use and is deprecated. Use "facility grounding system" or one of the components of the facility grounding system, as appropriate. *See also* **equipotential ground plane, facility grounding system, frequency, ground.**

high frequency (HF): 1. A frequency that lies in the range between 3 MHz (megahertz) and 30 MHz, the range having the alphabetic designator HF and the numeric designator 7. **2.** Usually, a radio, radar, or video frequency that is above 3 MHz (megahertz). *See* **extremely high frequency, lowest useful high frequency, superhigh frequency, tremendously high frequency, ultrahigh frequency, very high frequency.** *See also* **frequency, frequency spectrum designation, lowest usable frequency.** *Refer to* **Tables 1 and 2, Appendix B.**

High Frequency Adaptive Antenna Receiving System (HFAARS): A receiving antenna system that (a) suppresses signals that interfere with communications, (b) is used on Canadian Navy ships, and (c) consists of (i) four high frequency (HF) receivers that are matched in phase and frequency, (ii) digitizers, (iii) filters, and (iv) a programmable digital signal processing chip.

high frequency distribution frame (HFDF): A distribution frame that provides terminating and interconnecting facilities for the combined supergroup modulator output circuits and the combined supergroup demodulator input circuits that contain signals occupying the baseband spectrum. *See also* **baseband, distribution frame, frequency, group.**

high frequency omnidirectional radio range: *See* **very high frequency omnidirectional radio range.**

high frequency radio: A radio that (a) uses electromagnetic waves at frequencies above 3 MHz (megahertz) for communications purposes, (b) uses the higher frequencies to obtain greater propagation distance than are possible with the lower frequencies, and (c) usually depends on ionospheric reflection for long-distance communications. *See* **adaptive high frequency radio.** *Synonym* **shortwave radio.**

high gain antenna: An antenna with an antenna power gain that is achieved by confining its radiated power to a smaller solid angle than is obtained in other antennas. *See also* **antenna power gain, directive gain.**

high grade cryptographic system: *Synonym* **high grade cryptosystem.**

high grade cryptosystem: A cryptosystem that (a) provides long-lasting security, i.e., a cryptosystem that is inherently resistant to solution for a comparatively longer time than lower grade cryptosystems and (b) provides security almost indefinitely, i.e., for a very long time, regardless of the number of transmissions in which it is used. *Synonym* **high grade cryptographic system.**

high level control: In the hierarchical structure of a primary or secondary data transmission system, the conceptual level of control or processing logic that (a) is above the Link Layer and (b) controls Link Layer functions, such as device control, buffer allocation, and station management. *Note:* In the open systems architecture concept of a communications network, high level control is considered to be in the application, presentation, session, transport, and network layers. Low level control is considered to be in the data link and physical layers. *See also* **highlevel data link control, level, low level control, Open Systems Interconnection—Reference Model.**

highlevel data link control (HDLC): A Data Link Level protocol that provides for point-to-point transmission of a packet. *Note:* A subset of highlevel data link control (HDLC), known as "LAP-B," is the Layer 2 protocol for International Telegraph and Telephone Consultative Committee (CCITT) Recommendation X.25. *See also* **Advanced Data Communication Control Procedure, data, data transmission, level, link, Open Systems Interconnection—Reference Model, synchronous data link control, X series Recommendations.**

high level digital interface: An interface between two sets of station equipment, such as between data terminal equipment (DTE) and data circuit-terminating equipment (DCE), that operates at relatively high electrical voltages and currents. *Note:* An example of high level digital interface equipment is equipment that operates (a) at 60 V (volts) and 120 mA (milliamperes) direct current (dc) or (b) at 130 V and 20 or 60 mA, dc. *See also* **interface.**

high level language (HLL): 1. A computer programming language that (a) neither reflects nor depends on the structure of any one specific computer or any one specific class of computers, (b) requires that a statement in the language be interpreted, and (c) requires that corresponding intermediate, assembly, or machine language statements be compiled for use by a specific computer. *Note:* Examples of high level languages are Ada®, used in U.S. Department of Defense applications, and FORTRAN, BASIC, and COBOL, used in commercial applications. **2.** A computer programming language in which a notation scheme is used that (a) is convenient and readily understood by the users and the programmers and (b) is independent of the computer that will execute programs written in the language. **3.** A programming language that is somewhat independent of the limitations of a specific communications, computer, data processing, or control system, by the use of a compiler or an interpreter that converts the instructions of a high level language computer program to lower level language instructions, such as machine language instructions, that the system can execute. *Synonym* **high-order language.** *See also* **assembly language, compile, computer, computer language, computer oriented language, data description language, intermediate level language, language, low level language, machine language.**

high level protocol (HLP): A protocol that (a) is used in a computer distributed network to direct or control its communications, computer, and data processing resources for the purpose of accomplishing assigned tasks and (b) directs the communications, computing, data processing, and control operations in the system. *See also* **distributed network, protocol.**

highlighting: In display systems, the emphasizing of a specific display element, display group, or display image in the display space on the display surface of a display device. *Note:* Highlighting may be accomplished by various means, such as thickening lines, increasing luminous intensity above normal, increasing background intensity, converting from light on a dark background to dark on a light background, underscoring, surrounding with a halo, using different colors, using shading or cross hatching, or causing the highlighted element to blink regularly.

high loss fiber: *Synonym* **high loss optical fiber.**

high loss optical fiber: An optical fiber in which a propagating optical wave has a high energy loss per unit length of fiber. *Note 1:* Loss usually is measured in dB per kilometer at a specified wavelength. High loss usually is considered to occur at attenuation rates above 100 dB•km^{-1}. The attenuation primarily is due to scattering caused by metal ions and absorption caused by water, i.e., by OH ions, although all causes of loss are included in rating a given optical fiber, including Rayleigh scattering. *Synonym* **high loss fiber.** *See* **absorption, attenuation rate, low loss optical fiber, medium loss optical fiber, scattering.** *Refer to* **Fig. A-10 (49).**

highly reflective coating: A single or multilayer coating that is applied to an optical surface for the purpose of greatly increasing its reflectance over a specified range of wavelengths. *Note:* Examples of highly reflective coatings are (a) single films of aluminum or silver and (b) multilayers of at least two different dielectric materials when low absorption is important. Other parameters, such as incidence angle and incident radiant intensity, i.e., incident irradiance, are also significant in determining the reflectance produced by reflective coatings. *See also* **coated optics, reflectance.**

high-order language: *Synonym* **high level language.**

high-order mode: In an electromagnetic wave propagating in a waveguide, a mode (a) that corresponds to the larger eigenvalue solutions to the wave equations, regardless of whether it is a mode in a transverse electric (TE), transverse magnetic (TM), or transverse electromagnetic (TEM) wave, and (b) in which more than just a few whole wavelengths of the wave fit transversely in the guide and therefore can be supported by the guide. *Note:* A multimode optical fiber supports as many modes as the core diameter, numerical aperture, and wavelength permit. For the high-order modes, the normalized frequency, i.e., the V value, must be above 4 in order that the number of modes be sufficiently high for there to be high-order modes. In a fiber supporting a single mode, there are no high-order modes. Conceptually, however, if there are only two modes, one could be considered the low-order mode and the other the high-order mode. *See also* **low-order mode.** *Refer to* **Fig. P-17 (767).**

high-pass filter: A filter that passes frequencies above a given frequency and attenuates all other frequencies. *See also* **filter, frequency.**

high performance communications equipment: Communications equipment that (a) has the performance characteristics required in trunks or links, (b) is designed primarily for use in global and tactical communications systems, and (c) satisfactorily withstands electromagnetic interference when operating in a variety of network or point-to-point circuits. *Note:* Performance requirements for global and tactical high performance communications equipment may differ. *See also* **high performance equipment.**

high performance equipment: Equipment, such as communications, computer, data processing, and control equipment, that (a) has the performance characteristics required in trunks or links, (b) is designed primarily for use in global and tactical systems, and (c) satisfactorily withstands electromagnetic interference when operating in a variety of systems. *Note:* Performance requirements for global and tactical high performance equipment may differ. *See also* **low performance equipment.**

high probability of intercept receiver: A receiver that has a high probability of detecting any electromagnetic energy that illuminates its antenna within the frequency range that it is designed to cover. *See also* **low probability of intercept receiver.**

high selectivity: In communications systems, such as (a) continuously tuned radio receivers and (b) wavelength division multiplexed lightwave communications systems using optical fibers, the ability to select, separate, and amplify wanted signals and reject unwanted energy, such as undesired transmissions, crosstalk, noise, and interference. *See also* **high sensitivity.**

high sensitivity: 1. The capability of electronic equipment to accept weak incident signals and render them useful. **2.** The capability of electronic equipment to select, separate, and amplify wanted signals and reject unwanted energy, such as crosstalk, noise, and interference. *See also* **high selectivity.**

high speed Morse: The international Morse code that (a) is transmitted at speeds over 80 wpm (words per minute), (b) should be used only when time is critical, such as during search and rescue operations or military situations, and (c) should be used only when the higher probability of errors that it brings to transmission, reception, and interpretation has been considered.

high technology office: *Synonym* **office automation.**

high usage trunk group: A group of trunks for which an alternate route has been provided to absorb the relatively high rate of overflow traffic. *See also* **group, overflow, routing.**

highway: 1. A digital serial coded bit stream with time slots allotted to each call on a sequential basis. **2.** A bus that is capable of carrying a digital serial coded bit stream with time slots allotted to each call on a sequential basis. **3.** A common path or a set of parallel paths over which signals from more than one channel pass with separation achieved by time division. *See* **information superhighway.** *See also* **channel gate, time division multiplexing.**

highway system: *See* **intelligent vehicle highway system.**

HIPERLAN: high performance radio local area network, high-performance local area network.

Hiragana: The set of symbols used in one of the two Japanese phonetic alphabets in common use. *See also* **Katakana.**

hiss: 1. Noise in the audio frequency range, having sounds similar to the English pronunciation of a prolonged "s" as in "sister," "sh" as in "fish," and "z" as in "zoo." *Note:* These sounds are called the sibilant sounds. **2.** Noise that has no pronounced low frequency components. **3.** A rushing noise. *See also* **frequency, interference, noise.**

hit: 1. In communications systems, a transient disturbance to a communications system or a propagation medium. **2.** In information storage, retrieval, and control systems, a match of data to a prescribed criterion. **3.** In data processing, the situation that occurs when specified conditions are met, such as when two items of data are being compared and they are identical. *See* **amplitude hit, line hit, phase hit.** *See also* **dropout.**

hit-on-the-fly printer: *Synonym* **on-the-fly printer.**

HLL: high level language.

HLP: high level protocol.

hockel: In a cable, a loop or a kink that (a) forms when there is a twist that applies a torque to the cable, (b) usually occurs when the cable is being payed out from a spool or a reel and tension is applied to the cable, (c) may result in damage to the cable, and (d) can be prevented from forming by (i) applying a countertwist when the cable is being payed out or (ii) avoiding a twist when it is being wound on reels or spools. *Note:* Twist can occur when a cable is formed with biased or unbalanced twisted components or when the cable passes around idler pulleys and capstans when being wound.

hockey puck: A polishing fixture that (a) is used to facilitate the manual finishing of the endfaces of certain types of fiber optic connectors, (b) consists of the appropriate mating sleeve for the connector, which is mounted at right angles to, and in the center of, a disk of hard material, such as stainless steel, and (c) is used by (i) mounting in the fixture the unfinished connector that is secured to the fiber optic cable, (ii) allowing excess materials, such as fiber end, bead of adhesive material, and excess connector length, if present, to protrude from the opposite side of the disk, and (iii) grinding away the excess materials by manually sweeping to and fro, usually in a figure-8 pattern, with a piece of microfinishing film, which is supported by a rigid flat substrate, usually using two to four grades of microfinishing film with abrasive particles ranging in size from 0.3 μm (micron) to 15 μm, starting with the film with larger-size particles. *Note:* Various manufacturers use proprietary names to identify the hockey puck, but this popular term has become ubiquitous. *See also* **microfinishing film.**

Hocquenghem code: *See* **Bose-Chaudhuri-Hocquenghem code.**

Hogg horn antenna: An Earth station horn antenna that (a) has a somewhat higher antenna efficiency, (b) has a lower noise pickup, (c) usually is more bulky, and (d) usually is more expensive than an equivalent prime focus feed antenna or a Cassegrain antenna. *See also* **horn antenna.**

hold: *See* **call hold, mark hold, space hold.**

hold-in frequency range: 1. The range of frequencies over which a phase-locked loop can maintain lock-on. **2.** The range of frequency differences between the local oscillator or clock and the reference frequency of a phase-locked loop for which the local oscillator or clock will slowly change frequency in a direction that will reduce the frequency difference and, if not interrupted, will eventually reach the lock-on frequency and achieve phase lock. *See also* **capture range.**

holding: *See* **line holding.**

holding parameter: *See* **polarization-holding parameter.**

holding time: 1. The total length of time that a call occupies a trunk or a channel. *Note 1:* Holding time usually is measured in call•seconds. *Note 2:* An example of holding time is the elapsed time that a telephone call occupies equipment, measured from the time a demand for service is initiated until the restoration of an idle line condition. **2.** The time of a telephone call hold. *See also* **call hold, call•second, channel, trunk.**

hold mode: The operating mode of an analog computer during which integration is stopped and all variables are held at the value that they had when the mode was entered.

hold unit: *See* **track and hold unit.**

hole: In a semiconducting material containing a dopant that has one fewer electron for each atom than that required to complete the intrinsic bonding structure, a site at which an electron is missing to complete the bonding structure. *Note:* Initially, a hole is created by a dopant atom, but if an electron from a neighboring atom moves in to "fill" the hole, the neighboring atom will have a hole, thus the hole can be considered to have migrated, i.e., to have moved, especially under the influence of an applied electric field. *See* **carrier hole, feed hole.** *See also* **acceptor, donor.**

Hollerith code: A set of hole patterns that (a) is punched in a card, (b) corresponds to a set of alphanumeric characters, usually the 26 letters of the English alphabet, i.e., the Roman alphabet, the set of ten decimal digits, and a set of punctuation marks and special signs and symbols, such as the dollar sign, hyphen, comma, period, and question mark, and (c) is placed in a column on the card, each column corresponding to a character. *Note:* The original Hollerith code hole pattern was designed for the widely used standard 80-column, 12-row punched card. The columns on the card are numbered consecutively 1 through 80 from left to right. The rows are labeled X, Y, and 0 through 9, two holes being in the same column to represent one letter, usually an X or a Y punch and one of the numbered rows. The decimal digits are represented by a single hole in the row that corresponds to the digit. Other special combinations of two and three holes in a column are used to represent other signs and symbols. The card has 960 punch positions, and therefore has a capability of representing 2^{960}, or approximately 10^{280} different patterns. If every different pattern were to be placed on a different card, there would not be enough material matter in the known universe to make the 10^{280} cards that would be required, even if the known universe were solid. Avogadro's number is only 6.03×10^{23}. The potential of the 80-column card to represent data is not nearly used. *See also* **punch card.**

hollow chamber effect: *Synonym* **rain barrel effect.**

hologram: An in-depth, apparently three-dimensional image of great realism, produced by illuminating an object field with two interrelated coherent light beams, one directly from a light source and the other slightly delayed, thus giving a three-dimensional appearance. *Note:* The image can be recorded and re-created by using similar techniques. Damage to a spot on the holographic film results only in a slight reduction in resolution or brightness, not the loss of a picture element. A scanning technique permits a recording on film, which can later be recovered and displayed. The recording is rendered intelligible only by reversing the holographic process.

home computer: *Synonym* **personal computer.**

home fiber optic transceiver: A fiber optic receiver, including a photodetector, and a fiber optic transmitter, including a light source, located in the home or the office, that transmits lightwave signals to, and receives lightwave signals from, a fiber optic distribution box, via optical fibers. *Synonyms* **home fiber optic transmitter-receiver, home optical transceiver, home optical transmitter receiver.**

home fiber optic transmitter-receiver: *Synonym* **home fiber optic transceiver.**

home key: On a personal computer (PC) standard keyboard, a marked special key that, when actuated, causes a cursor or a pointer to return to the beginning of a line, file, spreadsheet, help screen, picture, or document, or to the top left corner of the screen. *Note:* In Lotus 1-2-3, pressing the home key causes the pointer to return to the upper left corner of the screen, to cell A1 of the spreadsheet, i.e., the Worksheet, to the left side of the edit line when in the edit mode, to the top of the list of help screens, to the first menu choice when in the menu mode, or to the top of a list of selected pictures. In Volkswriter, depressing the home key causes the cursor to move to the upper left corner of the current screen, and depressing it again causes it to move to the first position on line 1 of the current file. In WordPerfect, pressing the home key in conjunction with other keys causes the cursor to move to a number of different desired locations, such as the beginning of a document, the end of a document, and the beginning of a line. The home key is also used for other purposes, such as creating links to join characters that should not be separated or to form character strings that should not be broken by hard or soft returns. *See also* **end key.**

home optical transceiver: *Synonym* **home fiber optic transceiver.**

home optical transmitter receiver: *Synonym* **home fiber optic transceiver.**

homer: A homing station that serves as a radionavigation aid by making use of direction finding equipment. *See also* **direction finding, radionavigation aid.**

homing: 1. The moving of a mobile station that is directed, or directs itself, toward an electromagnetic, thermal, sonic, or other source of energy, whether primary or reflected, or follows a vector force field or a gradient of a scalar force field. **2.** In radio direction finding, the locating of a moving signal source by a moving direction-finding station that has a mobility advantage. **3.** The act of approaching a source of electromagnetic radiation in which the approaching vehicle is guided by a receiver with a directional antenna. **4.** Seeking, finding, intercepting, and engaging an object, i.e., a target, fixed or mobile, that may contain a signal source. *See* **active homing, dual homing, multiple homing, request homing, semiactive homing, split homing.**

homing adapter: A device that (a) produces audio or visual signals and (b) indicates the direction of a transmitting radio station with respect to the heading or the orientation of the device. *Note:* An example of a homing adapter is a device used with an aircraft radio receiver to indicate the direction from which received signals are coming with respect to the heading or with respect to the frame of the aircraft.

homing beacon: *Synonym* **radio beacon.**

homing guidance system: A guidance system that seeks a source of energy, such as an electromagnetic radiation source or an acoustic source, i.e., a sound source, including infrared sources and illuminated reflecting sources.

homing procedure: A procedure that is used to accomplish a rendezvous of stations, at least one of which is mobile. *Note 1:* An example of a homing procedure is a procedure used to enable an aircraft to reach a ship. The ship might use radar to determine the position of the aircraft and then issue heading instructions to the aircraft, or the aircraft might head in the direction of signals being received from the ship. Homing procedures make use of radio, radar, beacon, and other specialized direction finding (DF) equipment. *Note 2:* Types of homing procedures include those in which (a) the relatively fixed station, such as a ship or land station, to be homed on by a relatively faster mobile station, such as an aircraft, spacecraft, or land vehicle station, transmits a scheduled signal whose direction can be determined by the mobile station for homing purposes, this procedure being nearly always used for aircraft stations to home on ship stations or land stations, (b) the mobile station transmits a signal and the relatively fixed station obtains a bearing of the signal and transmits the proper bearing, course, or heading for the mobile station to arrive at the fixed station or at some predetermined location, and (c) a station to be homed on simply transmits a signal, such as a beacon signal or its call sign, and any mobile station desiring to home on it or obtain a bearing on it may do so with appropriate direction finding (DF) equipment. *See also* **bearing, direction finding, heading.**

homochronous: Pertaining to the relationship between two signals such that their corresponding significant instants are displaced by a constant time interval. *See also* **anisochronous, heterochronous, isochronous, mesochronous, plesiochronous, synchronous system.**

homodyne: Pertaining to the combining of two waves, such as electromagnetic waves, of the same frequency. *Note:* In fiber optics, one form of homodyning is based on the use of only one frequency, such as monochromatic light, to achieve detection. The light is launched into a length of optical fiber. The baseband signal, i.e., the parameter being detected, varies the length of the fiber, thus shifting the phase of the output wave. The output wave is combined with the original wave in a beam splitter and sent to a photodetector. Shifting the phase of one wave relative to the other causes the combined wave intensity at the photodetector to vary as the amount of cancellation and reinforcement of the two waves varies. Thus, the photocurrent amplitude will be a function of the phase shift, which in turn is a function of the baseband signal that is modulating the length of the fiber. If the relative phase shift varies through many cycles, the photodetector photocurrent frequency will be proportional to the rate of change of the phase shift. Calibration enables measurement of any physical parameter that changes the length of the fiber, such as a sound wave, pressure, force, magnetostriction, or displacement. In radio, another example of homodyning pertains to a system of suppressed carrier radio reception in which the receiver generates a voltage having the original carrier frequency and combines it with the incoming signal, i.e., zero beat reception. *See also* **heterodyne, homodyne detection, self-heterodyne.**

homodyne detection: Detection, i.e., demodulation, using techniques that depend on the mixing of two signals of the same frequency to obtain the baseband signal. *See also* **heterodyne, homodyne, self-heterodyne.**

homogeneous cladding: Cladding that consists of a homogeneous medium, where in particular the refractive index is constant throughout, i.e., is isotropic. *Note:* An optical fiber may have several homogeneous claddings, each with a different refractive index. *See also* **cladding, homogeneous medium, tolerance field.**

homogeneous computer system: A computer system that has architecture or layering similar to one or more other computer systems in order that similar corresponding layers readily or routinely can be interconnected or interrelated. *See also* **architecture, layer.**

homogeneous medium: In optical systems, a propagation medium that has light transmission parameters, such as the parameters of the constitutive relations, that are spatially constant in magnitude and direction, i.e., are not a function of space coordinates, although they may vary uniformly throughout the medium as a function of other parameters, such as time, temperature, pressure, humidity, and wavelength of a propagating electromagnetic wave. *Note:* Examples of homogeneous media are (a) glass with a constant refractive index throughout and (b) a vacuum with no objects within it. *See also* **constitutive relation, propagation medium.**

homogeneous multiplexed circuit: A multiplexed circuit in which all the information-bearer channels operate at the same data signaling rate (DSR).

homogeneous multiplexing: Multiplexing in which all of the information-bearer channels operate at the same data signaling rate (DSR). *See also* **heterogeneous multiplexing, multiplexing.**

homogeneous network: A network of similar or compatible computers, such as computers of one model manufactured by a given manufacturer.

homojunction: 1. A junction between two semiconductors that (a) differ in their doping levels, (b) differ in their electrical conductivities, and (c) do not differ in their atomic or alloy compositions. **2.** In a laser diode, a single junction, i.e., a single region of shift in doping from positive to negative majority carrier regions, or vice versa, and a change in refractive index at one boundary, and hence one energy level shift, one barrier, and one refractive index shift. *See also* **heterojunction.**

hook: *See* **off-hook, on-hook, verified off-hook.**

hook service: *See* **off-hook service.**

hook signal: *See* **off-hook signal.**

hop: 1. One excursion of a radio wave from the Earth to the ionosphere and back to the Earth. *Note:* The number of hops indicates (a) the number of reflections from the ionosphere and (b) one more than the number of reflections from the Earth. **2.** An excursion of a radio wave from the Earth to a geostationary satellite and back to the Earth. *Note:* The number of hops indicates the number of transmissions from the Earth and the number of transmissions from the satellite. **3.** A transmission from one geostationary satellite to another. *See also* **atmospheric duct, geostationary satellite, ionosphere, skip distance, skip zone, transmission.**

hop count: 1. The number of signal regenerating devices, such as gateways, bridges, routers, and repeaters, through which data must pass to reach their ultimate destination. **2.** In a data communications network, the number of legs or branches traversed by a packet between its source and its destination. *Note:* The hop count may be used to determine the "time-to-live" for some packets. **3.** In ionospheric transmission of radio waves, (a) the number of times a given radio wave is reflected back to Earth by the ionosphere or (b) 1 plus the number of times a given radio wave is reflected back to the ionosphere by the Earth and returned to the Earth.

hopper: In punched card machines, a receptacle for punch cards to be placed in, to be read, punched, sorted, or collated. *See* **frequency hopper.**

hopping: *See* **frequency hopping, mode hopping, time hopping.**

hopping spread spectrum: *See* **frequency hopping spread spectrum.**

horizon: *See* **radar horizon, radio horizon.**

horizon angle: In a vertical plane, the angle between (a) a horizontal line extending from the center of an antenna and (b) a line extending from the same point to the radio horizon. *Note:* A horizon angle usually is an angle of depression, i.e., a negative angle, because the radio horizon usually is lower that the antenna, particularly for antennas that are on ship masts, on aircraft, on towers, on high buildings, or on mountain tops. *See also* **antenna, horizontal line, k factor, path profile, radio horizon.**

horizon cover: The horizontal angle, i.e., the angle between two azimuths, that an electromagnetic wave source, such as a radio antenna, radar antenna, or light source, is capable of scanning or illuminating. *Synonym* **horizontal cover.** *See also* **vertical cover.**

horizon range: *See* **radio horizon range.**

horizontal: Pertaining to a direction that is perpendicular to a line through the center of the Earth. *See also* **vertical.**

horizontal cover: *Synonym* **horizon cover.**

horizontal line: **1.** A line that is perpendicular to a line through the center of the Earth. **2.** A line that is tangent to the Earth's average surface, i.e., a line that is perpendicular to a vertical line. *See* **horizon angle, vertical line.**

horizontally polarized electromagnetic wave: A uniform plane polarized electromagnetic wave in which the electric field vector is always and everywhere horizontal while propagating in space or in a material propagation medium. *Note:* The magnetic field vector may be inclined at any angle of elevation to the horizontal plane. If it is vertical, the wave will be propagated horizontally; if it is horizontal, the wave will be propagated vertically; and if it is inclined at 45°, the propagation will be at 45° to the horizontal, all depending on the antenna configuration or the time phasing of the electric and magnetic fields. Thus, if the wave is propagating vertically, the magnetic field will be horizontal. If the wave is propagating horizontally, the magnetic field will be vertical. However, in any case, for a horizontally polarized electromagnetic wave, the electric field vector must always and everywhere be horizontal regardless of the direction of the magnetic field vector or the direction of propagation. *See also* **vertically polarized electromagnetic wave.**

horizontal polarization: *See* **fixed horizontal polarization.**

horizontal redundancy check: *Synonym* **longitudinal redundancy check.**

horizontal resolution: In facsimile systems, the number of picture elements per unit distance in the direction of scanning or recording. *See also* **facsimile.**

horizontal return: On a typewriter or a printer, the relative movement between the print position and the data medium, usually paper mounted on a carriage, roller, or platen, in a direction opposite to that of the writing, usually back to the beginning of the current line or the next line. *Note:* If the horizontal return is to the beginning of the next line, an indexing operation is included in the horizontal return. *See also* **data medium.**

horizontal tabulation character: A format effector that causes the print position of a printer or the display position in the display space of a display device to move forward to the next of a series of predetermined positions along the same line. *Note:* The next predetermined position should be more than one character space away from the present position when the horizontal tabulation character is used. If the next desired position is only one space away, the space character may be used. *See* **format effector, space character.**

horizontal terminating block: In a switching center, a group of electrical terminals at which contact can be made with inside plant equipment. *See also* **combined distribution frame, inside plant, switching center.**

horizontal wraparound: On the display surface of a display device, the continuation of cursor movement from the rightmost print position, i.e., the last print position in a horizontal row or line, to the leftmost print position, i.e., the first print position, in (a) the next row or line, i.e., the row or the line below the current row or line, (b) the current row or line, or (c) the preceding row or line, i.e., the row or the line above the current row or line. *See also* **vertical wraparound.**

horn: **1.** In radio transmission systems, a waveguide section of increasing cross-sectional area used to radiate directly in the desired direction or to feed into a reflector that forms the desired beam. *Note 1:* Horns may have any of several expansion curves, i.e., longitudinal cross-section curves, such as elliptical, conical, hyperbolic, or parabolic curves. *Note 2:* By a combination of designs controlling the horn dimensions, the spacing of the reflector, and the reflector shape and dimensions, a very wide range of beam patterns may be formed. **2.** A portion of a waveguide in which at least one of the cross-sectional diameters is smoothly increased along the axial direction. *Note:* Horns are used as direct radiators or as feed devices for reflective antennas. **3.** In audio systems, a tube of increasing, often exponentially increasing, cross section for radiating or receiving acoustic waves. *Note:* The small or terminating end is usually provided with an impedance matching device or cross section as well as with an appropriate transducer. *See* **scalar feed horn.** *See also* **antenna, waveguide.**

horn antenna: An antenna that (a) has the shape of a tube with a circular or rectangular cross-sectional area that increases toward the open end through which radio waves are launched, (b) may vary in size from large Earth station antennas to small radio or microwave link antennas, and (c) may be an electrically or mechanically despun antenna. *See* **antenna, despun antenna, Earth station.**

horn feed system: *See* **five-horn feed system, four-horn feed system.**

horn gap switch: A switch that (a) is provided with arcing horns and (b) is used for interrupting large currents when disconnecting transmission power distribution lines, usually overhead lines, that are bearing high charging currents, i.e., are bearing heavy load currents.

horse: *See* **Trojan horse.**

host: In packet and message switched communications networks, the collection of hardware and software that makes use of packet or message switching to support user services, such as user-to-user, i.e., end-to-end communications, interprocess communications, and distributed data processing. *Note:* The term "host computer" is often used to refer to the host. *See* **bastion host.** *See also* **host computer.** *Refer to* **Figs. S-9 and S-11 (880), T-5 (1012).**

host computer: **1.** In a computer network, a computer that (a) provides users with various services and software, such as application program support, computation support, utility programs, programming languages, and database access, (b) usually performs network control functions, and (c) usually is located at a node in the network. **2.** In computer software engineering, a computer that is selected to perform specified functions on behalf of a program or a system, such as providing (a) access to programs or databases installed in the computer, (b) support for the development of software intended for another computer, and (c) processing capabilities to users in the network to which it is connected. *Note:* The term "host" by itself is often used to refer to the host computer. *Synonyms* **central computer, host machine, host processor.** *See also* **host.** *Refer to* **Fig. T-5 (1012).**

host computer interface: The interface that (a) is between a host computer and a network and (b) may be a front end processor. *See also* **front end processor.**

host interface: The interface that (a) is between (i) a host computer or a host processor and (ii) a data processing network and (b) may be a front end processor or a communications computer.

host language: In database management systems, a programming language that (a) has embodied within it a language that is used for some special purpose, such as COBOL, PL/1, Assembly Language, or FORTRAN, and (b) has a built-in data description language. *See also* **data description language, database management system.**

host machine: *Synonym* **host computer.**

host node: In a user application network, a node at which the host computer, i.e., the host processor, that controls a domain is located. *See also* **application network, domain, host computer, node, user.**

host processor: *Synonym* **host computer.**

hot bend test: In an optical fiber, fiber optic bundle, or fiber optic cable, a test to discover the effect on optical parameters, such as attenuation rate, dispersion, and the breaking point, during and after repeated alternate bend flexure at various elevated temperatures.

hot boot: *Synonym* **warm restart.**

hotline: **1.** A network-provided service feature in which a call is automatically originated to a preselected destination when the call originator goes off-hook, i.e., picks up, without any additional action by the user. *Note 1:* Hotlines cannot be used to originate calls other than to preselected destinations. *Note 2:* Various priority services that require dialing are not hotlines. **2.** A dedicated circuit that links two and only two terminals. **3.** A line that is monitored 24 hours a day, seven days a week, for providing special help, information, or services to callers in need. *See also* **circuit, dedicated circuit, leased circuit, off-hook service, permanent virtual circuit, point to point link.**

hotline circuit: A communications circuit that supports a hotline service. *See* **switched hotline circuit.** *See also* **circuit, hotline.**

hot standby: Pertaining to electronic equipment that is powered, ready for use, and usually connected, but is not in use. *See also* **cold standby.**

hour: A unit of time equal to 3600 seconds. *See* **busy hour, connections per circuit hour, frequency hour, group busy hour, network busy hour, switch busy hour, year worst hour.** *See also* **second.**

house cable: *Deprecated synonym for* **on-premises wiring.** Communications cable within a building or a complex of buildings. *Note:* House cable owned before divestiture by the Bell System and after divestiture by the Regional Bell Operating Companies will eventually be fully depreciated and will then belong to the customer. *See* **on-premises wiring.**

house cabling: *Synonym* **on-premises cabling.**

housekeeping: In a communications, computer, data processing, or control system, pertaining to operational hardware and software matters that do not contribute directly to the handling of user applications, i.e., to system applications, such as application programs, but do contribute directly to the operation of the system. *See also* **application program, application software.**

housekeeping information: *Synonym* **service information.**

housekeeping signal: *Synonym* **service signal.**

house wiring: *Synonym* **on-premises wiring.**

housing: A structure that (a) provides the outer protective covering for a component, device, or module and (b) usually serves to support and package the component. *See* **fiber optic splice housing, laser protective housing.**

h parameter: *See* **polarization-holding parameter.**

H plane bend: *Synonym* **H bend.**

HTML: HyperText Markup Language, hypertext metalanguage.

HTTP: hypertext transfer protocol, Hypertext Transfer Protocol.

Huffman coding: Coding used to compact data by representing the more common events with short codes and the less common events with longer codes. *Note:* Huffman coding is used in Group 3 facsimile.

hundred call•seconds (ccs): A unit of communications traffic or system utilization equal to 100 c•s (call•seconds). *Note:* The initial "c" in the abbreviation "ccs" is the Roman numeral for 100. *See* **call•second.** *See also* **hundred circuit•hours.**

hundred circuit•hours (cch): A unit of measure of use or time of a communications circuit or group of circuits. *Note 1:* If six circuits were in use for 168 hours, the total usage would be 1008 ch (circuit•hours). This would be equivalent to 10.08 cch (hundred circuit•hours). *Note 2:* The hundred circuit•hour (cch) unit could be used to specify any type of time, such as total use time for user traffic and outage time for any number of causes, such as time loss caused by preventive maintenance, time loss caused by corrective maintenance, and time loss caused by loss of power. *Note 3:* The initial "c" in the abbreviation "cch" is the Roman numeral for 100. *See also* **call•second, hundred call•seconds.**

hunting: 1. In a device, the seeking, and usually the finding, of a specified state, which (a) includes some oscillation about the specified position, significant instant, significant condition, or operating condition during the seeking, and (b) ceases when equilibrium is reached and oscillations about the specified state cease, with the device maintaining the specified state until disturbed. **2.** In telephone systems, the operation of a device or a circuit, such as a selector, to find and usually establish a connection with an idle circuit of a specified group. **3.** The continued failure to achieve a state of equilibrium, usually by alternately overshooting and undershooting the point of equilibrium. *Synonym* **ringing.** *See also* **failure, group, rotary hunting.**

h value: *See* **polarization-holding parameter.**

hybrid: 1. A functional unit in which two or more different technologies are combined to satisfy a given requirement. **2.** In electrical, electronic, fiber optic, or communications engineering, pertaining to a device, circuit, apparatus, or system made up of two or more components, which usually are used in different applications and are not usually combined to meet a given single requirement. *Note:* Examples of hybrids are (a) an electronic circuit that has both vacuum tubes and transistors, (b) a cable containing optical fibers and electrical wires, (c) a mixture of thin film and discrete integrated circuits, (d) a computer that has both analog and digital capabilities, and (e) a transformer, combination of transformers, or resistive network that provides paths to three branches, circuits *A, B,* and *C,* so arranged that *A* can send to *C* and *B* can receive from *C,* but *A* and *B* are effectively isolated. *See* **resistance hybrid.** *See also* **hybrid communications network, integrated optical circuit.** *Refer to* **Fig. N-8 (623).**

hybrid balance: A measure of the degree of balance between two impedances connected to two conjugate sides of a hybrid set. *Note:* The hybrid balance is expressed by the same relation as that used for return loss (L_r), which is given by the relation $L_r = |(Z_2 - Z_1)/(Z_2 + Z_1)|$, in which the impedances (Z_1 and Z_2) are toward the source and toward the load, respectively. *See also* **balanced, balanced line, impedance, return loss.**

hybrid cable: 1. A communications cable that contains (a) two or more different types of electrical conductors that bear electrical signals, (b) a mixture of signal-bearing electrical conductors and optical fibers, (c) two or more different types of optical fibers, or (d) any combination of these. **2.** A communications cable that contains (a) one or more signal propagation media and electrical power conductors, (b) one or more signal propagation media and optical power waveguides, or (c) a combination of these. *See also* **composite cable, propagation medium.**

hybrid coil: A single transformer that (a) effectively has three windings and is designed to be connected to four branches of a circuit so as to make them conjugate in pairs, and (b) primarily is used to convert between two-wire and four-wire communications circuits, such as when repeaters are introduced in a two-wire circuit. *Synonym* **bridge transformer.** *See also* **bridge, hybrid, hybrid set.**

hybrid communications network (HCN): A communications network that uses a combination of trunks, loops, or links, some of which are capable of transmitting only analog or quasi-analog signals and some of which are capable of transmitting only digital signals. *Synonym* **hybrid network.** *See also* **communications, quasi-analog transmission.**

hybrid communications system: A communications system that can accommodate both digital and analog signals.

hybrid computer: A computer that processes both analog and digital data and consists of both analog circuits and digital circuits.

hybrid connector: 1. A connector that (a) contains contacts for more than one type of service, (b) has various types of contacts, such as contacts for electrical power, optical power, coaxial cables, optical fibers or fiber optic cables for signals, and shielded or unshielded audio pairs, and (c) may contain active devices, such as repeaters, amplifiers, and signal conditioners. **2.** In fiber optics, a connector that contains optical fibers and electrical conductors.

hybrid coupler: 1. In antenna systems, a hybrid junction used as a directional coupler. *Note:* The coupling factor for a hybrid coupler usually is about 3 dB. **2.** In fiber optics, a fiber optic coupler with special features, such as the ability to couple optical signals and electrical power between hybrid cables or between a hybrid cable and either a fiber optic transmitter or a fiber optic receiver. *See also* **coupling, fiber optic coupler, fiber optics, hybrid cable, hybrid junction, waveguide.**

hybrid fiber optic circuit: An interconnection of optical fibers for transmission from point to point and optical integrated circuits (OICs) for logical decision and control operations. *See also* **integrated optical circuit, optical fiber.**

hybrid interface structure: In integrated services digital networks (ISDNs), an interface structure that uses both labeled and positioned channels. *See also* **labeled channel, positioned channel.**

hybrid junction: A waveguide or transmission line arranged such that (a) there are four ports, (b) each port is terminated in its characteristic impedance, and (c) energy entering any one port is transferred, usually equally, to two of the three remaining ports. *Note:* The hybrid junction is used as a mixing or dividing device. *See also* **matched junction, waveguide.**

hybrid mode: In electromagnetic waves, a mode in which components of both the electric and magnetic fields are in the direction of propagation. *Note:* In fiber optics, hybrid modes correspond to skew rays, i.e., nonmeridional rays, that spiral around the optical fiber axis. *See also* **meridional ray, mode, skew ray.**

hybrid network: *See* **hybrid communications network.**

hybrid ranging: In spread spectrum systems, ranging, i.e., the use of radar, in which short code sequences are combined with a constant-frequency synchronous tone to give a faster acquisition of objects, i.e., of targets, than is obtainable with longer sequences, but longer range than is obtainable with short sequences alone. *Note:* Spread spectrum hybrid ranging techniques are used in communications and radar systems to provide some security and to provide some antijamming capability. *See also* **spread spectrum.**

hybrid resistance: *Synonym* **resistance hybrid.**

hybrid routing: Routing in which numbering plans and routing tables are constructed and used to permit the collocation, such as location in the same area code, of switches using deterministic routing schemes with switches using flood search routing schemes. *Note 1:* The routing tables are constructed with no duplicate numbers so that direct dial service can be provided to all network subscribers. This may require the use of ten-digit numbers. *Note 2:* Hybrid routing is more difficult to construct and maintain than deterministic and flood search routing. *See also* **deterministic routing, flood search routing.**

hybrid set: Two or more transformers interconnected to form a network in which (a) there are four pairs of accessible terminals, (b) the four pairs of terminals may be connected to four impedances, and (c) the branches containing the four impedances may be made interchangeable. *See also* **bridge, four-wire terminating set, hybrid coil, line adapter circuit.**

hybrid spread spectrum: A combination of frequency hopping spread spectrum and direct sequence spread spectrum.

hybrid system: In communications, computer, data processing, and control systems, a system that accommodates both digital and analog signals. *See also:* **hybrid network.**

hybrid topology: *See* **network topology.**

hydrographic information: Information that pertains to (a) the measurement and the description of the physical features of bodies of water, such as oceans, seas, bays, lakes, rivers, and streams, (b) land areas adjacent to these bodies of water, and (c) the development and the use of these bodies of water, such as for navigation, water supply, recreation, and wildlife.

hydrographic message: A serially numbered broadcasted message that contains navigational and hydrographic information. *Synonym* **hydrolant.**

hydrolant: *Synonym* **hydrographic message.**

hydrological and meteorological fixed station (HMS): A fixed station used for the automatic transmission of either hydrological data or meteorological data, or both. *See also* **fixed station.**

hydrological and meteorological land station (HMS): A land station used for the automatic transmission of either hydrological data or meteorological data, or both. *See also* **land station.**

hydrological and meteorological mobile station (HMS): A mobile station used for the automatic transmission of either hydrological data or meteorological data, or both. *See also* **mobile station.**

hydrological and meteorological station (HMS): A fixed, land, or mobile station used for the automatic transmission of either hydrological data or meteorological data, or both.

hydrophone: A transducer that (a) transforms sound waves, i.e., pressure waves, in water into corresponding electrical or optical signals in a wire or an optical fiber and (b) uses one or more effects, such as piezoelectric, resistive, capacitive, inductive, or optical effects, that are produced by sound waves and therefore are capable of modulating a measurable quantity, such as an electric current, voltage, or electromagnetic wave, such as a radio wave or a lightwave. *See also* **transducer.**

hydroxyl ion absorption: The absorption of electromagnetic waves, particularly lightwaves, in optical glass caused by the presence of trapped hydroxyl ions remaining from water as a contaminant. *Note:* The hydroxyl (OH^-) ion can penetrate glass after product fabrication, such as while an optical fiber is being drawn, after it has been wound on a spool, after it has been cabled, during storage, and during operational use. Hydroxyl ion absorption is the major cause of attenuation in fiber optic transmission systems, which usually operate at a wavelength of 1.3 μm (microns). *Refer to* **Fig. H-1.** *Refer also to* **Fig. S-15 (931).**

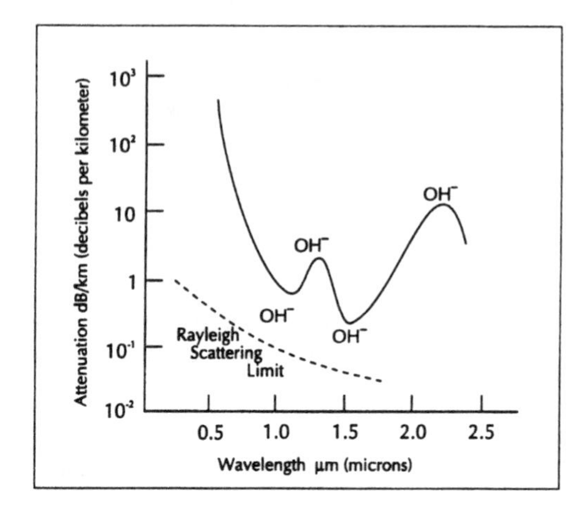

Fig. H-1. The attenuation rate versus wavelength for a low loss optical fiber, showing peaks and troughs, i.e., attenuation windows, caused by **hydroxyl ion absorption.**

hydroxyl ion overtone absorption: Absorption caused by the presence of hydroxyl (OH^-) ions, giving rise to fundamental vibrations and hence an absorption band centered at the 2.8-μm (micron) wavelength, with harmonics of the fundamental vibrations at 1.4, 0.97, and 0.75 μm, i.e., at the second, third, and fourth harmonics. *Note:* The OH^- ions originate primarily from water as a contaminant in optical glass. *See also* **absorption, harmonic, ion.**

hyperbolic navigation system: A navigation system in which electromagnetic waves are employed to establish a grid that (a) consists of intersecting wave fronts that can be used by a receiving station located in the grid to determine its position without transmitting any signals, and (b) is in the form of intersecting hyperbolas because two transmitters are used to establish the grid, and signal field strength varies inversely as the square of the distance from the antenna. *Note:* The signal power, i.e., the irradiance, varies inversely with distance from the antenna. *See also* **intermediate field, field strength, irradiance, navigation system.**

hyperbolic orbit: An orbit that (a) has the shape of a hyperbola, (b) is not a closed orbit, and (c) might be described as a hyperbolic trajectory. *Note:* Circular and elliptical orbits are closed orbits. Parabolic and hyperbolic orbits are not closed. *See also* **closed orbit, orbit, parabolic orbit.**

hypermedia: Computer-addressable documents and files that contain pointers for linking to multimedia information, such as text, graphic, video, or audio information, in the same or in other documents and files. *Note:* The using of hypertext links is known as navigating. *See also* **hypertext.**

hypertext: A system of coding that is used to (a) create hypermedia or (b) navigate among hypermedia in a logical sequence rather than a consecutive or ordered sequence, such as numerical or alphabetical order.

HyperText Markup Language: A language that permits conversion of standard Internet files into World Wide Web (WWW) format.

hypertext transfer protocol (HTTP): A protocol that provides file transfers for hypertext-based information between local and remote systems.

Hypertext Transfer Protocol (HTTP): In the World Wide Web (WWW), a protocol that provides file transfers for hypertext-based information among local and remote systems. *See also* **World Wide Web.**

hyphen: A data delimiter that allows two character strings to be separately distinguishable yet joined or associated. *See* **hard hyphen, soft hyphen.**

hyphen drop: In word processing, the suppression of a discretionary hyphen, i.e., a soft hyphen, that software places at the end of a line when the hyphenated word subsequently appears elsewhere and no longer requires a hyphen. *Note:* If a hyphen is always required for a given word and it also appears at the end of a line, a hyphen drop feature can result in elimination of the hyphen when the hyphenated word is moved to another place in respect to the right margin. If there is no hyphen drop feature, hyphens placed at the end of lines can result in words with unnecessary hyphens when the words subsequently appear elsewhere. Computer word processor software writers face a dilemma because of the ambiguity in the use of the hyphen. For example, if "online" appears at the end of a line of text, and automatic hyphenation shows "on-" at the end of a line and "line" on the next line, the reader will not know whether the proper spelling is "online" or "on-line". The most prevalent solution had been to install automatic hyphen drop, but hyphens deliberately placed in words by operators will not be dropped. Hyphens that are not to be dropped are called "hard hyphens." Hyphens placed at the end of a line by the software are called "soft hyphens." Thus, hard hyphens might be positioned anywhere in a line of text, will always remain in the text, and might serve as an end-of-the-line hyphen, whereas soft hyphens are placed only at the end of a line. Further complicating the issue is the myriad of sometimes conflicting rules governing the use of hyphens, with many exceptions to the rules. In some software systems, automatic hyphenation sometimes produces odd and awkward construction of text that might have to be manually edited. *See also* **data delimiter, hard hyphen, hyphen, soft hyphen.**

hypodermic fiberscope: A fiberscope that consists of (a) a light source for illumination, (b) groups of optical fibers with diameters on the order of 4 μm (microns) to 8 μm for forward transmission of the illumination for the object and backward transmission of the light reflected from the object, and (c) a device for displaying the reflected image of the subcutaneous tissue.

hysteresis loop: *See* **magnetic hysteresis loop.**

Hz: hertz.

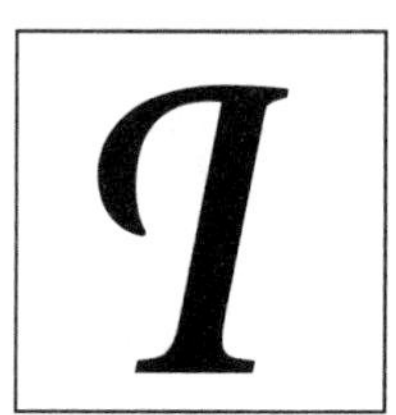

IAGC: instantaneous automatic gain control.

IATV: interactive television.

I band: The band of frequencies in the range from 8 GHz (gigahertz) to 10 GHz, consisting of ten numbered channels, each of which is 200 MHz (megahertz) wide. *Note:* The I band frequencies overlap parts of the old X band and H band.

IBM PC DOS: A personal computer (PC) disk operating system (DOS) that (a) was developed by the International Business Machines Corporation, (b) has sufficient software program control capability to manage basic operations on a PC, (c) performs operations such as copy, format, print, and display file names (DIR), (d) calls up application programs stored on disks operated by disk drives in the PC, (e) calls up files stored on disks, and (f) does not perform word processing or graphic operations. *See also* **application program, copy, disk drive, disk operating system, file names, format, personal computer, software.**

IC: integrated circuit.

ICA: International Communications Association.

icand: *Synonym* **multiplicand.**

Iceland spar: A transparent variety of the natural uniaxial crystal calcite that (a) displays a very strong double refraction and (b) chemically is calcium carbonate crystallized in the hexagonal rhombohedral crystallographic system. *Synonym* **calspar.**

ICI: incoming call identification.

ICMP: Internet Control Message Protocol.

icon: 1. A pictorial representation of an idea or a concept. *Note:* In computer operations, icons may be used in windows or menus on a monitor screen to represent commands, files, or options. **2.** In computer systems, a small, pictorial representation of an application software package, idea, or concept used in a window or a menu to represent commands, files, or options. *See also* **application software, monitor, menu, software package, window.**

ICS: integrated communications system.

ICW: interrupted continuous wave.

ID: identification, identifier.

ideal blackbody: *Synonym* **blackbody.**

identification (ID): Proof or evidence that indicates the identity of an entity. *Note:* Examples of identification are personal names, corporate names, station call signs, stripes on a zebra, fingerprints, pictures, signatures, passwords, blood type, and corporate logos. *See* **aircraft signature identification, automatic identification, automatic number identification, caller identification, configuration identification, equipment identification, incoming call identification, line identification, subscriber identification, terminal automatic identification, service identification, transit network identification, transmission identification.**

identification code: *See* **data network identification code.**

identification facility: A functional unit that is capable of (a) detecting a given entity, (a) examining or analyzing the given entity, and (c) identifying, classifying, or naming the entity. *See* **called line identification facility, calling line identification facility.** *See also* **functional unit.**

identification friend or foe (IFF): 1. In military systems, pertaining to equipment that (a) uses electromagnetic reception and transmission, (b) is in the hands of friendly forces, and (c) automatically responds to received signals by emitting signals that distinguish the friendly forces from hostile forces. **2.** Pertaining to the capability of determining the friendly or unfriendly nature of aircraft, ship, and ground systems by other aircraft, ship, and ground systems using electronic detection and signal processing equipment. *See also* **electronic warfare, recognition.**

identification friend-or-foe (IFF) personal identifier: The discrete identification friend-or-foe code assigned to a particular aircraft, ship, or other vehicle for identification by electronic means. *See also* **recognition.**

identification-not-provided signal: In a telephone or a telegraph system, a signal that is sent in response to a request for calling or called line identification when the corresponding facility is not provided in the source user network or in the destination user network.

identification request indicator: A signal that indicates that a request has been made for the identification of either the call originator or the call receiver. *See* **called line identification request indicator, calling line identification request indicator.**

identification signal: A sequence of characters that (a) is transmitted to data terminal equipment (DTE) and (b) identifies either the call originator or the call receiver, depending on whether the sequence of characters is sent or received. *See* **called line identification signal, calling line identification signal.**

identified outward dialing: *See* **automatic identified outward dialing.**

identified outward dialing lead: *See* **automatic identified outward dialing lead.**

identifier (ID): 1. In telecommunications and data processing systems, one or more characters used to identify, name, or characterize the nature, properties, or contents of a set of data elements. **2.** A string of bits or characters that names an entity, such as a program, device, or system, in order that other entities can call that entity. **3.** In computer programming languages, the name that (a) is assigned to an entity, such as the name of a variable, array, record, label, program, system, data element, or procedure, (b) usually consists of a unique permutation of alphanumeric characters, (c) usually cannot exceed a specified length, (d) is used in a program to call the entity when there is a need to do so, and (e) usually is placed in a label. *Note:* The label is attached to, is a part of, or remains associated with the information it identifies, and if it becomes disassociated from the information it identifies, the information may not be accessible. *See* **call identifier.** *See also* **identification, identity, label, pointer, programming language.**

identity: The set of known values or attributes that characterizes, i.e., that identifies, an entity. *Note 1:* The identity of an entity is established when the values or the attributes of the entity are detected or recognized, such as (a) when the peculiar keying habits of a telegraph operator are recognized, (b) when the voice of a particular radio operator is recognized, or (c) when a secret code or password is detected. *Note 2:* It may be said that "*id*entity" is the "ID" of an "entity." *See* **called line identity, calling line identity, country-network identity.** *See also* **attribute, identifier, value.**

identity gate: *Synonym* **EXCLUSIVE NOR gate.**

identity message: A message that identifies, i.e., that names, the calling or transmitting entity, such as a call originator, a transmitting station, or a beacon. *See* **calling line identity message, line identity message.**

identity operation: 1. The Boolean operation in which the result has an assigned specific value, such as 1, truth, go, or positive, when all the operands have a common Boolean value, such as all 1s, all 0s, all positive, or all negative. *Note:* An identity operation with only two operands is an equivalence operation. **2.** In logic operations, an operator that has the property that if P is a statement, Q is a statement, R is a statement, . . . , then P,Q,R, . . . is true if and only if all statements are true or all statements are false. *See also* **equivalence operation, EXCLUSIVE NOR gate, logical operator, statement.**

IDF: intermediate distribution frame.

idle channel: A channel that is momentarily not being used, though it is in a state of readiness to use at any instant and at electronic speeds. *Note 1:* Idle characters may be passing through the idle channel even though the channel is not being used. *Note 2:* Idle channels are selected for use in time assignment speech interpolation (TASI) systems.

idle channel loading: Meaningless signals, special codes, or noise deliberately applied to individual unused multiplexed channels to maintain a desired level of modulator loading. *See also* **channel, idle channel, level, modulator, multiplexing.**

idle channel noise: Noise that (a) is present in a communications channel when signals are not applied to the channel and (b) should only be measured under stated conditions and terminations in order for the measured value of the noise level to be significant. *See also* **channel, idle channel, noise.**

idle character: A control character that is sent when there is no information being sent or to be sent, i.e., a control character that is transmitted when no useful information is being transmitted. *See* **synchronous idle character.** *See also* **bit stuffing, character.**

idle condition: *Synonym* **idle state.**

idle line: *Synonym* **inactive line.**

idle line termination: An electrical network that (a) is controlled by a switch and (b) maintains a desired impedance at a trunk or line terminal when that trunk or line terminal is in the idle state. *Synonym* **inactive line termination.** *See also* **idle state, line, terminal impedance, trunk.**

idle node: *Synonym* **inactive node.**

idle state: The telecommunications service condition that exists whenever user messages are not being transmitted but the service is immediately available for use. *Synonym* **idle condition.** *See also* **availability, on-hook.**

idle station: *Synonym* **inactive station.**

idle time: 1. In a system, the duration, such as 4 hrs (hours), of an idle state. **2.** In a system, the calendar time interval, such as from 1:00 A.M. to 5:00 A.M. each day, that the system is in the idle state. *See also* **idle state.**

IDN: integrated digital network.

IDTV: improved definition television.

IDTV system: improved definition television system.

ier: *Synonym* **multiplier.**

IF: intermediate frequency.

I/F: interface.

if A then B gate: *Synonym* **B OR NOT A gate.**

IF amplifier: intermediate frequency amplifier.

if A then not B gate: *Synonym* **NAND gate.**

if B then A gate: *Synonym* **A OR NOT B gate.**

if B then not A gate: *Synonym* **NAND gate.**

IFF: identification friend or foe.

IFF personal identifier: identification friend or foe personal identifier.

IFOC: integrated fiber optic circuit.

IFR: instrument flight rule.

IFRB: International Frequency Registration Board.

IF repeater: 1. intermediate frequency repeater. 2. *Synonym* **heterodyne repeater.**

IFS: 1. ionospheric forward scatter. 2. *Synonym* **ionospheric scatter.**

IF . . . THEN statement: A computer program statement that indicates that should the condition specified after "IF" occur, such as a counter reaches a specified count, the computer is to perform the function or access the address indicated by that which is specified after "THEN."

ignore gate: *Synonym* **A IGNORE B gate.**

IIL: integrated injection logic.

I²L: integrated injection logic.

ILD: injection laser diode.

ILF: infralow frequency.

illegal character: A character, or a combination of bits, that is not valid in a given system according to specified criteria, such as with respect to a specified alphabet, a particular pattern of bits, a rule of formation, or a check code. *Note:* Examples of an illegal character might be (a) one of the bit patterns that go unused when decimal digits are encoded using four-bit strings, (b) in the American Standard Code for Information Interchange, one of the seven-bit patterns assigned to an unneeded foreign symbol, and (c) a specific bit pattern assigned the role of an illegal character in a given system. *Synonyms* **false character, forbidden character, improper character, unallowable character, unused character.**

illegal code: A code element, string, character, or symbol that appears, or purports to be, a proper element because it possesses features that are required of the proper elements of the code, but is in fact not a proper member of the defined set of codes, alphabet, or language of the code. *Note:* Examples of illegal codes might be (a) in a code in which all members must have five characters, designating the code XXXXX as an inadmissible code, or (b) one of the six four-bit bytes that are not used to represent a decimal digit. Should an illegal code occur in a system, appropriate gating circuits can be used to trigger a signal to alert the operator that a failure or malfunction has occurred in the system. *Synonyms* **forbidden code, improper code.** *See also* **code, gate, gating, gating circuit, string.**

illuminance: 1. In optical systems, the amount of light incident upon a unit area. *Note:* The SI unit (Système International d'Unités) for illuminance is lumens per square meter. **2.** In radio systems, such as radar, the amount of radiation upon a unit area, such as that of a target. *Note:* Radar illuminance usually is expressed in watts per square meter. *See also* **irradiance**.

illuminate: To cause electromagnetic energy to become incident upon a real or virtual surface.

illumination: Either (a) the act of causing electromagnetic energy to become incident upon a real or virtual surface or (b) the electromagnetic energy itself. *See* **fiber optic flood illumination, fiber optic spot illumination.** *See also* **illuminance, illuminate.**

illumination Cassegrain: *See* **uniform illumination Cassegrain.**

illumination detection: *See* **fiber optic illumination detection.**

illuminator: *See* **fiber optic illuminator.**

ILS: instrument landing system.

IM: intensity modulation, intermodulation.

IMA: integrated modular avionics.

image: In an optical system, a representation of an object, the representation being produced by means of light rays. *Note:* An optical element forms an image by collecting a bundle of light rays diverging from an object and transforming it into a bundle of rays that converge toward, or diverge from, another point. If the rays converge to a point, a real image of the object point is formed. If the rays diverge without intersecting each other, they appear to proceed from a virtual image. *See* **coded image, display image, double image, erect image, geometric image, ghost image, inverted image, latent image, real image, reflection image, reverted image, still image, virtual image, wire frame image.** *See also* **light ray, object, optical element, optical system.** *Refer to* **Fig. I-1.**

image antenna: A hypothetical mirror image antenna considered to be located as far below ground as the actual antenna is above ground. *Note 1:* The Earth's surface is considered to be an equipotential reflecting surface. *Note 2:* The image antenna is helpful in calculating electric field vectors, magnetic field vectors, and electromagnetic fields emanating from the real antenna, particularly in the vicinity of the antenna and along the ground. Each charge and each current in the real antenna has an image that may be considered as a source of radiation equal to, but differently directed from, its real counterpart. This enables field vector parameters to be calculated as though the image were a real antenna, much as the luminance from a light bulb image in a mirror must be used to calculate the luminance of a point on the real side of the mirror. *Note 3:* An image antenna can be considered to be on the opposite side of any equipotential plane surface, such as a grounded metal plate, much like the image of a light source in a perfect mirror. *See also* **antenna, ground plane.**

image aspect: The spatial orientation of an image, such as erect, normal, canted, inverted, reverted, rotated, or displaced, i.e., translated. *See also* **erect image, image, inverted image, reverted image.**

image brightness: In an optical system, the apparent brightness of an image as seen through an optical system. *Note:* Image brightness depends on many factors, such as (a) the brightness of the object and (b) the propagation medium, magnification, distortion, and diameter of the exit pupil of the optical system. *See also* **brightness, image, optical system.**

image compression: A reduction in the storage requirements for images. *Note:* The reduction is accomplished by digitizing the images and applying signal compression techniques on the digital data.

image dissector: 1. In fiber optic systems, a bundle of optical fibers that (a) has a tightly packed end on which an image may be focused, (b) has optical fibers that may be separated into groups, (c) allows each group to transmit its part of the image, and (d) allows each part to remain aligned and thus preserve the original image for presentation at a distance from the point of original focus. **2.** An optical system in which (a) an image is formed, and (b) the image is scanned to generate signals for transmission. *See also* **aligned bundle.**

image frequency: 1. In a device, a frequency that can be used to produce a desired frequency. **2.** In heterodyning, an undesired input frequency that produces the same intermediate frequency (IF) that is produced by the desired input frequency. *Note:* The effect of the image frequency demonstrates the mirror-like symmetry of signal and image frequencies about the beating oscillator frequency or the intermediate frequency, whichever is higher. *See also* **heterodyning, homodyning.**

image intensifier: A device that increases the luminance of a low intensity image. *See also* **image, luminance.**

image inverter: A device or an optical system that inverts an image, i.e., rotates the image 180° in its own plane.

image jump: The apparent displacement of an object caused by an optical system, such as caused by a prismatic condition or by refraction in the optical system. *See also* **object, prism, refraction.**

image method: An analytical method that (a) is used to analyze and calculate electric field distribution patterns, i.e., radiation patterns, emanating from a radiating source, (b) assumes that a given charge distribution produces a field in the region outside a plane-bounded conducting region as though there were a mirror image of the same charge distribution inside the plane-bounded conducting region, but of opposite polarity and inverted as though the distribution were a mirror image of the actual distribution, (c) uses both the real charge distribution and the image charge distribution to vectorially contribute to the calculation of the magnitude and direction of the real electric field at a given point, and (d) holds for line and point charge distributions, such as occur on a vertical straight antenna above the Earth's surface or above a ground plane. *See also* **electric field, radiation pattern.** *Refer to* **Fig. I-1.**

image persistence: In a video display, the continued appearance of (a) images from earlier video frames, (b) images that remain after their objects that have subsequently moved, and (c) images of objects that have left the scene of action but continue to remain in the display area, i.e., remain in the received video imagery. *Note:* Image persistence is caused by (a) the persistence of the phosphor used in the screen of the video display device and (b) the time required for an illuminated area on the screen to discharge its electrostatic charge.

image plane: The plane (a) in which an image lies or is formed, (b) that is perpendicular to the optical axis of an optical system, such as a lens, and (c) on which a real

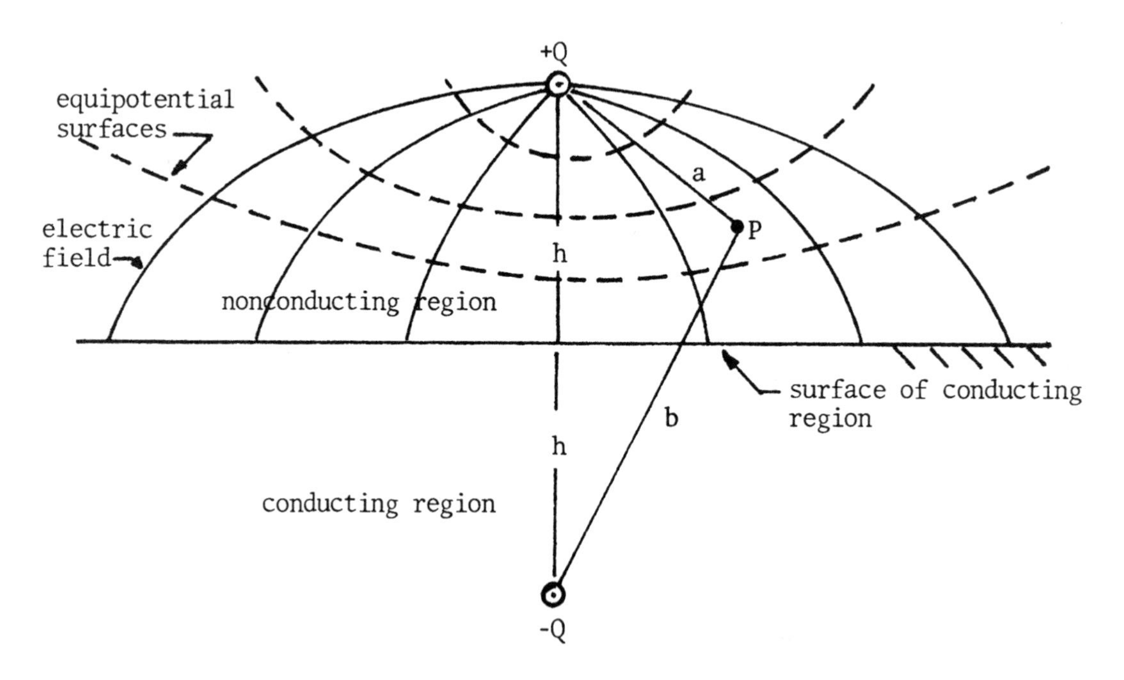

Fig. I-1. In the **image method,** the electric field at point P in the nonconducting region may be calculated as though the electronic charge $-Q$ also existed and the conducting region did not exist. The electric field at P is the result of the vector addition of the field caused by $+Q$ and the equal but opposite charge $-Q$ and the distances a and b according to the relation $\mathbf{E} = (Q/4\pi\epsilon_0 r^2)\mathbf{a}_r$ where Q is the electronic charge, ϵ_0 is the electric permittivity of the intervening space, r is the distance from the charge to the point P, and $\mathbf{a}_r$ is the unit vector in the direction of the line from the charge to the point P.

visible image is formed by a converging lens if a reflective surface, such as a screen, is placed in the plane. *See also* **image, lens, optical system, real image.**

image point: 1. A particular point on a real or virtual image. **2.** In optical systems that are capable of forming images, any point or location, coordinate position, or place in the image display space on the display surface of a display device that is also on an image displayed on that display surface. *See also* **coordinate position, display surface, image, optical system.**

image processing: The performing of operations on an image by an optical system, such as performing image dissection, image formation, image restoration, image filtering, or wavefront control by various processes such as phase conjugation, aperture tagging, wavefront compensation, and image sharpening. *See also* **aperture tagging, compander, compressor, expander, filtering, image, image dissector, image restoration, image sharpening, optical system, phase conjugation, wavefront compensation, wavefront control.**

image quality: The extent to which an image resembles the object from which it is made. *Note:* Characteristics of a lens or optical system that affect image quality primarily include resolving power and contrast. Aberrations contribute to poor image quality. Errors of construction and defects in material adversely affect image quality. Characteristic effects of aberrations on image quality make it possible to distinguish between their effects and those of accidental errors of workmanship, such as nonspherical surfaces, poor polish, scratches, pits, decentering, defects on cementing, and scattered light, all of which contribute to deterioration of the image. Defects in glass, such as bubbles, stones, striae, crystalline bodies, cloudiness, strain, seeds, chicken wire, and opaque minerals, also produce poor image quality. *See also* **aberration, bubble, chicken wire, image, object, seed, stone, striae.**

image rejection ratio (IRR): In heterodyning, the ratio of (a) the intermediate frequency signal level produced by the desired input frequency to (b) the level produced by the image frequency. *Note 1:* The image rejection ratio usually is expressed in dB. *Note 2:* When the ratio is measured, the input signal levels of the desired and image frequencies must be equal for the measurement to be meaningful. *See also* **circuit noise level, frequency, image frequency.**

image replacement: *Synonym* **image substitution.**

image restoration: The conversion of a distorted image to a less distorted image using image processing techniques, usually by reversing the processes that caused the distortion. *Note:* Selection of the most look-alike image from a database of images is not image restoration. *See also* **image processing.**

image rotator: A device or a system that rotates an image in a controlled manner and in a desired direction. *Note:* Examples of image rotators are (a) aligned fiber optic bundles that may be twisted along their length and (b) computer programs that cause rotation of images on the display surface of a display device, such as a monitor screen. *See also* **aligned fiber optic bundle, image, monitor.**

imagery: The representation of objects that are electronically or optically reproduced by any means on any data medium, such as on film and electronic display screens.

imagery system: *See* **medical diagnostic imagery system.**

image sharpening: 1. In image processing by an optical system, increasing of spatial light intensity gradients of an image on a display surface in order to increase the contrast at boundaries of components of the image, to more clearly distinguish individual object points from which the image was made. *Note:* Image sharpening may be accomplished by image restoration techniques, such as improved object illumination, improved focusing, precise wavefront control, and filtering. **2.** In image restoration, a method of wavefront control in which transition gradients from light to dark regions are increased in order to increase the clarity of the elements of an image. *See also* **filtering, illumination, image restoration, wavefront control.**

image storage space: In display systems and computer graphics, storage locations that are dedicated to storing coded images, used to store coded images, or occupied by coded images. *Synonym* **coded image space.** *See also* **computer graphics, display system, image.**

image substitution: The real-time exchanging of data displayed on a television screen during a broadcast with different data that are more relevant or more desirable for a local audience. *Note 1:* In most image substitution systems, the replacement image will fit the original space and adjust the viewing angle during panning. *Note 2:* An example of image substitution is the seamless replacement of advertisements on the walls of a stadium during a foreign sports event with advertisements or messages of more value to a domestic audience, such as replacing the foreign ad with an ad for a local product or changing a tobacco ad with a message on the hazards of smoking. *Synonym* **image replacement.**

imaging: The sensing of real objects, or representations of objects, followed by their representation by any means, such as mechanical, chemical, optical, or electronic means, on a data medium, such as a film, screen, plotter bed, cathode ray tube, or thin film electroluminescent (TFEL) display panel. *Note 1:* Imaging may be accomplished by controlling coded images with computer programs that enable manipulation of the images, such as to rotate, invert, expand, shrink, explode, implode, adjust, change, and color images. *Note 2:* Aligned bundles of optical fibers may be used for imaging. *Note 3:* Imaging photon-counting detector systems can produce images of objects by scanning and counting photons over a shorter time period than even the fastest photographic films, such as 20 photons/ms (milliseconds) for 2 minutes compared to 14 hours of exposure on the ASA 400 photographic film to produce the same quality image of the same object illuminated at an extremely low irradiance level. *See also* **computer program, image, object.**

imaging system: A display system capable of forming images of objects. *Note:* Examples of imaging systems are lens systems, fiber optic systems, computer controlled cathode ray tube (CRT) or fiberscope terminals, film systems, video tape systems, and thin film electroluminescent (TFEL) systems.

imbedded system: *See* **embedded system.**

IMD: intermodulation distortion.

imitation: *See* **signal imitation.**

imitative electronic deception: *See* **electronic deception.**

immediate access storage: Storage in which the access time is negligible compared to other operation times in the system to which it is connected. *Synonym* **instantaneous storage.** *See also* **access time, operation, storage.**

immediate address: In software, an operand, rather than an address of an operand, that is placed in the address part of an instruction word. *See* **address part, instruction word, operand, software.**

immediate data: In software, data that are contained in an instruction word rather than in a storage location specified by an address in the instruction word. *See also* **address, data, instruction word, software, storage location.**

IMMEDIATE message: In military operations, a message that (a) has a precedence designation higher than PRIORITY but lower than FLASH messages, (b) relates to situations that gravely affect the security of national or allied forces or the populace, and (c) requires immediate delivery to the addressee. *See also* **addressee, IMMEDIATE precedence, precedence, precedence designation.**

IMMEDIATE precedence: In military operations, the precedence designation that (a) is higher than PRIORITY but lower than FLASH, (b) is reserved for messages relating to situations that gravely affect the security of national or allied forces or the populace, and (c) when assigned to messages, requires that the messages be delivered immediately to the addressee. *See also* **addressee, IMMEDIATE message, precedence designation.**

immersion effect: *See* **fluid immersion effect.**

IMP: interface message processor.

impact point: *See* **drop zone impact point.**

impact printer: A printer that prints characters by using type to mechanically press a specially treated medium, such as carbon ribbon, carbon sheets, or inked ribbon, against a data medium, such as paper or cards. *See also* **carbon ribbon, data medium, inked ribbon, printer, type.**

impact test: In the testing of devices and components, such as transmitters, receivers, coaxial cables, connectors, optical fibers, fiber optic bundles, and fiber optic cables, a test to measure the ability of the device or the component to (a) withstand impact loading, (b) withstand breaking under an impact tensile load, or (c) withstand an impact on being dropped from given heights to hard surfaces, such as steel plate or concrete, without loss of design capabilities or with measurable loss of design capabilities.

impedance: The total passive opposition offered by a circuit to the flow of electric current when a direct current (dc) or alternating current (ac) voltage is applied to one or more components in the circuit. *Note 1:* The impedance is caused by a combination of resistance, inductive reactance, and capacitive reactance. *Note 2:* Impedance depends on frequency, the materials of which the circuit components are composed, the geometric configuration of the components, and the way they are interconnected. *Note 3:* The magnitude of an impedance consisting of lumped circuit elements, i.e., lumped parameters, connected in series is given by the relation $Z = [R^2 + (X_L - X_C)^2]^{1/2}$, and the vector value, i.e., the phasor value, is given by the relation $Z\angle\Theta = R + jX_L - jX_C$ where Z is the impedance, Θ is the phase angle, equal to arctan $(X_L - X_C)/R$, R is the resistance, X_L is the inductive reactance, equal to $2\pi fL$, where f is the frequency and L is the inductance, and

X_C is the capacitive reactance, equal to $1/2\pi fC$, where C is the capacitance. Usually R is in ohms, L is in henries, C is in farads, and f is in hertz. *See* **characteristic impedance, transimpedance, wave impedance.** *See also* **lumped parameter.** *Refer to* **Fig. T-8 (1024).**

impedance discontinuity: *See* **optical impedance discontinuity.**

impedance matching: The connection of an additional impedance to an existing one in order to accomplish a specific effect, such as balance a circuit or reduce reflection on a transmission line. *Note:* Examples of impedance matching are (a) to terminate a transmission line in its characteristic impedance in order to prevent reflections and (b) to adjust the impedance of a device in order to obtain maximum power transfer from a power source to which the device is connected. The matched impedance must include that of the load. *See also* **antenna matching, balancing network, building out, iterative impedance, loading, L pad.**

implant: *See* **ion implant.**

implementation: A physical realization of a concept in a less abstract form, i.e., a more concrete form. *Note:* Examples of implementation are (a) the creation of a computer program based on a flowchart, (b) the construction of prototype hardware from a detailed design, (c) the preparation of a machine language program from an assembly language program, and (d) the adoption and the distribution of a communications standard from a draft.

implementation conformance statement: *See* **protocol implementation conformance statement.**

implementation conformance statement pro forma: *See* **protocol implementation conformance statement pro forma.**

implementation phase: In the lifecycle of a system, the period during which implementation of an aspect or a portion of the system occurs. *See* **implementation.**

improper character: *Synonym* **illegal character.**

improper code: *Synonym* **illegal code.**

improved definition television (IDTV): Television transmitters and receivers that (a) are built to satisfy performance requirements over and above those required by the National Television Standards Committee (NTSC) standard, (b) remain within the general parameters of NTSC standard emissions, and (c) are improvements in conventional system performance and picture definition that do not make existing receivers and other equipment obsolete, i.e., the improvements are compatible improvements. *Note 1:* Improved definition television (IDTV) improvements may be made at the television (TV) transmitter or the receiver. *Note 2:* Examples of improved definition television (IDTV) improvements include enhancements in encoding, digital filtering, scan interpolation, interlaced line scanning, and ghost cancellation. *Note 3:* Improved definition television (IDTV) improvements must allow the television (TV) signal to be transmitted and received in the standard 4:3 aspect ratio. *Synonym* **enhanced quality television.** *See also* **advanced television, distribution quality television, extended definition television, high definition television.**

improved definition television (IDTV) system: A television (TV) system in which the characteristics of improved definition television (IDTV) are used. *See also* **advanced television system, enhanced definition television (EDTV) system, high definition television (HDTV) system.**

improvement: *See* **angle modulation noise improvement, preemphasis improvement.**

improvement factor: *See* **frequency modulation improvement factor.**

improvement threshold: *See* **frequency modulation improvement threshold.**

impulse: A surge of electrical, magnetic, or electromagnetic energy that (a) is of short duration, (b) usually is nonrepetitive, and (c) may be in the form of a step function, i.e., an instantaneous transition from one significant condition to another. *Synonym* **surge.** *See also* **pulse, shock wave, significant condition.**

impulse dialing: *Synonym* **loop disconnect pulsing system.**

impulse excitation: The production of a current in a circuit by impressing a voltage for a relatively short period compared with the duration of the current produced. *Note:* The impulse excitation usually produces current and voltage transients and oscillations. *Synonym* **shock excitation.** *See also* **impulse response, impulse response function.**

impulse noise: **1.** Noise that (a) consists of random occurrences of energy spikes with random amplitude and bandwidths, (b) can cause transmission errors in a data channel, and (c) primarily is caused by disturbances that have abrupt surges of short duration in electric current or voltage. **2.** In telephone circuits, noise that (a) is characterized by transient disturbances that are separated in time by quiescent intervals and (b) usually does not have systematic phase relationships. *Note:* The same disturbance source may produce impulse noise in one system and random noise in a different system. *See also* **noise, random noise.**

impulse response: 1. The amplitude of current or voltage output versus time of a device or a circuit in response to an impulse input. **2.** In a device, such as transmission equipment or an amplifier, the response, such as the current output amplitude as a function of time, after a Dirac-delta function, i.e., a step function, such as a step voltage or step optical signal, is applied to its input. **3.** For a device, the mathematical function that describes the output waveform that results when the input is excited by a unit impulse. *Note:* The impulse response is the inverse Laplace or Fourier transform of the transfer function, i.e., the output function is determined by the transfer function operating upon the input function. *See also* **impulse, impulse excitation, impulse response function, random noise.**

impulse response function: The function of time, $h(t)$, that describes the response of an initially relaxed system after a Dirac-delta function, i.e., a step function, is applied at time $t = 0$. *Note:* The root mean square (rms) duration of the impulse response may be used to characterize a component or a system by means of a single parameter rather than a function. The rms value of the impulse response function, σ_{rms}, is given by the relation

$$\sigma_{rms} = \{(1/M_0)\int_{-\infty}^{+\infty}(T - t)^2 h(t)dt\}^{1/2}$$

where

$$M_0 = \int_{-\infty}^{+\infty} h(t)dt$$

and

$$T = (1/M_0)\int_{-\infty}^{+\infty} th(t)dt.$$

The impulse response may be obtained by deconvolving, i.e., by deriving, the input waveform from the output waveform, or as the inverse Fourier transform of the transfer function. *See also* **impulse excitation, impulse response, root mean square deviation, root mean square pulse broadening, root mean square pulse duration, spectral width.**

impurity absorption: In traveling or standing waves in propagation media, such as glass, the absorption of light energy from the waves by foreign elements, such as iron, copper, cobalt, vanadium, chromium, hydroxide, and chloride ions. *Note:* Impurity absorption results in attenuation. *See* **propagation medium, standing wave, traveling wave.**

impurity level: In a material, an electron energy level that (a) is outside the normal energy level of the intrinsic material, (b) is caused by the presence of impurity atoms in the material, and (c) is capable of making an insulator semiconducting. *Note:* The impurity atom may be a donor or an acceptor. If a donor, the impurity induces electronic conduction through the transfer of an electron to the conduction band. If an acceptor, the impurity can induce hole conduction through the acceptance of an electron from the valence band. *See also* **acceptor, conduction band, donor, electron, hole, semiconductor, valence band.**

IN: intelligent network.

inactive: In communications, computer, data processing, and control systems, pertaining to (a) a lack of readiness to perform a function and (b) an unconnected, unenergized, not loaded, or not ready to run component, subsystem, or system.

inactive character: A character that (a) is transmitted in the information transfer phase of an information transfer transaction, such as a call, (b) usually is a stuffing character, and (c) does not represent any information. *See* **call, information transfer phase, information transfer transaction, stuffing character.**

inactive code word: A code word that (a) has been placed in use and (b) subsequently either has been replaced by another code word that has the same meaning or has had its use and meaning discontinued altogether. *See also* **code word.**

inactive file: In communications, computer, data processing, and control systems and databases, a file that (a) has been retired or removed from the system or the database, (b) has been placed in an archival file from which it is not currently, automatically, or readily available, and (c) is no longer intended to be accessed, but is still retrievable, i.e., is not erased or discarded. *See* **archiving, database, file.** *See also* **active file.**

inactive line: In a communications system, a line that is currently not available for use. *Synonym* **idle line.** *See also* **active line.**

inactive line termination: *Synonym* **idle line termination.**

inactive link: In a communications network or a telemetry system, a link that is not connected or is not available for connection to the network or the system. *Synonym* **idle link.** *See also* **active link.**

inactive node: In a communications network, a node that is unenergized, disconnected, or not available to the network. *Synonym* **idle node.** *See also* **active node.**

inactive optics: *Synonym* **fixed optics.**

inactive program: 1. In communications, computer, and data processing systems, a computer program that is not loaded and ready to be executed. **2.** A system development program that once was active but has been temporarily or permanently discontinued. *See also* **active program.**

inactive station: In communications, computer, data processing, and control systems, a station that is currently unable to enter or accept messages to and from the system to which it is or has been connected. *Synonym* **idle station.**

in-band noise power ratio (IBNPR): In multichannel equipment, the ratio of (a) the mean noise power measured in any channel with all channels loaded with white noise to (b) the mean noise power measured in the same channel with all channels but the measured channel loaded with white noise. *See also* **noise, white noise.**

in-band signaling: Signaling that uses frequencies or time slots that are within the bandwidth of the information channel. *See also* **bandwidth, channel, channel-associated signaling, common channel signaling, frequency, out-of-band signaling, signal, single frequency signaling.**

incandescence: The emission of light by thermal excitation, which brings about energy level transitions that produce (a) sufficient quantities of photons with sufficient energies to render the source of radiation visible or the radiation itself visible and (b) infrared radiation.

incidence: The act of falling upon or affecting, such as the action of a ray of light, a radio wavefront, or a rubber ball impinging upon a surface. *Note:* The ray is in the direction of propagation, and the wavefront is perpendicular to the direction of propagation. The ray and wavefront directions are used to determine the incidence angle. *See* **grazing incidence, normal incidence.** *See also* **incidence angle, incident ray.**

incidence angle: 1. In optics, the angle between the incident ray and the normal to a reference surface, such as a reflecting or refracting surface. **2.** In wave transmission, such as electromagnetic, sound, and water wave transmission, the angle between the direction of propagation of the wave and the normal to a reference surface, such as the core-cladding interface in an optical fiber, an aircraft fuselage, the surface of the ionosphere, a mountainside, or a shoreline. *Synonyms* **angle of incidence, incident angle.** *See also* **critical angle, incidence, ray, total internal reflection.** *Refer to* **Fig. L-9 (536), S-13 (919), S-18 (951).**

incidence sounding: *See* **oblique incidence sounding.**

incidental radiation device: A device that (a) radiates radio frequency (rf) energy during its operation and (b) is not intentionally designed to generate radio frequency energy. *Note:* Examples of incidental radiation devices are radio receivers, radar receivers, fiber optic receivers, and certain direction finding equipment. *See also* **restricted radiation device.**

incident angle: *Synonym* **incidence angle.**

incident ray: A ray that falls upon, or strikes, a reference surface, such as that of on object. *Note:* The incident ray is said to be incident to a surface. *See also* **incidence, incidence angle, ray.**

inclination: *See* **orbital inclination.**

inclination angle: For an Earth satellite orbit, the angle determined by (a) the plane containing the orbit and (b) the Earth's equatorial plane.

inclination of an orbit: *Synonym* **orbital inclination.**

inclined orbit: An orbit that (a) is a nonequatorial orbit and (b) may be (i) circular or elliptical, (ii) synchronous or asynchronous, and (iii) direct or retrograde. *See also* **direct orbit, equatorial orbit, geostationary orbit, geosynchronous satellite, polar orbit, retrograde orbit, satellite, synchronous orbit.**

inclusion: In a material, such as optical glass or plastic, a particle of foreign or extraneous matter. *Note 1:* Examples of inclusions are air bubbles, specks of dust, seeds, striae, hydroxyl ions, and iron ions. *Note 2:* In optical fibers, dopants used to achieve desired refractive index profiles are not considered as inclusions. *Note 3:* In dielectric waveguides, such as optical fibers and slab dielectric waveguides, inclusions cause scattering and absorption of lightwaves. *See also* **absorption, dopant, optical fiber, scattering, seed, slab dielectric waveguide, striae.**

inclusion gate: *Synonyms* **A OR NOT B gate, B OR NOT A gate.**

inclusive NOR gate: *Synonym* **NOR gate.**

inclusive OR gate: *Synonym* **OR gate.**

incoherent: In optics, pertaining to the coherence of lightwaves in which the coherence degree is significantly less than 0.88. *See also* **coherence degree, coherent.**

incoherent bundle: *A deprecated synonym for* **unaligned bundle.**

incoherent frequency shift keying (IFSK): Frequency shift keying (FSK) in which the instantaneous frequency is shifted between two discrete values, called the "mark" and "space" frequencies, without regard for phase discontinuity. *See also* **coherent frequency shift keying, frequency shift keying.**

incoherent moving target indicator: A moving target indicator (MTI) that does not make use of the Doppler effect to obtain the radial component of velocity of the object, i.e., the target, with respect to the radar site. *Note:* In the radar set, return signals, i.e., echoes, from objects are subtracted from the next echo signal, using a suitable temporary storage circuit. *See also* **coherent moving target indicator, Doppler effect, moving target indicator.**

incoherent radiation: Radiation that has a low coherence degree. *See also* **coherence, coherence degree, incoherent.**

incoming call identification (ICI): A network feature that allows an attendant to identify visually the type of service or trunk group associated with a call directed to the attendant position. *See also* **calling line identification facility, service feature.**

incorrect bit: A received bit that has a binary value that is the complement of the bit that was transmitted, i.e., a bit that is a wrong bit and not the bit that was intended or transmitted. *See also* **bit, complement, incorrect block.**

incorrect block: A block that (a) is successfully delivered to the intended destination user and (b) has one or more incorrect bits, additions, or deletions. *See also* **block, block error probability, block transfer failure, error, incorrect bit.**

increment: *See* **radio frequency channel increment.**

incremental compaction: Data compaction accomplished by specifying only the initial value and all subsequent changes. *Note:* An example of incremental compaction is the storing or the transmitting of a line voltage followed only by the deviations from the initial value. Thus, instead of transmitting the values 102, 104, 105, 103, 100, 104, and 106, only the values 102, +2, +1, −2, −3, +4, and +2, or only the values 100, +2, +4, +5, +3, 0, +4, and +6 need be sent, depending on the system used. Transmitting only the initial and incremental values requires much less time and space than transmitting the absolute values.

incremental magnetic permeability: In a material, the limit of the ratio of (a) the change in magnetic flux density, ΔB, produced to (b) the change in magnetic field strength, ΔH, required to produce the ΔB at a given point in the material, expressed by the relation $\Delta_{inc} = dB/dH$, where Δ_{inc} is the incremental magnetic permeability and dB/dH is the rate of change of magnetic flux density with respect to the magnetic field strength, i.e., the slope of the magnetization curve of the material. *See also* **absolute magnetic permeability, field strength, relative magnetic permeability.**

incremental phase modulation (IPM): In spread spectrum systems, phase modulation in which one binary code sequence is shifted with respect to another, usually to conduct a synchronizing search, i.e., a search to discover if the two sequences are the same, and perhaps thereby to enable two data streams to be synchronized.

incremental quantum efficiency: *Synonym* **differential quantum efficiency.**

incremental vector: *Synonym* **relative vector.**

increment size: The spatial distance between adjacent addressable points within a display space. *Note:* Examples of increment sizes are (a) the distance between centers of adjacent optical fibers in the faceplate of a fiberscope and (b) the minimum displacement that can be drawn by a plotter, i.e., the plotter step size. *See also* **addressable point, display space.**

indefinite call sign: 1. A call sign that represents a group of facilities, commands, authorities, activities, or units rather than one of these. **2.** In radio communications, a call sign that does not identify a station and that is used in the callup signal or in a message that has the station call sign encrypted in the text.

indefinite component life: The life of a component that is expected to remain in a satisfactory operational and serviceable condition during the entire life of the system of which it is a part. *See* **component, system.**

independent: *See* **bit-sequence-independent, code-independent.**

independent clock: A clock in a communications network timing subsystem that uses precise free-running clocks at the nodes for synchronization purposes. *Note:* Variable storage buffers installed to accommodate variations in transmission delay between nodes are made large enough to accommodate small time, i.e., small phase, departures among the nodal clocks that control transmission. Traffic occasionally is interrupted to reset the buffers. *See also* **clock, node, synchronized clock.**

independent product validation: *Synonym* **product validation.**

independent sideband: A sideband in a sideband transmission system in which the separate sidebands are used to convey independent or different information. *See* **sideband.**

independent sideband carrier: *See* **two-independent-sideband carrier.**

independent sideband (ISB) transmission: Double sideband transmission in which (a) each sideband carries different information, i.e., transmission in which the modulated frequencies on the opposite sides of the carrier frequency are related separately to two sets of modulating frequencies, i.e., by two separate baseband signals, and (b) the carrier may be transmitted, suppressed, or partially suppressed. *Synonym* **twin sideband transmission.** *See* **two-independent-sideband transmission.** *See also* **double sideband transmission, sideband transmission.**

independent system validation: *Synonym* **system validation.**

index: 1. A list of identifiers of the records in a file together with keys to identify the location of the records. **2.** To prepare a list of identifiers. **3.** A symbol used to identify a particular item in an array of similar items. *Note:* Examples of indexes are the subscripts in x_1, x_2, and x_3. **4.** To move a data medium to a prescribed position. **5.** To move a machine part to a predetermined position or by a predetermined distance or angle. *See* **absolute refractive index, absorption index, articulation index, code line index, cooperation index, group index, key-word-in-context index, key-word-out-of-context index, modulation index, radiation transfer index, refractive index, relative refractive index, surface refractive index.**

index contrast: *See* **refractive index contrast.**

index difference: *See* **equivalent step index refractive index difference.**

index dip: *Synonym* **refractive index dip.**

index flash: In visual signaling systems, a single panel that (a) indicates the top of numerals or words formed with signaling panels and (b) is placed first and removed last when signaling from ground to air. *See also* **panel.**

index gradient: *Synonym* **refractive index gradient.**

indexing crosstalk: Crosstalk that is coupled from one carrier circuit to another carrier circuit. *See also* **coupling, crosstalk, interaction crosstalk.**

index matching cell: *Synonym* **refractive index matching cell.**

index matching gel: *Synonym* **refractive index matching gel.**

index matching material: *Synonym* **refractive index matching material.**

index of cooperation: *Synonym* **cooperation index.**

index of refraction: *Synonym* **refractive index.**

index optical fiber: *See* **graded index optical fiber, step index optical fiber.**

index optical waveguide: *See* **graded index optical waveguide.**

index power leveling: *See* **power law refractive index power leveling.**

index profile: *See* **equivalent step index profile, parabolic refractive index profile, power law refractive index profile, refractive index profile.**

index profile parameter: *Synonym* **refractive index profile parameter.**

index register: In communications, computer, data processing, and control systems, a register that is used to perform counting and storing operations requiring that quantities be modified by prescribed amounts, usually in a consecutive manner, such as to (a) modify the address of an operand during the execution of an instruction, (b) control the execution of a loop, (c) control the use of an array, (d) support the use of a lookup table, and (e) serve as a pointer. *See* **register.**

index template: *See* **four-concentric-circle refractive index template.**

in-dialing: *See* **network in-dialing.**

indicating group: *See* **address indicating group, international address indicating group, national indicating group.**

indicating group allocation: *See* **address indicating group allocation.**

indicating radiobeacon: *See* **satellite emergency position-indicating radiobeacon.**

indication: *See* **automatic charge indication, automatic date and time indication, automatic duration indication, charge indication, clear indication, duration indication.**

indicator: 1. A device that (a) may be set into a prescribed state based on the result of a previous process or on the occurrence of a specified condition, (b) emits a signal announcing the existence of the prescribed state or the occurrence of the specified condition, and (c) may be used to determine a selection from among a group of alternatives, such as specify the routing for a message or the status of an instrument. **2.** A symbol, signal, or group of symbols or signals, that serves to identify a specific item, such as (a) a message from a specified source, (b) the specific cryptosystem key used to encrypt or authenticate a message or a transmission, or (c) a specific signal unit from other signal units for checking purposes. *See* **alarm indicator, alternative routing indicator, called line identification request indicator, calling line identification request indicator, coherent moving target indicator, collective routing indicator, destination code indicator, drop zone indicator, end of message indicator, error indicator, figure indicator, glide slope indicator, incoherent moving target indicator, letter indicator, message alignment indicator, mobile unit routing indicator, mode indicator, Morse code indicator, moving target indicator, negative indicator, overflow indicator, panel signaling indicator, panoramic indicator, planned position indicator, radar planned position indicator, range-height indicator, routing indicator, signal unit indicator, special purpose routing indicator, start of message indicator, user class indicator.**

indicator allocation: *See* **routing indicator allocation.**

indicator assignment: *See* **routing indicator assignment.**

indicator delineation map: *See* **routing indicator delineation map.**

indicator delineation table: *See* **routing indicator delineation table.**

indicator shutter: *See* **fallen clearing indicator shutter.**

indirect address: An address of a storage location that contains an operand address that is not necessarily a direct address.

indirect clock control: In digital data transmission, the use of a standard clock rate, such as 4, 8, or 128 times the modulation rate, that is higher than 2 times the modulation rate used in direct clock control. *Synonym* **indirect control.** *See also* **clock, direct clock control, modulation.**

indirect control: *Synonym* **indirect clock control.**

indirect mode: In a radionavigation system, the operation or the use of the several transmitters of the radionavigation service simply as radio beacons for direction finding (DF) rather than as part of a radionavigation system such as LORAN or hyperbolic navigation systems.

indirect wave: A wave, such as a lightwave, radio wave, or sound wave, that arrives at a point in a propagation medium by reflection or scattering from surrounding objects or surfaces, such as a mountain for sound waves, a steel structure for radio waves, and a refractive index interface surface for lightwaves, rather than directly from the source. *Note:* In fiber optic transmission, indirect waves may arrive at a point later than the direct waves and, by combining with the direct waves, will cause delay distortion. In television systems, an indirect wave may cause a secondary image, i.e., a double image, to appear on the screen, the second image usually being caused by the indirect wave. *See also* **direct wave.**

individual call sign: In a radio communications system or net, a call sign that identifies a single station. *See also* **collective call sign, net call sign.**

individual line: A line that connects a user to a switching center, such as a central office (C.O.). *See also* **leased circuit, loop.**

individual normal magnification: The apparent magnification that (a) is produced by a magnifier, such as a lens or a mirror, (b) usually is different from the absolute magnification, and (c) depends on the extent of the myopia or the hyperopia, i.e., hypermetropia, possessed by the individual. *See also* **absolute magnification, lens, magnification, mirror.**

individual reception: In the broadcasting satellite service, the reception of emissions from a space station by simple domestic installation, especially installations with small antennas. *See also* **broadcasting satellite service, space station.**

individual trunk: A single trunk that directly connects two switching centers, i.e., without intervening switching centers. *See also* **switching center, trunk.**

induced attenuation: The difference between (a) the attenuation in a component or a system, or the attenuation rate, caused by external factors, such as environmental changes, mechanical treatment, and immersion in fluids, and (b) the attenuation, or the attenuation rate, measured before the external factors were applied to the component or to the system. *Note:* When fiber optic components, such as optical fibers and fiber optic connectors, splices, and couplers, are tested, the induced attenuation, or the induced attenuation rate, is determined by measuring the attenuation, or the attenuation rate, before and after the application of external factors. *Synonym* **change in transmittance, transmittance change.** *See also* **fiber optic dosimeter.**

induced modulation: *See* **mechanically induced modulation.**

induced optical conductor loss: *See* **connector induced optical conductor loss.**

inductance: The ability of a configuration of conducting material to impede the flow of alternating current (ac) by inducing in itself an electromotive force (emf), i.e., a voltage, that is in opposition to the applied electromotive force that produces the alternating current. *Note 1:* The unit of inductance is the henry. *Note 2:* The inductance of a configuration is a function of the shape, i.e., the configuration, of the material, and not directly a function of the resistance or the capacitance, though these are usually present in a practical inductance. *Note 3:* The inductive reactance of an inductance is given by the relation $X_L = 2\pi fL$ where the inductive reactance, X_L, will be in ohms if the frequency, f, is in hertz, and the inductance, L, is in henries. *Note 4:* In a pure inductance, i.e., one of zero resistance and zero capacity, the current will lag the applied voltage by 90° (electrical), i.e., the current will be in phase quadrature with the applied ac voltage. *Note 5:* Many turns of wire wound on an iron core, such as are found in electric relays, transformers, motors, and generators, have a high inductance. *See also* **induction, inductive reactance.** *Refer to* **Table 3, Appendix B.**

induction: The generation of an electromotive force (emf), i.e., a voltage, by the time rate of change of magnetic flux, given by the relation $e = Nd\phi/dt = -Ldi/dt$ where e is the emf generated in a closed circuit through which there is a time rate of change of magnetic flux linkages, $d\phi/dt$ is the time rate of change of magnetic flux linkages, N is the number of conductor turns in the circuit, L is the inductance, a function of geometry and materials, and di/dt is the time rate of change of electric current in the circuit. *See* **mathematical induction.** *See also* **inductance, inductive reactance.**

induction field: *Synonym* **intermediate field.**

inductive assertion method: In communications, computer, data processing, and control systems, a proof of correctness method in which (a) certain assertions are made, (b) a set of relevant theorems is developed, and (c) the theorems are then proved to be true. *Note:* An example of the inductive assertion method is to (a) make a set of computer program input and output assertions, (b) develop a set of theorems relating to the satisfaction of the input and output assertions, and (c) run the program to prove that the theorems are true by examining the output that results from the input.

inductive coupling: 1. The induction of signals or noise in one conductor by a time rate of change of magnetic flux linkages that are created by electrical currents in another conductor. **2.** The transfer of energy from one circuit to another by mutual inductance between the circuits. *Note 1:* The coupling may be deliberate and desired, such as in an antenna coupler, or may be undesired, as in power line inductive coupling into a telephone line. *Note 2:* Inductive coupling favors transfer of lower frequencies, whereas capacitive coupling favors transfer of higher frequencies. *See also* **capacitive coupling, inductance, induction.**

inductive reactance: The opposition to the flow of alternating current (ac) that is produced by closed circuit configurations of conductors, such as straight wires and coils, when an alternating voltage is applied to the circuit. *Note 1:* The changing current, i.e., the alternating current (ac), produces a changing magnetic field. The changing flux linkages induce an electromotive force (emf), i.e., a voltage, in the circuit that opposes the flow of current producing the flux linkages. This creates the opposition to the flow of current produced by the applied alternating voltage. *Note 2:* If there is only inductive reactance in a circuit, i.e., there is no resistance and no capacitive reactance, the inductive reactance is the ratio of the applied voltage to the resultant current, i.e., the current is given by the relation $I = V/X_L$, where V is the applied sinusoidal voltage and X_L is the inductive reactance. I will be in amperes if X_L is in ohms and f is in hertz. *Note 3:* The inductive reactance is given by the relation $X_L = j2\pi fL$ where the inductive reactance, X_L, will be in ohms if the frequency, f, is in hertz and the inductance, L, is in henries; j is the quadrature operator given by the relation $j = (-1)^{1/2}$. L is a function of geometry and materials. *Note 4:* In a pure inductance, i.e., one of zero resistance and zero capacity, the current will lag the applied voltage by 90° (electrical), i.e., the current will be in phase quadrature with the applied sinusoidal ac voltage. *See also* **capacitive reactance, impedance, inductance, induction, resistance.**

industrial radio frequency (rf) heating equipment: Equipment in which a radio frequency (rf) generator, such as an rf oscillator, is used to control the generation of rf energy for industrial heating operations by inducing electrical currents in materials in manufacturing and production processes, such as for annealing copper tubing or extruding metals forced through dies.

industrial, scientific, and medical (ISM) application: The use of radio frequency energy by equipment or appliances designed to locally generate and use radio frequency energy for industrial, scientific, medical, domestic, or similar purposes, excluding use in telecommunications.

industrial, scientific, and medical (ISM) radio frequency (rf) equipment: Radiation devices that use electromagnetic waves for industrial, scientific, or medical (ISM) purposes, including the transfer of energy by radio, though not necessarily used, nor intended to be used, for communications purposes. *Note:* Examples of industrial, scientific, and medical (ISM) radio frequency (rf) equipment are devices that develop rf energy for induction heating, microwave heating, diathermy, and infrared heating.

industry standard: A voluntary, industry-developed standard that (a) establishes requirements for products, practices, services, or operations, or (b) the document containing the standard. *See also* **de facto standard.**

inertia: The characteristic of matter, i.e., of mass, to remain at rest or to remain at a constant velocity, i.e., constant speed and constant direction, unless acted upon by an outside force. *Note 1:* An applied outside force produces acceleration, a vector quantity. A component of applied force in the direction of motion produces a change in speed. A component of force transverse to the direction of motion produces a change in the direction of motion. The force **f**, mass *m*, and acceleration **a** are related in Newtonian mechanics by the relation $\mathbf{f} = m\mathbf{a}$. The relation applies as long as the speeds are very small compared to the speed of light. Speeds greater than 0.1 of the speed of light must have relativistic corrections. For curvilinear motion at constant speed, the acceleration **a** is replaced by $\mathbf{v}^2/\mathbf{r}$ where **v** is the speed and **r** is the radius of curvature, thus producing the relation $\mathbf{f} = m\mathbf{v}^2/\mathbf{r}$. *Note 2:* Inertia is used in inertial guidance systems, in impact fuzes, and to keep satellites in desired orbits. *See also* **inertial guidance system.**

inertial guidance system: A guidance system that is based on the principle of inertia. *Note:* An accelerometer, such as a spring-loaded mass, and integration can be used to obtain velocity and distance, provided that the initial conditions are known. *See also* **inertia, integration, noninertial guidance system.**

inferior position flaghoist: A flaghoist signaling message that is to be read after a superior position flaghoist message. *See also* **flaghoist signaling, superior position flaghoist.**

infinite bus: 1. An hypothetical electrical power bus that has a constant voltage regardless of the current it supplies, has a zero resistance, and can supply an infinite current on demand. **2.** A hypothetical optical power bus that can supply an infinite amount of optical power. *See also* **power bus.**

infix notation: Notation in which algebraic expressions are formed by prescribed unambiguous rules in which the operators are dispersed among the operands. *Note:* Examples of infix notation are $(a - b)/(a + b)$ where the / is the division operator, and $ax^2 + bx + c$ where the multiplication operator is implied. *See also* **operand, operator, parenthesis-free notation, prefix notation, postfix notation.**

infix operator: 1. An operator in an algebraic expression constructed with infix notation. **2.** An explicit or implicit operator located between operands. *See also* **infix notation.**

infobahn: *Synonym* **information superhighway.**

information: 1. The meaning that a human assigns to data by means of the known conventions used in their representation. *Note:* For information transfer to occur, the sender and the receiver must assign the same meaning to the data transferred. **2.** Data that have been processed and formulated by automated or manual means to satisfy a knowledge requirement for use by a decision maker. **3.** In communications and teleprocessing systems, any transmission of signals that usually are in the form of data or text messages. **4.** In military intelligence, unprocessed data that may be used in the production of intelligence. *See* **cryptographic information, cryptoinformation, exempt information, hydrographic information, overhead information, perishable information, protected information, protocol control information, raw radar information, record information, security information, sensitive information, service information, signaling information, system overhead information, user information, user overhead information.**

information addressee: The organization, activity, or person deemed by the message originator to require the information in a message but not required to take the action specified in the message. *See also* **action addressee, exempted addressee, information, message originator.**

information bearer channel (IBC): 1. A channel that (a) is capable of transmitting all the information required for communications, such as user data, synchronizing sequences, and control signals and (b) may operate at a higher signaling rate than that required for user data alone. **2.** A basic communications channel with the necessary bandwidth but without enhanced or value-added services. *See also* **channel, circuit, communications.**

information bit: A bit that (a) is generated by a data source and delivered to a data sink and (b) is not used by the communications system. *See* **user information bit.**

information block: *See* **user information block.**

information channel: The propagation media and the intervening equipment, such as the modulator, demodulator, error control equipment, and backward channel, that are used for the transfer of information in a given direction between two stations. *See* **backward channel, propagation medium.**

information content: 1. The information that is represented by a given set of characters. *Note:* An example of information content is the meaning assigned to the data in a message. **2.** In information theory, a measure of the information that (a) is conveyed by the occurrence of an event of definite probability and (b) is represented by $I(x_i)$, for the event x_i, usually expressed as the logarithm of the reciprocal of the probability, $p(x_i)$, that the particular event will occur, and given by the relation $I(x_i) = \log [1/p(x_i)] = -\log p(x_i)$. *Note:* If two characters, *x* and *y*, among others, can occur in a message, if the probability of *x* occurring in the message is 0.5, and if the probability of *y* occurring in the message is 0.25, then the information content of *x* is $\log_2 2 = 1$ shannon, $\log_{10} 2 = 0.301$ hartley, and $\ln 2 = 0.693$ nats (natural units) because these are the logarithms of 1/0.5, i.e., of 2, and the information content of *y* is 2 shannons, 0.602 hartley, and 1.386 nats because these are the logarithms of 1/0.25, i.e., of 4. *See* **conditional information content, joint information content, mutual information content, natural unit of information content, signaling information content, transinformation content.** *See also* **information, information theory, logarithm.**

information content binary unit: *Synonym* **shannon.**

information content decimal unit: *Synonym* **hartley.**

information content natural unit: *Synonym* **natural unit of information content.**

information content Napierian unit: *Synonym* **natural unit of information content.**

information descriptor: *See* **receiver information descriptor, transmitter information descriptor.**

information distribution system: *See* **multifunction information distribution system.**

information entropy: The mean value of the information content that is conveyed by the occurrence of any one of a finite number of mutually exclusive and jointly exhaustive events of definite probability of occurrence. *See* **conditional information entropy.**

information exchange: *See* **Advanced Project for Information Exchange, Defense Network Security Information Exchange.**

information facility: In a communications system, a facility that enables a user to send a predetermined address from a data station and gain access to information regarding available data communications services, such as directory inquiry services, charge information services, and fault reporting services. *Synonym* **inquiry facility.** *See also* **facility.**

information feedback: The return of received data to the source, usually for the purpose of checking the accuracy of transmission by comparison with the data originally transmitted. *See also* **echo check, feedback, forward error correction.**

information field: In data transmission, a field that (a) contains user information and (b) contains data that are not interpreted at the link level. *See also* **data transmission, Open Systems Interconnection—Reference Model, user information bit.**

information highway: *Synonym* **information superhighway.**

information link: *See* **tactical digital information link.**

information management system (MIS): A system used to manage the information required to operate an organization, enterprise, endeavor, or other system. *See* **corporate information management.**

information measure: 1. An indication of the amount or the volume of information that is represented by a given set of data in a particular situation or environment. *Note:* Coding of the information must be taken into account in determining the amount or the volume of information. **2.** In information theory, a function of the probability of occurrence of an event or a sequence of events out of a set of possible events. *Note:* Examples of information measure are (a) the probability of occurrence of a specified member of a set, (b) the probability of occurrence of a specified character or a specified word in a given position in a message, and (c) the probability of occurrence of a specified bit pattern in a bit stream. *See also* **information content, information theory.**

information payload capacity: *See* **payload rate.**

information processing: *Synonym* **data processing.**

information processing center (IPC): A facility that (a) is staffed and equipped for processing and distributing information and (b) may be centralized or may be geographically distributed.

information processing equipment: *See* **Federal Information Processing equipment.**

information processing system: *See* **Federal Information Processing system.**

information protocol: *See* **Common Management Information Protocol.**

information rate: 1. *Synonym* **character mean entropy. 2.** *See* **average information rate, system information rate.**

information resource: The information or its representation, i.e., the data, that an entity, such as a person, organization, program, endeavor, enterprise, facility, or system, has available for its use, such as for operations, decision making, problem solving, or data processing. *See* **data, information.**

information retrieval (IR): Actions, methods, procedures, and principles applied to the recovery of data from storage for specific use, including (a) organizing and indexing the data, (b) storing the data in a database, (c) developing search strategies for subsequent search, (d) maintaining the database in which the data are stored, and (e) obtaining the data from the database. *Synonym* **information storage and retrieval.** *See also* **data, database, index, retrieval.**

information security: The protection of information against unauthorized disclosure, transfer, modification, or destruction, whether accidental or intentional. *See also* **communications security.**

information separator: In communications, computer, data processing, and control systems, a control character that (a) is used to delimit data units in a hierarchical arrangement of data and (b) does not necessarily indicate the unit of data that it separates. *Synonym* **separating character.** *See also* **data delimiter.**

information server: *See* **Wide Area Information Server.**

information service: *See* **common management information service.**

information source: *Synonym* **source user.**

information station: *See* **traveler information station.**

information storage and retrieval: *Synonym* **information retrieval.**

information superhighway: The totality of interconnected communications facilities that (a) includes all transmission facilities, such as telephone, videophone, interactive video, telegraph, broadcast radio, worldwide cellular radio, broadcast television, facsimile, teleconferencing, and digital data facilities, (b) provides a myriad of propagation media, such as wire, coaxial cable, fiber optic cable, microwave, and satellite services, (c) is capable of handling multimedia information, such as voice, video, and digital data, (d) provides high capacity multichannel capabilities at the loop level, allowing each user to be connected to all other users, and (e) allows local area network (LAN) interconnection, remote access data processing, and computer-to-computer interconnection. *Note 1:* The information superhighway makes use of a coordinated open systems infrastructure consisting of (a) interconnected satellite, wire cable, coaxial cable, fiber optic cable, microwave, and radio links, (b) switching centers and interconnection facilities, and (c) applicable protocol suites. *Note 2:* Communications experts anticipate that by the year 2000, the photonic information superhighway will carry information at teraflop rates, i.e., billions of operations per second, by the year 2005 at thousand teraflop rates, and by the year 2010 at a million teraflop rates, i.e., at 10^{18} operations per second. *Note 3:* With universal interconnection, the information superhighway should provide secure, broadband, interactive, digital, analog, audio, video, imaging, and data services to all homes, workplaces, schools, and hospitals. *Note:* The information superhighway extends worldwide, including communications networks that (a) provide hundreds of channels directly to each user, (b) enable networking of computers and databases, (c) allow direct personal multimedia communications, such as via telephone, electronic mail, photophone, and teleconferencing, and (d) provide access to hundreds of channels of education, information, and entertainment. *Synonyms* **infobahn, information highway.** *See also* **channel, database, internet, multimedia, network.**

information system: 1. An automated or manual system that (a) is composed of people, machines, or methods and (b) is organized to collect, process, transmit, and disseminate data that represent user information. **2.** In communications, computer, data processing, and control systems, any interconnected system, subsystem, or component thereof, that is used in the handling of data or information, such as the acquisition, storage, manipulation, management, movement, control, display, switching, interchange, transmission, or reception of data, voice, or video signals, including software, firmware, and hardware. *See* **automated information system, electronic flight information system.**

information system architecture: 1. The physical configuration, logical structure, formats, protocols, and operational sequences for (a) handling data in, and

(b) controlling the configuration and operation of, an information handling system. **2.** The documentation that contains descriptions of the principles, physical configuration, functional organization, operational procedures, and data formats used as the basis for the design, construction, modification, and operation of an information system. *See also* **computer architecture, network architecture.**

information systems network: *See* **Defense Information Systems Network.**

information systems security (INFOSEC): 1. The protection of information systems against (a) unauthorized access to, or modification of, information in storage, processing, or transit, (b) the interruption or denial of service to authorized users, and (c) the provision of service to unauthorized users, including the measures necessary to detect, document, and counter such threats, intrusions, and penetrations. **2.** The means of protecting telecommunications systems, automated information systems, and the information in them. *See also* **automated information systems security, communications security.**

information systems security initiative: *See* **multilevel information systems security initiative.**

information technology: The branch of technology devoted to (a) the study and the application of knowledge regarding information and its processing and (b) the development and the use of the hardware, software, firmware, and procedures associated with the processing.

information theory: The branch of learning devoted to the study of measures and properties of information. *See also* **communication theory.**

information transfer: 1. The moving of data from one location to another, such as from one point in a communications system to another point in the system. *Note:* For information transfer to occur, the sender and the receiver must assign the same meaning to the data transferred. **2.** The moving of messages containing user information from a source to a sink. *Note:* The information transfer rate may not be equal to the transmission modulation rate. *See also* **information bearer channel, information transfer transaction, isochronous burst transmission.**

information transfer phase: In communications, computer, data processing, and control systems, the phase (a) that is the second of the three phases of an information transfer transaction and (b) during which user information is transferred between the source user and the destination user, i.e., between the call originator and the call receiver or from the message originator data terminal equipment (DTE) to the addressee DTE. *See also* **access phase, disengagement phase, information transfer transaction.**

information transfer transaction (ITT): A coordinated sequence of user and communications system actions that (a) causes information present at a source user data terminal equipment (DTE) or end instrument to become present at a destination user DTE or end instrument and (b) consists of three consecutive phases, i.e., the access phase, the information transfer phase, and the disengagement phase. *See also* **access phase, disconnect, disconnect signal, disengagement attempt, disengagement phase, information transfer, information transfer phase.**

INFOSEC: information systems security.

infralow frequency (ILF): A frequency that lies in the range from 300 Hz (hertz) to 3 kHz, the range having the alphabetic designator ILF and the numeric designator 3. *See also* **frequency, frequency spectrum, frequency spectrum designation.** *Synonym* **ultralow frequency.**

infrared (IR): The region of the electromagnetic spectrum bounded by the long-wavelength extreme of the visible red spectrum, i.e., visible red at about 0.75 μm (microns), and the shortest microwaves, i.e., radio frequency (rf) at about 100 μm, i.e., about 0.1 mm (millimeters). *Note 1:* Infrared (IR) uses include military and commercial fiber optic systems at 1.31 μm (microns), night vision devices, missile homing devices, night photographic systems, surveillance systems, and detection devices. *Note 2:* Infrared (IR) is not visible to the human eye. Persons are cautioned against looking into the end of a connected optical fiber. The light cannot be seen, and the radiation can seriously damage the retina. *Note 4:* The frequency of infrared (IR) is lower and the wavelength longer than those of visible red. The IR region of the spectrum is often divided into three regions, (a) the near IR region, 0.75 μm (micron) to 3 μm, (b) the middle IR region, 3 μm to 30 μm, and (c) the far IR region, 30 μm to 100 μm. *See* **far infrared, middle infrared, near infrared.** *See also* **frequency, light, microwave.** *Refer to* **Fig. A-9 (49).**

infrared absorption: In a material, such as optical fiber glass, absorption that (a) is caused by the motion of thermally excited ionic bonds, i.e., electric dipoles, that interact with the electric and magnetic fields of propagating lightwaves, resulting in a transfer of energy from the fields to the structure, and (b) is particularly significant at wavelengths greater than 1.5 μm (microns), which is not in the visible region. *See also* **absorption, bond frequency, infrared.**

infrared angle tracker: In an airborne infrared (IR) receiver, a device that gives (a) the look angle to the IR radiation source, such as an aircraft, a missile, a missile launch, a fire, a factory, an electric power generating plant, or the Sun, (b) the look angle change rate, i.e., the angular velocity, and (c) the radiation intensity of the received infrared radiation. *See also* **infrared, infrared point source flash detector, look angle.**

infrared band: The band of electromagnetic wavelengths that (a) lies between the extreme end of the visible part of the frequency spectrum, i.e., about 0.75 µm (microns), and the shortest microwaves, about 1000 µm, and (b) may be divided into three regions, (i) the near infrared, about 0.75 µm to 3 µm, (ii) the middle infrared, about 3 µm to 30 µm, and (iii) the far infrared, about 30 µm to 1000 µm. *See also* **frequency spectrum designation, infrared.**

infrared detector: A device that (a) detects infrared (IR) radiation, (b) indicates the intensity of the detected IR radiation, and (c) may create a visible image of the source of the IR radiation. *Synonym* **infrared sensor.** *See* **background-limited infrared detector.** *See also* **infrared, infrared band, infrared viewer.**

infrared detector dark resistance: The electrical resistance of the photoconductive infrared (IR)-sensitive material of an IR detector when no IR radiation is incident upon it. *See also* **infrared, infrared detector.**

infrared device: A device in which radiation in the infrared region of the electromagnetic spectrum is used. *Note:* Infrared (IR) radiation has a wavelength slightly longer and a frequency slightly lower than radiation in the visible region of the electromagnetic spectrum. *See* **active infrared device, passive infrared device.** *See also* **infrared.**

infrared jamming electronic countercountermeasure: A measure that reduces the effectiveness of active and passive infrared (IR) devices by the use of IR antijamming techniques, such as by (a) changing wavelengths, (b) increasing scanning rates to produce higher modulation frequencies, (c) using frequency modulation (FM) rather than amplitude modulation (AM), (d) reducing bandwidths, and (e) using squelch and multifrequency systems. *See also* **infrared, infrared detector.**

infrared point source flash detector: In an infrared (IR) receiver that accepts and displays IR energy, a detector warning device that gives (a) an indication of a point source of IR radiation, such as an aircraft, missile launch, or fire and (b) the azimuth and elevation angles at which the source of IR lies with respect to the location of the detector. *See also* **infrared angle tracker.**

infrared radiation (IR): Electromagnetic radiation that lies in the infrared (IR) band. *See also* **infrared, infrared band.**

infrared radiation suppression: The reduction of infrared radiation intensity by means of shielding, baffling, temperature reduction of operational devices, use of nonthermal devices, i.e., cool devices, cooling, and flame reduction.

infrared saturation jamming: Infrared (IR) jamming in which a high intensity IR source is used to (a) drive the guidance and control system of the object being jammed into the nonlinear portion of its operating characteristic, (b) deprive the object of effective guidance information, i.e., useful guidance information, and (c) "blind" the object with high intensity IR radiation. *See also* **infrared, infrared band, infrared detector.**

infrared sensor: 1. *Synonym* **infrared detector. 2.** *See* **photoconductive infrared sensor.**

infrared spin-scan radiometer: *See* **fiber optic visible/infrared spin-scan radiometer.**

infrared transmission: The transmission of messages or data using infrared (IR) radiation, i.e., radiation with a wavelength just above and a frequency just below that of the visible region of the electromagnetic spectrum. *See also* **infrared, infrared detector.**

infrared viewer: A device for viewing infrared (IR) radiation, particularly from the end of an optical fiber carrying infrared radiation from a laser source, which would be invisible without the viewer. *Note:* Direct viewing of infrared radiation, such as from the end of an optical fiber or a laser, can cause damage to the retina. *See also* **infrared, infrared band, infrared detector, ultraviolet viewer.**

inherent transparency: Transparency in which special transmission control characters are not needed. *See also* **transparency.**

inhibiting signal: A signal that prevents the occurrence of an event. *Note:* An inhibiting signal may be used to (a) disable an AND gate, thus preventing any signals from passing through it as long as the inhibiting signal is present, (b) prevent transmission on a line, and (c) prevent a ringing signal from reaching a telephone that is off-hook.

inhomogeneous optical fiber: 1. An optical fiber with a refractive index that is not constant throughout the fiber with respect to any spatial coordinates. **2.** An optical fiber with a refractive index profile that is not constant throughout the length of the fiber. *See also* **optical fiber, refractive index, refractive index profile.**

initial condition mode: In an analog computer, the operating mode during which the integrators are inoperative and the initial conditions are set or being set. *Synonym* **reset mode.**

initial warning radar: A radar that (a) is used to obtain the first indication of the existence of an object, i.e., a target, (b) usually is a long range radar that also may be an early warning radar, and (c) is not an initial warning radar if the warning is not in time to permit a preplanned effective reaction. *See also* **early warning radar, radar.**

initialize: To set a system or a device to a specified starting state or condition. *Note:* Examples of initialize are to (a) set a counter to zero, (b) set a bank of switches to specific on or off positions, (c) store zeros or other specific starting data in a set of storage locations or registers, and (d) set starting addresses in a computer program. *See also* **restore.**

initiative: *See* **multilevel information systems security initiative.**

injection fiber: *Synonym* **launching fiber.**

injection laser: *Synonym* **semiconductor laser.**

injection laser diode (ILD): 1. A laser diode that (a) uses a forward-biased semiconductor junction as the active medium to generate coherent light, (b) produces stimulated emissions of coherent light that are generated at a p-n junction, (c) operates as a laser producing a monochromatic light modulated by insertion of carriers across a p-n junction of semiconductor materials, and (d) has the narrower spatial and spectral width emission characteristic needed for the higher data signaling rate (DSR) systems than the light-emitting diodes (LEDs). *Note:* LEDs are useful for the large diameter, large numerical aperture optical fibers that operate at lower frequencies. *See also* **active laser medium, chirping, laser, laser diode, spectral width, superradiance.** *Refer to* **Fig. I-2.**

injection locked laser: A laser that has a peak spectral wavelength, i.e., a peak intensity emission wavelength, controlled by the injection of a separate optical signal from a different source or from an optical signal reflected from an external mirror.

injection logic: *See* **integrated injection logic.**

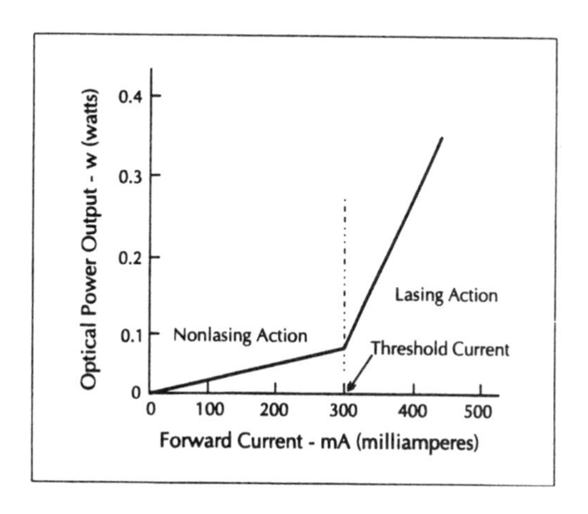

Fig. I-2. Output optical power versus forward input current for a typical **injection laser diode** light source.

injection optical fiber: *Synonym* **launching optical fiber.**

ink character recognition: *See* **magnetic ink character recognition.**

inked ribbon: A ribbon that (a) usually may be used repeatedly, (b) usually is made of cloth, and (c) is impregnated with an ink that will adhere to paper when pressed against the paper by the type of an impact printer to form characters on the paper. *See also* **carbon ribbon.**

inking: In display systems, the drawing of a line from one display position to another in the display space on the display surface of a display device, leaving a trail behind the display writer, in the manner of a pen drawing a line on paper. *Note:* Inking may be accomplished by manual control or by automatically directed control of the display writer. *See also* **display device, display position display space, display surface, display system, display writer.**

ink jet printer: 1. A printer with a printing mechanism that prints characters by deflecting droplets of ink on a record medium. **2.** A nonimpact printer that prints characters by projecting, i.e., by squirting, a jet of ink on a record medium. *See* **impact printer, record medium.**

ink jet recording: Recording in which ink is deposited directly upon the record medium. *See also* **facsimile, record medium.**

ink vapor recording: Recording in which vaporized ink particles are deposited directly upon the record medium. *See also* facsimile, record medium.

inline data processing: Executing data processing jobs in the same sequence in which they arrive at a data processing system, i.e., on a first in, first out (FIFO) basis, without their being sorted or prioritized at or by the system. *See also* **job, remote job entry.**

inline package: *See* **dual inline package.**

inline package switch: *See* **dual inline package switch.**

inline recovery: In communications, computer, data processing, and control systems, the immediate resumption of a sequential process by the system after an interruption without aborting the process or resorting to the execution of a different process. *Note:* An example of inline recovery is error recovery in which the affected process, such as the execution of a computer program, is resumed from a point in the program ahead of, i.e., prior to, the point at which the error occurred. *Synonym* **online recovery.**

Inmarsat System: International Maritime Satellite System.

innovative real time antenna modeling system (IRAMS): A programmed modeling tool that (a) simulates, in real time, a variety of antenna types, including the effects of changing aperture shapes, illumination, and electronic steering, and (b) is designed to meet the requirements of typical battlefield air-to-air or air-to-ground engagements.

inphase component life: In a system, the life of a component that can be placed into the maintenance or service cycle of the system in which an identical component is a part. *See also* **out-of-phase component life.**

input: 1. In a device, process, or channel, a point that accepts data. **2.** In a communications, computer, data processing, and control system, (a) pertaining to a device, process, point, or channel through which data may be entered into the system, (b) the data that are entered, or (c) the act of entering the data. *Note:* The data may be any representation of information, such as signals, bit strings, pulse strings, impulses, and modulated waves. **3.** The voltage, current, or power that is entered into a circuit or device. **4.** An access port, such as terminals and contacts, that is used to enter signals into a device or a system. **5.** In a computer, the data that are transferred into the computer or into its storage unit. **6.** Any data that are to be processed or operated upon. *Note:* Examples of inputs are (a) the operands in an arithmetic or logical operation, such as a Boolean operation, and (b) the augend and addend entered into an adder. **7.** The state, or the sequence of states, or the signals that occur or are made to occur on a specific channel. **8.** A device, or set of devices, that is used for inserting data into another device or set of devices. **9.** The state of the terminals at which signals are entered into a combinational circuit. **10.** The data obtained by a transmitter for transmission or the data received by a receiver. *Note:* When devices are in tandem, the output of one is the input to the next one in succession. **11.** A state, or a sequence of states, of a point that accepts data. **12.** A stimulus, such as data, a signal, or interference, that enters a functional unit, such as a communications system, a computer, or a computer program. *See* **maximum receiver input.** *See also* **combinational circuit, input data, output.**

input aperture: In an optical system, the aperture that (a) is at the point of entrance to an optical system and (b) usually is the aperture of the objective. *See also* **aperture, entrance pupil, objective, optical system, output aperture.**

input area: The set of storage locations reserved or used for input data. *See also* **input data, storage location.**

input data: 1. Data that are received or are to be received by a device or a computer program. **2.** Data to be processed.

input device: *Synonym* **input unit.**

input impedance: The ratio of the voltage to the current at the input terminals of a device or a circuit, such as an antenna, a transmission line, or an amplifier, when all sources of voltage inside the device or the circuit are shorted or nonexistent. *See also* **impedance.**

input noise temperature: *See* **effective input noise temperature.**

input-output channel: In a computer, a channel that handles the transfer of data between internal memory and peripheral equipment. *Synonym* **I/O channel.**

input-output controller (IOC): A functional unit that controls one or more input-output channels. *Synonym* **I/O controller.**

input-output device: *Synonym* **input-output unit.**

input-output (I-O, I/O) unit: In communications, computer, data processing, and control systems, equipment that enters data into, or extracts data from, a system, such as a terminal, channel, port, computer, storage unit, buffer, communications network, or front end processing system. *Note:* The data may be in the form of signals, pulses, impulses, or significant conditions. *Synonyms* **input-output device, I/O unit, I-O unit.** *See also* **terminal.**

input power: *See* **optical receiver maximum input power.**

input program: A computer program that organizes and controls an input process. *See also* **computer program.**

input protection: For analog input channels, protection against overvoltages that may be applied between any two input connectors or between any input connector and ground.

input station: *See* **data input station.**

input unit: A functional unit that can introduce data into a system, such as a communications, computer, data processing, or control system. *Note:* Examples of input units are facsimile transmitters, telephone set transmitters, keyboards, mice, styli, pointers, telegraph transmitters, and microphones. *Synonym* **input device.** *See also* **functional unit, input-output unit, output unit.**

insert: In a fiber optic connector, the portion of the connector that contains the precision alignment components. *See* **aerial insert, drop and insert.**

insertion: *See* **zero bit insertion.**

insertion gain (IG): The power gain that results from the insertion of an active device, such as an amplifier or a repeater, into a system, such as a transmission line, i.e., the ratio of (a) the power delivered to the insertion point in the system following the device to (b) the power delivered to that same point before insertion. *Note 1:* The insertion gain usually is expressed in dB. *Note 2:* The insertion gain expressed in dB is positive, i.e., the ratio is greater than unity. If not, an insertion loss is indicated. *Note 3:* For there to be a positive insertion gain, the inserted device must be an active device, i.e., a device that provides power. *See also* **active device, dB, gain, insertion loss.**

insertion loss (IL): 1. The power loss that results from the insertion of a device, such as a delay line, connector, or coupler, into a system, such as a transmission line, i.e., the ratio of (a) the power delivered to the insertion point in the system following the device to (b) the power delivered to that same point before insertion. *Note 1:* The insertion loss usually is expressed in dB. *Note 2:* The insertion loss expressed in dB is negative, i.e., the ratio is less than unity. If not, an insertion gain is indicated. *Note 3:* Insertion loss usually occurs from the insertion of a passive device because it consumes power. **2.** In a fiber optic system, the total optical power loss caused by insertion of an optical component, such as a connector, splice, or coupler. *Note:* In fiber optic systems, insertion losses may be attributed to many causes, such as absorption, scattering, diffusion, leaky waves, dispersion, microbends, macrobends, reflection, and lateral radiation. **3.** At the entrance to a waveguide in optical transmission systems, the optical power that is lost for any reason, such as Fresnel reflection, packing fraction, limited numerical aperture, axial misalignment, lateral offset, longitudinal displacement, initial scattering, and diffusion. **4.** In a fiber optic transmitter for an optical link, the radiant power lost at the point where the light source output is coupled into the optical fiber, i.e., is inserted into the fiber optic pigtail. *See also* **launch loss, transmission loss.** *See also* **dB, gain, insertion gain.**

insertion loss versus frequency characteristic: 1. A plot of the insertion loss as a function of frequency. **2.** The amplitude transfer function characteristic of a system or a component as a function of frequency. *Note:* The insertion loss versus frequency characteristic may be plotted as the amplitude versus frequency response, which may be (a) the absolute gain, loss, amplification, or attenuation versus frequency or (b) the ratio of (i) any one of these quantities at particular frequencies to (ii) their value at a specified reference frequency, i.e., the response curve may be normalized. *Synonym* **amplitude frequency response.** *See also* **amplitude distortion, frequency, transfer function.**

insertion sort: A sort in which each item in a set is placed one at a time into its proper position in the sorted set according to the sort criterion. *See also* **sort criterion.**

inside call: *Synonym* **internal call.**

inside plant: 1. All the equipment installed in a communications facility, including the main distribution frame (MDF) and all the equipment extending inward therefrom, such as central office equipment, cabling, teletypewriters, MDF protectors, and grounding systems. **2.** In radio and radar systems, all communications-electronics (C-E) equipment that (a) is installed in buildings and (b) is not located outside of buildings, such as on poles, laid on the ground, buried, or otherwise exposed to the weather. **3.** In wire, cable, and fiber optic systems, fixed ground communications-electronics (C-E) equipment that extends inward from the perimeter of a building or from the main distribution frame (MDF). *Note:* Examples of equipment that composes the inside plant are central office equipment, switching equipment, teletypewriters, switchboards, display devices, and data terminal equipment, all inside a building. Wire cables, buried, on the ground, or on poles, and equipment at user premises compose the outside plant. *See also* **facility, on premises wiring, outside plant.**

inside vapor deposition (IVD): In the production of optical fiber preforms, the deposition of dopants on the inside of a rotating glass tube, called a "bait tube," which is then sintered to produce a doped layer of higher-refractive-index glass on the inside. *Note:* A solid optical fiber is then pulled from the tube as part of the fiber drawing process. *See also* **bait tube, dopant, optical fiber, preform.**

inside vapor phase oxidation (IVPO) process: In the production of optical fibers, a chemical vapor phase oxidation (CVPO) process in which (a) dopants are burned with dry oxygen mixed with a gas fuel to form an oxide stream, i.e., a soot stream, (b) the soot stream is deposited on the inside of a rotating glass tube, called a "bait tube," (c) the bait tube is then sintered to produce a doped layer of higher-refractive-index glass on the inside for forming the core of the optical fiber, and (d) a solid optical fiber is then pulled from the tube as part of the fiber drawing process. *See also* **bait tube, dopant, chemical vapor phase oxidation process, modified inside vapor phase oxidation process, optical fiber, outside vapor phase oxidation process.**

inside wire: *Synonym* **on-premises wiring.**

in-slot signaling: Signaling performed within the associated channel time slot. *See also* **bit robbing, channel time slot, out-slot signaling, signaling, signaling time slot.**

inspection: *See* **interface inspection, TEMPEST inspection.**

inspection lot: A selected group of produced units wherein (a) a controlled sample is to be drawn and inspected to determine conformance with acceptability criteria, (b) the controlled sample usually is taken at random from the units produced in a given production run in which all units were produced under the same conditions, (c) the number of units in the selected group of produced units and the number of units to be randomly selected from the group for testing depend on the number of units being acquired or purchased, and (d) the selected group usually differs from a collection of units designated as a group for other purposes, such as for production, storage, packaging, and shipment.

instability: *See* **frequency instability, phase instability, time instability.**

install: In communications, computer, data processing, and control systems, to record a computer program in a storage unit of a system so that the program can be called and used by the system. *Note:* In personal computers, application programs usually are stored on a hard disk ready to be called when needed. *See also* **application program, call, computer program.**

installation: In communications, computer, data processing, and control systems, an assembly of equipment that can perform specific functions. *See* **customer installation, radio installation.**

instant: A point in time, not necessarily with reference to a specific time scale. *Note:* An instant has no duration and might be considered as a boundary between past time and future time. *See* **decision instant, significant instant.**

instantaneous automatic gain control (IAGC): Automatic gain control of an amplifier in which the response time is equal to or less than the received pulse duration, i.e., the circuit is fast enough to act during the time that a pulse is passing through the amplifier. *Note:* In radar systems, instantaneous automatic gain control (IAGC) can be used as an electronic counter-countermeasure (ECCM) against long-pulse interference or jamming. IAGC is the portion of the radar system that automatically adjusts the gain of an amplifier for each pulse it receives in order to obtain a substantially constant output pulse peak amplitude with different input pulse peak amplitudes. *See also* **amplifier, amplitude, automatic gain control, electronic countercountermeasure, input, interference, jamming, radar, response time.**

instantaneous companding: A process that reduces the quantizing noise on pulse code modulated (PCM) signals by passing the analog signals through a compressor before quantizing. *Note:* At the receiver, the reverse process must be applied by using an expander after the decoding operation. *See also* **compressor, expander, pulse code modulation, quantizing noise.**

instantaneous storage: *Synonym* **immediate access storage.**

instantaneous transmission rate: In the processing of data by a transmission facility, the short-term rate at which the data are transmitted, given by the ratio of (a) the number of transmitted data units, such as bits, characters, or blocks, in a short period to (b) the duration of the short period, such as 50,000 bits in 25 milliseconds, to yield an instantaneous transmission rate of 2 mbps (megabits per second). *Note:* The instantaneous transmission rate may vary from 0 to the maximum possible rate. However, it usually is higher than the effective transmission speed, i.e., the average transmission rate. *Synonym* **instantaneous transmission speed.** *See also* **effective transmission rate, efficiency factor, throughput.**

instantaneous transmission speed: *Synonym* **instantaneous transmission rate.**

instantaneous value: The magnitude of a varying quantity, such as a varying voltage, current, or power, at a particular instant. *Note:* Instantaneous values may be expressed in appropriate parameters and units, such as a power level, such as 10 mW (milliwatts); a phase angle, such as 45°; a phase time, such as 1 μs (microsecond); a frequency, such as 1.544 MHz (megahertz); or a spatial position, such as 2 μm (microns), relative to a given reference. *See also* **value.**

instant distortion: *See* **significant instant distortion.**

instruction: In a computer programming language or in a computer program, an expression that (a) specifies one or more operations and (b) identifies the applicable operands. *Note:* An example of an instruction is "add A and B," which may be written as ADD A,B, as +A,B, as A + B, or as A,B+, depending on the languages and systems used. Instructions may specify the addresses of storage locations where the operands are stored rather than the operands themselves. Instructions may specify the address where the result of the operation should be stored and may also include the address of the next instruction to be executed. Instructions may specify that nothing be done, that the computer should stop executing instructions, or that the computer control unit should jump to a specified instruction rather than execute the next instruction in the sequence of instructions. *See* **computer instruction, exit instruction, four-address instruction, immediate instruction, jump instruction, macroinstruction, multiple instruction, n-address instruction, no-op instruction, one-address instruction, signal operation instruction, standing communications operating instruction, three-address instruction, transmission instruction, two-address instruction, zero-address instruction.** *See also* **instruction word, operand, operation.** *Refer to* **Fig. P-15 (763).**

instruction address: In the execution of a computer program, the address of an instruction word, i.e., the address that the computer must use to fetch an instruction word from storage, in contrast to the address part of an instruction word, which usually specifies the address of an operand. *Note:* The conventional computer is directed by an internally stored program. The computer must usually retrieve the instruction from storage and place it in a register where it may be analyzed, in order to execute the instruction. *See also* **computer program, instruction word.**

instructional station: *See* **aviation instructional station.**

instruction code: In computer programming languages and in communications, computer, data processing, and control systems, a code that (a) is used to represent the instructions that a system is capable of executing and (b) is used to prepare a program, such as a computer program, for execution. *Note:* The program, consisting of a sequence of instructions selected from the code that the system is capable of executing, is entered into the system, stored, and executed. Each item in the code is interpreted by the system, usually converted to a machine language code, and executed. The codes usually are written in a somewhat mnemonic common language, such as COBOL, FORTRAN, ADA®, or BASIC, in which, for example, MPY might be used to indicate multiplication. *See* **machine instruction code.** *See also* **code, computer program, instruction, instruction repertory.**

instruction passing: The forwarding or relaying of an instruction over a communications system, such as from station to station, point to point, or ship to ship, in order that the instruction may arrive at its ultimate destination for action.

instruction repertory: The complete list of instructions that a computer is capable of executing. *Note:* The instructions in a computer program are selected from the instruction repertory. *Synonym* **instruction repertoire.** *See also* **computer program, instruction, instruction code.**

instruction word: In a computer program, a computer word that (a) expresses an instruction and (b) usually consists of an operation part, one or more address parts, and perhaps other characters, such as flags, for exercising various options. *See also* **address part, computer word, instruction, operation part.**

instrument: *See* **binocular instrument, end instrument, monocular instrument, terminal instrument, user end instrument.**

instrumentation signals intelligence: *See* **foreign instrumentation signals intelligence.**

instrument flight rule (IFR): At airports, a rule that governs the pilot's decisions when performing instrumented runway approach and landing, such as, for a given guidance system at a specific airport, the altitude at which the pilot must have the runway in sight in order to safely land the aircraft.

instrument landing system (ILS): 1. A radionavigation system that (a) provides aircraft with horizontal and vertical guidance just before and during landing and (b) at certain fixed points indicates the distance to the reference point of landing. **2.** A system of radio navigation that (a) is intended to assist aircraft pilots in landing, (b) provides lateral and vertical guidance, and (c) may include indications of distance from the optimum point of landing. *See also* **instrument landing system localizer, localizer.**

instrument landing system (ILS) glide path: The glide path followed by an aircraft during landing in which (a) an instrument landing system (ILS) is used to guide the aircraft, and (b) a glide path indicator and a localizer are used to indicate the vertical and horizontal deviations, respectively, from its optimum path of descent along the axis of the runway selected for landing.

instrument landing system (ILS) glide path indicator (GPI): In an instrument landing system (ILS), a vertical guidance path indicator that indicates the vertical deviation of the aircraft from its optimum path of descent along the axis of the runway selected for landing.

instrument landing system localizer: In an instrument landing system (ILS), a horizontal guidance indicator that indicates the horizontal deviation of the aircraft from its optimum path of descent along the axis of the runway selected for landing.

insulated: In electric circuits, the state that exists at a point that does not have a path to another point, except via the circuit, for other than negligible currents because of separation from other conducting points by extremely high resistance materials, such as plastic, rubber, fabric, tar, or other dielectric materials that are nonconductors. *Note:* An example of an insulated conductor is a wire wrapped in plastic or heavily coated with a plastic bonded to the wire. However, electrostatic, magnetic, electromagnetic, and magnetostatic coupling may still occur between insulated conductors in spite of their being insulated, in which case they are not isolated. *See* **isolated.**

insulation blocking: In a cable, such as a coaxial, twisted pair, or fiber optic cable, the ability of the outer covering, such as a jacket, sheath, or insulation, to withstand elevated temperatures without sticking to itself on adjacent turns or layers when coiled or wound on spools, reels, or bundpacks.

integer: Zero, one of the natural numbers, or one of the negated natural numbers, i.e., a whole number between minus and plus infinity, including 0. *Note:* Examples of integers are −6, −1, 0, 8, 19, and 544. *Synonym* **integral number.** *See also* **natural number.** *Refer to* **Fig. P-15 (763).**

integral: **1.** In the calculus, the result of the process of integration. *Note:* An integral results from the integration of an integrand. For example, the integral of $ax^2 + bx + c$ is $ax^3/3 + bx^2/2 + cx$ when integrated from 0 to x. The derivative of the integral yields the original integrand. **2.** Pertaining to a whole number, i.e., an integer. **3.** Pertaining to a component that is a permanent part of a system. *See also* **derivative, integer, integrand, integration.**

integral lens light-emitting diode (ILLED): A light-emitting diode (LED), such as a side emitting, i.e., edge emitting, high radiance double heterostructured diode, that is coupled to a multimode optical fiber via a photolithographically etched and regrown converging microlens structure on the emitting surface of the semiconductor diode. *Note:* Typical brightness is 75 $W \cdot sr^{-1} \cdot cm^{-2}$ (watts per steradian per square centimeter). *See also* **converging lens, light-emitting diode, optical fiber.**

integral number: *Synonym* **integer.**

integrand: In the calculus, the mathematical function, i.e., the operand, upon which the process of integration operates. *Note:* For example, if the integrand is $ax^2 + bx + c$, the result of integration, i.e., the integral, is $bx^3/3 + x^2/2 + cx$ when integrated from 0 to x. The derivative of the integral yields the original integrand used to obtain the integral. *See also* **derivative, integral, integration.**

integrated chip: *See* **large-scale integrated chip.**

integrated circuit (IC): An electronic circuit that consists of many individual circuit elements, such as transistors, diodes, resistors, capacitors, inductors, and other active and passive semiconductor devices, formed on a single chip of semiconducting material and mounted on a single piece of substrate material. *Synonym* **microcircuit.** *See* **application-specific integrated circuit, large-scale integrated circuit, lithium niobate optical integrated circuit, monolithic integrated circuit, optical integrated circuit.** *See also* **chip, circuit, wafer.** *Refer to* **Fig. I-3.** *Refer also to* **Figs. C-3 (128), P-14 (762), P-15 (763).**

integrated circuit filter-coupler-switch-modulator: *See* **optical integrated circuit filter-coupler-switch-modulator.**

integrated communications adapter: An adapter that provides for direct connection of multiple communications lines to a single system, such as to a computer or a signal processing unit. *See also* **adapter.**

integrated communications system (ICS): **1.** A communications system created by the consolidation of two or more existing originally autonomous communications systems into a single system with none of the original autonomy remaining, i.e., the complete inter-

Fig. I-3. Technicians performing maintenance on ion implant equipment in an advanced computer **integrated circuit** microchip manufacturing plant. (Courtesy Intel Corporation.)

connection and interoperation of the formerly separate systems. **2.** A communications system that transmits analog and digital traffic over the same switched network.

integrated database: A database from which redundant data have been removed or into which they have not been entered, i.e., a database in which files have been consolidated, correlated, and interrelated in such a manner that redundant data have been eliminated. *See also* **database.**

integrated digital network (IDN): A network in which both digital transmission and digital switching are performed. *See also* **Integrated Services Digital network.**

integrated end instrument: *Synonym* **integrated station.**

integrated fiber optic circuit: A fiber optic circuit, such as a fiber optic data link and a fiber optic bus, that is embedded, used in, or connected to, nonoptical circuits, such as wire circuits, coaxial cable circuits, microwave links, and radio links. *See also* **optical circuit, optical integrated circuit.**

integrated fiber optic communications system: A communications system or network that consists of fiber optic links connected to other types of links, such as wire and microwave links.

integrated injection logic (I^2L, IIL): An optical integrated circuit (OIC) that (a) contains injection laser diodes, waveguides, switches, gates, and related circuitry and (b) uses less power, is more stable, is cheaper, is more easily mass-produced, and is two orders of magnitude faster than transistor to transistor logic (TTL). *See also* **integrated optical circuit transistor.**

integrated management system: In personal computer (PC) and microcomputer systems, a software package that combines two or more application programs into one computer program. *Note:* An example of an integrated management system is a program that combines a word processing program, a spreadsheet program, a database manager, a graphics program, and a communications program into one program. *See also* **application program, computer program, database manager, personal computer, software package.**

integrated modem: A modem that has been embedded in, i.e., is an integral part of, the system or the device with which it operates. *See also* **modem, standalone modem.**

integrated modular avionics (IMA): The branch of science and technology devoted to the study, design, development, and application of avionic systems that (a) perform aircraft functions, such as communications, navigation, and identification (CNI), (b) are modular in construction for reconfiguration, reliability, and maintenance, and (c) may be integrated into a single system.

integrated optical spectrum analyzer: An integrated optical circuit that (a) performs electromagnetic spectral analyses by executing coherent optical signal processing operations, such as Fourier transform analyses, Fourier transform integration, and related correlation and convoluted functions, and (b) usually consists of monolithic structures of alloy compositions, such as gallium arsenide, that may be formed by liquid phase epitaxial techniques, energized molecular beam processes, or chemical vapor deposition processes. *See also* **Fourier analysis, integrated optical circuit.**

integrated optics: The branch of science and technology that (a) is devoted to the interconnection of miniature optical components via optical waveguides on transparent dielectric substrates, such as lithium niobate and titanium-doped lithium niobate, to make optical integrated circuits (IOCs), (b) includes the design, development, and operation of circuits that apply the technology of integrated electronic circuits produced by planar masking, etching, evaporation, and crystal film growth techniques to microoptic circuits on the dielectric substrate, and (c) combines electronic circuitry and optical waveguides for performing various communications, switching, and logic functions, including lightwave amplification, gating, modulation, generation, detection, filtering, multiplexing, processing, coupling, and storing. *See also* **optical integrated circuit.**

integrated receiver: *See* **monolithic integrated receiver, positive-intrinsic-negative field-effect transistor integrated receiver.**

integrated resource manager: *See* **functionally integrated resource manager.**

integrated services digital network (ISDN): A digital network in which (a) the same time division multiplexed switches and digital transmission paths are used to establish connections for different services, such as voice, video, and data services that include electronic mail and facsimile, and (b) the type of connection is often specified, such as switched connection, nonswitched connection, exchange connection, and Integrated Services Digital Network (ISDN) connection. *Note:* An example of an integrated services digital network (ISDN) is a network in which fiber optic cables are used for local loops, local area networks (LANs), metropolitan area networks (MANs), and long haul trunks between switching centers. One single-mode optical fiber in the loop to home or office could handle all forms of signals and many different user end instruments because of the wideband capability of the fiber, such as handle a group of (a) 20 analog video channels, with 100-Mbps (megabit per second) frequency shift keyed (FSK) digital data signaling with 200 MHz (megahertz) spacing between channels, a carrier/noise ratio of 16 dB, a bit error ratio (BER) of 10^{-9}, and modulation of a laser source with a 4-GHz signal at frequencies between 2 GHz (gigahertz) and 6 GHz, or (b) integrated services that can be all of (i) 50 analog broadcast-quality video channels (2 GHz bandwidth), (ii) 4 HDTV channels, frequency modulation (FM) (800 MHz), (iii) 4 switched video channels, digital (800 MHz), (iv) 25 digital audio channels (200 MHz), and (v) voice, data, and other services channels (200 MHz). The bandwidths needed for each service group sum to 4 GHz. *See also* **communications, electronic mail, integrated digital network.**

Integrated Services Digital Network: *See* **Broadband Integrated Services Digital Network.**

integrated services digital network-one (ISDN-1): A network that (a) handles simultaneous voice, video, data, and graphic communications traffic and (b) provides local and long distance service over a single copper telephone line.

integrated station: **1.** A station that provides integrated services, such as voice, video, and digital data. **2.** At a given site, a collection of terminals, one or more of which must be an integrated terminal. **3.** A user end instrument or terminal device (a) in which a telephone and one or more other devices, such as video display devices, keyboards, and printers, are combined into a single unit, (b) that is used over a single circuit, and (c) that may require more than one channel, such as several wire pairs or a fiber optic cable. *See also* **integrated system, integrated terminal.**

integrated system: **1.** A communications system that transfers analog and digital traffic over the same switched network. **2.** A system in which the hardware and software components have been effectively interconnected by means of appropriate interfaces. *See also* **communications system, embedded system, hybrid communications network, integrated station, integrated terminal, network, system integration.**

integrated terminal: A terminal that (a) provides integrated services, such as voice, video, and digital data, and (b) has the combined capability of a group of two or more different types of devices, such as (i) a telephone, video monitor, computer, and facsimile machine, (ii) a telephone and a video display unit, (iii) a telephone, keyboard, and printer, or (iv) a computer, video display unit, and a printer, each group integrated into a single functional unit and connected to a network via a circuit, such as a loop. *See also* **integrated station, integrated system.**

integrated voice/data terminal (IV/DT): *See* **integrated station.**

integrating circuit: An electrical circuit (a) that has one output and one or more inputs, and (b) in which the output is the time integral of all of the inputs. *Refer to* **Fig. I-4.**

integrating motor: A motor that has a constant ratio of (a) an output shaft rotational speed, i.e., an angular velocity, to (b) an input signal voltage amplitude, such that the total angle that the motor shaft turns with respect to a datum, i.e., an origin, is proportional to the time integral of the input signal.

integrating network: An electrical network or circuit that (a) produces an output waveform that is the time integral of the input waveform and (b) is used in signal processing, such as for producing sawtooth waves from square waves.

integration: 1. In communications, computer, data processing, and control systems, the combining of separate hardware and software components into a single functional unit or system. **2.** In the calculus, a process of summation in which an infinite number of infinitesimally small values are added to produce the whole. *Note:* An example of integration is the process, when applied to a mathematical function curve, that obtains the area that lies between the curve and an axis and between defined limits of the coordinates of that axis. The process of integration operates on a mathematical function, i.e., the integrand, to yield a result, i.e., the integral. For example, the integration of the function ax^2 between the limits of 3 and 6 yields $63a$, i.e.,

$$\int_3^6 ax^2dx = a(6^3 - 3^3)/3 = 63a$$

The integral is exact because the limits are given. The $63a$ represents the area under the curve between the limits of $x = 3$ and $x = 6$. An integrator can perform integration. *See* **radio and wire integration, system integration.** *See also* **differentiation, integrand, integral, integrator.**

integration device: *See* **radio-wire integration device.**

integration testing: An orderly progression of testing in which modules are tested and combined with others, and the combination is tested, the process then being repeated until the entire system has been integrated.

integrator: A device that performs integration. *Note:* In analog systems, an example of an integrator is a device that has an output analog variable that is the integral of an input analog variable with respect to time or another input variable. *See* **frequency modulation integrator, summing integrator.** *See also* **differentiator, integral, integrand, integrating circuit, integration.**

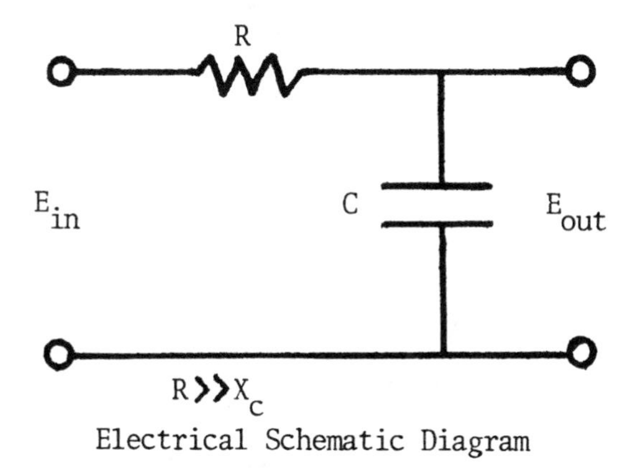

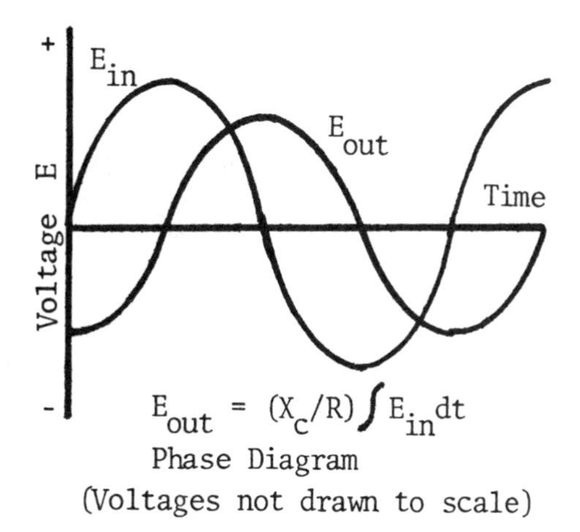

Fig. I-4. In the **integrating circuit,** the phase angle of the output voltage lags the phase angle of the input voltage by nearly 90° (electrical).

integrator feedback: *See* **video integrator feedback.**

integrity: In communications, computer, data processing, and control systems, the extent to which the validity, accuracy, completeness, precision, and other desirable qualities of data and software are maintained. *Note:* One way to ensure the integrity of a computer program is to store it in a read-only memory (ROM). Another way to preserve integrity is to allow data to be entered into a database from many authorized sources, but to permit only those persons responsible for maintaining the database to change or delete data. *See* **bit count integrity, bit integrity, character count integrity, character integrity, data integrity, service integrity, system integrity.** *See also* **confinement, protection.**

intelligence: 1. Information that (a) is gathered from various sources and (b) is of use, confirmed, processed, and made available on a timely basis. **2.** In military systems and operations, information that (a) is derived from input information from one or more sources, (b) usually is confirmed by multiple independent sources, (c) is sorted, organized, and processed for use by military and other interested units, and (d) is disseminated to those units that have a need for the information. *Note:* Examples of intelligence are (a) tactical or military intelligence, (b) order of battle, (c) naval intelligence, (d) air intelligence, (e) technical intelligence, and (f) strategic intelligence. *See* **artificial intelligence, closed source intelligence, communications intelligence, electronic intelligence, electronics intelligence, electrooptical intelligence, foreign instrumentation signals intelligence, photographic intelligence, signal intelligence, technical intelligence.**

intelligence communications: Communications methods, systems, and equipment that are used for the transmission of messages that contain foreign military or security information.

intelligent display device: 1. In display systems, a display device that contains a computer, a processor, or another arithmetic or logic unit that is capable of performing certain local operations within itself, such as to generate characters, move display elements, display groups, or display images by shifting device coordinates, storing data in a buffer, controlling scrolling, or controlling the scissoring of data in its buffer storage. **2.** A display device that (a) is directly supported by a computer that is dedicated to, and usually a part of, the display device it supports, and (b) performs certain operations locally.

intelligent network (IN): A network that (a) allows functional units to be flexibly distributed at a variety of nodes on and off the network and (b) allows the network architecture to be modified to control the network-provided services. *Note 1:* Intelligent networks (INs) are envisioned to provide advanced services, such as (a) distributed call processing across multiple network modules, (b) real time authorization code verification, (c) one-number service, and (d) flexible private network services, including (i) reconfiguration by the customer, (ii) traffic analyses, (iii) service restrictions, (iv) routing control, and (v) data on call histories. Levels of proposed intelligent network (IN) development are:

> **IN/1:** An intelligent network that provides services that (a) allow increased customer control and (b) can be provided by centralized switching equipment that serves a large customer base.
>
> **IN/1+:** An intelligent network that provides services using centralized switching equipment, such as access tandems that serve a large customer base.
>
> **IN/2:** An advanced intelligent network that extends the distributed IN/1 architecture to allow service independence. Traditionally, service logic has been localized at individual switching centers. IN/2 architecture provides flexibility in the placement of service logic, using advanced techniques to manage the distribution of both network data logic and network service logic across multiple IN/2 modules.

Note 2: Intelligent networks (INs) are envisioned for North America. *See* **advanced intelligent network.**

intelligent peripheral (IP): 1. A functional unit that (a) has advanced performance features, (b) is capable of performing logic functions, and (c) may be used most efficiently when locally accessed. **2.** An intelligent network (IN) feature that provides specialized telecommunications capabilities required by IN/2 service logic programs. *See also* **intelligent network.**

intelligent terminal: 1. In computer, communications, data processing, and control systems, a terminal that can execute (a) specified sets of instructions or procedures or (b) certain fixed or variable programs without intervention by the user during execution of the instructions, procedures or programs. **2.** In computer systems, a programmable terminal. *Synonyms* **programmable terminal, smart terminal.**

intelligent vehicle highway system (IVHS): A highway traffic management system that embraces various aspects of highway traffic management, such as automatic vehicle identification, in-vehicle identification, in-vehicle navigation, and traffic management communications systems.

intelligibility: The capability of being understood. *Note:* Intelligibility does not imply the recognition of a particular voice. *See* **voice intelligibility.** *See also* **recognition, signal to noise ratio.**

intelligible crosstalk: 1. Crosstalk that gives rise to intelligible signals. **2.** Crosstalk from which information can be derived. **3.** Crosstalk that consists of signals or sounds that can be interpreted by a person or a machine. *See also* **crosstalk, far-end crosstalk, interference, near-end crosstalk, unintelligible crosstalk.**

intensifier: *See* **image intensifier.**

intensity: In an electromagnetic wave, such as a radio wave, radar wave, microwave, or lightwave, the square of the electric field strength. *Note:* Intensity is proportional to irradiance. It is equal to irradiance only when the constant of proportionality is factored in. "Intensity" should not be used in place of the term "irradiance," unless only ratios are considered so that the constant of proportionality drops out. *See* **field intensity, luminous intensity, mean spherical intensity, peak radiant intensity, radiant intensity, traffic intensity.** *See also* **field strength, irradiance, radiometry.**

intensity modulation (IM): 1. In optical communications systems, modulation in which the optical power output of a source is varied in accordance with some characteristic of the modulating signal, i.e., the information-bearing signal. *Note:* In intensity modulation, there are no discrete upper and lower sidebands in the usually understood sense of these terms. However, the output of a light source does have a spectral width. The envelope of the modulated optical signal is an analog of the modulating signal in that the instantaneous power of the envelope is an analog of the characteristic of interest in the modulating signal. Recovery of the modulating signal is by direct detection rather than by heterodyning as in radio receivers. **2.** Variation of the power level of an optical carrier in accordance with a modulating signal. *Note 1:* Intensity modulation (IM) can be achieved in several ways, such as (a) direct modulation of a light source and (b) modulation of the optical carrier after launching. *Note 2:* Intensity modulation of an optical carrier is similar to amplitude modulation of a radio frequency (rf) carrier. *See also* **amplitude modulation, fiber optics, modulation.**

interaction crosstalk: Crosstalk caused by coupling between carrier and noncarrier circuits. *Note:* If the interaction crosstalk is, in turn, coupled to another carrier circuit, the crosstalk is called "indexing crosstalk." *See also* **coupling, crosstalk, indexing crosstalk.**

interactive: 1. Pertaining to an application or a mode of operation of a system in which each entry by an operator brings forth a response from the system, and usually vice versa. **2.** Pertaining to an application or mode of operation of a system in which entries and responses between the operator and the system occur in the same manner as in a dialogue, or conversation, between two people, i.e., pertaining to a conversational mode. *Note:* Personal computers (PCs) usually are operated in a conversational mode, consisting of declarative, interrogative, and imperative statements on the part of the PC and the operator. An airline reservation system usually is operated in an interactive mode because the operator either (a) enters an inquiry for seats available on a given flight and needs an immediate answer in order to sell a ticket or make a reservation or (b) enters data concerning seats sold or canceled so the database can be updated immediately. *See also* **conversational, conversational mode, noninteractive.**

interactive data transaction (IDT): A message that (a) is transmitted via a data channel and (b) requires a reply in order for communications and related work to proceed. *See also* **data transmission.**

interactive display system: A display system that (a) can be used in a conversational mode, i.e., a user can enter display commands or inquiries, and the system can respond as in a conversation by changes in display elements, display groups, or display images in the display space on the display surface of the display device that is a part of the system, and (b) usually is supported by a computer or data processing system, computer programs, a menu, display commands, and a display device, such as a cathode ray tube (CRT), fiberscope, gas panel, liquid crystal display panel, or light-emitting diode (LED) panel. *See also* **conversational mode.**

interactive graphics: The branch of science and technology devoted to the study and the application of display devices, their components, and associated equipment used in an interactive mode, i.e., in a mode similar to a conversation between two persons. *See also* **display device, interactive mode, passive graphics.**

interactive mode: In communications, computer, data processing, and control systems, a manner of operating a system in such a way that a sequence of alternating entries and responses takes place between the system and a user similar to a dialogue between two persons, usually so that the user can alter or interact with the system via an input-output unit, such as a display system that is supported by a computer or data processing system with a cursor, light pen, stylus, keyboard, or mouse, as well as computer programs, a menu, display commands, and a display device, such as a cathode ray tube (CRT), fiberscope, gas panel, liquid crystal display panel, or light-emitting diode (LED) panel. *See also* **conversational mode.**

interactive service (IS): In integrated services digital networks (ISDNs), a telecommunications service that provides for the bidirectional exchange of information among users or among users and hosts. *Note:* Interactive services are grouped into conversational services, message services, and retrieval services. *See also* **duplex operation.**

interactive television (IATV): Television that (a) allows the viewer to influence the television display using a hand-held remote control device and (b) that has a programming box in the television set to control the display. *Note:* Examples of interactive television (IATV) include viewer participation in game shows and controlling actions on the screen similar to video games.

interblock gap: An area on a data medium used to indicate the end of a block or a physical record. *Note:* Examples of interblock gaps are the gaps between blocks on magnetic tapes and disks. *See also* **band, block, data medium, magnetic tape.**

intercept: 1. In a telephone network, (a) to stop a telephone call directed to an improper, disconnected, or restricted telephone number and (b) to redirect that call to an attendant or to a recording. **2.** To gain possession of messages (a) usually without the knowledge or consent of the originator or the addressee, (b) usually without delaying or preventing the transmission, and (c) with or without authorization. **3.** The acquisition of a transmitted signal with the intent of delaying or preventing that signal from reaching the intended destination user. *See* **communications intercept, willful intercept.** *See also* **electronics intelligence, electronic warfare, monitoring.**

intercept communications method: Communications in which a station transmits a message to another station that repeats the message, thus transmitting it to a third station that (a) is the addressee station, (b) has no need to transmit the message, (c) may not transmit at all, (d) cannot receive messages directly from the station that originated the message, or (e) may transmit the message to yet another station. *See also* **broadcast communications method, intercepted terminal, receipt communications method, relay communications method.**

intercepted terminal: A terminal that has messages destined for it intercepted because it cannot accept messages. *See also* **intercept communications method, terminal.**

intercepting: 1. In communications system operations, the deliberate reception by a terminal or station of a call or a message that was intended for another terminal or station. **2.** The routing of a call or a message to another terminal or station when the message cannot be received by the terminal or the station for which it was intended, such as when the end instrument has been disconnected or is out of order, i.e., is in the outage state. *See also* **end instrument, outage state, station, terminal.**

intercepting trunk: A trunk to which a call to a specific call receiver is connected by an attendant when the call cannot be completed, such as when the call is to a vacant, i.e., unused, number, to a changed number, to an out of order line, or to a disconnected end instrument. *See also* **call, call receiver, completed call, end instrument, intercepting, trunk.**

interception: *See* **airborne interception, changed address interception, low probability of interception.**

intercept point: *See* **third-order intercept point.**

intercept radar: *See* **airborne intercept radar, track-while-scan airborne intercept radar.**

intercept receiver: A receiver that (a) is designed to detect electromagnetic waves, (b) provides visual or aural indication of the received emissions that occur within the particular portion of the electromagnetic spectrum to which it is tuned, (c) may be of fixed or variable frequency, and (d) has a frequency band that is narrow compared with a panoramic receiver. *See also* **panoramic receiver.**

intercept tape storage: 1. In communications, computer, data processing, and control systems, temporary storage on punched paper or magnetic tape for en route message traffic that is destined for nonoperating channels or for backlogged channels, i.e., for saturated circuits. **2.** In the Automatic Digital Network (AUTODIN), the providing of temporary storage on magnetic tape for message traffic destined for nonoperating or backlogged channels. *See also* **Automatic Digital Network.**

interchange: *See* **electronic data interchange.**

interchangeability: A condition that exists when two or more items possess functional and physical characteristics such that (a) they are equivalent in performance and durability, (b) they are capable of being exchanged one for the other without alteration of the items themselves or of adjoining items, except for adjustment, and (c) they need not be selected for fit and performance. *See also* **commonality, compatibility, interoperability.**

interchangeable connector: A connector that (a) shares common installation geometry with and (b) has the same transmission capabilities as another connector.

interchange circuit: 1. A circuit that (a) is between data terminal equipment (DTE) and data circuit-terminating equipment (DCE) for the purpose of exchanging data and signaling information, (b) may transmit many types of signals, such as control signals and timing signals, and (c) performs many types of functions, such as common return and circuit isolation functions. **2.** A circuit that provides for the interchange of traffic between an element of one communications system and an element of another system. *Note:* Examples of interchange circuits are (a) a circuit between military and nonmilitary communications systems, (b) a circuit between a wire communications system and a fiber optic communications system, and (c) a circuit between wire and fiber optic portions of the same system. *See also* **circuit, data circuit-terminating equipment, data terminal equipment.**

interchange code: *See* **extended binary coded decimal interchange code.**

intercharacter interval: In asynchronous transmission, the time interval that (a) is between the end of the stop signal of one character and the beginning of the start signal of the next character, (b) may be of any duration, and (c) has a signal sense that is always the same as the sense of the stop element, i.e., both are always a 1 or a mark signal, or both are always a 0 or a space signal. *See also* **asynchronous transmission, character, character interval.**

interco: In the international signal code, the proword that is used to precede, announce, or introduce the use of the code. *See also* **international signal code, proword.**

intercom: 1. Audio equipment that (a) allows a closed user group of persons to converse directly with one another without resorting to the use of an outside telephone network, and (b) serves members of a group that (i) are not within direct audio range and (ii) belong to a group that have something in common, such as belonging to the same organization, office, or crew. **2.** A distributed microphone and speaker system that enables personnel to talk to each other within an aircraft, ship, or office. **3.** A dedicated voice communications service in a specified user environment. *Synonyms* **intercommunications equipment, interphone.**

intercommunications equipment: *Synonym* **intercom.**

interconnect facility: In a communications network, one or more communications links that (a) are used to provide local area communications service among several locations, (b) usually form a node in the network, (c) may include network control and administrative circuits as well as the primary traffic circuits, (d) may use any medium available, and (e) may be redundant. *See also* **facility, interconnection, link, local area network, network.**

interconnect feature: *See* **fiber optic cable interconnect feature.**

interconnection: 1. The linking together of interoperable systems. **2.** The linkage used to join two or more communications functional units, such as systems, networks, links, nodes, and circuits. *See* **comparably efficient interconnection, fiber optic interconnection, Open Systems Interconnection.** *See also* **interconnect facility, link.** *Refer to* **Fig. L-4 (506).**

interconnection box: *See* **fiber optic interconnection box.**

interconnection protocol security option: *See* **revised interconnection protocol security option.**

interdiction: In communications system security, the denying of the use of system resources to a user or a group of users.

interest: *See* **community of interest.**

interexchange carrier (IXC): Any service provider offering inter-LATA telecommunications services. *See also* **local access and transport area.**

interface (I/F): 1. In a system, a shared boundary. *Note:* An example of an interface is the boundary between two subsystems or two devices. **2.** A shared boundary between two functional units, defined by specific attributes, such as functional characteristics, common physical interconnection characteristics, and signal characteristics. **3.** The specification of the connection of two devices having different functions. *Note:* The interface description usually includes (a) the type, quantity, and function of the interconnecting circuits, (b) the type, form, and content of signals to be interchanged, and (c) mechanical details of plugs, sockets, and pin numbers. **4.** A point of communications between two or more processes, persons, or other physical entities. **5.** A boundary or a point common to two or more similar or dissimilar command and control systems, subsystems, or other entities through which or at which data flow takes place. **6.** The point of interconnection between user terminal equipment and commercial communications service facilities, i.e., the point between the data terminal equipment (DTE) and the data circuit-terminating equipment (DTE) on the customer premises. **7.** To interrelate two or more entities at a common point or a shared boundary. **8.** To interrelate two or more dissimilar circuits or systems. **9.** In layered systems, the set of protocols that is applicable to the interaction between any two layers within a single system, such as a communications, computer, data processing, or control system. **10.** A connection that is needed to complete an operation. *See* **access unit interface, application program interface, basic rate interface, Basic Rate Interface, data transmission interface, electrical interface, Electronic Industries Association interface, framed interface, gateway interface, high level digital interface, host computer interface, multimedia interface, net radio interface, network interface, network terminating interface, optical interface, primary rate interface, radio-wire interface, R interface, T interface, transparent interface, user interface, user-system interface, virtual device interface.** *See also* **bridge, brouter, data circuit-terminating equipment, data terminal equipment, front-end processor, gateway, interface message processor, medium interface connector, medium interface point, network interface device, optical junction, radio and wire integration device, router.** *Refer to* **Figs. D-1 (203), E-4 (314), L-8 (536), L-9 (536), P-15 (763), R-10 (836), S-18 (951).**

interface bus: In a system, such as a computer, data processing, or control system, a bus that interconnects various components, such as microprocessors, storage units, arithmetic and logic units, and input-output registers. *Note:* In a personal computer (PC), the interface bus serves as the data path and attaches the random access memory (RAM), read-only memory (ROM), keyboard, display monitor, disk storage, printer, and any other devices that are connected to the PC, including all circuit boards and modems, to one another so that they may all be controlled by the program in the microprocessor.

interface capability: *See* **direct interface capability.**

interface connector: *See* **medium-interface connector.**

interface device: *See* **network interface device.**

interface equipment: *See* **process interface equipment.**

interface for computer environments: *See* **portable operating system interface for computer environments.**

interface functionality: In telephony, the characteristic of interfaces (a) that allows them to support transmission, switching, and signaling functions that are identical to those used in the enhanced services provided by the carrier, and (b) in which, as part of the carrier comparably efficient interconnection (CEI) offering, the carrier must make available standardized hardware and software interfaces that are able to support transmission, switching, and signaling functions identical to those used in the enhanced services provided by the carrier.

interface inspection: Inspection conducted to ensure that an interface is constructed in accordance with interface specifications.

interface message processor (IMP): In packet-switched networks, a processor-controlled switch used to route packets to their proper destination. *See also* **interface, node, packet switching, switch.**

interface payload: In integrated services digital networks (ISDNs), the part of the bit stream through a framed interface used for telecommunications services and for signaling purposes. *See also* **framed interface, payload.**

interface point: 1. *Synonym* **point of interface. 2.** *See* **medium interface point.**

interface processor: 1. In communications systems, a processor that serves as the interface between a network and another processor or terminal. **2.** A processor that is used to control the flow of data into or out of a network. *Synonym* **communications computer.**

interface protocol: *See* **Serial Line Interface Protocol.**

interface rate: The gross bit rate across an interface after all processing is completed. *Note:* An example of an interface rate is the actual bit rate at the boundary between the physical layer and the physical propagation medium. *See also* **propagation medium.**

interface requirement: A requirement of a hardware, software, firmware, or database element with which a system or component must interface, or that sets forth constraints on formats, timing, connection, voltage levels, or conditions that must be met by the two systems or components being joined so that the output of one can serve as the input to the other in either one or both directions.

interface specification: A specification that meets the interface requirements for systems or components.

interface standard: A standard (a) that describes (i) the functional characteristics, such as code conversion, line assignments, and protocol compliance, and (ii) describes the physical characteristics, such as electrical, mechanical, and optical characteristics, necessary to allow the exchange of information between two or more usually different systems or pieces of equipment, (b) that allows command and control functions to be performed using communications and computer systems, and (c) that may include operational characteristics and acceptable performance levels. *See* **fiber distributed data interface standard.**

interface structure: *See* **positioned interface structure.**

interface surface: In the transmission media of optical and fiber optic systems, the surface over which a discontinuity occurs in the medium property, characteristic, or parameter, such as the refractive index, electrical permittivity, electrical conductivity, magnetic permeability, or their gradients. *Note:* Interface surfaces exist in step index optical fibers, fiber optic couplers, optical integrated circuits (OICs), and surfaces of lenses, prisms, and back-surfaced mirrors, or wherever there is an abrupt change in those parameters of the medium that affect the transmission of a lightwave propagating in the medium. For example, a glass lens in air has interface surfaces with the air. If the lens is immersed or embedded in a transparent material with the same refractive index, and the material "wets" the surface, the lens will have no effect on the lightwave passing from the material to the lens, i.e., in effect, the lens is not there, and there is no interface surface. In the expression for the reflection coefficient, if $n_1 = n_2$, and, because of Snell's law, the incidence angle and the refraction angle are equal, the reflection coefficient reduces to 0, and the transmission coefficient becomes unity, i.e., there is no interface. *See also* **propagation medium.**

interface testing: Testing conducted to ensure that an interface performs in accordance with specifications, such as in communications systems, to ensure that data integrity is maintained when the data are passed from one system or component to another. *Note:* Interface testing is of great importance in interconnecting networks, computers, and databases, such as occurs in the Internet. *See also* **data integrity.**

interference: 1. In signal transmission systems, energy that (a) is extraneous, (b) may be from natural or man-made sources, (c) interferes with the reception of desired signals, and (d) is distinguished from noise by usually consisting of a narrower band of frequencies emanating from a limited number of sources and thus may be somewhat coherent, whereas noise usually consists of a wider band of frequencies from a large number of sources and usually is incoherent. **2.** The effect of unwanted energy caused by one or more emissions, radiation, or inductions upon reception in a radiocommunications system, manifested by performance degradation, misinterpretation, or loss of information that could be extracted in the absence of such unwanted energy. **3.** In optics, the interaction of two or more beams of coherent or partially coherent light. *Note:* Interference may be constructive or destructive. Constructive interference increases the field strength of an electromagnetic wave in the original direction. Destructive interference decreases the field strength in the original direction and increases lateral emanations and absorption. **4.** The phenomenon that occurs when two or more coherent waves are superimposed so that they form beats in time or patterns in space. **5.** Noise induced in a system by another system. *See* **accepted interference, adjacent channel interference, antenna interference, antiinterference, cochannel interference, common mode interference, conducted interference, differential mode interference, electromagnetic interference, harmful interference, intersymbol interference, mutual interference, permissible interference, radio frequency interference, single frequency interference, single tone interference.** *Synonym* **unintentional interference.** *See also* **blanketing, coordination area, interference filter, static.**

interference emission: 1. Emission that results in a signal being propagated into and interfering with the proper operation of equipment. **2.** Emission that causes an electrical signal to enter into and interfere with the proper operation of electrical or electronic equipment. *Note:* The frequency range of interference emission includes the entire electromagnetic spectrum. *See also* **electromagnetic interference, electromagnetic interference control, interference.**

interference filter: An optical filter that (a) selectively transmits some wavelengths and either reflects or absorbs other wavelengths, (b) usually consists of one or more thin layers of dielectric or metallic materials, (c) operates by using interference effects, and (d) is wavelength-sensitive because of the interference effects that occur between the incident and reflected waves at the thin film boundaries. *Synonym* **multilayer filter.** *See also* **dichroic filter, interference, optical filter.**

interference limit: 1. In radio transmission, the maximum permissible interference level as specified by recognized authorities, such the International Special Committee on Radio Interference. **2.** In radio transmission, the maximum permissible interference level as specified by the International Special Committee on

Radio Interference. *Synonym* **limit of interference.** *See also* **interference.**

interference pattern: In images reproduced by display systems, a defect that (a) usually has the form of regularly spaced curved or straight lines, (b) is superimposed on the display image, (c) is caused by recurring interference signals, and (d) usually is man-made. *See also* **interference.**

interference report: Information concerning interference, such as (a) a description of the nature or the type of interference that is being received, (b) the known or suspected source of the interference, (c) the transmitter or the receiver being interfered with, (d) the receiving station that is experiencing the interference, (e) the time and the date the interference is experienced, and (f) other relevant information. *See also* **interference, mutual interference.**

interference suppression and blanking: *See* **pulse interference suppression and blanking.**

interferometer: An instrument in which the interference of coherent electromagnetic waves is used to make measurements, such as determining (a) the accuracy of optical surfaces by means of Newton's rings, (b) the length of optical paths between surfaces, and (c) the values of physical variables, such as temperature, pressure, strain, electric currents, magnetic fields, linear displacement, linear velocity, linear acceleration, angular displacement, angular velocity, and angular acceleration, by means of fiber optic interferometric sensors. *See* **Fabry-Perot interferometer, speckle interferometer, Twyman-Green interferometer.** *See also* **coherency, coherent, interference, interferometric sensor.**

interferometric sensor: In fiber optics, a sensor in which (a) a single coherent narrow spectral width beam is divided into two or more beams, (b) the individual beams are subjected to the various environments or stimuli that are being sensed, (c) the beams are then recombined with the original beam so that cancellation and enhancement vary as the phases of the waves are shifted relative to each other, resulting in a modulation of irradiance incident upon a photodetector, (d) optical fibers are used to convey the light beams, (e) usually there is a sensing leg and a reference leg, the outputs of which are combined to develop the intensity variation, and (f) the sensing leg is exposed to the physical variable being sensed, while the reference leg is protected from the stimulating physical variable. *Note:* The interferometric sensor will still perform properly if both legs are equally exposed to another physical variable not being sensed, such as both legs being exposed to the same temperature, pressure, and magnetic field strength variations while electric field strength is being sensed. *See also* **Sagnac fiber optic sensor.**

interferometry: The branch of science and technology devoted to (a) the study and the measurement of the interaction of waves, such as electromagnetic and sound waves, and (b) the design, development, and use of instruments in which electromagnetic and sound wave interference is used to measure physical variables, such as temperature, pressure, strain, electric currents, magnetic fields, and linear displacement, velocity, and acceleration, as well as angular displacement, velocity, and acceleration. *Note 1:* The interaction of the waves can produce various spatial, time, and frequency domain energy distribution patterns. *Note 2:* Interferometric techniques are used to measure refractive index profiles of dielectric waveguides, such as optical fibers. *Note 3:* Coherent lightwaves can be made to interact so as to produce various spatial, temporal, and frequency domain energy distribution patterns that can be used for various purposes, such as measuring thickness of materials, locating discontinuities in optical fibers, and sensing physical variables. *See* **slab interferometry, transverse interferometry.** *See also* **interferometer.**

interframe time fill (ITF): In digital data transmission, the sequence of bits that (a) is transmitted between frames and (b) does not include bits stuffed within a frame. *See also* **frame, packet format.**

interior communications: *Synonym* **internal communications.**

interlaced scanning: In raster-scanned video displays, scanning in which all odd-numbered scanning lines are first traced consecutively, followed by consecutively tracing the even-numbered scanning lines, each of which is traced between a pair of odd-numbered scanning lines except for the last even line. *Note 1:* The pattern created by tracing the odd-numbered scanning lines is called the "odd field," and the pattern created by tracing the even-numbered scanning lines is called the "even field." Each field contains half the information content, i.e., half of the total number of pixels, of the complete video frame. *Note 2:* Image flicker is less apparent in an interlaced scanning display than in a noninterlaced scanning display, because the rate at which successive fields occur in an interlaced display is twice that at which successive frames would occur in a noninterlaced display containing the same number of scanning lines and having the same frame refresh rate. *Synonym* **interlacing.** *See also* **refresh.**

interlacing: *Synonym* **interlaced scanning.**

inter-LATA: Pertaining to communications between local access and transport areas (LATAs). *See also* **local access and transport area.**

interleaving: 1. In data processing, the arranging of the members of one sequence of items or events so that they alternate with members of one or more other sequences of items or events, each original sequence retaining its identity, i.e., its ordering sequence. **2.** In communications, the transmission of pulses from two or more digital sources in time division sequence over a single path. **3.** In digital data error detection and correction, the use of error-correcting codes to reduce the number of undetected error bursts by (a) reordering code symbols before transmission in such a manner that any two successive code symbols are separated by $I-1$ symbols in the transmitted sequence, where I is the degree of interleaving, and (b) upon reception, the reordered code symbols are again reordered into their original sequence, thus effectively spreading, i.e., randomizing, the errors in time to enable more thorough correction by a random error-correcting code. *See also* **digital multiplexer, error control, time division multiplexing.**

interlock: To prevent another action or event from taking place until after a current action or event is completed. *Note:* An example of an interlock is to prevent a ringing signal to be placed on a user line while a call is in progress or while a handset is off-hook. *See also* **deadlock.**

interlock code: In a communications system, a code that (a) is sent in the forward direction, (b) in some circumstances, is sent in the backward direction, and (c) indicates that a closed user group is engaged in a call. *See also* **backward direction, closed user group, forward direction.**

intermateable connector: *See* **fully intermateable connector, mechanically intermateable connector.**

intermediate distribution frame (IDF): In a central office (C.O.) or private branch exchange (PBX), a frame that (a) cross-connects the user line cable to the user line circuit, (b) may serve as a distribution point for multipair cables from the main distribution frame (MDF) or the combined distribution frame (CDF) to individual cables connected to equipment in areas remote from these frames, and (c) usually is connected between the MDF and the switching equipment or switching circuits. *See also* **circuit, combined distribution frame, distribution frame, main distribution frame.**

intermediate element: In a network, a line-unit-line termination (LULT) or a line-unit-network termination (LUNT).

intermediate equipment: At a data station, the auxiliary equipment, such as an interchange circuit, that (a) is inserted between the data terminal equipment (DTE) and either the data circuit-terminating equipment (DCE) or the signal conversion equipment, (b) performs certain additional functions before modulation or after demodulation, and (c) has input and output circuits and signals that conform to the established standards for the interface between the DTE and the DCE. *See also* **data circuit-terminating equipment, data station, data terminal equipment, interchange circuit.**

intermediate field: In an electromagnetic wave developed by an antenna or a source of electromagnetic radiation, the field (a) in which the electric field strength varies primarily as the inverse square of the distance, and significantly as the inverse distance and the inverse cube of the distance from the antenna or the source, (b) that exists between the near field and the far field, (c) that lies in the intermediate region at distances from the antenna between approximately 0.1 and 1.0 of the wavelength, for an antenna equivalent length that is small compared to this distance, and (d) that has smooth transitions (i) between the near field and the intermediate field itself and (ii) between the intermediate field itself and the far field, i.e., there are no distinct boundaries between the several fields. *Note 1:* The differential tangential component of the electric field vector produced by an ideal Hertzian dipole antenna, which most dipole antennas can be considered to be at the distances involved, is given by the relation $dE_\theta = (Idl\,(\sin\theta)/4\pi\varepsilon)[(-\omega\,(\sin\omega t)/rv^2) + (\cos\omega t)/r^2v) + (\sin\omega t)/\omega r^3)]$ where dE_θ is the differential tangential component of the electric field vector, I is the antenna current, dl is the elemental length of antenna bearing the current I, θ is the angle between the antenna direction of dl and the line from the dipole center to the point of measurement, π is approximately 3.1416, ε is the electric permittivity of the intervening medium, ω is the angular velocity equal to $2\pi f$ where f is the frequency, t is time, r is the distance from the center of the dipole antenna to the point of measurement, and v is the propagation velocity equal to $(\mu\varepsilon)^{-1}$ where μ is the magnetic permeability. Note that for the expression in brackets, the terms are for the radiation field, the induction field, and the electrostatic field, in that order. Note that at great distances, i.e., large values of r, the radiation field, i.e., the far field, is dominant. Closer to the antenna, the intermediate field is dominant, and just outside the antenna the near field is dominant. *Note 2:* The near and far fields become dominant at the edges of the intermediate field. *Synonyms* **Fresnel field, induction field.** *Deprecated synonyms* **intermediate field region, intermediate zone.** *See also*

electromagnetic radiation, far field radiation pattern, far field, near field. *Refer to* **Table 9, Appendix B.**

intermediate field radiation pattern: A radiation pattern in which (a) the electric and magnetic field strengths of an electromagnetic wave primarily vary inversely as the square of the distance from the source, and (b) other contributions to the field strengths, such as those that vary inversely as the cube of the distance or inversely as the distance, are less significant. *See also* **diffraction, electromagnetic radiation, far field, far field radiation pattern, intermediate field, intermediate field region, near field, near field diffraction pattern, near field radiation pattern, radiation pattern.**

intermediate field region: In an electromagnetic wave, the spatial volume (a) that is occupied by the intermediate field, (b) that lies between the near field region and the far field region, and (c) in which the field strength depends primarily on the existence of moving charges in the antenna, i.e., on currents. *Note:* The induction field primarily determines the field strength that varies as $1/d^2$ where d is the distance from the antenna, but with some contributions from the adjacent near field and far field. **2.** In radio communications, the region between (a) the near field region of an antenna and (b) the far field region. *Note:* The center of the intermediate field region usually is considered to be at a radius from the antenna equal to (a) twice the square of the antenna aperture length divided by (b) the wavelength. *Synonyms* **Fresnel region, induction field region.** *Deprecated synonym* **intermediate zone.** *See also* **antenna, effective Earth radius, far field, far field region, intermediate field, k-factor, near field, near field region, path clearance, path profile, path survey, propagation path obstruction.** *Refer to* **Table 9, Appendix B.**

intermediate format: *See* **quarter common intermediate format.**

intermediate frequency (IF): 1. A frequency to which a signal is shifted as an intermediate step in transmission or reception. *Note:* An intermediate frequency (IF) is used in heterodyning. **2.** In a heterodyne circuit, such as is used in a radio receiver, the frequency that (a) is produced when the frequency of a local oscillator is mixed with the audio frequency (AF) modulated incoming radio frequency (rf) signal, (b) usually is higher than the AF and lower than the rf, and (c) can be more effectively processed by the receiver circuits than either the AF or the rf. *See also* **heterodyne, heterodyne receiver.**

intermediate frequency (IF) amplifier: An amplifier in the intermediate frequency stage of a heterodyne radio receiver. *See:* **linear logarithmic intermediate frequency amplifier.** *See also* **intermediate frequency.**

intermediate frequency (IF) gain control: *See* **receiver intermediate frequency gain control.**

intermediate frequency (IF) repeater: *Synonym* **heterodyne repeater.**

intermediate language (IL): In communications, computer, data processing, and control systems, a language into which a source program or statement is translated prior to interpretation or further translation. *Note:* As an analogy, English is an intermediate language if a statement in Swahili is translated into English that is then translated into German or Dutch.

intermediate layer protocol: In a layered open communications system, a protocol resident in the middle layers of the system, such as in layers that correspond to Layer 3, Layer 4, and Layer 5 of the Open Systems Interconnection—Reference Model.

intermediate level language: In communications, computer, data processing, and control systems, a programming language that (a) is less machine-oriented than a machine language, (b) is not so machine-independent as a common language, such as Ada®, COBOL, or FORTRAN, (c) contains macros that are less powerful than common language macros, (d) usually is the object language of a root compiler, (e) has to be converted to machine language before execution, and (f) has features that lie between those of a machine language and a common language. *Note:* Examples of intermediate level languages are assembly languages, such as PL/1, BASIC, and FORTRAN. *Synonym* **intermediate order language.** *See also* **artificial language, common language, high level language, low level language, machine-dependent language, machine-independent language, machine language, natural language, object language, root compiler.**

intermediate node: In a network, a node that is directly connected to at least two other nodes by means of branches and therefore is not an end point node. *See also* **end point node, node.**

intermediate order language: *Synonym* **intermediate level language.**

intermediate switchboard: In telephone switchboard operations, any switchboard that is situated or connected between the switchboard that is serving a call originator and the switchboard that is serving the call receiver.

intermediate system: A system that provides an Open Systems Interconnection—Reference Model (OSI-RM) Network Layer relay function in which data received from one corresponding network entity are forwarded to another corresponding network entity. *See also* **Open Systems Interconnection—Reference Model.**

intermediate zone: *Deprecated synonym for* **intermediate field** *or for* **intermediate region.**

intermittent timing transmission: *Synonym* **random bit stream signaling.**

intermodal dispersion: 1. *Deprecated synonym for* **multimode distortion. 2.** *See also* **modal dispersion.**

intermodal distortion: *Synonym* **multimode distortion.**

intermodulation (IM): The modulation of the frequency components of an electromagnetic wave, resulting in the production of (a) frequencies corresponding to the sums and the differences of integral multiples of the input fundamental frequencies and (b) all the harmonics above the fundamental frequencies, such as are produced by (i) mixing and (ii) using nonlinear circuits or nonlinear elements of a system. *Note:* Intermodulation occurs more in frequency division multiplexed systems and wavelength division multiplexed systems than in time division multiplexed systems. Each frequency modulates all the others. *See* **radio frequency intermodulation.** *See also* **frequency, heterodyne, modulation.**

intermodulation distortion (IMD): Nonlinear distortion characterized by the appearance of frequencies, in the output, equal to the sums and the differences of integral multiples, i.e., of harmonics, of the component frequencies present in the input. *Note:* The harmonics that are present in the output usually are not considered to be part of the intermodulation distortion. They describe the shape of the output signal. When the harmonics are included as part of the distortion, a statement to that effect should be made. *See* **modulation frequency intermodulation distortion.** *See also* **composite two-tone test signal, distortion, frequency, heterodyne.**

intermodulation noise: In a transmission path or device, noise that (a) is generated during modulation and demodulation and (b) results from nonlinear characteristics in the path or the device. *See* **equipment intermodulation noise, path intermodulation noise.** *See also* **noise.**

intermodulation noise loss: The power loss that (a) is the difference between the actual output power and the theoretical output power of a wanted signal from a nonlinear circuit and (b) is attributable to the total power loss that is caused by unwanted intermodulation products that are generated by nonlinearities in the circuit.

intermodulation product: In an output signal produced by intermodulation, the frequencies equal to the sums and the differences of integral multiples of the frequencies present in the input signal. *Note:* Intermodulation products are caused by nonlinear circuits.

internal absorptance: In a propagation medium, the value (a) that is the ratio of (i) lightwave energy absorbed between the entrance and emergent surfaces of the medium, i.e., absorbed inside the medium, to (ii) the lightwave energy that has penetrated the entrance surface of the medium, (b) that does not include the effects of interreflections between the two surfaces, and (c) that is numerically equal to unity minus the internal transmittance. *See also* **internal transmittance, propagation medium, value.**

internal bias: In start-stop teletypewriter receiving mechanisms, bias that will have the same effect on the operating margin as an external bias applied to the receiver. *Note:* Internal bias may be a marking bias or a spacing bias. *See also* **bias, bias distortion, margin, marking bias, spacing bias.**

internal call: A call placed within a private branch exchange (PBX) or local switchboard, i.e., not through a central office (C.O.) in a public switched network. *Synonym* **inside call.** *See also* **external call.**

internal communications facility: An electrical, acoustical, or mechanical communications system that interconnects various operational compartments or spaces within a specified area, such as in a ship, aircraft, spacecraft, train, or motor vehicle, or in a building. *Note:* Examples of internal communications facilities are annunciators, intercoms, and telewriters. *Synonym* **interior communications.** *See also* **annunciator, convoy internal communications, intercom, telewriter.**

internal hard disk: A hard disk that (a) is mounted and operated inside the outer case of a computer, (b) usually is permanently installed at a shop or a factory, and (c) has a higher capacity than a diskette. *See also* **hard drive.**

internal label: A label that (a) can be read by a machine, i.e., is machine-readable, and (b) is recorded on the same medium as the set of data to which it refers. *See also* **machine-readable, machine-readable medium, medium.**

internal memory: In a computer, the set of storage spaces that (a) is directly accessible by a processor without the use of the computer input-output channels, i.e., is integral to the computer and not in a separate storage unit outside the computer, and (b) usually includes several types of storage, such as main storage, cache memory, and special registers, all of which can be directly accessed by the processor. *Synonym* **internal storage.** *See also* **main storage, random access memory, read-only memory.**

internal message distribution center: A communications service center that (a) provides communications services within an entity, such as an organization, area, building, headquarters, office, or factory, (b) contains support facilities, such as message distribution equipment, (c) delivers incoming messages or internal messages (i) automatically, such as by electronic means, (ii) semiautomatically, such as by pneumatic tubes or telewriters, or (iii) manually, such as by a periodic pickup and delivery messenger service, a mail delivery system, and the use of special envelopes, and (d) effects delivery by recognition of address designators, addressees, message subject matter, subject matter discriminators, and descriptors.

internal optical density: A measure of the degree of opacity of a propagation medium to lightwaves, such that the higher the optical density, the higher the opacity is, and the more lightwaves are attenuated in the medium, this value being numerically equal to the logarithm to the base 10 of the reciprocal of the internal transmittance. *Synonym* **transmission factor.** *See also* **internal transmittance, propagation medium.**

internal photoeffect: *Synonym* **internal photoelectric effect.**

internal photoeffect detector: *Synonym* **internal photoelectric effect detector.**

internal photoelectric effect: The changes in the characteristics of a material that occur when incident photons are absorbed by the material and excite the electrons in the various energy bands in the molecules composing the material. *Note 1:* Characteristic changes include changes in (a) electrical conductivity, i.e., the photoconductive effect, (b) electric potential development, i.e., the photovoltaic effect, and (c) photosensitivity. *Note 2:* Changes in emissivity, i.e., the photoemissive effect, are not an internal photoelectric effect. *Note 3:* Electrons may move from a valence band to a conduction band when the material is exposed to radiation, thereby increasing their mobility. The material includes the intrinsic material and the impurities, including dopants. The internal photoelectric effect includes both intrinsic and extrinsic effects. *Synonym* **internal photoeffect.** *See* **extrinsic internal photoelectric effect, intrinsic internal photoelectric effect.** *See also* **conduction band, external photoelectric effect, incidence, valence band.**

internal photoelectric effect detector: A device (a) in which incident photons raise electrons from a lower to a higher energy state, resulting in an altered state of the electrons, holes, or electron-hole pairs generated by the transition, which is then detected, and (b) which may be used as a photodetector in a fiber optic receiver. *Synonym* **internal photoeffect detector.** *See also* **internal photoelectric effect.**

internal reflection: In an optical element in which an electromagnetic wave is propagating, a reflection at an outside surface toward the inside such that a wave that is incident upon the surface is reflected wholly or partially back into the element itself. *Note:* Optical fibers depend on total internal reflection at the core-cladding interface for successful transmission of lightwaves, by confining to the core sufficient transmitted energy all the way to the end of the fiber. *See* **total internal reflection.** *See also* **incidence, optical element, reflection, total reflection.** *Refer to* **Fig. I-5.** *Refer also to* **Figs. C-9 (190), S-13 (919), S-18 (951).**

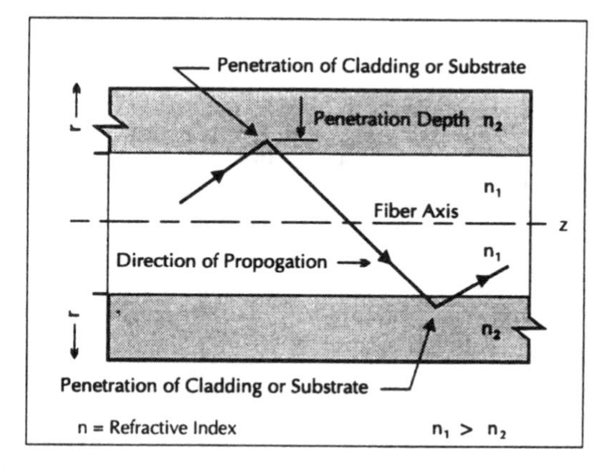

Fig. I-5. Internal reflection of a light ray in a slab dielectric waveguide, showing some penetration of the ray into the cladding.

internal reflection angle: In an internal reflection, the angle between the incident ray and the normal to the surface at the point of incidence. *See also* **critical angle, internal reflection, reflection angle, total internal reflection.**

internal reflection sensor: *See* **frustrated total internal reflection sensor.**

internal security audit: A security audit of a system that is conducted by persons who are responsible to the organization or the agency that operates the system, i.e., by persons responsible for the implementation of security measures and security policy for the system. *See also* **security audit.**

internal sort: A sort performed entirely with the use of internal memory or main storage only. *See also* **external sort, internal memory, main storage.**

internal storage: *Synonym* **internal memory.**

internal transmittance: In a propagation medium, the value (a) that is the ratio of (i) the lightwave energy transmitted between the entrance and emergent surfaces of the medium, i.e., the lightwave energy that reaches the emergent surface of the medium, to (ii) the lightwave energy that has penetrated the entrance surface of the medium, (b) that does not include the effects of interreflections between the two surfaces, and (c) that is numerically equal to unity minus the internal absorptance. *See also* **internal absorptance, transmission medium, value.**

internally stored program computer: *Synonym* **stored program computer.**

international address indicating group: An address indicating group (AIG) that (a) is assigned for international use, (b) is assigned to a nation in blocks, (c) allows each nation to allocate its assigned block as appropriate, and (d) may be permanent or temporary. *See also* **address indicating group, national address indicating group.**

international aeronautical emergency frequency: The nominal 121.5 MHz (megahertz) frequency that is used internationally, i.e., used worldwide, for emergency messages by aircraft stations, ship stations, and aeronautical land stations that are primarily concerned with safety and regulation of travel along international routes and lanes, particularly international civil air routes and international shipping lanes. *Note:* The emergency frequency band lies in the range from 118 MHz to 132 MHz. *See also* **emergency message.**

International Atomic Time (TAI) scale: The time scale established by the International Time Bureau (BIH) of the Bureau International des Poids et Mesures (BIPM) on the basis of atomic clock data supplied by cooperating institutions. *Note 1:* International Atomic Time (TAI) differs from Coordinated Universal Time (UTC) by an integral number of seconds. *Note 2:* The abbreviations "TAI" and "BIH" are a result of translation from the official international names, which are written in French. TAI stands for Temps Atomique International. BIH stands for Bureau International de l'Heure. *See also* **Coordinated Universal Time.**

International Atomic Time (TAI): The time expressed on the International Atomic Time (TAI) scale. *See also* **Coordinated Universal Time, International Atomic Time (TAI) scale.**

international broadcasting station (IBS): A broadcasting station that (a) uses frequencies allocated to the broadcasting service between 5950 kHz (kilohertz) (5.95 MHz) and 26,100 kHz (26.1 MHz) and (b) transmits directly to the general public in foreign countries.

international call sign: A call sign that is assigned in accordance with the rules of the International Telecommunication Union. *Note:* International calls signs are intended to permit a receiver to identify a radio station. *See* **military ship international call sign, ship international call sign.** *See also* **military call sign.**

international congestion signal: In an international communications system, a signal that (a) is sent in the backward direction and (b) indicates the failure of a call setup attempt, i.e., an unsuccessful access attempt, caused by network congestion, saturation, or contention that is being encountered in the international network or in the destination national network. *See also* **access attempt, backward direction, congestion, contention, saturation.**

international cooperation index: *Synonym* **diametral cooperation index.**

international frequency register: *See* **Master International Frequency Register.**

International Frequency Registration Board (IFRB): A permanent organization of the International Telecommunication Union (ITU) that implements frequency assignment policy and maintains the Master International Frequency Register (MIFR).

international index of cooperation: *Synonym* **diametral cooperation index.**

International Maritime Satellite (Inmarsat) System: A worldwide satellite communications system that (a) is designed for mobile communications via Earth stations and transportable terminals from remote locations, such as from (i) ship and ground vehicle terminals and (ii) portable briefcase-size satellite data link telephones and terminals that contain laptop computers and printers, and (b) provides various types of service, such as M service for voice, data, and facsimile and C

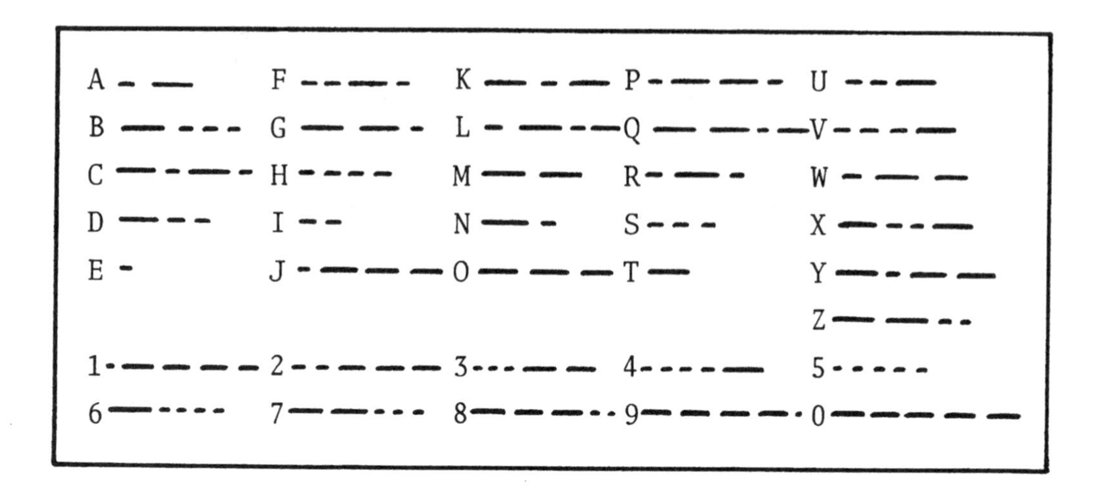

Fig. I-6. The **international Morse code.**

service for low cost data linkage, including Telex, E mail, and limited low rate facsimile.

international Morse code: A code (a) in which specific groupings of dots and dashes are used to represent letters and numerals and (b) that is especially used in wire telegraph, radio telegraph, and visual communications systems. *See also* **dash, dot, numerals.** *Refer to* **Fig. I-6.** *Refer also to* **Fig. T-3 (991).**

International Organization for Standardization (ISO): An international organization that (a) consists of member bodies that are the national standards bodies of most of the countries of the world, (b) is responsible for the development and publication of standards in various technical fields, after developing a suitable consensus, (c) is affiliated with the United Nations, and (d) has its headquarters at 1, rue de Varembé, Geneva, Switzerland. *Note:* Member bodies of the International Organization for Standardization (ISO) include, among others, the American National Standards Institute (ANSI), Association Française de Normalisation (AFNOR), the British Standards Institution (BSI), and the Deutsches Institut für Normung (DIN). ISO standards may be obtained from ANSI, 1430 Broadway, New York, NY 10018.

International Radio Consultative Committee (CCITT): A predecessor of International Telecommunication Union-R (ITU-R).

international radio telegraph distress frequency: The nominal 500 kHz (kilohertz) frequency that (a) is used internationally, i.e., is used worldwide, for distress radio telegraph messages that are sent by ship stations, aircraft stations, and survival craft stations when requesting assistance, (b) usually is used to request assistance through the stations of the maritime mobile service and the aeronautical mobile service, and (c) is used only for distress calls, distress traffic, distress messages, urgency messages, and safety signals. *Note:* The distress frequency range is from 405 kHz to 535 kHz. *See also* **aeronautical mobile service, distress calls, distress messages, distress traffic, maritime mobile service.**

international radio telegraph lifecraft frequency: The nominal 8364 kHz (kilohertz) frequency that (a) is used internationally, i.e., is used worldwide, for radio telegraph distress messages sent by lifeboat, lifecraft, and survival craft stations that usually are equipped to transmit on frequencies that are between 4000 kHz and 27.5 MHz (megahertz), (b) is used to request assistance through stations of the aeronautical mobile service and the maritime mobile service, and (c) is used only for establishing and conducting search and rescue communications. *See* **search and rescue communications, search and rescue frequency, survival craft station.**

international radio telephone distress frequency: The frequency that (a) is used internationally, i.e., is used worldwide, for radio telephone distress messages sent from ship stations, aircraft stations, and survival craft stations that usually are equipped to transmit on frequencies that are between 1605 kHz and 4000 kHz, (b) nominally is 2182 kHz (kilohertz), (c) is used to request assistance through stations of the aeronautical mobile service and the maritime mobile service, and (d) is used for distress calls, distress messages, distress traffic, urgency signals, urgency messages, and safety signals. *See also* **aeronautical mobile service, distress calls, distress messages, distress traffic, maritime mobile service, safety signals.**

international signal code: A signaling code, and its related systems and procedures for international use, that (a) consists of signs, symbols, letters numerals, prowords, and control signals that have been accepted for use by international agreement, (b) is used particularly when communicating with ship stations, and (c) has code groups in messages that should be preceded by the special pennant that identifies the code or by the proword interco. *See also* **code, interco, pennant, proword, ship station, signaling.**

International System of Units (Système International d'Unités) (SI): A consistent system of units that (a) is based on the meter, kilogram, second, and coulomb (MKSC), (b) includes units dependent on MKSC units, and (c) is based on agreements that were reached by the General Conference on Weights and Measures. *See* **metric system.**

International Telecommunication Union (ITU): A civil international organization established to promote standardized telecommunications on a worldwide basis. *Note:* The ITU-R and the ITU-T are committees of the International Telecommunication Union (ITU). The headquarters of the International Telecommunication Union (ITU) is in Geneva, Switzerland. While older than the United Nations, the ITU is recognized by the United Nations as the specialized agency for telecommunications. *Refer to* **Table 1, Appendix B.**

International Telecommunication Union-R (ITU-R): The Radiocommunications Sector of the International Telecommunication Union (ITU) that (a) studies technical issues relating to radiocommunications and (b) has some regulatory powers. *Note:* A predecessor of the International Telecommunication Union-R (ITU-R) was the International Radio Consultative Committee (CCIR), also of the ITU.

International Telecommunication Union-T (ITU-T): The Telecommunications Standardization Sector of the International Telecommunication Union (ITU) that (a) studies technical, operating, and tariff questions relating to telecommunications, (b) prepares and issues recommendations on these questions, and (c) has the goal of standardizing telecommunications on a worldwide basis. *Note:* The International Telecommunication Union-T (ITU-T) performs the standards-setting activities of the former International Telegraph and Telephone Consultative Committee (CCITT) and the former International Radio Consultative Committee (CCIR), both also of the ITU.

International Telecommunication Union (ITU) regulations: Telecommunications standards approved and published by the International Telecommunication Union (ITU).

International Telecommunication Union (ITU) world region: A subdivision of the world area for purposes of communications, particularly for distress signaling. *Note:* The Western Hemisphere is roughly Region 2. The South Pacific, including China, East Indies, Australia, New Zealand, and southern Japan, is roughly Region 3, and the remainder of the world, including Eurasia, Africa, and the North Atlantic, is Region 1.

International Telegraph Alphabet 1 (ITA-1): An alphabet that (a) consists of a two-condition five-unit code, (b) is used in Baudot synchronous telegraph systems, and (c) is specified by Article 16 of the Telegraph Regulations of the International Telecommunication Union, Geneva, Switzerland, 1958.

International Telegraph Alphabet 2 (ITA-2): An alphabet that (a) consists of a two-condition five-unit code, (b) is used in tape relay systems, (c) is the set of correspondences between (i) impulses or patterns of holes at each row on tape and (ii) the letters, figures, i.e., numerals, special signs and symbols, operating signals, and machine function that the patterns of holes represent, and (d) is specified in International Telegraph and Telephone Consultative Committee (CCITT), International Telecommunication Union (ITU) Recommendation F.1. *Note:* The early five-unit telegraph alphabet was developed circa 1895. *Synonyms* **teleprinter code, teletype code, teletypewriter code.**

International Telegraph Alphabet 3 (ITA-3): An alphabet that (a) consists of a two-condition seven-unit constant ratio code and (b) is specified in International Telegraph and Telephone Consultative Committee (CCITT), International Telecommunication Union (ITU) Recommendation S.13, 1972 or in International Radio Consultative Committee (CCIR), International Telecommunication Union (ITU) Recommendation 342-2, 1970.

International Telegraph Alphabet 4 (ITA-4): An alphabet that (a) consists of a two-condition six-unit code, (b) is composed of two code combinations corresponding to permanent positive and negative polarity in order that the multiplexed channel can be operated in a switched network, and (c) is used for time-division-multiplexed synchronous telegraph systems.

International Telegraph Alphabet 5 (ITA-5): An alphabet (a) in which a different seven-binary-digit pattern is assigned to each different letter, numeral, special sign and symbol, and control character, (b) that is used for effecting information interchange, (c) on which international agreement has been reached, (d) that is a result of a joint agreement between the International Telegraph and Telephone Consultative Committee (CCITT) of the International Telecommunication Union (ITU) and the International Organization for Standardization (ISO), (e) that is published as CCITT Recommendation V3 and as ISO 646, (f) that has been adopted by the North Atlantic Treaty Organization (NATO) for military use, (g) in which 12 of the seven-binary-digit patterns are unassigned for use in a given country that may have unique language requirements, such as monetary symbols and the tilde, umlaut, circumflex, and dieresis, (h) that has been adapted in the United States as the American Standard Code for Information Interchange (ASCII), published by the American National Standards Institute (ANSI), (i) that actually includes eight-digit binary patterns each consisting of seven primary binary digits and a parity check bit, and (j) that includes upper and lower case letters, ten decimal numerals, special signs and symbols, diacritical marks, data delimiters, and transmission control characters.

International Telegraph and Telephone Consultative Committee (CCITT): A predecessor of International Telecommunication Union-T (ITU-T).

International Teleprinter Network (TELEX): A worldwide teleprinter exchange service. *See also* **teleprinter exchange service.**

international temperature scale (ITS): A temperature scale that (a) is based on the Kelvin temperature scale, (b) is defined by international agreement, the latest of which is ITS-90, and (c) is used primarily to apply corrections, via published tables, to the Kelvin temperature scale, particularly at extremely high and low temperatures. *See also* **Kelvin temperature scale.**

International Time Bureau (BIH): The organization that is responsible for establishing and maintaining the International Atomic Time (TAI) scale. *Note:* The abbreviations "TAI" and "BIH" are a result of translation from the official international names, which are written in French. TAI stands for Temps Atomique International. BIH stands for Bureau International de l'Heure. *See* **International Atomic Time.**

international visual signal code: A code that (a) has been adopted by many nations for international visual communications, (b) contains combinations of letters that are assigned to words, phrases, and sentences, (c) transmits the letters by signaling systems designed to transmit internationally standardized letters by visual means, such as (i) by hoisting internationally standardized alphabet flags and pennants, (ii) by flashing lights, such as heliograph and searchlight, using the international Morse code, and (iii) by wigwag using the international Morse code. *See also* **code, flags, pennant, signaling, wigwag.**

internet: *Synonym* **internetwork.**

Internet: The formal collection of networks and gateways that (a) includes among others, the Military Network (MILNET) and the National Science Foundation Network (NSFNET), (b) uses the Transmission Control Protocol/Internet Protocol (TCP/IP) protocol suite, (c) functions as a single, virtual network, (d) provides global connectivity, (e) provides three levels of network services, identified as (i) less reliable, connectionless packet delivery, (ii) more reliable, full-duplex bit stream delivery, and (iii) application-level services, such as E-mail, and (f) is used internationally. *See also* **internetwork, National Science Foundation network.** *Refer to* **Fig. S-11 (880).**

Internet access: An access to Internet using Internet addresses, protocols, and servers. *See also* **Internet, Internet address, Internet protocol, server.**

Internet address: In the internet protocol, the fixed-length addresses that (a) identify the hosts of sources and destinations and (b) provides for fragmentation and reassembly of long datagrams, if necessary, for transmission through small-packet networks. *Synonym* **Internet protocol address.**

Internet Assistant: A Microsoft™ program written in HyperText Markup Language that is a combination of a file conversion tool and a World Wide Web (WWW) browser. *See also* **browser, HyperText Markup Language, Internet.**

Internet Control Message Protocol: An Internet protocol that (a) reports datagram delivery errors, (b) is a key part of the Transmission Control Protocol/Internet Protocol (TCP/IP) protocol suite, and (c) serves as a basis for the packet internet groper (PING). *See also* **Internet.**

internet groper: *See* **packet internet groper.**

Internet Protocol (IP): A U.S. Department of Defense (DoD) standard protocol that (a) is designed for use in interconnected systems of packet-switched computer communications networks, (b) provides for transmitting blocks of data, called "datagrams," from sources to destinations, where sources and destinations are hosts identified by fixed-length addresses, and (c) provides for fragmentation and reassembly of long datagrams, if necessary, for transmission through small-packet networks. *See also* **block, communications, protocol.**

Internet protocol address: *Synonym* **Internet address.**

Internet Protocol Suite: *See* **Transmission Control Protocol/Internet Protocol Suite.**

internetwork: An informal collection of packet-switching networks that (a) is interconnected by gateways, (b) uses protocols allowing it to function as a single, large, virtual network, (c) is a worldwide interconnection of networks, (d) consists of an interconnection of individual personal computers and computer networks each of which belongs to a public or private entity, such as a person, corporation, university, government agency, or laboratory, (e) is unregulated, (f) does not belong to any particular organization or individual, and (g) uses existing telecommunications facilities to establish interconnections. *Note:* When written in initial upper case, "Internet" refers specifically to the connected Internet and the Transmission Control Protocol/Internet Protocol (TCP/IP) protocols it uses. *Synonym* **internet.** *See also* **Internet.**

internetwork connection: *Synonym* **gateway.**

internetworking: The process of interconnecting a number of individual networks to provide a path from one network to another network. *Note:* The interconnected networks may be different types. However, each network is distinct, with its own addresses, internal protocols, access methods, and administration. *See also* **bridge, communications, gateway.**

Internetwork Packet Exchange (IPX): A proprietary local area network (LAN) protocol.

interoffice signaling: *See* **common channel interoffice signaling.**

interoffice trunk: A direct trunk between central offices (COs) with no intervening COs.

interoperability: 1. The ability of communications systems and components to function together. *Note:* The degree of interoperability should be defined in referring to specific cases. Interface devices may be placed between systems or components in order to achieve the desired degree of interoperability. **2.** The ability of systems, units, or forces to (a) provide services to and accept services from other systems, units, or forces and (b) use the services so exchanged to enable them to operate effectively together. **3.** The condition achieved among communications-electronics systems or items of communications-electronics equipment when information or services can be exchanged directly and satisfactorily among them and their users. **4.** The ability to exchange data in a prescribed manner and the processing of such data to extract intelligible information that can be used to control and coordinate operations. *Note:* Internet and the information superhighway are realizable because of communications systems interoperability. *See also* **commonality, compatibility, interchangeability, mobile service, mobile station, portability, transportability.**

interoperability standard: 1. A standard, or the published documentation, that establishes engineering and technical requirements that are to be used in the design of systems, units, or forces intended to operate directly together. *Note:* An example of an interoperability standard is the 1.31-μm (micron) wavelength used for fiber optic data links.

interoperation: The use of interoperable systems, units, or forces.

interphone: *Synonym* **intercom.**

interpolation: *See* **analog speech interpolation, digital speech interpolation.**

interposition trunk: 1. A connection between two positions of a large switchboard so that a line on one position can be connected to a line on another position. **2.** Connections terminated at test positions for testing and patching between testboards and patch bays within a technical control facility. *See also* **patch bay, tie trunk, trunk.**

interpret: 1. In the execution of a computer program, to translate and to execute each source language statement in the computer program before translating and executing the next statement or instruction. **2.** To represent coded characters in such a manner that they can be read by humans. *Note:* Examples of interpret are (a) to print at the top of a punch card column the character represented by the holes in the column and (b) to print on the card the characters magnetically recorded on the card, such as to print the figures on a transit system fare card that are stored in a magnetic stripe on the card.

interpreter: 1. A computer program that interprets. **2.** A computer program that enables a computer to (a) translate and execute each instruction or statement of a program written in a given language, such as a program written for a given computer that will enable the computer to execute a sequence of instructions written in an intermediate language, such as BASIC or FORTRAN, and (b) pick up each stored program instruction in turn and translate it into one or more machine language instructions for immediate execution. **3.** A device that

(a) accepts data on a data medium in machine-readable form that cannot readily be read by humans and (b) represents the data for human reading on the same data medium. *Note:* An example of an interpreter is a device that reads bar codes and prints the corresponding numerals beside the code. *See* **card interpreter.** *See also* **bar code, computer program, data medium, instruction, interpret, machine-readable, numeral.**

interrogating: In communications, computer, data processing, and control systems, the transmitting and the receiving of messages in the form of a question or in a form that requires a response. *Note:* Examples of interrogating are (a) the use of a signal or combination of signals intended to trigger a response and (b) the requesting by a master station that a slave station indicate its identity. *See also* **polling.**

interrogating station: *See* **oceanographic data interrogating station.**

interrogation: 1. The triggering, by a signal or a combination of signals, of a response. **2.** The requesting, by a station or device, that another station or device identify itself or give its status. *See also* **master station, polling.**

interrogator: In satellite communications, a pulse transmitter that is used exclusively for exciting a transponder.

interrupt: A suspension of a process, such as the execution of a computer program, (a) caused by an event external to that process and (b) performed in such a way that the process can be resumed. *Synonym* **interruption.**

interrupted continuous wave (ICW): Modulation in which there is on-off keying of a continuous wave (cw). *See also* **continuous wave, modulation.**

interrupted isochronous transmission: *Synonym* **isochronous burst transmission.**

interruption: *Synonym* **interrupt.**

interruption rate: In a telephone system in which direct current (dc) is used for control signals, i.e., for signaling, the rate at which a dc source is interrupted for signaling purposes.

interruption switch: In an electric power distribution system, a switch that (a) has the capability of opening, i.e., of interrupting, a circuit that is conducting an electric current and (b) usually has an interrupting capacity and rating, usually specified in amperes, before opening, and a voltage rating to ground or between its contacts, specified in volts, after opening, i.e., a switch that can be opened or closed under load within specified voltage and current limits. *See also* **disconnecting switch.**

interrupt signaling: *See* **digital carrier interrupt signaling.**

intersatellite service: A radiocommunications service that provides links between Earth artificial satellites.

intersection: In direction finding (DF), triangulation in which the location of a source of radiation, such as a transmitting radio antenna or an observed forest fire, is determined by (a) obtaining bearings of the source of radiation from at least two known locations, (b) plotting the two or more locations from which the bearings are taken on a map, (c) plotting the bearings from those locations on the map, and (d) determining the location on the map at which the bearings cross. *Note 1:* If three bearings are taken, the bearings will form a small triangle, the size depending on the precision with which the bearings are taken. *Note 2:* The smaller of the bearing intersection angles should be as large as possible, for increased precision. If the source is on, or nearly on, the line joining the two locations from which the bearings are taken, the location of the source cannot be obtained by intersection. *See* **bearing intersection, direction finding, space intersection, triangulation.** *See also* **resection.** *Refer to* **Fig. I-7.** *Refer also to* **Fig. R-11 (848).**

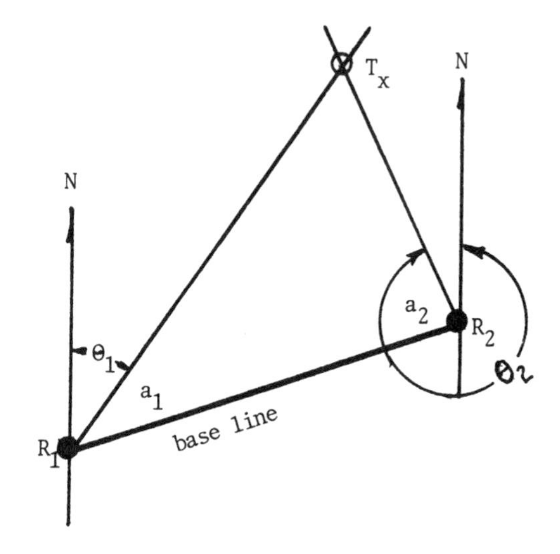

Fig. I-7. The **intersection** method of triangulation, in which receivers R_1 and R_2 of known location are used to determine the azimuths θ_1 and θ_2 that are bearings with true north, or their angles a_1 and a_2 with the baseline, to determine the position of the transmitter T_x on a map. The transmitter operator must know where the receivers are and the receiver operators must report the true north bearings, usually with (a) 180° added if the bearing is less than 180° or (b) 180° subtracted if the bearing is greater than 180°.

intersection gate: *Synonym* **AND gate.**

intership radio teletypewriter: A ship-to-ship radio communications system in which a teletypewriter is used for the transmission and the reception of messages.

interstitial defect: In the somewhat ordered array of atoms and molecules in a material, such as that used to make optical fibers and lenses, an extra atom or molecule that (a) is inserted or trapped in the space between the atoms and the molecules of a normal array and (b) can cause scattering, diffusion, heating, absorption, and resultant attenuation of signals or light rays. *See also* **scattering center, vacancy defect.** *Refer to* **Figs. S-8 (874), V-1 (1060).**

interswitch trunk: A trunk between switching nodes. *See also* **trunk.**

intersymbol interference: 1. Extraneous energy (a) that is in a signal representing a string of symbols during one or more keying intervals, and (b) that interferes with the reception of the signal during another keying interval. **2.** The disturbance caused by extraneous energy from a signal in one or more keying intervals that interferes with the reception of the signal in another keying interval. **3.** In digital transmission systems, distortion of received signals that results in the spreading and consequent overlap in space and time of individual pulses to the degree that the receiver cannot reliably distinguish between changes of state, i.e., cannot reliably distinguish between individual signal elements. *Note 1:* In fiber optic systems, intersymbol interference can occur when dispersion causes pulse broadening, i.e., spreading in time and space along the optical fiber, as it propagates, resulting in pulse overlap that might be so large that a photodetector can no longer distinguish clear boundaries between pulses. When this occurs to a sufficient degree in digital data systems, the bit error ratio (BER) may become excessive. If the spread is too much, either (a) the data signaling rate (DSR) has to be reduced so as to provide more time, or space, between pulses, or (b) the dispersion has to be reduced. *Note 2:* At a certain threshold, intersymbol interference will compromise the integrity of the received data. *Note 3:* Intersymbol interference attributable to the statistical nature of quantum mechanisms sets the fundamental limit to receiver sensitivity. *Note 4:* Intersymbol interference may be measured by eye patterns. *See also* **attenuation-limited operation, bit error ratio, distortion, distortion-limited operation, eye pattern.**

intertoll trunk: 1. A channel between two toll offices. **2.** In a telephone system, a trunk between toll offices that are in different exchanges. *See also* **exchange, toll office, trunk.**

interval: *See* **bit interval, character interval, intercharacter interval, Nyquist interval, protection interval, service time interval, significant interval, unit interval.**

interval error: *See* **time interval error.**

interval modulation: *See* **pulse interval modulation.**

interval theoretical duration: *See* **significant interval theoretical duration.**

interworking: The communicating by data terminal equipment (DTE) that belongs to one user service with a DTE that belongs to a different user service class. *See also* **data terminal equipment, service class, user service class.**

intra-LATA: Pertaining to within the boundaries of a local access and transport area (LATA). *See also* **inter-LATA, local access and transport area.**

intramodal dispersion: *Synonym* **material dispersion.**

intramodal distortion: Distortion caused by dispersion, such as material, profile, or waveguide dispersion, of a specific propagating mode. *Note 1:* An example of intramodal distortion is the distortion resulting from dispersion of group velocity of a propagating mode. *Note 2:* Intramodal distortion is the only form of modal distortion that can occur in a single-mode optical fiber. *See also* **dispersion, distortion, multimode distortion.**

intranodal: In a communications network, pertaining to any event, such as a transmission or a message exchange, that takes place entirely within the equipment at a single node. *See also* **node.**

intraoffice trunk: A trunk connection within the same central office (C.O.), i.e., the same switching center. *See also* **trunk.**

intrinsic absorption: In lightwave propagation media, the absorption of light energy from a traveling wave or a standing wave by the medium itself, causing attenuation as a function of many factors, such as distance, material properties, mode, and frequency, i.e., wavelength. *Note 1:* Intrinsic absorption is primarily caused by (a) charge transfer bands in the ultraviolet region and (b) vibration or multiphonon bands in the near infrared region, particularly if they extend into the region of wavelengths used in fiber optic communications systems, such as the 0.7-μm (micron) to 1.3-μm region of the optical spectrum. *Note 2:* Intrinsic absorption usually occurs in optical fibers and integrated optical circuits (OICs) made of glass, silica, or plastic. *See also* **absorption, attenuation, infrared, mode,**

multiphonon band, optical spectrum, propagation medium, standing wave, traveling wave, ultraviolet.

intrinsic coupling loss: In fiber optic systems, the insertion loss caused by optical fiber parameter mismatches, such as fiber dimension, refractive index profile parameter, and mode field diameter mismatches, when two dissimilar fibers are joined. *Synonym* **intrinsic joint loss.** *See also* **extrinsic coupling loss.**

intrinsic internal photoelectric effect: An internal photoeffect that depends only on the basic material rather than on any dopants or inherent impurities. *See also* **dopant, internal photoelectric effect.**

intrinsic joint loss: *Synonym* **intrinsic coupling loss.**

intrinsic-negative field-effect transistor integrated receiver: *See* **positive-intrinsic-negative field-effect transistor integrated receiver.**

intrinsic-negative photodiode: *See* **positive-intrinsic-negative photodiode.**

intrinsic-negative photodiode coupler: *See* **positive-intrinsic- negative photodiode coupler.**

intrinsic noise: In a transmission path or device, the noise that (a) is inherent to the path or the device and (b) does not result from modulation. *Note:* Examples of intrinsic noise are (a) thermal noise in a resistor, (b) domain rotation noise in a ferromagnetic material, and (c) microphonic noise in an amplifier, i.e., noise caused by vibration of the amplifying device. *See also* **noise.**

intrinsic quality factor (IQF): A measure of the quality of a multimode optical fiber referenced to the intrinsic coupling loss. *Note:* The intrinsic quality factor (IQF) can be used as an alternative to specifying precise refractive index profile requirements, such as values for core diameter, numerical aperture, concentricity error, and core noncircularity, by requiring optical fibers to meet an average intrinsic coupling loss. The use of the IQF allows simultaneous parameter deviations to compensate for one another in terms of measured intrinsic coupling loss.

intrinsic semiconductor: A pure semiconductor that (a) has not been doped with any dopant material and (b) is pure crystal with neither positive dopant, i.e., p-type or acceptor dopant, nor negative dopant, i.e., n-type or donor dopant, in it. *Note:* An example of an intrinsic semiconductor is the material in a positive-intrinsic-negative (PIN) diode between the two positive and negative dopant materials that make up the junction. *See also* **acceptor, donor, dopant, junction, semiconductor.**

intrusion: *See* **electromagnetic intrusion.**

intrusion-resistant communications cable: A communications cable that (a) provides substantial physical protection, electrical isolation, and optical isolation, (b) reduces the possibility of clandestine coupling to the wires or optical fibers in the cable that contain information-bearing signals, and (c) is used in conjunction with protective measures and specialized equipment to detect slight changes in the physical or electrical state of the cable to provide visible or audible indications at a central control point of attempts at intrusion. *Synonym* **alarmed cable.**

invariant routing: In communications system operations, the routing of messages such that messages from a given originating office or source user are always sent to a fixed set of destination offices or destination users. *See also* **destination office, destination user, originating office, routing, source user, user.**

inverse filtering: In image restoration, filtering in which (a) the distorted image is processed in a manner that reverses the transformations that caused the original distortion, (b) if S is an operator that maps g into h, in which h is a distorted image of the object g, then the inverse filter is the operator S^{-1} that operates on h to form a less distorted or more true image of g, (c) the operator S is the point spread function, and (d) S^{-1} performs the reverse. *Note:* For example, if the elements of an image of an object were shifted in one direction proportionally by S, then S^{-1} operating on the distorted image would shift the image elements, such as the display elements, display groups, and display images, proportionally in the opposite direction, thus obtaining the original, undistorted, or at least less distorted, image of the object. *See* **pseudoinverse filtering.** *See also* **display element, display group, display image, image, image restoration, object, operator, point spread function.**

inverse square law: In electromagnetic wave propagation, the irradiance, i.e., the power per unit area in the direction of propagation, of a spherical wavefront varies inversely as the square of the distance from the source, assuming that there are no losses caused by absorption or scattering. *Note:* For example, the power radiated from a point source, i.e., an omnidirectional isotropic antenna, or from any source at very large distances from the source compared to the size of the source, must spread itself over larger and larger spherical surfaces as the distance from the source increases. The inverse square law is also applicable to diffused and incoherent radiation. *See also* **geometric spreading.**

inversion: *See* **bit inversion, phase inversion, population inversion.**

inversion code: *See* **modified alternate mark inversion code.**

inversion signal: *See* **alternate mark inversion signal.**

inversion violation: *See* **alternate mark inversion violation.**

inverted directory: In telephone system operations, a listing (a) that consists of telephone numbers in numerical order, (b) in which each number is followed by the name of the user, such as a customer, subscriber, person, equipment, place, or organization, and (c) that is used primarily by system operating and administrative personnel. *Synonym* **reverse directory.** *See also* **telephone directory.**

inverted file: A file in which another of the data items in the records is used to sequence the records instead of the one originally used to create the file. *Note 1:* An inverted file can only be an inversion of another file. *Note 2:* Examples of inverted files are (a) an inverted directory, (b) a list of social security numbers in numerical order, each number being followed by the corresponding person's name, and (c) a list of the three-letter codes for international airports in alphabetical order, each code being followed by the name and the location of the corresponding airport. *See also* **inverted directory.**

inverted image: In an optical system, an image, either real or virtual, that has a vertical orientation opposite to that of the object, i.e., an image that is upside-down compared to the orientation of the object. *Note:* The image obtained directly on the retina is an inverted image. However, the brain interprets the image as being erect. The image obtained on the retina with the aid of an optical system is said to be inverted, in contrast to erect, when the orientation of the image is upside-down in comparison to the object as seen with the unaided eye. The image formed on the film of a single-lens camera is an inverted image. *See also* **erect image, image, object, optical system.**

inverted position: In frequency division multiplexing (FDM), a position of a translated channel in which an increasing signal frequency in the untranslated channel causes a decreasing signal frequency in the translated channel. *See also* **down-converter, erect position, frequency division multiplexing, frequency translation, up-converter.**

inverter: 1. In electrical engineering, a device for converting direct current (dc) into alternating current (ac), in contrast to a rectifier that converts ac to dc. **2.** In computer and communications systems, a device or a circuit, such as a transformer or a NEGATION gate, that inverts the polarity of a signal or a pulse. **3.** In analog systems, a device that has an output analog variable that is equal in magnitude to its input analog variable but is of opposite algebraic sign. *Deprecated synonym* **negation circuit.** *Refer to* **Fig. L-6 (529).**

invitation: In communications system operations, a contact by a station to allow another station to transmit a message if the contacted station has a message to transmit.

inward dialing: *See* **direct inward dialing.**

inward wide area telephone service (INWATS): A network-provided service feature that allows users to receive calls from within specified rate areas without a charge to the call originator. *Synonym* **eight-hundred (800) service.** *See also* **call.**

INWATS: inward wide area telephone service.

I/O: input-output.

I-O: input-output.

IOC: input-output controller.

I/O channel: *Synonym* **input-output channel.**

I/O controller: *Synonym* **input-output controller.**

I/O device: *Synonym* **input-output device.**

ion: An electrically charged particle that (a) usually is formed when there is an excess or a deficiency of electrons forming a parent atom or molecule, (b) when stationary, forms a static charge, and (c) when moving, constitutes an electric current. *Note 1:* Examples of ions are (a) a sodium atom with the single electron in its outer shell, i.e., energy level, missing, the ion thus being a positive ion, and (b) a chlorine atom that has its outer shell completed with an extra electron, the ion thus being a negative ion. An electronic bond, i.e., an ionic bond, is formed when the sodium atom electron fills the outer shell of the chlorine atom, forming sodium chloride, i.e., table salt. *Note 2:* Positive and negative ions continue to exist even though they are bound in pairs and groups in solid materials. In liquids and gases, ions can become dissociated and can wander about the fluid. *See also* **hydroxyl ion absorption, ion absorption.**

ion absorption: Absorption of lightwave energy by ions embedded in the propagation medium. *See* **hydroxyl ion absorption.** *See also* **ions, propagation medium.**

ion density: The concentration of ions in materials or free space, given as the number of ions per unit volume. *See also* **ionosphere.**

ion exchange process: In the manufacture of graded index optical fibers, a process in which the fiber is fabricated by replacing undesirable ions with desirable ions in order to achieve the desired composition and structure of the material used to fabricate the fiber. *Synonym* **ion exchange technique.** *See also* **chemical vapor deposition technique, double crucible method, graded index profile.**

ion exchange technique: *Synonym* **ion exchange process.**

ionic double refraction: *See* **magnetoionic double refraction.**

ion implant: To control the addition of special ions to semiconducting materials in the manufacture of integrated circuit microchips, i.e., large-scale integrated (LSI) circuit chips. *See also* **diffusion, dopant.** *Refer to* **Fig. I-3 (457).**

ionization: The production of ions by the removal of electrons from, or addition of electrons to, atoms or molecules. *Note 1:* Ionized layers of the stratosphere cause reflection, refraction, scattering, and absorption of electromagnetic waves that are incident upon them. Ionospheric reflection enables radio communications over greater distances around the Earth's surface than can be achieved by ground waves. *Note 2:* Dissociation of ions in fluids allows conduction of electric current, such as in gas lamps and storage batteries. *See also* **ionosphere.**

ion laser: A laser in which ionized gases, such as argon, krypton, and xenon, are used as the lasing material to produce lasing action.

ionosphere: 1. The region of the upper atmosphere that (a) is electrically conducting, (b) is greater than approximately 50 km above the Earth's surface, and (c) has free electron and ion densities sufficient to affect radio wave propagation. **2.** The part of the atmosphere in which reflection or refraction of electromagnetic waves occurs. **3.** The part of the atmosphere, extending from about 70 km (kilometers) to 500 km, in which ions and free electrons exist in sufficient densities to reflect or refract electromagnetic waves. *Note 1:* Below about 50 km above the Earth's surface, free electrons and ions recombine rapidly. Above about 50 km they persist for a time long enough to affect propagating radio waves. *Note 2:* A radio wave propagating vertically from the Earth will be reflected at an altitude where the free-electron density and the frequency of the radio waves are related by the relation $N_e = 1.24 \times 10^4 f_2$ where N_e is the free-electron density in number of electrons per cubic centimeter, and f is the frequency of the radio wave in megahertz. *Note 3:* Free-electron density increases with altitude up to approximately 350 km. Therefore, the higher the frequency of the radio wave is, the higher the altitude where reflection or refraction takes place up to the critical frequency, approximately 10 MHz (megahertz), with vertical propagation. Frequencies above the critical frequency propagate through the ionosphere without significant reflection or refraction. *See also* **critical frequency, D layer, D region, E layer, E region, F layer, F region, reflecting layer.**

ionosphere sounder: A device that transmits signals used to determine ionospheric conditions.

ionospheric absorption: Absorption that occurs as a result of interaction between (a) an electromagnetic wave and (b) gas molecules and ions in the ionosphere. *See also* **absorption, ionosphere.**

ionospheric disturbance: In the D region of the ionosphere, an increase in the ionization, i.e., an increase in ion densities, that (a) is caused by solar activity and (b) results in greatly increased radio wave absorption. *See also* **ionosphere, sudden ionospheric disturbance.**

ionospheric forecast: *See* **long-term ionospheric forecast, short-term ionospheric forecast.**

ionospheric forecasting: The forecasting of ionospheric conditions and the preparation and the distribution of electromagnetic wave propagation data that are derived from the forecasting. *See also* **ionosphere, ionospheric message format, long-term ionospheric forecast, short-term ionospheric forecast.**

ionospheric forward scatter: *Synonym* **forward propagation ionospheric scatter.**

ionospheric layer: One of the several ionization layers of the ionosphere. *See also* **D layer, E layer, F layer, ionosphere.**

ionospheric message: A message that contains information on the condition of the ionosphere, usually prepared from ionospheric forecasting. *Note:* An example of an ionospheric message is a message that contains a series of dates on which propagation disturbances will begin, to continue over the next 30 days, followed by a date up to which a 7-day daily forecast is given, followed by a seven-digit code in which each digit represents one of nine propagation conditions that range from "useless" to "excellent," followed by the time-date group. An example of an ionospheric message might be 1-2 Aug, 10-12 Aug, up to 7 Aug 2455666, DTG 020001Z. *See also* **ionospheric forecasting, ionospheric message format, long-term ionospheric forecast, short-term ionospheric forecast.**

ionospheric message format: A format used for preparing ionospheric messages in which specific conventions and codes are used to express ionospheric conditions obtained in ionospheric forecasting. *See also* **format, ionospheric message, long-term ionospheric forecast, short-term ionospheric forecast.**

ionospheric reflection: A bending by reflection and refraction of electromagnetic waves back toward the Earth as they strike successive ionized layers of the ionosphere. *Note:* The amount of bending and the skip distance depend on the extent of penetration, which is a function of frequency, incidence angle, direction of polarization of the wave, the height of the ionosphere, and ionospheric conditions, such as the ion density. *See also* **ion density, reflection, refraction, skip distance.**

ionospheric scatter: The propagation of radio waves by scattering as a result of irregularities or discontinuities in the ionization of the ionosphere. *Note:* Ionospheric scatter may be in the forward direction, i.e., forward scatter, or in the backward direction, i.e., backscatter. *See also* **backscatter, backward propagation ionospheric scatter, forward propagation ionospheric scatter, forward scatter.**

ionospheric sounding: Sounding that (a) provides real time high-frequency propagation data, (b) uses a basic system that consists of a transmitter and a receiver, (c) measures the time delay between transmission and reception, (d) uses radio wave propagation velocity that allows calculation of ionospheric layer heights, (e) may be vertical incident sounding using a collocated transmitter and receiver that directs a range of frequencies vertically to the ionosphere and measures the times and values of the reflected returned signals to determine the ionosphere layer heights, (f) is also used to determine the critical frequency, and (g) may use oblique sounding that uses a transmitter at one end of a circuit and a receiver, usually with an oscilloscope-type display, i.e., an ionogram, at the other end, such that the transmitter emits a stepped or swept frequency signal that is displayed or measured at the synchronized receiver, the measurement then being converted to a time delay to determine the ionospheric layer height while the ionogram shows layer height as a function of frequency. *See also* **sounding.**

ionospheric turbulence: A permanent ongoing turbulence or disturbance in the ionosphere that (a) has the effect of scattering incident electromagnetic waves and (b) results in irregularities in the composition of the ionosphere that change with time, causing changes in radio wave propagation, such as reflection properties, skip distance, fading, local intensification, and distortion of incident radio waves.

ion overtone absorption: *See* **hydroxyl ion overtone absorption.**

IOS: International Organization for Standardization. *Properly abbreviated as* **ISO.**

IP: intelligent peripheral, Internet Protocol.

IPC: information processing center.

IPM: incremental phase modulation.

IPX: Internetwork Packet Exchange.

IQF: intrinsic quality factor.

IR: information retrieval, infrared.

IRAMS: innovative real time antenna modeling system.

irradiance: 1. In an electromagnetic wave, the power flow rate, a vector quantity, that propagates in a specific direction at a particular point in a propagation medium, expressed as the energy per unit time passing through a point per unit cross-sectional area that is perpendicular to the direction of propagation of the wave. The power density diminishes with distance from the source, the attenuation being caused by absorption, dispersion, diffusion, deflection, reflection, scattering, and diffraction, as well as geometric spreading. *Note:* Irradiance usually is expressed in watts per square meter. **2.** A measure of the power that may be attributed to a particular frequency band, usually expressed in watts per hertz. **3.** Radiant power per unit area incident upon a surface. *Note 1:* Irradiance is usually expressed in watts per square meter but may also be expressed in joules per cubic meter and lumens per square meter. All three-dimensional concepts give rise to the colloquial term "power density." *Note 2:* The radiant power in a beam is the total power obtained by integrating the irradiance over the cross-sectional area of the beam. If the radiant power in a beam is distributed uniformly over the cross-sectional area of the beam, the irradiance is the radiant power divided by the cross-sectional area, i.e., the average irradiance. *Note 3:* Because the radiant power is propagating in the form of an electromagnetic wave with a velocity given by the relation $v = c/n$ where c is the speed of light in a vacuum and n is the refractive index of the material at the point at which the speed is being considered, the irradiance can also be expressed in terms of the energy per unit volume of vacuum or propagation medium, expressed as joules per cubic meter. *Note 4:* Irradiance can be increased at a point by causing convergence of waves, i.e., of rays, such as might be accomplished by means of a convex lens, concave mirror, or reflecting antenna, i.e., a dish. A positive lens only a few centimeters in diameter and less than a centimeter thick can increase the irradiance of solar radiation, i.e., sunlight, when focused at a point on the surface of a combustible material, to a level high enough to ignite the material, but the radiant power that enters the lens and that is incident on the surface are about the same. *Note 5:* It may be said that "irradiance" is "incident radiation (ir)" on a surface, rather than "radiance" radiating from a source of radiation, such as a light-emitting diode, a laser, a radio antenna, or the exit end of an energized optical fiber. *Synonym* **power density.** *Deprecated synonyms* **power density, radiant flux density.** *See* **spectral irradiance.** *See also* **geometric spreading, incidence, intensity, irradiance, noise power density, optical energy density, optical power density, propagation medium, radiance, radiant power, radiation, radiometry, reflection, spectral irradiance.** *Refer to* **Table 5, Appendix B.**

irradiation: The product of the irradiance and the time, i.e., the time integral of the irradiance, equivalent to the radiant energy incident per unit area during a given time interval. *Note 1:* An example of irradiation of 100 $J \cdot m^{-2}$ (joules per square meter) is obtained when an irradiance of 25 $W \cdot m^{-2}$ (watts per square meter) is continuously incident for 4 s (seconds). *Note 2:* Irradiation is a measure of incident radiation and is not a measure of energy absorption. *See also* **irradiance, rad, radiance.**

irrational number: A real number that (a) is not expressible as a quotient of two integers, i.e., cannot be obtained by dividing one integer by another nonzero integer, (b) is neither an imaginary nor a complex number, and (c) is not a rational number. *Note:* Examples of irrational numbers are (a) $\sqrt{2} = 1.414 \ldots$, $3^{1/2} = 1.732 \ldots$, $e = 2.718 \ldots$, and $\pi = 3.1416. \ldots$ *See also* **integer, rational number.**

irrelevance: In information theory, the conditional information entropy measured as the occurrence of specific messages at a message sink, given the occurrence of specific messages at the message source connected to the message sink by a specified channel. *Synonyms* **prevarication, spread.** *See also* **channel, conditional information entropy, information theory, message sink, message source.**

ISA: industry standard architecture: In communications, computer, data processing, and control systems, conventional systems architecture in which components meet industrial standards.

ISB: independent sideband.

ISDN: Integrated Services Digital Network.

ISDN-1: integrated services digital network-one.

ISM: industrial, scientific, and medical.

ISO: International Organization for Standardization. *The proper abbreviation for all languages is* **ISO.**

isochrone: In the propagation of electromagnetic waves, such as radio waves and lightwaves, a line on a map or a chart joining points that have a given, i.e., a constant, propagation time from the transmitter.

isochrone determination: Radiolocation in which a position line is determined by the difference in propagation times of signals from the same source but propagating along different paths. *See also* **path, propagation time delay, radiolocation.**

isochronous: 1. In a periodic signal, pertaining to transmission in which the time interval separating any two corresponding transitions is equal to the unit interval or to a multiple of the unit interval. **2.** Pertaining to data transmission in which corresponding significant instants of two or more sequential signals have a constant phase relationship with each other. *Note:* "Isochronous" and "anisochronous" are characteristics, while "synchronous" and "asynchronous" are relationships. *See also* **anisochronous, asynchronous transmission, heterochronous, homochronous, isochronous burst transmission, isochronous distortion, isochronous modulation, isochronous signal, mesochronous, plesiochronous.**

isochronous burst transmission (IBT): In a data network, transmission performed by interrupting, at controlled intervals, the data stream that is being transmitted. *Note 1:* Isochronous burst transmission enables communications between data terminal equipment (DTE) and data networks that operate at dissimilar data signaling rates, such as when the information bearer channel rate is higher than the DTE output data signaling rate (DSR). *Note 2:* In isochronous burst transmission, the binary digits are transferred at the information-bearer channel rate. The transfer is interrupted at intervals in order to produce the required average data signaling rate (DSR). The interruption is always for an integral number of digit periods. *Note 3:* Isochronous burst transmission has particular application where envelopes are being transferred between data circuit-terminating equipments (DCEs), and only the bytes contained within the envelopes are being transferred between the DCE and the data circuit-terminating equipment (DTE). *Synonyms* **interrupted isochronous transmission.** *Deprecated synonym* **burst isochronous transmission.** *See also* **information transfer, isochronous.**

isochronous demodulation (DM): Demodulation in which the time interval separating any two significant instants is equal to the unit interval or to an integral multiple of the unit interval. *See also* **demodulation, integral, isochronous, modulation, plesiochronous, significant instant.**

isochronous distortion: In a digital system, the difference between (a) the measured modulation rate and (b) the theoretical modulation rate, measured as the ratio of (i) the maximum measured difference between the actual and theoretical intervals that separate any two significant instants of modulation or demodulation to (ii) the unit interval, such that the difference is measured without regard to sign, the significant instants occur at significant condition transition points in the signal, and significant instants may not necessarily be consecutive. *Note 1:* Isochronous distortion usually is expressed in percent. The result of the measurement should be accompanied by an indication of the limited period during which the measurement was made. *Note 2:* For a prolonged modulation or demodulation, it is useful to consider the probability that an assigned value of the degree of isochronous distortion will be exceeded. *See also* **degree of isochronous distortion, demodulation, distortion, isochronous, modulation, significant condition, significant instant, unit interval.**

isochronous distortion ratio (IDR): In data transmission, the ratio of (a) the absolute value of the maximum measured difference between the actual and the theoretical intervals separating any two significant instants of modulation or demodulation to (b) the unit interval. *Note 1:* The significant instants need not necessarily be consecutive. *Note 2:* The isochronous distortion ratio is usually expressed in percent, i.e., percent isochronous distortion. *Note 3:* The result of the measurement of the isochronous distortion ratio (IDR) should be annotated by an indication of the performance measurement period, which usually is limited. For a prolonged modulation or demodulation measurement period, the probability that an assigned value of the IDR will be exceeded should be considered. *See also* **distortion, isochronous distortion.**

isochronous modulation (IM): Modulation in which the time interval separating any two significant instants is equal to the unit interval or to a multiple of the unit interval. *See also* **demodulation, isochronous, modulation, plesiochronous.**

isochronous restitution: Telegraph restitution in which the time interval that separates any two significant instants is theoretically equal to the unit interval or an integral multiple of the unit interval. *See also* **significant instant, significant interval, telegraph restitution, unit interval.**

isochronous signal: A signal in which the time interval separating any two significant instants is equal to the unit interval or to an integral multiple of the unit interval. *Note 1:* Variations in the time intervals are constrained within specified limits because there is no way of generating exactly equal unit intervals. *Note 2:* "Isochronous" is a characteristic, i.e., pertains to an entity, while "synchronous" is a relationship between entities. *See also* **integral, isochronous, signal, significant instant, unit interval.**

isochronous transmission: Transmission of a signal (a) in which there is always an integral number of unit intervals between any two significant instants, and (b) that is characterized by (i) a constant pulse repetition rate, (ii) a constant time interval, or multiples thereof, between electric field strength transitions or voltage transitions, and (iii) gating by a controlled clock. *See also* **field strength, integral, pulse repetition rate, significant instant.**

isolated: 1. A condition of an entity such that it cannot be influenced by any event that occurs outside of itself, such as pressure, shock, vibration, force fields, or corrosive environments. **2.** In electric circuits and communications systems, the state of a point such that its state cannot be changed or influenced by any stimulus from outside the circuit or the system to which it is connected, i.e., the point is not influenced by any outside or environmental electrostatic, magnetic, magnetostatic, or electromagnetic field, and is, therefore, also insulated, i.e., it is inaccessible to electric currents and voltages as well as fields from the outside. *Note:* The isolated state is somewhat achievable by means of nonconducting insulators and certain types of conducting shields, such as metal containers. *See also* **insulated.**

isolation: In a fiber optic coupler, the extent to which optical power from one signal path is prevented from reaching another signal path. *Note:* Path A is considered to be isolated from path B if the amount of optical power from path B coupled into path A is zero or extremely small compared to the power levels.

isolator: *See* **fiber optic isolator, optical isolator, waveguide isolator.**

isothermal region: *Synonym* **stratosphere.**

isotropic: 1. Pertaining to a material with properties, such as density, electrical conductivity, electric permittivity, magnetic permeability, chemical composition, and refractive index, that do not vary with distance or direction, i.e., their spatial derivatives are zero in all directions, and (b) in which an electromagnetic wave propagating in the material will be affected in the same way regardless of the direction of propagation and regardless of the direction or type of polarization of the wave. **2.** Pertaining to a material with magnetic, electrical, or electromagnetic properties that do not vary with the direction of static or propagating magnetic, electrical, or electromagnetic fields within the material. *Note:* Isotropy is exhibited in materials that have a uniform symmetrical molecular structure, such as is found in most glasses and plastics. Thus, the polarization of electromagnetic waves in isotropic materials is not affected by the direction of propagation. *See also* **anisotropic, birefringence, birefringent medium, isotropic.**

isotropic antenna: A theoretical antenna, such as a light source or radio antenna, that (a) radiates with equal irradiance, i.e., with equal power density, or equal field intensity, in all directions or (b) receives equally in all directions, i.e., has the same aperture in all directions. *Note 1:* An isotropic antenna can only be approximated in actual practice. However, it is a convenient conceptual reference for comparing and expressing the directional properties of actual antennas, such as radio antennas and optical sources. *Note 2:* An isotropic antenna has a hypothetical directive gain of unity in all directions. *Note 3:* The ratio of (a) the power gain when referred to an isotropic antenna to (b) that referred to a standard half-wave dipole antenna is equivalent to 2.15 dB. *See also* **antenna, directive gain, half-wave dipole antenna, light source.**

isotropic gain: *Synonym* **absolute gain.**

isotropic propagation medium: Pertaining to a material medium in which properties and characteristics of the medium are everywhere the same in reference to a specific phenomenon, such as a medium that has electromagnetic properties, e.g., the refractive index, at each point that are independent of the direction of propagation and polarization of a wave propagating in the medium.

isotropic space loss: At a specified frequency, the ratio of (a) the available power from a loss-free standard receiving antenna to (b) the power radiated by an identical transmitting antenna when the propagation medium is free space, i.e., there is no atmospheric attenuation, scattering, reflection, or absorption. *See also* **free space, propagation medium.**

ITA: international telegraph alphabet.

item: *See* **configuration item, data item, menu item.**

item control: *See* **configuration item control.**

iterative: Pertaining to a calculation that (a) obtains a desired result by repeating a cycle or a sequence of operations, each cycle bringing the result closer to the true value, (b) usually is used when a function contains the result of the calculation, (c) has the result obtained in each cycle placed back in the function where it is cited, and (d) has the process repeated until the function is precisely satisfied to a desired precision, or until further repetition does not produce a change in the result.

iterative impedance: At a pair of terminals of a four-terminal network, the impedance that will terminate the other pair of terminals in such a manner that the impedance measured at the first pair is equal to this terminating impedance. *Note:* The iterative impedance of a uniform transmission line is the same as the characteristic impedance. *See also* **characteristic impedance, impedance, impedance matching.**

ITS: Institute for Telecommunication Sciences, international temperature scale.

ITU: International Telecommunication Union.

ITU-R: International Telecommunication Union-R.

ITU-T: International Telecommunication Union-T.

IVD: inside vapor deposition.

IVDT: integrated voice data terminal.

IVHS: 1. intelligent vehicle highway system. 2. *Formerly,* **Intelligent Vehicle Highway Society;** *currently,* **Intelligent Transportation Society of America.**

IVPO: inside vapor phase oxidation.

IVPO process: *See* **inside vapor phase oxidation process.**

IXC: interexchange carrier.

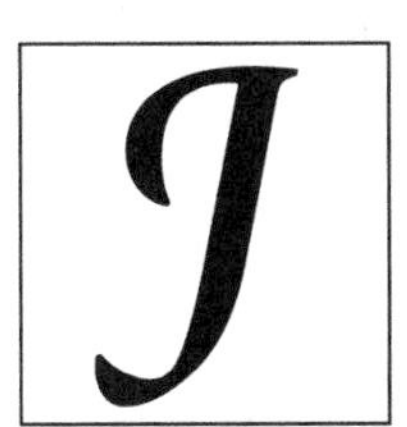

jabber: In a data station of a local area network (LAN), transmissions that continue beyond the time interval allowed by the operating protocol.

jack: In a telephone switchboard, a socket, hole, or receptacle (a) that is mounted on the face of the switchboard and (b) into which a plug can be inserted to complete a connection. *See* **connection, modular jack, monitor jack, registered jack.**

jacket: In a cable, a tough, usually fluid-resistant, layer of flexible material, such as plastic, that (a) is applied over the propagation medium, (b) protects the medium during spooling, storage, shipping, payout, installation, and service, and (c) may be in several separate layers, such as often occurs in communications cables. *Note:* Fiber optic cables often have an inner jacket surrounded by armor or strength members over which an outer jacket is applied. *See* **bundle jacket, cable bundle jacket, cable jacket, fiber optic cable jacket, optical fiber jacket.** *Refer to* **Fig. C-6 (142).**

jacketed cable: *See* **tight-jacketed cable.**

jacket leak: A measure of the imperviousness of a cable jacket, such as a jacket on a fiber optic, electrical, or hybrid cable, determined by one or more techniques, such as (a) submerging the cable in water and applying a vacuum, (b) submerging the cable in water and applying pressure, and (c) submerging the cable in a gas, applying a vacuum, and detecting the gas.

jackfield: A group of terminals that is used for making connections to jacks. *See* **channel patching jackfield, patching jackfield.**

jam: 1. To deliberately prevent the successful operation of a communications system by interfering with the transmission of signals in the system by using other signals to cause interference. **2.** To cause a halt to the dynamic interaction of moving parts of a mechanical system by a forced interference among the fitting of the parts. **3.** The binding caused by a forced interference of the mechanical parts of a system. *Note:* In a mechanical system or mechanism, a jam can be removed by various means, such as by removing the interference fit, by lubrication, and by adjustment. *See* **antijam, paper jam.** *See also* **jammer, jamming.**

jammer: A transmitting system that is designed specifically for jamming purposes. *See* **airborne radar jammer, automatic search jammer, automatic spot jammer, barrage jammer, coherent repeater jammer, deception jammer, direct noise amplification jammer, expendable jammer, locked pulse radar jammer, manual spot jammer, multipurpose jammer, noise jammer, preset jammer, reflecting jammer, repeater jammer, search and lock jammer, spot jammer, sweep jammer, swept jammer, swept repeater jammer.** *See also* **jamming.**

jammer analyzer: A signal processing device that (a) is used to determine specific characteristics of received jamming signals, such as the identity of the source, the type of modulation, and the power level of a received jamming signal, and (b) may be used to control an antijamming device. *See also* **antijamming, jamming, signal processing.**

jammer area coverage: The geographical area in which a jammer is capable of producing an effective jamming signal.

jammer modulation: The specific modulation of the carrier wave that is transmitted by a jammer. *Note:* Examples of jammer modulation are random noise modulation, locked pulse modulation, swept frequency carrier wave modulation, frequency-hopped modulation, and spread spectrum modulation. Frequency-hopped modulation and spread spectrum modulation require a variable carrier frequency. *See also* **carrier, jamming, modulation.**

jammer out-tuning: Changing the tuning of a radio, radar, or video receiver so as to reduce the effects of jamming at a particular frequency. *Note:* Out-tuning may be accomplished by selecting a faster tuning rate than a spot jammer can follow or by tuning to a portion of the frequency spectrum that the jammer cannot reach. *See also* **frequency spectrum, jamming, spot jammer, tuning.**

jammer polarization: The electromagnetic wave polarization that is used in a particular jammer. *See also* **jammer, polarization.**

jammer transmitter: The part of a jammer that (a) generates, amplifies, and sends the jamming signal to a transmitting antenna, (b) may also include an antenna, (c) usually is tunable to the frequency of the signal that is being jammed, and (d) may perform one or more different types of jamming, such as spot, barrage, or frequency hopping jamming. *See also* **frequency hopping, jammer, jamming, tuning.**

jamming: The deliberate introduction of interference into a communications channel with the intent of preventing error-free reception of transmitted signals, usually by radiating, reradiating, or reflecting signals, with the object of impairing the use or reducing the effectiveness of electronic devices, thus obliterating or obscuring the information that is contained in the signals. *Note:* Lightwaves confined to optical fibers are difficult to jam except by direct access to the core of the fibers with a compatible or coherent lightwave source. *See* **antielectronic jamming, antijamming, barrage jamming, communications jamming, electronic jamming, escort jamming, formation radar jamming, infrared saturation jamming, mechanical jamming, noise jamming, periodic waveform jamming, pulse modulation radar jamming, reflective jamming, saturation jamming, selective jamming, self-protection jamming, sequential jamming, side-lobe jamming, slow sweep noise modulated jamming, spoof jamming, spot jamming, spot noise jamming, spot off-frequency jamming, standoff jamming, support jamming, sweep jamming, synchronous jamming, unmodulated continuous wave jamming.** *See also* **electronic countermeasures, electronic warfare, interference, susceptibility.** *Refer to* **Fig. J-1.**

jamming antenna: An antenna that (a) receives jamming signals from a jammer transmitter, (b) is used for jamming purposes, and (c) has a design that includes the frequency, directivity, gain, polarization, power, and beam pattern requirements of the jamming signal it transmits. *See also* **beam pattern, directivity, frequency, gain, jammer, jamming, jamming signal, polarization, power, beam pattern.**

jamming azimuth: The angle between (a) a reference direction, such as true, grid, or magnetic north, and (b) the direction that a jamming signal is transmitted by a jammer antenna. *See also* **jamming bearing, jammer, jamming signal.**

jamming bearing: The angle, measured by an observer with equipment at a given point, between (a) a reference direction, such as true, grid, or magnetic north, and (b) the direction that a jamming signal is observed to be coming from, i.e., the back azimuth of a jamming signal. *See also* **back azimuth, jamming azimuth, jamming signal.**

jamming electronic countercountermeasure: *See* **infrared jamming electronic countercountermeasure.**

jamming margin: The level of interference, i.e., the level of jamming, that a system is able to accept and still maintain a specified level of performance, such as to maintain a specified bit error ratio (BER) even though the signal to noise (S/N) ratio is decreased. *See* **antijamming margin.** *See also* **antijamming measure, bit error ratio, interference, jamming, signal to noise ratio.**

jamming measure: *See* **antijamming measure.**

jamming modulation: The modulation used by a jammer, such as noise, pulse, repeater, carrier wave, amplitude, tone, or multitone modulation. *See also* **jammer, modulation.**

jamming power: The radio frequency (rf) power that a jammer can develop. *Note:* Examples of jamming power are (a) the total emitted power, (b) the power emitted at a specific frequency, (c) the power within a specified frequency band, or (d) the power at a given point in space, expressed in terms of the irradiance or

Fig. J-1. The AN/GRC-206(v)5 ruggedized military communications suite covers the 2 MHz to 400 MHz frequency band. It provides frequency agility communications in the presence of **jamming.** (Courtesy Magnavox Electronic Systems Company, Fort Wayne, IN.)

the field strength. *See also* **field strength, irradiance, jammer.**

jamming radar display: *See* **spot-barrage jamming radar display, spot-continuous-wave jamming radar display.**

jamming report: A report that (a) is prepared by a transmitting station, such as a radio, radar, or television station and (b) contains information on detected jamming, such as the identification of the frequency, channel, or band that is being jammed, the type of jamming signal, the time and the duration of the jamming, the jamming signal strength, the jammer bearing, the antijamming measure that was or is being taken by the reporting station, the effect of the jamming signal on the reporting station transmissions, and the reaction of the jamming source to the antijamming measure that was taken. *See also* **antijamming, antijamming measure, jammer bearing, jamming.**

jamming signal: A modulated or unmodulated carrier signal that (a) is transmitted for the purpose of electronic jamming and (b) may be used for other purposes, such as to transmit information, provide guidance signals, or serve as a homing beacon. *See also* **carrier, electronic jamming, homing beacon, jamming.** *Refer to* **Fig. J-2.**

jamming to signal ratio: At any point, the ratio of (a) the jamming signal power to (b) the station transmitted power at that point, i.e., the reciprocal of the signal to noise ratio, in which the signal is the useful or intelligence-bearing transmitted or received signal, and the jamming power is the noise. *Note 1:* The measurement conditions of the powers and the point of measurement, such as the point in space or in the circuits of a system, such as a radio, radar, beacon, or video system, should be indicated. *Note 2:* An example of the jamming to signal ratio is the ratio of the jamming power to the information-bearing signal power at the terminal of a receiving antenna. *See also* **signal to noise ratio.**

jam signal: A signal that carries a bit pattern sent by a data station to inform the other stations that they must not transmit. *Note 1:* In carrier-sense multiple access with collision detection (CSMA/CD) networks, the jam signal indicates that a collision has occurred. *Note 2:* In carrier-sense multiple access with collision avoidance (CSMA/CA) networks, the jam signal indicates that the sending station intends to transmit. *Note 3:* "Jam signal" should not be confused with "electronic jamming." *See also* **carrier-sense multiple access with collision avoidance, carrier-sense multiple access with collision detection.**

Fig. J-2. The AN/VRC-83 VHF/UHF vehicular radio set provides reliable communications in the presence of **jamming signals.** (Courtesy Magnavox Electronic Systems Company, Fort Wayne, IN.)

J band: The band of frequencies that (a) lies in the range from 10 GHz (gigahertz) to 20 GHz and (b) consists of ten numbered channels, each channel being 1 GHz wide. *Note:* The J band overlaps parts of the obsolete X band, K band, and M band. Frequencies in the range from 5.30 GHz (gigahertz) to 8.20 GHz overlap parts of the old C band, G band, X band, and H band, which are also obsolete.

JCL: job control language.

JCLAS system: joint computer-aided acquisition and logistics support system.

jerkiness: In video systems that usually display continuous full motion, the quality of the displayed motion such that the motion appears as a rapid sequence of distinct still motion images, such as a sequence of distinct "snapshots."

jet control: *See* **pulse-jet control.**

jet printer: *See* **laser jet printer.**

jinking: The flying of an aircraft in a path in which there is a continuous change in altitude, heading, i.e., azimuth, and sometimes air or ground speed.

jitter: 1. Referenced to a given instant, a point in a transmission system, or a fixed time frame, abrupt and spurious temporal or spatial variations of signal elements, such as variations in (a) pulse duration, (b) intervals between pulses, or (c) amplitude, frequency, or phase of successive pulses. *Note 1:* Jitter must be qualified, such as time, amplitude, frequency, or phase jitter. *Note 2:* The jitter form must be specified, such as pulse-delay-time jitter or pulse duration jitter. *Note 3:* In jitter measurements, the time-related or amplitude-related variation must be specified, such as average, rms, peak to peak, or maximum measurements. *Note 4:* The low frequency cutoff for jitter is 1 Hz (hertz), i.e., jitter frequency is above 1 Hz. Jitter, swim, wander, and drift are in the order of decreasing frequency. *Note 5:* Sources of jitter include signal-pattern-dependent laser turn-on delay, noise induced at a gating turn-on point, gating hysteresis, and variations in signal delay that (a) accumulate on a data link and (b) are caused by cable vibration. **2.** Rapid or jumpy undesired movement of display elements, display groups, or display images about their normal or mean positions on the display surface of a display device. *Note:* Examples of jitter are (a) movement of display images on the fiber optic faceplate of a fiberscope when the image source at the objective end of the aligned bundle of optical fibers is vibrated relative to the aligned bundle, and (b) movement of images on the screen of a cathode ray tube (CRT) when the bias voltage on the deflecting plates oscillates at, say, 5 to 10 Hz. If the jitter rate is above 15 Hz, or if the persistence of the screen is long, the jitter will cause the image to appear blurred because of the persistence of vision. *See* **longitudinal jitter, phase jitter, pulse jitter, time jitter.** *See also* **drift, fall time, phase perturbation, swim, time jitter, wander.**

jittered pulse repetition rate: A random, quasirandom, or pseudorandom variation of the pulse repetition rate of a transmitting system, such as a radio, radar, or video transmitter. *Note:* When performed deliberately as an electronic countercountermeasure (ECCM), a jittered pulse repetition rate provides some discrimination against repeater jammers because of the difficulty they may experience in following the jittered pulse repetition rate. *See also* **electronic countercountermeasure, jitter, pulse repetition rate, repeater jammer.**

job: 1. In communications, computer, data processing, and control systems, the data that (a) define a specific amount of work that a system is to perform, and (b) usually include the data needed for the work to be accomplished, such as computer programs, computer routines, files, special instructions, control procedures, and instructions to the operating system. **2.** In computing and data processing, a unit of work that (a) is defined by a user, (b) is accomplished by a computer or data processing system, (c) is identified by a label, and (d) usually includes a set of computer programs, files, and control statements to the operating system. **3.** A specific amount of processing that a data processing system is to perform to obtain a specified result. *See also* **computer program, job control language, master file, operating system, remote job entry.**

job control language (JCL): A problem-oriented language that is used to (a) express the statements that describe a job, (b) identify and describe jobs to a computer, (c) describe the requirements of a job, usually to an operating system, or (d) control the execution of computer programs that are specified for jobs. *See also* **job, operating system, problem-oriented language.**

job entry: In communications, computer, data processing, and control systems, to enter a job into a system in such a manner that it can be identified by the operating system and executed by the system. *See* **remote job entry.** *See also* **job, job control language.**

job recovery control file: *Synonym* **backup file.**

Johnson noise: *Synonym* **thermal noise.**

join gate: *Synonym* **OR gate.**

joint: For optical fibers and fiber optic cables, a splice or a connector. *See* **fiber optic rotary joint, multifiber joint.**

joint communications: The common use of communications facilities by two or more military services that belong to the same nation. *See also* **combined communications.**

joint computer-aided acquisition and logistics support (JCLAS) system: For the U.S. Department of Defense, an integrated system that (a) provides for acquiring, creating, improving, and managing logistic technical information, (b) links participating sites via local area networks (LANs) and wide area networks (WANs), (c) provides advanced imaging and display support, and (d) is designed to reduce voluminous and costly paperwork in the process of acquisition and maintenance of military systems.

joint denial gate: *Synonym* **NOR gate.**

jointed optical fiber: An optical fiber (a) that consists of several lengths joined in sequence, such as an optical fiber 10 km (kilometers) long consisting of ten 1-km lengths, (b) in which the joints may be accomplished by splices or connectors, and (c) in which the joints usually introduce insertion losses, such as reflection, absorption, and scattering losses. *See also* **connector, coupling loss, joint, insertion loss, optical fiber, splice.**

joint information content: In information theory, the information content that is conveyed by the occurrence of two events of definite joint probability. *See also* **information content, information theory.**

joint multichannel trunking and switching system: In military communications, the composite multichannel trunking and switching system formed from (a) assets of the U.S. Armed Services, (b) assets of the Defense Communications System (DCS), (c) assets controlled by the Joint Chiefs of Staff, and (c) other available systems that provide an operationally responsive, survivable, communications system, preferably in a mobile transportable recoverable configuration, for the joint force commander in an area of operations.

Joint Tactical Information Distribution System (JTIDS): An advanced information distribution system that (a) provides secure integrated communications, navigation, and identification (ICNI) capability for application to military tactical operations and (b) allows direct communications among the armed services, including ground, air, and sea forces, during tactical operations, i.e., during military operations.

Joint Telecommunications Resources Board (JTRB): The board required to be established by Section 2(b)(3) of Executive Order No. 12472 to assist the Director of the Office of Science and Technology in the exercise of assigned nonwartime emergency telecommunications functions.

joint uniform telephone communications precedence system (JUTCPS): A precedence system of the U.S. armed services that (a) is used in various communications systems, such as the Automatic Voice Network (AUTOVON), the Automatic Digital Network (AUTODIN), and the Automatic Secure Voice Communications System (AUTOSEVOCOM), and (b) applies to messages and calls. *See also* **call, message, precedence, precedence designation.**

journal: 1. In data processing operations, a chronological record of the operations that may be used to reconstruct a previous or an updated version of a file. *Synonym* **log. 2.** In database management systems, the record of all stored data items that have values changed as a result of processing and manipulation of the data.

JOVIAL: A procedure-oriented computer programming language that (a) is derived from ALGOL, (b) controls data by the byte or by the bit, and (c) has had specific application in command and control systems, such as satellite maneuver control, and has also been used in commercial and scientific applications. *Note:* JOVIAL is derived from "Jule's (Schwartz) own version of the international algorithmic language." Jovial was developed by the System Development Corporation. *See also* **ALGOL, programming language.**

joystick: 1. A manually operated lever that is used to interact with a system, such as a computer and display system, to perform various functions, such as (a) control the movement of display elements, display groups, or display images to other new display positions in the display space on the display surface of a display device, (b) generate or enter coordinate data, (c) select items from a menu, such as symbols and characters, (d) indicate a selection from a set of options or choices, and (e) indicate a display position. **2.** In control systems, a manually operated lever that is used to control the movement of a device or a system, such as an aircraft, a vehicle, or a remote-controlled mobile vehicle. *See also* **control ball, mouse.**

JTIDS: Joint Tactical Information Distribution System.

JTRB: Joint Telecommunications Resources Board.

Julian calendar: The calendar (a) that was introduced by Julius Caesar in Rome in 46 B.C., (b) in which the year consisted of 365 days, every fourth year having 366 days, (c) that established the 12-month year, (d) in which each month had 30 or 31 days, except February, which had 28 days, and (e) in which for the 366-day years, i.e., the leap years, February had 29 days, as in the Gregorian calendar of today. *See also* **Gregorian calendar.**

Julian date: The sequential-day count, reckoned consecutively from the epoch beginning January 1, 4713 B.C. *Note:* The Julian date for January 1, 1996, is 2,449,083. *Note:* In modern times, the tendency is to use January first of each year as the reference for determining "Julian date." "Day of year" should be used for this purpose rather than "Julian date." *See also* **day of year.**

jumbo group: In telecommunications carrier telephony, a grouping of six mastergroups. *Synonym* **supermastergroup.** *See* **channel jumbo group.** *See also* **carrier, group, telecommunications carrier, telephony.**

jumbo optical fiber: A relatively large optical fiber that (a) has a core diameter on the order of 500 μm (microns) or larger, (b) is used to replace fiber optic bundles in cables, and (c) can transmit sufficient optical power for (i) signaling, such as ringing, electronic circuit operation, switching, and repeater operation, and (ii) user information transmission. *Synonym* **macrofiber.** *See also* **core, fiber optic bundle, optical fiber, ringing, signaling.**

jump: 1. In computer and data processing systems, a departure from the normal, explicit, or declared sequence in which instructions are to be executed by a computer under the control of a computer program. **2.** To provide an electrical conducting path from one point to another. *See* **conditional jump, image jump, phase jump, unconditional jump.** *See also* **computer program, cross connection, jump instruction.**

jumper: 1. *Synonym* **cross connection. 2.** *See* **fiber optic jumper.**

jumper wire: A wire that makes a direct electrical connection between two points in order that they be maintained at the same electrical potential with respect to a reference, i.e., a wire that assures that no potential difference occurs between the points it connects. *See also* **level, potential difference.**

jump instruction: In a computer program, an instruction that specifies a location in the program other than the normal, implicit, declared, or very next instruction in the sequence of instructions in the program. *See also* **computer program, jump.**

junction: 1. An optical interface. **2.** In a semiconducting material, such as that used in a transistor, diode, or laser, the contact interface between two regions of different semiconducting material. *Note 1:* Examples of junctions are the interfaces between (a) n-type materials, which are negatively doped and therefore donor materials, and p-type materials, which are positively doped and therefore acceptor material, (b) n-type materials and intrinsic materials, and (c) p-type materials and intrinsic materials. *Note 2:* At a junction, a voltage, i.e., an electrical potential gradient or energy barrier, occurs because of the migration of electric charges (holes or electrons) to the junction and the accumulation of unlike charges on either side of it. *See* **double heterojunction, fiber optic junction, heterojunction, homojunction, hybrid junction, matched junction, optical junction, positive-negative junction, series T junction, single heterojunction.**

junction call: In telephone switchboard operation, a call that requires that a connection be made between two or more switchboards that are connected by trunks. *Synonym* **trunk call.** *See also* **call, switchboard.**

junction point: *Synonym* **node.**

junction stage: In the electrical layout of a system, a stage that (a) is a characteristic point that is identifiable and (b) is accessible to a stage in the same or another system. *Note:* Examples of junction stages are (a) a stage of voice frequency channels, (b) an intermediate frequency (IF) stage in a radio receiver, (c) a stage of groups of channels, such as primary groups and secondary groups, and (d) a radio frequency stage in an amplifier that may have several stages of amplification. *See also* **channel, intermediate frequency, primary channel, radio frequency, secondary channel, voice frequency channel.**

Jupiter system: A Raven-based system that (a) is a communications management system that (a) uses a 486-type computer, (b) uses a Windows-based man-machine interface, (c) has a minimum of 8 Mb (megabytes) of random access memory (RAM), and (d) is developed by Siemens Plessey Defence Systems.

justification: 1. The movement and alignment of data with respect to a specified fixed reference. *See also* **justify. 2.** *Synonym* **bit stuffing.**

justify: 1. To shift the contents of a register or a field so that the significant character at the specified end of the data is at a particular position in the register. **2.** To align text horizontally or vertically so that the first and last characters of every line, or the first and last line of the text, are aligned with their corresponding margins. *Note 1:* In English, text may be justified left, right, or both. Left justification is the most common. *Note 2:* The last line of a paragraph usually is only left justified. **3.** To align lines of text on any designated character position. *See* **full justify, left justify, right justify.** *See also* **register.**

JUTCPS: joint uniform telephone communications precedence system.

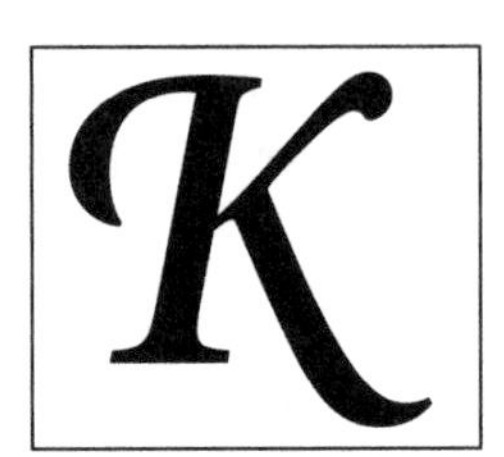

k: **1.** kilo. **2.** The symbol used to represent Boltzmann's constant. **3.** The symbol used to represent the wave number of an electromagnetic wave. **4.** In tape relay system operations, the message prosign that (a) is used to indicate an invitation to transmit, (b) is placed at the end of a transmission, i.e., the end of a message, and (c) indicates that a response is implied, necessary, requested, or expected, depending on the context of the transmitted message. *See also* **Boltzmann's constant, K, kelvin, tape relay, wave number.**

K: **1. kelvin.** **2.** 2^{10} or 1024, as in 200K bytes of storage, where the 1024 is equal to 2^{10}. *Note 1:* A 64K memory would mean 65,536 bytes of storage capacity. *Note 2:* When used alone, as in 273 K (kelvin) or as a suffix to signify an amount, the upper case, K, is used. **3.** 1000, as in (a) 22K resistor, meaning 22,000 Ω (ohms) and (b) $500K, meaning $500,000. *Note:* When used as a prefix, the lower case, k, is used, such as in (a) 50 kHz (kilohertz) for frequency, meaning 50,000 Hz (hertz), (b) 2 km (kilometers) for distance, meaning 2000 m (meters), (c) 3 kg (kilograms) for weights or mass, meaning 3000 grams, and (d) 4 kbps or $kb \cdot s^{-1}$ (kilobits per second) for signaling rates, meaning 4000 bits per second. *Note:* The "K" as a suffix and the "k" as a prefix are an abbreviation of the prefix "kilo," derived from the Greek *khilioi,* meaning "thousand." *See also* **kelvin.**

Kalman filter: A computational algorithm that processes measurements to deduce an optimum estimate of the past, present, or future state of a linear system by using (a) a time sequence of measurements of the system behavior, (b) a statistical model that characterizes the system and measurement errors, and (c) initial condition information. *See also* **digital filter.**

Kalman filtering: In image restoration, recursive filtering that (a) uses a linear time-variant discrete-time inverse filter function, (b) provides a least mean square error estimate of a discrete-time signal based on noise measurements or observations, (c) makes use of an adaptive algorithm, such as a computer program or segment, suitable for direct evaluation by a digital computer, and (d) becomes Wiener filtering if restricted to be time-invariant. *See also* **algorithm, computer program, digital computer, discrete, filter, filtering, image restoration.**

Kanji: A set of ideographic symbols or characters used in Japanese alphabets. *See also* **Hiragana, Katakana.**

Karnaugh map: A table (a) that allows the mapping of Boolean functions, (b) that is an orderly extension of the Venn diagram, (c) that consists of rows and columns numbered in sequence from left to right and top to bottom with numerals in Gray code, *i.e.,* reflected binary, 00, 01,11,10 for a 4 × 4 table, (d) in which a 1 is placed at each row and column intersection for each term present in a given Boolean function, and a 0 is placed in intersections at each row and column intersection for each term not present in the function, (e) in which each intersection of a row and a column represents the Boolean AND function, i.e., logical multiplication, (f) in which adjacent 1s represent terms that can readily be reduced to a simplified Boolean OR function, i.e., logical addition, (g) in which nonadjacent terms cannot be reduced by logical addition, and (h) which permits the immediate elimination of redundant variables in Boolean functions. *See also* **Boolean function, Gray code, logical addition, logical multiplication, Venn diagram, wraparound.** *Refer to* **Fig. K-1.**

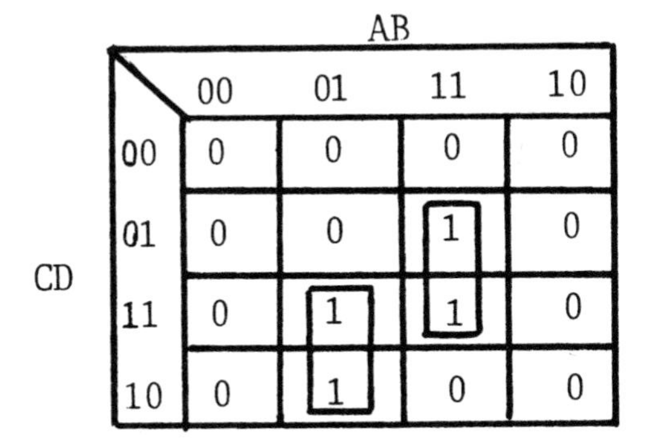

Fig. K-1. A **Karnaugh map** of the function $F = \Sigma(6, 7, 13, 15$, i.e., 0110, 0111, 1101, 1111) as indicated with the 1s in the appropriate squares. The numerals also represent Boolean functions. For example, (a) 1111 represents A AND B AND C AND D, and (b) 1101 represents A AND B AND NOT C AND D.

Katakana: The set of symbols used in one of the two Japanese phonetic alphabets in common use. *See also* **phonetic alphabet.**

K band: The band of frequencies that (a) are in the range from 20 GHz (gigahertz) to 40 GHz and (b) consists of ten numbered subbands, each of which is 2.00 GHz wide. *Note:* The K band overlaps parts of the obsolete P band and K band that consisted of frequencies in the range from 10.90 GHz (gigahertz) to 36.00 GHz, which was between the old X band and Q band, overlapping parts of the old X band and P band. The old K band was divided into 12 subbands, each designated by a distinctive letter. There were other subdivisions of the old K band that differed in the United States and the United Kingdom.

kbps: kilobits per second.

KDC: key distribution center.

keeping: *See* **station keeping.**

kelvin (K): The SI (Système International d'Unités) unit interval of thermodynamic temperature that (a) is equal to 1/273.16 of the triple point of water and (b) was adopted at the 13th General Conference on Weights and Measures (CGPM) 1967. *See* **Kelvin temperature scale.**

Kelvin (K) temperature scale: In the International System of Units (SI), the kelvin (K) is the fraction 1/273.16 of the thermodynamic temperature of the triple point of water. *Note 1:* The temperature 0 K is called "absolute zero," which is equivalent to −273.16°C (Celsius) or −459.6°F (Fahrenheit). *Note 2:* In citing temperatures on the Kelvin temperature scale, the symbol ° is not used. Thus, the temperature 280 K is read "280 kelvin." *Note 3:* The degree Celsius is called "an interval of one kelvin," rather than "one degree kelvin." *Note 4:* The Kelvin temperature scale was formerly called "the absolute temperature scale." *Note 5:* Room temperature of 20°C or 68°F is equivalent to 293 K. A temperature of 0°C is equal to 273.16 kelvins. To convert Celsius degrees to Fahrenheit degrees, use the relationship $F = 9C/5 + 32$, and to convert Fahrenheit degrees to Celsius degrees, use the relationship $C = 5(F - 32)/9$. A temperature of −40°F and −40°C may be written simply as −40° because at that temperature both scale readings are the same. *Note 6:* A temperature of 0 K (−273.16°C or −459.6°F) is called "absolute zero." Temperature is defined as the average kinetic energy of a moving molecule. Absolute zero coincides with the minimum theoretical molecular activity, i.e., thermal energy of matter, which occurs when the kinetic energy theoretically is 0. *Note 7:* For high precision measurements at extremely high and low temperatures, special temperature scales are defined by international agreement. The latest of these agreements is the international temperature scale of 1990 (ITS-90), which became effective January 1, 1990. Published tables delineate the corrections that must be made to the Kelvin temperature scale, and also to previous international temperature scales, for conformity to ITS-90. *Note 8:* The kelvin is the unit of measure, a noun. The adjective is Kelvin, such as in "Kelvin temperature scale." *See also* **kelvin, metric system.**

Kendall effect: In a facsimile system recorded copy, a spurious pattern, or other distortion, that (a) is caused by unwanted modulation products arising from the transmission of a carrier signal and (b) appears in the form of a rectified baseband that interferes with the lower sideband of the carrier. *Note:* The Kendall effect primarily occurs when the single sideband width is greater than half of the facsimile carrier frequency. *See also* **facsimile.**

Kennelly-Heaviside layer: *Synonyms* **E layer, E region.**

kernel: A module of a computer program that forms a logical entity or performs a unit function. *Note:* A special case of a kernel is the most vulnerable portion of code in a secure operating system. *See* **security kernel.**

Kerr cell: A substance, usually a liquid, that (a) has a refractive index that changes in direct proportion to the square of an applied electric field strength, (b) usually is configured so as to be part of the optical path of another system, and (c) provides a means of modulating the light in the optical path. *Note:* The Kerr cell is used to modulate light passing through a material. The modulation depends on the rotation of the polarization plane caused by an applied electric field. The amount of rotation controls the fraction of a beam that can pass through a fixed polarizing filter. *See also* **Kerr effect, Pockels cell, Pockels effect.**

Kerr effect: The creation of birefringence in a material that is not usually birefringent by subjecting the material to an electric field. *Note:* The degree of birefringence, i.e., the difference in refractive indices for light of orthogonal polarizations, is directly proportional to the square of the applied electric field strength. Thus, the Kerr effect is a polarization plane rotation that can be used with polarizers to modulate light, such as to modulate the irradiance of light incident upon a photodetector. *See also* **Kerr cell, Pockels cell, Pockels effect.**

Kerr effect sensor: A fiber optic birefringent sensor in which (a) the phase of the induced ordinary ray can be advanced or retarded relative to the phase of the extraordinary ray when an electric field is applied, (b) the electric field causes double refraction, i.e., causes an incident beam of light to be divided into two beams that propagate at different phase velocities, resulting in a rotation of the polarization plane, (c) the amount of phase shift is directly proportional to the square of the applied voltage, and (d) the direction of shift depends on the polarity of the applied voltage. *See also* **birefringence, birefringent sensor, electric field, extraordinary ray, incidence, ordinary ray, polarization plane.**

key: 1. To manually enter a representation of a character, such as a digit, letter, or punctuation mark, into a device, such as to push a button on a keyboard or actuate a switch. **2.** A device that, when actuated, connects a signal source to a transmission line. *Note:* The signal source usually is a code generator, and the device usually is a manually operated contact that is actuated by a finger. **3.** A set of characters within a set of data that (a) represents information about the data in the set and (b) identifies the set, as does a label, though the label is outside the set. **4.** Information, usually represented by a sequence of pseudorandom binary digits, used initially to set up, and periodically to change, the operations performed by cryptoequipment for the purpose of encrypting or decrypting electronic signals or for determining electronic countercountermeasure patterns, such as those used in frequency hopping or spread spectrum systems. *Note 1:* Given key may be used to produce other key. *Note 2:* In communications security (COMSEC), the term "key" has replaced the terms "cryptovariable," "key variable," "keying variable", and "variable," all of which are deprecated in this sense. *See* **absolute key, actual key, alternative key, anticlash key, arrow key, attention key, backspace key, back-tab key, calculate key, control key, cryptokey, dialing key, edit key, eject key, end key, enter key, escape key, function key, go-to key, graphic key, help key, home key, latch-down key, Morse key, name key, nonescaping key, no-print key, num-lock key, page key, pass key, pointer-movement key, print key, query key, repeat key, scroll-lock key, tab key, table function key, traffic encryption key, window key, word key.**

keyboard: 1. An input device that (a) is used to enter data by manual depression of keys and (b) causes the generation of the selected code element. **2.** An input device that (a) has an array of keys that are mounted on a panel, (b) has each key connected to a different member of a set of pulse code generating sources, (c) has each key marked with a letter, numeral, symbol, or control character, (d) is so configured that, when a key is depressed, the source that generates the particular pulse pattern that corresponds to the actuated key is connected to the outgoing line, i.e., the depression of each key is equivalent to the selection and the generation of a code that represents a character, and (e) has keys that may be used singly or in combination for additional codes. **3.** A set of keys systematically arranged for ease of selection of individual keys that are used to control equipment, such as telegraph transmitters, data storage devices, computers, computer input devices, and communications equipment. *See* **data keyboard, function keyboard, minipad keyboard, motorized keyboard, sawtooth keyboard, storage keyboard, telegraph keyboard.** *See also* **code, control character, data source, pulse code, source, telegraph transmitter.**

keyboard perferator: A perferator that (a) is provided with a keyboard and (b) is used to control the perforation of data media, such as paper tape. *See also* **data medium, keyboard.**

keyboard punch: *Synonym* **keypunch.**

keyboard send-receive teletypewriter: Equipment that (a) consists of a combined teletypewriter transmitter and a teletypewriter receiver, (b) enters data into a communications system only by means of a keyboard, and (c) furnishes data from a system only by means of a printer. *See* **keyboard, teletypewriter receiver, teletypewriter transmitter.**

key card: 1. A card that contains a pattern of punched holes or magnetic characters that represent the key for a cryptosystem during a cryptoperiod. **2.** A card (a) on which a code is recorded and (b) that, when inserted into a reader, will actuate a device, such as a lock to a vault or a door to a controlled area. *See also* **code, controlled area, cryptoperiod, cryptosystem.**

key distribution center (KDC): In communications security (COMSEC), a facility that generates and distributes key in electrical form. *See also* **cryptology, cryptomaterial, key.**

key encryption system: *See* **private key encryption system, public key encryption system.**

key escrow cryptography: Cryptography devoted to the design, development, production, and application of combined hardware and software systems to provide secure communications, such as a key escrow system that (a) is based on public key cryptography with private key held by escrow agents and public key held in communications products, (b) provides communications protection with available cryptographic algorithms, and (c) does not provide protection to classified algorithms. *See also* **key, private key, public key, encryption.**

key generator: A device that produces key in electronic form for use in communications security. *See also* **communications security, cryptokey, key.**

keying: The generating of signals by interrupting or modulating a steady signal or carrier, i.e., by controlling a parameter of a carrier, such as the amplitude, polarity, phase, or frequency. *See* **binary phase-shift keying, chroma keying, coherent frequency shift keying, differential phase-shift keying, differentially coherent phase-shift keying, double frequency shift keying, frequency shift keying, Gaussian minimum shift keying, incoherent frequency shift keying, low level keying, multiple frequency shift keying, multiple phase-shift keying, narrow shift frequency shift keying, phase exchange keying, phase-shift keying, phase reverse shift keying, quadrature phase-shift keying, rekeying, remote keying, two-tone keying, wide shift frequency shift keying.** *See also* **carrier.**

keying frequency: *See* **maximum keying frequency.**

keying head: A device that reads, i.e., that detects, hole patterns on tape and converts the patterns into electrical pulse trains that (a) represent the same information that the hole patterns represent, (b) usually are formed as a group of parallel bits for each row of holes, and (c) may be transmitted in parallel or may be serialized for transmission on a single line, in which case the pulse train in time sequence still constitutes a character with appropriate data delimiters, such as start and stop elements. *See also* **pulse, pulse train, start element, stop element.**

keying material: In cryptosystems, cryptomaterial that (a) describes the arrangements and the settings of cryptoequipment that is used in encryption and decryption, (b) describes the code sequences, messages, and signals that are used for command, control, and authentication, (c) indicates the code sequences that can be used in messages, and (d) contains key lists and instructions for the use of cryptomaterials that are changed as often as may be required in order to maintain security. *See also* **authentication, code, command, control, cryptokey, cryptomaterial, cryptosystem, key, security, signal.**

keying off: To halt the process of keying on. *See also* **keying on.**

keying on: To initiate, and continue to dispatch, a signal that (a) is generated by actuating, i.e., by depressing, a key or a switch so that power flow occurs, and (b) usually is used to cause another device to perform a specific function until the key is released, such as (i) closing a circuit and thereby causing a transmitter to transmit a modulated carrier signal for as long as the key is held depressed or (ii) causing a telegraph transmitter to send a dot or a dash until the key is released. *See also* **carrier, dash, dot, function, key, modulation, signal, switch, telegraph.**

keying variable: *Deprecated synonym for* **key.**

key list: A publication that (a) contains the keys for a cryptosystem and (b) specifies the keys and the periods when they are to be in use, i.e., the cryptoperiods. *See also* **cryptosystem, key.**

key management: In cryptosystems, i.e., encryption systems, the generation, storage, protection, transfer, loading, use, and destruction of key used by cryptoequipment for the purpose of encrypting or decrypting electronic signals. *See also* **cryptoequipment, cryptosystem, encryption, key.**

key management system: *See* **electronic key management system.**

keypad: A set of keys that are (a) mounted on a keyboard, (b) usually are placed together in one area of the keyboard, and (c) are dedicated to specific purposes, such as inserting a set of functions, inserting numbers, or moving a cursor. *Refer to* **Fig. K-2.**

key panel: On a console for a system, a set of keys used to enter data or functions into the system. *See also* **function key, function keyboard, keyboard.**

key point compaction: *See* **slope-keypoint compaction.**

key protection: *See* **lock and key protection.**

key pulsing: The sending of telephone calling signals in which the digits are transmitted by operation of a pushbutton key set or push-pull keys. *Note:* Key pulsing usually used by telephone system users and private branch exchange (PBX) operators is dual tone multifrequency signaling (DTMS). Each pushbutton causes generation of a unique pair of tones. In military systems, pushbuttons are also provided for additional signals, such as precedence. *See also* **dual tone multifrequency signaling, pulse, pulsing.**

keypunch: A device that (a) is actuated by a keyboard and (b) punches holes in a data medium, such as punch cards or paper tape. *Synonym* **keyboard punch.** *See also* **card punch, keyboard, punch card.**

key routing: In communications system operations, message routing that is based on a key in a fixed heading in the message and that is processed in message management facilities to determine the actual destination office, destination user, or message addressee. *See also* **key, message, routing.**

key set: A multiline or multifunction user terminal device. *See also* **key telephone system, private branch exchange.**

key stream: A sequence of symbols, or their electrical or mechanical equivalents, that (a) is produced in a machine or semiautomatic cryptosystem and (b) is combined with plain text to produce cipher text, control transmission security processes, or produce key. *See also* **ciphersystem, cipher text, cryptosystem, encrypt, key.**

key telephone system (KTS): In a local environment, terminals and equipment that (a) provide immediate access, from all terminals, to a variety of telephone services without attendant assistance and (b) may interface with the public switched telephone network. *See also* **key set, private branch exchange.**

key telephone unit (KTU): A unit of equipment, such as a key set, terminal, or Electronic Industries Association (EIA) interface, in a key telephone system. *Note:* Examples of key telephone units are line adapter circuits, line-unit-line terminations, and line-unit-network terminations. *See also* **key telephone system.**

key variable: *Deprecated synonym for* **key.**

keyword: 1. In information retrieval systems, a word use to search for and find a record or a file. **2.** A signif-

Fig. K-2. The AN/PRC-113 VHF/UHF manpack radio provides reliable communications in the harsh battlefield and jamming signals environment. The user operates the set by using the simple front panel **keypad.** (Courtesy Magnavox Electronic Systems Company, Fort Wayne, IN.)

icant word descriptive of the contents of a body of data or text. *Synonym* **descriptor.**

keyword in context (KWIC) index: An index for a set of documents in which the words in the string of keywords used in the index are cited along with the text in which they occur in each document. *Note 1:* The documents might be identified by a numeral. *Note 2:* The keywords in the index might be listed in alphabetical order and justified. *See also* **index, justify, keywords, string, words.**

keyword out of context (KWOC) index: An index for a set of documents in which the words in the string of keywords used in the index are taken out of the text in which they occur in each document. *Note 1:* The documents might be identified by a numeral. *Note 2:* The keywords in the index might be listed in alphabetical order and justified. *See also* **index, justify, keywords, string, words.**

k factor: 1. In tropospheric radio propagation, the ratio of (a) the effective Earth radius to (b) the actual Earth radius, the ratio being about 4/3. **2.** In ionospheric radio propagation, a correction factor that (a) is applied in calculations related to curved layers, (b) is a function of distance and the real height of the point of ionospheric reflection of a radio wave, and (c) takes into account the effects of ducting. *See also* **ducting, effective Earth radius, Fresnel region, Fresnel zone, horizon angle, ionosphere, path clearance, path profile, path survey, propagation path obstruction.**

kHz: kilohertz.

kilo (k): The SI (Système International d'Unités) prefix that indicates multiplication by 10^3, i.e., by 1000. *Note:* An example of the use of "kilo" is in 10 kHz (kilohertz), equivalent to 10^4 Hz. *Note:* The prefix "kilo" is derived from the Greek *khilioi,* meaning thousand. *See also* **k, K, metric system.**

kilobits per second (kbps): A unit of signaling rate equivalent to 10^3 bps (bits per second), i.e., 1000 bps. *See also* **signaling rate.**

kilobytes: 1. One thousand bytes. **2.** In storage capacity designation, 1024 bytes of storage capacity. *Note:* A 64-kilobyte memory implies exactly 65,536 bytes of storage capacity, derived from 2^{16}, rather than 64,000 bytes of storage capacity, derived from 64×1000. *See also* **byte, storage capacity.**

kilohertz (kHz): A unit of frequency that denotes 10^3 Hz (hertz), i.e., 1000 Hz. *See also* **frequency, hertz.**

kilometer (km): A unit of distance or length that denotes 10^3 m (meters), i.e., 1000 m. *See also* **meter.**

kilopulses per second (kpps): A unit of pulse repetition rate that denotes 10^3 pps (pulses per second), i.e., 1000 pps. *See also* **pulse repetition rate.**

kit: *See* **fiber optic breakout kit.**

km: kilometer.

knife edge effect: In electromagentic wave propagation, the transmission, i.e., the redirection, of electromagnetic waves or beams into the line of sight (LOS) shadow region caused by the diffraction that occurs because of an obstacle in the path of the waves, such as a sharply defined mountain top, building edge, leaf edge, or razor edge. *Note:* The uncut portion of an electromagnetic wave or beam continues, but the edges spread into the volume that would have been occupied by the wave had it not been cut, i.e., that the wave or the beam would have filled if the knife edge were not there, giving rise to a diffraction pattern such that the knife edge acts as a new source from which the wave or the beam emanates. Thus, the knife edge effect causes the transmission of electromagnetic waves, such as radio, radar, and video waves into the line of sight shadow region by means of the diffraction that occurs at the edge of an obstacle. The deflection of a light beam caused by a diffraction grating is an example of a multiplicity of knife edge effects on many portions of the beam. *See also* **diffraction.**

knife edge test: *See* **Foucalt knife edge test.**

K patch bay: A patching facility designed for patching and monitoring balanced digital data circuits at data signaling rates (DSRs) up to 1 Mbps (megabits per second). *Synonym* **K type patch bay.** *See also* **digital primary patch bay, D patch bay, facility, M patch bay, MM patch bay.**

kpps: kilopulses per second.

kT: noise power density.

KTS: key telephone system.

KTU: key telephone unit.

K type patch bay: *Synonym* **K patch bay.**

KWIC index: keyword in context index.

KWIC: keyword in context.

KWOC: keyword out of context.

KWOC index: keyword in context index.

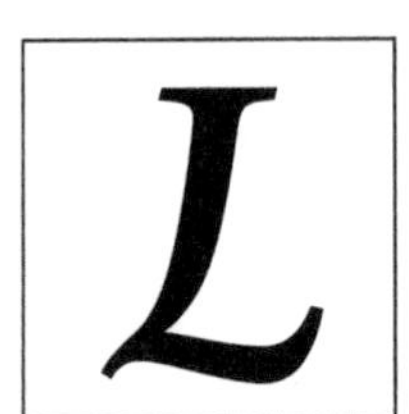

L: The symbol for electrical inductance, usually expressed in henries.

label: **1.** An identifier within or attached to a set of data items. **2.** One or more characters that (a) are within or attached to a set of data elements and (b) represent information about the set, including its identification. **3.** In communications, information within a message used to identify specific system parameters, such as the particular circuit with which the message is associated. *Note:* Messages that do not relate to call control should not contain a label. **4.** In computer programming languages, an identifier that names a statement. **5.** An identifier that indicates the sensitivity of the attached information. **6.** For classified information, an identifier that indicates (a) the security level of the attached information or (b) the specific category in which the attached information belongs. **7.** In some personal computer (PC) spreadsheet programs, any cell entry that begins with a letter or a label prefix. *See* **data label, internal label, long label, repeating label.** *See also* **identifier, label prefix.**

labeled channel: In integrated services digital networks (ISDNs), a time-ordered set of all block payloads that have labels containing the same information, i.e., containing the same identifiers. *See also* **hybrid interface structure.**

labeled interface structure (LIS): In integrated services digital networks (ISDNs), an interface structure that provides telecommunications services and signaling by means of labeled channels. *See also* **interface, labeled channel, signaling.**

labeled multiplexing: In integrated services digital networks (ISDNs), multiplexing by concatenation of the blocks of the channels that have different identifiers in their labels.

labeled statistical channel: In integrated services digital networks (ISDNs), a labeled channel in which the block payloads or the duration of each successive block is random. *See also* **block, labeled channel, payload.**

label prefix: In personal computer spreadsheet programs, a special character that (a) is used to indicate that the cell entry being entered is a label and not a value, and (b) in some programs, must be used on labels that begin with a number or begin with certain symbols, such as +, −, $, (, #, or @. *See also* **cell, computer program, entry, label, personal computer, spreadsheet, value.**

lambert: A unit of luminance equal to $10^4/\pi$ ca•m^{-2} (candela per square meter). *Note 1:* The SI (Système International d'Unités) unit of luminance is the lumen per square meter. *Note 2:* 4π lumens of light emanate from 1 candela.

Lambertian distribution: A radiance distribution that is uniform in all directions. *Synonym* **uniform Lambertian distribution.**

Lambertian radiator: A radiator or source of radiation in which the radiation is distributed angularly according to Lambert's cosine law. *Synonym* **Lambertian source.** *See also* **Lambert's cosine law.**

Lambertian reflector: A reflector of radiation in which the radiation is distributed angularly according to Lambert's cosine law. *See also* **Lambert's cosine law.**

Lambertian source: *Synonym* **Lambertian radiator.**

Lambert's cosine law: The radiance of perfectly diffusing, emitting, plane surfaces, known as "Lambertian radiators," "Lambertian sources," or "Lambertian reflectors," is (a) dependent upon the cosine of the angle between the viewing direction and the normal to the surface, (b) is given by the relation $N = N_o \cos \theta$, where N is the radiance, i.e., the radiant intensity, usually expressed in watts per square meter per steradian, i.e., $W{\bullet}m^{-2}{\bullet}sr^{-1}$, N_o is the radiance normal, i.e., perpendicular, to an emitting surface, and θ is the angle between the viewing direction and the normal to the surface being viewed, and (c) is a maximum in the direction of the normal to the surface and decreases in proportion to the cosine of the angle between the normal to the surface and the viewing direction. *Note:* Lambertian radiators are characterized by diffuse emission. *Synonyms* **cosine emission law, Lambert's emission law.**

Lambert's emission law: *Synonym* **Lambert's cosine law.**

Lambert's law: In the transmission of electromagnetic radiation, such as light, when propagating in a scattering or absorptive medium, the internal transmittance is given by the relation $T_2 = T_1^{d_2/d_1}$ where T_2 is the unknown internal transmittance of a given thickness d_2, and T_1 is the known transmittance of a given thickness d_1. *See also* **Beer's law, Bouger's law.**

lamp: In communications, computer, data processing, and control systems, a source of light that (a) is used to indicate the on or off, closed or open, or connected or disconnected condition of a circuit, (b) is mounted near the circuit, usually on a panel in front of the circuit, and (c) may have a label. *See* **answer lamp, cord lamp, signaling lamp.** *See also* **label.**

LAN: local area network.

landing: *See* **ground controlled approach landing.**

landing aid: *See* **radio landing aid.**

landing marker: *See* **no-landing marker.**

landing point marker: *See* **forbidden landing point marker.**

landing system: *See* **instrument landing system.**

landing system glide path: *See* **instrument landing system glide path.**

landline: A line that (a) is placed in areas on land and inland waterways, (b) includes conventional twisted-pair lines, coaxial cables, and fiber optic cables used in overhead, direct buried, underground, and microwave applications, and (c) does not include satellite, radio, and undersea transmission systems. *See also* **line.**

landline facility: A communications facility that (a) makes use of aerial, direct buried, and underground transmission lines, i.e., transmission cables, (b) includes land microwave systems, and (c) does not include satellite, radio, and undersea transmission systems. *See also* **line.**

landline service: A service provided by a landline facility. *See also* **landline facility, line.**

land mobile radio base station: *See* **airport land mobile radio base station.**

land mobile radio station: *See* **airport land mobile radio station.**

land mobile satellite service (LMSS): A mobile satellite service in which mobile Earth stations are on land. *See also* **Earth station.**

land mobile service (LMS): A mobile service (a) between base stations and land mobile stations or (b) between land mobile stations.

land mobile station (LMS): A mobile station in the land mobile service capable of surface movement, usually within the geographical limits of a country or a continent.

Landsat: A geographic satellite mapping system that creates photographic maps of sections of the Earth.

landscape mode: In facsimile systems, the mode of operation in which the scanning lines are parallel to the longer dimension of a rectangular object, i.e., a rectangular original. *See also* **landscape orientation.**

landscape orientation: **1.** In computer graphics, the orientation of a page in which the longer dimension is horizontal. **2.** An orientation of western world printed text on a page such that the lines of text are parallel to the long dimension of the page. *Note:* If the page contains an image, such as a picture, and the page is viewed in the normal manner, the long dimension of the page is parallel to the interpupillary distance of the viewer, i.e., the interpupillary line. If a person is standing or sitting in an upright position, the long dimension of the page and the interpupillary distance are parallel to the Earth's horizon. Thus, the orientation is with respect to the viewer, not with respect to the horizontal or the vertical directions of the Earth. Once a page is printed or a picture is created with a landscape orientation, it matters not how the page or picture is oriented with respect to the Earth. *See also* **landscape mode, portrait orientation.** *Refer to* **Fig. L-1.**

land station (LS): In the mobile service, a station that (a) is not intended to be used while in motion and (b) usually communicates with, on a secondary basis, with fixed stations or other land stations in the same category. *See* **aeronautical multicom land station, aeronautical telemetering land station, aeronautical utility land station, flight telemetering land station, hydrological and meteorological land station, radiolocation land station, radionavigation maintenance-test land station, radionavigation operational-test land station, radionavigation land station, radionavigation test land station, radiopositioning land station, telemetering land station.**

language: A set of characters, conventions, and rules that is used for representing and conveying information among persons and machines. *Note 1:* The rules of a language are used to (a) combine characters into larger groups, such as words, phrases, and sentences, that have assigned meanings and (b) combine the groups into messages to achieve specific and perhaps profound meanings. *Note 2:* A language may be natural or artificial. *See* **algorithmic language, application-oriented language, artificial language, body language, command language, computer language, computer oriented language, data description language, extensible language, general purpose language, high level language, host language, HyperText Markup Language, intermediate level language, job control language, low level language, machine language, metalanguage, natural language, object language, plain language, problem-oriented language, procedure-oriented language, programming language, query language, requirements specification language, sign language, source language, special purpose language, stratified language, symbolic language, unstratified language, user language.** *See also* **alphabet, character.** *Refer to* **Fig. M-1 (584).**

language code: In a communications system, an address digit that permits an originating attendant, i.e., originating operator, to request the assistance of an attendant in a desired language on an international call.

language override: *See* **plain language override.**

language processor: A computer program that performs tasks, such as translating and interpreting, required for processing a specified programming language, such as FORTRAN, COBOL, or a specific application language, such as a requirements specification language, a design language, or an air traffic controller language. *Note 1:* Examples of language processors are a FORTRAN processor and a COBOL processor. *See also* **computer program, interpret, processor, programming language, translation.**

LAP-B: Link Access Procedure B.

LAP-D: Link Access Procedure D.

lapping: In optical transmission systems, the use of tangential coupling in order to transmit lightwaves from

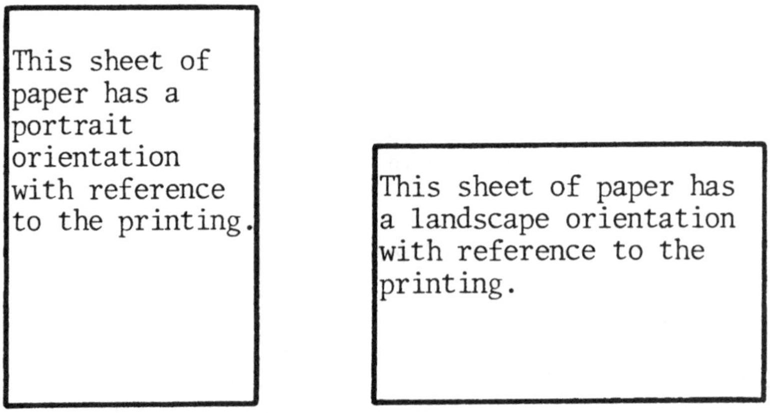

This sheet of paper has a portrait orientation with reference to the printing.

This sheet of paper has a landscape orientation with reference to the printing.

Fig. L-1. Portrait orientation and **landscape orientation** of a sheet of paper.

one optical element to another, such as from one optical fiber to another. *See also* **optical element, optical fiber.**

laptop: Pertaining to a personal computer that (a) is small enough to hold and operate on one's lap, (b) may be battery-powered for use anywhere, and (c) usually is operated in its own carrying case. *See also* **personal computer.**

laptop computer: A personal computer (PC) that (a) is small enough to fit in a standard attache case, (b) may be operated without a desk or support structure, (c) usually may be powered by a local ac power source or by internal batteries, and (d) may have a built-in modem. *See also* **modem, palmtop computer.**

large optical cavity (LOC) diode: An injection laser diode in which (a) the p-n junction is placed between two heterojunctions, (b) the optical cavity for lasing action is wide, the wider cavity having a higher refractive index than the material on either side of the cavity, and (c) the output beam is wider and the output power is greater than that of a light emitting diode (LED). *See also* **heterojunction, injection laser diode, lasing light-emitting diode, optical cavity, p-n junction.**

large-scale integrated (LSI) chip: An integrated circuit that (a) has many circuit elements that are formed from a single piece of semiconducting material and (b) is used for performing logic functions, i.e., Boolean functions. *Note:* An example of a large-scale integrated (LSI) chip is several thousand interconnected transistors and other circuit elements that form combinational circuits on a 1 sq. in. (square inch) silicon chip. *See also* **Boolean functions, circuit element, chip, combinational circuit, integrated circuit, gate, semiconductor.** *Refer to* **Figs. C-3 (128), I-3 (457), P-14 (762), P-15 (763), P-16 (764), W-1 and W-2 (1078).**

laser: 1. *Synonym* **light amplification by stimulated emission of radiation. 2.** A device that produces a high irradiance, narrow spectral width, coherent, highly directional, i.e., near zero divergence, beam of light by stimulating electronic, ionic, or molecular transitions to higher energy levels so that when the electrons, ions, or molecules fall back to lower energy levels they produce a stream of photons of various discrete energy levels. *Note 1:* The word "laser" is derived from "light amplification by stimulated emission of radiation." *Note 2:* The lasing action is produced by population inversion. *Note 3:* The release of radiated energy can be controlled in time and direction so as to generate an intense highly directional narrow beam of spatially or temporally coherent electromagnetic energy, i.e., the electromagnetic fields at every point in the beam are uniquely and specifically definable and predictable. *Note 4:* The coherence degree of laser radiation exceeds 0.88. The beam emitted by certain lasers is suitable for launching optical power into optical fibers. The beam can be modulated, or the laser itself can be modulated. *See* **atmosphere laser, atomic laser, distributed feedback laser, gas laser, glass laser, injection locked laser, ion laser, liquid laser, mixed gas laser, molecular laser, multiline laser, multimode laser, optical fiber laser, Q switched repetitively pulsed laser, semiconductor laser, solid state laser, soliton laser, Surface Emitting Distributed Feedback laser, telecommunications transverse excited atmosphere laser, thin film laser, tunable laser.** *See also* **light amplification by stimulated emission of radiation.** *Refer to* **Fig. L-2.** *Refer also to* **Fig. I-2 (451).**

laser basic mode: The primary or lowest-order fundamental transverse propagation mode for the lightwave emitted by a laser. *Note:* The energy emitted by a laser (a) usually has a Gaussian distribution, i.e., a bell curve distribution in space, and (b) all the lightwave energy is in a single beam, i.e., in a single lobe with no side lobes. *See also* **lobe, mode, side lobe, transverse electromagnetic wave.**

laser beam: A collimated, highly directional bundle of monochromatic light rays with zero or nearly zero divergence and exceptionally high irradiance, emitted from materials that are undergoing lasing action. *See also* **irradiance, lasing, monochromatic, ray bundle.**

laser chirp: An abrupt shift of the center wavelength of a laser during individual pulses. *Note:* Laser chirp is caused by laser instability, particularly during modulation by pulses. *See also* **chirping.**

laser connector: An active connector that uses a laser as the active semiconducting device to convert an incoming electrical signal to an optical signal. *See also* **active connector, optical signal.**

laser diode (LD): A semiconductor junction diode that (a) consists of positive and negative carrier regions with a p-n junction, i.e., a transition region, (b) emits electromagnetic energy, i.e., quanta of energy, at optical frequencies when injected electrons under forward bias recombine with holes in the vicinity of the junction, (c) is made of certain materials, such as gallium arsenide, that have a high probability of radiative recombination producing emitted light, rather than heat, at a frequency suitable for optical waveguides, such as optical fibers and optical integrated circuits (OICs), and (d) has polished ends that reflect and trap light to produce laser action, cause increased excitation, overcome losses, and produce increased emission. *Note:* Typical characteristics of a GaAlAs-type laser

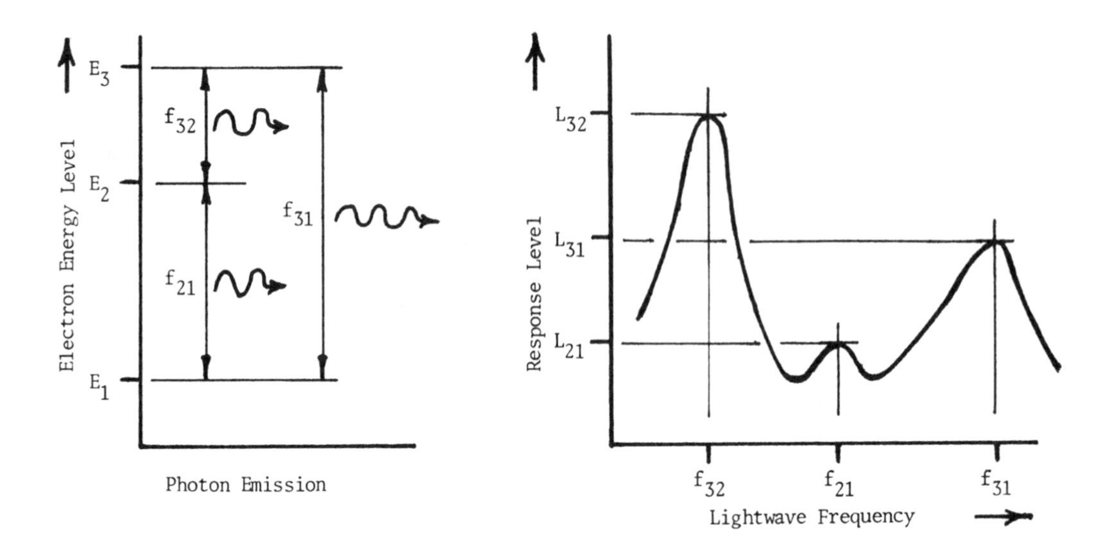

Fig. L-2. In a **laser,** electrons that were excited to higher energy levels return to lower levels and emit photons (left) at energy levels that correspond to the energy difference between the two levels producing the spectral response (right) when the energies are summed over their number per unit time for each level. For example, photon f_{32} is emitted when electrons drop from E_3 to E_2.

diode for fiber optic communications applications are (a) 0.80 μm (micron) wavelength, (b) 1 mW (milliwatt) to 40 mW optical power output, (c) 10 mA (milliampere) to 500 mA input current, (d) 2 V (volt) input voltage, (e) 1% to 20% power efficiency, (f) 1 gm (gram) of weight, (g) 10^4 hr (hours) to 10^7 hr operating life, (h) 10 by 35 degree divergence, and (i) direct modulation by the drive current. Typical characteristics of an InGaAsP-type laser diode for fiber optic communications applications are (a) 1.1 μm (micron) to 1.6 μm wavelength, (b) 1 mW (milliwatt) to 10 mW optical power output, (c) 20 mA (milliampere) to 200 mA input current, (d) 1.5 V (volt) input voltage, (e) 1% to 20% power efficiency, (f) 1 gm (gram) of weight, (g) up to 10^5 hr (hours) operating life, (h) 10 by 30 degree to 20 by 40 degree divergence, and (i) direct modulation by the drive current. The efficiency is the electromagnetic (optical) power output divided by the electrical power input. *Synonyms* **diode laser, semiconductor diode laser, semiconductor laser diode.** *See* **carrier, injection laser diode, junction, optical waveguide, radiative recombination, stripe laser diode.**

laser diode (LD) coupler: A coupler that (a) couples light energy from a laser diode (LD) source to an optical fiber or cable at the transmitting end of a fiber optic data link and (b) may consist of an optical fiber pigtail epoxied to the coupler. *See also* **coupler, coupling, data link, fiber optic coupler.**

laser element: *See* **thin-film laser element.**

laser emulsion storage: Storage in which (a) the data storage medium is a photosensitive surface that has been exposed to a controlled laser beam in minute areas, and (b) the laser beam can be modulated by a Kerr cell to produce the desired data pattern. *See also* **Kerr cell, laser.**

laser fiber optic transmission system: A transmission system that (a) consists of one or more laser transmitters and connected fiber optic cables, (b) limits radiation to fiber optic cables during normal operation, (c) may require a special tool to effect disconnection, especially if connections form part of the protective housing, and (d) may require the use of mechanical beam attenuators at connectors as part of safety requirements for fiber optic transmission systems. *See also* **connector, fiber optic cable, fiber optic connector.**

laser frequency switch: A switch that enables selection of laser output frequency, i.e., output wavelength, by electronically driving an electrooptic crystal in order to produce changes in the laser resonant cavity length, thus enabling control of laser frequency. *Note:* An example of a laser frequency switch is driving a deuterated ammonium dehydrogen phosphate crystal inside of a dye laser cavity with low voltage pulses that cause time-dependent variations in the refractive index of the electrooptic element to produce the changes in the resonant cavity optical length that will change the laser operating frequency. *See also* **dye laser, electroooptic, refractive index, resonant cavity, switch.**

laser hazard: In a laser, a feature that causes, or has the potential of causing, an injury. *Note:* An example of a laser hazard is the injury to the eye that can be caused by the radiation from a gallium arsenide laser that emits invisible infrared radiation from a glass window on its top. Precautions include observation only by indirect methods, such as through the use of a filter, a shield, or an infrared or ultraviolet viewer. *See also* **filter, infrared viewer, laser, ultraviolet viewer.**

laser head: A module that (a) contains an active laser medium, a resonant cavity, and other components within one enclosure and (b) does not necessarily include a power supply. *See also* **laser medium, module.**

laser intelligence (LASINT): Technical and intelligence information that (a) is derived from laser systems and (b) is a subcategory of electrooptical intelligence (ELECTROOPTINT).

laser jet printer: A dot matrix printer that (a) uses ink jets from a cartridge, toner, electrophotographic techniques, and light from a laser source to form and fix images with shapes and sizes that are selectable under program control, and (b) forms various images, including characters, that may be colored by mixing inks and that may be formed on paper or transparencies. *Note:* There are many manufacturers and types of laser jet printers. *See also* **dot matrix printer, image.**

laser line width: 1. In the operation of a laser, the frequency range over which a specified amount, such as 90%, of the laser beam energy is distributed. **2.** In the operation of a laser, the difference between the lower and upper frequencies at which the laser beam energy per hertz is down to a specified level, such as the 3-dB-down level, i.e., 50% of its midband or maximum level. *See also* **spectral width.**

laser medium: *Synonym* **active laser medium.**

laser protective housing: For a laser, a protective housing that (a) prevents human exposure to laser radiation in excess of an allowable, established, or statutory emission limit, and (b) has parts that can be removed or displaced, or, if not interlocked, may be secured in such a way that removal or displacement of the parts requires the use of special tools. *See also* **laser hazard, radiation, hazard of electromagnetic radiation to personnel.**

laser pulse duration: The duration of the burst of electromagnetic energy emitted by a pulsed laser. *Note:* Laser pulse duration usually is measured at the half-power points. *Synonyms* **laser pulse length, laser pulse width.** *See also* **half-power point, pulse.**

laser pulse length: *Synonym* **laser pulse duration.**

laser pulse width: *Synonym* **laser pulse duration.**

laser service connection: An access point in a laser-to-fiber-optic transmission system that (a) is designed for service and safety and (b) should require a special tool for disconnection.

laser sonar: A ranging system or a communications system (a) in which modulated laser beams are fired into a propagation medium to launch sound waves or phonons for detection at other places for communications or for reflection from objects for ranging, and (b) that usually is used underwater or underground. *See* **long-range laser sonar, short-range laser sonar.** *See also* **communications system, modulation, object, phonon, ranging, reflection, propagation medium.**

lasing: A phenomenon that (a) occurs in a specially prepared material, such as a uniformly doped semiconductor crystal that has free-moving, highly mobile, loosely coupled electrons that respond to the application of resonant frequency-controlled energy coupled to the electrons, (b) as a result of resonance and the imparting of energy collision or close approach, allows electrons to be raised to highly excited energy states, and (c) when the electrons move to lower energy states, results in the release of quanta of high energy electromagnetic radiation in a narrow beam of coherent high irradiance lightwaves. *Note 1:* Lasing action takes place in a laser. *Note 2:* Lasing action can take place in certain (a) semiconductors, (b) crystals, (c) gases, such as helium, neon, argon, krypton, and carbon dioxide, and (d) liquids, such as organic and inorganic dyes. *See also* **coherency, irradiance, laser, radiation, semiconductor.** *Refer to* **Fig. I-2 (451).**

lasing condition: *See* **prelasing condition.**

lasing medium: *Synonym* **active laser medium.**

lasing threshold: The lowest excitation level at which laser output is dominated by stimulated emission rather than by spontaneous emission. *See also* **spontaneous emission, stimulated emission.**

LASINT: laser intelligence.

last in, first out (LIFO): A queuing discipline in which entities at a location, such as a file, leave in an order reversed from that in which they arrived; i.e., in a last in, first out (LIFO) file, service is offered to the entity that has been in the file the shortest time. *Note:* An example of a last in, first out file is an in-box operated in such a manner that items placed on top are removed from the top. Files in which LIFO is used include cellars, pushdown files, and spike files. *Synonym* **pushdown.** *See also* **buffer; first in, first out; pushdown storage; queue traffic.**

LATA: local access and transport area.

latch-down key: On a keyboard, a key that when operated remains in its position until released. *Note:* Examples of latch-down keys on a standard keyboard are (a) the capslock key that causes letter keys to remain upper case keys, (b) the num lock key that converts a cursor pad, i.e., arrow keys, to a number pad, and (c) the ins key that allows characters to be inserted in a line with or without moving existing text to the right, depending on its position. *See also* **key, keyboard.**

latching: A switching action that requires a specific force or signal to place a switch in a position where it remains until another actuating force or signal is applied. *See* **nonlatching.**

latching switch: 1. A switch that requires a specific signal to place the switch in a position where it remains until another actuating force or signal is applied. **2.** In fiber optics, a switch that will selectively transfer optical signals from one optical fiber to another when an actuating force or signal is applied, and will continue to transfer signals after the actuating force or signal is removed until another actuating force or signal is applied. *See* **nonlatching switch.**

late distortion: Distortion that is caused by the delayed arrival of parts of a signal. *Note:* Signal parts are transmitted in a given sequence, each part at a specific instant in the signal time slot. A signal part that arrives after the proper instant of its time slot causes late distortion. *Synonym* **positive distortion.** *See also* **distortion, early distortion, signal, time slot.**

latency: *See* **ring latency.**

latent image: An image that exists on sensitized material after exposure but before development. *Note:* Examples of latent images are (a) electrostatic images produced by xerographic copying machines, (b) images produced by camera on photographic film when a frame is exposed and the image is not yet developed, and (c) images produced on sensitized paper but not yet developed as a blueprint. *See also* **electrostatic, image, xerographic recording.**

lateral communications: Communications between facilities that are at the same operational level or echelon of command. *See also* **facility, lateral tell.**

lateral displacement loss: *Synonym* **lateral offset loss.**

lateral magnification: The ratio of (a) the linear size of an image to (b) the linear size of the object when an enlarging lens is used. *See also* **enlarging lens, image, object.**

lateral offset: In a fiber optic joint, such as a butted joint of two optical fibers, an offset, in a radial direction, of the optical fiber axes. *Synonym* **transverse offset.**

lateral offset loss: In fiber optic systems, a power loss caused by transverse deviation, i.e., lateral or sideways deviation, from optimum alignment, at a connector or a splice, such as at a light source to an optical fiber joint, a fiber to fiber joint, or a fiber to photodetector joint. *Note 1:* The transverse direction is perpendicular to the optical fiber axis and may be in any radial direction. *Note 2:* The amount of lateral offset loss depends on (a) the extent of offset between the optical fiber axes, (b) the relative diameters of the respective cores and cladding, (c) the direction of propagation if the diameters are not the same, (d) the angular misalignment, and (e) the longitudinal displacement. *Synonyms* **lateral displacement loss, transverse displacement loss, transverse offset loss.** *See also* **angular misalignment loss, longitudinal displacement loss.**

lateral tell: Information transfer that occurs between facilities that are at the same operational level or echelon of command. *See also* **facility, lateral communications.**

latitude: *See* **celestial latitude, terrestrial latitude.**

launch angle: 1. The angle with the normal at which a light ray emerges from a surface. **2.** The beam divergence at an emitting surface, such as that of a light-emitting diode (LED), laser, lens, prism, or optical fiber endface. **3.** At an endface of an optical fiber, the angle between an input ray and the optical fiber axis. *Note:* If the endface of the optical fiber is perpendicular to the fiber axis, the launch angle is equal to the incidence angle when the ray is external to the fiber and the refraction angle when initially inside the fiber. *Synonym* **departure angle.** *See also* **acceptance angle, deviation angle, exit angle, incidence angle, launch numerical aperture, optical fiber, refraction angle.** *Refer to* **Fig. L-3.**

launch condition: When light rays exit from the surface of a material, such as the surface of a light source

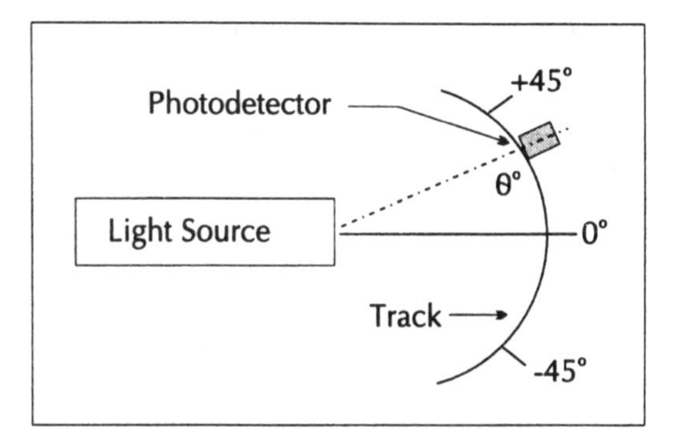

Fig. L-3. A **launch angle** plotter. The light source could be a laser, a light-emitting diode (LED), or the exit endface of an optical fiber.

or the exit endface of an optical fiber, a parameter that describes the geometric relationships between the ray and the surface, the optical characteristics of the interfaces and paths, the characteristics of the rays, and other related parameters, such as exit angle, launch angle, wavelength, refractive index contrast, launch numerical aperture, launch loss, and spectral width.

launching: *See* **single mode launching.**

launching optical fiber: 1. An optical fiber pigtail attached to a light source used to enable splicing to another component, such as a coupler or an optical fiber. *Note 1:* A launching fiber may be used in conjunction with a light source to excite the modes of another fiber in a particular fashion. *Note 2:* Launching optical fibers are often used in test systems to improve the precision and the repeatability of the measurements. **2.** An optical fiber used in conjunction with a source to excite the modes of another fiber in a particular fashion. *Note:* Launching optical fibers are most used in test systems to improve the precision of measurements. *Synonym* **injection optical fiber.** *See also* **mode, pigtail.**

launch loss: The radiant power that (a) is lost at the point where the radiant power output from a light source is coupled into an optical waveguide, such as an optical fiber pigtail attached to the light source, and (b) can be caused by aperture mismatch, angular misalignment, longitudinal displacement, lateral offset, and Fresnel reflection. *See also* **insertion loss.**

launch numerical aperture (LNA): The numerical aperture of an optical system used for coupling power into an optical waveguide. *Note 1:* The launch numerical aperture (LNA) may differ from the stated numerical aperture (NA) of a final focusing element for various reasons, such as when the element is underfilled or the focus, i.e., the focal length, is other than that for which the element is designed. *Note 2:* The launch numerical aperture (LNA) is one of the parameters that determine the initial distribution of power among the modes of an optical waveguide, such as an optical fiber. *See also* **acceptance angle, coupling, launch angle, numerical aperture.** *Refer to* **Fig. N-21 (641).**

launch spot: A spot of light (a) that is produced by a light beam from an optical element, such as from a lens or the exit endface of an optical fiber, when the beam is incident upon a transverse surface, i.e., a screen perpendicular to the optical axis of the element, (b) whose diameter is dependent upon the departure angle, i.e., the launch angle, from the element and the distance between the transverse surface and the element, and (c) whose size can be used to determine the numerical aperture of the optical element. *Note:* To determine the numerical aperture (NA), the optical element is aligned to provide maximum irradiance on the screen. The numerical aperture is given by the relation $NA = D/(D + 4d^2)^{1/2}$ where NA is the numerical aperture, D is the spot diameter, and d is the distance from the element, such as the optical fiber exit endface, to the screen. If $D/2d$ is less than 0.25, the numerical aperture of the element is given by the approximate relation $NA \approx D/2d = r/d$ where r is the spot radius and r/d is the tangent of the launch angle. *See also* **numerical aperture, XX/YY restricted launch.** *Refer to* **Fig. N-21 (641).**

law: *See* **a-law, Beer's law, Biot-Savart law, Boltzmann's emission law, Bouger's law, Brewster's law, encoding law, inverse square law, Lambert's cosine law, Lambert's law, Planck's law, radiance conservation law, reflection law, Richardson's law, Shannon's law, Snell's law.**

law enforcement training center: *See* **Federal Law Enforcement Training Center.**

law refractive index power leveling: *See* **power law refractive index power leveling.**

law refractive index profile: *See* **power law refractive index profile.**

lay: *Synonym* **lay length.**

layer: 1. In radio wave propagation, a stratum in the ionosphere that has a specified characteristic, such as a stratum in which the variation of free electron density with height attains a maximum value. **2.** In telecommunications networks with open systems architecture, a group of related functions that are performed in a given level in a hierarchy of groups of related functions. *Note:* In specifying the functions for a given layer, the assumption is made that the functions that are specified for the layers below the given layer are performed for the given layer, except for the lowest layer. In the lowest layer, such as the Physical Layer in the Open Systems Interconnection—Reference Model, the functions are performed for the layers above. *See* **barrier layer, D layer, E layer, F layer, end to end transport layer, Heaviside layer, ionospheric layer, reflecting layer.** *See also* **ionosphere, layered system, Open Systems Interconnection—Reference Model.** *Refer to* **Fig. L-4.** *Refer also to* **Fig. P-19 (771).**

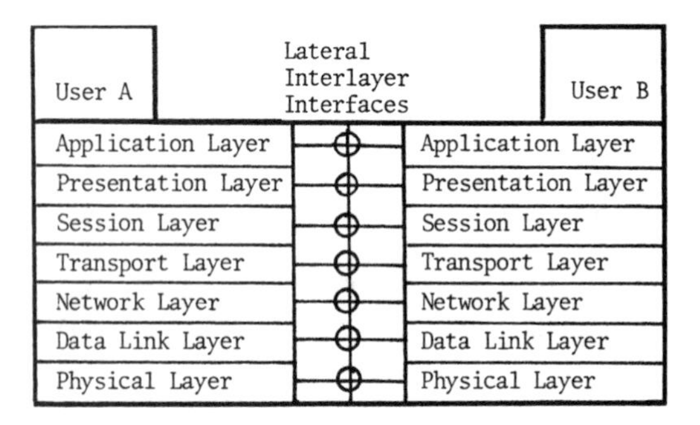

Fig. L-4. Each **layer** of a network or a station in the open systems architecture concept may be laterally associated through connections and protocols with peer layers of other networks or stations in order that the networks may be controlled and the user end-instruments may be interconnected.

layered protocol: In open systems network architecture, a protocol that is distributed over two or more layers of a network.

layered system: A system in which components are grouped in a hierarchical arrangement in such a manner that lower layers provide functions and services that support the functions and the services of higher layers. *Note:* Various systems, particularly communications systems and computing systems, of ever increasing capability can be built by changing, adding, or superimposing layers one above the other, or by changing, adding, or superimposing layers below to improve capability, each layer using the facilities below and supporting the layers above. The concept particularly allows for using the components and protocols that are in place while still improving the capability of the overall system. *See also* **layer, open systems architecture, Open Systems Interconnection—Reference Model.** *Refer to* **Fig. P-19 (771).**

layer protocol: In network architecture, the protocol that is applicable to the group of functions of a layer. *See also* **layer, network architecture, protocol.** *Refer to* **Fig. P-19 (771).**

lay length: For a given component in a helically wound cable, i.e., a cable in which the components, such as wires, strength members, or optical fibers, are twisted or spun around each other and consequently around the axis of the resulting cable, the longitudinal distance, i.e., the distance along the cable, required for the component to complete one revolution about the cable axis. *Note 1:* Different elements of the cable, such as wires, optical fibers, and strength members, usually have different lay lengths, ranging from a few centimeters to infinity, in which case there is no twist. Lay lengths are adjusted according to the stress/strain characteristics of the components and to allow for the pressure applied to inner components by outer components when the cable is under tension. *Note 2:* The lay length may be expressed in meters or meters per revolution. *Note 3:* The physical length of the component for each lay length depends on the lay length and the radial distance of the component from the axis of the cable. *Note 4:* In many fiber optic cables, the lay length is shorter than in metallic cables of similar diameter to avoid overstressing the fibers during the pulling associated with spooling, payout, and installation. *Note 5:* The total length of a cable is equal to the number of wraps times the lay length. The wraps should not be confused with the twists given twisted metallic pairs. Pairs of optical fibers are not twisted. *Synonym* **lay, pitch.** *See also* **cable component.**

layout: *See* **file layout.**

LBO: line buildout.

LCD: liquid crystal display.

LD coupler: laser diode coupler.

LD: laser diode.

LDDC dialing system: long-distance direct current dialing system.

LDM: limited distance modem.

LEA laser: longitudinally excited atmosphere laser.

lead: *See* **automatic identified outward dialing lead.**

leader: *See* **magnetic tape leader, tape leader.**

lead signaling: *See* **E and M lead signaling.**

lead time: *See* **procurement lead time.**

leak: *See* **carrier leak, jacket leak.**

leakage loss: *See* **light leakage loss.**

leaky bucket counter: A counter that (a) is incremented by unity each time an event occurs and (b) is periodically decremented by a fixed value.

leaky mode: In an optical waveguide, such as an optical fiber, a mode that (a) has an evanescent field in which the field strength decays monotonically for a finite distance in the transverse direction outside the core, but becomes oscillatory and thus conveys power everywhere beyond that finite distance, (b) in physical optics, corresponds to a leaky ray in geometric optics, (c) experiences attenuation, even if the waveguide is perfect in every respect, (d) usually is a high-order mode, (f) results in power loss in propagating signals, and (g) satisfies the relation $[(n_a k)^2 - (m/a)^2]^{1/2} \leq \beta \leq n_a k$ where n_a is the refractive index at the core-cladding interface surface, i.e., at the distance a from the optical axis where a is the core radius, k is the free-space wave number, given by the relation $k = 2\pi/\lambda$, where λ is the wavelength, m is the azimuthal index of the mode, and β is the imaginary part (phase term) of the axial propagation constant. *Synonym* **tunneling mode.** *See also* **bound mode, cladding mode, fiber optics, leaky ray, mode, optical fiber, radiation mode, unbound mode.**

leaky ray: In an optical waveguide, such as an optical fiber, a ray (a) for which geometric optics would predict total internal reflection at the core-cladding boundary, but which escapes from the core into the cladding because of microbends in the boundary and because the boundary is curved, i.e., the cross section of the core may be round or elliptical, (b) that results in power loss in propagating signals, (c) that, in geometric optics, corresponds to a leaky mode, i.e., a tunneling mode, in physical optics, and (d) at radial distance r from the optical axis, that satisfies the relation $n_r^2 - n_a^2 \leq \sin^2 \theta_r$ and the relation $\sin^2 \theta_r \leq [n_r^2 - n_a^2]/[1 - (r/a)^2 \cos^2 \Phi_r]$ where θ_r is the angle the ray makes with the optical axis of the waveguide, n_a is the refractive index at the distance a from the optical axis, n_r is the refractive index at the radial distance r from the optical axis, a is the core radius, and Φ_r is the azimuthal angle of the projection of the ray on the transverse plane. *Synonym* **tunneling ray.** *See also* **bound mode, cladding ray, guided ray, leaky mode, unbound mode.**

leaky wave: In a waveguide, such as an optical fiber, an electromagnetic wave that (a) is coupled or transferred to a propagation medium outside the waveguide, such as from inside to outside the cladding of optical fibers, (b) is no longer guided because it is no longer coupled to modes inside the waveguide, (c) usually stems from an incident wave that has a large skew component at entry into the waveguide, (d) becomes detached and radiates away from the guide after a short distance, i.e., a few wavelengths, of propagation within the guide, (e) will have escaped by the time the equilibrium length, i.e., the equilibrium modal power distribution length, is reached, (f) usually is a high-order mode, and (g) usually is in the cutoff frequency region. *Note:* In optical fiber waveguides, low-order modes usually are bound to the core and therefore are not leaky waves. Unless coupling to other guides is desired, efforts usually are made to reduce leaky waves to a minimum so that optical power is not lost, especially in long-haul communications links. *Refer to* **Figs. L-8 (536), P-17 (767).**

leap second: An occasional adjustment of one second, added to, or subtracted from, Coordinated Universal Time (UTC) to bring it into approximate synchronism with Universal Time (UT-1), which is the time scale based on the rotation of the Earth. *Note 1:* Adjustments, when required, are made at the end of the last minute of June 30, or preferably, December 31, Universal Time (UT), so that UTC never deviates from UT-1 by more than 0.9 second. *Note 2:* The adjustment is made so that the last minute of the day on which an adjustment is made has 61 or 59 seconds. *See also* **Coordinated Universal Time, International Atomic Time.**

learning: *See* **machine learning.**

leased circuit (LC): Dedicated common carrier facilities and channel equipment used by a network to furnish exclusive private line service to a specific user or a specific group of users. *Note:* In a leased circuit, privacy is complete, except perhaps for accidental crosstalk or a faulty connection. *See also* **dedicated circuit, hot line, individual line, party line, private line.**

leased circuit data transmission service: A communications service in which a data circuit of a public data network is made available to a user or a group of users for their exclusive use. *Note:* When there are only two sets of data circuit-terminating equipment (DCE), the leased circuit data transmission service is known as a point to point facility. When there are more than two, it is known as a multipoint facility. *See also* **data circuit, facility, point to point, multipoint, public data network.**

Leased Interfacility National-airspace Communications System (LINCS): A backbone digital voice and data communications network that handles air traffic routing and control for the Federal Aviation Administration.

least privilege: In computer and communications security, the granting of the minimum access authorization that is necessary for the performance of required or assigned tasks. *See also* **communications security.**

least time principle: *Synonym* **Fermat's principle.**

LEC: local exchange carrier.

LED: light-emitting diode.

LED coupler: light-emitting diode coupler.

left circular polarization: *Synonym* **left-hand circular polarization.**

left-hand circular polarization: Circular polarization of an electromagnetic wave in which the electric field vector rotates in a counterclockwise direction, as seen by an observer looking in the direction of propagation of the wave. *Synonym* **left circular polarization.** *See also* **right-hand circular polarization.**

left-hand helical polarization: Helical polarization of an electromagnetic wave in which (a) the electric field vector rotates in a counterclockwise direction as it advances in the direction of propagation, as seen by an observer looking in the direction of propagation, and (b) the tip of the electric field vector advances like a point on the thread of a left-hand screw, the normal screw being a right-hand screw, when entering a fixed nut or a tapped hole. *Synonym* **left helical polarization.** *See also* **right-hand helical polarization.**

left-hand polarized electromagnetic wave: An elliptically or circularly polarized electromagnetic wave, in which (a) the electric field vector, observed in a fixed plane normal to the direction of propagation while looking in the direction of propagation, rotates in a left-hand direction, i.e., in a counterclockwise direction, and (b) the direction of propagation is the same as the forward direction of a left-hand screw when being screwed into a fixed nut. *Synonyms* **anticlockwise polarized wave, counterclockwise polarized wave, left-hand polarized wave, left polarized wave.** *See also* **circular polarization, right-hand polarized electromagnetic wave.**

left-hand polarized wave: *Synonym* **left-hand polarized electromagnetic wave.**

left helical polarization: *Synonym* **left-hand helical polarization.**

left justification: The alignment of sets of stored data, such as lines of text or data in a register, such that the alignment reference is at the left end of the storage area, such as the page or the register. *See also* **left justify, right justification.**

left justify: 1. To shift data in a storage medium so that the left-hand end of the data is at a particular position relative to a given reference. *Note:* Examples of left justify are (a) to adjust the position of a word or a numeral in a register so that the left end of the data is at the left end of the register and (b) to adjust the words on a printed page so that the left margin is flush, i.e., so that all the beginnings of the lines of text are in a straight line and are at a fixed distance from the left-hand edge of the page. **2.** To align data in or on a medium to fit the left end positioning constraints of a required format. *See also* **position, register, right justify, shift.**

left polarized wave: *Synonym* **left-hand polarized electromagnetic wave.**

leg: 1. A part of an end-to-end route or path, such as a path from user to user via several networks and nodes within networks. *Note:* Examples of legs are several sequential microwave links between two switching centers or a transoceanic cable between two shore communications facilities, each connected to a node in a national network. **2.** A connection from a specific node to an addressable entity. *See also* **node.**

legend: *See* **accounting legend.**

length: *See* **available line length, average block length, back focal length, block length, borescope overall length, borescope tip length, borescope working length, coherence length, dispersion-limited length, effective antenna length, electrical length, equilibrium length, equivalent focal length, focal length, front focal length, lay length, nonequilibrium modal power distribution length, optical path length, path length, polarization beat length, register length, scanning length, scanning line usable length, sheath length, total line length, total scanning length, word length, wavelength.**

length discrimination: *See* **packet length selection.**

length equation: *See* **password length equation.**

length parameter: *See* **password length parameter.**

lens: An optical element that (a) has curved surfaces, (b) is made of one or more pieces of a material, such as glass or plastic, that is transparent to the rays passing through, (c) is capable of forming an image, either real or virtual, of an object, i.e., of the source of the rays, (c) has at least one curved surfaces that is convex or concave, (d) usually is spherical but sometimes may be aspheric, and (e) usually has two opposite polished major nonparallel surfaces, at least one of which is convex or concave in shape so that they serve to change the amount of convergence or divergence of the transmitted rays. *See* **achromatic lens, aplantic lens, bitoric lens, Cartesian lens, collective lens,**

compound lens, concentric lens, converging lens, condensing lens, corrected lens, cylindrical lens, dielectric lens, divergent meniscus lens, diverging lens, electron lens, field lens, finished lens, microcylindrical lens, plane lens, planoconcave lens, planoconvex lens, SELFOC® lens, single lens, tapered lens, telephoto lens, thick lens, thin lens, zoom lens. *See also* **convergence, divergence, image, object, optical element, real image, virtual image.**

lens coupling: In optical waveguides, the transfer of electromagnetic energy from source to guide, or from guide to guide, by means of a lens placed between the source and a sink. *Note:* Coupling loss can be reduced to the combined packing fraction, angular misalignment, longitudinal displacement, and lateral offset losses. *See also* **lens, sink, source.**

lensed connector: *Synonym* **expanded beam connector.**

lens light-emitting diode (LED): *See* **integral lens light-emitting diode.**

lens measure: A mechanical device for measuring surface curvature in terms of dioptric power. *See also* **dioptric power.**

lens speed: The property of a lens that affects the illuminance of the image it produces. *Note:* Lens speed may be specified in terms of aperture ratio, numerical aperture, T stop, or F number. *See also* **aperture ratio, F number, illuminance, image, lens, numerical aperture, T stop.**

lens system: Two or more lenses arranged to work in conjunction with one another.

lens watch: A dial depth gauge graduated in diopters. *See also* **diopters.**

LEOW: low cost exploitation operator workstation.

Le Système International d'Unités (SI): The International System of Units (SI).

letter: A character or a symbol that (a) is a member of an alphabet, (b) is used to represent a speech sound, and (c) in strings, is used to compose words to which meaning can be assigned. *See* **signal letter.**

letter case: 1. In the teletypewriter code for tape relay systems, the machine function that produces or interprets a hole pattern on tape or a pulse pattern on wires or optical fibers as one of the 26 letters of the English alphabet. *Note:* The letter case is obtained by actuating a special function key, usually marked LTRS. **2.** One of the groups into which the characters, particularly the letters of a character set, are placed. *Synonym in teletype systems* **lower case.** *See also* **figure case, function key, lower case, upper case.**

letter indicator: A panel signaling indicator that (a) consists of a single panel appropriately placed and (b) means that the following message is to be read as letters, i.e., the following panels, coded to represent numerals, are to be read as letters, such that the numerals 1–26 are to be read as the letters A–Z when the letter indicator is present. *See also* **panel, panel signaling indicator.**

letter shift: In teletypewriter operations, a case shift that results in the translation of signals into another group of characters, a group that was being used or interpreted before the shift. *See also* **case shift, escape character, figure shift, letter shift signal.**

letter shift signal: The signal, sent by a telegraph transmitter, that conditions a telegraph receiver to interpret all subsequent received signals as the letter case, which is a group of signals that primarily represent letters and machine functions, rather than the figure case that was in use before the letter shift signal occurred. *See also* **figure case, letter case, machine function.**

level: 1. In a circuit or a system, the absolute or relative voltage, current, or power at a particular point. **2.** In a hierarchical system, a tier or a layer. **3.** Pertaining to a tier or a layer of a hierarchical system, such as in "link level protocol" and "high level computer language." *See* **ambient noise level, average picture level, carrier level, carrier noise level, channel noise level, circuit noise level, cross level, digital voltage level, energy level, facsimile signal level, link level, optical power level, power level, priority level, quantization level, reference white level, relative transmission level, signal level, single sideband equipment reference level, standard telegraph level, transmission level, zero relative level.** *Refer to* **Fig. P-8 (737).**

level alignment: In a transmission system, the adjustment of the power, voltage, or current levels at various points in the system in order to prevent overloading, overdriving, or underdriving of the components in the system. *Note:* An example of level alignment is the adjustment of the voltage levels in single data links and data links in tandem in order to ensure that all received signals are at the specified levels above threshold levels to ensure acceptable response. *See also* **data link, level, response, tandem, threshold.**

level assignment: *See* **priority level assignment.**

level code: *See* **modified nonreturn to zero level code.**

level coding: *See* **biphase level coding.**

leveling: *See* **power law refractive index power leveling, power leveling.**

level operation: *See* **balanced link level operation.**

level point: *See* **transmission level point, zero dBm transmission level point.**

level transition: *See* **energy level transition.**

lever: *See* **optical lever.**

LF: line feed, low frequency.

LF character: line feed character.

LHOTS: long-haul optical transmission set.

liaison circuit: *See* **external liaison circuit.**

librarian: *See* **software librarian.**

library: *See* **electronic library, private library, software library, system library.**

LIDAR: 1. light detection and ranging. 2. A system that uses electromagnetic frequencies in the optical spectrum for detection and range determination of objects, i.e., targets, in a manner similar to radar.

life: *See* **component life, functioning life, indefinite component life, in-phase component life, out-of-phase component life, shelf life.**

lifecraft frequency: *See* **international radiotelegraph lifecraft frequency.**

lifecycle: In communications, computer, data processing, and control systems, the entire set of stages assumed by an entity during the period of its existence, usually from the early conceptual and definition stage to the final disposition and disposal stage, including stages such as the design, prototype, fabrication, test, installation, operational, and salvage stage. *See* **software lifecycle, system lifecycle.**

LIFO: last in, first out.

lifter: *See* **bridge lifter.**

light: 1. The region of the electromagnetic spectrum that (a) can be perceived by human vision, (b) is designated the visible spectrum, (c) nominally includes the wavelength range from 0.4 μm (microns) to 0.8 μm, and (d) includes all the colors, from red to violet, to which the normal human retina responds. **2.** The region of the electromagnetic spectrum (a) that is used in optical systems, such as those using lasers, optical fibers, and infrared night vision, homing, and photographic systems, (b) in which the basic optical techniques and procedures used in the visible spectrum systems are used, (c) is considered to extend from the near ultraviolet region of about 0.3 μm (micron), through the visible region, and into the mid-infrared region to about 3 μm, and (d) has been called the near visible region. *Note:* The 1.31-μm (micron) wavelength used in fiber optic systems is considered to be within the realm of lightwaves, though it is not within the visible spectrum and hence cannot be detected by the human eye but can cause damage to the retina. The optical spectrum is considered to extend even farther than the near visible region, i.e., from the beginning of vacuum ultraviolet at 0.001 μm to the end of far infrared at 100 μm. *See* **coherent light, collimated light, convergent light, divergent light, monochromatic light, polarized light, pyrotechnic light, space coherent light, speed of light, time coherent light, ultraviolet light, white light.** *See also* **frequency spectrum designation, infrared, lightwave communications, optical spectrum, ultraviolet, wavelength.**

light absorption: The conversion of light into other forms of energy upon traversing a propagation medium, thus attenuating the transmitted light beam. *Note:* Energy absorbed (A), reflected (R), scattered (S), and transmitted (T) obeys the law of conservation of energy, given by the relation $A + R + S + T = 1$, when the energies are normalized with respect to the input energy. *See also* **attenuation, beam, light, propagation medium.**

light adaptation: The ability of the eye to adjust itself to a change in the intensity of light, i.e., the irradiance level. *See also* **accommodation, irradiance.**

light amplification by stimulated emission of radiation (laser): Amplification of light by a device that produces an intense, coherent, directional beam of optical radiation by stimulating electronic, ionic, or molecular transitions to higher energy levels so that when the electrons, ions, or molecules return to lower energy levels they emit energy. *Note 1:* The release of radiated energy can be controlled in time and direction so as to generate an intense highly directional narrow beam of spatially or temporally coherent electromagnetic energy, i.e., the electromagnetic fields at every point in the beam are uniquely and specifically definable and predictable. *Note 2:* The coherence degree of laser radiation at the laser exceeds 0.88. *See also* **active laser medium, coherence, coherence degree, laser, laser diode, optical cavity.**

light analyzer: For incident light, a polarizing element that can be rotated about its optical axis to (a) control the amount of transmission, i.e., the transmission coefficient, of incident plane polarized light, or (b) determine the polarization plane of the incident light. *See also* **incidence, optical axis, plane polarization, polarization plane, transmission coefficient.**

light antenna: A system of reflecting and refracting components arranged to guide or direct a beam of light.

light attenuation: In a propagating lightwave, a reduction in the lightwave energy, usually by absorption, reflection, and scattering. *Note 1:* Light attenuation by absorption results in the conversion of the lightwave energy into other forms of energy when propagating a given distance in a propagation medium, thus reducing the transmitted lightwave energy, i.e., by reducing the transmittance. *Note 2:* When lightwaves propagate in a material medium, such as the atmosphere or glass, for a given distance, they undergo absorption, radiation, reflection, scattering, and transmission. Energy absorbed (A), reflected (R), scattered (S), and transmitted (T) obeys the law of conservation of energy, given by the relation $A + R + S + T = 1$ when the energies are normalized with respect to the input energy. Light energy absorbed, reflected, and scattered contributes to light attenuation. *Note 3:* Although geometric spreading contributes to attenuation, geometric spreading does not result in conversion of optical energy to other forms. *See also* **attenuation, propagation medium.**

light button: *Synonym* **virtual push button.**

light communications system: *See* **flashing light communications system.**

light conduit: *Synonym* **unaligned bundle.**

light current: *Synonym* **photocurrent.**

light detection and ranging (LIDAR): The use of electromagnetic frequencies in the optical spectrum for detection and range determination of objects, i.e., targets, in a manner similar to radar. *See also* **optical spectrum.**

light detector: *Synonym* **photodetector.**

light duty connector: In fiber optics, a connector that is designed and intended for use inside of an interconnection box, i.e., a distribution box or cabinet. *See also* **fiber optic interconnection box.**

light-emitting diode (LED): A semiconductor device, such as a positive-negative (p-n) junction, that (a) produces incoherent radiation by spontaneous emission under suitable operational conditions, such as when biased in the forward direction, (b) operates in a manner similar to that of an injection laser diode, i.e., by injecting electrons and holes across a semiconductor junction, (c) has about the same total optical output power and operational electric current densities as the injection laser diode, (d) is simpler and cheaper, has a lower data signaling rate (DSR), has a lower tolerance requirement, and is more rugged than an injection laser diode, (e) has a spectral width about ten times that of an injection laser diode, and (f) has a launch angle greater than that of an injection laser diode. *Note:* A typical peak spectral power output of a gallium arsenide light-emitting diode (LED) source occurs at 0.901 μm (microns) with a spectral width of about 0.050 μm. An aluminum arsenide LED source operates at about 0.820 μm with a 10 MHz (megahertz) wide spectrum. Both operate at about 1 mW (milliwatt) of spectral power output with a driving current of about 50 mA (milliamperes). *See* **edge-emitting light-emitting diode, front-emitting light-emitting diode, superluminescent light-emitting diode, surface-emitting light-emitting diode.** *See also* **electroluminescence, data signaling rate, incoherent, injection laser diode, source, spectral width, spontaneous emission.** *Refer to* **Fig. L-5.** *Refer also to* **Fig. O-3 (673).**

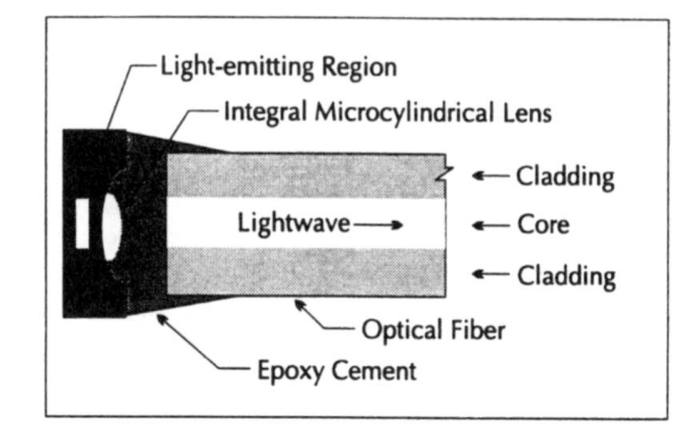

Fig. L-5. A **light-emitting diode** (LED) source, showing the butt coupling of optical output power through an integral microcylindrical lens into the core of an optical fiber.

light-emitting diode coupler: A coupler that (a) couples light energy from a light-emitting diode (LED) source to an optical fiber or fiber optic cable at the transmitting end of a fiber optic data link and (b) may be a fiber optic pigtail epoxied to the LED. *See also* **fiber optic cable, fiber optic data link, fiber optic pigtail, light-emitting diode, source.**

light guide: An assembly of optical elements, such as optical fibers, lenses, and optical waveguides, mounted and finished in a component that is used to transmit light from one or more points to one or more other points. *Synonym* **lightguide.** *See* **ultraviolet light guide.** *See also* **light pipe, optical element, optical fiber.**

lightguide: *Synonym* **light guide.**

light gun: *Synonym* **light pen.**

lighting: *See* **highlighting.**

light leakage loss: Light energy loss in a light transmission system, such as a light conduit, fiber optic cable, fiber optic connector, or optical integrated circuit (OIC), caused by any means of escape, such as imperfections at core-cladding boundaries, breaks in jackets, and sharp bends in optical fibers. *See also* **critical radius, microbends.**

lightning: The sudden discharge of an electrical voltage, i.e., of static electricity, that (a) builds up between clouds and between clouds and the Earth and (b) usually is accompanied by thunder.

lightning down conductor: The conductor that connects the air terminal, i.e., overhead ground conductor, to the earth electrode subsystem. *See also* **air terminal, arrester, earth electrode subsystem, facility grounding system, fault protection subsystem, ground, lightning protection subsystem.**

lightning protection subsystem: The collection of components used to protect a facility from the effects of lightning, usually including air terminals, lightning down conductors, the earth electrode subsystem, air gaps, arresters, and their interconnections. *See also* **air terminal, arrester, earth electrode subsystem, facility, lightning.**

light pen: A stylus that (a) usually is hand-held, (b) is a photodetector, a light source, or both, and (c) may be used to (i) interact with a computer through a specially designed monitor screen, (ii) identify display elements, display groups, or display images in the display space on the display surface of a display device, (iii) detect the light generated within an aiming symbol, (iv) provide or generate coordinate data in a display space, (v) make selections from a menu of items, (vi) indicate one or more choices from selectable alternatives, and (vii) illuminate specific optical fibers that terminate on the faceplate of an aligned bundle of optical fibers of a fiberscope in order to energize or enable specific combinational circuits in a digital system. *Synonyms* **light gun, selector pen.** *See also* **aiming symbol, aligned bundle, combinational circuit, coordinate data, display elements, fiberscope, gate, light source, menu, photodetector, stylus.**

light pencil: In optics, a narrow ray bundle diverging from a point source or converging toward an image point. *See also* **image point, point source, ray bundle.**

light pen detect: *Synonym* **light pen detection.**

light pen detection: In display systems, the sensing of light, using a light pen, from a display element, display group, or display image in the display space on the display surface of display device. *Synonyms* **light pen detect, light pen hit, light pen strike.** *See also* **light pen.**

light pen hit: *Synonym* **light pen detection.**

light pen strike: *Synonym* **light pen detection.**

light pipe: A passive optical element, such as (a) an optical fiber, (b) a slab dielectric waveguide, (c) an aligned or unaligned bundle of optical fibers, or (d) a hollow tube with a reflecting inner wall that conducts light from one place to another. *See also* **aligned bundle, light guide, optical element, optical fiber.**

light quantity: The product of luminous power and time, i.e., the integral of luminous flux with respect to time. *See also* **luminous power.**

light ray: In geometric optics, the path that (a) is described by a succession of tangents at each point in the direction of propagation of light energy, (b) is perpendicular to the wavefront, as defined in physical optics, (c) represents the lightwave itself, and (d) for plane polarized light, is in the same direction as the Poynting vector at each point. *See* **ray.** *See also* **geometric optics, isotropic medium, optical ray, physical optics, Poynting vector.** *Refer to* **Figs. I-5 (470), L-8 and L-9 (536).**

light-repeating ship: In visual signaling systems, a ship station that is designated to repeat flashing light signals that originate from other ships in order to relay messages to other ships that may be the destination addressees of the messages represented by the signals or that may repeat the signals. *See also* **flashing light ship signaling, ship station, visual signaling system.**

light request: In visual signaling systems, a special message that requests the activation of certain navigational lights for specific purposes, such as identification, message exchange, course indication, and safety.

lights: *See* **northern lights, plugging out lights, southern lights.**

light source: A device that produces or emits lightwaves. *Note:* Examples of light sources are light-emitting diodes (LEDs), lasers, and lamps. *See* **borescope light source, fiber optic light source.** *See also* **lamp, light-emitting diode, lightwave.** *Refer to* **Figs.**

A-3 (6), F-2 (339), I-2 (451), L-3 (505), L-5 (511), M-3 (590), O-3 (673), R-3 (799), T-6 (1013).

light susceptibility: *See* **ambient light susceptibility.**

light transmission: *See* **colored light transmission, flashing light transmission.**

light typewriter: By international convention, a typewriter that weighs less than 10 kg (kilograms).

light valve: *Synonym* **optical switch.**

light velocity: The speed and the direction of monochromatic lightwaves, i.e., the phase velocity. *Note:* The speed of light in a vacuum is 299,792.5 km•s^{-1} (kilometers per second). In a vacuum, the velocity of electromagnetic waves of all frequencies, i.e., all wavelengths, is the same. The phase velocity in a propagation medium is given by the relation $v_p = c_0/n$ where v_p is the phase velocity, c_0 is the light velocity in a vacuum, and n is the refractive index of the propagation medium relative to that of a vacuum. *Synonym* **velocity of light.** *See also* **group velocity, phase velocity, propagation medium.**

lightwave: An electromagnetic wave (a) that has a wavelength within or near the visible spectrum, usually including the ultraviolet, visible, and near infrared regions of the electromagnetic spectrum and (b) has a wavelength ranging from about 0.3 µm (microns) to 3 µm, i.e., about 300 nm (nanometers) to 3000 nm, which is a part of the optical spectrum. *See also* **electromagnetic wave, optical spectrum, visible spectrum.** *Refer to* **Figs. L-2 (502), P-2 and P-3 (708), P-9 (739), S-8 (874).**

lightwave communications: The branch of science and technology devoted to the study, development, and application of equipment that uses electromagnetic waves in or near the visible region of the electromagnetic spectrum for communications purposes, including photodetectors, photoconverters, optical integrated circuits (OICs), and related devices used for generating, processing, transmitting, and receiving lightwaves. *Note:* Optical communications is oriented toward optical equipment and systems, whereas lightwave communications is oriented toward the signal being processed. However, in practice, the terms are considered synonymous. *Synonym* **optical communications.** *See also* **light, visible spectrum, visual communications system.**

lightwave spectrum analyzer: A device that (a) is capable of determining the existence of, and measuring the energy levels of, various wavelengths in a light beam and (b) has a tunable filter that can examine each portion of the optical spectrum in the beam. *Note:* One type of lightwave spectrum analyzer has (a) a slit for admitting the incident beam and excluding stray ambient light, (b) some lenses, (c) a diffraction grating that is tuned by a stepper motor, (d) more lenses, and (e) another slit. A Fabry-Perot interferometer at the output of the diffraction grating will improve the resolution. Further improvements can be made by using a lightwave synthesizer, i.e., a tunable local oscillator with a sufficiently narrow bandwidth.

limit: *See* **accommodation limit, interference limit.**

limited: *See* **bandwidth-limited, diffraction-limited, dispersion-limited, power-limited.**

limited distance modem (LDM): A modem that may be used on privately owned circuits, such as local area networks (LANs), to enhance signals or increase data signaling rates (DSRs) over short distances.

limited infrared detector: *See* **background-limited infrared detector.**

limited operation: *See* **attenuation-limited operation, bandwidth-limited operation, bit-rate•length product-limited operation, detector-noise-limited operation, dispersion-limited operation, distortion-limited operation, quantum-limited operation, quantum-noise-limited operation, thermal-noise-limited operation.**

limited protection: Short-term communications security that is applied to the electromagnetic or acoustic transmission of unclassified information that (a) warrants a degree of protection against simple analysis and easy exploitation and (b) does not warrant protection to the extent needed for security of classified information. *See also* **communications security, Data Encryption Standard.**

limited protection voice equipment: Equipment that provides limited protection of voice information for purposes of security and privacy. *See also* **limited protection, privacy, security.**

limited scanning: In facsimile transmission, scanning that (a) is at a rate that is at integral multiples of the nominal scanning pitch, such as double or triple the scanning pitch, (b) is used to shorten the transmission time, and (c) may reduce the picture quality.

limiter: A device that (a) operates on a characteristic of an output signal, such as the voltage, current, power, frequency, or phase, (b) automatically prevents the characteristic from exceeding a specified value, and (c) in normal operation, yields a linearly proportional output for varying instantaneous inputs below a certain value, where, for inputs above this value, the output is at a constant peak value. *See* **band limiter,**

bandpass limiter, bend limiter. *See also* **clipper, compressor, limiting, peak limiting, signal, value.**

limiter circuit: A circuit of nonlinear elements that restricts the excursion of an electrical variable, such as the voltage, current, power, frequency, or phase, in accordance with specified criteria. *See also* **automatic gain control, circuit.**

limiter hardness: A measure of the extent to which a limiter restricts the excursion of its output signal in comparison with excursions of the input signal. *Note:* In an amplifier with a limiter, if the gain is severely reduced for large input signal excursions at high input signal levels, i.e., it provides a sudden, sharp limit or maximum at the output, it is a hard limiter. If the gain is not severely reduced for large input signal excursions, at high input signal levels, i.e., it provides a less sudden, less sharp limit or maximum at the output, it is a soft limiter. If the gain is constant for all values of input signals, there is no limiting function being performed. *See also* **limiter.**

limiting: 1. In a device, the adjusting of a characteristic of the output of the device by (a) preventing the characteristic from exceeding a predetermined value, as in hard limiting or (b) adjusting the transfer function of the device so that the transfer function depends on the instantaneous or integrated output level, as in soft limiting. **2.** The preventing of a parameter of an output signal, such as the voltage, current, power, frequency, or phase, from exceeding a specified value. *See* **hard limiting, soft limiting.** *See also* **limiter, peak limiting, value.**

limiting resolution angle: From a point of observation or measure, the angle that (a) is subtended by two distant points or parallel lines that are just far enough apart to permit an optical system to distinguish them as separate and (b) is inversely proportional to the resolving power. *Note:* The ability of an optical device to resolve two points or lines is called "resolving power." *See also* **resolution, resolving power.**

limit of interference: *Synonym* **interference limit.**

lincompex: linked compression and expansion.

LINCS: Leased Interfacility National-airspace Communications System.

line: 1. A device or equipment, such as a transmission line or loop, used for transferring signals from one point to another. **2.** In communications systems, the portion of a data circuit that (a) is external to data circuit-terminating equipment (DCE) and (b) connects the DCE to a data switching exchange (DSE), i.e., to a central office (C.O.), or to one or more other DCEs. **3.** A communications channel. **4.** A connection or a channel between (a) data circuit-terminating equipment (DCE) at one data station and (b) a DCE and data terminal equipment (DTE), such as a user terminal or end instrument, at another data station. **5.** A path, i.e., a trace, of a moving spot on the display surface in the display space of a display device, such as a cathode ray tube (CRT) or a liquid crystal display (LCD). *Note:* An example of a line is a television (TV) raster line. **6.** In facsimile, television, wirephoto, and telephoto systems, an element on an object, such as a document, that is scanned. **7.** A horizontal sequence of symbols or groups of symbols, such as a row of characters on a printed page. **8.** A mark, path, or trace drawn between two points. *See* **access line, acoustic delay line, active line, artificial transmission line, available line, balanced line, baseline, call forwarding busy line, common trunk line, delay line, digital subscriber line, direction finding baseline, direct line, display line, dual use access line, electromagnetic delay line, exchange line, fiber optic delay line, fiber optic line, flowline, format line, Goubau line, hidden line, hot line, inactive line, individual line, landline, long-distance line, long line, magnetostrictive delay line, marked access line, network access line, nonloaded line, nonswitched line, offline, online, open wire line, party line, phase alternation by line, position line, private line, scan line, single wire line, special grade access line, spectral line, switched line, terminated line, transmission line, unbalanced line, uniform transmission line.** *See also* **data circuit-terminating equipment, data station, data terminal equipment.** *Refer to* **Figs. D-1 (203), T-3 (991).**

line adapter circuit: A circuit that (a) is used at the station end of a user access line and (b) interfaces with a four-wire telephone. *See also* **four-wire circuit, four-wire terminating set, hybrid set, two-wire circuit.**

linear analog control: *Synonym* **linear analog synchronization.**

linear analog synchronization: Synchronization in which the functional relationships used to obtain synchronization are of simple proportionality, i.e., are directly proportional. *Synonym* **linear analog control.** *See also* **synchronization.**

linear array: *See* **uniform linear array.**

linear circuit: A circuit that (a) consists of linear components, i.e., components in which the ratio of voltage to current is a constant regardless of their magnitudes, (b) has a constant ratio of voltage to current at the input, i.e., a constant input impedance, (c) has a constant ratio of voltage to current at the output, i.e., a constant output impedance, and (d) has a constant transfer impedance, all at a given frequency. *Note 1:* A linear circuit usually will remain linear only over a specific range of input voltages. *Note 2:* The impedances of a linear circuit will vary with frequency, depending upon the magnitude and the distribution of resistances and the capacitive and inductive reactances in the circuit. *Synonym* **linear network.** *See also* **linearity.**

linear combiner: 1. A diversity combiner in which the combining consists of simple addition of two or more signals. **2.** A diversity combiner that adds outputs from two or more receivers. *See also* **diversity combiner, maximal ratio combiner.**

linear device: A device (a) that has an output that is directly proportional to the input and (b) in which the output has no components that are not present in the input, such as no new wavelengths or modulation frequencies. *Note 1:* The behavior of a linear device can be described by (a) a simple transfer function, such as given by the relation $y = mx + b$ where y is the output, x is the input, and m and b are constants, or (b) by an impulse response function. *Note 2:* An example of a linear device is a linear amplifier. *Synonym* **linear element.** *See also* **nonlinear device.**

linear diversity combining: Diversity combining in which the outputs of two or more receivers that operate in a diversity mode are combined in an adder. *See also* **dual diversity combining.**

linear element: *Synonym* **linear device.**

linear feedback shift register: A shift register that (a) has modulo-two feedback, (b) generates pseudorandom binary sequences, and (c) is amenable to mathematical theory and information theory in much the same ways as are linear networks and linear circuits. *See also* **nonlinear feedback shift register.**

linearity: 1. The property of a system in which, if input signals X and Y result in system output $S(X)$ and $S(Y)$, respectively, the input signal $X+Y$ will result in the output $S(X+Y) = S(X) + S(Y)$, where S is the system transfer function. *Note:* Linearity exists in a circuit when $S(X+Y) = S(X) + S(Y)$. For example, if S is a gain of 5, X is 2, and Y is 3, then $S(X)$ is 10, $S(Y)$ is 15, $S(X) + S(Y)$ is 25, and $S(X + Y)$ is 5(2 + 3), 5(5), or 25, in which case the circuit is linear. **2.** Pertaining to a relationship, over a designated range, between two parameters, i.e., two variables, such that the ratio of the two parameters is a constant. *Note 1:* The parameters might be the input and output values of a device, the relationship might be their ratio, and the values might be signal amplitudes, frequencies, and phase shifts. *Note 2:* If the gain of a device is constant over all values of input signal amplitudes, frequencies, and phase shifts at all times, the device is considered linear. If the input/output ratio of only one of the parameters is a constant, then it is linear only with respect to that parameter and nonlinear with respect to the other parameters. If the input/output ratio of a specified parameter is constant only over a specified range of the input parameter, the device is considered linear only over that range and nonlinear outside that range. *See* **modulation-demodulation linearity.** *See also* **fidelity, nonlinear distortion.**

linear logarithmic intermediate frequency (IF) amplifier: An intermediate frequency (IF) amplifier with a gain characteristic that (a) varies linearly as a function of lower input signal amplitudes, frequency changes, or phase shifts and (b) varies logarithmically as function of higher input signal amplitudes, frequency changes, or phase shifts. *Note:* The linear logarithmic intermediate frequency (IF) amplifier may be used to increase the dynamic range of a radar receiver in order to maintain a constant false alarm rate at the output of the receiver. *See also* **constant false alarm rate, linearity.**

linearly polarized (LP) mode: In an electromagnetic wave with linear polarization, a mode that (a) is propagating in a weakly guiding medium, such as a weakly guiding optical fiber, and (b) has small electric and magnetic field components in the direction of propagation, i.e., the longitudinal direction, compared to the magnitude of the transverse components. *Note:* Optical fibers used in telecommunications systems are usually weakly guiding fibers. *See also* **linear polarization, mode, weakly guiding optical fiber.**

linear network: *Synonym* **linear circuit.**

linear optimization: *Synonym* **linear programming.**

linear polarization: 1. Electromagnetic wave polarization in which the electric field vector, **E,** or the magnetic field vector, **H,** maintains a fixed spatial direction and varying magnitude. *Note:* Historically, the orientation of a polarized electromagnetic wave has been defined in the optical regime by the orientation of the electric vector and in the radio regime by the magnetic vector. **2.** In electromagnetic wave propagation, polarization such that the tip of the electric field vector describes a straight line segment in a fixed plane normal to the direction of propagation. **3.** The polarization of a uniform plane polarized electromagnetic wave in which two arbitrary sinusoidally varying rectangular electric components of the electric field vector are exactly in phase, i.e., their relative phase angle is zero, although their magnitudes may differ, depending on the orientation of the electric field vector with reference to a coordinate system. *See also* **circular polarization, electric field, electric vector, elliptical polarization, plane polarization, right-hand circular polarization.**

linear predictive coding (LPC): 1. The digital encoding of analog signals in which (a) the value of the signal at each sample time is predicted to be a linear function of the past values of the quantized signal, and (b) a single level or multilevel sampling system may be used. *Note:* Linear predictive coding (LPC) is related to adaptive predicting coding (APC), in that both use adaptive predictors. However, in LPC more prediction coefficients are used to permit use of lower information bit rates, such as 2.4 kbps (kilobits per second) to 8.4 kbps, than in APC, thus requiring a more complex processor. **2.** Coding that consists of a digital bit stream that is modulated by an analog signal, such as an analog voice signal. **3.** Coding that is produced by a vocoder that is smaller than a channel vocoder and that has an improved voice quality. *Note:* Solid state electronic components are used. LPC is sensitive to channel errors and requires hardware performance improvement at high frequencies. *See also* **adaptive predictive coding, code, code excited linear prediction coding, level.**

linear programming (LP): In operations research, programming in which the maximum or the minimum of a linear function of variables that are subject to linear constraints is located or determined. *Synonym* **linear optimization.**

linear receiver: A receiver that (a) has an output signal that varies in direct proportion to its input signal, i.e., has a constant transfer function, and (b) has the constant transfer function regardless of changes in the frequency, phase, or amplitude of the input signal. *Note:* The transfer function is the output signal value divided by the input signal value. *See also* **linearity, value.**

linear refractive index profile: In the core of a dielectric waveguide, such as an optical fiber, a refractive index profile (a) in which the refractive index varies uniformly with distance from the waveguide axis, i.e., from a given value at the axis to a given value at the core-cladding interface surface, and (b) that is produced when the refractive index profile parameter is unity, equal to exactly 1. *Synonym* **uniform refractive index profile.** *See also* **parabolic refractive index profile, power law refractive index profile, profile parameter, radial refractive index profile.**

linear scattering: *See* **nonlinear scattering.**

linear sequence code: The sequence of binary digits produced by a spread spectrum code sequence generator that uses only linear addition combinational circuit elements, such as modulo-two adders. *See also* **combinational circuit, spread spectrum code sequence generator.**

linear topology: *See* **network topology.**

line balance: The degree of electrical similarity of the two conductors of a transmission line, such as a twisted pair or open wire pair. *Note:* Improved precision of line balance reduces pickup of extraneous disturbances and interference of all kinds, including crosstalk. With proper balancing, (a) a disturbance in one conductor produces a cancellation of the disturbance by the other conductor, and (b) a disturbance that equally affects both conductors is canceled. *See also* **balanced, balanced line.**

line buildout (LBO): *Synonym* **building out.**

line character: *See* **new line character.**

line circuit: *See* **switched hotline circuit.**

line code: 1. A code chosen for use within a communications system for transmission purposes. *Note:* A line code may differ from the code generated at a user terminal and thus may require translation. **2.** A sequence of symbols that represent binary data for transmission purposes. *Note 1:* Examples of line codes are Manchester, return to zero, and block codes. *Note 2:* Line codes are used to recover precise timing and may be used to detect errors. **3.** A table of equivalences between (a) a set of digits that are generated by a data processing system or a data processing system component, such as a terminal, and (b) the pulse patterns that are selected to represent that set of digits for transmission on a line. *Note:* Consideration is given to the characteristics of the propagation medium in choosing a line code. *See* **optical line code.** *See also* **code, code conversion, propagation medium, translator, transmission line.**

line determination: *See* **radio position line determination.**

line disconnection: The interruption of the continuity or the transmission capability of a line. *Note 1:* Examples of line disconnection are physically removing a plug from a jack, opening relay contacts, and disabling a logic gate or a combinational circuit. *Note 2:* Line disconnection usually is performed at a private branch exchange (PBX) or a switching center, either by an attendant or by automatic equipment. *See also* **combinational circuit, gate, line.**

line driver: A digital or analog amplifier used to enhance transmission reliability over extended distances.

line editing: At a display station in a display system, editing in which data (a) are displayed, (b) can be accessed, and (c) can be entered into storage only one line at a time, usually by using a special device, such as a keyboard, cursor, stylus, or light pen, to insert commands. *See also* **full screen editing.**

line equipment: *See* **offline equipment, online equipment.**

line feed: 1. The movement, i.e., the displacement, of the paper on a page printer from one line of printing to the next in the vertical direction, i.e., in the downward or upward direction, without horizontal movement, i.e., without right or left, movement. **2.** A machine function that controls the vertical movement, i.e., downward or upward movement, of paper in a printer to allow line-by-line printing.

line feed character: A format effector that enables or causes the print or display position to move to the corresponding position on the next line. *See also* **display position, format effector, print position.**

line filter balance: A network designed to maintain phantom group balance when only one side of the group is equipped with a carrier system. *Note:* The line filter balance configuration usually is simpler than the filter that it balances because it balances the phantom group only for voice frequencies. *See also* **balanced line, filter, network, phantom group.**

line function: In a formatted message, the purpose or the use that is made of a particular line in the heading, text, or ending of the message.

line group: 1. A group of telecommunications lines that can be activated and deactivated as a unit. **2.** A group of lines connected to the same switch in such a manner that a call to a busy line is directed to another line in the group, as in line grouping. *See also* **call, line, line grouping.**

line grouping: The connection of a group of users with a common interest to one switch such that (a) the line to each user is grouped with all the others in such a manner that an incoming call to a busy line is routed to a free line of the group in a preferred sequence, (b) a call that is destined for a particular user is directed to that user, (c) if the user is busy, the call is directed to the next free line in the preferred sequence, (d) if all the lines in the group are busy, a busy signal is returned to the call originator, unless preemption or precedence is invoked, and (e) if a high precedence call is directed to a line in the group and that line is busy, preemption usually is not invoked unless all other lines in the group are also busy. *See also* **call, call originator, coordinate, line, precedence, preemption, switch, user.**

line hit: On a line, a sudden short-term interference that introduces noise into the connected circuits. *See also* **hit, interference, line.**

line holding: A continuation of line seizure. *See also* **line, line seizure.**

line identification: 1. The identity of a line. **2.** An identification of a line that is furnished by a communications network at the request of two users that are connected.

line identification facility: 1. A facility, i.e., a network-provided service feature, that enables a network to furnish a call receiver with the identity of the call originator. **2.** A facility that enables a network to determine the identity of a call originator, a call receiver, or both. *See* **called line identification facility, calling line identification facility.**

line identification request indicator: An indicator, i.e., a signal, that (a) is sent in the forward direction, indicating whether or not the called line identity should be included in the response message, or (b) is sent in the backward direction, indicating whether or not the calling line identity should be sent forward in a calling line identity message. *See* **called line identification request indicator, calling line identification request indicator.** *See also* **backward direction, forward direction.**

line identification signal: A signal, i.e., a sequence of characters, that (a) is transmitted to calling data terminal equipment (DTE), i.e., a call originator, to permit identification of the called line or (b) is transmitted to the called DTE, i.e., call receiver, to permit identification of the calling DTE, i.e., the call originator. *See* **called line identification signal, calling line identification signal.** *See also* **call originator, call receiver, data terminal equipment.**

line identity: Information that (a) is sent in the backward direction, consisting of a number of signals indicating the address of the called line, or (b) is sent in the forward direction, consisting of a number of signals indicating the address of the calling data terminal equipment (DTE), i.e., the call originator. *See* **called line identity, calling line identity.** *See also* **backward direction, data terminal equipment, forward direction.**

line identity message: A message that (a) is sent in the forward direction, indicating the identity of the calling data terminal equipment (DTE), i.e., the call originator, or (b) is sent in the backward direction, indicating the identity of the called DTE, i.e., the call receiver. *See* **called line identity message, calling line identity message.**

line impedance: The impedance that (a) is presented by a communications line to an input signal, (b) is a function of the line physical characteristics, i.e., the materials of construction, their geometric shape, their environment, and the frequency composition of the applied signal, (c) is the measured value at the line input terminals, (d) is measured as the ratio of the applied vector voltage to the obtained vector current, and (e) usually is expressed in ohms and a phase angle, i.e., a phasor angle. *See also* **impedance, line, vector current, vector voltage.**

line index: *See* **code line index.**

line interface protocol: *See* **Serial Line Interface Protocol.**

line length: 1. In a communications system, the physical length of a transmission line, i.e., the distance between the beginning and the end of the line. *Note:* The line length may be expressed in any suitable distance unit, such as feet, meters, kilometers, or statute miles. **2.** In a communications or power transmission line, the electrical length of the line, i.e., the phase difference between the ends of the line. *Note:* The line length usually is expressed in electrical degrees or radians. *See* **available line length, scanning line length, total line length, total scanning line length.**

line level: The signal level at a particular point in a communications line. *Note:* The line level usually is expressed as a number of dB up or down from the level at a reference point, such as the level at the beginning of the line, which might be designated as the zero transmission level point. *See also* **dB, level, signal level, zero transmission level point.**

line load control: A network-provided service feature that allows selective denial of call origination to certain lines when excessive demands for service are required of a switching center. *Note:* Line load control may be accomplished by delaying the dial tone on certain groups of lines according to specified criteria. Each line may be assigned a line load control designation, such as (a) a precedence level of temporary denial of a dial tone when circuits are overloaded, (b) a higher precedence level of denial of a dial tone when circuits are still overloaded, or (c) a still higher precedence level such that a dial tone is never denied. *Synonym* **line traffic load control.** *See also* **adaptive routing, avoidance routing, classmark, directionalization, loading, minimize, service feature, traffic load.**

line load control designation: In a communications system, one of the set of precedences that may be used to determine the sequence, i.e., the priority order, in which specific groups of users will have their communications service degraded in order to provide full services to users with more essential needs. *Note:* Line load control is used to provide a degree of automatic traffic overload protection during saturation conditions or when facilities are limited. *See also* **automatic traffic overload protection, precedence, saturation.**

line lock: A connection (a) that is established between a remote station and a host computer for the duration of transmission of an inquiry message from the station and the reception of a reply at the station, and (b) during which the line is not available to other stations. *See also* **host computer, reception, remote station, reply.**

line loop: *Synonym* **loop.**

line-modified: *See* **phase alternation by line—modified.**

line of sight (LOS): In communications, a direct propagation path that does not go below the radio horizon. *Note:* A path that is found by atmospheric or ionospheric reflection or refraction alone is still considered line of sight regardless of the geometric shape of the path. The path need not be a straight line. *See* **electronic line of sight.** *See also* **path, propagation, radio horizon.**

line of sight (LOS) equation: *See* **radar line of sight equation.**

line of sight (LOS) link: 1. In radar, radio, video, and microwave systems, a link in which line of sight conditions exist between the transmitting and receiving antennas. *Note:* A path that is found by atmospheric or ionospheric reflection or refraction alone is still considered line of sight regardless of the geometric shape of the path. **2.** In optical transmission systems, a direct unguided beam transmission path from point to point. *Note:* Examples of a line of sight link are (a) a microwave link between two towers and (b) a laser beam link between an Earth station and a satellite station. **3.** In visual communications, a direct visual path between a transmitter and a receiver, such as between signal lamps on two ships and between a heliograph and an observer.

line of sight (LOS) propagation: **1.** Propagation of an electromagnetic wave in which the direct ray from the transmitter to the receiver is unobstructed. *Note 1:* A path that is found by atmospheric or ionospheric reflection or refraction alone is still considered line of sight regardless of the geometric shape of the path. *Note 2:* Line of sight (LOS) propagation applies mostly at or above very high frequency (VHF). **2.** Electromagnetic wave propagation in the atmosphere in such a manner that the field intensity, i.e., the irradiance, decreases because of the spreading of the energy of the wave over larger areas according to the inverse square law with relatively minor effects caused by the composition and the structure of the atmosphere. *Note:* Line of sight propagation is considered to be unavailable when any ray from the transmitting antenna or the light source refracted by the atmosphere encounters any opaque object, such as the Earth, a mountain, or a building, that prevents the ray from proceeding directly to the receiving antenna. A path that is found by atmospheric refraction alone is considered a line of sight path regardless of the geometric shape of the path, i.e., the path need not be a straight line. *See also* **direct ray, electronic line of sight, free space loss, loss, propagation, radio horizon, refraction, shadow loss.**

line period: *See* **scanning line period.**

line rate: *See* **optical line rate, reading line rate, recording line rate, scanning rate.**

line recovery: *See* **offline recovery, online recovery.**

line residual equalizer: An electrical network or circuit that reduces the attenuation and the frequency distortion that remains in a propagation medium after other equalizers have been adjusted to their optimum condition. *See also* **attenuation, frequency distortion.**

line route map (LRM): In signal communications operations, a map or an overlay that shows (a) the actual routes and types of construction of transmission lines, such as wire lines, coaxial cables, and optical cables, and (b) the locations of stations and switching equipment, such as central offices (COs), switching centers, switchboards, and telegraph stations along the lines. *See also* **circuit.**

line seizure: In a communications system, the prevention by a user of further use of a line by another user or by an attendant, i.e., an operator.

line side: At a station, the portion of a device that faces toward the transmission path, i.e., that is connected to facilities external to the station, such as connected to a channel, loop, or trunk, rather than to switches, patch panels, test bays, and supervisory equipment inside the station. *See also* **equipment side, local side.**

line signaling: *See* **free line signaling.**

line slope: The rate of change, with respect to frequency, of the attenuation of a transmission line over the frequency spectrum. *Note:* Usually the attenuation is greater at high frequencies than at low frequencies. *See also* **attenuation.**

line source: **1.** An optical source that emits one or more beams, each with an extremely narrow spectral width, i.e., emits one or more monochromatic beams, rather than beams having wide spectral widths or a continuous spectrum with many wavelengths. *See also* **monochromatic.** **2.** An optical source that (a) emits one or more beams, each with a cross section that is extremely narrow compared to its length, and (b) has an active area, i.e., an emitting area, that is narrow compared to its length, such as that of an injection laser diode, and thus emits a spatially narrow beam of light. *See also* **coherent, spectral width.**

line spectrum: In optics, an emission or absorption spectrum consisting of one or more spectral lines of extremely narrow spectral width, as opposed to a continuous spectrum. *See also* **monochromatic, spectral line, spectral width.**

line speed: *Synonym* **data signaling rate.**

line temperature compensating equalizer: An equalizer network that is used with a transmission line to compensate for changes in attenuation and frequency distortion that are caused by changes in the temperature of the line. *See also* **attenuation, equalization, frequency distortion, network.**

line termination: At the network end, the physical termination of the loop. *See* **idle line termination.** *See also* **digital subscriber line, network termination.**

line-to-line correlation: In facsimile systems, the correlation of object information from scanning line to scanning line. *Note:* Line-to-line correlation is used in two-dimensional encoding. *See also* **facsimile.**

line traffic coordinator (LTC): **1.** In a switching center, the computer or the processor that is designated and used to control and coordinate traffic. **2.** In an Automatic Digital Network (AUTODIN), or Defense Digital Network (DDN), switching center, the processor that controls traffic on a line. *See also* **loading, traffic.**

line traffic load control: *Synonym* **line load control.**

line-unit-line termination (LULT): A termination that is on the network side of a loop that does not terminate on the common carrier equipment.

line-unit-network termination (LUNT): A termination that is on the customer side of a loop that does not terminate on the customer premises.

line usable length: *See* **scanning line usable length.**

line verification: *Synonym* **busy verification.**

line weighting: *See* **F1A line weighting, one forty four line weighting.**

line width: *See* **nominal line width, spectral line width.**

link: 1. The communications facilities between adjacent nodes of a network. **2.** A portion of a circuit connected in tandem with, i.e., in series with, other portions. **3.** A radio path between two points, called a "radio link." **4.** Communications facilities between two points. **5.** A conceptual circuit, i.e., a logical circuit that (a) is between two users of a network and (b) enables the users to communicate, even when different physical paths are used. *Note 1:* In all cases, the type of link, such as data link, downlink, duplex link, fiber optic link, line of sight link, point to point link, radio link, or satellite link, should be identified. *Note 2:* A link may be simplex, half-duplex, or duplex. **6.** In a computer program, a part, such as a single instruction or address, that passes control and parameters between separate portions of the computer program. **7.** In hypertext, the logical connection between discrete units of data. *See* **active link, control link, cross site link, data link, dedicated data link, downlink, feeder link, fiber optic link, fiber optic data link, functional signaling link, inactive link, line of sight link, logical data link, long-haul optical link, multiplex link, multipoint link, point to point link, radio link, repeatered fiber optic link, repeaterless fiber optic link, satellite optical link, signaling data link, signaling link, studio to transmitter link, tactical digital information link, television fiber optic link, uplink.** *See also* **hypertext.** *Refer to* **Figs. F-2 (339), F-3 (342), T-6 (1013).**

Link Access Procedure B (LAP-B): The Data Link Layer protocol as specified by International Telegraph and Telephone Consultative Committee (CCITT) Recommendation X.25 (1989). *See also* **high level data link control.**

Link Access Procedure D: A link protocol used in Integrated Services Digital Networks. *See also* **high level data link control.**

linkage: *See* **call sign linkage.**

link bay: *See* **U link bay.**

link control: The control of (a) activities and functions, such as monitoring, of a link and (b) transmissions, i.e., traffic, over a link. *See* **basic mode link control, high level data link control, synchronous data link control.**

link control sublayer: *See* **logical link control sublayer.**

linked compression and expansion (lincompex): Companding in which data links are used to interconnect the functional units of a compander. *See also* **compander, companding, data link, functional unit.**

link encryption: In a link of a communications system, encryption operation such that all data passing over the link are encrypted. *See* **multiplex link encryption.** *See also* **bulk encryption, link.**

link engineering circuit: A voice or data communications circuit that (a) serves as a transmission link between adjacent communications facilities that are interconnected by a data link and (b) is used only for coordination and control of link activities and functions, such as circuit monitoring and traffic control. *Synonym* **link orderwire.** *See also* **circuit, engineering orderwire, facility, link, orderwire circuit.**

link escape character: *See* **data link escape character.**

link establishment: *See* **automatic link establishment.**

linking protection (LP): In the linking function of adaptive high frequency (HF) radio automatic link establishment (ALE), protection that (a) is intended to protect the establishment of unauthorized links or the unauthorized manipulation of legitimate links and (b) is provided through an authorization process. *Note:* Linking protection may be provided by the use of cryptographic authentication in high frequency (HF) automatic link establishment (ALE) signaling.

Link Layer: 1. *Deprecated term for* **Data Link Layer. 2.** *See* **Open Systems Interconnection—Reference Model.**

link level: In the hierarchical structure of a primary or secondary station, the conceptual level of control or data processing logic that controls the data link. *Note:* The link level functions provide an interface between the station high-level logic and the data link. Link level functions include (a) transmit bit injection and receive bit extraction, (b) address and control field interpretation, (c) command response generation, transmission, and interpretation, and (d) frame check sequence computation and interpretation. *See* **Open Systems Interconnection—Reference Model, primary station, secondary station.**

link level operation: *See* **balanced link level operation.**

link noise temperature: *See* **equivalent satellite link noise temperature.**

link orderwire (LOW): *Synonym* **link engineering circuit.**

link protocol: A set of rules that (a) relate to data communications over a data link and (b) define data link parameters, such as transmission code, transmission mode, control procedures, and recovery procedures. *See also* **code, link, protocol.**

link quality analysis (LQA): In a radio link, the overall process by which measurements of signal quality are made, assessed, and analyzed. *Note 1:* In link quality analysis (LQA), signal quality is determined by measuring, assessing, and analyzing link parameters, such as bit error ratio (BER), the ratio of (a) signal plus noise plus distortion to (b) noise plus distortion (SINAD), and multipath (MP). *Note 2:* The measurements, assessments, and analyses are stored at, and exchanged between, stations for use in making decisions about automatic link establishment (ALE). *Note 3:* For adaptive high frequency (HF) radio, link quality analysis (LQA) is automatically performed and usually is based on analyses of bit error ratios (BERs) and on SINAD. *See also* **automatic link establishment, multipath, signal plus noise plus distortion to noise plus distortion ratio.**

link repeater: *See* **pulse link repeater.**

link subsystem: *See* **space-ground link subsystem.**

link threshold power: In satellite communications systems, the minimum Earth station transmitter carrier power that is necessary to maintain the received signal at the demodulator above a threshold level. *Note:* The received signal demodulator is at the destination addressee Earth station, not in the satellite. The system conditions under which the threshold is specified must also be defined. *See also* **threshold.**

lip sync: lip synchronization.

lip synchronization: The synchronization of audio signals with corresponding video signals so that there is no noticeable lag or lead time between them.

liquid core optical fiber: An optical fiber that (a) consists of optical glass, quartz, or silica tubing filled with a higher refractive index liquid and (b) usually has attenuation rate troughs in the attenuation rate versus wavelength characteristic curve less than 8 dB•km^{-1} (decibels per kilometer) at wavelengths of 1.090 μm (microns), 1205 μm, and 1.280 μm. *Note:* The refractive index core is a liquid that is pumped in after pulling or drawing of the optical fiber. Tetrachloroethylene is used as a liquid core in optical fibers. *See also* **attenuation rate.**

liquid crystal display (LCD): A display device that (a) creates characters by means of the action of electrical signals on a matrix of liquid cells that become opaque when energized and (b) may be viewed by reflected or transmitted light. *See* **active matrix liquid crystal display.**

liquid laser: A laser in which the lasing medium, i.e., the active medium, is in liquid form, such as organic dyes and inorganic solutions. *Note:* Dye lasers are commercially available and are often called "organic dye" or "tunable dye" lasers. *See also* **laser.**

LISP: A computer programming language that is primarily used for list processing. *Note:* LISP is derived from "list processing."

listening depth: The depth at which a submarine is able to receive messages over a specified communications system or network, such as the submarine operations and distress net. *Note:* Listening depth varies with many factors, such as weather conditions, the signal strength at the surface, the type of antenna, the signal to noise ratio, and operator experience. *Synonym* **reception depth.**

listening mode: In a communications network, a mode of operation in which a station does not send or receive messages, but does monitor messages in the network. *See also* **receiver watch.**

listening silence: 1. In a network, the condition of a station when (a) it is only receiving a transmission, (b) it is standing by ready to receive a transmission, and (c) it is not transmitting. **2.** In a network, the condition of a station when neither a receiver nor a transmitter is turned on, in order to eliminate radiation by transmitters and receivers.

listening station: A radio station on listening watch that has been designated to listen for a specific transmission from another station, usually a mobile station. *See* **listening watch, mobile station, overdue report.**

listening watch: 1. A radio communications watch in which (a) an operator is required to maintain a continuous receiver watch that is established for the reception of messages that are addressed to, or are of interest to, the listening station, and (b) a message log is usually kept. **2.** In radio telephone ciphony net operations, the period during which equipment and an operator at a station are ready to receive a message, should there be one. *Note:* Listening watches that are established in the cipher mode at directed times are usually conducted in the cipher-standby condition. At other times, listening watches may be conducted in the plain mode. *Synonym* **loudspeaker watch.** *See also* **copy watch, guard watch, radio communications cover, receiver watch.**

liter: The SI (Système International d'Unités) unit of volumetric measure, equal to the volume of a cube measuring 10 cm (centimeters), i.e., 0.1 m (meter) on each edge, or equal to 10^{-3} m^3 (cubic meter), which is the volume occupied by 1 kg (kilogram) of water at a temperature of 4°C and a pressure of 760 mm (millimeters) of mercury, i.e., 1 atmosphere. The liter is equivalent to about 1.056 U.S. quarts.

literal: 1. A representation that is explicit and is to be taken or understood at its face value. *Note:* Examples of literals are (a) command verbs FIND, LIST, and SEEK following the dot prompt in dBase software systems used in personal computers (PCs) and (b) COPY, ERASE, or FORMAT in PC-DOS software. **2.** In a source program, an explicit representation of an item that must remain unaltered during any translation of the program. *Note:* An example of a literal is the word STOP in the instruction "If a = 1, print STOP". Thus, the literal is not to be interpreted and therefore must remain unaltered during any translation. STOP is not a command. STOP is what is to be printed.

literal cipher equipment: Cipher equipment that (a) accepts symbols usually used in a language, such as letters of the alphabet, space characters, punctuation marks, and numerals 0 through 9, and (b) produces encrypted text that usually consists only of letters of the alphabet. *See also* **cipher, cipher equipment, encrypt.**

literal cryptosystem: A cryptosystem that is designed for literal communications, i.e., communications in which the plain text characters are primarily letters. *Note:* In some systems, other symbols that are usually used in a plain language, such as the numerals 0 through 9 and punctuation marks, are also used in encrypted text.

lithium niobate integrated circuit: An integrated circuit that performs filtering, coupling, switching, and modulation of lightwaves on a lithium niobate chip.

LLC: logical link control.

lm: lumen.

LNA: launch numerical aperture.

load: 1. The power consumed by a device or a circuit while performing its function. **2.** A power-consuming device connected to a circuit. **3.** To connect a power-consuming device to a circuit. **4.** To enter data or programs into storage or into the working registers of a computer or data processing system. **5.** In a database, to insert data into prescribed locations that previously contained no data. **6.** To place a machine-readable data medium on or into a sensing device. *Note:* Examples of to load are (a) to place a magnetic tape reel on a tape drive, (b) to place cards into the card hopper of a card punch or reader, and (c) to place a stack of bank checks in a magnetic ink reader hopper. *See* **critical technical load, demand load, dummy load, noncritical technical load, nonoperational load, nontechnical load, operational load, scatter load, station load, tactical load, traffic load.**

load capacity: In pulse code modulation (PCM), the level of a sinusoidal signal that has positive and negative peaks that coincide with the positive and negative virtual decision values of the encoder. *Note:* Load capacity usually is expressed in dBm0. *Synonym* **overload point.** *See also* **pulse code modulation.**

loaded diffused optical wavegude: *See* **strip-loaded diffused optical waveguide.**

loaded origin: The address of the initial storage location of a computer program usually when it is loaded into main storage. *See also* **load, main storage.**

loader: 1. A routine that reads data into main storage. **2.** In computer programming, software, such as a computer program, routine, subroutine, or software tool, that transfers an object program from secondary, auxiliary, or external storage into primary, main internal, or direct access storage prior to execution of the software. **3.** A computer program that brings data or other programs into storage in order that the data may be processed by the computer or the programs may be used by the computer. *See* **absolute loader, bootstrap loader.**

load factor: In power distribution systems, the ratio of (a) the average load over a designated period to (b) the peak load occurring during the period. *See also* **diversity factor, load.**

loading: 1. The adding of traffic to a communications system or component. **2.** The insertion of impedance into a circuit to change the characteristics of the circuit, such as to increase the real or reactive power in the circuit. *See also* **impedance matching. 3.** In multichannel telephony systems, the power that is required as a function of the number of channels. *Note:* The power may be the equivalent mean power or the peak power. **4.** The equivalent power of a multichannel group or a composite signal referred to the zero transmission level point (0TLP). **5.** In multichannel communications systems, the insertion of white noise or equivalent dummy traffic at a specified level to simulate system traffic and thus enable analysis of system performance. **6.** In multichannel telephony systems, the load, i.e., the power level, imposed by the busy hour traffic. *Note 1:* The loading may be expressed as (a) the equivalent mean power and the peak power as a function of the number of voice channels or (b) the equivalent power of a multichannel complex or a signal composite referred to the zero transmission level point (0TLP). *Note 2:* Loading is a function of the number of channels and the specified voice channel mean power. *See* **activity loading, idle channel loading, system loading.** *See also* **erlang, line load control, line traffic coordinator, loading characteristic, medium power talker, traffic intensity.**

loading characteristic: For the busy hour in multichannel telephone systems, a plot of (a) the equivalent mean power and the equivalent peak power as a function of (b) the number of voice channels. *Note:* The equivalent power of a multichannel signal, referred to the zero transmission level point (0TLP), is a function of the number of channels and has for its basis a specified voice channel mean power. *See also* **channel, loading, medium power talker.**

loading coil: A coil that (a) does not provide coupling with any other circuit, (b) is inserted in a circuit to increase its inductance, (c) when inserted along wire pairs, reduces the attenuation of the higher voice frequencies up to the cutoff frequency of the low-pass filter formed by (i) the inductance of the coils and the wires and (ii) the capacitance between the wires, (d) when inserted along wire pairs, usually is inserted at regular intervals along a line, and (e) attenuates the signal at a rapid rate above the cutoff frequency. *Note 1:* A common application of loading coils is to improve the voice frequency response characteristics of twisted wire pairs. When connected across a twisted pair at regular intervals, loading coils, in concert with the distributed resistance and capacitance of the pair, form an audio frequency filter that improves the high frequency audio response of the pair. *Note 2:* When loading coils are in place, signal attenuation increases rapidly for frequencies above the audio cutoff frequency. Thus, when the pair is used to support applications that require higher frequencies, such as in carrier systems, loading coils are not used.

load module: A computer program that is suitable for loading into main storage and is then ready for execution. *See also* **loading, main storage.**

load on call: In the execution of computer programs and program segments, an arrangement in which the programs and program segments remain resident in external, secondary, or auxiliary storage, such as disk storage, until they are called into internal, primary, or main storage, usually by the linkage editor when an entry point they contain is called. *See also* **call, computer program, entry, linkage editor, segment.**

load resistance: An electrical resistance that is used to accept electrical current from a source or to cause a voltage drop when it is conducting a current. *Note:* Load resistance is used to (a) adjust stability and sensitivity in active devices, such as amplifiers, photodetectors, transmitters, receivers, and electric power generators, and (b) adjust impedance of passive devices, such as transmission lines and antennas. *See also* **resistance.**

load test: A test of a device under various conditions of loading. *See* **tensile load test.** *See also* **loading.**

lobe: 1. An identifiable part of a radiation pattern. **2.** In a polar coordinate diagram and in each direction, the locus of points at which the power density, i.e., the irradiance, is a specified fraction, such as one-half, of the peak power density at the central axis of the radiation pattern of an emitter, i.e., an antenna or a source of radiation. *Note:* The central axis of the lobe is the line from the antenna or the source to a point at a particular range, i.e., the range for which a lobe is drawn, where the power density, i.e., the irradiance, is a maximum. **3.** A pair of channels between a data station and a lobe attaching unit, one channel for sending and one for receiving signals by the attached data station. *See* **antenna lobe, main lobe, side lobe.** *See also* **lobe attaching unit, radiation pattern.**

lobe attaching unit: In a ring network, a functional unit used to connect and disconnect data stations to and from the ring without disrupting network operations.

lobe blanking: *See* **side lobe blanking.**

lobe cancellation: *See* **side lobe cancellation.**

lobe jamming: *See* **side lobe jamming.**

lobe on receive only (LORO): Passive scanning in which a steady beam, usually of the nonscanning type, is used to illuminate an object, i.e., a target, while a separate receive antenna is scanned either conically by lobe switching or by unidirectional beams. *Note:* Lobe on receive only (LORO) permits an improved electronic countercountermeasure capability by reducing interference.

lobe switching: *See* **beam lobe switching.**

lobing: In radar systems, making use of the variation of energy levels of different parts of a lobe while illuminating an object, i.e., a target. *See* **paired lobing, sequential lobing.** *See also* **conical scanning.**

LOC: large optical cavity.

local: In communications, computer, data processing, and control systems, pertaining to an entity that is applicable to a specific area of consideration and is not applicable beyond that area, such as (a) in computer programming, pertaining to what is defined on one subdivision of a computer program and is not used in any other subdivision of that or any other program, (b) in personal computer (PC) systems, pertaining to a setting, definition, or condition that applies to only one part of a program and not to any other part, (c) at a node in a network, pertaining to a protocol that applies only to that node and not to any other station in the network, and (d) on a spreadsheet, pertaining to a setting that affects only one cell, or one set of cells, such as the column width, for one column on the spreadsheet.

local access and transport area (LATA): Under the terms of the Modification of Final Judgment, the geographical area within which a Bell Operating Company is permitted to provide exchange telecommunications and exchange access services after divestiture by the American Telephone and Telegraph (AT&T) Company. *See also* **divestiture, exchange area, inter-LATA, intra-LATA.**

local area network (LAN): A communications system that (a) lies within a limited geographical area, (b) has a specific user group, (c) has a specific topology, (d) is not a public switched telecommunications network, such as a metropolitan area network (MAN) or a wide area network (WAN), but may be connected to one, (e) usually is restricted to relatively small areas, such as rooms, buildings, clusters of buildings, industrial plants, military installations, ships, and aircraft, (f) uses fairly high data signaling rates (DSRs), (g) is not subject to public telecommunications regulations, and (h) usually is connected to a switching center or central office (C.O.) for outside communications via other networks, such as telephone and data networks. *See also* **metropolitan area network, network topology, wide area network.**

local area signaling service: *See* **custom local area signaling service.**

local battery: 1. In telegraphy, the source of power that actuates the telegraphic station recording instruments, as distinguished from the source of power that furnishes current to the line. **2.** In telephony, a system in which each telephone has its own source of power. *See also* **common battery.** *Refer to* **Fig. G-4 (404).**

local call: 1. A call that uses a single switching facility, such as a central office (C.O.) or a switchboard. **2.** A call within a local charging area. *See also* **call, long-distance call, number call, person to person call, station to station call.**

local central office: *Synonym* **central office.**

local clock: A clock (a) that is located in close proximity to an associated facility, such as a communications station, central office (C.O.), or node, and (b) that might be a remote clock relative to some other facility. *See also* **clock, external timing reference.**

local distribution system: In communications, a system that serves a group of users that have a fixed, local association, such as all the users in a single suite in an office building and served by a single private branch exchange (PBX).

local end: 1. The point at the end of a telegraph transmission line at which (a) components, such as the telegraph key, telegraph repeater, control units, and power sources, are located, and (b) transmission quality can be measured. **2.** The components, such as the telegraph key, telegraph repeater, control units, and power sources, located at the end of a telegraph transmission line.

local engineering circuit: 1. In a communications system, a communications circuit that is between (a) a technical control facility and (b) selected terminal or repeater facilities within the system. **2.** In multichannel radio communications systems, usually a handset connection at the radio station location. *Synonym* **local orderwire.** *See also* **handset, technical control facility.**

local exchange carrier (LEC): A local organization, such as a telephone company, that (a) is a communications common carrier, (b) provides local voice-grade telecommunications services, (c) provides service in a specified geographic area, and (d) is under regulation.

local exchange loop: An interconnection between customer premises equipment and public telephone central office (C.O.) equipment. *See also* **access line, drop, loop.**

localizer: A directional radio beacon that provides an aircraft station with an indication of its lateral position relative to a specific runway centerline while landing. *See also* **instrument landing system.**

localizer station: A radionavigation land station in the aeronautical radionavigation service that (a) provides signals for the lateral guidance of aircraft stations with respect to a runway centerline and (b) uses an instrument landing system localizer.

local line: *Synonym* **loop.**

local loop: *Synonym* **loop.**

local measured service: *Synonym* **measured rate service.**

local net: In radio net operations, the radio net to which a station belongs and within which it operates. *See also* **radio net.**

local office: *Synonym* **central office.**

local orderwire: *Synonym* **local engineering circuit.**

local oscillator: In a piece of electronic equipment, an oscillator that (a) usually is used as a source of frequencies for mixing with other frequencies that the equipment is handling, (b) usually is a crystal-controlled fixed-frequency oscillator, though it may be tunable, (c) usually is an inherent part of radar, radio, or television equipment, particularly receivers, and (d) produces output used for many purposes, particularly for local use. *Note:* Examples of local oscillators are (a) in a radio heterodyne receiver, an oscillator that produces a frequency that is mixed with the incoming radio frequency (rf) to produce the intermediate frequency (IF) that is to be processed in the receiver circuits, and (b) in a television receiver, an oscillator that will produce the sweep frequencies for raster scanning purposes. *See* **stabilized local oscillator.** *See also* **heterodyning, intermediate frequency, oscillator, radio frequency.**

local oscillator off: A radar operating condition such that (a) the local oscillator is turned off in order to (i) reduce the effects of jamming and (ii) provide an electronic countercountermeasure (ECCM) during barrage jamming, (b) barrage jamming cannot be seen on the radar scope, (c) objects, i.e., targets, that are on the jammer azimuth can be seen, (d) objects that are not on the jammer azimuth cannot be seen, and (e) other arrangements must be made, such as using an automatic azimuth switch or another receiver display system, for objects not on the jammer azimuth. *See also* **barrage jamming, electronic countercountermeasure, local oscillator.**

local oscillator tuning: The adjusting of the frequency of a local oscillator in electronic equipment, such as a radar, radio, or television receiver, for various purposes, such as signal tuning for improved reception, reduction of interference, and reduction of jamming effects. *See also* **local oscillator, tuning.**

local record: In communications systems, a display or hard copy of a transmitted message made on a receiver that (a) is associated with or connected to the transmitter and (b) is at the transmitting station.

local service area: The geographical area in which the telephones located in the area can be connected to one another without incurring a toll charge. *See also* **toll charge.**

local ship-shore station: A shore-based station that (a) is equipped with low-powered transmitters, (b) usually operates at low frequency (LF) or medium frequency (MF), (c) occasionally operates at very high frequency (VHF) or ultrahigh frequency (UHF), and (d) provides communications with ships in and around local areas, such as ports, harbors, bay areas, and estuaries. *See also* **frequency spectrum designation.**

local side: At a station, the portion of a device that faces toward the station facilities, rather than toward the transmission path, i.e., that is connected to switches, patch panels, test bays, and supervisory equipment in the station rather than to facilities external to the station, such as a channels, loops, or trunks. *See also* **equipment side, line side.**

local switching center (LSC): A switching center to which customer lines are connected, and that provides connections to trunks to and from other switching centers.

local telecommunications service: *See* **consolidated local telecommunications service.**

locating: In display systems, generating coordinate data corresponding to specific locations, in the display space on the display surface of a display device, by using a cursor, cross hairs, or a stylus guided by a manual control device, such as a control ball, thumb wheel, or joy stick. *Synonym* **positioning.** *See also* **control ball, cursor, display device, display space, display surface, display system, joystick, stylus, thumb wheel.**

location: *See* **radiolocation, storage location.**

location land station: *See* **radiolocation land station.**

location loss: *See* **receiver location loss.**

location mobile station: *See* **radiolocation mobile station.**

location station: *See* **radiolocation station.**

locator: In display systems, a device that (a) is used as an input device for computer and communications systems and (b) provides outputs that represent coordinate positions.

locator beacon: *See* **personnel locator beacon.**

locator transmitter: *See* **emergency locator transmitter.**

LOC diode: large optical cavity diode.

lock: *See* **angle break lock, false lock, frequency lock, line lock, transfer lock.**

lock and key protection: Protection that makes use of a password or key that is matched to an access requirement in order to obtain information from a system or to use a system.

lock code: A code used in a system, such as a communications, computer, data processing, or control system, to provide protection against improper use of components in the system, such as storage areas, files, input-output channels or devices, communications channels, and switching systems.

locked envelope: *See* **permanently locked envelope.**

locked laser: *See* **injection locked laser.**

locked loop: *See* **digital phase-locked loop, phase-locked loop.**

locked pulse radar jammer: A repeater-jammer that returns an "echo" to the jammed receiver that is different from the normal echo in range, bearing, and number of echoes, in that the pulses that are returned by the jammer are (a) locked to the radar transmitter that is being jammed, (b) timed to arrive at the jammed radar at the same time as the normal echo to give the impression of a larger target, (c) timed to arrive early, late, and on time to give the impression of many targets, (d) timed and powered to enter side lobes to give wrong bearings, and (e) made to appear as a normal echo from specific types of targets to give the impression that these targets exist when in fact they do not exist. *See also* **jammer.**

locker: *See* **flag locker.**

lock-in frequency: In a closed-loop system, a frequency at which the system can acquire and track a signal. *See also* **lock-in range.**

lock-in range: **1.** The range of frequencies between the minimum and the maximum lock-in frequencies. **2.** The dynamic range within which a closed-loop system can acquire and track a signal. *See also* **lock-in frequency.**

locking: In a communications system, the function of a code extension character that has the characteristic that the change in the interpretation of characters following the character applies to all coded representations that follow, or all coded representations in a given class, until the next appropriate code extension character occurs. *See also* **code extension character, locking code extension character.**

locking code extension character: A code extension character that indicates that the character change that is signaled by the character applies to all the succeeding characters that follow until the next appropriate code extension or escape character, i.e., it does not apply to just the one character that follows or to a specific number of characters that follow. *See* **code extension character, nonlocking code extension character.**

lock jammer: *See* **search and lock jammer.**

lock key: *See* **num-lock key, scroll lock key.**

lock loop: *See* **phase lock loop.**

lock on: The fixation of a radar tracking antenna on a specific object, i.e., target. *Note 1:* The tracking is automatically maintained by feeding the return signal (echo) from the object (target) being tracked through a control feedback loop so as to optimize the amplitude of the echo. *Note 2:* The lock-on condition in a radar system indicates that the system is continuously tracking an object in range, azimuth, and elevation.

lockout: **1.** In a telephone circuit controlled by two voice-operated devices, the inability of one or both users to get through, either because of excessive local circuit noise or because of continuous speech from either or both users. **2.** In mobile communications systems, an arrangement of control circuits such that only one receiver at a time can feed the system. *Synonym* **receiver lockout system.** **3.** In telephone systems, treatment of a user line or trunk that is in trouble or in a permanent off-hook condition, by automatically disconnecting the line from the switching equipment. **4.** In public telephone systems, a process that denies an attendant or other users the ability to reenter an established connection. *See also* **call spill over, head on collision, not ready condition.** **5.** An arrangement for restricting access to all or part of a computer system. *See also* **protection.**

lockup: *See* **scan-stop lockup.**

log: 1. *Synonym* **journal.** 2. *See* **aeronautical station master log, aeronautical telecommunications log, antilog, circuit log, operator log, position log, ship deck log, ship radio log, station log.**

logarithm: A number that consists of a characteristic and a mantissa that, when associated with a base, can be used to represent another number, i.e., an antilog. *Note:* The decimal, i.e., base 10, logarithm of the number 496 is approximately the numeral 2.6954817, where the 2 is the characteristic and the remaining numerals are the mantissa. The logarithm of 496 to the natural base $e \approx 2.71828$ is 6.2065759 and to the base 2 is 8.9541963. The logarithm of 1000 is 3 because $1000 = 10^3$. Logarithms are convenient because (a) adding them is equivalent to multiplication, (b) they can be used for expressing large numbers, and (c) they can be used as the exponent in a floating-point number. *See* **natural logarithm.** *See also* **antilog, base, characteristic, mantissa.**

logarithmic companding: Companding in which the transmitted signal is a logarithmic function of the amplitude of the input signal to the transmitting compressor portion of the compander. *See also* **amplitude, companding, compander, compressor, function, logarithmic function.**

logarithmic fast time constant (log FTC): In radar systems, a combination that (a) consists of a logarithmic function and a fast, i.e., a short, time constant, (b) is produced by a logarithmic, i.e., a nonlinear, intermediate frequency (IF) amplifier that has an output that is fed into a fast time constant (FTC) circuit, and (c) is effective in removing variations in output noise level, particularly in video systems, that might be caused by various forms of interference, such as spot noise jamming, wideband noise, and slow-sweep noise-modulated jamming. *See also* **fast time constant, noise, slow-sweep noise-modulated jamming, spot noise jamming, wideband noise jamming.**

logarithmic function: A nonlinear function, i.e., a function that (a) contains variables that have exponents other than unity and (b) may contain logarithms of variables. *Note:* As an example of a logarithmic function, the quadratic function mx^2 may be expressed as the logarithmic function $\log(mx^2) = \log m + 2\log x$.

logarithmic intermediate frequency (IF) amplifier: *See* **linear logarithmic intermediate frequency (IF) amplifier.**

logarithmic receiver: In radar, radio, video, and microwave systems, a receiver that (a) has a large dynamic range of automatic gain control (AGC), (b) provides protection against saturation by strong interference or jamming signals, and (c) is useful in communications systems against many forms of interference, such as precipitation, clutter, chaff, and spot jamming. *See also* **automatic gain control, dynamic range, interference, jamming, saturation.**

log FTC: logarithmic fast time constant.

logging: *See* **data logging.**

logic: An interconnected set of combinational circuits, such as AND, OR, NOR, NAND, and NEGATION gates, that (a) usually consists of interconnected transistors and other elements mounted on a semiconductor integrated circuit chip and (b) is used to perform various decision and control functions in a computer. *See* **fixed logic, formal logic, integrated injection logic, programmable logic.** *See also* **combinational circuit.** *Refer to* **Fig. P-15 (763).**

logical addition: The execution of the EXCLUSIVE OR function, i.e., the addition of binary digits such that a 1 is produced if either digit is a 1 and carry digits are ignored. *See also* **EXCLUSIVE OR, EXCLUSIVE OR gate, logical sum.**

logical channel: In a communications system operating in the packet mode, a means of two-way simultaneous transmission through a data link that consists of associated transmit and receive channels. *Note:* A number of logical channels may be derived from a data link by packet interleaving, i.e., by time division multiplexing. Several logical channels may exist on the same data link. *Synonym* **virtual circuit.** *See also* **data link, packet, packet mode.**

logical circuit: *Synonym* **virtual circuit.**

logical connection: In a communications network, a path that (a) is established between a source user and a destination user and (b) that has virtual circuits in it. *See also* **destination user, source user, virtual circuit.**

logical data link: A data link that is created by means of virtual circuits. *Note:* An example of a logical link is a data link that is established by time division multiplexing or by interleaving packets on a physical link, i.e., a real hardware link. *See also* **data link, logical circuit, packet, time division multiplexing.**

logical diagram: *Synonym* **logic diagram.**

logical expression: An expression that (a) can be evaluated as whether true or false and (b) consists of (i) logical operators, such as AND, OR, and NOT, (ii) operands, and (iii) usually paired parentheses. *Note:* Examples of logical expressions are (a) A + B, indicating A OR B, (b) AB, indicating A AND B, and

(c) A(A + B), indicating A AND (A OR B), which reduces to A + B, indicating A OR B. *Note:* The logical expression AB will have the value TRUE only if A and B are both true. *See also* **AND gate, Boolean function, gate, operand, operator, OR gate.**

logical file: *Synonym* **file.**

logical link control (LLC) sublayer: In a local area network/metropolitan area network (LAN/MAN) system, the part of the link level that (a) supports medium-independent data link functions and (b) uses the services of the medium access control sublayer to provide services to the network layer. *See also* **link, medium, Open Systems Interconnection—Reference Model.**

logical multiplication: The execution of the AND function, i.e., the multiplication of binary digits such that a 1 is produced if, and only if, all digits are 1s. *Note:* In logical multiplication, as in decimal system multiplication, if one or more of the factors is a 0, the result is a 0. *See also* **AND, AND gate.**

logical operation: An operation performed according to the rules of logic, such as the rules of the Boolean algebra. *Note 1:* Logical operations are in contrast to arithmetic operations. *Note 2:* Logical operations are performed in computers through the use of gates, such as AND, OR, and NEGATION gates. *See also* **arithmetic operation.**

logical operator: An operator that can be used in logical expressions to indicate the operation to be performed on the operands. *Note:* Examples of logical operators are (a) the AND in the logical expression A AND B and (b) the NOT in the logical expression NOT A. In the logical expression AB, the AND logical operator is implied. In the logical expression A OR B, the OR is the logical operator, and the expression might be written as A + B, where the + sign is the logical operator OR. *See also* **Boolean function, gate, logical expression, logical operation, operand, operation, operator.**

logical record: A record that (a) is independent of its physical environment, such as the data media in which it is recorded or the storage in which it is placed, (b) may have portions of the record located in different physical records, and (c) may have several records, or parts of records, placed in one physical record or storage area. *Note:* If the parts of a logical record are distributed among several physical records or storage areas, a pointer or an identifier must be placed at or in each part, in order that all the parts can be located. *See also* **data medium, record, storage area.**

logical signaling channel: **1.** A logical channel that provides a signaling path within an information channel. **2.** A logical channel that provides a signaling path within a physical channel. *See also* **signaling path.**

logical station (LS): A data station that is created by multiplexing and that is located at the end of a logical data link or at the end of a virtual circuit. *See also* **virtual circuit, logical data link.**

logical storage: A concept of the storage layout of a system, such that to a user the system behaves as though its storage areas and addresses were actually physically organized that way, but they are not, for the real storage is organized differently. *See also* **address, storage, storage area.**

logical sum: A sum of two binary numbers in which carry digits are ignored, i.e., a sum that is produced by an EXCLUSIVE OR gate. *Note:* The logical sum of 110101 and 011011 is 101110. The addend, augend, and sum must have the same number of digits. *See also* **EXCLUSIVE OR gate, logical addition.**

logical topology: In a network, the connection configuration that reflects the network functions, use, or implementation without reference to the physical interconnection of elements. *See also* **physical topology.**

logic array: *See* **programmable logic array.**

logic device: In communications, computer, data processing, and control systems, a device that executes logic operations, i.e., operations that follow the rules of symbolic logic, such as the Boolean operations of AND, OR, and NEGATION. *Note:* Communications, computer, data processing, and control systems use logic devices to perform assigned functions automatically. Logic devices may be constructed and interconnected as large-scale integrated (LSI) circuits, i.e., as chips, and as optical integrated circuits (OICs). *Refer to* **Fig. P-15 (763).**

logic diagram: A diagram that (a) is a graphic representation of (a) the interconnected logic, i.e., the interconnected decision-making logic combinational circuits, of systems, such as communications, computer, data processing, and control systems, (b) may include various functional components, such as gates and control flip-flops in symbolic form, counters, registers, encoders, decoders, storage units, and their interconnecting signal lines, (c) has special sets of symbols to express the logic events and show the flow path of data and decision-making control signal paths through the system, and (d) does not show supporting elements, such as (i) power supplies, illuminating wiring, disconnect switches, grounding, and air conditioning,

(ii) circuit elements, such as tubes, transformers, transistors, resistances, capacitors, and inductors, and (iii) clock pulse distribution networks for signal timing and shaping. *Synonym* **logical diagram.** *See also* **combinational circuit, gate.** *Refer to* **Fig. L-6.**

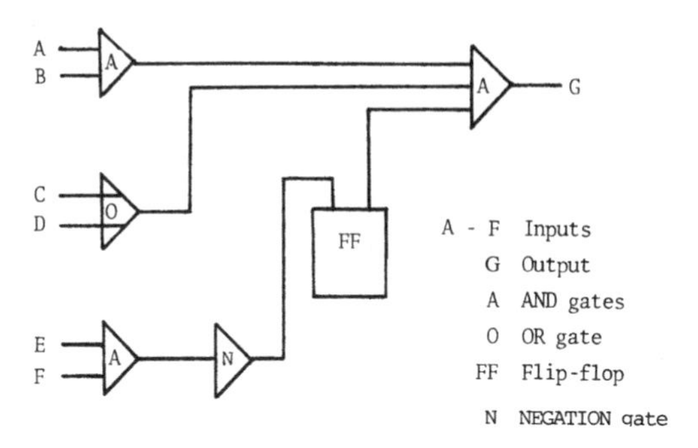

Fig. L-6. A **logic diagram** with six inputs and one output, three AND gates, one OR gate, a NEGATION gate, and a flip-flop. The diagram represents a combinational circuit that might be a special control unit.

logic element: *See* **combinational logic element.**

logic product gate: *Synonym* **AND gate.**

logic sum gate: *Synonym* **OR gate.**

logic variable: *Synonym* **switching variable.**

log in: *Synonym* **log on.**

log-in: *Synonym* **log-on.**

logistics support: *See* **joint computer-aided acquisition and logistics support system.**

log off: At a terminal, to close or end a period of operation of a communications, computer, data processing, or control system, i.e., to end a session, in a prescribed manner. *Synonym* **log out**

log-off: The closing or the ending of a session, i.e., the termination of a period of terminal operation, by a user. *Synonym* **log-out.**

log on: At a terminal, to open or begin a period of operation of a communications, computer, data processing, or control system, i.e., to start a session, in a prescribed manner. *Synonym* **log in.** *See also* **password.**

log-on: The opening or the beginning of a session, i.e., the beginning of a period of terminal operation, by a user. *Synonym* **log-in.** *See* **remote log-on.**

log out: *Synonym* **log off.**

log-out: *Synonym* **log-off.**

log periodic (LP) antenna: A wideband, multielement, unidirectional, narrow beam antenna (a) that has impedance and radiation characteristics that are regularly repetitive as a logarithmic function, i.e., an exponential function, of the excitation frequency, (b) that has length and spacing of antenna elements that increase logarithmically from one end of the antenna to the other, (c) that has a frequency response curve, i.e., response characteristic, that is repeated at frequencies that are equally spaced in the frequency spectrum, and (d) in which the value of the frequency spacing in the spectrum is determined by the logarithm of the ratio that is used to determine the physical length and the physical spacing of the antenna elements. *Synonym* **log periodic array.** *See also* **antenna element, frequency response curve, frequency spectrum.**

long board: A printed circuit (PC) board that extends for the full length of the cabinet in which it is installed. *Note:* An example of a long board is a printed circuit board that extends from (a) the plug-in side of a personal computer (PC) cabinet where it is fastened in a female connector to (b) a slot on the opposite side where it is also fastened. *See also* **printed circuit board, short board.**

long call: *Synonym* **long local call.**

long-distance call: **1.** A call that (a) is to a call receiver that is outside of the area served by the central office (C.O.) that serves the call originator, (b) usually is subject to a per call charge, and (c) may be an inter-LATA or intra-LATA call. **2.** In telephone system operation, a call on a long-distance line. **3.** A call in which a trunk that connects two switching centers is used, i.e., a call to a call receiver at the far end of one or more trunks from the end that the call originator is calling from. **4.** A call to a call receiver outside a specified area or not within a specified group of switching centers. *Synonym* **toll call.** *See also* **call, call originator, call receiver, long-distance line, switching center, trunk.**

long-distance direct current (LDDC) dialing system: A long-distance telephone dialing system in which direct current is used for transmission and signaling.

long-distance line (LDL): A line that, because of its electrical length, its physical length, its traffic load, or the number of connections that it has, requires one or more trunks between switching centers or exchanges.

long-haul communications (LHC): **1.** In public switched networks, pertaining to circuits that span large distances, such as the circuits in interlocal access and transport area (inter-LATA), interstate, and international communications systems. **2.** Communications that (a) are among users on a national or worldwide basis, (b) are over longer distances than are tactical communications, (c) have higher levels of users, such as the National Command Authority, (d) have more stringent performance requirements, such as higher quality circuits, (e) have long distances between users, including worldwide distances, (f) have high traffic volumes and densities, (g) have large switches and trunk cross sections, (h) have fixed and recoverable assets and (i) usually pertain to the U.S. Defense Communications System. *See also* **communications, tactical communications.**

long-haul communications network (LHCN): A communications network that (a) handles communications traffic over long distances, such as nationwide or worldwide traffic and (b) is characterized by long-distance trunks between towns and cities, large-size switching centers and central offices (COs), high quality equipment for high fidelity and high definition analog (voice) and digital data transmission, integrated services digital (data) networks (ISDN), high transmission capacity, and automatic switching for handling calls and messages without operator assistance. *Note:* It is anticipated that by the year 2000 over 600 billion voice circuit•kilometers will be on fiber optic long-haul and short-haul networks and only 16 billion on microwave, 160 billion on satellite, and 3 billion on coaxial cable. *See also* **short-haul communications network.**

long-haul fiber optic link: A fiber optic link that transmits lightwave signals over fiber optic cables over long distances, such as between telephone distribution frames or switching centers. *Synonyms* **long-haul optical link, long-haul fiber optic transmission set.** *See also* **data link, distribution frame, fiber optic cable, fiber optic link, switching center.**

long-haul fiber optic transmission set: *Synonym* **long-haul fiber optic link.**

long-haul optical link: *Synonym* **long-haul fiber optic link.**

longitude: *See* **celestial longitude, terrestrial longitude.**

longitudinal balance: **1.** The electrical symmetry of the two wires of a pair, such as a twisted pair or an open wire pair, with respect to ground. **2.** The measure of the difference in impedance of the two sides of a circuit. *Note:* Longitudinal balance occurs when the impedance of each side of a pair of wires is equal. **3.** In a parallel pair of wires, the electrical symmetry of the two wires with respect to ground such that the two wires cannot be distinguished from each other by any electrical test, such as an impedance or waveform distortion for identically transmitted waves. *See also* **balanced, balanced line.**

longitudinal compression sensor: *See* **fiber optic longitudinal compression sensor.**

longitudinal displacement attenuator: A fiber optic attenuator that (a) exploits longitudinal displacement loss to reduce the optical power level when inserted inline in the optical fiber path, (b) is used to prevent saturation or overdrive of components, such as prevent the saturation of a fiber optic receiver, and (c) should be placed inline near the fiber optic transmitter. *See also* **longitudinal displacement loss.**

longitudinal displacement loss: **1.** In fiber optic systems, a power loss caused by the longitudinal spacing, i.e., lengthwise spacing parallel to the optical fiber axis, at a connector or a splice, such as at a light source to optical fiber junction, a fiber to fiber junction, or a fiber to photodetector junction. *Note:* In fiber optic connectors, longitudinal displacement is needed to prevent optical fiber endface-to-endface contact that might cause damage to the endfaces when making and breaking connections with a connector. **2.** The optical power loss that occurs when an optical signal is transferred from one optical fiber to another that is axially aligned with it, i.e., there is no angular misalignment and no lateral offset of the axes of the fibers, but the endfaces of the fibers are separated in the axial direction. *Note 1:* The longitudinal displacement allows light from the source fiber to spread after leaving the fiber endface. When the rays strike the endface of the receiving fiber, some rays will enter the cladding and be lost, assuming that the core is not large enough to capture all the rays. *Note 2:* A longitudinal displacement loss may occur between a light source, such as a laser, and an optical fiber. The loss usually does not occur between an optical fiber and a photodetector because the sensitive area of the photodetector usually is larger than that of the fiber cross section. *Deprecated synonym* **gap loss.** *See also* **angular misalignment loss, gap loss, lateral offset loss.**

longitudinal jitter: In facsimile transmission, the effect caused by irregular scanning speed. *Note:* The speed irregularity may occur from many causes, such as irregular rotation of the drum or the helix which causes slight waviness or breaks in the lines of the reproduced image, lines that were straight on the scanned object.

longitudinally excited atmosphere (LEA) laser: A gas laser (a) in which the electric field excitation of the active medium is longitudinal to, i.e., in the direction of, the flow of the active medium and (b) that operates in a gas pressure range lower than that required for transverse excitation. *See also* **active medium, electric field, gas laser, transverse excited atmosphere laser.**

longitudinal parity check: A parity check performed on a serial sequence of binary digits that (a) are recorded on a single track of a data medium, such as a magnetic disk, tape, drum, or card, or (b) arrive in serial sequence on a single wire or line. *See also* **parity check, transverse parity check.**

longitudinal redundancy check (LRC): In a bit stream, a redundancy check in which (a) the check is based on the formation of a block check following preset rules, (b) the check formation rule applied to blocks is also applied to characters, and (c) the check is made on serial bit patterns. *Note 1:* When the longitudinal redundancy check (LRC) is based on a parity bit applied to characters and blocks in a single data stream, the LRC can only detect, with limited certainty, whether or not there is an error. It cannot correct the error. Detection cannot be guaranteed because an even number of errors in the same character or block will escape detection, whether odd or even parity is used. *Note 2:* When two-dimensional arrays of bits are used to represent characters or blocks, such as might occur in synchronized parallel data streams, a transverse redundancy check (TRC) may also be used. When a longitudinal redundancy check (LRC) and a transverse redundancy check (TRC) are combined, individual erroneous bits can be corrected. *See also* **block check, error control, transverse redundancy check.**

longitudinal resolution: 1. In a facsimile transmitter, the dimension along a scanning line of the smallest recognizable detail of an image that is reproduced by the shortest signal transmitted by the transmitter under specified conditions. *Note 1:* In phototelegraphy, the transmission resolution includes the dimensions and the effective luminance of the finest detail in question. *Note 2:* In document facsimile telegraphy, assuming that the contrast of the finest detail is adequate, the transmitted signal must correspond to either nominal black or nominal white. **2.** In a facsimile receiver, the dimension along a scanning line of the smallest recognizable detail of the image that is produced by the shortest signal capable of actuating the receiver under specified conditions. *Note 1:* In phototelegraphy, the longitudinal and transverse dimensions of the finest detail correspond to those of the picture element. *Note 2:* In document facsimile telegraphy, the longitudinal resolution is determined by the length of the line that is produced by the shortest nominal black or shortest nominal white signal that is capable of actuating the receiver. The transverse resolution is equal to the width of the scanning pitch. *Synonym* **longitudinal definition.** *See also* **document facsimile telegraphy, facsimile receiver, facsimile transmitter, image, nominal black, nominal white, phototelegraphy, picture element, scanning line, transverse resolution.**

longitudinal voltage: A voltage induced or appearing along the length of a propagation medium, such as a pair of parallel wires strung on poles, wires of a twisted pair, or coaxial cables. *Note 1:* Longitudinal voltages may effectively be eliminated by differential amplifiers or receivers that respond to voltage differences between the wires. Induced longitudinal voltages at low frequencies, such as power line frequencies, can be greatly reduced by twisting parallel wires, such as twisted wire pairs. *Note 2:* A longitudinal voltage is a common mode voltage applied to propagation media.

long label: In computer system spreadsheets, i.e., worksheets, an entry in a cell that extends beyond the width, i.e., the length, of the cell. *Note:* In most spreadsheet application software, such as Lotus 1-2-3, the entire label will be displayed on the control panel when the pointer is placed at the cell containing the long label, and on the spreadsheet, i.e., the worksheet, when there is space to the right of the long label cell. Thus, the entire label is stored even though it is truncated on the spreadsheet.

long line: 1. A physical conductor that is used for long-distance communications purposes, such as (a) open wire systems, (b) underground, buried, and overhead wires, (c) fiber optic cables, (d) coaxial cables, and (e) submarine cables, but not local connections. *Note:* A one-mile trunk between switching centers and a transatlantic cable are examples of long lines. A 20-mile-long loop to a user end instrument is not considered under the concept of long lines. Long lines are used in long-haul communications systems. They constitute the transmission elements of the long-distance telephone networks, and they may also contain radio relay systems, radio and microwave links, and satellite communications systems when these are integrated with the landline facilities. **2.** A transmission line in a long-distance communications network. *Note 1:* Examples of long lines are cross-country trunks, trunks between central offices, transoceanic submarine cables, satellite links, microwave links, and radio relay links. *Note 2:* Local connections are not considered as long lines. *Note 3:* There are many different types of long lines. *See also* **landline facilities, long-haul communications.** *Refer to* **Fig. R-12 (853).**

long local: *Synonym* **long local call.**

long local call: A call in which the foreign exchange service is used. *Synonyms* **long call, long local.** *See also* **foreign exchange service.**

long range aid to navigation (loran): 1. *Synonym* **long range radionavigation system. 2.** *See also* **very high frequency omnidirectional radio range.**

long range aid to navigation (loran) system: *Synonym* **long range radionavigation system.**

long range electronic navigation (loran): *Synonym* **long range radionavigation system.**

long range laser sonar: A sonar device for long range applications in which a laser is used for forming separate and identifiable beams, i.e., for forming a pulse-coded (PC) frequency-modulated (FM) wave.

long range radar: Radar that is capable of determining the range and direction, i.e., the range, azimuth, and elevation angle, of objects, i.e., of targets, at great distances compared to normal radar equipment. *See also* **early warning radar.**

long range radio aid to navigation (loran) system: *Synonym* **long range radionavigation (loran) system.**

long range radionavigation (loran) station: A land-based transmitter (a) that transmits the pulses used in the long range radionavigation (loran) system, and (b) in which the transmissions are synchronized with the transmissions of other loran stations. *See also* **long range radionavigation system.**

long range radionavigation (loran) system: A navigation system (a) that consists of (i) a fixed radio transmitter and (ii) a mobile radio receiver capable of selecting one hyperbolic line of position from information that is contained in one channel, i.e., received from one transmitter, (b) that is used to determine the position of the receiver, (c) that, for point position fixing, requires information from at least two channels in order to provide an intersection of at least two hyperbolas, i.e., at least two crossing lines, using specially prepared charts, (d) that produces a triangle at the intersection of three hyperbolas if transmissions from three transmitters are used, thus pinpointing the position with greater precision, (e) that uses the time difference of reception of pulse transmissions from two or more stations that are fixed, and (f) in which the transmissions of all stations are synchronized. *Note:* "Loran" is derived from long range radionavigation or long range aid to navigation. *Synonyms* **long range aid to navigation, long range aid to navigation system, long range electronic navigation, long range radio aid to navigation system.** *See also* **loran station.**

long-term ionospheric forecast: A forecast that (a) usually covers a three-month or longer period of ionospheric conditions and (b) usually is distributed via printed pamphlet, E-mail, Internet, or Bulletin Board to interested persons or organizations.

long-term security: In cryptographic communications systems, the protection that results from the use of a cryptosystem that will protect encrypted communications traffic from successful cryptanalysis for as long as the traffic has intelligence value. *See also* **cryptanalysis, cryptosystem.**

long-term stability: In an oscillator, the oscillator frequency as a function of time when the oscillator frequency is always measured under the same environmental conditions, such as the same applied voltage, load, and temperature. *Note:* Long-term frequency changes are caused by changes in the oscillator elements that determine frequency, such as crystal drift, inductance changes, and capacitance changes. *See also* **oscillator.**

longwave: Pertaining to electromagnetic waves, i.e., radio frequency (rf) waves, with a frequency that is below the medium frequency (MF) range, i.e., below 300 kHz (kilohertz), which corresponds to wavelengths greater than 1000 m (meters). *See also* **middlewave, shortwave.**

long wavelength: In fiber optic communications systems, pertaining to electromagnetic radiation wavelengths greater than 1 μm (micron). *Note:* In fiber optic communications systems, wavelength is standardized at 1.31 μm (microns).

LO off: local oscillator off.

look-ahead-for-busy (LFB) information: Information on network resources available for supporting higher precedence calls. *Note 1:* Available resources include idle circuits and circuits used for lower precedence calls. *Note 2:* Look-ahead-for-busy information may be used to make call path reservations.

look angle: The angle between a direction of sight and a reference direction. *Note:* Examples of look angles are (a) the angle between the direction from an aircraft to a radiation source and the longitudinal axis of the aircraft and (b) the angle between a radar antenna axis and true north.

look through: In electronic warfare systems, the irregular interruption of a jamming signal for short periods, usually extremely short periods, in order to allow for monitoring of the signal being jammed.

lookup table: An arrangement, usually a multidimensional array, of data such that (a) on the basis of a multiple entry into the table, a single or unique location in the array will permit an identification of a particular item in the array, (b) a single entry usually permits a set of items to be selected, such as a row of items, (c) another entry permits the selection of another set of items, such as a column of items, and (d) an item common to both sets can then be selected. *Note 1:* Examples of lookup tables are code tables, telephone system routing tables, logarithm tables, trigonometric tables, the periodic table of chemical elements, between-cities mileage tables, and transportation system time tables. *Note 2:* Lookup tables may assume many forms, such as (a) row and column, i.e., Cartesian, form, (b) polar form, and (c) chart form in which intersecting curved lines are followed to coordinate positions, such as are used in Smith charts for transmission line analyses and steam charts on which various lines of constant entropy, enthalpy, temperature, and pressure are drawn.

loop: 1. A communications channel from either a switching center or an individual message or packet distribution point to a user terminal or a user end instrument. *Note:* In telephone systems, the loop may consist of a pair of wires or a fiber optic cable to a user telephone. In a fiber optic loop, a fiber optic cable may connect user end instruments, such as telephones, data terminals, computers, and television sets, to a communications network, usually via multiplexing equipment to a network central office (C.O.) or switching center. *Synonyms* **local line, local loop, subscriber line, subscriber loop, user line.** *See also* **local exchange loop. 3.** Go and return conductors of an electrical circuit, i.e., a closed circuit. **4.** A closed path under measurement in a resistance test. **5.** A type of antenna extensively used in direction finding equipment and in ultrahigh frequency (UHF) reception. **6.** In a computer, a sequence of instructions that may be executed iteratively while certain conditions prevail or a specified event occurs. *Note:* In some implementations, no test is made until the loop has been executed once. The number of times a loop is to be repeated may be set in a descending counter. Each time the loop is executed, the counter value is reduced by 1 until the number in the counter reaches 0 or some prescribed number, at which time the iterative process is terminated. The repetition of the loop can also be terminated when an ascending counter reaches a specified value. *Synonyms* **line loop, local line, local loop, subscriber line.** *See* **conditioned loop, Costa's loop, digital phase-locked loop, feedback loop, fiber optic loop, fixed loop, ground loop, local exchange loop, magnetic hysteresis loop, phase-locked loop, squaring loop, switched loop, test loop, unrepeatered loop.**

loop back: 1. Pertaining to the performance of transmission tests of access lines from the serving switching center, i.e., tests that usually do not require the assistance of personnel at the served terminal. **2.** Pertaining to the performance of tests between not necessarily adjacent nodes or stations, in which two lines are used, with the testing being done at one node or station and the two lines interconnected at the distant node or station. *See also* **back-to-back connection, loop, loop test.**

loop carrier: *See* **digital loop carrier.**

loop check: *Synonym* **echo check.**

loop dialing: *Synonym* **loop disconnect pulsing system.**

loop disconnect pulsing system: An extension of the loop disconnect system that (a) is a system for telephone signaling, (b) allows numbers to be dialed into an automatic telephone exchange as a series of direct current (dc) pulses, (c) uses a standard pulse repetition rate of 10 pps (pulses per second), and (d) in a loop circuit, makes use of break pulses generated by a cam-operated dial spring switch. *Synonyms* **impulse dialing, loop dialing.** *See also* **break, loop disconnect system.**

loop disconnect system: A telephone signaling system that is used with manual exchanges and in which a handset rest switch, such as a cradle switch, hook switch, or gravity switch, applies a short circuit to the line paired cable when the telephone is not in use. *See also* **handset, signaling.**

looped dual bus: Pertaining to a distributed-queue dual-bus (DQDB) scheme in which the head of bus functions for both buses are at the same location.

looped tape storage: Storage in which (a) continuous closed loops of tape, such as magnetic or punched tape, are used to store data, (b) reading or recording usually may be accomplished forward or backward for faster access, (c) if there are many loops, they may be placed side by side in a bin, and each loop can have its own read/write head or one head can move from loop to loop on a rail, and (d) if the tapes are long, they may be loosely folded back and forth in a container or they may be wound off and on two reels. *Note:* The old eight-track magnetic tape cartridge may be considered as looped tape storage. The magnetic tape cassette, both audio and video, is not looped tape storage because it has to be rewound after reaching the end. The soundtrack tapes used to record combat aircraft sounds were looped tape storage. They were spliced with one twist, so that the tape would record on both sides with one head.

loop filter: In a phase-locked loop, a filter located between (a) the phase detector, i.e., the time discriminator, and (b) the voltage-controlled oscillator, i.e., the phase shifter.

loop frame: A collection of data or messages sent around a loop network during loop transmission. *See also* **loop transmission, network topology.**

loop gain: 1. The total usable power gain of a carrier terminal or two-wire repeater. *Note:* The maximum usable gain is determined by, and may not exceed, the losses in the closed path. **2.** The product, when expressed in ratios, or the sum, when expressed in dB, of the gain values acting on a signal passing around a closed path, i.e., around a closed loop. *See also* **gain, loop.**

loop multiplexer: *See* **fiber optic loop multiplexer.**

loop network: *See* **network topology.**

loop noise: The noise contributed by one or both loops of a circuit to the total circuit noise. *Note:* Whether the loop noise is for one or both loops should be stated, implied, or understood. *See also* **loop, noise.**

loop noise bandwidth: *See* **closed loop noise bandwidth.**

loop start: A supervisory signal transmitted by a telephone or a private branch exchange (PBX) in response to completing the loop path. *See also* **ground start, loop.**

loop test: In communications systems, a test in which (a) operating conditions are determined, faults are detected, and faults are located, (b) closed circuits or closed loops are used, and (c) various signal parameters and circuit conditions are measured. *Note:* Examples of closed loop tests are (a) a test of a loop by applying a short circuit, i.e., by forming a loop, at a remote end of the loop or by closing the loop with a known impedance, (b) a test to locate a fault in the insulation of a conductor when the conductor can be arranged to form part of a closed loop, (c) a test of the coupling of a signal from the transmitting path back to the local receiving path, which may include the whole of the transmitter and the receiver, and (d) in a satellite system, a test of the loop through the satellite transponder and return to Earth, the test also including the physical links at appropriate points in the transmitting and receiving equipment. *See also* **loop, loop back.**

loop transfer function: *See* **closed loop transfer function.**

loop translator: *See* **test loop translator.**

loop transmission: Multipoint transmission in which (a) all the stations in the network are serially connected in one closed loop, (b) there are no cross connections, (c) the stations serve as regenerative repeaters, forwarding messages around the loop until they arrive at their destination stations, and (d) any station can introduce a message into the loop by interleaving it with other messages.

loose buffer: *See* **buffer.**

loose cable structure: *Synonym* **loose tube cable.**

loose tube cable: A fiber optic cable in which (a) one or more optical fibers are fitted loosely in a tube, (b) one or more such tubes may be used to form a single cable, (c) the tubes may be filled with a seal or a gel to protect the fiber, and (d) within certain limits, stresses applied to the cable are not transferred to the fiber. *Synonym* **loose cable structure.** *See also* **tight-jacketed cable.**

loose tube splicer: A tube (a) with a square hole used to splice two optical fibers, (b) in which curved fibers are made to seek the same corner of the square hole, thus holding them in alignment until an index-matching epoxy, already in the tube, cures, thus forming a lowloss butted joint. *Synonym* **square tube splicer.** *See also* **precision sleeve splicer, tangential coupling.** *Refer to* **Fig. L-7.**

loose tube splicing: Splicing in which (a) a tube with a square hole is used to splice two optical fibers, and (b) the ends of the usually curved fibers are made to seek the same corner of the square hole, thus holding them in alignment until an index-matching epoxy, already in the tube, cures, thus forming a lowloss butted joint.

loran: long range aid to navigation, long range electronic navigation, long range radionavigation.

Lorentzian optical fiber: An optical fiber (a) that has a refractive index profile defined by the relation $n_r{}^2 = n_1{}^2/[1 + 2\Delta(n_1{}^2/n_2{}^2)(r/a)^2]$ for $r \leq a$ where Δ is given by the relation $\Delta = (n_1{}^2 - n_2{}^2)/2n_1{}^2$ where n_1 is the maximum refractive index of the core, which is at the fiber

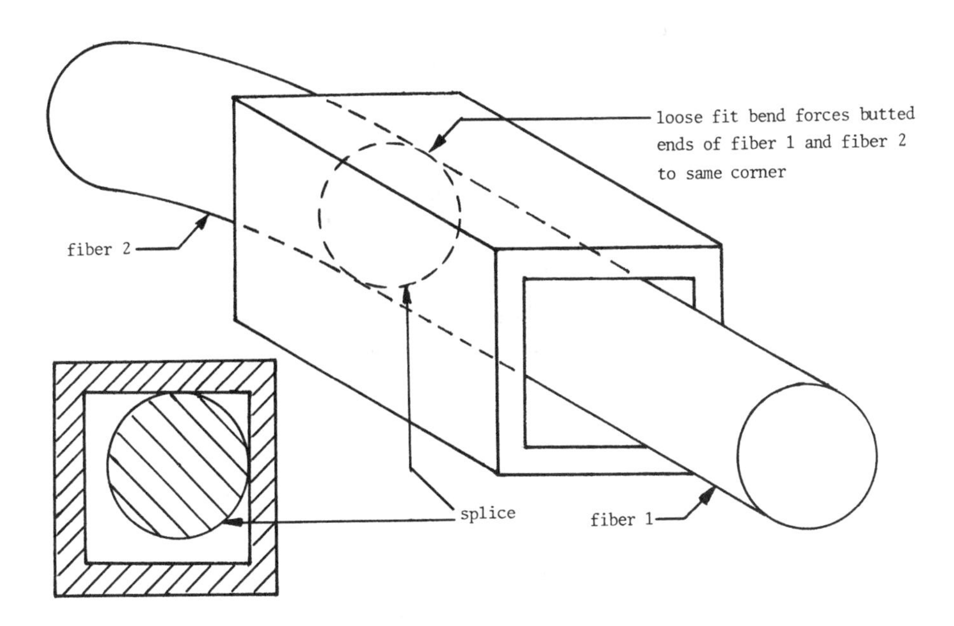

Fig. L-7. A **loose tube splicer** for butt-coupling optical fibers.

axis, i.e., at $r = 0$, and n_2 is the refractive index of the homogeneous cladding, i.e., the refractive index of the cladding is given by the relation $n_r = n_2$ for $r \geq a$ where r is the radial distance from the optical axis of the fiber and a is the radius of the core, and (b) in which skew rays entering the fiber at angles less than the acceptance angle, thus satisfying the critical angle criterion for total internal reflection, will travel a helical path within the fiber. *Note:* A skew ray is a ray that is neither a meridional ray nor an axial ray. If the value of Δ is appropriately selected, all the helical rays will arrive at the end of the fiber at the same time. However, this profile will then not permit the meridional rays and the axial rays to arrive at the end at the same time as the helical rays.

LORO: lobe on receive only.

LOS: line of sight.

LOS propagation: line of sight propagation.

loss: 1. In communications systems, the amount of electrical or optical attenuation in a circuit. **2.** The power consumed in an electrical or optical circuit or component. **3.** The energy dissipated without accomplishing useful work or purpose. *Note:* Loss usually is expressed in watts or dB. **4.** In directed signal and power transmission, the difference between power or energy that is dispatched and that which is received. *Note:* An example of loss in a fiber optic data link is the signal power that is absorbed, reflected, scattered, or radiated from a fiber optic cable between the transmitter and the receiver. Thus, the loss is the transmitted power minus the received power. *See* **absorption loss, angular misalignment loss, antenna dissipative loss, aperture to medium coupling loss, area loss, balance return loss, bending loss, bridging loss, connector-induced optical conductor loss, connector insertion loss, coupler loss, coupler excess loss, coupling loss, curvature loss, extrinsic coupling loss, fiber optic cable facility loss, free-space loss, Fresnel reflection loss, gap loss, insertion loss, intermodulation noise loss, intrinsic coupling loss, isotropic space loss, lateral offset loss, light leakage loss, microbend loss, mismatch of core radii loss, modal loss, net loss, packing fraction loss, numerical aperture loss, path loss, receiver location loss, reference loss, reflection loss, refractive index profile mismatch loss, return loss, scattering loss, shadow loss, source-coupler loss, source to optical fiber loss, splice loss, statistical fiber optic cable facility loss, transmission loss, via net loss.** *See also* **attenuation, circuit, dB, power.** *Refer to* **Figs. L-8** and **L-9.**

loss budget: In a link, such as a fiber optic link, the distribution of the total loss among the components of a system that can be tolerated in order that the received signal be above the threshold of sensitivity of the receiver, including allowance for a power margin. *Note:* The loss is distributed among the components of the system, such as cables, couplers, and splices, in an optimum fashion so as to design for minimum cost at tolerable bit error ratios (BERs) for digital data systems.

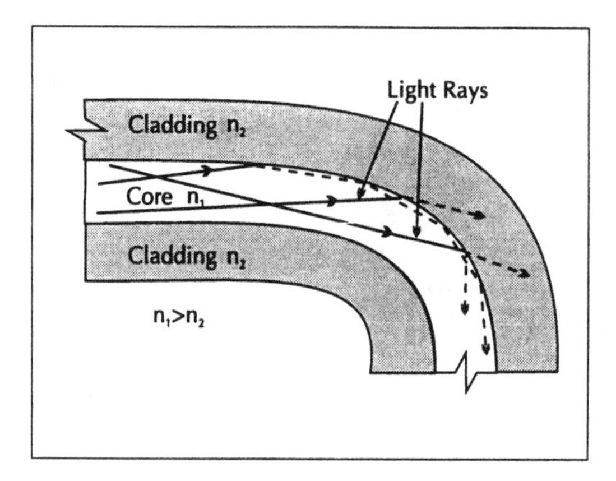

Fig. L-8. A curvature **loss** that occurs in light rays propagating in an optical fiber, which is bent so that at the bend (a) many of the rays strike the core-cladding interface surface at angles less than the critical angle and (b) the surface waves, i.e., evanescent waves and leaky waves, at the outside curve of the bend cannot travel faster than the speed of light and therefore become uncoupled with modes inside the core and radiate away. Thus, an optical fiber bent to the critical radius will glow at the bend if the light is in the visible spectrum.

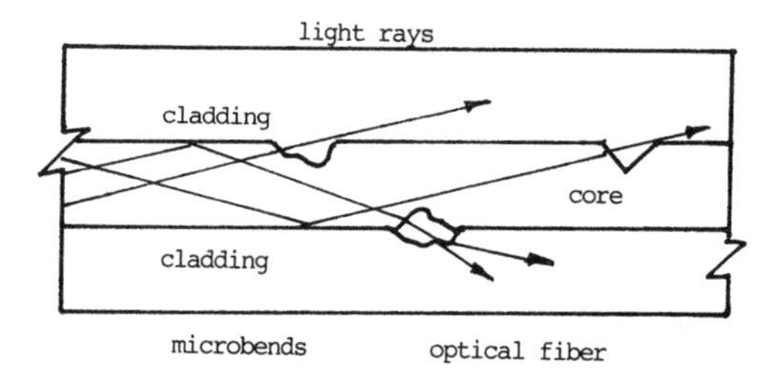

Fig. L-9. A microbend **loss** that occurs in light rays propagating in an optical fiber. The loss occurs because, at the microbends, some rays encounter a core-cladding interface surface such that the incidence angle is less than the critical angle at that microscopic point of incidence.

Required transmitted power, receiver sensitivity, intervening losses, and safety margins are all considered in the distribution, i.e., in the budgeting, of the losses.

loss budget constraint: In the design of a fiber optic station/regenerator section, a constraint (a) such that the terminal/regenerator system gain is equal to or larger than the sum of all of the optical power losses in the optical fiber path between terminal/regenerator spliced interfaces and (b) is given by the relation $G > L$ where G is the terminal/regenerator system gain given by the relation $G = P_T - P_R - P_D - R_P - M - U_{wdm} - l_{sm}U_{sm} - N_{con}U_{con}$ where P_T is the transmitter power, P_R is the receiver sensitivity (power), P_D is the dispersion power penalty, R_P is the reflection power penalty, M is the overall safety power margin, U_{wdm} is the worst case value of all losses associated with wavelength division multiplexing equipment at both ends, l_{sm} is the fiber length in the stations, U_{sm} is the worst case end of life loss (dB/km) of the single mode cable at both stations, N_{con} is the number of fiber optic connectors within the stations (inside plants), and U_{con} is the loss (dB) per connector. *Note:* G is used to overcome all the losses, L, in the optical cable facility (outside plant) of the station/regenerator section, given by the relation $L = l_t(U_c + U_{cT} + U_\lambda) + N_s(U_S + U_{ST})$ where l_t is the total sheath, i.e., jacket, length of spliced fiber optic cable (km), U_c is the worst case end of life cable attenuation rate (dB/km) at the transmitter nominal central wavelength, U_λ is the largest increase in cable attenuation rate that occurs over the transmitter central wavelength range, U_{cT} is the effect of temperature on the end of life cable attenuation rate at the worst case temperature conditions over the cable operating temperature range, N_s is the number of splices in the length of the cable in the fiber optic cable facility, including the splice at the station facility on each end and allowances for cable repair splices, U_S is the loss (dB/splice)) for each splice, and U_{ST} is the maximum additional loss (dB/splice) caused by temperature variation. L must be equal to or less than the fiber optic terminal/regenerator system gain, G, for the station/regenerator section to operate satisfactorily. *See also* **optical fiber global attenuation rate characteristic, statistical loss budget constraint.**

loss characteristic: *See* **optical fiber global loss characteristic.**

loss curve: *See* **spectral loss curve.**

loss distance: *See* **transmission loss distance.**

lossless half-wave dipole antenna: A theoretical half-wave dipole antenna that (a) has a directive gain of 2.15 dB over an isotropic antenna and (b) when used as a gain reference, has an antenna power gain of 0 dB. *See also* **antenna power gain, directive gain, half-wave dipole antenna, isotropic antenna.**

loss probability: *See* **block loss probability.**

loss ratio: *See* **block loss ratio.**

loss test set: *See* **optical loss test set.**

loss time: *See* **environmental loss time.**

loss variation: *See* **net loss variation.**

loss versus frequency characteristic: *See* **insertion loss versus frequency characteristic.**

loss window: *See* **minimum loss window.**

lossy medium: A propagation medium in which a significant amount of the energy of a propagating electromagnetic wave is absorbed per unit distance traveled by the wave. *Note:* In optical fiber cladding, lossy material may be used to absorb the energy that has leaked outside the optical fiber core. *See also* **absorption coefficient, Bouger's law, propagation medium.**

lost bit: A bit not delivered to the user within a specified maximum end-to-end bit transfer time. *See also* **block, block misdelivery probability, block transfer failure, error, misdelivered bit.**

lost block: A block not delivered to the user within a specified maximum end-to-end block transfer time. *See also* **block, block misdelivery probability, block transfer failure, error, misdelivered block.**

lost call: Except for calls in which the call receiver is busy, a call that has not been completed. *See also* **blocking, call, unsuccessful call.**

lost time: In facsimile transmission systems, the portion of the scanning line period that cannot be used for picture signal transmission. *Note:* In a drum transmitter, the lost time is the dead sector scanning time. *See also* **dead sector, facsimile picture signal, scanning time.**

lot: *See* **inspection lot.**

loudspeaker: In communications, a device that (a) converts electrical energy into sound energy, usually by accepting analog electrical signals that represent voice or sound and converting them into sound signals that can be heard a considerable distance from the device, the distance depending on the volume setting and the shape of the loudspeaker, and (b) usually consists of (i) an electromagnet, (ii) a vibrating diaphragm attached to an armature that is vibrated by the variations of electric current in the electromagnet, and (iii) perhaps a horn to direct the sound waves.

loudspeaker watch: *Synonym* **listening watch.**

loupe: *Synonym* **magnifier.**

low birefringence optical fiber: An optical fiber in which (a) the polarization beat length is 50 m (meters) or greater, (b) the beat length is given by the relation $L_B = 2\pi/(\beta_x - \beta_y)$ where β_x and β_y are the propagation constants of the two polarized waves caused by the birefringence of the optical fiber, (c) the propagation constant difference is small, and (d) the intrinsic birefringence is negligible, so that birefringence effects induced in the fiber by external forces can be calculated directly from the polarization state of the output waves. *Note 1:* Low birefringence applications include Faraday effect electric current sensors, magnetic field sensors, and fiber optic isolators. *Note 2:* Two approaches have been used to obtain very low birefringence optical fibers. One is done by maintaining the geometrical circular symmetry of the fiber cross section to a very high degree and using carefully selected dopant materials to reduce the transverse stress in the fiber. The other is done by rapidly spinning the preform while pulling the fiber. The spin pitch must be much less than the fiber beat length that would be obtained if the preform were not spun. *See also* **birefringence, high birefringence optical fiber.**

low cost exploitation operator workstation (LEOW): An image processing and manipulation system that allows an operator to perform various operations on images, such as edit images, remove grid lines from digitized maps, combine digitized photographs and maps, measure images, process text, produce graphics, and store images in transmission format.

low detectability electromagnetic wave transmission: A communications transmission system that provides an enhanced signal to noise ratio at the intended receiver, while at the same time it exhibits a low or less than unity signal to noise ratio, i.e., a negative dB value, at potential narrowband clandestine intercept receivers.

lower case: **1.** In an alphabet, pertaining to the set of letters that are characterized by being smaller than and shaped differently from the corresponding upper case letters, i.e., smaller than the capital letters. *Note:* Except for the initial letter of the first word and the "E" in "English," this sentence consists of lower case letters of the English alphabet. **2.** In teletypewriter transmission, the letter case. *See also* **figure case, letter case, upper case.**

lower frequency ground: **1.** *Deprecated term.* **2.** *See* **facility grounding system.**

lower layer protocol: In a layered open communications system, a protocol resident in the bottom layers of the system, such as in layers that correspond to Layer 1 and Layer 2 of the Open Systems Interconnection—Reference Model.

lower sideband: The sideband that has a lower frequency than the carrier frequency and the upper sideband. *Note:* The lower sideband frequency is lower than the upper sideband if there is no carrier, i.e., if the carrier is suppressed.

lowest-order transverse mode: The lowest frequency in an electromagnetic wave that can propagate in a given waveguide that can support more than one transverse electric (TE) mode or more than one transverse magnetic (TM) mode, the limitation in the number of modes being determined by the boundary conditions and the geometrical shape of the waveguide as well as the frequency, i.e., the wavelength. *Note:* Solution of Maxwell's equations with the boundary conditions for a rectangular waveguide operating in the TM mode yields the relation $\omega^2\mu\varepsilon > [(m\pi/a)^2 + (n\pi/b)^2]$ where ω is the angular velocity ($\omega = 2\pi f$ where f is the frequency), μ is the magnetic permeability of the material in the waveguide, ε is the electric permittivity of the material in the waveguide, a and b are the cross-section dimensions, and m and n are the whole numbers (eigenvalues) that satisfy and provide solutions to Maxwell's equations. The solutions to Maxwell's equations for transverse magnetic (TM) and transverse electric (TE) modes of propagation yield modes identified as TM_{mn} and TE_{mn}, except that for TM modes neither m nor n can be zero, because these conditions result in a zero electric and magnetic field. Therefore, the lowest order mode for TE is TE_{10} and the lowest order mode for TM is TM_{11}. The TE_{10} mode, called the "dominant mode," is obtained by designing a waveguide with a width to depth ratio of about 2, using an operating frequency above the cutoff frequency, given by the relation $f_c = v/2a$ where v is the velocity of light in the medium of the guide, but below the next higher cutoff frequency. In fiber optic waveguides, the dimensional parameters relate to those of the optical fiber. In a circular-cross-section optical fiber, the core diameter corresponds to the a and b for the rectangular waveguide. The variables for a rectangular electromagnetic metal waveguide relate one for one to the variables in the waveguide parameter (V value) equation for optical fibers, namely, that the number of modes is a function of the normalized frequency, i.e., the V number or the V value, determined by the relation $V = (2\pi a/\lambda_o)(n_1^2 - n_2^2)^{1/2}$ where a is the optical fiber core diameter, λ_o is the free-space wavelength, and n_1 and n_2 are core and cladding refractive indices. When both equations are solved for wavelength, it can be shown that the modes in both waveguides are functions of the dimensions of the guides, their refractive indices, the velocity of light, and the launch conditions. For the circular-cross-section optical fiber, the governing dimension is the diameter, or the radius, of the core. Thus, there is no fundamental difference between the equations for determining supportable modes in electromagnetic rectangular filled or hollow metal waveguides operating at radio, microwave, and video frequencies and the equations governing the number of supportable modes in optical fibers operating at optical frequencies. *See also* **dominant mode, low order mode, mode conversion.**

lowest usable frequency (LUF): 1. The lowest frequency in the high frequency (HF) band at which the received electric field strength or the irradiance is sufficient to provide the required signal to noise ratio during a specified period, such as 0100 to 0200 UTC (Universal Coordinated Time) on 90% of the undisturbed days of the month. **2.** In radio frequency (rf) electromagnetic wave transmission, such as is used in radio, video, microwave, and satellite communications systems, the lower limit of the frequencies that (a) can be used with good results between two specified locations, (b) includes propagation by reflection from the regular ionized layers of the ionosphere, and (c) is a median frequency that is applicable to 90% of the days of a month, as opposed to 50% for the maximum usable frequency (MUF). *Synonym* **lowest useful frequency.** *See also* **frequency, ionosphere, maximum usable frequency.**

lowest usable high frequency (LUHF): The lowest frequency in the high frequency (HF) band at which the electric field intensity at a receiver is sufficient to provide the required signal to noise ratio on 90% of the undisturbed days of the month. *Synonym* **lowest useful high frequency.**

lowest useful frequency: *Synonym* **lowest usable frequency.**

lowest useful high frequency: *Synonym* **lowest usable high frequency.**

low frequency (LF): Frequencies that are in the range from 30 kHz (kilohertz) to 300 kHz, the range having the alphabetic designator LF and the numeric designator 5. *See* **extremely low frequency, infralow frequency, very low frequency.** *See also* **frequency, frequency spectrum designation.** *Refer to* **Table 1, Appendix B.**

low frequency cutoff: The lowest electromagnetic frequency a waveguide with given dimensions and constructed of given materials is capable of supporting. *Note 1:* The low frequency cutoff occurs because of dimensional and boundary conditions. *Note 2:* The lowest frequency corresponds to the longest wavelength the guide can support. *Note 3:* In fiber optics, the low frequency cutoff is the frequency of the longest wavelength an optical fiber is capable of supporting. In essence, the core diameter has to be at least the order of magnitude of the wavelength of the mode propagating in the core, or else the core cannot transmit, i.e., cannot support, the mode. For a single mode optical fiber operating at the standard wavelength of 1.31 μm (microns), the diameter is about 6 μm.

low level control: In communications systems, control functions that (a) occur at or below the data link layer in the hierarchical structure of a communications station or network, (b) include various data processing functions, such as bit stuffing, i.e., bit injection, bit destuffing, i.e., bit extraction, address interpretation, response generation, and frame check sequence computation, and (c) in the open system architecture concept of a communications network, usually occur in the data link layer and the physical layer. *See also* **high level control, Open Systems Interconnection—Reference Model.**

low level keying: *Synonym* **low level signaling.**

low level language: A computer programming language that consists of codes that (a) are directly executable by a specific computer or class of computers, (b) are used to construct macros, (c) usually are the object codes of a compiler, and (d) are not convenient or readily understood by a user or a programmer, but are used for design, maintenance, and engineering purposes. *Note:* An example of a low level language is a machine language. *Synonym* **low order language.** *See also* **computer oriented language, high level language, intermediate level language, machine language.**

low level modulation: Modulation of a signal at a point in a system or a device, such as a radio transmitter, where the power level is low compared with the output power. *See also* **modulation.**

low level protocol (LLP): In the open systems architecture concept of a communications system, a protocol that (a) is at the lower layers of the system, such as the data link layer and the physical layer, and (b) places particular emphasis on the communications processes and the transmission operations. *See also* **high level protocol, Open Systems Interconnection—Reference Model.**

low level signaling: 1. On signal lines, the use of low levels of voltage, such as voltages that are between the limits of positive and negative 6 V (volts). **2.** In telegraph systems, the keying of electrical currents and voltages in which the currents and voltages are at the lowest possible values, usually not more than 6 V (volts) between keying contacts. *Note 1:* Purposes of low level signaling include (a) a reduction of the possibility of unauthorized detection of digital data and telegraph transmissions, (b) a reduction of crosstalk and interference with other channels, and (c) a reduction of high frequency harmonics that are caused by the sudden interruption of currents and voltages, because sharp-edged waveforms are equivalent to the existence of high frequency components in the signals. *Note 2:* Typical low level signaling values are 2 V (volts) at 70 μA (microamperes) and 6 V at 1 mA (milliampere). *Synonym* **low level keying.** *See also* **level, signal.**

lowloss FEP-clad silica optical fiber: An optical fiber that (a) consists of (i) a pure fused silica core with a refractive index of 1.458 and (ii) a perfluoronated ethylene-propylene (FEP), a commercial polymer, cladding with a refractive index of 1.338, (b) has a transmission loss of less than 2 to 3 dB/km (decibels per kilometer), and (c) has an ultraviolet transmission capability of less than 360 dB/km at 0.546 μm (micron) wavelength.

low loss optical fiber: Compared to a high loss optical fiber, an optical fiber that (a) has a low energy loss, from all causes, per unit length of optical fiber, usually considered to be below 20 dB/km (decibels per kilometer), though this figure is decreasing as time goes by, and (b) causes attenuation of a propagating wave primarily by (i) scattering caused by metal ions, (ii) scattering caused by the intrinsic fiber material, i.e., Rayleigh scattering, and (iii) absorption caused by water in the OH^- radical form, i.e., in the hydroxyl ion form. *See also* **high loss optical fiber, ultralow-loss optical fiber.** *Refer to* **Figs. A-10 (49), H-1 (430).**

low order language: *Synonym* **low level language.**

low order mode: In an electromagnetic wave propagating in a waveguide, a mode (a) that corresponds to the lesser, i.e., the first few, eigenvalue solutions to the wave equations, regardless of whether it is a mode in a transverse electric (TE), transverse magnetic (TM), or transverse electromagnetic (TEM) wave, and (b) in which only a few whole wavelengths of the wave fit transversely in the guide and therefore can be supported by the guide. *Note:* A multimode optical fiber supports as many modes as the core diameter, numerical aperture, and wavelength permit. For the waveguide to support only low order modes, the normalized frequency, i.e., the V parameter or V value, must be less than 4 in order that the number of modes be sufficiently small that there will be no high order modes. In a single mode fiber there are no high order modes. Conceptually, however, if there are only two modes, one could be considered the low order mode and the other the high order mode. Also, an optical fiber operating in single mode at a wavelength of 1.31 μm (microns) may operate in multimode if the wavelength is reduced to, say, 0.85 μm. *See also* **high order mode, low order mode, lowest order transverse mode.** *Refer to* **Fig. P-17 (767).**

low-pass filter: A filter that passes all frequencies below a specified frequency with little or no attenuation, but highly attenuates higher frequencies. *See also* **bandpass filter, filter, frequency, high-pass filter, band stop filter.**

low performance communications equipment: 1. In communications systems, equipment that (a) has imprecise characteristics, (b) usually does not meet communications system reliability requirements, and (c) has insufficiently exacting characteristics to permit its use in trunks or links. *Note:* Low performance communications equipment may be used in loops if it meets loop performance requirements. **3.** Tactical ground and airborne communications equipment that (a) has size, weight, and complexity characteristics that must be kept to a minimum and (b) is used in communications systems that have components with similar minimum performance characteristics. *See also* **high performance communications equipment.**

low performance equipment: 1. In communications, computer, data processing, and control systems, equipment that (a) has imprecise characteristics, (b) usually does not meet system reliability requirements, and (c) in communications, has insufficiently exacting characteristics to permit its use in trunks or links. *Note:* Low performance equipment may be used in loops if it meets loop performance requirements. **3.** Tactical ground and airborne equipment that (a) has size, weight, and complexity characteristics that must be kept to a minimum and (b) is used in systems that have components with similar minimum performance characteristics. *See also* **high performance equipment.**

low power communications device: A communications device (a) from which radiation levels of electromagnetic energy for communications purposes are restricted, (b) that is used for the transmission of signals, and (c) in which conducted or guided radio frequency techniques are not used. *Note 1:* Low power signals include (a) signals that represent intelligence of any nature, including signs, writing, images, and sounds, or (b) control signals. *Note 2:* Examples of low power communications devices include wireless microphones, phonograph oscillators, radio-controlled garage door openers, and radio-controlled models, such as model airplanes, ships, and ground vehicles.

low probability of detection (LPD): Pertaining to a signal, such as a radio frequency (rf) signal, in which there is a low probability of unauthorized detection of the presence of the signal. *See also* **low probability of interception.**

low probability of interception (LPI): Pertaining to a signal, such as a radio frequency (rf) signal, in which there is a low probability of unauthorized interception and extraction of intelligence contained in the signal. *See also* **low probability of detection.**

low probability of intercept receiver: A receiver in which (a) a directional antenna is used, (b) variable frequency tuning is used, (c) incident electromagnetic power is not detected unless the frequency, direction, and polarization of the receiver are in proper coincidence with the transmitter at the instant of transmission, (d) the ratio of (i) the time during which the transmission illuminates the receiver to (ii) the time taken by the receiver to cover the complete frequency band and spatial sector of search of the antenna is greater than unity, and (e) a sampling process is used. *See also* **high probability of intercept receiver.**

low speed Morse: The international Morse code transmitted at speeds that are lower than normal speeds, such as speeds of less than 18 words per minute. *Note:* Maritime and aeronautical radio organizations tend to limit signaling speeds to less than 18 words per minute, i.e., to low speed Morse.

low voltage enclosed air circuit breaker switchgear: Switchgear that (a) is designed for circuit interruption and closure at low voltages, usually not exceeding 600 V (volts, rms) for alternating current (ac) circuits and not exceeding 750 V for direct current (dc) circuits, (b) has air circuit breakers contained in individual compartments with controlled access, (c) is controlled remotely from front panels, i.e., from control consoles, (d) has only dead front switchgear assemblies, i.e., no power is available at the front panel, and (e) has all power at the back panel or remote from the control area.

LP: linear programming, linearly polarized, linking protection.

L pad: A circuit that consists of two discrete, i.e., lumped, electrical components, one series component, such as an inductor, and one shunt component, such as a capacitor, with respect to the line. *Note 1:* The components can be fixed or variable. If variable, the impedance of the L pad can be varied, or both components can be adjusted simultaneously, so as to maintain a constant impedance. *Note 2:* In an electrical schematic diagram, the orientation of the components and their connection are such that the diagram resembles the upper case inverted letter "L" with a component on each leg. *See also* **gain, impedance matching.**

LP mode: linearly polarized mode.

LP_{01} mode: 1. fundamental linearly polarized (LP) mode. 2. *See* **fundamental mode.**

LPC: linear predictive coding.

LPD: low probability of detection.

LPE: liquid phase epitaxy.

LPI: low probability of interception.

LQA: link quality analysis.

LRC: longitudinal redundancy check.

LSB: least significant bit, lower sideband.

LSI: large-scale integration.

LSI chip: large-scale integrated chip.

LSI circuit: large-scale integrated circuit.

LTC: line traffic coordinator.

Lucal code: *Synonym* **modified reflected binary code.**

LUF: lowest usable frequency, lowest usable high frequency, lowest useful frequency.

Lukasiewicz notation: *Synonym* **prefix notation.**

lumen (lm): The SI (Système International d'Unités) unit of luminous power that corresponds to $1/4\pi$ of the total luminous power emitted by a source that has a candlepower of 1 candela, thus being equal to the luminous power issuing from 1/60 cm^2 (square centimeter) of opening of the international standard source, and is included in a solid angle of 1 steradian. *See also* **candela, luminous power.**

lumen•hour (lm•hr): The quantity of light energy equal to 1 lm (lumen) flowing for 1 hr (hour).

lumen•second (lm•s): The quantity of light energy equal to 1 lm (lumen) flowing for 1 s (second).

lumerg: The centimeter-gram-second (cgs) unit of light energy equal to 10^{-7} lm•s (lumen•second).

luminance: 1. In a light source, (a) the luminous intensity emitted in a given direction by an infinitesimal area of the source divided by (b) the projection of that area of the source upon a plane perpendicular to the given direction. *Note 1:* Luminance is usually stated as luminous intensity per unit area or optical power, i.e., luminous power, emitted per unit solid angle per unit projected area upon which the power is incident, the areas being perpendicular to the direction in which the lightwave is propagating. *Note 2:* "Luminance" is the proper photometric term for the obsolete term "brightness," which is a relative effect on the human retina. **2.** The amount of light per unit area emitted or reflected from a surface. *Note:* The SI (Système International d'Unités) unit for luminance is candelas per square meter. *See also* **luminous intensity, luminous power.**

luminance compensation: In the transmission of pictures, whether by facsimile, phototelegraph, radiophoto, or fiber optic systems, compensation introduced at the receiver by the record medium to reproduce the image so as to exactly reproduce the luminance range of the object. *See also* **image, luminance, luminance range, object, record medium.**

luminance range: The difference between the lowest luminous intensity and the highest luminous intensity that is obtained directly from a light source or reflected by an object. *See also* **luminous intensity.**

luminance range compression: 1. In photographic transmission systems, such as facsimile, radiophoto, telephoto, and fiber optic systems, the decrease in the luminance range of the signals in the propagation medium, the display image, or the recorded copy below the luminance range of the object, such as the document or the picture being transmitted, i.e., the luminous range of the image is smaller than that of the object. *Note:* Luminance range compression could be accidental or deliberate. **2.** The ratio of (a) the luminance range of the signals in the propagation medium or in the display image to (b) the luminance range of the object such that the luminance range of the signals in the propagation medium or in the display image is smaller than that of the object. *See also* **luminous range, object, propagation medium.**

luminance range expansion: 1. In photographic transmission systems, such as facsimile, radiophoto, telephoto, and fiber optic systems, the increase in the luminance range of the signals in the propagation medium, the display image, or the recorded copy over and above the luminance range of the object, such as the document or the picture being transmitted, i.e., the luminous range of the image is larger than that of the object. *Note:* Luminance range expansion could be accidental or deliberate. **2.** The ratio of (a) the luminance range of the signals in the propagation medium or in the display image to (b) the luminance range of the object such that the luminance range of the signals in the propagation medium or in the display image is larger than that of the object. *See also* **luminous range, object, propagation medium.**

luminance temperature: The temperature of an ideal blackbody that would have the same luminance as the source for which the luminance temperature is desired for some narrow spectral region. *See also* **blackbody.**

luminance threshold: *See* **absolute luminance threshold.**

luminescence: In certain materials, such as radium, the emission of electromagnetic radiation that (a) is in excess of that attributable to the thermal state of the material and the emissivity of its surface, (b) is characteristic of the particular material, (c) occurs without outside stimulation, and (d) is restricted to certain wavelengths of the electromagnetic spectrum. *See* **electroluminescence, superluminescence.** *See also* **emissivity, fluorescence, phosphorescence.**

luminescent diode: *See* **superluminescent diode.**

luminosity: The ratio of (a) the luminous power to (b) the radiant power in a sample of light. *Note:* Luminosity may be expressed in units of lumens per watt of radiant energy. *Synonym* **luminous radiation efficiency.** *See also* **luminous power, radiant power.**

luminosity curve: The curve obtained by plotting luminous efficiency as a function of the wavelength of a lightwave. *See* **absolute luminosity curve.** *See also* **luminous efficiency.**

luminous density: In a lightwave, the luminous energy per unit of spatial volume occupied by the lightwave.

luminous efficiency: In a light source, the ratio of (a) the luminous power, i.e., optical power, emitted by the source to (b) the power consumed by the source, such as a laser, light-emitting diode (LED), incandescent bulb, or fluorescent bulb. *Note:* Luminous efficiency usually is expressed in lumens per watt.

luminous emittance: The total luminous power emitted by a unit area of a source, in contrast to a point or line source. *Note:* Luminous emittance usually is expressed in lumens per unit area. *See also* **luminous power.**

luminous energy: 1. Luminous power integrated over a period. **2.** The light energy contained in a given spatial volume occupied by luminous power. *See also* **luminous power.**

luminous intensity: The ratio of (a) the luminous power emitted by a light source, or an element of the source, in an infinitesimally small cone about a given direction, to (b) the solid angle of that cone. *Note:* Luminous intensity usually is stated as luminous power, i.e., optical power, emitted per unit solid angle, i.e., in lumens per steradian. *See also* **luminous power, steradian.**

luminous power: 1. Radiant power in the visible spectrum. **2.** The quantity that (a) specifies the capacity of the radiant power, i.e., optical power, from a light source to produce the attribute of visual sensation known as brightness, (b) is radiant power evaluated with respect to luminous efficiency of radiation, and (c) unless otherwise specified, pertains to the standard photopic observer. *Note:* Luminous power is expressed in lumens. A 1 candela light source emits 4π lumens of luminous power. *See also* **candela, lumen, luminous energy, radiant power, visible spectrum.**

luminous radiation efficiency: *Synonym* **luminosity.**

luminous range compression: 1. In lightwave, facsimile, or photographic transmission systems, the reduction in the luminance range of the signals in the propagation medium or in the display image from the luminance range of the object, such as the document or the picture being transmitted. **2.** The difference between (a) the luminance range of the signals in the propagation medium or in the display image and (b) the luminance range of the object, such that the luminance range of the signals in the propagation medium or in the display image is smaller than that of the object. *See also* **luminous range, object, propagation medium.**

luminous reflectance: The ratio of (a) the luminous power reflected by an object, i.e., reflected by the incident surface of the object, to (b) the incident luminous power. *See also* **luminous transmittance.**

luminous responsivity: In an optical source, such as a photodiode, the output optical flux per unit of input current, such as 0.10 $lm \cdot mA^{-1}$ (lumens per milliampere). *Refer to* **Figs. P-6 (721), S-16 (932).**

luminous transmittance: The ratio of (a) the luminous power transmitted by an object, i.e., that crosses the incident surface into the object, to (b) the incident luminous power. *See also* **luminous reflectance.**

lumped circuit element: A circuit element that (a) is considered to be located at a single point in the circuit in which it is connected, (b) ideally allows an electrical signal to propagate instantly through the element, (c) if linear, allows the use of ordinary differential equations to model the circuit in which the element, and other elements, are connected, (d) is considered to accept input wavelengths that are large compared to the dimensions of the element, and (e) usually does not radiate energy. *Note:* If the propagation time through a circuit element is appreciable, the element is considered to be distributed rather than lumped. *See also* **distributed circuit element.**

lumped parameter: Pertaining to electrical circuits in which (a) the circuit elements, i.e., circuit components, such as resistances, capacitances, inductances, semiconductor junctions, thermocouples, photosensitive materials, and connectors, are each considered to be located at a single point in the circuit in which they are connected, (b) an electrical signal is considered to propagate instantly through each element, (c) ordinary differential equations can be used to model the circuit if the resistance, capacitance, and inductance elements are linear, (d) the input wavelength is large compared to the dimensions of the elements, and (e) usually no energy is considered to be lost by radiation. *Note:* If the propagation time through a circuit element is appreciable, the circuit element is considered to be distributed rather than lumped. *See also* **circuit element, distributed circuit element, distributed parameter, lumped circuit element.**

lunar eclipse: An eclipse of the Moon by the Earth as the Earth, i.e., an obstruction, moves between (a) the Sun, i.e., the source, and (b) the Moon, i.e., a reflector. *See also* **eclipse, solar eclipse.**

lux: The SI (Système International d'Unités) unit of illuminance, equal to 1 lum•m^{-2} (lumen per square meter) of surface normal to the direction of propagation.

lynx: A World Wide Web (WWW) browser that (a) provides a character-based user interface to hypertext-based information and (b) displays only character-based portions of the hypertext information. *See also* **Internet, World Wide Web.**

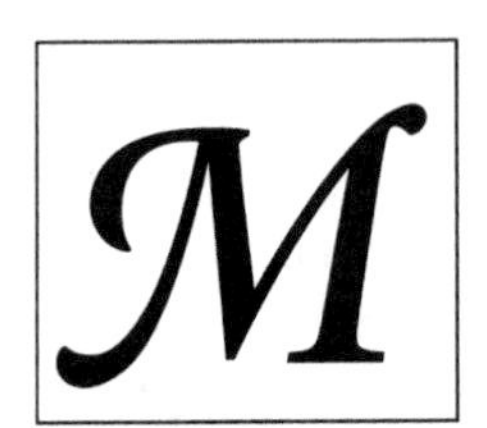

m: 1. meter. 2. As a prefix, milli.

M: 1. mega. 2. 10^6, i.e., one million. *Note 1:* $500M means five hundred million dollars, 4 Mbps or 4 Mb•s^{-1} means 4 x 10^6 bits per second, 4 M bytes means 4 megabytes (of storage capacity or of data), and 4 MHz means 4 megahertz (a frequency). *Note 2:* The "M" is derived from "mega," the prefix meaning million. *Note 3:* To mean million as a stand-alone character, the upper case, M, must be used. The stand-alone lower case, m, is for meter; as a prefix, milli.

MAA: maximum acceptance angle.

MAC: medium access control. *See* **medium access control sublayer.**

mach: *Synonym* **mach number.**

machine: *See* **abstract machine, dictation machine, optical fiber pulling machine, finite state machine, Turing machine.**

machine code: 1. Any code that a machine, such as a communications, computer, data processing, or control system, can recognize, sense, read or interpret. **2.** A set of coded instructions that a machine can sense, interpret, and execute. *Note:* Machine languages, machine functions, and machine instructions are described or written in machine code. *Synonym* **computer code.** *See* **absolute machine code, relocatable machine code.** *See also* **machine instruction, machine language, machine function.**

machine function: 1. An operation that is performed by equipment. **2.** The signal that causes equipment to perform an operation. *Note:* Examples of machine functions are shift, carriage return, search, indent, and save.

machine-independent: In telecommunications, computer, data processing, and control systems, pertaining to operations, procedures, computer programs, and processing that do not depend upon specific hardware for their successful execution.

machine instruction: An instruction that (a) is written in a machine language and (b) can be executed directly by the machine for which it was designed without translation or interpretation. *See also* **computer instruction, machine language.**

machine instruction code: For a given machine, the code used to write instructions for execution by that or an identical machine. *See also* **instruction code, machine instruction.**

machine language: 1. A language (a) that need not be modified, translated, or interpreted before it can be used by the processor for which it was designed, and (b) in which the operation codes and addresses used in instructions (i) can be directly sensed by the arithmetic and control unit circuits of the processor for which the language is designed, (ii) are usually used by computer designers rather than computer users, and (iii) are usually not suited for human use. *Note:* Instructions written in an assembly language or a high level language must be translated into machine language before they can be executed by a processor. **2.** A computer oriented language that (a) has instruction codes that are executed directly by a machine, i.e., a computer, (b) usually is based on or is a part of the internal electronic circuitry, i.e., the internal hardware, of a computer, (c) a programmer need not necessarily be aware of, (d) is an artificial language, (e) is designed and written for a specific computer, and (f) is used by the computer to compile an executable program from the high level language instructions cited by the application programmer. *Note:* Machine languages usually are used by computer designers rather than computer users. They usually are not suited for use by programmers who prepare application programs. *See also* **assembly language, compile, computer, computer language, computer oriented language, high level language.**

machine learning: The ability of a device to improve its performance based on the results of its past performance.

machine oriented language: *Synonym* **computer oriented language.**

machine-readable: Pertaining to the characteristic of being able to be sensed by a device, usually by a device that has been designed and built specifically to perform sensing or reading functions. *Note:* Machine-readable data are data that have been stored on tapes, cards, diskettes, drums, and other media that can be automatically sensed.

machine-readable medium: 1. A medium capable of storing data in a form that can be accessed by an automated sensing device. **2.** A medium that (a) usually is recorded on a data medium and (b) can convey data to a given sensing device. *Note:* Examples of machine-readable media are (a) magnetic disks, cards, tapes, and drums on which data have been recorded, (b) punched cards and punched paper tapes, (c) optical disks on which data have been recorded, and (d) paper on which magnetic ink characters have been recorded. *Synonym* **automated data medium, data medium.**

machine system: *See* **man-machine system.**

machine translation (MT): The translation of natural languages by programmed computers.

machine word: *Synonym* **computer word.**

mach number: A number used to express the speed of a body moving in a given medium relative to the medium in which the body is moving, given by the relation $M = v_b/v_s$ where M is the mach number, v_b is the actual speed of the body relative to the medium in which it is moving, and v_s is the speed sound would travel in the same medium under the same conditions, such as temperature, pressure, and humidity. *Note 1:* When expressed as $v_b = Mv_s$, the mach number expresses the number of times the speed of sound that a body is traveling in a given medium at the stated conditions. *Note 2:* The speed of a body traveling at 2160 ft/s (feet per second) in dry air at 0°C at sea level would be moving at mach 2 because the speed of sound in air under those conditions is 1080 ft/s. There is about a 1 ft/s increase in the speed of sound in air at sea level for each kelvin above 0°C. Thus, the body would have to move faster in warmer air if it were to remain at mach 2. *Note 2:* Speeds less than mach 0.8 are subsonic. Speeds between mach 0.8 and mach 1.2 are transonic. Speeds above mach 1.2 are supersonic. Speeds above mach 10 have been called hypersonic. *Synonym* **mach.**

Mach-Zehnder fiber optic sensor: An interferometric sensor in which (a) an electromagnetic wave, such as a lightwave, is split in half, and (i) the halves each propagate around half a loop in opposite directions, one half via a beam splitter and a fixed mirror and the other half via a movable mirror and a beam splitter, and (ii) both halves recombine in an optical fiber or on the sensitive surface of a photodetector where their relative phases can enhance or cancel each other, (b) the movable mirror is used to modulate the resultant irradiance, i.e., the field intensity, at the photodetector, (c) displacements as short as 10^{-13} m (meter) can be measured, and (d) optical fibers may be used for the light paths.

macro: *Synonym* **macroinstruction.**

macrobend: 1. In an optical fiber, a macroscopic deviation of the fiber axis from a straight line. *Note 1:* Macrobends are distinguished from microbends. *Note 2:* In optical fibers, macrobend radii must be greater than both the minimum bend radius, i.e., the radius less than which the fiber will break, and the critical radius, i.e., the radius less than which substantial radiation will occur. **2.** A relatively large radius bend in an optical fiber, such as might be found in a splice organizer tray or a fiber optic cable that has been bent. *Note:* A macrobend will not cause a significant radiation loss if it is of sufficiently large radius, the loss depending on the radius and the type of optical fiber. Single mode (SM) fibers have a low numerical aperture (NA), usually less than 0.15, and thus are more susceptible to bend losses than other types. Usually an SM fiber will not tolerate a critical radius (CR) of less than 6 cm (centimeters). Certain specialized SM fibers can tolerate a shorter CR without appreciable loss. A graded index multimode (MM) fiber having a core diameter of 50 μm (microns) and an NA of 0.20 will usually tolerate a CR of not less than 3.8 cm. The fibers usually used in customer premises applications (62.5-μm core) usually have a relatively high NA (0.27) and can tolerate a bend radius less than an inch (2.5 cm). **2.** In an optical waveguide, a macroscopic deviation of the optical axis from a straight line. *See also* **curvature loss, macrobending, microbending, microbend loss.**

macrobend attenuation: In a dielectric waveguide, such as an optical fiber, the power lost to radiation at a macrobend. *Note:* Evanescent waves, cladding modes, and high order leaky modes in the core become decoupled from the core propagating modes because of phase shifts and velocity differences at the bend. For macrobend radii less than the critical radius, most of the propagating optical power is radiated laterally from the waveguide. *See also* **curvature loss, dielectric waveguide.**

macrobending: In an optical fiber, the creating of macroscopic deviations of the fiber axis from a straight line. *Note 1:* Macrobending is distinguished from microbending. *Note 2:* In optical fibers, macrobend radii must be greater than both the minimum bend radius, i.e., the radius less than which the fiber will break, and the critical radius, i.e., the radius less than which substantial radiation will occur. *See also* **curvature loss, microbending, microbend loss.**

macrobending loss: *Synonym* **curvature loss.**

macrobend loss: *Synonym* **curvature loss.**

macrofiber: *Synonym* **jumbofiber.**

macrogenerator: A computer program that (a) is written with macroinstructions in a source language and (b) replaces them with specific sequences of instructions in the same source language. *See also* **computer program, macroinstruction, source language.**

macroinstruction: A computer instruction that (a) can be replaced by a specific sequence of instructions in the same language, (b) usually provides the parameters that are required for its own execution, and (c) may also designate the machine language instructions for performing a specified task. *Synonym* **macro.** *See also* **computer instruction, machine language.**

macroprocessor: A processor that (a) allows a programmer to define and use macroinstructions and (b) usually is a portion of an assembler. *See also* **assembler, macroinstruction, processor, programmer.**

magic cube: A three-dimensional storage unit that (a) is shaped like a cube about 0.5 in (inch) on an edge, (b) is capable of storing 1 Tb (terabit, 10^{12} bits) of data, or the equivalent of about 2.4 million floppy disks, (c) has a random access capability, (d) consists of optical integrated circuits (OICs), (e) may be incorporated into computer boards, and (f) is sponsored by the U.S. Air Force.

MAGNA*Phone*®M: A satellite telephone, i.e., a satphone, produced by Magnavox.

magnetic: 1. Pertaining to the phenomenon that causes certain materials, some of which are found in nature and some of which have been synthesized by humans, to attract or repel one another independently of their electric attraction or repulsion and independently of their gravitational attraction. **2.** Pertaining to a force field that (a) exhibits polarity and (b) can be created by electric currents consisting of moving electric charges, such as electrons and ions. *Note:* Strong magnetic fields can be created by passing an electric current through many turns of wire wrapped around an iron core. *See* **transverse magnetic.** *See also* **core, magnetic field.**

magnetic bearing: 1. A bearing with respect to the local measured magnetic direction, such as might be measured with a magnetic compass in the field. **2.** A bearing with respect to magnetic north, which differs from true north by the declination stated on maps and charts of the area. *Note:* Magnetic north is not spatially or temporally constant.

magnetic card: A card with a magnetizable layer on which data can be stored and from which stored data can be retrieved. *See also* **band, magnetic tape.**

magnetic card storage: Storage (a) in which data may be stored on magnetic coatings placed on cards that are selected and moved manually or by electromechanical devices in order to store data and retrieve the stored data, and (b) that usually consists of magnetic material coated on a nonmagnetic substrate in the form of a card, such as a plastic, fiberboard, laminate, or cardboard card.

magnetic circuit: 1. The complete closed path taken by magnetic lines of flux. **2.** A region of ferromagnetic material, such as the core of a transformer or a solenoid, that contains essentially all of the magnetic flux. *Note:* Every line of magnetic flux must close on itself. If a low reluctance path, i.e., high permeability path, is provided, the flux will follow that path, i.e., that circuit.

magnetic core: A configuration of magnetizable material, such as iron, ferrite, or iron oxide, (a) that is placed in proximity to current-carrying conductors such that the polarity of its magnetization can be reversed by selection of appropriate conductors, (b) in which the direction of magnetization, or its reversal, can be used to store and read binary data and perform logic operations, and (c) that may come in various shapes, such as wires, tapes, toroids, rods, thin films, slugs, beads, core-ropes, and annular rings.

magnetic core storage: A storage unit that (a) consists of separated pieces of magnetic materials, such as iron, iron oxide, and ferrite, in such shapes as wires, tapes, toroids, beads, rods, or thin films, (b) stores a binary digit in each piece or domain, and (c) stores and reads data by selectively magnetizing and reversing the magnetic polarity of the pieces with electric currents. *See also* **magnetic, magnetic core, storage unit.**

magnetic delay line: A delay line that consists of a length of magnetic material such that magnetic pulses introduced at one end emerge at the other end after an elapsed transit time. *See also* **delay line, magnetic, pulse, transit time.**

magnetic disk: A flat circular plate that (a) has a magnetizable surface on one or both sides of which data can be stored and from which stored data can be retrieved, (b) stores data in the form of magnetized spots or as an analog signal, and (c) is rotated so that data may be stored and read by one or more appropriately positioned recording heads, writing heads, or read/write heads.

magnetic disk unit: A device that contains magnetic disks, a disk drive, one or more magnetic heads, and associated controls.

magnetic drum: A right circular cylinder with a magnetizable surface on which data can be stored and from which stored data can be retrieved.

magnetic drum unit: A device that contains a magnetic drum, the mechanism for moving it, magnetic heads, and associated controls.

magnetic field: A field that (a) is created by an electric charge that is in motion, and (b) can be identified and quantified (i) by the force exerted on a magnetic pole placed in it (attraction or repulsion effect), (ii) by the electromotive force, i.e., voltage, it induces in a conductor moved relative to and transverse to it, provided that the magnetic field produces magnetic flux and the number of flux linkages that wrap around the circuit in which the current is flowing changes with respect to time (electric generator effect), or (iii) by the force exerted on an electric charge moving transverse to it (electric motor effect). *See also* **Biot-Savart law, field, magnetic, magnetic field force.**

magnetic field component: In an electromagnetic wave, the part of the wave that (a) consists of a time-varying magnetic field, (b) interacts with, and interchanges its energy with, an electric field, and (c) thereby gives rise to the propagation of a field of force or energy in a direction perpendicular to both fields. *Note:* Reflection, refraction, and transmission that occur at an interface between two different media depend on (a) the directions of the magnetic field component and the electric field component relative to the interface surface and (b) the electric permittivities, magnetic permeabilities, and electrical conductivities of the propagation media on both sides of the interface. *See also* **electric field component, Maxwell's equations, wave equation.**

magnetic field force: **1.** The force of attraction or repulsion between two magnetic poles, given, in vector notation, by the relation $\mathbf{F} = (\mu m_1 m_2 / r^2)\mathbf{a_r}$ where $\mathbf{F}$ is the vector force, μ is the magnetic permeability of the medium in which the magnetic poles are embedded, m_1 and m_2 are the magnetic pole strengths, r is the distance between the poles, and $\mathbf{a_r}$ is a unit vector in the direction joining the two poles. **2.** The force exerted on a magnetic pole or on an electric current-carrying wire properly oriented and situated in a magnetic field, given by the relation $d\mathbf{F} = \mathbf{B} \times i d\mathbf{s}$, where $d\mathbf{F}$ is the elemental force on an elemental piece of wire perpendicular to the magnetic field, $\mathbf{B}$ is the magnetic flux density, i is the electric current in the elemental piece of wire, and $d\mathbf{s}$ is the elemental length of wire. *Note:* The flux density is given by the relation $\mathbf{B} = \mu\mathbf{H}$, where $\mathbf{B}$ is the flux density, usually in gauss, μ is the magnetic permeability, and $\mathbf{H}$ is the magnetic field strength, usually in oersteds. *See also* **magnetic field strength, magnetic flux density.**

magnetic field strength: The intensity, i.e., the amplitude or the magnitude, of a magnetic field at a given point. *Note 1:* The magnetic field strength usually may be expressed in root mean square (rms) or direct current (dc) ampere•turns per meter, in oersteds, or as the magnetic flux density, in gauss or lines per square meter, divided by the magnetic permeability. *Note 2:* Instantaneous magnetic field strength is a gradient, and therefore is a vector quantity, because it has both magnitude and direction at the spatial position where the magnetic field strength is determined. *Note 3:* The magnetic flux density, $\mathbf{B}$, is given by the relation $\mathbf{B} = \mu\mathbf{H}$ where μ is the magnetic permeability, and $\mathbf{H}$ is the magnetic field strength. *See also* **electric field strength.**

magnetic field vector: **1.** At a point, a vector that represents the magnitude and the direction of the magnetic field strength at that point. **2.** In an electromagnetic wave, such as a radio wave or lightwave, the vector that represents the instantaneous magnetic field strength in magnitude and direction at any point in a medium in which the wave is propagating. *See also* **magnetic field, magnetic vector.**

magnetic flux: The magnetic lines of force that emanate from a magnet, i.e., magnetized material, or from a solenoid that develops lines of force because of the ampere•turns, i.e., the magnetomotive force, of its winding. *See also* **magnetic field, magnetomotive force.**

magnetic flux density: 1. The number of lines of magnetic flux passing through a unit area perpendicular to the direction of the magnetic lines, i.e., $\mathbf{B} = \mathbf{\Phi}/A$ where $\mathbf{B}$ is the magnetic flux density vector, i.e., the number of magnetic lines per unit area, a point function, $\mathbf{\Phi}$ is the magnetic flux vector, and A is the magnitude of area, perpendicular to the flux, through which the flux is passing. **2.** At a point, (a) the magnetic permeability of the material at the point times (b) the magnetic field strength at the point, i.e., $\mathbf{B} = \mu\mathbf{H}$ where $\mathbf{B}$ the magnetic flux density vector, μ is the magnetic permeability, and $\mathbf{H}$ is the magnetic field strength vector. **3.** At a point in a closed magnetic path, the magnetomotive force (mmf) divided by the product of (a) the magnetic reluctance, R, of the magnetic path over which the mmf is exerted and (b) the cross section of the path at the point, given by the relations $B = \phi/A = mmf/AR$ where B is the magnetic flux density, ϕ is the total flux in the path, a constant for all points in a closed path, A is the cross section of the path at the point, mmf is the total magnetomotive force applied to the entire closed path, and R is the reluctance of the closed path, such as the closed iron core of a transformer. *Note 1:* The variables are scalar values, i.e., magnitudes, because the path is a closed magnetic path. *Note 2:* The magnetomotive force (mmf) is proportional to the ampere•turns applied to the path. The reluctance is directly proportional to the permeability of the material forming the closed path, directly proportional to the length of the path, and inversely proportional to the cross section of the path. If the path is segmented with different cross sections for different lengths, the magnetic reluctance for each segment is evaluated separately. The segments are added if they are in series. For segments in parallel, their reciprocals are added to obtain the reciprocal of the parallel combination. *Note 3:* Magnetic circuit, i.e., magnetic path, analysis is similar to electric circuit analysis, where the magnetomotive force corresponds to the voltage, the magnetic flux to the current, and the magnetic reluctance to the impedance. *See also* **magnetic field strength, magnetic reluctance, magnetomotive force.**

magnetic head: An electromagnet that is capable of reading, writing, or erasing data in the form of minute magnetized areas, i.e., minute magnetized spots, on a magnetizable surface, such as on a magnetic tape, disk, drum, or card.

magnetic hysteresis loop: A plot of the relationship between (a) the magnetic flux density obtained in a magnetic material and (b) the applied magnetic field strength, i.e., the applied magnetomotive force (mmf) in which the material is placed, which is (i) first applied in one direction, and usually increased until saturation of the material is reached, (ii) then reduced to zero, (iii) then increased in the opposite direction, and usually increased until saturation occurs, and (iv) finally reduced to zero again, at which point some magnetization will remain in the material. When the cycle is repeated and plotted, a closed loop is observed. *Note:* Each time the applied magnetomotive force (mmf) is zero, some magnetization will remain, i.e., there will be residual magnetism or remanence. A reversed mmf is required to bring the magnetization to zero. Also, when the mmf is decreasing, the magnetization plot does not retrace the increasing mmf plot, and thus forms a loop. *See also* **magnetic flux density, magnetomotive force, residual magnetism.**

magnetic ink character recognition (MICR): The reading of characters that (a) are printed with ink that contains particles of magnetic material and (b) can be read by a magnetic reading head and by persons. *Note:* A set of magnetic ink characters have been standardized for use on bank checks for automatic reading of bank fund transfer routing and accounting information.

magnetic mode: *See* **transverse magnetic mode.**

magnetic permeability: A material medium parameter that (a) defines the magnetic characteristic of the medium, (b) serves as the constant of proportionality in the constitutive relation between the applied magnetic field strength and the resulting magnetic flux density, specifically, by the relation $\mathbf{B} = \mu\mathbf{H}$ where μ is the magnetic permeability, $\mathbf{B}$ is the magnetic flux density, and $\mathbf{H}$ is the magnetic field strength, such that if $\mathbf{B}$ is in gauss and $\mathbf{H}$ is in oersteds, μ will be in SI units (Système International d'Unités), and (c) serves as the constant of proportionality in the equation expressing the force of attraction of unlike magnetic poles a given distance apart. *Note:* Except for ferrous metals, which have a high magnetic permeability, and a few metals, which have a much lower permeability, but nevertheless a permeability greater than unity, the magnetic permeability of most materials, e.g., dielectric materials such as glass and plastic, is unity. Thus, magnetic fields are practically unaffected by the glass and other dielectric materials from which optical fibers are made. Therefore, the magnetic permeability of glass does not contribute to the refractive index of glass. For electrically nonconducting media, i.e., propagation media in which the electrical conductivity is zero, the relative velocities of electromagnetic waves, such as lightwaves, propagating within them, are given by the relation $v_1/v_2 = [(\mu_2\varepsilon_2)/(\mu_1\varepsilon_1)]^{1/2}$ where v is the velocity, and μ and ε are the magnetic permeability and electric permittivity, respectively, of media 1 and 2. If medium 1 is free space, the above equation becomes the relation $c/v = [(\mu\varepsilon)/(\mu_0\varepsilon_0)]^{1/2}$ where c is the velocity of light in a vacuum, v is the velocity in a material medium, the nonsubscripted values of the constitutive relations are for the material propagation medium and the subscripted values are for free space. However, c/v is the definition of the refractive index of a medium relative to free space. From this is obtained the relation $n = (\mu_r\varepsilon_r)$ where n is now a dimensionless number and relative to free space. The magnetic permeability and electric permittivity are also usually given relative to free space, or air. As stated above, for dielectric materials, μ_r is approximately 1. Hence, for the glass and plastics used to make optical fibers and other dielectric waveguides, the refractive index, relative to a vacuum, or air, is given by the relation $n \approx \varepsilon_r^{1/2}$ where ε_r is the electric permittivity of the propagation medium and the subscript denotes that the value is relative to a vacuum. Refractive indices are usually given relative to a vacuum, or relative to 1.0003 for air near the Earth's surface and less as the air gets thinner in the upper atmosphere and outer space. In absolute units for free space, the electric permittivity is $\varepsilon = 8.854 \times 10^{-12}$ F•m^{-1} (farads per meter), while the magnetic permeability is $4\pi \times 10^{-7}$ H•m^{-1} (henries per meter). The reciprocal of the square root of the product of these values yields approximately 2.998×10^8 m/s (meters per second), the velocity of light in a vacuum. Because the refractive index for a propagation medium is defined as the ratio between the velocity of light in a vacuum and the velocity of light in the medium, the ratio becomes unity when the propagation medium is a vacuum. *See* **refractive index, relative magnetic permeability.**

magnetic recording: 1. Recording data on a magnetic data medium with the use of magnetic heads. **2.** The result obtained from recording data on a magnetic data medium with the use of magnetic heads. *See* **nonreturn to zero change on ones magnetic recording.** *See also* **data, data medium, magnetic, magnetic head.**

magnetic reluctance: The resistance of a path to the establishment, i.e., to the flow, of magnetic flux, given (a) by the relationship $R = mmf/\phi$, where R is the magnetic reluctance of the magnetic path that is followed by the magnetic flux, mmf is the magnetomotive force that is applied to the path and is causing the magnetic flux, and ϕ is the total magnetic flux in the path, and (b) by the relation $R = L/\mu A$ where R is the magnetic reluctance, L is the length of the path, μ is the magnetic permeability of the material from which the path is constructed, and A is the cross-sectional area of the path. *Note:* Reluctances in a magnetic path may be connected and combined like resistances in an electric circuit. *Synonym* **reluctance.** *See also* **magnetic flux, magnetomotive force.**

magnetic rod storage: Magnetic storage in which data are stored in small slugs or rod-shaped pieces of magnetizable material that (a) function much like magnetic cores in a magnetic core storage unit and (b) need to be strung on wires like the annular ring cores. *Synonym* **magnetic slug storage.** *See also* **magnetic core, magnetic storage, storage unit.**

magnetic slug storage: *Synonym* **magnetic rod storage.**

magnetic storage: Storage in which the magnetic properties of materials are used to store data, usually in the form of magnetized area or spots on a continuous surface coated with a magnetizable material, such as magnetic tapes, drums, disks, cards, and thin films, or as individual pieces of magnetized material, such as magnetic cores, slugs, and core ropes. *See also* **magnetic.**

magnetic storm: A perturbation of the Earth's magnetic field. *Note:* Magnetic storms (a) are caused by solar disturbances, (b) usually last for brief periods, such as several days, and (b) are characterized by large deviations from the usual value of at least one component of the Earth's magnetic field. *See also* **sudden ionospheric disturbance.**

magnetic tape: 1. A tape with a magnetizable surface on which data can be stored and from which stored data can be retrieved. **2.** A tape or a ribbon of suitable material, such as plastic, impregnated or coated with magnetic or other material on which information may be represented in the form of either magnetically polarized spots for digital recording or an analog signal for voice and video recording. *Note 1:* Typical packing densities for magnetic tapes are 800 bpi (bits per inch) and 1600 bpi on each of several tracks. The bits for each character usually are in parallel tracks, such that the packing density for each track is the same as the character density for each tape. *Note 2:* Many modes of recording are used, such as nonreturn to zero (NRZ) and return to zero (RZ). *See also* **band interblock gap, magnetic card, magnetic disk, phase encoded recording.**

magnetic tape cassette: A container that (a) usually is made of plastic, (b) has two reels that can be driven for paying out and taking up magnetic tape so that the tape can be used for reading and writing digital or analog data without removal from the container, (c) has permanently built-in magnetic tape, the ends of which are fastened to the hub of the reels and which must be rewound each time the complete tape is used, i.e., the tape is not in a loop, as it is in loop bins and eight-track cartridges where the tape can be used continuously and need not be rewound, though it may be moved backward or forward, (d) may be inserted and removed from reading and recording equipment, and (e) is available for voice and video recording and reading for up to several hours. *See also* **video cassette, video cassette recorder.**

magnetic tape drive: A mechanism that (a) controls the movement of magnetic tape for reading, writing, and erasing data while the tape is moved past a magnetic head and (b) usually holds two reels for paying out and taking up the magnetic tape. *See also* **magnetic head, magnetic tape.**

magnetic tape handler: A device that (a) receives, interprets, and executes control signals, (b) has two or more spindles for mounting reels for paying out, taking up, or copying magnetic tapes, and (c) has magnetic tape drives and brakes for pulling magnetic tape from a payout reel, past one or more magnetic heads, usually via capstans and idler pulleys. *See also* **magnetic head, magnetic tape, magnetic tape drive.**

magnetic tape leader: The portion of magnetic tape that (a) precedes the beginning of tape mark and (b) is used to thread the tape on a tape handler or on a reel. *See also* **beginning of tape mark, magnetic tape, magnetic tape handler, magnetic tape trailer.**

magnetic tape storage: Storage in which magnetic tape is used as the data medium to store data. *See also* **data, data medium, magnetic tape, storage, store.**

magnetic tape trailer: The portion of magnetic tape that (a) follows the end of tape mark and (b) is used to retain the tape on the reel or on the tape handler. *See also* **end of tape mark, magnetic tape, magnetic tape handler, magnetic tape leader.**

magnetic thin film storage: Storage (a) in which data are stored by magnetic recording on a film of molecular thickness on the order of 1 μm (micron), i.e., 10^{-6} m (meter), and (b) that may be used for combinational circuits as well as storage elements. *See also* **combinational circuits, storage, thin film.**

magnetic vector: A representation that specifies the direction and the amplitude of magnetic flux, such as (a) the magnetic flux established by the magnetic field in an electromagnetic wave, (b) the magnetic flux produced by magnetic heads, electric motors, electric generators, and electromagnets, and (c) the Earth's magnetic flux. *See also* **electric vector, magnetic field, magnetic field vector, Poynting vector.**

magnetic wave: A wave that (a) is launched by a source, such as a coil of wire bearing an electric current, that produces a magnetic field, (b) does not travel very far from the source, (c) consists of closed lines of magnetic flux that emanate from the source, (d) when varied in magnitude or reversed in polarity, may be used to induce currents in coils, such as in transformers and ignition coils, and (e) is not necessarily associated with an electric wave and not necessarily part of an electromagnetic wave. *See* **transverse magnetic wave.** *See also* **electric wave, electromagnetic wave.**

magnetic wire storage: Storage in which (a) the data medium is a length of magnetizable wire on which data may be stored as magnetized spots along the wire, and (b) the wire is wound on spools and moved past a magnetic head. *See also* **data medium, storage.**

magnetism: The phenomenon that (a) is exhibited by a field of force, i.e., a magnetomotive force, (b) is produced by a magnet or by an electric current, and (c) exerts a force on magnetic poles or a torque on certain shapes of nonmagnetized ferromagnetic material. *See* **residual magnetism.**

magneto exchange: *Synonym* **magneto switchboard.**

magnetohydrodynamics (MHD): The branch of science and technology devoted to the study and the application of the flow of electrically conducting fluids, such as metallic vapors and blood, that are under the influence of electric, magnetic, and electromagnetic fields. *See also* **electric field, electromagnetic field, magnetic field.**

magnetoionic double refraction: For a linearly polarized wave entering the ionosphere, the combined effect of the Earth's magnetic field and atmospheric ionization, such that (a) the wave is split into two components, called the "ordinary wave" and the "extraordinary wave," and (b) the component waves follow different paths, experience different attenuations, have different phase velocities, and usually are elliptically polarized in opposite senses. *See also* **birefringence, ionosphere.**

magnetomotive force (mmf): The driving force that (a) creates a magnetic field strength at a point in free space and in materials, (b) is caused by electric current, (c) is proportional to the ampere•turns presented by a conductor carrying an electric current, and (d) is not dependent upon the magnetic permeability of the surrounding material. *Note:* An example of magnetomotive force is that which is produced by a long coil of wire, given by the relation $mmf = 0.4\pi NI$ where mmf is the magnetomotive force, N is the number of turns in the coil, and I is the electric current in the coil in amperes. The magnetic flux and the magnetic field strength produced by the coil depend on the shape and the magnetic permeability of the material in which the coil is embedded. *See also* **magnetic field strength, magnetic flux, magnetic permeability.**

magnetooptic: Pertaining to the changing of the refractive index of a material by changing a magnetic field applied to the material. *Note 1:* By controlling the refractive index of a material, the magnetooptic effect can be used to modulate a lightwave. *Note 2:* In certain materials, the magnetooptic effect can be used to rotate the polarization plane of an electromagnetic wave passing through the material. *See also* **Faraday effect, magnetooptic effect.**

magnetooptic effect: The effect produced by the interaction of a magnetic field with an electromagnetic wave passing through the field or the interaction of an electromagnetic wave with matter under the influence of a magnetic field. *Note 1:* The most important magnetooptic effect having application to optical communications is the Faraday effect, in which the polarization plane is rotated under the influence of a magnetic field parallel to the direction of propagation. This effect may be used to modulate a lightwave. *Note 2:* The amount of polarization plane rotation, i.e., the angular displacement, is given by the relation $\theta = AHL$ where θ is the angular displacement, usually expressed in radians, A is the constant of proportionality depending on the propagation medium and the units used, H is the applied magnetic field strength, usually expressed in oersteds, in the direction of propagation, and L is the distance the electromagnetic wave is in the magnetic field, usually expressed in meters. **2.** In a plane polarized lightwave propagating in a medium, the rotation of the polarization plane brought about by subjecting the medium to a magnetic field, i.e., a Faraday rotation. *Note:* The magnetooptic effect can be used to modulate a lightwave in a material because many parameters at interfaces, such as the reflection and transmission coefficients, acceptance angles, critical angles, refraction angles, transmission modes, and propagation velocities, are dependent on the direction of propagation of the lightwave relative to the interface surface. *See also* **electrooptic effect, Faraday effect.**

magnetooptic (m-o) modulator: A modulator that uses the magnetooptic effect to modulate a lightwave carrier. *See also* **carrier, lightwave, magnetooptic effect, modulate.**

magnetooptics: The branch of science and technology devoted to the study and the application of the interaction between (a) magnetic fields and (b) electromagnetic waves that are in the optical region of the electromagnetic spectrum. *See also* **acoustooptics, electrooptics, optooptics, photonics.**

magnetostatic field: The field of force that (a) is produced by moving electric charges, i.e., an electric current that is constant or varying slowly, (b) has a strength that can be measured by the force exerted on a magnetic pole, a solenoid, or a conductor bearing a direct current (dc) placed within the field, (c) usually is considered not to vary with time, or to vary slowly with respect to events of interest occurring within it, such as less than 400 Hz (hertz), and (d) is uncoupled from any other field, such as an electric field as in an electromagnetic wave. *Synonym* **staticmagnetic field.**

magnetostriction: The phenomenon exhibited by some materials, such as ferromagnetic materials including iron, cobalt, nickel, and certain alloys, in which dimensional changes occur when the material is subjected to a magnetic field, usually becoming longer in the direction of the applied field and shorter in the transverse direction. *Note 1:* The applied magnetic field tends to rotate and thus line up magnetic domains in the material, thus causing the material to elongate. *Note 2:* Magnetostriction can be used to launch a shock or sound wave each time the field is applied or changed, possibly giving rise to phonons that could influence energy levels in the atoms of certain materials, such as semiconductors and lasers, and thereby serve as a modulation method. Along with photon or electric field excitation, the phonon energy could provide threshold energy to cause electron energy level transitions, causing photon absorption or emission. *Note 3:* If an optical fiber is coated with a bonded magnetostrictive material, the length of the fiber can be modulated by varying the applied magnetic field. Hence, a coherent lightwave exiting from the end can be made to shift phase. An interferometer can be used to detect the phase shift. Therefore, the output of a photodetector can be modulated by the applied magnetic field that is causing the magnetostriction, and thus the strength of the magnetic field can be measured. The magnetostrictive effect can also be used to apply pressure to an optical fiber, causing changes in the refractive index, which, in turn, can be used as a modulation method and again as a magnetic field sensor, i.e., a magnetometer. *See also* **phonon, magnetic field.**

magnetostrictive delay line: A delay line that (a) operates on the principle of magnetostriction, i.e., consists of a material that changes dimensions when a magnetic field is applied, and (b) may be used to store data by (i) inserting a magnetic field pulse that causes a short length to increase at one end, thus generating a mechanical pulse or shock wave that travels the length of the line, and (ii) sensing the propagated mechanical pulse at the other end by means of an appropriate sensor, which may be magnetic or microphonic. *See also* **delay line, magnetic field, magnetostriction, microphonics.**

magneto switchboard: A manually operated telephone exchange in which the users and the operators call and clear by means of hand-cranked magnetos that serve as the source of signaling power, i.e., the source of calling and clearing power. *Note:* The signaling is called "ring" and "ring off." *Synonym* **magneto exchange.**

magnetron: *See* **fixed magnetron.**

magnification: The relationship between the linear dimension of an image and the corresponding dimension of the object when the image is larger than the object. *Note 1:* If an image is four times the linear size of the object, the magnification is expressed as 4X, 4:1, 400%, or 300% larger. If the magnification is 1.2:1, the magnification is 120% or 20% larger. *Note 2:* Usually magnification pertains to a given optical system. *See* **absolute magnification, angular magnification, borescope magnification, individual normal magnification.** *See also* **image, magnifying power, object, optical system.**

magnifier: An optical system that forms a magnified virtual image of an object placed near its front focal point. *Note:* Magnifier magnification usually ranges from 3X to 20X. *Synonyms* **loupe, magnifying glass, simple microscope.** *See also* **front focal point, magnification, optical system, virtual image.**

magnifying power: In an optical system, (a) a measure of the ability of the optical system to make an object appear larger than it appears to the unaided eye, (b) the diameter of the entrance pupil divided by the diameter of the exit pupil, (c) for a telescopic system, the focal length of the eyepiece, or (d) the tangent of an angle in the apparent field divided by the tangent of the corresponding angle in the true field. *Note:* If an optical element or system has a magnifying power of 2X, the object will appear twice as wide and twice as high as the object appears with the unaided eye. *See also* **entrance pupil, exit pupil, eyepiece, focal length, magnification, optical system.**

magnitude: In any domain, the size or the value of an entity. *Note 1:* Examples of magnitude are (a) in the frequency domain, a given value of frequency, a frequency shift, or a frequency band, (b) in the time domain, a given time or time interval, (c) in the spatial domain, a given displacement, distance, area, or volume, and (d) in electrical quantities, a given value of the current, voltage, or power. *Note 2:* In electrical quantities, the amplitude is a special case of magnitude. *See* **pulse magnitude.** *See also* **domain, value.**

mail: *See* **electronic mail.**

mailbox facility: A facility in which a message from an originating user is stored until the destination user requests delivery of that message. *Synonym* **mailbox-type facility.** *See also* **delayed delivery facility, electronic mail, store and forward.** *Refer to* **Fig. S-22 (977).**

mailbox-type facility: *Synonym* **mailbox facility.**

mail transfer protocol: *See* **Simple Mail Transfer Protocol.**

main beam: *Synonym* **main lobe.**

main distribution frame (MDF): A distribution frame (a) on which, on one part, the permanent outside lines entering a facility terminate, and on another part, cabling, such as customer line multiple cabling and trunk multiple cabling, terminates, (b) that is used for cross-connecting (i) any permanent outside line with (ii) any desired terminal of the multiple cabling or any other outside line, (c) that usually holds the central office (C.O.) protective devices, (d) that functions as a test point between a line and the C.O., (e) that, in a private exchange, performs functions similar to those performed by the distribution frame in a C.O., (f) that may have outside lines that include radio channels or circuits, (g) that usually holds all outside lines and their protective devices on the vertical side and all connections to central equipment that may be assigned to particular outside lines on the horizontal side, and (h) that, in communications facilities other than telephone facilities, has all station lines and equipment terminate on the vertical side and all patch fields on the horizontal side. *Synonym* **main frame.** *See also* **circuit, com-**

bined distribution frame, distribution frame, intermediate distribution frame.

main exchange: An exchange that (a) is connected to one or more satellite exchanges and (b) has one or more attendant positions, i.e., operator positions. *See also* **position, satellite exchange.**

mainframe: A high capacity, high performance, high speed, and usually larger computer to which other computers and terminals are connected to share its resources and computing power. *See also* **central processing unit, host, host computer, main distribution frame.**

main frame: *Synonym* **main distribution frame.**

main lobe: In an antenna radiation pattern, the lobe that (a) contains the maximum irradiance, i.e., contains the direction with the maximum radiation intensity, and (b) usually contains the largest amount of radiated power compared to any other lobe. *Note 1:* The horizontal radiation pattern usually is specified in describing the main lobe. The width of the main lobe usually is specified as the angle between the points where the power has fallen 3 dB below the maximum value. The vertical radiation pattern is also of interest and may be similarly specified. *Note 2:* Usually there is only one main lobe in an antenna radiation pattern, whereas there may be several side lobes and back lobes. *Note 3:* For a light source that launches a lightwave into an optical fiber, the main lobe is directed into an endface of the optical fiber. *Synonyms* **main beam, main lobe.** *See also* **antenna, antenna lobe, back lobe, directive gain, side lobe.**

main memory: *Synonym* **main storage.**

main station: A user instrument that (a) has a distinct unique number designation, (b) is connected to a loop, (c) is used for originating calls, and (c) is used to answer incoming calls from the central office (C.O.), i.e., the exchange, to which it is connected via the loop. *Note:* Examples of main stations are telephone sets, facsimile terminals, and private branch exchanges (PBXs). *See also* **call, end instrument, extension terminal, terminal.**

main storage: 1. In a computer, program addressable storage from which instructions and other data can be loaded directly into registers for subsequent execution or processing. *Note 1:* Main storage includes the total program addressable execution space that may include one or more storage devices. *Note 2:* "Main storage" usually refers to large and intermediate computers, whereas "memory" usually refers to microcomputers, minicomputers, and calculators. **2.** The part of internal storage into which instructions and other data must be loaded for execution or processing. **3.** The part of a processing unit in which programs are run. *Synonyms* **main memory, primary storage.** *See also* **auxiliary storage, internal memory, random access memory, read only memory.**

maintainability: 1. A characteristic of equipment or system design and installation that is expressed as the probability that the equipment or the system will be retained in, or restored to, a specified condition within a given period, when the maintenance is performed (a) in accordance with prescribed procedures and (b) with specified available resources. **2.** The ease with which maintenance of a functional unit can be performed in accordance with prescribed procedures. *See also* **availability, failure, mean time between failures, mean time to repair, mean time to service restoral.**

maintaining optical fiber: *See* **polarization-maintaining optical fiber.**

maintaining parameter: *See* **polarization-maintaining parameter.**

maintenance: 1. The actions taken (a) to retain materiel and equipment in, or to restore it to, a specified condition, and (b) that include inspection, testing, servicing, classification as to serviceability, repair, rebuilding, and reclamation. **2.** In military systems, the supply and repair actions taken to keep a force in condition to carry out its mission. **3.** The routine recurring work required to keep a facility, such as a plant, building, structure, ground facility, or utility system, in such condition that it may be continuously used (a) at its original or designed capacity and efficiency and (b) for its intended purpose. **4.** An activity that (a) is intended to restore or retain a functional unit in a state in which the unit can perform its required functions and (b) includes keeping the unit in the specified state by performing activities such as tests, measurements, replacements, adjustments, and repairs. *See* **adaptive maintenance, corrective maintenance, emergency maintenance, file maintenance, preventive maintenance, software maintenance.**

maintenance control circuit (MCC): 1. In communications, computer, data processing, and control systems, a circuit used by maintenance personnel for coordination. *Note:* A maintenance control circuit (MCC) is not available to operations or technical control personnel. **2.** In microwave systems, a voice circuit used by maintenance personnel over microwave links for coordination. *Note:* A maintenance control circuit (MCC) is not available to operations or technical control personnel. *See also* **circuit, engineering orderwire, link, maintenance, orderwire circuit.**

maintenance panel: A panel, such as (a) a connection, display, or control panel or (b) a control console, that is used by maintenance personnel for their interaction with the equipment that is being maintained. *See also* **panel.**

maintenance plan: In communications, computer, data processing, and control systems, a plan that (a) identifies the managerial, administrative, and technical procedures that will be used to maintain hardware, software, and firmware products in an operative condition and (b) includes various topics, such as software tools, resources, facilities, equipment, spare parts, schedules, costs, and risks. *Note:* A maintenance plan usually is documented, in which case the document itself may be called the "maintenance plan." *See also* **firmware, hardware, software.**

maintenance software: A computer program, routine, subroutine, or tool that supports the maintenance of hardware and firmware, such as by identification and reporting of devices that have failed, are not operating satisfactorily, are no longer connected to the system, or have lost power.

maintenance test land station: *See* **radionavigation maintenance test land station.**

major: In optics, a piece of glass to which a piece of glass with a different refractive index will be fused to make a multifocal lens. *See also* **lens, multifocal, refractive index.**

majority decoder: A decoder in which the output is based on the majority of the code values on the several input lines. *See also* **pseudo bit error ratio.**

majority operator: A logical operator having the property that if P is a statement, Q is a statement, R is a statement . . . , then MAJORITY P,Q,R . . . is true if more than half of the statements are true, false if half or fewer are true.

major lobe: *Synonym* **main lobe.**

major relay station: In tape relay systems, a tape station that (a) is connected to two or more trunk circuits and (b) that provides alternate routing in order to meet communications system performance requirements. *See also* **tape relay.**

make busy circuit: *See* **pilot make busy circuit.**

male connector: The half of a connector that usually has protruding elements or contacts that can be inserted into the cavity of a female connector to establish a connection. *See also* **female connector, gender changer.**

malfunction: *Synonym* **failure.**

MAMI: modified alternate mark inversion.

MAN: metropolitan area network.

managed object: 1. In a network, an abstract representation of network resources that are managed. *Note:* A managed object may represent a physical entity, a network service, or an abstraction of a resource that exists independently of its use in management. **2.** In telecommunications management, a resource within the telecommunications environment that may be managed through the use of operation, administration, maintenance, and provisioning application protocols.

management: *See* **accounting management, communications system management, configuration management, corporate information management, data management, frequency management, performance management.**

management complex: *See* **administrative management complex, customer management complex.**

management information protocol: *See* **Common Management Information Protocol.**

management information service: *See* **common management information service.**

management information system (MIS): 1. An organized assembly of resources and procedures required to collect, process, and distribute data for use in decision making. **2.** An information system that (a) is used to assist in the performance of management functions, such as controlling inventories, funds, and task performance in an organization, and (b) may be a computer, communications, data processing, or control system. *See also* **information management system.**

management network: *See* **telecommunications management network.**

management protocol: *See* **Simple Network Management Protocol.**

management system: *See* **database management system, electronic key management system, flight management system, information management system, integrated management system.**

manager: *See* **functionally integrated resource manager.**

Manchester code: The code obtained as a result of Manchester encoding. *See also* **Manchester encoding.**

Manchester encoding: The encoding, i.e., the combining, of data and clock signals to form a single self-synchronizing data stream such that (a) each encoded bit contains a transition at the midpoint of a bit period, (b) the direction of transition determines whether the bit is a 0 or a 1, (c) the first half is the true bit value, and (d) the second half is the complement of the true bit value. *See* **differential Manchester encoding.** *See also* **alternate mark inversion signal.**

mandrel: *See* **optical fiber mandrel.**

mandrel wrapping: In fiber optics, the wrapping of several turns of optical fiber around a mandrel approximately 1 cm in diameter in order to filter out excess higher-order modes that are launched into the optical fiber. *Note:* Mandrel wrapping enables mode filtering that is used to simulate or obtain equilibrium mode distribution without having to assume an equilibrium length. *See also* **critical radius, equilibrium length, equilibrium mode distribution.**

maneuvering signals: A signal that (a) is transmitted to stations on board ships, aircraft, spacecraft, land vehicles, or other mobile units for the purpose of controlling or directing their movement and (b) may be transmitted by any means of communication, such as voice, radio, heliograph, flaghoist, colored lights, or hand signaling.

Mangin mirror: A divergent meniscus lens in which (a) the second surface with respect to incident light, i.e., the convex surface, is silvered, and (b) spherical aberration can be corrected for any given position of the image by carefully choosing the radii of the surfaces of the mirror. *See also* **divergent meniscus lens, image, spherical aberration.**

manipulative communications deception (MCD): Communications deception by (a) controlling the insertion of misleading information into a communications system for the purpose of presenting a false traffic pattern, (b) misleading communications monitors by transmitting false data over communications circuits and channels, (c) using communications system operators or users to originate the deception traffic, and (d) altering or simulating electromagnetic communications radiation in order to falsify the information that can be obtained from analysis of the radiation. *See also* **communications deception, traffic.**

manipulative electronic deception (MED): The use of electromagnetic radiation to falsify the information that can be obtained from analysis of the radiation, including the alteration, simulation, reradiation, absorption, or reflection of electromagnetic radiation to accomplish deception. *See* **electronic deception.**

man-machine system: A system in which the functions of a human operator and a machine are integrated.

manpack ultrahigh frequency terminal: *See* **enhanced manpack ultrahigh frequency terminal.**

mantissa: The fractional part of a logarithm. *Note:* If 2.544 is a base 10 logarithm, the .544 is the mantissa. The mantissa is always less than 1. The number, i.e., the antilog, that has a base 10 logarithm of 2.544 is $10^{2.544} \approx 349.945$. The 2 is the characteristic. *Note:* The mantissa is often erroneously called "the exponent," which is proper for floating point numbers. In proper parlance, the base 10 logarithm of 349.945 is about 2.544, in which the 2 is the characteristic and the .544 is the mantissa. *See also* **antilog, characteristic, logarithm.**

manual answering facility: A facility that is used to establish a call only if the call receiver indicates a readiness to receive the call by means of a manual operation or a manually generated signal.

manual calling facility: A facility that (a) is manually operated by a call originator, (b) accepts selection signals from the call originator at an undefined, i.e., a nonspecific, data signaling rate (DSR), and (c) accepts signaling characters generated at the data terminal equipment (DTE) or at the data circuit-terminating equipment (DCE). *See also* **data circuit-terminating equipment, data signaling rate, data terminal equipment, facility, selection signal, signaling.**

manual exchange: An exchange in which the connections between incoming and outgoing lines, i.e., between outside lines, are manually controlled by an attendant, i.e., an operator. *See also* **exchange.**

manual relay system: A communications system in which messages are relayed manually from point to point. *Note:* Examples of manual relay systems are (a) tape relay systems in which messages received at a data station are torn off the receiver and manually moved to a unit for transmission to a final station or to another intermediate station to be relayed farther, and (b) hand flag transmission systems, such as semaphore, wigwag, or Morse flag signaling systems, in which messages are relayed from ship to ship or mountain top to mountain top. *See also* **hand flag transmission, torn tape relay system.**

manual spot jammer: A spot jammer (a) that can concentrate its transmitted power at a given frequency or channel by manually selecting the frequency and (b) in which the frequency usually can be manually varied continuously over a wide spectrum. *See also* **spot jammer.**

manual switch storage: Storage in which data are stored in the form of arrays of set switches, such as toggle switches, that are manually set and usually used to enter data, control data flow, or represent instructions.

manual telegraphy: Telegraph operation in which (a) the signal elements are individually keyed by an online operator from a knowledge of the code, and (b) the signals are immediately transmitted without storage at the transmitter. *See also* **key, signal element.**

manufacturer automation protocol (MAP): A procedure for automating manufacturing processes. *See also* **computer aided manufacturing.**

manufacturing: *See* **computer aided manufacturing.**

map: 1. A graphic representation of the relative spatial positions, temporal occurrences, or physical conditions of various entities, such as (a) the relative position of terrain features and installations, (b) the relative reading of two time scales, and (c) the corresponding readings of Centigrade and Fahrenheit temperature scales. **2.** To transform one form of representation of an entity into another form in such a manner that the transformation can be reversed and the original representation recovered. *Note 1:* Perfect mapping will not result in a loss of information. *Note 2:* An example of mapping is to transform one code into another code. *See* **gnomic map, Karnaugh map, line route map, orthodromic map, routing indicator delineation map, thimble map.**

MAP: manufacturer automation protocol.

mapped buffer: A display buffer in which each storage location in the buffer has a corresponding display position on the display surface of a display device, i.e., a character stored at a location in the buffer will be displayed in a corresponding position on the screen.

margin: 1. In communications systems, the maximum degree of signal distortion that can be tolerated without affecting the restitution, i.e., without incorrect signal interpretation. **2.** The allowable error rate, deviation from normal, or degradation of performance of a system. **3.** The allowance, such as the space along the edges of this page, that is made to ensure that data in or on a data medium do not exceed the limits of the medium. **4.** In a budget, such as a power budget, that which is added to the budget after all requirements have been accounted for in order to allow for possible cumulative tolerances, biases, and statistical variations among the elements of the budget. *See* **antijamming margin, design margin, fade margin, jamming margin, net margin, nominal margin, optical system power margin, power margin, radio frequency power margin, singing margin, start-stop margin, synchronous margin, synchronous receiver margin, system power margin, telegraph receiver margin, theoretical margin, system power margin.** *See also* **distortion, signal distortion.**

marginal check: *Synonym* **marginal test.**

marginal test: A test or a check in which certain operating conditions, such as voltage, frequency, data signaling rate (DSR), and signal to noise (S/N) ratios, are varied about their nominal values in order to detect and locate incipient faults in a system. *Synonym* **marginal check.**

margin bell: A machine function that (a) controls a bell, i.e., an audible signal, that indicates that the end of a printing or typing line is being approached and (b) is used on manual input and output devices, such as manually operated typewriters, keyboards, teletypewriters, and printers. *See also* **machine function.**

marine broadcast station (MBS): A coastal station that makes scheduled broadcasts of time, meteorological, and hydrological information. *Synonym* **maritime broadcast station.** *See also* **coastal station.**

marine radiobeacon station: A radiobeacon navigation land station that (a) is in the maritime radionavigation service, (b) transmits for the benefit of ships at sea, and (c) emits signals that are intended to enable a ship station to determine its bearing, i.e., its direction, in relation to the radiobeacon navigation land station. *Synonym* **maritime radiobeacon station.**

marine utility station (MUS): In the maritime mobile service, a station that (a) consists of one or more handheld radiotelephone units licensed under a single authorization, and (b) can be operated while being handcarried. *Synonym* **maritime utility station.**

maritime air communications (MAC): Communications systems, procedures, operations, and equipment that are used for message traffic between aircraft stations and ship stations in the maritime service. *Note:* Commercial, private, naval, and other ships are included in maritime air communications.

maritime air control authority: An organization that is responsible for the control of air operations that are performed by air and sea units.

maritime air radio organization: A worldwide air radio communications system in which high frequency (HF) international Morse code is used for controlling and reporting maritime patrol aircraft operations.

maritime air telecommunications organization: An organization for satisfying the communications requirements for the control and the reporting of maritime patrol aircraft operations and for the transmitting of air safety messages.

maritime broadcast communications net: A communications net that is used for international distress calling, including (a) international lifeboat, lifecraft, and survival craft high frequency (HF), (b) aeronautical emergency very high frequency (VHF), (c) survival ultrahigh frequency (UHF), (d) international calling and safety very high frequency (VHF), (e) combined scene of search and rescue, and (f) other similar and related purposes. *Note:* Basic international distress calling is performed at either medium frequency (MF) or high frequency (HF). *Synonym* **maritime distress communications net.** *Refer to* **Figs. S-3 and S-4 (869).**

maritime broadcast station: *Synonym* **marine broadcast station.**

maritime distress communications net: *Synonym* **maritime broadcast communications net.**

maritime mobile satellite service (MMSS): A mobile satellite service in which (a) mobile Earth stations are on board ships, and (b) survival craft stations and emergency position-indicating radiobeacon stations may also participate. *See also* **Earth station.**

maritime mobile service (MMS): A mobile service (a) that is between coast stations and ship stations, between ship stations, or between associated on board communications stations, and (b) in which survival craft stations and emergency position-indicating radiobeacon stations may also participate.

maritime patrol air broadcast net: A broadcast net that is used to control maritime aircraft patrol units and to transmit information to aircraft stations.

maritime patrol air reporting net: A communications net that (a) establishes two-way communications links between aircraft patrol units and their controlling maritime headquarters, and (b) may also be used for communications between aircraft stations and ship stations.

maritime radiobeacon station: *Synonym* **marine radiobeacon station.**

maritime radionavigation satellite service (MRNSS): A radionavigation satellite service in which Earth stations are on board ships. *See also* **Earth station.**

maritime radionavigation service (MRNS): A radionavigation service intended for the benefit and the safe operation of ships. *Refer to* **Figs. S-3 and S-4 (869).**

maritime ship-shore communications net: A communications net that is primarily used for long-distance ship-shore communications, such as for command and control communications between ship stations and shore stations.

maritime tactical air operations communications net: A communications net that is used for coordination of maritime air-related purposes, such as maritime air movements, helicopter units, maritime air patrol reporting, joint air support, and air homing.

maritime telecommunications system: *See* **automated maritime telecommunications system.**

maritime utility station: *Synonym* **marine utility station.**

mark: 1. A detectable symbol in or on a data medium. **2.** In binary communications, one of the two significant conditions of encoding, the other being the space. *Synonyms* **marking pulse, marking signal.** *See also* **space. 3.** A symbol or symbols that indicate the beginning or the end of a field, word, or data item in a file, record, or block. **4.** In telegraphy, one of the two significant conditions of encoding, the other condition being the space. *See* **beginning of tape mark, document mark, end of tape mark.** *See also* **code element, marking bias, modulation, neutral operation, pulse, signal transition, significant condition.**

mark active: In telegraph systems, pertaining to a signal in which (a) a mark is represented by a pulse or an on condition, such as (i) a direct current (dc) is flowing, (ii) a fixed frequency tone is sounding, or (iii) a circuit is closed and something is happening in it, and (b) a space is represented by no pulse or an off condition, such as (i) no current is flowing, (ii) the fixed frequency tone is not sounding, or (iii) a circuit is open and nothing is happening in it.

mark coding: *See* **biphase mark coding.**

marked access line: In telephone systems, a line that (a) has been marked for special access control, (b) when marked, is used to provide a preemption capability to a switchboard operator, such as that of a private branch exchange (PBX), (c) may be marked with a precedence level, such as priority, immediate, or flash, so that the operator can use the line marked for the corresponding level of precedence of a call that is being placed, and (d) in some systems, can only have a priority marking.

marker: In display systems, a symbol that (a) can be displayed in the display space on the display surface of a display device, (b) has a recognizable shape, and (c) usually has an obvious center or index point, such as a circle, square, caret, bracket, arrow, hourglass, or underscore. *See* **drop zone boundary limit marker, forbidden landing point marker, no landing marker, range marker, time marker.** *See also* **cursor.**

marker beacon: In the aeronautical radionavigation service, a transmitter that vertically transmits a distinctive radiation pattern to provide position information to aircraft. *See* **fan marker beacon, Z marker beacon.**

marker beacon station: *See* **aeronautical marker beacon station.**

mark hold: 1. The transmission of a steady mark or signal during a normal line condition when there is no traffic. **2.** In a telegraph system, a no-traffic condition in which a steady mark is transmitted. *See also* **line, mark, mark active, space hold.**

marking: *See* **multiple peg marking, position marking.**

marking bias: The uniform lengthening of the duration of all marking signal pulses at the expense of all spacing pulses. *See also* **bias, bias distortion, mark, pulse duration, spacing bias.**

marking end distortion: *See* **end distortion.**

marking panel: A sheet of material that is displayed on the ground for visual signaling to an aircraft. *See also* **visual signaling.**

marking pulse: *Synonym* **mark.**

marking signal: *Synonym* **mark.**

mark inversion code: *See* **modified alternate mark inversion code.**

mark inversion signal: *See* **alternate mark inversion signal.**

mark inversion violation: *See* **alternate mark inversion violation.**

Markov chain: A model of a sequence of events in which the probability of occurrence of a given event depends entirely upon the event that immediately precedes it.

mark scanning: 1. The automatic optical sensing of marks that usually are recorded manually on a data medium. **2.** Scanning performed during mark sensing. *See also* **data medium, mark, mark sensing.**

mark sensing: The automatic sensing of marks that (a) usually are made manually and (b) are detected by various means. *Note:* An example of mark sensing is detecting positioned spots made on paper. The spots are made with a lead pencil. The spots, usually in rows and columns, are detected by passing an electric current between contacts pressed on the paper. The lead pencil marks serve as a conductor. If there is a mark at a specified location, current will flow, indicating that a mark has been made. If there is no mark, current will not flow at the designated position. Mark sensing can be used to record answers and count correct answers to multiple-choice or true-false questions. The answers can be read automatically and the correct answers identified by the marks being at the positions corresponding to the correct answers. The score is computed automatically by adding the correct answer signals in a counter.

mark signal: 1. In telegraph and teleprinter operations, the signal that produces the inactive condition in a teleprinter, i.e., the signaling condition that produces a stop signal when using the International Telegraph Alphabet 2 (ITA-2). **2.** The signal that corresponds to an on key or transmit condition. *See also* **International Telegraph Alphabet 2, on key, stop signal.**

mark-space signaling: *See* **digital mark-space signaling, double current mark-space signaling.**

markup language: *See* **HyperText Markup Language.**

***m*-ary code:** *Synonym* ***n*-ary code.**

***m*-ary digit:** *Synonym* ***n*-ary digit.**

***m*-ary digital signal:** *Synonym* ***n*-ary digital signal.**

***m*-ary information element:** *Synonym* ***n*-ary information element.**

***m*-ary signaling:** *Synonym* ***n*-ary signaling.**

maser: 1. An oscillator that (a) depends on molecular interaction with electromagnetic waves and (b) generates microwaves, i.e., superhigh frequency (SHF) (3 GHz (gigahertz) to 30 GHz) radio waves, by means of microwave amplification by stimulated emission of radiation. *See* **optical maser.** *See also* **laser, microwave amplification by stimulated emission of radiation.**

mask: **1.** In communications systems, to obscure, cover up, obliterate, or otherwise prevent information from being derived from a signal. *Note:* The masking usually is performed by means of another signal, such as by noise, jamming, static, or other forms of interference. Thus, to mask is not the same as to erase or to delete. **2.** In computing and data processing systems, a pattern of characters that is used to retain or eliminate portions of another pattern of characters.

masked threshold: The level at which an indistinguishable signal becomes distinguishable from other signals or noise. *Note:* In acoustics, the masked threshold usually is expressed in dB.

masking: **1.** The use of transmitters to hide or obliterate a particular transmission, particularly to obscure the source, purpose, or information value of a transmission. **2.** The blocking or the preventing of a signal from reaching a receiver by means of deflection, reflection, absorption, dispersion, refraction, attenuation, knife edge effect, or any combination of these means caused by an intervening object. **3.** The obliteration of transmission in certain directions by means of a directed jamming signal. **4.** In computing and data processing, the use of a pattern of characters, such as a string of characters, to retain or eliminate portions of another pattern of characters. *See* **fault masking, radar masking.** *See also* **absorption, attenuation, deflection, dispersion, diffraction, knife edge effect, refraction, string.**

mass storage: Storage that (a) has a very large capacity and (b) usually is peripheral storage. *See also* **peripheral, storage, storage capacity.**

master: *See* **stencil master.**

master clock: A device that generates periodic, precisely spaced signals, i.e., timing signals, that are used for such purposes as timing, regulation of the operations of a processor, or generation of interrupts. *See* **Department of Defense master clock.** *See also* **clock, coordinated clock, Coordinated Universal Time, reference clock, slave clock, timing signal.**

master file: A relatively permanent file of essential data used as the primary or main file in support of a given job or set of jobs. *See also* **job.**

master frequency generator: In frequency division multiplexing (FDM), equipment that is used to provide system end-to-end carrier frequency synchronization and frequency accuracy of tones. *Note 1:* Various types of oscillators are used in master frequency generators, such as (a) a master oscillator as an integral part of a multiplexer set, (b) a submaster oscillator, i.e., a slave oscillator, as an integral part of a multiplexer set, (c) an external master oscillator that has extremely accurate and stable frequency and amplitude characteristics, and (d) an oscillator that has a continuously variable frequency capability for frequency-modulated systems. *Note 2:* The following types of oscillators are employed in the Defense Communications System FDM master frequency generators:

Type 1 - A master carrier oscillator as an integral part of the multiplexer set.

Type 2 - A submaster oscillator equipment or slave oscillator equipment as an integral part of the multiplexer set.

Type 3 - An external master oscillator equipment that has extremely accurate and stable characteristics.

Synonyms **master frequency oscillator, master oscillator.** *See also* **carrier, frequency, frequency division multiplexing, multiplexer set, oscillator.**

master frequency oscillator: *Synonym* **master frequency generator.**

mastergroup: In frequency division multiplexing of telephone channels, a grouping of supergroups, i.e., in the United States and Canada, a grouping of 10 supergroups or 600 voice channels. *Note:* Some mastergroups that are used in terrestrial telephone systems consist of 300 or 900 channels that are formed by multiplexing 5 or 15 supergroups, respectively. *See* **channel mastergroup, group.**

Master International Frequency Register (MIFR): A master list that (a) consists of all international frequency registrations of transmitting stations throughout the world and (b) is maintained by the International Frequency Registration Board (IFRB), a permanent organization of the International Telecommunication Union (ITU). *Note:* In each country or group of countries, specific assignments within national frequency allocations are made by their respective national frequency management organizations, such as, in the United States, the Federal Communications Commission (FCC) for non-federal government organizations and the National Telecommunications and Information Administration (NTIA) for federal government organizations. *See also* **assigned frequency, authorized frequency, frequency, frequency band allocation, frequency band allotment.**

master log: *See* **aeronautical station master log.**

master net control station (MNCS): A control station that (a) directs operational aspects of a network,

(b) directs the net control station, and (c) controls network equipment, usually for a unique or special communications system that includes common user or dedicated facilities. *See also* **net control station, network.**

master oscillator: 1. In a communications system, an oscillator that is used for frequency control. *Note:* Examples of master oscillators are (a) an oscillator that establishes the carrier frequency for the output signal of a transmitter and (b) an oscillator that provides or controls modulator frequencies for a channel, or groups of channels, in a frequency-modulated communications system or in a frequency division multiplexing system. **2.** *Synonym in frequency division multiplexing* **master frequency generator.**

master-slave timing: In a communications system, timing in which one station or node supplies the timing reference for all other interconnected stations or nodes. *See also* **democratically synchronized network, hierarchically synchronized network, mutual synchronization, oligarchically synchronized network, timing signal.**

master-slave timing system: In a communications system, a timing system in which one station or node supplies the timing reference for all other interconnected stations or nodes. *See also* **master-slave timing.**

master station: 1. In a data network, the station that (a) is designated by the control station to ensure data transfer to one or more slave stations, and (b) controls one or more data links of the data communications network at any given instant. *Note:* The assignment of master status to a given station is temporary and is controlled by the control station according to the procedures set forth in the operational protocol. Master status usually is conferred upon a station so that it may transmit a message, but a station need not have a message to send to be designated as the master station. **2.** In navigation systems using precise time dissemination, a station that has the clock used to synchronize the clocks of subordinate stations. **3.** In basic mode link control, the data station that has accepted an invitation to ensure a data transfer to one or more slave stations. *Note:* At a given instant, there can be only one master station on a data link. *See also* **contention, control station, data communications, data transmission, interrogation, network, primary station, secondary station, slave station, tributary station.**

matched cladding: In a dielectric waveguide, cladding composed of a single homogeneous layer of dielectric material.

matched junction: In a waveguide, a junction that (a) has four or more ports, i.e., arms, (b) has all ports except one terminated in the proper impedance, and (c) does not reflect energy from the junction when fed along the unterminated port. *See also* **hybrid junction, junction.**

match gate: *Synonym* **EXCLUSIVE NOR gate.**

matching: *See* **antenna matching, impedance matching.**

matching cell: *See* **refractive index matching cell.**

matching gel: *See* **refractive index matching gel.**

matching material: *See* **refractive index matching material.**

matching voice coder: *See* **pattern matching voice coder.**

material: *See* **communications security material, isotropic material, keying material, nonlinear optical (NLO) material, optically active material, polymeric nonlinear optical material, refractive index matching material, third-order nonlinear material.**

material absorption: 1. *Synonym* **absorption. 2.** *See* **bulk material absorption.**

material dispersion: 1. The dispersion, i.e., the broadening of the duration and the spatial width of an electromagnetic pulse, caused by the variation in the refractive index of a propagation medium as a function of the wavelength of the electromagnetic wave propagating in the medium. *Note:* Material dispersion contributes to group delay distortion, along with waveguide delay distortion and multimode group delay spread. **2.** In fiber optics, the broadening, i.e., the lengthening of the duration and the spatial width of a light pulse as it propagates in a dielectric waveguide. *Note 1:* The broadening is caused by the different velocities of the different wavelengths in the pulse, i.e., in the spectral width of the light. *Note 2:* Material dispersion should not be called "spectral" or "intramodal" dispersion because all dispersion is spectral, and material dispersion affects all modes. *Note 3:* Material dispersion is a part of the total dispersion of an electromagnetic pulse in a waveguide, such as an optical fiber, because the refractive index of the material used to make the waveguide is dependent on the wavelength of the electromagnetic wave propagating in the guide. As the wavelength is increased, and frequency is decreased, material dispersion decreases. At high frequencies, the rapid interaction of the electromagnetic field with the waveguide material renders the refractive index even more dependent upon frequency. Because each frequency experiences a different refractive index, some frequencies will propagate faster than others, so that they arrive at the end of the fiber at different times, making the pulse at the end of a path broader in space and time than at the beginning. Thus, the refractive index for a material is not a constant over all wavelengths. *Note 4:* In pure silica, the basic material from which most common telecommunications-grade optical fibers are made, material dispersion is minimum at wavelengths in the vicinity of 1.27 μm (microns). *Note 4:* Use of the term "chromatic dispersion" is discouraged although some workers in the field consider it to be a synonym. *Synonyms* **chromatic dispersion, intramodal dispersion.** *Deprecated synonym* **chromatic dispersion.** *See also* **dispersion, dispersion coefficient, profile dispersion, propagation medium, waveguide dispersion.**

material dispersion coefficient: *Synonym* **dispersion coefficient.**

material dispersion parameter: *Synonym* **dispersion coefficient.**

material dispersion wavelength: *See* **zero material dispersion wavelength.**

material scattering: 1. In an electromagnetic wave, scattering that is attributable to the intrinsic properties of the materials through which the wave is propagating. *Note 1:* Examples of material scattering are atmospheric scattering, ionospheric scattering, and Rayleigh scattering. *Note 2:* In an optical fiber, material scattering is caused by the materials used to fabricate the fiber, including the dopants used to create the refractive index profile. **2.** The electromagnetic wave power that is scattered per unit volume of the basic material, i.e., the propagation medium, through which the wave is propagating. *Note 1:* Material scattering follows a Rayleigh distribution, which is characteristic of a propagation medium that has a refractive index that changes over short distances compared to the wavelength of the propagating electromagnetic wave, such as a lightwave or a radio wave. *Note 2:* Material scattering losses are usually expressed in dB per kilometer. *Synonym* **bulk material scattering.** *See also* **ionospheric scattering, Rayleigh scattering, scattering, waveguide scattering.**

mathematical check: A check based on a mathematical relationship, such as a product, sum, or remainder. *Synonym* **arithmetic check.** *See also* **modulo-n check, product, sum.**

mathematical induction: Proving that a statement concerning terms based on natural numbers not less than N is valid by showing that if the statement is valid for the term based on N and that if it is also valid for an arbitrary value greater than N, then it is also valid for the term based on $N + 1$.

mathematical model: A representation of a process, device, concept, or other phenomenon by means of an algebraic expression.

mathematical morphology (morph or MORPH): A method of analyzing and transforming images by (a) generating and using structuring elements, i.e., operator sequences and parameter settings in the form of templates, and (b) synthesizing complex features using a fixed set of operators. *Note:* Mathematical morphology (MORPH) may be used in target identification, imagery, pattern recognition, and artificial intelligence.

matrix: 1. An array of items arranged in rows and columns so they may be manipulated in accordance with specified rules. *Note:* Matrix algebra contains the rules for the formation of matrices and their manipulation. **2.** Two sets of equidistant parallel conductors, each set running perpendicular to the other and in a separate plane, so that combinational circuits can be connected where the wires cross over each other. *Note:* The matrix is used in switching systems. **3.** An array of any number of items and dimensions. *Note:* The dimensions of a matrix usually are orthogonal, such as in a Cartesian coordinate system. *See* **coupler transmittance matrix, dot matrix, path quality matrix, route matrix, switching matrix, transmittance matrix.**

matrix liquid crystal display: *See* **active matrix liquid crystal display.**

matrix printer: *See* **dot matrix printer.**

matrix storage: Storage in which the storage elements are arranged so that access to any location requires the use of two or more rectangular coordinates, i.e., Cartesian coordinates. *Note:* Examples of matrix storage are (a) cathode ray tube (CRT) storage and magnetic core storage. *See also* **cathode ray tube storage, magnetic core storage, storage.**

maximal code sequence: In spread spectrum systems, a binary digit code sequence that is the longest that can be generated by a given spread spectrum code sequence generator before the sequence is repeated. *Note:* The longest code sequence for a single linear binary-code-sequence feedback shift-register generator is given by the relation $2n - 1$, where n is the number of stages, i.e., number of flip-flops, in the register of the generator. *See also* **spread spectrum code sequence generator.**

maximal ratio combiner: A diversity combiner in which (a) the signals from each channel are added together, (b) the gain of each channel is made proportional to the root mean square (rms) signal and inversely proportional to the rms noise in that channel, and (c) the same proportionality constant is used for all channels. *Synonym* **ratio squared combiner.** *See also* **diversity combiner, selective combiner.**

maximal-ratio-square diversity combining method: A method for combining the signal power outputs of two or more receivers that are operating in a diversity mode in which the combined output power squared equals the sum of the squares of the input signal powers, the net result being that the root mean square (rms) output power is the rms value of the input powers. *See also* **diversity combiner, diversity reception, diversity transmission.**

maximum acceptance angle (MAA): In an optical fiber, the maximum angle between the longitudinal axis of the fiber and an incident light ray at the fiber endface in order for the ray to be totally (internally) reflected at the core-cladding interface inside the fiber. *Note:* For an optical fiber, the sine of the maximum acceptance angle (MAA) is the numerical aperture (NA), given by the relation $MAA = \sin^{-1} NA$, where *MAA* is the maximum acceptance angle and $NA = (n_1^2 - n_2^2)^{1/2}$ where *NA* is the numerical aperture, n_1 is the fiber core refractive index at the point on the fiber endface at which the NA is being determined, and n_2 is the cladding refractive index. For a graded-index fiber, because the NA depends on the refractive index, and the refractive index varies with distance from the center of the fiber, the NA varies with the distance from the center of the fiber, and therefore the true MAA also depends on the maximum refractive index found on the fiber endface, which is at the center.

maximum access time: The maximum allowable waiting time between initiation of an access attempt and access. *See also* **access, access attempt, access time.**

maximum block transfer time: The maximum allowable waiting time between (a) the instant that a block transfer attempt is initiated and (b) the instant that a successful transfer of the block occurs. *See also* **block, block transfer rate, minimum cable cutoff wavelength, successful block transfer.**

maximum cable cutoff wavelength: The maximum wavelength at which a cabled waveguide, such as a cabled optical fiber, will support a mode. *Note:* The optical fiber will not support any wavelength longer than the maximum cable cutoff wavelength. The operating wavelength range, as determined by the transmitter nominal central wavelength and the transmitter spectral width, must be less than the maximum cutoff wavelength and greater than the minimum cutoff wavelength to ensure that the cable operates entirely in the fiber single mode region, i.e., as a single mode fiber. Thus, if the wavelength is too short, the fiber will operate in multimode. If the wavelength is too long, the fiber will not support any mode. Usually, the highest value of cabled fiber cutoff wavelength occurs in the shortest cable length. A criterion that will ensure that a system is free from high cutoff wavelength problems is given by the relation $\lambda_{ccmax} > \lambda_{tmin}$ where λ_{ccmax} is the maximum cutoff wavelength and λ_{tmin} is the minimum value of the transmitter central wavelength range, which is also caused by the worst case variations introduced by manufacturing, temperature, aging, and any other significant factors determined when the cable is operated under standard or extended operating conditions. *See also* **fiber cutoff wavelength, minimum cable cutoff wavelength.**

maximum call area: *Synonym* **maximum calling area.**

maximum calling area: The geographical area that (a) is within the calling limits permitted to a particular access line and (b) is based on the requirements for the particular line. *Note 1:* Maximum calling area limits are imposed for network control purposes and operating efficiency. *Note 2:* In the maximum calling area a user can place a call without special authorization or arrangements. *Synonym* **maximum call area.** *See also* **call.**

maximum contrast: The ratio of (a) the luminance of the brightest portion of an object or an image to (b) the darkest portion, i.e., the range of minimum to maximum luminance. *See also* **image, luminance, object.**

maximum detection range: In radio reception, the range that (a) corresponds to the minimum field strength that a signal can have and still be separated from noise and restored to the form in which it was transmitted by a specified transmitter that is transmitting under specified conditions, such as power, direction, and modulation, (b) depends upon such factors as the transmitting power, propagation conditions, and characteristics of the receiver, and (c) can be considered only in terms of a transmitter and receiver pair, not in terms of either transmitter or receiver alone. *See also* **field strength, threshold.**

maximum disengagement time: The maximum allowable waiting time between (a) the instant a disengagement attempt is initiated and (b) the instant a disengagement success is obtained. *See also* **disengagement**

attempt, disengagement success, disengagement time.

maximum input power: *See* **optical receiver maximum input power.**

maximum justification rate: *Synonym* **maximum stuffing rate.**

maximum keying frequency: In facsimile systems, the frequency numerically equal to (a) the spot speed divided by (b) twice the *x* dimension of the scanning spot. *Note:* If the spot speed is expressed in a distance unit per second and the *x* dimension is expressed in the same distance unit, the maximum keying frequency will be in hertz. *Synonym* **fundamental scanning frequency.** *See also* **facsimile, frequency, x dimension of scanning spot.**

maximum keying rate: *Synonym* **maximum keying frequency.**

maximum modulating frequency (MMF): In a facsimile transmission system, the highest picture frequency required for the system. *Note:* The maximum modulating frequency and the maximum keying frequency are not necessarily equal. *See also* **facsimile, frequency.**

maximum optical reflection: In a fiber optic transmitter, the maximum percent of total output optical power reflected back into the transmitter that it can accommodate and still maintain its stated performance.

maximum pulse rate (MPR): In optical waveguides and metallic conductors in which pulse dispersion limits the pulse repetition rate, i.e., the data signaling rate (DSR), for a given waveguide, the pulse rate, P_m, that (a) is just sufficient to create a specified bit error ratio (BER) and (b) that is given by the relation $P_m = AL^{-\Upsilon}$ where A is determined by line characteristics that fix the value of γ and L is the length of the waveguide. *Note:* Usually $\Upsilon = 1/2$ for single mode glass optical fibers. For multimode fiber guides, $1/2 < \Upsilon < 1$. For wires, $\Upsilon = 2$.

maximum receiver input: In an fiber optic receiver terminal/regenerator facility of a fiber optic station/regenerator section, the maximum value of the input optical power (dBm) to the receiver at the line side of the receiver module fiber optic connector or splice when operated under standard or extended operating conditions that the receiver will accept and still not exceed a specified bit error ratio (BER). *Note 1:* The maximum receiver input usually is expressed in dBm. *Note 2:* If the receiver input power should exceed the maximum receiver input, the section is overdesigned, or an optical attenuator must be inserted.

maximum stuffing rate (MSR): In a bit stream, the maximum rate at which bits can be inserted into, or deleted from, the stream. *Synonym* **maximum justification rate.** *See also* **binary digit, bit stuffing, destuffing, nominal bit stuffing, overhead bit.**

maximum theoretical numerical aperture: *See* **numerical aperture.**

maximum transceiver dispersion: The worst case dispersion, caused by a given length of optical fiber between members of a transmitter and receiver pair, that can be accommodated by the pair to meet the bit rate, i.e., data signaling rate, and bit error ratio (BER) specified by the manufacturer, when operated under standard or extended operating conditions. *Note:* For a given length of optical fiber, the maximum transceiver dispersion usually is expressed in $ps \cdot nm^{-1}$ (picoseconds/nanometer), i.e., picoseconds of increased pulse duration per nanometer of spectral width of the source.

maximum usable frequency (MUF): In radio transmission using reflection from the regular ionized layers of the ionosphere, the upper limit of the frequencies that can be used for transmission between two points at a specified time. *Note:* The maximum usable frequency (MUF) is a median frequency applicable to 50% of the days of a month, as opposed to 90% cited for the lowest usable high frequency (LUF) and the optimum traffic frequency (FOT). *See also* **frequency, frequency spectrum designation, lowest usable high frequency, optimum traffic frequency.**

maximum user signaling rate (MUSR): The maximum rate at which binary information can be transferred in a given direction between users over the telecommunications system facilities dedicated to a particular information transfer transaction, under conditions of continuous transmission and no overhead information. *Note 1:* The maximum user signaling rate usually is expressed in bits per second. *Note 2:* For a single channel, the maximum user signaling rate is given by the relation $SCSR = (1/T)\log_2 n$ where $SCSR$ is the single channel signaling rate in bits per second, T is the minimum time interval in seconds for which each level must be maintained, and n is the number of significant conditions of modulation of the channel. *Note 3:* In the case where an individual end-to-end telecommunications service is provided by parallel channels, the parallel channel signaling rate is given by the relation

$$PCSR = \sum_{i=1}^{m} \left(\log n_i\right) / T_i$$

where *PCSR* is the total signaling rate for m channels, m is the number of parallel channels, T_i is the minimum interval between significant instants for the ith channel, and n_i is the number of significant conditions of modulation for the ith channel. *Note 4:* In the case where an end-to-end telecommunications service is provided by tandem channels, the end-to-end maximum user signaling rate is the lowest of the maximum user signaling rates among the component channels. *See also* **channel, data signaling rate, effective speed of transmission, Shannon's law, signal, throughput.**

Maxwell's equations: A group of basic equations, in either integral or differential form, that (a) describe the relationships between the properties of electric and magnetic fields, their sources, and the behavior of these fields at material media interfaces, (b) express the relationships among electric and magnetic fields that vary in space and time in material media and free space, and (c) are fundamental to the propagation of electromagnetic waves in material media and free space. *Note 1:* Maxwell's equations are the bases for deriving the wave equation that expresses the electric and magnetic field vectors in an electromagnetic wave in a propagation medium, such as radio waves in free space and lightwaves in an optical fiber. *Note 2:* Maxwell's equations, in differential form, are given by the relations

$$\nabla \times \mathbf{E} = -\partial \mathbf{B}/\partial t$$

$$\nabla \times \mathbf{H} = \mathbf{J} + \partial \mathbf{D}/\partial t$$

$$\nabla \bullet \mathbf{B} = 0$$

$$\nabla \bullet \mathbf{D} = \rho$$

where (a) **E**, **H**, **B**, and **D** are the vectors of the electric field strength, i.e., electric field intensity (volts per meter), the magnetic field strength, i.e., magnetic field intensity (amperes per meter or oersteds), the magnetic flux density (webers per square meter or gauss), and the electric flux density or electric displacement (coulombs per square meter), respectively, (b) **J** is the electric current density (amperes per square meter), (c) ρ is the electric charge density (coulombs per cubic meter), and $(\partial)t$ is time. The ∇ is the "del" space derivative operator, expressing differentiation with respect to all distance coordinates, the $\times$ being the curl or cross product operator, resulting in a vector quantity, and the • being the divergence or dot product operator, resulting in a scalar quantity. The partial derivatives are with respect to time. These equations are used in conjunction with the constitutive relations to obtain useful practical results, given actual distributions of electrical charge and current in real media and given the boundary conditions. *Note 3:* Maxwell's equations are valid only when the field and current vectors are single-valued, bounded, and continuous functions of position and time and have continuous derivatives. Wave equations are the solutions of Maxwell's equations, given the boundary conditions. For optical fibers, the propagation medium is charge-free and electrically nonconducting. Thus, $\rho = 0$ and $\mathbf{J} = 0$, which result in simplified solutions to Maxwell's equations. *See also* **constitutive relation, wave equation.**

MAYDAY: The distress signal that (a) is used in voice communications, (b) is the first part of a distress call, (c) is repeated three times, and (d) is followed by the call sign of the mobile station in distress. *Note:* The call sign is also repeated three times. The distress message then follows the call signs.

M band: The band of frequencies in the range from 60 GHz (gigahertz) to 100 GHz, consisting of ten numbered channels that are each 4 GHz wide. *Note:* The old M band consisted of frequencies in the range from 10 GHz to 15 GHz, overlapping parts of the old X band and the old P band.

Mbps: megabits per second.

MCA: maximum calling area.

MCC: maintenance control circuit.

MCM: multicarrier modulation.

MCVD: modified chemical vapor deposition.

MCVD process: *See* **modified chemical vapor deposition process.**

MDAS: multiple database access system.

MDF: main distribution frame.

MDIS: medical diagnostic imagery system.

meaconing: A system for receiving radiobeacon signals and retransmitting them on the same frequency to confuse navigation and cause inaccurate bearings to be obtained by beacon users, such as aircraft stations, ship stations, and mobile ground stations.

mean circuit activity: The ratio of (a) the total time to (b) the time that a circuit is occupied by necessary signals, such as speech, data, and control signals. *Note 1:* A typical value of the mean circuit activity is 25%. With the greater incentive to use high cost international circuits more intensively, 37% is the typical value for trans-Atlantic links. *Note 2:* Mean circuit activity is an important factor in the use of time assignment speech interpolation (TASI) and in the use of digital speech interpolation systems for long-distance communications trunks. *See also* **time assignment speech interpolation.**

mean entropy: *See* **character mean entropy.**

mean information content: *Synonym* **character mean entropy.**

mean power: In radio, video, radar, and microwave transmission systems, the power that is supplied to the antenna transmission line by a transmitter during normal operation, averaged over a time that is sufficiently long compared with the period of the lowest frequency, i.e., the longest wavelength, that is encountered in the modulation. *Note:* A time of 0.1 s (seconds) during which the mean power is the greatest usually is selected as the measurement period. *See* **radio transmitter mean power.**

mean power of the talker volume distribution: The mean power talker volume less a conversion constant, taken as 3.9 dB by some authorities, 1.4 dB by others, and 2.9 dB in the U.S. Department of Defense, to convert from volume units (vu) to dB referenced to 1 mW, i.e., 0 dBm. *See also* **mean power talker.**

mean power talker: A talker that (a) represents the average of the powers of a group of talkers measured at a given location or in a given environment and (b) is the value of an ordered set of values below which and above which there are an equal number of values. *Synonym* **mean volume talker.** *See also* **mean power of the talker volume distribution, medium power talker, value.**

mean repair time: *Synonym* **mean time to repair.**

mean solar day: *Synonym* **civil day.**

mean spherical intensity: The average value of the radiant intensity of an electromagnetic source of radiation, such as a light source, with respect to all directions. *See also* **radiant intensity, value.**

mean square pulse duration: *See* **root mean square pulse duration.**

mean time: *See* **Greenwich Mean Time (GMT).**

mean time between failures (MTBF): 1. For a particular interval, (a) the total functioning life of a population of an item divided by (b) the total number of failures within the population during the performance measurement period. *Note:* Mean time between failures (MTBF) may be determined for time, cycles, kilometers, events, or other measure of life units. **2.** For a particular item in a population of items and during a specified performance measurement period, (a) the total functioning life of the population of the items divided by (b) the total number of failures within the population during the measurement period. *Note 1:* The total functioning life of a population of the items is the product of the total number of items in the population during the measurement period and the duration of the measurement period. Thus, if the number of items is 1000 and the measurement period is 100 s (seconds), the total functioning life is 10^5 item•s. *Note 2:* When computing mean time between failures (MTBF), any measure of life units may be used, such as time, cycles, kilometers, and events. If a total of 1000 events, such as burnouts, rotations, cycles, or kilometers, occur in a population of items during a measurement period of 100 seconds and there are a total of 10 failures among the entire population, the mean time between failures (MTBF) for each item is the functioning life, i.e., (1000)(100) or 10^5, divided by the number of failures, i.e., 10, to produce a MTBF of $(1000)(100)/10 = 10^4$ s (seconds) for each of the items, i.e., burnouts, rotations, cycles, or kilometers. **3.** In a system consisting of a large number items and over a long measurement period, the measurement period divided by the number of failures that have occurred during the measurement period. *See also* **availability, downtime, failure, maintainability, performance measurement period, reliability.**

mean time between outages (MTBO): In a system, the mean time between component failures that (a) result in loss of system continuity or unacceptable degradation of performance and (b) are not automatically bypassed. *Note:* The mean time between outages is expressed as $MTBO = MTBF/(1 - FFAS)$, where $MTBF$ is the nonredundant mean time between failures and FFAS is the fraction of failures for which the failed equipment is automatically bypassed. *See also* **availability, downtime, failure, failure fraction automatically switched, reliability.**

mean time to failure (MTTF): In a population of similar items, the mean time that an item functions successfully between (a) the initial instant that it is placed in operation and (b) the instant of first failure, i.e., the infant mortality period averaged over the population of items. *See also* **bathtub curve.**

mean time to repair (MTTR): In systems maintenance, (a) the total corrective maintenance time, i.e., the total time devoted to maintenance, divided by (b) the total number of corrective maintenance actions during a given period, such as a month, a year, or to date since placed in service. *Synonym* **mean repair time.** *See also* **availability, downtime, maintainability, reliability.**

mean time to service restoral (MTTSR): The mean time to restore service, i.e., to restore to an acceptable operational capability, following system failures that result in (a) a service outage or (b) unacceptable operational capability. *Note:* The time to restore service is all the time between (a) the instant that a failure occurs and (b) the instant that service is restored, including fault detection, fault location, and fault correction time, i.e., the actual time to repair. *See also* **availability, downtime, failure, maintainability, reliability.**

mean transinformation content: The average of the transinformation content that (a) is conveyed by the occurrence of any one of a finite number of mutually exclusive and jointly exhaustive events, given the occurrence of another set of mutually exclusive events, (b) is equal to the difference between the entropy of one of the two sets of events and the conditional information entropy of this set relative to the other, and (c) is given by the relation $MTC = H_x - H_{x/y} = H_y - H_{y/x}$ where H_x is the entropy of the message at the source, $H_{x/y}$ is the equivocation, H_y is the entropy of the message at the sink, and $H_{y/x}$ is the irrelevance. *Note:* In the transmission of one message, the difference between the entropy at the message source and the equivocation is equal to the difference between the entropy at the message sink and the irrelevance. *Synonym* **average transinformation content.** *See* **character mean transinformation content.** *See also* **conditional information entropy, equivocation, irrelevance, sink, source, transinformation content.**

mean volume talker: *Synonym* **mean power talker.**

measure: *See* **add-on security measure, addressability measure, electronic warfare support measure, information measure, lens measure, repeatability measure.**

measured rate service: Telephone service for which charges are made in accordance with the use of the line. *See also* **call, flat rate service, postalized rate, tariff.**

measurement: *See* **central wavelength measurement.**

measurement period: *See* **performance measurement period.**

measurement range: In an optical time domain reflectometer (ODTR), the length of optical waveguide, such as an optical fiber, that lies between the minimum length and the maximum length for which the ODTR can make an attenuation measurement, such as measure the distance to, and the insertion loss caused by, a fiber optic connector. *Note:* Measurement ranges usually are from several centimeters to over a hundred kilometers, operating at typical wavelengths of 0.85 µm (micron), 1.3 µm, and 1.55 µm. *See also* **distance resolution, dynamic range.**

measurement tolerance: *See* **phase measurement tolerance.**

measurement unit: *See* **noise measurement unit.**

measuring equipment: *See* **angle measuring equipment, distance measuring equipment, telegraph error rate measuring equipment.**

mechanical jamming: Jamming that is accomplished by using devices, such as chaff, that reradiate or reflect signals to produce extraneous signals, such as clutter, that tend to (a) saturate or degrade radar systems and (b) confuse and deceive radar operators. *See also* **chaff, clutter.**

mechanically despun antenna: An antenna that (a) is mounted on an articulating platform and (b) is driven to maintain a constant antenna pointing direction. *Note:* Examples of despun devices are (a) an antenna that is mounted on gimbals on a spinning satellite and constantly points toward the Earth and (b) a gun barrel that is mounted in a tank running over rough terrain and always points toward the fixed object, such as a terrain feature or a target, at which it was last aimed.

mechanically induced modulation (MIM): 1. In a signal propagating in an optical fiber, desirable or undesirable modulation caused by mechanical means, such as by vibrating, mechanical switching, microbending, macrobending, or squeezing an optical fiber. *Note 1:* Examples of mechanically induced modulation are (a) modulation of a lightwave in a violin, in which the strings are optical fibers, (b) modulation of a lightwave in a critical radius macrobend fiber optic sensor for measuring displacement, (c) modulation of a lightwave in a microbend sensor or a refractive index sensor for measuring pressure, (d) modulation of a lightwave in an optical fiber coated with a magnetostrictive material for measuring magnetic field strength, (e) modulation of a lightwave in a metallic coated optical fiber for measuring electric currents, and (f) speckle noise created in a multimode fiber by an imperfect splice or an imperfectly mated connector, or both, such that mechanical disturbance ahead of the joint will introduce changes in the modal structure, resulting in variations in joint loss. *Note 2:* The mechanically induced modulation may appear in many different forms, such as amplitude, phase shift, wavelength, and polarization plane rotation. **2.** In a lightwave propagating in an optical fiber, the production of undesirable modulation, such as noise, caused by physical distortion of the fiber.

mechanically intermateable connector: **1.** A connector that (a) is mechanically mateable with another connector, (b) does not create mechanical damage when mated or demated, and (c) is mated and demated without regard to attenuation properties. **2.** In fiber optic systems, a connector that will mate with another connector from a different source, i.e., a different manufacturer, without mechanical damage and without regard to optical properties. *See also* **interchangeable connector.**

mechanical optical switch: A manually or electromechanically operated optical switch that (a) either allows or prevents lightwave propagation in a given optical circuit, such as a fiber optic circuit, or selects alternate paths, (b) has switching times on the order of tens of milliseconds, and (c) is practical for light path coupling and decoupling in the same manner as electromechanical microswitches. *See also* **optical switch, path.**

mechanical splice: A fiber optic splice that (a) is accomplished by fixtures or materials, such as loose tubes and optical cements, rather than by thermal fusion and (b) may have refractive index matching material applied between the joined optical fiber ends. *Note 1:* The optical fibers may be secured by mechanical means or with an optical adhesive, i.e., optical cement. *Note 2:* When the fibers are secured by mechanical means, the gap between the endfaces usually is filled with a matching gel to reduce reflections. *See also* **fusion splice, matching gel, refractive index matching material, optical fiber splice.**

mechanics: *See* **celestial mechanics.**

mechanism: *See* **access control mechanism, bandwidth balancing mechanism.**

mechatronics: The branch of science and technology devoted to the design, development, fabrication, testing, and use of systems or devices in which mechanical and electronic components are integrated, usually the electronic components for control, sensing, and decision making and the mechanical components for motion and force development, such as a mouse, a light pen, and an electronic computer directed autopilot. *See also* **electromechanics, light pen, mouse.**

MED: manipulative electronic deception.

MediaCom™: Software and service that (a) was developed and is licensed by Bellcore, and (b) allows direct voice, voice messaging, facsimile, electronic mail, and access to electronic services, such as by means of a personal computer (PC), simultaneously over a single integrated services digital network (ISDN) line.

mediation function: A function that (a) is performed within a telecommunications management network (TMN) that routes or operates on information passing between network elements and network operations, and (b) can be shared among network elements, mediation devices, and network operation centers. *Note:* Examples of mediation functions are communications control, protocol conversion, data handling, communications of primitives, processing including decision making, and data storage functions. *See also* **telecommunications management network.**

media transmission: *See* **multiple media transmission.**

medical diagnostic imagery system (MDIS): An imagery system that (a) is a combined communications and picture archiving system, (b) connects large medical treatment facilities to smaller teleradiology installations, and (c) can digitize, process, store, and distribute images, such as X rays and stored films. *See also* **teleradiology.**

medical diathermy equipment: Equipment that (a) generates and transmits radio frequency electromagnetic waves for medical therapeutic or diagnostic purposes and (b) can create interference in communications equipment, particularly at receiving ends where the signal to noise ratios are low. *See also* **interference, radio frequency, signal to noise ratio.**

medical rf equipment: *See* **industrial-scientific-medical rf equipment.**

medium: **1.** In communications, computer, data processing, and control systems, a path along which a signal is confined and propagated. *Note:* Examples of media are wire pairs, coaxial cables, buses, and waveguides, such as optical fibers. **2.** In optics, any substance or vacuum through which light can propagate. **3.** In data recording, the material on which data are recorded. *Note:* Examples of media are paper sheets, paper tapes, punch cards, magnetic tapes, and magnetic disks. *See* **active laser medium, anisotropic propagation medium, birefringent medium, data medium, dispersive medium, empty medium, homogenous medium, isotropic propagation medium, lossy medium, machine-readable medium, nondispersive medium, propagation medium, record medium, sensitized medium, virgin medium.**

medium access control (MAC) sublayer: In a communications network, the part of the layer that supports topology-dependent functions and uses the services of the physical layer to provide services to the logical link control sublayer. *See also* **layer, link, Open Systems Interconnection—Reference Model.**

medium access unit (MAU): In a communications system, the equipment that adapts or formats signals for transmission over the propagation medium. *Note:* An example of a medium access unit (MAU) is a fiber optic transmitter that accepts an electrical signal at its input and converts it to an optical signal at its output. *See also* **propagation medium.**

medium coupling loss: *See* **aperture to medium coupling loss.**

medium frequency (MF): A frequency that lies in the range from 300 kHz (kilohertz) to 3000 kHz, i.e., to 3 MHz (megahertz), the range having the letter designation MF and a numerical designation of 6. *See also* **frequency, frequency spectrum designation.** *Refer to* **Table 1, Appendix B.**

medium interface connector (MIC): In communications systems, such as local area network/metropolitan area network (LAN/MAN) systems, the connector at the interface point between the bus interface unit and a terminal, i.e., at the medium interface point. *See also* **bus, local area network, medium interface point, metropolitan area network, terminal.**

medium interface point (MIP): In communications systems, such as local area network/metropolitan area network (LAN/MAN) systems, the location at which the standards for the interface parameters between a terminal and the bus interface unit are implemented. *See also* **interface, local area network, medium interface connector, metropolitan area network, network interface device.**

medium loss optical fiber: An optical fiber that has a medium level optical signal power loss per unit length of fiber, i.e., attenuation rate. *Note 1:* The attenuation rate usually is expressed in dB/km (decibels/kilometer) at a specified wavelength and includes all intrinsic losses. *Note 2:* In medium loss optical fiber, attenuation in amplitude of a propagating wave is caused primarily by (a) scattering caused by metal ions, (b) absorption caused by water in the form of the hydroxyl ion, and (c) Rayleigh scattering. The attenuation does not include extrinsic losses, such as high order mode loss caused by launch conditions, fiber optic connector losses at fiber-to-fiber interfaces, and curvature losses. *See also* **ultralow-loss optical fiber.** *Refer to* **Fig. A-10 (49).**

medium power talker: In a log normal, i.e., Gaussian or bell curve, distribution of talkers, a hypothetical talker that has a volume, i.e., power level, that lies at the medium power of all talkers that determine the volume distribution at the point of interest. *Note 1:* Medium power talker values are expressed in dB. *Note 2:* When the distribution of talker volume follows a log normal curve, the mean and the standard deviation can be used to compute the medium power talker. *Note 3:* The talker volume distribution follows a log normal curve, and the medium power talker is uniquely determined by the average talker volume. *Note 4:* The medium power talker volume, V, is given by the relation $V = V_0 + 0.115\sigma^2$ where V_0 is the average of the talker volume distribution in volume units (vu), and σ is the standard deviation of the distribution. *Synonym* **medium volume talker.** *See also* **loading, loading characteristic, mean power talker, power.**

medium volume talker: *Synonym* **medium power talker.**

meet me conference call: A conference call in which the operator notifies each conferee of the number to call at a specified time, prior to which the operator sets up all the necessary bridging circuits so that when the conferees place their calls, they will all be connected together, including the conference originator, i.e., the user requesting the call. *See also* **conference call, multiple call.**

mega (M): A prefix that indicates multiplication by 10^6, i.e., by a million. *Note:* An example of the use of "mega" is in 10 MHz (megahertz), equivalent to 10^7 Hz.

megabits per second (Mbps) signaling rate: A unit of data signaling rate (DSR) equal to 10^6 b•s^{-1}, i.e., one million bits per second. *See also* **bit, data signaling rate.**

megabyte (Mb): A unit of data or storage capacity equal to 10^6 Mb, i.e., one million bytes. *See also* **byte, data, storage capacity.**

megabytes per second (Mbps): A unit of data signaling rate (DSR) equal to 10^6 Mb•s^{-1}, i.e., one million bytes per second. *See also* **byte, data signaling rate.**

megahertz (MHz): A unit of frequency equal to 10^6 Hz, i.e., one million hertz. *See also* **frequency, hertz, metric system.** *Refer to* **Table 4, Appendix B.**

megaphone: A hand-held device that (a) is spoken into, (b) directs speech in the direction of the longitudinal axis, (c) usually has the shape of a horn, (d) may (i) be only voice-powered, (ii) contain a built-in power source, amplifier, and loudspeaker, or (iii) be powered by an external source, and (e) is used to project speech at increased volumes in desired directions. *See also* **horn.**

member: *See* **strength member.**

memory: 1. In a processing unit, the addressable storage space used to execute instructions. **2.** In calculators, microcomputers, and minicomputers, the main storage. *See* **cache memory, compact disk read-only memory, flash memory, random access memory, read-only memory, virtual memory.** *See also* **auxiliary storage, main storage, read-only storage, register.**

memory bounds checking: *Synonym* **storage bounds checking.**

memory card: A smart card that is either (a) a free-access card or (b) a secure-access card. *See also* **free-access card, secure-access card, smart card.**

memory disk: *See* **random access memory disk.**

meniscus: A lens that has a convex surface and a concave surface. *Synonym* **concavoconvex lens.** *See* **divergent meniscus lens.**

menu: 1. In display systems, a displayed list of items, usually commands, from which a user may select one or more items. *Note 1:* An example of a menu is a displayed list of options from which a user selects actions to be performed. *Note 2:* In a computer system, such as a personal computer (PC) system, menu items, i.e., menu choices, usually are displayed and then may be entered into the system by picking one of the choices on the menu, such as by moving a pointer, indicator, or cursor to the item and then performing an action, such as (a) depressing the enter key, (b) depressing a letter key that identifies the choice, (c) pressing the actuate switch on a mouse i.e., clicking a mouse, or (d) energizing a stylus, such as a light pen. **2.** In communications, computer, data processing, and control systems, a displayed list of options from which an operator or a user selects actions to be performed. *See* **command, command menu, nested command menu.** *See also* **indicator, light pen, menu item, menu pointer, mouse, picking, pointer, stylus.** *Refer to* **Fig. S-22 (977).**

menu item: In a communications, computer, data processing, or control system, one of the choices on a displayed menu that can be selected by an operator, i.e., a user, of the system. *Note:* In most communications, computer, data processing, and control systems, menu items, i.e., menu choices, are displayed, selection is made, and certain actions are taken by the user to enter the choice, usually a command, into the system. *See also* **menu, menu pointer, picking.**

menu pointer: In communications, computer, data processing, and control systems, the pointer that is used to for picking, i.e., for selecting a menu item. *Note:* Examples of menu pointers are reverse video bars and specially shaped displayed symbols, such as symbols shaped like arrow heads, hour glasses, closed brackets, underscores, squares, circles, and specified keys on a keyboard. *See also* **menu, menu item, picking, pointer.**

MERCAST: merchant ship broadcast system.

Mercator projection: A projection of the Earth's surface such that (a) the lines of longitude are made parallel lines, (b) the lines of latitude are parallel lines, (c) each set is perpendicular to the other set, (d) the poles are lines across the top and the bottom of the map, and (e) the polar areas, such as Siberia, Alaska, the Arctic Ocean, and Antarctica, appear oversized compared to equatorial areas of the same actual size. *Note:* A Mercator projection need not be of the entire Earth's surface. The smaller the area that is projected, the less the distortion at the edges.

merchant ship broadcast system (MERCAST): A maritime shore-to-ship broadcast system in which the ocean areas are divided into primary broadcast areas, each covered by a high powered shore radio station that broadcasts simultaneously on one medium frequency (MF) and one or more high frequencies (HF) for routing messages to oceangoing ships. *Note:* In some instances, ship broadcast coastal radio stations may repeat the messages.

mercury storage: Storage in which (a) the acoustic properties of mercury, such as ultrasonic sound wave transmission between electrosonic transducers, are used to form delay lines, and (b) the delay lines may be used open-ended for temporary storage or close-looped for continuous storage.

merge: To combine two or more sets of items into one set by (a) preserving the ordering in the original sets and (b) adhering to a sequencing criterion, such as merging one, two, or three at a time from each set. *See also* **merge-sort, sort.**

merge-sort: 1. A sort in which the items in two sets are merged, and the resulting set is sorted according to some criterion. *Note:* If the sequence R,B,P,S were merged one at a time, with the sequence Y,T,A,B, the result might be R,Y,B,T,P,A,S,B. If this sequence were sorted alphabetically, the result would be A,B,B,P,R,S,T,Y. **2.** A sort in which the items in a set are divided into two subsets, the items in each subset are sorted according to some ordering criterion, and the resulting sorted subsets are merged according to some sequencing criterion. *Note:* The final merged set may have to be sorted. *See also* **merge, sort.**

merging: The combining of two or more ordered sets of items into one ordered set while preserving the original ordering of the items that existed in the sets before they were combined. *See also* **merge, merge-sort, sort.**

meridian: *See* **celestial meridian.**

meridional plane: 1. Any plane that contains the optical axis of an optical system. **2.** In an optical fiber with a circular cross section, a plane that contains the optical fiber axis. **3.** Any plane that contains an Earth meridian. *See also* **celestial meridian, optical axis, optical fiber axis, optical system.**

meridional ray: In an optical fiber, a ray that passes through the fiber axis. *Note:* Along with axial rays, meridional rays lie in meridional planes, i.e., planes that contain the optical fiber axis. *See also* **axial ray, geometric optics, numerical aperture, optical fiber axis, paraxial ray, skew ray.** *Refer to* **Fig. R-5 (817).**

merit figure: *See* **optical fiber merit figure.**

Mersenne code sequence generator: A maximal code sequence generator (a) in which the length of the code sequence is a prime number of code elements, such as bits, (b) in which the code sequence is significant when cross correlation is a major factor of consideration, and (c) that is used in spread spectrum transmission and reception systems. *See also* **maximal code sequence generator, spread spectrum, spread spectrum code sequence generator.**

MESHnet: A vehicular secure communications system that (a) enables tactical field forces to route voice and data communications from a variety of information systems between division headquarters and subunits, (b) is in use by the Canadian Army, (c) is compatible with U.S. and U.K. counterparts, such as U.S. mobile subscriber equipment and the U.S. Army tactical command and control system, (d) operates in synchronous and asynchronous modes, and (e) consists of (i) a user control device, (ii) a network access unit, and (iii) a portable data terminal with liquid crystal display.

mesh network: *See* **network topology.**

mesh topology: *See* **network topology.**

mesochronous: Pertaining to the relationship between two signals such that their corresponding significant instants occur at the same average rate. *See also* **anisochronous, heterochronous, homochronous, isochronous, plesiochronous.**

message: 1. A thought or an idea expressed briefly in a plain, coded, or secret language, that (a) is prepared in a form suitable for transmission by any means of communication, (b) usually has three parts, (i) an indication of an addressee, (ii) the data that represent the information being communicated, and (iii) an indication of the sender or the originator, i.e., there must be a header, body, and ending, and (c) may consist of one or more units, i.e., may be a one-unit message or a multiunit message. **2.** In telecommunications, record information expressed in plain or encrypted language and prepared in a specified format for transmission by a telecommunications system. **3.** An arbitrary amount of information that has a defined, or implied, beginning and end. **4.** Information and control bit sequences that are transferred as an entity. **5.** In alphabetic telegraphy and in data transmission, a group of characters and control signal sequences that (a) are arranged at the transmitting station and (b) are transferred as an entity from a transmitter to a receiver. **6.** In information theory and in communication theory, an ordered arrangement of characters that are intended to convey information. *Deprecated synonyms* **cable, communication, line, signal, telegram, transmission, wire.** *See* **access control message, address message, amplifying switch message, book message, calling line identity message, clear message, codress message, direction finding message, distress message, drop message, dual precedence message, emergency message, error message, exclusive message, fallout message, fixed format message, FLASH message, FLASH OVERRIDE message, formal message, fix message, general message, hydrographic message, IMMEDIATE message, line identity message, multiaddress message, multiple part message, multiunit message, orderwire message, plaindress message, preburst message, PRIORITY message, pro forma message, refile message, response message, ROUTINE message, safety message, service message, signal message, single call message, single precedence message, straggler message, switch message, variable format message, warning message.** *See also* **body, communication theory, ending, header, information theory, signal, signal message.**

message accounting: *See* **automatic message accounting, centralized automatic message accounting.**

message address: In a message, the address part, usually consisting of one or more identifiers for the originator, action addressees, information addressees, and exempted addressees. *See* **codress message address.**

message alignment: In the placing of messages, or their parts, in or on a medium, the maintaining of specified temporal and spatial relationships (a) among the various data media that are used and (b) among the messages and the various parts within them. *See also* **medium.**

message alignment indicator (MAI): In a signal message, data that are placed between the user part of the message and the message transfer part of the message to identify the boundaries between the parts of the signal message. *See also* **signal message.**

message authentication: A communications security measure that (a) establishes the authenticity of a message by means of an authenticator that (i) is within the message and (ii) is derived from certain predetermined elements of the message itself, (b) is adopted to protect a communications system against the acceptance by a receiving station of a fraudulent message, and (c) usually is placed in effect by appropriate authority. *See also* **authenticator, communications security.**

message available at action office time: The time that a message is received by the action office, i.e., by the office of the action addressee. *See also* **action addressee, action office.**

message available for delivery time: The time that a received message is ready for distribution by the receiving communications center to the action addressee. *See also* **action addressee, communications center.**

message backlog: In a communications facility, all of the messages or data that are awaiting transmission or processing. *Synonym* **message queue.** *See also* **communications facility.**

message broadcast: An electronic mail (E-mail) conference capability in which data terminals are used. *Note:* Message broadcast control can be accomplished by the user or by the network. *See also* **electronic mail.**

message buffering: The placing of messages in a spooler for transmission (a) at an optimum rate, such as the maximum data signaling rate (DSR), and (b) at an appropriate time, such as when all higher precedence messages have been sent, when channels are available, or when operators are present. *See also* **data signaling rate, spooler.**

message center: The part of a communications center that is responsible for (a) the acceptance and the transmission of outgoing messages from message originators and (b) the reception and the delivery of incoming messages to message addressees. *See* **telecommunications center.**

message character: *See* **end of message character.**

message code: *See* **prearranged message code.**

message dispatch time: The date, including the time, that a message is dispatched (a) to an addressee by the receiving communications center, (b) to a receiving communications center by a sending communications center, or (c) to a sending communications center by a message originator. *See also* **date, date-time group.**

message distribution center: *See* **internal message distribution center, switching and message distribution center.**

message drafter: The person or the machine that composes a message for approval and release by the message originator or by the message releasing officer. *See* **message originator, message releasing officer.**

message element: Any part of a message. *Note:* Examples of message elements are accounting symbols, address designators, routing indicator groups, address groups, date-time groups, address indicating groups, call signs, originator designators, texts written by the message drafter or message originator, and endings. *See also* **accounting symbol, address designator, address group, address indicating group, date-time group, ending, call sign, message originator, originator designator, routing indicator group.**

message exchange: *See* **automatic message exchange.**

message feedback: The sending of received messages back to the sending end for comparison with the original messages that have been stored there, to check the accuracy of transmission of the messages.

message file time: **1.** The time that a message is brought into a message center for processing **2.** In automated remote entry systems, the time that a message is released to a communications system for transmission. *Synonym* **time of file.**

message format: A predetermined or prescribed temporal or spatial arrangement of the parts of a message that (a) is recorded in or on a data medium and (b) usually is transmitted or placed in storage. *Note:* Messages prepared for electrical transmission usually are composed on a printed blank form with spaces for each part of the message and for administrative entries. *See* **ionospheric message format, radiotelegraph message format.** *See also* **format line.**

message handling system (MHS): In the International Telegraph and Telephone Consultative Committee (CCITT) X.400 Recommendations, the family of services and protocols that provides the functions for global electronic mail transfer among local mail systems. *See also* **electronic mail, protocol.**

message header: *Synonym* **message heading.**

message heading: The message part or parts that (a) precede the text, i.e., the message body, temporally or spatially arranged according to established conventions, and (b) may include several message elements, such as address groups, routing indicators, action addressee designators, information addressee designators, exempted addressee designators, prosigns, prowords, clear indicators, date-time groups, originator designators, special instructions, and protocol symbols. *Note:* Several message heading elements may be combined into a message preamble. *Synonym* **message header.** *See also* **message element, message text, preamble.**

message indicator: *See* **start of message indicator.**

message mode: A mode of operating a communications network in which message switching is used rather than packet switching. *See also* **message switching, packet switching.**

message originator: The entity, such as the commander, authority, agency, agency head, person, or organization, that (a) is the authority under which a message is drafted, approved, and released for transmission and (b) is responsible for the actions of the message drafter and the message releasing officer. *See also* **message drafter, message releasing officer.**

message part: 1. In radio telephone system messages, one of the three major subdivisions of a message, namely the heading, the text, and the ending. *Note:* Each message part may have separate components, and each component may have certain message elements and contents. **2.** In cryptosystems, text that results from the separation of a long message into several shorter messages of different lengths as a transmission security measure. *Note:* Message parts usually are prepared in such a manner as to appear unrelated externally. Statements that identify the parts for assembly at reception are encrypted in the texts. *See also* **ending, message element, message heading, text.**

message preparation time: 1. From a communications system operator point of view, the time required for a message to be processed for entry into a communications system, including (a) for manually operated systems, the time during which data media, such as tape, are prepared and made ready for transmission, and (b) for automatic terminals, such as remote entry terminals, the time during which the message is read by the automatic message processing equipment. **2.** From a message originator point of view, the time that is required to write a message on a data medium, such as on a message form, or to type a message on a remote terminal, for acceptance by a communications system message center or by a communications system. *See also* **data medium, message center, message originator, remote terminal.**

message processing system: *See* **automatic message processing system.**

message processing time: The total time between (a) the instant the text of a message is completed by the message originator and (b) the instant the addressee commences to read the message. *Note:* The message processing time may be divided into (a) for the drafter, the time from the message release time to the message file time at a communications facility, (b) for the communications system operators, the time from the message file time at the originating communications facility to the available for delivery time at the receiving addressee communications facility, including the processing at both the originating and terminating communications facilities in addition to the total transmission time, and (c) for the destination addressee, the time from the available for delivery time at the receiving addressee communications facility to the time of receipt by the addressee. *See also* **available for delivery time, addressee, message drafter, message file time, message release time, transmission time.**

message processor: *See* **interface message processor.**

message protocol: *See* **Internet Control Message Protocol.**

message queue: *Synonym* **message backlog.**

message readdressal: The adding of a new addressee to a message by the message originator, by the initial addressee, or by another authority, without changing the addressees or text of the previously transmitted message. *Synonym* **double heading.** *See also* **addressee, message originator.**

message register lead: At a private branch exchange (PBX), one of the terminal equipment leads solely used for receiving direct current (dc) message register pulses from a central office (C.O.) so that message unit information usually recorded only at the C.O. is also recorded at the PBX.

message relay: To send, i.e., to retransmit, a message that has been received.

message release time: The time, i.e., the instant, at which an authorized message releasing officer authorizes a message for transmission. *See also* **message releasing officer.**

message releasing officer: The person who has the authority to authorize the dispatch of messages for, on behalf of, or in the name of the message originator, such as an organization, command, person, or agency. *See also* **message originator.**

message routing: In communications system operation, the forwarding of messages, including path selection, propagation medium selection, dispatch, transmission, propagation, and reception. *See also* **path, propagation medium.**

message sequencing: *See* **address message sequencing.**

message service: A switched service that (a) is furnished to the general public, (b) is not a private line service, and (c) includes exchange switched services and all switched services provided by interexchange carriers and completed by local telephone company access services, except when these services are provided by other means. *Synonym* **message toll service.** *See also* **carrier, private line, switch.**

message sink: *Synonym* **destination user.**

message source: The part of a communications system from which a message is considered to have originated. *See* **stationary message source.**

message switching: 1. In a central office (C.O.), i.e., in a switching center, (a) the handling of message traffic either from local users or from other switching centers and (b) the storing and forwarding of the message traffic through the office. **2.** The handling of message traffic through a switching center, either from local users or from other switching centers, in which (a) either (i) a connection is established between the message originating station and the message destination station or (ii) messages are forwarded, or stored and forwarded, through the center, (b) the message header determines the routing and the destination, and (c) there is no dial through. *See also* **circuit switching, message routing, packet switching, queue traffic, store and forward, store and forward switching center, switching center, switching system, traffic.**

message switching network: In telephone systems, a network of central offices (COs), i.e., switching centers, in which each CO (a) handles call and message traffic either from local users or from other switching centers and (b) may store and forward message traffic through the CO. *See also* **circuit switching, message routing, message switching, packet switching, queue traffic, store and forward, store and forward switching center, switching center, switching system, traffic.** *See also* **packet switching network.**

message switching node: In a message switching network, a node at which a central office (C.O.), i.e., a switching center, is located. *See also* **message switching, node, packet switching node.**

message system: *See* **electronic message system.**

message text: The message part that represents the thought or the idea that the message originator wants to send. *See also* **external header, message header.**

message time objective: The overall handling time (a) that communications systems personnel attempt to achieve for a message, and (b) that is measured between (i) the instant that a message is accepted for dispatch at a communications facility and (ii) the instant of available for delivery time or the instant of receipt by the addressee. *Note:* An example of a message time objective is 10 min (minutes) or less for a FLASH message. *See also* **available for delivery time.**

message toll service: *Synonym* **message service.**

message transfer part: The part of a common channel signaling system that transfers signal messages and performs associated functions, such as error control and signaling link security functions. *See also* **common channel signaling, communications security, error control, signal message.**

message unit: A unit of measure for charging calls, based on various parameters, such as the length of call, distance called, time of day, day of week, and holidays. *See also* **tariff.**

message verification: The determination of the validity of all or certain portions of a message when (a) the recipient believes that the text that was received is not what the message originator intended to send, (b) verification of the entire message requires repetition of the entire message, and (c) verification of portions of the message requires repetition of only the doubted portions. *See also* **message originator.**

message weighting: *See* **C message weighting.**

messaging service: In integrated services digital networks (ISDNs), an interactive telecommunications service that provides for information interchange among users by means of store and forward, electronic mail, or message handling functions.

metalanguage: A language that is used to specify, i.e., to describe, other languages as well as itself. *See also* **object language, target language.**

metal-clad switchgear: Electrical power and communications equipment that (a) is placed in an indoor or outdoor metal structure that contains switching equipment and other associated equipment, such as circuit breakers, measuring instruments, transformers, buses, and connectors that are insulated and placed in separate grounded metal containers, (b) usually has circuit breakers that (i) are equipped with self-coupling disconnecting devices and (ii) are arranged with a position-changing mechanism for moving the breaker vertically or horizontally in sequence from the connected position to a test or disconnected position, and (c) has interlocks to ensure the proper and safe sequence of operations during insertion or withdrawal of removable elements. *See also* **bus, circuit breaker, connector, ground, switching equipment, transformer.**

metallic circuit: An electrical circuit in which (a) metallic conductors are used and (b) the ground or earth does not form a part. *See also* **balanced line, circuit, four-wire circuit, metallic circuit current, two-wire circuit.**

metallic circuit current: An electric current that (a) flows in a metallic circuit, and (b) when the circuit consists of a pair of wires, flows in one direction in one of the wires and in the opposite direction in the other wire. *See also* **metallic circuit.**

metallic voltage: A voltage between metallic conductors, as opposed to a voltage between a metallic conductor and ground. *See also* **phantom circuit, virtual circuit.**

metal oxide semiconductor (MOS): A semiconductor that (a) consists of metal oxide, (b) may be doped to make positive-negative (p-n) junctions and positive-intrinsic-negative (PIN) junctions, and (c) is used to make diodes and transistors for conventional and integrated circuits. *See also* **room-temperature metal-oxide semiconductor.**

meteor burst communications: **1.** Communications by the propagation of radio signals reflected by ionized meteor trails. **2.** Radio communications that (a) operate at low data rates, (b) are effective at ranges from 800 km (kilometers) to 2000 km, and (c) make use of the 10^9 meteors that (i) enter the Earth's atmosphere each day, (ii) are trapped by the Earth's gravitational field, and (iii) have impact energies that produce ionization in the form of long thin paraboloids that are used in radio wave propagation. *See also* **ionosphere, ionospheric disturbance, ionospheric reflection, skip distance.**

meteorological aids service: A radio communications service used for meteorological, including hydrological, observations and explorations.

meteorological broadcast: *See* **routine meteorological broadcast, special meteorological broadcast.**

meteorological fixed station: *See* **hydrological and meteorological fixed station.**

meteorological land station: *See* **hydrological and meteorological land station.**

meteorological mobile station: *See* **hydrological and meteorological mobile station.**

meteorological organization: *See* **World Meteorological Organization.**

meteorological satellite Earth station: An Earth station that is in the meteorological satellite service. *See also* **Earth station, meteorological satellite service.**

meteorological satellite service: **1.** An Earth exploration satellite service for meteorological purposes. **2.** In satellite communications, a telemetering space service in which the results of meteorological observations are made by instruments that are on Earth satellites, and the results are transmitted to Earth stations by the space stations on the satellites. *See also* **Earth station, space station, space telemetering Earth station.**

meteorological satellite space station: A telemetering space station that (a) is in the meteorological satellite service and (b) is located on an Earth satellite. *See also* **Earth station, meteorological satellite service.**

meteorological warning: *See* **emergency meteorological warning.**

meter (m): **1.** The SI unit (Système International d'Unités) of length equal to 1,650,763.73 times the wavelength of the spectral line of orange light emitted when a gas consisting of the pure krypton isotope of mass number 86 is excited by an electrical discharge. *Note:* The meter was originally established by Napoleonic scientists as one ten-millionth of the distance between an Earth pole and the equator, i.e., 10^{-7} of that distance along the surface of the Earth. Later, the standard international meter was the distance between two fine lines engraved on a platinum bar held at the International Bureau of Weights and Measures near Paris, France. **2.** A device that measures a parameter value and indicates the result of the measurement. *See* **fiber optic blood flow meter, frequency meter, par meter, phase meter, photoconductive meter, photometer, photovoltaic meter.**

metering device: *See* **telephone traffic metering device.**

metering pulse: A pulse sent periodically over a telephone line to determine the duration of a call and the toll charges by counting the number of such pulses sent between (a) the instant that the call is completed, i.e., the call receiver answers, and (b) the instant that the call is finished, i.e., the instant that the call originator or the call receiver hangs up. *See also* **call originator, call receiver, completed call, finished call, hang up, pulse, toll.**

method: *See* **actuation method, alternative test method, broadcast communications method, fiber optic test method, frequency code method, Fresnel reflection method, image method, inductive assertion method, intercept communications method, maximal-ratio-square diversity combining method, Monte Carlo method, radiotelegraph operating method, receipt communications method, reference test method, relay communications method, refracted ray method, transverse scattering method.**

metric radar: A radar set that operates in the 1 m (meter) to 10 m wavelength band, i.e., that has a frequency that lies in the range that (a) extends from 30 MHz (megahertz) to 300 MHz, (b) has the letter designation VHF (very high frequency), and (c) has the numerical designation 8. *See also* **frequency, frequency spectrum designation, wavelength.**

metric system: A decimal system of weights and measures based on the meter, the kilogram, and the second, i.e., on the MKS system. *Note:* The latest version of this system uses the International System of Units (Système International d'Unités). *Refer to* **Table 3, Appendix B.**

metric wave: An electromagnetic wave with a wavelength that is between 1 m (meter) and 10 m, i.e., that has a frequency that lies in the range that (a) extends from 30 MHz (megahertz) to 300 MHz, (b) has the letter designation VHF (very high frequency), and (c) has the numerical designation 8. *See also* **frequency, frequency spectrum designation, wavelength.**

metropolitan area network (MAN): A network that usually (a) covers an area larger than a local area network (LAN) and smaller than a wide area network (WAN), (b) interconnects two or more LANs, (c) covers an entire metropolitan area, such as a large city and its suburbs, (d) operates at a higher speed than a WAN, (e) crosses administrative, i.e., political, boundaries, and (f) uses multiple access methods. *Note:* The Washington, DC, metropolitan area network (MAN) includes parts of the 301 (Maryland) and 703 (Virginia) area codes and all of the 202 (District of Columbia) area code. *See also* **communications, local area network, medium interface connector, medium interface point, wide area network.**

MF: medium frequency.

MFD: mode field diameter.

MFJ: Modification of Final Judgment.

MFSK: multiple frequency shift keying.

MHD: magnetohydrodynamics.

MHS: message handling system.

MHz: megahertz.

MIC: medium interface connector.

Michelson fiber optic sensor: A high resolution interferometric sensor in which (a) an electromagnetic wave, such as monochromatic light, is split, one-half reflected from a fixed mirror and back through the splitter to a photodetector, the other half passed directly through the splitter to a movable mirror that reflects it back to the splitter where it is reflected to the same photodetector, (b) the two waves can enhance or cancel each other, thereby modulating the irradiance, i.e., the light intensity, at the photodetector in accordance with an input signal in the form of a displacement of the movable mirror, such as might be produced by a sound wave or a pressure, strain, or temperature variation, (c) if an optical fiber is used, the ends of the fiber form the reflecting surfaces, (d) moving one end relative to the other produces the same effect as moving the mirror, and (e) displacements less than 10^{-13} m (meter) can be measured when narrow spectral width, i.e., monochromatic, laser sources are used.

MICR: magnetic ink character recognition.

microbend: In an optical waveguide, such as an optical fiber, a small deviation, dent, or recurring variation at intervals of a few millimeters, either in the core-cladding interface of the guide or in sudden deviation of the optical axis of the guide, resulting in sharp curvatures in the core-cladding interface surface or local displacements of the longitudinal axis. *Note 1:* Undulated core-cladding interface surfaces with large numbers of microbends per unit length are often created by the trauma that occurs during the cabling process in manufacturing fiber optic cables, in applying coatings, in winding cables on spools and reels, in passing cables around capstans, and in packaging, shipping, and installing cables. Microbends can be created in fiber optic cables under hydrostatic pressure, particularly when buffering is inadequate or strength members are improperly positioned. Simply bending a cable can introduce microbends because of stress differentials imposed on the fiber at the inside and outside radii of the bend. *Note 2:* The size of microbends may vary from a fraction of a micron to several microns in the direction of the optical axis or a fraction of a micron in the depth of a transverse dent. *Note 3:* A waviness of the optical axis, up to several millimeters in wavelength, is also considered to be microbends. *Note 4:* Lightwaves striking a microbend will escape from the core if the critical angle at the microbend becomes less than that required for total (internal) reflection. *Note 5:* Microbends can cause significant or excessive radiative loss and intermodal coupling. *Note 6:* Microbends are distinguished from microcracks. *See also* **critical radius, curvature loss, macrobending, microbending, microcrack, minimum bend radius.**

microbending: In an optical waveguide, the introduction of minute curves with small radii in the optical axis of the guide or the core-cladding interface surface, at which incidence angles of light rays propagating in the guide are, or tend to become, less than the critical angle so that some of the radiant power in the ray escapes from the waveguide core into the cladding and out of the guide. *Note:* When microbends in an optical fiber axis are large enough, they may be considered macrobends, such as will occur when the cable containing the fiber is bent around a curve in a raceway, duct, or trench, possibly resulting in a curvature loss. Care must be taken in installing fiber optic cables so that (a) microbending is not excessive, and (b) the macrobend radius is greater than both the minimum bend radius and the critical radius. *See also* **critical radius, curvature loss, macrobend, microbend, minimum bend radius.**

microbending loss: *Synonym* **microbend loss.**

microbend loss: In an optical fiber, the optical power loss caused by a microbend, i.e., a small bend, kink, or abrupt discontinuity in the direction of the optical axis of fibers, in the core-cladding interface surface, or both, usually caused by cabling, wrapping the fiber around drums and capstans, and winding the fiber on reels. *Note:* The microbend losses result from (a) the coupling of guided modes among themselves and among the radiation modes, and (b) rays escaping from the core into the cladding at a microbend because the critical angle requirement for total (internal) reflection at the core-cladding interface is violated. *Synonym* **microbending loss.** *See also* **core, cladding, critical angle, curvature loss, microbend, total internal reflection.** *Refer to* **Fig. L-9 (536).**

microbend sensor: A fiber optic sensor in which (a) an optical fiber is passed between toothed plates, i.e., between serrated plates, (b) a force or pressure is applied to the plates thereby squeezing the fiber, causing microbends in the fiber, (c) an electromagnetic wave, such as a lightwave, propagating in the fiber escapes at the bends, ejecting light into the cladding at each point at which the incidence angle at the core-cladding interface is greater than the critical angle for total (internal) reflection, (d) the attenuation of the main beam in the fiber increases, because of the loss of light into the cladding as pressure is applied to the plates, (e) the resulting light is detected by a photodetector that decreases its output as the pressure increases, and (f) the output signal of the photodetector is a function of the applied force or pressure.

microcircuit: 1. *Synonym* **integrated circuit. 2.** *See also* **chip, large-scale integrated circuit, wafer.**

microcode: A sequence of microinstructions that is fixed in storage, is not program addressable, and performs specific detailed processing functions.

microcomputer: A computer (a) in which the processing unit is a microprocessor and (b) that usually consists of a microprocessor, a storage unit, an input channel, and an output channel, all of which may be on one chip. *See also* **computer, microinstruction, microprogram.**

microcrack: In optical fibers, a minute crack or a partial break, that (a) usually is along the outer surface of the cladding, but may be deep enough to enter the core, (b) may become enlarged by exposure to adverse environmental conditions, such as moisture, bending, twisting, tensile stress, thermal gradients, vibration, shock, and corrosive atmospheres, and (c) will reflect, refract, and scatter incident light, usually causing the light to escape from the fiber. *Note 1:* Tensile testing of optical fibers under tension, i.e., tensile proofing, enlarges microcracks. If the cracks become sufficiently large, the fiber ruptures under the applied tension and must be spliced. However, proofing is done anyway, even if splicing is required, to reduce the possibility of a fiber rupturing (a) when it is not under tension or (b) when it is under small tensile stresses. *Note 2:* Microcracks are distinguished from microbends. *See also* **microbend.**

microcrack model: *See* **Griffith microcrack model.**

microcylindrical lens: An extremely small lens that (a) usually is made by an etching and crystal-growing process, (b) is attached to a light source or a photodetector on a monolithic basis, such as by attaching a cylindrical lens on an integral lens light-emitting diode, and (c) is used to assist in controlling the divergence of a beam in one dimension. *Synonym* **beam, cylindrical microlens, divergence, integral lens light-emitting diode, lens.**

microdot: A microform that (a) consists of an extremely small piece of film or print as the data medium, (b) is in the form of a period or a dot, such as a period at the end of a sentence or a dot on the letter "i," (c) contains microimages, such as printed graphics or pages of text, and (d) has no means for determining location other than unattached instructions. *See also* **data medium, microform, microimage.**

microelectronic circuit: An electronic circuit mounted on a single chip. *See also* **chip, large-scale integrated circuit, wafer.**

microfiche: A microform (a) that consists of film in the form of separate sheets as the data medium, (b) that contains microimages usually arranged in a grid pattern of rows and columns for location of microimages by means of Cartesian coordinates, and (c) in which each sheet usually has a title that can be read without magnification. *See also* **data medium, magnification, microimage, microform.**

microfilm: A microform that (a) consists of film as the data medium, (b) is in the form of a roll or a strip, (c) contains microimages in a sequential arrangement that is either in a cine oriented mode or in a comic strip oriented mode, rather than in rows and columns as on microfiche, and (d) has a leader with a title that can be read without magnification. *See also* **cine oriented mode, comic strip oriented mode, data medium, microform, microimage.** *See* **computer output microfilm.**

microfinishing film: A film that (a) consists of a substrate of dimensionally stable material, such as plastic, with a coating of carefully graded abrasive or polishing powder, i.e., particles, that has dimensions on the order of 1 μm (micron) or less, thus resembling extremely fine sandpaper, and (b) is used commercially to (i) shape and polish machined parts and (b) finish and polish the endfaces of fiber optic connectors. *See also* **hockey puck.**

microform: Any data medium that contains microimages. *Note:* Examples of microforms are microfiche, microfilm, microimage chips, and microdots. *See also* **data medium, microimage.**

microform reader: **1.** A device that has a blowback ratio large enough to permit microimages to be viewed and read by the unaided eye. **2.** A device that can scan and sense data directly from microimages and dispatch the data to another device. *Note:* Examples of microform readers are microfiche readers, microfilm readers, and fiber optic fiberscope enlargers. *See also* **blowback ratio, microimage.**

microform reader-copier: A device that can perform the functions of a microform reader and in addition produce a hard copy enlargement of selected microimages. *Synonym* **microform reader-printer.** *See also* **microform, microform reader.**

microform reader-printer: *Synonym* **microform reader-copier.**

micrographics: The branch of science and technology devoted to the methods and the techniques that are used for converting data to or from a microform or that are used to process data that are on a microform. *See also* **microform.**

microimage: An image that, under ordinary light, requires magnification to be interpreted by the eye.

microimage chip: A microform that consists of a small piece of material on which an image is recorded that requires the use of a high-powered optical system to render the image legible to the human eye. *Note:* An example of a microimage chip is a copy of the Torah, Bible, and Koran all together on one side of a 1-in.^{2} surface of photofilm. *See also* **image, microform, optical system.**

microinstruction: **1.** A basic, i.e., an elementary, machine instruction. **2.** An instruction that controls data flow and instruction execution sequencing in a processor at a more fundamental level than machine instructions. *Note:* Functions, such as individual machine instructions, may be implemented by microprograms consisting of a series of microinstructions that contain microcodes. *See also* **macroinstruction, microcode, microcomputer, microprogram.**

microlens: An extremely small lens that (a) usually is formed by epitaxial crystal growth and (b) is grown on or bonded to the device that forms part of an optical system. *Note:* Examples of microlenses are lenses bonded to the end of fiber optic pigtails (a) to launch light rays into optical fibers or focus rays on the active surface of photodetectors, (b) to focus light rays in fiber optic connectors, and (c) to focus light rays emitted by a light-emitting diode (LED). *See* **microcylindrical lens.** *See also* **lens, light-emitting diode, optical system, pigtail.**

micrometer (μm): *Synonym* **micron.**

microminiaturization: In electrical engineering, the technology of constructing electrical circuits and devices such that (a) a large number of circuit elements, such as transistors, diodes, resistors, capacitors, inductors, storage elements, and combinational circuits, are contained in extremely small packages, (b) equipment performance is improved over conventional circuitry, and (c) fabrication, manufacturing, and interfacing techniques are developed. *Note:* Examples of microminiaturization are the technology of large-scale integrated (LSI) circuits and optical integrated circuits (OICs).

micron (μm or μ): One millionth of a meter, i.e., 10^{-6} m (meters). *Note:* The SI (Système International d'Unites) unit is the micrometer. The micron, instead of the nanometer, which is widely used in the optics scientific community, is widely used by the fiber optic technology community to express the wavelengths of light and the geometry of optical fibers, such as the core, cladding, and mode field radii and diameters. This tends to simplify wavelength and geometric comparisons. The wavelengths of light used in fiber optics is on the order of 1 μm (micron), which makes the micron a convenient unit in fiber optics for expressing optical wavelengths and describing the geometry of optical fibers. *Synonym* **micrometer.**

microphone: 1. A device that converts variations in sound pressure, i.e., sound waves, into corresponding electric currents or voltages. **2.** A sound wave to electrical wave transducer. **3.** A device that converts analog sound signals into analog electrical signals.

microphonics: The generation, or the variation, of electrical currents or voltages in a circuit element or component caused by its own mechanical vibration, such as (a) a conductor vibrating in a magnetic field in such a manner that the magnetic flux linkages change has a voltage induced in it, and (b) a capacitor that changes its capacity as a result of vibration alters the current in the circuit in which it is connected with an applied voltage. *Note:* Microphonics may be undesirable when it introduces noise into a circuit and desirable when used to induce signals in circuits, such as in microphones, telephone transmitters, and vibration sensors.

microprocessor: 1. A central processing unit (CPU) mounted on a single chip, i.e., on an integrated circuit. **2.** An arithmetic, logic, or control unit constructed on a single large-scale integrated (LSI) chip. **3.** A processor that executes microcodes or microinstructions. **4.** A processor that is constructed of microelectronic circuits. *See also* **arithmetic-logic unit, central processing unit, computer, large-scale integrated chip, microinstruction.**

microprocessor card: A smart card that (a) contains a microprocessor, (b) provides one or more functions, such as cryptographic algorithm management and secret code management that may be used in such areas as banking and pay television, and (c) may be a user-specific application-oriented computer-programmed card. *See also* **smart card.**

microprogram: 1. A sequence of microinstructions that are in a special storage where they dynamically can be accessed to perform various functions. **2.** In communications, computer, data processing, and control systems, a sequence of microinstructions, i.e., a sequence of elementary instructions, that (a) usually corresponds to a single operation, (b) usually is held in main storage for immediate execution in response to a command, and (c) usually is initiated by the introduction of an instruction into a specific register of the computer. *Note:* Each microprogram usually corresponds to a single specific computer operation, such as (a) find a specified file, (b) place it in main storage, and (c) display the first page of the file on a screen, each operation controlled by a single microprogram. *See also* **computer instruction, microinstruction.**

microscope: An optical system, usually consisting of (a) one or more objective lenses that can be indexed for various magnifying powers, (b) one or more eyepieces, for one or more magnifying powers, that can be interchanged, (c) a means for illuminating the object, such as a pivoted mirror or light source, and (c) a means for holding and adjusting the position of the object or the objective along the optical axis to obtain a focused image on the retina of the eye of the observer. *See* **electron microscope, scanning electron microscope.** *See also* **eyepiece, focus, magnifier, magnifying power, objective, optical axis, optical system.**

microsecond: A unit of time equal to 10^{-6} s (seconds), i.e., one millionth of a second. *See also* **second.** *Refer to* **Table 4, Appendix B.**

microsoftware: Software designed and packaged, usually on diskettes, for installation and use on a personal computer (PC). Examples of microsoftware are WordPerfect, Volkswriter, Lotus 1-2-3, Framework, Harvard Total Project Manager, dBASE III, Dr. Halo, and MS Windows, all registered trademarks. *Note:* Microsoftware has popularly been called simply "software." *See also* **diskettes, personal computer, software.**

microwave (mw): 1. An electromagnetic wave that has a wavelength between 0.003 m (meter) and 3 m, thus covering frequencies from very high frequency (VHF) to extremely high frequency (EHF), i.e., from 0.1 GHz (gigahertz) to 100 GHz, or 10^8 Hz (hertz) to 10^{11} Hz. **2.** The portion of the electromagnetic spectrum corresponding to wavelengths between 0.003 m (meter) and 3 m. **3.** Pertaining to equipment and communications in the 0.003-m (meter) to 3-m band. *Note 1:* Microwave transmitters usually operate at wavelengths from 0.01 m to 0.3 m, i.e., from 1 GHz to 30 GHz, or 10^9 Hz to 30 × 10^9 Hz. *Note 2:* Microwave transmissions may easily be directionalized into a narrow beam, i.e., the beams have a high directive gain. *Note 3:* Microwave equipment is used in long-haul telephone networks that have (a) satellite links and (b) links from tower to tower using receivers, repeaters, and transmitters that relay signals between the towers, called "microwave relay towers." **4.** Pertaining to radio frequency wavelengths between 0.003 m (meter) and 3 m, that are so short they exhibit some of the properties of light and therefore may be easily concentrated into a beam. *Note:* By comparison with microwave, the optical spectrum extends from 0.3 μm (micron) to 30 μm, corresponding to frequencies from 0.01 PHz to 1 PHz, i.e., 10^{13} Hz to 10^{15} Hz. *See also* **frequency, frequency spectrum designation.** *Refer to* **Fig. R-12 (853).**

microwave amplification by stimulated emission of radiation (maser): 1. Amplification of microwaves by a device in which molecular interaction with electromagnetic radiation is used to achieve amplification of an input signal. *Note:* The electromagnetic interaction, rather than electric or magnetic interaction, in the maser results in very low noise. **2.** A device in which molecular interaction with electromagnetic radiation is used to achieve amplification of an input signal. *Note:* The electromagnetic interaction, rather than electric or magnetic interaction, in the amplifier, i.e., in the maser, results in very low noise. *See also* **laser.**

microwave auxiliary station: *See* **fixed microwave auxiliary station.**

microwave circulator: A lossless junction that is used to couple an antenna to a transmitter in one direction and to a receiver in the other direction. *Note:* An example of a microwave circulator is a diplexer. *See also* **coupling, diplexer, junction.**

microwave landing system (MLS): A precision runway approach landing system that enables all-weather landing of aircraft by riding a microwave beam for vertical and lateral control all the way to touchdown. *Note:* Microwave landing system (MLS) equipment is in three categories:

Category 1 equipment allows flight glideslope approach down to 200 feet above the runway. If the runway is in sight, the aircraft is permitted to land. If not, the pilot must abort and try again.

Category 2 equipment allows flight glideslope approach down to 100 feet above the runway.

Category 3 equipment allows flight glideslope approach from 50 feet above the runway with some visibility to a hands-off automatic landing with zero ceiling and visibility.

microwave relay tower: In long-haul telephone networks, such as those that have long links across mountains, deserts, and bodies of water, a high tower on which there is microwave equipment, including receivers, repeaters, and transmitters, that relays signals between the towers. *Note:* In regions of long periods of sunshine, the microwave equipment may be solar-powered. Batteries power the equipment during cloudy periods.

middle infrared: In the electromagnetic spectrum, pertaining to the region that lies between the longer-wavelength end of the near infrared region and the shorter-wavelength end of the far infrared region, i.e., from about 3 μm (microns) to 30 μm. *Note:* The near infrared is included in the near visible spectrum, but the middle infrared and far infrared regions are not included in the near visible region. The far infrared extends to the shorter-wavelength end of the radio wave region, which is also the longer-wavelength end of the optical spectrum, i.e., to about 100 μm. None of the infrared, i.e., the near, middle, or far, is included in the visible spectrum. *See also* **far infrared, near infrared.**

MIDS: multifunction information distribution system.

Mie scattering: In a propagation medium, such as in the glass of an optical fiber, scattering of an electromagnetic wave by particles with a size about equal to or greater than the wavelength. *Note:* Mie scattering is caused by the differences in refractive index between the particles and the propagation medium. *See also* **Rayleigh scattering.**

MIFR: Master International Frequency Register.

mile: *See* **geographical mile, nautical mile, radar mile, route mile, sheath mile, statute mile.**

mileage: In telecommunications, a specified distance used in tariff calculations, i.e., toll charge calculations. *Note:* Mileage is locally defined and often refers to airline distance rather than actual communications system route miles. *See also* **measured rate service, postalize, route mile, tariff.**

military call sign: A call sign that is used by and for units that are under military control. *See also* **call sign, international call sign.**

military common emergency frequency (MCEF): A frequency that (a) is used by all military units that are equipped to operate at that frequency or in the band in which that frequency lies, and (b) is also used internationally by survival craft stations and survival craft equipment.

military communications: The means that (a) are used by military commanders to exercise command and control over forces and areas that may be widely distributed, (b) are used to facilitate and expedite the transfer of instructions, orders, and information among persons and machines in military units, and (c) support administration, intelligence, operations, logistics, and related activities that are essential to the exercise of command and to mission accomplishment.

military communications system: A communications system that is established and used exclusively by military organizations for the conduct of military communications. *See* **strategic military communications system, tactical military communications system.** *See also* **military communications.**

military fixed communications network: A military communications network (a) that consists of fixed stations, (b) in which each station, usually a land station, can communicate with a number of stations in other areas by means of point-to-point nets, each net usually consisting of two stations that relay messages when stations are not in direct contact, (c) in which the net stations handle traffic into and out of local areas, and (d) in which the stations relay messages over the military fixed communications worldwide network that they form or are a part of. *See also* **civil fixed communications network, fixed station, land station, message, military communications, military communications system, traffic.**

military ship international call sign: An international call sign that is used (a) in military communications systems and (b) by and for ships that are under military control. *See also* **call sign, international call sign, military communications system.**

milliliter (ml): A unit of volume equal to (a) one thousandth of a liter, i.e., 10^{-3} l (liter), (b) a cubic centimeter, and (c) about 0.033 fl. oz. (fluid ounces). *See also* **liter.**

millimeter (mm): A unit of distance equal to one thousandth of a meter, i.e., 10^{-3} m (meters). *See also* **meter.**

millimicron: *Synonym* **nanometer.**

millimicrosecond: *Synonym* **nanosecond.**

millisecond: A unit of time equal to one thousandth of a second, i.e., 10^{-3} s (second). *See also* **second.**

milliwatt: A unit of power equal to one thousandth of a watt, i.e., 10^{-3} W (watt). *See* **digital milliwatt.** *See also* **watt.** *Refer to* **Table 4, Appendix B.**

MILNET: 1. military network. 2. *See also* **Internet.**

Milstar: An advanced communications satellite system that (a) provides secure communications among mobile tactical units, (b) is immune to electronic warfare jamming, and (c) does not compromise the location of using units. *See also* **jamming.**

minicomputer: A computer that (a) contains a microprocessor, (b) is smaller than a large-scale main frame computer, i.e., central processing unit, but not necessarily of small capacity, (c) usually executes programs in read-only memory (ROM) that also serves as main storage or internal memory, and (d) usually runs specific applications in communications, computer, data processing, and control systems. *Note:* Computers have been classified into (a) microcomputers that are constructed of microcircuits, (b) minicomputers that are small in size but not necessarily small in capacity, and (c) main frames, i.e., central processing units, computers that are large in size and have a high capacity. All of these are based on size. These distinctions are becoming less valid as computer capabilities increase with advances in technology. The microcomputer of today can be more powerful and more versatile than the minicomputer of yesterday, and even more powerful than the main frame computer of a few years ago. *See also* **central processing unit, computer, microcomputer.**

minimize: In communications systems, a condition in which usual message and telephone traffic is drastically reduced so that messages associated with an actual or simulated emergency are not delayed. *See also* **directionalization, line load control, precedence.**

minimum bend radius (MBR): As the radius of curvature of an optical waveguide, such as an optical fiber or fiber optic cable, is decreased from a straight line, the radius at which the waveguide, usually an optical fiber, will break. *Note 1:* Care must be taken in laying, installing, and operating fiber optic cables so as not to bend them at radii less than that of the minimum bend radius (MBR). After installation, i.e., during operation, the bend in cables must be less than the critical radius (CR) so that optical power is not lost at the bends. *Note 2:* The minimum bend radius (MBR) is of particular importance in the handling of optical fibers and fiber optic cables. The MBR varies with different fibers and cables. The manufacturer should specify the radius below which fibers and cables should not be bent, usually a value two or three times the MBR, for both the short term during installation and the long term during operation. Usually the MBR for long term is greater than the MBR for short term. *Note 3:* The MBR is a function of tensile stresses, such as occur when the optical fiber or cable is bent around a sheave or a capstan, particularly when the fiber or the cable is under tension. *Note 4:* If an MBR is not specified or known, a safe rule of thumb is to assume a long-term low-stress radius not less than 15 times the cable diameter. *See also* **critical radius, curvature loss.**

minimum cable cutoff wavelength: The minimum wavelength at which a cabled waveguide, such as a cabled optical fiber, will support a single mode. *Note:* For wavelengths shorter than the minimum cable cutoff wavelength, the waveguide will support more than one mode. The operating wavelength range, as determined by the transmitter nominal central wavelength and the transmitter spectral width, must be less than the maximum cutoff wavelength and greater than the minimum cutoff wavelength to ensure that the cable operates entirely in the single mode region. Usually, the highest value of cabled cutoff wavelength occurs in the shortest cable length. Worst case variations that are introduced by manufacturing, temperature, aging, and any other significant factors determine the minimum cable cutoff wavelength when the cable is operated under standard or extended operating conditions. *See also* **fiber cutoff wavelength, maximum cable cutoff wavelength.**

minimum discernible signal: In radar systems, the lower limit of power level that (a) can be useful to a radar receiver and (b) is determined by (i) the signal to noise ratio at the output of the receiving antenna and (ii) the signal characteristics and capability of the antenna and receiver system. *See also* **radar sensitivity.**

minimum dispersion slope: In a multimode optical fiber, the rate of change of material dispersion with respect to wavelength at the minimum dispersion wavelength. *Note:* In silica-based optical fibers, the minimum dispersion wavelength, and hence the minimum dispersion slope, occurs at about 1.3 μm (microns). *See also* **dispersion, material dispersion.**

minimum dispersion window: 1. In the attenuation versus wavelength characteristic of an optical fiber, the window, i.e., the trough, at which material dispersion is relatively small compared to the material dispersion at other wavelengths. *Note 1:* In silica-based optical fibers, the minimum dispersion window occurs at a wavelength of about 1.3 μm (microns). *Note 2:* The minimum dispersion window may be shifted toward the minimum loss window, i.e., about 1.55 μm (microns), by the addition of dopants during manufacture of the optical fiber. **2.** In single mode optical fibers, the window, or in the case of doubly or quadruply clad fibers, the windows, at which material and waveguide dispersion cancel one another, resulting in extremely low, near zero, dispersion and hence an extremely wide modulating bandwidth over a very narrow range of optical wavelengths.

minimum distance code: A binary code (a) in which the signal distance does not fall below a specified value, (b) that can be interpreted under noise conditions, and (c) in which a great many code words, i.e., binary combinations or binary patterns, are (i) unusable, (ii) invalid for transmission, and (iii) recognized as erroneous words when they are received. *Note:* An example of a minimum distance code is a code with a minimum signal distance of 3, constructed so that at least three bits must be changed to convert one valid code word to another. *See also* **binary code, code, code word, signal distance.**

minimum interval: In data transmission, the duration of the shortest significant interval when the durations of the significant intervals are not all multiples of the unit interval. *See also* **significant interval, unit interval.**

minimum loss window: In an optical fiber, the transmission window at which the attenuation coefficient is at or near the theoretical minimum, i.e., the quantum-limited, Rayleigh scattering, window. *Note 1:* If, on a single graph, the losses from various mechanisms are plotted as a function of wavelength, the minimum loss window occurs in the vicinity of the wavelength at which the Rayleigh-scattering attenuation curve and the infrared absorption curve intersect. *Note 2:* For silica-based optical fibers, the minimum loss window occurs at about 1.55 μm (microns).

minimum mean square error filtering: *Synonym* **minimum mean square error restoration.**

minimum mean square error (MMSE) restoration: Image restoration in which (a) an estimate, f_e, is made of the actual original light intensity distribution, f_i, of an image, (b) the error in the estimate is defined by the relation $e = f_o - f_i$, and (c) the total error of estimation is made a minimum over the entire ensemble of all possible images. *Synonyms* **minimum mean square error filtering, Wiener filtering.**

minimum picture interval: The minimum time between the television pictures that have been selected for encoding. *Note:* International Telegraph and Telephone Consultative Committee (CCITT) Recommendation

H.221 values for picture intervals are 1/29.97, 2/29.97, 3/29.97, and 4/29.97 s/p (seconds per picture).

minimum shift keying: *See* **Gaussian minimum shift keying.**

minipad keyboard: A small keyboard that (a) has a few keys compared with the standard keyboard, such as those used with computer, communications, and data processing stations and terminals, (b) usually is attached to the standard keyboard or is a part of it, (c) is used to add additional keying capabilities, and (d) in most personal computer keyboards is an integral part of the keyboard. *See also* **keyboard.**

minor relay station: In tape relay communications systems, a relay station that cannot provide an alternate route in either direction.

minuend: A number from which another number, the subtrahend, is subtracted to obtain a difference. *Note:* If the minuend is larger than the subtrahend, and both are positive numbers, the difference will be a positive number. If the subtrahend is larger than the minuend, and both are positive numbers, the difference will be a negative number.

minus lens: *Synonym* **diverging lens.**

minus ones complement: *See* **radix minus ones complement.**

MIP: medium interface point.

mirror: 1. A usually flat, optically ground, and polished surface on a reflecting material, or a transparent material that is coated to make it reflecting, used for reflecting light. *Note:* A beam-splitting mirror has a lightly deposited metallic coating that transmits a portion of the incident light and reflects the remainder. **2.** A smooth, highly polished plane or curved surface (a) that reflects light, (b) that usually has a thin coating of silver or aluminum on glass that constitutes the actual reflecting surface, and (c) that, when applied to the front face of the glass, is a front surface mirror and, when applied to the back, is a back surface mirror. *See* **back surface mirror, dichroic mirror, front surface mirror, Mangin mirror, off-axis paraboloidal mirror, partial mirror, triple mirror.**

mirroring: In display systems, the reflecting of all or part of a display element, display group, or display image in such a manner that (a) the element, group, or image appears to have been rotated 180° about a line in the plane of the display surface, (b) each point on the image, which now assumes the role of an object, appears to have been translated perpendicularly to a corresponding point on the image on the opposite side of the line as far from the line as it originally was, (c) the line may be considered as the "plane of the mirror" projected on the display surface, and (d) the element, group, or image will be reversed with respect to the direction perpendicular to the line. *See also* **display system, image, translation.**

mirror sensor: *See* **optical cavity mirror sensor.**

MIS: management information system, metal-insulator-semiconductor.

misalignment loss: *See* **angular misalignment loss.**

misdelivered bit: A bit received at a sink other than the one intended by the source. *See also* **bit, bit misdelivery probability, lost bit, sink, source.**

misdelivered block: A block received by a sink other than the one intended by the source. *See also* **block, block misdelivery probability, lost block, sink, source.**

misdelivery probability: *See* **bit misdelivery probability, block misdelivery probability.**

misdelivery ratio: *See* **bit misdelivery ratio, block misdelivery ratio.**

MISFET: metal insulator semiconductor gated field effect transistor.

mismatch loss: *See* **refractive index profile mismatch loss.**

mismatch of core radii loss: In an optical fiber, a loss of signal power introduced by a splice or a connector in which the radii of the cores of the two fibers that are joined or connected are not equal. *Note:* The mismatch of core radii loss usually is expressed in dB.

MISSI: multilevel information systems security initiative.

missile control: *See* **guided missile control.**

mission bit stream: 1. In a device at any point in a transmission system, the output rate, i.e., the input line rate less any overhead rate generated by the device. **2.** The total user information bits that are being passed through a communications channel. *Note:* Though framing, stuffing, control, and service bits are not mission bits, i.e., user information bits, they are a part of the bit stream. *See also* **bit stream, bit stream transmission, service bits, stuffing, transmission.**

mission traffic: The traffic for which a communications system was designed, constructed, and installed. *See also* **traffic.**

mistake: An action that is performed by a person and that produces an unintended result. *Note:* Examples of mistakes are the use of a wrong call sign in a message header, the misspelling of a word, or the use of a wrong circuit or channel to transmit a message. *See also* **error, failure, fault.**

MIVPO: modified inside vapor phase oxidation process.

mixed-base numeration system: A numeration system in which (a) positional notation is used, (b) each digit position is assigned a weight that is not the same for each position, and (c) there is not a fixed ratio between the weights of adjacent positions as in the conventional decimal and pure binary numeration systems. *Note:* If the bases of successive positions are *a, b,* and *c,* then the numeral 496 would represent $4a + 9b + 6c$. If the bases *a, b,* and *c* are 10, 5, and 1, respectively, the numeral 496 would be a mixed-base numeral equivalent to decimal 91. If the bases were 100, 10, and 1, the decimal numeral would be 496, but would not be a mixed-base numeral because there is a fixed ratio between the weights assigned to the adjacent digit positions. *See also* **numeration system, positional numeration system, pure binary numeration system, weight.**

mixed cell address: A cell address that consists of both an absolute cell address and a relative cell address. *See also* **absolute cell address, cell address, relative cell address.**

mixed gas laser: An ion laser in which a mixture of gases, such as argon and krypton, is used as the active laser medium, i.e., lasing medium. *See also* **active laser medium, ion laser, laser.**

mixed net: In radiotelephone communications systems, a net in which (a) one or more stations are equipped with ciphony or compatible cryptokey capabilities, (b) one or more other stations are not equipped with ciphony or compatible cryptokey capabilities, and (c) some stations may have had ciphony or compatible cryptokey capabilities but no longer have the capability to communicate in the cipher mode. *See also* **ciphony, cipher mode, cryptokey, net, radio telephone communications.**

mixer: *See* **fiber optic mixer, turnaround mixer.**

mix gate: *Synonym* **OR gate.**

mixing: *See* **optical mixing.** *See also* **heterodyning.**

mixing box: *See* **optical mixing box.**

mixing rod: *See* **optical mixing rod.**

MLP: multilevel precedence.

MLPP: multilevel precedence and preemption.

MLS: microwave landing system.

mm: millimeter.

MMI: multimedia interface.

MM patch bay: A patching facility designed for patching and monitoring of digital data circuits that operate at data signaling rates (DSRs) exceeding 3 Mbps (megabits per second). *See also* **data signaling rate, D patch bay, M patch bay, patch bay.**

MMSE: minimum mean square error.

MMSE filtering: minimum mean square error filtering.

MNCS: master net control station.

mnemonic: 1. Pertaining to the memory or to the quality of being easily remembered. *Note:* For example, pertaining to the use of symbols or combinations of symbols that are specifically chosen to aid the memory of a person when recall becomes necessary. Examples of mnemonic codes include the use of "tiger" to help one remember that **t**rees, **g**rass, and **r**agweed pollinate in that order each season and the use of "GI" for "government issued," which has been extended to pertain to U.S. soldiers. **2.** A group of letters, such as a word, abbreviation, acronym, or title that (a) is specifically designed to serve as an aid to its recall as well as serve to identify an entity and (b) makes use of prior training or experience to aid in its recall. *Note 1:* Examples of mnemonics are pulse train, bit stream, DIV, MPY, SUB, ADD. *Note 2:* Mnemonics in one language usually are not mnemonics in other languages. Thus, SUB, a mnemonic for "to subtract" in English would have to be written as ABN for *abnehmen*, which means "to subtract" in German.

mnemonic code: A code that can be remembered comparatively easily and that aids its user in recalling the information it represents. *Note:* Examples of mnemonic codes are MPY for multiply, NOTAM for Notice to Airmen, and ROY G. BIV for remembering the sequence of colors as they appear in the dispersion of sunlight by a prism or as they appear in a rainbow, i.e., **r**ed, **o**range, **y**ellow, **g**reen, **b**lue, **i**ndigo, **v**iolet. A Mnemonic code in one natural language usually is not a mnemonic code in another language. Mnemonic codes are widely used in computer programming and communications system operations to specify instructions. The systems can be designed to be responsive to arbitrary sets of symbols, but the mnemonic codes are used to help programmers and operators remember the

German Instruction	Mnemonic Code	Schlüssel Code	Mnemonic Code	English Instruction
Wagenrücklauf	WRL	0 0 1	CR	Carriage Return
Rechtsverschiebung	RVS	0 1 0	RS	Right Shift
Linksverschiebung	LVS	0 1 1	LS	Left Shift
Druckbefehl	DBF	1 0 0	PR	Print
Lochkartenlesen	LKL	1 0 1	RC	Read Cards
Sprungbefehl	SBF	1 1 0	JP	Jump
Weiterschalten	WSN	1 1 1	CF	Count Forward

Fig. M-1. The usefulness of a **mnemonic code** depends on the language that is familiar to the user.

codes that represent instructions and commands, such as upon entering the letter group SUB, a computer may be designed to execute a subtraction instruction. *Refer to* **Fig. M-1.**

mo: magnetooptic, magneto-optic.

mobile: The characteristic of an item such that (a) it is permanently mounted on a carrier, such as a vehicle, aircraft, or ship, (b) if a source of power is required for operation, it usually has a self-contained power source, and (c) it usually can be operated while the carrier is in motion. *See also* **portable, transportable.**

mobile communications: *See* **global system for mobile communications.**

mobile communications equipment: Communications equipment that (a) is installed in or on a vehicle, (b) can be operated while the vehicle is in motion, and (c) is considered to be transportable rather than mobile if the equipment (i) is mounted, loaded, or installed in or on a vehicle and (ii) can be operated only when the vehicle is stationary. *Note:* A mobile communications unit can transmit and receive messages and signals while it is moving. *See also* **transportability, transportable.**

mobile Earth station (MES): An Earth station in the mobile satellite service intended to be used while in motion or during halts at unspecified points. *See also* **Earth station.**

mobile phone service: *See* **Advanced Mobile Phone Service.**

mobile phone system standard: *See* **Advanced Mobile Phone System standard.**

mobile radio base station: *See* **airport land mobile radio base station.**

mobile radio station: *See* **airport land mobile radio station.**

mobile radiotelephone frequency: *See* **maritime mobile radiotelephone frequency.**

mobile satellite communications (MSATCOM): Communications systems for mobile stations that (a) use the uplinks and downlinks of satellite systems, (b) allow global point-to-point communications, and (c) allow communications while the station is in motion.

mobile satellite service (MSS): A radiocommunications service between (a) mobile Earth stations and one or more space stations, (b) space stations used by this service, or (c) mobile Earth stations by means of one or more space stations. *Note:* Mobile satellite service may also include feeder links necessary for its operation. *See also* **Earth station.**

mobile satellite (off-route) service: *See* **aeronautical mobile satellite (off-route) service.**

mobile satellite (route) service: *See* **aeronautical mobile satellite (route) service.**

mobile service (MS): A radiocommunications service between mobile and land stations, or between mobile stations. *See* **aeronautical mobile service, cellular mobile service, land mobile service, maritime mobile service.** *See also* **compatibility, interoperability, portability, transportability.**

mobile station (MS): A station in the mobile service intended to be used while in motion or during halts at unspecified points. *See* **aeronautical multicom mobile station, aeronautical telemetering mobile station, aeronautical utility mobile station, flight telemetering mobile station, hydrological and meteorological mobile station, radiolocation mobile station, radionavigation mobile station, radiopositioning mobile station, telemetering mobile station.** *See also* **compatibility, interoperability, portability, transportability.**

Mobile Subscriber Equipment (MSE) System: A mobile robust ruggedized communications system that can be rapidly deployed for use under adverse conditions, such as for battlefield communications in U.S. Army-wide tactical situations and for coordinating emergency relief efforts by fire, police, and hospitals in hurricane disaster areas.

mobile telecommunications system: *See* **Universal Mobile Telecommunications System.**

mobile unit routing indicator: A message routing indicator that is assigned to a mobile station, such as a ship station, aircraft station, or mobile land station. *See also* **indicator, message routing, mobile station.**

mobility: *See* **equipment mobility.**

modal dispersion: 1. In the propagation of an electromagnetic wave or a pulse in a waveguide, the changes introduced by the guide in the relative magnitudes of the frequency components of the wave or the pulse. *Note:* The guide is capable of supporting or introducing only a fixed number of frequencies, depending on its geometry and material parameters, such as permeability, permittivity, and conductivity. **2.** In fiber optics, the spreading of a light pulse in an optical fiber caused by the different paths taken by the various modes. *Synonyms* **intermodal dispersion, mode dispersion.** *See also* **multimode distortion.**

modal distortion: *Synonym* **multimode distortion.**

modal distribution: 1. In an optical waveguide, such as an optical fiber, operating at a given peak wavelength, (a) the number of modes supported by the waveguide and (b) the distribution of propagation time differences among the various modes. **2.** In an optical waveguide operating at multiple peak spectral wavelengths simultaneously, such as occurs in wavelength division multiplexing (WDM), the distribution of the separations of wavelengths among the modes being supported by the waveguide. *See also* **fiber optics, frequency, mode, mode volume, multimode optical fiber, single mode optical fiber.**

modal loss: 1. In a mode of an electromagnetic wave propagating in an optical waveguide, especially in an open waveguide, such as an optical fiber, energy loss caused by certain anomalies inside and outside the waveguide, such as obstacles outside the waveguide, abrupt changes in direction of the waveguide, or discontinuities in the structure of the waveguide. **2.** A loss of power from a mode in an electromagnetic wave propagating in a waveguide that occurs when radiant power is transferred to high order modes and leaky modes that enter the cladding where they may be absorbed, scattered, or radiated away from the waveguide. *See also* **fiber optics, loss, curvature loss, microbend loss, mode, speckle pattern.**

modal noise: In an optical waveguide, such as an optical fiber, noise generated by a combination of (a) mode-dependent optical losses, (b) fluctuations in the distribution of optical energy among the guided modes, (c) fluctuations in the relative phases of the guided modes, and (d) the effects of differential mode attenuation. *Synonym* **speckle noise.** *See also* **mode, mode partition noise, speckle pattern.**

modal partition noise: *Synonym* **mode partition noise.**

modal power distribution length: 1. *Synonym* **equilibrium length. 2.** *See* **nonequilibrium modal power distribution length.**

modal power distribution: In an electromagnetic wave propagating in a waveguide, the distribution of power among the various modes. *Note:* In a single mode fiber, all the propagating optical power is in the one mode. *See* **equilibrium length, modal power distribution length, nonequilibrium modal power distribution.**

mode: 1. In a waveguide or a cavity, one of the various possible patterns of propagating or standing electromagnetic fields. *Note 1:* The field pattern of a mode depends on the wavelength, refractive index, and cavity or waveguide geometry. In guided electromagnetic waves, each mode is a particular condition or arrangement of the waves in the waveguide. Different propagating wavelengths correspond to different propagating modes. *Note 2:* Each mode is characterized by a different frequency, polarization, electric field strength, or magnetic field strength. *Note 3:* In geometric optics, each path taken by a ray in a waveguide corresponds to a different mode. *Note 4:* Optical fibers can support many modes, the number being given by the relation $N = 2\pi^2 a^2(n_1 - n_2{}^2)/\lambda^2$ where N is the number of modes, a is one-half the core diameter, n_1 and n_2 are the refractive indices of the core and cladding respectively, and λ is the free space wavelength of the lightwave launched into the fiber, i.e., not the wavelength inside the fiber. Single mode fiber diameters range from 2 µm (microns) to 11 µm, but the number of modes that a given fiber can support depends on the core radius, the wavelength, and the numerical aperture, which depends on core and cladding refractive indices. Thus, a fiber operating in single mode can be made to operate in multimode if the wavelength is sufficiently reduced. **2.** Any electromagnetic field distribution that satisfies Maxwell's equations and the applicable boundary conditions, such as the boundary conditions imposed by a waveguide. **3.** In data communications, a protocol used to transfer data from switch to switch or from switch to terminal. *Note:* In military switches, such as in the Automatic Digital Network (AUTODIN), modes I, II, III, IV, and V are used. **4.** In statistics, the highest peak in a probability density function. **5.** In identification friend or foe (IFF), the number or the letter that refers to the specific pulse spacing of the signals transmitted by an interrogator. *See* **access mode, add mode, adjust text mode, Asynchronous Transfer Mode, bound mode, cine oriented mode, cipher mode, cladding guided mode, cladding mode, coasting mode, comic strip oriented mode, compute mode, controlled security mode, conversational mode, coupled mode, cutoff mode, data mode, dedicated mode, degenerate waveguide mode, deterministic transfer mode, dial mode, direct mode, dominant mode, edit mode, evanescent mode, fundamental mode, high order mode, hold mode, hybrid mode, indirect mode, initial condition mode, interactive mode, landscape mode, laser basic mode, leaky mode, linearly polarized mode, listening mode, lowest order transverse mode, low order mode, message mode, monitor mode, nontransparent mode, packet mode, packet transfer mode, passive mode, plain mode, propagation mode, radiation mode, ready mode, satellite mode, set mode, static test mode, synchronous transfer mode, transfer mode, transverse electric mode, transverse electromagnetic mode, transverse magnetic mode, tracking mode, transparent mode, unbound mode.** *Refer to* **Figs. E-4 (314), G-2 (399), P-17 (767), R-10 (836), S-18 (951).**

mode attenuation: *See* **differential mode attenuation.**

mode conversion: In a guided electromagnetic wave, such as a lightwave, the transfer of some or all of the electromagnetic power from one mode to another mode. *Note:* Mode conversion (a) can be intentional, such as when inducing high order cladding modes for transferring energy to another fiber or for microbend sensing, or (b) can be unintentional, such as when transferring energy to high order modes that escape from the guide and are radiated away, causing signal attenuation.

mode coupling: In an electromagnetic wave propagating in a waveguide, such as an optical fiber, the exchange of power among modes. *Note:* In an optical fiber, mode coupling reaches statistical equilibrium after the equilibrium length has been traversed. *See also* **coupling, equilibrium length, mode, mode scrambler, multimode optical fiber**.

mode delay: *See* **differential mode delay.**

mode dispersion: *Deprecated synonym for* **multimode distortion.**

mode distortion: *Synonym* **multimode distortion.**

mode field diameter (MFD): 1. A measure of distribution of the irradiance, i.e., the optical power intensity, across the endface of a single mode optical fiber. **2.** For Gaussian statistical distributions, i.e., bell curve distributions, of power and energy among the modes of an electromagnetic wave propagating in a single mode optical fiber, the diameter at which the electric and magnetic field strengths are reduced to $1/e$ of their maximum values, which is equivalent to the diameter at which the radiant power is reduced to $1/e^2$ of the maximum power because the power is proportional to the square of the electric field strength or the magnetic field strength. *Note:* The mode field diameter can be seen to a limited extent when the light exiting an optical fiber is projected on a screen, i.e., launched from the fiber and incident upon a screen.

mode filter: A device used to select, reject, or attenuate a certain mode or modes in an electromagnetic wave. *See* **core mode filter.** *See also* **dichroic filter, fiber optics, filter, mode.**

mode hopping: The transfer of radiant power from one mode to another. *Note:* Mode hopping occurs in lasers. *Synonym* **mode jumping.**

mode indicator: In communications, computer, data processing, and control systems, a word or a symbol that (a) is displayed on a control panel, (b) specifies the particular state or mode that the system is in, and (c) informs the operator which group of commands or menu items are currently applicable. *Note:* In Lotus 1-2-3, the mode indicators are displayed in the upper right corner of the screen, whereas other software personal computer programs may place them in other corners of the screen. In Lotus 1-2-3, the mode indicators are:

Indicator	**System Mode**
READY	Waiting for a command, value, or label
VALUE	Entering a number or a formula
LABEL	Entering a label
EDIT	Editing and entry
POINT	Pointing to a Worksheet (spreadsheet) entry
MENU	Selecting a menu item
HELP	Residing in the help facility
ERROR	Waiting for actuation of the ESC or ENTER key
WAIT	Calculating and cannot process commands
FIND	Performing a /Data Query Find operation

mode interference: *See* **common mode interference.**

mode jumping: *Synonym* **mode hopping.**

model: A representation of a real world process, device, or concept. *See* **debugging model, error model, Griffith microcrack model, mathematical model, reliability model.**

mode launching: *See* **single mode launching.**

mode link control: *See* **basic mode link control.**

modem: 1. A device that modulates and demodulates signals. *Note 1:* Modems are primarily used for converting digital signals into quasi-analog signals for transmission over analog communications channels and for reconverting the quasi-analog signals into digital signals. *Note 2:* Many additional functions may be added to a modem to provide for user service and control features. **2.** Signal conversion equipment that (a) includes both a modulator and a demodulator, (b) is a combination of equipment that (i) changes the type of modulation, (ii) modulates an incoming signal, (iii) demodulates an incoming signal, (iv) converts digital signals into quasi-analog signals usually for transmission, or (v) converts quasi-analog signals into digital signals usually for further processing, (c) usually is part of a terminal installation, (d) is connected to a data channel, (e) is used for the transmission and the reception of data at either end or both ends of a circuit or a channel, (f) may include clocks and signal generators, (g) usually does not contain error control equipment, (h) may be considered as a signal processor, and (i) may perform many additional functions to provide user services or communications system control features. *Synonyms* **modulator-demodulator, signal conversion equipment.** *See* **fiber optic modem, integrated modem, narrowband modem, stand-alone modem, wideband modem.** *See also* **acoustic coupler, data circuit-terminating equipment, input/output device, peripheral equipment, quasi-analog signal.** *Refer to* **Fig. M-2.** *Refer also to* **Figs. B-9 (96), C-7 (157).**

mode mixer: *Synonym* **mode scrambler.**

modem patch: 1. To electrically connect paths of a circuit by using back-to-back modems. **2.** Two modems connected back to back used to connect paths of a circuit. **3.** A patch that electrically connects two points, two paths, or two conductors by means of two modems that are connected back to back. *See also* **modem, patch.**

mode operation: *See* **packet mode operation.**

mode optical fiber: *See* **multimode optical fiber.**

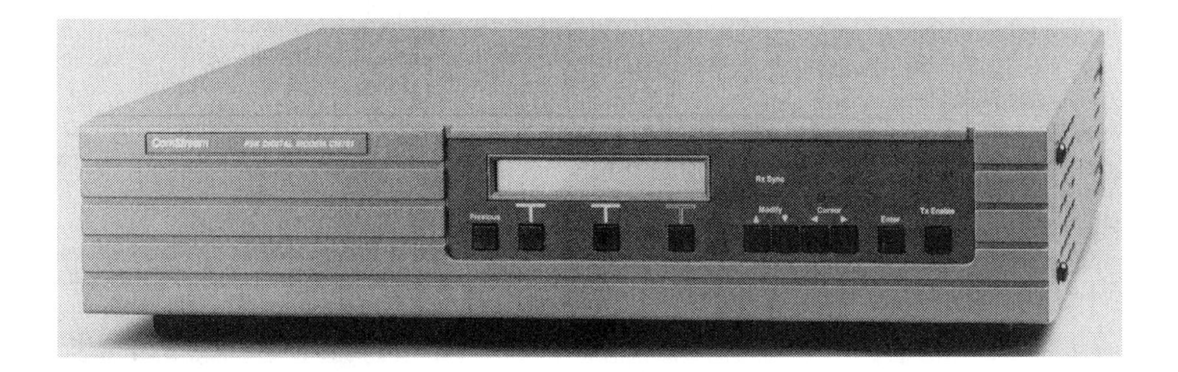

Fig. M-2. A satellite communications single-channel-per-carrier (SCPC) **modem.** (Courtesy ComStream Company.)

mode parameter: *Synonym* **normalized frequency.**

mode partition noise: In an optical link, phase jitter caused by the combined effects of mode hopping in the optical source and intramodal distortion. *Note:* Mode hopping causes random wavelength changes that affect the group velocity of the mode. The variation of group velocity creates the mode partition noise. *Synonym* **modal partition noise.** *See also* **modal noise.**

mode scrambler: 1. A device that (a) induces mode coupling in a waveguide and (b) is frequently used to produce a modal distribution that (i) is independent of source characteristics or (ii) meets specific requirements. **2.** In fiber optic systems, a device for inducing mode coupling in an optical fiber. **3.** In fiber optic systems, a device composed of one or more optical fibers in which strong mode coupling occurs. *Synonym* **mode mixer.** *See also* **coupling, fiber optics, mode, mode coupling, multimode optical fiber.**

mode simulator: *See* **equilibrium mode simulator.**

Mode S radar: A secondary surveillance radar that (a) is used as a primary air/ground digital message communicator to transmit Data Link information between aircraft and ground stations and (b) is deployed at airports and en route sites along air lanes. *See also* **Data Link.**

mode stripper: *See* **cladding mode stripper, fiber optic mode stripper.**

mode terminal: *See* **packet mode terminal.**

mode volume: The number of bound modes that an electromagnetic waveguide is capable of supporting. *Note:* For optical fibers with a step index profile, the mode volume, M_{si}, is given by the relation $M_{si} = V^2/2$ where M_{si} is the mode volume and V is the normalized frequency. For optical fibers with a power law refractive index profile, the mode volume, M_{gi}, is given by the relation $M_{gi} = (V^2/2)(g/(g + 2)$ where M_{gi} is the mode volume, V is the normalized frequency, and g is the refractive index profile parameter. *Note:* For single mode optical fibers, the normalized frequency, V, is given by the relation $V \leq 2.405$. For graded index multimode optical fibers, $V > 5$. *See* **effective mode volume.** *See also* **normalized frequency.**

Modification of Final Judgment (MFJ): The 1982 antitrust suit settlement agreement, i.e., "Consent Decree," entered into by the United States Department of Justice and the American Telephone and Telegraph Company (AT&T) that, after modification and upon approval of the United States District Court for the District of Columbia, required the divestiture of the Bell Operating Companies from AT&T.

modified alternate mark inversion (MAMI) code: A T-carrier alternate mark inversion (AMI) line code in which bipolar violations may be deliberately inserted to maintain system synchronization, depending on signal patterns. *Note 1:* The clock rate of an incoming T-carrier signal is extracted from its bipolar line code. The T-carrier was originally developed for voice applications. When voice signals are digitized for transmission via T-carrier, synchronization may readily be maintained because of the nature of the digitized signals. However, when used for the transmission of digital data, the conventional alternate mark inversion (AMI) line code may not have sufficient marks, i.e., sufficient 1s, to permit recovery of the incoming clock, resulting in a loss of synchronization. This occurs when there are too many consecutive zeros in the user data being transported. To prevent loss of synchronization when a long string of zeros is present in the user data, deliberate bipolar violations are inserted into the line code in order to create a sufficient number of marks to maintain synchronization. The receive terminal equipment recognizes the bipolar violations and removes from the user data the marks attributable to the bipolar violations. *Note 2:* The exact pattern of bipolar violations that is inserted and transmitted in any given case depends on the line rate and the polarity of the last valid mark in the user data prior to the unacceptably long string of zeros. *Note 3:* The number of consecutive zeros that can be tolerated in user data depends on the data rate, i.e., the level of the line code in the T-carrier hierarchy. The North American T1 line code (1.544 Mbps (megabits per second)) does not use bipolar violations. The European T1 line code (2.048 Mbps) may use bipolar violations when eight or more consecutive zeros are present. This line code is called the "bipolar with eight-zero substitution (B8Zs) code." In all levels of the European T-carrier hierarchy, the patterns of bipolar violations that are used differ from those used in the North American hierarchy. At the North American T2 rate (6.312 Nbps), bipolar violations are inserted if six or more consecutive zeros occur. This line code is called the "bipolar with six-zero substitution (B6Zs) code." At the North American T3 rate (44.736 Mbps), bipolar violations are inserted if three or more consecutive zeros occur. This line code is called the "bipolar with three-zero substitution (B3ZS) code."

modified alternate mark inversion (MAMI) signal: A signal that (a) does not meet the definition of an alternate mark inversion (AMI) signal and (b) deviates from an AMI signal in accordance with a set of rules. *See also* **alternate mark inversion signal, signal.**

modified chemical vapor deposition (MCVD) process: A modified inside vapor phase oxidation (IVPO) process for production of optical fibers in which (a) a burner travels along the glass tube optical fiber preform, (b) soot particles are created inside the tubing rather than in the burner flame as in the outside vapor phase oxidation (OVPO) process, (c) chemical reactants, such as oxygen, dopants, and silicon tetrachloride, are caused to flow through the rotating glass tube at a pressure of about 1 atmosphere, gauge, (d) high temperature causes the formation of oxides, in the form of a soot and a glassy deposit, on the inside surface of the tube, (e) the deposit determines the refractive index profile of the glass and forms the core, and (f) the tube is drawn into a solid fiber by an optical fiber pulling machine. *Synonym* **modified inside vapor phase oxidation process.** *See also* **optical fiber pulling machine, chemical vapor phase oxidation process.**

modified clear message: A message that contains combinations of clear text and encrypted text.

modified inside vapor phase oxidation process (MIVPO): *Synonym* **modified chemical vapor deposition process.**

modified nonreturn to zero (MNRZ) level code: A variation of the basic nonreturn to zero code in which the binary states corresponding to 0 and 1 are represented as voltage transitions, rather than as voltage levels as in the nonreturn to zero level codes, such as the nonreturn to zero mark (NRZM) code and the nonreturn to zero space (NRZS) code. *Note:* The basic nonreturn to zero (NRZ) code should be called the "nonreturn to zero level (NRZL) code" because the binary states corresponding to 0 and 1 are signal levels rather than signal transitions.

modified reflected binary code: A code that (a) is formed by adding an extra even parity bit to the rightmost position of each reflected binary word, (b) is used because of its error detection properties in some arithmetic operations, and (c) requires two to three times as many combinational circuits as the conventional binary adder. *Synonym* **Lucal code.** *See also* **combinational circuit, Gray code, reflected code.**

modular: 1. Pertaining to the design concept in which interchangeable units are used to create a functional end product. **2.** Pertaining to the composition of a system by means of assembled and interconnected modules, each of which usually can be individually replaced.

modular decomposition: The designing of a system by (a) subdividing it into separate components or modules, (b) designing each component or each module separately with due consideration being given to the interfaces between them, and (c) interconnecting the components or the modules. *See also* **component, hierarchical decomposition, interface, module.**

modularity: The extent to which a system or a functional unit is composed of discrete interconnected hardware or interrelated software components or modules. *See also* **component, functional unit, module.**

modular jack: A device that conforms to the *Code of Federal Regulations,* Title 47, part 68, which defines the size and the configuration of all units that are permitted for connection to the public exchange facilities.

modular programming: The developing of computer programs or operating systems as a collection of components or modules. *See also* **component, computer program, operating system, module.**

modular spread spectrum code sequence generator: In spread spectrum systems, a spread spectrum code sequence generator in which each flip-flop in the code sequence register is followed by a modulo-two adder. *See also* **code sequence register, flip-flop, spread spectrum code sequence generator.**

modulate: 1. To vary any property of a wave for the purpose of transferring information. **2.** To vary a parameter, such as amplitude, phase, frequency, pulse position, or pulse duration, of a wave usually for the purpose of transferring information. *Note 1:* A wave can be modulated by superimposing another wave or by varying a physical parameter to which the wave is sensitive, such as by varying attenuation in an optical fiber or controlling the output of a laser by varying the driving voltage. *Note 2:* Modulating in an uncontrolled or random manner creates noise or interference. **3.** To vary a characteristic of a carrier in accordance with an information-bearing signal. *See also* **carrier, demodulate, demodulation, modulation.**

modulated jamming: *See* **slow sweep noise modulated jamming.**

modulated signal: *See* **pulsed frequency modulated signal.**

modulating signal: The information-bearing signal that is used to modulate a carrier. *See also* **carrier, modulation.**

modulation: 1. The process, or the result of the process, of varying a characteristic of a carrier in accordance with an information-bearing signal. **2.** A controlled variation of any property of a wave for the purpose of transferring information. **3.** The controlled variation of a parameter, such as amplitude, phase, frequency, pulse position, or pulse duration, of a wave usually for the purpose of transferring information. *Note 1:* Modulation can be accomplished by superimposing another wave or by varying a physical parameter to which the wave is sensitive, such as by varying attenuation in an optical fiber or controlling the output of a laser by varying the driving voltage. *Note 2:* Uncontrolled or random modulation is considered to be noise or interference. *Note 3:* Examples of modulation are (a) variation of the amplitude or the frequency of a carrier in accordance with an analog signal, such as a voice or video signal, (b) variation of the irradiance, i.e., the intensity, of a beam from a light source, such as a laser, in accordance with an intelligence-bearing electronic signal applied to the source, and (c) variation of the radiant power at a point in a waveguide, such as an optical fiber, in accordance with a physical variable being sensed or measured, such as in a microbend or Sagnac sensor. *See* **absorption modulation, absorptive modulation, adaptive differential pulse code modulation, amplitude modulation, analog intensity modulation, analog modulation, angle modulation, balanced amplitude modulation, balanced modulation, binary modulation, conditioned diphase modulation, constant current modulation, continuously variable slope delta modulation, continuous phase modulation, cross modulation, delay modulation, delta modulation, delta sigma modulation, demodulation, differential modulation, differential pulse code modulation, differential trellis coded modulation, digital frequency modulation, digital modulation, digital phase modulation, direct sequence modulation, double modulation, electrooptic phase modulation, external optical modulation, final modulation, fixed reference modulation, frequency code modulation, frequency hopper direct sequence modulation, frequency modulation, full modulation, incremental phase modulation, intensity modulation, intermodulation, isochronous modulation, jammer modulation, jamming modulation, low level modulation, mechanically induced modulation, multilevel modulation, nonreturn to zero (change) modulation, nonreturn to zero (change on ones) modulation, percentage modulation, phase modulation, polarization modulation, pulse amplitude modulation, pulse code modulation, pulse duration modulation, pulse interval modulation, pulse modulation, pulse position modulation, pulse time modulation, quadrature amplitude modulation, quadrature modulation, reference modulation, start-stop modulation, subband adaptive differential pulse code modulation, suppressed clock pulse duration modulation, synchronous demodulation, telegraph modulation, trellis coded modulation, wavelength modulation, 100% modulation.** *See also* **carrier, demodulate, demodulation, modulate.** *Refer to* **Fig. M-3.** *Refer also to* **Figs. N-17 (635), N-18 (636), P-8 (737).**

modulation deception: *See* **radar scan rate modulation deception.**

modulation-demodulation linearity: A measure of the presence or the absence of harmonics of the frequency of the modulating signal in the signal that results from the modulating or demodulating process. *See also* **harmonic, modulating signal.**

modulation factor: **1.** In amplitude modulation, the ratio of (a) the peak variation actually used, i.e., the max-

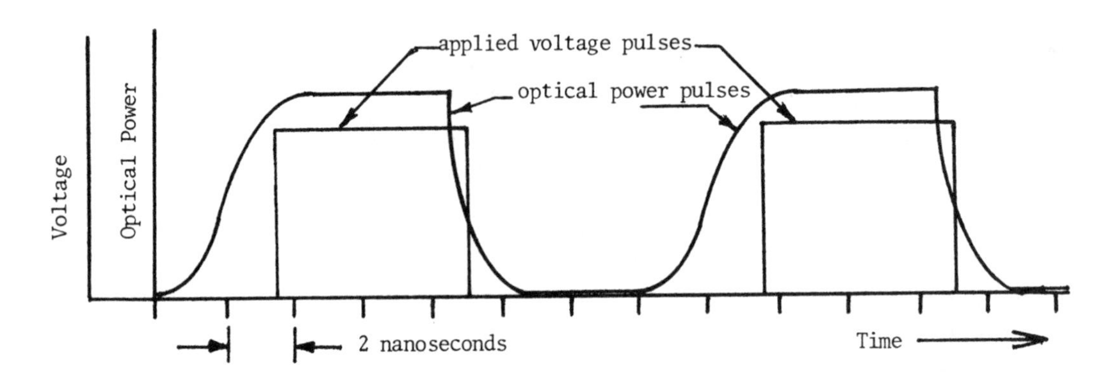

Fig. M-3. Optical power output pulses produced by **modulation** of a light source, such as a laser or a light-emitting diode (LED), with applied electrical pulses.

imum peak-to-trough value that occurs in a given signal, to (b) the maximum design variation, i.e., the maximum peak-to-trough value that the system is designed to allow. *Note:* In conventional amplitude modulation (AM), the maximum design variation is considered to be that for which the instantaneous amplitude of the modulated signal reaches zero. When zero is reached, the modulation factor is 100%. **2.** In an amplitude-modulated electromagnetic wave, the ratio of (a) the difference between the maximum amplitude and the minimum amplitude to (b) the maximum amplitude. *Note:* As the modulation factor is increased to the point where the modulated signal value reaches zero, the modulation factor is unity, i.e., the modulation is 100%. Attempts at modulation beyond this point will result in signal distortion. **3.** In a frequency-modulated electromagnetic wave, the ratio of (a) half the difference between the maximum and minimum frequencies to (b) the unmodulated sine wave carrier frequency. *Synonym* **modulation depth.** *See also* **amplitude modulation, modulation, modulation index, percentage modulation, unbalanced modulator.**

modulation frequency harmonic distortion: A constituent of nonlinear distortion that (a) consists of the production of sinusoidal components that have frequencies that are integral multiples of the fundamental frequency of the modulating signal, and (b) occurs particularly in the modulation frequency response of equipment to a sinusoidal modulated carrier input. *See also* **frequency response, modulating signal, modulation, nonlinear distortion.**

modulation frequency intermodulation distortion: A constituent of nonlinear distortion that consists of the production of sinusoidal components, i.e., intermodulation products, that have frequencies that are linear combinations, i.e., sums and differences, of the frequencies of the sinusoidal components of the modulating signal. *See also* **intermodulation products, modulating signal, nonlinear distortion.**

modulation improvement factor: *See* **frequency modulation improvement factor.**

modulation improvement threshold: *See* **frequency modulation improvement threshold.**

modulation index: In angle modulation, the ratio of (a) the frequency deviation of the modulated signal to (b) the frequency of a sinusoidal modulating signal. *Note:* The modulation index is numerically equal to the phase deviation in radians. *See also* **angle modulation, deviation ratio, frequency modulation, modulation, modulation factor, phase deviation, phase modulation.**

modulation integrator: *See* **frequency modulation integrator.**

modulation multiplexing: *See* **pulse code modulation multiplexing.**

modulation noise: *See* **equivalent pulse code modulation noise.**

modulation noise improvement: *See* **angle modulation noise improvement.**

modulation product: 1. One or more of the new frequencies that are obtained when one signal modulates another. **2.** In a modulator, the entire output product, including signal and noise, that results from modulation. *See also* **modulation, modulator.**

modulation radar jamming: *See* **pulse modulation radar jamming.**

modulation rate: 1. The rate at which a carrier is varied to represent the information in a digital signal. *Note:* Modulation rate and information transfer rate are not necessarily the same. **2.** A measure of how many individual signal elements are transmitted per unit time. *Note:* The term "modulation rate" can be used to describe either the capacity of a transmission channel or the rapidity of passage of significant instants of a signal. **3.** For modulated digital signals, the reciprocal of the unit interval of the modulated signal, measured in seconds. *Note:* The modulation rate is expressed in baud. *See also* **baud, bit rate, bits per second, data signaling rate, modulation, unit interval.**

modulation significant condition: A significant condition, such as a voltage, current, frequency, or phase, that (a) is assumed by a device and (b) corresponds to the quantization values of the characteristics that are chosen for modulation. *See also* **mark, modulation, restitution, significant condition, value.**

modulation suppression: In the reception of an amplitude-modulated (AM) signal, a reduction in the modulation factor, i.e., modulation depth, of a wanted signal, caused by the presence at the detector of a stronger unwanted signal. *See also* **amplitude modulation, signal to noise ratio, suppressed carrier transmission.**

modulation threshold effect: *See* **frequency modulation threshold effect.**

modulation threshold extension: *See* **frequency modulation threshold extension.**

modulation transfer function: 1. In an optical fiber in which an electromagnetic wave is propagating, that portion of the total optical fiber transfer function that is attributable to the modulation. *Note:* Although the optical fiber transfer function cannot extend to zero, the modulation transfer function can. **2.** In optics, the function that (a) describes the modulation of the image of a sinusoidal object as the frequency increases, and (b) usually is described by means of a graph or an equation. *Synonym* **sine wave response.** *See also* **frequency, image, modulation, object, optical fiber transfer function.**

modulator: A device that performs modulation, such as imposing a signal on a carrier, i.e., that converts a modulating signal into a modulation product. *Note:* In transmission systems, the modulation product signal is transmitted to a receiver that demodulates the signal to recover the original intelligence that was represented by the modulating signal. *See* **acoustooptic modulator, demodulator, electrooptic modulator, fiber optic modulator, magnetooptic modulator, optical integrated circuit filter-coupler-switch-modulator, optical phase modulator, telegraph modulator, thin film optical modulator, unbalanced modulator.** *See also* **carrier, modulating signal, modulation, modulation product.** *Refer to* **Fig. F-2 (339).**

modulator-demodulator: *Synonym* **modem.**

module: 1. An interchangeable item that contains components. *See also* **component. 2.** In the electrical circuits of communications, computer, data processing, and control systems, an interchangeable hardware plug-in unit that contains one or more circuit elements or components. *Note:* Examples of modules are fiber optic transmitters, oscillators, preamplifiers, repeaters, hard drives in personal computers (PCs), and printed circuit (PC) boards. **3.** In computer programming, a program unit that is discrete and identifiable and therefore can be treated as a unit, such as with respect to compiling, combining with other modules, and loading. *See* **multichip module.** *See also* **computer language, loading, overlay, patching, program patch.**

modulo: In mathematics, pertaining to the remainder after division of any integer by another specified integer. *Note:* For example, 26 modulo 8 is 2. *See* **division, integer.**

modulo-n arithmetic: Conventional arithmetic (a) that is performed in the usual manner, except that modulo values are used instead of the normal operands as in the conventional arithmetic, (b) in which the value of n may be any integer, (c) in which no number that is operated upon is allowed to become larger than $n - 1$, (d) in which all numbers are cyclic, (e) in which a number is congruent with another number modulo-n if the remainder after dividing each number by n is the same, and (f) that is used in checking systems for data storage, retrieval, and transmission in communications, computer, data processing, and control systems. *Note 1:* For example, the product of 59 and 38, modulo 5, is 2 because (a) 59 modulo 5 is 4, (b) 38 modulo 5 is 3, (c) 4×3 is 12, and (d) 12 modulo 5 is 2. Also, (a) 59×38 is 2242 and (b) 2242 modulo 5 is 2. *Note 2:* The numbers 5 and 23 are congruent modulo-9 and modulo-6, but not congruent modulo-8 and modulo-4. *See also* **modulo-n check, integer.**

modulo-n check: A check for verifying a computation, though with limited certainty, by (a) repeating the arithmetic operations to be checked using modulo-n arithmetic and (b) comparing the results with the results obtained using the original operands and conventional operations. *See also* **cyclic, modulo-n arithmetic.**

modulo-n counter: A counter in which the number that is represented by the counter reverts to 0 in the sequence of counting one at a time after reaching the maximum value of $n - 1$. *See also* **counter, modulo-n arithmetic, modulo-n check, value.**

modulo-two adder: *Synonym* **EXCLUSIVE OR gate.**

modulo-two addition: The addition of binary quantities or signals that (a) is performed without a carry action and (b) is accomplished by an EXCLUSIVE OR gate, i.e., an anticoincidence gate, that produces a 1 when either input is a 1 and that produces a 0 when both inputs are 0 and when both inputs are 1. *Note:* In normal binary addition, including carries, 11010 + 11100 = 110110. In modulo-two, addition of the same operands produces 001100.

modulo-two sum gate: *Synonym* **EXCLUSIVE OR gate.**

modulus counter: A counter (a) that produces an output pulse after a certain number, or multiple of this number, of input pulses is applied, and (b) in which the total count possible is based on the number of stages, i.e., digit positions. *Note 1:* A binary modulo-8 counter with three flip-flops, i.e., three stages, will produce an output pulse, i.e., display a 1, after eight input pulses have been counted, i.e., applied. This assumes that the counter started in the zero condition. The only output is that count of 1 each time the counter has counted 8. At the ninth pulse, the output is 0, and it remains at 0 until eight pulses have been counted again, i.e., it remains 0 for seven counts and remains 1 for every eighth count. *Note 2:* The output of the modulus counter can be used as input to a counter that will count the number of times the modulus counter has completed a cycle. *See also* **counter, flip-flop.**

molded case circuit breaker: A circuit breaker that (a) is assembled as an integral part in a supporting and enclosing housing of insulating material and (b) usually is used in systems in which ratings of 600 V (volts) or less may be used. *See also* **circuit breaker.**

molecular laser: A gas laser that has an active laser medium, i.e., a lasing medium, that is a molecular substance, i.e., a compound, such as carbon dioxide, hydrogen cyanide, or water vapor. *See also* **active laser medium, atomic laser.**

molecular stuffing (MS) process: A process of making graded index (GI) optical fibers in perhaps five broad steps, such as glass melting, phase separation, leaching, dopant introduction, and consolidation.

moment: *See* **current moment.**

monadic operator: An operator that can operate on only one operand. *Note:* Examples of monadic operators are negation, exponential, inverse, square root, sine, and log operators.

monitor: 1. In communications, to place emissions, such as radio, radar, video, optical, microwave, or sonar transmissions, under continuous surveillance, by detecting, measuring, recording, and interpreting the emissions. **2.** Software or hardware that is used to scrutinize and then display, record, supervise, control, or verify the operations of a system. *Note:* The many uses of monitors include indicating significant departures from the norm, determining levels of utilization of particular functional units, enabling human-computer interaction, and displaying data in a file. **3.** Software, hardware, or firmware that observes, tracks, traces, supervises, controls, or verifies the operations of a system. **4.** In communications, computer, data processing, and control systems, including personal computer (PC) systems, a display device that (a) enables interaction between the user, i.e., the system operator, and the system, via a keyboard or a mouse, and (b) displays data selected by the user. *Note:* Monitors usually contain cathode ray tube (CRT), liquid-crystal, or gas panel technology. *See* **reference monitor, software monitor, traffic monitor, variation monitor.** *See also* **computer, display device, keyboard, mouse, personal computer.**

monitoring: 1. The detecting of the presence of signals, such as electromagnetic radiation, sound, and visual signals, and usually the measurement of the signals. **2.** The act of listening, carrying out surveillance on, or recording the emissions of a station for the purpose of (a) improving transmissions, procedures, standards, and security, or (b) for maintaining a reference. **3.** In military communications systems, the act of listening, carrying out surveillance on, or recording of emissions for intelligence purposes. *See* **silent monitoring.** *See also* **electronic reconnaissance, intercept.** *Synonym* **radiological monitoring.**

monitor jack: A jack used to access communications circuits to observe signal conditions without interrupting the services being provided by the circuits. *See also* **bridging connection, circuit, monitor key.**

monitor key: A key used to access communications circuits to observe signal conditions without interrupting the services. *See also* **bridging connection, circuit, monitor jack.**

monitor mode: In network control systems, a mode of operation in which the host, control station, server, or other monitoring station is immediately notified when any of a given set of significant events occurs, such as an attention signal, a disconnect, or other unusual change in status on the line or in the network. *See also* **control station, host, monitoring, server.**

monitor printer: A printer that prints all the messages that are transmitted over the circuit to which it is connected.

monitor program: A computer program that is capable of observing, tracking, tracing, supervising, controlling, or verifying the operations of a system, such as a communications, computer, data processing, control, telemetering, or teleprocessing system. *See also* **supervise, trace, track, tracking, verify.**

monochromatic: In optics, pertaining to a single wavelength of electromagnetic radiation or to a single pure color. *Note:* Radiation is never perfectly monochromatic, i.e., it never consists of only one wavelength. In practice, a "single" wavelength source is at best a narrow band of wavelengths, i.e., radiation always has a finite spectral width. *See also* **coherent, line source, spectral width.**

monochromatic light: Electromagnetic radiation that (a) is in the visible or near visible spectrum and (b) ideally has only one frequency or wavelength. *Note:* Production of high radiance, high intensity, high energy light at a single wavelength, i.e., with a zero spectral width, is not practical at present. Light with narrow spectral widths, such as that produced by lasers, is considered monochromatic light, i.e., it consists of one color. *See also* **monochromatic radiation.**

monochromatic radiation: Electromagnetic radiation that ideally has one frequency or wavelength, usually in the visible frequency spectrum. *Note:* Electromagnetic radiation that has a very narrow spectral width is considered monochromatic radiation. A single line in a spectroscope is considered monochromatic radiation. *See also* **monochromatic light, spectroscope.**

monochromator: 1. An instrument for selecting narrow portions, i.e., selecting specific spectral lines, from an optical spectrum. **2.** In optics, an instrument that isolates narrow portions of the optical spectrum. *See also* **optical spectrum, spectral line.**

monochrome: Pertaining to a single color, though brightness may vary. *Note:* Many display device screens used in communications, computer, data processing, and control systems are of the monochrome type, such as green or amber characters on a dark or black background. *See also* **brightness, color.**

monocord switchboard: In telephone systems, a switchboard that (a) usually is a small magneto switchboard, (b) has loops such that each loop principally is a cord, jack, drop, and two-way level-type key, (c) is used to make conference connections by plugging the originating user plug or cord into the first call receiver jack, the first call receiver jack into the second call receiver jack, and so on, until all the conferees are connected, and (d) that usually has 4 to 12 cords, particularly in military and field installations. *See also* **cord circuit, jack, loop, magneto switchboard, plug, plug-ended cord.**

monocular instrument: An optical instrument that (a) has one optical axis, (b) has one eyepiece that makes use of, or allows the use of, one eye at a time, and (c) has no requirement to provide for convergence in order to accommodate two eyes. *Note:* Examples of monocular instruments are hand-held telescopes, magnifying glasses, monocles, some microscopes, and jeweler loupes. *See also* **convergence, eyepiece, loupe, magnifier.**

monofiber cable: *See* **monofilament cable.**

monofilament cable: A fiber optic cable that contains a single strand of optical fiber. *Synonym* **monofiber cable.** *See also* **fiber optic cable, optical fiber.**

monolayer: *See* **coplanar self-assembled monolayer.**

monolithic integrated circuit: An integrated circuit in which the substrate on which all circuit elements are mounted or constructed is an active material, usually a semiconductor, such as doped silicon, to create the active junctions. *See also* **circuit element, dopant, integrated circuit, junction, semiconductor, substrate.**

monolithic integrated receiver: An optical receiver in which (a) a positive-intrinsic-negative (PIN) photodiode is combined with a monolithic transimpedance amplifier in a single housing, and (b) the packaging is performed in such a way that the performance of the combination is better than that of the individual components packaged separately. *See also* **positive-intrinsic-negative field-effect transistor integrated receiver.**

monolithic technology: In electronics, a technology in which all circuit elements, such as transistors, diodes, resistors, and capacitors, are mounted or formed on one piece of substrate material. *Note:* The circuit elements may be formed by various methods, such as vapor deposition, etching, bonding, dipping, spraying, or combinations of these. *See also* **circuit element, monolithic integrated circuit, substrate, thin film circuit.**

monomode optical fiber: *Synonym* **single mode optical fiber.**

monopole radiated power: *See* **effective monopole radiated power.**

monopulse antenna feed system: An antenna feed system that (a) produces two or more partially overlapping antenna lobes, (b) has two or more feeds and one reflector, (c) has corresponding antenna beams with a small angular displacement between them, (d) has sum and difference channels in a tracking receiver that compares the amplitude or the phase of the antenna output signals, and (e) is used to deduce directional information in tracking radars in which an error signal to control the antenna is derived from the difference of signals received from the offset antenna beams. *See also* **antenna feed, antenna feed system, antenna lobe, beam, lobe.**

monopulse tracking system: *See* **pseudomonopulse tracking system.**

monorail double-heterojunction diode: A laser diode with a double heterojunction and a shift and return to level of the refractive index profile that has the spatial plot appearance of a square wave or step function on both sides of the junction. *See also* **double heterojunction, junction, laser diode, refractive index profile, square wave.**

monostable circuit: A circuit that has one stable state and an infinite number of unstable states, such that when the circuit is forced into an unstable state and the forces are removed, it will revert to the stable state, even though it may pass through a series of unstable states to reach the stable state. *Note:* An example of a monostable circuit is a capacitor with a resistance connected across the terminals. As long as no voltage is applied to the terminals, the capacitor will remain discharged and stable. If a voltage is applied to the terminals, the capacitor will charge to the level of the applied voltage and remain charged as long as the voltage is applied. When the voltage is removed, the capacitor will discharge through the resistor and return to the state it was in before the voltage was applied. *See also* **capacitor, circuit, resistor.**

Monte Carlo method: Any method based on pure chance. *Note:* Examples of the Monte Carlo method are (a) methods for obtaining approximate solutions to numerical problems by the use of random numbers, (b) random walk methods, and (c) procedures that use random number sequences to calculate an integral by (i) choosing a random number, (ii) calculating the integral from the integrand, (iii) choosing another number, larger or smaller, that will bring the chosen number closer to the actual solution, and (iv) repeating the process until the solution is reached. *See also* **integral, integrand.**

MORPH: mathematical morphology.

morpheme: In a language, an element that (a) indicates the relationship between semantemes in the language and (b) that contains no smaller meaningful part. *Note:* Examples of morphemes are the words but, and, if, or, with, not. The words forthcoming, always, overdone, telephone would not be considered as morphemes. *See also* **language, semanteme.**

morphology: *See* **mathematical morphology.**

Morse: 1. Samuel F. B. Morse, the person credited with inventing the telegraph. **2.** Pertaining to systems, devices, and techniques that make use of the dot and dash code and the type of telegraph equipment invented by Samuel F. B. Morse. *See* **high speed Morse, low speed Morse.** *See also* **code, dash, dot, telegraph.** *Refer to* **Figs. I-6 (472), T-3 (991).**

Morse code: A two-condition telegraph code in which (a) characters are represented by groups of dots and dashes, (b) each unique group represents a letter or a digit, (c) the groups are separated by spaces to form words, (d) usually the words are separated by slightly longer spaces to form sentences, and (e) sentences are separated by the longest spaces. *See also* **code, dash, dot, international Morse code, telegraph.** *Refer to* **Figs. I-6 (472), T-3 (991).**

Morse code indicator: In visual signaling systems, a panel signaling indicator that (a) consists of two panels that form an open V, and (b) when the panels are placed above the index flash, indicates the use of Morse code. *See also* **index flash, Morse code, panel, panel signaling, panel signaling indicator, visual signaling system.**

Morse continuous wave: Pertaining to telegraph operation in which (a) the signals are formed in accordance with the Morse code, and (b) the two basic methods of operation are (i) the direct current (dc) interruption and (ii) the carrier wave, usually a sinusoidal electromagnetic wave, interruption to form the dots and the dashes of the code. *Note:* Direct current (dc) interruption is suited for use in open and twisted pair wire lines and the carrier wave interruption is suited for radio and coaxial cable transmission. *Synonyms* **Morse cw, Morse CW.** *See also* **carrier wave, dash, dot, Morse code.**

Morse cw or CW: *Synonym* **Morse continuous wave.**

Morse flag signaling: Visual signaling in which (a) the Morse code is used, (b) a dot is indicated by raising both arms vertically overhead, (c) a dash is indicated by spreading both arms horizontally to the sides, (d) dot-dash separation is indicated by placing both arms before the chest, (e) separation of letters is indicated by lowering both arms to a 45° angle with the ground, (f) repeat or erase is indicated by circular motions of arms and hands overhead, and (g) the end of transmission is indicated by a quick vertical movement of arms in front of the body, *See also* **dash, dot, Morse code, semaphore signaling, visual signal, visual signaling, wigwag signaling.**

Morse key: A device that (a) can be used to form Morse code telegraph signals by manual operation, (b) consists of a spring-loaded contact that when depressed makes a connection and when released breaks the connection, (c) may be connected to other circuits and devices, such as spark suppressors and relays, to improve performance and extend useful life, and (d) is used to modulate a carrier by interrupting the carrier or by turning the carrier on and off for the dots, dashes, and spaces. *Synonym* **telegraph key.** *See also* **carrier, dashes, dots, modulation, relay, telegraph.**

Morse printer: A device that accepts Morse code signals and prints the characters that correspond to the signals. *See also* **Morse code, signal.**

Morse telegraphy: Telegraphy, i.e., telegraph system operation, that (a) forms signals in accordance with the Morse code, (b) includes the transmission and the reception of international Morse code signals, (c) includes the use of Morse keys and the semiautomatic and automatic devices associated with them, and (d) includes any form of alphabetic telegraphy using the Morse code. *See also* **Morse code, international Morse code, Morse key, signal, telegraphy.**

MOS: metal oxide semiconductor.

MOSAIC: 1. A security software capability that (a) provides limited security for sensitive information, such as proprietary, copyrighted, personal, and patent information, but not for security classified information, (b) is used in connection with a personal identification number and a cryptographic card, (c) resides in a user work station, (d) permits encryption of electronic mail, and (e) includes a digital signature for authentication and proof of delivery in a manner similar to registered mail. **2.** A portable World Wide Web (WWW) browser that provides a user with a graphical interface to hypertext-based information. *See also* **browser, hypertext, World Wide Web.**

mosaic telegraphy: Telegraphy, i.e., telegraph system operation, in which the patterns that form the characters are made up of units that are transmitted as separate signal elements. *Note:* An example of mosaic telegraphy is alphabetic telegraphy. *See also* **alphabetic telegraphy, signal, signal element, telegraphy.**

MOSFET: metal oxide semiconductor field effect transistor.

mosquito noise: In video systems, noise that, in displays, (a) is manifested by distortion that may be seen around the edges of moving images, (b) is characterized by (i) moving artifacts around the edges or (ii) blotchy noise patterns superimposed over the images, and (c) is similar in sight and sound to one or more mosquitoes flying around a person's head and shoulders.

motherboard: 1. A printed circuit (PC) board (a) into which another printed circuit board, a daughterboard, is plugged, (b) that provides access for the daughterboard to the system bus without requiring an additional port, i.e., pluggable slot, and (c) that is often used in microcomputer, minicomputer, and personal computer (PC) systems to add capacity or expand system capability, especially when physical space is extremely limited. **2.** In personal computer (PC) systems, the primary computer printed circuit (PC) board. *See also* **bus, daughterboard, microcomputer, minicomputer, personal computer, port, printed circuit board.**

motion compensation: Interframe coding that (a) is used to compress motion of video images, (b) uses an algorithm to examine a sequence of image frames to measure the difference from frame to frame in order to send motion vector information, and (c) allows only the picture changes between frames to be sent, thereby conserving bandwidth. *See also* **bandwidth, frame, image.**

motion operation rate: *See* **full motion operation rate.**

motion response degradation: The deterioration of motion video quality, usually full-motion video quality, resulting in a loss of perceived spatiotemporal resolution.

motion video: Video imagery that conveys the appearance of movement.

motor: *See* **integrating motor.**

motorboating: Singing that occurs in a circuit at a very low frequency, usually as a result of feedback and often at power frequencies, i.e., 50 Hz (hertz) to 400 Hz. *See also* **feedback, singing.**

motorized keyboard: A keyboard in which the energy that is required to move the combination bars into the position that is selected by depression of a key is derived from a motor that drives the instrument of which the keyboard is a part. *See also* **key, keyboard.**

mount: *See* **altazimuth mount, nonorthogonal antenna mount, X-Y mount.**

mounting cement: In optical systems, an adhesive that (a) is used to hold optical elements in their mounts and (b) that may be a thermoplastic, thermosetting, or chemical hardening material. *See also* **optical element, optical system.**

mouse: 1. A device that puts positional data into a computer. **2.** A computer input hand-held device that (a) generates the coordinates of a position indicator that is displayed on the screen of a display device, such as a computer monitor, (b) is operated by being moved on a flat surface, (c) usually has a control ball for rolling on the flat surface, and (d) has one or more switches for selecting options displayed on a screen. **2.** A hand-held device that (a) may be moved in contact with a flat surface in order to generate coordinate data for input to a computer, (b) may be used with a tablet, (c) may have an index mark and a light source, (d) usually is so small that it can be made to move about with a minimum of inertia so as to permit the following of curves on paper placed on the tablet, (e) can be made to perform the functions of a joystick, and (f) can be used to control or perform display operations, such as translating, i.e., moving display elements, rubber banding, scissoring, and tracking. *Note:* For example, if the mouse has a light source, and the flat surface is the faceplate of an aligned bundle of optical fibers, the selection and the illumination of a specific fiber, or group of contiguous fibers, can result in the generation of digital coordinate data that correspond to the position of the illuminated fiber or fibers by sending an enabling signal to a set of combinational circuits that will generate the digital coordinate data. *See also* **aligned bundle, combinational circuit, control ball, coordinate data, display element, enabling signal, faceplate, joystick, rubber banding, scissoring, tablet, track, tracking.**

m out of n code: A binary code that (a) has a fixed weight assigned to each digit position, (b) has n-digit words in which exactly m digits are the same, and (c) has word patterns consisting of a string of binary digits that may form a character or a group of characters. *Note:* An example of an m out of n code is a 2 out of 5 code in which every 5-bit word or byte has exactly two 1s and three 0s. If this rule is violated, perhaps because of a circuit malfunction, an error is indicated. The error detector need only be a modulo-n counter in which $n = 3$. *See* **binary code, binary digit, digit position, modulo-n counter, string, weight.**

move command: In communications, computer, data processing, and control systems, a command that (a) transfers data from one or more storage locations to one or more other storage locations and (b) usually does not allow the data to remain in the original location. *See also* **command, copy command, storage location.**

movement key: *See* **pointer movement key.**

movement service: *See* **ship movement service.**

moving grating sensor: A fiber optic sensor in which (a) two optical fibers are separated by a small gap in which a pair of gratings is placed in such a way that one is fixed and the other is movable, (b) both gratings consist of alternate opaque and transparent parallel elements of equal width, (c) when the transparent elements of both gratings line up, light is transmitted from the light source fiber to the photodetector fiber, (d) when the opaque elements of one grating line up with the transparent elements of the other, the light transmission from fiber to fiber is zero, (e) starting from a lineup of opaque elements covering transparent elements, and consequently zero transmission, as the movable grating starts to uncover the transparent elements, light begins to pass through the gratings, causing the transmitted amount of light to increase to the maximum when the transparent elements line up, (f) as the movable grating continues to be moved in the same direction, the amount of light decreases to zero, rising again as the grating continues to move in the same direction, until the end of its movement, (g) the number of times peak transmission is reached is proportional to the distance moved, (h) the frequency of the output signal is proportional to the speed of movement, and (i) the acceleration of the movable grating is proportional to the time rate of change of output signal frequency. *Note:* A sound wave impinging on the movable grating will be reproduced as an electronic output signal of the photodetector. *See also* **fiber optic sensor.**

moving target indicator (MTI): A radar display that shows only objects, i.e., only targets, that are in motion above a specified threshold. *Note 1:* In the radar set, return signals, i.e., echoes, from objects are subtracted from the next return signal, i.e., the echo signal, using a suitable temporary storage circuit. *Note 2:* There are two basic types of moving target indicators (MTIs), incoherent and coherent. Both systems are used to remove ground clutter and return signals, i.e., echoes, from slow moving objects. Both systems depend on a pulse-to-pulse comparison and subsequent cancellation of fixed or slow-moving object echoes. Cancellation can take place in either the intermediate frequency (IF) or the video stage. Coherent MTI, however, makes use of Doppler information from an object in a clutter environment, thus providing subclutter visibility that is not available in incoherent MTI. The radar displays only those objects that have a radial component of velocity in relation to the radar site. A change in radial distance from the radar site, i.e., in range, causes a shift in frequency, amplitude, or phase between two adjacent echoes from the object. In some MTI radars, the echo signal is fed into two channels. One of these is a normal amplifier. It amplifies the signal to video level. The other channel is a delay circuit. The outputs are compared with a new return signal to determine any amplitude, frequency, or phase change in a discriminator. The output of the discriminator presents to the radar display unit only those objects in which changes have occurred that are above a given threshold. *See* **coherent moving target indicator, discriminator, Doppler effect, incoherent moving target indicator.**

M patch bay: A patching facility designed for patching and monitoring digital data circuits at data signaling rates from 1 Mbps (megabits per second) to 3 Mbps. *See also* **D patch bay, patch bay.**

MPSK: multiple phase-shift keying.

MRS: multirole system.

MS: molecular stuffing.

MSATCOM: mobile satellite communications.

M sequence: A sequence of binary digits, i.e., bits, that can be generated with an M-stage linear shift register, where M can be any arbitrary integer, and having the property that, if binary 1s are fed into it, a pseudorandom binary sequence of $2^M - 1$ bits will be generated before the shift register returns to its original state and the output sequence is repeated. *See also* **integer, shift register, spread spectrum code sequence generator.**

MSE System: Mobile Subscriber Equipment System.

MSP: molecular stuffing process.

MT: machine translation.

MTBF: mean time between failures.

MTBO: mean time between outages.

MTI: moving target indicator.

MTM: McLintock theater model.

MTSR: mean time to service restoral.

MTTF: mean time to failure.

MTTR: mean time to repair.

MTTSR: mean time to service restoral.

mudbox: Equipment that is rugged enough to withstand adverse environments. *Note:* A mudbox is expected to operate unsheltered on the ground.

MUF: maximum usable frequency, maximum useful frequency.

mu-law: Pertaining to a standard algorithm used for pulse code modulation encoding of pulse amplitude modulation samples of signals used in digital voice communications. *Synonym* **µ law.** *Note:* The mu law algorithm is mainly used in North America and Japan.

mu-law algorithm: A standard analog signal compression algorithm that (a) is used in the digital communications systems of the North American digital hierarchy to optimize, i.e., to modify, the dynamic range of analog signals prior to their digitization, and (b) because the wide dynamic range of speech is not suitable for efficient linear digital encoding, reduces the dynamic range of the analog signal, thereby increasing the coding efficiency, resulting in a higher signal-to-distortion ratio than that obtained by linear encoding for a given number of bits.

muldem: multiplexer/demultiplexer.

multiaddress calling: A network-provided service feature that permits a user to nominate more than one addressee for the same data. *Note:* The network may sequentially or simultaneously perform multiaddress calling. *See also* **call, conference call, group alerting and dispatching system, selective calling, service feature.**

multiaddress calling facility: A facility or a process that (a) permits a user to use more than one address for the same message or data, (b) allows a call originator to call more that one call receiver or data station at the same time, (c) allows the communications system to establish the connections sequentially or simultaneously, perhaps at the option of the user, (d) uses the same call procedure as for a direct call, (e) allows the use of a special code or codes to designate all the required destinations or to indicate the individual full or abbreviated address of each destination user, and (f) may also be used with delay working procedures. *See also* **abbreviated address, address, code, delay working, direct call, facility.**

multiaddress instruction: *Synonym* **multiple address instruction.**

multiaddress service: A communications service that (a) provides for the delivery of messages that are destined for more than one addressee, (b) differs from broadcast service in that a multiaddress message need not be transmitted simultaneously to all addressees, and (c) is addressee-selective, i.e., only the addressees are to receive the messages, and no other user or station will. *See also* **broadcast service.**

multiband radar: Radar that (a) transits at more than one frequency simultaneously, (b) uses only one antenna, (c) allows for many sophisticated forms of video processing, (d) provides for improved performance, (e) overcomes some forms of interference, and (f) as an electronic countermeasure, imposes a severe burden on a jammer by requiring that the jammer jam all the frequencies, i.e., all the channels, at the same time. *See also* **channel, electronic countermeasure, interference, jam, jammer.**

multicarrier modulation (MCM): Modulation in which the data to be transmitted are divided into several interleaved bit streams, and these bit streams, are used to modulate two or more carriers. *Note:* Multicarrier modulation (MCM) is a form of frequency division multiplexing (FDM).

multicast: **1.** In a network, the simultaneous transmission of data, usually in packet form, to a selected set of destinations. *Note:* Some networks, such as Ethernet, support multicast by allowing a network interface to belong to one or more multicast groups. **2.** In a network, to simultaneously transmit data, usually in packet form, to a selected set of destinations in the network. **3.** To transmit information to a select group of stations, users, or devices without obtaining acknowledgment of receipt of the transmission.

multicast address: A routing address (a) that is used to simultaneously address all the computers in a group, and (b) that usually identifies a group of computers that share a common protocol, as opposed to a group of computers that share a common network. *Note:* Multicast address also applies to radio communications. *Synonym in Internet protocol* **class d address.**

Multicast Backbone: In Internet, a subset that (a) is capable of multicasting, i.e., distributing, specified packets to all addressees that request them, (b) operates interactively with users, and (c) handles audio, video, and digital data. *See also* **Internet, multicasting, packet.**

multicasting: In a data network, the simultaneous transmission of identical data to a specified group of recipients in the network. *Note:* Multicasting transmission may be confirmed or unconfirmed.

multicast packet: A packet that is addressed and delivered to more than one individually identified addressees or to all the stations on a specified network. *See also* **packet, unicast packet.**

multichannel: Pertaining to communications that (a) usually are full duplex communications on several channels and (b) may be accomplished by multiplexing, such as by time division, frequency division, phase division, or space division multiplexing.

multichannel-bundle cable: In fiber optic systems, two or more single-channel single-bundle cables, all in one outside jacket. *Note:* Each single-channel single-bundle cable has a jacket, which, in a multichannel-bundle cable, might be called an inside jacket. *See also* **cable, jacket, single-channel single-bundle cable, single-channel single-fiber cable.**

multichannel cable: In fiber optic systems, two or more cables combined in a unitizing element, such as a single jacket, harness, strength member, or cover. *See also* **cable, harness, jacket, strength member.**

multichannel optical analyzer: A device that analyzes the components or the content of electromagnetic radiation in the optical spectrum. *Note:* An example of a multichannel optical analyzer is a vidicon tube with a sensing area and an electron gun for reading signals that represent an image focused on its screen. *See also* **electromagnetic radiation, focus, image, screen, signal.**

multichannel single-fiber cable: In fiber optic systems, two or more single-fiber cables, all in one outside jacket. *See also* **cable, jacket.**

multichannel trunking and switching system: *See* **joint multichannel trunking and switching system.**

multichannel voice frequency telegraph (MCVFT): Telegraph in which two or more carrier current signals are used that have frequencies that are within the voice frequency range, i.e., in the range between 300 Hz (hertz) and 3000 Hz. *See also* **carrier, telegraph, voice frequency.**

multichip module: A module that (a) consists of two or more interconnected chips, (b) is capable of performing 10^{10}, i.e., 10 billion, flops (floating point operations per second, and (c) supports a wide range of applications ranging from intelligent systems management to imaging.

multicom land station: *See* **aeronautical multicom land station.**

multicom mobile station: *See* **aeronautical multicom mobile station.**

multicom service: *See* **aeronautical multicom service.**

multicoupler: *See* **antenna multicoupler.**

Multidimensional Application and Gigabit Internetwork Consortium (MAGIC): A consortium that (a) consists of government agencies, such as the U.S. Army and the U.S. Geological Survey, and other organizations, such as SRI International, Minnesota Supercomputer Center, Lawrence Berkeley Laboratory, Sprint, and the University of Kansas, and (b) is dedicated to the development and operation of a high-performance network research test bed to be used for geographically dispersed computing, storage, and display components.

multidither: *See* **frequency multidither.**

multidrop network: A network configuration in which one or more intermediate nodes are on a path between end point nodes. *See also* **end point node, intermediate node, network configuration, node, path.**

multielement dipole antenna: An antenna that (a) consists of an arrangement of a number of dipole antennas and (b) may produce various directivity patterns by varying the arrangement of the dipoles and the way they are driven. *See also* **antenna, dipole antenna, directivity pattern.**

multifiber cable: A fiber optic cable (a) that contains two or more optical fibers, and (b) in which each fiber provides a separate transmission channel. *Synonyms* **multifilament cable, multiple fiber cable.** *See also* **fiber optic cable, optical fiber, transmission channel.**

multifiber cable assembly: *See* **cable assembly.**

multifiber joint: A fiber optic splice or connector that is designed to (a) mate two multifiber cables and (b) provide simultaneous optical alignment of all individual fibers. *Note:* Optical coupling between the aligned optical fibers may be achieved by various techniques, such as proximity butting, the use of lenses, and the use of refractive index matching materials. *See also* **fiber optic connector, fiber optic splice, optical fiber, refractive index matching material.**

multifilament cable: *Synonym* **multifiber cable.**

multifocal: In optics, pertaining to a system or a component, such as a lens or a lens system, that has, or is characterized by, two or more foci. *See also* **focus, lens, optics, principal focus point.**

multiframe: In pulse code modulation (PCM) systems, a set of consecutive frames in which (a) each frame, and the position of each frame, can be identified by reference to a multiframe alignment signal for the group of consecutive frames, and (b) the multiframe alignment signal does not necessarily occur, in whole or in part, in each and every frame. *See also* **distributed frame alignment signal, frame, frame alignment signal, multiframe alignment signal, pulse code modulation, signal.**

multiframe alignment signal: A frame alignment signal that serves as a timing signal, i.e., a reference signal, for a specified group of consecutive frames. *See also* **frame, frame alignment signal, multiframe.**

multifrequency (MF) signaling: Signaling in which a combination of frequencies is used. *Note:* An example of multifrequency (MF) signaling is signaling (a) that uses combinations of two out of six (MF 2/6) or two out of eight (MF 2/8) voiceband frequencies to indicate telephone address digits, precedence ranks, and line or trunk busy, and (b) in which (i) MF 2/6 uses frequencies of 700, 900, 1100, 1300, 1500, and 1700 Hz (hertz), and (ii) MF 2/8 and dual tone multifrequency (DTMF) use 697, 770, 852, 941, 1209, 1336, 1447, and 1633 Hz. Thus, each separate digit is represented by two different frequencies, each frequency being one of a possible six or a possible eight in these cases. A maximum of 36 or 64 possible codes may be obtained for these cases. For the case of eight frequencies, two at a time but different frequencies for each code element, there are 56 possible codes. If permutations are not permitted, but only combinations, and the frequencies must be different, only 28 variations are available. For six frequencies, the corresponding values are 30 and 15. *Synonym* **multifrequency pulsing.** *See also* **dual tone multifrequency signaling, frequency.**

multifrequency pulsing: *Synonym* **multifrequency signaling.**

multifunction board: A printed circuit (PC) board that adds two or more distinct capabilities to a communications, computer, data processing, or control system. *Note:* Though a multifunction board may add several functions to a personal computer (PC) where physical space and ports, i.e., pluggable slots, are at a premium, the individual functions may not be so powerful as that provided by a single-function add-in board. *See also* **add-in board, printed circuit board.**

multifunction information distribution system (MIDS): An interaircraft communications system that (a) provides location information on other aircraft, (b) has secure voice channels, and (c) accepts input information concerning location of hostile aircraft from existing friendly airborne early warning aircraft.

multilayer filter: *Synonym* **interference filter.**

multilevel information systems security initiative (MISSI): A U.S. Department of Defense program to (a) provide security products and services to support defense communications systems, such as the message, packet, electronic mail, and file transfer systems within the defense message system (DMS), the integrated tactical and strategic digital network (ITSDN), and multilevel security systems, (b) handle multilevel security classified information from unclassified to top secret, (c) provide for compartmentation, (d) provide support to the multilevel security (MLS) needs for the U.S. armed services communications systems architectures, and (e) ensure that transmissions are authentic, were originated by validated sources, are received only by validated recipients, and are not made available to unauthorized persons and systems.

multilevel modulation: *Synonym* **n-ary signaling.**

multilevel precedence (MLP): Pertaining to communications systems and components that can handle more than one precedence designation. *See* **automatic multilevel precedence, precedence designation.**

multilevel precedence and preemption (MLPP): A precedence and preemption scheme, i.e., a priority scheme, (a) for assigning one of several precedence levels to specific calls or messages so that the system handles them in a predetermined order and time frame, (b) for gaining controlled access to network resources in which calls and messages can be preempted only by higher-priority calls and messages, (c) that is recognized only within a predefined domain, and (d) in which the precedence level of a call outside the predefined domain usually is not recognized. *See also* **domain.**

multilevel precedence capability: In automatically switched telephone networks, the capability of handling traffic at more than one precedence designation level. *Note:* An example of a multilevel precedence capability is the capability of handling flash override, flash, immediate, priority, and routine traffic, all in the same network and all within the specified precedence level definition limits. Usually a preemption procedure is used. *See also* **automatically switched network, precedence designation, preemption, traffic.**

multiline laser: A gas laser that is a multimode laser, i.e., that emits radiation at two or more different wavelengths. *See also* **gas laser, laser, multimode laser.**

multilink: **1.** In a communications system, pertaining to a multiplicity of data links. **2.** A group of data links that are obtained by means of multiplexing, i.e., multiplexed data links. **3.** In a data network, pertaining to a branch that consists of two or more data links. *See also* **branch, data link, data network, multiplexing, node.**

multilink operation: In packet-switched networks, the simultaneous use of multiple links for the transmission of different segments of the same message unit. *Note:* Multilink operation increases the effective rate of message transmission and requires special procedures for multiplexing and demultiplexing control.

multimedia: **1.** A combination of various means of information representation, such as audio, video, digital, analog, lightwave, microwave, smoke, and body language. **2.** A combination of various modes of transmission, such as radio, telephone, television, telegraph, facsimile, wirephoto, intercom, and loudspeaker. **3.** Pertaining to any communications system that uses two or more signal propagation technologies, such as wire, radio, microwave, or fiber optic cable. **4.** Pertaining to any information representation scheme using two or more formats, such as audio, text, graphics, and video. **5.** Pertaining to a single communications path that supports two or more information formats, such as audio, text, graphics, facsimile, and video. **6.** Pertaining to any communications station that uses two or more different types of user end instruments, such as (a) telephones, modems, and computers or (b) facsimile machines, teletypewriters, and television receivers. *See also* **multiple media transmission, multitasking.**

multimedia interface (MMI): Pertaining to an interface system that allows communications equipment to interface with various existing communications systems, propagation media, and transmission systems, such as radio, telephone, and telegraph systems that handle video, voice, and digital data.

multimegabit data service: *See* **switched multimegabit data service.**

multimeter: A test instrument that (a) is used for measuring electric voltages, currents, and resistances that may lie within a number of different ranges, (b) has a range selection capability to obtain measurement precision, and (c) must be operated with caution so as not to allow the indicator to go off scale, usually by starting a measurement with the scale that measures the largest unit. *Synonym* **volt-ohm milliammeter.**

multimode dispersion: *See* **optical multimode dispersion.**

multimode distortion: In an optical waveguide, the distortion that (a) results from differential mode delay, (b) is a result of the spread in time of a pulse because the velocity of propagation is not the same for all modes in an optical signal, and (c) is not a result of a dispersive mechanism, i.e., is not a form of dispersion. *Note 1:* Multimode distortion in multimode step-index optical fibers may be compared to multipath propagation of radio signals. The direct signal is distorted by the arrival of the reflected signals a moment later. In a step-index optical fiber, rays taking more direct paths through the fiber core, i.e., those undergoing fewer reflections at the core-cladding boundary, traverse the length of the fiber sooner than those rays that undergo more reflections, resulting in signal distortion at the end of the fiber. *Note 2:* Multimode distortion limits the bandwidth of a given multimode optical fiber. A typical step-index fiber with a 50-μm (micron) core would be limited to about 20 MHz (megahertz) for a 1-km (kilometer) length, i.e., a bandwidth•length product of 20 MHz•km. *Note 3:* Multimode distortion may be considerably reduced, but not completely eliminated, by the use of a core having a graded refractive index. The bandwidth•length product of a typical off-the-shelf graded index multimode fiber having a 50-μm (micron) core may be over 1 GHz•km (gigahertz•kilometer), with over 3 GHz•km fibers having been produced. *Note 4:* Because of its similarity to dispersion in its effect on optical signals, multimode distortion is sometimes incorrectly referred to as "intermodal dispersion," "modal dispersion," or "multimode dispersion." Such usage is incorrect because multimode distortion is not truly a dispersive effect. Dispersion is a wavelength-dependent phenomenon, particularly because of the spectral width of a pulse, whereas multimode distortion may occur to a single wavelength. *Synonyms* **intermodal distortion, modal distortion.** *See also* **coherence area, coherence length, coherence time, distortion, distortion-limited operation, fiber optics, mode.**

multimode facility: In communications systems, a facility that is capable of handling more than one transmission mode, such as telephone, telegraph, radio, and facsimile. *See also* **facility, transmission mode.**

multimode fiber: *Synonym* **multimode optical fiber.**

multimode group delay: *Synonym* **differential mode delay.**

multimode group delay spread: In an electromagnetic wave propagating in a waveguide, the variation in group delay time, caused by differences in group velocity, among bound propagating modes even at a single frequency. *Note:* The differences in arrival times of the leading and trailing edges of a pulse at the end of the waveguide, as compared to the sending end, are caused by the different propagation delays of the different modes. The modes can be considered as different optical paths of different optical path lengths. In an optical fiber, it is possible for photons or waves that propagate along the optical fiber axis of the core to arrive at the end sooner than those that follow a helical path through the core, thus causing the pulse duration at the end of the fiber to be increased. If the pulse duration is too great, intersymbol interference, i.e., an overlapping of consecutive pulses, will occur. The received pulse duration can be reduced if the refractive index profile of the core is arranged so that light rays taking a helical path along the outer edges of the core propagate through a lower-refractive-index material, and hence travel faster in the longer path than axial rays propagating along the optical axis or in a helical path closer to the axis of the core, in the higher-refractive-index material, so that all rays arrive at the end of the fiber at the same time. Actual propagation delay is of little consequence, as long as all rays of a given pulse arrive at the end of the fiber at the same time. Zero-dispersion fibers are being made. *Synonym* **differential mode delay.** *See also* **pulse duration.**

multimode laser: A laser that emits radiation containing two or more modes, i.e., two or more different wavelengths. *See also* **laser, multiline laser.**

multimode operation: In analog systems, the use of a common circuit or a single propagation medium for both analog and digital data, such as voice, binary coded data, facsimile, and Morse code, all on one circuit though not necessarily at the same time. *Note:* Simultaneous transmission in more than one transmission mode at the same time can be accomplished by using multiplexing techniques. *See also* **analog data, circuit, digital data, mode, multiplexing, propagation medium, transmission mode.**

multimode optical fiber: An optical fiber that supports the propagation of more than one bound mode at a given operating wavelength. *Note 1:* A multimode optical fiber may be either a graded index (GI) fiber or a step index (SI) fiber. *Note 2:* The number of modes that an optical fiber will support depends on the core diameter, the numerical aperture (NA), and the wavelength. *Synonym* **multimode fiber.** *See also* **cladding mode, coupled modes, fiber optics, modal distribution, modal noise, mode, mode scrambler, mode volume, optical fiber, single mode optical fiber.**

multimode waveguide: A waveguide that can support more than one mode. *Note:* Because different wavelengths constitute different modes and the number of modes is also dependent on the numerical aperture (NA) and the core diameter, a given "multimode waveguide" might support only one mode and therefore could be called a "single mode waveguide" if the operating wavelength were long enough, and, conversely, a given "single mode waveguide" might support several modes and therefore could be called a "multimode waveguide" if the operating wavelength were short enough. *See also* **mode.**

multinode network: In communications systems, a network in which users may be interconnected through more than one node. *See also* **network, node.**

multipaired cable: A paired cable that has two or more pairs of conductors, such as two or more twisted pairs. *See also* **paired cable, twisted pair.**

multiparty line: *Synonym* **party line.**

multipass sort: A sort in which items in a set are examined or passed over two or more times before all the items are sorted. *Note:* An example of a multipass sort is a sort in which items in internal or main storage are sorted and moved to external or auxiliary storage in batches, and then the batches are merged.

multipath: The propagation phenomenon that results in radio signals reaching the receiving antenna by two or more paths. *Note 1:* For radio, video, and microwave transmissions, causes of multipath include atmospheric ducting, ionospheric reflection and refraction, and reflection from terrestrial objects, such as mountains and buildings. *Note 2:* Multipath effects include constructive reinforcement, phase shifting, and destructive cancellation of the signal. *Note 3:* In facsimile and television transmission, multipath causes jitter and ghosting. *Note 4:* For lightwaves in dielectric waveguides, the different paths taken by the lightwaves may also be considered as a form of multipath. Causes include refractive index and entrance condition variations. *See also* **propagation, Rayleigh fading.**

multipath fading: In the propagation of electromagnetic waves, including radio waves and lightwaves, fading caused by multipath in which (a) energy is dissipated in each of the paths, causing distortion, and (b) signal reinforcement and cancellation at the destination, because of the differences in arrival times for waves in the paths that have different path lengths, also cause distortion. *Synonym* **multiple path fading.** *See also* **distortion, fading, multipath, path length, Rayleigh fading..**

multiphonon band: The group or the range of acoustic, elastic, or vibrational frequencies associated with a number of phonons with energies lying in a given range, the range usually containing phonons of many different frequencies and therefore different energies. *See also* **frequency, phonon.**

multiple: In communications, computer, data processing, and control systems, pertaining to a system of wiring such that a circuit, a line, or a group of lines is accessible at a number of points. *Synonym* **multipoint.** *See also* **circuit.**

multiple access: 1. The connection of a user end instrument such that (a) the instrument is connected to two or more central offices (COs), i.e., switching centers, by separate lines, and (b) the instrument has a single message routing indicator or telephone number. **2.** In satellite communications, the capability of a communications satellite to function as a portion of a communications link between more than one pair of satellite terminals concurrently, usually by passing a number of separately modulated carriers, perhaps from different Earth stations, through a single satellite transponder. *Note:* The three types of multiple access presently used with communications satellites are code division, frequency division, and time division multiple access. *See* **code division multiple access, demand assignment multiple access, frequency division multiple access, pulse address multiple access, satellite switched time division multiple access, space division multiple access, spread spectrum multiple access, time division multiple access.** *See also* **alternate routing, dual access, dual homing, extension facility, multiple homing, satellite.**

multiple access rights terminal: A terminal that may be used for more than one user service class. *Note:* An example of a multiple access rights terminal is a terminal in which different groups of users have access rights to certain classes of data and not to other classes of data. *See also* **terminal, user service class.**

multiple access with collision avoidance: *See* **carrier sense multiple access with collision avoidance.**

multiple access with collision detection: *See* **carrier sense multiple access with collision detection.**

multiple address instruction: A computer program instruction that has two or more address parts. *Note:* Examples of multiple address instructions are four-address instructions, three-address instructions, and two-address instructions. *Synonym* **multiaddress instruction.** *See also* **address part, computer program, four-address instruction, instruction, three-address instruction, two-address instruction.**

multiple address message: A message (a) that is destined for two or more addressees, (b) in which each addressee usually is informed of all addressees, and (c) in which each addressee usually is informed by the message originator as an action, information, or exempted addressee. *See also* **action addressee, exempted addressee, information addressee, message originator.**

multiple beaming: In radar systems, the transmission of two or more beams from an antenna, each in a different direction at any given instant. *See also* **beam, conical beam split antenna.**

multiple-bundle cable: A number of jacketed bundles placed together in a common, usually cylindrical, outer jacket. *See also* **bundle, cable, fiber optic bundle, jacket.**

multiple-bundle cable assembly: A multiple-bundle cable that has been terminated and is ready for installation. *See also* **bundle, fiber optic bundle, fiber optic cable, fiber optic terminus.**

multiple call: 1. In telephone switchboard operation, a call from one user to several other users. *Synonym* **bunched call. 2.** In radio telegraph transmission, a call in which (a) two or more different call signs are transmitted, (b) call signs may be individual station, net, or collective call signs, and (c) each call sign may be transmitted more than once when operating conditions render transmissions difficult. *See also* **call sign, collective call sign, conference call.**

multiple call message: A message that contains routing indicators that require segregation at intermediate relay stations. *See also* **relay station, routing indicator, routing line segregation.**

multiple domain network: A network that has two or more host nodes. *See also* **domain, host node, network.**

multiple fiber cable: *Synonym* **multifiber cable.**

multiple fiber cable assembly: A fiber optic cable that (a) contains two or more optical fibers, (b) is terminated, and (c) is ready for installation. *See also* **fiber optic cable, optical fiber, terminus.**

multiple frequency shift keying (MFSK): Frequency shift keying (FSK) in which (a) multiple codes are used in the transmission of digital signals, and (b) the coding schemes use multiple frequencies that are concurrently or sequentially transmitted. *See also* **frequency, frequency shift keying, keying.**

multiple homing: 1. In telephone systems, the connection of a terminal facility so that it can be served by one or several switching centers. *Note:* Multiple homing may use a single directory number. **2.** In telephone systems, the connection of a terminal facility to more than one central office (C.O.), i.e., one switching center, by separate access lines. *Note:* Separate directory numbers are applicable to each central office or switching center accessed. *See also* **alternate routing, dual homing, multiple access, split homing.**

multiple lines: *See* **common address multiple lines.**

multiple media transmission: Transmission in which more than one type of propagation medium, such as a combination of fiber optic cable, wireline, and radio, is used to transfer information from one point to another. *See also* **multimedia, propagation medium.**

multiple-part message: A message in which parts are transmitted separately.

multiple-path fading: *Synonym* **multipath fading.**

multiple peg marking: In telephone switchboard operation, a system of color-coded pegs, some with numbers, that are used to indicate the status of the jack into which a peg is inserted. *Note:* Examples of multiple peg markings are (a) plain black to indicate a spare line, no telephone connected, or not in use and (b) plain red to indicate out of order. *See also* **jack, peg count.**

multiple phase-shift keying (MPSK): Keying, i.e., modulation, in which there are as many phase states, i.e., phase positions or significant conditions, of a carrier as there are digital information input code elements to represent. *Note:* In a quaternary code, i.e., a code in which there are four significant conditions, each perhaps representing one of the digits 0, 1, 2, and 3, there are four different phase positions of the carrier, usually 90° (electrical) apart. *See also* **carrier, keying, significant condition.**

multiple routing: Message routing in which more than one destination address is specified in the header. *See also* **header.**

multiple selection signal code: *See* **multiple selection signal code.**

multiple spot scanning: In facsimile systems, scanning simultaneously performed by two or more scanning spots, each one analyzing its fraction of the total scanned area of the object. *See also* **facsimile, object, scanning.**

multiplex (MUX): 1. To use a common propagation medium to provide for two or more channels. **2.** Pertaining to the use of a single or common channel or path in order to make two or more channels, such as by (a) sharing the time of the channel, i.e., time division multiplexing, or (b) superimposing many frequencies at the same time, i.e., frequency division multiplexing, in order that many signal sources and sinks may communicate during a given period. **3.** To use one channel for connecting two or more source and sink pairs. *See also* **channel, demultiplex, multiplexer, multiplexing, sink, source, propagation medium.**

multiplex aggregate bit rate: In a time division multiplexer, the bit rate that is equal to the sum of the input channel data signaling rates (DSRs) available to the user plus the rate of the overhead bits required, given by the relation $MABR = R(\Sigma n_i + H)$, summed from $i = 1$ to $i = m$, where $MABR$ is the multiplex aggregate bit rate, R is the pulse repetition rate of the frame of the output channel, n_i is the number of bits per multiplex frame associated with the ith channel, m is the maximum number of input channels to the multiplexer, including nonworking channels, equipped channels, or both, and H is the number of overhead bits per multiplex frame of the output channel. The number of bits in the multiplex frame is assumed to be constant. *See also* **binary digit, bit rate, channel, data signaling rate, time division multiplexing.**

multiplex baseband: In frequency division multiplexing, the frequency band occupied by the aggregate of the signals in the line interconnecting the multiplexing and the radio or line equipment. *See also* **baseband, multiplexing.**

multiplex baseband receive terminal: The point in the baseband circuit that is nearest to the multiplex equipment and from which connection usually is made to the radio, video, optical, or line baseband receiving terminals or intermediate facility. *See also* **baseband, facility.**

multiplex baseband transmit terminal: The point in the baseband circuit that is nearest to the multiplex equipment and from which connection usually is made to the radio, video, optical, or line baseband transmitting terminals or intermediate facility. *See also* **baseband, facility.**

multiplexed circuit: *See* **homogeneous multiplexed circuit.**

multiplexer (MUX): 1. A device that performs multiplexing. *Note 1:* The term "multiplexing equipment" is often used to refer to the set of equipment that performs both multiplexing and demultiplexing. *Note 2:* A multiplexer may be combined with pulse code modulation (PCM) equipment. **2.** A device for creating additional channels from a given set of equipment. **3.** A device for multiplexing two or more channels. *Synonym* **multiplexing equipment.** *See* **data multiplexer, digital multiplexer, fiber optic demultiplexer (active), fiber optic demultiplexer (passive), fiber optic loop multiplexer, fiber optic multiplexer (active), fiber optic multiplexer (passive), optical demultiplexer (active), optical demultiplexer (passive), optical multiplexer (active), optical multiplexer (passive), thin film optical multiplexer, trunk group multiplexer.** *Refer to* **Figs. B-9 (96), C-7 (157).**

multiplexer-demultiplexer (muldem): A device that combines the functions of multiplexing and demultiplexing of digital signals. *Note:* A multiplexer-demultiplexer (muldem) is in contrast to a modulator-demodulator (modem). A muldem is in the digital domain, whereas a modem is in the analog domain.

multiplexer filter: *See* **fiber optic multiplexer-filter, fiber optic rod multiplexer-filter.**

multiplex hierarchy: In frequency division multiplexing, the rank of occupied frequency bands occupied. *Note:* The ranks are (a) 12 channels equal 1 group, (b) 5 groups (60 channels) equal 1 supergroup, (c) 5 supergroups (300 channels) equal 1 mastergroup (International Telegraph and Telephone Consultative Committee (CCITT) standard), (d) 10 supergroups (600 channels) equal 1 mastergroup (U.S. standard), and (e) 6 U.S. mastergroups equal 1 jumbo group. *See* **digital multiplex hierarchy.** *See also* **frequency division multiplexing, group, multiplexing, time division multiplexing.**

multiplexing (MUXing): 1. In communications systems, the creating of data links, circuits, or channels by (a) using a given set of physical equipment, (b) increasing the capacity of the equipment to carry data, and (c) using various means to increase the capacity, such as time sharing, frequency division, time division, space division, and phase shifting. *Note:* In electrical communications, the two basic forms of multiplexing are time division multiplexing (TDM) and frequency division multiplexing (FDM). In optical communications, what corresponds to FDM is referred to wavelength division multiplexing (WDM). **2.** The combining of two or more information channels onto a common propagation medium. **3.** Using a single transmission channel for two or more user signals. *See* **asynchronous time division multiplexing, color division multiplexing, frequency division multiplexing, heterogeneous multiplexing, homogeneous multiplexing, optical multiplexing, pulse code modulation multiplexing, space division multiplexing, statistical multiplexing, statistical time division multiplexing, synchronous time division multiplexing, time division multiplexing, wavelength division multiplexing.** *See also* **digital multiplex hierarchy, group, multiplex hierarchy, time sharing.** *Refer to* **Figs. S-14 (927), T-6 (1013), W-3 (1082).**

multiplexing equipment: 1. *Synonym* **multiplexer. 2.** *See* **pulse code multiplexing equipment.**

multiplex link: A link that (a) enables data terminal equipment (DTE) to have several access channels to a network over a single circuit, (b) performs interleaving, such as packet interleaving, byte interleaving, and bit interleaving, and (c) is established by any form of multiplexing. *See also* **byte, link, multiplexing, packet.**

multiplex link encryption: 1. Encryption in which a single cryptographic device, i.e., a single piece of cryptoequipment, is used to encrypt the data in all of the channels within a multiplex link. **2.** Encryption in which a single cryptographic device is used to encrypt all of the data in a multiplexed link. *See also* **cryptoequipment, encrypt, encryption, multiplexing, multiplex link.**

multiplex operation: Operation of a communications path, such as a circuit, channel, or link, in such a manner that there is simultaneous transmission of two or more messages in either direction or both directions over the transmission path. *See also* **channel, circuit, link.**

multiplex set: *See* **engineering circuit multiplex set.**

multiplex structure: *See* **frame multiplex structure.**

multiplicand: In multiplication, the number that (a) is to be multiplied by another number, the multiplier, (b) usually is on hand or in a register first, and (c) in a binary digital computer multiplier, is shifted and added according to the digits in the multiplier until the product is obtained. *Synonym* **icand.** *See also* **multiplication, multiplier.**

multiplication: An arithmetic process in which a number, the multiplicand, is added to itself one fewer time than another number, the multiplier, to produce a result, the product. *See* **avalanche multiplication, logical multiplication, quarter squares multiplication.** *See also* **multiplicand, multiplier, product.**

multiplier: 1. In multiplication, the number that determines how many times another number, the multiplicand, is to be added to itself to produce a result, i.e., the product. *Note:* In a binary digital computer, the multiplier digits determine whether the shifted multiplicand is to be added or not to produce partial products and ultimately the final product. *Synonym* **ier. 2.** A device that performs multiplication. *See* **analog multiplier, quarter squares multiplier.** *See also* **multiplicand, multiplication, multiplier, product, shift.**

multiply connected star network: *See* **network topology.**

multipoint: 1. Pertaining to the ability to interconnect three or more devices, such as sets of data circuit-terminating equipment (DCE), without intervening switchboards or switching centers, in order to obtain satisfactory operation. *See also* **point to point. 2.** *Synonym* **multiple.**

multipoint access: Access in which more than one terminal is supported by a single network termination. *See also* **network termination, terminal.**

multipoint circuit: A circuit that interconnects three or more separate points. *See also* **broadcast operation, circuit.**

multipoint connection: 1. A connection between two data stations via one or more other intermediate stations. **2.** A connection that (a) is among three or more data stations in order to accomplish data transmission among them and (b) may include switching facilities. *See also* **data station, data transmission, facility, switching.**

multipoint distribution service (MPDS): A one-way domestic public radio service rendered on microwave frequencies from a fixed station that (a) usually transmits in an omnidirectional radiation pattern and (b) transmits to multiple receiving facilities located at fixed points.

multipoint grounding system: Equipment bonded together and also bonded to the facility grounding system at the point nearest the equipment. *See also* **earth electrode subsystem, facility grounding system, ground.**

multipoint link: In communications systems, a data link that interconnects three or more terminals or data stations. *See also* **data link, data station, link, terminal.**

multiport coupler: *See* **fiber optic multiport coupler, optical multiport coupler.**

multiport optical coupler: *Synonym* **optical multiport coupler.**

multiport repeater: In digital networks, an active device (a) in which there is a multiplicity of input-output (I-O) ports, (b) in which a signal introduced at the input of any port appears at the output of every port, (c) that usually performs regenerative functions, such as amplifying, timing, and reshaping signals, and (c) depending on the application, that may be designed not to repeat a signal back to the port from which it originated.

multiposition switchboard: In telephone switchboard operations, a switchboard that is built to accommodate more than one operator and is therefore equipped with more than one position. *See also* **position, switchboard.**

multiprocessing: 1. Simultaneous or concurrent processing by two or more processors of a multiprocessor. **2.** The simultaneous execution of two or more computer programs or sequences of instructions by a computer. **3.** The parallel processing of two or more computer programs simultaneously. *See also* **computer program, multiprocessor, multiprogramming, parallel processing, time sharing.**

multiprocessor: A computer that has two or more processors that have common access to a main storage. *See also* **main storage.**

multiprogramming: The interleaved execution, i.e., the simultaneous or concurrent execution, of two or more computer programs by a single processor such that operators and programmers obtain the impression that their program is the only one being executed by the processor. *See also* **computer program, multiprocessing, time sharing.**

multipurpose jammer: 1. A jammer that is capable of simultaneously jamming many frequencies throughout a wide band of frequencies. **2.** A jammer that is capable of simultaneously combining two or more electronic countermeasures (ECMs), such as barrage jamming and deception jamming. **3.** A jammer that is capable of accomplishing two or more different types of jamming, such as spot jamming and synchronous jamming. *See also* **jammer, jamming.**

multirefracting crystal: A transparent crystalline substance that is anisotropic with respect to the velocity of light propagating within it in different directions, i.e., the refractive index is different in different directions.

multirole system (MRS): A battlefield switching system that (a) has a 16-port switch, (b) carries 30 channels in each port, (c) uses application-specific integrated circuits, and (d) is a Siemens Plessey Defence Systems product.

multisatellite link (MSL): A radio link that (a) is between a transmitting Earth station and a receiving Earth station through two or more communications satellites, without any intermediate Earth station and (b) consists of one uplink, one or more satellite-to-satellite links, and one downlink. *See also* **downlink, Earth station, uplink.**

multitasking: 1. Pertaining to a computer operating system capability that allows two or more different sets of functions to be performed by the operating system. **2.** The concurrent performance or interleaved execution of two or more tasks. *Synonym* **concurrent operation.** *See also* **operating system.**

multiterminal computer: A computer that (a) has two or more remote terminals connected to it, (b) can simultaneously support the terminals such that the operations at each terminal are not significantly delayed because of the connection and the use of the other terminals, and (c) usually has a response time that is so short for all terminals that each terminal operator is not aware that other operators are using the computer concurrently. *See also* **remote terminal, response time.**

multiunit message: A message in which more than one signal unit is used when it is transmitted. *See also* **signal unit.**

multivibrator: A relaxation oscillator that consists of two stages that are coupled so that the input of each stage is derived from the output of the other. *See also* **oscillator, relaxation oscillator.**

Murray code: *Synonym* **International Telegraph Alphabet 2.**

mutation: *See* **program mutation.**

mutual information content: *Synonym* **transinformation content.**

mutual interference: Interference, i.e., noise, that is obtained, created, or induced in one device when it is operating in conjunction with another device, and vice versa. *Note:* An example of mutual interference is the exchange of energy between two antennas that are radiating simultaneously, particularly in the same frequency band and in proximity, because of the electrostatic, magnetostatic, magnetic, and electromagnetic coupling that occurs between them. *See also* **interference.**

mutually synchronized network: A network in which transmission and reception of signals are synchronized by having each clock in the network exert some degree of control on all other clocks in the network. *See also* **clock, democratically synchronized network, hierarchically synchronized network, master-slave timing, mutual synchronization, oligarchically synchronized network, synchronization.**

mutual synchronization: 1. Synchronization in which the frequency of the clock at a particular node is controlled by a weighted average of the timing on all signals received from neighboring nodes. **2.** A timing subsystem in which (a) directed control is not used, and (b) the frequency of the clock at a particular node is controlled by some weighted average of all the timing signals received from other nodes. *See also* **democratically synchronized network, hierarchically synchronized network, master-slave timing, mutually synchronized network, node, oligarchically synchronized network, synchronization.**

MUX: multiplex, multiplexer, multiplexing.

mw: microwave.

Mylar: A polyester material, made by DuPont, often used in film form as a base for magnetic coatings for making magnetic tape. *See also* **magnetic tape.**

myocardium simulator: *See* **fiber optic myocardium simulator.**

myriametric wave: An electromagnetic wave that lies in the range that (a) extends from 3 kHz (kilohertz) to 30 kHz, (b) has a letter designator of VLF (very low frequency), and (c) has a numeric designator of 4. *See also* **frequency spectrum designation, radio frequency.**

n: 1. In optics, the symbol for refractive index, usually implying the refractive index of a material relative to that of a vacuum. **2.** In computers and communications, the symbol for an arbitrary integer. **3.** In semiconductors, the abbreviation for "negative," as in "n-type dopant" or "n-type semiconductor." *See also* **n-ary, refractive index.**

NA: numerical aperture.

NACISO: NATO Communications and Information Systems Agency.

***n*-address instruction:** A computer program instruction that has *n* address parts, *n* being a whole positive number, usually 1, 2, 3, or 4. *See also* **address part, computer program, instruction.**

***n*-adic Boolean operation:** A Boolean operation in which there are exactly *n* operands, where *n* is a whole positive number. *Note:* An example of an *n*-adic Boolean operation in which $n = 3$, i.e., a triadic Boolean operation, is A(B + C), read as A AND (B OR C). *See also* **Boolean operation.**

nadir: In radio communications systems, a point on a line between the Earth's center and an observer located farther from the Earth's center than the point. *Note:* The nadir is opposite the zenith.

nailed-up circuit: 1. A circuit semipermanently established through a circuit switching facility for point-to-point connection. **2.** *Deprecated synonym for* **permanent virtual circuit** *and for* **dedicated circuit.**

NAK: negative acknowledge character.

NAK attack: In communications security systems, a security penetration technique that (a) makes use of the negative acknowledge transmission control character and (b) capitalizes on a potential weakness in a system that handles asynchronous transmission interruption in such a manner that the system is in an unprotected state against unauthorized access during certain periods.

name: *See* **file name.**

named range: In communications, computer, data processing, and control systems, a specified group of storage locations that has been given a name, so that the data in these locations can be called in a function, moved elsewhere in storage or on a spreadsheet, or printed as a single unit. *See also* **storage location.**

name key: In personal computer systems, a key that, when depressed while the system is in an applicable mode, will cause the system to display the names that have been assigned to groups of data, such as the file names that have been assigned to files.

name server: *See* **domain name server.**

name system: *See* **Domain Name System.**

naming: 1. In communications systems, the selection and the assignment of identifiers, i.e., titles, to an addressee for purposes of message delivery. *Note:* Examples of naming are the assignment of (a) "Personnel Coordinator" to a person, (b) "Room 912" to a place, and (c) "Traffic Control" to a function, for purposes of message or mail delivery. *See also* **identifier.** **2.** In a display system, the assignment of a name or a label to a segment, display file, computer program, or other software used for operating the display system. *See also* **computer program, display file, display system, function, segment, software.**

NAND element: *Synonym* **NAND gate.**

NAND gate: A binary logic combinational circuit or device that is capable of performing the NAND operation, i.e., the negative AND operation, such that the NAND of A, B, C . . . is false if all the statements are true and true if any one or more of the statements are false. *Note:* The NAND gate behaves like the AND gate when the output of the AND gate is simply negated. In terms of binary notation with 0s and 1s as input digits, the NAND gate will yield an output of 0 when all the inputs are 1 and a 1 when at least one input is a 0. *Synonyms* **alternate denial gate, dispersion gate, if A then NOT B gate, if B then NOT A gate, NAND element, negative AND gate, nonconjunction gate, NOT AND gate, NOT both gate, Sheffer stroke gate.** *See also* **combinational circuit, gate.** *Refer to* **Fig. N-1.**

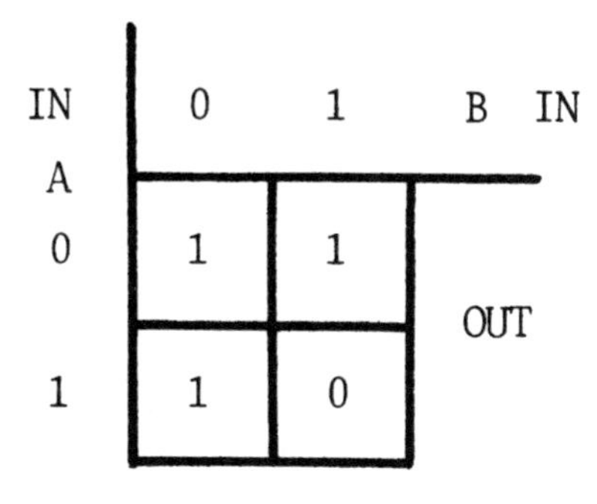

Fig. N-1. A truth table that shows the input and output digits of a **NAND gate.**

nanomechanics: The branch of science and technology devoted to the design, development, and production of electronic and optical integrated circuits that (a) usually constitute entire operating systems, such as computers, controllers, and sensors, (b) operate at rates higher than 1 Gflop (10^9 floating point operations per second), (c) have circuit elements less than 1 μm (micron) in size, and (d) are mounted on a single chip substrate.

nanometer (nm): One billionth (U.S.) of a meter, i.e., 10^{-9} m (meter) or 10 Å (angstroms). *Note:* Though a large part of the scientific community still uses nanometers to express the wavelength of light, the trend in the fiber optics engineering and industrial community is toward the use of the micron (μm), 1 millionth of a meter or 1000 nm (nanometer). The micron is more convenient because of the numbers that occur when indicating spatial relationships in optical fibers. The optical wavelengths used in fiber optics are on the order of 1 μ (micron). In fiber optics, (a) optical fiber core and cladding diameters, (b) cladding, core, and reference tolerance areas and fields, (c) optical fiber joint longitudinal displacement and lateral offset, (d) optical thin film thicknesses, and (e) many other dimensional aspects of optical waveguides are also expressed in microns, which simplifies comparisons between optical wavelengths and optical waveguide geometry, such as comparing mode field diameters and fiber core diameters and comparing fiber mode volumes using formulas that contain both geometric values and wavelengths. Many of the equations and formulas contain wavelengths and geometric values in numerators and denominators. Calculations and estimates are simplified when geometric values and wavelength values are expressed in the same units. *Refer to* **Table 4, Appendix B.**

nanosecond: One thousandth of a millionth of a second, i.e., 10^{-9} s (seconds). *See also* **second.** *Refer to* **Table 4, Appendix B.**

narrative traffic: **1.** Traffic that consists of plain or encrypted messages written in a natural language and transmitted in accordance with standard formats and procedures. *Note:* Examples of narrative traffic include (a) messages that are placed on paper tape, transmitted via a teletypewriter (TTY), and, on reception, converted back to a printed page on another teletypewriter or teleprinter, and (b) messages that are printed on a sheet of paper, transmitted via optical character recognition (OCR) equipment, and, on reception, converted back to a printed page on a printer. **2.** Messages usually prepared in accordance with standardized procedures for transmission via optical character recognition equipment or teletypewriter. *Note:* In contrast to data pattern traffic, narrative messages contain additional message format lines. *See also* **record traffic, traffic.**

narrowband: **1.** Pertaining to a group or a range of frequencies that (a) is within a relatively restricted part of the electromagnetic frequency spectrum, (b) has limiting frequencies that are relatively close together, the concept of "closeness" being dependent upon the position in the electromagnetic frequency spectrum, and (c) usually is in a band that is less than 0.1% of the median or operating frequency. *Note:* At a median frequency of 100 kHz, a band of frequencies 5 kHz wide would be considered wide. The same 5-kHz band at 100 MHz would be considered narrow. **2.** A group or a range of frequencies that is in a band that is less than 0.1% of the median, operating, or assigned frequency.

narrowband modem: 1. A modem in which a narrow bandwidth is used, or that has a narrow bandwidth capability. *Note:* The bandwidth may range from 3 kHz (kilohertz) to 20 kHz. **2.** A modem in which the modulated output signal frequency spectrum is limited to that which can be wholly contained within, and faithfully transmitted through, a voice channel with a nominal 4-kHz bandwidth. *Note:* Modems usually used with high frequency (HF) systems are limited to operation over a voice channel with a nominal 3-kHz bandwidth. *See also* **channel, modem, narrowband radio voice frequency, wideband modem.**

narrowband radio voice bandwidth (NBRVB): In narrowband radio, the nominal 3-kHz (kilohertz) bandwidth allocated for single-channel radio that provides a transmission path for analog and quasi-analog signals.

narrowband radio voice frequency (NBRVF): In narrowband radio, the midband or center frequency of the nominal 3-kHz (kilohertz) bandwidth allocated for single-channel radio that provides a transmission path for analog and quasi-analog signals.

narrowband signal: 1. A signal that has a spectral content, i.e., a frequency spectrum, that is limited to a narrow band. **2.** Any analog signal, or analog representation of a digital signal, with a frequency spectrum that is limited to that which can be contained within, i.e., represented within, a voice channel with a nominal 4-kHz (kilohertz) bandwidth. *Note:* Narrowband radio uses a voice channel with a nominal 3-kHz bandwidth. *See also* **bandwidth, channel, narrowband, narrowband radio voice frequency, signal, wideband.**

narrow beam antenna: An antenna that (a) has a high directive gain, i.e., its radiated energy is confined to a small solid angle, and thus has a narrow main lobe, (b) usually has comparatively small side lobes, (c) is less subject to interference because of its directional capability, (d) is more difficult to jam than wide beam antennas or omnidirectional antennas, and (e) in radar systems, is well suited for tracking. *See also* **directive gain, main lobe, side lobe, wide beam antenna.**

narrow beam radar: A radar that (a) emits a signal that has its energy confined to a small solid angle or to a small angle in elevation or azimuth, (b) uses a narrow beam antenna, and (c) has an antenna radiation pattern with relatively small side lobes compared to the narrow main lobe, such as that from a spotlight or the light from a parabolic reflector. *See also* **narrow beam antenna, radiation pattern, wide beam radar.**

narrow shift frequency shift keying: In radiotelegraph communications systems, frequency shift keying that (a) is accomplished by the modulating signal, (b) is small compared to the frequencies being used, (c) is used to distinguish the mark from the space, such as to distinguish 1s from 0s, dots and dashes from spaces, or tone from silence. *Note:* An example of narrow shift frequency shift keying is a shift of +425 Hz (hertz) for the mark and a shift of −425 Hz for the space, both with reference to the assigned frequency. *See also* **frequency shift keying, wide shift frequency shift keying.**

NARSAP: National Advanced Remote Sensing Application Program.

***n*-ary: 1.** Pertaining to a selection, choice, or condition that has *n* possible different values or states. *Note 1:* The *n* is an integer greater than 1. *Note 2:* A prefix may be substituted for the *n*, such as in binary, ternary, and quaternary. **2.** Pertaining to a fixed-radix numeration system having a radix of *n*. *Note:* Except for 0 and 1, the *n* may be any number, such as 2, 8, 10, or *e*, the base of natural logarithms, approximately equal to 2.718.

***n*-ary code:** A code (a) that has *n* significant conditions, where *n* is an integer greater than 1, and (b) in which the integer substituted for *n* indicates the specific number of significant conditions, i.e., quantization states, in the code. *Note 1:* An example of an *n*-ary code is an 8-ary code, i.e., an octonary code, that has eight significant conditions and can convey three bits per code symbol, such as the digits 0 through 7. *Note 2:* A prefix may be substituted for the *n*, such as in binary, ternary, and quaternary. *Synonym* ***m*-ary code.** *See also* **binary digit, code, level, significant condition.**

***n*-ary digit:** A digit or a member selected fron an *n*-ary code. *Note:* Examples of *n*-ary digits are (a) a 1 in a binary code consisting of 0 and 1, i.e., a binary digit, (b) a 3 in a quaternary code consisting of 0, 1, 2, and 3, i.e., a quaternary dight, and (c) a ternary digit.

***n*-ary digital signal:** A digital signal in which a signal element may assume *n* discrete states, where *n* may be any integer greater than 1. *Note:* An example of an *n*-ary digital signal is a ternary digital signal element that could have any one of three stable states, significant conditions, or levels, such as +V, 0, and −V, where V could be any value of voltage, such as 6 V (volts). *Synonym* **m-ary digital signal.** *See also* **significant condition, value.**

***n*-ary information element:** An information element that enables the representation of *n* distinct states, conditions, or levels where *n* may be any integer great than 1. *Note:* Examples of *n*-ary information elements are (a) a ternary information element, i.e., data element, that has three and only three possible data items, such as masculine, feminine, and neuter as data items for the data element *gender,* and (b) in a ternary numeration system, one of the three possible digits. *Synonym* ***m*-ary information element.**

***n*-ary signaling:** Signaling in which each digital data symbol has *n* significant conditions, i.e., *n* different states. *Note 1:* The *n* may be any integer greater than 1. *Note 2:* When *n*-ary signaling is used, more than two states are usually implied. In *n*-ary signaling there are multiple decision thresholds and significant conditions. *Note 3:* A prefix may be substituted for the *n*, such as in binary signaling, ternary signaling, and quaternary signaling. *Synonym* **multilevel modulation.** *See also* **decision threshold, level, modulation, signal, significant condition.**

nat: natural unit of information content.

NATA: North American Telecommunications Association.

national address indicating group: An address indicating group (AIG) that is assigned for use within a specific nation. *See also* **international address indicating group.**

National Advanced Remote Sensing Application Program (NARSAP): A program that (a) makes use of aerial photography and photogrammetry and (b) develops tools to allow authorized users to consolidate multispectral imagery to aid in environmental monitoring, land management, or scientific research, such as allowing the coordinated use of existing sensors to aid in the solution of problems, e.g., soil erosion, crop blight, pollution plumes, and thermal changes for seismic analysis.

National Command Authorities (NCA): The President of the United States and the U.S. Secretary of Defense or their duly deputized alternates or successors.

National Communications System (NCS): 1. The organization that (a) was established by Section 1(a) of Executive Order No. 12472 to assist the President, the National Security Council, the Director of the Office of Science and Technology Policy, and the Director of the Office of Management and Budget in the discharge of their national security emergency preparedness (NS/EP) telecommunications functions, and (b) consists of both the telecommunications assets of the entities represented on the Committee of Principals and an administrative structure consisting of the Executive Agent, the Committee of Principals, and the Manager. **2.** The telecommunications system that results from the technical and operational integration of the separate telecommunications systems of the executive branch departments and agencies having significant telecommunications capabilities.

National Coordinating Center (NCC): The telecommunications joint industry and federal government operation established by the National Communications System (NCS) to assist in the initiation, coordination, restoration, and reconstitution of national security or emergency preparedness (NS/EP) telecommunications services and facilities.

National Electric Code® (NEC): The standard that (a) governs the installation and the use of electrical wire, cable, and fixtures in buildings, (b) is developed and maintained by the National Electric Code® (NEC) Committee of the American National Standards Institute (ANSI), (c) is sponsored by the National Fire Protection Association (NFPA), (d) is adopted by the U.S. government, and (e) is identified as ANSI/NFPA 70-1990.

national information infrastructure (NII): Within a nation, (a) the totality of communications, computer, and information systems and components and (b) their interconnection and gateways to international systems. *Note:* Internet is an early model that describes the national information infrastructure (NII).

National Science Foundation network (NSFnet): The network that (a) serves as the backbone network for Internet, (b) provides high speed switching, network capacity, and compatibility with the synchronous optical network (SONET), (c) enables T-1 links to meet broadband requirements, (d) enables users to access T-3 networks from T-1 links, (e) uses asynchronous transfer mode (ATM), and (f) automatically adjusts application speed to available network bandwidth. *See also* **asynchronous transfer mode, Internet, synchronous optical network.**

National Security or Emergency Preparedness (NS/EP) telecommunications: Telecommunications services that are used to (a) maintain a state of readiness and (b) respond to and manage any local, national, or international event or crisis that causes, or could cause, (i) injury or harm to the population,

(ii) damage to or loss of property, (iii) degradation of or a threat to the national security or emergency preparedness posture of the United States, or (d) any combination of these. *See also* **priority level, priority level assignment, private National Security or Emergency Preparedness (NS/EP) telecommunications services, public National Security or Emergency Preparedness (NS/EP) telecommunications services, service identification, service user, Telecommunications Service Priority service, Telecommunications Service Priority system.**

National Security or Emergency Preparedness (NS/EP) telecommunications service: *See* **private National Security or Emergency Preparedness (NS/EP) telecommunications service, public switched National Security or Emergency Preparedness (NS/EP) telecommunications service.**

national switching equipment congestion signal: In international communications, a signal that (a) is sent in the backward direction and (b) indicates that the call setup attempt was unsuccessful because of the congestion encountered at a switching center in a national network, i.e., that the access attempt ended in access failure. *See also* **access, access attempt, access failure, backward direction, call setup, congestion, network, switching center.**

National Television Standards Committee (NTSC) standard: The North American standard for the generation, transmission, and reception of television signals (a) in which the 525-line picture is the standard, whereas the European phase alternation by line (PAL) system uses the 600-line picture, (b) in which the picture information is transmitted in amplitude-modulated (AM) signals, and the sound information is transmitted in frequency-modulated (FM) signals, (c) that is compatible with International Telegraph and Telephone Consultative Committee (CCITT) Standard M, and (d) that is also used in Central America, a number of South American countries, and some Asian countries, including Japan. *See also* **phase alternation by line, phase alternation by line modified, système electronique coleur avec memoire, teleconference, television.**

NATO Communications and Information Systems Agency (NACISO): The agency of the North Atlantic Treaty Organization (NATO) that (a) serves as the procurement and contracting agent for communications and information systems and (b) is responsible for development of a viable command, control, and communications infrastructure that (i) can support military operations against known possible adversaries, (ii) can provide a basic capability to meet unknown adversaries, (iii) can integrate communications with non-NATO coalition forces, (iv) focuses on mobile communications to support quick reaction strike forces, (v) keeps NATO command and staff informed of available communications capabilities, and (vi) notifies NATO command and staff of the extent to which a specific planned operation can be given adequate communications support with available communications resources.

natural frequency: 1. The frequency at which a circuit or a device will freely oscillate after the application of an impulse. **2.** For an antenna, the lowest frequency at which the antenna resonates without the addition of any inductance or capacitance. *See also* **antenna, antenna matching, circuit, frequency.**

natural language: A language in which the symbols, conventions, and rules for its use are based on current usage and were not specifically or explicitly prescribed before use of the language. *See also* **artificial language, language.**

natural logarithm (ln): A logarithm to the base $e \approx 2.71828$. *See also* **logarithm.**

natural number: 1. Any of the numbers obtained by repeatedly adding 1 to a number that was obtained in the same way, including 1 itself. *Note:* Examples of natural numbers are 1, 2, 3, . . . *n*, where *n* is a whole number. Thus, 0 is not a natural number. **2.** Any number obtained by adding 1 to a number that was obtained in the same way, including 0. *Note:* Examples of natural numbers are 0, 1, 7, 13, 23, and 1776.

natural unit of information content (nat): A unit of logarithmic measure of information content that is expressed as a Napierian logarithm, i.e., as a natural logarithm, a logarithm to the base $e \approx 2.71828$. *Note:* The decision content of a character set of eight characters and the information content of a character from the set of eight characters is $\ln 8 \approx 2.079$ nats (natural units of information content). *See also* **decision content, information content.**

nautical mile (nmi): 1. A measure of the distance equal to one minute of arc on the Earth's surface. *Note 1:* The United States has adopted the International Nautical Mile equal to 1852 m (meters) or 6076.11549 ft (feet) or about 1.15 smi (statute miles). **2.** The fundamental unit of distance used in navigation. *Note 2:* The nautical mile is frequently confused with the geographical mile, which is equal to 1 min of arc on the Earth's equator or about 6087.15 ft. *Note 3:* The U.S Department of Defense (DoD) and the U.S. Department of Commerce (DoC) adopted the International Nautical Mile in 1954. *See also* **statute mile.**

navigate: In cyberspace, to browse and search among multimedia via browsers, servers, and networks, such as Internet. *See* **browser, cyberspace, Internet, multimedia.**

navigating: In cyberspace, the using of hypertext links. *See also* **cyberspace, hypermedia, multimedia, navigate.**

navigation: *See* **long range aid to navigation, radar navigation, radionavigation, short range aid to navigation.**

navigation aid: *See* **radionavigation aid.**

navigation and ranging: *See* **sound navigation and ranging.**

navigation service: *See* **maritime radionavigation service.**

navigation system: A system that (a) enables the determination of present location with respect to known objects, (b) provides a means for determination of a direction, course, bearing, or azimuth for a mobile unit to reach a desired location, and (c) usually makes use of radio transmitters, receivers, and directional antennas. *See also* **beacon navigation system, hyperbolic navigation system, terminal guidance navigation system.**

NBH: network busy hour.

***n*-bit byte:** A byte that consists of *n* binary dights, i.e., *n* bits, where *n* can be any positive integer. *Note:* In communications, computer, data processing, and control systems, *n* usually is not greater than 16 or less than 2. The eight-bit byte is the most common because characters, particulary the American Standard Code for Information Interchange (ASCII) characters, require seven bits, allowing for 128 characters, and one additional bit in each byte for parity checking. *See also* **byte, character.**

***n*-body problem:** A class of problems in classical celestial mechanics that treats the relative motion of an arbitrary number, *n*, of point masses that (a) are only under their mutual gravitional attraction, (b) are released under a set of initial conditions, i.e., initial velocities and initial distances of separation, and (c) have no forces acting on them other than their own inertial forces and their forces of gravitational attraction. *Note:* The *n*-body problems are encountered in satellite systems. *See also* **celestial mechanics, restricted circular three-body problem, three-body problem, two-body problem.**

NBRVF: narrowband radio voice frequency.

NCA: National Command Authority.

NCC: National Coordinating Center.

NCS: 1. National Communications System. 2. nomadic communications service. *See* **personal communications service.**

near end crosstalk: Crosstalk that is propagated in a disturbed channel in the direction opposite to the direction of propagation of a signal in the disturbing channel. *Note:* The terminals of the disturbed channel, at which the near end crosstalk is present, and the energized terminal of the disturbing channel, are usually near each other. *See also* **crosstalk, far end crosstalk, intelligible crosstalk, interference.**

near field: In a propagating electromagnetic wave, the field (a) that lies within the near field region, (b) that is primarily caused by the electric charge distribution in the antenna, (c) that varies approximately as $1/d^3$ where d is the radial distance from the antenna, and (d) in which the transit time of an electromagnetic wave from the antenna to a point in the field is negligible compared to the transit time to the far field. *Note:* The differential tangential component of the electric field vector produced by an ideal Hertzian dipole antenna, which most dipole antennas can be considered to be at the distances involved, is given by the relation $dE_\theta = (Idl\ (\sin\theta)/4\pi\varepsilon)[-\omega(\sin\omega t)/rv^2 + (\cos\omega t)/r^2v) + (\sin\omega t)/\omega r^3]$ where dE_θ is the differential tangential component of the electric field vector, I is the antenna current, dl is the elemental length of antenna bearing the current I, θ is the angle between the antenna direction of dl and the line from the dipole center to the point of measurement, π is approximately 3.1416, ε is the electric permittivity of the intervening medium, ω is the angular velocity equal to $2\pi f$ where f is the frequency, t is time, r is the distance from the center of the dipole antenna to the point of measurement, and v is the propagation velocity equal to $(\mu\varepsilon)^{-1}$ where μ is the magnetic permeability. Note that for the expression in brackets, the terms are for the radiation field, the induction field, and the electrostatic field, in that order. Note that at great distances, i.e., large values of r, the radiation field, i.e., the far field, is dominant. Closer to the antenna the intermediate field is dominant, and just outside the antenna the near field is dominant. *Synonyms* **electrostatic field, refractive field.** *See also* **far field, intermediate field, near field region.** *Refer to* **Table 9, Appendix B.**

near field diffraction pattern: 1. The diffraction pattern of an electromagnetic wave that (a) is close to a source, such as an antenna, light-emitting diode, aperture, or endface of an optical fiber, as distinguished from a far field diffraction pattern and (b) is observed up to a distance, *d*, from the source, given by the relation $d < s^2/\lambda$ where *s* is a characteristic dimension of the source, such as half the length or width, and λ is the wavelength. *Note 1:* If the source is a uniformly illuminated circle, then *s* is the radius of the circle. *Note 2:* The near field diffraction pattern in the output plane is called the "near field radiation pattern." **2.** In fiber optic systems, the region close to a source or an aperture. **3.** A diffraction pattern for an electromagnetic wave observed in the near field region. *Synonym* **Fresnel diffraction pattern.** *See also* **diffraction, far field diffraction pattern, far field region, intermediate field region, near field region.**

near field pattern: *Synonym* **near field radiation pattern.**

near field radiation pattern: 1. In the output plane of a source, a radiation pattern in which (a) the electric and magnetic field strengths of an electromagnetic wave vary inversely as the cube of the distance from the source and (b) other contributions to the field strength, such as those that vary inversely as the square of the distance or inversely as the distance, are negligible. **2.** For an optical fiber, the radiation pattern that describes the distribution of radiant emittance as a function of position in the plane of the exit face of the fiber. *Synonym* **near field pattern.** *See also* **diffraction, electromagnetic radiation, far field radiation pattern, intermediate field radiation pattern, intermediate field region, near field diffraction pattern, radiation pattern.**

near field region: In an electromagnetic wave, the spatial volume (a) that is occupied by the near field, (b) that lies close to an aperture or an antenna, where the electric charge distribution in the antenna, hence the electrostatic field, primarily determines the field strength, which varies primarily as $1/d^3$, where *d* is the distance from the aperture or the antenna, and (c) in which the propagation times from the aperture or the antenna to points within the volume are negligible compared to propagation times to points in the far field region. *Note:* If the antenna has a maximum overall dimension that is not large compared to the wavelength, this field region may not exist. For an antenna focused at infinity, the radiating near field is referred to as the "Fresnel region." *Synonyms* **electrostatic field region, near zone, refractive field region.** *See also* **far field region, intermediate field region, near field.**

near field scanning: In fiber optics, measuring the refractive index profile of an optical fiber by (a) scanning the entrance endface of the fiber with a light source and (b) measuring the point-by-point radiance, i.e., the radiant emittance, at the exit endface. *See also* **refracted ray method.**

near field template: *See* **four-concentric-circle near field template.**

near full motion operation: In video equipment operation, such as television, operation that unconditionally provides a decoded video frame update rate greater than or equal to 6.0 fps (frames per second), but less than the update rate required to describe full motion operation. *Note:* In North America, full motion television (TV) operates at 30 fps. *See also* **full motion operation.**

near infrared: Pertaining to the region of the electromagnetic spectrum that lies between the longer-wavelength end of the visible spectrum and the shorter-wavelength end of the middle infrared region, i.e., from about 0.8 µm (micron) to 3.0 µm. *Note:* The near infrared is included in the near visible region, but the middle infrared and far infrared regions are not included in the near visible region. The far infrared extends to the shorter-wavelength end of the radio wave region, which is also the longer-wavelength end of the optical spectrum, i.e., to about 100 µm (microns). None of the infrared regions, i.e., the near infrared, middle infrared, and far infrared regions, is included in the visible spectrum.

near real time delay: The delay that (a) usually is caused by transmission and automated processing and (b) occurs between the occurrence of an event and the display of the data describing the event at some other location. *See also* **real time.**

near vertical incidence skywave: In radio propagation, a wave that (a) is reflected from the ionosphere at a nearly vertical angle and (b) is used in short range communications to reduce the area of the skip zone and thereby improve reception beyond the limits of ground wave reception.

near zone: *Synonym* **near field region.**

NEC: National Electric Code®.

necessary bandwidth: For a given class of emission, the frequency bandwidth that (a) is just sufficient to ensure the transmission of information at the rate and with the quality required under specified conditions and (b) includes emissions useful for the adequate functioning of the receiving equipment, such as the emission corresponding to the carrier of reduced carrier systems. *Note:* Annex J of *NTIA Manual of Regulations and Procedures for Federal Radio Frequency Management* contains formulas that may be used to calculate necessary bandwidth. *See also* **bandwidth, bandwidth compression, carrier, frequency, nominal bandwidth, occupied bandwidth, spurious emission, spurious response, suppressed carrier transmission.**

negate: 1. To perform the function of a NEGATION gate. **2.** To perform the operation of negation, i.e., the logic NOT operation, such that if A is a true statement, then the statement NOT A is false, and vice versa. *Note:* Negation is one of the *n*-adic Boolean operations in which *n* is 1, i.e., a monadic Boolean operation.

negation circuit: *Deprecated synonym for* **inverter.**

NEGATION gate: In a two-state variable system, such as a binary digital system, a device that is capable of producing the state other than the input state at any given instant, i.e., a binary logic combinational circuit or device that is capable of performing the NOT operation, such that if A is a statement that is true, then NOT A is false, and vice versa. *Note:* In practice, the NEGATION gate converts an input 1 to an output 0 and an input 0 to an output 1. *See also* **combinational circuit.** *Refer to* **Fig. N-2.** *Refer also to* **Figs. B-6 (78), L-6 (529).**

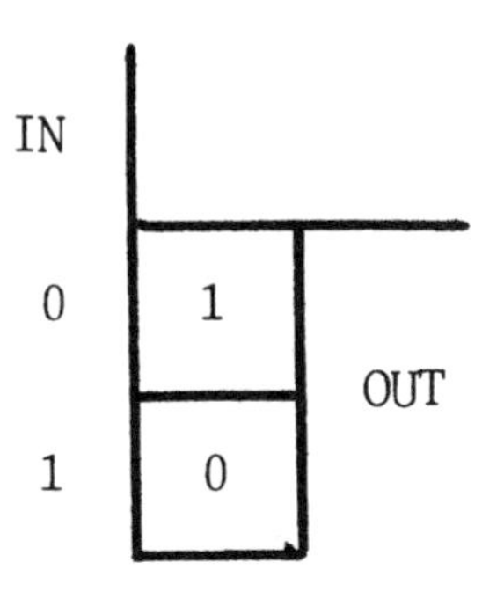

Fig. N-2. A truth table that shows the input and output digits of a **NEGATION gate.**

negative acknowledge: In signal transmission error control that relies on repeating any message received with detected errors, pertaining to the return signals that report the errors. *See also* **error control.**

negative acknowledge character (NAK): A transmission control character that (a) is sent by a station as a negative response to the station with which the connection has been set up, (b) in binary synchronous communications protocol, is used to indicate that an error was detected in the previously received block and that the receiver is ready to accept retransmission of the erroneous block, and (c) in multipoint systems, is used as the not-ready reply to a poll. *See also* **acknowledge character, character, compelled signaling, control character.**

NEGATIVE A IGNORE B gate: A two-input binary logic combinational circuit or device that is capable of performing the logic operation of NEGATIVE A IGNORE B, such that if A is a statement and B is a statement, then the result of logical NEGATIVE (A IGNORE B) is true when A is false and false when A is true, i.e., (a) the output is the inverse of A and independent of B, (b) performance is the same as an A IGNORE B gate that has its output negated, and (c) the operation usually is temporary and controllable, such as a NOR gate, which, upon signal, is temporarily converted to the NEGATIVE A IGNORE B gate. *Note:* True and false may be represented by 1 and 0, or vice versa, and 0 and 1 may be represented by the significant conditions of a signal. *See also* **combinational circuit, gate, significant condition.** *Refer to* **Fig. N-3.**

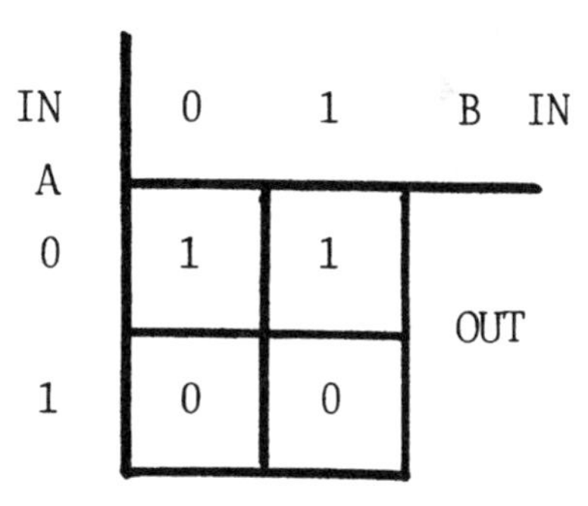

Fig. N-3. A truth table that shows the input and output digits of a **NEGATIVE A IGNORE B gate.**

NEGATIVE B IGNORE A gate: A two-input binary logic combinational circuit or device that is capable of performing the logic operation of NEGATIVE B IGNORE A, such that if A is a statement and B is a statement, then the result of logical NEGATIVE B IGNORE A is true when B is false and false when B is true, i.e., (a) the output is the inverse of B and independent of A, (b) performance is the same as a B IGNORE A gate that has its output negated, and (c) the operation usually is temporary and controllable. *Note:* The NEGATIVE B IGNORE A gate is the same as the NEGATIVE A IGNORE B gate except that the variables are reversed or simply renamed. If the line identities are to remain unchanged, only the terminals need to be reversed. *Note:* True and false may be represented by

1 and 0, or vice versa, and 0 and 1 may be represented by the significant conditions of a signal. *See also* **combinational circuit, gate.** *Refer to* **Fig. N-4.**

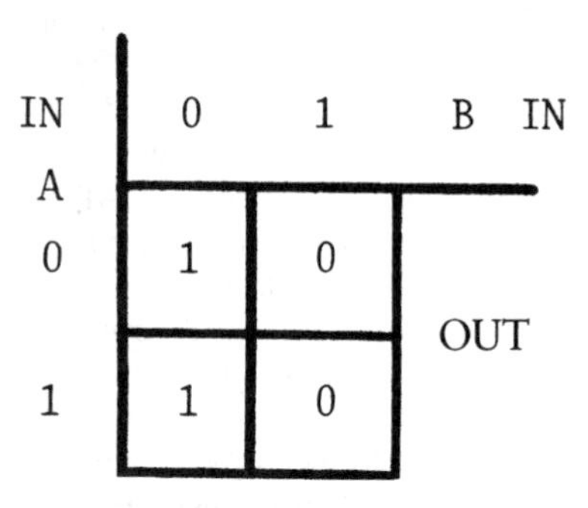

Fig. N-4. A truth table that shows the input and output digits of a **NEGATIVE B IGNORE A gate.**

NEGATIVE B IMPLIES A gate: *Synonym* **B AND NOT A gate.**

negative clearing: In double supervisory working of telephone switchboards, clearing in which (a) an indicator, such as a light disk or a signal lamp, associated with a plug that is in use, turns on, i.e., appears, when a connection is set up, and (b) when a user clears, i.e., hangs up, the associated indicator is turned off or disappears, thus giving a negative indication of the clearing. *See also* **clearing, double supervisory working, positive clearing.**

negative distortion: *Synonym* **early distortion.**

negative entry: The entry of a number along with an associated negative indicator. *See also* **negative indicator.**

negative feedback: The return of a fraction of the output signal of a circuit, device, or system to the input in such a manner that (a) it reduces the amplitude of the input signal, usually by entering the input signal and the feedback signal into a subtractor before entry into the circuit, device, or system, (b) it has the effect of reducing the overall gain of the circuit, device, or system including the subtractor, although the intrinsic forward gain is not affected, (c) distortion is reduced, (d) linearity is improved, and (e) stability is improved. *Note 1:* Negative feedback is used in animate and inanimate systems. *Note 2:* The phase of the signal fed back, relative to the input or the output, must also be taken into consideration. It also has an effect on the overall transfer function of the system. *See also* **feedback, positive feedback.**

negative field-effect transistor integrated receiver: *See* **positive-intrinsic-negative field-effect transistor integrated receiver.**

negative indicator: An indicator or an operator used to ensure that an associated number is treated as a negative number in any process or operation in which it may occur. *Note:* If a number, such as an augend, is to be added to another number, such as the addend, it will be subtracted from the augend if a negative indicator, such as a minus sign, parentheses, or a red color, is present. If a number with a negative indicator is subtracted from a positive number, the numbers will be added, and the result will be a positive number. If two numbers that both have negative indicators are multiplied or divided, the product or the quotient will be positive. If a negative indicator is not present, the number will be treated as a positive number even though a positive indicator, such as a + sign, is not present.

negative justification: *Synonym* **destuffing.**

negative lens: *Synonym* **diverging lens.**

negative meniscus: *Synonym* **divergent meniscus lens.**

negative OR gate: *Synonym* **NOR gate.**

negative photodiode: *See* **positive-intrinsic-negative photodiode.**

negative photodiode coupler: *See* **positive-intrinsic-negative photodiode coupler.**

negative pulse stuffing: *Synonym* **destuffing.**

negative scanning shift: In facsimile transmission sytems, (a) a relative shift of the scanning device with repect to the document being scanned when the surface of the document is scanned along the lines from right to left and from top to bottom, and (b) in a drum scanning apparatus, helicoidal scanning toward the right.

Negotiations Archive: *See* **Automated Recourse to Electronic Negotiations Archive.**

neither NOR gate: *Synonym* **NOR gate.**

***N*-entity:** In the Open Systems Interconnection—Reference Model (OSI-RM), an active element in the *N*th layer that (a) interacts directly with the elements, i.e., the entities, of the layer immediately above or below the *N*th layer, (b) is defined by a unique set of rules, i.e., a unique syntax, and information formats, including data and control formats, and (c) performs a defined set of functions. *Note 1:* The *N* refers to any one of the seven layers of the OSI-RM. *Note 2:* In an existing layered open system, the N may refer to any given layer in the system. *Note 3:* Layers are conventionally numbered from the lowest, i.e., the physical layer, to the highest, so that the $N + 1$ layer is above the Nth layer and the $N - 1$ layer is below. *See also* **format, function, Open Systems Interconnection—Reference Model, protocol, syntax.**

NEP: noise equivalent power.

neper (Np): A unit used to express gain, loss, and relative values, given by the relation $Np = \ln(x_1/x_2)$, where Np is the value in nepers, x_1 is the measured value, x_2 is the reference value, and ln is the natural logarithm, i.e., logarithm to the base $e \approx 2.71828$. *Note 1:* One neper (Np) ≈ 8.686 dB, where 8.686 ≈ 20/ln10. *Note 2:* Nepers are often used to express voltage and current ratios, whereas dB usually are used to express power ratios. *Note 3:* Like the dB, the Np (neper) is a dimensionless unit. *Note 4:* The decineper, equal to 0.1 Np (neper), and the centineper, equal to 0.01 Np, are also used. *Note 5:* The International Telegraph and Telegraph Consultative Committee (CCITT) and the International Radio Consultative Committee (CCIR) Recommendations recognize both units. *See also* **dB.**

nest: 1. To build or place a structure or structures of one kind into a structure of the same kind. *Note:* Examples of nesting are (a) to incorporate a loop, the nested loop, into another loop, the nesting loop, (b) to incorporate one subroutine intact within another subroutine, and (c) to form an algebraic expression in parentheses, then place this expression in parentheses within other expressions inside parentheses, and so on, such as in the expression $a = b(c + d(e - f(g + h)))$, in which the inner parenthetical expressions are nested within the outer ones. To avoid multiple parentheses, braces, { }, and brackets, [], are often used as data delimiters for the nested items, such as loops, subroutines, and algebraic expressions. **2.** *Synonym* **pushdown storage.** *See also* **algebraic expression, data delimiter, loop, nested command menu, subroutine.**

nested command menu: A command menu that is called by a command in another command menu. *Synonym* **nested command list.** *See also* **command menu.**

net: 1. *Synonym* **communications net. 2.** *See* **airborne early warning net, alternate communications net, answering net, calling net, ciphony protected net, command net, computer net, control net, coordination net, direct working local net, directed net, distress communications net, fiber optic net, free net, giganet, ground-air net, local net, maritime amphibious communications net, maritime broadcast communications net, maritime distress communications net, maritime patrol air broadcast net, maritime air reporting net, maritime ship-shore communications net, maritime tactical air operations communications net, mixed net, opeations net, Petri net, radio net, relay net, reporting net, ship-air net, ship-shore net, ship to air net, ship to ship net, ship to shore net, shore net, Telnet, triangulation net, warning net, working net.** *See also* **communications network, network.**

net authentication: Authentication in which a net control station authenticates itself, and all other stations in the net systematically establish their authenticity. *See also* **authentication, net control station.**

net call sign: In a radio net, a call sign that uniquely identifies each station in the net. *See also* **call sign, collective call sign, individual call sign.**

NETCONSTA: *Synonym* **net control station.**

net control station (NCS): A station that (a) performs net control functions, such as controlling traffic and enforcing operational discipline, (b) controls network operational discipline within a given net, (c) is responsible to the master net control station (MNCS), and (d) assumes control of the overall communications system, or portion of the system, as directed by the MNCS. *Synonym* **NETCONSTA.** *See also* **master net control station, net operation, network management.**

net gain: In communications, computer, data processing, and control systems, the overall gain of a transmission circuit. *Note 1:* Net gain is measured by applying a test signal at an appropriate power level at the beginning, i.e., at the input end, of a circuit and measuring the power delivered at the other end, i.e., at the output end. The net gain is calculated by taking ten times the logarithm of the ratio of the output power to the input power, i.e., $10 \log(P_o/P_i)$. The test signal must be chosen so that its power level is within the usual operating range of the circuit being tested. *Note 2:* The net gain expressed in dB may be positive or negative. If the net gain expressed in dB is negative, it is also called the "net loss." *Note 3:* If the net gain is expressed as a ratio, and the ratio is less than unity, a net loss is implied. *See also* **circuit, dB, gain, net loss, standard test signal, standard test tone, transmission level.**

net glossary: An organized cross-referenced list of communications network functional titles of all nets, such as police, fire, rescue, medical emergency, army, navy, air force, coast guard, marine, maritime, air navigational, meteorological, and forestry nets, available for use or contact by a person or an organization, usually along with addresses, frequencies, and telephone numbers.

net loss: The overall loss in a transmission circuit. *Note 1:* The net loss is measured in a manner identical to that of measuring net gain. *Note 2:* If the calculated net gain, expressed in dB, is negative, it is the net loss. If it is expressed as a ratio, and the ratio is equal to or greater than unity, there is no net loss, and there is a net gain. If the ratio is less than unity, there is a net loss, calculated by subtracting the ratio from unity. *See also* **dB, loss, net gain, net loss variation, via net loss.**

net loss variation: The maximum change in net loss that occurs in a specified portion of a communications system during a specified period. *See also* **loss, net loss.**

net margin: In telegraph operation, the margin of the telegraph apparatus when the modulation rate at the input of the apparatus is the standard rate. *See also* **margin, modulation rate.**

net operation: The operation of an organization of stations that (a) is capable of direct communications on a common channel or frequency, (b) allows ordered conferences among participants who have common interests, common information needs, or similar functions to perform, (c) is characterized by adherence to standard formats, (d) adheres to signal operation instructions (SOIs), and (e) is responsive to a common supervisory station, called the "net controller" or the "net control station," that permits access to the net and maintains net discipline. *See also* **communications net, polling, signal operation instruction.**

net radio interface (NRI): An interface between (a) single-channel radio stations that are usually in a net and (b) switched communications systems. *See also* **communications system, interface, radio and wire integration.**

netting: In radiotelegraph communications, the tuning of the frequency of two or more stations, such as when (a) a net is established, (b) stations join a net, or (c) a net is being operated. *Note:* Netting is accomplished under the direction of the net control station. *See* **radar netting.**

netting station: *See* **radar netting station.**

network: **1.** An organization of stations capable of intercommunication but not necessarily on the same channel or circuit. **2.** Two or more interrelated circuits. **3.** A combination of switches, terminals, and circuits in which transmission facilities interconnect the user end instruments directly, i.e., there are no switching, control, or processing centers. **4.** A combination of switches, terminals, and circuits that serves a given purpose. **5.** A combination of circuits and terminals serviced by a single processing center or switching center. **6.** In network topology, an interconnection of nodes by means of branches. **7.** An interconnection of three or more communicating entities, such as terminals, data terminal equipment (DTEs), facsimile machines, and telephones via one or more nodes that usually contain switching equipment. **8.** For a circuit, an interconnection of passive or active electronic components that serves a given purpose. *See* **active network, advanced intelligent network, Advanced Research Projects Agency Network, administrative network, air defense command-control-communications network, analog network, application network, asynchronous network, Automatic Digital Network, Automatic Secure Voice Communications Network, automatic voice network, Automatic Voice Network, balancing network, blocking network, branching network, Broadband Integrated Services Digital Network, centralized computer network, civil communications network, civil fixed telecommunications network, Clos network, common user network, communications network, computer network, data network, data transfer network, decentralized computer network, deemphasis network, Defense Data Network, Defense Information Systems Network, democratically synchronized network, despotically synchronized network, differentiating network, digital network, distributed network, distributed-queue dual-bus network, equivalent network, fiber optic data transfer network, fixed network, fully connected network, heterogeneous network, hierarchical computer network, hierarchically synchronized network, hierarchical network, homogeneous network, hybrid communications network, integrated digital network, integrated services digital network, International Teleprinter Network, internetwork, local area network, long-haul communications network, loop network, mesh network, message switching network, metropolitan area network, military fixed communications network, multidrop network, multinode network, multiple domain network, mutually synchronized network, neural network, nodal network, oligarchically synchronized network, packet switching network, passive network, preemphasis network, public data network, public switched network, public switched telephone network, radiotelephone network, satellite network, seamless network, shaping network, short-haul communications network, single domain**

network, single node network, slotted ring network, star network, subnetwork, switched network, synchronous data network, synchronous fiber optic network, synchronous network, tandem tie trunk network, telecommunications management network, telecommunications network, teletypewriter network, token bus network, token ring network, transparent network, tree network, value-added network, virtual network, weighting network, wide area network. *See also* **branch, channel, circuit, network topology, node, switching center, switching equipment, terminal.** *Refer to* **Figs. B-9 (96), L-4 (506), N-5 through N-15 (622 and 623), S-10 and S-11 (880).**

network access line: A line that connects the communications facility of an organization to a communications network. *Note:* An example of a network access line is a tie line to a private branch exchange (PBX).

network administration: The set of network management functions that (a) provide support services, (b) ensure that the network is used efficiently, (c) ensure that prescribed service quality objectives are met, and (d) include various operating functions, such as network address assignment, routing protocol assignment, routing table configuration, and directory service configuration.

network analog: An analog or a system that usually is not a network and that is used to simulate, represent, or serve as a model for a network, such as an electrical network. *Note:* Examples of network analogs are (a) a digital computer program that, when executed, simulates the performance of an electrical network, communications network, railroad network, or any other type of network and (b) a set of equations, usually differential equations, that represent an electrical network and that can be used to analyze or predict the performance of the real electrical network, such as the differential equation that represents the instantaneous voltages in a series electrical circuit, given by the relation $e = iR + Ldi/dt + q/C = Rdq/dt + Ldq^2/dt^2 + q/c$ where e is the applied voltage, i is the instantaneous current equal to dq/dt, R is the resistance, L is the inductance, di/dt is the instantaneous rate of change of current with respect to time, q is the instantaneous charge on the capacitor, and C is its capacity. The equation can be solved and the performance of the circuit analyzed and predicted, such as when the applied voltage, e, is constant, sinusoidal, or exponential.

network architecture: 1. The design principles, physical configuration, functional organization, operational procedures, and data formats used as the bases for the design, construction, modification, and operation of a communications network. *See* **open network architecture.** *See also* **Open Systems Interconnection—Reference Model. 2.** The structure of an existing communications network, including the physical configuration, facilities, operational structure, operational procedures, and data formats in use. *See also* **centralized operation, distributed control, distributed network, distributed switching, network connectivity.**

network automatic identification: The automatic identificaton of a station or of a user in which the answer back code is provided by the network. *See also* **answer back code.**

network busy hour (NBH): The busy hour calculated for an entire network. *See* **busy hour.** *See also* **switch busy hour, traffic load.**

network configuration: 1. The arrangement of the components that compose a network. *Note:* Examples of network configurations are (a) the network topology, such as the arrangenment of the nodes and the branches of a network, (b) the interconnection of switching centers and connecting trunks, and (c) in open systems interconnection, the layering of protocols, hardware, software, and functions. **2.** A schematic representation of the elements of a network. *See also* **network topology.** *Refer to* **Figs. C-7 (157), N-16 (624), S-9 and S-11 (880), T-5 (1012).**

network congestion: In a communications network, the condition that exists when the network (a) cannot handle the traffic load imposed upon it, (b) has reached saturation, (c) cannot complete additional calls, and (d) informs call originators by messages or signals that trunks are busy. *See also* **contention, saturation.**

network connectivity: 1. The interconnection of the nodes of a network in terms of (a) the switching equipment, trunks, and circuit terminations and (b) their locations, paths, and quantities. **2.** The topological description of a network that specifies the interconnection of the nodes of a network in terms of (a) the switching equipment, trunks, and circuit terminations and (b) their locations, paths, and quantities. *See also* **distributed network, network, network architecture, node.**

network control (NC): In a communications network, the control and the implementation of functions that (a) prevent or eliminate degradation of any part of the network, (b) initiate immediate response to demands that are placed on the network, (c) respond to changes in the network to meet long range requirements, and (d) include (i) immediate circuit utilization actions, (ii) control of circuit quality, (iii) control of equipment performance, (iv) development of procedures for immediate repair, restoration, or replacement of facilities or equipment, (v) liaison with network users and with representatives of other networks, and (vi) the rendering of advice and assistance in network use. *Note:* Network control at the system level usually is called "system control." *See* **operational network control.** *See also* **network control center, network management, system control.**

network control center: In a command, control, and communications network, a location where (a) the network main control function is performed, (b) a switching center, a processing center, or both, may be located for use by the network and local users, and (c) usually a network node is located. *See also* **network control.**

network control channel: In a communications network, a channel between nodes that carries information that (a) concerns the status of the network and (b) is used in the excercise of network control. *See also* **channel, network control, node.**

network controller: In communications systems, a device usually controlled by a programmed computer, that directs the operation of the data links of a communications network. *See also* **computer program, data link.**

network control phase: In a call, a data call, or an information transfer transaction, the period (a) during which network control signals are exchanged between data terminal equipment ((DTE) and the network and (b) that includes the three phases of an information transfer transaction, i.e., the access phase, the information transfer phase, and the disengagement phase. *See also* **call, data call, data terminal equipment, information transfer transaction, network.**

network data collection: In communications systems, (a) the gathering of data from a group of addressees, (b) assembling the data within the communications system, and (c) delivering the data in the form of messages to specified addressees.

network delay: In communications, the time between (a) the instant that the leading edge of a signal or the beginning of the first character of a data unit is entered into a network and (b) the instant that the trailing edge of the signal or the last character of the data unit enters the receiver. *Note:* Network delay (a) includes the transit time and the message length time and (b) does not include network access time and waiting time, such as waiting for a token or waiting for a receiver to indicate that it is ready to receive. *Synonym* **delay time.**

network element (NE): In Integrated Services Digital Networks (ISDNs), a piece of telecommunications equipment that provides support or services to the user. *See also* **network management.**

Network File System (NFS): A proprietary distributed file system that (a) is widely used by Transmission Control Protocol/Internet Protocol vendors, (b) allows different computer systems to share files, and (c) uses User Datagram Protocol (UDP) for data transfer.

network identification: *See* **transit network identification.**

network identification code: *See* **data network identification code.**

network identity: *See* **country network identity.**

network indialing (NID): *Synonym* **network inward dialing.**

network interface: **1.** The point of interconnection between a user terminal and a private or public network. **2.** The point of interconnection between the public switched network and a privately owned terminal. *Note:* The *Code of Federal Regulations,* Title 47, part 68, stipulates the interface parameters. **3.** The point of interconnection between (a) two networks, (b) a network and a portion of another network, or (c) portions of two networks. *See also* **divestiture, entrance facility, gateway, interface, network, network terminating interface, registered jack, service termination point.**

network interface device (NID): **1.** A device that performs interface functions, such as code conversion, protocol conversion, and buffering, required for communications into and out of a network. **2.** A device primarily used within a local area network (LAN) to allow a number of independent devices, with varying protocols, to communicate with each other. *Note 1:* The network interface device (NID) converts each device protocol into a common transmission protocol. *Note 2:* The transmission protocol may be chosen so as to directly accommodate a number of the devices used within the network without the need for protocol conversion for those devices by the network interface device (NID). **3.** A device used to access a bus or a network. *Note:* An example of a network interface device (NID) is a modem. *Synonym* **network interface unit.** *See also* **local area network, medium interface point, network.**

network interface unit (NIU): *Synonym* **network interface device.**

network inward dialing (NID): **1.** In an automatically switched telephone network, a network-provided service feature in which a call originator may dial directly to an extension number at the call receiver facility without human intervention. **2.** In automatically switched telephone networks, the capability of dialing an extension number at the call receiver installation from the call originator end instrument and obtaining a ring and an answer (a) without human intervention altogether or (b) at least without human intervention at the call receiver switching center, switchboard, or exchange. *Synonyms* **direct inward dialing, network indialing.** *See also* **network, network outward dialing, service feature.**

network layer: In open systems architecture, the layer that (a) provides the functions, procedures, and protocol that are needed to control the transfer of data in a specific transmission facility or system and (b) masks the routing and switching characteristics of the data link layer and the physical layer below from the layers above. *Note:* In a fiber optic network, the data link and physical layers consist of fiber optic and electronic components and systems. *See also* Open Systems Interconnection—Reference Model (OSI-RM).

Network Layer: *See* **Open Systems Interconnection—Reference Model.**

network management: The execution of the set of functions (a) that is required for controlling, planning, allocating, deploying, coordinating, and monitoring the resources of a telecommunications network, (b) that includes the performance of various functions, such as initial network planning, frequency allocation, predetermined traffic routing to support load balancing, cryptographic key distribution authorization, configuration management, fault management, security management, performance management, and accounting management, and (c) that does not include (i) customer elements, such as customer end instruments, and (ii) functions performed under system management. *See also* **accounting management, configuration management, fault management, network control, performance management, security management.** *Refer to* **Figs. C-7 (157), N-16 (624).**

network management protocol: *See* **Simple Network Management Protocol.**

network manager: In network management, the entity that initiates requests for management information from managed systems or receives spontaneous management-related information from managed systems.

network outward dialing (NOD): In an automatically switched telephone network, a network-provided service feature in which a call originator may dial directly all network user numbers by dialing through the local switching center, switchboard, or exchange and obtaining a call receiver ring and answer (a) without human intervention at all or (b) without human intervention at the call originator switching center, switchboard, or exchange. *Synonyms* **direct outward dialing, network outdialing.** *See also* **network interface device, network inward dialing, service feature.**

network security: The collective measures that are taken to protect a network, including its hardware, software, and operators, against (a) unauthorized access, (b) accidental or unauthorized willful interference with operations, (c) accidental or unauthorized willful destruction of network facilities and equipment, and (d) accidental or willful bodily harm or injury to operating personnel.

network security information exchange: *See* **Defense Network Security Information Exchange.**

network service: *See* **complementary network service, special network service.**

network simulator: A computer program that (a) enables the animated simulation and display of interconnected systems and components, (b) indicates network performance data, such as response times, queuing data, message delivery, message loss, and component use, (c) enables analysis of embedded or distributed components, such as computers and communications networks, and (d) indicates the effects of changing parameters and protocols.

network slowdown: In communications network operations, reduced operation in which network controls limit the volume of new calls, new data, or new messages accepted by the network. *Note 1:* If priority precedence procedures are being used, these may be invoked during the slowdown. *Note 2:* Network slowdown is invoked when network congestion occurs, such as when buffer availability drops below a threshold or demand exceeds traffic capacity.

network standard: *See* **Synchronous Optical Network standard.**

network supplement: In tape relay communications networks, the alteration of the format of a message or a group of messages by adding control or information symbols to the message. *Note:* Network supplements should be avoided as much as possible.

network terminal number (NTN): In the International Telegraph and Telephone Consultative Committee (CCITT) X.121 format, the sets of digits that compose the complete address of the data terminal end point, i.e., user end instrument. *Note:* For a network terminal number (NTN) that is not part of a national integrated numbering format, the NTN is the 10 digits of the International Telegraph and Telephone Consultative Committee (CCITT) X.25 14-digit address that follow the Data Network Identification Code (DNIC). When part of a national integrated numbering format, the NTN is the 11 digits of the CCITT X.25 14-digit address that follow the DNIC.

network terminating interface (NTI): The point where the network-service-provider responsibility for service begins or ends. **2.** *Synonym* **demarcation point.** *See also* **customer premises equipment, entrance facility, interface, network interface, plant.**

network terminating unit: A simplified form of data circuit-terminating equipment (DCE) that terminates a circuit in a specialized data network. *See also* **data circuit-terminating equipment, data network.**

network termination: 1. Network equipment that performs functions that (a) are necessary for network implementation of integrated services digital network (ISDN) access protocols and (b) are essential for network-provided service features. **2.** At the customer end, the physical termination of the customer digital line (DSL). *See also* **digital subscriber line, Integrated Services Digital Network, line termination, protocol, terminal.**

network termination 1 (NT1): In Integrated Services Digital Networks (ISDNs), a functional group that (a) is equivalent to Layer 1, i.e., the Physical Layer, of the ISO Open Systems Interconnection—Reference Model (OSI-RM), (b) is associated with the proper physical and electromagnetic termination of the network, and (c) performs various network functions, such as Layer 1 line transmission terminations, maintenance, performance monitoring, timing, power transfers, Layer 1 multiplexing, and interface terminations.

network termination 2 (NT2): In Integrated Services Digital Networks, (ISDNs), a functional group that (a) is equivalent to Layer 1 and higher of the CCITT Recommendation X.200 Reference Model, (b) provides NT2-type functions, (c) includes equipment, or combinations of equipment, such as private branch exchanges (PBXs), local area networks (LANs), and terminal controllers, and (d) performs various network functions, such as (i) Layer 2 and Layer 3 protocol handling functions and (ii) Layer 2 and Layer 3 multiplexing, switching, concentration, maintenance, and interface termination functions.

network topology: 1. The branch of science and technology that is devoted to the study and analyses of the configuration and properties of networks. **2.** The specific physical, i.e., the real, or the specific logical, i.e., the virtual, arrangement of the elements of a network. *Note 1:* Two networks have the same topology if their connection configuration is the same, although the networks may differ in physical interconnections, distance between nodes, transmission rates, and signal types. *Note 2:* The basic types of network topologies are bus, linear, mesh, star, tree, and ring topologies. These networks may be interconnected in various combinations giving rise to various hybrid network topologies. *Refer to* **Figs. N-5 through N-16.**

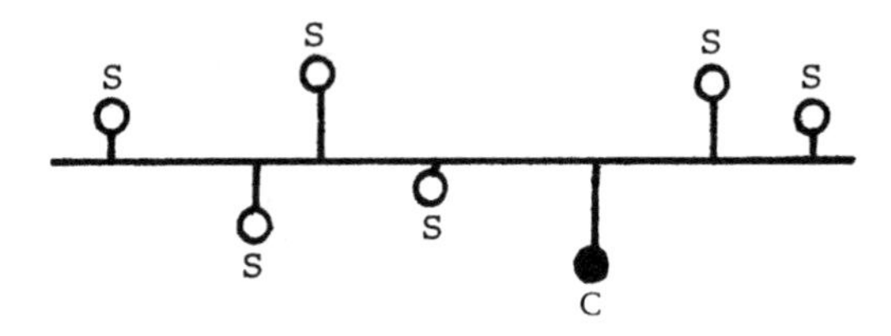

Fig. N-5. A **network topology** for a bus network with a control station.

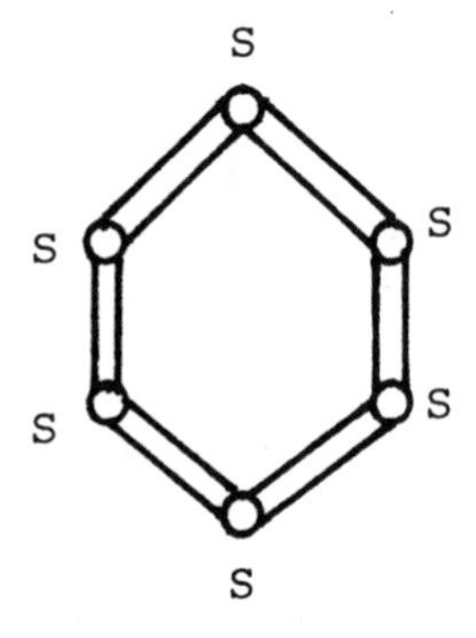

Fig. N-6. A **network topology** for a dual-ring network, such as might be used for a token-passing network.

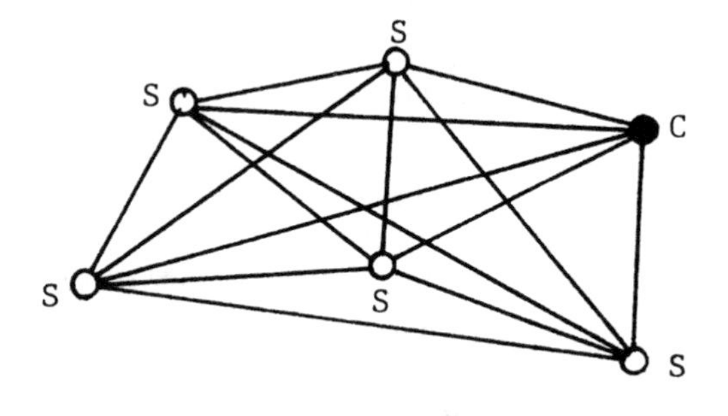

Fig. N-7. A **network topology** for a fully connected network with a control station.

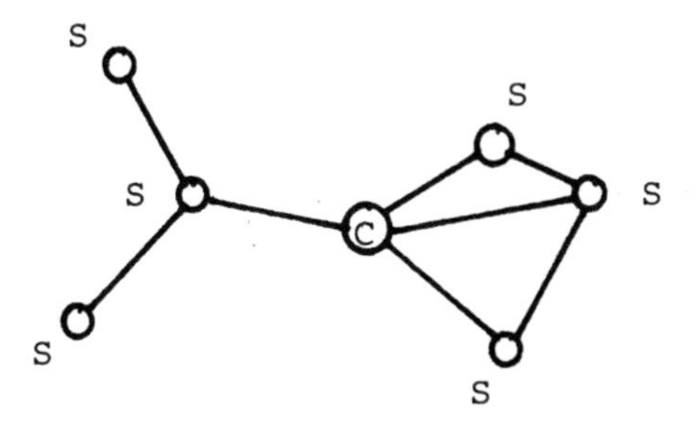

Fig. N-8. A **network topology** for a hybrid network with a control station.

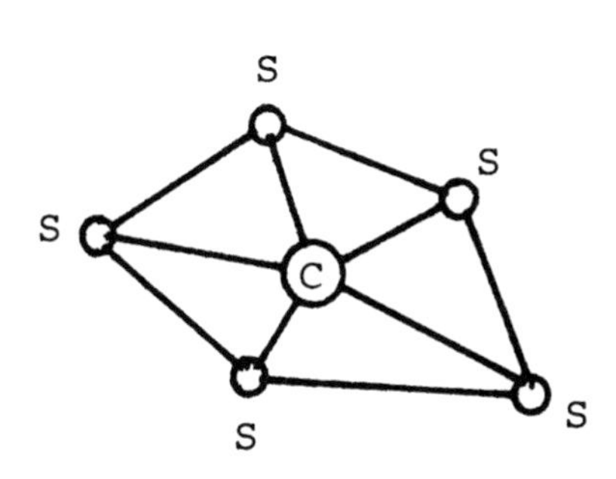

Fig. N-9. A **network topology** for a mesh network with a control station.

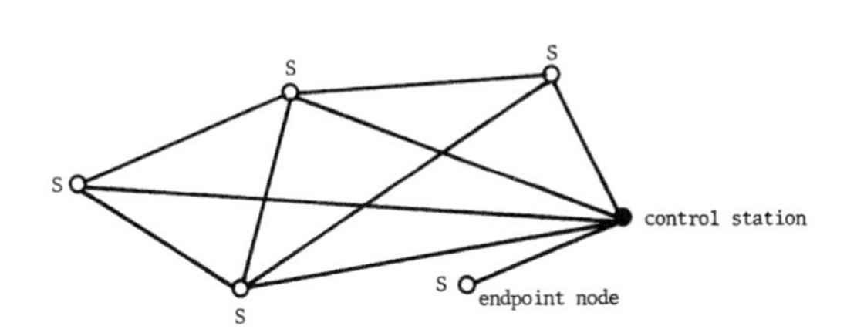

Fig. N-10. A **network topology** for a mesh network with a control station and an end-point node.

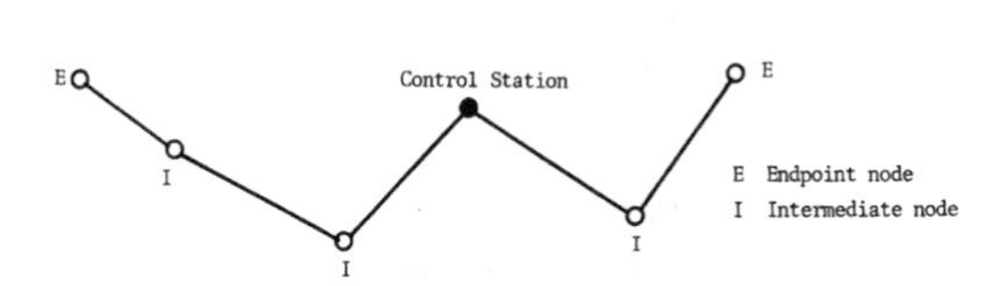

Fig. N-11. A **network topology** for a linear network with a control station and end-point nodes.

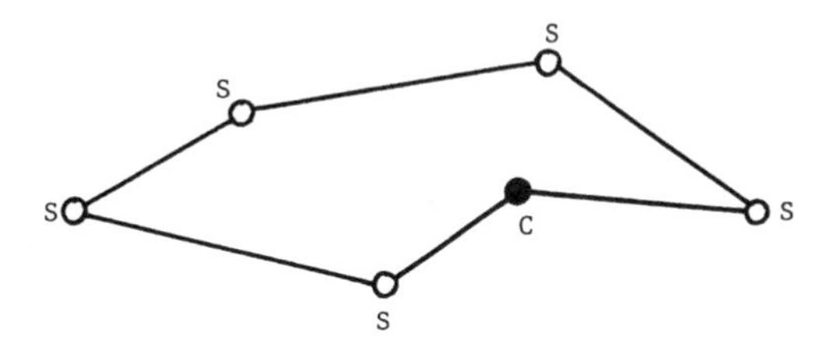

Fig. N-12. A **network topology** for a ring network with a control station.

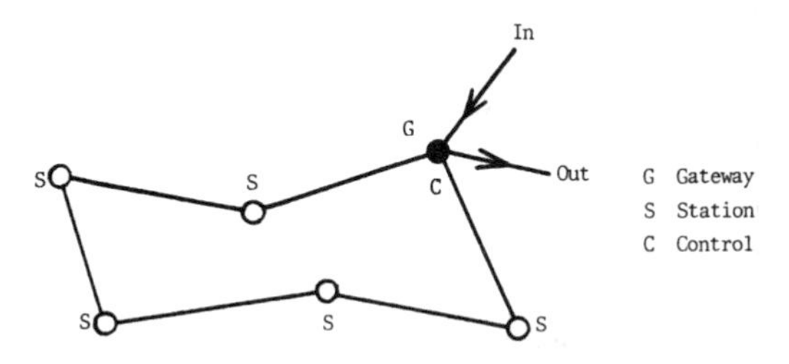

Fig. N-13. A **network topology** for a ring network with a control station that also serves as a gateway to another network.

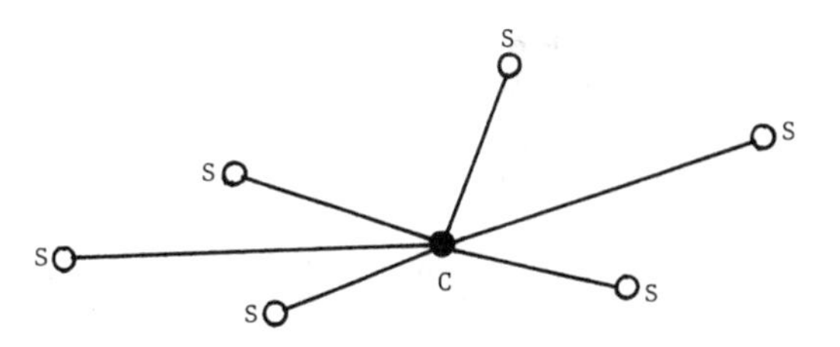

Fig. N-14. A **network topology** for a star network with a control station.

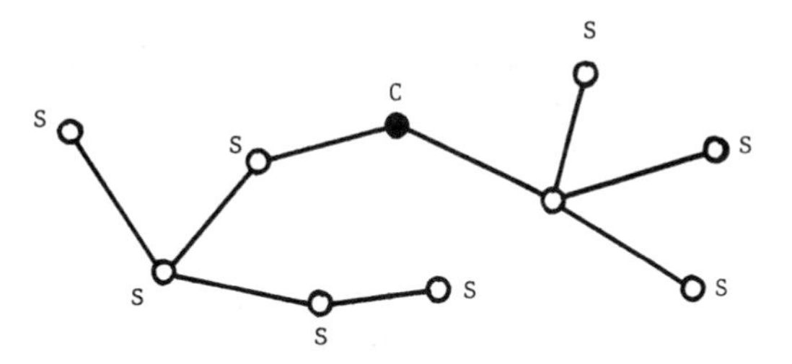

Fig. N-15. A **network topology** for a tree network with a control station.

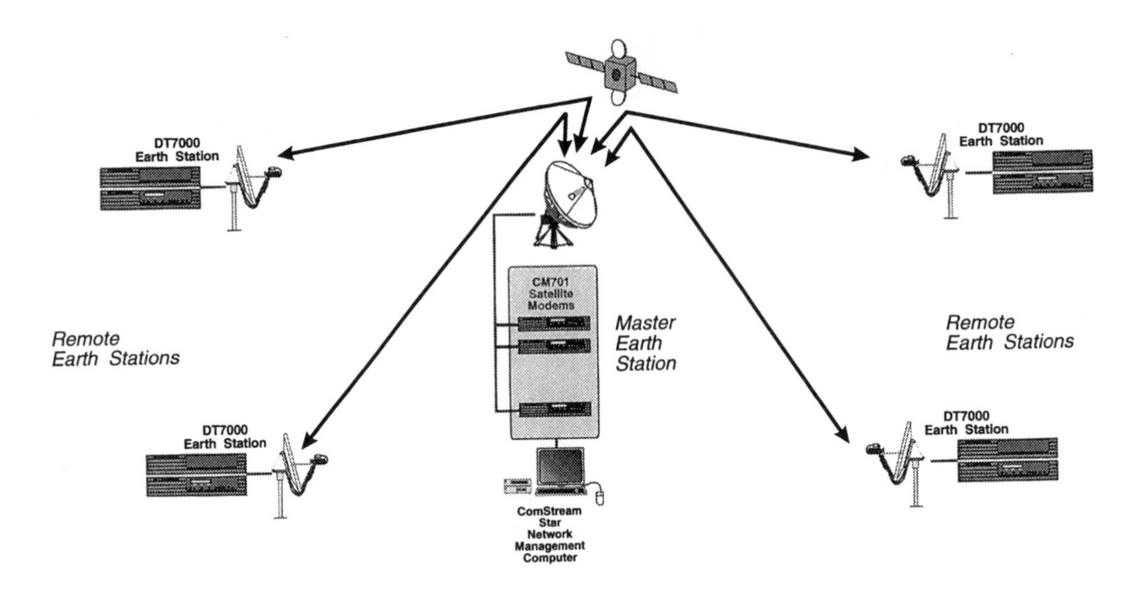

Fig. N-16. A **network topology** for a typical single-channel-per-carrier (SCPC) satellite star network. (Courtesy ComStream Company).

Note 3: Examples of network topologies are:

bus network: 1. A network in which all nodes are serially connected and, except for those at the ends, all nodes are capable of transmitting in, and receiving from, two and only two directions. *Note 1:* All nodes may communicate with all other nodes on the bus because intermediate nodes act as repeaters or passive transparent nodes. *Note 2:* The failure of a single branch linking any two nodes will result in the isolation of a minimum of one node from the rest of the network. **2.** A network in which all nodes, i.e., all stations, are connected together by a single bus.

distributed star coupled network: *Synonym* **star-mesh network.**

fully connected mesh network: A mesh network in which all nodes are directly connected to every other node. *Note:* In a fully connected mesh network, the number of branches is given by the relation $N_{br} = n(n - 1)/2$ branches where N_{br} is the number of branches, and n is the number of nodes.

hybrid network: A network in which at least two basic network topologies are combined to form a single network. *Note 1:* Instances can occur where two basic network topologies, when connected together, can still retain the basic network character, and therefore not be a hybrid network. Thus, a tree network may be connected to a tree network, and the resulting network may still be a tree network, depending on how the two tree networks are connected. Therefore, a hybrid network accrues only when two basic networks are connected, and the resulting network topology fails to meet one of the basic topology definitions. *Note 2:* Examples of hybrid networks are (a) two star networks connected together such that the central node of one is connected to the noncentral end of a branch of the other and (b) several star networks with their central nodes connected by a mesh network, i.e., a multiply connected star network. *Note 3:* A hybrid topology always accrues when two different basic network topologies are connected. *Synonym* **hybrid topology.**

linear network: A network in which there are exactly two end-point nodes, any number of intermediate nodes, and only one path between any two nodes. *Synonym* **linear topology.**

loop network: A network in which (a) each node is directly connected to two and only two nodes, both of which are adjacent to it, (b) one and only one path connects all nodes, (c) there are no end-point nodes, and (d) one or more nodes may serve as a port or a connection to other networks. *See also* **ring network.**

mesh network: 1. A network in which there are at least two nodes that have two or more paths between them. *Note:* A fully connected mesh network is a special case of a mesh network. **2.** A network configuration in which there is more than one path between any two nodes and thus there are no end-point nodes, that is, no nodes connected to only one branch of the network. *Synonym* **mesh topology.** *See* **fully connected mesh network.**

multiply connected star network: *Synonym* **star-mesh network.**

ring network: 1. A network in which every node has exactly two branches connected to it. **2.** A network configuration that consists of a closed loop propagation medium. *Synonym* **ring topology.**

single node network: A network in which all stations, user end instruments, or other devices are interconnected via one node. *Note:* Examples of single node networks are star networks and tree networks. *Note 2:* The functions of the single node may be geographically distributed.

star-mesh network: 1. A mesh network in which one or more nodes serve also as the central node of a star network. Inversely, a star-mesh network is a communications network configuration in which the central nodes of two or more star networks are connected together by a bus, such as by a partially connected or fully connected mesh network bus. **2.** A mesh network in which a large number of local loops or data links are connected to mesh nodes that serve as network control, processing, and switching centers. **3.** A hybrid network in which two or or more star networks have their nodes connected together by a mesh network. *Synonyms* **distributed star coupled bus, distributed star coupled network, multiply connected star network.**

star network: 1. A radial (starlike) configuration of communications network nodes such that (a) each end-point node is connected via a single branch directly to the central node that serves as a central distribution node, and (b) each end-point node can communicate with any other node only via the central node and two branches. *Note 1:* The central node serves as a distribution node to all the end-point nodes. The central node may rebroadcast all received messages to all end-point nodes, including the originating node, or to selected end-point nodes, depending on whether the central node is an active or a passive node. *Note 2:* The failure of a transmission line or channel linking any end-point node to the central node will result in the isolation of that end-point node from all others. *Note 3:* If the central node is passive, the originating node must be able to tolerate the reception of an echo of its own transmission, delayed by the two-way transmission time, i.e., to and from the central node, plus any delay generated in the central node. An active star network, i.e., a star network with an active central node, may have means to prevent echo-related problems. **2.** The interconnection of nodes such that there is a direct path between each node and a central node that serves as a control node. The central node is connected directly to all end-point nodes, which usually are the locations of equipment, such as a data station, ship system, computer, display device, switchboard, or other user end instrument. *Synonym* **star topology.**

tree network: 1. A network in which nodes are interconnected via a series of one or more branches such that there exists a unique path between any two nodes in the network. *Note:* Tree networks cannot contain any closed loops. **2.** A network that, from a purely topological viewpoint, resembles a star network in that individual peripheral nodes are required to transmit to and receive from one other node only, toward a central node, and are not required to act as repeaters or regenerators. *Note 1:* The function of the central node may be distributed. *Note 2:* As in the conventional star network, individual nodes may thus still be isolated from the network by a single point failure of a transmission path to the node. *Note 3:* A single point failure of a transmission path within the distributed node will result in separating two or more stations from the rest of the network. *Synonym* **tree topology.**

network utility: An internetwork administrative signaling mechanism in the call control procedure between packet switching public data networks.

neural network: A network (a) that consists of a simulated interconnection of many nodes via many branches, such as might occur in a real communications network, (b) that accepts real netwok experience data that adjust the weights assigned to the capabilities of the branches, (c) that is trained, i.e., is adjusted, by algorithms in accordance with the experience data it receives, (d) in which, as training data continue to be accepted, algorithms continue to adjust the interconnection weights so that the network outputs approximate more and more closely the desired response for a given set of input data, and (e) in which, if new experience data are accepted, the algorithms readjust the network accordingly.

neutral: 1. The alternating current (ac) power system conductor that (a) is intentionally grounded on the supply side of the service disconnect and (b) provides a current return path for ac currents. *Note:* In a single-phase alternating current (ac) power supply circuit that has three wires and a green safety wire, the neutral is the low potential, i.e., the white wire. The safety ground (green) conductor should not be used to provide a return path for currents, except during fault conditions. The high potential wires usually are red and black. *Note:* In a three-phase alternating current (ac) "Y," i.e., wye, power distribution system, the neutral is the low potential fourth wire that conducts currents only when the "Y" is unbalanced or during a fault condition. **2.** In an electric circuit, pertaining to a point that is not polarized relative to other points in the circuit. *Note:* An example of a neutral point is the common point of a balanced three-phase star, i.e., three-phase "Y," alternating current (ac) power distribution system. *See also* **common return, facility grounding system, fault protection subsystem, ground, neutral ground.**

neutral direct current telegraph system: A telegraph system in which (a) current flows during marking intervals and no current flows during spacing intervals for transmission of signals over a line, and (b) the direction of current flow is irrelevant. *Note:* In polar telegraph systems, the signal alternates between positive and negative voltages, and thus the current flow reverses on the line during transitions between marks and spaces. *Synonyms* **single current system, single current transmission system, single Morse system.** *See also* **communications system.**

neutral ground: An intentional ground applied to the neutral conductor or neutral point of a circuit, transformer, machine, apparatus, or system. *See also* **facility grounding system, fault protection subsystem, ground, neutral.**

neutral operation: A method of teletypewriter operation in which (a) marking signals are formed by current pulses of one polarity, either positive or negative, and (b) spacing signals are formed by reducing the current to zero or nearly zero. *See also* **mark, polar operation, space.**

neutral relay: A relay in which the movement of the armature does not depend upon the direction of the current in the circuit controlling the armature. *See also* **polar operation, relay.**

neutral telegraphy: *Synonym* **single current telegraphy.**

neutral transmission: *Synonym* **single current signaling.**

neutron bombardment resistance: In optical transmission systems, the ability of an optical element to continue to perform its designed function when subjected to high energy neutrons. *Note:* An optical fiber subjected to a 3×10^9 n•cm^{-2} (neutrons per square centimeter) beam can result in an increase in attenuation rate of 0.0005 dB•km^{-1} (decibel per kilometer) per 10^8 n•cm^{-2} (incident), resulting in a 5% to 10% increase in scattering losses, although a 50% recovery occurs 5 min (minutes) after cessation of bombardment.

new address signal: *See* **redirected to new address signal, redirect to new address signal.**

new customer premises equipment: All customer premises equipment not in service or in the inventory of a regulated telephone utility as of December 31, 1982. *See also* **customer premises equipment, embedded customer premises equipment.**

new-line (NL) character: A format effector that causes the print or display portion of a printer to move to the first print or display position on the next line. *See also* **format effector, line, position.**

newton (N): The SI (Système International d'Unités) unit of force such that 1 N (newton) produces an acceleration of 1 m•s^{-2} (meter per second square) when acting on a 1-kg (kilogram) mass.

Newton's fringes: *Synonym* **Newton's rings.**

Newton's rings: The series of rings, bands, or fringes that are formed when (a) two clean polished surfaces are placed in contact with a thin air film between them, and (b) reflected, usually chromatic, beams of light from the two adjacent surfaces interfere with each other, causing alternate cancellation and reinforcement of light as the distance between the surfaces are multiples or submultiples of the wavelength. *Note:* By counting Newton's rings from the point of actual contact of two surfaces, the departure of one surface from the other can be determined. The regularity of the rings maps the regularity of the distance between the two surfaces. This is the usual method of determining the fit of a surface under test to a standard surface of a test glass. *Synonyms* **Newton's bands, Newton's fringes.** *See also* **beam, interference, map.**

NF: noise figure.

NFS: Network File System.

***N*-function:** In the Open Systems Interconnection—Reference Model, a defined action that (a) is performed by an *N*-entity and (b) may be (i) a single action, i.e., a primitive function, or (ii) a set of actions. *See also* **function, *N*-entity, *N*-layer, Open Systems Interconnection—Reference Model.**

nibble: *Deprecated.* Obsolete term for part of a byte, usually half of a byte. *See* **byte.**

NID: network interface device, network inward dialing.

night effect: The variations in the state of polarization of electromagnetic waves that (a) are reflected by the ionosphere, (b) sometimes result in errors in direction finding (DF) bearings, and (c) are most frequent and most pronounced at nightfall. *See also* **direction finding, electromagnetic wave, ionosphere, polarization.**

Night Hawk: In a network, a multilevel security (MLS) router that (a) receives data passing over the network from a variety of internal and external sources, (b) processes the data by acting as a firewall, blocking all data, (c) uses a sophisticated identification and labeling system to determine the security classification of the information, such as whether the information relates to routine missions or to classified contingency missions, permits or denies access accordingly, and disseminates all incoming information in accordance with the classification or other criteria. *See also* **router.**

NII: national information infrastructure.

nine-hundred (900) service: A network-provided service feature in which the call originator is charged for access to information on a charge-per-call or charge-per-unit time basis, the charge basically being furnished to the call receiver who furnishes the information. *Synonym* **900 service.**

nines complement: The complement of a decimal number obtained by (a) taking each digit of the decimal number, place by place, (b) subtracting it from 9, and (c) recording the results of each subtraction in the corresponding position, disregarding any carries, and ensuring that there are as many digits in the nines complement as there are in the original number unless the lead zeros, if any, are eliminated. *Note:* The nines complement of 496544 is 503455, and vice versa. Either number is the nines complement of the other. A practical application of the nines complement is in the excess-three binary coded decimal numeration system. Decimal 3 is represented in excess-three code as 0110. If this is stored in a bank of four flip-flops, the opposite sides of the flip-flops would be 1001, which represents 6, the nines complement of 3, the binary numerals being in the excess-three code. Thus, each decimal digit and its nines complement are immediately available, which has many practical applications and advantages in performing calculations. The nines complement is a radix-minus-one complement where the radix is 10. Also, the largest four-digit binary numeral is 1111, which corresponds to decimal 15. The difference between 9 and 15 is 6, the nines complement of 3, the basis of the excess-three code. *See also* **numeration system, radix.**

niobate integrated circuit: *See* **lithium niobate integrated circuit.**

NIU: network interface unit.

NL: new-line character.

nm: nanometer.

NMCS: National Military Command System.

nmi: nautical mile.

NNI: network-network interface

NOD: network outdialing, network outward dialing.

nodal clock: At a particular node in a network, the principal clock or the alternate clock that provides the timing reference for all major functions at that node.

nodal point: *Synonym* **node.**

nodal switching center: A switching center that (a) is within a telephone network of an integrated communications system, (b) uses circuit switching techniques, and (c) contains communications electronic equipment designed to interconnect terminals with other circuits in the network. *See also* **integrated communications system, switching center.**

node: 1. In network topology, a terminal of any branch of a network or an interconnection common to two or more branches of a network. *Synonyms* **junction point, nodal point.** *See* **end-point node.** *See also* **branch, network. 2.** In a switched network, one of the switches that (a) form the network backbone and (b) may also include patching and control facilities. **3.** In a data network, a point at which one or more data transmission lines are interconnected. **4.** In a data network, the location of a data station where one or more functional units interconnect transmission lines or data links. *Synonyms* **junction point, nodal point, vertex. 5.** A technical control facility (TCF). *See also* **communications, extension facility, extension terminal, interface message processor, network 6.** A point in a standing wave, i.e., a stationary wave, at which the amplitude is a minimum. *Note:* In electrical systems, the node should be identified as a current node or a voltage node. Nodes occur in acoustic waves, such as when echoes occur, and in elastic waves, such as in vibrating strings. *Synonym* **null. 7.** In a communications, computer, data processing, control, and telemetry network, a functional unit, such as a station, terminal installation, terminal, termination, communications computer, or repeater, connected to the network. *See* **active node, adjacent node, antinode, data processing node, end-point node, host node, inactive node, intermediate node, message switching node, packet switching node.** *See also* **standing wave ratio.**

nodeless network: A communications network in which all connections are direct between all users. *See also* **communications network.**

noise: 1. Within a given frequency band used in a communications system, unwanted or disturbing energy introduced into the communications system from man-made and natural sources. **2.** A disturbance that (a) affects a signal and (b) may alter the information carried by the signal. **3.** Random variations of one or more characteristics of any entity, such as voltage, current, or data. **4.** A random signal of known statistical properties, such as amplitude, distribution, and spectral density. **5.** A disturbance that tends to, or does, interfere with the proper operation of a device or a system. *Note:* In video systems, noise has been given various descriptive names, such as snow, rain, stripes, and running rabbits. *See* **acoustic noise, atmospheric noise, background noise, black noise, blue noise, cosmic noise, equipment intermodulation noise, equivalent pulse code modulation noise, external noise, feeder echo noise, galactic noise, idle channel noise, impulse noise, intermodulation noise, intrinsic noise, Johnson noise, loop noise, modal noise, mosquito noise, notched noise, path intermodulation noise, phase noise, pink noise, polarization noise, pseudorandom noise, quantizing noise, quantum noise, random noise, reference noise, shot noise, sky noise, solar noise, spot noise, thermal noise, triangular random noise, total channel noise, white noise.** *See also* **antenna noise temperature, circuit noise level, precipitation static.**

noise amplification jammer: *See* **direct noise amplification jammer.**

noise bandwidth: *See* **closed loop noise bandwidth.**

noise current: 1. The electrical current caused by noise voltage. **2.** An interfering and unwanted electrical current in a device, circuit, or system. **3.** In optical communications systems, a root mean square (rms) component of the photodetector electrical output current that occurs when signal power is not applied. *See also* **noise voltage.**

noise density: *See* **carrier to receiver noise density.**

noise equivalent power (NEP): 1. At a given data signaling rate (DSR) or a given modulation frequency, operating wavelength, and effective noise bandwidth, the radiant power input that (a) produces a signal to noise ratio of unity at the output of a given photodetector and (b) usually is measured with a blackbody radiation source at 500 kHz (kilohertz) and a bandwidth of 1 Hz or 5 Hz. *Note:* The modulation rate varies with the type of detector and usually is between 10 Hz (hertz) and 1000 Hz. It is essential that the measurement conditions be specified in indicating a value of noise equivalent power (NEP). NEP is only useful for comparison when all measurement conditions, such as frequency, modulation rate, bandwidth, detector area, and temperature, are specified. **2.** The minimum detectable power per square root bandwidth. *Note:* The noise equivalent power (NEP) has the units of watts/square root hertz, $W \bullet Hz^{-1/2}$ on $W \bullet s^{1/2}$. The total NEP usually is expressed in watts, milliwatts, microwatts, or picowatts. In this sense, the NEP is a misnomer because power should be expressed in power units, such as watts. **3.** In photodetectors used in fiber optic systems, the radiant power that produces a signal to dark current noise ratio of unity. *Note 1:* The noise equivalent power (NEP) is a common parameter in specifying detector performance. It is useful for comparison only if modulation frequency, bandwidth, detector area, and temperature are specified. *Note 2:* The noise equivalent power (NEP) measurement is valid only if the dark current noise dominates the noise level. **4.** In optics, the root mean square (rms) value of optical power required to produce unity rms signal to noise ratio. *See also* **bandwidth, blackbody, dark current, detectivity, noise, photodetector, signal to noise ratio, specific detectivity.**

noise factor: *Synonym* **noise figure.**

noise figure (NF): 1. In a device, such as a transducer or an amplifier, the ratio of (a) the output noise power to (b) the portion of the output noise power stemming from thermal noise at the input at standard noise temperature, usually 290 K (kelvin). *Note:* The noise figure usually is expressed in dB. **2.** The ratio of (a) the output noise power to (b) the noise power that would remain if the device itself did not introduce noise. *Note 1:* In heterodyne systems, the output noise includes spurious noise from image frequency transformations. *Note 2:* In heterodyne systems at standard noise temperature, the portion of output noise power stemming from thermal noise at the input includes that which appears via the principal frequency transformation and excludes that which appears via the image frequency transformations. **3.** The amount of noise added by signal-handling equipment to the noise existing at its input, equal to the signal to noise ratio (S/N) at the input divided by the S/N ratio at the output. **4.** A measure of the deterioration of the signal to noise power density ratio that occurs in the process of amplification, equal to the actual noise power at the output divided by the noise power that is calculated by assuming that the device itself is noiseless, i.e., always greater than unity, given by the relation $NF = (T_n + 290)/290$ where NF is the noise figure and T_n is the effective input noise temperature. *Note:* Unless otherwise stated, the input noise is assumed to be that which would be caused by a matched resistor at the input at room temperature, i.e.,

290 K (kelvin). *Synonym* **noise factor.** *See also* **effective input noise temperature, noise, noise power density, thermal noise.**

noise improvement: *See* **angle modulation noise improvement.**

noise jammer: A jammer that emits a signal in which the power is randomly distributed over a wide band of frequencies.

noise jamming: Jamming in which (a) a carrier wave is modulated by noise, (b) noise at the desired output frequencies is amplified and radiated without a carrier, or (c) in radar and communications systems, a signal is transmitted that has many different and usually random frequencies, such as white noise or spot noise. *See also* **carrier, jamming, spot noise, white noise.**

noise level: The noise power, usually relative to a reference level. *Note 1:* Noise level usually is expressed in dB for relative power and picowatts, or dBpW, for absolute power. A suffix is added to denote a particular reference or specific qualities of the measurement. *Note 2:* Examples of noise level measurement units are dBa, dB(F1A), dBa(HA1), dBa0, dBm, dBm(psoph), dBm0p, dBp, dBrn, dBrnC, dBrnC0, dBrn(f_1-f_2), dBrn(144-line), pW, pWp, and pWp0. *See* **ambient noise level, carrier noise level, channel noise level, circuit noise level.** *See also* **noise, noise weighting, signal to noise ratio.**

noise-limited operation: *See* **detector noise-limited operation, quantum-noise-limited operation, thermal-noise-limited operation.**

noise loss: *See* **intermodulation noise loss.**

noise measurement unit (NMU): A unit of noise power measurement, either relative or absolute. *Note:* The dB is the most common unit for noise power measurements. The picowatt is also used. To obtain an absolute unit when expressing noise levels in dB, a suffix is added to indicate the reference level. A suffix is also added to indicate a particular measuring instrument or other specific qualities or conditions of the measurement. Examples of noise measurement units include dBa, dBa(F1A), dBa(HA1), dBa0, dBm, dBm0, dBm(Psoph), dBrn, dBrn(144-line), dBrnC, dBrn(f_1-f_2), pW, pWp, and pWp0. *See also* **noise weighting.**

noise-modulated jamming: *See* **slow sweep noise-modulated jamming.**

noise plus distortion ratio: *See* **signal plus noise plus distortion to noise plus distortion ratio.**

noise power: 1. Interfering and unwanted power in an electrical device, circuit, or system. **2.** The power generated by a random electromagnetic process. **3.** The power that is developed by unwanted electromagnetic waves from all sources in the output of a device, such as a transmission channel or an amplifier, expressed as a function of the noise voltage squared and the equivalent source resistance. **4.** In a system or a device, the total noise of electromagnetic waves that have frequencies within the passband of the system or the device, including crosstalk, distortion, and intermodulation products. **5.** In the acceptance testing of radio transmitters, the mean power supplied to the antenna transmission line by a radio transmitter when loaded with noise having a Gaussian amplitude versus frequency distribution rather than a constant distribution of amplitude. **6.** In information theory, a disturbance that does not represent any information in a message or from a signal source. *See* **flat noise power, received noise power.** *See also* **black noise, blue noise, crosstalk, distortion, intermodulation product, noise, noise equivalent power, pink noise, white noise.**

noise power density (NPD): The noise power in a bandwidth of 1 Hz (hertz), i.e., the noise power per hertz at a point in a noise spectrum. *Note:* The noise power density of the internal noise that is contributed by a receiving system to an incoming signal is expressed as the product of Boltzmann's constant, k, and the equivalent noise temperature, T_n. Thus, the noise power density is often expressed simply as kT. *Synonym* **kT.**

noise power ratio: *See* **carrier to noise power ratio, in-band noise power ratio, single sideband noise power ratio.**

noise ratio: *See* **carrier to noise ratio, energy to noise ratio, signal plus noise to noise ratio, signal to noise ratio.**

noise ratio per bit: *See* **signal to noise ratio per bit.**

noise resistance: In a circuit, device, equipment, or system, the capability of preventing or reducing the development of noise within and at the output of the circuit, device, equipment, or system. *See* **equivalent noise resistance.**

noise strobe: 1. In radar systems, a scan for the purpose of determining noise levels as a function of azimuth and elevation. **2.** A display of noise that is obtained by a radar strobe pulse. **3.** In electronic warfare systems, a scanning of the frequency spectrum for noise levels. *See also* **azimuth, elevation, frequency spectrum, noise, radar strobe.**

noise suppression: **1.** The reduction of the noise power level in a device, circuit, or system. **2.** The automatic reducing of the noise output of a receiver during periods when a carrier is not being received. *See also* **squelch circuit.**

noise temperature: At a pair of terminals in a network, the temperature of a passive system that has an available noise power per unit bandwidth at a specified frequency equal to that of the actual terminals of the network. *Note:* The noise temperature of a simple resistor is the actual thermal temperature of that resistor. The noise temperature of a diode in operation may be many times the actual thermal temperature of the diode. *See* **antenna gain to noise temperature, antenna noise temperature, effective input noise temperature, effective noise temperature, equivalent noise temperature, equivalent satellite link noise temperature, front end noise temperature, receiving system noise temperature, standard noise temperature.**

noise voltage: **1.** Interfering and unwanted voltage in an electronic device, circuit, or system. **2.** In optical communications systems, a root mean square (rms) component of the photodetector electrical output voltage. *Note 1:* The noise voltage is incoherent with the signal radiant power. *Note 2:* Noise voltage usually is measured with the signal power removed. *See also* **noise current.**

noise weighting: A specific amplitude versus frequency characteristic that permits a measuring set to give numerical readings that approximate the interfering effects to any listener using a particular class of telephone instrument. *Note 1:* Noise weighting measurements are made in lines terminated by either the measuring set or the class of instrument. *Note 2:* The noise weightings usually were established by agencies concerned with public telephone service. They were based on characteristics of specific commercial telephone instruments, and they represented successive stages of technological development. The coding of commercial apparatus appears in the nomenclature of certain weightings. The same weighting nomenclature and units are used in military versions of commercial noise measuring sets. *See also* **dBa, dBm0p, dBrn, dBrnC, dBrn(f_1-f_2), dBrn(144-line), noise measuring unit, psophometric voltage, weighting network.**

noise window: A trough, i.e., a dip, in the noise frequency spectrum characteristic of a device, such as a transmitter, receiver, channel, or amplifier, from external sources or internal sources. *Note:* The noise window usually is represented as a band of lower amplitude noise in a wider band of higher amplitude noise.

noisy black: In facsimile or other display systems, such as television (TV), a nonuniformity in the black area of an object, such as a document, picture, or image. *Note 1:* In scanning of a black area, noisy black causes signal level changes that constitute noise. *Note 2:* An example of noisy black is a signal that is supposed to represent a black area on the object, but that has a noise content that causes the creation of white spots on the display surface or the record medium. *See also* **noisy white.**

noisy white: In facsimile or other display systems, such as television (TV), a nonuniformity in the white area of an object, such as a document, picture, or image. *Note 1:* In scanning of a white area, noisy white causes signal level changes that constitute noise. *Note 2:* An example of noisy white is a signal that is supposed to represent a white area on the object, but that has a noise content that causes the creation of black spots on the display surface or the record medium. *See also* **noisy black.**

no-landing marker: *Synonym* **forbidden landing point marker.**

nomadic communications service (NCS): A communications service that (a) provides access to a switched telecommunications system from any point, either worldwide or within a specified area, (b) provides communication from stationary or moving stations, and (c) usually provides multimedia services. *See also* **personal communications service.**

no-operation instruction: *Synonym* **no-op instruction.**

no-print key: A key that is used to prevent a printer, teletypewriter, or typewriter from printing or typing a character.

nominal band center frequency: The arithmetic or geometric mean of the lower and upper limits of a frequency band. *Note 1:* The arithmetic mean of the frequency band is given by the relation $f_{ca} = (f_l + f_u)/2$ where f_l is the lower limit and f_u is the upper limit of the frequency band. The arithmetic mean usually is used when constant-bandwidth filters, such as 50 Hz (hertz), are specified. *Note 2:* The geometric mean is given by the relation $f_{cg} = (f_l \times f_u)^{1/2}$. The geometric mean usually is used for filters of constant-percentage bandwidth, such as one-third octave band filters.

nominal bandwidth: The widest band of frequencies, inclusive of guard bands, assigned to a channel. *See also* **bandwidth, frequency, necessary bandwidth, occupied bandwidth, radio frequency bandwidth.**

nominal bit stuffing rate: The rate at which stuffing bits are inserted or deleted when both the input and output bit rates are at their nominal values. *See also* **binary digit, bit stuffing, destuffing, maximum stuffing rate.**

nominal black: In display systems, such as facsimile, television, and radar, a characteristic or a value, such as a signal level or a frequency, that corresponds to the darkest black that can be represented or transmitted by the system. *See also* **nominal white, value.**

nominal central wavelength: In an optical transmitter, the wavelength (a) at which the effective optical power is concentrated, such as the wavelength of the peak mode or a power weighted measurement, and (b) that lies within the total range of transmitter wavelengths caused by worst case factors, such as manufacturing tolerances, temperature variations, and aging, when operating under standard or extended conditions. *See also* **ultralow-loss fiber**.

nominal linewidth: In facsimile systems, the average separation between centers of adjacent scanning or recording lines. *See also* **facsimile.**

nominal margin: In communications systems, the minimum value that is set for the effective operating margin of a device, such as a receiver, detector, transducer, or transmitter, while operating under standard adjustment and environmental conditions. *See also* **margin, value.**

nominal value: The assigned, specified, or intended value of a quantity. *Note:* Allowable tolerances in nominal values should be specified.

nominal white: In display systems, such as facsimile, television, and radar, a characteristic or a value, such as a signal level or a frequency, that corresponds to the brightest white that can be represented or transmitted by the system. *See also* **nominal black, value.**

nonassociated common channel signaling: Common channel signaling in which the signaling channel serves one or more trunk groups, at least one of which terminates at a point other than the signal transfer point at which the signaling channel terminates. *See also* **associated common channel signaling, common channel signaling, signal transfer point.**

nonblocking: Pertaining to a communications system in which 100% of all access attempts, i.e., all calls, are completed. *See* **blocking probability, completed call.**

nonblocking switch: 1. A switch that has enough paths across it that a call can always reach an idle line without encountering a busy condition on the way. **2.** A switching network or system in which any idle outlet can always be reached from any given inlet under all traffic conditions. *See also* **busy, call, idle line, path, switch, switching center, switching matrix, traffic.**

noncentralized operation: In communications systems, a control discipline for multipoint data communications links in which transmission may be between tributary stations or between network control stations and tributary stations. *See also* **multipoint, multipoint circuit, multipoint connection, multipoint link, network control station, tributary station.**

noncircularity: 1. *Synonym* **ovality. 2.** *See* **cladding noncircularity, core noncircularity, reference surface noncircularity.**

noncoherent bundle: *Synonym* **unaligned bundle.**

nonconjunction gate: *Synonym* **NAND gate.**

noncritical technical load: Of the total technical load at a facility during normal operation, the part that is not required for synchronous operation and operation of automatic switching equipment. *See also* **critical technical load, load, technical load.**

nondestructive cursor: A cursor that does not erase the symbols or the characters at its current display position or at any of the display positions over which it passes. *See also* **cursor, display position, erase.**

nondestructive storage: Storage that does not, or may not, have its contents erased at a location when data are read from that location. *Note:* Examples of nondestructive storage are (a) card, punch card, double core, and permanent storage and (b) magnetic tape, disk, drum, and card storage. *See also* **destructive storage, nonvolatile storage, volatile storage.**

nondirectional beacon: A radionavigation aid that (a) consists of transmitters, receivers, and antennas for direction finding and radio determination, (b) usually has a range that varies with the power output of the transmission source and the sensitivity of a receiver, but rarely exceeds 500 km (kilometers), (c) provides a single bearing, and (d) enables resection and radio determination from two beacons. *Note 1:* Short ranges and fairly large resection angles are necessary to achieve location precision. *Note 2:* Nondirectional implies omnidirectional, meaning all directions, not "no" directions. *See also* **directional antenna, direction finding, radio determination, radionavigation aid, resection.**

nondisjunction gate: *Synonym* **NOR gate.**

nondispersive medium: In the propagation of electromagnetic waves in material media, a propagation medium in which (a) the phase velocity does not vary with frequency, and (b) all the different wavelengths of a pulse will arrive at a given point at the same time, i.e., the phase velocity and the group velocity are equal. *Note:* All materials are dispersive to some extent. *See also* **dispersive medium, propagation medium.**

nonequality gate: *Synonym* **EXCLUSIVE OR gate.**

nonequilibrium length: *Synonym* **nonequilibrium modal power distribution length.**

nonequilibrium modal power distribution: In an electromagnetic wave propagating in a multimode waveguide, such as a multimode optical fiber, the distribution of radiant power among modes such that (a) the fraction of the total propagating power in each mode changes, i.e., the total power is continuously redistributed among the modes, as the wave propagates along the waveguide, and (b) the fractional power distribution among the modes will continue to change until the equilibrium length is reached, after which the fractional power in each mode will remain constant. *Note:* In an electromagnetic wave propagating in a waveguide, the total power will diminish because of scattering and absorption.

nonequilibrium modal power distribution length: The distance in a waveguide, such as an optical fiber, between the entrance endface and the beginning of equilibrium modal power distribution, i.e., the point at which the modal radiant power fractional distribution no longer changes with distance along the guide. *Note:* The nonequilibrium modal power distribution length and the equilibrium length are the same because the end of nonequilibrium and the beginning of equilibrium occur at the same point. *Synonym* **nonequilibrium length.** *See* **equilibrium length.**

nonequivalence gate: *Synonym* **EXCLUSIVE OR gate.**

nonerasable memory: *Synonym* **read-only memory.**

nonescaping key: A key that permits a character to be typed or printed without subsequent change in print position. *See also* **key.**

nongeostationary satellite: A satellite that does not meet all the conditions for a geostionary satellite, i.e., a satellite that is not at the proper altitude, is in an orbit that is not in the equatorial plane, is not moving in the same direction and in synchronism with the Earth's rotation, and does not remain fixed with respect to all points on the Earth, but might meet some of the conditions, such as (a) remain in synchronism with the Earth's rotation but not in the equatorial plane or (b) be at the altitude of a geostationary satellite but in a retrograde orbit. *See also* **geostationary satellite, orbit, retrograde orbit, satellite.**

nonimpact printer: A printer that prints characters without using type (a) to mechanically press specially treated materials, such as an inked or carbon ribbon or a sheet of carbon paper against plain paper or (b) to apply pressure to a data medium, such as a sheet of special chemically treated pressure-sensitive paper. *Note:* An example of a nonimpact printer is an ink jet printer. *See also* **carbon ribbon, data medium, inked ribbon, ink jet printer, printer, type.**

noninertial guidance system: A guidance system that does not make use of the principle of inertia, e.g., it does not doubly integrate the output of an accelerometer to obtain distance, but uses other means of guidance, such as (a) command, beam riding, and homing guidance systems, which are subject to interference and jamming, and (b) preprogrammed and logging systems, which are not subject to interference and jamming. *See also* **inertia, inertial guidance system.**

noninteractive: Pertaining to a mode of operation of a system in which there is no interaction between the system and an operator, except perhaps for support functions, such as to install a program, insert initial data, remove output results on a data medium, initialize variables, and apply power. *See also* **conversational, conversational mode, interactive.**

nonlatching: Pertaining to a switch that requires continuous application of an activating force or signal to cause the switch to remain in the position that it is in. *See also* **latching.**

nonlatching switch: 1. A switch that will remain in a given position only as long as a force or energy is applied. **2.** In fiber optics, a switch that (a) selectively transfers optical signals from one optical fiber to another when an actuating force or signal is applied, (b) continues to transfer signals only as long as the actuating force or signal is continuously applied, and (c) when the actuating force or signal is removed, reverts to its original position. *See also* **latching switch.**

nonlinear device: A device that (a) has an output that is not a linear function of its input and (b) operates such that, if the output is represented by the variable y and the input by the variable x, then the output is related to the input by a nonlinear function, such as by the relation $y = ax^2 + bx + c$ where a, b, and c are constants. *Note:* An example of a nonlinear device is a device in which the output electric field, voltage, or current is not linearly proportional to the input electric field, voltage, or current. Such a device might have a resistance (varistor) that is a function of the current within it such that, for the relation $e = ir$ where e is the instantaneous output voltage, i is the instantaneous current, and r is a function of i, such as $r = ai + b$, giving rise to the relation $e = ai^2 + bi$, new wavelengths or modulation frequencies would be generated by the device. The behavior of a nonlinear device can be described by a transfer function or an impulse response function. *See also* **linear device.**

nonlinear distortion: 1. Distortion caused by a deviation from a linear relationship between specified input and output parameters, such as current, voltage, and power, of a system or a component. **2.** Distortion that is a deviation from a linear relationship between specified signal parameters as the signal passes through equipment or system components. *Note:* Examples of nonlinear distortion are the distortion that occurs because (a) the gain or the loss in a circuit is not constant over all values of signal amplitude or all values of frequency, and (b) the constitutive relation constants change with frequency, electric current density, electric field strength, magnetic field strength, and direction of propagation of electromagnetic waves with respect to the propagation medium. *See also* **constitutive relation, distortion, linearity, propagation medium.**

nonlinear feedback shift register: A shift register in which (a) the feedback elements are not necessarily reducible to modulo-two functions, and (b) multiple feedback loops and special networks are used to create the nonlinear feedback shift register. *See also* **linear feedback shift register, modulo-n, shift register.**

nonlinearity: *See* **phase nonlinearity.**

nonlinear material: *See* **third-order nonlinear material.**

nonlinear optical (NLO) material: In fiber optics, a transparent material, such as a special glass, a polymer, or an organic material, that displays nonlinear properties, such as generate higher harmonic frequencies when energized with optical signals. *Note:* The generation of second and higher harmonic frequencies implies switching speeds of less than 100 femtoseconds. The process is also practically lossless and thus can be repeated in tandem many times. An exit wave can be frequency-mixed, and output beams can be considerably different from input beams. At the exit point of an optical integrated circuit (OIC) waveguide, different signals can be switched to different ports. *See* **polymeric nonlinear optical material.**

nonlinear scattering: 1. Scattering of electromagnetic radiation in which wavelengths other than the original wavelength are generated, such as occurs in Raman scattering and Brillouin scattering. **2.** Direct conversion of a photon from one wavelength to one or more other wavelengths. *Note 1:* In an optical fiber, nonlinear scattering usually is not significant below the threshold irradiance for stimulated nonlinear scattering. *Note 2:* Examples of nonlinear scattering are Raman scattering and Brillouin scattering. *See also* **photon, scattering.**

nonloaded cable: *Synonym* **nonloaded line.**

nonloaded line: A transmission line, such as a coaxial cable, overhead cable, open wire, or twisted pair, that has no intentionally added impedance, such as inductive reactance in the form of loading coils. *Synonym* **nonloaded cable.** *See also* **coaxial cable, impedance, inductive reactance, loading coil, open wire, transmission line, twisted pair.**

nonlocking code extension character: A code extension character that has the characteristic that the change in interpretation that is to be made of the characters that follow applies only to a specified number of the coded representations, i.e., characters, that follow the code extension character, usually only to the character that immediately follows it. *See also* **code extension character.**

nonoperational load: In communications, computer, data processing, and control systems, (a) the administrative, support, and housing power or (b) the administrative, support, and housing electrical equipment. *Synonym* **utility load.** *See also* **operational load.**

nonorthogonal antenna mount: In radar systems, an antenna mount (a) in which two axes of rotation are not at right angles to each other, and (b) that has advantages for particular situations where fast tracking may be required. *See also* **antenna, track.**

nonprecedence call: 1. In telephone system operations, a routine precedence call in networks in which a precedence system is used. **2.** A call in a telephone system in which precedences are not assigned to calls. **3.** A call to which a precedence has not been assigned. *See also* **call, precedence, precedence call.**

nonradiative recombination: In an electroluminescent diode in which electrons and holes are injected into the p-type and n-type regions by application of a forward bias, i.e., a forward voltage, the recombination of injected minority carriers with the majority carriers in such a manner that the energy released upon recombination (a) is not radiated in the form of electromagnetic waves, (b) results in heat, which is dissipated primarily by conduction and some thermal radiation, and (c) in light-emitting diodes (LEDs), does not contribute to light energy for optical use, such as energizing optical fibers or driving optical integrated circuits (OICs). *See also* **radiative recombination.**

nonreflective star coupler: A fiber optic coupling device that enables signals in one or more optical fibers to be transmitted to one or more other fibers by entering the input signal into an optical fiber volume without an internal reflecting surface so that the diffused signals pass directly to the output fibers on the opposite side of the fiber volume for propagation away in one or more of the output fibers. *Note:* The optical volume is a shaped piece of optical fiber material that allows transmission from two or more inputs to two or more outputs. *Synonym* **transmitting star coupler.** *See also* **optical fiber, reflective star coupler, star coupler.**

nonresonant antenna: *Synonym* **aperiodic antenna.**

nonreturn to zero (NRZ): Pertaining to a signal that represents a sequence of digits, such as binary, ternary, or *n*-ary digits, in which the signal level does not return to a zero amplitude, or to a value that represents the zero level, between the digits, though zero may be a significant condition that represents a 0 or a 1, and the signal might pass through zero during transition from one significant condition to another. *Note 1:* If the varying parameter is a phase shift, the signal will not return to a zero phase shift between digits. If the parameter is an electric current or an electric field strength, it will not return to a zero value between digits. *Note 2:* The four basic forms of nonreturn to zero (NRZ) data, or signals that represent information, are NRZ change, NRZ level, NRZ mark, and NRZ space. *Note 3:* Nonreturn to zero (NRZ) data in binary form have successive 1s and 0s following one another. The signal does not return to the zero level between digits, but remains at the significant condition of the previous bit for as long as the bit does not change. If a positive voltage represents a 1 and a negative voltage represents a 0, the voltage will remain positive for as long as there is a series of consecutive 1s, go negative at the first occurrence of a 0, remain negative for as long as there are 0s, go positive at the next occurrence of a 1, and so on for the length of the sequence of digits. *See also* **significant condition, value.** *Refer to* **Figs. N-17, N-18.**

nonreturn to zero (NRZ) binary code: A code in which (a) 1s are represented by one significant condition and 0s are represented by another, (b) transitions are direct from one stable state to another, and (c) there is no neutral, zero, or rest condition between the stable states, such as a zero amplitude in amplitude modulation (AM), zero phase shift in phase-shift keying (PSK), or midfrequency in frequency shift keying (FSK). *Note 1:* Nonreturn to zero (NRZ) codes are in contrast with Manchester codes and return to zero codes. *Note 2:* For a given data signaling rate (DSR), i.e., bit rate, the nonreturn to zero (NRZ) code requires only one-half the bandwidth required by the Manchester code. *See also* **bipolar signal, code, duobinary signal, Manchester encoding, nonreturn to zero mark code, nonreturn to zero space code, change on ones code, return to zero code.**

nonreturn to zero change (NRZ-C): Pertaining to a binary signal stream in which (a) a signal level, i.e., a significant condition, transition occurs whenever there is a change from the state that represents a 1 to the state that represents a 0, and a transition in the reverse direction occurs when the state that represents a 0 changes to the state that represents a 1, (b) as long as the digits remain the same, there are no transitions, and (c) the transitions represent the data rather than the stable states, i.e., the significant conditions. *See also* **significant condition.**

nonreturn to zero change on ones (NRZ-1) code: *Synonym* **nonreturn to zero mark (NRZ-M).**

nonreturn to zero change on ones magnetic recording: A nonreturn to zero change magnetic recording in which 1s are represented by a change in magnetic condition, such as polarized magnetization in a given direction, and 0s are represented by no change in the magnetic condition.

nonreturn to zero change on zeros (NRZ-0) code: *Synonym* **nonreturn to zero space (NRZ-S) code.**

nonreturn to zero (NRZ) coding: Coding in which electrical or lightwave pulses do NOT return to the zero-level significant condition of the signal parameter, such as power, current, voltage, phase, or light intensity, between consecutive pulse on-conditions. *Note:* The binary digits 111001 would be represented by (a) an on-condition for three time intervals, (b) an off-condition for two time intervals, and (c) an on-condition for one time interval. There would be no returning to the off-condition between, or during the passing of, the binary 1s or between, or during the passing of, the binary 0s. *See also* **pulse, signal parameter, significant condition.**

nonreturn to zero level (NRZ-L): Pertaining to a bit stream in which (a) the signal level of the signal parameter, such as the voltage, power, current, phase, frequency, or light intensity, remains in one state, say high, for as long as there are just 1s in the bit stream that is being represented, and, in this case, low for as long as there are 0s in the bit stream, or 0 could be

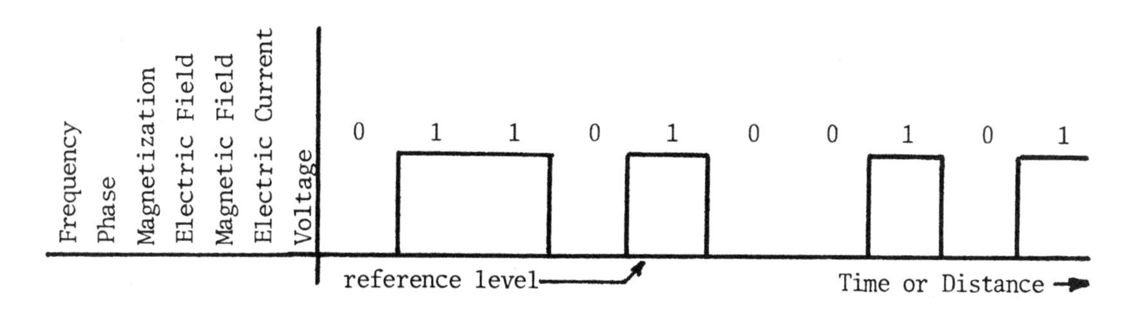

Fig. N-17. In **nonreturn to zero (change) modulation,** the physical variable, i.e., the parameter, that represents the data will transition to the opposite significant condition only when the digit in the bit stream changes.

high and 1 could be low, (b) the level shift corresponds exactly to the binary code, (c) for long strings of 1s and for long strings of 0s, there are direct current (dc) and low frequency components and perhaps little energy at high frequencies, which causes the data to degrade rapidly under high-pass filtering in which the low frequency components of the signal cannot get through, (d) signal integrity is retained only under low-pass filtering, and (e) in order to pass data, the transmission channel must be able to accept and transmit low frequencies, even down to dc. *See also* **bit stream, level, low-pass filter, signal.**

nonreturn to zero level code: *See* **modified nonreturn to zero level code.**

nonreturn to zero mark (NRZ-M): 1. Pertaining to a code in which 1s are represented by a change in a significant condition, and 0s are represented by no change. **2.** Pertaining to a code used in a binary signal stream in which a signal parameter, such as electric current or voltage, undergoes a change in a significant condition or level every time that a 1 occurs, but when a 0 occurs, it remains the same, i.e., no transition occurs. *Note:* The transitions could also occur only when 0s occur and not when 1s occur. If the significant condition transition occurs on each 0, it is called "nonreturn to zero space (NRZ-S)." The signals are interchangeable. It is only necessary to maintain proper accounting in the electronic circuitry when distinguishing between NRZ-M and NRZ-S. *Synonyms* **conditioned baseband representation, differentially encoded baseband, nonreturn to zero change on ones (NRZ-1) code, NRZ-B.**

nonreturn to zero (change) modulation: The modulation of a signal parameter that is used in a bit stream in which one of the binary digits, say the 1, is represented by a transition from a specified signal significant condition, such as magnetization, electric field strength, magnetic field strength, electric current, voltage, or amplitude, to another given significant condition, and the other binary digit, in this case the 0, is represented by a transition from the given back to the specified condition. *Note:* The two conditions may be a nonzero value of a signal parameter and a zero value, two nonzero values, two polarities, or oppositely directed fields. This mode is called "change modulation" because the signal condition changes only when the bits in a bit stream change from a 0 to a 1 or from a 1 to a 0. Thus, only the change is significant. It indicates that the digit has changed. *See also* **bit stream, modulation, signal parameter, significant condition.** *Refer to* **Fig. N-17.**

nonreturn to zero (change on ones) modulation: In a bit stream, modulation of a signal parameter in which one of the binary digits, say the 1, is represented by a change in the signal significant condition, and the other binary digit, the 0 in this case, is represented by the absence of a change, i.e., by no change. *Note 1:* The signal parameter that changes significant condition might be the magnetization, electric field strength, magnetic field strength, electric current, voltage, phase, or frequency. *Note 2:* Nonreturn to zero (change on ones) modulation has also been called "mark modulation" or "space modulation" because only the 1s (mark) or only the 0s (space) are explicitly recorded. *See* **bit stream, modulation, signal parameter, significant condition.** *Refer to* **Fig. N-18.**

nonreturn to zero space (NRZ-S): 1. Pertaining to a code in which 0s are represented by a change in a significant condition, and 1s are represented by no change. **2.** Pertaining to a code used in a binary signal stream in which a signal parameter, such as electric current or voltage, undergoes a change in a significant condition or level every time that a 0 occurs, but when a 1 occurs, it remains the same, i.e., no transition occurs. *Note:* The transitions could also occur only when 1s occur and not when 0s occur. If the significant condition transition occurs on each 1, it is called "nonre-

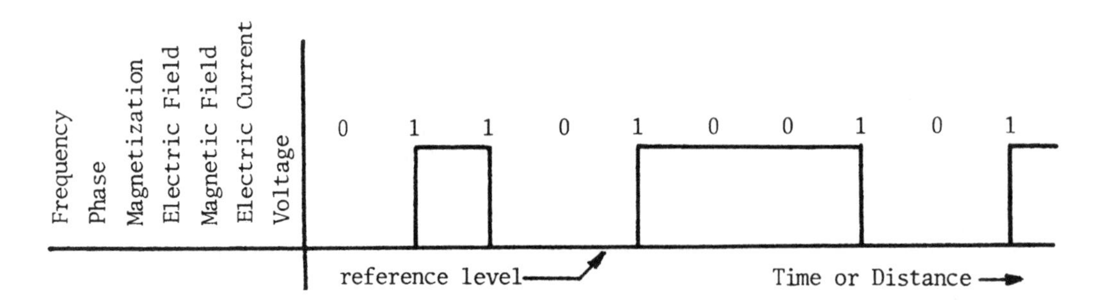

Fig. N-18. In **nonreturn to zero (change on ones) modulation,** the physical variable, i.e., parameter, that represents the data will transition to the opposite significant condition each time a binary 1 occurs in the bit stream.

turn to zero mark (NRZ-M)." The signals are interchangeable. It is only necessary to maintain proper accounting in the electronic circuitry when distinguishing between NRZ-S and NRZ-M. *Synonym* **nonreturn to zero change on zeros (NRZ-0) code.**

nonshifted optical fiber: *Synonym* **dispersion unshifted optical fiber.**

nonstore through cache: Pertaining to a mode of computer and data processing systems operation in which store or write instructions cause data to be temporarily placed in cache storage rather than placed in main storage immediately. *See also* **cache storage.**

nonswitched line: In a communications network, a line that does not require switching equipment to be connected and therefore cannot be be connected by dialing. *Note:* An example of a nonswitched line is a line that is permanently connected to a station. *See also* **dialing, line, switching equipment.**

nonsynchronous data transmission channel: *Synonym* **asynchronous data transmission channel.**

nonsynchronous network: *Synonym* **asynchronous network.**

nonsynchronous satellite: *Synonym* **asynchronous satellite.**

nonsynchronous system: *Synonym* **asynchronous system.**

nonsynchronous transmission: *Synonym* **asynchronous transmission.**

nontechnical load: Of the total operational load at a facility during normal operation, the part used for support purposes, such as general lighting, heating, air conditioning, and ventilating equipment. *See also* **air conditioning, operational load, plant, technical load.**

nontransparent mode: A mode of operating a data transmission system in which control characters are treated and interpreted as such, rather than simply as data or text bits in a bit stream. *See also* **bit stream, control character.**

nonuniformly distributive coupler: A fiber optic coupler in which (a) optical power input to each input port is not distributed equally among the output ports, and (b) the optical power input to each output port is not distributed equally among the input ports. *See also* **uniformly distributive coupler.**

nonvolatile storage: Storage in which the contents are not permanently lost when power is removed, but become available again when power is restored. *Note:* Examples of nonvolatile storage are magnetic tape, disk, drum, and card storage. *See also* **storage, volatile storage.**

NOR gate: A binary logic combinational circuit or device that is capable of performing the logic operation of NOR, i.e., negative OR, such that if A is a statement, B is a statement, C is a statement, . . . , the result of NOR A, B, C, . . . is false if at least one of the statements is true, and true only if all the statements are false. *Note:* The behavior of the NOR gate is the same as that of an OR gate, i.e., inclusive OR gate, with its output negated, i.e., inverted, by sending the output through a NEGATION gate. *Synonyms* **inclusive NOR gate, joint denial gate, negative OR gate, neither NOR gate, nondisjunction gate, rejection gate, zero match gate.** *See also* **combinational circuit, gate.** *Refer to* **Fig. N-19.**

normal distribution: In a population of entities, a distribution of attributes that follows a Gaussian distribution. *See also* **Gaussian distribution.** *Refer to* **Fig. P-13 (759).**

normal emergence: In optics, a condition in which a ray emerges perpendicular to the surface of a propaga-

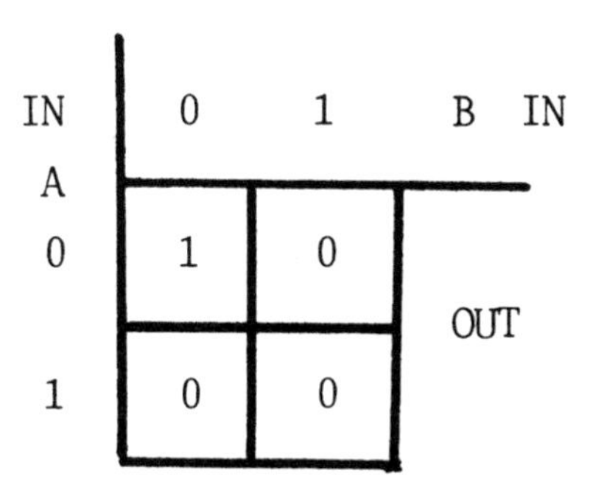

Fig. N-19. A truth table that shows the input and output digits of a **NOR gate.**

tion medium. *See also* **emergence, grazing emergence, propagation medium.**

normal incidence: Incidence that (a) is in a direction that is at an angle of 90° with a surface, i.e., is perpendicular to the surface at the point of incidence, or (b) is perpendicular to a plane tangent to the surface at the point of incidence. *Note:* A lightwave, radio wave, soundwave, or water wave that strikes a surface at an angle of 90° to the surface is a normal wave, i.e., a wave at normal incidence. *Refer to* **Fig. S-18 (951).**

normalized: *Synonym* **fractional.**

normalized detectivity: *Synonym* **specific detectivity.**

normalized error: The ratio of (a) an observed or measured value to (b) the true, specified, theoretically correct, or calculated value, both usually expressed in the same units. *See also* **absolute error, relative error.**

normalized frequency (V): 1. In an optical fiber, a dimensionless quantity, V, given by the relation $V = (2\pi a/\lambda)(n_1^2 - n_2^2)^{1/2}$ or $V = 2\pi a(NA)/\lambda$ where V is the normalized frequency, a is the core radius, λ is the wavelength in vacuum, i.e., the wavelength of the source, n_1 is the maximum refractive index of the core, n_2 is the refractive index of the homogeneous cladding, and NA is the numerical aperture. *Note 1:* In multimode operation of an optical fiber having a power law refractive index profile, the approximate number of bound modes, i.e., the mode volume, is given approximately by the relation $M_p \approx (V^2/2)[g/(g + 2)]$ where M_p is the mode volume, V is the normalized frequency greater than 5, and g is the profile parameter. *Note 2:* For a step index fiber, the mode volume is given approximately by the relation $M_s \approx V^2/2$ where M_s is the mode volume and V is the normalized frequency. For single mode operation, $V \leq 2.405$. The larger the normalized frequency, the larger the mode volume, and consequently the larger the number of modes that can be supported. *Synonyms* **mode parameter, V number, V parameter, V value. 2.** The ratio of (a) an actual frequency, such as a transmitted or a measured frequency, and (b) a reference frequency. **3.** The ratio of (a) an actual frequency, such as an instantaneous frequency and (b) its nominal or midrange value. *See also* **bound mode, core, frequency, mode volume, parabolic profile, power law index profile.** *Refer to* **Fig. R-10 (836).**

normalized offset: An offset divided by the associated nominal value. *Synonyms* **fractional offset, relative offset.**

normalized power: 1. In electric circuits, the mean or average value of the voltage squared of a waveform, divided by a pure resistance, that is defined as a voltage as a function of time and dissipated in the pure resistance, given by the relation $P_n = (1/RT)\int[v(t)]^2dt$, integrated from 0 to T, where P_n is the normalized power, R is the pure resistance into which the power is dissipated, T is the period over which the average is taken, usually the period of a periodic wave, and $v(t)$ is the voltage as a function of time. *Note:* The normalized power will be in watts if the voltage is in volts, the resistance is in ohms, and the time units for the function and the period are compatible. **2.** In electric circuits, the mean or average value of the voltage squared of a waveform that is defined as a voltage as a function of time and dissipated in a 1-Ω (ohm) resistor. *Note:* The dimension of normalized power is volts squared. It is numerically equal to watts when the power is dissipated in a 1-Ω (ohm) resistor.

normal magnification: *See* **individual normal magnification.**

northern lights: *Synonym* **aurora borealis.**

NOT: The logical operator that has the property that if P is a statement, then NOT P is true if P is false and false if P is true. *Note:* The NOT operation is performed by the NEGATION gate. *See also* **logical operator, NEGATION gate.**

not accepted signal: *See* **call not accepted signal.**

NOTAL: An indicator that is placed after a statement or a reference in a message heading to indicate that the statement is neither intended for nor needed by all of the addressees of the message. *Note:* NOTAL is derived from "not all." *See also* **heading, indicator.**

not AND gate: *Synonym* **NAND gate.**

notation: *See* **binary coded decimal notation, binary notation, infix notation, parenthesis-free notation, postfix notation, prefix notation.**

notation one: *See* **abstract syntax notation one.**

not both gate: *Synonym* **NAND gate.**

notch: **1.** In a frequency spectrum that primarily consists of white noise, though not necessarily of uniform amplitude, i.e., of constant amplitude, a narrow band of the spectrum in which the amplitudes of the waves are low or zero, i.e., they essentially are not present. **2.** The trough in the frequency spectrum produced by a notch filter that is fed with white noise. *See also* **frequency spectrum, notched noise, notch filter, white noise.**

notch antenna: An antenna that (a) forms a radiation pattern by means of a notch or a slot in a radiating surface, (b) has characteristics similar to those of a properly proportioned metal dipole antenna, and (c) may be evaluated with techniques similar to those used to evaluate a dipole antenna. *See also* **dipole antenna, radiation pattern.**

notched card: *See* **edge notched card.**

notched noise: Noise that (a) is distributed over a wide frequency spectrum, (b) has one or more narrow bands of frequencies removed, and (c) usually is used for testing devices or circuits. *See also* **frequency spectrum, noise, notch.**

notch filter: *Synonym* **band stop filter.**

notice: *See* **go ahead notice, stop notice.**

Notice to Airmen (NOTAM): In aviation communications, messages that are used to provide timely information for conditions that are essential to flight operations. *Note:* Notices to Airmen (NOTAMs) are controlled by the National Flight Data Center. *See also* **aeronautical broadcast station.**

not if then gate: *Synonym* **A AND NOT B gate.**

not obtainable signal: In a communications system, a signal that (a) is sent in the backward direction and (b) indicates that a call cannot be completed because of any of several reasons, such as the call destination number or line is not in use, is in a different user group, or is in a different user class. *See also* **backward direction, call, user class, user group.**

not ready condition: At the data terminal equipment/data circuit-terminating equipment (DTE/DCE) interface, a steady state condition indicating that the DCE is not ready to accept a call request signal or that the DTE is not ready to accept an incoming call. *Note:* The not ready condition may be controlled or uncontrolled. *See also* **call restriction, data circuit-terminating equipment, data terminal equipment, data transfer request signal, onhook, onhook signal.**

not ready signal: *See* **controlled not ready signal.**

Np: neper.

NRI: net radio interface.

NRZ: nonreturn to zero.

NRZ-B: *Synonym* **nonreturn to zero mark.**

NRZ-C: nonreturn to zero change.

NRZ coding: nonreturn to zero coding.

NRZ-L: nonreturn to zero level.

NRZ-M: nonreturn to zero mark.

NRZ-S: nonreturn to zero space.

NRZ-0: nonreturn to zero change on zeros.

NRZ-1: nonreturn to zero change on ones.

NSEP: National Security and Emergency Preparedness.

NS/EP: National Security and Emergency Preparedness.

***n*-sequence:** A pseudorandom binary sequence of n bits that (a) is the output of a linear shift register and (b) has the property that, if the shift register is set to any nonzero state and then cycled, a pseudorandom binary sequence of a maximum of $n = 2^m - 1$ bits will be generated, where m is the number of stages, i.e., the length or the number of bit positions in the register, before the shift register returns to its original state and the n-bit output sequence repeats. *Note:* The register may be used to control the sequence of frequencies for a frequency hopping spread spectrum transmission system. *See also* ***n*-sequence generator, register, shift register.**

***n*-sequence generator:** A bit pattern generator that usually consists of (a) a linear shift register, i.e., a row of flip-flops connected in tandem, (b) one or more feedback loops in which the condition of selected flip-flops is gated back to other selected flip-flops in the register, (c) an input at one end to accept the stepping pulse, and (d) an output at the other end from which the pseudorandom sequence of bits emerges. *Note 1:* The state of the register at any moment is determined by the initial state, i.e., the initial setting, of the register, the feedback loop configuration, and the stepping pulse input. *Note 2:* The n-sequence may be changed by changing (a) the number of flip-flops in the register, (b) the initial setting of the register, (c) the configuration of feedback loops, or (d) any combination of these. *Note 3:* An operator usually only changes the initial setting of the register. The manufacturer sets the other parameters in the hardware. *Note 4:* The *n*-sequence generator is used in spread spectrum systems. *See also* **bit pattern, feedback, *n*-sequence, pseudorandom number sequence.**

NTI: network terminating interface.

NTN: network terminal number.

NTSC: National Television Standards Code, National Television Standards Committee.

NTSC standard: National Television Standards Committee standard.

n-type semiconductor: A semiconducting material, such as silicon or germanium, (a) that has been doped with minute amounts of donor-type material, i.e., with material having a valence that allows extra relatively free electrons to wander about the lattice structure of the semiconducting crystal, all other atomic bonds being complete, (b) in which the free electrons can move readily from atom to atom under the influence of an applied electric field, albeit a relatively weak field, thus constituting an electric current, and (c) in which the dopant, i.e., the doping material "donates" electrons to this current, constituting a stream of negatively charged particles, hence the name n-type semiconducting material. *See also* **p-type semiconductor.**

nuclear hardening: *See* **fiber optic nuclear hardening, optical fiber nuclear hardening.**

nuclear hardness: 1. A measure of the extent to which the performance of a system will degrade in a given nuclear environment. **2.** The physical attributes of a system or component that will allow survival in an environment that includes nuclear radiation, electromagnetic impulses (EMI), and electromagnetic pulses (EMP). *Note 1:* The hardness measure may be in terms of either susceptibility or vulnerability. *Note 2:* Hardness usually is measured in terms of the effects of nuclear environmental conditions, such as overpressure, peak velocities, energy absorbed, and electrical stress, on the characteristics of the system or component, such as its electrical conductivity, molecular structure, or tensile strength. *Note 3:* Hardness is achieved through design specifications and is verified by test and analysis techniques. *Note 4:* After exposure to a nuclear environment, the extent of performance degradation is measured, such as outage time, data lost, and equipment damage. The environment must be defined, such as radiation levels, overpressure, peak velocities, energy absorbed, and electrical stress. **3.** The physical attributes of a system or a component that will allow a defined degree of survivability in a given environment created by a nuclear weapon.

NUL: null character.

null: 1. A dummy letter, symbol, or code group inserted into an encrypted message to delay or prevent its solution, or to complete encrypted groups for transmission or for transmission security purposes. **2.** In an antenna radiation pattern, a point at which the radiance of a transmitting antenna or the response sensitivity of a receiving antenna is either zero or near zero relative to the maximum radiance of the main beam. *Note:* A null often has a narrow directivity angle compared to that of the main beam. Thus, the null is useful for several purposes, such as radio navigation and suppression of interfering signals in a given direction. **3.** In a standing wave in a cavity or on a transmission line, pertaining to a point at which the current, voltage, or power is zero or near zero. *Synonym in this sense:* **node.** *See also* **node.**

null (NUL) character: 1. In transmission systems, a control character (a) that is used to accomplish media-fill stuffing or a time-fill stuffing in storage devices or in data transmission lines, and (b) that may be inserted and removed from a series of characters without affecting the meaning of the series. *Note:* The null character may affect the control of equipment or the format of messages. **2.** In cryptosystems, a letter that is inserted into an encrypted message in order to (a) delay or prevent its solution, (b) complete encrypted groups for tranmission, or (c) provide for transmission security. *See also* **cryptosystems, encryption, transmission control charcter.**

NULL gate: A device that (a) produces a signal, or a sequence of signals, that represents strings of 0s in a given system, using specified convention or representation, for as long as the power is on and the gate is enabled, (b) may be controlled or uncontrolled by an input control on-off signal, i.e., an enabling or inhibiting signal, and (c) is the inverse of the GENERATOR gate. *See also* **gate, GENERATOR gate, string.** *Refer to* **Fig. N-20.**

null string: A string in which the number of entities in the string is reduced to zero. *See also* **string.**

number: 1. The designation of a quantity of entities, represented as a numeral. *Note:* A telephone number may be represented by the numerals (804) 933-8615, or the number of days in a year may be represented by the numerals 365, CCCLXV, or 101101101. **2.** A mathematical entity or abstraction used to indicate a quantity or an amount of units, such as the number of

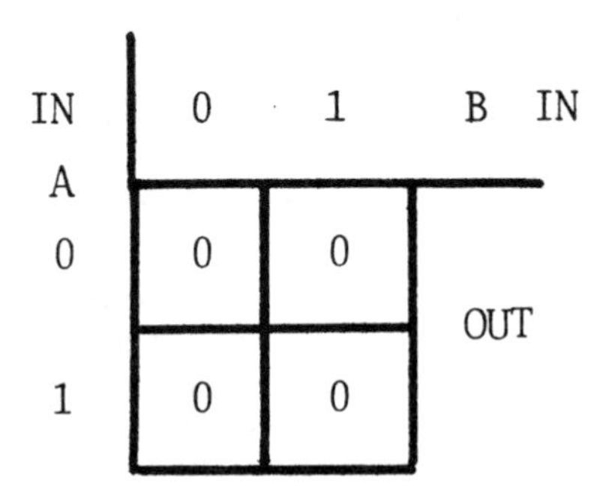

Fig. N-20. A truth table that shows the input and output digits of a **NULL gate.**

wires in a cable, feet in a statute mile, digits in a hand, or miles per hour. **3.** In communications systems, an identifier, such as a string of numerals, assigned to an end instrument. *See* **binary number, F number, irrational number, mach number, natural number, network terminal number, prime number, random number, rational number, ringer equivalency number, significant condition number, spot number, telephone number, T number, wave number.** *See also* **end instrument, identifier, numeral, string.**

number call: A call that (a) is to a user end instrument number rather than to the user name, (b) speeds service, (c) enables an attendant, if any, to proceed to other business as soon as anyone at the number answers, (d) allows the call originator to seek the proper person at the destination number, i.e., at the called number, if necessary, and (e) in automatically switched systems, does not require attendant assistance. *Note:* A number call often is called a station to station call, in contrast to a person to person call. *See also* **local call, person to person call, station to station call.**

number crunching: Performing calculations, such as computing the value of an algebraic expression when numbers are substituted for the operands, i.e., the arguments, parameters, or variables, in the expression. *Note 1:* Examples of number crunching are (a) the calculation of the value of the geometric series $a(1 + r + r^2 + r^3 + \ldots ar^n)$ when values are given to a, r, and n and (b) evaluating the value of a matrix of numbers. *Note 2:* In communications systems management, number crunching is performed in route selection, traffic and statistical analyses, and service charge calculations.

number generator: *See* **pseudorandom number generator.**

number identification: *See* **automatic number identification.**

numbering plan area: *See* **data numbering plan area.**

numeral: The graphic or discrete design, symbol, or set of designs or symbols used for the representation of a number. *Note:* The number of days in a year might be represented as (a) three hundred and sixty-five (words in the English language), (b) dreihundert fünf und sechzig (words in the German language), (c) 365 (an Arabic numeral in the decimal numeration system), (d) CCCLXV (a Roman numeral), or (e) 101101101 (a numeral in the pure binary numeration system). *See* **binary numeral, ship radio telephone numeral.** *See also* **decimal, number, numeration system, pure binary numeration system.**

numeral flag: In visual communications systems, a flag that (a) is one of a set of ten different flags, (b) is used to represent one of the numerals from 0 through 9, (c) is square, (d) has a pattern of colors to represent the numerals, and (e) has a different color pattern on each flag to represent the numerals. *See also* **flag, numeral, numeral pennant.**

numeral pennant: In visual connunications systems, a pennant that (a) is one of a set of ten different pennants, (b) is used to represent one of the numerals from 0 through 9, (c) is elongated and tapered, (d) is hung with the base of the taper at the halyard, (e) has a pattern of colors to represent the numerals, and (f) has a different color pattern on each flag to represent the numerals. *See also* **numeral, numeral flag, pennant, halyard.**

numeral sign: In semaphore signaling, a flag position that (a) is used before and after each group of numerals, or groups of mixed letters and numerals, in the text of a message, (b) is used to indicate that the group of mixed numerals and letters, or numerals only, is to be considered and recorded as a single group, and (c) is particularly important when codes are being used, enabling each code group to be separately identifiable for decoding or deciphering. *See also* **decipher, decode, flag position, numeral, semaphore, semaphore signaling.**

numeration system: A notation system for representing numbers. *Synonyms* **number representation system, number system.** *See* **decimal numeration system, mixed base numeration system, positional numeration system.**

numerical analysis: The study and application of methods of obtaining useful quantitative solutions to problems that have been expressed in mathematical form, in which numbers may be substituted for the parameters and the variables used in equations, including the evaluation of the effects of errors and assumptions. *Note:* Numerical analysis may be used to obtain the solutions to the differential equations that describe the motion of bodies subjected to applied forces and fields.

numerical aperture (NA): 1. The product of (a) the sine of half the vertex angle of the largest cone of meridional rays that can enter or leave an optical system or element and (b) the refractive index of the medium in which the cone is located. *Note:* The numerical aperture (NA) usually is measured with respect to an object or image point and will vary as that point is moved. **2.** For an optical fiber in which the refractive index decreases monotonically from the optical fiber axis to the core-cladding interface, the maximum numerical aperture, NA_{max} is given by the relation $NA_{max} = (n_1^2 - n_2^2)^{1/2}$ where n_1 is the refractive index at the fiber axis, and n_2 is the refractive index of the innermost homogeneous cladding. *Note 1:* The refractive indices are relative to the refractive index of the medium from which or to which the rays enter or leave the fiber. *Note 2:* The numerical aperture (NA) is a point function across the endface of the core, depending on the refractive index at the point. In a step index fiber, the NA does not vary across the endface of the core because the refractive index is constant from point to point. In a graded index fiber, the refractive index decreases with increased radial distance from the fiber axis. Therefore, the NA is a maximum at the center and decreases with increased distance from the fiber axis. *Note 3:* For optical fibers, the numerical aperture (NA) is a measure of the light-accepting property of a fiber, inasmuch as only the light that propagates an appreciable distance, such as propagates beyond the equilibrium length, inside the core can be considered to have been accepted by the fiber. Thus, the aperture of an unaided fiber is limited to an acceptance cone, namely, the cone within which all rays will undergo total (internal) reflection when inside the core. *Note 4:* Typical numerical apertures for optical fibers range from 0.25 to 0.45. *Note 5:* Loose terms, such as "openness," "light gathering ability," "angular acceptance," and "acceptance cone," have been used to describe the numerical aperture (NA) of an optical fiber. *Note 6:* One way to measure the numerical aperture of an optical fiber is to measure the launch spot size produced by an illuminated fiber. The spot of light produced by a light beam from the exit endface of an optical fiber is adjusted for maximum irradiance on a transverse screen, i.e., a surface perpendicular to the fiber axis, by aligning the fiber to be normal to the surface. The diameter of the spot is dependent upon the departure angle, i.e., the launch angle, from the fiber endface and the distance from the fiber endface to the screen. The numerical aperture is given by the relation $NA = D/(D + 4d^2)^{1/2}$ where D is the spot diameter, and d is the distance from the fiber exit endface to the screen. If $D/2d$ is less than 0.25, the numerical aperture of the element is given by the approximate relation $NA \approx D/2d$. The NA is basically defined as the sine of the acceptance angle. However, the ratio of (a) the spot diameter to (b) twice the distance from the fiber endface to the screen, which is the same as the ratio of the radius of the spot to the distance to the screen, is the tangent of the acceptance angle. For the smaller angles, the sine and the tangent are nearly the same. *See* **launch numerical aperture.** *See also* **acceptance angle, fiber optics, focus depth, launch spot, meridional ray, optical fiber, radiation angle, radiation pattern.** *Refer to* **Fig. N-21.** *Refer also to* **Figs. A-6 (24), R-10 (836).**

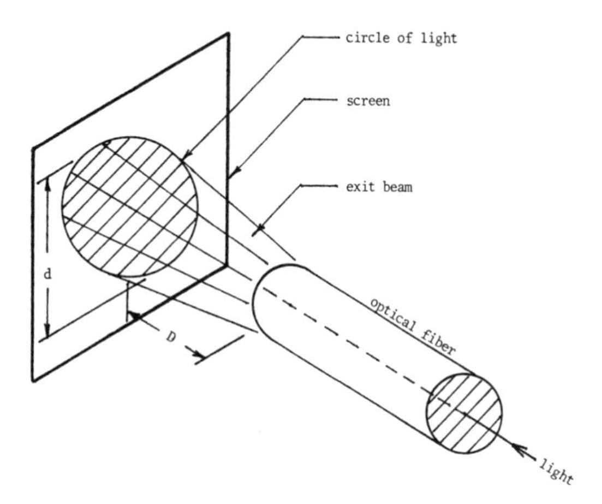

Fig. N-21. Measurement of the **numerical aperture (NA)** for an optical fiber using $NA = d/2D$ where d is the diameter of the spot of light formed on a screen held perpendicular to the fiber axis and caused by the beam launched from the fiber exit endface, and D is the distance between the endface and the screen.

numerical aperture loss: A loss of optical power that occurs at a splice or a pair of mated connectors when numerical aperture (NA) of the "transmitting fiber" exceeds that of the "receiving fiber," even if the cores have precisely the same diameter and are perfectly in angular alignment with zero lateral offset. *Note 1:* The higher numerical aperture (NA) of the transmitting fiber means that it emits a larger cone of light than the receiving fiber is capable of accepting, resulting in a coupling loss. *Note 2:* In the opposite case of numerical aperture (NA) mismatch, where the transmitting fiber has a lower NA, no NA loss occurs because the receiving fiber is capable of accepting light from any bound mode of the transmitting fiber. *See also* **numerical aperture.**

numerical control: The control of a device, process, or system in which use is made of numeric data, usually introduced in real time. *See also* **numeric data, real time.**

numeric character set: A character set that contains digits and that may contain control characters, special characters, and the space character, but not letters. *See also* **character set, control character, digit, letter, space character, special character.**

numeric character subset: A character subset that contains digits and that may contain control characters, special characters, and the space character, but not letters. *See also* **character set, character subset, control character, digit, letter, space character, special character.**

numeric code: A code in which a numeric character set is used to represent data. *See also* **code, numeric character set.**

numeric coded character set: A character set that is used to code characters that may be (a) numeric characters, (b) nonnumeric characters, such as letters, special signs and symbols, the space character, and control characters, or (c) both numeric and nonnumeric characters. *Note:* Examples of numeric coded character sets are (a) the character set of a binary coded decimal (BCD) code and (b) the character set of the International Telegraph Alphabet 5 (ITA-5), from which the American Standard Code for Information Interchage (ASCII) is derived, and in which patterns of binary digits are used to represent the characters, each character being represented by a binary numeral. *See also* **character set, numeric character set.**

numeric data: Data that are represented entirely by numerals. *See also* **numerals.**

numeric keypad: On a standard computer keyboard, a keypad, usually located at the right end of the main keyboard, that may be used as number keys when the NUM LOCK key is in the appropriate position. *Note:* Actuating the NUM LOCK key alternates the keypad between being (a) a numeric key pad and (b) being a cursor or pointer keypad. *See also* **keypad.**

numeric representation: A discrete representation of data in which only numerals are used.

numeric word: A group of digits that (a) represent data to which meaning has been assigned, and (b) consists of digits and perhaps the space character and other special chracters, such as the parenthesis, decimal point, and comma. *See also* **digit.**

NUM LOCK key: On a computer standard keyboard, a key that (a) is used to alternate a keypad between being a numeric keypad and being a cursor keypad and (b) causes the keypad to change its function each time it is depressed. *See also* **cursor keypad, key, keyboard, keypad, numeric keypad.**

***n*-unit code:** A code in which the signals or the groups of digits that represent coded items, such as characters, have the same number of signal elements or digits, namely *n* elements or digits, where *n* may be any positive nonzero integer. *Note:* An example of an *n*-unit code is the 7-unit code, 8-unit with parity, in the American Standard Code for Information Interchange (ASCII) code. Each character is represented by a pattern of seven binary digits. The units may also be characters or other special signs.

***N*-user:** In the International Organization for Standardization (ISO) Open Systems Interconnection—Reference Model (OSI-RM), the identification of a user of a given layer, designated by the letter *N*. *Note 1:* The *N*-user is an $N + 1$ entity that uses the services of the *N*-layer, and below, to communicate with another $N + 1$ entity. *Note 2:* If *N* identifies a specific or a reference layer, the $N + 1$ layer is the layer above the *N* layer and the $N - 1$ layer is the layer below the *N* layer. Thus, the $N + 2$ layer is two layers above the *N* level, and so on. *See also* **Open Systems Interconnection—Reference Model (OSI-RM).**

nutation: The nodding of the axis of a rotating body, such as a spinning top or a satellite, about the cone of precession. *Note:* The effect is caused by impulses or rearrangement of the masses of a spinning body relative to its axis of rotation. Special equipment for damping nutation usually is fitted to satellites that use spin stabilization in order to accomplish specific purposes, such as (a) in communications satellites, to control footprint location and Earth coverage, (b) in meteorological satellites, to control camera optical axis orientation, and (c) in space exploration satellites, to control the direction of telescope gaze. *See also* **communications satellite, Earth coverage, footprint, satellite.**

nu value: *Synonym* **Abbe constant.**

NVIS: near vertical incidence skywave.

NXX code: In the North American direct distance dialing numbering plan, a central office (C.O.) code of three digits that designates a particular C.O. or a given 10,000-line unit of customer lines. *Note:* In the NXX code, N is any of the decimal digits 2 through 9, and X is any of the decimal digits 0 through 9. *See also* **access code, code.**

Nyquist bandwidth: *Synonym* **Nyquist sampling rate.**

Nyquist criterion: In order to faithfully reproduce a signal with high fidelity using sampling techniques in which a signal is sampled, transmitted in discrete pulses, and reconstituted at the receiving end, the sampling rate must be at least twice the highest significant frequency present in the signal being sampled. *Synonym* **sampling criterion.** *Refer to* **Fig. S-1 (866).**

Nyquist interval: In a given signal, the maximum time that (a) is between regularly spaced samples that will permit the signal waveform to be completely determined, and (b) is equal to the reciprocal of twice the highest frequency in the sampled signal. *Note 1:* If the bandwidth, i.e., the highest significant frequency in the signal being sampled, is in hertz, the Nyquist interval will be in seconds. If the frequency is in megahertz, the Nyquist interval will be in microseconds. Thus, for a frequency of 5 MHz, the Nyquist interval will be 0.1 μs (microsecond), i.e., a sample must be taken not greater than every 0.1 μs for the Nyquist criterion to be satisfied. *Note 2:* In practice, when analog signals are sampled for the purpose of data processing or digital transmission, the sampling rate must be higher than that calculated from the Nyquist interval in accordance with the Nyquist theorem, because of quantization errors introduced by the digitizing process. The required sampling rate is then determined by the accuracy and the precision of the digitizing process. *See also* **Nyquist rate, Nyquist sampling rate, Nyquist theorem, sampling rate, signal sampling.**

Nyquist noise: *Synonym* **thermal noise.**

Nyquist rate: *Synonym* **Nyquist sampling rate.**

Nyquist sampling rate (NSR): 1. The reciprocal of the Nyquist interval. **2.** The minimum sampling rate needed to faithfully reproduce a sampled waveform. *Note 1:* If f is the highest frequency in the sampled waveform, the Nyquist sampling rate, N_s, is given by the relation $N_s = 2f$. If f is in hertz, the sampling rate will be in samples per second. *Note 2:* The time interval, T_S, between samples is given by the relation $T_S = 1/N_s = 1/2f = 0.5/f$ where N_s is the Nyquist sampling rate, and f is the highest frequency in the sampled waveform. *Synonyms* **Nyquist bandwidth, Nyquist rate.** *See also* **Nyquist interval, Nyquist theorem, sampling rate.**

Nyquist theorem: An analog signal waveform may be uniquely and precisely reconstructed from samples taken of the waveform at equal time intervals, provided the sampling rate is equal to, or greater than, twice the highest significant frequency in the analog signal. *Synonym* **sampling theorem.** *See also* **Nyquist interval, Nyquist sampling rate, sampling rate, signal sampling.**

OAA: open avionics architecture.

object: 1. In an optical system, the figure (a) that consists of natural or artificial structures or targets or a real or virtual image formed by another optical system, and (b) that is viewed through or imaged by the optical system. *Note:* In the field of optics, an object should be considered as an aggregation of points, each of which has its own radiance when energized or illuminated. *In radar, synonym* **target. 2.** In facsimile systems, the item that is scanned by the scanning spot in a facsimile transmitter. *Note:* The object usually is a hard-copy document. *In facsimile, synonym* **original.** *In facsimile, deprecated synonyms* **original copy, subject, subject copy.** *See also* **image, object point, optical system, record copy, recorded copy, virtual image.**

object code: The code produced, usually from a source code, by hardware or software. *Note:* Examples of object code are (a) the output code from an assembler and (b) a compiler that can be executed as machine code or that can be processed to produce executable machine code. *Synonym* **target code.** *See also* **assembler, code, compiler, machine code.**

objective: In an optical system, the optical component, such as a lens or a prism, that receives light directly from the object and forms the first or primary image. *Note 1:* In simple cameras without added lenses, such as zoom or telescopic lenses, the image formed by the objective is the final image. *Note 2:* In telescopes, microscopes, range finders, binoculars, and other special instruments, the image formed by the objective usually is further processed by other optical components, such as in magnification by a magnifier or by an eyepiece. *See also* **eyepiece, image, lens, magnifier.**

object language: 1. In a translation process, the language into which another language, i.e., the source language, is translated. **2.** A language that is specified or described by a metalanguage. **3.** In computing, data processing, and communications systems, a language into which statements are translated. *Note:* Translators, assemblers, and compilers prepare object language programs, usually machine language programs, from source language programs, usually high level language programs written by programmers. *Synonym* **target language.** *See also* **language, metalanguage, source language, statement.**

object plane: In an optical system, the plane that contains the object points lying within the field of view. *See also* **object, object point, optical system.**

object program: 1. An assembled, compiled, or translated computer program that is ready for execution and loading into a computer. **2.** A computer program that is produced by an assembler, compiler, or translator. **3.** A computer program that is to be produced by an assembler, compiler, or translator. *Synonym* **target program.** *See also* **assemble, compile, load, source program, translate.**

oblate spheroid: The surface that is generated by spinning an ellipse about its minor axis. *Note:* The shape of the Earth's surface approximates that of an oblate spheroid, i.e., it bulges at the equator. The primary distortion of an oblate spheroid is a pear shape, where the southern hemisphere is slightly larger than the northern hemisphere. The Earth's polar radius is about 6357 km (kilometers). The Earth's equatorial radius is about 6378 km. These dimensions are considered in placing communications satellites in orbit. *See also* **communications satellite.**

oblique incidence sounding: The transmission of an electromagnetic wave obliquely to a surface and back again, particularly to the ionosphere and back. *Note:* The incidence angle with the normal to the surface of the ionosphere that is used in ionospheric sounding, i.e., measuring the height of the ionosphere, is on the order of 20° to 40°. *See also* **electromagnetic wave, incidence angle, ionosphere, sounding, transmission.**

obliquity: *See* **elliptic obliquity.**

obstruction: *See* **propagation path obstruction.**

OCC: other common carrier.

occluder: In optics, a device that completely or partially limits the amount of light reaching the eye.

occultation: An eclipse of a celestial body by the Moon. *Note:* A special case of occultation is the solar eclipse, in which the Sun is eclipsed by the Moon as observed from a point on Earth.

occupancy: For equipment, such as a circuit or a switch, the ratio of (a) the actual time the equipment is in use to (b) the available time during a performance measurement period, usually 1 hour. *Note 1:* Occupancy usually is expressed in percent. *Note 2:* Occupancy may be plotted versus time of day if the performance measurement period is short enough. *Synonym* **usage.**

occupied bandwidth: The width of a frequency band such that (a) below the lower frequency limit and above the upper frequency limit the mean powers emitted are each equal to a specified percentage, B/2, of the total mean power of a given emission, where B/2 should be taken as 0.5%, unless otherwise specified by the International Radio Consultative Committee (CCIR) for the appropriate class of emission, (b) the band contains 99% of the total radiated power, and (c) any discrete band in which the radiated power is 0.25% of the total radiated power is included. *Note 1:* The power outside the occupied bandwidth is represented by B, the percentage of the total mean power of the given emission. *Note 2:* In some cases, such as in multichannel frequency division multiplexing (FDM) systems, use of the 0.5% limits may lead to difficulties in the practical application of the definitions of occupied bandwidth and necessary bandwidth. In such cases, a different percentage may prove useful. *See also* **bandwidth, frequency, necessary bandwidth, nominal bandwidth, rf bandwidth.**

occurrence time: *Synonym* **time of occurrence.**

oceanic display and planning system (ODAPS): A computer-based programmable oceanic air traffic control system that (a) accepts and processes data from various sources, such as the aeronautical telecommunications network (ATN) and the automatic dependent surveillance (ADS) system, and (b) performs various traffic control functions, such as flight plan processing, high frequency (HF) position report processing, conflict prediction and assessment, and air traffic situation display. *See also* **automatic dependent surveillance system.**

oceanographic data interrogating station (ODIS): A station in the maritime mobile service used to initiate, modify, or terminate functions of equipment directly associated with an oceanographic data station, including the functions of the station itself.

oceanographic data station (ODS): A station that (a) is in the maritime mobile service, (b) is used for transmission of oceanographic data, and (c) is located on a ship, buoy, or other sensor platform.

OCR: optical character reader, optical character recognition.

octal: 1. Pertaining to eight. *Note:* An octal numeration system has a radix of eight and uses eight different digits, such as 0 through 7, inclusive. Each octal digit can be represented by a binary numeral that consists of three binary digits, called a "triad." Thus, 101 would represent octal 5. **2.** Pertaining to a numeration system in which the radix is 8. *Note:* Each digit position to the left of a given position has a weight of eight times that of the given position, i.e., the weights from right to left are 1, 8, 64, 512, . . . etc. The octal number 544 is equal in value to the decimal number 356. *See also* **numeration system, radix.** *Refer to* **Fig. C-2 (121).**

octave: *See* **frequency octave.**

octave band: The interval between two specified frequencies, usually the nominal band center frequencies, in two given bands of frequencies. *Note:* The octave band usually is calculated by taking $\log_2$, or $3.322 \times \log_{10}$, of the ratio of the nominal band center frequencies of the two bands.

octet: A byte that (a) consists of eight binary digits and (b) usually is operated upon as a single entity. *Note:* An octet is used in the International Telegraph Alphabet 5 (ITA-5), from which the American Standard Code for Information Interchange (ASCII) is derived, to represent each character. There are 128 characters, which require seven bits. The eighth bit is the parity bit. *Synonym in modern usage* **byte.** *See also* **binary digit, byte.**

octet alignment: 1. The configuration of a field composed of an integral number of octets such that if the number of bits in the field is not divisible by eight, usually zero bits are added to either (a) the first octet, for left justification, or (b) the last octet, for right justification. **2.** The temporal and spatial positioning of an octet with respect to a time frame or a spatial frame of reference. *See also* **binary digit, byte, octet.**

octet timing signal: A signal that (a) identifies each octet in a contiguous sequence of serially transmitted octets and (b) usually identifies the first bit in each octet.

OD: optical density.

ODAPS: oceanic display and planning system.

odd-even check: *Synonym* **parity check.**

odd parity: Parity in which the number of 1s, or marks, in a group of binary digits, or in a group of mark and space signaling elements, such as those that compose a character or a word, is always maintained as an odd number, including the parity bit itself. *Note:* The 1s in a group, such as a seven-bit byte in the American Standard Code for Information Interchange (ASCII) coded character set, are counted. If the number is even, a 1 bit is appended to the byte, making it an octet with an odd number of 1s. If the count is odd, a 0 is added to the byte, making it still an octet with an odd number of ones. During processing or transmission to various points in a system, such as a communications, computer, data processing, or control system, the 1s in each byte are counted. If the count is even, an error will be indicated. However, if the errors are compensating, no error will be indicated, though there may be two errors. It is highly unlikely that compensating errors will continue to occur for more that a few contiguous bytes, in which case the error signal will be sent. The same process can be applied to other groups of bits, such as frames, blocks, and encrypted groups. *See* **byte, even parity, mark, parity, parity check, signaling element, space.**

OFC: optical fiber, conductive.

OFCP: optical fiber, conductive, plenum.

OFCR: optical fiber, conductive, riser.

off-axis optical system: An optical system that (a) has an optical aperture axis that is not coincident with the mechanical center of the aperture, (b) is primarily used (i) to avoid obstruction of the primary aperture by secondary optical elements, instrument packages, or sensors, and (ii) to provide ready access to instrument packages or sensors at the focus, and (c) has an engineering tradeoff of an increase in image focusing aberrations. *See also* **aberration, aperture, focus, optical system.**

off-axis paraboloidal mirror: A paraboloidal mirror that has a paraboloidal surface through which the optical axis does not pass. *See also* **optical axis, paraboloidal mirror.**

off-frequency jamming: *See* **spot off-frequency jamming.**

off-hook: 1. In telephone operations, the conditions that exist, or pertaining to the conditions that exist, when the receiver or the handset of a telephone instrument, or another piece of equipment that is similarly operated, is removed from its cradle, hook, switch, or rest position, i.e., is changed from its rest or nonuse condition. **2.** One of the two possible signaling states, such as (a) tone or no tone, (b) ground connection or battery connection, and (c) on or off, that is opposite to the on-hook condition. *Note:* If off-hook pertains to one state, on-hook pertains to the other. **3.** The active state, i.e., the closed-loop state, of a private line loop or a private branch exchange (PBX) loop. **4.** An operating state of a communications link in which data transmission is enabled either for voice and data communications or for network signaling. *See* **verified off-hook.** *See also* **dial tone, loop, on-hook, open circuit.**

off-hook service: 1. The automatic establishment of a connection between specified users when the call originator removes the receiver or the handset from its switch, rest, cradle, or hook, i.e., goes off-hook. **2.** A telephone service in which (a) a single action on the part of one user establishes a connection directly and immediately with another user or set of users, (b) two signaling states are used, such as (i) tone or no tone, (ii) ground or battery connection, or (iii) on or off, and (c) lifting a handset off its cradle, hook, switch, or rest position immediately causes a prearranged number, or group of numbers, to ring. *Synonym* **automatic signaling service.** *See also* **dedicated circuit, handset, hot line, permanent signal, permanent virtual circuit.**

off-hook signal: In telephone switching, a signal that (a) indicates seizure of a line or a loop, request for service, or a busy condition, and (b) usually is generated and transmitted when a handset is removed from its cradle, hook, switch, or rest position. *See also* **call, call control signal, dial tone, seizure signal, signal.**

office: *See* **central office, destination office, end office, local central office, originating office, tandem office, toll office, tributary office.**

office automation: The organization and the integration of several applications of computers and networks to accomplish office work. *Note:* Examples of office automation are (a) word processing, machine dictation, and document preparation, (b) computer-based information storage, information retrieval, and data banking, (c) communications using electronic mail, electronic message service, and teleconferencing, (d) computer-based financial accounting and electronic funds transfer, (e) management information systems, and (f) computerized commercial transactions. *Synonym* **computer-based office work.**

office class: *Synonym* **class of office.**

office classification: Prior to divestiture, the numbers that (a) were assigned to offices according to their hierarchical function in the U.S. public switched telephone network, and (b) were assigned as follows:

Class 1: Regional Center (RC).

Class 2: Sectional Center (SC).

Class 3: Primary Center (PC).

Class 4: Toll Center (TC), with operators; otherwise, Toll Point (TP).

Class 5: End Office (EO) (local central office).

Note: Any one office handles traffic from one center to two or more centers lower in the hierarchy. Since divestiture, these designations have become less firm. *See also* **central office, class of office, network, switching center.**

officer: *See* **message releasing officer, senior ship radio officer.**

officer of the watch: *See* **radio officer of the watch.**

office terminal: *See* **customer office terminal.**

office trunk: *See* **cross-office trunk.**

offline: 1. In communications, computer, data processing, and control systems, pertaining to the operation of a functional unit when it is (a) not under the direct control of the system with which it is associated, (b) not available for immediate use on demand by the system, (c) able to be independently operated, and (d) in need of human intervention to bring the unit online. **2.** Pertaining to a condition of devices or subsystems associated with a system such that (a) they are not connected to, do not form a part of, and are not subject to the same controls as the operational system with which they are associated, and (b) they may be independently operated. **3.** Pertaining to equipment that is disconnected from a system, is not in operation, and usually has the power off. *See also* **extension facility, online.**

offline cryptographic operation: Cryptosystem operation in which (a) the processes of encryption and transmission, or reception and decryption, are performed in separate steps rather than automatically and simultaneously with transmission and reception, (b) encryption and decryption usually is performed as a self-contained operation, similar to the use of hand machines that are not connected to the line, and (c) the encryption process is not associated with a transmission mode, and thus the resulting cryptogram can be transmitted by any means. *See also* **cryptogram, cryptosystem, decryption, encryption, line.**

offline equipment: Equipment that is in an offline status. *Note:* Examples of offline equipment are (a) in data transmission, equipment that provides data to, or accepts data from, intermediate, bulk, or peripheral storage and not directly from a transmission line, (b) in communications systems, equipment that is not connected to a transmission line such that signals cannot pass to or from the equipment and the line without operator intervention, and (c) in computing and data processing systems, equipment that is not under the direct or continual control of a computer or a data processing system. *See also* **offline, online, online equipment.**

offline recovery: 1. The recovery of nonprotected message traffic by use of an offline processor, central processing unit (CPU), or computer. **2.** The resumption of a process after an interruption and after diagnosing and removing the cause of the interruption while other processes are being executed. *Note:* An example of offline recovery is debugging a computer program while other programs are being run. *See also* **monitoring, offline, run.**

offline storage: 1. Storage that is not under the control of a processing unit. **2.** Storage that is not under the immediate control of a system, subsystem, or component, such as a central processing, arithmetic, logic, or control unit of a communications, computer, data processing, or control system. *Note:* Examples of offline storage are data media, such as punched cards, magnetic tape, and magnetic cards, that contain data and are stored in cabinets and cannot be accessed automatically by a system.

offnet calling: 1. Calling in which telephone calls that originate or pass through private switching systems in transmission networks are extended to stations in the public switched telephone system. **2.** In communications, calling in which calls that originate from, or are destined for, users in nonnetwork entities, are handled by a given communications network. *Note:* Examples of offnet calling are (a) handling a call from a call originator in a private local telephone network to a call receiver in a military network, and (b) using a private or commercial telephone network to handle a call from a user in a military network to a user in the same military network but at a great distance away. *See also* **call, network.**

offnet extension: Relative to a given communications network, such as a telephone or telegraph network, the ability of a central office (C.O.), i.e., an exchange or switching center, to extend an incoming call into another network. *Note:* The provision of offnet extensions from outside the net call originators may not be within the mission requirements of the local net. Offnet extension accommodation rests with the local net authority.

off-premises extension (OPX): An extension telephone, private branch exchange (PBX) station, or key system station that is located on property that is not contiguous with that on which the main telephone, PBX station, or key system station, respectively, is located. *See also* **extension facility, on-premises extension.**

off-route service: *See* **aeronautical mobile satellite (off-route) service.**

offset: *See* **channel offset, common return offset, frequency offset, lateral offset, normalized offset, phase offset.**

offset loss: *See* **lateral offset loss.**

offset prime focus feed: A specialized arrangement for coupling transmitter output signals to an antenna.

offset ratio: *See* **optical fiber/coating offset ratio.**

offset voltage: *See* **common return offset voltage.**

off the air: 1. In radio communications systems, pertaining to a station that is completely shut down, i.e., that is not transmitting any signal, not even an unmodulated carrier. **2.** In a radio station, pertaining to a particular source of modulation, such as a specific microphone, that is disconnected, i.e., is no longer capable of modulating the carrier. *Note:* The carrier may continue unmodulated, or it may be modulated by another signal source. *See also* **off-the-air monitoring, on the air.**

off-the-air monitoring: 1. In radio net operations, monitoring of the transmissions of stations in the net that (a) is performed by the net control station, particularly to check the quality of their transmissions, and (b) usually is performed during periods when the net control station is not transmitting. **2.** The act of a station listening to its own transmissions, by receiving the signal that it has transmitted from its transmitting antenna, in order to discover the quality of the signal that it is transmitting to other stations or that it is broadcasting. *Note 1:* Off-the-air monitoring requires that the received signal must have traveled through the air some distance from the transmitting antenna and not be a signal that is tapped on its way to the transmitting antenna internal to the station or in the antenna transmission line or feeder. The monitoring distance should be such that direct inductive or capacitive coupling between the transmitting antenna and the monitor antenna does not occur, i.e., the receiver should be in the far field region, though the intermediate field region might suffice. *Note 2:* Monitoring transmissions by means of delayed transmission or delayed broadcasting for purposes of editing transmissions or ensuring that erroneous or illegal transmissions do not occur is not off-the-air monitoring. *See also* **monitoring, net, net control station.**

off the shelf: 1. Pertaining to equipment already manufactured and available for delivery from stock. **2.** In military procurement, pertaining to an item that (a) has been developed and produced in accordance with military or commercial standards and specifications, (b) is readily available for delivery from an industrial source, and (c) may be procured without change to satisfy a military requirement.

off-the-shelf hardware: Equipment that has already been produced, is available, and is ready for use.

off-the-shelf item: 1. Equipment already manufactured and available for delivery from stock. **2.** In military procurement, an item that (a) has been developed and produced in accordance with military or commercial standards and specifications, (b) is readily available for delivery from an industrial source, and (c) may be procured without change to satisfy a military requirement.

OFN: optical fiber, nonconductive.

OFNP: optical fiber, nonconductive, plenum.

OFNR: optical fiber, nonconductive, riser.

OFTF: optical fiber transfer function.

ohm: The SI (Système International d'Unités) unit of electrical resistance of material or free space (a) that is equivalent to a resistance such that 1 A (ampere) of electric current will flow through the material or the space between two points when 1 V (volt) of potential difference is applied across the points, and (b) that is a function of the material and its geometrical shape, i.e., its geometry. *See also* **potential difference, resistance.**

old telephone service: *See* **plain old telephone service.**

oligarchically synchronized network: A synchronized network in which the timing of all clocks is controlled by a few selected clocks. *See also* **democratically synchronized network, hierarchically synchronized network, master-slave timing, mutually synchronized network, mutual synchronization.**

OLTS: optical loss test set.

Omega: A global radionavigation system (a) that provides position information by measuring the phase difference between signals radiated by a network of eight transmitting stations distributed worldwide, (b) in which the transmitted signals time share transmission on frequencies of 10.2, 11.05, 11.33, and 13.6 kHz (kilohertz), (c) in which the transmissions are coordinated with Coordinated Universal Time (UTC) United States Naval Observatory (USNO), and (d) that provides a time reference.

omnidirectional antenna: 1. An antenna that has a radiation pattern that is nondirectional. **2.** An antenna that has a receiving pattern that is nondirectional. **3.** An antenna that has a radiation pattern that is nondirectional in azimuth, i.e., the radiance is equal in every horizontal direction. *Note:* The vertical radiation pattern may be of any shape. **4.** An antenna that has a receiving pattern that is nondirectional in azimuth, i.e., the threshold of sensitivity is equal in every horizontal direction. *Note:* The vertical receiving pattern may be of any shape. *See also* **antenna, antenna lobe, directional antenna, radiation pattern, threshold, unidirectional antenna.**

omnidirectional radio range: A radionavigation aid that creates an infinite number of paths in space throughout 360° of azimuth and elevation angle. *Synonym* **omnirange.** *See* **high frequency omnidirectional radio range.** *See also* **azimuth, elevation angle, radionavigation aid.**

omnidirectional range station (ORS): In the aeronautical radionavigation service, a radionavigation land station (a) that provides the direct bearing, i.e., omnibearing, of that station from an aircraft, (b) that furnishes the aircraft with the back azimuth of the aircraft from the station so that personnel in the aircraft get what they want, i.e., get the azimuth of the station from the aircraft, and (c) at which station personnel have to obtain the bearing on the aircraft, either add 180° if it is less than 180° or subtract 180° if it is greater than 180°, and send the result to the aircraft. *See also* **radionavigation land station.**

omnidirectional signaling system: A signaling system in which signals are transmitted with nearly equal radiance, field strength, or irradiance in all directions, or at least in a given plane, such as the horizontal plane. *Note:* Examples of omnidirectional signaling systems are (a) a flashing light transmission system in which a point source of light is used, such as a light bulb, and (b) a radio transmission system in which a single dipole antenna is used. *See also* **dipole, flashing light transmission, irradiance, radiance, signaling.**

ONA: open network architecture.

on black: *See* **phasing on black.**

on-board communications station (OCS): A low-powered mobile station in the maritime mobile service intended for communications (a) on board a ship, (b) between a ship and its lifeboats and life rafts during lifeboat drills or operations, (c) within a group of vessels being towed or pushed, and (d) in line handling and mooring operations. *See also* **mobile station.**

on-call patching service: A network-provided service feature that (a) establishes a connection between authorized users, and (b) provides a temporary path between them in order to fulfill an unanticipated or special communications request. *See also* **path.**

on-demand call tracing: In a specific call, call tracing that is performed on request (a) particularly during the information transfer phase, (b) during the disengagement phase, (c) after the call is finished, or (d) after the circuits are disconnected. *See also* **call, finished call, information transfer phase.**

one-address instruction: A computer program instruction in which the address part contains only one address, such as the address of an operand. *See also* **four-address instruction, multiple address instruction, three-address instruction, two-address instruction.**

one forty four line weighting: In telephone systems, a noise weighting that is used in a noise-measuring set to measure noise on a line that would be terminated by an instrument with a No. 144 receiver, or by a similar instrument. *Note:* The meter scale readings are in dBrn(144-line). *Synonym* **144-line weighting.** *See also* **C message weighting, flat weighting, F1A line weighting, HA1 receiver weighting, line, noise, noise measurement unit, noise weighting.**

one forty four receiver weighting: In telephone systems, a noise weighting that is used in a noise measuring set to measure noise across the receiver of an instrument equipped with a No. 144 receiver or similar receiver. *Note:* The meter scale readings are in dBrn(144-receiver). *Synonym* **144-receiver weighting.** *See also* **C message weighting, flat weighting, HA1 receiver weighting, line, noise, noise measurement unit, noise weighting.**

one gate: *Synonym* **OR gate.**

one hundred percent modulation: Modulation in which the magnitude of the modulating signal produces the full range of modulated signal, from zero to the maximum or reference value. *Note:* Examples of one hundred percent modulation are (a) in an amplitude-modulated signal, the situation that occurs when the modulation envelope just reaches the zero baseline at one point in the cycle, and (b) in a phase-modulated signal, the situation that occurs when the modulation signal causes the phase position of the modulated carrier to just reach the zero phase shift position. *Synonym* **100% modulation.**

one-operator ship: In radiotelegraph communications, a ship station that has only one radiotelegraph operator. *Note:* Messages for a specific one-operator ship station are received and held for that station during its nonoperational periods and transmitted by other ship or shore stations to that station during its operational periods. In ship stations, the guard watch can be rotated, particularly if the ships are in the same convoy or in the immediate ocean area. *See also* **guard watch.**

ones complement: The binary numeral obtained by taking each digit of another binary numeral, the complement of which is desired, subtracting it from 1, digit place by digit place, i.e., subtracting it from a string of 1s, and writing the digits down one by one, such that (a) no carries are considered in the process, (b) there are as many digits in the complement as in the numeral from which it was derived, provided that leading zeros are not suppressed, and (c) either numeral is the complement of the other. *Note 1:* The ones complement of a numeral, and only a pure binary numeral, can be obtained simply by reversing all digits, changing all the 1s to 0s and all the 0s to 1s. Also, when a bank of flip-flops is used to represent a binary numeral by using one side of the flip-flops, the other side represents the ones complement. A practical application is in subtraction, which may be performed by adding the minuend to the ones complement of the subtrahend, adding a 1 at the least significant digit position of the result, propagating all carries, and then deleting the most significant digit of the result. A ones complement is a radix-minus-one complement in a numeration system in which the radix is 2, i.e., the pure binary numeration system. *Note 2:* The ones complement of 101101 is 010010. *See* **radix minus ones complement.** *See also* **minuend, numeral, numeration system, pure binary, subtraction, subtrahend.**

one-way communication: Communication in which information is always transferred in only one preassigned direction. *Note 1:* One-way communication is not necessarily constrained to one transmission path. *Note 2:* Examples of one-way communications systems are broadcast stations, one-way intercom systems, and wireline news services.

one-way only channel: A channel in which information can be transferred in one, and only one, fixed direction and cannot be reversed. *Synonym* **unidirectional channel.** *See also* **channel, simplex circuit, simplex operation.**

one-way only operation: *Synonym* **simplex operation.**

one-way operation: *Synonym* **simplex operation.**

one-way reversible operation: *Synonym* **half-duplex operation.**

one-way trunk: 1. In communications systems, a trunk that (a) is between switching centers, (b) is used for traffic in one fixed preassigned direction, (c) usually is used to collect originating traffic from a particular switching center for transmission to a particular destination, and (d) at the traffic originating end, is known as an "outgoing" trunk, and at the other end, is known as an "incoming" trunk. **2.** A trunk between two switching centers over which traffic may be originated from one preassigned location only. *Note:* The traffic may consist of two-way communications. The term "one-way" refers only to the constraint on the location that may originate a call. *See also* **trunk.**

one-wire line: *Synonym* **single-wire line.**

on-hook: 1. In telephone systems, the condition that exists during the time the receiver or the handset is resting on its switch or is in its cradle. *See also* **dial tone. 2.** One of two possible signaling states, such as (a) tone or no tone and (b) ground connection or battery connection. *Note:* If on-hook pertains to one state, off-hook pertains to the other. **3.** The idle state, i.e., open loop, of a customer or private branch exchange (PBX) user loop. **4.** An operating state of a communications link in which data transmission is disabled by a high impedance to the network. *Note:* During the on-hook condition, the link is sensitive to ringing signals. *See also* **off-hook, open circuit.**

on-hook signal: In telephone switching systems, a signal that (a) indicates a disconnect, an unanswered call, or an idle condition and (b) usually is transmitted when the switch on an end instrument is allowed to remain undisturbed in its cradle, hook, switch, or rest position. *See also* **call, call control signal, idle state, not ready condition, signal.**

online: **1.** In communications, computer, data processing, and control systems, pertaining to the operation of a functional unit when it is (a) under the direct control of the system with which it is associated, (b) available for immediate use on demand by the system, (c) not able to be independently operated, and (d) not in need of human intervention for the system to access the unit. **2.** Pertaining to a condition of system-associated devices or subsystems such that (a) they are connected to, form a part of, and are subject to the same controls as the operational system with which they are associated, and (b) they may not be independently operated. **3.** Pertaining to equipment that is connected to a system, is in operation, and has the power on. *See also* **extension, extension facility, offline, on-premises extension.**

online computer system: A computer system (a) that is a part of, or is embedded in, a larger entity, such as a communications system and (b) that functions within that entity to obtain data from the entity, perform operations on that data, and provide the results to that entity. *Note:* An online computer system usually interacts in real or near real time with the entity in which it is embedded and with its users. *See also* **multiprocessing, time sharing.**

online cryptographic operation: Cryptosystem operation in which (a) cryptoequipment is used that is directly connected to a line, (b) encryption and transmission, reception and decryption, or both, are single processes, (c) operation includes the automatic encryption associated with a particular transmission system in which messages are encrypted and passed directly to a line to operate decryption equipment at one or more distant stations, and (d) the user may not be aware of the encryption-decryption process, but the user is aware of the security classification of the messages. *See also* **cryptosystem, decryption, encryption, offline cryptographic operation.**

online equipment: Equipment that is in an online status. *Note:* Examples of online equipment are (a) in data transmission, equipment that provides data to, or accepts data from, intermediate, bulk, or peripheral storage directly via a transmission line, (b) in communications systems, equipment that is connected to a transmission line such that signals can pass to or from the equipment and the line without operator intervention, and (c) in computing and data processing systems, equipment that is under the direct or continual control of a computer or a data processing system. *See also* **offline, offline equipment, online.**

online recovery: *Synonym* **inline recovery.**

online storage: **1.** Storage that is under the control of a processing unit. **2.** Storage that is under the direct and immediate control of a system, subsystem, or component, such as a central processing, arithmetic, logic, or control unit of a communications, computer, data processing, or control system. *Note:* Examples of online storage are data media, such as punched cards, magnetic tape, and magnetic cards, that automatically can be accessed by a system, such as a communications, computer, data processing, or control system. *See also* **access, offline storage, storage.**

on-premises cabling (OPC): Customer-owned metallic or optical fiber communications transmission lines that (a) are installed within or between buildings, (b) include horizontal cabling and backbone cabling, (c) extend from the external network interface to the user workstation areas, and (c) include the total communications cabling, to transport current or future data, voice, video, local area network (LAN), and image information. *Deprecated synonym* **house cabling.** *See also* **on-premises wiring.**

on-premises extension: An extension telephone, private branch exchange (PBX) extension station, or key system extension station located on property that is contiguous with that on which the main telephone, PBX, or key system is located. *See also* **extension facility, off-premises extension, online.**

on-premises wiring (OPW): Customer-owned metallic or optical fiber communications transmission lines, such as twisted pair, coaxial cables, and fiber optic cables, that (a) are installed within or between buildings, (b) include horizontal wiring and backbone wiring, (c) extend from the external network interface to the user workstation areas, and (c) include the total communications wiring, to transport current or future data, voice, video, local area network (LAN), and image information. *Synonyms* **inside wire, premises wiring.** *Deprecated synonym* **house wiring.** *See also* **on-premises cabling.**

on the air: 1. In radio communications systems, pertaining to a station that is transmitting a carrier, whether or not the carrier is modulated. **2.** In a radio station, pertaining to a particular source of modulation, such as a specific microphone, that is connected, i.e., is capable of modulating the carrier. *See also* **off the air.**

on-the-fly printer: An impact printer that has type slugs that continue moving during the time that they press an inked or carbon ribbon against a data medium, such as paper. *Synonym* **hit-on-the-fly printer.** *See also* **carbon ribbon, data medium, impact printer, inked ribbon, type slug.**

on white: *See* **phasing on white.**

opacity: A quality of a material that allows it to prevent the conduction of lightwaves by means of maximum absorption. *See also* **transparency.**

opaque: 1. A quality of a material that makes it impervious to light so that its luminous transmittance is zero. **2.** A substance that is impervious to light applied to transparent or translucent substances. **3.** To make impervious to light. *See also* **luminous transmittance, translucent, transparent.**

opcode: *Synonym* **operation code.**

open: 1. In electric circuitry, pertaining to a circuit that is incomplete, and in which electric current thus cannot flow. **2.** In communications, pertaining to a circuit that is available for transmission. **3.** In display systems, to prepare a specific sequence of graphic or display commends in order that the segment may accept additional commands. *See also* **close.**

open avionics architecture (OAA): The structure of a system that (a) is an open system, (b) consists of avionic components, and (c) usually is constructed of modular components. *See also* **avionics, open system, system.**

open circuit (OC): 1. An electrical path that (a) closes on itself and (b) contains a nearly infinite impedance. *Note 1:* In theory, an idealized open circuit will conduct 0 current with the application of a finite voltage. *Note 2:* An open circuit may be intentionally open, as with a switch, or may be unintentionally open, as in a severed cable. **2.** In communications, a circuit available for use. *Synonym in this sense* **available line.** *See also* **circuit, off-hook.**

open dual bus (ODB): A dual bus in which the head of bus functions for the two buses are at different locations. *See also* **dual bus.**

open network architecture (ONA): Network architecture that (a) covers the basic facilities and services that a communications carrier provides and (b) allows all users of the network to interconnect to specific network functions and interfaces on an unbundled equal-access basis. *Note:* In the Federal Communications Commission (FCC) Computer Inquiry III, the open network architecture (ONA) concept consists of three integral types of components, namely, (a) basic serving arrangements (BSAs), (b) basic service elements (BSEs), and (c) complementary network services. *See also* **network architecture.**

open shop operation: In computer and data processing system operations, the operation of a computing facility or center such that computer time is made available to the job-originating users who work outside the facility or center and who also write their own programs. *See also* **closed shop operation, job.**

open-source intelligence (OSCINT): Intelligence derived by gathering, processing, and analyzing information that (a) usually is derived or obtained from sources that make the information available to the public, (b) usually is available via multimedia, (c) is not security-classified by the source, (d) is not proprietary, except for copyright, (e) is not derived from sensitive contracting, and (f) is not obtained by clandestine or covert means. *Note:* Examples of open-source intelligence (OSCINT) are intelligence reports derived from newspapers, radio, television, journals, and magazines. *See also* **closed-source intelligence.**

open system: A system with characteristics that comply with specified, publicly maintained, readily available standards and that therefore can be connected to other systems that comply with these same standards. *Refer to* **Fig. L-4 (506).**

open system architecture (OSA): The structure, configuration, or model of a communications or distributed data processing system that (a) enables system description, design, development, installation, operation, improvement, and maintenance to be performed on a hierarchical structure or layering of functions, (b) allows each layer to provide a set of accessible functions that can be controlled and used by the functions in the layer above it, (c) enables each layer to be implemented without affecting the implementation of other layers, and (d) provides for alteration of system performance without disturbing the large investment in existing equipment, procedures, and protocols, by making alterations at higher or lower levels. *Note 1:* Examples of independent alterations include (a) converting from wire to optical fibers at a physical layer without affecting the data link layer or the network layer except to provide more traffic capacity and (b) altering the operational protocols at the network layer without altering the physical layer. *Note 2:* Open systems architecture may be implemented by using the Open Systems Interconnection—Reference Model (OSI-RM) as a guide but designing the system to meet performance requirements. *Refer to* **Figs. L-4 (506), P-19 (771).**

open systems: *See* **Corporation for Open Systems.**

open systems interconnection (OSI): 1. The interconnection of communications, computer, data processing, and control systems, usually using standard procedures. **2.** A logical architecture, i.e., a logical structure, for a communications network in which (a) operations are standardized by the International Organization for Standardization (ISO), the standard being a seven-layer network architecture for defining network protocol standards to enable any open systems interconnection (OSI) compliant device, such as a computer or a user end instrument, to communicate with any other OSI compliant device, (b) the OSI protocol specifications are the lowest level of abstraction within the OSI standards scheme, (c) each OSI protocol specification operates at a single layer, and (d) each protocol specification defines the primitive operations and permissible responses required for the exchange of information between peer processes in communicating systems, to carry out all or a subset of the services defined within the OSI service definitions for that layer. *Note:* The Open Systems Interconnection—Reference Model (OSI-RM) is an abstract description of the digital communications between application programs running in distinct systems. The model uses a hierarchical structure of seven layers, including the (1) physical (lowest), (2) data link, (3) network, (4) transport, (5) session, (6) presentation, and (7) application (highest) layers. Each layer performs value-added service at the request of the adjacent higher layer and, in turn, requests more basic services from the next lower layer. The OSI service definitions are at the next lower level of abstraction below that of the OSI reference model. The OSI service definitions for each layer define the layer abstract interfaces and the facilities provided to the user of the service, independent of the mechanism used to accomplish the service. The OSI systems management function is the collection of functions in the application layer that are related to the management of various resources and their status across all layers of the OSI architecture. *See also* **Open Systems Interconnection—Reference Model, Open Systems Interconnection—Service Definition, Open Systems Interconnection—Systems Management.**

Open Systems Interconnection (OSI): Pertaining to a logical structure for communications networks standardized by the International Organization for Standardization (ISO). *Note:* Adherence to the standard enables any Open Systems Interconnection (OSI) compliant system to communicate with any other OSI compliant system for an efficient and meaningful exchange of information.

Open Systems Interconnection (OSI) architecture: Communications system architecture that adheres to the set of International Organization for Standardization (ISO) standards that relates to open systems architecture.

Open Systems Interconnection (OSI)—Protocol Specification: An Open Systems Interconnection (OSI) protocol specification that (a) operates at a single layer, (b) defines the primitive operations and permissible responses required to exchange information between peer processes in communicating systems, (c) enables the execution of all or a subset of the services defined within the Open Systems Interconnection (OSI)—Service Definitions for the layer, and (d) is at the lowest level of abstraction within the Open Systems Interconnection—Reference Model (OSI-RM).

Open Systems Interconnection—Reference Model (OSI-RM): A model that (a) may be used as a reference to describe the architecture of a communications system and the operational processes within the system, (b) consists of a hierarchical structure of seven layers, each containing a set of specific functions, (c) identifies the value-added services performed at each layer, and (d) identifies the set of functions in each layer that responds to requests for services from the adjacent higher layer and, in turn, requests more basic services from the adjacent lower layer. *Note 1:* In a graphical representation of the layers, Layer 7 (Application Layer) is at the top and is the highest layer. Layer 1 (Physical Layer) is at the bottom and is the lowest layer. *Note 2:* The seven layers of the Open Systems Interconnection—Reference Model (OSI-RM) are:

(7) Application Layer: Layer 7, the highest of the seven layers, contains a set of functions that (a) directly interfaces the application processes, (b) performs common applications services for the applications processes, (c) requests services from the Presentation Layer, and (d) provides communications services that perform semantic conversion between associated application processes. Examples of common applications services performed by Application Layer functions include the implementation of virtual file, virtual terminal, job transfer, and job manipulation protocols.

(6) Presentation Layer: Layer 6 contains a set of functions that (a) responds to service requests from the Application Layer, (b) requests services from the Session Layer, and (c) relieves the Application Layer from being required to resolve syntactical differences in data representation within the end user systems. An example of a Presentation Layer function is the conversion of an Extended Binary Coded Decimal Interchange Code (EBCDIC) text file to an American Standard Code for Information Interchange (ASCII) text file.

(5) Session Layer: Layer 5 contains a set of functions that (a) responds to service requests from the Presentation Layer, (b) requests services from the Transport Layer, (c) manages the dialogue between end user application processes, (d) provides for either duplex or half-duplex operation, and (e) establishes checkpointing, adjournment, termination, and restart procedures.

(4) Transport Layer: Layer 4 contains a set of functions that (a) responds to service requests from the Session Layer, (b) requests services from the Network Layer, (c) provides transparent data transfer between end users, and (d) relieves the upper layers from being required to provide reliable and cost-effective data transfer.

(3) Network Layer (NL): Layer 3 contains a set of functions that (a) responds to service requests from the Transport Layer, (b) requests services from the Data Link Layer, (c) provides the functional and procedural means for transferring variable-length data sequences from a source to a destination via one or more networks while maintaining the service quality requested by the Transport Layer, and (d) performs network routing, flow control, data segmentation and desegmentation, and error control functions.

(2) Data Link Layer (DLL): Layer 2 contains a set of functions and protocols that (a) responds to service requests from the Network Layer, (b) requests services from the Physical Layer, (c) provides the functional and procedural means to transfer data among network entities, and (d) detects, and possibly corrects, errors that may occur in the Physical Layer. Examples of Data Link protocols are (a) high level data link control (HDLC) and Advanced Data Communication Control Procedure (ADCCP) protocols for point-to-point and switched networks and (b) logical link control (LLC) protocols for local area networks (LANs).

(1) Physical Layer (PL): Layer 1, the lowest of the seven layers, contains a set of functions that (a) performs services requested by the Data Link Layer, (b) establishes and terminates connections to communications media, (c) participates in the processes in which the communications system resources are effectively shared among multiple users, such as contention resolution and flow control, and (d) converts the representation of data in user equipment to corresponding signals to be transmitted over a communications channel. *Refer to* **Fig. L-4 (506).**

Open Systems Interconnection—Service Definition (OSI-SD): For each of the seven layers of the Open Systems Interconnection—Reference Model (OSI-RM), a definition of (a) each of the abstract interfaces within the layer and with adjacent layers and (b) the services and the facilities provided to their user independent of the mechanisms used to accomplish them. *Note:* Open Systems Interconnection (OSI)—Service Definition is the next level of abstraction below that of the Open Systems Interconnection—Reference Model (OSI-RM).

Open Systems Interconnection—Systems Management (OSI-SM): In the Application Layer of the Open Systems Interconnection—Reference Model (OSI-RM), the set of functions related to the management and the status of various resources identified in all layers of the Open Systems Interconnection—Reference Model (OSI-RM).

open waveguide: 1. An all-dielectric waveguide in which (a) electromagnetic waves are guided by a refractive index gradient so that the waves are confined to the guide by refraction or reflection from the outer surface of the guide or from surfaces within the guide, and (b) the electromagnetic waves propagate, without radiation, within the waveguide, although evanescent waves coupled to internal waves may travel in the space immediately outside the waveguide. *Note:* Examples of open waveguides are optical fibers, slab dielectric waveguides, and waveguides in optical integrated circuits. **2.** A waveguide (a) that does not have electrically conducting walls, (b) in which electromagnetic waves are guided only by the refractive index profile, (c) in which the waves are confined to the guide by refraction and reflection at the outer surface of the guide and the dielectric interface surfaces within the guide, and (d) that is constructed only with dielectric materials, such as glasses and plastics. *Synonym* **all-dielectric waveguide.** *See also* **closed waveguide, optical fiber, optical waveguide.** *Refer to* **Fig. P-17 (767).**

open wire: 1. Individual electrical conductors that (a) usually are solid wires separately supported with insulators on poles or towers above the surface of the Earth, (b) may be insulated or uninsulated, and (c) may be used for signal and power transmission. *Note:* Open wires for signal transmission are rapidly being eliminated in many areas, including urban areas and along railroad lines. They are being replaced with above- and below-ground sheathed cables, especially fiber optic cables, and with radio nets. Open wire lines are still being installed for long-distance high voltage power transmission and for rural low voltage power distribution. **2.** A wire that has an electrical resistance of such large magnitude that its successful use as an electrical path or connection is precluded. *Note:* Examples of open wires are (a) a wire with a break in it and (b) a wire that is not effectively connected to an end terminal. *See also* **twin cable, twisted pair, two-wire circuit.**

open wire line: A line that (a) usually consists of uninsulated solid wire fastened to insulators with pieces of the wire itself, (b) is mounted on poles or towers to traverse open country, such as along roads and railroad tracks, and (c) is used for signal or power transmission. *Note:* Open wire line insulators usually are made of glass, ceramic, or plastic. *See* **line, open wire.**

operand: An entity on which an operation is performed. *Note:* Examples of operands are (a) the a and b in the expression $a + b$, (b) the θ in $\sin \theta$, and the A and B in the Boolean logic operation A OR B. *See also* **operation.**

operate mode: *Synonym* **compute mode.**

operating agency: In communications system operations, the agency or the authority that is responsible for, or that controls, the operation of a communications network or system. *See* **telecommunications private operating agency.**

operating bounce time: *See* **switch operating bounce time.**

operating condition: *See* **extended operating condition, standard operating condition.**

operating instruction: *See* **standing communications operating instruction.**

operating method: *See* **radiotelegraph operating method.**

operating signal: In communications systems, a group of characters used to convey instructions, orders, requests, reports, or other information to facilitate communications. *Note:* An example of an operating signal is a group of three letters that are sometimes followed by a numeral and that may begin with a 0 or a Z. They are used as brevity codes to express various stereotyped phrases that are encountered in the daily conduct of communications. In voice systems, they may be simple prosigns or prowords, such as OVER, ROGER, OUT, SOS, MAYDAY, and TENFOUR.

operating space: *Synonym* **display space.**

operating system (OS): 1. In a computer, an integrated collection of routines that (a) services the sequencing and processing of programs, (b) provides a central control to users and programs, (c) performs input, output, accounting, resource allocation, and storage assignment functions, and (d) performs other system-related functions. **2.** In communications, computer, data processing, and control systems, the totality of (a) all software, such as executive, call, loading, and recovery routines, (b) priorities, (c) associated data, (d) arrangements for control of (i) available hardware, storage, and central processing unit time allocation, (ii) queuing, input-output, receipt, and transmission operations, and (iii) data flow, and (e) arrangements for multiprocessing by both parallel processing and multiprogramming, all to permit total system operation. *Note:* In a personal computer (PC), the operating system is a set of fundamental instructions that controls operations, such as the disk operating systems IBM DOS and MS DOS. **3.** The software that (a) controls the execution of computer programs in a communications, computer, data processing, or control system, (b) may provide supporting services, such as resource allocation and scheduling, input-output control, and data management, and (c) may be partially or completely implemented as firmware. *See* **disk operating system, general purpose operating system.** *Refer to* **Fig. P-18 (769).**

operating system interface for computer environments: *See* **portable operating system interface for computer environments.**

operating time: 1. The time between (a) the instant of occurrence of a specified input condition to a system and (b) the instant of completion of a specified operation. **2.** In communications, computer, data processing, and control systems, the time between (a) the instant a request for service is received from a user and (b) the instant of final release of all facilities by the user or either of two users. **3.** In communications systems conference calls, the time between (a) the instant a request for service is received from one of a group of concurrent users and (b) the instant all but one of the users have released all facilities. *See* **switch operating time, system operating time.**

operation: 1. The use a device, equipment, or system. *Note:* Examples of operations are (a) the action of obtaining a result from one or more operands in accordance with a rule, (b) a program step that includes the execution of an instruction in a computer program to produce a result, (c) the action that is performed by a combinational circuit to produce an output, and (d) the insertion of a switchboard plug into a jack to connect a line or create a circuit. **2.** A defined action that, when applied to any permissible combination of known entities, produces a new entity. *Note:* An example of an operation is the addition of 5 and 3 to obtain 8. The numbers 5 and 3 are the operands, the number 8 is the result, and the plus sign is the operator, indicating that the operation performed is addition. **3.** In a computer, a program step, usually specified by a part of an instruction word, that is undertaken or executed by the computer. *Note:* Examples of operations are (a) addition, multiplication, extraction, comparison, shift, and transfer, and (b) calculation of the values of sin θ for given values of θ. *See* **arithmetic operation, asynchronous operation, attenuation-limited operation, automatic digital network operation, automatic operation, auxiliary operation, balanced link level operation, bandwidth-limited operation, binary arithmetic operation, bit by bit asynchronous operation, bit-rate•length product-limited operation, bit synchronous operation, Boolean operation, broadcast operation, centralized operation, closed shop operation, coherent pulse operation, concurrent operation, conference operation, continuous operation, cross-polarized operation, detector noise-limited operation, diplex operation, dispersion-limited operation, distortion-limited operation, drop channel operation, equivalence operation, fail safe operation, half-duplex operation, identity operation, multilink operation, multimode operation, multiplex operation, n-adic Boolean operation, net operation, neutral operation, noncentralized operation, offline cryptographic operation, online cryptographic operation, open shop operation, packet mode operation, parallel operation, polar operation, push to talk operation, push to type operation, quantum-limited operation, quantum-noise-limited operation, repetitive operation, reversible operation, sequential operation, simplex operation, slip-free operation, speech plus duplex operation, synchronous cryptooperation, synchronous operation, thermal-noise-limited operation, unattended operation, unidirectional operation.** *See also* **instruction, operand.**

operational amplifier: In analog computing systems and in communications systems, a high gain amplifier that is connected to external elements in order to perform specific operations or functions, such as multiplication by a constant.

operational control: *See* **change in operational control.**

operational evaluation (OPEVAL): Evaluation of system performance and capabilities based on complete system operation in the real environments in which the system is to be used. *See also* **technical evaluation.**

operational load: In a communications system, the total power requirement for communications facilities. *See also* **nonoperational load.**

operational network control: In a communications network, the control and the implementation of functions that (a) prevent or eliminate degradation of any part of the network, (b) provide response to changes in the network to meet long range requirements, (c) provide immediate response to demands that are placed on the network, and (d) provide for efficient operation of the network, such as (i) perform immediate circuit utilization actions, (ii) continuously control circuit quality, (iii) continuously control equipment performance, (iv) develop procedures for immediate repair, restoration, or replacement of facilities and equipment, (v) conduct continuous liaison with network users and with representatives of other networks, and (vi) provide advice, assistance, and instruction in network operation to operating personnel and in network utilization to users.

operational reliability: In communications, computer, data processing, and control systems, the reliability of a system, including hardware, software, and firmware, when operating in the environment for which it was designed to be used, rather than in a specified or test environment. *See also* **reliability, specified reliability, test reliability.**

operational research: *Synonym* **operations research.**

operational service period: 1. A period during which a telecommunications service remains in an operational state. *Note:* The operational state must be defined in accordance with specified criteria. **2.** A performance measurement period, or succession of performance measurement periods, during which a telecommunications service remains in an operational service state. *Note:* An operational service period begins at the beginning of the performance measurement period in which the telecommunications service enters the operational service state, and ends at the beginning of the performance measurement period in which the telecommunications service enters the nonoperational service state. The end of the operational service state and the beginning of the immediately following nonoperational service state occur at the same instant. *See also* **operational service state, performance measurement period.**

operational service state (OSS): During any performance measurement period of a communications, computer, data processing, or control system, a service condition that existed when the calculated values of specified performance parameters were equal to or better than their associated outage thresholds. *See also* **operating time, operational service period, outage, outage threshold, performance measurement period, performance parameter.**

operational test land station: *See* **radionavigation operational test land station.**

operational threshold: *See* **system operational threshold.**

operation code (opcode): 1. In communications systems, a code that (a) is used for immediate and rapid communications during written and oral exchanges of information among operating, maintenance, and management personnel, (b) consists of words and phrases, with prearranged or assigned meanings, that are capable of being pronounced or spelled, usually both, and (c) usually are contractions of terms, practically all of which are in this dictionary. *Note:* Examples of operation codes are "opstate" for operational service state, "zeezero" or "zeesubzero" for the characteristic impedance of a transmission line, "disc" for the disconnect command, "opcode" for operation code, and "metrodata" for meteorological data. **2.** A code that is used to write operations that a system is to execute. *Note:* An example of an opcode is the instruction code for preparing or creating a computer program. *See also* **instruction code.**

operation instruction: *See* **signal operation instruction.**

operation net: In tactical or strategic military communications systems, a communications net that is used primarily for the exchange of messages that are in support of tactical or strategic military operations.

operation part: In a computer program instruction word, the part that contains the operator or that specifies the operation that is to be performed with or on the operands that either (a) are in the instruction word or (b) are in the storage locations specified by the addresses that are in the address part of the instruction word. *See also* **computer program, instruction word, address part, operator.**

operation rate: *See* **full motion operation rate.**

operations analysis: *Synonym* **operations research.**

operations communications net: *See* **maritime tactical air operations communications net.**

operations per second: *See* **floating point operations per second.**

operations research (OR): 1. The design, study, and use of models for the solution of complex problems arising in systems operations in order to optimize the use of available resources, including analytical methods, such as mathematical, analog, and simulation methods, for solving these problems. *Note:* Examples of systems that may be modeled are (a) a communications, computer, data processing, or control system, (b) a railroad, busline, airline, or highway network to analyze traffic patterns to ensure optimal use of resources, (c) the economy of a country to effectively use available resources, and (d) a battle scenario to analyze operations and make optimal use of military capabilities, minimize losses to friendly forces, and maximize losses to hostile forces. **2.** The design of models for complex problems concerning the optimal allocation of available resources and the application of mathematical methods for solving those problems. **3.** The analytical study of military problems undertaken to provide responsible commanders and staff agencies with a scientific basis for deciding on actions to improve military operations. *Synonyms* **operational research, operations analysis.** *See also* **system analysis.**

operations security (OPSEC): 1. In military systems, the analyzing of friendly actions attendant to military operations and other activities to (a) identify those actions that can be observed by adversary intelligence systems, (b) determine indicators that hostile intelligence systems might obtain that could be interpreted or pieced together to derive critical information in time to be useful to the hostile forces, and (c) select and execute measures that eliminate, or reduce to an acceptable level, the vulnerabilities of friendly actions to hostile-force exploitation. **2.** The denying to potential adversaries information about friendly capabilities and intentions, by identifying, controlling, and protecting unclassified evidence of the planning and the execution of sensitive activities.

operations security (OPSEC) assessment: The process of analyzing information sources and indicators associated with operations and related activities to evaluate and improve the effectiveness of an organization in protecting its critical information resources from real and potential adversaries, including the use of techniques such as (a) identification of critical information that requires protection, (b) identification of information and indicators that can be observed or obtained by adversaries that could be interpreted, analyzed, or assembled to derive critical information in time to be useful to adversaries, (c) selection and recommendation of measures that eliminate or reduce the vulnerability of friendly actions or information, and (d) elimination or reduction of the use of friendly actions or information by adversaries. *See also* **operations security, operations security survey.**

operations security (OPSEC) survey: An on-site examination of an operation or an activity to determine if there are vulnerabilities that would permit adversaries to obtain and exploit critical information during the planning, preparation, execution, and postexecution phases of any operation or activity. *See also* **operations security, operations security assessment.**

operations service: *See* **port operations service, space operation service.**

operations service element protocol: *See* **remote operations service element protocol.**

Operations Support System (OSS): The command and control decision-aid computer and database system that (a) provides U.S. Navy command centers with a hybrid information-management command, control, communications, computers, and intelligence (C^4I) capability, (b) captures the functions from a variety of other (C^4I) systems, (c) integrates the functions into a single cohesive uniform user-friendly system, and (d) has an integrated database that supports command center functions, such as information display, message transfer, platform-position tracking, operations planning, and capabilities analysis.

operations system (OS): In network management, a system that processes telecommunications-management information and that supports, controls, and performs various telecommunications-management functions, including the performance of surveillance and testing functions to support customer access maintenance.

operation table: A table that (a) consists of a list of permissible combinations of operators and all possible values of operands and (b) lists the results for each of the combinations of operators. *See* **Boolean operation table.**

operation time: 1. In communications, computer, data processing, and control systems, the time required to interpret and execute a specified operation. *Note:* Examples of operation times are (a) the time required to select a storage location and store a number and (b) the time required to add two numbers, including the access time for obtaining the operands, performing the addition, and storing the results. **2.** In a computer, the time required to execute a computer program instruction. **3.** In a computer, the instant at which the execution of an instruction starts. *See also* **instruction.**

operator: 1. In telephone switchboard operation, the person at a switchboard position who performs the functions that are required to handle telephone traffic and inquiries, such as to make connections between call originators and call receivers via trunks or loops, set up multiple calls, and complete call tickets. **2.** In communications, computer, data processing, and control systems, a person who interacts with a system to ensure successful operation of the system. *Note:* Typical operators are telephone operators, telegraph operators, teletype operators, facsimile system operators, and computer operators. **3.** In the description of an operation or a process, such as an arithmetic or logical operation, that which (a) performs an action on an operand or (b) indicates an action to be performed on an operand to obtain a result. *Note:* Examples of operators are (a) the + sign in the arithmetic operation $a + b$, (b) the OR in the Boolean operation A OR B, which might be performed by an OR gate, (c) the differentiating operator d/dt for taking the time derivative of a function of time, (d) the exponent in the expression x^2, and (d) the • in the expression for the dot product of **A** and **B**, i.e., **A • B,** in the vector algebra. *See also* **Boolean operation, call originator, call receiver, switchboard, switchboard position, system operator.**

operator-assisted call: A call that is established with the help of a person, i.e., an operator, attendant, or supervisor, rather than only with automatic switching equipment and related equipment.

operator delay: In a communications system, the time from (a) the instant that operator assistance is requested by a user to (b) the instant that the operator commences to process the requested service. *See also* **operator-assisted call.**

operator log: In a communications system, a chronological record of events that relate to the operation of a particular system, subsystem, or component, such as a line, circuit, group, trunk, link, exchange, station, radio transmitter, radio receiver, or radar set.

operator period station: In communications systems, a station that is not operated on a continuous 168-hours-per-week basis, such as operation by a single-operator period watch or operation only during daylight hours. *See also* **constant watch station, single-operator period watch, two-operator period watch.**

operator ship station: *See* **one-operator ship station.**

operator signaling: *See* **remote operator signaling.**

OPEVAL: operational evaluation.

OPSEC: operations security.

optic: *See* **electrooptic, fiber optic, magnetooptic.**

optic active connector: *See* **fiber optic active connector.**

optical: 1. Pertaining to the field of optics. **2.** Pertaining to eyesight. **3.** Pertaining to systems, devices, or components that generate, process, and detect lightwaves or light energy, such as lasers, lens systems, optical fibers, fiber optic bundles, fiber optic cables, and photodetectors. **4.** Pertaining to the optical spectrum, i.e., the lightwave region, of the electromagnetic spectrum, i.e., the region in which the techniques and the components used in the visible spectrum also apply to the region extending somewhat beyond the visible region into the ultraviolet and infrared regions, corresponding to wavelengths between 0.001 µm (micron) and 100 µm. **5.** Pertaining to the range of wavelengths of optical radiation, which is the electromagnetic spectrum within the wavelength region extending from the vacuum ultraviolet at 0.001 µm (micron) to the far infrared at 100 µm, i.e., from about 1 nm (nanometer) to 0.1 mm (millimeter), which lies between the region of transition from radio waves and the transition to X rays. *See also* **optical spectrum.**

optical adaptive technique: *See* **coherent optical adaptive technique.**

optical analyzer: *See* **multichannel optical analyzer.**

optical attenuator: In optical communications, a device used to reduce the power level of an optical signal. *Note 1:* In optical attenuators used in fiber optic systems, the attenuation may depend on (a) the modal distribution in the optical signal, (b) the absorptive, reflective, and scattering characteristics of the attenuator material, and (c) the use of special coatings and films. *Note 2:* One type of optical attenuator uses a filter that consists of a metal film evaporated into a sheet of glass to obtain the attenuation. The filter might be tilted to avoid reflection back into the source, such as a laser or fiber optic cable output. *Note 3:* The basic types of optical attenuators are fixed, stepwise variable, and continuously variable. *See* **continuously variable optical attenuator, stepwise variable optical attenuator.** *See also* **fiber optic attenuator.**

optical axis: 1. In an optical waveguide, the longitudinal geometric axis of symmetry. *Note:* The optical axis of an optical fiber of circular cross section is the locus of all points at the centers of cross-sectional circles, i.e., the central longitudinal axis of the core. **2.** In a lens element, the straight line that passes through the center of curvature of the lens surfaces. **3.** In an optical system consisting of two or more elements, the line formed by the coincident principal axes of the series of optical elements. *See also* **bandwidth, fiber optics, optic axis, optical element, optical fiber, optical fiber axis.** *Refer to* **Fig. N-21 (641).**

optical beamsplitter: *See* **beamsplitter.**

optical blank: A casting that (a) consists of a molded optical material, such as silica glass, (b) has the desired geometry for grinding, polishing, and reshaping into desired optical and mechanical properties, and (c) in the case of some optical fiber manufacturing processes, can be drawn to an optical fiber, after adding dopants, in accordance with the final optical and mechanical specifications for the fiber. *See also* **optical fiber blank, optical fiber preform.** *Refer to* **Fig. P-20 (776).**

optical branching device: A device that has three or more ports and shares input light among its ports in a given fashion without changing the input signal other than distributing its energy among the ports. *Note:* Types of branching devices include unidirectional, bidirectional, symmetric, and asymmetric branching devices.

optical cable: *Synonym* **fiber optic cable.**

optical cable assembly: *Synonym* **fiber optic cable assembly.**

optical cable facility: *Synonym* **fiber optic cable facility.**

optical cable facility loss: *Synonym* **fiber optic cable facility loss.**

optical cable interconnect feature: *Synonym* **fiber optic cable facility.**

optical cable jacket: *Synonym* **fiber optic cable jacket.**

optical cavity: A region, i.e., a geometric space, in vacuum or material media, bounded by two or more reflecting surfaces, referred to as mirrors or as cavity mirrors, that have elements that are aligned to provide multiple reflections of lightwaves, i.e., produce standing waves of high irradiance, i.e., high intensity, at certain wavelengths, depending on the material of construction and the dimensions of the region or the space. *Note 1:* Standing waves might be obtained in a ruby crystal laser with two plane or spherical mirrors, forming a resonant cavity. The cavity is the portion of the crystal that lies between the mirrors. The molecules in the cavity can be excited by an inert gas lamp which causes the molecules in the cavity to generate and emit a narrow beam of monochromatic light of high irradiance, i.e., high electric field strength or high radiant power, in the direction of the crystal axis. *Note 2:* The resonator in a laser is an optical cavity. *Synonym* **resonant cavity.** *See also* **active laser medium, laser.**

optical cavity diode: *See* **large optical cavity diode.**

optical cavity mirror sensor: A fiber optic sensor in which (a) operation is based on changes in the optical reflection coefficient caused by changes in pressure, temperature, strain, or other imposed stimuli, (b) an arrangement of mirrors causes light to be reflected back to a photodetector via a partial mirror through which the incident beam has passed from the light source, (c) the mirrors are positioned in such a way that one of the reflecting surfaces is at the critical angle of the incident light in the optical fiber core, and (d) the slightest variation in mirror position, caused by the applied stimulus to be sensed, causes the reflected light to pass into or out of total (internal) reflection, which changes the amount of light sent back to the photodetector.

optical cement: A permanent and transparent adhesive capable of withstanding extremes of temperature. *Note:* Examples of optical cement are Canada balsam and synthetic adhesives, such as methacrylates, caprinates, and epoxies.

optical character reader (OCR): A device used for optical character recognition (OCR).

optical character recognition (OCR): The machine identification of printed characters through the use of light-sensitive devices. *Note:* Optical character recognition equipment includes optical character readers and facsimile scanners. *See also* **character, magnetic ink character recognition.**

optical chopper: A device that interrupts a light beam. *Note:* Examples of optical choppers are (a) rotating disks with radial slots through which a collimated beam must pass on to a photodetector to produce a signal with a pulse repetition rate proportional to the angular rotation rate of the disk, and (b) objects on a conveyor belt that interrupts a light beam passing across the belt. Optical fibers can be used to bring the light to and from the optical chopper, i.e., to a rotating disk or any other moving element.

optical circuit: A circuit that includes an optical component, such as a fiber optic link, an optical isolator, or an optical path component. *See also* **circuit, component, optical path component.**

optical circuit filter-coupler-switch modulator: *See* **integrated optical circuit filter-coupler-switch modulator.**

optical combiner: **1.** In fiber optics, a passive optical device in which (a) radiant power from two or more input optical fibers is distributed among one or more output fibers, (b) the input fibers may each insert light with a different wavelength, and (c) the output fiber, or fibers, will contain all the input wavelengths. **2.** A passive optical device in which radiant power from two or more input ports is distributed among a smaller number of output ports.

optical communications system: A communications system that makes use of optical systems and components, such as (a) light sources, (b) propagation media that include open waveguides, such as optical fibers, slab dielectric waveguides, and optical integrated circuits (OICs), (c) photodetectors, and (d) perhaps display systems and nonoptical communications systems, such as twisted pair, coaxial cable, and radio systems. *Note:* Visual signaling systems, such as those that use heliographs, semaphores, ship pennants, ground panels, flags, arms and hands, wig-wag. and smoke, are not considered as optical communications systems. *See also* **visual communications system.**

optical computer: A computer in which internal operations, such as storage, arithmetic and logic, and control functions, are performed by using optical paths, optical switches, optical integrated circuits (OICs), and optical images. *Note:* An optical computer might have an optical modulator tube that stores optical information as charge patterns while performing arithmetic functions, and makes images in memory consisting of picture elements in the form of the on-off state of light sources.

optical conduction: The propagation or the transmission of lightwaves through a vacuum or a material medium in which they usually are guided or confined. *Note 1:* In an optical data link, lightwaves are guided from a light source via a fiber optic cable to a photodetector with minimal loss of light energy, the losses being caused by absorption, dispersion, deflection, reflection, scattering, or diffusion. The intelligence carried by lightwaves modulated at the source can be recovered at the end of the cable. *Note 2:* Care must be taken in using the terms "conductor," "conduction," and "conducting" in fiber optics because these terms are applied to materials that conduct electric currents, such as semiconducting materials and metals, whereas optical materials are dielectric materials which are referred to as "nonconductors." Thus, optical fibers are poor "conductors" of electric currents but good "conductors" of lightwaves. In fiber optics, the terms "guided," "transmitted," and "propagated" are preferred over the term "conducted," thus avoiding ambiguous statements, such as "Optical fibers are good conductors." *See also* **optical conductor.**

optical conductor: 1. A transparent material that offers a low optical attenuation rate to the transmission of lightwaves. *Note:* Examples of optical conductors are certain glasses, plastics, and crystals. **2.** An optical waveguide. *Note:* The term "optical conductor" is deprecated. "Conductor" should be reserved for electrical, sonic, and thermal systems rather than optical systems. *See also* **optical conduction.**

optical conductor loss: *See* **connector induced optical conductor loss.**

optical connector: 1. *Synonym* **fiber optic connector. 2.** *See* **fiber optic connector, receiver fiber optic connector, transmitter fiber optic connector.**

optical connector variation: *Synonym* **fiber optic connector variation.**

optical contact: A condition in which (a) two sufficiently clean and close-fitting surfaces adhere with no reflection at the interface, and (b) the resulting joint is as strong as the body of the joined items, such as optical waveguides. *See also* **fiber optic terminus, terminus.**

optical countermeasure: 1. A measure or an action that is taken to deliberately prevent the use of the optical spectrum, i.e., the portion of the electromagnetic spectrum extending from the far infrared through the ultraviolet. **2.** A measure or an action that is taken to deliberately prevent the use of the visible spectrum, i.e., the portion of the electromagnetic frequency spectrum extending from about 0.40 μm (micron) to 0.75 μm. *See also* **electromagnetic spectrum, optical spectrum, visible spectrum.**

optical coupler: A device that interconnects optical conductors, such as optical fibers, slab dielectric waveguides, and optical integrated circuit (OIC) components. *See also* **fiber optic coupler, optical directional coupler.**

optical data processing: 1. The performing of digital data processing operations, such as execution of Boolean functions, using a combination of optical and electronic elements. **2.** The performing of analog data processing operations that model or emulate actual systems represented by mathematical functions and relationships governing the behavior of light passing through optical elements, such as by making use of diffraction patterns, interference effects, reflection effects, and wavefront transformations produced by variously shaped optical elements. *Note:* An application of optical data processing is in the use of Fourier transform optical wave analysis, which can be used to perform ocean wave analyses, antenna pattern analyses, phased array signal processing, spread spectrum systems analyses, folded spectrum applications, filtering, image processing, and word pattern recognition. *See also* **Boolean function, optical element.**

optical demultiplexer (active): An active optical device that (a) uses optical components and other components, such as electrical and acoustic components, to accept signal streams or messages that have been previously multiplexed into a single channel, and separates the streams or messages and places them in individual independent channels so the messages can be used immediately or be separately routed to their ultimate destination, and (b) might consist of an electrically operated switching device that separates individual wavelengths so as to cause each color in the polychromatic light to be incident upon a separate photodetector, in order that the intelligence-bearing signal that modulated each color in the optical transmitter can be recovered by the optical receiver. *See also* **fiber optic demultiplexer (active), fiber optic multiplexer (active), optical multiplexer (active).**

optical demultiplexer (passive): A passive optical device that (a) uses only optical components to accept signal streams or messages that have been previously multiplexed into a single channel, and separates the streams or messages and places them in individual independent channels so the messages can be used immediately or be separately routed to their ultimate destination, (b) might consist of a prism that can accept polychromatic light resulting from the multiplexing action of a mixer and separate the individual wavelengths so as to cause each color in the polychromatic light to be incident upon a separate photodetector, in order that the intelligence-bearing signal that modulated each color in the optical transmitter can be recovered by the optical receiver. *See also* **fiber optic**

demultiplexer (passive), fiber optic multiplexer (passive), optical multiplexer (passive).

optical density (OD): An inverse measure of the transmittance of an optical element. *Note 1:* Optical density (OD) is expressed by the relation $\log_{10}(1/T)$, where T is the transmittance. *Note 2:* The higher the optical density (OD), the lower the transmittance, and vice versa. *Note 3:* Ten times the optical density (OD) is equal to the transmission loss expressed in dB. Thus, an OD of 0.3 corresponds to a transmission loss of 3 dB. *Note 4:* When the transmittance approaches unity, the optical density (OD) is a more useful value, such as in dealing with transparent materials. *Note 5:* The analogous term, $\log_{10}(1/R)$, where R is the reflectance, is called the "reflection density." *Note 6:* The sum of the reflectance and the transmittance is unity because, at an interface, the reflected radiant power plus the transmitted radiant power equals the incident radiant power. There is no power or energy loss in crossing an interface. *Synonym* **transmittance density.** *See also* **reflectance, reflection density, transmission loss, transmittance.**

optical detector: A transducer that converts optical input signals to output signals, such as electrical, sound, other optical, chemical reaction, or mechanical signals. *See* **video optical detector.** *See also* **optoelectronic, photodetector.**

optical detector type: The characterization of an optical detector by identifying its performance or construction. *Note:* Examples of optical detector types are the positive-intrinsic-negative (PIN) photodiode, avalanche photodiode (APD), germanium (Ge) photodiode, silicon (Si) photodiode, and superluminescent photodiode.

optical device: *See* **active optical device, Bell integrated optical device, passive optical device.**

optical directional coupler: An optical coupler that (a) is used with dielectric waveguides, such as optical fibers, fiber optic bundles, fiber optic cables, slab dielectric waveguides, and optical integrated circuit (OIC) waveguides, (b) operates on lightwaves propagating in one or both directions, (c) may be controlled or uncontrolled, and (d) may be used for tapping, directional control of propagation, switching, dropinserting, monitoring, branching, mixing, and wavelength multiplexing. *Note 1:* Optical directional couplers are used in fiber optic communications systems, such as cable television (CATV) and fiber optic data links, for (a) optical fiber measurements and (b) combining or splitting optical signals at desired ratios by insertion into a transmission line. *Note 2:* An example of an optical directional coupler is a four-port unit with precise connectors at each port to enable inputs to be coupled and transmitted via multiple output ports. *See also* **fiber optic coupler.**

optical disk: A flat circular disk that (a) is coated with a photosensitive substance on which binary digits in the form of light and dark spots may be stored by scanning the disk with a narrow beam from a modulated light source, such as a laser, as the disk is rotated past the radially moving beam and developing the photosensitive material, and (b) can be read with a reflected or transmitted light beam as it is rotated. *Note:* Currently, optical disks are read-only storage devices that can store on the order of 10^{13} bytes, i.e., 10 Tb (terabytes) for a 30-cm (centimeter)-diameter disk, with 10 Mbps (megabits per second) recording rates and bit error rates (BERs) less than 10^{-7}. *See also* **optical disk storage.**

optical disk storage: Photooptic storage in which the photosensitive surface is placed on a substrate shaped in the form of a flat circle that may be rotated for recording and reading as a permanent storage device. *See also* **optical disk, permanent storage device, photooptic storage, reading, recording, substrate.**

optical dispersion: *See* **dispersion.**

optical dispersion attenuation: The attenuation of a signal in an optical waveguide in which each frequency component of a launched pulse is attenuated in such a manner that the shorter wavelengths in the spectral width of the pulse, i.e., the higher frequencies, are attenuated more than the lower frequencies, giving rise to a form of distortion. *Note:* The optical dispersion attenuation factor, D_{af}, is given by the relation $D_{af} = e^{-df}$ where D_{af} is the optical dispersion attenuation factor, d is an empirically determined constant, dependent on material and geometry, and f is a significant frequency component of the attenuated signal.

optical distortion: An aberration of spherical surface optical systems caused by variation in magnification with distance from the optical axis. *See also* **optical axis, optical system.**

optical dynamic range: In an optical communications system, the dynamic range required of an optical signal in order to maintain a specified bit error ratio (BER).

optical element: A component or a part of an optical system. *Note:* Examples of optical elements are the whole of, and parts or segments of, lenses, prisms, mirrors, display devices, light-emitting diodes, photodetectors, optical fibers, fiber optic cables, and fiber optic bundles.

optical emitter: A source of optical radiant power, i.e., a source of electromagnetic radiation, in the optical spectrum, i.e., the visible and near visible region of the frequency spectrum. *See also* **electromagnetic radiation, frequency spectrum, optical spectrum, visible spectrum.**

optical end-finish: The surface condition at the endface of an optical waveguide. *See also* **hockey puck, microfinishing film, optical conductor, optical fiber end finish.**

optical energy density: In a light beam propagating through a unit area normal to the direction of maximum power gradient, the amount of optical energy contained in a unit volume of material or free space within the beam, usually expressed in joules per cubic meter, which corresponds to an irradiance level in watts per square meter because the electromagnetic waves in the beam are propagating with the speed of light. *Note:* The time rate of energy flow across a transverse unit area of a beam is the optical energy density multiplied by the speed, joules/meter^3 × meters/second, which is joules/meter^2-second. But joules/second is watts. Therefore, watts/meter^2 remain as the equivalent to joules/meter^3, which implies that the optical energy density and the irradiance, i.e., the optical power density, are equivalent for a propagating electromagnetic wave.

optical fiber: A single discrete filament-shaped transparent dielectric material that (a) guides light, (b) usually consists of glass or plastic, (c) usually has a cylindrical core and one or more claddings on the outside, (d) usually has a round cross section, and (e) may have a special-purpose noncircular cross section, such as an elliptical, rectangular, planar, or slotted cross section. *Note 1:* The refractive index of the core of an optical fiber must be higher than that of the cladding for the light launched into the fiber to remain captured by, and to propagate within, the fiber. If the incidence angle of light rays at the core-cladding interface exceeds a certain angle called the "critical angle," rays headed out of the core will be reflected back into the core. *Note 2:* Optical fibers, singly and in bundles, may be used for signal transmission, sensing, illumination, display, image transmission, and endoscopy. *Note 3:* The lightwaves in an optical fiber may be modulated with an information-bearing signal. *Note 4:* Light that enters one end of an optical fiber within its acceptance cone emerges at the other end diminished by losses that are dependent upon many factors, such as the length, absorption, scattering, microbends, and various transfers of energy from coupled to uncoupled modes. *Note 5:* Each optical fiber is a separate channel that may be multiplexed. In an aligned bundle, each fiber carries a picture element. *Synonyms* **fiber lightguide, light pipe, optical fiber waveguide.** *See* **active optical fiber, advanced tactical optical fiber, all-glass optical fiber, all-plastic optical fiber, all-silica optical fiber, circularly symmetric optical fiber, compensated optical fiber, deeply depressed cladding optical fiber, dispersion flattened optical fiber, low loss optical fiber, multimode optical fiber, overcompensated optical fiber, polarization-maintaining optical fiber, radiation hardened optical fiber, SELFOC® optical fiber, self-focusing optical fiber, SEL® optical fiber, single mode optical fiber, tapered optical fiber, triangular cored optical fiber, undercompensated optical fiber.** *See also* **absorption, acceptance cone, buffer, cladding, cladding mode, core, coupled mode, critical angle, fiber optic cable, fiber optics, microbend, optical waveguide, picture element, scattering, ultralow-loss optical fiber.** *Refer to* **Fig. O-1.** *Refer also to* **Figs. A-2 (5), A-3 (6), A-6 (24), A-9 (49), C-4 (123), D-4 (264), G-2 (399), H-1 (430), L-3 (436), L-5 (511), L-7 (535), L-8 and L-9 (536), N-21 (641), P-1 (695), P-11 (749), P-12 (751), P-20 (776), R-1 (796), R-9 (835), S-14 (927), S-15 (931), S-18 (951), T-6 (1013), T-9 (1037), W-3 (1082).**

optical fiber absorption: In an optical fiber, the optical power attenuation caused by absorption in the fiber core and cladding material, a loss usually evaluated by measuring the optical power emerging from the end of successively shortened known lengths of the fiber. *See also* **absorption, attenuation, optical power.**

optical fiber acoustic sensor: *Synonym* **fiber optic acoustic sensor.**

optical fiber active connector. *Synonym* **fiber optic active connector.**

optical fiber amplifier: An amplifier that (a) amplifies an optical signal directly without the need to (i) convert it to an electrical signal, (ii) amplify it electrically, and then (iii) reconvert it back to an optical signal, (b) consists of a specially prepared optical fiber, usually a fiber doped with erbium, (c) uses the fiber to carry the optical signal, (d) is optically pumped, usually with a high-powered laser, to provide the power for amplification of the optical signal, (e) is well suited for a wide variety of digital and analog applications mainly because neither optical-electrical conversion nor electrical amplification takes place, (f) does not require extraordinary frequency control, i.e., wavelength control of the pumping optical source, (g) intensifies the signal by means of Raman amplification, and (g) is relatively simple. *Note:* One type of optical fiber amplifier uses an erbium-doped optical fiber that is optically pumped with a laser having a high-powered continuous output at an optical frequency

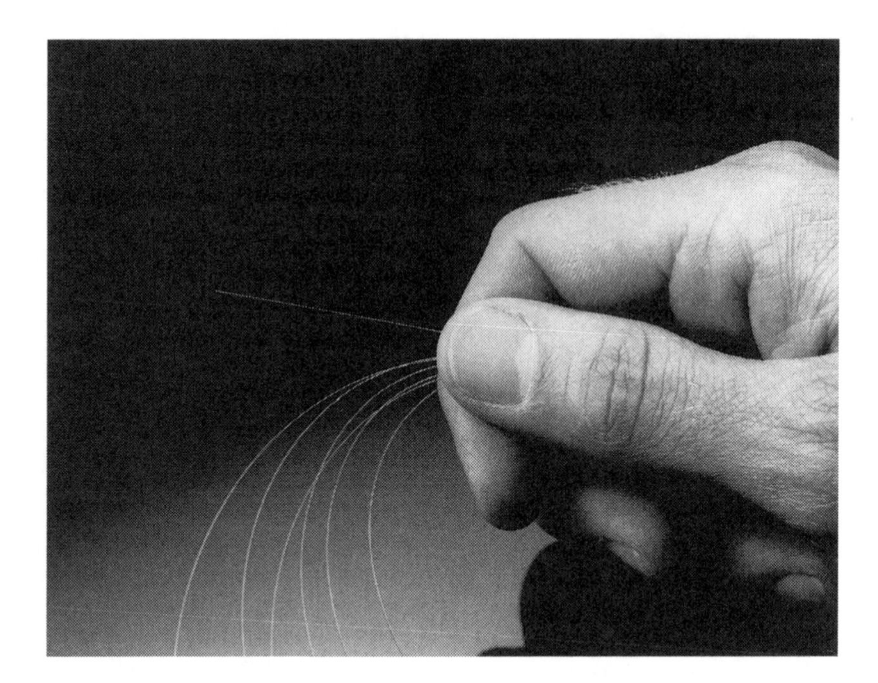

Fig. O-1. Uncabled **optical fiber.** (Courtesy Siecor Corporation, Hickory, NC.)

slightly higher, i.e., a wavelength slightly lower, than that of the communications optical signal. *Synonyms* **fiber amplifier, Raman amplifier.** *Refer to* **Fig. P-20 (776).**

optical fiber axis: 1. The locus of the core centers of an optical fiber. **2.** In an optical fiber, the line connecting the centers of the circles that circumscribe the core and that determine the tolerance field. *Synonym* **fiber axis.** *See also* **tolerance field.** *Refer to* **Fig. N-21 (641), R-1 (796), R-10 (836).**

optical fiber bandwidth: The lowest frequency at which the magnitude of the modulation transfer function of a given length of optical fiber decreases to a specified level, usually to one-half of the zero modulation frequency value. *Note 1:* Optical fiber bandwidth depends on the modulation transfer function of the fiber and not on the optical frequency transfer function. *Note 2:* The optical fiber bandwidth is a measure of the fiber's information-carrying capacity at a specified optical wavelength. The bandwidth is limited by modal distortion and material dispersion in multimode optical fibers, and by material dispersion and waveguide dispersion in single mode fibers. However, graded index optical fibers can be made with zero dispersion at specific operating wavelengths. *Note 3:* The bit-rate•length (BRLP) product is a measure of the information-carrying capacity of optical fiber. The BRLP is limited by dispersion, the modulation scheme, jitter, and the desired data signaling rate (DSR). Another measure of the information-carrying capacity of an optical fiber is the full-width at half-maximum (FWHM) point. *See also* **bandwidth•distance factor, modulation transfer function.**

optical fiber blank: A cylindrically shaped piece of glass of high purity, such as glass with a transition metal composition of 10 to 50 ppb (parts per billion, i.e., 10^{-9}), used to make a preform from which optical fibers are pulled. *Note:* In the doped deposited silica (DDS) process for making optical fibers, the glass and dopants are deposited on a mandrel mounted on a lathe using an inside vapor phase oxidation (IVPO) process or an outside vapor phase oxidation (OVPO) process to control the refractive index profile of the fiber. After sufficient glass has been deposited on the mandrel to form the optical fiber blank, it is removed from the lathe. The mandrel is then removed, and the blank is consolidated into a dense glass preform. The fiber is subsequently pulled from the preform. *See also* **optical blank, optical fiber mandrel, optical fiber preform.**

optical fiber buffer: A protective material bonded to an optical fiber over the outside cladding for various purposes, such as (a) preserving fiber strength, (b) inhibiting cabling losses, (c) protecting against mechanical damage, such as is caused by microbending, macrobending, cabling, and spooling, (d) protecting against harsh environments, such as moisture and corrosive atmospheres, and (e) providing compatibility with fiber and cable manufacturing processes. *Note:* Optical fiber buffer application methods include dipping, extrusion, spraying, and electrostatic methods. Buffer materials include fluoropolymers, Teflon™, Kynar™, polyurethane, and many others. *See also* **optical fiber coating.** *Refer to* **Fig. C-1 (102).**

optical fiber bundle: *Synonym* **fiber optic bundle.**

optical fiber cable: *Synonym* **fiber optic cable.**

optical fiber cladding: In an optical fiber, a light-conducting material that (a) surrounds the core of the fiber and (b) has a lower refractive index than the core material. *See also* **optical fiber, optical fiber core.**

optical fiber coating: Special materials bonded to optical fibers to make them sensitive to a specific physical variable that is to be sensed, such as magnetostrictive materials to measure magnetic field strength, metals to measure electric current, and thermally sensitive materials for measuring temperature. *Note:* In one application of optical fiber coatings, a length of the optical fiber with the bonded material is exposed to the physical variable to be sensed. A similar length of fiber is isolated from the physical variable being sensed. A coherent lightwave is sent through both via suitable couplers. The phase shift caused by the variation of the physical variable is measured by interferometric techniques. Both fibers may be subjected to a physical variable not being sensed as long as the exposure is the same for both, such as both subjected to the same temperature or pressure when magnetic field strength is being sensed. *See also* **coating, optical fiber buffer.**

optical fiber communications: *Synonym* **fiber optic communications.**

optical fiber concentrator: *Synonym* **fiber optic concentrator.**

optical fiber concentricity error: In an optical fiber, the distance between the center of the two concentric circles that specify the cladding diameter and the center of the two concentric circles that specify the core diameter. *Note:* The optical fiber concentricity error is used in conjunction with tolerance fields to specify optical fiber core and cladding geometry. *Synonyms* **concentricity error, core eccentricity.**

optical fiber, conductive (OFC): The designation given by the National Fire Protection Association (NFPA) to interior fiber optic cables that (a) contain at least one electrically conductive but non-current-carrying component, such as a metallic strength member, armor, overarmor, or vapor barrier, and (b) are not certified for use in plenum or riser applications.

optical fiber, conductive, plenum (OFCP): The designation given by the National Fire Protection Association (NFPA) to interior fiber optic cables that (a) contain at least one electrically conductive but non-current-carrying component, such as a metallic strength member, armor, overarmor, or vapor barrier, and (b) are certified for use in plenum applications.

optical fiber, conductive, riser (OFCR): The designation given by the National Fire Protection Association (NFPA) to interior fiber optic cables that (a) contain at least one electrically conductive but non-current-carrying component, such as a metallic strength member, armor, overarmor, or vapor barrier, and (b) are certified for use in riser applications.

optical fiber connector set: The total of the fiber optic connector parts required to provide demountable coupling between two or more fiber optic cables. *See also* **fiber optic cable, fiber optic connector, fiber optic splice.**

optical fiber core: In an optical fiber, the central portion that (a) is surrounded by the cladding and (b) has a higher refractive index than the cladding material. *See also* **optical fiber, optical fiber cladding.**

optical fiber core diameter: In an optical fiber, the diameter of the boundary between the fiber core and the fiber cladding. *See also* **optical fiber, optical fiber cladding, optical fiber core.**

optical fiber count: The number of optical fibers in a fiber optic bundle, given by the relation $N \approx 0.907 (D/d - 1)^2$ where N is the optical fiber count, D is the bundle diameter, and d is the fiber diameter. *See also* **fiber optic bundle, optical fiber.**

optical fiber coupler: *Synonym* **fiber optic coupler.**

optical fiber crosstalk: In an optical fiber, the exchange of optical power between a core and the cladding, between the cladding and the ambient surrounding, or between layers with different refractive indexes. *Note:* Optical fiber crosstalk usually is undesirable because differences in path length and propagation time usually result in dispersion, thus reducing transmission distances. Therefore, attenuation is deliberately introduced into the cladding by making it a lossy medium. *See also* **attenuation, dispersion, lossy medium, refractive index.**

optical fiber cutoff wavelength: 1. In a short, uncabled, single mode optical fiber with a specified radius of curvature, the wavelength at which the presence of the second-order mode of a propagating electromagnetic wave introduces a measurable attenuation increase when compared to a wave in a fiber with a differential mode attenuation that is not changing at that wavelength. *Note:* A cabled optical fiber cutoff wavelength (FCW) is a more functional value for cutoff wavelength than the uncabled FCW because the cutoff wavelength depends on length, bend, and cabling. The cabled FCW usually is lower than the uncabled FCW. **2.** The wavelength greater than which a particular mode ceases to be a bound mode. *Note:* In a single mode optical fiber, concern is with the cutoff wavelength of the second-order mode. **3.** The wavelength at which the normalized frequency is equal to 2.4.

optical fiber cutting tool: A special cutting tool designed to prepare the ends of optical fibers for splicing or connecting by bending the fiber, and at a spot under tension, using a sharp-edged blade to initiate a microcrack, which fractures the fiber and produces a smooth, flat cleavage. *See also* **microcrack.**

optical fiber delay line: *Synonym* **fiber optic delay line.**

optical fiber demultiplexer: *Synonym* **fiber optic demultiplexer.**

optical fiber demultiplexer (active): *Synonym* **fiber optic demultiplexer (active).**

optical fiber demultiplexer (passive): *Synonym* **fiber optic demultiplexer (passive).**

optical fiber-detector coupling: In fiber optic transmission systems, the transfer of optical signal power from a transmission optical fiber to a photodetector for conversion to an electrical signal. *Note:* Many fiber optic photodetectors have an optical fiber pigtail for connection by means of a splice, or a have a fiber optic connector for connection to a transmission fiber optic cable. *See also* **fiber optic connector, photodetector, pigtail, splice.**

optical fiber diameter: The outside diameter of an optical fiber, usually including the core, all cladding, and any bonded materials not usually removed when making a connection, such as hermetic seals and special buffers. *See also* **buffer, cladding, core.**

optical fiber dimensional stability: In an optical fiber, the extent to which the physical dimensions, particularly the diameter and the length, remain constant under stress and temperature changes that occur during pulling, winding, storage, shipping, payout, cabling, installation, operation, and maintenance. *See also* **cabling, pulling machine.**

optical fiber dispersion: In an optical pulse propagating in an optical fiber, the increase in pulse duration as a function of distance as the pulse travels along the fiber, the increase being caused by (a) material dispersion, caused by the wavelength dependence of the refractive index, (b) modal dispersion, caused by the different group velocities of the different modes, and (c) waveguide dispersion, caused by the wavelength dependence of the propagation constant of the mode on that particular wavelength. *See also* **material dispersion, modal dispersion, pulse duration, waveguide dispersion.**

optical fiber distribution panel: An optical fiber connector panel for splicing or connecting distribution cables and jumper cables, fiber by fiber, to the optical transmitters and optical receivers.

optical fiber drawing: The controlled pulling of an optical fiber, in a melted or softened state, through an aperture, causing it to elongate and thus reduce its diameter as it cools. *See also* **optical fiber pulling machine.** *Refer to* **Fig. D-4 (264).**

optical fiber drawing machine: *Synonym* **optical fiber pulling machine.**

optical fiber end finish: The surface condition on the endface of an optical fiber. *See also* **end finish, hockey puck, microfinishing film, optical end finish.**

optical fiber global attenuation rate characteristic: A plot that shows (a) the attenuation rate as a function of transmitter wavelength for an optical fiber, (b) the peaks and the valleys caused by the different absorption, scattering, and reflection levels at different wavelengths, (c) the useful wavelength regions for the fiber, and (d) typical values of attenuation rate rather than worst case values so that the plot can be used to determine how the fiber will respond to new applications and to changes made in existing systems. *Note:* Designers ensure that optical fibers operate at a wavelength at which the attenuation rate is the lowest, such as the 1.31 μm (microns) used as the standard wavelength. Attenuation rates in dB per kilometer are usually plotted at 23°C for wavelengths about 20% on either side of the transmitter nominal central wavelength. In designing a fiber optic station/regenerator section, allowance must be made for increases in typical attenuation rates because of transmitter spectral width, the effect of temperature at the worst case temperature conditions, and aging. *Synonyms* **fiber global-loss characteristic, fiber optic global attenuation rate characteristic.** *See also* **attenuation rate, loss budget constraint, fiber optic cable facility loss.**

optical fiber hazard: A feature of a fiber optic system that has the potential of causing injury. *Note 1:* Examples of optical fiber hazards are (a) fiber optic cables that have been treated with a special flame-retardant material that is also caustic and can cause skin lesions, (b) the output of high radiation lasers that can cause eye injury even though the radiation is invisible, (c) fragments of fibers that can cause eye damage, and (d) grinding and polishing residues suspended in air that can cause lung damage if inhaled. *Note 2:* Precautions against optical fiber hazards include the use of gloves, goggles, masks, ventilators, filters, and frequent washing with warm water and soap during handling and processing operations. Particular precaution must be taken in regard to viewing radiation from the end of an energized optical fiber because the radiation may not be in the visible spectrum of the electromagnetic spectrum, thus giving the impression that the fiber is not energized or the appearance that the radiation is at a lower irradiance level than it actually is.

optical fiber interconnection box: *Synonym* **fiber optic interconnection box.**

optical fiber jacket: The material that covers the buffered or unbuffered optical fiber. *See also* **fiber optic cable jacket, sheath.**

optical fiber junction: An interface formed by butting the endfaces of two optical fibers to allow direct fiber-to-fiber optical transmission. *See also* **fiber optic connector, fiber optic splice.**

optical fiber laser: A laser in which the lasing medium is in the form of an optical fiber for direct splicing to an optical fiber. *Synonym* **fiber laser.**

optical fiber link: *Synonym* **fiber optic link.**

optical fiber mandrel: A cylindrically shaped piece of glass of high purity, such as glass with a transition metal composition on the order of 10 to 50 ppb (parts per billion), i.e., on the order of 10^{-9}, used to make an optical fiber blank which is used to make an optical fiber preform. *Note:* In the doped deposited silica (DDS) process for making optical fibers, the glass and dopants are deposited on the mandrel mounted on a lathe by using an inside vapor phase oxidation (IVPO) process or an outside vapor phase oxidation (OVPO) process to control the refractive index profile of the fiber. After sufficient glass has been deposited, the blank is removed from the lathe. The mandrel is then removed, and the blank is consolidated into a dense glass preform. The fiber is subsequently pulled from the preform with an optical fiber pulling machine under precisely controlled melt, tension, flow, and diameter control conditions. *See also* **optical fiber blank, optical fiber preform, optical fiber pulling machine.**

optical fiber merit figure: A figure that indicates the ability of an optical fiber to handle high frequency signals over given distances with specified distortion limits and bit error ratios (BERs). *Note:* Examples of optical fiber merit figures are (a) the pulse repetition rate times the fiber length when pulse dispersion is the specified limiting factor in signal distortion rather than attenuation, and (b) a figure of merit, F_m, for multimode fibers which is given by the relation $F_m = PL^s$ where F_m is the optical fiber merit, P is the pulse repetition rate, L is the length of the fiber, and s varies between 0.5 and 1, depending on fiber characteristics, such as the refractive index profile, type of glass, and dimensions.

optical fiber multiplexer (active): *Synonym* **fiber optic multiplexer (active).**

optical fiber multiplexer (passive): *Synonym* **fiber optic multiplexer (passive).**

optical fiber, nonconductive (OFN): The designation given by the National Fire Protection Association (NFPA) to interior fiber optic cables that (a) do not contain any electrically conductive components, such as a metallic strength member, armor, overarmor, or vapor barrier, and (b) are not certified for use in plenum or riser applications.

optical fiber, nonconductive, plenum (OFNP): The designation given by the National Fire Protection Association (NFPA) to interior fiber optic cables that (a) do not contain any electrically conductive components, such as a metallic strength member, armor, overarmor, or vapor barrier, and (b) are certified for use in plenum applications.

optical fiber, nonconductive, riser (OFNR): The designation given by the National Fire Protection Association (NFPA) to interior fiber optic cables that (a) do not contain any electrically conductive components, such as a metallic strength member, armor, overarmor, or vapor barrier, and (b) are certified for use in riser applications.

optical fiber nuclear hardening: Design allowances made to prevent or reduce the effects of high-energy radiation and bombardment, such as gamma radiation and neutron bombardment, which cause some optical fibers to depart from normal operating parameters, e.g., to darken and to increase attenuation. *Note:* Optical devices, such as light sources, light-emitting diodes (LEDs), lasers, and couplers also need to be hardened to prevent malfunction when exposed to high radiation high radiation levels. *Synonym* **fiber nuclear hardening.** *See also* **fiber optic nuclear hardening.**

optical fiber organizer: *Synonym* **fiber and splice organizer.**

optical fiber pattern: *Synonym* **fiber optic pattern.**

optical fiber pigtail: A short length of optical fiber (a) that extends from and is permanently attached to a component, such as a fiber optic light source, coupler, connector, repeater, tap, mixer, or photodetector, (b) to which an optical fiber can be spliced, (c) that is used to facilitate connecting the component to another optical fiber or component, such as for coupling optical power or signals between the component and a transmission line, and (d) that performs a function similar to that of an electrical pigtail, which is a length of wire extending from an electrical circuit element, such as a resistor, transistor, capacitor, diode, or inductor, used for connecting the element to another circuit element or terminal. *Note:* An optical fiber pigtail on a light source is also called a "launching fiber."

optical fiber polarizer: An optical fiber that (a) usually has a specially shaped cross section, (b) produces a single well-defined polarized mode by selecting one of the two polarized modes from a lightwave propagating in a single mode fiber, (c) extinguishes the unwanted polarization, (d) transmits the other polarization, (e) when spliced into a conventional single mode system, eliminates the need for collimation, and (f) reduces sensitivity to various disturbances, such as vibration. *Synonym* **fiber optic polarizer.**

optical fiber preform: A specially shaped piece of optical material from which optical fibers may be pulled, drawn, or rolled. *Note 1:* Examples of optical fiber preforms are (a) a solid rod made with a higher-refractive-index glass than the tube into which it is slipped to be heated and pulled or rolled into a cladded optical fiber, and (b) four lower-refractive-index rods surrounding a higher-refractive-index rod heated and drawn into a cladded fiber. *Note 2:* The drawing process results in an optical fiber many times longer than the preforms. For round solid preforms, the length of the drawn optical fiber is equal to the square of the ratio of the preform and fiber radii times the length of the preform, resulting in a fiber tens of kilometers long from a single preform. *Note 3:* In the doped deposited silica (DDS) process for drawing optical fiber, a cylindrically shaped piece of glass of high purity, such as glass with a transition metal composition of 10 to 50 ppb (parts per billion or 10^{-9}), is used to make a fiber optic blank, which is used to make the preform. The glass and dopants are deposited on a mandrel mounted on a lathe. An inside vapor phase oxidation (IVPO) process or an outside vapor phase oxidation (OVPO) process is used to control the refractive index profile of the optical fiber. After sufficient glass has been deposited, the blank is removed from the lathe. The mandrel is then removed, and the blank is consolidated into a dense glass preform. The fiber is subsequently pulled from the preform. *Synonym* **preform.** *See also* **chemical vapor deposition technique, ion exchange technique, optical fiber blank, optical fiber mandrel, optical fiber pulling machine.** *Refer to* **Fig. P-20 (776).**

optical fiber pulling machine: A device that (a) draws an optical fiber from a preform heated to fusion temperature, (b) operates at increased temperatures to provide for ductility and fusion of dopants to control the refractive index profile, (c) controls the cross section of the fiber as the cross section decreases and the length increases as the fiber is drawn, (d) controls cooling and hardening of the drawn fiber, and (e) winds the fiber on a spool, reel, drum, or bundpack. *Synonym* **optical fiber drawing machine.** *See also* **bundpack, dopant, preform, refractive index.** *Refer to* **Fig. P-12 (751).**

optical fiber radiation damage: In lightwaves propagating in an optical fiber, the increased attenuation (a) that is caused by increased losses from absorption, diffusion, scattering, or deflection at specific sites created by exposure to high energy bombardment, such as by gamma rays, cosmic rays, or high energy neutrons, (b) in which the degradation can be catastrophic or graceful, and (c) in which the effects of radiation on optical fibers are cumulative, and for the most part irreversible, though there have been indications of considerable recovery after exposure to short bursts of high energy radiation.

optical fiber ribbon: *Synonym* **fiber optic ribbon.**

optical fiber scattering: In lightwaves propagating in an optical fiber, scattering caused primarily by impurities, such as hydroxyl and iron ions, and the intrinsic material from which the fiber is made, i.e., Rayleigh scattering. *See* **configuration scattering.** *See also* **Rayleigh scattering, scattering.**

optical fiber sensor: *See* **coated optical fiber sensor, distributed optical fiber sensor, flat coil optical fiber sensor.**

optical fiber source: *Synonym* **fiber optic source.**

optical fiber splice: A connection that (a) is between two optical fibers, made by joining an end of one fiber to an end of another fiber, (b) consists of the fiber joint and a fiber splice housing, and (c) usually is contained within an interconnection box or a fiber optic cable splice closure. *Note:* Types of optical fiber splices include fusion, ultraviolet cured or bonded, mechanical, rotary mechanical, and ribbon splices. *See also* **fiber optic splice.**

optical fiber strain: The percentage or the per-unit elongation, i.e., the strain, of an optical fiber when subjected to a given stress or force. *Note:* Optical fiber fracture usually occurs at about 1.0%. Operating strains are held to less than 0.2%.

optical fiber tensile strength: The maximum load that a short piece of optical fiber can sustain without breaking, usually calculated by the relation $T_s = Er/r_{min}$ where T_s is the safe tension stress, E is Young's modulus, i.e., the modulus of elasticity, which for glass usually is about 10^7 lb/in^2 (pounds per square inch) (7×10^9 kg•m^{-2}) (kilograms per square meter) or 70×10^9 N•m^{-2} (newtons per square meter), r is the radius of the fiber, and r_{min} is the minimum bend radius before breaking. *Note:* The calculated optical fiber tensile strength, T_s, is approximate and must be adjusted for probabilistic increases in occurrences of fractures and microcracks in longer fibers.

optical fiber transfer function (OFTF): The transformation that an optical fiber performs on an electromagnetic wave that enters it, such that if the input signal composition is known, and the transfer function of the fiber is known, the output signal can be determined. *Note:* An example of an optical fiber transfer function (OFTF) is the relation $OFTF = e^{-af}$ where a is a constant for the fiber and f is a frequency component of the input signal, i.e., the input electromagnetic wave.

optical fiber trap: An optical fiber that (a) breaks easily when stressed, (b) can be placed on approaches to areas to be secured, such as fences and fields, (c) can signal the location of a break and so cannot be cut without detection, (d) can be used to detect the presence of trespassers, and (e) is almost invisible, particularly in dimly lit areas. *Synonym* **security optical fiber.**

optical fiber video trunk: *Synonym* **fiber optic video trunk.**

optical fiber waveguide: *Synonym* **optical fiber.**

optical filter: **1.** A device that selectively either transmits or blocks a specified range of optical wavelengths. **2.** An optical element that alters the spectral composition of lightwaves that are incident upon it. *Note:* The transmitted light emanating from an optical filter (a) might have the radiant power of certain wavelengths in the incident light reduced or wholly removed, (b) will not contain wavelengths that are not in the incident wave, (c) will have an optical power level less than the input optical power level, and (d) may have a different polarization composition from the incident wave. *See* **tunable optical filter.**

optical frequency division multiplexing: *Synonym* **wavelength division multiplexing.**

optical glass: *See* **bulk optical glass.**

optical harness: An assembly that (a) consists of a number of multifiber cables or jacketed bundles placed together in an array that contains branches, (b) usually is installed within other equipment, and (c) is mechanically secured to that equipment.

optical harness assembly: An optical harness that is terminated and ready for installation. *See also* **optical harness.**

optical heterodyning: *Synonym* **optical mixing.**

optical impedance discontinuity: In an optical waveguide, such as an optical fiber or slab dielectric waveguide, an abrupt spatial variation in a parameter of the guide, such as the refractive index, absorption coefficient, scattering coefficient, geometric configuration, electric permittivity, magnetic permeability, or electrical conductivity. *Note:* Optical impedance discontinuities are not only a function of the intrinsic physical properties of the material from which the guide is constructed, but they are also a function of other extrinsic parameters, such as the frequency or wavelength of an electromagnetic wave propagating in the waveguide, the temperature, and the applied pressure.

optical integrated circuit (OIC): A circuit, or a group of interconnected circuits, that (a) may be either monolithic or composed of hybrid circuits, (b) usually is mounted on semiconductor or dielectric substrates, i.e., on chips, (c) is used for signal processing, and (d) consists of miniature solid state active and passive optical, electrical, electrooptic, or optoelectronic components, such as (i) light-emitting diodes (LEDs), (ii) lasers, (iii) optical filters, (iv) photodetectors, (v) optical memories, (vi) fiber amplifiers, (vii) logic gates, such as AND, OR, NAND, NOR, and NEGATION gates, and (viii) thin film optical waveguides. *Synonym* **integrated optical circuit.** *See also* **hybrid.** *Refer to* **Figs. R-8 (834), S-12 (917).**

optical integrated circuit (OIC) filter-coupler-switch-modulator: Two or more optical waveguides fabricated on a minute piece of material, such as lithium niobate, (a) that has light-propagating characteristics, (b) in which energy interaction can be varied in order to perform the four major functions performed by a radio receiver, namely, filtering, coupling, switching, and modulating, and (c) in which electrodes that apply a voltage across a section of a waveguide alone, or across a section of a waveguide shared by two or more waveguides, control the performance of the various functions. *See also* **coupler, electrode, filter, modulator, optical waveguide, switch.**

optical interface: In a fiber optic communications link, a point at which an optical signal is passed from one equipment or medium to another without conversion to an electrical signal.

optical isolator: 1. At a point in a transmission path, a device that (a) prevents light energy from propagating in a given direction and (b) prevents reflections from proceeding toward the source of the incident light. **2.** A two-port optical device that attenuates optical radiation in one direction more than in the opposite direction. *Note:* An example of an optical isolator is the Faraday optical isolator, in which the magnetooptic effect is used to obtain the isolation. *See also* **fiber optic isolator, optoisolator.**

optical junction: In an optical system, an optical interface, i.e., a physical interface, between optical components. *Note:* Examples of optical junctions are interfaces between light sources and optical fibers, fibers and other fibers, fibers and detectors, beams in air and prisms or lenses, fibers and lenses, lenses and other lenses, lenses and light sources, and lenses and photodetectors. *See also* **fiber optic junction, optical system.**

optical length: *See* **optical path length.**

optical lever: A means of detecting and measuring small angular displacements of an object by (a) reflecting a beam of light from the object, such as from a smooth reflecting surface of the object or from a mirror or a prism attached to the object, and (b) allowing the reflected beam to form a spot of light on a scale to measure the angular displacement.

optical line code: Sequences of optical pulses suitably structured to permit data transfer over a fiber optic link. *See also* **fiber optic link.**

optical line rate: The data signaling rate (DSR), at a given wavelength, that appears at the interface, i.e., the boundary, between an optical fiber and transmission equipment. *Note:* The optical line rate usually is expressed in bits per second.

optical link: *Synonym* **fiber optic link.**

optical loss test set (OLTS): A device (a) that consists of a stabilized light source, such as a laser or an incoherent light-emitting diode (LED), and a calibrated optical power meter and (b) that is used (i) to measure the optical power loss between two points by launching a known amount of power at one of the points and comparing it with the amount of power received at the other point, (ii) to measure insertion losses, and (iii) to measure attenuation rates. *See also* **optical time domain reflectometry.**

optically active material: A material that (a) rotates the polarization of the light that passes through it, (b) exhibits different refractive indices for left-hand and right-hand circular polarizations of the light, and (c) demonstrates circular birefringence. *See also* **birefringence, left-hand polarization, polarization, right-hand polarization.**

optical maser: A source that (a) produces monochromatic and coherent electromagnetic radiation by the synchronous and cooperative emission of optically pumped ions introduced into a crystal host lattice, a gas, or a liquid, the atoms of which are excited in a discharge tube, and (b) produces a lightwave with a sharply defined wavelength, i.e., with a narrow spectral width, that propagates in a high power, highly directional beam, i.e., a beam with near zero divergence. *See also* **coherence.**

optical material: *See* **nonlinear optical material, polymeric nonlinear optical material.**

optical mixing: 1. The production of a lightwave that contains all the wavelengths that are in all the input lightwaves. *Note:* An example of optical mixing is the production of dichromatic or polychromatic radiation from two or more monochromatic lightwaves by any of various means, such as the use of an optical mixing box, a mixing rod, or sets of mirrors. **2.** The mixing of two coherent lightwaves to obtain a beat frequency, as in an interferometer. **3.** The mixing of two pulse-modulated optical signals to obtain a beat frequency equal to the difference between the pulse repetition rates. **4.** The production of a pulsating electrical current proportional to the difference in wavelength between two interfering monochromatic lightwaves incident upon the same photodetector. *Synonym* **optical heterodyning.** *See also* **interference, photodetector.** *Refer to* **Fig. O-2.** *Refer also to* **Fig. P-9 (739).**

optical mixing box: A device that (a) consists of a piece of transparent material, (b) receives signals of several different optical wavelengths, or frequencies, i.e., lightwaves entering the box are usually a group of monochromatic waves, each of a different wavelength

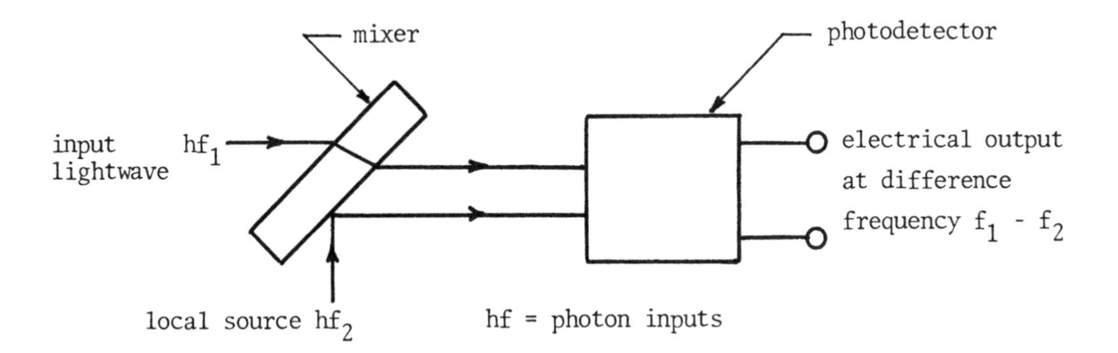

Fig. O-2. Optical mixing by means of a partially reflecting and partially transmitting mirror, i.e., an optical mixer in the form of a dichroic filter or mirror, and a photodetector. The input lightwaves must be coherent to obtain the difference frequency.

and each having been modulated separately, (c) mixes them to produce dichromatic or polychromatic lightwaves for dispatch via one or more output ports for transmission elsewhere and subsequent separation into the constituent wavelengths, i.e., demodulation, to produce the original intelligence introduced by the modulation of each of the different wavelengths, and (d) has reflective inner surfaces, except at the ports, to accomplish the mixing. *See also* **fiber optic coupler, demodulation, modulation.** *Refer to* **Fig. P-9 (739).**

optical mixing rod: An optical mixing box that has the general shape of a right circular cylinder, usually with pigtails for entrance and exit ports. *See also* **optical mixing box, pigtail.** *Refer to* **Fig. P-9 (739).**

optical modulation: *See* **external optical modulation.**

optical modulator: *See* **thin film optical modulator.**

optical multimode dispersion: Dispersion of the various frequencies contained in the pulses in an optical waveguide caused by mode scrambling that occurs when two or more transmission modes are supported by the same guide. *Note:* Optical multimode dispersion is greatly reduced in graded index (GI) optical fibers and somewhat reduced by using a monochromatic light source, such as a laser. *See also* **dispersion, fiber optic multimode dispersion, graded index optical fiber, mode, monochromatic.**

optical multiplexer (active): A device (a) that uses optical components and other types of components, such as electrical and acoustic components, to create data links, circuits, or channels to increase the transmission capacity of a system, such as by using space division, wavelength division (optical), frequency division (baseband), and time division multiplexing schemes, either singly or in combination, and (b) in which additional energy is required to operate the device, over and above that contained in the lightwaves being multiplexed. *Note:* Examples of active optical multiplexers are optoelectronic multiplexers and electrooptic multiplexers. *See* **thin film optical multiplexer.** *See also* **fiber optic demultiplexer (active), fiber optic multiplexer (active), optical demultiplexer (active), optical multiplexer (passive).**

optical multiplexer (passive): A device that (a) uses only optical components to create data links or channels to increase the transmission capacity of a system, such as by using space division or wavelength division multiplexing schemes, and (b) does not require energy to operate it, other than that contained in the lightwaves being multiplexed. *Note:* An example of a passive optical multiplexer is an optical mixer. *See* **thin film optical multiplexer.** *See also* **fiber optic demultiplexer (passive), optical demultiplexer (passive), fiber optic multiplexer (active), fiber optic multiplexer (passive), optical multiplexer (active).**

optical multiplexing: Multiplexing of lightwaves by (a) mixing different fixed wavelengths, i.e., wavelength division multiplexing (WDM), (b) mixing different baseband modulation frequencies, i.e., frequency division multiplexing (FDM), (c) sharing the time of a channel, i.e., time division multiplexing, (d) using two or more waveguides in a single cable, i.e., space division multiplexing (SDM), or (e) any combination of these. *Note:* Optical multiplexing may be active, requiring external power, or passive, using only the power contained in the lightwaves that are being multiplexed.

optical multiport coupler: An optical coupler (a) that has at least three ports, (b) in which at least one port is an input port, and (c) in which at least one port is an output port. *Synonym* **multiport optical coupler.**

optical network standard: *See* **synchronous optical network standard.**

optical path component: A usually passive component that (a) has only one function, namely, to convey an optical signal through a point or from one point to another, i.e., simply to provide a path for the signal, (b) ideally does not introduce propagation delay, distortion, attenuation, or noise, or in any way change or operate upon the optical signals that are passing through it, and (c) does not perform signal processing functions, such as amplifying, repeating, and shaping signals. *Note 1:* Examples of optical path components are dielectric waveguides, optical fibers, fiber optic cables, splices, connectors, termini, penetrators, rotary joints, distribution frames, splice trays, and raceways. *Note 2:* Optical multiplexers, demultiplexers, filters, and attenuators are examples of optical components that are not optical path components because they operate upon the signals. *See also* **passive optical device.**

optical path length (OPL): For a light ray passing through a medium, the integral of *nds* integrated from 0 to *S*, i.e., given by the relation $L = \int_0^S n\,ds$ where *L* is the optical path length, *S* is the physical length of the path, i.e., the spatial length, such as the measured length in meters, and *n* is the refractive index at the point along the path where *ds,* an element of length, is taken. *Note 1:* For a light ray passing through an isotropic medium, i.e., a medium in which the refractive index is constant, the path length becomes $L = nS$, where *n* is the refractive index and *S* is the spatial distance the ray travels along the path. *Note 2:* In a lightwave propagating along a path, the optical path length between two points along the path is proportional to the difference in phase of the lightwave between the two points. *Synonym* **optical length.** *See also* **optical thickness, physical path length.** *Refer to* **Fig. G-2 (399).**

optical phase modulator: An optical device that controls or varies the phase of a lightwave relative to a fixed reference or relative to another lightwave in accordance with an information-bearing signal. *Note:* An example of an optical phase modulator is one that uses the same Sagnac interferometric principle used in fiber optic gyroscopes. Signals are kept amplified, coherent, and in phase prior to entering the Sagnac optical fiber loop. *Synonym* **fiber optic phase modulator.**

optical plastic: *See* **bulk optical plastic.**

optical power: The radiant power of electromagnetic waves in the optical spectrum portion of the electromagnetic frequency spectrum. *See* **transmitter optical power.** *See also* **electromagnetic radiation, electromagnetic wave, frequency spectrum, optical spectrum, radiant power, visible spectrum.** *Refer to* **Fig. I-2 (451), L-5 (511), M-3 (590), O-3.** *Refer to* **Table 5, Appendix B.**

optical power budget (OPB): In an optical transmission system, the distribution of the available power that (a) is required for transmission within specified distortion limits or bit error ratios (BERs), (b) is based on the difference between the light source output power and the threshold power sensitivity of the photodetector, (c) is distributed among the intervening components, such as fiber optic cables, connectors, splices, couplers, multiplexers, demultiplexers, attenuators, and mixers, and (d) usually includes a safety margin for component loss tolerances and an allowance for splicing breaks that occur in the fiber optic cable after installation. *Note 1:* In a fiber optic link, the optical power budget usually is in dB for each optical path component in the link from light source to photodetector. Optical power budgets usually are about 30 dB from light source to photodetector. *Note 2:* The optical power launched into an optical fiber depends on many factors, such as (a) the type of source, such as light-emitting diode (LED) or laser diode, (b) the core diameter, and (c) the numerical aperture (NA). A power budget is specified for a particular fiber and for a particular length of that fiber.

optical power density: In a propagating or standing electromagnetic wave, the irradiance expressed as energy per unit volume occupied by the wave, such as joules per cubic meter, rather than as watts per square meter incident upon a transverse real or notional surface. *See also* **irradiance.**

optical power efficiency: The ratio of (a) the electromagnetic output power emitted by an optical source to (b) the electrical input power to the source.

optical power output: The power in the optical portion of the electromagnetic spectrum emitted by a device. *Note:* Optical power output may be expressed in total watts, total lumens per second, or watts per unit solid angle per unit of projected area of the source in a given direction, i.e., the radiance. *Refer to* **Fig. O-3 (673).**

optical protective coating: A coating that (a) is applied to a buffered or unbuffered optical surface, such as that of an optical fiber or lens, primarily for protecting the surface from mechanical abrasion, chemical corrosion, or both, (b) usually is bonded to the surface, and (c) may have special optical qualities, such as transparency, opacity, and refractive index. *Note:* An important class of protective coatings consists of evap-

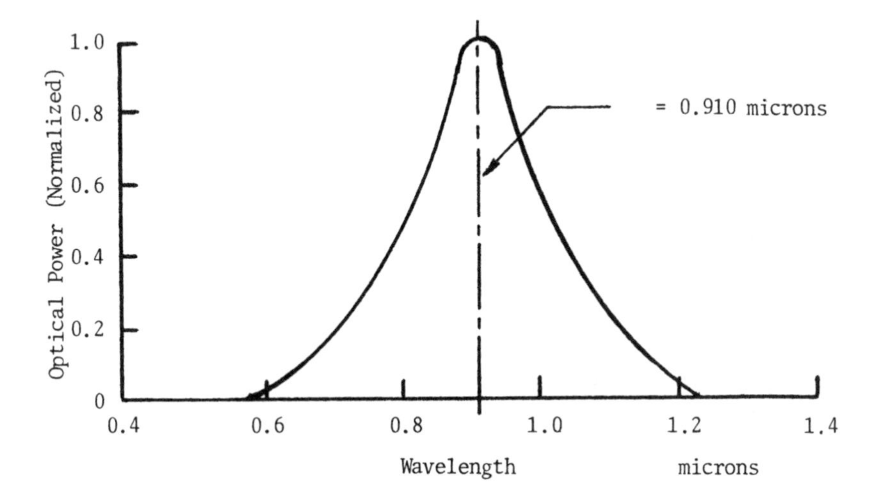

Fig. O-3. Normalized **optical power output** of a typical light-emitting diode (LED) source.

orated thin films of titanium dioxide, silicon monoxide, or magnesium fluoride. A thin layer of silicon monoxide may be added to protect an aluminized surface to prevent corrosion. *See also* **opacity, transparency.**

optical pulse: A pulse of power or energy in the optical portion of the electromagnetic spectrum. *See also* **optical spectrum, pulse.**

optical pulse broadening: *See* **optical pulse stretching.**

optical pulse compression: In the transmission of pulses in optical fibers, the reduction of pulse duration of certain frequency-modulated pulses arriving at the end of the fiber when longer wavelengths emitted later catch up with shorter wavelengths emitted earlier. *See also* **pulse duration.**

optical pulse distortion: An unintentional change in a property that characterizes an optical pulse, such as its shape, amplitude, phase, wavelength, position, or duration.

optical pulse lengthening: *See* **optical pulse stretching.**

optical pulse stretching: The increasing of optical pulse duration caused by transmission through a series of optical path components, such as a length of optical fiber or a series of fibers connected in tandem. *Note 1:* The amount of optical pulse stretching usually is given by the empirical relation $T_2 = T_1 (L_2/L_1)^k$ where T_2 is the pulse duration at the end of a fiber length L_2, T_1 is the pulse duration at the end of fiber length L_1, L_2 is the overall system fiber length including L_1, and k is a factor that is dependent upon the system length, L_2, and the refractive index profile. The exponent k usually lies between 0.7 and 0.9 for a given optical fiber. *Note 2:* Pulse stretching is caused primarily by dispersion. *Synonym* **optical pulse broadening, optical pulse lengthening, optical pulse widening.** *See also* **optical path component, pulse duration.**

optical pulse suppressor: A device that attenuates an optical pulse, such as 1 km (kilometer) of optical fiber. *Note:* Optical pulse suppressors were originally conceived to compensate for the dead zones of optical time domain reflectometers (OTDRs) so that measurements could be made beginning at the entrance face of an optical fiber, a connector, or a pigtail.

optical pulse widening: *See* **optical pulse stretching.**

optical radiation: Electromagnetic radiation at wavelengths in the optical spectrum, i.e., wavelengths between the transition to X rays and the transition to radio waves, i.e., between (a) about 1 nm (nanometer) and about 0.1 mm (millimeter) or (b) about 0.001 μm (micron) and 100 μm.

optical range (OR): 1. The direct line-of-sight distance between two points at given altitudes above the surface of the Earth. **2.** The range that is achieved by a straight line from one point above the Earth's surface to another point above the Earth's surface and tangent to the Earth's surface at one point in between.

optical ray: In geometric optics, a representation of the direction of propagation of electromagnetic radiation as defined by directional parameters. *Note:* A ray is perpendicular to a wavefront and in the direction of the Poynting vector at each point in a propagation medium. *See also* **geometric optics, Poynting vector.**

optical reader: A device that (a) is capable of sensing, interpreting, and transmitting encoded handwritten, typed, or printed characters by being able to distinguish their shape as a distinct pattern of light areas, i.e., reflective areas, and dark areas, i.e., nonreflective (absorptive) areas, for each character, on a data medium. *Note 1:* Examples of optical readers are (a) the reader in optical character recognition equipment used for computer input and (b) the device that reads the data on an optical disk. *Note 2:* Scanning and transmitting an image, such as in facsimile, radiophoto, and phototelegraphy systems, is not considered as optical reading. *See also* **optical scanner.**

optical receiver: A device that (a) detects optical signals, (b) converts them to electrical signals, and (c) processes the electrical signals for subsequent use. *See also* **fiber optic receiver, fiber optics, photodetector.**

optical receiver maximum input power: The maximum input optical power (dBm) coupled to an optical receiver, as measured by using specified standard procedures, from the line side of the receiver fiber optic connector when operated under standard or extended operating conditions, that the receiver will accept and still (a) not exceed the manufacturer-specified bit error ratio (BER) in digital systems or (b) operate in a linear nonsaturated mode in analog systems.

optical reflection: *See* **maximum optical reflection.**

optical regenerative repeater: *Synonym* **optical repeater.**

optical regenerator: *Synonym* **optical repeater.**

optical repeater: An optoelectronic device or module that receives an optical signal, amplifies it, or, in the case of a digital signal, also reshapes, retimes, or otherwise reconstructs it, and retransmits it as an optical signal representing the same information as the input signals. *Note:* The optical repeater is inserted at a point in a propagation medium, such as a long fiber optic cable or a stage in an optical integrated circuit (OIC), to overcome the effects of attenuation and dispersion. *Synonyms* **optical regenerative repeater, optical regenerator.** *See also* **fiber optic repeater, fiber optics.**

optical repeater power: 1. In optical systems, the power required to operate an optical repeater. *Note:* Optical repeater power may be delivered to the repeater by various means, such as (a) an electrical conductor in the cable that also contains the optical waveguides, i.e., a hybrid cable, (b) a separate electrical wire or cable, (c) a solar cell, (d) a local battery, primary or secondary, such as a seawater battery, (e) an electrical power system local power outlet, (f) an optical power cable separate from the optical signal cable or a fiber optic bundle in the same cable with the signal waveguides, or (f) a tap of the signal power contained in the signals in the optical fibers without disturbing the information content of the signals and without excessive attenuation of the signals.**2.** The electrical power or the optical power delivered to an optical repeater. **3.** The optical power output of an optical repeater. *See also* **fiber optic repeater, optical repeater, signal path.**

optical rotation: 1. The angular displacement of the polarization plane of an electromagnetic wave, such as a lightwave propagating in a propagation medium. **2.** The azimuthal displacement of the field of view achieved through the use of a rotating prism. *See also* **electromagnetic wave, polarization plane.**

optical scanner: A device in which (a) a light source emits a beam that illuminates objects, such as images, marks, codes, and characters, (b) the beam sweeps over the objects, and (c) a reflection sensor develops digital or analog electrical signals that represent the images, marks, codes, and characters that are sensed. *Note:* Examples of optical scanners are (a) devices that are held like a writing pen and moved across marks on a display surface, (b) oscillating scanning heads that scan an entire page, line by line, (c) hand-held devices that emit a beam that is swept across coded patterns, such as bar codes on labels and packages, and (d) flat surface scanners that emit a beam from within that is reflected back through glass to read patterns, such the scanners used at grocery and merchandise checkout counters and shipping centers. *See also* **optical reader.**

optical scanning: The scanning of objects, such as images, marks, bar codes, and letters, in which a light source illuminates the objects, and a reflection sensor develops coded electrical signals that represent the objects sensed. *See also* **optical scanner.**

optical sensitivity: *See* **receiver optical sensitivity.**

optical serrodyne: A modulator in which the phase of a coherent lightwave is modulated by controlling the transit time.

optical signal: A signal that (a) contains optical power and (b) is transmitted in an optical waveguide, such as an optical fiber, slab dielectric waveguide, or optical integrated circuit (OIC). *See* **two-level optical signal.** *See also* **optical power, visual signal.**

optical source: 1. A device that provides radiant power in the optical portion of the electromagnetic spectrum, usually by converting other forms of power, such as electrical power, to radiant power. *Note:* The two most common optical sources used in fiber optic systems are light-emitting diodes (LEDs) and laser diodes. **2.** Test equipment that generates a stable optical signal for the purpose of making optical transmission loss measurements. *Note:* Optical sources are used in interferometry and optical time domain reflectometry (OTDR). *See* **standard optical source, video optical source.** *See also* **interferometry, optical spectrum, optical time domain reflectometry, source efficiency**.

optical source type: Characterization of a light source by identifying its (a) specific features, such as a laser, a light-emitting diode (LED), or a superluminescent light-emitting diode (SLED), (b) compositional material, such as InGaA, or (c) generic structure, such as edge emitting and surface emitting.

optical space division multiplexing (OSDM): The use of independent optical fibers contained in a single fiber optic cable to provide parallel optical paths for separate channels. *See also* **channel, fiber optic cable, path.**

optical spectrum: In the electromagnetic spectrum, the range of wavelengths extending from (a) 0.001 μm (micron), 1 nm (nanometer), or 10 Å (angstroms) to (b) 100 μm or 0.1 mm (millimeter), i.e., from the vacuum shortest ultraviolet wavelength at 1 nm (nanometer) to the longest far infrared wavelength at 0.1 mm (millimeter). *Note 1:* The visible spectrum lies within the optical spectrum. The optical spectrum lies between the shortest wavelengths of radio, i.e., microwaves, and the longest wavelengths of X rays. No formal spectral limits for the optical spectrum are nationally or internationally recognized. *Note 2:* The optical spectrum is of interest in the sense that techniques for handling lightwaves and components designed for the human visible spectrum also apply to the spectrum somewhat beyond the human visible region, i.e., the optical spectrum is "visible" with respect to the components, though not in its entirety to humans. *Note 3:* The standard fiber optic wavelength of 1.31 μm (microns) is outside the visible spectrum but inside the optical spectrum. *See also* **electromagnetic spectrum, emissivity, infrared, light, visible spectrum.**

optical spectrum analyzer: *See* **integrated optical spectrum analyzer.**

optical speed: 1. In lens systems, the reciprocal of the square of the f-number, which is inversely proportional to twice the numerical aperture. **2.** In an optical fiber, a number that is directly proportional to the numerical aperture squared, which is equal to the difference of the squares of the refractive indices of the core and the cladding. *See also* **f-number, lens system, numerical aperture.**

optical splice: *Synonym* **fiber optic splice.**

optical splitter: A device that extracts a portion of the optical signal power propagating in a dielectric waveguide, such as an optical tap or a partial mirror, and diverts that power to a destination different from that of the remaining portion. *Synonym* **beamsplitter.** *See also* **fiber optic splitter.**

optical station facility: *Synonym* **fiber optic station facility.**

optical station/regenerator section: *Synonym* **fiber optic station/regenerator section.**

optical storage: Storage in which data are stored on a photosensitive surface. *Synonym* **photooptic storage.** *See also* **laser emulsion storage, optical disk storage.**

optical surface: In an optical system, an identifiable geometric surface, which usually occurs at a boundary or a discontinuity of optical quality or characteristic of a material, such as an abrupt change of refractive index, absorptive quality, transmittance, and vitrification. *Note:* Examples of optical surfaces are reflecting and refracting surfaces of optical elements, such as mirrors, lens surfaces, and core-cladding interface surfaces in optical fibers. *See also* **optical element, optical system.**

optical switch: A switch that (a) enables signals in optical fibers and optical integrated circuits (OICs) to be selectively switched from one circuit or path to another, or (b) performs logic operations with these signals by using electrooptic effects, magnetooptic effects, or other effects or other methods that operate on transmission modes, such as transverse electric versus transverse magnetic modes, the type and the direction of polarization, or other characteristics of electromagnetic waves. *Note:* Examples of optical switches are multimode achromatic electrooptic waveguide switches and optical polarization switches. *Synonyms* **light valve, photonic switch.** *See* **mechanical optical switch, thin film optical switch.** *See also* **flip switch.**

optical system: A group of interrelated parts and equipment that (a) accepts, processes, and utilizes electromagnetic waves in the optical spectrum, (b) transmits and guides the waves for the most part in material media, such as lenses, prisms, optical fibers, filters, multiplexers, couplers, connectors, mixers, and various waveguides, and (c) accomplishes such functions as communications, illumination, inspection (endoscopy), sensing, and display. *Note:* Examples of optical systems are fiber optic links, telemetry systems, optical gyrocompasses, telescopes, microscopes, and aligned bundles of optical fibers. *See* **off-axis optical system.**

optical system power margin: A system power margin for an optical system, such as a fiber optic link. *See also* **system power margin.**

optical taper: An optical fiber that (a) has an increasing or decreasing diameter that is a linear function of its length in a longitudinal direction, and (b) can be used (i) to increase or decrease the size of an image in an aligned bundle and (ii) as a transition fiber to splice fibers with different diameters. *Synonyms* **fiber optic taper, tapered optical fiber.**

optical tapoff: *Synonym* **optical tapping.**

optical tapping: The removal or the extraction of some of the radiant power, i.e., the optical power, from electromagnetic waves, such as lightwaves, that are propagating in an optical waveguide, such as an optical fiber, fiber optic bundle, fiber optic cable, slab dielectric waveguide, or waveguide in an optical integrated circuit (OIC). *Note:* Optical tapping usually is accomplished by using connectors, couplers, taps, and branches. *Synonym* **optical tapoff.**

optical thickness: 1. The physical thickness times the refractive index of an optical element consisting of an isotropic propagation medium. **2.** The total optical path length through (a) an optical element, such as an optical fiber, optical filter, or lens system, or (b) a series of elements. *See also* **optical path length.**

optical thin film: A thin layer of an optical propagation medium, usually deposited or formed on a substrate, with a geometric shape and refractive quality that enable control, transmission, and guidance of lightwaves for specific purposes, usually to accomplish logical functions by forming optical switches and gates. *Note:* An example of an optical thin film is a thin layer of transparent material used to guide lightwaves in an optical integrated circuit (OIC). *See also* **optical integrated circuit.** *Refer to* **Figs. P-2 and P-3 (708), R-8 (834), S-12 (917).**

optical thin-film storage: Storage that (a) stores data in the form of light and dark spots, and (b) usually is read by a scanning light beam or by movement of the film past an optical reading head. *See also* **optical disk.**

optical time domain reflectometer (OTDR): A measurement device used to characterize an optical fiber by (a) launching an optical pulse into the fiber and measuring the resulting light that is scattered and reflected back to the input, i.e., back to the reflectometer, as a function of time. *Note:* In an optical fiber, the optical time domain reflectometer (OTDR) can be used to (a) estimate the optical attenuation coefficient as a function of distance, (b) identify defects, (c) determine localized losses, such as connector and splice insertion losses, (d) measure fiber length, and (e) determine the distance from the reflectometer to a break. *See also* **fiber optics, optical time domain reflectometry.**

optical time domain reflectometry (OTDR): The field of science and technology devoted to the characterization of an optical fiber, in which an optical pulse is transmitted through the fiber and the resulting light that is scattered and reflected back to the input is measured as a function of time. *Note:* Optical time domain reflectometry (OTDR) is useful in (a) estimating the optical attenuation coefficient as a function of distance, (b) identifying defects, (c) determining localized losses, such as connector and splice insertion losses, (d) measuring optical fiber length, and (e) measuring the distance to a break in a fiber. *Synonym* **backscattering technique.** *See also* **fiber optics, optical time domain reflectometer.**

optical transceiver: *Synonym* **fiber optic transceiver.**

optical transimpedance: In an optical transmission system, the ratio of (a) the output voltage at the photodetector to (b) the input current at the light source. *See also* **fiber optic transimpedance, transimpedance.**

optical transmission: The transmission of an optical signal in an optical propagation medium for the purpose of transferring information from one point to another. *See also* **optical signal, optical transmitter, propagation medium, visual transmission.**

optical transmittance: *Synonym* **transmittance.**

optical transmitter: 1. A light source that (a) is capable of being internally modulated or that emits light that is externally modulated, and (b) is coupled to a propagation medium, such as an optical fiber or an optical integrated circuit (OIC). **2.** A device that (a) accepts an electrical signal as its input, (b) processes this signal, and (c) uses the result of the processing to modulate an optoelectronic device, such as a light-emitting diode (LED) or an injection laser diode, to produce an optical signal capable of being transmitted via an optical propagation medium. *See also* **fiber optics, fiber optic transmitter, optical link, optical transmission.**

optical video disk (OVD): A disk on whose surface digital data may be recorded at high packing densities in concentric circles, or in a spiral, by using a laser beam to record spots that are read by means of a reflected laser beam of lower irradiance, i.e., lower intensity, than the recording intensity. *Note:* Up to 10^{13} bits may be recorded on a single disk at rates of 1 Mbps (megabits per second), making optical video disks suitable for many hours of television programming playback.

optical waveguide: Any structure that guides the flow of optical energy along a path parallel to its axis and, at the same time, contains the energy within, or adjacent to, its surface. *Note 1:* An optical waveguide is not necessarily an all-dielectric waveguide. *Note 2:* In communications systems, optical waveguides may be (a) optical fibers or (b) thin-film deposits used in optical integrated circuits (OICs), both used to transmit signals. *See* **diffused optical waveguide, graded index optical waveguide, single mode optical waveguide, step index optical waveguide, thin film optical waveguide.** *See also* **fiber optic waveguide, optical fiber.** *Refer to* **Figs. P-2 and P-3 (708).**

optical waveguide connector: A device that (a) transfers optical power from one optical waveguide to another and (b) can be connected and disconnected repeatedly.

optical waveguide coupler: 1. A device that distributes optical power to two or more ports. **2.** A device that couples optical power between a waveguide and either a light source or a photodetector. *See also* **fiber optic coupler.**

optical waveguide splice: A permanent joint that couples optical power from one waveguide to another. *See also* **fiber optic splice.**

optical waveguide termination: The point in an optical waveguide at which optical power propagating along the waveguide leaves the waveguide and continues in a nonwaveguide mode of propagation. *Note:* In most applications, care must be taken to prevent leakage, losses, and reflection at a termination.

optic amplifier: *See* **fiber optic amplifier,**

optic arteriovenous oximeter: *See* **fiber optic arteriovenous oximeter.**

optic attenuator: *See* **fiber optic attenuator.**

optic axis: In an anisotropic medium, a direction of lightwave propagation in which orthogonal polarizations have the same phase velocity. *See also* **optical axis, optical fiber axis.**

optic blood flow meter: *See* **fiber optic blood flow meter.**

optic borescope: *See* **fiber optic borescope.**

optic branching device: *See* **fiber optic branching device.**

optic breakout cable: *See* **fiber optic breakout cable.**

optic breakout kit: *See* **fiber optic breakout kit.**

optic bundle: *See* **fiber optic bundle.**

optic cable: *See* **bidirectional fiber optic cable, fiber optic cable, fiber optic cable core, special fiber optic cable, station fiber optic cable, TEMPEST proofed fiber optic cable.**

optic cable assembly: *See* **fiber optic cable assembly.**

optic cable driver: *See* **fiber optic cable driver.**

optic cable facility: *See* **fiber optic cable facility.**

optic cable facility loss: *See* **fiber optic cable facility loss, statistical fiber optic cable facility loss.**

optic cable feed through: *See* **fiber optic cable feed through.**

optic cable pigtail: *See* **fiber optic cable pigtail.**

optic choledochoscope: *See* **fiber optic choledochoscope.**

optic communications: *See* **fiber optic communications.**

optic concentrator: *See* **fiber optic concentrator.**

optic connector: *See* **fiber optic connector, fixed fiber optic connector, free fiber optic connector.**

optic connector variation: *See* **fiber optic connector variation.**

optic coupler: *See* **fiber optic coupler.**

optic cross connection: *See* **fiber optic cross connection.**

optic data link: *See* **fiber optic data link.**

optic data transfer network: *See* **fiber optic data transfer network.**

optic delay line: *See* **fiber optic delay line.**

optic demultiplexer (active): *See* **fiber optic demultiplexer (active).**

optic demultiplexer (passive): *See* **fiber optic demultiplexer (passive).**

optic display device: *See* **fiber optic display device.**

optic distribution frame: *See* **fiber optic distribution frame.**

optic distribution system: *See* **protected fiber optic distribution system.**

optic dosimeter: *See* **fiber optic dosimeter.**

optic drop: *See* **fiber optic drop.**

optic-electronic device: *Synonym* **optoelectronic device.**

optic endoscope: *See* **fiber optic endoscope.**

optic facility: *See* **station fiber optic facility.**

optic filter: *See* **fiber optic filter.**

optic flood illumination: *See* **fiber optic flood illumination.**

optic gyroscope: *See* **fiber optic gyroscope.**

optic illumination detection: *See* **fiber optic illumination detection.**

optic illuminator: *See* **fiber optic illuminator.**

optic interconnection: *See* **fiber optic interconnection.**

optic interconnection box: *See* **fiber optic interconnection box.**

optic interface device: *See* **fiber optic interface device.**

optic isolator: *See* **fiber optic isolator.**

optic jumper: *See* **fiber optic jumper.**

optic light source: *See* **fiber optic light source.**

optic link: *See* **fiber optic link, repeatered fiber optic link, repeaterless fiber optic link, television fiber optic link.**

optic mixer: *See* **fiber optic mixer.**

optic modem: *See* **fiber optic modem.**

optic mode stripper: *See* **fiber optic mode stripper.**

optic modulator: *See* **acoustooptic modulator, electrooptic modulator, fiber optic modulator.**

optic multiplexer (active): *See* **fiber optic multiplexer (active).**

optic multiplexer (passive): *See* **fiber optic multiplexer (passive).**

optic multiport coupler: *See* **fiber optic multiport coupler.**

optic myocardium stimulator: *See* **fiber optic myocardium stimulator.**

optic network: *See* **synchronous fiber optic network.**

optic patch panel: *See* **fiber optic patch panel.**

optic penetrator: *See* **fiber optic penetrator.**

optic photodetector: *See* **fiber optic photodetector.**

optic power: *See* **fiber optic repeater power.**

optic probe: *See* **fiber optic probe.**

optic receiver: *See* **fiber optic receiver.**

optic repeater power: *See* **fiber optic repeater power.**

optic ribbon: *See* **fiber optic ribbon.**

optic ringer: *See* **fiber optic ringer.**

optic rotary joint: *See* **fiber optic rotary joint.**

optic rotor: *See* **fiber optic rotor.**

optics: The branch of science and technology devoted to the study of (a) the nature and the properties of electromagnetic radiation in the optical spectrum, with most of the emphasis placed on the visible and near visible spectra, (b) the application of the optical radiation in useful systems, and (c) the phenomena of vision. *See* **acoustooptics, active optics, coated optics, crystal optics, electron optics, electrooptics, fiber optics, fixed optics, geometric optics, integrated optics, physical optics, ultraviolet fiber optics, woven fiber optics.**

optic scrambler: *See* **fiber optic scrambler.**

optic section: *See* **station/regenerator fiber optic section.**

optic sensor: *See* **fiber optic sensor, Sagnac fiber optic sensor.**

optic splice: *See* **fiber optic splice.**

optic splice tray: *See* **fiber optic splice tray.**

optic splitter: *See* **fiber optic splitter.**

optic spot illumination: *See* **fiber optic spot illumination.**

optic supporting structure: *See* **fiber optic supporting structure.**

optic switch: *See* **fiber optic switch.**

optic telecommunications cable: *See* **fiber optic telecommunications cable.**

optic telemetry system: *See* **fiber optic telemetry system.**

optic terminus: *See* **fiber optic terminus.**

optic test method: *See* **fiber optic test method.**

optic test procedure: *See* **fiber optic test procedure.**

optic transceiver: *See* **fiber optic transceiver.**

optic transimpedance: *See* **fiber optic transimpedance.**

optic transmission system: *See* **fiber optic transmission system.**

optic transmitter: *See* **fiber optic transmitter.**

optimum traffic frequency: *Synonym* **optimum transmission frequency.**

optimum transmission frequency (FOT): 1. The highest frequency that is predicted to be available for skywave transmission over a particular path at a particular hour for 90% of the days of the month. *Note 1:* The abbreviation "FOT" is derived from "frequency of optimum traffic." *Note 2:* The optimum transmission frequency (FOT) usually is just below the maximum usable frequency (MUF). **2.** The most effective frequency for ionospheric reflection of radio waves between two specified points. *Note:* In the prediction of useful frequencies, the optimum transmission frequency is usually taken as 15% below the monthly median value of the maximum usable frequency (MUF) for the specified time and path. *Synonyms* **frequency of optimum traffic, frequency of optimum transmission, ionospheric reflection, optimum traffic frequency, optimum working frequency.** *See also* **frequency, path.**

optimum working frequency: *Synonym* **optimum transmission frequency.**

option: *See* **revised interconnection protocol security option.**

optoacoustics: *Synonym* **acoustooptics.**

optoacoustic transducer: A device that converts audio frequency-modulated lightwaves into a sound tone by causing a crystal to vibrate at the audio modulation frequency of the lightwave. *See also* **acoustooptic transducer.** *See also* **modulation.**

optocoupler: *Synonym* **fiber optic link.**

optoelectrical cable: A cable that contains optical waveguides, such as optical fibers, and electrical conductors.

optoelectronic: 1. Pertaining to devices that (a) contain both optical and electronic components, (b) usually convert optical power or energy into electrical power or energy, such as the conversion of an optical signal into an electrical signal, or vice versa, representing the same information, (c) must have at least one electrical port, and (d) respond to, emit, or modify optical radiation, or use optical radiation for internal operation. **2.** Pertaining to (a) the class of devices that function as electrical to optical or optical to electrical transducers or (b) instruments in which such devices are used. *Note 1:* Examples of optoelectronic devices commonly used in fiber optic communications systems are photodiodes, light-emitting diodes (LEDs), injection laser diodes, and optical integrated circuit (IOC) elements. *Note 2:* "Electro-optical" and "electrooptical" are often erroneously used as synonyms for "optoelectronic." *See also* **electrooptic, electrooptic effect, optical detector, optical source.**

optoelectronic detector: 1. A device that detects radiation and uses the influence of light to form an electrical signal. **2.** A device that (a) accepts an optical signal, (b) converts the optical signal into an electrical signal that has the same information content as the optical signal, and (c) emits the electrical signal. **3:** A device that detects or measures optical power by using incident light on a photosensitive surface to generate or control an electrical current by the use of the photoconductive, photoemissive, or photovoltaic effect. *See also* **information content, signal.**

optoelectronic device: 1. A device that (a) accepts and uses, (b) modifies and emits, and (c) neither accepts nor emits, but uses within for its internal operation, coherent or incoherent electromagnetic radiation in the visible, infrared, or ultraviolet regions of the electromagnetic frequency spectrum, i.e., in the optical spectrum. *Note:* The wavelengths currently handled by optoelectronic devices range from approximately 0.3 µm (micron) to 30 µm. **2.** An electronic device that (a) is associated with light and (b) usually serves as a light source, a processor of lightwaves, or a photodetector. *Synonym* **optic-electronic device.** *See also* **light.**

optoelectronic directional coupler: A directional coupler in which the coupling function is electronically controllable, usually with a photodetector that permits an electronic circuit to be driven by the coupler by tapping some of the passing optical power. *See also* **directional coupler, optical power.**

optoelectronic receiver: A receiver that accepts an optical signal, usually from a fiber optic cable, and converts it to an electrical signal for further processing or use. *See also* **fiber optic cable, receiver, signal.**

optoelectronics: The combination of pure and applied electronics, optics, and electromagnetics. *Synonym* **optronics.** *See also* **electrooptics.**

optoelectronic transmitter: 1. A transmitter that accepts an optical signal and converts it to an electrical signal that represents the same information as the optical signal. **2.** A transmitter in which the electrooptic effect is used in its operation.

optoisolator: An optical transmission link that (a) consists of an optical line driver, an optical waveguide, and a photodetector, (b) provides electrical isolation between two communications systems or components, and (c) serves as an electrical insulator as well as an optical transmission line. *See also* **fiber optic isolator, optical isolator.**

optooptics: The branch of science and technology devoted to the use of optical power to control the generation, transmission, reception, and processing of optical signals and optical power. *Note:* Optooptics is being applied in optical integrated circuits (OICs) using nonlinear optical materials to perform modulation, multiplexing, and switching functions. *See also* **acoustooptics, electrooptics, magnetooptics, photonics.**

optronic library: A database that (a) consists of data stored in a combination of optical and electronic circuits, (b) may be fixed or programmable, and (c) is usually mounted on integrated circuit (IC) chips for high capacity and transportability. *Note:* An optronic library might contain and process all the data required to operate a passenger aircraft, including flight scheduling, flight path data, weather data, airport configuration data, passenger listings, announcements, ticketing data, and meal menus.

optronics: *Synonym* **optoelectronics.**

OPX: off-premises extension.

OR: 1. Operations research. 2. The Boolean operator for the logic operation performed by an OR gate, i.e., a combinational circuit, that, for two or more input lines, an output signal is obtained if there is a signal on at least one input line, and no output signal is obtained if there are no input signals. *Note 1:* The statement A OR B may be written as A + B or as A v B. *Note 2:* The OR operator produces the logical sum. *See* **EXCLUSIVE OR.** *See also* **Boolean function, Boolean operation, Boolean operator, combinational circuit, logical sum, OR gate.**

ORBCOMM: A personal communications system based on a constellation of satellites that enables national and international point-to-point communications, particularly for areas outside cellular communications areas.

orbit: A path that (a) is relative to a specified reference, (b) is described by the center of mass of an object in space, such as a satellite, and (c) primarily is subjected to natural forces, mainly the force of gravity. *Note:* Examples of orbits are the paths of the Earth around the Sun, the Moon around the Earth, and an Earth communications satellite around the Earth. *See* **closed orbit, direct orbit, equatorial orbit, geostationary orbit, geostationary satellite orbit, hyperbolic orbit, inclined orbit, parabolic orbit, phased orbit, polar orbit, randomly phased orbit, retrograde orbit, synchronous orbit.**

orbital inclination: The angle determined by the plane containing an orbit and the Earth's equatorial plane. *Synonym* **inclination of an orbit.**

orbit determination: In space, the describing of the past, present, or predicted position of an object, such as a satellite, in terms of orbital parameters, such as (a) distances from the Earth's center, (b) angular position with respect to a reference, (c) linear and angular velocity, and (d) linear and angular acceleration, all as a function of time.

OR circuit: *Synonym* **OR gate.**

order: 1. To arrange or place items relative to each other spatially or temporally according to a specified set of rules. *Note:* Examples of order are to arrange a set of items in alphabetical order, numerical order, or chronological order or according to size, length, or weight. **2.** The arrangement of items according to a specified set of rules. *See* **absolute order, diffraction grating spectral order, relative order.** *See also* **order of diversity.**

order channel: *Synonym* **engineering channel.**

order circuit: *Synonym* **engineering circuit.**

ordering group: *See* **centralized ordering group.**

order intercept point: *See* **third-order intercept point.**

order nonlinear material: *See* **third-order nonlinear material.**

order of diversity: 1. The number of independently fading propagation paths or frequencies, or both, used in diversity reception. **2.** The number of signals simultaneously detected and combined through the use of (a) space-, frequency-, or time-division multiplexing, (b) modulation characteristics, (c) polarization, or (d) a combination of these. *Synonym* **diversity order.** *See also* **diversity reception, dual diversity combining, quadruple diversity combining.**

orderwire: The channel or the path in a communications system used for signals to control, or direct the control, of system operations. *Note 1:* (a) An analog voice orderwire may operate at 300 Hz (hertz) to 400 Hz, (b) a digital orderwire, i.e., a data orderwire, may operate at 16 kbps (kilobits per second), and (c) a telemetering orderwire may operate at 2 kbps. *Note 2:* In a fiber optic cable, the orderwire may be (a) a metallic conductor, (b) the same optical fiber used for user communications, or (c) another fiber. *See* **acquisition and tracking orderwire, remote orderwire.** *See also* **datawire, engineering channel, engineering circuit.**

orderwire circuit: *Synonym* **engineering circuit.**

orderwire complex: Equipment specifically designed for carrying orderwire traffic, i.e., engineering circuit traffic, as opposed to one designed for carrying mission traffic, i.e., user traffic. *Note:* The orderwire complex may use channels that are multiplexed with user traffic channels. *Synonym* **orderwire multiplex.** *See also* **multiplexing, orderwire circuit.**

orderwire message (OWM): A message that (a) is exchanged among technical control facilities for the purpose of controlling network operations, (b) usually is generated by and for any combination of operators and automated controllers, and (c) is not a message generated by or for users of the network. *See also* **orderwire circuit.**

orderwire multiplex: *Synonym* **orderwire complex.**

orderwire multiplex set: *See* **engineering orderwire multiplex set.**

ordinary ray: In a doubly refracting crystal, the ray that (a) has an isotropic velocity and (b) obeys Snell's law of refraction at the crystal surface. *See also* **extraordinary ray.**

ored: *See* **wire ored.**

OR element: *Synonym* **OR gate.**

organic semiconductor: A semiconductor that (a) may be used to fabricate light-emitting diodes (LEDs) that emit light in different colors, including red, blue, green, white, and purple, (b) transmits data photonically, (c) may be fabricated on silicon wafers, (d) may be used for flat panel displays and photonic communications between high speed chips, and (e) may be used to fabricate photodetectors that respond to the light produced by the LEDs. *See* **chip, wafer.**

organization: Persons and equipment assembled, grouped, associated, or arranged in a specified manner for a specified purpose. *Note:* Examples of organizations are the International Organization for Standardization, the American National Standards Institute, the Institute for Telecommunications Science, AT&T, SPRINT, Southern Bell, and Northern Telecom. *See* **area communications organization, maritime air radio organization, maritime air telecommunications organization, World Meteorological Organization.**

organization chart: *See* **radio organization chart.**

organizer: *See* **fiber and splice organizer, splice organizer.**

OR gate: A device that is capable of performing the logical OR, i.e., the INCLUSIVE OR operation, namely, that if P is a statement, Q is a statement, R is a statement . . . , then the OR of P, Q, R . . . is true if at least one statement is true, and false only if all statements are false. *Note:* P OR Q OR R is often written as P + Q + R or as P v Q v R. The OR gate can serve as a buffer because all input signals pass through the gate without having an input signal on any input line pass back out on any other input line. *Synonyms* **alternation gate, buffer gate, disjunction gate, inclusive OR gate, join gate, logic sum gate, mix gate, one gate, OR circuit, OR element, OR unit, positive OR gate, union gate.** *See also* **OR.** *Refer to* **Fig. O-4.** *Refer also to* **Figs. B-5 (77), B-6 (78), L-6 (529).**

orientation: 1. The spatial positioning or facing of an object or an image with respect to a frame of refer-

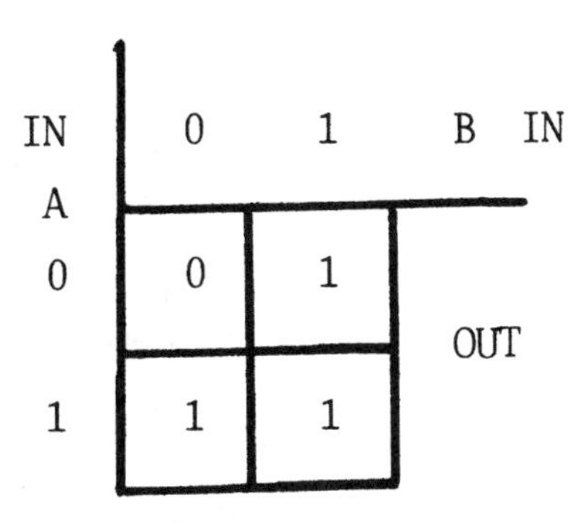

	0	1
0	0	1
1	1	1

Fig. O-4. A truth table that shows the input and output digits of an **OR gate.**

ence, such as a coordinate system or a reference direction. *Note:* If the front of an object were facing north, that position would be an orientation for that object. **2.** In synchronous transmission systems, a systematic phase difference between the rotation of the receiving distributor and the rotation of the transmitting distributor in order to take into account the propagation time of the signals or the response of the receiving device. *See* **landscape orientation, portrait orientation.**

orientation range: In data transmission systems, the phase difference that lies between the limits, in either direction, of the orientation of receiving and transmitting distributors that is compatible with the correct translation of signals.

oriented data transfer protocol: *See* **connection oriented data transfer protocol.**

oriented language: A language that is used, or intended to be used (a) in a specific area of interest, (b) by a specialized group of people, and (c) for a specific purpose. *See* **application-oriented language, computer-oriented language, problem-oriented language, procedure-oriented language.**

origin: *See* **computer program origin, loaded origin.**

original: *In facsimile systems, synonym for* **object.**

original copy: *In facsimile, deprecated synonym for* **object.**

originating office: In a communications system, the switching center or the central office (C.O.) at which messages or calls are entered into the system. *Note:* Examples of originating offices are (a) the central office (C.O.) from which a call originator places a call, and (b) the message center at which a message originator places a message for dispatch. *See also* **call, call originator, central office, message originator, switching center.**

originating user: The initiator, i.e., the originator, of a particular information transfer transaction. *Note 1:* The originating user may be either the source user or the destination user. *Note 2:* Examples of originating users are call originators and message originators. *See also* **access originator, communications source.**

originator: *See* **access originator, disengagement originator, message originator.**

originator to recipient speed of service: *Synonym* **speed of service.**

ORSA: operations research/systems analysis.

orthodromic map: A map that is made by projecting all great circles as straight lines even through the angles between them are correct at only one point, namely, the point of contact, i.e., the Earth tangent point, while at other points a distorted angular scale must be used. *Note:* An orthodromic map is made from a gnomonic projection formed by projecting from the center of the Earth onto a plane that touches the Earth's surface at the tangent point. This has the disadvantage that only a limited portion of the Earth's surface can be shown on any one map. A map that covers an area of 100 km on a side has noticeable distortion of shapes and angles at the edges. The scale or the representative fraction increases so that on the map objects appear farther apart than they really are according to the scale at the center of the map. A modification of a purely orthodromic map, i.e., a Mercator projection, of the Earth shows the lines of longitude as parallel lines and the lines of latitude as parallel lines, each set perpendicular to the other. The poles would be lines across the top and bottom of the map. This would make polar areas, such as Siberia, Alaska, the Arctic Ocean, and Antarctica, appear oversized compared to equatorial areas of the same actual size.

orthogonal: 1. Pertaining to being at right angles or mutually perpendicular. *Note:* Two vectors are orthogonal when the integral of their scalar product, i.e., dot product, throughout space always and everywhere is zero because the scalar product of two vectors is the product of their magnitudes times the cosine of the angle between them. If they are at right angles, i.e., are orthogonal, the angle is 90°, and the cosine of 90° is zero. Their vector product would be a vector equal to the product of their magnitudes and in a direction perpendicular to the plane of the vectors. *See also* **Poynting vector. 2.** The property of being independent and mutually exclusive.

orthogonal antenna mount: *See* **nonorthogonal antenna mount.**

orthogonal multiplex: 1. The combining of two or more digital signals that have mutually independent pulses in order to avoid intersymbol interference. **2.** Time division multiplexing in which pulses with orthogonal properties are used so as to avoid intersymbol interference. *See also* **intersymbol interference, multiplexing.**

orthogonal signals: A pair of signals that, at least theoretically, are mutually noninterfering. *Note:* Frequency-modulated signals and amplitude-modulated signals are orthogonal to each other. Thus, orthogonality of signals is relative and not an intrinsic property of a single or given signal.

orthomode transducer: A device that (a) is part of an antenna feed and (b) serves to combine or separate orthogonally polarized signals. *See also* **polarization.**

OR unit: *Synonym* **OR gate.**

oscillating sort: A sort of a set of items in which subsets are sorted and merged alternately until one sorted set is formed.

oscillator: In electronic systems, a device or a circuit that (a) produces an ideally stable alternating voltage or current at a specified frequency or frequencies, (b) has an output wave shape that may be of any shape, though usually the first stage output is in the shape of a single frequency sine wave, (c) subsequently may perform signal processing, such as clipping and differentiating, (d) may convert the sinusoidal wave to a rectangular wave, a sawtooth wave, or a series of pulses or spikes that become the final output, (e) may produce multiples and submultiples of the original frequency, and (f) may produce a frequency that is fixed, stepwise-variable, or continuously variable. *See* **crystal oscillator, quartz oscillator, relaxation oscillator.**

oscillator off: *See* **local oscillator off.**

oscillator tuning: Adjusting the frequency of an oscillator to a desired value, usually by adjusting one or more of the components in the oscillator circuit. *See* **local oscillator tuning.**

oscilloscope: A laboratory instrument that (a) is used to display representations of waveforms that are encountered in electronic and electrical circuits, (b) usually consists of a cathode ray tube with accessible terminals for access to electrostatic or magnetic, or both, electron beam deflection components, (c) accepts signals on the beam deflectors to move the electron beam that impinges on a screen, enabling a waveform to be displayed, (d) usually has a sawtooth, i.e., a linearly rising sweep voltage with an adjustable repetition rate or a rate that is triggered by, i.e., is in synchronism with, the waveform being observed, thus producing a time scale, (e) accepts the signal to be displayed on the vertical deflector, and (f) usually can accept signals on both deflectors to observe Lissajou figures, such as circles, ellipses, lemniscates, and cardioids.

OSCINT: open-source intelligence.

OSDM: optical space division multiplexing.

OSI: Open Systems Interconnection.

OSI-RM: Open Systems Interconnection—Reference Model.

OSS: operations support system.

OTAM: over the air management (of automated high frequency (HF) network nodes).

OTAR: over the air rekeying (of cryptographic equipment).

OTDR: optical time domain reflectometer, optical time domain reflectometry.

OTF: optimum traffic frequency, optimum transmission frequency.

other common carrier (OCC): A specialized common carrier, a domestic or international record carrier, or a domestic satellite carrier engaged in providing services authorized by the U.S. Federal Communications Commission. *See also* **common carrier, divestiture, resale carrier, specialized common carrier.**

OUT: In voice communications systems, the procedure word used to indicate that (a) the transmission is ended, i.e., the message is ended, (b) a response is neither requested nor expected, and (c) the calling station, i.e., the call originator, may not be ready to receive a response. *See* **time out.** *See also* **procedure word.**

outage: 1. A communications system operational or service condition in which (a) a user is completely deprived of service by the system, and (b) for a particular system or a given situation, the system service condition is below a defined threshold of acceptable performance. **2.** In a system, the operational condition of a component that cannot perform the function it is assigned or the function for which it was designed and installed. *See also* **continuous operation, degraded service state, failure, operational service state, outage state, performance measurement period, performance parameter.**

outage duration: *Synonym* **outage period.**

outage period: 1. The period between (a) the instant that an outage starts, and (b) the instant that restored service starts. **2.** In a communications system, a performance measurement period, or a succession of performance measurement periods, that (a) is the period during which a communications service remains in the outage state, (b) begins at the beginning of the performance measurement period in which the communications service enters the outage state and ends either (i) at the beginning of the performance measurement period in which the communications service leaves the outage state or (ii) on expiration of the service time interval, whichever comes first, and (c) may occur on equipment other than communications equipment, such as computing, data processing, control, and radar equipment. *Synonym* **outage duration.** *See also* **performance measurement period.**

outage probability: The probability that the outage state will occur within a specified period. *Note:* The outage probability is indicated by the outage ratio. *See also* **outage ratio, performance measurement period.**

outage ratio (OR): The ratio of (a) the sum of all the outage periods during a performance measurement period to (b) the performance measurement period. *Note:* The outage ratio is an indicator of the outage probability. *See also* **performance measurement period.**

outages: *See* **mean time between outages.**

outage state: In communications, computer, data processing, and control systems, a system condition in which (a) a user is completely deprived of service or use regardless of the cause of the outage, and (b) the system is in neither the operational service state nor the degraded service state. *See also* **degraded service state, operational service state, outage.**

outage threshold: *Synonym* **system operational threshold.**

outdialing: *See* **network outdialing.**

outer space: The space that is beyond the Earth's atmosphere. *See also* **deep space.**

outgoing access: 1. Access by a user in one network to a user in another network via the networks to which they are connected. **2.** The capability of a user in one network of communicating with a user in another network via the networks to which they are connected. *See* **closed user group with outgoing access.** *See also* **access, gateway, network, port.**

out-of-band emission (OBE): Emission at frequencies immediately outside the necessary bandwidth, including frequencies resulting from the modulation process, but excluding spurious emissions. *See also* **emission.**

out-of-band signaling (OBS): 1. Signaling via a different channel from that used for primary information transfer. *Note:* Either frequency division multiplexing (FDM) or time division multiplexing (TDM) may be used for creating channels for out-of-band signaling. **2.** Signaling using a portion of the channel bandwidth provided by the propagation medium, such as the carrier channel, but denied to the speech or intelligence path by filters. *Note:* This mode of signaling results in a reduction of the effective available bandwidth. **3.** Signaling in which bits or bandwidths over and above those allocated for normal traffic are used, although the transmissions are still within the established frequency channels for frequency division multiplexing or time slots for time division multiplexing. *Note:* Examples of out-of-band signaling are signaling in which (a) frequencies that are within the guard band between channels are used, (b) bits that are other than information bits are used in a digital system, and (c) only a portion of the channel bandwidth is provided by the propagation medium, such as the carrier channel, and the speech or intelligence path is denied by filters, resulting in a reduction of the effective bandwidth. *See also* **bandwidth, channel, channel-associated signaling, common channel signaling, effective bandwidth, frequency, in-band signaling, propagation medium, retrieval, signal, signaling.**

out-of-frame alignment time: In a data transmission system, the period that consists of (a) the time during which frame alignment is effectively lost, (b) the time needed to detect the loss of frame alignment, and (c) the reframing time. *See also* **frame, frame alignment, reframing time.**

out of order signal: In a communications system, a signal that is sent in the backward direction indicating that a call cannot be completed because the called terminal or the called terminal access line (a) is in the outage state, (b) has been taken out of service, or (c) has a fault that renders it unserviceable. *See also* **backward direction, outage.**

out-of-phase component life: In a given system, the life of a component that cannot be placed into the maintenance or service cycle with the components that have an in-phase component life. *See also* **in-phase component life.**

outpulsing: The transmitting of digital address information over a trunk from one switching center or switchboard to another. *See also* **pulse, pulse train, pulsing, system signaling and supervision.**

output: 1. In an entity, such as a circuit, component, device, equipment, subsystem, system, terminal, or channel, from or through which data are emitted or extracted. **2.** Data retrieved, extracted, or removed from a functional unit, such as an amplifier, repeater, circuit, network, or computer, usually after some processing. *Note:* The output may be a state, or a sequence of states, of a terminal on a functional unit, the states representing the information retrieved, extracted, or removed from the functional unit. **3.** A terminal or a channel used to retrieve, extract, or remove data from a functional unit. **4.** In a computer, pertaining to a device, process, or channel used in the extraction of data produced by the computer or by any of its components. **5.** That which is extracted from an entity. *Note:* Examples of output are (a) the electric current, voltage, or power extracted from a circuit, (b) the optical power emitted by a laser, (c) the sound energy emitted by a speaker or a loudspeaker, (d) the data transferred from internal storage to external storage, (e) data that have been processed and removed from the processor, (f) the state or the sequence of states or signals that occur on a channel at the receiving end, (g) a device or a collective set of devices that are used for taking data out of another device, (h) the channel that indicates the result of an operation by a combinational circuit, such as an AND gate, and (i) the data obtained from a transmitter or from the end of a communications line, channel, or link. *See* **optical power output, power output.** *See also* **input.**

output angle: *Synonym* **radiation angle.**

output aperture: In an optical system, the aperture at the point of exit from the system, usually equal to the aperture of the final optical component, such as an eyepiece lens or the end of an optical fiber, from which the image or the exit beam is projected for viewing or for input to another device. *Note:* The output aperture of an optical fiber can be measured by projecting the exiting beam on a screen and taking the ratio of (a) half the beam diameter at the screen to (b) the distance to the screen. *See also* **antenna aperture, aperture, input aperture.**

output area: The set of storage locations reserved or used for output data.

output data: 1. Data that are obtained or are to be obtained from a device. **2.** Data that have been processed and are extracted from a device, such as a computer or a data processing system. *See also* **input data.**

output device: *Synonym* **output unit.**

output power: *See* **rated output power.**

output program: A computer program that organizes and controls the output process of a communications, computer, data processing, control, or telemetering system.

output rating: 1. The power that is available at the output terminals of a device, such as a transmitter, when connected to the normal load or its equivalent. **2.** Under specified ambient conditions, the power that can be delivered by a device without overheating (a) over a long period of continuous operation or (b) over a specified operational duty cycle, i.e., operation with specified on and off periods. *See also* **output, peak power output, power, rated output power.**

output unit: In a system, a functional unit that (a) obtains output from the system, such as a communications, computer, data processing, or control system, and (b) furnishes the output in some form to another entity. *Note:* Examples of communications output units are printers, display devices, loudspeakers, and facsimile transmitters. *Synonym* **output device.** *See* **input-output unit.** *See also* **input unit, output.**

outside call: *Synonym* **external call.**

outside plant: 1. In telephone systems, all equipment located between (a) the main distribution frame of a switching facility and (b) either (i) the entrance to another switching facility or (ii) the premises of the serviced user equipment. *Note:* Examples of outside plant equipment include all cables, poles, ducts, conduits, wires, repeaters, and load coils. **2.** In telephony, all equipment, such as cables, conduits, ducts, poles, towers, repeaters, and repeater huts, between (a) a demarcation point in a switching facility and (b) a demarcation point in another switching facility or at a customer premises. *Note:* Demarcation points usually are easily identifiable and accessible points, such as distribution frame terminals, cable heads, or microwave transmitter terminals. **3.** In U.S. Department of Defense (DoD) communications systems, the portion of intrabase communications equipment extending from the main distribution frame outward to user end instruments or to the terminal connections for these instruments. *See also* **access point, aerial cable, direct buried cable, facility, inside plant, underground cable.**

outside vapor deposition (OVD): In the production of optical fibers, (a) the depositing of materials in vapor or soot form on the outside of optical fiber blanks and (b) the sintering of the blanks to make optical fiber preforms with the desired refractive index profile. *Note:* Optical fibers are then pulled from the preforms, buffered, and wound on spools, reels, or bundpacks for subsequent cabling. *See also* **cabling, optical fiber, optical fiber preform, optical fiber pulling machine.**

outside vapor phase oxidation (OVPO) process: A chemical vapor phase oxidation (CVPO) process for the production of optical fibers, in which the soot stream and the heating flame are deposited on the outside surface of a rotating glass rod. *See also* **chemical vapor phase oxidation process, inside vapor phase oxidation process.**

out-slot signaling: Signaling performed in digit time slots that are not within the channel time slot. *See also*

bit robbing, channel time slot, in-slot signaling, signaling, signaling time slot.

out station: In a radio net, a station that (a) is listening and not transmitting and (b) usually responds to a call in the alphabetical order of the call signs of the stations in the net. *See also* **radio net.**

out tuning: *See* **jammer out tuning.**

outward dialing: *See* **automatic identified outward dialing, direct outward dialing, network outward dialing.**

outward dialing lead: *See* **automatic identified outward dialing lead.**

ovality: 1. The attribute of an optical fiber in which the cross section of the core or the cladding deviates from a perfect circle. **2.** In an optical fiber, the degree of deviation, from perfect circularity, of the cross section of the core or the cladding. *Note 1:* As a first approximation, the fiber cross section is assumed to be an ellipse. The ovality may be calculated from the relation $2(a - b)/(a + b)$, where a is the major axis length, and b is the minor axis length. If $a = b$, the cross section is a circle, and the ovality is 0. The ovality may be expressed in percent. *Note 2:* Ovality may also be calculated by the use of tolerance fields for the core and for the cladding, each consisting of two concentric circles, within which the cross-section boundaries must lie. *Synonym* **noncircularity.** *See also* **cladding noncircularity, concentricity error, core noncircularity, tolerance field.**

OVD: optical video disk, outside vapor deposition.

OVER: In voice communications systems, the procedure word used to indicate that (a) the transmission is ended, i.e., the message is ended, (b) a response is requested or expected, (c) the calling station, i.e., the call originator, is ready to receive the response, and (d) the called station, i.e., the call receiver, should transmit immediately. *See also* **procedure word.**

overall length: *See* **borescope overall length.**

overarmor: An additional cover, jacket, or sheath placed over a cable, such as a fiber optic cable, to (a) provide additional strength and protection against harsh environments and (b) allow special uses, such as (i) a tether cable from ship to a buoy or ship to ship, (ii) a shore-to-ocean interface, (iii) a ground cable where abrasive abuse is expected, and (iv) a riser from ground to pole top in an aerial insert. *Note 1:* An additional cover or jacket may be placed over the overarmor for ease of handling and marking. *Note 2:* Overarmor usually is braided for flexibility and usually is made of metal or a high strength polymer. *Note 3:* Gel-filled smooth metal tubing that serves as the jacket over loose optical fibers is not considered overarmor. *Refer to* **Fig. C-1 (102).**

overcoat: In a buffered optical fiber, an additional protective coating that (a) is placed over the buffer, i.e., over the undercoat, and (b) usually is applied by spraying or dipping. *See also* **overarmor.**

overcompensated optical fiber: An optical fiber with a refractive index profile adjusted so that the high order propagating modes arrive ahead of the low order modes. *Note:* In an uncompensated optical fiber, the higher order modes arrive after the lower order modes. *Note:* Higher order modes have higher eigenvalues in the solution of the wave equations, higher frequencies, and shorter wavelengths than lower order modes. *See also* **compensated optical fiber, undercompensated optical fiber.**

overdue report: 1. In communications systems, a report that was expected by a listening station from a mobile station at an earlier specified time, but has not yet been received. **2.** A report that was expected by a listening station at a time when a mobile station was expected to reach a specified position, but, that time having passed, has not yet been received. *See also* **listening station, mobile station.**

overfill: In coupling optical power from a source, such as a laser, a light-emitting diode (LED), or an optical fiber, to a receiver, such as another fiber or a photodetector, the situation that occurs when the useful diameter of the receiver, such as the core of a fiber or the sensitive surface of a photodetector, is smaller than the diameter, at the surface of the receiver, of the cone of light emanating from the source. *Note 1:* In coupling optical fibers, overfill occurs when (a) the spot of light incident on the endface of the fiber is larger than the core diameter, (b) the launch angle of the source exceeds the acceptance angle of the fiber, or (c) both. *Note 2:* In fiber optic communications systems testing, overfill in both numerical aperture (NA) and mean diameter, i.e., core diameter and spot size, usually is required in order to make proper comparisons among components. *See also* **fill, underfill.**

overflow: 1. The generation of potential traffic that exceeds the capacity of a communications system or subsystem. **2.** A count of telephone call access attempts made on groups of busy trunks or access lines. **3.** Traffic handled by auxiliary equipment. **4.** Traffic that exceeds the capacity of the switching equipment and is therefore lost. **5.** On a particular route, excess traffic that is offered to another route, i.e., an alternate route. **6.** In computers and data processing systems, the part of a given set of data, such as a sum or a set of characters, that exceeds the capacity of the location or cell designated for its storage, or pertaining to the condition that caused the excess data. *Note:* In many systems, arrangements are made to capture and store overflow in a special storage area. *See* **arithmetic overflow, traffic overflow.** *See also* **high usage trunk group, storage area.**

overflow indicator: An indicator, such as an audio or visual signal, indicating that an overflow has occurred. *Note:* Examples of overflow indicators are (a) a visual display on the display register of a calculator indicating that an overflow has occurred, (b) a statement displayed on the screen of a monitor indicating that the last line has been printed on a page, and another sheet of paper must be inserted into the printer, and (c) a visual signal on a switchboard, coupled with an audio signal, indicating that a trunk is saturated. *See also* **overflow, saturation.**

overflow register: 1. In a telephone system, a device that records the number of access attempts that are made to contact busy lines or busy trunks. **2.** In a computer, a register in which an overflow may be stored, should one occur. *See also* **access attempt, busy.**

overhead bit: 1. In a message, a bit other than a user information bit. **2.** A bit that (a) is contained in a message, (b) is not an information bit, and (c) may be part of user overhead or part of system overhead. *Note:* Examples of overhead bits are check bits, stuffing bits, and communications system control bits. *Synonym* **overhead communications bit.** *See* **delivered overhead bit.** *See also* **binary digit, front end processing, maximum stuffing rate, overhead information, service bit, system overhead, user information bit, user overhead.**

overhead block: *See* **delivered overhead block.**

overhead communications bit: *Synonym* **overhead bit.**

overhead information: In a digital communications system, information that (a) is transferred across the functional interface between (i) a user and the system, (ii) a user end instrument and the system, and (iii) any two functional units, (b) may be system overhead information or user over head information, and (c) is used for directing or controlling the transfer of user information. *Note:* Overhead information that is originated by the user is user overhead information and is not system overhead information. Overhead information that is generated by or within the system and not delivered to the user is system overhead information. Thus, the user throughput is reduced by both overheads, while system throughput is reduced only by system overhead. *See also* **block transfer efficiency, check bit, check character, check digit, delivered overhead bit, delivered overhead block, heading, overhead bit, service bit, system overhead information, user overhead information.**

overhead rate: In a transmission system, the bit rate at an interface corresponding to the bits required to operate the system, usually consisting of bits for error detection and correction, synchronization, buffer stuffing, control, and other system purposes. *Note 1:* The interface may be between a user end instrument and the system or between functional units within the system. *Note 2:* Overhead bits representing information generated by the user are not considered as system overhead bits, whereas overhead bits generated within the system and not delivered to the user are considered as system overhead bits. Thus, user throughput is reduced by both overheads, while system throughput is reduced only by system overhead. *Note 3:* The interface rate is the payload rate plus the system overhead rate. *See also* **interface rate, overhead bit, payload rate, system overhead information, user overhead information.**

overlap: In facsimile transmission systems, a defect in reproduction by a facsimile receiver when the width of the scanning line is greater than the scanning pitch. *See also* **scanning line, scanning pitch.**

overlap tell: In radar systems, the transfer of data to an adjacent radar facility concerning objects, i.e., concerning targets, that are detected in the area of responsibility of the adjacent radar facility.

overlay: 1. In a computer program, a segment of the program that is not permanently stored in internal or main storage. **2.** During the execution of a computer program, one of several segments of the program that occupies the same area of main storage, one segment at a time. **3.** To repeatedly use the same areas of internal storage during different stages of the execution of a computer program. **4.** In the execution of a computer program, to load part of the computer program in a storage area that was occupied by parts of the program that are not currently needed or are no longer needed. **4.** A transparent film used to superimpose data on a map. *Note:* The data on the overlay might indicate the location of communications facilities, supply depots, or military units. *See* **form overlay, radarscope overlay.** *See also* **computer program, internal storage, patching.**

overlay path: All the segments of a particular computer program in a given overlay structure. *See also* **computer program, overlay segment, overlay structure.**

overlay program: A computer program that has two or more control segments that can be stored in the same storage area, i.e., the same overlay region, at different times during execution of the program. *See also* **computer program, overlay region, program segment.**

overlay region: The area in main storage in which computer programs or overlay segments can be stored independently of overlay paths in other areas, such that only one path within the area can be in internal or main storage at a time. *See also* **computer program, internal storage, overlay path, overlay segment.**

overlay segment: A portion of a computer program that may be executed without the entire computer program necessarily being stored in internal or main storage at any one time. *See also* **computer program, internal storage, main storage.**

overlay structure: A graphic representation that shows all the segments of an overlay program and how the overlay segments are arranged to use the same overlay region, i.e., a designated storage area in internal or main storage, at different times. *Synonym* **overlay tree.** *See also* **overlay program, overlay region, program segment, storage area.**

overlay tree: *Synonym* **overlay structure.**

overload: A load, placed on a device, that is greater than the device is capable of handling, so that the device cannot perform the functions for which it was designed. *Note:* Examples of overloads are (a) traffic entered into a communications system that is greater than the traffic capacity of the system, (b) for analog inputs, voltage levels above which an analog to digital converter cannot distinguish a change, and (c) in electrical circuits, an electrical current that will result in damage from overheating.

overload point: *Synonym* **load capacity.**

overmodulation: 1. Modulation in which the instantaneous level of the modulating signal exceeds the value necessary to produce 100% modulation of the carrier, resulting in distortion of the modulator output signal and distortion of the recovered modulating signal. **2.** Modulation in which the mean level of the modulating signal is such that the peak value of the modulating signal exceeds the value necessary to produce 100% modulation of the carrier, resulting in distortion of the modulator output signal and distortion of the recovered modulating signal. *See also* **distortion, modulation, percentage modulation.**

over-over mode: *Synonym* **two-way alternate communication.**

override: 1. For a person or another system, to take control of a system, such as to manually take control of an automatic system. **2.** For a user other than the current user of the circuit, to enter or seize a busy circuit, i.e., an occupied circuit. *Note:* Examples of override are (a) the seizure of a circuit by an attendant after a busy verification, (b) the seizure of a circuit by a user with a higher precedence level than the current user of the circuit, (c) a condition in which a third user enters or seizes an active circuit, and (d) manually taking control of an automatic switch. *See also* **flash override precedence, precedence, seizing.**

override precedence: *See* **flash override precedence.**

overrun: 1. The loss of data that occurs when a receiver is unable to accept the data at the arriving data signaling rate (DSR). **2.** In systems management, the situation that occurs when (a) an actual cost exceeds the projected cost, or (b) the spending rate exceeds the available rate.

overshoot: 1. In the transition of any parameter from one value to another, the temporary value of the parameter that exceeds the final or target value. *Note:* Overshoot occurs when the transition is from a lower value to a higher value. When the transition is from a higher value to a lower value, the situation is called "undershoot." **2.** In an amplifier, the increased amplitude of a portion of a nonsinusoidal wave caused by the particular characteristics of the circuit. *Note 1:* Overshoot may be used to decrease the response time of a signal by decreasing the rise time or decreasing the fall time, or both. Overshoot causes distortion of the signal. *Note 2:* Overshoot and undershoot may be removed, i.e., may be damped, in various ways, such as by using limiting circuits, clippers, and circuit elements that filter overshoot and undershoot oscillation, i.e., ringing, that tends to occur at the natural frequency of the circuit whenever a signal level transition to another significant condition occurs. *See also* **response time. 3.** In radio transmission, the result of an unusual ionospheric or atmospheric condition that causes radio signals to be received where they are not intended. *See also* **ducting, natural frequency, undershoot.**

over-the-air rekeying: Changing traffic encryption key or transmission security key in remote cryptoequipment by sending new key directly to the remote cryptoequipment over the communications path it secures. *See also* **encryption, key.**

over-the-horizon radar: A radar system that (a) makes use of atmospheric reflection and refraction to extend its range of detection beyond the range of the line of sight, and (b) may be either a forward scatter or a backscatter system.

overtime: *Synonym* **overtime period.**

overtime period: 1. On a time-charged telephone call, the period that (a) begins at the end of a specified period, such as 1 min (minute), after the call receiver goes off-hook, i.e., the call is answered, and (b) ends when either the call originator or the call receiver goes on-hook (hangs up), i.e., when the call is finished. *Note:* The specified period is usually one minute, in which case the overtime period is the period that begins one minute after the call receiver goes off-hook and ends when the call originator or the call receiver goes on-hook. **2.** On a time-charged telephone call, a length of time equal to the length of the call minus a specified period, such as one minute, in which case the overtime period is the length of the call minus one minute. **3.** In Wide Area Telephone Service (WATS), pertaining to the occupancy time of circuits in excess of the time, in each period, specified in a contract. *Synonym* **overtime.** *See also* **off-hook, on-hook, tariff.**

overtone: Of a sinusoidal wave, an integral multiple of the frequency of the wave, other than the fundamental itself. *Note 1:* In counting or labeling overtones, (a) the first overtone is twice the frequency of the fundamental and thus corresponds to the second harmonic, (b) the second overtone is three times the frequency of the fundamental and thus corresponds to the third harmonic, and (c) so on for the higher overtones. *Note 2:* Use of the term "overtone" usually is confined to acoustic waves, especially in applications related to music, rather than communications systems. *Note 3:* When referring to electrical and electromagnetic waves, the term "harmonic" usually is used, rather than "overtone." *See also* **harmonic.** *Refer to* **Fig. A-10 (49).**

overtone absorption: *See* **hydroxyl ion overtone absorption.**

overwrite: To store or record data in a storage location or data medium in such a manner that the data that had been stored or recorded there are lost. *See also* **data medium, storage location.**

OVPO: outside vapor phase oxidation.

OVPO process: *See* **outside vapor phase oxidation process.**

oxidation process: *See* **axial vapor phase oxidation process, chemical vapor phase oxidation process, inside vapor phase oxidation process, modified inside vapor phase oxidation process, outside vapor phase oxidation process.**

PA: personal agent, personal assistant.

PABX: private automatic branch exchange.

pacing: Communications systems operation in which the receiving station controls the effective data transmission rate of the sending station. *Note:* Pacing is used to prevent undesirable system conditions, such as overrun, congestion, saturation, collision, and contention.

pack: *See* **disk pack.**

package: *See* **dual inline package, program package, software package.**

package switch: *See* **dual inline package switch.**

packet: **1.** In data communications, a sequence of binary digits that (a) includes data, control signals, and possibly error control signals, (b) is transmitted and switched as a composite whole, (c) is arranged in a specific format, such as a header part and a data part, (d) may consist of several messages or may be part of a single message, (e) is used in asynchronous switched systems, and (f) usually is dedicated to one user for one session. **2.** A group of frequencies or pulses, such as those that compose a photon, group of photons, group of phonons, or high frequency sonic pulse. *See* **encapsulated packet, multicast packet, trace packet, unicast packet.** *See also* **binary digit, burst switching, format, packet switching, protocol, protocol data unit.**

packet assembler-disassembler (PAD): A functional unit that (a) enables data terminal equipment (DTE) not equipped for packet switching to access a packet-switched network, and (b) may be colocated with data terminal equipment (DTE). *See* **packet assembly-disassembly facility, packet-switched data transmission service.**

packet assembly-disassembly facility: A facility that (a) is accessible to users, (b) enables data terminal equipment (DTE) not equipped for the packet-switching mode of operation to access a packet-switched network, and (c) is placed at one or more locations within the network. *See also* **packet assembler-disassembler, packet-switched data transmission service.**

packet exchange: *See* **Internetwork Packet Exchange.**

packet flow control: In packet-switching communications systems, the control of the transfer rate of data in packets between two specified points in a data network. *Note:* An example of packet flow control is the control of the data transfer rate between data terminal equipment (DTE) and a data switching exchange (DSE) or between two DTEs. *See also* **packet switching, data transfer rate.**

packet format: The structure of user data and transmission control data in a packet. *Note:* The size and the content of the various fields in a packet are defined by a set of rules that are used to construct the packet. *See also* **format, protocol.**

packet internet groper (ping): **1.** In Transmission Control Protocol/Internet Protocol (TCP/IP), a protocol function that tests the ability of a computer to communicate with a remote computer by sending a query and receiving a confirmation response. **2.** In Transmission Control Protocol/Internet Protocol (TCP/IP), to use a protocol function that tests the ability of a computer to communicate with a remote computer by sending a query and receiving a confirmation response.

packet length selection: A user facility in which data terminal equipment (DTE) is used to select a certain maximum user data field length from a specified set of fixed lengths. *See also* **data, data terminal equipment, field.**

packet mode (PM): A mode of operating a communications network in which packet switching is used rather than message switching.

packet mode operation: A mode of operating a communications network in which packet switching is used. *See also* **communications network, packet switching.**

packet mode terminal: Data terminal equipment (DTE) that can control, format, transmit, and receive packets. *See also* **mode, packet switching, terminal.**

packet sequencing: Controlling the order in which packets are processed by a communications system. *Note:* An example of packet sequencing is ensuring that packets are delivered to the receiving data terminal equipment (DTE) in the same sequence in which they were received by the sending DTE, or in the same sequence in which they were transmitted by the sending DTE. *See also* **packet, packet switching.**

packet switch: In a data network that provides packet-switched data transmission service, the collection of hardware and software that (a) is used to implement network procedures, such as routing, resource allocation, and error control, and (b) provides access to network switching services via a network host interface. *See also* **data network, error control, host interface, packet switching, routing.** *Refer to* **Fig. C-7 (157).**

packet-switched data transmission service (PSDTS): A service that (a) provides for the transmission of data in the form of packets and (b) may provide for the assembly and the disassembly of data packets. *See also* **data, data transmission.**

packet switching: The routing and the transferring of data by means of addressed packets so that (a) a channel is occupied only during the transmission of the packet, and (b) upon completion of the transmission of the packet, the channel is made available for the transfer of other traffic. *Note 1:* A packet may contain several messages, or abnormally long messages may require several packets. Messages are user units. Packets are communications system units. Hence, they can overlap each other or be made up of each other. *Note 2:* Packet mode operation potentially increases the channel traffic capacity. *Note 3:* In some communications networks, the data may be formatted into one or more packets by data terminal equipment (DTE) or by other equipment within the network, such as a packet assembler/disassembler (PAD). The packets are formed for various purposes, such as transmission and multiplexing. *See* **fast packet switching.** *See also* **channel, circuit switching, compelled signaling, connectionless mode transmission, data, interface message processor, message switching, packet, packet mode terminal, packet assembler/disassembler, packet switching network, public switched network, switching system.**

packet switching network (PSN): A network that transmits data in the form of packets. *Note:* The packet transmitted by the network may require format conversion at a gateway. *See also* **burst switching, data, packet, packet switching.**

packet switching node: In a packet switching network, a node at which data switches and equipment for controlling, formatting, transmitting, routing, and receiving data packets are located. *Note:* In the Defense Data Network (DDN), the equipment at a packet switching node usually is configured to support up to thirty-two X.25 56-kbps (kilobits per second) host connections, as many as six 56-kbps interswitch trunk (IST) lines to other packet switching nodes, and at least one Terminal Access Controller (TAC). *See also* **message switching node, packet switching.**

packet transfer mode (PTM): A mode of information transfer that allows dynamic sharing of network resources among many users by means of packet transmission and packet switching. *See also* **connection.**

packing: *See* **channel packing.**

packing density (PD): 1. In data storage systems, the number of storage cells per unit length, area, or volume of storage media. *Note 1:* Examples of packing density are (a) the number of bits or characters stored per unit length, such as bits per inch (bpi or b/in.) of magnetic tape, and (b) the number of bits stored per unit length of track on an optical disk. *Note 2:* A storage cell usually contains a single data unit, such as a bit, byte, character, word, block, or frame. *Note 3:* In a linear arrangement of entities, the packing density is numerically equal to the reciprocal of the pitch. *Synonym* **recording density. 2.** In a fiber optic bundle,

(a) the end cross-sectional total core area of all the optical fibers in the bundle divided by (b) the total cross-sectional area of the bundle, usually within the ferrule, including jacket, strength member, buffer, core, cladding, and interstitial areas, i.e., including the total cross section within the outer jacket and the jacket itself. *Note:* In a fiber optic bundle, the packing density varies with various factors, such as the size of the optical fibers, the core areas relative to the total fiber area, and the pressure applied to the fiber optic bundle. *See also* **bit density, packing fraction, pitch.**

packing fraction (PF): In a fiber optic bundle, the ratio of (a) the aggregate fiber cross-sectional core area to (b) the total cross-sectional area of the bundle. *Note 1:* For round optical fibers tightly packed in a round bundle, the packing fraction is given by the relation $PF = N(d/D)^2$ where PF is the packing fraction, N is the number of fibers, a is the diameter of the core, and D is the diameter of the bundle or the whole assembly, including jacketing. *Note 2:* The packing fraction varies with many factors, such as the overall size of the optical fibers, core cross-sectional areas relative to fiber cross-sectional areas, the geometric or spatial distribution of the fibers, and the tightness of packing. *See also* **core, fiber optics.**

packing fraction loss: The optical power loss incurred because the packing fraction is less than unity. *Note:* The packing fraction loss usually is expressed in dB. *See also* **fiber optics.**

PACVD: plasma activated chemical vapor deposition.

PACVD process: plasma activated chemical vapor deposition process.

pad: 1. A device, consisting of fixed resistors, that attenuates signals by a fixed amount with negligible distortion. *Note:* The pad is called an "attenuator" if the resistance is adjustable. *See* **L pad.** *See also* **attenuation, fixed attenuator, network. 2.** To character fill, i.e., to add null characters, usually zeros or blank spaces, to a data unit, such as a word, block, file, or record, to accomplish a specific purpose, such as to obtain constant block lengths. *See also* **character fill.**

PAD: packet assembler-disassembler.

padding: 1. In a communications system, the appending of redundant bits to a bit stream in order to accomplish a specific purpose, such as to bring the bit rate up to a specified value or to maintain an idle state when there are no data to transmit. **2.** In a communications system, the extraneous text that is added to a message for a specific purpose, such as to conceal its beginning, ending, or length. *See also* **bit rate, bit stream.**

page: 1. In tape relay communications systems, a single sheet of paper that contains, or has the capacity to contain, not more than a prescribed number of lines of printed material, such as 20 lines. **2.** In a storage system for a computer, a fixed-length block of data that has an address and that can be transferred as a unit between storage units or to and from other units and storage. **3.** In computer systems, a fixed number of lines of text identified by data delimiters. **4.** In a document, one side of a sheet that (a) usually is made of paper, (b) usually is bound or associated with other sheets, (b) has a fixed size, and (d) may be numbered. *See also* **data delimiter.**

page frame: 1. A storage area that is capable of holding a volume of data equivalent to the contents of one page. **2.** A frame the size of a page. *See also* **frame, storage area.**

page key: On a standard keyboard, a key used as a control device to process one page at a time, such as to print a page, display the next page, display the previous page, store a page, or retrieve a page. *See also* **key.**

page printer: A device that prints an array of characters that are arranged on a page.

pager: A device used to obtain the attention of a person in a certain area or within a group of persons. *See* **radio pager.** *See also* **paging.**

pages: *See* **white pages.**

page swapping: Transferring a page from main storage to auxiliary storage and vice versa.

page turning: *Synonym in this sense* **paging.**

paging: 1. A one-way communications service, from a base station to mobile or fixed receivers, that provides signaling or information transfer by various means, such as tone, tone-voice, tactile, and visual readout. *See* **radio paging.** *See also* **radio pager, signaling. 2.** Obtaining the attention of a person in a given area or within a group of persons. *See also* **page, pager. 3.** A time-sharing arrangement in which pages are transferred from main storage to auxiliary storage and vice versa. **4.** The transfer of pages between main and auxiliary storage. *Synonym in this sense* **page turning. 5.** Storage allocation in which the storage is divided into page frames. *See also* **auxiliary storage, main storage, page frame.**

paging rule: 1. In tape relay communications systems, a rule that governs the identification and the formatting of messages that consist of two or more pages.

2. In computer systems, a rule that pertains to the transfer of pages between storage units, between storage and other units, or between other units. *See also* **page.**

pain threshold: In audiometry, the sound amplitude or frequency above which the sound causes a person to feel pain in the ear receiving the sound. *See also* **audiometry, level, sound.**

pair: *See* **Darlington pair, shielded pair, shielded twisted pair, symmetrical pair, twisted pair.**

paired cable: A cable that (a) is made up of one or more separately insulated twisted pairs of conductors, i.e., of lines, and (b) is arranged so that the pairs are not arranged with others to form quads. *Synonym* **twisted pair cable.** *See also* **cable, quadded cable, shielded pair, twisted pair.**

paired disparity code: A code in which some or all of the characters are represented by two sets of digits of opposite disparity that are used in sequence so as to minimize the total disparity of a longer sequence of digits. *Note 1:* An alternate mark inversion signal is an implementation of a paired disparity code. *Note 2:* The digits may be represented by disparate physical quantities, such as two different frequencies, phases, voltage levels, magnetic polarities, or electrical polarities, each one of the pair representing a 0 or a 1. *Synonym* **alternating code.** *See also* **alternate mark inversion signal, disparity, disparity code.**

paired lobing: In radar systems, an extension of sequential lobing in which radar echo signals that are received by the upper pair or the right-hand pair of four horns symmetrically placed about the radar antenna axis are combined and compared with the combined signals that are received by the lower pair or the left-hand pair of horns, respectively. *See also* **conical scanning, radar echo, sequential lobing.**

pair gain: Pertaining to a user transmission system that serves a number of users with a smaller number of wire pairs by using concentrating equipment, multiplexing equipment, or both. *See also* **concentrating equipment, multiplexing.**

pair gain system: A transmission system that uses concentrators or multiplexers so that fewer wire pairs can provide service to a given number of customers. *See also* **concentrator, multiplexing.**

pairing: *See* **bit pairing.**

PAL: phase alternation by line.

PAL-M: phase alternation by line (modified).

palmtop computer: A personal computer that (a) is small enough to be held in one hand and operated with the other hand, (b) usually is used for on-the-spot data entry, such as point-of-sale data entry, logging crop yields on the fly, and executing selected functions, (c) usually is internal-battery- or solar-powered, and (d) may be pocketed when not in use. *See also* **laptop computer.**

PAM: pulse amplitude modulation.

PAMA: pulse-address multiple access.

pancratic lens: *Synonym* **zoom lens.**

panel: 1. In visual signaling systems, a piece of background-contrasting material, such as cloth, canvas, paper, plastic, wood, grass, sod, or sand, that (a) should be at least 2 ft by 6 ft, (b) is laid or created on the ground, (c) is of contrasting shape or color, (d) is used for sending messages to aircraft, (e) may be used for position marking or signaling, and (f) usually is displayed in groups in accordance with a prearranged code to convey messages. **2.** In communications, computer, data processing, and control systems, a surface on which items may be mounted. *Note:* Examples of panels are switchboards, instrument boards, and consoles. *See* **control panel, fiber optic patch panel, gas panel, maintenance panel, marking panel, optical fiber distribution panel, patch panel, touch panel.** *See also* **panel signaling, signaling panel code.**

panel code: A prearranged code that makes use of panels that are designed for visual communications from ground to aircraft. *Synonym* **surface code.** *See* **panel, signaling panel code.**

panel display: *See* **flat panel display.**

panel signaling: In visual communications systems, signaling by means of panels that are laid flat on the ground for communications with aircraft by means of a limited prearranged code. *Note:* Panels may be made of flexible strips, pieces of wood, or markings on snow, ice, earth, grass, or sand. A standard panel that is used for ground-to-air signaling is a 2-ft by 6-ft rectangle, usually made of cloth or plastic. *See also* **panel, panel code.**

panel signaling indicator: In a standard panel signaling system, one of the four standard panel displays, namely, the index flash, the letter indicator, the Morse code indicator, and the figure indicator, i.e., the numeral indicator, which is used to indicate the kind of character that is being represented by the remainder of the display in which the indicator occurs. *See also* **figure indicator, index flash, letter indicator, Morse**

code indicator, panel, panel signaling, signaling panel code.

panel signaling sign: A panel display that is used to indicate prearranged messages from the ground to aircraft. *See* **special panel signaling sign.** *See also* **panel signaling vocabulary.**

panel signaling vocabulary: In visual communications systems, a set of correspondences between (a) numerals, i.e., figures, and (b) messages that may be sent from ground stations to aircraft stations by means of panel arrangements, which are to be read as numerals that correspond to the messages being sent. *Note:* An example of the use of the panel signaling vocabulary is, on the ground, to display the index flash alone and then the numeral panel that indicates 5. The aircraft station reads the numeral 5 and looks up the meaning of 5 in the prearranged vocabulary. The 5 might mean a request for medical aid. *See also* **index flash, panel, panel code, panel signaling indicator.**

panning: 1. On a computer monitor screen, to shift an entire display image laterally, i.e., horizontally. *Note:* The panning direction is orthogonal to the scrolling direction. **2.** In video technology, using a camera to horizontally scan objects. **3.** In antenna systems, successively changing the azimuth of a beam of radio frequency (rf) energy over the elements of a given horizontal region, or the corresponding process in reception.

panoramic adapter: In radar systems, an attachment that is designed to operate with a search receiver to provide a visual presentation, on an oscilloscope screen, of a band of frequencies that extends above and below the center frequency to which the search receiver is tuned.

panoramic indicator: In radar systems, auxiliary equipment that is used with a receiver and presents a visual indication of all signals that are contained within the frequency coverage of the associated receiver.

panoramic receiver: In radar systems, a receiver that (a) has a very wide frequency coverage, (b) has an integral or auxiliary panoramic indicator, (c) usually tunes automatically, (d) rapidly sweeps the frequency spectrum, (e) has a visual presentation of picked-up signals, (f) usually displays a spike or a pip on a frequency baseline, i.e., on the abscissa axis, and (g) usually is used for general search and monitoring of transmission, such as for jammer control. *See also* **automatic receiver, jammer.**

paper: *See* **sensitized paper.**

paper feed: The controlled movement of paper in a device, such as a facsimile transmitter, facsimile receiver, duplicating machine, printer, typewriter, calculator, or cash register.

paper jam: A condition in which a paper feed mechanism fails to function, causing the paper to be (a) caught, trapped, and held stationary when it should be moving, (b) possibly damaged, and (c) overprinted by successive lines.

paper throw: The movement of paper in a device, such as a facsimile transmitter, facsimile receiver, duplicating machine, printer, typewriter, calculator, or cash register, at an excessive line feed rate such that the designated normal line spacing or scanning pitch is exceeded.

par: peak to average ratio.

parabolic antenna: An antenna that consists of (a) a parabolic reflector and (b) a radiating or receiving element at or near the focus of the parabola. *Note:* If the reflector is in the shape of a paraboloid, the antenna is called a "paraboloidal antenna." Cylindrical paraboloids and partial paraboloids are also used in parabolic antennas. In one type of parabolic antenna, the reflector surface is parallel to the longitudinal axis of the radiating or receiving element. It has a parabolic cross section, i.e., the intersection of the reflector surface and a longitudinal plane is a parabola with the radiating element at the focus. It emits a thin beam that spreads out like a hand-held fan. Another type of parabolic antenna has a paraboloidal reflector of the shape that is generated by revolving a plane parabola about its longitudinal axis, producing a shape that looks like the nose of a jet plane, also with the radiating or receiving element at the focus. It emits a narrow pencil-like beam of circular transverse cross section, similar to that of a spotlight. *See also* **antenna.** *Refer to* **Fig. R-2 (799).**

parabolic index profile: *Synonym* **parabolic refractive index profile.**

parabolic orbit: An orbit that has the shape of a parabola and therefore is not a closed orbit. *Note:* Circular and elliptical orbits are closed orbits. Like a parabolic orbit, a hyperbolic orbit is also not closed. *See also* **hyperbolic orbit, orbit.**

parabolic refractive index profile: In round optical fibers, a refractive index profile in which the refractive index of the core is a quadratic function of the distance from the optical fiber axis, i.e., the variation is such that the refractive index within the core at any distance from the axis, r, is given by the relation $n_r = n_1(1 - br^2)^{1/2}$ where n_1 is the refractive index at $r = 0$, i.e., at the axis, r is the radial distance from the axis, and b is a constant given by the relation $b = 2\Delta/a^2$ where a is the value of r for which the refractive index becomes uniform, i.e., the radius of the core, and Δ is given by the relation $\Delta = (n_1^2 - n_2^2)2n_1^2$ where n_1 again is the refractive index at $r = 0$, i.e., at the axis, and n_2 is the refractive index at the outer edge of the core, i.e., at $r = a$. *Note:* For most optical fibers, the refractive indices n_1 and n_2 are nearly equal. Therefore, Δ is also given by the relation $\Delta = 1 - (n_2/n_1)$ where the parameters are as defined above. Because $n_r = n_2$ at $r = a$, the parabolic refractive index profile is also given by the equivalent relation $n_r = n_1(1 - 2\Delta(r/a)^2)^{1/2}$. Another form of the equation, approximately the same as those above, is given by the relation $n_r^2 = n_1^2(1 - 2\Delta(r/a)^2)$ where the parameters are as defined above. *Synonyms* **parabolic index profile, parabolic profile, quadratic profile, quadratic refractive index profile.** *See also* **graded index profile, linear refractive index profile, mode volume, multimode optical fiber, power law refractive index profile, profile parameter, radial refractive index profile, step index profile.** *Refer to* **Fig. P-1.**

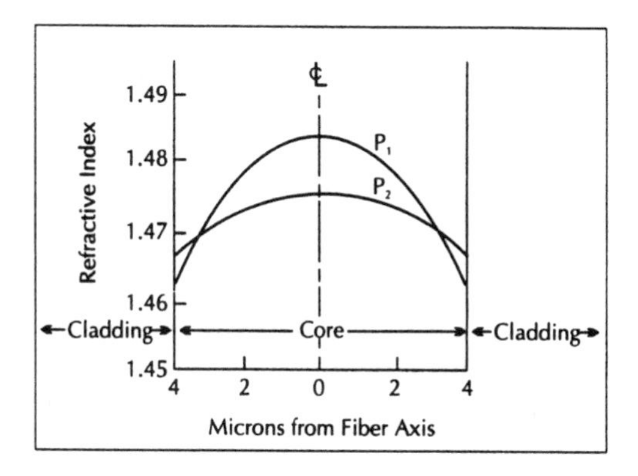

Fig. P-1. Parabolic refractive index profiles of two different optical fiber cores that have an 8-μm diameter.

paraboloidal mirror: A concave mirror (a) that has the form of a paraboloid of revolution, (b) that is free from chromatic aberration, and (c) in which all axially directed parallel light rays are focused at the focal point of the paraboloid without spherical aberration and, conversely, all light rays emanating from a source at the focal point are reflected as a bundle of parallel rays without spherical aberration, provided that they do not miss the mirror. *Note:* If a paraboloidal mirror consists of only a portion of a paraboloidal surface through which the axis does not pass, it is called an "off-axis paraboloidal mirror." *See* **off-axis paraboloidal mirror.**

parallel: *See* **declination parallel.**

parallel adder: A digital adder that concurrently adds both the digits in all the corresponding digit positions of the augend and the addend at one time, developing partial sums and carries, if any, in appropriate registers at digit positions where they are to be added again simultaneously at the next pulse cycle, again developing a partial sum and carries, if any, until no more carries are generated, i.e., the carry register reaches zero, in which case the register that has been holding the partial sums now holds the final arithmetic sum, the addition operation is completed, and the adding process is halted. *See also* **adder, digital.**

parallel computer: 1. A computer that has multiple arithmetic units or multiple logic units that are used to accomplish parallel processing, i.e., the concurrent or simultaneous execution of two or more processes in a single unit. **2.** A computer that has parallel circuits to accept, transmit within, process within, and emit all the bits of a computer word simultaneously, such as parallel adders, parallel circuits with as many conductors in each path as there are bits in the computer word, registers in which all the bits of a word are stored and retrieved simultaneously, and parallel ports. *See also* **arithmetic-logic unit, parallel adder, parallel port.**

parallel conversion: *See* **serial to parallel conversion.**

parallel operation: In communications, computer, data processing, and control systems, an operating mode in which operations are performed concurrently in one or more devices. *Note:* Examples of parallel operation are to concurrently (a) store data while computing, (b) read while adding, and (c) transfer data over several wires or optical fibers. *See also* **sequential operation.**

parallel port: A port (a) through which data are passed two or more bits at a time, such as all the bits of an eight-bit byte at a time, (b) that requires as many electrical wires, optical fibers, or channels as the number of bits that are to be handled simultaneously, (c) that is faster than a serial port that requires only one wire, fiber, or channel, and (d) that usually is used with printers and other parallel transmission devices. *See also* **serial port.**

parallel processing: Pertaining to the concurrent or simultaneous execution of two or more processes in a single unit, such as printing and calculating at the same time.

parallel search storage: Storage that may have one or more of its storage locations in one or more of its parts queried simultaneously.

parallel storage: Storage in which all the elements that compose digits, characters, or words may be transferred in or out simultaneously or concurrently. *Note:* The storage may be parallel by byte, i.e., all the bits in a byte are parallel, but the bytes and words are serial. *See also* **byte.**

parallel to serial conversion: The conversion of a spatial distribution of signal states that represent information into a corresponding time sequence of signal states that represent the same information. *Note:* In a parallel to serial converter, each of the parallel input signals requires a separate channel. Only one output channel is needed. *Synonyms* **dynamicization, serialization.** *See also* **parallel to serial converter.**

parallel to serial converter: 1. A digital device that converts a group of simultaneous inputs on two or more parallel channels, often constituting a specific data unit, such as a byte or a word, into corresponding time-sequenced signal elements on a single channel. **2.** A device that converts a spatial distribution of signal states that represent data into a corresponding time sequence of signal states such that each of the parallel input signals requires a separate channel, i.e., a parallel port, while the serial output requires only a single channel, i.e., a serial port. *Synonyms* **dynamicizer, serializer.** *See also* **channel, parallel port, serial port, serial to parallel converter.**

parallel transmission: 1. The simultaneous transmission of the signal elements of a data unit. *Note:* Examples of parallel transmission are (a) the transmission of a signal in which each signal element is characterized by a combination of three out of twelve frequencies that are simultaneously transmitted over a three-wire channel, and (b) the simultaneous transmission of the eight bits of an American Standard Code for Information Interchange (ASCII) character on eight separate wires or separate optical fibers. **2.** In digital communications, the simultaneous transmission of related signal elements over two or more separate paths. *Note:* Parallel transmission protocols, such as those used for computer ports, have been standardized by standards bodies, such as the American National Standards Institute (ANSI). *See also* **serial transmission.**

paramagnetic: Pertaining to a material with a magnetic permeability that is only slightly greater than that of air or that of a vacuum, i.e., slightly greater than unity. *Note:* Examples of materials that exhibit paramagnetic effects are nickel, cobalt, and chromium. The material is attracted by a magnet with a force that is weak compared with that of a ferromagnetic material, such as iron and some forms of iron oxide. Like a ferromagnetic material, a length of paramagnetic material tends to align its longest axis parallel to the lines of magnetic flux, i.e., tends to align itself with the magnetic flux. *See also* **diamagnetic, ferromagnetic, magnetic flux.**

parameter: 1. In an entity, a characteristic or an attribute that can have a finite set of usually specific values, each of which identifies a variation of the entity. *Note:* Examples of parameters are (a) in a communications channel, the set of specific signaling rates and (b) in a fiber optic link, the set of wavelengths that may be used. **2.** In an algebraic expression, a value that is held constant for some purpose but can be varied for other purposes. *Note:* In the equation for a straight line in Cartesian coordinates, $y = mx + b$, usually y is the dependent variable and x is the independent variable, while m and b are considered as parameters that can assume a set of values that give rise to (a) a set of straight lines either with different slopes for a fixed value of b and various values of m or (b) a set of parallel lines with a fixed value of slope m and various values of y-intercept b. Thus, m and b are parameters. If one is fixed and the other is varied, the result is a set of parametric equations. *See* **application parameter, communications parameter, distributed parameter, global fiber parameter, lumped parameter, password-length parameter, performance parameter, polarization-holding parameter, material dispersion parameter, polarization-maintaining parameter, profile dispersion parameter, refractive index profile parameter, signal parameter, value, v parameter, wave parameter.** *See also* **attribute, value**

parametric amplifier (paramp): An amplifier that (a) has a very low noise level, (b) has a main oscillator that is tuned to the input signal frequency, (c) has another pumping oscillator of a different frequency that periodically varies the parameters, i.e., the capacitance or the inductance, of the main oscillator circuit, and (d) enables amplification of the input signal by making use of the energy from the pumping action. *Note:* Paramps with a variable-capacitance main oscillator semiconductor diode are used in radar tracking and communications Earth stations, Earth satellite stations, and deep-space stations. The noise temperature of paramps cooled to the temperature of liquid helium, about 5 K (kelvin), is in the range of 20 K to 30 K. Paramp gains are about 40 dB.

parametric oscillator: *See* **optical parametric amplifier.**

paramp: parametric amplifier.

parasitic element: In an antenna system, a directive element that (a) is coupled to the driven element only by the transmitted fields, and (b) therefore is not connected to a radio transmitter or receiver either directly or via a feeder. *Synonym* **passive element.** *See also* **antenna, feeder, field, reflective array antenna.**

parasitic emission: In a communications system in which one or more electromagnetic sources are used, electromagnetic radiation, such as lightwaves, radio waves, microwaves, X rays, or gamma rays, from one or more of the sources that is not harmonically related, i.e., is not coherent, with the transmitted carrier. *Note:* Parasitic emissions usually are caused by undesired oscillations or energy level transitions in the sources. *See also* **energy level transition.**

paraxial ray: In optical systems, a ray that (a) is in a group or a bundle, (b) approaches the chief ray of the group or the bundle as its limiting position, (c) is nearly parallel with the optical axis, (d) is a ray in the sense of Gaussian or first-order optics, and (e) for computation, makes an angle, θ, with the optical axis, that is small enough for $\sin \theta$ to be practically equal to $\tan \theta$, where either or both can be replaced by θ in radians. *See also* **axial ray, light ray.**

parenthesis-free notation: Notation in which algebraic expressions are delimited by rules and conventions other than the use of parentheses or symbols that perform a function similar to that performed by parentheses. *Note:* Using parentheses, the multiplication of two binomials might conventionally be written as $(a + b)(c - d)$. In parenthesis-free notation, the equivalent expression might be written as $ab + cd - \times$, which is read as "fetch a and b and add them, fetch c and d and subtract d from c, then multiply the sum and difference. Note that d is subtracted from c because d follows c. Also, the expression ab does not imply multiplication as in a conventional algebraic expression. Parenthesis-free notation is convenient for computer programming because the computer must first fetch the operands and then be told what to do with them. If a computer were to read the expression $a + b$ in sequence, it would be told to perform an operation before it had fetched or read the second operand. *See also* **postfix notation, prefix notation.**

parity: In binary coded data, a condition that is maintained such that, in any permissible coded expression, the total number of 1s, or 0s, is always odd or always even. *Note 1:* Parity is used in error-detecting and error-correcting codes. *Note 2:* In the American Standard Code for Information Interchange (ASCII) code or in the International Telegraph Alphabet 5 (ITA-5) code as usually implemented, seven bits are used to represent each character, and one bit is used as a parity check bit. Usually odd parity is used. *See also* **block parity, even parity, odd parity.** *Refer to* **Fig. E-3 (310).**

parity bit: A bit that (a) is appended to an array of binary digits to make the sum of all the binary digits, i.e., the result obtained by adding all the 1s, including the appended bit, i.e., the parity bit, always odd or always even, (b) makes the number of 1s in each array always odd or always even when the data are generated, (c) usually is associated with a character, word, or block for the purpose of providing a means for checking the accuracy of transmission, storage, and retrieval, and (d) has a value of 0 or 1, depending on the previous sequence and whether the system operates on odd parity or even parity. *Note:* In using the American Standard Code for Information Interchange (ASCII), which is the U.S. implementation of the International Telegraph Alphabet 5 (ITA-5), the eighth bit is appended to each character so as to always have an odd or always an even number of 1s. Usually odd parity is used. *See* **check bit, check digit, even parity, odd parity, parity.**

parity check: A check that tests whether (a) the number of 1s in an array of binary digits is odd or even, or (b) the number of 0s in an array of binary digits is odd or even, including the parity bit. *Note:* In odd parity, the number of 1s is maintained odd including the parity bit. In even parity, the number of 1s is maintained even including the parity bit. Odd parity is standard for synchronous transmission, and even parity is standard for asynchronous transmission. *Synonym* **odd-even check.** *See also* **block parity, check bit, check digit, code, error control, parity, parity bit, parity check coding.**

parity check coding: Error-detecting coding (a) that consists of the addition of an extra bit, a parity bit, at the end of each binary coded array, such as a byte, character, word, or block, that is transmitted or stored, and (b) in which the extra bit is always chosen so as to make the number of 1s in each code group always odd or always even, including the parity bit. *Note:* In odd parity, if the code group representing a four-bit byte were 1011, it would become 10110. The parity bit is always included in the count. In even parity, if the initial count of 1s is odd, a 1 is added to make the total count even, in which case the 1011 would become 10111. *See also* **parity, parity bit.**

par meter: A meter that (a) is used to determine the performance characteristics of telephone channels, (b) measures, calculates, and displays the ratio of the peak power level to the time-averaged power level in a circuit, i.e., the peak to average ratio (p/a r), (c) is used as a quick means to identify degraded telephone channels, (d) is very sensitive to envelope delay distortion, (e) may be used for idle channel noise, nonlinear distortion, and amplitude distortion measurements, and (f) may be used for determining the peak to average ratio for many signal parameters, such as voltage, current, power, frequency, and phase. *Note:* The term "par" is derived from "peak to average ratio." *See also* **grade of service, peak to average ratio.**

parse: To decompose a larger computer program unit, such as a block, line, command phrase, or instruction word, into smaller or more elementary subunits, such as parts of lines, syllables, characters, or bytes.

Parseval's theorem: An extension of the superposition theorem (a) that is applied to aperiodic wave shapes for spectral irradiance, i.e., for spectral power density, and (b) in which, in the periodic case, the various signals are considered orthogonal over the period, but in the aperiodic case, the interval of orthogonality extends over the entire time axis from minus to plus infinity. *See also* **orthogonal, orthogonal signals.**

part: *See* **address part, message part, message transfer part, operation part, user part.**

partial coherence: In an electromagnetic wave, or waves, coherence such that the electromagnetic fields at two points in space or two instants of time have a low statistical level of correlation, i.e., a coherence degree below 0.88 but not 0.

partial degradation: A condition in which a system (a) does not fail completely, i.e., catastrophically, (b) continues to provide partial service, (c) operates with a reduced capability, and (d) reaches the new reduced operating condition suddenly rather than gracefully, i.e., gradually. *See also* **catastrophic degradation, graceful degradation.**

partial mirror: **1.** A surface that (a) simultaneously transmits and reflects a significant portion of incident electromagnetic radiation, such as a beam of lightwaves, and (b) usually consists of a substrate, such as glass, with a layer of reflective coating, such as a metal, so thin that it reflects approximately half of the incident radiation and transmits approximately half, with negligible absorption. **2.** A mirror that (a) reflects a portion of the incident optical power, such as 50%, and transmits the remaining portion, (b) may be directional, such that it reflects all or none of the incident light in one direction, and reflects a portion in the other direction, (c) may use an extremely thin metallic film on glass or plastic to reflect all the incident light from one direction and transmit practically all incident light from the other direction, and (d) may be used in optical mixers and couplers. *Note:* If two persons, A and B, are on opposite sides of such a mirror, A might see B, but B cannot see A. If the mirror is constructed with precise dimensions, such as a half or quarter wavelength thick, with a thin reflective and transmitting coating on both sides of the glass, it can be used as a beam splitter, combiner, or interferometer to mix two coherent lightwaves from different directions that are incident on opposite sides of the mirror. Depending on the nature of the reflective and transmitting coatings on both sides of the transparent substrate, the output of the mirror would then consist of either (a) two coherently mixed beams, each component of the mixture down 3 dB from its original power level, or (b) one coherently mixed beam. *See also* **beam splitter.** *Refer to* **Fig. O-2 (671).**

partial response signal: A signal that has been processed in such a manner that though there is a controlled amount of intersymbol interference inherent in the signal design, there are desirable spectral properties at the band edges so that it is less susceptible to filter degradation. *See also* **band edge absorption, filter, intersymbol interference.**

partial sum gate: *Synonym* **EXCLUSIVE OR gate.**

partial tone reversal: In facsimile transmission, a defect in which the scanning output signal does not have a smooth transition from white to black, but goes from white to black and then toward white again, or vice versa, like a damped oscillation. *Note:* One cause of partial tone reversal is the incorrect setting of black and white limits in the subcarrier frequency modulator. *See* **damped oscillation, frequency modulation, subcarrier.**

particle: *See* **alpha particle, radioactive particle.**

part message: *See* **multiple part message.**

party: *See* **alternate party, busy party.**

party line: In telephone systems, an arrangement in which (a) several user end instruments, usually telephones, are connected to the same loop, (b) all users are assigned the same number, (c) individual users are distinguished by different ringing signals, such as a different number of rings or a different combination of long and short rings, and (d) privacy is limited, and congestion often occurs. *Note:* Party lines primarily remain only in rural areas where loops are long. *Synonym* **multiparty line.** *See also* **private line.**

Part 68: The section of Title 47 of the *Code of Federal Regulations* governing (a) the direct connection of telecommunications equipment and customer premises wiring with the public switched telephone network and certain private line services, such as (i) foreign exchange lines at the customer premises end, (ii) the station end of off-premises stations associated with private branch exchange (PBX) and Centrex® services, (iii) trunk to station tie lines at the trunk end only, and (iv) switched service network station lines, i.e., common control switching arrangements, (b) the direct connection of (i) all PBX and similar systems to private line services for tie trunk type interfaces, (ii) off-premises station lines, and (iii) automatic identified outward dialing (AIOD) and message registration, and (c) the rules that provide the technical and procedural standards under which direct electrical connection of customer-provided telephone equipment, systems, and protective apparatus may be made to the nationwide network without causing harm and without a requirement for protective circuit arrangements in the service provider networks.

pascal (P): A unit of sound pressure, equal to 1 $N \cdot m^{-2}$ (newton per square meter). *Note 1:* The micropascal, µP, equal to 10^{-6} $N \cdot m^{-2}$ (newtons per square meter), is the standard unit of sound pressure. *Note 2:* The minimum human ear sensitivity level at 1 kHz (kilohertz) is 20 µP (micropascals). For underwater acoustics, the reference pressure level is 1 µP. Sound pressures are expressed in dB referred to these levels. *See also* **dBp.**

PASCAL: A block-structured general purpose computer programming language that is primarily used for logical, string, set, and numerical processing.

pass: *See* **bandpass, bypass, sort pass.**

passband: The band of frequencies, between specified limits, that a circuit, such as a filter or a telephone circuit, is capable of transmitting. *Note 1:* All frequencies below the lower limit and above the upper limit are not transmitted, i.e., are not allowed to pass. *Note 2:* The limiting frequencies are those at which the transmitted power level decreases to a specified level, usually 3 dB below the maximum level, as the frequency is decreased or increased from that at which the transmitted power is a maximum. *Note 3:* The difference between the limits is the passband bandwidth, usually expressed in hertz. *Note 4:* All types of equipment and circuits have a passband. *See also* **stopband.**

passing: *See* **instruction passing, token passing.**

passive communications satellite: A communications satellite that only reflects signals from one Earth station to another, or from several Earth stations to several others. *See also* **active communications satellite.**

passive detection: In communications systems, the detection of electromagnetic radiation by using the energy that is emitted by the source, i.e., the energy that is in the radiation being detected, without emitting radiation to perform the detection.

passive detection system: A system that (a) does not radiate electromagnetic energy, (b) has the capability of determining the existence, the location of the source, and the type of the radiation, such as the modulation and polarization, of received electromagnetic radiation, (c) uses triangulation to determine the location of the source, and (d) has a range that depends on many factors, such as detector sensitivity, location of both the source and the detector, and directivity patterns of antennas of the source and the detector. *See also* **active detection system, directivity pattern, sensitivity, triangulation.**

passive detection tracking: In radar systems, the detection and the tracking of radar or jammers by (a) performing triangulation on signals that are received at two or more locations, (b) combining azimuth data on jamming or radar strobes from several receiving stations to obtain intersections that indicate the position of jammers, and (c) in some cases, obtaining an indication of the jammer or radar average power output. *See also* **azimuth, jammer, strobe, triangulation.**

passive device: A device that (a) does not contain a source of energy and (b) has an output that is a function of present or past inputs. *Note:* Examples of passive devices are electrical resistors, electrical capacitors, diodes, optical fibers, cables, wires, glass lenses, and certain filters. *See also* **active device.**

passive element: *Synonym* **parasitic element.**

passive filter: A filter that does not require power to perform its function. *See also* **active filter.**

passive flight phase: In satellite launching, the part of the spacecraft flight in which (a) propulsion engines are shut off, and (b) the trajectory depends on (i) the final state of the preceding powered flight phase and (ii) natural forces, such as atmospheric drag and gravitational forces. *See also* **powered flight phase.**

passive graphics: In display systems, the field of application and study in which display devices, their components, and associated equipment are used in a passive mode, i.e., without changing the displayed data and without human interaction with the displayed data. *Note:* Examples of passive graphics are reading data and transporting data, such as occurs in the use of plotters, microfilm readers and recorders, and outdoor fixed signs and billboards. *See also* **interactive graphics.**

passive homing: Homing in which (a) a mobile station tracks an object, i.e., a target, (b) the mobile station does not radiate energy, (c) the radiation from the object being tracked is used to guide the mobile station to the object, (d) the object being tracked may be fixed or mobile, and (e) triangulation is not used. *Note:* Examples of passive homing are (a) missile to target homing in which the missile tracks the target without the use of radiation from itself or its launcher and uses radiation from the target, such as radio waves, lightwaves, infrared radiation, or reflected sunlight, to reach the object, (b) an aircraft station that uses direction finding to guide its way to an aircraft carrier station or land station, and (c) a submarine station that tracks the sound waves emanating from a target ship propeller or engines without launching sonar pulses. *See also* **active homing, direction finding, homing, mobile station, semiactive homing.**

passive infrared device: A device that (a) can detect, measure, or determine the direction of a source of infrared radiation and (b) does not itself contain a source of infrared radiation. *See also* **active infrared device, infrared radiation.**

passive mode: 1. A mode of operation of a device or system, such as a communications, computer, data processing, or control system, that does not allow a user to alter or interact with the system. **2.** In display systems, a mode of operation of a display device in such a manner that a user (a) can call for data to be displayed and (b) cannot alter the displayed data, i.e., cannot interact with a display element, display group, or display image in the display space of the device. *See also* **interactive mode.**

passive network: A network that does not include a power source. *See also* **active network, network.**

passive optical device: A device (a) that operates with, or performs specific operations on, electromagnetic waves having wavelengths that are in the optical spectrum, i.e., in the optical region of the electromagnetic spectrum, usually within the visible and near visible spectra, and (b) in which operations are performed by the propagation media through which the waves pass without the use of input energy other than that contained in the waves themselves. *Note:* Examples of passive optical devices are optical couplers, splitters, mixers, filters, attenuators, gratings, lenses, prisms, and all-optical multiplexers and demultiplexers. Fiber optic cables, splices, and connectors are also passive optical devices, but they perform no specific operation on lightwaves other than to transport them between and through points. *See also* **active optical device, optical path component.**

passive optical fiber branching device: *Synonym* **optical fiber branching device.**

passive optical processor: A device that (a) performs specific operations on electromagnetic waves having wavelengths that are within or near the visible spectrum of the electromagnetic spectrum, i.e., are in the optical spectrum, and (b) uses only the energy that is contained in the waves themselves and the characteristics of the material media through which the waves pass to perform the operations. *Note:* Examples of passive optical processors are optical couplers, splitters, mixers, filters, attenuators, gratings, lenses, and prisms. *See also* **electromagnetic spectrum, optical spectrum, visible spectrum.**

passive optics: *Synonym* **fixed optics.**

passive repeater: A repeater that (a) is used to route signals, such as a microwave signal, past or around a point, such as an obstruction or a fault. *Note:* Examples of passive repeaters are (a) a pair of antennas connected and oriented so as to turn a microwave through an angle of 90°, (b) a flat reflector used as a mirror, and (c) a bypass around a failed node or a failed active repeater in a network.

passive satellite: In satellite communications, a satellite that (a) retransmits received signals after only amplification and signal shaping prior to retransmission and (b) does not perform complex functions, such as code translation, frequency translation, and link switching. *Note:* The Moon is a passive satellite and Echo I has been a passive satellite in that they only have served as reflectors. *See also* **active satellite.**

passive sensor: 1. A sensor that performs its sensing function without input energy other than the energy contained in that which is being sensed. **2.** In the Earth exploration satellite service or in the space research service, an instrument used to obtain information from the reception of radio waves of natural origin. **3.** A sensor that does not emit radiation to perform its sensing function, such as finding targets by maintaining a listening watch for signals from them. *See also* **active sensor, sensor.**

passive sonar: Sonar that (a) depends on the reception of sound generated by a source, such as the sound source on board an aircraft, ship, or submarine, and (b) does not launch a soundwave and depend on its reflection. *See also* **sonar.**

passive station: On a multipoint connection or a point-to-point connection using basic mode link control, a tributary station waiting to be polled or selected. *See also* **node, terminal.**

passive tracking: In radar and sonar systems, the tracking of a moving object that is accomplished by (a) depending on the energy emitted by the object being tracked and (b) not radiating energy, such as sonar or radar pulses, of its own to obtain a reflected signal from the object being tracked. *Note:* Examples of passive tracking are tracking by using (a) infrared radiation from the tracked object, (b) sunlight reflected by the object, and (c) the sound energy from the object.

passive wiretapping: The monitoring or the recording of data that are being transmitted in a communications line or circuit. *Note:* Passive wiretapping may be authorized or clandestine.

password: 1. In communications, computer, data processing, and control systems, a string of characters, such as a word or group of words, that permits access to otherwise inaccessible data, information, or facilities. *See also* **log in. 2.** A secret word or distinctive sound used to reply to a challenge. **3.** A protected, private character string that is used to authenticate an identity. *See also* **authenticate, character, communications security.** *Refer to* **Fig. P-18 (769).**

password-length equation: An equation that determines an appropriate password length, L, that provides an acceptable probability, P, that a password will be guessed in its lifetime. *Note:* The password length, M, is given by the relation $M = (\log S)/(\log N)$ where M is the password length in number of characters, S is the size of the password space, and N is the number of characters available. The password space, i.e., the number of unique algorithm generated passwords, S, is given by the relation $S = LR/P$ where L is the maximum lifetime that a password can be used to log in to a system, R is the number of guesses per unit of time, and P is the probability that a password can be guessed in its lifetime. *See also* **log in, password-length parameter.**

password-length parameter: A basic parameter affecting the password length needed to provide a given degree of security. *Note 1:* Password-length parameters are related by the expression $P = LR/S$ where P is the probability that a password can be guessed in its lifetime, L is the maximum lifetime that a password can be used to log in to a system, R is the number of guesses per unit of time, and S is the password space, i.e., the number of unique algorithm-generated passwords. *Note 2:* The degree of password security is determined by the probability that a password can be guessed in its lifetime.

patch: 1. To connect circuits together temporarily. *Note 1:* In communications, patches may be made by means of a cord, i.e., by means of a cable, known as a "patch cord." *Note 2:* In automated systems, patches may be made electronically. *See also* **circuit, connection, patch bay, patching. 2.** To make a temporary or expedient modification of a computer program in order to locate and correct an error. **3.** To modify a program or a routine in a rough, temporary, crude, or expedient way. **4.** A computer program modification at the object program level made by replacing a portion of existing machine language code with a modification of that portion of the machine language code. *See* **computer program patch, machine code, machine language, modal patch, modem patch, object program, smart patch, telephone patch.**

patch and test facility (PTF): At a facility, such as a central office (C.O.), i.e., a switching center, equipment that (a) functions under the technical supervision of a designated technical control facility (TCF) and (b) performs supporting functions, such as (i) quality control checks and tests on equipment, links, and circuits, (ii) troubleshooting, (iii) activation, changing, and deactivation of circuits, and (iv) technical coordination and reporting. *See also* **circuit, facility, technical control facility.**

patch bay: An assembly of hardware that (a) is arranged so that a number of circuits, usually of the same or a similar type, appear on jacks for monitoring, interconnecting, and testing purposes, (b) is used at many locations, such as technical control facilities, patch and test facilities, and central offices (COs), such as telephone exchanges and switching centers, and (c) is often used for special purposes, such as direct current (dc), voice frequency (VF), group, coaxial, equal level, and digital data circuits. *See* **coaxial patch bay, digital circuit patch bay, digital primary patch bay, direct current patch bay, D patch bay, equal level patch bay, group patch bay, K patch bay, MM patch bay, M patch bay, voice frequency primary patch bay.** *See also* **circuit, interposition trunk, patch, patch panel, secure voice cord board.**

patching: 1. The temporary connection of circuits by direct connection, by (a) the use of cords with plugs inserted into appropriate jacks, (b) the use of a patch bay, (c) interconnection of telephone and radio links, or (d) electronic means. *See also* **circuit, connection, overlay, patch, patch bay. 2.** The inserting of a program patch in a computer program, routine, or subroutine. **3.** The inserting of a program, routine, or subroutine into an existing computer program. *See also* **patch.**

patching jackfield: In a communications system, a group of jacks that (a) terminate permanent input and output connections to channeling equipment or (b) provide monitoring points to allow circuits to be set up by means of plug-terminated cord circuits to allow the connection of monitoring equipment. *See* **channel patching jackfield.**

patching service: *See* **on-call patching service.**

patch panel: 1. In a communications system, a segment of a patch bay. **2.** In a communications system, a board, panel, or console that (a) is equipped with a jackfield mounted on its surface and (b) enables connection to various points in the system for patching, monitoring, and testing via the jacks. *See* **fiber optic patch panel.** *See also* **jackfield patch bay.**

path: 1. In communications systems and network topologies, a route that (a) lies between any two points, such as between (i) any two nodes of a network, (ii) two telephones, (iii) a control panel and a motor controller of a ship propulsion system, (iv) a remote data input terminal and a computer, or (v) two switching centers, (b) usually consists of several branches and several nodes that lie en route between the two points, and (c) has a beginning point and an end point. **2.** In radio communications, the route that (a) lies between a transmitter and a receiver and (b) may consist of two or more concatenated links. *Note:* Examples of radio paths are line-of-sight routes and ionospheric routes. **3.** In a computer program, the logical sequence of instructions in the program that are to be executed by a computer. **4.** In database management systems, a series of physical or logical connections between records or segments that usually require the use of pointers. *See* **access path, data path, engineering path, instrument landing system glide path, overlay path, signaling path, virtual path.** *See also* **database, pointer.**

path attenuation: In communications, the power losses that (a) occur in a wave or a signal propagating between a transmitter and a receiver and (b) may be caused by many effects, such as free-space loss, refraction, reflection, aperture-propagation-medium coupling loss, and absorption. *Note:* Path attenuation usually is measured in dB with reference to the transmitted power. *Synonym* **path loss.**

path clearance: 1. In microwave line-of-sight (LOS) communications, the perpendicular distance from the radio beam axis to obstructions such as trees, buildings, or terrain. *Note:* The required path clearance usually is expressed, for a particular k-factor, as some fraction of the first Fresnel zone radius. **2.** In electromagnetic wave propagation, the distance between the nearest point on a wave path and a propagation path obstruction. *Note:* If the clearance exceeds 0.6 of the first Fresnel zone radius, the obstruction does not cause a significant loss and thus may not be considered to be an obstruction. *See also* **effective Earth radius, Fresnel zone, k-factor, path profile, path survey, propagation path obstruction.**

path component: *See* **optical path component.**

path fading: *See* **multiple path fading.**

pathfinder: In radar systems, a set of equipment that is used for navigating or homing to reach an objective or a position when lack of visibility precludes accurate visual navigation. *Note:* The lack of visibility may be caused by various factors, such as precipitation, fog, smoke, trees, walls, buildings, blinding light, and darkness.

path intermodulation noise: Noise in a transmission path that combines with a signal by modulating it, such as by mixing with the signal and distorting it. *Note:* The noise results from the nonlinear characteristics of the path. These nonlinear characteristics create noise from the signal energy. In addition, the different frequencies in the signal and the noise are affected differently by the propagation media of the path, such as by absorption, refraction, diffraction, dispersion, reflection, and scattering. These effects cause more noise, which further modulates the signal. *See* **intermodulation noise, modulation, propagation medium.**

path length: 1. The actual physical distance, i.e., the geometric distance, measured along a specified route between two points. **2.** In electromagnetic transmission, such as radio, radar, television, and lightwave transmission, the distance obtained as the product of the propagation velocity and the transit time between two points. *Note:* The propagation velocity will depend on the extent to which the frequencies of an electromagnetic wave are affected by the refractive indices of the propagation media that constitute the path, including slowing down and speeding up of the wave, ducting, bending by refractive index gradients, and knife edge effects. **3.** In the propagation of waves, such as electromagnetic waves and sound waves, the length that is proportional to the time that a wave or a signal takes to travel from one point to another. *Note 1:* The path length is directly proportional to the physical length and directly proportional to the integrated or effective refractive index of the path. *Note 2:* If two electromagnetic waves travel between the same two points via different paths, one path physically longer than the other, but the longer physical path has a lower refractive index allowing the wave to travel faster, and therefore arrive at the destination point earlier, than the slower-moving wave taking the shorter physical path, the wave arriving earlier had the shorter path length although its physical path was longer. *See* **optical path length, transmission path length.** *See also* **multipath, refractive index.**

path loss: In a propagating electromagnetic wave, the attenuation of the wave that occurs between a transmitter and a receiver. *Note 1:* Path loss usually is expressed in dB. *Note 2:* Path losses include losses caused by refraction, reflection, diffraction, scattering, aperture-medium coupling, absorption, and free-space spreading, i.e., geometric spreading, losses. *Synonym* **path attenuation.** *See also* **loss, shadow loss, transmission loss.**

path obstruction: *See* **propagation path obstruction.**

path profile: A graphic representation that (a) shows the physical features of a propagation path in the vertical plane containing both end points of the path, showing the Earth surface and including trees, buildings, knife edges, and other features that may obstruct the radio signal, and (b) is drawn either with (i) an effective Earth radius simulated by a parabolic arc, in which case the ray paths are drawn as straight lines, or (ii) a flat Earth surface, in which case the ray paths are drawn as parabolic arcs. *See also* **effective Earth radius, Fresnel zone, horizon angle, k-factor, path clearance, propagation path obstruction, smooth Earth.**

path quality analysis (PQA): In a communications path, an analysis that (a) includes an overall evaluation of (i) the component quality measures, (ii) the individual link quality measures, and (iii) the aggregate path quality measures, and (b) is performed by evaluating communications parameters, such as bit error ratios (BERs), signal plus noise plus distortion to noise plus distortion ratios (S+N+D)/(N+D), and spectral distortions.

path quality matrix: A data bank that (a) contains path quality analysis information used to support path selection and to support routing determination and (b) in adaptive radio automatic link establishment (ALE), usually contains path quality data on (i) reachable destinations via directly reachable relays for multilink paths and (ii) directly reachable destinations for single link paths.

path survey: The gathering, assembling, and recording of geographical and environmental data that (a) particularly concern the existing or potential paths that a signal may use and (b) are needed to design a radio communications system, especially a microwave system. *See also* **Fresnel zone, k-factor, path clearance, path profile, propagation path obstruction.**

patrol air broadcast net: *See* **maritime patrol air broadcast net.**

patrol air reporting net: *See* **maritime patrol air reporting net.**

pattern: *See* **acceptance pattern, bit pattern, diffraction pattern, directivity pattern, equilibrium radiation pattern, eye pattern, far field diffraction pattern, far field radiation pattern, fiber optic pattern, frame synchronization pattern, Fraunhofer diffraction pattern, Fresnel diffraction pattern, interference pattern, near field diffraction pattern, near field radiation pattern, optical fiber pattern, radiation pattern, receiver pattern, receiving pattern, speckle pattern, test pattern.**

pattern-matching voice coder: A voice coder in which (a) the short-term speech frequency spectrum is compared with the spectral data that are stored at specific addresses, and (b) the appropriate address is then transmitted to a receiver that locates the same pattern that is stored at the receiving speech synthesizer. *See* **vocoder.**

pattern plotter: *See* **radiation pattern plotter.**

pattern recognition (PR): The sensing and the interpretation of signal shapes and components of complex wave forms. *Note 1:* Pattern recognition may be applied to signals and waveforms from any source, including (a) transmitted signals, (b) reflections from moving, spinning, oscillating, or vibrating objects, such as jet engine exhaust, marine engine noise, or speech vibrations, and (c) galactic noise. *Note 2:* Pattern recognition may be accomplished by finding the nearest match in a library of previously stored signals. If the signals are digitized, stored digital patterns may be used in the match test. The stored patterns are identified with various waveforms from known specific sources, thus permitting identification, i.e., permitting recognition, of the source.

pattern-sensitive fault: In a communications, computer, data processing, or control system, a fault, malfunction, or error that occurs only when a particular pattern of binary digits is entered, stored, or generated in the system.

paulin signal: *See* **visual paulin signal.**

Pawsey stub: A device for connecting an unbalanced coaxial feeder to a balanced antenna. *See also* **balun, feed.**

PAX: private automatic exchange. *See* **private branch exchange.**

payload: In a set of data, such as a data field, block, or stream, being processed or transported, the part that (a) represents user information and user overhead information, (b) may include additional information, such as network management and accounting information when requested by the user, and (c) does not include system overhead information. *See also* **system overhead, user overhead.**

payload module: The portion of a payload that completely occupies one or more channels.

payload rate: In a transmission system, the interface rate minus the overhead rate. *Synonym* **information payload capacity.** *See also* **interface rate, overhead rate.**

PBER: pseudo bit error ratio.

PBX: private branch exchange.

PBX tie trunk: private branch exchange tie trunk.

PBX trunk: private branch exchange trunk.

PC: carrier power, personal communications, personal computer, power of a carrier.

PCB: power circuit breaker, printed circuit board.

PC DOS: personal computer disk operating system.

PCM: pulse code modulation, pulse code multiplexing.

PCM multiplexing equipment: pulse code modulation multiplexing equipment.

PCM noise: pulse code modulation noise.

PCS: personal communications service, personal communications system, plastic-clad silica.

PCS fiber: plastic-clad silica fiber.

PCSS: personal communications system service.

PD: photodetector.

PDM: pulse duration modulation.

PDN: public data network.

PDS: protected distribution system.

PDU: protocol data unit.

PE: phase encoded.

peak: *See* **absorption peak.**

peak busy hour: *Synonym* **busy hour.**

peak emission wavelength: For a source of optical power, i.e., of an optical emitter, the wavelength at which the source emits the maximum value of optical power, i.e., the spectral line that has the greatest power.

peak envelope power (pX, PX): In a radio transmitter, the average power supplied to the antenna transmission line by the transmitter during one radio frequency cycle at the highest crest of the modulation envelope taken under normal operating conditions.

peak limiting: 1. The limiting of the absolute instantaneous value of a signal parameter by preventing it from exceeding a specified value. **2.** Limiting in which (a) the positive or negative, or both, instantaneous values of a signal parameter or (b) the maximum or minimum, or both, instantaneous values of a signal parameter are prevented from exceeding a specified value. *See also* **attack time, clipper, compandor, compressor, expander, limiter, limiting, pulse code modulation.**

peak power output (PPO): During transmission of an electromagnetic wave, the output power averaged over the cycle that has the maximum peak value that can occur. *See also* **output rating, peak envelope power, rated output power.**

peak radiance wavelength: In a light source, the wavelength at which the radiance in a given direction is a maximum. *See also* **radiance.**

peak radiant intensity: The maximum value of radiant intensity of a lightwave. *See also* **lightwave, radiant intensity, value.**

peak signal level: 1. In a transmission path, the maximum instantaneous signal power, voltage, or current at any point. **2.** At a given point in a transmission path, the maximum instantaneous signal power, voltage, or current that occurs during a specified period. *See also* **level, signal.**

peak spectral power: The maximum radiance, i.e., optical power, emitted by a light source, usually occurring at a specific wavelength or within a specific range of wavelengths. *Note:* A gallium arsenide light-emitting diode (LED) emits its peak spectral power at a wavelength of 0.910 μm (micron).

peak spectral wavelength: The wavelength at which a light source emits its peak spectral power, i.e., its maximum total optical power. *Note:* A gallium arsenide light-emitting diode (LED) emits its peak spectral wavelength of 0.910 μm (micron) when operating at a peak diode current of 50 mA (milliamperes) at 1 mW (milliwatt) of total optical power output with about 0.3 mW of the optical power output coupled to a fiber optic pigtail. *See also* **fiber optic pigtail, optical power, peak spectral power.**

peak to average ratio (p/ar) The ratio of (a) the instantaneous peak value, i.e., maximum magnitude, of a signal parameter to (b) its time-averaged value. *Note:* Peak to average ratio can be determined for many signal parameters, such as voltage, current, power, frequency, and phase. *See also* **output rating.**

peak to peak value: 1. The algebraic difference between the extreme values of a varying quantity. *Note:* An example of peak to peak value is the difference between the most negative value and the most positive value of a polarized signal. **2.** The absolute value of the difference between the maximum and the minimum magnitudes of a varying quantity. *See also* **output rating.**

peak value: The maximum value that a varying value may have with respect to a given reference. *Note:* Examples of a peak value are (a) the value of the crest of a sinusoidal wave with reference to the zero value and (b) the maximum value of the significant condition of a pulse with reference to −6 V (volts). *See also* **value.**

peak wavelength: In a source of electromagnetic radiation, the wavelength at which the radiant power, i.e., the radiance or the radiant intensity, is a maximum. *See also* **irradiance, radiance, spectral line, spectral width.**

peer: In open systems architecture, a member of a peer group. *See also* **peer entity, peer group.** *Refer to* **Fig. L-4 (506).**

peer entity: In open systems architecture, one member of a set of entities (a) in one layer of a given system or (b) in the equivalent layer of another system. *See also* **open systems architecture, open systems interconnection.**

peer group: In open systems architecture, a group of functional units in a given layer of a network such that all the functions performed by the functional units extend throughout the system at the same layer. *Note:* The peer group layer usually extends across all the nodes and branches of a network. *See also* **open systems architecture, open systems interconnection.** *Refer to* **Fig. L-4 (506).**

peg count: The number of seizures of, or the number of attempts to seize, specific equipment, such as telephone trunks, access lines, and switches, during a specified period of time. *See also* **seizing, traffic usage recorder.**

peg count register: A register that is used on telephone trunks to record the number of calls that are attempted, i.e., the number of access attempts. *See also* **access attempt, call, trunk.**

peg marking: *See* **multiple peg marking.**

pel: 1. In a facsimile system, the smallest discrete scanning line sample containing only black/white information, i.e., not containing gray shading information. *See also* **facsimile. 2.** In a television (TV) picture, the smallest area that is capable of being represented by a signal element that passes through the TV system. **3.** The smallest element that can be distinguished vertically and horizontally within a black-and-white image telemetry system. *See also* **elemental area, facsimile, picture element, pixel, scanning.**

pen: *See* **light pen, sonic pen.**

penalty: *See* **dispersion power penalty, reflection power penalty.**

pencil: *See* **light pencil, voltage pencil.**

pending signal: *See* **call pending signal.**

penetration: In communications, computer, data processing, and control systems, a successful unauthorized access to a system. *Note:* The penetration may be deliberate or accidental. **2.** The successful bypassing of the security mechanisms of a cryptographic system or an automated information system. **3.** The passage through a partition, wall, hull, bulkhead, tank, equipment housing, or enclosure by a propagation medium, such as a cable or a wire. **4.** In fiber optic systems, the passage through a partition, wall, hull, tank, bulkhead, or other obstacle, usually by means of a fiber optic penetrator or continuous fiber optic cable. *See* **technological penetration.** *See also* **cable, fiber optic cable, fiber optic penetrator.** *Refer to* **Fig. I-5 (470).**

penetration profile: In system security, a delineation of the activities that are required to effect a penetration of a system. *See also* **communications system security, penetration.**

penetration reaction: *See* **real time penetration reaction.**

penetration signature: In system security, the description of a situation or a condition in which a penetration of a system could occur, or the set of events that indicates that a penetration is in progress or has occurred. *Note:* Most security systems are designed in such a manner that a penetration signature will reveal the fact that a penetration is in progress.

penetration test: **1.** In system security, to test the ability of a system to withstand penetration and to identify system weaknesses against penetration. **2.** In communications system security, a test that consists of attempting access at various points and measuring (a) the access successes without detection and (b) the detection successes. *See also* **access success, communications system security.**

penetrator: *See* **fiber optic penetrator.**

pennant: In visual communications systems, an elongated and tapered flag on which colored patterns are used to represent words, letters, and numerals. *Note:* Pennants are hung with the base to the halyard. *See* **numeral pennant, special pennant.**

penta prism: A prism that has the unique property of being able to divert a light beam 90° in the principal plane even if the beam does not strike the endfaces exactly with normal incidence. *See also* **beam, normal incidence, prism.**

PEP: peak envelope power. The usual abbreviation for peak envelope power is **"PX"** or **"pX."**

percentage modulation: **1.** In angle modulation, 100 times the fraction of a specified reference modulation. **2.** In amplitude modulation, the modulation factor expressed in percent. *Note:* The percentage modulation may also be expressed in dB below 100 percent modulation. *See also* **modulation, modulation factor, overmodulation.**

percent break: In telephone dialing, 100 times the ratio of (a) the open circuit time to (b) the sum of the open and closed circuit times allotted to a single dial pulse cycle. *See also* **dial signaling.**

percent modulation: *See* **one hundred percent modulation.**

perfluoronated ethylene propylene (FEP): A commercial polymer used for making the cladding in optical fibers. *See also* **cladding, optical fiber.**

perforated tape: A tape in which patterns of holes are punched to represent information.

perforated tape retransmitter: An automatic retransmitter that consists of a reperforator that feeds a tape directly into an automatic transmitter.

perforation: *See* **chadless perforation.**

perforator: In telegraph systems, a device that is used for punching patterns of holes to represent data in paper tape for transmitting the data via a tape transmitter. *Note:* When the punching of holes is automatically controlled by incoming signals, the device is called a reperforator, i.e., a receiving perforator or a receiver-perforator. *See* **keyboard perforator, printing perforator, tape perforator.**

perforator tape: *Synonym* **punch tape.**

performance evaluation: The assessment of the capability of a system, subsystem, equipment, or component to reach the operating objectives in actual use under the operating conditions for which it was designed.

performance management: In communications network management, (a) a set of functions that evaluate and report information on the performance of communications equipment and the effectiveness of network performance, and (b) a set of various subfunctions, such as gathering statistical information, maintaining and examining historical logs, determining system performance under natural and artificial conditions, and altering system operational modes.

performance measurement period: In communications, computer, data processing, and control systems, the period during which performance parameters are measured. *Note:* A performance measurement period is determined by required confidence limits and may vary as a function of the observed parameter values. System user time is divided into consecutive performance measurement periods to enable measurement of user information transfer reliability. *See also* **acceptance test, operational service period, outage, outage probability.**

performance parameter: In a system, a quantity that (a) has a numerical value and (b) characterizes an aspect of a particular system capability or attribute. *Note:* Examples of performance parameters include a peg count, a ratio of accesses to access attempts, mean time between failures, and outage probability. *See also* **acceptance test, communications, operational service state, outage, system operational threshold.**

performance requirement: A requirement that defines a capability that a system or a component must possess, such as speed, accuracy, frequency, and stability. *See also* **performance parameter, requirement.**

performance specification: A specification that sets forth a performance requirement for a system or a component. *Synonym* **requirement specification.** *See also* **performance parameter, performance requirement, specification.**

periapsis: In a satellite orbit, the point that is closest to the gravitational center of the primary body about which the satellite is orbiting. *Note:* In Earth-based satellite systems, the periapsis is called the "perigee." *See also* **geostationary orbit, perigee, satellite.**

perigee: In an orbit of a satellite orbiting the Earth, the point that is closest to the gravitational center of the Earth. *See also* **apogee, apogee altitude, geostationary orbit, satellite, perigee altitude.**

perigee altitude: The altitude of the perigee above a specified reference surface that represents the surface of the Earth, i.e., the vertical distance from the perigee to a specified reference surface serving to represent the surface of the Earth. *See also* **apogee, apogee altitude, perigee.**

perimeter protection system (PPS): A system, such as a distribution of field disturbance sensors or optical fibers, installed around a facility to detect unauthorized entry or exit. *Note:* A field disturbance sensor might consist of buried fiber optic cables in which leaky modes are propagating.

period: 1. Elapsed time. **2.** In a regularly recurring phenomenon, such as a swinging pendulum or a sequence of equally spaced pulses of equal duration, the time required for each complete occurrence of the phenomenon. **3.** In a recurring wave, the reciprocal of the frequency. *See* **access period, antenna rotation period, constant failure period, distress period, early failure period, operational service period, outage period, overtime period, performance measurement period, satellite period, scanning line period, scanning period, synodic period, wearout failure period.** *See also* **frequency, periodic.**

periodic: Pertaining to a phenomenon, such as a recurring wave, an oscillating pendulum, or a vibrating elastic string, that (a) repeats itself during time intervals that are equally spaced, (b) may be a real mechanical, electrical, or magnetic phenomenon or may be a notional phenomenon, and (c) has an associated frequency or repetition rate. *Note 1:* The frequency of a periodic phenomenon is the reciprocal of the time interval that is required for one whole cycle of the repeating phenomenon. If the period is in seconds, the frequency will be in hertz. If the period is in microseconds, the frequency will be in megahertz. If the period is in nanoseconds, the frequency will be in gigahertz, and so on. *Note 2:* A signal is periodic if and only if any selected value is the same as a value taken a full cycle time later, given by the relation $v(t + T_c) = v(t)$ where $v(t)$ is a given signal parameter, such as voltage, current, or power as a function of time, T_c is the smallest constant period of repetition, i.e., is the fundamental period, and t is time, which may vary from minus infinity to plus infinity. *See also* **frequency, period, signal.** *Refer to* **Fig. P-4 (712).**

periodically distributed thin-film optical waveguide: A slab dielectric waveguide that is formed by evaporating, growing, etching, cementing, depositing, masking, or otherwise placing thin films of transparent dielectric material on substrates. *Note:* An example of a periodically distributed thin-film optical waveguide is a waveguide with a periodically varying width deposited on an optical integrated circuit (OIC) chip so that only selected modes are propagated while other modes are eliminated because of the geometric shape. *See also* **dielectric.** *Refer to* **Figs. P-2, P-3.**

periodic antenna: An antenna that has an approximately constant input impedance over a narrow range of frequencies. *Note:* An example of a periodic antenna is a dipole array antenna. *Synonym* **resonant antenna.** *See also* **aperiodic antenna, impedance, log periodic antenna.**

periodic waveform jamming: Jamming in which (a) periodic waveforms, such as saw-toothed waves, i.e., triangular waves, or boxcar waves, i.e., square waves, are used to modulate jamming transmitters, (b) usually amplitude or frequency modulation is used, (c) in radar systems, the jamming signal level is raised in the receiver, thereby obliterating echoes reflected from objects, i.e., from targets, or thereby producing error signals that indicate that there is a target when, in fact, there is none, (d) signals will often produce an appearance on the display that is less dense than noise, and (e) mixed noise and periodic signals can be used to modulate either the amplitude or the frequency of a jamming oscillator. *See also* **jamming.**

period processing: In a computer or data processing system, processing in which (a) various levels of sensitive information are processed at distinctly different times, and (b) the system is purged of all information from one processing session or run before transitioning to the next session or run when there are different users with different authorizations during each session or run. *See also* **communications security, periods processing.**

periods processing: In a data processing system, processing in which (a) various levels of classified or un-

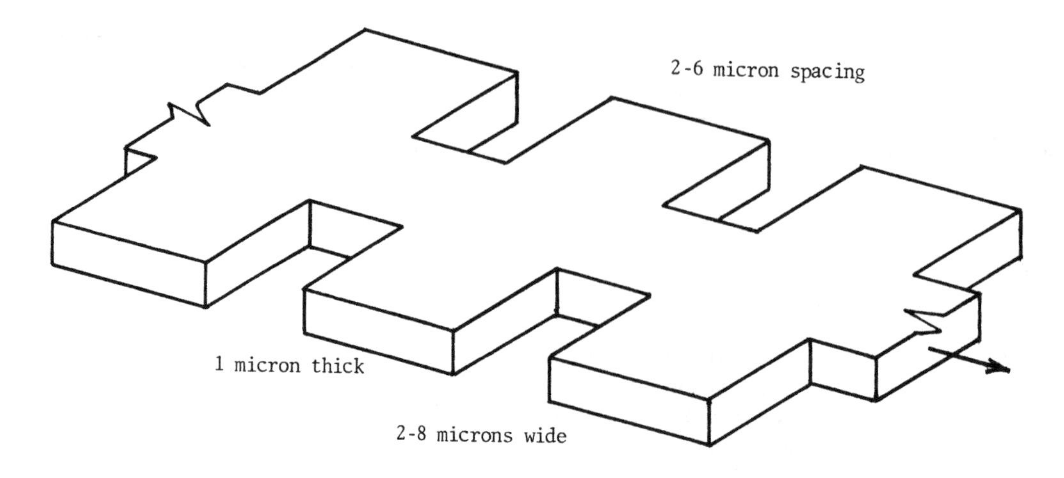

Fig. P-2. A slab dielectric **periodically distributed thin-film optical waveguide.** The arrow indicates the direction of a propagating electromagnetic wave. The guide will transmit lightwaves that have the wavelength that matches the spacing.

classified information are processed at distinctly different times, i.e., different periods, and (b) the system must be purged of all information from one processing period before transitioning to the next period when there are different users with differing authorizations. *See also* **communications security, period processing.**

period station: *See* **operator period station.**

period watch: *See* **single-operator period watch, two-operator period watch.**

peripheral: *See* **computer peripheral.**

peripheral device: 1. *Synonym* **peripheral equipment. 2.** *See* **computer peripheral device.**

peripheral equipment: In a communications, computing, data processing, or control system, online or offline equipment that (a) is associated with, but distinct from, the system, (b) provides the system with additional capabilities, (c) may be used independently from the system with which it is associated, and (d) may be shared by several users. *Note 1:* In a communications system, peripheral equipment is equipment that may or may not be connected to a circuit or a network. *Note 2:* In a computing or data processing system, peripheral equipment is distinct from the central processing unit (CPU) and is accessed via system input/output (I/O) channels. *Note 3:* Examples of peripheral equipment include printers, plotters, modems,

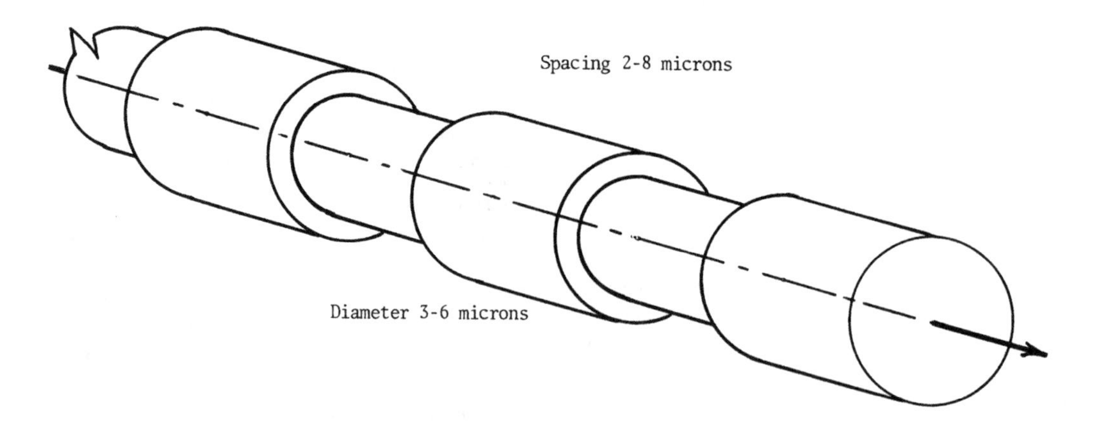

Fig. P-3. A circular-cross-section **periodically distributed thin-film optical waveguide.** The arrow indicates the direction of a propagating electromagnetic wave. The guide will transmit lightwaves that have the wavelength that matches the spacing.

auxiliary storage, and monitors. *Synonym* **peripheral device.** *See also* **keyboard, modem, terminal.**

peripheral node: *Synonym* **endpoint node.**

peripheral strength member fiber optic cable: A fiber optic cable that contains optical fibers that are surrounded by a group of high-tensile-strength materials, such as stranded or solid contrahelical or longitudinal steel wires, nylon or Kevlar strands, steel or copper tubes, or other material, with a crush-resistant jacket, sheath, or overarmor on the outside of the cable. *See also* **central strength member fiber optic cable.** *Refer to* **Fig. C-1 (102).**

peripheral transfer: The transfer of data from one piece of peripheral equipment to another, such as from a magnetic tape unit to a printer. *See also* **magnetic tape unit, peripheral equipment, printer.**

periscope antenna: An antenna consisting of (a) an active transmitting portion oriented to produce a vertical radiation pattern and (b) a flat or parabolically curved passive reflector mounted above the active transmitting portion to direct the beam in a horizontal path toward the receiving antenna. *Note:* The periscope antenna allows increased terrain clearance, short transmission lines, and location of the active equipment at or near ground level for ease of maintenance. The periscope antenna is often used in microwave relay antenna arrangements in which the feed and the antenna are at the base of a tower, and an angled passive reflector is at the top, aimed to reflect signals to the next tower. *See also* **antenna, microwave.**

permanent bond: A bond not expected to require disassembly for operational or maintenance purposes. *See also* **connection.**

permanent call tracing: Call tracing that is performed on all calls that are made through specified equipment, such as a given line, trunk, switchboard, exchange, or switching center. *See also* **call tracing.**

permanent circuit: In communications systems, a circuit that is permanently installed for use under all conditions.

permanent echo: In radar systems, return signals, i.e., echoes, that (a) are dense and fixed returns that are caused by the Earth's surface features or other fixed or moving objects, such as swaying trees, and (b) are distinguishable from ground clutter by being reflected from identifiable locations rather than large areas. *See also* **ground clutter.**

permanently locked envelope: In signal transmission, an envelope that is always separated by a number of bits that correspond to an integral number of envelopes. *See also* **bit, envelope, integral, signal, transmission.**

permanent signal (PS): In telephone systems, an extended off-hook condition not followed by dialing. *See also* **off-hook.**

permanent storage: Storage in which stored data are nonerasable. *See also* **read-only storage.**

permanent virtual circuit (PVC): A virtual circuit (a) that is used to establish a long-term association between data terminal equipments (DTEs), (b) in which the long-term association is identical to the data transfer phase of a virtual call, and (c) that eliminates the need for repeated call setup and clearing. *Deprecated synonym* **nailed-up circuit.** *See also* **circuit, hot line, off-hook service, virtual call capability, virtual circuit.**

permanent voice call sign: A call sign that is permanently assigned to a station or an organization, and that is to be used for voice transmissions. *Note:* An example of a permanent voice call sign is RED FOX permanently assigned to a radio net station. *Refer to* **Fig. P-5 (718).**

permeability: *See* **absolute magnetic permeability, incremental permeability, magnetic permeability, relative magnetic permeability.**

permissible interference: Observed or predicted interference that complies with quantitative interference and sharing criteria contained in International Radio Consultative Committee (CCIR) Recommendations, in the Radio Regulations (RRs), or in special agreements made in accordance with the RRs. *See also* **accepted interference, interference.**

permittivity: *See* **electric permittivity, relative electric permittivity.**

permutation table: In communications security systems, such as cryptosystems, a table of data items that is used to (a) systematically construct code groups for encoding messages and (b) correct garbles that occur in coded or encrypted text. *See also* **cryptosystem, encrypt, garble.**

permuter: In secure communications systems, a device that is used in cryptoequipment to change the order in which the contents of a shift register are used in various nonlinear combining circuits. *See also* **cryptoequipment, secure communications, shift register.**

Perot fiber optic sensor: *See* **Fabry-Perot fiber optic sensor.**

persistence: *See* **image persistence.**

personal agent (PA): A software-driven system that (a) scans incoming data, such as electronic mail, (b) identifies data of interest in accordance with keywords, such as news articles on news wire lines, and (c) informs the user of the data or routes the data to other users or to storage. *Synonym* **personal assistant.**

personal assistant: *Synonym* **personal agent.**

personal call: *Synonym* **private call.**

personal communications (PC): A communications capability that (a) enables a person to communicate with others from any location, (b) usually provides multimedia services, and (c) usually is based on a device that (i) is highly portable, (ii) is lightweight, (iii) is the size of an attache case or smaller, and (d) is tied to broadband integrated services digital networks (BISDNs). *Note:* Examples of personal communications (PC) include (a) the interconnection of personal computers via Internet and (b) cellular telephone systems. *See also* **personal computer.**

personal communications service (PCS): 1. A communications service that (a) is provided by a personal communications system (PCS) and (b) provides a communications capability between individuals via portable communications terminals and handsets. **2.** A wireless communications service that (a) is supported by ground and satellite communications systems, (b) is provided to persons on the move, and (c) usually includes hand-held computers that allow users to (i) communicate with each other from any location, (ii) access electronic messages, (iii) access multimedia information systems and databases, and (iv) transmit and receive information via facsimile. *Synonyms* **nomadic communications service, personal communications system service.**

personal communications system: A family of mobile or portable separate radio services for individuals and businesses that (a) is integrated with a variety of national networks, (b) provides advanced voice paging, two-way acknowledgment paging, and digital data services, (c) may have fixed or broadcast capability if part of the portable/mobile service, (d) is divided into 50 kHz paired and unpaired channel categories nationwide, (e) has five 50/50 kHz (kilohertz) unpaired, three 50/12.5 kHz paired, and two 50 kHz unpaired channels, and (f) operates in the range from 1850 MHz (megahertz) to 1990 MHz, a total of 140 MHz of assigned frequencies. *See also* **cellular telephone, nomadic communications service, paging.**

personal communications system service: *Synonym* **personal communications service.**

personal computer (PC): A self-contained, stand-alone computer, usually with sufficient internal and external storage capacity to run a variety of software programs. *Note:* A typical personal computer configuration might consist of (a) a computer unit or a central processor with high speed internal storage and one or more disk drives, (b) one or more input devices, such as a keyboard, mouse, light pen, or joy stick, and (c) one or more output devices, such as a printer, display device, or communications link. Personal computers usually are designed to accept computer programs on diskettes, i.e., on floppy or flexible disks, on disk cartridges, or on hard disks. The programs usually are transferred, i.e., booted, to internal storage by the disk operating system (DOS), such as IBM PC DOS or MS DOS. Data files and other operating software features usually are stored on disks or implemented by firmware or hardware. Special operating features may be added with plug-in printed circuit (PC) boards on which special circuitry and chips are mounted. *See also* **computer, personal communications.** *Refer to* **Figs. C-7 (157), V-4 (1067).**

Personal Computer Memory Card International Association (PCMCIA): An association devoted to the development and the standardization of plug-in memory cards for personal computers, such as the PCMCIA Type 2 card that (a) once installed, communicates using a carrier-sense multiple-access with collision avoidance (CSMA/CA) scheme and (b) uses radio rather than cable as the propagation medium.

personal identifier: *See* **identification friend or foe personal identifier.**

personal sign: In communications systems, a call sign that (a) is composed of one or more letters, usually personal initials, and (b) is used when endorsing station records and messages to indicate the responsible operator or supervisor.

personal telecommunications (PT): Telecommunications in which each user (a) is provided access to telecommunications services that may be obtained during personal mobility, (b) is provided a user-defined set of subscriber services, (c) may initiate and receive calls, (d) is assigned a unique, personal, network-independent number that transcends multiple networks in order to reach any fixed, movable, or mobile terminal irrespective of geographic location, and (e) obtains service subject only to the limitations that are imposed by (i) terminal and network capabilities and (ii) restrictions imposed by the network provider. *See* **Universal Personal Telecommunications.**

personnel locator beacon: In search and rescue communications, a beacon that (a) is capable of providing homing signals and (b) is used to enable the location of personnel who are in distress. *See also* **beacon, homing, search and rescue communications, signal.**

personnel security: In communications security, procedures that are used to ensure that persons who have requested access, or have a need for access, to classified or sensitive information have the required authority, need to know, and clearance for that information. *See also* **communications security.**

person to person call: A call that requires operator intervention, even in automatically switched systems, in order to verify that the call receiver who was explicitly requested by the call originator is, in fact, on the line. *See also* **call, line, local call, number call, station to station call.**

perturbation: In communications, computer, data processing, and control system circuits or lines, a distinct disturbance, surge, or spike that (a) usually is in the form of a sudden shift and return in the value of a parameter, such as voltage, current, power, phase, frequency, or polarization, (b) usually is identifiable and of short duration, and (c) usually is a form of interference, though deliberate insertion may be performed for test purposes. *See* **phase perturbation.** *See also* **value.**

peta (P): A prefix that indicates multiplication by 10^{15}, i.e., by (a) a thousand million million or (b) in the United States, by a thousand trillion. *Note:* An example of the use of "peta" is in 10 PHz (petahertz), equivalent to 10^{16} Hz.

petahertz (PHz): A unit of frequency equal to 1 thousand trillion hertz, i.e., 10^{15} Hz, about the frequency of visible violet light. *See also* **hertz.** *Refer to* **Table 4, Appendix B.**

Petri net: An abstract model of information flow in a computer, communications, data processing, or control system showing the static and dynamic properties of the system. *Note:* A Petri net usually is represented as a network having nodes that are connected by arcs with markings indicating the dynamic properties of the network branches.

PF: packing fraction.

PFM: pulse frequency modulation.

phantom circuit (PC): A third circuit derived from two suitably arranged pairs of wires, called "side circuits," with each pair of wires being a circuit in itself and at the same time acting as one conductor of the third circuit. *Note:* The side circuits are coupled to their respective drops by center-tapped transformers, usually called "repeat coils." The center taps are on the line side of the side circuits, and current from the phantom circuit is split evenly by the center taps. This arrangement cancels crosstalk from the phantom circuit to the side circuits. *See also* **circuit, logical circuit, physical circuit, side circuit, simplex circuit, simplexed circuit, simplex signaling, virtual circuit.**

phantom group: Three circuits that are derived from simplexing two physical circuits to form a phantom circuit. *See also* **circuit, line filter balance.**

phase: **1.** A distinguishable state of a phenomenon. **2.** The instant at which a specified function occurs in a sequential list of functions. **3.** In a periodic function or phenomenon, such as the electric field vector in an electromagnetic wave, the instant, event, or position at which a specified significant parameter of the function occurs relative to a given time reference or relative to a significant instant, event, or position in another function or phenomenon. *Note 1:* Phase can be identified by the time of its occurrence, elapsed from a specified reference correctly called "phase time," but frequently abbreviated to "phase." If the phenomenon is periodic, such as a sine wave or a square wave, the phase can be identified either by angle or by time, both measured from a specified reference, depending on the dimensions assigned to the reference period. The period may be divided into 360° (electrical degrees), 2π radians, or time units that are fractions of the period, such as nanoseconds, for a period of 1 μs (microsecond). A transient phenomenon may also be divided into phases. *Note 2:* For a sinusoidal function, phase may be represented (a) in polar coordinates by $A\angle\Theta$ where A is the magnitude and Θ is the phase angle, and (b) in Cartesian coordinates by $a + jb$ where a is the real component, b is the imaginary component, i.e., the quadrature component, and the phase angle $\Theta = \tan^{-1} b/a$. *Note 3:* An example of a phase relationship is the angular displacement between the peak values of two sinusoidal waves of the same frequency. Thus, when lightwaves propagate in an optical fiber, waves of different wavelength are shifted in phase over given lengths of the fiber by different amounts because of their different propagation velocities, giving rise to distortion of signals because the signals usually consist of more than one optical frequency, i.e., more than one wavelength. *See* **access phase, assembly phase, disengagement phase, information transfer phase.** *Refer to* **Figs. D-2 (233), I-4 (459), P-4 (712).**

phase alternation by line (PAL): The television (TV) signal standard that calls for 625 lpf (lines per frame) and 50 Hz (hertz), 220 V (volts, rms) primary power. *Note:* Phase alternation by line (PAL) is used in the United Kingdom, much of the rest of western Europe, several South American countries, some Middle East and Asian countries, several African countries, Australia, New Zealand, and other Pacific island countries. *See also* **National Television Standards Committee standard, phase alternation by line (modified), système electronique coleur avec memoire.**

phase alternation by line—modified: *Synonym* **phase alternation by line (modified).**

phase alternation by line (modified) (PAL-M): A modified version of phase alternation by line (PAL) such that there are 525 lpf (lines per frame) and 50 Hz (hertz), 220 V (volts, rms) primary power rather than 625 lpf (lines per frame) and 50 Hz (hertz), 220 V (volts, rms) primary power as in PAL. *Note:* Phase alternation by line (modified) (PAL-M) is used in Brazil. *Synonym* **phase alternation by line—modified.** *See also* **National Television Standards Committee standard, phase alternation by line, système electronique couleur avec memoire.**

phase-amplitude distortion: In a nonlinear system, distortion of an input signal that (a) occurs because the phase shift introduced by the system is different for different amplitudes of the input signal, and (b) is measured while the system is operated under steady state conditions and usually with a sinusoidal input wave form. *Note 1:* An amplifier introduces some phase shift, somewhat like a delay line. The phase shift should be constant for all frequencies, and their amplitudes, in the input signal. If the phase shift is different for different amplitudes, the signal will be distorted. This distortion is phase-amplitude distortion. *Note 2:* Nonlinearity of frequency versus gain is another cause of distortion. *See also* **distortion, nonlinear distortion, steady state condition.**

phase angle: In a periodic wave, the angle that corresponds to a point on the wave relative to a point on a frame of reference. *Note 1:* Units of measure of phase angle include angular degrees, electrical degrees, radians, wavelengths, and time, such as nanoseconds, when the frequency is known. *Note 2:* The frame of reference for a phase angle usually is a Cartesian coordinate system on which a temporal or spatial plot of the wave may be made such that the magnitude of the wave is the ordinate, and the unit of measure of the phase angle is the abscissa. *Note 3:* For one cycle of the wave, the abscissa is usually divided into 360°, 2π radians, or time units. *Note 4:* When a full cycle of the wave is plotted on the frame of reference, the reference point may be any specified point on the wave, such as a point where the wave crosses the abscissa axis of the frame of reference, a significant instant, or the leading edge of a timing pulse. On a time plot, the leading edge is on the left. On a spatial plot with distance increasing toward the right, the leading edge is in the direction of propagation, which is to the right. *Note 5:* The value of a phase angle of a point on the wave is the point on the abscissa that corresponds to the point on the wave. *Note 6:* The phase angle of a quantity is usually written as $A\angle\theta$ where A is the magnitude, and θ is the phase angle relative to a specified reference. *Refer to* **Fig. P-4.** *Refer also to* **Figs. D-2 (233), I-4 (459), Q-1 (786).**

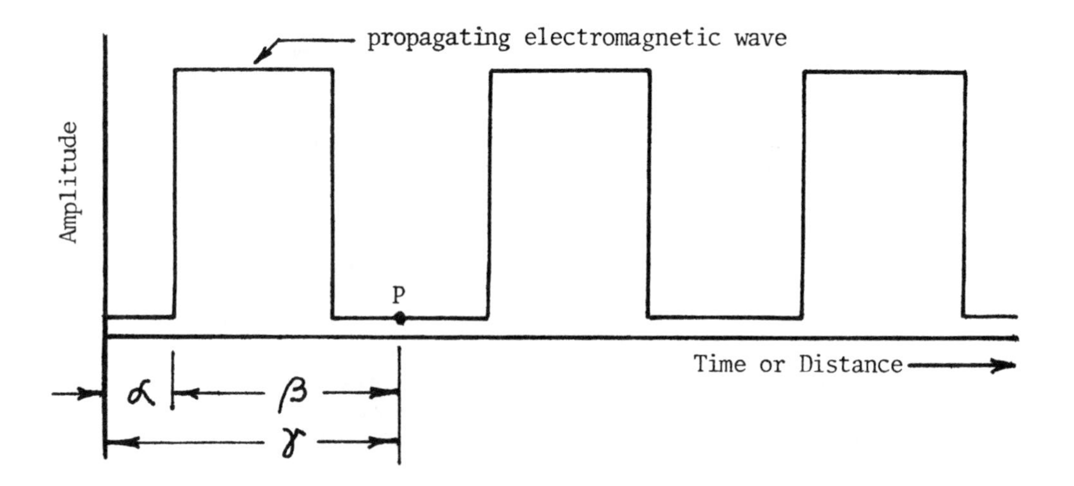

Fig. P-4. Plot in which α is the absolute **phase angle** of a transition at a significant instant of a periodic wave relative to a fixed spatial or temporal reference, β is the relative phase angle of a point P on the wave relative to the significant instant, and γ is the absolute phase angle of the point P relative to the fixed reference. The wave need not be a square wave. All phase angles may be expressed in electrical degrees, radians, or time units.

phase bandwidth: In a network or a device, the width of the continuous frequency band over which the phase versus frequency characteristic does not depart from linearity by more than a stated amount. *See also* **bandwidth, frequency, phase.**

phase change coefficient: At a point in a transmission line and at a given frequency, the phase change per unit length, i.e., the phase shift per unit length, β, given by the relation $\beta = d\phi/dl$ or $d\phi = \beta dl$ where β is the phase change coefficient, $d\phi$ is the differential phase shift, and dl is the differential distance over which $d\phi$ occurs and, if the transmission line is uniform, $\beta = B/L$ or $B = \beta L$ where B is the total phase shift over the length of the line, β is the phase change coefficient, and L is the length of the line. *See also* **transmission line, uniform transmission line.**

phase-coherence: The state in which two signals maintain (a) a fixed phase relationship with each other or (b) a fixed phase relationship with a third signal that can serve as a reference for each of the two signals. *See also* **coherent, frequency, frequency synthesizer, phase.**

phase-coherent: Pertaining to two signals that maintain (a) a fixed phase relationship with each other or (b) a fixed phase relationship with a third signal that can serve as a reference for each of the two signals. *See also* **coherent, frequency, frequency synthesizer, phase.**

phase constant: *Synonym* **phase term.**

phased array: An arrangement of any type of antennas in which (a) the signal feeding each antenna can be varied in such a way that radiation is reinforced in a desired direction and suppressed in undesired directions, (b) rapid scanning in azimuth and elevation can be accomplished by signal feed control to each antenna, and (c) signal variation among the antennas in the array usually is a variation in relative phase or magnitude among them. *Note:* An example of a phased array is an antenna array of dipoles in which the phase of the signal that is fed to each dipole is varied in such a way that beams can be formed, shaped, and directed very rapidly in azimuth and elevation without requiring physical movement of any of the antennas in the array. *See also* **antenna, multielement dipole antenna, reflective array.**

phase delay: In the transmission of a single-frequency wave from one point to another, the delay of the point in the wave that identifies its phase. *Note:* Phase delay may be expressed in any convenient unit, such as (a) seconds, degrees, radians, or fractions of a wavelength, such as 10^{-6}, or (b) the ratio of the total phase shift in seconds to the frequency in hertz. *See also* **absolute delay, coherent, delay equalizer, phase.**

phase departure: **1.** In a wave propagating in a transmission line or waveguide, a phase deviation from a specified value. **2.** In a wave propagating in a transmission line or waveguide, an unintentional deviation from the nominal phase value. *See also* **coherent, error, phase, phase jump.**

phase detector: A circuit or an instrument that detects the difference in phase between corresponding points on two signals. *Note:* If signals are synchronous, the phase difference between them is constant. *See also* **phase.**

phase deviation: In phase modulation, the maximum difference between the instantaneous phase angle of the modulated wave and the phase angle of the carrier. *Note:* For a sinusoidal modulating wave, the phase deviation, expressed in radians, is equal to the modulation index. *See also* **angle modulation, carrier (cxr), differential phase-shift keying, frequency modulation, modulation index, phase, phase difference, phase-frequency distortion, phase inversion, phase modulation, phase shift.**

phase diagram: A graphic representation of the phase relationships between two or more waveforms. *Note 1:* A phase diagram may be represented as a "vector" diagram, i.e., a phasor diagram, or as a magnitude plotted as a function of time. *Note 2:* Each "vector" on the phase diagram is represented by a line in a given direction. The length of the line represents the magnitude of the "vector," and the direction of the line represents the phase, indicated by an angle from a specified reference direction; hence such lines are often called "phasors." Each plotted "vector" may be identified as $A\angle\Theta$ where A is the magnitude, and Θ is the phase angle relative to the specified reference. *Note 3:* These "vectors" indicate magnitude and phase, not the magnitude and spatial direction of a vector quantity. The lines on a phase diagram have the appearance of vectors that represent the magnitude and spatial direction of a vector quantity, and thus they are often called "vectors." *See also* **coherent, phase, phase deviation, phase modulation, phasor.**

phase difference: **1.** The time or the angle by which one wave leads or lags another. **2.** The difference in phase angle between specified points on a given wave or between specified points, one on each of two waves. **3.** The relative phase of two points on the same wave or on different waves. *See also* **coherent, phase, phase deviation, phase modulation, phase offset.**

phase distortion: *Synonym* **delay distortion.**

phased orbit: An orbit in a system that uses a group of satellites in which the orbit position of an individual satellite has a fixed or controlled relationship with one or more other satellites in the group. *See* **randomly phased orbit.** *See also* **orbit.**

phase encoded (PE) recording: Binary recording on magnetic media, such as magnetic disks, tapes, and cards, in which a 1 is represented by a magnetic flux reversal to the polarity of the interblock gap, and a 0 is represented by a magnetic flux reversal to the polarity opposite to that of the interblock gap, both when recording in the forward direction. *See also* **code, magnetic tape, phase, phase flux reversal.**

phase equalizer: *Synonym* **delay equalizer.**

phase exchange keying: In the phase modulation of an electromagnetic wave, a transition from one phase-significant condition to another that is achieved by reducing the amplitude of the carrier component of one phase to zero while replacing it with the carrier component of the other phase. *Note:* The phase and the amplitude of the output signal are determined at any instant by the vector sum of the two components. In the case of a transition between two phase-significant conditions that are separated by 180°, one intermediate state is the exact cancellation of the two components. This results in an amplitude of zero. Phase exchange keying (a) is a form of balanced amplitude modulation and (b) can be implemented by presenting a shaped baseband signal to a balanced modulator that is excited by the desired carrier. *Synonym* **phase exchange signaling.** *See also* **balanced amplitude modulation, phase keying, significant condition.**

phase exchange signaling: *Synonym* **phase exchange keying.**

phase flux reversal: In phase encoded recording, a magnetic flux reversal written at the nominal midpoint between successive 1 bits or between successive 0 bits to establish proper magnetic polarity. *See also* **phase, phase encoded recording.**

phase-frequency characteristic: A Cartesian coordinate plot, i.e., an x-y plot, in which the phase shift introduced by a circuit or a device is plotted as the dependent variable, i.e., the ordinate or y-axis value, and the input frequency as the independent variable, i.e., the abscissa or x-axis value. *Note 1:* A circuit has phase linearity if the phase shift for each frequency in the input signal is proportional to that frequency, i.e., the phase-frequency characteristic (phase versus frequency characteristic) is a straight line through the origin and with proper positive slope. Distortion will not occur because of phase shift. The phase relationships in the input signal will be preserved by the circuit. *Note 2:* A circuit has phase nonlinearity if the phase shift for each frequency in the input signal is not proportional to that frequency, i.e., the phase-frequency characteristic (phase versus frequency characteristic) is not a straight line through the origin and with proper positive slope. Distortion will occur because of the different phase shifts for the different frequencies in the input signal. The phase relationships in the input signal will not be preserved by the circuit. *See also* **phase-frequency distortion, phase linearity, phase nonlinearity.**

phase-frequency distortion: 1. Distortion in which the phase-frequency characteristic is not linear over the frequency range of interest, i.e., the phase shift that is introduced by a circuit, component, or system is not linear for all frequencies of input signal. **2.** Distortion that occurs when the phase-frequency characteristic (phase versus frequency characteristic) either (a) is not a straight line with the proper positive slope in the frequency range of interest or (b) has a zero frequency intercept that is not zero or is not an integral multiple of 2π radians, or (c) both. *Note: See also* **error, frequency, phase, phase deviation, phase-frequency characteristic, phase linearity, phase modulation, phase nonlinearity.**

phase-frequency equalizer: A circuit or a network that compensates for phase-frequency distortion within a specified frequency band. *See also* **delay equalizer, group delay.**

phase front velocity: In an electromagnetic wave, such as a radio wave or a lightwave, the velocity of a specified point on a wavefront, the wavefront being a surface described by the locus of all points at which the same significant instant occurs in the wave, such as the surface that describes the locus of all points at which the peak value of the electric field vector is a constant. *See also* **electromagnetic wave.**

phase hit: 1. In a transmission system, a momentary perturbation that is caused by a sudden phase shift in a signal. **2.** In a transmission system, a sudden phase shift caused by (a) a disturbance generated in the system or (b) a disturbance from outside the system. *See also* **perturbation, phase shift.**

phase instability: The fluctuation of the phase of a wave relative to a reference. *Note:* Phase instability is often due to unknown causes. *See also* **phase depar-**

ture, phase deviation, phase jitter, phase perturbation.

phase interference fading: The variation in signal amplitude produced by the interaction of two or more signal elements with different relative phases. *See also* **diurnal phase shift, fading, flutter, interference, ionosphere, phase.**

phase inversion: The production of a phase difference of 180° or π radians. *Note:* Phase inversion usually is produced by the inversion of a symmetrical periodic signal, resulting in a change in sign. A signal represented by $f(t) = Ae^{j\omega t} = A(\cos \omega t + j \sin \omega t)$, after phase inversion, becomes $f_1 = Ae^{j(\omega t + \pi)} = -Ae^{j\omega t} = A((\cos (\omega t + \pi) + j\sin (\omega t + \pi)) = -A(\cos \omega t + j\sin \omega t)$ where A is the magnitude, ω is the angular velocity equal to $2\pi f$ where f is the frequency, and t is time. The vector sum of $f(t)$ and $f_1(t)$ is always zero. *See also* **phase, phase perturbation.**

phase jitter: 1. A rapid or repeated phase perturbation that results in the intermittent shortening or lengthening of signal elements. **2.** Relative to a reference, an undesired, rapid, usually small shifting of the phase of a signal. *Note 1:* Phase jitter may be random or cyclic. *Note 2:* The phase departure in phase jitter usually is smaller, but more rapid, than that of phase perturbation. Phase jitter may be expressed in degrees, radians, or seconds. Phase jitter usually is random. However, if cyclic, phase jitter many be expressed in hertz as well as in degrees, radians, or seconds. *See also* **jitter, phase, phase perturbation.**

phase jump: A sudden phase change in a signal. *See also* **phase, phase departure.**

phase keying: In signaling systems, keying in which digital information is transmitted by a representation of each of the possible signaling states that can be assumed by the phase of a carrier. *Note:* In a practical system of limited bandwidth, an abrupt transition from one phase-significant condition to another cannot be achieved while maintaining constant amplitude. *See also* **amplitude, keying, phase, phase-shift keying, phase exchange keying, significant condition.**

phase linearity: In the phase-frequency characteristic, i.e., the phase versus frequency characteristic, of a circuit or a device, the condition that exists when the phase shift for each frequency in the input signal introduced by a circuit is directly proportional to that frequency in the input signal, i.e., the phase versus frequency characteristic is a straight line through the origin and with a positive slope. *Note:* When phase linearity occurs, the input signal to a circuit will not be distorted because the phase shift of the different frequencies in the input signal will be proportional to frequency. All frequencies will preserve the phase relationship they had at the input. *See also* **frequency, phase, phase-frequency characteristic, phase nonlinearity.**

phase-locked loop (PLL): 1. An electronic circuit that (a) controls an oscillator so that it maintains a constant phase angle relative to a signal from a reference signal source, (b) is used in situations in which signals that are shifted in phase with respect to one another can maintain a fixed phase relationship, (c) in spread spectrum systems, is used to cause an oscillator internal to the feedback loop to oscillate at the incoming carrier frequency, (d) has a feedback circuit, i.e., a servoloop, that uses the output of a phase-sensitive detector, via a low-pass filter, to control the frequency of its own reference signal, (e) has a feedback loop that is damped to permit tracking of the carrier phase changes at the input, but not tracking of the modulation changes, and (f) provides a low noise threshold. **2.** A circuit for synchronizing a variable local oscillator with the phase of a transmitted signal. *Note:* Phase-locked loops (PLLs) are widely used in space communications for coherent carrier tracking and threshold extension, bit synchronization, and symbol synchronization. *See also* **acquisition, carrier synchronization, Costas loop, digital phase locked loop, false clock, loop, phase, threshold.**

phase measurement tolerance: The maximum allowable difference between a measured phase value and the actual phase value. *See also* **error, phase, value.**

phase meter: An instrument that may be used for measuring and indicating the difference in phase between two alternating quantities that have the same frequency.

phase modulation (PM): Angle modulation in which the phase angle of a carrier is caused to depart from its reference value by an amount proportional to the instantaneous value of the modulating signal. *See also* **angle modulation, carrier, continuous phase modulation, deviation ratio, differential phase-shift keying, frequency modulation, modulation index, phase, phase deviation, phase difference, phase-frequency distortion.**

phase modulator: *See* **optical phase modulator.**

phase noise: Rapid, short-term, random fluctuations in the phase of a wave, caused by time domain instabilities in circuit elements, such as those of an oscillator. *Note:* Phase noise power, N_ϕ, in dB relative to carrier power (dBc) on a 1-Hz (hertz) bandwidth, is given by the relation $N_\phi = 10\log 0.5(S_\phi)$ where N_ϕ is the phase noise power, and S_ϕ is the normalized spectral density of the phase fluctuations. *See also* **spectral density.**

phase nonlinearity: In the phase-frequency characteristic, i.e., phase versus frequency characteristic, of a circuit or a device, the condition that exists when the phase shift for each frequency in the input signal introduced by a circuit is not directly proportional to that frequency in the input signal, i.e., the phase versus frequency characteristic is not a straight line with a positive slope, or, if it is a straight line with a positive slope, it does not pass through the origin. *Note:* When phase nonlinearity occurs, the input signal to a circuit will be distorted because the phase shift of the different frequencies in the input signal will not be proportional to frequency. All frequencies will not preserve the input phase relationship they had at the input. *See also* **frequency, phase, phase-frequency characteristic, phase linearity.**

phase offset: The algebraic phase difference between a given signal and a specified reference signal. *See also* **phase difference.**

phase oxidation process: *See* **chemical vapor phase oxidation process, inside vapor phase oxidation process, modified inside vapor phase oxidation process.**

phase perturbation: 1. Relative to a reference, an undesired, usually rapid, shifting of the phase of a signal. *Note 1:* The phase perturbation may be random, cyclic, or a combination of both. *Note 2:* The phase departure in phase perturbation usually is larger, but less rapid, than that of phase jitter. Phase perturbation may be expressed in degrees, radians, or seconds. A cyclic phase perturbation many be expressed in hertz as well as in degrees, radians, or seconds. **2.** The cause of a rapid shifting of the relative phase of a signal. The phase shift may be random, cyclic, or a single shift. Other signal parameter perturbations may also occur along with a phase perturbation, such as superimposed noise and spikes caused by propagation media anomalies, propagation media abrupt changes, and interference. The phase perturbation may be expressed in electrical degrees or radians. A cyclic perturbation, i.e., a recurring perturbation, of any sort may be expressed in hertz. Phase perturbation may be negligible in a given transmission system. *See also* **error, jitter, phase, phase jitter.**

phase quadrature: *See* **quadrature.**

phase reverse keying: Phase-shift keying in which the phase of a sinusoidal carrier is switched plus or minus 90° (electrical) by the transitions of a digital data stream. *See also* **carrier, phase-shift keying.**

phase-sensitive detector-demodulator: A demodulator (a) in which a local oscillator reference voltage at the received carrier frequency is applied as a parallel input to a diode balanced circuit, with the phase of the reference voltage in quadrature with the carrier signal that is applied as a differential input, (b) in which the output, for phase differences of less than 90° (electrical) between a modulated carrier input and the reference, is proportional to the sine of the phase difference angle, and (c) that acts as an amplitude detector if the reference voltage is arranged to be in phase with the received carrier. *See also* **carrier, demodulator, quadrature.**

phase shift (PS): 1. On a waveform, a change in phase of a point with respect to a reference. **2.** The change in phase of a periodic signal with respect to a reference. *See* **compensating phase shift, diurnal phase shift.** *See also* **differential phase-shift keying, frequency shift, phase, phase deviation, phase-shift keying.**

phase shift coefficient: *Synonym* **phase change coefficient.**

phase-shift keying (PSK): 1. In digital transmission, angle modulation in which the phase of the carrier is discretely varied in relation to (a) a reference phase or position or (b) the phase of the previous signal element, in accordance with the data being transmitted. **2.** In a communications system, the representing of characters, such as bits, quaternary digits, or octal digits, by shifting the phase of a carrier wave with respect to a reference, by an amount corresponding to each symbol being encoded. *Note 1:* Examples of phase-shift keying are, when encoding bits, (a) the phase shift could be 0° for encoding a 0 and 180° for encoding a 1, and (b) the phase shift could be −90° for encoding a 0 and +90° for encoding a 1, thus making the representations for 0 and 1 a total of 180° degrees apart in both examples. *Note 2:* In phase-shift keying (PSK) systems designed so that the carrier can assume only two different phase angles, each change of phase, i.e., each signal element, represents one bit of information, i.e., the bit rate equals the modulation rate. Thus, each bit uses 2π radians of the carrier. If the number of recognizable phase angles is increased to four, then two bits, i.e., quaternary digits, of information can be encoded into 2π radians of the carrier. Likewise, eight phase angles can encode three bits, i.e., octal digits, in each 2π radians of the carrier. *Synonym* **phase-shift signaling.** *See* **differentially coherent phase-shift keying, differential phase-shift keying, filtered symmetric differential phase-shift keying, quadrature phase-shift keying.** *See also* **carrier, digital modula-**

tion, digital phase modulation, double frequency shift keying, frequency modulation, frequency shift keying, keying, modulation, phase, phase shift.

phase-shift signaling: *Synonym* **phase shift keying.**

phase slew: *See* **clock phase slew.**

phase telegraphy: *See* **relative phase telegraphy.**

phase term: In the propagation of an electromagnetic wave in a uniform waveguide, such as an optical fiber or a metal waveguide, the parameter that (a) is the imaginary part of the propagation constant for a particular mode, (b) indicates the phase change per unit distance of the wave at any point along the waveguide, and (c) is represented by h in the expression for the exponential variation of the electric field strength of the propagating wave, given by the relation $E = E_0e^{-pz} = E_0e^{-ihz-az}$ where E_0 is the initial electric field strength at $z = 0$, p is the propagation constant, $p = h + a$, where a is the attenuation term caused by absorption and scattering, z is the distance from where E_0 is taken and in the direction of propagation, and i is the complex operator, $i = (-1)^{1/2}$, identifying the quadrature phase component of the exponent. *Note 1:* The phase term usually is expressed in radians per unit length. *Note 2:* The phase term h is dependent upon the electric permittivity and magnetic permeability of the material filling the guide, the frequency of the electromagnetic wave, and the modal characteristics of the wave. *Note 3:* The phase term may also be applied to the magnetic field, in which case H may be substituted for E, and H_0 may be substituted for E_0. *Synonym* **phase constant.** *See also* **attenuation term, mode, phase, propagation constant.**

phase tolerance: The maximum permissible departure of the phase of a signal from a desired or assigned value.

phase velocity: In a uniform plane polarized electromagnetic wave propagating in a medium, the velocity of propagation of the wave, i.e., the velocity an observer would have to move to make the wave characteristic appear to remain constant in phase in a given propagation medium, given by (a) the product of the free space wavelength, i.e., the source wavelength, and the frequency of the wave divided by (b) the refractive index of the medium. *Note 1:* In free space, the refractive index is unity, and the group and phase velocities are equal. *Note 2:* In SI (Système International d'Unités) units, the refractive index of free space may be considered to be unity because the refractive indices of materials are normalized with respect to that of a vacuum. Phase velocity can only be applied to a single-frequency wave, such as an unmodulated carrier wave from a single-frequency or monochromatic source. For a given frequency of an electromagnetic wave, which remains constant across material media interfaces, the speed is less in material media than in free space. In nondispersive media, i.e., media that do not cause dispersion, the phase velocity and the group velocity are equal. *See also* **coherence time, group index, group velocity, phase, propagation constant.**

phasing: In facsimile transmission, the ensuring of the coincidence between a point on the scanning field at the receiver with the corresponding point at the transmitter so as to ensure that the positioning of the picture or data on the record medium is correct. *See also* **record medium.**

phasing on black: In facsimile transmission, phasing that (a) is between the transmitter and the receiver, (b) is ensured by a phasing signal that corresponds to nominal black, and (c) is accomplished by a short interruption, corresponding to nominal black, that is transmitted during lost time. *See also* **lost time, nominal black, phasing on white, phasing signal.**

phasing on white: In facsimile transmission, phasing that (a) is between the transmitter and the receiver, (b) is ensured by a phasing signal that corresponds to nominal white, and (c) is accomplished by a short interruption, corresponding to nominal white, that is transmitted during lost time. *See also* **lost time, nominal black, phasing on white, phasing signal.**

phasing signal: In facsimile transmission, a signal that (a) is sent by a transmitter to the corresponding receiver, (b) ensures that the proper phase relationship exists between the receiver and the transmitter, and (c) ensures that the receiver scanner is at the corresponding point on the record medium as the transmitter scanner is on the object being scanned. *See also* **object, record medium.**

phasor: In electric circuits bearing alternating sinusoidal voltages and currents, a representation, as vectors, of the complex impedances, voltages, and currents, such as (a) the magnitude and the phase angles of the voltages across each lumped impedance relative to the current in a series circuit or (b) the phases of the current in each leg of a parallel circuit relative to the applied voltage. *Note:* The phasors must satisfy Kirchoff's laws in a given circuit.

phasor impedance: *Synonym* **vector impedance.**

phon: In acoustics, a unit of subjective loudness level equal to the sound pressure level in dB compared to that of an equally loud standard sound. *Note 1:* The accepted standard is a 1 kHz (kilohertz) pure sine wave tone or narrowband noise centered at 1 kHz. For example, a sound that is judged to be as loud as a 40 dB, 1-kHz tone has a loudness level equal to 40 phons. *Note 2:* The minimum perceptible sound level of the normal human ear in air corresponds to a sound pressure of 20 μPa (micropascal) or 20 $\mu N \bullet m^{-2}$ (micronewtons per square meter) at 1 kHz (kilohertz). Thus, 40 phons would be 40 dB above that minimum perceptible level of the normal human ear.

phonation: The opening and the closing of the vocal folds or cords between 100 and 200 times per second, emitting puffs of air that constitute voice.

phone: 1. *Synonym* **telephone. 2.** The voice operation mode in radiotelephone communications. *See* **hydrophone, videophone.**

phone patch: *Synonym* **telephone patch.**

phone service: *See* **Advanced Mobile Phone Service.**

phone system standard: *See* **Advanced Mobile Phone System standard.**

phonetic alphabet: 1. A list of standard common words used to identify voiced letters in a message transmitted by radio or telephone. *Note:* The following are the internationally adopted words, listed in alphabetical order, for each letter in the English alphabet: **A**lpha, **B**ravo, **C**harlie, **D**elta, **E**cho, **F**oxtrot, **G**olf, **H**otel, **I**ndia, **J**uliet, **K**ilo, **L**ima, **M**ike, **N**ovember, **O**scar, **P**apa, **Q**uebec, **R**omeo, **S**ierra, **T**ango, **U**niform, **V**ictor, **W**hiskey, **X**-ray, **Y**ankee, **Z**ulu. *See also* **alphabet. 2.** A special alphabet in which each character represents a specific sound that is pronounceable by a person. *Note:* The symbol θ might be used to represent the sound of the TH in such words as thing, through, thread, and theta, and not for the sound of the TH in words such as the, these, those, and them. The initial vowel sound in the words southern and brother would be represented by the same symbol. Thus, words spelled one way with these symbols would be pronounced the same way by everyone who has learned to use the phonetic alphabetic. Some of these alphabets have been standardized, are used in voice, enunciation, and language instruction in schools, and can be found in the better dictionaries. *Refer to* **Fig. P-5.**

phoneticised: Transliterated by using the phonetic alphabet. *See also* **alphabet, phonetic alphabet, transliterated.**

phonon: An acoustic energy packet (a) that is similar in concept to a photon because its energy is a function of the frequency of vibration of the sound source that produced it, (b) streams of which form a sound wave, depending on whether one speaks in terms of the quantum theory or the wave theory of sound, just as is the case for light, (c) has an energy usually less than 0.1 eV (electron•volts), an order of magnitude less

Letter	Word	Pronunciation	Letter	Word	Pronunciation
A	Alpha	AL FAH	N	November	NOH VEM BER
B	Bravo	BRAH VOH	O	Oscar	AHS KAH
C	Charlie	CHAR LEE	P	Papa	PAH PAH
D	Delta	DELL TAH	Q	Quebec	KAY BECK
E	Echo	ECK OH	R	Romeo	ROH MEE OH
F	Foxtrot	FAHKS TRAHT	S	Sierra	SEE AIR AH
G	Golf	GAHLF	T	Tango	TANG GO
H	Hotel	HOH TELL	U	Uniform	YOU NEE FORM
I	India	IN DEE AH	V	Victor	VICK TAH
J	Juliett	JEW LEE ETT	W	Whiskey	WISS KEE
K	Kilo	KEE LOH	X	Xray	ECKS RAY
L	Lima	LEE MAH	Y	Yankee	YANG KEE
M	Mike	MIKE	Z	Zulu	ZOO LOO

Fig. P-5. The **phonetic alphabet** used for spelling words or citing letters of the alphabet in voice transmission systems. Although the phonetic alphabet is used internationally, the associated pronunciation aid shown is primarily for use in the English-speaking world. Other languages use the same phonetic alphabet but use a different pronunciation aid so that the phonetic alphabet sounds the same in all languages. For example, the German pronunciation aid for Yankee would be JANG KIE and for Uniform would be U NIE FORM.

than the energy of a photon, (d) when dealing with band-gap energies in semiconductors, has energy that is not negligible, and with appropriate coupling, can contribute appreciably to the movement of electrons into higher energy levels on a statistical basis, such that when these excited electrons move to lower energy levels, photon emission occurs, and (e) the existence of which can affect the spectral composition of radiation emitted by a light source. *Note:* When photon and phonon energies interact in semiconductors used in communications systems, undesirable behavior of the system can occur.

phonon absorption: In a material, the conversion of light energy to vibrational energy. *Note:* Phonon absorption determines the fundamental, i.e., quantum limit of attenuation, i.e., minimum attenuation, in silica-based glasses in the far infrared region of the electromagnetic spectrum.

phonon band: 1. The group of acoustic, elastic, or vibrational frequencies associated with or that are contained in a phonon or a group of phonons. **2.** The frequency composition of a phonon or a group of phonons. *See also* **frequency, phonon.**

phosphorescence: The emission of electromagnetic radiation by a material (a) during stimulation from an outside source of energy and (b) after stimulation from an outside source of energy ceases. *Note:* Phosphorescence may continue for a time ranging from an extremely short time to an extremely long time after stimulation ceases. *See also* **fluorescence, luminescence.**

PHOTINT: photographic intelligence.

photocell: A device that produces an output electric current, voltage, or power proportional to incident irradiance, i.e., optical power, integrated over the illuminated area of a surface that is sensitive because of photoconductive, photoemissive, or photovoltaic effects. *See also* **irradiance, photoconductive effect, photoemissive effect, photovoltaic effect.**

photoconduction: In a material, an increase in the electrical conduction capability that results from the absorption of electromagnetic energy.

photoconductive: Pertaining to the capability of changing electrical conductivity as a result of exposure to electromagnetic radiation in the optical spectrum. *See also* **photoemissive, photovoltaic.**

photoconductive cell: A device that detects or measures electromagnetic radiation intensity, i.e., irradiance, by variation of the conductivity of a substance caused by absorption of the radiation. *Synonyms* **photoresistive cell, photoresistor.** *See also* **absorption.**

photoconductive device: A device that makes use of photoconductivity, such as a photoconductive cell. *See also* **photoconductive cell, photoconductivity.**

photoconductive effect: The phenomenon, an internal photoelectric effect, in which some nonmetallic materials exhibit a marked increase in electrical conductivity upon absorption of photon energy. *Note 1:* The photovoltaic effect is also an internal photoelectric effect. *Note 2:* Photoconductive materials include (a) gases that have been ionized by the energy in photons and (b) certain crystals. They are used in conjunction with semiconductor materials that are ordinarily poor electrical conductors but become distinctly conducting when subjected to photon absorption. The incident photons excite electrons into the conduction band where they move more freely, resulting in increased electrical conductivity. The increase in conductivity occurs because of the additional free carriers generated when photon energies are absorbed in energy transitions. The rate at which free carriers are generated and the length of time that they remain in conducting states, i.e., their lifetime, determines the amount of change in conductivity produced by the incident stream of photons. *Note 3:* The increase in conductivity can be used to measure the irradiance of incident optical power. *See also* **internal photoelectric effect, photoemissive effect, photoelectromagnetic effect, photovoltaic effect.**

photoconductive film: A film of material that has an electrical-current-carrying capacity that is enhanced when it is illuminated by electromagnetic radiation, particularly with radiation in the visible and near visible spectra, i.e., the optical spectrum of the electromagnetic frequency spectrum.

photoconductive gain factor: The ratio of (a) the number of electrons per second flowing through a circuit containing a cube of semiconducting material with sides of unit length to (b) the number of photons per second of incident electromagnetic radiation absorbed in the cube.

photoconductive infrared sensor: An infrared electromagnetic radiation detector that makes use of the photoconductive effect, i.e., a device that makes use of a material in which the electrical conductivity is proportional to the frequency or the intensity of incident infrared electromagnetic waves. *See also* **electrical conductivity, infrared, photoconductive effect, radiation.**

photoconductive meter: An exposure meter in which a battery supplies power through a photoconductive cell to an electrical-current-measuring device, such as a milliammeter or a microammeter, to measure the intensity of electromagnetic radiation of specific frequencies incident upon its active surface. *See also* **photoconductive cell.**

photoconductive photodetector: A photodetector that (a) makes use of photoconductivity in its operation, (b) detects the presence of electromagnetic radiation in the optical spectrum of the electromagnetic frequency spectrum by changing its electrical resistance in accordance with the intensity of the incident radiation, (c) has an output current proportional to the intensity of the incident radiation, i.e., the irradiance, (d) has a voltage power source, such as a battery, to provide the current, and (e) has an instrument for measuring and indicating the current, but with a scale calibrated in irradiance units, such as watts per square meter. *See also* **photoconductivity, photodetector.**

photoconductivity: 1. In some nonmetallic materials, particularly semiconductors, the increase in electrical conductivity caused by the increase in free carriers generated when photon energy is absorbed by the material. **2.** The conductivity of a material when subjected to optical radiation, i.e., electromagnetic waves in the optical spectrum. *See also* **carrier, electrical conductivity, photoconductive effect, photoelectric effect.**

photoconductor: A material, usually a nonmetallic solid, in which the electrical conductivity increases when it is exposed to electromagnetic radiation, particularly in the optical spectrum. *See also* **electrical conductivity.**

photocurrent: In a material medium, the electrical current that results from the absorption of electromagnetic radiation in the optical spectrum. *Note 1:* In a strict sense, photocurrents result from photovoltaic or photoemissive effects rather than from an increase in conductivity, i.e., from the photoconductive effect, because then the actual source of current is a power source external to the photodetector. However, because the electric current is a function of incident electromagnetic radiation, the term may be applied in any case. *Note 2:* Photocurrents occur in photodiodes and other photodetectors. *Note 3:* Photocurrents may be enhanced by internal gain caused by interaction among ions and photons under the influence of applied fields, such as occurs in avalanche photodiodes (APDs). *Note 4:* Photocurrents may be calculated from the relation $I_{ph} = \Gamma eP/hf$ where I_{ph} is the photocurrent, Γ is the carrier collection quantum efficiency, e is the electron charge, P is the incident optical power, h is Planck's constant, and f is the frequency of the incident radiation. *Synonym* **light current.** *See also* **dark current, photodiode.**

photodarlington: A device that (a) consists of a combination of a photodetector and a Darlington pair transistor circuit and (b) detects optical signals and produces amplified electrical versions of the signals.

photodetector (PD): 1. A device that detects the presence of radiation in the optical spectrum. **2.** A transducer that accepts an optical signal and produces an electrical signal that represents the same information as in the optical signal. **3.** A device that extracts the information in a modulated optical carrier and converts the resultant data into electrical signals. *Note 1:* Examples of photodetectors are avalanche photodiodes, fiber optic photodetectors, photoelectromagnetic photodetectors, photoemissive photodetectors, and photovoltaic photodetectors. *Note 2:* The two main types of photodetectors are the photodiode (PD) and the avalanche photodiode (APD). *See* **field effect transistor photodetector.** *See also* **optical detector.** *Refer to* **Figs. A-3 (6), F-2 (339), L-3 (505), P-7 (723), R-3 (799), T-6 (1013).**

photodetector responsivity: The ratio of (a) the root mean square (rms) value of the output current or voltage of a photodetector to (b) the rms value of the incident optical power. *Note:* In most cases, photodetectors are linear in that their responsivity is independent of the irradiance, i.e., the power density or intensity, of the incident radiation. Thus, the photodetector output in amperes or volts is proportional to the incident optical power in watts. Differential responsivity applies to small variations in input optical power. Photodetectors are square law detectors in that they respond to the irradiance of the incident electromagnetic wave, i.e., to the square of the electric field strength associated with the incident wave, which is proportional to the optical power.

photodetector signal to noise ratio: The signal to noise ratio for a photodetector, given by the relation $S/R_{ph} = (N_p/B)(1 - e^{-hf/kT})$ where S/R_{ph} is the photodetector signal to noise ratio, N_p/B is the number of incident photons per unit of bandwidth, h is Planck's constant, f is the frequency of the incident photons, k is Boltzmann's constant, and T is the absolute temperature.

photodiode (PD): A photodetector that (a) consists of a semiconductor diode, (b) uses photoconductive effects, (c) requires a source of voltage, (d) produces photocurrent by absorbing light and increasing electron mobility, (e) detects optical power, and (f) converts optical signals to electrical signals. *Synonym*

diode photodetector. *See* **avalanche photodiode, germanium photodiode, photocurrent, PIN photodiode, silicon photodiode.** *Refer to* **Fig. P-6.** *Refer also to* **Fig. S-16 (932).**

Fig. P-6. A high-speed silicon fiber optic **photodiode** with integral fiber optic pigtail. Typical operating values are a maximum of 100 μA reverse-bias dark current, a continuous photocurrent density of 5 mA•mm^{-2}, a foreward current of 10 mA, a responsivity of about 0.5 A•W^{-1}, a luminous responsivity of 0.12 lm•mA^{-1}, and a quantum efficiency of about 85%, all values applied and measured at a wavelength of 0.83 μm.

photodiode coupler: *See* **avalanche photodiode coupler, positive-intrinsic-negative photodiode coupler.**

photoeffect: *See* **external photoeffect, extrinsic internal photoeffect, internal photoeffect, intrinsic internal photoeffect.**

photoeffect detector: *See* **external photoeffect detector, internal photoeffect detector.**

photoelastic: 1. Pertaining to the property of a material in which the elasticity depends upon the intensity of incident electromagnetic radiation, i.e., the incident irradiance. **2.** The property of a material in which the mechanical elasticity, i.e., the coefficient of elasticity or the modulus of elasticity (Young's modulus), expressed as stress/strain, at a point in the material is a function of the instantaneous incident electromagnetic power density, i.e., irradiance, usually expressed as watts per square meter or joules per cubic meter, at the point, such that an increase in power density results in a decrease in the modulus of elasticity, i.e., the material becomes more pliable or softer. *Note:* The power density may be expressed as incident energy per unit area per unit time. *See also* **irradiance, photoelastic effect.**

photoelastic effect: 1. That property of a material in which Young's modulus—i.e., the modulus of elasticity, the mechanical elasticity, or the coefficient of elasticity (stress/strain)—at a point in the material is a function of the instantaneous electromagnetic irradiance at the point such that an increase in irradiance results in a decrease in the modulus of elasticity, i.e., the material becomes softer. **2.** The effect that changes in elasticity have on electromagnetic waves propagating in a material medium. *Note 1:* When an electromagnetic wave is propagating in a material or free space, the irradiance of the wave is (a) the instantaneous electromagnetic power passing through per unit cross-sectional area, usually expressed in watts per square meter, (b) the electromagnetic energy contained in a unit volume of the radiation, usually expressed in joules per cubic meter at the point in the material, or (c) the electromagnetic energy incident per unit area per unit time, usually expressed as joules per square meter per second. These are identical units of irradiance. *Note 2:* The photoelastic effect is a point function.

photoelastic sensor: A fiber optic sensor (a) in which the photoelastic effect is used for sensing based on the change in refractive index or the birefringence that is produced when stress is applied to a transparent material, especially transparent plastics, (b) in which the several lightwaves caused by the changes in refractive indices reinforce or cancel one another so that light patterns indicate points of stress, and (c) that produces a signal indicating the resultant modulation of light from specific points being subjected to stress in a mechanical system.

photoelectric effect: 1. In a material, the changes in the electrical characteristics caused by photon energy absorption, including (a) the liberation of electrons and other electrical charge carriers in the material, (b) their increased flow as a result of increased conductivity, (c) their aggregation at material interfaces to produce a voltage, or (d) any combination of these, as a result of photon energy absorption exciting the material. *Note:* The energy of the incident individual photon is given by the relation $E_p = hf$ where h is Planck's constant, and f is the frequency of the photon, i.e., the frequency of the incident radiation. If the incident radiation consists of a mixture of frequencies within a given range, the energy level of each photon will be determined by one of the frequencies in the range. The electrons may be (a) released into a vacuum or into another material or (b) may circulate within a closed circuit in the form of a photocurrent. The excited material may be a solid, liquid, or gas. Thus, photoconductive, photoelectromagnetic, photoemissive, and photovoltaic effects are all photoelectric effects. *See* **internal photoelectric effect.** *See also* **photoconductive effect, photoelectromagnetic effect, photoemissive effect, photon, photovoltaic effect, Planck's constant.**

photoelectric effect detector: *See* **internal photoelectric effect detector.**

photoelectric equation: The equation that relates photon energy, the work function of a material, such as certain metals and oxides, and the kinetic energy of an electron emitted by the material when a photon strikes it, given by the relation $E_{ph} = hf = W + (1/2)mv^2$ where E_{ph} is the photon energy, h is Planck's constant, f is the frequency associated with the incident photon, W is the work function of the material the photon strikes, m is the mass of the electron, and v is the velocity with which the electron is ejected from the material. *Note:* If the emitted electron energy is zero, i.e., the electron has negligible velocity, the photon has just sufficient energy to overcome the work function. The frequency corresponding to this condition is the threshold frequency. If the energy of every photon in a stream of incident photons is less than that of the work function, the photons cannot cause an electron to be emitted no matter (a) how many photons bombard the material, expressed as the number of photons, (b) how long a time the electrons continue to bombard, in time units such as hours, and (c) at what rate they bombard the material, in photons per unit time, such as electrons per second. The only case in which electrons are emitted is that in which the energy of the individual photon is equal to or higher than the work function of the material. The work function of a material (a) is an intrinsic characteristic of the material and (b) can be measured and published for a given material. *See also* **threshold frequency, work function.**

photoelectromagnetic effect: The effect produced in a photoconductive material in which an electric potential difference is created as a result of the interaction of a magnetic field with the photoconductive material when (a) the material is subjected to incident electromagnetic radiation, such as lightwaves, (b) the incident radiation creates hole-electron pairs that diffuse into the material, and (c) the magnetic field causes the paired components to separate as they move in the same direction, but they constitute oppositely directed electric currents, resulting in the production of an electric potential difference across the material. *Note:* In most applications of the photoelectromagnetic effect, (a) light is made to fall on a flat surface of an intermetallic semiconductor located in a magnetic field that is parallel to the surface, (b) excess hole-electron pairs are created by the incident radiation and, with the magnetic field, produce a current flow through the semiconductor that is at right angles to both the light rays and the magnetic field, and (c) the current flow is caused by transverse forces acting on electrons and holes that are diffusing into the semiconductor from the surface. *Synonym* **photomagnetoelectric effect.**

photoelectromagnetic photodetector: A photodetector that makes use of the photoelectromagnetic effect, i.e., a photodetector that (a) requires an applied magnetic field, (b) accepts incident optical electromagnetic radiation, and (c) produces an electric current proportional to the electric field strength of the incident radiation.

photoemissive: Pertaining to the capability of emitting electrons as a result of exposure to electromagnetic radiation. *See also* **photoconductive, photovoltaic.**

photoemissive cell: A device that (a) detects radiant energy and (b) usually measures radiant energy by measuring the quantity of electrons emitted from a surface that is made of a photoemissive material, i.e., a material that displays the photoemissive effect. *See also* **photoemissive, photoemissive effect.**

photoemissive effect: In certain materials, the emission of electrons from the surface when the material absorbs photon energy. *Note 1:* The photons must have sufficient energy, given by the relation $E_{ph} = hf$ where E_{ph} is the photon energy, h is Planck's constant, and f is the frequency of the source, to overcome the work function of the material. If this condition is not met, increased irradiance and longer exposure time will not induce emission of free electrons from the surface. *Note 2:* The electrons may be emitted into a vacuum or into adjacent material. *Synonym* **external photoelectric effect.** *See also* **photoconductive effect, photoelectric equation, photoelectromagnetic effect, photon, photovoltaic effect, work function.** *Refer to* **Fig. P-7.**

photoemissive photodetector: A photodetector that makes use of the photoemissive effect. *Note 1:* An applied electric field is necessary to attract or collect the emitted electrons, thus producing an electric current. Were it not for the applied field, the density of emitted electrons would build up and create an electric field, or a space charge cloud, that would inhibit further emission. *Note 2:* The electrons may be emitted into a vacuum or into adjacent material, i.e., into proximate material.

photoemissive tube photometer: A photometer that (a) uses a tube made of a photoemissive material, (b) requires electronic amplification, (c) is highly accurate, and (d) is used primarily in laboratories. *See also* **photoemissive.**

photoemissivity: The property of a material that (a) causes it to emit electrons when electromagnetic

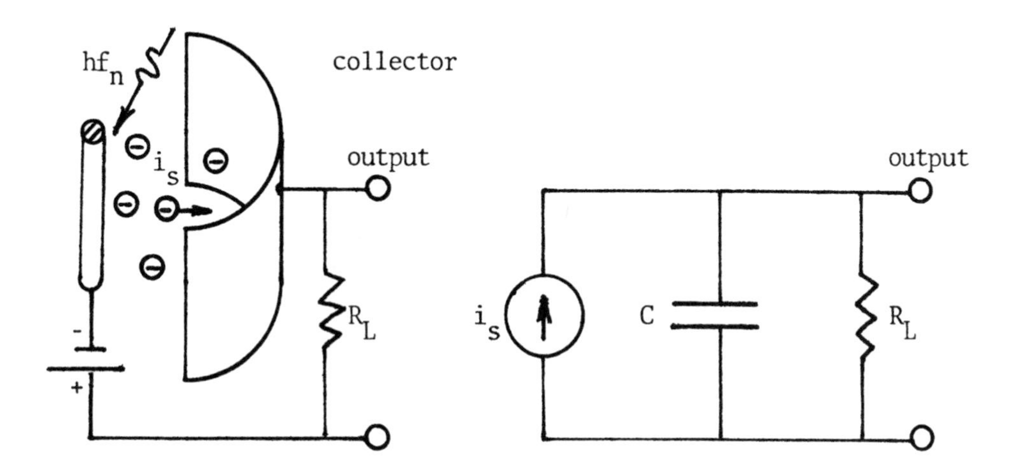

Fig. P-7. A **photoemissive effect** photodetector and its equivalent circuit.

radiation is incident upon it and (b) in practice, usually requires an applied voltage to collect the electrons and drive them through the external part of a circuit and return them to the material. *See also* **photoemissive effect.**

photographic intelligence (PHOTINT): Information that is obtained from photographs, or is on photographic media, and has been processed for use.

photography: The branch of science and technology that (a) is devoted to the creation of pictures, usually on hard copy, by exposing sensitive chemically treated substances, such as photographic paper and plastic, to an image composed of electromagnetic radiation, usually in the optical spectrum, which includes infrared and ultraviolet radiation, (b) includes activities related to the making of pictures, such as finding, creating, or traveling to or near objects of interest, and (c) includes still and motion pictures, though the subdivision of motion pictures tends to be called "cinematography," especially when the motion pictures relate to entertainment. *See also* **electromagnetic radiation, image, optical spectrum.**

photomagnetoelectric effect: *Synonym* **photoelectromagnetic effect.**

photometer: A device that measures and indicates irradiance levels, i.e., intensity levels, in the optical spectrum. *See* **photoemissive tube photometer.** *See also* **photovoltaic meter.**

photometry: 1. The branch of science and technology that (a) is devoted to the measurement of electromagnetic radiation, especially in the infrared, visible, and ultraviolet regions, i.e., the optical spectrum, of the electromagnetic spectrum and (b) includes the science and the technology devoted to light sources, photodetectors, and the effects of radiation on materials. *Note:* Although photometry was originally devoted to the effects of electromagnetic radiation on the human eye, the area of interest has been expanded to include the science and the technology devoted to light sources, photodetectors, and the effects of radiation on materials. **2.** The branch of science and technology that (a) is devoted to measurement of the effects of electromagnetic radiation on the eye, (b) is an outgrowth of the psychophysical aspects induced by the radiation, and (c) includes determination of visual effectiveness by considering the response to various levels of incident radiation, i.e., irradiance, and the sensitivity of the eye to various frequencies, i.e., various wavelengths. *See also* **irradiance, optical spectrum.**

photomultiplier: An electron tube that (a) multiplies the effect of incident electromagnetic radiation by accelerating primary emission electrons and using them to impinge upon other surfaces, i.e., secondary electrodes called "dynodes," knocking out additional electrons as secondary emission, which, in turn, impinge on additional dynodes, until a large electric current is produced for low incident radiation levels, (b) consists of an electron tube that contains a photocathode, one or more dynodes, and an output electrode, and (c) multiplies the electron flow emitted from the cathode by secondary emission from the dynodes by cascading the original electron emission. *See also* **photoemissive effect, primary emission, secondary emission.**

photon: A discrete packet, i.e., a quantum, of electromagnetic energy given by the relation $E_{ph} = hf$ where E_{hp} is the photon energy, h is Planck's constant, and f is the frequency, usually in the optical spectrum, of the source. *Note:* The photon energy will be expressed in joules if Planck's constant is given as 6.626×10^{-34} J•s (joule•seconds), and the frequency is given in hertz. The energy of a photon with a wavelength of 1.31 µm (micron), the standard wavelength for fiber optic systems, is approximately 1.6 eV (electron•volts). By comparison, a value for the energy of a typical phonon is somewhat less than 0.1 of that of the typical photon. Photons have some particle-like characteristics, such as that they can be counted and their individual energy measured. *See* **gamma photon.** *See also* **nonlinear scattering, Planck's constant.** *Refer to* **Figs. L-2 (502), O-2 (671).**

photon detector: A device that (a) responds to incident photons, i.e., a device that signals, with some reasonable probability of being correct, the absorption of a photon, i.e., a quantum of electromagnetic energy, and (b) contains a sensitive material that exhibits a change in property when it absorbs a photon, i.e., the material responds to photoemissive, photoconductive, photovoltaic, or photoelectromagnetic effects.

photonics: The branch of science and technology devoted to (a) the study, control, and use of photons and electromagnetic waves in the optical spectrum, i.e., lightwaves, including the generation, transmission, reception, and processing of optical signals and power, and (b) the use of electrical, optical, acoustic, or other means to control the photons and lightwaves. *See also* **acoustooptics, electrooptics, magnetooptics, optooptics.**

photonic switch: *Synonym* **optical switch.**

photon noise: Quantum noise attributed to (a) the discrete particle, i.e., the quantum, nature of light and (b) the streams of incident photons that interact with the molecular structure of materials such that the absorption of each photon produces a discrete contribution to the total photon noise. *See also* **dark current, quantum noise, shot noise, thermal noise.**

photooptic storage: *Synonym* **optical storage.**

photoresistive cell: *Synonym* **photoconductive cell.**

photoresistor: *Synonym* **photoconductive cell.**

photosensitive material: A material that changes its chemical composition when exposed to electromagnetic radiation. *Note:* An example of photosensitive material is the coating on unexposed photofilm used in cameras. *See also* **photosensitive recording.**

photosensitive recording: In facsimile systems, recording by exposing a material with a photosensitive surface to a signal-controlled light beam, spot, or image. *See also* **dark current, facsimile, photosensitive material, recording.**

phototelegraphy: In facsimile systems, transmission by telegraphic means in which (a) the objects are pictures, and (b) a photographic process at the receiver produces a recorded copy, i.e., a permanent image, of the picture that was scanned by the transmitter.

phototransistor: A transistor that (a) produces electrical output signals corresponding to input incident optical signals and (b) may be used as a photodetector in circuits, such as in photodarlington circuits.

phototronic photocell: *Synonym* **photovoltaic cell.**

photovoltaic: Pertaining to the capability of generating a voltage as a result of exposure to electromagnetic radiation, usually in the optical spectrum. *See also* **photoconductive, photoemissive, radiovoltaic.**

photovoltaic cell: A device that (a) detects or measures radiant energy by producing a voltage proportional to the incident radiation intensity, i.e., the irradiance, (b) can operate without an additional source of voltage because it develops a voltage, and (c) generates a potential difference at a junction, i.e., at a barrier layer, between two types of materials, upon absorption of radiant energy. *Synonym* **barrier layer cell, barrier layer photocell, boundary layer photocell, photronic photocell.** *See also* **photovoltaic effect.**

photovoltaic effect: In certain materials, the production of a potential difference, i.e., a voltage, at the junction of two dissimilar materials, such as a semiconductor p-n junction, when the materials absorb photon energy contained in incident radiation. *Note:* The electrical potential difference is caused by the diffusion of hole-electron pairs across the junction potential barrier, which the incident photons cause to shift or increase, leading to direct conversion of a part of the absorbed energy into usable electromotive force (emf), i.e., a usable voltage, proportional to the irradiance of the incident radiation. *See also* **photon, photoconductive effect, photoelectromotive effect, photoemissive effect, photovoltaic cell.**

photovoltaic meter: A photometer that (a) consists of a photovoltaic cell, (b) a sensitive current-measuring device, such as a microammeter, and (c) a calibrated scale on the current-measuring device that indicates the irradiance levels. *See also* **irradiance, photometer, photovoltaic cell.**

photovoltaic photodetector: A photodetector that (a) uses the photovoltaic effect, (b) does not require an external voltage source because it generates a voltage internally, and (c) is connected in a circuit with a voltage-, or current-, measuring device with a scale calibrated in units of irradiance. *See also* **photodetector, photovoltaic effect.**

physical circuit: An actual circuit that (a) is constructed with hardware, such as fiber optic cables, coaxial cables, fiber optic transmitters and receivers, microwave transmitters and receivers, switching equipment, modems, multiplexers, resistors, inductors, and capacitors, and (b) is used to (i) make direct connections as a real circuit, (ii) create circuits by multiplexing, such as time division and frequency division multiplexing, and (iii) create logical circuits, phantom circuits, and virtual circuits. *See also* **physical layer, physical network.**

physical frame: *See* **frame.**

physical layer: In open systems architecture, the layer that (a) provides the functions that are used or needed to establish, maintain, and release physical connections, (b) conducts signals between data terminal equipment (DTE), data circuit-terminating equipment (DCE), and switching equipment, i.e., between user end instruments and end-point nodes, between switching nodes and end-point nodes, or between switching nodes, and (c) is implemented with physical circuits that use fiber optic and electronic components. *See also* **Open Systems Interconnection—Reference Model, physical circuit, physical network.**

Physical Layer: *See* **Open Systems Interconnection—Reference Model.**

physical length: *Synonym* **physical path length.**

physical network: A network composed of physical circuits and components. *See also* **component, physical circuit, physical layer.**

physical optics: The branch of optics in which light is described as consisting of (a) a form of wave motion in which energy is propagated by electromagnetic radiation in the form of waves and wavefronts, i.e., as surfaces, rather than as rays as in geometric optics, (b) wavefronts that are perpendicular to the direction of propagation at every point, i.e., perpendicular to the Poynting vector, (c) wavefront planes that are defined by the electric and magnetic field vectors at any given point, and (d) wavefronts that are bent at the interface between two dissimilar media or may be curved in a medium in which the refractive index is a function of position. *Synonym* **wave optics.** *See also* **geometric optics, wavefront.**

physical path length: In the propagation of a lightwave, the geometric distance that (a) is traveled by a lightwave when propagating from point A to point B in a medium, (b) is independent of the refractive index of the medium, and (c) is dependent on the route taken by the lightwave when propagating from point A to point B. *Note:* The physical path length usually is expressed in meters. *Synonyms* **geometric length, physical length.** *See also* **optical path length.**

physical requirement: A requirement that defines a specific mechanical characteristic that a system or a system component must have, such as size, weight, shape, color, strength, elasticity, or odor. *See also* **requirement.**

physical security: In communications, computer, data processing, and control systems, security that results from all physical measures necessary to safeguard classified equipment, material, and documents from access thereto or observation thereof by unauthorized persons. *See also* **communications security, security.**

physical signaling sublayer (PLS): In a local area network (LAN) or a metropolitan area network (MAN) using open systems interconnection (OSI) architecture, the portion of the physical layer that (a) interfaces with the medium access control sublayer, (b) performs character encoding, transmission, reception, and decoding, and (c) performs optional isolation functions. *See also* **local area network, metropolitan area network.**

physical topology: In a communications system, the network topology and the network connectivity, including loops, circuits, and trunks, that interconnect real equipment, such as switches, transmitters, receivers, terminators, and end instruments. *See also* **logical topology, network connectivity, network topology.**

PHz: petahertz.

picket: A ship, aircraft, vehicle, or troop, or combinations of these, placed in a specific temporal and spatial relation to a larger group for a specific purpose, such as sounding an alarm, transmitting weather information, or alerting the larger group. *Note:* Formerly, a picket was only a detachment of soldiers or a single ship, such as a small outguard or a patrol boat, serving to protect an army or a fleet from surprise. Modern communications technology has resulted in the deployment of new kinds of pickets with new roles. *See* **direction finding picket, radar picket.**

picking: In display systems in which a stylus is used to select, identify, or point to a coordinate position in the display space on the display surface of a display device, the identifying and the selecting of display elements, display groups, or display images, or parts thereof. *Note 2:* Examples of styli used for picking include tracing needles attached to pantographs, light pens, sonic pens, tablet styli, voltage pencils, and cursors. *See also* **cursor, light pen, sonic pen, tablet stylus, voltage pencil.**

pick-off coupling: *Synonym* **tangential coupling.**

pickup: *See* **call pickup.**

picosecond (ps): One millionth of 1 millionth of a second, i.e., 10^{-12} s (second). *See also* **second.**

picowatt (pW): A unit of power equal to 10^{-12} W (watt), i.e., −90 dBm. *Note:* The picowatt usually is used for both weighted and unweighted noise measurements. In making noise measurements, the context in which the picowatt is used must be carefully observed for one fully to appreciate its significance. The type of measurement must be specified. *See also* **pWp.**

picowatt, psophometrically weighted: *See* **noise weighting.**

picowatt, psophometrically weighted, referred to a zero transmission level point (pWp0): *See* **noise weighting.**

picture black: In communications systems, such as facsimile, radiophoto, and wirephoto, pertaining to the signal that corresponds to the darkest part, i.e., the spot with the lowest illuminance, of the object being scanned. *See also* **picture white.**

picture element (PE): 1. In display systems, a display element that can be used to construct a display image, or a picture, in the display space on the display surface of a display device. *Note:* For a fiber optic aligned bundle, one picture element is transmitted by each fiber of the bundle that forms the picture on a faceplate. For a cathode ray tube, a picture element is the output from a single separate piece of the mosaic that forms the screen and that can be independently controlled by the electron beam. **2.** In a facsimile transmission system, (a) the area of the object that coincides with the scanning spot at a given instant or (b) the area of the finest detail that can be effectively reproduced on the record medium. *Note:* Both the mark and the space are considered as separate picture elements. *See also* **dot frequency, object, pel, pixel, record medium.**

picture facsimile telegraphy: Facsimile transmission in which the object is scanned, and the resulting signal is used to produce, on the record medium, an image that has shades of gray in addition to black and white.

picture frequency: In facsimile systems, a frequency generated by scanning an object, i.e., a frequency in the baseband signal. *Note:* Picture frequencies do not include frequencies that are in a modulated carrier. *See also* **baseband, facsimile, frequency.**

picture level: *See* **average picture level.**

picture white: In communications systems, pertaining to the signal that corresponds to the brightest part, i.e., the spot with the highest illuminance, of the object being scanned. *See also* **picture black.**

piecewise linear encoding: *Synonym* **segmented encoding.**

piezoelectric effect: 1. The physical property demonstrated by certain natural and synthetic crystals by which they are mechanically deformed under the influence of an electric field, usually increased in length in the direction of the applied field. **2.** In certain crystals, the effect of producing a voltage when an applied stress, either compression, expansion, or shear, is undergoing a change, and conversely, producing a stress by applying a voltage to the crystals. *Note:* The voltage is produced only when the stress is changing, and vice versa, not when the voltage, or the stress, is static. The output, voltage or stress, as a function of time, is proportional to the time derivative of the input, stress or voltage, respectively.

pigeon: *See* **carrier pigeon.**

piggyback entry: In communications, computer, data processing, and control systems, an unauthorized access to a system via the legitimate connection of an authorized user.

pigtail: 1. A short length of conductor, such as an optical fiber or wire that (a) is permanently fixed to a component and (b) is used to connect the component to another component. **2.** A short length of single-fiber cable that (a) usually is tight-buffered, (b) has an optical connector on one end, and (c) has a length of exposed fiber at the other end. *Note:* The exposed optical fiber of the pigtail is spliced to one fiber of a multifiber trunk cable, i.e., an arterial cable, to enable the multifiber cable to be "broken out" into individual single-fiber cables that may be connected to a patch panel, an input port, or an output port of a fiber optic receiver or transmitter. **3.** A short length of electrical conductor permanently affixed to a component, used to connect the component to another conductor. *See*

fiber optic cable pigtail, optical fiber pigtail. *See also* **launching fiber.** *Refer to* **Fig. P-6 (721).**

pilot: 1. In a communications system, a signal that (a) usually is a single frequency, such as a single-frequency tone, (b) is transmitted over the system for specific purposes, such as for supervisory, control, equalization, synchronization, continuity, and reference purposes, and (c) is not used to transmit user information. *Note 1:* Several different independent pilot frequencies may be used at the same or different times. *Note 2:* Most radio relay systems transmit their own pilots as well as transmit pilots belonging to the carrier system. **2.** In tape relay systems, instructions that appear in message format and are related to the transmission or the handling of the message in which they are contained. *See* **synchronizing pilot, tape relay pilot.** *See also* **carrier, carrier system, frequency, signal.**

piloted vehicle: *See* **remotely piloted vehicle, unretrievable remotely piloted vehicle.**

pilot make-busy (PMB) circuit: In a carrier system, a circuit in which a pilot is used to make trunks busy to the switching equipment for specific purposes, such as during a telephone system outage or during radio signal fading. *Note:* A busy signal is preferable to the transmission of an unintelligible signal or a completely erroneous signal. The busy signal, after a length of time or several access attempts, will prompt the user to use other means of communication. *See also* **access attempt, busy signal, circuit, fading, failure, outage.**

pilot regulation: *See* **two-pilot regulation.**

pilot tone: A pilot in the form of a single-frequency tone, i.e., an audible pilot. *See also* **pilot.**

PIN: positive-intrinsic-negative.

pincushion distortion: In the image of a straight-sided square or rectangular object, a distortion of the image in such a manner that the display element, display group, or display image has concave sides, i.e., sides that are bowed in. *See also* **barrel distortion.**

PIN diode: positive-intrinsic-negative diode.

ping: 1. In sonar, the audible underwater signal transmitted to locate objects by analyzing the echo. *Note:* The term "ping" is derived from the characteristic audible sound of an underwater sonar pulse. **2. packet internet groper.**

pink noise: 1. In electromagnetics, noise power with a spectral density that is inversely proportional to the noise frequency over a specified frequency range. *Note:* When noise power is plotted with respect to frequency, pink noise has a negative slope, white noise has a zero slope, and blue noise has a positive slope. **2.** In acoustics, noise in which the power in each octave is at the same level. *See also* **black noise, blue noise, level noise, white noise.**

PIN photodiode: positive-intrinsic-negative photodiode.

pip: *Synonym* **spike.**

pipe: *See* **light pipe, voice pipe.**

pipelining: In computer and communications systems operation, the fetching of the next instruction, or the setting up for the next operation to be performed, while the current instruction or operation is being performed. *See also* **instruction, operation.**

piping: In a tree file, the establishment and the use of paths to subdirectories and subsections for use by computer programs. *See also* **computer program, directory.**

piston: In a waveguide, a transverse longitudinally movable metallic plane surface that reflects essentially all the incident energy. *Synonym* **plunger.**

pitch: 1. In a linear arrangement of entities, the distance between adjacent entities, numerically equal to the reciprocal of the packing density. *Note:* If the bit packing density on a tape or disk is 1000 bpi (bits per inch), the pitch is 0.001 in. (inches). *Synonym in twisted elements* **lay length.** *See* **frame pitch, scanning pitch, track pitch. 2.** On a frequency scale, the relative position of a tone that (a) consists of a single frequency or a specified mixture of frequencies and (b) usually is of specified loudness or intensity as determined by a normal ear. *See also* **pitch detector.**

pitch detector: In a voice encoder, i.e., a vocoder, a circuit that (a) detects the overall volume and frequency of a sound, such as a speech sound or an audio frequency sound in electrical form, (b) transmits the frequency information to the receiving synthesizer, (c) modulates the pulse generator that provides the carrier input that is modulated by the signals in the spectral channels, and (d) thereby reconstitutes the sounds, usually voiced sounds, that were originally detected. *See also* **modulation, vocoder.**

pixel: In a facsimile system, the smallest discrete scanning line sample that can contain gray scale information. *See also* **elemental area, facsimile, gray scale, pel, picture element, scanning.**

PLA: programmable logic array.

plaindress message (PDM): A message in which (a) the addresses of the message originator and the message receiver are externally indicated, i.e., are not embedded in the text, (b) the address indicating group (AIG) number or its address group usually is on the "TO" line of the message address, (c) additional action addressees are on the "TO" line, (d) information addressees are on the "INFO" line, and (e) the message originator title may be on the "FROM" or "FM" line. *See also* **addressee, address indicating group.**

plain language: 1. The language that is used to (a) write plain text or speak in an intelligible language and (b) convey intelligible meaning in the language in which the text is written or the language is spoken, without using codes to deliberately hide or obscure the meaning. **2.** The intelligible language that (a) results from decrypted text, (b) is used to write intelligible text or speak in an intelligible language, (c) uses signals that usually have understood meanings, and (d) can be read, heard, and acted upon without the application of decoding or decryption devices. *See also* **decrypt, plain text.**

plain language address designator: An address designator that is written in plain language rather than in the language that results from encryption. *See also* **address designator, encryption, plain language.**

plain language override: In cryptosystems, a capability that (a) is used primarily in radiotelephone communications systems and (b) is built into certain cryptoequipment that permits operators to (i) listen for plain language signals in the frequency to which they are tuned for operations, (ii) receive signals in the guard channel when their equipment is set for reception of ciphony transmissions, and (iii) operate in the cipher standby condition. *Synonym* **plain text override.** *See also* **cipher, cryptosystem, plain language, plain text.**

plain mode: In communications systems, a nonsecure transmission mode of operating ciphony equipment in which voice transmissions are not encrypted, i.e., plain text is transmitted as spoken. *See also* **ciphony, encrypt, encryption, plain language, plain text.**

plain old telephone service (POTS): Telephone service that (a) provides voice and low speed digital data services only, (b) uses wire pairs, such as twisted pairs and open wires, in loops, and (c) cannot convey video, high speed digital, and broadband analog signals. *Note:* If plain old telephone service (POTS) wires to homes and offices were replaced with fiber optic cables, integrated multimedia services could be provided. *See also* **twisted pair.**

plain text: 1. Signals that represent text that are not encrypted and from which the information they represent can be extracted relatively easily. **2.** Unencrypted information in textual form. *Note:* Plain text may be printed, voiced, transmitted, and stored. *Synonym* **clear text.** *See also* **cipher text, cryptology, plain language.**

plain text override: In cryptosystems, a capability that (a) is used in radiotelephone and radio telegraph communications systems and (b) is built into certain cryptoequipment that permits operators to (i) monitor for plain text signals in the frequency to which they are tuned for operations, (ii) receive signals in the guard channel when their equipment is set for reception of ciphony transmissions, and (iii) operate in the cipher standby condition. *Synonym* **plain language override.** *See also* **cipher, cryptosystem, plain language, plain text.**

plan: *See* **Aeronautical Emergency Communications System Plan, aircraft communications standard plan, call sign allocation plan, communications plan, demand assignment access plan, frequency allotment plan, frequency assignment plan, maintenance plan, preassignment access plan, project plan, ship fitting communications standard plan, test plan.**

planar array: An antenna (a) in which all of the elements, including active and parasitic, are in one plane, (b) that provides a large aperture and may be used for directional beam control by varying the relative phase of each element, and (c) that may be used with a reflecting screen. *See also* **active element, antenna, parasitic element.**

planar dielectric waveguide. *Synonym* **slab dielectric waveguide.**

planar diffused waveguide: A waveguide that usually (a) is a thin-film slab dielectric waveguide, (b) is 1 µm (micron) to 10 µm thick, and (c) usually has a graded refractive index profile that is made by controlling the diffusion of dopants during its construction. *See also* **dielectric, dielectric waveguide, refractive index.**

plan area: *See* **data numbering plan area.**

planar waveguide: *Synonym* **slab dielectric waveguide.**

Planck's constant: A physical constant that (a) is approximately 6.626×10^{-34} J•s (joule•seconds) and (b) usually is designated by *h*. *Note:* Planck's constant is used in many relations that apply to electronic and fiber optic systems, including communications, control, sensing, telemetry, endoscopic, illumination, and imaging systems:

(a) The energy of a photon is given by the relation $E_{ph} = hf$ where f is the frequency of the radiation associated with the photon. The energy of a photon of 1 μm (micron) wavelength is about 1.2 ev (electron•volts). Electrons can only exist in certain energy levels. Therefore, only certain transitions between these levels can occur, giving rise to absorption and emission of only certain quanta of energy, as expressed by hf, where h is Planck's constant, and f is the frequency of the absorbed or emitted photon. Therefore, only certain discrete wavelengths of optical radiation exist in a given spectral line.

(b) Photon energy is given by the relations $E_{ph} = hf = hkc$ where h is Planck's constant, k is the wave number, and c is the velocity of light. The wave number turns out to be the number of wavelengths per unit distance in the direction of propagation.

(c) The work function of a material is defined according to the relation $W = hf_0$ where h is Planck's constant, and f_0 is the threshold frequency required to release an electron from the material.

(d) Photocurrents may be calculated from the relation $I_{ph} = \Gamma eP/hf$ where I_{ph} is the photocurrent, Γ is the carrier collection quantum efficiency, e is the electron charge, P is the incident optical power, h is Planck's constant, and f is the frequency of the incident radiation.

(e) The signal to noise ratio for a photodetector is given by the relation $S/R_{ph} = (N_p/B)(1 - e^{-hf/kT})$ where S/R_{ph} is the signal to noise ratio for the photodetector, N_p/B is the number of incident photons per unit of bandwidth, h is Planck's constant, f is the frequency of the incident photons, k is Boltzmann's constant, and T is the absolute temperature.

(f) The equation that relates photon energy, the work function of a material, such as certain metals and oxides, and the emitted electron energy, i.e., the kinetic energy with which an electron will be emitted by the material when the photon strikes it, is given by the relation $hf = W + (1/2)mv^2$ where h is Planck's constant, f is the frequency associated with the incident photon, W is the work function of the material the photon strikes, m is the mass of the electron, and v is the velocity with which the electron is ejected from the surface of the material into free space, assuming that there is no surface charge to cause attraction back into the material.

(g) In a photoemissive detector, the remaining energy of an electron that escapes from the emissive material when energy is imparted to it by an incident photon. The remaining energy is given by the relation $E_e = hf - eW$ where E_e is the remaining electron energy, h is Planck's constant, f is the photon frequency, e is the charge of an electron, and W is the work function of the emissive material.

(h) In quantum mechanics, the total energy, E, in a stream of photons of many frequencies during the time when the flow occurs, is given by the relation $E = h\Sigma f_i n_i t_i$ summed from $i = 1$ to $i = m$ where E is the total energy in the stream, f_i is the ith frequency, n_i is the number of photons of the ith frequency, t_i is the duration of the ith frequency summed over all the frequencies, photons, and time durations of each in a given light beam or beam pulse, and h is Planck's constant. *Synonym* **quantum of action.** *See also* **photon.**

Planck's law: The fundamental law of quantum theory that (a) has direct application in optical communications, i.e., lightwave communications, (b) describes the essential concept of the quanta of electromagnetic energy, i.e., describes the concept of the particle, granular, or corpuscular theory of electromagnetic radiation, particularly of light, and (c) states that the quantum of energy, E_{ph}, associated with an electromagnetic field, is given by $E_{ph} = hf$ where E_{ph} is the quantum of energy associated with a photon, h is a constant of proportionality, i.e., Planck's constant, and f is the frequency of the electromagnetic radiation, i.e., of the photon. *Note 1:* Planck's constant usually is given in joule•seconds and the frequency in hertz. Thus, the quantum of energy is usually given in joules. *Note 2:* The product of energy and time is sometimes referred to as the "elementary quantum of action." Hence, h is sometimes referred to as the "elementary quantum of action." *See also* **Planck's constant.**

plane: *See* **automatic radio relay plane, ecliptic plane, equipotential ground plane, focal plane, ground plane, image plane, meridional plane, object plane, polarization plane.**

plane electromagnetic wave: An electromagnetic wave (a) that predominates in the far field region of an antenna, (b) that has a wavefront that is essentially in a flat plane, and (c) in which all of the wavefront planes are parallel, i.e., the wave surfaces of constant phase are infinite parallel planes normal to the direction of propagation. *Note:* A uniform plane polarized electromagnetic wave, i.e., a plane wave, propagating in free space or in dielectric media has a characteristic impedance, i.e., the ratio of the electric and magnetic field strengths, given by the relation $Z = E/H$ where E and H (a) are the electric field and magnetic field strengths, (b) are orthogonal, and (c) are in a direction perpendicular to the direction of propagation, i.e., are perpendicular to the Poynting vector. In free space, the characteristic impedance of a plane electromagnetic wave is 377 ohms. From Maxwell's equations, the characteristic impedance of a plane wave in a linear, homogeneous, isotropic, dielectric, and electric-charge-free propagation medium is given by the relation $Z = (\mu/\varepsilon)^{1/2}$ where μ is the magnetic permeability and ε is the electric permittivity, where $\mu = 4\pi \times 10^{-7}$ H/m (henries per meter) and $\varepsilon = (1/36\pi) \times 10^{-9}$ F/m (farads per meter), from which 120π or 377 Ω (ohms) is obtained for free space. For dielectric media, the electric permittivity is the dielectric constant, which for glass, ranges from about 2 to 7. However, the refractive index is the square root of the dielectric constant. Thus, the characteristic impedance of dielectric materials is $377/n\Omega$, where n is the refractive index.

plane lens: A lens that (a) has no curved surfaces or has two curved surfaces that neutralize each other, (b) has no net refracting power, and (c) usually is used to isolate optical systems from their environment. *Note:* Examples of plain lenses are flashlight lenses, copier surface glass, house windows, light bulb envelopes, and auto windshields.

plane of polarization: *Synonym* **polarization plane.**

plane polarization: 1. In an electromagnetic wave, such as a lightwave or a radio wave, polarization of the electric and magnetic field vectors such that (a) they both remain in a plane that is perpendicular to the direction of propagation, (b) the orientation of the plane usually is constant with respect to distance coordinates or with respect to the longitudinal axis of a waveguide, (c) the orientation of the plane does not change with time, except perhaps slowly compared to the frequency, and (d) a slowly rotating source is capable of producing plane polarized waves at each instant of time. **2.** The polarization of a plane polarized electromagnetic wave. *See also* **electromagnetic wave, polarization.**

plane polarized electromagnetic wave: An electromagnetic wave in which (a) the electric field vector is contained in a plane that is perpendicular to a fixed direction of propagation so that successive planes in space remain parallel, and (b) wavefronts are planar rather than spherical. *See* **uniform plane polarized electromagnetic wave.** *See also* **electric field, electric vector, electromagnetic wave.**

plane wave: 1. A wave in which surfaces of constant phase are infinite parallel planes normal to the direction of propagation. **2.** An electromagnetic wave that predominates in the far field region of an antenna and that has a wavefront that essentially is in a plane. *Note 1:* The characteristic impedance of free space for a plane electromagnetic wave is given by $Z = E/H$, where E is the electric field strength in volts/meter and H is the magnetic field strength in oersteds. E and H are perpendicular, and the direction of propagation is perpendicular to both. *Note 2:* The characteristic impedance to a plane electromagnetic wave in a linear, homogeneous, isotropic, dielectric, electric-charge-free medium is given by the relation $Z = (\mu/\varepsilon)^{1/2}$ where μ is the magnetic permeability and ε is the electric permittivity. For free space, $\mu = 4\pi \times 10^{-7}$ H/m (henries per meter), and $\varepsilon = (1/36\pi) \times 10^{-9}$ F/m (farads per meter), from which 120π or 377 Ω (ohms) is obtained. *Note 3:* The refractive index is the square root of the dielectric constant. Thus, the characteristic impedance of dielectric materials to an electromagnetic plane wave is $377/n$ Ω, where n is the refractive index. *See also* **antenna, far field radiation pattern.**

plan format: *See* **communications plan format.**

planned position indicator (PPI): A map or a display that (a) usually is in the form of a polar diagram, (b) is in polar coordinates with the user or the observer assumed to be at the origin, and (c) indicates the position of an object relative to the observer or to other objects. *See* **radar planned position indicator.** *See also* **polar diagram.**

planning tool: *See* **architectures design, analysis, and planning tool.**

planoconcave lens: A lens with one plane and one concave surface. *See also* **lens.**

plant: In communications, especially in telephone systems, all the facilities and equipment used to provide communications services. *Note:* At a given facility, the plant usually is divided into the outside plant and the inside plant. The outside plant includes all equipment outside the central office (C.O.) or switching center, such as all poles, repeaters and their housing, ducts, cables, aerial inserts, and the inside portion of interfacility cables outward from the main distributing frame (MDF) in a C.O. or switching center. The inside plant includes the MDF and all equipment and facilities within a C.O. or switching center. *See also* **network terminating interface, nontechnical load.**

plasma activated chemical vapor deposition (PACVD) process: In the manufacture of graded index (GI) optical fibers, a chemical vapor deposition

(CVD) process in which (a) a series of thin layers of materials of different refractive indices are deposited on the inner wall of a glass tube as chemical vapors flow through the tube, and (b) a microwave resonant cavity is used to stimulate the formation of oxides by means of a nonisothermal plasma generated by the microwave cavity.

plasma panel: *Synonym* **gas panel.**

plastic: *See* **bulk optical plastic.**

plastic clad silica (PCS) fiber: An optical fiber that (a) has a silica core, i.e., a glass core, (b) has a plastic cladding, usually a soft silicone material, (c) usually has a higher optical attenuation rate and dispersion than all-glass fibers, and (d) usually has a step index profile. *See also* **attenuation rate, cladding, core, dispersion, step index profile.**

plastic fiber: *See* **all-plastic optical fiber.**

plastic optical fiber: *See* **all-plastic optical fiber.**

plate: *See* **type plate.**

platen: 1. On a printer or a typewriter, a cylindrical roller that guides and holds paper during printing. **2.** A transparent surface of a copying machine upon which material to be copied is placed.

plenum: In a building, an enclosure that (a) is created by components of the building, such as a space above a suspended ceiling or below an access floor, i.e., false floor, (b) is used for the movement of environmental air, and (c) is used for laying communications and power cables to reach equipment installed in open space, such as open office, laboratory, machinery, and equipment space. *Note:* Cables installed in plenums must meet applicable environmental and fire protection regulations. This may mean enclosing them in suitable ducts or using cables that have components, such as jackets and strength members, made of material that are resistant to open flame and are nontoxic at high temperatures.

plesiochronous: The relationship between two signals such that (a) their corresponding significant instants occur at nominally the same rate, (b) any variations are constrained within a specified limit, and (c) there is no limit to the phase difference that can accumulate between corresponding significant instants over a long period. *See also* **anisochronous, asynchronous transmission, heterochronous, homochronous, isochronous, isochronous modulation, mesochronous, synchronous.**

PLL: phase-locked loop.

PLM: *Abbreviation for the deprecated synonym* **pulse length modulation.** *See* **pulse duration modulation.**

plosive: An aperiodic speech sound that is created by an explosive action, i.e., a step function or a pulse of a sound that is created by the speech-forming mechanisms. *Note:* In English, plosives are the sounds made in naming, i.e., pronouncing, the letters B, K, P, and T in reciting the alphabet, or when they are used in words in which they are not silent.

plotter: 1. An output unit that presents data in the form of a two-dimensional graphic representation. **2.** A display device that (a) displays data on a display surface, usually on a point-by-point or line-segment-by-line-segment basis, (b) uses a stylus or a writing head placed at a coordinate position by a mechanical or electromechanical driving mechanism, and (c) makes a permanent recorded copy of an image on paper or film placed on a flat or cylindrical surface. *See* **acceptance pattern plotter, flatbed plotter, radiation pattern plotter.** *See also* **image, recorded copy, stylus, writing head.**

plotter bed: The display surface of a plotter.

plotter step size: The smallest displacement that can be drawn by a plotter, usually corresponding to the distance between addressable points within the display space of the plotter. *See also* **addressable point, displacement, display space, plotter.**

plow-in fiber optic cable: A fiber optic cable that (a) is designed specifically to be buried in direct contact with the earth, i.e., the ground, without any further preparation, treatment, or protection, and (b) usually is buried by slit-plowing a trench, paying the cable into the slit, and allowing the slit to close on the cable. *Note:* In plowing-in cross-country cables, special care is taken when obstacles are encountered, such as rock formations, driveways, road intersections, and buried utility lines, such as buried water, sewer, gas, and electric power lines. Certain obstacles may require an aerial insert. The location of the cable is clearly marked, and connection boxes are placed where appropriate. *See also* **aerial fiber optic cable, aerial insert, buried cable, underground cable.**

PLS: physical signaling sublayer.

plug: In telephone switchboard operation, a metal fitting that (a) has an insulated grip, (b) usually has two contacts, one at its tip and one on its shaft, (c) is connected to the conductors of a flexible cord circuit, and (d) is used to make a connection by inserting it into a jack. *See* **answering plug, calling plug.** *See also* **connection, cord circuit, jack.**

plugboard: A perforated board or panel on which an array of jacks is mounted, into which pins, plugs, or jumpers may be inserted in order to create circuits to control equipment in a desired manner. *Note:* Examples of plugboards are (a) in cryptoequipment, an electrical device that (i) may be used to introduce variations in cryptooperations and (ii) usually consists of a number of wires with plugs and an array of jacks into which the plugs may be inserted, (b) in analog computers, a panel with an array of jacks into which pins, plugs with wires attached, or jumpers may be inserted to connect the various components, such as summers, multipliers, integrators, and differentiators, that are required for solving a problem, and (c) in telephone switchboards, the panel that holds the jacks into which plugs connected to cord circuits are inserted. *See also* **analog computer, cryptoequipment, jumpers, plugs, switchboard.**

plug connector: *See* **tapered plug connector.**

plug-ended cord: A conductor that has a plug at each end and that may be used to establish a connection between two jacks into which they are inserted, such as for (a) connecting an incoming line, i.e., a calling line, to an outgoing line, i.e., a called line, on a telephone switchboard, and (b) connecting a summer and a differentiator by means of a panel of jacks in an analog computer. *See also* **jack, plug.**

plugging out lights: In telephone switchboard operation, an invariably prohibited action by an operator in which a call is handled by deliberately inserting a plug into the calling jack, thereby causing the call indicator light to go out, and then removing the plug without any attempt to complete the call. *See also* **call, calling jack, completed call, plug.**

plunger: *Synonym* **piston.**

plus lens: *Synonym* **converging lens.**

PL/1: A general purpose computer programming language that (a) is designed for use in a wide range of commercial and scientific computer applications and (b) is primarily used for numerical and logical processing operations, such as computation, character string handling, and record processing operations. *See also* **computer program.**

pm: phase modulation.

PM: phase modulation, post meridiem, preventive maintenance.

PMB: pilot-make-busy.

PMB circuit: pilot-make-busy circuit.

p-n junction: In a semiconductor device, such as a transistor or a diode, the interface that (a) is between p-type and n-type semiconductor material, thus creating holes, i.e., acceptor sites, on one side of the interface, and relatively free electrons, i.e., donor sites, on the other side of the interface, and (b) usually is assumed to be abruptly or linearly graded in its transition region across the interface from the p-type to the n-type material, i.e., from the positively doped to the negatively doped region. *Note:* Holes are less mobile than electrons. Both contribute to the total electric current. Intrinsic materials that may be doped for p-n junctions include germanium and silicon.

Pockels cell: A device in which (a) a material, usually a crystal, such as lithium niobate, is used that has a refractive index that changes linearly with an applied electric field, (b) the configuration is such that the crystal is part of an optical system, (c) provision is made for modulating the light in the optical path in which the material is placed, (d) the modulation depends on the rotation of the polarization plane of the light in a beam, the rotation being caused by an applied electric field, (e) a light beam is passed through a polarizer that transmits only certain parts of the beam, and (f) the parts of the beam that are transmitted depend on the orientation of the polarization plane of the part. *See also* **Kerr cell, Pockels effect, polarization plane.**

Pockels effect: In doubly refracting media, i.e., birefringent media, an increase in their usual or natural birefringence, caused by an applied electric field. *Note:* The change in birefringence is linearly proportional to the applied electric field strength. The electric field causes a rotation of the polarization plane of a plane polarized electromagnetic wave. When combined with polarizers, the effect can be used in active optical devices, such as modulators and switches. *See also* **birefringence, Kerr effect, Pockels cell, polarization plane.**

Pockels effect sensor: A fiber optic birefringent, i.e., doubly refracting, sensor in which (a) an applied electric field produces birefringence in a material that is not ordinarily a birefringent medium without the applied field, or enhances birefringence in a material that is inherently doubly refractive, (b) the voltage applied across the material causes plane polarized light propagating within it to be resolved into two orthogonal vectors, (c) the change in phase retardation between the two vectors (ellipticity) is directly proportional to the applied electric field strength, (d) a fiber optic polarizer analyzes the output beam, causing irradiance modulation, i.e., intensity modulation, of light at a photodetector, and (e) the applied electric field strength controls the output of the photodetector. *Note:* The Pockels effect is found in particular crystals that are capable of advancing or retarding the phase of the induced ordinary ray, relative to the phase of the extraordinary ray, when the electric field is applied. *See also* **birefringence, extraordinary ray, irradiance, modulation, ordinary ray, photodetector.**

POI: point of interface.

point: 1. A location of zero dimension. **2.** In designating the size of type font, one point is equal to 1/72 of an inch. *Note:* The point size of a font is measured from the top of an upper case letter, i.e., a capital letter, to the bottom of a descending lower case letter. The overall dimension of a 12-point font is 0.167 in. (inches). This would be the actual width of a line and does not include the spacing between lines. *See* **access point, addressable point, adjunct service point, available point, demarcation point, dew point, drop zone impact point, entrance point, entry point, exit point, first principal point, focal point, image point, object point, principal focus point, principal point, reentry point, reference point, rerun point, restart point, secondary control point, second principal point, service access point, service termination point, signal transfer point, singing point, subsatellite point, test point, third-order intercept point, transmission level point, T reference point, viewpoint, visibility point, V reference point, zero dBm transmission level point.**

pointer: 1. A function or operation indicator that (a) is under the direct control of a computer operator, (b) can have a solid line structure, and (c) is used to (i) indicate displayed information, (ii) highlight data, (iii) identify areas of interest, (iv) serve as a graphic display cursor, and (v) select icons. **2.** In computer graphics, a manually operated functional unit used to specify an addressable point. **3.** In computer programming, an identifier that indicates the location of specific data. **4.** In display systems, a manually operated functional unit used to conduct interactive graphic operations, such as selection of one member of a predetermined set or menu of display elements or indication of a coordinate position in the display space on the display surface of a display device, thus generating corresponding coordinate data. *See* **addressable point, cell pointer, identifier, menu pointer.**

pointer movement key: On computer standard keyboards, one of a set of keys that is used to move the pointer or the cursor to various points on the monitor screen. *Note 1:* Examples of pointer movement keys are the UP, LEFT, DOWN, RIGHT, HOME, END, PgUp, PgDn, and TAB keys. Pointer movement keys may be used in combination to accomplish special purposes, such as move the cursor to the beginning of a document, to the end of a document, and to the left or right end of a line. The set of pointer keys is usually located in a contiguous keypad at the right end of the keyboard, in coincidence with a numeric keypad. A special key, called the NUM LOCK key, is used to switch the keypad from a numeric keypad to a pointer movement keypad. *Note 2:* On a computer standard keyboard, pointer keys are designated as:

KEY	**FUNCTION: Moves Pointer:**
↑ [UP]	One row or line up
← [LEFT]	One column or space to the left
↓ [DOWN]	One row or line down
→ [RIGHT]	One column or space to the right
[HOME]	To upper left corner of screen
[END]	To end of line or screen
[PgUp]	Up by one page
[PgDn]	Down by one page
£	Left by one screen or page
¢	Right by one screen or page
−	To bottom of screen or screen down
+	To top of screen or screen up

pointing: In display systems, indicating or picking one of several symbols reserved for a specific use without using a keyboard to key in data, such as a label or coordinate data. *Note:* Examples of pointing are (a) touching a light pen to a displayed symbol to clear the screen of a cathode ray tube (CRT) or the faceplate of a fiberscope, (b) picking a symbol adjacent to a selected display image to indicate a response to a displayed query, (c) using a mouse to move a cursor to a displayed list of options and clicking the mouse to enter the selection, and (d) using the pointer movement keys to move the cursor to an item on a displayed list of options and pressing the ENTER key. *See also* **image, light pen, picking.**

pointing angle: In satellite communications systems, the elevation angle or the azimuth angle of an Earth station antenna when it is directed toward a satellite.

point of interface (POI): In a telecommunications system, the physical interface between the local access and transport area (LATA) access and interLATA functions. *Note:* The point of interface (POI) is used to establish the technical interface, the test points, and the points of operational responsibility. *Synonym* **interface point.** *See also* **local access and transport area.**

point of presence (POP): A physical layer within a local access and transport area (LATA) at which an interLATA carrier establishes itself for the purpose of obtaining LATA access, and to which the local exchange carrier provides access services. *Synonym* **presence point.** *See also* **local access and transport area.**

point of train (POT): In infrared transmission systems, a steady infrared light that is used to assist the transmitter in locating a receiving station and for keeping the transmitted light pointed in the proper direction for satisfactory reception.

point protocol: *See* **Point to Point Protocol.**

points: *See* **emission beam angle between half-power points.**

point source: A source of electromagnetic radiation such that (a) the source is so distant from a point of observation, measurement, or reception of the radiation that the wavefront of the radiation is essentially a planar surface, in relation to an antenna or a detector at the point, rather than a curved surface, regardless of the shape of the source, (b) the size or the shape of the source has no influence on the shape of the wavefront at the point of observation or measurement, and (c) the source need not necessarily radiate with equal radiance in all directions. *Note:* The wavefront from a point source in a uniform propagation medium is always and everywhere spherical, but when the distance from the point source is very large compared to a receiving aperture, the wavefront is practically that of a plane as it enters the aperture, such as that of a receiving antenna. *See also* **aperture, propagation medium, wavefront.**

point source flash detector: *See* **infrared point source flash detector.**

point to point: In communications systems, pertaining to the interconnection of two devices, particularly user end instruments, such as telephones, display devices, remote terminals, facsimile terminals, recorders, and personal computers, without short-term connection via intervening switchboards or switching centers.

point to point circuit: **1.** A circuit that is established between two terminals, including any intermediate patching or switching facilities connected on a long-term basis. **2.** A circuit that directly connects two terminals, data stations, or user end instruments.

point to point connection: In communications systems, a connection that (a) is established between only two terminals, nodes, or data stations and (b) may be (i) a direct connection without intervening switching facilities or (ii) via switching facilities in which the connections are established for long periods, i.e., are permanent or at least semipermanent.

point to point link: A data link that connects only two stations, terminals, or nodes. *See also* **dedicated circuit, hot line, link.**

Point to Point Protocol (PPP): A network protocol that, when installed in a computer, enables dialing into an Internet service provider.

point to point transmission: Transmission directly between two designated stations. *See also* **broadcast operation, dedicated circuit, hot line, leased circuit.**

Poisson distribution: A statistical distribution of values of a variable that differs from the normal bell curve (Gaussian) distribution only in that the variable cannot assume all values, but is limited to a particular and finite set of values. *Note:* Examples of Poisson distributions are (a) a distribution of groups of bits that can occur only in certain patterns, (b) distributions of pulses that can occur only in certain ways or in certain groups, and (c) the distribution that is encountered in counting the number of blood cells that pass through an orifice while it is suspended in a liquid. The cells pass through in clusters, with a finite number of cells in each cluster according to a Poisson distribution. The clusters are counted by a resistance change across the orifice as the cluster passes through it. The total number of cells is determined by the number of clusters counted and the statistical number of cells in each cluster.

polar diagram: A display on a relatively flat display surface of a display device in which distances or ranges from a reference, origin, or observation point are indicated as a straight line radial distance to an observed point, and headings or azimuth angles are indicated as angular displacements from a radial reference line. *Note:* In most radar polar diagrams, the observation point is considered to be in the center of the polar diagram. However, if the area of concern is within a sector of a circle, the origin might be located along one edge of the display surface. Ranges are measured by concentric circles that indicate radial distance to an object. A reference direction, such as true north or a line from the display device to the bow of a ship, is indexed to the screen for measuring azimuth. The arrangement is similar to that of a conventional polar coordinate system, in which the coordinates are ρ, the radial distance outward from the origin at the center, and θ, the angular displacement from a reference direction, i.e., a zero direction, usually a horizontal line drawn toward the right from the origin, similar to the positive x-direction in Cartesian coordinates. In Cartesian coordinates one usually reads to the right or the left along the x-axis and then up or down in the direction of the y-axis. In polar coordinates, one rotates a radial line about the origin from the zero reference direction through an angle θ and then moves radially outward from the origin a distance ρ. The angle θ corresponds to the azimuth of the object, and the distance ρ corresponds to the range to the object.

polar direct current telegraph transmission: Binary telegraph transmission in which (a) positive and negative direct currents (dc) represent the significant conditions, (b) the signal alternates between positive and negative voltages, and (c) the current reverses on the line during transitions between marks and spaces. *Synonym* **double current transmission.** *See also* **direct current signaling, telegraph.**

polarential telegraph system: A direct current (dc) telegraph system in which polar transmission is used in one direction, and differential duplex transmission is used in the other direction. *Note:* Two types of polarential systems, known as Type A and Type B, are in use. In half-duplex operation of the Type A polarential telegraph system, the direct current (dc) balance is independent of line resistance. In half-duplex operation of the Type B polarential telegraph system, the direct current (dc) is substantially independent of the line leakage. Type A is better for cable loops, where leakage is negligible but resistance varies with temperature. Type B is better for open wire, where variable line leakage is frequent. *See also* **communications system, half-duplex operation.**

polarimetric sensor: *Synonym* **polarization sensor.**

polariscope: A combination of a polarizer and an analyzer used to detect birefringence in materials placed between them or to detect polarization plane rotation caused by materials placed between them. *See also* **birefringence, polarization plane.**

polarization: In an electromagnetic wave, the time-varying direction and amplitude of the electric field vector. *Note 1:* The type of polarization may be identified by the figure traced, as a function of time, by the projection of the extremity of the electric field vector onto a fixed plane that is perpendicular to the direction of propagation. It may be traced in a clockwise sense, i.e., right-hand polarization, or a counterclockwise sense, i.e., counterclockwise polarization, as viewed in the direction of propagation. The sense of rotation is that of a right-hand screw rotated clockwise and viewed from behind the screw and facing the direction of forward motion of the tip of the screw as it is turned clockwise into a fixed nut or tapped hole. The rotation direction is obtained by rotating the electric vector into the magnetic vector through the smaller angle between them. The propagation direction at every instant is perpendicular to both the electric and magnetic fields, i.e., in the direction of the Poynting vector. The figure traced by the tip of the electric field vector is that of an ellipse. Circular polarizations are obtained when the ellipse becomes a circle. Linear polarizations are obtained when the ellipse collapses and becomes a straight line. *See* **circular polarization, elliptical polarization, left-hand circular polarization, left-hand helical polarization, plane polarization, right-hand circular polarization, right-hand helical polarization, rotating polarization, variable length polarization, variable polarization.** *See also* **depolarization, diversity reception, magnetooptic, Poynting vector.**

polarization beat length: For a plane polarized lightwave propagating in an optical fiber, the distance over which the polarization goes through one complete cycle of change caused by modal birefringence, such as (a) from linear polarization in one transverse direction, (b) through elliptical polarization to circular polarization, (c) then through linear polarization in the opposite direction from before, (d) then through elliptical polarization to circular polarizations, and (e) back to linear polarization in the original direction, i.e., the polarization plane rotates through an angle of 360° during one beat length. *Note:* The beat length for a single mode optical fiber is given by the relation $L_B = 2\pi/(\beta_x - \beta_y)$ where β_x and β_y are the propagation constants of the two polarized waves caused by the birefringence of the fiber. Beat lengths of single mode fibers are typically a few centimeters. *See also* **birefringence.**

polarization diversity: 1. Pertaining to changing the polarization of an electromagnetic wave, usually at the source of radiation, by changing the direction of the polarization plane or by changing the linear, circular, elliptical, or helical polarization of the wave. **2.** Pertaining to two or more types of polarization mixed in the same light beam or in the same transmission. **3.** Pertaining to transmission and reception of electromagnetic waves in which the same information signal is transmitted and received simultaneously using orthogonally polarized waves with independent propagation characteristics. *See also* **depolarization, diversity reception, dual diversity.**

polarization-holding optical fiber: *Synonym* **polarization-maintaining optical fiber.**

polarization-holding parameter: *Synonym* **polarization-maintaining parameter.**

polarization-maintaining (PM) optical fiber: An optical fiber (a) in which the polarization planes of lightwaves launched into the fiber are maintained during propagation with little or no cross coupling of optical power between the polarization modes, and (b) that has a flattened, elliptical, or rectangular core such that the direction of the polarization plane of a plane polarized lightwave entering the fiber remains fixed with respect to the direction of the optical axis of the fiber as the lightwave propagates along the fiber. *Note 1:* Launch conditions at the entrance of the fiber must be consistent with the direction of the transverse axis of the fiber cross section. *Note 2:* Cross sections of polarization-maintaining optical fibers range from elliptical to rectangular. *Note 3:* Polarization-maintaining optical fibers are used in special applications, such as in fiber optic sensing and interferometry. *Synonyms* **polarization-holding optical fiber, polarization-preserving optical fiber.**

polarization-maintaining parameter: In a polarization-maintaining (PM) high birefringence optical fiber, a parameter that (a) indicates the ability of the fiber to maintain the polarization of a lightwave launched into the fiber and (b) describes the average rate at which optical power is transferred from one excited mode to the other orthogonally polarized mode. *Note:* The relative cross-coupled power increases with the length of the fiber according to the relation $P_x/(P_x + P_y) = (1 - e^{-2hL})/2$ where P_x and P_y are the optical powers in the two orthogonal modes of the high birefringence fiber, h is the polarization-maintaining parameter, and L is the fiber length. A polarization-maintaining parameter value of 1×10^{-5} m^{-1} corresponds to 1% average cross-coupled power after 1 km (kilometer) of optical fiber. *Synonyms* **h-parameter, h-value, polarization holding parameter, polarization-preserving parameter.**

polarization modulation: The modulation of an electromagnetic wave such that the polarization of the wave, such as (a) the direction of polarization of the electric and magnetic field vectors or (b) the relative phasing of the electric and magnetic field vectors to produce changes in polarization angle in linear, circular, or elliptical polarization, is varied in accordance with a characteristic of an intelligence-bearing signal, such as a pulse-or-no-pulse digital signal. *Note:* Polarization modulation can be accomplished in waves propagating in waveguides. In dielectric waveguides, such as optical fibers, polarization shifts that are made in accordance with an input signal are a practical means of modulation. *See also* **modulation, polarization.**

polarization noise: Fluctuations of the direction of the polarization plane of an electromagnetic wave emanating from the end of an optical fiber. *Note 1:* Polarization noise is caused by variations in the direction of polarization of the input lightwave and from added fluctuations caused by longitudinal and transverse variations in the refractive index profile along the fiber. *Note 2:* Polarization noise above certain threshold levels will interfere with the operation of polarization-maintaining optical fibers. *Synonym* **birefringence noise.** *See also* **birefringence.**

polarization plane: At any point in a medium in which an electromagnetic wave is propagating, the plane determined by the two intersecting vectors that represent the direction of the electric and magnetic fields at that point. *Note 1:* The electric and magnetic field vectors lie in, and thus define, the polarization plane. *Note 2:* The direction of propagation of the wave is perpendicular to the polarization plane. *Note 3:* The polarization plane vector may be represented by the vector product of the electric and magnetic field vectors given by the relation $\mathbf{P} = \mathbf{E} \times \mathbf{H}$ where $\mathbf{P}$ is the Poynting vector, $\mathbf{E}$ is the electric field strength, and $\mathbf{H}$ is the magnetic field strength. *Note 4:* In geometric optics, the unit vector that represents the direction of the polarization plane is always and everywhere in the same direction as the Poynting vector, i.e., in the direction of the ray. *Note 5:* In physical optics, the wavefront at a point lies in the polarization plane at that point. *See also* **electromagnetic wave, polarization.**

polarization-preserving optical fiber: *Synonym* **polarization-maintaining optical fiber.**

polarization sensitivity: In an optical device, the change in a performance parameter, such as the transmittance of an attenuator or the power-splitting ratio of a coupler, per unit change in the angle of the polarization plane. *See also* **polarization.**

polarization sensor: In fiber optics, a sensor in which (a) the polarization plane of an electromagnetic wave, such as a lightwave in an optical fiber, is rotated by a longitudinal magnetic field applied by a coil wrapped around the fiber, (b) the total number of times the polarization plane is rotated when the fiber passes through the magnetic field is a function of the magnetic field strength and the distance that the fiber is in the field, (c) a change in the number of polarization plane rotations produced by a change in the electric current level in the coil can be detected by a photodetector placed after a polarization analyzer, the output of which is a maximum when the analyzer grating is parallel to the polarization plane and is zero when it is perpendicular, (d) a finite number of output signal peaks occurs for a finite change of current in the coil, (e) the instantaneous pulse repetition rate or the frequency of the output pulses is proportional to the rate of change of the current in the coil, and (f) the time derivative of an amplitude-modulated input electrical signal to the coil results in a frequency-modulated output signal from the photodetector. *Synonym* **polarimetric sensor.** *See also* **polarization plane.**

polarized antenna: *See* **single-polarized antenna.**

polarized electromagnetic wave: An electromagnetic wave in which the electric vectors, i.e., the electric fields, of all the propagating modes are in the same direction, and that direction is perpendicular to the direction of propagation. *Note:* Narrow slit sources, such as slot antennas, narrow aperture antennas, edge emitting diodes, and certain lasers can be made such that they emit polarized electromagnetic waves. Unpolarized radio waves and lightwaves, such as sunlight, which is unpolarized, can be polarized by passing the waves through a polarizer, such as a diffraction grating, which eliminates all the electric fields and their components that are transverse to the grating. Once polarized, radio waves will remain polarized unless they are scattered, assuming that they are not absorbed. Polarization can be maintained in optical fibers by launching a polarized wave into a flattened optical fiber, i.e., a polarization-maintaining optical fiber. *See* **horizontally polarized electromagnetic wave, left-hand polarized electromagnetic wave, plane polarized electromagnetic wave, right-hand polarized electromagnetic wave, uniform plane polarized electromagnetic wave, vertically polarized electromagnetic wave.** *See also* **polarization-maintaining optical fiber, polarizer.**

polarized light: A light beam that has its electric vector oscillating in a direction that does not change, unless the propagation direction changes, such as when a light ray is bent by a refractive index gradient or a radio wave is bent by a knife edge. *Note:* If the time-varying electric vector can be resolved into two perpendicular components that have equal amplitudes and that differ in phase by 1/4 wavelength, the light is circularly polarized. Circular polarization is obtained whenever the phase difference between two perpendicular components is any odd integral number of quarter wavelengths. If the electric vector is resolvable into two perpendicular components of unlike amplitudes that differ in phase by values other than 1/4, 1/2, 3/4, 1 . . . wavelengths, the light beam is elliptically polarized. *See also* **circular polarization, electric vector, elliptical polarization, polarization.**

polarized mode: *See* **linearly polarized mode.**

polarized operation: *See* **cross polarized operation.**

polarized return to zero modulation: The modulation of a binary signal in which (a) one of the binary digits, say the 1s, is represented by a signal significant condition, such as the magnetization direction, electric field intensity polarity, magnetic field intensity polarity, voltage polarity, or the electric current direction, in one sense, and the other binary digit, in this case the 0s, is represented by a signal significant condition in the opposite sense, and (b) for each binary digit, there is a time or spatial significant condition that represents a zero level of the signal, i.e., there is always a significant condition between all consecutive significant instants. *See also* **modulation, significant condition, significant instant.** *Refer to* **Fig. P-8.**

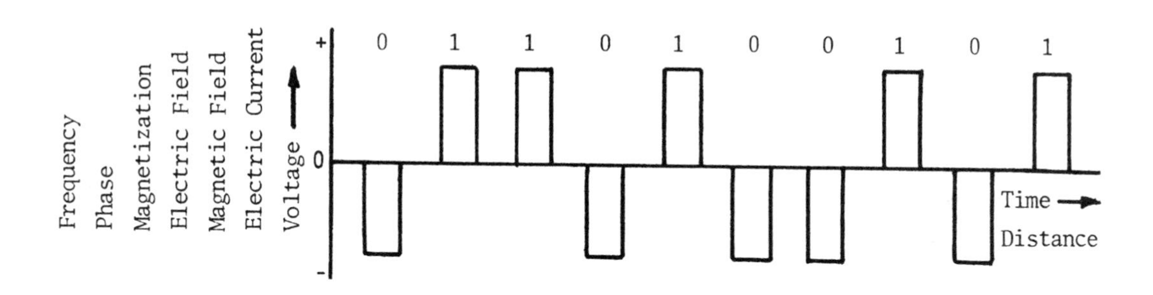

Fig. P-8. In signals that result from **polarized return to zero modulation,** the physical variable. i.e., parameter, returns to a zero level between the binary digits. An opposite, complementary, or other nonzero significant condition is used for the 0 and 1 digits.

polarizer: 1. An optical device that transforms unpolarized light, i.e., diffused or scattered light, into polarized light. **2.** An optical device that alters the type of polarization of polarized light. *See* **optical fiber polarizer.**

polar operation: Operation of a transmission line such that (a) marking signals are formed by current or voltage pulses of one polarity, and (b) spacing signals are formed by current or voltage pulses of equal magnitude but opposite polarity, i.e., bipolar signaling is used, such as using one polarity for one of the binary digits, say the 1s, and the other polarity for the other binary digit, in this case the 0s. *Note:* A positive pulse, represented by a current flowing in one direction in a circuit, could be used to represent a mark or a 1, and a negative pulse, represented by a current flowing in the opposite direction in the circuit, could be used to represent the space or a 0. The direction of electric current flow, or the polarity of a voltage, is not to be confused with the direction of information flow. Thus, a negative voltage pulse at the transmitting end that causes current to flow from the receiving end toward the transmitting end is nevertheless a pulse that travels toward the receiver and bears with it the information sent to the receiver. The phenomenon observed at the receiver would be one in which the polarity goes negative as current is withdrawn from the receiver, signifying the arrival of the pulse at the receiver. *See also* **bipolar signal, neutral operation, pulse.**

polar orbit: An orbit such that (a) it is in a plane with an inclination angle of 90°, i.e., an inclined orbit with an inclination angle of 90°, (b) a satellite in the orbit will pass directly over the North and South Poles, (c) the obit will remain stationary with respect to the Earth's rotation, and (d) the Earth will rotate under the polar orbit with one rotation every 24 hours while a satellite in a polar orbit will complete an orbit in about 90 minutes. *See also* **direct orbit, equatorial orbit, geostationary orbit, inclined orbit, orbit, retrograde orbit, satellite, satellite synchronous orbit.**

polar relay: A relay that consists of a solenoid, i.e., an electromagnet, and an armature such that it can actuate the armature in a direction of movement that depends upon the direction of current flow in the solenoid. *See also* **neutral relay, relay.**

polar transmission: *Synonym* **double current signaling.**

pole: *See* **celestial pole.**

policy: *See* **communications security policy, electromagnetic emission policy, emission policy.**

Polish notation: *Synonym* **prefix notation.**

poll: In a communications system, to interrogate a group of devices one at a time for a specific purpose, such as to prevent contention, determine operational status, determine readiness to receive or transmit messages, or invite to transmit. *See also* **blind, interrogating, lockout, select.**

polling: 1. A network control system in which the control station invites the tributary stations to transmit in the sequence specified by the control station. **2.** In multipoint or point-to-point communications, the process in which stations are invited, one at a time, to transmit. **3.** A sequential interrogation of devices for various purposes, such as avoiding contention, determining operational status, or determining readiness to send or receive data. *See also* **interrogating, network, response.**

polychromatic radiation: Electromagnetic radiation that consists of two or more frequencies or wavelengths. *Synonym* **dichromatic radiation.** *See also* **monochromatic radiation.**

polychromatism: *Synonym* **dichroism.**

polyline: In display systems, a display element, display group, or display image that consists of a set of connected lines, usually capable of being operated upon as a unit, such as by translating, rotating, and scissoring the set of lines all at once.

polymeric nonlinear optical material: A transparent polymer, that (a) functions as a high-order nonlinear material, (b) displays nonlinear optical properties when energized by lightwaves, (c) produces second- and higher-order harmonic frequencies when energized with single frequency lightwave signals, (d) allows switching speeds of less than 100 fs (femtoseconds), (e) allows lossless operations, i.e., operations that can be passively repeated in successive stages many times, (f) has an exit wave that can be frequency-mixed, (g) has output beams that can be considerably different from input beams, and (h) is being used to manufacture optical modulators, multiplexers, and switches. *Note:* At the exit point of an optical integrated circuit (OIC) waveguide made of polymeric nonlinear optical material, different signals can be switched to different ports.

polyphase sort: A sort of a set of items in which (a) subsets of sorted items that result from internal sorts are merged, and (b) the distribution of the sorted subsets in auxiliary or external storage is based on a specific algorithm, function, or rule, such as a Fibonacci series. *See also* **Fibonacci series, internal sort, merge, sort.**

polystep dither: In optical systems, a method of obtaining and tagging perturbations that consist of discrete changes made consecutively on a fixed time schedule. *See also* **frequency multidither.**

POP: point of presence.

population inversion: A redistribution of energy levels in a population of chemical elements such that, instead of there being more atoms with lower-energy-level electrons, there are fewer atoms with higher-energy-level electrons. *Note 1:* The increase in the total number of electrons in the higher excited states occurs at the expense of the energy in the electrons in the ground or lower state and at the expense of the resonant energy source, i.e., the pump. *Note 2:* Population inversion is not an equilibrium condition. Population inversion is brought about by, and must be maintained by, the pumping action of an energy source. When population inversion occurs, the probability of downward energy transitions, giving rise to radiation, is greater than the probability of upward energy transitions, giving rise to photon absorption. This results in a net output radiation level, thus producing stimulated emission, i.e., the laser action that occurs in a laser. *See also* **pump frequency.**

port: 1. In a communications network, a point at which signals can enter or leave the network en route to or from another network. *Note:* An example of a port is the point in a shipboard data transfer network (DTN) at which a ship to shore communications link can be connected. **2.** In a communications network, a place of access to a device or a network where energy may be supplied or withdrawn or where the device or network variables may be observed or measured. **3.** For a ship, a safe haven that (a) affords protection from adverse weather conditions, (b) allows loading and unloading operations to take place, and (c) allows various services to be obtained, such as maintenance, supply, and medical services. *See also* **bridge, brouter, data circuit-terminating equipment, gateway, input/output device, router, terminal.** *Refer to* **Fig. P-9.** *Refer also to* **Fig. B-9 (96).**

portability: 1. A measure of the ease with which an entity can be moved from one location to another. **2.** The ability to transfer data from one system to another without being required to re-create or reenter data descriptions or to significantly modify the application being transported. **3.** The ability of software of a system to run (a) on more than one type or size of computer or (b) under more than one operating system. *See also* **compatibility, interoperability, mobile service, mobile station, portable operating system interface for computer environments, transportability.** *Refer to* **Fig. S-7 (870).**

portable: The characteristic of an item such that it (a) can be carried by one or more persons, (b) if a source of power is required for operation, usually has a self-contained power source, (c) usually can be oper-

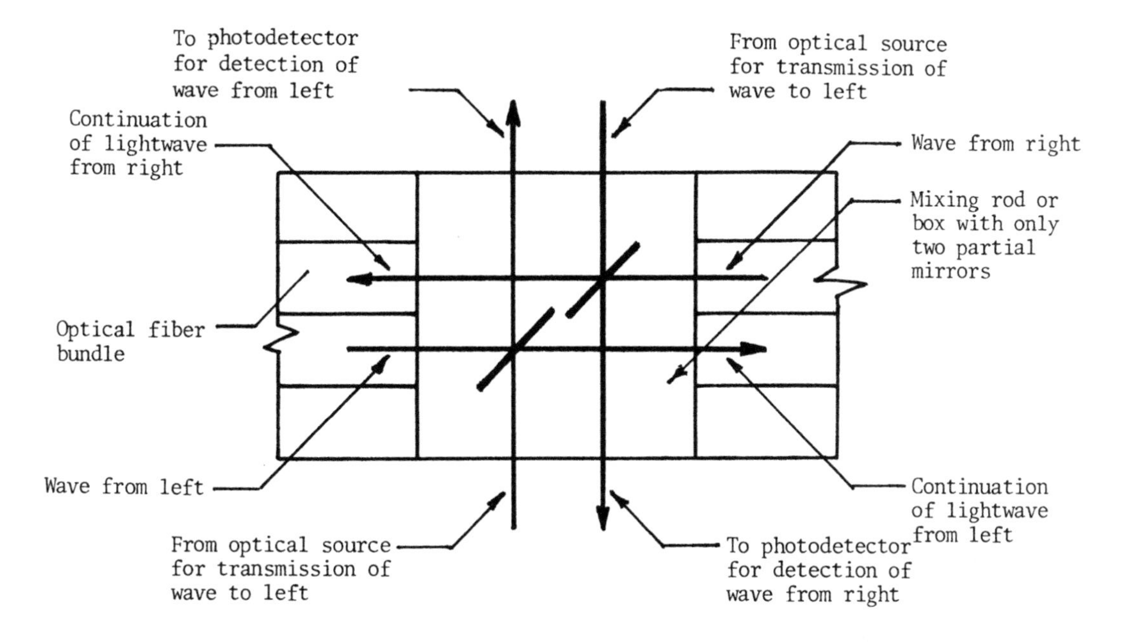

Fig. P-9. Connection of fiber optic bundles to an optical mixing rod, showing entry and exit **ports.**

ated while being carried or during brief halts at unspecified locations, and (d) may be operated from a carrier, such as a vehicle, aircraft, or ship, while in motion. *See* **air portable.** *See also* **mobile, transportable.**

portable operating system interface for computer environments (POSIX): A U.S. government information processing standard for a vendor-independent interface between an operating system and an application program, including operating system interfaces and source code functions.

portable station (PS): A radio station that (a) can be carried by one or more persons, (b) usually has a self-contained power source, (c) usually is operated while being carried or during brief halts at unspecified locations, and (d) may be operated from a vehicle while in motion. *Refer to* **Fig. P-10.** *Refer also to* **Fig. S-7 (870).**

port operations service: A maritime mobile service that (a) operates in or near a port between coast stations and ship stations, and between ship stations, (b) restricts messages to the handling, movement, and safety of ships and, in emergency, to the safety of persons, and (c) excludes public correspondence messages.

port radio transmission control: A special control that is imposed on a ship station while it is in port, such as sealing radio equipment, restricting all messages except distress messages, and obtaining permission from local consular authority to transmit.

portrait mode: In facsimile systems, the mode of operation in which the scanning lines are perpendicular to the longer dimension of a rectangular object, i.e., a rectangular original. *See also* **landscape orientation.**

portrait orientation: 1. In facsimile systems, the mode for scanning lines across the shorter dimension of a rectangular object, i.e., a rectangular original. **2.** In computer graphics, the orientation of a page in which the longer dimension is vertical. **3.** An orientation of printed text on a page such that the lines of text are perpendicular to the long dimension of the page. *Note:* If the page contains an image, such as a picture, and the page is viewed in the normal manner, the long dimension of the page will be perpendicular to the interpupillary distance of the viewer, i.e., the interpupillary line. If a person is standing, sitting in an upright position, lying on a side, or has the head cocked to the side, portrait orientation calls for the long dimension of the page to be perpendicular to the interpupillary distance. Thus, the orientation is with respect to the viewer, not with respect to the horizontal or vertical directions of the Earth or the Earth's horizon. Once a page is printed or a picture is created with a portrait orientation, it matters not how the page or the picture is oriented with respect to the Earth. *Synonym* **portrait mode.** *See also* **landscape orientation.** *Refer to* **Fig. L-1 (500).**

port station: A coast station in the port operations service.

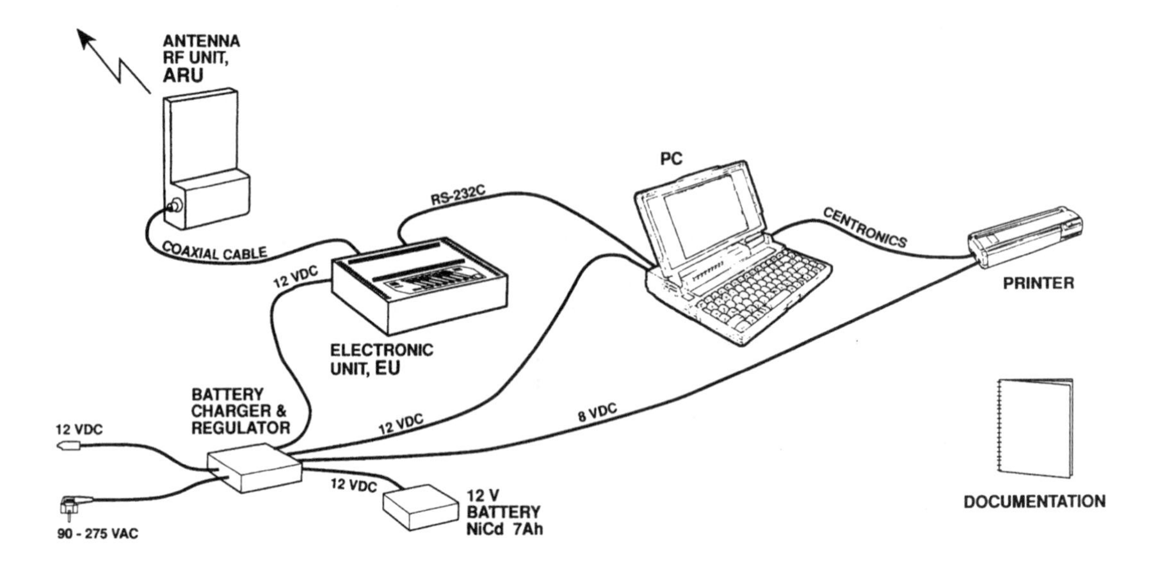

Fig. P-10. The Saturn C lightweight Inmarsat-C satellite communications **portable station** that (a) is designed for use in remote areas and (b) provides access to the International Telex® Network. (Courtesy Mackay Communications.)

position: **1.** In a character string, a location that may be occupied by a character and that may be identified by a serial number. **2.** In a communications system, a certain point, location, or piece of equipment, such as a transmitter distributor or a contact on it, the location of an operator at a switchboard or a control console, a service clerk in a message center, a tape cutting unit, a typing reperforator, or a place of work in a communications facility. *See* **attendant position, coordinate position, current position, dead reckoning position, digit position, display position, erect position, inverted position, print position, supervisor position, teletype position, typing position.** *See also* **point.**

positional numeration system: A numeration system in which a numeral is represented such that the total value contributed to the number by a digit depends upon its position in the numeral relative to the other digits, as well as its particular numerical value, i.e., weights are assigned to digit positions in the numeral. *See also* **digit position, numeration system, value, weight.**

position classification: In direction finding systems, an evaluation of the precision with which the location of a station is or can be measured by two or more radio direction finding stations. *Note:* Positions may be classified as (a) Class A, a position measurement estimated to be precise to within 5 nmi (nautical miles), (b) Class B, precise to within 20 nmi, or (c) Class C, precise to within 50 nmi. The position classification is specified by the direction finding station. *See also* **direction finding, evaluation, position, precision, radio direction finding station.**

positioned channel: A channel that occupies dedicated bit positions in an integrated services digital network (ISDN) framed data stream in ISDN user interfaces. *Note:* In integrated services digital networks (ISDNs), examples of positioned channels are the B, H, and D channels. *See also* **integrated services digital network.**

positioned interface structure: Within a framed interface, a structure in which positioned channels provide all user services and supervisory signaling. *See also* **Broadband Integrated Services Digital Network, framed interface, Integrated Services Digital Network, positioned channel.**

position flaghoist: *See* **inferior position flaghoist, superior position flaghoist.**

position-indicating radiobeacon station: *See* **emergency position-indicating radiobeacon station, satellite emergency position-indicating radiobeacon.**

position indicator: *See* **planned position indicator, radar planned position indicator.**

positioning: In display systems, using a display writer, such as a light pen, sonic pen, laser beam, electron gun, voltage pencil, plotter, stylus, or tablet stylus, to indicate the location, in the display space on the display surface of a display device, where a specific display element, display group, display image, character, or picture is to be placed. *Synonym* **locating.** *See also* **display device, display writer.**

positioning land station: *See* **radiopositioning land station.**

positioning mobile station: *See* **radiopositioning mobile station.**

positioning system: *See* **differential global positioning system.**

positioning time: *Synonym* **seek time.**

position line: In radiolocation, a line that is drawn on a map at a radio determination azimuth through a known point, such as through the location of the transmitting antenna or the location of the receiving antenna. *See also* **radiodetermination, radiolocation.**

position line determination: *See* **radio position line determination.**

position log: In a radio or radar station position, i.e., a workstation, a description of operating events that pertain to the position. *Note:* A position log might include the station identification, position identification, opening and closing time, date, stations communicated with, operating frequencies, tests performed, communications summaries, distress actions, safety actions, operating difficulties, maintenance problems, and repairs and adjustments effected. *See also* **aeronautical station master log.**

position marking: In a military operational situation, marking used for identifying personnel or equipment on the ground by elements on that ground, such as the location of the most advanced elements of an advancing friendly military unit or the location of an advancing forest fire-fighting or rescue unit. *Note:* In visual signaling systems, a panel exposed on the ground or a cloud of colored smoke may be used for position marking.

position modulation: *See* **pulse position modulation.**

position sensing: *See* **rotational position sensing.**

position system: *See* **traffic service position system.**

positive AND gate: *Synonym* **AND gate.**

positive clearing: In double supervisory working of telephone switchboards, clearing in which an indication, such as a light or a disk, associated with a plug in use (a) appears when a user clears a connected call and (b) remains displayed until the associated plug is disconnected. *See also* **double supervisory working, negative clearing.**

positive distortion: *Synonym* **late distortion.**

positive feedback: The return of a processed part of the output signal of a device or a system to the input in such a manner that the signal fed back (a) has an in-phase component relative to the output signal or relative to the input signal, (b) is additive to the input signal, i.e., the feedback signal is regenerative, (c) increases the gain of the device and increases the distortion of the signal, (d) decreases linearity and stability of the device, (e) can occur unintentionally or inadvertently, and (f) can be harmful because of the huge gain and sustained high amplitude oscillations that can occur. *See also* **feedback, negative feedback, regeneration, regenerative feedback.**

positive-intrinsic-negative (PIN) diode: A junction diode that (a) has a junction that consists of three semiconducting materials joined in sequence in the forward current-conducting direction such that (i) the first of the three materials is doped positive, i.e., is a p-type semiconductor, thereby creating holes, i.e., acceptor sites, (ii) the second is not doped, i.e., it is made of intrinsic material, and (iii) the third is doped negative, i.e., is n-type material, thereby creating free electrons, i.e., donor sites, and (b) is used extensively as a photodetector in fiber optic systems and optical integrated circuits (IOCs). *See also* **junction, photodiode, positive-negative (p-n) junction, semiconductor.**

positive-intrinsic-negative field-effect transistor (PIN-FET) integrated receiver: An optical receiver (a) that is formed by combining a positive-intrinsic-negative doped (PIN) photodiode and a field effect transistor (FET) in a single housing, and (b) in which the packaging is performed in such a way that the performance of the combination is better than that of the individual components connected and used as separate and discrete components. *Note:* "PIN" is an abbreviation for "positive-intrinsic-negative doped." *See also* **monolithic integrated receiver.**

positive-intrinsic-negative (PIN) photodiode: A photodiode that (a) has a large intrinsic region that is not doped or is lightly doped between (i) a positive-doped semiconducting region, i.e., an acceptor region, and (ii) a negative-doped semiconducting region, i.e., a free electron region, and (b) changes its electrical conductivity, i.e., its electrical resistance, in accordance with the intensity and the wavelength of incident light when photon energy is absorbed. *Note 1:* Photons absorbed in the intrinsic region create electron-hole pairs that are separated by an electric field, thus generating an electric current, i.e., a photocurrent. *Note 2:* "PIN" is an abbreviation for "positive-intrinsic-negative doped." *See also* **dopant.**

positive justification: *Synonym* **bit stuffing.**

positive lens: *Synonym* **converging lens.**

positive-negative (p-n) junction: In a semiconductor, i.e., a solid state device, the interface between (a) semiconducting material that has been doped positively, i.e., with an impurity material that produces holes, i.e., acceptor sites, on one side of the interface, and (b) semiconducting material that produces free electrons, i.e., donor sites, on the other side. *Note:* The junction usually is assumed to be abrupt or linearly graded with a high gradient in the transition region across the interface from a positive carrier (hole) dopant to a negative carrier (electron) dopant in the intrinsic semiconductor material, such as germanium or silicon.

positive OR gate: *Synonym* **OR gate.**

positive pulse stuffing: *Synonym* **pulse stuffing.**

positive scanning shift: In facsimile transmission systems, a relative shift of the scanning device with respect to the surface of the object being scanned when the object is scanned along the line from left to right, and where the scanning lines proceed from top to bottom. *Note:* In a drum facsimile transmitter, helicoidal scanning toward the left is a positive scanning shift.

POSIX: portable operating system interface for computer environments.

post: To place an item, such as a letter or a package, at a designated point, such as a mailbox or a postal service office, for pickup and delivery to an addressee. *See* **airborne command post.** *See also* **addressee, postal address.**

postal address: An address that is used for posting, i.e., for mailing. *Note:* The postal address for the International Organization for Standardization is 1211 Geneva 20, Switzerland and that for the American National Standards Institute is 1430 Broadway, New York, NY 10018.

postalize: In communications, to structure rates or prices so that they are not distance-sensitive but depend on other factors, such as call duration, type of service, and time of day. *See also* **extended area service, flat rate service, measured rate service.**

postalized rate: In a transmission service, a rate that is not dependent upon distance. *Note:* In international transmission service, a postalized rate is not dependent on the location within the country that originates the message. *See also* **extended area service, flat rate service, measured rate service.**

postal, telegraph, and telephone (PTT): In a given country, pertaining to the system, such as a network, common carrier, recognized private operating agency (RPOA), or post office, that provides communications services for the public. *Note:* Each country has a different arrangement for handling postal, telegraph, and telephone communications. *See also* **common carrier, divestiture.**

postal, telegraph, and telephone (PTT) organization: In a given country, the organization that (a) performs the communications functions, such as postal, telegraph, and telephone functions, for the country and (b) usually is either government-regulated or a governmental department, ministry, or agency that serves as the common carrier for the country. *Note 1:* Examples of postal, telegraph, and telephone organizations are the U.S Postal Service in the United States, the Post Office in the United Kingdom, the Bundespost in Germany, and the Nippon Telephone and Telegraph Public Corporation in Japan. *Note 2:* In the United States, separate organizations perform postal, telegraph, and telephone operations. *See also* **common carrier, divestiture.**

postamble: Characters that follow a specified group of characters. *Note:* An example of a postamble is a sequence of characters that is recorded at the end of each block of data on a phase encoded magnetic tape for the purpose of synchronization when the tape is read backward, i.e., in the reverse direction.

postdetection combiner: A circuit or a device that combines two or more signals after demodulation. *See also* **diversity combiner, maximal ratio combiner.**

postdetection filter: A filter that is used to filter the baseband output of a detector. *See also* **baseband, detector, filter.**

postdetector bandwidth: In the baseband channels for the signal output of a detector, the bandwidth that is determined by the bandpass of the postdetection filter. *See also* **bandwidth, baseband, passband, postdetection filter.**

postdetector signal to noise ratio: The ratio of the baseband signal power to the total noise power in the postdetector bandwidth. *Note:* The postdetector signal to noise ratio may be expressed as a ratio, as a percentage, or in dB. *See also* **baseband, postdetector bandwidth.**

postdevelopment review: *Synonym* **system follow-up.**

postfix notation: A notation system for forming mathematical expressions in which each operator is preceded by its operands and indicates the operation to be performed on the operands or on the intermediate results that precede it. *Note 1:* AB× indicates A is to be multiplied by B, AB+ indicates B is to be added to A, AB− indicates B is to be subtracted from A, and AB−C× indicates B is to be subtracted from A and the difference is to be multiplied by C. *Note 2:* AB does not imply multiplication as in conventional notation. *Note 3:* Postfix notation indicates the sequence in which operations might be performed by a computer, such as, for AB×, "fetch A and place it in a multiplicand register, fetch B and place it in a multiplier register, multiply the contents of the two registers." *Synonyms* **reverse Polish notation, suffix notation.** *See also* **prefix notation.**

postimplementation review: *Synonym* **system follow-up.**

postmortem dump: A dump that is made after a computer program has been executed, usually for debugging, auditing, or record-keeping purposes. *See also* **dump.**

postprocessor: A computer program that performs processing after other computation or processing is finished. *Note:* An example of a postprocessor is a computer program that converts data produced by a processor to the format of an emulated system. *See also* **computer program, emulate, format, preprocessor, processing, processor.**

postproduction processing: In broadband integrated services digital networks (B-ISDNs) applications, the processing of audio and video information after contribution and prior to final use. *See also* **broadband integrated services digital network, contribution, distribution.**

postselection time: In the establishment of a connection or the placement of a call, the time between (a) the end of transmission of selection signals and (b) the receipt of the call completed signal at the call originator data terminal equipment (DTE). *Note:* An example of postselection time is the time between the completion of dialing and the end of ringing, i.e., receipt of an off-hook signal. *See also* **call setup time.**

post-telegraph-telephone (PTT): The organization that (a) usually is a governmental department, administration, or ministry, (b) regulates or operates its national communications common carriers, and (c) includes services such as radio, television, mail, telephone, and telegraph services.

post-telegraph-telephone (PTT) authority: The authority that regulates or operates national communications common carriers, such as radio, television, mail, telephone, and telegraph services.

pot: *Synonym* **potentiomenter.**

POT: point of train.

potential: *See* **absolute potential, ground potential.**

potential difference: In electrical equipment, the difference between the absolute potential of one point and that of another point, i.e., the voltage between them. *Note 1:* Potential difference is relative, i.e., it can only exist between two points. *Note 2:* Potential difference usually is expressed in volts, or multiples and submultiples of volts, such as kilovolts and millivolts. *Note 3:* Examples of potential difference are (a) the voltage measured between the terminals of a battery, regardless of the absolute potential of either of the terminals, (b) the voltage between two conductors or lines, whether or not either is grounded, and (c) the voltage between a conductor and ground. *Note 4:* The potential difference can change at high frequencies, such as during high pulse repetition rates. *See also* **absolute potential, point, volt, voltage.** *Refer to* **Fig. E-1 (285).**

potentiometer: A spatially distributed electrical resistance in which (a) the resistive element usually is uniformly distributed, (b) there is a terminal on each end and a movable contact that slides along the resistive element in such a manner that the resistance from either end to the sliding contact can be varied, (c) the resistive element may be of any shape, but usually is round, (d) a sliding contact is mounted on a shaft so it may slide along the resistive element by being rotated, (e) usually a potential difference, i.e., a voltage, is applied across the ends of the resistive element, (f) a fraction of the voltage will occur between the sliding contact and either end of the resistive element, depending on (i) the fraction of resistance between either end and the contact and (ii) the amount of current passing through the contact, (g) if only a negligible amount of current is drawn through the contact, the fraction of the end-to-end voltage that appears at the sliding contact will be the same as the fraction of resistance between the sliding contact and the ends of the resistive element, and (h) adjustments of output voltage can be made for electrical loading. *Note:* The potentiometer has many applications in which voltage control is needed, such as volume control, bias control, and circuit balancing. *Synonyms* **pot, voltage divider.**

POTS: plain old telephone service.

powder: *See* **glass powder, starting powder.**

power: **1.** At a point in a system, circuit, or component, the rate of transfer or absorption of energy per unit time. *Note 1:* Power usually is expressed in watts. A modifier may be used, such as power(electrical), power(optical), or power(sound). *Note 2:* Whenever the power of a transmitted electromagnetic wave is indicated, an appropriate modifier should be used, such as peak envelope power, average power, mean power, instantaneous power, root mean square (rms) power, or carrier power. **2.** In optics, a measure of the ability of an optical element to bend or refract light. *Note 1:* Optical power is expressed in diopters. *Note 2:* A six-power (6X) optical instrument would make an object that is 600 m (meters) away appear to be only 100 m away. **3.** In mathematics, the number of times the base of a numeration system is to be multiplied by itself, including itself in the count. *Note:* If 10 is raised to the third power, the result is be 1000. **4.** In the transmission of an electromagnetic wave, the rate of transfer of energy, expressed in one of the following forms, according to the class of emission, using the arbitrary symbols indicated:

Peak envelope power (PX or pX)

Mean power (PY or pY)

Carrier power (PZ or pZ)

Note: For different classes of emission, the relationships between peak envelope power, mean power, and carrier power, under the conditions of normal operation and no modulation, are contained in International Radio Consultative Committee (CCIR) Recommendations, which may be used as a guide. For use in formulas, the symbol p denotes power expressed in watts and the symbol P denotes power expressed in dB relative to a reference level. *See* **absolute power, apparent power, average power, auxiliary power, bundle resolving power, carrier power, chromatic resolving power, disturbance power, effective isotropic radiated power, effective monopole radiated power, effective power, effective radiated power, equivalent isotropically radiated power, fiber optic repeater power, flat noise power, grating chromatic power, jamming power, link threshold power, luminous power, magnifying power, mean power, noise equivalent power, noise power, normalized power, optical power, optical receiver maximum input power, optical repeater power, peak envelope power, peak**

power, peak spectral power, primary power, prism chromatic resolving power, psophometric power, radiant power, radio transmitter mean power, rated output power, reactive power, received noise power, relative power, resolving power, signaling power, source to fiber coupled power, theoretical resolving power, transmitted power, transmitter optical power. *See also* **dB, power level.**

power balancing: In communications systems, the adjustment of individual radiated and nonradiated power levels in a multipoint network. *Note:* Examples of power balancing are (a) in a satellite communications system, achieving optimum performance of each component of the network by distributing the available power among the components, (b) in a radio network, adjusting the transmitted power, i.e., the radiated power, of each station for optimum reception and minimum interference among all stations, (c) in a satellite communications system, optimizing the available downlink power that is allocated to the individual accessing signals, depending on the number of accessing signals and their relative uplink strengths, the power levels being controlled by adjusting the transmitter power at the Earth stations, and (d) in a switched network, distributing the available power among the switches and trunks for optimum transmission and reception of signals. *See also* **power budget, power sharing.**

power budget: In a system, the allocation of available power among (a) the various functions that need to be performed, and (b) the various losses that must be sustained. *Note:* Examples of power budgets are (a) in a communications satellite, the allocation of available power among various functions, such as maintaining satellite orientation, maintaining orbital control, performing signal reception, and performing signal retransmission, and (b) in a fiber optic link, the allocation of power among the components of the link, such as source to optical fiber loss, fiber attenuation, connector losses, and fiber to photodetector loss. The total power budget for fiber optic links is about 30 dB to 40 dB. *Synonym* **system budget.** *See* **optical power budget, power balancing.**

power bus: A conductor, i.e., a bus, that (a) carries electrical or optical power to a set of devices or components. *Note 1:* An electrical bus usually (a) is maintained at a constant voltage level, usually by means of voltage regulators, (b) has a negligible, nearly zero resistance, and (c) is capable of delivering large currents compared to the currents required for the connected loads. *Note 2:* An optical bus has more than sufficient optical power to drive all the attached optical components at levels above their threshold operating levels. *See also* **bus, infinite bus.**

power circuit breaker (PCB): 1. The primary switch used to apply power to and remove power from equipment. **2.** A circuit breaker used on alternating current (ac) circuits rated in excess of 1500 V (volts). *See also* **protector.**

power density: 1. A measure of the power that may be attributed to a particular frequency band. *Note:* Power density usually is expressed in watts per hertz. *See* **noise power density, optical power density.** *See also* **optical energy density. 2.** *Deprecated synonym for* **irradiance.**

power distribution: *See* **core-cladding power distribution, equilibrium modal power distribution, nonequilibrium modal power distribution.**

power distribution length: *Synonym* **nonequilibrium modal power distribution length.**

powered: *See* **sound-powered.**

powered flight phase: In the launching of a satellite or spacecraft, the part of satellite or spacecraft flight during which propulsion engines are developing thrust. *Note:* In most cases powered flight ends when the required height and velocity have been reached, and the launch vehicle or booster is separated from the satellite or spacecraft. When launch conditions are such that the satellite cannot be propelled directly into the chosen orbit, the launch sequence consists of several powered flight phases, alternating with unpowered intervals with the engines off, until the proper height and velocity are reached. *See also* **passive flight phase.**

powered telephone: *See* **electrically powered telephone, sound-powered telephone.**

power efficiency: *See* **optical power efficiency, source power efficiency.**

power factor (PF): 1. In an alternating current (ac) power transmission and distribution system, the cosine of the phase angle between the voltage and the current at any point in the system. **2.** At any point in an alternating current (ac) power distribution system, the ratio of (a) the effective power, i.e., the real power in watts, to (b) the apparent power in volt•amperes. *Note 1:* When the load is inductive, i.e., the inductive reactance exceeds the capacitive reactance, such as is obtained when the load consists of motors and transformers, the power factor is a lagging power factor. When the load is capacitive, i.e., the capacitive reactance exceeds the inductive reactance, such as is obtained with capacitors and synchronous condensers, the power factor is a leading power factor. *Note 2:* Power factors other than unity cause increased power losses in power lines because of the increased currents required to deliver a given amount of effective power to a load, and, in effect, cause a reduction in power system capacity. Power companies encourage customers to maintain power factors within prescribed limits, usually requiring customers that have excessive inductive loads to pay a premium when (a) the apparent power in volt•amperes exceeds the effective power in kilowatts by a specified amount, or (b) the ratio of effective power to apparent power is less than a specified value. Because most loads are inductive, the power companies often give encouragement to customers that produce a leading power factor to bring the power factor in long lines closer to unity, thereby reducing the line currents with a consequent reduction in resistive losses.

power gain: *See* **antenna power gain.**

power gain of an antenna: *Synonym* **antenna gain.**

power law refractive index power leveling: At the end of an optical propagation medium, such as an optical fiber, adjusting the optical power delivered to a device, such as a photodetector, by adjusting the refractive index profile of the medium to avoid overdrive or signal distortions. *See also* **power level.**

power law refractive index profile: In round optical fibers, a refractive index profile in which the refractive index of the core is an exponential function of the distance from the fiber axis, i.e., the variation is such that the refractive index within the core at any distance from the axis, r, is given by the relation $n_r = n_1(1 - br^g)^{1/2}$ where n_1 is the refractive index at $r = 0$, i.e., at the axis, r is the radial distance from the axis, g is the refractive index profile parameter that determines the shape of the refractive index profile, and b is a constant given by the relation $b = 2\Delta/a^g$ where a is the value of r for which the refractive index becomes uniform, i.e., the radius of the core, and Δ is given by the relation $\Delta = (n_1^2 - n_2^2)/2n_1^2$ where n_1 again is the refractive index at $r = 0$, i.e., at the axis, and n_2 is the refractive index at the outer edge of the core, i.e., at $r = a$. For most optical fibers, the refractive indices n_1 and n_2 are nearly equal. Therefore Δ is also given by the relation $\Delta = 1 - (n_2/n_1)$ where the parameters are as defined above. Because $n_r = n_2$ at $r = a$, the power law refractive index profile is also given by the equivalent relation $n_r = n_1(1 - 2\Delta(r/a)^g)^{1/2}$. Another form of the equation, approximately the same as those above, is given by the relation $n_r^2 = n_1^2(1 - 2\Delta(r/a)^g)$ where the parameters are as defined above. The profile parameter, g, again defines the shape of the profile. The Δ is the refractive index contrast when the refractive index of the cladding is constant. If $g = 1$, a straight line results, i.e., a triangular variation of refractive index as a function of radial distance from the axis results. If $g = 2$, a parabolic index profile results. If the g is about 2.25, intermodal dispersion will be minimized or nearly eliminated for most fibers. As g approaches infinity, the core refractive index is constant and equal to n_1, which is the case for the step index fiber. *See also* **graded index profile, linear refractive index profile, mode volume, parabolic refractive index profile, profile parameter, radial refractive index profile, step index profile.**

power level: At any point in a transmission system, the ratio of (a) the power at that point to (b) some arbitrary amount of power chosen as a reference. *Note 1:* The power level usually is expressed either in dB referred to one milliwatt, i.e., dBm, or in dB referred to one watt, i.e., dBW. *Note 2:* The power level at a point may also be expressed in direct power units, such as milliwatts, watts, or kilowatts. *See also* **dB, level, power.**

power leveling: Adjusting the power that is delivered at the end of a propagation medium to a load that has a reactance that varies with frequency, particularly to avoid overloading, reflections, or distortion. *Note:* Power leveling usually is accomplished at the receiving end of transmission lines, cables, and waveguides. *See also* **distortion, frequency, load, power law refractive index power leveling, power level, propagation medium, reactance, reflection, transmission line.**

power-limited: In a communications system, pertaining to the situation in which the total traffic capacity of the system is limited by the power that is available for allocation to the signals being processed at the channels being operated. *Note 1:* In radio transmission, the power limitation is only on radio frequency (rf) energy availability. *Note 2:* The alternative to being power-limited is to be bandwidth-limited. This occurs when the total traffic capacity is limited by the frequencies that are available for use, i.e., by the available bandpass. *Note 3:* Communications system power limitation and bandwidth limitation are somewhat analogous to automatic data processing system input-output (bandwidth) limitation and processor (power) limitation. *See also* **bandwidth-limited.**

power margin: In designing communications systems and components, such as microwave, wire, coaxial cable, and fiber optic links, a power loss (dB) value used in the power budget to allow for unexpected losses and ensure that performance criteria are met, such as the bit rate, i.e., the data signaling rate (DSR), and the bit error ratio (BER). *Note:* The power margin should not include expected losses and degradations, such as laser aging, cable aging, reflections, repairs, coupling losses, connector losses, and cable losses. These known losses should be included in the appropriate parameters of the specific devices and in the power budget. *Synonyms* **power safety margin, safety power margin.** *See* **optical system power margin, radio frequency power margin, system power margin.** *See also* **power budget.**

power of the talker volume distribution: *See* **mean power of the talker volume distribution.**

power output: The power, such as electrical, optical, or sound power, emitted by a system, component, or device. *Synonym* **output power.** *See* **optical power output, peak power output.**

power output rating: The power, such as electrical, optical, or sound power, a system is designed to furnish under normal operating conditions. *See* **transmitter power output rating.**

PowerPC: A high performance microprocessor reduced instruction set computing (RISC) chip being designed, developed, produced, and marketed under a joint effort consisting of (a) Apple Computer, Incorporated, (b) the IBM Corporation, and (c) Motorola.

power penalty: *See* **dispersion power penalty, reflection power penalty.**

power point: *See* **full wave half power point.**

power safety margin: *Synonym* **power margin.**

power sharing: The distribution of power among the components of a system. *Note:* An example of power sharing in a satellite communications system is the fixed division of available satellite downlink power among the individual transponders or separate pathways through the satellite. The allocation of power among individual access signals is a variable, the distribution depending on the number of accessing signals and the relative uplink signal strengths, controlled by adjustment of transmitter power at the Earth stations. The controlled adjustment of distribution is called "power balancing." Usually only the radio frequency (rf) power is implied. *See also* **downlink, power balancing, uplink.**

power talker: *See* **medium power talker.**

Poynting vector: In an electromagnetic wave, the resulting vector that (a) is obtained from the cross product, i.e., the vector product, of the electric and magnetic field vectors, given by the relation $\mathbf{P} = \mathbf{E} \times \mathbf{H}$ where **P** is the Poynting vector, **E** is the electric field strength, and **H** is the magnetic field strength, (b) is obtained when the electric field vector is rotated clockwise into the magnetic field vector, as in conventional vector algebra, (c) represents the electromagnetic field at each and every point in material media and free space in which the wave propagates, (d) indicates the direction of the wave at every point, (e) together with propagation media parameters, such as the electrical conductivity, electric permittivity, and magnetic permeability, is proportional to the irradiance at each point, and (f) has the same direction as that of the endface of a right-hand screw when screwed into a fixed nut. *See also* **electromagnetic wave.**

PP fiber: polarization-preserving fiber.

PPI: planned position indicator.

ppm: parts per million.

PPM: pulse position modulation.

pps: pulses per second.

pragmatics: The branch of philosophy that deals with the testing of the value and the truth of ideas by their practical consequences or utility, including the study, i.e., the discipline, concerned with the practical relationships among symbols, strings, or groups of characters and their users or interpreters. *See also* **string, value.**

preamble: A segment of text that appears or occurs in space or time before a larger body of text. *Note:* Examples of preambles are (a) in computer and communications systems, a component of the heading of a message, including items such as the precedence designation, date-time group, and message instruction, (b) in spread spectrum systems, a binary code sequence that is used for acquiring synchronization, particularly synchronization of the spread spectrum code sequence generators, (c) in magnetic tape storage, a sequence of characters that is recorded at the beginning of each block of data on a phase encoded magnetic tape for the purpose of synchronization, and (d) the widely known preamble to the Constitution of the United States. In most instances, preambles usually are short compared to other character sequences or codes. *See also* **postamble.**

prearbitrated slot: In a communications network, a slot dedicated by the head of bus function for transferring isochronous service octets.

prearranged message code: A code that (a) is adopted for use by corresponding parties, (b) requires a special or technical vocabulary, and (c) usually is composed exclusively of groups of characters that represent complete or nearly complete statements or messages. *Note:* An example of a prearranged message code is the code "One if by land and two if by sea."

preassignment access plan (PAP): In satellite communications system operations, a fixed communications channel access plan, as opposed to a demand assignment access plan in which allocation of accesses or the number of channels per access is varied in accordance with the demand. *See also* **demand assignment access plan.**

preburst message: A message that indicates the effective forward wind and the downwind distance for a given geographical area, information that (a) is to be used in the event of a nuclear explosion in the area and (b) is of particular interest in regard to the effects, in the downwind area from ground zero, of a burst. *See also* **fallout message.**

precedence: 1. In communications systems, the level of urgency that an originator assigns to a message. *Note:* The precedence is assigned according to the estimated effect on life and property that would be caused if transmission of the message were delayed. **2.** A designation assigned to a message by the originator to indicate to communications personnel the relative order of handling and to the addressee the order in which the message is to be noted. *Note:* The descending order of precedence for military messages is FLASH OVERRIDE, FLASH, IMMEDIATE, PRIORITY, and ROUTINE. *See also* **classmark, flash message, immediate message, minimize, override, preemption priority message, routine message, seizing, special grade of service.** *Refer to* **Figs. T-1 and T-2 (983).**

precedence and preemption: *See* **multilevel precedence and preemption.**

precedence call: In telephone switchboard operation, a call that has a precedence above routine in networks in which a precedence system is in effect. *See also* **nonprecedence call.**

precedence capability: *See* **multilevel precedence capability.**

precedence designation: In communications systems, a designation that (a) is assigned to a message or a call by the originator to indicate the relative order in which communications personnel or equipment are to handle a message relative to other messages and (b) indicates an estimate of the degree of urgency and an estimate of the adverse effects on life and property that may occur if the message is delayed. *Note:* Precedence designators that are used in military communications systems include FLASH OVERRIDE, FLASH, IMMEDIATE, PRIORITY, and ROUTINE, in descending order of urgency. Some of these categories are used in public and private communications systems. *See also* **emergency category, flash override precedence, flash precedence, immediate precedence, precedence designator, priority precedence, routine precedence.**

precedence designator: A letter, group of letters, or word that is used to indicate the precedence designation of a message or call, i.e., used to indicate the precedence that is assigned to a message or call. *See also* **precedence designation.**

precedence message: *See* **dual precedence message, single precedence message.**

precedence responsibility: The responsibility for assigning the precedence to a message or a call, said responsibility usually resting with the originator.

precedence system: *See* **Joint Uniform Telephone Communications Precedence System.**

precedence telephone system: A telephone system that requires the use of precedence designations, i.e., precedence designators, on all calls or on all calls over certain trunks of the system. *See also* **precedence designation, precedence designator.**

precession: The effect that results from the application of an external torque to a spinning body, in which the gyroscopic forces cause the resultant displacement due to the torque to be orthogonal to the applied force and orthogonal to the vector representing the spinning mass. *Note 1:* The vector that represents the spinning mass is parallel to the axis of rotation and in the direction of a right-hand screw turning in the direction of spin. The effect occurs in all spinning masses from electrons to satellites. The effect must be taken into account in spin-stabilized communications satellites in order to maintain proper orientation of communications equipment, particularly antennas. *Note 2:* Common examples of precession are that (a) a top spinning on its point will tend to list, with the applied force of gravity, centered at the center of gravity of the top, supplying the torque that causes the axis of the top to precess as the top rotates, and (b) a free bicycle wheel spinning on its axle with one end of the axle held fixed and the other end left free, allowing gravity to apply the torque about the fixed end, will precess about the fixed end of the axle, assuming in both examples that the spin rate is sufficient to maintain equilibrium at least long enough for one to observe the precession.

precipitation attenuation: The loss of energy by an electromagnetic wave that (a) occurs during its passage through a volume of the atmosphere containing precipitation, such as fog, rain, snow, hail, or sleet, and (b) results from scattering, reflection, refraction, diffraction, and absorption. *Note:* A volume of falling snow or rain will appear as a reflection, i.e., a return signal, on a radar screen, indicating that at least a part of the signal is reflected, and perhaps an attenuated part penetrates the snow or rain volume and proceeds to an object, i.e., a target. *See also* **attenuation, loss, path loss, transmission loss.**

precipitation gage station: *See* **radar beacon precipitation gage station.**

precipitation static (p-static): Radio interference that (a) is caused by the impact of charged particles against an antenna and (b) may occur in a receiver during fog and certain weather conditions, such as snow, hail, rain, sleet, and dust storms, or a combination of these. *See also* **electromagnetic environment, noise.**

precise frequency: 1. A frequency that is maintained to the known precision of an accepted reference frequency standard. *Note:* Current frequency precision among international standards is approximately 1 part in 10^{13}. **2.** A frequency requirement precise to within one part in 10^{9}. *See also* **Coordinated Universal Time, DoD master clock, frequency, frequency accuracy, International Atomic Time, primary time standard, reference frequency, standard frequency and time signal service.**

precise time: 1. A time mark position that is accurately known with reference to an accepted reference time standard. *Note:* Current time precision among current international standards is approximately 1 part in 10^{13}. **2.** A time duration, i.e., a time interval, that is accurately known to an accepted reference standard. **3.** In U.S. military organizations for operational purposes, time precise to within 10 ms (milliseconds). *See also* **clock, Coordinated Universal Time, DoD master clock, International Atomic Time, primary time standard, reference frequency, standard frequency and time signal service.**

precise time and time interval (PTTI): 1. The branch of science and technology devoted to the study and the implementation of precision and accuracy in timekeeping and time information transfer. **2.** An instant of time, elapsed time, and time interval expressed to a specified tolerance, i.e., specified precision, with respect to an indicated time standard.

precision: 1. A measure of the ability to distinguish between nearly equal values. *Note:* Six-place numerals are more precise than four-place numerals. A digital clock that displays seconds is more precise than a clock that displays minutes. A six-place logarithmic or trigonometric table is more precise than a four-place table. However, the six-place table may have a typographical error in one of the entries, making the more precise table inaccurate. High precision merchandise, such as machines and tools, is made to smaller than usual tolerances, thereby reducing interference and loose fits of mating parts. **2.** The degree of discrimination or resolution with which a quantity is stated. *Note:* Four-digit decimal numerals discriminate among 10,000 possibilities. **3.** The degree of mutual agreement among a series of individual measurements, values, or results that are often expressed by the standard deviation. **4.** With respect to a set of independent devices of the same design, pertaining to the ability of these devices to produce the same value or result, given the same input conditions and operating in the same environment. **5.** With respect to a single device put into operation repeatedly without adjustments, pertaining to the ability to produce the same value or result, given the same input conditions and operating in the same environment. *See also* **accuracy, fidelity, reproducibility, resolution.**

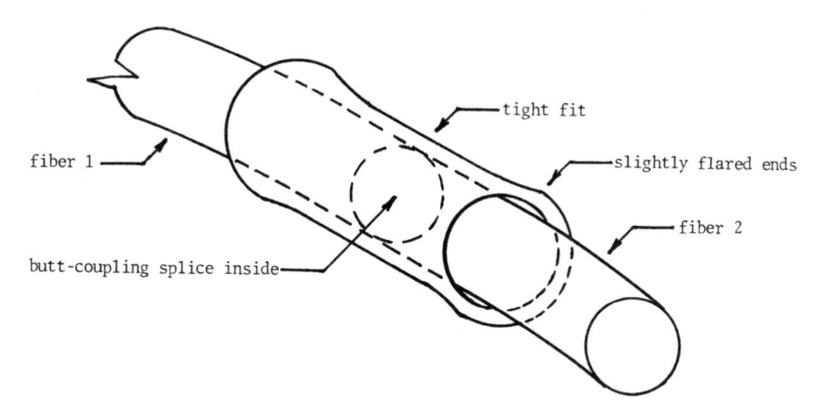

Fig. P-11. A **precision sleeve splicer** for butt coupling two optical fibers.

precision coding compaction: *See* **variable precision coding compaction.**

precision sleeve splicer: A round tube of suitable material, such as plastic, (a) that has a round hole with a diameter equal to the outer diameter of two optical fibers to be spliced, (b) that contains an index matching material, such as an epoxy, (c) into which two fibers may be inserted from opposite ends to form a butted joint, (d) the ends of which may be crimped, if necessary, to hold the fibers tightly, at least until the index matching material cures, and (e) in which the index matching material may be an optical cement. *See also* **loose tube splicer.** *Refer to* **Fig. P-11.**

precision sleeve splicing: The creating of a splice through the use of a round tube of suitable material, such as plastic, (a) that has a round hole with a diameter equal to the outer diameter of two optical fibers to be spliced, (b) that contains an index matching material, such as an epoxy, (c) into which two fibers may be inserted from opposite ends to form a butt coupling, (d) the ends of which may be crimped, if necessary, to hold the fibers permanently and tightly, at least until the index matching material cures, and (e) in which the index matching material may be an optical cement. *See also* **butt coupling, loose tube splicing, tangential coupling.**

precombining: The combining of multiplexed signals prior to the modulation of the carrier. *Synonym* **premodulation combining.** *See also* **combiner, combining.**

predetection combiner: A circuit or a device that combines two or more signals prior to demodulation of any of the signals and prior to the demodulation of the combined signal. *See also* **combiner, combining, diversity combiner.**

predetection combining: Signal combining in which all incoming diversity signals are brought into approximate phase coincidence before being combined, thus obtaining an improved signal from multiple radio receivers engaged in diversity reception. *See also* **diversity combiner, maximal ratio combiner, selective combiner.**

predetermined routing: In message transmission, the routing of messages in which specific instructions for routing are used and applied to the routing indicators of message headers in order to assign relay responsibility and to prevent the need for specific routing instructions for each message. *Note:* Alternative routing is a form of predetermined routing. *See also* **alternative routing, free routing, routing directory, routing indicator.**

prediction: *See* **frequency prediction, reliability prediction.**

prediction coding: *See* **code excited linear prediction coding.**

prediction publication: *See* **propagation prediction publication.**

prediction station: *See* **sounder prediction station.**

prediction table: In satellite communications systems, a table that (a) contains the longitude of subsatellite points, node times, and orbit inclinations at daily intervals for a particular satellite in order to allow pointing angles to be calculated for any point on the Earth's surface that is within the satellite's Earth coverage, and (b) is used in conjunction with an acquisition table that is appropriate for the satellite orbit in order to initiate communications via the satellite. *See also* **orbit, orbit inclination, subsatellite point.**

predictive coding: *See* **adaptive predictive coding, linear predictive coding.**

preemphasis: 1. In an electromagnetic wave of many frequencies at the input of a system, circuit, or device, the increasing of the magnitude of certain wave components with specific frequencies with respect to the magnitude of others in order to reduce or compensate for certain adverse effects on the signal, such as noise, in various parts of the system, circuit, or device. **2.** In a communications system, signal processing in which the magnitude of some frequency components are increased with respect to the magnitude of others in order to reduce adverse effects, such as frequency distortion and changes in frequency response, in subsequent parts of the system. **3.** In frequency-modulated (FM) transmission, the intentional alteration of the amplitude versus frequency characteristics of the signal to reduce adverse effects of noise in a communications system. *Note:* The higher frequency signals are emphasized, usually by increasing gain at higher frequencies, to produce a more equal modulation index for the transmitted frequency spectrum, and therefore a better signal to noise ratio (S/N) for the entire frequency range. *Synonym* **emphasis.** *See also* **deemphasis, frequency, noise, signal to noise ratio.**

preemphasis circuit: A circuit that modifies the gain-frequency, i.e., frequency response, of a system or a device, usually by increasing the relative amplitude of the higher frequency signal components in conjunction with a separate, but corresponding, decrease of these amplitudes in a deemphasis circuit, to restore the original signal and thereby improve the signal to noise ratio. *Note:* A deemphasis of the lower frequency components and their restoration would have similar effects. *See also* **deemphasis, deemphasis circuit, preemphasis improvement.**

preemphasis improvement: In the signal to noise ratio (S/N) of the high frequency end of the baseband, the S/N improvement that results from (a) passing the modulating signal at the transmitter through a preemphasis network, which increases the magnitude of the higher frequencies in signals, and then (b) passing the output of the discriminator at the receiver through a deemphasis network to restore the original signal power distribution among the frequencies of the baseband. *See also* **baseband, deemphasis, emphasis, frequency, noise, signal to noise ratio.**

preemphasis network: A network that (a) is inserted in a system in order to increase the magnitude of one range of frequencies with respect to another and (b) usually is used in frequency-modulated (FM) or phase-modulated (PM) transmitters to equalize the modulating signal drive power in terms of deviation ratio. *Note:* The receiver demodulation process makes use of a reciprocal network, called a "deemphasis network," to restore the original signal power distribution among the frequencies in the baseband signal. *See also* **frequency.**

preempting call: A call in which preemption is exercised by the call originator. *See* **multilevel precedence and preemption.**

preemption: **1.** The seizure of facilities that are being used to serve lower precedence traffic, such as calls, messages, and signals, in order to immediately serve higher precedence traffic. *Note:* Preemption usually is accomplished automatically. **2.** The reallocation of communications, computer, data processing, and control systems services and equipment from a lower precedence use to a higher precedence use. *Note:* In many systems that have a preemption capability, calls and messages can be preempted by a call originator while another call is in progress, i.e., during the information transfer phase. *See* **multilevel precedence and preemption.** *See also* **automatic preemption, call, precedence, seizing, traffic.**

preemption tone: In telephone systems, a distinctive tone that is used to indicate to connected users, i.e., customers, that their call has been preempted by a call of higher precedence. *Note:* An example of a preemption tone is a distinctive, steady, high pitch tone transmitted for three seconds or until the preempted user hangs up. *See also* **preemption.**

prefix: In a message, one of the parts of the header. *Note:* Examples of prefixes are (a) accounting information, such as prepaid, collect, and service, (b) group count, and (c) SVC, indicating that the message is a service message. In some radiotelephone system messages, the prefix consists only of the accounting information or the group count. *See also* **amplifying prefix, label prefix, service message.**

prefix-free code: *Synonym* **comma-free code.**

prefix notation: A notation system for forming mathematical expressions in which each operator (a) precedes its operands and (b) indicates the operation to be performed on the operands or on the intermediate results that follow it. *Note 1:* ×AB indicates A is to be multiplied by B, +AB indicates A is to be added to B, −AB indicates A is to be subtracted from B, and −AB×C indicates A is to be subtracted from B, and the difference is to be multiplied by C. *Note 2:* AB does not imply multiplication as in conventional notation. *Note 3:* Prefix notation indicates the sequence in which operations might be performed by a computer, such as, for ×AB, "set the multiplicand register, set the multiplier register, fetch A and place it in the multiplicand register, fetch B and place it in the multiplier register, perform the multiplication" or "multiply the operand in the A register by the operand in the B register." *Synonyms* **Lukasiewicz notation, Polish notation.** *See also* **postfix notation.**

preform: *See* **optical fiber preform.** *Refer to* **Fig. P-12.**

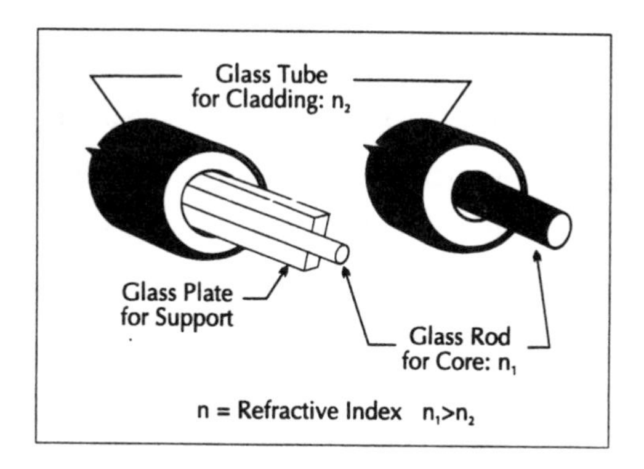

Fig. P-12. Two optical fiber **preforms,** (a) on the left, a central rod, supporting plate, and tubing and (b) on the right, a glass tube surrounding a glass rod. Optical fibers may be drawn from the preforms in a fiber pulling machine.

pregroup combining: In communications systems, assembling a number of narrowband channels, such as 40-kHz-wide telephone channels, into a specified frequency band such that, after pregroup translation, they may be formed with other pregroups into a standard group, such as an International Telegraph and Telephone Consultative Committee (CCITT) basic group, by frequency division multiplexing.

pregroup translation: In communications systems, the transposing, in frequency, of a pregroup of channels, such as telephone or data channels, in such a manner that they may be formed into a standard group, such as an International Telegraph and Telephone Consultative Committee (CCITT) basic group, by frequency division multiplexing.

prelasing condition: In a laser, an operating condition such that radiation primarily is spontaneous and incoherent.

preliminary call: In radio transmission, a call that (a) includes at least the identification of the calling station and the called station, (b) is designed to establish communications with a particular station, and (c) usually includes a request to the called station to reply, although the request may be implied by the recitation of the call signs. *See also* **call, call sign.**

preliminary call-up signal: In the schedule for an area broadcast station or a coastal radio station, one of the calling signals, such as the collective call sign, transmitting station call sign, and schedule serial numbers, that immediately precede the traffic list of messages to be transmitted. *See also* **area broadcast station, calling signal, call sign, coastal radio station, collective call sign, traffic list.**

premises equipment: *See* **customer premises equipment, embedded customer premises equipment.**

premises wiring: *Synonym* **on-premises wiring.**

premodulation combining: *Synonym* **precombining.**

Prentice's rule: The prism power at any point on a lens is obtained by multiplying the dioptric power by the distance in centimeters from the optical center. *See also* **dioptric power, lens, prism power.**

prepaid: 1. In a communications system, a prefix in a message indicating that the charges for the message were paid at the source, usually prior to transmission. **2.** Pertaining to a message for which the charges have been paid prior to transmission. *See also* **collect.**

preparation time: *See* **message preparation time.**

preprocessor: 1. A computer program that performs processing preliminary to other computation or processing. *Note:* Examples of preprocessors are (a) a computer program that examines a source program and effects changes before it is executed, and (b) a computer program that converts the data format used in a system to the format required by an emulator of that system. **2.** A processor that processes data and programs prior to their insertion into another processor, which usually is the central processing unit of a computer. *Note:* The preprocessor might change format, interpret codes, or effect program changes before execution of the program by the main processor. *See also* **computer program, emulate, emulator, processor, source program.**

preselection filter: A bandpass filter for a local oscillator and for frequency rejection in the signal path prior to down conversion in a receiver. *Synonym* **preselector filter.** *See also* **bandpass filter, down conversion, local oscillator.**

preselector filter: *Synonym* **preselection filter.**

presence: *See* **continuous presence, point of presence.**

presence point: *Synonym* **point of presence.**

presentation layer: In open systems architecture, a layer that (a) usually provides the functions, procedures, services, and protocols that are selected by the application layer, (b) may include data definition, data entry control, data exchange, and data display, and (c) is directly below the Application Layer and directly above the Transport Layer in the Open Systems Interconnection—Reference Model. *See also* **Open Systems Interconnection—Reference Model.**

Presentation Layer: *See* **Open Systems Interconnection—Reference Model.**

preset conference: A network-provided service feature that permits the automatic connection of a fixed group of users, or a closed user group with outgoing access, by keying a single directory number. *See also* **closed user group with outgoing access.**

preset jammer: A jammer (a) in which the frequency of the jamming transmitter is fixed before the transmitter is placed in operation, (b) that is most useful in airborne jamming operations where weight and space requirements may prohibit the use of operators or elaborate control equipment in flight, and (c) that usually is used in barrage jamming over a wide band, usually in overlapping series of frequency bands. *See also* **jammer, jamming.**

press to talk operation: *Synonym* **push to talk operation.**

press to type operation: *Synonym* **push to type operation.**

press traffic: In a communications system, messages that (a) contain text destined to be released to public information media, such as newspapers, magazines, radio stations, and television (TV) stations, (b) constitute user traffic, and (c) do not necessarily contain information originated or gathered by the news media, which is also user traffic. *Note:* An example of press traffic is a distributed announcement to the news media by the government of the development of a new weapon, a new public interest program, or a new communications system protocol. The transmission of the same information from one news media office to another is not press traffic, but is communications system user traffic.

prestore: In computer and data processing systems operations, to store required data before a computer program, routine, or subroutine that needs the data is entered for execution. *See also* **computer program.**

prestored string: In data communications and personal computer (PC) operations, a string of characters that is used often or repetitively and that can be called when needed. *Note:* Examples of prestored strings are (a) a string used in a log-in sequence that can be called up with only one or two characters or key strokes and (b) a stored macroinstruction. *See also* **call, macroinstruction, string.**

prevarication: *Synonym* **irrelevance.**

preventive maintenance (PM): 1. The care and the servicing of equipment and facilities in order to maintain the equipment and the facilities in satisfactory operating condition by providing for systematic inspection, detection, and correction of incipient failures either before they occur or before they develop into major defects. **2.** The systematic or prescribed maintenance intended to reduce the probability of failure of equipment and facilities. **3.** Maintenance, including tests, measurements, adjustments, and parts replacement, performed specifically to prevent failures from occurring or to increase the mean time between failures (MTBF). *See also* **corrective maintenance, fault, maintenance, mean time between failures.**

PRF: pulse repetition frequency.

PRI: primary rate interface.

primary: 1. In high level data link control (HDLC) procedures that are used in communications systems, the part of a data station that (a) supports the main control functions of the data link, (b) generates commands for transmission, (c) interprets responses, and (d) performs specific functions at the data link level, such as (i) initialization of control signaling, (ii) organization of data flow, and (iii) error control. **2.** *Synonym* **primary winding.** *See also* **high-level data link control.** *Refer to* **Fig. T-8 (1024).**

primary aeronautical station: A station in the aeronautical mobile service that (a) at any given instant, has the right to select and to transmit information to a secondary station, (b) has the responsibility to ensure information transfer, (c) is the only primary station on a data link or a radio net at one time, (d) has control of the data link or the radio net at a given instant, (d) is assigned primary status temporarily, (e) is governed by standard control procedures, (f) is conferred primary status so that it may transmit a message, although a station need not have a message to transmit to be selected as the primary station, and (g) is selected by the master net control station. *See also* **aeronautical mobile service, master net control station, secondary station.**

primary axis: In movable antenna systems, the axis that (a) has the ground, i.e., the earth, as a reference datum and (b) serves as the datum for the secondary axis, if any. *Note:* An example of a primary axis is the vertical axis about which an antenna may rotate to scan in azimuth. The secondary axis may scan in elevation. *See also* **secondary axis.**

primary channel (PC): 1. The channel that (a) is designated as a prime transmission channel and (b) is used as the first choice in restoring priority circuits. **2.** In a communications network, the channel that (a) has the highest signaling rate of all the channels that share a common interface, and (b) may support the transfer of information in one direction only, either direction alternately, or both directions simultaneously. *See also* **channel, secondary channel.**

primary clock: *Deprecated synonym for* **principal clock.**

primary coating: 1. In the manufacture of optical fibers, the optical fiber coating that (a) is in intimate contact with the outermost cladding surface, (b) is usually bonded to the cladding, and (c) is used to preserve and protect the integrity of the fiber outer surface. **2.** The plastic overcoat that (a) is in intimate contact with the cladding of an optical fiber, (b) is applied during the manufacturing process, (c) usually has an outside diameter of about 250 μm (microns) to 750 μm, (d) serves to protect the fiber from mechanical damage and chemical attack, (e) enhances optical properties of fibers by stripping off cladding modes, (f) where multiple fibers are used inside a single buffer tube or cable, suppresses cross coupling of optical signals from one fiber to another, i.e., suppresses crosstalk, and (g) should not be confused with a tight buffer or with the plastic cladding of a plastic-clad silica (PCS) fiber. *Synonyms* **primary polymer coating, primary polymer overcoat.** *See also* **cladding, optical fiber coating.**

primary control: *See* **central primary control.**

primary distribution system (PDS): An alternating current (ac) distribution system for supplying electrical power to the primaries of distribution transformers from generating stations or substation distribution buses. *See also* **facility, primary power, primary winding.**

primary emission: The emission of electrons from a material that (a) is exposed to electromagnetic radiation and (b) demonstrates photoemissivity, i.e., the photoemissive effect. *See also* **photoemissive effect, photoemissivity, photomultiplier, secondary emission.**

primary frequency (PF): 1. A frequency that is assigned for usual use on a particular circuit. **2.** The first-choice frequency that is assigned to a fixed or mobile station for radiotelephone communications. *See also* **secondary frequency.**

primary frequency standard: A frequency source that (a) meets national standards for precision and (b) operates without the need for calibration. *Note:* An example of a primary frequency standard is a cesium clock. *See also* **cesium clock, cesium standard, DoD master clock, frequency offset, frequency standard, standard frequency and time signal service.**

primary group: *Synonym* **basic group.** *See* **group.**

primary guard: In aeronautical communications, the principal radio guard responsible for maintaining communications with an aircraft station during flight.

primary guard station (PGS): In aeronautical communications, the station that holds primary guard responsibility for a particular flight of an aircraft station.

primary patch bay: *See* **digital primary patch bay.**

primary polymer overcoat: *Synonym* **primary coating.**

primary power: In communications, computer, data processing, and control systems, a reliable source of electrical power that usually supplies the main bus, such as the main bus of a station. *Note 1:* The primary power source may be a government-owned generating plant, a privately owned generating plant, or a public utility power system. *Note 2:* A Class A primary power source assures a continuous supply of alternating current (ac) electrical power. *See also* **auxiliary power, power, primary distribution system, primary substation, station battery.**

primary radar (PR): 1. A radio determination system based on the comparison of reference signals with radio signals reflected from the position to be determined, i.e., from the target. **2.** Radar that (a) uses only the reflection of its own signals from an object, i.e., from a target, and (b) does not depend on active signals that are transmitted by the object that it finds or tracks. *See also* **radar, radiodetermination, secondary radar.**

primary radiation: Radiation that is incident upon a material and produces secondary emission from the material. *See also* **radiation, secondary emission, secondary radiation.**

primary radiation source: A source that generates and emits electromagnetic waves. *Note:* Examples of primary radiation sources are energized radio transmitters, light bulbs, lasers, light-emitting diodes (LEDs), the Sun, and radioactive materials. *See also* **secondary radiation source.**

primary rate interface (PRI): An integrated services digital network (ISDN) interface standard (a) that is designated in North America as having 23B+D channels, (b) in which all circuit-switched B channels operate at 64 kbps (kilobits per second), and (c) in which the D channel also operates at 64 kbps. *Note:* The primary rate interface (PRI) combination of channels results in a digital speech interpolation (DSI) (T1) interface at the network boundary. *See also* **Integrated Services Digital Network.**

primary route: The predetermined path of a message from (a) its source, i.e., from its sending or originating station, to (b) a message sink, i.e., to a receiving, addressee, or destination station. *Note 1:* In telephone switchboard operations, the primary route is the route that is attempted first by the attendants or equipment when completing a call. *Note 2:* Alternatives to the primary route are based on network traffic conditions and supervisory policy. *See also* **alternative routing, completed call.**

primary service area: The service area of a broadcast station in which the groundwave is not subject to objectionable interference or objectionable fading.

primary ship-shore station: A shore-based ship-shore station that is responsible for providing the radio guard for any or all of the worldwide ship-shore frequencies that (a) may be changed from time to time and (b) are used by ship broadcast systems for direct access transmissions. *See also* **local ship-shore station, secondary ship-shore station, ship-shore station.**

primary spectrum: The main, first, or characteristic chromatic aberration of a simple nonachromatized lens or prism. *See also* **achromat, achromatic, chromatic aberration, lens, prism, secondary spectrum.**

primary station: In a communications network, the station that (a) controls balance in a data link, (b) generates commands and interprets responses, and (c) at the link level, ensures the (i) initialization of data and control information interchange, (ii) organization and control of data flow, (iii) performance of retransmission control, and (iv) performance of all recovery functions. *See also* **control station, data communications, link, master station, network, secondary station, slave station, tributary station.**

primary storage: *Synonym* **main storage.**

primary substation: Equipment that switches or modifies the characteristics of primary power, such as the voltage and the frequency. *See also* **primary power.**

primary time standard: A time standard that (a) is based on the International System of Units (Système International d'Unités) definition of the second and (b) does not require calibration. *Note:* The international second is based on the microwave frequency associated with the hyperfine ground-state levels of the cesium-133 atom in a magnetically neutral environment. Realizable cesium frequency standards use a strong electromagnet to deliberately introduce a magnetic field that overwhelms that of the Earth. Slight variations in the calibration of the electric current in the electromagnet introduce minuscule negligible frequency variations among different cesium oscillators. *See also* **cesium clock, cesium standard, coordinated clock, Coordinated Universal Time, DoD master clock, International Atomic Time, precise frequency, precise time, reference clock, reference frequency, second, standard frequency and time signal service.**

primary winding: 1. The winding in a transformer that is connected to a power source. **2.** In an electrical transformer, the winding that accepts the input alternating current (ac) power. *Synonym in this sense* **primary.** *See also* **transformer.**

prime focus feed: An antenna feed arrangement in which the radiating feed horn is located at the principal focus point of the main reflector. *Synonym* **front feed.** *See* **offset prime focus feed, principal focus point.**

prime number: A whole number that has no whole number divisors except 1 and itself, i.e., that when divided by a whole number other than 1 and itself will always produce a mixed number, i.e., a whole number and a fraction. *Note:* The first few prime numbers are 1, 2, 3, 5, 7, 11, 13, 19, and 23. Even numbers except 2, products of two or more whole numbers, mixed numbers, irrational numbers, repeating numbers, such as 777 or 333, and 0 are not prime numbers.

principal clock: Of a set of redundant clocks, the clock that is selected for normal use. *Note 1:* The principal clock may be selected because of a property that makes it a unique member of the set of clocks in a system, such as superior clock quality. *Note 2:* "Primary clock" should not be used because of possible confusion with "primary frequency standard." *Deprecated synonym* **primary clock.** *See also* **clock, Coordinated Universal Time, International Atomic Time, reference frequency.**

principal focus: *Synonym* **principal focus point.**

principal focus point: The point to which incident parallel rays of light converge, or from which they diverge, when they have been acted upon by a lens or a mirror. *Note:* A lens has a single point of principal focus on each side of the lens. A mirror has only one principal focus. A lens or a mirror has an infinite number of image points, real or virtual, one for each position of, or point on, the object. *Synonym* **principal focus.** *See also* **focal point, image, lens, mirror, object, point, principal point.**

principal point: In an optical system, any one of two or more significant points, i.e., definable points, in the system. *Note 1:* Examples of principal points are the geometric center of a lens, a focal point, and the position of a virtual image on the optical axis. *Note 2:* In a lens system, principal points are counted from the object, i.e., point zero, toward the objective lens. *See* **first principal point, second principal point.** *See also* **optical axis, optical system, point.**

principal ray: In an optical system and starting from the object, i.e., the object space, the ray that (a) is directed at the first principal point and hence (b) in the image space, intersects the optical axis at the second principal point. *Note:* In a single positive lens system, the object is at the zero principal point, the lens is at the first principal point, and the principal ray crosses the optical axis at the second principal point and creates the image at the third principal point, at which the image is focused. *See also* **first principal point, ray.**

principle: *See* **Fermat's principle, firewall principle.**

print command: 1. An instruction given by a communications, computer, data processing, or control system operator to print specific data on a specified printing device in a predetermined, fixed, or specified format. **2.** One of a set of commands given to a system by an operator, all the members of the set being related to the printing process. *Note:* Examples of print commands are the margin setting, font to use, beginning and ending of pages, character and line spacing, printer to use, and whether to use single-page or continuous-form paper. *See also* **command, instruction.**

print contrast ratio: On a printed page, the ratio of (a) the difference between (i) the maximum reflectance in the nonprinted region, R_n, found within a specified distance from an inspection area in the print region, and (ii) the reflectance of that inspection area in the print region, R_p, to (b) that maximum reflectance in the nonprinted region, given by the relation $(R_n - R_p)/R_n$ where R_n is the maximum reflectance in the nonprinted region a specified distance from an inspection area in the printed region, and R_p is the reflectance of that inspection area in the printed region. *See also* **reflectance.**

print control character: A control character that (a) is used to designate and effect operations associated with printing and (b) does not control the actual printing, i.e., that which is to be printed is controlled by the user. *Note:* Examples of print control characters are characters that control line spacing, new page alignment, paper feed, page ejection after the last line, carriage return, underlining, and font changing.

printed circuit (PC): An electrical circuit that (a) is formed on a substrate material, (b) usually is made by starting with a metal-clad surface that is coated with a photosensitive material, (c) usually is made by etching away undesired metal to form connections between circuit components that are also mounted on the board, and (d) is accessed by means of sliding surfaces mounted on the board, i.e., male connectors, that are slipped into sockets, i.e., female connectors, that are usually connected to a bus.

printed circuit board (PCB): The substrate and the mounted electronic components for a printed circuit, such as the printed circuit connections, integrated circuits, transistors, lumped components, and other electrical and mechanical components, such as necessary plugs, terminals, connectors, and guides.

printer: A device that (a) forms graphic characters and images on a data medium, such as paper, cardboard, tape, or plastic, (b) accepts data and print control characters to form the graphic characters, and (c) uses a writing medium, such as liquid ink, powdered ink, carbon ribbon, or inked ribbon, to form the characters on the data medium. *See* **bar printer, daisy wheel printer, direct printer, dot matrix printer, impact printer, ink jet printer, laser jet printer, monitor printer, Morse printer, nonimpact printer, on-the-fly printer, page printer, wheel printer.** *See also* **data medium, print control character.**

printing key: A key that, when actuated, causes a character to be printed or typed and that usually causes the paper carrier or the type carriage to move to the next print position. *See also* **key, print position, type.**

printing perforator: A perforator that, when perforating tape, also prints, on the tape at each hole pattern position, a graphic character, such as an alphanumeric character or a symbol that represents a control function, that corresponds to the hole pattern on the tape. *Note:* A card punch that prints, usually at the top of each column of a punch card, the graphic character represented by the hole patterns in the column is called a "card interpreter." *See also* **card punch, perforator, printing reperforator, punch card, punched card, reperforator.**

printing-recording: *See* **ink vapor printing-recording.**

printing reperforator: A reperforator that, when reading and perforating tape, also prints on the tape at each hole pattern position a graphic character, such as an alphanumeric character or a symbol that represents a control function, that corresponds to the hole pattern on the tape. *See also* **printing perforator.**

printing telegraphy: *Synonym* **alphabetic telegraphy.**

printing teletypewriter telegraphy (PTT): In communications systems, manual telegraph operation in which signals are (a) sent by means of a keyboard and (b) are recorded by a receiver, usually a teleprinter, in the form of printed characters. *See also* **printer, telegraph, telegraphy, teleprinter, teleprinting.** *Refer to* **Fig. P-10 (740).**

printout: 1. The output from a printer, such as printed or typed pages. **2.** To print or type graphic characters, such as text or control symbols. *See also* **printer.**

print position: The location or the point on a data medium where a graphic character may be placed or formed. *Synonym* **imprint position.** *See also* **character, data medium, graphics, point.**

print speed: The rate at which data units, such as characters, words, lines, and pages, are printed per unit of time, such as 600 characters per minute, two lines per second, or six pages per minute. *See also* **printer.**

print spooler: A storage device, such as a buffer storage, that is used specifically to hold data that are to be printed, such as those for an entire page. *Note:* When a print command is given, the spooler can receive and hold the data while the printer is printing character by character or line by line, thus allowing the computer to proceed to other tasks or allowing the operator to use the computer for other operations, such as editing the next page or performing calculations. *See also* **buffer storage, printing.**

print-through: A transfer of a recorded signal from one part of a data medium to another part of the data medium when these parts are brought into physical contact. *Note:* Print-through usually is undesired. *See also* **data medium.**

print wheel: A wheel that (a) has the type font mounted on its periphery instead of on a ball or on the ends of rods as on many typewriters, (b) is used on many impact printers, such as daisy wheel printers that have only one wheel and cylindrical wheel printers that have a wheel for each print position, (c) usually is used on printers connected to computers, and (d) is used when higher than dot matrix quality printing is needed. *See also* **daisy wheel printer, font, print position, type.**

priority: 1. Unless specifically qualified, the right to occupy a specific frequency that (a) is used for authorized purposes and (b) is free of harmful interference from stations of other organizations and agencies. **2.** In U.S. Department of Defense (DoD) record communications systems, one of the five precedence levels used to establish the time frame for handling a given message. **3.** In U.S. Department of Defense (DoD) voice communications systems, one of the precedence levels assigned to a customer telephone for the purpose of preemption of telephone services. *See* **frequency priority, restoration priority, with priority.** *See also* **precedence, precedence designation, PRIORITY message,**

priority facility: In a communications system, a facility that (a) enables a user to exercise precedence over the other users of the system and (b) handles priority or precedence designation arrangements for system services, such as calls, messages, and packet transfers.

priority level: The level that may be assigned to a National Security or Emergency Preparedness (NS/EP) telecommunications service specifying the order in which provisioning or restoration of the service is to occur relative to other NS/EP or non-NS/EP telecommunications services. *Note:* Authorized priority levels (a) for provisioning, from highest to lowest priority, are designated "E," "1," "2," "3," "4," and "5" and (b) for service restoral, from highest to lowest priority, are designated "1," "2," "3," "4," and "5." *See also* **National Security or Emergency Preparedness (NS/EP) telecommunications.**

priority level assignment: The priority level designated for the provisioning and the restoration of a particular National Security or Emergency Preparedness (NS/EP) telecommunications service. *See also* **National Security or Emergency Preparedness (NS/EP) telecommunications.**

priority message: A category of precedence that (a) is reserved for messages that require expeditious action by the addressees or furnish essential information for the conduct of operations in progress, and (b) is used when routine precedence will not suffice. *See also* **precedence, seizing.**

PRIORITY message: In military communications systems, a message that (a) has the next higher level of precedence above ROUTINE and (b) has an element of urgency that is not so great as that of a FLASH OVERRIDE, FLASH, or IMMEDIATE message. *See also* **precedence designation.**

priority precedence: The precedence designation that is reserved for calls and messages that (a) require expeditious action by the addressees, (b) concern the conduct of operations in progress, (c) relate to other important and urgent matters, and (d) require a precedence designation above that of routine precedence. *See also* **precedence, precedence designation, routine precedence.**

priority service: *See* **Telecommunications Service Priority Service.**

priority system: *See* **Telecommunications Service Priority System.**

priority system user: *See* **Telecommunications Service Priority System user.**

prism: A transparent body (a) that has at least two polished plane faces inclined with respect to each other, (b) from which light is reflected or through which light is refracted, such that when light is refracted and the refractive index exceeds that of the surrounding medium, the light is deviated, i.e., it is bent, toward the thicker part, and (c) that may be used to demultiplex a light beam containing many wavelengths by refracting and dispersing the beam into separate beams, each with a different wavelength, each in a different direction, and each incident upon a different photodetector. *See* **dispersion, penta prism.**

prism chromatic resolving power: An expression for prism resolving power in which (a) parallel rays of light are incident on a prism, (b) the prism is oriented at the angle of minimum deviation at wavelength λ, and (c) the entire height of the prism is used, so that the resolving power R, deduced on the basis of Rayleigh's criterion, is given by the relation $R = \lambda/\Delta\lambda = Bdn/d\lambda$ where R is the prism chromatic resolving power, λ is the wavelength of the light rays, B is the maximum thickness of the prism traveled by the light rays, n is the refractive index for the wavelength λ, and $dn/d\lambda$ is the rate of change of refractive index with respect to λ. *Note:* The rate of change of refractive index with respect to λ, $dn/d\lambda$, is often called the "prism dispersion," and the maximum thickness, B, is often called the "prism base length." *See also* **dispersion, resolving power.**

privacy: 1. In a communications system or network, the protection given to information to conceal it from persons within the system or the network. *Synonym* **segregation. 2.** In communications systems, short-term protection given to unclassified information, such as radio transmissions of law enforcement personnel, that requires safeguarding from unauthorized persons. **3.** In a communications system, the protection given to prevent unauthorized disclosure of the information in the system. *Note 1:* The required protection may be accomplished by various means, such as by communications security measures and by directives to operating personnel. *Note 2:* The limited protection given certain voice and data transmissions by commercial cryptoequipment is sufficient to deter a casual listener, but could not withstand a competent cryptanalytic attack. *See also* **authenticate, communications, cryptanalysis, cryptoequipment. 4.** The right of a person's self-determination as to the degree to which the person is willing to share with others personal information, i.e., private information about her- or himself, that is in the custody of others and that might be compromised by unauthorized exchange or release of such information. *Note:* It is the right of persons and organizations to control the collection, storage, and dissemination by others of information about themselves. This right is statutory, and it is above and beyond the professional ethical obligation of doctors, lawyers, and members of the clergy to protect personal information about their patients, clients, or congregation. Under the Information Privacy Act, privacy is a legal right.

privacy code: A code that (a) is used to protect the contents of a message from being understood when read by unauthorized persons, and (b) does not afford any security or protection against organized cryptanalysis. *See also* **cryptanalysis, protected information, protection, security.**

private aircraft station (PAS): An aircraft station that is on board an aircraft that is (a) owned and operated by a private person or a private organization, (b) usually is not for hire, and (c) is neither owned nor operated by a commercial air carrier, a commercial airline, or a government agency.

private automatic branch exchange (PABX): A private branch exchange (PBX) that (a) has automatic switching equipment and (b) usually does not have a manually operated switchboard. *Note:* "Private branch exchange (PBX)" is preferred over both "private automatic branch exchange (PABX)" and "private automatic exchange (PAX)," regardless of the degree of automation. *See also* **private branch exchange.**

private automatic exchange (PAX): *Synonym* **private automatic branch exchange.**

private branch exchange (PBX): 1. A privately owned exchange that (a) usually is not owned by the public switched network but usually has access to the network, (b) becomes a customer of the switched network when provided access to the network, and (c) may have a manual switchboard or an automatic switch. **2.** A switch that (a) serves a selected group of users and (b) is subordinate to a switch at a higher level of management and control. **3.** A private telephone switchboard that provides on-premises dial service and may provide connections to local and trunked communications networks. *Note:* A private automatic exchange (PAX), i.e., a private automatic branch exchange (PABX), has an automatic switch. "Private branch exchange (PBX)" is being used rather than "PABX" and "PAX" even when the exchange has an automatic switch, i.e., regardless of the degree of automation. *See also* **key set, private automatic branch exchange, key telephone system.** *See* **satellite private branch exchange.** *Synonym* **private exchange (PX).** *Refer to* **Figs. C-7 (157), T-5 (1012).**

private branch exchange tie trunk: A direct connection between two private branch exchanges (PBXs). *See also* **private branch exchange, trunk.**

private branch exchange trunk: In telephone system operation, a trunk that is used to interconnect a private branch exchange (PBX) with the network switching center that serves it. *See also* **private branch exchange, trunk.**

private call: In telephone systems operation, a call that is made by, on behalf of, or in the interest of the caller rather than on behalf of the organization or in the interest of the organization to which the telephone service is being provided. *Note:* Many organizations do not permit private calls either to be made or to be received from organizational telephones, except under prescribed circumstances, such as when they are less than one minute in length, when they are local calls, when they are registered and paid for by the caller, or when they result in a monthly long-distance charge of less than two dollars. *Synonym* **personal call.**

private exchange: *Synonym* **private branch exchange.**

private key encryption system: A cryptosystem in which (a) key is assigned to a designated user, link, or user group, (b) the key is used to both encrypt and decrypt messages, and (c) the security of the cryptosystem is compromised if the key is obtained by unauthorized persons. *Note:* The Data Encryption Standard (DES) is a well-known private key encryption system used to provide a degree of privacy, but it is not approved for national security purposes, such as for U.S. Department of Defense security classified information. *See also* **cryptomaterial, cryptosystem, key.**

private library: A software library that is user-owned and usually is not a part of a system library. *See also* **software library, system library.**

private line: In telephone systems, an arrangement in which (a) each user end instrument or set of instruments on a user premises is connected to a single separate loop, (b) no end instrument belonging to any other user is connected to that loop, (c) each user is assigned a unique number, and (d) privacy is complete, except perhaps for accidental crosstalk, a faulty connection, or eavesdropping. *See also* **leased circuit, party line.**

private National Security or Emergency Preparedness (NS/EP) telecommunications service: A National Security or Emergency Preparedness (NS/EP) telecommunications service that (a) is a non-common-carrier telecommunications service and (b) includes private line, virtual private line, and private switched network services. *See also* **public switched National Security or Emergency Preparedness telecommunications service.**

private operating agency: *See* **telecommunications private operating agency.**

privilege: *See* **least privilege.**

probability: *See* **access denial probability, added block probability, bit misdelivery capability, block error probability, blocking probability, block loss probability, block misdelivery probability, disengagement denial probability, outage probability, service probability.**

probability analysis compaction: Data compaction accomplished by specifying applicable analytical formulas that are characterized by expressing the distribution of a set of data by certain numbers, such as only the median value and the standard deviation. *Note:* An example of probability analysis compaction is the transmission of only the median value and perhaps those values that exceed one, two, or three standard deviations rather than the entire set of measurements. Probability analysis compaction is based on the as-

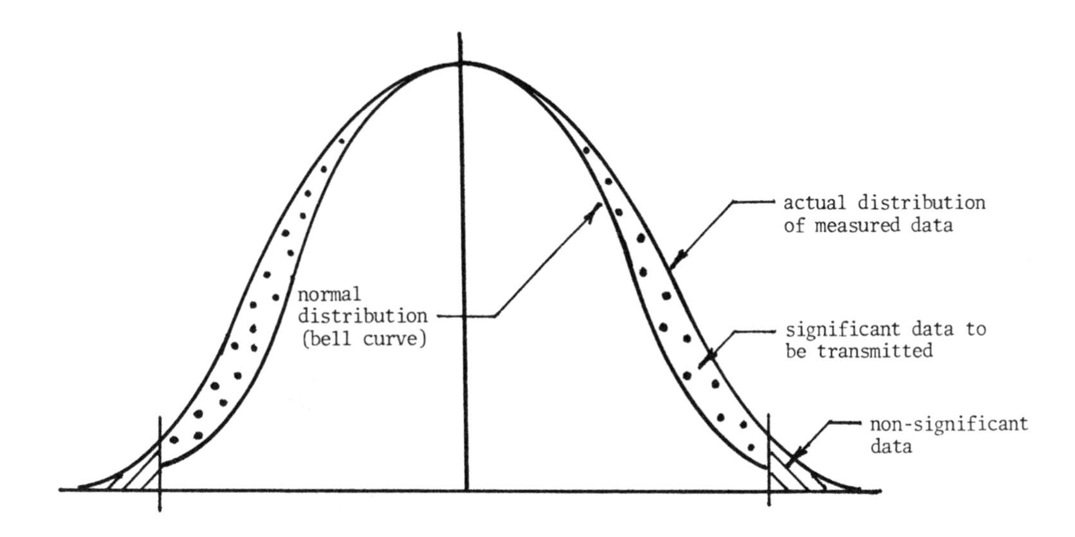

Fig. P-13. In **probability analysis compaction,** only data that are outside of a normal distribution, i.e., Gaussian distribution, of values need be stored or transmitted, thus saving space, time, equipment, and cost. Extreme values may be ignored.

sumption that only the large deviations are significant and may require attention, action, interpretation, or analysis. All the other values that entered into the determination of the median and the standard deviation need not be stored or transmitted. *Refer to* **Fig. P-13.**

probability of detection: *See* **low probability of detection.**

probability of interception: *See* **low probability of interception.**

probe: *See* **fiber optic probe.**

problem: *See* **benchmark problem, n-body problem, restricted circular three-body problem, three-body problem, two-body problem.**

problem-oriented language: A computer programming language (a) that is specifically designed to express problems for computer solution, and (b) in which the method of solution of the problems is not a primary factor. *See also* **application-oriented language, programming language.**

problem time: In an analog simulation, the time during which the simulation process takes place or is represented to take place. *Note:* In the ballistic trajectory of a missile, the problem time, usually in seconds, is the time that corresponds to each coordinate position of the missile, though the missile is not actually being flown. In the early days of electronic computers, as the ENIAC (Electronic Numerical Integrator and Calculator) calculated the ordinate values of the trajectory of a simulated round of ammunition fired from a gun under specified initial conditions and drag functions, one accumulator displayed the ordinate values, i.e., the missile altitude values, another displayed the abscissa, i.e., the range values, and a third displayed the time of flight, usually in seconds, that represented the accumulated increments of time that were being inserted into the ballistic equation to compute the ordinate and range values, i.e., the problem time. The ballistic equation of motion expressed position as a function of time. The values were calculated and displayed faster than the time required to fly the missile, the real time. The problem time ended when the altitude became negative. The problem time was slow enough to watch but faster than the real time. *See also* **real time, real time operation.**

procedure: In communications, computer, data processing, and control systems operation, the actions that are taken to accomplish a task or reach a desired result, such as the actions required to implement a protocol, control a communications network, or write a computer program. *See* **Advanced Data Communications Control Procedure, break-in procedure, call control procedure, circuit unavailability procedure, crossband radio telegraph procedure, data communications control procedure, fiber optic test procedure, flow control procedure, homing procedure, radiotelegraph procedure, radiotelephone procedure, recovery procedure, ship to shore radiotelephone procedure, simplex radio telegraph procedure, test procedure.**

Procedure B: *See* **Link Access Protocol B.**

Procedure D: *See* **Link Access Protocol D.**

procedure-oriented language: A computer problem-oriented programming language that facilitates expressing a procedure in the form of explicit algorithms. *Note:* Examples of procedure-oriented languages are FORTRAN, ALGOL, COBOL, and PL/1. *See also* **machine language.**

procedure sign (PROSIGN): One or more characters, such as letters numerals, or words, that are used to represent complete statements that are frequently used in communications. *Note 1:* The use of procedure signs (PROSIGNS) provides for shorter, more efficient, and more meaningful communications system operation. PROSIGNS are used to convey messages, i.e., user traffic, in contrast to PROWORDS, which are used to conduct communications operations. *Note 2:* Examples of procedure signs (PROSIGNS) are SOS to mean "help," AA to mean "all after," AB to mean "all before," and, in law enforcement, B&E to mean "breaking and entry." *Synonym* **prosign.** *See also* **procedure word.**

procedure word (PROWORD): In communications system operations, such as in voice radio and radiotelephone operations, a word that (a) has an extended procedural meaning, (b) is used relatively often when carrying on communications, and (c) enables communications to be effected more efficiently. *Note 1:* The use of procedure words (PROWORDS) provides for shorter, more efficient, and more meaningful communications system operation. PROWORDS are used to conduct communications operations, in contrast to procedure signs (PROSIGNS), which are used to convey messages, i.e., user traffic. *Note 2:* Examples of procedure words (PROWORDS) are "ROGER," "OVER," and "OUT." The PROWORD in voice systems is used in place of the procedure sign (PROSIGN) in telegraph systems, such as SOS to mean "help," AA to mean "all after," and AB to mean "all before." *Synonym* **proword.** *See also* **procedure sign.**

proceed to select: In communications systems operation, pertaining to a signal or an event in the access phase of a call that confirms the reception of a call request signal and advises the call originator data terminal equipment (DTE) to proceed with the transmission of the selection signals. *Note:* Examples of proceed to select pertain to a dial tone in a telephone system.

proceed to select signal: In a communications system, a signal that indicates that the system is ready to receive a selection signal. *Note:* An example of a proceed to select signal is a dial tone. *See also* **call selection signal.**

process: *See* **axial vapor phase oxidation process, cabling process, chemical vapor deposition process, chemical vapor phase oxidation process, concurrent process, doped deposited silica process, double crucible process, inside vapor phase oxidation process, ion exchange process, iterative process, modified chemical vapor deposition process, molecular stuffing process, outside vapor phase oxidation process, plasma activated chemical vapor deposition process, sequential process, software development process, thermographic process, vapor phase axial deposition process.**

process computer system: A system that (a) consists of a computer and process interface equipment and (b) monitors or controls technical processes, such as industrial production processes for the manufacture of paper, petroleum products, steel products, and automobiles. *See also* **process interface equipment.**

process control: Control of a process or processes in which a computer or specialized control equipment usually is used to regulate usually continuous operations or processes, such as industrial production processes for the manufacture of paper, petroleum products, steel products, and automobiles. *See* **automatic process control.**

process control equipment: Equipment that (a) measures the variables of a technical process, (b) directs the process according to control signals from the process computer system, and (c) provides appropriate signal transformation. *Note:* Examples of process control equipment include actuators, sensors, and transducers. *See also* **process computer system.**

process control system: A system that consists of (a) a computer, (b) process control equipment, and (c) possibly process interface equipment that may be separate equipment or may part of the computer, i.e., a special purpose computer. *See also* **process interface equipment.**

process gain: In a spread spectrum communications system, the signal improvement, such as the improvement in signal gain, signal to noise ratio, and signal shape, that is obtained by coherent band spreading, remapping, and reconstitution of the desired signal. *See also* **coherence, spread spectrum.**

processing: In communications, computer, data processing, and control systems, the performing or causing of events, such as a sequence of operations, that are intended to accomplish a desired purpose, produce a desired effect, or obtain a desired result. *Note:* An example of processing is performing a sequence of operations on data, such as storing, shifting, transmitting, receiving, deleting, adding, changing, translating, checking, converting, distributing, and collecting data. *See* **automatic data processing, background processing, batch processing, call processing, data processing, distributed data processing, distributed processing, foreground processing, front end processing, full screen processing, image processing, inline data processing, multiprocessing, optical data processing, parallel processing, postproduction processing, real time data processing, remote access processing, remote batch processing, signal processing, word processing.** *See also* **operation.**

processing center: In a switched communications network, a point (a) at which signal and data processing is accomplished for the network or for users of the network and (b) that usually is located at a node in the network. *See also* **node, point.**

processing equipment: *See* **automatic data processing equipment, Federal Information Processing equipment.**

processing gain: *See* **signal processing gain.**

processing node: *See* **data processing node.**

processing station: *See* **data processing station.**

processing system: *See* **automatic data processing system, automatic message processing system, distributed data processing system, Federal Information Processing system.**

processing terminal: *See* **data processing terminal.**

processing time: *See* **message processing time.**

processing unit: A functional unit that consists of one or more processors and their internal storage. *See* **central processing unit.**

process interface equipment: A functional unit that adapts process control equipment to the computer system in a process control system. *Synonym* **process interface system.**

process interface system: *Synonym* **process interface equipment.**

ecutes instructions and (b) consists of at least an instruction control unit and an arithmetic unit. *See* **front end processor, interface processor, interference message processor, language processor, microprocessor, passive processor, word processor.** *Refer to* **Figs. P-14 through P-16.**

processor unit: *See* **communications processor unit.**

process time: *See* **switching process time.**

procurement: The obtaining of personnel, services, information, supplies, and equipment.

procurement lead time: The time (a) that is between (i) the instant a procurement action is initiated and (ii) the instant the manufactured product that is acquired as the result of the action enters the supply system, (b) that consists of production lead time and administrative lead time, and (c) that does not include prototypes. *See also* **product.**

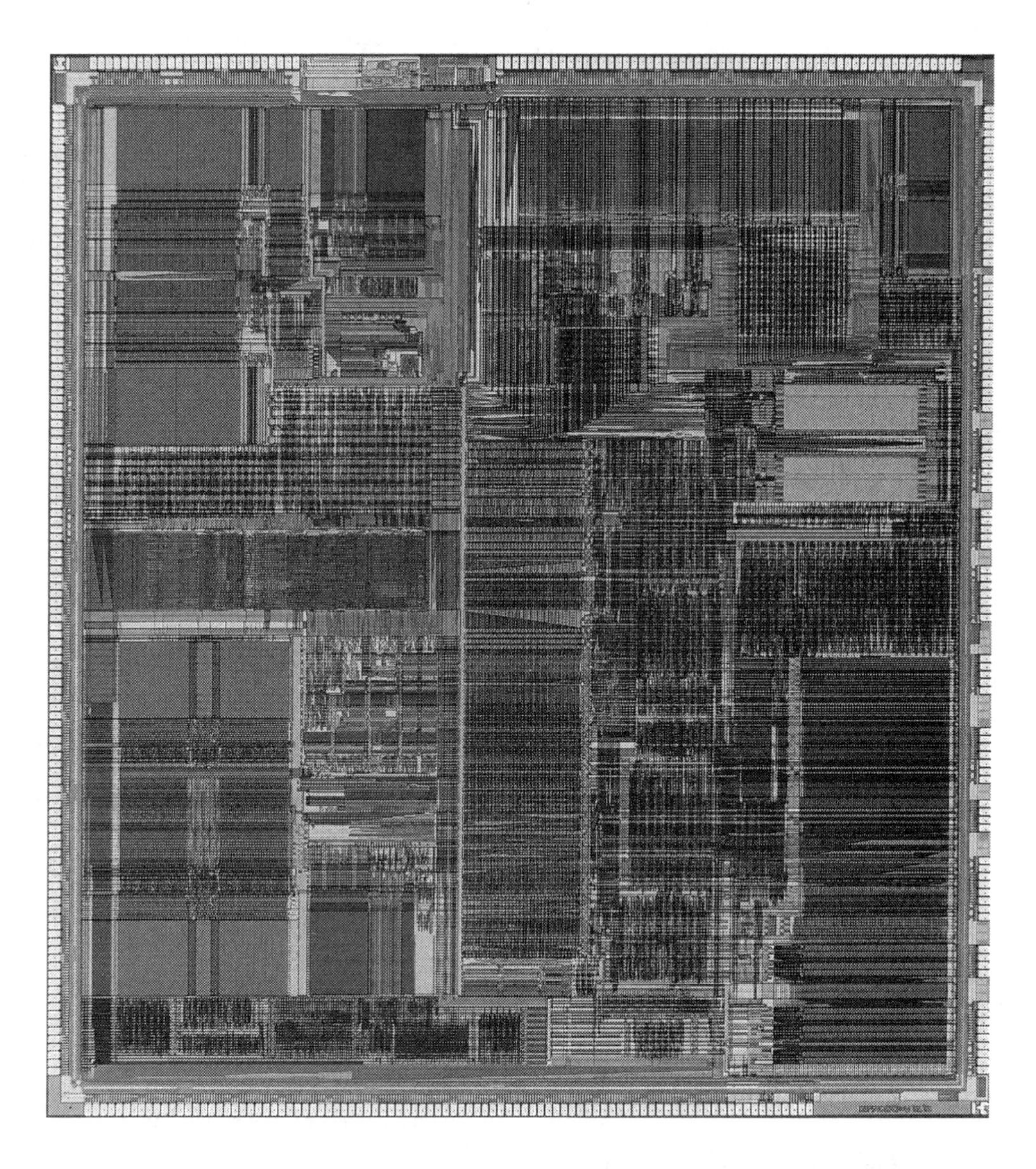

Fig. P-14. The high performance Intel Pentium™ **processor** manufactured on Intel's 0.6-μm (micron) four-layer metal BiCMOS advanced technology chip, at iCOMP™ Index 735/90 MHz (megahertz) and Pentium Processor at iCOMP Index 815/100 MHz, and containing 3.3 million transistors. (Courtesy Intel Corporation.)

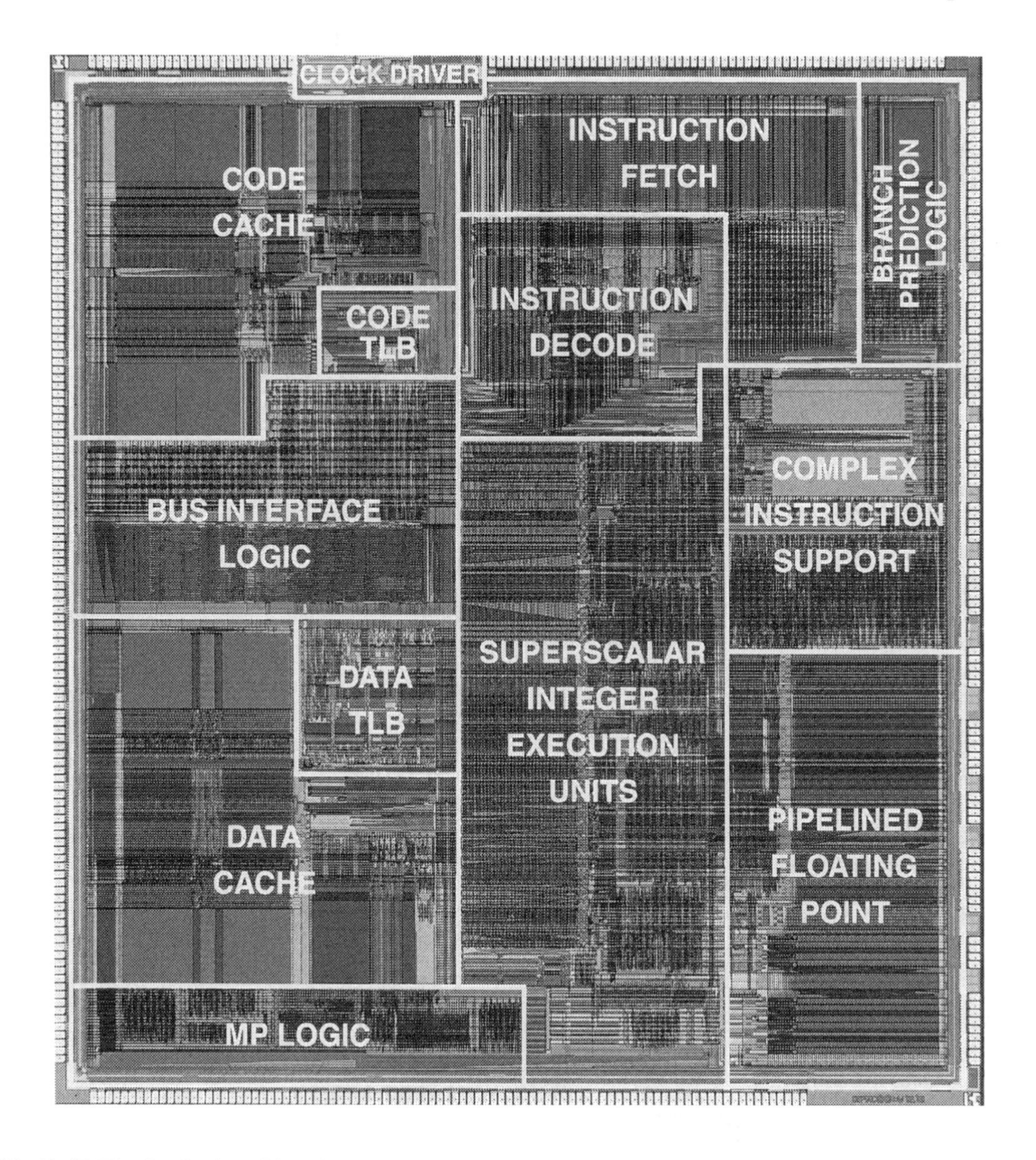

Fig. P-15. The distribution of functions on the Intel Pentium™ **processor** chip. (Courtesy Intel Corporation.)

product: 1. A result obtained after a sequence of operations, such as manufacturing operations. **2.** In arithmetic operations, (a) the result obtained when a number, the multiplicand, is multiplied by another number, the multiplier, or (b) the number obtained when a number, the multiplicand, is added to itself one fewer time than another number, the multiplier. *Note:* To multiply a number by 2, the number need only be added to itself once. In digital computers, binary numbers are multiplied by means of successive addition, which is performed by shifting and either adding or not adding, depending on the digits in the multiplier. *See* **bit-rate•length product, gain•bandwidth product, intermodulation product, modulation product, scalar product, type 1 product, type 2 product.** *See also* **multiplicand, multiplication, multiplier.**

production processing: *See* **postproduction processing.**

Fig. P-16. The Intel Pentium™ **processor** family at 60, 66, 90, and 100 MHz (megahertz). (Courtesy Intel Corporation.)

product validation: Validation of an item of hardware or software by an organization that is technologically and managerially independent from the organization responsible for developing the item. *Synonym* **independent product validation.**

profile: *See* **communications security profile, equivalent step index profile, functional profile, graded index profile, index profile, linear refractive index profile, parabolic refractive index profile, path profile, penetration profile, power law refractive index profile, radial refractive index profile, refractive index profile, standardized profile, step index profile, uniform index profile.**

profile dip: *Synonym* **refractive index dip.**

profile dispersion: In a dielectric waveguide, the dispersion of an optical pulse that occurs because the refractive index contrast, the profile dispersion parameter, and the profile parameter of the waveguide are different for each of the different wavelengths in the spectral width of the lightwave. *Note:* The refractive index contrast is dependent on the difference of nearly equal quantities, namely, the refractive indices of the core and the cladding. Thus, small variations in either of these makes the refractive index contrast vary by large percentages, seriously affecting the performance characteristics of the waveguide, including dispersion. *See also* **dispersion, material dispersion, profile dispersion parameter, profile parameter, refractive index contrast, refractive index profile, waveguide dispersion.**

profile dispersion parameter (P): In an optical fiber, the parameter that characterizes the part of the refractive index profile dispersion caused by a variation of the refractive index contrast with wavelength. *Note 1:* The profile dispersion parameter, P, is given by the relation $P = (n_1/N_1)(\lambda/\Delta)(d\Delta/d\lambda)$ where P is the profile dispersion parameter, n_1 is the maximum refractive index of the core, N_1 is the group index of the core, i.e., the group index corresponding to n_1, given by $N_1 = n_1 - \lambda(dn_1/d\lambda)$ where λ is the wavelength of the source, i.e., the free-space wavelength, and Δ is the refractive index contrast, given (a) by the relation $\Delta = (n_1{}^2 - n_2{}^2)/2n_1{}^2$ or (b) by the relation $(n_1 - n_2)/n_1$ for optical fibers in which the refractive indices are not different from each other by more than 1%. *Note 2:* The refractive index of the homogeneous cladding, n_2, is given by $n_2 = n_1(1 - 2\Delta)^{1/2}$. *Note 3:* In some texts, the profile dispersion parameter has been defined with the factor -2 in the numerator. *See also* **dispersion.**

profile mismatch loss: *See* **refractive index profile mismatch loss.**

profile optical fiber: *See* **uniform index profile optical fiber.**

profile parameter (g): In the power law index profile of an optical fiber, the parameter, g, that defines the shape of the refractive index profile. *Note:* The optimum value of the profile parameter, g, for minimum dispersion is about 2.25. *See* **refractive index profile parameter.** *See also* **power law index profile, refractive index profile.**

pro forma: *See* **protocol implementation conformance statement pro forma.**

pro forma message: A message that (a) is in a precise standard format and (b) has successive prescribed elements that are understood by prearrangement among the originator, addressee, and communications system operators. *Synonym* **proforma message.** *See also* **format.**

proforma message: *Synonym* **pro forma message.**

program: 1. To select a set of actions that is designed to achieve a specific result when the actions are performed, such as (a) to prepare, design, write, and check a computer program by selecting a sequence of instructions and then testing the set to determine if in fact it accomplishes its intended purpose, and (b) to select a sequence of operations in a communications system. **2.** The product that results when a set of actions is selected that, when performed, will achieve a specific result. *See* **active program, application program, computer program, communications program, conditional control program, control program, diagnostic program, Federal Communications Commission Registration Program, foreground program, graph**

accomplishes its intended purpose, and (b) to select a sequence of operations in a communications system. **2.** The product that results when a set of actions is selected that, when performed, will achieve a specific result. *See* **active program, application program, computer program, communications program, conditional control program, control program, diagnostic program, Federal Communications Commission Registration Program, foreground program, graph program, inactive program, input program, microprogram, monitor program, object program, output program, overlay program, reentrant program, registration program, resident program, reusable program, sort program, source program, structured program, subprogram, supervisory program, trace program, utility program.**

program analyzer: A software tool that may be used to provide information about a computer program. *See also* **computer program, software, software tool.**

program architecture: The structure, relationships, and arrangement of the components of a computer program, usually including the program interface and interface requirements for the operating environment, particularly the operating system. *See also* **computer program, interface, operating system.**

program block: An identifiable portion of a computer program, usually consisting of a group of related statements, used to accomplish a specific purpose, such as to delimit routines, specify storage allocation, delineate the applicability of labels, or segment other parts of the program for other purposes. *See also* **computer program, data delimiter, program segment, routine, statement.**

program disk: A portable disk (a) on which computer programs are stored, (b) that may be inserted and removed from computers, such as control computers and personal computers, and (c) that, when inserted, can be accessed by the computer. *See also* **computer program, diskette, personal computer.**

program extension: An enhancement made to an existing computer program to increase or improve its capability. *See also* **computer program.**

program interface: *See* **application program interface.**

program library: *Synonym* **software library.**

programmable: Pertaining to a device that can accept instructions that alter its basic functions.

programmable logic: A set of combinational circuits, i.e., combinational logic elements, in which (a) the interconnection of the circuits can be changed by means of a computer program, i.e., the combinational circuits are subject to program control, and (b) usually a system of enabling and disabling signals to the gates in the combinational circuits is used. *Synonyms* **field programmable gate array, variable logic.** *See also* **combinational circuit, disabling signal, enabling signal, fixed logic, gate.**

programmable logic array (PLA): An array of gates with interconnections that can be programmed to perform a specific logical function. *See also* **gate.**

programmable read-only memory (PROM): Storage in which data and computer programs may be stored, at which moment the storage becomes a read-only memory. *See also* **read-only memory.**

programmable terminal: 1. A terminal that has computational capabilities. **2.** A terminal that can be programmed to perform user-selected functions, such as checking, storing, and formatting data. **3.** An intelligent terminal that may be programmed to perform sets of functions by executing computer program instructions. *Synonym* **smart terminal.** *See also* **intelligent terminal.**

programmed check: A check performed by means of a computer program. *Note:* Examples of programmed checks are modulo-n checks and summation checks.

programmer: 1. The part of digital equipment that controls the timing and the sequencing of operations. **2.** A person who prepares computer programs, i.e., writes sequences of instructions for execution by a computer. *See also* **compile, computer, computer language, computer program.**

programming: The branch of science and technology devoted to the study, development, and application of sequences of instructions for execution in communications, computer, data processing, and control systems. *See* **automatic programming, bottom-up programming, computer programming, dual programming, dynamic programming, modular programming, multiprogramming, structured programming, top-down programming.** *See also* **computer program.**

programming flowchart: A flowchart that shows in a graphical form the sequence of operations in a computer program. *See also* **computer program, flowchart, operation.**

programming language: An artificial language that is used to generate or to express computer programs. *Note:* Examples of programming languages are COBOL, FORTRAN, ALGOL, BASIC, and Ada®.

programming system: One or more programming languages and the software necessary for using these languages with particular computers or automatic data processing equipment.

program module: A discrete, identifiable set of computer program instructions that can be called and executed as a unit by an assembler, compiler, editor, loader, subroutine, routine, or program. *See also* **call.**

program mutant: *Synonym* **program mutation.**

program mutation: **1.** A computer program that (a) is a deliberately altered version of another computer program and (b) is often used with test data to detect the ability of the program to detect the alteration. **2.** The creation of altered programs to evaluate the adequacy, propriety, or correctness of program test data. *Synonym* **program mutant.** *See also* **computer program.**

program origin: *See* **computer program origin.**

program package: In personal computer (PC) systems, a set of computer application programs that (a) usually is recorded on one or more diskettes, (b) will enable the application programs to be accessed and used by a PC under control of an operator and an operating system, and (c) may be accessed by booting up the computer with an access program, i.e., an operating system, such as MS DOS or IBM PC DOS, by (i) inserting the program package in a disk drive, (ii) calling for the package with the appropriate disk drive and proper code designation or file name, such as Lotus for Lotus 1-2-3, VW for Volkswriter, or WP for WordPerfect, and (iii) depressing the ENTER key so that the computer will access the program in the package, which may include word processing, spreadsheets, graphics, tutorials, databank management, calendars, office management, games, or any other of a host of possible application programs. *See also* **application program, computer program.**

program patch: *Synonym* **computer program patch.**

program protection: The actions and the measures taken to prevent unauthorized access or modification of a computer program, usually by means of applied internal and external controls. *See also* **security.**

program segment: In a computer program, a portion that (a) performs a distinct function, such as to read a set of data, provide a sequence of instructions, or perform a sort, and (b) usually can be identified, stored, and called separately from the program that it is a part of. *See also* **computer program.**

progressive conference call: A conference call in which the operator (a) obtains each conferee on the conference circuit and (b) either has the originator wait on the line or calls the originator back when all the conferees are on the line, depending on (i) the time required to obtain all or most of the conferees on the line and (ii) the wishes or the seniority of the originator. *See also* **conference call, meet me conference call.**

progress signal: *See* **call progress signal, waiting in progress signal.**

progress tone: *See* **call progress tone.**

project for information exchange: *See* **Advanced Project for Information Exchange.**

projection: *See* **flood protection, Mercator projection, spot protection.**

projection scanning: *See* **spot projection scanning.**

project plan: A plan, or the management document that describes the plan, that contains (a) the approach that will be taken in executing a project, (b) the work to be done, (c) the methods to be used, (d) the configuration management and quality assurance procedures to be followed, (e) the schedules to be met, (f) the project organization, (g) the resources to be used, including personnel, money, materials, information, and facilities required, and (h) any other unique and significant aspects of the project.

PROM: programmable read-only memory.

prompt: **1.** In interactive display systems, to display messages in the display space on the display surface of a display device to help the operator plan and execute subsequent operations. **2.** In communications, computer, data processing, and control systems, to inform an operator that the system is ready for the next input, such as command or a data element. **3.** In interactive display systems, a message displayed in the display space on the display surface of a display device to help the operator plan and execute subsequent operations. **4.** A cue or a reminder provided by a computer that alerts or guides the user to take some action. **5.** The message informing the user of a communications, computer, data processing, or control system that the system is ready for the next user action, such as a command or a message. *Note:* Prompts can occur during the execution of a command, such as a message to wait, or prior to execution, such as a message that the system is ready, such as the blinking A>, indicating that the system is ready to accept one of a set of legal commands, such as the DIR a:, a command in IBM PC DOS that directs the computer to display the file names of all the files on the diskette in disk drive A. **6.** Displayed or spoken data or text to assist a speaker, usually without the knowledge of the audience. *Note:* In some systems, the prompt is on a transparent data medium such that the data are visible to the speaker and not to the audience. *Synonym* **cue.**

proofed fiber optic cable: *See* **TEMPEST proofed fiber optic cable.**

propagation: **1.** The motion of waves within, through, or along a medium, such as the atmosphere, outer space, coaxial cables, and optical fibers. *Note:* For electromagnetic waves, propagation may occur in a vacuum and in material media. **2.** In communications, the progression of an electromagnetic wave or an acoustic wave. *Note:* Propagation usually is described in terms of phase or group velocity. *See* **anomalous propagation, line of sight propagation, scatter propagation, sporadic E propagation.** *See also* **backscattering, diffraction, direct ray, forward scatter, ionospheric scatter, multipath, propagation medium, refraction, scatter, tropospheric scatter.** *Refer to* **Figs. P-17, S-18 (951).**

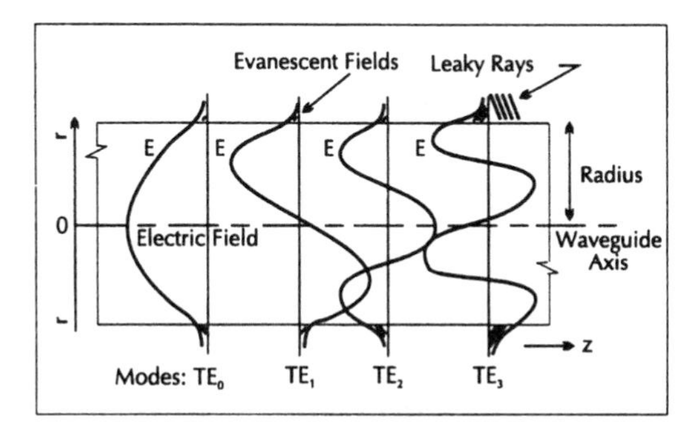

Fig. P-17. Lower-order **propagation modes** in a slab dielectric waveguide showing evanescent waves outside the guide but coupled to waves inside the guide, and leaky waves outside the guide but becoming uncoupled from the waves inside the guide. The leaky waves will radiate into space and usually occur with the higher-order modes.

propagation constant: **1.** In an electromagnetic wave propagating in a waveguide, such as an optical fiber or a metal pipe, the factor, *p,* (a) that is in the expression for the exponentially varying characteristics of the wave, given by the relation $e^{-pz} = e^{-ihz-az}$ where pz is a complex number, p is the propagation constant, $p = ih + a$, $i = (-1)^{1/2}$, h is the phase term, i.e., the imaginary part, and a is the attenuation term, i.e., the real part, and (b) that governs the propagation characteristics of the wave in the guide. *Note:* Dispersion occurs because the propagation constant is a function of frequency as well as a function of the materials of construction of the guide. **2.** A rating for a line or a medium through which an electromagnetic wave of a given frequency is propagating. **3.** For an electromagnetic field mode varying sinusoidally with time at a given frequency, the logarithmic rate of change, with respect to distance in a given direction, of the complex amplitude of any field component. *Note:* The propagation constant is given by the relation $e^{-pz} = e^{-ihz-az}$ where pz is a complex number, p is the propagation constant, $p = ih + a$, $i = (-1)^{1/2}$, h is the phase term, i.e., the imaginary part, and a is the attenuation term, i.e., the real part, and p governs the propagation characteristics of the wave in the guide. *See* **axial propagation constant, transverse propagation constant.** *See also* **attenuation term, Bouger's law, phase term.**

propagation delay: **1.** In an optical device, such as a fiber optic transmitter, receiver, cable, or coupler, the delay between (a) the instant the leading edge of an input signal enters the device and (b) the instant the leading edge of the corresponding output signal emerges from the device. **2.** The time required for a signal to travel from one point to another. *Note:* Examples of propagation delay are (a) the time required for the leading edge of a pulse to travel from the input end to the output end of a transmission line, such as a wire, coaxial cable, or an optical fiber, (b) in an optical receiver, the time between (i) the instant the leading edge of an optical input pulse is incident on the photodetector and (ii) the instant the leading edge of the corresponding electrical output pulse emerges from the receiver, and (c) the time required for a signal to propagate from one Earth station to another via a satellite station, about 0.33 s (seconds) for a geostationary Earth satellite. *Synonyms* **propagation time delay, transit time, transmit time.** *See also* **block transfer time, transmission, transmission time.**

propagation ionospheric scatter: *See* **backward propagation ionospheric scatter, forward propagation ionospheric scatter.**

propagation medium: **1.** Any material substance that can be used for the propagation of signals, usually in the form of modulated radio, light, or acoustic waves, from one point to another. *Note:* Examples of propagation media are (a) optical fibers, cables, or bundles, (b) wires, including coaxial cables, twisted pairs, and open wires, (c) slab dielectric waveguides, (d) the sea-atmosphere interface, and (e) atmospheric ducts. Examples of materials that are used to produce or serve as propagation media are metals for electrical current signals, glass and other dielectric materials for lightwave signals, and air for sound wave signals, with the possible exception that vacuum is considered as a propagation medium for electromagnetic wave signals, such as light, radio, video, and microwave signals. Except for electromagnetic transmission in vacuum and the atmosphere, propagation media usually are shaped, i.e., have a specific form, in order to guide the energy in a signal from a point of dispatch to a point of reception. **2.** In telecommunications, the transmission path along which a signal propagates, such as a wire pair, coaxial cable, waveguide, optical fiber, or radio path. **3.** Any material item or substance, such as an optical fiber, a fiber optic cable, a fiber optic bundle, a wire, a dielectric slab, water, or air, that can be or is used for the propagation of signals, usually in the form of modulated radio, light, or acoustic waves, from one point to another. *Note:* By extension, free space is also considered a propagation medium for electromagnetic waves, although it is not a material medium. *Synonym* **transmission medium.** *See* **anisotropic propagation medium, isotropic propagation medium.** *See also* **circuit, communications, data medium, link, loop, transmission channel.** *Refer to* **Fig. S-21 (967).**

propagation mode: 1. In radio transmission, the mode used by radio signals to travel from a transmitting antenna to a receiving antenna, such as by ground wave, sky wave, direct wave, ground reflection, or scatter. **2.** One of the possible electric and magnetic field configurations in which electromagnetic energy propagates in a waveguide, duct, or transmission line, i.e., an allowable electromagnetic field condition, distribution, or configuration that can exist in a waveguide relative to the direction of propagation of the wave in the guide. *Note:* Each mode has a factor, i.e., an eigenvalue, that defines the propagation constant but not the attenuation term in the wave equation for the discrete mode. The entire field can be described in terms of these modes. Bends and discontinuities in a waveguide or a propagation medium may lead to mode conversion, i.e., the transfer of energy from mode to mode, but not necessarily to energy that would radiate away from the waveguide. The number of modes a waveguide can support is dependent upon the dimensions of the guide, the wavelength, and the refractive indices of the propagation media. With proper choice of waveguide size, refractive index, and operating wavelength, single mode transmission can be achieved. In an open waveguide, i.e., a dielectric waveguide such as an optical fiber, evanescent fields are established in a transverse plane. These modes are guided by the gradient of the refractive index. *Synonym* **transmission mode.** *See also* **modal loss, mode, mode volume, propagation medium, transmission, wave equation.** *Refer to* **Fig. P-17.**

propagation path obstruction: A man-made or natural physical feature that lies near enough to a radio path to affect path loss, exclusive of reflection effects. *Note:* A propagation path obstruction may lie to the side, or even above the path, although usually it will lie below the path. Ridges, cliffs, buildings, and trees are examples of propagation path obstructions. If the clearance from the nearest anticipated path position, over the expected range of Earth radius k-factor, exceeds 0.6 of the first Fresnel zone radius, the feature usually is not considered as a propagation path obstruction. *See also* **effective Earth radius, Fresnel zone, k-factor, path clearance, path profile.**

propagation prediction publication: A publication that contains charts, guides, tables, and narrative information to enable stations to select optimum frequencies for radio transmissions, given relevant factors, such as distance, time of day, season, area of the Earth, and transmission power.

propagation time delay: *Synonym* **propagation delay.**

proportional spacing: The spacing occupied by printed characters according to their natural width. *Note:* The letter "m" may require three units of spacing escapement while the letter "i" may require only one. *See also* **escapement.**

proportional vector: In satellite communications systems, the part of a satellite command message that indicates the magnitude or the quantity of a function. *Note:* Examples of command vectors are (a) the number of degrees of angular movement of a satellite steerable antenna and (b) the number of thruster pulses required for a satellite orbital or orientation adjustment.

proprietary standard: A standard that (a) is prepared and documented by a private organization, (b) specifies equipment, practices, or operations unique to that organization, and (c) is copyrighted.

proration: 1. In communications systems, the proportional distribution or allocation of parameters, such as noise power and losses, among a number of tandem-connected items, such as equipment, cables, links, or trunks, in order to balance the performance of communications circuits. *Synonym* **budgeting.** **2.** In a telephone switching center, the distribution or allocation of equipment or components proportionally

access to (a) a system and (b) the data the system contains. **2.** In communications, computer, data processing, or control systems, to prevent the deliberate, inadvertent, or accidental modification of data or programs by the operator of a system or by the system itself. *Note:* In most personal computer programs, a protect feature may be applied or removed at the discretion of the operator. The protect feature may be applied to a range of cells in a spreadsheet so that their contents are not altered accidently, such as in Lotus 1-2-3 programs, the /Worksheet Global Protection Enable commands or the /Range Input commands must be given to allow the protect feature to be applied to a range of cells on a spreadsheet, i.e., a worksheet.

protected: *See* **copy protected, crash protected.**

protected distribution system (PDS): A communications system that includes (a) adequate acoustical, electrical, electromagnetic, and physical safeguards to allow its use for the transmission of unencrypted classified information and unencrypted sensitive national security information, (b) the user data terminal equipment (DTE), and (c) the interconnecting equipment. *Deprecated synonym* **approved circuit.** *Refer to* **Fig. P-18.**

protected fiber optic distribution system: A fiber optic distribution system that includes (a) adequate acoustical, electrical, electromagnetic, and physical security safeguards to permit its use for the transmission of unencrypted classified information and unencrypted sensitive national security information, (b) the user data terminal equipment (DTE), and (c) the interconnecting equipment. *See also* **approved circuit, secure line.**

protected field: A field in which data cannot be entered, changed, or deleted. *Note:* An example of a protected field is, on the display surface of a display device, a display field that cannot be changed. *See also* **display field, field, unprotected field.**

protected frequency: A frequency that is not to be deliberately jammed by friendly forces, particularly during a specified period. *See also* **guarded frequency, taboo frequency.**

protected information: Information that should not be disclosed to unauthorized persons. *Note:* Examples of protected information are security classified, personal confidential, private, and proprietary information. The protection is afforded by the steps that are taken to prevent unauthorized disclosure. *Refer to* **Fig. P-18.**

protected line: 1. In fiber optic and wire transmission systems, a line that is safe from unauthorized physical

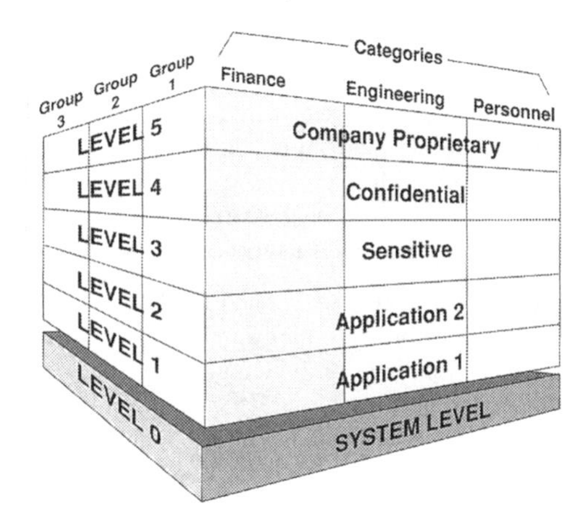

Fig. P-18. A database of **protected information,** the protection resulting from the use of operating system commands, labeling systems, rules tables, and password systems that reside at the System Level (Level 0) to provide for hacker immunity. (Courtesy Harris Computer Systems Corporation.)

access. **2.** In fiber optic and wire transmission systems, a line that cannot be damaged by normal environmental conditions. *See also* **secure line.**

protected storage: Storage whose contents cannot be changed by certain programs, particularly user programs, such as a user program to copy or modify the storage contents.

protected wireline distribution system: A wireline distribution system that includes (a) adequate acoustical, electrical, electromagnetic, and physical security safeguards to permit its use for the transmission of unencrypted classified information and unencrypted sensitive national security information, (b) the user data terminal equipment (DTE), and (c) the interconnecting equipment. *See also* **approved circuit, secure line.**

protection: 1. The prevention of unauthorized access to (a) communications, computer, data processing, and control systems and (b) the data the systems contain. **2.** In communications, computer, data processing, or control systems, the prevention of the deliberate, inadvertent, or accidental modification of data or programs by the operator of a system or by the system itself. *See* **communications protection, data-dependent protection, lock and key protection, program protection, read protection, read-write protection, write protection.** *See also* **confinement, integrity, lockout.**

control systems, the prevention of the deliberate, inadvertent, or accidental modification of data or programs by the operator of a system or by the system itself. *See* **communications protection, data-dependent protection, lock and key protection, program protection, read protection, read-write protection, write protection.** *See also* **confinement, integrity, lockout.**

protection interval (PI): 1. The period during which protection is afforded to hardware or software. **2.** In high frequency (HF) radio automatic link establishment (ALE), the period between changes in the time of day portion of the time-varying randomization data used for encrypting transmissions. *See also* **approved circuit, encrypt, encryption, protection, secure line.**

protection jamming: *See* **self-protection jamming.**

protection ratio: At a receiver input, the minimum value of the ratio of (a) the wanted signal to (b) the unwanted signal determined under specified conditions such that a specified reception quality of the wanted signal is achieved at the receiver output. *Note:* The protection ratio usually is expressed in dB.

protection subsystem: *See* **fault protection subsystem, lightning protection subsystem.**

protection system: *See* **perimeter protection system.**

protection voice equipment: *See* **limited protection voice equipment.**

protective coating: *See* **optical protective coating.**

protective covering: In fiber optics, a covering placed over a component, such as a fiber optic cable, splice, or connector, usually in the form of a wrapped tape or bonded coating designed to protect the device against the environment, such as humidity, abrasion, and bending. *Note:* Examples of protective coverings are wrapped tapes, bonded coatings, caps, dust covers, pipes, and tubing.

protective housing: *See* **laser protective housing.**

protector: In communications and control systems, a device that (a) is used to protect facilities and equipment from unusually high voltages or currents, (b) may be designed to operate on short-duration phenomena, on long-duration phenomena, or both, and (c) may contain arresters. *Note:* The duration of the protection should be specified. *See* **transient protector.** *See also* **air terminal, arrester, circuit breaker, fuse, power circuit breaker.**

protocol: 1. In communications, computer, data processing, and control systems, a set of formal conventions that govern the format and control the interactions between two communicating functional units. **2.** A set of formal semantic and syntactic rules that determines the behavior of functional units in achieving communications. *Note 1:* Protocols may govern portions of a network, types of service, or administrative procedures. *Note 2:* An example of a protocol is a data link protocol that specifies the methods whereby data communications over a data link are performed in terms of the particular transmission mode, control procedures, and recovery procedures. **3.** In layered communications systems architecture, a set of formal procedures that are adopted to facilitate functional interoperation within the layered hierarchy. *See* **address resolution protocol, Common Management Information Protocol, connection-oriented data transfer protocol, File Transfer Protocol, high level protocol, Internet Control Message Protocol, layered protocol, layer protocol, link protocol, low level protocol, Point to Point Protocol, remote operations service element protocol, Serial Line Interface Protocol, Simple Mail Transfer Protocol, Simple Network Management Protocol, transmission control protocol, Transmission Control Protocol, User Datagram Protocol, XMODEM protocol.** *See also* **communications, handshaking, link, N-entity, network, packet, packet format, Open Systems Interconnection—Protocol Specifications, Open Systems Interconnection—Reference Model.** *Refer to* **Fig. L-4 (506).**

protocol control information (PCI): 1. Information that (a) usually consists of the queries and the replies among communications equipment and (b) is used to determine the respective capabilities of equipment at each end of a communications link. **2.** For layered systems, information exchanged between entities of a given layer, via the service provided by the next lower layer, to coordinate their joint operation. *See also* **handshaking.**

protocol data unit (PDU): A data unit that (a) is transferred among peer entities of a network and (b) contains protocol information, such as control information and address information. *See* **command protocol data unit, response protocol data unit.** *See also* **Open Systems Interconnection—Reference Model.**

protocol hierarchy: In open systems architecture, the distribution of the network protocol among the various layers of the network. *See also* **open systems architecture, Open Systems Interconnection—Reference Model, protocol.** *Refer to* **Fig. P-19.**

protocol implementation conformance statement (PICS): A statement that (a) has been completed by an implementor of a standard and (b) indicates exactly which standard options the particular implementation

Protocol Layer	Functions
Application	Cost Accounting, Information Retrieval, Text Editing
Utility	Data Transfer, Terminal Support
User	Interprocess Communication
Access	Network Access Services, Security Control
End to End	Traffic Flow Control
Node to Node	Congestion Control, Routing
Link Control	Error Detection, Error Correction, Data Integrity

Fig. P-19. Protocol hierarchy showing typical functions in an open systems architecture concept.

supports. *See also* **protocol implementation conformance statement (PICS) pro forma.**

protocol implementation conformance statement (PICS) pro forma: A list of the requirements and options of a standard. *Note:* A protocol implementation conformance statement (PICS) pro forma usually is provided by the developers of the applicable standard. *See also* **protocol implementation conformance statement (PICS).**

Protocol/Internet Protocol Suite: *See* **Transmission Control Protocol/Internet Protocol Suite.**

protocol security option: *See* **revised interconnection protocol security option.**

Protocol Suite: *See* **Transmission Control Protocol/Internet Protocol Suite.**

protocol translator: A collection of hardware and software that is required or used to convert the high level protocols used in one network to those used in another network. *See also* **protocol.**

prototype: 1. A preproduction functional unit that usually is (a) the first of its type and (b) is used for the evaluation of design, performance, and production potential of the production unit. **2.** A model suitable for evaluation of design, performance, and production potential of the production unit.

provider: *See* **competitive access provider.**

provisioning: In telecommunications, the supplying of telecommunications services to a user, including all associated transmissions, cabling, and equipment. *Note:* In National Security or Emergency Preparedness (NS/EP) telecommunications services, "provisioning" and "initiation" are synonymous and include altering the state of an existing priority service or capability. *See also* **National Security or Emergency Preparedness (NS/EP) telecommunications.**

proword: *Synonym* **procedure word.**

PROWORD: procedure word.

proximity coupling: In fiber optics, the transfer of radiant energy, i.e., of optical power, from one optical fiber to another by stripping their cladding for a short distance and placing their cores close together, i.e., adjacent to each other for the stripped length. *Note:* The amount of optical power that is transferred by proximity coupling (a) can be controlled by the stripped length and the proximity of the fibers and (b) is also dependent upon the modal power distribution and the evanescent waves. *See also* **evanescent field coupling, evanescent mode, evanescent wave, modal power distribution.**

proximity fuze detonation: The detonation of a device, such as an artillery shell, a bomb, or a missile, that (a) contains a fuze that is activated by a radio signal that the device launched and that is reflected from an object, such as the ground or a target in space, (b) measures the time that elapses between the launching of the signal and the receipt of the echo from the target, (c) is calibrated in distance units from the target, and (d) is set for the desired distance. *Note:* The radio signal propagation velocity and the time introduced by the fuze train to the explosive are known and are taken into consideration in determining the setting.

PRR: pulse repetition rate.

ps: picoseconds.

PS: permanent signal.

psec: picoseconds.

pseudobinary code: A three-level signal code in which (a) a 0, or off bit, is represented by 0 volts and (b) a 1, or on bit, is represented by either a positive or a negative pulse, or vice versa, i.e., the representations for 0 and 1 may be reversed. *Note:* Pseudobinary codes include bipolar and duobinary codes. *Synonym* **pseudoternary code.**

pseudo bit error ratio (PBER): 1. A bit error ratio (BER) derived by a majority decoder that processes redundant transmissions. *Note:* In adaptive high frequency (HF) radio automatic link establishment (ALE), the pseudo bit error ratio (PBER) is determined by using the fraction of nonunanimous votes in a two-out-of-three majority decoder. **2.** In adaptive high frequency (HF) radio systems, a bit error ratio that (a) is derived by a majority decoder that processes redundant transmissions and (b) in adaptive high frequency radio automatic link establishment (ALE), is determined by the extent of error correction, such as by using the fraction of nonunanimous votes in the two-out-of-three majority decoder. *See also* **adaptive radio, automatic link establishment, bit error ratio, majority decoder.**

pseudocode: A code that (a) must be translated or interpreted before it can be used and executed in a computer program and (b) usually consists of a combination of artificial and natural language codes used for computer program design. *See also* **artificial language, computer program, interpret, natural language, translate.**

pseudoflaw: In security procedures, an apparent loophole that is deliberately planted in communications system operating procedures or computer programs as a trap for intruders who are attempting a penetration. *See also* **entrapment, penetration, security, security filter.**

pseudoinverse filtering: In image restoration, the use of inverse filtering in which (a) a nonunique inverse transformation function is used in lieu of a nonexistent exact unique transformation function, (b) images are restored to original form when the image distortion function is known, can be derived, or can be produced and used even if it is nonunique, and (c) ambiguity that may be produced can be removed by using other specific criteria for selecting the appropriate output image, i.e., criteria that were not used in the inverse transformation function. *See also* **image restoration, inverse filtering.**

pseudomonopulse tracking system: In radar systems, a signal tracking device in which a five-horn monopulse feed is used, and that produces tracking error signals that are analogous to conical scan difference patterns. *See also* **conical scanning, five-horn feed system.**

pseudorandom: Pertaining to a quality of randomness that (a) can be defined by some arithmetic process, (b) is sufficiently random and satisfactory for a given purpose, and (c) satisfies at least one of the standard statistical tests for randomness. *Note:* A signal sequence of binary digits determined by a specific algorithm is not truly random because it repeats itself after a period of time or a number of bits. It can be predicted by a receiver that is properly coded or controlled. Therefore, a generated pseudorandom bit sequence fails some test for randomness. A sequence of binary digits is truly random only when the next bit has an exactly 0.5 probability of being a 1 or a 0 for all time. Having the property of being produced by a definite calculation process and simultaneously satisfying one or more of the standard tests for statistical randomness are the basic criteria for pseudorandomness.

pseudorandom binary sequence (PRBS): A sequence of binary digits that meets at least one of the standard tests for randomness. *See also* **pseudorandom number sequence.** *Refer to* **Fig. C-2 (121).**

pseudorandom noise: Noise that satisfies one or more of the standard tests for statistical randomness. *Note 1:* Most noise is pseudorandom noise. *Note 2:* Although pseudorandom noise seems to lack any definite pattern, most noise contains a sequence of pulses that repeats itself, even after a long time or after a long sequence of pulses. *Note 3:* In spread spectrum systems, modulated carrier transmissions appear as pseudorandom noise to a receiver (a) that is not locked on the transmitter frequencies, or (b) that is incapable of correlating a locally generated pseudorandom code with the received signal. *Synonym* **pseudonoise.** *See also* **noise, white noise.**

pseudorandom number generator: 1. An analog or digital source of unpredictable, unbiased, and usually independent bits. **2.** In cryptosystems, a random bit generator used for several different functions, such as to generate keys or to start all the cryptoequipment at the same point in the key stream. *See also* **binary digit, data scrambler, descrambler, limited protection voice equipment, pseudorandom number sequence, scrambler.**

pseudorandom number sequence: A sequence of numbers that (a) appears to be random, (b) has been determined by some defined arithmetic process but is effectively a random number sequence for the purpose for which it is required, and (c) meets at least one of the standard tests for randomness, such as that the sequence does not repeat itself even after a long period or after a long sequence of numbers. *Note 1:* Although a generated pseudorandom number sequence may appear to lack any definite pattern, it will repeat itself after a long enough time interval or after a sufficiently long sequence of numbers. *Note 2:* Pseudorandom number sequences can be generated by algorithms that require a set of continuously executed arithmetic operations in which the operand for the next operation is the result of the previous set of operations, such as algorithms that generate the digits for the calculation of (a) the value of π, i.e., the ratio of the circumference of a circle and its diameter, or (b) the value of *e,* the base of the natural logarithms. *See also* **pseudorandom number generator, random number, spread spectrum.** *Refer to* **Fig. S-17 (940).**

pseudoternary code: *Synonym* **pseudobinary code.**

PSK: phase-shift keying.

PSN: public switched network.

psophometer: An instrument that (a) provides a visual indication of the aural effects of disturbing voltages of various frequencies and (b) usually incorporates a weighting network whose characteristics depend on the type of circuit under investigation, such as whether the circuit is used for high fidelity music or for normal speech. *See also* **dBm(psoph), noise, weighting network.**

psophometrically weighted dBm: *See* **dBm(psoph), dBm0p.**

psophometric electromotive force: The electromotive force (emf) that (a) permits the quantitative expression of the degree of interference that a disturbing emf from outside sources would have on a telephone conversation, and (b) is twice the psophometric voltage that would be measured across a resistance of 600 Ω (ohms) closing the circuit at the point of measurement either directly or by means of a transformer that matches the impedance of the circuit to 600 Ω, the sending end of the circuit being terminated by its matched impedance. *See also* **impedance, impedance matching, psophometric voltage.**

psophometric power: The power that is absorbed by a resistance of 600 Ω (ohms) from a source of psophometric electromotive force (emf). *Note:* The psophometric power, expressed in picowatts, i.e., 10^{-12} W (watt), is given by the square of the psophometric electromotive force (emf) expressed in millivolts, divided by 0.0024. *See also* **psophometric voltage.**

psophometric voltage: 1. Circuit noise voltage measured with a psophometer that includes an International Telephone Consultative Committee (CCIF-1951) weighting network. *Note 1:* "Psophometric voltage" should not be confused with "psophometric emf," i.e., the electromotive force (emf) in a generator or a line with 600 ohms of internal resistance. For practical purposes, the psophometric emf is twice the corresponding psophometric voltage. The situation is analogous to the fact that a source of emf, such as a battery, with constant internal resistance delivers its maximum power to a load when the load resistance is equal to the internal resistance of the source, in which case the power in the load is the same as the power consumed by the internal resistance of the source, and the total power developed by the source is twice that which is dissipated in the load. *Note 2:* Psophometric voltage readings, *V,* in millivolts, are commonly converted to dBm(psoph) by dBm(psoph) = $20 \log_{10}V - 57.78$. *Note 3:* The International Telephone Consultative Committee (CCIF) is a predecessor of the International Telegraph and Telephone Consultative Committee (CCITT). **2.** At a point in a telephone line, the voltage that (a) has a frequency of 800 Hz (hertz) and (b) if it replaced a disturbance voltage, would produce the same degree of interference with a telephone conversation as the disturbance voltage. *See also* **noise weighting, psophometric weighting.**

psophometric weighting: A noise weighting established by the International Telephone Consultative Committee (CCIF), the predecessor of the International Telegraph and Telephone Consultative Committee (CCITT), designated as CCIF-1951 weighting, for use in a noise measuring set or psophometer. *Note:* The shape of this characteristic is virtually identical to that of F1A weighting. The psophometer is, however, calibrated with a tone of 800 Hz, 0 dBm, so that the corresponding voltage across 600 Ω (ohms) produces a reading of 0.775 V (volt), i.e., $P = V^2/R$, which is evaluated as $(0.775)^2/600 = 1$ mW = 0 dBm. This introduces a 1-dBm adjustment in the formulas for conversion to dBa. *See also* **dBm(psoph), dBr, noise.**

PSTN: public switched telephone network.

psychological warfare communications: Communications methods, systems, and equipment that (a) are used for the transmission of messages that contain psychological warfare information, (b) include the use of communications media as well as the preparation and the distribution of information, and (c) are designed, by means of disseminated information, to (i) bring information to bear on specific populations,

(ii) influence their attitudes, and (iii) influence their behavior.

PT: payload type, personal telecommunications.

PTF: patch and test facility.

PTM: pulse time modulation.

PTT: printing teletypewriter telegraphy, post-telegraph-telephone.

PTT authority: post-telegraph-telephone authority.

PTTI: precise time and time interval.

p-type semiconductor: A semiconductor material, such as silicon or germanium, that has been doped with minute amounts of acceptor-type material, i.e., material with a chemical valence that creates molecular centers that (a) lack an electron to complete their energy shells and (b) attract an electron from their neighbors, leaving a hole among the neighbors. *Note:* Holes migrate when electrons, moving under the influence of an electric field, albeit a relatively weak field, fill the holes, thus constituting an electric current that conventionally is oppositely directed to the flow of electrons. The flow constitutes a positive electric current. Thus, the dopant creates holes, or positive centers, that trap electrons; hence the material is called "p-type material." The positive current constituted by the migrating holes must be added to the oppositely directed negative current of electrons in calculating the total electric current. *See also* **n-type semiconductor, semiconductor.**

publication: *See* **communications publication, propagation prediction publication.**

public correspondence: Any communication, such as a call or a message, that the offices and stations, such as central offices (COs) and radio stations, must accept for transmission, by reason of their being at the disposal of the public.

public data network (PDN): A network, established and operated by a telecommunications administration, a recognized private operating agency (RPOA), or a common carrier, for the specific purpose of providing data transmission services for the public. *See also* **common carrier, communications, data transmission, public data transmission service, public switched network, public switched telephone network.**

public data transmission service (PDTS): A data transmission service that (a) is established and operated by a telecommunications administration or by a recognized private operating agency and (b) uses a public data network. *Note:* The public data transmission service may include circuit-switched, packet-switched, and leased-circuit data transmission networks. *See also* **data transmission, public data network.**

public key encryption system (PKES): A cryptosystem in which (a) a pair of keys, known as the public key and the private key, can be assigned to, or generated by, a user, thereby permitting the user to send encrypted messages to, or receive messages from, other users without divulging the user secret private key and (b) the public key can be openly publicized without compromising the security of the system. *Note:* Uses of the public key encryption system include (a) to permit a user to encrypt a message, using the private key such that anyone can decrypt the ciphertext message, using a user public key, and thereby know that only the designated user could have generated the message, (b) to permit anyone to encrypt a message, using a user public key, such that only the designated user can decrypt the message using the corresponding private key, (c) to permit combined use by two participating users, A and B, such that only A can encrypt a message that can be decrypted only by B, and (d) to permit authentication and electronic signature verification. *See also* **communications security, Data Encryption Standard, limited protection, private key encryption.**

public relations communications: Communications methods, systems, and equipment that are used for the transmission of messages that are addressed to the public and that contain information about an organization and its operations deemed to be in the interest of the organization and of interest to the public.

public switched National Security or Emergency Preparedness (NS/EP) telecommunications service: A National Security and Emergency Preparedness (NS/EP) telecommunications service that (a) uses public switched networks and (b) may include both interexchange and intraexchange network facilities, such as switching systems, interoffice trunks, and customer loops. *See also* **private National Security or Emergency Preparedness telecommunications services.**

public switched network (PSN): A common carrier network that provides circuit switching for the general public. *Note:* Public switched network (PSNs) usually are telephone networks, but they could also be other types of networks, such as data networks and packet-switched networks. *See also* **circuit, communications, packet switching, public data network, public switched telephone network.**

public switched telephone network (PSTN): A domestic communications network (a) that usually is accessed by telephones, key telephone systems, private branch exchange (PBX) trunks, and data equipment, such as modems, and (b) in which completion of the circuit between the call originator and the call receiver requires network signaling in the form of dial pulses or in the form of multifrequency tones, such as dual tone multifrequency (DTMF) signals. *See also* **circuit, communications, public data network, public switched network, telephony.**

public utility commission (PUC): In the United States, a state regulatory body charged with regulating intrastate utilities, including telecommunications systems. *Note:* In some states this function is performed by public service commissions or state corporation commissions.

PUC: public utility commission.

puck: *See* **hockey puck.**

pull-in frequency range: The maximum frequency difference that can occur between the local clock and the reference clock in order that the phase-locked loop can lock the local clock to the reference clock.

pulling machine: *See* **optical fiber pulling machine.**

pulsating direct current: An electric current level that is changing at regular or irregular intervals but is always in the same direction, i.e., is not altering its polarity. *See also* **level.**

pulse: 1. A waveform that consists of (a) a short period of rapid transition from one significant condition to another, i.e., a short rise time, (b) a significant condition period that usually is longer than the transition period, and (c) thereafter a short period of rapid transition to the original significant condition, i.e., a short fall time. **2.** A temporal or spatial variation, usually characterized as a rise and a decay from a baseline or a reference, of the magnitude, such as amplitude, phase, frequency, or other parameter of a physical quantity, such as the electric field strength of an electromagnetic wave, the variation being short relative to the time or space schedule of interest. **3.** In the magnitude of a quantity, a transition to a different level, followed by a return to the original level, that occurs rapidly in relation to the time span of interest. *Note 1:* Examples of pulses are (a) a rapid increase followed by a rapid decrease in irradiance, i.e., in optical power or electric field strength, (b) a sudden shift and return in phase, (c) a sudden change and return in frequency, and (d) a change in polarization and a return to the original state that existed before the change occurred. The pulse magnitude is the extent to which a parameter deviates from a reference. *Note 2:* If the magnitude rapidly changes only from one level to another, the change is called a "step change" or a "step function" rather than a pulse. *Note 3:* A pulse may have a spatial width measured in microns, a temporal width measured in nanoseconds or picoseconds, a bandwidth measured in hertz, or a phase shift measured in radians. Workers in the field tend to use "pulse length" for spatial dimensions, "pulse duration" for temporal dimensions, "spectral width" for the wavelength content of a pulse, and "bandwidth" for the frequency composition of a pulse. *Note 4:* Pulses may be used to modulate a carrier or may be modulated by a baseband signal to represent digital data or discrete samples of analog signals. Pulses are usually coded or modulated and transmitted in streams at high data signaling rates (DSRs) or pulse repetition rates (PRRs) to represent and transport information. *See* **Gaussian pulse, optical pulse, radar pulse.** *Refer to* **Figs. M-3 (590), P-20.**

pulse address multiple access (PAMA): To access, by means of a communications satellite, several Earth terminals simultaneously, i.e., to receive signals from the Earth terminals and to amplify, translate, and relay these signals back to other Earth terminals, based on the address of each Earth terminal. *Note 1:* Each Earth terminal address is a unique combination of time and frequency slots. *Note 2:* Pulse address multiple access (PAMA) may be restricted by allowing only some of the Earth terminals access to the satellite at any given time. *See also* **dual access, multiple access.**

pulse amplitude: A measure of the extent to which a physical quantity, such as optical power, electric current, or voltage, used to represent a pulse, changes from a zero or other baseline value for the pulse duration, i.e., the magnitude of a pulse parameter, such as the electric field strength, voltage level, current level, or power level. *Note 1:* Pulse amplitude is measured with respect to a specified reference and therefore should be modified by qualifiers, such as "average," "instantaneous," "peak," and "root mean square," to indicate the particular meaning intended. *Note 2:* Pulse amplitude also applies to the amplitude of frequency-modulated (FM) and phase-modulated (pm) waveform envelopes but not to the frequency or phase shift. *Note 3:* A shift in frequency or phase is considered as a change in magnitude and not a change in amplitude. *Note 4:* Examples of pulse amplitudes are (a) the voltage of a radar pulse on a transmission line, (b) the momentary change in optical power output of a laser to represent a single pulse, and (c) the amount of the change in photocurrent of a photodetector from the dark current value to the light current value. *Synonym* **pulse height.** *See also* **pulse, pulse magnitude.**

pulse amplitude modulation (PAM): Modulation in which the amplitude of a pulsed carrier is varied in accordance with some characteristic of the modulating

Fig. P-20. At Bell Laboratories, a lathe heats a glass rod treated with the rare-earth element erbium in the first step toward making optical amplifiers for boosting light signals, such as light **pulses**, for long-distance telecommunications networks. The rod will be stretched until the diameter is about that of human hair. Courtesy of AT&T Archives.

band AM with an output equal to the average power in the pulse transmission. However, PAM is readily adaptable to time division multiplexing (TDM). *Synonym* **pulse height modulation.** *See also* **carrier, modulation, pulse, pulse frequency modulation.**

pulse broadening: 1. *Synonym* **pulse dispersion. 2.** *See* **root mean square pulse broadening.**

pulse carrier: A carrier wave that (a) consists of a series of pulses, usually of constant pulse duration, constant spacing, constant repetition rate, and, when not modulated, constant amplitude, and (b) usually is used as a subcarrier. *See also* **carrier, carrier wave.**

pulse code: In data representation, the table or the list of equivalences between the quantized value of a sample and the corresponding signal, pulse pattern, or pulse train that represents the corresponding character, numeral, or word.

pulse code modulation (PCM): 1. Modulation in which an analog signal is converted to a digital signal by means of coding. **2.** Modulation in which the pulse amplitude, duration, or position has a definite code with a defined meaning. **3.** Signal processing in which (a) the signal is sampled, (b) the magnitude of each sample with respect to a fixed reference is quantized, and (c) the quantized value is converted by coding to a digital signal. *Note:* Each element of information consists of different kinds or numbers of pulses and spaces. Pulse code modulation (PCM) requires a large bandwidth because of the large number of pulses that must be transmitted. Light pulses have the necessary bandwidths at the frequencies required. PCM as a data quantization and formatting technique is misnamed as

and (c) the quantized value is converted by coding to a digital signal. *Note:* Each element of information consists of different kinds or numbers of pulses and spaces. Pulse code modulation (PCM) requires a large bandwidth because of the large number of pulses that must be transmitted. Light pulses have the necessary bandwidths at the frequencies required. PCM as a data quantization and formatting technique is misnamed as such. The modulating function, i.e., the baseband signal, is sampled in terms of amplitude, and the observed discrete values are represented by a coded arrangement of pulses, often a binary representation of the discrete value that is observed. The presence or the absence of pulses, not their shape, amplitude, position, or spacing, determines the intelligence that is conveyed. The pulse train usually is transmitted to a distant point by using another time modulation or keying technique on the data link. PCM is often used in satellite and industrial telemetry, and in large-capacity, short-distance telephone circuits. The number of possible signal amplitudes that can be represented depends upon the number of pulse positions in which the sample amplitude lies at the moment of sampling. At the receiver, each group is converted into a signal of the amplitude depicted by the code. The succession of signals thus created is passed through a smoothing network that reproduces an approximation of the original signal. In the reception of pulse signals, it is necessary to determine only whether or not a pulse is present in each of the time positions in a group of pulses. This can be done faithfully unless the noise peaks are strong enough to introduce a pulse where none should be, or to obliterate a pulse that should be there. To offset this in a long link, the pulses can be regenerated by repeaters along the link at intermediate points. PCM is subject to a unique type of noise that is caused by small errors that are inherent in transmitting continuously varying signals as a series of discrete amplitude steps. This is known as "granular" or "quantizing" noise. To reduce this noise, additional amplitude steps can be used. However, additional steps require additional pulses in the pulse groups, and this increases the required bandwidth. Another way of reducing the granular noise is to compress the original signal before it is sampled and quantized, and then expand it again at the receiver. This combination of compressing and expanding is called "companding." The pulses for several different signals can be timed to permit time division multiplexing. The fidelity of reproduction depends upon the number of pulses in the sampling group. The more amplitude samples per pulse interval, the more closely the reconstructed signal will resemble the original signal that was sampled. PCM pulses are not affected by circuit noise, unless the noise is so strong that its peaks are greater than one-half of the amplitude of the signal pulses. This is so because one-half of the pulse amplitude usually is chosen as the threshold for deciding whether there is a pulse at a sample point, or there is no pulse at the sample point. If the sample is less than half, it is considered that there is no pulse. If it is greater than half, it is considered that there is a pulse. PCM enables transmission over digital links. *Synonym* **pulse coding.** *See* **adaptive differential pulse code modulation, differential pulse code modulation, subband adaptive differential pulse code modulation.** *See also* **balanced code, code conversion, frame, load capacity, pulse code modulated noise, modulation, multiframe, pulse code modulation multiplex equipment, peak limiting, signal sample.**

pulse code modulation (PCM) multiplexing: Multiplexing by deriving a single digital signal at a defined data signaling rate (DSR) from two or more analog channels by a combination of pulse code modulation and time division multiplexing. *Note 1:* The process is reversed for carrying out the inverse function, i.e., for demultiplexing. *Note 2:* In expressing pulse code modulation (PCM) multiplexing values, the relevant equivalent binary digit rate should be stated, such as 2048 kbps (kilobits per second) PCM multiplexing. *See also* **data signaling rate, pulse code modulation, time division multiplexing.**

pulse code modulation noise: *See* **equivalent pulse code modulation noise.**

pulse code multiplexing (PCM) equipment: Equipment for digitizing two or more analog signals and combining the resultant bit streams into a single aggregate signal by a combination of pulse code modulation and time division multiplexing. *See also* **channel, code, multiplexing, pulse code modulation.**

pulse coding: *Synonym* **pulse code modulation.**

pulse compression: A reduction of the pulse duration by using impedance matching filtering to discriminate against signals that do not correspond to the transmitted signal. *See* **optical pulse compression, radar pulse compression.** *See also* **pulse stretching.**

pulse compression radar: Radar in which transmitted signals and echoes are passed through an impedance matched narrow bandpass filter so as to reduce the pulse duration. *See also* **bandpass filter.**

pulse decay time: *Synonym* **fall time.**

pulsed frequency-modulated signal: In spread spectrum communications systems, a signal in which (a) a pulsed carrier is swept in frequency for the duration of the pulse, and (b) on reception of the pulsed frequency-modulated signal, a matched filter with a passband that varies in the same manner as the frequency of the transmitted pulse carrier is used to detect the signal. *Synonym* **chirp.** *See also* **pulse carrier, spread spectrum.**

pulse dispersion: 1. An increase in pulse duration that occurs during propagation. *Note 1:* Pulse dispersion may be specified by the impulse response, the root mean square pulse broadening, or the full-duration half-maximum pulse broadening. *Note 2:* The three basic mechanisms that cause pulse dispersion are material dispersion, waveguide dispersion, and multimode dispersion. Specific causes include surface roughness, presence of scattering centers, bends in guiding structures, deformation of guides, inhomogeneities in the propagation medium, spectral width of the pulse, multipath, and the variation of refractive index with wavelength **2.** Increasing the pulse width (spatial) or the pulse duration (temporal) of a pulse as a result of dispersion. *Note:* The distortion is caused by the spreading in time and space of an electromagnetic pulse, such as a telegraph, radio, radar, or optical pulse, propagating along a propagation medium or path, such as the atmosphere, an electrical transmission line, or an optical fiber. The pulse broadening, i.e., the dispersion, is caused by the variation of the propagation constant for each wavelength in the signal pulse, i.e., the different wavelengths that compose the pulse propagate at different speeds. This phenomenon limits the useful transmission bandwidth of the path, or for a given bandwidth, limits the data signaling rate (DSR), i.e., limits the pulse repetition rate that may be transmitted because of the intersymbol interference that occurs when the pulse duration at the receiving end increases to a point where the pulses begin to overlap each other upon arrival. For an optical fiber, the amount of spreading depends on the spectral width of the light, the refractive index profile of the fiber, the length of the fiber, and the duration of the launched pulse. Because of the dependence on length, a significant measure of optical fiber performance is the bandwidth•length product, i.e., the bit-rate•length product. Pulse dispersion may be specified by the impulse response or by the full-duration half-maximum pulse broadening. *Synonyms* **pulse broadening, pulse spreading.** *See also* **full-width half-maximum, impulse response, root mean square pulse broadening.**

pulse dispersion specification: The specification or the measurement of the pulse dispersion, i.e., of the pulse spreading, caused by a propagation medium, such as the atmosphere or an optical fiber. *Note:* The 50% dispersion time factor is given by the relation $\tau_{50} = (D_1 - D_2)^{1/2}/(L_1 - L_2)$ where τ_{50} is the 50% dispersion time factor, D_1 is the pulse duration at 50% of the maximum pulse amplitude of a test specimen, such as a length of optical fiber selected in accordance with a specified sampling scheme, D_2 is the pulse duration at 50% of the maximum pulse amplitude of a reference output wave form, L_1 is the length of the test specimen, and L_2 is the length of the reference. *Synonym* **pulse spreading specification.** *See also* **dispersion, pulse duration.**

pulse distortion: The difference between the shape of a transmitted pulse and the shape of that same pulse when it is received. *See* **optical pulse distortion.** *See also* **pulse shape.**

pulsed laser: *See* **Q switched repetitively pulsed laser.**

pulse Doppler radar: Pulsed radar in which the Doppler effect is used in order to obtain a direct measurement of the radial component of velocity of an object, i.e., a target, relative to the radar antenna. *See also* **Doppler effect.**

pulse duration: 1. The time between (a) the instant that a pulse starts and (b) the instant that the pulse ends. *Note:* The pulse usually is used to represent a binary digit. **2.** In amplitude modulation, the time interval between the points on the leading and trailing edges of a pulse at which the instantaneous value bears a specified relation to the peak pulse amplitude, such as the time between the full-wave half-power points in rise and fall or between the 10% values in rise and fall. **3.** In frequency or phase shift modulation, the time interval during which the frequency or the phase remains changed to represent a binary digit. **4.** In a pulse waveform, the time between a specified reference point on the first transition and a similarly specified point on the last transition. *Note:* The time between 10%, 50%, or $1/e$ of the peak pulse amplitude points is usually used, as in the root mean square (rms) pulse duration. The e is the base of the natural logarithms, approximately equal to 2.718. **5.** In radar, the duration of a pulse transmission, usually expressed in microseconds, i.e., the time the radar transmitter is energized during each cycle. *Note:* Because pulses have a duration, i.e., they have a temporal width, and they are propagating at a specific speed in a propagation medium, such as free space, the atmosphere, coaxial cables, or optical fibers, they also have a spatial width, i.e., they occupy a space in the propagation medium. However, the spatial width is rarely considered because primary interest is directed to events that occur at a point, which has no dimension, and therefore temporal matters are of prime concern, and duration becomes significant. *Deprecated synonyms* **pulse length, pulse width.** *See* **laser pulse duration, root mean square pulse duration.** *See also* **distortion, frequency, propagation medium, pulse, pulse fall time.**

pulse duration discrimination: The ability to distinguish variations in the duration of pulses or the spatial width of pulses. *Deprecated synonym* **pulse length discrimination.** *See also* **discriminator, pulse duration.**

pulse duration discriminator: A device that (a) measures the pulse duration of signals and passes only those pulses that have a duration that falls within a predetermined time interval specified by design tolerances, (b) offers good protection against long-pulse jamming, most types of low frequency modulation, and some forms of interference, and (c) affords little or no protection against short pulses and high frequency modulation. *See also* **pulse duration, jamming, modulation.**

pulse duration distortion: 1. In an optical fiber, the difference between (a) the pulse duration of an optical input pulse and (b) the duration of the corresponding optical output pulse at the distal end of the fiber. **2.** In a fiber optic transmitter, the difference between (a) the duration of an optical input pulse and (b) the duration of the corresponding electrical output pulse. **3.** In an optical receiver, the difference between (a) the pulse duration of an optical input pulse and (b) the duration of the corresponding optical output pulse. **4.** In an optical link, the difference between (a) the duration of an electrical pulse at the input of the transmitter and (b) the duration of the electrical pulse at the output of the receiver. *Deprecated synonym* **pulse length distortion.**

pulse duration modulation (PDM): Modulation in which the duration of a carrier pulse is varied in accordance with some characteristic, i.e., some attribute, of the modulating signal. *Note:* The modulating signal may vary the time of occurrence of the leading edge or the trailing edge of the carrier pulse, or both. In modulation, 0s and 1s may be represented by a carrier wave that is on or off for longer or shorter spans of time. However, in pulse duration modulation, a pulse is transmitted for each sampling of the baseband signal, i.e., the modulating signal. Pulse duration modulation yields the same signal to noise improvement as pulse position modulation. However, the average power that is required is greater in pulse duration modulation (PDM) than in pulse position modulation (PPM) because of the longer average duration of the pulses. PDM is rarely used for direct modulation of a radio carrier, but it has been applied in the intermediate processes of some PPM receiving devices. In these devices, the transformation usually is made by having the position of the short pulse control the starting time of the two longer pulses with a fixed terminating time. *Synonym* **pulse telegraph modulation.** *Deprecated synonyms* **pulse length modulation, pulse width modulation.** *See* **suppressed clock pulse duration modulation.** *See also* **modulation, pulse, pulse position modulation, pulse time modulation.**

pulse duty factor (PDF): In a pulse train, the ratio of the pulse duration to the pulse period. *Note:* The pulse duty factor is dimensionless. The pulse period is the time between corresponding points on consecutive pulses, such as the time between their leading edges. *See also* **duty cycle, pulse, pulse duration, pulse train.**

pulse edge tracking: Tracking that (a) is performed on the edge of a radar echo rather than on the peak or the center of the echo, (b) primarily is used to prevent capture of tracking circuits by fixed or slow-moving objects, such as towers, mountains, or chaff, (c) keeps automatic range tracking to a minimum range, allowing closer approaches to objects, i.e., to targets, for visual identification purposes, and (d) has the effectiveness of the tracking circuits improved by operator selection of the edge of the signal pulse for tracking control. *See also* **radar echo, track, tracking, tracking radar.**

pulse expanding: *Synonym* **pulse stretching.**

pulse fall time: The time that is required for a pulse amplitude instantaneous value to decrease from one specified fraction of its peak value, say 90% of its peak value, to another fraction of its peak value, say 10% of its peak value. *Synonym* **pulse decay time.** *See also* **fall time, pulse duration, pulse rise time.**

pulse frequency modulation (PFM): Modulation in which (a) the pulse repetition rate (PRR) of a pulsed carrier is varied in accordance with some characteristic of the modulating signal, and (b) bandwidth expansion occurs for digital data in which the carrier frequency is made to vary linearly with the time across the channel bandwidth during each data period, i.e., during the period that each bit of information is being transmitted. *Note 1:* In one form of pulse frequency modulation, binary states represented by 0 and 1 correspond to decreasing and increasing pulse repetition rates, respectively, or vice versa. The transmission method is that of frequency modulation (FM). The modulating waveform in the time domain is a succession of sawtooth waves, i.e., waves that have a slow linear rise time and a rapid fall time, or vice versa, with positive or negative slopes that correspond to the binary data. *Note 2:* Pulse frequency modulation has advantages for air-ground digital communications and geostationary satellite communications with marginal power budgets because it combats multipath reception and Doppler shifts. *Note 3:* In frequency code modulation, discrete frequencies are used. *Note 4:* In a certain sense, "pulse frequency modulation" is a misnomer because it is the pulse repetition rate (PRR) of a pulse carrier that is being modulated, which results in a change in the frequency spectrum of the pulses. *Synonym* **pulse repetition rate modulation.** *See also* **frequency code modulation, modulation, pulse carrier, pulse repetition rate.**

pulse half-duration: In a pulse, half of the full duration of the pulse, the full duration being the time between specified points on the pulse waveform, such as between 90%, half, or $1/e$ of maximum value points. *Note:* The e is the base of the natural logarithms, approximately equal to 2.718. *Deprecated synonyms* **pulse half-length, pulse half-width.**

pulse half-length: *Deprecated synonym for* **pulse half-duration.**

pulse half-width: *Deprecated synonym for* **pulse half-duration.**

pulse height: *Synonym* **pulse amplitude.**

pulse height modulation: *Synonym* **pulse amplitude modulation.**

pulse interference suppression and blanking (PISAB): In radar systems, removing or discriminating against interference by suppressing or blanking all video signals that are not synchronous with the radar pulse repetition rate (PRR), thereby making the radar less susceptible to random pulse interference and certain types of jamming. *See also* **blanking, discriminator, interference, jamming, pulse repetition rate.**

pulse interval modulation: Modified pulse position modulation in which the time between carrier pulses is changed in accordance with an intelligence-bearing variation in the modulating signal. *See also* **pulse carrier, pulse position modulation.**

pulse jet control: The control of thrusts that are applied to a satellite for orbital control, spin control, or orientation control, such that if the thrust of each pulse produces a fixed change in momentum, various values of momentum changes can be produced by changing the pulse jet repetition rate. *Note:* Thrusts may be applied to a spinning satellite once per revolution, or a continuously applied thrust may have its amplitude changed in abrupt bursts of varying length.

pulse jitter: The jitter of a pulse parameter, such as the jitter of the pulse duration, pulse time or space position, or pulse amplitude. *See also* **jitter.**

pulse length: *Deprecated synonym for* **pulse duration.**

pulse length discrimination: *Deprecated synonym for* **pulse duration discrimination.**

pulse length distortion: *Deprecated synonym for* **pulse duration distortion.**

pulse length modulation: *Deprecated synonym for* **pulse duration modulation.**

pulse link repeater: In E & M signaling, a device that (a) interfaces the signal paths of concatenated trunk circuits, (b) responds to a ground on the "E" lead of one trunk by applying −48 V (volt) direct current (dc) to the "M" lead of the connecting trunk, and vice versa, and (c) is a built-in, switch-selectable option in some commercially available carrier channel units. *See also* **circuit, E & M signaling, link.**

pulse magnitude: A measure of the extent to which the physical parameter, quantity, or phenomenon used to represent a pulse changes from a baseline or a reference for a short time, usually to represent digital data. *Note 1:* Examples of pulse magnitudes are (a) the amount of change in optical power output of a laser to represent a bit, (b) the extent of the phase shift of a monochromatic wave in phase-shift keying, and (c) the extent of a frequency change in a frequency-modulated system. *Note 2:* Often it is necessary to use modifiers, such as average, instantaneous, peak, and root mean square, to indicate the significance of the units or measure used to define the pulse magnitude. *See also* **pulse amplitude.**

pulse modulation: In the representation of digital data, modulation (a) in which the temporal characteristics or parameters, such as pulse duration, pulse position, or pulse magnitude, e.g., pulse amplitude, phase shift, or frequency change, of a carrier wave are changed in accordance with an intelligence-bearing signal, (b) that is used in radar, radio, telegraph, and facsimile systems to represent digital data, and (c) in fiber optic systems, that can be accomplished by (i) applying an intelligence-bearing electrical modulating signal directly to a light source or (ii) modulating the constant lightwave carrier output of the source after emission, such as by launching the lightwave carrier into an optical fiber and subjecting the fiber to a continuously varying or pulsating physical variable, e.g., pressure, sound, temperature, or interferometric variations, and thereby producing a pulsed carrier. *See also* **pulse magnitude, pulse repetition rate.**

pulse modulation radar jamming: Radar jamming that (a) causes a confusing pattern on a radar display device, (b) produces a pattern that may vary from a number of apparent echoes, often lying in radial lines, to continuous spiral lines that are similar to running rabbits, and (c) can have reduced effects through the use of circuits with variable time constants. *See also* **jamming.**

pulse operation: *See* **coherent pulse operation.**

pulse period: 1. The time between corresponding points on consecutive pulses, such as the time between their leading edges. **2.** The reciprocal of the pulse repetition rate (PRR). *See also* **pulse repetition rate.**

pulse position modulation (PPM): 1. Modulation in which the temporal positions of the pulses of a pulse carrier are varied with respect to a reference in accordance with some characteristic of the modulating signal without modifying the pulse duration. *Note:* In pulse position modulation (PPM), usually very narrow pulses deviate from uniformly spaced reference positions by time intervals that are proportional to the instantaneous amplitudes of a sampled input, usually an analog input. In PPM, one pulse is sent for each sampling of the input signal. The pulses all have the same pulse amplitude and duration, but the exact time, within a range, that each pulse is sent, i.e., its position with respect to a fixed time reference, depends on the magnitude of the input signal. When PPM is sent by amplitude modulation (AM) techniques, it has a higher signal to noise ratio than normal double sideband AM. This improvement, in terms of the current ratio, is proportional to the pulse shift in time and to the bandwidth that is used. This requires a greater bandwidth, however, than double sideband AM. The increase in bandwidth is proportional to the ratio of the maximum interval between pulses to the pulse duration. **2.** In a fiber optic transmission system, modulation that causes the arrival time of optical pulses at a detector to vary according to a signal impressed on a pulsed source, the detector output being a function of the arrival time with respect to a fixed time reference. *See also* **modulation, pulse, pulse carrier, pulse magnitude, pulse time modulation.**

pulse radar: Radar in which use is made of pulses, i.e., short bursts of radio frequency energy. *Note:* An example of pulse radar operation is to transmit pulses that (a) have a duration of 1 μs (microsecond), (b) are 5 μs apart, i.e., there is a 4-μs interval between pulses, with a pulse repetition rate (PRR) of 200 kpps (kilopulses per second), (c) have a radio frequency (rf) within each pulse of 10 MHz (megahertz), and (d) therefore have 10 cycles of rf energy in each pulse. *See also* **radar, radio frequency.**

pulse radar jammer: *See* **locked pulse radar jammer.**

pulse rate: 1. *Synonym* **pulse repetition rate. 2.** *See* **maximum pulse rate.**

pulse repetition frequency (PRF): *Deprecated synonym for* **pulse repetition rate.**

pulse repetition rate (PRR): The number of pulses that occur per unit time at a particular point in a propagation medium. *Note 1:* Pulse repetition rate (PRR) usually is expressed in pulses per second, such as 2 kpps (kilopulses per second). *Note 2:* In pulse-modulated systems, such as radar, pulse repetition rate (PRR) is not to be confused with the carrier frequency, i.e., the transmission frequency, which is the frequency of the radio frequency (rf) carrier wave, usually expressed in hertz, that is modulated by the pulses. The carrier frequency is always higher than the PRR. The modulating pulses usually modulate the carrier by turning it on and off at the PRR. *Note 3:* In most pulsed communications systems, the desired bit error ratios (BERs), dispersion, attenuation, available power, detector sensitivity, noise levels, and other factors limit the pulse repetition rate that a given transmission system can handle. *Synonym* **pulse rate.** *Deprecated synonym* **pulse repetition frequency.** *See* **Doppler pulse repetition rate, jittered pulse repetition rate, staggered pulse repetition rate.** *See also* **frequency, pulse, pulse radar.**

pulse repetition rate modulation: *Synonym* **pulse frequency modulation.**

pulse rise time: The time that is required for a pulse amplitude instantaneous value to rise from one specified fraction of its peak value, say 10% of its peak value, to another fraction of its peak value, say 90% of its peak value. *See also* **pulse fall time, rise time.**

pulse shape: A parameter or a characteristic of a pulse, such as its amplitude, duration, position, phase, or frequency, that (a) usually is expressed as a function of time, (b) usually is a voltage, current, power, or field strength as a function of time, and (c) may be displayed on the screen of a cathode ray tube (CRT) as the ordinate value, i.e., *y*-axis value, using a linear time scale on the abscissa, i.e., *x*-axis. *See also* **pulse.**

pulse shaping: *Synonym* **pulse regeneration.**

pulse spreading: *Synonym* **pulse dispersion.**

pulse spreading specification: *Synonym* **pulse dispersion specification.**

pulse stretching: Increasing the duration of a pulse. *Synonym* **pulse expanding.** *See* **optical pulse stretching.** *See also* **pulse compression, pulse duration, radar pulse compression.**

pulse string: *Synonym* **pulse train.**

pulse stuffing: The insertion of pulses into a stream of pulses to achieve a specific purpose, such as achieving synchronism between two digital communications systems, filling a buffer, or completing a fixed-length frame. *Synonym* **positive pulse stuffing.** *See also* **bit stuffing.**

pulse suppressor: *See* **optical pulse suppressor.**

pulse test: *See* **shuttle pulse test.**

pulse time modulation (PTM): Modulation in which the time of occurrence of some characteristic of the pulsed carrier is varied with respect to some characteristic of the modulating signal. *Note:* Pulse time modulation (PTM) includes pulse position modulation and pulse duration modulation (PDM). *See also* **modulation, pulse, pulse duration modulation, pulse position modulation.**

pulse train: In communications, computer, data processing, and control systems, a series of pulses that have similar characteristics. *Synonym* **pulse string.** *See also* **character string, pulse.**

pulse width: *Deprecated synonym for* **pulse duration.**

pulse width distortion: *Deprecated synonym for* **pulse duration distortion.**

pulse width modulation (PWM): *Deprecated synonym for* **pulse duration modulation.**

pulsing: In telephone systems, the transmission of address information to a central office (C.O.) or to a switching center by means of digital pulses. *Note:* Pulsing methods include multifrequency, rotary, and revertive pulsing. *See* **dial pulsing, key pulsing, multifrequency pulsing, revertive pulsing, wink pulsing.** *See also* **dial pulse, dial signaling, multifrequency signaling, pulse, pulse address multiple access, rotary dial.**

pump frequency: The frequency of an oscillator used to provide sustaining power to devices, such as lasers and parametric amplifiers, that require radio frequency (rf) sinusoidal or pulsed power. *Note:* Pumps usually are designed to provide resonant frequencies to raise power levels in the devices they pump energy into. *See also* **laser, population inversion, power, pumping.**

pumping: In electric circuits, the action of an oscillator that provides cyclic changes to control a reaction device. *Note 1:* Examples of pumping are (a) the action that provides the sustaining power that results in amplification of a signal by a parametric amplifier, and (b) the action that provides a laser with input power to sustain the optical power output. *Note 2:* Usually there must be some form of synchronization or relationship between the pump frequency and the frequency of the item being pumped for the pumping to be effective.

punch: *See* **card punch.**

punch card: A card in which one or more patterns of holes can be punched by a card punch to represent information, the holes being the data that represent the information stored on the card when the card is punched. *See also* **card punch, Hollerith card, punched card.**

punched card: A card that has one or more patterns of holes that have been punched by a card punch to represent information, the holes being the data that represent the information stored on the card. *See also* **card punch, Hollerith card, punch card.**

punched tape: Tape that has one or more patterns of holes that have been punched by a perforator to represent information, the holes being the data that represent the information stored on the tape.

punch tape: Tape that (a) has sprocket holes for feeding, (b) is used in a perforator, and (c) can be punched with hole patterns to represent characters. *Synonym* **perforator tape.** *See also* **perforator, punched tape.**

pupil: *See* **artificial pupil, entrance pupil, exit pupil.**

pure binary numeration system: A fixed-radix positional numeration system for representing numbers characterized by the arrangements of binary digits in sequence, with the understanding that consecutive digits are to be interpreted as coefficients of descending powers of base 2 as read from left to right, and the presence of a 1 at a digit position indicating that the weight of the digit position is to be added to obtain the total value of a number, and a 0 at a digit position indicating that the weight of the digit position is not to be added to obtain the total value of the number. *Note 1:* The pure binary numeral 1101.01 is equivalent to the total value of $1 \times 2^3 + 1 \times 2^2 + 0 \times 2^1 + 1 \times 2^0 + 0 \times 2^{-1} + 1 \times 2^{-2}$, which equals 8 + 4 + 0 + 1 + 0 + 0.25, which sums to decimal 13.25, the decimal value of the binary number. *Note 2:* The number of days in a non-leap year is represented in the pure binary numeration system as 101101101, i.e., 365 in the decimal numeration system. *Synonym* **binary notation.** *See also* **binary coded decimal, binary coded decimal (BCD) notation, binary digit, binary notation, code, Gray code, numeration system.** *Refer to* **Fig. G-3 (402).**

purity: *See* **spectral purity.**

pushbutton: *See* **virtual pushbutton.**

pushbutton dialing: Dialing that (a) uses pushbuttons or keys to actuate and connect audible tone oscillators to the line, each oscillator emitting a unique frequency that is used to create a signal to represent the digit or the symbol corresponding to the button, (b) is accomplished by selecting and transmitting digits, each represented by a different frequency or tone or group of frequencies or tones, (c) is performed by frequency division multiplexing rather than by the pulse counts produced by a rotary dial, (d) is an alternative to and is replacing rotary dialing on telephone handsets, (e) provides faster circuit selection, reduced telephone exchange time, and reduced line holding time compared to rotary dialing, and (f) usually uses tone or direct current (dc) signaling. *Synonym* **pushbutton signaling.** *See* **dial, dialing, dual tone multifrequency signaling, key.**

pushbutton dialing pad: A keyboard, usually a numeric keypad on a pushbutton telephone, each button, i.e., each key, of which is used to actuate an oscillator or oscillators to generate a tone that has a frequency composition corresponding to the digits 0 through 9 and two or three other symbols for special purposes. *See also* **dialing, pushbutton dialing, numeric keypad.**

pushbutton signaling: *Synonym* **pushbutton dialing.**

pushdown: *Synonym* **last in, first out.**

pushdown storage: Storage in which the most recently stored item is the next item to be retrieved. *Synonyms* **cellar, nest, running accumulator.** *See also* **last in, first out; stack.**

push to talk operation: In telephone or two-way radio systems, communications over a speech circuit in which the talker is required to keep a switch operated while talking. *Note:* In radio push to talk operation, the same frequency is used by both transmitters. *Synonym* **press to talk operation.** *See also* **circuit, conversational mode, half-duplex circuit.**

push to type operation: In telegraph or data transmission systems, communications in which the operator at a station must keep a switch operated in order to send messages. *Note 1:* In radio push to type operation, the same frequency is used by both transmitters. *Note 2:* Push to type operation is a derivative form of transmission. *Note 3:* Push to type operation may be simplex, half-duplex, or duplex operation. *Synonym* **press to type operation.** *See also* **half-duplex circuit.**

pushup: *Synonym* **first in, first out.**

pushup storage: Storage in which the item stored the longest is the next item to be retrieved, as occurs in a theater box-office queue. *See also* **first in, first out.**

PVC: permanent virtual circuit.

pW: picowatt.

PWM: *Abbreviation for the deprecated synonym* **pulse width modulation.** *See* **pulse duration modulation.**

pWp: picowatt, psophometrically weighted.

pWp0: picowatt, psophometrically weighted, referred to a zero transmission level point.

pX: peak envelope power, peak envelope power (of a transmitter).

PX: peak envelope power, private exchange. *See* **private branch exchange.**

p x 64: In video teleconferencing, pertaining to a family of International Telegraph and Telephone Consultative Committee (CCITT) Recommendations, where p is a nonzero positive integer that indicates the number of 64 kbps (kilobits per second) channels. *Note:* The p x 64 family includes International Telegraph and Telephone Consultative Committee (CCITT) Recommendations H.221, H.230, H.242, H.261, and H.320. These Recommendations form the basis for video telecommunications interoperability.

pyrotechnic code: In visual signaling systems, a prearranged code in which meanings are assigned to the various colors and arrangements of pyrotechnics.

pyrotechnic device: A device that (a) contains chemicals that produce smoke or brilliant lights when they are ignited and allowed to burn, (b) usually is fired with a launcher, such as a pistol, cannon, or self-contained rocket, and (c) may be used for various purposes, such as signaling, lighting, or obscuring vision. *Note:* Examples of pyrotechnic devices are fireworks, flares, rockets, Roman candles, and smoke bombs.

pyrotechnic distress signal: A distress signal that is transmitted by pyrotechnic means. *Note:* A single red pyrotechnic light or a succession of red pyrotechnic lights might indicate that the aircraft that launched them is in distress. Pyrotechnic lights of any color fired at short intervals might indicate that the ship that launched them is in distress.

pyrotechnic signal: In visual signaling systems, a signal that is produced by a device that generates light or smoke by means of chemical reaction, such as combustion. *See also* **pyrotechnic device.**

pyrotechnic smoke: Smoke that (a) is produced by means of an exothermic chemical reaction, such as combustion, (b) usually is white, yellow, black, brown, orange, or red, and (c) may be used for signaling, marking, or obscuring vision.

pyrotechnic transmission: The transmission of messages by means of pyrotechnic devices, such as flares, rockets, fire, and smoke. *Note:* Pyrotechnic communications rely heavily on prearranged coding of messages for recognition, interpretation, and decoding.

PZT: piezoelectric transducer.

q: The symbol for quantity of electronic charge, usually expressed in coulombs (C). *Note:* The voltage across a capacitor may be expressed as $v = q/C$, where v is the voltage in volts, q is the charge in coulombs, and C is the capacity in farads, and $dv/dt = (1/C)dq/dt = i/C$ where i is the current in amperes, and t is time in seconds.

Q: *Synonym* **Q-gain.**

QA: quality assurance.

QAM: quadrature amplitude modulation.

Q-band: The band that consists of frequencies in the range from 36 GHz (gigahertz) to 46 GHz. *Note 1:* The Q-band is now obsolete. It was divided into five subranges of 2 GHz each. The ranges were designated QA through QE. The Q-band also has consisted of frequencies in the range from 26 GHz through 40 GHz between the old K and V bands. *Note 2:* Frequency bands are no longer designated by letter. *See also* **frequency spectrum designation.**

QC: quality control.

Q-gain: An increase in the output power of an electrical device produced by obtaining resonance in a tuned circuit. *Synonym* **Q.** *Note 1:* The Q of a coil is given by the relation $Q = \omega L/R$ where Q is the Q-gain, $\omega = 2\pi f$ where π is approximately 3.1416 and f is the frequency in hertz, L is the inductance in henries, ωL is the inductive reactance in ohms, and R is the resistance in ohms. *Note 2:* The Q-gain usually is expressed in dB, based on the ratio of the output power of tuned and untuned circuits. The Q-gain often is expressed in nepers.

QOS: quality of service.

QPSK: quadrature phase-shift keying.

QST: A standard operating signal that is used as a suffix to a message to indicate that the charge for the message immediately preceding is to be ascertained.

Q-switch: A device that (a) prohibits pulsed laser emission until energy increases to a certain level in the active laser medium, (b) increases the pulse power level by shortening pulse duration while keeping the pulse energy constant, and (c) provides shorter and more intense pulses at higher pulse repetition rates than could be achieved by simply pulsing the active laser medium. *See also* **active medium, laser, pulse repetition rate.**

Q-switched repetitively pulsed laser: A solid state laser in which continuous emission is converted into pulses by a Q-switch. *See also* **laser, Q-switch.**

quad: A group of four wires composed of two twisted pairs twisted together. *Note:* The individual wires in the twisted pairs have a longer lay length, i.e., a longer pitch, than the two twisted pairs that form the quad. *See also* **four-wire circuit.**

quadded cable: A cable formed of multiples of quads, paired and separately insulated, and twisted together within an overall jacket. *See also* **cable, paired cable, spiral-four cable.**

quadratic refractive index profile: *Synonym* **parabolic refractive index profile.**

quadrature: 1. Separated by 90°, i.e., π/2 radians. **2.** Pertaining to the phase relationship between two periodic quantities that vary with the same period, i.e., with the same frequency or repetition rate, when the phase difference between them is one-fourth of their period, i.e., the spatial or electrical phase difference between them is 90°, i.e., π/2 radians. **3.** Pertaining to two spatial vectors, such as the electric and magnetic fields in a circularly polarized electromagnetic wave, that are perpendicular to each other. **4.** When an alternating voltage is applied to a pure inductance or a pure capacitance, pertaining to the phase angle between the applied voltage and the resulting current. *Note:* In an inductance, the current vector lags the voltage vector by 90°, i.e., π/2 radians, and in a capacitance the current vector leads the voltage vector by 90°, i.e., π/2 radians. In both cases the voltage and the current are in quadrature. **5.** Pertaining to the relationship between real and imaginary numbers, especially when plotted in Cartesian coordinates. **6.** In physical and geometric optics, pertaining to the relationship between wavefronts and rays. *Synonym* **phase quadrature.** *See also* **modulation, quadrature modulation.** *Refer to* **Fig. Q-1.**

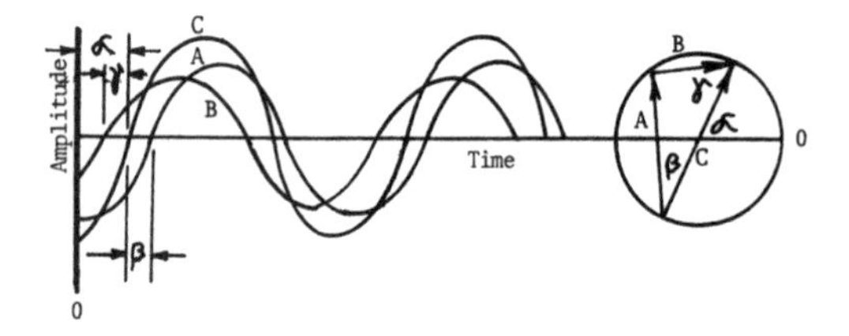

Fig. Q-1. For an arbitrary sine wave, **C,** also represented by the phasor **C** drawn as a diameter of a reference circle, the two **quadrature components, A** and **B,** are shown as lagging and leading wave components, respectively, also represented by the inscribed-phases **A** and **B.** In both diagrams, **A** and **B** are 90° (electrical) apart, the sum of **A** and **B** equals **C** in magnitude and phase at every instant of time, **C** has a frame of reference phase angle of α, **A** lags **C** by the angle β, whereas **B,** the other quadrature component of **C,** leads **C** by the angle γ. (β + γ = 90°.)

quadrature amplitude modulation (QAM): Quadrature modulation in which the two carriers are amplitude-modulated. *See also* **modulation, quadrature modulation.**

quadrature component: Either of two orthogonal components, i.e., components that are at right angles to each other, into which any vector quantity may be resolved such that (a) the vector sum of the two quadrature components, i.e., the resultant, is the original vector from which they were resolved, (b) the two components may vary in direction and magnitude, but they must sum to the original vector, (c) when plotted, and assuming that the vector is the diameter of a circle, the two quadrature components must be inscribed within either semicircle formed by the diameter, and (d) the components form a right triangle with the diameter as the hypotenuse. *See also* **orthogonal, vector.** *Refer to* **Fig. Q-1.**

quadrature modulation (QM): Modulation in which two carriers out of phase by 90° are modulated by separate signals. *See also* **modulation, quadrature amplitude modulation.**

quadrature phase-shift keying (QPSK): Phase-shift keying in which four different phase angles are used in each signal element, i.e., there are four phase states or positions in the time or frequency domains within a single period of a sinusoidal carrier wave. *Note:* In quadrature phase-shift keying (QPSK), the four angles are usually out of phase by 90°. *Synonyms* **quadriphase keying, quaternary phase-shift keying.** *See also* **keying, modulation, phase, phase-shift keying.**

quadriphase keying: *Synonym* **quadrature phase-shift keying.**

quadriphase modulation: Modulation in which (a) multiple phase-shift keying is used, (b) four phase significant conditions, i.e., four phase states, of a sinusoidal carrier wave are used to convey four digital data codes, such as 00, 01, 10, and 11, one for each significant condition of the carrier, and (c) the members of each adjacent pair of the four phase states are in quadrature with each other. *See also* **carrier, modulation, multiple phase-shift keying, phase-shift keying, quadrature, significant condition.**

quadriphase shift keying: Shift keying a sinusoidal carrier wave in which (a) 0s and 1s are represented by phase shifts that occur at 0°, +90°, −90°, and +180°, (b) the four phase significant conditions are used in the time or frequency domain within a period of the sinusoidal carrier wave, (c) the binary code assigned to each significant condition depends on the system that is used, (d) if all four significant conditions are assigned a different code, such as 00, 01, 10, and 11, there is no redundancy, and (e) other codes may be used in which there is redundancy with improved reliability, resistance to noise, and reduction of error. *See also* **shift keying, significant condition.**

quadruple diversity: Diversity transmission and reception (a) in which four independently fading signals are used, and (b) that may be accomplished through the use of space, frequency, angle, time, or polarization multiplexing, or combinations of these. *See also* **diversity reception, dual diversity, order of diversity.**

quadruple diversity combining: The simultaneous combining of, or selection from, four independently fading signals and their detection through the use of space, frequency, phase angle, time, modulation, or polarization characteristics, or combinations of these. *See also* **diversity reception, dual diversity combining, order of diversity.**

quadruplex system: A telegraph system that is arranged for the simultaneous independent transmission of two messages in each direction over a single circuit.

quadruply clad optical fiber: An optical fiber that has four claddings. *Note 1:* In a quadruply clad optical fiber, the core usually has a relatively very high refractive index, while the claddings, starting from the core radially outward, have relatively very low, high, low, and medium refractive indices. *Note 2:* Quadruply clad optical fibers usually are designed to operate in single mode. *See also* **cladding, deeply depressed cladding optical fiber, fiber optics, optical fiber.**

qualification testing: Formal testing that (a) is designed to demonstrate that the software and the hardware of a system meet specified requirements, (b) may be accomplished at any time during the life of a system, such as during prototype development, manufacturing, shipment, storage, installation, and operation, and (c) is conducted to determine the extent to which a system passes specified performance criteria. *See also* **acceptance testing, data validation, system testing.**

quality: *See* **image quality, light quantity, toll quality.**

quality analysis: *See* **path quality analysis.**

quality assurance (QA): 1. The actions taken to ensure that standards and approved procedures are adhered to and that delivered products or services meet performance requirements. **2.** The planned systematic activities necessary to ensure that a component, module, or system conforms to established technical requirements. **3.** In data systems, the policy, procedures, and systematic actions established in an enterprise for the purpose of providing and maintaining some degree of confidence in data integrity and accuracy throughout the lifecycle of the data, which includes input, update, manipulation, and output. *See also* **acceptance test, quality control, test and validation.**

quality control (QC): A management function in which control of the quality of raw materials, assemblies, produced materiel, and services is exercised for the purpose of preventing (a) production of defective materiel and (b) provision of faulty services. *See also* **grade of service.**

quality factor: *See* **intrinsic quality factor.**

quality matrix: *See* **path quality matrix.**

quality of service (QOS): 1. The quality specification of a communications channel or system. *Note:* Quality of service (QOS) may be quantitatively indicated by channel or system performance parameters, such as signal to noise ratio (S/N), bit error ratio (BER), message throughput rate, and call blocking probability. **2.** A subjective rating of telephone communications quality in which listeners judge transmissions by qualifiers, such as excellent, good, fair, poor, or unsatisfactory. *Synonym* **service quality.** *See also* **call, grade of service.**

quality of service (QOS) parameter set: A set of quality of service (QOS) requirements based on (a) a target value desired by the call originator, (b) the lowest-quality acceptable value agreeable to the call originator, (c) an available value that the network is willing to provide, and (d) a selected value that is agreeable to the call receiver. *Note:* In defining the lowest acceptable quality of service (QOS), the term "minimum" is used when referring to throughput and the term "maximum" is used in lieu of "lowest" when referring to transit delay.

quality television: *See* **distribution quality television.**

quantization: The conversion of a signal into discrete values such that (a) the continuous range of values of the signal is divided into nonoverlapping subranges, (b) a unique discrete value of the output is assigned to each subrange, (c) the subranges are not necessarily equal, and (d) whenever the signal value falls within a given subrange, the output has the corresponding discrete value. *Note:* For example, in pulse code modulation, when quantization is performed by sampling techniques, if the sampled signal value falls within a given subrange, the sample is assigned the corresponding discrete value. Thus, the exact sample values of a signal are converted to the nearest equivalents in a finite set of discrete values in order to permit digital encoding as a next possible step. *See also* **a-law, quantization level, signal, uniform encoding.**

quantization distortion: The distortion that results from a quantization process, including the errors in the output analog and digital signals. *Synonym* **quantizing distortion.**

quantization level: **1.** In a quantization process, the discrete digital value of the output assigned to a particular subrange of the analog input. **2.** In coded digital transmission, a signal level, i.e., a significant condition, that is used in quantization. *Note:* Signal quantization results in the transmission of several discrete signal levels. *Synonym* **quantizing level.** *See also* **a-law, digital to analog converter, digitize, digitizer, level, quantization, uniform encoding.**

quantization noise: Noise that (a) is caused by the error of approximation in a quantization process and (b) is solely dependent upon the particular quantization process that is used and the statistical characteristics of the quantized signal. *Synonyms* **granular noise, quantizing noise.** *See also* **noise, quantization.**

quantized feedback: **1.** In a feedback system, the digital signal that is fed back. *Note 1:* Several forms of analog to digital converters contain a quantized feedback loop following the basic A-D converter. *Note 2:* The quantized feedback is often processed before it is fed back to the input. **2.** A feedback system (a) in which the output digital signal is fed back to the input, (b) that may contain one of several forms of analog to digital converters that have a quantized feedback loop following the basic A-D converter, and (c) in which the quantized feedback is often processed before it is fed back to the input. *See also* **delta modulation, feedback, loop.**

quantized synchronization: *See* **amplitude quantized synchronization.**

quantizing distortion: *Synonym* **quantization distortion.**

quantizing level: *Synonym* **quantization level.**

quantizing noise: *Synonym* **quantization noise.**

quantum: In electromagnetics, a unit of electromagnetic energy equal in magnitude to *hf*, where *h* is Planck's constant and *f* is the frequency of the radiation. *Note 1:* A quantum of electromagnetic energy is released when an electron in an excited or radioactive chemical element moves from a higher to a lower energy level. *Note 2:* A photon is a quantum of electromagnetic energy in the optical spectrum. The lowest-energy photon would have the lowest frequency, i.e., the longest wavelength, at the extreme end of the far infrared region of what is considered to be the optical spectrum, 100 μm (micron), corresponding to a frequency of 3 THz (terahertz), or 3×10^{12} Hz (hertz). The highest-energy photon wavelength is 0.001 μ, which corresponds to 0.3 EHz (exahertz) or 0.3×10^{18} Hz. The energy of a 1-μm photon is about 1.2 eV (electron•volts).

quantum efficiency: **1.** In a quantum device, such as an optical source or detector, the ratio of (a) the countable elementary events at the output to (b) the countable elementary events at the input. *Note 1:* The quantum efficiency defines the device transfer function for the countable events. *Note 2:* Examples of quantum efficiency are (a) in an optical semiconductor source, such as a photodiode, the ratio of the number of photons emitted to the number of electrons applied by an input electrical pulse and (b) in a photodetector, the ratio of the number of electrons generated in the photocurrent to the number of photons applied by an input optical pulse. **2.** In a light source or a photodetector, the ratio of output quanta to input quanta. *Note:* Input and output quanta need not both be photons. *See* **differential quantum efficiency, response quantum efficiency.** *Refer to* **Fig. P-6 (721).**

quantum-limited operation: In the operation of a photodetector, the inability of the detector to measure the incident radiation levels below a threshold level because of fluctuations in the output current that are not caused by the incident photons, such as dark currents and noise. *See also* **dark current, noise, photodetector.**

quantum noise: **1.** Noise attributable to the discrete and probabilistic nature of physical phenomena and their interactions. *Note:* Quantum noise represents the fundamental limit of the achievable signal to noise ratio of an optical communications system. This limit is not reached in practical systems. **2.** Noise attributed to the discrete or particle nature of electromagnetic radiation, i.e., radiation in the form of streams of individual packets of energy, such as photons that occur in lightwaves, gamma rays, and cosmic rays. *Note:* The absorption of each photon produces a discrete contribution to the total noise. *See also* **dark current, photon noise, shot noise, thermal noise.**

quantum-noise-limited operation: **1.** In a device or a system, the condition that prevails when quantum noise at a point in the device or the system limits its performance or the performance of the device or the system to which it is connected. *Note:* An example of quantum-noise-limited operation is the condition that prevails in an optical link when quantum noise is the predominant mechanism that limits link performance. **2.** Operation in which the minimum signal that can be detected is determined by quantum noise. *See also* **at-**

tenuation-limited operation, bandwidth-limited operation, dispersion-limited operation, distortion-limited operation, shot noise.

quantum of action: *Synonym* **Planck's constant.**

quarter adder: A binary adder with two inputs that (a) produces a modulo-two sum, (b) does not produce a carry digit, and (c) performs the function of an EXCLUSIVE OR gate. *See also* **binary half-adder.**

quarter common intermediate format (QCIF): A video signal format defined in International Telegraph and Telephone Consultative Committee (CCITT) Recommendation H.261 that is characterized by 176 luminance pixels on each of 144 lines, with half as many chrominance pixels in each direction. *Note:* Quarter common intermediate format (QCIF) has one-fourth as many pixels as the full common intermediate format (CIF). *See also* **common intermediate format.**

quarter squares multiplication: Multiplication of analog or digital quantities based on the identity relation $xy = ((x+y)^2 - (x-y)^2)/4$ where x and y are the variables to be multiplied. *See also* **multiplier.**

quarter squares multiplier: An analog multiplier that (a) also may be programmed on a digital computer, (b) is based on the identity relation $xy = ((x+y)^2 - (x-y)^2)/4$ where x and y are the variables to be multiplied, and (c) consists of interconnected inverters, analog adders, and square law devices. *See also* **multiplier.**

quartet: A byte that consists of four binary digits, such as 1011. *Synonym* **tetrad.** *See also* **byte.**

quartz: *See* **fused quartz.**

quartz clock: A clock that contains a quartz crystal oscillator (XO) that determines the accuracy and the precision of the clock.

quartz oscillator (XO or QO): An oscillator in which a quartz crystal is used to stabilize the frequency. *Note:* The piezoelectric property of the quartz crystal results in a nearly constant output frequency, depending upon the crystal size, shape, and excitation.

quasi-analog signal: A digital signal that has been converted to a form suitable for transmission over a specified analog channel. *Note:* The specification of the analog channel should include frequency range, frequency bandwidth, signal to noise ratio (S/N), and envelope delay distortion. When this form of signaling is used to convey message traffic over dialup telephone systems, it is often referred to as voice data. A modem may be used for the conversion. *Synonym* **voice-data signal.** *See also* **modem, signal.**

quasi-analog transmission: Transmission in which (a) a modulator is used to modulate one or more voice frequency (VF) carriers, thus making a digital signal suitable for transmission over an analog voice circuit, and (b) a demodulator is used to recover the digital signal at the other end of the circuit. *See also* **hybrid communication network, modem.**

quasi-synchronous: Pertaining to the state of being partially synchronous, nearly synchronous, or synchronous in a limited respect. *Note:* A quasi-synchronous satellite is a satellite in an orbit with a height that is nearly but not exactly that required for geostationary synchronism so that the apparent longitudinal motion, i.e., the motion with respect to a meridian of longitude, of the satellite relative to the Earth is that of a slow drift in the equatorial plane. *See also* **geostationary satellite, orbit, synchronous.**

quaternary: **1.** Pertaining to a numeration system in which (a) the radix is four, (b) there are four different digits, such as 0, 1, 2, and 3, and (c) digit positions are weighted 256, 64, 16, 4, 1 . . . from left to right with the least significant digit on the right. **2.** Pertaining to an n-ary code in which n is 4, modulo-n arithmetic in which n is 4, or an n-ary digit in which n is 4.

quaternary operator: **1.** An operator that requires exactly four operands. **2.** An operator that operates on four operands.

quaternary phase-shift keying: *Synonym* **quadrature phase-shift keying.**

quaternary signal: A digital signal that has four significant conditions. *See also* **signal, significant condition.**

quench: *See* **source quench.**

query call: In adaptive high frequency (HF) radio, an automatic link establishment (ALE) call that requests responses from stations having connectivity to the destination, i.e., to the call receiver, specified in the call. *See also* **adaptive high frequency radio.**

query key: In communications, computer, data processing, and control systems, a key that will, when depressed, cause data in certain previously specified locations to be displayed in the display area on the display surface of a display device. *Note:* An example of a query key is a function key that causes the most recently specified operation to be repeated and the results to be displayed, even if new data have been entered in the spreadsheet. *See also* **display area, function key, key.**

query language: A language that is used to interrogate a system. *Note:* An example of a query language is a language that enables a user to (a) interact directly with a database management system and (b) retrieve, store, change, and store data. *See also* **database, database management, language.**

question file: *See* **frequently asked question file.**

queue: **1.** A collection of items, such as telephone calls, that is arranged in sequence. *Note:* Queues are used to store data or log events that occur at random times so that they may be serviced according to a prescribed discipline that may be fixed or adaptive. **2.** A list or string of items that (a) usually have accumulated over time and (b) are to be processed in some way and in some sequence, such as examined for a particular quality and selected from either end or at random. *See also* **buffer, queue traffic.**

queue traffic: **1.** In a store and forward switching center, the outgoing messages awaiting transmission at the outgoing line position. **2.** A series of calls waiting for service. *See also* **buffer; camp on; first in, first out; message switching; queuing; selective calling.**

queuing: The entering of elements into or the removing of elements from a queue. *See also* **buffer, fair queuing, queue traffic.**

queuing delay: **1.** In a switched network, the time between (a) the instant of completion of signaling by the call originator and (b) the instant of arrival of a ringing signal at the call receiver. *Note:* Queuing may be caused by delays at the originating switch, intermediate switches, or the call receiver servicing switch. **2.** In a data network, the sum of the delays between (a) the instant of completion of a request for service and (b) the instant of establishment of a circuit to the called data terminal equipment (DTE). **3.** In a packet-switched network, the sum of the delays encountered by a packet between (a) the instant of insertion into the network and (b) the instant of delivery to the addressee. *See also* **buffer.**

quick engagement clip: In visual signaling systems, a clip that enables rapid attachment and detachment of flags and pennants to and from a halyard.

quiescing: **1.** The bringing of a process or an operation to a halt by rejection of requests for new or additional work. **2.** Being ready but not performing.

quintet: A byte that is composed of five binary digits, such as 11001. *See also* **byte.**

quit command: In communications, computer, data processing, and control systems, a command to the system to leave the current mode or status and return to another, usually higher, command level or mode. *Note:* Examples of quit commands are (a) in WordPerfect, the command to EXIT or CANCEL and (b) in Lotus 1-2-3, the QUIT command menu item that will return the computer to (i) the disk operating system, such as IBM PC DOS or MS DOS, or (ii) to the Lotus Access System, depending on which one the operator started or booted from. *See also* **boot, command, disk operating system.**

quotient: The result obtained when the arithmetic process of division is performed, i.e., when a number, the dividend, is divided by another number, the divisor. *Note:* In binary digital computers, division is performed by counting the number of times the divisor can be subtracted from the dividend. When the difference becomes smaller than the divisor, the division is complete, and the number smaller than the divisor is the remainder. If 496 is to be divided by 50, it would be subtracted 9 times, where 9 is the quotient, and the remaining 46 is the remainder. However, as a mixed number, the entire $9^{23}/_{25}$ could be considered as the quotient, or as a decimal number the entire 9.92 could be considered as the quotient. *Note 2:* In binary digital computers, subtraction is performed by adding the ones complement of the subtrahend to the minuend, adding 1 to the least significant digit of the sum, and propagating all carries, if any. Ones complements of numbers are available on the opposite side of flip-flops when a number is stored in a bank of flip-flops, i.e., in a register.

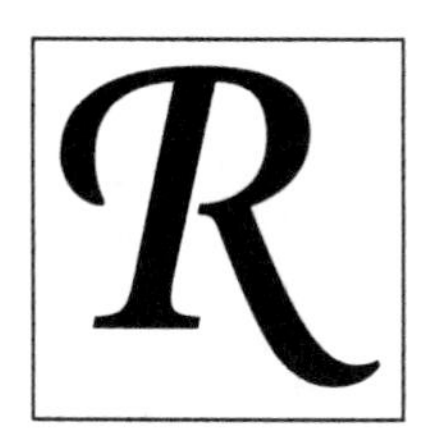

raceway: A linear enclosure, channel, or sequence of brackets used to hold and protect wires, cables, or buses. *Synonym* **wireway.** *See also* **bus.**

rack: A frame in which one or more pieces of equipment are mounted. *Note:* In the U.S Department of Defense (DoD), racks usually are vertical.

racon: *Synonym* **radar beacon.**

rad: radian, radiation absorbed dose.

radar: **1.** Radio detection and ranging. **2.** A radio detection and ranging (RADAR) system that (a) is used to determine the distance to and the direction of objects, i.e., of targets, by transmitting an electromagnetic wave and observing the reflected return, i.e., the echo, (b) includes the methods and equipment that use beamed and reflected electromagnetic energy for detecting, identifying, locating, and measuring characteristics of objects, such as their range, azimuth, elevation angle, velocity, and altitude, and (c) is used for various other purposes, such as navigation, homing, bombing, missile tracking, sounding, and mapping. **3.** A radiodetermination system based on the comparison of reference signals with radio signals reflected or retransmitted from the position to be determined. *Note:* "Radar" is derived from "radio detection and ranging." *See* **acquisition radar, airborne intercept radar, complex pulse radar, conical beam split radar, continuous wave radar, Doppler radar, early warning radar, frequency agile radar, guidance radar, initial warning radar, long range radar, metric radar, multiband radar, narrow beam radar, over the horizon radar, primary radar, pulse compression radar, pulse Doppler radar, pulse radar, search radar, secondary radar, side-looking airborne radar, tracking radar, track-while-scan airborne intercept radar, variable pulse recurrent frequency radar, wide beam radar.** *See also* **radiodetermination.**

radar altimetry area: Large and comparatively level terrain that has a defined elevation that can be used for determining the altitude of airborne equipment by the use of radar.

radar beacon (RACON): **1.** A radionavigation system that (a) transmits a pulsed radio signal with specific characteristics and (b) may transmit automatically or in response to a predetermined received signal. **2.** A receiver-transmitter that (a) is associated with a fixed navigational mark that, when triggered by a radar, automatically returns a distinctive signal that can appear on the display of the triggering radar, and (b) provides range, bearing, and identification information. **3.** A radionavigation transponder that is used to transmit, in response to a received signal, a pulsed radio signal with specific characteristics that enable determination of bearing and range of the transponder from the interrogator. *Note:* A radar beacon may also be used to identify the transponder itself. **4.** A pulsed radio signal with specific characteristics. **5.** A radio beacon in which radar is used to obtain bearing and range to an object, i.e., to a target, and to transmit this information by radio to another station. **6.** A receiver-transmitter combination that sends out a coded signal when triggered by the proper type of pulse, enabling determination of range and bearing information by the interrogating station or aircraft. *Note:* "RACON" is derived from "radar beacon." *Synonym* **racon.**

radar beacon precipitation gage station: A transponder station that (a) is in the meteorological aids service and (b) is used for telemetering precipitation data. *See also* **transponder.**

radar blind range (RBR): The range that corresponds to the situation in which a radar transmitter is on and hence the receiver must be off, so that the radar transmitted signal does not saturate, i.e., does not blind, its own receiver. *Note:* Radar blind ranges occur because there is a time interval between transmitted pulses that corresponds to the time required for a pulse to propagate to the object, i.e., to the target, and for its reflection to travel back. This causes an attempt to measure the range just as the radar transmitter is transmitting the next pulse. However, the receiver is off, so this particular range cannot be measured. The width of the range value that cannot be measured depends on the duration of the time that the radar receiver is off, which depends on the duration of the transmitted pulse. The return time interval could be coincident with the very next radar transmitted pulse, i.e., the first pulse following a transmitted pulse, or the second, or the third, and so on, giving rise to a succession of blind ranges. The blind ranges are given by the relation $r_m = mc/2fn$ where r_m is the blind range for a given value of m, m is a positive integer that has the values 1, 2, 3, 4, . . . , and that indicates which of the blind ranges is being determined, c is the velocity of electromagnetic wave propagation in a vacuum, approximately 3×10^8 m/s (meters per second), f is the radar pulse repetition rate, and n is the refractive index of the propagation medium (nearly 1 for air). *See also* **radar blind speed.**

radar blind speed (RBS): The magnitude of the radial component of velocity of an object, i.e., of a target, relative to a radar site, that cannot be measured by the radar unit. *Note:* Radar blind speeds occur because of the relationship between the transmitted pulse repetition rate (PRR) and the received pulse repetition rate, i.e., the radar return repetition rate. The Doppler pulse repetition rate is the difference between the transmitted and the received pulse repetition rates. For example, when the object is stationary, or moving with a constant radius from the radar site, the reflected PRR is the same as the transmitted PRR so that a net zero Doppler signal is indicated for the radial component of velocity, and the radial component is zero. If it happens that the Doppler PRR is the same as the transmitted PRR, i.e., the illuminating PRR, or it is a multiple of the transmitted PRR, a zero signal is also obtained, and hence the radar is blind to these speeds, one for each multiple of the transmitted pulse repetition rate. It is not the absolute magnitude of the speed of the object that is measured by Doppler radar, but only the radial component of the speed. The radial components of blind speeds, v_m, are given by the relation $v_m = m\lambda f/102$, where v is the blind speed in knots, m is the multiple of the radar pulse repetition rate and the number of the blind speed, i.e., a positive integer, 1, 2, 3, 4, . . . , for the first, second, third, fourth, and so on, blind speed, λ is the wavelength of the illuminating radar in centimeters, f is the transmitter pulse repetition rate in pps (pulses per second), and the 102 is a units conversion factor. *See also* **radar blind range, radar return.**

radar cross section (RCS): A measure of the extent to which an object, i.e., a target, reflects radar pulses. *Note:* The radar cross section of an aircraft can vary by a factor of over 100, depending on the aspect angle of the aircraft to the radar transmitter. Radar reflection, i.e., radar returns, off the nose of the aircraft usually represents the smallest radar cross section, while a broadside presentation to the signal produces the greatest cross section. Shape, surface roughness, and reflective material as well as orientation also affect the radar cross section. *See also* **radar reflectivity, radar return.**

radar deception repeater: In radar electronic warfare, a deception device that (a) samples an interrogating radar signal, (b) instantaneously stores its frequency, (c) subsequently repeats the signal after changing one or more of its characteristics, and (d) may consist of a receiver, a frequency memory device, a signal processor, a wideband amplifier, a pulse programmer, and a transmitter. *Note:* Some radar deception repeaters deceive by repeating either a part or all of the received signal, i.e., the radar return, and creating false signal characteristics in order to deceive, mislead, or confuse a computing system or operator. Other repeaters deceive by amplifying and directly repeating all portions of a received signal without altering its characteristics. These repeaters may deceive by creating a fictitious reflective area that does not exist in terms of the size represented by the transmitted signal magnitude. *See also* **radar return.**

radar diplexing: The simultaneous operating of radar sets on at least two different frequencies while using a single, i.e., a common, antenna.

radar display: A graphic display of a set of variables that are measured by means of radar, such as the azimuth of an object, i.e., of a target, the range to an object, or the signal strength received from an object, all in real time, delayed time, or predicted time. *See* **spot barrage jamming radar display, spot continuous wave jamming radar display.**

radar echo: 1. In radar systems, the electromagnetic energy that is received by a radar antenna after reflection from an object, i.e., a target. **2.** The deflection or the change in intensity of the display element on a display surface, such as a cathode ray tube (CRT) screen, produced after the electromagnetic energy reflected from an object, i.e., a target, is received by the radar antenna, amplified, and processed by the radar receiver. *See also* **radar return.**

radar height indicator: *Synonym* **range height indicator.**

radar homing beacon: A beacon that (a) uses radar to establish the bearing and distance from the beacon location to a mobile station, such as a ship station, aircraft station, or land mobile station, that is being tracked and (b) may be installed in any fixed or mobile station or platform. *See also* **beacon, bearing, radar beacon, track, tracking.**

radar horizon: The locus of points of tangency at which rays from a radar antenna are tangential to the Earth's surface. *Note:* On the open sea, the radar horizon is to the geographic horizon, i.e., it is nearly horizontal and circular, the radius depending on the height of the antenna above the sea surface, such as when it is on a mast. On land, the slope of the line to the radar horizon points varies with the terrain in each different direction from the antenna. It could vary from a positive slope to a negative slope, depending on the height of the antenna and the terrain in each direction to the horizon. *See also* **antenna height above average terrain, radio horizon range.**

radar information: *See* **raw radar information.**

radar intelligence (RADINT): Intelligence information derived from data collected by radar.

radar jammer: *See* **airborne radar jammer, locked pulse radar jammer.**

radar jamming: *See* **formation radar jamming, pulse modulation radar jamming.**

radar line of sight (LOS) equation: An equation that expresses the radar horizon range (RHR), given by the relation $RHR_s = (2h)^{1/2} + (2a)^{1/2} = 1.414(h^{1/2} + a^{1/2})$ where RHR_s is the radar horizon range in statute miles, h is the antenna height in feet, and a is the object critical altitude, i.e., the target critical altitude, in feet, below which the transmitter cannot illuminate the object, i.e., target. The RHR is also given by the relation $RHR_k = 4.12(h^{1/2} + a^{1/2})$ where RHR_k is the radar horizon range in kilometers, when h and a are in meters. The effective Earth radius, i.e., 4/3 times the actual Earth radius, is used in deriving these formulas. Second-order differentials are neglected. They contribute less than 0.1%. *See also* **radio horizon range.**

radar masking: On a radar display, the obliteration of echoes by any cause, such as (a) interference from local sources, (b) low signal to noise ratio, and (c) saturating the receiving antenna with large-amplitude wideband signals that may be from jamming, such as might be produced by white noise generators operating over a wide range of frequencies. *See also* **interference, jamming, white noise.**

radar mile: The time required for a radar pulse to travel 1 mile to an object, i.e., to a target, reflect, and return to the receiver. *Note:* A radar statute mile is 10.8 μs (microseconds). A radar nautical mile is 12.4 μs. The time for any other radar unit distance is readily determined, such as the radar meter or the radar kilometer.

radar navigation: The use of radar in navigation and pilotage. *See also* **radar, radiodetermination, radio direction finding, radio navigation aid.**

radar netting: The interconnection of several radar systems and their connection to a single center for coordination, direction, and control, so as to (a) provide integrated operations, (b) exchange raw radar information among systems, and (c) maintain radar coverage of a large sector, such as by providing for coverage of an area that might be uncovered if a radar system in the net experienced technical difficulties or otherwise became nonoperational. *See also* **raw radar information.**

radar netting station: A station that can receive data from radar tracking stations and exchange these data among other radar tracking stations. *See also* **track, tracking.**

radar picket: A radar system, usually mounted in a vehicle, ship, or aircraft, that is stationed in relation to an area, or to a group of mobile units, for the purpose of increasing the radar detection range for radar protection of the area or the group of mobile units. *Note:* Examples of radar pickets are radars on board (a) a ship to provide greater range in the direction of the area to be covered than can be achieved by land-based radar, (b) a ship distant from a fleet to extend the area of radar coverage for the fleet, and (c) an aircraft distant from an aircraft formation to extend radar coverage for the formation. *See also* **picket.**

radar planned position indicator: A planned position indicator for a radar system that (a) displays a map, usually in polar coordinates, of the area near or surrounding the system and (b) dynamically displays identified objects, i.e., identified targets, as echoes, i.e., pips, on the display surface at screen coordinates that correspond to the actual direction and range of the objects from the radar installation or from a specified origin. *See also* **planned position indicator, radar echo, screen coordinate.**

radar pulse: A pulse that (a) consists of a short burst of radio frequency (rf) electromagnetic energy, (b) is emitted in a controlled direction by a radar antenna usually at fixed time intervals, and (c) is used to obtain a reflection, i.e., a radar return, from an object, i.e., a target, for radio detection and ranging purposes. *See also* **pulse, radar, radar return.**

radar pulse compression: The decrease in the pulse duration of received radar pulses, i.e., radar echoes, that results from stretching, i.e., increasing the duration of, transmitted radar pulses, by using matched impedance filter techniques, thus permitting an increase in the average transmitted power without an increase in the peak power requirement with no loss in range resolution. *See also* **pulse duration, pulse stretching, radar resolution cell, range resolution.**

radar reflectivity: The characteristic of an object, i.e., a target, that causes it to reflect electromagnetic waves, such as radar pulses. *Note:* The reflectivity usually is expressed in units of equivalent area of a flat reflector, oriented normal to the direction of wave propagation, that would produce the same echo strength as the object for a given transmitted power. *See also* **electromagnetic wave, radar cross section, radar pulse, reflection, reflectivity, reflector.**

radar relay: The transmission of radar video signals to a display device.

radar repeater: A device that (a) has several display surfaces, such as cathode ray tube (CRT) screens, and (b) is fitted with facilities that enable it to display selected radar data from locations that usually are remote from the radar antenna. *See also* **cathode ray tube, display surface, radar, screen.**

radar resolution cell (RRC): The volume of space that (a) is occupied by a radar pulse, (b) is determined by the pulse duration and the horizontal and vertical beamwidths of the transmitting radar, (c) prevents the radar from distinguishing between two separate objects that lie within the same volume of space, (d) has a depth that remains constant regardless of its distance from the radar transmitting antenna, i.e., a depth that does not increase with range, the depth being given by the relation $D = 150d$ where D is the depth in meters and d is the radar pulse duration in microseconds, (e) has a horizontal width that increases with range, given by the relation $W = (HBW)(R/57)$ where W is the horizontal width of the volume of space occupied by the radar pulse, i.e., the width of the resolution cell, HBW is the horizontal beamwidth in degrees, and R is the range, and (f) has a vertical height that increases in range, given by the relation $H = (VBW)(R/57)$ where H is the vertical height of the cell, VBW is the vertical beamwidth in degrees, and R is the range. *Note:* The range, R, is the distance from the radar antenna to the reflecting object, i.e., to the target. The width and the height of the radar resolution cell (RRC) will be in the same units in which the range is given. For example, if the range is given in meters, the width and the height of the RRC will be in meters. The 57 merely converts degrees to radians. If the beamwidths are given in radian measure, the 57 is omitted. *See also* **beamwidth, radar pulse.**

radar return: A signal that has been transmitted by a radar transmitter and reflected back to the same or another radar receiver, i.e., a radar echo. *See also* **radar echo.**

radar scan: **1.** One complete revolution of a rotating search radar antenna. **2.** One complete cycle of a scan by a radar phased array antenna. **3.** The area or the angle that is covered by a radar receiving antenna.

radar scan rate modulation deception: Electronic deception that (a) is used against tracking radars and (b) uses modulation of the radar deception repeater or transponder output at or near the radar frequency. *Note:* A particular form of radar scan rate modulation deception that is effective against conical scan, track while scan (TWS), and lobe on receive only (LORO) tracking uses a square wave, i.e., on-off gating of the output state of a repeater at a variable frequency that includes the scan frequency of the radar being countered. As this modulation approaches the radar scan frequency, increasingly large error signals appear in the radar servo loops. This causes rapid random gyration of the antenna system. Frequently these error signals are sufficient to cause the radar to lose the object, i.e., the target, completely. Because the scan channel passbands of the conical scan radars are narrow, usually 1 Hz (hertz) to 4 Hz, the modulation in the countermeasure must be swept slowly if a maximum effect is to be realized. *See also* **conical scanning, electronic deception, lobe on receive only, radar deception repeater, servomechanism, track, tracking, track while scan airborne intercept radar.**

radarscope overlay: A transparent overlay for placement on a radar screen in order to identify, compare, and analyze radar echoes. *See also* **radar echo.**

radar screen: A display surface for a radar system. *See also* **display device, display space, display surface.**

radar sensitivity: 1. The response of a radar receiver to signals on its designed frequency. **2.** A measure of the ability of a radar receiver to amplify and make usable very weak signals, i.e., radar returns with very low field strength. *See also* **field strength, minimum discernible signal, radar return, response.**

radar shadow: 1. The region that is obscured from surveillance by a radar because of a natural or artificial obstruction in the path of the radar signal. **2.** The region of low intensity or zero intensity radar signal that is produced on the side of an object opposite to that of a radar transmitter.

radar shift: The movement of the origin of a radial display, i.e., of a polar diagram, away from the center of the display space on the display surface of a display device, such as a cathode ray tube (CRT) or a gas panel. *Note:* Radar shift enables efficient use of the radar display area on the display surface of a display device when only a sector is of interest, such as the runways of an airport. However, 360° of radar coverage may not be obtained. *See also* **display device, display space, display surface, polar diagram, radar display.**

radar signature: The detailed wave shape of the radar echo (a) that is received by a radar receiver and (b) that can be used to identify or distinguish among objects, i.e., among targets, such as aircraft, decoys, missiles with warheads, and chaff. *See also* **aircraft signature identification, radar echo.**

radar silence: A restriction against the use of radar, including time restriction and emission power level restriction. *See* **ship radar silence.**

radar spoking: In a radar system polar diagram, periodic flashes of the rotating time base that are sometimes caused by mutual interference. *See also* **polar diagram.**

radar station: An installation of radar equipment, including transmitter, receiver, signal processing and display equipment, and necessary ancillary and supporting equipment. *See* **surveillance radar station.**

radar strobe: 1. An intensified spot in the sweep of a deflection indicator that is used as a reference mark for ranging or expanding a display, i.e., a presentation. **2.** On a radar planned position indicator or B-scope, an intensified sweep that (a) may result from certain types of interference or (b) may be purposely applied as a bearing or heading marker. **3.** A line on a radar console oscilloscope screen that represents the azimuth data generated by the radar. *See also* **azimuth, bearing, console, display, heading, oscilloscope, radar planned position indicator, screen.**

radar system: *See* **distributed radar system.**

radar warning system: A warning system, such as an early warning system or an initial warning system, that is based on the use of radar. *See* **airborne radar warning system.** *See also* **early warning radar, initial warning radar.**

RADHAZ: electromagnetic radiation hazard.

radial: *See* **cardinal radial.**

radial distortion: In a lens system, an aberration characterized by (a) the imaging of an extra-axial straight line, i.e., a nonaxial straight line, as a curved line, or (b) unsymmetrical or irregular distortions of the image caused by imperfect location of optical centers or irregularity of optical surfaces. *Note:* There may be little or no effect on the resolving power of the system.

radially stratified fiber: *Synonym* **step index fiber.**

radial refractive index profile: In an optical fiber with a circular cross section, the refractive index described as a function of the radial distance from the optical fiber axis and the refractive index at the center, i.e., a function described by the relation $n_r = n_0 f(r)$ where n_r is the refractive index at a radial distance r from the center, n_0 is the refractive index at the center, and $f(r)$ is the function of r that expresses the refractive index at the distance r from the center. *Note:* Zero point symmetry, i.e., axial symmetry, of the refractive index profile is assumed. *See also* **linear refractive index profile, parabolic refractive index profile, power law refractive index profile.** *Refer to* **Fig. R-1.**

radian (rad): The SI (Système International d'Unités) supplementary unit of measure for plane angles. *Note 1:* 1 rad is equal to a central angle that subtends a circular arc equal in length to the radius of the circle. *Note 2:* 1 rad is equal to $360/2\pi$ degrees, approximately 57°17′44.6″, or approximately 57.29578°. *Note*

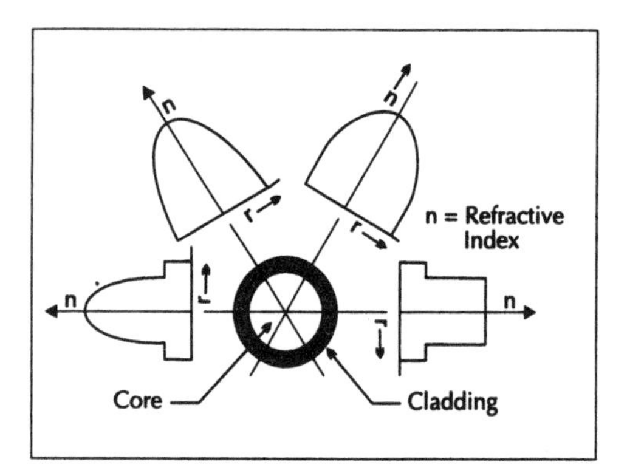

Fig. R-1. Graded and stepped **radial refractive index profiles** for four different optical fibers, the refractive index being a function of the radial distance from the optical fiber axis.

3: In a full circle there are 2π radians where $\pi \approx 3.1416$. *See* **metric system.**

radiance: In a given direction and at a point that is a given distance from a source, the radiant power per unit solid angle per unit of projected area of the source, as viewed from the given point, calculated as follows: (a) the radiant power, in a given direction and on a real, virtual, or notional surface, transmitted by an elementary beam passing through the given point and propagating in the solid angle containing the given direction, divided by (b) the product of the value of this solid angle, the cross-sectional area of that beam containing the given point, and the cosine of the angle between the normal to that area and the beam, the surface being any surface that can be considered to emit, intersect, or receive a beam. *Note 1:* Radiance usually is expressed in watts per steradian•(square meter) or watts•steradian^{-1}•(square meter)$^{-1}$, such as 2 W•sr^{-1}•m^{-2}. *Note 2:* Radiant power per unit area that is incident at a point on a real, virtual, or notional area is the irradiance, usually expressed in watts per square meter. *Deprecated synonym* **radiant flux.** *See also* **conservation of radiance, emissivity, irradiance, radiant energy, radiant intensity, radiometry, spectral radiance.** *Refer to* **Table 5, Appendix B.**

radiance conservation law: A passive optical device or optical system cannot increase the quantity given by the relation $Q = L/n^2$, where Q is the normalized radiance, L is the radiance of a beam, and n is the local refractive index. *Note 1:* The normalized radiance, Q, would be constant if losses, such as losses caused by absorption and scattering, were zero. *Note 2:* Radiance conservation has been called "conservation of brightness," "conservation of radiance," and the "brightness theorem." *Synonyms* **brightness conservation, brightness theorem, radiance conservation, radiance conservation law, radiance conservation principle.** *See also* **fiber optics, radiance.**

radiance conservation principle: *Synonym* **radiance conservation law.**

radiance wavelength: *See* **peak radiance wavelength.**

radiant efficiency: In an optical system, a measure of the percentage of the input power that is converted into useful optical power output, given by 100 times the ratio of (a) the forward useful radiant power from a source, i.e., the radiance integrated over the total forward solid angle from a source, to (b) the total power input to the source. *See also* **radiation efficiency.**

radiant emittance: 1. The radiant power emitted into a full sphere, i.e., 4π steradians, per unit area of a source. *Note:* Radiant emittance usually is expressed in watts per square meter. **2.** At a point on the surface of a source, the radiant power emitted per unit area of the source, i.e., (a) the radiant power leaving an element of the surface at the point divided by (b) the cross-sectional area of that element. *Synonym* **radiant exitance.** *See also* **emissivity, radiation mode, radiometry, spectral radiance, spectral width.** *Refer to* **Table 5, Appendix B.**

radiant energy: Energy that is emitted, transferred, or received via electromagnetic waves, i.e., the energy obtained from the time integral of radiant power. *Note 1:* Radiant energy usually is expressed in joules. *Note 2:* Radiant energy is not considered to use the motion or action of material matter to achieve its propagation, in contrast to elastic, kinetic, and some forms of potential energy. *See also* **radiance, radiant power.** *Refer to* **Table 5, Appendix B.**

radiant exitance: *Synonym* **radiant emittance.**

radiant flux: *Deprecated synonym for* **irradiance, radiance, radiant power.**

radiant intensity: Radiant power per unit solid angle emitted by a source. *Note:* Radiant intensity usually is expressed in watts per steradian. *See* **peak radiant intensity.** *See also* **radiance, radiant emittance, radiometry.** *Refer to* **Table 5, Appendix B.**

radiant power: The time rate of flow of radiant energy, i.e., electromagnetic energy. *Note 1:* If the radiant power is incident upon a surface, the amount of radiant power distributed per unit area of the surface at a point is the irradiance at the point, i.e., the power density. Thus, the radiant power in a beam is the total power obtained by integrating the irradiance over the cross-sectional area of the beam. If the radiant power in a beam is distributed uniformly over the cross-sectional area of the beam, the irradiance is the radiant power divided by the cross-sectional area, which is also the average irradiance. For an electromagnetic wave propagating in a vacuum or a material medium, the radiant power propagating per unit area in the direction of maximum power gradient is also the irradiance, which usually is expressed in watts per square meter. Because the wave is propagating with a velocity, v, given by the relation $v = c/n$ where c is the speed of light in a vacuum and n is the refractive index of the material at the point at which the speed is being considered, the irradiance can also be expressed in terms of the energy per unit volume of vacuum or material medium, expressed as joules per cubic meter, hence giving rise to the term "electromagnetic energy density." *Note 2:* A convex lens a few centimeters in diameter and a centimeter thick can increase the irradiance of solar radiation, i.e., sunlight, when focused on the surface of a combustible material, to a level high enough for combustion, but the total radiant power at the lens and at the combustible material is the same, the loss between the lens and the material usually being negligible. *Note 3:* Radiant power usually is expressed in watts, i.e., joules per second. *Note 4:* The modifier is often dropped, and "power" is used to mean "radiant power." *Deprecated synonyms* **flux, radiant flux.** *See also* **radiance radiometry.** *Refer to* **Table 5, Appendix B.**

radiant power rise time: In a light-emitting diode (LED) or a laser, the time required for the output radiation level to rise from 10% to 90% of the maximum level during the pulse, from the time of onset or median value of the leading edge of the driving pulse. *Note:* The radiant power rise time usually is expressed in nanoseconds or picoseconds. *Deprecated synonym* **flux rise time.**

radiant transmittance: The ratio of (a) the radiant power transmitted by an object to (b) the incident radiant power.

radiated power: The total electromagnetic power, i.e., the total radiant power, emitted by a source. *See* **effective isotropic radiated power, effective monopole radiated power, effective radiated power, equivalent isotropically radiated power.** *See also* **radiant power.**

radiating repeater: *See* **reradiating repeater.**

radiation: 1. Energy in the form of electromagnetic waves or photons, such as radio waves, lightwaves, gamma rays, and X rays. **2.** In radiocommunications, the emission of energy in the form of electromagnetic waves, or the electromagnetic waves themselves. **3.** From any source, the outward flow of energy in the form of electromagnetic waves. *See* **coherent radiation, Earth radiation, electromagnetic radiation, gamma radiation, incoherent radiation, infrared radiation, light amplification by stimulated emission of radiation, microwave amplification by stimulated emission of radiation, monochromatic radiation, optical radiation, polychromatic radiation, primary radiation, secondary radiation, solar radiation, spurious radiation, thermal radiation, time-coherent radiation, visible radiation.** *See also* **antenna, hazards of electromagnetic radiation to fuel, hazards of electromagnetic radiation to ordnance, hazards of electromagnetic radiation to personnel, radiation pattern, radiation scattering, spurious radiation, thermal radiation.** *Refer to* **Figs. R-2 and R-3 (799).** *Refer to* **Table 9, Appendix B.**

radiation absorbed dose (rad): 1. The amount of radiant energy absorbed by a material. **2.** The amount of radiant energy absorbed per unit mass of material. *Note 1:* The basic unit of measure of radiation energy absorption by a material is the rad. One rad (radiation absorbed dose) corresponds to an absorption of 0.01 J/kg (joules per kilogram) or 100 erg/g (ergs per gram) of absorbing material. *Note 2:* The absorbed radiant energy heats, ionizes, or destroys the material upon which it is incident, depending on the nature of the material. *Note 3:* The radiation absorbed dose (rad) should not be confused with irradiance, which is a measure of the incident energy per unit area per unit time, independent of the amount of energy that is absorbed. Much of the incident energy could be reflected or could pass through, i.e., be transmitted, without being absorbed. Thus, a person behind a glass window or a wooden wall could receive a lethal number of rads with no effect on the window or the wall. *Note 4:* 1 erg = 1 dyne•centimeter. 1 dyne = 1/980 of a gram of force, or about 10^{-5} N (newtons). 1 J (joule) = 10^7 ergs. *See also* **absorption, radiation hardness.**

radiation absorption: *Synonym* **absorption.**

radiation absorption factor: The ratio of (a) the amount of absorbed thermal energy to (b) the total radiant energy other than heat, i.e., the electromagnetic radiation other than heat, incident upon a given surface area. *Note:* For solar radiation, the radiation absorption factor is called the "solar absorption factor."

radiation angle: In fiber optics, half the vertex angle of the cone of light emitted at the exit face of an optical fiber. *Note:* The cone boundary usually is defined by the angle at which the far field irradiance has decreased to a specified fraction of its maximum value or as the cone within which there is a specified fraction of the total radiant power at any point in the far field. *Synonym* **output angle.** *See also* **acceptance angle, far field region, numerical aperture, optical fiber.**

radiation balance: *See* **Earth radiation balance.**

radiation control: *See* **electromagnetic radiation control.**

radiation damage: *See* **optical fiber radiation damage.**

radiation device: *See* **restricted radiation device.**

radiation effects on electronics: *See* **transient radiation effects on electronics.**

radiation efficiency: At a given frequency, the ratio of (a) the radiant power emitted by a radiator to (b) the power supplied to the radiator. *See also* **antenna, radiant efficiency, radiant emittance.**

radiation field: *Synonym* **far field.**

radiation-hardened optical fiber: An optical fiber made with core and cladding materials that recover their attenuation rate, or their attenuation term, within an acceptable period after exposure to a radiation pulse, such as an electromagnetic pulse (EMP). *See also* **attenuation rate, attenuation term, optical fiber.**

radiation hardness: The ability of a system, such as a fiber optic communications system, a fiber optic sensor, an electronic computer, a control system, or a data processing system, to function satisfactorily after exposure to a given irradiance level, i.e., radiant power, distributed over a given area of the system, for a specified time or after absorbing a given number of rads, i.e., radiation absorbed dose. *See also* **irradiance, radiant power, radiation absorbed dose.**

radiation hazard: *See* **electromagnetic radiation hazard.**

radiation mode: **1.** In an open waveguide, such as an optical fiber, a leaky mode, i.e., any mode that is not a bound mode. **2.** In an optical fiber, a mode that (a) has a component of radiant power in a direction transverse to the direction in which power is intended to be transferred or guided, i.e., transverse to the optical fiber axis, (b) has electric and magnetic fields that (i) are transversely oscillatory everywhere external to the waveguide and (ii) exist even at the limit of zero wavelength, (c) satisfies the relation $\beta \leq [(n_r k)^2 - (\mathring{a}/r)^2]^{1/2}$ where β is the imaginary part, i.e., the phase term, of the propagation constant, n_r is the refractive index at r, r is the radial distance from the optical fiber axis, k is the free space wavenumber equal to $2\pi/\lambda$ where λ is the source wavelength, and $\mathring{a}$ (an integer) is the azimuthal index of the mode at the point r, and (d) corresponds to a refracted ray in the terminology of geometric optics. *Note:* In some instances, the wavenumber, k, has been expressed as $1/\lambda$. *Synonym* **unbound mode.** *See also* **bound mode, cladding mode, fiber optics, leaky mode, mode, optical fiber, refracted ray.**

radiation pattern: **1.** A diagram that relates irradiance or field strength to direction relative to an antenna. *Note:* Radiation patterns (a) usually refer to planes or the surface of a cone containing the antenna and (b) usually are normalized to the maximum value of the irradiance or the electric field strength. **2.** For an antenna, a polar or spherical coordinate plot of the irradiance or the electric field strength in each radial direction from the antenna or the light source. *Note:* The plot usually is a three-dimensional surface plotted in spherical coordinates in which the length of a radius vector from the origin at the source of the radiation pattern to the surface is proportional to the irradiance or the electric field strength in the direction in which the vector is drawn. The lines of intersection of this surface with planes through the origin usually are displayed as antenna radiation patterns. The radiation pattern of a radio antenna is measured in the far field, i.e., the radiation field. This field is beyond the distance of $2D^2/\lambda$ where D is the radiating length of the antenna, and λ is the wavelength. For light sources, the wavelength is so short that there is only the far field, i.e., the radiation field. **3.** For the exit face of an optical fiber or a fiber optic bundle, the curve of the irradiance plotted as a function of the angle between the optical fiber axis or the fiber optic bundle and a normal to the surface on which the irradiance is being measured, i.e., the output irradiance versus direction of measurement relative to the optical axis. *Note:* The radiation pattern for a light source or a radio antenna usually is shown for the two principal planes, namely azimuth and elevation. **4.** The radiant emittance of a source as a function of direction. *Note:* The radiation pattern is represented graphically for the far field conditions in either horizontal or vertical planes. *Synonym* **directivity pattern.** *See* **equilibrium radiation pattern, near field radiation pattern.** *See also* **antenna, beamwidth, far field diffraction pattern, main beam, receiving pattern.** *Refer to* **Fig. R-2.** *Refer also to* **Fig. R-3.** *Refer to* **Table 9, Appendix B.**

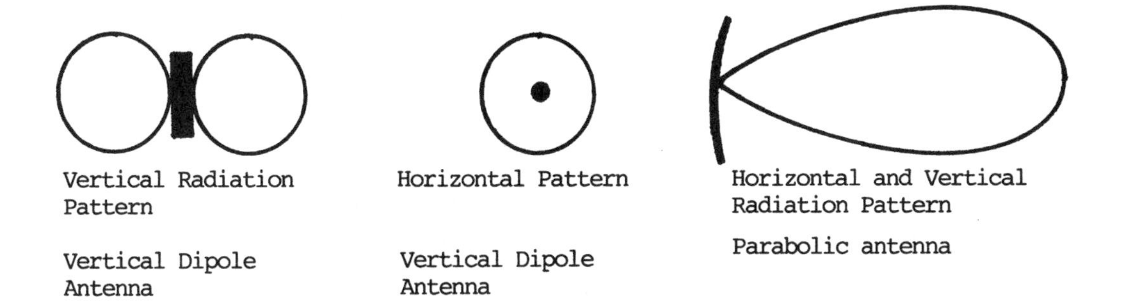

Fig. R-2. Radiation patterns show the distribution of electromagnetic energy radiated by an antenna as a function of direction.

radiation pattern plotter: A device that measures the light intensity, i.e., the irradiance, in various directions from a source relative to a reference line or plane on the source, such as the front emitting surface of a laser or a light-emitting diode (LED). *Refer to* **Fig. R-3.**

radiation resistance: The resistance that, if inserted in place of an antenna, would consume the same amount of power that can be radiated by the antenna. *See also* **antenna, antenna matching, dummy load.**

radiation scattering: The diversion of radiation, such as thermal, electromagnetic, and nuclear radiation, from its original path as a result of interaction or collisions with atoms, molecules, or larger particles in the atmosphere or other propagation media between the source of radiation, such as a light-emitting diode (LED), laser, or nuclear explosion, and a point some distance away. *Note 1:* As a result of radiation scattering, free neutrons and electromagnetic radiation, such as high frequency radio waves and gamma rays, will be received at a point from many directions instead of only from the direction of the source. *Note 2:* In an optical fiber, radiation scattering contributes to signal attenuation and pulse dispersion, i.e., pulse spreading or broadening. *See also* **absorption index, electromagnetic radiation, ionospheric scatter, propagation medium, pulse dispersion, radiation, scatter.**

radiation source: *See* **primary radiation source, secondary radiation source.**

radiation suppression: *See* **infrared radiation suppression.**

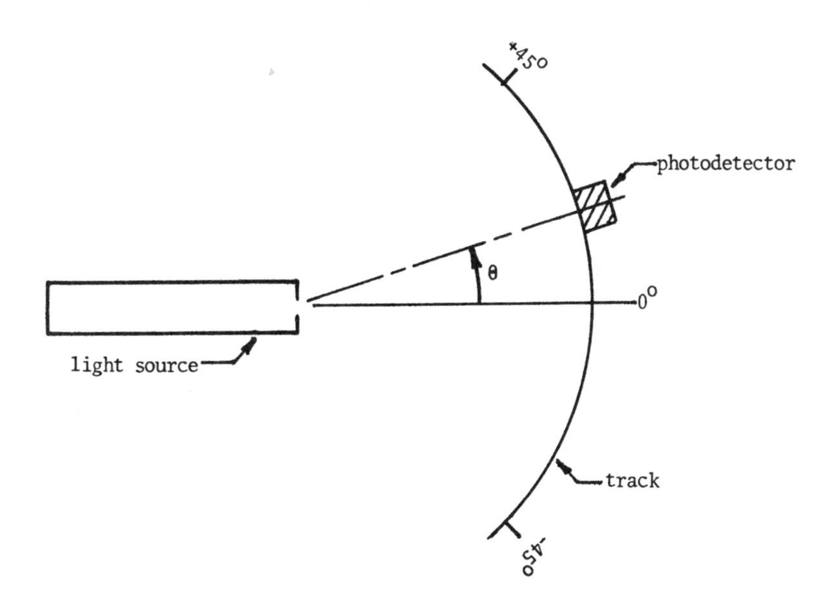

Fig. R-3. A **radiation pattern plotter** for measuring the radiation pattern of a particular light source.

radiation temperature: *See* **total radiation temperature.**

radiation transfer index: In a radiation transmission system, the 1/10th root of the overall transmission efficiency, given by the relation $RTI = \alpha^{1/10}$ where *RTI* is the radiation transfer index and α is the ratio of (a) the value of the radiation energy, radiation power density, optical power, irradiance, or number of photons at the end of a system for which the RTI is desired to (b) the value, expressed in the same units, at the beginning. *Note:* Because dB is defined as $dB = 10\log P_{in}/P_{out} = -10\log P_{out}/P_{in} = 10\log_{10}\alpha^{-1}$, it follows that $RTI = 10^{-dB/100}$ and that $dB = -100\log_{10}(RTI)$. *See also* **irradiance, transmission efficiency, value.**

radiative recombination: In an electroluminescent diode in which electrons and holes are injected into the p-type and n-type regions by application of a forward bias, the recombination of injected minority carriers with the majority carriers in such a manner that the energy released upon recombination results in the emission of photons of energy *hf,* where *h* is Planck's constant, and *f* is the frequency of the radiation, i.e., of the photon, which is approximately equal to the band gap energy. *Note:* Radiative recombination produces the light in a light-emitting diode (LED), which can be modulated for signaling purposes using optical fibers for transmission or optical integrated circuits (OICs) for switching. *See also* **electroluminescent diode, nonradiative recombination.**

radiator: A device that emits radiation. *Note:* Examples of radiators are (a) radio, radar, television, and microwave transmitter antennas, (b) light sources, such as lasers, light-emitting diodes, fluorescent bulbs, and incandescent bulbs, and (c) radioactive materials. *See* **Lambertian radiator.** *See also* **radiation.**

radii loss: *See* **mismatch of core radii loss.**

RADINT: radar intelligence.

radio: 1. Pertaining to the use of electromagnetic waves in the frequency band between 10 kHz (kilohertz) and 3000 GHz (gigahertz), i.e., 3 THz (terahertz). **2.** Communication by modulating, radiating, receiving, and demodulating electromagnetic waves. **3.** A device, or pertaining to a device, that transmits or receives electromagnetic waves that are in the frequency band between 10 kHz (kilohertz) and 3 THz (terahertz). *See* **adaptive high frequency radio, adaptive radio, analog radio, automated radio, combat net radio, high frequency radio, teleprinter on radio.** *See also* **electromagnetic wave, Hertzian wave, International Telecommunication Union–R, modulation, radio wave.** *Refer to* **Fig. R-4.**

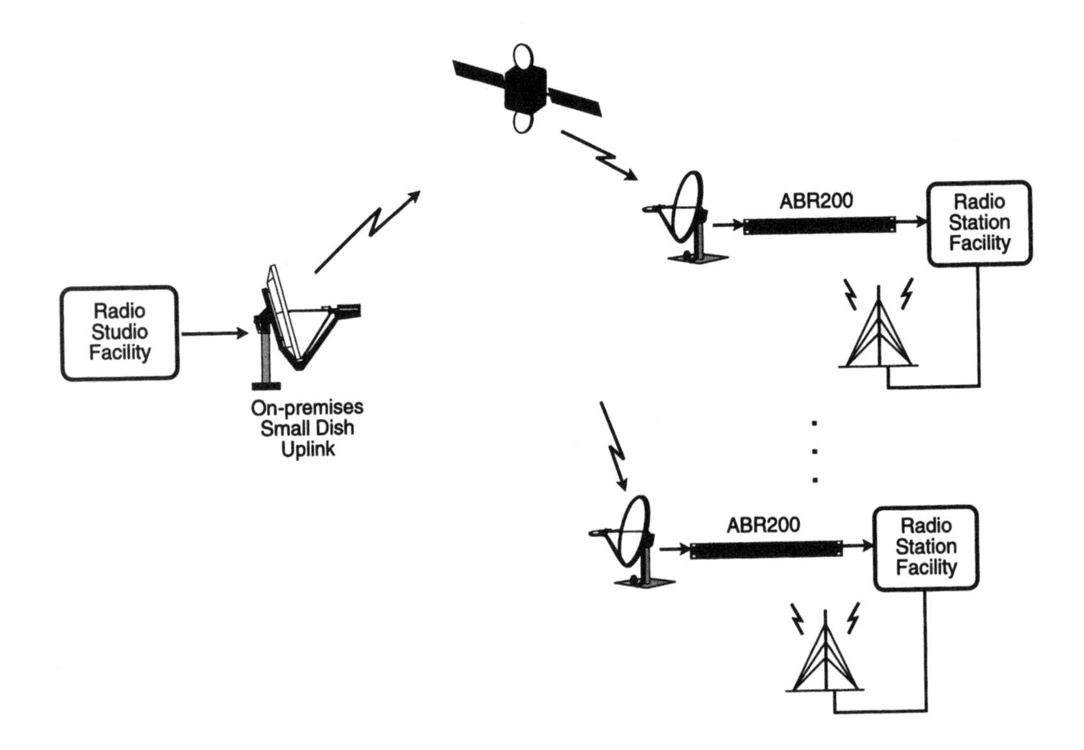

Fig. R-4. Radio program distribution via satellite. (Courtesy ComStream Company.)

radioactive atom: An atom in which electrons and other atomic components are undergoing energy band transitions that result in absorption or emission of high energy quanta. *See also* **absorption, emission, energy band, quantum.**

radioactive particle: A particle of matter that is attached to radioactive atoms and therefore seems to be radioactive itself.

radioactivity: The activity of radioactive atoms, such as the emission or the absorption of gamma radiation, X rays, or other high energy quanta. *See also* **gamma radiation, quantum, radioactive atom, X rays.**

radio altimeter: 1. Radionavigation equipment that (a) is on board an aircraft or a spacecraft and (b) is used to determine the height of the aircraft or the spacecraft above the Earth's surface or another surface. **2.** An instrument that (a) displays the distance between the aircraft datum and the Earth's surface vertically below as determined by a reflected radio/radar transmission and reception, (b) uses beamed and reflected electromagnetic energy to measure height above ground by means of a phase displacement between the transmitted radio signal and its echo reflected from the ground, and (c) makes use of the fact that the altitude in meters is 150 $m \cdot \mu s^{-1}$ (meters per microsecond) of phase displacement.

radio and wire integration: *Synonym* **radio-wire integration.**

radio and wire integration device: *Synonym* **radio-wire integration device.**

radio approach aid: Equipment that makes use of radio to determine the position of an aircraft with high accuracy for the time it is in the vicinity of an airfield until it reaches a position from which landing can be carried out by direct visual means. *See also* **ground controlled approach landing, ground controlled approach system, instrument landing system, radio.**

radio astronomy: Astronomy based on the reception of radio waves of cosmic origin.

radio astronomy service: A service in which radio astronomy is used. *See also* **radio astronomy.**

radio astronomy station (RAS): A station that (a) is in the radio astronomy service and (b) is always a receiving station.

radio autocontrol: The control of an object by means of radio signals that are launched from the object and reflected back to the object from other objects. *Note:* Examples of radio autocontrol are (a) control of the height above ground of an aircraft by ground reflection of signals originating from the aircraft, and (b) control of the flight path of an aircraft by reflection of signals from obstacles such as mountains and towers. *See also* **radio.**

radio baseband: The baseband of radio transmission of a radio station or network. *Note:* An example of a radio baseband is the frequency band that is available for all radio signaling, data, and telephone call channels in a radiotelephone system. *See* **baseband.**

radio baseband receive terminal: The point in the baseband circuit that is nearest to the radio receiver and from which connection usually is made to the multiplex baseband receive terminal or to an intermediate facility. *See also* **multiplex baseband receive terminal, radio baseband send terminal.**

radio baseband send terminal: The point in the baseband circuit that is nearest to the radio transmitter and from which connection usually is made to the multiplex baseband send terminal or to an intermediate facility. *See also* **multiplex baseband send terminal, radio baseband receive terminal.**

radio base station: *See* **airport land mobile radio base station.**

radiobeacon: 1. A transmitter that (a) emits a distinctive or characteristic electromagnetic signal that is used for the determination of bearings, courses, or location in radionavigation systems. **2.** The electromagnetic signal, or the device that transmits the signal, that is used to determine the bearing of the transmitter from a distant point using direction finding (DF). **3.** A radio/radar station that (a) determines the location an object by radar and (b) transmits bearing information to the object by radio. *Synonym* **homing beacon.** *See* **satellite emergency position-indicating radiobeacon.** *See also* **bearing.**

radiobeacon station: A station in the radionavigation service that emits an electromagnetic signal that enables a mobile station to determine its bearing in relation to the radio beacon station. *See* **aeronautical radiobeacon station, emergency position-indicating radiobeacon station, marine radiobeacon station.** *See also* **bearing.**

radio beam: In a directional antenna, a radiation pattern in which the energy of the transmitted radio wave is confined to a small angle in at least one dimension, i.e., the beamwidth is concentrated into a sector that is small in either azimuth, or elevation, or both. *See also* **antenna, antenna lobe, beamwidth, field intensity, field strength, radiation pattern.**

radio broadcast day: The scheduled period of time that a radio station transmits, usually during a single day. *Note:* An example of a radio broadcast day is the period from 6:00 A.M. to 9:00 P.M. each day that the station is actively broadcasting. Stations usually indicate the beginning and the end of their radio broadcast day. *Synonym* **radio day.**

radio channel: An assigned band of frequencies sufficient for radiocommunications. *Note 1:* The bandwidth of a radio channel depends upon the type of transmission and the frequency tolerance. *Note 2:* A channel usually is assigned for a specified radio service to be provided by a specified transmitter using a specified modulation scheme. The particular service performs a specific communications function, such as police, amateur, medical telemetry, aeronautical, broadcast, or military communications. *See also* **assigned frequency, channel, frequency.**

radio circuit: In radiotelegraph communications, a circuit between two stations that has at least one radio link, i.e., at least one segment of the connection path makes use of electromagnetic waves in free space. *See also* **free space, path, radio link.**

radio common carrier (RCC): A common carrier that (a) provides Public Mobile Service and (b) does not provide landline local central office (C.O.) telephone service. *Note:* Radio common carriers (RCCs) were formerly called "miscellaneous common carriers." *See also* **common carrier, radio.**

radiocommunications: 1. Communication by means of radio waves. **2.** The use of electromagnetic waves in the radio frequency (rf) spectrum, i.e., 10 kHz (kilohertz) to 3 THz (terahertz), of the electromagnetic spectrum in free space for communications purposes, i.e., the use of radio waves that are not guided between the transmitting and receiving antennas by installed artificial physical paths, such as wire, coaxial cables, metallic waveguides, and optical fibers. *See* **space radiocommunications, terrestrial radiocommunications.** *See also* **electromagnetic spectrum, free space, path, radio frequency.**

radiocommunications copy watch: To maintain a continuous copy watch while keeping a complete radio station log. *See also* **copy watch, radiocommunications cover, station log.**

radiocommunications cover: To maintain a continuous copy watch (a) while keeping a calibrated transmitter available but not necessarily ready for immediate use, and (b) in which maintaining a complete radio station log is optional. *See also* **copy watch, guard watch, listening watch, radiocommunications copy watch, station log.**

radiocommunications guard: 1. A radiocommunications station that is designated to (a) listen for and record transmissions and (b) handle traffic on a designated frequency for certain other units or organizations. **2.** To maintain a continuous copy watch along with a transmitter ready for immediate use. **3.** In radiocommunications, a ship station, aircraft station, or other mobile station that is designated to copy, cover, guard, or listen for transmissions and to handle traffic on a designated frequency for other mobile stations. *See also* **copy watch, cover watch, guard watch, listening watch.**

radiocommunications service (RCS): A service that (a) includes the transmission, emission, or reception of radio waves for specific telecommunications purposes and (b) relates to terrestrial radiocommunications. *See also* **space radiocommunications.**

radiocommunications system: A communications system in which radio is used as the propagation medium. *See also* **communications system, propagation medium, radio.**

radio control: The control of a device through the use of electromagnetic waves, such as the control of the movement of an aircraft, vehicle, missile, or other mobile unit, either manned or unmanned, from a radio station on the ground or in another mobile unit. *See also* **radio autocontrol.**

radioconverter: *See* **telegraph radioconverter.**

radio coverage diagram: A diagram that shows the area within which a radio station is broadcasting an effective signal strength in relation to a given standard. *Note:* An example of a radio coverage diagram is a polar plot, in each direction from the antenna, of the distance from the antenna at which the signal strength is equal to a specified value, i.e., it is the locus of all points at which the signal strength is equal to a specified value.

radio day: *Synonym* **radio broadcast day.**

radio deception: 1. The use of radio to deceive hostile forces. *Note:* Radio deception includes the use of radio to perform deceptive actions, such as sending false dispatches, using deceptive headings, and using the call signs of hostile forces. **2.** The use of radio transmission to deceive the recipient of the transmitted signals. *Note:* Examples of radio deception include the sending of false messages, deceptive headings, and misleading call signs by radio. *See also* **electronic countermeasures, electronic warfare.**

radio detection: The use of radio to detect the presence of an object by radiolocation without precisely determining the position of the object. *See also* **monitoring.**

radio detection and ranging: *See* **radar.**

radiodetermination: The determination of (a) the parameters of an object, such as its location, orientation, rotation, velocity, size, and shape, and (b) information relating to these parameters by means of the propagation properties of radio waves. *See also* **radio goniometry, radio range finding.**

radiodetermination satellite service: A radiocommunications service for the purpose of radiodetermination that includes the use of one or more space stations.

radiodetermination service: A service in which radio is used to determine (a) the parameters, such as position and velocity, of an object and (b) information relating to these parameters by means of the propagation properties of radio waves. *See also* **radiodetermination satellite service.**

radiodetermination station: A station in the radiodetermination service. *See also* **radiodetermination.**

radio direction finding (RDF): 1. Radio location in which only the direction to a station is determined by means of its emissions. **2.** Radiodetermination using the reception of radio waves for the purpose of determining the direction to a station or an object. *See also* **monitoring, radio fix, radio location.**

radio direction finding equipment: Radio receiving equipment that (a) is used to obtain bearings of transmitting stations and (b) has a receiver in the form of an automatic direction finder that automatically and continuously indicates the direction to a transmitter from the receiver. *See also* **direction finder, direction finding, radio.**

radio direction finding station: A radiodetermination station that (a) uses radio direction finding and (b) determines the direction to other stations by means of radio transmission from those stations. *See also* **direction finding, radiodetermination, radio direction finding.**

radio Doppler: The determination of the radial component of the relative velocity between a source of radio waves and another object by (a) measuring the difference in frequency that occurs between the radio waves that are emitted from the source and the waves that are reflected from the object, i.e., from the echoes that are returned from the object, and (b) calculating the radial component of velocity by using the frequency difference. *Note 1:* If the echo frequency is higher than the transmitted frequency, there is a component of velocity of the object toward the transmitter, and a velocity component is radially away from the transmitter if the echo frequency is lower. *Note 2:* If radio frequency (rf) pulses are used, the difference in transmitted and received pulse repetition rates determines the radial component of velocity. *See also* **echo, radio wave.**

radio equation: *See* **one-way radio equation.**

radio fadeout: *Synonym* **flutter.**

radio failure: The inability to receive messages, transmit messages, or both, by radio. *Note:* In the event of radio failure, other means of communication are used, such as wire, coaxial cable, optical fiber, or visual signaling, such as flag, pennant, panel, semaphore, heliograph, wig-wag, or pyrotechnic signaling. Aircraft experiencing radio failure fly a triangular pattern to the right if only the receiver is operating, and to the left if the receiver and the transmitter are both inoperative, in an attempt to establish radar contact.

radio field intensity: *Synonym* **field strength.**

radio fix: 1. The locating of a radio transmitter, such as might be on board a ship or an aircraft, by determining the intersection of bearings taken from two or more direction finding stations, the site of the transmitter being at the point of intersection. **2.** The location of a receiver, such as might be on board a ship or an aircraft, by means of resection, i.e., by determining the direction of radio signals coming to the receiver from two or more sending stations, the locations of which are known. *See also* **monitoring, radio direction finding, radionavigation.**

radio fixing aid: Equipment that makes use of radio to assist in the determination of the geographical location of transmitting or receiving stations, i.e., to obtain a radio fix.

radio frequency: In the electromagnetic spectrum, a frequency that (a) usually is associated with radio wave propagation and (b) is considered to lie between 3 kHz (kilohertz) and 3000 GHz (gigahertz), i.e., 3 Thz (terahertz). *Note:* Radio frequency (rf) nomenclature is as follows:

Frequency Designator	Frequency Range	Metric Subdivision
VLF	3–30 kHz	myriametric waves
LF	30–300 kHz	kilometric waves
MF	300–3000 kHz	hectometric waves
HF	3–30 MHz	decametric waves
VHF	30–300 MHz	metric waves
UHF	300–3000 MHz	decimetric waves
SHF	3–30 GHz	centimetric waves
EHF	30–300 GHz	millimetric waves
THF	300–3000 GHz	decimillimetric waves

See also **electromagnetic spectrum, frequency, frequency spectrum designation, radio, radio wave.**

radio frequency allotment (RFA): A designated frequency in an agreed-upon plan, adopted by a competent conference or a recognized authority, for use by one or more administrations for a terrestrial or space radiocommunications service in one or more identified countries or geographical areas and under specified conditions. *See also* **adaptive channel allocation, assigned frequency, assigned frequency band, authorized frequency.**

radio frequency assignment (RFA): 1. An authorization, given by an administration, or by the appropriate competent or recognized authority, for a radio station to use a radio frequency under specified conditions. **2.** The authorizing of a specific frequency, group of frequencies, or frequency band to be used at a certain location under specified conditions, such as bandwidth, power level, azimuth, duty cycle, and modulation. *Note:* The International Frequency Registration Board (IFRB), a permanent organization of the International Telecommunication Union (ITU), maintains a Master International Frequency Register (MIFR). The MIFR is a master list of all international frequency registrations. In each country or group of countries, specific assignments within national frequency allocations are made by their respective national frequency management organizations, such as, in the United States, the Federal Communications Commission (FCC) for nonfederal government organizations and the National Telecommunications and Information Administration (NTIA) for federal government organizations. *Synonym* **frequency assignment.** *See also* **assigned frequency, authorized frequency, frequency, frequency band allocation, frequency band allotment.**

radio frequency bandwidth: 1. A bandwidth that is within the radio frequency region of the electromagnetic spectrum. **2.** In a radio transmitter, the difference between the highest and lowest values of the emission frequencies in the region of the carrier frequency. *Note:* In single-channel emission, the radio frequency bandwidth is the region of the carrier frequency beyond which the amplitude of any frequency, such as those that result from modulation, those that are subcarrier frequencies, or those that result from distortion products, is less that 5%, i.e., −26 dB, of the rated peak output amplitude of the carrier or a single-tone sideband, whichever is greater. For multiplex emission, the radio frequency bandwidth is the same, except that the 5% applies to the subcarrier or a single-tone sideband of the carrier, whichever is greater. *Synonym* **rf bandwidth.** *See also* **electromagnetic spectrum, necessary bandwidth, nominal bandwidth, occupied bandwidth, radio frequency, value.**

radio frequency channel allotment (RFCA): A designated frequency channel in an agreed-upon plan, adopted by a competent conference or a recognized authority, for use by one or more administrations for a terrestrial or space radiocommunications service in one or more identified countries or geographical areas and under specified conditions. *See also* **adaptive channel allocation, assigned frequency, assigned frequency band, authorized frequency.**

radio frequency channel assignment: 1. An authorization, given by an administration, or by the appropriate competent conference or recognized authority, for a radio station to use a radio frequency channel under specified conditions. **2.** The authorizing of a specific radio frequency channel, a group of radio frequency channels, or frequency bands to be used at a certain location under specified conditions, such as bandwidth, power level, azimuth, duty cycle, and modulation. *See also* **frequency band allocation, frequency band allotment, assigned frequency, authorized frequency, frequency.**

radio frequency channel increment: The frequency separation between adjacent channels in a multichannel transmission system.

radio frequency combining: The combining of a number of multichannel trunks for different transmissions, such as voice and digital data transmissions, over a single wideband facility where the combining action occurs at radio frequencies, as opposed to video or audio frequencies. *See also* **combining, radio frequency.** *Synonym* **frequency division multiplex combining.**

radio frequency interference (RFI): 1. Interference that is generated or induced in electronic circuits and is in the radio frequency region of the electromagnetic spectrum. **2.** Electromagnetic phenomena that either directly or indirectly contribute to a degradation in performance of a system, such as a radio receiver. *See also* **radio frequency.**

radio frequency intermodulation: A constituent of nonlinear distortion that (a) consists of the occurrence of harmonics in the response of electrical components and (b) is caused by intermodulation distortion in the radio frequency (rf) states of a receiver. *See also* **harmonic, intermodulation, intermodulation distortion, nonlinear distortion.**

radio frequency power margin: An amount of transmitter power above that which is required to meet calculated values of desired signal levels at specified distances from the antenna. *Note:* The radio frequency (rf) power margin allows for uncertainties in (a) empirical components of the signal level prediction method, (b) terrain characteristics, (c) atmospheric conditions, and (d) equipment performance parameters. *See also* **design margin, fade margin.**

radio frequency pulse: A train, string, or packet of radio frequency (rf) waves that have an envelope in the shape of a pulse. *Note:* The number of waves in a radio frequency pulse is given by the relation $n = ft$ where n is the number of waves, i.e., of cycles, f is the radio frequency of the waves in the pulse, and t is the pulse duration. Thus, a 20-MHz (megahertz) pulse that has a duration of 2 μs (microseconds) has 40 wave cycles in the pulse, assuming that the frequency remains constant during the period of the pulse. *See also* **envelope, pulse, pulse shape, wave shape.**

radio frequency shift: A frequency shift, i.e., a frequency change, of a frequency in the radio frequency region of the electromagnetic spectrum. *See also* **electromagnetic frequency spectrum, frequency shift, radio frequency.**

radio frequency system: *See* **wired radio frequency system.**

radio frequency tight: Pertaining to a high degree of electromagnetic shielding effectiveness. *See also* **TEMPEST.**

radio goniometry: The determination of the relative direction of a distant object by means of its radio emissions, whether the emissions are independent, reflected, or automatically retransmitted. *See also* **direction finding, radiodetermination, radio range finding.**

radiogram: *Synonym* **radio telegram.**

radio guard: A ship, aircraft, or land radio station designated to (a) listen for and record transmissions and (b) to handle traffic on a designated frequency for specified units or for other stations. *See also* **monitoring, traffic.**

radio horizon: The locus of the points at which direct rays from an antenna are tangential to the Earth's surface. *Note 1:* On a spherical surface, the radio horizon is a circle. *Note 2:* The distance to the radio horizon is affected by (a) the height of the transmitting antenna, (b) the height of the receiving antenna, and (c) atmospheric refraction. *See also* **horizon angle, line of sight propagation, radio horizon range, smooth Earth.**

radio horizon range (RHR): The distance at which a direct radio wave can reach a receiving antenna of given height from a transmitting antenna of given height. *Note:* The radio horizon range in nautical miles, R, is given by the relation $R = 1.23\ (h_t^{1/2} + h_r^{1/2})$ where h_t and h_r are the heights of the transmitting and receiving antennas in feet. The radio horizon range, R, in nautical miles is also given by the relation $R = 2.23(h_t^{1/2} + h_r^{1/2})$ where h_t and h_r are the heights of the transmitting and receiving antennas in meters. The effective Earth radius, 4/3 times the actual Earth radius, is used in deriving the formulas. Second-order differentials are neglected. They are on the order of 0.1%. *See also* **radar horizon, radar line of sight equation, radio horizon.**

radio installation: In a radio station, all the equipment that (a) is required to make the station fully operational for use and (b) includes various equipment and components, such as transmitters, receivers, antennas, feeders, transducers, amplifiers, and modulation equipment as well as the structure or the container in which it is housed, such as a case, vehicle, container, or building.

radio landing aid: Radio equipment that is used to assist in the landing of an aircraft. *See also* **ground controlled approach landing, ground controlled approach system, instrument landing system, radio.**

radio link: Between two points, a connection that (a) is made by radio for communications purposes and (b) usually is part of a transmission path that also contains other propagation media. *See also* **point, propagation medium, radio.**

radiolocation: Radiodetermination used for purposes other than for radionavigation, such as for determining the position of a transmitter or a receiver by triangulation. *See also* **radiodetermination, radionavigation.**

radiolocation land station (RLLS): In the radiolocation service, a station not intended to be used while in motion. *See also* **radiolocation, radiolocation service.**

radiolocation mobile station (RLMS): In the radiolocation service, a station not intended to be used while in motion or during halts at unspecified points. *See also* **radiolocation, radiolocation service.**

radiolocation service: A radiodetermination service that also provides a radionavigation service.

radiolocation station: A station in the radiolocation service.

radio log: In a radio station, a continuing record that (a) may be chronological or structured, (b) describes significant events, important messages, and other matters that pertain to the communications operations of the station, and (c) may include on- and off-watch operator signatures, dates, call signs, schedules copied, date-time groups of messages sent, frequencies used, data on received messages, radio silence periods, and alarm signals used. *See also* **ship radio log.**

radio logical monitoring: *See* **monitoring.**

radiology: *See* **teleradiology.**

radiometer: An instrument that measures electromagnetic radiant intensity, i.e., measures irradiance. *See* **fiber optic visible/infrared spin-scan radiometer.** *See also* **irradiance.**

radiometric: Pertaining to the use of electromagnetic waves to measure physical parameters, such as the distance to an object, the height of the ionosphere, the refractive index profile of a dielectric waveguide, or the attenuation rate in an optical fiber. *See also* **electromagnetic wave, radiometry, sounding.**

radiometry: The branch of science and technology devoted to the measurement of electromagnetic radiation. *Note:* In lightwave communications and the use of optical fibers, primary interest is in radiometry rather than photometry, but not to the exclusion of photometry.

radionavigation: 1. Radiolocation intended for the determination of location, direction, or obstruction warning in navigation. *See also* **radio fix. 2.** Radiodetermination used for navigation and obstruction warning. *See also* **radiodetermination, radiolocation.**

radionavigation aid: 1. A radio facility that is used in navigation. **2.** Radio equipment that uses radio waves, i.e., electromagnetic waves between 10 kHz (kilohertz) and 3000 GHz (gigahertz), for purposes of assisting aircraft stations in air navigation and assisting ship stations in maritime navigation or pilotage.

radionavigation land station (RNLS): In the radionavigation service, a station not intended to be used while in motion or during halts at unspecified points.

radionavigation maintenance-test land station: A radionavigation land station and maintenance-test facility in the aeronautical radionavigation service that (a) is used as a radionavigation calibration station for the transmission of essential information in connection with the testing and the calibration of aircraft navigational aids, and (b) usually has receivers, transmitters, and interrogators that are used for maintenance and testing of aircraft communications equipment installed at predetermined surface locations during testing operations. *See also* **radionavigation operational-test land station.**

radionavigation mobile station (RNMS): In the radionavigation service, a station intended to be used while in motion or during halts at unspecified points.

radionavigation operational-test land station: A radionavigation land station and operational-test facility in the aeronautical radionavigation service that (a) is used as a radionavigation calibration station for the transmission of essential information in connection with the testing and the calibration of aircraft navigational aids, receiving equipment, and interrogators at predetermined surface locations during testing operations, and (b) is primarily used by pilots to check radionavigation systems in an aircraft prior to its takeoff. *See also* **radionavigation maintenance-test land station.**

radionavigation satellite Earth station: An Earth station in the radionavigation satellite service.

radionavigation satellite service (RNSS): 1. A radiodetermination satellite service used for radionavigation. *Note:* The radionavigation satellite service may provide feeder links. **2.** A service in which space stations on Earth satellites are used for the purpose of radionavigation, including the transmission or the retransmission of supplementary information that may be necessary for the operation of the radionavigation system. *See* **aeronautical radionavigation satellite service.**

radionavigation satellite space station: A space station that (a) is in the radionavigation satellite service and (b) is on board an Earth satellite.

radionavigation service: A radiodetermination service that provides radio aids for aeronautical, maritime, and land mobile navigation. *See* **aeronautical radionavigation service, maritime radionavigation service.**

radionavigation station: A station in the radionavigation service.

radio net: 1. An organization of radio stations that conduct direct communications on a common frequency. **2.** An organization of radio stations that broadcast common programming, not necessarily simultaneously, at different frequencies from different locations. *See also* **network.**

radio noise: *Synonym* **cosmic noise.**

radio officer: In an organization, the person responsible for managing radio operations for the organization. *See* **senior ship radio officer.**

radio officer of the watch: The radio officer or operator who is responsible for a specified radio watch for specified periods during the voyage of a ship.

radio operator: In radiocommunications system operation, a person who, during a radio watch, (a) operates radio transmission and receiving equipment for the transmission and the reception of messages, (b) makes minor repairs and adjustments to the equipment, and (c) maintains a radio log when required.

radio organization: *See* **maritime air radio organization.**

radio organization chart: A chart that illustrates interrelated fixed and mobile radio nets.

radio pager: A pocket-size radio receiver capable of alerting its wearer that there is a phone call, either from a displayed phone number or to a predesignated number. *Synonym* **beeper.**

radio paging: The use of a pocket-size radio receiver capable of alerting its wearer that there is a phone call, either from a displayed phone number or to a predesignated number. *Synonym* **beeping.**

radiophoto: A facsimile transmission system in which radio waves, i.e., waves in the radio frequency region of the electromagnetic spectrum, are used for the transmission of signals that are generated by a scanning transmitter and a recording receiver.

radiopositioning land station: A station in the radiolocation service that is not a radionavigation station and is not intended for operation while in motion. *See also* **radiopositioning mobile station.**

radiopositioning mobile station: A station in the radiolocation service that is not a radionavigation station and is intended for operation while in motion or during unspecified temporary halts while on board a vehicle. *See also* **radiopositioning land station.**

radio position line determination: The determination of a position line by means of radiolocation.

radio range (RR): 1. The distance from a transmitter at which the signal strength remains above the minimum usable level for a particular antenna and receiver combination. **2.** The maximum distance from a transmitter at which the signal level is above a specified value. **3.** A radio aid to air navigation that creates an infinite number of paths in space throughout a given sector or azimuth angle by various methods of transmission and reception of electromagnetic waves. **4.** A range in which a series of radio stations is used for aircraft or missile control or for flight path measurements. *See* **omnidirectional radio range, very high frequency omnidirectional radio range, visual-aural radio range.**

radio range finding: Radiolocation in which the distance to an object is determined by means of its radio emissions whether independent, reflected, or retransmitted on the same or another wavelength, i.e., frequency. *See also* **radiodetermination, radio goniometry, radio location.**

radio range station (RRS): A radionavigation station in the aeronautical radionavigation service that (a) provides radio equisignal zones for location determination by a receiver and (b) usually is at a fixed position on land, but may also be on board a ship, aircraft, spacecraft, or land vehicle.

radio receiver: A receiver that accepts radio frequency electromagnetic waves and selectively converts them, or makes them available for conversion, into intelligible and useful form.

radio recognition: 1. The radiodetermination of the character, individuality, or nature of a received radio signal and the drawing of conclusions concerning the transmitting station and its operators, based on signal analysis. **2.** In military systems, the determination by radio means of the friendly or hostile character or individuality of another. *See also* **identification friend or foe.**

radio recognition and identification: *Synonym* **identification friend or foe.**

radio relay: 1. The reception and the retransmission by a radio station of signals that are received either from another radio station or from a wire, fiber optic cable, microwave, coaxial cable, or other link of an integrated land line and radiocommunications system component. **2.** A terrestrial point-to-point communications system, such as a microwave relay communications system or a satellite communications system. *Note:* The siting of radio relay stations and the radio coverage diagrams of the antenna radiation patterns are arranged for minimum interference with satellite Earth stations. The analog and digital baseband arrangements are similar to satellite systems. Radio relay links may form part of the connection between an Earth station and a switching center. *See* **airborne radio relay, down the hill radio relay.**

radio relay aircraft: *See* **automatic radio relay aircraft.**

radio relay electronic countermeasure: A measure that is used to counter deliberate efforts to introduce interference, i.e., jamming, into radio relay circuits.

radio relay plane: *See* **automatic radio relay plane.**

radio relay system (RRS): 1. A point-to-point radio transmission system in which the signals are received, amplified, and retransmitted by one or more intermediate radio stations. **2.** A communications system that is used to perform a radio relay function. *See also* **drop and insert, drop channel operation, link, radio relay.**

radio service: Any of the functions and services provided by a radio station, radio net, or radio network, such as message transmission, broadcasting, multicasting, radionavigation signaling, radio direction finding, radiodetermination, radiolocation, radiobeaconing, and radio astronomy. *See* **Armed Forces Radio Service, basic exchange telecommunications radio service, rural radio service, terrestrial radio service.**

radio set: 1. An apparatus that consists of a transmitter, a receiver, an antenna, and related equipment and is used for the transmission and the reception of radio signals. **2.** A radio transmitter and an antenna. **3.** A radio receiver and antenna.

radio silence: In radiocommunications, an electronic systems operational condition in which, during a designated period, (a) designated transmitters are not transmitting, (b) all of certain specified radio equipment that is capable of emitting electromagnetic radiation, including certain receivers, is kept inoperative, (c) radio transmissions are not to be made, and (d) no electronic equipment is to be radiating. *Note:* Radio silence may be required to (a) reduce interference to emergency traffic, (b) prevent confusion to navigational aids or missile guidance, or (c) as a radiocommunications electronic countercountermeasure, prevent the opportunity to gather intelligence through interception and analysis of radio signals. *See* **ship radio silence.** *See also* **communications silence.**

radiosonde: In the meteorological aids service, an automatic radio transmitter that (a) usually is carried on an aircraft, free or tethered balloon, kite, or parachute, and (b) transmits meteorological data. *Note:* The data sent by radiosonde are obtained by sensors, converted by transducers, and telemetered to a ground station.

radiosonde station: A station that is in the meteorological service and uses radiosonde equipment.

radio station: An installation, assembly of equipment, location, or vehicle from which radio signals may be transmitted or received. *Note:* The vehicle may be a spacecraft, aircraft, ship, or land vehicle. *See* **airport land mobile radio station, ship broadcast area radio station, ship broadcast coastal radio station.** *Refer to* **Fig. R-4 (800).**

radio telecontrol: The control of mechanisms, devices, equipment, components, and systems by radio waves sent to them over greater distances than those that occur in radio control. *See also* **radio control.**

radiotelegram (RT): 1. A telegram that is to be sent, is being sent, or has been sent over at least one radio link. **2.** A telegram (a) that originated in, or is intended for, a mobile station or a mobile Earth station, and (b) that is transmitted on all or part of its route over the radiocommunications channels of the mobile service or the mobile satellite service. *Synonym* **radiogram.** *See also* **radiotelephone call.**

radiotelegraph communications: Communications in which telegraph codes are transmitted (a) by electromagnetic waves that (i) are in the radio wave region of the electromagnetic spectrum and (ii) are in waveguides or free space, and (b) not by electrical signals in wires and optical signals in optical fibers.

radiotelegraph distress frequency: The international distress and calling frequency for radiotelegraphy, such as 500 kHz (kilohertz) for use by ship stations, aircraft stations, or survival craft stations, for distress calls, distress messages, urgency signals, safety signals, and other distress and urgency calls. *See also* **distress calls, distress messages, safety signals, urgency signals.**

radiotelegraph lifecraft frequency: *See* **international radiotelegraph lifecraft frequency, radiotelegraph lifecraft frequency.**

radiotelegraph message format: The prescribed arrangement of the parts of a message that has been prepared for radiotelegraph transmission.

radiotelegraph operating method: One of three operating methods for passing messages from one station to another, i.e., (a) the broadcast communications method, (b) the intercept communications method, and (c) the receipt communications method. *Note:* The operating method that is used in a particular situation is determined by the operational requirements. *See also* **broadcast communications method, intercept communications method, receipt communications method.**

radiotelegraph procedure: A procedure that (a) is used in radiotelegraph communications systems, (b) is designed to provide a concise and definitive language whereby radiotelegraph communications may be conducted accurately, rapidly, and with the security that is obtainable in radio circuits, (c) is used in transmissions over radio and line circuits that use the international Morse code as well as other codes, and (d) when commercial messages are handled by noncommercial systems, usually is contained in the message heading or in the calling and routing instructions. *See* **cross band radiotelegraph procedure, simplex radiotelegraph procedure.**

radiotelegraph transmission: The transmission of telegraphic codes by means of radio. *See also* **code, telegraph, telegraphy, telephony.**

radiotelegraphy: **1.** The transmission of telegraph-code messages by means of radio. **2.** The branch of science and technology devoted to the design, development, manufacture, installation, operation, and maintenance of telecommunications systems and equipment that (a) make use of radio to effect communications and (b) enable the reproduction at a distance of any kind of information in any form, such as written, printed, and pictorial form. *See also* **radiophoto, telephony.**

radiotelemetry: Telemetry by means of radio waves. *See also* **radio wave, telemetry.**

radiotelephone: **1.** A telephone equipped with a radio transmitter and receiver that may be used to place calls to similar equipment or to the public switched telephone network. **2.** Pertaining to radiotelephone communications, services, or systems.

radiotelephone call: **1.** A telephone call in which at least one radio link is used. **2.** A telephone call (a) that originated in, or is intended for, a mobile station or a mobile Earth station, and (b) that is transmitted on all or part of its route over the radiocommunications channels of the mobile service or the mobile satellite service. *See also* **call, radio telegram.**

radiotelephone communications: Communications in which sound signals, such as voice, are transmitted by electromagnetic waves that (a) are in the radio wave region of the electromagnetic spectrum and (b) are in waveguides or free space, rather than by electrical signals in wires and optical signals in optical fibers.

radiotelephone distress frequency (RTDF): An international distress and calling frequency for mobile radiotelephone stations, survival craft, and emergency position indicating radio beacons. *Note:* A radio telephone distress frequency is 2180 kHz (kilohertz). *See* **international radiotelephone distress frequency.**

radiotelephone frequency: A carrier frequency that (a) is suitable for modulation by a voice frequency baseband signal, (b) is used in radiotelephone links, radiotelephone networks, and radio transmission systems to transmit signals that represent sound, and (c) is demodulated upon reception. *See* **convoy radiotelephone frequency, maritime mobile radiotelephone frequency.** *See also* **carrier, voice frequency.**

radiotelephone network: A group of radiotelephone stations that (a) operate on common frequencies and (b) support each other in a defined manner to ensure maximum dependability of communications for handling telephone traffic. *Note:* Examples of radiotelephone networks are (a) a cellular telephone network and (b) a group of radiotelephone aeronautical stations that (i) operate on common frequencies, (ii) conduct radio guard operations, and (iii) support each other to ensure maximum dependability of air-ground communications for handling air-ground traffic.

radiotelephone numeral: One of the ten numerals that are enunciated in voice telephone communications systems operations. *See* **ship radiotelephone numeral.**

radiotelephone procedure: A procedure that (a) is used in radiotelephone communications systems, (b) is designed to provide a concise and definitive language whereby radiotelephone communications may be conducted accurately, rapidly, and with the security that is obtainable in radio circuits, and (c) is used in transmissions over radio and line circuits to handle calls. *Note:* An example of a radiotelephone procedure is the call originator identification procedure. *See* **ship to shore radiotelephone procedure.**

radiotelephone service: *See* **air-ground radiotelephone service.**

radiotelephone station: A station in a radiotelephone system or network. *See also* **radiotelephone communications, radiotelephone network, radiotelephone system, station.**

radiotelephone system: A communications system in which (a) a telephone is used by the call originator or the call receiver, and (b) radio links are used in the communications path between them. *See also* **call originator, call receiver, communications system, path, radio link, telephone.**

radiotelephone transmission: The transmission of speech by means of modulated radio waves. *See also* **radiotelephony.**

radiotelephone voice system: A radiotelephone system that transmits sound signals, such as speech, by means of voice-frequency-modulated radio waves. *See* *also* **modulation, radiotelephone system, radio wave, voice frequency.**

radiotelephony: The branch of science and technology devoted to the design, development, manufacture, installation, operation, and maintenance of telecommunications systems and equipment that (a) make use of radio to effect communications and (b) enable the reproduction at a distance of sound signals, such as speech. *See also* **radiotelephone system, radiotelegraphy, telephony.**

radioteleprinter: A teleprinter used in association with radio circuits and radio networks.

radioteletype (RTTY or RATT): A communications system in which teletypewriters, perhaps teleprinters, and radio links are used. *See also* **teletype.**

radioteletypewriter (RTTY or RATT): A teletypewriter used in a radiocommunications system. *See also* **radiocommunications system, teletypewriter.**

radio Telex® call: 1. A Telex® call in which a radio link is used. **2.** A Telex® call (a) that originated in, or is intended for, a mobile station or a mobile Earth station, and (b) that is transmitted on all or part of its route over the radiocommunications channels of the mobile service or the mobile satellite service. *See also* **Telex®**.

radio transmission control: The control and the limitation that is imposed upon radio operation. *Note:* Examples of radio transmission control are (a) the control that limits transmissions to the absolute minimum essential to carry on communications traffic, and (b) the control that ensures that transmission procedures are precisely adhered to. *See* **port radio transmission control.**

radio transmission security: Security measures that are applied to radiocommunications systems and networks. *See* **ship radio transmission security.** *See also* **radio network, radiocommunications system, security.**

radio transmitter: A device used for the production, modulation, amplification, and emission of radio frequency energy that (a) is used for effecting communications, (b) is fed to an antenna or an antenna array, and (c) is coupled to free space as radio waves. *See also* **antenna, antenna array, feeder, free space, modulation, radio frequency, radio wave.**

radio transmitter mean power: During normal operation of a radio transmitting station, the average power supplied to the antenna transmission line by the transmitter during a period sufficiently long compared with that of the lowest frequency encountered in the modulation. *Note:* A period of 0.1 second, during which the mean power is greatest, usually is selected for mean power measurements. *See also* **antenna, directive gain, effective radiated power, radio transmitter.**

radiovoltaic: Pertaining to the production of a voltage that occurs when ionized nuclear alpha and beta particles create electron-hole pairs in the crystal lattice of a semiconductor material. *Note 1:* The effects of nuclear radiation on cadmium sulfide and cadmium telluride thin-film cells have been applied in this form of energy conversion for limited biomedical purposes. Power levels of 1 μW (microwatt) per cell have been achieved. *Note 2:* An alpha particle is a free helium nucleus, and a beta particle is a free electron. *See also* **photovoltaic.**

radio watch: An assigned period of duty of an operator at a radio station, with emphasis on monitoring reception of signals and taking appropriate action. *See also* **copy watch, cover watch, guard watch, listening watch.**

radio watch shift: *Synonym* **area broadcast shift.**

radio wave: 1. An electromagnetic wave with a frequency less than 3000 GHz (gigahertz), i.e., less than 3 THz (terahertz), and usually higher than 10 kHz (kilohertz). **2.** An electromagnetic wave of frequency arbitrarily lower than 3000 GHz, propagated in space without an artificial guide, i.e., without a man-made guide. *Synonym* **Hertzian wave.**

radio-wire integration (RWI): The combination and the interconnection of wire circuits and radio facilities. *Synonym* **radio and wire integration, wire-radio integration.**

radio-wire integration device: An interface device that permits interconnection and communications between wire circuits and radio facilities, such as interconnection of wire links and radio links in a radio-wire integrated communications system. *Synonym* **radio and wire integration device.** *See also* **radio-wire interface.**

radio-wire interface: An interface between a single-channel radio net and a switched communications system. *Synonyms* **net radio interface, wire-radio interface.** *See also* **radio-wire integration device.**

radius: *See* **critical radius, effective Earth radius, minimum bend radius.**

radius sensor: *See* **critical radius sensor.**

radix: The positive integer by which the weight of the digit position in a numeral of a given positional numeration system is multiplied to obtain the weight of the digit position with the next higher weight. *Note:* In the decimal numeration system, the radix of each digit is 10 because the weight of a given digit position has to be multiplied by 10 to obtain the weight of the next higher position. Thus, the weight of the fourth position to the left of the decimal point, i.e., the 1000s position, is obtained by multiplying the weight of the third position, i.e., the 100s position, by 10. In the pure binary numeration system, the radix is 2. In a biquinary code, the radix of each 5s position is 2. *See also* **integer, positional numeration system, pure binary numeration system.**

radix complement: 1. A number obtained by (a) subtracting each digit of a given number from one less than the radix of that digit position, one at a time, (b) adding 1 to the least significant digit of the result, and (c) propagating all carries that result when adding the 1. *Note 1:* The radix complement of decimal 496 is obtained by subtracting each digit from 9 to produce 503. When 1 is added to the 3, 504 is obtained, which is the radix complement of 496. *Note 2:* In most conventional numeration systems, when a number (a) is added to its radix complement, (b) all carries are propagated, and (c) the most significant digit of the sum is deleted, all zeros are obtained. Thus, when 504 is added to 496, 1000 is obtained, and deleting the 1 results in all 0s remaining. **2.** A number obtained by subtracting a given number, the radix complement of which is desired, from a number created by following 1 with a many 0s as there are digits in the given number and generating all the carries as in ordinary arithmetic subtraction. *Note:* If the given number is 544, subtract it from 1000, thus obtaining 456, the radix complement of 544. However, this method may not work too well in a given computer because the operand register may not be large enough to hold the that fourth digit, i.e., the 1. *See also* **radix minus ones complement.**

radix minus ones complement: A number obtained by subtracting each digit of a given number from one less than the radix of that digit position, one at a time. *Note:* The radix minus ones complement of decimal 496 is 503, obtained by subtracting each digit from 9. The radix minus ones complement of pure binary 1101 is 0010, obtained by subtracting each digit from 1, or simply by reversing each digit. *Note:* In a bank of flip-flops, a given number and its radix minus ones complement are both always present for use in arithmetic operations. To set a register to all 1s, add any number to its radix minus ones complement. To set it to all 1, add 1 to the result, and propagate all the carries. To subtract two numbers, add the radix minus ones complement of the subtrahend and add 1 to the result, propagating all the carries. The most significant digit has to be deleted if an additional digit position is created. *See also* **radix complement.**

radome: A cover that (a) protects (i) a radiator, such as a radio transmitter antenna, radar transmitter antenna, or light source or (ii) protects a radiation receiver from environmental damage, (b) is designed to avoid signal distortion, and (c) may introduce some undesirable distortion and attenuation of signals. *See also* **antenna, attenuation, distortion, radiator.**

railing: On a radar screen, the appearance of an image when radar pulse jamming occurs at high pulse repetition rates, such as 50 kHz (kilohertz) to 150 kHz, such that the jamming results in an image that resembles a rail or picket fence. *See also* **image, jamming, pulse repetition rate, radar, screen.**

rain barrel effect: The sound distortion that (a) occurs when telephone line equalization is excessive, i.e., the line is overcompensated, and (b) is similar to that obtained in talking into an empty, or partially filled, rain barrel or into a hollow chamber. *Synonyms* **echoing, echo chamber effect, hollow chamber effect.** *See also* **distortion, equalization, line.**

Rajchman selection switch: *Synonym* **core rope storage.**

RAM: random access memory.

Raman amplifier: *Synonym* **optical fiber amplifier.**

Raman scattering: The generation of many different wavelengths of light from a nominally single-wavelength source (a) by means of lasing action and interaction with molecules, thereby creating many different excited molecular energy levels that will produce photons of various energy levels, i.e., various wavelengths, when transitions to lower excited states occur, and (b) by the beating together of two frequencies, thus inducing dipole moments in molecules at the difference frequencies and thereby causing modulation of laser-molecule interaction, which, in turn, produces light at side frequencies, i.e., side wavelengths relative to the nominal wavelength. *See also* **Brillouin scattering, Mie scattering, Rayleigh scattering.**

random access memory (RAM): 1. A read/write, nonsequential access memory used for the storage of instructions or data. **2.** A high speed read/write memory with an access time that is essentially the same for all storage locations. *Note 1:* Random access memory (RAM) is characterized by a shorter access time than disk or tape storage. *Note 2:* Random access memory

(RAM) usually is volatile. *See also* **direct access storage, read-only memory, volatile storage.**

random access memory disk: 1. In personal computer (PC) systems, a computer program that (a) simulates a disk drive by using the direct access storage, i.e., the random access memory (RAM), (b) in most PC systems, is lost, along with the associated data, when the power is turned off, (c) is not lost when stored on real diskettes or hard disks, and (d) operates faster than, and avoids repeated use of, the real disk drives. **2.** Software that is built into a printed circuit (PC) board, i.e., into firmware, that simulates disk drives by using the computer random access memory, i.e., direct access storage, thus making the extra direct access storage appear as though it were a disk. *See also* **direct access storage, disk drive, diskette, firmware, personal computer, printed circuit.**

random access storage: *Synonym* **direct access storage.**

random binary sequence: *See* **pseudorandom binary sequence.**

random bit stream signaling: Signaling in which bit-intermittent transmission of signals on a unit interval basis is used without regard to the presence or the absence of a code or an alphabet. *Synonym* **intermittent timing transmission.** *See also* **signaling, unit interval.**

random conference: A conference that is established among communications network users at any moment upon their request to a network operator.

randomizer: 1. A device used to invert the sense, such as polarity or the representation of 0 and 1, of pseudorandomly selected bits in a bit stream to avoid long sequences of bits of the same sense. *Note:* The same randomizer selection pattern must be used on the receive terminal in order to restore the original bit stream. **2.** In cryptographic equipment operations, a random bit generator that starts all the cryptoequipment at the same point in the key stream. *See also* **binary digit, cryptoequipment, data scrambler, descrambler, key, limited protection voice equipment, scrambler.**

randomly phased orbit: In a system that consists of a group of satellites, an orbit in which the position of an individual satellite does not have a fixed or controlled time or spatial relationship to the position of others in the same group. *See also* **orbit, phased orbit, satellite.**

random noise: Noise that consists of a large number of transient disturbances with a statistically random time distribution. *Note:* An example of random noise is thermal noise. *See also* **impulse noise, noise.**

random number: 1. A number selected from a known set of numbers in such a way that each number in the set has the same probability of occurrence. **2.** A number obtained by chance. **3.** One of a sequence of numbers considered appropriate for satisfying certain statistical tests or believed to be free from conditions that might bias the result of a calculation. **4.** A number, such as a bit string, that satisfies at least one of the criteria for randomness. *See also* **pseudorandom number sequence.**

random scan: *Synonym* **directed scan.**

random signal: A sequence of pulses such that (a) each pulse does not have a fixed time relationship to any other pulse, and (b) a pulse's individual time of occurrence is unpredictable, although some rule of randomness may be broken, such as (i) the pulses must occur within a fixed minimum time interval between them, or (ii) there must be a fixed number of pulses in a given time interval. *See also* **pulse, pseudorandom, pseudorandom number sequence.**

range: 1. The maximum distance at which a signal from a source can be effectively received and used. **2.** In radiocommunications, the maximum distance between specified radio stations over which effective communications can be provided. **3.** In radar, the maximum distance at which an object, i.e., a target, can be detected or tracked. *Note:* Some of the factors that influence range are the transmitted power, the antenna gain, i.e., the antenna directivity, of the transmitting and receiving antennas, receiver sensitivity, atmospheric conditions, and, in the case of radar, the characteristics of the reflecting object, i.e., the target. **4.** The distance from a given location to a location of interest. **5.** The values between two values on the same scale. **6.** The difference between two limits on the same scale. **7.** All the possible values that a parameter may have. *See* **burn-through range, capture range, dynamic range, emission intercept range, frequency range, lock-in range, luminance range, maximum detection range, measurement range, named range, omnidirectional radio range, optical dynamic range, optical range, orientation range, radar blind range, radio horizon range, radio range, slant range, transmitter central wavelength range, very high frequency omnidirectional radio range, transmitter central wavelength range, visual-aural radio range.** *See also* **value.**

range advantage: In radar, the difference between (a) the maximum range at which a radar transmission can be intercepted by a search receiver and (b) the maximum detection range against the object using the search receiver. *Note:* The range advantage obtained by a small ship using direction-finding equipment is greater than that of a large ship, and that obtained by a submarine is greater than that of a frigate using similar equipment. *Note:* Even if the range advantage is small, it may still give earlier detection than would ordinarily be obtained without the advantage. Thus, an object with a small radar cross section has a range advantage over an object with a large radar cross section. *See also* **radar, range.**

range check: A check to determine if the value of a variable lies at or between lower and upper limits, i.e., is neither below a lower limit nor above an upper limit. *See also* **value, variable.**

range command: 1. In communications, computer, data processing, and control systems, a command that sets a specified range that a variable or a parameter can have. **2.** In personal computer systems, a command that affects a range of cells on a spreadsheet, such as to specify, to delete, or to protect a range of cells. *See also* **parameter, range, value.**

range compression: A reduction in the range of values that a variable may have. *Note:* Examples of range compression include reduction of the spectral width of a range of colors or reduction of the range of shades a color may have. *See* **luminance range compression.**

range designation: *See* **data signaling rate range designation.**

range designation of data signaling rates (DSRs): *Synonym* **data signaling rate (DSR) range designation.**

range discrimination: 1. A measure of the precision with which range can be measured. *Note:* A radar may have a high range discrimination, i.e., it can precisely measure range to, say, a few meters, depending on the depth of the radar resolution cell. It may have a poor bearing discrimination and a poor elevation discrimination, depending upon the width and the height of the radar resolution cell. A small azimuth or elevation angle tolerance corresponds to many meters at long ranges, i.e., at normal operating ranges. **2.** The ability of a transmitter-receiver to distinguish, separate, or identify on the basis of range, a given signal source from all other signals, noise, or interference. *Note:* In radar systems, range discrimination is obtained from the width of the range gate, i.e., from the depth of the resolution cell. *See also* **bearing discrimination, detection discrimination, discrimination, precision, radar resolution cell, range gate.**

range equation: *See* **self-screening range equation.**

range expansion: An increase in the range of values that a variable may have. *Note:* Examples of range expansion are an increase in the spectral width of a range of colors, or an increase in the range of shades a color may have. *See* **luminance range expansion.**

range finding: The determining or the measuring of the range from a given point to a point of interest. *See* **radar range finding.**

range gate: A radar circuit that measures the time between (a) the instant that a pulse from the radar transmitter is launched, and (b) the instant that the pulse echo from an object, i.e., from a target, is detected by the radar receiver. *Note:* The launched pulses and the echoes can be displayed on a screen. The screen distance between them can be adjusted so that when they are made coincident, the range in convenient units, such as meters, kilometers, or nautical miles, can be read directly. The gate time scale is adjusted until the pulses are coincident.

range gate stealing: In radar systems, the situation that occurs when a tracking radar is caused to lose range tracking of an object, i.e., a target. *Note:* Range gate stealing operations against a pulsed radar cause the radar to lose lock. If the gate stealing cycle is fast, the radar may not appear to hold lock or to break lock quickly. Against a slower cycle, such as several seconds, distinct indications of range gate capture and walkoff may be evident. Video dims or grass recedes as the jammer retransmitted pulses drive down the radar automatic gain control (AGC). The radar overtake indication decreases or does not increase fully as range gate walkoff occurs. The real object may become visible at shorter ranges. Break lock occurs when the jammer completes its transmission. *See also* **range gate, range gate walkoff, track, tracker, tracking.**

range gate walkoff: In radar systems, a continued increase in the range to an object, i.e., a target, that is displayed as an image on a radar range display, such as an A-scope or B-scope, in which the range gate plays a role in such a manner that the range to the object appears to increase to a maximum value as the range pulse gradually shifts out of the display area on the display surface of the display device, such as the radar screen. *See also* **range gate.**

range height indicator (RHI): A device that (a) scans an object, i.e., a target, by radar, (b) determines the range, or uses range data supplied by another source, (c) measures the elevation angle or depression angle to the target, (d) computes the difference in elevation between the device installation and the object, and

(e) displays the range and the height of the object. *Synonyms* **radar height indicator, radio height indicator.**

range laser sonar: *See* **short-range laser sonar.**

range marker: In radar systems, a single calibration pulse, i.e., a single blip, that (a) is fed to the time base of a radar radial display, i.e., polar diagram, and (b) causes a series of concentric circles, as the time base rotates, to be drawn on the plan position indicator screen. *Note:* The concentric circles remain long enough to be drawn again by the next calibration pulse. The concentric circles may be used to measure range because the concentric circles form a range scale along any radial line. *See also* **range resolution.**

range rate memory: A radar circuit that (a) is used to reject spurious signals that differ significantly from the echo signal obtained from the desired object, i.e., desired target, particularly in range closing rate signals and signals from randomly dispensed chaff, and (b) provides range closing rate and tracking, i.e., lock-on, stability during momentary loss of signal because of scintillation effects. *See also* **chaff, scintillation.**

range resolution: 1. In radar systems, the ability of radar equipment to measure and represent the separation of two reflecting objects, i.e., reflecting targets, on a similar bearing but at different ranges from the radar antenna. *Note:* The range resolution is determined primarily by the pulse duration that is used. The pulse duration determines the depth of the radar resolution cell. Range resolution does not change with range. The radar resolution cell depth, D, is given by the relation $D = 150d$ where D is the radar resolution cell depth in meters, and d is the radar pulse duration in microseconds. **2.** The precision with which the range to a given or selected signal source can be measured or determined, i.e., can be resolved. *Note:* Range resolution precision may be expressed as a tolerance in various terms, such as absolute distance, or as a percent of range. *See also* **radar resolution cell.**

range site: *See* **downrange site.**

range station: A station in a range used for the measurement of trajectories of moving objects. *Note:* Examples of range stations are (a) a station in a missile range for gathering position and time data, (b) an artillery range for proof firing of artillery weapons and munitions, and (c) a radio range station. *See* **omnidirectional range station, radio range station.**

ranging: The measuring of distance, usually from a known observation or reference point, to an object. *Note:* Examples of ranging are echo, intermittent, manual, navigational, explosive echo, optical, and radar ranging. *Synonym* **spotting.** *See* **hybrid ranging, light detection and ranging, sound navigation and ranging.**

raster: Within the display space on the display surface of a display device, a pattern of scanning lines that can be formed by sweeping a light beam or an electron beam capable of (a) being modulated in intensity in order to draw or write a picture as it sweeps, or (b) sensing the reflectivity or luminance of small areas or surfaces sequentially in order to read or sense an image. *Note:* An example of a raster is the pattern formed by an electron beam scanning the screen of a television (TV) camera or TV set. The pattern can be seen when the video signal is suppressed, and the screen is examined closely. *See also* **scanning.** *See also* **display device, display space, display surface, luminance, modulation, reflectivity, scanning.**

raster count: The total number of raster scanning lines within a display space on the display surface of a display device. *Note:* The raster count could be the raster density times the vertical width of the display space, the vertical width being measured perpendicular to the scanning line. In most display systems with rasters, the raster count is fixed, and the density changes with the vertical dimension of the display space, assuming that the scanning lines are horizontal. *See also* **display device, display space, display surface, raster, scanning.**

raster density: In display systems, the number of scanning lines per unit distance perpendicular to the scanning direction. *Note:* The raster density times the vertical dimension of the display space equals the raster count, assuming that the scanning lines are horizontal. *See also* **scanning.**

raster display: The generation of a display image on the display surface of a display device that makes use of a raster scan to create the image. *See also* **display device, display space, display surface, image, raster, scanning.**

raster scan: In display systems, the generating, recording, or displaying of display elements, display groups, or display images by means of a line-by-line sweep of a modulated writing element, such as an electron beam or a light beam, across the display space on the display surface of a display device. *Note:* Raster scans are performed (a) on television (TV) screens by sweeping a modulated electron beam horizontally and rapidly while moving the beam relatively slowly vertically, or (b) on a fiberscope faceplate that terminates an aligned bundle of optical fibers formed in a Cartesian coordinate system by illuminating the fibers in sequence in each horizontal row and moving vertically to the next row when the end of each row is reached. A wraparound is used to repeat the scan when the bottom of the display space is reached. Although raster scans are performed automatically by circuits built into the display device, the scans may also be directed by a computer program. *See also* **aligned bundle, display device, display element, display group, display image, display space, display surface, directed scan, faceplate, fiberscope, image, raster, scanning, wraparound.**

raster scanning: Scanning in which the motion of the scanning spot follows a raster. *See also* **raster, scanning.**

raster unit: The physical spatial distance between corresponding points on adjacent scanning lines, i.e., the distance between lines plus the width of one line. *Note:* The raster unit should be constant over the display space to avoid distortion. The raster unit times the raster count usually gives the vertical dimension of the display space, assuming that the scanning lines are horizontal. *See also* **display space, raster count, scanning.**

rate: *See* **actual transfer rate, average error rate, average information rate, average transinformation rate, average transmission rate, binary serial signaling rate, bit rate, block error rate, block transfer rate, calling rate, chip rate, clock rate, constant false alarm rate, data rate, data signaling rate, data transfer rate, Doppler pulse repetition rate, effective data transfer rate, effective transfer rate, effective transmission rate, error rate, failure rate, full motion operation rate, instantaneous transmission rate, interface rate, interruption rate, jittered pulse repetition rate, maximum pulse rate, maximum stuffing rate, maximum user signaling rate, modulation rate, multiplex aggregate bit rate, nominal bit stuffing rate, Nyquist rate, optical line rate, overhead rate, payload rate, pulse repetition rate, reading line rate, recording line rate, regeneration rate, repetition rate, sampling rate, scanning line rate, scan rate, staggered pulse repetition rate, symbol rate, system information rate, transient attenuation rate.**

rate characteristic: *See* **wavelength-dependent attenuation rate characteristic.**

rated output power: The power available at a specified output of a device under specified conditions of operation. *Note:* The rated output power may be further described, such as maximum rated output power or average rated output power. *See also* **dBW, output rating, peak envelope power, peak power output.**

rate efficiency: *See* **block rate efficiency.**

rate interface: *See* **basic rate interface, primary rate interface.**

rate•length product: *See* **bit rate•length product.**

rate measuring equipment: *See* **error rate measuring equipment.**

rate memory: *See* **range rate memory.**

rate range designation: *See* **data signaling rate range designation.**

rate transparency: *See* **data signaling rate transparency.**

rating: *See* **output rating, transmitter power output rating.**

ratio: *See* **access denial ratio, access success ratio, added block ratio, aperture ratio, aspect ratio, axial ratio, bit error ratio (BER), bit misdelivery ratio, block error ratio, block loss ratio, block misdelivery ratio, blowback ratio, call completion ratio, carrier to noise ratio, carrier to noise power ratio, common mode rejection ratio, compression ratio, deviation ratio, disengagement ratio, disengagement success ratio, diversity ratio, echo attenuation ratio, energy to noise ratio, fiber/coating offset ratio, front to back ratio, image rejection ratio, inband noise power ratio, jamming to signal ratio, outage ratio, photodetector signal to noise ratio, postdetector signal to noise ratio, protection ratio, pseudo bit error ratio, radiant power ratio, reduction ratio, reproduction ratio, residual error ratio, signal plus noise plus distortion to noise plus distortion ratio, signal plus noise to noise ratio, signal to crosstalk ratio, signal to noise ratio, single sideband noise power ratio, standing wave ratio, start-stop distortion ratio, voltage standing wave ratio.**

rational number: A real number, i.e., neither an imaginary number nor a complex number, that (a) is the quotient of an integer (dividend or numerator) divided by an nonzero integer (divisor or denominator), (b) is not necessarily a whole number, and (c) may be a repeating decimal. *Note:* Examples of rational numbers are (a) 4 obtained from 8 divided by 2, (b) 1.875 obtained from 15 divided by 8, (c) 16.666 . . . obtained from 100 divided by 6, and (d) about 1.0967742 obtained from 544 divided by 496. *See also* **integer, irrational number.**

ratio per bit: *See* **signal to noise ratio per bit.**

ratio squared combiner: *Synonym* **maximal ratio combiner.**

ratio square diversity combining method: *See* **maximal ratio square diversity combining method.**

ratio tester: *See* **bit error ratio tester.**

RATT: radio teletype, radio teletypewriter.

RATTY: radio teletype, radio teletypewriter.

Raven system: A tactical radio system that (a) carries voice, data, facsimile, and teleprinter communications, (b) may be manpacked or vehicular-mounted, (c) includes various electronic countercountermeasures (ECCM), such as built-in burst data transmission, frequency hopping, digital encryption, remote control, and frequency offset, (d) may be operated by remote control, and (e) is developed by Siemens Plessey Defence Systems.

raw radar information: Information that is obtained directly from a radar antenna, from the radar circuits that control radar displays, or from the radar displays.

ray: An infinitesimally narrow beam of electromagnetic radiation, such as a lightwave, that (a) is represented by a straight line drawn in the direction of propagation at a point in a propagation material medium or free space, where the line is drawn tangent to and in the direction of the line representing the path taken by the wave, (b) is used for analyses of lightwave behavior in the terminology of geometric optics, and (c) is in the direction of the Poynting vector and therefore perpendicular to wavefronts in the terminology of physical optics. *Note:* Examples of rays are light rays, X rays, gamma rays, cosmic rays, and possibly microwaves. *See* **axial ray, chief ray, direct ray, emergent ray, extraordinary ray, field ray, gamma ray, guided ray, helical ray, incident ray, leaky ray, light ray, meridional ray, optical ray, ordinary ray, paraxial ray, principal ray, reflected ray, refracted ray, skew ray, X ray.** *Refer to* **Figs. I-5 (470), L-8 and L-9 (536).**

ray bundle: **1.** A group of rays, such as ordinary rays, in which the rays differ from one another in some detailed respect, such as frequency, i.e., wavelength or color, or intensity, yet have one or more aspects in common, such as direction. *Note:* The rays in a bundle may be separated from each other by dispersion caused by (a) a prism, in which case a multifrequency bundle of rays incident on the prism is separated into distinct rays of different color, or (b) an optical fiber, which causes the different frequencies to arrive at the end of the fiber at different times. **2.** A group of light rays, such as the rays emanating from or reflected from a point, considered as such for some purpose or discussion. *Synonym* **bundle of rays.** *See also* **dispersion, ordinary ray, point, prism, ray.**

Rayleigh distribution: A mathematical statement of the frequency distribution of random variables, for the case in which two orthogonal variables (a) are independent, i.e., are uncorrelated or incoherent, and thus mutually exclusive, and (b) are usually distributed with the same variance. *Note:* In the propagation of electromagnetic waves, the Rayleigh distribution (a) affects scattering by material media, producing a lower limit to attenuation in the media, and (b) occurs because of the intrinsic molecular structural pattern of the propagation medium, such as (i) air and moisture in the atmosphere or (ii) the glass used in optical fibers. *See also* **propagation medium, Rayleigh fading, Rayleigh scattering.**

Rayleigh fading: In the transmission of electromagnetic waves, phase interference fading that (a) is caused by multipath and (b) is approximated by the Rayleigh distribution. *See also* **fading, fading distribution, multipath, Rayleigh distribution.**

Rayleigh scattering: In an electromagnetic wave propagating in a material medium, scattering caused by (a) the atomic or molecular structure of the medium and (b) spatial inhomogeneities, such as spatial variations in refractive index and distances between inhomogeneities, i.e., distances between scattering centers, that are small compared to the wavelength. *Note 1:* Rayleigh scattering losses vary as the reciprocal of the fourth power of the wavelength. *Note 2:* Ionospheric scattering is partly caused by Rayleigh scattering. *Note 3:* In material media, Rayleigh scattering sets a theoretical lower limit to electromagnetic wave attenuation as a function of wavelength. *Note 4:* The familiar light green color of soft-drink bottles is caused by Rayleigh scattering by distributed iron atoms that have not been removed from the glass. *See also* **Brillouin scattering, ionospheric scattering, material scattering, Mie scattering, Raman scattering, Rayleigh distribution, scatter, waveguide scattering.** *Refer to* **Figs. A-9 and A-10 (49), H-1 (430).**

ray method: *See* **refracted ray method.**

ray optics: *Synonym* **geometric optics.**

ray trajectory: The path or the course that (a) is taken by a light ray in a propagation material medium or a vacuum and (b) is perpendicular to the wavefront. *See also* **propagation medium, ray, wavefront.** *Refer to* **Fig. R-5.**

RBOC: Regional Bell Operating Company.

RC: reflection coefficient.

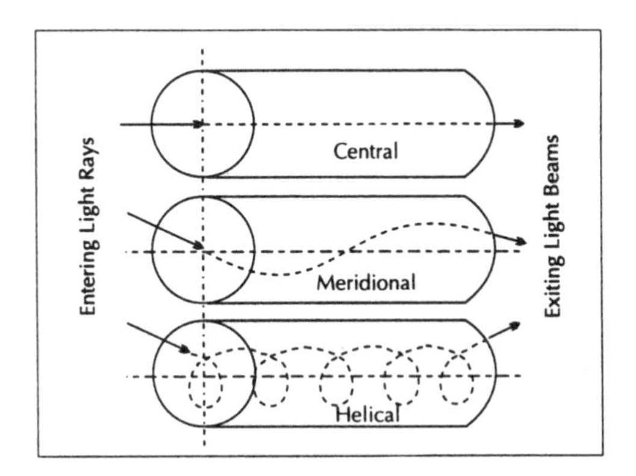

Fig. R-5. Ray trajectories in graded index optical fibers.

RCC: radio common carrier.

RDF: radio direction finding.

reactance: *See* **capacitive reactance, inductive reactance.**

reaction time: *See* **switching reaction time.**

reactive: *See* **volt•amperes reactive.**

reactive power: In steady state sinusoidal voltages and currents in a circuit, the product of the current, the voltage, and the sine of the phase angle between them. *Note:* Reactive power may be expressed in various units, such as vars (volt•amperes reactive), and mvars (millivolt•amperes reactive). *See also* **real power.**

read: 1. To obtain data from a source, such as a storage device or a storage medium. **2.** In voice communications, to receive and understand a message. **3.** In voice communications, a proword that is used in a query or a reply to indicate reception and understanding of a statement or a message. *Note:* An example of the use of "read" is in the transmitted expression "Do you read me?" which indicates "Did you receive and understand my message?" If the answer is affirmative, the reply is "I read you." *See also* **copy, proword, storage, storage medium.**

readability: 1. In communications systems operations, the extent to which the content of a message is understood. **2.** The ability to accept and interpret signals regardless of the means of communication or the propagation medium.

readable: *See* **machine-readable.**

read-around ratio: 1. The maximum number of times a specific location in a storage device can be interrogated before the data stored in neighboring locations are affected, i.e., before the data stored at those locations are lost. *Note 1:* Examples of read-around ratio are (a) in an electrostatic storage device, such as a cathode ray tube (CRT), the maximum number of times a specific location can be interrogated before spillover, i.e., migration of electrons into neighboring locations, is about to cause loss of data at those locations, and (b) in a nondestructive core storage unit, the maximum number of times a specific core can be interrogated before the magnetic condition of the core deteriorates because of magnetic fields that occur when neighboring cores are interrogated. *Note 2:* Before the read-around ratio is reached, the stored data have to be regenerated. In some storage devices, the data are restored at a location every time the location is read, i.e., is interrogated. *See also* **regenerative storage.**

readdressal: *See* **message readdressal.**

reader: A device capable of sensing data on a data medium and generating signals that represent the data. *See* **coupled reperforator tape reader, microform reader, optical reader, tape reader.** *See also* **magnetic ink character recognition, optical character recognition.**

reader-copier: *See* **microform reader-copier.**

read head: A magnetic head that only can read.

reading: The acquisition or interpretation of data from a source, such as a storage device or a data medium.

reading head: *See* **tape reading head.**

reading line frequency: *Synonym* **reading line rate.**

reading line rate: The scanning line rate during reading of an object, such as a document, scene, or picture. *Synonym* **reading line frequency.** *See also* **scanning line rate.**

read only: Pertaining to data that (a) only can be read and used and (b) cannot be operated upon in any manner, such as copied, printed, modified, or transferred. *Note:* A read-only microfilm database file can be read by the user but cannot be changed, copied, or removed from the database. Within copyright laws, the user can hand-copy the data or take notes.

read-only memory (ROM): A memory in which data usually only can be read. *Note:* Read-only memories (ROMs) are primarily used in microcomputers and minicomputers. *Synonym* **nonerasable memory.** *See* **compact disk read-only memory, programmable read-only memory.** *See also* **random access memory.**

read-only storage (ROS): 1. A storage device in which the contents usually cannot be modified. **2.** Storage in which (a) the contents cannot be changed during normal operation by a user, (b) the contents can only be changed by physical modification, (c) there is an implied restriction on the ability or the authority of persons to effect modification of the contents, and (d) changes to the contents cannot be effected by means of a computer program. *Note 1:* The contents of a read-only storage (ROS) may be modified only by a particular user or when operating under particular conditions. *Note 2:* An example of a read-only storage (ROS) is a storage device in which writing is prevented by a programmed or manual lockout. *Note 3:* Read-only storages (ROSs) are primarily used in large mainframe computers. **3.** In personal computer (PC) systems, internal direct access storage (a) from which data and programs can be retrieved, (b) into which new data and programs cannot be entered, (c) from which data and programs are not lost when power is turned off, i.e., the storage is nonvolatile, and (d) that is fixed, permanent, and nonerasable. *Synonyms* **fixed storage, nonerasable storage.** *See also* **erase, firmware, permanent storage, reading, storage.**

read protection: The protection that (a) is afforded to stored data by preventing the reading of the data from a data medium, such as a file, record, or set of data on a magnetic disk, and (b) is invoked to prevent unauthorized access and thereby prevent the reading of the data. *See also* **access, read.**

read station: The position in a reader at which data on a data medium are sensed, i.e., are read. *See also* **data medium, reader.**

read-write head: A magnetic head that can dynamically read data from, or write data on, magnetic data media, such as magnetic tapes, disks, drums, and cards. The head, a small electromagnet, reads by sensing magnetized spots and writes by creating magnetized spots on the data medium. *Note:* The level and the direction of magnetization of the spots depend on the type of modulation used. When the spots are sensed, because the head and the data medium move relative to one another, the output is the time derivative of the magnetization and must be interpreted accordingly. *See also* **data medium, magnetic head, modulation.**

read-write opening: *Synonym* **read-write slot.**

read-write protection: The protection that (a) is afforded to stored data by preventing the reading of the data from, and the writing of the data on, a data medium, such as a file, record, or set of data on a magnetic disk, and (b) is invoked to prevent unauthorized access and thereby prevent the reading or changing of the data. *See also* **access, read.**

read-write slot: An opening in the jacket of a diskette to allow access by the read/write heads. *Synonym* **read/write opening.**

ready condition: In communications system operation, a steady state condition at the interface between data terminal equipment (DTE) and data circuit-terminating equipment (DCE) that indicates that the DCE is ready to accept a call request signal or that the DTE is ready to accept an incoming call. *See also* **call not ready condition.**

ready for data signal: 1. A call control signal that is transmitted by the data circuit-terminating equipment (DCE) to the data terminal equipment (DTE) to indicate that the connection is available for data transfer between both DTEs. **2.** A signal that (a) is sent in the backward direction in the interexchange data channel indicating that all the succeeding exchanges in the connection have been through-connected, or a signal sent in the forward direction in the interexchange data channel indicating that all the preceding exchanges in the connection have been through-connected, (b) is sent by the user terminal, and (c) corresponds to the ready for data state at the user interface.

ready mode: In communications, computer, data processing, and control systems, a state or a condition in which a system is prepared to accept a command from the operator whether or not the system is still performing an operation or executing a prior command from the operator. *Note:* In some personal computer (PC) system programs, the ready mode is indicated by a message on the screen. *See also* **computer program.**

ready signal: *See* **controlled not ready signal.**

real address: The address of a physical storage location in a storage device.

real image: An image represented by actual rays of light that, when incident upon a reflecting surface, can be seen, or that, when incident upon a transmitting medium, can be transmitted a considerable distance. *Note:* An optical system, such as a camera, can convey light rays that produce a real image on a film, and an aligned bundle of optical fibers can convey light rays that produce a real image on a faceplate. *See also* **image.**

reality deck: *See* **virtual reality deck.**

reality system: *See* **virtual reality system.**

real power: *Synonym* **effective power.**

real storage: Actual physical storage, usually physical main storage or physical auxiliary storage. *Note 1:* Examples of real storage are (a) the hardware in a storage device that stores data, such as magnetic disks or arrays of magnetic cores, and (b) the locations on magnetic tape that are addressed for access by a read-write head. *Note 2:* Real storage contains only a part of the range of addresses that are available to a user. *See also* **virtual storage.**

real system: In communications, computer, data processing, and control systems, a system that (a) includes one or more actual items, such as hardware, firmware, software, peripheral equipment, and physical processes, (b) includes human operators, (c) has the means of internal communication to form an autonomous whole, (d) performs information processing, information transfer, or both, and (e) may be an implementation of a model or a simulation.

real time: 1. In communications, computer, data processing, and control systems, the actual calendar time during which a physical process occurs. **2.** In a communications, computer, data processing, or control system, pertaining to the transmission or processing of data in connection with a process outside the system according to the time requirements imposed by the outside system, such as (a) observing pictures from an endoscope, (b) displaying telemetered emergency medical data from a remote patient for immediate diagnosis and treatment, (c) transmitting live coverage of a news event, (d) operating a system in the conversational mode, i.e., the interactive mode, or (e) controlling a process while the process is in progress. **3.** Pertaining to the performance of a computation during the actual time that the related physical process occurs, in order that results of the computation can be used to guide the physical process. **4.** The absence of delay, except for the time required for electromagnetic transmission, between the occurrence of an event, or the transmission of data, and knowledge of the event, or reception of the data, at some other location, such as in radar tracking of a moving object, i.e., target. *See also* **absolute delay, near real time, transmission.**

real time computing: *Synonym* **real time data processing.**

real time data processing: Data processing in which the data processed originate from a process in real time, and the results of the data processing are used to influence or control the process, or related processes, while they occur. *Note:* Examples of real time data processing are (a) sensing the composition of the wood pulp slurry in a paper mill, computing the changes that are necessary, and altering the mix before the next step in the process, and (b) controlling the trajectory of a missile in flight by radar data obtained from the target and perhaps the missile as well. *Synonym* **real time computing.** *See also* **real time.**

real time delay: *See* **near real time delay.**

real time operation: 1. A mode of operation of a system in which the occurrence of an event at the input is followed by the occurrence of a corresponding event at the output with no delay that is apparent to the system user. *Note:* If the maximum absolute delay appropriate to each application is specified, and the maximum is not exceeded, the application usually is considered to be conducted in real time. **2.** In analog computers, operation in the compute mode during which the time scale factor is 1. *See also* **problem time, time scale factor.**

real time penetration reaction: In computer and communications security, a system response to a penetration attempt that is detected and diagnosed in time to take action to prevent the penetration. *See also* **communications security, security.**

real time transmission: Transmission of data with such little delay that the information is available in time to influence or control the process being monitored or controlled by the data.

real world coordinate: In display systems, an actual physical coordinate of a point on or in a real object, or part thereof, whose image is displayed in the display space on the display surface of a display device. *Note:* The user may select other more convenient coordinates for display or programming purposes, i.e., user coordinates, and the display device may have its own intrinsic coordinates inherent in the device, such as the 0 to 100 scale on a plotter bed, on a grid superimposed on a cathode ray tube (CRT) screen, i.e., screen coordinates, or numbered index markings on a thumbwheel. *See also* **device coordinate, display space, screen, user coordinate.**

reasonableness check: A test to determine whether a value conforms to specified criteria. *Note 1:* The reasonableness check can be used to prevent processing of questionable data. *Note 2:* Examples of reasonableness checks are checks made to ensure that (a) values lie within a specified range, (b) sudden large changes in values do not occur, (c) a given name is in a specified list of names, or (d) posted dates are within the time span of interest. *See also:* **wild point detection.**

rebroadcast: 1. A radio relay system in which a simplex broadcast radio at a relay station retransmits a received signal automatically or by manual operation of the equipment. **2.** To broadcast a program or a message that has been previously broadcast. *Note:* The rebroadcast may be by (a) the same station that originally broadcast the program or message, or (b) another station.

recalculation: In personal computer spreadsheet and database management programs, the execution of an indicated algebraic operation or set of operations in accordance with algebraic expressions after a change is made in any of the variables or parameters in any of the algebraic expressions identified as such on the spreadsheet, i.e., worksheet. *See also* **algebraic expression, parameter, spreadsheet, variable.**

recall: In telephone systems operations, to obtain the attention of an attendant after a call has been established. *See* **camp on with recall.** *See also* **recall signal.**

recall response signal: In telephone switchboard operations, after a user sends a recall signal, the signal that is sent to a user indicating that the user recall signal has been detected.

recall signal: In telephone switchboard operations, the signal that is used to bring in an attendant, i.e., an operator, after a call has been established. *Note:* An example of a recall signal is the signal generated by tapping on the hook or cradle switch of the telephone instrument and thereby "flashing" the attendant to get attention.

receipt: In communications system operations, a transmission made by a receiving station to indicate that a message has been satisfactorily received.

receipt communications method: 1. A method of communications system operation in which the receiving station is required to send a receipt to the transmitting station for each message that is received. *Note:* The receipt method of operation is used in an attempt to obtain certainty of reception. **2.** A method of communications in which each message is receipted by each addressee. *See also* **broadcast communications method, intercept communications method, relay communications method.**

receipt time: The date and the time at which a communications agency completes reception of a message that is transmitted to it by another communications agency. *Note:* An example of receipt time is the time and the date that a message is received from a communications circuit at a communications center.

receive: To accept an entity, such as a message, signal, or electromagnetic wave, that was sent or transmitted from another location. *See* **confirmation to receive.** *See also* **transmit.**

receive after transmit time delay: The time between (a) the instant of keying off of the local transmitter to stop transmitting and (b) the instant that the local receiver output has increased to 90% of its steady state value in response to a radio frequency (rf) signal from a distant transmitter. *Note 1:* The radio frequency (rf) signal from the distant transmitter must exist at the local receiver input prior to, or at the time of, keying off of the local transmitter. *Note 2:* Receive after transmit time delay applies only to half-duplex operation. *See also* **absolute delay, receiver attack time delay, transmit after receive time delay.**

received noise power: 1. At the receiving end of a circuit, channel, link, or system, the calculated or measured noise power within the bandwidth being used. **2.** At a receiving point, the measured or calculated absolute power of the noise. *Note:* The related bandwidth and the noise weighting must be specified. **3.** The value of noise power from all sources measured at the line terminals of a listener telephone set. *Note:* Either flat weighting or some other specific amplitude-frequency characteristic or noise weighting characteristic must be associated with the measurement and must be specified along with the measurement. *See also* **bandwidth, noise, noise power.**

received signal level (RSL): The signal level at a receiver input terminal. *Note 1:* The signal bandwidth and the established reference level must be specified. *Note 2:* The received signal level (RSL) usually is expressed in dB with respect to 1 mW (milliwatt), i.e., with respect to 0 dBm. *See also* **level, signal.**

receive only (RO): Pertaining to a device or a mode of operation in which messages can be received, but cannot be transmitted. *See* **lobe on receive only.**

receive-only teletypewriter: A teletypewriter that (a) can receive signals and print the data represented by the signals, and (b) has no keyboard and no tape transmitter and cannot transmit signals. *See also* **automatic send and receive teletypewriter, keyboard send-receive teletypewriter.**

receiver: 1. In communications systems, a device that (a) accepts signals from a transmitter or from a propagation medium, (b) interprets the data represented by the signals, and (c) furnishes the data, or the information represented by the data, to a user. **2.** The portion of a communications system in which electromagnetic signals are (a) converted into visible images or audible sounds or (b) accepted, processed, and furnished to another portion of the system. *See* **automatic receiver, continuous receiver, drum receiver, facsimile receiver, fiber optic receiver, high probability of inter-**

cept receiver, intercept receiver, jammer receiver, linear receiver, logarithmic receiver, low probability of intercept receiver, monolithic integrated receiver, optical receiver, optoelectronic receiver, panoramic receiver, positive-intrinsic-negative field-effect transistor integrated receiver, positive-intrinsic-negative integrated receiver, radio receiver, search receiver, superheterodyne receiver, telephone receiver, warning receiver. *Refer to* **Figs. R-6 and R-7.** *Refer also to* **Figs. B-9 (96), M-2 (587), S-2 (868), T-3 (991).**

Fig. R-6. A satellite digital audio **receiver**. (Courtesy ComStream Company.)

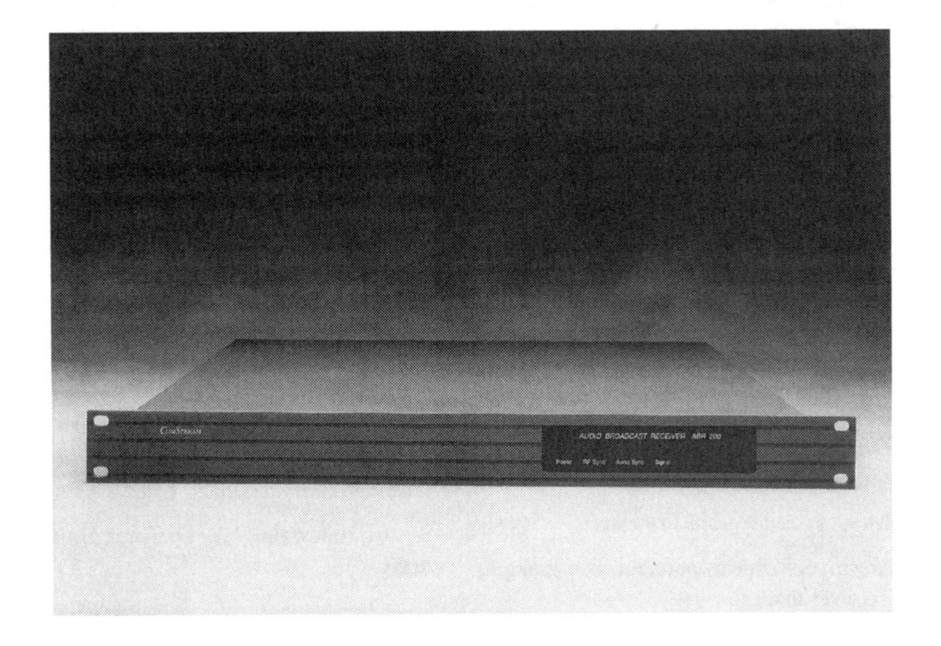

Fig. R-7. A multichannel digital audio **receiver**. (Courtesy ComStream Company.)

receiver aperture: *See* **facsimile receiver aperture.**

receiver attack time delay: The time interval from (a) the instant that a step radio frequency (rf) signal, at a level equal to the receiver threshold of sensitivity, is applied to the receiver input to (b) the instant that the receiver output amplitude reaches 90% of its steady state value. *Note:* If a squelch circuit is operating, the receiver attack time delay includes the time for the receiver to break squelch. *See also* **receiver release time delay, squelch circuit, squelching.**

receiver desensitization: The effect that occurs in a radio or video receiver when a high power carrier signal with nearly the same operating frequency as its useful signal enters the receiver, masks its useful signal, and thereby reduces its effective sensitivity.

receiver information descriptor: A unique descriptor from which information about a receiver can be determined, such as the manufacturer, terminal equipment association, system design application (single mode, multimode), performance specifications, detector type (APD, PIN photodiode), temperature controller, and manufacturer product change designation, i.e., the issue or the revision.

receiver input: *See* **maximum receiver input.**

receiver intermediate frequency gain control: The control of the intermediate frequency (IF) stage gain in a radio, video, microwave, or radar receiver.

receiver location loss: In satellite communications systems, a single parameter that is used in power-balancing calculations to account for path loss variation with Earth station location, as a function of slant range to the satellite and reduction of satellite antenna gain. *Note:* The antenna gain is measured from the on-axis maximum signal at the corresponding off-axis angle of the Earth station. The receiver location loss can be computed for these variables as a function of each station antenna elevation angle. The loss can be used for downlink design to establish the minimum signal power per access from the satellite for acceptable signal power above receiver threshold. It can also be used for uplink design in calculating the minimum Earth station transmitter power to achieve the minimum allowable downlink signal strength. *See also* **path loss, power balancing.**

receiver lockout system: *Synonym* **lockout.**

receiver margin: *See* **synchronous receiver margin, telegraph receiver margin.**

receiver maximum input power: *See* **optical receiver maximum input power.**

receiver noise density: *See* **carrier to receiver noise density.**

receiver optical connector: The optical connector (a) that is provided at the input of an optical receiver, (b) that is attached to the receiver pigtail, and (c) whose description should include (i) manufacturer, (ii) type, such as biconic or FC, (iii) model number, (iv) classification, such as multimode or single mode, and (v) mating connector model number.

receiver optical sensitivity: The worst case value of input optical power (dBm) coupled into a receiver, as measured on the line side of the receiver connector under specified standard or extended operating conditions, that is necessary to achieve the manufacturer-specified bit error ratio (BER) as measured under standard procedures. *Note 1:* The worst case value combines manufacturing, temperature, aging, extinction ratio, and rise-fall time variations in a worst case fashion. *Note 2:* The receiver sensitivity should not include power penalties associated with dispersion or reflection.

receiver release time delay: The time between (a) the instant of removal of radio frequency (rf) energy at the receiver input and (b) the instant that the receiver output is squelched. *See also* **receiver attack time delay, squelch circuit, squelching.**

receiver threshold: The minimum input carrier power to noise power ratio (C/N) at a receiver demodulator for an acceptable signal to noise ratio in the baseband output.

receiver watch: A watch at a radio or video receiver for a specified purpose, such as performing a copy watch, cover watch, guard watch, or listening watch. *See also* **copy watch, cover watch, guard watch, listening watch, watch.**

receiver weighting: *See* **HA1-receiver weighting, one forty four receiver weighting.**

receive teletypewriter: *See* **automatic send-receive teletypewriter, keyboard send-receive teletypewriter.**

receive terminal: *See* **multiplex baseband receive terminal, radio baseband receive terminal.**

receive time delay: *See* **transmit after receive time delay.**

receiving element: In an optical system, the accepting terminus of a junction of optical elements. *See also* **optical element, optical system, terminus.**

receiving end crossfire: Crossfire that (a) is introduced into one channel, usually from one or more adjacent channels, at the receiving end, i.e., the end remote from the transmitter and (b) is caused by signals transmitted by the transmitter. *See also* **crossfire.**

receiving pattern: At a receiver, the signal strength, i.e., the irradiance or field strength, of electromagnetic waves that are received above the threshold of sensitivity for specified frequencies and specified directions. *See also* **radiation pattern.**

receiving perforator: *Synonym* **reperforator.**

receiving system noise temperature: The temperature that corresponds to the measured noise voltage caused by thermal agitation of molecules and atoms in the input circuit of a receiver. *Note:* Noise temperature usually is given in kelvin, such as 300 K. Thermal noise is caused by the motion of conduction electrons in materials, such as wires, resistors, transistors, and diodes, because of their temperature. Temperature is a measure of the average kinetic energy of moving molecules. From the statistical theory of thermodynamics, the root mean square (rms) noise voltage, E, that is developed across a resistor is given by the relation $E = (4RkT\Delta f)^{1/2}$ where E is the rms voltage developed across the resistor in volts, R is the resistance of the resistor in ohms, T is the absolute temperature in kelvin (Celsius + ~273), k is Boltzmann's constant ($k \approx 1.38 \times 10^{-23}$ W•s•K^{-1} (watt•seconds per kelvin), and Δf is the bandwidth of the measuring system in hertz. *See also* **noise.**

reception: In communications systems, listening to, accepting, copying, collecting, recording, or viewing received signals. *Note:* The received signals usually are converted into useful form by the receiving equipment. *See* **community reception, diversity reception, exalted carrier reception, zero-beat reception.**

reception congestion: In a network, congestion that occurs at a receiving station. *Note:* Example of reception congestion are contention, saturation, delay, and a queue of messages or calls at a telephone switching center or a data switching exchange.

reception depth: *Synonym* **listening depth.**

reciprocal bearing: 1. The course that a direction finding (DF) station tells a requesting mobile station to steer to reach the direction finding (DF) station. *Note:* The direction finding (DF) station course computes the course to steer by adding 180° to the true bearing of the requesting station from the DF station if the bearing is less than 180°, and subtracting 180° if the bearing is greater than 180°. The course to steer is given without correction for wind or current and without regard for the actual destination of the requesting station. **2.** The opposite direction to a given bearing. *See also* **azimuth, back azimuth, reciprocal heading.**

reciprocal heading: The opposite direction to a given heading, i.e., a given course. *See also* **azimuth, back azimuth, reciprocal bearing.**

recirculating register: A register that (a) stores data by moving the data through the register positions in sequence such that the data emanating from one end are fed back to the other end, and (b) furnishes data from either (i) the end only or (ii) any or all positions in the register. *Note:* Recirculating registers have been constructed of (a) a bank of flip-flops connected in tandem, (b) a string of lumped capacitive and inductive elements, (c) delay lines that use the piezoelectric properties of fused quartz, and (d) delay lines that use the magnetostrictive properties of materials, such as ferrites and nickel wire. If the stored data can be read and statically retained at the same time, the register is said to provide nondestructive readout. However, most recirculating registers are dynamic in the sense that the data are moving all the time, and the readout is simply a gating action. *See also* **gate, read.**

reckoning: *See* **dead reckoning.**

reckoning position: *See* **dead reckoning position.**

recognition: 1. The determination, by any means, of the character, individuality, or nature of persons or objects, such as aircraft, ships, radiation patterns, sound signatures, or telegraph key operators. **2.** In military systems, such as those used in ground combat operations, the determination that an object is within a known category of objects, such as tanks, trucks, or artillery pieces, including their types and capabilities. *See* **character recognition, magnetic ink character recognition, optical character recognition, pattern recognition, radio recognition.** *See also* **authenticate, identification friend or foe, identification friend or foe personal identifier, intelligibility.**

recognition device: *See* **pattern recognition device.**

recognized private operating agency (RPOA): A private agency that operates a telecommunications system and that adheres to international communications conventions. *Note:* Recognized private operating agencies (RPOAs) usually are obliged by their national governments to adhere to Article 19 of the International Telecommunication Convention and the Regulations that are annexed thereto.

recombination: *See* **nonradiative recombination, radiative recombination.**

recommendation: *See* **V series Recommendation.**

reconditioned carrier reception: *Synonym* **exalted carrier reception.**

reconfiguration: 1. In communications, computer, data processing, and control systems, a rearrangement of the connections among the components of a system. **2.** Repair strategy in which failed or failing components of a system are switched out of the system and replaced by other components, usually that are ready to be switched in. *Note:* An example of reconfiguration is the switching out of a channel, trunk, or network and the switching in of another in its place, rather than rerouting traffic.

reconnaissance: *See* **electronic reconnaissance.**

reconstructed sample: An analog sample that (a) is generated at the output of a decoder when a specified character signal is applied at its input, and (b) has an amplitude that is proportional to the value of the corresponding encoded sample. *See also* **digital to analog converter.**

record: 1. A set of data treated as a unit. *Note 1:* Examples of records are (a) the medical data pertaining to one person, (b) all the data on an invoice, (c) all the data in a message, or (d) all the data in one row of a spreadsheet. *Note 2:* A usual data structure hierarchy, from bottom to top, is bits, bytes, characters, words, blocks, records, files, and databanks. **2.** To write data on a data medium, such as to write data on magnetic tape or on paper. *See* **call record, frequency record, local record, logical record, root record.** *See also* **database, file, write.**

record communications facility: 1. A communications facility that (a) is responsible for filing the original of a transmitted message and (b) usually is also the facility that accepted the message for transmission. **2.** A communications facility that handles record traffic. *See also* **record traffic.**

record communications: Communications in which a hard copy of the transmission, such as teletypewriter or facsimile transmission, is produced for the record. *See also* **communications, facsimile, hard copy, record traffic, teletypewriter.**

recorded copy: 1. A duplicate, of an original set of data, that has been placed in a storage medium. **2.** An original set of data that has been placed in a storage medium. **3.** In facsimile systems, a visible image of the object, i.e., of the original, in record form. *Note:* The recorded copy usually is authenticated and is filed for retention. *See also* **copy, facsimile, hard copy, object, recording.**

recorded spot: In facsimile transmission systems, the image that is left by the recording spot on the record sheet. *See also* **image, recording spot, record sheet, *x*-dimension of recorded spot, *y*-dimension of recorded spot.**

recorded spot *x*-dimension: *Synonym* ***x*-dimension of recorded spot.**

recorded spot *y*-dimension: *Synonym* ***y*-dimension of recorded spot.**

recorder: *See* **airborne direction finding recorder, facsimile recorder, siphon recorder, traffic image recorder.**

recorder signal: *Synonym* **busy signal.**

recorder warning tone (RWT): 1. A tone that (a) is periodically applied to a telephone line to indicate that at least one of the users is recording the conversation, (b) is required by law to be generated as an integral part of any recording device used for the purpose, (c) is required not to be under the control of the call originator, and (d) is recorded together with the conversation. *Note:* An example of a recorder warning tone during an information transfer transaction is a one-half second burst of 1400 Hz (hertz) applied to a telephone line every 15 seconds to indicate to the call receiver that the call originator is recording the conversation, or vice versa. *See also* **busy verification.**

record information: Information that (a) is in such a form that it is or can be registered, (b) can be retrieved, reproduced, transmitted, received, or preserved, (c) may be registered and stored in either temporary or permanent form, (d) may consist of various representations, such as text, graphic images, digital data, or sound recordings, and (e) is not volatile. *See also* **hard copy, record medium, volatile storage.**

recording: 1. Producing a copy of data on a record medium i.e., creating a recorded copy, in order to preserve the data. *Note:* Examples of recording are (a) the storing of a message on magnetic tape, (b) the capturing of an image on photographic film, (c) the placing of data on an optical disk or on a magnetic disk, tape, card, or drum, and (d) operating a recorder to capture a telephone conversation or message. **2.** In facsimile systems, the converting of the received electrical signals to an image on the record sheet, i.e., the creating of the recorded copy. **3.** The data on a record medium together with the record medium itself. *Note:* Examples of recordings are optical discs, photographic films, and magnetic disks, tapes, cards, and drums on which data have been recorded. *See* **black recording, call detail recording, direct recording, electrochem-**

ical recording, electron beam recording, electrolyte recording, electromechanical recording, electrophotographic recording, electrosensitive recording, electrostatic recording, electrothermal recording, fiberscopic recording, ink vapor recording, magnetic recording, nonreturn to zero magnetic recording, phase encoded recording, photosensitive recording, station message detail recording, streaming tape recording, tape recording, xerographic recording.

recording area: *Synonym* **film frame.**

recording density: *Synonym* **bit density.**

recording line frequency: *Synonym* **recording line rate.**

recording line rate: The scanning line rate during the recording, i.e., the writing of an image. *Synonym* **recording line frequency.** *See also* **scanning line period, scanning line rate.**

recording medium: *Synonym* **record medium.**

recording sheet: *Synonym* **record sheet.**

recording spot: In a facsimile recorder, the spot that is used to generate the recorded copy on the record sheet. *See also* **facsimile, recorded copy, recording, x-dimension of recorded spot, x-dimension of scanning spot, y-dimension of recorded spot, y-dimension of scanning spot.**

recording telegraphy: *See* **signal recording telegraphy.**

recording trunk: A trunk between telephone exchanges, such as between two telephone switching centers or private branch exchanges (PBXs), used only for communications between operators, i.e., attendants. *See also* **exchange, switching center, trunk.**

record medium: **1.** The physical medium on which information is stored in recoverable form. **2.** In facsimile systems, the physical medium on which the recorder forms an image of the object, i.e., creates the recorded copy. *Synonym* **recording medium.** *See also* **facsimile, hard copy, recorded copy, record information, record sheet.**

record sheet: In facsimile transmission systems, the data medium that is used to produce the recorded copy, a visible image of the object, i.e., of the original, in record form. *Note:* The record sheet in facsimile systems performs the same function as the record medium in other systems, such as magnetic tape, drum, and disk systems. *Synonym* **recording sheet.** *See also* **facsimile, object, recorded copy, record medium.**

record signal: In facsimile systems, the signal that creates the recorded copy on the record medium. *See* **start record signal, stop record signal.** *See also* **recorded copy, record medium.**

record traffic: **1.** Traffic that is recorded, i.e., record copies are made in permanent or quasi-permanent form, by the originator, the addressee, or both. **2.** Traffic that is permanently or semipermanently recorded in response to administrative procedures or public law. **3.** Messages that have been electrically transmitted and that must be received by the destination user in such a form as to permit permanent or semipermanent storage. *See also* **hard copy, narrative traffic, record copy.**

recoverable error: A fault, malfunction, error, or mistake that allows error recovery to be made. *See also* **error, error recovery, fault, malfunction, mistake.**

recovery: In a database management system, the procedures and the capabilities available for reconstruction of the contents of a database to a state that prevailed before the detection of processing errors and before the occurrence of a hardware or software failure that resulted in the destruction of some or all of the stored data. *See* **backward recovery, error recovery, inline recovery, offline recovery, timing recovery.** *See also* **database management system.**

recovery procedure: **1.** After a system failure, the actions necessary to restore the system to the system capability that existed before the failure. **2.** After an automated information system failure, the actions necessary to restore (a) the system data files to their condition before the failure, and (b) the system computational capability that existed before the failure. **3.** In data communications, a procedure in which a data station attempts to resolve conflicting or erroneous conditions that arise during the transfer of data, i.e., during an information transfer transaction. *Synonym* **recovery process.** *See also* **clear collision, error.**

recovery process: *Synonym* **recovery procedure.**

recovery routine: A computer routine that (a) is entered when incorrect results are obtained, (b) isolates the cause, such as error, fault, malfunction, or mistake, (c) assesses the significance, (d) indicates the courses of action, (e) attempts to make corrections, and (f) causes resumption of operation. *See also* **computer routine, error, fault, malfunction, mistake.**

recovery time: The time that is required by a receiver to return to a zero signal condition after receiving an input signal, such as an overdrive signal or a jamming pulse of saturation intensity.

rectifier: An electrical device that converts alternating current (ac) to direct current (dc) by presenting a high resistance in one direction and a low resistance in the opposite direction, with the net result that the current flows only in one direction. *Note 1:* Examples of rectifiers are vacuum tube diodes, crystal diodes, controlled gas-filled rectifiers (thyratrons), and semiconductor rectifiers, such as selenium rectifiers. *Note 2:* A low-pass filter, such as a shunt capacitor or a series inductor, or both, usually is used to filter the output of a rectifier, which would otherwise have a very high second harmonic. The amount of filtering depends on the use to be made of the rectifier direct current (dc) output. If the output is for a direct current (dc) power supply for communications equipment, practically all of the output ripple voltage may have to be removed to suppress noise. If the output is for a dc motor, the motor inductance may sufficiently reduce ripple currents. If the output is for charging a battery or for electroplating, a reasonable ripple voltage usually can be tolerated. *See also* **ripple voltage.**

recurrent frequency radar: *See* **variable pulse recurrent frequency radar.**

recursive filtering: In image restoration, inverse filtering in which the function that operates upon the distorted image, to produce a less distorted image, is based on statistical estimation and smoothing that use recursive functions which use a linear combination of previous estimates. *Note:* Assuming that an image is in Cartesian coordinates, if its desired to estimate $f(x+1,y+1)$ as a function of its immediate upper left neighbors and the value of the recorded image at those points, the estimate could be computed recursively by the relation $f(x+1,y+1) = a_1 f(x+1,y)+a_2 f(x,y+1)+ a_3 f(x,y)+a_4 g(x,y)$ where x and y are spatial coordinates in the two-dimensional image. *See also* **image, image processing, image quality, image restoration.**

recursive function: 1. A mathematical function that (a) is defined in terms of itself, or a modification of itself, and (b) usually includes itself, or a modification of itself, as an operand. *Note:* Examples of recursive functions are (a) $e^x + \int_{-\infty}^{x} e^x dx$, (b) Fibonacci$(N)$ = Fib$(N-1)$ + Fib$(N-2)$, and (c) factorial(N) = Nfact$(N-1)$. *See also* **Fibonacci series.**

recursive routine: A computer routine that (a) may be used as a subroutine of itself, (b) may call itself directly or be called by another subroutine, perhaps one that it itself has called, and (c) usually requires the keeping of records of the status of its unfinished business or applications so that it will complete or halt its execution, and, perhaps, so that program control will be returned to a nonrecursive program, lest the computer never stop. *See also* **computer routine.**

RED-BLACK concept: In secure communications systems, a concept in which (a) equipment items, such as electrical circuits, fiber optic circuits, modems, and transmitters, that handle plain language security classified information (RED) must be separated from items that handle unclassified or encrypted security classified information (BLACK), particularly because of the level of protection that must be afforded each category, (b) RED and BLACK designation is used to clarify specific criteria that relate to and differentiate such items and the areas in which they are located, (c) items are marked, i.e., color marked or completely colored, red or black to identify the proper level of protection to be given to them, (d) equipment with red marking must be given the higher level of protection, and (e) the RED and BLACK designations may change before and after encryption and before and after decryption. *See also* **BLACK designation, compromising emanation, RED designation, TEMPEST.**

RED designation: In secure communications systems, a designation that is applied (a) to all items, such as lines, wires, coaxial cables, fiber optic cables, modems, switching equipment, transmitters, and receivers, and (b) to all equipment within a terminal or switching center that carry plain language security classified information, including (i) all lines that are between the encrypted side of the online cryptoequipment and the user, i.e., the customer or subscriber end instruments or terminal equipment, (ii) the equipment that terminates plain language security classified processing equipment, and (iii) areas and spaces that contain these lines, equipment, interconnections, and auxiliary facilities. *See also* **RED-BLACK concept.**

red-green-blue (RGB): Pertaining to a color video signal transmission system in which (a) three separate signals are used to carry the red, green, and blue components of a color image, (b) there is no need to encode the image, (c) the resolution and picture clarity obtained are higher than specified by the National Television Standards Committee (NTSC) standard, and (d) three separate lines are required for connection, whereas systems that adhere to the NTSC standard require only one line.

redirected call: A call that had been connected but that had to be disconnected and reconnected to another call receiver.

redirected call indicator: Information that (a) is sent in the forward direction and (b) indicates that the call is a redirected call. *See also* **forward direction, redirected call.**

redirected to new address signal: In communications systems operations, a signal that (a) is sent in the backward direction and (b) indicates that a call has been redirected to an address other than the destination address selected by the call originator. *See also* **backward direction, redirected call.**

redirection: *See* **call redirection.**

redirection address: In communications systems operations, information that (a) is sent in the backward direction and (b) indicates that a call has been redirected to an address other than the destination address selected by the call originator. *See also* **backward direction, redirect to new address signal.**

RED signal: In cryptographic systems, a signal that contains security classified information that has NOT been encrypted. *See also* **BLACK, BLACK signal.**

reduced carrier: A carrier wave that (a) is transmitted at a power level that usually is between 6 dB and 32 dB, more often between 16 dB and 26 dB, below the peak envelope power level, and (b) usually is transmitted in order to achieve automatic frequency control or automatic gain control at the receiver. *See also* **full carrier, level, peak envelope power, suppressed carrier.**

reduced carrier single sideband emission: A single sideband emission in which the degree of carrier suppression enables the carrier to be reconstituted and to be used for demodulation. *See also* **full carrier single sideband emission, single sideband emission, suppressed carrier transmission.**

reduced carrier transmission (RCT): Amplitude-modulated (AM) transmission in which the carrier is transmitted at a power level below that required for demodulation but high enough to serve as a frequency reference at the receiver. *Note:* At the receiver, the carrier power level can be raised for demodulation. *See* **double sideband reduced carrier transmission.** *See also* **reduced carrier single sideband emission, suppressed carrier transmission.**

reduction: **1.** The relationship between the linear dimension of an image and the corresponding dimension of an object when the image is smaller than the object. *Note:* If an image is one-fourth the size of the object in linear dimension, not in area, the reduction may be expressed as 4X, 1 to 4, 1/4, 25%, or 75% smaller. **2.** In display systems, a reduced-size copy of a display element, display group, or display image. *See also* **image, object.**

reduction ratio: **1.** The ratio of (a) the distance between two points on an image of an object to (b) the distance between corresponding points on the object when the ratio is less than unity. **2.** The reciprocal of the number of times that the linear dimensions of an image of an object would have to be enlarged in order to be as large as the object. *Note:* If the image has to be enlarged to four times its linear size to be as large as the object, the reduction ratio is 1/4. If the enlargement is performed in steps, i.e., enlargements are made of enlargements to reach full size, the product of the reduction ratios is the total reduction ratio. Thus, if enlargements are 4/1, two enlargements would correspond to a reduction ratio of 1/16.

redundancy: **1.** In the transmission of data, the excess of transmitted message symbols over the amount required to convey the essential information in a noise-free circuit. *Note:* Redundancy may be introduced intentionally, such as by error detection and correction codes, or inadvertently, such as by oversampling a band-limited signal or by using inefficient code formats. **2.** In a communications system, surplus capability usually provided to improve the reliability and the quality of service, such as by the inclusion of multiple, alternate, or secondary system components to ensure continued operation in the event that the primary component fails. **3.** In information theory, the amount by which the decision content of a message or data exceeds the entropy, given by the relation $R = H_0 - H$ where R is the redundancy, H_0 is the decision content, and H is the entropy. *Note:* Usually messages can be represented with fewer characters by using suitable codes. **4.** Without loss of information content, the amount of decrease of the length of a message that may be accomplished by judicious coding, i.e., the code itself should not introduce additional redundancy. *See also* **continuous operation, data, data compression, decision content, entropy, frequency diversity, information content, space diversity.**

redundancy check: **1.** A check used to verify that any redundant hardware or software in a communications system is in an operational condition. **2.** A check in which one or more extra characters or binary digits are attached to data for the detection of errors. *See* **cyclic redundancy check, transverse redundancy check.** *See also* **operational service state, performance parameter, relative redundancy.**

redundant code: A code in which more signal elements than necessary are used to represent the intrinsic information content in a given amount of data, such as a message. *Note 1:* A five-unit code in which all the characters of the International Telegraph Alphabet 5 (ITA-5) are used in messages is not a redundant code. If only the digits, and not the alphabetic and other symbols, were used, the same code would be redundant. A seven-unit code in which only the signals that have four space digits and three mark digits are used is redundant. An eight-unit code in which one of the bits is a parity bit is redundant. *Note 2:* The redundancy may be used for error control. *See also* **error control, redundancy.**

REED: restricted edge emitting diode.

reed relay: A relay that (a) uses enclosed magnetically operated reeds as the contact members and (b) may be of the type that is mercury-wetted. *See also* **relay.**

reek: *See* **block reek.**

reel: *See* **tape reel.**

reeling condition: The condition, i.e., the stress or the stimulus, such as a tensile, thermal, bending, torsional, compressive, or humidity condition, imposed on a spooled item, such as an optical fiber, optical fiber bundle, fiber optic cable, coaxial cable, or twisted pair, during winding onto a spool, reel, or bundpack.

reencryption: The encryption of a message that is already encrypted, i.e., the encryption of a message that has been encrypted and transmitted, without necessarily having the message decrypted between the two encryption operations. *See also* **decrypt, encrypt.**

reenlargement: 1. The restoration of an image to a size larger than the size to which it had been reduced. **2.** The enlargement of an image that already had been enlarged, in order to make the image larger than it could be made in a single enlargement. **3.** The restoration of a microimage to a size that renders it legible to the normal unaided human eye. *See also* **microimage.**

reenterable program: *Synonym* **reentrant program.**

reentrant program: A computer program that (a) may be entered repeatedly, even before execution is completed, provided that none of its parameters or instructions is modified during execution, and (b) may be used simultaneously by two or more computer programs. *Synonym* **reenterable program.** *See also* **computer program.**

reentry point: In the execution of a computer program, the point, such as the address or label of an instruction, at which the computer program is continued after a temporary departure from the program, such as after completion of the execution of a subroutine called by the program. *See also* **computer program, point, reentrant program.**

reference: *See also* **external timing reference.**

reference antenna: An antenna that (a) may be real, virtual, or theoretical, and (b) has a radiation pattern that can be used as a basis of comparison for other antenna radiation patterns. *Note:* Examples of reference antennas are unit dipoles, half-wave dipoles, and isotropic or omnidirectional antennas. *See also* **radiation pattern.**

reference architecture: The structural features that describe the visibility points of distributed and open systems. *Note:* Examples of reference architecture are the interfaces and protocols that are necessary for the interconnection of the components of a system. *See also* **distributed system, interface, protocol, visibility point.**

reference black level: In television (TV) systems, the level corresponding to the specified maximum excursion of the luminance signal in the black direction. *See also* **luminance, reference white level, television.**

reference circuit: A hypothetical circuit of specified length and configuration with a defined transmission characteristic, primarily used (a) as a reference for measuring the performance of other circuits and (b) as a guide for the planning and engineering of circuits and networks. *Note:* Several types of reference circuits usually are defined, each with different configurations, to meet communications requirements for a wide range of distances. *See also* **circuit, reference system.**

reference clock (RC): 1. A clock that emits timing signals that are compared with timing signals of another clock. **2.** A clock, usually of high stability and accuracy, used to stabilize and time a group of mutually synchronized clocks of lower stability. *Note:* The failure of a reference clock does not necessarily cause loss of synchronism. *See also* **clock, coordinated clock, Coordinated Universal Time (UTC), DoD master clock, master clock, precise time.**

reference configuration: In integrated services digital networks (ISDNs), a combination and an arrangement of functional groups and reference points that reflect possible network topologies. *See also* **functional group, network topology, reference point, topology.**

reference coupling: Crosstalk coupling that is necessary to produce a specified signal level in the disturbed circuit when a signal of a specified level is inserted into the disturbing circuit. *Note:* An example of crosstalk coupling is the coupling that occurs when a 0-dBr signal is caused in the disturbed circuit when a test signal, such as a test tone, of 90 dBr is placed on the disturbing circuit. Both dBr values must be determined with the same weighting characteristic. *See also* **crosstalk, crosstalk coupling, level, noise weighting.**

reference edge: In a data carrier or data medium, the edge that is used to establish specifications or measurements for the layout of the data on the carrier or the medium. *Synonym* **guide edge.** *See also* **data carrier, data medium.**

reference frequency (RF): 1. A standard fixed frequency (a) from which operational frequencies may be derived or with which they may be compared, and (b) that may be used to (i) specify an assigned frequency, (ii) fix a characteristic frequency, and (iii) fix a carrier frequency. **2.** A frequency (a) that has a fixed and specific position with respect to the assigned frequency, and (b) that has a displacement, with respect to the assigned frequency, with the same absolute value and sign that the displacement of the characteristic frequency has with respect to the center frequency of the band occupied by the emission. *See also* **assigned frequency, carrier, characteristic frequency, frequency, precise frequency, precise time, primary time standard, principal clock, syntonization.**

reference level: *See* **single sideband equipment reference level.**

reference model: *See* **Open Systems Interconnection—Reference Model.**

reference modulation: In a modulating signal, the magnitude of the signal, such as the amplitude, maximum phase shift, or maximum frequency excursion, that is used as a standard of comparison for measurement, indication, or application of various forms of modulation. *Note:* The actual modulation magnitude can be expressed in percent of the reference modulation magnitude. If the reference modulation is 100%, then the concept of percent modulation, such as 25%, 50%, 75%, or 90% modulation, becomes useful to define, discuss, or represent the level of modulation. *See* **fixed reference modulation, modulation.**

reference monitor: 1. A device or a computer program that ensures that references maintain required values. **2.** In access control, a functional unit that mediates all accesses to objects. *See also* **access, access control, controlled access, restricted access.**

reference noise: The magnitude of circuit noise chosen as a reference for measurement. *Note 1:* Many different reference noise levels with a number of different weightings are in use. The proper parameters and measurement conditions must be stated. *Note 2:* An example of a reference noise is the magnitude of circuit noise that will produce a circuit noise meter reading, i.e., a psophometer reading, that is equal to that produced by 10^{-12} W (watt), i.e., 1 pW (picowatt) or −90 dBm, of electrical power at 1000 Hz (hertz) for noise calibrated in dBrn(144-line) or dBrnC. For psophometers that are calibrated in dBa(F1A), the reference noise is adjusted to −85 dBm. *See also* **dBa, dBa(F1A), dBa(HA1), dBa0, dBm(psoph), dBm0, dBmp0p, dBrn, dBrnC, dBrnC0, dBrn(f_1-f_2), dBrn(144-line), noise, noise weighting.**

reference point: In integrated services digital networks (ISDNs), a logical point that (a) is between nonoverlapping functional groups and (b) is designated as an interface when equipment is placed at the point. *See* **T reference point, V reference point.** *See also* **functional group, Integrated Services Digital Network (ISDN).**

reference subsystem: *See* **signal reference subsystem.**

reference surface: In the joining of two open waveguides, such as optical fibers, the surface that is used to contact the transverse alignment elements of a component, such as a connector or splice tube. *Note 1:* The reference surface usually is the outer surface of the outermost cladding. *Note 2:* Examples of reference surfaces for various optical fiber types are (a) the core-cladding interface, i.e., the fiber core, (b) the outer surface of the outermost cladding, and (c) the outer surface of the buffer. *Note 3:* In certain cases the reference surface may not be an integral part of the optical fiber. *See also* **optical fiber connector.**

reference surface center: 1. In the cross section of an optical fiber, the center of the circle that (a) best fits the outer limit of the reference surface and (b) may not be the same as the core and cladding centers. *Note:* The method of best fit must be specified. **2.** The center of the smallest circle into which the reference surface can be fitted. *See also* **cladding, core, reference surface.**

reference surface concentricity: *See* **core reference surface concentricity.**

reference surface noncircularity: The value obtained by dividing (a) the difference between the diameters of the two circles used to define the reference surface tolerance field by (b) the reference surface diameter. *See also* **reference surface, reference surface tolerance field.**

reference surface tolerance field: In the cross section of an optical fiber, the region between (a) the smallest circle concentric with the center of the reference surface circumscribed about the core area and (b) the largest circle, concentric with the first one, that fits inside the reference surface.

reference system: **1.** In communications, computer, data processing, and control systems, a group of related reference circuits. **2.** A hypothetical system of specified configuration with defined characteristics, primarily used (a) as a reference for measuring and evaluating the performance of system components and (b) as a guide for the planning and the engineering of system components, such as circuits and networks. *Note 1:* An example of a reference system is a group of related reference circuits. *Note 2:* Several types of reference systems may be defined, each with different configurations, to meet communications requirements for a wide range of distances and capabilities. **3.** In communications, computer, data processing, and control systems, a component, device, equipment, subsystem, or system selected and used as a reference for comparison with a given component, device, equipment, subsystem, or system. *See* **worldwide geographical reference system.** *See also* **reference circuit, system.**

reference test method (RTM): In fiber optics, a test method (a) in which a given characteristic of a specified class of fiber optic devices, such as optical fibers and fiber optic cables, connectors, photodetectors, and light sources, is measured strictly according to the definition of the given characteristic, and (b) that gives accurate and reproducible results relatable to practical use. *See also* **alternative test method.**

reference TLP: reference transmission level point.

reference transmission level point: The point in a transmission system that is selected and used as a reference point from which the signal voltage, current, or power level at other points may be measured or compared. *Note 1:* The comparison usually is expressed in dB, though it may also be expressed in nepers, in percent, or as a ratio. *Note 2:* The zero transmission level point (0TLP) or a channel input terminal power level might be chosen as the reference transmission level point. All gains and losses from that point are expressed in dB with reference to the specified reference point power level. Thus, if the input terminal is chosen as the reference transmission level point, and if the input terminal power level is 100 W (watt), and the output power at the end of any channel is 0.1 W, the output power level is −30 dB, i.e., 30 dB down from the reference transmission level point. *See also* **relative transmission level, transmission level point.**

reference volume: In the measurement of the volume, i.e., the level, magnitude, or intensity, of electrical signals, such as those that represent speech, music, noise, or sound, a level that produces a zero volume unit (VU) on a standard volume meter. *Note:* The sensitivity of the volume meter is adjusted so that the zero VU, i.e., the reference volume, is read when the meter is connected across a 600 Ω (ohm) resistor in which 1 mW (milliwatt) of electrical power at a frequency of 1000 Hz (hertz) is dissipated. *See also* **level, volume unit.**

reference white level: In television (TV), the level corresponding to the specified maximum excursion of the luminance signal in the white direction. *See also* **luminance, reference black level, television.**

refile: In a communications system, the conversion of a message that is prepared for transmission in accordance with the procedures for one communications system into a message that is prepared for transmission in accordance with the procedures for another system. *See* **commercial refile.**

refile message: A message that (a) is received at a communications center for filing, (b) was originally filed at another, originating, communications center, and (c) is to be dispatched over yet another communications system. *Note:* The filing time for refile messages is the date and the time that the message was received by the receiving communications center for filing. *See also* **filing, filing time, refile.**

reflectance: In optics, the ratio of reflected power to incident power. *Note 1:* Reflectance may be expressed in dB, in percent, or as the reflectance density, i.e., the logarithm to the base ten of the reciprocal of the ratio of reflected to incident power. *Note 2:* In communications, reflectance usually is expressed in dB. *Note 3:* The conditions under which the reflectance occurs should be stated, such as the spectral composition and polarization of the incident wave, the geometrical shape of the reflecting surface, and the composition of the propagation media on both sides of the interface. *Note 4:* Reflection from a smooth surface is called "specular reflection," whereas reflection from a rough surface is called "diffuse reflection." *See also* **Fresnel reflection loss, reflection.**

reflectance density: The logarithm to the base ten of the reciprocal of the reflectance. *See also* **optical density, reflectance.**

reflectance loss: *Synonym* **Fresnel reflection loss.**

reflected binary code: *See* **modified reflected binary code.**

reflected binary unit distance code: *Synonym* **Gray code.**

reflected code: A cyclic binary code such that (a) if the code words, i.e., numerals, are listed in a column, lines can be drawn that may be considered as mirrors, above and below which the code words appear to be in reflected pairs when the most significant digits are ignored, and (b) the most significant digit usually is a 0 above each mirror and a 1 below each mirror. *Note 1:* The reflected code list looks as though the part below the mirror is a mirror image of the part above the mirror, except for the most significant digits. *Note 2:* An example of a reflected code is the Gray code. *Note 3:* Reflected codes are used to avoid large signal distances between sequential codes, such as occurs when 1 is added to pure binary 11111 to obtain pure binary 100000. *See also* **Gray code.**

reflected ray: When an electromagnetic wave, such as a radio wave or a lightwave, is incident upon an interface surface between two different propagation media, such as a boundary between dielectric materials with different refractive indices, the ray that is turned back into the medium containing the incident ray. *Note 1:* If the refractive indices at the interface vary as a step function across the interface, i.e., there is a sharp boundary, spectral reflection takes place. The angle that the reflected ray makes with the normal to the interface surface, namely, the reflection angle, has the same value that the incident ray has with the normal, i.e., the incidence angle is equal to the reflection angle. The power of the reflected ray is equal to the power in the incident ray times the reflection coefficient, as determined by the refractive indices and the incidence angle, assuming that no power is absorbed at the interface, i.e., at the reflection surface, which usually is the case. *Note 2:* In an optical fiber, the reflected ray is the portion of a ray that is in the core and incident to the core-cladding interface that is returned to the core. For core-cladding interface incidence angles greater than the critical angle, all the optical power in the incident ray is contained in the reflected ray. For incidence angles equal to the critical angle, a refracted ray will propagate along the core-cladding interface surface. If there is a smooth transition of refractive index at the core-cladding interface of an optical fiber, such as in a graded index fiber, there will be some penetration of an incident ray a short distance into the cladding, and the incident ray will be returned to the core by successive or continuous reflection. *See also* **Goos-Haenchen shift, refracted ray.**

reflecting jammer: In radar system jamming, a passive device or object that (a) reflects electromagnetic radiation, i.e., produces echoes, and (b) serves to confuse radar systems. *Note:* Examples of reflecting jammers are (a) a large quantity of metallic or metallized strips or wires, such as window, chaff, or rope and (b) false target reflectors, such as corner reflectors and Luneberg reflectors, that create a reflection similar to a large target. The jammer reflects electromagnetic radiation in such a manner as to obscure other objects, to prevent their identification and to prevent successful operation of the radar system being jammed. *See also* **chaff, corner reflector, echo, jammer, jamming, radar.**

reflecting layer: In the ionosphere, a layer that has a free electron density sufficient to reflect radio waves. *Note 1:* The principal reflecting layers are the E, F_1, and F_2 layers in the daylight hemisphere. *Note 2:* A critical frequency is associated with the reflection caused by each layer. *See also* **critical frequency, D layer, E layer, F layer, ionosphere.**

reflecting loss: *Synonym* **reflection loss.**

reflection: 1. The abrupt change in direction of a wavefront at an interface between two dissimilar propagation media so that the wavefront returns into the medium from which it originated. **2.** Energy diverted back from the interface of two propagation media. *Note 1:* The reflection may be (a) specular, i.e., coherent, clear, or direct, for smooth surfaces, such as the surfaces of mirrors and highly polished metals, or (b) diffuse, i.e., scattered, for rough surfaces, such as ground glass or fogged glass. *Note 2:* After specular reflection, an image of the original source can be formed, but an image of the reflecting surface cannot be formed. After diffuse reflection, an image of the original source cannot be formed, but an image of the diffusely reflecting surface can be made. *See* **diffuse reflection, Fresnel reflection, half-reflection, internal reflection, ionospheric reflection, maximum optical reflection, specular reflection, total internal reflection, total reflection.** *See also* **critical angle, reflectance, reflectivity.** *Refer to* **Fig. S-13 (919).**

reflection angle: When a ray of electromagnetic radiation strikes a surface and is reflected in whole or in part by the surface, the angle between the normal to the reflecting surface and the reflected ray. *See also* **critical angle, ray, reflection.**

reflection coating: *See* **antireflection coating.**

reflection coefficient (RC): 1. In a wave incident upon a surface, such as a material discontinuity or mismatch, the ratio of (a) the amplitude of the reflected wave to (b) the amplitude of the incident wave. *Note:* For smooth surfaces, such as a mirror, the reflection coefficient may be near unity. At large incidence angles, i.e., near grazing incidence, rough surfaces have a large reflection coefficient, i.e., a reflection coefficient approaching unity. **2.** At any specified place in a transmission line between a power source and a sink, the complex ratio of the electric field strength of the reflected wave to that of the incident wave. *Note:* The magnitude of the reflection coefficient (RC) is given by the relation $RC = |(Z_1 - Z_2)/(Z_1 + Z_2)|$ and by the relation $RC = (SWR - 1)/(SWR + 1)$ where RC is the reflection coefficient, Z_1 is the impedance toward the source, Z_2 is the impedance toward the load, the vertical bars designate absolute magnitude, and SWR is the standing wave ratio. **3.** At oblique incidence and when the *electric* field vector of an incident plane polarized electromagnetic wave is parallel to the interface surface between dielectric propagation media of different refractive indices, the ratio of (a) the reflected electric field strength to (b) the incident electric field strength, given by the relation $R_e = (m_2 \cos A - m_1 \cos B)/(m_2 \cos A + m_1 \cos B)$ where m_1 and m_2 are the reciprocals of the refractive indices of the incident and transmitted media, respectively, and A and B are the incidence and refraction angles, respectively. *Note:* The relation is one of the Fresnel equations. The sum of the reflection coefficient and the transmission coefficient is not necessarily unity. **4.** At oblique incidence and when the *magnetic* field vector of an incident plane polarized electromagnetic wave is parallel to the interface surface between dielectric propagation media of different refractive indices, the ratio of (a) the reflected electric field strength to (b) the incident electric field strength, given by the relation $R_m = (m_1 \cos A - m_2 \cos B)/(m_1 \cos A + m_2 \cos B)$ where m_1 and m_2 are the reciprocals of the refractive indices of the incident and transmitted media, respectively, and A and B are the incidence and refraction angles, respectively. *Note:* The relation is one of the Fresnel equations. The sum of the reflection coefficient and the transmission coefficient is not necessarily unity. *See also* **Fresnel reflection loss, reflection loss, refractive index, refractive index contrast, refractive index profile, return loss, standing wave ratio, transmission coefficient.**

reflection density (RD): An inverse measure of the reflectance of an optical element. *Note 1:* Reflection density (RD) is expressed by $\log_{10}(1/R)$, where R is the transmittance. *Note 2:* The higher the reflection density (RD), the lower the reflectance, and vice versa. *Note 3:* Ten times the reflection density (RD) is equal to the reflection loss expressed in dB. For example, an RD of 0.3 corresponds to a reflection loss of 3 dB. *Note 4:* When the reflectance approaches unity, the reflection density (RD) is a more useful value, such as in dealing with nearly perfect mirrors. *Note 5:* The analogous term, $\log_{10}(1/T)$, where T is the transmittance, is called the "optical density." *See also* **optical density, reflectance, transmission loss, transmittance.**

reflection factor: The reciprocal of the scalar value of the reflection loss. *See also* **reflection loss.**

reflection image: An image formed by a reflecting surface. *Note:* An unwanted reflection image is a ghost image. *See also* **ghost image, image, reflection.**

reflection law: When a ray of electromagnetic radiation is reflected in whole or in part at a point in a reflecting surface, (a) the reflection angle is equal to the incidence angle, and (b) the incident ray, reflected ray, and normal to the surface are all in the same plane.

reflection loss: 1. In a wave of given frequency incident at a discontinuity or a mismatch of propagation media, such as a discontinuity in a transmission line, i.e., a junction of a source of power and a load, the ratio of the incident power to the reflected power, given by the relation $20 \log_{10}|(Z_1 + Z_2)/(4Z_1Z_2)^{1/2}|$ where the reflection loss is in dB, the vertical bars designate absolute magnitude, Z_1 is the impedance of the source of power including the line, and Z_2 is the impedance of the load. *Note 1:* The reflection loss usually is expressed in dB. *Note 2:* When the incident and reflected waves have opposite phases and appropriate magnitudes, a "reflection gain" may be obtained, i.e., a negative loss is obtained. *Note 3:* Any point may be considered as a junction of a source and a load. The source is on one side of the point, and the load is on the other. *See also* **echo attenuation, Fresnel reflection loss, loss, path loss, reflection coefficient, return loss, scattering loss, standing wave ratio.**

reflection method: *See* **Fresnel reflection method.**

reflection power penalty: In an optical receiver, the additional power that is required by the receiver (a) when the manufacturer-specified value of maximum optical reflection is introduced at the line side of the associated transmitter connector, to achieve the same bit error ratio (BER) that is obtained without the introduced reflection, and (b) when standard measurement procedures are used.

reflection sensor: *See* **frustrated total reflection sensor.**

reflective array antenna: An antenna, such as a billboard antenna, that (a) has active elements, i.e., the driven elements, that are situated at a predetermined distance from a surface designed to reflect the incident energy from the active antenna in a desired direction, (b) usually has many active elements working in conjunction with an electrically large reflection surface to produce a unidirectional beam, (c) may be used to increase antenna gain in desired directions or reduce radiation in unwanted directions, and (d) may contain parasitic elements as well as driven elements. *See also* **antenna, parasitic element, phased array, radiation pattern, reflector.**

reflective coating: A coating placed on an optical element to increase its reflectance. *See* **highly reflective coating.**

reflective jamming: Jamming in which reflectors are used to return false and confusing signals to the radar receiver that is being jammed. *Note:* An example of reflective jamming is jamming with a large quantity of metallic strips or wires, such as window, chaff, rope, or corner reflectors. *See also* **chaff, corner reflector, jamming, radar, victim emitter.**

reflective membrane optical scintillator (REMOS): A test bed that (a) is used to evaluate the performance of laser receivers, (b) creates realistic turbulence conditions, (c) affects laser beams before they strike the receiver, (d) simulates the conditions encountered by laser beams as they pass through unstable air or engine plumes, and (e) was developed by the U.S. Air Force.

reflective sensor: *See* **fiber optic reflective sensor.**

reflective star coupler: A fiber optic coupler in which signals in one or more optical fibers are transmitted to one or more other fibers by entering the signals into one end of an optical fiber, cylinder, or other piece of transparent material with a reflecting back surface, in order to reflect the diffused signals into output ports for additional transmission in one or more fibers. *See also* **fiber optic coupler, nonreflective star coupler, tee coupler.**

reflectivity: 1. The reflectance of the surface of a material so thick that the reflectance does not change with increasing thickness. **2.** The intrinsic reflectance of the surface of a material irrespective of other parameters or effects, such as the reflectance of the rear surface. *Note:* "Reflectivity" is no longer in common use. *See* **spectral reflectivity.** *See also* **reflectance.**

reflectometer: *See* **optical time domain reflectometer.**

reflectometry: *See* **optical time domain reflectometry.**

reflector: 1. In radio antenna systems, a set of electrical conductors or conducting surfaces that reflects radiant energy. *Note:* Reflectors come in many different sizes and shapes, the most common being the parabolic reflector, often called a "dish," used in radar, radio astronomy, and satellite communications systems. **2.** In optics, a surface with a high reflection coefficient at all incidence angles, such as a mirror. *See* **corner reflector, Lambertian reflector, trihedral corner reflector.** *See also* **parasitic element, passive repeater, reflective array antenna.**

refracted near field scanning method: *Synonym* **refracted ray method.**

refracted ray: 1. In a ray that is incident upon an interface surface between two different propagation media, the ray that emerges on the side of the interface opposite the side of incidence. *Note 1:* When the incident ray is in a higher refractive index material, such as the core of an optical fiber, the refracted ray is bent away from the normal and toward the interface surface. When the incident ray is in a lower refractive index material, the refracted ray is bent toward the normal and away from the interface surface. *Note 2:* In optical fibers, for incidence angles greater than the critical angle, there is no refracted ray. For incidence angles equal to the critical angle, the refracted ray will propagate along the core-cladding interface surface. *Note 3:* The amplitude of the refracted ray is given by the transmission coefficient, i.e., the refraction coefficient, and the amplitude of the incident ray. **2.** In an optical fiber, a ray that is refracted from the core into the cladding. *Note 1:* The direction of a refracted ray is given by the relation $[n_r^2 - n_a^2]/[1-(r/a)^2 \cos^2 \phi_r] \leq \sin^2 \Theta_r$ where r is the radial distance from the optical fiber axis, ϕ_r is the azimuthal angle of projection of the ray at r on the transverse plane, Θ_r is the angle the ray makes with the fiber axis, n_r is the refractive index at r, n_a is the refractive index at the core radius, and a is the core radius. *Note 2:* In optical fibers, refracted rays correspond to radiation modes. *See also* **cladding ray, guided ray, leaky ray, radiation mode, reflected ray, transmission coefficient.**

refracted ray method: A method for measuring the refractive index profile of an optical fiber by scanning the entrance face with the vertex of a high numerical aperture (NA) cone of light and measuring the change in the power of refracted rays, i.e., of unguided rays. *Synonym* **refracted near field scanning method.** *See also* **refracted ray, refraction.**

refracting crystal: A transparent crystalline substance in which the ratio of the refraction angle to the incidence or reflection angle is high, i.e., which has a high refractive index compared to glass. *See* **multirefracting crystal.** *See also* **refractive index.**

refraction: 1. The abrupt change of direction and speed of a soundwave, radiowave, or lightwave as it passes obliquely from one medium with a given refractive index to another medium with a different refractive index. **2.** The changing of direction and speed of a wavefront when it passes obliquely through a boundary between two dissimilar media, or while it propagates in a medium with a refractive index that is a certain continuous function of position, i.e., in a graded refractive index medium. *See also* **deviation angle, graded refractive index, propagation medium, refractive index, wavefront.** *Refer to* **Fig. S-13 (919).**

refraction angle: When an electromagnetic wave propagating in one propagation medium strikes a surface of another propagation medium and is wholly or partially transmitted into the new medium, the acute angle between the normal to the surface and the refracted ray. *See also* **propagation medium, refracted ray.** *Refer to* **Fig. S-13 (919).**

refraction coefficient: *Synonym* **transmission coefficient.**

refraction law: *Synonym* **Snell's law.**

refraction profile: *Synonym* **refractive index profile.**

refractive field: *Synonym* **near field.**

refractive field region: *Synonym* **near field region.**

refractive index (n): 1. In a refractive medium, i.e., a propagation medium other than a vacuum, the ratio of (a) the velocity of propagation of an electromagnetic wave in vacuum to (b) the velocity of propagation of the wave in the medium. **2.** At a point in a propagation medium and in a given direction, the ratio of (a) the velocity of light in vacuum to the magnitude of the phase velocity of a sinusoidal electromagnetic plane wave propagating in that direction. **3.** The ratio of the sines of the incidence and refraction angles as a light ray passes across an interface between two different propagation media. *Note 1:* If the refractive index for one material medium is known relative to a vacuum, i.e., approximately relative to air, the refractive index of the other can be determined. *Note 2:* The refractive index of a vacuum is defined as unity. Refractive indices, relative to a vacuum, are 1.000292 for air, 1.333 for water, 1.156 for crown glass, and 1.475 for silica glass (SiO_2). Refractive indices for materials are dimensionless because they are relative to the refractive index of a vacuum. In material media, the refractive index changes slightly with changes in wavelength, usually exhibiting a minimum value at a certain wavelength for each material. *Note 3:* The refractive index of a dielectric material is given by the relation $n = (\mu\varepsilon/\mu_0\varepsilon_0)^{1/2}$ where μ is the magnetic permeability of the material, ε is the electric permittivity of the material, and the subscripted values are for a vacuum. In absolute units for free space, the electric permittivity is $\varepsilon = 8.854 \times 10^{-12}$ F/m (farads per meter) and the magnetic permeability is $\mu = 4\pi \times 10^{-7}$ H/m (henries per meter). The reciprocal of the square root of the product of these values yields approximately 2.998×10^8 m•s^{-1} (meters per second), the speed of light in a vacuum. *Note 4:* Because the refractive index for a propagation medium is defined as the ratio of the velocity of light in a vacuum to the velocity of light in the medium, the ratio becomes unity when the propagation medium is a vacuum. Thus, the relative refractive index is given by the relation $n_r = (\mu\varepsilon)^{1/2}$ where the n_r is the refractive index relative to vacuum (or air). The refractive indices for the materials used in optical fibers are always given as relative to a vacuum and are therefore dimensionless. Also, except for ferrous materials, cobalt, and nickel, and some diamagnetic materials, the relative magnetic permeability of materials, especially dielectric materials, is approximately 1. For a vacuum it is 1. In dielectric materials, the conductivity is considered to be zero, and the resistivity is considered to be infinite. *Synonym* **index of refraction.** *See* **relative refractive index.** *See also* **electric permittivity, magnetic permeability, relative electric permittivity, relative magnetic permeability, wave impedance.** *Refer to* **Figs. R-8, R-9.** *Refer also to* **Figs. C-9 (190), I-5 (470), P-1 (695), R-1 (796), R-5 (817), R-10 (836).**

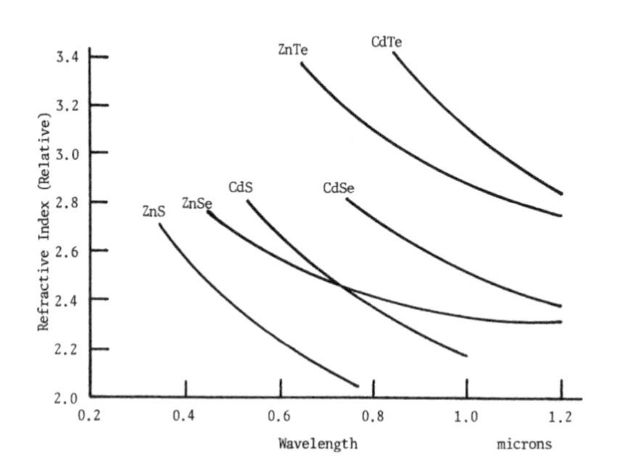

Fig. R-8. The **refractive index** as a function of wavelength for single-crystal optical thin films, each doped by diffusion of a specific dopant for use in optical integrated circuits.

refractive index contrast (Δ): A measure of the relative difference in refractive index across an interface surface between propagation media with different re-

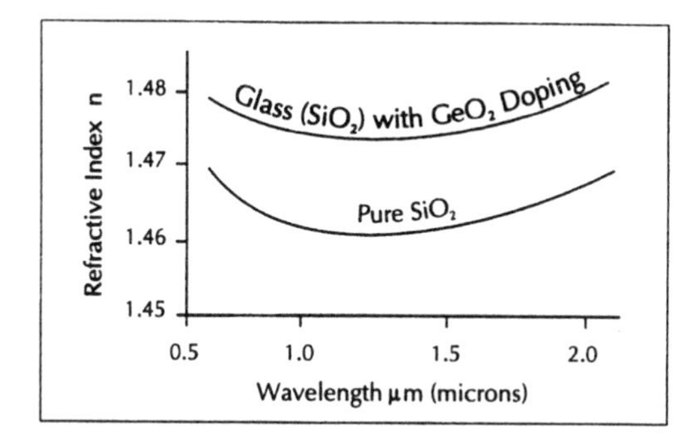

Fig. R-9. The variation of the **refractive index** with wavelength in different materials. The levels of the curves can be raised or lowered by the addition of different dopants and different concentrations of those dopants.

fractive indices, such as between the core and the cladding of an optical fiber. *Note 1:* The refractive index contrast is given by the relation $\Delta = (n_1^2 - n_2^2)/2n_1^2$. For refractive indices that do not differ by more than 1%, the refractive index contrast is given by the relation $\Delta \approx (n_1 - n_2)/n_1$ where Δ is the refractive index contrast, n_1 is the refractive index of the denser material, and n_2 is the refractive index of the less dense medium. For an optical fiber, n_1 is the maximum refractive index of the core, and n_2 is the refractive index of the homogeneous cladding. *Note 2:* The refractive index contrast is used in the expressions for the power law index profile. *Note 3:* For optical fibers, the refractive index contrast usually is less than 1%. *See also* **refractive index.**

refractive index difference: *See* **equivalent step index profile refractive index difference.**

refractive index dip: In an optical fiber, a decrease in the refractive index at the center of the core. *Note:* An index dip is caused by certain optical fiber manufacturing processes. *Synonyms* **index dip, profile dip.** *See also* **refractive index profile.**

refractive index gradient: The rate of change of refractive index with respect to distance in a propagation medium, i.e., the slope of the refractive index profile. *Note 1:* The refractive index gradient is expressed in units of reciprocal distance. The refractive index is dimensionless because it is assumed to be the refractive index relative to a vacuum. *Note 2:* Examples of refractive index gradients are (a) in a radial direction in the core of a graded index optical fiber, the rate of change of refractive index at a point with respect to distance, and (b) in a vertical direction in the atmosphere, (i) the difference in refractive index at two different altitudes divided by (ii) the difference in altitude. *Note 3:* A gradient is a vector point function. *See also* **ducting, refractive index, refractive index profile, relative refractive index, surface refractive index gradient.**

refractive index matching cell: A container that is filled with a liquid that has a refractive index equal to the refractive index of inserted butt-jointed optical fibers. *Note:* The reflection coefficient for normal incidence is given by the relation $(n_1 - n_2)/(n_1 + n_2)$. The reflected power at a butted joint of the inserted fibers is zero when $n_1 = n_2$, i.e., there effectively is no interface surface. *Synonym* **index matching cell.**

refractive index matching gel: A gel that (a) is a colloid in which the disperse phase has combined with the continuous phase to produce a semisolid material, as in a jelly such as petroleum jelly, (b) has selective refractive index properties that are the same as those of optical fibers, and (c) is used in fiber optic joints to minimize reflective losses at optical fiber endfaces. *Synonym* **index matching gel.** *See also* **fiber optic joint, gel, optical fiber, refractive index.**

refractive index matching material: 1. Transparent material, i.e., light-transmitting material, with a refractive index such that when it is used in intimate contact with other transparent materials, radiant power loss is reduced at interfaces by its reducing reflection, increasing transmission, avoiding scattering, and reducing dispersion. **2.** In optical fibers, a material with a refractive index nearly equal to the fiber core refractive index. *Note 1:* Index matching materials are used to reduce scattering, dispersion, and Fresnel reflections at an optical fiber endface. *Note 2:* Index matching materials may be liquids, gels, or cements. *Synonym* **index matching material.** *See also* **antireflection coating, Fresnel reflection, mechanical splice, refractive index.**

refractive index power leveling: *See* **power law refractive index power leveling.**

refractive index profile: 1. In a dielectric waveguide, the variation of refractive index in a cross section of the guide, i.e., in a direction transverse to the direction of wave propagation. **2.** In a round optical fiber, the refractive index as a function of distance from the optical fiber axis along any fiber diameter. *Note:* The refractive index profile of a round optical fiber is approximately symmetrical about the optical fiber axis, i.e., the profile has zero axis, i.e., zero point, symmetry. *Synonyms* **index profile, refraction profile.** *See* **linear refractive index profile, parabolic refractive index profile, power law refractive index**

profile, radial refractive index profile. *See also* **fiber optics, graded index profile, refractive index, step index profile.** *Refer to* **Fig. R-10.** *Refer also to* **Figs. G-2 (399), P-1 (695), R-1 (796).**

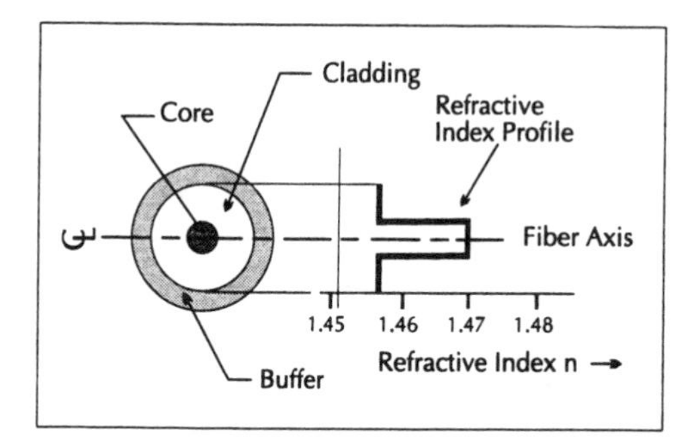

Fig. R-10. The **refractive index profile** for a step index (SI) fiber. The refractive index of the core is higher than that of the cladding, the change occurring abruptly at the core-cladding interface surface. The number of modes the fiber is capable of supporting depends on the core radius, the numerical aperture, and the wavelength, i.e., depends on the normalized frequency, *V.* For normalized frequencies over 2.4, the fiber will support $V^2/4$ modes.

refractive index profile mismatch loss: A loss of signal power introduced by a fiber optic splice in which the graded index profiles of the two coupled optical fibers are not the same. *See also* **graded index profile, refractive index.**

refractive index profile parameter: The exponent, *g,* in the several relations for the power law refractive index profile. *Synonym* **index profile parameter.** *See also* **power law refractive index profile.**

refractive index template: *See* **four-concentric-circle refractive index template.**

refractivity gradient: *Synonym* **refractive index gradient.**

reframing time: The time between (a) the instant that a valid frame alignment signal is available at the receiving data terminal equipment (DTE) and (b) the instant that frame alignment is established. *Note:* The reframing time includes the time required for replicated verification of the validity of the frame alignment signal. *Synonym* **frame alignment recovery time.** *See also* **frame.**

refresh: 1. To repeat the production of a display image on a display surface so that the image remains visible. **2.** In a storage device that requires regeneration in order that the stored data remain stored, to repeat the data store cycle. *See also* **regeneration.**

regeneration: 1. The coupling of a processed part of the output of an amplifier to its input. **2.** In computer graphics, a repetition of the sequence of events used to generate a display image from its representation in storage. **3.** In a repeater, the process in which digital signals are amplified, reshaped, retimed, and retransmitted. **4.** In a storage or display device, the restoration of stored or displayed data that have deteriorated. *Note:* In display devices, such as a cathode ray tube (CRT), displayed data must continually be regenerated in order that the data remain displayed. In storage devices in which data that are read are destroyed, the data must be regenerated and restored. **5.** In a destructive storage device, the restoration of data subsequent to reading. *See* **pulse regeneration, signal regeneration.** *See also* **closed loop transfer function, dejitterizer, destructive storage, feedback, positive feedback, read, refresh.**

regeneration rate: The rate at which the regeneration cycle is repeated in order that (a) in a display device, the display element, display group, or display image remain displayed, or (b) in a storage device that requires regeneration, the data remain stored, such as in an electrostatic storage tube. *Note:* The regeneration rate may be expressed in regeneration cycles per unit time, such as 50 times per second, or in hertz. *Synonym* **refresh rate.** *See also* **regeneration.**

regenerative feedback: Feedback in which the portion of the output signal that is returned to the input has a component that is in phase with the input signal. *See also* **positive feedback.**

regenerative repeater: A repeater in which digital signals are amplified, reshaped, retimed, and retransmitted. *Synonym* **regenerator.** *See also* **optical repeater, pulse, repeater.**

regenerative storage: Storage in which stored data must be continually restored in order that the data remain stored, whether or not the data are read. *See* **destructive storage, read, regeneration, regeneration rate.**

regenerative track: On a magnetic drum or disk, part or all of a track that has separately spaced pairs of read and write heads on the same track, thus allowing data recorded by the write head to be read by the read head a short time later, so as to produce a revolving data loop that can be used for fast-access storage. *Note:* The data in the loop can be changed by gating in new data at the write head. The length of the loop depends on the packing density and the track distance between the write head and the read head. *See also* **head, packing density, read, read head, track, write, write head.**

regenerator: *Synonym* **regenerative repeater.**

regenerator fiber optic section: *See* **fiber optic station/regenerator section.**

regenerator section: *See* **fiber optic station/regenerator section.**

regenerator system gain: *See* **statistical terminal/ regenerator system gain, terminal/regenerator system gain.**

region: A domain, such as the frequency, time, and spatial domain, with explicit or implied boundaries or limits. *See* **diffraction region, far field region, Fraunhofer region, Fresnel region, intermediate field region, International Telecommunication Union world region, near field region, overlay region.**

regional center: *See* **office classification.**

Regional Bell Operating Company (RBOC): A company (a) that is any one of the seven holding companies formed from the divestiture action of the American Telephone and Telegraph Company, i.e., AT&T, and (b) into which one or more of the Bell System local telephone companies were assigned.

register: 1. A device that (a) is accessible to a number of input circuits, (b) accepts and stores data, and (c) usually is used only as a device for temporary storage of data. **2.** A temporary memory device used to receive, hold, and transfer data, usually a computer word, to be operated upon by a processing unit. *Note:* Computers usually contain a variety of registers, such as (a) general purpose registers that perform a variety of functions, such as holding constants, holding operands, and accumulating results of arithmetic and logical operations, and (b) special purpose registers that perform special functions, such as holding the instruction being executed, holding the address of a storage location, holding data being retrieved from, or sent to, storage, and holding a count number. *See* **arithmetic register, bounds register, code sequence register, feedback shift register, index register, linear feedback shift register, Master International Frequency Register, nonlinear feedback shift register, overflow register, peg count register, recirculating register, shift register, storage register, traffic usage register.** *See also* **buffer, fetch protection, M sequence, read-only storage, permanent storage, random access memory, storage.**

registered jack (RJ): Any of the RJ series of jacks, described in the *Code of Federal Regulations,* Title 47, part 68, used to provide an interface to the public telephone network. *See also* **connection, interface, network.**

register generator: *See* **shift register generator.**

register length: In a register, the storage capacity, such as 12 decimal digits or 32 binary digits. *See also* **register, storage capacity.**

registration: 1. The positioning of an entity relative to a reference. **2.** In communications, the act of registering terminal equipment and protective circuits in accordance with the *Code of Federal Regulations,* including the assigning of numbers to the equipment and testing the equipment. *See also* **Federal Communications Commission Registration Program, circuit, terminal equipment.**

registration board: *See* **International Frequency Registration Board.**

registration program: *See* **Federal Communications Commission Registration Program.**

regression testing: In communications, computer, data processing, and control systems, subsystems, components, and equipment, selective retesting to (a) detect errors that may have been introduced during modifications (b) verify that modifications have not caused undesired adverse effects, or (c) verify that modified items still meet originally specified requirements.

regular reflection: *Synonym* **specular reflection.**

regular station: In a communications network, a station that, under normal conditions, is required to be in communication with a specified set of stations or is required to intercept messages from a specified set of stations. *Note:* An example of a regular station is a fixed station in an aeronautical communications network that is required to be in communication with aircraft stations. *See also* **fixed station, intercept communications method.**

regular transmission: *Synonym* **specular transmission.**

regulate: 1. In communications, computer, data processing, and control systems, to make changes that compensate for changing environmental conditions, thereby permitting continued normal operation. **2.** To control by holding values to specified values so that system changes do not occur, regardless of changing environmental conditions or applied perturbations, such as by using feedback or servomechanisms that react in real time. *See also* **feedback, real time, servomechanism.**

regulation: 1. In a system, compensation for changes in order to maintain constant conditions and values. *Note:* Examples of regulation are (a) adding heat to compensate for heat losses and maintain a constant temperature, (b) raising signal levels at a point to compensate for increased attenuation in signals transmitted to other points, and (c) raising an internally generated voltage to maintain a constant terminal voltage or constant voltage at the load as the load is increased. **2.** Controlling a parameter, such as controlling electric current flow to maintain a prescribed flow schedule. **3.** The variation of a parameter, such as the percentage by which voltage at a load drops from no-load to full-load, expressed by the relation $R = 100(V_0 - V_L)V_0$ where R is the regulation in percent, V_0 is the no-load voltage, and V_L is the full-load voltage. *See* **two-pilot regulation.** *See also* **parameter.**

rejection gate: *Synonym* **NOR gate.**

rejection ratio: *See* **common mode rejection ratio, image rejection ratio.**

rekeying: The changing of one or more keys that are used for either communications security (COMSEC) or transmission security (TRANSEC) functions. *See* **automatic remote rekeying, over the air rekeying.** *See also* **key.**

relateral tell: The relaying of messages between two facilities via a third facility when (a) it is absolutely necessary, (b) the communications environment between the two facilities has become untenable, such as lines have been storm damaged, or (c) a degraded service state exists between the two facilities. *See also* **degraded service state.**

relation: *See* **constitutive relation.**

relational database: A database in which the relationships among files, records, and other data items are explicitly specified such that the content of the files indicates the relationships among the files, in contrast to a database that has a hierarchical, linear, sequential, or shallow-network structure. *Note 1:* A data description language may be used to automatically create a data dictionary to store all the relationships between the various sets of data. *Note 2:* An example of a relational database is a file on cities that contains information on the population of each city and the name of the city with the next higher population. If all the cities with a population above a certain value needed to be identified, every file would not have to be retrieved to examine the population figure. The first city found with a population greater than the specified figure would automatically lead to all the other cities with a higher population. *Note 3:* A relational database can be manually operated or be computer-programmed. *See also* **database, data description language, data dictionary.**

relational operator: An operator that (a) operates on two or more operands, (b) yields a true or false value, and (c) usually is placed between the operands. *Note:* In the expression "4 ≥ 9," the 4 and 9 are operands, ≥ is the relational operator, and the value is false. In the expression "MART *is an initial word of* MARTIN," MART and MARTIN are operands, *is an initial word of* is the relational operator, and the value is true. In "Area *has the units of* time," "Area" and "time" are the operands, *has the units of* is the relational operator, and the value is false. *See also* **operand, operator, value.**

relations communications: *See* **public relations communications.**

relative address: In computer and data processing programming, an address that is expressed as a difference in relation to a base address. *See* **self-relative address.** *See also* **base address, difference.**

relative bearing: A bearing that is measured from a given reference direction or point, usually other than true north.

relative cell address: In a personal computer spreadsheet formula, a cell address that will change when copied, such that a copied formula that referred to a cell a certain number of rows and columns away will now refer to another cell that is the same relative distance away from its new location on the spreadsheet. *Note:* Relative cell addresses usually are not preceded by a special symbol, such as a $, as might be used to indicate an absolute cell address. *See also* **absolute cell address, cell, cell address.**

relative coordinate: In display systems, the coordinate of a point expressed as a displacement from a given addressable coordinate in the display space on the display surface of a display device or in an image storage space. *Note:* The relative coordinate indicates the coordinate distance between the given addressable point and some other addressable point in the particular coordinate system in which the points lie. *See also* **absolute coordinate, display surface, image storage space, relative coordinate data.** *Refer to* **Fig. P-4 (712).**

relative coordinate data: In display systems, such as computer interactive display terminals with cathode ray tube (CRT) screens or fiberscope faceplates on the ends of an aligned bundle of optical fibers, values that specify displacements from an actual coordinate, such as an absolute coordinate, user coordinate, or screen coordinate, in the display space on the display surface of a display device or an image storage space. *Note:* The coordinates specified by relative coordinate data may be (a) contained in a computer program, (b) stored in a storage unit, such as main storage, buffer storage, or auxiliary storage, or (c) printed on a hard copy, such as a sheet of paper. In an aligned bundle, the coordinate of a particular optical fiber may be given relative to a given fixed or specified fiber in the bundle. *Synonym* **relative data.** *See also* **absolute coordinate data, aligned bundle, fiber optic bundle, relative coordinate, screen coordinate, user coordinate.**

relative electric permittivity: The electric permittivity of a material, ε_r, usually represented as ε and implied to be relative to the absolute or incremental electric permittivity of free space, i.e., compared with that of a vacuum, ε_0. *Note 1:* The absolute or incremental electric permittivity of free space in SI units (Système International d'Unités) is 8.854×10^{-12} C•m^{-1}•V^{-1} (coulombs per square meter)/(volts per meter), which is the same as farads per meter. The refractive index for the silica glass of optical fibers is given approximately by the relation $n \approx \varepsilon^{1/2}$ where ε is the electric permittivity of the glass. Because it is relative to free space, a vacuum, or air, it is dimensionless for specific materials, about 2 or 3 for glass and up to 10 or 12 for the special dielectric materials used in capacitors to increase their capacity and provide for electrical separation of electrodes. *Note 2:* In common practice, the electric permittivity usually is given as the relative electric permittivity, but simply called the "electric permittivity" with the understanding that it is relative to that of a vacuum. *See also* **electric permittivity, refractive index.**

relative entropy: In information theory, the ratio of (a) the absolute entropy of a message or a set of data to (b) the decision content of the message or the set of data, expressed by the relation $H_r = H/H_0$ where H_r is the relative entropy, H is the absolute entropy, and H_0 is the decision content of the given quantity of information. *See also* **absolute entropy, decision content, message.**

relative error: The ratio of (a) an absolute error to (b) the true, specified, or theoretically correct value of the quantity that is in error. *See also* **absolute error, normalized error.**

relative level: *See* **zero relative level.**

relative magnetic permeability: The magnetic permeability of a material, μ_r, usually represented as μ and implied to be relative to the absolute or incremental magnetic permeability of free space, i.e., compared with that of a vacuum, μ_0. *Note 1:* The absolute or incremental magnetic permeability of free space in SI units (Système International d'Unités) is $4\pi \times 10^{-7}$ wb•A^{-1}•m^{-1} or H•m^{-1} (webers per square meter) per (ampere per meter), webers per ampere•meter, or henries per meter. For an optical fiber, the magnetic permeability is very nearly equal to that of free space. Thus, μ_r for an optical fiber is very close to unity, and because it is relative to free space, a vacuum, or air, it is dimensionless for specific materials. *Note 2:* In common practice, the magnetic permeability is usually given as the relative magnetic permeability and is simply called the "permeability" with the understanding that it is the magnetic permeability relative to that of a vacuum. *See also* **magnetic permeability, refractive index.**

relative offset: *Synonym* **normalized offset.**

relative order: In display systems, a display command that may be in a segment, display file, or computer program, or the instruction repertory thereof, that can cause a display device to interpret the data following the order as relative coordinate data rather than absolute coordinate data. *See also* **absolute coordinate data, display command, relative coordinate data, segment.**

relative phase telegraphy: Telegraphy in which signal coding is used that is dependent on the phase relation of an input signal to the previous digit, rather than on its absolute value. *See also* **differential encoding, phase.** *Refer to* **Fig. P-4 (712).**

relative power: A power level expressed in relation to a given absolute power level or reference power level. *Note:* Relative power may be expressed in any convenient power unit, such as dB, dBr, dBm, percent, or per unit when normalized. *See* **dB, dBr, dBm.**

relative redundancy: In information theory, the ratio of (a) the absolute redundancy to (b) the decision content of a given quantity of data, such as a message or a set of data, given by the relation $R_r = R/H_0 = (H_0 - H)/H_0$ where R_r is the relative redundancy, H_0 is the decision content, and H is the entropy of the given quantity of data. *See also* **absolute redundancy, decision content, entropy.**

relative refractive index: The refractive index of a substance relative to another substance. *Note 1:* If two glasses have refractive indices, i.e., refractive indices relative to a vacuum, of $n_1 = 2.100$ and $n_2 = 1.781$, then the relative refractive index of substance 1 relative to substance 2 is n_1/n_2, or 1.179. If substance 2 is air, the relative refractive index of substance 1 is about 2.100. *Note 2:* At an interface between materials with different refractive indices, the relative refractive index is more directly informative than the refractive index relative to a vacuum for the two materials, particularly with regard to reflection and refraction. *Note 3:* Refractive indices given in tables of refractive indices are relative to a vacuum and therefore are dimensionless. Thus, relative refractive indices are also dimensionless. *See also* **refractive index, Snell's law.**

relative spectral width (RSW): 1. The spectral width normalized by dividing (a) the root mean square deviation in which the independent variable is wavelength, λ, and $f(\lambda)$ is a suitable radiometric quantity, by (b) the central or nominal wavelength. **2.** The spectral width normalized by dividing (a) the root mean square deviation in which the independent variable is wavelength, λ, and $f(\lambda)$ is a suitable radiometric quantity, by (b) the full width at half maximum (FWHM), i.e., the difference between the wavelengths at which the magnitude drops to one-half of its maximum value. *Note 1:* The relative spectral width is given by the relation $(\Delta\lambda)\lambda$ where $\Delta\lambda$ is the spectral width determined by suitable criteria, and λ is the central or nominal wavelength. *Note 2:* The full width at half maximum (FWHM) may be difficult to obtain when the spectral composition envelope has a complex shape. *See also* **coherence length, dispersion, line spectrum, root mean square deviation, spectral width.**

relative transmission level (RTL): At a given point in a transmission system, the ratio of (a) the signal power at that point to (b) the signal power at some point chosen as a reference. *Note:* The relative transmission level (RTL) usually is determined by (a) applying a standard test tone at zero transmission level point (0TLP), or adjusted test tone power at any other point, and (b) measuring the gain or the loss to the location of interest. A distinction should be made between the standard test tone power and the expected power of the actual signal required as the basis for the design of the transmission system. *See also* **dBr, level, standard test tone, transmission, transmission level, transmission level point, zero transmission level point.**

relative unit: A unit that (a) has been normalized to a specified reference unit, (b) that usually is obtained by dividing an absolute unit by a reference unit, and (c) usually can be converted to an absolute unit by multiplying by the reference unit. *Note 1:* Examples of relative units are dB (relative to a specified power level at a specified point), refractive index (relative to the absolute refractive index of a vacuum), directive gain (relative to the power level of an isotropic radiator), and per unit (relative to a specified absolute unit, such as a kilovolt, so that all voltages may be expressed as a number to be multiplied by 1000 V (volts)). *Note 2:* Relative units are dimensionless. *Note 3:* Relative units can be converted to absolute units only when the reference unit is known. *See also* **absolute unit.**

relative vector: In display systems, a vector that (a) has a starting point usually specified by an absolute vector and usually the last reached point of the immediately preceding display element, and (b) has an end point specified by a displacement from the starting point. *Note:* Relative vectors usually are visually indicated in display areas on display surfaces, such as cathode ray tube (CRT) screens on computer monitors, fiberscope faceplates, light-emitting diode panels, and gas panels. *Synonym* **incremental vector.** *See also* **absolute vector, display element, vector.**

relaxation oscillator: A device in which a voltage in one stage is built up across a capacitor to a specified level, at which point the condenser is discharged through another stage that has a breakdown point, thus producing a pulsed output. *See also* **multivibrator.**

relay: 1. In communications systems, an intermediate station that accepts information from a station and passes the information to another station. **2.** An intermediate station that passes information between terminals or other relay stations. *Note:* The purpose of relaying may be to reach an area not covered by a given station, increase range, interconnect radio stations operating at different frequencies, or use different modulation methods. **3.** An electromechanical device that (a) enables a part in one circuit to control electrical currents or voltages in other circuits, (b) usually consists of at least a magnetic coil connected to one circuit that can control the movement of an armature fitted with contacts for opening and closing one or more other circuits connected to other contacts, and (c) using special arrangements of one or more coils and one or more spring-loaded armatures, can be used to (i) control high power circuits with low power signals, (ii) serve as a combinational circuit, i.e., a gate, and (iii) simply open and close circuits. **4.** At a station or a terminal in communications, computer, data processing, and control systems, to accept a message and forward it to another station or terminal, usually without changing its information content. *See* **airborne radio relay, down-the-hill radio relay, frame relay, message relay, neutral relay, polar relay, radar relay, radio relay, reed relay, satellite relay, side stable re-**

lay, tape relay, telegraph relay, torn tape relay, vibrating relay. *See also* **circuit, combinational circuit, gate, relateral tell, relay communications method, relay configuration, repeater.** *Refer to* **Figs. R-12 (853), T-3 (991).**

relay aircraft: *See* **automatic radio relay aircraft.**

relay circuit: *See* **voice-operated relay circuit.**

relay communications method: A method of communication in which a station transmits a message to another station that repeats the message, transmitting it to a third station that repeats it to a fourth, and so on, to the last station in the series or to the destination station. *Note:* Examples of the relay communications method are (a) the use of radio relay and (b) the use of a series of microwave links. *See also* **broadcast communications method, intercept communications method, radio relay, radio relay system, receipt communications method.**

relay configuration: An operating configuration in which (a) a circuit is established between two stations via an intermediate relay station, (b) two links are simultaneously used, and (c) the channel connections at the relay station are accomplished completely within the station. *See also* **circuit, radio relay, relay.**

relay electronic countercountermeasure: *See* **radio relay electronic countercountermeasure.**

relay equipment: *See* **telegraph automatic relay equipment.**

relay net: A communications network used for the relaying of message traffic. *See also* **relay.**

relay pilot: *See* **tape relay pilot.**

relay plane: *See* **automatic radio relay plane.**

relay satellite: A satellite that has a radio relay station on board, i.e., a satellite station that receives messages and retransmits or rebroadcasts them. *See also* **radio relay, satellite station.**

relay service station: *See* **cable television relay service station.**

relay station: A station, such as a radio station, microwave station, tape station, or telegraph station, (a) that accepts transmissions from other stations, (b) that retransmits the messages to other stations, and (c) in which most, if not all, of the traffic received is retransmitted and is not intended for use at the relay station. *See* **major relay station, minor relay station.**

relay switching: *See* **semiautomatic continuous tape relay switching, tape relay switching.**

relay system: *See* **automatic relay system, automatic tape relay system, manual relay system, radio relay system, semiautomatic relay system, torn tape relay system.**

relay tower: A tower or a mast that has a relay station, such as a microwave repeater or a radio relay transmitter, at the top.

relay working: In voice radio station operations, operation in which the operator at a radio relay station retransmits messages by voice. *See also* **radio relay.**

release: 1. In communications systems, to authorize the dispatch of a message. **2.** In communications security, to authorize the transfer of security classified information to another authority, organization, or foreign government. *See* **circuit release, switching release.**

released acknowledgment signal: *See* **circuit released acknowledgment signal.**

released loop: *Synonym* **switched loop.**

released signal: *See* **circuit released signal.**

release signal: *See* **switching release signal.**

release time: 1. The time after the end of an enabling signal, as in a vogad or echo suppressor, during which suppression continues. **2.** In the suppression of a signal in a device, such as a voice-operated gain adjusting device (vogad) or an echo suppressor, the time between (a) the instant that a received enabling signal ends and (b) the instant that suppression ceases. *Note:* The suppression continues after the enabling signal is received because of the time required for terminals to discharge before remanent inhibiting suppressor biases are removed. *See also* **echo suppressor. 3.** In a relay, the time interval between de-energization of the relay coil and the end of contact closure. **4.** The time between de-energization of a relay coil and the beginning of contact opening, i.e., end of closure, for a relay that usually is open when not energized. **5.** The time between de-energization of a relay coil and the beginning of contact closure, i.e., end of open, for a relay that usually is closed when not energized. *Note:* A usually open relay has a spring that holds the contacts open until the coil is energized. A usually closed relay has a spring that holds the contacts closed until the coil is energized. *See* **call release time, message release time, switch release time.** *See also* **relay.**

release time delay: *See* **receiver release time delay, transmitter release time delay.**

releasing officer: *See* **message releasing officer.**

reliability: **1.** The ability of an item to perform a required function under stated conditions for a specified period. **2.** The probability that a functional unit will perform required functions for a specified period under specified conditions. *Note:* An example of reliability is the probability that a message will arrive at its intended destination within a reasonable time and without undesired alteration of the text or loss of meaning. **3.** The achieved level of operational capability of existing hardware or software. *See* **operational reliability.** *See also* **acceptance test, availability, maintainability, mean time between failures, mean time between outages, mean time to repair, mean time to service restoral, reliability assessment.**

reliability assessment: **1.** The determination of whether existing hardware, firmware, or software has achieved a specified level of operational reliability. **2.** The determination of the level of operational reliability achieved by existing hardware, firmware, or software. *Synonym* **reliability evaluation.**

reliability data: Information necessary or used for reliability assessment of hardware or software at selected points in the lifecycle. *Note:* Examples of reliability data are (a) error data and time data for reliability models, (b) program attributes, such as complexity, and (c) programming characteristics, such as the program development techniques used. *See also* **reliability, reliability assessment.**

reliability estimation: **1.** A quantitative statement concerning hardware and software operational reliability based upon measurements taken on the existing hardware and software. **2.** A quantitative statement concerning the operational reliability of a given set of hardware and software based upon measurements taken on other existing hardware and software of the same type or class. *See also* **operational reliability, reliability assessment, reliability prediction.**

reliability evaluation: *Synonym* **reliability assessment.**

reliability growth: The improvement in hardware or software reliability that results from correcting faults or errors.

reliability model: A model used for reliability assessment, estimation, or prediction. *See also* **model.**

reliability prediction: A quantitative statement concerning future hardware or software operational reliability based upon many factors, such as size, complexity, functional requirements, past performance data, and reliability assessment. *See also* **operational reliability, reliability assessment, reliability estimation.**

relocatable machine code: Machine code that is loaded in storage such that its relative address is different from its absolute address. *See also* **absolute address, absolute machine code, machine code, relative address.**

relocate: In computer programming, to move a computer program, routine, subroutine, database, or parts thereof, and to adjust necessary references so that they can be used after being moved.

relock: In radar operations, to reobtain lock-on on an object, i.e., a target, on which tracking was lost. *See also* **lock-on, tracking, transfer lock.**

reluctance: *Synonym* **magnetic reluctance.**

remail site: An E-mail forwarding site. *Note:* A remailer at a remail site can convert return addresses to pseudonyms and render E-mail untraceable.

remanence: The magnetization, i.e., the magnetic polarization, that remains in a magnetized material after the applied magnetic field that magnetized the material is removed.

remapping: In spread spectrum systems, a process of correlation in which a spread spectrum signal is converted into a coherent narrowband signal, and undesired signals are converted into wider bandwidths. *See also* **coherence.**

remodulation: In transponders usually used in satellite links, line-of-sight radio links, and microwave relay links, the (a) down conversion of an incoming signal to an intermediate frequency (IF), (b) amplification of the IF signal, (c) demodulation of the IF signal to obtain the baseband signal, (d) modulation of a radio frequency (rf) carrier by the baseband signal, and (e) up conversion of the modulated rf signal for transmission. *See also* **baseband, down-converter, up-converter.**

remodulator: A demodulator that changes the form of modulation in a reverse manner from that of a conventional demodulator. *Note:* An example of a remodulator is a facsimile system device that changes an amplitude-modulated signal to an audio frequency-shift-modulated signal.

REMOS: reflective membrane optical scintillator.

remote access: **1.** Pertaining to communications with and among data processing facilities through the use of data links. **2.** A private branch exchange (PBX) service feature that allows a user at a remote location to access PBX features by telephone, such as access wide area telephone service (WATS) lines by telephone. *Note:* For remote access, individual authorization codes usually are required. *See also* **access, data link, private branch exchange, remote control equipment, service feature, Wide Area Telephone Service.**

remote access data processing (RADP): Data processing in which input-output functions are performed by devices that are connected to the data processing system by means of a communications link.

remote batch entry: The submitting of batches of data by using an input unit that has access to a computer via a data link. *See also* **remote job entry.**

remote batch processing: 1. Batch processing in which input and output units have access to a computer via a data link. **2.** Batch processing in which jobs are transmitted from a remote station to a computer input-output channel via a data link, i.e., by remote job entry, and the results are transmitted back to the remote station. *See also* **batch processing, data link, remote job entry.**

remote call forwarding: A network-provided service feature that allows calls to a remote call forwarding number to be automatically forwarded to any answering location designated by the call receiver. *Note 1:* Customers may have a remote forwarding telephone number in a central office (C.O.) without having any other local telephone service in that office. *Note 2:* Calls coming to the remote call forwarding number are automatically forwarded to any answering location desired by the user. *See also* **call forwarding.**

remote clock: 1. A clock remotely located from a particular facility, such as a communications station or node, with which it is associated. **2.** A clock remotely located from another clock with which timing signals are to be compared. *See also* **clock, DoD master clock, master clock, reference clock.**

remote control: 1. The control of a functional unit from a distant point by any means, such as by electrical, electronic, electromagnetic, sonic, or mechanical means. **2.** Radio transmitter control in which the control functions are performed electrically from a distance through the use of wire, coaxial cable, optical fiber, microwave, or other radio circuits. *Note:* Examples of remote control systems are (a) systems in which radio programs originate downtown or from a mobile unit while the transmitter is on the top of a mountain at the foot of a tower with the antenna at the top, and (b) systems in which a receiver in an aircraft and a transmitter on the ground can be used to control the flight of the aircraft.

remote control equipment: Equipment used for controlling, monitoring, or supervising the functions of a system from a distance. *See also* **access point, remote access.**

remote job entry: 1. In communications, computer, data processing, and control systems, operation in which computer programs, data, or control functions are entered into the system from a remote site, and results are obtained at the remote site through the use of communications links. **2.** In computer operations at a given site, a mode of operation that allows (a) the insertion of a job into the computer from a remote site and (b) the receipt of the output at the same or another remote site via a communications link. *See also* **remote access.**

remote keying: In cryptosystems, the electrical distribution and insertion of cryptographic key variables into the key generator from a distant point, i.e., through a communications link. *See also* **cryptographic information, cryptosystem, key, variable.**

remote log-on: At a terminal of a host computer, a log-on that allows a user to connect to another host computer via an internet work and to interact with that computer as if the terminal were directly connected to that computer. *See also* **internetwork.**

remotely piloted vehicle: A vehicle that is completely under remote control. *See* **unretrievable remotely piloted vehicle.** *See also* **remote control.**

remote master data circuit-terminating equipment: Data circuit-terminating equipment (DCE) that is capable of controlling other DCEs through the use of communications links. *See also* **data circuit-terminating equipment, remote control.**

remote operations service element (ROSE) protocol: A protocol that (a) provides remote operation capabilities, (b) allows interaction between entities in a distributed application, and (c) upon receiving a remote operations service request, allows the receiving entity to attempt the operation and report the results of the attempt to the requesting entity.

remote operator signaling: Signaling between a central office (C.O.) and a remote operator position that allows a remote operator to process operator calls through the use of communications links. *See also* **remote control, signaling.**

remote orderwire: An extension of a local engineering circuit, i.e., an orderwire, to a point that is more convenient for operators and maintenance personnel to perform required monitoring functions. *See also* **engineering circuit.**

remote rekeying: The encrypted transmission of key from a remote source. *See* **automatic remote rekeying.** *See also* **encryption, key, rekeying, source.**

remote reprogramming: *See* **automatic remote reprogramming.**

Remote Sensing Application Program: *See* **National Advanced Remote Sensing Application Program.**

remote trunk arrangement (RTA): An arrangement that permits the extension of traffic service position system (TSPS) functions to remote locations.

REN: ringer equivalency number.

reorder signal: *Synonym* **busy signal.**

reorder tone: A unique tone received by a calling station when communications equipment or facilities, such as switching paths and trunks, except the called terminal, i.e., the call receiver equipment, are not available for use during an access attempt. *See also* **access attempt.**

repair: The restoration of an item to serviceable condition by (a) correcting specific failures and malfunctions and (b) correcting conditions that caused the failures and the malfunctions. *See* **mean time to repair.**

repairing circuit: *See* **self-repairing circuit.**

repeatability: *See* **test repeatability.**

repeatability measure: In display systems, a measure of the degree of spatial coincidence obtained when a specific display element, display group, or display image is repeatedly generated at the same coordinate position. *See also* **test repeatability.**

repeated selection sort: A sort in which (a) a set of items is divided into subsets, (b) a selection sort is performed on each subset to form a second-level subset, (c) a selection sort is performed on this second-level subset, (d) the selected item is appended to the sorted set and is replaced by the next eligible item in the original subset, and (e) the process is repeated until all items are in the sorted set. *See also* **selection sort, sort.**

repeater. A device that (a) performs one or more signal processing functions on an input signal, such as to recover, filter, amplify, reshape, and retime the signal, depending on whether the signal is an analog or a digital signal, (b) may perform image processing and image restoration functions, (c) may be bidirectional or unidirectional, and (d) transmits the signal. *See* **fiber optic repeater, multiport repeater, optical repeater.** *See also* **regenerative repeater.**

repeater jammer: A jammer that (a) consists of a receiver and a transmitter and (b) amplifies, multiplies, and retransmits the signals that it receives for purposes of deception or jamming. *See* **coherent repeater jammer, swept repeater jammer.** *See also* **jammer.**

repeater power: *See* **fiber optic repeater power, optical repeater power.**

repeatered fiber optic link: A fiber optic link that has fiber optic repeaters along its cable to recover, amplify, and perhaps reshape and retime optical signals. *See also* **repeaterless fiber optic link.**

repeaterless fiber optic link: A fiber optic link, usually in a long-haul communications link, in which repeaters are not used because of the short length, the high bandwidth•distance factor, or the low attenuation rate of the optical fiber. *Note 1:* Repeaterless fiber optic links may be thousands of kilometers long. Optical links for short distances, such as for intercity or local area network (LAN) use in which repeaters were never needed or anticipated, are simply called "fiber optic links" rather than "repeaterless fiber optic links," even though they have no repeaters. *Note 2:* The term "repeaterless fiber optic link" was adopted because early optical fibers had such high attenuation and dispersion rates that optical repeaters had to be installed every few kilometers. Later, as attenuation and dispersion rates were reduced, even long distances did not require repeaters, i.e., repeaterless fiber optic links were installed in new installations, or the older repeatered fiber optic links were replaced. *See also* **repeatered fiber optic link.**

repeating character: In communications, computer, data processing, and control systems, a specific character that is placed in every character position in a specified storage area, storage cell, or data medium. *Note:* Examples of repeating characters are (a) a series of dashes or hyphens across a page and (b) a string of all 1s completely filling a specified storage area.

repeating coil: A voice frequency transformer that (a) consists of (i) a closed core, (ii) a pair of identical balanced primary windings i.e., line windings, that are connected to the line, and (iii) a pair of identical but not necessarily balanced secondary windings i.e., drop windings, that are connected to the drop, (b) is characterized by a low transmission loss at voice frequencies, (c) transfers voice voltages from one winding to another by magnetic induction, (d) matches the line and drop impedances, and (e) prevents direct conduction between the line and the drop. *See also* **balanced, drop, drop repeater.**

repeating label: In personal computer spreadsheets and storage devices, a label that is reproduced over and over again according to a specified criterion, such as for the entire length of a spreadsheet line or over a specified storage area or data medium. *See also* **data medium, label, spreadsheet, storage area.**

repeating ship: In visual communications systems, a ship that (a) is equipped to relay a message manually or automatically, (b) elects to relay a message to facilitate communications, or (c) is used for routing messages. *See* **light repeating ship.**

repeat key: 1. A specific key that, when actuated, enables other keys to repeat their functions as long as they are held depressed. **2.** Any key that repeats its function automatically for as long as it is held depressed. *Note:* Examples of repeat keys are the keys that (a) are on a personal computer standard keyboard operating with certain application programs and (b) repeat their functions for as long as they are held depressed, such as letter keys, the spacebar, and cursor movement keys operating in WordPerfect software. *See also* **key, keyboard.**

repeat request: *See* **automatic repeat request.**

reperforator: In teletypewriter systems, a device that (a) consists of a receiver and a perforator and (b) is used to punch hole patterns in a tape in accordance with arriving signals, thus permitting reproduction of the signals for retransmission. *Synonyms* **receiver perforator, receiving perforator.** *See* **printing reperforator.** *See also* **chad, chadless tape, chad tape, tape relay, torn tape relay.**

reperforator tape reader: *See* **coupled reperforator tape reader.**

reperforator-transmitter: A teletypewriter that (a) consists of a reperforator and a tape transmitter, each independent of the other, (b) is used as a relaying device, and (c) is suitable for temporary message queuing and for converting the signaling rate of incoming data to a different rate. *See also* **message backlog, signaling rate.**

repertory: *See* **instruction repertory.**

repertory dialer: A telephone set that stores a group of frequently called numbers and retransmits the dialing information to the central office (C.O.) by a single action. *See also* **call, card dialer, speed calling.**

repetition rate: 1. The number of occurrences of a repeating event per unit time. **2.** In telephone systems, (a) the number of times that users request a connection divided by (b) the period during which the requests were made, i.e., the number of access attempts per unit time measured during a specified period. *Note:* Repetition rates are used to assess the effectiveness of transmission over a telephone line during different periods. If the events are pulses, the repetition rate is the pulse repetition rate. *See* **Doppler pulse repetition rate, jittered pulse repetition rate, pulse repetition rate, staggered pulse repetition rate.**

repetitive operation: In analog computers, operation that (a) includes the automatic repetition of the solution of a set of equations with fixed combinations of parameters, such as initial conditions, coefficients, exponents, and constraints, (b) is often used to permit the display of apparently steady state solutions, and (c) is used to permit manual adjustment or optimization of one or more parameters in order to obtain certain desired results in the solution. *See also* **analog, analog computer.**

repetitively pulsed laser: *See* **Q-switched repetitively pulsed laser.**

reply: 1. Any response to a message or a signal. **2.** A message to the originator of a previous message that contained a question. **3.** The answer to a challenge in an identification procedure, such as identification friend or foe. *See* **situation awareness beacon with reply.** *See also* **challenge, challenge and reply.**

reply authentication: *See* **challenge and reply authentication.**

report: *See* **interference report, jamming report, overdue report, situation report, weather report.**

reporting net: A communications net that is designated for the free and rapid interchange of information, such as information that usually is associated with the maintenance of up-to-date plots, tactical situations, maps, picture coverage, displays, and related data, in contrast to command and control system information. *See* **maritime patrol air reporting net.**

representation: *See* **analog representation, coded representation, numeric representation.**

reproducibility: 1. A measure of the ease with which (a) an entity, such as an object, event, or image, can be reproduced, or (b) a known result can be obtained. *Note:* The precision of the reproduction or the result should be specified. **2.** With respect to a set of independent devices of the same design, the standard deviation of the values produced by these devices. **3.** With respect to a set of independent devices of the same design, pertaining to the ability of these devices to produce the same value or result, given the same input conditions and given the same operational environment. **4.** With respect to a single device put into operation repeatedly without adjustments, pertaining to the ability to produce the same value or result, given the same input conditions and operation in the same environment. *See also* **fidelity, precision, resettability, resolution.**

reproducible fault: In communications, computer, data processing, and control systems, a fault that will occur each time the set of conditions causing the fault occurs. *Note:* The conditions under which the fault occurs and the precision with which the fault occurs must be specified in determining the reproducibility of a fault.

reproduction ratio: The ratio of (a) a linear dimension, such as the distance between two points on an image, such as a recorded copy, to (b) the corresponding dimension on the object, such as the distance between corresponding points on the object. *See also* **image, recorded copy.**

reproduction speed: 1. In facsimile systems, the rate at which recorded copy is produced. *Note:* The reproduction speed usually is expressed as (a) as the area of recorded copy produced per unit time, such as square meters per second, or (b) the number of pages per unit time, such as pages per minute. **2.** In duplicating equipment, the rate at which copies are made. *Note:* The reproduction speed usually is expressed in pages per unit time, such as pages per second, minute, or hour.

reprogramming: *See* **automatic remote reprogramming.**

request: *See* **access request, automatic repeat request, correction request, disengagement request, light request, statistics on request, tracer request.**

request data transfer: A call control signal that (a) is sent by data terminal equipment (DTE) to data circuit-terminating equipment (DCE) to request the establishment of a data connection and (b) is used in switched and leased circuit service.

request data transfer (RDT) signal: A signal, sent by data terminal equipment (DTE) to data circuit-terminating equipment (DCE), requesting the establishment of a connection. *Synonym* **data transfer request signal.** *See also* **call control signal, data, data circuit-terminating equipment, data terminal equipment, signal.**

request fix: In radiotelephone direction finding, a request from a mobile station for a determination of position or direction by a direction finding station. *See also* **direction finding.**

request homing: In radiotelephone direction finding, a request from a mobile station for course information, i.e., bearing information, that will lead the requesting station to the direction finding station.

request indicator: *See* **called line identification request indicator, calling line identification request indicator.**

request repeat: *See* **automatic request repeat.**

request repeat (RQ) signal: A signal from a receiver to a transmitter asking that a message be transmitted again.

request repeat system: *Synonym* **automatic request repeat.**

request separator character: *See* **facility request separator character.**

request signal: *See* **clear request signal, data terminal equipment clear request signal, data transfer request signal, facility request signal.**

request time: *See* **call request time.**

request to send signal: A signal that is generated by a receiver in order to condition a remote transmitter to commence transmission. *See also* **clear to send signal.**

requirement: 1. A stated or implied (a) condition that a system or a component must meet or (b) capability that a system or a component must have. **2.** A statement that (a) describes the capabilities that the user expects a system or a component to have and (b) usually forms the basis for subsequent development of the system or the component, particularly the specifications. *See* **design requirement, electrical requirement, functional requirement, interface requirement, performance requirement, physical requirement.** *See also* **specification.**

requirements analysis: The reviewing, analyzing, studying, and verifying of user needs to arrive at a definition of system hardware or software requirements. *See also* **requirement.**

requirement specification: *Synonym* **performance specification.**

requirements phase: The period in the lifecycle of hardware or software during which the requirements, such as functional requirements and performance requirements, are defined and documented. *See also* **functional requirements, performance requirements.**

requirements specification: 1. A specification that (a) sets forth the requirements for a hardware or software system or component and (b) usually includes design requirements, electrical requirements, functional requirements, interface requirements, performance requirements, physical requirements, and standards. **2.** The preparing or the stating of requirements for system hardware or software. *See also* **requirement, requirements specification language, specification.**

requirements specification language: A language with special constructs and verification protocols used to specify, verify, and document requirements. *Note:* Examples of requirements specification language are (a) to clearly distinguish between calculated values and measured values, (b) to use "shall" where appropriate, rather than "will," (c) to express tolerances precisely in conventional format, and (d) to use metric units followed by English units in parentheses. *Note:* The requirements specification language should not be confused with the requirements specifications themselves. *See also* **requirements specification.**

reradiating repeater: A repeater that receives, amplifies, shapes, and radiates received signals without frequency translation. *See also* **frequency translation.**

reradiation: 1. Radiation, at the same or a different frequency, of energy received from an incident wave. **2.** Undesirable radiation of signals locally generated in a device, such as a radio receiver. *Note:* Reradiation might cause interference or reveal the location of the device. *See also* **interference.**

rering: In telephone system operations, a facility that is provided on trunk lines for recalling a distant operator or for ringing off where no automatic signaling is provided. *See also* **ring-off.**

rerouting: The restarting of route selection from the first point of routing control when congestion is encountered at some intermediate switching point in the connection to be established. *See also* **clear collision, congestion.**

rerun: 1. In communications systems, the retransmission by the transmitter of a received message. **2.** In tape relay communications systems, the retransmission by the transmitter of a message that was previously sent. *Note:* The rerun message usually is stored on a data medium or in a storage device for subsequent comparison of the rerun message. **3.** In computing and data processing systems, a repeat of a computer program run from its beginning. *Note:* The rerun may be necessary for various reasons, such as a false start, an interruption, or a need to effect a required change. *See also* **retransmission, run.**

rerun point: A point, location, or instruction in the sequence of instructions in a computer program at which all the data needed to rerun the program from the beginning are available. *See also* **computer program, instruction, point, restart point.**

RES: reserved (in asynchronous transfer mode).

resale carrier: A carrier that buys the services of a common or commercial carrier and retails or redistributes the services to the public. *See also* **common carrier, divestiture, other common carrier, specialized common carrier.**

resale service: The right of a buyer of basic telecommunications services, such as private lines, foreign exchanges, and wide area telephone services (WATS), to resell or share unused capacity. *Note:* Resale service rights are defined in the deliberations and rulings of the Federal Communications Commission (FCC).

rescue beacon: *See* **search and rescue beacon.**

rescue beacon frequency: *See* **search and rescue beacon frequency.**

rescue communications: 1. Scene of action communications that occur at a rescue scene. **2.** Communications that occur in connection with a rescue operation, including stations at the scene of action and at support or message relay locations. *See* **search and rescue communications.** *See also* **rescue control channel, scene of action communications.**

rescue control channel: The channel, i.e., the frequency, that is used at the scene of a rescue operation to control all activities and actions in the area intended to alleviate or remove the emergency condition, such as evacuation operations, emergency medical operations, and environmental control operations, such as those for fighting fire and weather conditions. *See also* **scene of action communications, search and rescue communications.**

rescue frequency: *See* **scene of air rescue frequency.**

rescue point: *Synonym* **restart point.**

research: *See* **operations research.**

research Earth station: *See* **space research Earth station.**

research service: *See* **space research service.**

research service space station: *Synonym* **space research space station.**

research space station: *See* **space research space station.**

research station: *See* **experimental research station.**

resection: Position-finding triangulation in which (a) an unknown location of a single receiver is determined by obtaining bearings from at least two radiation sources of known location, and (b) plotting the bearings through those locations on a map determines their point of intersection, that point being the location of the receiver. *See also* **direction finding, intersection, position finding, triangulation.** *Refer to* **Fig. R-11.**

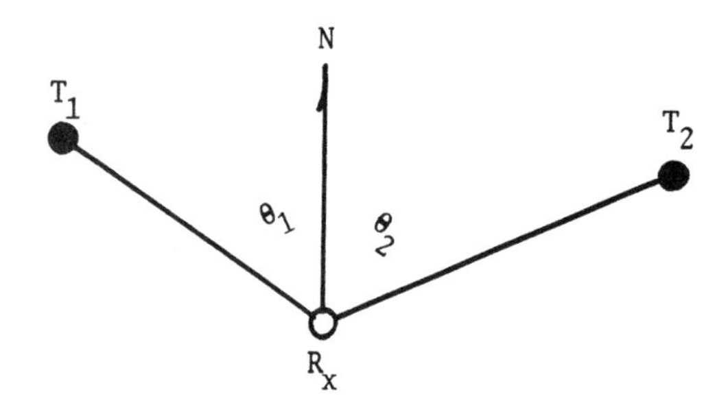

Fig. R-11. Resection may be used to determine the location of point R_x. By means of direction finding techniques, the bearings, i.e., the azimuths, are taken on two transmitters whose locations are known. On a map, the intersection of the back azimuths through the known locations determines the location of R_x.

reserved circuit service: In integrated services digital network (ISDN) applications, a telecommunications service in which (a) a communications path is established in response to a user-network signaling request, and (b) the service time is preset according to a user request.

reserved word: In programming languages, a word that (a) is defined by the programming language in which it is used and (b) cannot be changed by the user. *Note:* Ada® and COBOL have reserved words. FORTRAN does not have reserved words.

reset: 1. In a data processing system, to cause a counter or a register to take the state that corresponds to a previously specified value, such as the state of initial value or the state that represents zero. **2.** In a computer or a computer program, to cancel a previous setting, return to a global condition, or assume a default configuration. *See also* **set.**

reset mode: *Synonym* **initial condition mode.**

resettability: A measure of the ability to duplicate controllable conditions. *Note:* An example of resettability is the ability to reset the frequency controls of radio equipment so as to obtain the exact same frequency of transmission or tuning as was previously obtained. *See also* **reproducibility.**

resident: 1. Pertaining to computer programs that remain in a particular storage area. **2.** Data or programs that (a) are located in a specified storage area and (b) usually are placed where they are subject to call. *See also* **resident program.**

resident program: A computer program that (a) is stored in a specified storage area and (b) may be called by another computer program or routine. *See also* **storage area.**

residual equalizer: *See* **line residual equalizer.**

residual error rate: *A deprecated synonym for* **undetected error ratio.**

residual error ratio: 1. The ratio of (a) the sum of the data units, such as data elements, bits, characters, or blocks, that are received (i) incorrect but undetected and (ii) incorrect but uncorrected by the error control equipment to (b) the total number of data units received. *Note:* The residual error ratio is the error ratio that remains after error detection and correction have been accomplished. It signifies the error rate of the detection and correction equipment. **2.** The error ratio that remains after attempts at correction are made. *Synonym* **undetected error ratio.** *Deprecated synonyms* **residual error rate, undetected error rate.**

residual magnetism: The magnetization, i.e., magnetic polarization, that remains in a magnetized material after all attempts to remove the magnetization have been made. *Note:* An example of residual magnetization is the magnetization remaining from magnetic recording after attempts at degaussing. *See also* **degauss, remanence.**

residual modulation: *Synonym* **carrier noise level.**

resilience: *See* **system resilience.**

resistance: 1. A measure of the ability of an entity to withstand or survive environmental conditions or attack. **2.** A measure of the opposition to an electric current by a material or free space when a potential difference, i.e., a voltage or potential gradient, is applied between two points, given, for a resistor, in the Ohm's law relation $R = V/I$ where R is the resistance, V is the voltage applied across the resistor, and I is the current through the resistor. *Note:* The SI (Système International d'Unités) unit of resistance is the ohm. The reciprocal of the resistance is the conductance, usually expressed in mhos, which is ohm spelled backward. The resistance of metallic conductors causes attenuation of signal currents. In optical fibers, there are no electric currents. The fibers are made of dielectric materials. However, there is some interaction between lightwaves in optical fibers and the fiber material, which causes (a) attenuation by scattering and absorption and (b) minor heating, which, in certain respects, is analogous to resistance. *See* **crosstalk resistance, electromagnetic interference resistance, equivalent noise resistance, infrared detector dark resistance, load resistance, neutron bombardment resistance, radiation resistance.** *See also* **capacitance, conductance, dielectric, electrical resistivity, inductance, ohm, potential difference.** *Refer to* **Table 3, Appendix B.**

resistance coupling: The coupling of voltages or currents, such as those that represent signals or noise, from one conductor to another by means of the resistance or conductance between them.

resistance hybrid: A network of resistors (a) to which four branches of a circuit may be connected to make them conjugate pairs, (b) that is used to convert between two-wire and four-wire communications circuits, and (c) is required when repeaters are introduced in a two-wire circuit. *Synonym* **hybrid resistance.**

resistant communications cable: *See* **intrusion-resistant communications cable.**

resistor: An electric circuit element that has the capability of providing resistance to the flow of electric current when a potential difference, i.e., a voltage, occurs across it. *Note:* The value of the resistance, in ohms, provided by a resistor, is selected by the designer of the equipment in which the resistor is installed. *See also* **resistance.**

resolution: 1. The minimum difference between two discrete values that can be distinguished by a measuring device. *Note:* High resolution does not necessarily imply high accuracy, but it usually does imply high precision. **2.** The degree of precision to which a quantity can be measured or determined. **3.** A measurement of the smallest detail that can be distinguished by a sensor system under specific conditions. **4.** In display systems, a measure (a) of the ability of a device to distinguish between adjacent objects and display them as separate images, and (b) of the sharpness of a display image, i.e., the degree to which luminous intensity discontinuities across the display area of the image match those of the object. *Note:* Resolution usually is expressed in the number of points per unit area or the number of lines per unit length that are discernible as separate points or lines in the image. *See* **angular resolution, borescope target resolution, detection resolution, distance resolution, horizontal resolution, longitudinal resolution, range resolution, transverse resolution.** *See also* **accuracy, definition, display area, image, object, precision, resolving power.**

resolution angle: *See* **limiting resolution angle.**

resolution cell: *See* **radar resolution cell.**

resolution protocol: *See* **address resolution protocol.**

resolver: A device that accepts the coordinates of a point in polar coordinates and produces Cartesian coordinates of the same point, or vice versa, given by the relations $x = \rho \cos \theta$ and $y = \rho \sin \theta$ where x and y are the Cartesian coordinates of a point, and ρ and θ are the polar coordinates for conversion from polar coordinates to Cartesian coordinates, and by $\rho = (x^2 + y^2)^{1/2}$ and $\theta = \tan^{-1} y/x$ for conversion from Cartesian to polar coordinates.

resolving power: A measure of the ability of a lens or an optical system to form separate and distinct images of two objects with small angular separation. *Note:* The resolving power of an optical system is limited by diffraction at the aperture. An optical system cannot form a perfect image of a point. It produces instead a small disk of light, i.e., an Airy disk, surrounded by alternately dark and bright concentric rings. When two object points are at that critical separation from which the first dark ring of one diffraction pattern falls upon the central disk of the other, the points are just resolved, i.e., distinguished as separated, and the points are said to be at the limit of resolution. *See* **bundle resolving power, chromatic resolving power, grating chromatic resolving power, prism chromatic resolving power, theoretical resolving power.** *See also* **aperture, diffraction, resolution.**

resonance: In an electrical circuit, the oscillation that occurs when the inductive reactance and the capacitive reactance are of equal magnitude, causing electrical energy to oscillate between the magnetic field of the inductor and the electric field of the capacitor. *Note 1:* Resonance occurs because the collapsing magnetic field of the inductor generates an electric current in its windings that charges the capacitor, and then the discharging capacitor provides an electric current that builds the magnetic field in the inductor, and thus the process is repeated. *Note 2:* At resonance, the series impedance of the two elements is at a minimum, and the parallel impedance is at a maximum. Resonance is used for tuning and filtering because it occurs at a particular frequency for given values of inductance and capacitance. Resonance can be detrimental to the operation of communications circuits by causing unwanted sustained and transient oscillations that may cause noise, signal distortion, and damage to circuit elements. *Note 3:* As stated, at resonance the inductive reactance and the capacitive reactance are equal. Thus, $\omega L = 1/\omega C$, where $\omega = 2\pi f$, in which f is the resonant frequency in hertz, L is the inductance in henries, and C is the capacity in farads when standard SI (Système International d'Unités) units are used. Thus, the resonant frequency, f, is given by the relation $f = (1/2\pi)(LC)^{-1/2}$. *See also* **resonant cavity, resonant frequency.**

resonant cavity: **1.** A bounded region in a material medium, such as (a) a rectangular free space in a laser crystal, (b) a length of hollow tubing closed on both ends, or (c) two parallel walls, all of such dimensions that a standing wave, such as an electromagnetic, acoustic, or elastic wave, can be sustained and raised to high intensity by application of stimulation consisting of applied energy of an appropriate frequency, from the outside or in the inside of the cavity. *Note 1:* The stimulation frequency for a resonant cavity is determined by the dimensions of the cavity. The wavelength of the stimulation is a multiple or a submultiple of the width of the resonant cavity. The walls must be reflective if a standing wave is to be sustained. *Note 2:* Resonant cavities are used in some lasers in which they form part of the laser head. **2.** A geometric space that (a) may be empty or may contain a fluid or solid material, (b) is bounded by two or more reflectors, and (c) can reflect electromagnetic waves back and forth, thus producing standing waves of high intensity at certain wavelengths. *Note:* An example of a resonant cavity is a ruby crystal laser having two plane or spherical mirrors with the crystal itself between the mirrors. The molecules of the crystal can be excited by an inert gas lamp and thus generate and emit a narrow beam of monochromatic light of high power and high radiance, i.e., high optical power density or intensity, in the direction of the crystal axis. *See also* **laser head, optical cavity, radiance.**

resonant frequency: **1.** In an electrical circuit, the frequency at which circuit inductive reactance and capacitive reactance, both of which are functions of frequency and geometry, are equal, thus creating oscillation or interchange of energy between the magnetic field of the inductor and the electric field of the capacitor. *See also* **resonance.** **2.** In a resonant cavity, the frequency at which standing waves can be sustained by application of stimulation at an appropriate frequency. *See also* **resonance, resonant cavity.**

resource: *See* **information resource.**

resource allocation: In communications, computer, data processing, and control system operations, the assignment of various components and devices to a system or a user prior to or during operation.

resource controller: In satellite communications systems, a processor or a group of processors that control access to satellite payload communications resources within an individual satellite program.

resource manager: *See* **functionally integrated resource manager.**

resources board: *See* **Joint Telecommunications Resources Board.**

respond opportunity: *Synonym* **response opportunity.**

response: **1.** A reply to a query. **2.** In a data transmission response frame, the content of the control field that advises the primary station concerning the processing by the secondary station of one or more command frames from the primary station. **3.** The effect produced by an active or passive functional unit upon receipt of an input signal. **4.** The reaction of a functional unit to an input stimulus, such as a step function of voltage. *See* **impulse response, scene cut response, spectral response.** *See also* **polling, response time.** *Refer to* **Fig. L-2 (502).**

response degradation: *See* **motion response degradation.**

response frame: **1.** In data transmission, a frame that is transmitted by a secondary station. **2.** In high level data link control (HDLC) procedures, a frame that may be transmitted by a secondary station or by a combined station that has the local or transmitting combined station address. *See also* **combined station, frame, secondary station.**

response function: **1.** *Synonym* **transfer function.** **2.** *See* **impulse response function.**

response message: A message that (a) is sent in the backward direction and (b) contains (i) an indication of the called terminal line condition, (ii) an indication of a network condition, (iii) information relating to user and network facilities, and (iv) in the case of some user facilities, identifying information, such as addresses and authenticators. *See also* **authenticator, backward direction.**

response opportunity: In data transmission, the link-level logical control condition during which a given secondary station may transmit a response. *Synonym* **respond opportunity.** *See also* **secondary station, transit flow control.**

response protocol data unit (RPDU): In a communications network, a protocol data unit (PDU) transmitted by a logical link control (LLC) sublayer in which the PDU command/response (C/R) bit is a 1.

response quantum efficiency: **1.** Quantum efficiency that (a) is a ratio of (i) the number of countable output events to (ii) the number of incident photons that occur when energy in the form of electromagnetic waves, such as lightwaves, gamma radiation, X rays, and cosmic rays, is incident upon a material, (b) is often measured as electrons emitted per incident quantum, i.e., for lightwaves, per incident photon, (c) indicates the efficiency of conversion or utilization of optical energy, (d) is an indication of the number of events produced for each quantum incident on the sensitive surface of a photodetector, and (e) is a function of the wavelength, incidence angle, polarization, and other factors. **2.** In a material, a measure of the probability that an incident photon will trigger a measurable event. *Note:* Response quantum efficiency is an intrinsic quality of materials. It is a function of wavelength, incidence angle, and polarization of the incident electromagnetic field. It can be quantified as the number of electrons released, i.e., emitted, on the average, for each incidence angle and for each polarization of the incident electromagnetic wave. The electron-hole pairing that is caused by an incident photon is a complex probabilistic phenomenon that depends on the details of the energy band structure of the material. *See also* **differential quantum efficiency.**

response signal: A signal sent as a reply to a query. *See* **partial response signal, recall response signal.**

response test: A test performed to determine the adequacy of a response signal to perform a desired function. *See* **frequency response test.** *See also* **response signal.**

response time: **1.** The time that a functional unit takes to react to a given input. **2.** In a computer, the time between (a) the instant at which an operator at a terminal enters a request for a response from a computer and (b) the instant at which the first character of the response is received at that or another indicated terminal. **3.** In a data system, the time between the end of transmission of an inquiry message and the beginning of the receipt of a response message, measured at the inquiry originating station. **4.** In a circuit or a device, the time between (a) the instant that the input signal is 90% of its maximum value and (b) the instant that the output signal reaches 90% of its maximum value. *See also* **overshoot, turnaround time, value.**

response timer (T_K): In multilevel precedence and preemption, the device that controls the length of time the call receiver of the precedence call is allowed in order to accept the incoming precedence call. *Note:* The length of the time, in the form of a time out, usually is set in the range of 4 s (seconds) to 30 s.

response unit: *See* **audio response unit.**

responsibility: *See* **precedence responsibility.**

responsibility chain: *See* **visual responsibility chain.**

responsiveness: The ability of a functional unit to provide service within the required time frame. *Note:* "Timeliness" is sometimes used incorrectly to mean responsiveness.

responsivity: **1.** In an optical detector, such as a photodetector, the ratio of (a) the electrical output voltage, current, or power to (b) the optical input power. *Note 1:* Optical detector responsivity may be expressed in amperes, volts, or watts of output per watt of incident radiant power or per unit of irradiance, such as watts per square meter. *Note 2:* An example of optical detector responsivity is 3 μA/mW (microamperes per milliwatt) at a wavelength of 0.810 μm (micron). *Note 3:* Optical detector responsivity depends on the wavelength of the incident radiation. **2.** In a light source, such as a laser or a light-emitting diode (LED), the ratio of (a) the electrical input voltage, current, or power to (b) the optical output power. *Note 1:* Light source responsivity may be expressed in lumens or watts of radiant output per ampere, volt, or watt of input. *Note 2:* An example of light source responsivity is 3 lumens/mA (lumens per milliampere) at the design wavelength. *Note 3:* The light source responsivity will depend on the operating conditions, such as lasing conditions and driving voltage levels. *Deprecated synonym* **sensitivity.** *See* **luminous responsivity, photodetector responsivity, spectral responsivity.**

restart: The resumption of the execution of a computer program using the data recorded at a checkpoint. *See* **cold restart, warm restart.** *See also* **checkpoint.**

restartpoint: A point, place, or instruction, such as a restart instruction, in the sequence of instructions in a computer program at which all the data needed to continue execution of the program from the point are available. *Synonym* **rescue point.** *See also* **rerun point.**

restitution: In demodulation, a series of significant conditions determined by the decisions made according to the products of the demodulation process. *See* **isochronous restitution, start-stop restitution.** *See also* **demodulation, detection, modulation, significant condition.**

restoral: *See* **mean time to service restoral.**

restoration: **1.** To restore an entity to an original condition or to an equivalent-to-original condition. **2.** In a communications system, the action taken to repair and return to service one or more communications services, including repair of a damaged or impaired communications facility, that have a degraded quality of service or have a service outage. *Note:* Restoration may be accomplished by various means, such as patching, routing, substitution of component parts, or selecting other pathways. The means depend on (a) the nature of the impairment, (b) environmental conditions, (c) availability of resources, and (d) precedence requirements. Restoration is performed by or under the direction of the service vendor. *See* **channel restoration, circuit restoration, image restoration.**

restoration circuit: *See* **direct current restoration circuit.**

restoration priority: 1. In communications systems operations, the sequence in which communications services will be upgraded when system capacity becomes available after an extended outage or degraded service state period. *Note:* Certain access lines may have a high priority in obtaining restored or improved services when capacity becomes available. Other lines may have an intermediate or low priority. Restoration priorities become highly significant in the event of extended outages caused by power outages, natural disasters, and nuclear attack. **2.** In communications circuit restoration, the relative order of precedence assigned to each circuit to aid in determining which communications requirement will be continued, in case an outage occurs, by reassigning the functional communications assets of a lower-precedence-level communications requirement to a higher-precedence-level communications requirement. *See also* **degraded service state, outage, Telecommunications Service Priority service.**

rest pad: *See* **type bar rest pad.**

restricted access: *Synonym* **restricted access service.**

restricted access service: 1. A class of service in which users may be denied access to one or more system features or operating levels. **2.** A switching system operating condition in which certain users do not have access to one of more of the network service features because of the limitations of a specific switch, such as (a) the manner in which the switch is designed, wired, or programmed or (b) that the switch can only accept rotary dial pulses. *Synonyms* **restricted access, restricted service, restriction.** *See also* **access control, classmark, class of service, code restriction, controlled access, dial pulse, service feature.**

restricted area: In communications, computer, data processing, and control systems security, a physical space or area (a) that is intended for use or access only by authorized persons and (b) in which special control and security measures are enforced to prevent unauthorized entry. *Note:* Examples of restricted areas are (a) a locked radio room on board a ship or an aircraft, (b) an area containing cryptoequipment at a communications center, or (c) a spacecraft control room.

restricted channel (RC): In digital communications systems, a channel that has a useful capacity of only 56 kbps (kilobits per second), instead of the usual 64 kbps. *Note:* The restricted channel, currently common in North America, originally was developed to satisfy a ones-density limitation in T1 circuits. *See also* **T carrier.**

restricted circular three-body problem: A mathematical simplification of the three-body problem pertaining to three-body orbital calculations in which one of the three bodies is assumed to have a very small mass and is assumed to be under the forces of attraction of two bodies of very much larger mass that are moving in circular orbits about their common center of mass. *Note:* A problem of this kind is encountered in studies of the motion of an asteroid or a comet under the influence of the attraction of the Sun and Jupiter. The jovian orbit, to a first approximation, is considered to be circular. The problem is also directly related to investigations of the motion of spacecraft and satellites in the Earth-Moon system. No rigorous solution of the restricted circular three-body problem has been obtained on a fully analytical basis. In celestial mechanics, approximate methods have been developed that enable a sufficiently exact solution of the problem for most purposes, such as satellite orbital calculations, including geostationary Earth satellites. *See also* **n-body problem, three-body problem, two-body problem.**

restricted edge-emitting diode (REED): An edge-emitting light-emitting diode (ELED), i.e., a light-emitting diode (LED) in which light is emitted over a small portion of an edge, permitting improved coupling with dielectric waveguides, such as optical fibers and optical integrated circuits (OICs).

restricted radiation device: A device (a) in which the generation of radio frequency (rf) energy intentionally is incorporated into the design, and (b) that generates rf energy that is conducted along wires or is radiated, exclusive of transmitters, for which special provisions are given in the *Radio Regulations* of the International Telecommunication Union. *See also* **field disturbance sensor, incidental radiation device.**

restricted service: *Synonym* **restricted access service.**

restriction: 1. *Synonym* **restricted access service. 2.** *See* **call restriction, code restriction.**

restructuring: *See* **dynamic restructuring.**

retention: *See* **cable retention, optical fiber retention.**

reticle: A scale, indicator, or pattern placed in one of the focal planes of an optical instrument that (a) appears to the observer to be superimposed upon the field of view, (b) is used in various patterns to determine the center of the field of view or to assist in gauging distance, determining leads, or making a measurement, and (c) may (i) consist of fine wires or optical fibers mounted on a support at the ends or (ii) be a scrupulously polished and cleaned plane parallel plate of glass on which grooves have been etched.

retiming: Adjustment of the intervals between corresponding significant instants of a digital signal in reference to a timing signal.

retrace: In an oscillator operating in a stable condition at a specified test temperature, the difference between (a) the frequency at a specified time after oscillator turn-on and (b) the frequency immediately prior to oscillator turnoff. *Note:* The retrace is determined by turning the oscillator off for a specified period, maintaining the oscillator at the specified test temperature, and turning it on again. *See also* **oscillator.**

retransmission: 1. At an intermediate node of a communications network, the transmission of a received signal, usually performed automatically. **2.** A repeated transmission of a previously transmitted signal or message. *Note:* In contrast to multicasting, retransmission of a specific message must be to the same addressee as that of the received message. **3.** The transmission of data identical to that which was previously sent by a given transmitter. **4.** The transmission of received data by a repeater that had not transmitted the identical data before. **5.** The repetition of a message, signal, or other transmission that was previously transmitted, by the same or any other mode of transmission. *See* **automatic retransmission.** *See also* **rerun.** *Refer to* **Fig. R-12.**

retransmit: To transmit the same set of signals that was received, either by the original transmitter or by a transmitter at the receiver location, whichever is specified.

Fig. R-12. A long lines microwave radio relay tower at Martinsville, New Jersey in 1974, for reception and **retransmission** of signals. Courtesy of AT&T Archives.

retransmitter: *See* **automatic retransmitter, perforated tape retransmitter, tape retransmitter.**

retrieval: **1.** In common channel signaling, (a) the guarding against the loss of signaling information when a signaling link fails and changeover is initiated, and (b) the retransmission of lost or mutilated messages. **2.** In information processing, the recovering of data or information from storage. *See* **information retrieval.** *See also* **common channel signaling, link, out-of-band signaling.**

retrieval function: In data processing, a capability to (a) select and locate stored records with specified characteristics and (b) transfer those records to a work area for any required further processing by application programs. *See also* **application program.**

retrieval service: In integrated services digital network (ISDN) applications, an interactive telecommunications service that allows access to and retrieval of stored information, i.e., of information stored in a database.

retrodirective reflector: *Synonym* **triple mirror.**

retrograde orbit: Of a satellite orbiting the Earth, an orbit in which the projection of the satellite position on the Earth's equatorial plane revolves in the direction opposite to that of the Earth's rotation. *See also* **equatorial orbit, geostationary orbit, inclined orbit, polar orbit, synchronous orbit.**

return: *See* **carriage return, common return, ground return, horizontal return, radar return, sea return.**

return character: *See* **carriage return character.**

return circuit: *See* **ground return circuit, signal return circuit.**

return key: *Synonym* **enter key.**

return loss: **1.** In a fiber optic system, optical power that is reflected back toward the source by a component, such as a fiber optic splice, connector, coupler, attenuator, or rotary joint. *Note 1:* Return loss usually is expressed in absolute power units, such as microwatts, or in dB with reference to the incident optical power. *Note 2:* Return loss is measured by using optical time domain reflectometry (ODTR) and is used to determine the location of a component or a fault in a fiber optic cable or in a transmission line. Backscatter is included in the return loss measurement. **2.** The part of incident power that is not transmitted, absorbed, scattered, or radiated by a connected component or by a discontinuity in a waveguide or in a transmission line. **3.** At the junction of a transmission line and a terminating impedance, the ratio of (a) the amplitude of the reflected wave to (b) the amplitude of the incident wave. *Note 1:* The return loss usually is expressed in dB. *Note 2:* The amplitude of the return loss indicates the dissimilarity between the characteristic impedances of the line and the terminating impedance. *Note 3:* The return loss is directly proportional to the reflection coefficient, which is given by the relation $RC = |(Z_1 - Z_2)/(Z_1 + Z_2)|$ where RC is the reflection coefficient, Z_1 is the impedance toward the source, Z_2 is the impedance toward the load, and the vertical bars indicate absolute magnitude. *See* **balance return loss.** *See also* **echo, echo attenuation, echo suppressor, forward echo, loss, reflection coefficient, reflection loss.**

return signal: *See* **ground return signal.**

return to zero: **1.** Pertaining to a digital signal in which there is a signal significant condition, such as current, voltage, power, phase, or frequency, that represents zero, i.e., is in a rest condition, between significant conditions that are not zero. *Note:* The significant condition that represents zero may or may not represent an *n*-ary digit, depending on the coding used. **2.** Pertaining to a signal in which the signal level remains at the absolute zero level for a finite length of time in each bit period, such as (a) a signal that has two differently polarized levels and returns to the zero level between transitions from one polarized state to the other, or (b) a signal that reverts to zero during each bit period, such as returning to the zero level between each bit even in a series of consecutive 1s, i.e., mark conditions. **3.** The transition of a signal level to (a) the absolute zero signal level or (b) the significant condition that represents zero. **4.** Pertaining to a code that contains a zero value in its representation. *See* **nonreturn to zero.** *See also* **level, n-ary, nonreturn to zero, signal, significant condition, transition.** *Refer to* **Fig. P-8 (737).**

return to zero (RZ) binary code: A code for a digital signal that (a) has two information states, i.e., information significant conditions, and (b) returns to a rest state during a portion of the bit period. *Note:* The two information significant conditions are usually called (a) the "zero state" and the "one state" or (b) the "space" and the "mark." *See also* **alternate mark inversion signal, bipolar signal, code, duobinary signal.**

return to zero change: *See* **nonreturn to zero change.**

return to zero coding: Coding in which electrical or optical pulses are used that return to a significant condition that is zero or that represents zero, such as a power, current, voltage, phase, frequency, or luminous intensity condition that either has a zero value or that represents zero, between significant conditions that do not have a zero value or do not represent zero. *Note:* An example of return to zero coding is coding in which the binary numeral 111001 is represented by a voltage on-condition for three time intervals, each followed by an off-condition, (b) is off for two whole bit periods and a fraction of a bit period, (c) is on for one time interval, and (c) is off for at least a fraction of a bit period. There is a zero condition between consecutive digits, regardless of whether a digit is a 0 or a 1 or whether bipolar signaling is used. *See also* **significant condition.**

return to zero level code: *See* **modified nonreturn to zero level code.**

return to zero mark: *See* **nonreturn to zero mark.**

return to zero space: *See* **nonreturn to zero space.**

reusable program: A computer program that may be loaded into main storage once and then executed repeatedly, usually provided that the program parameters, such as addresses and instructions that were modified, are returned to their original states after each use. *See also* **address, computer program, instruction, main storage.**

reversal: *See* **partial tone reversal, phase flux reversal.**

reverse battery signaling (RBS): Loop signaling in which the battery and ground connections are reversed on the tip and the ring of the jack to give an "off-hook" signal when the call receiver answers, i.e., goes off-hook. *Note:* Reverse battery signaling may be used (a) either for a short period or until the call is finished to indicate that the call is a toll call, and (b) to provide toll diversion in a private branch exchange (PBX). *See also* **off-hook signal, on-hook signal, signal, toll diversion.**

reverse directory: *Synonym* **inverted directory.**

reverse keying: *See* **phase reverse keying.**

reverse Polish notation: *Synonym* **postfix notation.**

reverse video: In a display area, highlighting, changing, or interchanging the display image on a screen with its background, such as to change green characters on a black background to black characters on a green background. *See also* **display area, display image, highlighting, screen.**

reverse video bar: In a display area, a highlighted area that (a) has a rectangular shape in which characters are represented with dark lines on an illuminated background, (b) is often used to indicate certain areas on a screen, such as (i) a group of cells on a spreadsheet, (ii) a range of spreadsheet cells to be printed, or (iii) a selected menu item, (c) usually has a size that is adjusted to the size of the item that it is indicating or pointing to, and (d) in many personal computer software programs, is used to indicate or select a menu item or cell by being placed at the selected item with the arrow keys followed by depression of the enter key to initiate the desired action. *See also* **arrow key, display area, enter key, highlighting, key, pointing.**

reversible counter: **1.** A counter that can store a number that can be increased or decreased in accordance with a control signal. **2.** A counter that (a) has a finite number of possible states, (b) can represent a number that can be increased or decreased by unity or by a given constant, upon receipt of an appropriate signal, and (c) usually is capable of bringing the represented number to a specified value, such as zero. *See also* **counter.**

reversible operation: An operation (a) in which an operator operates on an operand to produce a result, and (b) that can be reversed in such manner that the original operand is obtained. *Note:* Examples of reversible operations are (a) addition that can be reversed by subtraction, (b) multiplication that can be reversed by division, (c) taking a square root that can be reversed by squaring, and (d) taking a reciprocal that can be reversed by taking the reciprocal again. Erasing a number and radio broadcasting are not reversible operations.

reverted: In optical systems, being turned the opposite way so that right becomes left, and vice versa, such as occurs when a plane mirror reflects an image.

reverted image: In an optical system, an image whose right side appears to be the left side, and vice versa.

revertive pulsing: In telephone networks, pulsing (a) in which the near end receives signals from the far end, and (b) that is used to control distant switching selections.

review: *See* **design review.**

review file: *See* **audit review file.**

revised interconnection protocol security option (RIPSO): A protocol security option that (a) is part of the Defense Network Security Information Exchange (DNSIX) and (b) defines identification labels for all data transmissions. *Note:* A network level module (a) verifies transmission labels, (b) provides identity tags for sources that do not support the protocol security option, and (c) prevents unauthorized devices from using an existing interconnection profile address or from posing as a network router. *See also* **Defense Network Security Information Exchange.**

revolution: *See* **cyberrevolution.**

revolver track: *Synonym* **regenerative track.**

rf: radio frequency.

RF: radio frequency.

rf bandwidth: radio frequency bandwidth.

rf equipment: radio frequency equipment.

rf heating equipment: radio frequency heating equipment.

rf power margin: radio frequency power margin.

rf tight: radio frequency tight.

RFI: radio frequency interference.

RGB: red-green-blue.

RHI: range-height indicator.

rhombic antenna: An antenna that (a) consists of wire radiators that form the sides of a rhombus, whose two halves are fed equally in opposite phase at an apex, (b) usually is terminated and unidirectional, and (c) when unterminated, is bidirectional. *See also* **antenna.** *Note:* A rhombus is an equilateral parallelogram.

RHR: radar horizon indicator, radio horizon indicator.

RI: routing indicator.

ribbon: *See* **carbon ribbon, fabric ribbon, fiber optic ribbon, inked ribbon.**

ribbon cable: A fiber optic cable in which the optical fibers are held in grooves and laminated within a flat semirigid strip of material, such as plastic, that positions, holds, and protects them. *Note 1:* Ribbon cables may be stacked to produce fiber optic cables with large numbers of optical fibers. Buffers, strength members, fillers, perhaps gels to inhibit liquid penetration, and jacketing usually are added to produce the final cable. *Note 2:* A ribbon cable may be either a tight jacketed or a loose tube fiber optic cable, depending on how tightly, or loosely, the optical fibers are held by the grooves in the ribbons. *See also* **fiber optic cable.**

Richardson's law: The basic law of thermionic emission, expressed by the Richardson-Dushman equation, i.e., the current density caused by thermal excitation in cathode material is given by the relation $J = AT^2e^{-bq/kT}$ where J is the current density, A is a material constant, T is the cathode absolute temperature, b is the work function of the material, q is the charge of an electron, and k is Boltzmann's constant, all in compatible units, such as coulombs, meters, amperes, joules, and kelvin, as appropriate. *Note:* The exponential function is steep.

Rician fading: Radio signal fading that (a) is fast fading or slow fading, (b) is not approximated by the Rayleigh distribution, (c) is influenced by the k-factor, (c) may be determined by the ratio of the direct signal power to the multipath signal power, and (d) includes a combination of white noise power and multiplicative, frequency-nonselective, correlated Rayleigh fading. *See also* **fading, fading distribution, k-factor, Rayleigh fading, white noise.**

right angle adapter: A connector that allows an optical fiber, a fiber optic bundle, or a fiber optic cable to enter or exit a fiber optic transmitter or receiver at a right angle, i.e., 90°, to the surface of the unit.

right-hand circular polarization: Circular polarization of an electromagnetic wave in which (a) the electric field vector rotates in a clockwise direction as the wave advances in the direction of propagation and as seen by an observer looking in the direction of propagation, and (b) in a waveguide, the electric field vector always remains perpendicular to the direction of propagation as well as perpendicular to the magnetic field vector. *See also* **circular polarization. right-hand helical polarization.**

right-hand helical polarization: Polarization of an electromagnetic wave in which (a) the electric field vector rotates in a clockwise direction as it advances in the direction of propagation and as seen by an observer looking in the direction of propagation, (b) the tip of the electric field vector advances like a point on the thread of a right-hand screw when entering a fixed nut or a tapped hole, thus describing a helix in the shape of the thread itself, and (c) in a waveguide, the electric field vector always remains skewed to the direction of propagation. *See also* **right-hand circular polarization.**

right-hand polarized electromagnetic wave: An elliptically or circularly polarized electromagnetic wave in which the electric field vector, observed in a fixed plane normal to the direction of propagation as one looks in the direction of propagation, rotates in a right-hand direction, i.e., in a clockwise direction. *Note:* The direction of propagation of the wave would be the same as the forward direction of a right-hand screw when being screwed into a fixed nut. *Synonyms* **clock-**

wise polarized electromagnetic wave, clockwise polarized wave. *See also* **circular polarization, left-hand polarized electromagnetic wave.**

right-hand rule: *Synonym* **Fleming's rule.**

right justification: The alignment of sets of stored data, such as lines of text or data in a register, such that the alignment reference is at the right end of the storage area, such as the page or the register. *See also* **left justification, left justify, right justify.**

right justify: 1. To shift data in a storage medium so that the right-hand end of the data is at a particular position relative to a given reference. *Note:* Examples of right justify are (a) to adjust the position of a word or a numeral in a register so that the right end of the data is at the right end of the register, and (b) to adjust the words on a printed page so that the right margin is flush, i.e., so that all the ends of the lines of text are in a straight line and are at a fixed distance from either the right-hand or the left-hand edge of the page. **2.** To align data in or on a medium to fit the right end-positioning constraints of a required format. *See also* **justify, left justify.**

rights terminal: *See* **multiple access rights terminal.**

right-through control: Control in a switched communications network in which the originating nodal switching center (NSC) (a) determines the route of a call to the destination NSC and thus to the destination terminal, and (b) sets up the connection all the way through to the call receiver end instrument.

rigid borescope: A portable fiber optic borescope that (a) has an aligned bundle of optical fibers in a stiff fiber optic cable, or a train of lenses mounted in a rod, with sufficient rigidity to cross unsupported gaps in a path to an objective, (b) usually has an eyepiece for viewing the image, (c) may have a controllable articulating tip on the objective end, i.e., the distal end, to enable inspection at various angles, such as up to 90° from the cable optical axis, (d) is hand-held, (e) is energized via a fiber optic cable from a light source, and (f) usually consists of a basic borescope unit and ancillary equipment required for operation. *See also* **flexible borescope, image, objective.**

rigid disk: *Synonym* **rigid diskette.**

rigid diskette: An assembly that (a) consists of a thin magnetic disk that is permanently enclosed in a rigid protective cover with openings for access by read-write heads, (b) has mechanical attachments that enable the disk to be rotated, (c) in which data may be recorded on either or both sides of the disk, (d) usually is used for storing data or computer programs for personal computers (PCs), and (e) often is used as a part of a software package. *Note:* A typical standard double-sided high-density rigid diskette measures 3.5 in. (inches) by 3.5 in. and holds 1.44 Mb (megabytes) when IBM-formatted. *Synonym* **rigid disk.** *See also* **diskette, floppy diskette, magnetic disk, read/write head, software package.**

ring: In telephone systems, a signal of specific duration and character that indicates to a call receiver that a call originator is engaged in an attempt to gain access to that call receiver end instrument. *See* **guard ring, rering.** *See also* **access attempt, circuit, ring network.**

ringaround: 1. The improper routing of a call back through a switching center already engaged in attempting to complete the same call. **2.** In secondary surveillance radar, the presence of false targets declared as a result of transponder interrogation by side lobes of the interrogating antenna. *See also* **alternate routing, routing.**

ringback: 1. In telephone switchboard operations, ringing a call receiver end instrument (a) by using the answering plug and (b) usually after getting the call receiver on the line after a booked call. *Note:* The calling plug is connected to the call receiver line and therefore is already in use. **2.** The signal used to alert the user who had placed a booked call and to get the user on the line to indicate the progress of the call, such as the call receiver is on the line, the call receiver is not available, the call cannot be completed, or the call must be delayed pending circuit availability.

ringback signal: 1. In telephone systems, a signal that (a) usually consists of an audio tone interrupted at a slow rate, and (b) is provided to a call originator to indicate that the call receiver instrument is receiving a ringing signal. *Note:* The ringback signal may be generated by the call receiver servicing switch or by the call originator servicing switch. **2.** A signal returned to a call originator to indicate that one of the types of delayed automatic calling is now ringing the call receiver end instrument. *See also* **call, call control signal.**

ringback tone: *Synonym* **ringing tone.**

ring connection: The connection of a set of terminals in which a single one-way ring cable goes from station to station in sequence, each terminal connected by a relatively short cable or T-coupler to the ring cable, which has no ends. *Refer to* **Figs. N-12 and N-13 (623).**

ring counter: A device (a) that consists of a serial loop of interconnected storage elements, such as flip-flops, only one of which can be in a specified stable state at any given time, (b) in which each consecutive applied pulse causes the next storage element in the loop to switch to the specified state, (c) in which the total count possible is equal to the number of storage elements in the loop, i.e., the number of stages, and (d) in which, when a selected storage element is in the specified state, the condition can be gated out to another ring counter, which can count the number of times the first loop has been completed, and so on, each ring counter serving as a digit position in a numeration system. *See also* **binary counter, modulus counter.**

ringdown: In telephone systems, signaling in which (a) telephone ringing current is sent over the line to operate a lamp and cause the drop of a self-locking relay, (b) manual operation is used, as differentiated from dialing, (c) a continuous or pulsed alternating current (ac) signal is transmitted over the line, and (d) there may not necessarily be a switchboard in use. *Note:* The term "ringdown" originated in magneto telephone signaling, in which cranking the magneto in a telephone set would not only "ring" its own bell but also cause a lever to fall "down" at the central office (C.O.) switchboard.

ringdown circuit: 1. In telephone systems, a circuit in which manually generated signaling power is used to perform ringdown. **2.** A telephone circuit in which signaling is manually applied but not necessarily manually generated. *See also* **ringdown, signaling.**

ringdown signaling: In telephone systems, the application of a signal to a line (a) to operate a line signal lamp or a supervisory signal lamp at a switchboard to alert an operator, or (b) to ring a call receiver end instrument. *See also* **ring, ringdown, signal.**

ringer: *See* **fiber optic ringer.**

ringer equivalency number: A number that (a) is determined in accordance with the *Code of Federal Regulations,* Title 47, part 68, and (b) that represents the ringer loading effect on a line. *See also* **loading.**

ringing: 1. *Synonym* **hunting. 2.** *See* **ac-dc ringing, bridged ringing.**

ringing signal: *Synonym* **ringing tone.**

ringing tone: 1. The tone that is received by a calling telephone to indicate that a called telephone is being rung. *Note:* The ringing tone is not generated by the called telephone. **2.** In telephone switchboard operations, an audible tone that indicates (a) that a dialed end instrument is ringing, (b) the circuit status, to inform the call originator that the required directory number is being called, (c) that the called line is not busy, (d) that all connections to the called line have been made, and (e) that all that remains for the call to be completed is for the call receiver to remove the handset from the hook or lift it off its cradle switch to answer, i.e., to pick up. *Synonyms* **audible ringing tone, ringback tone, ringing signal.**

ring latency: In a ring network, such as a token passing ring network, the time required for a signal to propagate once around the ring. *Note 1:* Ring latency may be measured in seconds or in bits at the data transmission rate. *Note 2:* Ring latency includes signal propagation delays in (a) the ring medium, (b) the drop cables, and (c) the data stations connected to the ring network. *See also* **network topology, token.**

ring network: *See* **slotted ring network, token ring network.** *See also* **network topology.**

ring-off: In telephone switchboard operations, an operation in which a user sends a clearing signal to the switchboard operator to indicate to the operator that the call is finished. *See also* **clearing, clearing signal, finished call, signaling.**

rings: *See* **Newton's rings.**

ring topology: *See* **network topology.**

ring transit time: *Synonym* **round-trip delay time.**

ring trip: In telephone system operations, a signal that (a) is sent to a switchboard by a user end instrument to indicate that the user has answered and to stop the ringing tone, (b) usually is a seizure signal, and (c) in the event of a failure to function, may result in the ringing tone's being coupled to the receiver and ringing in the call receiver's ear. *See also* **ringing tone, seizure signal, user end instrument.**

R interface: For a basic rate access in an integrated services digital network (ISDN) environment, the interfacing specifications covering pre-ISDN standards, such as Electronic Industries Association EIA-232C Standard. *See also* **Integrated Services Digital Network, S interface, T interface, U interface.**

ripple voltage: 1. The alternating component of a direct current (dc) voltage that is residually retained from the rectification of alternating current (ac) power, i.e., the generation of dc power from an ac power source. *Note:* Ripple voltages usually are reduced to a minimum with filters, particularly to reduce (a) the 60 Hz (hertz) from local alternating current (ac) power lines, (b) the 60 Hz from half-wave rectifiers, and (c) the 120 Hz from full-wave rectifiers. *Note:* A low-pass filter, such as a shunt capacitor or a series inductor, or both, usually is used to remove ripple voltages from the output of a rectifier, which would otherwise have very high first or second harmonics of the alternating current power source. The amount of filtering depends on the use to be made of the rectifier direct current (dc) output. If the output is for a direct current (dc) power supply for communications equipment, practically all of the output ripple voltage might have to be removed to suppress noise. If the output is for a dc motor, the motor inductance might sufficiently reduce ripple currents. If the output is for charging a battery or for electroplating, a reasonable ripple voltage usually can be tolerated. **2.** Interference, originating from the alternating component of direct current (dc) voltage, that is coupled into a circuit. *See also* **interference, rectifier.**

riser: 1. In communications systems, a cable that (a) runs vertically, usually inside a building, to carry signals to the various floors, (b) is connected to outside plant cables, usually in an underground vault, and (c) is cross-connected, spliced, or split off at each floor, usually in a distribution box, for horizontal distribution to user end instruments. **2.** In electric power systems, a cable that (a) runs vertically, usually inside a building, to carry power to the various floors, (b) is connected to distribution cables, usually in an underground vault, and (c) is connected to distribution cables or wires at each floor, usually in a power distribution closet on each floor for the larger buildings, for horizontal distribution to power outlets in each room. **3.** In telemetering systems, a cable that is used to telemeter data from lower to higher levels, or vice versa, such as from the ocean floor to the surface, up a mine shaft, or up a drilled well shaft. **4.** In an aerial insert, a cable that runs from underground to the top of a pole or tower, or vice versa. **5.** In radar displays, a display element that has appeared on and shortly thereafter suddenly disappears from the display surface.

rise time: 1. The time required for a signal level, such as the pulse amplitude, to increase from a specified value, usually near the lowest value for the case of a pulse, to another specified value, usually near the peak value. *Note:* The specified values for determining pulse rise time usually are 10% and 90% of the peak value. When other than 10% and 90% values are used, they should be specified. *Synonym* **pulse rise time. 2.** The time required for the magnitude of a pulse, such as the amplitude, the phase shift, the frequency shift, or any other parameter used to represent the pulse, to increase to near the peak value, usually 90% of the peak value. *Note:* Overshoot and oscillation, i.e., hunting, may need to be specified in determining rise time. *See also* **drift, fall time, jitter, root mean square pulse broadening, swim, wander.**

risk analysis: An analysis of system resources and their vulnerabilities used to establish expected loss from specified events, based on estimated or calculated probabilities of occurrence of those events.

RJ: registered jack.

RJE: remote job entry.

RLC: resistance inductance capacity.

rms: root mean square.

rms pulse duration: *See* **root mean square pulse duration.**

RO: receive only.

robbing: *See* **bit robbing.**

robustness: The extent to which hardware or software continues to perform correctly despite some violation of its specification, such as incorrect data input, incorrect operator commands, or environmental conditions that exceed original requirements. *See* **system robustness.**

rod: *See* **GRIN[R] rod, optical mixing rod, type rod.**

rod coupler: *See* **fiber optic rod coupler.**

rod-in-tube technique: A technique in which a rod placed in a tube is used as an optical fiber preform in the manufacture and the drawing of optical fibers. *See also* **preform.**

rod multiplexer filter: *See* **fiber optic rod multiplexer-filter.**

rod storage: *See* **magnetic rod storage.**

ROGER: In voice communications systems, the word used by the called station, i.e., the call receiver to (a) acknowledge that the calling station transmission is ended, i.e., the message is ended, and (b) to indicate that the transmission, i.e., the message, has been received. *Note 1:* ROGER may be followed by one or more transmissions, such as (a) a repeat message request, (b) another message to the calling station, (c) OVER if further messages are expected from the calling station, or (d) OUT if no further transmissions are expected from the calling station. *Note 2:* The use of the sequence "ROGER, OVER, and OUT" is improper.

rollback: 1. In the execution of a computer program, a programmed return to a prior point in the sequence of instructions, usually to a check point or a restart point. **2.** In a database management system, to return the database to a prior state or condition. *See also* **checkpoint,**

computer program, database, database management, restartpoint.

roll in: To transfer back to main or internal storage, data or computer programs that previously have been rolled out. *See also* **computer program, main storage, roll out.**

roll out: To transfer data or computer programs from main to auxiliary storage or from internal to external storage, for the express purpose of freeing main or internal storage areas for other use and usually to be rolled back in at a later time. *See also* **auxiliary storage, external storage, internal storage, main storage, roll in.**

roll-out-roll-in: In computer and automatic data processing system operation, pertaining to storage management in which data and computer programs are moved in and out of main or internal storage and auxiliary or external storage to make the most effective use of limited main or internal storage capacity. *See also* **auxiliary storage, external storage, internal storage, main storage.**

ROM: read-only memory.

roof filter: *Synonym* **roofing filter.**

roofing filter: A low-pass filter used to reduce unwanted higher frequencies. *Synonym* **roof filter.** *See also* **filter.**

room: *See* **entrance room, equipment room.**

room noise level: *Synonym* **ambient noise level.**

room temperature metal oxide semiconductor (RTMOS): A metal oxide semiconductor (MOS) that (a) operates at room temperature, (b) can be used to make ultrasmall chips, i.e., to make 0.1 μm (micron) chips, (c) may be fabricated by using a conventional silicon chip manufacturing processes, (d) operates at low voltage, such as 1.5 V (volts), compared to 3 V to 5 V for conventional integrated circuits, and (e) can operate up to about 120 GHz (gigahertz). *See also* **metal oxide semiconductor.**

root: The highest level of a hierarchy.

root compiler: In computer programming, a compiler that (a) has a machine-independent intermediate level language as an object language, and (b) when combined with a machine language code generator, forms a complete or full compiler. *See also* **code generator, compiler, intermediate level language, machine language, object language.**

root mean square: In a set of values, the square root of the sum of the squares of the values. *Note:* Taking the values of the voltages for each frequency that occur simultaneously in a circuit, such as the voltages in a square wave signal as determined by a Fourier analysis of the signal, squaring each of the voltages, summing the squares, and obtaining the square root of the sum provides a voltage that is considered the effective voltage for determining power and heat production in the circuit. *See also* **Fourier analysis, value.**

root mean square (rms) deviation: A single quantity, σ_{rms}, that (a) characterizes a function, $f(x)$, given by the relation

$$\sigma_{rms} = [(1/M_0)\int_{-\infty}^{+\infty}(x-M_1)^2 f(x)dx]^{1/2}$$

where $M_0 = \int_{-\infty}^{+\infty} f(x)dx$ and $M_1 = (1/M_0)\int_{-\infty}^{+\infty} f(x)dx$, and (b) in probability and statistics, is used where the normalization, M_0, is unity. *See also* **impulse response, root mean square, root mean square pulse broadening, root mean square pulse duration, spectral width.**

root mean square (rms) pulse broadening: The temporal root mean square (rms) deviation of the impulse response of a component or a system. *See also* **impulse response function, root mean square deviation, root mean square pulse duration, spectral width.**

root mean square (rms) pulse duration: Root mean square deviation in which the independent variable is time, and the pulse waveform, $f(t)$, is substituted for $f(x)$. *See also* **impulse response function, root mean square deviation, root mean square pulse broadening, spectral width.**

root record: In a hierarchically structured database, i.e., a tree-structured database, a record located at the first level, i.e., the top level, from which all second levels branch out. *Synonym* **base node record.** *See also* **database.**

root segment: The part of an overlay program that remains in storage during the entire execution of the overlay program. *See also* **overlay program.**

rope storage: *See* **core rope storage.**

rotary dial: A calling device that (a) is incorporated within a telephone set, and (b) when wound up and released, generates the direct current (dc) pulses required for establishing a connection in a telephone system. *See also* **dial pulse, dial signaling, dual tone multifrequency signaling.**

rotary dial service: A telephone service that is provided to users with telephones that have a rotary dial. *Note:* Usually, if rotary dial service is provided, outgoing call number selection cannot be made with pushbutton, i.e., dual-tone, telephones. *See also* **dial pulsing, rotary dial.**

rotary hunting: Hunting in which all the numbers in the hunt group are sequentially selected. *Note:* In modern electronic switching systems, the numbers in the hunt group are not necessarily sequentially selected. *See also* **hunting.**

rotary joint: *See* **fiber optic rotary joint.**

rotary switching: In telephone systems, electromechanical switching in which the selection mechanism consists of a rotating element using several groups of wipers, brushes, and contacts. *See also* **switching system.**

rotating: In graphics and display systems, turning a display element, display group, or display image about an axis perpendicular to the display space on the display surface of a display device. *See also* **display element, display space, translating, tumbling.**

rotating polarization: Polarization of an electromagnetic wave, such as a lightwave or a radio wave, in which the polarization plane rotates with an angular displacement or direction that is a function of the distance along the light ray or radio propagation path.

rotation: *See* **antenna beam rotation, antenna rotation, optical rotation.**

rotational delay: *Synonym* **search time.**

rotational position sensing: In rotational storage media, such as magnetic disks, magnetic drums, and optical disks, the locating of a given sector, a desired track, and a specific record by continuous comparison of the read-write head position with appropriate synchronization signals.

rotation period: *See* **antenna rotation period.**

rotator: *See* **image rotator.**

rotor: *See* **fiber optic rotor.**

round: In mathematics, to delete or omit one or more of the less significant digits of a numeral and to adjust the retained more significant digits in accordance with a prescribed rule, thus limiting precision but reducing the number of characters in the numeral. *Note 1:* The most usual rule is to round at any given position by adding 1 to the next higher significant digit if the digit in the lower significant position is 5 or greater, and drop all less significant digits if the numeral in the lower significant position is 4 or less. Thus, the numeral 496.5442 can be rounded to 496.544, 496.54, 496.5, 497, or 500. Care must be taken so that rounding does not produce a bias. If 544.47 is rounded to 544.5, the 544.5 should not be rounded to 545. The 544.47 should have been rounded to 544 if three significant digits are desired. *Note 2:* If there are many numerals to be processed, such as added or multiplied, and the less significant digits are evenly distributed, the error caused by rounding usually will be negligible. *See also* **truncate.**

round-trip: In satellite communications, the distance that (a) is measured from an Earth originating station through a satellite to a receiving Earth station and back via the satellite to the originating station, and (b) is used in computing the round-trip delay time.

round-trip delay time: 1. The time required for the transit of a signal over a closed circuit. *Note:* The round-trip delay time is significant in systems that require two-way interactive communications, such as voice telephone or acknowledge/negative-acknowledge (ACK/NAK) data systems where the round-trip time directly affects the throughput rate. The round-trip delay time may range from a few microseconds for a short line-of-sight (LOS) propagation radio system to many seconds for multiple-link circuits with one or more satellite links in tandem. Round-trip delay time includes the node delays as well as the propagation medium transit time. **2.** In a token-passing ring network, the time required for a signal transmitted by a station to return to that station. **3.** In radar systems, the time required for a transmitted pulse to reach a target and for the echo to return to the receiver. *See also* **circuit, propagation medium, turnaround time.**

route: 1. In communications systems operations, the geographical path that is followed by a call or a message over the circuits that are used in establishing a chain of connection. **2.** To determine the path that a message or a call is to take in a communications network. **3.** To construct the path that a call or a message is to take in a communications network in going from one station to another or from a source user end instrument to a destination user end instrument. **4.** In the Transmission Control Protocol/Internet Protocol (TCP/IP), the path that network traffic follows from its source to its destination. *Note:* In a Transmission Control Protocol/Internet Protocol (TCP/IP) internet, each IP datagram is routed separately. The route followed by a datagram may include many gateways and many physical networks. *See also* **call, message, traffic, user end instrument.**

route dialing: In the setting up of a trunk call over an automatically switched communications network, dialing in which either the call originator, i.e., the source customer, subscriber, or switchboard operator, is required to generate a directory number for each stage of the route to the destination user, i.e., the call receiver. *Note:* An example of route dialing is dialing by a switchboard attendant who dials each switching exchange along the route as each connection is made.

route dispersion: *Synonym* **route diversity.**

route diversity: The allocation of circuits between two points over more than one geographical or physical route with no geographical points in common. *Synonym* **route dispersion.** *See also* **alternate routing, avoidance routing, diversity routing.**

route map: *See* **line route map.**

route matrix: In communications network operations, a record that (a) indicates the interconnections between pairs of nodes in the network and (b) is used to select direct routes, select alternate routes, and create available route tables from point to point.

route mile: A distance of 1 mi (statute mile) along an installed cable. *Note:* A route mile requires more than a sheath mile of cable because of the cable consumed for purposes other than covering distance, such as slack, coupling, aerial inserts, offsets, and connection boxes along the route. *See also* **sheath mile.**

router: 1. In a communications network, a functional unit that (a) is used to interface networks, (b) provides switching among the networks, (c) reads all packets or messages transmitted by a network, (d) accepts only those addressed to another network, (e) buffers each accepted packet for retransmission to the other network, (f) does the same for the other network, (g) operates at the Network Layer (Layer 3) of the Open Systems Interconnection—Reference Model, (h) reads and filters packets or frames by putting them onto the correct link for their destination, and (i) may perform the functions of a bridge. **2.** In a communications network, equipment that automatically selects one of the several paths that network traffic will follow. *Note 1:* When used with the Transmission Control Protocol/Internet Protocol (TCP/IP), a router is used as an IP gateway that routes datagrams using IP destination addresses. *Note 2:* At the lowest level, a physical network bridge performs the functioning of a router because it chooses whether to pass packets from one physical wire to another. *Note 3:* Within a long-haul packet-switching network, each individual packet switch performs the functions of a router because it chooses routes for the individual packets. **3.** A functional unit that forwards packets using a specific protocol type from one logical network to another. *Note:* In the Open Systems Interconnection—Reference Model, a router receives Physical Layer signals, performs Data Link Layer and Network Layer protocol processing, and then sends the signals via appropriate Data Link Layer and Physical Layer protocols to another network. A router operates at the Network Layer because it uses Network Layer addresses. A router usually performs more functions than a bridge, and may contain flow control mechanisms as well as source routing or nonsource routing features. *See* **screening router.** *See also* **bridge, brouter, gateway, port.** *Refer to* **Figs. S-9 and S-11 (880).**

route selection: *See* **automatic route selection.**

route service: *See* **aeronautical mobile satellite (off-route) service, aeronautical mobile satellite (route) service.**

routine: *See* **computer routine, computer subroutine, supervisory routine.**

routine message: A message that (a) is sent by rapid means, such as wire, radio, and telegraph, and (b) is not of sufficient urgency to require a precedence higher than routine precedence. *See also* **precedence, routine precedence.**

ROUTINE message: In military communications systems, a message that has the lowest precedence level for military messages. *See also* **precedence designation.**

routine meteorological broadcast: A broadcast that (a) is periodic except in an emergency, (b) contains a meteorological forecast of weather conditions of a general nature that is useful to all stations, (c) usually is broadcast on ship broadcast systems by many nations, and (d) is also broadcast on fleet and merchant ship broadcast stations by certain nations. *See also* **emergency meteorological warning, special meteorological broadcast.**

routine precedence: The precedence designation that is used for all messages that require transmission by rapid means, such as wire, radio, and telegraph, but are not of sufficient urgency to justify a higher precedence. *See also* **precedence designation, routine message.**

routing: In a communications network, determining and prescribing the path or the method to be used for establishing connections and forwarding messages. *See* **affinity routing, alternate routing, avoidance routing, collective routing, destination routing, deterministic routing, diverse routing, dynamically adaptive routing, flood search routing, free routing, heuristic routing, invariant routing, key routing, multiple routing, predetermined routing, satura-**

tion routing, transaction routing. *See also* **high usage trunk group, ring around, spill forward.**

routing bulletin: *Synonym* **routing directory.**

routing chart: In telephone system operations, instructions on routing that are in a diagrammatic or tabular form.

routing diagram (RD): In a communications system, a diagram that (a) shows all links between all switchboards, exchanges, switching centers, and stations in the system, such as the links between primary relay, major relay, minor relay, and tributary stations as well as supplementary links, (b) is used to identify the stations and the links, and (c) is used to indicate tape relay routes, transfer circuits, refile circuits, radio links, operational status, line conditions, and other network information required for network operations and management.

routing directory (RD): In communications systems operations, an alphabetical listing of (a) all switchboards, exchanges, switching centers, and stations in a system and (b) the first, or primary, route for communications traffic. *Note:* An example of a routing directory (RD) for a telephone network is a list of (a) primary route switching centers, central offices (COs), and switchboards, (b) alternate route switching centers, COs, and switchboards adjacent to each primary route switching center, central office CO, and switchboard, and (c) additional information that may be appended to the listed entries, such as status, hours of operation, and outages. *Synonym* **routing bulletin.**

routing indicator (RI): 1. In the heading of a message, an address, or a group of characters, that specifies routing instructions for the transmission of the message to its final destination. *Note:* Routing indicators may also include addresses of intermediate points. **2.** In military communications systems, a group of letters assigned to indicate (a) the geographical location of a station and (b) a fixed headquarters of a command, activity, or unit at a geographical location. *Note:* In tape relay systems, the routing indicator includes the general location of a tape relay or tributary station to facilitate the routing of traffic over the tape relay networks. *See* **alternative routing indicator, collective routing indicator, mobile unit routing indicator, special purpose routing indicator.** *See also* **alternate routing, character, deterministic routing, flood search routing, hybrid routing.**

routing indicator allocation (RIA): The allocation of a block of routing indicators to an organization for distribution among the elements of the organization. *Note:* An example of routing indicator allocation is the allocation of a block of five-letter routing indicators, all beginning with the letters AA, to the First Field Army. *See also* **routing indicator assignment.**

routing indicator assignment (RIA): The assignment of a specific routing indicator, from the block of allocated routing indicators, to (a) a specific organizational element of the larger unit to which the block was allocated or (b) specific geographical areas. *Note:* An example of routing indicator assignment is the assignment of the five-letter routing indicator AABAC to the Headquarters of the XXIII Corps. Thus, the BAC indicates the Headquarters of the XXIII Corps, and the AA indicates the next higher echelon, such as the First Field Army to which the XXIII Corps is assigned. *See also* **routing indicator allocation.**

routing indicator delineation map: A map that shows the boundaries of each geographical area that has been assigned a letter, for use in preparing routing indicators for messages for destination users, i.e., for addressees, in those areas.

routing indicator delineation table: A table that shows preallocated worldwide routing indicators, each consisting of four to seven characters, with each character designating a specific part of a routing indicator, such as a letter for each nation or geographical area, a letter for each major relay station, or a letter for tributary stations.

routing line segregation: Routing in which (a) the basic routing line of the message heading is altered as the message passes through each relay station along the route, and (b) only those routing indicators that are pertinent to the onward transmission are left in the routing line of the standard message format. *See also* **multiple call messages.**

routing table: 1. A matrix, associated with a network control protocol, that gives the hierarchy of link routing at each node. **2.** A matrix associated with a network control protocol that gives the preferred onward line preference at each node in the network. *See also* **distributed control, queue traffic.**

row: In data processing, a horizontal arrangement of characters or character positions, such as a horizontal line of characters on a spreadsheet, a punched card, or a punched tape. *Note:* A row on a spreadsheet or punched card corresponds to a line on a typed or printed page. *See also* **column.**

row binary: Pertaining to the representation of binary numbers in which the successive digits are placed in a row, rather than in a column, usually with the least significant digit at the right end, such as using each row in a standard 80-column punch card to represent 80 consecutive bits, or to represent two 40-bit numbers. *Note:* If all the rows on a standard 12-row, 80-column card were used to store one pure binary numeral with 960 digits, all of the 2^{960} numerals from 0 to 2^{960} could not be punched, i.e., all possible different hole patterns could not be punched. *See also* **column binary.**

RQ: repeat request.

r region: *Synonym* **far field region.**

RSL: received signal level.

RSW: relative spectral width.

RTA: remote trunk arrangement.

RTI: radiation transfer index.

RTM: reference test method.

RTTY: radio teletypewriter.

rub: *Synonym* **crush.**

rubber banding: In graphics and display systems, a technique in which the ends of a set of straight lines are moved while the other ends remain fixed. *Note:* Several lines from the corners of the base of a projected image of a pyramid converge at the apex. In rubber banding, if the apex were moved, the ends at the apex would move with the apex, and the ends at the base would remain fixed. If one of the corners of the base were moved, the line ends converging there would move, and all other ends would remain fixed.

rubidium clock: A clock that contains a quartz oscillator stabilized by a rubidium standard. *See also* **rubidium standard.**

rubidium standard: A frequency standard (a) in which a specified hyperfine transition of rubidium-87 atoms is used to control the output frequency, (b) that consists of a gas cell that has an inherent long-term instability, and (c) that, as a consequence, is relegated to the status of a secondary standard. *See also* **primary frequency standard.**

rub-out character: *Synonym* **delete character.**

run: 1. The execution of one or more computer jobs or programs. **2.** In a communications system, the transmission of a complete message or the continuous transmission of a group of messages or packets. **3.** In systems maintenance, a complete sequence of testing operations. *See* **cable run, rerun.**

run-length encoding: In a facsimile system, encoding that uses redundancy reduction in which a set of consecutive picture elements having the same state, i.e., the same gray scale or color, is encoded into a single codeword. *See also* **code, facsimile.**

runner cut: On the surfaces of glass and other ground or polished materials, a curved scratch, such as might be caused by a grinding or polishing wheel.

running accumulator: *Synonym* **pushdown storage.**

running fix: A method of determining the location of a stationary signal source by means of direction finding (DF), by one mobile DF unit, from two or more locations at different times. *Note:* If the transmitter that is being located is fixed, i.e., is stationary, a running fix is readily obtained. If it is moving slowly, a running fix may also be obtainable to a reasonable accuracy if the DF station can move fast enough. If the mobile transmitter that is being located is fast-moving, obtaining a running fix is difficult, because the attempt to obtain a running fix reduces to a simple chase with radio contact rather than visual contact. If the pursuing station is fast enough, visual contact may result.

runway approach aid: Any system, device, or marking that is used to assist pilots in the landing of aircraft. *See also* **ground controlled approach landing, instrument landing system.**

rural radio service: A public radio service that (a) is provided by fixed stations on frequencies below 1000 MHz (megahertz) and (b) provides (i) public message communications service between a central office (C.O.) and users, i.e., customers, located in rural areas to which it is impractical to extend service via landlines, (ii) public communications service between landline COs and different exchange areas that are impractical to interconnect by any other means, or (iii) private line telephone, telegraph, or facsimile service between two or more points to which it is impractical to extend service via landlines.

rural subscriber station: A fixed station in the rural radio service used by a customer for communications with a central office (C.O.) station.

RWI: radio and wire integration, radio-wire integration, radio-wire interface.

RX: receive, receiver, receiving.

RZ: return to zero.

RZ coding: return to zero coding.

s: second.

SABER: situation awareness beacon with reply.

safety message: The message that follows the safety signal and that is sent at distress frequencies at certain times in certain world regions. *Note:* An example of a safety message is the message containing warnings of hazards, such as icebergs, ice, icing conditions, bad weather, earthquakes, wrecks, and fires, that is sent a few seconds before the 18th and the 48th minute after every hour in all International Telecommunication Union (ITU) regions.

safety power margin: *Synonym* **power margin.**

safety service: A radiocommunications service permanently or temporarily used for the safeguarding of human life and property.

safety signal: A signal used to indicate that a radio station is about to transmit a safety message, such as a message concerning navigational, aeronautical, or public safety. *Note 1:* Upon receipt of a safety signal, stations (a) monitor the safety message that follows and (b) usually do not make any transmissions that might interfere with the safety message. *Note 2:* Examples of safety signals are (a) in continuous wave (cw) transmissions, TTT sent three times and (b) in voice transmissions, the word "securité" (the French word for "security," pronounced "say-coor-ee-tay") pronounced three times.

safety traffic: The totality of messages that (a) pertain to safety, (b) contain emergency meteorological warnings, (c) are deemed to be in the safety category of emergency messages, and (c) usually take precedence over all traffic except distress messages and urgent messages. *See also* **distress message, emergency message, emergency meteorological warning, safety message, traffic, urgent message.**

Sagnac fiber optic sensor: An interferometric sensor in which (a) a lightwave is split and passed in opposite directions around the same rigid rotating optical fiber loop, by means of mirrors, to a single photodetector, (b) phase cancellation and enhancement change, and hence irradiance, i.e., light intensity, changes when the angular velocity is varied, thereby producing an output frequency that is proportional to the angular acceleration, and (c) the output signal from the photodetector can be integrated once to obtain angular velocity and twice for angular displacement. *See also* **irradiance, optical fiber, phase, photodetector.**

salesman algorithm: *See* **traveling salesman algorithm.**

SAM: coplanar self-assembled monolayer, serving area multiplex.

sample: *See* **reconstructed sample, signal sample.**

sample change compaction: Data compaction accomplished by (a) specifying a constant level, or an easily definable varying level, of a value of a parameter or a variable, and (b) also specifying the deviations from the level in discrete or continuous values. *Note:* An example of sample change compaction is the transmission or storage of one absolute value and a series of incremental changes to that value, rather than transmitting a series of absolute values. Thus, transmitting 54496+3+1−5 is more compact than transmitting 54496 54499 54500 54495. *See also* **compaction.**

sampling: *See* **signal sampling.**

sampling criterion: *Synonym* **Nyquist criterion.**

sampling frequency: *Deprecated synonym for* **sampling rate.**

sampling rate: The number of samples taken per unit time, i.e., in communication systems, the rate at which signals are sampled for subsequent use, such as for modulation, coding, conversion, and quantization. *Note:* The sampling rate usually is expressed as the number of samples taken per unit time in a specified channel or line. *Deprecated synonym* **sampling frequency.** *See also* **Nyquist rate, signal sampling.**

sampling signal: 1. The signal that is used to determine or measure the signal level of a given signal, i.e., its instantaneous magnitude, at equally spaced instants. *Note:* The measured signal may be somewhat satisfactorily reconstructed for most purposes from these samples, provided that at least two samples are taken during the period that corresponds to the highest frequency component of the signal being sampled, i.e., if the Nyquist theorem is applied and the Nyquist criterion is satisfied. **2.** The signal that is obtained from the process of sampling a given signal at equally spaced instants. *Note:* The reconstituted signal is the envelope of the signal that is obtained from the sampling process. The envelope is obtained by a detector that usually consists of an integrating circuit with a time constant that is properly adjusted to the sampling rate. *See also* **Nyquist theorem.** *Refer to* **Fig. S-1.**

sampling theorem: *Synonym* **Nyquist theorem.**

sampling time: The reciprocal of the sampling rate, i.e., the time between corresponding points on two adjacent, i.e., two consecutive, sampling pulses of the sampling signal. *See also* **sampling rate, sampling signal.**

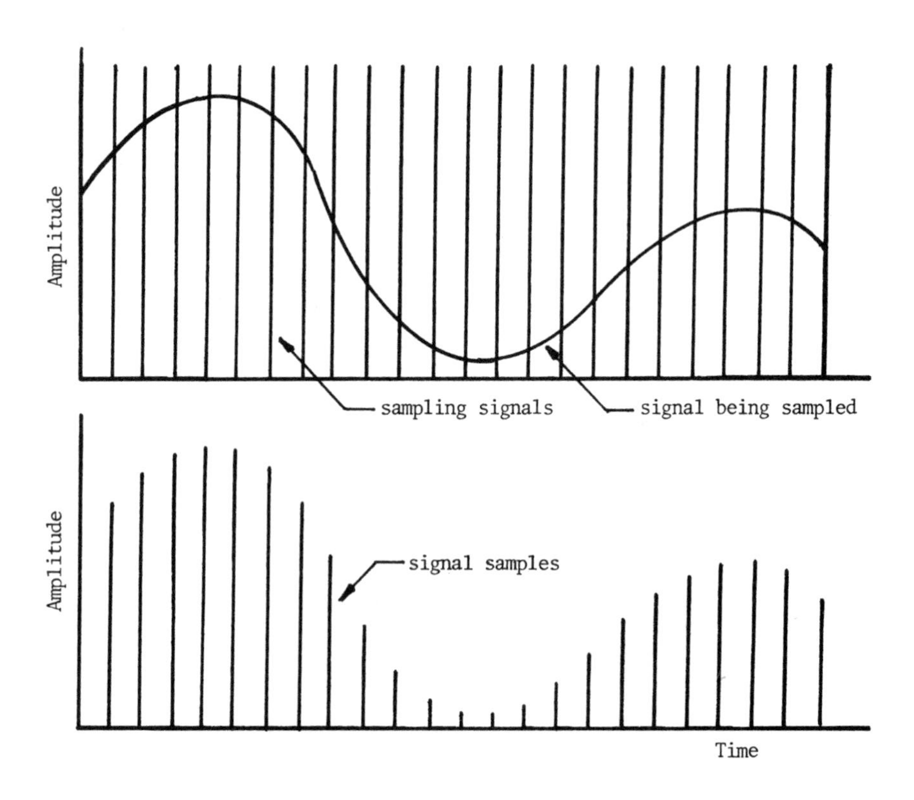

Fig. S-1. If the **sampling signal** (pulsed) and the signal being sampled (analog) are entered into an AND gate, the output will be the signal samples. The output samples can then be digitized, transmitted, detected, or otherwise processed.

sanitizing: In communications, computer, data processing, and control systems, the degaussing or overwriting of sensitive information on a data medium, such as magnetic tape or disks. *See also* **degauss.**

SARAH: search and rescue and homing.

SARBE: search and rescue beacon.

saros: *See* **cycle of saros.**

SAS: survivable adaptive system.

satellite: 1. A body that revolves around another body. **2.** One of two bodies revolving about their common center of gravity. **3.** A body that (a) revolves around another body of preponderant mass and (b) has a motion primarily and permanently determined by the force of gravitational attraction of that other body. **4.** An object or a vehicle that is orbiting, or is intended to be placed in orbit, around the Earth, the Moon, or any other celestial body. *Note:* A parent body and its satellite both revolve about their common center of gravity. The smaller body is considered to be the satellite of the larger. *See* **active communications satellite, active satellite, advanced communications technology satellite, asynchronous satellite, communications satellite, direct broadcast satellite, direct broadcast television satellite, Earth satellite, geostationary satellite, geosynchronous satellite, nongeostationary satellite, passive communications satellite, passive satellite, relay satellite, synchronous Earth satellite, synchronous satellite.** *Refer to* **Fig. W-4 (1087).**

satellite access: In satellite communications systems, the establishment of contact with a communications satellite space station. *Note:* An example of satellite access is access at the moment when an Earth station commences to use a satellite space station as a signal repeater, i.e., to use its transponder. Each radio frequency (rf) carrier that is relayed by a satellite space station at any time occupies an access channel. Accesses, i.e., channels, are distinguishable by various system parameters, such as frequency, time, or code, depending on the system configuration. *See also* **Earth station, multiple access.** *Refer to* **Fig. C-7 (157).**

satellite address: The identification code of the communications satellite space station for which a message or a transmission is intended.

satellite antenna: *See* **global beam satellite antenna.**

satellite attitude: The angle between (a) a satellite axis and (b) a reference line, such as the angle between an antenna axis and a line between the Earth's center and the satellite's center.

satellite beacon: A discrete radio frequency (rf) signal that (a) is radiated by a satellite for Earth station antenna tracking, and (b) is modulated by an identifying frequency tone.

satellite channel mode: In satellite communications, one of the four possible connections of receive and transmit antennas, namely, (a) Earth coverage for both, (b) Earth coverage for transmit and narrow beam for receive, (c) narrow beam for transmit and Earth coverage for receive, and (d) narrow beam for both.

satellite communications: A telecommunications service that is provided via one or more satellite relays and their associated uplinks and downlinks. *See also* **acquisition, acquisition time, active satellite, asynchronous satellite, communications satellite, demand assignment, despun antenna, direct orbit, downlink, Earth coverage, Earth terminal, Earth terminal complex, equatorial orbit, footprint, geostationary orbit, inclined orbit, link, multiple access, orbit determination, periapsis, perigee, polar orbit, pulse address multiple access, satellite, satellite relay, space subsystem, synchronous orbit, synchronous satellite, uplink, view.** *Refer to* **Fig. S-2.** *Refer also to* **Figs. C-7 (157), N-16 (624), P-10 (740), T-4 (994).**

satellite communications system: A communications system in which messages are relayed from Earth station to Earth station via (a) one or more communications satellite space stations with either a receiver-amplifier-transmitter or a reflector and (b) the propagation media between the stations. *Note:* The communications satellite space station may perform some signal processing. *See also* **communications satellite space station, propagation medium.** *Refer to* **Figs. S-3 through S-7.** *Refer also to* **Figs. B-9 (96), C-8 (158), M-2 (587), R-7 (821), T-5 (1012).**

satellite computer: 1. A computer that is online with, is under the control of, accepts jobs from, and furnishes results to another computer. **2.** An offline auxiliary computer used to perform certain tasks in conjunction with another computer. *See also* **job, offline, online.**

satellite Earth station: 1. *Synonym* **Earth station. 2.** *See* **communications satellite Earth station, meteorological satellite Earth station, radionavigation satellite Earth station.**

satellite Earth terminal: *Synonym* **Earth terminal.**

satellite emergency position-indicating radiobeacon: An Earth station that (a) is in the mobile satellite service and (b) emits signals intended to facilitate search and rescue operations.

Fig. S-2. A digital data **satellite communications** receiver. (Courtesy ComStream Company.)

satellite exchange: In a communications system, an exchange, such as a private branch exchange (PBX) or a Centrex™ system exchange, that (a) is not equipped with an operator position, i.e., an attendant position and (b) is associated with an attended exchange that provides attendant functions for the communications system. *Note:* The satellite exchange usually is not on a communications satellite. *See also* **exchange, position.**

satellite hop time: *Synonym* **satellite transit time.**

satellite link: A radio link that (a) is between a transmitting Earth station and a receiving Earth station via one satellite, (b) may be a link in an Earth surface communications system, and (c) usually consists of one uplink, one downlink, and a satellite station, such as a satellite relay. *See also* **downlink, Earth station, satellite relay, uplink.** *Refer to* **Figs. C-7 (157), P-10 (740), T-4 (994).**

satellite link noise temperature: *See* **equivalent satellite link noise temperature.**

satellite network: A satellite system, or a part of a satellite system, that consists of only one satellite and the cooperating Earth stations. *Refer to* **Fig. C-7 (157).**

satellite operation: *Synonym* **satellite private branch exchange (SPBX).**

satellite optical link: A beamed optical transmission channel that (a) usually consists of a laser source shining a lightwave in the form of a narrow beam on an avalanche photodetector, (b) operates between an Earth station and a satellite space station, and (c) has a source that is modulated in accordance with an intelligence-bearing signal. *See also* **Earth station, laser, satellite space station.**

satellite orbit: *See* **geostationary satellite orbit.**

satellite PBX: satellite private branch exchange.

satellite period: 1. The time between two consecutive passages of a satellite through a specified point on its orbit. **2.** The time between two consecutive passages of a satellite through a characteristic point on its orbit.

satellite point: *See* **subsatellite point.**

satellite private branch exchange (SPBX): A private branch exchange (PBX) or Centrex® System that (a) is not equipped with attendant positions and (b) is associated with an attended main PBX or Centrex® System. *Note:* The main exchange attendant provides attendant functions for the satellite private branch exchange (SPBX). *Synonym* **satellite operation.** *See also* **Centrex® service, private branch exchange.**

Fig. S-3. The above-deck equipment (ADE) and the below-deck equipment (BDE) of the Saturn MM Inmarsat M Marine Terminal for a **satellite communications system** dial-up link. A single coaxial cable connects the ADE and the BDE. (Courtesy Mackay Communications.)

Fig. S-4. The Saturn Mm Inmarsat M above-deck component (ADE) of the marine terminal in a **satellite communications system** environment. (Courtesy Mackay Communications.)

Fig. S-5. The ABB Nera Saturn Bp station that (a) provides a **satellite communications system** dial-up link to terrestrial networks for telephony, facsimile, data, and Telex™ traffic, (b) is a portable terminal for the Inmarsat B system, and (c) is compact, robust, and designed for use over a wide temperature range in rain and snow. (Courtesy Mackay Communications.)

Fig. S-6. The ABB Nera Saturn Bp station that provides a **satellite communications system** dial-up link to terrestrial networks for telephony, facsimile, data, and Telex™ traffic in a field environment. (Courtesy Mackay Communications.)

satellite relay: In a communications satellite space station, an active or a passive satellite repeater that relays signals between two Earth terminals. *Note:* An example of a satellite relay is a radio relay station on a

Fig. S-7. The Saturn MiniPhone™, a lightweight portable Inmarsat M **Satellite communications system** station that (a) provides access to the international dial-up telephone, facsimile, and data network, (b) has a 4.5 kbps (kilobits per second) digital coded voice, a CCITT Group III at 2400 bps facsimile, and a 2400 bps data rate capability, and (c) is briefcase-size and weighs about 9 kg. (Courtesy Mackay Communications.)

satellite orbiting a celestial body, such as the Earth, the Moon, or Mars. *See also* **radio relay station.**

satellite service: *See* **aeronautical mobile satellite (off-route) service, aeronautical mobile satellite (route) service, aeronautical radionavigation satellite service, amateur satellite service, broadcasting satellite service, communications satellite service, Earth exploration satellite service, fixed satellite service, meteorological satellite service, radionavigation satellite service.**

satellite space station: *See* **broadcasting satellite space station, communications satellite space station, meteorological satellite space station, radionavigation satellite space station.**

satellite switched time division multiple access (SS-TDMA): In a satellite communications system, multiple access that is obtained by the use of time division multiplexing with one or more of the switching facilities forming part of a communications satellite space station. *See also* **communications satellite space station, multiple access, time division multiplexing.**

satellite system: A space system using one or more artificial Earth satellites. *See also* **space subsystem, space system.**

satellite system for evaluation and test: *See* **Functional Advanced Satellite System for Evaluation and Test.**

satellite telephone (satphone): A portable, briefcase-size telephone that is capable of communication via satellite systems. *Synonym* **satphone.**

satellite transit time: In a satellite communications system, the time taken for a signal to propagate from one Earth station to another via a satellite link, approximately one-third of a second for geostationary Earth satellites. *Synonyms* **satellite hop time, transmit time.** *See also* **Earth station, geostationary Earth satellite, satellite link.**

satellite wide-area network (SWAN): A wide-area network (WAN) that is implemented with (a) very small aperture terminal (VSAT) antennas, (b) satellite and Earth stations, and (c) digital cellular telephone systems. *See also* **very small aperture terminal antenna, wide-area network.**

satphone: *Synonym* **satellite telephone.**

SATPHONE®: A satellite-supported portable telephone that (a) enables worldwide connection from virtually anywhere on Earth, (b) uses the Inmarsat-M satellite system, and (c) was developed by Alden.

sat routing: *Synonym* **saturation routing.**

saturation: 1. In a communications system, the condition in which a component, such as a line or a circuit, of the system has just reached its maximum traffic-handling capacity. *Note:* Saturation is equivalent to one erlang per component. When this occurs, the component is saturated. *See also* **busy hour, erlang, traffic load. 2.** The point at which the output of a linear device, such as a linear amplifier, deviates significantly from being a linear function of the input when the input signal is increased. *Note:* Modulation often requires that amplifiers operate below saturation in order to remain linear.

saturation jamming: The jamming of all transmitters in a specified area. *See* **infrared saturation jamming.**

saturation routing: In a switched telephone network, a routing procedure in which (a) a call selection signal is extended forward from each nodal switching center, in turn, on a broadcast basis, over all the internodal links in a part of, or in the whole of, a network, (b) the nodal switching center to which the call receiver terminal is connected acknowledges the call by means of

progressively to the call originator nodal switching center, each time over the link on which the call selection signal was first received, so that the best available route between the centers can be selected for routing the call. *Note:* Saturation routing can also be used in tape relay systems. *Synonym* **sat routing.** *See also* **backward direction, broadcast, routing, call selection signal.**

saturation signaling: In communications systems, signaling in which a route through a number of switching points to a destination user, i.e., destination customer, is determined by simultaneously interrogating all switching points on a progressive basis through the network until the destination user end instrument is found, and a ringing signal reaches it.

Savart law: *See* **Biot-Savart law.**

save area: An area of main storage used to store data pending immediate use, such as the contents of registers, a file being edited, a menu, or a current program module. *See also* **main storage, menu, program module, register.**

sawtooth keyboard: A manually operated keyboard in which the energy that is required to move the code bars into positions selected by the depression of a key is derived from the operator. *See also* **keyboard, position.**

scalar: Pertaining to a parameter, variable, or quantity that (a) only has magnitude, such as amplitude, phase, frequency, mass, time, speed, volume, or signal to noise ratio, and (b) does not have direction, such as area, velocity, electric field, angular momentum, or linear length. *Note:* A vector has both magnitude and direction. The magnitude of a vector is a scalar quantity, but not vice versa. The gradient of a scalar quantity is a vector, such as the derivative of the refractive index with respect to distance or the derivative of electric field strength with respect to distance. The scalar product, i.e., the dot product, of two vectors is a scalar quantity. *See also* **scalar product, vector.** *Refer to* **Figs. P-15 (763), T-7 (1024).**

scalar feed horn: A corrugated antenna feed horn that (a) has a radiation pattern nearly axially symmetric and independent of polarization, and (b) is used in some shipborne communications satellite Earth stations. *See also* **antenna, antenna feed, communications satellite Earth station, horn, horn antenna, radiation pattern.**

scalar product: A scalar quantity equal to the product of (a) the magnitudes of two vectors and (b) the cosine of the angle between them. *Synonym* **dot product.** *See* **vector product.**

scale: 1. To change the representation of a quantity by expressing it in units or by using a scale factor that will bring its value to within a specified range. **2.** In display systems, to enlarge or reduce the size of all or part of a display element, display group, or display image by multiplying the coordinates of the element, group, or image parts by a constant value that is greater than unity for enlargement and less than unity for reduction. *See* **Celsius temperature scale, coordinated time scale, Fahrenheit temperature scale, gray scale, International Atomic Time scale, international temperature scale, Kelvin temperature scale, time scale, uniform time scale.** *See also* **display element, enlargement, reduction, time scale factor.**

scan: 1. To sequentially examine, part by part. **2.** To routinely examine every reference or every entry in a file as part of a retrieval scheme. **3.** In electromagnetic or acoustic search, one complete rotation of an antenna. *Note:* The antenna rotation may provide a time base. **4.** In electronics intelligence (ELINT), the motion of an electromagnetic beam through space searching for a target. *Note:* Scanning may be accomplished by physical motion of the antenna, such as by rotation, or by beam switching, such as in phased array antennas. **5.** The path that is followed by the end of a moving beam. *Note:* Examples of a scan are the path that is traced (a) periodically by an electromagnetic wave exploring outer space, (b) by an electron beam that is writing on a cathode ray tube (CRT) screen, and (c) by a single sweep of a directional rotating antenna, such as a searchlight radar antenna. **6.** To sweep a beam about a point or about an axis. **7.** In a facsimile transmitter, to pass over recorded data with a reading head or light spot. *See* **circular scan, directed scan, flying spot scan, radar scan, raster scan, sector scan.**

scan airborne intercept radar: *See* **track while scan airborne intercept radar.**

scan line: 1. The line produced on a record medium by a single sweep of a scanner. **2.** In an imaging system, such as a facsimile, radio photo, or telephoto system, the area that is explored by (a) the scanning spot in one sweep from one side of the scanning field to the other, i.e., from one side of the object to the other, in a transmitter or (b) the recording spot in one sweep from one side of the recording area, i.e., the recorded copy, to the other in a receiver. *Synonym* **scanning line.** *See also* **object, recording spot, scanning spot.**

scanner: **1.** A device that performs a scanning operation. **2.** A device that (a) examines a spatial pattern, such as the object in a facsimile transmitter, one part after another, and (b) generates analog or digital signals corresponding to the pattern. *Note:* Scanners are often used in mark sensing, pattern recognition, character recognition, facsimile transmission systems, and television (TV). **3.** In facsimile systems, the part of the transmitter that systematically translates the densities of the scanned object, such as a document, into a signal for transmission. **4.** In radar, a device, such as a rotating directional antenna or a phased array antenna, that emits a beam to search for possible objects, i.e., targets, in space. *See* **flying spot scanner, optical scanner.** *See also* **facsimile, scanning, simple scanning.**

scanning: **1.** In communications systems, examining traffic activity to determine whether further processing is required. *Note:* Scanning usually is performed periodically. **2.** In communications systems, determining, analyzing, or synthesizing the light intensity values or equivalent attributes of parts of an object, such as a display image, picture element, or graphic character, usually to obtain a baseband signal for modulating a carrier for transmission. **3.** In television (TV), facsimile, and picture transmission, successively analyzing the colors and the densities of the object according to a predetermined pattern. **4.** Tuning a device through a predetermined range of frequencies in prescribed increments and at prescribed times. *Note:* Scanning may be performed at regular or random increments and intervals. **5.** Sweeping a beam, i.e., rotating a beam, about a point or about an axis. **6.** In radar, directing a beam of radio frequency (rf) energy successively over the elements of a given region. **7.** The sweeping of a region by a directional receiving antenna in a search for possible signals. *See* **conical scanning, dynamic scanning, electronic line scanning, frequency scanning, limited scanning, mark scanning, multiple spot scanning, optical scanning, simple scanning, solid state scanning, spot projection scanning.**

scanning density: **1.** In facsimile systems, the line advance in number of scanning lines per unit distance perpendicular to the scanning line, i.e., the number of scanning pitches per unit length. **2.** The reciprocal of the scanning pitch. *See also* **scanning line, scanning pitch.**

scanning direction: In facsimile transmitting equipment, the scanning of the plane, or the developed plane in the case of a drum transmitter, of an object, such as a message surface, along lines running from right to left commencing at the top, so that scanning commences at the top right-hand corner of the surface and finishes at the bottom left-hand corner. *Note 1:* The scanning direction is equivalent to scanning over a right-hand helix on a drum. *Note 2:* The orientation of the message on the scanning plane will depend upon its dimensions and is otherwise of no consequence in scanning, transmission, or reception. *Note 3:* In facsimile receiving equipment, scanning takes place from right to left and top to bottom, in the above sense, for "positive" reception and from left to right and top to bottom, in the above sense, for "negative" reception. *Note 4:* Scanning direction conventions are included in International Consultative Committee for Telegraph and Telephone (CCITT) Recommendations for phototelegraphic equipment. *See also* **facsimile, scanning.**

scanning electron microscope (SEM): An electron microscope that (a) has an electron beam that can scan an object in a regular or controlled pattern and (b) can be used to expose resist material in high resolution scanning patterns used in the fabrication of optical integrated circuit (OIC) components. *See also* **electron microscope, object, optical integrated circuit components, resolution, scanning.**

scanning field: In facsimile systems, the total of the areas that are actually explored by the scanning spot during the scanning of the object by the transmitter or during scanning of the record medium by the receiver. *See also* **object, record medium, scanning spot.**

scanning line: **1.** The path traversed by a scanning spot during a single line sweep. **2.** The area traversed by a scanning spot during single sweep, i.e., during a scanning period. *Note:* The area of a scanning line is equal to the product of the scanning line length and the *y*-dimension of the scanning spot in a transmitter or in a receiver. *Synonym* **scan line.** *See also* **scanning spot, x-dimension of scanning spot, y-dimension of scanning spot.**

scanning line frequency: *Synonym* **scanning line rate.**

scanning line length: In facsimile systems, the total length of a scanning line, equal to the spot speed divided by the scanning line rate. *Note:* The scanning line length usually is greater than the available line length. *See also* **available line length, dead sector, facsimile, scanning, spot speed.**

scanning line period: *Synonym* **scanning period.**

scanning line rate: In facsimile systems, the rate at which a fixed line perpendicular to the direction of scanning is crossed by a scanning spot. *Note:* The scanning line rate is equivalent to drum speed in some mechanical systems. *Synonym* **scanning line frequency.** *See also* **facsimile, frequency, scanning, stroke speed.**

scanning line usable length: In facsimile systems, the maximum length of the scanning line that can be achieved with a particular set of facsimile equipment. *See also* **scanning line.**

scanning period: 1. In facsimile systems, the time between the scanning instants of two corresponding points on two consecutive scanning lines during transmission or reception. **2.** In facsimile systems, the reciprocal of the scanning line rate. *Synonyms* **scanning line period, scan period. 3.** In a radar scanning pattern, the time taken by a radar scanner to scan from any arbitrary point back to the same point, such as to complete a scan pattern and return to a reference starting point. *Synonym* **scan period.**

scanning pitch: 1. The distance between the centers of consecutive scanning lines. **2.** The reciprocal of the scanning density. *See also* **facsimile, scanning, scanning density.**

scanning rate: *Synonym* **scanning speed.**

scanning shift: *See* **negative scanning shift, positive scanning shift.**

scanning speed: 1. In facsimile systems, the linear speed of the scanning spot in its movement over the object in the transmitter or over the record medium in the receiver. **2.** In facsimile and television (TV) systems, the rate of displacement of the scanning spot along the scanning line. *Synonym* **scanning rate.** *See also* **facsimile, scanning, scanning spot.**

scanning spot: 1. In facsimile transmission systems, the area on the object, i.e., the original, covered at any instant, i.e., illuminated instantaneously, by the pickup scanner in the transmitter. **2.** In facsimile reception systems, the smallest area on the record medium covered at any instant, i.e., illuminated instantaneously, by the recording scanner of the receiver to ensure the synthesis of the recorded copy. **3.** In phototelegraphy and photography, the part of the light-sensitive area that is exposed at a given instant. *See* ***x*-dimension of scanning spot, *y*-dimension of scanning spot.** *See also* **elemental area, facsimile, object, raster scanning, recorded copy, record medium, scanning.**

scanning spot *x*-dimension: *Synonym* ***x*-dimension of scanning spot.**

scanning spot *y*-dimension: *Synonym* ***y*-dimension of scanning spot.**

scanning technique: *See* **near field scanning technique.**

scan period: *Synonym* **scanning period.**

scan radiometer: *See* **fiber optic visible/infrared spin-scan radiometer.**

scan rate: 1. The rate of scanning an object or an image. *Note:* The scan rate may be expressed in various units, such as distance, characters, lines, display positions, area, and pages scanned per unit time. **2.** In radar systems, the angular speed of a rotating antenna or of a phased array beam. *Note:* The scan rate usually is expressed in revolutions per minute, radians per minute, or radians per second. *See also* **scanning rate, scanning speed.**

scan rate modulation deception: *See* **radar scan rate modulation deception.**

scan-stop lockup: In automatic link establishment (ALE) radios, the undesired condition in which the normal process of (a) scanning radio channels, (b) stopping on the desired channel, or (c) returning to scan is terminated by the equipment.

scatter: For a wave propagating in a material medium, to change the direction, frequency, or polarization of the wave when it encounters medium discontinuities, such as embedded impurities, abrupt discontinuities in the refractive index, and molecular structural discontinuities in the propagation medium, that have lengths or separation distances of the order of one wavelength of the propagating wave. *Note 1:* If the particles and the discontinuities are larger than one wavelength of the propagating wave, the diffusion or the deflection consists of specular reflection and refraction rather than scattering. *Note 2:* Scattering results in (a) a disordered or random change in the incident energy distribution, (b) an increase in the mode volume, i.e., number of modes, and (c) a decrease in coherency. *See* **backscatter, backward propagation ionospheric scatter, forward scatter, ionospheric scatter, tropospheric scatter.** *See also* **backscattering, propagation medium, scattering, spectral reflection.**

scattered seeds: In transparent propagation media, seeds that (a) usually are about 5 cm (centimeters) to 10 cm apart, though occasionally they are closer together or farther apart, and (b) usually are highly visible and coarse. *See also* **heavy seeding, propagation medium, seed, seeding.**

scattering: 1. The deflection of electromagnetic waves propagating in material media. *Note 1:* Scattering is caused by (a) the intrinsic molecular structure (Rayleigh scattering), (b) the existence of free and bound charges and the movement of free charges, (c) embedded particles, (d) trapped ions, such as hydroxyl and iron ions, and (e) discontinuities and inhomogeneities in the material media in which the radiation is propagating (Mie scattering). *Note:* Scattering is more of a random process than a specular process. **2.** In an electromagnetic wave propagating in a material medium, the waves that are deflected by (a) the intrinsic molecular structure (Rayleigh scattering), (b) the existence of free and bound charges and the movement of free charges, and (c) embedded particles, trapped ions, such as hydroxyl and iron ions, and discontinuities and inhomogeneities in the material media in which the radiation is propagating (Mie scattering). *See* **backscattering, Brillouin scattering, bulk material scattering, configuration scattering, nonlinear scattering, optical fiber scattering, radiation scattering, Raman scattering, Rayleigh scattering, transverse scattering, waveguide scattering.** *See also* **scatter.** *Refer to* **Fig. S-8.**

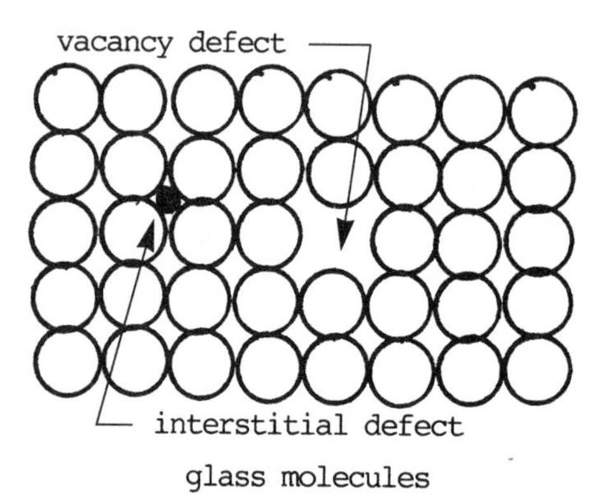

Fig. S-8. Scattering of transmitted lightwaves in glass is caused by interstitial defects, vacancy defects, and irregularity in the molecular structure, resulting in increased optical attenuation.

scattering center: A site in the microstructure of a propagation medium at which electromagnetic waves are scattered. *Note 1:* Examples of scattering centers are (a) vacancy defects, such as missing atoms or molecules in the somewhat orderly though amorphous structure of the propagation medium, (b) interstitial defects, (c) inclusions, such as a gas molecules, hydroxide ions, iron ions, and trapped water molecules, (d) microcracks or fractures in dielectric waveguides, and (e) abrupt changes in the refractive index, possibly caused by pressure points. *Note 2:* Scattering centers are frozen in the medium when it solidifies and may not necessarily cause Rayleigh scattering, which varies inversely as the fourth power of the wavelength. In glass optical fibers, there is a high attenuation band at 0.95 μm (micron), primarily caused by scattering and absorption by OH^- (hydroxide) ions. Iron selectively scatters green, giving rise to the green hue of certain soft-drink bottles. *See also* **Mie scattering, propagation medium, scatter.**

scattering coefficient: A measure of the attenuation that is caused by the scattering of radiant or sound energy as a result of its passage through a material medium. *Note 1:* The scattering coefficient usually is expressed in units of reciprocal distance and appears in the exponent of the expression that defines Bouger's law. *Note 2:* The sum of the scattering coefficient and the absorption coefficient is the attenuation coefficient. *Note 3:* Both parts of the attenuation coefficient, i.e., the scattering part and the absorption part, depend on impurities and the irregularity of the molecular structure of the intrinsic material that constitutes the propagation medium. *See also* **absorption coefficient, attenuation coefficient, attenuation constant, phase constant, Bouger's law, propagation constant.**

scattering cross section (SCS): The area on an incident wavefront, at a reflecting surface or medium, such as an object in space, through which will pass an amount of power that, if it were isotropically scattered from that point, would produce the same power at a given receiver as is actually provided by the entire reflecting surface. *Note:* The scattering cross section may be calculated by considering that when an electromagnetic wave is reflected from an object in space and a portion of the reflected power reaches a receiving antenna, it is a simple matter to calculate the power of a hypothetical isotropic transmitter located at the object in space that would produce the same irradiance at the receiving antenna. The power per unit area, i.e., the irradiance, of the original incident signal is known. Thus, it is again a simple matter to calculate the cross-sectional area at the object location through which the same amount of power passes. This area is the scattering cross section. *Synonym* **echo area.** *See also* **irradiance.**

scattering loss: In the propagation of electromagnetic waves in a material medium, such as lightwaves in an optical fiber or radio waves in the atmosphere, the power loss that is caused by (a) scattering of a transmitted wave within the material medium or (b) scattering of a reflected wave from the rough surface of a material medium, such as ground or pebbled glass. *See also* **loss, reflection loss, scatter, transmission loss.**

scattering method: *See* **transverse scattering method.**

scatter load: To place the control sections of a load module in nonadjacent areas of main storage, usually by using a loader. *See also* **loader, load module, main storage.**

scatter propagation: In electromagnetic wave propagation, the propagation achieved by means of scattering, both forward scattering and backward scattering, caused by irregularities or discontinuities in the ionization densities of the ionosphere and the physical properties of the troposphere. *Note:* Most of the irregularities and the discontinuities are caused by turbulence. *See also* **backscatter, backscattering, forward scatter, scattering.**

SCC: specialized common carrier.

scene cut: A change in video imagery where adjacent frames are not highly correlated, giving rise to an interruption or an irregularity to the displayed motion.

scene cut response: In video systems, the apparent impairments to the imagery associated with a scene cut.

scene of action communications: Communications methods, systems, and equipment used for the transmission of messages to, from, or at a specific location for the purpose of reporting special events, controlling operations, or describing activities at the location, such as air-sea rescue operations, an insurrection, disaster area events, or a sporting event.

scene of air-sea rescue frequency: A frequency used for communications between aircraft, surface vessels, and submarines in action at the scene of an air-sea rescue operation.

schedule: *See* **ship broadcast schedule, ship to shore schedule.**

scheduled time: The time in which a specific functional unit is intended to perform the function for which it was designed and built. *See also* **functional unit.**

scheduler: A computer program that controls the flow of jobs, such as scheduling, initiating, and terminating them. *See also* **computer program, job.**

schematic: 1. Pertaining to a diagram, drawing, or sketch that details the elements of a system, such as the elements of an electrical circuit or the elements of a logic diagram for a computer or communications system. **2.** A diagram, drawing, or sketch that details the elements of a system, such as the elements of an electrical circuit or the elements of a logic diagram for a computer or communications system. *See also* **circuit, system.**

Schottky effect: *Synonym* **shot noise.**

Schottky noise: *Synonym* **shot noise.**

science: *See* **computer science.**

scientific-medical radio frequency equipment: *See* **industrial-scientific-medical radio frequency equipment.**

scintillation: 1. In electromagnetic wave propagation, a small random fluctuation of the received electric field strength about its mean value. *Note:* Scintillation effects become more significant as the frequency of the propagating wave increases. **2.** In radar systems, the instability that occurs in a displayed radar echo. *Note:* Scintillation may be caused by (a) variations in propagation conditions in the propagation medium, (b) phase changes brought about by chaff elements, (c) vibration of tracked objects, i.e., targets, and (d) rapid fluctuations in the amplitude of the received signal. **3.** In satellite communications systems, the rapid fluctuations of a received signal amplitude usually caused by ionospheric and tropospheric turbulence. *Note:* Scintillations occur at typical communications satellite frequencies of 4 GHz (gigahertz) and 6 GHz. They are presumed to be caused by very dense and thick irregular layers in the F layer of the ionosphere, although there is no ionospheric absorption in this part of the frequency spectrum. They usually exist only in the early part of the evening. The scattering irregularities distort the phase of the wavefront as the front moves along the propagation path. The signal received by the satellite Earth station fluctuates with time. The scintillations are greatest near the equator and are significant between 30°N and 30°S latitude. They show a very strong diurnal peak about one hour after local sunset and seasonal peaks near the vernal and autumnal equinoxes. The amplitude of the scintillations usually are less than 4 dB with periods of 4 s (seconds) and 6 s between adjacent fades. *See also* **chaff, field intensity, field strength, F layer, ionosphere, propagation medium.**

scissoring: 1. In display systems, removing parts of display elements, display groups, or display images, in the display space on the display surface of a display device, that lie outside a window. **2.** In display systems, removing (a) display elements from display groups or display images and (b) display elements or display groups from display images that lie outside a window. *Synonym* **clipping.** *See also* **display device, display element, display group, display image, display space, display surface, window.**

SCOI: standing communications operating instruction.

scope: A device that displays images on a screen or on the retina of a person's eye, such as an endoscope, fiberscope, oscilloscope, microscope, sniperscope, or telescope. *Note 1:* In radar systems, the scope usually is a cathode ray tube (CRT) that displays the results of radar measurements. In electronic laboratories, the scope is a CRT oscilloscope for viewing signal wave shapes. *Note 2:* The word "scope" is used for brevity. The context should indicate the kind of scope. *See also* **image, screen.**

scramble: 1. In telephony, to make plain text or speech unintelligible to casual interpretation. **2.** In cryptography, to mix characters in a random or pseudorandom fashion so as to render the text unintelligible until it is restored by a reverse process. *Note:* A scrambled message does not provide the security of an encrypted message. *See also* **encryption, key, scrambler.** *Refer to* **Fig. C-2 (121).**

scrambler: A device that (a) transposes or inverts signals or otherwise encodes a message at the transmitter to make the message unintelligible at a receiver not equipped with an appropriately set descrambling device, (b) usually uses a fixed algorithm or mechanism, (c) produces an output signal sequence that is based on an input sequence and a transformation algorithm in such a manner that the output is unintelligible to those who do not have knowledge concerning the nature of the algorithm or the mechanism, and (d) provides communications privacy that is inadequate for U.S. Department of Defense (DoD), Central Intelligence Agency (CIA), and National Security Agency (NSA) security information, i.e., is inadequate for security classified traffic. *See* **data scrambler, descrambler, fiber optic scrambler, mode scrambler.** *See also* **limited protection voice equipment, randomizer.**

scrambling: In communications security, the process of altering, inverting, displacing, and substituting signals, such as line, radio, video, or optical signals, in order to make them unintelligible until they are unscrambled by properly set complementary circuits. *Note:* Scrambling does not provide the security of encryption. *See* **speech scrambling.** *See also* **encryption.**

scratch: In optics, a marking or a tearing of the surface of an optical element, such as a lens or a prism, that looks as though it was caused by a hard, sharp, or rough instrument or substance. *See also* **block reek, crush, runner cut, sleek.**

scratchpad storage: *Synonym* **temporary storage.**

screen: 1. In a telecommunications, computing, or data processing system, to examine entities that are being processed to determine their suitability for further processing. **2.** A nonferrous metallic mesh used to provide electromagnetic shielding. **3.** To reduce undesired electromagnetic signals and noise by enclosing devices in electrostatic or electromagnetic shields. **4.** A viewing surface, such as that of a cathode ray tube (CRT) or liquid crystal display (LCD). *See* **help screen, radar screen, split screen.** *See also* **filter, shield, shielding effectiveness.** *Refer to* **Fig. N-21 (641).**

screen coordinate: In display systems, a device coordinate that is one of the arbitrary coordinates in a coordinate system placed on the display surface of a display device. *Note:* A screen coordinate system might be a Cartesian (*x*-*y*) coordinate grid or a polar coordinate grid on a transparent sheet placed over a cathode ray tube screen. *See also* **device coordinate, display device.**

screening range equation: *See* **self-screening range equation.**

screening router: A router that (a) performs a routing function in accordance with specified criteria or protocols, (b) usually is a basic firewall component, and (c) when associated with a bastion host, provides an effective firewall. *See also* **bastion host, firewall, gateway, router.**

scroll: In a display device, to move the display window of the screen either vertically or horizontally to view the contents of a stored document. *Note:* Scrolling may be continuous or incremental. *See also* **display, window.**

scrolling: In display systems, the vertical (top to bottom or bottom to top) or horizontal (side to side) movement of all or part of a display image contained within a window or otherwise displayed in the display space on the display surface of a display device in such a manner that as new data appear at one edge of the window or display space, old data disappear at the opposite edge. *Note:* In personal computer (PC) systems, scrolling is accomplished in different ways, depending on the application program. In Lotus 1-2-3 Worksheet operations, i.e., spreadsheet operations, scrolling may be accomplished with the scroll lock key, and in WordPerfect operations, scrolling may be accomplished with various keys on the keypad at the right end of the standard keyboard. *See also* **display area, display image, display space, display surface, scroll lock key.**

scroll lock key: On a computer standard keyboard, a key that (a) is marked SCROLL LOCK, (b) when actuated, will cause the depressing of the arrow key to scroll the data on the screen in the direction of the arrow, and (c) when actuated again, will disable this function. *See also* **arrow key, key, keyboard.**

scrubbing: *Synonym* **sanitizing.**

SDLC: synchronous data link control.

SDM: space division multiplex, space division multiplexer, space division multiplexing.

SE: software engineering, systems engineering.

seamless image substitution: Image substitution in which there is little or no noticeable, distinctive, or unnatural boundary between (a) the remaining part of the original image and (b) the substituted image, i.e., the substitution is close to perfect, including the changing perspective during panning.

seamless network: A network that (a) consists of an interconnection of existing networks, (b) has transparent interfaces, gateways, routers, ports, and bridges, (c) may use any transmission medium, such as wire, fiber optic, or radio, (d) permits easy access from anywhere to anywhere at any time, and (e) handles data in any form, such as voice, data, video, digital, analog, or graphic. *See also* **network, transparency, transparent.**

search: *See* **binary search, dichotomizing search, electronic search, global search, tree search.**

search and lock jammer: A jammer that (a) has a transmitter that is automatically controlled by a receiver-transmitter tuning device, (b) when a transmission is detected, automatically tunes in and locks on to the frequency of that transmitter, (c) turns on the jammer transmitter, (d) after a certain length of time, ceases jamming and unlocks the receiver, and (e) commences a search for a frequency to jam. *See also* **jammer, jamming.**

search and rescue beacon (SARBE): 1. A beacon that is used by survivors of a distress situation to guide search and rescue units to the distress scene. **2.** A beacon that is used by search and rescue units as an aid in the conduct of search and rescue operations. *See also* **beacon.**

search and rescue beacon frequency: A frequency used by search and rescue beacons (SARBE). *Note:* One frequency used by search and rescue beacons (SARBE) and for search and rescue and homing systems (SARAH) is 234.5 MHz (megahertz).

search and rescue communications: Communications methods, systems, and equipment used for transmission of messages that (a) contain information about search and rescue operations, (b) are sent to or from search and rescue units or survivors, and (c) are primarily used for directing, coordinating, and supporting search and rescue operations.

search and rescue frequency: A radio frequency (rf) that is specially selected for and useful to mobile stations during search and rescue operations. *Note:* Examples of search and rescue frequencies are (a) 3023.5 kHz (kilohertz) and 5880 kHz for aeronautical mobile stations, (b) 8364 kHz for survival craft stations, and (c) 123.1 MHz (megahertz) for non-VHF equipped stations.

search jammer: *See* **automatic search jammer.**

searchlighting: In radar systems, orienting a radar beam such that the beam is centered on the object, i.e., on the target, being tracked in order to provide (a) a higher concentration of power for increased burn-through range and (b) a higher data rate from the object. *See* **burn-through range.**

searchlight sonar: Sonar that develops a sound wave in the form of a beam that resembles the beam produced by a radio parabolic antenna, spotlight, or flashlight. *Synonym* **single pulse carrier wave sonar.** *See also* **parabolic antenna, sonar.**

search radar: Radar that is specially designed to identify the existence of an object, i.e., a target, rather than accurately determine its range or velocity.

search receiver: A receiver that can be tuned over a relatively wide frequency band in order to detect, identify, or measure electromagnetic signals.

search routing: *See* **flood search routing.**

search time: In data processing systems, the time required for a storage device to locate a particular data element, record, or file.

sea rescue frequency: *See* **scene of air-sea rescue frequency.**

sea return: In radar systems operations, an electromagnetic wave or echo that is obtained by reflection from the sea. *See also* **echo.**

season: *See* **busy season.**

sea surveillance system: A system that (a) collects, reports, correlates, and presents information that supports and is derived from the task of sea surveillance and (b) is based primarily on monitoring radio and video transmissions.

sec: second.

SECAM: système electronique couleur avec memoire (electronic color system with memory).

second: The SI unit (Système International d'Unités), i.e., the International System of Units (SI) unit, for a time interval equal to 9,192,631,770 periods of the radiation corresponding to the transition between the two hyperfine levels of the ground state of the cesium-133 atom. *Note:* The mean solar second is 1/31,556,925.9747 of the tropical year for 1900 January 0 at 1200 hours. The sidereal second is 1/86,400 of a sidereal day. The January 0 comes from the fact that the Julian day begins at noon, usually called "1200 hours," when the sun is at its zenith. January 1 would imply the Gregorian calendar. December 31 would be the wrong day. *Synonym* **cesium clock second.** *See* **bits per second, call•second, leap second, lumen•second.** *See also* **cesium clock, cesium standard, Coordinated Universal Time, DoD master clock, International Atomic Time, sidereal day, weighted standard work second.**

secondary: In a communications system, the part of a data station that (a) executes data link control functions, such as supporting the main control functions of the data link, generating commands for transmission, interpreting responses, and performing specific functions at the data link level, such as (i) initialization of control signaling, (ii) organization of data flow, and (iii) error control, as required by the primary, and (b) in high level data link control (HDLC) procedures, usually supports the interpretation of received commands and generation of responses for transmission. *See also* **high-level data link control, primary.** *Refer to* **Fig. T-8 (1024).**

secondary aeronautical station: A station that (a) may be a land, ship, or mobile station, (b) is not a net control, guard, or standby station, (c) is in an aeronautical mobile service net, and (d) provides a communications service to aircraft stations.

secondary axis: In movable antenna systems, i.e., rotatable antenna systems, the axis that has the primary axis, rather than the ground, as a datum. *See also* **primary axis.**

secondary channel: In a system in which two channels share a common interface, the channel that has a lower data signaling rate (DSR) capacity than the other channel, i.e., primary channel. *See also* **auxiliary channel, channel, primary channel.**

secondary coating: A coating applied to the primary coating of an optical fiber for added protection during handling, cabling, installation, and use.

secondary control point: In high level data link control (HDLC) procedures, a point in a data network at which a secondary is located. *See also* **high level data link control, secondary.**

secondary emission: **1.** Particles or radiation, such as photons and alpha particles, Compton recoil electrons, delta rays, secondary cosmic rays, and secondary electrons, that are produced by the action of primary radiation on matter. **2.** The emission of electrons from a material caused by bombardment of the material by electrons that have been emitted by another material that (a) is exposed to electromagnetic radiation and (b) demonstrates photoemissivity, i.e., the photoemissive effect. *See also* **photoemissive effect, photoemissivity, photomultiplier, primary emission, primary radiation.**

secondary frequency (SF): **1.** In a radiotelephone network, a frequency that is assigned to an aircraft station as a second choice for air-ground communications. **2.** A frequency that is assigned for use on a particular radio communications circuit when the primary frequency becomes unusable. *See also* **primary frequency.**

secondary frequency standard: A frequency standard that (a) does not have sufficient inherent accuracy and (b) must be calibrated against a primary frequency standard. *Note:* Secondary standards include crystal oscillator and rubidium standards. A crystal oscillator depends for its frequency on its physical dimensions, which vary with fabrication tolerances and environmental conditions. Though a rubidium standard uses atomic transitions, it is not a primary standard because (a) it is in the form of a gas cell through which an optical signal is passed, (b) the gas pressure varies from various causes, such as temperature variations, and (c) the concentrations of the required buffer gases vary, all of which circumstances cause frequency variations, i.e., deviations from a reference, such as a cesium-133 standard. *See also* **primary frequency standard, primary time standard.**

secondary group: **1.** *Synonym* **supergroup.** **2.** *See* **group.**

secondary radar (SR): **1.** A radiodetermination system based on the comparison of reference signals with radio signals retransmitted from the position to be determined. **2.** Radar that (a) uses active signals that are transmitted by an object and (b) does not rely on reflection of its own signals. *See also* **primary radar, radar.**

secondary radiation: Radiation, such as photons, delta rays, X rays, gamma rays, and secondary cosmic rays, that is produced by the action of primary radiation on matter. *See also* **primary radiation, secondary emission.**

secondary radiation source: A source that emits, or appears to emit, electromagnetic waves but does not generate them. *Note:* Examples of secondary sources are images in a mirror, illuminated parabolic reflectors, illuminated radar targets, and the Moon. *See also* **primary radiation source.**

secondary service area: For a broadcast station, the service area (a) that is served by the skywave, (b) that is not subject to objectionable interference, and (c) in which the broadcast signal is subject to intermittent variations in strength.

secondary ship-shore station: A shore-based ship-shore station that (a) uses and guards frequencies other than those frequencies that are guarded by primary stations, and (b) is designed to provide communications with ship stations. *See also* **guard, guard ship, local ship-shore station, primary ship-shore station, ship station.**

secondary spectrum: In an achromatic lens, the residual chromatic aberration that (a) is primarily the longitudinal chromatic aberration, (b) is unlike the primary spectrum, and (c) causes the image formed in one particular color to lie nearest the lens, the images in all other colors being formed behind the first at distances that increase sharply toward both ends of the useful wavelength spectrum. *See also* **achromatic lens, chromatic aberration, primary spectrum.**

secondary station: 1. In a communications network, a station that (a) is responsible for performing unbalanced link level operations as instructed by the primary station, (b) interprets received commands, and (c) generates responses. *See also* **backward supervision, control station, data communications, link, master station, primary station, slave station, tributary station. 2.** In an aeronautical air-ground communications system, a station that is destined or planned to serve as the next guard station during a particular flight of an aircraft station. *Note:* For a particular flight, all stations, other than the primary guard station and the secondary station, are considered as standby stations. *Synonym* **en route guard station.** *See also* **guard station, primary guard, primary guard station, standby station.**

secondary status: In high level data link control (HDLC) operations, the current condition of a secondary station with respect to processing the series of commands that are received from the primary station. *See also* **primary station.**

secondary storage: *Synonym* **auxiliary storage.**

secondary time standard: A time standard that requires periodic calibration against a primary time standard. *See also* **primary time standard.**

secondary winding: 1. The winding in a transformer that is connected to a power load. **2.** In an electrical transformer, the winding that contains the output alternating current (ac). *Synonym in this sense:* **secondary.** *See also* **transformer.**

second principal point: In an optical system, the second significant point, i.e., definable point, in the system, counting in the direction established in going from the object toward the objective lens. *Note:* The center of the objective lens might be the first principal point, and the principal focus point of the objective lens might be the second principal point. *See also* **first principal point, object, objective lens, principal focus point.**

second window: In silica-base glasses, such as those used in optical fibers, the spectral window at approximately 1.3 μm (microns). *Note 1:* The second window is the minimum dispersion window in silica-based glasses. *Note 2:* The standard wavelength for fiber optic systems is 1.31 μm (microns). *See also* **spectral window.**

SECORD: secure voice cord board.

SECTEL: secure telephone.

section: *See* **fiber optic station/regenerator section.**

sector: A predetermined, addressable angular part of a track or a band on a magnetic drum or a magnetic disk. *See* **dead sector.**

sectorial harmonic function: A spherical harmonic function in which the sign is consistent along lines of constant longitude, but the magnitudes change with latitude. *See also* **spherical harmonic function.**

sectoring: *See* **soft sectoring.**

sector scan: 1. In a moving antenna, a scan in which the antenna oscillates through a specific angle. **2.** In a phased array antenna, a scan in which the emitted beam is made to scan within a specific angle.

sector sweep: To scan a specific solid angle vertically or horizontally with a transmitting or receiving antenna. *Note:* An example of a sector scan is to conduct a radar or direction finding (DF) search only in a small angle

phased array antenna, a scan in which the emitted beam is made to scan within a specific angle.

sector sweep: To scan a specific solid angle vertically or horizontally with a transmitting or receiving antenna. *Note:* An example of a sector scan is to conduct a radar or direction finding (DF) search only in a small angle for various reasons, such as (a) to prevent interference caused by search in the unswept sectors, (b) conserve power, (c) save time, or (d) avoid searching sectors in which there is no interest.

secure: Pertaining to the prevention of unauthorized use, access, operation, or interpretation, whereby, for example (a) secure messages cannot be interpreted or understood by unauthorized persons, (b) secure equipment has been fastened, stored, or turned off to prevent access, use, damage, or alteration by unauthorized persons or natural forces, (c) secure call signs cannot be linked by unauthorized persons to plain language address designators, and (d) secure circuits or networks cannot be accessed by unauthorized persons or systems. *Note:* A channel, circuit, or net can be secured by the use of online cryptoequipment for telegraph, data, facsimile, or voice operation. A circuit can be secured by closing it down. An area can be secured by closing and locking all doors and hatches leading to it. *See also* **communications security, secure communications, security.**

secure-access card: A smart card that (a) ensures secure read-write functions in memory, (b) may be a reusable or a nonreusable card, and (c) may be a token card, such as a telephone, movie, parking, identification, subscription, transaction, or pay-per-view television card.

secure communications: Communications that effectively are secured (a) by communications security (COMSEC) equipment or (b) by protected distribution systems. *Note:* Secure communications prevent hostile exploitation of systems and the data in them. *See also* **communications security equipment, firewall principle, protected distribution system.** *Refer to* **Figs. S-9 through S-11.**

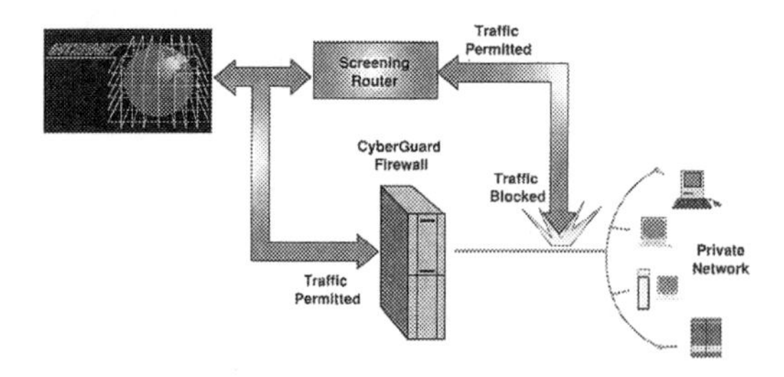

Fig. S-9. The interconnection of a firewall and a screening router to provide for a screened-host easy-to-implement gateway for effecting highly **secure communications.** (Courtesy Harris Computer Systems Corporation.)

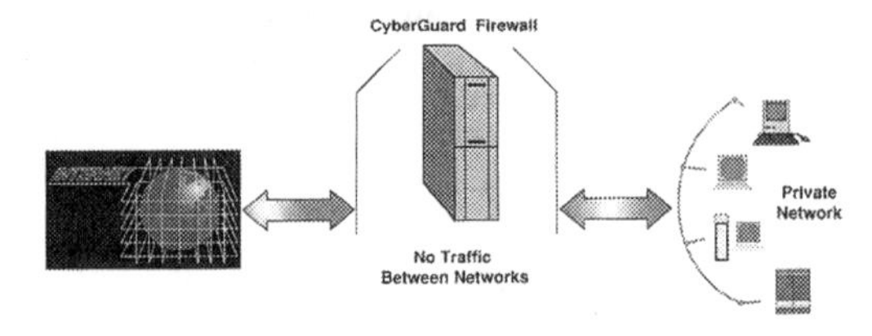

Fig. S-10. The interconnection of a firewall and a private network that provides (a) a dual-homed gateway, (b) a single point of control, (c) a small zone of risk, and (d) a **secure communications** system. (Courtesy Harris Computer Systems Corporation.)

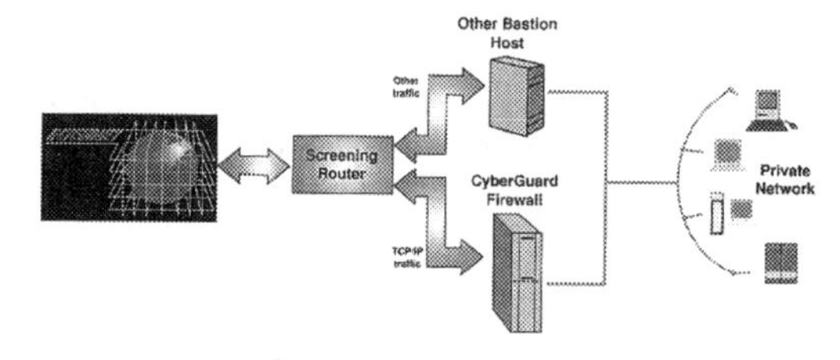

Fig. S-11. The interconnection of a firewall, screening router, and another bastion host in the form of a screened subnet to provide very **secure communications** while supporting gradual transition to the Internet. (Courtesy Harris Computer Systems Corporation.)

secure line: In optical or wire transmission systems, a line that (a) carries information-bearing signals that are coded so that they cannot be interpreted except by persons having knowledge of the code, (b) is protected from unauthorized physical access, and (c) is protected from damage by environmental conditions. *See also* **protected line.**

secure mobile antijam reliable tactical terminal (SMART-T): An extremely high frequency satellite antenna and terminal that (a) may be mounted on a heavy high-mobility multipurpose wheeled vehicle and (b) is designed for tactical operations. *Note:* An example of a secure mobile antijam reliable tactical terminal (SMART-T) is one developed by the Harris Corporation

with a prearranged code. *Note:* Telephones designated as "secure telephones" are not suitable for security classified information. *See also* **encryption.**

secure telephone service: *See* **Federal Secure Telephone Service.**

secure telephone unit (STU): A U.S. government-approved telecommunications terminal that protects the transmission of sensitive or classified information in voice, data, and facsimile systems. *See also* **secure communications.**

secure transmission: 1. In transmission security, transmission of data that can be interpreted only by the addressee. **2.** In spread spectrum systems, the transmission of binary coded sequences that represent information that can be recovered only by persons or systems that have the proper key for the spread spectrum code sequence generator, i.e., have a synchronized code sequence generator that generates a code sequence identical to that used for transmission. *Refer to* **Fig. S-9.**

secure voice: *Synonym* **encrypted voice.**

secure voice communications system: *See* **automatic secure voice communications system.**

secure voice cord board (SECORD): A desk-mounted patch panel that provides the capability for controlling (a) sixteen 50-kbps (kilobits per second) wideband or sixteen 2400-bps (bits per second) narrowband user lines and (b) five narrowband trunks to Automatic Voice Network (AUTOVON) or other Defense Data Network (DDN) narrowband facilities. *See also* **patch bay, wideband.**

security: 1. The condition achieved when designated information, materiel, personnel, activities, and installations are protected against espionage, sabotage, subversion, and terrorism, as well as against loss or unauthorized disclosure, including the measures necessary to achieve this condition and the organizations responsible for implementing the measures. **2.** Measures taken by a military unit, an activity, or an installation to protect itself against all acts designed to, or that may, impair its effectiveness. **3.** The condition that results from the establishment and the maintenance of protective measures that ensure a state of inviolability from hostile acts or influences. **4.** In the handling of security classified matter, the condition that prevents unauthorized persons from accessing information that is safeguarded in the interests of national security. **5.** The condition that results from measures taken to protect information, personnel, systems, components, and equipment from unauthorized persons, acts, or influences. *See* **add-on security, administrative security, automated information systems security, automatic secure voice communications, communications security, computer security, data security, electronic emission security, electronic security, electronics security, emanation security, end-to-end security, environmental security, information security, long-term security, operations security, personnel security, physical security, ship radio transmission security, short-term security, signal security, telephone circuit security, teleprocessing security, traffic flow security, transmission security.** *See also* **communications security material, firewall principle, security management.**

security assessment: *See* **operations security assessment.**

security audit: In communications, computer, data processing, and control systems, an audit that is designed to examine data security procedures for the purpose of evaluating their adequacy and compliance with established policy. *See* **external security audit, internal security audit.** *See also* **audit, data security.**

security classified: Pertaining to information that must be handled according to specific rules in order to prevent compromise, i.e., prevent unauthorized disclosure. *Note:* In the U.S. Department of Defense (DoD), one form of security classification of information is to categorize information into unclassified, confidential, secret, and top secret information. Unauthorized disclosure of confidential, secret, and top secret information affects the security of the United States.

security custodian: In communications security, the person responsible for the custody, accounting, handling, safeguarding, and required destruction of communications security material that is received from an issuing authority and is required to be kept on hand for use. *See also* **communications security material.**

security equipment: *See* **communications security equipment.**

security filter: 1. In communications security, the hardware, firmware, or software used to prevent access to specified data by unauthorized persons or systems, such as by (a) preventing transmission, (b) preventing forwarding messages over unprotected lines or circuits, and (c) requiring special codes for access to read-only files. **2.** In communications, computer, data processing, and control systems, a trusted subsystem that enforces security policy on the data that pass through them. *Refer to* **Fig. S-9.**

security information: *See* **communications security information.**

security information exchange: *See* **Defense Network Security Information Exchange.**

security initiative: *See* **multilevel information systems security initiative.**

security kernel: 1. In computer and communications security, the central part of computer or communications system hardware, firmware, and software that implements the basic security procedures for controlling access to system resources. **2.** A self-contained, usually small, collection of key security-related statements that (a) works as a part of an operating system to prevent unauthorized access to, or use of, the system and (b) contains criteria that must be met before specified computer programs can be accessed. **3.** The hardware, software, and firmware elements of a trusted computing base that implement the reference monitor concept. *See also* **communications security, key, operating system, statement.**

security management: In communications, computer, data processing, and control systems management, the set of functions (a) that protect systems, facilities, and equipment, such as communications networks and data processing systems, from unauthorized access by persons, acts, or influences, and (b) that includes many subfunctions, such as (i) creating, deleting, and controlling security services and mechanisms, (ii) distributing security-relevant information, (iii) reporting security-relevant events, (iv) controlling the distribution of cryptographic key, and (v) authorizing access, rights, and privileges. *See also* **security.** *Refer to* **Fig. S-9.**

security material: *See* **communications security material.**

security measure: *See* **add-on security measure.**

security mode: *See* **controlled security mode.**

security optical fiber: *Synonym* **fiber optic trap.**

security option: *See* **revised interconnection protocol security option.**

security policy: *See* **communications security policy.**

security profile: *See* **communications security profile.**

SEDFB laser: Surface Emitting Distributed Feedback laser.

seed: In transparent media, such as clear glass and plastic, a gaseous inclusion with an extremely small diameter. *See* **scattered seed.** *See also* **heavy seeding.**

seeding: *See* **error seeding, heavy seeding.**

seek: To selectively position the access mechanism of a direct access storage device.

seek time: The time required for the access arm of a direct access storage device to be positioned on the appropriate track. *Synonym* **positioning time.**

seepage: In communications, computer, data processing, and control systems, data that (a) are presumed to be controlled by security measures and (b) gradually, in piecemeal fashion, flow to unauthorized persons. *See also* **security, security management.**

segment: 1. In interactive display systems, a basic sequence of display commands or instructions in a display file that (a) enables the transmission and the display of display elements, display groups, or display images in the display space on the display surface of a display device, and (b) enables their manipulation through the performance of operations upon the elements, groups, and images. *See also* **display device, display element, display group, display image, display space, display surface. 2.** To divide a computer program, routine, or subroutine into identifiable and usually addressable parts. *See* **ground segment, overlay segment, program segment, root segment, space segment. 3.** In distributed queue dual bus (DQDB) networks, a protocol data unit (PDU) that (a) consists of 52 octets transferred between distributed queue dual bus (DQDB) layer peer entities as the information payload of a slot, (b) contains a header of 4 octets and a payload of 48 octets, and (c) is either a prearbitrated segment or a queued arbitrated segment. *See also* **distributed queue dual bus (DQDB) network, protocol data unit. 4.** In satellite communications, a ground segment. *See also* **ground segment.**

segmented encoding: Encoding in which an approximation to a smooth wave is obtained by a number of sequential linear segments. *Synonym* **piecewise linear encoding.** *See also* **code, encoding law.**

segregation: 1. *Synonym* **privacy. 2.** *See* **routing line segregation.**

seismic wave: A movement, such as a shock or a tremor, of a local Earth area that (a) travels through or across the surface of the Earth and (b) usually is initiated by a sudden physical disturbance, such as an earthquake, volcanic eruption, tidal wave, or fault slippage, that causes a movement of large masses of earth before energy contained in the traveling wave is expended sufficiently for it to subside.

seized condition: *See* **trunk seized condition.**

seizing: The temporary dedication of various parts of a communications system to a specific use, usually in response to a user request for service. *Note:* The parts seized may be automatically connected, such as by direct distance dialing (DDD), or may require human intervention. *See also* **dedicated circuit, override, peg count, precedence, preemption.**

seizure: *See* **line seizure.**

seizure signal: In telephone systems, a signal that (a) is used by the calling end of a trunk or a line to indicate a request for service and (b) may be used in various systems, such as telephone, facsimile, telegraph, tape relay, and radio systems. *See also* **access request, call control signal, off-hook signal, signal.**

selcall: selective calling.

select: In a communications system, to request one or more data stations to receive data, the stations usually being in a net or connected by a multipoint connection. *See also* **data station, multipoint connection, net.**

selecting: In a communications system, the requesting of one or more data stations to receive data, the stations usually being in a net or connected by a multipoint connection. *See also* **data station, multipoint connection, net.**

selection: 1. In telegraph receiver operation, the primary operation of telegraph translation in which the control character or the data character to be printed is chosen manually or automatically by interpreting the received signal. *Note:* Examples of selection are (a) the use of a received set of binary digits to select a letter from an alphabet and (b) a person, such as a telegraph operator, selecting a letter from an alphabet or a word from a vocabulary upon hearing a Morse code sequence that corresponds to the letter. *See also* **translation. 2.** The choosing of a line, circuit, set of lines, or set of circuits, the choice being based on the satisfaction of certain conditions or criteria. *See* **automatic route selection, packet length selection.** *See also* **call selection signal.**

selection character: *See* **end of selection character.**

selection position: *Synonym* **decision instant.**

selection signal: *See* **call selection signal.**

selection signal code: *See* **network multiple selection signal code.**

selection sort: A sort in which (a) the items of a set are examined to find items that meet certain criteria, (b) each item that meets the criteria is removed from the unsorted set and placed in the set of sorted items, and (c) the process is repeated until all items are in the sorted set. *See* **repeated selection sort.**

selection switch: *Synonym* **core rope storage.**

selection time: *See* **call selection time, postselection time.**

selective absorption: 1. In the propagation of radio waves through a material medium, absorption such that the medium absorbs all the frequencies contained in the wave or the beam of electromagnetic radiation except those that it reflects or transmits. *Note:* Some substances are transparent to waves of certain frequencies, allowing them to be transmitted, while absorbing waves of other frequencies. **2.** In optical systems, the absorption of all frequencies in a light ray except those that it transmits. *Note:* Some reflecting surfaces will absorb light of certain frequencies and reflect others. The color of a transparent object is the color that it transmits. The color of an opaque object is the color that it reflects, and all other colors are absorbed. Absorbed colors are not seen. The color that an opaque object reflects must be in the incident light, or else the object will appear black. *See also* **selective transmission.**

selective calling (selcall): 1. The ability of a transmitting station to specify which of several stations on the same line is to receive a message. **2.** Calling in which a transmitting station specifies which of several stations on the same line or net is to receive a message. *See* **digital selective calling.** *See also* **call, calling, conference call, multiaddress calling facility, party line, queue traffic, selective ringing.**

selective cavity: A radio frequency (rf) resonant cavity that (a) filters all but a narrow band of frequencies, (b) usually has an adjustable spatial width to obtain maximum selectivity for a specific frequency, i.e., wavelength, and (c) has a minimum spatial width of one-fourth of a wavelength of the operating frequency.

selective combiner: A diversity combiner that selects and uses the diversity signal that has the most desirable characteristics. *Note:* The selection process uses signal characteristics, such as signal amplitude, signal to noise ratio (S/N), or transition characteristics. *See also* **diversity combiner, maximal ratio combiner, predetection combining.**

selective dump: A dump of data only in certain specified storage locations or storage areas. *See also* **dump.**

selective fading: Fading in which the components of the received radio signal fluctuate, i.e., fade, independently of each other. *Note:* Selective fading may be caused by selective absorption and selective transmission. *See also* **fading, flat fading, flutter, selective absorption, selective transmission.**

selective jamming: Jamming in which (a) only certain frequencies are selected for jamming, (b) only certain transmitters are selected as victim emitters, (c) only certain sectors are selected for jamming, or (d) any combination of these. *See also* **electronic warfare, jammer, jamming, victim emitter.**

selective ringing: In a party line, ringing only the desired user instrument. *Note:* Without selective ringing, all the instruments on the party line will ring at the same time, selection being made by the number of rings, the spacing of rings, or the combination of long and short rings. *See also* **selective calling.**

selective transmission: 1. In the propagation of electromagnetic waves through a material medium, transmission such that the medium transmits all the frequencies contained in the wave or the beam of electromagnetic radiation except those that it reflects or absorbs. *Note:* Some substances are transparent to waves of certain frequencies, allowing them to be transmitted, while absorbing waves of other frequencies. **2.** In optical systems, the transmission of all frequencies in a light ray except those that it reflects or absorbs. *Note:* Some reflecting surfaces will absorb light of certain frequencies and reflect others. The color of a transparent object is the color it transmits. The color of an opaque object is the color it reflects. Absorbed colors are not seen. The color that an opaque object reflects must be in the incident light, or else the object will appear black. *See also* **selective absorption.**

selective transmittance: The variation of transmittance as a function of the wavelength of electromagnetic radiation transmitted through a substance. *See also* **transmittance.**

selective waveguide: *See* **frequency-selective waveguide.**

selectivity: A measure of the ability of a receiver to discriminate between a wanted signal on one frequency and unwanted signals on other frequencies. *See* **adjacent channel selectivity, channel selectivity, high selectivity.** *See also* **discriminator, frequency.**

selector: In an automatic telephone exchange, a device that establishes a request for one or more data stations to receive data. *See* **group selector.**

selector pen: *Synonym* **light pen.**

select signal: *See* **proceed to select signal.**

SELED: surface-emitting light-emitting diode.

self-adapting computer: 1. A computer that has the ability to change its performance or its capability in response to environmental changes, such as to run a different program if the weather changes. **2.** A computer that will modify itself in order to maintain continued unchanged performance in the face of environmental changes, such as a computer that will switch to alternate power sources and circuits in the face of rising temperatures so as not to allow a programmed course of action to be altered because of the rising temperature.

self-adjusting: In fiber optic communications networks, pertaining to networks in which (a) an adequate number of redundant fiber optic links are installed to accommodate optical signal streams in both directions, (b) automatic alternate routing is provided if a link fails, thereby eliminating the necessity of repairing fiber optic cables before service can be restored, and (c) a digital access and fiber optic cross-connect system usually is used to reroute optical signals. *Synonym* **self-healing.**

self-assembled monolayer: *See* **coplanar self-assembled monolayer.**

self-authentication: Authentication in which (a) a transmitting station, i.e., a calling station, establishes its own validity without the participation of the receiving station, i.e., the called station, (b) the calling station establishes its own authenticity, (c) the called station is not required to challenge the calling station, and (d) use is made of the technique only when one-time authentication systems are used to derive the authentication. *See also* **authentication, authenticator.**

self-checking code: *Synonym* **error-detecting code.**

self-delineating block: A block in which a bit pattern or a flag identifies the beginning or the end of the block, i.e., identifies the block end points. *See also* **block.**

self-demarcating code: A code in which (a) the symbols are so selected and arranged that the occurrence of false combinations by interaction of segments from two successive code groups is prevented, (b) data delimiters, i.e., data separators, between code groups are not needed, and (c) redundancy is increased. *See also* **code group, data delimiter, redundancy.**

self-evident code: A code in which the entries in the table of correspondences for encoding and decoding are fairly obvious in a particular language. *Note:* Examples of self-evident codes are brevity codes and mnemonic codes. *See also* **brevity code, mnemonic code.**

self-focusing optical fiber: An optical fiber that (a) is capable of focusing the lightwaves it propagates and its output at one or more points by precision control of geometry, refractive index profile, and other parameters at a given operating wavelength, (b) is wavelength-selective, and (c) in the form of optical fibers, rods, and lenses, is used in optical integrated circuits (OICs) and other fiber optic and optoelectronic components, such as fiber optic connectors, attenuators, switches, multiplexers, and sensors. *See also* **GRIN^R^ rod, SELFOC^R^ lens.**

self-healing: *Synonym* **self-adjusting.**

self-heterodyne: In fiber optics, pertaining to a fiber optic communications network in which the reference oscillator for coherent communications is derived from the same light source as that used for generating the signal that is detected. *See also* **coherence, homodyne.**

self-lasing optical fiber: *Synonym* **optical fiber laser.**

SELFOC^R^ lens: A cylindrically shaped lens, such as a short length of optical fiber, that (a) has a graded refractive index profile parameter that will cause all light rays entering within a given acceptance cone to be focused at one point for a given operating wavelength, (b) collimates and focuses a light beam emanating from a point source, (c) was jointly developed by Nippon Sheet Glass Company and Nippon Electric Company, Ltd., and (d) has a profile parameter of approximately 2, making the refractive index profile close to that of parabola. *See also* **GRIN^R^ rod, self-focusing optical fiber.**

SELFOC^R^ optical fiber: A self-focusing optical fiber produced by Nippon Electric Company and Nippon Sheet Glass Company.

self-organizing computer: A computer that can restructure itself so that it can perform a given job as efficiently as possible. *See also* **job.**

self-protection jamming: Jamming in which the equipment with a jamming capability is carried in the vehicle, aircraft, or ship that is being protected against jamming signals. *See also* **jamming, support jamming.**

self-relative address: A relative address that uses the address of the computer program instruction in which it appears as the base address. *See also* **base address, computer program, instruction, relative address.**

self-repairing circuit: A circuit that can ameliorate the effects of a failure in itself, by itself, i.e., without intervention by a person or another piece of equipment. *Note:* Self-repairing can be accomplished by automatic reconfiguration or by automatic connection to an installed spare part.

self-screening range equation: An equation that combines the two-way radar equation and the one-way radio equation to obtain an expression for the maximum detection range at which a reflected signal, i.e., an echo, that is present at a radar receiving antenna can be masked by a signal that is transmitted from a remote location. *Synonym* **burn-through range equation.** *See also* **burn-through range, two-way radar equation, one-way radio equation.**

self-synchronizing code: A code (a) in which the symbol stream formed by a portion of one code word, or by the overlapped portion of any two adjacent code words, is not a valid code word, (b) that permits the proper framing of transmitted code words, provided that no uncorrected errors occur in the symbol stream, and (c) in which external synchronization is not required. *Note:* Examples of self-synchronizing code words are found in High Level Data Link Control (HDLC) frames and Advanced Data Communication Control Procedure (ADCCP) frames. *See also* **code, comma-free code.**

self-testing circuit: A circuit that uses the data it is transmitting to perform tests on itself to ensure that it is transmitting properly, i.e., that all circuit components are functioning properly.

Sellmeier equation: An equation that expresses the group delay per unit length for an optical fiber, given by the relation $\tau(\lambda) = A + B\lambda^2 + C\lambda^{-2}$ where A, B, and C are empirical fit parameters, or by the relation $\tau(\lambda) = \tau_o + (S_o/8)(\lambda - \lambda_o^2/\lambda)^2$ where τ is the group delay per unit length of fiber, λ is the free-space wavelength, τ_o is the relative group delay minimum at λ_o, λ_o is the zero-dispersion wavelength, and S_o is the zero-dispersion slope. *Note:* These functional forms may be assumed when measured group delay data near the zero-dispersion wavelength of a dispersion-unshifted fiber are numerically fitted for the purposes of calculating the chromatic dispersion coefficient, given by the relation $D(\lambda) = d\tau(\lambda)/d\lambda = S_o\lambda(1 - \lambda_o^4/\lambda^4)/4$ where all the variables are as defined above. *See also* **chromatic dispersion coefficient, group delay.**

SEL® optical fiber: An optical fiber produced by Standard Electric Lorenz.

SEM: scanning electron microscope.

semanteme: In a language, an element that expresses a definite image or idea, such as a word or a part of a word. *Note:* Examples of semantemes are words of a natural language, data elements, data items, and code words to which meaning has been assigned. *Note:* When the same meaning is assigned to two or more different semantemes, communication difficulties do not occur. Difficulties occur in communications because of ambiguities among semantemes, i.e., two or more different meanings are assigned to the same semanteme. The story is told of a misunderstanding that occurred during World War II in the Pacific Theater of Operations. An ammunition supply depot commander reported that the ammunition he had received was faulty. He was instructed to refuse the entire lot, so he sent it back to the mainland with great loss of life and property. The original instruction meant that he should change the fuze assemblies on the rounds of ammunition. In those days "refuse" meant "to reject material" and "to change the fuses." Thus, the new words "fuze" and "refuze" were introduced and used in connection with ordnance. *See also* **morpheme.**

semantics: The relationships among characters or groups of characters and their meanings, independent of the manner of their interpretation and use.

semaphore: A shared variable used to synchronize concurrent processes in computer, communications, or data processing systems by indicating whether a specified action has been completed or a specified event has occurred. *See also* **variable.**

semaphore signaling: Visual signaling in which the positions of the hands, each holding a flag, are used to represent the letters of the alphabet, numerals, punctuation marks, and certain procedure words and prosigns that are used for the transmission of messages. *See also* **flag, Morse flag signaling, procedure word, prosign, wigwag signaling.**

semiactive homing: A missile-to-target guidance system in which the missile tracks radiation that is reflected from the target but is originated by a target-tracking ground or airborne radar. *See also* **active homing, passive homing.**

semiautomated tactical command and control system: In military tactical command and control, a machine-aided system in which human intervention is required in varying degrees to operate the system.

semiautomatic continuous-tape relay switching: Message switching in which incoming messages at a station are received on continuous printed or perforated tape and given onward electrical transmission according to routing requirements or instructions, i.e., routing indicators, through a pushbutton panel connection of a transmitter-distributor into the appropriate output channels. *See also* **routing indicator, semiautomatic relay system, tape relay station, torn tape relay.**

semiautomatic message switching: Message switching in which (a) an attendant performs message routing functions according to information contained in the messages, and (b) other operations are performed by automatic switching equipment.

semiautomatic relay system: A tape relay communications system in which operator intervention is required at a tape relay station to physically handle the message tapes without the rekeying of any of the messages, in order to route them to other stations according to established routing procedures or according to the routing indicators in the messages. *See also* **automatic relay system, routing indicator, semiautomatic continuous tape relay switching, torn tape relay.**

semiautomatic switching system: 1. In telephone systems, a switching system in which (a) attendants receive oral call instructions from users, and (b) the attendants complete the calls by automatic equipment. **2.** At tape relay intermediate stations, the manual routing or rerouting of taped messages without their being rekeyed. *See also* **private branch exchange (PBX).**

semiconductor: A material that (a) has a filled valence electron energy band separated by a finite band gap energy from a higher-energy conduction band, (b) is neither an insulator with large band gaps and small electronic mobilities nor a metallic conductor with extremely high conductivity, i.e., low resistivity, (c) possesses covalent bonding in which electron pairs are held tightly in the region between adjacent atoms or ions, and (d) is often grown in single crystals and sliced or cut, to form diodes or transistors, thus maintaining an ordered crystal lattice structure suitable for accepting positive or negative dopants to create junctions. *Note 1:* Examples of semiconductors are diamond, silicon, germanium, gray tin, tellurium, and selenium. *Note 2:* The band gap energy is the energy required to break an electron out of bonding with an atom or an ion. *See* **intrinsic semiconductor, n-type semiconductor, organic semiconductor, p-type semiconductor.** *See also* **band gap energy, dopant, energy band, junction, valence.**

semiconductor diode laser: *Synonym* **laser diode.**

semiconductor laser: A laser in which the lasing occurs at the junction of n-type and p-type semiconductor materials, such as occurs in an injection laser diode. *See also* **junction laser, laser, lasing, n-type semiconductor, p-type semiconductor, semiconductor.**

semiduplex operation: 1. Circuit operation that is simplex at one end and duplex at the other. *Note:* In radiocommunications, duplex operation and semiduplex operation usually require two frequencies, whereas simplex operation may use either one or two frequencies. **2.** Operation of a communications network in which a base station operates in a duplex mode with a group of remote stations operating in a half-duplex mode. *Note:* The terms "half-duplex" and "simplex" are used differently in wire and radio communications. *See also* **duplex operation, half-duplex operation, simplex operation.**

semielectronic switch: A switch that is composed of electromechanical relays and electronic control circuits. *See also* **electronic switching system, relay.**

semiotics: The branch of science and technology that (a) is devoted to the use of symbols and (b) is often considered to include syntax, semantics, and pragmatics. *See also* **pragmatics, semantics, symbol, syntax.**

sender: In communications systems, a device that (a) accepts address information from a register or routing information from a translator and (b) transmits the proper routing digits to a trunk or to local equipment. *Note:* Sender, register, and translator functions may be performed by a single unit. *See also* **address.**

sending end crossfire: 1. In wire communications systems, such as wire telephone, telegraph, teletype, and facsimile systems, the interfering electrical current in a channel that is caused when one or more adjacent channels transmit from the same end at which the interference is measured. **2.** At a station in a teletypewriter (TTY) system, the interfering current in a channel from one or more adjacent TTY channels transmitting from the station. *See also* **channel, interference, teletypewriter, wireline communications, wireline transmission.**

send-receive teletypewriter: *See* **automatic send-receive teletypewriter, keyboard send-receive teletypewriter.**

send signal: *See* **clear to send signal, request to send signal.**

send terminal: *See* **multiplex baseband send terminal, radio baseband send terminal.**

senior ship radio officer: The ship officer responsible for all matters pertaining to communications on board a ship, including (a) reporting to communications authorities before sailing, (b) establishing radio watch, (c) procuring and securing radio equipment, (d) ensuring that communications equipment is in operating condition, (e) informing the ship master concerning broadcast schedules, (f) establishing radio guard, (g) maintaining the radio log, (g) arranging for emergency communications, and (h) ensuring radio operational discipline. *See also* **secure.**

sense: *See* **heading sense.**

sense multiple access with collision avoidance: *See* **carrier sense multiple access with collision avoidance.**

sense multiple access with collision detection: *See* **carrier sense multiple access with collision detection.**

sensing: *See* **carrier sensing, mark sensing, rotational position sensing.**

sensitive: *See* **touch-sensitive.**

sensitive detector-demodulator: *See* **phase-sensitive detector-demodulator.**

sensitive information: 1. Information that (a) if lost, misused, accessed, or modified without authorization, could adversely affect (i) the national interest, (ii) the conduct of federal programs, or (iii) the privacy to which individuals are entitled under Title 5, U.S.C, Section 552A (The Privacy Act), and (b) has not been specifically authorized under criteria established by Executive Order or an Act of Congress to be kept secret in the interest of national defense or foreign policy. **2.** In computer and communications security, information that (a) requires a degree of protection and (b) must not be made available to anyone except authorized persons. *Note:* Examples of sensitive information are (a) private information, i.e., personal information, (b) proprietary information, (c) defense security information, and (d) national security information. *See also* **communications security.**

sensitivity: In input-output devices, such as amplifiers and communications system receivers, the minimum input signal level required to produce a specified output signal level that meets specified criteria, such as a specified signal to noise ratio (S/N), or to maintain a specified bit error ratio (BER). *Note 1:* The sensitivity may be expressed in power units, such as watts or dBm, or in electric field strength units, such as volts per meter. *Note 2:* The device input impedance must be stipulated. *Note 3:* Sensitivity is not the same as responsivity. *See* **high sensitivity, polarization sensitivity, radar sensitivity, receiver optical sensitivity.** *See also* **discriminator, responsivity, signal to noise ratio, spectral responsivity.**

sensitivity-time control: In radar systems, control that is performed by a circuit that reduces receiver sensitivity for the first few thousand meters of each scan and then gradually restores the sensitivity to normal. *See also* **radar, scan, scanning, sensitivity.**

sensitized medium: A medium that has been treated or coated with a substance that responds to a stimulus, such as pressure, heat, light, or another form of energy. *Note:* Examples of sensitized media are photographic paper, pressure-sensitive paper, or camera photofilm. *See also* **sensitized paper.**

sensitized paper: Paper that has been treated or coated with a substance that responds to a stimulus, such as pressure, heat, light, or another form of energy, so as to produce an image or a recorded copy, thus serving as a data medium. *See also* **data medium, image, recorded copy.**

sensor: 1. A device that (a) responds to a physical stimulus, such as heat, light, sound, pressure, magnetism, or motion, and (b) produces a signal. **2.** A device or a means for extending the natural senses. *Note 1:* Examples of sensors are devices that (a) measure or indicate terrain configurations or the presence of objects, or their motion, by means of energy that is emitted or reflected by the objects, (b) detect and measure physical variables, such as temperature, pressure, speed, humidity, force, weight, vibration, or acceleration, (c) detect and measure the intensity of wave phenomena, such as radio, sound, elastic, and lightwaves, including radioactivity, and (d) measure the presence and the concentration of chemicals, such as pollutants and irritants. *Note 2:* Most sensors are, in effect, somewhat like transducers in that they detect, measure, and indicate the presence of one form of energy by means of another form of energy. However, a transducer converts one form of energy to another without the intent to detect. **3.** Equipment that detects, indicates, or makes a recorded copy of objects and activities by means of energy or particles emitted, reflected, or modified by the objects or the activities. *Note 1:* Examples of sensors are thermometers, gyroscopes, barometers, tachometers, and speedometers. *Note 2:* Optical fibers are being used as the sensing element in various types of sensors. *See* **active sensor, alarm sensor, birefringence sensor, bobbin-wound sensor, coated fiber optic sensor, coated optical fiber sensor, critical angle sensor, critical radius sensor, distributed optical fiber sensor, Fabry-Perot fiber optic sensor, fiber optic acoustic sensor, fiber optic axial alignment sensor, fiber optic axial displacement sensor, fiber optic end-to-end separation sensor, fiber optic longitudinal compression sensor, fiber optic reflective sensor, fiber optic sensor, fiber optic strain-induced sensor, fiber optic tension sensor, fiber optic transverse compression sensor, field disturbance sensor, flat coil optical fiber sensor, frustrated total internal reflection sensor, interferometric sensor, Kerr effect sensor, Mach-Zehnder fiber optic sensor, Michelson fiber optic sensor, microbend sensor, moving grating sensor, optical cavity mirror sensor, passive sensor, photoconductive infrared sensor, photoelastic sensor, Pockels effect sensor, polarization sensor, Sagnac fiber optic sensor, sun sensor.** *See also* **recorded copy, transducer.**

sensor array: A spatial arrangement of sensors in one or more dimensions. *Note:* Examples of sensor arrays are (a) a spatial distribution of receiving antennas, (b) a distribution of optical fibers to form a perimeter warning system, and (c) several hydrophones spaced linearly to improve directive gain as applied to acoustic reception in an ocean acoustic telemetry system, for classifying, localizing, and tracking ships or other acoustic sources at sea. *See also* **directive gain.**

sensor system: *See* **distributed sensor system.**

sentinel: *Synonym* **flag.**

separate channel signaling: *See* **separate channel supervisory signaling.**

separate channel supervisory signaling: Signaling in which the whole or a part of one or more channels in a multichannel system is used to provide for supervisory and control signals for the user message traffic channels. *Note:* The same channels, such as frequency bands or time slots, that are used for supervisory signaling usually are not used for the user message traffic, i.e., mission traffic. *Synonym* **separate channel signaling.** *See also* **channel, common channel signaling, signaling.**

separating character: *Synonym* **information separator.**

separation sensor: *See* **fiber optic end-to-end separation sensor.**

separative sign: In visual communications systems, the semaphore flag positions that are used as a special character to separate words. *Note:* An example of a separative character is ii sent as a group. *See also* **flag, semaphore.**

separator: *See* **address separator, file separator, information separator.**

separator character: *See* **facility request separator character.**

septet: A byte composed of seven binary elements. *Synonym* **seven-bit byte.**

sequence: A set of entities that (a) has a defined beginning and ending, (b) is used for a specific purpose, such as controlling a spread spectrum system, and (c) has members that (i) are temporally or spatially consecutive and (ii) have a particular form or manifestation, such as an electrical pulse or a semaphore flag position. *See* **calling sequence, flag sequence, frame-check sequence, M sequence, pseudorandom sequence, pseudorandom number sequence.** *See also* **automatic sequential connection, bit-sequence independence, spread spectrum code sequence generator.**

sequence code: *See* **linear sequence code.**

sequence generator: *See* **Mersenne code sequence generator, spread spectrum code sequence generator.**

sequence independent: *See* **bit-sequence independence.**

sequence modulation: *See* **direct sequence modulation.**

sequence register: *See* **code sequence register.**

sequence spread spectrum: *See* **direct sequence spread spectrum.**

sequencing: *See* **address message sequencing, packet sequencing.**

sequential: Pertaining to the occurrence of events in time sequence, with no simultaneity, overlap, or concurrence, such as pertaining to a serial sequence of electrical pulses that occur at a point as a function of time. *See* **bit sequential.** *See also* **sequence.**

sequential access: *Synonym* **serial access.**

sequential circuit: 1. A device that has at least one output channel and one or more input channels, all characterized by discrete states, such that the state of each output channel is determined by the previous states of the input channels. **2.** A logic device that (a) has output values at any given instant that depend on a set of input values and the internal state at that instant, which depends on the immediately preceding input values and the preceding internal state, (b) has a storage capability to enable recall of the previous state and input values, and (c) may be regarded as a finite automaton because it can only assume a finite number of internal states. *Synonym* **sequential logic element.** *See also* **combinational circuit.**

sequential computer: A computer in which events, such as computer instruction execution, input, output, processing, and storage transfer operations, occur in time sequence with little or no simultaneity, concurrence, or overlap of events.

sequential connection: *See* **automatic sequential connection.**

sequential control: In computer operations, control in which computer program instructions are executed in an implicit and predetermined sequence until a different explicitly defined sequence is initiated, such as by a jump instruction. *See also* **jump instruction.**

sequential jamming: Jamming in which (a) specific frequencies, or frequency bands, are jammed in succession until all frequencies have been jammed during a given period, and (b) at the end of the period, the jamming sequence is repeated. *See also* **jammer, jamming.**

sequential lobing: In radar echo signal sensing, the sampling of received energy at fixed points located symmetrically about the antenna axis in the focal plane of the echo by sequentially switching to the fixed points to obtain one angular coordinate. *Note:* The error signal is determined by comparing the amplitude of the signals in opposite antenna feeds. Lobing permits optimizing the echo signal while tracking in order to remain on target and at the same time determine the precise radar angle. *See also* **conical scanning, echo, paired lobing, tracking.**

sequential logic element: *Synonym* **sequential circuit.**

sequential operation: A processing mode in which two or more operations are performed one after the other in time sequence rather than concurrently. *Note:* For example, when a source user to destination user connection is completed in a communications network, the connections are serial. Pulses emitted from one end of the connection may then be emitted sequentially, i.e., in time sequence, for the duration of the connection. *See also* **parallel operation.**

sequential process: A series of operations executed one at a time. *See also* **concurrent process.**

sequential transmission: *Synonym* **serial transmission.**

serial: 1. Pertaining to a process in which all events occur one after the other or are positioned in a single line and processed sequentially. *Note:* An example of serial transmission is transmission of the bits of a character according to the International Consultative Committee for Telegraph and Telephone (CCITT) V.25 protocol. **2.** Pertaining to the sequential or consecutive occurrence of two or more related activities in a single device or channel. **3.** Pertaining to the sequential processing of the individual parts of a whole, such as the bits of a character or the characters of a word, using the same facilities for successive parts. *See also* **parallel processing. 4.** An element or a group of elements within a series that is given a numerical or alphabetical designation for convenience in planning, scheduling, and control. *Note:* A convoy may be divided into several serials, each serial with its own identity and control. The links of a chain are spatially serial with respect to one another, but when moving over a sprocket, they are moving in temporal sequence, i.e., serially, with respect to the hub. Similarly, the bits in a bit stream are spatially separated in a waveguide and arrive serially in time with respect to one another at a data terminal. *See also* **bit stream, sequence.**

serial access: 1. Pertaining to the sequential or consecutive transmission of data into or out of a device, such as a computer, transmission line, or storage. **2.** A process in which data are obtained from a storage device or are entered into a storage device in such a way that (a) the process depends on the location of those data, and (b) the process depends on a reference to data previously accessed. *Synonym* **sequential access.** *See also* **access, data communications control procedure, direct access.**

serial access storage: Storage in which the access time depends upon the location of the data and on a reference to data that have been previously accessed. *See also* **access time, direct access storage, storage.**

serial adder: A digital adder that adds both the digits in corresponding digit positions of the augend and the addend, one digit position at a time, usually starting with the least significant position, and propagating carries, if any, to the next higher weighted digit position for inclusion there. *See also* **adder, addend, augend, digital adder, digit position, half-adder, weight.**

serial computer: *Synonym* **serial digital computer.**

serial data stream: A sequence of signals, such as pulses, that (a) are sent either one after the other without interruption, or separately, but not simultaneously, (b) may be either (i) serial by signal element and serial by character, in which case each character is sent element by element, such as bit by bit, or (ii) parallel by signal element and serial by character, in which case all the signal elements of a given character, such as all the bits of a given character, are sent at the same time on parallel channels, but the characters are sent one after the other, i.e., serially. *See also* **sequence.**

serial digital computer: 1. A digital computer in which the bits that compose each digit or each character are handled serially. *Note:* In a serial digital computer, an adder would add the bits of operands serially, i.e., one bit at a time from the least significant bit to the most significant bit. **2.** A digital computer in which the bytes representing characters are handled serially. *Note:* The bits that compose a character, such as a digit, letter, or control character, may be handled either (a) serially, i.e., on one wire, or (b) in parallel on as many wires as there are bits in the bytes, such as on four wires for sexadecimal digits or on eight wires for American Standard Code for Information Interchange (ASCII) characters. *Synonym* **serial computer.** *See also* **digital computer.**

serialization: *Synonym* **parallel to serial conversion.**

serializer: *Synonym* **parallel to serial converter.**

Serial Line Interface Protocol (SLIP): A network protocol that (a) may be installed in a computer, such as a main frame or a personal computer (PC), and (b) permits access to communications services, such as Internet, E-mail, Usenet, Telnet, FTP, and WWW (Worldwide Web). *See also* **protocol.**

serial port: A port (a) through which data are passed serially, i.e., one bit at a time, (b) that requires only one electrical wire or optical fiber to handle a set of bits, such as all the bits of a byte, therefore usually requiring less material for construction, (c) that is slower than a parallel port, which handles all the bits of a byte at the same time, and (d) usually is used with serial devices, such as modems and some printers.

serial signaling rate: *See* **binary serial signaling rate.**

serial sort: A sort of items in a storage unit in which only serial access to the items is possible, that is, the items must (a) be examined in the sequence in which they are stored, such as when they are stored on magnetic tape or on a set of tracks on a magnetic disk, (b) have the sort criteria applied, and (c) be placed in sorted position in the data medium. *Note:* A serial sort is overly time consuming and often requires an abundance of dynamic movement of the data medium because the same data have to be passed over many times.

serial to parallel conversion: Conversion by a device that (a) accepts signal elements in time sequence, i.e., one element at a time, and (b) converts the sequence into a spatial distribution among multiple parallel transmission output channels. *Note:* Each of the parallel output signal sequences requires a separate channel. *See also* **parallel to serial converter, sequence.**

serial to parallel converter: A digital device that accepts a time sequence of signal elements on a single input line and distributes them among multiple parallel output lines. Each of the parallel output signal sequences requires a separate channel. *Synonym* **staticizer.** *See also* **parallel to serial converter.**

serial transmission: The sequential transmission of the signal elements of a group that represents a data unit, such as a byte or a character. *Note:* The signal elements are transmitted in a temporal sequence over a single line rather than simultaneously over two or more lines as occurs in parallel transmission. The sequential elements in either serial or parallel transmission may be transmitted with or without interruption. *Synonym* **sequential transmission.** *See also* **data communications control procedure, data stream, parallel transmission.**

series: *See* **electrochemical series.**

series T junction: A three-port waveguide junction that has an equivalent circuit in which the impedance of the branch waveguide is predominantly in series with the impedance of the main waveguide at the junction.

serrodyne: *See* **optical serrodyne.**

server: 1. In a local area network (LAN) of computer workstations, such as a network of personal computers (PCs), the station that performs the operational, executive, and administrative functions for the network, such as recognizing authorized users, recording usage data, and maintaining available online equipment inventories. **2.** In a local area network (LAN) of interconnected computers, the computer that maintains control over the resources of the network, i.e., the entire system, including the data, hardware, and software. **3.** A network device that provides service to the network users by managing shared resources, particularly in the context of a client-server architecture for a local area network (LAN). *Note:* Examples of servers are (a) a printer service that provides hard copy printing services, such as color printing, so that each terminal on the network need not have its own printer, (b) a file server that provides storage and archival services to augment the limited storage capacity of terminals on the network, and (c) an application server that provides expensive application software services to terminals that use the software infrequently. **4.** An entity that provides access to a network, such as the Internet. *See* **client-server, domain name server, file server, Wide Area Information Server.**

service: 1. A function provided by one entity in support of or on behalf of another entity. **2.** In open system interconnection, a capability of a given layer, and the layers below it, that (a) is provided to the entities of the next higher layer and (b) for a given layer, is provided at the interface between the given layer and the next higher layer. **3.** In a communications system, a function or an operating feature of the system that meets user communications requirements. *See* **absent subscriber service, Advanced Mobile Phone Service, aerodrome control service, aeronautical fixed service, aeronautical mobile satellite (route) service, aeronautical mobile satellite (off-route) service, aeronautical mobile service, aeronautical multicom service, aeronautical radionavigation satellite service, aeronautical radionavigation service, air-ground radiotelephone service, airport control service, amateur satellite service, amateur service, approach control service, Armed Forces Radio Service, basic exchange telecommunications radio service, basic service, broadcasting satellite service, broadcasting service, cellular mobile service, centralized attendant service, Centrex® service, circuit switched data transmission service, commercial communications service, common management information service, common user communications service, common user service, communications satellite service, communications service, complementary network service, consolidated local telecommunications service, courier service, custom local area signaling service, datagram service, data service, data transmission service, dedicated communications service, dedicated service, demand service, direct dialing service, direction finding service, directory service, distribution service, Earth exploration satellite service, eight-hundred service, enhanced service, essential service, extended area service, facsimile service, Federal Secure Telephone Service, five-hundred service, fixed satellite service, fixed service, flat rate service, foreign exchange service, four-wire subset user service, general purpose user service, graphic service, inward wide area telephone service, landline service, land mobile service, leased circuit data transmission service, maritime radio navigation service, measured rate service, meteorological aids service, multiaddress service, nine-hundred service, off-hook service, on-call service, packet-switched transmission service, personal communications service, plain old telephone service, port operations service, private National Security or Emergency Preparedness (NS/EP) telecommunications service, public data transmission service, public switched National Security or Emergency Preparedness (NS/EP) telecommunications service, quality of service, radio astronomy service, radio determination service, radio location service, radionavigation satellite service, radionavigation service, resale service, reserved circuit service, restricted access service, retrieval service, rotary dial service, rural radio service, safety service, seven-hundred service, ship movement service, space operation service, space research service,**

space service, special grade of service, special network service, standard frequency service, switched multimegabit data service, teleaction service, telecommunications service, Telecommunications Service Priority service, teleprinter exchange service, Teletypewriter Exchange Service, teletypewriter service, terrestrial radio service, terrestrial service, transmission service, trunk grade service, universal service, user service, user to user service, voice-data service, wide area telecommunications service, wide area telephone service. *See also* **interface, layer, open system interconnection, Open Systems Interconnection—Reference Model.**

serviceability: 1. The operational state of a functional unit or a system. **2.** In communications, computer, data processing, and control systems, a measure of or the degree of achievement of the best technical performance of a service, together with the capability of rapid restoration of the service in the event of an interruption. *See also* **mean time to service restoral.**

service access point: 1. A physical point at which a circuit may be accessed. **2.** A point in a communications system at which a user may obtain service in person or by means of communications equipment. **3.** In an Open Systems Interconnection—Reference Model (OSI-RM) protocol layer, a point at which a designated service may be obtained. *See also* **access point, Open Systems Interconnection—Reference Model, point.**

service area: *See* **local service area, primary service area.**

service assistance: *See* **dial service assistance.**

service assistance switchboard: A switchboard that (a) is used to provide communications services to users and (b) usually is operated by an attendant. *See* **dial service assistance switchboard.**

service bit: A network overhead bit used for providing a network service, such as a request for repetition of a message or for a numbering sequence. *Note:* A service bit is not a check bit. *See also* **binary digit, check bit, front end processing, overhead bit, overhead information.**

service board: *See* **dial service board.**

service channel: 1. *Synonym* **engineering channel. 2.** *See* **transmission service channel.**

service circuit: *Synonym* **engineering circuit.**

service class: *Synonym* **class of service.**

service connection: *See* **laser service connection.**

service data unit (SDU): In layered communications systems, a set of data that (a) is sent by a user of the services of a given layer and (b) is transmitted to a peer service user semantically unchanged. *See also* **layer, semantics, Open Systems Interconnection—Reference Model, peer, peer group.**

service element: *See* **basic service element.**

service element protocol: *See* **remote operations service element protocol.**

service feature: In a communications network, an initially or additionally specified function of equipment that provides a service to the user in addition to the basic function of the equipment. *Note:* Modern telephone switches may be obtained with a wide variety of service features, such as call forwarding and call waiting. Many earlier service features are now basic services.

service grade: *Synonym* **grade of service.**

service identification: The information that uniquely identifies a National Security or Emergency Preparedness (NS/EP) telecommunications service to the service vendor and the service user. *See also* **National Security or Emergency Preparedness (NS/EP) telecommunications.**

service information: In a digital data transmission system, information that is added to a user message to enable the equipment associated with the message to function correctly, and possibly to provide ancillary facilities.

service integrity: The extent to which a service, such as a communications, computer, or data processing service, once initiated, is continued to be provided without being impaired, i.e., without a reduction in such attributes as quality, value, quantity, or strength.

service message: In communications systems, a brief, concise message between operating, supervisory, maintenance, administrative, or management personnel at communications centers or relay stations that pertains to any aspect of traffic handling, status of facilities, circuit conditions, or other matters that affect communications system operation, including circuit continuity checks, tracer actions, and error corrections.

service period: *See* **operational service period.**

service point: *See* **adjunct service point.**

service position system: *See* **traffic service position system.**

service priority service: *See* **Telecommunications Service Priority Service.**

service priority system: *See* **Telecommunications Service Priority System.**

service priority system user: *See* **Telecommunications Service Priority System user.**

service probability: **1.** The probability of obtaining a specified grade of service during a given period. **2.** The probability of obtaining a specified or higher grade of service during a given period. *See also* **grade of service, outage probability.**

service program: *Synonym* **utility program.**

service quality: *Synonym* **quality of service.**

service restoral: *See* **mean time to service restoral.**

service routine: *Synonym* **utility routine.**

services digital network: *See* **integrated services digital network.**

service seeking: In a communications or computer network control system, the interrogation of devices on a multipoint line to determine their readiness to transmit or receive data or messages.

service signal: A signal that enables data systems equipment and subsystems to (a) function correctly and (b) provide ancillary facilities. *Note:* Examples of service signals are (a) a signal indicating that certain equipment is not in service, and (b) a signal that is automatically transmitted by a communications network to a calling station that indicates that access failure has occurred because of equipment malfunction. *Synonym* **housekeeping signals.** *See also* **access failure, signal.**

service speed: *Synonym* **speed of service.**

service state: *See* **operational service state.**

service station: *See* **cable television relay service station.**

service termination point: **1.** Proceeding from a network toward a user terminal, the last point of service rendered by a commercial carrier under applicable tariffs. *Note 1:* The service termination point usually is on the customer premises. *Note 2:* The customer is responsible for equipment and operation from the service termination point to user end instruments. **2.** In a switched communications system, the point at which common carrier service ends and user-provided service begins, i.e., the interface point between the communications system equipment and the user terminal equipment, under applicable tariffs. *See also* **common carrier, customer premises equipment, network interface, network terminating interface, point, user.**

service time interval: In communications system operations, a performance measurement period, or a succession of performance measurement periods, (a) during which a communications service is scheduled to perform its normal function, (b) that includes operation time and outage time, and (c) that is scheduled by the calendar, such as 0600 to 2200 hours daily, except Sunday 1200 to 2200 hours. *See also* **performance measurement period.**

service unit: *See* **customer service unit, data service unit.**

service user: An individual or an organization, including a service vendor, that is provided a telecommunications service for which a priority level has been requested or assigned. *See also* **priority level.**

serving arrangement: *See* **basic serving arrangement.**

servomechanism: **1.** An automatic, usually electromechanical, device that uses feedback to control the spatial position of an element or a part. **2.** A control device in which one or more of the internal or external signals represent mechanical motion. *Note:* Examples of servomechanisms are (a) a device that controls the precise positioning of an elevator at a floor, (b) an autopilot, and (c) an antenna position controller that maximizes the received signal strength.

session: In communications, computer, data processing, and control systems, (a) a period of time, (b) a temporary grouping of equipment, or (c) a set of interactions among humans, among machines, or among humans and machines. *Note:* Examples of sessions are (a) the time interval between logging on and logging off a computer remote terminal, (b) a connection between two terminals that allows them to communicate for a period, (c) a temporary connection between a control panel and a ship or aircraft propulsion system controller to enable real time online control of the system or one of its subsystems, and (d) a conference telephone call. *See* **editing session, work session.** *See also* **session layer.**

session layer: In open system architecture, the layer that provides the functions and services that may be used to (a) establish and maintain connections among elements of a session, (b) maintain a dialogue of requests and responses between the elements of a session, and (c) terminate the session. *See also* **session, Open Systems Interconnection—Reference Model.**

Session Layer: *See* **Open Systems Interconnection—Reference Model.**

set: **1.** Any defined group of entities, usually defined according to specified criteria, such as all the protocols used in a given communications system. **2.** A finite or infinite number of objects, entities, or concepts that have a given property or properties in common. **3.** In a specified communications, computer, data processing, or control system, to put all or part of system hardware and software into a specified state. *See* **alphabetic character set, alphabetic coded character set, alphanumeric coded character set, character set, coded character set, coded set, code set, data set, engineering circuit multiplex set, four-wire terminating set, hybrid set, key set, numeric character set, numeric coded character set, optical fiber connector set, optical loss test set, radio set, telegraph set, television set, tone wedge set, user set.**

set mode: The operating mode of an analog computer during which the coefficients or the equations, terms, or parameters are entered, such as (a) the potentiometers are set for division, (b) gains are set for multiplication factors, and (c) summers are set for constants of integration. *See also* **analog computer, compute mode, parameter, potentiometer.**

settling time: In systems that require changes to different frequencies, such as spread spectrum frequency shift keying and wavelength-modulated fiber optic transmission systems, the time required for the frequency of a multifrequency device to change to, and stabilize at, a new operating frequency.

set top box: *See* **television set top box.**

setup: **1.** In communications, computer, data processing, and control systems, the assembly of individual components, their interconnection, and the adjustments needed for their successful operation, usually for a given run or operational period. **2.** The preparation of a system to perform an operation, job, or step. **3.** The actual arrangement of data and components of a system, (a) including the performance of preoperational tasks, such as mounting reels of tape, inserting disks, loading programs, loading data, and stacking cards in a hopper, and (b) usually not including maintenance, actual runs, performing diagnostic tests, removing data media, and disconnecting components. *See also* **data medium, job, loading, operation, run, step.**

setup string: In communications, computer, data processing, and control systems, a series of characters used to place, i.e., to set, a component, device, or system in a specified condition. *Note:* An example of a setup string is a sequence of characters that is used to initialize a printer. *See also* **initialize, sequence, string, set, setup.**

setup time: *See* **call setup time.**

seven-bit byte: *Synonym* **septet.**

seven-hundred service (700 service): A personal telephone service that allows call receivers to receive, via a single number, telephone calls at various locations, such as home, office, or car, from call originators that use the same common carrier. *Synonym* **700 service.**

sexadecimal: *Synonym* **hexadecimal.**

sextet: A byte composed of six binary elements, such as six binary digits. *Synonym* **six-bit byte.**

SF: single frequency, store and forward.

SF signaling: single frequency signaling.

SF signaling system: single frequency signaling system.

SF signal unit: single frequency signal unit.

SFTS: standard frequency and time signal. *See also* **standard frequency and time signal service.**

shading: In display systems, to change one or more attributes, such as tone, color, shade of color, intensity, and composition, of a specific area or part of a display element, display group, or display image in the display space on the display surface of a display device. *See also* **attribute, display device, display element, display group, display image, display space, display surface.**

shadow area: *Synonym* **blind area.**

shadow loss: **1.** In a radio signal, the attenuation caused by obstructions in the propagation path. **2.** In an antenna, the reduction in the effective aperture of the antenna caused by the masking effect of antenna parts, such as (a) the feed masking the radiation from a parabolic reflector and (b) supporting structures masking radiation from a billboard antenna. *See also* **line of sight propagation, path loss.**

shannon: The unit of information derived from the occurrence of one of two equiprobable, mutually exclusive, and mutually exhaustive events. *Note 1:* A bit may, with perfect formatting and source coding, contain one shannon of information. However, the information content of a bit usually will be less than a shannon. *Note 2:* The decision content of a character set of 8 characters is 3 shannons, i.e., $\log_2 8 = 3$. *Synonym* **information content binary unit.** *See also* **decision content.**

Shannon's law: The theoretical maximum rate at which error-free digits can be transmitted over a bandwidth-limited channel in the presence of noise, is given by the relation $C = W\log_2(1 + S/N)$ where C is the channel capacity in bits per second, W is the bandwidth in hertz, and S/N is the signal to noise ratio. *Note:* Error correction coding can improve the communications performance relative to uncoded transmission, but no practical error correction coding system exists that can even closely approach the theoretical performance limit given by Shannon's law. *See also* **erlang, loading, maximum user signaling rate, throughput.**

shape: *See* **pulse shape, wave shape.**

shaped pulse: *See* **Gaussian-shaped pulse.**

shaped reflector Cassegrain: *Synonym* **uniform illumination Cassegrain.**

shape factor: In a filter, a measure of how rapidly the filter frequency response curve falls or rises as a function of frequency. *Note:* The shape factor usually is expressed in dB per hertz, dB per kilohertz, or percent per kilohertz. *See also* **filter, frequency, frequency response curve.**

shaping: *See* **signal shaping.**

shaping circuit: *Synonym* **shaping network.**

shaping network: A network, i.e., an electrical circuit, such as a pulse regeneration circuit, that (a) is inserted into a circuit and (b) improves or modifies the wave shape of signals or pulses that are passing through the circuit. *Synonym* **shaping circuit.** *See also* **circuit, pulse regeneration, pulse shape.**

shared disk: In communications, computer, data processing, and control systems, a magnetic or optical disk that may be used for information storage concurrently by two or more systems.

sharing: *See* **frequency-geographical sharing, frequency sharing, frequency-time sharing, geographical sharing, power sharing, time sharing.**

sharpening: *See* **image sharpening.**

sheath: **1.** In communications and power cables, the outermost cover that (a) protects the cable components from damage and (b) may serve as a shield. *Note 1:* The sheath may have markings that identify the type of cable. *Note 2:* Overarmor may be placed over the sheath for added protection in extremely harsh environments, such as aerial insert risers, direct buried cable river crossings, shore-to-ocean transitions, and ship-to-shore or ship-to-ship tethers. **2.** *Synonym* **fiber optic cable jacket.** *See also* **optical fiber jacket, shield, shielded pair.**

sheath length: The length of a cable when wound on a spool and ready to be installed. *Note:* Sheath length may be expressed in any convenient unit of length, such as feet, meters, or statute miles. *See also* **sheath mile.**

sheath mile: A 1 mi (statute mile) actual physical length of a cable. *Note 1:* Sheath miles are often expressed in average route miles. *Note 2:* A route mile requires more than a sheath mile of cable because of cable consumed for purposes other than covering forward distance, such as slack, coupling, aerial insert risers, offsets, and connection boxes along the route. *See also* **route mile.**

shelf life: The maximum length of time before use that a system, component, device, or material will remain in serviceable condition and not significantly deteriorate so as to affect the operational behavior that it would have had, had it been placed in operation when it was made.

shell: In a computer environment, an operating system command interpreter, i.e., the part of an operating system that reads an input that specifies an operation and performs, directs, or controls the specified operation, such as (a) to permit a user to switch among application programs without terminating any of them, and (b) to obtain its input from different sources, such as a user terminal, a file, or a server. *See also* **operating system, operation.** **2.** In a fiber optic connector, the primary connector housing. *See* **backshell.**

SHF: superhigh frequency.

shield: **1.** A housing, screen, sheath, or cover that substantially reduces the coupling of electric and magnetic fields into or out of circuits. **2.** A housing, screen, sheath, or cover that prevents the accidental contact of objects or persons with parts or components operating at hazardous voltage levels. *See* **ferrous shield.** *See also* **screen.**

SHIELD: silicon hybrids with infrared extrinsic long-wavelength detector.

shielded enclosure: An area, room, or box that is capable of preventing, usually by attenuation or reflection, the entry or the exit of electromagnetic fields, sound waves, or both, that originate inside or outside of the enclosed space. *Note:* Necessary openings in the shielded enclosure, such as doors, air vents, and electrical or optical feedthroughs, are especially designed to maintain the shielding. Electromagnetic shielding material usually is sheet metal or wire screen. Soundproofing material usually is a thick layer of honeycombed, fibrous, nonelastic, nonmetallic material, such as pressed straw or styrofoam.

shielded twisted pair: A two-wire transmission line surrounded by a sheath of conductive material that (a) protects it from the effects of external fields, (b) confines internal fields to the line, and (c) provides additional protection against physical damage. *See also* **paired cable, quadded cable.**

shielding: **1.** Using shields to reduce the undesirable effects on circuits caused by electromagnetic, electric, and magnetic fields originating outside the circuit. **2.** Using shields to prevent the escape of electromagnetic fields from electrical and electronic circuits **3.** Using shields to prevent the accidental contact of objects or persons with parts or components operating at hazardous voltage levels. **4.** The material used to create a shield. *See also* **cladding, radio frequency tight, screening, shield, shielding effectiveness.**

shielding effectiveness: A measure of the attenuation in the electromagnetic, electric, or magnetic field strength caused by a shield inserted between the field source and the point of measurement. *Note:* Shielding effectiveness usually is expressed in dB. *See also* **screen, shield.**

shift: **1.** The movement of some or all of the characters of a word by the same number of character positions in the direction of a specified end of a word. **2.** In radar, to move the origin of a radial display away from the center of a screen, such as that of a cathode ray tube (CRT). **3.** In radar, the amount the origin of a radial display is moved away from the center of a screen, such as that of a cathode ray tube (CRT). **4.** To change the frequency or the phase of a signal. **5.** A change in the frequency or the phase of a signal. **6.** In teletype systems, a machine function that controls the positioning of certain components of teletypewriter-teleprinter equipment to permit the printing of upper case or lower case characters as required, i.e., to permit a case shift. **7.** In radio communications for merchant ships, the process that a convoy or independent ship station uses to change its radio watch from one ship broadcast area or subarea to the next. **8.** In communications, computer, data processing, and control system operations, the period during which a person is on duty. *See* **area broadcast shift, arithmetic shift, carrier shift, case shift, circular shift, compensating phase shift, data shift, diurnal shift, Doppler shift, end-around shift, facsimile frequency shift, figures shift, frequency shift, Goos-Haenchen shift, letter shift, negative scanning shift, phase shift, positive scanning shift, radar shift, radio frequency shift, signal frequency shift, subcarrier frequency shift.** *Refer to* **Fig. S-17 (940).**

shifted optical fiber: *See* **dispersion-shifted optical fiber, dispersion-unshifted optical fiber.**

shift frequency shift keying: *See* **narrow shift frequency shift keying.**

shift-in character: A code extension character that (a) is used to terminate a series of a certain type of character, i.e., from a certain character set, that has been introduced by the shift-out character, and (b) makes effective again the characters of the character set that was in effect before the shift-out character occurred, usually by a return to the normal, standard, base, or default set, such as a return to Roman font after the shift-out character introduced italics. *Note:* The terms "in" and "out" usually are with reference to the most prevalent set or the default set of characters. *See also* **code extension character, shift-out character.**

shift keying: The switching from one condition, mode, or state to another. *See* **binary phase-shift keying, coherent frequency shift keying, differentially coherent phase-shift keying, differential phase-shift keying, double frequency shift keying, filtered symmetric differential phase-shift keying, frequency shift keying, Gaussian minimum shift keying, incoherent frequency shift keying, multiple frequency shift keying, multiple phase-shift keying, narrow shift frequency shift keying, phase-shift keying, quadrature phase-shift keying, wide shift frequency shift keying.**

shift-out character: A code extension character that introduces another or an alternate character set. *Note 1:* An example of a shift-out character is a character that introduces the use of italics in lieu of Roman font until a shift-in character occurs, at which point there is a return to the Roman font. Other shift-out characters are boldface, underline, double underline, strike out, outline, and shadow. *Note 2:* The terms "in" and "out" usually are with reference to the most prevalent set or the default set of characters. *See also* **code extension character, shift-in character.**

shift register: **1.** A register in which shifts are performed. **2.** A storage device that (a) may serially move an ordered set of data as a unit into a discrete number of storage locations and (b) may be configured so that stored data may be (i) moved in one or more directions, (ii) entered and stored from multiple inputs, and (iii) grouped into arrays in order to perform more complex data operations. *See* **nonlinear feedback shift register, linear feedback shift register.** *See also*

data, n-sequence, register, shift, storage register. *Refer to* **Fig. S-17 (940).**

shift register generator: *Synonym* **spread spectrum code sequence generator.**

shift signal: *See* **letter shift signal.**

ship: *See* **constant watch ship, guard ship, light-repeating ship, repeating ship.**

ship-air net: A communications net that is used to transmit messages from ship stations to aircraft stations and aircraft stations to ship stations. *See also* **ship to air net.**

ship alarm signal: A signal that automatically is transmitted by radio in the event of a ship distress or emergency. *Note:* Standard alarm signals for merchant ships are (a) an audio alarm consisting of twelve 4 s (second) dashes, with 1 s intervals, at 500 kHz (kilohertz) and (b) an alarm at 2182 kHz consisting of two audio tones of different pitch transmitted for a period of 30 s to 60 s.

ship broadcast area radio station: A shore radio station that transmits messages to ship stations in a given long-range broadcast area. *See also* **ship broadcast area radio station.**

ship broadcast coastal radio station: A shore radio station that (a) transmits messages to ship stations in local coastal waters and (b) usually does not transmit at times and frequencies in which long-range ship broadcast area radio stations schedule their transmissions. *See also* **ship broadcast area radio station.**

ship broadcast schedule: The schedule of transmissions by a ship broadcast area radio station or a ship broadcast coastal radio station that usually consists of the call tape, the preliminary call-up signal, the traffic list, the messages on the list, and the sign-off transmission after the last message. *See also* **call tape, preliminary call, preliminary call-up signal, traffic list.**

ship broadcast service: *See* **merchant ship broadcast service.**

ship call sign: A radio call sign that (a) is assigned to a ship, such as merchant ships, miscellaneous government-owned ships, and private personally owned ships and (b) uses the characters in the ship call sign as the ship international call sign.

ship collective call sign: A call sign that is used by or assigned to a group of ships. *Note:* An example of a ship collective call sign is a call sign that is assigned to all the ships registered in Panama or all the ships in a specific convoy.

ship communications: Communications methods, systems, or equipment used for the transmission of messages that (a) contain information about ship operations, such as messages that are sent to or from ships or shore communications facilities for handling matters relating to maritime operations, such as coordinating ship movements, defining shipping routes, providing escort, and handling distress and rescue operations, and (b) do not contain private or personal messages except in an emergency.

ship deck log: A log that may be chronological, subject-structured, or a combination of both, covering significant events that (a) usually are not classified and (b) occur on board a ship during a voyage. *Synonym* **ship log.** *See also* **ship radio log.**

ship Earth station: In the maritime mobile satellite service, a mobile Earth station on board a ship. *See also* **maritime mobile service.**

ship emergency transmitter: On board a ship, a transmitter to be used exclusively on a distress frequency for distress, urgency, or safety purposes.

ship fitting communications standard plan: A table of standard communications equipment and components that are recommended as minimum fit for ships of various classes and functions.

ship international call sign: The call sign that (a) is unencrypted, (b) is assigned to each ship, and (c) uniquely identifies the ship and, in this sense, is equivalent to the name of the ship. *See* **military international call sign.**

ship log: *Synonym* **ship deck log.**

ship movement service (SMS): In the maritime mobile service, a safety service that is (a) between coast stations and ship stations, (b) between ship stations, (c) restricted to messages relating to the movement of ships, (d) not a port operations service, and (e) not used for messages that are of a public correspondence nature.

ship net: *See* **ship to ship net.**

ship radar silence: A restriction against the use of ship radar equipment to maintain radar silence for such purposes as (a) reduction of interference, (b) avoidance of confusion of launched missiles, (c) prevention of missiles homing on radar transmission, and (d) prevention of ship identification or discovery from radar signal analysis, i.e., radar signature analysis. *See also* **homing, interference, radar signature.**

ship radio log: In a ship station, a continuing record that (a) may be chronological or structured, (b) describes significant events, important messages, and other matters that pertain to the communications operations of the station, usually for each voyage, and (c) that may include on and off watch operator signatures, dates, call signs, schedules copied, date-time groups of messages sent, frequencies used, data on received messages, radio silence periods, distress signals received and the response, alarm signals used, and other significant events concerning ship safety traffic. *See also* **radio log, safety traffic, ship station.**

ship radio officer: *See* **senior ship radio officer.**

ship radio silence: A restriction against the use of ship radio equipment, to maintain radio silence for such purposes as (a) reduction of interference, (b) avoidance of confusion of launched missiles, (c) prevention of missiles homing on radio transmission, and (d) prevention of ship identification or discovery from radio signal analysis. *See also* **homing, interference, radio silence, signal analysis.**

ship radiotelephone numeral: 1. One of the special pronunciations of the decimal numerals that are used over maritime mobile voice radiotelephones for international understanding. *Note:* The ship radiotelephone numerals are unaone, bissotwo, terrathree, kartefour, pantafive, soxisix, setteseven, oktoeight, novenine, and nadazero. **2.** A ship radiotelephone directory number for routing telephone calls to a specific ship.

ship radio transmission security: Communications security measures that are taken to ensure that a ship station is operated in accordance with the rules and the procedures prescribed in security regulations. *See also* **communications security, procedure, ship station.**

ship-shore communications: Communications methods, procedures, or equipment that are used for the transmission of messages that (a) contain information about ship-shore operations, such as coordinating, directing, and conducting operations that include sea, ground, and air units, (b) are sent from shore, i.e., land stations, to ship stations and from ship stations to land stations, i.e., communications are bidirectional, and (c) are sent between fixed and mobile stations.

ship-shore communications net: *See* **maritime ship-shore communications net.**

ship-shore net: A radio net that is used to send messages from ship stations to land stations, i.e., shore stations, and from land stations to ship stations. *See also* **ship to shore net.**

ship-shore radio teletypewriter: A teletypewriter that is used for sending or receiving messages in ship to shore, shore to ship, and ship-shore radio nets.

ship-shore station: A station that (a) on land, can send messages to and receive messages from ships or (b) on ships, can send messages to and receive messages from land. *See* **local ship-shore station, primary ship-shore station, secondary ship-shore station.**

ship signaling: *See* **flashing light ship signaling.**

ship station: 1. In the maritime mobile service, a mobile station that (a) is on board a ship that is not permanently moored and (b) is not a survival craft station. **2.** The assigned spatial position of a ship in a convoy. **3.** An assigned duty position, location, or function on board a ship. *Note:* Examples of ship stations are senior radio officer, radio operator, radio room, and pilot station. *See also* **one-operator ship station.**

ship to air net: A communications net that is used to transmit messages from ship stations to aircraft stations. *See also* **ship-air net.**

ship to ship net: A communications net that is used to transmit and receive messages among individual ships whether or not they are in a convoy.

ship to shore net: A communications net that is used by ship stations for passing messages to land stations, i.e., shore stations, for onward relay or distribution. *See also* **ship-shore net.**

ship to shore radiotelephone procedure: The procedure used to place telephone calls from ships to telephone systems on shore. *Note:* Ship to shore radiotelephone procedures are described in Article 33, *International Telecommunications Regulations.*

ship to shore schedule: A list of call signs, answering frequencies, watch hours, and calling frequencies for each ship to shore station in a maritime communications system. *Note 1:* For example, for the ship to shore radio station Halifax, Nova Scotia, the call sign is VCS, one answering frequency is 4293.5 kHz (kilohertz), operational hours are 0000 to 1000 UTC, and the calling band is from 4177 kHz to 4187 kHz. *Note 2:* Ship to shore schedules are subject to change.

shock excitation: *Synonym* **impulse excitation.**

shock test: *See* **thermal shock test.**

shock wave: A sudden or instantaneous transition from one pressure level to another. *Note:* Examples of shock waves are (a) the sound from subsonic, transonic, supersonic, and hypersonic missiles, aircraft, and spacecraft in the atmosphere, i.e., sonic booms, (b) the sound from an imploded incandescent light bulb, (c) the initial sound wave from a lightening strike, and (d) the sound wave launched by a nuclear explosion.

SHORAN: short range aid to navigation.

shore communications: *See* **ship-shore communications.**

shore net: A radio net that consists of stations on land but is used to handle messages that are destined for ship stations at sea. *See* **ship-shore net, ship to shore net.** *See also* **ship station.**

shore radiotelephone procedure: *See* **ship to shore radiotelephone procedure.**

shore radioteletypewriter: *See* **ship-shore radioteletypewriter.**

shore station: In maritime communications, a land station. *See* **local ship-shore station, primary ship-shore station, secondary ship-shore station.**

short address: *See* **abbreviated address.**

short board: A printed circuit (PC) board that (a) is not as long as the cabinet in which it is installed and (b) usually is supported only by being fastened at the plug-in side of the cabinet and by being inserted into its female connector. *See also* **long board, printed circuit.**

short-haul communications network: A communications network that (a) handles communications traffic usually over distances of less than 20 km (kilometer), (b) usually is within an area covered by a single telephone switching center or central office (C.O.), such as metropolitan areas, large cities, or an entire county, (c) is characterized by not having long-distance trunks between towns and cities or more than one switching center or C.O., and (d) like long-haul communications networks, may have high quality equipment for high fidelity and high definition analog (voice) and digital transmission, integrated services digital networks (ISDN), high transmission capacity, and automatic switching for handling calls and messages without operator assistance. *Note:* It is anticipated that by the year 2000 over 600 billion voice circuit•kilometers will be on fiber optic long-haul and short-haul networks and only 16 billion on microwave, 160 billion on satellite, and 3 billion on coaxial cable. *See also* **long-haul communications network.**

short-haul toll traffic: Message toll traffic between nearby points, usually less than 50 miles apart.

short-range aid to navigation (SHORAN): A short-range radionavigation system that (a) is used to obtain precise fixes for mobile receivers, (b) consists of a pulse transmitter, a pulse receiver, and two transponder beacons at fixed points, and (c) with a properly equipped receiving station, can obtain a fix based on the directions and phase relationships of received signals, with the assistance of appropriate charts and tables. *Synonyms* **short-range navigation, short-range radionavigation.**

short-range laser sonar: A sonar device for short ranges in which a controlled laser beam is used to produce separate identifiable beams and thus produce code pulses of a carrier wave. *See also* **carrier, laser, sonar.**

short-range navigation: *Synonym* **short-range aid to navigation.**

short-range radionavigation: *Synonym* **short-range aid to navigation.**

Shortstop: A battlefield electronic system that (a) automatically detonates incoming munitions, such as mortar and artillery shells, (b) provides for omnidirectional or directed coverage, and (c) may be either a lightweight manpack, a vehicle-mounted, or a ground-tripod-mounted system.

short-term ionospheric forecast: A 1-wk (week) or shorter forecast of ionospheric conditions that is distributed in broadcast or wire message form. *Note:* Sudden, very short duration ionospheric disturbances cause immediate fading and are unpredictable. Certain ionospheric storms can be forecast. *See also* **long-term ionospheric forecast.**

short-term security: The protection that is obtained from using a cryptosystem that will protect communications traffic from successful cryptanalysis for up to approximately one month, the length of time varying with each cryptosystem, depending on the cryptomaterials used, the key used, the principle used, and the encrypted communications traffic volume. *See also* **cryptanalysis, cryptomaterial, encrypt, key.**

short title: 1. An abbreviated form of a proper name. *Note:* Examples of short titles are INTELSAT for International Telecommunications Satellite Consortium, CINCPAC for Commander-in-Chief Pacific Command, and SUNOCO for Sun Oil Company. **2.** An identifying combination of letters and numerals that is assigned to communications security material for brevity. *Synonym* **abbreviated title.** *See also* **communications security.**

shortwave: Pertaining to electromagnetic waves with a frequency above the medium frequency (MF) range, i.e., above 3 MHz (megahertz), corresponding to wavelengths that are less than 100 m (meters). *See also* **long wavelength, short wavelength.**

short wavelength: In fiber optic communications, a wavelength less than 1 µm (micron), i.e., less that 10^{-6} m (meter). *See also* **shortwave.**

shortwave radio: *Synonym* **high frequency radio.**

shot effect: *Synonym* **shot noise.**

shot noise: 1. The noise caused by the fluctuation in the electric current when charge carriers pass through a surface at statistically independent times. *Note:* In an optical system, shot noise is the mean square noise current expressed in amperes squared rather than noise power expressed in watts. **2.** In a photodetector, the noise produced by current fluctuations that are caused by the discrete nature of charge carriers and random emission of charged particles from an emitter, given by the relation $MSSNS = 2qIb$ where $MSSNS$ is the mean square shot noise current, q is the electronic charge of each particle, I is the average photocurrent, and b is the bandwidth. *Note:* The average photocurrent contains contributions from the signal current, background radiation-induced photocurrent, and dark current. The mean square shot noise current (ampere square) is converted to noise power (watts) using the equivalent resistance of the photodetector and its output circuit. Shot noise current would reduce to zero if the magnitude of an individual charge tended to zero. This indicates that the discrete nature of the charge is the underlying cause of shot noise. If there were no dark current and background radiation on the detector so that the only contribution to the average photocurrent was caused by the optical signal, the resulting shot noise current density would produce noise at the lower limit of detector noise, which leads to quantum-limited sensitivity. The quantum limit to optical sensitivity is caused by the granularity, i.e., the particle nature, of light. Thus, the minimum energy increment of an electromagnetic wave is hf, i.e., the energy of a photon. The shot noise is then the quantum noise given by the relation $QN = hfb$ in the limit when photocurrent is caused only by the optical signal, where QN is the quantum noise, h is Planck's constant, f is the frequency, and b is the bandwidth. *Synonyms in this case* **photon noise, quantum noise. 3.** Noise, i.e., random surges of voltage, current, electrical power, or optical power, caused in electronic and optoelectronic devices by random variations in the number and velocity of electrons emitted by heated cathodes, photosensitive surfaces subjected to incident radiation, and p-n junctions in semiconductors. *Note:* The random noise occurs because electric currents consist of discrete electric charges, and optical power consists of a stream of discrete photons, both of which are capable of irregular and erratic movements, such as clustering as well as reinforcing and canceling one another. Also, shot noise from the electronic circuits in a light source can produce equivalent surges in optical power output, which will produce corresponding surges, i.e., noise, in photodetector electrical power, current, and voltage output. *Synonyms* **shot effect, Schottky effect, Schottky noise.** *See also* **quantum noise.**

shutter: *See* **fallen clearing indicator shutter.**

shuttle pulse test: The obtaining and viewing of pulse shapes that would occur at regular intervals in a long optical fiber, without pulsing the long fiber, by (a) taking a short sample of the long fiber, (b) equipping the short length with a reflector at both ends, (c) pulsing the fiber through one reflector at the near end, (d) observing the pulses through the reflector at the far end after each $2N - 1$ transits where N is an integral number of transits, and (e) thereby measuring optical properties, such as mode coupling and dispersion, that would occur in a long fiber by using short lengths of test fibers.

SI: Système International, Système International d'Unités. *See* **International System of Units.**

SI character: shift-in character.

SID: sudden ionospheric disturbance.

side: *See* **equipment side, line side, local side.**

sideband: 1. In an amplitude-modulated carrier, the frequency band that (a) is located above or below the carrier frequency and (b) results from the modulation process. *Note:* The frequencies above the carrier frequency are called the "upper sideband," and those below the carrier frequency are called the "lower sideband." **2.** The spectral energy distributed above and below the carrier frequency of a modulated electromagnetic wave. *Note:* In conventional amplitude modulation (AM) transmission, both sidebands are present. Transmission in which one sideband is present is called "single-sideband transmission." *See* **double sideband, independent sideband, lower sideband, single sideband, upper sideband, vestigial sideband.** *See also* **amplitude modulation, modulation, sideband transmission, single sideband suppressed carrier transmission.**

sideband emission: *See* **reduced carrier single sideband emission, single sideband emission, two-independent-sideband carrier.**

sideband equipment reference level: *See* **single sideband equipment reference level.**

sideband noise power ratio: *See* **single sideband noise power ratio.**

sideband suppressed carrier: *See* **double sideband suppressed carrier.**

sideband suppressed carrier transmission: *See* **double sideband suppressed carrier transmission.**

sideband transmission: Amplitude-modulated (AM) transmission in which (a) one or more sidebands are used to convey information, (b) the frequencies above the carrier frequency are called the "upper sideband," and those below the carrier frequency are called the "lower sideband," (c) the upper and lower sidebands may carry the same or different information, (d) the carrier and either sideband may be suppressed independently, and (e) in conventional AM transmission, both sidebands carry the same information, and the carrier is present. *See* **compatible sideband transmission, double sideband transmission, independent sideband transmission, single sideband transmission, two-independent-sideband transmission, vestigial sideband transmission.** *See also* **amplitude modulation, band, carrier, double sideband suppressed carrier transmission, frequency, modulation, reduced carrier transmission, sideband, single sideband suppressed carrier transmission, suppressed carrier transmission.**

sideband voice: *See* **single sideband voice.**

side circuit: Either of the two circuits used to derive a phantom circuit. *See also* **circuit, phantom circuit.**

side lobe: In a directional antenna or light source radiation pattern, a lobe that (a) is in any direction other than that of the main lobe, (b) usually is smaller than the main lobe, (c) usually is adjacent to the main lobe or another side lobe, (d) has a much lower power density than the main lobe at a given range although in a different direction, and (e) in lasers, light-emitting diodes (LEDs), and narrow beam directional antennas, is such that as much energy as possible is confined to the main lobe, thus making side lobes negligible. *See also* **antenna, antenna lobe, main beam, main lobe.**

side lobe blanking: Radar operation in which (a) the signal strengths between an omnidirectional antenna and a normal radar antenna are compared, (b) the omnidirectional antenna, plus receiver, has slightly more gain than the side lobes of the normal antenna but less gain than the main lobe, (c) in electronic countercountermeasures, effective off-azimuth jamming and deception is removed, (d) spoofjamming signals at false azimuths can be removed, and (e) there is a disadvantage caused by the turning off of receivers where there could be objects of interest, i.e., targets of interest. *See also* **azimuth, deception jammer, gain, jamming, main lobe, omnidirectional antenna, side lobe, side lobe cancellation, spoofjamming.**

side lobe cancellation: In a radar system, anti-interference and antijamming operations in which (a) the same antenna and receiver configuration is used as in side lobe blanking, except that a gain matching and canceling process occurs, (b) extraneous signals that enter the side lobe of the main antenna are canceled while the main echoes from the object, i.e., from the target, remain uncanceled, (c) cancellation is obtained on the order of 20 dB against a single noise jammer, and (d) the radar system sensitivity is reduced by the same amount. *See also* **side lobe blanking.**

side lobe jamming: Jamming in which (a) the jamming signal is received through a side lobe of a receiving antenna, (b) an attempt is made to obliterate the desired signal that is received through the main lobe of the receiving antenna, and (c) in radar, the operator can be confused about the real azimuth of the jammer because of the injection of multiple signals, i.e., multiple strobes. *See also* **jamming, radar strobe, side lobe.**

side-looking airborne radar: In an aircraft, high resolution airborne radar in which (a) the radar main lobe is directed at right angles to the direction of flight or at right angles to the longitudinal axis of the aircraft, and (b) a presentation of terrain or moving longitudinal objects is produced. *See also* **radar, resolution.**

sidereal day: **1.** The actual mean period of one Earth rotation, equal to about 86,164 s (seconds), more precisely equal to 23 h (hours), 56 min. (minutes), and 4.095 s. **2.** The time between two successive transits of a fixed star past a given meridian on Earth. *Note:* Because of the star distance, the Earth-star line can be considered as a fixed line in space for a period as short as 1 day. *See also* **second, synodic period.**

side stable relay: A polar relay that remains in the last signaled contact position. *See also* **polar relay.**

sidetone: In a telephone set, the sound, consisting of the voice of the speaker and the background noise, coming from the telephone receiver. *Note:* The sidetone volume usually is suppressed relative to the transmitted volume. *Synonym* **telephone sidetone.** *See also* **feedback.**

Siecor cable: A fiber optic cable produced by Siecor Optical Cable, Inc.

sifting sort: *Synonym* **bubble sort.**

sight: *See* **line of sight.**

sight equation: *See* **radar line of sight equation.**

sighting: An actual direct visual contact with an object by a person rather than by means of an instrument, such as radar, sonar, sound, echo, reflection, or radio. *See also* **echo, radar, radio, reflection, sonar, sound.**

sight link: *See* **line of sight link.**

sight propagation: *See* **line of sight propagation.**

SIGINT: signal intelligence, signals intelligence.

sigma: In telephone systems, a group of telephone wires, usually the majority or all of the wires that constitute a line or a local loop, that are treated as a unit for various purposes, such as computing noise levels, arranging connections for the measurement of noise, and establishing electrical current balance.

sigma modulation: *See* **delta-sigma modulation.**

sign: In communications systems, a unique indicator of (a) the identity of an entity, such as a radio station, or (b) the content of a coded message. *See* **aircraft call sign, answering sign, attention sign, call sign, collective call sign, front sign, indefinite call sign, individual call sign, international call sign, military call sign, military ship international call sign, net call sign, numeral sign, permanent voice call sign, personal sign, procedure sign, separative sign, ship call sign, ship collective call sign, ship international call sign, special panel signaling sign, visual call sign.**

signal: 1. A representation of information conveyed by a carrier. **2.** Detectable transmitted energy that can be used to carry information. **3.** A time-dependent variation of a characteristic of a physical phenomenon used to convey information. **4.** A transmitted electrical impulse. **5.** A message that (a) consists of one or more symbols, such as letters, words, characters, signal flags, visual displays, or special sounds, with prearranged meanings, and (b) is conveyed or transmitted by any means, such as visual, acoustical, or electrical means. **6.** The code or the pulse that represents intelligence, a message, or a control function conveyed over a communications system. **7.** A visual or mechanical action to which meaning is assigned, such as puffs of smoke, sunlight flashed from a mirror, displays, motion of flags or pennants, or motion of a cable, wire, or rope. *See* **access barred signal, acknowledgment signal, address signal, alarm signal, alternate mark inversion signal, analog signal, anisochronous signal, answer signal, asynchronous signal, attention signal, backward signal, baseband signal, black signal, bunched frame alignment signal, busy signal, call accepted signal, call control signal, called line identification signal, call failure signal, calling line identification signal, calling signal, call not accepted signal, call pending signal, call progress signal, call selection signal, call-up signal, call waiting signal, camp on busy signal, character signal, chrominance signal, circuit released acknowledgment signal, circuit released signal, clear confirmation signal, clear forward signal, clear request signal, clear signal, clear to send signal, coherent signal, composite two-tone test signal, composite video signal, compound signal, connection signal, connection in progress signal, continuity failure signal, controlled not ready signal, data circuit-terminating equipment clear signal, data circuit-terminating equipment waiting signal, data terminal equipment clear request signal, data terminal equipment waiting signal, data transfer request signal, digital signal, disabling signal, disconnect signal, disjoint signal, distress signal, distributed frame alignment signal, duobinary signal, duress signal, electronic signal, emergency signal, enabling signal, erasure signal, error signal, facility request signal, facsimile picture signal, figure shift signal, forward signal, frame alignment signal, function signal, ground return signal, ground to air visual body signal, identification not provided signal, inhibiting signal, international congestion signal, isochronous signal, jamming signal, letter shift signal, maneuvering signal, marking signal, mark signal, minimum discernible signal, modulating signal, multiframe alignment signal, n-ary digital signal, narrowband signal, national circuit group congestion signal, national switching equipment congestion signal, octet timing signal, off-hook signal, on-hook signal, operating signal, orthogonal signal, out of order signal, partial response signal, permanent signal, phasing signal, preliminary call-up signal, proceed to select signal, pulsed frequency-modulated signal, pyrotechnic distress signal, pyrotechnic signal, quasi-analog signal, quaternary signal, random signal, ready for data signal, re-call signal, response signal, redirected to new address signal, redirect to new address signal, request data transfer signal, request to send signal, response signal, ringback signal, safety signal, sampling signal, seizure signal, selection signal, service signal, ship alarm signal, space signal, standard test signal, standard frequency and time signal, start record signal, start signal, stop record signal, supervisory signal, switching release signal, synchronous signal, symmetric signal, synchronization correction signal, synchronizing signal, system blocking signal, telegraph signal, terminal engaged signal, ternary signal, test signal, timing signal, two-level optical signal, two-level signal, undesired**

signal, unipolar signal, video signal, visual paulin signal, visual signal, waiting in progress signal, white signal, wideband signal. *See also* **message.** *Refer to* **Figs. E-2 (291), S-1 (866).**

signal alarm: *See* **emergency signal alarm.**

signal amplitude: The physical magnitude of a signal that (a) is measured from a zero level or a baseline, (b) is expressed in quantitative values, such as (i) field strength in volts per meter, irradiance in watts per square meter, or energy in joules per cubic meter for electromagnetic waves, (ii) volts, amperes, or watts for electrical signals in wires, (iii) optical power in watts for optical signals in optical fibers, and (iv) pascals for sound wave intensity, and (c) is not a phase, frequency, or wavelength measurement, though the signal has these parameters also. *Note:* In referring to a signal, the amplitude, phase, frequency, and wavelength all have a magnitude.

signal analysis: The study and the analysis of received signals, such as radio, radar, and sonar signals, in order to determine their technical characteristics, their tactical or strategic use, or the character or nature of their sources. *See also* **signature, signature analysis.**

signal center: In military communications, a combination of signal communications facilities that (a) is operated by a military unit in the field and (b) usually consists of a communications center, a telephone switching central, and appropriate means of signal communications. *See also* **communications, communications center.**

signal code: *See* **international signal code, international visual signal code, network multiple selection signal code.**

signal communications: The means that are used to convey information from one person, place, or machine to another, except by direct, unassisted conversation or by written correspondence through the mail.

signal communications axis (SCA): 1. The line or the route containing the starting position and probable future locations of the command post of a military unit. **2.** The main route along which messages are relayed or sent to and from tactical and service units during a military situation or exercise. *Synonym* **axis of signal communications.**

signal communications center (SCC): An agency (a) that is responsible for the receipt, transmission, and delivery of messages, and (b) that may include a message center, transmitting facilities, and receiving facilities. *Note:* Transmitting, receiving, and relay stations are not necessarily located in a communications center, nor are they necessarily signal communications centers in themselves, but facilities for their remote control may terminate in a signal communications center.

signal compression: 1. In audio frequency systems, the reduction of the dynamic range of a signal by controlling the range as a function of the inverse relationship of its instantaneous value relative to a specified reference level, i.e., instantaneous values of the input signal that are low, relative to the reference level, are increased, and those that are high are decreased. *Note 1:* Signal compression usually is expressed in dB. *Note 2:* Signal compression usually is accomplished by separate devices called "compressors." Signal compression is used for many purposes, such as for improving signal to noise ratios (S/N), preventing overload of succeeding elements of a system, or matching the dynamic ranges of two or more devices. *Note 3:* Signal compression may be a linear or nonlinear function of the signal level across the frequency band of interest, and essentially may be instantaneous or have fixed or variable delay times. *Note 4:* Signal compression always introduces distortion, which usually will not be objectionable if the compression is limited to a few dB. **2.** In facsimile systems, a process in which the number of pels scanned on the object is larger than the number of encoded bits of picture information transmitted. *See also* **compaction, compandor, compression ratio, compressor, data compaction, data compression, expander, instantaneous value, level.**

signal contrast: 1. A ratio or a difference between any two specified signals. **2.** In facsimile, the ratio of (a) the level of the white signal to (b) the level of the black signal. *Note:* The signal contrast usually is expressed in dB, though it may be expressed in other units, such as a number, percent, or nepers. The signals must be expressed in the same units, such as volts, watts, frequency, or electrical degrees, in calculating the signal contrast. *See also* **facsimile, signal to noise ratio.**

signal conversion equipment: *Synonym* **modem.**

signal converter: A device in which the input and output signals are formed according to the same code but not necessarily according to the same type of electrical modulation. *See* **analog to digital converter, parallel to serial converter, serial to parallel converter.** *See also* **code, signal.**

signal delay: *See* **absolute signal delay.**

signal distance: 1. The number of digit positions in which the corresponding digits of two binary numbers, or words, of the same length are different. *Note 1:* The signal distance between 1011101 and 1001001 is two. *Note 2:* The concept can be extended to other notation systems. For example, the signal distance between 2143896 and 2233796 is four, between 544 and 496 is three, and between "tones" and "roses" is two. *Synonym* **Hamming distance.** *See also* **unit distance code. 2.** A measure of the difference between a given signal and a reference signal. *Note:* For analog signals, the signal distance is the root mean square difference between the given signal and a reference signal over a symbol period.

signal distortion: 1. In a signal, any departure from specified input-output relationships of signal elements over a range of frequencies, amplitudes, or phase shifts during a time interval. **2.** In start-stop teletypewriter signaling, the shifting of the significant instants, i.e., transition points, of the signal pulses from their proper positions relative to the beginning of the start pulse. *Note:* The magnitude of the signal distortion usually is expressed in percent of the theoretical attribute of the pulse. *See* **distortion, end distortion, teletypewriter signal distortion.** *See also* **pulse, signal.**

signal droop: In an otherwise essentially flat-topped rectangular pulse, the distortion characterized by a decline, i.e., a negative slope, of the top of the pulse as a function of time.

signal element: In a signal, a part usually distinguished by its nature, magnitude, duration, transition, and relative position, or by some combination of these. *Note:* Examples of signal elements are (a) parts of a telegraph signal, (b) single pulses in a group of pulses, (c) parts of a pulse, (d) transitions from one signal condition to another, (e) significant conditions between consecutive significant instants, (f) significant instants, (g) voltage changes, (h) current changes, (i) frequency shifts, (j) frequencies in frequency shift keying or dual tone multifrequency signaling, and (k) binary elements. *See also* **binary digit, signal, signal transition, significant condition, significant instant, unit interval.**

signal format: A structure for the transmission of logical states that constitutes a method of transferring information from one point to another.

signal frequency shift: In a facsimile system, the numerical difference between the frequencies that correspond to a white signal and those corresponding to a black signal at any point in the system. *See also* **black signal, frequency shift, white signal.**

signal generator: An instrument that produces a specific form of signals, such as electrical voltages and currents, electromagnetic waves, and optical pulses, all of known characteristics, such as amplitude, frequency, phase, wave shape, or timing, for any purpose, such as information representation, transmission, modulation, testing, or measurement. *See also* **generator, signal.**

signal group: A group of characters used for a specific signaling purpose. *Note:* Examples of signal groups are call signs, prosigns, basic groups, and address designators. *See also* **spare signal group.**

signal imitation: 1. The detection, recognition, identification, duplication, and transmission of signals. **2.** The false operation of a receiver by elements of speech or code patterns that occur in communications traffic.

signaling: 1. The use of signals for controlling communications. **2.** In a communications network, the information exchange that (a) is required for establishing and controlling connections, (b) is required for managing the network, and (c) does not include user information transfer. **3.** The sending of a signal from a given end of a circuit to inform a user at the opposite end that the given end is about to send a message to the user. **4.** The transmitting of signals. **5.** The signals that are transmitted. *See* **associated common channel signaling, baseband signaling, binary signaling, biphase signaling, channel-associated signaling, common battery signaling, common channel interoffice signaling, common channel signaling, composite signaling, compelled signaling, confirmation signaling, control signaling, dial signaling, digital carrier interrupt signaling, digital mark space signaling, digital signaling, direct current signaling, double current signaling, dual tone multifrequency signaling, E and M lead signaling, E and M signaling, en bloc signaling, flaghoist signaling, flashing light ship signaling, free line signaling, frequency change signaling, frequency exchange signaling, functional signaling, in-band signaling, low level signaling, Morse flag signaling, multifrequency signaling, n-ary signaling, out-of-band signaling, out-of-slot signaling, panel signaling, random bit stream signaling, remote operator signaling, reverse battery signaling, ringdown signaling, saturation signaling, semaphore signaling, separate channel signaling, simplex signaling, single current signaling, single frequency signaling, sound signaling, speech digit signaling, speech-plus signaling, supervisory signaling, system signaling, telephone circuit signaling, visual engaged signaling, visual signaling, wigwag signaling.** *See also* **signal.**

signaling data link: A data link that (a) is used only for signaling purposes, (b) is not used for the transfer of user information, (c) usually consists of a combination of two data channels that operate together in a single signaling system, and (d) has data channels that operate in opposite directions and at the same data signaling rate. *See also* **channel, data link.**

signaling exchange: *See* **common battery signaling exchange.**

signaling indicator: *See* **panel signaling indicator.**

signaling information: In a communications system, information that is used by the system, including its operators, to maintain control of transmissions, such as calls and messages, while they are being handled by the system. *Note:* An example of signaling information is the information represented by transmission control signals, call control signals, synchronizing signals, and address designators.

signaling information content: In a call or a message, the information that (a) is provided or added by a communications system, (b) is related to a call, message, system, or user action, such as a call control procedure or a system management function, (c) is part of a call or a message, and (d) is not entirely user information. *Note:* Address designators and routing information are often a signaling function not provided by the user. Message alignment, synchronizing, and service signals are not part of the signaling information content of a message. They are not in a message. They are required for proper system functioning for every message.

signaling information field: In a signal unit, such as a message or a packet, the fixed-length field that (a) holds signaling information, (b) does not necessarily exist in all signal units, (c) in a given system, usually consists of a fixed number of bits, such as 40 bits in some systems, and (d) may contain signaling information that pertains to the whole or a part of a message. *See also* **signaling information, signaling information content, signal unit.**

signaling lamp: A light source that (a) easily can be flashed for signaling and (b) usually can be changed in intensity, angle of coverage, or direction when required. *See also* **signaling.**

signaling link: In a signaling system, a functional unit that consists of one signaling data link and associated control functions. *See* **functional signaling link.** *See also* **data link, signaling.**

signaling panel code: In visual signaling systems, a code that enables numerals, letters, and special signs to be transmitted by means of panel signaling. *Note:* An example of a signaling panel code is a panel code that corresponds to international Morse code in that a dot is represented by a panel with its long axis at right angles to the line of code, and a dash is represented by a panel that is placed parallel to the line of code. The Morse code indicator and the index flash are used to indicate the orientation of the character and therefore the direction of reading. Some Morse code characters become different characters when read in reverse, such as A(•−) becomes N(−•) when read in reverse. Other letters are symmetrical and therefore are the same regardless of their orientation, such as E(•), I(••), O(−−−), and T(−−). *See also* **index flash, international Morse code, Morse code indicator, panel, panel code, panel signaling.**

signaling path: In a transmission system, a path that (a) is used in the performance of management and operational functions, such as system control, synchronization, checking, and transmission of service signals, (b) is not used for the data, messages, or calls of the users, i.e., mission traffic, and (c) may be used to transfer power for various system needs. *See also* **engineering channel, path, service signal, signal.**

signaling power: In a transmission system, the electrical, optical, or sound power that (a) is used to obtain attention at a receiving station for purposes of control or for indicating that a connection is desired, and (b) may be provided by the transmitting station or the receiving station. *Note:* Signaling power may be obtained from an unmodulated electrical carrier used to drive an electroacoustic transducer or an unmodulated optical carrier used to drive an optoacoustic transducer to produce a signaling tone. *See also* **carrier, ringing tone, transducer.**

signaling rate: *See* **binary serial signaling rate, data signaling rate, maximum user signaling rate, range designation of data signaling rate.**

signaling rate range designation: *See* **data signaling rate range designation.**

signaling rate transparency: *See* **signaling rate transparency.**

signaling service: *See* **custom local area signaling service.**

signaling sign: *See* **special panel signaling sign.**

signaling system: *See* **directional signaling system, omnidirectional signaling system, single frequency signaling system.**

Signaling System No. 7: A common channel signaling system that (a) is defined by the International Telegraph and Telephone Consultative Committee (CCITT) Recommendations Q.771 through Q.774, and

(b) is a prerequisite for implementation of the Integrated Services Digital Network (ISDN). *See also* **common channel signaling, Integrated Services Digital Network.**

signaling time slot: A time slot that (a) starts at a particular phase or instant in each frame and (b) is allocated to the transmission of supervisory and control data, i.e., transmission of signaling information. *See also* **data, digit time slot, frame, signal, signaling information, time slot.**

signaling unit: *See* **direct current signaling unit.**

signaling vocabulary: *See* **panel signaling vocabulary.**

signal intelligence: *Synonym* **signals intelligence.**

signal intercept from low orbit (SILO): A tracking beacon that (a) uses global positioning system (GPS) signals to determine the location of an object, (b) retransmits GPS location data with a unique identification code, (c) uses an ultrahigh frequency (UHF) carrier, (d) can receive signals from space, airborne, or ground systems, and (e) is about the size of a video cassette. *See also* **Global Positioning System.**

signal letter: A letter in the set of letters of an international call sign that is used in visual signaling. *See* **international call sign, visual signaling.**

signal level: 1. A measure of the power of a signal at a specified point in a communications system. **2.** The magnitude of a signal parameter or element, such as the magnitude of the electric field strength, voltage, current, power, phase shift, or frequency. *Note:* The signal level may be expressed in absolute or relative quantities, such as dB, percent, millivolts, hertz, volts per meter, microamperes, oersteds, electrical degrees, watts per square meter, watts, or lumens per square meter. *See also* **dBm, level, signal, signal magnitude, transmission level point.**

signal level compensator: In wire communications systems, an automatic gain control used to adjust the incoming signal level before the signal enters the receiving equipment. *See also* **automatic gain control.**

signal message: In communications systems, a message, i.e., an assembly of signaling information, that (a) includes associated message alignment indicators and service information, (b) pertains to a call or a message, and (c) is transferred via the message transfer part. *See also* **message, message alignment indicator, message transfer part, service information, signal.**

signal/noise ratio: *Synonym* **signal to noise ratio.**

signal operating instruction: *Synonym* **signal operation instruction.**

signal operation instruction (SOI): One of a series of orders that (a) are used for the technical control and coordination of signal communications activities of an organization, (b) usually are detailed and usually cover day-to-day operations related to communications, and (c) may also contain instructions for the regular changing of cryptoprocedures and encryption codes, i.e., key. *Synonyms* **communications operation instruction, signal operating instruction.** *See also* **key.**

signal parameter: Any of the characteristics or the attributes of a signal, such as the frequency, phase, pulse amplitude, pulse duration, polarization type, modulation type, strength, power, irradiance, or polarity. *See also* **signal.**

signal plus noise plus distortion to noise plus distortion ratio (SINAD): 1. The ratio of (a) total received power, i.e., the received signal plus noise plus distortion power, to (b) the received noise plus distortion power. **2.** The ratio of (a) the recovered audio power, i.e., the original modulating audio signal power plus noise power plus distortion power, from a modulated radio frequency carrier to (b) the residual audio power, i.e., noise plus distortion powers, remaining after the original modulating audio signal is removed. *Note:* The signal plus noise plus distortion to noise plus distortion ratio (SINAD) usually is expressed in dB.

signal plus noise to noise ratio ((S+N)/N): At a given point in a communications system, the ratio of (a) the amplitude of the desired signal plus the noise to (b) the amplitude of the noise at the given point. *Note 1:* The signal plus noise to noise ratio ((S+N)/N) usually is expressed in dB. *Note 2:* The signal plus noise to noise ratio ((S+N)/N) may also be represented as 1 + S/N where S/N is the signal to noise ratio. *See also* **channel noise level, noise, point, signal.**

signal processing: The performance of operations on signals, such as detection, shaping, converting, coding, and time positioning, that results in their transformation into other forms, such as other wave shapes, power levels, and coding arrangements, usually by means of electronic circuits or optical devices, such as lens systems, waveguides, antennas, detectors, rectifiers, pulse compressors, pulse expanders, pulse generators, nonlinear circuits, clocks, and gates. *See also* **automatic error detection, signal, signal regeneration.**

signal processing gain (SPG): 1. The ratio of (a) the signal to noise ratio (S/N) of a processed signal to (b) the signal to noise ratio of the unprocessed signal.

Note: Signal processing gain usually is expressed in dB. **2.** In a spread spectrum communications system, the signal gain, signal to noise ratio, signal shape, or other signal improvement obtained by coherent band spreading, remapping, and reconstitution of the desired signal.

signal recording telegraphy: Telegraphy in which the received signal elements are recorded automatically for subsequent translation by an operator. *See also* **signal elements, telegraphy.**

signal reference subsystem: The portion of a facility grounding system that (a) provides reference planes, such as ground return circuits, for all of the signal paths in the facility and (b) is isolated from other circuits, especially isolated from circuits that carry fault, lightning discharge, and power distribution currents. *See also* **earth electrode subsystem, facility grounding system, fault protection subsystem, signal.**

signal regeneration: 1. Signal processing, i.e., signal transformation, that restores a signal so that it conforms to its original characteristics. **2.** To the extent practical, the restoration of a signal to an original configuration, shape, or position in time or space. *Note:* Signal regeneration may include several operations, such as amplification, clipping, clamping, differentiation, integration, and clocking, i.e., timing. *See also* **dejitterizer regeneration, signal, signal processing.**

signal return: *Synonym* **signal return circuit.**

signal return circuit: A current-carrying return path from a load back to the signal source, i.e., back to the low side of the closed loop energy transfer circuit between a source-load pair. *Synonym* **signal return.** *See also* **loop back, ringback signal.**

signal sample: The value of a particular characteristic, such as the amplitude, frequency, phase, or polarity, of a signal at a chosen instant. *Note:* An example of a signal sample is the instantaneous voltage level of a given signal. If a relatively wide voltage pulse and an extremely narrow clock pulse are both entered into a two-input AND gate, the output can be arranged to be the amplitude of the voltage pulse at the instant of the clock pulse, i.e., the output will be an instantaneous sample of the input pulse. *See also* **AND gate, pulse code modulation, signal, signal sampling, significant instant.**

signal sampling: The process of obtaining a sequence of instantaneous values of a particular signal characteristic, usually at regular time intervals. *Note:* An example of signal sampling is obtaining the instantaneous phase in phase-shift keying transmission. *See also* **sampling rate, signal, signal sample.**

signal security: 1. The combination that includes all of communications security (COMSEC) and all of electronic security (ELSEC). **2.** The protection that results from all measures that are designed to deny unauthorized persons information of value that might be derived from interception and analysis of radiation or emission from any electrical, electronic, or electromagnetic system, component, or equipment or from related documentation. *See also* **communications security, electronic security, electronic warfare, emission, protected distribution system, protection, radiation, TEMPEST.**

signals intelligence (SIGINT): A category of intelligence information that consists, either individually or in combination, of all communications intelligence, electronics intelligence, and foreign instrumentation signals intelligence, however transmitted. *Note:* Signals intelligence is an intelligence activity that usually is performed by using communications-electronics systems at the national or strategic level, though it may also be performed at the tactical level. *Synonym* **signal intelligence.** *See* **foreign instrumentation signals intelligence.** *See also* **communications intelligence, electronics intelligence, electronic warfare.**

signal support: In an organization, the provision of personnel, equipment, or services for the establishment, maintenance, and operation of required communications systems. *Note:* Signal support may be provided by (a) an integral part of the organization being supported or (b) a separate organization or unit on a demand basis.

signal to crosstalk ratio (SCR): At a specified point in a circuit, the ratio of (a) the power of the wanted signal to (b) the power of the unwanted signal from another channel. *Note 1:* The signals are adjusted in each channel so that they are of equal power relative to the zero transmission level point (0TLP). *Note 2:* The signal to crosstalk ratio usually is expressed in dB. *See also* **circuit, crosstalk, relative transmission level, signal, signal to noise ratio, zero transmission level point.**

signal to noise ratio (SNR): 1. At a given point in a system, the ratio of the signal power to the noise power. **2.** At a given point in a system, the ratio of (a) the amplitude of the desired signal to (b) the amplitude of the noise. *Note 1:* The signal to noise ratio (SNR) usually is expressed in dB and in terms of peak values for impulse noise and root-mean-square values for random noise. *Note 2:* Both the signal and the noise should be defined to avoid ambiguity, such as peak signal to peak noise ratio. **3.** At a point in the domain of cyberspace, such as that of the Internet, and according to a set of criteria, the ratio of (a) wanted, useful, and relevant information to (b) the unwanted, useless, and irrelevant information that enters and remains at or passes through the point. *Note:* The signal to noise ratio in this sense is analogous to the ratio of (a) needed, wanted, or requested mail to (b) unwanted junk mail. *See also* **carrier to noise ratio, carrier to receiver noise density, channel noise level, detectivity, FM improvement factor, FM improvement threshold, FM threshold extension, intelligibility, modulation suppression, noise, noise level, preemphasis, preemphasis improvement, sensitivity, signal.**

signal to noise ratio per bit (SNR/bit): The ratio given by the relation SNR/bit = E_b/N_0, where E_b is the signal energy per bit, and N_0 is the noise energy per hertz of noise bandwidth.

signal transfer point (STP): In a common channel signaling network, a switching center that provides for transfer from one signaling link to another. *Note:* In nonassociated common channel signaling, the signal transfer point need not be the point through which the call, which is associated with the signaling being switched, passes. *See also* **common channel signaling, link, nonassociated common channel signaling, signal, switching center.**

signal transformation: *Synonym* **signal shaping.**

signal transition: 1. In a signal, a change from one significant condition to another that occurs at a significant instant. **2.** In the modulation of a carrier, a change from one significant condition to another in order to represent the information contained in the modulating signal. *Note 1:* Examples of signal transitions are (a) a change from one electric current, voltage, or power level to another, (b) a change from one optical power level to another, or (c) a change from one frequency, wavelength, or phase to another. *Note 2:* Signal transitions are used to separate, i.e., to delimit, the significant conditions that are used to create signals that represent information, such as 0 and 1. **3.** In teletypewriter systems, the change from one significant condition, i.e., one signaling condition, to another, such as the change from "mark" to "space" or from "space" to "mark." *See also* **mark, signal, significant condition, significant instant, space.**

signal unit (SU): The group of bits or pulses that (a) is encompassed by a single error check process or covered by a specified set of error check bits, and (b) includes the check bits. *Note:* In most communications systems, all the information that is transferred by a system is formatted in signal units. *See also* **check bit, format.**

signal unit indicator: An indicator that (a) is in a specific field within a signal unit, and (b) is used to identify or distinguish signal units from one another. *See also* **field, indicator, signal unit.**

signature: 1. In communications-electronics, the complete baseband signal received from a source, such as an infrared source, a radio transmitter, a radar transmitter, or a target. *Note:* Signatures usually are analog signals that may be analyzed to (a) indicate the nature of the source, (b) locate the source, and (b) assist in recognition and identification of the source. **2.** The attributes of an electromagnetic wave that (a) has been reflected from or transmitted through an object, and (b) contains information indicating the attributes of the object. *See* **electromagnetic signature, penetration signature, radar signature, spectrum signature, target signature.**

signature analysis: An analysis that (a) is designed to yield information concerning a radiation source or a reflector of radiation, (b) is based on the measured characteristics of a received electrical, electromagnetic, or acoustic signal, and (c) is based on the frequency composition, amplitude, phase, direction, polarization, distortion, shape, and other features, characteristics, and attributes of the received signal. *See also* **spectrum analysis, spectrum signature.**

signature identification: *See* **aircraft signature identification.**

sign bit: In a specified digit position of a number, a bit that indicates whether the number is a positive or a negative number. *Note:* A 0 in a specified position in a number might indicate that the number is a negative number. *See also* **digit position.**

significance: *Synonym* **weight.**

significant backlog: In a communications system, a traffic backlog of such volume that (a) messages are being delayed at one or more points in the system, such as at a receiver, a transmitter, a message processing unit, a service section, a cryptographic section, a store and forward switching center, a booked call position, or a transceiver, (b) buffers are being filled, and (c) users are postponing calls and messages because of busy signals and delayed transmission. *Note:* Insufficient circuit capacity, personnel, or processing equipment or an excessive traffic volume may cause a significant backlog. Communications system saturation prevents overloading. A significant backlog is the result. *See also* **buffer, point, traffic.**

significant condition: In the modulation of a carrier, one of the discrete values of the signal parameter chosen to represent information. *Note 1:* The length or the duration of a significant condition occurs between successive significant instants. *Note 2:* Examples of significant conditions are (a) electric current, voltage, and power levels, (b) optical power levels and wavelengths, and (c) carrier frequencies, wavelengths, and phases. A significant condition may be the state or the condition of a device or a signal that corresponds to a quantized value of a characteristic chosen for modulation. A voltage level of 8 mV (millivolts) might represent a digit value that corresponds to a 1, and a voltage level of 0 might represent a digit value of 0, both conditions thus serving to represent numerals in the pure binary numeration system. Often, half values are used as the threshold of decision for 0s and 1s. In this case, a voltage of 4 mV or above would be interpreted as a 1, and a voltage below 4 mV would be interpreted as a 0. *Note 3:* Significant conditions are used to create signals that represent information, such as 0 and 1. *Note 4:* The significant conditions are recognized by an appropriate device. Each significant condition is determined at the instant that the device assumes a condition or a state usable for performing a specific function, such as recording, processing, or gating. *Note 5:* Transitions between significant conditions usually are not considered as significant conditions, except when coding is based on the direction of transition or on whether or not there is a transition. *See* **modulation significant condition.** *See also* **modulation, signal transition, significant instant, value.** *Refer to* **Figs. N-17 (635), N-18 (636), P-8 (737).**

significant condition number: In a signal coding system, such as is used in telegraph or digitized voice transmission, an integer that indicates the number of different significant conditions that are used to form or characterize the signal, such as in modulation, restitution, or quantization. *Note:* Examples of significant condition numbers are (a) 3 in the alternate mark inversion signal and in the bipolar coding signal, (b) 2 in the binary nonreturn to zero signal, and (c) 4 in the quadriphase modulation signal. The significant condition number must be greater than 1 in order to transfer information.

significant digit: In a numeral, a digit that is needed for a given purpose. *Note:* Examples of significant digits are (a) digits that must be kept to maintain a given accuracy or a given precision, (b) least significant digits, and (c) most significant digits.

significant instant: **1.** In a signal, such as that obtained in the modulation of a carrier, the instant at which successive significant conditions of the signal begin or end. *Note:* Each significant instant occurs when an appropriate device assumes a significant condition or state usable for performing a specific function, such as recording, processing, gating, or representing digits. **2.** In a time plot of a signal, such as a time plot of a pulsed electromagnetic wave, an instant at which a particular type of a usually repetitive event occurs, such as a transition to a different electric field strength, power level, polarization, frequency, or phase in a modulated wave. *Note:* In some coding schemes, the code is based on the transition that occurs at a significant instant rather than on the significant condition, such as (a) whether or not there is a transition or (b) the direction of transition. *See also* **decision instant, signal element, signal transition, significant condition, significant interval, start-stop distortion.** *Refer to* **Fig. P-4 (712).**

significant instant distortion: In modulation, demodulation, or transmission, the ratio of (a) the maximum temporal or spatial displacement, expressed algebraically, of the actual significant instant from the ideal or desired significant instant to (b) the unit interval, the displacement being considered positive when the actual significant instant occurs in time or space after the ideal instant. *Note:* Significant instant distortion usually is expressed as a percentage of the unit interval. *See also* **significant condition, significant instant, unit interval.**

significant interval: The period between two consecutive significant instants.

significant interval theoretical duration: In a signal, such as a telegraph code, digital data code, or modulated wave, the exact duration that is prescribed or desired for a significant interval.

sign language: A body language (a) in which the letters, words, phrases, and sentences of a natural language are formed by various positions of parts of the body, particularly the fingers, hands, arms, and head, (b) that usually is more precise than most body languages, and (c) that usually is not as precise as most natural spoken and written languages. *See also* **body language, natural language.**

silence: *See* **communications silence, cone of silence, electronic silence, listening silence, radio silence, ship radar silence, ship radio silence.**

silence zone: *Synonym* **skip zone.**

silent monitoring: To listen, unnoticed, to a telephone conversation. *Note:* Federal statutes limit silent monitoring. *See also* **eavesdropping.**

silica: Silicon dioxide (SiO_2) that (a) occurs in crystalline or amorphous form, (b) occurs naturally in impure forms, such as quartz and sand, (c) is the most common basic raw material of which glass is made, and (d) is purified and doped for making most communications optical fibers. *See* **fused silica, vitreous silica.**

silica cladded optical fiber: *See* **doped silica cladded optical fiber.**

silica graded optical fiber: *See* **doped silica graded optical fiber.**

silica optical fiber: *See* **all-silica optical fiber, hard clad silica optical fiber, lowloss FEP-clad silica optical fiber, plastic clad silica optical fiber.**

silica process: *See* **doped deposited silica process.**

silicon photodiode: A photodiode that (a) is silicon-based, (b) is either a PN or PIN photodiode, (c) is used for direct detection of optical wavelengths shorter than 1 μm (micron), and (d) has a greater bandgap and thus is quieter, i.e., generates less noise, than the germanium-based photodiodes that are usually used for wavelengths greater than 1 μm.

SILO: signal intercept from low orbit.

simple buffering: The assigning of buffer storage for the duration of the execution of a computer program. *See also* **buffer.**

Simple Mail Transfer Protocol (SMTP): The Transmission Control Protocol/Internet Protocol (TCP/IP) standard protocol that (a) allows for transfer of electronic mail messages, (b) specifies how two systems are to interact, and (c) specifies the format of messages used to control the transfer of the electronic mail. *See* **electronic mail, protocol.**

simple microscope: *Synonym* **magnifier.**

Simple Network Management Protocol (SNMP): The Transmission Control Protocol/Internet Protocol (TCP/IP) standard protocol that (a) is used to manage and control IP gateways and the networks to which they are attached, (b) uses IP directly, bypassing the masking effects of TCP error correction, (c) has a direct access to IP datagrams on a network that may be abnormally operating, thus requiring management, (d) defines a set of variables that the gateways must store, and (e) specifies that all control operations on the gateways are a side effect of fetching or storing those data variables, i.e., operations that are analogous to writing commands and reading statuses. *See also* **common management information protocol, gateway, protocol.**

simple scanning: In facsimile systems, scanning using only one scanning spot at a time. *See also* **facsimile, scanner, scanning, scanning spot.**

simplex: In communications, computer, data processing, and control systems, pertaining to a singular arrangement, condition, or state, such as one way only, one wire only, or one antenna transmitting at one carrier frequency.

simplex circuit: A circuit that provides transmission in one and only one direction. *Note:* Formerly, simplex circuits provided transmission in both directions but only in one direction at a time. This type of circuit is now called a "half-duplex circuit." *See also* **circuit, duplex circuit, duplex operation, half-duplex circuit, half-duplex operation, one-way only channel, phantom circuit, simplexed circuit, simplex operation.**

simplex operation: 1. Primarily in telephony, operation in which transmission occurs in one and only one preassigned direction. *Note:* Duplex operation may be achieved by simplex operation of two or more simplex circuits or channels. *Synonym* **one-way operation. 2.** Primarily in radio, operation in which transmission is conducted alternately in each direction of a telecommunications channel, such as by means of manual control. *Note:* In radio communications, duplex operation and semiduplex operation require two frequencies, whereas simplex operation may use one or two frequencies. *Synonym* **two-way alternate communications.**

simplex radiotelegraph procedure: A radiotelegraph network operational procedure in which (a) a calling station, such as a ship station, calls another station, such as a shore station, (b) the called station answers, and (c) the calling station transmits its message, all on the same frequency. *See also* **cross band radiotelegraph procedure.**

simplex signaling (SX): Signaling in which (a) two conductors are used for a single channel, (b) a center-tapped coil, or the equivalent, is used at both ends of a circuit, including user circuits, (c) a one-way signaling scheme suitable for intracentral-office signaling may be used, and (d) the two legs may be connected to duplex, i.e., full duplex, signaling circuits that then function like composite signaling (CX) circuits with E and M lead control. *See also* **E and M lead signaling, signal, simplex operation.**

simplex transmission: *Synonym* **two-way alternate transmission.**

SIMSCRIPT: A computer programming language that is primarily used for digital simulation. *See also* **digital, simulation.**

simulate: To represent certain features of the behavior of a physical or abstract system by the behavior of another system. *Note 1:* Examples of simulation are (a) to use delay lines to represent propagation delay and phase shift caused by an actual transmission line, (b) to represent a physical phenomenon by means of operations performed by an analog or digital computer, (c) to represent the operations of a computer by those of another computer, (d) to represent the economy of a country by using computer hardware and software, (e) to represent the operations of a communications system, transportation system, traffic control system, or an airline, and (f) to represent war games. *Note 2:* The simulator may imitate only a few of the phenomena, operations, or functions of the unit it simulates. *See also* **emulate.**

simulation: The representation of the features, characteristics, parameters, or behavior of a physical or abstract system by the behavior of another system. *See also* **emulate, emulator, simulate, simulator.** *Refer to* **Fig. T-10 (1041).**

simulative electronic deception: *See* **electronic deception.**

simulator: A functional unit that can imitate the behavior of the hardware or the software of a physical or abstract system. *Note:* Examples of a simulator are (a) a programmed computer that can represent the behavior of a communications network, the economy or ecology of a country, or a set of war games, or (b) a device that can be controlled by an operator who responds to instrument displays that represent the actual flight of an aircraft. *See* **equilibrium mode simulator.** *See also* **computer simulation, emulator, functional unit.**

simultaneous: Pertaining to the occurrence of two or more events at the same instant.

simultaneous communications: *See* **two-way simultaneous communications.**

simultaneous direction finding: Direction finding (DF) in which (a) a control station of a net requests addressees to obtain specified information on an identified transmission at a precise instant and to report the information in a formatted message, and (b) the stations in the net report to the control station in alphabetical order of their call signs. *See also* **control station, direction finding.**

SINAD: signal plus noise plus distortion to noise plus distortion ratio.

SINCGARS: Single-Channel Ground and Airborne Radio System.

sine condition: The property of a lens in which the sine of the maximum angular opening of the axial bundle of refracted rays is no longer inversely proportional to the F number for a given refractive index. *See also* **aperture ratio, aplanatic lens, F number, ray, ray bundle, refraction, refractive index.**

sine junction gate: *Synonym* **A AND NOT B gate.**

sine wave: A wave that (a) has instantaneous values that correspond to the relation $A \sin \theta = A \sin \omega t$ where A is the peak amplitude, i.e., the maximum amplitude, θ is the angular displacement, ω is the angular velocity, i.e., $\omega = 2\pi f$ where π is approximately 3.1416 and f is the frequency, and t is time, (b) has smooth transitions from one point to another, such as from peak positive values to peak negative values, and vice versa, (c) unless specially contrived, is generated by electrical oscillators, and (d) may be constructed graphically by plotting the magnitude of the projection, on a diameter of a circle, of the radius vector of the circle drawn from the origin to a point on the circle as the point moves around the circle, such that, in Cartesian coordinates, the angular displacement of the radius vector is plotted as the independent variable, i.e., the x-axis coordinate, and the length of the projection as the dependent variable, i.e., the y-axis coordinate. *See also* **Fourier analysis, value.**

sine wave object: An object that has a sinusoidal variation of luminance, having the advantage that the image will have a sinusoidal variation of illuminance, and the only effect of degeneration by a lens system will be to decrease the modulation amplitude in the image relative to that in the object. *See also* **luminance, image, object.**

sine wave response: *Synonym* **modulation transfer function.**

singing: An undesired self-sustained audio oscillation in a circuit. *Note:* Singing usually is caused by positive feedback, excessive gain, or unbalance of a hybrid termination. *Note:* When singing occurs at a very low frequency, it is called "motorboating." *See also* **circuit, motorboating, singing margin, singing point, unbalanced line.**

singing device: *See* **antisinging device.**

singing margin: The difference in power levels between the singing point and the operating gain of a system or a component.

singing point: In the gain of a system, the threshold point at which additional gain will cause sustained self-oscillation, i.e., singing. *See also* **singing.**

singing suppression circuit: A circuit or a component that (a) is inserted into another circuit, (b) prevents or reduces the tendency for singing to occur, (c) may limit the gain in the circuit in which it is inserted, and (d) usually provides negative feedback. *See also* **anti-singing device, volcas circuit.**

single band: Pertaining to communications equipment, such as radio transmitters and receivers, that can operate in only one frequency band. *See also* **band.**

single bundle cable: *See* **single-channel single-bundle cable.**

single call: In radiotelegraph transmission, a call in which only one call sign precedes the call sign of the calling station. *See also* **call sign.**

single call message: A message that will not require reprocessing, such as routing line segregation, at any relay station. *See also* **relay station, routing line segregation.**

single call sign calling: In radiotelegraph and radiotelephone operations, calling in which (a) communications are established and conducted such that subordinate station call signs are used exclusively, and (b) prior agreement has been reached by participating entities, such as organizations, stations, networks, and countries. *See also* **call sign.**

Single-Channel Ground and Airborne Radio System: A military radio communications system that allows communications (a) between ground and air forces, usually to coordinate battlefield operations, (b) horizontally between battlefield units, and (c) vertically between command echelons.

single-channel single-bundle cable: A fiber optic cable that consists of a single fiber optic bundle with buffers and strength members, all inside a protective cover, such as a jacket. *See also* **fiber optic bundle, fiber optic cable, jacket.**

single-channel single-fiber cable: A fiber optic cable that consists of a single optical fiber with buffers and strength members, all inside a protective cover, such as a jacket. *See also* **fiber optic cable, jacket, optical fiber.**

single clear: In telephone switchboard operation, a single steady supervisory signal that may be received during double supervisory working, and that indicates that a user has cleared at one end of a circuit but not at the other end. *See also* **clear, double supervisory working, supervisory signal.**

single current signaling: Direct current (dc) telegraphy in which (a) on-off working is used, (b) the battery voltage is applied to the line in the mark condition, and electrical current flows to operate the receive relay, (c) no voltage or current is applied in the space condition, (d) both transmitter and receiver are biased to the off condition by mechanical or electrical means, (e) operation is restricted to slow signaling speeds as the line takes time to charge and discharge, (f) noise and interference affect the receiver during the space condition, (g) bias of the receive relay needs frequent adjustment to balance varying line conditions, (h) because the line current must overcome the bias, the relay sensitivity is low, and (i) unidirectional currents are used. *Synonyms* **neutral telegraphy, neutral transmission, single current telegraphy, single current transmission, unipolar telegraphy, unipolar transmission.**

single current system: *Synonym* **neutral direct current telegraph system.**

single current telegraphy: *Synonym* **single current signaling.**

single current transmission: *Synonym* **single current signaling.**

single current transmission system: *Synonym* **neutral direct current telegraph system.**

single domain network: A network that has one and only one host computer. *See also* **host computer, network.**

single-ended control: *Synonym* **single-ended synchronization.**

single-ended synchronization: Synchronization (a) that is used between two locations, and (b) in which phase error signals used to control the clock at a location are derived from a comparison of the phase of the incoming signals and the phase of the internal clock at the same location. *Synonym* **single-ended control.** *See also* **bilateral synchronization, clock, double-ended synchronization, synchronization, unilateral synchronization system.**

single frequency conversion: *Synonym* **single frequency translation.**

single frequency interference: **1.** Interference caused by a single frequency source. *Note 1:* The interference caused by the single frequency source may have other frequencies and may also appear in many channels. *Note 2:* Examples of single frequency interference are (a) the interference in a transmission channel induced by a 60-Hz (hertz) power source that may have harmonics that contribute to the interference, and (b) the interference caused in a transmission channel by a direct current (dc) power supply that has a ripple voltage in addition to the dc. **2.** In a circuit, device, or system, such as a telephone line, interference that contains only a narrow band of frequencies such that they appear, sound, or are measured as one frequency. *See also* **frequency, interference, ripple voltage, single tone interference.**

single frequency (SF) signaling: In telephone communications, signaling in which dialing or supervisory signals are conveyed with one or more specified single frequencies. *Note 1:* The signals usually are used in in-band signaling for long-haul communications. *Note 2:* The Defense Data Network (DDN) transmits direct current (dc) signaling pulses or supervisory signals, or both, over carrier channels or cable pairs on a four-wire basis using a 2600-Hz signal tone. Single frequency (SF) signaling units convert the signaling pulses or supervisory signals into signal tones, or vice versa. *See also* **frequency, in-band signaling, signal, signaling.**

single frequency signaling system: In telephone communications, a system in which (a) single frequency signaling is used, and (b) the conversion of pulses into tones, or vice versa, is done by single frequency signal units. *Note:* A system could transmit direct current (dc) signaling pulses or supervisory signals, or both, over carrier channels or cable pairs on a four-wire basis using a 2600-Hz (hertz) signal tone. *See also* **single frequency signaling.**

single harmonic: **1.** One of the harmonics of the fundamental frequency, i.e., of the first harmonic. **2.** A sinusoidal wave of only one frequency, which would be the fundamental frequency, i.e., the first harmonic, given by the relation $A = A_m \sin \omega t$ where A is the instantaneous value of the single frequency sine wave, A_m is the maximum amplitude, i.e., the peak value, of the wave, ω is the angular velocity equal to $2\pi f$ where π is approximately 3.1416 and f is the frequency, and t is time. *See also* **fundamental frequency, harmonic, sine wave.**

single harmonic distortion: In a signal, the ratio of (a) the power of a given harmonic, such as the second, third, or fourth harmonic, to (b) the power of the fundamental frequency, i.e., the first harmonic. *Note:* Single harmonic distortion is measured at the output of a device under specified conditions and usually is expressed in dB or percent. *See also* **distortion, frequency, fundamental frequency, harmonic distortion, total harmonic distortion.**

single heterojunction: In a laser diode, a junction that (a) performs two energy level shifts and two refractive index shifts and (b) provides increased confinement of radiation direction, improved control of radiative recombination, and reduced nonradiative (thermal) recombination. *See also* **laser diode, radiative recombination.**

single lens: A lens composed of only one piece of optical material, such as glass or plastic. *See also* **compound lens.**

single-line repeater: A telegraph repeater that uses a pair of cross-coupled polar relays inserted in series with a circuit to insert additional power into the signal.

single mode: *See* **mode.**

single mode fiber: *Synonym* **single mode optical fiber.**

single mode launching: The insertion of an electromagnetic wave into a waveguide in such a manner that (a) only one propagation mode is coupled into, and hence transmitted by, the guide, (b) various parameters, such as incidence angle, beam diameter, skew ray angle, and source to waveguide longitudinal displacement, are controlled, and (c) propagation of the mode depends on waveguide dimensions, wavelength, and refractive indices of the material constituting the guide.

single mode optical fiber: An optical fiber in which only one bound mode, i.e., the lowest-order bound mode, can propagate at a given wavelength, numerical aperture, and core radius. *Note 1:* The lowest-order bound mode may be a pair of orthogonally polarized electric and magnetic fields. *Note 2:* To support one mode, the core radius must be less than twice the wavelength of the source of radiation, and the numerical aperture must be adjusted accordingly. *Note 3:* In step index optical fibers, single mode operation occurs when the normalized frequency, V, is less than 2.405. For power law profiles, single mode operation occurs for a normalized frequency, V, less than approximately $2.405[(g+2)/g]^{1/2}$, where g is the profile parameter. *Note 4:* If appropriate conditions are met, the orthogonal polarizations will not be associated with degenerate modes. *Synonyms* **monomode fiber, monomode optical fiber, single mode fiber.** *See* **dispersion-unshifted**

single mode optical fiber. *See also* **bound mode, multimode optical fiber, normalized frequency, profile parameter, step index optical fiber.** *Refer to* **Fig. S-15 (931).**

single mode optical waveguide: An optical waveguide that is capable of supporting the propagation of only one mode at a given wavelength. *Note:* An optical waveguide designed to operate in single mode at a given wavelength may support more than one mode if operated at shorter wavelengths. *See also* **optical waveguide, single mode, multimode, single mode optical fiber.**

single Morse system: *Synonym* **neutral direct current telegraph system.**

single node network: *See* **network topology.**

single-operator period watch: A communications watch in which (a) the duration of the watch is based on availability of only one operator at a station, (b) if there are several stations in contact, the watch can be rotated among them, (c) messages that are intended for a given station can be held by the other stations when the operator of the given station is off duty, and (d) the messages can then be relayed when the operator is on duty. *See also* **communications watch, continuous communications watch, one-operator ship, two-operator period watch, watch.**

single optical fiber: An optical fiber that is optically isolated from, but may be combined with, other optical fibers to form fiber optic cables, aligned bundles, unaligned bundles, and fiber optic faceplates. *See also* **aligned bundle, fiber optic cable, optical fiber.**

single-polarized antenna: An antenna (a) that radiates or receives radio waves with a specific polarization, and (b) in which the desired sense of polarization is maintained only for certain directions or within the major portion of the radiation pattern. *See also* **antenna.**

single position magneto switchboard: A magneto switchboard that is operated by one operator without a supervisor. *See also* **magneto switchboard, position.**

single precedence message: A message in which (a) the same precedence is applicable to all addressees, i.e., to both action addressees and information addressees, and (b) only one precedence designator is needed. *See also* **dual precedence message, precedence, precedence designator.**

single pulse carrier wave (CW) sonar: *Synonym* **searchlight sonar.**

single sideband (SS or SSB): Pertaining to amplitude modulation that (a) primarily is used in carrier telephony and high frequency (HF) radio to increase transmission efficiency, i.e., power efficiency, (b) is used to increase electromagnetic spectrum utilization in terms of the total number of channels available in a given bandwidth, (c) uses only one sideband for transmission while the other sideband and the carrier are suppressed, and (d) although proposed for the uplink and the downlink of satellite systems, has had limited use in satellite systems. *See also* **amplitude modulation, electromagnetic spectrum, sideband.**

single sideband (SSB) emission: An amplitude-modulated emission with only one sideband. *See* **full carrier single sideband emission, reduced carrier single sideband emission, suppressed carrier single sideband emission.** *See also* **carrier, double sideband reduced carrier transmission, double sideband suppressed carrier transmission, double sideband transmission, sideband transmission.**

single sideband (SSB) equipment reference level: The power of one of two equal tones that, when used together to modulate a transmitter, cause it to develop its full rated peak power output. *See also* **level, peak power output, rated power output, reference circuit, sideband transmission.**

single sideband noise power ratio: The ratio of (a) the output power, measured with a notch in, to (b) the output power, measured with the notch out. *Note 1:* Measurements are made in which (a) notched noise is used, (b) power is in the notch bandwidth, and (c) power is measured at the output of the device for which the single sideband (SSB) noise power ratio is being determined. *Note 2:* The input power must be sufficient to maintain a constant total system mean noise power output. *See also* **noise, notch, notched noise.**

single sideband suppressed carrier (SSB-SC) transmission: Single sideband transmission in which (a) the carrier is suppressed, and (b) the carrier power level is suppressed so that it is insufficient for signal demodulation.

single sideband transmission: Sideband transmission in which (a) only one sideband is transmitted, and (b) the carrier may be suppressed.

single sideband voice: A mode of voice radio transmission in which one sideband is transmitted, while the carrier and the other sideband are suppressed. *See also* **carrier, sideband.**

single simplex operation: *Synonym* **two-way alternate communications.**

single tone interference: An undesired discrete frequency, i.e., a single harmonic, that (a) appears in a transmission channel and (b) appears in the channel regardless of the nature of the source. *See also* **channel, frequency, interference, single frequency interference.**

single-wire line: A communications line that has only one wire, with the ground, i.e., earth, serving as the return part of the circuit. *Synonym* **one-wire line.**

sink: 1. In communications, electrical, and optical circuits, an absorber of energy. *Note:* A resistor may serve as a sink for real power. A capacitor or an inductor may serve as a sink for reactive power. *See also* **load. 2.** In communications systems, a device that receives signals, such as user data and control signals, from a source. **3.** In communications systems, the part of the system in which messages are considered to be received. *See also* **communications sink, data circuit-terminating equipment, data sink, data terminal equipment, destination user, source, source user.**

sink user: *Synonym* **destination user.**

S interface: For basic rate access in an Integrated Services Digital Network (ISDN) environment, a user-to-network interface reference point that (a) is characterized by a four-wire, 144-kbps (kilobits per second) (2B+D) user rate, (b) serves as a universal interface between ISDN terminals or terminal adapters and the network channel termination, (c) allows a variety of terminal types and customer networks, such as private branch exchanges (PBXs), local area networks (LANs), and controllers, to be connected to the network, and (d) operates at 4000 48-bit fps (frames per second), i.e., 192 kbps, with a user portion of 36 bpf (bits per frame), i.e., 144 kbps. *See also* **Integrated Services Digital Network, R interface, T interface, U interface.**

sinusoidal function: A mathematical function that expresses the amplitude of the projected line segment formed on the diameter by a point that is moving in a circle, or, if the speed of the moving point is constant, the amplitude of the projection as a function of time of the uniformly rotating radius vector, and given by the relation $A\sin\theta = A\sin\omega t$ where A is the peak amplitude, i.e., the maximum amplitude, θ is the angular displacement, ω is the angular velocity, i.e., $\omega = 2\pi f$ where π is approximately 3.1416 and f is the frequency, and t is time, and also given by functions of such a function. *Note:* Electromagnetic waves and alternating currents (ac) usually are sinusoidal functions, or at least can be represented as a summation of such functions. *See also* **Fourier analysis.**

siphon recorder: A recorder in which the recording stylus is fed with ink via a fine siphon. *See also* **stylus.**

site: *See* **downrange site, remail site.**

site link: *See* **cross-site link.**

situation awareness beacon with reply (SABER): A global positioning system (GPS) that (a) provides a two-way communications capability, (b) uses ultrahigh frequency (UHF), (c) acquires GPS signals, (d) tags the positional information with a unique beacon identification code for rebroadcast via satellite-to-ground stations, and (e) provides the acquired information for relaying to command center terminals for real time monitoring. *See also* **global positioning system.**

situation report: A report or a message that is used to describe an existing or ongoing situation, incident, or action. *See also* **scene of action communications.**

SI unit: Système International unit.

six-bit byte: *Synonym* **sextet.**

six-zero substitution: *See* **bipolar with six-zero substitution.**

size: *See* **increment size, plotter size, spot size.**

skew: 1. In parallel transmission, the difference in arrival time of bits transmitted at the same time. **2.** For data recorded on multichannel magnetic tape, the difference in reading time of bits recorded in a single transverse row. **3.** In facsimile systems, the angular deviation of the received frame from rectangularity caused by asynchronism between the scanner and the recorder. **4.** In a nonrotating object undergoing translation, the angle between (a) the longitudinal axis of the object and (b) the direction of motion of the object. *Note:* Skew is numerically expressed as the tangent of the deviation angle. **4.** In facsimile systems, the angle between the scanning line, or recording line, and the perpendicular to the paper path. **5.** In lens systems and optical waveguides, such as optical fibers, pertaining to a light ray that is not parallel to the optical axis. *See also* **facsimile, optical axis.**

skew ray: In a multimode optical fiber, a bound ray, i.e., a totally internally reflected ray, that travels in a helical path along the fiber and thus (a) is not parallel to the fiber axis, (b) does not lie in a meridional plane, and (c) does not intersect the fiber axis. *See also* **axial ray, geometric optics, hybrid mode, meridional ray, paraxial ray.**

skim: In optical elements, streaks of dense seeds with accompanying small bubbles or inclusions. *See also* **heavy seeding, inclusion, optical element, scattered seed, seed.**

skin antenna: An antenna that can be conformally placed on a surface, such as the outer surface of an aircraft, spacecraft, ground vehicle roof, shelter roof, or ship mast.

skin effect: The tendency of alternating current (ac) to flow near the surface of a conductor, thereby (a) restricting the current to a small part of the total cross-sectional area and (b) increasing the resistance to the flow of current. *Note:* Skin effect is due to the inductance of the conductor, which causes an increase in the inductive reactance, especially at high frequencies. The inner filaments experience an inductive reactance with all the surrounding filaments, their reactance thus being higher than that of the outer filaments. Thus, the current tends toward the lower-reactance filaments, i.e., the outside filaments. At high frequencies, the circumference is a better measure of resistance than the cross-sectional area. The depth of penetration of current at high frequencies can be very small compared to the diameter. Skin effect must be taken into account in designing antennas and metallic waveguides. *See also* **inductive reactance.**

skip: To pass over or ignore. *Note:* Examples of skip are (a) to pass over one or more instructions in a computer program without executing them, (b) to pass over one or more locations in a data medium, such as to perform one or more line feed operations in sequence, (c) to pass over recorded text without reading it, and (d) to have an electromagnetic wave pass over a skip zone. *See also* **skip zone.**

skip distance: In a transmitted wave reflected from the ionosphere and in a given direction from the transmitter, the distance between the transmitter and the closest point of return of the wave to the Earth. *See also* **hop, ionosphere, skip zone.**

skip zone: Within the transmission range of a transmitter, a ring-shaped region, i.e., a washer-shaped region, bounded (a) on the inside by the locus of the farthest points reached by the ground wave and (b) on the outside by the locus of the nearest points at which the reflected sky waves return to the Earth. *Note:* In the skip zone, signals from the transmitter are (a) below the reception threshold of receivers, (b) are not received, or (c) have such a low signal to noise ratio (S/N) that clear reception is impossible. *Synonyms* **silent zone, zone of silence.** *See also* **direct ray, ionosphere, skip distance.**

sky noise: The total noise that (a) is received by an antenna from sources that are outside the Earth, i.e., from nonterrestrial sources, and (b) is not from man-made sources, such as satellites, missiles, and spacecraft, even though they are outside the Earth.

skywave: 1. A radio wave that travels vertically upward from the antenna. *Note:* A sky wave may be returned to the Earth from the ionosphere. **2.** A radio wave that descends upon the Earth at a high elevation angle, usually above 30° with the Earth's surface. *See also* **direct ray, ionosphere, surface wave.**

skywave management for automatic robust transmission network (SMART NET): A network that provides (a) multimedia message handling, (b) service for worldwide networks, and (c) automatic data and voice communications beyond the line of sight (LOS) without the aid of relays or satellites. *Note:* The skywave management for automatic robust transmission network (SMART NET) was developed and built by Rockwell International.

slab dielectric optical waveguide: A slab dielectric waveguide that (a) operates in the optical spectrum of the electromagnetic spectrum, (b) consists of rectangular layers or ribbons of materials of differing refractive indices that support one or more lightwave transmission modes, (c) guides the waves by confining certain modes in the highest refractive index medium, the lower-index media serving as cladding, jacketing, and surrounding media, and (d) is used in optical integrated circuits (OICs) for geometrical convenience, in contrast to optical fibers in fiber optic cables used for long-distance transmission. *See also* **optical fiber, optical spectrum, refractive index, slab dielectric waveguide.** *Refer to* **Fig. P-2 (708).**

slab dielectric waveguide: An electromagnetic waveguide that (a) consists of a dielectric propagation medium of rectangular cross section (b) has a width and a thickness, refractive indices, and a wavelength that determine the modes the guide will support and hence transmit over a long distance, i.e., more than a hundred wavelengths or beyond the equilibrium length, (c) may be cladded, protected, distributed, and electronically controllable, (d) may be used in various applications, such as in optical integrated circuits (OICs) for geometrical convenience, in contrast to the round optical fibers used in fiber optic cables for long-distance transmission, (e) usually is a polarization-maintaining waveguide, and (f) operates on the same principle as optical fibers. *Synonym* **planar waveguide.** *See* **doubly cladded slab dielectric waveguide.** *See also* **core area.** *Refer to* **Fig. S-12.** *Refer also to* **Figs. I-5 (470), P-2 (708), P-17 (767).**

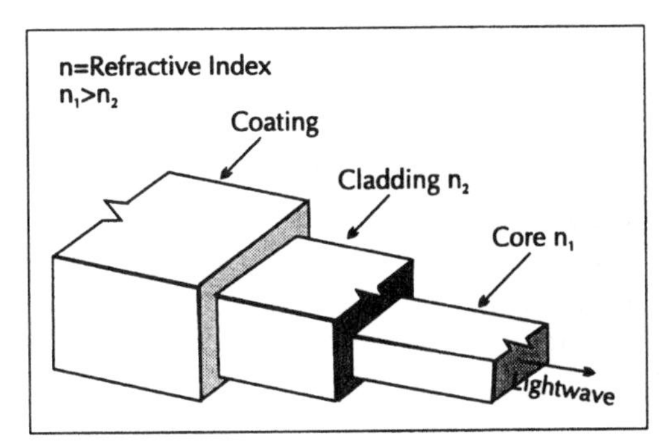

Fig. S-12. A **slab dielectric waveguide** that (a) is a few microns wide and thick, and (b) might be formed by an etching or chemical vapor deposition process on an optical integrated circuit (OIC) chip.

slab interferometry: Measurement of the refractive index profile of a dielectric waveguide, such as an optical fiber or a slab dielectric waveguide, by using (a) a thin sample, i.e., a short length, that has its faces perpendicular to the axis of the waveguide, and (b) an interferometer that scans across an endface. *Note:* The index profile is calculated from the interference patterns obtained by the interferometer. *Synonyms* **axial interference microscopy, axial slab interferometry, slice interferometry.** *See also* **interferometer, transverse interferometry.**

slant range: The line of sight distance between two points, not at the same level, i.e., not at the same height or altitude, relative to a specific datum. *Note 1:* The line of sight distance from a radar antenna to an object, i.e., a target, is the slant range. *Note 2:* The slant range and the elevation angle are used to calculate the horizontal and vertical distances from an antenna to a target. *See also* **antenna, horizon angle, line of sight propagation.**

slave clock: A clock that is coordinated with, i.e., controlled by, a master clock. *Note 1:* Slave clock coordination usually is achieved by phase-locking the slave clock signal to a signal received from the master clock. *Note 2:* To adjust for the transit time of the signal from the master clock to the slave clock, the phase of the slave clock may be adjusted with respect to the signal from the master clock so that both clocks are in phase. Thus, the time markers of both clocks, at the output of the clocks, occur simultaneously at their respective locations. *See also* **master clock.**

slave station: 1. In a data network, the station that (a) is selected and controlled by a master station, and (b) usually can only call, or be called by, a master station. **2.** In navigation systems that employ precise time dissemination, a station that has a clock that is synchronized by a master station. *See also* **control station, master station, primary station, secondary station, tributary station.**

slave timing system: *See* **master-slave timing system.**

SLD: superluminescent diode.

SLED: superluminescent light-emitting diode.

sleek: In optical elements, a polishing scratch without visible conchoidal fracturing at the edges of the scratch. *See also* **optical element.**

sleeve splicer: *See* **precision sleeve splicer.**

sleeve splicing: *See* **precision sleeve splicing.**

slew: *See* **clock phase slew.**

slewing: 1. Rotating a directional antenna or transducer rapidly, usually in a horizontal or vertical direction or both. **2.** Changing the frequency or the pulse repetition rate of a signal source. **3.** Changing the tuning of a receiver, usually by sweeping through many or all frequencies.

slice: *See* **time slice.**

slice interferometry: *Synonym* **slab interferometry.**

sliding window: A variable-duration window that allows a sender to transmit a specified number of data units before an acknowledgment is received or before a specified event occurs. *Note:* An example of a sliding window in packet transmission is one in which, after the sender fails to receive an acknowledgment for the first transmitted packet, the sender "slides" the window, i.e., resets the window, and sends a second packet. This process is repeated for a specified number of times before the sender interrupts transmission. *Synonym* **acknowledgment delay period.**

slip: In a sequence of transmitted symbols, a signal phase shift, i.e., a signal positional displacement, that causes (a) the loss of one or more symbols or (b) the insertion of one or more extraneous symbols. *Note:* Slips usually are caused by inadequate synchronization of the two clocks controlling the transmission and reception of the signals that represent the symbols. *See* **bit slip, digital slip, frame slip.** *See also* **clock.**

slip-free operation: Operation of a communications system with sufficient phase locking (a) to prevent slip or (b) to prevent the overflowing or emptying of buffers. *See also* **phase.**

slope: In a transmission line, the rate of change of attenuation with respect to frequency over the frequency spectrum. *Note 1:* The slope usually is expressed in dB per hertz. *Note 2:* In metallic transmission lines, the slope and the attenuation, and consequently the attenuation rate, usually are greater at high frequencies than at low frequencies. *See* **chromatic dispersion slope, line slope, minimum dispersion slope, zero-dispersion slope.** *See also* **attenuation, frequency, slope equalizer.**

slope equalizer: A device or a circuit used to achieve a specified slope in a metallic transmission line. *See also* **attenuation, frequency, slope.**

slope facility: *See* **glide slope facility.**

slope indicator: *See* **glide slope indicator.**

slope-keypoint compaction: Data compaction accomplished by stating (a) a specific point of departure, (b) a direction or a slope of departure, (c) the maximum deviation from a prescribed specific value, and (d) a new point and a new slope. *Note:* An example of slope-keypoint compaction is the storage or the transmission of a slope and one point on a straight line instead of storing and transmitting a large number of values, i.e., of points, on the line.

slope station: *See* **glide path slope station.**

slot: In a distributed queue dual bus (DQDB) network, a protocol data unit (PDU) that (a) consists of 53 octets used to transfer segments of user information, (b) has the capacity to contain a segment of 52 octets and a 1-octet access control field, and (c) may be a prearbitrated (PA) slot or a queued arbitrated (QA) slot. *See* **channel time slot, digit time slot, frame alignment time slot, prearbitrated slot, signaling time slot, time slot.** *See also* **cell, data unit, octet, protocol.**

slot antenna: A radiating element formed by a slot in a conducting surface or in the wall of a waveguide. *See also* **antenna.**

slot signaling: *See* **out-of-slot signaling.**

slotted ring network: A ring network that allows unidirectional data transmission between data stations by transferring data in predefined slots in the transmission stream over one propagation medium such that the data return to the originating station. *See also* **network topology.**

slot time: In communications systems using carrier sense multiple access networks with collision detection (CSMA/CD), the length of time that a transmitting station waits before trying to transmit again after a collision, which varies from station to station. *Note:* When implementing the CSMA/CD protocol, a transmitting data station that detects another signal while transmitting, stops sending, and it may then send a jam signal. *See also* **carrier sense multiple access with collision detection.**

slowdown: *See* **network slowdown.**

slow-sweep noise-modulated jamming: Jamming in which the radar screen displays bright "walking" strobes that vary a few degrees in azimuth and fluctuate in intensity. *See also* **jamming. strobe.**

slow time scale factor: *Synonym* **extended time scale factor.**

slug: *See* **type slug.**

slug matrix storage: *Synonym* **magnetic rod storage.**

smart antenna: A phased-array antenna that (a) has no moving parts, (b) is computer-controlled, and (c) can be programmed for automatic adjustment of beam frequency, forming, steering, and polarization.

smart card: A small plastic credit-card-size device that (a) usually contains digital data and an embedded integrated circuit (IC) or computer chip capability, (b) has applications in various areas, such as public pay telephone, bank account access, building and computer network security, parking meter, and point-of-sale payment systems, (c) is used to improve transaction speed, financial management, buying pattern analysis, system security, transportation management, theater marketing, retail promotion, and telecommunications access, (d) is an improvement over magnetic stripe and holographic cards in terms of security, simplicity, and flexibility, and (e) may be either (i) a memory card, such as a free access or a secure access card, or (ii) a microprocessor card. *Note:* Smart cards preclude the use of cash and check payment for each transaction. *See also* **free-access card, memory card, microprocessor card, security access card.**

SMART NET: skywave management for automatic robust transmission network.

smart patch: A piezoceramic sensor that (a) detects vibration, (b) inhibits the vibration by generating frequencies that counter it, (c) detects strain levels five orders of magnitude below those detected by conventional strain gauges, (d) is lightweight, and (e) may be used for improving (i) stability of structures, (ii) precision control of telescopes, (iii) structural fault detection, (iv) automobile and aircraft vibration reduction, (v) nonintrusive fluid sensing, and (vi) artillery aiming. *See also* **piezoelectric.**

SMART-T: secure mobile antijam reliable tactical terminal.

smart terminal: *Synonym* **programmable terminal.**

SMDR: station message detail recording.

Smith chart: A chart that is used (a) to establish a set of real and complex coordinates for the transmission line generalized voltage reflection coefficient and (b) to aid in transmission line performance analysis, such as to determine impedance, phase, and frequency values at various fractions of a wavelength toward a sinusoidal source of power, from a load of given impedance connected to a uniform line of given characteristic impedance. *See also* **characteristic impedance, load, reflection coefficient, uniform transmission line.**

SMNP: Simple Network Management Protocol.

smoke: *See* **pyrotechnic smoke.**

smooth: To reduce rapid fluctuations. *Note:* Examples of smoothing are (a) to decrease sudden changes in data values, (b) to reduce sudden or large changes in the magnitude of an electrical current, (c) to eliminate voltage surges, (d) to remove ripple voltages from a direct current (dc) source, and (e) to draw a line with gentle curves that best matches a set of plotted points.

smooth Earth: 1. Idealized surfaces, such as water surfaces or very level terrain, having radio horizons that are not formed by prominent ridges or mountains but are determined solely as a function of antenna height above ground and the effective Earth radius. **2.** A theoretical Earth that is perfectly round and has no surface irregularities, such as mountains, valleys, and canyons. *Note:* The smooth Earth concept is used for theoretical calculations related to radio wave transmission, such as line-of-sight and radio horizon calculations. *See also* **antenna height above average terrain, path profile, radio horizon.**

S/N: signal to noise ratio.

S+N: signal plus noise.

snapshot dump: A dump that (a) is both a dynamic dump and a selective dump and (b) may be made at several places during the execution of a computer program. *See also* **computer program, dump, dynamic dump, execute, selective dump.**

sneak current: 1. In a communications circuit, an unwanted, steady, slightly higher than normal current that presents no immediate danger, but may cause improper operation of, or damage to, the circuit. **2.** In an electric power, communication, or control circuit, an electrical current that (a) is coupled from other circuits into wire lines, (b) may be too weak to cause immediate trouble or damage, and (c) can produce harmful effects, such as heating and corrosion, if allowed to continue over long periods.

Snell's law: When an electromagnetic wave, such as a light ray, (a) passes into a higher refractive index medium, its path is deviated toward the normal to the surface at the point of incidence, i.e., the refraction angle is less than the incidence angle, (b) when the wave passes into a lower refractive index medium, its path is deviated away from the normal, i.e., the refraction angle is greater than the incidence angle, (c) the incidence angle is equal to the reflection angle, and (d) the relationship among the incidence angle, the refraction angle, and the refractive indices on both sides of the interface surface is given by $\sin\theta_1/\sin\theta_2 = n_2/n_1$ where θ_1 is the incidence angle, θ_2 is the refraction angle, n_2 is the refractive index of the medium into which the ray enters after passing through the interface, and n_1 is the refractive index of the medium containing the incidence ray, all angles being with respect to the normal to the surface at the point of incidence. Thus, the ratio of the sines of the incidence and refraction angles is a constant, provided that there is a refracted ray. *See also* **critical angle, incidence angle, reflection angle, refraction angle, refractive index, ray.** *Refer to* **Fig. S-13.**

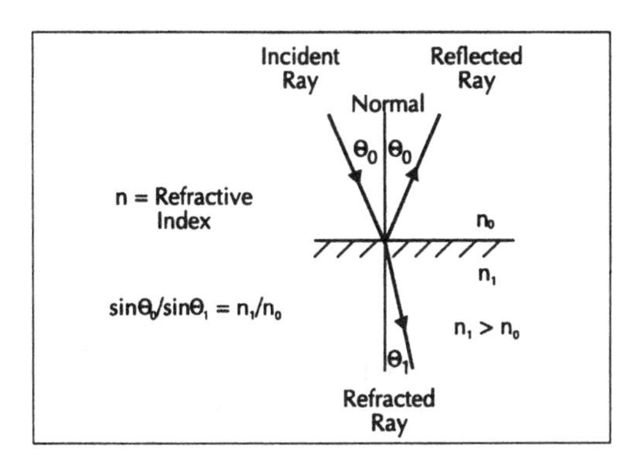

Fig. S-13. Reflection and refraction in accordance with **Snell's law.** The incidence and reflection angles are equal. The ratio of the sines of the incidence and refraction angles equals the inverse ratio of the refractive indices of the media.

(S+N)/N: signal plus noise to noise ratio.

SNOBOL: A computer programming language that is used primarily for string handling. *See also* **computer programming, string.**

snow: 1. In video display systems, noise that (a) is distributed uniformly on the display surface, such as that of a television (TV) or radar screen, (b) has the appearance of a uniform distribution of fixed or moving spots, mottling, cross hatching, or speckling, and (c) usually is caused by random noise on an intensity-modulated signal in a display device, such as a cathode ray tube (CRT). **2.** On radar display screens, an effect that (a) is caused by sweep jamming and (b) has a cross-hatched or speckled appearance.

SNR: signal to noise ratio.

soft: *See* **fail soft.**

soft copy: 1. A nonpermanent display image. **2.** A display element, display group, or display image that (a) is not permanent, (b) is not separable from the display space on the display surface of a display device, (c) usually is not readily transportable in a physical sense, except when the display device itself is transported or except in the sense that the display elements, groups, or images can be transmitted as signals inside the display device or transmitted as signals over communications lines, in which case they are sill soft copy, and (d) usually is volatile, i.e., is lost when the power is removed from the device that maintains it. *Note:* Examples of soft copy are (a) images on cathode ray tube (CRT) screens or liquid crystal display (LCD) surfaces and (b) facsimile pulse trains. *See also* **hard copy, liquid crystal display.**

soft hyphen: A hyphen that (a) is placed at the end of a line of text only to improve right justification without using proportional spacing to achieve complete right justification, and (b) is generated by using rules for word hyphenation that are built into word processing software packages. *Note:* When automatic hyphenation computer programs are used, some programs will remove these soft hyphens if the text is reformatted after editing, as necessary, and add new ones in the reformatted text as appropriate. *See also* **hard hyphen, justify, software package.**

soft limiting: 1. In a device, limiting in which (a) the transfer function of the device is dependent upon its instantaneous or integrated output level, and (b) the output waveform is therefore distorted but not hard-limited or clipped. **2.** Limiting with appreciable variation in output in the range where the output is limited, i.e., is controlled, when the input varies widely. *See also* **clipper, clipping, hard limiting, level, limiting, transfer function.**

soft sectoring: On magnetic disks, magnetic drums, and optical disks, the identification of sector boundaries by using recorded information. *See also* **hard sectoring.**

software: 1. A set of programs, procedures, and related documentation pertaining to the operation of communications, computer, data processing and control systems. *Note:* Examples of software are protocols, computer programs, compilers, assemblers, translators, routines, subroutines, algorithms, logic diagrams, manuals, routing tables, routing matrices, and circuit diagrams. **2.** In a display system, the segments, display files, computer programs, and documents required for the operation of the display system. *See* **application software, demo software, maintenance software, microsoftware, system software, utility software, virtual environment software.** *See also* **computer, computer program, display file, firmware, hardware, segment.**

software accessibility: A measure of the extent to which software facilitates selective use of its parts so as to obtain maximum programming efficiency in terms of time and use of resources in a given application.

software adaptability: A measure of the ease with which communications, computer, data processing, and control systems software can be altered to satisfy differing system constraints or user needs.

software development process: The process in which (a) user requirements are translated into software requirements, (b) software requirements are transformed into computer programs, and (c) programs are tested, documented, and certified for operational use. *See also* **computer program, software, software lifecycle.**

software engineering (SE): The branch of science and technology that is devoted to the design, development, and use of software. Design criteria include accuracy, efficiency, flexibility, comprehensibility, maintainability, time, and cost. *See* **computer aided software engineering.** *See also* **software.**

software engineering technology: *See* **computer aided software engineering technology.**

software librarian: The person responsible for establishing, controlling, operating, and maintaining a software library. *See also* **software, software library.**

software library: A library that (a) consists of a controlled, usually organized, collection of software and related documentation designed to aid in software development, use, and maintenance, and (b) may be a master library for all the software at an installation or may be a specialized software library, such as a software development library, a software production library, a software repository for application software, or a combination of these. *Synonym* **program library.** *See also* **system library.**

software lifecycle: The sequence of events that (a) starts when a software requirement is conceived, defined, or specified and ends when the software is no longer available for use, and (b) usually has several phases, such as a requirements phase, design phase, implementation phase, test phase, installation phase, checkout phase, operational phase, maintenance phase, and retirement or archival phase. *See also* **software development process.**

software maintenance: Maintenance performed on software after its delivery to the user, such as the correction of faults, incorporation of enhancements or improvements, and revision of accompanying documentation, such as user manuals.

software monitor: A software tool that (a) is executed concurrently with a computer program and (b) contains detailed information about the execution of the program. *See also* **computer program, software tool.**

software package: A set of software that (a) consists of one or more computer programs, an instruction manual for their use, and other related information, such as service and updating information, utility programs, and tutorial programs, (b) usually can handle one or more specific classes of application, such as word processing, graphics, database management, project management, tabulation, or calculation, and (c) usually is recorded on a diskette if the software is to be used on a personal computer (PC).

software system: *See* **command, control, communications, and intelligence reusable software system.**

software tool: Software, such as a computer program, routine, subroutine, program block, or program module, that can be used to develop, test, analyze, or maintain a computer program or its documentation. *Note:* Examples of software tools are automated software verification routines, compilers, program maintenance routines, bootstraps, program analyzers, and software monitors.

software verification: *See* **automated software verification.**

SOH: start of heading character.

SOI: signal operating instruction, signal operation instruction, signal operations instruction.

solar absorption factor: The radiation absorption factor when the Sun is the source of the radiation. *See also* **radiation absorption factor.**

solar array: An assembly of solar cells used for the production of electrical power by means of the photovoltaic effect. *Note:* An example of a solar array is a group of photocells that cover most of the external surfaces of a satellite to provide electrical power for satellite control and communications. *See also* **photovoltaic effect.**

solar day: *See* **true solar day.**

solar eclipse: An eclipse of the Sun by the Moon as the Moon moves between an observer, such as a person or a photodetector, and the Sun, thus obscuring the Sun. *See also* **eclipse.**

solar flare: A powerful eruption of radiation from the Sun that (a) is associated with sun spots, (b) usually lasts 15 min (minutes) to 20 min, (c) in rare instances may last up to a few hours, and (d) produces intense radiation that reaches the Earth about 26 h (hours) after the eruption. *See also* **radiation.**

solar noise: Electromagnetic radiation noise received from the Sun. *Note:* Solar noise can interfere with radio, television, microwave, and lightwave communications. All solar radiation is electromagnetic. The infrared and visible wavelengths are perhaps the most beneficial; the higher frequencies perhaps cause genetic changes. Solar noise is by far the most prominent radio noise source. At 1 GHz (gigahertz) the solar noise temperature is approximately 2.8×10^5 K (kelvin). This decreases in nearly logarithmic manner to 6×10^3 K at 30 GHz. When an antenna beamwidth is less than 0.5°, which is the angle subtended by the Sun from the Earth, the solar noise temperature is given by the relation $T_{sn} = (1.96/f) \times 10^{14}$. This relation is obtained from $T_{sn}/290 = (675/f) \times 10^9$ where T_{sn} is the solar noise temperature in kelvin and f is the frequency in hertz. *See also* **standard noise temperature.**

solar radiation: Radiation that (a) includes the totality of electromagnetic radiation from the Sun, (b) is in the range from low frequency (LF) to the frequency of cosmic rays, the highest known frequency, and (c) is primarily in the infrared, visible, and ultraviolet regions of the frequency spectrum. *See also* **cosmic ray, frequency spectrum, infrared, ultraviolet, visible spectrum.**

solar wind: An electrically charged gas, i.e., a plasma, that (a) in addition to light and heat, is radiated by the Sun, (b) penetrates the Earth's magnetic field, (c) creates auroras, i.e., the aurora borealis and the aurora australis, or the northern and southern lights, (d) may disrupt radio communications, radionavigation, and radio control systems, and (e) may damage radar, microelectronic circuits, and electronically controlled electric power supplies, causing wide-area power blackouts. *See also* **aurora, Sunspot.**

sole user: Pertaining to a communications system in which specific communications facilities or components within the system are provided for use by one user, or a relatively small number of users, rather than for use by essentially all of the users in the area in which the system is located.

sole user circuit: An exclusive user circuit that (a) is set up between a user and one or more other users on a full-time basis for a defined period, and (b) is provided with automatic restoration of connection in the event of a failure. *See also* **exclusive user circuit, failure, switched hotline circuit.**

solid angle: In solid geometry, a volumetric angle at a point subtended by a surface that lies within a perimeter on a unit sphere. *Note:* Solid angles are expressed in steradians. A full sphere consists of 4π sr (steradians). *See* **unit solid angle.**

solid-state circuit: An electrical circuit that (a) is composed entirely of solid materials and (b) for operation, depends on the control of electric, electromagnetic, and magnetic phenomena in solid materials. *Note:* Examples of solid-state circuits are (a) circuits composed entirely of resistors, crystal diodes, transistors, solid conductors, and semiconductors and (b) a large-scale integrated (LSI) circuit made of solid semiconductors.

solid-state laser: A laser in which the active laser medium is a solid material, such as glass, crystal, or semiconductor material, rather than a fluid. *See also* **active laser medium, laser, semiconductor, semiconductor laser.**

solid-state scanning: Scanning that (a) is performed all or in part by electronic commutation of an array of solid state elements, (b) may be used in facsimile systems, phased array transmitter antennas, and distributed receiver antennas, and (c) does not make use of moving elements. *See also* **facsimile, scanning.**

solid-state storage: Storage in which the storage elements consist of large-scale integrated (LSI) semiconductor elements that are accessed by LSI circuits enabling fast access to storage locations, all mounted on a single chip. *Note:* Optical disks and magnetic disks, tapes, drums, and cards are not considered to be solid-state storage. *See also* **chip, large-scale integrated circuit, solid-state circuit.**

soliton: An optical pulse that (a) has a special shape, spectral content, spectral distribution, and power level and (b) uses nonlinear effects in waveguides essentially to eliminate dispersion, even over long distances.

soliton laser: A mode-locked 1.4 μm (micron) to 1.6-μm wavelength laser that has an output pulse duration and shape that are controlled by a single-mode polarization-maintaining optical fiber incorporated into its feedback loop. *See also* **feedback loop, polarization-maintaining optical fiber, single mode.**

SOM: start of message.

sonar: sound navigation and ranging. *See* **active sonar, laser sonar, long range laser sonar, passive sonar, searchlight sonar, short range laser sonar.**

sone: A unit used to define the intensity of loudness sensation. *Note:* One sone corresponds to a loudness level of 40 phons. For loudness levels of 40 phons or greater, the loudness level, *P,* in phons and the loudness level, *S,* in sones, are given by the relation $\log_{10}S = 0.030(P - 40)$. This relation is derived from the relation $S = 2^{(P - 40)/10}$ and $\log_{10}2 \approx 0.30$. *See also* **phon.**

sonet: synchronous optical network.

SONET: Synchronous Optical Network (Standard).

SONET standard: Synchronous Optical Network standard.

sonic equipment: *See* **ultrasonic equipment.**

sonic pen: A stylus that (a) may be used to provide coordinate data in the display space on the display surface of a display device, (b) uses an acoustic wave, i.e., a sound wave, to couple energy to the display surface to indicate position, (c) produces an acoustic wave that is highly directive for accurate positioning, and (d) enables a specific location to be identified when energized. *See also* **coordinate data, display device.**

sonobuoy: In sound navigation and ranging (sonar) systems, a device (a) that is used to detect sound waves, such as those produced by ships and submarines, (b) that, when activated, relays information by radio or undersea cable, (c) that may be active or passive, and (d) that may be directional or omnidirectional.

soot: The dopant material, such as germanium tetrachloride or titanium tetrachloride, that is deposited on a glass substrate, pattern, plug, or slug, and is followed by heating, oxidation, and drawing processes to form an optical fiber with a controlled step or graded refractive index profile in its cross section. *See also* **chemical vapor phase oxidation process, dopant, graded index profile, optical fiber, step index profile.**

soot process: *Synonym* **chemical vapor phase oxidation process.**

sort: To separate a group of entities into sets according to some rule, key, characteristic or property, such as to separate (a) mixed mail into groups according to zip code, (b) a pile of invoices according to make of car, (c) personnel records according to city of employment, and (d) words according to the number of characters in each word, and usually, after separation, to arrange the items in each group in accordance with a specified criterion, such as sequencing in alphabetical or numerical order. *See* **bubble sort, exchange sort, external sort, insertion sort, internal sort, merge sort, multipass sort, oscillating sort, polyphase sort, repeated selection sort, selection sort, serial sort, tournament sort.**

sort-pass: During sorting, a single processing of all items, such as data items, to be sorted, for the purpose of decreasing the number of strings in which the items are sorted and increasing the number of items in each string that remain after a pass.

sort program: A computer program that, when executed, sorts items of data. *See also* **computer program, sort.**

SOS: The distress signal used in Morse code, carrier wave (cw) modulation, or radiotelegraph communications systems.

SOSIC: silicon on sapphire integrated circuit.

SOSTEL: solid state electronic logic.

sound: 1. A vibrational disturbance of the molecular structure of a fluid or elastic solid, usually in the range of 30 Hz (hertz) to 20 kHz (kilohertz). **2.** A sensation detected by the ear. *See* **voiced sound, unvoiced sound.** *See also* **sound wave.**

sound communications: The use of sound waves to transmit messages, such as in transmitting signals by whistle, siren, horn, sounder, or other device, in which (a) the duration of the sound signal can be controlled, (b) the international Morse code can be used, and (c) special meanings can be assigned to other sound formations, such as (i) a single blow of a whistle to initiate or halt an action, (ii) a series of horn blasts to indicate that a ship is leaving a pier, (iii) two blasts of a railroad whistle to indicate that a train is about to depart from a station, and (iv) the use of a percussion device to send Morse code, with one stroke for a dot and two rapid strokes for a dash. *See also* **sounder, sound signaling, sound waves.**

sounder: A telegraph receiving instrument in which Morse code signals are converted into sound signals differentiated by various coding arrangements, such as (a) intervals between two diverse sounds, (b) two different audio frequencies, or (c) audio frequency dots and dashes. *See* **echo sounder.** *Refer to* **Figs. G-4 (404), T-3 (991).**

sounder prediction station: A station equipped with an ionosphere sounder for real time monitoring of ionospheric phenomena or for obtaining data for the prediction of propagation conditions. *See also* **ionosphere, sounder.**

sounder station (SS): A station equipped with an ionosphere sounder used for the real time selection of frequencies for communications circuits. *See also* **ionosphere, sounder.**

sounding: 1. The measuring of geophysical parameters using radiometric devices and techniques. *Note:* The measurements are usually made to optimize the performance of communications systems. **2.** The determination of distance to an object, such as a mountain or a building, or to a surface, such as the bottom of the sea, by launching a pulsed sound wave, measuring the time taken for an echo to return, and calculating the distance by using the measured time and the known speed of sound in the propagation medium. **3.** In automated adaptive high frequency (HF) systems, the broadcasting of a station address to permit stations receiving a transmitted signal to measure link quality from the sending station. *See* **air sounding, automatic sounding, echo sounding, ionospheric sounding, oblique incidence sounding.** *See also* **propagation medium, radiometric.**

sound navigation and ranging (sonar): A sonic device that primarily is used for the detection and the location of underwater objects by reflecting sound waves from the objects or by interception of sound waves from an underwater, surface, or above-surface sound source. *Note:* Sound navigation and ranging (sonar) systems operate with sound waves in the same way that radar and radio direction finding equipment operate with electromagnetic waves, including use of the Doppler effect, radial component of velocity measurement, signature analysis, and triangulation. *See also* **radar.**

sound-powered: Pertaining to a device, such as a microphone or a telephone, that (a) derives power by converting acoustic energy into electrical power, (b) is operated by using the derived power, and (c) is not aided by additional external power.

sound-powered telephone: A telephone in which (a) the operating power for voice transmission is derived from the speech input, and (b) no other source of power is used except for signaling, such as using a magneto to operate an attention device, e.g., a bell or a buzzer. *Synonyms* **speech-powered telephone, voice-**

powered telephone. *See also* **electrically powered telephone.**

sound signaling: 1. Signaling in which modulated sound waves are used for transmission of supervisory signals between a transmitter and a receiver. *Note:* Sound signaling might make use of various devices, such as whistles, sirens, bells, and buzzers, to obtain attention or send coded signals. **2.** Using modulated sound waves to send messages by various means, such as Morse code or prearranged coded sound patterns. *See also* **sound communications.**

sound subcarrier: *See* **stereophonic sound subcarrier.**

sound subchannel: *See* **stereophonic sound subchannel.**

sound wave: 1. An acoustic wave that is audible to the normal human ear. **2.** A longitudinal wave that (a) consists of a sequence of longitudinal pressure pulses or elastic displacements of the material, whether gas, liquid, or solid, in which the wave propagates, (b) in gases, consists of a sequence of condensations (dense gas) and rarefactions (less dense gas) that travel through the gas, (c) in liquids, consists of a sequence of combined elastic deformation and compression waves that travel though the liquid, and (d) in solids, consists of a sequence of elastic compression and expansion waves that travel though the solid. *Note:* The speed of a sound wave in a material medium is determined by the temperature, pressure, and elastic properties of the medium. Sound in air propagates at 1087 ft/s (feet per second) or 332 m/s (meters per second) at 0°C. In air, its speed increases about 2 ft/s or 0.6 m/s for each kelvin above 0°C. *See also* **acoustic wave, elastic wave.**

source: 1. A generator of power. *Note:* Examples of sources are devices from which electronic signals or electromagnetic radiation emanates. **2.** In communications, a device that transmits signals, such as user information signals or control signals, to a sink. **3.** In communications, the part of a system from which messages are considered to originate. *See* **borescope light source, communications source, data source, fiber optic source, fiber optic light source, Lambertian source, light source, line source, message source, optical fiber source, optical source, point source, primary radiation source, secondary radiation source, standard optical source, standard source, stationary message source, video optical source, virtual source, YAG/LED source.** *See also* **data terminal equipment, destination user, sink, source user.**

source chirp: The signal from a source that emits a varying frequency, usually from shorter to longer wavelengths, i.e., higher to lower frequency, during a pulse time. *Note:* Source chirp results in pulse compression in optical fibers because the longer wavelengths of light emitted later will catch up with the shorter wavelengths emitted earlier. The distribution of radio or optical frequencies in a source chirp is analogous to the distribution of audio frequencies in the "chirp" of a bird.

source code: The code furnished to hardware or software from which the object code is produced. *Note:* An example of a source code is the code furnished to an assembler or a compiler to produce the executable machine code. *See also* **assembler, compiler, execute, hardware, machine code, software.**

source-coupler loss: In a fiber optic data link, the optical power loss that occurs between the light source and the device or the material that couples the light source energy from the source to the fiber optic cable that connects to the sink. *See also* **fiber optic cable, fiber optic data link, optical power, sink, source.**

source-coupling efficiency: The maximum efficiency with which the optical power, i.e., the integrated irradiance, luminous intensity, or optical power density of a Lambertian radiator, such as a light-emitting diode (LED), is coupled to an optical waveguide, such as an optical fiber, given by the relation $\Gamma_{sc} = A_f(NA)^2/2A_s n_0^2$ where Γ_{sc} is the source coupling efficiency, A_f is the cross-sectional area of the fiber core, NA is the numerical aperture, which is the sine of the acceptance angle or half of the apex angle of the acceptance cone, and dependent upon the refractive indices of the core and the cladding, A_s is the cross-sectional area of the source-emitted light beam, and n_0 is the refractive index of the propagation medium between the light source and the fiber endface.

source data: Original data that (a) are provided by a user for entry into a data processing system or a computer, (b) are usually contained in or are available to a source program, and (c) may first have been provided via a source document. *See also* **source document, source program.**

source document: A document that contains the source data that are to be made available to or are to be entered into a data processing system. *Note:* Examples of source documents are sales tickets, bills of lading, travel vouchers, or examination answer sheets.

source efficiency: In an electrically powered optical source, the ratio of (a) the optical power output to (b) the electrical power input. *Synonym* **source power efficiency.** *See also* **optical source.**

source-fiber coupling: In fiber optic transmission systems, the transfer of optical signal power emitted by a light source into an optical fiber. *Note:* The coupling is dependent upon many factors, particularly the geometry of the light source and the optical fiber. Many optical sources used in fiber optic systems have a fiber optic pigtail for connection by means of a splice or a connector to a transmission optical fiber. The source to pigtail coupling is permanent and is made as efficient as possible. *See also* **source to fiber coupled power.**

source intelligence: *See* **closed source intelligence.**

source language: 1. In a translation process, the language from which another language, i.e., the target language, is derived. **2.** In computing, data processing, and communications systems, a language from which statements are translated. *Note:* Translators, assemblers, and compilers prepare target language programs, usually machine language programs, from source language programs, usually high level language programs written by programmers. *See also* **intermediate language, target language.**

source power efficiency: *Synonym* **source efficiency.**

source program: 1. A computer program written in a source language, usually by a programmer. *Note:* An example of a source program is a program that serves as the input to an assembler, compiler, or translator. **2.** A computer program that must be assembled, compiled, or translated before it can be executed by a computer. *See also* **source language, target program.**

source quench: Congestion control in which a computer experiencing data, message, or packet traffic congestion sends a request back to the traffic source telling it to stop transmitting.

source to optical fiber coupled power: The optical power that is coupled from a light source into the interior of an optical fiber, given by the relation $P_c = P_0(\theta_a/\theta_e)T$ where P_c is the source to optical fiber coupled power, P_0 is the source optical power output, θ_a is the acceptance solid angle, i.e., the angle of collection, of the fiber, θ_e is the emission solid angle, i.e., the beam divergence, of the light source, and T is the transmission coefficient of the optical system, i.e., of the source to fiber interface. *Note:* In cases of angular misalignment or lateral offset, the θ_a and the θ_e must be replaced by the source and fiber cross-sectional areas. *See also* **acceptance angle, optical fiber, optical power, source.**

source to optical fiber loss: The total optical power lost when the output power of a source is coupled to an optical fiber, equal to the difference between (a) the optical power output of the source and (b) the optical power transferred into the interior of the optical fiber. *Note:* Components of the source pigtail to optical fiber loss include angular misalignment loss, lateral offset loss, longitudinal displacement loss, core and cladding diameter mismatch loss, and refractive index mismatch loss.

source type: *See* **optical source type.**

source user: The user that provides the information to be transferred to a destination user during a particular information transfer transaction. *Synonym* **information source.** *See also* **access originator, call originator, communications source, destination user, information transfer transaction, sink, source.**

southern lights: *Synonym* **aurora australis.**

space: 1. In display systems, any reference system, such as a coordinate system, in which a display element, display group, or display image can be described, located, defined, displayed, or operated upon. **2.** A machine function that is used to control the horizontal movement of the carriage of a teletypewriter or a teleprinter and is required to advance the paper laterally without printing a character on the page, the movement being obtained by use of the space bar, corresponding tape perforation pattern, or space signal. **3.** In Morse code, the designation given to one of the two significant conditions in the code, the other condition being designated as the mark condition. **4.** In binary communications and binary modulation, one of the two significant conditions of encoding, the other condition being specified as the mark condition. **5.** The time interval between the code elements, characters, and words in the Morse code. *Note:* The unit time interval is the dot. The space (a) between code elements may be one unit interval, (b) between characters two or three unit intervals, and (c) between words six or seven unit intervals. *Synonyms* **spacing pulse, spacing signal.** *See* **address space, backspace, controlled space, cyberspace, display space, dead space, deep space, free space, image storage space, nonreturn to zero space, outer space, unshift on space, virtual address space, virtual space, work space.** *See also* **code element, dot, mark, modulation, neutral operation, signal transition.**

space-active: Pertaining to a telegraph signal in which (a) a space is represented by a pulse, such as a direct current (dc) condition or a constant frequency signal or tone, and (b) a mark is represented by an open circuit, i.e., no current or no tone.

space character: A character that (a) usually is represented by a blank site in a series of graphic characters, (b) has a function equivalent to that of a format effector that causes the print or display position to move one position forward without producing, printing, or displaying any graphic character, (c) usually has a function that is equivalent to that of an information separator, i.e., a data delimiter, such as separating words and sentences, and (d) is not a control character. *See also* **horizontal tabulation character.**

space coding: *See* **biphase space coding.**

space coherence: *Synonym* **coherence.**

space-coherent light: In the wavefront of a lightwave or when a lightwave is incident upon a real or notional surface, light such that the amplitude, phase, and time variation are predictable and correlated. *Note:* The surface usually is in a plane perpendicular to the direction of propagation. Spatial noncoherence refers to a random and unpredictable state of the phase over a surface perpendicular to the direction of propagation. *See also* **coherent light, time-coherent light, wavefront.**

space communications system: A group of cooperating Earth stations and space stations that (a) provides a given space service and (b) may use objects in space, such as satellites and spacecraft, for reflection or retransmission, i.e., relaying, of radio or video signals. *See also* **Earth station, space service, space station.**

spacecraft: **1.** A man-made vehicle intended to go beyond the major portion of the Earth's atmosphere. **2.** Any type of space vehicle, including Earth satellites or deep space probes, whether manned or unmanned. **3.** An artificial body that is capable of traveling beyond the Earth's atmosphere but is not captured in orbit around a parent body. *Note 1:* Examples of spacecraft are (a) a space vehicle to the Moon, (b) a lunar landing module, and (c) a deep space probe to Venus or Mars. *Note 2:* In the United States, a satellite is an orbiting body that cannot be maneuvered while in orbit. A spacecraft is a body that has the ability to maneuver, whether controlled from the ground, by a computer on board, or by personnel on board. Thus, in the United States, a spacecraft can be fully or partially in orbit around a parent body, but it must be capable of maneuvering into or out of an orbit. A spacecraft has the option of being a satellite. A satellite has no options, except to remain in orbit, get lost in space, perhaps intersect the Earth's surface, or self-destruct. *See also* **satellite.**

spaced aerial diversity: *Synonym* **spaced antenna diversity.**

spaced antenna diversity: Pertaining to signal transmission or reception, or both, in which (a) spatially separated antennas are used, (b) both are used when the transmitted information is the same, (c) the antennas usually are 2 or 3 wavelengths apart but may be up to 50 wavelengths apart to overcome the adverse effects of tropospheric scatter or to make use of tropospheric scatter for propagation purposes, and (d) the effects of fading, especially flat and selective fading, are minimized by making use of common polarization. *Synonym* **spaced aerial diversity.**

space diversity: **1.** Pertaining to signal transmission and reception in which (a) two or more separate and independent propagation media or paths are used for transmitting the same information, (b) on reception, the signals from the separate media or paths are combined, (c) the effects of fading, transmission loss, delay, interference, and other effects that tend to degrade a signal are overcome, and (d) diversity is accomplished through the use of spatial separation. **2.** A method of transmission or reception, or both, in which (a) the effects of fading are minimized by the simultaneous use of two or more antennas located a number of wavelengths apart, and (b) the antennas use different transmission paths or polarization. *See also* **antenna, cross-polarized operation, diversity reception, dual diversity, spaced antenna diversity.**

space division (SD): In communications systems, the use of space to obtain channel separation. *Note:* Examples of space division include (a) a telegraph or telephone switching system in which every connection requires an individual line through the switching center, thereby limiting the number of connections between stations to the number of physical lines between them, (b) a cable consisting of a bundle of twisted pairs, and (c) 144 optical fibers in a single fiber optic cable. *See also* **frequency division, time division.**

space division multiple access (SDMA): Satellite communications in which (a) communications traffic capacity is divided into spot beam areas in which all of the available frequency band is allocated for use on a spatial separation basis, and (b) communications operations include the use of complex routing and switching within the communications satellite itself. *See also* **spot beam.**

space division multiplex: Pertaining to the separation of channels by means of spatial separation of conductors or waveguides. *Note:* Examples of space division multiplexed channels are (a) separate optical fibers in a fiber optic cable, and (b) a large number of twisted pairs in an electrical cable. Each of the space division multiplexed channels may be independently time and frequency division multiplexed.

space division multiplexing (SDM): Multiplexing in which separate channels are obtained by spatially separating conductors and waveguides, such as wires, optical fibers, and coaxial cables. *Note 1:* Examples of space division multiplexing are (a) using separate optical fibers for each channel and placing many fibers in a single cable, and (b) placing many metallic twisted pairs in a single cable. *Note 2:* Each space division multiplexed channel may be further multiplexed by time division multiplexing or frequency division multiplexing, i.e., wavelength division multiplexing, or both. Phantom circuits may also be used in some cases. *See* **space division optical multiplexing.** *Refer to* **Fig. S-14.**

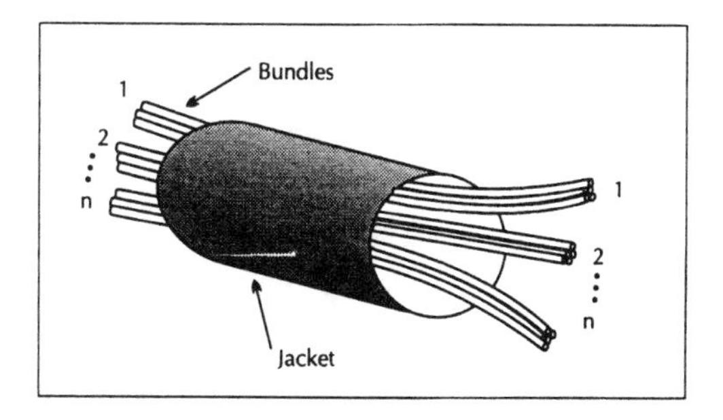

Fig. S-14. Space division multiplexing by maintaining separate fiber optic bundles, or separate optical fibers, in a fiber optic cable.

space division switching: In switched networks, switching in which (a) single transmission path-routing determination is accomplished in a switch by using a physically separated set of matrix contacts or cross points, (b) the interconnections of input and output terminations are established and maintained with electrical or optical continuity, and (c) channels are physically separated during transmission or during the information transfer phase of an information transfer transaction. *See also* **information transfer transaction, switching system.**

space-ground link subsystem: In satellite communications systems, an integrated set of (a) satellite tracking and telemetry equipment on Earth and (b) command equipment consisting of communication functional units in satellites, space probes, and spacecraft.

space hold: In telegraph operations when no traffic is being transmitted, a transmission condition in which a steady space is transmitted. *See also* **mark hold, space.**

space intersection: The use of geometry, including triangulation and trilateration, in geodetic and satellite ephemeris calculations. *See also* **ephemeris, triangulation, trilateration.**

space loss: *See* **free space loss, isotropic space loss.**

space operation service: A radiocommunications service that (a) is concerned exclusively with the operation of spacecraft, and (b) performs various functions, such as space tracking, space telemetry, and space telecommand, these functions usually being provided within the service in which the space station is operating.

space optical communications: *See* **free space optical communications.**

space radiocommunications: Radiocommunications that includes the use of objects in space, such as one or more space stations or one or more reflecting satellites. *See also* **radiocommunications service.**

space research Earth station: An Earth station in the space research service.

space research service: A space service in which objects in space, such as spacecraft, space probes, and satellites, are used for scientific or technological research purposes.

space research space station: A space station in the space research service. *Synonym* **research service space station.**

space segment: In a satellite communications system, the parts of the system that are used for the launch, maintenance, tracking, telemetry, command, and control of the communications satellites. *Note:* The space segment does not include (a) the terrestrial systems that are concerned with the payload or the primary function of the satellite communications system, (b) the mission of the communications satellite, (c) the communications Earth stations, and (d) the traffic handling equipment. These are parts of the ground segment. *See also* **ground segment.**

space service (SS): A radiocommunications service that (a) is provided (i) between Earth stations and space stations, (ii) between space stations, and (iii) between Earth stations when signals are retransmitted by space stations, (b) may accomplish transmissions via reflection from objects in space, (c) does not include transmissions by reflection or scattering by the ionosphere, and (d) does not include transmissions entirely within the Earth's atmosphere.

space signal: 1. In telegraph, teletype, and teleprinter operations, the signal that (a) initiates the active condition in a teletypewriter or a teleprinter, and (b) usually is used as the signaling condition that precedes the production of a start signal in the International Telegraph Alphabet 2 (ITA-2). **2.** In telegraph, teletype, and teleprinter operations, the signal that (a) corresponds to the code combination that causes the printing position to be advanced by one character pitch, and (b) does not cause the printing of a graphic character. **3.** The signal that corresponds to an off-key or nontransmit condition. *See also* **mark signal.**

space signaling: *See* **digital mark-space signaling.**

space station: A station that (a) is in the space service, (b) is located on an object that is beyond, is intended to go beyond, or has been beyond, the major portion of the Earth's atmosphere, (c) may be located on a satellite in orbit, a powered spacecraft in space, a deep space probe in deep space probing a planet, or a ballistic missile in flight, and (d) is not on board an aircraft. *See also* **terminal.** *See* **broadcasting satellite space station, communications satellite space station, meteorological satellite space station, radionavigation satellite space station, space research space station, space telecommand space station, space telemetering space station, space tracking space station.**

space subsystem: In satellite communications, the portion of the satellite link that is in orbit.

space system: A group of cooperating Earth stations or space stations that uses space radiocommunications. *See also* **Earth station, space station.**

space telecommand: The use of radiocommunications for the transmission of signals to a space station to initiate, modify, or terminate functions of equipment on an object in space, such as a satellite or a spacecraft, including the space station itself.

space telecommand Earth station: An Earth station that transmits messages and instructions for space telecommand purposes. *See also* **Earth station, space telecommand.**

space telecommand space station: A space station that receives messages and instructions for space telecommand purposes. *See also* **space station, space telecommand.**

space telemetering: The use of telemetry for the transmission, from a space station, of results of measurements made in a spacecraft, including those relating to the functioning of the spacecraft itself. *See also* **space telemetry, telemetry.**

space telemetering Earth station: An Earth station that receives space telemetering signals or messages from space stations.

space telemetering space station: A space station that transmits space telemetering signals or messages to Earth stations.

space telemetry: The transmission of signals between space stations and Earth stations for the purpose of (a) controlling the making of measurements in a space station, (b) controlling measurements made by a space station, (c) transmitting the results of the controlled and uncontrolled measurements to one or more Earth stations, and (d) transmitting measurements relating to the operation of the spacecraft itself. *See also* **Earth station, space station, space telemetering, telemetry.**

space tracking: Determining the orbit, velocity, or instantaneous position of an object in space by means of radiodetermination, excluding primary radar, for the purpose of following the movement of the object. *See also* **radiodetermination.**

space tracking Earth station: An Earth station that transmits or receives messages or signals that are used for space tracking. *See also* **Earth station, space tracking.**

space tracking space station: A space station that transmits, or receives and retransmits, messages or signals that are used for space tracking. *See also* **space station, space tracking.**

spacing: *See* **proportional spacing.**

spacing bias: The uniform lengthening of all spacing signal pulses at the expense of all marking signal pulses. *See also* **bias, bias distortion, marking bias, mark signal, space, space signal.**

spacing end distortion: *Synonym* **end distortion.**

spacing pulse: *Synonym* **space.**

spacing signal: *Synonym* **space.**

spar: *See* **Iceland spar.**

spare: An individual part, subassembly, or assembly supplied for the maintenance or the repair of systems or equipment.

spare signal group (SSG): A signal group that (a) is not used on a regular basis, (b) is held as a spare in the event it is needed, and (c) is used by a ship station for the duration of a voyage, a land station for a specified period, or an aircraft station for the duration of a flight. *Note:* Examples of spare signal groups are indefinite call signs, collective call signs, radio distinguishing groups, suffixes to radio distinguishing groups, and key lists for encryption of call signs. *See also* **signal group.**

spatial application: In display systems and imagery, an application that (a) requires a high spatial resolution capability and (b) may occur at the expense of reduced temporal positioning precision, i.e., may result in increased jerkiness. *Note:* Examples of spatial application are (a) the requirement to display small characters and (b) the requirement to resolve fine detail in still video, freeze-frame video, or motion video that contains very limited motion, i.e., very slow motion. *See also* **display system, imagery, temporal application.**

spatial coherence: *Synonym* **coherence.**

spatial degree: A unit of measure of a plane angle in space, of such magnitude that 360 spatial degrees constitute a full plane rotation of a line about a point on the line. *See also* **electrical degree.**

spatially coherent: *Synonym* **coherent.**

spatially coherent radiation: *Synonym* **coherent.**

spatial resolution: *Synonym* **distance resolution.**

spatter: In applying evaporative coatings on optical elements, such as lenses, prisms, and mirrors, small chunks of material that fly from the hot crucible onto the optical element surface and adhere there. *See also* **optical element.**

speaker channel: *Synonym* **engineering channel.**

speaker circuit: *Synonym* **engineering circuit.**

special character: A graphic character in a character set that is not a letter, not a numeral, and not a space character. *Note:* Examples of special characters are hyphens, commas, slashes, question marks, and equal signs.

special fiber optic cable: A fiber optic cable with one or more features intended to meet one or more specific environmental conditions or performance requirements not found in standard or regular general-purpose production-quality fiber optic cables. *Note:* Examples of special fiber optic cables are cables with (a) a flat cross section, (b) a heavy armor covering, i.e., overarmor, (c) an abnormally high tensile strength, and (d) optical fibers that have extremely small or extremely large diameters. *Synonym* **special optical cable.**

special flag: In visual communications systems, a square flag on which a specific pattern of colors is used to represent a specific word, which may constitute an entire message, such as "Quarantine."

special grade access line: 1. An access line that is capable of providing a special grade of service. **2.** In the Automatic Voice Network (AUTOVON), an access line specially conditioned, usually by providing amplitude and delay equalization, to give it characteristics suitable for handling special services, such as reducing data signaling rates (DSRs) to a rate between 600 bps (bits per second) and 2400 bps. *See also* **Automatic Voice Network, equalization, line, special grade of service.**

special grade of service: 1. A network-provided service feature in which specially conditioned trunks and access lines are used to provide special services to users on demand, such as (a) the required transmission capability for secure voice, secure data, and secure facsimile transmission, (b) high fidelity transmission, (c) high data signaling rates (DSRs), and (d) multimedia transmission for (i) analog and digital data and (ii) audio and video signals. **2.** In the Automatic Voice Network (AUTOVON), a network-provided service feature in which specially conditioned interswitch trunks and access lines are used to provide secure voice, data, and facsimile transmission. *See also* **Automatic Voice Network, classmark, grade of service, precedence.**

special grade trunk: A trunk that has the capability to provide a special grade of service, such as to provide interference-free, uninterrupted, dedicated connection service for voice, data, or facsimile transmission, above that which is provided for ordinary voice transmission. *See also* **special grade of service.**

special grade user: A communications system user that is provided with a special grade of service. *See also* **special grade of service, user.**

special handling designator: A repeated letter that (a) usually is repeated five times, (b) immediately follows the security classification letter, and (c) is used as a designator for messages that require handling different from the handling of regular messages. *See also* **security classified.**

special interest group: *Synonym* **community of interest.**

specialized common carrier (SCC): A common carrier that (a) provides unique or limited services, special types of services, or a special grade of service or (b) serves limited markets, businesses, or areas. *See also* **common carrier, divestiture, other common carrier, resale carrier, special grade of service.**

special meteorological broadcast: Complete meteorological information that is placed on a special broadcast or a facsimile broadcast and copied by stations that are concerned. *See also* **broadcast, copy, emergency meteorological warning, routine meteorological broadcast.**

special network service: A network-provided service feature in which a network is provided that (a) affords privacy of service within a specified community of interest and (b) denies network users access to any of the general network users. *See also* **closed user group.**

special optical cable: *Synonym* **special fiber optic cable.**

special panel signaling sign: A special panel or pennant display that is used to indicate special prearranged messages from ground stations to air stations. *Note:* Examples of special panel signaling signs are panels that indicate special prearranged messages, such as "land in this direction," "pick up message here," "message drop not recovered," "unable to proceed," or "you may not proceed." *See also* **display, panel, panel code, panel signaling.**

special pennant: In a visual communications system, a pennant of specified shape with a specified color pattern that is used to represent a particular word or to represent a special panel signaling sign. *See also* **pennant.**

special purpose computer: A computer that can only operate on a restricted class of problems.

special purpose language: A computer programming language that can be used for programming the solution of a given class of problems or for a given class of applications, such as engineering design, communications network control, simulation, or numerical control of machines or machine tools. *See also* **application software.**

special purpose routing indicator: A message routing indicator designed and used to accomplish a special purpose, such as to indicate an emergency relocation site, an alternate command post, or an alternate control station. *See also* **message routing, routing indicator.**

special service: A radiocommunications service that (a) is provided exclusively to meet the specific needs of a community of interest, (b) is not open to general public correspondence, and (c) is not in one of the defined categories of radiocommunications services. *See also* **community of interest.**

specification: **1.** An essential technical requirement for items, materials, or services, including (a) the procedures to be used to determine whether the requirement has been met, and (b) the requirements for preservation, packaging, packing, and marking. **2.** The defining of a requirement, detailing of a design, or developing of a performance criterion, including determining and obtaining the necessary information for performing these actions and preparing a document. **3.** The document that contains the essential technical requirements for items, materials, or services, including (a) the procedures to be used to determine whether the requirement has been met, and (b) the requirements for preservation, packaging, packing, and marking. *Note:* To be of value, specifications must be clear, precise, and accurate. *See* **design specification, functional specification, interface specification, performance specification, pulse dispersion specification, requirements specification.** *See also* **design objective, standards.**

specification language: *See* **requirements specification language.**

specific detectivity: For a photodetector, a figure of merit used to characterize detector performance, equal to (a) the ratio of the square root of the product of the photosensitive region and the effective noise bandwidth to (b) the noise equivalent power (NEP), given by the relation $D^* = (A\Delta f)^{1/2}/NEP$ where D^* is the specific detectivity, A is the area of the photosensitive region of the detector that is exposed to radiation, Δf is the effective noise bandwidth, and NEP is the noise equivalent power. The relation may also be written as $D^* = D(A\Delta f)^{1/2}$ where D is the detectivity, i.e., the reciprocal of the noise equivalent power, and the other variables are as before. *Note 2:* The higher the figure of merit, the better the performance. *Synonyms* **normalized detectivity, D star.** *See also* **detectivity, noise equivalent power.**

specific integrated circuit: *See* **application-specific integrated circuit.**

specified reliability: The reliability of hardware or software in a specified environment rather than in its operational or test environment. *See also* **operational reliability, test reliability.**

specimen: *See* **test specimen.**

speckle effect: A mottling effect due to color distortion caused by interference in light transmission produced by the propagation medium itself, such as an optical fiber. *See also* **interference, propagation medium.**

speckle interferometer: An interferometer in which (a) a laser illuminates a diffusely reflecting surface that is viewed through the interferometer, (b) the optical system forms easily visible speckle patterns on the surface of the image, (c) interference with a reference beam causes the light and dark speckles to reverse shade when the object moves one-fourth of the wavelength along the optical axis, and (d) if the object is vibrated in the direction of the optical axis, the image washes out at various vibrating places and remains intact at the vibrational nodes, i.e., where the vibrational amplitude is zero. *See also* **interference, interferometer, interferometry, speckle pattern.**

speckle noise: *Synonym* **modal noise.**

speckle pattern: In optical systems, a field intensity pattern produced by the mutual interference of partially coherent beams that are subject to minute temporal and spatial fluctuations. *Note:* In a multimode fiber, a speckle pattern results from a superposition of mode field patterns. If the relative modal group velocities change with time, the speckle pattern will also change with time. If differential mode attenuation (DMA) occurs, modal noise results. *See also* **fiber optics, modal noise, mode, multimode optical fiber.**

spectral absorptance: The absorptance of electromagnetic radiation by a material, evaluated at one or more wavelengths.

spectral absorption coefficient: In the transmission of electromagnetic waves in a propagation medium, the attenuation term, a, in the relation $I = I_0 e^{-ax}$ where I is the radiance or the irradiance, i.e., the optical power density or intensity of radiation at point x, and I_0 is the initial radiance or irradiance at the point from which x is measured, when the attenuation is caused only by absorption and not by any other cause, such as refraction, diffraction, scattering, dispersion, diffusion, divergence, or geometric spreading. *Note 1:* The spectral absorption coefficient usually is expressed in units of reciprocal distance and appears in the exponent of the expression that defines Bouger's law. *Note 2:* The sum of the spectral absorption coefficient and the spectral scattering coefficient is the spectral attenuation term. *Note 3:* Both parts of the attenuation term, i.e., the absorption part and the scattering part, depend on impurities and the irregularity of the molecular structure in the intrinsic material of the propagation medium. *See also* **absorption coefficient, attenuation term, Bouger's law, propagation medium.**

spectral attenuation rate characteristic: For an optical fiber, the attenuation rate plotted as a function of wavelength. *Refer to* **Fig. S-15.**

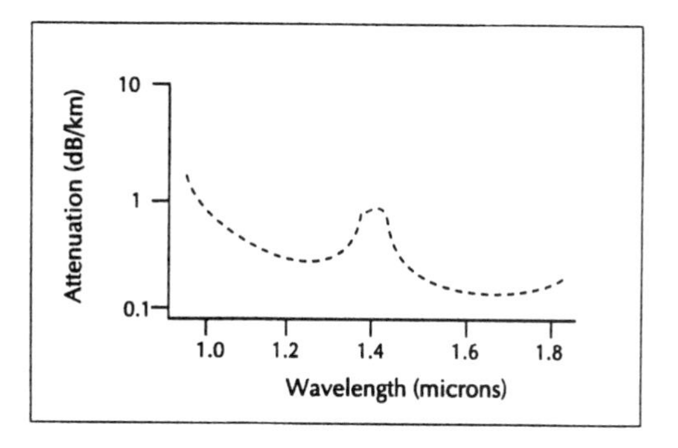

Fig. S-15. The **spectral attenuation rate characteristic** curve for a typical single mode optical fiber. A hydroxyl ion (OH–) absorption peak occurs at about 1.4 μm, at which the attenuation rate is usually less than 1.0 dB•km^{-1}.

spectral bandwidth: *Synonym* **spectral width.**

spectral density: 1. In a specified bandwidth of radiation consisting of a continuous frequency spectrum, (a) the total power in the specified bandwidth divided by (b) the specified bandwidth. *Note:* The spectral density usually is expressed in watts per hertz. **2.** In a finite bandwidth of electromagnetic radiation consisting of a continuous or distributed spectrum of frequencies, the optical power distribution expressed in watts per hertz of the given bandwidth.

spectral dispersion: *Synonym* **material dispersion.**

spectral emittance: The radiant emittance plotted as a function of wavelength.

spectral irradiance: The irradiance per unit wavelength interval at a given nominal median wavelength, calculated as (a) the irradiance contained in an elementary range of wavelengths at a given wavelength divided by (b) the range of wavelengths. *Note:* Spectral irradiance usually is expressed as power per unit area, such as watts per square meter, per unit wavelength interval, such as nanometers of wavelength interval, i.e., W•m^{-2}•nm^{-1}. *Refer to* **Table 5, Appendix B.**

spectral line: 1. In spectroscopy, a group of contiguous optical wavelengths that has an extremely narrow spectral width such that the group appears as a single line on a spectroscope as though there were a single wavelength. **2.** Electromagnetic radiation with an extremely narrow spectral width. *Note 1:* Only certain discrete wavelengths of optical radiation exist in a given spectral line. *Note 2:* Spectral lines are observed and measured with a spectroscope. *See also* **line source, line spectrum, monochromatic, spectral line width.** *Refer to* **Table 5, Appendix B.**

spectral line width: 1. In spectroscopy, the difference between the limiting wavelengths of a spectral line. **2.** A measure of (a) the distribution of wavelengths, i.e., the wavelength composition of the optical radiation, in a spectral line and (b) the difference between the shortest wavelength and the longest wavelength in the spectral line. *See also* **spectral line, spectral width.**

spectral loss curve: In a lightwave propagating in an optical waveguide, such as an optical fiber, a curve that shows attenuation as a function of wavelength. *Note:* Spectral loss curves should be normalized before meaningful comparisons of different optical waveguides can be made. *See also* **fiber optics.**

spectral order: *See* **diffraction grating spectral order.**

spectral power: *See* **peak spectral power.**

spectral purity: The degree to which a beam or a signal is coherent. *See also* **coherence, coherence degree, coherent.**

spectral radiance: At a given wavelength and in a given wavelength interval, the radiance per steradian per unit area per unit wavelength interval, i.e., per unit of spectral width. *Note:* The spectral radiance may be calculated as (a) the radiance contained in an elementary range of wavelengths at a given central wavelength divided by (b) the range of wavelengths. It may be expressed in watts per steradian•(square meter)•nanometer, i.e., $W{\bullet}sr^{-1}{\bullet}m^{-2}{\bullet}nm^{-1}$, or any other convenient set of equivalent units, such as microwatts, angstroms, or microns. *See also* **radiance, radiant emittance, radiometry.** *Refer to* **Table 5, Appendix B.**

spectral reflectivity: The reflectivity of a surface evaluated as a function of wavelength. *See also* **reflectivity.**

spectral response: 1. In photodetectors, the responsivity as a function of the wavelength of incident electromagnetic radiation, usually indicated as a plot of (a) the electrical current or power output divided by the incident optical power input on the ordinate, as a function of (b) the wavelength of the incident radiation on the abscissa, i.e., the responsivity is plotted as a function of wavelength of incident radiation. *Note 2:* The spectral response shows the wavelength dependence of the detector. *Note 3:* Photodetector spectral response may be expressed in various units, such as amperes/$watt_{optical}$, amperes/(lumen•second), or coulombs/lumen, as a function of incident wavelength. **2.** In light sources, the responsivity, indicated as the optical power output as a function of driving electrical current or power input, plotted as a function of output wavelength. *Note:* Light source spectral response usually is plotted as $watts_{optical}$/ampere, lumens/(ampere•second), or lumens/ coulomb on the abscissa and output wavelength on the ordinate. *See also* **responsivity.** *Refer to* **Fig. S-16.** *Refer also to* **Fig. L-2 (502).**

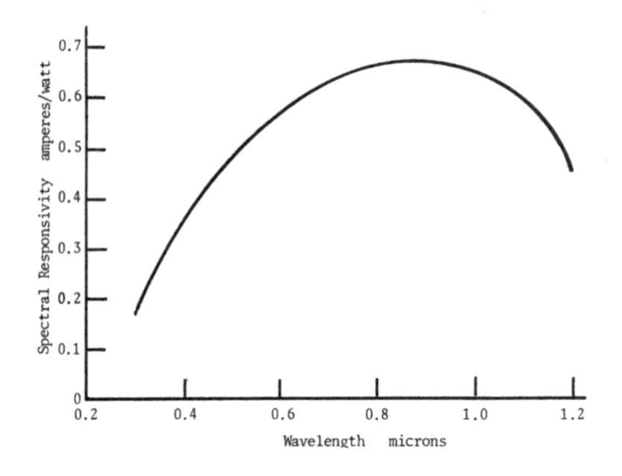

Fig. S-16. A **spectral response** curve for a typical photodiode.

spectral responsivity (SR): The responsivity per unit of incident wavelength interval at a given nominal or median wavelength, such as watts/ampere•nanometer at 1.31 μm (microns). *See also* **responsivity, spectral response.**

spectral transmittance: Transmittance evaluated at one or more wavelengths.

spectral wavelength: *See* **peak spectral wavelength.**

spectral width (SW): In electromagnetic radiation, such as radio waves or lightwaves, the wavelength interval in which the power level is greater than a specified fraction of its maximum value, such as 0.50 of the maximum power level, i.e., 3 dB, or 0.707 of the maximum current or voltage level for radio waves and the irradiance half-power points for lightwaves that are used to bound the band of frequencies in the wavelength interval. *Note 1:* The spectral width may be specified (a) as a special case of root mean square deviation in which the independent variable is wavelength, λ, and $f(\lambda)$ is a suitable radiometric quantity, or (b) as the full width at half maximum (FWHM), i.e., the difference between the wavelengths at which the magnitude drops to one-half of its maximum value. The latter specification may be difficult to apply when the spectral quantity has a complex shape. *Note 2:* The relative spectral width, $(\Delta\lambda)\lambda$, is frequently used, where $\Delta\lambda$ is obtained according to the specifications in Note 1. *Note 3:* The spectral width of a laser is an order of magnitude less than that of a light-emitting diode (LED). Optical fiber wavelengths are on the order of 1 μm (micron). Spectral widths of LEDs are on the order of 20 to 50 nm (nanometers), whereas those of lasers are on the order of 1 or 2 nm. Dispersion of an optical pulse in an optical fiber might be expressed in nanoseconds per microsecond of launched pulse per nanometer of spectral width of the light from the source. The difference of the electromagnetic frequencies corresponding to the two limiting wavelengths would express the same spectral width in hertz. If the spectral width of a laser is 1 nm, this would correspond to a spectral bandwidth of 3×10^{11} Hz. If the nominal wavelength is 1 μm, the nominal frequency is 3×10^{14} Hz. The ratio of width to nominal frequencies is 10^{-3}, and the ratio of width to nominal wavelengths is also 10^{-3}. *Synonym* **spectral bandwidth.** *See* **relative spectral width.** *See also* **coherence length, dispersion, impulse response function, line spectrum, monochromatic, root mean square deviation, root mean square pulse broadening, root mean square pulse duration, spectral linewidth.**

spectral window: In a given propagation medium or material, a wavelength region of relatively high transmittance, surrounded by regions of low transmittance, i.e., a trough in a plot of optical attenuation, caused by the material, when plotted versus wavelength. *Note:* A major spectral window in the optical spectrum occurs between 1.1 μm (microns) and 1.6 μm, with minor spectral windows at 1.1, 1.3, and 1.55 μm. A wavelength of 1.31 μm is a standard wavelength used in fiber optic communications systems. The 0.8-μm to 0.9-μm region is also favorable for transmission and is currently in use because components operating at this wavelength are readily available. To achieve lower losses, reduced attenuation rates, particularly for long-haul communications, and longer wavelengths, operating deeper in the infrared may be required. For short-haul communications, such as local area networks (LANs), networks in urban areas, loops, and networks on board ships and aircraft, use of the far infrared region of the optical spectrum may not be necessary. *Synonym* **transmission window.**

spectrometer: A spectroscope that (a) is equipped with an angle scale, (b) is used to measure the angular deviation of radiation of different wavelengths, and (c) is used to measure angles between surfaces of optical elements.

spectroscope: An instrument that disperses radiation into its component wavelengths and allows observation or measurement of the wavelength composition of the resultant spectrum.

spectrum: A continuous range of a group, or a group of ranges, of frequencies of waves that have something in common, such as (a) all the frequencies in a string of similar, equally spaced, rectangular pulses, (b) all the frequencies in a beam of visible light, (c) a band of radio frequencies (rf), or (d) all the frequencies in a sound wave, such as a musical chord. *See* **continuous spectrum, electromagnetic spectrum, direct sequence spread spectrum, frequency hopping spread spectrum, frequency spectrum, frequency spectrum designation, line spectrum, optical spectrum, primary spectrum, secondary spectrum, spread spectrum, visible spectrum, visual spectrum.** *See also* **characteristic frequency, frequency.**

spectrum analysis: An analysis that shows (a) the distribution of energy or energy density contained in a set of electromagnetic waves of more than one frequency as (b) a function of frequency. *See also* **signature analysis, spectrum signature.**

spectrum analyzer: *See* **integrated optical spectrum analyzer, lightwave spectrum analyzer.**

spectrum code sequence generator: *See* **spread sprectrum code sequence generator.**

spectrum designation: *See* **frequency spectrum designation.**

spectrum designation of frequency: *Synonym* **frequency spectrum designation.**

spectrum signature: 1. The frequencies that make up the waves emanating from a particular source, such as (a) the frequencies (or wavelengths) in lightwaves emanating from a light-emitting diode (LED), and (b) the frequencies contained in a sound wave emanating from a given ship. **2.** In a device, the pattern of radio signal frequencies, amplitudes, and phases, that (a) characterizes the output of the device, (b) tends to distinguish the device from other devices, and (c) may be used to identify the device. **3.** In electronic equipment, a collection of the spectral characteristics of emanations, transmissions, or reflections from the equipment that are measured, calculated, or estimated. *Note:* Spectrum signatures may be obtained from any radiative equipment, such as radio and radar transmitters and receivers, antennas, and direction finding (DF) equipment. Some of the electromagnetic characteristics that may be evident in a spectrum signature are bandwidth, power output, sensitivity, selectivity, modulation, spurious radiation and responses, intermodulation, cross modulation, antenna beam pattern, and dynamic range. **4.** The electromagnetic wave that is obtained as an echo from an object in space. *Note:* The wave shape contains information or evidence concerning the nature of the object. The echo may be analyzed as a function of the amplitudes of its frequency components as a function of time, and used to identify the nature or the motion of the reflector when this information is coupled with other information from other sources. *See also* **signature analysis, spectrum analysis.**

specular reflection: Reflection from a smooth curved or planar surface, such as that of a mirror or polished silver, so that there is negligible diffusion, as Snell's laws of reflection and refraction are microscopically obeyed over a uniformly directed surface. *Note 1:* Specular reflection results in clear and sharp images, such as are obtained from plane, concave, or convex high-grade mirrors. *Note 2:* Coherence can be maintained during specular reflection. *Synonym* **regular reflection.** *See also* **diffuse reflection.**

specular transmission: The propagation of lightwaves in a propagation medium such that (a) diffusion attenuation is negligible, (b) smooth changes in refractive indices result in the refraction of rays at the microscopic level, and (c) in aligned bundles, clear images are preserved during propagation. *Note:* Specular transmission results in improved performance of optical waveguides, such as optical fibers, bundles, and cables. *Synonym* **regular transmission.** *See also* **propagation medium.**

speech: *See* **coded speech.**

speech digit signaling: Signaling (a) in which digit time slots primarily used for encoded speech are used for signaling, and (b) that is an option in networks based on the Integrated Services Digital Network (ISDN). *See also* **Integrated Services Digital Network, signaling.**

speech interpolation: *See* **analog speech interpolation, digital speech interpolation.**

speech plus: Pertaining to a circuit that was designed and used for speech transmission, but to which other uses, such as digital data transmission, facsimile transmission, telegraph, or signaling superimposed on the speech signals, have been added by means of multiplexing. *See also* **multiplexing.**

speech plus duplex operation: Operation in which (a) speech and telegraphy, duplex or simplex, are transmitted simultaneously over the same circuit, and (b) mutual interference is eliminated by the use of filters. *Note:* The circuit usually is used for the transmission of voice signals. *See also* **circuit, composite circuit, duplex operation.**

speech plus signaling: Pertaining to equipment that permits the use of part of a speech band for signaling. *See also* **circuit, composite circuit, signal.**

speech plus telegraph: An arrangement of equipment that permits the use of part of a speech band, i.e., voice frequency band, for the transmission of telegraph signals for data transmission.

speech-powered telephone: *Synonym* **sound-powered telephone.**

speech power unit: *Synonym* **volume unit.**

speech scrambler: In communications systems operations, a device that converts speech signals into unintelligible form before transmission in order to obtain some measure of privacy in the event of interception, such as casual overhearing, by unauthorized persons. *See also* **speech scrambling.**

speech scrambling: In communications systems operations, (a) the converting of speech signals into unintelligible form before transmission and (b) the restoring of the signals to intelligible form at reception, in order to obtain some measure of privacy in the event of interception, such as casual overhearing, by unauthorized persons. *See also* **speech synthesizing.**

speech synthesizer: In communications systems operations, a device that (a) accepts digital or analog data, (b) develops intelligible speech that corresponds to the input data, and (c) does not resort to recorded sounds or simply perform speech scrambling in reverse. *See also* **speech scrambling.**

speech synthesizing: In communications systems operations, (a) the accepting of digital or analog data and (b) the developing of intelligible speech that corresponds to the input data without resorting to recorded sounds or without simply performing speech scrambling in reverse. *See also* **speech scrambling.**

speed: *See* **drum speed, lens speed, optical speed, print speed, radar blind speed, reproduction speed, scanning speed, service speed, spot speed, transmission speed.**

speed buffer: *Synonym* **data buffer.**

speed calling: A service feature that enables a switch or a station to store certain telephone numbers and dial them automatically when a short code, such as a one-, two-, or three-digit code, is entered. *See also* **abbreviated dialing, card dialer, repertory dialer, service feature.**

speed calling 8: A network-provided service feature that (a) gives the call originator fast, accurate, one-touch dialing of eight phone numbers, and (b) may be used on all phones connected to the line.

speed dialing: Dialing at a speed greater than the normal 10 pps (pulses per second). *See also* **abbreviated dialing, pulse, pulsing.**

speed Morse: *See* **high speed Morse.**

speed of light: The speed of an electromagnetic wave in free space, i.e., a vacuum, which is precisely 299,792,458 $m \cdot s^{-1}$ (meters per second). *Note 1:* The speed of light figure is precise because, by international agreement, the meter is defined in terms of the speed of light, represented by the symbol *c*. *Note 2:* The speed of a wave, such as a lightwave, radio wave, or sound wave, is equal to the product of the wavelength and the frequency, whether the wave is propagating in free space or in a material medium. *Note 3:* In a material medium, the speed of a lightwave is lower than in free space. Consequently, the wavelength is also shorter in the material medium than in free space. In a vacuum, the speed of electromagnetic waves of all frequencies, i.e., all wavelengths, is the same. The phase velocity in a propagation medium is given by the relation $v_p = c_0/n$ where v_p is the phase velocity, c_0 is the light velocity in a vacuum, and n is the refractive index of the propagation medium relative to that of a vacuum. *See also* **group velocity, phase velocity, propagation medium.** *Refer to* **Fig. L-8 (536).**

speed of service: 1. From the point of view of the communications system user, i.e., the customer, the time between (a) the instant that a message is released by the message originator and (b) the instant that the message is received by the addressee, i.e., the originator to recipient service speed. *Synonym* **originator to recipient speed of service. 2.** From the point of view of the communications system operator, the time between (a) the instant that a message is entered into a communications system, i.e., the filing time, and (b) the instant that the message is received at the destination addressee communications facility, i.e., the message available for delivery time. *Synonym* **service speed.** *See also* **effective speed of transmission, efficiency factor, filing time, message available for delivery time, throughput.**

speed-up tone: *Synonym* **camp-on-busy tone.**

spelling table: *Synonym* **syllabary.**

sphere: *See* **celestial sphere.**

spherical harmonic function: An orthogonal mathematical function in a spherical coordinate system that (a) is analogous to a Fourier series in a rectilinear coordinate system, (b) may be used to define the shape of the Earth and points on its surface and to describe the frequency components of electromagnetic waves as a function of time and distance when propagating over the surface of the Earth, and (c) is divided into three classes of harmonic functions, (i) zonal harmonics which have values that are constant along parallels of latitude, (ii) sectorial harmonics which have values of consistent sign along lines of constant longitude but with magnitudes that change with latitude, and (iii) tesseral harmonics which maintain a consistent sign with a tessera, but change sign and magnitude as functions of both latitude and longitude. *See also* **Fourier series, tessera.**

spherical intensity: *See* **mean spherical intensity.**

spheroid: *See* **oblate spheroid.**

spherometer: An instrument used to measure precisely the radius of curvature of surfaces.

spike: 1. An extremely short pulse usually plotted in the time domain. **2.** An extremely narrow band of frequencies plotted in the frequency domain. **3.** A bright spot that represents an object in space and that appears on a radar planned position indicator (PPI) screen. *Synonym in radar and sonar* **pip.**

spike file: A filing device that (a) consists of a spike on a pedestal or a stand, (b) is used to store documents by impaling them with the spike, and (c) operates on a last in, first out basis. *See* **last in, first out.**

spill forward: In automatic switching, the transfer of full control of a call to the succeeding central office (C.O.) by sending forward the complete telephone address of the call receiver. *See also* **adaptive channel allocation, distributed switching, routing.**

spill forward feature: A network service feature in which (a) an intermediate central office (C.O.) acts on incoming trunk service treatment indicators, (b) the intermediate C.O. assumes routing control of calls from the originating C.O., and (c) the call completion rate is increased by offering the call to more trunk groups than are available at the originating C.O. *See also* **routing service feature.**

spillover: 1. In an antenna feed, the part of the radiated energy from the feed that does not impinge on the reflectors and hence is spilled into the back and side lobes. **2.** In alternating current (ac) signaling on multilink connections, the part of a signal that passes from one link section to another before the connection between the sections is broken. *See also* **antenna feed.**

spin-scan radiometer: *See* **fiber optic visible-infrared spin-scan radiometer.**

spin stabilization: Control of the stability of a free body in which (a) the attitude of the free body, such as a satellite, missile, bullet, or artillery round, is stabilized by spinning about its axis of maximum moment of inertia, (b) gyroscopic forces are developed that maintain stability, and (c) forces that tend to change the moment of inertia vector that defines the attitude will be resisted.

spiral-four cable: A quadded cable with exactly four conductors. *Synonym* **star quadded cable.** *See also* **cable, quadded cable.**

splice: 1. To permanently join physical media that conduct or transmit power or communications signals. **2.** A device that joins conducting or transmitting media. **3.** A completed joint of conducting or transmitting media. *See* **cable splice, fiber optic splice, fusion splice, mechanical splice, optical fiber splice, optical waveguide splice, waveguide splice.** *See also* **connector.** *Refer to* **Fig. P-11 (749).**

splice closure: At a splice in a cable, an encasement that (a) usually is made of tough plastic, (b) covers the exposed area between the spliced cables, i.e., the area where the jackets have been stripped back to expose the individual propagation media, such as individual optical fibers or wires, that are spliced, (c) usually contains a means to maintain continuity of the tensile strength of the cable, (d) may contain a means to provide electrical continuity of metallic armor, or provide external connectivity to such armor for electrical grounding, (e) in the case of fiber optic cables, may contain a splice organizer to facilitate the splicing process and protect exposed optical fibers from mechanical damage, (f) is sealed at the seams and the points of entry of the spliced cables, (g) may be filled with a material to further retard the entry of moisture, and (h) usually is weatherproof and may be directly buried. *See* **cable splice closure.**

splice housing: *See* **fiber optic splice housing.**

splice loss: In fiber optic systems, a loss of optical power at a splice. *Note 1:* Losses at splices are (a) intrinsic losses in the material of the splice, such as losses in the glass and index matching materials of the splice, and (b) extrinsic losses that depend on the method or the device being used to create the splice, such as losses caused by angular misalignment, lateral offset, longitudinal displacement, differences in core diameter (larger to smaller), and refractive index mismatch at interfaces. *Note 2:* Splice loss caused by a single fiber optic splice can vary from a small fraction of a dB to several dB. *Synonym* **splicing loss.** *See also* **insertion loss.**

splice organizer: In fiber optic communications systems, a device that (a) facilitates the splicing or the breaking out of fiber optic cables, (b) provides a means to separate and secure individual optical fibers or fiber optic pigtails, (c) provides a means to secure (i) mechanical splices or (ii) protective sleeves used in fusion splices, and (d) has a means to contain the slack fiber that remains after splicing is completed. *See* **optical fiber and splice organizer.**

splicer: A device used to create a splice. *See* **loose tube splicer, precision sleeve splicer.** *See also* **splice.**

splice tray: *See* **fiber optic splice tray.**

splicing: The creating of a splice. *See* **field splicing, fusion splicing, precision sleeve splicing.** *See also* **splice.**

splicing loss: *Synonym* **splice loss.**

split homing: The connection of a terminal to more than one central office (C.O.), i.e., switching center, by separate access lines, each having separate directory numbers. *Note:* Split homing is a survivability feature of a communications system. *See also* **access, dual homing, multiple homing.**

split screen: On a display device, display space that has been divided into two or more areas, so that each area can display different portions of the same file or portions of different files. *Note 1:* The split screen excludes the data lying between the portions of the file or files being displayed and includes the desired data in the two or more windows afforded by the split screen. *Note 2:* Examples of split screens are screens that display different portions of a spreadsheet, database, graph, or picture that are too far apart in storage to be viewed or displayed simultaneously as a single image on a single screen.

splitter: *See* **beam splitter, fiber optic splitter.**

splitting: *See* **call splitting, storage splitting.**

spoiling: The reduction or the nullification of the value of a radio network as a direction finding (DF) radionavigation aid by suitably siting a synchronized transmitter that was installed to mutually add to the service area of coverage of the radio network. *See also* **direction finding, radionavigation aid.**

spoking: *See* **radar spoking.**

spontaneous emission: 1. Electromagnetic radiation, such as radiation from a light-emitting diode (LED) or an injection laser operating below the lasing threshold, emitted when the energy level of an internal quantum system, such as electron energy levels in a population of molecules, drops to a lower energy level without regard to the simultaneous presence of similar radiation. **2.** The emission of electromagnetic radiation, such as light, that does not have an amplitude, phase, frequency, time, or other relationship with an applied signal and is therefore a spurious, random, or noise-like form of radiation. **3.** In a quantum mechanical system, radiation emitted when the internal energy of the system drops from an excited level to a lower level without regard to the simultaneous presence of similar radiation. *Note:* Examples of spontaneous emission include radiation from light-emitting diodes (LEDs) and from injection lasers operating below the lasing threshold. *See also* **injection laser diode, light amplification by stimulated emission of radiation, light-emitting diode, stimulated emission, superradiance.**

spoof: 1. To deliberately induce a user of a system, such as a communications, computer, data processing, or control system, to act incorrectly in relation to the system. **2.** To deliberately cause a component of a system to perform incorrectly. *Note:* Examples of spoofing are (a) in radar systems electronic countermeasures (ECM), causing an operator to believe that a target or targets are at a particular location when they are in fact somewhere else, and (b) causing the operator to believe that there are many objects, i.e., targets, when there are in fact no more than two. *See* **anti-spoof.** *See also* **electronic countermeasure.**

spoofing: 1. In communications security (COMSEC) applications, the interception, alteration, and retransmission of a cipher signal or data in such a way as to mislead the receiver. **2.** In automated information systems applications, an attempt to gain access to an automated information system by posing as an authorized user.

spoof jamming: The radiation of a signal that will appear, or be received, as a radar echo signal when there is no object from which the received signal could have been reflected. *See also* **radar echo.**

spool: 1. In communications, computer, data processing, and control systems, to perform two or more on-line operations on peripheral equipment. *Note 1:* Examples of spooling are (a) to use auxiliary storage as a buffer storage to reduce processing delays in transferring data between peripheral equipment and the processors of a computer or data processing system and (b) to read input data streams and write output data streams concurrently with computer program execution. *Note 2:* "Spool" is derived from the expression "simultaneous peripheral operation on line." **2.** To wind or unwind materials, such as wire, cable, optical fiber, or tape, on spools, reels, or bobbins. *See also* **spooler, spooling.**

spooler: 1. A buffer storage between a system component, such as a main frame or the internal memory of a computer, and a peripheral device, used to enable the component to continue to perform other operations while the peripheral device, such as a printer or auxiliary storage, is transmitting data to or receiving data from the component. *Note:* An example of a spooler is a buffer between a computer and a printer that allows the computer to continue processing while the printer is printing. **2.** A program segment or subroutine that controls spooling. *See* **print spooler.** *See also* **spool, spooling.**

spooling: 1. The use of auxiliary storage as buffer storage to reduce processing delays in transferring data between peripheral equipment and the processors of a computer. *Note:* "Spooling" is derived from the expression "simultaneous peripheral operation on line." **2.** The reading of input data streams and the writing of output data streams concurrently under computer program control. **3.** The winding or the unwinding of materials, such as wire, cable, optical fiber, or tape, on spools, reels, or bobbins. *See also* **spool, spooler.**

sporadic E: Irregular scattered patches of relatively high density ionization that (a) seasonally develop within the E region of the ionosphere, (b) reflect and scatter radio frequencies (rf) up to 150 MHz (megahertz), (c) are a regular daytime occurrence over the equatorial regions, (d) are common in the temperate latitudes in late spring, early summer, and to a lesser degree early winter, (e) at high latitudes, can occur along with aurora-associated disturbed magnetic conditions, and (f) can sometimes support reflections for distances up to 2400 km (kilometers) at radio frequencies (rf) up to 150 MHz. *See also* **critical frequency, D layer, D region, E layer, E region, F layer, F region, ionosphere, reflecting layer, sporadic E propagation.**

sporadic E propagation: Radio wave propagation by means of reflection from irregular ionization densities appearing at heights of about 90 km to 120 km above the Earth's surface. *Note:* The maximum frequency reflected from this layer can be much greater than that from the usual E layer. Close to the equator, sporadic E propagation essentially is a daytime phenomenon. In the auroral zones, it is most prevalent at night. The effect causes electromagnetic waves to forward-scatter more readily, thus increasing the range of very high frequency (VHF) communications because of the increased ionization, i.e., ion density, of the E layer of the ionosphere. Sporadic E propagation is more common in summer, during daylight and early evening hours, than in winter. *See also* **forward scatter, ionosphere, propagation, sporadic E.**

spot: *See* **launch spot, recorded spot, recording spot, scanning spot, sunspot, x-dimension of recorded spot, x-dimension of scanning spot, y-dimension of recorded spot, y-dimension of scanning spot.**

spot barrage jamming radar display: A radar screen display that consists of (a) a bright strobe or wedge appearing at jammer azimuth at low jam power, (b) noise patterns, (c) bright strobes at other azimuths, and (d) a dark wedge at jammer azimuth at high jam power. *See also* **azimuth, jammer, jamming, strobe.**

spot beam: In satellite communications systems, a narrow beam from a satellite station antenna that illuminates, with high irradiance, a limited area of the Earth by using parabolic reflectors rather than omnidirectional antennas or Earth coverage antennas.

spot continuous wave jamming radar display: A radar screen display that consists of (a) a bright strobe or wedge appearing at jammer azimuth at low jam power, (b) a dark wedge at jammer azimuth, and (c) bright strobes at other azimuths at high jam power. *See also* **azimuth, display, jammer, radar, screen, strobe.**

spot illumination: *See* **fiber optic spot illumination.**

spot jammer: A jammer that (a) directs its jamming energy against a single narrow bandwidth, thereby increasing the chance of saturating a specific receiver, (b) usually is tunable over a wide frequency band, (c) usually jams only one transmission at a time, and (d) tunes accurately, has a tunable directional antenna, and is thus less likely than a barrage jammer to cause interference with other transmissions. *See* **automatic spot jammer, manual spot jammer.**

spot jamming: Jamming in which (a) a specific channel, frequency, or transmitter is jammed, (b) all the jamming energy is concentrated in a narrow band of frequencies, (c) jamming signals are tunable over a wide band of frequencies, and (d) a directional antenna is used for maximum gain and reduction of interference with friendly transmissions. *Note:* Spot jamming is the most widely used form of jamming. *See also* **electronic warfare.**

spot noise: Noise that is confined to a specific, usually narrow, portion of the electromagnetic frequency spectrum.

spot noise figure: The available output noise power per unit bandwidth divided by the portion of output noise power that is caused by a matched input termination at the standard noise temperature of 290 K (kelvin). *See also* **noise figure, termination.**

spot noise jamming: Jamming in which (a) a single radar is jammed by using a noise bandwidth that is slightly greater than the bandwidth of the radar that is being jammed, (b) the jamming effectiveness is higher than in other forms of jamming, and (c) effectiveness is the greatest when one spot noise jammer is assigned to each radar to be jammed. *Note:* Care must be taken in spot noise jamming implementation. If the jammers are in separate aircraft, the effect would be to give each radar to be jammed a clear view of one jamming aircraft. This would negate the jamming effectiveness quickly. The problem of having a multiplicity of radars to be jammed is more likely than this, and its solution is not simple. The large numbers of emitters at diverse frequencies in modern communications systems make spot noise jamming difficult to implement. Nevertheless, spot noise jamming is the most common form of jamming, primarily because it causes minimum interference with other communications systems.

spot number: A number assigned to a frequency. *Note:* Spot numbers are assigned throughout the frequency spectrum in communications systems operations to quickly identify transmitters and frequencies.

spot off-frequency jamming: Jamming in which the jamming frequency is near, but not at, the frequency that is being jammed. The effect is to reduce the effectiveness of the system being jammed in an attempt to evade jamming by moving off the frequency being jammed.

spot projection scanning: In facsimile systems, optical scanning in which (a) a scanning spot is moved across the object, and (b) the scanning spot size is determined by the illuminated area of the spot on the object. *See also* **facsimile, object, scanning.**

spot size: **1.** On the screen of a cathode ray tube (CRT), the size of the electron spot used to generate a displayed image. *Note:* The spot size is larger than the diameter of the electron beam because of the spillover of electrons into adjacent areas of the screen near the spot. The spot size is a function of the ability of the tube to focus the electron beam, as well as of the electron gun aperture. **2.** In facsimile systems, the diameter or the dimensions of the scanning or recording spot. *See also* **recording spot, scanning spot, x-dimension of recording spot, y-dimension of recording spot, x-dimension of scanning spot, y-dimension of scanning spot.**

spot speed: In facsimile systems, the speed of the scanning or recording spot along the available line. *Note:* The spot speed is usually measured on the object or on the recorded copy in a convenient unit, such as meters per second. *See also* **facsimile, signal, synchronizing.**

spread: 1. *Synonym* **irrelevance. 2.** *See* **Doppler spread, multimode group delay spread.**

spreading: *See* **bandspreading, geometric spreading.**

spreadsheet: A representation of a set of rows and columns, as in a ledger, in which (a) the intersections are addressed locations, i.e., cells, in which data, mathematical formulas, or instructions may be placed, (b) each location has a unique address so that its contents can be represented in formulas and instructions placed at other locations, (c) the display shows the results of a calculation rather than the formula or the instruction placed in it, and (d) if an operand identified by an address in a formula is changed, the new result will be displayed at the location in which the formula is placed. *Note:* For personal computers, spreadsheet programs are used to create, edit, store, and print spreadsheets. The Lotus 1-2-3 spreadsheet, i.e., Worksheet, has 2047 rows, each designated by a numeral from 1 to 2047, and 256 columns, each designated by a letter or a pair of letters, from A through Z, AA through AZ, BA through BZ, and so on, all the way to IV. *See also* **column, row.**

spreadsheet command: In personal computer (PC) spreadsheet operations, one of the set of commands that pertains to the spreadsheet as a whole rather than to a specific location, i.e., a specific cell. *Note:* Examples of spreadsheet commands are formatting, column insertion, column deletion, column width, configuration, recalculation, print, status, and other global commands. *See also* **command, global, spreadsheet.**

spread spectrum (SS): 1. Pertaining to signal coding and transmission in which a sequence of different electromagnetic wave frequencies is used in a pseudorandom manner to transmit a given signal, the instantaneous frequency being selected by a pseudorandom pulse generator that gates the available frequencies in a coded sequence. *Note:* A spread spectrum system requires the use of a bandwidth greater than would be required without the use of spread spectrum transmission in order to allow for all the different frequencies that occur during transmission. A signal to noise ratio improvement is obtained, as well as some antijamming capability and signal security. The receiver must use an identical synchronized pseudorandom pulse generator to unscramble the received signal. **2.** Pertaining to signal structuring in which direct sequence spread spectrum transmission or frequency hopping, or a hybrid of these, is used for multiple access or multiple functions. *Note:* The use of spread spectrum systems (a) usually decreases the interference to other receivers, (b) achieves privacy, and (c) increases the immunity of spread spectrum receivers to noise and interference. Spread spectrum systems usually make use of a sequential noise-like signal structure to spread the usually narrowband information signal over a relatively wide band of frequencies. The receiver correlates the signals to retrieve the original information signal, i.e., the baseband signal. **3.** A transmitted sequence of fixed frequencies that are selected in a pseudorandom manner in accordance with a code. *Note 1:* The duration of each frequency may be fixed or may be varied in accordance with a code. *Note 2:* The transition between frequencies usually is rapid compared to the duration of each frequency. *Note 3:* Spread spectra are used in frequency hopping and spread spectrum transmission systems. **4.** Pertaining to telecommunications signal structuring and transmission in which a signal is transmitted in a bandwidth considerably greater than the frequency content of the original information. *Note:* Spread spectrum transmission uses a sequential noise-like signal structure to spread the usually narrowband information over a relatively wide band of frequencies. The receiver uses the frequency coding structure of the transmitter to correlate the signals and retrieve the original information signal. *See also* **antijam, direct sequence spread spectrum, frequency hopping, frequency hopping spread spectrum, pseudorandom number sequence, spread spectrum code sequence generator.** *Refer to* **Fig. C-2 (121).**

spread spectrum code sequence generator (SSCSG): A generator (a) that produces a pseudorandom sequence of 0s and 1s, (b) that consists of a register, i.e., a bank of series-connected flip-flops with each output connected to the next flip-flop in the series, (c) in which a selection is made of individual flip-flop outputs that are gated back to the input of the first flip-flop in the bank, (d) in which the pseudorandom sequence can be changed by selecting different flip-flops for the feedback loop and by changing the initial states of the flip-flops, (e) in which the output is the sequence of binary pulses corresponding to the states of the last flip-flop in the bank, and (f) in which the output sequence of pulses will regularly repeat itself after a specific number of output pulses. *Synonym* **shift-register generator.** *See* **modular spread spectrum code sequence generator.** *See also* **feedback shift register, frequency hopping generator.** *Refer to* **Fig. S-17.** *Refer also to* **Fig. L-6 (529).**

spread spectrum multiple access (SSMA): In satellite communications systems, code division multiple access (CDMA) in which (a) all carriers that are radiated from Earth stations occupy the entire bandwidth of the satellite transponder, not specific frequencies within the bandwidth, (b) a special code is inserted into each carrier that enables it to be identified by the addressed ground terminal, and (c) some of the commonly used coding methods are phase modulation, frequency hopping, and phase-shift keying. *See also* **bandspreading technique, code division multiple access, maximal code sequence.**

spurious emission: Emission on a frequency or frequencies that (a) are outside the necessary bandwidth and (b) are at a level that may be reduced without affecting the corresponding transmission of information. *Note:* Spurious emissions include harmonic emissions, parasitic emissions, intermodulation products, and frequency conversion products. Out-of-band emissions are not included. *See also* **bandwidth, emission, frequency, necessary bandwidth, spurious response, stray current, susceptiveness.**

spurious radiation: 1. Any unintentional emission. **2.** In communications, computer, data processing, and control systems, any emission outside the intended communications bandwidth. *See also* **frequency, radiation, spurious emission.**

spurious response: 1. Any response other than the desired response of a device, especially of a receiver or a transducer. **2.** Any undesired response to frequencies

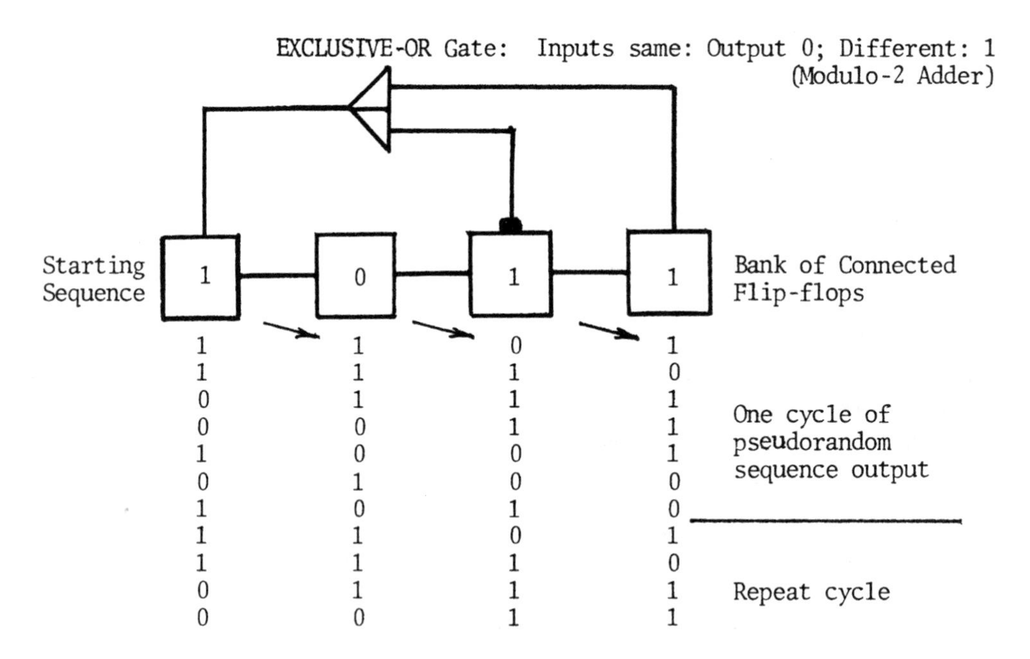

Fig. S-17. The output sequences of the **spread spectrum code sequence generator** are the stepped states of the flip-flops in the shift register. The states are used to gate different frequencies, each flip-flop of the register controlling whether or not a specific frequency is to be transmitted. The horizontal rows of binary digits indicate the set of consecutive states of the register. Each vertical column is derived from the column on its left as the register is stepped, i.e., is shifted, to the right. The leftmost column is determined by the output of the EXCLUSIVE OR gate. The output of the EXCLUSIVE OR gate is determined by the outputs of the rightmost flip-flop and the second flip-flop from the right that has the feedback tap. The starting sequence, the feedback tap locations, and the number of flip-flops can be changed to obtain other pseudorandom number sequences for frequency control. When a cycle is competed, it is repeated. The number of flip-flops, feedback taps, and gates determines the code sequence and its length. The starting sequence is not in the cycle. Some pseudorandom code sequences do not occur.

outside the intended communications bandwidth of a device. **3.** A response in the receiver intermediate frequency (IF) bandpass stage produced by an undesired emission in which the fundamental frequency, or harmonics above the fundamental frequency, of the undesired emission mixes with the fundamental or a harmonic of the receiver local oscillator. *See also* **bandwidth, error, intermediate frequency, response, spurious emission.**

square law: *See* **inverse square law.**

square law device: A nonlinear device that has output signal values that are proportional to the square of the input signal values. *See also* **linear, signal, values.**

square law index profile: *Synonym* **parabolic index profile.**

squares multiplication: *See* **quarter squares multiplication.**

squares multiplier: *See* **quarter squares multiplier.**

square tube splicer: *Synonym* **loose tube splicer.**

square wave: A wave that (a) has two significant conditions, i.e., two levels of amplitude, that change from one condition to the other in a relatively short time compared to the wavelength, (b) when the instantaneous amplitude is plotted versus time or distance, has a rectangular shape, (c) when differentiated, produces alternate polarity spikes, and (d) when integrated, produces a triangular waveshape. *See also* **significant condition, differentiation, integration.** *Refer to* **Fig. P-4 (712).**

squaring loop: A Costas loop that includes a square law device or circuit. *See also* **Costas loop, square law device.**

squelch: A circuit that (a) suppresses the audio output of a receiver, (b) is activated in the absence of sufficiently strong desired input signals, and (c) excludes undesired lower-power input signals that are at or near the frequency of the desired signals. *See also* **circuit, FM improvement threshold, noise suppression.**

squelch circuit: A circuit that reduces background noise in the absence of desired input signals. *See also* **noise suppression.**

squelching: The reducing or the eliminating of the noise in a radio receiver when no carrier signal is present.

sr: steradian.

SRD: superradiant diode. *See* **superluminescent light-emitting diode.**

SS: single sideband.

SSB: single sideband.

SSB-SC: single sideband suppressed carrier.

SSB-SC transmission: single sideband suppressed carrier transmission.

SSMA: spread spectrum multiple access.

SS-TDMA: satellite switched time division multiple access.

stability: The extent to which a specified property, characteristic, or parameter of a substance, device, or apparatus varies with time or with varying environmental conditions. *See* **carrier frequency stability, clock stability, dimensional stability, frequency stability, frequency standard stability, optical fiber dimensional stability, wavelength stability.**

stabilization: *See* **spin stabilization.**

stabilized local oscillator: A local oscillator with a fixed, constant, i.e., nondrifting, output frequency. *Note:* A stable local oscillator is required in the superheterodyne radar receiver in a moving object, i.e., moving target, tracking system.

stable relay: *See* **side stable relay.**

stable state: A state or a condition in which a system will remain unless it is disturbed by some outside force. *Note:* A flip-flop is in a stable state when it is in either of its states before the application of a suitable pulse that could cause it to assume the opposite state, and an Egyptian pyramid is in a stable state in the absence of Earth tremors.

stack: A list or a string of items that is operated upon in such a manner that the item that has been in the list or the string the least time is the next item to be retrieved or moved out, as in the case of a pile of dishes depressed into a counter, i.e., a cellar, or as in a spike file. *See also* **last in, first out; pushdown storage.**

stage: *See* **junction stage.**

stagger: In facsimile systems, a periodic error in the position of the recorded spot along the recorded line. *See also* **facsimile.**

staggered pulse repetition rate: The variation of the time interval between radar pulses in a predetermined manner. *Note:* A staggered pulse repetition rate (a) may eliminate blind speeds associated with fixed pulse repetition rate (PRR) radar systems, and (b) may serve to counter repeater deception as an electronic countercountermeasure (ECCM).

staging: In computer operating systems, the transferring of data from an offline, low priority device to an online, high priority device, or vice versa. *See also* **operating system.**

stand-alone: Pertaining to a device, program, or system that operates independently of another device, program, or system, such as pertaining to (a) a computer that is not served by communications facilities, (b) a modem that is separate from the unit with which it operates, (c) an emulator that is not controlled by a control program, and (d) a word processor that does not depend on other equipment or resources and is operated by one operator at a time.

stand-alone modem: A modem that is separate from the unit with which it operates, and to which it is connected by a cable or a bus. *Synonym* **external modem.** *See also* **integrated modem.**

standard: 1. An agreement reached on products, practices, or operations and formally approved by nationally or internationally recognized industrial, professional, trade, or governmental bodies. **2.** The document that contains the agreements reached on products, practices, or operations and formally approved by nationally or internationally recognized industrial, professional, trade, or governmental bodies. *Note:* Standards usually do not include de facto and proprietary "standards." **3.** An exact value, a physical entity, or an abstract concept, established and defined by authority, custom, or common consent, to serve as a reference, model, or rule in measuring quantities or qualities, establishing practices or procedures, or evaluating results. **4.** A fixed quantity or quality. *See* **Advanced Mobile Phone System standard, ANSI/EIA/TIA-568 cabling standard, cesium standard, data encryption standard, Data Encryption Standard, de facto standard, Electronic Industries Association standard, encryption standard, federal communications standard, Federal Information Processing Standard, fiber distributed data interface standard, industry standard, National Television Standards Committee standard, primary frequency standard, primary time standard, proprietary standard, rubidium standard, secondary frequency standard, secondary time standard, Synchronous Optical Network standard, system standard, time standard.**

standard accuracy: *See* **frequency standard accuracy.**

standard frequency and time signal (SFTS): 1. A time-controlled radio signal broadcast at scheduled intervals on a number of different frequencies by government-operated radio stations in the United States and other countries. **2.** A carrier frequency and time signal emitted in allocated bands in conformity with International Consultative Committee for Radio (CCIR) Recommendation 460. *Note:* In the United States, standard frequency and time signals (SFTS) are broadcast by the U.S. Naval Observatory and the National Institute of Standards and Technology, formerly the National Bureau of Standards. *Synonym* **standard frequency-time signal.** *See also* **Department of Defense master clock, frequency, precise frequency, precise time, primary frequency standard, primary time standard, signal, standard frequency and time signal service.**

standard frequency and time signal satellite service: A radiocommunications service that (a) uses space stations on Earth satellites for the same purpose as it uses those of the standard frequency and time signal service, and (b) may include feeder links. *See also* **standard time and frequency signal service.**

standard frequency and time signal (SFTS) service: 1. A radiocommunications service that (a) serves scientific, technical, and other purposes, (b) transmits specified frequencies or time signals, or both, of specified high precision, and (c) is intended for reception by the general public. **2.** A radio transmission service that broadcasts standard frequency and time signals. *Note 1:* In the United States, the standard frequency and time signal (SFTS) service is called the "standard time and frequency signal (STFS) service." *Note 2:* In the United States, standard time and frequency signals (SFTS) are broadcast by the U.S. Naval Observatory and the National Institute of Standards and Technology, formerly the National Bureau of Standards. *See also* **Coordinated Universal Time, Department of Defense master clock, frequency, precise frequency, precise time, primary frequency standard, primary time standard, signal, standard frequency and time signal.**

standard frequency and time signal station: A station in the standard frequency and time signal service.

standard frequency broadcast: *See* **time signal standard frequency broadcast.**

standard frequency service: A radio communications service that (a) transmits electromagnetic waves of specified standard frequencies of known high accuracy and precision, and (b) is intended for general use, primarily for scientific and technical purposes that require a high degree of frequency stability. *See also* **accuracy, frequency stability, precision, stability.**

standard frequency station: A station in the standard frequency service.

standard frequency-time signal: *Synonym* **standard frequency and time signal.**

standardized profile: In open systems, a profile that specifies one or more interoperable interconnection stacks that are intended to cover one or more specific functional areas. *Note:* Examples of standardized profiles are the International Organization for Standardization (ISO) standardized profiles (ISPs) and the North Atlantic Treaty Organization (NATO) standardized profiles (NSPS).

standard noise temperature: Approximately room temperature, i.e., 290 K (kelvin), used as a reference for noise measurements and for comparison of noise levels in electronic components, particularly in circuits with high gain and low signal to noise ratios. *See also* **solar noise.**

standard operating condition: In the design of a fiber optic station/regenerator section, a nominal environmental condition over which specified parameters must maintain their stated performance ratings, such as terminal input voltage range, room ambient temperature range, relative humidity range, electrical interfaces, and optical line rate. *See also* **extended operating condition.**

standard optical source: A reference optical source to which emitting and detecting devices are compared for calibration purposes. *Note:* In the United States, standard optical sources must be traceable to the National Institute of Standards and Technology (NIST), formerly the National Bureau of Standards (NBS). *Synonym* **standard source.** *See also* **reference frequency.**

standard plan: *See* **aircraft communications standard plan, ship fitting communications standard plan.**

standard precision: *See* **frequency standard precision.**

standard source: *Synonym* **standard optical source.**

standard telegraph level (STL): The power per individual telegraph channel required to yield the standard composite signal level, i.e., operating signal level. *Note:* For a composite signal level of −13 dBm at 0 dBm transmission level point (0TLP), the standard telegraph level (STL) would be −25.0 dBm for a 16-channel voice frequency carrier telegraph (VFCT) terminal computed from $STL = -(13 + 10 \log_{10} n)$, where n is the number of telegraph channels and the STL is in dBm. In this example, $n = 16$. *See also* **level.**

standard test signal: A single frequency sinusoidal signal at a standardized level used for various purposes, such as (a) testing circuit peak power transmission capability, (b) measuring the total harmonic distortion of circuits or parts of a circuit, (c) adjusting the level of individual links and circuits in tandem, and (d) aligning circuits. *Note 1:* One standard test signal is designed for use at the 600-Ω (ohm) audio portions of a circuit. Usually it has a power level of 1 mW (milliwatt) at a frequency of 1 kHz (kilohertz) and is applied at the zero level transmission reference point (0TLP). If 1 mW exceeds the limiting or linear signal range of the circuit under test, a power of −10 dBm at 0 dB level may be used to measure the gain or the loss. A corresponding decrease in test power at the receiver must be allowed to indicate the true gain or loss and hence the true level. *Note 2:* Standardized test signal levels and frequencies are listed in MIL-STD 188-100 and in the *Code of Federal Regulations,* Title 47, part 68. *Synonym* **standard test tone.** *See also* **circuit, frequency, level, reference frequency, total harmonic distortion, transmission level.**

standard test tone: *Synonym* **standard test signal.**

standard test tone power: One milliwatt (electrical), i.e., 0 dBm, at 1 kHz (kilohertz).

standard time and frequency signal (STFS): *See* **standard frequency and time signal (SFTS).**

standard time and frequency signal (STFS) service: *See* **standard frequency and time signal (SFTS) service.**

standard work second: *See* **weighted standard work second.**

standby: **1.** In computer and communications systems operations, pertaining to a power-saving condition or the status of operation of equipment that is ready for use but not in use. *Note:* An example of a standby condition is a radio station operating condition in which the operator can receive but not transmit. **2.** Pertaining to a dormant operating condition or state of a system or equipment that permits complete resumption of operation in a stable state within a short time. **3.** Pertaining to a nonoperating set of equipment that is placed in operation only when equipment in use becomes inoperative, such as a spare set. *See* **cold standby.**

standby station: **1.** A station in a standby status. **2.** An aircraft station, ship station, or land station that (a) is not designated as the primary guard station or as the secondary station for another station, and (b) usually remains in a standby status until called upon to assume the role of a guard station or a secondary station. *See* *also* **guard station, primary station, secondary station, standby.**

standing communications operating instruction (SCOI): An instruction that pertains to the operation of a communications system on a regular or daily basis. *Note 1:* Examples of standing communications operating instructions are hours of operation, frequencies, power levels, and standby notices. *Note 2:* Certain instructions in the standing communications operating instruction (SCOI), such as those that pertain to cryptographic communications, cryptokeying material, and key, are changed as required to maintain security.

standing wave (SW): 1. A wave that (a) exists in a spatial dimension, such as on a transmission line, in a resonant cavity, or on a vibrating string with fixed ends, (b) does not propagate in the ordinary sense of a propagating wave, (c) is characterized by a crest that increases, decreases, and reverses direction or polarity while remaining at the same point on the line, in the cavity, or on the string, and (d) is further characterized by a crest amplitude that has maxima and minima along a spatial dimension, with the distance between the maxima, or any other corresponding points, usually being one-half of the wavelength in absolute terms or a whole wavelength when polarity is taken into account. **2.** In a transmission line or waveguide, a wave (a) in which a distribution of current, voltage, or field strength is formed by two sets of waves propagating in opposite directions, one wave transmitted at one end of the line and one reflected from the other end, and (b) that is characterized by a series of maxima and minima along the line. *Note 1:* In a transmission line, the reflected wave usually occurs as a result of impedance mismatch at the end of the line. Standing waves are undesirable in transmission lines. Lines are terminated in their characteristic impedance so that there are no reflections and thus no standing waves. *Note 2:* Standing waves are induced in optical cavities, such as in lasers, in order to pump the laser to a high optical power level. *Synonym* **stationary wave.** *See also* **optical cavity, traveling wave.**

standing wave ratio (SWR): In a uniform transmission line, the ratio of (a) the amplitude of a standing wave at an antinode to (b) the amplitude at a node. *Note 1:* The standing wave ratio, SWR, in a uniform transmission line is given by the relation $SWR = (1+RC)/(1-RC)$ where RC is the reflection coefficient. *Note 2:* The standing wave ratio (SWR) increases as the reflection coefficient increases, approaching infinity as the reflection coefficient approaches unity. The SWR is limited by the reflection coefficient, the input power, and the losses in the line. The reflection coefficient may be reduced to near zero, and consequently the SWR to near unity, by terminating the line in its characteristic impedance. *See also* **antinode, characteristic impedance, node, reflection coefficient, reflection loss.**

stand-off jamming: Jamming that is accomplished from an aircraft station, ship station, or land station that is close enough to a group being protected, or to the transmitters being jammed, to accomplish effective jamming, but that does not accompany the group being protected. *See also* **jamming.**

star: *See* **specific detectivity.**

star connection: Connection of a set of terminals in which each terminal is individually connected to a central location by a separate cable extending all the way to a central distribution box or a star coupler. *See* **nonreflective star coupler, reflective star coupler.**

star coupler: 1. In fiber optic systems, a coupler that (a) distributes optical signal power from one input port to two or more output ports, (b) combines optical power from two or more input ports and distributes it to a larger number of output ports, or (c) combines optical power from two or more input ports and distributes it to a smaller number of output ports. **2.** In fiber optic systems, a passive device in which signal power from one or several input optical fibers is distributed among a larger number of output optical fibers. *See* **reflective star coupler.** *See also* **network topology, optical combiner, tee coupler.**

star-mesh network: *See* **network topology.**

star network: *See* **network topology.**

star quadded cable: *Synonym* **spiral-four cable.**

STARS: stored terrain access and retrieval system.

start: *See* **ground start, loop start.**

start code: *See* **transmitter start code.**

starting frame delimiter: A specified bit pattern that indicates the start of a transmission frame. *See also* **flag sequence.**

starting powder: 1. A raw glass powder that (a) is used for producing optical fibers, and (b) as the first step in the process of producing optical fibers, must be purified. **2.** Pure glass powder used for making preforms. *See also* **optical fiber, preform.**

start key: *Synonym* **enter key.**

start message: *Synonym* **go ahead notice.**

start notice: *Synonym* **go ahead notice.**

start of heading character (SOH): A transmission control character used as the first character of a message header to indicate the start of the message heading. *See also* **character, control character.**

start of message indicator: An indicator that (a) is used to activate automatic message switching equipment and (b) is required on messages that pass into or through automatic switching systems to indicate the start of a message. *Note:* An example of a start of message indicator is the four-letter group ZCZC that is used in some tape relay communications systems. *Synonym* **start of transmission indictor.** *See also* **indicator.**

start of text character (STX): A transmission control character that (a) precedes text and (b) may be used to terminate the message heading. *Synonym* **STX character.** *See also* **character, control character.**

start of transmission indicator: *Synonym* **start of message indicator.**

start pulse: *Synonym* **start signal.**

start recording signal: *Synonym* **start record signal.**

start record signal: In facsimile systems, a signal used for starting the creation of the recorded copy, i.e., starting the process of converting the electrical signal to an image on the record medium, i.e., the record sheet. *Synonym* **start recording signal.** *See also* **facsimile, recording, signal.**

start signal: 1. In start-stop transmission, a signal at the beginning of a character that prepares the receiving device for the reception of the code elements. *Note:* A start signal is limited to one signal element usually having the duration of a unit interval. **2.** A signal that prepares a device to receive data or to perform a function. *Synonym* **start pulse.** *See also* **A condition, control character, start-stop transmission, stop signal.**

start-stop character: A character that includes one start signal at its beginning and one or two stop signals at its end.

start-stop distortion: The highest absolute value of individual distortion affecting the significant instants of a start-stop modulation. *Note:* The degree of distortion of a start-stop modulation, or demodulation, usually is expressed as a percentage. Distinction can be made between the degree of late, i.e., positive, distortion and the degree of early, i.e., negative, distortion. *See also* **distortion, signal element, significant instant, start-stop modulation.**

start-stop distortion ratio (SSDR): 1. In asynchronous data transmission, the ratio of (a) the absolute value of the maximum measured difference between (i) the actual and (ii) the theoretical intervals that separate any consecutive significant instants of modulation or demodulation, measured from the significant instant of the immediately preceding start element, to (b) the unit interval. *Note 1:* The start-stop distortion ratio (SSDR) usually is expressed in percent, i.e., percent start-stop distortion. *Note 2:* Distinction can be made between late start-stop distortion, i.e., positive start-stop distortion, and early start-stop distortion, i.e., negative start-stop distortion. **2.** In start-stop teletypewriter signaling, the ratio of (a) the highest absolute value of the amount of shifting of the transition points of the signal pulses from their proper positions relative to the beginning of the start pulse to (b) the ideal unit pulse duration. *Synonym* **degree of start-stop distortion.** *See also* **decision instant, distortion, end distortion, pulse duration, significant instant, start-stop distortion.**

start-stop equipment: Equipment that is used in a telegraph system in which start-stop modulation is used. *See also* **start-stop modulation.**

start-stop margin: In start-stop modulation, the maximum amount of overall start-stop distortion that is compatible with the correct translation by the start-stop equipment of all the character signals that appear (a) singly, (b) at the maximum allowable speed, or (c) at the standard modulation rate. *See also* **distortion, start-stop distortion, start-stop modulation, translation.**

start-stop modulation: 1. Modulation in which the time of occurrence of the bits within each character, or block of characters, relates to a fixed time frame, but the start of each character, or block of characters, is not related to this fixed time frame. **2.** Modulation that (a) results from the synchronous modulation of the signal elements that constitute a character, or block of characters, with an undefined significant interval between the consecutive characters, or blocks of characters, (b) permits transmission of single characters, or groups of characters, that are separated by time intervals of random length, (c) uses the start signals to prepare the receiving equipment to receive the character, or group of characters, immediately following, and (d) uses the stop signal to place the equipment in a rest condition so as to prepare it for receipt and interpretation of the next start signal. *See also* **asynchronous communications system, binary digit, modulation, significant interval.**

start-stop restitution: A characteristic that results from isochronous restitution of the signal elements that (a) constitute a set of data, such as a character or a block, and (b) have an undefined significant interval between consecutive data units, such as characters, blocks, and frames. *See also* **isochronous restitution, signal element, significant interval.**

start-stop system: *Synonym* **asynchronous communications system.**

start-stop teletype distortion: *Synonym* **teletypewriter signal distortion.**

start-stop transmission: 1. Asynchronous transmission in which a start pulse and a stop pulse are used for each symbol. **2.** Signaling in which each group of code elements corresponding to an alphanumeric character is (a) preceded by a start signal that serves to prepare the receiving mechanism for the reception and the registration of a character, and (b) followed by a stop signal that serves to bring the receiving mechanism to rest in preparation for the reception of the next character. *See also* **asynchronous operation, asynchronous transmission, code, pulse.**

start-stop transmission system: An asynchronous data transmission system in which each group of code elements corresponding to an alphanumeric character is (a) preceded by a start signal that serves to prepare the receiving mechanism for the reception and the registration of a character, and (b) followed by a stop signal that serves to bring the receiving mechanism to rest in preparation for the reception of the next character. *See also* **asynchronous operation, asynchronous transmission, code, pulse.**

start-stop TTY distortion: *Synonym* **teletypewriter signal distortion.**

state: *See* **degraded service state, excited state, ground state, idle state, operational service state, outage state, stable state.**

state diagram: A directed graph, network, or diagram that (a) has nodes that correspond to the internal states of a system, (b) has edges or branches that correspond to transitions from one state to another, and (c) may be used for describing a system in terms of past, present, and future states and state changes.

statement: 1. In programming languages, a language construct that represents a set of declarations or a step in a sequence of actions. **2.** In computer programming, a symbol string or another arrangement of symbols. **3.** In computer programming, a meaningful expression or a generalized instruction, represented in a source language. *Note:* "Statement" has often been erroneously used as a synonym for "instruction," and vice versa. *See* **IF . . . THEN statement, protocol implementation conformance statement.**

statement pro forma: *See* **protocol implementation conformance statement pro forma.**

statement verb: In a statement, the word that describes the function or the action to be performed when the computer program that contains it presents the statement for execution. *See also* **computer program, execution, statement.**

static: In communications systems, the interference caused by natural electrical disturbances in the atmosphere. *Note:* Examples of static are the noises that are heard from a radio receiver and that are caused by lightning or northern lights. *See* **precipitation static.** *See also* **interference.**

static analysis: The evaluation of the performance of a functional unit before it is actually operated or run. *Note:* Examples of static analyses are (a) the evaluation of a computer program without execution of the program and (b) evaluation of a protocol before it is implemented. *See also* **dynamic analysis.**

static analyzer: In computer programming, a software tool that aids in the evaluation of computer programs without actually executing the programs. *Note:* Examples of static analyzers are syntax checkers, compilers, cross-reference generators, flow characters, and standards enforcers. *See also* **dynamic analyzer, software tool.**

static binding: Binding that (a) is performed prior to execution of a computer program and (b) usually is not subject to change during program execution. *See also* **binding, computer program, dynamic binding, execution.**

static buffering: Allocating buffer storage to incoming or outgoing data or to a job, computer program, or routine before or at the beginning of its execution, rather than as needed during execution. *See also* **buffer storage, computer program, dynamic buffering, job.**

static data: In a storage device, data that remain stored and fixed in time and space, do not have to be regenerated to be retained, and do not have to be moved to be read or removed. *Note 1:* An example of static data are data in a register that consists of a group of storage devices, such as flip-flops or magnetic cores used to store a number of bits. The bits may be entered into or removed from the storage devices in parallel or serially. *Note 2:* Examples of data that are not static are (a) data stored in cathode ray tubes (CRT) or liquid crystal display (LCD) devices because of their volatility, and (b) data stored on magnetic tapes, disks, cards, or drums because the data media have to be moved to be read. *See also* **data medium, serial to parallel converter.**

static dump: A dump that is made (a) during or after the execution of a computer program, (b) usually when the program is halted or interrupted, and (c) usually under the control of a supervisory program rather than the program being executed. *See also* **computer program, dump, supervisory program.**

static electricity: An accumulation of positive or negative electric charges (a) on a body that has the capacity to store them, (b) that causes the buildup of an electrical potential difference, i.e., a voltage, with respect to other bodies, (c) that is discharged as soon as the voltage is high enough for an available path to conduct the charges off the body, and (d) that discharges as an electric current, with the current and the voltage producing surges of electromagnetic energy that (i) can cause interference in communications systems, (ii) can damage electronic circuitry, detonate explosives, set fires, harm living tissue, or cause equipment to malfunction, and (iii) can produce harmful effects, such as electric polarization caused by high electric fields, magnetization due to high magnetic fields caused by high discharging currents, and heating caused by magnetically induced currents or conduction currents during discharging. *Note:* Effective grounding of equipment is essential in order to reduce the effects of static electricity. *See also* **electromagnetic radiation hazard, lightning.**

staticizer: *Synonym* **serial to parallel converter.**

static magnetic field: *Synonym* **magnetostatic field.**

static storage: Storage in which the data medium used for storage does not move to accomplish reading and writing. *Note:* Examples of static storage are magnetic core storage, cathode ray tube (CRT) storage, and flip-flop storage. *See also* **data medium.**

station: 1. In communications systems, an assembly of equipment composed of data terminal equipment (DTE), data circuit-terminating equipment (DCE), and the interface between them. **2.** In high level data link control (HDLC) operation, a primary station, a secondary station, or a combined station. **3.** In radio communications, one or more transmitters, receivers, or accessory equipment, or a combination thereof, that (a) is necessary at one location for carrying on a radiocommunications service or the radio astronomy service, and (b) is classified by the service in which the transmitters, receivers, or accessory equipment permanently or temporarily operate. *See* **active station, aeronautical advisory station, aeronautical broadcast station, aeronautical Earth station, aeronautical fixed station, aeronautical marker beacon station, aeronautical multicom land station, aeronautical multicom mobile station, aeronautical radio beacon station, aeronautical station, aeronautical telemetering land station, aeronautical telemetering mobile station, aeronautical utility land station, aeronautical utility mobile station, aircraft Earth station, aircraft station, airport land mobile radio base station, airport land mobile radio station, altimeter station, area broadcast station, aviation instructional station, balanced station, base station, broadcasting satellite space station, broadcasting station, cable television relay service station, called station, calling station, coast Earth station, coast station, combined station, communications satellite Earth station, communications satellite space station, configurable station, constant watch station, control station, data input station, data processing station, data station, directly connected station, dual access tributary station, Earth station, emergency position-indicating radiobeacon station, experimental export station, experimental research station, experimental station, experimental testing station, fixed Earth station, fixed microwave auxiliary station, fixed station, flight telemetering land station, flight telemetering mobile station, flight test station, glide path slope station, hydrological and meteorological fixed station, hydrological and meteorological land station, hydrological and meteorological mobile station, inactive station, inquiry station, integrated station, international broadcasting station, land station, listening station, localizer station, local ship-shore station, logical station, main station, major relay station, marine broadcast station, marine radio beacon station, master net control station, master station, meteorological satellite Earth station, meteorological satellite space station, minor relay station, mobile Earth station, mobile station, net control station, oceanographic data interrogating station, oceanographic data station, omnidirectional range station, one-operator ship station, operator period station, primary ship-shore station, primary station, primary substation, private aircraft station, radar beacon station, radar beam precipitation gage station, radar netting station, radio astronomy station, radio beacon station, radio determination station, radio direction finding station, radiolocation land station, radiolocation station, radionavigation mobile station, radionavigation operational test land station, radionavigation satellite Earth station, radio navigation**

satellite space station, radionavigation station, radiopositioning land station, radiopositioning mobile station, radio range station, radiosonde station, radio telephone station, read station, regular station, relay station, rural subscriber station, secondary aeronautical station, secondary ship-shore station, secondary station, ship broadcast area radio station, ship broadcast coastal radio station, ship Earth station, ship station, shore station, slave station, space research space station, sounder prediction station, sounder station, space station, space telecommand Earth station, space telecommand space station, space telemetering Earth station, space telemetering space station, space tracking Earth station, space tracking space station, standard frequency station, standby station, surface station, surveillance station, survival craft station, telemetry command station, telemetering fixed station, telemetering land station, telemetering mobile station, terminal station, terrestrial station, tracking telemetry command station, transportable station, transport station, traveler information station, tributary station, way station, workstation.

stationary direct current: Direct current (dc) that does not vary in magnitude, i.e., is steady rather than pulsating. *See also* **direct current, pulsating direct current.**

stationary information source: *Synonym* **stationary message source.**

stationary message source: A message source that transmits messages each of which has a probability of occurrence independent of its occurrence time. *Synonym* **stationary information source.**

stationary satellite: *Synonym* **geostationary satellite.**

stationary wave: *Synonym* **standing wave.**

station battery: Within a facility, a separate battery power source that (a) provides for all significant communications requirements for direct current (dc) input power, (b) usually is centrally located, (c) provides dc power for radio and telephone equipment, and (d) may provide power for emergency lighting and controls for the facility equipment. *See also* **auxiliary power, facility, primary power.**

station cable: *See* **fiber optic station cable.**

station call: *See* **station to station call.**

station clock: In a station, the principal clock, or the alternate clock, that (a) provides the timing reference at the station, (b) controls all station communications equipment requiring time control, and (c) may also be used to provide timing or frequency signals for other equipment. *See also* **clock, frequency synthesizer, principal clock.**

station designator: *See* **address designator.**

station equipment: *Synonym* **customer premises equipment.**

station fiber optic cable: In a fiber optic station/regenerator section, the fiber optic cable that (a) connects the optical transmitter or receiver fiber optic pigtail to the splice on the fiber optic cable facility side, i.e., the line or outside-plant side, of the fiber optic station facility, (b) includes the cable to, and within, the cable interconnect feature, (c) is used within a building environment to connect the outside-plant fiber optic cable to the fiber optic system terminal equipment, and (d) may provide the optical signal path in the form of fiber optic patch panels that allow rearrangement of paths to the outside-plant optical fibers. *Synonyms* **fiber optic station cable, optical station cable, station optical cable.**

station fiber optic facility: The part of a fiber optic station/regenerator section that lies within a station, i.e., the inside plant, including the fiber optic transmitter, station fiber optic cable, connectors, and cable interconnect features. *Synonyms* **fiber optic station facility, station optical facility.**

station keeping: 1. In satellite communications systems, the controlling of a satellite to ensure that it maintains a desired orbit. *Note:* Examples of station keeping are (a) maintaining a nominally geostationary satellite within a given region relative to the Earth, and (b) using data relays from satellites to control the relative position of other satellites. **2.** Maintaining an assigned position of a ship relative to other ships in a convoy. *Note:* In ship convoy communications, each station, i.e., each location or position, in a convoy may be assigned an internal convoy call sign that is relinquished to any ship that is assigned to occupy that station, i.e., that position, in the convoy.

station load: The total electrical power that a station consumes. *See also* **disconnect switch, load, power, technical load.**

station log: A chronological record of significant events that occur at a station, such as operating schedules, message traffic conditions, outages, message handling difficulties, delays, and names of operators.

station master log: *See* **aeronautical station master log.**

station message detail recording (SMDR): A record of all calls originated or received by a switching system. *Note:* Station message detail recordings (SMDRs) usually are generated by a computer. *See also* **activity factor, audit trail.**

station/regenerator section: *See* **fiber optic station/regenerator section.**

station side: *Synonym* **equipment side.**

station symbol: A letter group that is used to indicate a class or a type of transmitting station. *Note:* The letters "BCI" might be used to designate an international broadcasting station.

station to station call: In an automatically switched telephone network, a long-distance call in which (a) the call originator enters, i.e., dials, the number of the call receiver and obtains a connection through the use of automatic equipment, (b) there is no human intervention, and (c) at least two exchanges are used, the calling and the called exchange. *See also* **local call, number call, person to person call.**

statistical fiber optic cable facility loss: In a single mode fiber optic station/regenerator section, the mean loss in the fiber optic cable facility, i.e., in the outside plant, given by the relation $\mu_L = l_t(\mu_c + \mu_{cT} + \mu_{S\lambda}) + N_s(\mu_S + \mu_{ST})$ with the standard deviation given by the relation $\sigma_L = \{l_t l_R(\sigma^2_{cmx} + \sigma^2_{cT}) + N_S(\sigma_S^2 + \sigma_{ST}^2)\}^{1/2}$ where l_t is the total sheath, i.e., jacket, length of spliced fiber optic cable in kilometers, l_R is the reel length of fiber optic cable in kilometers, μ_c is the mean end-of-life cable attenuation rate in decibels per kilometer (dB/km) at the transmitter nominal central wavelength, $\mu_{S\lambda}$ is the increase in mean cable attenuation rate above μ_c measured at the wavelength within the transmitter central wavelength range at which the largest mean plus two sigma loss occurs, μ_{cT} and σ_{cT} are the mean and the standard deviation of the effect of temperature on cable loss (dB/km) at the worst case temperature conditions over the expected cable operating temperature, σ_{cmx} is the standard deviation of cable loss (dB/km) determined at the wavelength at which the largest mean plus two sigma cable loss occurs, including an allowance for uncertainty in cable loss measurements, N_s is the number of splices in the length of the cable in the fiber optic cable facility, including the splice at the station facility on each end and allowances for cable repair splices, μ_{ST} and σ_{ST} are the mean and the standard deviation of the effect of temperature on splices (dB/splice) at the worst case temperature conditions over the cable operating temperature range, which must be specified for underground, buried, and aerial applications, and μ_S and σ_S are the mean and the standard deviation of splice loss (dB). *Synonym* **statistical optical cable facility loss.**

statistical loss budget constraint: In the design of an optical station/regenerator section using statistical methods, the constraint imposed by the relation $\mu_{G\text{-}L} > 2\sigma_{G\text{-}L}$ where $\mu_{G\text{-}L} = \mu_G - \mu_L$ and $\sigma_{G\text{-}L}{}^2 = \sigma_G{}^2 + \sigma_L{}^2$ where μ_G is the statistical (mean) terminal/regenerator system gain in dB, μ_L is the statistical (mean) fiber optic cable facility loss (dB), σ_G is the standard deviation of the statistical terminal/regenerator system gain (dB), and σ_L is the standard deviation of the fiber optic cable within the fiber optic cable facility of the fiber optic station/regenerator system section, assuming that distributions are Gaussian, splice losses are uncorrelated with fiber losses, cable losses are uncorrelated from reel to reel, loss allowance for transmitter wavelength drift is correlated in different reels, averages and sigmas are representative over time, sample sizes are sufficient to warrant the use of Gaussian statistical theory, and product distributions are reasonably Gaussian in shape. *See also* **loss budget constraint.**

statistical multiplexing: Multiplexing in which channels are established on a statistical basis. *Note 1:* An example of statistical multiplexing is the making of connections according to the probability of need. *Note 2:* Statistical multiplexing is based on prior usage, i.e., on service demand history. *See also* **channel, multiplexing.**

statistical optical cable facility loss: *Synonym* **statistical fiber optic cable facility loss.**

statistical terminal/regenerator system gain: In a single mode fiber optic station/regenerator section, the transmitter optical power less the receiver power sensitivity and the optical power requirements of the terminal or regenerator station facilities, such as the dispersion power penalty, the reflection power penalty, the overall safety margin, and only the power losses in cables and connectors within the station facility at both ends. *Note:* The mean statistical terminal/regenerator system gain is given by the relation $\mu_G = P_T - P_R - P_D - R_P - M - \mu_{WDM} - l_{SM}\mu_{sm} - N_{con}\mu_{con}$ and the standard deviation is given by the relation $\sigma_G = (\sigma^2_{PT} + \sigma^2_{PR} + \sigma^2_{WDM} + \sigma^2_{PD} + N_{con}\sigma^2_{con})^{1/2}$ where P_T and σ_{PT} are the mean and the standard deviation of the transmitter power, the optical power in dBmW (decibels referred to 1 milliwatt) into the single mode station fiber optic cable on the line side of the transmitter unit fiber optic connector or splice, specified as P_{T1} and σ_{PT1} for standard operating conditions or P_{T2} and σ_{PT2} for extended operating conditions, and P_R and σ_{PR} are the mean and the standard deviation of the receiver sensitivity, the input optical power (dBm) to the fiber optic receiver on the line side of the receiver connector or splice, specified as P_{R1} and σ_{PR1} for standard operating conditions or P_{R2} and σ_{PR2} for extended operating conditions, that are necessary to achieve the manufacturer-specified bit error ratio (BER) and that include the performance degradations caused by manufacturing variations introduced by temperature and aging drifts, the maximum transmitter power penalty resulting from the use of a transmitter with a worst case extinction ratio, and the maximum transmitter power penalty resulting from use of a transmitter with a worst case rise and fall time, but do not include the power penalty associated with dispersion; and P_D and σ_{PD} are the mean and the standard deviation of the dispersion power penalty (dB), R_P is the reflection power penalty (dB), M is the overall safety margin (dB), μ_{WDM} and σ_{WDM} are the mean and the standard deviation of the all-inclusive loss (dB) associated with wavelength division multiplexing equipment at both ends, including the effects of temperature, humidity, and aging, l_{SM} is the total length (km) and μ_{SM} is the mean attenuation rate (dB/km) for the single mode station/regenerator fiber optic cable, N_{con} is the number of single mode connectors, not including the transmitter and receiver unit connectors, and μ_{con} and σ_{con} are the mean and the standard deviation of the single mode connector insertion loss. *See also* **statistical fiber optic facility loss.**

statistical time division multiplexing: Time division multiplexing in which connections to communications circuits are made on a statistical basis, such as a service demand history basis. *See also* **circuit, multiplexing.**

statistics on request: Upon request of a user, certain data that are furnished concerning the distribution of calls that are made under defined headings, such as international calls, national calls, calls to certain users, total calls, and other related information. *Note:* Information concerning the nature or the content of the subject matter of the calls is not furnished.

status: *See* **basic status, global status, secondary status.**

status channel: A channel that is used to indicate whether a group of bits is (a) for data exchange between users or (b) for communications system or transmission control purposes.

status signaling: Signaling that is used to determine the progress of a call. *Note:* Examples of status signaling are dial tone, ringing tone, busy signal, and number-unobtainable tone.

statute mile: A unit of distance equal to 5280 ft (feet), 1760 yds (yards), about 1.609 km (kilometers), about 0.869 nmi (nautical mile). *See also* **nautical mile.** *Note:* In some countries, these numbers may be written as 1,609 km (kilometers), 0,869 nmi (nautical miles), or 5 280 ft (feet). Care must be taken in using a comma, as it serves as a decimal point in many countries, such as France and Germany.

steady state condition: 1. In a communications circuit, a condition in which some specified characteristic of a parameter, such as a value, rate, periodicity, or amplitude, exhibits only negligible change over an arbitrarily long period. **2.** In an electrical circuit, a condition (a) that occurs after all initial transients or fluctuating conditions have damped out, and (b) in which currents, voltages, or fields remain essentially constant, or oscillate uniformly, without changes in characteristics, such as changes in amplitude, frequency, or wave shape. *In fiber optics and modal power distribution, synonym for* **equilibrium modal power distribution.** *See also* **circuit, equilibrium modal power distribution.**

stealing: *See* **velocity gate stealing, range gate stealing.**

steer: *Synonym* **vector.**

steering: *See* **beam steering.**

stencil duplicating: Duplicating in which ink passes through perforations in a stencil master to form a recorded copy, i.e., an image, on the record medium, such as paper. *See also* **stencil master.**

stencil duplicator: A duplicating device that has one or more revolving drums that use the stencil duplicating process, which produces multiple copies from a stencil master mounted on a drum, through which ink passes to a sheet of a recording medium, such as paper, on each revolution of the drum. *Note:* An example of a stencil duplicator is a mimeograph machine.

stencil master: A sheet of material that carries an image, in the form of perforations, of the object, i.e., the original, to be copied. *See* **image, object.**

step: 1. In communications, computer, data processing, and control systems, to cause the execution of one operation. **2.** An operation performed by a device, component, or system.

step by step switch (SXS): In a switching system, a switch composed of components that select the links from switching stage to switching stage, one after the other, independently of the state of following stages. *See also* **switch.**

step by step (SXS) switching system: An automatic dial telephone system in which calls go through the switching equipment by a succession of switches that move one step at a time, from stage to stage, each step being made in response to the dialing of a number. *See also* **crossbar switch, electronic switching system, switching center.**

step function: 1. A sudden transition from one significant condition to another without consideration of a return to the original condition. *Note 1:* The step function might be that of any parameter, such as a voltage, current, power, frequency, or phase. *Note 2:* A step function might be used as a stimulus to measure the response of a circuit or a device. **2.** A pulse of theoretically infinite duration. **3.** In a mathematical function, a sudden shift in the value of a variable or a parameter. *See also* **significant condition, value.**

step index fiber: *Synonym* **step index optical fiber.**

step index optical fiber: An optical fiber that (a) has a uniform homogeneous isotropic refractive index core and a uniform isotropic refractive index cladding, with the core refractive index being greater than the cladding refractive index, and (b) may have more than one cladding, each usually with a different refractive index which is lower than that of the core. *Note:* For the step index optical fiber, the profile parameter, *g*, is infinite. *Synonyms* **radially stratified optical fiber, step index fiber.** *See also* **fiber optics, graded index optical fiber, profile parameter, step index profile.** *Refer to* **Fig. S-18.** *Refer also to* **Figs. C-9 (190), R-1 (796).**

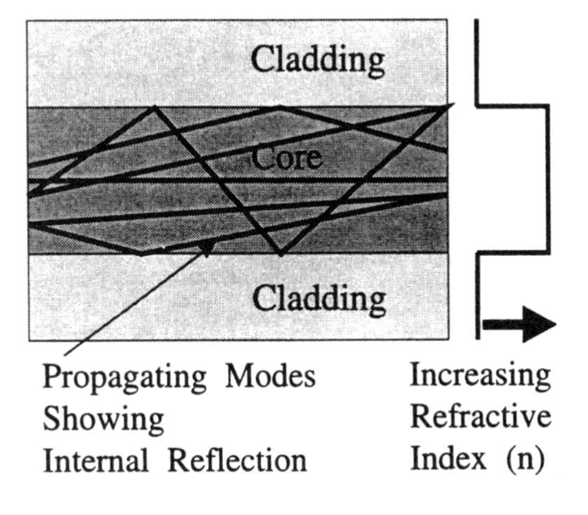

Fig. S-18. Propagating modes and refractive index profile in a **step index optical fiber,** showing the total internal reflection that occurs because the incidence angle at the core-cladding interface is greater than the critical angle, all angles being measured with respect to the normal at the interface surface at the point of incidence.

step index optical waveguide: A dielectric waveguide that has a core of uniform refractive index and one or more claddings, usually each with a different uniform refractive index.

step index profile: In an optical fiber, a refractive index profile characterized by (a) a uniform refractive index within the core, (b) a sharp decrease, with respect to radial distance, in refractive index at the core-cladding interface, and (c) a uniform refractive index in one or more claddings. *Note 1:* The step index profile corresponds to a power law index profile with the profile parameter approaching infinity. *Note 2:* The step index profile is used in most single mode optical fibers and some multimode fibers. *See* **equivalent step index profile.** *See also* **critical angle, graded index profile, multimode optical fiber, normalized frequency, profile parameter, refractive index.**

step index profile refractive index difference: *See* **equivalent step index profile refractive index difference.**

stepped: *See* **bit-stepped, character-stepped.**

stepped start-stop transmission system: A start-stop transmission system in which the start signals occur at regular intervals. *See also* **asynchronous communications.**

step size: *See* **plotter step size.**

stepwise variable optical attenuator: A device that attenuates the irradiance or the radiance of lightwaves in discrete steps, each of which is selectable by some means, such as by changing sets of filter cells, rather than over a continuously variable range. *Note:* An example of a stepwise variable optical attenuator is one in which fixed-attenuation cells of 0, 1, 2, 6, and 10 dB (decibels) are used three at a time so that all attenuations from 0 to 14 dB are achievable in steps of 1 dB. The attenuator may be inserted into an optical link to control the irradiance, i.e., the intensity of light at the receiver photodetector. *See also* **continuously variable optical attenuator.**

steradian (sr): A unit of solid angle measure equal to the solid angle at the center of a sphere subtended by an area equal to the radius squared on the surface of the sphere. *Note 1:* 1 sr (steradian) is $1/4\pi$ of the total solid angle of a sphere, i.e., there are 4π sr in each and every sphere. *Note 2:* The solid angle subtended by a cone of half-angle A is $SA = 2\pi(1 - \cos A)$ where SA is the solid angle in steradians. *See also* **unit solid angle.**

stereophonic crosstalk: An undesired signal that occurs (a) in the main channel from modulation of the stereophonic channel or (b) in the stereophonic channel from modulation of the main channel.

stereophonic sound subcarrier: A subcarrier within the frequency-modulated (FM) broadcast baseband used for transmitting signals for stereophonic sound reception of the main broadcast program service.

stereophonic sound subchannel: The band of frequencies from 23 kHz (kilohertz) to 99 kHz containing sound subcarriers and their associated sidebands.

STFS: standard time and frequency signal, standard time-frequency signal. *See* **standard frequency and time signal, standard frequency and time signal service, standard frequency and time signal station.**

stick: *See* **joystick.**

still image: Nonmoving visual information, i.e., fixed images. *Note:* Examples of still images are a single graph, a drawing, and a picture. *See also* **facsimile, freeze frame, full motion, near full motion.**

stimulated emission: 1. Electromagnetic radiation, such as radiation from an injection-locked laser operating above the lasing threshold, emitted when the energy level of an internal quantum system, such as electron energy levels in a population of molecules, drops to a lower energy level when induced to do so by the simultaneous presence of radiation of suitable wavelength. **2.** In a laser, the emission of light caused by a signal applied to the laser such that the response is directly proportional to, and in phase coherence with, the electromagnetic field of the stimulating signal, this coherency between applied signal and response being the key to the operation and the usefulness of the laser. **3.** In a quantum mechanical system, radiation emitted when the internal energy of the system drops from an excited level to a lower level when induced by the presence of radiant energy at the same frequency. *Note:* An example of stimulated emission is the radiation from an injection laser diode operating above the lasing threshold. *See also* **injection laser diode, light amplification by stimulated emission of radiation, spontaneous emission.**

stimulated emission of radiation: *See* **light amplification by stimulated emission of radiation, microwave amplification by stimulated emission of radiation.**

stimulator: *See* **fiber optic myocardium stimulator.**

STL: standard telegraph level, studio to transmitter link.

stone: In the glass and plastic used in optical systems, a small, solid, trapped opaque particle that causes deflection, dispersion, diffusion, reflection, scattering, and absorption in optical elements, such as lenses and optical fibers.

stop: *See* **aperture stop, field stop, hard stop, T stop.**

stopband: The band of frequencies, between specified limits, that a circuit, such as a filter or telephone circuit, is not capable of transmitting; i.e., all frequencies above the stopband lower limit and below the stopband upper limit are not transmitted, i.e., cannot pass. *Note 1:* The limiting frequencies are those at which the transmitted power level increases to a specified level, usually 3 dB below the maximum level, as the frequency is decreased or increased from that at which the transmitted power is a minimum. *Note 2:* The difference between the limits is the stopband bandwidth, usually expressed in hertz. *Note 3:* Not all types of equipment and circuits have a stopband. Most equipment passes either low frequencies or high frequencies, or may pass both, without a stopband. *Note 4:* In the case of a stopband filter, if the stopband is a single frequency or a very narrow band of frequencies, the filter is called a "notch filter." *See also* **passband.**

stop character: *See* **start-stop character.**

stop code: In word processing systems, the code for an instruction that causes an operation, such as reading, writing, printing, or transferring, to halt. *See also* **instruction.**

stop distortion: *See* **start-stop distortion.**

stop element: A stop signal that (a) consists of one signal element and (b) has a duration equal to or greater than a specified minimum value. *See also* **signal element, stop signal, value.**

stop filter: *See* **band-stop filter.**

stop lockup: *See* **scan-stop lockup.**

stop message: *Synonym* **stop notice.**

stop notice: In tape relay communications systems, a service message to a relay or to a tributary station requesting that the operator stop transmitting over specified channels. *Synonym* **stop message.** *See also* **go ahead notice.**

stop pulse: *Synonym* **stop signal.**

stop recording signal: *Synonym* **stop record signal.**

stop record signal: In facsimile systems, a signal used for stopping the process of converting the electrical signal to an image on the record medium, i.e., of creating a recorded copy. *Synonym* **stop recording signal.** *See also* **facsimile, signal.**

stop restitution: *See* **start-stop restitution.**

stop signal: 1. In start-stop transmission, a signal that (a) is at the end of a character, (b) prepares the receiving device for the reception of a subsequent character, and (c) usually is limited to one signal element of any duration equal to or greater than a specified value. *Synonym* **stop pulse. 2.** A signal to a receiving mechanism to wait for the next signal. *See also* **control character, overhead bit, start signal.**

stop transmission: *See* **start-stop transmission.**

stop transmission system: *See* **start-stop transmission system, stepped start-stop transmission system.**

storage: 1. The retention of data in a device that holds the data. **2.** The action of placing data into a device that holds the data. **3.** A functional unit into which data can be placed, in which the data can be retained, and from which the data can be retrieved. *Note:* The means for storing the data, i.e., for representing the data in storage, include chemical, mechanical, electrical, magnetic, and optical means. *Synonym* **memory, store.** *See* **acoustic storage, associative storage, auxiliary storage, buffer storage, capacitor storage, cathode ray tube storage, changeable storage, circulating storage, core storage, core-rope storage, cryogenic storage, dedicated storage, delay line storage, destructive storage, direct access storage, disk storage, dynamic storage, erasable storage, external storage, fixed storage, flip-flop storage, free storage, immediate access storage, intercept tape storage, magnetic storage, magnetic card storage, magnetic rope storage, magnetic tape storage, magnetic wire storage, magnetic thin-film storage, main storage, manual switch storage, mass storage, matrix storage, mercury storage, nondestructive storage, offline storage, online storage, optical disk storage, optical thin-film storage, parallel search storage, parallel storage, permanent storage, photooptic storage, primary storage, protected storage, pushdown storage, pushup storage, read-only storage, real storage, regenerative storage, secondary storage, serial access storage, solid state storage, static storage, temporary storage, unaddressable storage, unprotected storage, virtual storage, volatile storage, word-organized storage.** *See also* **erase, fetch protection, read-only storage, register.**

storage allocation: The assignment of storage units, storage areas, and storage locations to specified data or programs, special purposes, specific users, or particular systems. *See also* **dynamic storage allocation.**

storage area: A part or a subdivision of storage.

storage bounds checking: Checking (a) that tests the results of a computer program designed to determine whether or not there is access to storage outside of the authorized limits, and (b) that is used to ensure that unauthorized access to specified storage areas does not occur. *Synonym* **memory bounds checking.**

storage capacity: The amount of data that can be contained in a storage device. *Note:* Storage capacity may be expressed or specified in bits, bytes, characters, words, or other application-related units of data, usually in eight-bit bytes, such as a 1.44-Mb (megabyte) storage capacity.

storage cell: 1. An addressable storage unit. **2.** The smallest subdivision of storage into which a unit of data can be entered, stored, and retrieved. *Synonym* **storage element.**

storage center: In a communications system, a centralized storage unit for a store and forward data transmission system.

storage compaction: Increasing the efficiency of utilization of the storage capacity of a storage unit, by using more efficient coding schemes rather than by relocating fragmented data or programs that are scattered in storage into contiguous areas, thus filling in unused spaces. *See also* **compaction, data compaction, data compression.**

storage compression: Data compression accomplished by increasing the efficiency of use of the storage capacity of a storage unit, such as by increasing the ratio of the volume of data stored to the storage capacity. *Note:* An example of storage compression is the relocation of fragmented data or computer programs that are scattered in storage into contiguous areas, thus filling unused spaces. *See also* **storage compaction.**

storage cycle: The sequence of events that is required each time data are transferred to or from a storage unit. *See also* **storage unit.**

storage device: *Synonym* **storage unit.**

storage disk: *Synonym* **data disk.**

storage element: *Synonym* **storage cell.**

storage location: A storage area that (a) is uniquely and explicitly specified by an address, and (b) usually has a specified fixed capacity, such as the capacity to store a single computer word. *See also* **storage area.**

storage cycle: The sequence of events that is required each time data are transferred to or from a storage unit. *See also* **storage unit.**

storage device: *Synonym* **storage unit.**

storage disk: *Synonym* **data disk.**

storage element: *Synonym* **storage cell.**

storage location: A storage area that (a) is uniquely and explicitly specified by an address, and (b) usually has a specified fixed capacity, such as the capacity to store a single computer word. *See also* **storage area.**

storage register: A register used to store data rather than instructions or control signals. *See also* **instruction, register.**

storage space: *See* **image storage space.**

storage splitting: The partitioning or the dividing of the storage capacity of a storage unit into separate storage areas that (a) may be independently addressed and (b) may be assigned to and accessed by specific computer programs. *See also* **computer program, storage area, storage capacity, storage unit.**

storage tube: A cathode ray tube (CRT) that (a) retains a display image on its screen for a period longer than is required for a regeneration cycle, and (b) can scan the image and transmit the resulting signals to an amplifier for recording or transmission by radio, wire, fiber optic cable, or another propagation medium, sequentially or simultaneously, such as when a picture on the tube screen is (i) sequentially scanned and transmitted serially or (ii) transmitted as a whole image over a fiber optic aligned bundle and displayed on the faceplate of a fiberscope at the receiving end. *See also* **aligned bundle, display image, fiber optic bundle, propagation medium, regeneration.**

storage unit: In communications, computer, data processing, and control systems, the part in which data and computer programs may be stored and from which they may be retrieved, other parts being the arithmetic-

Fig. S-19. The Personal Information Center has a speakerphone, caller ID, and the capability to **store** 200 names along with various items of personal information, such as telephone numbers, addresses, birthdates, and notes. (Courtesy AT&T Archives.)

next routing point or to the addressee, i.e., the ultimate destination. *See also* **electronic mail, message switching.**

store and forward data transmission (SAFDT): 1. Data transmission (a) in which data and messages are temporarily stored at one or more intermediate points in a route prior to their delivery to the next point in the route or to the ultimate destination in the route, (b) that is used to relay message traffic, (c) in which a message or a packet is sent from a terminal station of a network via a computer-controlled switching center, and (d) in which messages or packets that are to be transmitted are assembled into a special format and stored ready for transmission at a suitable time. *Note 1:* An advantage of store and forward data transmission is its usefulness in error control, in which handshaking usually is used. Disadvantages of store and forward data transmission are the cost of storage and the delay in transmission caused by addressing and storage time. *Note 2:* Store and forward data transmission is used to relay traffic in which a message or a packet is sent from an originating station of a network via a computer-controlled switching center. The switching center computer accepts, analyzes, stores, and retransmits each received message in accordance with procedures contained in the programs. **2.** Usually at a user request, the recording by a communications network of a message for subsequent automatic forwarding to the message addressee. *Synonym* **store and forward transmission.** *See also* **store, store and forward switching center, data transmission.**

store and forward switching center: 1. A message switching center in which data or messages that are destined for other switching centers, or for users connected to a given switching center, are temporarily stored until forwarded to their ultimate destination. *Note:* The storage duration depends on traffic volume, transmission capacity, message priority, and availability of channels. **2.** A message switching center in which a message is accepted from the originating user when it is offered, held in a physical storage, and forwarded to the destination user in accordance with (a) the priority placed on the message by the originating user and (b) the availability of an outgoing channel. *See also* **circuit switching unit, message switching, store, store and forward data transmission, switch, switching system.**

store and forward transmission: *Synonym* **store and forward data transmission.**

stored program computer: A computer that (a) can synthesize and internally store instructions, (b) is controlled by the internally stored instructions, i.e., internally stored programs, and (c) can subsequently execute those instructions.

stored terrain access and retrieval system (STARS): An avionics system for aircraft that stores and displays terrain features so that support processing can (a) mask, i.e., overlay, an aircraft position symbol and pop-up threats, (b) provide terrain-aided navigation data, (c) perform route planning, replanning, and ranging, (d) provide radar-coordinated navigation data, and (e) provide ground-collision avoidance information.

storm: *See* **magnetic storm.**

STP: signal transfer point, standard temperature and pressure.

STRADIS: Structured Analysis, Design, and Implementation of Information System.

strain: *See* **optical fiber strain.**

strain-induced sensor: *See* **fiber optic strain-induced sensor.**

strap: *Synonym* **cross connection.**

strapping: *See* **cross strapping.**

strategic military communications system: A communications system that (a) usually is global in nature, (b) is operated on either a common user or a special purpose basis, (c) may be confined to a specified area, (d) may be limited to a particular type of traffic, (e) usually is configured such that interoperation with other strategic systems is possible when desired or required, (f) has equipment that is procedurally compatible with other strategic communications systems in order to facilitate efficient communications traffic interchange, and (g) has a worldwide routing plan as a prerequisite to operations.

stratified language: A language that is not usable as a metalanguage. *Note:* Examples of stratified languages are FORTRAN and BASIC. *See also* **metalanguage.**

stratosphere: The portion of the Earth's atmosphere that (a) lies between the troposphere and the ionosphere, (b) has a temperature that is practically constant in a vertical direction, i.e., is an isothermal region, (c) is free from clouds, except for occasional dust clouds, (d) is free from active convection, (e) has an altitude at its base that varies in regular fashion with latitude and with the seasons over the Earth as a whole, and that fluctuates irregularly from day to day, and (f) that extends from (i) 10 to 13 km (6 to 8 mi) to (ii) about 50 km (30 mi) above the Earth's surface. *Synonym* **isothermal region.** *See also* **atmosphere, ionosphere, tropopause, troposphere.**

stray current: Current through a path other than the intended path. *See also* **spurious emission.**

stream: *See* **bit stream, data stream, key stream, mission bit stream, serial data stream.**

streamer: *Synonym* **streaming tape drive.**

streaming tape drive: 1. A magnetic tape unit especially designed to make a nonstop dump or restore data on magnetic disks without stopping at interblock gaps. **2.** A magnetic tape unit that (a) makes a nonstop dump, i.e., transmits all of the contents to another storage medium without stopping, or (b) records data received from another storage medium without stopping at interblock or interrecord gaps. *Note:* Streaming tape drives are often used for bulk transfer of data between tape and disk storage. *Synonym* **streamer.**

streaming tape recording: Recording on magnetic tape in which continuous tape motion is maintained without stopping and starting within the interblock or interrecord gaps.

street address: The address used for the physical geographic location of an organization. *Note:* The street address of the International Organization for Standardization is 1, rue de Varembe, Genève, Suisse, i.e., 1 Varembe Street, Geneva, Switzerland.

strength: *See* **dielectric strength, electric field strength, field strength, magnetic field strength, optical fiber tensile strength, tensile strength, twist strength.**

strength member: A component of a cable, such as a communications, power, fiber optic, or metallic cable, that protects the elements of the cable from excessive tensile, compression, torsional, and bending stresses during spooling, payout, installation, and operation.

strength member fiber optic cable: *See* **central strength member fiber optic cable, peripheral strength member fiber optic cable.**

stressed environment: In radiocommunications, an environment that is under the influence of extrinsic factors that degrade communications systems integrity. *Note 1:* A stressed environment exists when (a) the benign communications medium is disturbed by natural or man-made events, such as an intentional nuclear burst, (b) the received signal is degraded by natural or man-made interference, such as jamming signals or co-channel interference, or (c) an interfering signal can reconfigure the network. *Note 2:* In military systems, in a stressed environment, where an adversary poses a threat to successful communications, radio signals may be encrypted in order to deny the adversary an intelligible message, traffic flow information, network information, or automatic link establishment (ALE) control information.

stretching: *See* **optical pulse stretching, pulse stretching.**

stria: In optical materials, particularly glass, a defect that (a) consists of a sharply defined streak of material with a slightly different refractive index from that of the main body of the material, (b) usually causes wave-like distortions in the images of objects seen through the material, exclusive of similar distortion caused by variations in thickness or curvature, and (c) usually is due to temperature variation or poor mixing of ingredients, which causes the optical density, i.e., the refractive index, to vary from place to place in the material.

string: A sequence of elements that (a) usually consists of entities of the same type, such as characters, (b) has a beginning and an end identified by data delimiters, and (c) is treated as a unit. *See* **alphabetic string, bit string, character string, null string, prestored string, pulse string, setup string, symbol string, unit string.** *See also* **data delimiter, pulse train.**

strip: *See* **bunching strip.**

stripe laser diode: A laser diode usually fabricated as a multiheterostructured monolithic element with deposited metallic stripes, for electrical conductors to create necessary excitation, electric fields, and contacts.

striping: On a flowchart, placing a line across the upper part of a symbol on the flowchart to indicate that a detailed description of what the symbol represents is given elsewhere in the document containing the flowchart. *See also* **flowchart.**

strip-loaded diffused optical waveguide: A three-dimensional optical waveguide, constructed from a two-dimensional diffused optical waveguide upon whose surface a dielectric strip of lower-refractive-index material has been deposited, thus confining the electromagnetic fields of the propagating mode to the vicinity of the strip, and hence achieving a three-dimensional waveguide. *See also* **diffused optical waveguide, mode, optical waveguide, refractive index.**

stripper: *See* **cladding mode stripper, fiber optic mode stripper.**

strobe: *See* **noise strobe, radar strobe.**

strobe triangulation: Triangulation in which a jammer is located by plotting the azimuths of the jammed sector strobes of two or more remotely located radars that are jammed simultaneously by the jammer. *See also* **jammer, triangulation.**

stroke: A straight line or an arc that is used as a segment of a graphic character.

stroke character generator: A character generator that generates characters composed of adjacent or connected strokes for presentation in the display space on the display surface of a display device, usually as alphanumeric characters. *See also* **display device, dot matrix character generator.**

stroke edge: In character recognition, the line of discontinuity between a side of a stroke and the background, obtained by averaging, over the length of the stroke, the irregularities resulting from printing and detection.

stroke speed: In facsimile systems, the rate at which a fixed line perpendicular to the direction of scanning is crossed in one direction by a scanning or recording spot. *Note 1:* The stroke speed usually is expressed as a number of strokes per minute. When the system scans in both directions, the stroke speed is twice this number. *Note 2:* In most conventional mechanical systems, the stroke speed is equivalent to drum speed. *See also* **facsimile, scanning line frequency.**

stroke width: In character recognition, the distance, measured perpendicularly to the stroke centerline, between the two stroke edges.

structure: *See* **control structure, fiber optic supporting structure, frame multiplex structure, hierarchical structure, overlay structure, positioned interface structure, tree structure.**

Structured Analysis, Design, and Implementation of Information System (STRADIS): A computer aided software engineering (CASE) tool that (a) is used for analyzing, designing, and implementing information systems and (b) is being used by the North Atlantic Treaty Organization (NATO).

structured program: A computer program that (a) is constructed by using a basic set of controls, each one having one entry point and one exit point, (b) usually consists of a set of control structures that includes (i) a sequence of two or more operations, (ii) a conditional branch to one of two or more operations, and (iii) a repetition of a sequence of operations, and (c) usually is made up of a hierarchy of program modules with but one entry point and one exit point, control being passed down through the structure without jumping or branching back to higher levels of the structure.

structured programming: Programming in which (a) programs are organized and coded, (b) a hierarchy of modules is used, (c) each module has a single entry and a single exit point, (d) control is passed downward through the structure without unconditional branches to higher levels of the structure, and (e) three types of control flow are used, namely, sequential, test, and iteration.

STU: secure telephone unit.

stub: *See* **Pawsey stub.**

studio to transmitter link (STL): A communications link used for the transmission of broadcast material from a studio to the transmitter. *Note:* The studio to transmitter link (STL) may be a microwave, radio, or landline link. The studio may be downtown and the transmitter on a mountaintop at the foot of a tower, with the antenna at the top of the tower.

stuffer: A device that appends redundant bits to an asynchronous bit stream to bring the bit rate up to a standard constant value that subsequently may be handled by synchronous equipment or equipment with a different transmission speed. *Note:* An example of a stuffer is a time division multiplexer that assembles the data streams that are in a number of asynchronous channels into a single stream in an asynchronous channel. *See also* **padding, stuffing character.**

stuffing: 1. In a fiber optic cable, the material used for such functions as mechanical spacing, core wrapping, and optical fiber buffering. **2.** The addition of data items, elements, or codes to a given set of data for the purpose of creating a specified amount of data rather than for the purpose of adding information. *See* **bit stuffing, destuffing, pulse stuffing.**

stuffing character: A character that (a) is used in isochronous transmission and (b) is added to transmitted data to bring the character rate up to a standard constant value that may be subsequently handled by synchronous equipment or equipment with a different transmission rate. *Note:* An example of a stuffing character is a character that is used to take account of differences in clock speed. *See also* **clock speed, stuffer.**

stuffing process: *See* **molecular stuffing process.**

stuffing rate: *See* **maximum stuffing rate, nominal bit stuffing rate.**

stunt box: A device that controls the performance of specified operations, such as the nonprinting functions of a printer at a terminal.

STX: start of text.

STX character: *Synonym* **start of text character.**

stylus: A pointer that (a) provides coordinate data when a cursor is positioned in the display space on the display surface of a display device, and (b) controls operations such as (i) the movement of display elements, display groups, or display images, (ii) entering coordinate data, (iii) making selections from a menu, and (iv) indicating display positions. *Note:* Examples of styli are light pens, sonic pens, tablet styli, and voltage pencils. *See also* **display device, menu, pointer.**

SU: signal unit.

SUB: substitute, substitute character.

subband: A subdivision of a given frequency band. *Note:* An example of a subband is a subdivision of the speech frequency spectrum into channels of differing bandwidth such that each subdivision makes an equal contribution to speech intelligibility. This has particular application in vocoders. *See also* **band, frequency spectrum.**

subband adaptive differential pulse code modulation (SB-ADPCM): Modulation in which (a) an audio frequency band is split into two subbands, i.e., a higher and a lower band, and (b) the signals in each subband are encoded by using adaptive differential pulse code modulation (ADPCM). *See also* **adaptive differential pulse code modulation, differential pulse code modulation, pulse code modulation.**

subcarrier: A carrier that modulates another carrier, which can be used to modulate another carrier, which can be used to modulate another carrier, and so on, so that there can be several levels of subcarriers, i.e., several intermediate subcarriers. *See* **stereophonic sound subcarrier.** *See also* **carrier, frequency.**

subcarrier frequency shift: The conveying of information by modulating an audio frequency carrier that is then used to modulate a radio frequency (rf) carrier for radio transmission. *Note:* When only two discrete steps of subcarrier frequency shift are used, the subcarrier frequency shift is called "two-tone keying." *See also* **audio frequency, carrier, modulation, two-tone keying.**

subchannel: *See* **stereophonic sound subchannel.**

SUB character: substitute character.

subject: *In facsimile, deprecated synonym for* **object.**

subject copy: *In facsimile, deprecated synonym for* **object.**

subjunction gate: *Synonym* **A AND NOT B gate.**

sublayer: 1. In layered systems, a subdivision of a layer. **2.** In the Open Systems Interconnection—Reference Model, a subdivision of a layer. *See* **logical link control sublayer, medium access control sublayer, physical signaling sublayer.**

submarine cable: *Synonym* **submersible cable.**

submerged cable: A cable that is (a) immersed in a body of water, such as a river, harbor, strait, lake, sea, or ocean, or (b) is immersed in an autoclave during testing. *See also* **cable.**

submersible cable: An electrical power or communications cable, such as a fiber optic cable, designed to convey power or messages while immersed in a body of water, such as a river, harbor, strait, lake, sea, or ocean. *See also* **cable.** *Refer to* **Fig. S-20.**

subnet address: In an Internet Protocol (IP) address, an extension that allows users in a network to use a single IP network address for two or more physical subnetworks. *Note:* The three parts of the Internet Protocol (IP) address are the host, the subnet, and the network addresses. Inside the subnetwork, hosts and gateways divide the local portion of the IP address into a host and a subnet address. Outside of the subnetwork, routing continues as usual by dividing the destination address into a network portion and a local portion.

subnetwork: A collection of equipment and physical transmission media that forms an autonomous whole and that can be used to interconnect systems for purposes of communications. *Refer to* **Fig. S-11 (880).**

subprogram: 1. A computer program that is invoked by another computer program. **2.** In FORTRAN, a program module that is headed or called by a function, subroutine, program, or block data statement. *See also* **computer program, FORTRAN.**

subreflector: An intermediate antenna reflector that (a) forms part of the propagation path to and from other or main reflectors of an antenna, and (b) is used to produce a desired directive gain. *See also* **directive gain.**

subroutine: *See* **computer subroutine.**

subsatellite point: At any given instant, the point on the Earth's surface where the Earth's surface is intersected by a line from a satellite to the Earth's center. *Note:* To an observer at the subsatellite point, the satellite is at the zenith, i.e., is directly overhead.

subscriber: *Synonym* **customer.**

subscriber identification: *Synonym* **user identification.**

Fig. S-20. A (larger) TAT 7 copper coaxial cable and a (smaller) TAT 8 AT&T SL (Submarine Lightwave) **submersible cable** that has more than double the number of circuits, is half the size, and is one-third the weight. (Courtesy AT&T Archives.)

subroutine: *See* **computer subroutine.**

subsatellite point: At any given instant, the point on the Earth's surface where the Earth's surface is intersected by a line from a satellite to the Earth's center. *Note:* To an observer at the subsatellite point, the satellite is at the zenith, i.e., is directly overhead.

subscriber: *Synonym* **customer.**

subscriber identification: *Synonym* **user identification.**

subscriber line: 1. *Synonyms* **line loop, loop. 2.** *See* **digital subscriber line.**

subscriber loop: *Synonym* **loop.**

subscriber service: *See* **absent subscriber service.**

subscriber set: *Synonym* **user set.**

subscriber station: *See* **rural subscriber station.**

subscriber terminal equipment: *See* **data subscriber terminal equipment.**

subset: *See* **alphabetic character subset, alphanumeric character subset, character subset, numeric character subset.**

subsidence: 1. In the atmosphere, a descending motion of air that (a) usually occurs over a broad area, (b) occurs most often in the upper atmosphere, and (c) is often revealed by cloud layers. **2.** A sinking of a part of the Earth's crust, the sinking being caused by various factors, such as underground excavations, underground water removal, rock shifts, erosion, cave collapse, and settling.

substation: *See* **distribution substation, primary substation.**

substitute character (SUB): A control character that replaces a character that is recognized to be invalid or in error, or that cannot be represented on a given device. *Note:* An example of a substitute character is an

materials are deposited by evaporation or are bonded by cementing or etching.

subsystem: In a system, a part that (a) is an identifiable component and (b) may also be a system in its own right. *See* **communications subsystem, earth electrode subsystem, fault protection subsystem, lightning protection subsystem, signal reference subsystem, space-ground link subsystem, space subsystem.**

subtracter: A device that forms the algebraic or arithmetic difference between two quantities, i.e., the minuend and the subtrahend, that are presented as inputs. *See* **adder-subtracter.** *See also* **comparator.**

subtrahend: A number that is subtracted from another number, the minuend, to obtain their difference. *Note:* If the minuend is larger than the subtrahend, and both are positive numbers, the difference will be a positive number. The operands, i.e., the minuend and subtrahend, may be scalar or vector quantities.

subvoice-grade: Pertaining to a system, such as a circuit, line, or channel, that (a) has a bandwidth narrower than that of a voice-grade circuit and (b) usually is a subchannel of a voice-grade line. *See also* **bandwidth, voice grade, voice-grade circuit.**

subvoice-grade channel: A channel that (a) has a bandwidth narrower than that of a voice-grade circuit and (b) usually is a subchannel of a voice-grade line. *See also* **bandwidth, voice grade, voice-grade circuit.**

success: *See* **access success, block transfer success, disengagement success.**

successful block delivery: The transfer of a nonduplicate user information block between the source user and the intended destination user. *Note:* Successful block delivery includes the delivery of correct and incorrect blocks. *See also* **block, block transfer failure, successful block transfer.**

successful block transfer: Between the source user and the intended destination user, the transfer of a correct, nonduplicate, user information block that (a) occurs at the moment when the last bit of the transferred block crosses the functional interface between the telecommunications system and the intended destination user data terminal equipment (DTE) or system interface unit, and (b) must occur within a defined maximum block transfer time after initiation of a block transfer attempt. *See also* **block, block transfer attempt, block transfer failure, block transfer time, maximum block transfer time, successful block delivery.**

successful disengagement: *Synonym* **disengagement success.**

success ratio: *See* **access success ratio, disengagement success ratio.**

suction feeder: In a duplicating machine, a device that removes data media, such as individual sheets of paper, from the top or the bottom of a stack, one item at a time and conveys the item into the machine for processing, such as printing.

sudden ionospheric disturbance (SID): In the D region of the ionosphere, an abnormally high ionization density that (a) is caused by an occasional sudden outburst of ultraviolet light from the Sun, i.e., from a sudden solar flare, and (b) results in a sudden increase in radio wave absorption that is most severe in the upper medium frequency (MF) and lower high frequency (HF) ranges of the frequency spectrum. *See also* **ionosphere, ionospheric disturbance, magnetic storm.**

suffix: *See* **amplifying suffix.**

suffix notation: *Synonym* **postfix notation.**

suite: *See* **Transmission Control Protocol/Internet Protocol Suite.**

sum: The result obtained when two or more quantities, such as numbers or vectors, are added. *See* **checksum, logical sum.**

sum check: *Synonym* **summation check.**

summation check: **1.** A check in which (a) the sum of the individual digits of a numeral is formed, and (b) the sum usually is compared with a previously computed value or with a specified value. **2.** A comparison of checksums on the same data on different occasions, at different places, or on different representations of the data in order to verify data integrity. *Note:* The digits in the number 106242370 sum to 25. The number may be stored or transmitted as 106242370:25. At any time or place, if a summation of the left part does not sum to 25, an error has been made. However, if the sum is 25, it is not absolutely certain that an error has not been made because there could be compensating errors such that the sum remains 25 even though there are errors. *Synonym* **sum check.** *See also* **checksum.**

summer: In analog systems, a device that has an output analog signal, i.e., an analog variable, that is equal to the sum, or a weighted sum, of two or more input analog signals, i.e., analog variables. *Synonym* **analog adder.** *Refer to* **Fig. T-7 (1024).**

summing integrator: In analog systems, a device that has an output analog signal, i.e., analog variable, that is the integral, or weighted sum, of input analog signals, i.e., analog variables, with respect to time or with respect to another input analog signal, i.e., analog variable. *See also* **integral.**

sun sensor: In satellite communications systems, a satellite component that (a) senses the time or the incidence angle of solar radiation and (b) provides an electrical signal that may be used to compute the satellite position or attitude, i.e., orientation.

sunspot: 1. In the photosphere, i.e., the disk, of the Sun, a dark marking that (a) lies between 30°N and 30°S latitude on the Sun disk, (b) has a temperature that is about 2000 K lower than that of the surrounding photosphere, (c) lasts from a few hours to several months, depending on the size, (d) has a size that varies from a few hundred kilometers wide to several times the size of the Earth, (e) is periodic, with about 11 years between successive maxima and minima, and (f) is associated with intense solar activity and interference with radio and video communications. **2.** A focused spot of light that is an image of the sun and that often reaches the kindling temperature of some substances. **3.** A focused spot of light from a large incandescent lamp that is provided with a filter to imitate the light from the Sun in color cinematography.

sunspot activity: 1. The electromagnetic effects that are produced by sunspots and that cause interference, disturbances, and anomalies in radio and television transmission. **2.** The occurrence of sunspots, their growth, and their disappearance, and the degradation of radio wave ionospheric propagation caused by the magnetic storms the sunspots induce.

sunspot cycle: A cycle of about 11 years in which (a) sunspot activity increases and decreases, the peak activity being marked by intense magnetic fields and magnetic storms on Earth, (b) the higher communications frequencies that are propagated almost entirely by ionospheric reflection are seriously degraded during severe sunspot activity, i.e., every 11 years, (c) a disturbance of the solar surface appears as a relatively dark center, i.e., an umbra, surrounded by a less dark area, i.e., a penumbra, and (d) the sunspots usually occur in groups, are relatively short-lived, and with few exceptions, occur in the region between 30°N and 30°S latitude on the Sun disk.

supercomputer: A computer with an extremely large direct access storage capacity, such as 200 Mb (megabytes), low storage access time, such as 10 ps (picoseconds), ultrahigh switching speeds, such as 1 ps, and fast operational speeds, such as 100 Gflops (gigafloating point operations per second), and perhaps using optical storage rather than magnetic storage.

superencryption: The encryption of encrypted information, i.e., the further encryption of already encrypted text, for increased security or a higher level of privacy of messages. *Note:* Superencryption occurs automatically when a message encrypted offline is transmitted over a secured circuit, or when information encrypted by the originator is multiplexed onto a communications trunk in which all transmissions are encrypted, i.e., are bulk-encrypted. *See also* **encrypt.**

superfiber: An optical fiber that has the best attributes of all the specialty fibers, such as high strength, long shelf life, precision polarization control, precision dispersion shifting, moisture resistance, high level radiation hardness, low attenuation rate, benign cabling, and low cost. *See also* **optical fiber.**

superframe: *See* **extended superframe.**

supergroup: A group of telephone voice channels that usually consists of 60 voice channels, five groups of 12 channels each, occupying a frequency band from 312 kHz (kilohertz) to 552 kHz. *Synonym* **secondary group.** *See* **channel supergroup, group, through supergroup.**

supergroup distribution frame (SGDF): In frequency division multiplexing (FDM), the distribution frame that provides terminating and interconnecting facilities for group modulator output, group demodulator input, supergroup modulator input, and supergroup demodulator output circuits of the basic supergroup frequency spectrum between 312 kHz (kilohertz) and 552 kHz.

supergroup equipment: *See* **through supergroup equipment.**

superhacker: A specially skilled, expert, usually felonious hacker, i.e., an überhacker. *Synonym* **überhacker.** *See also* **hacker.**

superheterodyne receiver: A receiver that operates on the principle of mixing the received signal with a local oscillator output signal in order to improve selectivity and amplification.

superhigh frequency (SHF): A frequency that lies in the range from 3 GHz (gigahertz) to 30 GHz, i.e., in the range from 3×10^9 Hz (hertz) to 30×10^9 Hz, or in wavelengths from 1 to 10 cm (centimeters), the range having the alphabetic designator SHF and the numeric designator 10. *See also* **frequency spectrum designation.** *Refer to* **Table 1, Appendix B.**

superhighway: *See* **information superhighway.**

superior position flaghoist: In visual communications systems, a flaghoist signaling message that (a) is to be read ahead of others, and (b) is placed higher or more forward than an inferior position flaghoist message. *See also* **flaghoist signaling, inferior position flaghoist.**

superluminescence: The amplification of spontaneous emission in a gain medium that (a) is characterized by moderate spectral line narrowing and directionality, (b) usually is distinguished from lasing action by the absence of positive feedback, and (c) does not have a well-defined modal distribution of optical power. *Synonyms* **superluminescent effect, superradiance.** *See also* **laser, luminescence, spontaneous emission, stimulated emission.**

superluminescent effect: *Synonym* **superluminescence.**

superluminescent LED: *Synonym* **superluminescent light-emitting diode.**

superluminescent light-emitting diode (SLED): A light-emitting diode (LED) in which there is stimulated emission with amplification but insufficient feedback for oscillations to build up to achieve lasing action. *Synonym* **superluminescent LED.** *See also* **light-emitting diode, spontaneous emission, stimulated emission.**

supermaster group: 1. *Synonym* **jumbo group. 2.** *See* **group.**

supermicro: 1. Pertaining to an improved microcomputer that handles many users and processes data at extremely high speeds. **2.** An improved microcomputer that handles many users and processes data at extremely high speeds. *See also* **microcomputer.**

superradiance: *Synonym* **superluminescence.**

supervise: In telephone switchboard operations, the action in which an operator, i.e., an attendant, puts a receiver across a line to ascertain whether a call has been satisfactorily completed or whether a call is finished. *See also* **completed call, finished call.**

supervision: *See* **backward supervision, forward supervision, system supervision.**

supervisor: 1. In telephone switchboard operations, a person who (a) is an assistant to the chief supervisor, (b) performs some of the duties of the chief supervisor, (c) directs, assists, and instructs the operators in the performance of their duties, and (d) if the senior supervisor, takes charge of the exchange in the absence of the chief supervisor. *See also* **chief supervisor. 2.** In computer programming, a supervisory program. *See also* **supervisory program.**

supervisor position: The position, i.e., the place, at a communications station or center, such as a telephone exchange, occupied by the supervisor of communications operations, and at which special equipment is located for performing supervisory functions.

supervisory control: The use of characters or signals to automatically actuate equipment or indicators. *See also* **character, signal, supervisory signals.**

supervisory program: 1. A computer program that (a) usually is part of an operating system, (b) controls the execution of other computer programs, and (c) regulates the flow of work in a data processing system. **2.** A computer program that (a) usually is part of an operating system, (b) controls the execution of routines, subroutines, and other programs, and (c) regulates various functions, such as work scheduling, input-output operations, and error control actions. *See also* **control station. 3.** A computer program that (a) allocates computer component space and (b) schedules computer events, usually by task queuing and system interrupts. *Note:* Control of the system is returned to the supervisory program frequently enough to ensure that demands on the system are being met. *Synonyms* **executive program, supervisor.**

supervisory routine: A computer routine that (a) allocates computer component space, and (b) schedules computer events, usually by task queuing and system interrupts. *Note:* Control of the system is returned to the supervisory program frequently enough to ensure that demands on the system are being met.

supervisory signal: In a particular connection, a signal that (a) indicates or controls the various operating states of the circuits or circuit combinations in the connection, (b) usually indicates the progress of a call, and (c) includes signals such as off-hook, clear, and recall signals. *See also* **forward busying, engineering circuit, orderwire, signal, supervisory control.**

supervisory signaling: Signaling that (a) is associated with the initiation of a call and its termination by a user or by a switching center, and (b) includes various signals, such as off-hook, clear, and recall signals. *See* **separate channel supervisory signaling.**

supervisory working: *See* **double supervisory working.**

supplement: *See* **network supplement.**

supplementary heading: In the readdressing of messages, a heading that (a) is placed in front of the preamble to the message that is being readdressed, (b) usually includes all procedural lines as required, and (c) omits all parts of the message heading that originally preceded the preamble. *See also* **heading, preamble.**

support: *See* **joint computer-aided acquisition and logistics support system, signal support, system support.**

supporting structure: *See* **fiber optic supporting structure.**

support jamming: Jamming in which the jamming equipment (a) is carried in a vehicle, aircraft, spacecraft, satellite, or ship that is not part of the group that is being protected, (b) may be land-based, and (c) may be either in a stand-off jamming role or in an escort jamming role. *See also* **escort jamming, jamming, self-protection jamming, stand-off jamming.**

support measure: *See* **electronic warfare support measure.**

support software: *Synonym* **system software.**

support system: *See* **global decision support system, operations support system.**

suppressed carrier: A carrier transmitted at a power level more than 32 dB below the peak envelope power, but preferably more than 40 dB below the peak envelope power. *See* **double sideband suppressed carrier, reduced carrier, single sideband suppressed carrier.** *See also* **full carrier, reduced carrier.**

suppressed carrier single sideband emission: A single sideband emission in which the carrier is reduced to a power level below that required for demodulation and therefore is not intended to be used for demodulation.

suppressed carrier transmission: Amplitude-modulated (AM) transmission in which the carrier is reduced to a minimum power level, below that required for demodulation. *Note 1:* When the carrier is suppressed, either or both of the sidebands may be transmitted with increased power. *Note 2:* The carrier is not intended to be used for demodulation. *Note 3:* Suppressed carrier transmission is a special case of reduced carrier transmission. *Note 4:* The carrier power must be restored by the receiving station to permit demodulation. *See* **double sideband suppressed carrier transmission, single sideband suppressed carrier transmission.** *See also* **carrier, modulation suppression, reduced carrier transmission, sideband transmission, single sideband transmission, transmission.**

suppressed carrier transmission system: *See* **double sideband suppressed carrier transmission system.**

suppressed clock pulse duration modulation: Modulation that is similar to pulse duration modulation, except that clock information is removed, thus reducing the required bandwidth.

suppression: *See* **infrared radiation suppression, modulation suppression, noise suppression, zero suppression.**

suppression and blanking: *See* **pulse interference suppression and blanking.**

suppression circuit: *See* **singing suppression circuit.**

suppressor: *See* **echo suppressor, half-echo suppressor, optical pulse suppressor.**

surf: 1. To switch from channel to channel, medium to medium, network to network, or database to database, seeking that which might prove to be of interest, remaining at a point only for as long as one's interest is maintained, and switching to another point when that interest is lost, all while not seeking anything in particular. **2.** To browse in cyberspace via the information superhighway, such as via Internet or the World Wide Web. *See* **channel surf.**

surface: *See* **display surface, emitting surface, equipotential surface, interface surface, optical surface, reference surface.**

surface center: *See* **reference surface center.**

surface code: *Synonym* **panel code.**

surface concentricity: *See* **core reference surface concentricity.**

surface detection equipment: *See* **airport surface detection equipment.**

Surface Emitting Distributed Feedback Laser: A laser that (a) instead of mirrors, has gratings that eliminate optical damage and beam divergence inherent in reflector-type lasers, (b) allows for a large emitting area without the use of facets, the primary cause of laser and light-emitting diode (LED) failure, (c) may be used in a variety of applications, such as medical laser systems, materials processing, laser igniters, laser radar, and military optoelectronic countermeasures, and (d) is developed by Hughes.

surface-emitting light-emitting diode (SLED or SELED): A light-emitting diode (LED) that (a) emits radiation normal to the plane of the junction, i.e., perpendicular to the surface plane rather than from the edges of the surface plane, (b) has a lower output radiance and a lower coupling efficiency to an optical fiber or an optical integrated circuit (OIC) than the edge-emitting LEDs and the injection lasers have, and (c) along with edge-emitting LEDs, provides several milliwatts of optical power in the 0.8-μm (micron) to 1.3-μm wavelength range at drive currents of 100 mA (milliamperes) to 200 mA, whereas diode lasers at these currents provide several tens of milliwatts of optical power. *Synonyms* **front-emitting light-emitting diode, Burrus diode, Burrus light-emitting diode.** *See also* **edge-emitting light-emitting diode, injection laser.**

surface mirror: *See* **back surface mirror, front surface mirror.**

surface noncircularity: *See* **reference surface noncircularity.**

surface refractive index: The refractive index of the Earth's atmosphere at the Earth's surface, usually calculated from observations of pressure, temperature, and humidity at the surface. *Note:* The gradient of the surface refractive index is (a) the difference between the refractive index at the surface and the refractive index at a given altitude, such as 100 m (meters), 500 m, or 1 km (kilometer), divided by (b) the given altitude. *Synonym* **surface refractivity.** *See also* **ducting, refractive index, refractive index gradient.**

surface refractive index gradient: The value of the refractive index gradient at specific points near the Earth's surface, determined by (a) measuring pressure, temperature, and humidity at the points, (b) calculating the refractive index at those points using the measurements, and (c) dividing (i) the difference between the refractive index at a given altitude and the index at the surface by (ii) the given altitude. *Note:* Surface refractive index gradients are determined at various locations and altitudes, such as 100 m (meters), 1 km (kilometer), or 1000 ft (feet), depending on the need for radio wave propagation path evaluation and transmission system performance requirements. *Synonym* **surface refractivity gradient.** *See also* **ducting, refractive index, surface refractive index.**

surface refractivity: *Synonym* **surface refractive index.**

surface refractivity gradient: *Synonym* **surface refractive index gradient.**

surface station: A station that is on or near the surface of the Earth, such as a land station, a ship station, a mobile vehicular station, a fixed station, and, in satellite communications, an aircraft station because it operates in the Earth's atmosphere and relatively close to the surface of the Earth compared to a satellite orbit, outer space, or deep space.

surface surveillance device: *See* **airport surface surveillance device.**

surface to air communications: One-way communications from surface stations to aircraft stations. *See also* **surface station.**

surface tolerance field: *See* **reference surface tolerance field.**

surface wave: 1. An electromagnetic wave that (a) propagates along the interface surface between different propagation media in accordance with the geometrical shape of the surface and the properties of the propagation media near that surface, such as the electric permittivity, magnetic permeability, and electrical conductivity that define the refractive indices of the media, (b) is guided by the interface surface between the two different media, such as by a refractive index gradient in the media, (c) has electromagnetic field components that may exist throughout space but become negligible within a short distance from the interface, (d) propagates along the interface surface without radiating away from the interface surface, (e) has energy that is not converted from the surface wave field to another form of energy, and (f) does not have a component directed normal to the interface surface. *Note:* Examples of surface waves are (a) evanescent waves in optical fiber transmission that travel along the core-cladding interface or the cladding-air interface, and (b) ground waves in radio transmission that propagate close to the surface of the Earth, the Earth having one refractive index and the atmosphere another, so that they constitute an interface surface. **2.** An electromagnetic wave that propagates relatively close to the Earth's surface, which serves as an interface between two media with different refractive indices, the ground below the surface and the atmosphere above the surface, the surface thus serving as a guide of the wave. *Note 1:* Waves below the surface tend to become trapped because the refractive index of the ground is higher than that of the atmosphere. *Note 2:* Surface waves include direct waves, indirect waves, and ground waves, but not skywaves. *See also* **direct wave, ground wave, indirect wave, refractive index, skywave.** *Refer to* **Fig. L-8 (536).**

surfing: 1. Switching from channel to channel, medium to medium, network to network, or database to database seeking that which might prove to be of interest, remaining at a point only for as long as one's interest is maintained, and switching to another point when that interest is lost, all while not seeking anything in particular. **2.** Browsing in cyberspace via the information superhighway, such as via Internet or the World Wide Web. *See also* **channel surf, surf.**

surge: *Synonym* **impulse.**

surge protector: *Synonym* **transient protector.**

surge suppressor: *Synonym* **arrester.**

surveillance: *See* **air surveillance, electromagnetic surveillance, electronic surveillance, exploratory electromagnetic surveillance.**

surveillance device: *See* **airport surface surveillance device.**

surveillance radar: *Synonym* **acquisition radar.**

surveillance radar station: A radionavigation land station or ship station in the aeronautical radionavigation service that uses radar to display the presence of objects within its range.

surveillance system: *See* **Amazon surveillance system, automatic dependent surveillance system.**

surveillance unit: *See* **automatic dependent surveillance unit.**

survey: *See* **path survey.**

survivability: 1. In an entity, such as a system, subsystem, equipment, process, or procedure, a property that (a) provides a defined degree of assurance that the entity will continue to function during and after a natural or man-made disturbance, such as an earthquake or a nuclear burst, and (b) is qualified by specifying (i) the range of conditions over which the entity will survive, (ii) the minimum acceptable level of postdisturbance functionality, and (iii) the maximum acceptable outage duration. **2.** In communications, computer, data processing, and control systems, the capability of a system to resist any interruption or disturbance of service, particularly that caused by warfare, fire, earthquake, harmful radiation, or other physical or natural catastrophe rather than by electromagnetic interference or crosstalk. *Synonym* **survivable operation.** *See* **communications survivability, electromagnetic survivability.** *See also* **communications system, continuous operation, endurability.**

survivable adaptive system (SAS): A digital ruggedized tactical wireless communications system that provides (a) communications for high mobility stations, (b) fast installation, (c) high flexibility for changing military situations, (d) multimedia capabilities, (e) high capacity, i.e., high transmission rates and throughput, and (f) support to the digitized battlefield. *See also* **digitized battlefield.**

survivable operation: *Synonym* **survivability.**

survival craft station: A mobile station that (a) is in the maritime mobile service, (b) is intended solely for survival purposes, and (c) is located on survival equipment, such as lifeboats and life rafts. *See also* **maritime mobile service.**

susceptibility: 1. The extent to which a device, equipment, or a system is open to effective degradation caused by one or more inherent weaknesses and environmental conditions. *Note:* An example of susceptibility is the degree to which a communications system is susceptible to interference from sunspot activity. **2.** The extent to which electronic equipment is adversely affected by radiant energy, such as radiation from local transmitters or jammers. **3.** In psychological warfare, the vulnerability of a target audience to particular forms of psychological operations approach. **4.** In electronic warfare, the extent to which electronic equipment is affected by electromagnetic energy radiated by hostile forces equipment, such as jamming transmitters. *See* **ambient light susceptibility.** *See also* **electronic countercountermeasure, electronic jamming, electronic warfare, jamming.**

susceptiveness: In telephone systems, the extent to which circuits pick up noise and low frequency energy by induction from power systems. *Note:* Susceptiveness depends on the mutual inductance between the telephone lines and the power lines, i.e., the higher the mutual inductance, the greater the susceptiveness. Efforts are made to reduce the mutual inductance by various methods, such as achieving telephone circuit balance, using wire and connection transpositions to cancel induced voltages and currents, increasing wire spacing between power lines and telephone lines, and increasing isolation from ground and ground currents. *See also* **crosstalk, spurious emission.**

SWAN: satellite wide area network.

swap: In time-sharing computer systems, to interchange data or jobs, such as pages or programs, between main storage and external storage or between other storage units.

swapping: *See* **page swapping.**

sweep: 1. In electronic systems, the pattern of light or marking on the face of a cathode ray tube (CRT) that is caused by the predetermined deflection and modulation of the electron beam. *Note:* The deflection voltage versus time that is placed on one of the pairs of deflecting plates, or one of the deflecting coils, of the cathode ray tube (CRT) enables the display of the voltage that is applied to the other pair of plates, or to the other deflecting coil. The sweep voltage has the waveshape of a sawtooth, i.e., it has a slow linear rise and a sudden drop. When only the sweep is applied, a single straight line appears as a time baseline. If magnetic deflection is used, the same principles apply to the deflecting current. **2.** To vary the frequency of a transmitter continuously over a wide frequency range. **3.** To tune a receiver continuously over a wide frequency range. *See* **antenna sweep, sector sweep.**

sweep acquisition: In a radio receiver, acquisition in which the frequency of the local oscillator is slowly swept past the reference in order to ensure that the pull-in range is reached. *See also* **pull-in frequency range.**

sweep jammer: A jammer that (a) electronically sweeps a narrow frequency band of jamming energy over a wide band, and (b) emits a jamming signal that consists of a carrier wave, modulated or unmodulated, whose frequency is continuously varied within a given bandwidth. *See also* **jammer, jamming.**

sweep jamming: Jamming in which (a) the frequency of the jamming signal is continuously varied within a specific bandwidth, (b) a narrow frequency band of jamming energy is repeatedly swept over a relatively wide frequency band, and (c) the sweep rate is such as to be on any given frequency only long enough to accomplish its jamming task, returning to that frequency again before expiration of the jammed circuit's recovery time. *Note 1:* Sweep jamming combines the advantages of both spot and barrage jamming by rapidly electronically sweeping a narrow band of jamming signals over a broad frequency spectrum. *Note 2:* The disadvantage of sweep jamming is its high susceptibility to electronic countercountermeasures (ECCM).

sweep noise-modulated jamming: *See* **slow sweep noise-modulated jamming.**

swept jammer: A jammer in which (a) the frequency is varied continuously over specific frequency bands, and (b) the bands usually are just above or below the frequencies to be jammed. *See also* **jammer, jamming.**

swept repeater jammer: A sweep jammer that transmits only on those frequencies that (a) are within its bandwidth, and (b) are specifically selected because they are in use by the systems that are being jammed. *See also* **jammer, jamming, sweep jammer.**

swim: Relative to jitter, a slow, graceful, undesired movement of display elements, groups, or images about their mean position on a display surface, such as the screen of a cathode ray tube (CRT) or the optical fiber faceplate of a fiberscope. *Note 1:* Swim is so slow it usually can be followed by the human eye, whereas jitter is so rapid it usually appears as a blur. Swim has a larger amplitude than jitter and usually pertains to images, whereas jitter usually pertains to images and pulses. *Note 2:* Jitter, swim, wander, and drift have increasing periods of variation in that order. *See also* **drift, jitter, wander.**

switch: In communications systems, a mechanical, electromechanical, or electronic device for making, breaking, or changing the connections in or among circuits. **2.** In communications systems, to transfer a connection from one circuit to another. *Note:* "Switch" and "switching center" are not synonymous. **3.** In a computer program, a conditional instruction and a flag that is interrogated by the instruction. **4.** In a computer program, a parameter that (a) controls branching and (b) is bound prior to the branch point's being reached. *Synonym* **switchpoint. 5.** In computer programming, a functional unit, such as toggles and conditional jumps, used to make selections, i.e., exercise options. *See* **analog switch, crossbar switch, digital switch, disconnect switch, dual inline package switch, electronic switch, ferrite switch, fiber optic switch, fuse disconnecting switch, horn gap switch, interrupt switch, laser frequency switch, latching switch, mechanical optical switch, nonblocking switch, nonlatching switch, optical switch, packet switch, Q switch, semielectronic switch, step by step switch, tandem switch, thin film optical switch, trunk switch, waveguide switch.** *Refer to* **Fig. S-21 (967).**

switchboard: 1. Equipment used to perform switching operations manually. **2.** A manually operated set of equipment at a telephone exchange on which the various circuits from users and from other exchanges are terminated in order to enable operators to establish communications among users (a) on the same exchange, (b) on the same and other exchanges, and (c) on other exchanges. *See* **cordless switchboard, dial service assistance switchboard, field switchboard, monocord switchboard, multiposition switchboard, service assistance switchboard, single position magneto switchboard.** *See also* **cord circuit, private branch exchange.**

switch bounce time: The switch operating bounce time or the switch release bounce time.

switch busy hour: The busy hour for a single switch. *See also* **busy hour, erlang, group busy hour, network busy hour, traffic load.**

switched: *See* **failure fraction automatically switched.**

switched circuit: In a communications network, a circuit that temporarily may be established at the request of one or more of the connected stations. *See also* **circuit, switching system.**

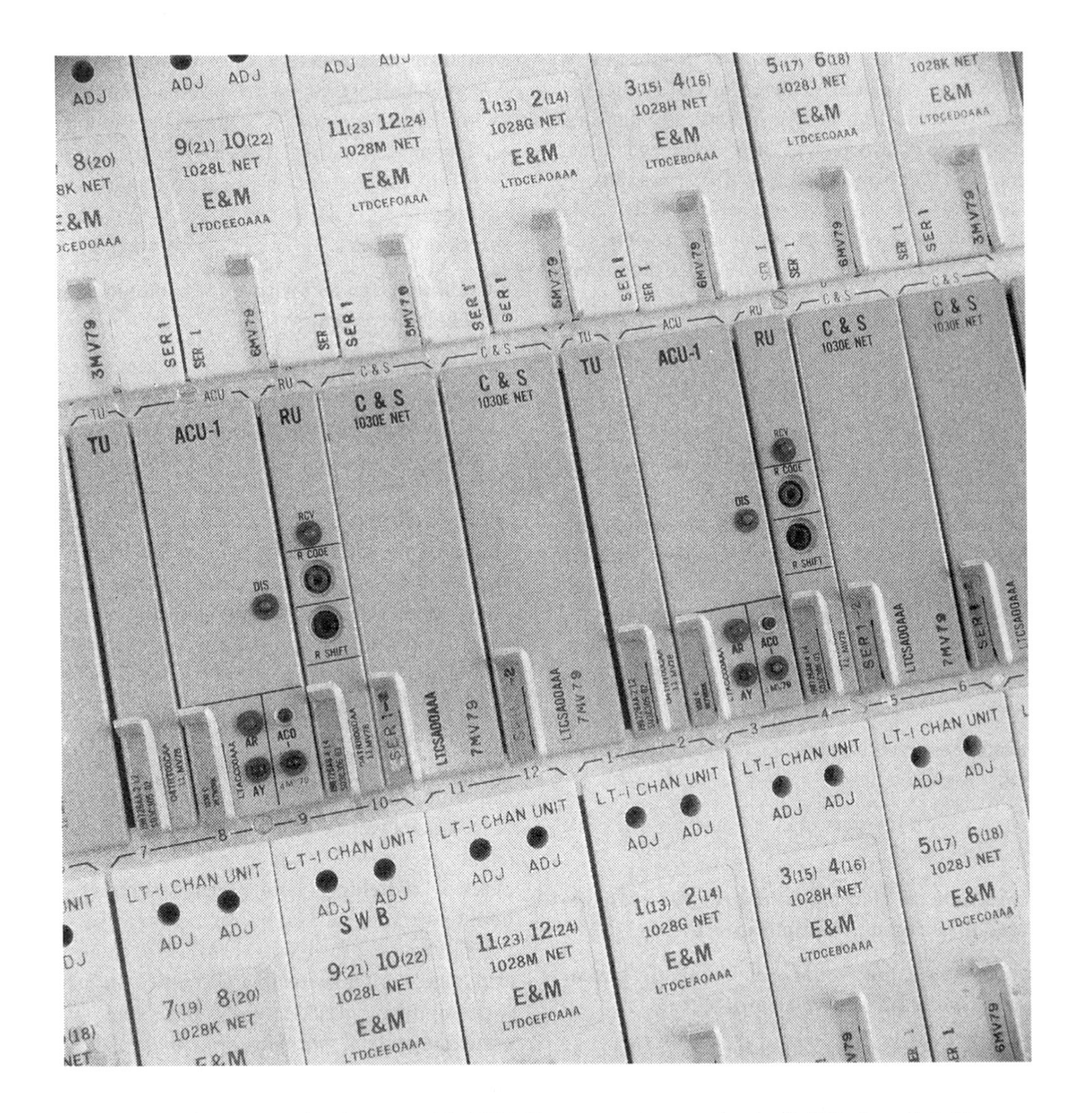

Fig. S-21. Channel and control units, part of the LT-1 Connector, that couples the Bell System digital long-distance **switches** with analog propagation media at St. Louis Missouri in 1979. (Courtesy AT&T Archives.)

(c) on other exchanges. *See* **cordless switchboard, dial service assistance switchboard, field switchboard, monocord switchboard, multiposition switchboard, service assistance switchboard, single position magneto switchboard.** *See also* **cord circuit, private branch exchange.**

switch bounce time: The switch operating bounce time or the switch release bounce time.

switch busy hour: The busy hour for a single switch. *See also* **busy hour, erlang, group busy hour, network busy hour, traffic load.**

switched: *See* **failure fraction automatically switched.**

switched circuit: In a communications network, a circuit that temporarily may be established at the request of one or more of the connected stations. *See also* **circuit, switching system.**

switched connection: *See* **circuit switched connection.**

switched data transmission service: *See* **circuit switched data transmission service, packet switched data transmission service.**

switched hot line circuit: An exclusive-user preprogrammed circuit set up as required between a user and one or more other users, in which user action is limited to going off-hook to establish a connection within service that provides local-area-network-like (LAN-like) performance and the features provided by metropolitan area networks (MANs) and wide area networks (WANs). *Note:* Currently switched multimegabit data service (SMDS) operates at 1.544 Mbps (megabits per second) or 44.736 Mbps. These are the T1/DS1 and DS3 rates, respectively, over switched fiber optic networks. **2.** A data service that (a) provides high speed packet-switched data service, (b) interconnects computers, computer networks, and local area networks (LANs), and (c) is provided over wide geographical areas.

switched network: 1. A communications network in which (a) any user may be connected to any other user, (b) message, circuit, or packet switching is performed, and (c) control devices are used. *Note:* An example of a switched network is the public switched telephone network. **2.** A network that provides switched communications services. *See* **public switched network.** *See also* **network, public switched telephone network, switched multimegabit data service, switching system.**

switched repetitively pulsed laser: *See* **Q-switched repetitively pulsed laser.**

switched telephone network: *See* **public switched telephone network.**

switched time division multiple access: *See* **satellite switched time division multiple access.**

switchgear: *See* **low voltage enclosed air circuit breaker switchgear, metal clad switchgear.**

switching: 1. In communications networks, the controlling or the routing of signals in circuits to transmit data between specific points in the networks. **2.** In computers, the controlling or the routing of signals in circuits to execute logical or arithmetic operations. *Note:* Switching may be performed electronically or optically, in structural levels ranging from (a) gating in the microstructure of electronic or optical integrated circuits on chips to (b) selective coupling of trunks and loops in the switches of communications network switching centers. *See* **analog switching, automatic message switching, automatic switching, beam lobe switching, burst switching, circuit switching, digital data switching, digital switching, distributed switching, fast packet switching, message switching, packet switching, rotary switching, semiautomatic continuous tape relay switching, semiautomatic message switching, space division switching, tape relay switching, time division switching.** *See also* **gate.**

switching and message distribution center: In a communications system, an installation that (a) has switching equipment to connect communications equipment for routing messages or packets toward their ultimate destinations, i.e., their addressees, and (b) distributes messages to addressees in the local area served by the installation. *See also* **routing.**

switching arrangement: *See* **common control switching arrangement.**

switching center: 1. In communications systems, an installation or a facility (a) in which switches are used to interconnect communications circuits on a circuit switching, message switching, or packet switching basis, and (b) that usually is located at a node in a network. *In telephone systems, synonyms* **switching exchange, switching facility.** *Note:* "Switching center" and "switch" are not synonymous. **2.** In satellite communications systems, the interface with existing terrestrial communications systems. **3.** A center of an integrated telegraph and telephone network (a) in which store and forward message switching is used, and (b) that consists of message switching equipment capable of terminating dedicated point-to-point circuits and setting up connections via nodal switching centers of communications system switched telegraph and telephone networks. *See* **circuit switching center, gateway switching center, local switching center, nodal switching center, store and forward switching center.** *See also* **central office, data switching exchange, digital switching, end office, information superhighway, packet switching, step by step switching system, switching system.**

switching equipment: Communications equipment that is used to effect the onward transmission of information through connections of circuits, loops, channels, or trunks. *See* **automatic switching equipment, dial switching equipment.**

switching equipment congestion signal: *See* **national switching equipment congestion signal.**

switching exchange: *Synonym* **switching center.**

switching facility: *Synonym* **switching center.**

switching matrix: In telephone system automatic switching exchanges, a connection matrix that consists of (a) relays connected to crossbars or solid state circuits connected to cross buses and (b) lines between incoming and outgoing circuits.

switching network: *See* **message switching network, packet switching network.**

switching node: *See* **message switching node, packet switching node.**

switching process time: The time from (a) the instant at which a switching center receives the call receiver complete address, i.e., complete number, to (b) the instant of commencement of the status signal, such as a ringing tone, busy signal, out of service signal, or service disconnect signal, from the call receiver end instrument.

switching reaction time: In telephone systems, the time between (a) the establishment of a circuit condition and (b) the receipt of a signal that indicates the condition.

switching release signal: In telephone systems, a signal sent to a switchboard to be used for disconnecting circuits and restoring them to the idle state. *See also* **idle state, switchboard.**

switching system: 1. A communications system that consists of switching centers and their interconnecting signal propagation media. **2.** Any mechanical, electrical, electromechanical, electronic, or optical system that processes an input signal and delivers an output signal. *See* **automatic switching system, common control switching system, electronic switching system, joint multichannel trunking and switching system, semiautomatic switching system, step by step switching system, tactical automatic digital switching system.** *See also* **circuit switching, communications system, message switching, packet switching, switching center.**

switching time: The time between (a) the instant of initiation of the switch actuation method, such as application of a signal or a force field, and (b) the instant of the settling of the active output of the switch to within 10% of the nominal steady state transmittance, given by the relation $T_S = T_O + T_B$ where T_S is the switching time of the switch, T_O is the switch operating time, and T_B is the switch bounce time. *See also* **transmittance.**

switching trunk: *See* **toll switching trunk.**

switching unit: *See* **circuit switching unit.**

switching variable: A variable that may assume only a finite number of possible values or states. *Note:* An example of a switching variable is an unspecified character of a character set, such as a character set consisting of 0 and 1.

switch message: A message that contains information obtained by listening to radio broadcasts, such as the frequency, call sign, and output power level of a detected transmitter. *See* **amplifying switch message.**

switch modulator: *See* **integrated optical circuit filter coupler switch modulator.**

switch operating time: In a switch, the time between (a) the instant that an applied actuating force or signal is raised to a specified value and (b) the instant that the transmittance has changed by 90% of its maximum value. *See also* **switching time, transmittance.**

switchpoint: *In computer programming, synonym* **switch.**

switch release time: In a switch, the time between (a) the instant that an applied actuating force or signal ceases and (b) the instant that the transmittance has changed by 90% of its maximum value. *See also* **switching time, transmission.**

switchroom: 1. In a telephone exchange, a room in which a switchboard, as distinct from other equipment, is situated. **2.** A room in which switching equipment is installed.

switch storage: *See* **manual switch storage.**

switch train: A set of switches that (a) is used to establish a connection from a call originator to a call receiver, and (b) usually consists of mechanical, electromechanical, or solid state devices or a combination of these.

SWR: standing wave ratio.

SX: simplex signaling.

SXS: step by step switch, step by step switching system.

syllabary: In a code book, a list of individual letters, combinations of letters, permutations of letters, or syllables, that (a) are accompanied by their equivalent code groups and (b) are used for spelling out words or proper names that are not present in the vocabulary of a code. *Synonym* **spelling table.** *See also* **code group.**

syllabic companding: The action of a compandor that is just fast enough to correct the power level of individual syllables of speech. During the enunciation of a syllable, the gain of the expander and the loss of the compressor are proportional to the mean power talker. *See also* **mean power talker.**

syllable: A character string or a bit string within a word. *See also* **bit string, character string.**

symbol: A representation of a concept, usually where agreement has been reached about the concept that is being represented. *Note:* Examples of symbols are

(a) a small circle of light on a radar screen used to represent an object, i.e., a target, in space, (b) a letter of an alphabet that is used, together with others, to form words to which meaning has been assigned, (c) a flag used to represent a nation or a country, and (d) a red cross. *See* **accounting symbol, aiming symbol, station symbol, tracking symbol.**

symbolic address: In communications, computer, data processing, and control systems, an address expressed in a convenient form, such as mnemonic form, as an aid to computer programmers and system operators.

symbolic language: A computer programming language used to express addresses and instructions with symbols and expressions convenient to humans rather than to machines, such as ADD for addition, SUB for subtraction, and MPY for multiplication, rather than terminology, such as X for addition, Y for subtraction, and Z for multiplication, that may be perfectly suitable for a machine. *Note:* A symbolic language is often related to a natural language, such as English, and to an artificial language, such as a computer programming language, in the sense that (a) in the symbolic language, expressions representing the same concept in different natural languages are different, such as SUB suffices as a symbol for subtraction in English, but in German ABN (abnehmen) suffices for subtraction, i.e., the same operation but different natural languages are represented, and (b) symbols usually represent entities, such as operands and operators, in computer programming languages used by computer programmers. *See also* **artificial language, natural language, symbol.**

symbolic logic: The logical discipline in which (a) valid arguments, i.e., valid parameters or operands, and (b) operations on the arguments are dealt with by using an artificial language designed to avoid the ambiguities and logical inadequacies of natural languages. *See also* **artificial language, natural language.**

symbol rate: In a communications system, the number of symbols transmitted from, received at, or passing through a point per unit of time, usually expressed in bits, bytes, or characters per second.

symbol string: A string that consists only of symbols.

symmetrical binary code: A code (a) that is derived from a binary code, (b) in which the sign of the quantized value, positive or negative, is represented by one digit, (c) in which the remaining digits constitute a binary code that represents the magnitude of the value, and (d) in which the weight of each digit and the use that is made of the symbols 0 and 1 in the various digit positions are specified.

symmetrical channel: A channel in which the send and receive circuits have the same data signaling rate (DSR). *See also* **channel, data signaling rate.**

symmetrical double heterojunction diode: *Synonym* **four-heterojunction diode.**

symmetrical pair: A balanced two-conductor transmission line that has equal conductor resistances per unit length, equal impedances from each conductor to earth, and equal impedances to other lines. *See also* **balanced line.**

symmetrical pair cable: 1. A cable that contains one symmetrical pair. **2.** A multipaired cable that contains two or more symmetrical pairs. *See also* **balanced line, symmetrical pair.**

symmetric binary channel: A channel that (a) transfers messages that consist of binary characters and (b) has the characteristic that the conditional probability of inadvertently, i.e., of wrongly, changing any one binary character to the other binary character is equal for both characters.

symmetric difference gate: *Synonym* **EXCLUSIVE OR gate.**

symmetric differential phase-shift keying: *See* **filtered symmetric differential phase-shift keying.**

symmetric optical fiber: *See* **circularly symmetric optical fiber.**

symmetric signal: 1. A signal that has a quality of symmetry with respect to a reference. *Note 1:* For example, if a symmetric signal is represented with its amplitude on the vertical axis and time on the horizontal axis, the portion to the left of the vertical axis is the mirror image of the portion to the right of the vertical axis, whether the signal is upright or inverted. *Note 2:* A symmetric signal may have one or more of many kinds of symmetry, such as x-axis symmetry, y-axis symmetry, zero-point, i.e., origin, symmetry, and arbitrary line symmetry. **2.** A signal that has both negative and positive polarities over a long period in such a manner that the net positive flow of current equals the net negative flow of current, thus preventing long-term accumulation of charges and hence drifting of voltage and baseline levels.

symmetry: *See* **bidirectional symmetry.**

SYN: synchronous.

synaptic antenna: An antenna that (a) consists of a three-dimensional grid of radio frequency (rf) elements, such as conducting segments joined by electrooptical computer-controlled switches, and (b) may be controlled to adjust for frequency tuning, beam polarization, beam forming, beam steering, and impedance matching.

SYN character: synchronous idle character.

synchronism: 1. The state of being synchronous. **2.** For repetitive events with the same, multiple, or submultiple repetition rates, a relationship among the events such that a significant instant of one event bears a fixed time relation with a corresponding instant in another event or other events. *Note:* Synchronism is maintained when there is a fixed, i.e., a constant, phase relationship among a group of repetitive events. **3.** The simultaneous occurrence of two or more events at the same instant on the same coordinated time scale. *See* **bit synchronism.** *See also* **synchronous network.**

synchronization: 1. The attaining of synchronism. **2.** The obtaining of a desired fixed relationship among corresponding significant instants of two or more signals. *Note:* Examples of synchronization are (a) in facsimile transmission, the establishment of equal scanning line frequencies at the transmitter and the receiver, and (b) in spread spectrum systems, the obtaining of timing agreement between a transmitter and a receiver so that, coupled with the same spread spectrum binary sequence generator, the information content of the received signal can be recovered. *See* **amplitude quantized synchronization, analog synchronization, bilateral synchronization, bit synchronization, carrier synchronization, double-ended synchronization, frame synchronization, linear analog synchronization, mutual synchronization, single-ended synchronization.** *See also* **acquisition time, bit synchronous operation, mutually synchronized network, synchronization code, synchronous data link control, synchronous data network, unilateral synchronization system.**

synchronization acquisition time: The time required for a receiving circuit to acquire bit synchronism after application of the first digital signal pulse to the circuit terminals. *See also* **bit synchronism, bit synchronous operation.**

synchronization bit: A binary digit used to achieve or maintain synchronism. *Note:* Synchronization bits are used in digital data streams. Synchronization pulses are used in analog signals. The bits and the pulses may have the same shape when the dependent variable used to represent the bit, such as the current, voltage, or power, is plotted versus an independent variable, such as time or distance. *See also* **binary digit, bit synchronization, character, digit, timing signal.**

synchronization code: In digital systems, a sequence of digital symbols introduced into a signal to achieve or maintain synchronism. *See also* **frame synchronization, synchronization, synchronous data network.**

synchronization correction signal: A signal that is used for correcting or recovering synchronization in a synchronous system after synchronism has been lost.

synchronization pattern: *See* **frame synchronization pattern.**

synchronization pulse: A pulse that (a) usually is one of a series of pulses and (b) is used to achieve or maintain synchronism. *Note:* Synchronization pulses are used in analog signals. Synchronization bits are used in digital data streams. The bits and the pulses may have the same shape when the dependent variable used to represent the pulse, such as the current, voltage, or power, is plotted versus an independent variable, such as time or distance. *See also* **pulse, timing signal.**

synchronization system: *See* **unilateral synchronization system.**

synchronized clock: A clock that has an output signal that is synchronized with another clock, i.e., both clocks are synchronized to the same timing signal reference source. *See also* **independent clock.**

synchronized network: *See* **democratically synchronized network, despotically synchronized network, hierarchically synchronized network, mutually synchronized network, oligarchically synchronized network.** *See also* **master-slave timing.**

synchronized transmitter group: Two or more transmitters that (a) have their transmitted signals synchronized, (b) usually are modulated or keyed from a single signal source, (c) operate at exactly the same frequency, (d) are not necessarily in the same phase, (e) may completely mask another transmitter, and (f) may direct a signal in a specific direction by adjusting the phase relations among them.

synchronizing: 1. Achieving and maintaining synchronism. *See also* **gating. 2.** In facsimile systems, achieving and maintaining predetermined speed relations between the scanning spot and the recording spot within each scanning line. *See also* **facsimile.**

synchronizing code: *See* **self-synchronizing code.**

synchronizing pilot: In frequency division multiplexing (FDM), a reference frequency used for achieving and maintaining synchronization of the oscillators of a carrier system or for comparing the frequencies or the phase of the currents generated by those oscillators.

See also **carrier, frequency division multiplexing, pilot, timing signal.**

synchronizing signal: A signal that is used for achieving and maintaining synchronism. *Note:* Examples of synchronizing signals are (a) in facsimile systems, the signal that maintains predetermined speed relations between the scanning spot and the recording spot within each scanning line, (b) in digital data synchronous transmission systems, the signal that is used to maintain synchronism between the transmitter and the receiver during data transfer, and (c) in a synchronized transmitter group, the signal that is used to maintain synchronism among the transmitters of the group. *See also* **facsimile, signal, spot speed, timing signal.**

synchronous: 1. Pertaining to events that occur at the same time or at the same rate, such as pertaining to the dependence upon the simultaneous occurrence of specific events, such as the simultaneous occurrence of two timing signals, i.e., timing pulses, from two clocks. **2.** Pertaining to events that occur with a regular or predictable time relationship among them. *Note:* "Isochronous" and "anisochronous" pertain to characteristics. "Synchronous" and "asynchronous" pertain to relationships. **3.** For repetitive events with the same, multiple, or submultiple repetition rates, a relationship among the events such that a significant instant of one event (a) bears a fixed time relationship to a corresponding instant in another event, or (b) maintains a time relationship, to a corresponding instant in another event, that remains within defined limits within the smallest period of repetition of either event. *See* **quasisynchronous.** *See also* **asynchronous transmission, bit by bit asynchronous operation, frame alignment time slot, frame duration, framing.**

synchronous communications: *See* **binary synchronous communications.**

synchronous correction: In a synchronous system, the correction of the phase relationship between a signal in one device, such as a receiver, and a signal in another device, such as a transmitter. *Note:* Synchronism is maintained by the use of a phase signal proportional to the phase difference between the two signals that are to remain synchronized. This signal is used to cause the device with the advanced phase signal to slow down or shift back, or to cause the device with the delayed or lagging phase signal to speed up or shift forward, thus correcting the phase difference between the signals to the desired value, such as zero or exact quadrature. If there is continuous slippage in phase, i.e., the frequencies are different, another signal is developed to cause one of the frequencies to either increase of decrease until the frequencies are the same, and the phase lock circuit can maintain synchronism at the desired phase relationship.

synchronous counter: A counter in which (a) all stages of the counter are simultaneously moved to their next state, (b) the clock pulses, or the input pulses to be counted, initiate all the changes simultaneously, and (c) the total time taken to count one pulse is much less than that of an equivalent asynchronous counter in which the output state of each stage is dependent on the previous stage's having changed, i.e., the flip-flops do not all change state synchronously because it takes a finite time for changes to propagate from stage to stage through the counter.

synchronous cryptooperation: Online operation of a cryptosystem such that (a) terminal cryptoequipment timing systems keep the equipment in step, and (b) synchronism of the system is independent of the traffic passing on the applicable channel. *See also* **cryptology.**

synchronous data link control (SDLC): 1. Control of a synchronous transmission over data links in a data network. **2.** A bit-oriented protocol for the control of synchronous transmission over data links in a data network. *See also* **Advanced Data Communication Control Procedure, binary synchronous communications, data, data link, data network, data transmission, link, network, synchronization.**

synchronous data network: A data network in which (a) synchronism is achieved and maintained (i) between data circuit-terminating equipment (DCE) and the data switching exchange (DSE) and (ii) between DSEs, and (b) the data signaling rates (DSRs) are controlled by timing equipment within the network. *See also* **data circuit-terminating equipment, data switching exchange, link, synchronization, synchronization code.**

synchronous data transmission channel: A data channel in which timing signals are transferred with the traffic between the data terminal equipment (DTE) and the data circuit-terminating equipment (DCE). *See also* **data channel, data circuit-terminating equipment, data terminal equipment, traffic.**

synchronous demodulation: Phase-sensitive angle demodulation in which (a) the local oscillator is synchronized, i.e., is locked, in phase, and therefore in frequency, to the distant transmitter, (b) allowance is made for Doppler shifts in frequency caused by any radial component of relative velocity between the transmitter and the receiver, and (c) allowance is made for phase delays during transmission, i.e., for transit time. *See also* **angle modulation, demodulation, local oscillator, transit time.**

synchronous digital computer: A digital computer in which the execution of each instruction of a computer program is initiated by selected equally spaced signals from a master clock. *See also* **asynchronous digital computer, computer program, instruction, master clock.**

synchronous Earth satellite: A satellite that (a) orbits about the center of the Earth in a circle in exact synchronism with the rotation of the Earth, though not necessarily in the equatorial plane, and (b) maintains a fixed longitude but may oscillate back and forth in the sky. *Note:* If a synchronous Earth satellite is at the proper height, i.e., proper altitude, and proper speed in the equatorial plane, it will always remain at a fixed point above the equator, and hence will be a geostationary satellite. Three geostationary synchronous Earth satellites can provide full Earth coverage, except in the polar regions because of the low elevation angles of the satellite when viewed from the polar regions, where the satellite appears stationary but is barely above the horizon. *See also* **Earth coverage, geostationary satellite.**

synchronous equatorial satellite height: *Synonym* **geostationary satellite height.**

synchronous fiber optic network: A fiber optic communications network that (a) consists of fiber optic data links and nodes, (b) makes use of synchronous transmission, (c) operates with internal synchronism, and (d) may operate in synchronism with other networks. *Note:* An example of a synchronous fiber optic network is a network in which the Synchronous Optical Network (SONET) standard is implemented. *Synonym* **synchronous optical network.** *See also* **data link, Synchronous Optical Network (SONET) standard, synchronous transmission.**

synchronous height: The orbital altitude of a satellite at which the period of rotation of the satellite about the parent body equals that of the parent body about its axis. *Note:* The synchronous height for an Earth geostationary satellite is about 19,200 nmi (nautical miles), i.e., about 35,600 km (kilometers), above the equator (1 nmi = 1.852 km). *See also* **geostationary satellite.**

synchronous idle character (SYN): A transmission control character used in synchronous transmission systems to provide a signal from which synchronism, or synchronous correction, may be achieved between sets of data terminal equipment (DTE), particularly when no other character is being transmitted.

synchronous jamming: Jamming in which signals are used that are modulated at an exact multiple of the basic pulse repetition rate of the system, such as a radar system or a communications system, that is being jammed. *See also* **jamming, pulse repetition rate.**

synchronous margin: In a start-stop transmission system, the maximum value of the margin that is obtained by adjusting the modulation rate of the input signals to the most favorable value with respect to the timing characteristics of a receiver. *See also* **margin, modulation, start-stop transmission system, value.**

synchronous network: A network in which clocks are controlled so as to run, (a) ideally, at identical rates, or (b) at the same mean rate with limited but constant relative phase displacement, within a limited tolerance. *Note:* Ideally, the clocks are synchronous, but they may be mesochronous in practice. By common usage, such mesochronous networks are frequently considered to be synchronous. *See also* **clock, data transmission, mesochronous, synchronism.**

synchronous operation: Operation such that each event in the operation, or the operation itself, either occurs as a result of action by a clock or occurs regularly or predictably with respect to the occurrence of an action or an event in another process or system. *See* **asynchronous operation, bit synchronous operation.**

synchronous optical network: *Synonym* **synchronous fiber optic network.**

Synchronous Optical Network (SONET) standard: A fiber optic communications network interface standard that (a) was developed by the Exchange Carriers Standards Association (ECSA) and approved by the American National Standards Institute (ANSI), (b) describes synchronous 2.46-Gbps (gigabits per second) fiber optic transmission systems, (c) is based on a data signaling rate (DSR) of 51.840 Mbps (megabits per second), called "OC1," i.e., "Optical Carrier 1," (d) consists of a hierarchy of data signaling rates (DSR) based on multiples of OC1, up to and including OC48, which has a data signaling rate (DSR) of 2.48832 Gbps (gigabits per second), (e) has a set of optical transmission rates and formats established so that operating telephone companies can synchronize public telephone networks and efficiently interconnect high speed fiber optic links using an eight-bit byte format, rather than a single-bit-stream format, that will enable the telephone networks to be efficiently synchronized to an eight-bit frame so that optical signals at, say, 45 Mbps need not be demultiplexed down to, say, 1.5 Mbps for routing signals around the network, and (f) permits data to be transmitted at any data signaling rate on any line. *See also* **Fiber Distributed Data Interface standard, synchronous optical network.**

synchronous orbit: The path traversed by a synchronous satellite. *Note 1:* A satellite in a synchronous orbit traverses the entire orbit in a period equal to the average rotational period of the Earth. *Note 2:* A synchronous orbit that is circular and lies in the equatorial plane is a geostationary orbit. The geostationary orbit is about 19,200 nmi (nautical miles) above the equator (1 nmi = 1.825 km). If the satellite orbital motion is synchronized with the rotation of the Earth but not in the equatorial plane, the satellite remains at a fixed longitude but not above a fixed point. *See also* **direct orbit, equatorial orbit, geostationary orbit, inclined orbit, polar orbit, retrograde orbit, satellite.**

synchronous receiver margin: In a synchronous receiver, the margin determined by the extent of isochronous distortion. *See also* **isochronous distortion, margin.**

synchronous satellite: A satellite that (a) rotates about the Earth in synchronism with the Earth's rotation and (b) always remains at a fixed longitude but not necessarily above a fixed point. *Note:* A geostationary satellite is necessarily a synchronous satellite, but a synchronous satellite is not necessarily a geostationary satellite. *See also* **geostationary orbit, geostationary satellite, satellite, synchronism, synchronous.**

synchronous signal: A member of a set of signals such that between any two significant instants there is an integral number of unit intervals, as when keeping in step with a clock or a trigger signal. *See also* **asynchronous signal, integral, significant instant, unit interval.**

synchronous system: A system in which events, such as signals, occur in synchronism. *Note:* Examples of synchronous systems are (a) a transmitter and a receiver that operate with a fixed time relationship, (b) a telegraph or data system in which isochronous modulation or restitution is used, and in which the transmitting and receiving instruments operate continuously with a constant phase relationship, and (c) an alphabetic telegraphy system in which synchronous transmission is used. *See also* **alphabetic telegraphy, communications system, framing, homochronous, mesochronous, restitution, synchronous transmission.**

synchronous TDM: synchronous time division multiplexing.

synchronous time division multiplexing (TDM): Multiplexing in which timing is obtained from a clock that controls both the multiplexer and the channel source. *See also* **asynchronous time division multiplexing, time division multiplexing.**

synchronous transfer mode: In the transport layer of a broadband integrated services digital network (B-ISDN), a transfer mode in which time division multiplexing (TDM) and switching are used across the user-network interface. *See also* **Integrated Services Digital Network.**

synchronous transmission: 1. Digital data transmission in which the time interval between any two similar significant instants in the overall bit stream is always an integral number of unit intervals. *Note:* "Isochronous" and "anisochronous" are characteristics. "Synchronous" and "asynchronous" are relationships. **2.** Digital data transmission in which the time of occurrence of each signal that represents a bit is related to a fixed time base. **3.** Data transmission in which the time of occurrence of a specified significant instant, such as the leading edge of a start signal, in each unit of data, such as a byte, character, word, block, or frame, occurs with a specified time relationship to a preceding signal on the channel in accordance with a specified timing pulse or in accordance with a specified time frame. *Note:* Fiber optic communications systems may be operated in both synchronous and asynchronous transmission modes. *See also* **asynchronous transmission, bit by bit asynchronous operation, frame alignment time slot, frame duration, framing, significant instant, time of occurrence.**

syncopated code: In spread spectrum systems, a code that (a) is produced by a spread spectrum binary code sequence generator and (b) has asynchronous bit timing, i.e., the pulses in the code do not occur at regular or fixed periods. *See also* **spread spectrum code sequence generator.**

synodic period: The time of revolution of any satellite, including the Earth and the Moon, with reference to the body it revolves about. *Note:* The average sidereal day period of the Moon about the Earth is 27 1/3 days, compared to an average synodic period, or lunar month, of 29 1/2 days. *See also* **sidereal day.**

syntax: 1. In a given language, the relationships among characters or groups of characters, independent of their meanings or the manner of their interpretation and use. **2.** The structure of expressions in a language. *Note 1:* In artificial languages, such as computer programming languages, the syntax is prepared first and is used to generate the language. There are no syntax violations and no ambiguities. *Note 2:* In natural languages, such as languages developed and spoken by humans even before being written, the syntax is prepared after the language is in use. The syntax is loose, there are syntax violations, and there are ambiguities. **3.** The rules governing the structure of a language. *Note:* Examples of syntax in the English language are that (a) the letter "q" cannot begin a word unless it is followed by the letter "u," and (b) a + sign or a − sign must always be associated with a number or with a symbol that can be evaluated as a number. Thus, a language can be completely described by listing its symbols, their uses, their meanings, and the syntax that establishes their permissible arrangements. If the rules are never violated and nothing is permitted unless explicitly allowed by the rules, the language is said to be "ruly." Artificial languages, such as computer programming languages, tend to be ruly because syntax is seldom if ever violated. Natural languages tend to be "unruly" because syntax is often violated. **4.** The relationship among symbols. *See* **abstract syntax.** *See also* **semanteme, semantics.**

syntax notation one: *See* **abstract syntax notation one.**

synthesizer: *See* **frequency synthesizer, speech synthesizer.**

synthesizing: *See* **frequency synthesizing, speech synthesizing.**

syntonization: The setting of the frequency of one oscillator equal to that of another. *See also* **reference frequency.**

SYSGEN: system generation.

system: **1.** In communications, computers, information processing, and control, a group of persons, equipment, and methods that (a) is organized to accomplish a set of specific functions and (b) includes the equipment and facilities needed to ensure proper operation and maintenance. **2.** Any organized assembly of resources and procedures united and regulated by interaction or interdependence to accomplish a set of specific functions. *See* **access system, active detection system, adaptive communications system, adaptive system, advanced television system, airborne radar warning system, aircraft control warning system, air defense communications system, air-ground worldwide communications system, Amazon surveillance system. antenna feed system, asynchronous communications system, asynchronous system, authentication system, automated information system, automated maritime telecommunications system, automated tactical command and control system, automatic data processing system, automatic dependent surveillance system, automatic error detection and correction system, automatic message processing system, automatic relay system, automatic switching system, automatic tape relay system, automatic telephone system, autonomous system, beacon navigation system, beam riding guidance system, centralized dictation system, cipher system, ciphony communications system, code-dependent system, code-independent system, Com21™ system command and control system, command guidance system, common control switching system, communications subsystem, communications system, continuous time system, crossbar system, data system, decimal numeration system, differential global positioning system, digital access and cross connect system, digital transmission system, directional signaling system, discrete time filtering system, discrete time system, disk operating system, display system, distributed data processing system, distributed radar system, distributed sensor system, diversity system, Domain Name System, Doppler system, double sideband suppressed carrier transmission system, electronic flight information system, electronic key management system, electronic message system, electronic switching system, embedded system, Emergency Broadcast System, end system, enhanced definition television system, error-correcting system, error-detecting system, expert system, facility ground system, facsimile system, Federal Information Processing system, Federal Telecommunication System, fiber optic communications system, fiber optic telemetry system, fiber optic transmission system, filtering system, five-horn feed system, flight management system, frequency and amplitude system, general purpose operating system, global decision support system, grandfathered system, ground controlled approach system, group alerting and dispatching system, guidance system, Hell system, heterogeneous computer system, high definition television system, high grade cryptosystem, homing guidance system, homogeneous computer system, hybrid communications system, hybrid system, hyperbolic navigation system, imaging system, improved definition television system, inertial guidance system, information management system, information system, instrument landing system, integrated communications system, integrated fiber optic system, integrated system, intelligent vehicle highway system, interactive display system, joint multichannel trunking and switching system, Joint Uniform Telephone Communication**

Precedence System, Jupiter system, key telephone system, layered system, laser fiber optic transmission system, lens system, local distribution system, long-distance direct current dialing system, loop disconnect system, loop disconnect pulsing system, management information system, management system, man-machine system, manual relay system, master-slave timing system, medical diagnostic imagery system, merchant ship broadcast system, metric system, mixed base numeration system, monopulse antenna feed system, multifunction information distribution system, multirole system, National Communications and Information System, Network File System, neutral direct current telegraph system, noninertial guidance system, numeration system, off-axis optical system, omnidirectional signaling system, open system, operating system, operations support system, operations system, optical communications system, optical system, pair gain system, passive detection system, perimeter protection system, personal communications system, polarential telegraph system, positional numeration system, precedence telephone system, primary distribution system, private key encryption system, process computer system, process control system, programming system, protected distribution system, protected fiber optic distribution system, protected wireline distribution system, pseudomonopulse tracking system, public key encryption system, pure binary numeration system, quadruplex system, radio communications system, radio relay system, radiotelephone system, radiotelephone voice system, Raven system, reference system, satellite communications system, sea surveillance system, semiautomated tactical command and control system, semiautomatic relay system, semiautomatic switching system, Signaling System No. 7, single frequency signaling system, space-ground link subsystem, space system, start-stop transmission system, step by step switching system, stepped start-stop transmission system, strategic military communications system, switching system, synchronous system, tactical automatic digital switching system, tactical communications system, tactical military communications system, teleautography system, Telecommunications Service Priority System, telecommunications system, telegraph system, telemetry system, telephone system, television system, terminal guidance navigation system, terrain avoidance system, terrestrial system, terrestrial telecommunications system. threat warning system, torn tape relay system, traffic service position system, trusted computer system, turnkey system, unilateral synchronization system, Universal Mobile Telecommunications System, virtual reality system, virtual system, visual communications system, Wheatstone automatic system, wideband communications system, wideband system, wired radio frequency system, worldwide geographical reference system. *Refer to* **Fig. S-22.** *Refer also to* **Figs. V-4 (1067), W-4 (1087).**

system abbreviation: *See* **aeronautical communications system abbreviation.**

system analysis: A systematic investigation of a real or planned system to determine (a) the functions of the system, (b) how system functions relate to each other, and (c) how system functions relate to other systems. *Synonym* **systems analysis.** *See also* **operations research.**

system architecture: *See* **information system architecture, open system architecture, telecommunications system architecture.**

system blocking: *Synonym* **access denial.**

system blocking signal: A control message that (a) is generated within a telecommunications system to indicate temporary unavailability of system resources required to complete a requested access, and (b) is part of system overhead information. *See also* **blocking, overhead information, signal.**

system budget: *Synonym* **power budget.**

system consolidation: *See* **communications system consolidation.**

system control: In communications, computer, data processing, and control systems, the control and the implementation of a set of functions that (a) prevent or eliminate degradation of any part of the system, (b) initiate immediate response to demands that are placed on the system, (c) respond to changes in the system to meet long range requirements, and (d) may include various subfunctions, such as (i) immediate circuit utilization actions, (ii) continuous control of circuit quality, (iii) continuous control of equipment performance, (iv) development of procedures for immediate repair, restoration, or replacement of facilities and equipment, (v) continuous liaison with system users and with representatives of other systems, and (vi) provision of advice and assistance in system use. *See also* **network control.**

system controller: 1. In communications, computer, data processing, and control systems, a person at a technical control facility who is responsible for maintaining quality control of the system functions, such as the switching functions that are performed by commu-

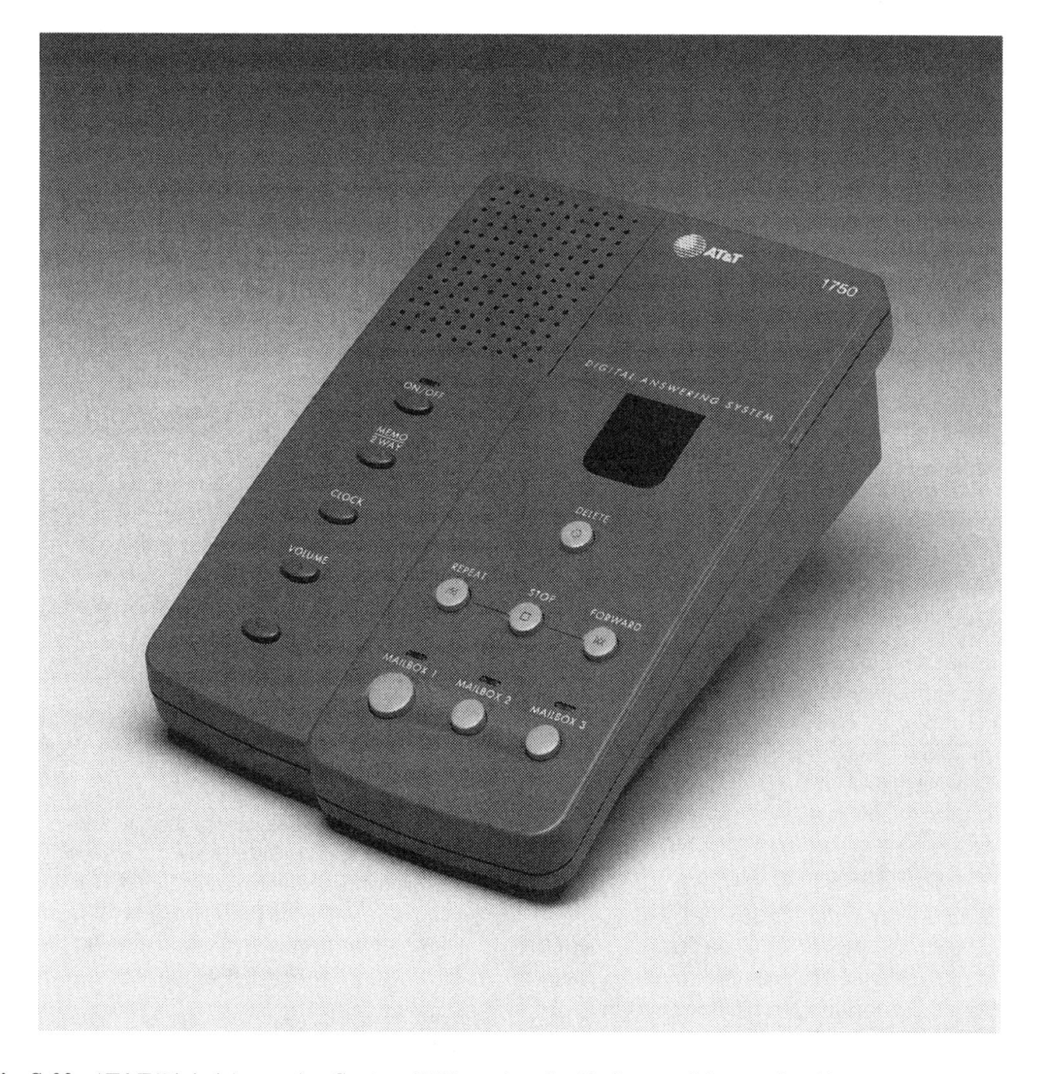

Fig. S-22. AT&T Digital Answering **System** 1750, equipped with three mail boxes, time/day announcement, and voice menu for directing call originator through remote commands. (Courtesy AT&T Archives.)

meet long range requirements, and (d) may include various subfunctions, such as (i) immediate circuit utilization actions, (ii) continuous control of circuit quality, (iii) continuous control of equipment performance, (iv) development of procedures for immediate repair, restoration, or replacement of facilities and equipment, (v) continuous liaison with system users and with representatives of other systems, and (vi) provision of advice and assistance in system use. *See also* **network control.**

system controller: 1. In communications, computer, data processing, and control systems, a person at a technical control facility who is responsible for maintaining quality control of the system functions, such as the switching functions that are performed by communications equipment. **2.** Equipment at a technical control facility that controls the functioning of switching and supporting equipment. *See also* **technical control facility.**

system design: 1. The defining of the hardware and software architecture, components, modules, interfaces, and data for a system to satisfy specified requirements. **2.** The preparation of an assembly of methods, procedures, or techniques united by regu-

système electronique couleur avec memoire (SECAM) (electronic color system with memory): A television signal standard that calls for 625 lines per frame and 50-Hz (hertz), 220-V (volt) primary power. *Note:* Système electronique couleur avec memoire (SECAM) is used in France, eastern European countries, countries of the former Soviet Union, and some African countries. *See also* **National Television Standards Committee (NTSC) standard, phase alternation by line, phase alternation by line modified.**

Système International d'Unités (SI): *See* **International System of Units.**

system engineering: A systematic sequence of operations, including requirements analysis, system description, design, development, prototype fabrication, testing, production, packaging, shipping, installing, operating, maintaining, and improving real or planned systems. *See* **communications system engineering.** *See also* **operations research, system, system analysis.**

system environmental file: A compilation of information concerning the environment in which a system is required to operate, such as the location, height above ground, identity, frequency assignment, and operational characteristics, and the equipment characteristics of surrounding radiating electronic equipment, that may be potentially important to compatibility studies for system design, development, and operation.

system failure transfer: In the event of the catastrophic failure of a central office (C.O.), the transfer of the C.O. trunks and the interoffice trunks to predetermined stations to allow incoming and outgoing calls to be completed.

system followup: The study of the effects and the performance of a system after it has reached a stabilized state of operational use.

system for evaluation and test: *See* **Functional Advanced Satellite System for Evaluation and Test.**

system for mobile communications: *See* **global system for mobile communications.**

system gain: *See* **statistical terminal/regenerator system gain, terminal/regenerator system gain.**

system generation (SYSGEN): The process of (a) selecting optional parts of an operating system, and (b) creating a particular operating system tailored to the requirements of a data processing installation.

system information rate: In optical transmission systems, the bit repetition rate internal to equipment that, when acted upon by the line coding algorithm, results in the optical line rate. *Note:* The system information rate usually is expressed in bits per second.

system integration: The progressive linking and testing of system components to (a) merge their functional and technical characteristics into a comprehensive and interoperable system, and (b) allow data existing in disparate systems to be shared or accessed across functional or system boundaries. *See also* **embedded system, integrated system.**

system integrity: 1. In a system, the condition that exists when mandated operational and technical parameters are within prescribed limits. **2.** The quality of a system, such as a communications, computing, or data processing system, that performs intended functions in an unimpaired manner. **3.** The assurance that (a) a communications, computer, data processing, or control system is logically designed, correctly operating, reliable, complete, and adequately protected in all hardware and software aspects, and (b) data integrity is maintained. *See also* **data integrity.**

system interface: *See* **user-system interface.**

system interface for computer environments: *See* **portable operating system interface for computer environments.**

system library: A controlled and organized collection of software that (a) is directly and automatically available to a communications, computer, data processing, or control system, (b) can be accessed for use, and (c) may be incorporated into other computer programs by reference. *See also* **software library.**

system lifecycle: The course of developmental changes through which a system passes from its conception to the termination of its use and subsequent salvage. *Note:* A system lifecycle might include the phases and the activities associated with the analysis, acquisition, design, development, test, integration, operation, maintenance, and modification of the system.

system loading: In a frequency division multiplexed (FDM) transmission system, the absolute power level of the composite signal transmitted in one direction. *Note 1:* The absolute power level is referenced to the zero transmission level point (0TLP). *Note 2:* The composite signal contains signaling, speech, and digital signals. *See also* **level, loading, transmission level point, zero transmission level point.**

system management: In a communications network, management extended to include customer equipment, such as private branch exchanges (PBXs) and user end instruments. *See* **communications system management.** *See also* **network management.**

system noise temperature: *See* **receiving system noise temperature.**

system operating time: The time during which a system (a) is capable of being operated, and (b) produces accurate results in accordance with system standards. *See also* **system downtime, system uptime.**

system operational threshold: For a supported performance parameter of a system, the value that establishes the minimum operational service performance level for the parameter. *Note:* A measured parameter value worse than the system operational threshold indicates that the system is in an outage state. The parameter is in a worse-than-threshold state when it is outside a specified range, i.e., is above a specified maximum threshold value or below a specified minimum threshold value. *Synonym* **outage threshold.** *See also* **performance parameter.**

system operator: In communications, computer, data processing, and control systems, the person who (a) ensures satisfactory operation of a system, (b) usually inserts input and accepts output, (c) reacts in accordance with system requirements, (d) usually interacts with the system, and (e) performs only minor maintenance at the control site, attendant position, workstation, or control console. *See* **telecommunications system operator.**

system overhead information: In a communications system message, the part of overhead information that (a) is added by the system and (b) is not delivered to the destination user, i.e., to the addressee. *Note:* Examples of system overhead information are check characters, routing instructions, and certain address groups. *See also* **overhead information, user overhead.**

system plan: *See* **Aeronautical Emergency Communications System Plan.**

system power margin: The difference between available power and the power needed to overcome system losses and still satisfy the minimum, i.e., threshold, input power requirements of the receiver, i.e., the load, sink, or user. *Note:* System power margin includes the effects of component aging. *See* **optical system power margin.** *See also* **optical power budget, power budget, threshold.**

system programmer: A programmer who (a) prepares and maintains the operating system, (b) usually controls the use of the system, and (c) improves system productivity. *See also* **operating system, programmer.**

system reliability: The probability that a system, including all hardware, firmware, and software, will satisfactorily perform the task for which it was designed or intended for a specified time and in a specified environment. *See also* **reliability.**

system resilience: The ability of a system, such as a communications, computer, data processing, control, or weapon system, to continue to function despite the existence of faults in its component subsystems or parts. *Note:* System performance may be diminished or degraded until the faults are corrected. *See also* **system robustness.**

system robustness: The measure or the extent of the ability of a system, such as a computer, communications, data processing, control, or weapons system, to continue to function despite the existence of faults in its component subsystems or parts, i.e., the measure or the extent of system resilience. *Note:* System performance may be diminished or otherwise altered; i.e., the system may continue to operate in a diminished or degraded state until the faults are corrected. *See also* **system resilience.**

systems: *See* **Corporation for Open Systems.**

systems analysis: *Synonym* **system analysis.**

system saturation: *See* **communications system saturation.**

system signaling: In transmission systems, signaling (a) that is used to perform system supervision, including operational and control functions, such as system control, addressing, routing, error detection and correction, level control, priority control, traffic control, message accountability, and overhead information generation and transfer functions, and (b) that is compatible throughout the system. *See also* **centralized operation, circuit, distributed control, distributed switching, network architecture, outpulsing.**

systems interconnection reference model: *See* **Open Systems Interconnection—Reference Model.**

systems network: *See* **Defense Information Systems Network.**

system software: **1.** Software that (a) is part of a system, such as part of the operating system, (b) is application-independent, and (c) supports the running of user application software. **2.** Software that is designed for use by a specific communications, computer, data processing, or control system, or for use by a family of such systems, to facilitate the operation and the maintenance of other software designed for specific applications. *Note:* Examples of system software are operating systems, compilers, assemblers, translators, and utility programs. *Synonym* **support software.**

systems security: *See* **automated information systems security.**

systems security initiative: *See* **multilevel information systems security initiative.**

system standard: 1. One of the minimum interoperability or performance characteristics of communications systems or subsystems. *Note:* Mandatory system standards are based on empirical data derived from proof of concept testing of the system or the subsystem. **2.** In a system, the specific characteristics that (a) are necessary to permit interoperation, and (b) are not dictated by electrical or optical performance requirements. *Note:* Examples of communications system standards are (a) the values of the center frequency for telegraph channels and for test tones and (b) the value 1.31 μm (microns) for the operating wavelength of fiber optic systems. *See* **Advanced Mobile Phone System standard.** *See also* **design objective.**

system supervision: 1. In communications systems, the use of signals and techniques to perform system management functions, such as system control, addressing, routing, error detection and correction, level control, priority control, traffic control, and message accountability. **2.** In transmission systems, the use of system signaling to (a) perform system operational and control functions, such as system control, addressing, routing, error detection and correction, level control, priority control, traffic control, and message accountability, and (b) generate overhead information and transfer instructions that may be contained in system overhead portions of messages. *See also* **centralized operation, circuit, distributed control, distributed switching, network architecture, outpulsing.**

system support: 1. The continued provision of services and material necessary for the use and the improvement of a system during its lifecycle. **2.** The continued provision of services and material necessary for the use and the improvement of a system after it becomes operational.

system testing: Testing that (a) is conducted to verify that a system meets its specified performance and functional requirements, and (b) includes testing of software and hardware with all system components connected in an integrated system. *See also* **acceptance testing, qualification testing.**

system test time: The part of operating time during which a system, or a functional unit within the system, is tested for proper operation. *Note:* In a computer, the system test time includes the time for testing programs belonging to the operating system. *See also* **operating system.**

system user: *See* **Telecommunications Service Priority System user.**

system validation: In a system, evaluation of hardware or software, or both, that is performed (a) to ensure compliance with requirements, and (b) at the end of the development process or during system testing and performance evaluation in accordance with approved plans, procedures, and standards. *Synonym* **independent system validation.** *See* **qualification testing.** *See also* **performance evaluation, system testing.**

T

T: tera (10^{12}). *See* **metric system.**

tab key: On a keyboard, such as a printer, computer, or communications keyboard, a key that, when depressed, will cause the device under control of the keyboard to step forward a specified number of increments, such as positions, columns, screens, or storage locations, such as (a) in a personal computer, cause the computer to display a new screen of columns to the right on a spreadsheet or (b) on a printer or typewriter, cause the printer or typewriter to step forward along a line to a preset position. *Note:* The action of the tab key might be equivalent to actuating a space bar a requisite number of times, or it might be equivalent to actuating the right arrow key a requisite number of times. *See also* **backtab key.**

table: *See* **Boolean operation table, garble table, lookup table, operation table, permutation table, prediction table, routing indicator delineation table, routing table, teletypewriter garble table, truth table.**

table function key: In personal computer system spreadsheet operations, a key that initiates an operation pertaining to a specific cell in the spreadsheet. *See also* **spreadsheet.**

tablet: A flat surface over which a drawing or a graph may be placed such that, when it is used with a stylus or a mouse, coordinate data can be generated such that the coordinate data correspond to the position of the stylus or the mouse on the surface. *See also* **coordinate data, mouse, stylus.**

taboo frequency: A frequency of such importance to friendly forces that jamming or deliberate interference is prohibited. *Note:* Examples of taboo frequencies may include frequencies used for communications, command and control, intelligence, missile control, early warning air defense, civil defense, search and rescue operations, and emergencies. *See also* **protected frequency, guarded frequency.**

tabulation character: *See* **horizontal tabulation character, vertical tabulation character.**

tabulator: A device that reads data recorded on a data medium or stored in a storage unit, such as punched cards, magnetic tape, or auxiliary storage, and prints lists, tables, or totals in designated formats.

TACK: In flaghoist signaling, the voice equivalent for the tackline prosign, a 2-m (meter) length of halyard that is used as a separator between flags or pennants that, if not separated, would convey a different meaning. *See also* **prosign, tackline.**

tackline: In visual communications systems, a length of halyard about 2 m (meters) long that is used in flaghoist signaling as a separator between flags or pennants that, if not separated, would convey a different meaning. *Note:* The voice prosign for the displayed tackline is TACK. *See also* **flag, flaghoist signaling, halyard, pennant, TACK.**

TACSIM: tactical simulation.

tactical air operations communications net: *See* **maritime tactical air operations communications net.**

tactical automatic digital switching system (TADSS): Any military transportable digital switching system designed for rapid deployment in support of tactical forces. *See also* **digital, switching system, transportable.**

Tactical Automatic Digital Switching System (TADSS): A specific operational military transportable store and forward message switching system that may be rapidly deployed in support of tactical forces. *See also* **switching system.**

tactical command and control (C^2) system: The equipment, communications, procedures, and personnel essential to a commander for planning, directing, coordinating, and controlling tactical operations of assigned forces pursuant to assigned missions. *See* **automated tactical command and control system.**

tactical communications: Within tactical forces, communications in which information of any kind, especially orders and decisions, is conveyed from one command, person, or place to another, usually by means of electronic equipment, including communications security equipment, which is organic to the tactical forces. *Note:* Tactical communications do not include communications provided (a) to tactical forces by the Defense Communications System (DCS), (b) to nontactical forces by the DCS, (c) to tactical forces by nontactical military commands, or (d) to tactical forces by civil organizations. *See also* **combat net radio, communications, communications security, long-haul communications, tactical load, TRI-TAC equipment.** *Refer to* **Figs. S-5 and S-6 (869).**

tactical communications system (TCS): A communications system that (a) is used within, or in direct support of, tactical forces, (b) is designed to meet the requirements of changing tactical situations and varying environmental conditions, (c) provides securable communications, such as secure voice, data, and video transmissions, among mobile users, (d) is used in the exercise of command and control within and among tactical forces, and (e) usually requires extremely short installation times, usually on the order 1 hour, in order to meet frequent relocation requirements. *Note:* Based on different requirements of the multichannel trunking networks, a distinction is made between (a) tactical communications systems that require extremely short installation times, such as from less than a minute to an hour, necessitated by relocation requirements that are sometimes frequent, and (b) tactical communications systems for which longer periods are available for installation and commencement of operation. Tactical communication systems usually are composed of mobile or transportable communications equipment and components that are assigned as organizational equipment in accordance with tables of organization and equipment (TO&E) in the hands of the personnel of the tactical unit. *See also* **communications system, TRI-TAC equipment.** *Refer to* **Figs. T-1, T-2.** *Refer also to* **Figs. S-5 and S-6 (869).**

tactical data information link (TADIL): A standardized communications link that (a) is approved by the U.S. Joint Chiefs of Staff, (b) is suitable for the transmission of digital information, (c) uses a rigidly controlled protocol, (d) provides communications support to tactical units, and (e) is characterized by standardized message formats and transmission characteristics.

tactical data information link A (TADIL–A): A netted link in which (a) one unit acts as a net control station and interrogates each unit by roll call, i.e., by polling, (b) each interrogated unit transmits its data to the net, (c) as a result, each unit receives all the information transmitted, and (d) there is a direct transfer of data, and no relaying is needed.

tactical data information link B (TADIL–B): A point-to-point data link between two units that provides for simultaneous transmission and reception of data, i.e., provides for duplex operation.

tactical load: For the host service tactical forces, the power that (a) constitutes the total power requirements for communications, including the requirements for weapons, detection, command and control systems, and related support functions, and (b) is a part of the operational load. *See also* **load, operational load, tactical communications.**

tactical military communications system: A communications system that is established and used exclusively by military organizations for the conduct of tactical military communications, i.e., for the support of military units during field operations, such as during field exercises, troop movements, and combat. *See also* **military communications, military communications system, strategic military communications system.**

tactical optical fiber: *See* **advanced tactical optical fiber.**

TADIL: tactical data information link, tactical digital information link.

Fig. T-1. Two sets in a network of up to 30 subscriber sets in a two-wire no-exchange no-online-operator **tactical communications system** with built-in intelligence, used for conference, broadcast, group call, collective call, precedence, and call/forward/transfer capabilities of the AWITEL® (Albis Wire Telephone Communication System with Distributed Intelligence). (Courtesy Siemens-Albis AG, Switzerland.)

Fig. T-2. The AWITEL® (Albis Wire Telephone Communication System with Distributed Intelligence) **tactical communications system** in operation with tactical equipment. (Courtesy Siemens-Albis AG, Switzerland.)

TADIXS-B: Tactical Data Information Exchange System—Broadcast.

TADSS: tactical automatic digital switching system, Tactical Automatic Digital Switching System.

tag: *Synonyms* **flag, label.**

tagging: *See* **aperture tagging.**

tag image file format (TIFF): A file format that (a) is used to store an image and (b) makes use of the particular data structure of the file.

TAI: International Atomic Time (Temps Atomique International).

tail circuit: A leased or privately owned communications line that (a) extends from the end of a major transmission link, such as a microwave link, satellite link, or local area network (LAN), to the end user location, and (b) is a part of a user-to-user connection. *See also* **circuit, loop.**

tailing: In transmission systems, the excessive prolongation of the transition of a signal from one signal condition to another. *Note 1:* Tailing usually is caused by a defect that delays the decay of a signal. *Note 2:* Examples of tailing are (a) in facsimile systems, an irregular shifting of the recorded copy when there is a sudden transition in the signal level, such as a sudden transition from black to white, and (b) in video systems, a smearing of an image at the boundary where there is a sudden transition from light to dark. *Note 3:* Tailing does not apply to the reproduction of details that are smaller than the picture element and that may be deliberately prolonged at the transmitter. *Synonym* **hangover.** *See also* **facsimile, underlap.**

tailoring: *See* **display tailoring.**

takeoff angle: *Synonym* **departure angle.**

talbot: In the meter-kilogram-second (MKS) system of units, a unit of luminous energy equal to 10 million lumergs and equal to 1 L•s (lumen•second).

talk: *See* **crosstalk.**

talker: *See* **mean power talker, medium power talker.**

talker volume distribution: *See* **mean power of the talker volume distribution.**

Talon: A vehicle license plate recognition, imaging, and identification system developed by Racal Radio Limited.

tandem: 1. Pertaining to an arrangement or sequencing of networks, circuits, or links, in which the output terminals of one network, circuit, or link are connected directly to the input terminals of another network, circuit, or link. *Note:* An example of a tandem connection is several microwave links connected in series. **2.** The connection of a group of entities in series, as in a chain. *Synonym* **concatenation, catenation, chain.** *See also* **link.**

tandem center: In communications systems, an installation in which switching equipment connects trunks to trunks and does not connect any loops. *See also* **end office, exchange, extension facility, switching center, tandem tie trunk network, trunk.**

tandem data circuit: A data circuit that contains more than two sets of data circuit-terminating equipment (DCE) in series, i.e., in tandem. *See also* **data circuit, tandem.**

tandem exchange: *Synonym* **tandem switch.**

tandem office (TO): A central office (CO) used as an intermediate switching point for traffic between COs.

tandem switch: A manual or automatic switch that connects the output terminals of the circuits of one trunk to the input terminals of the circuits of another trunk, thereby connecting both trunks in series, i.e., in tandem. *Synonym* **tandem exchange.**

tandem tie trunk network (TTTN): A network arrangement that (a) permits sequential connection of tie trunks between private branch exchange (PBX) locations by using tandem operation, and (b) permits two or more dial tie trunks to be connected together at a tandem center location to form a through connection. *See also* **switching center, tandem center, tie trunk, trunk.**

tangent: 1. In a right triangle, the ratio of (a) the length of the side opposite a given acute angle to (b) the length of the shorter side of the acute angle. *Note:* The length of the hypotenuse is irrelevant. **2.** From a point outside a curve or a surface, such as that of a circle, ellipse, parabola, curved line, or sphere, a straight line that (a) touches the curve or the surface at one and only one point and (b) does not cross the curve or the surface.

tangential coupling: The coupling of signals from one optical fiber to another by placing or fusing the core of a fiber containing a signal in its proximity with another fiber core for a short distance, to allow some of the signal to leak or spill over to the other fiber. *Note:* The degree of coupling is determined by the core-to-core spacing and the length of the coupling. Tangential coupling makes use of the evanescent waves that are bound to the waves in the optical fiber but are traveling on the outside of the fiber. *Synonym* **pick-off coupling.** *See also* **butt coupling, evanescent field coupling, evanescent wave, loose tube splicer.**

tap: 1. To draw energy from a circuit. **2.** To remove a part of the signal energy from a line, such as a wire, a coaxial cable, or an optical fiber. *Note:* Tapping may be authorized for purposes of establishing a signal line, or it may be clandestine, though clandestine tapping of optical fiber is difficult to accomplish and easily detected. **3.** The physical connection to a signal line that removes part of the signal energy in the line. *Note:* Examples of taps are branches, couplers, and induction coils. **4.** To monitor, with or without authorization, the information that is being transmitted in a communications circuit. **5.** In fiber optics, a device for extracting a portion of the optical signal from an optical fiber.

tape: *See* **call tape, carriage control tape, chadless tape, chad tape, computer tape, magnetic tape, punched tape, punch tape, Wheatstone tape.**

tape drive: *See* **magnetic tape drive, streaming tape drive.**

tape handler: *See* **magnetic tape handler.**

tape leader: The portion of tape that (a) precedes the beginning of tape mark and (b) is used to thread the tape on a tape handler or reel. *See* **magnetic tape leader.**

tape loop storage: *Synonym* **looped tape storage.**

tape mark: *See* **beginning of tape mark, end of tape mark.**

tape perforator: A device that records signals on paper tape by creating patterns of holes that (a) correspond to the signals and (b) are punched in accordance with a predetermined code.

taper: *See* **optical taper.**

tape reader: A device with a tape reading head that (a) senses hole patterns or cuts in a perforated tape, and (b) produces corresponding electrical signals in accordance with a predetermined code. *See* **coupled reperforator tape reader.** *See also* **magnetic head, tape reading head.**

tape reading head: A device in an automatic transmitter that explores a tape and produces electrical signals that correspond to the hole patterns that are punched in the tape in accordance with a predetermined code. *See also* **magnetic head, tape reader.**

tape recording: 1. The recording of audio or video signals on magnetic tape. **2.** The recorded audio, radio, or video signals on magnetic tape, including the tape itself. *See* **streaming tape recording.**

tapered lens: A lens with a cross section that has a greater edge thickness, i.e., a greater diameter, on one side than on the other.

tapered optical fiber: An optical fiber that has a cross section that increases or decreases with distance along the fiber axis. *Synonym* **optical taper.**

tapered plug connector: A fiber optic connector with a single optical fiber that uses a refractive index matching material between the fiber ends to reduce or eliminate fiber interface loss, i.e., gap loss. *See also* **gap loss, refractive index matching material.**

tape reel: A cylinder with flanges on which tape, such as magnetic or paper tape, may be wound between the flanges. *See also* **tape spool.**

tape relay: Pertaining to the retransmission of teletypewriter (TTY) traffic from one channel to another. *Note 1:* In tape relay operation, messages arriving on an incoming channel are recorded on perforated tape. Then the tape either is (a) directly and automatically fed into an outgoing channel, or (b) manually transferred to an automatic transmitter for transmission on an outgoing channel. *Note 2:* The reception and the retransmission of messages in tape form in teletype-teleprinter systems may be accomplished via manual, semiautomatic, or fully automatic relay stations. *See* **torn tape relay.** *See also* **automatic tape relay system, chadless tape, chad tape, reperforator.**

tape relay pilot: Instructions that (a) are placed in a message, such as on format line 1 of a standard message, and (b) pertain to the transmission or the handling of the message.

tape relay switching: At a station, message switching accomplished by moving perforated tapes from incoming tape perforators to outgoing tape readers, i.e., tape transmitters. *See* **semiautomatic continuous tape relay switching.**

tape relay system: *See* **automatic tape relay system, torn tape relay system.**

tape retransmitter: A device that (a) automatically records on tape, such as magnetic or paper tape, messages arriving on incoming lines, and (b) retransmits the messages on outgoing lines. *See* **perforated tape retransmitter.**

tape spool: A cylinder without flanges on which tape, such as magnetic or paper tape, may be wound. *See also* **tape reel.**

tape storage: *See* **looped tape storage, magnetic tape storage.**

tape trailer: The portion of tape that (a) follows the end of tape mark and (b) is used to retain the tape on a tape handler or reel. *See* **magnetic tape trailer.**

tapoff: *See* **optical tapoff.**

TARE: telegraph automatic relay equipment.

target: *Synonym* **object.**

target code: *Synonym* **object code.**

target indicator: *See* **coherent moving target indicator, moving target indicator.**

target language: *Synonym* **object language, object program.**

target resolution: *See* **borescope target resolution.**

target signature: The characteristic pattern of a target, i.e., of an object, that (a) is displayed or recorded by detection and identification equipment, (b) usually is an electromagnetic or sonic pattern, (c) may be actively launched by or reflected from a target, and (d) may be subjected to signature analysis. *See also* **signature analysis.**

tariff: The rates or the charges for a specific unit of equipment, facility, or type of service, such as a service provided by a telecommunications common carrier. *Note:* Tariffs usually are contained in published schedules. *See also* **access charge, common carrier, measured rate service, message unit, mileage, overtime period.**

TASA: Telefónica de Argentina.

TASI: time assignment speech interpolation.

tasking: *See* **multitasking.**

tau jitter: *Synonym* **dither.**

Tbps: terabits per second.

TC: thermocouple.

T carrier: A generic designator for any of several digitally multiplexed telecommunications carrier systems. *Note 1:* The designators for T carrier in the North American digital hierarchy correspond to the designators for the digital signal (DS) level hierarchy. *Note 2:* T carrier systems originally transmitted only digitized voice signals. Applications also include digital data transmission. *Note 3:* If an "F" precedes the "T," an optical fiber cable system is indicated at the same data signaling rates (DSRs). *Note 4:* The North American and Japanese T-carrier hierarchies are based on multiplexing 24 voice frequency channels and multiples thereof. The European hierarchy is based on multiplexing 30 voice frequency channels and multiples thereof. *Refer to* **Tables 6, 7, 8, 10, Appendix B.**

TCB: trusted computing base.

TCF: technical control facility.

T coupler: *Synonym* **tee coupler.**

TCP: transmission control protocol.

TCP/IP Suite: Transmission Control Protocol/ Internet Protocol Suite.

TCS: trusted computer system.

TCU: teletypewriter control unit.

TDC: theater deployable communications.

TDC system: theater deployable communications system.

TDM: time division multiplexing.

TDMA: time division multiple access.

TDTG: true date-time group.

TE: transverse electric.

tea laser: *Synonym* **transversely excited atmosphere laser.**

TECHEVAL: technical evaluation.

technical area: In communications, computer, data processing, and control systems, an area in which temperature, humidity, or access is controlled because it contains equipment, such as communications, computing, control, and support equipment, that requires such controls.

technical control facility (TCF): A physical plant, or a designated and specially configured part thereof, that (a) contains the equipment necessary for ensuring fast, reliable, and secure exchange of information, (b) usually includes (i) distribution frames and associated panels, jacks, and switches and (ii) monitoring, test, conditioning, and orderwire equipment and engineering circuits, and (c) allows telecommunications systems control personnel to (i) exercise operational control of communications paths and facilities, (ii) make quality analyses of communications and communications channels, (iii) monitor operations and maintenance functions, (iv) recognize and correct deteriorating conditions, (v) restore disrupted communications,

(vi) provide requested on-call circuitry, and (vii) take or direct such actions as may be required and practical to provide effective telecommunications services. *See also* **communications center, engineering circuit, facility, patch and test facility.**

technical control hubbing repeater: *Synonym* **data conferencing repeater.**

technical evaluation (TECHEVAL): Evaluation of system performance and capability based on system design and component test results that usually are obtained under controlled conditions and environments. *See also* **operational evaluation.**

technical intelligence: Information pertaining to foreign technology, particularly information on foreign military equipment, such as weapons, aircraft, ships, munitions, and vehicles.

technical load: The portion of the operational load required for communications, tactical operations, and ancillary equipment, including necessary lighting, air conditioning, or ventilation required for full continuity of communications. *See* **critical technical load, noncritical technical load.** *See also* **load, operational load, station load, tactical load.**

technical monitoring: The monitoring of signals at selected points in a communications system, such as the received and transmitted signals at various points in an Earth station.

technical vulnerability (TV): In information systems, a hardware, software, or firmware weakness, or design deficiency, that leaves a system open to assault, harm, or unauthorized exploitation, either externally or internally, thereby resulting in an unacceptable risk of information compromise, information alteration, or service denial. *See also* **electromagnetic vulnerability.**

technique: *See* **bandspreading technique, coherent optical adaptive technique, cutback technique, near field scanning technique, rod in tube technique.**

technological penetration: In communications, computer, data processing, and control systems security, a penetration that is effected by circumventing or nullifying hardware or software access controls rather than by subverting persons, such as operators, attendants, programmers, clerks, administrators, users, maintenance personnel, and janitorial personnel. *See also* **penetration.**

technology: *See* **computer aided software engineering technology, information technology, monolithic technology.**

technology satellite: *See* **advanced communications technology satellite.**

technology transfer: The transfer of government technical information, particularly information resulting from space and military systems research and development programs, for industrial and commercial exploitation and application. *Note:* Examples of technology transfer include transfer of (a) the results of early research and development information on electronic computers for military systems, (b) technology of communications systems developed under military and space programs, and (c) information on special materials developed in the federal space programs.

TED: trunk encryption device.

tee coupler: 1. A passive coupler that has three ports, i.e., a three-port star coupler. **2.** In fiber optics, a coupler that (a) connects three ports, (b) operates in such a manner that (i) optical power into any one port can be distributed to the other two ports, or (ii) power into any two ports can be combined and sent out on the third port, and (c) may be combined with another similar coupler to form a single unit for input and output signals in both directions of propagation. *Synonyms* **bifurcation connector, T coupler.** *See also* **coupler, coupling, nonreflective star coupler, reflective star coupler, star coupler.**

TEK: traffic encryption key.

telautography system: A telegraph system, primarily intended for the immediate transmission and reproduction of handwritten messages, in which the transmitted signals (a) represent the positions of a succession of points on a document, and (b) are to be reproduced, by the receiver, in the form of a copy of the handwritten message. *Synonym* **telewriting system.** *See* **telewriter.**

teleaction service: In Integrated Services Digital Network (ISDN) implementations, a telecommunications service that (a) uses very short messages with very low data transmission rates between the user and the network, (b) is computer-based, (c) usually operates at a low data rate, and (d) provides the user, consumer, or operator with conveniences that save time and effort in managing various everyday fragmented and miscellaneous transactions, such as merchandise checkout and inventory control, electronic funds transfer, automatic teller machine operation, lottery transactions, home and office security and surveillance, remote meter reading, and remote control of home and office appliances and equipment, such as radio-controlled garage doors.

telecom center: *Synonym* **telecommunications center.**

telecom control: *Synonym* **telecontrol.**

telecommand: The use of telecommunications for the transmission of signals to initiate, modify, or terminate functions of equipment at a distance. *See* **space telecommand.**

telecommand space station: *See* **space telecommand space station.**

telecommunication: Any transmission, emission, or reception of signs, signals, writings, images, sounds, or any other representation of information of any nature by sonic, electrical, electronic, or electromagnetic means, such as by wire, radio, video, or visual systems. *Note:* Telecommunications transmissions usually are made over long distances, often worldwide distances, such as intercity, interstate, and international distances. *See* **National Security and Emergency Preparedness telecommunications.** *See also* **automatic data processing, communications.**

telecommunications: Pertaining to the science and the technology devoted to the transmission, emission, or reception of signs, signals, writing, images, or sounds that represent information or intelligence of any nature by any communications system, such as wire, radio, or optical communications systems. *See* **personal telecommunications, Universal Personal Telecommunications.**

telecommunications access: *See* **virtual telecommunications access.**

telecommunications administration: The governmental agency responsible for overseeing telecommunications systems in a given country.

telecommunications cable: *See* **fiber optic telecommunications cable.**

telecommunications carrier: An organization that (a) provides telecommunications services, (b) may be a private organization or a government agency, and (c) may serve the general public or a select group of users. *Note:* Examples of telecommunications carriers include the Teletypewriter Exchange Service (TWX®), the International Teleprinter Network (TELEX®), a common carrier, and the AUTOVON network of the U.S. Department of Defense.

telecommunications center (TC): A communications facility that (a) usually serves more than one organization or terminal, (b) accepts, transmits, receives, processes, and distributes incoming and outgoing messages, and (c) performs other telecommunications services and support functions as required. *Synonym* **telecom center.** *See also* **communications center, facility, technical control facility.**

telecommunications circuit: A circuit that (a) provides a means of two-way communication between two points and (b) consists of a go and a return channel, each with a forward and a backward direction capability. *See also* **backward direction, forward direction.**

telecommunications conference: A conference among persons who are (a) remote from one another and (b) linked by a telecommunications system, such as a television, telephone, or telegraph system. *Note:* Examples of telecommunications conferences are conferences in which conference calls or closed circuit television is used. *See also* **teleconference, teleconferencing.**

telecommunications device for the deaf: *See* **teletypewriter/telecommunications device for the deaf.**

telecommunications facility: 1. At a location in a telecommunications system, the aggregate of equipment, such as radios, telephones, teletypewriters, facsimile equipment, cables, and switches, used for providing telecommunications services, such as data, voice, and video services. **2.** In data transmission, the aggregate of equipment, such as radios, telephones, teletypewriters, facsimile equipment, cables, switches, and ancillary, auxiliary, and support equipment, used for various modes of transmission, such as digital data, audio, and video transmission, including distribution systems and communications security facilities, as used and defined in the Federal Resources Management Regulations.

Telecommunications Industries Association (TIA): An association that (a) is accredited by the American National Standards Institute (ANSI), (b) is concerned with matters of interest to telecommunications organizations, especially telecommunications equipment manufacturers, and (c) develops telecommunications standards, such as standards for personal communications services (PCS). *See also* **personal communications service.**

telecommunications laser: A laser that (a) is used as a carrier wave source, and (b) has an output that can be modulated to transmit data in free space over long distances, such as between towers on mountains or between Earth stations and satellite stations. *See also* **Earth station, free space, laser.**

telecommunications laser beam: A laser beam that (a) is generated by a telecommunications laser, (b) has a fixed frequency, i.e., fixed wavelength, (c) is used as a carrier wave, and (d) may be modulated to transmit information between the source station and a detector receiver.

telecommunications log: *See* **aeronautical telecommunications log.**

telecommunications management network (TMN): A network that interfaces with a telecommunications network at several points in order to receive information from, and to control the operation of, the telecommunications network. *Note:* Telecommunications management networks (TMNs) may use parts of the managed telecommunications network to provide for the TMN communications. *See also* **engineering circuit, orderwire.**

telecommunications network: All the lines and the equipment that (a) provide telecommunications services among geographic locations that are widely distributed, and (b) consist of at least end instruments, loops, switching centers, and trunks. *See* **civil telecommunications network, military telecommunications network.** *See also* **communications network, telecommunications system.**

telecommunications organization: *See* **maritime air telecommunications organization.**

telecommunications private operating agency (TPOA): A private agency that (a) operates a telecommunications service, facility, network, or system, and (b) generates signals that may interfere with the operation of systems operated by other agencies.

telecommunications radio service: *See* **basic exchange telecommunications radio service.**

telecommunications resources board: *See* **Joint Telecommunications Resources Board.**

telecommunications security: Communications security applied to telecommunications systems, facilities, equipment, and components. *See also* **communications security.**

telecommunications service: A service that (a) includes a specified set of information transfer capabilities, (b) is provided to a group of users by a telecommunications system, (c) holds the user responsible for the information content of the transmitted messages, and (d) is responsible for the acceptance, transmission, and delivery of the messages. *See* **consolidated local telecommunications service, public switched National Security and Emergency Preparedness (NS/EP) telecommunications service, wide area telecommunications service.**

Telecommunications Service Priority (TSP) Service: A regulated service provided by a telecommunications provider, such as an operating telephone company or a carrier, for National Security and Emergency Preparedness (NS/EP) telecommunications. *Note:* The Telecommunications Service Priority (TSP) service replaced Restoration Priority (RP) service in September 1990. *See also* **National Security and Emergency Preparedness telecommunications.**

Telecommunications Service Priority (TSP) System: A system that provides a means for telecommunications users to obtain priority treatment from service providers for the National Security and Emergency Preparedness (NS/EP) telecommunications requirements. *Note:* The Telecommunications Service Priority (TSP) System replaced the Restoration Priority (RP) system in September 1990. *See also* **National Security and Emergency Preparedness telecommunications.**

Telecommunications Service Priority (TSP) System user: Any individual, organization, or activity that interacts with the National Security and Emergency Preparedness (NS/EP) Telecommunications Service Priority (TSP) System.

telecommunications standard: *See* **federal telecommunications standard.**

telecommunications system: A system that (a) is used for communications among points separated by distances so great that communications equipment or systems are required, (b) has capabilities beyond those of a single communications system, (c) may be composed of several communications systems, (d) is delimited by a set of functional interface points that allow access to the system by users, (e) performs the basic communications system functions of acceptance, transmission, and delivery of messages, or handling of calls, and (f) performs functions over global distances. *See* **automated maritime telecommunications system, Federal Telecommunications System, terrestrial telecommunications system, Universal Mobile Telecommunications System.** *See also* **communications system, telecommunications network.**

telecommunications system architecture (TSA): In a telecommunications system, the overall plan and configuration governing the capabilities of functional units and their interactions, including the configuration, integration, standardization, lifecycle management, and definition of protocol specifications among these functional units.

telecommunications system management. Communications system management applied to telecommunications systems, facilities, equipment, and components. *See* **communications system management.**

telecommunications system operator (TSO): The organization, person, or operator of the system that provides telecommunications services to users.

telecommunications union: *See* **International Telecommunication Union.**

telecommunications union world region: *See* **International Telecommunication Union world region.**

telecommuting: Using telecommunications equipment, such as personal computers, remote terminals, communications networks, modems, data stations, telephone lines, radio links, and television systems, at a place of business to conduct business in a manner similar to what would be done if the contacted persons were present at the place of business, in lieu of their commuting to the site in person, such as to conduct business from the office, home, or car via a personal computer, modem, and telephone line or cellular link.

teleconference: 1. A conference among persons and machines remote from one another but linked by a telecommunications system. *Note 1:* The teleconference may be supported by any telecommunications equipment, such as telephone, video, facsimile, teletype, or radio equipment. *Note 2:* The teleconference may be conducted on a live real time or an interactive basis. **2.** The live exchange of information among persons and machines remote from one another but linked by a telecommunications system. *Note:* The teleconference is supported by a telecommunications system that provides audio, video, and data services by one or more real time means, such as telephone, radio, and television (TV). *Synonym* **telepresence.** *See* **video teleconference.** *See also* **telecommunications conference.** *Refer to* **Fig. V-4 (1067).**

teleconferencing: The act of conducting a teleconference. *See* **video teleconferencing.** *See also* **teleconference.**

teleconferencing unit: *See* **video teleconferencing unit.**

telecontrol: The use of a telecommunications system to control a remote system, equipment, or device. *Synonym* **telecom control.** *See* **radio telecontrol.**

telegram: 1. Information in written, printed, or pictorial form that is transmitted by telegraphy for delivery to the addressee. *Note 1:* Telegrams include radio telegrams, i.e., radiograms. *Note 2:* "Telegraph" and "telegraphy" have the same general meaning as defined by the 1979 General Worldwide Administrative Radio Conference Convention. **2.** The document or the spoken message containing the information that was sent, transmitted, received, and delivered by a telegraph service, including any reproduction of the message made in the course of transmission or delivery to the addressee. *See* **radio telegram.**

telegraph: A telecommunications system in which coded signals are used. *Note:* "Telegraph" and "telegraphy" have the same general meaning as defined by the 1979 General Worldwide Administrative Radio Conference Convention. *See* **multichannel voice frequency telegraph, speech-plus telegraph, voice frequency telegraph.** *See also* **code, polar direct current telegraph transmission, radio telegraphy.** *Refer to* **Figs. G-4 (404), T-3.**

telegraph alphabet: A convention that indicates the correspondences between telegraph characters and the coded signals that represent them.

telegraph alphabet 1: *See* **International Telegraph Alphabet 1.**

telegraph alphabet 2: *See* **International Telegraph Alphabet 2.**

telegraph alphabet 3: *See* **International Telegraph Alphabet 3.**

telegraph alphabet 4: *See* **International Telegraph Alphabet 4.**

telegraph alphabet 5: *See* **International Telegraph Alphabet 5.**

telegraph and telephone consultative committee: *See* **International Telegraph and Telephone Consultative Committee.**

telegraph automatic relay equipment (TARE): In a telegraph communications network, a store and forward device that has connections with telegraph terminals or with other relay equipment that are fixed, i.e., are permanently connected on a point-to-point basis.

telegraph channel: A channel that allows one-way transmission of telegraph signals. *See also* **telegraph circuit.** *Refer to* **Fig. T-3.**

telegraph circuit: A pair of telegraph channels that permits two-way simultaneous transmission between two points. *See also* **telegraph channel.** *Refer to* **Fig. T-3.**

telegraph code: 1. A group of rules and conventions according to which telegraph signals used to represent a message are formed, formatted, transmitted, and translated in a given system. **2.** The signals that result

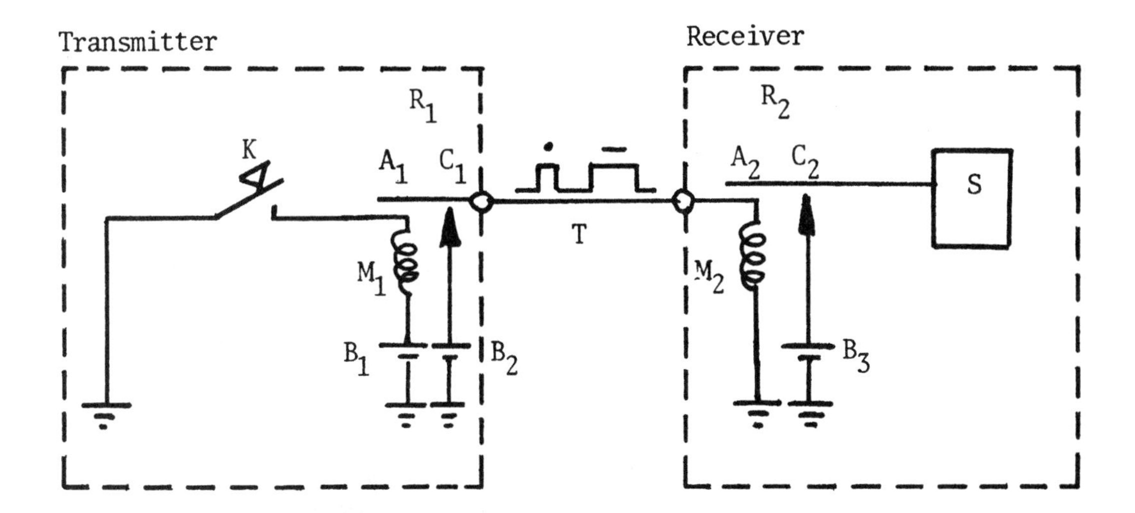

Fig. T-3. A telegraph channel, or one-half of a complete **telegraph circuit.** At the transmitter, depressing key K connects local battery B_1 to magnetic coil M_1, which attracts spring-loaded armature A_1, closing contact C_1 of relay R_1. This connects local battery B_2 to the transmission line T for as long as key K is depressed, say, for the length of a dot followed by the length of a dash for the Morse code for the letter "A." When a pulse arrives at the receiver, coil M_2 is energized, which attracts the armature A_2, closing contact C_2 of relay R_2. This connects local battery B_3 to sounder S for the duration of the pulse. A telegraph station usually has both a transmitter and a receiver, and perhaps a repeater. Two-way alternate communication can be accomplished on a one-wire line. Two-way simultaneous communication, i.e., duplex communication, requires two independent telegraph channels.

from the application of the rules and the conventions used to describe how a message should be formed, transmitted, and translated in a given system.

telegraph concentrator: Switching equipment that enables connections to be established and disestablished as required between existing circuits and telegraph sets. *Note:* In a given telegraph system, if the number of circuit terminations exceeds the number of telegraph sets, there may be a need for a concentrator.

telegraph conversation: An exchange, in real time, of messages between two telegraph users or operators.

telegraph demodulator: A demodulator that converts a received telegraph signal into another form for specific use, such as for input to an audio device, tape perforator, or printer.

telegraph discriminator: A discriminator that converts frequency shift telegraph transmission signals to direct current (dc) signals for such uses as telegraph relaying, audio signaling, tape perforation, printing, or sounder operation.

telegraph distortion: Distortion of telegraph signals, during either modulation or demodulation, i.e., during restitution, that usually is caused when significant intervals do not have their exact theoretical duration. *See also* **significant interval.**

telegraph distress frequency: *See* **international radiotelegraph distress frequency.**

telegraph error rate measuring (TERM) equipment: Test equipment that (a) is used to conduct loop tests in a telegraph data transmission system by substituting the test equipment for the normal telegraph transmit-and-receive equipment at the loop terminals, (b) generates telegraph test signals for transmission, and (c) compares the test signals with the restored output from the loop to measure the data error rate. *See also* **loop.**

telegraph keyboard: A device (a) that consists of an assembly of keys that is used to control the operation of a power source, such as a telegraph transmitter, and (b) in which each key, when actuated, causes the transmission of a different graphic character, space character, or transmission control character.

telegraph level: *See* **standard telegraph level.**

telegraph lifecraft frequency: *See* **international radiotelegraph lifecraft frequency.**

telegraph message format: *See* **radiotelegraph message format.**

telegraph modulation: The succession in time of significant conditions that correspond to a transmitted telegraph signal, such as when a high frequency carrier wave is modulated by an audio frequency signal, and both the carrier and the modulating signal are interrupted to form the dots and the dashes of a Morse code. *See also* **significant condition.**

telegraph modulator: A modulator controlled by a telegraph signal, such as a signal that represents (a) the dots and the dashes of the Morse code, (b) the code of the International Telegraph Alphabet Number 2, or (c) the code of the International Telegraph Alphabet Number 5, i.e., the American Standard Code for Information Interchange (ASCII) seven-bit code.

telegraph radioconverter: A device that accepts audio or intermediate frequency telegraph signals from a radio and converts, i.e., translates, them into rectangular pulses of constant amplitude that are capable of operating a telegraph receiver, i.e., a telegraph recorder. *Synonym* **telegraph radio recorder, telegraph recording unit.**

telegraph radio recorder: *Synonym* **telegraph radioconverter.**

telegraph receiver margin: The maximum value of signal distortion that (a) is compatible with correct translation, restitution, conversion, or demodulation by an end instrument, such as a teleprinter, telegraph receiver, or tape perforator, and (b) is determined when the signal arrives at the receiving end instrument under the most unfavorable conditions of frequency composition and distortion.

telegraph recording unit: *Synonym* **telegraph radioconverter.**

telegraph relay: 1. Store-and-forward telegraph transmission in which messages are first stored and then transmitted when the destination addressee or the next storage point in the network is available to receive them. **2.** An electromagnet that (a) has (i) a moving spring-loaded armature, (ii) a circuit making and breaking contact, and (iii) four terminals, two to the electromagnet coil and two to the circuit contact, (b) is designed to transfer signals in one circuit, to which the coil is connected, to another circuit, to which the contact is connected, and (c) is used primarily for repeating telegraph signals.

telegraph set: The set of equipment and parts that constitute a telegraph transmitter and a telegraph receiver, both usually to be used at one station.

telegraph signal: A signal that (a) is used to represent all or part on one or more telegraph messages, and (b) could range from a single dot, dash, or bit to an entire sequence of dots, dashes, or bits that represent the characters of a complete message.

telegraph system: A communications system that (a) transmits and receives messages, (b) uses telegraph equipment and components, and (c) consists of at least two telegraph sets, one at each of two locations. *See* **neutral direct current telegraph system, polarential telegraph system.** *See also* **telegraph set.**

telegraph transmission: *See* **polar direct current telegraph transmission, radio telegraph transmission.**

telegraph transmitter: A device that generates and transmits telegraph signals over a telegraph channel. *See also* **telegraph channel, telegraph signal.** *Refer to* **Fig. G-4 (404), T-3 (991).**

telegraph word: A group of contiguous signals that represent a group of characters that are separated from other groups of characters by more spaces than are used to separate the characters from each other within the group. *Note 1:* Usually a telegraph word is considered to consist of six characters, i.e., five actual characters and a space character, which is about the average word length in English. Encrypted text usually is divided into groups of five characters each and a space between the groups. *Note 2:* Telegraph traffic capacity usually is expressed in words per minute.

telegraphy: 1. The information transmission that (a) uses a signal code, and (b) includes any process that usually provides long-distance transmission, reception, and reproduction of information. *Note:* "Telegraph" and "telegraphy" have the same general meaning as defined by the 1979 General Worldwide Administrative Radio Conference Convention. **2.** A system of transmission that (a) is used for the communication of information represented by symbols, such as letters and numerals, (b) uses symbols represented by a preestablished set of signals in the form of a code, (c) is used primarily for record communications, though it can also be used on a nonrecorded basis between operators, such as when the Morse code is used, (d) includes any and all processes that provide for the transmission and the reproduction at a distance of documentary matter, (e) is a telecommunications system, and (f) does not include pulse-code modulated (PCM) telephony. *See* **alphabetic telegraphy, automatic telegraphy, document facsimile telegraphy, facsimile telegraphy, four-frequency diplex telegraphy, frequency shift telegraphy, manual telegraphy, Morse telegraphy, mosaic telegraphy, picture facsimile telegraphy, printing teletypewriter telegra-**

phy, radio telegraphy, signal recording telegraphy, two-condition telegraphy.

telemetering: 1. The sensing or the measuring of a physical parameter or variable, such as temperature, pressure, wind velocity, or flow rate, and the transmitting of the measurement over long distances via a telecommunications link. **2.** The use of telecommunications facilities for transmitting, displaying, and recording data at a distance from the point at which the data were sensed or measured. *Note:* Usually, telemetering is accomplished by automatic telecommunications systems, such as radio communications in a fixed or mobile service. *See* **space telemetering, space telemetering Earth station.**

telemetering fixed station: 1. A fixed station at a source of information used for telemetering the information from the source to a receiving station. **2.** A fixed station that receives telemetered information from a source of information. *See also* **fixed station, telemetering.**

telemetering land station (TLS): 1. In aeronautics, an Earth-based station used in the flight-testing of manned or unmanned aircraft, missiles, or their major components. **2.** In flight telemetry, an Earth-based station used for telemetering to (a) a balloon, (b) a booster or a rocket, excluding a booster or a rocket in orbit about the Earth or in deep space, or (c) an aircraft, excluding a station used in the flight-testing of an aircraft. **3.** In surface telemetry, a fixed, Earth-based station that telemeters data to another Earth-based station. *See also* **aeronautical telemetry mobile station, flight telemetering land station, space telemetry, telemetry.**

telemetering mobile station: A mobile station whose emissions are used for telemetering information from a source of information to a receiver. *See* **aeronautical telemetering mobile station, flight telemetering mobile station.** *See also* **telemetering.**

telemetering space station: *See* **space telemetering space station.**

telemetering station: A fixed, land, or mobile station that transmits data from a source data generator to a receiver.

telemetry: 1. The branch of science and technology devoted to the study, development, and implementation of methods for (a) measuring the values of variables, such as pressure, temperature, humidity, radiation, and sound levels, (b) transmitting the results of the measurements by means of a telecommunications system, and (c) interpreting, analyzing, indicating, displaying, recording, or using the information that is obtained. **2.** The use of telecommunications for automatically indicating or recording measurements at a distance from the measuring instrument. *See also* **radio telemetry, space telemetry.**

telemetry command: An instruction or control signal sent from one point to another by means of telemetered signals. *Note:* In satellite communications systems, a telemetry command may be sent from an Earth station satellite control facility to a satellite station to adjust its orbit, or a telemetry command may be sent from a satellite station to an Earth station for relaying to another satellite station.

telemetry command station: An Earth station that (a) receives telemetered data from satellites, (b) transmits instructions, i.e., commands, to satellites, (c) is separate from communications stations, (d) operates on a frequency and performs functions that are different from those of communications stations, (d) tracks and maintains satellites in orbit, (e) sets up access channels, and (f) with others, may be controlled from a central station, usually one that has extensive computer facilities. *See* **tracking telemetry command station.**

telemetry system: A system, usually consisting of (a) a sensor, i.e., a data source, that measures the absolute values and the variations of a physical variable, such as acceleration, pressure, temperature, fluid flow, humidity, radiation, sound, smoke, flame, or toxic fumes, (b) a data sink that receives, converts, displays, or otherwise processes the data, such as a photodetector and an indicator, and (c) a data link that connects the source to the sink. *Note:* In most telemetry systems, the sensor is at a remote or inaccessible location, and the sink is at another location, where the data can be interpreted, analyzed, displayed, recorded, or used. In some systems, signals can be sent from the sink, or another location, to (a) reset, calibrate, test, or otherwise control the sensor, or (b) control the sensor vehicle, such as a satellite, radio-controlled vehicle, or radio-controlled aircraft. *See* **fiber optic telemetry system.**

telephone: 1. Pertaining to a communications system or device that makes use of sound and the magnetic, electrical, or electromagnetic representation of sound to transmit messages, i.e., handle calls. **2.** A user end instrument (a) that is used to transmit and receive voice or tone-coded messages via an interconnecting network, and (b) that consists of (i) a transmitter for converting sound waves into electrical signals, (ii) a receiver for converting electrical signals into sound waves, (iii) a dialing or selection device, such as a rotary dial or a pushbutton panel, (iv) signaling circuits, and (v) switches. *Synonym* **phone.** *See* **analog**

telephone, cellular telephone, electrically powered telephone, satellite telephone, sound-powered telephone, underwater telephone. *Refer to* **Fig. T-4.**

Fig. T-4. The Saturn Bt transportable Inmarsat B satellite **telephone,** which provides voice and data communications between areas where other communications are poor or nonexistent, into the international switched telephone and data network. (Courtesy Mackay Communications.)

telephone call: A call that (a) begins when a call originator goes off-hook, i.e., picks up, and initiates selection signals by using a rotary dial or pushbuttons, and (b) ends when the call receiver or the call originator goes on-hook, i.e., hangs up. *See* **radiotelephone call.** *See also* **call, call originator, call receiver, dialing, off-hook, on-hook.**

telephone circuit security: Communications security that (a) is applied to a telephone circuit, (b) is a measure of the extent to which a telephone circuit is immune from monitoring, i.e., from tapping, by unauthorized persons or equipment on an effective basis, and (c) when a maximum, implies nonsusceptibility to unauthorized monitoring of calls.

telephone circuit signaling: In a telephone system, signaling that (a) provides the information necessary to promote the orderly transfer of call traffic, (b) informs users of the progress of a call, (c) provides signals to equipment that controls the progress of a call, and (d) may be in one of the three categories of signaling in the processing and progress of a call, i.e., supervisory, control, and status signaling. *See also* **signaling.**

telephone connection: *Synonym* **telephone cord circuit.**

telephone consultative committee: *See* **International Telegraph and Telephone Consultative Committee.**

telephone cord circuit: A device that (a) consists of one or more plugs and their associated flexible conducting cords, and (b) enables the connection of telephone circuits, i.e., enables the connection of loops to each other, loops to trunk circuits, and trunk circuits to each other. *Synonym* **telephone connection.**

telephone directory: An alphabetical listing of the names of users, including persons, processes, equipment, places, organizations, and services, of a telephone system, each name being followed by a telephone number and perhaps a town and street address. *See also* **street address, telephone number.**

telephone distress frequency: *See* **international radiotelephone distress frequency, radiotelephone distress frequency.**

telephone exchange: In a telephone system, an exchange that is primarily used for switching calls between other exchanges. *See also* **exchange, switching center.**

telephone exchange director: A telephone exchange register and translator that (a) accepts all or part of a call receiver number in the form of strings of dialed pulses, and (b) controls the setting up of a call by transmitting the strings of pulses used for the code, tandem, numerical, and final selectors in succession, for establishing a connection to a telephone circuit or trunk. *See also* **selector.**

telephone frequency: The band of frequencies used in telephone systems to carry voice signals, ranging from about 300 Hz (hertz) to 3400 Hz. *See also* **audio frequency, voice frequency.**

telephone number: The number that is assigned to a telephone user, i.e., customer or subscriber, for routing telephone calls. *Note:* For example, the telephone number for the Institute for Telecommunication Sciences is (303) 497-3000.

telephone patch: 1. In a telephone network, a temporary connection between circuits or systems. **2.** A connection that (a) is between the air-ground communications link and the local or long-distance telephone system, and (b) usually is made at a specific radiotelephone station. *Synonym* **phone patch.**

telephone receiver: The part of a telephone end instrument that (a) serves as a receiver of electrical signals, (b) converts the signals into sound signals, and (c) is placed near the ear for listening, or after amplification, may be connected to a loudspeaker. *See also* **end instrument, telephone, telephone transmitter.**

telephone service: A service provided by a common carrier telephone network or by a user. *See* **Federal Secure Telephone Service, inward wide area telephone service, plain old telephone service, wide area telephone service.** *See also* **service feature.**

telephone service class: A type of telephone service in which (a) certain classes of users are authorized to conduct certain types of business and are allowed access to certain networks, and (b) the class of user, the type of business, and the networks that are accessed determine the telephone service class. *See also* **class of service.**

telephone sidetone: In a telephone, the transmission and the reproduction of undesirable sounds through a local path from the transmitting transducer to the receiving transducer of the telephone while a person is talking or while other sounds are entered into the transmitter. *Synonym* **sidetone.**

telephone system: A communications system that is set up for the transmission of sounds, such as speech and coded sound signals. *See* **automatic telephone system, key telephone system, precedence telephone system.** *Refer to* **Figs. T-1 and T-2 (983).**

telephone traffic metering device: A device that (a) may be connected to telephone circuits and switches, and (b) is used to automatically record or count the occurrence or frequency of specific events, such as circuit uses, access attempts, and contacts to busy lines.

telephone transmission: *See* **radio telephone transmission.**

telephone transmitter: The part of a telephone that (a) serves as a transmitter of electrical signals, (b) converts the sound signals it receives into the electrical signals that it transmits, (c) is spoken into by the user or accepts other tones or sound waves, and (d) is similar to a microphone. *See also* **end instrument, microphone, telephone, telephone receiver.**

telephone unit: *See* **key telephone unit, secure telephone unit.**

telephony: 1. The branch of science and technology devoted to (a) the long-distance sensing, transmission, reception, and reproduction of sounds, such as speech and tones that represent digits for signaling, (b) transmission via any medium, such as wire, optical fibers, microwave, or radio, (c) the analog representation, transmission, and reception of sounds, (d) the digitalization of analog signals for transmission and their reconversion to analog form on reception, and (e) the development and the use of signaling systems. **2.** Telecommunications systems, equipment, and components set up for the transmission and the reception of signals originating from sounds, such as speech or coded sound signals. *See* **cordless telephony, radio telephony, voice telephony.** *See also* **message, public switched telephone network.** *Refer to* **Figs. S-5 through S-7 (869, 870).**

telephoto: 1. Pertaining to pictures that are (a) taken at the object location, (b) converted to coded signals, (b) transmitted via a communications system to a location usually a long distance from the object, and (d) reconstituted as pictures at the distant location. **2.** Pertaining to pictures transmitted via a telecommunications system. *See also* **wirephoto.**

telephoto lens: An objective lens optical system that (a) consists of a converging lens, i.e., a positive lens, and a diverging lens, i.e., a negative lens, separated from each other, each having such power and separation that the back focal length of the entire system is small in comparison with the equivalent focal length, and (b) is used for producing large images of distant objects with no need for a cumbersome length of the optical system. *See also* **convergent lens, divergent lens, lens, objective.**

telepresence: *Synonym* **teleconference.**

teleprinter: 1. A device that (a) receives data in the form of electrical signals, (b) prints the data represented by the signals, (c) is similar to a typewriter without a keyboard, and (d) does not transmit data. **2.** A device that (a) consists of a keyboard transmitter and a printing receiver, (b) operates in accordance with a start-stop transmission system, the start-stop elements defining the beginning and the end of each character, (c) is a printing telegraph instrument that has a signal-actuated mechanism for automatically printing messages that are received in the form of electrical signals, (d) has a keyboard similar to that of a typewriter, and (e) is used to receive and send messages. *See* **radio teleprinter.** *See also* **teletypewriter.**

teleprinter code: *Synonym* **International Telegraph Alphabet 2.**

teleprinter exchange service: A commercial communications service that (a) provides teleprinter communications services on the same basis as telephone service, (b) operates through a central switchboard to various stations within the same city, to other cities, or via worldwide networks, to other countries, (c) is limited to specific users of the service, and (d) enables users to communicate directly and temporarily among themselves by means of start-stop system equipment that makes use of public data transmission services or public data networks. *Note:* Examples of teleprinter exchange services are the Teletypewriter Exchange Service (TWX®) and the International Teleprinter Network (TELEX®). *Synonym* **teletypewriter exchange service.**

teleprinter network: *See* **International Teleprinter Network.**

teleprinter on multiplex: A teleprinter connected to multiplexed circuits. *Note:* Originally the teleprinter on multiplex converted the five-unit teleprinter code, i.e., the International Telegraph Alphabet 2, into a seven-unit code, i.e., the International Alphabet 5, thus permitting automatic error detection at the receiving end by using the extra bits for automatic error recovery. *See also* **error recovery.**

teleprinter on radio: A teleprinter connected to radiotelegraph circuits. *Note:* Originally the teleprinter on radio converted the five-unit teleprinter code, i.e., the International Telegraph Alphabet 2, into a seven-unit code, i.e., the International Alphabet 5, thus permitting automatic error detection at the receiving end by using the extra bits for automatic error recovery. *See also* **error recovery.**

teleprinting: Pertaining to a communications system that (a) transmits and receives messages in the form of signals, (b) uses teleprinters and associated equipment, (c) usually includes the use of perforated tape, manual keyboards, and printed pages, and (d) may use radio, wire, or fiber optic transmission systems. *Synonym* **teletypewriting.**

teleprocessing: **1.** The combining of telecommunications and computer operations that interact for the automatic processing, reception, and transmission of signals that represent information, such as analog and digital data. **2.** The overall function of an information transmission system in which telecommunications, computing, automatic data processing, and person-machine interface equipment are combined and controlled to operate as an integrated whole. **3.** Data processing by means of a combination of computers and sets of data terminal equipment connected by telecommunications systems, such as distributed data processing using communications systems. **4.** Remote access data processing in which certain input-output functions at different locations are connected by communications facilities.

teleradiology: The application of telecommunications systems to the practice of radiology, such as (a) transmission of X rays from remote or smaller medical facilities for real time diagnosis or diagnostic confirmation at larger medical centers, (b) transmission of archived X rays to smaller facilities for study and comparison, and (c) submission of X rays for archival storage. *See also* **medical diagnostic imagery system.**

telescopic aerial: *Synonym* **telescopic antenna.**

telescopic antenna: An antenna that can be extended or retracted, i.e., lengthened or shortened, by being constructed in the form of overlapping concentric cylinders that are able to slide inside of each other and make electrical contact with each other. *Note:* Submarines usually are fitted with telescopic high frequency (HF) and very high frequency (VHF) antennas that allow the transmission and the reception of signals at periscope depth by raising the antenna above the sea surface. Portable radios and motor vehicles may be equipped with telescopic antennas for convenience and for adjusting for wavelength. *Synonym* **telescopic aerial.**

teleseminar: *Synonym* **teletraining.**

teletex: An international store-and-forward, essentially error-free, communications service that (a) is defined by the International Telegraph and Telephone Consultative Committee (CCITT), (b) has a data signaling rate (DSR) of 2400 bps (bits per second) over switched telephone networks, and (c) has a communications protocol that supports the CCITT Group 4 facsimile service.

teletext: A one-way information service in which (a) a subscriber can receive data on a video display device, (b) data are transmitted to the subscriber display device over a common carrier channel, and (c) a proprietary video adapter unit is required for reception. *Synonym* **videotext.** *See also* **viewdata.**

teletraining: Training that (a) is accomplished at a distance through the use of telecommunications facilities, (b) may be accomplished on a point-to-point basis or on a point-to-multipoint basis, and (c) may assume many forms, such as a teleseminar, a teleconference, or an electronic classroom. *Synonyms* **distance training, electronic classroom, teleseminar.**

teletype (TTY): Pertaining to the general class of transmitting and receiving equipment that is used in telegraph systems, particularly keyboard, printing, tape perforating, tape reading, transmitting, and receiving equipment. *See* **radioteletype.** *See also* **Teletype.**

Teletype®: The trademark of The Teletype Corporation. *Note:* "Teletype" has been adopted to apply to the general class of transmitting and receiving equipment that is used in telegraph systems, particularly keyboard, printing, tape perforating, tape reading, transmitting, and receiving equipment. The common abbreviation for Teletype is TTY. *See* **radioteletype.** *See also* **teletype.**

teletype code: *Synonym* **International Telegraph Alphabet 2.**

teletype marking pulse: The significant condition of modulation that results in an active selected operation in a teleprinter or in the receiver portion of a teletypewriter. *See also* **modulation, significant condition.**

teletypewriter (TTY): A printing telegraph instrument that (a) has a signal-actuated mechanism for automatically printing received messages, (b) may have a keyboard similar to that of a typewriter for sending messages by manually depressing the keys, (c) may have a paper tape perforator for recording received messages, and (d) may have a paper tape reader to send messages recorded on perforated paper tape. *Note 1:* Radio circuits carrying teletypewriter (TTY) traffic are called "RTTY circuits" or "RATT circuits." *Note 2:* "Teletype" is a trademark of The Teletype Corporation. *See also* **radio teletypewriter, sending end crossfire, teleprinter.**

teletypewriter code: *Synonym* **International Telegraph Alphabet 2.**

teletypewriter communications: Communications in which teletypewriters, teletype circuits, radioteletypewriters, and related communications equipment are used. *Note:* "RATT" is used to designate teletypewriter communications over a radio link. "TTY" designates teletypewriter communications over other than radio links, such as open wire, twisted pair, fiber optic cable, coaxial cable, and microwave links.

teletypewriter control unit (TCU): A device that controls and coordinates operations between teletypewriters and message switching centers.

teletypewriter exchange (TTX) service: *Synonym* **teletypewriter service.**

Teletypewriter Exchange Service® (TWX): A teleprinter exchange service. *See also* **teleprinter exchange service.**

teletypewriter garble table: A table that (a) consists of a matrix of characters, and (b) is used as an aid in the interpretation of garbled text that may result from gains or losses of holes in tape or from picked-up or dropped pulses on circuits that are connected to a teletypewriter. *See also* **garble, garble table.**

teletypewriter network: A communications network that consists of a group of interconnected teletypewriter stations.

teletypewriter position: The physical space, electrical connections, and terminals that are used by a teletypewriter, a teleprinter, or both, at a station in a network for point-to-point, ground-air, ship-shore, or ship-to-ship teletype operations.

teletypewriter service: A service in which teletypewriters and teleprinters are connected as end instruments to a central office (C.O.) for access to other teletypewriters and teleprinters. *Note:* Commercial teletypewriter services are provided by the Bell operating companies, independent telephone networks, the Teletypewriter Exchange Service® (TWX), and the International Teleprinter Network® (TELEX). *Synonym* **teletypewriter exchange service.**

teletypewriter signal distortion: In a teletypewriter (TTY) system, the shifting of signal pulse transitions from their proper positions, i.e., from their proper significant instants, relative to the beginning of the start pulse. *Note:* The magnitude of the distortion usually is expressed in percent of an ideal unit pulse duration. *Synonym* **start-stop teletype distortion.** *See also* **cyclic distortion, distortion, end distortion, marking bias, signal, significant instant, spacing bias.**

teletypewriter/telecommunications device for the deaf (TTY/TDD): A telecommunications device for the deaf (TDD) in which teletypewriter (TTY) principles and technology are used.

teletypewriter telegraphy: *See* **printing teletypewriter telegraphy.**

teletypewriting: *Synonym* **teleprinting.**

television (TV): 1. Telecommunications used for the transmission of transient images of fixed or moving objects. *Note 1:* The picture signal usually is accompanied by the sound signal. *Note 2:* In North America, television (TV) signals are generated, transmitted, received, and displayed in accordance with the National Television Standards Committee (NTSC) standard. **2.** A system or a mode of telecommunication in which (a) an object, such as a picture, scene, or display image, is scanned to produce an analog signal that has an instantaneous value that corresponds to the light intensity at a scanned spot on the object, (b) the signal is used to modulate a carrier, (c) control signals, such as synchronizing signals, are superimposed, (d) audio signals are added, (e) the modulated carrier is transmitted to a receiver that uses the signal to modulate and control a display device that reconstructs and displays an image of the scanned object on a screen, (f) the screen may be part of a display device, such as a cathode ray tube (CRT), gas panel, light-emitting diode panel, or fiberscope, (g) the scanning rate is sufficiently rapid so as to enable the transmission and the reception of signals that represent changing pictures, scenes, display images, or objects in real time and online, (h) electromagnetic waves in free space, coaxial cables, and fiber optic cables may be used, and (i) satellite communications may be used for live real time worldwide coverage of ongoing events. *See* **advanced television, distribution quality television, extended definition television, freeze frame television.** *See also* **National Television Standards Committee (NTSC) standard.** *Refer to* **Fig. F-3 (342).**

television broadcast day: The scheduled period in which a television station transmits, usually during a single day. *Note:* An example of a television broadcast day is the period from 6:00 A.M. to 9:00 P.M. each day that the station is actively broadcasting. Stations usually indicate, i.e., announce, the beginning and the end of their television broadcast day, often along with their call sign, transmission power, location, and frequency, i.e., channel number, all as authorized by the Federal Communications Commission (FCC).

television broadcast translator: A translator that converts television programs broadcasted on a given channel, i.e., on a given frequency, to one or more other channels. *See also* **translator.**

television fiber optic link: A video signal link that (a) accepts an electrical video signal from a television camera, (b) converts it into an optical signal, (c) transmits it over a fiber optic cable, and (d) reconverts or reconstitutes it back to an electrical signal for use in a television monitor or a television set. *Synonym* **TV fiber optic link.**

television imagery: Imagery that (a) is acquired by a television (TV) camera, and (b) is (i) recorded on a video data medium on-site or (ii) telemetered electronically in real time and displayed by a television set. *See also* **imagery, telemetering, television set.**

television relay service station: *See* **cable television relay service station.**

television satellite: *See* **direct broadcast television satellite.**

television set: A device that dynamically displays pictures generated from received video signals consisting of modulated electromagnetic waves that arrive at the receiver antenna via free space or cable. *Synonym* **TV set.** *See also* **television.**

television set top box: An electronic or fiber optic device that (a) is connected to a television set, (b) controls various performance aspects of the television set, (c) performs various specific functions, such as (i) searches for desired channels, (ii) converts digitized signals into sound signals and video signals, i.e., picture signals, (iii) allows interaction with the set, and (iv) sends signals back down the cable to the signal source, (d) usually is placed on top of or near the television set, and (e) usually has a remote control. *Synonym* **TV set top box.** *See also* **television set.**

television system: 1. A system that (a) consists of one or more television cameras, transmitters, and television sets, and (b) generates, transmits, and displays video signals. **2.** Pertaining to a mode of television operation, such as a high definition television system. *See* **advanced television system, enhanced definition television system, high definition television system, improved definition television system.** *See also* **system, television.**

telewriter: The equipment that (a) is used in teleautography, in which the manually controlled movements of a stylus, i.e., a pen, over a plane surface electrically control similar movements of a writing pen at the receiving end, and (b) primarily is used for point-to-point rapid transmission of messages and sketches, such as between a machine shop and a design office or between two laboratories in the same or nearby buildings. *See also* **telautography system.**

telewriting system: *Synonym* **telautography system.**

TELEX: International Teleprinter Network.

Telex®: 1. The International Teleprinter Network. **2.** A telecommunications service in which teletypewriters and teleprinters are interconnected via automatic exchanges, central offices (C.O.s), and switching centers. *See also* **teleprinter exchange service, teletypewriter service.** *Refer to* **Figs. P-10 (740), S-5 and S-6 (869).**

Telex® call: *See* **radio Telex® call.**

tell: *See* **back tell, forward tell, lateral tell, overlap tell, relateral tell.**

telling: *See* **track telling.**

Telnet: The Transmission Control Protocol/Internet Protocol (TCP/IP) standard network virtual terminal protocol that (a) is used for remote terminal connection service, and (b) allows a user at one site to interact with systems at other sites as if that user terminal were directly connected to computers at those other sites.

TEM: transverse electromagnetic, transverse electromagnetic mode.

TEM mode: transverse electromagnetic mode.

TE mode: transverse electric mode.

temperature: A measure of the average kinetic energy of a moving atom or molecule of a substance, such that absolute zero temperature occurs when all molecular or atomic motion ceases. *Note 1:* In common terms, temperature is an indication of hotness or coldness of a substance using a defined scale. *Note 2:* The SI (Système International d'Unités) unit of temperature is the kelvin, which is one hundredth of the difference between the temperatures of the freezing point and the boiling point of water at sea level, with absolute zero at about −273.16 K. If the aforementioned points are 0 and 100 respectively, the scale is called "Celsius" or "centigrade." *See* **ambient temperature, antenna gain to noise temperature, antenna noise temperature, color temperature, effective input noise temperature, equivalent noise temperature, equivalent satellite link noise temperature, fictive temperature, front end noise temperature, luminance temperature, noise temperature, receiving system noise temperature, standard noise temperature, total radiation temperature.** *See also* **Celsius temperature scale, Fahrenheit temperature scale, Kelvin temperature scale.**

temperature-compensating equalizer: *See* **line temperature-compensating equalizer.**

temperature scale: *See* **Celsius temperature scale, Fahrenheit temperature scale, international temperature scale, Kelvin temperature scale.**

TEMPEST: Pertaining to the investigation, study, and control of compromising emanations, including conducted and radiated emanations, from electrical and electronic equipment, particularly communications, computer, data processing, and control systems and connected or related equipment. *Note:* "TEMPEST" is often used as a synonym for "compromising emanations," as in "TEMPEST test," "TEMPEST inspection," and "TEMPEST proof." *See also* **compromising emanations, electromagnetic emission control, electronic emission security, electronics security, electronic warfare, emanations security, RED/BLACK concept.**

TEMPEST inspection: An inspection designed to provide the means for conducting facility evaluations of emanations from electrical and electronic equipment, particularly communications, computer, data processing, and control systems and equipment, to determine the adequacy of emanation control measures, such as measures to control installation radiation levels.

TEMPEST proofed fiber optic cable: Fiber optic cable that has been tested and certified to meet U.S. National Security Agency and military standards in regard to emission levels of optical radiation.

TEMPEST test: A test that (a) may be conducted in a laboratory, in a production plant, or at an operational site, (b) is designed to determine the nature and the extent of conducted or radiated signals from electronic or electrical equipment, particularly from communications, computer, data processing, and control systems equipment, and (c) includes (i) detection and measurement of the signals, (ii) analyses of the signals to determine the extent to which they contain compromising information, i.e., information that should not be disclosed without authorization, and (iii) correlation of the detected signals, their strength, and the extent to which they could be detected by unauthorized persons with specialized detection equipment.

template: *See* **four-concentric-circle near field template, four-concentric-circle refractive index template.**

temporal application: In display systems and imagery, an application that (a) requires a high temporal resolution capability and (b) may occur at the expense of reduced spatial positioning precision, i.e., may result in reduced jerkiness. *Note:* An example of temporal application is the requirement to distinguish accurately such items as facial expressions and lip movements in display systems, such as videophone, television, slow-motion video, and full-motion video systems. *See also* **display system, imagery, spatial application.**

temporal coherence: *Synonym* **time coherence.**

temporally coherent light: *Synonym* **time-coherent light.**

temporally coherent radiation: *Synonym* **time-coherent radiation.**

temporary circuit: A circuit that (a) is required for a limited, usually short period, such as a few hours or days, (b) is kept in operation for a period longer than required for a single call or single message transmission, such as occurs in a switched connection, and

(c) is not kept in operation as long as a permanent circuit or connection. *See also* **permanent circuit.**

temporary storage: 1. In computer and data processing systems, storage reserved for intermediate results, usually during the execution of a computer program. **2.** In communications, computer, data processing, and control systems, storage that (a) is used to hold data ready for use, (b) holds data for a short time, such as during the execution of a particular series of operations, (c) usually is volatile, and (d) usually is cleared after the series of operations is completed and the final results are placed in more permanent storage. *Synonyms* **scratchpad storage, working storage.**

TEM wave: transverse electromagnetic wave.

tens complement: The complement of a decimal number, obtained by (a) adding 1 to the least significant digit of its nines complement and propagating the carries only as far as they go, or (b) subtracting if from a number that consists of a 1 followed by as many 0s as there are digits in the number whose complement is desired. *Note:* The nines complement of the decimal numeral 496544 is 503455, obtained by subtracting each digit from nine, to which 1 is added to produce 503456, the tens complement of 496544. The tens complement of 496544 may also be obtained by subtracting it from 1000000. Thus, the sum of a number and its tens complement will always be a number that consists of a 1 followed by as many zeros as there are digits in the number whose tens complement is desired. The sum of 496544 and its tens complement, 503456, is 1,000,000.

tensile load test: A test used to determine the characteristics of an entity, such as a wire, coaxial cable, optical fiber, fiber optic bundle, fiber optic cable, or connector, when subjected to elevated tensile stresses. *Note:* Tensile load tests may measure breaking points or transmission parameter changes, such as attenuation, dispersion, or elongation, under elevated tensile stresses.

tensile strength: A measure of the ability of an entity to withstand tensile stresses. *See* **optical fiber tensile strength.**

tensiometer: A gauge that (a) is used to measure tension, such as might be encountered in laying, paying out, or threading a cable, and (b) usually is a direct readout strain gauge. *See also* **threading.**

tension sensor: *See* **fiber optic tension sensor.**

tera (T): A prefix that indicates multiplication by 10^{12}, i.e., by (a) a million million or (b) in the United States, by a thousand trillion. *Note:* An example of the use of "tera" is in 10 THz (terahertz), equivalent to 10^{13} Hz.

terahertz (THz): 1. A unit that denotes a million million hertz, i.e., 10^{12} Hz (hertz). **2.** A frequency, cycle, or repetition rate of 10^{12} Hz (hertz). *Note:* Optical and visible frequencies are approximately 10^{15} Hz (hertz), i.e., 1 PHz (petahertz), for a wavelength of 0.3 μm (micron) in the region of ultraviolet light. The visible spectrum is from about 0.4 μm to about 0.8 μm. *See also* **frequency, frequency spectrum designation, metric system.** *Refer to* **Table 4, Appendix B.**

term: attenuation term, phase term.

terminal: 1. A device capable of sending, receiving, or sending and receiving, information over a communications channel. **2.** A communications facility that constitutes a point of origin or termination of a circuit or a channel. **3.** A collection of hardware and software that enables contact with, entry into, or exit from a system, such as a communications, computer, data processing, or control system or network. **4.** A point, such as a jaw, post, screw, or joint, at which an electrical connection can be made. *See* **computer terminal, customer office terminal, data processing terminal, enhanced manpack ultrahigh frequency terminal, extension terminal, intelligent terminal, intercepted terminal, multiple access rights terminal, multiplex baseband receive terminal, multiplex baseband transmit terminal, packet mode terminal, radio baseband receive terminal, radio baseband send terminal, satellite Earth terminal, user terminal.** *See also* **bit synchronous operation, called line identification facility, call release time, data circuit-terminating equipment, data terminal equipment, end instrument, input/output device, main station, passive station, peripheral equipment, port, virtual terminal.** *Refer to* **Figs. S-3 and S-4 (869).**

terminal access controller (TAC): A host computer that (a) accepts terminal connections, usually from dial-up lines, and (b) allows the user to invoke Internet remote log-on procedures, such as those of Telnet. *See also* **packet switching network.**

terminal adapter: An interface device that (a) is used at the "R" reference point in an Integrated Services Digital Network (ISDN) environment, (b) allows connection of a non-ISDN terminal at the physical layer to communicate with an ISDN network, and (c) usually will support standard RJ-11 telephone connection plugs for voice and RS-232C, V.35, and RS-449 interfaces for data. *See also* **Integrated Services Digital Network.**

terminal automatic identification: Automatic identification in which the answerback code is provided by the receiving terminal itself. *See also* **answerback code.**

terminal complex: *See* **Earth terminal complex.**

terminal emulator: An emulator (a) that enables a peripheral device, such as a personal computer, to operate with a host computer as though the peripheral device were just another input-output terminal or remote station of the host computer, and (b) that assists the host computer in the transfer of data files to and from the peripheral device.

terminal endpoint functional group (TEFG): A functional group that includes functions broadly belonging to Layer 1 and higher layers of the CCITT Recommendation X.200 Reference Model. *Note 1:* The functions of terminal endpoint functional group (TEFGs) are performed on various types of equipment, or combinations of equipment, such as digital telephones, data terminal equipment (DTEs), and integrated workstations. *Note 2:* Examples of functions performed by terminal endpoint functional groups (TEFGs) are protocol handling, maintenance, interface, and connection functions.

terminal engaged signal: A signal that (a) is sent in the backward direction and (b) indicates that a call cannot be completed because the called terminal connecting the user end instrument is engaged with another call, i.e., is busy. *See also* **backward direction, busy signal, completed call, end instrument.**

terminal equipment (TE): 1. An assembly of communications equipment used to transmit or receive signals on a channel, circuit, or link. **2.** In radio relay systems, equipment used at points where data are inserted or derived, as distinct from equipment only used to relay a reconstituted signal. **3.** Telephone and telegraph switchboards and other centrally located equipment at which wire and fiber optic circuits are terminated. **4.** Communications equipment that (a) constitutes data stations, (b) is at each end of a data link, and (c) is used to permit the stations to accomplish the mission for which the link was established. **5.** In a communications network, the facilities at a station. **6.** The equipment that constitutes a point of origin or termination of a line, circuit, channel, or data link. *See* **data subscriber terminal equipment, data terminal equipment, grandfathered terminal equipment.**

terminal exchange: In telephone switchboard operations, the exchange, i.e., the switchboard, to which a call receiver is directly connected, such as, in a series of exchanges connected in tandem, the last exchange or switchboard through which a call passes before reaching the call receiver end instrument. *Synonym* **terminal switchboard.**

terminal guidance navigation system: A navigation system that (a) uses a device that seeks, points to, or homes on a moving or stationary source of energy, such as a source of electromagnetic or sound energy, (b) establishes the direction from its present position to the source of radiation, (c) automatically steers the vehicle in which it is located in any desired direction with reference to the direction toward the source, and (d) if the device is in a missile, may direct itself to move in a course that will intercept the source.

terminal impedance: 1. At the output terminals of transmission equipment, such as a line, the impedance measured when no signals are being transmitted and the load is disconnected. **2.** At the output terminals of a device, the ratio of the voltage to the current with the load connected. **3.** The impedance that is connected across the output terminals of a device, i.e., the load impedance. *Note:* The impedance is always the complex impedance. Even a "pure" resistance and a "perfectly tuned" circuit have some reactive element, though perhaps a negligible one. *See also* **idle line termination, impedance.**

terminal instrument: A telecommunications device that provides a point of origin or termination of a circuit or a channel. *Note:* Examples of terminal instruments are (a) data terminal equipment (DTE), (b) a private branch exchange (PBX), and (c) a user end instrument, such as a telephone, teletypewriter, telegraph transmitter, telegraph receiver, facsimile transmitter, facsimile receiver, fiberscope, personal computer, or modem.

terminal port: In a data network, the functional unit at a node through which data can enter or leave the network. *Note:* The terminal port may be located at a node common to two or more networks, or it may be a designated station in another network for the purpose of transferring traffic between the networks. *Synonym* **transfer station.** *See also* **brouter, bridge, gateway, router.**

terminal/regenerator system gain: In a single mode fiber optic transmission system, the gain that (a) is represented by the transmitter optical power less the receiver sensitivity (power) and the optical power requirements of the terminal or regenerator station facilities, such as the dispersion power penalty, the reflection power penalty, the overall safety margin, and only the power losses in cables and connectors within the station facility at both ends, (b) is given by the relation

$G = P_T - P_R - P_D - R_P - M - U_{wdm} - l_{sm}U_{sm} - N_{con}U_{con}$ where P_T is the transmitter power, P_R is the receiver power sensitivity, P_D is the dispersion power penalty, R_P is the reflection power penalty, M is the overall safety margin, U_{wdm} is the worst case value of all losses associated with wavelength division multiplexing equipment at both ends, l_{sm} is the optical fiber length in the stations, U_{sm} is the worst case end of life loss (dB/km) of the single mode cable at both stations, N_{con} is the number of fiber optic connectors within the stations (inside plants), and U_{con} is the loss (dB) per connector, (c) is used to overcome all the losses, *L,* in the fiber optic cable facility, i.e., the outside plant, which lies between the two stations of the fiber optic station/regenerator section, and (d) must be equal to or greater than *L* for the section to function satisfactorily, i.e., the relation $G > L$ must hold, where *L,* the sum of all the losses in the fiber optic cable facility of the section expressed in dB, is given by the relation $L = l_t(U_c + U_{cT} + U_\lambda) + N_s(U_S + U_{ST})$ where l_t is the total sheath length, i.e., the total jacket length, of spliced fiber optic cable (km), U_c is the worst case end of life cable attenuation rate (dB/km) at the transmitter nominal central wavelength, U_λ is the largest increase in cable attenuation rate that occurs over the transmitter central wavelength range, U_{cT} is the effect of temperature on the end of life cable attenuation rate at the worst case temperature conditions over the cable operating temperature range, N_s is the number of splices in the length of the cable in the fiber optic cable facility, including the splice at the fiber optic station facility on each end and allowances for cable repair splices, U_S is the loss (dB/splice) for each splice, and U_{ST} is the maximum additional loss (dB/splice) caused by temperature variation. P_T and P_R are expressed in dB referenced to a common power unit, such as milliwatts, because they are single-ended in regard to optical power, but all the losses that contribute to *L* are double-ended and thus are simply expressed in dB. *See* **statistical terminal/regenerator system gain.** *See also* **fiber optic cable facility loss.**

terminal repeater: A repeater used at the end of a trunk or a line. *See also* **repeater.**

terminal station: 1. Receiving equipment and associated multiplex equipment that are used at the ends of a communications link. **2.** The last station in a sequence of stations used in routing a message from the message originator to the addressee, i.e., the station beyond which a given message is not routed.

terminal switchboard: *Synonym* **terminal exchange.**

terminated line: A line in which (a) the total impedance that is connected across its far end is equal to the characteristic impedance of the line, (b) no standing waves are present, and (c) no reflections occur when a signal is entered at the near end. *Note:* The total impedance connected across the far end of the line usually consists of the load impedance and another impedance connected so that the combination is equal to the characteristic impedance of the line.

terminating block: *See* **horizontal terminating block, vertical terminating block.**

terminating equipment: *See* **data circuit-terminating equipment.**

terminating interface: *See* **network terminating interface.**

terminating set: *See* **four-wire terminating set.**

termination: 1. The load connected to a transmission line, circuit, or device. *Note:* For a uniform transmission line, if the termination impedance, i.e., the impedance of the load at the end of the line, is equal to the characteristic impedance of the line, wave reflections from the end of the line will not occur. **2.** In waveguides, the point at which energy propagating in the waveguide continues in a nonwaveguide propagation mode into a load. *See* **idle line termination, line termination, optical waveguide termination.** *See also* **characteristic impedance, terminal impedance.**

termination efficiency test: In an optical fiber, fiber optic bundle, or fiber optic cable, a test designed to measure or evaluate the loss in transmitted optical power.

termination point: *See* **service termination point.**

terminus: A device that (a) is used to terminate an electrical conductor or dielectric waveguide, such as a coaxial cable or an optical fiber, and (b) provides a means of positioning and holding the conductor or the waveguide within a connector. *Note:* In electrical systems, a contact may perform the function of a terminus. In fiber optics, use of the term "contact" is deprecated. *See* **fiber optic terminus.**

ternary signal: A signal that can assume, at any given instant, one of three significant conditions of the power level, the phase position, the pulse duration, the frequency value, or another parameter. *Note:* Examples of ternary signals are (a) a pulse that can have either a positive, zero, or negative voltage value at any given instant, (b) a sine wave that can assume phase positions of 0°, 120°, or 240° of electrical position with reference to a clock pulse or relative to the three phase positions, and (c) a carrier wave that can assume any one of three different frequencies, depending on three different modulation signal significant conditions. *See also* **significant condition, three-condition coding.**

terrain: *See* **antenna height above average terrain.**

terrain avoidance system: An electronic system that (a) provides a continuous indication of the relative distance or height of an aircraft from the terrain in front of and directly beneath the aircraft, (b) is used to assist the pilot or the autopilot in taking a proper course, including azimuth and altitude, such as to avoid crashing into a mountainside or flying too close to the ground, and (c) operates similarly to radar.

terrestrial flight telephone system (TFTS): An airborne telephone service that enables telephone calls to be made from aircraft to ground terminals and user end instruments via ground stations rather than satellites. *Note:* Terrestrial flight telephone system (TFTS) implementation might (a) provide standard equipment and service across national boundaries, (b) use time division multiple access (TDMA), (c) use time division multiplexing (TDM), (d) provide automatic ground-station call handoff, (e) use digital modulation, and (f) operate in the 1.6 GHz (gigahertz) to 1.8 GHz band. *See also* **handoff.**

terrestrial latitude: The angle (a) that lies between (i) the line from a given point on the Earth's surface to the Earth's center and (ii) the line from the equator, at the same longitude as the given point, to the Earth's center, (b) that is the central angle, at the Earth's center, measured from the equatorial plane to a line from the Earth's center through the point on the Earth's surface whose latitude is desired, (c) that is measured north or south from the equatorial plane in a longitudinal plane, i.e., in a meridian plane, and (d) that is designated as 0° at the equatorial plane and 90° at the poles.

terrestrial longitude: The angle (a) that is measured from the Greenwich, England, meridian along the Earth's equator, or in a plane parallel to the equatorial plane and along a parallel of constant latitude, east or west, to the meridian through the given point whose longitude is desired, (b) that is never greater than 180°, such that 180° West longitude is the same as 180° East longitude, (c) that usually is expressed in degrees, minutes, and seconds of angles east or west, and (d) that also may be expressed in hours, minutes, and seconds of time because Earth rotation is a constant, and thus time and angular displacement are directly proportional, i.e., are interchangeable through a constant of proportionality with an allowance for the 24-hour clock versus the limit of 180°.

terrestrial radiocommunications: 1. Radio communications in which transmissions are made from transmitters on Earth directly to receivers on Earth via ground waves, ionospheric reflection, line-of-sight transmission, scattering, ducting, and radio relay. *Note 1:* Terrestrial radiocommunication includes ship, aircraft, and land stations. *Note 2:* Terrestrial radiocommunications does not include space radiocommunications and radio astronomy. **2.** Radiocommunications other than space radiocommunications and radio astronomy.

terrestrial radio service: A radio service other than a space service or the radio astronomy service.

terrestrial service: A radio service other than a space service or the radio astronomy service.

terrestrial station: A station in a terrestrial radio service.

terrestrial system: In communications, a ground-based communications system that makes use of cables, land lines, tower-to-tower microwave relays, radio land stations, aircraft stations, and ship stations rather than satellite systems.

terrestrial telecommunications system: In telecommunications, a ground-based telecommunications system that (a) makes use of cables, land lines, tower-to-tower microwave relays, radio land stations, aircraft stations, and ship stations, and (b) is not a satellite, radio astronomy, or space radiocommunications system.

tessera: A curvilinear rectangle. *Note:* An example of a tessera is an area of the Earth's surface bounded by lines of constant longitude and latitude.

tesseral harmonic function: A spherical harmonic function, i.e., a trigonometric function, that has values that maintain a consistent sign within a tessera but change sign and magnitude as functions of both latitude and longitude.

test: 1. A set of measurements of the performance of a functional unit that are made to establish conformity with specifications or design objectives. **2.** Physical measurements that (a) are taken to verify conclusions obtained from mathematical modeling and analysis, (b) are taken for the purpose of developing mathematical models, and (c) usually include validation of the results. *See* **acceptance test, attenuation test, benchmark test, communications test, conformance test, engaged test, Foucault knife edge test, frequency response test, Functional Advanced Satellite System for Evaluation and Test, hot bend test, impact test, load test, loop test, marginal test, penetration test, shuttle pulse test, TEMPEST test, tensile load**

test, termination efficiency test, thermal shock test, vibration test.

test and validation: To perform tests, make measurements, and analyze data for (a) the verification of conclusions obtained from modeling, simulation, and analyses, or (b) the development of models and simulators. *See also* **acceptance test, quality assurance, validation.**

test antenna: An antenna of known performance characteristics used to determine the transmission characteristics (a) of equipment, such as antennas, transmission lines, transmitters, receivers, reflectors, and connectors, and (b) of associated propagation paths. *See also* **antenna.**

test bed: 1. A test environment, including the hardware, instrumentation, tools, emulators, simulators, computer programs, and other supporting software necessary for testing a system or its components. **2.** A prototype of the ultimate production system, operating in the same environment that the production system is being designed to operate in.

test board: A switchboard that (a) contains test equipment, such as signal sources, response indicators, and volt-ohm meters, and (b) is arranged in such a manner that connections can be made to telephone lines, circuits, trunks, exchanges, and central office equipment for test purposes, particularly via lines from the test points to the switchboard. *See also* **test points.**

test center: In communications, computer, data processing, and control systems, a location from which various system and component tests can be conducted. *See also* **patch and test facility, technical control facility, test point.**

test chart: *Synonym* **test pattern.**

tester: *See* **bit error ratio tester.**

test facility: *See* **patch and test facility.**

testing: The operation of a system to determine whether system performance in a given environment is suitable for a specific purpose. *See* **acceptance testing, bottom-up testing, integration testing, interface testing, qualification testing, regression testing, system testing, top-down testing.**

testing station: *See* **experimental testing station.**

test land station: *See* **radionavigation maintenance test land station, radionavigation operational test land station.**

test loop: In a communications system, a loop that (a) is formed by temporarily linking complementary equipment, such as modulator-demodulator, transmitter-receiver, or input-output devices, (b) may be used to test certain parts of a communications system, such as remote parts that may be isolated while local units are being tested, and (c) enables testing to take place without the tester's actually going to remote sites or distant locations with the test equipment.

test loop translator: A translator that is used in a test in which (a) the translator is fed with a sample of a transmitted signal that it translates to another frequency, (b) the new frequency is then fed to a receiver that is being tested, and (c) the receiver that is being tested also receives the originally transmitted frequency in a test loop configuration, thus enabling analysis of the performance of the transmitter and the receiver. *See also* **translator, turnaround mixer.**

test method: *See* **alternative test method, fiber optic test method, reference test method.**

test mode: *See* **adjust text mode, static test mode.**

test pattern: In display systems, a conventional or standard drawing with precise characteristics, usually with lines at various angles and spacings, used as a reference for evaluating the quality of images, the resolving power of imaging systems, or the distortion introduced by optical or other means of image processing, such as by spherical aberration, inverse filtering, or recursive filtering. *Synonym* **test chart.** *See also* **eye pattern, inverse filtering, recursive filtering, spherical aberration.**

test plan: A management plan, or the document in which the plan is described, that (a) consists of the testing approach that will be taken to determine whether a system or a component meets specified requirements, and (b) usually includes (i) the equipment required to perform the test, (ii) the location of the test, (iii) the resources required to perform the test, (iv) a detailed description of the steps of the actual test to be performed, (v) the schedules to be met, (vi) the reporting requirements, (vii) the evaluation criteria for judging the results of the test, and (viii) any other significant matters related to the conduct of the test, such as the personnel designated to conduct it.

test point: In a communications, computer, data processing, or control system, a point, i.e., a terminal, that (a) enables access to signals, (b) may be used to check levels, detect faults, locate faults, or isolate faults, and (c) may be used to perform troubleshooting. *See also* **fault, level, point, terminal.**

test procedure: Detailed instructions that (a) call for the setup, operation, and evaluation of the results of one or more tests of a system, equipment, or a component, and (b) usually is contained in the test plan. *See* **fiber optic test procedure.** *See also* **test plan.**

test reliability: The reliability of hardware or software in a test environment rather than in an operational, artificial, or hypothetical environment. *See also* **operational reliability.**

test repeatability: In a given test, a measure of the extent to which the test produces the same results each time it is performed.

test set: optical loss test set.

test signal: A signal used for testing the ability of equipment to effectively handle signals. *Note:* In radiotelegraph systems, an example of a test signal is a signal that is transmitted for testing transmitters and receivers. One standard test signal consists of not more than three groups of three V's followed by the transmitter call sign and terminated by the letters AR. *See* **composite two-tone test signal, standard test signal.**

test specimen: An item selected, usually using a specified sampling scheme, for testing.

test station: *See* **flight test station.**

test time: *See* **system test time.**

test tone: A tone sent at a predetermined level and frequency through a transmission system for test purposes, such as for (a) facilitating measurements and (b) aligning gains and losses in the system. *See* **standard test tone.** *See also* **standard test signal, standard test tone.**

TETRA: trans-European trunked radio system.

tetrad: *Synonym* **quartet.**

TE wave: transverse electric wave.

text: In a message, the words that (a) are used to describe the thought or the idea that is desired to be communicated from the message originator to the addressee, and (b) consist of a sequence of characters, words, phrases, and sentences. *Note:* In a radiotelephone system message, the text may be preceded and followed by the prosign BREAK. In a radiotelegraph and tape relay system message, the text may be preceded and followed by the prosign BT. *See* **cipher text, hypertext, message text, plain text, video text.** *See also* **prosign.**

text character: *See* **end of text character, start of text character.**

text processing: *Synonym* **word processing.**

TFEL: thin film electroluminescent.

TFTS: terrestrial flight telephone system.

TG: telegraph.

TGM: trunk group multiplexer.

THAAD: theater high altitude area defense.

THD: total harmonic distortion.

theater-area command: A military organization that (a) usually is operated and controlled by national authorities, (b) usually is designed and organized so that it may be transferred to an international organization at the appropriate time, and (c) is responsible for military operations, functions, and activities, including communications activities. *Note:* The theater director of communications-electronics (C-E) is responsible to the commander of the theater-area command.

theater deployable communications system: A switched communications system that (a) provides land-based forces with tactical voice and data communications capabilities, including wired and wireless circuits, packet switching, message switching, inter- and intrabase transmission and multiplexing, and technical control, and (b) uses personal communications system (PCS) equipment.

theater director of communications-electronics: An officer who (a) is appointed by the theater-area command, and (b) is responsible for coordinating communications-electronic (C-E) activities in the theater or the area.

theorem: four-color theorem, Nyquist's theorem, Parseval's theorem, sampling theorem.

theoretical duration: *See* **significant interval theoretical duration.**

theoretical margin: The margin calculated (a) from the design objectives, specifications, or construction data of a device, such as a telegraph receiver or a radio transmitter, and (b) under the assumption that the device is operating under ideal conditions. *See also* **margin.**

theoretical resolving power: The maximum possible resolving power determined by diffraction, frequently measured as an angular resolution and given by the relation $A = 1.22\lambda/D$ where A is the limiting resolution angle in radians, λ is the free-space wavelength of the light at which the resolution is determined, and D is the effective diameter of the aperture. *See also* **aperture.**

theory: *See* **electromagnetic theory, information theory.**

thermal blanket: In satellite systems, such as satellite communications systems, an insulating material that (a) is used in the construction of a satellite, and (b) is placed on the outside surface or around internal equipment for protection of the satellite and onboard equipment against environmental conditions, such as direct sunlight.

thermal noise: In a conductor, the noise that (a) is generated by thermal agitation of charge carriers, such as electrons and ions, (b) is given by the relation $P = kT\Delta f$ where P is the noise power in watts, k is Boltzmann's constant in joules per kelvin, T is the conductor temperature in kelvins, and Δf is the bandwidth in hertz, and (c) per hertz, is uniformly distributed throughout the electromagnetic frequency spectrum, depending only on Boltzmann's constant and the temperature. *Note:* Thermal noise was first described by Nyquist in 1928. *Synonyms* **Johnson noise, Nyquist noise.** *See also* **bandwidth, effective input noise temperature, noise, noise figure, thermal radiation.**

thermal-noise-limited operation: The operation of a photodetector in which the minimum detectable signal is limited by the thermal noise of the detector, the load resistance, and the amplifier noise.

thermal radiation: 1. Electromagnetic emission that (a) is emitted from a source as a consequence of its temperature, and (b) may consist of ultraviolet, visible, and infrared radiations, such as those produced by an incandescent bulb or a nuclear explosion. **2.** Electromagnetic emission that (a) is radiated energy extracted from the thermal excitation of atoms and molecules, (b) is electromagnetic radiation in the infrared region of the frequency spectrum, and (c) when incident upon living organisms, is sensed by living organisms as heat. *Note:* Excess exposure to infrared radiation can cause dehydration and severe damage to tissues even though the burn is slow. Thermal radiation from heat sources, such as industrial chimneys and aircraft exhaust, can be used as a homing source of radiation. The frequency of the infrared portion of the electromagnetic frequency spectrum is just below that of the visible spectrum. *See also* **radiation, thermal noise.**

thermal shock test: A measurement of performance changes when equipment and components are subjected to sudden and large changes in temperature. *Note:* In an optical fiber, fiber optic bundle, or fiber optic cable, changes in parameters, such as attenuation, pulse dispersion, and tensile strength, are measured when sudden and large changes in temperature are applied.

thermocouple: A loop made of two thermojunctions connected together, the resultant current in the loop being a function of the temperature difference between the two junctions. *Note:* If the current is calibrated, the temperature difference between the two junctions can be measured by using a calibrated thermocouple.

thermoelectric effect: The effect produced in a conducting circuit constructed of two wires of different metals fused to each other at the ends in such a manner that (a) if a temperature difference is maintained between the junctions, an electric current will flow in the circuit, and (b) the energy for maintaining the current is derived from the thermal sources maintaining the temperature difference.

thermographic process: In a document-copying machine, a duplicating process based on the effect of infrared radiation on heat-sensitive material, usually contained on paper.

thermojunction: A junction that consists of two dissimilar metals bonded together, thereby forming an interface across which a voltage is developed when the junction is heated. *Note:* If two such junctions are held at different temperatures and are connected together by electrical conductors, thus forming a thermocouple, an electric current will be maintained that is somewhat proportional to their temperature difference. *See also* **thermocouple.**

thermoneutral: Pertaining to a chemical reaction in which (a) there is no overall energy change, and (b) the reaction is neither endothermal nor exothermal, i.e., the reaction neither requires heat to occur nor produces heat when it occurs.

thermoplastic: Pertaining to the property of a solid material such that the material softens, melts, or fuses when heated and rehardens when cooled. *Note:* Optical glass and optical fibers usually are thermoplastic. Special glasses and ceramics of a refractory type are not thermoplastic. *See also* **thermosetting.**

thermoplastic cement: An adhesive that has a viscosity that decreases, i.e., an adhesive that becomes softer, as the temperature is raised. *Note:* Canada balsam, resin, and pitch are common thermoplastic optical cements. *See also* **thermosetting cement.**

thermosetting: Pertaining to the property of a material that does not soften, melt, or fuse when heated, but often becomes even harder as the temperature is raised. *See also* **thermoplastic.**

thermosetting cement: An adhesive that (a) permanently sets or hardens at a certain high temperature, (b) remains hard when cooled, and (c) will not soften when reheated. *Note:* Methacrylate is a common thermosetting optical cement. *See also* **thermoplastic cement.**

THF: tremendously high frequency.

thick lens: A lens that has an axial thickness so large that the principal points and the optical center do not coincide at a single point on its optical axis. *See also* **optical axis, principal point, thin lens.**

thickness: *See* **optical thickness.**

thimble map: A computer program that (a) directs a computer printer to substitute other characters for those that the printer is set up to print, such as a program sequence that causes the printer to print a British pound sign (£) in lieu of the brace ({), and (b) takes into account the position of each substituted character on the print thimble.

thin film: A film of a material that (a) has a thickness measured in numbers of atomic or molecular layers, (b) usually is created by evaporative or etching techniques, and (c) may be on the order of 10^{-9} m (meters) thick. *See* **optical thin film.** *Refer to* **Fig. S-12 (917).**

thin-film circuit: A circuit constructed by creating films of material on an inert material substrate, where (a) films are on the order of 10^{-9} m (meter), i.e., a millimicron, thick, (b) circuit elements are created by control of geometry and material composition, (c) films may be formed by various techniques, such as evaporative deposition, dipping, spraying, etching, and doping, and (d) solid state semiconducting crystals are also used. *Note:* If the film is of magnetic material, the thin film may be used for data storage or for control. Large-scale integrated (LSI) circuits usually consist of thin films. The circuits are used primarily to perform combinational logic operations, such as those performed by logic gates. Thin-film circuits have an extremely high circuit element packing density, such as 10,000 logic elements or storage elements per square inch. *Refer to* **Figs. P-2 and P-3 (708), R-8 (834).**

thin-film laser (TFL): A laser that (a) is constructed by thin-film deposition techniques on a substrate for use as a light source, (b) usually is used to drive thin-film optical waveguides, and (c) usually is used in optical integrated circuits (OICs).

thin-film laser element: In a thin-film optical circuit, such as an optical circuit in an optical integrated circuit (OIC), a circuit element that consists of a laser. *See also* **optical integrated circuit, thin-film laser.**

thin-film optical modulator (TFOM or TOM): A modulator that (a) consists of multilayered films of materials of different optical or electrical characteristics, (b) modulates transmitted light by using electrooptic, electroacoustic, or magnetooptic effects to obtain signal modulation, and (c) usually is used as a component in optical integrated circuits (OICs).

thin-film optical multiplexer (TFOM or TOM): A multiplexer that (a) consists of multilayered films of material of different optical or electrical characteristics, (b) is capable of multiplexing transmitted light by using electrooptic, electroacoustic, or magnetooptic effects to obtain signal multiplexing, and (c) usually is used as a component in optical integrated circuits (OICs).

thin-film optical switch (TFOS or TOS): A switch that (a) consists of multilayered films of material of different optical characteristics, (b) switches transmitted light by using electrooptic, electroacoustic, or magnetooptic effects to obtain signal switching, (c) usually supports only one propagation mode, i.e., operates in single mode, and (d) usually is used as a component in optical integrated circuits (OICs). *See also* **mode.**

thin-film optical waveguide (TFOW or TOW): A slab dielectric waveguide that (a) consists of multilayered films of material of different optical characteristics, (b) uses lower refractive index material on the outside to serve as the substrate as well as the cladding, (c) is capable of guiding transmitted light in single mode, and (d) usually is used as a component in optical integrated circuits (OICs). *Note:* Thin-film optical waveguides (TOWs) usually operate from laser sources for effective launching of lightwaves into the thin films. Various combinations of thin-film waveguides, lasers, modulators, switches, directional couplers, filters, and other components may be light-coupled to form optical integrated circuits (OICs), which may be optically or electrically powered to perform logical operations. *See* **periodically distributed thin-film optical waveguide.** *Refer to* **Figs. P-2 and P-3 (708).**

thin-film storage: *See* **magnetic thin-film storage, optical thin-film storage.**

thin-film waveguide: 1. A thin film of semiconducting or dielectric material that (a) guides electric or electromagnetic waves, such as pulses, (b) usually is mounted on a substrate to form a chip, and (c) is used to perform arithmetic and logic operations. **2.** In fiber optics, a transparent dielectric film that (a) is bounded by lower index materials, (b) is capable of guiding electromagnetic waves in the optical spectrum, and (c) usually operates in single mode. *See* **periodically**

distributed thin-film waveguide. *See also* **optical fiber, optical waveguide.**

thin lens: A lens that has an axial thickness so small that the principal points, the optical center, and the vertices of its two surfaces can be considered as coinciding at a single point on its optical axis. *See also* **optical axis, principal point, thin lens.**

third cosmic velocity: The minimum radial velocity that (a) an object must acquire at the Earth's surface to overcome the attraction of the Earth, Moon, and Sun, and thus to just leave the solar system with negligible velocity, and (b) has a magnitude of about 16.7 km•s^{-1} (kilometers per second).

third harmonic nonlinear material: *Synonym* **third-order nonlinear material.**

third-order intercept point: In a radio receiver, a point that (a) is an extrapolated convergence, and therefore is not directly measurable, of intermodulation distortion products in the desired output, (b) indicates how well the receiver performs in the presence of strong nearby signals, and (c) is tested by using two frequencies that fall within the first intermediate frequency mixer passband, usually using test frequencies that are about 20 kHz (kilohertz) to 30 kHz apart, i.e., the difference between the two test frequencies is about 20 to 30 kHz.

third-order nonlinear material: In fiber optics, a transparent material, such as a special glass, polymer, or organic material, that generates a third harmonic frequency when energized with optical signals. *Note:* The generation of third harmonic frequencies means switching speeds of fewer than 100 fs (femtoseconds). The generation process is also lossless and thus can be repeated many times in sequence. An exit wave from third-order nonlinear material can be frequency-mixed, and output beams can be considerably different from input beams. At the exit point of an optical integrated circuit (OIC) waveguide, different signals can be switched to different ports. *Synonym* **third harmonic nonlinear material.**

thrashing: A computer or data processing system operating condition in which little useful work is accomplished because of the excessive time spent in transferring data and computer programs among storage units and areas, such as excessive paging. *See also* **paging.**

threading: In installing communications cables, the pulling of a cable around, over, under, and through various obstacles in order to use a single piece of cable from one distant point to another without making splices to make the connection. *Note:* An example of threading is connecting, via a single continuous unspliced cable, a distribution frame terminal to an end termination in a different building.

threat: 1. In any system, such as a telecommunications, computing, data processing, or control system, a circumstance or an event that can harm the system. *Note:* The harm may be in any form, such as unauthorized (a) destruction of the system or components, (b) disclosure of data in the system, (c) data modification, or (d) use of services. **2.** In communications, computer, data processing, and control systems, (a) the technical and operational capability of a natural or man-made entity to detect, exploit, destroy, subvert, or otherwise reduce the capability of the system, and (b) the demonstrated, presumed, or inferred intent or capability of that entity to conduct or cause such activity. *See* **threat warning system.**

threat warning system: A system that (a) provides warning of the existence of a potential or actual threat, such as a storm, a hurricane, or military attack, and (b) may be of any type, such as a radar, infrared, electrooptical, perimeter, sonar, or visual system. *Note:* An example of a threat warning system is an electronic system installed on board an aircraft for warning against ground, sea, or air attack. *See also* **threat.**

three-address instruction: A computer program instruction that has an address part that contains three addresses, such as the addresses of two operands and the address for the results of the operation specified in the operation part. *Note:* A computer instruction, written as 23 544 496 906 might indicate that the operation, 23, is that of addition, 544 is the address of the addend, 496 is the address of the augend, and 906 is the address at which the sum is to be stored.

three-bit byte: *Synonym* **triplet.**

three-body problem: 1. In celestial mechanics, the problem that concerns the motion of a small body, usually, but not necessarily, of negligible mass, relative to, and under the gravitational influence of, two other finite point masses that are also moving. **2.** The motion of three bodies that (a) are free to move under each other's attractive forces according to Newton's law of universal gravitation, (b) have no other forces acting on them, (c) are regarded as material particles with large distances between them compared to their size, i.e., there are no collisions, and (d) are given an initial velocity. *Note:* Most often, in dealing with the three-body problem, it is necessary to determine the motion of two bodies relative to the third. The motion of all three bodies is encountered in studying the motion of the Moon or other Earth satellites, such as communications satellites, under the influence of the Earth and the Sun, or in studying the motion of stars in three-star systems. *See also* **n-body problem, restricted circular three-body problem, two-body problem.**

three-condition coding: Pertaining to a signaling system code or signal, such as a telegraph or data transmission signal, in which there are exactly three significant conditions. *Note:* The significant conditions may be those of any of a number of signal parameters, such as voltage level, frequency, phase position, or polarity. *See also* **significant condition, ternary signal.**

three-way calling: 1. A network-provided service feature that allows a user to add a third user during a call without the assistance of an attendant. **2.** A network-provided service feature that lets the call originator get other call receivers, in two different places, on the phone at the same time. *See also* **call, conference call, multiple.**

three-zero substitution: *See* **bipolar with three-zero substitution.**

threshold: 1. In telecommunications, the minimum signal level that can be detected by a system or a sensor. **2.** A value used to denote predetermined conditions, such as those pertaining to the volume of message storage used in a message switching center. **3.** The minimum value of a parameter that will activate a device. **4**. The minimum value a stimulus may have to create a desired effect. *See* **absolute luminance threshold, fault rate threshold, fault threshold, frequency modulation improvement threshold, hearing threshold, lasing threshold, outage threshold, pain threshold, receiver threshold, system operational threshold.** *Refer to* **Figs. F-4 (356), M-3 (590).**

threshold current: In a laser, the driving current that corresponds to the lasing threshold. *See also* **lasing threshold.**

threshold effect: *See* **frequency modulation threshold effect.**

threshold extension: A change in the value of a threshold. *Note:* An example of threshold extension is, in a receiver, the lowering of the noise threshold by decreasing the receiver bandwidth. In an angle modulation receiver, this enables signals to be recovered from input that contains a large amount of noise. *See* **frequency modulation threshold extension.** *See also* **angle modulation, threshold, value.**

threshold frequency: In optoelectronics, the frequency of incident radiation below which there is no photoemissive effect in the illuminated material. *Note:* Below the threshold frequency, electrons are not emitted because the incident photons do not have sufficient energy to overcome the work function of the material, no matter how high the radiant intensity is or how long the material is irradiated. *See also* **frequency, frequency tolerance, work function.**

threshold power: *See* **link threshold power.**

throttle: *See* **upload throttle.**

through: *See* **dial through.**

through control: *See* **right through control.**

through group: A group of 12 voice frequency channels that is routed through a repeater as a unit and in which there is no frequency translation. *See also* **frequency, group, through-group equipment.**

through-group equipment: In carrier telephone transmission, equipment that accepts multiplexed signals from a group receiver output and attenuates or amplifies them to the proper signal level for insertion, without frequency translation, at the input of group transmitter equipment. *See also* **group, translation.**

throughput: 1. The volume of traffic per unit time, such as bits, characters, blocks, messages, or calls per second, that can pass through all or part of a communications system when the system, or part, is operating at optimum capacity, usually at saturation. *Note:* Operational throughput usually is less than the theoretical maximum throughput. **2.** A measure of the performance of a system during a given period, such as the number of calls or jobs completed in a given day or month. *See also* **binary digit, block, block transfer efficiency, block transfer rate, data communications, data transfer rate, effective data transfer rate, effective speed of transmission, efficiency factor, job, maximum user signaling rate, Nyquist rate, Shannon's law, speed of service.**

through supergroup: A group of 60 voice frequency channels (a) that is routed through a repeater as a unit, and (b) in which no frequency translation is performed. *See also* **frequency, group, through group.**

through-supergroup equipment: In carrier telephone transmission, equipment that accepts multiplexed signals from a supergroup receiver output and attenuates or amplifies them to the proper signal level for insertion, without frequency translation, at the input of supergroup transmitter equipment. *See also* **group.**

throw: *See* **paper throw.**

thumb wheels: In display systems, a set of captive wheels that (a) are often mounted in the surface panel of a display console, (b) may be used for manipulating a cursor, a set of cross hairs, or a plotter stylus on the display surface of a display device, and (c) are rotated through a radial angle that corresponds to changes in coordinate data. *See also* **coordinate data.**

THz: terahertz.

TIA: Telecommunications Industry Association.

TIBS: Tactical Information Broadcast Service.

ticket: *See* **call ticket.**

ticketed call: A call for which a record is made of certain relevant information, such as the time and the date it was placed, its duration, the call originator and call receiver numbers, and the name or the initials of the operator.

TIE: time interval error.

tie line: *Synonym* **tie trunk.**

tie trunk: A telephone line that directly connects two private branch exchanges (PBXs). *Synonym* **tie line.** *See* **private branch exchange tie trunk.** *See also* **interposition trunk, tandem tie trunk network, trunk.**

tight: *See* **radio frequency tight.**

tight jacketed cable: In fiber optics, a fiber optic cable in which the coated optical fibers are not free to assume their own position within the jacket but are completely constrained by the jacket.

tilt: *See* **antenna electrical beam tilt.**

time: 1. An epoch, i.e., the designation of an instant on a selected time scale, such as an astronomical or atomic time scale. *Note:* "Time" is used in the sense of time of day, usually expressed as a numeral that represents the hour, minute, and possibly second of the day, starting at midnight. **2.** On a time scale, (a) the interval between two events or (b) the duration of an event. *See* **access denial time, access time, acquisition time, active time, assembly time, attack time, available time, block transfer time, calendar time, call filing time, call release time, call request time, call selection time, call setup time, chip time, coherence time, computer time, Coordinated Universal Time, data transfer time, disengagement time, downtime, dwell time, environmental loss time, filing time, flux rise time, Greenwich Civil Time, Greenwich Mean Time, group delay time, guard time, holding time, idle time, International Atomic Time, lost time, maximum access time, maximum block transfer time, maximum disengagement time, message file time, message preparation time, message release time, operating time, operation time, out of frame alignment time, postselection time, precise time, problem time, procurement lead time, pulse fall time, pulse rise time, real time, receipt time, recovery time, reframing time, release time, response time, rise time, round-trip delay time, sampling time, satellite transit time, scheduled time, search time, seek time, settling time, slot time, switch bounce time, switching process time, switching reaction time, switching time, switch operating time, switch release time, system downtime, system operating time, system test time, synchronization acquisition time, total delay time, transmission time, transit time, turnaround time, Universal Time, variable time.** *See also* **coordinated time scale, primary time standard, standard frequency and time signal service, time scale.**

time ambiguity: A situation in which two or more different times or time measurements are obtained under given conditions.

time assignment speech interpolation (TASI): An analog switching technique (a) in which an additional user is switched onto a temporarily idled channel when the original user has stopped speaking, and in which, when the first user resumes speaking, that (original) user, in turn, is switched to a channel that happens to be idle, and (b) that is used on long frequency division multiplexed (FDM) links to improve voice transmission efficiency. *See also* **adaptive routing, digital speech interpolation, frequency division multiplexing, link.**

time availability: *Synonym* **circuit reliability.**

time between failures: *See* **mean time between failures.**

time between outages: *See* **mean time between outages.**

time block: In communications, an arbitrary grouping of several consecutive hours of a day, usually for a specific reason and during a particular season, during which it is assumed that propagation data are statistically homogeneous and therefore somewhat predictable during the several hours.

time bureau: *See* **International Time Bureau.**

time chart: *Synonym* **time conversion chart.**

time code: A code that (a) is used for the identification and the transmission of time signals, and (b) has a specific format. *See also* **code, Coordinated Universal Time.**

time code ambiguity: The shortest time interval between successive repetitions of the same time code value. *Note:* Examples of time code ambiguity are (a) in a time code in which the century is the most slowly changing field, a time code ambiguity of 100 years, (b) for a digital clock in which hours and minutes up to a maximum of 11:59 are displayed, a time code ambiguity of 12 hours, and (c) for a digital clock in which hours and minutes up to a maximum of 23:59 are displayed, a time code ambiguity of 1 minute. *See also* **time code resolution.**

time code precision: *Synonym* **time code resolution.**

time code resolution: The period between two successive time code states as determined by the most rapidly changing symbol position within the time code. *Note:* An example of time code resolution is a resolution of 1 minute for a digital clock that only displays hours and minutes. *See also* **time code ambiguity.**

time coherence: In waves, such as acoustic and electromagnetic waves, pertaining to the existence of a correlation (a) between the phases of two waves, (b) between the phases of one wave at one point in space at two or more instants of time, and (c) between phase relationships such that they can be calculated and predicted everywhere over a time period. *Synonym* **temporal coherence.**

time-coherent light: Lightwaves in which the amplitude, phase, polarization, or other characteristic of the lightwaves at any point in space and time can be predicted at that point, given the values of the same characteristics at any previous time at that same point.

time-coherent radiation: Electromagnetic radiation in which the amplitude, phase, polarization, or other characteristic of the radiation at any point in space and time can be predicted at that point, given the values of the same characteristics at any previous time at that same point. *Synonym* **temporally coherent radiation.**

time constant: A constant that indicates the time required for a system or a circuit to change from one state or condition to another, contained in the relation $A_t = A_0 e^{-t/a}$ where A_t is the value of the state at time t, A_o is the value of the state at time $t = 0$, a is the time constant, and t is the time that has elapsed from the start of the exponential decay. *Note 1:* When $t = a$, $A_t/A_o = 1/e$, or about 0.37, and the system has changed about 63% toward its new value in one time constant. A system is considered to have completely changed its state after the elapse of three time constants, which corresponds to a 95% change in state. Thus, an electrical capacitor, C, is 95% discharged through a resistor, R, after the elapse of $3RC$ seconds, where R is in ohms and C is in farads, because the discharge is given by the relation $q_t = q_0 e^{-t/RC}$, where q represents the charge. *Note 2:* Time constants usually are expressed in seconds, such as 3.5×10^{-6} second, i.e., 3.5 µs (microseconds). *See* **fast time constant, logarithmic fast time constant.**

time control: *See* **sensitivity time control.**

time conversion chart: A table that enables (a) time expressed in one mode at one location to be expressed in the same or another mode at another location, and (b) the determination of time differences between the locations. *Synonym* **time chart.**

time delay: *See* **near real time delay, receive after transmit time delay, receiver attack time delay, receiver release time delay, transmit after receive time delay, transmitter attack time delay, transmitter release time delay.** *Note:* These terms are intrinsically redundant in the sense that delay can only be with respect to time. However, they are defined as such by recognized standards bodies.

time delay distortion: *Synonym* **delay distortion.**

time-derived channel: A channel obtained by time division multiplexing a line, a circuit, or another channel. *See also* **frequency-derived channel, time division multiplexing.**

time difference: *See* **clock time difference.**

time diversity: **1.** Pertaining to transmission in which a signal is sent through the same channel more than once. **2.** Pertaining to transmission and reception in which the same information, though it may be represented by different signals, is transmitted and received at the same frequency more that once, but not simultaneously. *See also* **time diversity transmission.**

time diversity transmission: Transmission in which identical signals are sent over the same channel at different times. *Note:* Time diversity transmission is often used in systems subject to burst error conditions. The time spacing is adjusted to be longer than an error burst. *See also* **frequency diversity, space diversity, time diversity.**

time division (TD): In communications, the use of time in sharing the use of a system or a device when the device can only handle one user at a time. *Note:* Examples of time division are (a) to use a single channel by separating the available time into time slots that may be used to create additional channels, (b) to divide the time of a switch among a multiplicity of users, and (c) to divide the use of equipment, such as facsimile, transmission, and computing equipment, among several users, one at a time. *See also* **frequency division, space division, time division multiplexing.**

time division demultiplexing: Reversing the process of time division multiplexing such that messages, conversations, packets, blocks, or other units of data that time-shared one channel are each assigned separate channels so that they can each be routed to different destinations, i.e., a form of serial to parallel conversion.

time division duplexing: Duplex communication accomplished by the simultaneous transmission and reception of two signals, each representing different information, over a common path by using different time intervals for each signal. *See also* **duplex communication.**

time division multiple access (TDMA): In a telecommunications system, the allocation of unique time slots to the different users of a common channel. *Note:* Time division multiple access (TDMA) may be used in multiple or broadcast transmission. It is used extensively in satellite systems, local area networks (LANs), physical security systems, and combat net radio systems. *See also* **channel time slot, frequency division multiple access, multiple, multiplexing.** *Refer to* **Fig. T-5.**

time division multiplex equipment: Communications equipment that (a) allows several relatively slow messages to be transmitted through one channel, (b) includes complementary demultiplexing equipment at the distant end to obtain the individual messages, (c) may include a matched computer to which the complete multiplexed signal may be fed for demultiplexing, and (d) may use channel time sharing, such as (i) scanning, i.e., sampling, in which each possible input is connected in turn until all inputs have been scanned, and the sequence is then repeated, (ii) polling, in which a code is used to ask for a message that is ready to be transmitted now, and the one unit that is ready is connected and sends its address to the serving communications center, i.e., serving exchange, and, if two or more units are ready at the same time, a priority is arranged, and (iii) addressing remote devices that then send back a ready signal or a not-ready signal.

time division multiplexing (TDM): In a single channel operating in a given frequency spectrum and in a propagation medium connecting two or more points, the derivation of two or more concurrent channels from the single channel by assigning discrete time intervals, in sequence, to each of the derived channels, i.e., a form of parallel to serial conversion. *Note 1:*

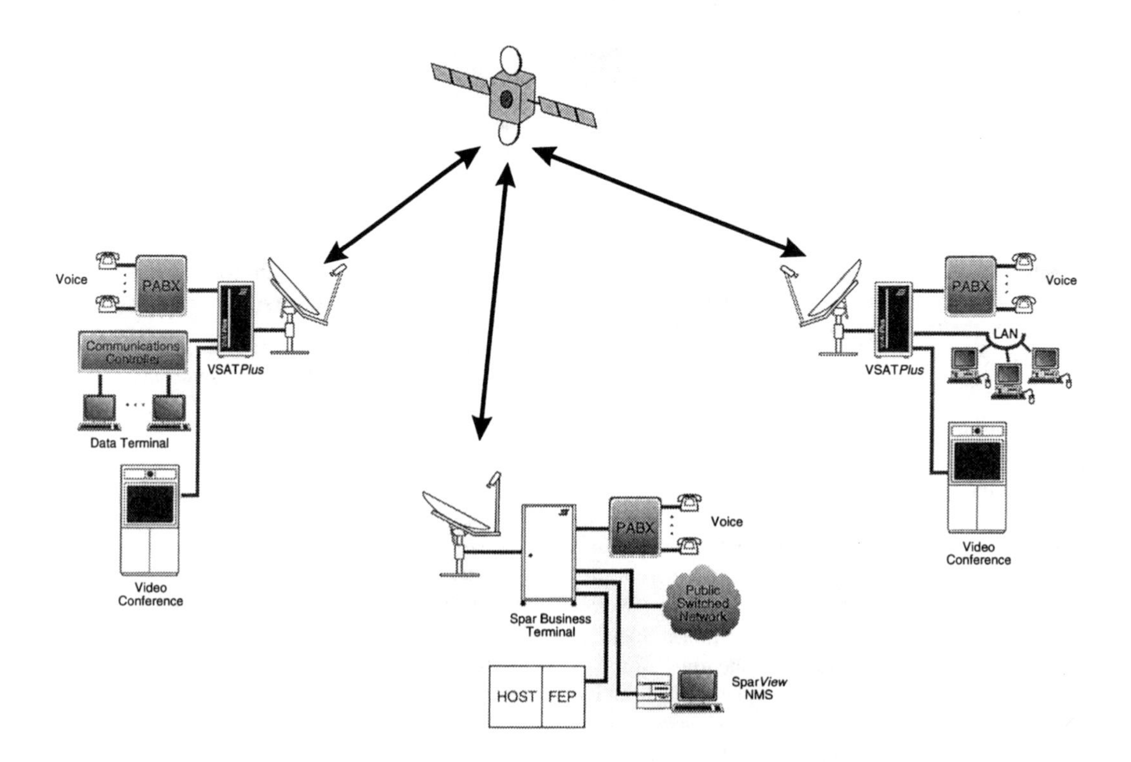

Fig. T-5. A typical mesh **time division multiple access (TDMA)** communications satellite network configuration. (Courtesy ComStream Company.)

The entire available frequency spectrum can be used by the derived channels during their assigned time intervals. *Note 2:* Time division multiplexing (TDM) systems usually use pulse transmission. The multiplexed pulse train is the interleaved pulse trains of the derived channels. The derived channel pulses may be individually analog- or digitally modulated. Thus, individual signals from separate sources share the time of a circuit by being assigned time slots on the circuit. In fiber optic data links a single optical fiber can share its time among a large number of sources. If each source is assigned 1 μs (microsecond) out of each ms (millisecond) of time on the channel, 1000 sources can be accommodated by the single fiber. Of course, the individual sources can be frequency division multiplexed, giving rise to many more channels. In an optical fiber, the lightwaves in the fiber can be wavelength division multiplexed, i.e., a different-color lightwave can be used for each channel. *See* **asynchronous time division multiplexing, synchronous time division multiplexing.** *See also* **channelization, concentrator, digit time slot, frequency-derived channel, frequency division multiplexing, highway, multiplex aggregate bit rate, multiplex hierarchy, multiplexing, time sharing.** *Refer to* **Fig. T-6.**

time division switching: 1. The switching of time division multiplexed (TDM) channels by shifting the data from one time slot to another in the TDM frame. **2.** In a switching system, an arrangement in which (a) the various connections share a common path but are separated in time, (b) each connection is permitted to use the common path, in sequence, for a short period, and (c) the interconnection of input and output terminations is established repetitively for short time intervals such that the elements of the switching matrix may be time-shared by a number of terminations. *See also* **digital switching, switching system, termination, time division multiplexing.**

time domain reflectometer: *See* **optical time domain reflectometer.**

time domain reflectometry: *See* **optical time domain reflectometry.**

time error (TE): After a period following perfect synchronization between a real clock and an ideal uniform time scale, the time difference between that clock and the time scale. *Synonym* **time interval error.**

time fill: *See* **interframe time fill.**

time filtering system: *See* **discrete time filtering system.**

time frame: A specified period, based on two or more events or significant instants, using time as a basis for measurement or reference. *See also* **significant instant.**

time-frequency signal: *See* **standard frequency-time signal.**

time gated direct sequence spread spectrum: *Synonym* **direct sequence spread spectrum.**

time group: *See* **date-time group, true date-time group.**

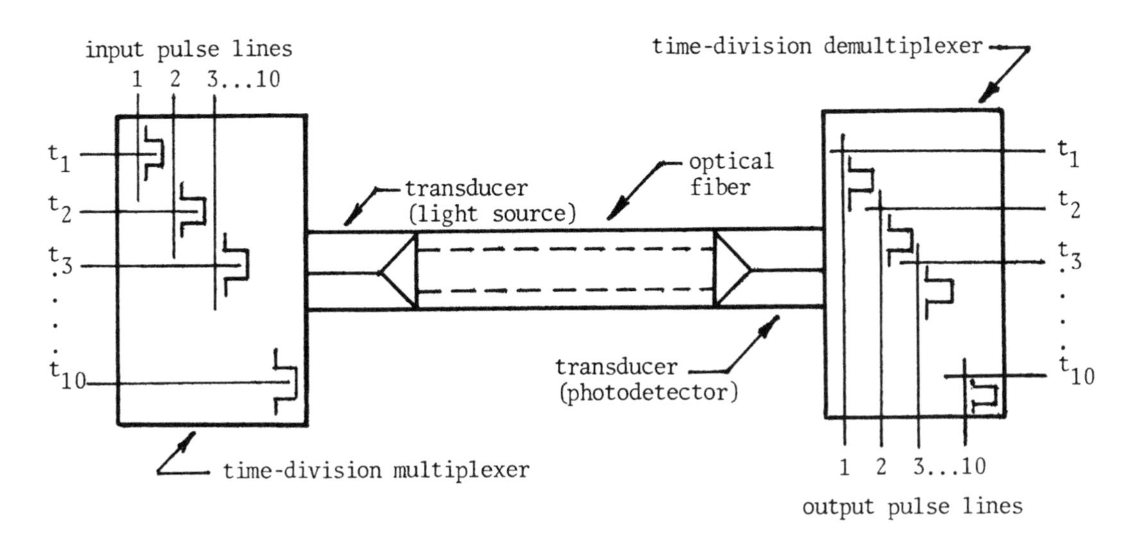

Fig. T-6. Time division multiplexing on a fiber optic link.

time guard band: A time interval left vacant on a channel to provide a margin of safety against intersymbol interference in the time domain between sequential operations, such as detection, integration, differentiation, transmission, encoding, decoding, or switching. *See also* **band, frequency guard band, time division multiplexing.**

time hopping: The shifting of the time of occurrence of a pulse in accordance with a given signal, such as a signal from a code sequence generator. *See also* **code sequence generator, time of occurrence.**

time indication: *See* **automatic date and time indication.**

time instability: The fluctuation of the time error caused by the instability of a real clock.

time interval: *See* **precise time and time interval, service time interval.**

time interval error (TIE): The time difference between a real clock and an ideal uniform time scale, after a time interval following perfect synchronization between the clock and the scale. *Synonym* **time error.**

time jitter: Short-term variation or instability in the duration of a specified time interval or in the time of occurrence of a specific event, such as a signal transition from one significant condition to another, where, when viewed with reference to a fixed time frame, the signal transition moves back and forth on the time axis. *See also* **drift, jitter, time of occurrence, swim, wander.**

time marker: A reference signal, often periodically repeated, enabling the correlation of specific events with a given time scale. *Note:* An example of a time marker is a signal used for establishing synchronization.

time modulation: *See* **pulse time modulation.**

time objective: *See* **message time objective.**

time of arrival: *See* **estimated time of arrival.**

time of file: *Synonym* **message file time.**

time of occurrence: The date of an event, i.e., the instant that an event occurs, with reference to a specified time scale. *Note:* The date is a particular instant on a particular time scale. *Synonym* **occurrence time.** *See also* **Coordinated Universal Time, time scale.**

time operator: *See* **real time operator.**

time-out: **1.** A network parameter related to an enforced event designed to occur at the end of a predetermined elapsed time. **2.** In a system, a specified period that is allowed to elapse before a specified event takes place unless another specified event occurs first, such that, in either case, the period is terminated when either event takes place. *Note 1:* A time-out condition can be canceled by the receipt of an appropriate time-out cancellation signal. *Note 2:* Examples of time-outs are (a) the 20 seconds in which a ringing signal might be allowed to occur before automatic disconnect occurs, unless the call originator hangs up first, and (b) the length of time in which a dial tone is allowed to occur before a loud attention signal is transmitted to an off-hook telephone to alert the user that the telephone is off-hook. After a time-out, the attention signal is discontinued, and the line goes dead until the telephone is placed on-hook. **3.** An event that occurs at the end of a predetermined period that began at the occurrence of another specified event. *Note:* The time-out can be prevented, i.e., can be inhibited, by an appropriate signal. *See also* **call control signal.**

timer: *See* **response timer.**

time scale: **1.** A time measuring system used to relate the passage of temporal events that have occurred since a selected epoch. *Note:* Time scales are graduated in (a) intervals, such as seconds, minutes, hours, days, months, and years, and (b) submultiples of a second, such as milliseconds, nanoseconds, and picoseconds. **2.** Time coordinates placed on the abscissa (*x*-axis) of Cartesian coordinate graphs used for depicting phenomena that are a function of time, such as waveshapes at a spatial point. *See* **coordinated time scale, International Atomic Time scale, uniform time scale, variable time scale.** *See also* **Coordinated Universal Time, epoch, International Atomic Time, time of occurrence.**

time scale factor: **1.** A multiplier used to transform the real time of occurrence of an event or a problem into system time, such as that of a telecommunications system or a computer. **2.** In analog computers, a number used as a multiplier to transform problem time to computer time. **3.** In simulation, the ratio of computer time to corresponding problem, real, actual, clock, or calendar time, or any other designated measure of time on a time scale. *See* **extended time scale factor, fast time scale factor.** *See also* **computer time, problem time, real time.**

time sharing: **1.** The interleaving of two or more independent processes being performed by one functional unit. **2.** Pertaining to the interleaved use of computer time that enables two or more users to execute programs concurrently. **3.** A mode of operation of a device that provides for its use by two or more entities at different times, though there is the appearance of simultaneous operation. **4.** A distribution of calendar time among two or more pieces of equipment because simultaneous operation of the pieces of equipment would cause unacceptable conditions, such as (a) overloading of a circuit and (b) excessive interference, such as when two radio stations operate at the same frequency and in proximity. *See* **frequency time sharing.** *See also* **frequency sharing, multiprocessing, multiprogramming, online computer system.**

time signal standard frequency broadcast: A broadcast, such as a radio broadcast or a television broadcast, that furnishes (a) information about standard time in specified areas, (b) information about standard frequencies, (c) standard timing signals, and (d) standard frequency signals for the synchronization and the calibration of equipment.

time slice: An interval of time that (a) is designated for the performance of a specific task or operation and (b) may be used for such purposes as (i) preventing monopolization of time beyond a fixed limit for a given message or user, (ii) organizing time sharing, and (iii) allowing multiple channels to use a single propagation medium, i.e., for multiplexing. *Note:* Examples of time slices are (a) a segment of time allocated to a specific job or (b) in time division multiplexing, an interval of time assigned to a channel. *See also* **time sharing, time slot.**

time slot: 1. A time interval that can be recognized and uniquely defined. **2.** A period (a) during which specific, usually controlled actions occur or are scheduled to occur, and (b) that has a defined beginning and a specific duration or a defined beginning and a defined ending. *Note 1:* An example of a time slot is a specified elapsed time after a beginning of frame signal. *Note 2:* A time slot may be defined and timed by a clock, i.e., a count of clock pulses. *See* **digit time slot.** *See also* **time slice, time sharing.**

time standard: A stable device that (a) emits signals at equal intervals such that their count may be used as a clock, and (b) may be used to calibrate various devices, such as oscillators, clocks, transmitters, receivers, and interferometers. *See* **primary time standard, secondary time standard.** *See also* **clock, Coordinated Universal Time, DoD master clock, master-slave timing.**

time system: *See* **continuous time system, discrete time system.**

time tick: A time mark output of a clock system.

time to failure: *See* **mean time to failure.**

time to repair: *See* **mean time to repair.**

time to service restoral *See* **mean time to service restoral.**

time transmission: *See* **real time transmission.**

time zone: An adopted relation of local time to Greenwich Mean Time (GMT), Coordinated Universal Time (UTC), or Universal Time (UT) for a particular geographical area such that (a) local time usually is an integral number of hours ahead of or behind GMT or UTC as far as clock time is concerned, (b) each local time is assigned an identifying letter, (c) the GMT, UTC, or UT at Greenwich, England, is identified by the letter "Z" (ZULU), (d) in many international communications networks, ZULU time (GMT), (UTC), or (UT) is used on all messages rather than local standard or daylight saving time, and (e) local time areas are about 1000 miles wide at the equator, though this varies around the world according to (i) land mass distributions, (ii) natural boundaries, and (iii) national borders.

timing extraction: *Synonym* **timing recovery.**

timing recovery: The derivation of a timing signal from a received signal. *Synonym* **timing extraction.** *See also* **signal.**

timing reference: *See also* **external timing reference.**

timing signal: 1. The output of a clock. **2.** A signal used to synchronize interconnected equipment. *See* **octet timing signal.** *See also* **clock, master-slave timing, signal, synchronization bit, synchronization pulse, synchronizing pilot.**

timing system: *See* **master-slave timing system.**

timing tracking accuracy: A measure of the ability of a timing synchronization system to minimize the clock difference between a master clock and any slaved clock. *See also* **clock, clock difference, synchronization.**

T interface: For basic rate access in an Integrated Services Digital Network (ISDN) environment, a user-to-network interface reference point that (a) is characterized by a four-wire, 144-kbps (kilobit per second) (2B+D) user rate, (b) accommodates the link access and transport layer function in the ISDN architecture, (c) is located at the user premises, (d) is distance-sensitive to the servicing network terminating equipment, and (e) functions in a manner analogous to that of the Channel Service Units (CSUs) and the Data Service Units (DSUs). *See also* **Integrated Services Digital Network, R interface, S interface, U interface.**

tip length: *See* **borescope tip length.**

title: *See* **short title.**

titling: In communications, computer, data processing, and control systems, the placing of headings on data units, such as (a) files, (b) messages, (c) graphs, (d) pictures, (e) rows or columns on a spreadsheet, so that they will remain in place when the rest of the spreadsheet is scrolled, (f) graphs, using one of the graph menu options of a personal computer program, and (g) control data.

T junction: *See* **series T junction.**

TLM: telemetry.

TLP: transmission level point.

TM: transverse magnetic.

TM mode: transverse magnetic mode.

TM wave: transverse magnetic wave.

T-number: The equivalent F-number of a fictitious lens that (a) has a circular opening, (b) has 100% transmittance, (c) gives the same central illumination as the actual lens being considered, and (d) is given by the relation $T_n = L/D = (L/2)(\pi/AT)^{1/2}$ where T_n is the T-number, L is the equivalent focal length, D is the diameter of the T-stop, π is about 3.1416, A is the area of the entrance pupil, and T is the transmittance of the lens system. *See also* **entrance pupil, F-number, transmittance, T-stop.**

TOD: time of day.

toggle: A functional unit that has two stable states and that changes to the opposite state and remains there each time the same command is given to it; e.g., on a personal computer keyboard, depressing the CAPS LOCK key shifts from upper to lower case letters or vice versa, depressing the INSERT key changes from normal to insert state or vice versa, or depressing the SCROLL LOCK key changes to scrolling and back. Many power switches are toggles, pushed once to turn on and once again to turn off. *Note:* Toggle switches need a special and separate indication of which state the switch is in. Knife switches and conventional wall switches are not toggles because they have to be pulled or pushed, in and out or up and down, to be reversed so that the actuating commands are not the same, and their state is indicated by the position of the switch. *See also* **flip-flop.**

token: In communications system protocols, such as local area network (LAN) protocols, a group of bits that (a) serves as a symbol of authority, (b) is passed among data stations, and (c) is used to indicate that the station with the symbol is temporarily in control of the propagation medium. *See also* **token passing.** *Refer to* **Fig. N-6 (622).**

token bus network: A bus network in which token passing is used. *See also* **local area network, network topology, token.**

token passing: In communications systems, such as a local area network (LAN), a control protocol in which (a) a token is passed from station to station, and (b) the station with the token is temporarily in control of the network, i.e., is the only station allowed to transmit. *See also* **local area network, protocol, token.** *Refer to* **Fig. N-6 (622).**

token ring network: A ring network in which unidirectional data transmission occurs between data stations by means of token passing over one propagation medium such that the transmitted data return to the transmitting station. *See also* **network topology.**

tolerance: In an entity, the permissible range of variation of some characteristic from its nominal value. *See* **clock tolerance, diameter tolerance, fault tolerance, frequency tolerance, phase measurement tolerance, phase tolerance.**

tolerance area: *See* **cladding tolerance area, core tolerance area.**

tolerance band compaction: *See* **fixed tolerance band compaction, variable tolerance band compaction.**

tolerance field: 1. The region between two curves, such as circles or rectangles, used to specify the tolerance on component size. **2.** In optical fiber cladding cross sections, the annular region between the two concentric circles of diameter $D+\Delta D$ and $D-\Delta D$, in which (a) the first circle, i.e., the larger circle, is the smallest circle that circumscribes the outer surface of the homogeneous cladding, and (b) the second circle, i.e., the smaller circle, is the largest circle that fits within the outer surface of the homogeneous cladding. **3.** In optical fiber core cross sections, the annular region between the two concentric circles of diameter $d+\Delta d$ and $d-\Delta d$ in which the first circle, i.e., the larger circle, is the smallest circle that circumscribes the core area, and (b) the second circle, i.e., the smaller circle, is the largest circle that fits within the core area. *Note 1:* The first circle, i.e., the larger circle, may be used as the center of the concentric circles for the cladding and the core. If the pairs of circles for the core and the cladding are determined independently, the distance between the centers of the two concentric pairs, i.e., the core pair and the cladding pair, defines the offset between core and cladding, i.e., the core-cladding offset or concentricity error. The width of the annulus bounded by the cladding circles defines the ovality of the cladding. The width of the annulus bounded by the core circles defines the ovality of the core. *Note 2:* The cladding circles usually are not concentric with the core circles. *See* **reference surface tolerance field.** *See also* **cladding, concentricity error, core, homogeneous cladding.**

toll: In public switched telephone systems, a charge that (a) is made for a connection to a central office (C.O.) or a user end instrument that is beyond a call originator exchange boundary, and (b) usually is based on various factors, such as distance, the number of exchanges used, duration of the call, time of day, locations of call originator and call receiver, type or class of service, and extent of use.

toll call: *Synonym* **long-distance call.**

toll diversion: A network-provided service feature (a) in which users are denied the ability to place long-distance calls without the assistance of an attendant, (b) that affects the entire switching system, and (c) that does not discriminate between individual subscribers. *See also* **call, call restriction, calls barred facility, restricted access, service feature.**

toll office (TO): A central office (C.O.) primarily used for supervising and switching long-distance calls. *See also* **toll switching trunk.**

toll quality (TQ): In telephone systems, the voice quality resulting from the use of a nominal 4-kHz (kilohertz) telephone channel. *Note:* Toll quality (TQ) may be quantized in terms of a specified bit error ratio (BER). *See also* **voice grade.**

toll restriction: *See* **classmark.**

toll switching trunk: A trunk that connects one or more end offices to a toll office as the first stage of concentration of long-distance calls. *Note:* Attendant assistance or participation may be an optional function. In U.S. common carrier telephone service, a toll office designated "Class 4C" is an office where assistance in completing incoming calls is provided in addition to other traffic. A toll office designated "Class 4P" is an office where attendants handle only outbound calls, or where switching is performed without human intervention. *See also* **end office, switching system, toll office.**

toll traffic: *See* **short-haul toll traffic.**

tone: An audible sound, usually of a given fixed frequency. *See* **call progress tone, dial tone, disabling tone, engaged tone, go ahead tone, number unobtainable tone, pilot tone, preemption tone, recorder warning tone, reorder tone, ringing tone, sidetone, test tone.**

tone control aperture: In photographic receiving systems, such as fiber optic, facsimile, or wirephoto systems, an aperture that permits controlled variation of the amount of light reaching the light-sensitive recording material in relation to the magnitude of the received signals and the attributes of the material. *See also* **aperture.**

tone delay: *See* **dial tone delay.**

tone dialing: *Synonym* **pushbutton dialing.**

tone diversity: In voice frequency telegraph (VFTG) transmission systems, the use of two channels to carry the same information. *Note:* Tone diversity usually is achieved by twinning the channels of a 16-channel voice frequency telegraph (VFTG) to obtain 8 channels with dual diversity. *See also* **diversity reception, dual diversity.**

tone interference: *See* **single tone interference.**

tone keying: *See* **two-tone keying.**

tone multifrequency signaling: *See* **dual tone multi-frequency signaling.**

tone reversal: *See* **partial tone reversal.**

tone test signal: *See* **composite two-tone test signal.**

tone wedge: A member of a tone wedge set. *See also* **tone wedge set.**

tone wedge set: A set of wedge-shaped optical elements (a) each of which has a slightly different absorptive and diffusive capability or optical density, with the set ranging in small steps from completely transparent to completely opaque, and (b) that can be inserted into an optical system to control tone or intensity from black to white through all shades of gray. *See also* **optical element.**

tool: In some computer languages, a small program usually executed as a shell command. *Note:* In other computer languages, such as BASIC, a small program or program segment is called a "utility." *See* **architectures design, analysis, and planning tool; optical fiber cutting tool; software tool.**

top box: *See* **television set top box.**

top-down: In communications, computer, data processing, and control systems, pertaining to an approach to system design, development, or fabrication that starts with the highest level of component and progressively proceeds through lower level components until the system is completed, such as pertaining to top-down designing, top-down programming, and top-down testing. *See also* **bottom-up.**

top-down designing: Designing a system by identifying its major components, dividing them into their lower level components, and then repeating the process until a designated level of detail is achieved. *See also* **bottom-up designing.**

top-down programming: The creating of software, such as a computer program or an operating system, by using a top-down approach, i.e., by writing the highest level programs, such as executive routines and control programs first, making reference to lower level routines and subroutines, then writing them, until the software is completed. *See also* **bottom-up programming.**

top-down testing: Testing hierarchically organized hardware or software by using a top-down approach, i.e., by testing whole higher level systems, then progressively proceeding to lower level subsystems and parts until the entire system and all its parts are tested. *Note:* If lower level components fail or are not available during system tests, they may have to be simulated so that the top-down test can proceed. *See also* **bottom-up testing.**

topology: *See:* **network topology, physical topology.**

torn tape relay: To relay incoming messages on paper tape by tearing the tape from receivers, usually at the end of a message or a sequence of messages, and placing the tape on one or more transmitters. *See also* **torn tape relay system.**

torn tape relay system: An antiquated tape relay system in which perforated tape from a receiver is manually transferred by an operator to the appropriate transmitter by (a) tearing off a section of tape bearing one or more messages, (b) processing the tape in accordance with operating instructions, (c) hand-carrying the tape to one or more transmitters according to routing instructions, and (d) inserting the leading end of the tape into the tape reading head of the transmitters. *See also* **relay, reperforator, semiautomatic continuous tape relay switching, semiautomatic relay system, tape relay.**

total: *See* **hash total.**

total acceptance angle: Two times the acceptance angle, i.e., twice the acceptance angle. *See also* **acceptance cone.**

total channel noise: Noise that (a) consists of the sum of random noise, intermodulation noise, and crosstalk, and (b) does not include impulse noise because different techniques are required for its measurement. *See also* **channel, channel noise level, noise.**

total delay time: The time between (a) the instant that a device, such as a user end instrument, sensor, ship transmitter, or aircraft system, has a signal sequence or a message ready to transmit and (b) the instant that the message passes completely through the destination network data circuit-terminating equipment/data terminal equipment (DCE/DTE) interface, including the source message length time, waiting time, transit time, and sink message length time. *Note:* Message length time is a responsibility of the user and the network. Waiting time and transit time are responsibilities of the network. *See also* **differential mode delay, group delay, propagation delay time.**

total diffuse reflectance: *Synonym* **diffuse reflectance.**

total emissivity: The ratio of (a) the radiation intensity of a given surface to (b) that of an ideal blackbody at the same temperature. *See also* **blackbody.**

total harmonic distortion (THD): In a signal, the ratio of (a) the sum of the powers in all the harmonic frequencies above the fundamental frequency to (b) the power of the fundamental frequency. *Note 1:* The total harmonic distortion (THD) usually is expressed in dB, though it is often also expressed in nepers, in percent, as a ratio, or as a fraction. *Note 2:* Measurements for calculating the total harmonic distortion (THD) are made at the output of a device under specified conditions. *Note 3:* The harmonic frequencies do not include the fundamental frequency, i.e., the input frequency. The distortion is caused by the nonlinearity of the component or the system. *See also* **distortion, frequency, harmonic distortion, single harmonic distortion, standard test signal.**

total internal reflection: In fiber optics, reflection that occurs when an electromagnetic wave propagating in a material medium, such as the glass core of an optical fiber, with a higher refractive index medium strikes the interface surface of a lower refractive index medium, such as the cladding of an optical fiber or air, at an incidence angle greater than or equal to the critical angle. *Note 1:* At angles greater than the critical angle, all of the incident power is reflected, and there is no refracted wave. At the critical angle, the incident ray will propagate along the interface surface. *Note 2:* In optical fibers, total internal reflection allows signal transmission over long distances by confining certain modes to the core because light rays in the core that approach the core-cladding interface at angles greater than the critical angle are totally reflected back, i.e., internally, into the core. *See also* **internal reflection, total reflection.** *Refer to* **Figs. C-9 (190), E-4 (314), S-18 (951).**

total internal reflection angle: *Synonym* **critical angle.**

total internal reflection sensor: *See* **frustrated total internal reflection sensor.**

total line length: In facsimile systems, the spot speed divided by the scanning line frequency, i.e., the repetition rate. *Note:* The total line length may be greater than the length of the available line. *See also* **available line, facsimile, spot speed.**

total radiation temperature: The temperature at which a blackbody radiates a total amount of electromagnetic radiation equal to that radiated by the body whose total radiation temperature is being considered. *See also* **blackbody.**

total reflection: Reflection that occurs when an electromagnetic wave propagating in a higher refractive index medium strikes the interface surface of a lower refractive index medium at an incidence angle greater than or equal to the critical angle. *Note 1:* At angles greater than the critical angle, all of the incident power is reflected, and there is no refracted wave. At the critical angle, the incident ray will propagate along the interface surface. *Note 2:* In optical fibers, total reflection allows signal transmission over long distances by confining certain modes to the core because light rays in the core that approach the core-cladding interface at angles greater than the critical angle are totally reflected back, i.e., internally, into the core. *See also* **incidence angle, critical angle, fiber optics, internal reflection, leaky mode, reflection coefficient, total internal reflection.**

total reflection sensor: *Synonym* **frustrated total reflection sensor.**

total scanning line length: In facsimile systems, (a) the product of the scanning speed and the scanning line period, or (b) the sum of (i) the available line length and (ii) the product of the scanning speed and the lost time. *See also* **available line length, lost time, scanning line period, scanning speed.**

touch panel: A human-to-machine interface device that allows the user to control a functional unit by touching selected points on a screen or a panel. *Note:* Touch panels are especially useful for menu-driven systems.

touch-sensitive: Pertaining to a control unit that allows a user to interact with a system, such as a telecommunications, computer, or data processing system, by touching an area on the surface of the unit with a finger, a pencil, or another object. *Note:* Most microwave ovens have a touch-sensitive control unit.

tournament sort: A repeated selection sort in which each subset has not more than two items in it.

tower: *See* **control tower, relay tower.**

TPOA: telecommunications private operating agency.

trace: 1. A visible reference line or a one-time signal line that (a) appears in the display space on the display surface of a display device, such as a cathode ray tube (CRT), light-emitting diode panel, gas panel, or fiberscope faceplate, and (b) usually is caused by a signal or a sweep oscillator. **2.** A record of a sequence of events, such as a record of the execution of a computer program, including the sequence in which the instructions were executed. **3.** A record of all or specific classes of computer instructions or program events that occur during the execution of a computer program. **4.** To record a sequence of events as they occur. **5.** To follow a record of events for a specific purpose, such as to discover where an error occurred. **6.** To reproduce a figure or a drawing by placing a transparent or translucent medium over it and following the lines with a writing instrument, such as a pen, pencil, or paint brush. **7.** In a telephone call, to determine the call originator or to determine the number and the address of the end instrument that was used to originate the call. *See* **dark trace.**

trace packet: In a packet switching network, a unique packet that causes a report of each stage of its progress to be sent to the network control center from each system element it reaches. *See also* **audit trail.**

trace program: A computer program that checks another computer program (a) by displaying the sequence in which instructions are executed, and (b) usually by displaying the results obtained from executing the instructions.

tracer: A software tool used to produce a trace. *See also* **software tool, tool, trace.**

tracer action: In communications, computer, data processing, and control systems, particularly telephone and telegraph systems, the action taken by operating personnel at one station to determine the cause, source, or nature of errors introduced into messages or operations by personnel or equipment at other stations. *Note:* Examples of tracer actions are investigations to determine (a) the reason for the delay in delivery or the nondelivery of a message and (b) the number of a call originator line or end instrument.

tracer request: A request from a user or an operator of a communications system to conduct a tracer action. *Note:* Examples of tracer requests are requests to determine (a) the reason for the delay or the nondelivery of a message, (b) the source of an incoming call, or (c) the charges for a finished call. *See also* **finished call, trace, tracer action.**

tracing: *See* **call tracing, on demand call tracing.**

track: 1. On a data medium, the path associated with a single read-write head as data move past the head. *Note:* Examples of tracks are (a) the ring-shaped portions of the surface of a magnetic drum or disk, and (b) the line of positions on a perforated tape, magnetic tape, or magnetic card that can be accessed by one fixed read-write head. **2.** To maintain contact with a moving object. *Note:* Examples of tracking are (a) to maintain radar contact with a moving target, (b) to maintain direction finding (DF) contact with a moving aircraft station or mobile land station, and (c) with a cursor, to maintain contact with a moving display element on the display surface of a display device. *See* **address track, clock track, regenerative track.**

track and hold unit: In analog systems, a device that has an analog output variable that is either (a) the input variable or (b) a sample of the input variable sampled by the action of an external signal operating such that, when tracking, the device follows the input analog variable and, when holding, the device holds the value of the analog input variable at the instant of switching. *Synonym* **track and store unit.**

track and store unit: *Synonym* **track and hold unit.**

track ball: *Synonym* **control ball.**

track correlation: In radar systems, the correlation of radar tracking data with other available data for identification purposes.

track density: The number of tracks per unit length, measured in a direction perpendicular to the tracks. *Note:* The reciprocal of the track density is the track pitch. *See also* **track, track pitch.**

tracker: *See* **infrared angle tracker.**

tracking: 1. In display systems, following or determining the position of a moving display writer by using (a) a stylus, such as a light pen, tablet stylus, or voltage pencil, or (b) a pointer, such as a control ball, joystick, mouse, or digitizer moving a display surface symbol, such as a cursor or an aiming circle. **2.** In radar, maintaining radar contact with a moving target. *See* **aided tracking, passive tracking, pulse edge tracking, space tracking.** *See also* **tracking symbol.**

tracking accuracy: *See* **timing tracking accuracy.**

tracking Earth station: *See* **space tracking Earth station.**

tracking error: 1. The deviation of a dependent variable with respect to a reference function. **2.** The difference between (a) the actual position of a tracked object and (b) the position measured and indicated by the tracking system. *See also* **track, tracking.**

tracking mode: An operational mode during which a system is operating within specified movement limits relative to a reference. *Synonym* **tracking phase.** *See also* **coasting mode, frequency tolerance.**

tracking orderwire: *See* **acquisition and tracking orderwire.**

tracking phase: *Synonym* **tracking mode.**

tracking radar: Radar that (a) tracks objects by means of lock-on, (b) usually receives data from scanning radar usually located in ship stations, aircraft stations, and land stations, (c) uses the data to obtain a lock on the moving object, and (d) in a weapon system, provides data needed for target intercept guidance. *See also* **lock-on, track, tracking.**

tracking space station: *See* **space tracking space station.**

tracking symbol: A symbol that (a) is visible in the display space on the display surface of a display device, and (b) can be moved and used to indicate the position of a pointer, i.e., a device capable of generating coordinate data, such as a stylus, joystick, control ball, or mouse. *Note:* Display surfaces include cathode ray tube screens, fiberscope faceplates, light-emitting diode panels, gas panels, and plotter beds.

tracking system: The totality of equipment used for tracking, including the operator. *See* **pseudomonopulse tracking system.**

tracking telemetry command station: An Earth station that (a) determines satellite positions and (b) provides telemetry and command facilities.

track pitch: The distance between the centerlines of adjacent tracks. *Note:* The reciprocal of the track pitch is the track density. *See also* **track, track density.**

track telling: In command, control, and communications, the communicating of radar air surveillance data between command and control systems and facilities.

See also **back tell, forward tell, lateral tell, relateral tell.**

track-while-scan airborne intercept radar: An airborne intercept radar that (a) tracks an object, i.e., a target, while it also scans for other objects, (b) searches and tracks in the same mode, (c) emits a signal that does not intermittently illuminate a jamming aircraft during search, (d) does not constantly illuminate a jamming aircraft during lock-on, and (e) emits a signal that at all times appears to be searching or intermittently illuminating.

traffic: 1. The information moved over a communications channel. **2.** In communications systems, the system control and user information that (a) is moved in the form of data represented by signals over a communications path, and (b) includes (i) the spatial totality of all the messages in a communications system or network, (ii) the flow rate of messages at a point in a network, channel, trunk, or other path or group of paths, (iii) message intelligence, such as voice, telegraph, or other data for which the communications system was established, and (iv) routine and special messages between system operators and maintenance personnel even though they usually are carried in channels separate from the channels that carry user information, i.e., information for which the system was established. **3.** The sum of the lengths of all the messages transmitted by a communications system during a specified period. *Note:* Traffic may be expressed in convenient units, such as call•seconds, character•seconds, or message•seconds. *See* **clear traffic, communications traffic, distress traffic, mission traffic, narrative traffic, press traffic, queue traffic, record traffic, safety traffic, secure traffic, short-haul toll traffic, urgency traffic, welfare traffic.** *See also* **busy hour, call•second, communications, erlang.** *Refer to* **Figs. B-10 (101), S-5 and S-6 (869), S-9 through S-11 (880).**

traffic analysis: 1. In a communications system, the analysis of traffic rates, volumes, densities, capacities, and patterns specifically for system performance improvement. **2.** The analysis of the external characteristics of tactical and strategic information and communications systems and components to obtain useful information, such as information that may be used (a) for drawing inferences of intelligence value, (b) as an aid to cryptoanalysis, (c) as a guide to efficient intercept operations, and (d) as a basis for communications deception. **3.** The analysis of the communications-electronics (C-E) environment for use in the design, development, and operation of new communications systems.

traffic capacity: The maximum traffic per unit of time that a given telecommunications system, subsystem, or device can carry under specified conditions. *See also* **busy hour, call•second, communications, erlang, traffic load.**

traffic control communications: *See* **air traffic control communications.**

traffic coordinator: *See* **line traffic coordinator.**

traffic encryption key (TEK): Key used to encrypt plain text, superencrypt text, or decrypt cipher text.

traffic engineering: The determination of the numbers and the kinds of circuits and the quantities of related terminating and switching equipment required to meet current and anticipated traffic loads throughout a communications system. *See also* **design margin, design objective, operational load, via net loss.**

traffic flow security: 1. The protection resulting from features, inherent in some cryptoequipment, that conceal the presence of valid messages on a communications circuit. *Note:* Traffic flow security usually is achieved by causing the circuit to appear busy at all times. **2.** Measures used to conceal the presence of valid messages in an online cryptosystem or a secure communications system. *Note:* Traffic flow security may be achieved by several methods, including (a) encrypting sender and receiver addresses, (b) causing the circuit to be busy at all times by sending dummy messages, and (c) more commonly, sending a continuous encrypted signal, whether or not user traffic is being transmitted. *See also* **cryptology, electronic warfare.**

traffic frequency: *See* **optimum traffic frequency.**

traffic intensity: In communications systems, the average occupancy of a facility during a specified period, i.e., the ratio of (a) the time during which a facility is continuously or cumulatively occupied to (b) the time during which the facility is available for occupancy. *Note 1:* Traffic intensity usually is measured in erlangs during the busy hour. *Note 2:* A traffic intensity of 1 erlang means continuous occupancy of a facility during the specified period, whether or not information is transmitted. Thus, if a telephone connection is established between two user end instruments and the connection remains for two hours, then even if the users are not talking, or talked only a part of the time, the traffic intensity for those telephones for that two-hour period is 1 erlang. If the telephones were connected for one-half hour during the two-hour period, the traffic intensity for those telephones is 0.25 erlang. *Synonym* **call intensity.** *See also* **busy hour, call-second, erlang, traffic load.**

traffic list: A list containing a serial number followed by the appropriate call sign and date-time group for each message to be transmitted by a station. *Note:* A station might transmit the traffic list prior to transmission of the messages identified in the traffic list in the order in which they are identified on the list. A broadcast serial number might consist of station letters or figures followed by a serial number from the series of numbers assigned consecutively to individual messages by an area broadcast station during a month.

traffic load: The total traffic carried by a trunk or a trunk group during a specified time interval. *See also* **busy hour, connections per circuit hour, erlang, group busy hour, level alignment, line load control, saturation, switch busy hour, traffic capacity, traffic intensity.** *Refer to* **Fig. B-10 (101).**

traffic metering device: *See* **telephone traffic metering device.**

traffic monitor: In a communications network, a network feature that provides basic statistics on the amount and the type of traffic handled by the network or a component of the network. *See also* **monitoring, service feature, station message detail recording, switch, traffic load, traffic usage recorder.**

traffic overflow: **1.** In a communications system, a condition in which (a) the traffic offered to a portion of the system exceeds its capacity, and (b) the excess traffic may be blocked or may be provided with alternate routing. **2.** The excess traffic that is blocked or provided with alternate routing when traffic offered to a portion of a communications system exceeds the capacity of that portion of the system. *See also* **alternate routing, overflow.**

traffic overload protection: *See* **automatic traffic overload protection.**

traffic register: In a communications system, a register that records specified events that occur in the system, such as (a) the number of call attempts that did not result in access, (b) the number of calls handled by a given trunk, (c) the number of ringing signals sent to a call receiver end instrument, that number being used to initiate a time-out signal, or (d) the number of abandoned calls. *See also* **register.**

traffic service position system (TSPS): A stored-program electronic system, i.e., a computerized system, that (a) is associated with one or more toll, i.e., long-distance, switching systems, and (b) at one location, provides centralized traffic service position functions for several central offices (COs).

traffic unit: *Synonym* **erlang.**

traffic usage recorder: A device for measuring and recording the amount of traffic, particularly telephone, telegraph, data, and facsimile traffic, carried by a line, a trunk, a switch, a trunk group, or several groups of switches or trunks. *See also* **monitoring, peg count, traffic monitor.**

traffic usage register: A device that is used to indicate, record, or both, the usage time of system components, such as the total cumulative time of usage, over an extended period, of communications system components, such as lines, circuits, trunks, and switches, particularly for telephone, telegraph, and facsimile equipment. *Note:* Usage time is recorded in convenient units, such as hundred call second (CCS) units.

traffic volume: *See* **communications traffic volume.**

trail: *See* **audit trail.**

trailer: *See* **magnetic tape trailer.**

train: *See* **pulse train, switch train.**

trajectory: *See* **ray trajectory.**

transaction: *See* **information transfer transaction, interactive data transaction.**

transaction routing: In communications systems operations, the routing of messages individually according to processed keys, destination names, or other data placed in the message heading by the originator, source user, or originating office.

transceiver: **1.** The combination of transmitting and receiving equipment (a) that is in a common housing, and (b) that (i) usually is portable or mobile, (ii) uses common circuit components for both transmitting and receiving, and (iii) usually provides half-duplex operation. **2.** A device that performs telecommunications transmitting and receiving functions within one chassis. *See* **facsimile transceiver, fiber optic transceiver, home fiber optic transceiver.**

transceiver dispersion: *See* **maximum transceiver dispersion.**

transcoding: The direct digital-to-digital conversion of signals from one encoding scheme, such as voice Linear Predictive Coding 10 (LPC-10), to a different encoding scheme without returning the signals to analog form. *Note:* The transcoded signals, *i.e.*,the digital representations of analog signals, may be any digital representation of any analog signal, such as voice, facsimile, or quasi-analog signals.

transducer: A simple device or component that converts energy from one form to another, usually with such fidelity that if the original energy represents information, the transformed energy represents the same information, its time derivative, or its time integral. *Note:* Examples of transducers are (a) microphones that convert sound waves to corresponding electrical currents, (b) modulated lasers that convert electrical currents to modulated lightwaves, (c) photodetectors that convert modulated lightwaves to electrical currents, and (d) piezoelectric crystals that produce voltages proportional to the time derivatives of the pressure waves that are applied to them and conversely produce pressure waves proportional to the time derivatives of the applied voltages. *See* **acoustooptic transducer, optoacoustic transducer, orthomode transducer.** *See also* **interface, optoelectronic.** *Refer to* **Fig. T-6 (1013).**

transfer: In communications systems, to send data from one location and to receive the data at another location. *See* **block transfer, call transfer, information transfer, peripheral transfer, successful block transfer, system failure transfer, technology transfer.**

transfer attempt: *See* **block transfer attempt.**

transfer characteristic: In a system, subsystem, or equipment, an intrinsic parameter that enables full or partial determination of the output when the input is known. *Note:* If the input, all the transfer characteristics, and their relationships are known, the output can be fully determined. *See also* **transfer function.**

transfer circuit: **1.** A circuit that transfers message traffic from (a) a system that is operated by one nation or group of nations to (b) a system that is operated by another nation or group of nations. **2.** A circuit used to transfer a message from one network to another via a terminal port, gateway, or bridge. *Note:* An example of a transfer circuit is a circuit used to transfer a message from a network that serves the originator, i.e., the source user, to a network that serves the message addressee, i.e., the destination user, perhaps via intermediate networks. *See also* **bridge, brouter, gateway, router.**

transfer failure: *See* **block transfer failure.**

transfer function: **1.** A mathematical statement that describes the transfer characteristics of a system, subsystem, or equipment. **2.** The relationship between the input and the output of a system, subsystem, or equipment in terms of the transfer characteristics. *Note 1:* When the transfer function operates on the input, the output is obtained. Given any two of these three entities, the third can be obtained. *Note 2:* Examples of simple transfer functions are voltage gains, reflection coefficients, transmission coefficients, and efficiency ratios. An example of a complex transfer function is envelope delay distortion. *Note 3:* For an optical fiber, the transfer function may be the ratio of the output optical power to the input optical power as a function of modulation frequency. *Note 4:* For a feedback circuit, the transfer function, T, is given by the relation $T = e_o/e_i = G/(1 + GH)$ where e_o is the output, e_i is the input, G is the forward gain, and H is the backward gain, i.e., the fraction of the output that is fed back and combined with the input in a subtractor. Whether the feedback is positive or negative depends on the sign, or phase, of H. If H is considered positive, i.e., in phase, the feedback is negative because of the subtractor, resulting in stability. If H is negative, the subtractor will make the feedback positive, resulting in instability. If $GH = -1$, the system will be completely unstable and possibly self-destruct if there are no limiting conditions. In mathematics, it produces a pole. *Note 5:* For a linear system, the transfer function and the impulse response are related through the Fourier transform pair, commonly given by the relations $H(f) = \int_{-\infty}^{+\infty} h(t)e^{i\omega t}dt$ where $H(f)$ is the ratio of output optical power to the input optical power as a function of modulation frequency, f is the frequency, t is time, and $h(t)$ is given by the relation $h(t) = \int_{-\infty}^{+\infty} H(f)e^{-\omega t}df$ where $\omega = 2\pi f$. $H(f)$ is often normalized to $H(0)$ and $h(t)$ is normalized to the relation $h_n(t) = \int_{-\infty}^{+\infty} h(t)dt$ which, by definition, is $H(0)$. *Synonyms* **baseband response function, frequency response function, response function.** *See* **closed loop transfer function, complex transfer function, fiber optic bundle transfer function, modulation transfer function, optical fiber transfer function.** *See also* **insertion loss versus frequency characteristic, transfer characteristic.** *Refer to* **Fig. T-7.**

transfer impedance: In a device or a circuit, the ratio of (a) the input voltage to (b) the output current. *Synonym* **transimpedance.**

transfer index: *See* **radiation transfer index.**

transfer lock: To change lock-on from one object, i.e., from one target, to another when a tracking device, such as a radar, sonar, optical tracker, or direction finder, is tracking an object. *See also* **lock-on, relock, track, tracking.**

transfer mode: In an integrated services digital network (ISDN), the transmitting, multiplexing, and switching of data. *See* **Asynchronous Transfer Mode, circuit transfer mode, deterministic transfer mode, packet transfer mode, synchronous transfer mode.**

transfer network: *See* **data transfer network, fiber optic data transfer network.**

transfer part: *See* **message transfer part.**

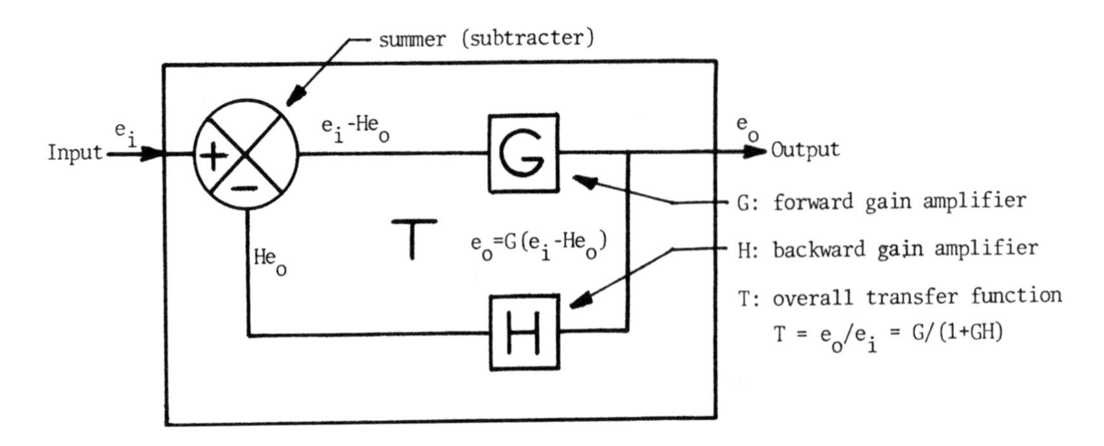

Fig. T-7. The overall **transfer function,** *T,* for the feedback amplifier is $T = G/(1 + GH)$ where G is the forward gain, and H is the backward gain, i.e., the feedback factor, both of which may be real or complex functions, i.e., scalar or vector functions.

transfer phase: *See* **data transfer phase.**

transfer point: *See* **signal transfer point.**

transfer protocol: *See* **connection oriented data transfer protocol, File Transfer Protocol, Simple Mail Transfer Protocol.**

transfer rate: *See* **actual transfer rate, block transfer rate, data transfer rate, effective data transfer rate, effective transfer rate.**

transferred information content: *Synonym* **transinformation.**

transfer request signal: *See* **data transfer request signal.**

transfer signal: *See* **request data transfer signal.**

transfer station: *Synonym* **terminal port.**

transfer success: *See* **block transfer success.**

transfer time: *See* **block transfer time, data transfer time, maximum block transfer time.**

transfer transaction: *See* **information transfer transaction.**

transformer: An electrical device that converts voltage, current, or impedance to either a higher or a lower level. *See also* **level.** *Refer to* **Fig. T-8.**

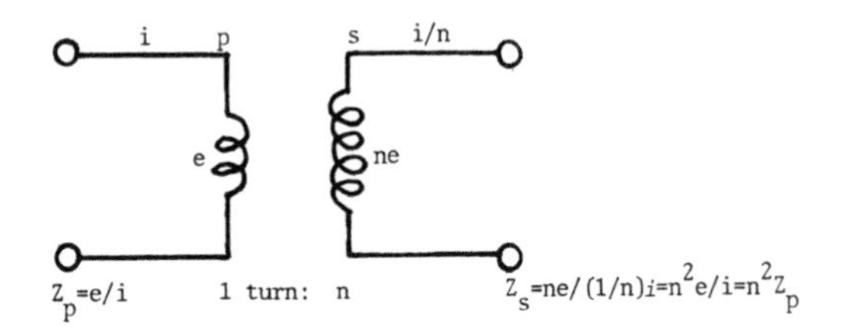

Fig. T-8. In this simple **transformer,** the turns ratio of primary to secondary is $1{:}n$. Thus, the voltage ratio is n, the current ratio is $1/n$, and the impedance ratio is n^2. All ratios are primary to secondary. For example, if the primary to secondary turns ratio is 10:1, i.e., there are 10 turns on the primary for each turn on the secondary, the turns ratio, n, is 0.1 and the transformer is called a step-down transformer, i.e., it reduces the voltage to 10%.

transient: 1. In communications, computer, data processing, and control systems, a temporary change in the value of a physical parameter or variable, such as (a) the current, voltage, power, or acoustic level, (b) frequency, or (c) phase. *Note 1:* Transients may be expressed in various units, such as (a) volts, (b) amperes, (c) watts, (d) volt•seconds, (e) ampere•seconds, i.e., coulombs, (f) watt•seconds, i.e., joules, (g) hertz, and (h) degrees or radians, all as a function of time. *Note 2:* Examples of transients are (a) a short-duration high amplitude surge or spike of electrical current or voltage in a wire line and (b) a sudden short-duration surge of optical power level in an optical fiber. **2.** In a wire line or an optical fiber, a change in a steady state power level, i.e., a change from one level to another, and usually back to the original level. *See* **dynamic variation.**

transient attenuation rate: 1. In a multimode dielectric waveguide, such as an optical fiber, a change (a) that occurs in the attenuation rate with distance along the guide between the entrance face and the equilibrium length, i.e., the equilibrium modal power distribution length, and (b) that can be an increase or a decrease in the attenuation rate with distance. *Note:* Within the equilibrium length, the attenuation rate changes with distance because of the over- or underexcitation of the lossy high order propagating modes. Beyond this length, the attenuation rate does not change with distance along the guide. Thus, this change, i.e., the "transient" condition, occurs as a function of distance. **2.** At a point between the entrance face of a multimode dielectric waveguide, such as an optical fiber, and the equilibrium length, i.e., the equilibrium modal power distribution length, a temporary change in the attenuation rate at the point, usually occurring during the rise and fall time of optical pulses or with temporal changes in environmental conditions. *Note:* The transient attenuation rate occurs as a function of time. The attenuation rate is a point function along an optical fiber.

transient protector: A device that will prevent transients from reaching equipment connected to a power or communications line. *Note:* An example of a transient protector is a device that filters or absorbs voltage surges, line noise, radio frequency interference (RFI), electromagnetic interference (EMI), and other forms of interference in a line before they reach connected terminal receiving or transmitting equipment. *Synonym* **surge protector.**

transient radiation effect on electronics (TREE): A temporal effect that (a) is caused by radiation, and (b) occurs particularly on active electronic elements, such as transistors and integrated circuits. *Note:* An example of a transient radiation effect on electronics is a radiation-induced change in the parameters of a transistor used in a combinational logic circuit so as to cause undesirable voltage, current, or power fluctuations or operating points and thus produce noise, interference, erroneous results, or system failure. *See also* **combinational circuit.**

transillumination: Illumination that (a) results from light passing through, rather than reflecting from, an element to be viewed, and (b) may be used to illuminate console panels or in indicators using edge-lighting or back-lighting techniques on clear, translucent, fluorescent, or laminated materials.

transimpedance: *See* **fiber optic transimpedance, optical transimpedance.**

transimpedance amplifier: *See* **photodetector transimpedance amplifier.**

transinformation content: The difference between (a) the information content that can be conveyed by the occurrence of an event and (b) the conditional information content that can be conveyed by the occurrence of the same event but given the occurrence of another event. *Note:* Seeing a bolt of lightning strike the ground bears a certain amount of information, such as magnitude and direction. Hearing the resultant thunder seconds later is indicative of the distance to the point of strike. This additional information is the transinformation content because it is dependent on the original event, the lightening strike. Hearing the thunder without seeing the strike also bears less information than both seeing the strike and hearing the thunder. This is also transinformation content for the two events. *Synonyms* **mutual information content, transferred information content, transmitted information content.** *See* **character mean transinformation content, mean transinformation content.**

transinformation rate: *See* **average transinformation rate.**

transistor: A solid state electronic device that (a) consists of a semiconducting material, such as germanium or silicon, (b) has two junctions, (c) has three regions alternately doped positive-negative-positive (p-n-p) or negative-positive-negative (n-p-n), one outer region serving as the emitter, or source of charge currents, the other outer region as the collector, and the intervening region as the base, (d) operates in such a manner that a large reverse bias is placed on the base-collector junction, thus allowing small signals applied between the base and the emitter to control larger currents and power in the base-collector circuit and thereby serve as a signal level amplifier, the power being supplied by a power source in the collector circuit, (e) can perform most of the functions that are performed by vacuum tubes, (f) has a low power dissipation loss, (g) can operate at extremely high frequencies, (h) is packaged in various ways, such as in individual plug-in units, mounted on a printed circuit board, or fabricated as an integrated circuit by various processes, such as etching and evaporation, to form large-scale integrated (LSI) circuits on chips, and (i) is capable of performing various functions, such as (i) amplifying electrical signals, (ii) mixing signals, (iii) modulating and demodulating signals, (iv) gating, (v) buffering, (vi) storing, (vii) shifting, (viii) shaping, (ix) switching, such as in communications, computer, data processing, and control systems in the areas of transmission, multiplexing, and signal processing, and (x) Boolean logic functions, such as AND, OR, NOR, NAND, and NEGATION. *See* **phototransistor.**

transistor integrated receiver: *See* **positive-intrinsic-negative field-effect transistor integrated receiver.**

transistor photodetector: *See* **field-effect transistor photodetector.**

transistor-transistor logic (TTL): A combinational circuit that (a) performs logic operations, such as gating operations, (b) consists of transistors that are connected directly to each other in such a manner that the output of one or more transistors is used as the input to one or more other transistors with few or no intervening circuit elements, and (c) is used in large-scale integrated (LSI) circuits. *See also* **combinational circuit, gate.**

transit: An eclipse of the Sun by a planet as the planet moves between (a) an observer or a photodetector and (b) the Sun.

transit delay: Between two given points in an integrated services digital network (ISDN), the sum of (a) the time between (i) the instant that the first bit of a data unit, such as a frame or a block, passes the first given point and (ii) the instant that bit passes the second given point and (b) the transmission time required by the length of the data unit. *See also* **frame.**

transition: 1. The changing from one specified condition to another specified condition, such as (a) from one current level to another, (b) from one state to another, such as from the solid state to the liquid state, (c) from a slab dielectric waveguide cross section to a round optical fiber cross section, and (d) in facsimile, from a white area to a black area or vice versa. **2.** In a signal, the changing from one significant condition to another. *Note:* Examples of transitions in a signal are (a) the changing from one voltage level to another in a square wave telegraph signal, (b) the shifting from one phase position to another in phase-shift keying, and (c) the translation from one frequency to another. *See* **energy level transition, signal transition.** *See also* **significant condition.** *Refer to* **Figs. N-17 (635), N-18 (636), P-4 (712).**

transition frequency: In an atomic system, the frequency associated with the difference between two discrete energy levels given by the relation $f_{2,1} = (E_2 - E_1)/h$ where $f_{2,1}$ is the frequency associated with the difference between two energy levels, E_2 and E_1 are the two energy levels, $E_2 \rangle E_1$, and h is Planck's constant. *Note:* If a transition from E_2 to E_1 occurs, a photon with frequency $f_{2,1}$ is likely to be emitted. *Refer to* **Fig. L-2 (502).**

transition optical fiber: An optical fiber of such geometric shape at input and output ends that it enables coupling between elements of different cross-sectional shape. *Note:* Examples of a transition optical fiber are (a) a fiber used to couple a thin-film optical waveguide to a round optical fiber, the transition fiber having a smooth transition from a rectangular to a round cross section, and (b) a tapered fiber used to couple two fibers with different core and cladding diameters.

transition zone: *Synonym* **intermediate field region.**

transit network identification: A network service feature that specifies the sequence of networks used to establish, or partially establish, a virtual circuit.

transit time: 1. *Synonym* **propagation time delay. 2.** *See* **satellite transit time.**

translating: In a display system, moving a display element, display group, or display image from one coordinate position to another without rotating the element, group, or image about a point within the system or relative to a point in the coordinate system in the display space on the display surface of the display device. *See also* **rotating, tumbling.**

translating program: *Synonym* **translator.**

translation: 1. In communications, changing the frequency without altering the baseband data or type of modulation. *Note:* In satellite communications systems, up-converters and down-converters are used to translate communications signals in satellite and Earth stations. **2.** In telegraph systems, reestablishing the text of a message from the received telegraph signal, including recording of the text. **3.** The conversion of text from one language to another. **4.** In computers and data processing, the conversion of a source program in one computer programming language to another program in another programming language that will produce the same results when executed. *Note:* The computer programming language translation occurs from the source language to the object language. *In a spatial sense, synonym* **displacement.** *See* **frequency translation, pregroup translation, single frequency translation.** *See also* **transliteration.**

translation equipment: *See* **channel translation equipment.**

translator: A device that converts information from one system of representation into equivalent information in another system of representation. *Note:* Examples of translators are (a) in telephone systems, a device that converts dialed digits into call routing information, (b) a computer program that translates a given language into another language, such as from one programming language into another programming language, (c) a computer program that translates a program written in one language to an equivalent program written in another language, (d) in communications systems, a device that changes the frequency of a signal without altering the baseband data or the type of modulation, (e) in radio and television broadcasting, a repeater station that receives a primary station signal, amplifies it, shifts it in frequency, and rebroadcasts it, and (f) a device that converts one frequency to another. *In computer programming, synonym* **translating program.** *See* **address translator, television broadcast translator, test loop translator.** *See also* **frequency translation, transponder.**

transliterate: To convert the characters of one alphabet to the corresponding characters of another alphabet. *Note:* Transliteration usually is performed on a one-for-one basis and is reversible.

transliteration: The letter-by-letter or the letter-by-letter-group conversion of text from the letters of one alphabet to the corresponding letters of another alphabet without changing the meaning but perhaps changing the pronunciation, such as (a) exchanging letters of the Roman alphabet for letters of the Greek alphabet, (b) exchanging Cyrillic or Hebrew letters for Roman letters, and (c) exchanging binary code patterns that are assigned to letters of a given alphabet, such as changing from the extended binary coded decimal interchange code (EBCDIC) alphabet to the American Standard Code for Information Interchange (ASCII) alphabet. *Synonym* **alphabet translation.** *See also* **translation.**

transmission: 1. The dispatching, for reception elsewhere, of a signal, a message, or another form of information representation. **2.** The propagation of a signal, a message, or another form of information representation, by any means, such as by telegraph, telephone, radio, television, or facsimile via any medium, such as wire, coaxial cable, microwave, optical fiber, heliograph, semaphore, or horns. **3.** In communications systems, a series of data units, such as blocks, messages, or frames. **4.** The transfer of energy from one location to another, such as the transfer of (a) electrical energy via conductors, (b) optical energy via optical fibers, and (c) acoustic energy via material media. *Note:* The transfer of energy may be for communications or power transfer purposes. *See* **analog transmission, anisochronous transmission, asynchronous transmission, batched transmission, bidirectional symmetrical transmission, bidirectional transmission, biternary transmission, bit stream transmission, black facsimile transmission, blind transmission, burst transmission, bursty transmission, byte serial transmission, colored light transmission, compatible sideband transmission, connection oriented mode transmission, data transmission, diffuse transmission, digital voice transmission, directional transmission, diversity transmission, double current transmission, double sideband reduced carrier transmission, double sideband transmission, duplex transmission, electrical transmission, electromagnetic wave transmission, flaghoist transmission, flashing light transmission, full carrier transmission, half-duplex transmission, hand flag transmission, independent sideband transmission, infrared transmission, isochronous burst transmission, isochronous transmission, loop transmission, low detectability electromagnetic wave transmission, multiple media transmission, parallel transmission, point to point transmission, polar direct current transmission, pyrotechnic transmission, quasi-analog transmission, radiotelegraph transmission, radiotelephone transmission, real time transmission, reduced carrier transmission, secure transmission, selective transmission, serial transmission, sideband transmission, single sideband suppressed carrier transmission, single sideband transmission, specular transmission, start-stop transmission, store and forward data transmission, suppressed carrier transmission, synchronous transmission, time diversity transmission, two-independent-sideband transmission, unidirectional transmission, vestigial sideband transmission, victim transmission, visual transmission, white facsimile transmission, white transmission, wireline transmission.** *See also* **transfer, transmit, two-independent-sideband carrier.** *Refer to* **Fig. W-3 (1082).**

transmission authentication: The establishment of the authority of the originator or the source of transmission that includes (a) self-authentication, station authentication, and message authentication and (b) procedures that enable stations to indicate and convey the authenticity of their own transmission.

transmission block: 1. A group of bits or characters that (a) is transmitted as a unit and (b) usually contains coding for error control purposes. **2.** In data transmission, a group of records sent, processed, or recorded as a unit. *Note:* A transmission block usually is terminated by an end of block character (EOB), end of transmission block character (ETB), or end of text character (EOT or ETX). *See also* **binary digit, character.**

transmission block character: *See* **end of transmission block character.**

transmission channel: All of the transmission facilities between the input to a channel at an initiating node and the output from the channel at a terminating node. *Note 1:* A transmission channel plus two loops constitutes a circuit. *Note 2:* In telephony, transmission channels may have various bandwidths, such as nominal 3-kHz (kilohertz), nominal 4-kHz, or nominal 48-kHz-group bandwidths. *Note 3:* "Transmission channel" is not to be confused with the more general term "channel." *See* **asynchronous data transmission channel, data transmission channel, synchronous data transmission channel.** *See also* **bandwidth, channel, information bearer channel.**

transmission character: *See* **end of transmission character.**

transmission circuit: *See* **data transmission circuit.**

transmission coefficient (TC): 1. The ratio of (a) the transmitted field strength to (b) the incident field strength of an electromagnetic wave when it is incident upon an interface surface between media with two different refractive indices. *Note 1:* At oblique incidence and when the electric field component of an incident plane polarized electromagnetic wave is parallel to the interface surface, the transmission coefficient is given by the relation $T_e = 2m_2 \cos A/(m_2 \cos A + m_1 \cos B)$ where m_1 and m_2 are the reciprocals of the refractive indices of the incident and transmitted media, respectively, and A and B are the incidence and refraction angles, respectively. This is one of the Fresnel equations. *Note 2:* At oblique incidence and when the magnetic field component of an incident plane polarized electromagnetic wave is parallel to the interface surface, the transmission coefficient is given by the relation $T_m = 2m_2 \cos A/(m_1 \cos A + m_2 \cos B)$ where m_1 and m_2 are the reciprocals of the refractive indices of the incident and transmitted media, respectively, and A and B are the incidence and refraction angles, respectively. This is another Fresnel equation. *Note 3:* The sum of the reflection coefficient and the transmission coefficient is not necessarily unity. **2.** At a discontinuity in a transmission line, the ratio of (a) the amplitude of the complex transmitted wave to (b) the amplitude of the incident wave. **3.** The probability that a portion of a communications system, such as a line, circuit, channel, link, or trunk, will meet specified performance criteria. *Note:* The value of the transmission coefficient is directly related to the quality of the line, circuit, channel, link, or trunk. **4.** At a refractive interface, the ratio of (a) the amplitude of the transmitted wave to (b) the amplitude of the incident wave. **5.** A number that indicates the probable performance of a portion of a transmission circuit. *Note:* The value of the transmission coefficient is directly related to the quality of the line, circuit, channel, link, or trunk. *Synonym* **refraction coefficient.** *See also* **Fresnel reflection loss, path loss, reflection coefficient, transmission loss.**

transmission command: *See* **asynchronous transmission command.**

transmission control: *See* **port radio transmission control, radio transmission control.**

transmission control character: A character used to control or facilitate the transmission of data over communications networks, usually for data communication control over data links between sets of data terminal equipment (DTE) at two separate stations. *Note:* Examples of transmission control characters are ACK, DLE, ENQ, EOT, ETB, ETX, NAK, SOH, STX, SYN, and WRU. *Synonyms* **communication control character, data communication control character, data transmission control character.** *See also* **control character.**

transmission control protocol (TCP): A protocol used for controlling host-to-host transfer in a communications system. *Note:* Examples of transmission control protocols are (a) the Transmission Control Protocol in the Internet Protocol (IP) suite and (b) the U.S. Department of Defense Standard Transmission Control Protocol. *See also* **host, host computer, transfer.**

Transmission Control Protocol (TCP): 1. In the Internet Protocol (IP) suite, a network control protocol that controls host-to-host transmissions over packet switching communications networks. **2.** The U.S. Department of Defense standard protocol intended for use as a reliable host-to-host protocol in a packet-switched computer/communications network, corresponding closely to the Open Systems Interconnection—Reference Model (OSI-RM) Layer 4, i.e., the Transport Layer. *See also* **control character, host, host computer, Open Systems Interconnection—Reference Model, protocol.**

Transmission Control Protocol/Internet Protocol (TCP/IP) Suite: A set of protocols used for transmission control in Internet.

transmission efficiency: 1. The ratio of the power (a) at the output of a functional unit to (b) the power at the input. *Note:* For passive devices, the transmission efficiency will always be less than unity. **2.** The ratio of (a) the number of good or correct data units, such as bits, bytes, characters, or words, received at a point to (b) total number of the same units that were transmitted to the point.

transmission ending: In radiotelegraph and radiotelephone transmission, a sign or a symbol used to indicate the end of a message or a transmission. *Note:* Examples of endings are the prosigns (a) K for an invitation to transmit, (b) AR for end of transmission in a radio or tape relay data transmission system, (c) ROGER for a radiotelephone voice transmission receipt of message, and (d) OVER for radiotelephone voice invitation to transmit.

transmission factor: *Synonym* **internal optical density.**

transmission frame: A data structure that (a) begins and ends with data delimiters, and (b) consists of fields predetermined by a protocol for the transmission of user data and control data.

transmission frequency: In radar, the radio or video frequency of the cycles that (a) are repeated within each transmitted pulse duration, and (b) usually occur at a rate several orders of magnitude higher than the pulse repetition rate. *Note:* The pulse repetition rate is the number of radar pulses that occur per unit time. Each pulse is transmitted for its entire duration at a certain transmission radio frequency (rf) or frequencies.

transmission function: *See* **end of transmission function.**

transmission group: *See* **digital transmission group.**

transmission identification: In a message heading, a required identification that (a) provides a means of maintaining continuity of traffic, (b) is used to identify messages on a channel between two stations, and (c) may consist of (i) only a station designator and a channel designator or (ii) a start of message indicator, station and channel designators, a figures shift, numeral characters, and a letters shift, especially for stations that transmit directly into fully automatic relay stations.

transmission instruction: In radiotelegraph communications, an instruction that (a) usually consists of prosigns, call signs, address designators, and operating signals that are required for routing, relaying, and delivering of messages, and (b) in direct communication between stations, may not be required.

transmission interface: *See* **data transmission interface.**

transmission layer: In a layered system, the set of functions and protocols that pertain to the transfer of data between geographically distinct locations. *See also* **Open Systems Interconnection—Reference Model.**

transmission level: **1.** At a point in a transmission system, the absolute or relative current, voltage, or power level. *Note:* The absolute power level usually is expressed in dBm. The relative power level usually is expressed in dB relative to a specified reference level, such as the zero transmission level point (0TLP). **2.** At any given point in a transmission system, the power that should be or is measured at that point when a standard test signal, such as a 0-dBm 1000-Hz (hertz) test signal, is transmitted at some point chosen as a reference point. *Note 1:* The transmission level usually is measured in dBm. *Note 2:* The measured absolute power level of a point depends on system design and the test signal. It is a measure of (a) the design gain, i.e., the nominal gain, at 1000 Hz, of the system between the chosen reference point, i.e., the zero transmission level point (0TLP), and the point at which the absolute power is being measured, and (b) the deviation from the design gain. *See* **relative transmission level.** *See also* **dBr, level, net gain, net loss, standard test signal, standard test tone.**

transmission level point (TLP): A point in a transmission system at which the ratio of (a) the power of a test signal to (b) the power of the test signal at a reference point is specified. *Note 1:* The power ratio usually is expressed in dB. *Note 2:* A zero transmission level point (0TLP) is an arbitrarily established point in a communications circuit to which all levels at other points in the circuit are referred. *Note 3:* The power level measured at a point in a circuit is so closely associated with the point that the power level and the point are used interchangeably. For example, a point where a reading of −16 dBm is obtained would be a "−16 transmission level point," often abbreviated as −16 TLP. This point could be designated as the zero transmission level, i.e., 0TLP. *See* **zero dBm transmission level point, zero transmission level point.** *See also* **dBr, level, relative transmission level, system loading.**

transmission level reference point: *Synonym* **reference transmission level point.**

transmission line: **1.** A line that (a) physically connects separated points and (b) is capable of transferring signals or power between the points. *Note:* Examples of communications transmission lines are open wires, twisted pairs, coaxial cables, overhead cables, and fiber optic cables. Examples of power transmission lines are high voltage cross-country power lines and underground power buses. **2.** The material medium or structure, such as a propagation medium, that forms all or part of a path from one place to another for directing the transmission of energy, usually in the form of currents and waves, such as electric currents, magnetic fields, acoustic waves, and electromagnetic waves. *Note 1:* Examples of transmission lines include wires, optical fibers, coaxial cables, rectangular closed waveguides, and dielectric slabs. *Note 2:* Transmission lines are primarily used for the transfer of signals or electric power. *See also* **propagation medium.** *Refer to* **Fig. T-3 (991).**

transmission loss: In the transmission of a signal from one point to another, the decrease in power level that occurs (a) within a component, (b) from the output of one component to the input of another component, or (c) from one point to another in a propagation medium. *Note:* Transmission loss usually is expressed in dB. The number of dB of transmission loss is given by the relation $10 \log_{10} (P_o/P_i)$ where P_o/P_i is the ratio of the output power at one point to the input power at another point, such as the output power divided by the input power of a transmission line, a connector, or another component. The transmission loss will be expressed as a negative number and labeled as a loss because P_o/P_i is less than unity, i.e., less than 1. There will always be a transmission loss when only passive elements are between the points of measure. If there are active elements and the ratio of P_o/P_i is greater than unity, the value of $10 \log_{10} (P_o/P_i)$ will be positive, and a transmission gain, in dB, will be indicated. *See also* **attenuation, insertion loss, net loss, optical density, path loss, reflection, scattering loss, transmittance.**

transmission loss distance: The distance in which a specified power loss or attenuation occurs for an electromagnetic signal of specified wavelength propagating in a given propagation medium. *Note 1:* Examples of transmission loss distances might be 0.01 km (kilometer) for ordinary window glass and 100 km for the purified silica glass used in optical fibers, both for a 5-dB loss at a wavelength of 0.85 µm (microns). *Note 2:* Transmission loss distance for a given material may be expressed in kilometers per dB at a specified wavelength.

transmission medium: *Synonym* **propagation medium.**

transmission mode: The manner of transmission for a system that makes use of one or more propagation media. *Note:* Examples of transmission modes are telephone, radio, television, microwave, telegraph, facsimile, wirephoto, radiophoto, semaphore, wigwag, heliograph, flaghoist signaling, panel signaling, and body language. *See also* **propagation medium, propagation mode.**

transmission rate: In a communications system, the rate at which data units, such as characters, words, and messages, are (a) sent from a point, (b) received at a point, or (c) pass through a point in the system. *Note:* Transmission speeds may be measured or expressed in any convenient unit, such as bits per second (bps), characters per minute (cpm), words per minute (wpm), and messages per hour or per day. *Synonym* **transmission speed.** *See* **average transmission rate, effective transmission rate, instantaneous transmission rate.** *See also* **transmission time.**

transmission security: 1. Communications security that results from all measures designed to protect transmissions from interception and exploitation by means other than cryptanalysis. **2.** The protection of information to prevent its unauthorized disclosure during transmission. **3.** The result obtained from the protection given to prevent the unauthorized disclosure of information during transmission. *See* **ship radio transmission security.** *See also* **communications security.**

transmission security key (TSK): Key used in the control of transmission security processes, such as in frequency hopping and spread spectrum systems. *See also* **key, transmission security.**

transmission service: A service that provides the means for transmitting messages from one point to another, such as from one user end instrument to another user end instrument. *See* **circuit switched data transmission service, data transmission service, leased circuit data transmission service, packet-switched data transmission service, public data transmission service.**

transmission service channel: In video systems, the one-way video signal transmission path between two designated points. *See also* **video, video signal.**

transmission speed: *Synonym* **transmission rate.**

transmission system: 1. In communications, computer, data processing, and control systems, a system that effects the transfer of energy that represents information from point to point. **2.** In power systems, a system that effects the transfer of energy that does not represent information from point to point. *See* **digital transmission system, double sideband suppressed carrier transmission system, fiber optic transmission system, laser fiber optic transmission system, start-stop transmission system, stepped start-stop transmission system.** *See also* **transfer.**

transmission time: 1. The time required for a message to pass a point in a transmission system. *Note:* The transmission time is a function of the length of the message and the data signaling rate (DSR), i.e., the signaling speed. **2.** The time between (a) the start of transmission of a message at the message source, such as an originating station, and (b) the receipt of the end of the message at the message sink, such as the destination station. **3.** In facsimile systems, the time between (a) the start of picture signals and (b) the detection of the end of the transmission character (EOT) by the receiver for a single object, such as a single document. *See also* **facsimile, propagation time delay, transmission speed.**

transmission window: *Synonym* **spectral window.**

transmissivity: In optics, the transmittance of a unit length of material, at a given wavelength, excluding the reflectances at the surfaces of the material, i.e., the internal transmittance per unit thickness of a nondiffusing substance, such as clear glass, plastic, or crystal. *Note:* The term is no longer in common use. *See also* **transmittance.**

transmit: To move an entity (a) from one place toward another, such as to broadcast radio waves, to dispatch data via a propagation medium, and to send data from data circuit-terminating equipment (DCE), (b) with the intent that it be received at another location, (c) in communications systems, without including propagation and reception, so that the process is not communication, and (d) in a manner similar to emission. *See also* **emit, receive.**

transmit after receive time delay: The time between (a) the instant that radio frequency (rf) energy at the local receiver input is removed and (b) after the local transmitter is automatically keyed on, the instant that the transmitted rf signal amplitude has increased to a specified level, usually 90% of its steady state value. *Note:* High frequency (HF) transceiver equipment usually is not designed with an interlock between the receiver squelch and the transmitter on-off key. The transmitter can be keyed on at any time, whether or not a signal is being received at the receiver input. *See also* **transmitter attack time delay, transmitter release time delay.**

transmit flow control: In data communications systems, the control of the rate at which data are transmitted from a terminal so that the data can be received by another terminal. *Note 1:* Transmit flow control may occur between data terminal equipment (DTE) and a switching center, via data circuit-terminating equipment (DCE), or between two DTEs. The transmission rate may be controlled because of network or DTE requirements. *Note 2:* Transmit flow control can occur independently in the two directions of data transfer, thus permitting the transfer rates in one direction to be different from the transfer rates in the opposite direction. *See also* **control character, data terminal equipment, data transfer time, data transmission, respond opportunity, transfer.**

transmittance: The ratio of the transmitted power to the incident power when an electromagnetic wave, such as a lightwave, is incident upon an interface surface between propagation media with different refractive indices. *Note 1:* In communications, transmittance usually is expressed in dB. In optics, transmittance usually is expressed as optical density, either in percent or as the base 10 logarithm of the reciprocal of the transmittance. *Note 2:* The conditions under which the transmittance, and the associated reflectance, occurs should be stated, such as the wavelength composition and polarization of the incident wave, the geometrical distribution of the reflecting surface or surfaces, and the composition of the media on both sides of the reflecting surface. *Note 3:* Transmittance was formerly called "transmission." *See* **coupler transmittance, optical transmittance, spectral transmittance.** *See also* **optical density, transmissivity.**

transmittance change: *Synonym* **induced attenuation.**

transmittance density: *Synonym* **optical density.**

transmittance matrix: *See* **coupler transmittance matrix.**

transmittancy: The ratio of (a) the transmittance of a solution, consisting of a solvent and a solute, to (b) that of an equal thickness of the solvent alone.

transmitted information content: *Synonym* **transinformation content.**

transmitted power: The energy per unit time, i.e., the power, such as electrical, optical, or acoustic power, propagating in a direction normal to a specified surface, such as (a) the cross section of a communications cable, (b) the surface of a fictitious sphere completely surrounding an antenna, and (c) a unit area in air or in free space. *Note 1:* Transmitted power can vary with time, and the specified cross-sectional area can change. Thus, transmitted power can be measured or expressed in many forms, such as (a) the peak envelope power, (b) the power in a given direction, (c) the power averaged over time, (d) the power averaged over an area or a solid angle, (e) the total carrier power delivered to an antenna, (f) the total power radiated and integrated over all directions, (g) the power limited to a specified portion of a frequency spectrum or bandwidth, (h) the power emanating per unit area per unit solid angle from a source, i.e., the radiance, and

(i) the power incident per unit area, i.e., the irradiance. *Note 2:* Transmitted power usually is expressed in watts, watts per square meter, or watts per steradian. *See also* **irradiance, radiance.**

transmitter (XMITTER): A device that (a) is capable of sending signals, and (b) usually is capable of generating and modulating signals for various purposes, such as communications, command, control, and telemetering purposes. *Note:* Examples of transmitters are (a) combinations consisting of a microphone, amplifier, and modulator for sending electromagnetic waves, (b) microphones in telephones, and (c) modulated lasers used in fiber optic links. *See* **automatic numbering transmitter, automatic telegraph transmitter, drum transmitter, electrooptic transmitter, facsimile transmitter, fiber optic transmitter, flat bed transmitter, jammer transmitter, optical transmitter, optoelectronic transmitter, radio transmitter, reperforator-transmitter, ship emergency transmitter, telegraph transmitter, telephone transmitter, tunable transmitter.** *Refer to* **Figs. I-7 (476), R-11 (848), T-3 (991).**

transmitter attack time delay: The time that (a) is between (i) the instant that a transmitter is keyed on and (ii) the instant that the transmitted radio frequency (rf) signal amplitude has increased to a specified level, usually 90% of its key-on steady state value, and (b) does not include the time required for automatic antenna tuning. *See also* **transmit after receive time delay.**

transmitter calibration factor: In satellite communications, a parameter used in uplink power balancing calculations that (a) are used to determine the minimum Earth station transmitter power required on each communications satellite link in order to achieve minimum downlink signal strength for reception, (b) take into account all system variables that influence power levels, and (c) usually use the output power of an Earth station as the reference transmission level point. *See also* **reference transmission level point.**

transmitter central wavelength range: In fiber optic communications, the total allowed range of fiber optic transmitter central wavelengths that (a) is the worst case variation, (b) is caused by many factors, such as manufacturing tolerances, temperature variations, and aging under standard or extended operating conditions, (c) is often given by the relation $\Delta\lambda = \lambda_{tmax} - \lambda_{tmin}$ where $\Delta\lambda$ is the transmitter central wavelength range, λ_{tmax} is the maximum transmitted wavelength in the range, and λ_{tmin} is the minimum transmitted wavelength in the range, and (d) has limits at the points where the optical power level per hertz is down 3 dB. *See also* **fiber optic receiver, fiber optic transmitter.**

transmitter fiber optic connector: The fiber optic connector that (a) is provided at the output of a fiber optic transmitter and (b) is attached to the transmitter pigtail. *Note:* The transmitter fiber optic connecter description should include (a) the name of the manufacturer, (b) the connector type, such as biconic or FC, (c) the model number, (d) the design application, such as multimode or single mode, and (e) the mating connector model number.

transmitter group: *See* **synchronized transmitter group.**

transmitter information descriptor: In fiber optics, a unique descriptor from which information about a transmitter can be determined, such as the (a) manufacturer, (b) terminal equipment association, (c) system design application, such as single mode or multimode, (d) operating wavelength, (e) output power level, (f) source type, such as laser or light-emitting diode (LED), (g) temperature controller, according to its Federal Communications Commission (FCC) classification, such as Class I or Class II, and (h) manufacturer product change designation, such as original issue or revision.

transmitter link: *See* **studio to transmitter link.**

transmitter mean power: *See* **radio transmitter mean power.**

transmitter optical power: The worst case minimum value of optical power coupled into a fiber optic cable on the line side of the fiber optic transmitter connector under specified standard or extended operating conditions using standard measurement procedures. *Note 1:* The transmitter optical power usually is expressed in dBm. *Note 2:* The worst case minimum value combines manufacturing, temperature, and aging variations in a worst case manner. *See also* **fiber optic transmitter.**

transmitter power output rating: The power output of a radio transmitter under stated conditions of operation and measurement. *Note:* Power output ratings may be made by using different measurement criteria, such as those used to measure peak envelope power, peak power, mean power, carrier power, noise power, and stated intermodulation level. *See also* **output rating.**

transmitter release time delay: The time between (a) the instant that a transmitter is keyed off and (b) the instant that the transmitted radio frequency (rf) signal amplitude has decreased to a specified level, usually

10% of its key-on steady state value. *See also* **transmitter attack time delay.**

transmitter start code: A code that (a) is sent by a switching center, (b) is used for polling all stations, and (c) if a polled station has no message to transmit, triggers the station to respond with an answer back code. *See also* **call directory code.**

transmitter: *See* **emergency locator transmitter.**

transmit terminal: *See* **multiplex baseband transmit terminal.**

transmit time: 1. *In transmission systems, synonym* **propagation time delay. 2.** *In satellite systems, synonym* **satellite transit time.**

transmit time delay: *See* **receive after transmit time delay.**

transmitting element: The radiating terminus at an optical junction.

transmitting star coupler: *Synonym* **nonreflective star coupler.**

transmultiplexer (XMUX): Equipment that transforms (a) signals, such as group or supergroup signals, derived from frequency division multiplexing equipment to (b) time division multiplexed signals having the same structure as those derived from primary or secondary pulse code modulation (PCM) multiplexing equipment, and vice versa. *See also* **frequency, frequency division multiplexing, pulse code modulation, time division multiplexing.**

transparency: 1. The property of an entity that allows another entity to pass through it without appreciably altering either of the entities, such as the property of a thin sheet of clear glass when subjected to incident lightwaves. *Note:* In the case of lightwave transmission, a high transparency implies little or no absorption or scattering of the lightwaves. Zero transparency implies opacity, i.e., complete absorption, reflection, or backscattering. **2.** In telecommunications networks, the property that allows end-to-end transmission of signals without changing their form or information content. **3.** The quality of a data communications system or device that uses a bit-oriented link protocol that does not depend on the bit sequence structure used by the data source. **4.** An image fixed on a clear base by means of a photographic printing, chemical, or other process, especially adaptable for viewing by transmitted light. *See* **data circuit transparency, data signaling rate transparency, inherent transparency.** *See also* **code-independent data communication, commonality, communications, compatibility, Integrated Services Digital Network, interoperability, opacity, transparent network.**

transparent: 1. Pertaining to the property of a material in which electromagnetic waves, usually in the visible spectrum, are propagated without attenuation, except attenuation caused by divergence, i.e., geometric spreading, and thus pertaining to a material with an extinction coefficient of zero. *Note:* Most optical elements, such as lenses, prisms, and optical fibers, are intended to be transparent. **2.** Pertaining to transparency. *See also* **Bouger's law, code-independent, seamless network, transparency.**

transparent interface: An interface that allows the connection and the operation of a system, subsystem, or equipment with another system, subsystem, or equipment without modification of system characteristics or operational procedures on either side of the interface. *See also* **commonality, interface, interoperability, mechanically intermateable connectors.**

transparent mode: A mode of operating a data transmission system in which all the bits are transmitted as bits in a bit stream without distinction, including special purpose characters, such as control characters, which are not interpreted as such.

transparent network: In telecommunications, a network that (a) allows end-to-end, i.e., source-user-to-destination-user, transmission of signals without changing the form or the information content of the signals, and (b) uses a bit-oriented link protocol that does not depend on the bit sequence structure used by the data source. *See also* **transparency.**

transponder: 1. A receiver-transmitter combination that receives, amplifies, and retransmits a signal on a different frequency. **2.** A device that automatically transmits a predetermined message in response to a predefined received signal. *Note:* Transponders are used in identification friend or foe (IFF) systems and air traffic control (ATC) systems. **3.** A receiver-transmitter that will generate a reply signal upon proper interrogation. **4.** A radio or radar transmitter-receiver that (a) automatically retransmits signals when the proper signals are received, (b) depends on some frequency memory or timing procedure to effect on-frequency retransmissions, (c) usually accepts the electronic challenge of an interrogator and automatically transmits an appropriate reply, (d) in a satellite communications system, may consist of receiver and transmitter relay equipment, (e) usually has an output power, frequency, and direction different from those of the input, and (f) may serve as a deception device that receives a signal and returns another signal of a predetermined nature, such as by transmitting high amplitude signals to create multiple false targets on a radar screen. *See also* **identification friend or foe.**

transportability: 1. In communications, the quality of equipment, devices, systems, and associated hardware that permits their being moved from one location to another to interconnect with local equipment or networks. *Note:* Transportability implies the use of standardized components, such as standardized plugs and transmission media. **2.** The capability of material to be moved by towing, self-propulsion, or carrier via any means, such as by rail, highway, waterway, pipeline, ocean, or air. *See also* **commonality, compatibility, interchangeability, interoperability, mobile service, mobile station, portability.**

transportable: 1. Pertaining to equipment that (a) may be carried or moved from location to location and (b) usually is not operated while it is being moved. **2.** The characteristic of an item such that it (a) must be packaged for transport on a carrier, such as a vehicle, aircraft, or ship, and (b) must be unloaded from the carrier, unpacked, assembled, connected, and powered for operation. *See also* **portable, mobile.**

transportable communications equipment: Fixed communications equipment that (a) may be easily transported, (b) is rapidly assembled and disassembled, (c) may be operated in a changing, moving, or tactical environment, and (d) usually is not operated while being moved.

transportable station: A station that (a) can be transferred to various fixed locations and (b) is not intended to be used while being transferred.

transport layer: In open systems network architecture, the layer that provides functions and facilities for the actual movement of data among network elements, such as among user end instruments, land vehicles, ships, aircraft, spacecraft, subsystems, control panels, and network nodes. *Note:* Microwave, radio, coaxial cable, satellite, wire, and fiber optic systems can be used to implement the transport layer and the physical layer. *See* **end to end transport layer.** *See also* **layer, Open Systems Interconnection—Reference Model.**

Transport Layer: *See* **Open System Interconnection—Reference Model.**

transport station: In distributed systems, the protocols, procedures, processes, and mechanisms that execute the data transfer protocols of the transport layer. *See also* **transport layer.**

transposition: 1. In data transmission, a transmission defect in which, during one character period, one or more signal elements are changed from one significant condition to the other, and an equal number of elements are changed in the opposite sense. *See also* **code, translator. 2.** In outside plant configuration, an interchange of spatial positions of the several conductors of a circuit between successive lengths. *Note:* Transposition usually is used to achieve inductive cancellation and thus reduce interference in communications circuits.

transputer: A computer chip that consists of arrays of interconnected transistorized logic gates, such as several million gates made from many more millions of transistors. *See also* **combinational circuit, gate.**

transverse compression sensor: *See* **fiber optic transverse compression sensor.**

transverse displacement loss: *Synonym* **lateral offset loss.**

transverse electric (TE): Pertaining to an electromagnetic wave in which the electric field vector is perpendicular to the direction of propagation and therefore, in a waveguide, such as an optical fiber, always perpendicular to the longitudinal axis of the guide, regardless of the direction of the magnetic field vector.

transverse electric (TE) mode: In an electromagnetic wave propagating in a waveguide, such as (a) a rectangular closed waveguide, (b) a dielectric waveguide, e.g., an optical fiber, or (c) a thin-film waveguide, a mode in which the electric field vector is normal to the longitudinal axis of the waveguide, and the magnetic field vector is not necessarily normal to the longitudinal axis of the waveguide. *Note:* In an optical fiber, transverse electric and transverse magnetic modes correspond to meridional and axial rays. *See also* **meridional ray, mode, transverse magnetic mode.**

transverse electric (TE) wave: An electromagnetic wave propagating in a propagation medium, including free space, in such a manner that the electric field vector is directed entirely transverse, i.e., perpendicular, to the general forward direction of propagation. *Note:* In a transverse electric (TE) wave propagating in a waveguide, the electric field vector is entirely perpendicular to the longitudinal axis of the waveguide, but the magnetic field may have a longitudinal component. In optical systems, the propagation medium usually is a waveguide with a uniform cross section in the longitudinal direction. *See also* **transverse magnetic wave.**

transverse electromagnetic (TEM) mode: In an electromagnetic wave propagating in a waveguide, a mode that has its electric and magnetic field vectors perpendicular to the longitudinal axis of the waveguide. *Note 1:* A transverse electromagnetic (TEM) mode might not exist in a given electromagnetic wave in a given waveguide. A transverse electric (TE) or a transverse magnetic (TM) mode might exist, but usually not both. *Note 2:* In an electromagnetic wave, whether in free space or in a waveguide, the instantaneous direction of propagation is always perpendicular to both the electric and the magnetic field vectors, i.e., in the direction of the Poynting vector. *See also* **mode, Poynting vector.**

transverse electromagnetic (TEM) wave: An electromagnetic wave in which (a) the electric field vector and the magnetic field vector, although varying in time at every point in the space occupied by the wave, whether in free space or in a material propagation medium including that of a waveguide, are contained in a local plane at that point, (b) the orientation of the plane at that point is independent of time, and (c) the orientation of the local plane is different for different points in space or material media, the exception to this being the special case of a uniform plane polarized electromagnetic wave, in which all the polarization planes are parallel, and the direction of propagation is perpendicular to these planes.

transverse interferometry: The measurement of the refractive index profile of a dielectric waveguide, such as an optical fiber, by (a) placing a short length of the guide in an interferometer, (b) illuminating it in a direction transverse to the optical axis, and (c) usually using a computer to interpret the interference patterns obtained. *See also* **interferometer, interferometry, slab interferometry, transverse scattering.**

transverse jitter: In facsimile transmission systems, an effect that (a) is caused by the irregularity of the scanning pitch and (b) results in concurrent overlap and underlap in the recorded copy. *See also* **recorded copy, scanning pitch.**

transverse magnetic (TM): Pertaining to an electromagnetic wave in which the magnetic field vector is perpendicular to the direction of propagation and therefore, in a waveguide, such as an optical fiber, always perpendicular to the longitudinal axis of the guide, regardless of the direction of the electric field vector.

transverse magnetic (TM) mode: In an electromagnetic wave propagating in a waveguide, such as (a) a rectangular closed waveguide, (b) a dielectric waveguide, e.g., an optical fiber, or (c) a thin-film waveguide, a mode in which the magnetic field vector is normal to the longitudinal axis of the waveguide, and the electric field vector is not necessarily normal to the longitudinal axis of the waveguide. *Note:* In an optical fiber, transverse electric and transverse magnetic modes correspond to meridional and axial rays. *See also* **meridional ray, mode, transverse magnetic mode.**

transverse magnetic (TM) wave: An electromagnetic wave propagating in a propagation medium, including free space, in such a manner that the magnetic field vector is directed entirely transverse, i.e., perpendicular, to the general forward direction of propagation. *Note:* In a transverse magnetic (TM) wave propagating in a waveguide, the magnetic field vector is entirely perpendicular to the longitudinal axis of the waveguide, but the electric field may have a longitudinal component. In optical systems, the propagation medium usually is a waveguide with a uniform cross section in the longitudinal direction. *See also* **transverse electric wave.**

transverse mode: *See* **lowest-order transverse mode.**

transverse offset: *Synonym* **lateral offset.**

transverse offset loss: *Synonym* **lateral offset loss.**

transverse parity check: A parity check performed on a group of binary digits recorded on parallel tracks of a data medium, such as a magnetic disk, tape, drum, or card.

transverse propagation constant: In a waveguide, such as a closed rectangular waveguide, and an optical fiber, the propagation constant evaluated in a direction perpendicular to the longitudinal axis of the guide. *Note:* The transverse propagation constant for a given mode can vary with the transverse coordinates, i.e., with the distance from the longitudinal axis of the guide. *See also* **propagation constant.**

transverse redundancy check (TRC): In synchronized parallel data streams, a redundancy check (a) that is based on the formation of a block check following preset rules, (b) in which the check formation rule applied to blocks is also applied to characters, and (c) in which the check is made on parallel bit patterns, i.e., each bit of a character is in a separate data stream with separate streams for the check bits. *Note 1:* When the transverse redundancy check (TRC) is based on a parity bit applied to characters and blocks in parallel data streams, the TRC can only detect, with limited certainty, whether or not there is an error. It cannot correct the error. Detection cannot be guaranteed because an even number of errors in the same character or block will escape detection, whether odd or even parity is used. *Note 2:* When two-dimensional arrays of bits are used to represent characters or blocks, such as might occur in synchronized parallel data streams, a longitudinal redundancy check (LRC) may also be used. When an LRC and a transverse redundancy check (TRC) are combined, individual erroneous bits can be corrected. *Synonym* **vertical redundancy check.** *See also* **block check, error control, longitudinal redundancy check.**

transverse resolution: 1. In a facsimile receiver, the dimension that (a) is perpendicular to a scanning line and (b) is the smallest recognizable detail of the image produced by the shortest signal capable of actuating the facsimile receiver under specified conditions. **2.** In phototelegraphy, the dimension that (a) is perpendicular to the scanning line, (b) defines the finest detail of the image produced, (c) corresponds to the transverse dimension of a picture element, (d) is equal to the width of the scanning pitch, and (e) includes the dimensions and the luminance of the finest detail that can be handled by the system. *Note:* The contrast of the finest detail is represented by a signal that corresponds to a nominal black or a nominal white in digital systems in which only two signal levels are used.

transverse scattering: The scattering that (a) is observed in examining the far field irradiance pattern and (b) is used to measure the refractive index profile of an optical fiber or preform by using coherent light to illuminate the fiber or the preform transversely to its axis. *Note:* A computer usually is required to accurately interpret the pattern of the scattered light. *See also* **scattering, transverse interferometry.**

transverse scattering method: A method for measuring the refractive index profile of an optical waveguide, such as an optical fiber or preform, by (a) illuminating the guide coherently and transversely to its longitudinal axis, (b) examining the far field irradiance pattern, and (c) using a computer to interpret the pattern of scattered light obtained. *See also* **slab interferometry, transverse interferometry.**

trap: *See* **optical fiber trap, wave trap.**

TRAP: Tactical Receive Equipment and Related Applications.

trap door: In communications, computer, data processing, and control systems security, a breach that (a) is intentionally created in a system for the purpose of collecting, delaying, or destroying data, and (b) is made for unauthorized purposes.

trapped electromagnetic wave (TEW): An electromagnetic wave that enters a layer of material that is surrounded on both sides by a layer of material of a lesser refractive index such that, if the wave is traveling parallel or nearly parallel to the surfaces of the layers, and thus the incidence angles between the direction of propagation of the wave and the normal to the surfaces, are greater than the critical angle, i.e., the angles are grazing the surfaces, total reflection will occur on both sides and hence trap the wave. *Note:* Dielectric slabs, optical fibers, and layers of air of different density, temperature, or moisture content can serve as an electromagnetic wave trap, thus confining the wave to a given direction of propagation or to a given point.

trapped mode: *Synonym* **bound mode.**

trapped ray: *Synonym* **guided ray.**

traveler information station (TIS): A radio station in the local government radio service used to broadcast noncommercial voice information pertaining to traffic conditions, road conditions, hazards, directions, lodging, rest stops, service stations, local points of interest, and other information of interest or possible use to travelers.

traveling salesman algorithm: 1. In a distribution of geographic locations, an algorithm that (a) determines the best path from any point to another and (b) is used to determine optimal routes in various applications, such as (i) communications network routes among nodes, (ii) cable routing within buildings, and (iii) road routes without duplication. **2.** In a network, an algorithm that (a) enables determination of the shortest or optimum path from a given node to another via various intervening nodes, and (b) allows for weighting of branches for factors other than distance, such as difficulty, reliability, cost, and efficiency. *Note:* Traveling salesman algorithms are available that can handle thousands of nodes. *See also* **branch, network, network topology, node.**

traveling wave (TW): A wave that (a) propagates in a propagation medium, including free space, (b) has a velocity determined by the launching conditions and the physical properties of the medium, and (c) may be a longitudinal or transverse wave. *Note 1:* A traveling wave is not a wave that is reduced to a standing wave by reflections from a distant boundary. *Note 2:* Examples of traveling waves are radio waves propagating in free space, lightwaves propagating in optical fibers, water waves on the surface of the ocean, and seismic waves.

traveling wave tube (TWT): An electron tube in which an electron beam interacts with a guided electromagnetic wave, resulting in amplification at ultrahigh frequencies. *Note:* The traveling wave tube amplifier (TWTA) can be used as the final power amplifier of a typical transmitter in satellite and Earth stations.

tray: A relatively flat usually rectangular container for housing or mounting electrical or optical circuit components and their interconnecting cables, wires, or optical fibers. *See* **fiber optic splice tray.**

TREE: transient radiation effects on electronics.

tree file: A file that is divided into subsections that are each further subdivided.

tree network: *See* **network topology.**

tree search: In a tree structure, a search in which it is possible to decide, at each step, which part of the tree may be rejected without a further search in the rejected part.

tree structure: A hierarchical structure in which (a) a given node is an ancestor of all the lower level nodes to which the given node is connected, (b) the root node, i.e., the base node, is an ancestor of all the other nodes, and (c) there is one and only one path from any point in the structure to any other point in the structure. *See also* **network topology.**

tree topology: *See* **network topology.**

T reference point: In Integrated Services Digital Network (ISDN) implementations, the conceptional reference point that divides NT2 and NT1 reference points in a particular ISDN arrangement.

trellis coded modulation (TCM): Modulation that (a) is a modification of continuous phase modulation (CPM), (b) improves performance without changing bandwidth, (c) when filtered, does not have the constant envelope of CPM, and (d) uses expanded signal sets that make it useful in the design of spectrally efficient communications systems. *See* **differential trellis coded modulation.** *See also* **continuous phase modulation, modulation.**

tremendously high frequency (THF): A frequency in the range from 300 GHz (gigahertz) to 3000 GHz, the range having an unofficial American alphabetical designator of THF and an International Telecommunication Union (ITU) numerical designator of 12. *See also* **frequency, frequency spectrum designation.** *Refer to* **Table 1, Appendix B.**

triad: *Synonym* **triplet.**

trial: *See* **acceptance trial.**

triangular-cored optical fiber: An optical fiber that consists of a core with a cross section shaped like a triangle with bowed convex sides placed in a hollow jacket or cladding tube such that only the vertices of the core touch the inner walls of the jacket or cladding, and air surrounds the core to perform the function of the cladding. *Refer to* **Fig. T-9.**

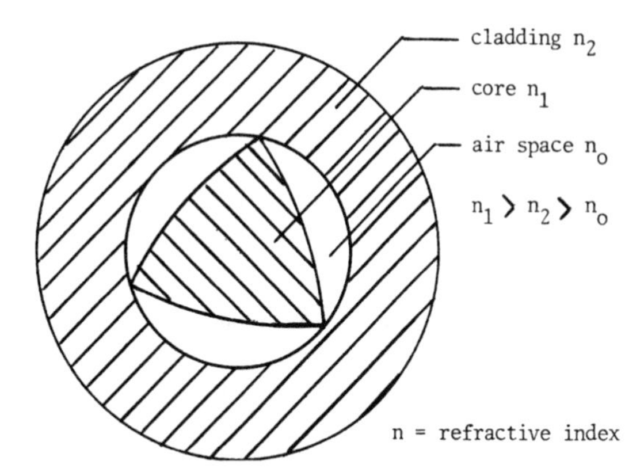

Fig. T-9. Cross section of a **triangular-cored optical fiber** with airspace between the core and the cladding.

triangular random noise: Random noise that has a frequency distribution in which the noise power per unit bandwidth is proportional to the square of the frequency.

triangulation: 1. Direction finding by either intersection or resection. **2.** The obtaining of a fix on a transmitter by plotting, on a map, the bearings of the transmitter from the plotted positions of the stations that are taking and reporting the bearings, and determining the location of the intersection of those bearings. *Note:* Instead of a map, tables of longitude and latitude can be used, or the intersection point can be calculated by using the known positions and the bearings. The intersection is a point that is common to two or more sectors that are defined by the bearings that are taken. If bearings are taken from three or more stations, the intersections will define an area whose size depends on the precision of the bearing measurements. **3.** A land survey in which accurate surveying and mapping is obtained from systematic point-to-point angular and distance measurements, from which triangles are constructed and checked for closure. **4.** In satellite communications systems, the use of Earth station pointing angles and the accurately measured distance between fixed Earth stations for obtaining antenna azimuth and elevation angles to satellite stations. *See* **strobe triangulation.** *See also* **trilateration.** *Refer to* **Figs. I-7 (476), R-11 (848).**

triangulation net: A special communications net that (a) handles triangulation information, (b) reports the bearings of a source of radiation, moving or fixed, from two or more stations so that the position of the source may be determined, (c) has stations with known positions, (d) reports the positions of its stations if they are mobile, (e) reports the times at which the bearings are taken for use in connection with moving sources, and (f) if the traffic volume is large, may be separate from other communications and reporting nets. *See also* **bearing, triangulation.**

tributary office: A central office (C.O.) that (a) is located outside the central office, i.e., outside the exchange, in which a toll center is located, and (b) has a rate center that is different from its toll center.

tributary station: **1.** In a communications network, a station other than the control station. **2.** On a multipoint connection or a point-to-point connection using basic mode link control (BMLC), any data station other than the control station. *See also* **control station, data communication, master station, primary station, secondary station, slave station.**

trigger circuit: A circuit that has a total of two or more stable and unstable states, at least one being stable, such that desired transitions from one state to another, i.e., one significant condition to another, can be accomplished by the application of a suitable pulse. *See also* **flip-flop.**

trihedral corner reflector: **1.** A device that (a) consists of three metallic surfaces or screens perpendicular to one another, and (b) is used in radio detection and ranging (radar) to produce, by means of multiple reflections from smooth surfaces, a radar return of greater magnitude than might be expected from an object with the same physical size. **2.** A reflector that (a) consists of three mutually perpendicular intersecting conducting flat surfaces, and (b) reflects an electromagnetic wave in a direction opposite to that of the incident wave. *Note:* Trihedral corner reflectors are often used to create chaff, radar targets, markers, and decoys. **3.** A passive optical mirror that (a) consists of three mutually perpendicular flat, intersecting reflecting surfaces, and (b) returns an incident light beam in a direction opposite to that of the incident beam. *See also* **chaff, corner reflector antenna, dihedral corner reflector.**

trilateration: Land surveying in which (a) accurate mapping is obtained from systematic point-to-point distance measurements, (b) a minimum of three different measurements usually are made for each point that is surveyed, (c) the distances between points have to match in the form of triangles, (d) in satellite communications, a satellite ranging transponder may be used to assist in accomplishing the task of locating points, and (e) points on different continents can be surveyed. *See also* **triangulation.**

trim effect: In a crystal oscillator, a combination of (a) a degradation of frequency-temperature stability and (b) an increase in frequency offset. *Note:* The trim effect results in a frequency change that produces a rotation and a distortion of the initial frequency-temperature characteristic. *See also* **frequency offset.**

trip: *See* **ring trip, round trip.**

trip delay time: *See* **round-trip delay time.**

triple mirror: A mirror that (a) consists of three reflecting surfaces, mutually at right angles to each other, arranged like the inside corner of a cube, (b) may be constructed of solid glass, in which case the transmitting face is normal to the diagonal of the cube, (c) may consist of three plane mirrors supported in a precisely constructed metal framework, (d) has a constant deviation of 180° for all incidence angles, and (e) reflects an incident ray of light from any angle back parallel to itself but with an offset depending on the location of the point of incidence. *Synonyms* **corner cube reflector, retrodirective reflector.** *See also* **corner reflector.**

triple precision: In computer programming, precision characterized by the use of three computer words to represent a number in accordance with the required precision.

triplet: A byte composed of three bits. *Synonyms* **three-bit byte, triad.**

tri-services tactical equipment (TRI-TAC-E): In military communications systems, equipment that (a) accommodates the transition from the manual and analog systems currently being used to fully automated digital systems, and (b) provides message switching, circuit switching for voice communications, secure voice terminals, digital facsimile systems, and digital voice terminals. *Synonym* **TRI-TAC equipment.** *See also* **tactical communications.**

TRI-TAC: tri-services tactical. *See* **tactical communications.**

TRI-TAC equipment: tri-services tactical equipment.

TRIXS: Tactical Reconnaissance Intelligence Service.

Trojan horse: In communications, computer, data processing, and control systems security, a computer program that apparently or actually is useful and also contains a trap door. *See also* **trap door.**

tropopause: The transition region that (a) lies between the stratosphere and the troposphere, and (b) usually occurs at an altitude of about 8 km (kilometers) to 14 km in polar and temperate zones and about 17 km in the tropics, i.e., the torrid zone. *See also* **atmosphere, ionosphere, stratosphere, troposphere.**

troposcatter: *Synonym* **tropospheric scatter.**

troposphere: 1. The lower layers of the Earth's atmosphere, in which (a) the change of temperature with height is relatively large, and (b) clouds form, convection is active, and mixing is continuous and more or less complete. **2.** The layer of the Earth's atmosphere (a) that lies between the surface and the stratosphere, (b) in which temperature usually decreases with altitude, and (c) in which about 80% of the total Earth air mass is contained. *Note:* The thickness of the troposphere varies with season and latitude. It usually is 16 km (kilometers) to 18 km thick over tropical regions and less than 10 km thick over the poles. *See also* **atmospheric duct, ionosphere.**

tropospheric duct: *Synonym* **atmospheric duct.**

tropospheric scatter: In electromagnetic wave propagation, scatter that (a) results from irregularities or discontinuities in the physical properties of the troposphere, (b) is used in transhorizon communications at frequencies from about 350 MHz (megahertz) to 8400 MHz, and (c) results from a propagation mechanism not fully understood, though it includes several distinguishable but changeable mechanisms, such as (i) propagation by means of random reflections, (ii) scattering from irregularities in the electric permittivity, i.e., the dielectric constant, and its gradient, (iii) irregularities in density of the troposphere, (iv) smooth Earth diffraction, and (v) diffraction over isolated obstacles, such as knife edge diffraction. *Note:* The electrical permittivity, magnetic permeability, and electrical conductivity at a point determine the refractive index at that point. These three parameters are subject to change (a) along a radio propagation path and (b) with time. The changes, absolute values, and gradients of these three parameters also contribute to tropospheric scatter. *Synonym* **troposcatter.** *See also* **backscattering, ionospheric scatter, propagation, scatter.**

tropospheric wave: A radio wave component that results from reflection of a ground wave at a place of abrupt change in the electric permittivity, i.e., dielectric constant, or its gradient, in the troposphere. *Note 1:* The ground wave may be altered so that the new components, i.e., the tropospheric waves, arising from reflection are distinguishable from the other components. *Note 2:* The electrical permittivity, along with the magnetic permeability and the electrical conductivity at a point, determine the refractive index at that point. These three parameters are subject to change (a) along a radio propagation path and (b) with time. The changes, absolute values, and gradients of these three parameters influence the magnitude and the direction of a tropospheric wave. *See also* **ground wave, sky wave.**

trouble desk facility: In an automatically switched telephone network, a facility that (a) consists of test boards and associated circuit patch bays and (b) is used to perform tests, such as busy, talking, listening, signaling, transmission, and noise tests.

troubleshoot: To locate and eliminate bugs in hardware or software, such as to identify, locate, and remove an error in a computer program or a fault in a computer, line, or circuit. *See also* **bug, error, fault.**

true bearing: At a point, a bearing with respect to true north, i.e., the direction to the North Pole or parallel to a meridian of longitude. *See also* **bearing, magnetic bearing.**

true date-time group (TDTG): The original date and time group that (a) is assigned to a message for identification purposes, (b) is not necessarily the date-time group that appears in the message external heading, and (c) remains identified with a message regardless of the number of times it is transmitted, encrypted, or addressed.

true field: 1. In an optical system or instrument, the view field. **2.** *In an instrument, synonym* **view field.** *See also* **view field.**

true power: *Synonym* **effective power.**

true solar day: The time that (a) is between two successive transits of the Sun across a meridian, i.e., a longitude, on Earth, (b) has a duration that varies with the Earth's rotation rate and time of year as the Earth moves in its orbit around the Sun, and (c) has a mean of 86,404 s (seconds), or about 24.001 h (hours). *See also* **civil day, second.**

TrueWave™: Pertaining to an ultrahigh capacity optical fiber that (a) was developed by the American Telephone and Telegraph (AT&T) Bell Laboratories, and (b) is capable of transmitting up to 80 Gbps (gigabits per second), about 16 times the data signaling rate (DSR) of conventional standard optical fibers, equivalent to about 1 million voice frequency channels simultaneously on a single fiber, which is achieved primarily by reduction of interchannel interference in wavelength division multiplexed (WDM) signals. *See also* **wavelength division multiplexing.**

truncate: To delete less significant digits to the right of a decimal point in a numeral. *Note 1:* The numeral 496.5442 can be truncated to 496.544, 496.54, 496.5, 496. The 496 cannot be truncated any further without losing significance. *Note 2:* In some computer programs, truncated values can be stored for possible significant accumulations. *See also* **round.**

truncated binary exponential backoff: In carrier sense multiple access with collision avoidance (CSMA/CA) networks and in carrier sense multiple access with collision detection (CSMA/CD) networks, the algorithm used to schedule retransmission after a collision, such that the retransmission is delayed by an amount of time derived from the slot time and the number of attempts to retransmit. *See also* **carrier sense multiple access with collision avoidance, carrier sense multiple access with collision detection.**

truncation: 1. The deletion or the omission of a leading or a trailing portion of a character string or a bit string in accordance with specified criteria. **2.** In data transmission, the deliberate removal of intelligence signal components without loss of message intelligibility. *Note:* Examples of truncation are (a) the removal of digits from overflow or shift operations on a computer, resulting in a less precise representation, (b) the removal of signals that are outside a limited bandwidth where noise or distortion renders them unreliable, (c) an acceptable reduction in signal bandwidth to obtain a reduction of transmitted power needed to achieve a required signal to noise ratio, and (d) an acceptable reduction in signal bandwidth to avoid band crowding and consequent interference from signals in the adjacent frequencies, i.e., extending the width of the guard band. **2.** In computers and data processing, the removal of the ends of words and numerals. *Note 1:* Examples of truncation are (a) the removal of the two least significant digits of the numeral 38.02926 to obtain 38.029 and (b) the removal of the last seven letters of the word "literature" so as to obtain "lit," as in the case of a student studying "English lit." *Note 2:* Truncation can result in (a) reduced precision, (b) errors, (c) loss of information, and (d) ambiguity. *See* **voice frequency truncation.**

truncation error: 1. An error caused by truncation. *Note:* The truncation error is calculated as the difference between the truncated value and the untruncated value. **2.** An error that occurs in the process of truncation, such as truncating the three least significant digits in a number when only two should have been truncated.

trunk: 1. In a communications network, a single transmission channel between two points that are switching centers or nodes. **2.** A circuit between switching equipment, as distinguished from circuits between a central office (C.O.) and data terminal equipment (DTE), such as loops. *Note 1:* Trunks may be used to interconnect switches, such as major, minor, public, and private switches, to form networks. *Note 2:* Examples of trunks are (a) optical light sources, fiber optic cables, photodetectors, and associated equipment forming data links, (b) connecting circuits between selectors of different rank in an automatic switching exchange, (c) circuits connecting two parts of a manual exchange, (d) connections between the exchange equipment of two switching centers, and (e) in satellite communications systems, baseband multiplexed radio links between two switching centers or between satellite and Earth stations, i.e., uplinks and downlinks. *See* **central office trunk, common trunk, cross office trunk, fiber optic video trunk, individual trunk, intercepting trunk, interoffice trunk, interposition trunk, interswitch trunk, intertoll trunk, intraoffice trunk, one-way trunk, private branch exchange tie trunk, recording trunk, special grade trunk, tie trunk, toll switching trunk.** *See also* **channel, circuit, tandem center, tandem tie trunk network, transmission channel.**

trunk answer from any station: A network-provided service feature that allows a user to dial a special code and answer an incoming call when the feature is activated. *See also* **service feature.**

trunk arrangement: *See* **remote trunk arrangement.**

trunk call: *Synonym* **junction call.**

trunk circuit: In telephone switchboard operations, a circuit that directly connects two exchanges or switchboards.

trunk delay working: A mode of telephone switchboard operation in which all calls carried over specified trunks are booked calls. *See also* **booked call.**

trunk encryption device (TED): A bulk encryption device that (a) is used to provide secure communications over a wideband digital transmission link, and (b) usually is located between the output of a trunk group multiplexer and a wideband radio or cable facility. *See also* **bulk encryption, link encryption.**

trunk exchange: In telephone switchboard operations, an exchange or a switchboard primarily used to handle trunk calls.

trunk free condition: In telephone systems operations, a condition (a) that is represented by a signal sent in the forward direction in the interexchange channel, (b) in which the circuit has been released by the sending exchange, i.e., the circuit is idle, is available for a new call, or is waiting for release by the other exchange, and (c) that is indicated by a clearing signal at the user interface, data terminal equipment (DTE), or end instrument. *See also* **end instrument, forward direction.**

trunk-grade service: The communications grade of service provided by a trunk network under varying communications traffic conditions. *See also* **grade of service.**

trunk group: **1.** Two or more trunks of the same type between two given points. **2.** A specified combination of trunks between a pair of switching facilities. *See* **high usage trunk group.** *See also* **group.**

trunk group multiplexer (TGM): A time division multiplexer (TDM) that combines individual digital trunk groups into a higher data signaling rate (DSR) bit stream for transmission over wideband digital communications links. *See also* **group, link, multiplexing.**

trunk hunting: *See* **hunting.**

trunking and switching system: *See* **joint multichannel trunking and switching system.**

trunk line: *See* **common trunk line.**

trunks busy: *See* **all trunks busy.**

trunk seized condition: In telephone systems operations, a condition (a) that is represented by a signal sent in the forward direction in the interexchange channel, (b) in which the circuit has been seized by a user, an attendant, or system equipment, and (c) that is indicated by a connection in progress signal at the destination user interface, data terminal equipment (DTE), or end instrument.

trunk switch: In a communications network, a switching exchange that interconnects trunk channels in the network.

trusted computer base (TCB): Within a computer system, all of the protection mechanisms, including hardware, firmware, and software, that are used to enforce a unified security policy. *Note:* The use of a computer system to correctly enforce a unified security policy depends on the mechanisms within the trusted computer base (TCB) and the correct input of parameters related to the security policy, such as user clearances, administrative classification designators, and priority designators. *See also* **automated information systems security, computer, data security.**

trusted computer system (TCS): An automated information system, including all of the hardware, firmware, and software, that (a) has undergone sufficient benchmark validation and testing, as well as acceptance and user testing, to meet the user requirements for reliability, security, and operational effectiveness with specified performance characteristics, and (b) primarily is intended for simultaneously processing various levels of sensitive and classified information without danger of compromise. *See also* **automated information systems security, computer, data security.** *Refer to* **Fig. T-10.**

Fig. T-10. The Night Hawk family of real time **trusted computer systems** for simulation and data acquisition applications. (Courtesy Harris Computer Systems Corporation.)

truth table: **1.** An operation table for a logic operation. **2.** A table that describes a logic function by listing all possible combinations of input values and indicating, for each combination, such as each Boolean function, the output value. *Refer to* **Fig. T-11.** *Refer also to* **Figs. A-7 (34), B-6 (78), E-7 (316), N-1 (609), N-19 (637), O-4 (681).**

TSK: transmission security key.

TSP: Telecommunications Service Priority, traffic service position.

TSPS: traffic service position system.

TSP system: Telecommunications Service Priority system.

T stop: The equivalent of a perfectly transmitting circular opening, given by the relation $T = (\pi/A)(D/2)^2$ or by the relation $D = 2(TA/\pi)^{1/2}$ where D is the diameter of the circular opening, T is the transmittance of the lens system, A is the area of the entrance pupil of the objective, and π is about 3.1416.

TTL: transistor-transistor logic.

TTTN: tandem tie trunk network.

TTY: teletype, teletypewriter.

TTY/TDD: teletypewriter/telecommunications device for the deaf.

tube: *See* **bait tube, cathode ray tube, dark trace tube, storage tube, traveling wave tube.**

tube photometer: *See* **photoemissive tube photometer.**

tube splicer: *See* **loose tube splicer.**

tumbling: In display systems, modification of the movement of an object in which (a) the object, i.e., an image of the object projected on a flat display surface, is modified so that it appears to be turning about an axis that is not perpendicular to the display surface, and (b) the modification usually is performed on a two-dimensional representation of a three-dimensional object, enabling the viewer to obtain a better understanding of the object's form. *Note:* If the object appears to be turning about a fixed axis perpendicular to the display surface, the object is considered to be rotating, and if the perpendicular axis is translating while the object is rotating, the object is considered to be rolling with or without slip, depending on the rate of translation, the rate of rotation, and the radius of the object. *See also* **rotating, translating.**

tunable laser: A laser that (a) has a spectral emission wavelength, i.e., the wavelengths that make up its spectral width, that can be varied, (b) may have a spectral range that (i) can be varied continuously across a broad spectral range, such as in organic dye or optical parametric oscillator lasers, or (ii) is only stepwise-tunable, such as in carbon dioxide or other molecular lasers, and (c) has an emission that can be tuned to any one of several nominal wavelengths, i.e., one of several spectral lines. *See also* **spectral width.**

tunable optical filter: An optical filter that (a) is capable of tuning to specific optical frequencies, i.e., wavelengths, (b) may be used for multiplexing and demultiplexing lightwaves, (c) may consist of a thin film of a liquid crystal between two glass mirrors separated by about 1 μm (micron), and (d) is operated by applying a voltage across the liquid crystal, changing the physical properties of the crystal and thereby allowing only a particular wavelength to pass through both mirrors, the wavelength depending on the magnitude of the applied voltage, the separation between the mirrors, and the thickness of the liquid crystal.

tuning: Adjusting the parameters and the components of a circuit so that it resonates at a particular frequency,

P	Q	PQ	P+Q	$\overline{PQ}$	$\overline{P+Q}$	$\overline{P}$	$\overline{Q}$	$\overline{P}\,\overline{Q}$	$\overline{P}+\overline{Q}$
Values		AND	OR	NAND	NOR	NOT P	NOT Q	*	#
0	0	0	0	1	1	1	1	1	1
0	1	0	1	1	0	1	0	0	1
1	0	0	1	1	0	0	1	0	1
1	1	1	1	0	0	0	0	0	0

* Examination of the respective columns shows that $\overline{PQ} = \overline{P}+\overline{Q}$

Examination of the respective columns reveals also that $\overline{P}\,\overline{Q} = \overline{P+Q}$

Note: $\overline{P}\,\overline{Q}$ is read as NOT P AND NOT Q; $\overline{P}+\overline{Q}$ is read as NOT P OR NOT Q

Fig. T-11. A **truth table** that (a) shows Boolean variables and functions and (b) can be used to prove identities among Boolean functions.

or so that the current or the voltage is either maximized or minimized at a specific point in the circuit. *Note:* Tuning usually is accomplished by adjusting the capacitance or the inductance, or both, of elements that are either connected to or in the circuit being tuned. *See* **jammer out-tuning, local oscillator tuning.**

tunnel: A virtual point-to-point link that (a) connects two points in two different networks, such as local area networks (LANs), islands, and Ethernet-like networks, (b) usually has end points that are workstations, such as personal computers (PCs), and (c) transfers Internet Protocol (IP) multicast encapsulated packets that are handled like normal unicast packets by the intervening routers and subnets. *See also* **unicast packet.**

tunneling mode: *Synonym* **leaky mode.**

tunneling ray: *Synonym* **leaky ray.**

turboboard: *Synonym* **accelerator board.**

turbulence: *See* **ionospheric turbulence.**

Turing machine: A mathematical model of a device that (a) changes its internal state, (b) reads from, writes on, and moves a potentially infinite tape, (c) can be programmed, and (d) satisfies the minimal requirements for a computer, all in accordance with its present state, thereby constituting a model for computer-like behavior.

turnaround mixer: In satellite communications systems, an Earth station component that (a) accepts a sample of an uplink signal, (b) changes it in frequency by using a frequency translation process, and (c) passes it through the station receiver for test purposes. *See also* **test loop translator.**

turnaround time: 1. The time between (a) the submission of a job, task, or computer program to a data processing center and (b) the return of completed results. **2.** The time required to reverse the direction of transmission, usually including the propagation time, line effects time, modem time, and machine reaction time. **3.** In a half-duplex circuit, the time required to reverse the direction of transmission from transmit to receive or vice versa. *See also* **half-duplex circuit, response time, round-trip delay time.**

turning flag: In visual signaling systems, a large flag that may be flown to indicate that a moving ship convoy is about to execute a turn, i.e., a change in heading or course. *Note:* A large E flag might be used to indicate a simultaneous turn by all ships at a 45° angle to starboard, and an I flag for a similar turn to port. As in most flaghoist signaling systems, the command is to be executed by all ships simultaneously when the flag is hauled down.

turnkey: Pertaining to procurement that (a) includes contractual actions at least through the system, subsystem, or equipment installation phase, and (b) may include follow-on contractual actions, such as testing, training, logistical, and operational support. *Note:* Precise definitions of the types of allowable contractual features are contained in the Federal Acquisition Regulations (FARs).

turnkey system: A system that (a) is supplied by a source or a vendor to a user, (b) is in a ready-to-run condition, (c) is installed, set up, and tested by the source or the vendor at the user site, and (d) usually is supported by the vendor in various areas, such as logistics, training, and consumable supplies.

tutorial: A body of basic information concerning a specific subject. *Note:* Examples of tutorials are (a) a seminar on the fundamentals of fiber optics, (b) in communications, computer, data processing, and control systems, a body of fundamental background information concerning operation of a system, and (c) in personal computer system operations, fundamental information recorded on diskettes that can be called up and displayed to inform the user on how to use the accompanying software.

TV: television.

TV fiber optic link: *Synonym* **television fiber optic link.**

TV set: *Synonym* **television set.**

TV set top box: *Synonym* **television set top box.**

twin cable: A cable (a) that consists of two insulated conductors laid parallel, and (b) in which the conductors either are attached to each other by the insulation or are bound together with a common covering. *See also* **cable, open wire.**

twinplex: Pertaining to a frequency shift keyed (FSK) carrier telegraphy system in which (a) four unique tones, i.e., two pairs of tones, are transmitted over a single transmission channel, such as one twisted pair, and (b) one tone of each pair represents a "mark" and the other a "space." *See also* **frequency shift keying.**

twinplex telegraphy: *Synonym* **four-frequency diplex telegraphy.**

twin sideband transmission: *Synonym* **independent sideband transmission.**

twist: In a telephone transmission line, a change in the response characteristic as a function of temperature.

twisted pair: Two metal separately insulated conductors that (a) have been twisted together with a short lay length, i.e., a short pitch, (b) are made of stranded or solid wire, (c) may have a strength member, and (d) usually are bound in small, medium, or large numbers in sheathed cables. *Note:* The twist keeps the pairs together for identification and provides for a balanced line. *See* **shielded twisted pair.** *See also* **balanced line.**

twisted-pair cable: *See* **paired cable.**

twisted ring counter: *Synonym* **feedback shift register.**

twist strength: The ability of an optical fiber, fiber optic bundle, or fiber optic cable to withstand alternate torsional flexing without breaking fibers or reducing transmission capability.

two adder: *See* **modulo-two adder.**

two addition: *See* **modulo-two addition.**

two-address instruction: A computer program instruction in which the address part contains exactly two addresses, such as the addresses of two operands. *See also* **four-address instruction, multiple address instruction, one-address instruction, three-address instruction.**

two-body problem: The problem in celestial mechanics that is concerned with (a) the relative motion of two point masses that are influenced only by their mutual gravitational attraction and their initial velocities relative to each other or relative to a fixed frame of reference, such as the undisturbed motion of a parent body and its satellite, (b) motion that usually is elliptical with one body at one of the foci of the ellipse, (c) the situation that occurs in conic sections, including circles, ellipses, parabolas, hyperbolas, and straight lines according to Kepler's and Newton's laws of motion, and (d) deviations, called "perturbations," caused by the attraction of other bodies. *See also* **n-body problem, restricted circular three-body problem, three-body problem.**

two-condition telegraphy: Telegraphy or data transmission in which only two significant conditions are used in forming the transmitted signals, such as in (a) a simple on and off signal, (b) a current or no current signal, (c) two different frequencies to represent 0 and 1, and (d) positive and negative currents with no zero current condition. *See also* **significant condition.**

two-frequency half-duplex: Pertaining to a communications circuit with a two-way simultaneous traffic capacity in which different frequencies are used for the two directions of transmission. *Synonym* **two-frequency simplex.**

two-frequency simplex: *Synonym* **two-frequency half-duplex.**

two-independent-sideband carrier: A carrier wave that (a) has two sidebands, each of which is independent of the other even though they are sidebands of the same carrier, and (b) may have a center frequency of the two independent sidebands that is not the same as the carrier frequency.

two-independent-sideband transmission: Transmission of a radio wave in which there are two sidebands that (a) are independent of each other, (b) may be separately modulated, and (c) use the same carrier.

two-level optical signal: A modulated lightwave that is at either one of two different intensities or power density levels at any given instant, such as the optical power being either on or off at any given instant.

two-level signal: A modulated carrier that is at either one of two different levels, such as voltage, current, irradiance, optical power density, or electric field strength levels, at a point at any given instant. *Note:* Examples of two-level signals are (a) an on-off voltage telegraph signal and (b) a lightwave carrier that is turned on and off depending on whether a 1 or a 0 is being transmitted at any given instant.

two-operator period watch: A communications watch for a duration that is based on the availability of two operators at one station. *See also* **continuous communications watch, single-operator period watch.**

two out of five code: A binary coded decimal notation in which (a) each decimal digit is represented by a binary numeral consisting of five binary digits, of which two are of one kind, called "ones," and three are of the other kind, called "zeros," and (b) the usual weights assigned to the digit positions are 0-1-2-3-6, except that "zero" is represented as 01100.

two-pilot regulation: In frequency division multiplexed (FDM) systems, the use of two pilot frequencies within a band so that the change in attenuation caused by twisting conductors can be detected and compensated by a regulator. *See also* **frequency, frequency division multiplexing, pilot.**

two-sample deviation: The square root of the Allan variance. *See also* **Allan variance.**

two-sample variance: *Synonym* **Allan variance.**

twos complement: The complement of a pure binary number obtained by adding 1 to the least significant digit of its ones complement and propagating the carries only as far as they go. *Note:* The ones complement of the binary number is obtained by changing each 0 to a 1 and each 1 to a 0. Leading 0s may be removed. Thus, the ones complement of 1011000 is 0100111. When 1 is added, 0101000 is obtained, the twos complement of 1011000. The twos complement of a number also may be obtained by subtracting the number from a number consisting of a 1 followed by as many 0s as there are digits in the number and truncating the most significant digit, i.e., the left-most digit of the result. *See also* **tens complement.**

two-source frequency keying: *Synonym* **frequency exchange signaling.**

two-tone keying: 1. In telegraphy systems, keying using a transmission path that consists of two channels in the same direction, one for transmitting the "space" for binary modulation, the other for transmitting the "mark" for the same modulation. **2.** Keying in which the modulating wave causes the carrier to be modulated with a single tone for the "mark" and modulated with a different single tone for the "space." *See also* **frequency exchange signaling, frequency shift keying, keying, voice frequency telegraph.**

two-tone telegraph: *Synonym* **two-tone keying.**

two-tone test signal: *See* **composite two-tone test signal.**

two-way alternate communication: Operation of a communications system in such a manner that data or information may be transferred in both directions, such as from source to sink and vice versa, one direction at a time. *Note:* Two-way alternate communications traffic usually is handled on a half-duplex or two-way simplex link, i.e., a link in which one user claims the channel in order to transmit, and the other user cannot interrupt and transmit until the first user has switched over. Lightweight mobile and portable equipment sometimes consists of a transmitter-receiver in which some circuits are used for both transmitting and receiving, requiring the use of a push-to-talk switch that is not pushed to listen. In radiotelegraph and data transmission systems, two-way alternate communication may consist of either the use of a single frequency, time slot, or code address for transmission and another frequency, time slot, or code address for reception, or the use of the same frequency, time slot, or code address for both transmission and reception. In wire telegraph systems, simplex operation may be used over either a half-duplex circuit or a neutral direct current circuit. *Synonyms* **either-way communication, over-over communication, over-over mode, simplex communication, simplex operation, simplex transmission, single simplex operation.** *See also* **duplex circuit, duplex operation, half-duplex circuit, half-duplex operation, one-way communication, one-way only channel, phantom circuit, simplex circuit, simplex signaling.**

two-way alternate operation: *Synonym* **half-duplex operation.**

two-way radar equation: An equation that expresses the peak or average signal power that is returned to a radar antenna terminal upon reflection from an object, i.e., a target, as a function of circuit and propagation medium parameters, transmitted power, antenna gain, wavelength, object area, and range. *See also* **one-way radio equation, self-screening range equation.**

two-way simultaneous communication: A mode of operation of a communications system in which data may be transferred in both directions, i.e., between source and sink, at the same time. *Synonym* **both-way simultaneous communication, duplex communication.** *Refer to* **Fig. T-12.**

two-way simultaneous operation: *Synonym* **duplex operation.**

two-wire circuit (TWC): 1. A communications circuit formed by two metallic conductors insulated from each other. **2.** In contrast to a four-wire circuit, a circuit in which one line or channel is used for communications in both directions. *See also* **balanced line, circuit, four-wire circuit, line adapter circuit, metallic circuit, open wire.**

TWT: traveling wave tube.

TWX®: teletypewriter exchange service.

Twyman-Green interferometer: An interferometer in which an observer sees a contour map of an emergent electromagnetic wavefront based on the wavelength of the light that is entering the system.

type: 1. A raised or embossed character that (a) is on an element, bar, or rod, (b) usually is on a typewriter or an impact printer, and (c) is used to make an imprint. **2.** A class to which an entity may be said to belong. *See* **optical detector type, optical source type.** *See also* **impact printer, imprint.**

type ball: A spherical type element that (a) has types on its surface, (b) is rotated on a typewriter or printer carriage to key-selected print positions, (c) presses a ribbon against paper to print or remove a character,

Fig. T-12. The MAGNASTAR digital airborne radiotelephone that provides **two-way simultaneous communications** (multiplexed) for aircraft passengers and crew. (Courtesy Magnavox Electronic Systems Company, Fort Wayne, IN.)

and (d) is available in various fonts. *See also* **print position.**

type bar: 1. A bar-shaped device that (a) has one or more type slugs mounted on it, (b) usually is mounted with its long axis vertical on most impact printers, and (c) usually is moved vertically to position the slugs for printing characters. **2.** On a typewriter, a lever that (a) has a type slug mounted on the free end and (b) is pivoted so that it can press an inked ribbon against paper and print a character.

type bar fulcrum: A rod on which type bars are pivoted.

type bar guide: A guide, such as a groove, track, slit, or slot, that guides a moving type bar to and from print positions.

type bar rest pad: A device that holds type bars in the rest position, i.e., the position held before and after their movement to and from the print position. *See also* **print position.**

type carrier: A device, such as a type ball, bar, cylinder, disk, plate, rod, slug, or wheel, on which one or more types, or type slugs, are mounted.

type cylinder: A device that (a) is cylindrically shaped, (b) has types or type slugs mounted on its curved surface, and (c) prints characters when an inked ribbon and paper are pressed against the types or slugs as they rotate into print position. *See also* **print position.**

type disk: A device that (a) is disk-shaped, (b) has types or type slugs mounted on its flat surface, and (c) prints characters when an inked ribbon and paper are pressed against the types or slugs as they rotate into print position. *See also* **print position.**

type font: A given size or style of type, such as pica, elite, italic, orator, 10-point Bodoni Modern, Courier, Times Roman, and script.

type plate: A device that (a) is rectangularly shaped, (b) has types or type slugs mounted on its flat surface, and (c) prints characters when an inked ribbon and paper are pressed against the types or slugs as they rotate into print position. *See also* **print position.**

type rod: A rod-shaped device with types or type slugs mounted on it so that it can move them into print position. *See also* **print position.**

type slug: A small piece of hard material, usually metal, on which one or more types, i.e., embossed characters, are formed.

typewriter: A machine that prints characters on paper or similar material when keys corresponding to the characters are depressed by an operator. *See* **heavy typewriter, light typewriter, teletypewriter.**

type 1 product: A classified or controlled cryptographic item endorsed by the National Security Agency (NSA) for securing classified and sensitive U.S. government information when appropriately keyed. *Note:* Type 1 products are simply products that (a) do not include information, key, services, or controls, (b) contain security classified National Security Agency algorithms, (c) are available to approved users, and (d) are subject to export restrictions. *See also* **encryption.**

type 2 product: An unclassified cryptographic piece of equipment, assembly, or component that (a) is endorsed by the National Security Agency (NSA) for use in telecommunications and automated information systems for the protection of national security information, (b) contains security classified NSA algorithms that distinguish them from products containing unclassified data algorithms, (c) are simply products and do not contain information, key, services, or controls, (d) may not be used for security classified information, but contain the security classified NSA algorithms, and (e) are subject to export restrictions. *See also* **encryption.**

type 3 algorithm: A cryptographic algorithm that (a) has been registered by the National Institute of Standards and Technology, (b) is used for protecting unclassified, sensitive, or commercial information, and (c) has been published as a Federal Information Processing Standard. *See also* **algorithm**

type 4 algorithm: An unclassified cryptographic algorithm that (a) has been registered by the National Institute of Standards and Technology, (b) is used for protecting unclassified, sensitive, or commercial information, and (c) has not been published as a Federal Information Processing Standard. *See also* **algorithm.**

typing alternative: In communications, computer, data processing, and control systems, a capability of software in which macroinstructions are assigned to certain combinations of keys to serve as a substitute for keying the entire set of instructions to accomplish the same effect, such as in personal computer (PC) systems, (a) pressing the ALT key with another key to duplicate entries on a spreadsheet and (b) depressing the CONTROL and a function key to move or copy blocked text.

typing position: The print position on a typewriter. *See also* **print position.**

T1 carrier: *See* **T carrier.**

T1C carrier: *See* **T carrier.**

T2 carrier: *See* **T carrier.**

T3 carrier: *See* **T carrier.**

T4 carrier: *See* **T carrier.**

u: **1.** A prefix that (a) occasionally is used to indicate "micro," as in microwatts, microamperes, and micrometers, (b) has the value of one millionth, i.e., 10^{-6}, and (c) usually is only used when µ is not available in the printing mechanism. **2.** A symbol sometimes used in lieu of µ or µm to indicate micrometers or microns, i.e., 10^{-6} m (meter). *Note:* The symbol um is also used to indicate microns.

überhacker: *Synonym* **superhacker.**

UDP: User Datagram Protocol.

UHF: ultrahigh frequency. *See also* **frequency spectrum designation.**

U interface: For basic rate access in an Integrated Services Digital Network (ISDN) environment, a user-to-network interface reference point that is characterized by the use of a two-wire loop transmission system that (a) conveys information between the four-wire user-to-network interface, i.e., the S/T reference point, and the servicing central office (C.O.), (b) is located in the servicing C.O., and (c) is not as distance-sensitive as a service using a T interface. *See also* **basic rate access, Integrated Services Digital Network, R interface, S interface, T interface.**

ULF: ultralow frequency. *See also* **frequency spectrum designation.**

U-link bay: A rack or a panel of connectors that (a) enables connecting or breaking into any signal line that enters or leaves a station and (b) is used for test purposes.

ultrafiche: Microfiche that contains images that (a) have an extremely small reduction ratio, such as less than 1/50, and (b) require a high blowback ratio to make them legible to the unaided eye. *See also* **blowback ratio, microfiche, reduction ratio.**

ultrahigh frequency (UHF): A frequency that is in the range from 300 MHz (megahertz) to 3000 MHz, the range having the American designator UHF and the International Telecommunication Union numerical designator 9. *See also* **frequency spectrum designation.** *Refer to* **Table 1, Appendix B.**

ultrahigh frequency terminal: *See* **enhanced manpack ultrahigh frequency terminal.**

ultralow frequency (ULF): *Synonym* **infralow frequency.**

ultralow-loss (ULL) optical fiber: **1.** Optical fiber that has an extremely low attenuation rate, a rate caused only by Rayleigh scattering. **2.** Optical fiber that has a potential attenuation rate less than 0.01 dB/km (decibel/kilometer), perhaps as low as 0.002 dB/km, the attenuation rate being limited only by Rayleigh scattering. *Note 1:* Ultralow-loss (ULL) optical fibers operate at a window in the 2-µm (micron) to 5-µm region of the optical spectrum, i.e., in the near infrared and middle infrared regions of the electromagnetic spectrum. *Note 2:* Certain halogen glasses, such as zirconium fluoride glasses, may be used to produce ultralow-loss (ULL) fibers. *Note 3:* Ultralow-loss (ULL) fibers may be used in repeaterless links over long distances, such as transoceanic distances. *Refer to* **Figs. A-9 and A-10 (49).**

ultrasmall aperture terminal (USAT) antenna: An antenna that (a) is less than 1 m (meter) in size, (b) transmits and receives signals, (c) has a small aperture, and (d) may be used in interactive networks, satellite Earth stations, and global positioning systems (GPS). *See also* **Earth station, global positioning system, very small aperture terminal antenna.**

ultrasonic equipment: Equipment that (a) generates radio frequency energy and (b) excites or drives an electromechanical transducer for the production of high frequency sound waves for various purposes, such as industrial, scientific, medical, and communications purposes.

ultraviolet (uv): The portion of the electromagnetic spectrum in which the longest wavelength is just below the visible spectrum, extending from approximately 0.004 μm (micron), i.e., 4 nm (nanometers), to approximately 0.4 μm, i.e., 400 nm. *Note 1:* Some scientists place the lower limit of ultraviolet (uv) at values between 1 nm (nanometer) and 40 nm, 1 nm being the upper wavelength limit of X rays. The 400-nm limit is the lowest-visible-wavelength, i.e., the highest-visible-frequency, violet. *Note 2:* Because the frequency of ultraviolet radiation is higher than that of the visible region, the rays, i.e., the photons, can kill some living organisms, such as some germs and bacteria, and can have a damaging effect on human cells if exposure time and irradiance are sufficiently high. Attenuation rates in glass are higher for the shorter ultraviolet wavelengths because of the higher frequencies. Attenuation rates are lower for the infrared because of the lower frequencies; hence the interest in the use of infrared rather than ultraviolet in long-haul fiber optic communications systems. The standard wavelength for fiber optic systems is 1.31 μm (microns), a wavelength well into the infrared region of the electromagnetic spectrum, which starts at about 0.8 μm, the upper wavelength limit of the visible spectrum. *See also* **infrared, light, ultraviolet fiber optics.** *Refer to* **Fig. A-10 (49).**

ultraviolet fiber optics: Fiber optics in which ultraviolet (uv) light is used in optical components designed to handle uv light. *Note:* Ultraviolet fiber optic applications include medical engineering, medicine, measurements, materials testing, sterilization, photochemistry, genetics, security, and secondary emission (fluorescence) applications. *See also* **ultraviolet.**

ultraviolet light: Radiation in the ultraviolet region of the electromagnetic spectrum, i.e., electromagnetic waves that have wavelengths between 1 nm (nanometer) and 400 nm, or between 0.001 μm (micron) and 0.4 μm. *Note:* The lightwave region of the electromagnetic spectrum lies between 0.3 μm (micron) and 3 μm, which is the region in which the techniques and components designed for the visible spectrum, 0.4 μm to 0.8 μm, also apply. Thus, only the narrow region between 0.3 μm and 0.4 μm might be considered ultraviolet light, and the remaining portion of the ultraviolet region, from 0.001 μm to 0.3 μm, might be considered ultraviolet radiation. *See also* **ultraviolet fiber optics.**

ultraviolet light guide: Special optical elements in various geometric shapes that (a) transmit light in the ultraviolet region of the electromagnetic spectrum, such as wavelengths from 0.1 μm (micron) to 0.4 μm, i.e., wavelengths less than those of visible violet light and therefore frequencies higher than those of visible light, and (b) are used in medicine, biochemistry, microscopy, physiology, and medical engineering.

ultraviolet viewer: A device for viewing ultraviolet (UV) radiation, particularly from the end of an optical fiber carrying ultraviolet radiation, which would be invisible to the human eye without the viewer. *Note:* Direct viewing of ultraviolet (uv) radiation, such as from the end of an optical fiber or a laser, can cause damage to the retina. *See also* **ultraviolet, infrared viewer.**

um: A symbol sometimes used in lieu of μ or μm to indicate micrometers or microns, i.e., 10^{-6} m (meter). *Note:* The symbol u is also used (a) to indicate microns, i.e., micrometers, and (b) as a prefix to mean micro, as in us (microseconds), especially when a Greek font is not available.

UMTS: Universal Mobile Telecommunications System.

unaligned bundle: A fiber optic bundle (a) in which the spatial coordinates of each fiber do not bear the same relationship to each other at the two ends of the bundle, (b) that cannot be used to transmit a picture because the image will be destroyed during transmission as the optical fibers in the bundle change their relative positions, thus scrambling the image as each fiber carries its picture element to a relative image point different from the corresponding relative object point, (c) that usually is in a single sheath or jacket, (d) in which the optical fibers are randomly placed, (e) that is considered as a single transmission medium, path, or channel, (f) that usually is used simply as a means of guiding light, with no concern for spatial relationships among the fibers, and (g) that may be used for carrying optical power or optical communications pulses. *Synonym* **light bundle.** *Deprecated synonyms* **incoherent bundle, noncoherent bundle.** *See also* **aligned bundle, bundle, fiber optic bundle, image, optical fiber, picture element.**

unallowable character: *Synonym* **illegal character.**

unattended operation: The operation of a system or a device without a human operator, such as the automatic transmission and reception of messages without human intervention, e.g., by an attendant or a maintenance crew. *See also* **automatic switching, station to station call.**

unavailability: 1. A measure of the extent to which a system, a subsystem, or equipment is not operable and is not in a committable state at the start of a mission, when the mission is called for at an unknown, i.e., a random, point in time. *Note 1:* The conditions determining operability and committability must be specified. *Note 2:* Expressed mathematically, unavailability is 1 minus the availability. **2.** The ratio of (a) the total time that a functional unit is not capable of being used during a performance measurement period to (b) the length of the period. *Note:* If a unit is not capable of being used for 68 hours in a week, the unavailability is $68/168 \approx 0.405$. *See also* **availability, performance measurement period.**

unavailability procedure: *See* **circuit unavailability procedure.**

unbalanced line: A transmission line, such as (a) a coaxial cable, (b) a single wire and a ground return, and (c) two wires of different shape and size, in which the magnitudes of the voltages on the two conductors are not equal with respect to ground. *Note:* In a coaxial cable, because the inner conductor is shielded by the outer conductor, their voltages cannot be equal with respect to ground unless they are both grounded, in which case they cannot be used for signals except as a ground return. *See also* **balanced, balanced line, common return offset, ground return circuit, line, singing, unbalanced wire circuit.**

unbalanced modulator: A modulator in which the modulation factor is different for the alternate half-cycles of the carrier. *Synonym* **asymmetrical modulator.** *See also* **balanced, carrier, modulation, modulation factor.**

unbalanced wire circuit: A wire circuit in which the two sides are inherently electrically dissimilar. *Note:* Examples of unbalanced wire circuits are (a) a coaxial cable, (b) a single wire and a ground return, (c) two wires of different size or material, and (d) two identical wires that take different paths. *See also* **ground return circuit, unbalanced line.**

unbound mode: *Synonym* **radiation mode.** *See also* **bound mode.**

unbundling: The separating of individual tariffed offerings and services that are associated with a specific element in the comparably efficient interconnection (CEI) or open network architecture (ONA) tariff from other tariffed basic service offerings. *Note:* Unbundling is based on the Federal Communications Commission (FCC) Computer III Inquiry and is defined in FCC *Report and Order*, dated June 16, 1986. *See also* **basic service element, basic serving arrangement.**

unconditional jump: In a computer program, a jump that (a) takes place when the program instruction that specifies it is executed, and (b) does not require that any other condition be specified or met prior to or during execution. *See also* **conditional jump, jump.**

uncoordinated bearing: A bearing that is determined by a single direction-finding station.

undercompensated optical fiber: An optical fiber that has a refractive index profile adjusted so that the high-order propagating modes arrive at the end of the fiber after the low-order modes, as in the case of the uncompensated fiber, i.e., a fiber designed such that no deliberate effort is made to compensate for dispersion. *Note:* Higher-order modes propagate faster than lower-order modes in a propagation medium with a given refractive index. In an undercompensated fiber they are made to travel longer paths in a higher refractive index medium and hence are delayed so they arrive at the end after the lower-order modes, which are made to travel shorter paths in lower refractive index material. Higher-order modes have higher eigenvalues in the solution of the wave equations. Higher-order modes also have higher frequencies and shorter wavelengths than lower-order modes. *See also* **compensated optical fiber, overcompensated optical fiber, wave equation.**

underfill: In coupling optical power from a source, such as a laser, a light-emitting diode (LED), or an optical fiber, to a receiver, such as another fiber or a photodetector, the situation that occurs when the useful diameter of the receiver, such as the core of a fiber or the sensitive surface of a photodetector, is larger than the diameter, at the surface of the receiver, of the cone of light emanating from the source. *Note:* In coupling optical fibers, underfill occurs when the spot of light incident on the endface of the fiber is smaller than the core diameter, and the launch angle of the source is smaller than the acceptance angle of the fiber. *See also* **fill, overfill.**

underflow: In a buffer, the condition in which the buffer is not filled to capacity. *See* **arithmetic underflow, overflow.**

underground cable: A communications cable that can be placed under the surface of the Earth in a duct system that isolates it from direct contact with the earth. *Note:* An underground cable is distinguished from a direct buried cable that is in direct contact with the earth, i.e., the soil. *See also* **cable, direct buried cable, outside plant.**

underlap: In facsimile systems, a defect that occurs when the width of the scanning line is less than the scanning pitch. *See also* **facsimile, tailing.**

undershoot: 1. In the transition of any parameter from one value to another, the condition that occurs when a temporary value is less than the final or target value. **2.** In the transition of any parameter from one value to another, the amount that a temporary value is less than the final or target value. **3.** In the transition of any parameter from one value to another, the temporary value of the parameter when it is less than the final or target value. *See also* **overshoot, value.**

underwater telephone: A telephone that (a) can be used to communicate underwater, (b) may use sound, electric, or electromagnetic transmission, and (c) usually uses frequency-modulated (FM) carriers.

undesired signal: In communications, computer, data processing, and control systems, a signal that produces degradation of performance in the operation of the systems, subsystems, and equipment. *See also* **distortion, interference, signal, spurious emission.**

undetected error rate: *A deprecated synonym for* **undetected error ratio.**

undetected error ratio (UER): The ratio of (a) the number of data units, such as bits, unit elements, characters, or blocks, incorrectly received and undetected to (b) the total number of data units sent. *Note:* The numerator and the denominator of the undetected error ratio (UER) must be expressed in the same data units so that the ratio will be dimensionless. *Synonym* **residual error ratio.** *Deprecated synonyms* **residual error rate, undetected error rate.** *See also* **binary digit, bit error ratio, block, character, error, error control.**

undisturbed day: A day in which sunspot activities or ionospheric disturbances do not interfere with radio communications.

undulator: A receiver used in Morse telegraph systems in which (a) a two-position stylus feeds ink to a continuously moving paper tape in conformity with a two-significant-condition incoming signal, thus creating a legible record, (b) the receiver records high speed Morse signals as a continuous ink line on a paper tape driven at constant speed, (c) a pen moves side to side across the tape for mark and space signals, and (d) the long and short undulations can be converted to text by trained operators. *See also* **Morse telegraphy.**

unformatted diskette: A diskette that (a) has not been subjected to the format process, and (b) has only a surface of unmagnetized magnetic material.

UNI: user-network interface.

unicast packet: A packet that is addressed and delivered to a single addressee. *See also* **multicast packet, packet.**

UNICOM station: *Synonym* **aeronautical advisory station.**

unidirectional antenna: 1. An antenna that (a) radiates a wave in one direction only, at any given moment, and (b) must be rotated or reoriented to radiate in a different direction. **2.** An antenna that is fixed and radiates in one and only one direction. *See also* **directional antenna, omnidirectional antenna.**

unidirectional channel: *Synonym* **one-way-only channel.**

unidirectional coupler: 1. A directional coupler that has terminals or connections for sampling waves in only one direction of transmission. **2.** In fiber optic transmission systems, a directional coupler that (a) permits signals to propagate in one and only one direction, or (b) has terminals or connections for sampling waves in one direction of transmission. *See also* **bidirectional coupler, directional coupler.**

unidirectional operation: Operation in which data are transmitted from a transmitter to a receiver in only one direction. *See also* **half-duplex operation.**

unidirectional transmission: Transmission between terminals, one of which can serve only as a transmitter and the other only as a receiver.

uniform density lens: A layered lens, or blank, in which (a) one layer is a clear material and the other is an absorptive material, (b) the clear portion is surfaced to the desired curvature, (c) the thickness of the tinted layer remains constant, and (d) the result is a lens with the same shade, i.e., the same transmittance, at the center as at the periphery.

uniform encoding: Analog to digital conversion in which all of the quantization subrange values are equal except for the highest and lowest quantization steps. *Synonym* **uniform quantizing.** *See also* **analog encoding, code, quantization, quantization level, signal.**

uniform illumination Cassegrain: Pertaining to an antenna reflector shape in which the field pattern law is linear across the effective antenna area. *Synonym* **shaped reflector Cassegrain.** *See also* **Cassegrain antenna.**

uniform index profile: In a dielectric waveguide, such as an optical fiber, a uniform linearly decreasing refractive index from the longitudinal axis radially toward the outside; i.e., the refractive index is inversely proportional to the distance from the center of the guide. *See also* **optical transmission, propagation medium.**

uniform Lambertian distribution: A Lambertian distribution that is constant over a specified surface. *See also* **Lambertian distribution.**

uniform linear array: An antenna that consists of a relatively large number of usually identical elements arranged (a) in a single line, i.e., they are collinear, or (b) in a plane with uniform spacing, and usually with a uniform feed system. *See also* **antenna.**

uniformly distributed constant amplitude frequency spectrum: *Synonym* **white noise.**

uniformly distributive coupler: A fiber optic coupler in which (a) the optical power input to each input port is distributed equally among the output ports, and (b) the optical power input to each output port is not distributed equally among the input ports. *See also* **nonuniformly distributive coupler.**

uniform plane polarized electromagnetic wave: A plane polarized electromagnetic wave in which the electric and magnetic field vector amplitudes are independent of the coordinate direction perpendicular to the direction of propagation, i.e., they are independent of the transverse coordinates, but are dependent on the longitudinal coordinates in the direction of propagation. *See also* **plane polarized electromagnetic wave.**

uniform quantizing: *Synonym* **uniform encoding.**

uniform refractive index profile: *Synonym* **linear refractive index profile.**

uniform spectrum random noise: *Synonym* **white noise.**

uniform time scale: A time scale that has equal intervals, i.e., that has equally spaced time markings.

uniform transmission line: A transmission line (a) that has electrical properties, i.e., resistance, inductance, and capacitance per unit length, that are constant throughout the length of the line, (b) in which the relation $dz/ds = 0$ where dz is the elemental change in impedance over an elemental length of the line, and ds is the elemental length of the line at any point in the line, and (c) in which the impedance, and therefore the voltage to current ratio, does not vary with distance along the line, if the line is terminated in its characteristic impedance. *Note 1:* Examples of uniform transmission lines are coaxial cables, twisted pairs, and identical single wires at constant height above ground, all of which have no changes in geometry, materials, or construction along their length. *Note 2:* If a line has discontinuities in its electrical properties along its length, special provision can be made for preventing reflections of waves in the line, such as providing termination in a characteristic impedance at points of discontinuity, thus enabling the line to operate as, and appear to be, a uniform transmission line. *Note 3:* Signal attenuation along the length is a function of the length of the line. However, the impedance, and therefore the voltage to current ratio, remains the same at every point on the terminated line. *Refer to* **Fig. C-6 (142).**

unilateral control system: *Synonym* **unilateral synchronization system.**

unilateral synchronization system (USS): A synchronization control system in which (a) signals at two locations are synchronized, and (b) the clock at one location controls the clock at a second location, but the clock at the second location does not control, i.e., does not influence, the clock at the first location. *Synonym* **unilateral control system.** *See also* **bilateral synchronization, clock, double-ended synchronization, single-ended synchronization, synchronization.**

unintelligible crosstalk: Crosstalk that consists of unintelligible signals, hence signals from which information cannot be derived. *See also* **crosstalk, intelligible crosstalk, interference.**

unintentional interference: *Synonym* **interference.**

union: *See* **International Telecommunication Union.**

union gate: *Synonym* **OR gate.**

unipolar: Pertaining to the transmission of signals with one polarity, i.e., either positive or negative voltage signals, but not both. Usually a mark is represented by an electric current in one direction, i.e., of one polarity, and a space is represented by the absence of current. *Note:* Unipolar signaling is used in single current signaling. *See also* **bipolar.**

unipolar signal: A two-significant-condition signal in which one of the conditions is represented by the presence of a positive or a negative voltage, but not both in the same line, and the other condition is represented by no voltage or current. *Note:* The current flow can be in either direction; hence the voltage polarity can be of either sign that remains fixed. *See also* **alternate mark inversion signal.**

unipolar telegraphy: *Synonym* **single current signaling.**

unipolar transmission: *Synonym* **single current signaling.**

unique key: Key held only by one piece of cryptoequipment and its associated distribution center. *See also* **cryptoequipment, encryption, key.**

unit: *See* **absolute unit, added data unit, airborne direction finding unit, answer back unit, arithmetic-logic unit, arithmetic unit, audio response unit, automatic calling unit, automatic dependent surveillance unit, central processing unit, circuit switching unit, command protocol data unit, communications processor unit, control unit, customer service unit, data service unit, dead zone unit, DX signaling unit, expedited data unit, field replaceable unit, functional unit, input-output unit, input unit, key telephone unit, medium access unit, message unit, noise measurement unit, output unit, protocol data unit, raster unit, relative unit, response protocol data unit, secure telephone unit, service data unit, signal unit, storage unit, teletypewriter unit, track and hold unit, video teleconferencing unit, visual display unit, volume unit, X unit.**

unit code: *See* **n-unit code.**

unit code alphabet: *See* **n-unit code alphabet.**

unit disparity binary code: In a line, a signal sequence in which (a) there is a difference of only unity, i.e., a difference of 1, between (i) the quantity of bits that represent 1 and (ii) the quantity of bits that represent 0, and (b) the long-term direct current (dc) component in the line is minimized to nearly zero.

unit distance code: An unweighted code that changes at only one digit position when going from one number to the next in a consecutive sequence of numbers, i.e., the signal distance is always 1. *Note 1:* Use of one of the many unit distance codes can minimize errors at symbol transition points in converting analog quantities into digital quantities. *Note 2:* An example of a unit distance code is the Gray code. *Note 3:* A decimal numeration system is not a unit distance code because in passing over multiples of 10 the signal distance is not 1, such as from 0999 to 1000, where the signal distance is 4. *See also* **signal distance.**

unit element: In the representation of a character, a signal element that has a duration, i.e., a length or a width, equal to the unit interval. *See also* **character, code.**

unit indicator: *See* **signal unit indicator.**

unit interface: *See* **access unit interface.**

unit interval: In an isochronous transmission system signal, the shortest significant interval, i.e., the shortest interval between consecutive significant instants. *Note:* The theoretical durations of significant intervals usually are whole multiples of the unit interval. The time interval between corresponding significant instants usually is equal to, or is a whole multiple of, the unit interval. *See also* **character interval.**

unit of information content: *See* **natural unit of information content.**

unit routing indicator: *See* **mobile unit routing indicator.**

unit solid angle: A steradian, equal to the solid angle that (a) intercepts $1/4\pi$ of the surface of a sphere and (b) has its vertex, i.e., its apex at the center of the sphere. *Note:* 4π steradians of solid angle intercept an entire surrounding sphere. Solid angles are measured in terms of the fraction of the surface of the surrounding sphere intercepted by the sidewalls of the solid angle, in contrast to degrees or radians between the two sides of plane angles, in which 360° and 2π rad (radians) intercept a complete circle. Solid angles are of interest in coupling a light beam confined to a given solid angle, thus specifying the divergence of the rays of the beam, into an optical fiber. *See also* **aperture ratio.**

unit string: A string that consists of only one entity, such as one bit or one letter. *See also* **bit string, character string, string.**

unit vector: A vector that (a) has the numerical magnitude of unity, i.e., of 1, (b) indicates direction only, (c) usually is associated with a coefficient that expresses the magnitude of a vector, (d) is expressed in a coordinate system, such as Cartesian, cylindrical, or spherical coordinates, and (e) may be expressed in the form of a set of components in the specific coordinate system. *Refer to* **Fig. I-1 (436).**

Universal Mobile Telecommunications System (UMTS): A European system that is an attempt to unify and standardize low-power short-distance radio transmissions, such as cellular, cordless, low-end wireless, local area network, private mobile radio (PMR), and paging systems.

Universal Personal Telecommunications (UPT): A service that (a) provides users access to telecommunications services while personally mobile, (b) allows each user to participate in a user-defined set of subscriber services, (c) enables each user to initiate and receive calls, (d) assigns to each user a unique, personal, network-independent number that transcends multiple networks, in order to reach any fixed, movable, or mobile terminal irrespective of geographic location, and (e) may be accessed subject only to the limitations imposed by (i) terminal and network capabilities and (ii) restrictions imposed by the network provider.

universal receiver-transmitter: In data communications, a circuit or a device that (a) is used in asynchronous, synchronous, or synchronous-asynchronous applications, (b) provides the logic for the reception of data in serial fashion and the transmission of the same data in parallel fashion, or vice versa, and (c) usually is a full-duplex circuit.

universal service: A service concept in which (a) basic local telephone service is made available at an affordable price to all persons within a country or a specified jurisdictional area, and (b) in some cases, certain other telecommunications and information services may also be made available.

Universal Time (UT): 1. The basis for coordinated dissemination of time signals, counted from 0000 at midnight. **2.** In celestial navigation applications, the time that gives the exact rotational orientation of the Earth, obtained from Coordinated Universal Time (UTC) by applying increments determined by the U.S. Naval Observatory. **3.** The official civil time of the United Kingdom. *Note: Formerly called* **Greenwich Mean Time.** *See* **Coordinated Universal Time.** *See also* **International Atomic Time.**

UNIX™: **1.** A computer operating system. **2.** A portable, multiuser, time-shared operating system that supports process scheduling, job control, and a programmable user interface. *Note 1:* Many proprietary operating systems are based on UNIX and are colloquially referred to as "UNIX," but are not necessarily interoperable. *Note 2:* Most UNIX-based operating systems are compliant with the portable operating system interface for computer environments (POSIX). *See also* **operating system, portable operating system interface for computer environments.**

unmodulated continuous wave jamming: Jamming that (a) consists of an unmodulated carrier wave (CW) signal, (b) may be used for various types of jamming, such as radar, communications, direction finding, and guided missile control jamming, (c) can be used for spot jamming or sweep jamming, (d) has virtually no bandwidth, and (e) appears similar to spot noise on a radar screen.

unnumbered command: In a data transmission, a command that does not contain sequence numbers in the control field. *See also* **control character, unnumbered response.**

unnumbered response: In a data transmission, a response that does not contain sequence numbers in the control field. *See also* **response, unnumbered command.**

unobtainable tone: *See* **number unobtainable tone.**

unprotected field: A field in which data can be entered, changed, or deleted. *Note:* On a display surface of a display device, an example of an unprotected field is a display field in which the operator can change the data displayed in the field. *See also* **field, protected field.**

unprotected storage: Storage whose content can be modified by computer programs. *Note:* An example of unprotected storage is a diskette that has a file that can be copied onto another diskette.

unrepeated loop: *Synonym* **unrepeatered loop.**

unrepeatered loop: A loop that does not make use of repeaters to amplify signals that propagate within it. *Synonym* **unrepeated loop.**

unretrievable remotely piloted vehicle: In communications systems, a mobile station that (a) is on board a vehicle that is controlled by signals from another station, (b) is not retrieved either to be used again or to retrieve any of the onboard data, and (c) transmits data that are relayed, retransmitted, or telemetered to another station that is not necessarily the controlling station.

unshifted fiber: *See* **dispersion-unshifted fiber.**

unshifted single mode fiber: *See* **dispersion-unshifted single mode fiber.**

unshift on space: A telegraph receiver operation in which (a) on receiving the space signal, the receiver is caused to shift from the figures case to the letters case, (b) the printing position is advanced by one character pitch, and (c) no character is printed.

unstratified language: A language that can be used as its own metalanguage. *Note:* Most natural languages, such as English, French, German, Spanish, and Russian, are unstratified languages. *See also* **metalanguage, natural language.**

unsuccessful call: *Synonym* **unsuccessful call attempt.**

unsuccessful call attempt: A call attempt that does not result in the establishment of a connection, i.e., does not result in access. *Synonym* **unsuccessful call.** *See also* **lost call.**

unused character: *Synonym* **illegal character.**

unvoiced sound: A sound created by that part of the human speech production organs where the sounds are generated, by forcing air through a movable constriction of the vocal tract without adding a sound from a vibrating vocal cord. *Note:* Examples of unvoiced sounds are whispers and hissing sounds. *See also* **voiced sound.**

unwind: In computer programming, to describe explicitly, in full, and without the use of modifiers, all the instructions that are used in the execution of a loop. *See also* **loop.**

up-converter: A device that (a) translates signals at lower frequencies to signals at higher frequencies, (b) usually does not change the type of modulation, (c) does not change the information content of the signals, and (d) is the converse of a down-converter. *Note:* In satellite communications systems, a satellite or Earth station may use an up-converter to translate a number of baseband trunks to a higher carrier frequency, usually without changing the modulation method or the occupied bandwidth. *Synonym* **up-translator.** *See also* **down-converter, erect position, frequency, frequency translation, inverted position, occupied bandwidth.**

update: *See* **display update.**

uplink: **1.** The portion of a communications link used for the transmission of signals from an Earth terminal to a satellite or to an airborne platform. *Note:* An uplink is the converse of a downlink. **2.** Pertaining to data transmission from a data station to the head end. *See also* **downlink, link, satellite.** *Refer to* **Fig. B-9 (96).**

upload: To transfer data from a local file to a central computer. *Note:* An example of upload is to transfer a file from a diskette on a personal computer (PC) to a host computer internal storage. *See also* **download.**

upload throttle: A system, device, procedure, or protocol that (a) controls the transfer of data from a local file to a host main frame, and (b) ensures that all host-main-frame data communications parameters and conditions are met, such as data signaling rates, the number of characters per message, time delays, character receipt delays, and character delays, for the transfer to be successfully completed.

upper-layer protocol: In a layered open communications system, a protocol resident in the higher layers of the system, such as in layers that correspond to Layer 6 and Layer 7 of the Open Systems Interconnection—Reference Model.

upper sideband: The sideband that has a higher frequency than the carrier frequency. *See also* **lower sideband, sideband.**

upright position: *Synonym* **erect position.**

upstream: **1.** The direction opposite the direction of data flow. **2.** In a network, the direction opposite the direction of data flow along a specified bus, *i.e.,* the direction toward the head of bus function. *See also* **downstream.**

UPT: Universal Personal Telecommunications.

uptime: The time during which a functional unit is fully operational. *See also* **available time, downtime.**

up-translator: *Synonym* **up-converter.**

urgency signal: A signal that (a) is used to indicate that the calling station has a message to transmit that concerns the safety of persons and equipment, (b) usually is transmitted by an aircraft station, a ship station, or another mobile station, and (c) has priority over all other communications traffic except distress traffic. *Note:* Examples of urgency signals are (a) the carrier wave (cw) urgency signal XXX transmitted three times and (b) the voice urgency signal PAN (pronounced pahn) transmitted three times. *See also* **distress message, distress signal, urgency message.**

urgency traffic: All messages that (a) contain information concerning the safety of persons and the protection of equipment, and (b) take precedence over all traffic except distress traffic. *Note:* Examples of urgency traffic are (a) messages from mobile stations with persons on board who will be in need of medical assistance in the near future, and (b) messages to mobile stations warning them of imminent danger.

usable frequency: *See* **lowest usable frequency.**

usable length: *See* **scanning line usable length.**

usable line length: *Synonym* **available line.**

usage: *Synonym* **occupancy.**

usage recorder: *See* **traffic usage recorder.**

usage register: *See* **traffic usage register.**

USAT antenna: ultrasmall aperture terminal antenna.

useful high frequency: *See* **lowest useful high frequency.**

useful line: *Synonym* **available line.**

user: 1. In communications, computer, data processing, and control systems, an entity, such as a person, an organization, or a system, that uses the services of a communications, computer, data processing, or control system. *Note:* Examples of users are (a) a person who uses an applications network to process or exchange data, (b) a computer that uses a communications network to transfer a program to another computer, (c) the transducer of a sensor that uses a radio link to telemeter data to a receiving station, (d) a call originator, (e) a call receiver, (f) a message originator, and (g) a facsimile transmitter that uses a communications channel to transmit the contents of a document. **2.** In telecommunications, an entity, such as a person, an organization, or a system, that uses the services of a telecommunications system for the transfer of information. *Note:* There are a source user and a destination user in each information transfer transaction. **3.** In automated information systems, a person, a process, or equipment that accesses an automated information system by direct connection via a terminal or by indirect connection via preparation of input data or receipt of output data. **4.** In communications security (COMSEC), a person who (a) is required to use COMSEC material in the performance of assigned duties, and (b) is responsible for safeguarding the COMSEC material. **5.** A component, equipment, or a subsystem that makes use of the system of which it is a part. **6.** In a telephone system, a customer or a subscriber of the services provided by the system. *See* **Automatic Digital Network user, Automatic Voice Network user, busy user, called user, common user, destination user, end user, originating user, sole user, service user, source user, special grade user, Telecommunications Service Priority System user.** *See also* **access originator, automated information system, communications sink, communications source, terminal.** *Refer to* **Figs. U-1, U-2.** *Refer also to* **Fig. C-5 (140).**

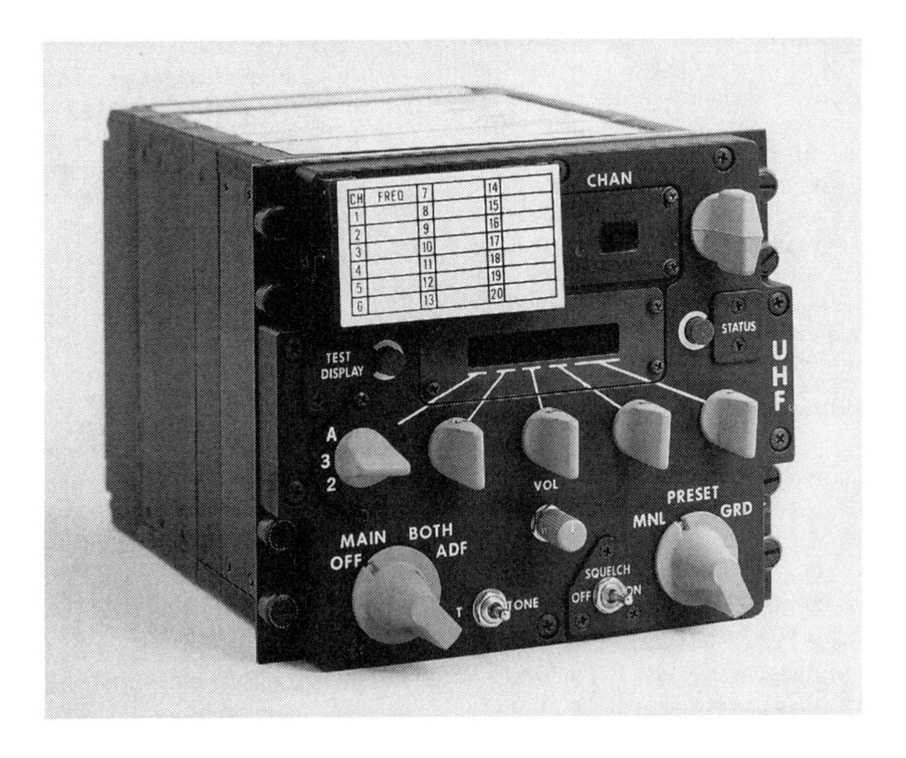

Fig. U-1. The AN/ARC-164 UHF radio's modular slice construction allows **users** to install field modulations to attain additional capabilities. (Courtesy Magnavox Electronic Systems Company, Fort Wayne, IN.)

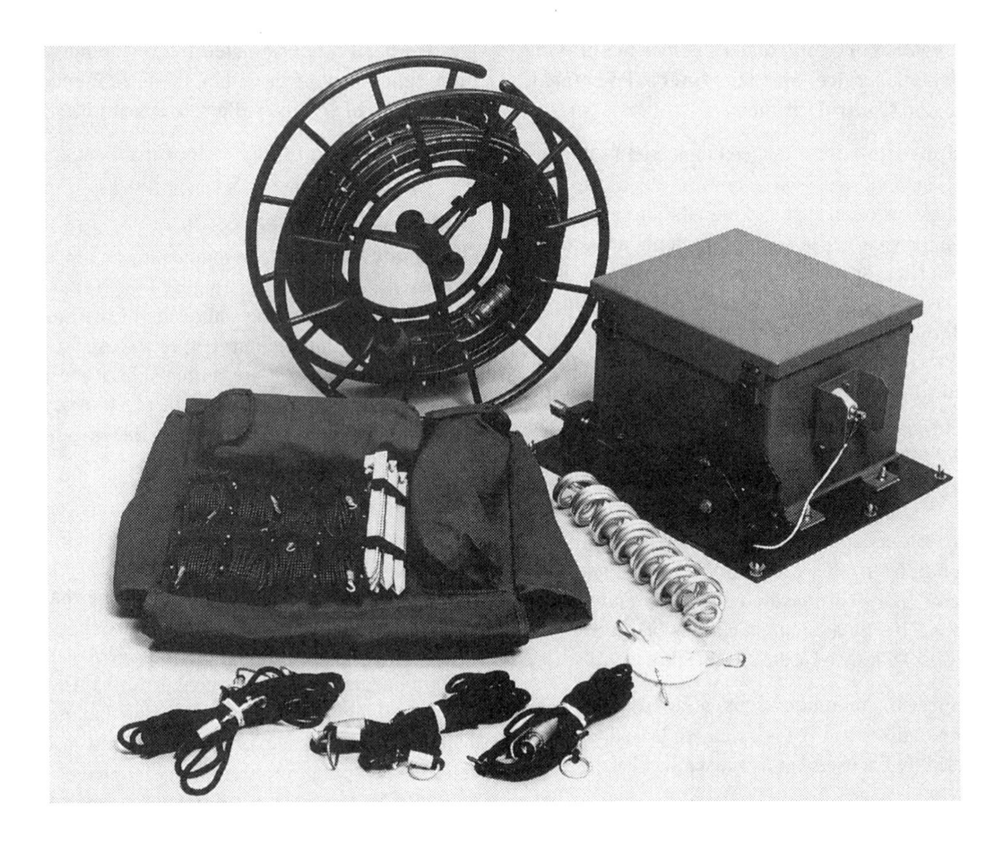

Fig. U-2. The AN/ARC-222's remote control unit provides **user**-friendly system operation. (Courtesy Magnavox Electronic Systems Company, Fort Wayne, IN.)

user circuit: *See* **common user circuit, exclusive user circuit, sole user circuit.**

user class indicator: In a communications system, information that (a) is sent in the forward direction and (b) indicates the class of service of the source user. *See also* **class of service, forward direction, source user.**

user communications: *See* **common user communications.**

user-computer interface: The manner or the means by which (a) a user and a computer communicate with each other, (b) control of the computer is exercised by the user, and (c) user-computer interaction may occur in many areas, such as (i) information presentation, displays, displayed information, formats, and data elements, (ii) command forms and languages, (iii) input devices and techniques, (iv) dialog, interaction, and transaction forms, (v) timing and pacing of operations, (vi) feedback and error diagnosis, (vii) prompting and queuing, (viii) job performance, and (ix) decision making. *See also* **human-machine interface, user-system interface.**

user coordinate: In display systems, a coordinate that (a) is expressed in a coordinate system that is convenient for a user, (b) is independent of device coordinates, (c) usually is in a virtual space that is specified by the user, and (d) enables image spatial manipulation operations to be performed, such as translating, rotating, isolating, scaling, scissoring, scrolling, tumbling, mirroring, rubberbanding, rolling, and zooming. *See also* **device coordinate, real world coordinate.**

user data: In layered systems, data transferred between entities of a given layer on behalf of the entities of the next higher layer for which the entities of the given layer are providing services. *See also* **user information.**

User Datagram Protocol (UDP): In the Internet Protocol (IP) suite, a standard, low-overhead, connectionless, host-to-host protocol that (a) is used over packet-switched computer communications networks, (b) allows an application program on one computer to send a datagram to an application program on another computer, and (c) differs from the Transmission

Control Protocol (TCP) in that TCP does not provide connectionless service. *See also* **Internet Protocol, Transmission Control Protocol.**

user end instrument: A device connected to the terminals of a circuit and used to convert electrical, optical, acoustic, or other types of signals into usable information, or vice versa, such as a telephone, control panel, computer, or remote data terminal. *Note 1:* Data transfer networks in (a) land-based communications systems, (b) aircraft communications and control systems, (c) shipboard communications and control systems, and (d) local area networks (LANs), including optical fiber communications systems, are designed to allow communication among users by means of their end instruments, such as intercoms, telephones, data terminals, and personal computers (PCs). *Note 2:* Operational systems and equipment, such as computers, aircraft propulsion, ship propulsion systems, and sensor systems, may be considered as user end instruments connected by a communications or control network. *Refer to* **Figs. D-1 (203), L-4 (506).**

user facility: In communications, a facility that (a) is on a user premises and (b) is operated, leased, owned, or protected by the user. *See also* **user service.**

user group: *See* **closed user group.**

user group with outgoing access: *See* **closed user group with outgoing access.**

user identification: In an automatically switched telephone network, the number used by the equipment to identify a user end instrument. *Note:* An example of user identification is a three-digit area code, followed by a three-digit exchange number, followed by a four-digit line or private branch exchange (PBX) number, and perhaps followed by an extension number for each single user end instrument or each group of user end instruments connected to the PBX. *See also* **user, user end instrument.**

user information: Information transferred across the functional interface between a source user and a telecommunications system for delivery to a destination user. *Note:* In telecommunications systems, user information includes user overhead information. *See also* **delivered overhead bit, delivered overhead block, destination user, interface, overhead bit, overhead information, source user.**

user information bit: A bit transferred from a source user to a telecommunications system for delivery to a destination user, including bits that are introduced into a message by the user to represent user overhead information. *Note:* User information bits do not include the overhead bits originated by, or having their primary functional effect within, the telecommunications system. *See also* **binary digit, overhead bit.**

user information block: A block that contains at least one user information bit. *See also* **block.**

user interface: An interface between a user and another entity, such as a communications, computer, data processing, or control system, such that (a) all parts of the entity that are on the same side of the interface as the user are within the purview of the user, and (b) all parts of the entity that are on the side of the interface opposite the user are not within the purview of the user. *See* **graphical user interface.** *See also* **interface, user, user-system interface.**

user language: 1. In a communications, computer, data processing, or control system, a language that (a) is used by the user of the system and (b) is acceptable to the system. **2.** A natural language that is used by persons in a specific field, such as engineering, medicine, law, or divine revelation, for developing or using a special purpose language. *Note:* An example of a user language is an engineering language, or a mathematical language, or both in concert, used to develop a computer programming language, such as FORTRAN. *See also* **artificial language, metalanguage, natural language.**

user line: *Synonym* **loop.**

user network: *See* **common user network.**

user overhead: In a communications system message, the part of overhead information, such as the name and the address of the addressee of the message, that is provided by the source user. *See also* **overhead information.**

user part: A functional part of a common channel signaling system that (a) transmits messages to, and receives messages from, the message transfer part, and (b) may be different for different systems. *Note:* In telephone and data communications systems, each user part is specific to a particular use of the signaling system. Thus, a user may be furnished a dial tone, a call receiver busy signal, or a call receiver ringing tone, but the user may not be furnished an intermediate trunk not available signal or a supervisory signal. *See also* **message transfer part.**

user service: One of a set of services and facilities, each of which can be made available by a network to a user on demand. *Note 1:* User services are provided to the user by a system or an organization, such as a public communications system, a private corporation, or a government agency. Some user services may be available on a per call basis, and others may be provided for an agreed period at the request of the user and with the agreement of the system operators. *Note 2:* Examples of network-provided user services are call hold, calling line identification facility, and call waiting. *Synonym* **user facility.** *See* **common user service, four-wire subset user service, general purpose user service.**

user service class: *Synonym* **class of service.**

user set: The communications equipment, exclusive of connecting lines, that is located at, used by, owned by, leased by, or protected by a user. *Note:* Examples of user sets are private branch exchanges (PBXs) and end instruments, such as telephones, teletypewriters, television sets, radio sets, modems, facsimile machines, and personal computers (PCs). *Synonyms* **customer set, subscriber set.**

user signaling rate: *See* **maximum user signaling rate.**

user-system interface: 1. An interface that includes aspects of information system design that affect and effect user participation in information handling transactions. **2.** In telecommunications, the manner in which, or the means by which, a user can access and communicate with a system, especially for exchanging messages with other users. *See also* **human-machine interface, user-computer interface.**

user terminal: An input-output device with which a user may (a) communicate with a system, such as a communications, computer, data processing, or control system, or (b) communicate with another user via a communications system.

user-to-user service: In communications systems, a service in which (a) two or more users are mutually connected continuously over an extended period, and (b) it appears to the users, or each group of users currently in direct communication with each other, that they are the only users connected to the system.

UT: Universal Time.

UTC: Coordinated Universal Time.

UTC(*i*): 1. Coordinated Universal Time (*i*). 2. Coordinated Universal Time (UTC) as kept by the "*i*" laboratory, where "*i*" is any laboratory cooperating in the determination of UTC. *Note:* In the United States, the official UTC is kept by the U.S. Naval Observatory and is referred to as UTC(USNO). *See also* **Coordinated Universal Time.**

utility: *See* **network utility.**

utility commission: *See* **public utility commission.**

utility land station: *See* **aeronautical utility land station.**

utility load: *Synonym* **nonoperational load.**

utility mobile station: *See* **utility mobile station.**

utility program: A computer program or an operating system software item that is in general support of the processes of a computer. *Note:* Examples of utility programs are program modules, program segments, diagnostic programs, trace programs, and programs used to perform routine tasks, i.e., perform everyday tasks, such as copying data from one storage location to another. *Synonym* **service program.** *See also* **tool.**

utility routine: *Synonym* **service routine.**

utility software: In communications, computer, data processing, and control systems, software that performs a support function for the application software, for the operating system, or for system users.

uv: ultraviolet.

UV: ultraviolet.

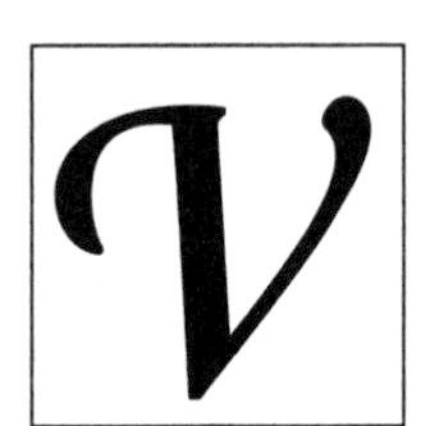

vacancy defect: In the somewhat ordered array of atoms and molecules in optical fiber material, a site at which an atom or a molecule is missing in the array. *Note:* A vacancy defect can serve as a scattering center, causing diffusion, heating, absorption, and resultant attenuation. *See also* **interstitial defect.** *Refer to* **Fig. V-1.**

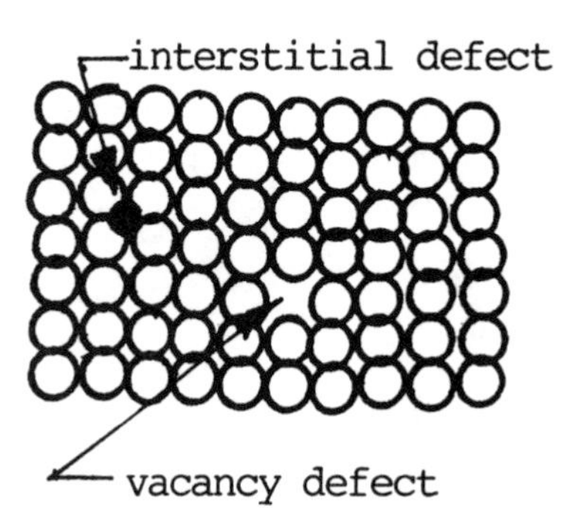

Fig. V-1. The molecular structure of glass, such as that of an optical fiber, showing a **vacancy defect** and an interstitial defect, such as an embedded impurity molecule. The molecular alignment in glass is not as uniform as in crystals. These defects cause scattering, resulting in an increase in optical attenuation.

vacuum bed: In a document copier, a mesh or perforated area of the machine on which a document is held in place by means of a vacuum applied during exposure.

VAD: vapor phase axial deposition.

VAD process: vapor phase axial deposition process.

valence band: In a semiconductor, the range of electron energy that (a) is lower than that of the conduction band, and (b) is possessed by electrons that are held bound to an atom of the material, thus reducing conductivity for electric currents even under the influence of an applied electric field, i.e., an applied voltage. *Note:* When electron energies are raised, such as by thermal excitation or by photon bombardment, electrons at the higher energy levels of the valence band are raised to the lower energy levels of the conduction band, thus leaving holes in the atoms from which they came and that remain in the valence band. In a conducting material, such as copper, aluminum, silver, gold, or lead, the valence band energy level is lower than in nonconductors, i.e., in insulators, thus allowing the electrons to be more free to move from atom to atom as an electric current under an applied field, i.e., an applied voltage.

validation: 1. The testing of a system, a subsystem, or equipment to determine whether it meets specified performance requirements. *Note:* The tests are made to ensure that the item tested achieves the proper levels of accuracy, precision, and relevancy. **2.** The checking of data for correctness or for compliance with applicable criteria, such as standards, rules, and conventions. **3.** In automated information systems security, the portion of the test and validation, i.e., test and evaluation (T&E), process that includes development of the specialized security test and evaluation procedures, tools, and equipment needed for the acceptance of an automated information system for joint usage. *Note 1:* Joint usage includes usage by two or more organizations, such as government agencies and commercial organizations. *Note 2:* Validation may include final development, evaluation, and testing preparatory to acceptance by security test and evaluation personnel. **4.** The use of physical measurements taken to verify conclusions obtained from mathematical modeling and analysis or taken for the purpose of developing mathematical models. *See* **data validation, product validation, system validation, test and validation.** *See also* **automated information system, automated information systems security, trusted computer system.**

value: 1. An occurrence of an attribute. *Note:* Examples of values are red for the attribute color, rough for the attribute smoothness, rose for flower, noon for time of day, and 106-24-2370 for Social Security number. **2.** A quantity assigned to a constant, variable, parameter, symbol, or character. *Note:* Examples of attributes are (a) the number 10 when applied to the number of milliwatts of power level and (b) in personal computer spreadsheet programs, cell entries that are numbers or formulas. **3.** The net worth of an entity. *See* **assumed value, data value, peak to peak value.** *See also* **label.** *Refer to* **Fig. P-13 (759).**

value added carrier: An organization that sells the services of a value added network. *See also* **enhanced service.**

value added network (VAN): A network that (a) uses the communications services of other commercial carriers, and (b) offers enhanced and new telecommunications services through the use of hardware, firmware, and software. *See also* **enhanced service, network.**

VAN: value added network.

Van Allen belt: Two concentric regions that (a) contain particles ionized by radiation, (b) are above the Earth's equator, (c) consist of protons, electrons, ions, and accompanying radiation at extremely high energy levels, (d) contain particles that have been captured by the Earth's magnetic field, (e) contain particles that have a quantity, density, and radiation energy level that fluctuates with solar activity, (f) have a more energetic inner belt that has a maximum radiation level and particle density at about 3000 km (kilometers) above the equator, and (g) have an outer belt that has a maximum radiation level and particle density at about 16,000 km. *See also* **ionosphere.**

vapor deposition: outside vapor deposition.

vapor deposition process: *See* **chemical vapor deposition process, modified chemical vapor deposition process, plasma activated chemical vapor deposition process.**

vapor phase axial deposition (VAD) process: A chemical vapor deposition (CVD) process in which dopants in vapor form are axially deposited on substrates, such as glass tubes, to make a preform from which optical fibers may be pulled.

vapor phase oxidation process: *See* **axial vapor phase oxidation process, chemical vapor phase oxidation process, inside vapor phase oxidation process, modified inside vapor phase deposition process, outside vapor phase oxidation process.**

vapor printing recording: *See* **ink vapor printing recording.**

vapor recording: *See* **ink vapor recording.**

var: volt•amperes reactive.

variable: 1. A quantity that can assume or be assigned any of a given set of possible values. **2.** In computer programming, a character, or a group of characters, that (a) refers to a value or corresponds to an address, and (b) usually is specified and qualified, such as may be elemented, structured, arrayed, subscripted, superscripted, and pointered. *See* **analog variable, switching variable.** *See also* **value.** *Refer to* **Figs. F-4 (356), N-17 (635), N-18 (636), P-8 (737), T-11 (1042), V-2 and V-3 (1064).**

variable format message: A message (a) in which transmission control characters are neither deleted when the message is received nor inserted when it is transmitted, and (b) that is intended for terminals and stations that have similar characteristics. *See* **fixed format message.** *See also* **format, transmission control character.**

variable function generator: In an analog computing system, a function generator in which the function it generates may be set or changed by the user before or during computation. *See also* **function, function generator.**

variable length buffer: A buffer into which data may be entered at one rate and removed at another rate without changing the data sequence. *Note:* Most first in, first out (FIFO) storage devices are variable length buffers in that the input rate may be variable while the output rate is constant, or the output rate may be variable while the input rate is constant. Various clocking and control systems are used to allow control of underflow or overflow conditions. *See also* **buffer, data, elastic buffer.**

variable logic: *Synonym* **programmable logic.**

variable optical attenuator: *See* **continuously variable optical attenuator, stepwise variable optical attenuator.**

variable polarization: In an electromagnetic wave, polarization that can be varied at a given point in space and time by any of various means, such as by (a) controlling the source or (b) operating on the wave while it is propagating, e.g., by (i) applying electric or magnetic fields or (ii) changing the parameters of the propagation medium. *See also* **polarization, propagation medium.**

variable precision coding compaction: Data compaction accomplished by (a) allowing the precision required of a set of data to vary with a given parameter, such as the magnitude of an evaluated function, time, or the value of the independent variable, (b) allowing the coding to vary with requirements, and (c) sending only a few signals at certain times, enabling those to be compacted and thus reduce storage and transmission requirements. *Note:* An example of variable precision coding compaction is that used in the countdown of a rocket launch. Timing signals and checking signals every 15 minutes are sufficient several days or hours before a launch. As time passes and events become critical, intervals between timing signals and between checking signals must become shorter, until at launch, fractions of a second are very significant and critical. The alternative without compaction would be to send checking signals and timing signals at fraction-of-a-second intervals throughout the hours, days, weeks, and perhaps months before the launch.

variable pulse recurrent frequency radar: A radar that (a) has a pulse repetition rate (PRR) that can be varied by a small amount about its nominal operating rate, usually about ±10%, (b) can use the variation to reduce interference from other sources of radiation, such as adjacent radars, (c) can use the variation to escape from some forms of jamming, and (d) provides some protection against deception by using the variation as an electronic countercountermeasure (ECCM).

variable slope delta modulation: *See* **continuously variable slope delta modulation.**

variable time (VT): 1. Pertaining to an event, such as a pulse, that occurs in a time interval whose duration can be either controlled or allowed to vary at random. **2.** Pertaining to an event such that (a) it occurs at a given instant, and (b) the time of occurrence is controlled. *See also* **time of occurrence.**

variable time scale: A time scale that can be changed during use, such as during the execution of a computer program, usually by changing the time scale factor. *See also* **time scale, time scale factor.**

variable tolerance band compaction: Data compaction accomplished by specifying one or more discrete bands to indicate the operational limits of a quantity, parameter, or value. *Note:* An example of variable tolerance band compaction is the transmission of data such that a prior arrangement is made in which a specified variable can be assumed to be always within a given range, and hence transmission of its value need not be made, except when the variable is greater or less than the values of the band limits, in which case a transmission is made for only as long as the variable remains outside the limits of the band, i.e., as long as the variable remains abnormal, thus reducing transmission and storage requirements that would be necessary if the absolute value of the parameter were to be transmitted at regular intervals. This arrangement must be extremely reliable or fail-safe because no transmission will be interpreted as "in band" when in reality there is no signal because the system is malfunctioning. *See also* **value.**

variance: *See* **Allan variance.**

variant: 1. One of two or more cipher or code symbols that have the same plain text equivalent. **2.** One of the several plain text meanings that are represented by a single code group. *Synonym* **alternative.** *See also* **code, cryptology, encode.**

variation: *See* **dynamic variation, fiber optic connector variation, net loss variation.**

variation monitor: A device that (a) senses deviations of any measured variable, such as voltage, current, frequency, and optical fiber diameter, and (b) is capable of initiating a programmed action, such as transferring to other power sources, transferring to other control devices, adjusting a control device, halting operations, or sounding an alarm, when programmed limits of the measured variable are exceeded, i.e., the departure from the nominal value exceeds a specified value. *Note:* An example of a variation monitor is a device that (a) monitors a voltage source, (b) disconnects the source if the voltage level changes by a specified amount, (c) reconnects the load to another voltage source, (d) turns on a red light, and (e) sounds a buzzer. Thus, proper power level continuity is assured, and attendants are notified of the event. *See also* **alarm sensor, reasonableness check, telemetry.**

V band: 1. An obsolete designation for a band of frequencies in the range from 46 GHz (gigahertz) to 56 GHz. *Note:* The V band was divided into five subbands, designated VA to VE, each subband being 2 GHz wide. **2.** An obsolete designation for frequencies in the range from 40 GHz (gigahertz) to 60 GHz between the old Q and O bands.

VC: virtual circuit.

VCI: virtual channel identifier

VCR: video cassette recorder.

VDU: visual display unit.

vector: 1. Pertaining to a parameter or a variable that (a) possesses both magnitude and direction, and (b) may be a function of time. *Note:* Examples of vector quantities are velocity, electric field, displacement, area, and refractive index gradient. Quantities that are not vector quantities are speed, mass, volume, signal to noise ratio, and frequency. These have magnitude only and therefore are scalar quantities. **2.** In direction finding (DF), communications, and navigation, to head or steer in a given direction. *Synonym* **steer.** *See* **absolute vector, electric field vector, electric vector, magnetic field vector, magnetic vector, Poynting vector, proportional vector, relative vector, unit vector.** *Refer to* **Figs. I-1 (436), Q-1 (786), T-7 (1024).**

vector current: The current obtained when the vector voltage is applied to the vector impedance. *Refer to* **Fig. Q-1 (786).**

vector generator: A functional unit, such as a display device or a computer program, that (a) generates directed line segments and (b) may connect them in sequence using a curve generator to generate curves for presentation as a display element, display group, or display image in the display space on the display surface of a display device, such as on a screen, fiberscope faceplate, light-emitting diode panel, gas panel, or surface of a plotter.

vector impedance: Impedance expressed in vector form, a form in which the resistance and the reactances are in quadrature. *Note 1:* The reactances depend on the frequency and whether the circuit element is an inductor, a capacitor, or a combination of both. The vector impedance in Cartesian coordinates may be expressed by the relation $z = r + jx$ where z is the vector impedance, r is the resistance, j is the quadrature designator, and x is the reactance. *Note 2:* Impedances, resistances, and reactances usually are expressed in ohms, per unit, or normalized values. *Note 3:* Capacitive reactance is given by the relation $X_C = 1/j\omega C$ where X_c is the capacitive reactance, j is the quadrature designator, $\omega = 2\pi f$ where π is about 3.1416, f is the frequency in hertz, and C is the capacity in farads, in which case the capacitive reactance will be in ohms. *Note 4:* Inductive reactance is given by the relation $X_L = j\omega L$ where X_L is the inductive reactance, j is the quadrature indicator, $\omega = 2\pi f$ where π is about 3.1416, f is the frequency in hertz, and L is the inductance in henries, in which case the inductive reactance will be in ohms. *Note 5:* The vector impedance may be expressed in polar form as $Z\angle\theta$ where Z is the magnitude of the impedance, and θ is the phase angle where $\tan\theta$ may be determined, in a series circuit, by the ratio of (a) the inductive reactance minus the capacitive reactance divided by (b) the resistance. *Note 6:* When plotted in Cartesian or polar coordinates, the vector impedance has both magnitude and direction and thus is called a "vector," though it is not of the same nature as a spatial vector, such as might represent a body in motion or a gradient. *Synonym* **phasor impedance.** *See also* **impedance.**

vector processor: *Synonym* **array processor.**

vector product: A vector quantity, **C**, equal to the product of (a) the magnitudes of two vectors, **A** and **B**, and (b) the sine of the angle between them, in which **C** has a direction perpendicular to the plane of the two vectors and directed such that if vector **A** is rotated about **C**, as an axis, through the smaller angle it makes with **B** in a right-handed, i.e., clockwise direction, the direction of **C** will be in the direction of a right-hand screw. *Synonym* **cross product.** *See also* **scalar product.**

vector voltage: An alternating sinusoidal voltage of given frequency with an associated phase angle, relative to a given reference, at that frequency. *Note 1:* The vector voltage may be represented (a) in Cartesian coordinates, as $v = a + jb$ where a is the in-phase component, i.e., the real component, j is the quadrature indicator, and b is the quadrature component, i.e., the imaginary component, and (b) in polar coordinates, as $V\angle\theta$ where V is the magnitude, and θ is the phase angle. Also $\tan\theta = b/a$ and $V = (a^2 + b^2)^{1/2}$. *Note 2:* Vector voltages may be an operand in the vector algebra with other scalar and vector quantities, such as vector currents, to produce scalar and vector results, such as real and reactive power. *Refer to* **Fig. Q-1 (786).**

vee value: *Synonym* **Abbe constant.**

vehicle: *See* **remotely piloted vehicle, unretrievable remotely piloted vehicle.**

vehicle harness: An assembly of interconnected equipment that (a) is on board a vehicle and (b) is used to provide specified communications functions, such as to control a radio and an intercom system.

vehicle highway system: *See* **intelligent vehicle highway system.**

Veitch diagram: A diagram that (a) shows, in a compact form, the information contained in a truth table, (b) has its variables expressed in a pure binary numeration system sequence, and (c) is a variation of the Karnaugh map, which has its variables in a Gray code sequence, thus permitting pairing of expressions, i.e., of literals, such as *ABCD,* which is an AND function, making the Karnaugh map more convenient to use. *See also* **Karnaugh map, pure binary numeration system, truth table, Venn diagram.** *Refer to* **Fig. V-2.**

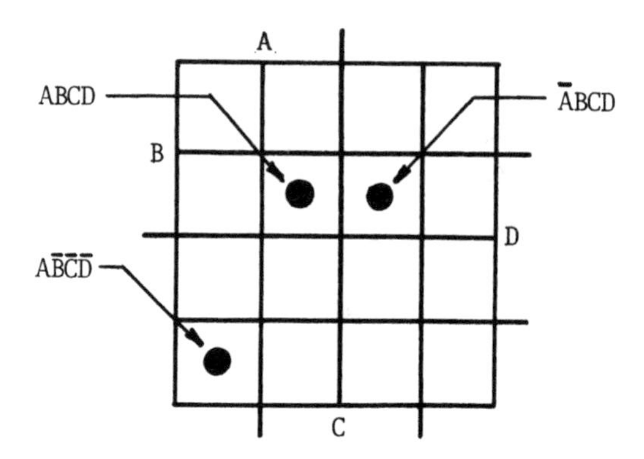

Fig. V-2. A **Veitch diagram** for four variables. The Boolean functions represented are (a) A AND B AND C AND D, (b) NOT A AND B AND C AND D, and (c) A AND NOT B AND NOT C AND NOT D.

velocity: The speed and the direction of an entity with respect to a reference point and a reference direction. *See* **electromagnetic wave velocity, group velocity, light velocity, phase front velocity, phase velocity, third cosmic velocity.**

velocity gate stealing: In Doppler radar systems, causing the radial velocity gate to lose its ability to accurately determine the radial component of velocity of a tracked object, i.e., a tracked target. *See also* **Doppler effect.**

velocity of light: *Synonym* **light velocity.**

Venn diagram: A graphic diagram that (a) shows allowable sets as closed areas, such as circles, (b) uses the relative positions of the closed areas to show relationships among variables or members of the sets represented by the closed areas, (c) may have closed areas that (i) overlap other areas, (ii) may be wholly separate from all other areas, or (iii) may be completely contained in other areas, and (d) is used to facilitate, describe, and illustrate relationships of the members of the sets, such as whether (i) all members of a set are also members of other sets, (ii) some members of a set are members of other sets, (iii) all members of a set are not members of any other set, and (iv) some members of a set are not members of any other set. *Refer to* **Fig. V-3.**

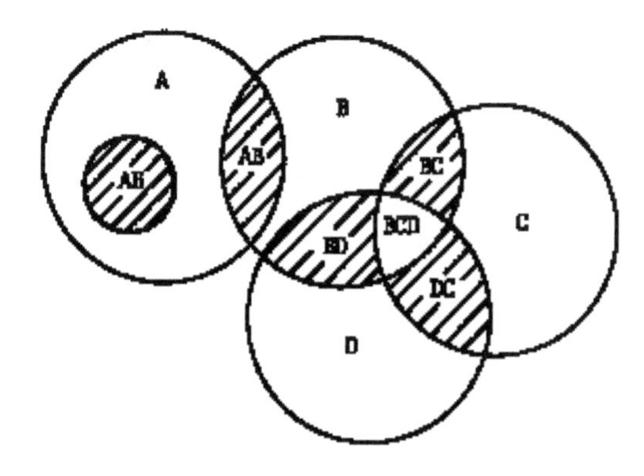

Fig. V-3. A **Venn diagram,** a graphic illustration of set relationships. For example, if Circle *A* represents the entire population of electronic equipment and Circle *B* represents the entire population of optical equipment, then the common area *AB,* defined as the Boolean function A AND B, would represent the entire population of optoelectronic equipment. If Circle *E* represents the entire population of electromagnetic wave transmitters, then all electromagnetic wave transmitters are electronic, according to the diagram. If there were some transmitters that were not electromagnetic wave transmitters, such as mechanical sound sources, then Circle *E* would not be wholly engulfed by Circle *A*. The area identified as *BCD* represents a population of entities that have all three features represented by the Circles *B, C,* and *D.*

verb: *See* **statement verb.**

Verdet constant: The constant of proportionality, i.e., the *V*, in the equation that defines the Faraday effect, which (a) is given by the relation $\theta = VB\int ds = V\mu H\int ds$ where θ is the rotation angle of the polarization plane of plane polarized light propagating in a medium, *V* is the Verdet constant, *B* is the magnetic flux density in the medium, μ is the magnetic permeability of the medium, *H* is the magnetic field strength at the point in the medium at which the path length differential, *ds*, is taken, and the integrand is the dot (scalar) product, i.e., the component of *H* in the direction of *ds*. *Note:* If the magnetic field is parallel to the propagation path, such as occurs when an optical fiber is parallel to a magnetic field, the magnitude of the rotation angle is directly proportional to the product of the path length and the magnetic flux density, or, for nonferrous substances such as glass, the product of the path length and the magnetic field strength because the magnetic permeability of glass is unity. Typical units for the Verdet constant are degrees/centimeter•tesla, degrees/meter•oersted, or radians/meter•tesla, depending on the units used for the variables in the solution integral of the Faraday equation for the magnetooptic effect. *See also* **magnetooptic effect.**

verification: *See* **automated software verification, busy verification, message verification.**

verified off-hook: In telephone systems, a network-provided service feature in which (a) a device is inserted on each end of a transmission circuit for verifying supervisory signals, and (b) a connection is automatically established between specified users when the call originator removes the receiver or the handset from its switch, i.e., goes off-hook. *See also* **circuit, dedicated circuit, hot line, point to point link, ringdown.**

verifier: A device that checks the correctness of transcribed data, usually by comparing with a second transcription of the same data or by comparing a retranscription with the original data.

verify: To determine whether an operation, such as the transcription of data, the proper functioning of a telephone circuit, or keypunching, is being or has been performed correctly.

vernal equinox: The line of intersection of the celestial equatorial plane and the ecliptic plane, at which the Sun passes from the Southern Hemisphere into the Northern Hemisphere, bringing spring to the Northern Hemisphere and autumn to the Southern Hemisphere on or about March 21. *See also* **autumnal equinox, ecliptic plane, equatorial plane.**

vertex: In an optical system, the point of intersection of the optical axis with any surface in the system. *See also* **node.**

vertex angle: At either end of an optical fiber, the solid angle formed by the extreme bound meridional rays accepted by the fiber or emerging from it, i.e., the angle formed by the largest cone of light accepted by the fiber or emerging from it. *Note:* The vertex angle is equal to twice the acceptance angle. *See also* **acceptance angle, meridional ray.**

vertical: Pertaining to a direction of a line through the center of the Earth. *See also* **horizontal.**

vertical cover: The vertical angle, i.e., the elevation angle, that an electromagnetic wave emitter, such as an antenna, is (a) capable of scanning or (b) capable of illuminating.

vertical coverage diagram: A graphic pattern, described in terms of altitude and range, that shows the distribution of power of a radio or radar signal in a vertical plane. *Note:* If the transmitter antenna is considered to be at the origin or the center of a polar coordinate system, then the length of a line from the origin to a perimeter of the pattern, i.e., to any point on the pattern, is directly proportional to the power per unit solid angle in the direction in which the line is drawn. Thus, the pattern is usually drawn as the locus of the points where the irradiance is a constant or is one-half of the maximum value for the given range. The diagram may be drawn for any vertical plane. It usually is measured empirically. It may also be calculated from theoretical and empirical functions of known parameters.

vertical line: A line that passes through the center of the Earth, i.e., a line that connects the zenith to the nadir. *See* **horizontal line, nadir, zenith.**

vertically polarized electromagnetic wave: A uniform plane polarized electromagnetic wave in which (a) the electric field vector is always and everywhere vertical, and hence the polarization plane is always vertical, (b) the magnetic field vector is always and everywhere horizontal, and hence the wave propagates in a horizontal direction, (c) the direction of propagation, i.e., the azimuth, of the wave is determined by the direction of the magnetic field as determined by the antenna, and (d) the wave cannot have a propagation component in the vertical direction. *See also* **horizontally polarized electromagnetic wave.**

vertical polarization: *See* **fixed vertical polarization.**

vertical redundancy check (VRC): *Synonym* **transverse redundancy check.**

vertical tabulation (VT) character: A format effector that causes the print or display position to move to the corresponding position on the next series of predetermined lines. *See also* **format effector, print position.**

vertical terminating block: In communications systems, a group of terminals at which contact can be made to permanent outside lines entering a station. *See also* **combined distribution frame.**

vertical wraparound: On the display surface of a display device, the continuation of cursor movement from the bottom position in a vertical column to the top position in the right adjacent column or from the top position in a vertical column to the bottom position in the left adjacent vertical column. *See also* **horizontal wraparound, wraparound.**

very high frequency (VHF): A frequency that lies in the range from 30 MHz (megahertz) to 300 MHz, the range having an American designator of VHF and an International Telecommunication Union designator of 8. *See also* **frequency spectrum designation.** *Refer to* **Table 1, Appendix B.**

very high frequency omnidirectional radio range: A very high frequency (VHF) radio-beacon short-range air navigation system that can provide a pilot with bearing information and left-right track signals from a selected ground station that (a) operates between 108 MHz (megahertz) and 118 MHz, (b) is a line of sight (LOS) station, (c) operates over distances at which signals can be received that are a function of altitude as well as transmitted power, (d) transmits two signals, one fixed and one rotating, i.e., scanning, (e) operates such that a receiver compares the phase of the signals and produces an indication of the magnetic bearing of the radio range station, (f) transmits a unique identification code, such as a three-letter code, and (g) usually is equipped with distance-measuring equipment. *See also* **long-range aid to navigation.**

very-large-scale integrated circuit (VLSIC): A large-scale integrated circuit (LSIC) that has a packing density and a speed capability an order of magnitude higher than in a conventional LSIC.

very low frequency (VLF): A frequency that lies in the range from 3 kHz (kilohertz) to 30 kHz, the range having an American designator of VLF and an International Telecommunication Union designator of 4. *See also* **frequency spectrum designation.** *Refer to* **Table 1, Appendix B.**

very small aperture terminal (VSAT) antenna: An antenna that (a) ranges in size from 1 m (meter) to 3 m, (b) transmits and receives signals, (c) has a small aperture, and (d) may be used in interactive networks, satellite Earth stations, and global positioning systems (GPS). *See also* **Earth station, global positioning system, ultrasmall aperture terminal (USAT) antenna.**

VES: virtual environment software.

vestigial sideband: In an amplitude-modulated signal, the small fraction of a sideband that is transmitted along with the whole or part of the carrier and the whole other sideband, thus allowing a mode of transmission in which all or part of the original carrier and a fraction of the other sideband are available at the receiver to assist in demodulation, thus making demodulation easier than in single sideband transmission. *See also* **single sideband transmission, vestigial sideband transmission.**

vestigial sideband (VSB) transmission: Double sideband transmission in which (a) one sideband, (b) all or part of the carrier, and (c) only a portion of the other sideband are transmitted. *See also* **carrier, double sideband transmission, frequency, sideband, sideband transmission.**

VF: voice frequency.

VFCT: voice frequency carrier telegraph.

VFCTG: voice frequency carrier telegraph. *See* **voice frequency telegraph.**

VF primary patch bay: *See* **voice frequency primary patch bay.**

VHF: very high frequency.

via net loss (VNL): Pertaining to circuit performance prediction and description that allows circuit parameters to be predetermined and the circuit to be designed to meet established criteria by analyzing actual, theoretical, and calculated losses. *See also* **circuit, design objective, traffic engineering.**

vibrating relay: A telegraph relay that (a) has additional windings that are excited so as to produce a certain regular oscillation of the armature between its stops in the absence of line current in the main winding, and (b) has increased sensitivity.

victim emitter: A source of electromagnetic energy that (a) is placed under electronic surveillance or (b) is considered for being placed under electronic surveillance for such purposes as monitoring, interception, deception, jamming, frequency control, net discipline, electronic emission control, securing electronic intelligence, obtaining order of battle information, and obtaining target information. *See also* **victim frequency.**

victim frequency: 1. A frequency that is intended to be jammed. **2.** A frequency that is undergoing interference. **3.** The frequency of a victim emitter. *See also* **victim emitter.**

victim transmission: Transmission that is (a) intercepted for intelligence purposes, (b) selected for electronic countermeasure (ECM) action, or (c) transmitted by a victim emitter.

video: 1. In radio transmission and display systems, pertaining to image information signals. **2.** Pertaining to the sections of a television system that carry television signals, in either unmodulated or modulated form. **3.** In communications systems, pertaining to a capability to transmit or process signals having frequencies that lie within the part of the frequency band employed for the representation of pictures that change in accordance with events occurring in real time. *Note:* Video frequency bands range from 100 kHz to several megahertz and are generally used for television transmission. **4.** In radio detection and ranging (radar) systems, pertaining to the demodulated radar signal that is applied to a radar display device. **5.** Pertaining to the bandwidth, i.e., the data signaling rate (DSR), necessary for the transmission of real time television pictures. *Note:* In practice, the baseband bandwidth required to meet the National Television Standards Committee standard for transmission of television pictures, not including the audio carriers, is about 5 MHz (megahertz). *See* **motion video.** *See also* **television, video signal.** *Refer to* **Fig. V-4.**

video amplitude limiting: The preventing of a video input signal from driving input circuits above a fixed power or voltage level. *Note:* In radar systems, video amplitude limiting is used to prevent the physical deterioration of the range-rate memory circuit performance that is caused by the greater amplitude of echo signals from chaff when dispensed from an aircraft with a relatively small reflecting surface.

video bar: *See* **reverse video bar.**

video buffer: A buffer that is used to store data to be displayed on a screen. *Note:* In a personal computer (PC) system, an area in random access memory (RAM) where software writes what is to be put on the display screen, using an adapter circuit that reads this video buffer many times per second and places its contents in readable form on the screen. *See also* **buffer.**

video cassette: A cassette that (a) contains a storage medium for video signals, such as magnetic tape or a

Fig. V-4. The AT&T Vistium™ Personal **Video** System combines desktop videoconferencing with collaboration, allowing users to see each other on their personal computer screens as they work on the same document. (Courtesy AT&T Archives.)

compact disk (CD), (b) can be inserted into a video recorder for one or more hours of video signal recording, and (c) can be played back many times by using a television set or a similar device. *See also* **video cassette recorder.**

video cassette recorder (VCR): A recorder that (a) accepts, holds, and drives a video signal storage medium, such as magnetic tape or a compact disk (CD), in a video cassette and (b) usually can be set to record and play back television programs. *See also* **camcorder, video cassette.**

video codec: *See* **codec.**

videoconference: *Synonym* **video teleconference.**

videoconferencing: *Synonym* **video teleconferencing.**

video discrimination circuit: A circuit that requires or forces a reduction in the frequency band pass of the video amplifier stage in which it is used. *Note:* In radar systems, when interference is experienced, the video discrimination circuit materially improves reception. A normal radar signal has a steep, i.e., a sharp, wavefront that may be partially or totally deteriorated by certain types of jamming. The video discrimination circuit may permit salvage of an otherwise lost signal if the signal is not completely deteriorated, i.e., if the interference level is lower than the desired signal.

video disk: *See* **digital video disk, optical video disk.**

video display terminal: *Synonym* **visual display unit.**

video display unit: *Synonym* **visual display unit.**

video filter: A filter that processes, i.e., either removes or passes, frequencies in the bands used in video systems, such as frequencies in the very high frequency (VHF) and ultrahigh frequency (UHF) bands.

video frequency (VF): 1. The frequency band that extends from approximately 100 MHz (megahertz) to several gigahertz. **2.** Pertaining to frequencies in the very high frequency (VHF) and ultrahigh frequency (UHF) ranges.

video graphic board: A printed circuit (PC) board that enables a computer to (a) control the display of images on the display surface of a display device, such as a personal computer (PC) monitor, (b) plot graphs, (c) control the display of diagrams, (d) allow interactive operation between the display device and the computer operator, and (e) control monochrome displays, polychrome displays, or both. *Synonym* **video graphic card.**

video graphic card: *Synonym* **video graphic board.**

video integrator feedback: Radar signal reception in which (a) received signals are recirculated, i.e., fed back, through a delay line, (b) in-phase synchronous signals are added, (c) the recirculation reduces the response to asynchronous signals, such as noise, interference, and jamming, (d) a limiter is used prior to feedback, and (e) pulsed signals, particularly spikes that result from noise and jamming, are materially reduced. *See also* **feedback, interference, jamming, limiter, noise.**

video optical detector: In a video signal transmission system, a device that converts (a) an optical video signal in an optical propagation medium, such as an optical fiber, to (b) an electrical video signal in an electrical propagation medium, such as a coaxial cable or wire.

video optical source: In a video signal transmission system, a device that converts (a) an electrical video signal in an electrical propagation medium, such as a coaxial cable or wire, to (b) an optical video signal in an optical propagation medium, such as an optical fiber.

videophone: A communications terminal that (a) is coupled to an imaging device that enables the call receiver or the call originator, or both, to view one another as on television, if they so desire, (b) has a video teleconference capability, (c) is usually configured as a small, desktop unit, designed for one operator, and (d) is a single, integrated unit.

video signal: A signal that (a) usually is used to transmit changing pictorial information, (b) in television, includes a sound signal, (c) is the generic signal between a transmitter and a receiver, and (c) has a bandwidth that depends upon the manner of transmission, such as high-definition television (HDTV), slow scan, full scan, or digitized full scan television (TV). *Note:* Traditionally, video bandwidth implied a bandwidth of about 4 MHz (megahertz). This varies, depending upon the application. *See* **composite video signal.** *See also* **signal, television.**

video teleconference (VT): 1. A teleconference in which video signals are used. **2.** A two-way electronic communications system that permits two or more persons in different locations to engage in the equivalent of face-to-face audio and visual communications. *Note:* Video teleconferences may be conducted as if all of the participants were in the same room. *Synonym* **videoconference.** *See also* **video signal.** *Refer to* **Figs. T-5 (1012), V-4.**

video teleconferencing: The conducting of a video teleconference. *Synonym* **videoconferencing.**

video teleconferencing unit (VTU): Equipment (a) that performs video teleconference functions, such as (i) coding and decoding of audio and video signals and (ii) multiplexing of video, audio, data, and control signals, and (b) usually does not include input/output (I/O) devices, cryptographic devices, network interface equipment, network connections, or the communications network to which the unit is connected. *Refer to* **Fig. T-5 (1012).**

video trunk: *See* **fiber optic video trunk.**

view: In satellite communications, the ability of a satellite Earth terminal to access a satellite, having it sufficiently above the horizon and clear enough of other obstructions that it is within a free line of sight from the satellite Earth terminal. *Note:* A pair of satellite Earth terminals has a satellite in mutual view when both simultaneously have unobstructed line of sight contact with the satellite. *See also* **footprint, horizon angle, line of sight propagation, satellite.**

viewdata: A telecommunications service in which a user can (a) access a remote database via a common carrier channel, (b) request data, and (c) receive requested data on a video display unit over a separate channel. *Note:* The access, request, and reception usually are via common carrier broadcast channels. *Synonym* **videotext.** *See also* **teletext.**

viewer: *See* **infrared viewer, ultraviolet viewer.**

view field: The maximum cone or fan of rays that are (a) passed through an aperture and (b) measured at a given vertex. *See* **borescope view field.** *See also* **aperture, ray, true field, vertex.**

viewing angle: *See* **borescope axial viewing angle.**

viewpoint: 1. The point from which a real object is seen, based on the image obtained on the retina and interpreted by the brain. **2.** In display systems, the origin from which angles and scales are used to map real world coordinates of an object into the user coordinates or device coordinates of display elements, display groups, or display images in the image storage space or in the display space on the display surface of a display device. **3.** In display systems, the location from which objects are considered to be viewed for the purpose of creating display elements, display groups, or display images in the image storage space or in the display space on the display surface of a display device. *See also* **device coordinate, display device, display element, display group, display image, display space, display surface, image, object, real world coordinate, user coordinate.**

viewport: *Synonym* **window.**

vignetting: In an optical element, the loss of light rays that (a) occurs when a light ray bundle passes through a given area, such as a pupil or an aperture, and (b) is caused by blocking of some of the light rays by various mechanisms, such as absorption, reflection, refraction, diffraction, scattering, and deflection. *See also* **aperture, optical element, pupil, ray bundle.**

village: *See* **electronic village.**

violation: *See* **alternate mark inversion violation.**

virgin medium: In communications, computer, data processing, and control systems, a data medium in or on which data can be recorded but have never been recorded. *Note:* Examples of virgin media are an unmarked sheet of paper, paper tape with no holes or with only sprocket holes, new magnetic tape on which data have never been recorded, or an unformatted diskette. *See also* **data medium.**

virtual address: The address of a storage location in virtual storage. *See also* **virtual storage.**

virtual address space: In virtual storage, the virtual address space, i.e., virtual storage area, assigned to a specific job or task, such as a batched job, a terminal job, a system task, or a command-initiated task. *See also* **job, storage area, virtual address space, virtual storage.**

virtual call: A call in which (a) the capabilities of either a real or a virtual circuit are used, (b) use is made of a virtual call capability, and (c) all or any part of the resources of the circuit is used for the duration of the call.

virtual call capability: A network-provided service feature in which (a) a call establishment procedure and a call disestablishment procedure determine the period of communication between data terminal equipment (DTE) in which user data are transferred by the network in the packet mode of operation, (b) end-to-end transfer control of packets within the network is required, (c) data may be delivered to the network by the call originator before the call access phase is completed, but the data are not delivered to the call receiver if the call attempt is unsuccessful, (d) the network delivers all the user data to the call receiver in the same sequence in which the data are received by the network, and (e) multiaccess DTEs may have several virtual calls in progress at the same time. *See also* **call, data terminal equipment, network, permanent virtual circuit, virtual call, virtual circuit.**

virtual call facility: A communications facility that has a virtual call capability.

virtual carrier: The location in the frequency spectrum that carrier energy would occupy if carrier energy were present. *Synonym* **virtual carrier frequency.** *See also* **carrier, frequency, permanent virtual circuit.**

virtual carrier frequency: *Synonym* **virtual carrier.**

virtual circuit (VC): 1. A communications system arrangement in which data from a source user are passed to a destination user over various real circuits during a single period of communication, i.e., during a single information transfer transaction. *Note:* Virtual circuits usually are established on a per call basis and are disestablished when the call is terminated. However, a permanent virtual circuit can be established as a dedicated link. **2.** In a communications system operating in the packet mode, a means of two-way simultaneous transmission through a data link that consists of associated transmit and receive channels. *Note:* A number of virtual channels may be derived from a data link by packet interleaving, i.e., by time division multiplexing. *Synonym* **logical circuit, virtual connection.** *See* **permanent virtual circuit.** *See also* **circuit, data, data circuit, information transfer transaction, network, phantom circuit, physical circuit, virtual call capability.**

virtual circuit capability: A network-provided service feature in which a user is provided with a virtual circuit. *Note:* The virtual circuit capability is not necessarily limited to packet mode transmission. An analog signal may be converted to a digital signal and then be routed over the network via any available route.

virtual connection: 1. *Synonym* **virtual circuit. 2.** A connection to a virtual circuit.

virtual device interface (VDI): A loadable computer program, such as an IBM PC DOS device driver, that (a) acts as an interpreter between a graphics application program and output devices, such as printers, plotters, and monitors, (b) can be used by software developers to write application programs that communicate with a virtual device rather than with specific hardware, and (c) in turn, sends commands to the specific hardware drivers that directly control the devices.

virtual disk drive: A direct access storage device that can be identified and used as a disk drive, but does not physically exist as an independent disk drive.

virtual environment software (VES): Computer software than can (a) be stored in a computer with a display device, (b) create a virtual environment on the display surface, and (c) support a virtual reality system. *See also* **display device, virtual reality system.**

virtual height: In the ionosphere, the height of an ionized layer determined by (a) measuring the time interval between the instant that a vertical signal is transmitted and the instant that an ionospheric echo is received, (b) estimating the speed of electromagnetic wave propagation in the atmosphere, considering the refractive index, (c) calculating the distance traveled by the transmitted wave and its echo, and (d) dividing this distance in half. *See also* **air sounding, ionosphere.**

virtual image: An image that (a) is formed at a point from which a divergent light ray bundle appears to proceed when the rays have a given divergence but no real physical point of intersection, (b) is formed at a distance that is inversely proportional to the divergence of the rays, (c) cannot be focused on a screen because there is no physical intersection of rays, i.e., there is no real image that can be focused on a screen, (d) is of a real object, (e) is always produced by a negative lens or a convex mirror, and (f) is produced by a positive lens when the object is located within its focal length. *See also* **focal length, image, lens, ray bundle, real image.**

virtual memory: In a computer system, a memory that (a) appears to, i.e., is available to, the operating programs running in the central processing unit (CPU), and (b) may be smaller, equal to, or larger than the real memory present in the system. *See also* **central processing unit, virtual storage.**

virtual network (VN): A network that (a) provides virtual circuits and (b) is established by using the facilities of a real network. *See also* **virtual circuit.**

virtual path: 1. In a network, a path between a data source and a data sink that uses one or more virtual circuits. **2.** In a network, a path that (a) lies between a data source and a data sink, (b) may be created by various physical circuit configurations without adding additional physical facilities to create the path, and (c) may be used for the transmission of messages. *Synonyms* **logical route, virtual route.**

virtual pushbutton: A display element or a display group on the display surface of a display device used, in connection with a stylus, such as a light pen or a cursor, as though it were a function key. *Synonym* **light button.**

virtual reality deck: The interconnected components that (a) are used to create and display virtual reality presentations, and (b) consist of (i) sensor input data terminals, (ii) input data from sensors geographically distributed, (iii) image processing equipment, and (iv) display equipment.

virtual reality system: A simulation system in which (a) an apparent three-dimensional computer-driven image, such as a landscape, spacecraft, or satellite, is displayed in the display space of a display device, (b) a user can interact with the image as if the user were in the image, such as (i) moving about in the image and (ii) grasping, moving, and releasing items in the image, and (c) a user can guide the orientation of the image. *See also* **display system, image.**

virtual route: *Synonym* **virtual path.**

virtual source: A source that is apparent rather than real when observed or measured at a distance from the real source. *Note:* Examples of virtual sources are (a) the main focal point of an antenna, (b) the focal point of a light source, (c) the focal point of an array or a distribution of radiators, (d) an image of a light source in a mirror, and (e) the focal point of an array of sound sources.

virtual space: In display systems, a space in which display elements, display groups, and display images are defined in user coordinates that may be independent of device coordinates but may be related to real world coordinates.

virtual storage: Storage space that may be regarded as addressable main storage by the user of a computer system, in which virtual addresses are mapped into real addresses though the physical storage area itself actually is not in the main storage area but in other storage areas, such as auxiliary, buffer, peripheral, or bulk storage areas. *Note:* The size of virtual storage is limited by the addressing scheme of the computer system and by the amount of auxiliary storage available, not by the actual number of main storage locations. Thus, in a computer, when the random access memory runs out of storage space, a computer program can make the other storage areas appear as main storage. However, time delays will be encountered in accessing these areas during operation. *See also* **main storage, real storage, virtual memory.**

virtual system: 1. In telecommunications, computer, information processing, and control systems, a system that (a) provides virtual networks or virtual circuits, or both, and (b) is established from or constructed with the facilities of a real system. **2.** A functional simulation of a system, such as a computer or a storage unit and its associated devices.

virtual telecommunications access: Access of a telecommunications network in which a set of computer programs is used to control communications among terminals and application programs.

virtual terminal (VT): In open systems, an applications service that (a) allows host terminals on a multiuser network to interact with other hosts regardless of terminal type and characteristics, (b) allows remote log-on by local area network (LAN) managers for the purpose of management, (c) allows users to access information from another host processor for transaction processing, and (d) serves as a backup facility.

visibility: In open systems architecture, the extent to which the functions and the operations in a layer below a given layer can be detected by the functions of the layer above, and vice versa. Usually layers mask each other, particularly when they are not adjacent. *See also* **layer, open system.**

visibility point: A point in a system, such as a communications, computer, data processing, or control system, at which a contact or a connection can be made. *Note:* Examples of visibility points are plugs, binding posts, terminals, connectors, and sockets.

visible infrared spin scan radiometer: *See* **fiber optic visible infrared spin scan radiometer.**

visible radiation: Electromagnetic radiation with wavelengths in the visible spectrum. *See also* **visible spectrum.**

visible spectrum: The portion of the electromagnetic frequency spectrum (a) to which the human retina is sensitive, (b) by which humans see, (c) that is considered to extend from a wavelength of about 400 nm (nanometers) to about 800 nm, i.e., 0.4 μm (micron) to 0.8 μm in air, which corresponds to a frequency band of 3.75 to 7.50×10^{14} Hz (hertz), and (d) that lies between the ultraviolet and the infrared regions, the visible wavelengths being longer than the ultraviolet and shorter than the infrared. *Note:* Fiber optic transmission systems operate at wavelengths slightly longer than the visible wavelengths, such as 1.31 μm (microns), a near infrared standard wavelength. *See* **light, optical spectrum.**

visual: *See* **audiovisual.**

visual-aural radio range: A radionavigation aid in an aeronautical or maritime navigation system that has four radio range legs, one pair of which provides a visual indication of range from the stations or a reference while the other pair provides an aural indication of range at the mobile station.

visual body signal: *See* **ground to air visual body signal.**

visual call sign: A communications system call sign transmitted by a means visible to the eye that may be aided by hand-held optical instruments, such as binoculars or telescopes. *Note:* Visual call signs may be transmitted by various means, such as semaphore, wigwag, flashing light, flaghoist, panel, heliograph, and pyrotechnics.

visual communications system (VCS): A communications system based on the use of direct observation by the human eye, such as systems that use heliographs, sun flash, semaphores, pennants, panels, arm and hand signals, smoke signals, flares, and flags. *Note:* Though visual communications systems do effect communications, sometimes over great distances, such as ship to ship, ship to aircraft, ground to air, and over large areas of a battlefield, they usually are not considered to be telecommunications systems.

visual display unit (VDU): A device that has a display screen and that usually is equipped with a keyboard. *Note:* Examples of visual display units (VDUs) are cathode ray tubes, light-emitting diode (LED) arrays, liquid crystal arrays, and plasma panels. *Synonyms* **monitor, video display terminal, video display unit.**

visual emergency: *See* **ground-air visual emergency.**

visual engaged signaling: In telephone switchboard operation, signaling that (a) consists of a visual indication that a line is no longer busy, and (b) when used, means that the operator need not perform the engaged test. *See also* **engaged test.**

visual paulin signal: In visual signaling systems, a signal that (a) is sent by means of patterns made of folded tarpaulins or sails, (b) usually is sent from surface to air, i.e., from ground to air or from sea surface to air, (c) in an emergency, may be sent by "tarpaulins" improvised with other materials, and (d) is primarily used by survivors in distress situations. *Synonym* **visual sail signal, visual tarpaulin signal.**

visual responsibility chain: In visual communications systems, the sequence of relaying or repeating stations that have visual contact and along which messages can be sent and relayed.

visual sail signal: *Synonym* **visual paulin signal.**

visual signal: A signal that can be seen by the eye and may be aided by a hand-held optical instrument, such as a pair of binoculars or a telescope. *Note:* Examples of visual signals are signals sent by the use of hands, flags, panels, smoke, heliograph, semaphore, and wigwag. *See also* **optical signal.**

visual signal code: *See* **international visual signal code.**

visual signaling: Signaling (a) in which communications are based on the use of direct observation by the human eye, such as systems that use heliographs, sun flash, flashing lights, semaphores, pennants, panels, arm and hand signals, smoke signals, flares, and flags, (b) in which communications may be effected over long and short distances, such as ship to ship, ship to aircraft, ground to air, mountain top to mountain top, and over large areas of a battlefield, (c) that with a few exceptions, such as flashing lights and flares, may only be performed during daylight hours, and (d) in which signals are not considered to be telecommunications signals. *See also* **signaling.**

visual spectrum: The bands of color produced when polychromatic light, such as white light, is decomposed into its wavelength components by means of dispersion. *Note:* Examples of visual spectra are (a) the rainbow produced by sunlight passing through a cloud of water droplets, forming a lens or a prism, suspended in air, and (b) the spread of colors produced by a prism, beveled plate glass, or a cut diamond when illuminated by white light or sunlight.

visual tarpaulin signal: *Synonym* **visual paulin signal.**

visual transmission: Transmission in which (a) signals are visible to the eye, and (b) messages may be sent and relayed by various signaling systems, such as semaphore, wigwag, heliograph, pyrotechnic, flaghoist, panel, paulin, and body signaling systems. *See also* **electrical transmission, optical transmission.**

vitreous silica: Glass that consists of almost pure silicon dioxide (SiO_2). *Synonym* **fused silica.** *See also* **fused quartz.**

VLF: very low frequency.

VNL: via net loss.

V number: *Synonym* **normalized frequency.**

vocabulary: *See* **panel signaling vocabulary.**

vocoder: *Synonym* **voice coder.**

vodas: *Synonym* **voice-operated device antising.**

vogad: *Synonym* **voice-operated gain-adjusting device.**

voice: In communications systems, pertaining to a capability or a characteristic used to detect, transmit, or process signals with frequencies that lie within the frequency band used for the representation of speech. *Note:* Voice frequency bands range from about 300 Hz (hertz) to about 3400 Hz. *See* **digitized voice, encrypted voice, single sideband voice.**

voice band: The 300 Hz (hertz) to 3400 Hz electrical wave band that represents the sound wave frequency band used in telephone equipment for the transmission of voice and sound signals that represent data. *Synonym* **voice frequency band.**

voice call sign: *See* **permanent voice call sign.**

voice coder (vocoder): A device (a) that usually consists of a speech analyzer and a speech synthesizer, (b) in which the analyzer circuitry converts analog speech waveforms into narrowband digital signals, (c) in which the synthesizer circuitry converts digital signals into artificial speech sounds, (d) that, for communications security (COMSEC) purposes, may be used in conjunction with a key generator and a modulator-demodulator to transmit digitally encrypted speech signals over narrowband voice communications channels, (e) that is used to reduce the bandwidth requirements for transmitting digitized speech signals, and (f) that may translate the frequency of incoming signals from one portion of the frequency spectrum to another portion. *Synonym* **vocoder, voice-operated coder.** *See* **pattern matching voice coder.** *See also* **code, communications security, key.**

voice communications system: *See* **automatic secure voice communications system.**

voice cord board: *See* **secure voice cord board.**

voice-data service: A communications systems service that handles messages (a) in voice form, i.e., in analog form, and (b) in data form, i.e., in digital form. *See also* **data service.**

voice-data signal: *Synonym* **quasi-analog signal.**

voice-data tributary station: *Synonym* **dual access tributary station.**

voiced sound: Sound that is created by the speech organs, in which sounds are generated by the vibration of vocal cords. *See also* **unvoiced sound.** *Refer to* **Fig. P-5 (718).**

voice encoder: *See* **channel voice encoder.**

voice equipment: *See* **limited protection voice equipment.**

voice frequency (VF): A frequency that (a) is within the audio range, i.e., the voice band, and (b) is used for the transmission of speech signals. *Note:* In telephony, the usable voice frequency band extends from about 300 Hz (hertz) to about 3400 Hz. The bandwidth allocated for a single voice frequency transmission channel usually is 4 kHz (kilohertz), including guard bands. *See also* **audio frequency, frequency, guard band, voice band, voice frequency channel.** *Refer to* **Fig. P-5 (718).**

voice frequency band: *Synonym* **voice band.**

voice frequency carrier telegraph (VFCTG or VFCT): *Synonym* **voice frequency telegraph.**

voice frequency channel: A channel that (a) can carry analog and quasi-analog signals and (b) operates at voice frequencies, i.e., frequencies from about 300 Hz (hertz) to about 3400 Hz. *See also* **channel, frequency, voice frequency.**

voice frequency primary patch bay: A patching facility (a) that provides the first appearance of local user voice frequency (VF) circuits in the technical control facility (TCF), (b) that allows patching, monitoring, and testing for all VF circuits, and (c) in which signals have various levels and signaling schemes, depending on the user terminal equipment. *See also* **circuit, patch bay.**

voice frequency telegraph (VFTG or VFT): A telegraph system in which (a) one or more direct current (dc) telegraph channels are multiplexed into a composite nominal 4 kHz voice frequency channel for further processing through a transmission line or a radio network, (b) one or more carrier frequency currents are used that are within the voice frequency range, and (c) the transmission equipment conveys telegraph signals by voice-frequency tones. *Note:* "TG" is the abbreviation for "telegraph." *Synonym* **voice frequency carrier telegraph.** *See* **multichannel voice frequency telegraph.** *See also* **channel, frequency, two-tone keying.**

voice frequency truncation: The suppression of the higher frequencies of speech, i.e., frequencies above about 3400 Hz, or the suppression of lower frequencies of speech, i.e., frequencies below about 300 Hz, in speech circuits and audio circuits. *Note:* Voice frequency truncation (a) may occur as a consequence of circuit limitations, or (b) may be an intended process to limit the bandwidth and thereby eliminate the noise in the upper and lower bandwidths. *Synonym* **speech spectrum truncation.**

voice grade: In the public regulated telecommunications services, a service grade that (a) does not require any specific signaling or supervisory scheme and (b) is described in part 68, Title 47 of the *Code of Federal Regulations (CFR)*. *Note:* Voice grade service does not imply any specific signaling or supervisory procedure or scheme. *See also* **subvoice grade channel.**

voice grade circuit (VGC): A circuit that has the frequency response and attenuation characteristics needed to effectively handle voice frequency signals. *See* **conditioned voice grade circuit.** *See also* **attenuation, frequency response, voice frequency, voice-operated gain-adjusting device.**

voice-graphics tributary station: *Synonym* **dual access tributary station.**

voice intelligibility: In a telephone communications system, the capability of understanding speech without necessarily implying the ability to associate a particular articulation with a particular person. *Refer to* **Fig. P-5 (718).**

voice network: *See* **Automatic Voice Network.**

Voice of America: A communications system that (a) sends messages in many languages to many countries, usually via radio and Internet, (b) is intended to inform audiences concerning world, national, and local news events, (c) provides health and environmental information, and (d) is supported by private and public contributions and by derived income.

voice-operated device antising (vodas): A device that (a) is used to prevent overall voice frequency (VF) singing in a two-way telephone circuit by ensuring that transmission can occur in only one direction at any given instant, and (b) is operated by sound that attenuates oscillations of any type or from any source, such as resonance, interference, ringing, singing, or noise. *Synonym* **vodas.** *See also* **singing.**

voice-operated gain-adjusting device (vogad): A device that has a substantially constant output amplitude over a wide range of input amplitudes. *Synonym* **vogad.** *See also* **clipper, compandor, compressor, expander, peak limiting, voice grade circuit.**

voice-operated relay circuit (vox): A relay circuit that (a) consists of an acoustoelectric transducer and a keying relay, (b) is connected so that the keying relay is actuated when sound energy, such as speech, above a certain threshold level is sensed by the transducer, and (c) eliminates the need to push to talk in the operation of some transmitters by using the sound energy in the speech of the operator to turn on the transmitter. *Synonyms* **voice-operated transmit/voice-operated relay circuit, vox.** *See also* **attack time, push to talk operation.**

voice-operated transmit/voice-operated relay circuit (vox): *Synonym* **voice-operated relay circuit.**

voice pipe: In aural communications systems, a hollow tube that is used to conduct voice signals from one place to another, such as from the bridge to the engine room of a ship, from the entrance of a building to an apartment, or from a machine operator to a supervisor's office.

voice-plus circuit: *See* **composited circuit, conference call.**

voice-powered telephone: *Synonym* **sound-powered telephone.**

voice service: A communications service that (a) is designed to transmit and receive voiced sounds, i.e., voiced calls, such as the transmission and the reception provided by a telephone system, (b) provides analog voice transmission and reception, (c) may provide digitized voice transmission and reception, and (d) may provide tone signaling and digital signaling, such as in facsimile and supervisory signaling.

voice telephony: Pertaining to a communications system established for the transmission of sounds, such as speech.

voice transmission: The transmission of speech signals in analog or digital form. *See* **digital voice transmission.**

voice unit: *Synonym* **volume unit.**

volatile storage: A storage device in which the contents are lost when power is removed. *Note:* Examples of volatile storage are electrostatic storage, acoustic delay line storage, electromagnetic delay line storage. *See also* **data volatility, nonvolatile storage.**

volatility: *See* **data volatility.**

volcas circuit: A voice-operated loss-control echo-singing suppression circuit. *See also* **antisinging device, singing suppression circuit.**

volt: The SI (Système International d'Unités) unit of electric potential difference, i.e., of voltage, that may exist between two points, i.e., the amount of electromotive force required to drive an electric current of 1 A (ampere), i.e., 1 $C{\bullet}s^{-1}$ (coulomb per second), through a resistance of 1 Ω (ohm). *Note:* The coulomb can be expressed as equivalent to an electric charge produced by 6.2418×10^{18} electrons, and the ohm can be expressed in terms of material, geometry, temperature, humidity, and pressure.

voltage: The measured, indicated, or existing potential difference, expressed in volts, including multiples and submultiples of volts, such as kilovolts, megavolts, millivolts, and microvolts, between two points. *See* **common mode voltage, common return offset voltage, disturbance voltage, noise voltage, psophometric voltage, ripple voltage, vector voltage.** *See also* **potential difference, volt.**

voltage divider: *Synonym* **potentiometer.**

voltage drop: *See* **distribution voltage drop.**

voltage level: *See* **digital voltage level.**

voltage pencil: A stylus that can be (a) used to identify coordinate positions in the display space on the display surface of a display device, (b) used to select symbols and operate from a source of voltage, and (c) detected at specific locations on the display surface. *See also* **stylus.**

voltage standing wave ratio (VSWR): In a transmission line, the ratio of (a) the maximum voltage to (b) the minimum voltage in a standing wave pattern. *Note:* The voltage standing wave ratio (VSWR) is used as a measure of impedance mismatch between the transmission line and its load. The higher the VSWR, the greater is the mismatch. If there is no standing wave, there is no mismatch, i.e., there is no reflected wave to interact with the incident wave to create a standing wave, and all the power arriving at the load is absorbed by the load. *See also* **standing wave ratio.**

voltaic: *See* **photovoltaic, radiovoltaic.**

voltaic effect: *See* **photovoltaic effect.**

voltaic photodetector: *See* **photovoltaic photodetector.**

volt•ampere (VA): In steady state sinusoidal voltages and currents in a circuit, a unit consisting of the product of the voltage in volts and the current in amperes. *Note:* 1 VA (volt•ampere) is equal to (a) 1 V times 1 A, (b) 2 V times 0.5 A, (c) 0.1 V times 10 A, and so on, usually in root mean square (rms) or direct current (dc) values. *See also* **apparent power, reactive power, real power, volt•amperes reactive.**

volt•amperes: *Synonym* **apparent power.**

volt•amperes reactive (VAR): In a circuit, a unit of the reactive power, i.e., the power in phase quadrature with the real power. *Note 1:* The square root of the sum of the squares of the reactive power and the real power is equal to the volt•amperes (VA). *Note 2:* Reactive power requires real current. In order to reduce resistive losses, i.e., I^2R losses where I is the current and R is the resistance, the transmission efficiency can be improved by inserting reactance, capacitive reactance if the current is lagging the voltage in phase or inductive reactance if the current is leading, so as to obtain a unity power factor, in which case the current and the voltage are in phase, and the supplied current is a minimum for the effective power load, i.e., the real power load. *See also* **real power, volt•ampere.**

volt-ohm-milliameter: *Synonym* **multimeter.**

volume: 1. A portion of data, together with the data carrier, that can be handled conveniently as a unit. *Note:* Examples of volumes include the data on a reel of magnetic tape, disk, diskette, or magnetic card. **2.** A defined portion of a storage area that is accessible to a single read-write head on a storage device, such as a magnetic disk, drum, card, or tape storage unit. *See* **communications traffic volume, effective mode volume, mode volume, reference volume.**

volume distribution: *See* **mean power of the talker volume distribution.**

volume unit (vu): The unit of measurement of the power of an audio frequency signal, as measured by a volume unit (vu) meter. *Note 1:* The volume unit (vu) meter, or its modern equivalent, is built and used in accordance with American National Standard C16.5-1942. It has a scale and specified dynamic and other characteristics used to obtain correlated readings of speech power that are necessitated by the rapid fluctuation in the level of voiced signals. *Note 2:* In using the volume unit (vu) meter to measure sine wave test tone power, 0 VU = 0 dBm = 1 mW. The reference 0 VU (volume unit) is 1 millivolt of a steady state rms sinusoidal electrical wave into a 600 Ω (ohm) resistive load. An acceptable working amplitude is between −10 VU and −30 VU. Values above 0 VU are too loud, and values below −55 VU are too soft, i.e., too weak. Readings will be a function of the calibration and the frequency response of the meter. Synonyms **speech power unit, voice unit.** *See also* **transmission level point.**

vox: voice-operated relay circuit, voice-operated transmit/voice-operated relay circuit.

V parameter: *Synonym* **normalized frequency.**

VPI: virtual path identifier.

V reference point: In an Integrated Services Digital Network (ISDN) environment, the interface between the line termination and the exchange termination, i.e., the local office termination.

VSAT antenna: very small aperture terminal antenna.

VSB: vestigial sideband.

V series Recommendation: A member of a set of telecommunications protocols and interfaces defined in International Telegraph and Telephone Consultative Committee (CCITT) Recommendations. *Note:* Some of the more common V series Recommendations are:

V.21	300-bps Modem
V.22	1200-bps Modem
V.23	600/1200-bps Modem
V.25	Autodial/Autoanswer
V.26	2.4-kbps Modem
V.26*bis*	1.2/2.4-kbps Modem
V.27	4.8-kbps Modem
V.27*bis*	2.4/4.8-kbps Modem with Automatic Equalization
V.27*ter*	2.4/4.8-kbps Modem
V.29	9.6-kbps Modem (Point to Point Only)

See also **X series Recommendation.**

VSWR: voltage standing wave ratio.

VT: variable time.

vu: volume unit (abbreviation).

VU: volume unit (unit of measure).

vulnerability: 1. The characteristics of a system that allow it to degrade, i.e., to reduce its capacity to perform its mission, usually as a result of some form of assault. **2.** The extent to which a system will degrade when subjected to a specified set of environmental conditions. *See also* **electromagnetic vulnerability, technical vulnerability.**

V value: *Synonym* **normalized frequency.**

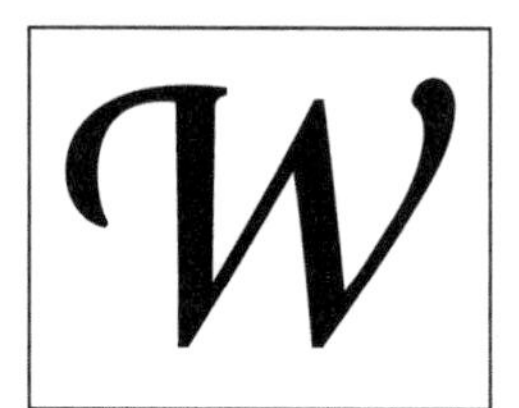

W: **watt.**

wafer: A thin slice of a suitable semiconductor, such as a silicon or germanium crystal, on which microcircuits are constructed by using various techniques, such as diffusion, etching, and evaporative deposition of dopants. *Note:* Thousands of microcircuits may be constructed on a single wafer. Individual and groups of microcircuits may be separated by scoring and breaking the wafer into individual chips or "dice." *See also* **chip, large-scale integrated circuit, microcircuit.** *Refer to* **Figs. W-1 and W-2.** *Refer also to* **Fig. C-3 (128).**

WAIS: **Wide Area Information Servers.**

waiting: *See* **call waiting.**

waiting in progress signal: In a communications system, a signal that is sent in the backward direction to indicate that (a) the call receiver has the connect when free facility, (b) the call receiver is busy at the moment, and (c) the call has been placed in a queue.

waiting signal: *See* **call waiting signal, data circuit-terminating equipment waiting signal, data terminal equipment waiting signal.**

walking code: A binary code that (a) gives the appearance of walking bit by bit through a register, one bit for each number change, and (b) has a pattern that is advanced by a count of 1 by (i) shifting the number to the left with the bit that is shifted out complemented and fed in at the right or (ii) shifting the number to the right with the bit that is shifted out complemented and fed in at the left. *Note:* Relatively simple counter circuits can be made to use the walking code. *Note:* An example of a three-bit unit-distance walking code is 000,001,011,111,110,100. *Synonym* **creeping code.**

WAN: **wide area network.**

wand: **1.** A hand-held device that (a) usually is long and narrow like an orchestra conductor baton, (b) contains (i) a light source and a photodetector or (ii) a magnetic reading head, (c) can be used to detect the nature of images or shapes printed or magnetically recorded on a surface, such as bar codes or magnetic ink recordings, and (d) usually contains circuits that convert the images or data scanned into electrical signals for transmission to an interpreting or display device. **2.** To scan with a hand-held device that will sense the images that are scanned. **3.** Pertaining to a hand-held device that (a) is used to scan a surface to detect images on it, and (b) is particularly suitable for reading bar codes when the object containing the bar code cannot be moved across the face of a fixed bar code reader, such as those used at store checkout counters. *See also* **wanding.**

wander: Relative to jitter and swim, long-term random variations of the significant instants of a digital signal from their ideal positions. *Note 1:* Wander variations are those that occur over a period greater than 1 s (second). *Note 2:* Jitter, swim, wander, and drift have increasing periods of variation in that order. *See also* **drift, jitter, significant instant, swim.**

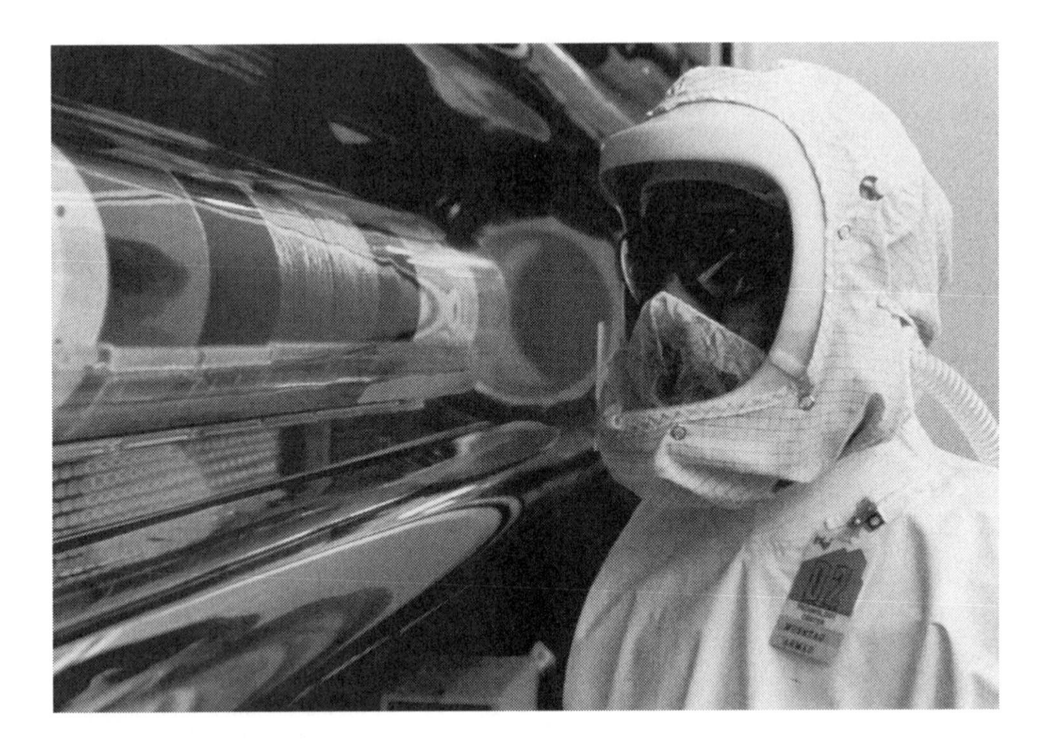

Fig. W-1. Furnaces glow to bake **wafers** to over 1000°C. (Courtesy Intel Corporation.)

Fig. W-2. The use of robotic equipment (behind panels) to handle silicon **wafers** in areas where strong chemicals are used to etch patterns into the surface of the wafers. (Courtesy Intel Corporation.)

wanding: Passing the tip of a wand over recorded data in order to read the data. *See also* **wand.**

warfare: *See* **electronic warfare.**

warfare communications: *See* **psychological warfare communications.**

warfare support measure: *See* **electronic warfare support measure.**

warm boot: *Synonym* **warm restart.**

warm restart: The restarting of equipment, after a usually unintentional, sudden, or flash closedown or shutdown, that allows reuse of previously retained initialized input data, programs, and output queues. *Note:* A warm start or restart cannot occur if initial data, programs, and files are not retained after closedown. *Synonyms* **warm boot, warm start.** *See also* **cold restart.**

warm start: *Synonym* **warm restart.**

Warner exemption: A statutory exemption pertaining to the acquisition of telecommunications systems that meet the exclusionary criteria of the Warner Amendment, Public Law 97-86, 1 December 1981, which is also known as the "Brooks Bill." *Note:* The use of Federal Telecommunications System (FTS) 2000 by U.S. government agencies is mandatory when telecommunications systems are required. However, the Warner Amendment excludes the mandatory use of FTS 2000 in instances related to maximum security.

warning: **1.** The content of a message that contains information concerning an impending danger, such as a natural disaster, a hostile force, bad weather, or a power outage. **2.** In a communications system, an advance notice that a possible error in transmission has occurred. *See* **airborne early warning, emergency meteorological warning.**

warning message: **1.** A message that contains information concerning an impending danger, such as a natural disaster, a hostile force, bad weather, or a power outage. **2.** A message or an indication that a possible error has been detected.

warning net: A communications network used for the transmission of warnings of impending dangers, such as natural disasters, hostile forces, bad weather, or power outages. *See* **airborne early warning net.**

warning radar: *See* **early warning radar, initial warning radar.**

warning receiver: A receiver that (a) usually is an intercept receiver, and (b) has the primary function of warning the user or the operator of impending danger.

warning system: *See* **airborne radar warning system, aircraft control warning system, threat warning system.**

warning tone: *See* **recorder warning tone.**

watch: A session of duty at a radar, radio, or video station with emphasis on reception of signals. *See* **communications watch, continuous communications watch, copy watch, cover watch, guard watch, lens watch, listening watch, radio communications copy watch, receiver watch, single-operator period watch, two-operator period watch, World Weather Watch.**

watch ship: In radiotelegraph communications at sea, a ship station that is on listening watch, is transmitting, or is receiving messages. *See* **constant watch ship.**

watch station: *See* **constant watch station.**

WATS: wide area telecommunications service, wide area telephone service.

watt: A unit of power equal to 10^7 ergs, 10^7 dyne•cm (dyne•centimeters), 1 A (ampere) at 1 V (volt) (dc or rms), 1/746 horsepower, or 1 J•s^{-1} (joule per second). *Note:* A dyne is 1/980g (gram) of force.

wave: A movement that (a) may be to and fro, up and down, or an increase and a decrease in the value of a parameter, usually a physical parameter, such as that of (i) material matter, as in water surface waves and acoustic waves, and (ii) force fields, as in the electric and magnetic fields of an electromagnetic wave, (b) tends to propagate, though it can also be a standing wave, (c) may be periodic or aperiodic, (d) may be simply a single surge, *i.e.,* a transient occurrence, and (e) usually is considered to be a periodic, propagating phenomenon, such as a radio wave, a lightwave, or a sound wave. *See* **acoustic wave, audio frequency wave, carrier wave, centimeter wave, clockwise polarized wave, complementary wave, continuous wave, counterclockwise polarized wave, decametric wave, decimetric wave, decimillimetric wave, direct wave, elastic wave, electric wave, electromagnetic wave, evanescent wave, ground wave, guided wave, hectometric wave, Hertzian wave, horizontally polarized electromagnetic wave, indirect wave, interrupted continuous wave, leaky wave, left-hand polarized electromagnetic wave, lightwave, magnetic wave, metric wave, millimetric wave, Morse continuous wave, myriametric wave, plane electromagnetic wave, plane polarized electromagnetic wave, plane wave, polarized electromagnetic wave, radio wave, right-hand polarized**

electromagnetic wave, seismic wave, shock wave, short wave, skywave, sound wave, standing wave, surface wave, transverse electric wave, transverse electromagnetic wave, transverse magnetic wave, trapped electromagnetic wave, traveling wave, tropospheric wave, uniform plane polarized electromagnetic wave, vertically polarized electromagnetic wave.

wave analyzer: A device capable of measuring the energy contained in electromagnetic waves in specified or selected frequency bands.

wave equation: The equation that (a) is derived from Maxwell's equations, the constitutive relations, and the vector algebra, and relates the electromagnetic field of an electromagnetic wave time and space derivatives to the propagation medium electric permittivity and magnetic permeability in a region without electrical charges or currents, i.e., in a charge free dielectric propagation medium, such as that of free space and optical fibers, (b) has solutions that yield the electric and magnetic field strengths everywhere in the medium in which the wave is propagating, and (c) is given by the relation $\nabla^2\mathbf{H} - \mu\varepsilon\partial^2\mathbf{H}/\partial t^2 = 0$ or by the relation $\nabla^2\mathbf{E} - \mu\varepsilon\partial^2\mathbf{E}/\partial t^2 = 0$ where **E** is the electric field strength, **H** is the magnetic field strength, ε is the electric permittivity, μ is the magnetic permeability, usually equal to 1, $\partial^2/\partial t^2$ is the second partial derivative with respect to time, and ∇ is the vector spatial derivative operator. *Note:* When the wave equation for uniform plane waves is transformed into Cartesian coordinates, the scalar equations for the electric and magnetic field strengths in the transverse directions (x and y for a wave propagating in the z direction) are obtained in terms of propagation distance and propagation medium parameters, namely, the electric permittivity and magnetic permeability. These are the Helmholtz equations. They are ordinary differential equations. Their solutions are simple exponential functions with exponents that are the propagation constants. For example, the relation $(d^2/dz^2)\mathrm{E}_x - \Gamma^2 E_x = 0$ is one of the Helmholtz equations derivable from the wave equation. A solution of this Helmholtz equation is given by the relation $E_x = E_{xo}e^{-\Gamma z}$ where $\Gamma = -\omega^2\mu\varepsilon$, μ being the magnetic permeability, ε the electric permittivity, and $\omega = 2\pi f$, where f is frequency. In transparent dielectric media, $\mu = 1$. Thus, the relative refractive index is given by the relation $n_r = \varepsilon_r^{1/2}$ where ε_r is the electric permittivity relative to a vacuum, or approximately to air. The wave equation, the derived Helmholtz equations, and their solutions apply to electromagnetic waves propagating in free space and optical waveguides. Their eigenvalues and boundary conditions give rise to many possible modes, of which only a limited number can be supported by a given optical waveguide, such as an optical fiber or planar waveguide.

waveform: The representation of a signal as a plot of amplitude versus time or amplitude versus distance. *See also* **wavefront.**

waveform jamming: *See* **periodic waveform jamming.**

wavefront: 1. In a wave, the surface that is defined by the locus of points that have the same phase, i.e., that have the same path length from the source. **2.** In an electromagnetic wave, the surface that (a) is defined by the locus of points that have the same phase, i.e., that have the same path length from the source, (b) at a point in the wave, is perpendicular to the ray, (c) is the plane in which the electric and magnetic field vectors lie at every point in the wave, (d) is represented by a vector that is in the direction of propagation, i.e., the Poynting vector, (e) for parallel rays, such as rays in the far field region and collimated rays, practically is a plane, (f) for rays diverging from a point, or converging toward a point, practically is a sphere, and (g) for rays with varying divergence or convergence, has other shapes, such as elliptical, parabolic, and hyperbolic surfaces, depending on the nature of, and distance from, the source. *Note:* The electric and magnetic field vectors at a point in space define a plane, that plane being tangent to the wavefront surface at that point. *See also* **geometric optics, path length, physical optics, Poynting vector, ray.**

wavefront compensation: In image restoration, wavefront control in which distortion of the original wavefront is removed by the introduction of coherent compensatory electromagnetic waves, by (a) creating an error map of local deviation or tilt from ideal spherical waves and (b) using these data to correct the distortion.

wavefront control: In a propagation medium, such as an optical system, performing operations to manipulate the shape of the wavefront of an electromagnetic wave, usually in the visible and near visible region of the frequency spectrum and usually with the intent of obtaining clear images of illuminated objects, i.e., of obtaining a spherical wavefront, by various methods, such as phase conjugation, aperture tagging, wavefront compensation, and image sharpening. *See also* **aperture tagging, image sharpening, phase conjugation, wavefront.**

wavefront optics: The branch of optics devoted to the study, design, development, and production of devices that function according to the phase of electromagnetic waves.

waveguide: A material medium that (a) confines, supports, and guides the energy of a propagating wave, such as an electromagnetic, acoustic, or elastic wave, along a prescribed, usually narrow, and controllable path, (b) usually consists of a configuration of materials, and (c) usually affects the waves in some way during their passage along the medium. *Note 1:* For radio waves, such as radio frequency (rf) and microwaves, waveguides may consist of (a) coaxial cables or hollow metallic conductors, usually rectangular, elliptical, or circular in cross section, which may be filled with a solid or gaseous dielectric material, (b) ionized layers of the stratosphere, and (c) refractive surfaces of the troposphere. *Note 2:* For lightwaves, waveguides may consist of a solid dielectric filaments, such as optical fibers, usually circular in cross section. In optical integrated circuits (OICs), waveguides may consist of thin-film dielectric materials. *Note 3:* For acoustic waves, waveguides may consist of air-filled hollow tubes, refractive surfaces of the troposphere, or refractive layers of seawater of different density. *See* **circular dielectric waveguide, closed waveguide, dielectric waveguide, diffused optical waveguide, fiber optic waveguide, frequency selective waveguide, graded index optical waveguide, grooved waveguide, multimode waveguide, open waveguide, optical waveguide, periodically distributed thin-film waveguide, planar diffused waveguide, single mode optical waveguide, slab dielectric optical waveguide, slab dielectric waveguide, step index optical waveguide, strip-loaded diffused optical waveguide, thin-film optical waveguide.** *See also* **optical fiber.** *Refer to* **Figs. C-4 (133), C-6 (142), E-2 (291), I-5 (470), P-2 and P-3 (708), P-17 (767), S-12 (917).**

waveguide connector: *See* **optical waveguide connector.**

waveguide coupler: *See* **optical waveguide coupler.**

waveguide delay distortion: In an electromagnetic wave, such as a lightwave, propagating in a waveguide, the distortion in received signals due to the differences in group delay time for each wavelength or each propagation mode, causing (a) dispersion, i.e., an increase in pulse duration or (b) distortion of analog signals, as the wave propagates along the waveguide. *See also* **pulse duration.**

waveguide dispersion: In a dielectric waveguide, dispersion of an optical pulse that occurs because of the dependence of the phase and group velocities on wavelength as a consequence of the geometric properties, such as cross-sectional dimensions and numerical apertures, of the waveguide. *Note 1:* For dielectric waveguides that are circular in cross section, such as optical fibers, waveguide dispersion depends on the ratio of a/λ where a is the core radius, and λ is the wavelength. Waveguide dispersion is of importance only in single mode fibers, caused by the dependence of the phase and group velocities on core radius, numerical aperture, and wavelength. *Note 2:* Practical single mode fibers may be designed so that material dispersion and waveguide dispersion cancel one another at the wavelength of interest, giving rise to a trough in the dispersion versus wavelength curve at that point. *See also* **dispersion, distortion, intramodal distortion, material dispersion coefficient, multimode distortion, profile dispersion, profile dispersion parameter.**

waveguide isolator: A passive attenuator that (a) has transmission losses that are much greater in one direction than in the other, (b) causes absorption of end reflections, backscatter, and noise, and (c) may be constructed of various materials with nonreciprocal, i.e., with directional, properties, such as some ferrite materials.

waveguide mode: A mode of a wave propagating in a waveguide. *See* **degenerate waveguide mode.**

waveguide scattering: Scattering that (a) is caused by variations in the dimensions and the refractive index profile of a waveguide, and (b) does not include scattering by the intrinsic material from which the waveguide is constructed, such as Rayleigh, Brillouin, and Mie scattering. *See also* **Brillouin scattering, material scattering, Mie scattering, nonlinear scattering, Rayleigh scattering, scatter.**

waveguide splice: A permanent joint between the transmission elements of two waveguides so that signals may pass from one waveguide to the other with minimal loss. *See* **optical waveguide splice.**

waveguide switch: A mechanically, electromechanically, electrically, magnetically, or electromagnetically controlled device that can stop or divert the propagation of electromagnetic energy at a specific point in a waveguide.

waveguide termination: *See* **optical waveguide termination.**

wave impedance: In an electromagnetic wave propagating at a point in a propagation medium, the ratio of (a) the electric field strength to (b) the magnetic field strength at the point and the instant of observation, such that if the electric field strength is expressed in volts per meter and the magnetic field strength is expressed in ampere•turns per meter, the wave impedance will have the units of ohms. *Note:* The characteristic impedance of a plane polarized electromagnetic wave in free space is 377 ohms. From Maxwell's equations, the characteristic impedance in a linear, homogeneous, isotropic, dielectric, and electric-charge-free propagation medium is given by the relation $Z = (\mu/\varepsilon)^{1/2}$ where $\mu = 4\pi \times 10^{-7}$ H/m (henries per meter) and is the magnetic permeability, and $\varepsilon = (1/36\pi) \times 10^{-9}$ F/m (farads per meter) and is the electric permittivity, from which relation 120π ohms, i.e., 377 ohms is obtained for free space. For many dielectric media, such as glass, the electric permittivity, i.e., the dielectric constant, ranges from about 2 to 7. However, the refractive index is the square root of the electric permittivity. The characteristic impedance of dielectric materials is $377/n$, where n is the refractive index. *Note 2:* Although the ratio is called the "wave impedance," it is also the impedance of the free space or the material medium. *See also* **field strength, impedance, refractive index.**

wavelength: The distance between points of corresponding phase of two consecutive cycles of a wave, such as between two consecutive crests, which (a) for sinusoidal waves is given by the relation $\lambda = v/f$ where λ is the wavelength, v is the propagation velocity, and f is the frequency, and (b) for pulsed waves is given by the relation $\delta = v/r$ where δ is the wavelength, v is propagation velocity, and r is the pulse repetition rate. *Note 1:* Wavelength is applicable to any form of wave, such as electromagnetic waves, acoustic waves, elastic waves, and water waves. *Note 2:* Wavelength determines the characteristics of the wave, such as (a) the color of visible light, (b) the ability to propagate in a specific waveguide, (c) the ability to penetrate various media, such as wire mesh, layers of the ionosphere, and layers of the atmosphere, and (d) the ability to reflect from specific surfaces. *Note 3:* In an electromagnetic wave in free space, $\lambda f = c$ where λ is the wavelength, f is the frequency, c is the speed of light in vacuum, about 3×10^8 m•s^{-1} (meters per second). In a material medium, $\lambda f = c/n$ where n is the refractive index of the propagation medium. *See* **cable cutoff wavelength, central wavelength, critical wavelength, cutoff wavelength, long wavelength, maximum cable cutoff wavelength, minimum cable cutoff wavelength, nominal central wavelength, optical fiber cutoff wavelength, peak emission wavelength, peak radiance wavelength, peak spectral wavelength, peak wavelength, short wavelength, zero dispersion wavelength, zero material dispersion wavelength.** *See also* **electrical length, frequency, light, phase velocity.** *Refer to* **Figs. H-1 (430), O-3 (673), R-8 (834), R-9 (835), W-3.**

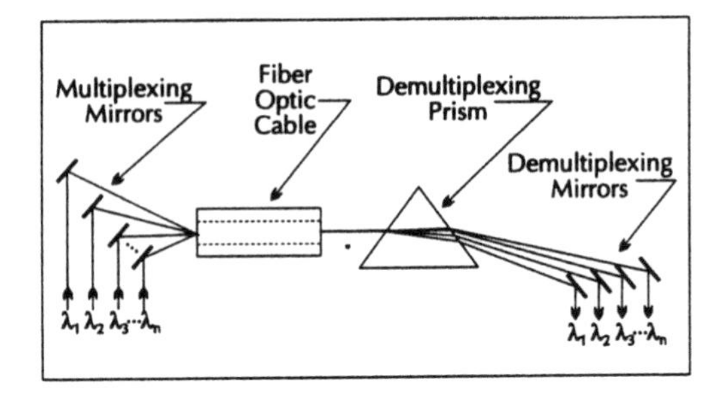

Fig. W-3. Wavelength division multiplexing by transmitting many different wavelengths in the same optical fiber and demultiplexing the resulting signal by means of dispersion.

wavelength-dependent attenuation rate characteristic: In an optical waveguide, such as an optical fiber, a plot that (a) shows the variation of (i) the attenuation rate, usually expressed in dB per kilometer, as a function of (ii) the wavelength of the energizing light source, (b) usually also shows the minimum, nominal, and maximum wavelength in the spectral width of the source, and (c) is used to determine the increase in attenuation rate that has to be allowed for in the design of a fiber optic station/regenerator section. *Note:* The increase in attenuation rate to be used in a design is the maximum value, which may occur at the minimum or the maximum wavelength, or anywhere in between, relative to the nominal central wavelength.

wavelength division multiplexing (WDM): 1. Multiplexing in which two or more signals that have different wavelengths are simultaneously transmitted in the same direction over one waveguide. *Note 1:* The multiplexed signal is demultiplexed at the far end of the waveguide. *Note 2:* Wavelength division multiplexing (WDM), though similar to frequency division multiplexing (FDM), is defined separately (a) because the wave parameters and the multiplexing and demultiplexing equipment, such as microwave transmitting and receiving antennas, light sources, photodetectors, and optical multiplexers and demultiplexers, are described, designed, and measured in terms of their ability to handle wavelengths below and frequencies above the radio frequency (rf) range, (b) because WDM may be discussed in terms of colors as well as wavelengths or frequencies, and most important, (c) because optical frequencies are so high that frequency division multiplexed (FDM) baseband pulse patterns can be time division multiplexed (TDM) and combined with space division multiplexing (FDM). Thus, WDM, FDM, TDM, and SDM can all be used concurrently. *Note 3:* Use of the term "wavelength division multiplexing (WDM)" also avoids confusion with the possible use of the term "frequency division multiplexing (FDM)" in assembling baseband signals that are to be transmitted over a fiber optic link at each of the many possible operating optical wavelengths. **2.** In optical systems, such as fiber optic networks, planar optical waveguides, and integrated optical circuits (OICs), the deriving of two or more simultaneous, continuous channels from the optical medium that connects two points by using separate wavelengths for each of the individual channels. *Note 1:* The different wavelengths might be considered as different colors, each of which can be separately modulated and demodulated. *Note 2:* Wavelength division multiplexing and frequency division multiplexing are similar. In radio systems, frequency is the major concern in the sense that most equations, tests, operations, transmissions, and receptions are described in terms of frequency. In optical systems, most relations and descriptions are described in terms of wavelength. Wavelength and frequency are inversely proportional. *Synonym* **optical frequency division multiplexing.** *See also* **fiber optics, frequency division multiplexing, modulation, multiplexing, time division multiplexing.** *Refer to* **Fig. W-3.**

wavelength measurement: *See* **central wavelength measurement.**

wavelength modulation (WM): The changing of the wavelength of an electromagnetic wave, either by direct input to the source or externally after launching, in accordance with the instantaneous value of an input signal.

wavelength range: *See* **transmitter central wavelength range.**

wavelength stability: In an optical source during a specified period, the maximum deviation of the peak wavelength from its mean value.

wave number: A number that (a) is the reciprocal of the wavelength of a single frequency sinusoidal wave, such as a single frequency uniform plane polarized electromagnetic wave or monochromatic lightwave, and (b) expresses the number of wavelengths per unit distance in the direction of propagation. *Note:* The wave number is used for waves in or near the visible spectrum because wavelength is more readily measured than frequency, but it is frequency, proportional to wave number, that is directly related to energy. For example, photon energy is given by the relations $PE = hf = hkc$ where h is Planck's constant, k is the wave number, and c is the velocity of light. *See also* **wave parameter.**

wave object: *See* **sine wave object.**

wave parameter: In periodic waves, such as electromagnetic waves, the parameter that is given by the relation $p = 2\pi/\lambda$ where π is approximately 3.1416, and λ is the wavelength. *See also* **wave number.**

wave ratio: *See* **standing wave ratio, voltage standing wave ratio.**

waveshape: In an electromagnetic wave, the dependent wave parameter, such as electric voltage, current, field strength, or power, specified as a function of an independent domain parameter, such as time, frequency, or spatial coordinates. *Note 1:* Examples of waveshapes are (a) a voltage square wave plotted as a function of time and (b) the envelope of a modulated carrier. *Note 2:* Waveshapes may be of any arbitrary shape, such as sinusoidal, rectangular, triangular, sawtoothed, spiked, plane polarized, elliptically polarized, or arbitrary transient.

wave transmission: *See* **electromagnetic wave transmission, low detectability electromagnetic wave transmission.**

wave trap: A device that (a) is used to exclude unwanted wave components of a signal, such as unwanted frequency components, noise, or interference, and (c) usually is tunable to permit selection of the unwanted and interfering components of signals.

wave velocity: The speed and the direction of propagation of a wave. *See* **electromagnetic wave velocity.**

way station: In a multipoint circuit of a communications system, a station between two other stations, the other two stations being the only other stations to which it is connected. *Note:* The term "way station" originated from the stations for (a) changing horses and perhaps riders along the route of the Pony Express, (b) the stagecoach lines between major towns and cities, used for changing horses or for a night's lodging, and (c) the early rail, telegraph, and telephone lines, used for lodging and for relaying messages and calls, all based on the expression, "It's a station along the way," rather than a station at a hub or at the end of a line.

WDM: wavelength division multiplexing.

weakly guiding fiber: An optical fiber in which the refractive index contrast is small, usually less than 1%. *Note:* As the refractive index contrast gets smaller, the critical angle at the core-cladding interface gets larger, and there are more leaky rays and hence less guidance. When the contrast is zero, there is no interface, the critical angle becomes 90°, there is no reflection, all the rays eventually leak, and hence there is no guidance. *See also* **refractive index, refractive index contrast.**

wear-out failure period: In a device with a large population of components, the third and final failure period, which (a) is characterized by a rising failure rate above that of the constant failure period, and (b) is caused by the increased rate at which deterioration causes components to reach the end of their designed useful lives. *See also* **bathtub curve, component life, constant failure period, early failure period.** *Refer to* **Fig. B-3 (71).**

weather communications: Communications activities, systems, and equipment devoted to the transmission of messages that (a) contain information about meteorological conditions, and (b) are sent to and from weather forecasting and weather information disseminating services, such as broadcasting, multicasting, and telemetering services, for the purpose of collecting and disseminating weather information.

weather report: A message that (a) contains information on local weather that actually is being experienced at the time and the place from which the message is sent, (b) usually is in a specific format, such as the weather code form developed by the World Meteorological Organization (WMO), and (c) is transmitted in accordance with security measures as required. *See also* **World Meteorological Organization.**

weather watch: *See* **World Weather Watch.**

web: *See* **World Wide Web.**

wedge: *See* **tone wedge.**

wedge set: *See* **tone wedge set.**

weight: In the mathematics of positional numeration systems, the factor that a value, represented by a character in a digit place, is multiplied by in order to obtain its additive contribution to the total value represented by a real number. *Note:* If the weights in a pure binary number are 8, 4, 2, and 1, the additive contributions for the digits of the binary number 1101 are 8+4+0+1, and the sum is equal to 13, the decimal equivalent of the binary number. In other numeration systems, the positions of numbers have other weights. *See also* **basis weight, Hamming weight.**

weighted standard work second (WSWS): In the measurement of communications systems operational performance, a unit that (a) is used to express the relative time required to perform various kinds of functions, such as call establishment, information transfer, and call disestablishment, and (b) is weighted to reflect applicable degrees of waiting-to-serve time, such as the time between the instant that the first digit of a called number is received by the system and the instant that the last digit is received.

weighting: **1.** The application of an adjustment to a measurement to compensate for specified conditions, such that if the adjustment is not made, the measurement will be biased, distorted, misleading, and perhaps useless. **2.** The adjusting of the charactistics of a circuit to compensate for specified effects. *See* **C message weighting, flat weighting, F1A line weighting, frequency weighting, HA1 receiver weighting, noise weighting, one-forty-four line weighting, one-forty-four receiver weighting, psophometric weighting.**

weighting network: A network that (a) has a loss versus frequency of the applied voltage characteristic with a predetermined shape, and (b) is used for improving or correcting transmission characteristics and making noise measurements. *See also* **noise weighting.**

welfare traffic: Communications traffic that consists of messages of a humanitarian nature, whether personal or official.

wet-mateable connector: A connector that can be mated and demated underwater, usually many times and under high pressure, without deleterious effects on the quality of the connection. *Note:* An example of a wet-mateable connector is a fiber optic connector that can be connected and disconnected many times on the ocean bottom.

Wheatstone automatic system: In communications, a Morse system in which (a) signals are automatically transmitted, (b) a previously prepared perforated tape is used for transmission, and (c) a perforated tape for automatic printing, or an inked tape for subsequent interpretation by an operator, is used for reception.

Wheatstone tape: A tape that (a) is used for automatic transmission and reception of the international Morse code, (b) for transmission, provides for two-unit perforation, i.e., two holes perforated transversely for a dot and two holes perforated diagonally for a dash, and (c) for reception, is drawn through an ink recorder, in which a pen draws a continuous inked line such that dots and dashes are indicated by transverse fluctuations of the inked line, and is read by a trained operator. *Synonym* **Boehme tape.**

wheel: *See* **print wheel, thumb wheel.**

wheel printer: A printer that (a) has the type slugs mounted or embossed on the outside of the rim of a disk or a wheel, (b) has a wheel for each print position in a line of print, (c) has the wheels keyed to a shaft, thus forming a cylinder that is rotated at high speed, and (d) sends a signal to a magnetically driven hammer to press paper against a ribbon to print the character, which is printed when it comes opposite the print position on the line to be printed. *See* **daisy wheel printer.**

whisper facility: A telephone system facility that enables whispered speech to be received at a power level nearly equal to that of normal speech.

white: 1. Pertaining to electromagnetic radiation in which there is a uniform power density, i.e., distribution of energy per hertz, throughout a broad spectrum of frequencies. **2.** The sensation obtained when the retina is exposed to a uniform distribution of electromagnetic energy in the visible region of the spectrum, i.e., when there is a balanced mixture of all colors incident on the retina, similar to the sensation obtained from average noon sunlight on a clear day. *See* **noisy white, nominal white, picture white.** *See also* **white light.**

white area: In radio transmission, the area or the population that does not receive (a) interference-free primary service from an authorized amplitude modulation (AM) station, or (b) a signal electric field strength of at least 1 $mV \cdot m^{-1}$ (millivolt per meter) from an authorized frequency modulation (FM) station. *See also* **blanketing.**

white facsimile transmission: 1. In an amplitude-modulated facsimile system, transmission in which the maximum transmitted power corresponds to the minimum density, i.e., the white area, of the object. **2.** In a frequency-modulated facsimile system, transmission in which the lowest transmitted frequency corresponds to the minimum density, i.e., the white area, of the object. *See also* **black facsimile transmission, facsimile, object.**

white level: *See* **reference white level.**

white light: Electromagnetic waves having a spectral distribution, i.e., energy distribution, among the wavelengths such that the color sensation to the average human eye is identical to that of average noon sunlight on a clear day. *See also* **white.**

white noise: Noise that has a frequency spectrum that is continuous and fairly uniform over a specified frequency range, i.e., that has a power per hertz that varies little over a specified frequency range. *Synonyms* **additive white Gaussian noise, uniformly distributed constant amplitude frequency spectrum, uniform spectrum random noise.** *See also* **black noise, blue noise, continuous spectrum, frequency, in-band noise power ratio, pink noise, pseudorandom noise, white light.**

white pages: 1. A hard copy telephone directory listing subscriber names, addresses, and telephone numbers. *Note:* "White pages" are in contrast to the "yellow pages" that list businesses and organizations according to type of business or service. **2.** An electronic information database that contains user names and their associated network addresses, in the manner of a telephone directory. *Note:* Electronic "white pages" usually contain additional information, such as office location, phone number, and mailstop.

white recording: 1. In an amplitude-modulated facsimile system, recording in which the maximum received power corresponds to the minimum density, i.e., the white area, of the object. **2.** In a frequency-modulated facsimile system, recording in which the lowest received frequency corresponds to the minimum density, i.e., the white area, of the record medium. *See also* **black facsimile transmission, black recording, facsimile, object, record medium.**

white signal: In facsimile systems, the signal that (a) results from scanning a minimum density area, i.e., the white area, of the object, (b) has a magnitude that corresponds to a signal parameter value, such as amplitude, frequency, or phase shift, and (c) corresponds to the lowest signal parameter value. *See also* **facsimile, object, white facsimile transmission.**

who are you (WRU) character: A transmission control character used for (a) switching-on an answer back unit in the station with which a connection has been established, (b) triggering the receiving unit to transmit an answer back code to the terminal that transmitted the original signal, and (c) initiating a response that might include station identification, an indication of the type of equipment that is in service, and the status of the station. *Note 1:* The who are you (WRU) character corresponds to the seven-bit code assigned to the WRU transmission control character. *Note 2:* The receiving unit may be any data terminal equipment (DTE), such as a telegraph unit, telephone set, or facsimile device. *Synonym* **WRU signal.**

wicking: In a cable, such as a coaxial, multipaired, or fiber optic cable, an indication of the amount of liquid, usually water, absorbed by one portion of the cable when an adjacent portion is immersed in water. *Note:* Wicking is measured in various ways, such as (a) by the weight of water absorbed in a specified time when a specified length is immersed vertically to a specified depth, and (b) by the height to which a dye solution is raised vertically by capillary or soaking action when a specified length is immersed vertically to a specified depth in the dye solution.

Wide Area Information Server: A distributed text searching system that (a) uses the protocol standard ANS Z39.50 to search databases on remote computers, (b) finds libraries that are most often found on the Internet, (c) allows users to discover and access information resources in the network without regard to their physical location, and (d) has software that uses the client-server model.

wide area network (WAN): A physical or logical network that (a) provides communications services to a larger number of independent users than are usually served by a local area network (LAN), and (b) is usually spread over a larger geographic area than that of a LAN. *Note 1:* Users may include physical networks, such as Integrated Services Digital Networks (ISDNs), X.25 networks, and T1 networks, all of which may serve as the physical or logical wide area network (WAN) environment. *Note 2:* A metropolitan area network (MAN) is a wide area network (WAN) that serves all the users in a metropolitan area. In telephone systems, a WAN may include all or parts of several area codes. WANs may be countrywide or worldwide. *See also* **local access and transport area, local area network, metropolitan area network, network.**

wide area telecommunications service (WATS): 1. In a local access and transport area (LATA), a telecommunications service for a fixed rate that is measured by distance and time, such as by zones and minutes. **2.** A telecommunications service, such as telephone, digital data, and analog data service, over a geographical area larger than that of a local area, such as an area served by a single central office (C.O.). *See also* **local access and transport area.**

wide area telephone service (WATS): 1. A network-provided long-distance service, i.e., toll service, in which (a) dial telephone telecommunications are provided between a given user station and stations within specified geographical rate areas, (b) a signal access line is provided between the user station, i.e., customer station, and the serving central office (C.O.), (c) each access line may be arranged for outward service (OUT-WATS), inward service (IN-WATS), or both, and (d) the service is provided at interLATA, i.e., between local access and transport area, and intraLATA fixed rates determined by geographical zones, duration of call, and hours of the day, day of the week, or holidays. **2.** A telephone service that (a) does not have individual call toll charges and (b) extends over a geographical area that includes a community of interest, such as the Washington, DC metropolitan area that covers all of the District of Columbia and parts of northern Virginia and southern Maryland, embracing all of the 202 area code and parts of the 703 and 301 area codes. *Synonym* **extended area telephone service.** *See also* **local access and transport area, local area network, metropolitan area network, wide area network.**

wideband (WB): 1. Pertaining to the property of a communications facility, equipment, circuit, channel, or system in which the transmitted bandwidth is greater than 0.1% of the midband frequency. *Note:* "Wideband" has many meanings depending upon application. At audio/telephone frequencies, a bandwidth exceeding 4 kHz is considered wideband. At high frequency (HF) radio frequencies (3–30 MHz), a bandwidth must be much larger than 3 kHz to be considered wideband. **2.** In communications security systems, pertaining to a bandwidth exceeding that of a nominal 4-kHz telephone channel. **3.** Pertaining to the property of a circuit that has a bandwidth wider than normal for the type of circuit, frequency of operation, and type of modulation. **4.** In commercial telephone systems, pertaining to the property of a circuit that has a bandwidth greater than 4 kHz. **5.** Pertaining to a signal that occupies a broad frequency spectrum. *Note:* "Wideband" is often used only to distinguish between "wideband" and "narrowband," where both terms are subjectively defined relative to the context. *Synonym* **broadband.** *See also* **bandwidth, communications system, frequency, frequency spectrum designation, narrowband, narrowband signal.**

wideband channel: In voice communications systems, a channel that has a bandwidth equivalent to 12 or more voice-grade channels.

wideband communications system (WBCS): A communications system that (a) handles, uses, or requires a wide range of frequencies, (b) usually provides numerous communications channels in which frequency, time, and space division multiplexing are used, (c) usually includes multichannel systems, such as multichannel telephone cable, tropospheric scatter, and line of sight (LOS) microwave systems, (d) has increased traffic capacity, and (e) occupies a larger part of the frequency spectrum than narrowband systems. *Refer to* **Fig. W-4.**

wideband modem: 1. A modem in which the modulated output signal frequency spectrum can be broader than that which can be wholly contained within, and

Fig. W-4. Telstar I, the first active telecommunications satellite. Telstar supported a **wideband communications system.** (Courtesy AT&T Archives.)

nel telephone cable, tropospheric scatter, and line of sight (LOS) microwave systems, (d) has increased traffic capacity, and (e) occupies a larger part of the frequency spectrum than narrowband systems. *Refer to* **Fig. W-4.**

wideband modem: 1. A modem in which the modulated output signal frequency spectrum can be broader than that which can be wholly contained within, and faithfully transmitted through, a voice channel with a nominal 4-kHz (kilohertz) bandwidth. *Note:* The 4 kHz (kilohertz) is an arbitrary value. Other bandwidths, such as 20 kHz, have been proposed as the limiting value between wideband and narrowband. **2.** A modem that has a bandwidth capability that is greater than that of a narrowband modem. *See also* **bandwidth, modem, narrowband modem, narrowband signal.**

wideband signal: 1. A voice frequency signal that has a spectral content greater that 4 kHz (kilohertz) wide. **2.** A radio frequency (rf) signal that occupies a frequency band greater that 0.1% of the midband frequency. *Note:* At a nominal midband frequency of 10 MHz (megahertz), a signal with a bandwidth greater than 10 kHz (kilohertz) would be a wideband signal. However, at a nominal midband frequency of 1 GHz (gigahertz), a signal with a bandwidth of 10 kHz, or even 100 kHz, would be a narrowband signal. **3.** An analog signal, or an analog representation of a digital signal, that has an essential spectral content that is greater than that which can be contained within a voice channel with a nominal 4 kHz (kilohertz) bandwidth. *See also* **bandwidth, narrowband, wideband.**

wideband system: 1. A system in which a bandwidth greater than a specified bandwidth is used. *Note:* An example of a wideband system is a system that consists of a group of voice frequency channels that use a bandwidth greater than 4 kHz (kilohertz), but whose bandwidth is often considered to be greater than 20 kHz. **2.** A system with a multichannel bandwidth greater that 0.1% of the midband frequency of opera-

because the radar display will be more nearly saturated, though the radar will not be as accurate in tracking as the narrow beam antenna, and (c) wastes a large amount of energy when a specific object, i.e., a specific target, is being tracked.

wide beam radar: A radar that emits a signal that (a) has its energy spread over a large solid angle, and (b) usually has a radiation pattern with relatively large side lobes. *Note:* Wide beam radar may be likened to a floodlight rather than a spotlight. Each has its advantage for a particular application.

wide shift frequency shift keying: In radiotelegraph communications, frequency shift keying (FSK) in which a frequency shift of a large number of hertz is used to (a) distinguish marks from spaces, such as dots and dashes distinguished from spaces, or (b) to distinguish tone from silence. *Note:* If a frequency shift of +425 Hz is used for the mark, a shift of −425 Hz is used for the space, and the assigned nominal frequency is 3 kHz (kilohertz), the frequency shift will be a wide shift because the shift is greater that 10% of the assigned nominal frequency of 3 kHz. *See also* **narrow shift frequency shift keying.**

wide web: *See* **World Wide Web.**

width: *See* **bandwidth, column width, nominal line width, relative spectral width, spectral line width, spectral width, stroke width.**

Wiener filtering: *Synonym* **minimum mean square error restoration.**

wigwag signaling: Visual signaling in which a flag is waved to the right of the sender in a 90° arc to represent a dot and in a similar motion to the left to represent a dash, in order to transmit international Morse code. *See also* **Morse flag signaling, semaphore signaling.**

wild point detection: The detection of invalid data that fail to meet specific criteria by testing data samples. *Note:* When invalid data are detected at a point in a system, the data passing through that point can be eliminated from further processing. *See also* **reasonableness check.**

wildcard character: A character that may be substituted for any of a defined subset of all possible characters, such as a string of American Standard Code for Information Interchange (ASCII) characters. *Note 1:* Examples of wildcard characters are (a) the "?" used in high frequency (HF) radio automatic link establishment (ALE) in which the "?" may be substituted for any one of the 36 characters "A" through "Z" and "0" through "9," and (b) the "*" and the "?" used in computers to substitute for any one or more of the American Standard Code for Information Interchange (ASCII) characters. *Note 2:* The string of one or more characters that the wildcard character represents must be specified.

willful intercept: In a communications system, to deliberately intercept messages that (a) were intended for other stations, and (b) could not be received by these stations for various reasons, such as equipment malfunction, operator unavailability, power failure, or weather conditions.

WIN: Worldwide Military Command and Control System Intercomputer Network.

Winchester drive: A magnetic disk drive that (a) is often used in personal computer (PC) systems and in microcomputers, (b) usually includes a controller, motor drive mechanism, disk holding mechanism, and the recording head assembly, and (c) may be built in or contained in removable cartridges, each with a storage capacity usually greater than 40 Mb (megabytes).

wind: *See* **solar wind.**

window: 1. A spatial or temporal break in an ongoing phenomenon. *Note:* Examples of windows are (a) a temporary improvement in a bad weather pattern, (b) a temporary cessation of barrage jamming, and (c) a low attenuation region, i.e., a trough, in an attenuation versus wavelength characteristic of an optical fiber. **2.** In fiber optics, a band of wavelengths at which an optical fiber is sufficiently transparent for practical use in communications applications. *See* **spectral window. 3.** In imagery, a portion of a display surface in which display images pertaining to a particular application can be presented. *Note:* Different applications may be simultaneously displayed in different windows on the same screen. *Synonym* **viewport.** *See* **glass window, minimum dispersion window, minimum loss window, noise window, sliding window, spectral window.** *Refer to* **Fig. H-1 (430).**

windowing: In video display systems, the sectioning of the display surface, such as the screen of a television (TV) set or computer monitor, into two or more separate areas for the purpose of displaying images from different sources. *Note:* In windowing, one window could display fixed or changing data, another motion video images from a remote site, and another graphic information.

window key: 1. In personal computer (PC) systems, a key that enables an operator to view a previously determined window so that the system can quickly display data stored in another storage area, such as another portion of a spreadsheet, database, or graph. **2.** A key that enables an operator to switch a pointer from one window to another on a split screen. *See also* **split screen, window.**

wink: 1. In telephone switching systems, a single supervisory pulse. **2.** A short off period for a light source that is on for a much longer period. **3.** One cycle of a wink pulsing signal. *See also* **wink pulsing.**

wink pulsing: In telephone switching systems, recurring pulsing in which the off-condition is relatively very short compared to the on-condition. *Note:* On key-operated telephone instruments, the hold position, i.e., the hold condition, of a line is often indicated by wink pulsing the associated lamp at 120 pulses per minute. The lamp is off during 6% of the pulse period and is on 94% of the period, i.e., the lamp is off for 30 ms (milliseconds) and on for 470 ms. *See also* **blinking, flicker, pulse, pulsing.**

wire: A metallic conductor of electric current, usually long compared to its width or diameter. *Note:* In addition to twisted pairs and open wirelines, coaxial cables and metallic hollow or dielectric-filled waveguides are considered as wirelines because of the presence of metals and their dependence on high conductivity. However, fiber optic cables, the atmosphere, vacuum, semiconductors, dielectric materials, and similar transmission media are not considered as wirelines. *See* **chicken wire, field wire, jumper wire, open wire.** *See also* **message, wireline.**

wire cable: 1. A wrapped and sheathed bundle of separately insulated wires, usually copper wires, used to form many separate channels for analog and digital signals. **2.** A cable in which wires are used to achieve space division multiplexing. *See also* **fiber optic cable, space division multiplexing.**

wire circuit: *See* **four-wire circuit, two-wire circuit, unbalanced wire circuit.**

wired radio frequency (rf) system: A system in which (a) restricted radiation devices are used, and (b) radio frequency (rf) energy is conducted or guided along wires or in cables, such as electric power lines, telephone twisted-pair wirelines, and coaxial cables.

wire frame image: In the display space on the display surface of a display device, an image that (a) shows all major lines of an object, including hidden lines, as though the object itself were made only of wires that represent edges, and (b) if hidden lines are not indicated as such, shows the shape of the object, though the view or the orientation of the object may be ambiguous.

wire integration: *See* **radio and wire integration.**

wire interface: *See* **radio-wire interface.**

wireless local area network (WLAN): A local area network (LAN) in which radio or lightwave transmission is used to interconnect stations, rather than wire or coaxial cable. *See also* **local area network.**

wireline: Wire paths in which metallic conductors are used to provide electrical connections between the components of a system, such as a communications or control system. *Note:* In addition to twisted pairs and open wirelines, coaxial cables and metallic hollow or dielectric-filled waveguides are considered as wirelines because of the presence of metals and their dependence on high conductivity. However, fiber optic cables, the atmosphere, vacuum, semiconductors, dielectric materials, and similar transmission media are not considered as wirelines. *See also* **wireline communications.**

wireline common carrier (WCC): A common carrier that provides telephone services via landlines and central offices (COs).

wireline communications: The use of a physical path, i.e., a propagation medium, made of metallic conductors that are connected between terminals or other components of a communications system. *See also* **fiber optic communications.**

wireline distribution system: *See* **protected wireline distribution system.**

wireline transmission: Transmission in which (a) signals, such as telephone, telegraph, teletypewriter, facsimile, data, intercom, closed circuit television, and control signals, are conducted in metallic conductors, and (b) transmission is by means of electric currents that require the movement of electric charges, i.e., of electrons, to build up voltages. *Note:* In addition to twisted pairs and open wirelines, coaxial cables and metallic hollow or dielectric-filled waveguides are considered as wirelines because of the presence of metals and their dependence on high conductivity. However, fiber optic cables, the atmosphere, vacuum, semiconductors, dielectric materials, and similar transmission media are not considered as wirelines. *See also* **wireline.**

wire-ored: 1. The connection of several conductors to a single point so that a signal on any conductor is transmitted to all other conductors, and signals on all conductors are transmitted to any one conductor. **2.** The connection of several conductors with incoming signals, i.e., input conductors, to a single point in such a manner that all signals are transmitted away from the point by a single conductor, i.e., an output conductor, whether or not arrangements are made to prevent signals on any incoming conductors from traveling on other incoming conductors. *Note:* Signals may be prevented from being conducted away from a point by means of (a) a buffer, such as a diode for electrical systems, or (b) a dichroic filter, for fiber optic systems.

wirephoto (WP): A facsimile transmission system in which wirelines with electrical currents are used for the transmission of signals from a scanning transmitter to a recording receiver. *See also* **wireline, wireline transmission.**

wire-radio integration: *See* **radio and wire integration.**

wire storage: *See* **magnetic wire storage.**

wiretapping: *See* **active wiretapping, passive wiretapping.**

wireway: *Synonym* **raceway.**

wiring: 1. The installing of wirelines in a structure, such as a building, ship, vehicle, or aircraft. **2.** The wirelines in a structure, such as a building, ship, vehicle, or aircraft. *See* **on-premises wiring.** *See also* **wirelines.**

with priority: A precedence designator that (a) is honored by (i) commercial and common carrier communications agencies, (ii) national communications systems, and (iii) recognized private operating agencies, and (b) indicates a degree of precedence to be provided to messages. *See also* **precedence.**

WLAN: wireless local area network.

WM: wavelength modulation.

WMO: World Meteorological Organization.

wooding: Pertaining to the undesirable obstruction of an antenna beam by other equipment, such as a ship superstructure, a radio tower, or an aircraft frame.

W optical fiber: A doubly cladded optical fiber that (a) has two layers of concentric cladding, (b) has a core that has the highest refractive index, (c) has an inner cladding that has the lowest refractive index, and (d) has several advantages, such as reduced bending losses, over conventionally cladded fibers that have (a) a single-stepped refractive index, (b) a graded refractive index core and a single uniform index cladding, and (c) a graded refractive index all the way from the fiber axis to the outside surface.

word: A string of data units, such as characters, bytes, and bits, that is considered, processed, or handled as a unit, i.e., as a single entity, for some purpose. *Note 1:* Meaning usually may be assigned to a word to represent information. *Note 2:* In telegraph communications, six character intervals are considered to be a word in computing traffic capacity in words per minute, which is computed by (a) multiplying the data signaling rate (DSR) in baud by 10 and (b) dividing the resulting product by the number of unit intervals per character. Thus, if the modulation rate is 50 baud, i.e., 50 unit intervals per second, and there are 5 unit intervals per character, the traffic capacity or traffic rate is (50)(10)/5 or 100 wpm (words per minute) at an implied 6 c/w (characters per word). The combination of 50 unit intervals per second and 5 unit intervals per character is equivalent to 10 characters per second, or 600 cpm (characters per minute), which is equivalent to 100 wpm at 6 c/w, as obtained previously. *See* **alphabetic word, code word, computer word, inactive code word, instruction word, keyword, numeric word, procedure word, reserved word, telegraph word.** *See also* **baud, binary digit, bit string, block, byte, character.**

word key: In word processing, a control key used to process text only one word at a time.

word length: The number of data units, such as binary digits, bytes, or characters, in a word. *See also* **binary digit, character.**

word-organized storage: Storage in which data can be entered, stored, and retrieved only in words, such as computer words.

word processing: 1. Using a system, such as a personal computer (PC) system, to manipulate text. *Note:* Examples of word processing functions include entering, editing, rearranging, sorting, storing, retrieving, displaying, and printing text. **2.** The automated processing of text by machines or software specifically designed for that purpose, primarily as a means of improving the efficiency and the effectiveness of business communications, especially in document preparation and handling. *Synonym* **text processing.**

word processor: 1. A computer program that enables a computer to execute word processing operations. **2.** The hardware and applicable software used to process data. *See also* **word processing.**

word wrap: In word processing, the automatic movement of the entire last word that does not fit on a line or row to the next line or row so that the operator can key continuously without concern for end-of-line operations, such as margin control, hyphenation, carriage return, or indexing.

work factor: In communications, computer, data processing, and control systems security, the effort or the time that can be expected to be expended to overcome a protective measure and effect a penetration with specified expertise and resources.

work file: A file that is used to hold data needed only for the duration of a job or between phases of a job. *See also* **job.**

work function: The amount of energy required to free an electron from an atom in a given material, usually measured in electron•volts. *Note:* If a photon has more energy than the work function of the material it strikes, it will be capable of freeing an electron from that material. The freed electron will possess kinetic energy equal to the difference between the energy of the photon and the material work function. The work function of a material is given by the relation $W = hf_0$ where W is the work function, h is Planck's constant, and f_0 is the threshold frequency required to release an electron from the material. If the frequency of the incident photon is above the threshold frequency required to overcome the work function, the freed electron will emerge from, or move within, the material, possessing kinetic energy. *See also* **photoelectric equation.**

working: *See* **delay working, direct working, double supervisory working, relay working, trunk delay working.**

working diameter: *See* **borescope working diameter.**

working frequency: 1. A specific frequency used during a given situation or operation. *Note:* An example of a working frequency is the frequency used during all or part of an emergency period, such as during a search-and-rescue operation, after contact has been established on the emergency calling frequency. **2.** The frequency that a station uses to transmit its messages to another station, particularly in crossband radiotelegraph procedures. *See also* **answering frequency, calling frequency, crossband radiotelegraph procedure.**

working length: *See* **borescope working length.**

working net: In ship-shore communications, the communications net that usually is used by ship stations for the transmission of messages to shore stations.

working storage: *Synonym* **temporary storage.**

work second: *See* **weighted standard work second.**

work session: A session that (a) begins with a successful log-on, (b) is devoted to accomplishing useful operations by an operator of an interactive system, and (c) ends with a log-off. *See also* **interactive system.**

worksheet: *Synonym* **spreadsheet.**

work space: In computer and data processing systems, the portion of main storage that is used by a computer program for the temporary storage of data. *See also* **main storage.**

workstation: 1. In automated systems, such as computer, data processing, communications, and control systems, the input, output, display, and processing equipment that provides the operator-system interface. **2.** Input, output, display, and processing equipment that constitutes a stand-alone system, i.e., a system that does not require external access. *Note:* Examples of workstations are (a) a personal computer (PC) with associated software, monitor, keyboard, printer, and perhaps a mouse, (b) an inquiry station, and (c) a communications system operator position.

world coordinates: In display systems, device-independent coordinates that may be used to map user coordinates.

World Meteorological Organization (WMO): A specialized agency of the United Nations that seeks to promote, support, and encourage stations for meteorological and geophysical observations, including meteorological satellite space stations. *Note:* The World Meteorological Organization (WMO) was founded in 1950. In 1951, it took over the International Meteorological Organization that was established in 1878. WMO headquarters is in Geneva, Switzerland. *See also* **World Weather Watch.**

world region: *See* **International Telecommunication Union world region.**

World Time: 1. *Synonym* **Coordinated Universal Time. 2.** *Synonym* **Greenwich Mean Time.**

World Weather Watch (WWW): An international worldwide system that (a) issues weather reports on a worldwide basis, (b) makes use of meteorological satellite stations, and (c) is operated by the World Meteorological Organization (WMO) of the United Nations.

worldwide communications system: *See* **air-ground worldwide communications system.**

worldwide geographic reference system: A geographic location reference system in which the Earth is divided into (a) twenty-four 15° longitudinal zones, lettered A through Z, omitting I and O, eastward from

Greenwich, England, (b) 12 latitude zones of 15° each, lettered A through M, omitting I, lettered northward from the South Pole, and (c) and 1° units, lettered eastward and northward from each southwest corner of each resulting 15° zone.

World Wide Web (WWW): An international virtual-network-based information service that (a) is composed of Internet host computers that provide online information in a specific hypertext format, (b) has servers that provide hypertext metalanguage (HTML) formatted documents using the hypertext transfer protocol (HTTP), (c) enables access to information using a hypertext browser, such as Mosaic, Viola, or Lynx, (d) has no existing hierarchy, and (e) enables the same information to be found by many different approaches.

worm: A computer virus capable of disrupting a computer program. *See also* **computer program.**

worst case chromatic dispersion over wavelength range: *Synonym* **worst case dispersion coefficient.**

worst case dispersion coefficient: In a single mode fiber optic station and fiber optic cable facility, the worst case end-of-life dispersion coefficient that corresponds to the greater absolute value obtained by comparing (a) the difference between the lower limit of the transmitter light source central wavelength range and the lower limit of the fiber zero dispersion wavelength range and (b) the difference between the upper limit of the transmitter light source central wavelength range and the upper limit of the fiber zero dispersion wavelength range. *Note 1:* The ranges are caused by manufacturing tolerances, temperature and humidity variations, and aging. *Note 2:* The worst case dispersion coefficient usually is expressed in picoseconds per nanometer•kilometer ($ps \cdot nm^{-1} \cdot km^{-1}$), where the picoseconds are a measure of the increase in pulse duration, the nanometers are a measure of the pulse spectral width, and the kilometers are a measure of the length of the optical fiber. *Synonym* **worst case chromatic dispersion over wavelength range.**

worst hour: *See* **year worst hour.**

worst hour of the year: *Synonym* **year worst hour.**

wound sensor: *See* **bobbin-wound sensor.**

woven fiber optics: The application of a combination of textile weaving and fiber optic techniques for various purposes, such as producing special effect devices, e.g., reproducible image guides, image dissectors, color image panels, high speed alphanumeric display surfaces, and distributed fiber sensors.

W-profile optical fiber: *Synonym* **doubly clad optical fiber.**

wrap: An indication of the satisfactory or acceptable completion of a task or an operation. *See* **core wrap, word wrap.**

wraparound: In the use of display systems, the extension or the continuation of an operation from a conceived last position in an image to a conceived first position in the image. *Note:* Examples of wraparound are (a) the continuation of a read operation or a cursor movement from the last character position in a display buffer storage or an image storage space to the first position, (b) the automatic moving of a cursor from the lower right corner of a display space to the upper left corner, and (c) the movement of data in a register such that as the data move out one end, they move in at the opposite end. *See* **horizontal wraparound, vertical wraparound.**

wrapping: In open systems architecture, the use of a network to connect two other networks, thus providing an increased interaction capability between the two connected networks. *Note:* Recurring application of wrapping usually results in a hierarchical communications network structure. *See also* **open systems architecture.**

write: 1. To make a permanent or temporary record of data in a storage device or on a data medium. *Note:* The transfer of a block of data from internal storage to external storage implies writing when viewed from the external storage, and implies reading when viewed from the internal storage. Writing always implies recording, i.e., creating, a record or a recorded copy. **2.** In communications, computer, data processing, and control systems, to permanently or temporarily record data in or on a storage medium, in a storage unit, or on a display surface. *See* **overwrite.**

write cycle time: The minimum time between the starts of successive write cycles of a storage device that has separate reading and writing cycles.

write head: A magnetic head capable of writing only.

write protection: The protection that (a) is afforded to stored data by preventing the writing of data, such as data items, files, records, or additions thereto, on a data medium, such as a magnetic disk, tape, or drum, and (b) is invoked to prevent unauthorized or inadvertent changing of the data already stored. *See also* **access, read.**

write protection label: On a diskette, a removable label, tab, or sticker used to prevent the writing of data on the diskette. *Synonym* **write protect tab.**

write protect tab: *Synonym* **write protection label.**

writer: *See* **display writer.**

write through: In display systems, the capability of a display device, such as a cathode ray tube (CRT), storage tube, or light-emitting diode panel, to superimpose another image on an image that has not yet disappeared after its regeneration has ceased.

writing bar: In facsimile transmission systems, a part that (a) is used in some continuous receivers in conjunction with a helix, and (b) consists of a rectilinear rib that determines the position of the scanning line on the record medium. *Synonyms* **chopper bar, writing edge.**

writing edge: *Synonym* **writing bar.**

WRU character: who are you character.

WRU signal: *Synonym* **who are you character.**

WWMCCS: Worldwide Military Command and Control System.

WWW™: World Weather Watch, World Wide Web.

wye coupler: In fiber optics, a coupler with one input port and two output ports. *Synonym* **Y coupler.** *See also* **fiber optic coupler.**

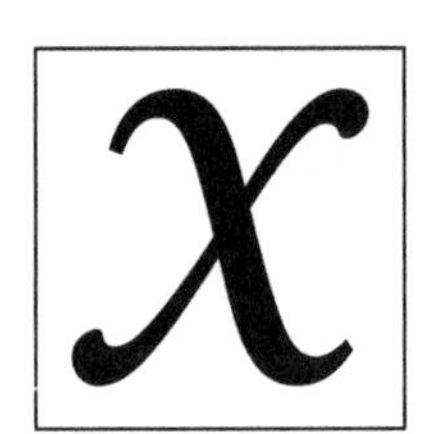

X band: **1.** The band of frequencies that (a) consisted of the range from 5.2 to 10.9 GHz (gigahertz), between the old S band and K band and overlapping with parts of the old C, G, J, H, and M bands in the United States, (b) was divided into 12 subbands, designated XA through XK, and (c) was also composed of frequencies in the range from 8.2 to 12.4 GHz, between the old J and P bands, also in the United States. *Note 1:* In the United Kingdom, the X band consisted of frequencies in the range from 7.0 GHz (gigahertz) to 12.0 GHz, overlapping part of the old C band. *Note 2:* Frequency band designation by letter is obsolete. **2.** In surface warning radar, the band of frequencies that (a) was on the order of 10 GHz (gigahertz) and (b) ranged from 5.2 GHz to 10.9 GHz.

***x*-dimension of recorded spot:** In facsimile systems, the effective recorded spot dimension measured in the direction of the recorded line; i.e., the effective recorded spot dimension is the largest center-to-center spacing between recorded spots, and gives minimum peak-to-peak variation of density of the recorded line. *Note 1:* The *x*-dimension of the recorded spot implies that the response to a constant density of the object is a succession of discrete recorded spots. *Note 2:* The numerical value of the *x*-dimension of the recorded spot depends upon the type of system. *Synonym* **recorded spot *x*-dimension.** *See also* **facsimile, maximum keying frequency, recording, *y*-dimension of recorded spot.**

***x*-dimension of scanning spot:** In facsimile systems, the distance between the centers of adjacent scanning spots measured in the direction of the scanning line on the object. *Note:* The numerical value of the *x*-dimension of the scanning spot depends upon the type of system. *Synonym* **scanning spot *y*-dimension.** *See also* **facsimile, scanning, *y*-dimension of the scanning spot**.

xerographic recording: Recording by action of a light spot on an electrically charged photoconductive insulating surface where a latent image is developed with a resinous powder. *See also* **recording.**

X-modem protocol: A data transfer protocol that (a) is in the public domain, (b) permits protocol matching between systems, (c) detects errors introduced during transmission, and (d) causes retransmission until (i) the transmitted data are correctly received, (ii) error-free transmission is assured, and (iii) much of the conventional file transfer protocol and handshaking is eliminated.

XO: crystal oscillator.

XOFF: An abbreviation for the American Standard Code for Information Interchange (ASCII) transmission control character meaning "transmitter off."

XON: An abbreviation for the American Standard Code for Information Interchange (ASCII) transmission control character meaning "transmitter on."

X ray: A ray of electromagnetic waves with wavelengths between 0.01×10^{-10} m (meter) and 100 ×

10^{-10} m, and therefore a frequency of 0.03×10^{18} Hz (hertz) to 300×10^{18} Hz. *Note:* The energy of a photon is given as Planck's constant times the frequency. Thus, the energy of an X-ray photon is sufficient to deeply penetrate solids, ionize gases, and destroy or change molecular structures, such as those of living tissue. X rays can penetrate directly through a substance with minimal or negligible attenuation, except for highly dense substances, such as lead and gold. Thus, X rays must be used with great care, such as when producing X-ray photos. They are used to produce images of materials they pass through in transit to sensitive film. The X-ray photos thus produced indicate variations in transmittance of the material through which they pass. *See also* **optical spectrum, visible spectrum.**

X series Recommendation: A member of a set of telecommunications protocols and interfaces defined in International Telegraph and Telephone Consultative Committee (CCITT) Recommendations and listed in X.24. *Note:* Some of the common X series Recommendations are:

X.21	Physical Layer Interface
X.21*bis*	Physical Layer Interface
X.25	DTE to DCE Interface
X.28	PAD Access to X.25
X.29	PAD Access to DTE
X.75	Internetwork Access for X.25 Networks (Gateway)
X.400	Addressing for E-mail

See also **V series Recommendation.**

XT: crosstalk.

XU: X unit.

X unit: A unit of wavelength for X rays or gamma rays, given by the relation 1 XU = 1.00202 Å (angstroms) or about 10^{-10} m (meters).

XX/YY restricted launch: In the launching of a light beam from a light source into an optical fiber, a launch with (a) a launch spot size equal to XX percent of the fiber numerical aperture (NA) and (b) a source aperture equal to YY percent of the fiber (NA). *See also* **launch spot, numerical aperture.**

x-y mount: A variation of the basic altitude over azimuth antenna mount, with the primary axis parallel to the Earth's surface for improved zenith tracking, such as might be used in telescope mounts and satellite tracking mounts. *See also* **altitude over azimuth.**

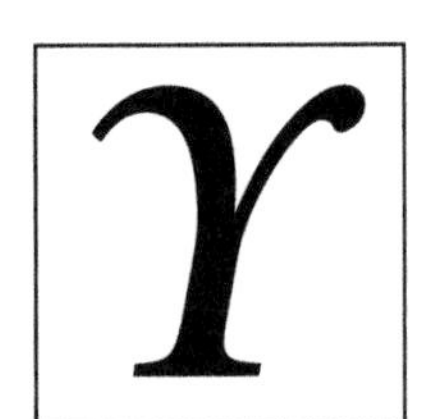

Yagi antenna: An antenna that (a) consists of at least two elements, and often three elements, such as (i) a radiator that usually is a dipole, (ii) a reflector that is mounted behind the dipole, and (iii) a director mounted in front of the dipole, and (b) is constructed to improve the directional gain.

YAG/LED source: A laser light source that (a) is used for fiber optic transmission, (b) consists of a neodymium (Nd) yttrium garnet (YAG) crystal laser, (c) usually is pumped by a light-emitting diode (LED), (d) emits a narrow spectrum, about 0.5 nm (nanometer) wide at an operating wavelength of 1064 nm or 1.064 μm (microns) at peak intensity, and (e) is inefficient and bulky. *Synonym* **yttrium aluminum garnet source.**

yarn: *See* **aramid yarn.**

Y coupler: *See* **wye coupler.**

***y*-dimension of recorded spot:** In facsimile systems, the effective recorded spot dimension measured perpendicular to the recorded line; i.e., the effective recorded spot dimension that (a) is the largest center-to-center spacing between recorded spots, (b) is measured perpendicular to the recorded line, and (c) gives minimum peak-to-peak variation of density perpendicular to the recorded line. *Note:* The *y*-dimension of the recorded spot implies that the response to a constant density of the object is a succession of discrete recorded spots. *Synonym* **recorded spot *y*-dimension.** *See also* **facsimile, maximum keying frequency, recording, scanning, *x*-dimension of recorded spot.**

***y*-dimension of scanning spot:** In facsimile systems, the distance between the centers of adjacent scanning spots measured perpendicular to the scanning line on the object. *Note:* The numerical value of the *y*-dimension of the scanning spot depends upon the type of system. *Synonym* **scanning spot *y*-dimension.** *See also* **facsimile, scanning, *x*-dimension of scanning spot.**

year: *See* **day of year.**

year worst hour: 1. In electromagnetic transmission systems, the hour of the year during which the median noise over any radio path is a maximum. **2.** The hour of the year during which the greatest transmission loss occurs. *Note:* Usually the greatest loss and the most noise occur during the same hour. *Synonym* **worst hour of the year.** *See also* **path loss, transmission loss.**

y-mount: *See* **x-y mount.**

yttrium aluminum garnet/light emitting diode (YAG/LED) source: *See* **YAG/LED source.**

yttrium aluminum garnet source: *Synonym* **YAG/LED source.**

Z: **Zulu Time.**

Zeeman effect: The splitting of electromagnetic radiation into its component frequencies, that is, the splitting of spectral wavelengths, i.e., of spectral lines, by an applied magnetic field.

Zehnder fiber optic sensor: *See* **Mach-Zehnder fiber optic sensor.**

zenith: In radio communications systems, a point on an extension of a line from the center of the Earth to an observer. *Note 1:* The zenith is farther from the Earth's center than the observer is. *Note 2:* The zenith is opposite the nadir. *See also* **nadir.**

zero: *See* **nonreturn to zero, return to zero.**

zero address instruction: An instruction in a computer program that has no address part because either the address is implicit, or no address is needed to execute the instruction.

zero-beat reception: In suppressed carrier radio transmission and reception, homodyning in which the receiver generates a voltage having the original carrier frequency and combines it with the incoming signal. *See also* **homodyne, homodyning.**

zero binary code: *See* **return to zero binary code.**

zero bit insertion: Bit stuffing in which bit-oriented protocols are used to ensure that (a) six consecutive "one" bits never occur between the two flags that define the beginning and the end of a transmission frame, (b) when five consecutive 1 bits occur in any part of the frame other than the beginning and the end flag, the sending station inserts an extra 0 bit, and (c) when the receiving station detects five 1 bits followed by a 0 bit, it removes the extra 0 bit, thereby restoring the bit stream to its original value.

zero crossing counter: A counter that (a) counts, records, or indicates the number of times that a radar input signal instantaneous amplitude value exceeds, i.e., crosses, the detection level threshold that is established for detection, (b) makes use of a constant false alarm rate (CFAR), (c) consists of a wideband limiting intermediate frequency (IF) amplifier followed by a counter that indicates the detection of an object, i.e., a target, when the rate of crossing falls below a predetermined value, (d) operates in such a manner that the limiting action of the wideband amplifier and the random nature of the noise permit the use of a fixed threshold detection level that is independent of the power or the amplitude of the input signal and independent of the total input noise power if the noise spectrum is flat, and (e) establishes a false alarm rate in the absence of an object, i.e., a target, by a counter at the output of the threshold circuit of the detector.

zero dBm transmission level point (0 dBm TLP): In a communications system, a point at which the reference level is 1 mW (milliwatt), i.e., 0 dBm. *Note:* The real power level of the communications traffic is not necessarily 0 dBm. It usually is below the reference level. The reference is for system design and test purposes. *Synonym* **zero transmission level point.**

zero dispersion: Pertaining to a waveguide in which (a) an electromagnetic wave is propagating such that all modes in the wave have the same end-to-end propagation delay, (b) the refractive index is different for each of the various wavelengths contained in the spectral width of the source, (c) the refractive index profile of the guide is adjusted so that the geometric path taken by each mode is different, the faster modes taking longer geometric paths and the slower modes shorter geometric paths, resulting in all modes arriving at the far end at the same time, i.e., all modes having the same path length, and (d) the optical path length, i.e., the product of the geometric distance and the refractive index, of all the modes is the same, which reduces the signal distortion by waveguide dispersion to nearly zero. *Note 1:* The geometric distance is the actual measured spatial distance over which a given mode propagates. *Note 2:* If the proper operating wavelength is chosen, material dispersion can also be reduced to nearly zero, resulting in a zero dispersion waveguide at a particular wavelength. In a single mode optical fiber, there is a singular case in which the total dispersion is identically equal, or nearly equal, to zero, i.e., the $ps•nm^{-2}•km^{-1}$ (picoseconds per square nanometer•kilometer) = 0, where the picoseconds are a measure of the increase in pulse duration, the nanometers are a measure of the pulse spectral width, and the kilometers are a measure of the length of the optical fiber. For an appropriate refractive index profile, zero dispersion will occur at a specific wavelength. The wavelength at which zero dispersion occurs will vary with manufacturing tolerances, temperature, humidity, and aging. *See also* **pulse duration, spectral width, zero material dispersion wavelength.**

zero dispersion slope: In an optical fiber, the slope of the global fiber dispersion characteristic at the wavelength at which zero dispersion occurs, i.e., the value of the chromatic dispersion slope at the zero dispersion wavelength of the waveguide, usually expressed by the relation $S_0 = S(\lambda_0)$ where S_0 is the zero dispersion slope, $S(\lambda)$ is the chromatic dispersion slope as a function of wavelength, and λ_0 is the zero dispersion wavelength of the waveguide. *Note 1:* The zero dispersion slope usually is expressed in $ps•nm^{-2}•km^{-1}$ (picoseconds per square nanometer•kilometer). *Note 2:* The zero dispersion slope is used to calculate the upper and lower limiting values of the global dispersion coefficient. The zero dispersion slope will vary with manufacturing tolerances, temperature variations, humidity variations, and aging. *See also* **chromatic dispersion slope, material dispersion coefficient, zero dispersion wavelength.**

zero dispersion wavelength (ZDW) (λ_0): In a single mode optical fiber, the wavelength at which material dispersion and waveguide dispersion cancel one another, i.e., zero or nearly zero dispersion occurs, because of an appropriate refractive index profile and the matching operating wavelength. *Note:* The zero dispersion wavelength will have a maximum and a minimum value among optical fibers and over time because of manufacturing tolerances, temperature variations, humidity variations, and aging. *See also* **dispersion, material dispersion coefficient.**

zero-fill: 1. To character-fill with a representation of the character that denotes zero. **2.** To fill unused storage locations with character strings, each of which represents 0. *Note:* The character string that represents 0 is not necessarily a character string that consists of all binary 0s. *See also* **character-fill, zeroize.**

zeroize: To restore a storage unit, or the contents of a storage unit, such as a register, to a condition in which all 0s are stored. *Note:* "All 0s" does not necessarily imply all binary 0s. *See also* **zero-fill.**

zero level code: *See* **modified nonreturn to zero level code.**

zero level decoder: A decoder that yields an analog level of 0 dBm at its output when the input is the digital milliwatt signal. *Note:* The digital milliwatt signal is a 1-mW (milliwatt) 1000-Hz (hertz) sine wave. *See also* **digital milliwatt.**

zero match gate: *Synonym* **NOR gate.**

zero material dispersion wavelength: In an optical waveguide, the electromagnetic wavelength at which there is no material dispersion. *Note:* The zero material dispersion wavelength usually occurs at that point in the electromagnetic frequency spectrum at which the electronic band edge absorption ceases, in a region starting from where the ultraviolet ceases and infrared absorption begins. The electronic band edge absorption occurs in the visible spectrum and is primarily caused by oxides of silicon, sodium, boron, calcium, germanium, and other elements, and by hydroxyl ions. For zero dispersion in an optical waveguide, the refractive index profile must be such that the guide is compensated for zero dispersion at the operating wavelength. *See also* **zero dispersion, zero dispersion wavelength.**

zero relative level (ZRL): In communications circuits, the voltage and impedance reference point for power ratios. *Note:* In a transmission line, a voltage level of 0.775 V (volt) in a 600 Ω (ohm) line, i.e., an absolute power level of 1 mW, is standard. Power is given as $P = V^2/R$ where V is voltage, and R is resistance. Thus, $P = (0.775)^2/600 = 1$ mW. On the dB scale, 1 mW (milliwatt) is equivalent to 0 dBm, which may also be considered as 0 dBr, i.e., zero relative level. *See also* **zero transmission level point.**

zero substitution: *See* **bipolar with eight-zero substitution, bipolar with six-zero substitution, bipolar with three-zero substitution.**

zero suppression: The elimination of nonsignificant zeros from a numeral. *Note:* Nonsignificant zeros include (a) zeros to the left of the leftmost nonzero digit in the integral part of a numeral and (b) zeros to the right of the rightmost nonzero digit in the fractional part of a numeral. In a decimal numeral, the integral part is to the left of the decimal point, real or implied, and the fractional part is to the right of the real decimal point. In binary numerals, the point is the binary point.

zero transmission level point (0TLP): In a communications system, a point chosen as the reference level for measuring and expressing all relative transmission levels. *Note 1:* The absolute level at the zero transmission level point (0TLP) is measured and usually is not 0 dBm, i.e., not 1 mW (milliwatt). *Note 2:* The point chosen for the zero transmission level point (0TLP) usually is a conveniently accessible point, such as the input of a transmission line. For instance, if the input point to a transmission line is chosen as the zero transmission level point (0TLP) and has a 100 mW (milliwatt) absolute power level, and a measurement at a point *A* somewhere along the line has a power level of 1 mW, then the power level at point *A* is −20 dBa0 or −20 dBr, i.e., 20 dB below 0TLP. *Synonym* **zero dBm transmission level point.** *See also* **dBa0, dBr, dBrnC0, level, relative transmission level, signal level, transmission level.**

ZIP code: A decimal number that (a) is assigned to each U.S. Postal Service office in the United States, (b) is used to expedite the transfer, delivery, and forwarding of mail, (c) is equivalent to the address designator and address indicating group that are used on messages for electrical transmission, and (d) currently consists of five essential decimal-digit numerals plus four digits for local destinations. *Note:* Street names, house numbers, box numbers, and addressee names are significant for mail delivery, but the city, town, and state names are redundant when the five-digit ZIP code is given.

zip cord: In fiber optic communications, a two-fiber cable that (a) includes both loose buffer and tight buffer designs, and (b) consists of two single-fiber cables having their jackets conjoined by a strip of jacket material so that the two fiber optic cables can be easily separated by slitting or tearing the two jackets apart, to allow for the installation of optical connectors on the separated cables, in an arrangement similar to common electrical lamp cord in which the two conductors can be torn apart for wiring the lamp socket or a wall plug.

Z marker beacon: A beacon that (a) is identical to the fan marker beacon except that it is installed, as part of a four-course, i.e., a four-legged, radio range, at the intersection of the four range legs, (b) radiates vertically in order to indicate to aircraft stations that they are exactly directly over the range station, or the extent to which they are on either side of it, (c) usually is not keyed for identification, and (d) is so named because (i) in conventional Cartesian coordinate systems, the *z*-axis is vertical, while the *x*-axis and the *y*-axis are horizontal, and (ii) when over the beacon, the aircraft station is at the zenith with respect to the beacon. *See also* **fan marker beacon.**

zonal harmonic function: A spherical harmonic trigonometric function that has values that are constant along a parallel of latitude.

zone: A defined region or area in a given domain, such as time, space, frequency, temperature, velocity, pressure, or any other measurable or defined quantity. *Note:* Examples of zones are the world geographic reference system zones, time zones, speed zones, landing zones for aircraft or parachutists, regions of the radiation field of an antenna, and layers of the ionosphere. *See* **auroral zone, communications zone, control zone, dead zone, drop zone, Fresnel zone, skip zone, time zone, transmission zone.**

zone boundary limit marker: *See* **drop zone boundary limit marker.**

zone exchange: In telephone systems, an exchange that (a) is used as the main switching center for a certain number or group of exchange centers in a defined area, and (b) has access, directly or indirectly, to all other main switching centers.

zone impact point: *See* **drop zone impact point.**

zone indicator: *See* **drop zone indicator.**

zone of silence: *Synonym* **skip zone.**

zoom back: In display systems, the creating of a smaller image of an object, thus giving the appearance of moving away from the object because the optical angle of intercept required to place the image of the object on the retina is decreased. *See also* **image, object, zoom in.**

zoom in: In display systems, the creating of a larger image of an object, thus giving the appearance of moving closer to the object because the optical angle of intercept required to place the image of the object on the retina is increased. *See also* **zoom back.**

zoom lens: A lens in an optical system that (a) has components that move in such a way as to change the focal length while maintaining a fixed image position, and (b) can vary the image size while the optical system remains in a fixed position. *Synonym* **pancratic lens.**

zooming: In display systems, the scaling of that part of a display element, display group, or display image that lies within a window in such a manner as to give the appearance of movement toward or away from a point or an object of interest. *Note:* Unless otherwise stated, to zoom in implies the appearance of moving closer to an object, i.e., to enlarge the image. To zoom back implies the reverse. *See also* **zoom back, zoom in.**

Z Time: 1. Zulu Time. 2. *Synonym* **Coordinated Universal Time.**

Zulu Time (Z): 1. *Synonym* **Coordinated Universal Time. 2.** *Formerly a synonym for* **Greenwich Mean Time.**

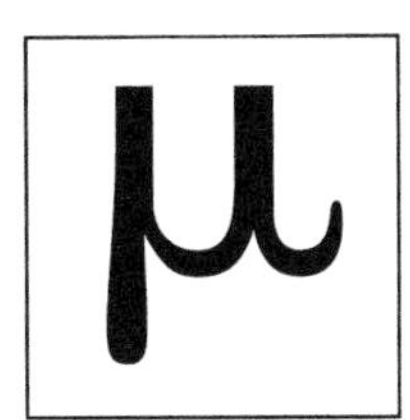

0TLP: zero transmission level point.

100% modulation: *See* **one hundred percent modulation.**

144-line weighting: *Synonym* **one forty four line weighting.**

144-receiver weighting: *Synonym* **one forty four receiver weighting.**

144-line weighting: In telephone systems, a noise weighting used in a noise measuring set to measure noise on a line that would be terminated by an instrument equipped with a No. 144-receiver or a similar instrument. *Note:* The meter scale readings are in dBrn (144-line). *See also* **noise, noise measurement unit, noise weighting.**

144-receiver weighting: In telephone systems, a noise weighting used in a noise measuring set to measure noise across the receiver of an instrument equipped with a no. 144-receiver or a similar instrument. *Note:* The meter scale readings are in dBrn (144-receiver). *See also* **noise, noise measurement unit, noise weighting.**

500 service: *Synonym* **five hundred service.**

700 service: *Synonym* **seven hundred service.**

800 service: *Synonym* **eight hundred service.**

900 service: *Synonym* **nine hundred service.**

Δ–β switch: delta-beta switch.

Δ–Σ modulation: delta-sigma modulation.

μ: **1.** A prefix that (a) is used to indicate "micro," as in μW (microwatts), μA (microamperes), and μV (microvolts), and (b) has the value of one millionth, i.e., 10^{-6}. **2.** A symbol often used to indicate micrometers or microns, i.e., 10^{-6} m (meter). *Note:* The symbol μm (microns, micrometers) is the preferred symbol used to indicate 10^{-6} m (meters).

μ law: *Synonym* **mu-law.**

μm: A symbol often used to indicate micrometers or microns, i.e., 10^{-6} m (meter). *Note:* The symbol μ is also used to indicate microns. However, μ is also used as a prefix for micro, such as μW (microwatts) and μA (microamperes).

APPENDIX A

*Acronyms and Abbreviations**

a	atto (10^{-18})
A	ampere
Å	angstrom
AAR	automatic alternate routing
AARTS	automatic audio remote test set
AAS	aeronautical advisory station
AB	asynchronous balanced [mode] (Link Layer OSI-RM)
ABCA	American, British, Canadian, Australian armies
abs	absolute
ABS	aeronautical broadcast station
ABSBH	average busy season busy hour
ac	alternating current
AC	absorption coefficient
	access charge
	access code
ACA	automatic circuit assurance
ACC	automatic callback calling
ACCS	associated common-channel signaling
ACCUNET	AT&T switched 56-kbps service
ACD	automatic call distributing

*Alphabetized letter by letter. Uppercase letters follow lowercase letters. All other symbols are ignored, including superscripts and subscripts. Numerals follow Z, and Greek letters are at the very end.

	automatic call distributor
ac-dc	alternating-current direct-current (ringing)
ACE	automatic cross-connection equipment
ACF/VTAM	advanced communications facility (VTAM)
ACK	acknowledge character
	acknowledgment
ACK/NAK	acknowledgment/negative acknowledgment
ACS	airport control station
	asynchronous communications system
ACSE	association service control element
ACTS	advanced communications technology satellite
ACU	automatic calling unit
AD	addendum
A/D	analog to digital
A-D	analog to digital
ADAPT	architectures design, analysis, and planning tool
ADC	analog-to-digital converter
	automatic digital control
ADCCP	Advanced Data Communications Control Procedure (ANSI)
ADH	automatic data handling
ADP	automatic data processing
	automatic data processor
ADPCM	adaptive differential pulse code modulation
	Association for Data Processing and Computer Management
ADPE	automatic data processing equipment
ADPS	automatic data processing system
ADPSSO	automatic data processing system security officer
ADS	automatic dependent surveillance
ADSP	AppleTalk Data Stream Protocol
ADSU	automatic dependent surveillance unit
ADU	automatic dialing unit
ADX	automatic data exchange
AECS	Aeronautical Emergency Communications System
AECSP	Aeronautical Emergency Communications System Plan
AEF	aircraft emergency frequency
	aeronautical emergency frequency
AES	aeronautical Earth station
AF	audio frequency
AFC	area frequency coordinator
	automatic frequency control

AFI	address and format identifier (ISO)
AFNOR	Association Française de Normalisation (French standards body)
	Association Française des Normes (French Standards)
AFP	AppleTalk Filing Protocol
AFRN	Armed Forces Radio Network
AFRS	Armed Forces Radio Service
AFS	aeronautical fixed service
	aeronautical fixed station
AG	antenna gain
AGC	automatic gain control
AGE	aerospace ground equipment
AGRS	air-ground radiotelephone service
AGWCS	air-ground worldwide communications system
AHF	adaptive high frequency
AHF radio	adaptive high frequency radio
AI	absorption index
	artificial intelligence
AIG	address indicating group
	address indicator group
AIM	amplitude intensity modulation
AIN	advanced intelligence network
AIOD	automatic identified outward dialing
AIS	automated information system
AISS	automated information systems security
AJ	antijamming
AL	Application Layer
ALAP	AppleTalk Link Access Protocol
ALBAM	air-land battle assessment model
ALC	automatic level control
	automatic load control
ALE	automatic link establishment
ALGOL	Algorithmic Language
ALMRBS	airport land mobile radio base station
ALMRS	airport land mobile radio station
ALOHA	an experimental radio broadcast network
ALU	arithmetic and logic unit
	arithmetic and logical unit
A/m	amperes per meter
AM	amplitude-modulated
	amplitude modulation

	amplitude modulator
	angle modulation
	ante meridiem
AMA	automatic message accounting
AMC	administrative management complex
AME	amplitude modulation equivalent
	automatic message exchange
AMI	alternate mark inversion
	alternate mark inversion (signal)
AMLCD	active matrix liquid crystal display
AM/PM/VSB	amplitude modulation/phase modulation/vestigial sideband
AMPS	Advanced Mobile Phone Service
	automatic message processing system
AMPSSO	automated message processing system security officer
AM(R)S	aeronautical mobile (route) service
AMS	aeronautical mobile service
	aeronautical multicom service
AMS(OR)S	aeronautical mobile-satellite (off-route) service
AMS(R)S	aeronautical mobile-satellite (route) service
AMSS	aeronautical mobile-satellite service
AMTS	automated maritime telecommunications system
ANA	article numbering association
ANI	automatic number identification
ANL	ambient noise level
	automatic noise limiter
ANMCC	Alternate National Military Command Center
ANS	American National Standard
ANSI	American National Standards Institute (formerly ASA)
ANT	antenna noise temperature
AP	anomalous propagation
	applications processor
APC	adaptive predictive coding
	adaptive predictive coding
	automatic phase control
apd	avalanche photodiode
a.p.d.	avalanche photodiode
APD	avalanche photodetector
	avalanche photodiode
APEX	Advanced Project for Information Exchange
API	application program interface

API/CS	application program interface/communications service
APK	amplitude phase-shift keying
APL	A Programming Language
	Array Programming Language
	average picture level
APPC	Advanced Program-to-Program Communications (IBM)
APPN	Advanced Peer-to-Peer Networking
APSK	amplitude phase-shift keying
APT	Automatically Programmed Tools
ARENA	Automated Recourse to Electronic Negotiations Archive
ARM	asynchronous response mode (Link Layer OSI-RM)
ARN	air reporting net
ARNS	aeronautical radionavigation service
ARNSS	aeronautical radionavigation-satellite service
ARP	address resolution protocol
	Address Resolution Protocol (standard)
ARPA	Advanced Research Projects Agency (now DARPA)
ARPAnet	Advanced Research Projects Agency Network
ARPANET	Advanced Research Projects Agency Network
ARQ	automatic recovery quotient
	automatic repeat-request
	automatic request for transmission
ARR	airborne radio relay
ARS	automatic route selection
	automatic route selector
ARTIC	a realtime interface coprocessor
ARU	audio response unit
AS	abstract syntaxes
	aeronautical station
ASA	American Standards Association (now ANSI)
ASAS	all-source analysis system
ASC	AUTODIN Switching Center
	axis of signal communications
ASCII	American Standard Code for Information Interchange
ASE	automatic switching equipment
ASIC	application specific integrated circuit
ASN.1	Abstract Syntax Notation 1 (ISO)
ASOCnet	Army Special Operations Command Network
ASP	adjunct service point
	Aggregated Switch Procurement

	AppleTalk Session Protocol
ASR	automatic send and receive
ASRT	automatic send/receive terminal
ASS	amateur satellite service
ASTM	American Society for Testing and Materials
AT	asynchronous transmission
ATACS	Army Tactical Communications System (U.S.)
ATB	all trunks busy (signal)
ATC	air traffic control
	air traffic controller
ATDM	asynchronous time-division multiplexing
ATE	automatic test equipment
ATF	advanced tactical (optical) fiber
ATM	asynchronous transfer mode
	Asynchronous Transfer Mode (standard)
	automatic teller machine
	automatic tracking mechanism
ATOF	advanced tactical optical fiber
ATOW	acquisition and tracking orderwire
ATV	advanced television
au	astronomical unit
AU	access unit
AUI	access unit interface
	attachment unit interface
AUTODIN	Automatic Digital Network (U.S. DoD)
AUTOSEVOCOM	Automatic Secure Voice Communications Network
AUTOVON	Automatic Voice Network (U.S. DoD)
AVD	alternate voice/data
AVPO	axial vapor phase oxidation
AWG	American Wire Gauge
AWGN	additive white Gaussian noise
AWM	appliance wiring material
AWSIM	air warfare simulation
AZ	azimuth
b	bit (binary digit)
	byte
B	bel
	byte
balun	balanced-to-unbalanced

BAS	basic activity set (OSI Session)
basecom	base communications
BASIC	beginner's all-purpose symbolic instruction code
BB	baseband
BBC	back-to-back connection
	British Broadcasting Company
BBLAN	baseband local area network
BBS	bridge-to-bridge station
BCC	block check character
BCD	binary coded decimal
BCH	Bose-Chaudhuri-Hochquenghem (code)
BCI	bit count integrity
BCS	B-channel circuit-switched
	basic combined set (OSI Session)
BCVT	basic class virtual terminal (OSI)
Bd	baud
BEL	bell (transmission control character)
bell	BEL character
BEP	block error probability
BER	basic energy reduction
	bit error rate (*use* bit error ratio)
	bit error ratio
	block error ratio
BERT	bit error ratio test
	bit error ratio tester
BES	base Earth station
BETRS	Basic Exchange Telecommunications Radio Service
BEX	broadband exchange
BFICC	British Facsimile Industry Consultative Committee
BFL	back focal length
BIGFON	experimental German video service
BIH	Bureau International de l'Heure
	International Time Bureau
BIOD	Bell integrated optical device
BIOS	basic input-out system
bis	second version of a recommendation (CCITT)
BISDN	broadband integrated services digital network
B-ISDN	Broadband Integrated Services Digital Network
bi-sync	binary synchronous
	binary synchronous communication

	binary synchronous communications
bit	binary digit
BIT	built-in test
BITE	built-in test equipment
BIU	basic information unit (IBM/SNA)
	bus interface unit
BLER	block error rate
BLERT	block error rate test
	block error rate tester
	block error rate testing
BLP	block loss probability
BLR	block-loss ratio
BMLC	Basic Mode Link Control
BMMG	British Minicomputer Manufacturing Group
BMR	bit misdelivery ratio
	block-misdelivery ratio
BNF	Backus-Naur form
BNR	Bell Northern Research
BOB	break-out box
BOC	Bell Operating Company
BOCs	Bell Operating Companies
BOP	Bit-Oriented Protocol
bpi	bits per inch
	bytes per inch
BPI	bits per inch
	bytes per inch
bps	bits per second
BPS	B-channel packet-switched
BPSK	binary phase-shift keying
BR	bit rate
BRA	basic rate access
BRI	basic rate interface
	Basic Rate Interface (ISDN)
BRL	balance return loss
BRLP	bit-rate•length product
b/s	bits per second
b•s^{-1}	bits per second
BS	broadcasting service
BSA	basic serving arrangement
BSC	binary synchronous communication

BSE	basic service element
b/sec	bits per second
b•sec^{-1}	bits per second
BSGL	branch systems general license
BSI	British Standards Institution
BSO	bit-synchronous operation
BSS	basic synchronized set (OSI Session)
	broadcasting-satellite service
BT	British Telecom
BTI	British Telecom International
BTN	billing telephone number
BTR	block transfer rate
BW	bandwidth
BWDF	bandwidth•distance factor
B3ZS	bipolar with three-zero substitution
B6ZS	bipolar with six-zero substitution
B8ZS	bipolar with eight zero substitution
c	centi (10^{-2})
c	symbol for the speed of light in vacuum
C	coulomb (electric charge)
C	symbol for capacitance (electric circuit)
C^2	command and control
	command and control system (tactical)
C^3	command, control, and communications
	command, control, and communications system (tactical)
C^4	command, control, communications, and computers
CA	collision avoidance
CAA	collinear antenna array
CAC^2	combined arms command and control
CACS	centralized alarm control system
CAD	computer aided design
	computer aided dispatching
	computer assisted design
CAD/CAM	computer aided design/computer aided manufacturing
CALS	continuous acquisition and lifecycle support
CAM	computer aided manufacturing
	computer assisted manufacturing
CAMA	centralized automatic message accounting
CAN	cancel character (control character)

CAP	competitive access provider
	customer administration panel
CARS	cable television relay service
	cable television relay station
CAS	centralized attendant services
	channel-associated signaling
CASE	common application service elements
	computer-aided software engineering
	computer-aided system engineering
	computer-assisted software engineering
	computer-assisted system engineering
CATV	cable television
	cable TV
	community antenna television
CBEMA	Computer and Business Equipment Manufacturers Association (Sponsor ANSI X3, Information Systems Standards Committee)
CBMS	computer-based messaging system (NIST)
CBR	Carson bandwidth rule
CBS	common-battery signaling
	corps battle simulation
CBX	computer branch exchange
	computerized branch exchange (IBM)
cc	cubic centimeter
CC	country code
CCA	carrier-controlled approach
CCC	call control character
	command and control center
	communications control center
CCCCP	Committee on Computer-to-Computer Communication Protocols
CCD	charge-coupled device
CCEB	Combined Communications-Electronics Board
cch	hundred circuit•hours
CCH	connections per circuit-hour
	hundred circuit-hours
CCI	character count integrity
CCIF	International Consultative Committee for Telephone
CCIR	International Radio Consultative Committee
CCIS	common channel interoffice signaling
CCIT	International Consultative Committee for Telegraph (now CCITT)
CCITT	Comité Consultatif International Télégraphique et Téléphonique

	International Consultative Telegraph and Telephone Committee
C^3CM	command, control, and communications countermeasures
CCR	commitment, concurrency, and recovery
ccs	hundred call•seconds (c = 100, Roman)
CCS	clear confirmation signal
	common channel signaling
	hundred call-seconds (C = 100, Roman)
CCSA	common control switching arrangement
CCSS No.7	common channel signaling system No. 7 (CCITT)
CCTV	closed-circuit television
CCW	cable cutoff wavelength
cd	candela
CD	collision detection
CD ROM	compact disk read-only memory
CDC	chromatic dispersion coefficient
CDF	combined distribution frame
	cumulative distribution function
CDM	color-division multiplexing
	conditioned diphase modulation
CDMA	code division multiple access
	code-division multiple-access (spread spectrum)
CDPSK	coherent differential phase-shift keying
CDR	call detail recording
CDS	chromatic dispersion slope
CDT	control data terminal
CDU	central display unit
C-E	communications-electronics
CEC	Committee of European Communities
CEI	comparably efficient interconnection
CELP	code-excited linear prediction (coding)
CEM	common equipment module (Ferranti)
CEN	Comité Européen de Normalization
CENLEC	European Committee for Electrotechnological Standardization
CEP	circular error probability
	circular error probable
CEPT	Conference Européen des Administrations des Postes et des Télécommunications
CERN	European Nuclear Research Center
CES	coast Earth station

CF	critical frequency
CFE	contractor-furnished equipment
CFR	Cambridge fast ring
	Code of Federal Regulations
CFSK	coherent frequency shift keying
cgs	centimeter-gram-second (system of units)
CHIEF	customs handling of import and export freight
CHILL	CCITT high-level language (telephone exchange use)
CHINAPAC	China national packet switching network
CHOP	change in operational control
ChR	channel reliability
C^3I	command, control, communications, and intelligence
C^4I	command, control, computers, communications, and intelligence
CI	computer industry
	cooperation index
CIAS	circuit inventory and analysis system
CIC	content indicator code
CICS	customer information control system
CIF	common interface format
	common intermediate format
CIFAX	ciphered facsimile
CIM	computer-integrated manufacturing
	corporate information management
CINC	commander-in-chief
CINCLANT	Commander-in-Chief, Atlantic
CINCPAC	Commander-in-Chief, Pacific
ciphony	ciphered telephony
CiR	circuit reliability
CIS	communications and information systems
C/kT	carrier-to-receiver noise density
cl	centiliter
CL	connectionless
CL-mode	connectionless mode (ISO)
CLASS	custom local area signaling services
CLID	calling line identification
CLIP	calling line identification presentation
CLIR	calling line identification restriction
CLIS	calling line identification signal
CLIST	command list
CLR	clear request (connection)

cm	centimeter
CM	configuration management
CMA	centralized message accounting
CMC	communications management configuration
	customer management complex
CMDR	command reject
CMI	coded mark inversion
CMIP	Common Management Information Protocol
CMIS	common management information service
	common management information system
CMOS	complementary metal oxide semiconductor
	complementary metal oxide substrate
CMOT	CMIS management over TCP
CMP	communications management processor
CMRR	common-mode rejection ratio
CMT	connectionless mode transmission
CMW	compartmented mode workstation
C/N	carrier to noise ratio
CNL	carrier noise level
CNM	communications network management
	communications network manager
CNR	carrier to noise ratio
	combat net radio
CNS	complementary network services
CO	central office
	central office (telephone network)
	connection-oriented
C.O.	central office
COAM	customer owned and maintained equipment
COAT	coherent optical adaptive technique
coax	coaxial cable
COBOL	Common Business Oriented Language
codec	coder-decoder
COG	centralized ordering group
CO-LAN	central office local area network
COM	computer output microform
COMINT	communications intelligence
COMJAM	communications jamming
compandor	compressor-expander
COMPUSEC	computer security

COMSAT	Communications Satellite Corporation
COMSEC	communications security
CONEX	connectivity exchange
CONUS	continental United States
	Continental United States
COS	Corporation for Open Systems (U.S.)
COT	customer office terminal
CO/TP	Connection-Oriented Transaction Protocol
CPE	customer premises equipment
cpi	characters per inch
CPI	common physical interface (ISDN)
cpm	characters per minute
	counts per minute
CP/M	control program for microcomputers
CPP	Cable Patch Panel (DEC)
cps	characters per second
	counts per second
	cycles per second (*deprecated for* frequency)
CPS	call progress signal
CPU	central processing unit
	communications processor unit
CR	call request (ISO)
	carriage return (character)
	channel reliability
	character recognition
	circuit reliability
	connection request (ISO)
	contrast ratio
CRADA	cooperative research and development agreement
CRC	cyclic redundancy check
CREG	concentrated range extension with gain
CRITCOM	critical intelligence communications
CRITICOM	Critical Communications Intelligence Network
CROM	control read-only memory
CRSS	C^3I reusable software system
CRT	cathode ray tube
c/s	cycles per second (*deprecated for* frequency)
CS	carrier system
	communications survivability
CSC	circuit-switching center

	common signaling channel
CSCINT	closed source intelligence
CSDN	circuit-switched data network
CSE	communications security equipment
CSM	communications security material
	communications system management
	communications systems management
	Communications and Systems Management
CSMA	carrier-sense multiple-access
CSMA/CA	carrier-sense multiple-access w/collision avoidance
	CSMA with collision avoidance
CSMA/CD	carrier-sense multiple-access w/collision detection
	CSMA with collision detection
CSNET	computer science network (U.S. research)
CST	compatible sideband transmission
CSU	channel service unit
	circuit switching unit
	customer service unit
CTCA	Canadian Telecommunications Carriers Association
CTD	clock time difference
CTL	critical technical load
CTM	circuit transfer mode
CTS	clear to send
	clear to send (EIA RS-232)
CTX	Centrex®
	clear to transmit
CUG	closed user group (X.25)
CVD	chemical vapor deposition
CVPO	chemical vapor phase oxidation (process)
CVSD	continuously variable slope delta (modulation)
cw	carrier wave
	composite wave
	continuous wave
CW	carrier wave
	composite wave
	continuous wave
CWSS	communications wiring system
CX	composite signaling
CX-mode	connection mode (ISO)
cxr	carrier

d	deci (10^{-1})
D*	D star (specific detectivity)
da	deka (10)
D/A	digital to analog
D-A	digital to analog
DACS	digital access and cross-connect system
DAD	draft addendum
DAMA	demand assignment multiple access
DAP	Data Access Protocol (DECnet)
DAR	dynamically adaptive routing
DARPA	Defense Advanced Research Projects Agency (U.S.)
DASS	digital access signaling system
DASS2	digital access signaling system two
DATAPAC	a public packet-switched network (Canada)
DATEX-L	digital circuit-switched service (Germany)
DATEX-P	public packet-switched network (Germany)
dB	decibel
dBa	weighted noise power in dB referred to –85 dBm
DBA	directed broadcast address
dBa(F1A)	noise power measured with F1A-line weighting
dBa(HA1)	noise power measured with HA1-receiver weighting
dBa0	noise power measured at zero transmission level point
dBc	dB relative to carrier power
DBC	dense binary code
DBI	decibel above isotropic
dBm	dB referred to 1 milliwatt (0 dBm = 1 milliwatt)
dBm(psoph)	noise power in dBm with psophometric weighting
dBm0	noise power in dBm referred to or measured at 0TLP
dBm0p	noise power in dBm0 measured with psophometric weighting
	noise power in dBm0 measured with C-message weighting
DBMS	database management system
dBmV	dB referred to 1 millivolt across 75 ohms
	decibel referred to 1 millivolt (signal level)
dBnW	dB referred to 1 nanowatt (0 dBnW = 1 nW)
DBP	Deutsche Bundespost (German Communications)
dB(psoph)	noise power in dBm measured with psophometric weighting
dBpW	dB referred to 1 picowatt (0 dBpW = 1 pW)
dBr	dB referred to a given reference
dBrn	dB above reference noise
dBrnC	noise power in dBrn measured with C-message weighting

dBrnC0	noise power in dBrnC referred to or measured at 0TLP
dBrn(f_1–f_2)	flat noise power in dBrn
dBrn(144-line)	noise power in dBrn measured with 144-line weighting
DBS	direct broadcast satellite
DBTVS	direct broadcast television satellite
dBv	decibels relative to 1 volt peak-to-peak
dBV	dB relative to 1 V (volt)
dBW	dB referred to 1 W (watt) (0 dBW = 1 W)
	decibels referred to 1 watt (0 dBW = 1 W)
dBx	dB above reference coupling
	decibels above reference coupling
dB0	noise power in dBm referred to or measured at 0TLP
DB-25	EIA RS-232 connector
dBμW	dB referred to 1 microwatt (0 dBμW = 1 μW)
dc	direct current
DC	direct current
	double crucible (process)
DCA	Defense Communications Agency
	document content architecture (IBM)
DCC	data country code
	digital cross-connect
DCD	data carrier detect
DCE	data circuit-terminating equipment
DCF	doubly clad fiber
DCI	diametral cooperation index
DCOF	doubly clad optical fiber
DCPSK	differentially coherent phase-shift keying
DCS	Defense Communications System
DCTN	Defense Commercial Telecommunications Network
DCWV	direct-current working volts
DDCMP	Digital Data Communications Message Protocol (DEC)
DDD	direct distance dialing
DDL	data description language
DDM	distributed data management
DDN	Defense Data Network (U.S.)
DDP	Datagram Delivery Protocol
DDR	disengagement denial ratio
DDS	dataphone digital service (AT&T)
	digital data service
	direct dialing service

	doped deposited silica
DE	data element
DEB	differentially encoded baseband
DEC	Digital Equipment Corporation
DECCA	digital positioning system
DECnet	Digital Equipment Corporation network architecture
DECT	Digital European Cordless Telecommunications
	digital European cordless telephone
	digital European cordless telephony
DEL	delete character (control character)
DELNI	DEC Local Network Interface
demarc	demarcation point
demod	demodulator
DEMOD	demodulator
DEMPR	DEC Multiport Repeater
demux	demultiplex
	demultiplexed
	demultiplexer
	demultiplexing
DEPS	departmental entry-processing system
dequeue	double-ended queue
DES	data encryption standard
	Data-Encryption Standard (NBS, now NIST)
	double-ended synchronization
detem	detector/emitter
DETEM	detector/emitter
DF	direction finder
	direction finding
	distress frequency
DFB	distributed feedback (laser)
DFM	digital frequency modulation
DFSK	double frequency-shift keying
DGPS	differential global positioning system
DGT	Direction Générale des Télécommunications (France)
DHCF	Distribution Host Command Facility
DIA	Defense Intelligence Agency
	document interchange architecture (IBM)
DID	direct inward dialing
DIN	Deutsche Industrie Normenausschuss (Germany)
	Deutsches Institut für Normung (German Standards)

	Deutsches Institut für Normung (German standards body)
	German Industrial Standards Committee
	German Institute for Standardization (German standards body)
DINA	direct noise amplification (jammer)
diopt	diopter
DIP	dual in-line package
	dual in-line package (switch)
DIS	Draft International Standard
DISA	Defense Information Systems Agency
DISC	disconnect
	disconnect command
DISH	data interchange for shipping (U.K.)
DISN	Defense Information Systems Network
DISNET	Defense Integrated Secure Network
DISOSS	distributed office support system
DIW	D-inside wire
D/L	downlink
DLA	Defense Logistics Agency
DLC	digital loop carrier
DLCI	data link connection identifier
DLE	data link escape
	data link escape character
DLL	Data Link Layer
DM	delay modulation
	differential modulation
	delta modulation
DMA	Defense Mapping Agency
	differential mode attenuation
	direct memory access
	direct memory access (data transfer)
DME	distance measuring equipment
DMI	digital multiplexed interface (data transfer)
DMS	data management system
	Defense Message System
	digital multiplexed switch (Northern Telecom)
DMT	digital multiplexed trunk
DMW	digital microwave
DN	digital network
	directory number
DNA	DEC network architecture

	Defense Nuclear Agency
DNIC	data network identification code
	Data Network Identification Code (CCITT)
DNPA	data numbering plan area
DNS	Domain Name System
DNSIX	Defense Network Security Information Exchange
DO	design objective
DoC	Department of Commerce (U.S.)
DoD	Department of Defense (U.S.)
DOD	direct outward dialing
DoDD	Department of Defense Directive (U.S.)
DoDI	Department of Defense Instruction (U.S.)
DoDISS	DoD Index of Specfications and Standards (U.S)
DoD-STD	Department of Defense Standard
DOS	disk operating system
DOV	data over voice
dp	draft proposal (ISO)
DPBX	data private branch exchange
DPCM	differential pulse-code modulation
dpIS	draft proposed international standard (ISO)
DPLL	digital phase-locked loop
DPM	digital phase modulation
	dual-precedence message
DPNSS	digital private network signaling system
DPPB	digital primary patch bay
DPRS	domestic public radio services
DPSK	differential phase-shift keying
DQDB	distributed queue, dual bus (MAN)
	distributed-queue dual-bus (network)
DQTV	distribution-quality television
DRAM	dynamic random access memory
DRS	disengagement request signal
DS	digital signal
	direct support
	dispersion-shifted (optical fiber)
	duplex signaling
DSA	dial service assistance
	Defense Security Agency
	directory service agent
DSAP	destination service access point (IEEE 802)

DSB	Defense Science Board
	dial service board
	double-sideband (transmission)
DSB-RC	double-sideband reduced carrier
DSB-SC	double-sideband suppressed carrier (transmission)
DSC	digital selective calling
DSCS	Defense Satellite Communications System
DSE	data switching equipment
	data switching exchange
DSI	digital speech interpolation
DSL	digital subscriber line
	digital subscriber loop (AT&T)
DSM	delta-sigma modulation
	direct-spread modulation
DSN	Defense Switched Network
DSP	digital signal processing
DSR	data sampling rate
	data signaling rate
	digital signaling rate
	disengagement-success ratio
	domain specific part (ISO addressing)
DSRI	destination station routing indicator
DSS	direct station selection
DSSCS	Defense Special Service Communications System
DSSS	direct-sequence spread spectrum
DSTE	data subscriber terminal equipment
DSU	data service unit
DS0	digital signal 0
DS1	digital signal 1
DS1C	digital signal 1C
DS2	digital signal 2
DS3	digital signal 3
DS4	digital signal 4
DS-1	multiplexed telephone signal (1.544 Mb/s)
DS-3	multiplexed telephone signal (44.736 Mb/s)
DTCM	differential trellis coded modulation
DTE	data terminal equipment
	data terminal equipment (CCITT)
DTG	date-time group
DTI	Department of Trade and Industry (U.K.)

DTMF	dual-tone multifrequency
	dual-tone multifrequency (signaling)
DTN	data transfer network
	data transmission network
DTP	data transfer process
DTR	data terminal ready
	data transfer rate
DTRS	data transfer request signal
DTU	data tape unit
	data transfer unit
	digital transmission unit
	direct to user
DU	data unit
	dispersion-unshifted (optical fiber)
DVD	digital video disk
DVM	data/voice multiplexed
	data/voice multiplexer
	data/voice multiplexing
DX	direct current signaling
	duplex
D3	Type 3 fiber optic connector
D4	Type 4 fiber optic connector
D(λ)	chromatic dispersion coefficient
e	base of natural logarithms, ~2.718
	symbol of the charge of an electron.
E	exa (10^{18})
EA	effective availability
EAN	electronic article numbering
EAR	echo attenuation ratio
EARN	European academic and research network
EAS	extended area service
EBCDIC	extended binary-coded decimal interchange code (IBM)
EBE	embedded base equipment
EBO	embedded base organization
EBS	Emergency Broadcast System
EBX	electronic branch exchange
EC	Earth coverage
	Earth curvature
	echo canceling

	electrooptic coefficient
ECC	electronically controlled coupling
	error-correcting code
ECCM	electronic countercountermeasure
ECM	electronic countermeasure
ECMA	European Computer Manufacturing Association
ECPE	embedded customer-premises equipment
ECSA	Exchange Carriers Standards Association
ECU	European currency unit
EDACS™	Enhanced Digital Access Communications System
EDC	error detection and correction
	error-detecting code
EDI	electronic data interchange
EDP	electronic data processing
EDTR	effective data transfer rate
EDTV	extended-definition television
EEC	electromagnetic emission control
	European Economic Community (Common Market)
EEP	electromagnetic emission policy
EER	effective Earth radius
EESS	Earth exploration-satellite service
EFIS	electronic flight information system
EFL	equivalent focal length
EGC	equal gain combiner
EGP	exterior gateway protocol
	Exterior Gateway Protocol (DARPA)
EHF	extremely high frequency
EHz	exahertz (10^{18} Hz)
EIA	Electronic Industries Association
EIN	European Information Network
eirp	effective isotropically radiated power
	equivalent isotropically radiated power
EKMS	electronic key management system
el	elevation
ELECTRO-OPTINT	electro-optical intelligence
ELED	edge-emitting light-emitting diode
ELF	extremely low frequency
ELINT	electromagnetics intelligence
	electronics intelligence
ELSEC	electronics security

ELT	emergency locator transmitter
EM	end-of-medium (control character)
E & M	ear and mouth (receiver/transmitter leads)
	ear and mouth (signaling)
E & M leads	Ear and Mouth receive and transmit leads
e-mail	electronic mail
e-Mail	electronic mail
E-mail	electronic mail (Federal standard)
E-MAIL	electronic mail
EMC	electromagnetic compatibility
EMCON	emission control
EMD	equilibrium mode distribution
EME	electromagnetic environment
emf	electromotive force (voltage)
EMI	electromagnetic interference
	electromagnetic interference (control)
EMP	electromagnetic pulse
EMPD	equilibrium modal-power distribution
EMR	electromagnetic radiation
	electromagnetic radiation (hazard)
e.m.r.p.	equivalent monopole radiated power
EMS	electronic message system
EMSEC	emanations security
emu	electromagnetic unit
EMUG	European MAP Users Group
EMUT	enhanced manpack ultrahigh frequency terminal
EMV	electromagnetic vulnerability
emw	electromagnetic wave
EMW	electromagnetic warfare
	electromagnetic wave
E_b/N_0	signal-energy-per-bit per hertz of thermal noise
ENA	extended network addressing
ENQ	enquiry
	enquiry character (control character)
ENT	equivalent noise temperature
ENVOY 100	Telecom Canada electronic mail system
EO	end office (network)
EOD	end of data (signal)
EOF	end of file
EOL	end of line

EOM	end of message
	end of message (control character)
EOP	end of output
	end of program
EOS	end of selection character
EOT	end of tape
	end of text
	end of text (control character)
	end of transmission
	end of transmission character
	end of transmission character (control)
EOW	engineering orderwire
EP	Emergency Preparedness
	emulator package
EPRM	erasable programmable read-only memory
EPROM	electrically programmable read-only memory
	erasable programmable read-only memory
EPSCS	enhanced private switched communications system
EQ	equalizer
EQTV	extended-quality television
ERC	electromagnetic radiation control
ERL	echo return loss
e.r.p.	effective radiated power
ERP	effective radiated power
ES	Earth station
	end system (network)
	enhanced service
	expert system
ESC	escape (control character)
	escape character (control)
ESD	electrostatic discharge
ESF	extended superframe
ESI	equivalent step index (profile)
ESM	electronic warfare support measure
ESP	enhanced service provider
ESPRIT	European Strategic Precompetitive Research Programme for Information Technology
ESS	electronic switching system
	Electronic Switching System (AT&T)
ESTELLE	electronic format for protocol description (ISO)

	formal format for protocol description (ISO)
esu	electrostatic unit
ET	Earth terminal
	exchange transmission
ETA	estimated time of arrival
ETB	end of transmission block
ETB	end of transmission-block character (control)
ETC	Earth terminal complex
ETD	estimated time of departure
ETR	effective transmission rate
	external timing reference
ETS	effective transmission speed (*deprecated, use* ETR)
ETSI	European Telecommunications Standards Institute
ETX	end of text
	end of text (control character)
eV	electron volt
EW	electronic warfare
EWSM	electronic warfare support measure
EXCSA	Exchange Carriers Standards Association
f	femto (10^{-15})
f	symbol for frequency
F	frequency
FAA	Federal Aviation Agency
	frequency assignment authority
$FAADC^2$	forward area air defense command and control
facs	facsimile
FADU	file accessible data unit (ISO)
FAI	functional address indicator
FAL	file access list
FAQ	frequently asked question
FAQ file	frequently asked question file
facs	facsimile
FAS	frame alignment signal
FATAM	file application, transfer, access, and management
fax	facsimile
FAX	facsimile
FC	fiber-optic connector
	functional component
FCC	Federal Communications Commission (U.S.)

FCCRP	Federal Communications Commission Registration Program
FCS	frame check sequence
	Frame Check Sequence (CRC)
FCS	frequency-change signaling
FCW	fiber cutoff wavelength
FD	frequency division
FDDI	fiber distributed data interface
	Fiber Distributed Data Interface (ANSI Standard)
FDDI-2	Fiber Distributed Data Interface 2
FDHM	full-duration half-maximum
FDM	frequency-division multiplex
	frequency-division multiplexed
	frequency-division multiplexer
	frequency-division multiplexing
FDMA	frequency-division multiple access
	frequency-division multiplexed access
FDX	full duplex
FE	format effector
FEC	far end crosstalk
	forward error correction
	front end computer
FECC	Federal Emergency Communications Coordinators
FED-STD	federal standard
FEMA	Federal Emergency Management Agency
FEN	feeder echo noise
FEP	fluorinated ethylene propylene
	front end processor
	perfluorinated ethylene propylene
FES	frequency-exchange signaling
FET	field effect transistor
FEXT	far-end crosstalk
FFL	front focal length
FGS	facility grounding system
FHSS	frequency-hopping spread spectrum
FIFO	first in, first out
FIP	Federal Information Processing
FIPS	Federal Information Processing Standard
FIPS-Pub	Federal Information Processing Standard (Publication)
FIR	finite impulse response
FIRM	functionally integrated resource manager

FIRMR	Federal Information Resources Management Regulation
FISINT	foreign instrumentation signal intelligence
FLETC	Federal Law Enforcement Training Center
flop	floating point operation
flops	floating point operations
	floating-point operations per second
FLTSATCOM	Fleet Satellite Communications
FM	frequency-modulated
	frequency modulating
	frequency modulation
	frequency modulator
	full modulation
FMIF	frequency modulation improvement factor
FMS	flight management system
FO	fiber optic
	fiber optics
FOC	fiber optic combiner
	fiber optic connector
	final operational capability
	full operational capability
FOCC	fiber optic cable component
FOG	fiber optic gyroscope
FOL	fiber optic link
FOM	fiber optic modem
FOMS	fiber optic myocardium stimulator
FOT	frequency of optimum traffic
	frequency of optimum transmission
	optimum traffic frequency
	optimum transmission frequency
FOTM	fiber optic test method
FOTP	fiber optic test procedure
FOTS	fiber optic transmission system
FPD	flat panel display
FPGA	field programmable gate arrays
FPI	functional process improvement
FPIS	forward propagation ionospheric scatter
fpm	feet per minute
fps	feet per second
	frames per second
fp/s	foot•pounds per second

FPS	fast packet switching
	fault protection subsystem
	focus projection and scanning
FRENA	frequency and amplitude (system)
FRM	fixed-reference modulation
FRMR	frame reject
FSC	frequency spectrum congestion
FSD	frequency spectrum designation
FSDPSK	filtered symmetric differential phase-shift keying
FSF	fully separate facility
FSK	frequency shift keying
FSL	facsimile signal level
FSS	fixed-satellite service
	fully separate subsidiary
FST	frequency-shift telegraphy
FSTS	Federal Secure Telephone System
FT	fiber optic T-carrier
FTAM	file transfer, access, and management
	File Transfer, Access, and Management (ISO)
FTC	fast time constant
	fast time constant (logarithmic)
FTP	file transfer protocol
	File Transfer Protocol (DARPA)
FTS	Federal Telecommunications System
FTS2000	Federal Telecommunications System 2000
FTSC	Federal Telecommunications Standards Committee
FTT	fault-tracing time
FU	functional unit
FWHM	full width at half maximum
	full-width half-maximum (pulse)
	full-width half-maximum (spectral power)
FX	fixed service
	foreign exchange (telephone connection)
	foreign exchange service
FYDP	Five Year Defense Plan
g	acceleration of gravity
	profile parameter
G	giga (10^9)
GAT	go-ahead tone

GBH	group busy hour
Gbps	gigabits per second (10^9 bps)
GCA	ground controlled approach (landing)
	ground controlled approach (system)
GCT	Greenwich Civil Time (UTC)
GDSS	global decision support system
GDF	group distribution frame
GFC	general flow control
GFE	government-furnished equipment
GFI	general format identifier
GGP	gateway-to-gateway protocol
	Gateway-to-Gateway Protocol (DARPA)
GHz	gigahertz
GI	graded-index (optical fiber refractive index)
GIP	graded-index profile
GKS	Graphical Kernel System
GMSK	Gaussian minimum shift keying
GMT	Greenwich Mean Time (UTC)
GOS	grade of service
GOSIP	Government Open Systems Interconnection Profile
	Government OSI Profile
GPI	glide-path indicator
GPIRS	global positioning inertial reference system
GPS	global positioning system
	Global Positioning System
GRC	ground-return circuit
GRIN	graded-index (optical fiber refractive index)
GRWSIM	ground warfare simulation
GS	group switch (Ericson)
GSA	General Services Administration
GSI	glide slope indicator
GSTN	general switched telephone network
G/T	gain to noise-temperature ratio (antenna)
GTMOSI	general teleprocessing monitor for OSI
GTP	Government Telecommunications Program
GTS	Government Telecommunications System
GUI	graphical user interface
GUS	Guide to the Use of Standards
h	hecto (10^2)
	hour

h	symbol for Planck's constant
HBC	half-baud coding
HCN	hierarchical computer network
	hybrid communications network
HCS	hard clad silica (optical fiber)
HD	half-duplex (circuit)
	half-duplex (operation)
HDLC	high level data link control
	High Level Data Link Control (ISO)
HDSL	high bit-rate digital subscriber line
HDTV	high definition television
	high definition TV
HDX	half-duplex (circuit)
	half-duplex (operation)
HEC	header error check
	heading error check
HE_{11} mode	fundamental hybrid mode (optical)
HEMP	high altitude electromagnetic pulse
HERF	hazards of electromagnetic radiation to fuel
HERO	hazards of electromagnetic radiation to ordnance
HERP	hazards of electromagnetic radiation to personnel
hf	photon energy (Planck's constant × frequency)
HF	high frequency
HFAARS	High Frequency Adaptive Antenna Receiving System
HFDF	high frequency distribution frame
HIPERLAN	high-performance local area network
	high performance radio local area network
HLL	high level language
HLP	high level protocol
HM	Her Majesty's (U.K.)
	His Majesty's (U.K.)
	heterogeneous multiplexing
HMS	hydrological and meteorological station
HOC	highest outgoing channel
HRC	hybrid ring control
HSN	hierarchically synchronized network
HTML	hypertext metalanguage
HTTP	hypertext transfer protocol
	Hypertext Transfer Protocol (standard)
HV	high voltage

h value	polarization-holding parameter
Hz	hertz (frequency unit)
I	information
I	symbol for electric current
IA	International Alphabet (CCITT)
IAGC	instantaneous automatic gain control
IASA	integrated AUTODIN system architecture
IATV	interactive television
IA1	International Alphabet No. 1 (CCITT)
IA2	International Alphabet No. 2 (CCITT)
IA3	International Alphabet No. 3 (CCITT)
IA4	International Alphabet No. 4 (CCITT)
IA5	International Alphabet No. 5 (CCITT)
IBC	information-bearer channel
	integrated broadband communications
IBCN	integrated broadband communications network
IBDN	Integrated Building Distribution Network (Northern Telecom)
IBM	International Business Machines (Corporation)
IBNPR	in-band noise power ratio
IBS	international broadcasting station
	international broadcasting system
IBT	isochronous burst transmission
I & C	installation and checkout
IC	integrated circuit
ICA	integrated communications adapter
	International Communications Association
icand	multiplicand
ICI	incoming call identification
	interface control information (ISO)
ICL	International Computers, Ltd.
ICMP	Internet Control Message Protocol
ICNI	Integrated Communications, Navigation, and Identification
ICS	integrated communications system
ICW	interrupted continuous wave
ID	identification
	identifier
IDDD	International Direct Distance Dialing
IDF	intermediate distribution frame
IDM	isochronous demodulation

IDN	integrated digital network
IDP	internetwork datagram protocol
	Internetwork Datagram Protocol (Internet)
IDR	isochronous distortion ratio
IDT	interactive data transaction
IDTV	improved definition television
IDU	interface data unit (ISO)
IEC	International Electrotechnical Commission
IEE	Institution of Electrical Engineers (UK)
IEEE	Institute of Electrical and Electronics Engineers (U.S.)
ier	multiplier
I/F	interface
IF	intermediate frequency
IFF	identification friend or foe
IFIP	International Federation for Information Processing
IFOC	integrated fiber optic circuit
IFR	instrument flight rule
IFRB	International Frequency Registration Board
IFS	ionospheric forward scatter
IG	insertion gain
IHL	Internet header link
IIL	integrated injection logic
IILS	Information Industry Liaison Committee
IIR	infinite impulse response
	infinite impulse response (filter)
IKBS	intelligent knowledge-based system
IL	insertion loss
	intermediate language
I^2L	integrated injection logic
ILD	injection laser diode
ILF	infralow frequency
ILLED	integral lens light-emitting diode
ILS	instrument landing system
I & M	installation and maintenance
IM	intensity modulation
	intramodulation
	isochronous modulation
IMA	integrated modular avionics
Inmarsat	International Maritime Satellite (System)
IMD	intermodulation distortion

IMP	interface message processor
	Interface Message Processor (DARPA)
IMS	information management system
IN	intelligence network
	intelligent network
INFOSEC	information systems security
INS	inertial navigation system
	Inertial Navigation System
	integrated network system
	Integrated Network System (NTT)
INSP	Internet Name Server Protocol
INT	interchange
	internal
	international
INTAP	Interoperability Technology Association for Information Processing (Japan)
INTELSAT	International Telecommunications Satellite Consortium
INWATS	incoming wide area telephone service
	inward wide area telephone service
IN/1	intelligent network concept 1
IN/1+	intelligent network concept 1 plus
IN/2	intelligent network concept 2
I/O	input/output (data, device, channel)
	input/output (devices, data, terminals, channels)
IOC	initial operational capability
	input-output controller
	integrated optical circuit (OIC is preferred)
IOS	*improper abbreviation for* ISO
IP	intelligent peripheral (device)
	Internet Protocol (ISO and U.S. DoD)
	internetwork protocol
IPA	intermediate power amplifier
IPC	information processing center
	interpersonal communication
	interpersonal communications
IPDS	intelligent printer data stream (IBM)
IPDU	internet protocol data unit
ipm	impulses per minute
	interference prediction model
	internal polarization modulation

	interruptions per minute
IPM	incremental phase modulation
	Institute of Practical Mathematics (Germany)
	interference prediction model
	internal polarization modulation
	interpersonal messaging (X.400)
ips	inches per second
	interruptions per second
IPSS	international packet-switched network (U.K.)
IPX	Internet Packet Exchange
	internetwork packet exchange
IQF	intrinsic quality factor
IR	information retrieval
	infrared
IRAC	Interdepartment Radio Advisory Committee (U.S.)
IRAMS	innovative real time antenna modeling system
IRC	Interagency Radio Committee
	international record carrier
IRF	industrial radio frequency
IRQ	interrupt request
IRR	image rejection ratio
IS	information superhighway
	interactive service
	international standard (ISO)
ISA	industry standard architecture
ISB	independent-sideband (transmission)
ISDN	integrated services digital network
	Integrated Services Digital Network (Standard)
ISI	intersymbol interference
ISLU	integrated services line unit (AT&T)
ISM	industrial, scientific, and medical (applications)
ISN	information systems network (AT&T)
ISO	International Organization for Standardization
ISPBX	integrated services PBX (British Telecom)
ISUP	ISDN user part
ITA	International Telegraph Alphabet
ITA-1	International Telegraph Alphabet Number 1
ITA-2	International Telegraph Alphabet Number 2
ITA-3	International Telegraph Alphabet Number 3
ITA-4	International Telegraph Alphabet Number 4

ITA-5	International Telegraph Alphabet Number 5
ITAB	Information Technical Advisory Board
ITC	International Teletraffic Congress
ITF	interframe time fill
ITI	Industrial Technology Institute (MAP)
ITS	Institute for Telecommunication Sciences
	international temperature scale
ITSO	International Telecommunications Satellite Organization
IT&T	International Telephone and Telegraph (Company)
ITT	International Telephone and Telegraph (Company)
	information-transfer transaction
	invitation to tender
ITU	International Telecommunication Union
ITV	interactive television
IVD	inside vapor deposition
	integrated voice/data
IV/DT	integrated voice/data terminal
	integrated voice-data terminal
IVHS	intelligent vehicle highway system
	Intelligent Vehicle Highway Society
IVPO	inside vapor phase oxidation (process)
IXC	interexchange carrier
	interexchange character

JANAP	Joint Army-Navy-Air Force Publication
JANET	Joint Academic Network (U.K.)
JCL	job control language
JCLAS	joint computer-aided acquisition and logistics support
JCS	Joint Chiefs of Staff
JCSPubl	Joint Chiefs of Staff Publication No. 1 (Glossary)
JCS1	Joint Chiefs of Staff Publication No. 1 (Glossary)
JEDI	joint electronic data interchange
JES	job entry subsystem
JPL	Jet Propulsion Laboratory
JSC	Joint Steering Committee (now JTSSG)
JTC^{3}A	Joint Tactical Command, Control, and Communications Agency
JTIDS	Joint Tactical Information Distribution System
JTM	job transfer and manipulation
JTRB	Joint Telecommunications Resources Board
JTSSG	Joint Telecommunications Standards Steering Group

JUTCPS	joint uniform telephone communications precedence system
k	Boltzmann's constant
k	kilo (10^3) (prefix)
	kelvin
	kilo-ohm (10^3 ohms, 1000 ohms)
	1000 (suffix)
	1024 (storage)
K	coefficient of absorption
kbps	kilobits per second (data signaling rate unit)
KDC	key distribution center
KDD	Kokusai Denshin Denwa Company, Ltd. (Japanese international carrier)
KDR	keyboard data recorder
KDT	keyboard display terminal
kg	kilogram (unit of mass)
kgms	kilogram-meter-second (*deprecated, use* MKS)
kHz	kilohertz (1000 Hz, unit of frequency)
km	kilometer (unit of distance)
kpps	kilopulses per second
KSR	keyboard send/receive device
*k*T	Boltzmann's constant times temperature
	noise power density
KTS	key telephone system
KTU	key telephone unit
KWIC	keyword in context (index)
KWOC	keyword out of context (index)
L	inductance (electric circuit)
LAA	locally administered address
LAN	local area network
LANFLT	Atlantic Fleet (U.S.)
LAP	link-access procedure
LAPB	link access procedure, balanced (X.25)
LAP-B	Data Link Layer protocol (CCITT X.25 (1989))
LAPD	link access procedure D
	link access procedure for D channel (ISDN)
LAP-D	link access procedure for D channel (ISDN)
LAPM	link access procedure, modem
laser	light amplification by stimulated emission of radiation

LASINT	laser intelligence
LAT	local area terminal
	Local Area Transit (DEC protocol)
LATA	local access and transport area
	local access and transport area (U.S. telephone)
LBO	line buildout
LC	leased circuit
LCD	liquid crystal display
LCN	logical channel number
LD	laser diode
	long distance
LDD	limited distance dialing
LDDC	long-distance direct current (dialing)
LDL	long-distance line
LDM	limited distance modem
LEA	longitudinally excited atmosphere (laser)
LEC	local engineering circuit
	local exchange carrier
LED	light-emitting diode
LEN	low-entry networking
LEOW	low-cost exploitation operator workstation
LF	line feed
	line feed (character)
	low frequency
LFB	look-ahead-for-busy (information)
LGN	logical group number
LHC	long-haul communications
LHCN	long-haul communications network
LHOTS	long-haul optical transmission set
LI	length indicator (OSI)
LIC	lowest incoming channel
LIDAR	light detection and ranging
LIF	logarithmic intermediate frequency (amplifier)
LIFO	last in, first out
LIM	line-access module (Ericson)
LINCS	Leased Interfacility National-airspace Communications System
LIS	labeled interface structure
LISP	list processing (programming language)
LL	link level
LLC	logical link control

	logical link control (sublayer)
LLIF	linear LIF (amplifier)
LLP	low level protocol
l/m	lines per minute
lm	lumen
LME	layer management entity
LMF	language media format
lm•hr	lumen•hour
lm•s	lumen•second
LMS	land mobile service
	land mobile station
LMSS	land mobile-satellite service
LNA	launch numerical aperture (optical)
LNMF	local network management function
LOC	large optical cavity (diode)
LOF	lowest operating frequency
	lowest operational frequency
log FTC	logarithmic fast time constant
loran	long range aid to navigation
	long range electronic navigation
LORAN	long range aid to navigation
	long range electronic navigation
LORO	lobe on receive only (radar)
LOS	line of sight
LOTOS	formal format for protocol description (ISO)
LOW	link orderwire
LP	linear programming
	linearly polarized (wave)
	linking protection
	log periodic
	log periodic (antenna)
	log periodic (array)
LP_{01}	fundamental mode (optical)
LPA	linear power amplifier
LPC	linear predictive coding
LPD	low probability of detection
lpi	lines per inch
LPI	low probability of interception
lpm	lines per minute
LPP	Lightweight Presentation Protocol

lps	lines per second
LPS	lightning protection subsystem
LQA	link quality analysis
L-R	left-to-right
LR	long route
LRC	longitudinal redundancy check
LRM	line route map
l/s	lines per second
LS	land station
	logical station
	logical storage
LSAP	link service access point (OSI)
LSB	least significant bit
	lower sideband
LSC	local switching center
LSE	local system environment
	Local System Environment (OSI)
LSI	large-scale integrated (chip)
	line status identification
LSIC	large-scale integrated circuit
LTC	line traffic coordinator
LU	logical unit (IBM/SNA)
LU 6.2	logical unit (program-to-program) (SNA)
LUF	lowest usable frequency
	lowest useful frequency
LUHF	lowest usable high frequency
	lowest useful high frequency
LULT	line-unit-line termination
LUNT	line-unit-network termination
LV	low voltage
LW	long-wave
LX	local exchange
LZW	Lempel-Ziv-Welsh (algorithm)
m	meter
	milli (10^{-3})
	minute
	mega (10^6) (prefix, as in megahertz)
M	mega (abbreviation, as in MHz and Mbps)
	thousand (suffix, as in 100M)

M	symbol for mutual inductance (electric circuit)
MAA	maximum acceptance angle
MAC	maritime air communications
	medium access control (LAN)
	medium access control (sublayer)
MACOM	major command
MAI	message alignment indicator
MAMI	modified alternate mark inversion
MAN	metropolitan area network
MAP	manufacturer automation protocol
	Manufacturing Automation Protocol (General Motors)
maser	microwave amplification by stimulated emission of radiation
MAU	medium access unit
	medium attachment unit (802.3)
	multistation access unit (802.5)
Mb	megabyte
Mbps	megabits per second (data signaling rate)
MBR	minimum bend radius
MBS	marine broadcast station
MCA	maximum calling area
MCC	maintenance control circuit
MCD	manipulative communications deception
MCEB	Military Communications-Electronics Board (U.S.)
MCEF	military common emergency frequency
MCM	multicarrier modulation
MCS	Master Control System
MCVD	modified chemical vapor deposition (process)
MCVFT	multichannel voice frequency telegraph
MCW	modulated continuous wave
MCXO	microcomputer-compensated crystal oscillator
MDAS	multiple database access system
MDC	material dispersion coefficient
MDF	main distribution frame
MDIS	medical diagnostic imagery system
MDN	managed data network
MDS	multipoint distribution service
MDT	mean downtime
MED	manipulative electronic deception
MEECN	Minimum Essential Emergency Communications Network
Megastream	2.048 Mbps British Telecom system

MERCAST	merchant ship broadcast system
Mercury	Mercury Communications Ltd. (Cable & Wireless subs.)
MES	mobile Earth station
MF	medium frequency
	multifrequency
	multifrequency (signaling)
MFD	mode field diameter
MFJ	Modification of Final Judgment
MFS	modem fault simulator
MFSK	multiple frequency shift keying
MHD	magnetohydrodynamics
MHF	medium high frequency
MHS	message-handling service
	message-handling system
MHW	magnetohydrodynamic wave
MHz	megahertz (unit of frequency)
mi	mile (statute)
MIB	management information base
MIC	medium interface connector
	microphone
	microwave integrated circuit
	minimum ignition current
	monolithic integrated circuit
	mutual interface chart
MICR	magnetic ink character reader
	magnetic ink character reading
	magnetic ink character recognition
MIDS	multifunction information distribution system
MIFR	Master International Frequency Register
Milnet	military network
MIL-NET	military network
MIL-STD	military standard
MIL-STD-XXX	Military Standard (Number)
MIM	mechanically induced modulation
Minitel	videotex-based service (France)
	videotex-based terminal (France)
MIP	medium interface point
mips	million instructions per second
MIS	management information system
MISFET	metal insulator semiconductor field effect transistor

MISSI	multilevel information systems security initiative
MIVPO	modified inside vapor phase oxidation (process)
MKS	meter-kilogram-second (system of units)
ML	multilevel
MLP	multilevel precedence
MLPP	multilevel precedence and preemption
MLS	microwave landing system
mm	millimeter
MM	multimode optical fiber
mmf	magnetomotive force (voltage)
MMF	maximum modulating frequency
MMFS	Manufacturing Message Format Standard (MAP)
MMI	man-machine interface
	multimedia interface
MMS	manufacturing message service
	maritime mobile service
MMSE	minimum mean square error
MMSS	maritime mobile satellite service
MMW	millimeter wave
MNCS	master net control station
MNP	Microcom Networking Protocol
MNRZ	modified nonreturn to zero (code)
mo	magnetooptic
	magneto-optic
MOD	modulator
modem	modulator-demodulator
modified AMI	modified alternate mark inversion (code)
mol	mole (molecular mass in grams)
MOP	Maintenance Operation Protocol
MORPH	mathematical morphology
MOS	metal oxide semiconductor
MOSFET	metal oxide semiconductor field effect transistor
MOTIS	message-oriented text interchange system (ISO)
MPDS	multipoint distribution service
MPR	maximum pulse rate
MPSK	multiple phase-shift keying
MRNS	maritime radionavigation service
MRNSS	maritime radionavigation satellite service
MRS	multirole system
ms	millisecond (10^{-3} s)

MS	message store (U.K.)
	mobile service
	mobile station
	molecular stuffing (process)
MSATCOM	mobile satellite communications
MSAU	multistation access unit
MSB	most significant bit
MSE	Mobile Subscriber Equipment (System)
msec	millisecond (10^{-3} s)
MSK	minimum-shift keying
MSL	multisatellite link
MSN	multiple subscriber number
MSR	maximum stuffing rate
MSS	mobile satellite service
MT	machine translation
MTA	message transfer agent
MTBF	mean time between failures
MTBM	mean time between maintenance
MTBO	mean time between outages
MTBPM	mean time between preventive maintenance
MTF	modulation transfer function (optical fiber)
MTI	moving target indicator
MTL	message transfer layer
MTM	McLintock theater model
MTP	message transfer part
MTS	message transfer service
MTSR	mean time to service restoral
MTTF	mean time to failure
MTTR	mean time to repair
MTTSR	mean time to service restoral
MUF	maximum usable frequency
	maximum useful frequency
muldem	multiplexed/demultiplexed
	multiplexer/demultiplexer
	multiplexing/demultiplexing
MUS	marine utility station
MUSR	maximum user signaling rate
MUX	multiplex
	multiplexed
	multiplexer

	multiplexing
MUXing	multiplexing
MVS	multiple virtual storage (an operating system)
mw	microwave
MW	medium wave
	microwave
	middle wave
MWI	message waiting indicator
MWV	maximum working voltage
M(λ)	material dispersion coefficient
n	nano (10^{-9})
n	refractive index
N	newton (force)
N	symbol for force
N_0	sea level refractivity (refractive index)
	spectral noise density
NA	numerical aperture (waveguides, antennas)
NACISO	NATO Communications and Information Systems Agency
NACK	negative acknowledge
NACSEM	National Communications Security Emanation Memorandum
NACSIM	National Communications Security Information Memorandum
NAK	negative acknowledge character (control)
	negative acknowledgment (transmission control)
NAND	negative AND (gate)
NAPLPS	North American presentation-layer protocol syntax
NARSAP	National Advanced Remote Sensing Application Program
NASA	National Aeronautics and Space Administration
NATA	North American Telecommunications Association
NATO	North Atlantic Treaty Organization
NAU	network addressable unit (IBM/SNA)
NAVSTAR	Navigational Satellite Timing and Ranging
NAVSTAR-GPS	NAVSTAR Global Positioning System
NBFM	narrowband frequency modulation
NBH	network busy hour
NBP	NetBIOS Protocol
NBRVB	narrowband radio voice bandwidth
NBRVF	narrowband radio voice frequency
NBS	National Bureau of Standards (U.S.) (now NIST)
NBSV	narrowband secure voice (system)

NC	network connection
	network control
NCA	National Command Authorities
	National Command Authority
NCC	network control center
	National Coordinating Center
NCCF	network communications control facility
NCP	NetWare Core Protocol
	network control program
NCS	National Communications System
	net control station
NCSC	National Communications Security Committee
NCTE	network channel-terminating equipment
N+D	noise-plus-distortion
NDC	National Destination Code
NDCS	network data control station
	network data control system
NDRENET	Norwegian Defence Research Establishment network
NE	network element
NEACP	National Emergency Airborne Command Post
NEC	National Electric Code®
NEP	noise equivalent power
NES	noise equivalent signal
NET	Normes Européennes de Télécommunications
NetBIOS	network basic I/O system
NETCONSTA	net control station
NEXT	near-end crosstalk
NF	noise figure
	normalized frequency
NFAR	network file access routine
NFS	network file system
	Network File System (Standard)
NFSP	NetWare File Service Protocol
NIC	network interface card
NICE	network information and control exchange
NICS	NATO Integrated Communications System
NID	network information database
	network indialing
	network interface device
	network inward dialing

NII	national information infrastructure
NIOD	network inward/outward dialing
NIST	National Institute of Standards and Technology (formerly NBS)
NIU	network interface unit
NL	Network Layer
	new line
	new-line (control character)
NLDM	network logical data manager
NLO	nonlinear optical (material)
nm	nanometer (10^{-9} meter)
NMA	network management architecture
NMAR	network management and access routine
NMC	network monitoring and control
NMCS	National Military Command System
nmi	nautical mile
NMI	nonmaskable interrupt
NMM	network management module
NMU	noise measurement unit
NNBS	NetView Network Billing System
NNI	network-network interface
	network to network interface
NOD	network outdialing
	network outward dialing
NOR	negative OR (gate)
Nordforsk	Nordic Cooperative Organization for Applied Research
NORDUNET	Nordic University Network
NOS	network operating system
NOTAL	not all (message indicator)
NOTAM	Notice to Airmen
Np	neper
NPA	numbering plan area
NPD	noise power density
NPDA	network problem determination application
NPDN	Nordic Public Data Network
NPDU	network protocol data unit (ISO)
NPR	noise power ratio
NPSI	NCP packet-switching interface
NR	Nyquist rate
N(R)	receive sequence (ACK) number
NRI	net radio interface

NRM	network resource manager
	normal response mode
NRZ	nonreturn to zero
NRZB	nonreturn to zero, mark
NRZ-B	nonreturn to zero, mark
NRZI	nonreturn to zero, inverted
NRZ-I	nonreturn to zero, inverted
NRZM	nonreturn to zero, mark
NRZ-M	nonreturn to zero, mark
NRZS	nonreturn to zero, space
NRZ-S	nonreturn to zero, space
NRZ1	nonreturn to zero, change on 1s
NRZ-1	nonreturn to zero, one
ns	nanosecond (10^{-9} second)
N(S)	send sequence number
NSA	National Security Agency
NSAP	network service access point (OSI)
NSC	National Security Council
NSDU	network service data unit (OSI)
nsec	nanosecond (10^{-9} second)
NS/EP	National Security Emergency Preparedness
NSFnet	National Science Foundation Network
NSP	Network Services Protocol
NSR	Nyquist sampling rate
NSTAC	National Security Telecommunications Advisory Committee
NT	network termination
NTDS	Naval Tactical Data System
NTE	network termination equipment (British Telecom)
NTELOS	NetView Traffic Engineering Line Optimization System
NTI	network terminating interface
NTIA	National Telecommunications and Information Administration (U.S.)
NTN	network terminal number
NTO	network terminal operation
	network terminal operator
	network terminal option
NTSC	National Television Standards Committee
NTSC STD	NTSC standard
NTT	Nippon Telegraph and Telephone
NT1	network terminator, Type 1 (ISDN)

NT2	network terminator, Type 2 (ISDN)
NUL	null character
NVIS	near vertical incidence skywave
NVIS	near vertical incident skywave
NXX code	North American direct distance dialing numbering system
OA	office automation
OAA	open avionics architecture
OBCS	onboard communications station
OBE	out-of-band emission
OBS	out-of-band signaling
OC	open circuit
OCC	other common carrier
OCR	optical character reader
	optical character reading
	optical character recognition
OCS	onboard communications station
OCU	orderwire control unit
OCVCXO	oven-controlled/voltage-controlled crystal oscillator
OCXO	oven-controlled crystal oscillator
OD	optical density
	outside diameter
ODA	office document architecture
ODAPS	oceanic display and planning system
ODB	open dual bus
ODETTE	European motor industry EDI project
ODIF	office document interchange format
ODIS	oceanographic data interrogating station
ODS	oceanographic data station
OECD	Organization for Economic Cooperation and Development
OFC	optical fiber, conductive
OFCP	optical fiber, conductive, plenum
OFCR	optical fiber, conductive, riser
OFN	optical fiber, nonconductive
OFNP	optical fiber, nonconductive, plenum
OFNR	optical fiber, nonconductive, riser
OFTEL	Office of Telecommunications
OFTF	optical fiber transfer function
OIC	optical integrated circuit
OLTS	optical loss test set

O & M	operations and maintenance
	organization and methods
OMAP	operations and maintenance applications part
OMB	Office of Management and Budget (U.S.)
ONA	open network architecture
	Open Network Architecture (Bell)
OPB	optical power budget
OPC	on-premises cabling
opcode	operation code
OPEVAL	operational evaluation
OPL	optical path length
opm	operations per minute
OPMODEL	operations model
OPSEC	operations security
OPSEC-A	operations security (OPSEC) assessment
OPSEC-S	operations security (OPSEC) survey
OPW	on-premises wiring
OPX	off-premises extension
OR	off-route aeronautical mobile service
	operations research
	optical range
	outage ratio
O/R	originator/recipient (X.400)
ORS	omnidirectional range station
ORSA	operations research/systems analysis
OS	operating system
	operations system
OSA	open systems architecture
OSCINT	open-source intelligence
OSD	optical space division multiplexing
OSHA	Occupational Safety and Health Administration
OSI	open switching interval
	open systems interconnection
	Open Systems Interconnection (ISO)
OSIA	Open Systems Interconnection architecture
OSIE	Open Systems Interconnection—Environment (ISO)
OSINet	Open Systems Interconnection Network (U.S. NIST)
OSI-RM	Open Systems Interconnection—Reference Model (ISO)
OSI-SD	Open Systems Interconnection—Service Definition
OSI-SM	Open Systems Interconnection—Systems Management

OSI-TP	Open Systems Interconnection—Transaction Processing
OSNS	open systems network support
OSRI	originating station routing indicator
OSS	operational service state
	Operations Support System
OSSN	originating station serial number
OTAM	over-the-air management (automated HF network nodes)
OTAR	over-the-air rekey
	over-the-air rekeying
	over-the-air rekeying (of cryptographic equipment)
OTDR	optical time domain reflectometer
	optical time domain reflectometry
OTF	optimum transmission frequency
OUTWATS	outgoing wide-area telephone service
OVD	optical video disk
	outside vapor deposition
OVPO	outside vapor phase oxidation (process)
OW	orderwire (signaling circuit)
OWM	orderwire message

p	pico (10^{-12})
P	peta (10^{15})
P	profile dispersion parameter
	symbol for power (acoustic, electrical, and optical)
Pa	pascal (1 Pa = 1 N•m^{-2})
PA	personal agent
PABX	private automatic branch exchange (*use* PBX)
PACVD	plasma activated chemical vapor deposition
PACX	private automatic computer exchange
PAD	packet assembler/disassembler
PAL	phase alternate line
	phase alternation by line
PAL-M	phase alternation by line—modified
PAM	pulse amplitude modulation
PAMA	pulse-address multiple access
PAP	preassignment access plan
	Printer Access Protocol
par	peak-to-average ratio
p/ar	peak-to-average ratio
p/a-r	peak-to-average ratio

P/AR	peak-to-average ratio
PAR	performance analysis and review
	positive acknowledgment plus retransmission
paramp	parametric amplifier
PARAMP	parametric amplifier
PARC	Palo Alto Research Center (Xerox)
par meter	peak-to-average ratio meter
PAS	private aircraft station
	profile alignment system
PASL	primary area switch locator
PAX	private automatic exchange
PBER	pseudo-bit error ratio
PBX	private branch exchange
PC	carrier power (radio transmitter)
	personal communications
	personal computer
	phantom circuit
	primary channel
	printed circuit
PCB	power circuit breaker
	printed circuit board
PC DOS	personal computer disk operating system
PCI	protocol control information
	Protocol Control Information (ISO)
PCM	plug compatible module
	process control module
	pulse code modulation
	pulse code modulator
	pulse code multiplexing
PCMCIA	Personal Computer Memory Card International Association
PCS	personal communications service
	personal communications system
	plastic clad silica (optical fiber)
PCSR	parallel channels signaling rate
PCSS	personal communications system service
PD	packing density
	photodetector
	photodiode
PDF	pulse duty factor
PDM	plain-dress message

	pulse delta modulation
	pulse duration modulation
PDN	public data network
PDP	profile dispersion parameter
PDS	power distribution system
	Premises Distribution System (AT&T)
	primary distribution system
	program data source
	protected distribution system
PDT	programmable data terminal
PDTS	public data transmission service
PDU	protocol data unit
	Protocol Data Unit (ISO standard)
PE	phase encoded (recording)
	picture element
	protocol emulator
pel	picture element
PEM	peripheral equipment module (Ferranti)
PEP	Packet Exchange Protocol
	peak envelope power
pF	picofarad (10^{-12} farad)
PF	packing fraction (fiber optics)
P/F	poll/final (link protocol)
PF	power factor
	primary frequency
PFM	pulse-frequency modulation
PGI	parameter group identifier (ISO)
PGS	primary guard station
PH	packet handler
PHOTINT	photographic intelligence
PHY	Physical Layer Protocol (FDDI)
PHz	petahertz (10^{15} Hz)
PI	parameter identifier (ISO)
	protection interval
PIC	plastic insulated cable
PICS	protocol implementation conformance statement
PIDB	peripheral interface data bus (AT&T)
PIN	positive-intrinsic-negative (semiconductor junction)
PINFET	PIN photodiode field effect transistor
ping	packet internet groper

PING	Packet Internet Groper (Internet)
PIRS	positioning inertial reference system (global)
PISAB	pulse interference suppression and blanking
PIV	peak inverse voltage
PKES	public key encryption system
PL	Physical Layer
	Presentation Layer
PL/1	programmable language 1
	Programming Language 1
PLA	programmable logic array
PLL	phase-locked loop
PLM	pulse-length modulation (*use* PDM)
PLN	private line network
PLR	pulse link repeater
PLS	physical signaling sublayer
pm	phase modulation
PM	mean power
	packet mode
	phase modulation
	polarization-maintaining (optical fiber)
	polarizing maintaining (fiber)
	post meridiem
	preventive maintenance
	pulse modulation
PMB	pilot make-busy (circuit)
PMD	physical medium-dependent specification (FDDI)
p-n	positive-negative (semiconductor junction)
POI	point of interface
POP	peak power output
	point of presence (networking)
POSI	Promoting Conference for OSI (Japan)
POSIX	portable operating system interface for computer environments
POST	power-on self-test
pot	potentiometer
POT	point of train
POTS	plain old telephone service
P-P	peak-to-peak
P/P	point-to-point
PP	polarization-preserving (optical fiber) (*use* PM)
PPI	planned position indicator

ppm	parts per million
PPM	parts per million
	pulse position modulation
PPO	peak power output
PPP	Point to Point Protocol
pps	pulses per second
PPS	perimeter protection system
PPSN	pilot packet-switched network (U.K.)
PQA	path quality analysis
PR	pattern recognition
	primary radar
	pulse rate
PRA	primary rate access
PRBS	pseudorandom binary sequence
PRF	pulse repetition frequency (*use* PRR)
PRI	primary rate interface
PRM	pulse-rate modulation
PROFS	Professional Office System (IBM)
PROM	programmable read-only memory
PROSIGN	procedure sign
PROWORD	procedure word
PRR	pulse repetition rate
PRSL	primary area switch locator
PS	packet switch
	packet switched
	packet-switching
	permanent signal
PS/2	Personal System/2
ps	picosecond (10^{-12} s)
PS	permanent signal
	phase shift
	portable station
PSDN	packet-switched data network
PSDTS	packet-switched data transmission service
psec	picosecond (10^{-12} s)
psi	pounds per square inch (pressure)
PSK	phase-shift keying
PSN	packet switching network
	public switched network
	packet switching node

PSPDN	packet-switched public data network
PSS	packet-switched stream
p-static	precipitation static
P-static	precipitation static
PSTN	public switched telephone network
PT	payload type
	personal telecommunications
PTF	patch and test facility
PTI	packet type indication
PTM	packet transfer mode
	pulse-time modulation
PTT	post, telephone, and telegraph
	postal, telegraph, and telephone
	printing teletypewriter telegraphy
	push to talk (operation)
PTTC	paper-tape transmission code
PTTI	precise time and time interval
PU	physical unit (IBM/SNA)
	power unit
PUC	public utility commission
PUP	PARC universal packet (XNS)
PVC	permanent virtual circuit
	Permanent Virtual Circuit (X.25)
	polyvinyl chloride (insulation)
	polyvinyl chloride (plastic)
pW	picowatt (10^{-12} W)
PWM	pulse-width modulation (*use* PDM)
pWp	picowatt (pW), psophometrically weighted
pWp0	pWp relative to a 0TLP
pX	peak envelope power, in dB
PX	peak envelope power, in watts
	private exchange (*use* PBX)
PXP	Packet Exchange Protocol
pY	mean power, in dB
PY	mean power, in watts
pZ	carrier power, in dB
PZ	carrier power, in watts
PZT	piezoelectric transducer
q	symbol for electronic charge
QA	quality assurance

QAM	quadrature amplitude modulation
QBE	query by example
QC	quality control
QCIF	quarter common intermediate format
QM	quadrature modulation
QMR	qualitative material requirement
QO	quartz oscillator
QOS	quality of service
QPSK	quadrature phase-shift keying
QRC	quick reaction capability
QSTAG	Quadripartite Standardization Agreement
R	route service
R	resistance (electric circuit)
RACE	Research in Advanced Communications Technology for Europe
racon	radar beacon
RACON	radar beacon
rad	radian
	radiation absorbed dose
radar	radio detection and ranging
RADC	Rome Air Development Center
RADHAZ	electromagnetic radiation hazard
RADINT	radar intelligence
RADP	remote access data processing
RAM	random access memory
	reliability, availability, and maintainability
RAS	radio astronomy station
RATT	radio teletype (system)
	radio teletypewriter
RBOC	regional Bell operating company
	Regional Bell Operating Company
RBR	radar blind range
RBS	radar blind speed
	reverse-battery signaling
RbXO	rubidium-crystal oscillator
RC	receive clock
	reference clock
	reflection coefficient
	resource controller
	restricted channel

RCC	radio common carrier
RCP	real time control program
RCS	radar cross section
	radiocommunications service
RCT	reduced-carrier transmission
RCVR	receiver
RD	received data
	reflection density
R & D	research and development
RD	routing diagram
	routing directory
RDF	radio direction finding
	request data forwarding
RDI	restricted digital information
RDT	request data transfer
	request data transfer (signal)
RDT&E	research, development, test, and evaluation
RDTF	radio-telephone distress frequency
REA	Rural Electrification Administration
REED	restricted edge-emitting diode
REJ	reject (link protocol command)
REMOS	reflective membrane optical scintillator
REN	ringer equivalency number
RENAN	ISDN pilot project (France) (French philosopher)
REP	reply
RES	reserved (asynchronous transfer mode)
rf	radio frequency
	range finder
RF	reference frequency
RFA	radio frequency allotment
	radio frequency assignment
RFC	request for commands
RFCA	radio frequency channel allotment
RFI	radio frequency interference
RFP	request for proposal
RFQ	request for quotation
RFS	real file store
	real file system
RG/U	grade of coaxial cable
RGB	red-green-blue (television)

RH	relative humidity
RHI	range height indicator
RHR	radar horizon range
	radio horizon range
RI	ring in
	ring indicator
	routing indicator
RIA	routing indicator allocation
	routing indicator assignment
RIP	refractive index profile
RIPSO	revised interconnection protocol security option
RISC	reduced instruction set chip
RJ	registered jack
RJE	remote job entry
RLC	resistance inductance capacity
	resistive-inductive-capacitive (circuit)
RLLS	radiolocation land station
RLMS	radiolocation mobile station
RLSD	receive line signal detect
RM	reference model (OSI) (ISO)
rms	root mean square
RNLS	radionavigation land station
RNMS	radionavigation mobile station
RNR	receiver not ready (link protocol command)
RNSS	radionavigation satellite service
RO	read only
	receive only
	ring off
	ring out
ROC	required operational capability
ROM	read-only memory
ROS	read-only storage
	remote operation service
ROSE	remote operations service element
	Remote Operations Service Element (protocol)
ROW	remote orderwire
RPC	remote procedure call
RPDU	response protocol data unit
rpm	revolutions per minute
RPM	rate per minute

RPOA	recognized private operating agency (common carrier)
rps	revolutions per second
RQ	repeat request
	request repeat
RR	radio range
	receiver ready (link protocol command)
	repetition rate
RRC	radar resolution cell
RRS	radio range station
	radio relay system
RSA	Rivest, Shamir, Adleman
RSL	received signal level
rss	root-sum-square
R/S/T	ISDN reference points
RSW	relative spectral width
RT	radiotelegram
R/T	real time
RTA	remote trunk arrangement
RTI	radiation transfer index
RTL	relative transmission level
RTM	reference test method
RTMOS	room temperature metal oxide semiconductor
RTMP	Routing Table Maintenance Protocol
RTS	reliable transfer service
	request to send
	Request to Send (EIA RS-232)
RTSE	reliable transfer service element
RTTY	radio teletype (system)
	radio teletypewriter
RTU	remote terminal unit
RTX	request to transmit
RVA	reactive volt-ampere (*deprecated, use* VAR)
RWI	radio and wire integration
	radio-wire integration
	radio-wire interface
RWT	recorder warning tone
RX	receive
	receiver
	receiving
RZ	return to zero (code)

RZI	redundant zero insertion
s	second
SAA	systems applications architecture (IBM)
SABER	situation awareness beacon with reply
SABM	set asynchronous balanced node
SABME	set asynchronous balanced node extended
SAFDT	store-and-forward data transmission
SAM	self-assembled monolayer (coplanar)
	serving area multiplex
SAP	service access point (OSI)
SAPI	service access point identifier
SARAH	search and rescue and homing
SARBE	search and rescue beacon
SARM	set asynchronous response mode
SAS	survivable adaptive system
SASE	specific application service element (OSI)
SAT	Société Anonyme Télécommunications (Telecommunications, Ltd.)
SATCOM	satellite communications
SB-ADPCM	subband adaptive differential pulse code modulation
SC	subcommittee (ISO)
SCA	signal communications axis
SCC	signal communications center
	specialized commercial carrier
	specialized common carrier
SCCP	signaling connection control part
SCE	service creation environment
SCF	service control facility
SCOI	standing communications operating instruction
SCP	service control point
SCPC	single channel per carrier
SCR	semiconductor controlled rectifier
	Signal Corps radio (U.S. Army)
	Signal Corps Radio (number)
	signal to crosstalk ratio
	silicon-controlled rectifier
SCS	scattering cross-section
SCSR	single-channel signaling rate
SD	space division
SDCD	secondary data carrier detect

SDE	submission and delivery entity (X.400)
SDL	specification and description language (CCITT)
SDLC	synchronous data link control
SDM	space-division multiplex
	space-division multiplexer
	space-division multiplexing
SDN	software-defined network
SDU	service data unit
	Service Data Unit (OSI)
SE	software engineering
	systems engineer
	systems engineering
sec	second
SECAM	electronic color system with memory
	Séquential à Mémoire (Color TV Standard, France)
	système electronique couleur avec mémoire
SECORD	secure voice cord board
SECTEL	secure telephone
SEDFB	Surface Emitting Distributed Feedback (laser)
SELED	surface-emitting light-emitting diode
SEM	scanning electron microscope
SER	satellite equipment room (DEC)
SETAMS	systems engineering, technical assistance, and management services
SEVAS	Secure Voice Access System
SF	secondary frequency
	single frequency (signaling)
	store and forward
S-F	store and forward
SFT	system fault tolerance
SFTS	standard frequency and time signal
SG	study group
SGDF	supergroup distribution frame
S/H	sample and hold
S&H	sample and hold
SHA	sidereal hour angle
SHF	superhigh frequency
SHIELD	silicon hybrids with infrared extrinsic long-wavelength detector
SHORAN	short-range aid to navigation
SI	International System of Units

	Système International d'Unités
SID	sudden ionospheric disturbance
SIGCOMM	special interest group on data communications (ACM)
SIGINT	signals intelligence
SILO	signal intercept from low orbit
SINAD	signal-plus-noise-plus-distortion to noise-plus-distortion ratio
SINCGARS	Single-Channel Ground and Airborne Radio System
SIP	step-index profile
SITRRO	simplification of international trade procedures
SL	Session Layer
SLA	service-level agreement
SLC	subscriber loop carrier
	system life cycle
SLD	superluminescent diode
	superluminescent LED
SLED	superluminescent light-emitting diode
SLI	service logic interpreter
SLIC	subscriber line interface card
SLL	superluminescent LED
SLP	service logic program
SLR	service-level reporter
SM	single-mode (optical fiber)
SMA	type connector (optical or rf)
SMART NET	skywave management for automatic robust transmission network
SMART-T	secure mobile antijam reliable tactical terminal
SMB	server message block
SMDR	station message-detail recording
SMDS	switched multimegabit data service
SMS	ship movement service
SMSA	standard metropolitan statistical area
SMT	station management (FDDI)
SMTP	Simple Mail Transfer Protocol (ARPA/DOD)
S/N	signal-to-noise ratio
SN	subscriber number
SNA	systems network architecture (IBM)
SNADS	SNA distribution service (IBM)
SNAP	Subnet Access Protocol
S+N+D	signal-plus-noise-plus-distortion
SNDCP	Subnet Dependent Convergence Protocol (ISO)
SNI	SNA network interconnect

SNICP	Subnet Independent Convergence Protocol (ISO)
SNMP	Simple Network Management Protocol
(S+N)/N	signal plus noise to noise ratio
SNR	signal to noise ratio
SNR/bit	signal to noise ratio per bit
(S/N)+1	signal plus noise to noise ratio
SOH	start of header
	start of heading character
SOI	signal operating instruction
	signal operation instruction
	signal operations instruction
SOM	start of message
sonar	sound navigation and ranging
SONET	synchronous optical network
SOP	standard operating procedure
SOR	start of record
SOS	distress signal
SOSIC	silicon on sapphire integrated circuit
SOSTEL	solid state electronic logic
SOW	statement of work
S/P	serial-to-parallel
SPAG	Standards Promotion and Applications Group (Europe)
SPDU	session protocol data unit
SPG	signal processing gain
SPP	Sequence Packet Protocol
SQL	structured query language
sr	steradian
SR	secondary radar
S/R	send and receive
SR	spectral responsivity
SRD	superradiant diode (*See* SLED)
SREJ	selective reject
SRTS	secondary request to send
SS	selection signal
	single sideband
	sounder station
	space service
	spread spectrum
S/S	start/stop (character, asynchronous)
SSAP	session service access point (ISO)

SSB	single sideband
	single sideband (transmission)
SSB-SC	single-sideband suppressed carrier
	SSB-suppressed carrier (transmission)
SSCP	system services control point (IBM)
SSCSG	spread spectrum code sequence generator
SSDR	start-stop distortion ratio
SSG	spare signal group
SSMA	spread spectrum multiple access
SSN	station serial number
SSP	service switching point
SSRT	subsecond response time
SSUPS	solid-state uninterruptible power system
SS7	signaling system 7
STALO	stabilized local oscillator
STANAG	Standardization Agreement (NATO)
STARS	stored terrain access and retrieval system
STAT MUX	statistical multiplexer
STD	standard
	subscriber trunk dialing
	Subscriber Trunk Dialing (Europe)
STE	Signaling Terminal Equipment (CCITT)
STELLA	satellite interconnection of LANs (Europe)
STFS	standard time and frequency signal
	Standard Time and Frequency Signal (service)
STL	standard telegraph level
	studio-to-transmitter link
STM	synchronous transfer mode
STP	signal transfer point
	standard temperature and pressure
STRADIS	Structured Analysis, Design, and Implementation of Information System
STS	synchronous transfer signal
STU	secure telephone unit
STX	start of text
	start of text character
SU	signal unit
SUB	substitute
	substitute character
SVC	switched virtual call

SW	short-wave
	spectral width
	standing wave
SWAN	satellite wide-area network
SWR	standing wave ratio
SX	simplex signaling
SXS	step-by-step
	step-by-step switch
	step-by-step switching
	step-by-step switching system
SYN	synchronization
	synchronize
	synchronizer
	synchronizing
	synchronous
	synchronous idle character
SYSGEN	system generation
S(λ)	chromatic dispersion slope
t	symbol for time
T	tera (10^{12})
T_K	response time
	response timer
TA	Technical Advisory (Bellcore)
	terminal adapter (ISDN)
TAC	terminal access controller
TACSIM	tactical simulation
TADIL	tactical data information link
TADIL–A	tactical data information link—A
TADIL–B	tactical data information link—B
TADIXS–B	Tactical Data Information Exchange System—Broadcast
TADS	teletypewriter automatic dispatch system
TADSS	tactical automatic digital switching system
	Tactical Automatic Digital Switching System
TAI	International Atomic Time (Temps Atomique International)
TARE	telegraph automatic relay equipment
TASA	Telefónica de Argentina
TASI	time-assignment speech interpolation
TAT	trans-Atlantic telecommunication (cable)
Tbps	terabits per second

TC	technical committee (ISO)
	telecommunications center
	thermocouple
	toll center
	transmission coefficient
	transmit clock
	transport connection
TCAM	telecommunications access method (IBM)
TCAP	transaction capabilities application part
TCB	trusted computing base
TCC	telecommunications center
TCC	telephone country code (CCITT)
TCCF	Tactical Communications Control Facility
TCF	technical control facility
TCM	time-compression multiplexing
	trellis coded modulation
TCP	transmission control protocol
	Transmission Control Protocol (U.S. DoD)
TCP/IP	Transmission Control Protocol/Internet Protocol
TCS	tactical communications system
	trusted computer system
TCTS	trans-Canada telephone system
TCU	teletypewriter control unit
TCVXO	temperature-compensated voltage-controlled crystal oscillator
TCXO	temperature-controlled crystal oscillator
TC97	ISO Technical Committee for Information Systems
TC 97/SC 1	Subcommittee for Vocabulary, TC 97, ISO
TD	time delay
	time division
	transmitted data
	transmitter distributor
TDC	theater deployable communications
TDCS	theater deployable communications system
TDD	telecommunications devices for the deaf
TDI	trade data interchange
TDM	time-division multiplexed
	time-division multiplexer
	time-division multiplexing
TDMA	time-division multiple access
TDR	time-domain reflectometer

	time-domain reflectometry
TDT	total delay time
TDTG	true date-time group
TE	terminal endpoint
	terminal equipment
	time error
	transverse electric
TE mode	transverse electric mode
TECHEVAL	technical evaluation
TED	trunk encryption device
TEF	Teflon®
TEFG	terminal endpoint functional group
TEI	terminal endpoint identifier
TEK	traffic encryption key
telco	telephone company
TELENET	a public packet-switched network (U.S.)
Telex®	The International Teleprinter Network
TELEX	International Teleprinter Network
TELNET	Teletypewriter Network Protocol (DARPA)
TEM	transverse electromagnetic
TEM mode	transverse electromagnetic mode
TEM wave	transverse electromagnetic wave
TEMPEST	compromising emanations
ter	third version of a recommendation (CCITT)
TERM equipment	telegraph error rate measuring equipment
TETRA	trans-European trunked radio system
TEW	trapped electromagnetic wave
TE1	Terminal Equipment Type 1 (ISDN terminal)
TE2	Terminal Equipment Type 2 (non-ISDN terminal)
TFL	thin-film laser
TFOM	thin-film optical modulator
	thin-film optical multiplexer
TFOS	thin-film optical switch
TFOW	thin-film optical waveguide
TFTP	Trivial File Transfer Protocol
TFTS	terrestrial flight telephone system
TGM	trunk group multiplexer
THAAD	theater high altitude area defense
THD	total harmonic distortion
THF	tremendously high frequency

THz	terahertz (10^{12} H)
TIA	Telecommunication Industries Association
TIBS	Tactical Information Broadcast Service
TIE	time interval error
TIFF	tag image file format
TIMS	transmission impairment measuring set
TIP	terminal interface processor
TIS	traveler information station
TL	Transport Layer
TLM	telemetry
TLP	transmission level point
TLS	telemetering land station
TLV	type, length, value (OSI)
TM	transverse magnetic
TM mode	transverse magnetic mode
TM wave	transverse magnetic wave
TMN	telecommunications management network
TO	tandem office
	timeout (in protocols)
	toll office
TOD	time of day
TOM	thin-film optical modulator
	thin-film optical multiplexer
TOP	Technical and Office Protocol
TOS	thin-film optical switch
	type of service (DARPA)
TOW	thin-film optical waveguide
TP	toll point
	Transport Protocol (ISO NIST)
TPDU	transport protocol data unit
TPOA	telecommunications private operating agency
TQ	toll quality
TR	Technical Reference
	transfer request
TRADACOMS	trading data communications
TRANSEC	transmission security
TRANSPAC	public packet-switched network (France)
TRAP	Tactical Receive Equipment and Related Applications
TRC	transverse redundancy check
TREE	transient radiation effects on electronics

TRF	tuned radio frequency
TRI-TAC	tri-services tactical
TRITAC-E	tri-services tactical equipment
TRIXS	Tactical Reconnaissance Intelligence Service
TRMS	transmission resource management system
TRN	token ring network
TROPO	tropospheric scatter
TSA	telecommunications system architecture
TSAP	transport service access point
TSDU	transport service data unit
TSI	time slot interchanger
TSK	transmission security key
TSM	telecommunications system management
TSO	telecommunications system operator
TSP	telecommunications service priority
TSPS	traffic service position system
TSR	telecommunications service request
	terminate and stay resident
TTL	transistor-transistor logic
TTTN	tandem tie trunk network
TTX	teletypewriter exchange (service)
	telex
TTXAU	teletex access unit
TTY/TDD	telecommunications device for the deaf
TTY	teletypewriter
TV	technical vulnerability
	television
TW	traveling wave
TWA	two-way alternate
TWC	two-wire circuit
TWS	two-way simultaneous
TWT	traveling wave tube
TWTA	traveling wave tube amplifier
TWX®	Teletypewriter Exchange Service
TX	transit exchange
	transmit
TYMNET	a public packet-switched network (U.S.)
T1	1.544-Mbps transmission system
UA	unnumbered acknowledgment
	user agent (in message systems)

UAE	user agent entity (in message systems)
UAL	user agent layer
UART	universal asynchronous receiver transmitter
UDI	unrestricted digital information
UDP	user datagram protocol
	User Datagram Protocol (DARPA)
UER	undetected error ratio
UHF	ultrahigh frequency
UI	unnumbered information
UL	Underwriters Laboratories
U/L	uplink
ULF	ultralow frequency
ULL	ultralow-loss (fiber)
um	micron, micrometer (10^{-6} m, when μ is not available)
UMTS	Universal Mobile Telecommunications System.
UNI	user-network interface
UNIVERSE	U.K. project connecting LANs via satellites
UNIX™	computer operating system
UNTDI	United Nations trade data interchange
UPS	uninterruptible power supply
UPT	Universal Personal Telecommunications
US	microsecond (10^{-6}, when μ is not available)
USART	universal sync, async, receiver, transmitter
USAT	ultrasmall aperture terminal (antenna)
USB	upper sideband
USDA	U.S. Department of Agriculture
USDJ	U.S. Department of Justice
USFJ	U.S. Forces, Japan
USFK	U.S. Forces, Korea
USNO	U.S. Naval Observatory
USOC	universal service order code
USS	unilateral synchronization system
USTA	United States Telephone Association
UT	Universal Time
UTC	Coordinated Universal Time
UTC(*i*)	Coordinated Universal Time (*i*)
UTP	unshielded twisted pair
uv	ultraviolet
UV	ultraviolet

V	normalized frequency
	number (normalized frequency)
	reference point (Integrated Services Digital Network)
	volt
VA	value-added (service)
	value-added (network service)
	volt-ampere
	volt•ampere
	volt•amperes
VAD	vapor phase axial deposition
VAD process	vapor phase axial deposition process
VADS	value-added and data services
VAN	value-added network
VAR	value-added reseller
	volt•ampere, reactive
	volt-amperes, reactive
VARISTAR	variable resistor
VARS	volt•amperes, reactive
VAT	value-added tax
VC	virtual call (X.25)
	virtual circuit
VCI	virtual channel identifier
VCO	voltage-controlled oscillator
VCR	video cassette recorder
VCS	visual communications system
VCXO	voltage-controlled crystal oscillator
V/D	voice/data
Vdc	volts, direct-current
VDI	virtual device interface
VDT	video display terminal (CRT)
VDU	video display unit
	visual display unit
VES	virtual environment software
VF	video frequency
	voice frequency
VFCT	voice frequency carrier telegraph
VFCTG	voice frequency carrier telegraph
VFDF	voice frequency distribution frame
VFO	variable-frequency oscillator
VFS	virtual file store

	virtual file system
VFT	voice frequency telegraph
VFTG	voice frequency telegraph
VGC	voice-grade circuit
VHF	very high frequency
VLF	very low frequency
VLSI	very large-scale integration (of semiconductors)
VLSIC	very large-scale integrated circuit
V/m	volts per meter (electric field gradient)
VN	virtual network
VNL	via net loss
VNLF	via net loss factor
V number	normalized frequency
vocoder	voice-coder
vodas	voice-operated device antising
vogad	voice-operated gain-adjusting device
volcas	voice-operated loss-control and echo-signaling device
VOM	volt-ohm meter
vox	voice-operated relay circuit
	voice-operated relay device
V parameter	normalized frequency.
VPI	virtual path identifier
VRC	vertical redundancy check
VSAT	very small aperture terminal
	Very Small Aperture Terminal (satellite)
VSAT antenna	very small aperture terminal antenna
VSB	vestigial sideband
	vestigial sideband (transmission)
VSM	vestigial sideband modulation
VSWR	voltage standing wave ratio
VT	variable time
	vertical tabulation
	video teleconference
	virtual terminal
VTAM	Virtual Telecommunications Access Method (IBM)
VT character	vertical tabulation character
VTP	Virtual Terminal Protocol
VTS	virtual terminal service
VTU	video teleconferencing unit
VTX	videotex

vu	volume unit (abbreviation)
VU	volume unit (unit of measure)
V value	normalized frequency
W	watt
WADS	wide area data service
WAIS	Wide Area Information Servers
WAN	wide area network
WARC	World Administrative Radio Conference
WATS	wide area telecommunications service
	wide area telephone service
WAWS	Washington Area Wideband System
WB	wideband
WBCS	wideband communications service
	wideband communications system
WCC	wireline common carrier
WD	wavelength division
	working draft
WDM	wavelength-division multiplexed
	wavelength-division multiplexer
	wavelength-division multiplexing
WG	working group (ISO SC subgroup)
WIN	WWMCCS Intercomputer Network
WITS	Washington Integrated Telecommunications System
WM	wavelength modulation
WMO	World Meteorological Organization
WORM	write once, read many times
WP	wirephoto
wpm	words per minute
wps	words per second
WRU	who are you character
	who-are-you (control character)
WSWS	weighted standard work second
WT	World Time
WU	Western Union
wv	working voltage
WVDC	working voltage, direct current
WWDSA	worldwide digital system architecture
WWMCCS	World Wide Military Command and Control System
WWW	world weather watch

	World Wide Web
X	reactance (electric circuit)
XDR	exchange data representative
XID	exchange identification
XMIT	transmit
XMITTER	transmitter
XMODEM	file transfer protocol, micro to mainframe
XMSN	transmission
XMTD	transmitted
XMTR	transmitter
XMUX	transmultiplexer
XNS	Xerox Network Systems (Xerox protocol)
XO	crystal oscillator
XOFF	control character to stop information flow
	transmitter off (transmission-control character)
X-OFF	transmitter off
XON	control character to start information flow
	transmitter on (transmission-control character)
X-ON	transmitter on
XT	crosstalk
XTAL	crystal
XU	X unit
YAG	yttrium aluminum garnet
Z	impedance (electric circuit)
	Zulu Time
Z_0	characteristic impedance
ZBTSI	zero byte time slot interchange
ZD	zero defects
ZDW	zero-dispersion wavelength
ZIP	Zone Information Protocol
ZRL	zero relative level
Z Time	Zulu Time
0TLP	zero transmission level point
2B1Q	two binary, one quaternary
4B/5B	four bits encoded into five bits (FDDI)
4B3T	four bits in three ternary signals

5ESS	Number 5 Electronic Switching System (AT&T)
Δ	delta modulation
	refractive index contrast
Δ-β switch	delta-beta switch
Δ-Σ modulation	delta-sigma modulation
ε	electrical permittivity (epsilon)
λ	wavelength (lambda)
λ_0	zero-dispersion wavelength
λ_{cf}	fiber cutoff wavelength
λ_{co}	cutoff wavelength
μ	magnetic permeability (mu)
	micro (10^{-6}, prefix, as in µA, µF, and µm)
	micrometer (10^{-6} m)
µA	microampere (10^{-6} A)
µF	microfarad (10^{-6} F)
µ-law	mu-law
µm	micron (10^{-6} m)
	micrometer (10^{-6} m)
µPa	micropascal (10^{-6} Pa, sound pressure unit)
µs	microsecond (10^{-6} s)
µV	microvolt (10^{-6} V)
µW	microwatt (10^{-6} W)
ρ	electric charge density (rho)
σ	electrical conductivity (sigma)
Ω	ohm (omega)

APPENDIX B

Tables

Table 1: **Frequency Ranges and Designators**

Frequency Ranges	United States Designators	ITU Designators*
30–300 Hz	ELF (Extremely Low Frequency)	2**
300–3000 Hz	ULF (Ultra Low Frequency)	3**
3–30 kHz	VLF (Very Low Frequency)	4
30–300 kHz	LF (Low Frequency)	5
300–3000 khz	MF (Medium Frequency)	6
3–30 Mhz	HF (High Frequency)	7
30–300 Mhz	VHF (Very High Frequency)	8
300–3000 Mhz	UHF (Ultra High Frequency)	9
3–30 Ghz	SHF (Super High Frequency)	10
30–300 Ghz	EHF (Extremely High Frequency)	11
300–3000 Ghz	THF*** (Tremendously High Frequency)	12

*The eight obsolete microwave frequency bands, designated by the letters C, L, S, X, K, Q, V, and W, that represent the frequency ranges from 225 MHz to 100 GHz and were used to describe radar bands, have no official status and are deprecated.

**An extrapolation that is not designated by the International Telegraph Union (ITU) as part of the radio spectrum.

***The term "THF" is not in the *Radio Regulations* of the International Telecommunication Union (ITU).

Table 2: Higher Frequency Ranges and Extension Designators

Frequency Ranges	ITU Designators*
3–30 Thz	13
30–300 Thz	14
300–3000 Thz	15
3–30 Phz	16
30–300 Phz	17
300–3000 Phz	18
3–30 Ehz	19
30–300 Ehz	20
300–3000 Ehz	21

*An extension of the ITU designators

THz = Terahertz (10^{12} Hz)

PHz = Petahertz (10^{15} Hz)

EHz = Exahertz (10^{18} Hz)

Table 3: The Metric System of Units

Quantity	Base Units	Abbreviation
length	meter	m
mass	kilogram	kg
time	second	s
electric current	ampere	A
temperature	kelvin	K
amount of substance	mole	mol
luminous intensity	candela	cd
Supplementary Units		
plane angle	radian	rad
solid angle	steradian	sr

Certain derived terms have been standardized, such as the "hertz" (2π radians per second), formerly one cycle per second.

Table 4: **Prefixes Used with Metric Units**

Unit	Abbrev.	Value
exa	E	10^{18}
peta	P	10^{15}
tera	T	10^{12}
giga	G	10^{9}
mega	M	10^{6}
kilo	k	10^{3}
hecto	h	10^{2}
deka	da	10^{1}
—	—	10^{0}
deci	d	10^{-1}
centi	c	10^{-2}
milli	m	10^{-3}
micro	μ	10^{-6}
nano	n	10^{-9}
pico	p	10^{-12}
femto	f	10^{-15}
atto	a	10^{-18}

Examples

Term	Abbreviation	Meaning
megahertz	Mhz	10^{6} hertz
picofarad	pF	10^{-12} farads
nanosecond	ns	10^{-9} seconds

Note: Additional examples of metric units and units used in the meter-kilogram-second system of units are (a) electric charge, coulombs, C; (b) energy, joules (watt-seconds), J; (c) force, newtons, N; (d) power, watts, W; (e) voltage, volts, V; (f) capacity, farads, F; (g) resistance, ohms, Ω; (h) inductance, henries, h; and (i) magnetic field strength, oersteds, H.

Table 5: Radiometric Terms

Term	Symbol	Quantity	Unit
radiant energy	Q	energy	joule (J)
radiant power ***Synonym* optical power**	ϕ	power	watt (W)
irradiance	E	power incident per unit area irrespective of angle	$W \bullet m^{-2}$
spectral irradiance	E_λ	irradiance per unit wavelength interval at a given wavelength	$W \bullet m^{-2} \bullet nm^{-1}$
radiant emittance ***Synonym* radiant exitance**	W	power emitted (into a full sphere) per unit area	$W \bullet m^{-2}$
radiant intensity	I	power per unit solid angle	$W \bullet sr^{-1}$
radiance	L	power per unit solid angle per unit projected area	$W \bullet sr^{-1} \bullet m^{-2}$
spectral radiance	L_λ	radiance per unit wavelength interval (at a given central wavelength)	$W \bullet sr^{-1} \bullet m^{-2} \bullet nm^{-1}$

Table 6: T-Carrier Hierarchy for North America

Designator† (DS Level)	DSR Mbps* No. of Channels
T0 (DS 0)	0.064
T1 (DS 1)	1.544 24
T1C	3.152 48
T2 (DS 2)	6.312 96
T3 (DS 3)	44.736 672
T4 (DS 4)	274.176 4032
T5 (DS 5)	400.352 5760

*Data signaling rate, megabits/second.

†See Table 10.

Table 7: **T-Carrier Hierarchy for Japan**

Level†	Rate mbps* No. of Channels
0	0.064
1	1.544 24
2	6.312 96
3	32.064 480
4	97.728 1440
5	565.148 7680

*Data signaling rate, megabits/second.

†See Table 10.

Table 8: **T-Carrier Hierarchy for Europe (CEPT)**

Level†	DSR Mbps* No. of Channels
1	2.048 30
2	8.448 120
3	34.368 480
4	139.268 1920

*Data signaling rate, megabits/second.

†See Table 10.

Table 9: Near, Intermediate, and Far Electromagnetic Fields

Near	Intermediate	Far
Near field	Intermediate field	Far field
Near field region	Intermediate field region	Far field region
Electrostatic field	Induction field	Radiation field
Near zone	Intermediate zone	Far zone
Varies $1/r^3$	Varies $1/r^2$	Varies $1/r$
Fresnel region	—	Fraunhofer region
Fresnel zone	—	Fraunhofer zone
*$E_\theta \propto (\sin \omega t)/\omega r^3$	$E_\theta \propto (\cos \omega t)/r^2 v$	$E_\theta \propto -\omega (\sin \omega t)/r v^2$
*$E_r \propto (\sin \omega t)/\omega r^3$	$E_r \propto (\cos \omega t)/r^2 v$	—
*$r \langle 2A^2/\lambda$	$r \langle \sim\lambda/6$	$\sim\lambda/6 \rangle r \rangle 2A^2/\lambda$
**Adjacent to antenna	Short distances away	Great distances

*v = propagation velocity = $1/(\mu\varepsilon)^{1/2}$
μ = magnetic permeability
ϵ = electric permittivity
$\omega = 2\pi f$
f = frequency
r = radial distance from antenna
λ = wavelength
A = antenna aperture

**Distances compared to wavelength and antenna size.

Table 10: T-Carrier Characteristics

A T-carrier is the generic designator for any of several digitally multiplexed telecommunications carrier systems.

1. The designators for T-carrier in the North American digital hierarchy correspond to the designators for the digital signal (DS) level hierarchy.
2. T-carrier systems were originally designed to transmit digitized voice signals. Current applications also include digital data transmission.
3. Tables 6, 7, and 8 list the designators and rates for current T-carrier systems.
4. If an "F" precedes the "T," a fiber optic cable system is indicated at the same rates.
5. T-carrier hierarchies for North America and Japan are based on multiplexing 24 voice-frequency (VF) channels and multiples thereof. The hierarchy for Europe is based on multiplexing 30 voice-frequency (VF) channels and multiples thereof.
6. *See also* **carrier system, digital signal, digital transmission system, digroup.**

APPENDIX C
Bibliography

Abebe, M., C. A. Villarruel, and W. K. Burns. Reproducible fabrication method for polarization preserving single-mode fiber couplers. *Journal of Lightwave Technology* **6** (7): 1191–1198 (July 1988).

Akinniyi, A. R. and J. S. Lehnert. Characterization of noncoherent spread-spectrum multiple access communications. *IEEE Transactions on Communications* **42** (1): 139–148 (January 1994).

Alexander, S. B. Design of wide-band optical heterodyne balanced mixer receivers. *Journal of Lightwave Technology* **LT-5** (4): 523–537 (April 1987).

American National Standards Institute, *American National Dictionary for Information Systems,* Washington, DC: Computer and Business Equipment Manufacturers Association, 1995.

Anagnostou, M. E. and E. N. Protonotarios. Analysis of a buffered TDM system with a general arrival process. *IEEE Transactions on Communications* **42** (2/3/4): 1752–1757 (February/March/April 1994).

Andonovic, I., L. Tančevski, M. Shabeer, and L. Bazgaloski. Incoherent all-optical code recognition with balanced detection. *Journal of Lightwave Technology* **12** (6): 1073–1080 (June 1994).

Atternas, L. and L. Thylen. Single-layer antireflection coating of semiconductor lasers: Polarization properties and the influence of the laser structure. *Journal of Lightwave Technology* **7** (2): 426–430 (February 1989).

Bauch, H. Transmission systems for the BISDN. *IEEE Magazine, Lightwave Telecommunications* **2** (3): 31–36 (August 1991).

Benedetto, S. and P. T. Poggiolini. Multilevel polarization shift keying: Optimum receiver structure and performance eval. *IEEE Transactions on Communications* **42** (2/3/4): 1174–1186 (February/March/April 1994).

Bertolotti, M., P. Mascuilli, and C. Sibilia. Mol numerical analysis of nonlinear planar waveguide. *Journal of Lightwave Technology* **12** (5): 899–908 (May 1994).

Besse, P. A., M. Bachmann, H. Melchior, et al. Optical bandwidth and fabrication tolerances of multimode interference couplers. *Journal of Lightwave Technology* **12** (6): 1004–1009 (June 1994).

Bialkowski, M. E. and S. T. Jellett. Analysis and design of a circular disk 3-dB coupler. *IEEE Transactions on Microwave Theory and Technology* **42** (8): 1437–1442 (August 1994).

Bisdikian, C. and A. N. Tantawy. A mechanism for implementing preemptive priorities in DQDB subnetworks. *IEEE Transactions on Communications* **42** (2/3/4): 834–839 (February/March/April 1994).

Bordon, E. E. and W. L. Anderson. Dispersion-adapted monomode fiber for propagation of nonlinear pulses. *Journal of Lightwave Technology* **7** (2): 353–357 (February 1989).

Braagaard, C., B. Mikkelsen, T. Durhuus, and K. E. Stubkjar. Modeling the DBR laser used as wavelength conversion device. *Journal of Lightwave Technology* **12** (6): 943–951 (June 1994).

Brandt-Pearce, M. and B. Aazhang. Multiuser detection for optical code division multiple access systems. *IEEE Transactions on Communications* **42** (2/3/4): 1801–1810 (February/March/April 1994).

Brosson, P. Analytical model of a semiconductor optical amplifier. *Journal of Lightwave Technology* **12** (1): 49–54 (January 1994).

Bruno, W. M. and W. B. Bridges. Powder core dielectric channel waveguide. *IEEE Transactions on Microwave Theory and Technology* **42** (8): 1524–1532 (August 1994).

Buchholtz, F., K. P. Koo, and A. Dandridge. Effect of external perturbations on fiber-optic magnetic sensors. *Journal of Lightwave Technology* **6** (4): 507–512 (April 1988).

Burton, R. S., T. E. Schlesinger, and Michael Munowitz. An investigation of the modal coupling of simple branching semiconductor ring laser. *Journal of Lightwave Technology* **12** (5): 754–759 (May 1994).

Calero, J., S.-P. Wu, C. Pope, et al. Theory and experiments on birefringent optical fibers embedded in concrete structures. *Journal of Lightwave Technology* **12** (6): 1081–1091 (June 1994).

Capmany, J. and J. Cascón. Discrete time fiber optic signal processors using optical amplifiers. *Journal of Lightwave Technology* **12** (1): 106–117 (January 1994).

Chao, S.-C., W.-H. Tsai, and M.-S. Wu. Extended Gaussian approximation for single mode graded-index fibers. *Journal of Lightwave Technology* **12** (3): 392–395 (March 1994).

Chen, K. L. and D. Kerps. Coupling efficiency of surface-emitting LEDs to single-mode fibers. *Journal of Lightwave Technology* **LT-5** (11): 1600–1604 (November 1987)

Chen, W.-T., P. R Sheu, and J.-H. Yu. Time slot assignment in TDM multicast switching systems. *IEEE Transactions on Communications* **42** (1): 149–165 (January 1994).

Cheng, Y. H., T. Okoshi, and O. Ishida. Performance analysis and experiment of a homodyne receiver insensitive to both polarization and phase fluctuations. *Journal of Lightwave Technology* **7** (2): 368–374 (February 1989).

Chiang, K. S. Stress-induced birefringence fibers designed for single-polarization single-mode operation. *Journal of Lightwave Technology* **7** (2): 436–441 (February 1989)

Chu, P. L. and T. Whitebread. An on-line fiber drawing tension and diameter measurement device. *Journal of Lightwave Technology* **7** (2): 255–262 (February 1989).

Cohen, R. and A. Segall. Multiple logical token-rings in a single high-speed ring. *IEEE Transactions on Communications* **42** (2/3/4): 1712–1721 (February/March/April 1994).

———. An efficient priority mechanism for token-ring networks. *IEEE Transactions on Communications* **42** (2/3/4): 1769–1777 (February/March/April 1994).

Corvaja, R. and L. Tomba. Crosstalk interference in PSK coherent optical systems. *Journal of Lightwave Technology* **12** (4): 670–678 (April 1994).

Culshaw, B. *Optical Fibre Sensing and Signal Processing*. London: Peter Peregrinus Ltd., 1984.

Dallal, Y. E. and S. Shamai. Coding for DPSK noisy phase channels. *IEEE Transactions on Communications* **42** (2/3/4): 927–940 (February/March/April 1994).

Darcie, T. E. Differential frequency-division multiplexing for lightwave networks. *Journal of Lightwave Technology* **7** (2): 314–322 (February 1989).

Del Bimbo, A. and E. Vicario. Transport measurements over an ethernet LAN. *IEEE Transactions on Communications* **42** (2/3/4): 1486–1489 (February/March/April 1994).

Djuknic, G. M. and D. L Schilling. Performance analysis of an ARQ transmission scheme for meteor burst communications. *IEEE Transactions on Communications* **42** (2/3/4): 268–271 (February/March/April 1994).

Dong, L., T. A. Birks, M. H. Ober, and P. St. J. Russell. Intermodal coupling by periodic microbending in dual-core fibers—Comparison of experiment and theory. *Journal of Lightwave Technology* **12** (1): 24–27 (January 1994).

D'Orazio, A., M. DeSario, V. Petruzzelli, and F. Prudenzano. Leaky mode propagation in planar multilayer birefringent waveguides: Longitudinal dielectric tensor configuration. *Journal of Lightwave Technology* **12** (3): 453–462 (March 1994).

Drögenmüller, K. A compact optical isolator with a plano-convex YIG lens for laser-to-fiber coupling. *Journal of Lightwave Technology* **7** (2): 340–346 (February 1989).

Dziong, Z. and L. G. Mason. Call admission and routing in multi-service loss networks. *IEEE Transactions on Communications* **42** (2/3/4): 2011–2022 (February/March/April 1994).

Eguchi, M. and M. Koshiba. Accurate finite element analysis of dual-mode highly elliptical-core fibers. *Journal of Lightwave Technology* **12** (4): 607–613 (April 1994).

Ehrhardt, A., M. Eiselt, G. Grosskopf, et al. Semiconductor laser amplifier as optical switching gate. *Journal of Lightwave Technology* **11** (8): 1287–1295 (August 1993).

Engle, A. G., Jr., N. I. Dib, and L. P. B. Katehi. Characterization of a shielded transition to a dielectric waveguide. *IEEE Transactions on Microwave Theory and Technology* **42** (5): 847–854 (May 1994).

Fang, J. Absorbing boundary conditions applied to model wave propagation in microwave integrated circuits. *IEEE Transactions on Microwave Theory and Technology* **42** (8): 1506–1513 (August 1994).

Fawer, V. A coherent spread-spectrum diversity receiver with AFC for multipath fading channels. *IEEE Transactions on Communications* **42** (2/3/4): 1300–1311 (February/March/April 1994).

Feldhauer, T., A. Klein, and P. W. Baier. A low cost method for CDMA and other applications to separate non-orthogonal signals. *IEEE Transactions on Communications* **42** (2/3/4): 881–883 (February/March/April 1994).

Friedrich, N. T. Fiber optic trends: Linking LANs with fiber: Cost and other factors. *Photonics Spectra:* 89–92 (September 1988).

Frigo, N. J. A model of intermodal distortion in non-linear multicarrier systems. *IEEE Transactions on Communications* **42** (2/3/4): 1216–1222 (February/March/April 1994).

Ganz, A. and Y. Gao. Time-wavelength assignment algorithms for high performance WDM star based systems. *IEEE Transactions on Communications* **42** (2/3/4): 1827–1836 (February/March/April 1994).

Ganz, A, Y. Gong, and B. Li. Performance study of a low earth-orbit satellite system. *IEEE Transactions on Communications* **42** (2/3/4): 1866–1871 (February/March/April 1994).

Giulianelli, L. C. and A. B. Buckman. Fiber optic circuits for direct phase conversion with two outputs in quadrature. *Journal of Lightwave Technology* **11** (7): 1263–1265 (July 1993).

Glance, B., I. P. Kaminov, and R. W. Wilson. Applications of the integrated waveguide router. *Journal of Lightwave Technology* **12** (6): 957–762 (June 1994).

Golovchenko, E. A. and A. N. Pilipetski. Acoustic effect and the polarization of adjacent bits in soliton communication lines. *Journal of Lightwave Technology* **12** (6): 1052–1056 (June 1994).

Grandhi, S. A., R. Vijayan, and D. J. Goodman. Distributed power control in cellular radio systems. *IEEE Transactions on Communications* **42** (2/3/4): 226–228 (February/March/April 1994).

Haas, Z. and M. A. Santoro. A multimode filtering scheme for improvement of the bandwidth-distance product in multimode fiber systems. *Journal of Lightwave Technology* **11** (7): 1125–1131 (July 1993).

Hayashi, Y., N. Ohkawa and D. Yanai. Optical non-repeatered transmission system employing high output power EDFA boosters. *Journal of Lightwave Technology* **11** (8): 1369–1376 (August 1993).

Hecht, J. *Understanding Fiber Optics.* Indianapolis, IN: Howard W. Sams & Company, 1987.

Horiguchi, M., K. Yoshino, M. Shimizu, et al. Erbium-doped optical fiber amplifiers pumped in the 660- and 820-nm bands. *Journal of Lightwave Technology* **12** (5): 810–820 (May 1994).

How, H., J. B. Thaxter, and C. Vittoria. Generation of Gaussian-like electromagnetic pulses. *IEEE Transactions on Microwave Theory and Technology* **42** (1): 68–72 (January 1994).

Hsu, K. and C. M. Miller. Theory and measurements of speed-of-light effects in long cavity fiber Fabry-Perot scanning interferometers. *Journal of Lightwave Technology* **11** (7): 1204–1208 (July 1993).

Huang, Y.-F. and S.-L. Lai. Regular solution of shielded planar transmission lines. *IEEE Transactions on Microwave Theory and Technology* **42** (1): 84–91 (January 1994).

Hui, J. Y. and T. Renner. Queueing analysis for multicast packet switching. *IEEE Transactions on Communications* **42** (2/3/4): 723–731 (February/March/April 1994).

Hwifets, S. A. and S. A. Kheifets. Longitudinal electromagnetic fields in an aperiodic structure. *IEEE Transactions on Microwave Theory and Technology* **42** (1): 108–117 (January 1994).

Ilić, I., R. Scarmozzino, R. M. Osgood, et al. Modeling multimode-input star couplers in polymers. *Journal of Lightwave Technology* **12** (6): 996–1003 (June 1994).

Inoue, K. Experimental study on channel crosstalk due to fiber four-wave mixing around the zero-dispersion wavelength. *Journal of Lightwave Technology* **12** (6): 1023–1028 (June 1994).

International Organization for Standardization, *Information Processing Vocabulary. International Standard 2382.* New York: American National Standards Institute, 1995.

Jalali, B., L. Naval, and A. F. J. Levi. Si-based receivers for optical data links. *Journal of Lightwave Technology* **12** (6): 930–935 (June 1994).

Kaasila, V.-P. and A. Mammela. Bit error probability of a matched filter in a Rayleigh fading multipath channel. *IEEE Transactions on Communications* **42** (2/3/4): 826–828 (February/March/April 1994).

Kahrizi, M., T. K. Sarkar, and Z. A. Marićević. Dynamic analysis of a microchip line over a perforated ground plane. *IEEE Transactions on Microwave Theory and Technology* **42** (5): 820–825 (May 1994).

Kang, S. and T. R. Fischer. Trellis excitation speech coding at low bit rates. *IEEE Transactions on Communications* **42** (2/3/4): 1902–1910 (February/March/April 1994).

Kapany, N. S. *Fiber Optics.* New York: Academic Press, 1967.

Kasper, B. L. and J. C. Campbell. Multi-gigabit/s avalanche photodiode lightwave receivers. *IEEE Journal of Lightwave Technology* **LT-5** (10): 1138–1142 (October 1987).

Kazovsky, L. G. Phase- and polarization-diversity coherent optical techniques. *Journal of Lightwave Technology* **7** (2): 279–292 (February 1989).

Kiasaleh, K. An all optical coherent receiver for self-homodyne detection of digitally phase modulated optical signals. *IEEE Transactions on Communications* **42** (2/3/4): 1496–1500 (February/March/April 1994).

Kipp, R. A. and C. H. Chan. Complex image method for sources in bounded regions of multilayer structures. *IEEE Transactions on Microwave Theory and Technology* **42** (5): 860–865 (May 1994).

Kitagawa, T., K. Hattori, Y. Hibino, and Y. Ohmori. Neodymium-doped silica-based planar waveguide lasers. *Journal of Lightwave Technology* **12** (3): 436–442 (March 1994).

Kumar, A., R. L. Gallawa, and I. C. Goyal. Modal characteristics of bent dual mode planar optical waveguides. *Journal of Lightwave Technology* **12** (4): 622–624 (April 1994).

Kumar, A., R. K. Varshney, and R. K. Sinha. Scalar modes and coupling characteristics of eight-port waveguide couplers. *Journal of Lightwave Technology* **7** (2): 293–296 (February 1989).

Kusyk, R. G., W. A. Krzymien, and T. E. Moore. Analysis techniques for the reduction of jitter caused by SONET pointer adjustments. *IEEE Transactions on Communications* **42** (2/3/4): 2036–2039 (February/March/April 1994).

Kuwabara, T., Y. Mitsunaga, and H. Koga. Calculation method of failure probabilities of optical fiber. *Journal of Lightwave Technology* **11** (7): 1132–1138 (July 1993).

Lachs, G., S. M. Zaidi, and A. K. Singh. Sensitivity enhancement using coherent heterodyne detection. *Journal of Lightwave Technology* **12** (6): 1036–1041 (June 1994).

Lau, K. Y. Short-pulse and high-frequency signal generation in semiconductor lasers. *Journal of Lightwave Technology* **7** (2): 400–419 (February 1989)

Lavoie, P., D. Haccoun, and Y. Savaria. A systolic architecture for fast stack sequential decoders. *IEEE Transactions on Communications* **42** (2/3/4): 324–335 (February/March/April 1994).

Lee, J.-S. and S.-Y. Shin. On the validity of the effective index method for rectangular dielectric waveguides. *Journal of Lightwave Technology* **11** (8): 1320–1324 (August 1993).

Lee, W. W. S. and E. K. N. Yung. The input impedance of a coaxial line fed probe in a cylindrical waveguide. *IEEE Transactions on Microwave Theory and Technology* **42** (8): 1468–1473 (August 1994).

Levitt, B. K., U. Cheng, A. Polydoros, and M. K. Simon. Optimum detection of slow frequency-hopped signals. *IEEE Transactions on Communications* **42** (2/3/4): 1990–2000 (February/March/April 1994).

Li, S. and H. Sheng. Discrete queueing analysis of multimedia traffic with diversity of correlation and burstiness properties. *IEEE Transactions on Communications* **42** (2/3/4): 1339–1351 (February/March/April 1994).

Liany, G. C., R. S. Whithers, B. F. Cole, and N. Newman. High-temperature superconductivity devices on sapphire. *IEEE Transactions on Microwave Theory and Technology* **42** (1): 34–40 (January 1994).

Louri, A. and H. Sung. An optical multi-mesh hypercube: A scalable optical interconnection network for massively parallel computing. *Journal of Lightwave Technology* **12** (4): 679–683 (April 1994).

Lui, C.-L. and K. Feher. Pilot symbol aided coherent M-ary PSK in frequency-selective fast Rayleigh fading channels. *IEEE Transactions on Communications* **42** (1): 54–62 (January 1994).

Lupu, V. and L. B. Milstein. Performance analysis of a coherent frequency hopped spread-spectrum system with multipath channel equalization in the presence of jamming. *IEEE Transactions on Communications* **42** (2/3/4): 1325–1338 (February/March/April 1994).

Lüsse, P., P. Stiuve, J. Schüle, and H.-G. Unger. Analysis of vectorial mode fields in optical waveguides by a new finite difference method. *Journal of Lightwave Technology* **12** (3): 463–468 (March 1994).

Madisetti, V. K. A fast spotlight-mode synthetic aperture radar imaging system. *IEEE Transactions on Communications* **42** (2/3/4): 873–876 (February/March/April 1994).

Mahlke, G. and P. Gössing. *Fiber Optic Cables.* Chichester, England: John Wiley & Sons, Ltd., 1987.

Mahmoud, S. F. and A. M. Kharbat. Transmission characteristics of a coaxial optical fiber line. *Journal of Lightwave Technology* **11** (11): 1717–1721 (November 1993).

Mak, H.-M. and H. Yanagawa. High extinction directional coupler switches by compensation and elimination methods. *Journal of Lightwave Technology* **12** (5): 899–908 (May 1994).

Marcuse, D. *Light Transmission Optics.* New York: Van Nostrand Reinhold Company, 1982.

Mason, L. J. Estimation of fine-time synchronization error in FH FDMA SATCOM systems using the early-late filter technique. *IEEE Transactions on Communications* **42** (2/3/4): 1254–1263 (February/March/April 1994).

Meier-Hellstern, K. S., G. P. Pollini, and J. J. Goodman. Network protocols for the cellular packet switch. *IEEE Transactions on Communications* **42** (2/3/4): 1235–1244 (February/March/April 1994).

Mileant, A. and S. Hinedi. Satellite communications overview of arraying techniques for deep space communications. *IEEE Transactions on Communications* **42** (2/3/4): 1856–1865 (February/March/April 1994).

Miller, S. L. Code sequence analysis of direct sequence code-division multiple-access with M-ary FSK modulation. *IEEE Transactions on Communications* **42** (2/3/4): 829–833 (February/March/April 1994).

Miyao, Y. A dimensioning scheme in ATM networks. *Proceedings of Networks* (92), Kobe (1992).

Mohrdiek, S., H. Burkhard, and H. Walter. Chirp reduction of directly modulated semiconductor lasers at 10 Gb/s by strong CW light injection. *Journal of Lightwave Technology* **12** (3): 418–424 (March 1994).

Mohrdiek, S. *Theory of Dielectric Optical Waveguides.* New York: Academic Press, 1974.

Nagano, N., T. Susaki, M. Soda, et al. Monolithic ultrabroadband transimpedance amplifiers using AlGaAs/GaAs heterojunction bipolar transistors. *IEEE Transactions on Microwave Theory and Technology* **42** (1): 2–10 (January 1994).

Nagata, H., N. Niyamoto, T. Saito, and R. Kaizu. Reliable jacket stripping of optical fibers. *Journal of Lightwave Technology* **12** (5): 727–729 (May 1994).

Nam, S. H. and C. K. Un. Performance analysis of a broadcast star network with priorities. *IEEE Transactions on Communications* **42** (2/3/4): 1785–1794 (February/March/April 1994).

Narasimha, M. J. A recursive concentrator structure with applications to self-routing switching networks. *IEEE Transactions on Communications* **42** (2/3/4): 896–898 (February/March/April 1994).

Narayan, S. and P. J. McLane. Noise detected, fractionally spaced equalization for analog cellular data modems. *IEEE Transactions on Communications* **42** (2/3/4): 1270–1276 (February/March/April 1994).

Norimatsu, S. and K. Iwashita. Damping factor influence on linewidth requirements for optical PSK coherent detection systems. *Journal of Lightwave Technology* **11** (7): 1226–1233 (July 1993).

Osman, N., M. Koshiba, and R. Kaji. A comprehensive analysis of multilayer channel waveguides. *Journal of Lightwave Technology* **12** (5): 821–826 (May 1994).

Otoshi, T. Y. Maximum and minimum return losses from a passive two-port network terminal terminated with a mismatched load. *IEEE Transactions on Microwave Theory and Technology* **42** (5): 787–792 (May 1994).

Pagnoux, D., J.-M. Blondy, P. Di Bin, P. Faugeras, and P. Facq. Azimuthal far-field analysis for the measurement of effective cutoff of wavelength in single-mode fibers—Effects of curvature, length, and index profiles. *Journal of Lightwave Technology* **12** (3): 385–391 (March 1994).

Panajotav, K. P. Polarization properties of a fiber-to-asymmetric planar waveguide coupler. *Journal of Lightwave Technology* **12** (6): 983–988 (June 1994).

Pant, D. K., Ar. Ali, and D. W. Langer. Computing the eigen-modes of lossy field-induced optical waveguides. *Journal of Lightwave Technology* **12** (6): 1015–1022 (June 1994).

Paris, D. T. and F. K. Hurd. *Basic Electromagnetic Theory.* New York: McGraw-Hill Book Company, 1969.

Personick, S. D. *Fiber Optics.* New York: Plenum Press, 1985.

Petersen, J. and T. Gillen. The idle time distribution of a system D/D/1. *IEEE Transactions on Communications* **42** (2/3/4): 854–856 (February/March/April 1994).

Poole, C. D. and D. L. Favin. Polarization-mode dispersion measurements based on transmission spectra through a polarizer. *Journal of Lightwave Technology* **12** (6): 917–929 (June 1994).

Prade, B. and J. Y. Vinet. Guided optical waves in fibers with negative dielectric constant. *Journal of Lightwave Technology* **12** (1): 6–18 (January 1994).

Pratt, W. R. *Laser Communications Systems.* New York: John Wiley & Sons, 1969.

Ramaswami, R. and K. N. Sivarajan. A packet switched multihop lightwave network using subcarrier and wavelength division multiplexing. *IEEE Transactions on Communications* **42** (2/3/4): 1198–1211 (February/March/April 1994).

Ramseier, S. Bandwidth-efficient correlative trellis-coded modulation schemes. *IEEE Transactions on Communications* **42** (2/3/4): 1595–1605 (February/March/April 1994).

Rivlidou, F.-N. Queueing two-dimensional models for cellular mobile systems. *IEEE Transactions on Communications* **42** (2/3/4): 1505–1511 (February/March/April 1994).

Rizzoli, V., F. Mastri, and D. Masotti. Performance prediction and optimization of a coherent phase modulated low noise analog optical link operating at microwave frequencies. *IEEE Transactions on Microwave Theory and Technology* **42** (5): 801–806 (May 1994).

Saley, A. A. M. and J. Stone. Two-stage Fabry-Perot fibers as demultiplexers in optical FDMA LAN's. *Journal of Lightwave Technology* **7** (2): 323–330 (February 1989).

Schmidt, F. An adaptive approach to the numerical solution of Fresnel's wave equation. *Journal of Lightwave Technology* **11** (9): 1425–1434 (September 1993).

Seikai, S., K. Kusunoki, and S. Shimokado. Experimental studies on wavelength division bidirectional optical amplifiers using Er^{3+}-doped fiber. *Journal of Lightwave Technology* **12** (5): 849–854 (May 1994).

Shami (Shitz), S. and A. Dembo. Bounds on the symmetric binary cutoff rate for dispersive Gaussian channels. *IEEE Transactions on Communications* **42** (1): 39–53 (January 1994).

Shaw, L., M. Ahmadi, M. Siddigui, et al. Monolithic V-band frequency converter chip set development using 0.2 μm InGaAs/GaAs pseudomorphic HEMT technology. *IEEE Transactions on Microwave Theory and Technology* **42** (1): 11–17 (January 1994).

Sheikh, A., Y.-D. Yao, and S. Cheng. Through-put enhancement of direct-sequence spread spectrum packet radio networks by adaptive power control. *IEEE Transactions on Communications* **42** (2/3/4): 884–890 (February/March/April 1994).

Shigesawa, H., M. Tsuji, P. Lampariello, et al. Coupling between different leaky-mode types in stub-loaded leaky waveguides. *IEEE Transactions on Microwave Theory and Technology* **42** (8): 1548–1560 (August 1994).

Shimada, J. I., O. Ohguchi, and K. E. Stubkjaer. Focusing characteristics of a wide-striped laser diode integrated with a microlens. *Journal of Lightwave Technology* **12** (6): 936–942 (June 1994).

Shin, J.-D. and H. F. Taylor. Dielectric mirror embedded optical fiber couplers. *Journal of Lightwave Technology* **12** (1): 68–73 (January 1994).

Simon, M. K. A simple evaluation of DPSK error probability performance in the presence of bit timing error. *IEEE Transactions on Communications* **42** (2/3/4): 263–267 (February/March/April 1994).

Sochacka, M. Optical fibers profiling by phase-stepping transverse interferometry. *Journal of Lightwave Technology* **12** (1): 19–23 (January 1994).

Surette, M. R., D. R. Hjelme, and A. R. Mickelson. An optically driven phased array antenna utilizing heterodyne techniques. *Journal of Lightwave Technology* **11** (9): 1500–1509 (September 1993).

Svinnset, I. Non-blocking ATM switching networks. *IEEE Transactions on Communications* **42** (2/3/4): 1352–1358 (February/March/April 1994).

Thijs, P. J. A., T. v. Dongen, L. F. Tiemeijer, and J. J. M. Binsma. High performance $\lambda = 1.3$ μm InGaAsP-InP strained-layer quantum well lasers. *Journal of Lightwave Technology* **12** (1): 28–37 (January 1994).

Thirstrup, C., P. N. Robson, P. L. K. Wa, et al. Well hetero nipi waveguides. *Journal of Lightwave Technology* **12** (3): 425–429 (March 1994).

Tomita, N., H. Hakasugi, N. Atobe, et al. Design and performance of a novel automatic fiber line testing system with OTDR for optical subscriber loops. *Journal of Lightwave Technology* **12** (5): 717–726 (May 1994).

Trifunović, V. and B. Jakanović. Review of printed Marchand and double Y baluns: Characteristics and applications. *IEEE Transactions on Microwave Theory and Technology* **42** (8): 1454–1462 (August 1994).

U.S. Department of Defense. *Military Specifications, Fiber Optic Component Series* (1989).

———. *MIL-STD-2196(SH) Glossary, Fiber Optics* (12 January 1989).

U.S. General Services Administration, *Federal Standard 1037C: Telecommunications: Glossary of Telecommunication Terms.* Federal Supply Service Bureau, Specification Section, 470 East L'Enfant Plaza, SW, Suite 800, Washington, DC 20407.

Unger, C. and W. Stöcklein. Investigation of microbending sensitivity of fibers. *Journal of Lightwave Technology* **12** (4): 591–596 (April 1994).

Ury, I. Optical communications. *Microwave Journal:* 24–35 (April 1985).

Valdimarsson, E. Blocking in multirate interconnection networks. *IEEE Transactions on Communications* **42** (2/3/4): 2028–2035 (February/March/April 1994).

van der Mark, M. B. and L. Bosselaar. Noncontact calibration of optical fiber cladding diameter using exact scattering theory. *Journal of Lightwave Technology* **12** (1): 1–5 (January 1994).

van Deventer, M. O. and A. J. Boot. Polarization properties of stimulated Brillouin scattering in single-mode fibers. *Journal of Lightwave Technology* **12** (4): 585–590 (April 1994).

van Deventer. T. E., L. P. B. Katehi, and A. C. Cangellaris. Analysis of conductor losses in high-speed interconnects. *IEEE Transactions on Microwave Theory and Technology* **42** (1): 78–83 (January 1994).

Weik, M. H. *Communications Standard Dictionary, 2nd Ed.* New York: Van Nostrand Reinhold Company, 1989.

———. *Fiber Optics Standard Dictionary, 2nd Ed.* New York: Van Nostrand Reinhold Company, 1989.

———. *Standard Dictionary of Computers and Information Processing, 2nd Ed.* Rochelle Park, NJ: Hayden Book Company, 1977.

Weik, M. H.. C. M. Davis, et al. *Fiberoptic Sensor Technology Handbook.* Alexandria, VA: Optical Technologies and Dynamic Systems, Inc., 1986.

West, R. H., H. Buker, E. J. Friebele, H. Henschel, and P. B. Lyons. The use of optical time domain reflectometers to measure radiation-induced losses in optical fibers. *Journal of Lightwave Technology* **12** (4): 614–621 (April 1994).

Wieselthier, J. E., C. M. Barnhart, and A. Ephremides. A neural network approach to routing without interference in multihop radio networks. *IEEE Transactions on Communications* **42** (1): 166–177 (January 1994).

Wong, P. C. and T.-S. P. Yum. Design and analysis of a pipeline ring protocol. *IEEE Transactions on Communications* **42** (2/3/4): 1153–1161 (February/March/April 1994).

Wu, X.-D. and K. Chang. Novel active FET circular patch antenna arrays for quasi-optical power combining. *IEEE Transactions on Microwave Theory and Technology* **42** (5): 766–771 (May 1994).

Yu, J., D. Yevick, and D. Weidman. A comparison of beam propagation and coupled-mode methods: Application to optical fiber couplers. *Journal of Lightwave Technology* **12** (5): 790–795 (May 1994).

Yu, J., P.-H. Zongo, and P. Facq. Refractive-index profile influences on mode coupling effects at optical fiber splices. *Journal of Lightwave Technology* **11** (8): 1270–1273 (August 1993).

Yum, T. S. P. and M. S. Chen. Multicast source routing in packet-switched networks. *IEEE Transactions on Communications* **42** (2/3/4): 1212–1215 (February/March/April 1994).

———. Dynamic channel assignment in integrated services cable networks. *IEEE Transactions on Communications* **42** (2/3/4): 2023–2027 (February/March/April 1994).

Zaccarin, D. and M. Kavehrad. Performance evaluation of optical CDMA systems using non-coherent detection and bipolar codes. *Journal of Lightwave Technology* **12** (1): 96–105 (January 1994).

Zhang, P.-G. and D. Irvine-Halliday. Measurement of beat length in high-birefringent optical fiber by way of magnetooptic modulation. *Journal of Lightwave Technology* **12** (4): 597–600 (April 1994).

Zheng, W., O. Hultín, and R. Rylander. Erbium-doped fiber splicing and splice loss estimation. *Journal of Lightwave Technology* **12** (3): 430–435 (March 1994).

Zvonar, Z. and D. Brady. Multiuser detection in single-path fading channels. *IEEE Transactions on Communications* **42** (2/3/4): 1729–1739 (February/March/April 1994).